The ARRL
Antenna Book

FOR RADIO COMMUNICATIONS

ARRL The National Association for Amateur Radio®

Twenty-Fifth Edition

- **Volume 1: Antenna Fundamentals — Chapters 1 – 5**
- **Volume 2: Basic Antenna Types and MF and HF Antennas — Chapters 6 – 14**
- **Volume 3: VHF, UHF, and Microwave Antennas and Specialty Applications — Chapters 15 – 24**
- **Volume 4: Building and Maintaining Antenna Systems and Index — Chapters 25 – 29**

Editor
Ward Silver, NØAX

Production Editor
Mark Derks, KC1RVQ

Production
Chris Dunham
Jodi Morin, KA1JPA
David F. Pingree, N1NAS

Editorial Assistant
Maty Weinberg, KB1EIB

Contributors to the Twenty-Fifth Edition

Alan Applegate, KØBG
Jim Breakall, WA3FET
Jim Brown, K9YC
Wyatt Dirks, ACØRA
Frank Donovan, W3LPL
James Duffey, KK6MC
Peter East
Doug Grant, K1DG
Rus Healy, K2UA
Matt Heere, N3NWV
Wolfgang Herrmann
Chuck Higgins
Martin Huml, OK1FUA
Dale Hunt, WB6BYU
Justin Johnson, GØKSC
Al Katz, K2UYH
Hal Kennedy, N4GG
Pete Kobak, KØBAK
Steve Kostro, N2CEI
Sean Kutzko, KX9X
Greg Lapin, N9GL
Dan Layne, ADØCY
Roy Lewellan, W7EL
Fred Maas, KT5X
Richard Marsden
Eduard Mol
Dave Mueller, N2NL
Mario Natali, IØNAA
Ron Ochu, KOØZ
Dave Olean, K1WHS
Steve Olney, VK2XV
Charles Osborne, K4CSO
Jonathan Pettingale
Alex Pettit, KK4VB
John Portune, W6NBC
Bruce Randall, NT4RT
Whitman Reeve
Rick Rosen, K1DS
Rich Russell, ACØUB
Charles Scharlau, NZØI
Tom Schiller, N6BT
Rudy Severns, N6LF
Lowell Silver, WW7SAR
Andrea Slack, K2EZ
Jim Smith, KKØU
John Stanley, K4ERO
Steve Stearns, K6OIK
Patrick Stoddard, WD9EWK
Chip Sufitchi, N2YO
Dave Typinkski, AJ4CO
Ray Uberecken, AAØL
Paul Wade, W1GHZ
Bill Wortman, N6MW
Ken Wyatt, KX7KW

Printed in the USA

ISBN: 978-1-62595-175-5 Hardcover
978-1-62595-176-2 Softcover
978-1-62595-183-0 Four-Volume Set

Kindle eBook Editions
ISBN: 978-1-62595-177-9 — Volume 1
ISBN: 978-1-62595-179-3 — Volume 2
ISBN: 978-1-62595-180-9 — Volume 3
ISBN: 978-1-62595-182-3 — Volume 4

Twenty-Fifth Edition, Second Printing

We strive to produce books without errors. Sometimes mistakes do occur, however. When we become aware of problems in our books (other than obvious typographical errors), we post corrections on the ARRL website.
If you think you have found an error, please check **www.arrl.org/arrl-antenna-book-reference** for corrections.
If you don't find a correction there, please let us know by sending an email to **pubsfdbk@arrl.org**.

Table of Contents

A detailed Table of Contents is included at the beginning of each chapter.
See the **Project Index** for a complete list of projects.

VOLUME 2

Basic Antenna Types and MF and HF Antennas

VOLUME 3

VHF, UHF, and Microwave Antennas and Specialty Applications

VOLUME 4

Building and Maintaining Antenna Systems and Index

Foreword

The antenna system is what all amateur stations have in common — it is where signals begin and end their propagation journeys. Little is more important to the effectiveness of an amateur radio station. Even the simplest equipment can perform well if connected to a good antenna. Conversely, the latest and greatest gear won't perform their wonders without one. That's why the antenna system is the subject of more ARRL publications than any other topic.

You might have noticed the word "system" in the preceding paragraph. Beginning at the transceiver, the antenna system includes all the cables, tuners, switches, instruments, and supporting structures, as well as the antenna itself. Each part of the antenna system is addressed in this book so you can select and use the best components and materials.

The *Antenna Book* is organized to take you from the basics all the way through design, building, and measuring your antenna system. The book begins with a section on fundamentals: the basic concepts of antennas, the familiar dipole and monopole that make up so many antennas, and how antennas interact with the ground under and around them. Even the path between antennas is covered in the material on radio wave propagation. A chapter on computer modeling of antennas puts these ideas into practice. Then the book dives right into the various forms of antennas, building from the fundamentals to understand the principles of long wires, phased arrays, dipole arrays, large and small loops, and frequency-independent antennas.

A set of chapters then presents antennas for the widely used medium- and high-frequency (MF and HF) bands. This includes general-purpose antennas, directional antennas, and how to design an effective HF antenna system for your terrain and operating styles. A following section tackles antennas for the VHF, UHF, and microwave spectrum where so many amateurs operate throughout their amateur radio careers, including everything from terrestrial mobile stations to satellite and "moonbounce." Next is a set of chapters for popular portable "on the air" operating, mobile and maritime stations, stealth and limited size installations, and antennas just for receiving from MF through outer space. Finally, the components and techniques that make up the full antenna system get an in-depth treatment, too.

Even if some of the topics sound familiar from previous editions, you'll find that the book has been comprehensively updated by the largest set of expert contributors for any edition. Here are just a few examples of their significant new work:

- More fundamentals, especially of monopoles and ground system effects
- Propagation at VHF/UHF along with solar and ionospheric effects
- New set of OWA Yagi designs from Justin Johnson, GØKSC, and Jim Breakall, WA3FET
- Shunt-feeding is explained for using your tower on the MF and lower HF bands
- Receiving antenna designs for MF/HF operating, direction-finding, and amateur radio astronomy
- New VHF/UHF slot antenna theory and design, plus portable satellite antenna systems
- Portable and ultra-light antenna systems for operating away from home
- Best practices for construction, building, and installing antenna systems safely

The supplemental material available to readers has also been significantly updated. Along with the software and referenced articles, there are more projects and designs than ever. You'll find many interesting items in the book's online information.

Back in 1939, it became clear that antennas deserved their own handbook. When that very first *Antenna Book* was created, it was a recognition of the importance of the antenna to the success of amateur radio and many of those original designs are still familiar. The book you are reading today extends the long tradition of antenna application and innovation in amateur radio. It has been a pleasure to bring it to you!

73,
Ward Silver, NØAX
Lead Editor

The ARRL Antenna Book
Online Material — (print edition only)

A wealth of additional material for this edition of *The ARRL Antenna Book* is available with the online material. As a purchaser of the print edition, you are entitled to download this material. If you purchased the book directly from ARRL through **arrl.org** you should have received an email with directions for downloading the content and a link. If you purchased *The ARRL Antenna Book* from another retailer, lost, or never received the email with the link to the online material, go to **arrl.org/antennabook25download**. You must log in to the ARRL website or register an account with **arrl.org/shop** to access the content.

Searchable Edition of *The ARRL Antenna Book*

The online material includes a print-replica, PDF version of this edition of *The ARRL Antenna Book*. Using *Adobe Reader*, you can view, print, or search the entire book.

Supplemental Files for Each Chapter

The online material contains supplemental information for most chapters of this book. This includes articles from *QST, QEX,* and other sources; material from previous editions of *The ARRL Antenna Book*; tables and figures in support of the chapter material; and files that contain information to build and test the projects provided in the chapters. The supplemental information is arranged in folders for each chapter.

ARRL Antenna Modeling Files

A set of *EZNEC* modeling files representing many of the antennas discussed in *The ARRL Antenna Book*. The models are grouped in folders named for the type of antenna they represent. Requires *EZNEC* antenna modeling software (**eznec.com**), not supplied.

Propagation Prediction Tables

Propagation-prediction tables generated by Dean Straw, N6BV, for more than 240 different transmitting locations throughout the world, including 42 locations in the USA. Each file is a PDF for viewing and printing using Adobe *Acrobat Reader*.

Companion Software

The following software is also included with the online material:

• *HFTA* (*HF Terrain Assessment for Windows*) — A ray-tracing program designed to evaluate the effect of foreground terrain on the elevation pattern of up to four multi-element HF monoband Yagis in a stack. See the **HF Antenna System Design** chapter in this book for details about the theory behind ray tracing and diffraction in *HFTA*.

• *TLW* (*Transmission Line for Windows*) — A sort of "Swiss Army Knife" for transmission line and antenna tuner calculations.

• *YW* (*Yagi for Windows*) — A special-purpose program, designed strictly for evaluation of monoband Yagis. It has the advantage of running many times more quickly than general-purpose programs such as *NEC,* but it has some attendant limitations.

Visit **arrl.org/arrl-antenna-book-reference** for instructions on installing the software with contemporary operating systems.

About ARRL

We're the American Radio Relay League, Inc. — better known as ARRL. We're the largest membership association for the amateur radio hobby and service in the US. For over 100 years, we have been the primary source of information about amateur radio, offering a variety of benefits and services to our members, as well as the larger amateur radio community. We publish books on amateur radio, as well as four magazines covering a variety of radio communication interests. In addition, we provide technical advice and assistance to amateur radio enthusiasts, support several education programs, and sponsor a variety of operating events.

One of the primary benefits we offer to the ham radio community is in representing the interests of amateur radio operators before federal regulatory bodies advocating for meaningful access to the radio spectrum. ARRL also serves as the international secretariat of the International Amateur Radio Union, which performs a similar role internationally, advocating for amateur radio interests before the International Telecommunication Union and the World Radiocommunication Conference.

Today, we proudly serve nearly 151,000 members, both in the US and internationally, through our national headquarters and flagship amateur radio station, W1AW, in Newington, Connecticut. Every year we welcome thousands of new licensees to our membership, and we hope you will join us. Let us be a part of your amateur radio journey. Visit www.arrl.org/join for more information.

ARRL The National Association for Amateur Radio®

225 Main Street
Newington, CT 06111-1400 USA
Tel: 860-594-0200
FAX: 860-594-0259
Email: membership@arrl.org

www.arrl.org

Common Schematic Symbols Used in Circuit Diagrams

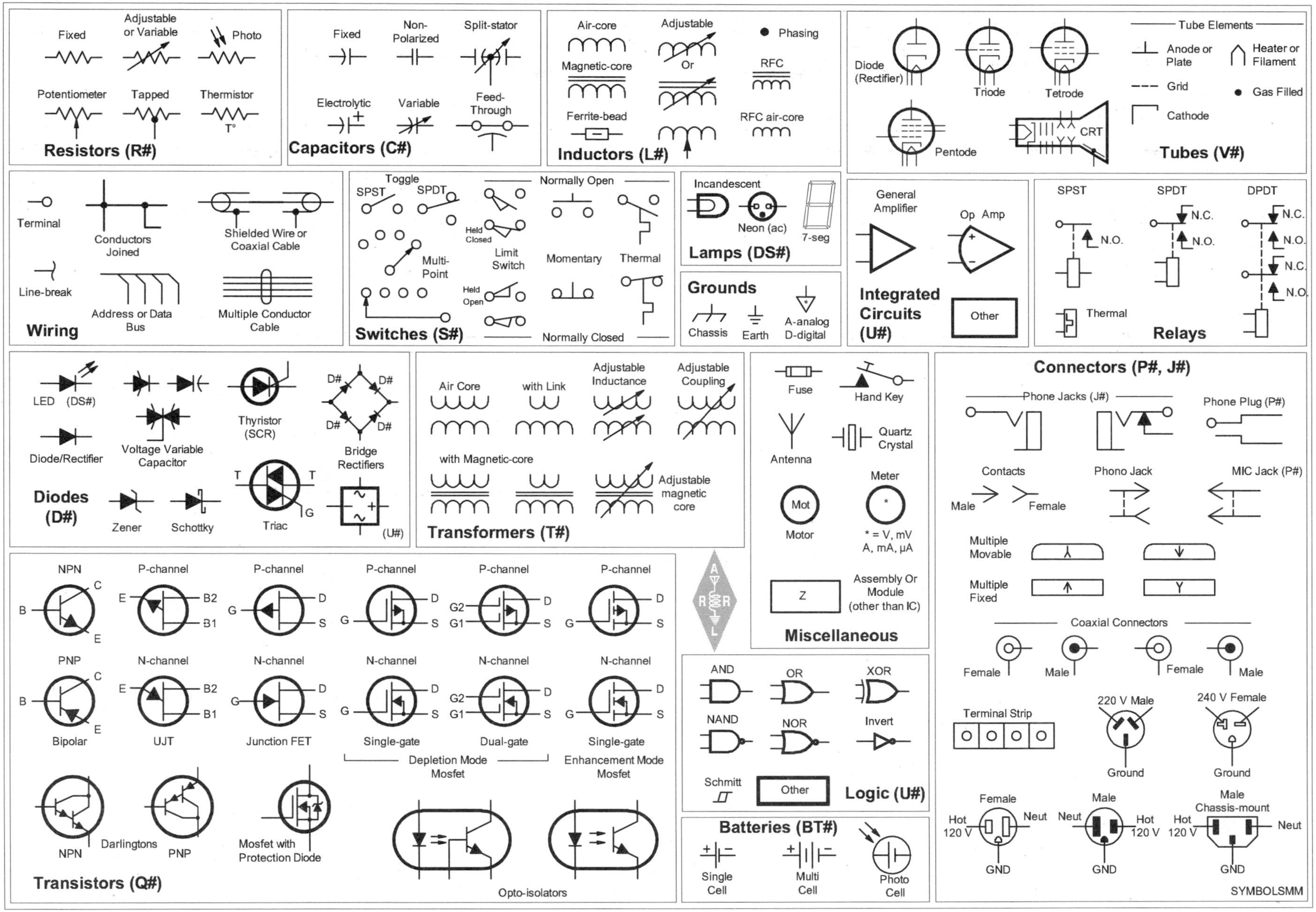

TABLE OF CONTENTS

Chapter 1 — Online Material

Articles

- "Hands-On Radio: Experiment #178 — Maxwell's Equations — Grad, Div, and Curl," Ward Silver, NØAX
- "Hands-On Radio: Experiment #179 — Maxwell's Equations — The Wave Emerges," Ward Silver, NØAX
- "Radio Mathematics" — supplemental information about math used in radio and a list of online resources and tutorials about common mathematics
- "Why an Antenna Radiates" by Kenneth Macleish, W7TX

Chapter 1

Antenna Fundamentals

Where does the word "antenna" come from? As related by Dr. Ulrich Rohde, N1UL, the term originated with Guglielmo Marconi during early radio tests in 1895 during which he used wire "aerials" attached to a vertical tent pole. The aerial wire then ran down the pole to the transmitter. In Italian, a tent pole is known as "l'antenna central" and so the pole with the wire became simply, "l'antenna." In the beginning of radio, antennas were attached directly to generators and transmitters and were considered part of a common assembly. It wasn't until after 1900 that antennas began to be regarded as separate elements of the system, independent of the transmitter or receiver.

While there are an enormous variety of antennas, they share basic characteristics, and all are designed to radiate and receive electromagnetic waves. In this chapter, we begin by defining what an electromagnetic wave is and how it is described. We then define the most important characteristics of an antenna — impedance, directivity, and polarization — and show how those characteristics are measured and displayed. For information on how waves propagate and related concepts, see the **Radio Wave Propagation** chapter or the ARRL's *Here to There: How Radio Waves Propagate*.

1.1 Introduction to Electromagnetic Fields and Waves

1.1.1 E and H Fields

In 1820 Hans Oersted discovered that a current flowing in a wire would deflect the needle of a nearby compass. We attribute this effect to a magnetic or H-field, which at any given location is denoted by the letter **H**. The magnetic field's amplitude is expressed in A/m (Amperes/meter) along with a direction. (Direction can also be expressed as some value of phase with respect to a reference.) Because a magnetic field has *both* amplitude and direction, it is a *vector*. Symbols representing a vector are printed in boldface.

Figure 1.1 shows a typical experimental arrangement that demonstrates the presence of a magnetic field. The shape of the magnetic field is roughly shown by the distribution of the iron filings. This field distribution is very similar to that around a vertical antenna.

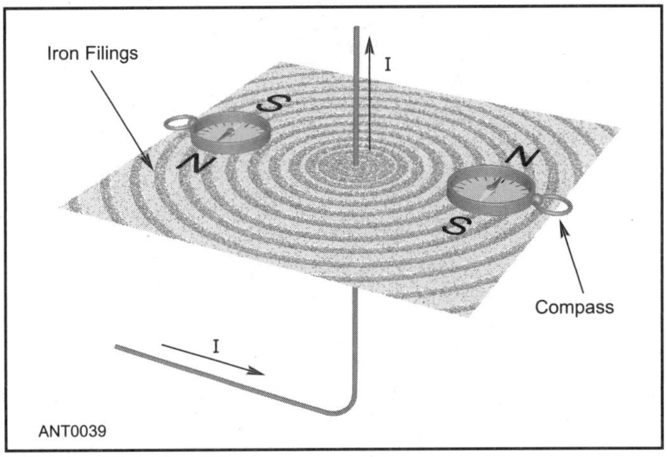

Figure 1.1 — Visualization of a magnetic field. The magnetic lines of force that surround a conductor with an electric current flowing in it are shown by iron filings and small compass needles. The needles point in the direction of the magnetic or H-field. The filings give a general view of the field distribution in the plane perpendicular to the conductor.

A compass needle (a small magnet itself) will try to align itself parallel to **H**. As the compass is moved around the conductor, the orientation of the needle changes accordingly. The orientation of the needle gives the direction of **H**. If you attempt to turn the needle away from alignment you will discover a torque trying to restore the needle to its original position. The torque is proportional to the strength of the magnetic field at that point. This strength is called the *field intensity* or amplitude of **H** at that point. If a larger current flows in the conductor, the amplitude of **H** will increase in proportion. Currents flowing in an antenna also generate an H-field.

An antenna will also have an electric or E-field, which can be visualized using a parallel-plate capacitor, as shown in **Figure 1.2**. If we connect a battery with a dc potential across the capacitor plates there will be an electric field **E** established between the plates, as indicated by the lines and directional

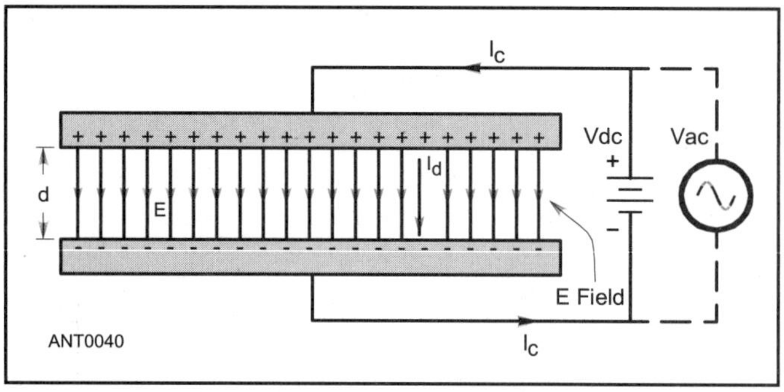

Figure 1.2 — Visualization of an electric field, $E=V_{dc}/d$. When the dc source is replaced with an ac source there will be a displacement current (I_d) flowing between the capacitor plates.

Math Tutorials

You will encounter a fair amount of intermediate level mathematics in this book. If you would like to brush up on your math skills or learn about an unfamiliar topic, a list of free online math tutorials is included with the online material for this book and on the ARRL website under "Math Tutorials" at **arrl.org/tech-prep-resource-library**.

arrows between the plates. (Like **H**, the electric field also has an amplitude and direction and so is a vector as well.) The magnitude of vector **E** is expressed in V/m (volts per meter), so for a potential of V volts and a spacing of d meters, E = V/d in units of V/m. The amplitude of **E** will increase with voltage and/or a smaller separation distance (d). In an antenna, there will be ac potential differences between different parts of the antenna and from the antenna to ground. These ac potential differences establish the electric field associated with the antenna.

1.1.2 Conduction And Displacement Currents

If we replace the dc voltage source in Figure 1.2 with an ac source, an ac current will flow in the circuit. In the conductors between the ac source and the capacitor plates, current (I_c) flows, because of the movement of charge, usually electrons. But in the space between the capacitor plates (particularly in a vacuum) there are no charge carriers available to carry a conduction current. Nonetheless, current still flows in the complete circuit, and we attribute this to a *displacement current* (I_d) flowing between the capacitor plates to account for the continuity of current in the circuit. Displacement and conduction currents are two different phenomena but they both represent current, just two different kinds. Some observers prefer to call conduction currents "currents" and displacement currents "imaginary currents." That terminology is OK, but to account for the current flow in a closed circuit with capacitance you must keep track of both kinds of current, whatever you call them. The accepted convention is to use the term "displacement current."

1.1.3 Electromagnetic Fields And Waves

This section is a description of the fields that make up electromagnetic waves and some basic properties of waves. For a more complete discussion of how radio waves propagate, see the **Radio Wave Propagation** chapter.

ELECTROMAGNETIC FIELDS

An electromagnetic field is created when the potential energy stored in an electric field or magnetic field changes. The changing electric and magnetic fields create *electromagnetic waves* (what we call radio waves) that propagate through space carrying both electric and magnetic energy. The electric and magnetic fields in the wave vary with time in a sinusoidal pattern. The potential energy is shared between the electric and magnetic fields making up the electromagnetic field.

The *field strength* of an electromagnetic wave can be measured either by the electric field (volts/meter) or the magnetic field (amps/meter). Usually, the wave's field strength is stated only in volts/meter since that is easier to measure than the magnetic field. If we multiply the electric and magnetic field strengths, we have power per unit area:

$$\frac{V}{m} \times \frac{A}{m} = \frac{watts}{m^2}$$

MAXWELL'S EQUATIONS

The basic theory of electromagnetic fields was established by James Maxwell in 1860 – 1864. The behavior of the fields is described by the four equations known today as *Maxwell's equations.* (This form of the equations was actually produced by Oliver Heaviside in his work prior to 1890.) The existence of the electromagnetic waves predicted by Maxwell in 1864 was demonstrated by Heinrich Hertz in 1886. For more information about the experiment, see the References and Bibliography listing at the end of this chapter for G.S. Smith's article analyzing Hertz's experiments.

A complete treatment of the equations is provided by Bob Zavrel, W7SX, in his book *Antenna Physics: An Introduction, Second Ed.* (see the References and Bibliography section of this chapter). A summary treatment of the equations in a pair of Hand-On Radio columns by Ward Silver, NØAX, is included in the online material, as well. These include an explanation of the vector calculus concepts of gradient, divergence, and curl, as well as illustrating how waves are created by moving electric charge.

ELECTROMAGNETIC WAVES

An electromagnetic wave, as the name implies, is composed of both an electric field and a magnetic field that vary with time. Electric and magnetic fields that do not change with time, such as those created by a dc current or voltage, are called *electrostatic fields*. The fields of a radio wave are created by an ac current in an antenna, usually having the form of a sine wave. As a result, the fields in a radio wave vary in the same sinusoidal pattern, increasing and decreasing in

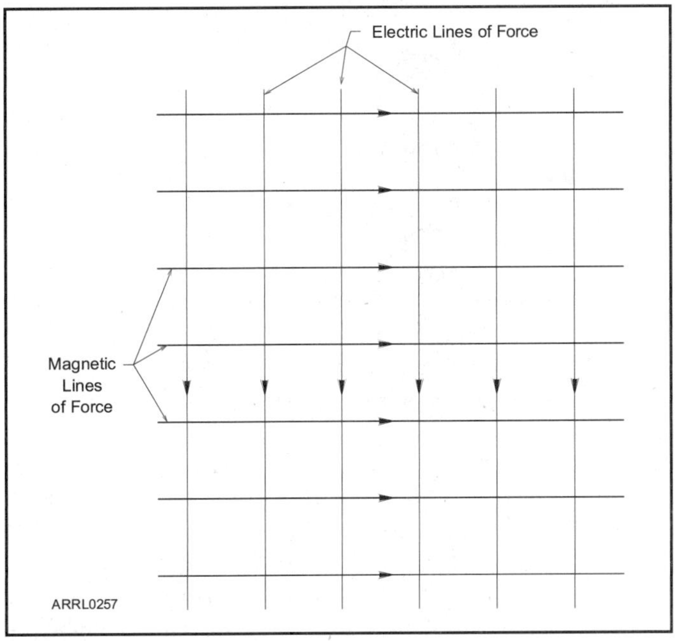

Figure 1.3 — Representation of electric and magnetic lines of force in an electromagnetic wavefront. Arrows indicate the instantaneous directions of the fields for a wavefront in a wave traveling toward you, out of the page. Reversing the direction of one of the fields would also reverse the direction of the wave.

strength and reversing direction with the same frequency, f, as the ac current. It is the movement of electrons — specifically the acceleration and deceleration as the ac current moves back and forth — that creates the electromagnetic wave.

The two fields of the electromagnetic wave are oriented at right angles to each other as shown by **Figure 1.3**. The term "lines of force" in the figure means the direction in which a force would be felt by an electron (from the electric field) or by a magnet (from the magnetic field). The direction of the right angle from the electric field to the magnetic field, clockwise or counterclockwise, determines the direction the wave travels, as illustrated in the figure.

Figure 1.3 represents a wavefront of the electromagnetic wave; a flat surface or plane moving through space on which the electric and magnetic fields have a constant value. To an observer moving in the same direction as the wave and at the same speed, the strength of the fields does not change. To that observer, the electric and magnetic field strengths are fixed, as in a photograph.

Just as an ac voltage is made up of an infinite sequence of instantaneous voltages, each slightly larger or smaller than the next, an infinite number of wavefronts make up a *propagating wave*, one behind another like a deck of cards. The direction of the wave is the direction in which the wavefronts move. The fields on each successive wavefront have a slightly different strength so as they pass a fixed location, the detected field strength changes as well.

To an observer staying in one place, such as a stationary receiving antenna, the electric and magnetic fields of the wave appear to oscillate as the wave passes. That is, the fields create forces on electrons in the antenna that increase and decrease in a sine wave pattern. Some of the energy in the propagating wave is transferred to the electrons as the forces from the changing fields cause them to move. This creates a sine wave current in the antenna with a frequency determined by the rate at which the field strength changes as the wave passes.

Figure 1.4 is a drawing of what would happen if we could suddenly freeze all of the wavefronts in the wave and take measurements of the electric and magnetic field strengths in each. In this example, the electric field is oriented vertically and the magnetic field horizontally. (Each of the vertical lines in the electric field can be thought of as representing an individual wavefront.) All of the wavefronts are moving in the direction indicated — the whole set of them moves together at the same speed. As the wave — the set of wavefronts — moves past the receiving antenna, the varying field strengths of the different wavefronts is perceived as a continuously changing wave. What we call a "wave" is made up of this entire group of wavefronts moving through space.

One more important note about electromagnetic waves: The electric and magnetic fields are *coupled*, that is they are both aspects of the same entity, the electromagnetic wave. They are not perpendicular electric and magnetic fields that simply happen to be in the same place at the same time! The fields cannot be separated, although the energy in the wave can be detected as electric or magnetic force. The fields are created as a single entity — an electromagnetic wave — by the motion of electrons in the transmitting antenna.

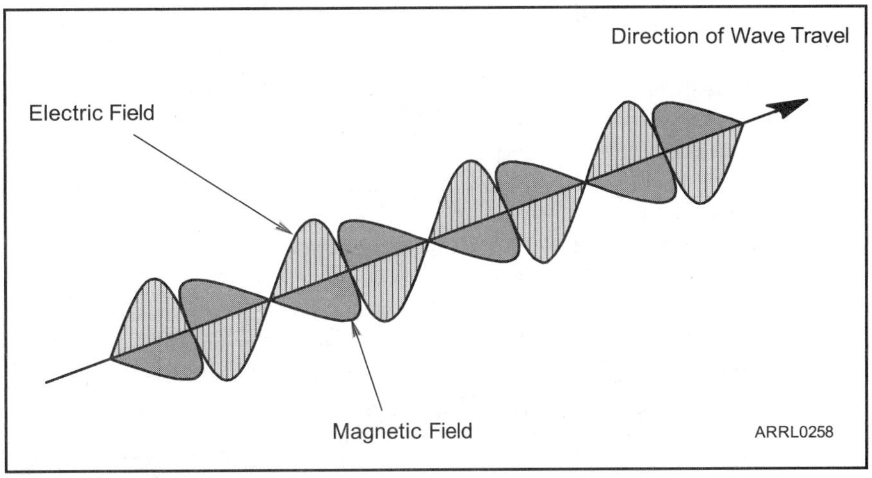

Figure 1.4 — Representation of the magnetic and electric field strengths of an electromagnetic wave. In the diagram, the electric field is oriented vertically and the magnetic field horizontally.

SPEED OF PROPAGATION AND WAVELENGTH

Because the velocity of wave propagation is so great, we tend to ignore it. Only 1⁄7 of a second is needed for a radio wave to travel around the world — but in working with antennas and transmission lines the time factor is extremely important. We can hardly discuss antenna theory or performance at all without involving travel time, consciously or otherwise.

The speed at which electromagnetic waves travel is called the *velocity of propagation*. It is determined by the permittivity (ε) and permeability (μ) of the medium in which the wave is traveling. This is commonly referred to as the "speed of light" and represented by *c*.

$$c = \frac{1}{\sqrt{\varepsilon\mu}}$$

The speed of light is highest in the vacuum of free space, approximately 300 million or 3×10^8 meters per second and the special symbols e_0 and μ_0 are used. It is often more convenient to remember this value as 300 m/μs (the actual value is 299.7925 m/μs).

It is also useful to know a radio wave's *wavelength* — the distance traveled during one complete cycle of a wave. Since one complete cycle takes 1/f and the velocity of a wave is the speed of light, c, the wavelength, λ, is thus:

$$\lambda = c / f$$

In free space

$$\lambda = 299.7925 \times 10^6 / f$$

where λ is the free-space wavelength in meters.

More convenient approximate formulas for use at radio frequencies are:

λ in meters = 300 / f in MHz, and

λ in feet = 983.6 / f in MHz

The ratio between the wave's velocity in a specific medium and that of free space is called the medium's *velocity factor* (VF) and is a value between 0 and 1. If the medium is air, the reduction in velocity of propagation can be ignored in most discussions of propagation at frequencies below 30 MHz. In the VHF range and higher, temperature and moisture content of the medium have increasing effects on the communication range, as will be discussed in the **Radio Wave Propagation** chapter. In materials such as glass or plastic the wave's velocity can be quite a bit lower than that of free space. For example, in polyethylene (commonly used as a center insulator in coaxial cable), the velocity of propagation is about 2⁄3 that in free space. In distilled water (a good insulator) the speed is about 1⁄9 that of free space.

PHASE OF WAVES

There will be few pages in this book where phase, wavelength and frequency do not enter the discussion. It is essential to have a clear understanding of their meaning in order to understand the design, installation, adjustment or use of antennas, matching systems or transmission lines in detail. In essence, *phase* means *time*. When something goes through periodic variations as an alternating current does, corresponding instants in succeeding periods are *in phase*.

It is important to distinguish between phase and *polarity*. Polarity is simply a convention that assigns a positive and negative direction or convention. Reversing the leads on a feed line reverses a signal's polarity but does not change its phase.

Phase is a relative measure of time within and between waveforms. The points A, B, and C in **Figure 1.5** are all in phase. They are corresponding instants at intervals of 1 λ. The distance between A and B or between B and C is one wavelength. This is a conventional view of a sine wave alternating current, with time progressing to the right. It also represents the *instantaneous* value of intensity of the traveling fields, if distance is substituted for time in the horizontal axis. The field-intensity distribution follows the sine curve, in both amplitude and polarity, corresponding exactly to the time variations in the current that produced the fields. Remember that this is an *instantaneous* picture of the many wavefronts similar to Figure 1.4.

Waves used in radio communication may have frequencies from about 10,000 to several hundred billion Hz. Suppose the frequency is 30 MHz. One cycle, or period, is completed in 1/30,000,000 second. The wave is traveling at 300,000,000 meters per second through the air, so it will move only 10 meters during the time that the current is going through one complete period of alternation. The electromagnetic field 10 meters away from the antenna is caused by the current that was flowing one period earlier in time. The field 20 meters away is caused by the current that was flowing two periods earlier, and so on.

If each period of the current is simply a repetition of the one before it, the currents at corresponding instants in each period will be identical. The fields caused by those currents will also be identical. As the fields move outward from the antenna they become more thinly spread over larger and larger spherical surfaces centered on the antenna. The field amplitudes decrease with distance from the antenna but they do not

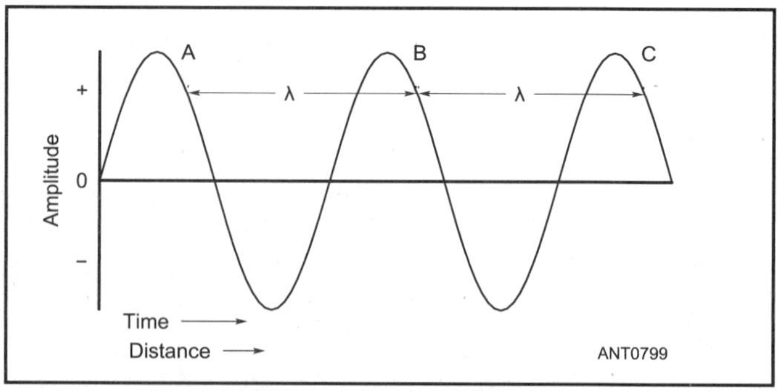

Figure 1.5 — The instantaneous amplitude of both fields (electric and magnetic) varies sinusoidally with time as shown in this graph. Since the fields travel at constant velocity, the graph also represents the instantaneous field intensity along the wave path. The distance between two points of equal phase such as A – B and B – C is the wave's wavelength.

lose their identity with respect to the instant of the period at which they were generated. They are, and they remain, in phase. In the example above, on the spherical surfaces separated by intervals of 10 meters measured outward from the antenna, the phase of the waves at any given instant is identical.

These spherical surfaces are the wavefronts described earlier. When the sphere is so large that the surface is essentially flat, the wavefront is called a *plane wave*. On every part of this surface, the wavefront, the wave has the same phase. The wavelength is the distance between two wavefronts having the same phase at any given instant. This distance must be measured perpendicular to the wave fronts along the line that represents the direction of travel.

WAVE POLARIZATION

A wave like that in Figure 1.3 is said to be *polarized* in the direction of the electric lines of force. The polarization here is vertical, because the electric lines are perpendicular to the surface of the Earth. If the electric lines of force are horizontal, the wave is said to be horizontally polarized. Horizontally and vertically polarized waves may be classified generally under *linear polarization*. Linear polarization can be anything between horizontal and vertical. In free space, "horizontal" and "vertical" have no meaning, since the reference of the seemingly horizontal surface of the Earth has been lost. *Instantaneous polarization* is the polarization of the wave at the stationary observer at a specific instant in time.

In many cases the polarization of waves is not fixed, but rotates continually, sometimes at random. When this occurs the wave is said to be *elliptically polarized*. A gradual shift of polarization in a medium is known as *Faraday rotation*. For space communication, *circular polarization* is commonly used to overcome the effects of Faraday rotation. A circularly polarized wave rotates its polarization through 360 degrees as it travels a distance of one wavelength in the propagation medium. The direction of rotation as viewed from the transmitting antenna defines the direction of circularity — right-hand (clockwise) or left-hand (counterclockwise). Linear and circular polarization may be considered as special cases of elliptical polarization.

IMPEDANCE OF FREE SPACE

Like all mediums in which electromagnetic waves travel, free space has a permittivity (ε_0) and permeability (μ_0). (The subscript 0 indicates free space.) These determine the speed at which the waves travel as shown in a previous section. The ratio of these two properties also determines the ratio of the E and H fields. Similarly to the ratio of voltage to current in a circuit, the ratio of E and H is the impedance of the medium in which the wave is traveling which is free-space in this case:

$$\frac{E}{H} = \sqrt{\frac{\mu_0}{\varepsilon_0}} = Z_0$$

The values of ε_0 and μ_0 result in a value of approximately 377 Ω for Z_0, the *characteristic impedance of free space*.

FIELD STRENGTH AND POWER DENSITY

A measurement of the strength of the wave at a distance from the transmitting antenna is its field strength (or *field intensity*). It is measured as the voltage between two points lying on an electric line of force in the plane of the wave front. The standard measure for field strength is the voltage developed in a wire that is 1 meter long, expressed as volts per meter. (If the wire were 2 meters long, the voltage developed would be divided by two to determine the field strength in volts per meter.)

The voltage in a wave is usually low so the measurement is made in millivolts or microvolts per meter. The voltage goes through time variations like those of the current that caused the wave. It is measured like any other ac voltage — in terms of the RMS value or, sometimes, the peak value. It is fortunate that amateurs do not customarily measure actual field strength as the equipment required is elaborate. We need to know only the effect of an adjustment so relative measurements are satisfactory. These can be made easily with home-built equipment.

The relationship between field intensity and power density is like that for voltage and power in ordinary circuits. They are related by the impedance of free space, which is approximately 377 Ω. A field intensity of 1 volt per meter is therefore equivalent to a power density of:

$$P = \frac{E^2}{Z} = \frac{1\ (\text{volt}/\text{m})^2}{377\ \Omega} = 2.65\ \text{mW}/\text{m}^2$$

The energy from a propagated wave decreases with distance from the source. This decrease in strength is caused by the spreading of the wave energy over ever-larger spherical surfaces as the distance from the source increases. In free space, the field intensity of the wave varies inversely with the distance from the source, once in the radiating far field of the antenna. If the field strength at 1 mile from the source is 100 millivolts per meter, it will be 50 millivolts per meter at 2 miles, and so on. Because of the relationship between voltage and power, the power density varies with the *square* of the field intensity, or inversely with the square of the distance. If the power density at 1 mile is 4 mW per square meter, then at a distance of 2 miles it will be 1 mW per square meter.

In practice, attenuation of the wave energy may be much greater than the inverse-distance law would indicate. The wave does not travel in a vacuum and the receiving antenna seldom is situated so there is a clear line of sight. The Earth is spherical and the waves do not penetrate its surface appreciably, so communication beyond visual distances must be by some means that will bend the waves around the curvature of the Earth. These means involve additional energy losses that increase the path attenuation with distance above that for the theoretical *free-space attenuation* or *spreading loss* in a vacuum. Additional causes of attenuation are discussed in the chapter on **Radio Wave Propagation**.

1.2 Basic Antenna Behavior

1.2.1 Reciprocity In Receiving And Transmitting

Many of the properties of an antenna used for reception are the same as its properties in transmission. It has the same directive pattern in both cases and delivers maximum signal to the receiver when the signal comes from a direction in which the antenna has its best response. The impedance of the antenna is the same, at the same point of measurement, in receiving as in transmitting. This is the principle of *reciprocity*.

1.2.2 Antenna Polarization

An antenna's polarization is defined to be that of its electric or E-field, in the direction where the field strength is maximum. For example, if a half-wavelength dipole is mounted horizontally over the Earth, the electric field is strongest perpendicular to its axis (that is, at right angles to the wire) and parallel to the Earth. Thus, since the maximum electric field is horizontal, the polarization in this case is also considered to be horizontal with respect to the Earth. If the dipole is mounted vertically, its polarization will be vertical. See **Figure 1.6**. Note that if an antenna is mounted in free space, there is no frame of reference and hence its polarization is indeterminate.

Antennas composed of several elements arranged so that their axes lie in the same or parallel directions have the same polarization as that of any one of the elements. For example, a system composed of a group of horizontal dipoles is horizontally polarized. If both horizontal and vertical elements are used in the same plane and radiate in phase, however, the polarization is the resultant of the contributions made by each set of elements to the total electromagnetic field at a given point some distance from the antenna. In such a case the resultant polarization is still linear but is tilted between horizontal and vertical.

In directions other than those where the radiation is maximum, the resultant wave, even for a simple dipole, is a combination of horizontally and vertically polarized components. The radiation off the ends of a horizontal dipole is actually vertically polarized, albeit at a greatly reduced amplitude compared to the broadside horizontally polarized radiation — the sense of polarization changes with compass direction.

CIRCULAR POLARIZATION

If vertical and horizontal elements in the same plane are fed out of phase (where the beginning of the RF period applied to the feed point of the vertical element is not in time phase with that applied to the horizontal), the resultant polarization is elliptical. See **Figure 1.7**. Circular polarization is a special case of elliptical polarization. The wave front of a circularly polarized signal appears (to a stationary observer) to rotate every 90° between vertical and horizontal, making a complete 360° rotation once every period. Circular polarization is frequently used for space communications and is discussed further in the chapter **Antennas for Space Communications**.

1.2.3 Near And Far Fields

The region very close to the antenna is called the *reactive near-field* region. The term "reactive" refers to the fact that the E and H fields are in quadrature (90 degrees out of phase) as are V and I in a reactance. Just as a reactance does not consume any real power, since E and H are 90 degrees out of phase there is no time-averaged power flow through space which is assumed to be resistive and doesn't add any phase shift of its own. In the reactive near field, the antenna stores energy as though it is a lumped-constant inductor or capacitor. (The near field is sometimes called the "induction field," meaning that this is the region in which energy is stored.)

The strength of the reactive near field decreases in a complicated fashion with increasing distance from the antenna. Beyond the reactive near field, the antenna's radiated field is divided into two other regions: the *radiating near field* and the *radiating far field*. Historically, the terms *Fresnel* and *Fraunhöfer* fields have been used for the radiating near and far fields, but these terms have been largely supplanted by the more descriptive terminology used here. Even inside the reactive near-field region, both radiating and reactive fields

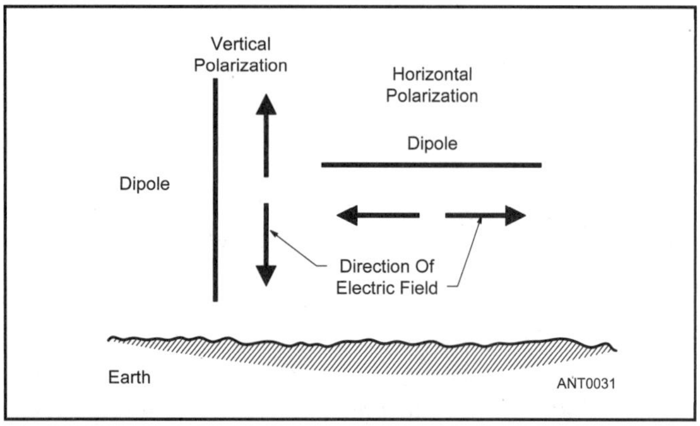

Figure 1.6 — Vertical and horizontal polarization of a dipole above ground. The direction of polarization is the orientation of the electric field in the direction of maximum field strength.

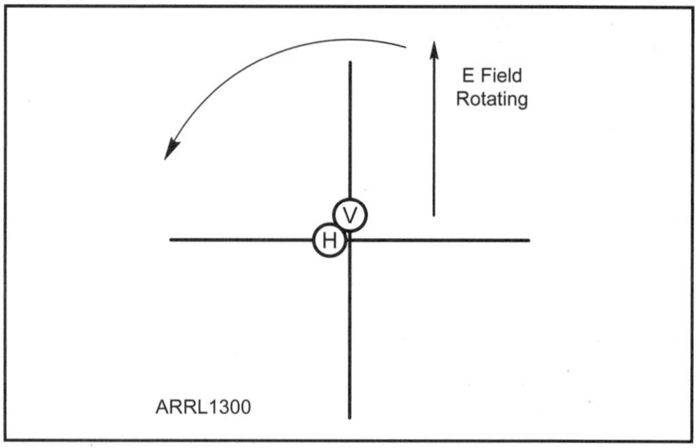

Figure 1.7 — Two crossed dipoles in the same plane and fed 90 degrees out of phase will produce a circularly polarized wave.

coexist although the reactive field predominates very close to the antenna.

As you get farther from the antenna (thus the name "far field"), the equation describing the fields shifts to a form in which the E and H fields are in phase so time-averaged power is not zero, and power flows away from the antenna as an electromagnetic wave. In the far field, the shape of the antenna's *radiation pattern* (see the sections "Antenna Directivity and Gain" and "Antenna Radiation Patterns") does not change with distance. The amplitude of the radiation changes but not the shape of the pattern.

Because the boundary between the fields is not a precise distance, experts debate where one field begins and another leaves off but the boundary between the radiating near and far fields is generally accepted as:

$$D = \frac{2L^2}{\lambda}$$

where L is the largest dimension of the physical antenna expressed in the same units of measurement as the wavelength, λ. Remember, many specialized antennas do not follow this equation's rule of thumb exactly. **Figure 1.8** depicts the three fields surrounding a simple dipole antenna. (An analysis of why this boundary is difficult to determine is provided in the article by Abdallah et al. in the Bibliography.)

Throughout the rest of this book, we will discuss mainly the radiating far fields, those forming the propagating electromagnetic waves and which will simply be referred to as the "far field." In the far field, intensity is inversely proportional to the distance, and that the electric and magnetic components, although perpendicular to each other in the wave front, are in phase as defined earlier. (Remember that power density is proportional to the square of intensity so it is inversely proportional to the square of distance.) The total energy is equally divided between the electric and magnetic fields. Beyond several wavelengths from the antenna these are the only fields we need to consider. For the reader interested in learning more about the various regions of the antenna field, see Sections 2.2.4 on "Field Regions" and 4.4 on "Region Separation" in *Antenna Theory* by Balanis (see the References and Bibliography).

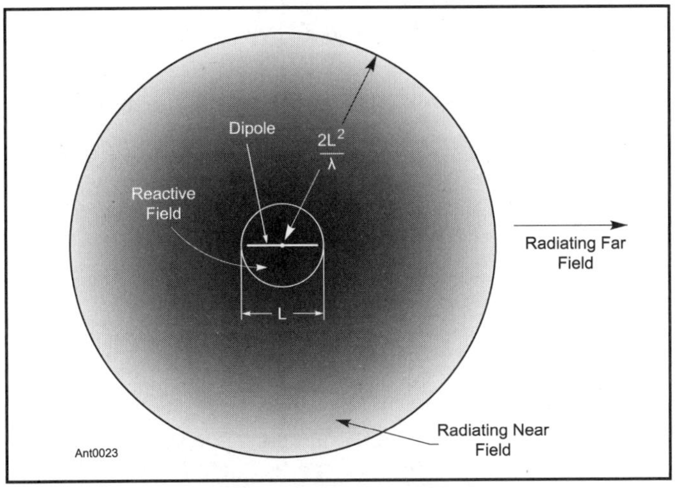

Figure 1.8 — The fields around a radiating antenna. Very close to the antenna, the reactive field dominates. Within this area mutual impedances are observed between the antenna and any other antennas or conductors. Outside the reactive near field, the radiating near field dominates up to the distance shown where L is the length of the largest dimension of the antenna. Beyond the near/far field boundary lies the radiating far field, where power density varies as the inverse square of radial distance.

1.2.4 Radiation Resistance and Efficiency

The power supplied to an antenna is dissipated in two ways: radiation of electromagnetic waves and heat losses in the wire and nearby conductors and dielectric material. The radiated power is what we want, the useful part, but it represents a form of "loss" just as much as the power used in heating the wire or nearby dielectrics is a loss. In either case, the dissipated power is equal to I^2R.

In the case of heat losses, R is a real resistance. In the case of radiation, however, R is a "virtual" resistance, which, if replaced with an actual resistor of the same value, would dissipate the power actually radiated from the antenna. This resistance is called the *radiation resistance* to the radiated electromagnetic wave, also called the *radiation reaction*. For the interested reader, a plain language explanation of radiation resistance is provided in Section 3.7 "Radiation Resistance and Efficiency" in Blake and Long's *Antennas* (see References and Bibliography). Rudy Severns, N6LF, has also contributed "Some Thoughts on Radiation Resistance, Parts 1 and 2" in the online material.

The total power in the antenna is therefore equal to $I^2(R_r+R)$, where R_r is the radiation resistance and R represents the total of all the loss resistances. Radiation efficiency is defined as:

$$\eta = \frac{R_r}{R_r + R}$$

where loss resistance (R) is calculated from (or normalized to) the same point at which R_r is determined. As R is reduced, the antenna's radiation efficiency increases.

It is important to understand that radiation resistance and the resistive part of feed point impedance are two different things. Radiation resistance is a property of the antenna and its environment, regardless of how power is applied. Feed point impedance varies with the location of the feed point and represents the ratio of applied voltage to the resulting current.

In antennas that are not electrically small compared to the signal wavelength, the power lost as heat in the conductor does not exceed a few percent of the total power supplied to the antenna. Expressed in decibels, the loss is less than 0.1 dB. The RF loss resistance of copper wire even as small as #14 AWG is very low compared with the radiation resistance of an antenna that is reasonably clear of surrounding objects and for which ground losses are low. You can therefore assume that the ohmic loss in a reasonably well located antenna is negligible and that the total resistance shown by the antenna (the feed point resistance) is radiation resistance. As a radiator of electromagnetic waves, such an antenna is a highly efficient device.

For antennas that are electrically small, incorporate loading or tuning circuits, are close to the ground (horizontal antennas), or rely on the ground as a path for current (vertical antennas), the power losses can be substantial. In these cases, such as small loops, mobile antennas, and vertical monopoles, it is important to reduce loss resistance wherever possible by using high-quality materials, ensuring that connections are secure and low-loss, and providing the lowest-loss current path practical.

1.2.5 Current And Voltage Distribution

When power is fed to an antenna, the current and voltage vary along its length. The current is minimum at the ends, regardless of the antenna's length. The current does not actually reach zero at the current minima, because of capacitance at the antenna ends. Insulators, loops at the antenna ends and support wires all contribute to this capacitance, which is called the *end effect*. The opposite is true of the RF voltage. That is, there is a voltage maximum at the ends.

In the case of a half-wave antenna there is a current maximum at the center and a voltage minimum at the center as illustrated in **Figure 1.9**. The curved lines in the figure are not waveforms and do not represent the phase relationship of voltage and current. The curves only show the amplitude of the voltage and current at any point along the antenna. This is called the antenna's voltage and current *distribution*. The pattern of alternating current and voltage minima ¼ wavelength apart repeats every ½ wavelength along a linear antenna as shown in Figure 1.6B. The phase of the current and voltage are inverted in each successive half-wavelength section.

1.2.6 Feed Point Impedance

Since amateurs are free to choose our operating frequencies within assigned bands we need to consider how the feed point impedance of a particular antenna varies with frequency within a particular band or even in several different bands if we intend to use one antenna on multiple bands. Impedance is simply the ratio of voltage to current. *Feed point impedance* is the ratio of voltage to current at the point where power is supplied to the antenna. There are two components of feed point impedance: *self impedance* and *mutual impedance*.

SELF IMPEDANCE

As you might expect, self impedance is what is measured at the feed point terminals of an antenna located completely free from the influence of any other conductors.

As you can see from Figure 1.9, self impedance is high for a feed point at the ends of a dipole where voltage is high and current is low. In the center of the dipole, feed point impedance is low, and it takes an intermediate value between those points. In the longer antenna of Figure 1.9B, there are several high-impedance and low-impedance points.

Where the current and voltage are exactly in phase, the self impedance is purely resistive with zero reactance and the antenna is *resonant*. (Amateurs often use the term "resonant" rather loosely, usually referring to a minimum value of SWR or an impedance of 50 Ω. Resonance has nothing to do with

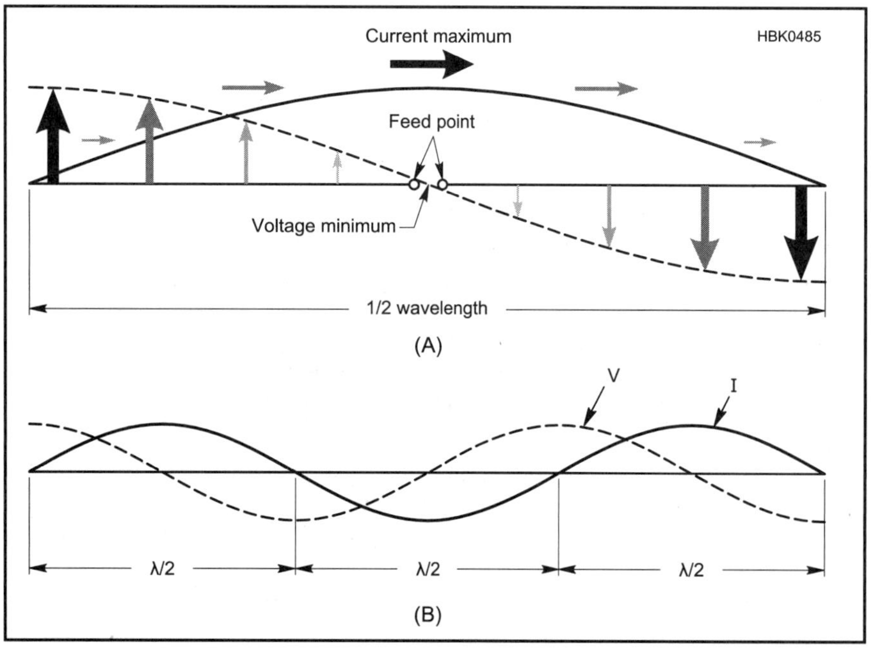

Figure 1.9 — The amplitude of voltage and current along a half-wavelength dipole (A) and along an antenna made from a series of half-wavelength sections. The curves show how voltage and current are distributed along the antenna.

Is Resonance Required?

An antenna need not be resonant in order to be an effective radiator. There is, in fact, nothing magic about having a resonant antenna, provided of course that you can devise some efficient means to feed the antenna. Many amateurs use non-resonant (even random-length) antennas fed with open-wire transmission lines and antenna tuners. They radiate signals just as well as those using coaxial cable and resonant antennas and as a bonus can usually be used on multiple frequency bands. It is important to consider an antenna and its feed line as a system in which all losses should be kept to a minimum.

the value of the impedance, only that it is purely resistive.)

Except at the one frequency where it is exactly resonant, the current in an antenna has a different phase compared to the applied voltage. In other words, the antenna exhibits a feed point impedance that is not just a pure resistance. The feed point impedance is composed of either capacitive or inductive reactance in series with a resistance.

MUTUAL IMPEDANCE

Mutual, or *coupled*, impedance is due to the parasitic effect of nearby conductors located within the antenna's reactive near field. This includes the effect of ground, which is a lossy conductor, but a conductor nonetheless. Mutual impedance is defined using Ohm's law, just like self impedance. However, mutual impedance is the ratio of voltage in one conductor, divided by the current in another (coupled) conductor. Mutually coupled conductors can distort the pattern of a highly directive antenna, as well as change the impedance seen at the feed point. Mutual impedance and coupling will be considered in detail in the chapters **Phased Arrays** and **HF Beam Antennas**.

1.3 Antenna Directivity and Gain

From IEEE Std 145-2013, IEEE Standard for Definitions of Terms for Antennas, the *directivity* of an antenna measures the degree to which the radiation emitted is concentrated in a single direction. It is the ratio of the radiation intensity in a given direction from the antenna to the radiation intensity averaged over all directions. It is the property of radiating or receiving more strongly in some directions than in others.

1.3.1 The Isotropic Radiator

Before we can fully describe practical antennas, we must first introduce a completely theoretical antenna, the *isotropic radiator*. Envision an infinitely small antenna at a point located in free space and completely removed from anything else around it. Then consider an infinitely small transmitter feeding this infinitely small, point-size antenna. You now have an isotropic radiator.

The uniquely useful property of this theoretical point-source antenna is that it radiates equally well in all directions. An isotropic antenna favors no direction at the expense of any other. In other words, it has absolutely no directivity. The isotropic radiator is useful for comparison with actual antenna systems.

You will find later that real, practical antennas all exhibit some degree of directivity. The radiation from a practical antenna never has the same intensity in all directions and may even have zero radiation in some directions. The fact that a practical antenna displays directivity (while an isotropic radiator does not) is usually desirable. The directivity of a real

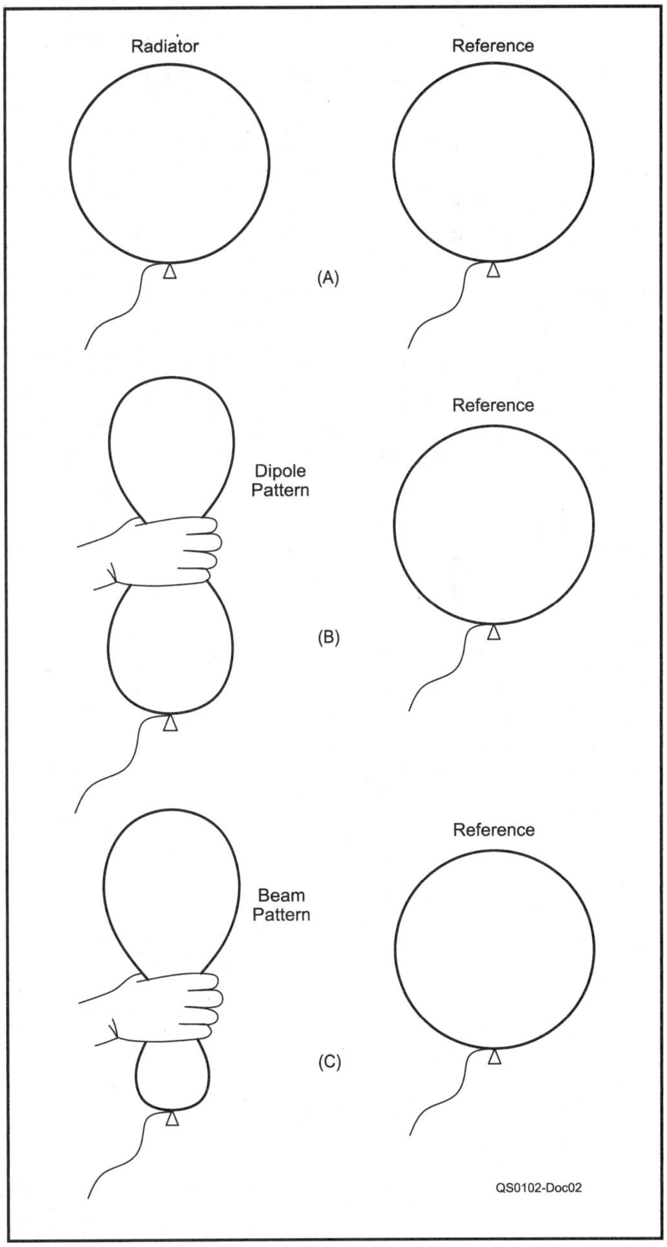

Figure 1.10 — Demonstrating antenna pattern gain with balloons. Take a balloon, blow it up so that it is roughly spherical in shape and declare that it is a radiation pattern from an isotropic radiator. Next, blow up another balloon to the same size and shape — this will be the "reference" antenna (A). Then, squeeze the first balloon in the middle to form a sort of figure-8 shape and a dipole pattern and compare the maximum size to that of the reference "antenna" (B). The dipole can be seen to have some "gain" over the reference isotropic. Next, squeeze the lower end of the dipole balloon to create a shape with more volume toward the top, demonstrating the sort of pattern a beam antenna creates (C).

antenna is often carefully tailored to emphasize radiation in particular directions. For example, a receiving antenna that favors certain directions can discriminate against interference or noise coming from other directions, thereby increasing the signal-to-noise ratio for desired signals coming from the favored direction.

1.3.2 Directivity And Gain

Let us now examine directivity more closely. As mentioned previously, all practical antennas, even the simplest types such as dipoles, exhibit directivity. Here's another picture that may help explain the concept of directivity. **Figure 1.10A** shows a balloon blown into a spherical shape. This represents a "reference" isotropic source described previously. Squeezing the balloon in the middle in Figure 1.10B produces a dipole-like figure-8 pattern with peak levels at top and bottom that are larger than the reference sphere. Compare this with Figure 1.10C. Next, squeezing the bottom end of the balloon produces a pattern that gives even more "gain" compared to the reference. Note that the volume of the balloon, representing the total amount of power radiated by the antenna, does not increase when squeezed — its shape is only rearranged.

Free-space directivity can be expressed quantitatively by comparing the three-dimensional pattern of the antenna under consideration with the perfectly spherical three-dimensional pattern of an isotropic antenna. For an isotropic antenna, the field strength (and thus power per unit area, or power density) is the same everywhere on the surface of an imaginary sphere having a radius of many wavelengths and centered on the antenna. For a directive antenna radiating the same total power as an isotropic antenna and surrounded by the same sphere, the directivity results in greater power density at some points on the sphere and less at others. The ratio of the maximum power density to the average power density taken over the entire sphere (which is the same as from the isotropic antenna under the specified conditions) is the numerical measure of the directivity of the antenna.

$$D = \frac{P}{P_{av}}$$

where

D = directivity

P = power density at its maximum point on the surface of the sphere

P_{av} = average power density

P_{av} is equivalent to the power density from a loss-free isotropic radiator with the same radiated power. D can be expressed in dB with respect to the directivity of a reference antennas as 10 log [D/D_{ref}].

GAIN

The *gain* of an antenna is related to its directivity, and directivity is related to the shape of the directive pattern. Because directivity is based solely on the shape of the directive pattern, it does not take into account any power losses that may occur in an actual antenna system. To determine gain, these losses must be subtracted from the power supplied to the antenna. The loss is normally a constant percentage of the power input, so the antenna gain is:

$$G = k\frac{P}{P_{av}} = kD$$

Introduction to the Decibel

The power gain and pattern measurements such as front-to-back ratio of an antenna system are usually expressed in decibels (dB). The decibel is a practical unit for measuring power ratios because it is more closely related to the actual effect produced at a distant receiver than the power ratio itself. One decibel represents a just-detectable change in signal strength, regardless of the actual value of the signal voltage. A 20-decibel (20 dB) increase in signal, for example, represents 20 observable steps in increased signal.

The power ratio 100 to 1 corresponds to 20 dB. The number of decibels corresponding to any power ratio is equal to 10 times the common logarithm of the power ratio, or

$$dB = 10 \log_{10} \frac{P_1}{P_2}$$

If the voltage ratio is given, the number of decibels is equal to 20 times the common logarithm of the ratio. That is,

$$dB = 20 \log_{10} \frac{V_1}{V_2}$$

When a voltage ratio is used, both voltages must be measured across the same value of impedance. Unless this is done the decibel figure is meaningless, because it is fundamentally a measure of a power ratio.

The main reason a decibel is used is that successive power gains expressed in decibels may simply be added together. Thus a gain of 3 dB followed by a gain of 6 dB gives a total gain of 9 dB. In ordinary power ratios, the ratios must be multiplied together to find the total gain.

A reduction in power is handled simply by subtracting the requisite number of decibels. Thus, reducing the power to ½ is the same as subtracting 3 decibels. For example, a power gain of 4 in one part of a system and a reduction to ½ in another part gives a total power gain of 4 × ½ = 2. In decibels, this is 6 – 3 = 3 dB. A power reduction or loss is simply indicated by including a negative sign in front of the appropriate number of decibels.

When P_2 and V_2 are some fixed reference value, a letter is added to "dB" to indicate "decibels with respect to" the reference value. This allows absolute values of power and voltage to be expressed in dB, as well. You will often encounter dBm (P_2 = 1 mW) and dBμV (V_2 = 1 μV) in amateur radio.

For more information about the decibel, read "A Tutorial on the Decibel" available at **arrl.org/files/file/A%20Tutorial%20on%20the%20Decibel%20-%20Version%202_1%20-%20Formatted.pdf**.

where

G = gain (expressed as a power ratio)

D = directivity

k = efficiency (power radiated divided by power input) of the antenna

P and P_{av} are as above

For many of the antenna systems used by amateurs, the efficiency is quite high (the loss amounts to only a few percent of the total). In such cases the gain is essentially equal to the directivity. The more the directive diagram is compressed — or, in common terminology, the sharper the lobe — the greater the power gain of the antenna. This is a natural consequence of the fact that as power is taken away from a larger and larger portion of the sphere surrounding the radiator, it is added to the volume represented by the narrow lobes. Power is therefore concentrated in some directions at the expense of others.

A simple approximation for gain over an isotropic radiator can be used, but only if the side lobes in the antenna's pattern are small compared to the main lobe and if the resistive losses in the antenna are small.

$$G \approx \frac{41,253}{H_{3dB} \times E_{3dB}}$$

Where H_{3dB} and E_{3dB} are the half-power points, in degrees, (i.e., the beamwidth, see below) for the H and E-plane patterns. When the radiation pattern is complex, numerical integration must be employed to give the actual gain.

BEAMWIDTH

Beamwidth is a measure of the width of the major lobe (or lobes) of the plotted pattern. The width is expressed in degrees at the half-power or –3 dB points. Beamwidth is often used as a general index of directivity and, therefore, gain. This information provides only a general idea of relative gain, rather than an exact measure. This is because an absolute measure involves knowing the power density at every point on the surface of a sphere, while a single diagram shows the pattern shape in only one plane of that sphere. It is customary to examine at least the E-plane and the H-plane patterns before making any comparisons between antennas.

dBi and dBd

Directivity and gain are usually expressed in decibels and are always calculated with reference to a *standard antenna* — usually an isotropic radiator or a dipole. Because the isotropic radiator has no directivity, meaning no variation of gain in any direction, it is an unambiguous standard and is frequently used as the comparison for gain measurements.

When the reference for directivity or gain is the isotropic radiator, they are expressed as dBi, with the "i" indicating "isotropic." When the reference is a dipole, dBd is used with the "d" indicating "dipole." The dipole's direction of maximum radiation, broadside to the antenna, is used as the reference. When the reference antenna is an isotropic antenna, then $D_{dBi} = 10 \log [D]$. If the reference antenna is a dipole, then $D_{dBd} = 10 \log [D / 1.64]$.

Because the dipole in free space has some gain compared to an isotropic radiator, the dipole's gain can be expressed as 2.15 dBi in its direction of maximum radiation. Gain in dBi can be converted to dBd by subtracting 2.15 dB. Gain in dBd can be converted to dBi by adding 2.15 dB.

FREE-SPACE GAIN

It is also important to know if the reference antenna gain was obtained with the antenna in free space. This is called *free-space gain*. If the antenna is located over ground or some other reflecting surface, the reflections will modify its radiation pattern. The reflected energy adds to the direct radiation, increasing the gain of the antenna in some directions. This additional gain is referred to as *ground gain* and can be as much as 6 dB at certain heights and elevation angles. Because the height above ground and the type of ground affect the amount of ground gain, the best comparison standard is either dBi or dBd in free space. For more information about ground reflections, see the chapters on **Effects of Ground** and **HF Antenna System Design**.

RECEIVE DIRECTIVITY FACTOR

Receive directivity factor (*RDF*) is the difference between an antenna's peak gain and gain averaged over all directions. (See the **Receiving Antennas** chapter and the entry for Ordy in the Bibliography.) RDF is measured in dB and is used to determine how well an antenna will reject noise or interference from directions away from the main lobe.

MEAN EFFECTIVE GAIN (dB-MEG)

dB-MEG is similar to RDF in that MEG measures gain in both the E- and H-plane over the whole hemisphere above ground or a ground plane. This provides an idea of how the antenna responds to signals coming in from all elevations and azimuths. It is defined as the ratio of average power received by the antenna to the sum of the average powers that would had been received by two reference antennas, vertically and horizontally polarized, respectively. MEG is most useful in VHF/UHF mobile environments where signals are heavily reflected and received with random polarizations. See the Reference entry under A. Glazunov for a more thorough explanation of MEG.

1.3.3 Antenna Aperture And Gain

An antenna's *effective area* defines its *effective aperture*. Aperture is a way of thinking about how much power an antenna receives from incoming signals in specific directions. It is defined as an area so you might think of an antenna's aperture as showing "how big" it is with respect to incoming radio waves.

Imagine a dipole antenna on the surface of sphere that surrounds a signal source. The sphere is so large that we can assume the waves arriving at the dipole from the source are flat plane waves with a certain power density, P_R, as defined earlier in W/m^2. Assuming the plane wave and the dipole have the same polarization, if the dipole receives some amount of power, P_0, in watts from the wave, its effective area or effective aperture, A_e, in square meters, is:

$$A_e = \frac{P_0}{P_R}$$

For example, say that the plane wave has a power density of 100 μW/m² at the antenna, which receives 10 μW. The antenna's effective area is 10 μW / 100 μW/m² = 0.1 m².

We cannot assume the aperture of an antenna is directly related to its physical dimensions. A linear wire antenna may have an effective aperture that is much greater than its very small physical area while a dish or horn antenna may have a slightly smaller effective aperture than its physical area.

Note that an antenna's aperture is only defined in a specific direction. Going back to the dipole, if it is rotated so that it is no longer aligned with the incoming wave's E field, it will receive less power and will have a smaller effective aperture. At the dipole's fundamental frequency, its maximum aperture will be broadside to the antenna but much smaller to either end.

Aperture also depends on frequency and wavelength. From Zavrel's explanation of aperture in Chapter 2 of *Antenna Physics: An Introduction* (see the References and Bibliography), "a 7 MHz dipole is considerably larger than a 144 MHz dipole. Intuitively we think that a much bigger structure will couple more power into…(or) from free space. If our intuition is correct, we should arrive at different aperture values for different frequencies." He goes on to show that the aperture of an antenna "*increases at a rate equal to the square of wavelength, or the reciprocal of the square of the frequency*" and that the aperture of an isotropic antenna is:

$$A_{iso} = \frac{\lambda^2}{4\pi}$$

Since gain is the ratio of power received by the antenna under consideration to an isotropic radiator, we have the fundamental equation relating gain and aperture:

$$G = \frac{P_0}{P_{iso}} = \frac{A_e}{A_{iso}} = \frac{4\pi A_e}{\lambda^2}$$

and 10 log (G) is gain in dBi.

Since antennas are reciprocal, they have the same aperture and therefore, gain, for receiving and transmitting are the same when certain conditions are met. One such condition is that both antennas must work into load impedances matched to their own impedances, so that maximum power is transferred in both cases. In the receiving case, the antenna is the source of power delivered to the receiver, rather than the load for a source of power (as in transmitting). Maximum possible output from the receiving antenna is obtained when the load to which the antenna is connected is the same as the feed point impedance of the antenna. We then say that the antenna is *matched* to its load.

In addition, the comparison antenna should be oriented so it gives maximum response to the signal used in the test. That is, it should have the same polarization as the incoming signal and should be placed so its direction of maximum gain is toward the signal source.

1.4 Antenna Radiation Patterns

The directivity of an antenna is directly related to the pattern of its radiated field intensity in free space. A graph showing the actual or relative field intensity at a fixed distance as a function of the direction from the antenna system, is called a *radiation pattern*. Since we can't actually see electromagnetic waves making up the radiation pattern of an antenna, we can consider an analogous situation.

Figure 1.11 represents a flashlight shining in a totally darkened room. To quantify what our eyes are seeing we might use a sensitive light meter like those used by photographers, with a scale graduated in units from 0 to 10. We place the meter directly in front of the flashlight and adjust the distance so the meter reads 10, exactly full scale. We also carefully note the distance. Then, always keeping the meter the same distance from the flashlight and keeping it at the same height above the floor, we move the light meter around the flashlight, as indicated by the arrow, and take

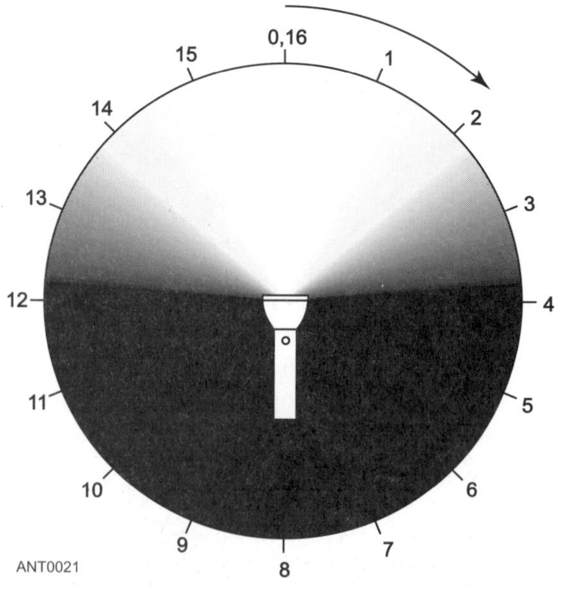

Figure 1.11 — The beam from a flashlight illuminates a totally darkened area a shown here. Readings taken with a photographic light meter at the 16 points around the circle may be used to plot the radiation pattern of the flashlight.

light readings at a number of different positions.

After all the readings have been taken and recorded, we plot those values on a sheet of polar graph paper, like that shown in **Figure 1.12**. The values read on the meter are plotted at an angular position corresponding to that at which each meter reading was taken. Following this, we connect the plotted points with a smooth curve, also shown in the figure. When this is finished, we have completed a radiation pattern for the flashlight.

Antenna radiation patterns can be constructed in a similar manner. Power is fed to the antenna under test and a field-strength meter indicates the amount of signal. We might wish to rotate the antenna under test, rather than moving the measuring equipment to numerous positions about the antenna. Since the pattern while receiving is the same as that while transmitting (see the section on Reciprocity later in this chapter), a source antenna fed by a low-power transmitter illuminates the antenna under test and the signal intercepted by the antenna under test is fed to a receiver and measuring equipment. Additional information on the mechanics of measuring antenna patterns is contained in the chapter **Antenna and Transmission Line Measurements**.

Some precautions must be taken to assure that the measurements are accurate and repeatable and one of the most important is to prevent mutual coupling between the source and receiving antennas that may alter the pattern you are trying to measure. This sort of mutual coupling can occur in the near field very close to the antenna under test and can be inductive or capacitive in nature. Later, in the chapters dealing with arrays of antennas, you will find that mutual coupling between elements can be put to good use to purposely shape the radiated pattern. When making pattern measurements, however, for accurate measurement of radiation patterns we must place our measuring instrumentation at least several wavelengths away from the antenna under test.

1.4.1 Types Of Radiation Patterns

In the example of the flashlight, the plane of measurement was at one consistent height above the floor. In **Figure 1.13A**

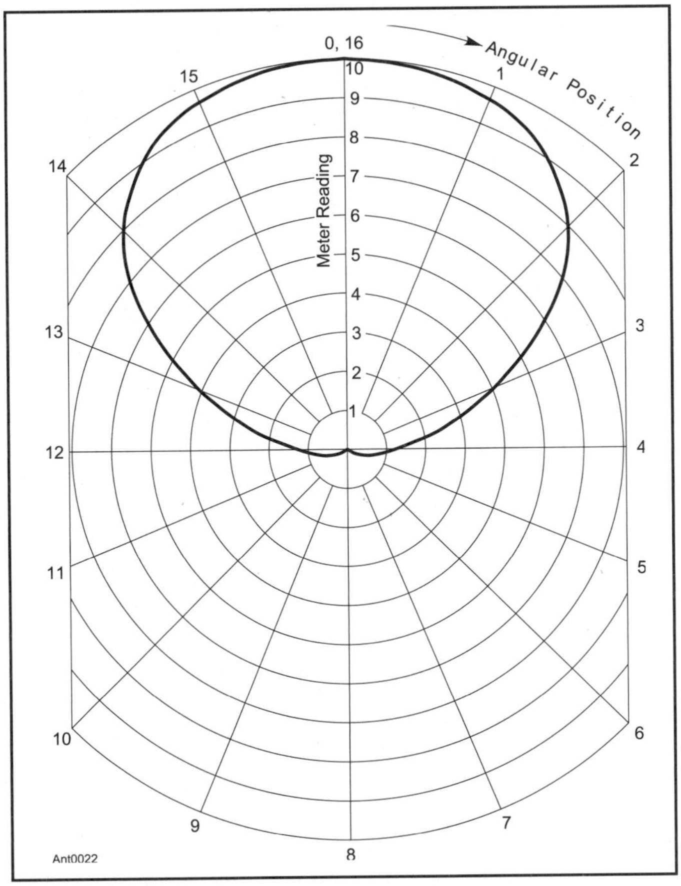

Figure 1.12 — The radiation pattern of the flashlight in Figure 1.7. The measured values are plotted and connected with a smooth curve.

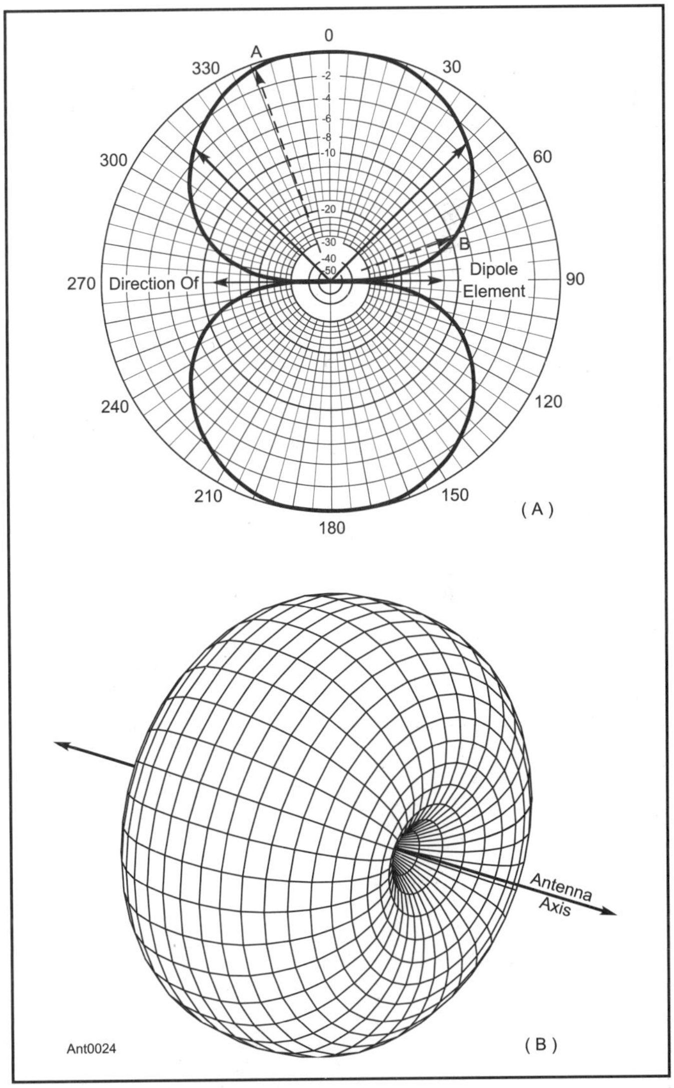

Figure 1.13 — Radiation patterns of a half-wavelength dipole in free-space. At A, the pattern in the plane containing the wire axis. The length of each dashed-line arrow represents the relative field strength in that direction, referenced to the direction of maximum radiation at right angles to the wire's axis. The arrows at approximately 45 degrees and 315 degrees are the half-power or –3 dB points. At B, a wire grid representation of the solid pattern for the same antenna.

a similar radiation pattern is shown for a half-wavelength dipole (see the **Dipoles and Monopoles** chapter) in free space, measured in a single plane containing the antenna wire. The antenna is located at the exact center of the plot with its orientation specified by the two-headed arrow. The antenna radiates best broadside to the wire axis and hardly at all off the ends of the wire.

Radiation patterns are graphic representations of an antenna's directivity. Shown in polar coordinates (see the math tutorial reference in the online information for this book for information about polar coordinates), the angular scale shows direction and the scale from the center of the plot to the outer ring. The smooth line in the shape of a figure 8 shows the relative strength of the antenna's radiated signal at each angle.

The pattern in **Figure 1.14** shows both *nulls* (angles at which a pattern minimum occurs) and *lobes* (radiation at angles between nulls). The *main lobe* is the lobe with the highest amplitude unless noted otherwise and unless several plots are being compared, the peak amplitude of the main lobe is placed at the outer ring as a reference point. The peak of the main lobe can be located at any angle. All other lobes are *side lobes* which can be at any angle, including to the rear of the antenna. In addition to the labels showing the main lobe and nulls in the pattern, the so-called *half-power* points on the main lobe are shown. These are the angles at which the power is one-half of the peak value in the main lobe.

Actually, the pattern for any antenna is three-dimensional and therefore cannot be represented by a single-plane drawing. The total radiation pattern of an antenna in free space would be found by measuring the field strength at every point on the surface of an imaginary sphere having the antenna at its center. The information so obtained would then be used to construct a solid figure, where the distance from a fixed point (representing the antenna) to the surface of the figure is proportional to the field strength from the antenna in any given direction. Figure 1.13B shows a three-dimensional solid representation of the radiation pattern of a half-wave dipole.

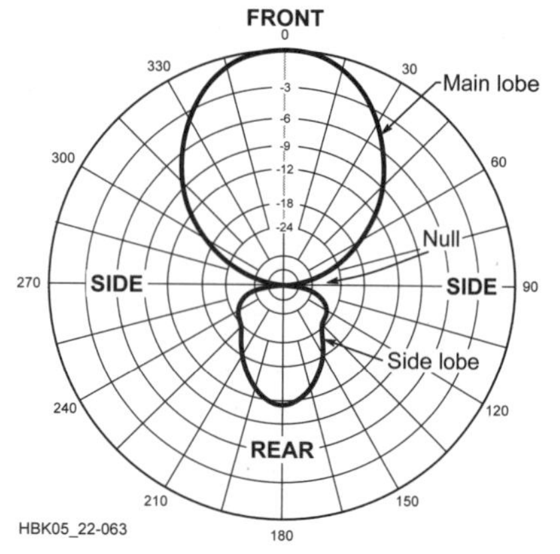

Figure 1.14 — Azimuthal pattern of a typical three-element Yagi beam antenna in free space. The Yagi's boom is aligned along the 0 degrees to 180 degrees axis and the beam's elements are in the plane of the pattern.

Figure 1.13A can be thought of as a cross-section of the solid pattern through the axis of the antenna. Two such diagrams, one in the plane containing the straight wire of a dipole and one in the plane perpendicular to the wire, can convey a great deal of information. After a little practice and with the exercise of some imagination, the complete solid pattern can be visualized with fair accuracy from inspection of the two planar diagrams, provided of course that the solid pattern of the antenna is smooth such as for simple antennas like the dipole of Figure 1.13.

AZIMUTH AND ELEVATION PATTERNS

When a radiation pattern is shown for an antenna mounted over ground rather than in free space, we automatically gain two frames of reference: *an azimuth angle* and an *elevation angle*. The azimuth angle is usually referenced to 0 degrees in the direction of maximum radiation from the antenna or it could be referenced to True North for an antenna oriented in a particular compass direction.

The elevation angle is referenced to the horizon at the Earth's surface, where the elevation angle is 0 degrees. Of course, the Earth is round but because its radius is so large, it can in this context be considered to be flat in the area directly under the antenna. An elevation angle of 90 degrees is directly above the antenna (the *zenith*) and the angles then reduce back to 0 degrees toward the horizon directly behind the antenna. (Professional antenna engineers often describe an antenna's orientation with respect to the point directly overhead — using the zenith angle, rather than the elevation angle. The elevation angle is computed by subtracting the zenith angle from 90 degrees.)

Figure 1.14 is an *azimuthal* or *azimuth pattern* that shows the antenna's gain in all horizontal directions (azimuths) around the antenna. As with a map, 0 degrees is at the top and bearing angle increases clockwise. (This is different from polar plots generated for mathematical functions in which 0 degrees is at the right and the angle increases counter-clockwise.)

Figure 1.15 is an *elevation pattern* that shows the same antenna's directivity but this time at all vertical angles. In this case, the horizon at 0 degrees is located to both sides of the antenna and the zenith (directly overhead) at 90 degrees. The plot shown in Figure 1.15 assumes the presence of ground

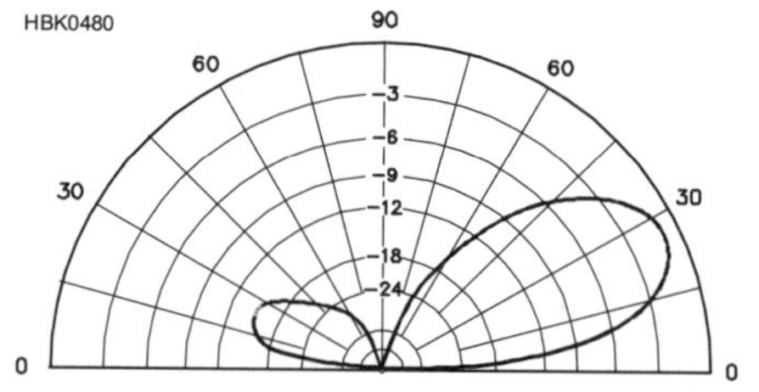

Figure 1.15 — Elevation pattern of a three-element Yagi beam antenna placed ½ wavelength above ground. The Yagi's boom lies on the 0 degree – 0 degree axis and the elements are perpendicular to the page on the same axis.

(the horizontal axis from 0 degrees to 0 degrees). The ground reflects or blocks radiation at negative elevation angles, making below-surface radiation plots unnecessary. In free space, the plot would include the missing semicircle with –90 degrees at the bottom. Without the ground reference, the term "elevation" has little meaning, however.

For amateur work, relative values of field strength (rather than absolute values) are quite adequate in pattern plotting. In other words, it is not necessary to know how many microvolts per meter a particular antenna will produce at a distance of 1 mile when excited with a specified power level. (This is the kind of specification that AM broadcast stations must meet to certify their antenna systems to the FCC.)

Regardless of whether the data is collected by measurements, simulated by computer software, or calculated from theoretical equations, it is common to normalize the plotted values so the field strength in the direction of maximum radiation coincides with the outer edge of the chart. That way, on a given system of polar coordinate scales the shape of the pattern is not altered by proper normalization, only its size. (See the section "Coordinate Scales for Radiation Patterns" later in this chapter for information about how radiation pattern scales are determined.)

E- AND H-PLANE PATTERNS

You'll also encounter *E-plane* and *H-plane radiation patterns*. These show the antenna's radiation pattern in the plane parallel to the E field or H field of the antenna. For antennas with horizontal elements, the E field is in the horizontal plane, so the E-plane radiation pattern is the same as an azimuthal pattern in the plane of the antenna. The H field is perpendicular to the E field, so the H-plane pattern is in a plane perpendicular to the E-plane pattern. If the E-plane pattern is an azimuthal pattern, then the H-plane pattern will be an elevation pattern.

It's important to remember that the E plane and H plane do not have a fixed relationship to the Earth's surface. For example, the E-plane pattern from a horizontal dipole is an azimuthal pattern but if the same dipole is oriented vertically, the E-plane pattern becomes an elevation pattern. For this reason, most E- and H-plane radiation patterns are created with the antenna in free space.

POLAR COORDINATES

It is often helpful to consider the radiation pattern from an antenna in terms of polar coordinates, rather than trying to think in purely linear horizontal or vertical coordinates. The reference axis in the polar system shown in **Figure 1.16** is vertical to the Earth under the antenna. The zenith angle is usually referred to as θ (Greek letter theta) and the azimuth angle is referred to as ϕ (Greek letter phi). Instead of zenith angles, most amateurs are more familiar with elevation angles, where a zenith angle of 0 degrees is the same as an elevation angle of 90 degrees, straight overhead.

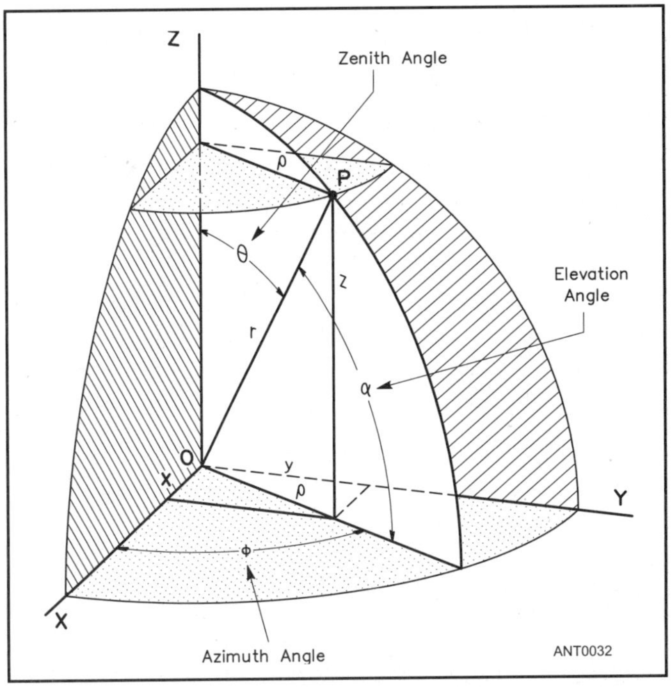

Figure 1.16 — Diagram showing polar representation of a point P lying on an imaginary sphere surrounding a point-source antenna. The various angles associated with the coordinate system are shown referenced to the x, y, and z-axes.

COORDINATE SCALE SYSTEMS FOR RADIATION PATTERNS

A number of different coordinate scales or grid systems are in use for plotting antenna patterns. Polar coordinate systems use an amplitude scale that is one of three types: linear, logarithmic, and modified logarithmic. Azimuth or elevation angle is shown directly on the outer circle. Cartesian coordinates or rectangular grid systems generally use a logarithmic vertical scale (dB) with angle (azimuth or elevation) in degrees on the horizontal scale.

A very important point to remember is that the shape of a pattern (its general appearance) is highly dependent on the system used for the plotting. This is illustrated in **Figure 1.17**, where the radiation pattern for a beam antenna is presented using three polar coordinate systems discussed in the paragraphs that follow.

Linear Coordinate Systems

The polar coordinate system in Figure 1.17A uses a linear amplitude scale. Equally spaced divisions could be used but to correspond to the usual signal level definitions, concentric circles are spaced to represent a decibel response, referenced to 0 dB (maximum amplitude) at the outer edge of the plot. In these plots the minor lobes with peaks more than 15 dB or so below the main lobe disappear completely because of their small size. Linear coordinate scales are not common.

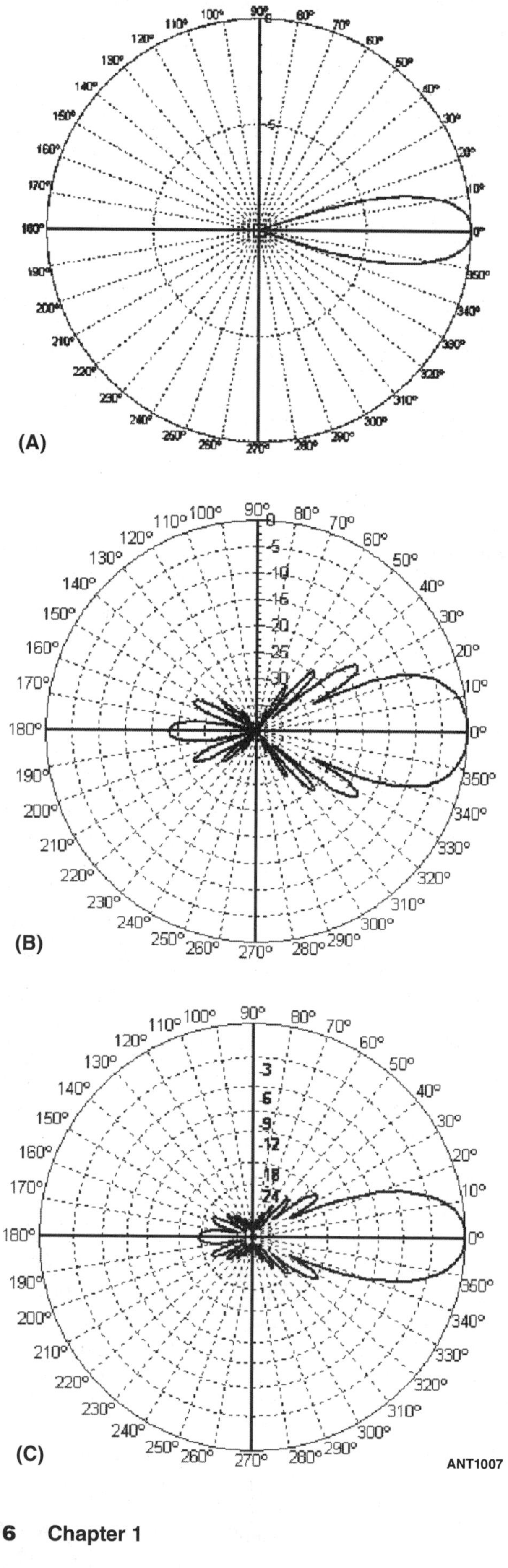

Logarithmic Coordinate System

Another coordinate system is the logarithmic scale, where the concentric grid lines are spaced according to the logarithm of the signal amplitude. See Figure 1.17B. The scale circles are graduated in steps of a consistent dB value. In that sense, the logarithmic grid might be termed a linear-log grid, one having linear divisions calibrated in decibels.

This grid enhances the appearance of the minor lobes. If the intent is to show the radiation pattern of an array supposedly having an omnidirectional response, this scale will also enhance that appearance. An antenna having a difference of 8 or 10 dB in pattern response around the compass appears to be closer to omnidirectional on this scale than on any of the others.

ARRL Log Coordinate System

The modified logarithmic scale in Figure 1.17C used by the ARRL has a system of concentric circles spaced according to the logarithm of 0.89 times the value of the signal voltage. In this grid, minor lobes that are 30 and 40 dB down from the main lobe are distinguishable. Such lobes are of concern in VHF and UHF work. The spacing between plotted points at 0 dB and –3 dB is significantly greater than the spacing between –20 and –23 dB, which in turn is significantly greater than the spacing between –50 and –53 dB.

For example, the scale distance covered by 0 to –3 dB is about 1/10 of the radius of the chart. The scale distance for the next 3-dB increment (to –6 dB) is slightly less, 89% of the first, to be exact. The scale distance for the next 3-dB increment (to –9 dB) is again 89% of the second. The scale is thus constructed so that the innermost two circles represent –36 and –48 dB and the chart center represents –100 dB. This spacing corresponds generally to the relative significance of such changes in antenna performance.

To construct this pattern in software, the distance of a gain point from the center of the circle = radius × $0.89^{(-g/3)}$, where radius is the radius of the pattern's outer ring and g is the normalized gain in dB (g = 0 dB when g = maximum gain). When g is an integer multiple, n, of –3 dB, the distance is 0.89n, which is equivalent to n times 89%. (Thanks to VA2EW for this formula.)

Antenna pattern plots in this book are generally made on the modified-log scale similar to that shown in Figure 1.17C. Patterns plotted by modeling software such as *EZNEC* may use a similar but somewhat different scale.

Figure 1.17 — Radiation pattern plots for a high-gain Yagi antenna on different grid coordinate systems. At A, the pattern on a linear-power dB grid. Notice how details of side lobe structure are lost with this grid. At B, the same pattern on a grid with constant 5 dB circles. The side lobe level is exaggerated when this scale is employed. At C, the same pattern on the modified log grid used by ARRL. The side and rearward lobes are clearly visible on this grid. The concentric circles in all three grids are graduated in decibels referenced to 0 dB at the outer edge of the chart. The patterns look quite different, yet they all represent the same antenna response!

Rectangular Coordinate Systems

While polar coordinates show the direction of radiation pattern features directly as azimuth angles, a rectangular grid or cartesian coordinates as in **Figure 1.18** makes it easier to display low-level minor lobes and other pattern structure. Multiple patterns in polar coordinates can be difficult to compare, particularly at low amplitudes close to the center of the plot. The vertical axis shows signal strength on a logarithmic scale calibrated in dB and the azimuth or elevation angle on the horizontal axis. The rectangular grid also makes comparison of antenna patterns much easier since the low-level features are not compressed into the area around the origin. Rectangular grids are very common in professional literature.

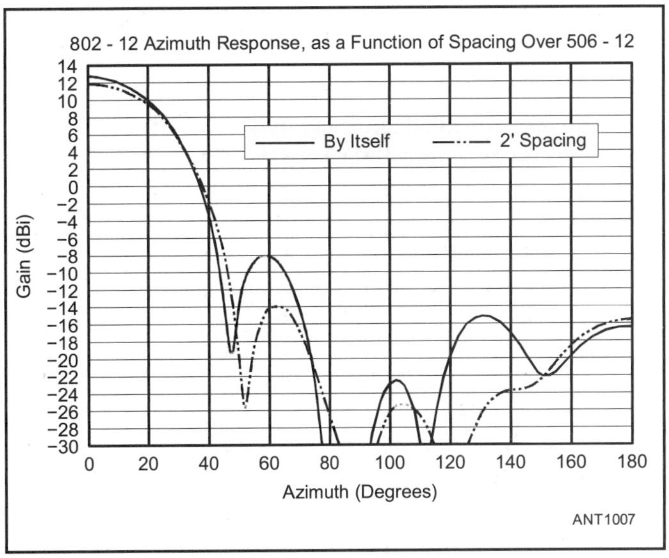

Figure 1.18 — The rectangular azimuthal patterns of two VHF Yagi antennas. The vertical axis is calibrated in dB and the horizontal axis shows azimuth angle in degrees. This example shows how a rectangular plot allows easier comparison of minor lobes and structure in the antenna patterns away from the main lobe.

1.4.2 Radiation Pattern Metrics

Given the basic radiation pattern and scales, it becomes easy to define several useful measurements or metrics by which antennas are compared by using their azimuthal patterns. Beamwidth is such a measurement and was defined in a previous section.

To reiterate from the previous sections, because an isotropic antenna has equal gain in all directions it is often used as a reference for gain measurements. Gain with respect to an isotropic antenna is stated as dBi. Gain with respect to a dipole's maximum gain, broadside to the antenna is stated as dBd. The dipole's maximum gain is 2.15 dB greater than that of the isotropic antenna. To convert from gain given as dBi to dBd, subtract 2.15 dB. Note that to specify gain in dBd, the dipole must be at the same effective height as the antenna being specified. Alternately, the free-space values for gain could be used. Be sure to state clearly which set of values are used.

Next to gain, the most commonly used metric for directional antennas is the *front-to-back ratio (F/B)* or just "front-to-back." This is the difference in dB between the antenna's gain in the specified "forward" direction and in the opposite direction. The front-to-back ratio of the antenna in Figure 1.14 is about 11 dB. *Front-to-side ratio* is also used and is the difference between the antenna's "forward" gain and gain at right angles to the forward direction. This assumes the radiation pattern is symmetric and is of most use to antennas such as Yagis and quads that have elements arranged in parallel planes. The front-to-side ratio of the antenna in Figure 1.14 is more than 30 dB. Because the antenna's rearward pattern can have large amplitude variations, the *front-to-rear ratio* is sometimes used. Front-to-rear uses the average of rearward gain over a specified angle, usually the 180-degree semicircle opposite the direction of the antenna's maximum gain, instead of a single gain figure at precisely 180 degrees from the forward direction.

In Figure 1.14, the antenna's beamwidth is about 54 degrees, since the pattern crosses the –3 dB gain scale approximately 27 degrees to either side of the peak direction. Antenna patterns with comparatively small beamwidths are referred to as "sharp" or "narrow."

An antenna with an azimuthal pattern that shows equal gain for all azimuths is called *omnidirectional.* This is not the same as an isotropic antenna that has equal gain in all directions, both azimuth and elevation.

1.5 Other Antenna Characteristics

1.5.1 Antenna Bandwidth

The *bandwidth* of an antenna refers generally to the range of frequencies over which the antenna can be used to obtain a specified level of performance. The bandwidth can be specified in units of frequency (MHz or kHz) or as a percentage of the antenna's design frequency.

Popular amateur usage of the term antenna bandwidth most often refers to the 2:1 SWR (standing wave ratio) bandwidth, such as, "The 2:1 *SWR bandwidth* is 3.5 to 3.8 MHz" or "The antenna has a 10% SWR bandwidth" or "On 20 meters, the antenna has an SWR bandwidth of 200 kHz." (Standing wave ratio or SWR is discussed in the **Transmission Lines** chapter.) Other specific bandwidth terms are also used, such as the *gain bandwidth* (the bandwidth over which gain is greater than a specified level) and the *front-to-back ratio bandwidth* (the bandwidth over which front-to-back ratio is greater than a specified level).

As operating frequency is lowered, an equivalent bandwidth in percentage becomes narrower in terms of frequency range in kHz or MHz. For example, a 5% bandwidth at 21 MHz is 1.05 MHz (more than wide enough to cover the whole band) but at 3.75 MHz only 187.5 kHz! Because of the wide percentage bandwidth of the lower frequency bands (160 meters is 10.5% wide, 80 meters is 13.4% wide) it is difficult to design an antenna with enough bandwidth to cover the whole band.

It is important to recognize that SWR bandwidth does not always relate directly to gain bandwidth. Depending on the amount of feed line loss, an 80-meter dipole with a relatively narrow 2:1 SWR bandwidth can still radiate a good signal at each end of the band, provided that an antenna tuner is used to allow the transmitter to load properly. Broadbanding techniques, such as fanning the far ends of a dipole to simulate a conical type of dipole, can help broaden the SWR bandwidth.

Q OF ANTENNAS

As with circuits, antennas can also be considered to have Q which affects their SWR bandwidth. In an antenna, Q is a measure of how much energy is stored in the near field compared to how much is radiated during each cycle. Higher values of Q can result from low radiation resistance, as is the case for electrically small antennas, and result in reduced antenna bandwidth.

High-Q antennas often have high voltage and currents due to the high stored energy, just as high-Q LC circuits have high voltages and circulating currents. This is discussed in detail in the paper by Yaghjian and Best, "Impedance, Bandwidth and Q of Antennas," listed in the References and Bibliography section of this chapter. Because of the high stored energy, high-Q antennas can create hazardous E and H field strengths in their near field. When using power levels of more than a few watts with a high-Q antenna, be cautious about people being close to the antenna.

1.5.2 Antenna Factor

When measuring field strength for RF exposure or EMI testing, it is important to understand that the voltage measured at the output terminals of an antenna is not the actual field intensity due to actual antenna gain, aperture characteristics, and loading effects. To convert between field strength and output voltage, *antenna factor* (*AF*) is used.

Antenna factor is defined as the ratio of the electric field, E, in V/m or µV/m, to the voltage, V, in V or µV, generated by the field across the terminals of an antenna. Antenna factor has units of m^{-1} or "per meter" and is calculated as AF = E / V. (E and V are RMS values, not instantaneous values.) If using dB units, AF = E (in dBV/m) – V (in dBV). dBµV/m and dBµV can also be used but both quantities must use the same units. To measure the magnetic field, H, the field strength is in units of A/m and the resulting antenna factor is in units of A/(V-m).

For a 50 Ω load on the antenna:

$$AF = \frac{\sqrt{377 P_D}}{\sqrt{50 P_D A_e}} = \frac{9.73}{\lambda \sqrt{G}}$$

where

$A_e = (\lambda^2 G)/4\pi$ = the antenna's effective aperture in meters

P_D = Power density in W/m^2 at the antenna

G = the antenna gain (a linear ratio, not dB)

λ = wavelength in meters

1.5.3 Frequency Scaling

Any antenna design can be scaled in size for use on another frequency or on another amateur band. The dimensions of the antenna may be scaled as follows:

$$D = \frac{f1}{f2} \times d$$

where

D = scaled dimension

d = original design dimension

f1 = original design frequency

f2 = scaled frequency (frequency of intended operation)

From this equation, a published antenna design for, say, 14 MHz can be scaled in size and constructed for operation on 18 MHz, or any other desired band. Similarly, an antenna design could be developed experimentally at VHF or UHF and then scaled for operation in one of the HF bands. For example, an element of 39.0 inches length at 144 MHz would be scaled to 14 MHz as follows: D = 144/14 × 39 = 401.1 inches, or 33.43 feet.

To scale an antenna properly, all physical dimensions must be scaled, including element lengths, element spacings, boom diameters and element diameters.

Lengths and spacings may be scaled in a straightforward manner as in the above example, but element diameters are often not as conveniently scaled. For example, assume a 14 MHz antenna is modeled at 144 MHz and perfected with ⅜-inch cylindrical elements. For proper scaling to 14 MHz, the elements should be cylindrical, of 144/14 × ⅜ or 3.86 inches diameter. From a realistic standpoint, a 4-inch diameter might be acceptable, but cylindrical elements of 4-inch diameter in lengths of 33 feet or so would be quite unwieldy (and quite expensive, not to mention heavy). Choosing another, more suitable diameter is the only practical answer and will result in changes to the antenna's performance. To correct for the different electrical sizes, the design will have to be remodeled or recalculated and adjusted, using actual sizes.

DIAMETER SCALING

Simply changing the diameter of dipole type elements during the scaling process is not satisfactory without making a corresponding element-length correction. This is because changing the diameter results in a change in the length/diameter (l/d) ratio from the original design, and this alters the corresponding resonant frequency of the element. The element length must be corrected to compensate for the effect of the different diameter actually used.

To be more precise, however, the purpose of diameter scaling is not to maintain the same resonant frequency for the element, but to maintain the same ratio of self-resistance to self-reactance at the operating frequency — that is, the Q of the scaled element should be the same as that of the original element. This is not always possible to achieve exactly for elements that use several telescoping sections of tubing.

TAPERED ELEMENTS

Rotatable beam antennas are usually constructed with elements made of metal tubing. The general practice at HF is to taper the elements with lengths of telescoping tubing. The center section has a large diameter, but the ends are relatively small. This reduces not only the weight, but also the cost of materials for the elements. Tapering of HF Yagi elements is discussed in detail in the chapter on **HF Beam Antennas**.

LENGTH CORRECTION FOR TAPERED ELEMENTS

The effect of tapering an element is to alter its electrical length. Two elements of the same length, one cylindrical and one tapered but with the same average diameter as the cylindrical element, will not be resonant at the same frequency. The tapered element must be made longer than the cylindrical element for the same resonant frequency.

A procedure for calculating the length for tapered elements has been worked out by Dave Leeson, W6NL, from work done by Schelkunoff at Bell Labs and is presented in Leeson's book, *Physical Design of Yagi Antennas*. On the *ARRL Antenna Book* website is a subroutine called *EFFLEN.FOR*. It is written in Fortran and is used in the *SCALE* program to compute the effective length of a tapered element. The algorithm uses the Leeson-Schelkunoff algorithm and is commented step-by-step to show what is happening. Calculations are made for only one half of an element, assuming the element is symmetrical about the point of boom attachment.

Also, read the documentation in SCALE.PDF for the *SCALE* program, which will automatically do the complex mathematics to scale a Yagi design from one frequency to another, or from one taper schedule to another. (Both *SCALE* and *EFFLEN.FOR* are available for download from **arrl.org/antenna-book-reference**.)

1.5.4 Effective Radiated Power (Erp)

When evaluating total station performance, accounting for the effects of the entire system is important, including antenna gain. This allows you to evaluate the effects of changes to the station. Transmitting performance is usually computed as *effective radiated power* (*ERP*). ERP is calculated with respect to a reference antenna system — usually a dipole but occasionally an isotropic antenna — and answers the question, "How much power does my station radiate as compared to that if my antenna was a simple dipole?" Effective isotropic radiated power (EIRP) results when an isotropic antenna is used as the reference. If no antenna reference is specified, assume that ERP uses a dipole as the reference antenna. (The abbreviation EDRP is sometimes used to clarify that a dipole is the reference.) ERP is most often used by amateurs when working with repeater stations, so the examples here show typical calculations in that setting.

ERP calculations begin with the *transmitter power output* (*TPO*). (This is assumed to be the output of the final power amplification stage if an external power amplifier is used.) Then the *system gain* of the entire antenna system including the antenna, the transmission line, and all transmission line components is applied to TPO to compute the entire station's output power.

System Gain = Transmission Line Loss – Transmission Components Loss + Antenna Gain

There is always some power lost in the feed line and often there are other devices inserted in the line, such as a filter or an impedance-matching network. In the case of a repeater system, there is usually a duplexer so the transmitter and receiver can use the same antenna and perhaps a circulator to reduce the possibility of intermodulation interference. These devices also introduce some loss to the system. The antenna system then usually returns some gain to the system. (See the **Transmission Lines** chapter for information on feed line loss.)

ERP = TPO × System Gain

Since the system gains and losses are usually expressed in decibels, they can simply be added together, with losses written as negative values. System gain must then be converted back to a linear value from dB to calculate ERP.

$$\mathrm{ERP} = \mathrm{TPO} \times \log^{-1}\left(\frac{\text{System Gain (dB)}}{10}\right)$$

It is also common to work entirely in dBm and dB until the final result for ERP is obtained and then converted back to watts. (dBm represents "decibels with respect to one mW" such that 0 dBm = 1 mW, 30 dBm = 1 W, and so forth.)

ERP (in dBm) = TPO (in dBm) + System Gain (in dB)

Suppose we have a repeater station that uses a 50 W transmitter and a feed line with 4 dB of loss. There is a duplexer in the line that exhibits 2 dB of loss and a circulator that adds another 1 dB of loss. This repeater uses an antenna that has a gain of 6 dBd. Our total system gain looks like:

System gain = –4 dB + –2 dB + –1 dB + 6 dBd = –1 dB

Note that this is a loss of 1 dB total for the system from TPO to radiated power. The effect on the 50 W of TPO results in:

$$\mathrm{ERP} = 50\ \mathrm{W} \times \log^{-1}\left(\frac{\text{system gain (dB)}}{10}\right) =$$

$$50 \times \log^{-1}(-0.1) = 50 \times 0.79 = 39.7\ \mathrm{W}$$

This is consistent with the expectation that with a 1 dB system loss we would have somewhat less ERP than transmitter output power.

As another example, suppose we have a transmitter that feeds a 100 W output signal into a feed line that has 1 dB of loss. The feed line connects to an antenna that has a gain of 6 dBd. What is the effective radiated power from the antenna? To calculate the total system gain (or loss) we add the decibel values given:

System gain = – 1 dB + 6 dBd = 5 dB

and

$$\mathrm{ERP} = 100\ \mathrm{W} \times \log^{-1}\left(\frac{\text{system gain (dB)}}{10}\right) =$$

$$100 \times \log^{-1}(0.5) = 100 \times 3.16 = 316\ \mathrm{W}$$

The total system has positive gain, so we should have expected a larger value for ERP than TPO. Keep in mind that the gain antenna concentrates more of the signal in a desired direction, with less signal in undesired directions. So, the antenna doesn't really increase the total available power. If directional antennas are used, ERP will change with direction.

Example: What is the effective radiated power of a repeater station with 150 W transmitter power output, 2 dB feed line loss, 2.2 dB duplexer loss, and 7 dBd antenna gain?

System gain = –2 dB – 2.2 dB + 7 dBd = 2.8 dB

$$\mathrm{ERP} = 150\ \mathrm{W} \times \log^{-1}\left(\frac{\text{system gain (dB)}}{10}\right) =$$

$$150 \times \log^{-1}(0.28) = 150 \times 1.9 = 285\ \mathrm{W}$$

Example: What is the effective radiated power of a repeater station with 200 W transmitter power output, 4 dB feed line loss, 3.2 dB duplexer loss, 0.8 dB circulator loss, and 10 dBd antenna gain?

System gain = –4 – 3.2 – 0.8 + 10 = 2 dB

$$\mathrm{ERP} = 200\ \mathrm{W} \times \log^{-1}\left(\frac{\text{system gain (dB)}}{10}\right) =$$

$$200 \times \log^{-1}(0.2) = 200 \times 1.58 = 317\ \mathrm{W}$$

What is the effective isotropic radiated power of a repeater station with 200 W transmitter power output, 2 dB feed line loss, 2.8 dB duplexer loss, 1.2 dB circulator loss, and 7 dBi antenna gain?

System gain = –2 – 2.8 – 1.2 + 7 = 1 dB

$$\mathrm{ERP} = 200\ \mathrm{W} \times \log^{-1}\left(\frac{\text{system gain (dB)}}{10}\right) =$$

$$100 \times \log^{-1}(0.1) = 200 \times 1.26 = 252\ \mathrm{W}$$

1.5.5 Ionospheric Propagation and Reciprocity

In long-distance transmission and reception via the ionosphere, the relationship between receiving and transmitting may not be exactly reciprocal. This is because the waves do not always follow exactly the same paths at all times and may vary the amount of time it takes for a transmission to travel between two stations. Also, when more than one ionospheric layer is involved in the wave travel (see the chapter **Radio Wave Propagation**), it is sometimes possible for reception to be good in one direction and poor in the other over the same path.

Wave polarization usually shifts in the ionosphere. The tendency is for the arriving wave to be elliptically polarized, regardless of the polarization of the transmitting antenna. Vertically polarized antennas can be expected to show no more difference between transmission and reception than horizontally polarized antennas. On average, however, an antenna that transmits well in a certain direction also gives favorable reception from the same direction, despite ionospheric variations.

1.6 References and Bibliography

Balanis, C., *Antenna Theory, Analysis and Design* (Harper & Row, 1982).

Blake L, and Long, M., *Antennas: Fundamentals, Design, Measurement*, 3rd Edition (Scitech Publishing, 2009).

Collin, R., *Antennas and Radiowave Propagation* (McGraw-Hill College, 1985).

Collin R., and Zucker, F., *Antenna Theory, Part 1 & 2*, (McGraw-Hill, 1969).

Elliott, R., *Antenna Theory and Design* (Prentice Hall, 1981).

Glazunov, A., Molisch, A., Tufvesson, F., "Mean effective gain of antennas in a wireless channel," *IET Microwaves Antennas & Propagation*, Apr. 2009, pp. 214 – 227.

Hansen, R., "Antennas," *Reference Data for Engineers*, 9th Edition, Van Valkenburg M., and Middleton M., Eds., (Newnes/Butterworth-Heinemann, 2002).

Johnson R., and Jasik H., *Antenna Engineering Handbook*, 2nd d. (New York: McGraw-Hill, 1984).

R. C. Johnson, *Antenna Engineering Handbook*, 3rd Edition McGraw-Hill, 1993).

Jordan E., and Balmain, K., *Electromagnetic Waves and Radiating Systems*, 2nd Edition (Prentice-Hall, 1968). Chapters 10 – 17 on antennas and propagation.

King, R., *Theory of Linear Antennas* (Harvard Univ. Press, 1956).

Kraus, J., *Electromagnetics* (McGraw-Hill).

Kraus, J., *Antennas*, 2nd Edition (McGraw-Hill, 1988).

Laport, E., *Radio Antenna Engineering* (McGraw-Hill, 1952).

Leutzelschwab C., et al, *Here to There: How Radio Waves Propagate* (ARRL, 2023).

Moxon, L., *HF Antennas for All Locations* (RSGB, 1982), pp. 109 – 111.

Ordy, G., **seed-solutions.com/gregordy/Amateur%20 Radio/Experimentation/RDFMetric.htm**

Ramo S., and Whinnery, J., *Fields and Waves in Modern Radio* (Wiley, 1953).

Rothammel's Antenna Book, 13th Edition., A. Krischke, DJØTR, Ed. (DARC Verlag, English translation, 2019).

Schelkunoff, S., *Advanced Antenna Theory* (Wiley, 1952).

Schelkunoff S., and Friis, H., *Antennas: Theory and Practice* (Wiley, 1952).

Silver, W., NØAX, "Hands-On Radio: Experiment #178 — Maxwell's Equations — Grad, Div, and Curl," *QST*, Nov. 2017, pp. 77 – 78.

Silver, W., NØAX, "Hands-On Radio: Experiment #179 — Maxwell's Equations — The Wave Emerges," *QST*, Dec. 2017, pp. 59 – 60.

Smith, G., "Analysis of Hertz's Experimentation Crucis on Electromagnetic Waves," IEEE *Antennas and Propagation Magazine*, Oct. 2016, pp. 96 – 108.

Terman, F., *Radio Engineers' Handbook*, 1st Edition. (McGraw-Hill, 1943).

Terman, F., *Radio Engineering*, 3rd Edition (McGraw-Hill, 1947).

Weeks, W., *Antenna Engineering* (McGraw-Hill, 1968).

Welch, G., *Wave Propagation and Antennas* (D. Van Nostrand Co, 1958), pp. 180 –182.

Yaghjian A., and Best, S., "Impedance, Bandwidth and Q of Antennas," Digest of IEEE Antennas and Propagation Society Symposium, Columbus, Ohio, Vol. 1, June 2003, pp. 501 – 504.

Zavrel, R., "Radiation Resistance, Feed Point Impedance and Mythology," *QEX*, Mar./Apr. 2014, pp. 32 –35.

Zavrel, R., *Antenna Physics: An Introduction*, 2nd Edition (ARRL, 2020).

Radio Broadcast Ground Systems, available from Smith Electronics, Inc, 8200 Snowville Rd, Cleveland, OH 44141.

Wiley Electrical and Electronics Engineering Dictionary (Wiley - IEEE Press, 2004)

TABLE OF CONTENTS

Chapter 2

Dipoles and Monopoles

Dipoles and monopoles are not only popular antennas, they are the basic elements from which most antennas used by amateurs are constructed, including beams. This chapter explores the basic characteristics of these antennas in support of the specific designs the reader will encounter later in the book. Material from previous editions is augmented by contributions from the dipole and vertical chapters of the 5th edition of *ON4UN's Low-Band DXing*. Additional material is adapted from Robert Zavrel, Jr., W7SX's ARRL book, *Antenna Physics: An Introduction*, 2nd edition.

2.1 Dipoles

The *dipole* is a fundamental antenna — in its most common form it is about one-half wavelength long at the frequency of use. (See **Figure 2.1**.) Many types of more complex antennas are constructed from elements that approximate dipoles. Its name, di- meaning *two* and -pole meaning *electrical polarity*, describes the opposite voltages on each half of the antenna, creating two electrical halves.

Figure 2.2 shows how the magnitudes of voltage and current vary along a half-wave dipole. The dipole has opposite voltages on either side of center, reaching a maximum at each end. The current at each end of a dipole is approximately zero. If the dipole is one-half wavelength long, the current is maximum at the center. This is the *distribution* of voltage and current. The distribution of current, in particular, determines how a dipole radiates.

If the dipole is *resonant*, the instantaneous waveforms of the voltage and current are exactly in phase. This means that the antenna's feed point impedance is purely resistive with no reactance. At resonance the stored energies in the electric and magnetic fields around the antenna are exactly equal.

2.1.1 Effects of Conductor Diameter

The physical length of a resonant ½-λ antenna will not be exactly equal to the half wavelength of a radio wave of that frequency in free space, but depends on the thickness of the conductor in relation to the wavelength as shown in **Figure 2.3**. **Table 2.1** gives resonant lengths for dipoles in free space, made of #12 AWG bare copper wire. If thinner wire is used, the resonant length will be a few percent longer, and vice versa.

An additional shortening effect occurs with wire antennas

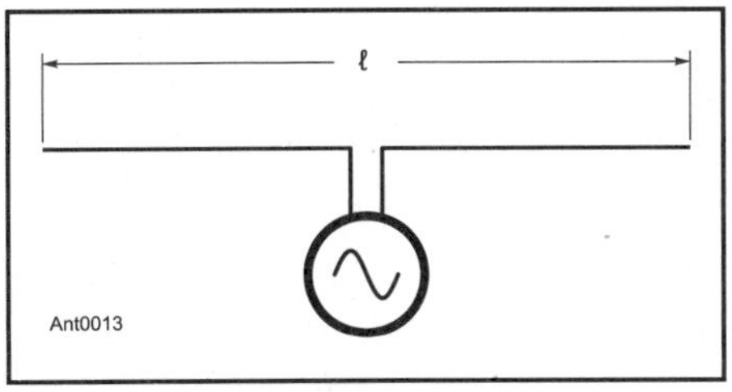

Figure 2.1 — The center-fed dipole antenna. It is assumed that the source of power is directly at the antenna feed point, with no intervening transmission line. Although λ/2 is the most common length for amateur dipoles, the length of a dipole antenna can be any fraction of a wavelength.

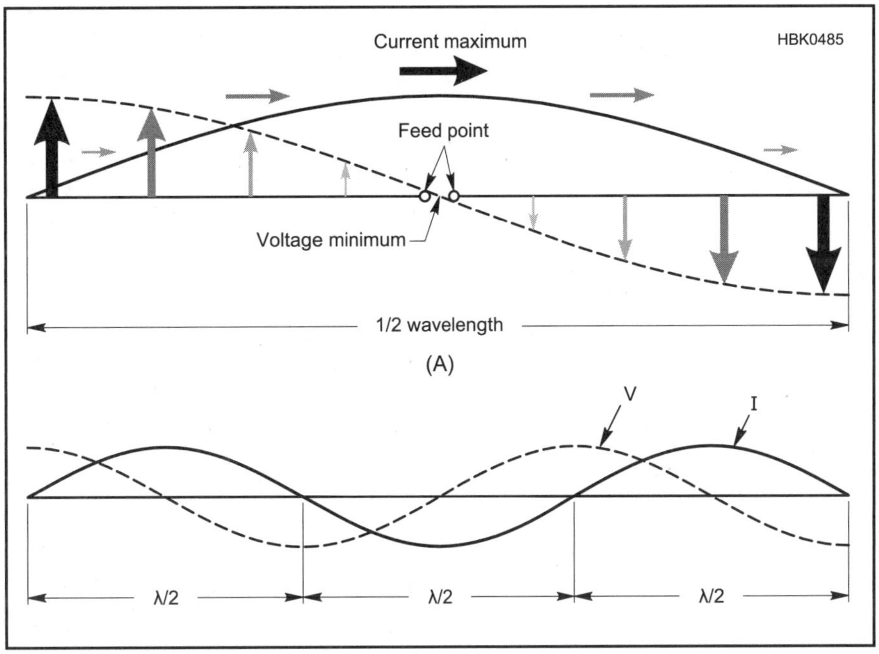

Figure 2.2 — The current and voltage distribution along a half-wave dipole (A) and for an antenna made from a series of half-wave sections (B).

Table 2.1
Resonant λ/2 Dipole Lengths in Free Space

Freq (MHz)	*½λ_{fs} (ft)*	*L (ft)*	*L (m)*
1.82	270.3	263.8	80.4
3.6	136.7	133.0	40.5
3.85	127.8	124.4	37.9
5.35	92.0	89.5	27.3
7.15	68.8	66.9	20.4
10.1	48.7	47.3	14.4
14.15	34.8	33.7	10.3
18.1	27.2	26.3	8.0
21.2	23.2	22.5	6.9
24.9	19.8	19.1	5.8
28.3	17.4	16.8	5.1
51	9.6	9.3	2.8

Dipole constructed from #12 AWG bare copper wire
½λ_{fs} is the free-space wavelength computed as 492/f(MHz)

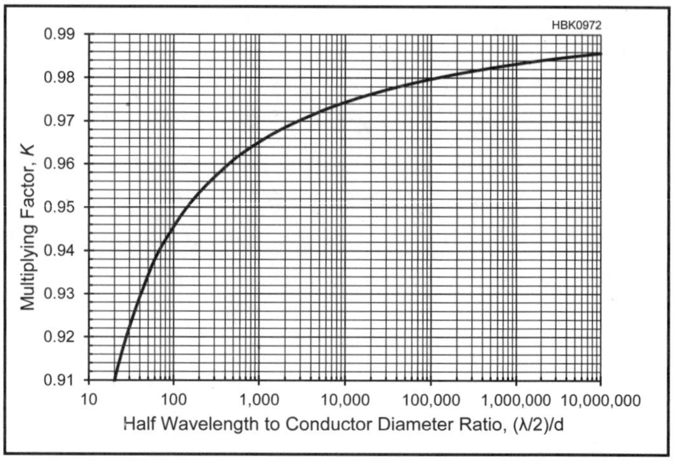

Figure 2.3 — Effect of antenna diameter on length for half-wavelength resonance in free-space, shown as a multiplying factor, K. The thicker the conductor relative to the wavelength, the shorter the physical length of the antenna at resonance. For antennas over ground, additional factors affect the antenna's electrical length.

supported by insulators at the ends (and at the feed point) because of the capacitance added to the system by the loops of wire through the insulators. This shortening is called *end effect.*

K FACTOR

Based on the standard value for the speed of light, the wavelength, λ, of an electromagnetic wave in free-space at frequency, *f*, is given by:

$$\lambda(\text{in feet}) = \frac{983.571}{\text{f(in MHz)}} \tag{1a}$$

and for one-half wavelength:

$$\frac{\lambda}{2}(\text{in feet}) = \frac{491.786}{\text{f(in MHz)}} \tag{1b}$$

A dipole with a length of exactly one-half of a free-space wavelength λ/2 has an impedance of 73.079 + *j*42.515 Ω, essentially independent of wire diameter. The dipole must be shortened to achieve resonance where reactance is equal to zero. As the diameter of a conductor increases, its capacitance per unit length increases and inductance per unit length decreases. This increases the ratio of stored electric field energy to magnetic field energy, with the result of lowering the frequency at which the dipole is resonant. Stated another way: For a given frequency, the larger a conductor's diameter, the lower the frequency at which a dipole made of that conductor is resonant.

A dipole's resonant length is given by the product of the free-space half-wavelength and a constant, K:

$$L_{\text{Resonant}}(\text{in feet}) = K \times \frac{491.786}{\text{f(in MHz)}} \quad \text{or} \tag{2}$$

$$L_{\text{Resonant}}(\text{in inches}) = K \times \frac{5901.43}{\text{f(in MHz)}}$$

where K is a constant between zero and one, 0 < K < 1, that depends on the dipole's "thickness." There are several ways to express dipole thickness. Antenna builders prefer to specify dipole thickness in terms of the half-wavelength-to-diameter ratio (λ/2)/d because a builder generally knows d and wants to calculate L. (Antenna theorists use the parameter Ω = 2 × ln (2L/d), where ln( ) is the natural logarithm.) The graph in **Figure 2.3** satisfies this need and allows dipole resonant length to be determined for a given physical thickness and frequency. Most half-wavelength dipoles at HF typically have (λ/2)/d ratios in the range of 2500 to 25,000 with values of K from 0.97 to 0.98.

History of the K Factor Graph

All *ARRL Handbooks* and *Antenna Books* from 1947 through 1996 published a graph of the K factor and dipole resonant resistance. The graphs were based on an early but approximate solution of Hallén's equation due to R.W.P King, F.G. Blake, and C.W. Harrison. Separately, the RSGB published a graph of the K factor based on R.A. Smith (1949) that was based on Schelkunoff's induced EMF (IEMF) equations. However, the ARRL and RSGB graphs did not agree.

In 1996, L.B. Cebik, W4RNL, pointed out that the ARRL's 70-year old K factor graph did not agree with *NEC2* calculations. So, from 1998 to 2018, the ARRL published a revised K factor graph that was based on *NEC2* modeling. Steve Stearns, K6OIK, discovered that although the original graph had overestimated *K*, the revised graph underestimated it! (See the discussion of Method of Moments accuracy in the **Antenna Modeling** chapter.) Consequently, the ARRL published a second revised graph beginning with the 2019 *ARRL Handbook*. The new revised graph, which is shown here in Figure 2.3 and based on Eq 3, is the result of a re-analysis of Schelkunoff and the IEMF method and has been checked against authoritative theory, approximations, and data published by C.T. Tai, R.S. Elliott, and the late R.C. Hansen. — *Contributed by Steve Stearns, K6OIK*

While K can be determined exactly from the induced EMF method (see the Bibliography entry for Hansen), K6OIK gives a simpler formula for K:

$$K = 1 - \frac{0.225706}{\ln\left(\frac{\lambda/2}{d}\right) - 0.429451} \quad (3)$$

Example 1: A half-wavelength dipole for 7.2 MHz has an uncorrected length of 491.786 / 7.2 = 68.3 feet. If it is made from #12 AWG wire (0.081 inch diameter), it has a $(\lambda/2)/d$ ratio of:

$$\frac{491.786}{7.2}\text{(ft)} \times \frac{12 \text{ in/ft}}{0.081 \text{ in}} = 10{,}119$$

From Figure 2.3 or Eq 3, a $(\lambda/2)/d$ ratio of 10,119 gives K = 0.974. Thus, by Eq 2, the resonant length of the half-wavelength dipole is 0.974 × 68.3 = 66 feet 7 inches.

It should be understood that K is not a velocity factor because it is unrelated to waves or the speed of wave travel. Rather, K arises because if a dipole is exactly one-half wavelength long, the stored energies in the electric and magnetic fields are not exactly equal. A dipole must be shortened to obtain equality and resonance.

The graph shown in Figure 2.3 and Eq 3 were determined by Steve Stearns, K6OIK, who evaluated and compared theoretical and numerical methods for calculating dipole and monopole impedance. (See the Bibliography entries for Stearns, Tai and Long, Elliott, and Schelkunoff.) If better accuracy than the graph is needed, the formula in Eq 3 should be used.

DIPOLE LENGTH FORMULAS

For wire dipoles for frequencies at and below 10 MHz and that are made of common wire sizes and are installed at heights of ⅛ to ¼ λ, the traditional "4-6-8" formula is:

$$L(\text{in feet}) = \frac{468}{f(\text{in MHz})} \quad (4)$$

This formula is an approximation that assumes the resonant length is 5% less than a half wavelength, i.e. K = 0.95, independent of wire thickness. 0.95 is outside the range of 0.97 to 0.98 that is typical for HF wire antennas. The next example illustrates the traditional formula.

Example 2: A half-wave dipole for 7200 kHz (7.2 MHz) is 468/7.2 = 65.0 feet. Note the traditional formula results in a dipole length that is 1 foot 7 inches too short compared to the more accurate calculation in Example 1 for the case of #12 AWG wire.

Example 3: Find the resonant length of a half-wave dipole antenna at 50.1 MHz, if the dipole is made of ½-inch diameter tubing. A half wavelength in space is 491.786 / 50.1 = 9.82 ft, and the ratio of half wavelength to conductor diameter (changing wavelength to inches) is (9.82 ft × 12 in/ft) / 0.5 in = 235.6. From Figure 2.3 or Eq 4, K = 0.955 for this ratio. The resonant length of the dipole is therefore 0.955 × 9.82 = 9.376 ft or 9 feet 4.5 inches. Calculated directly in inches, the resonant length is 0.955 × 5901.43 / 50.1 = 112.5 in.

The discussion and examples above are for antennas in free space. The effect of ground is not included. For single-wire HF antennas above ground, the free-space dipole length formulas can be inaccurate, but antenna modeling software can be used to come up with more accurate lengths. (See the **Antenna Modeling** chapter.) However, at VHF and UHF where the wavelength is short and antennas are several wavelengths above ground, the free-space dipole length formulas are accurate and useful.

2.1.2 Radiation Patterns and Effects of Ground

The radiation pattern of a dipole antenna in free space is strongest at right angles to the wire as shown in **Figure 2.4**,

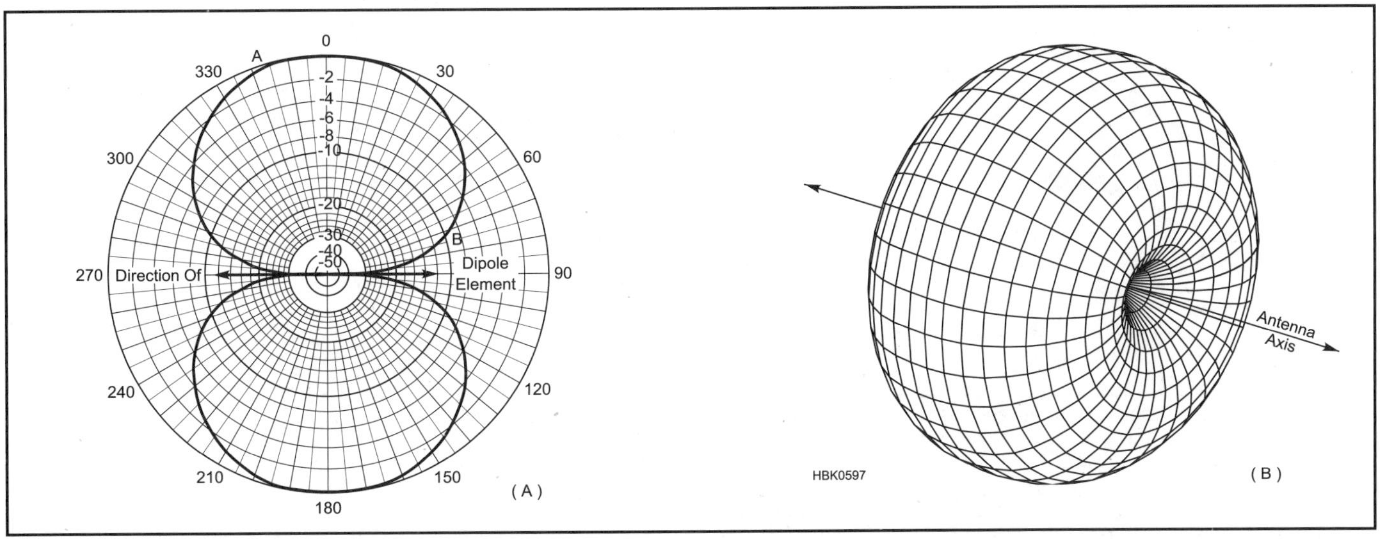

Figure 2.4 — Response of a dipole antenna in free space in the plane of the antenna with the antenna oriented along the 90° to 270° axis (A). The full three-dimensional pattern of the dipole is shown at (B). The pattern at A is a cross-section of the three-dimensional pattern taken directly through the axis of the antenna.

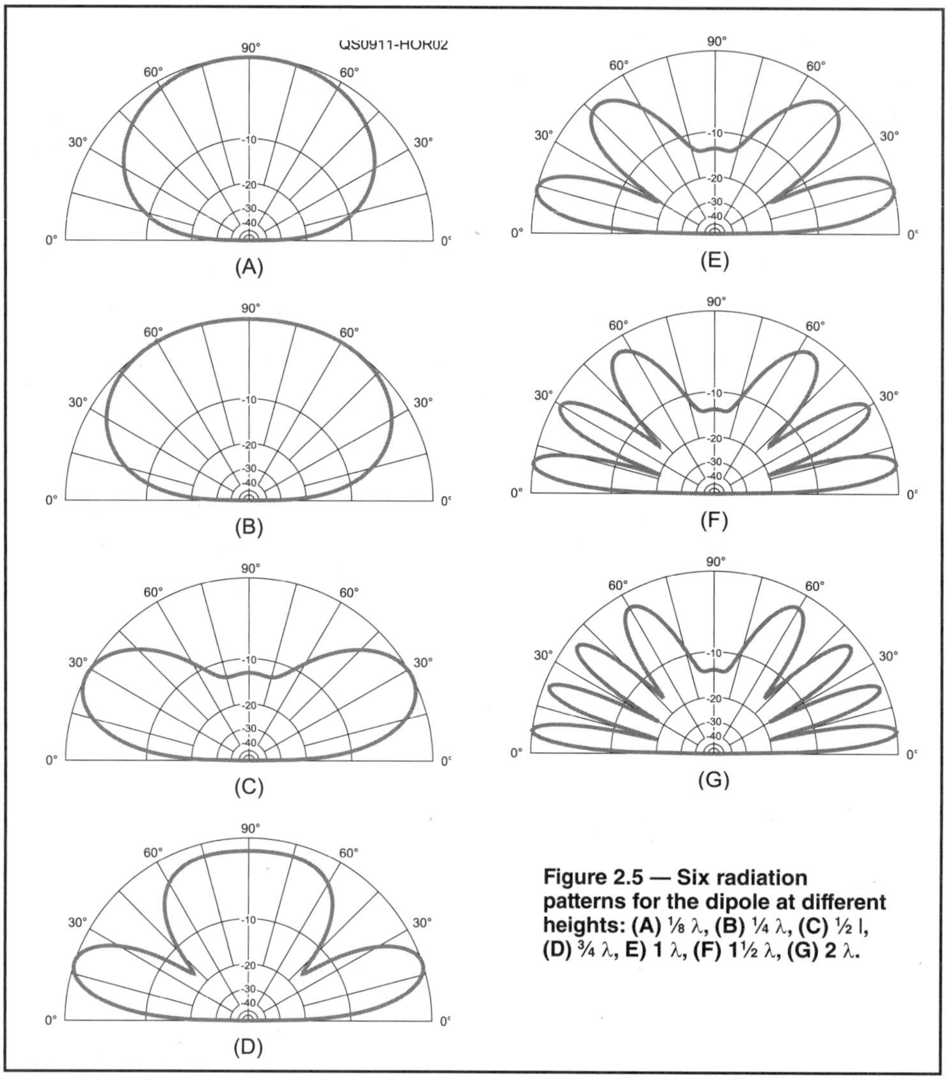

Figure 2.5 — Six radiation patterns for the dipole at different heights: (A) ⅛ λ, (B) ¼ λ, (C) ½ l, (D) ¾ λ, E) 1 λ, (F) 1½ λ, (G) 2 λ.

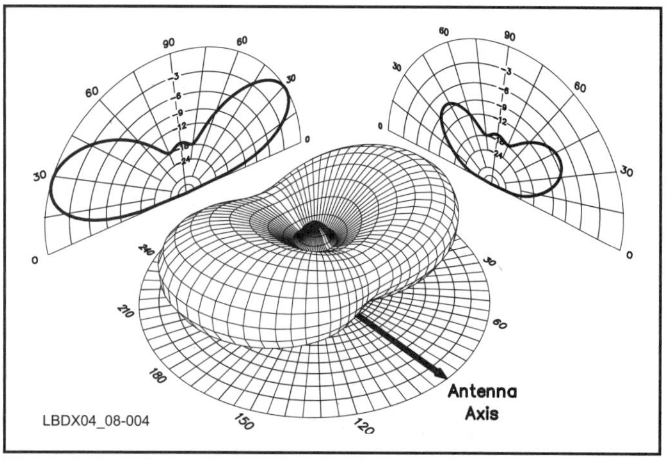

Figure 2.6 — Three-dimensional representation of the radiation patterns of a half-wave dipole, ½ λ above ground.

a free-space radiation pattern. The dipole in free space has a gain of 2.15 dBi.

In an actual installation, the figure-8 pattern is less directive due to reflections from ground and other conducting surfaces. As the dipole is raised to ½ λ or greater above ground, nulls off the ends of the dipole become more pronounced. Sloping the antenna above ground and coupling to the feed line tend to distort the pattern slightly.

As a horizontal dipole is brought closer to ground, reflections from the ground combine with the direct radiation to create lobes at different angles as shown in **Figure 2.5**. In addition, the directivity of the dipole also changes with height. For example, **Figure 2.6** shows the dipole's three-dimensional pattern at a height of ½ λ. The deep null along the axis of the wire in Figure 2.6 is filled in with a substantial amount of radiation.

Figure 2.7 shows the radiation pattern for dipoles at different heights above ground and at four different elevation angles from 15° to 60°. You can see that for low heights (the H = ¼ λ figure) the dipole becomes almost omnidirectional at elevation angles of 60° and higher.

The type of ground under the dipole also affects the radiation pattern. **Figure 2.8** illustrates what happens over two different types of ground; very poor soil (desert) and saltwater. These two types of ground represent the extremes of what amateurs are likely to encounter and most installations will be somewhere in between these two examples.

For antennas over ground, the height above ground also affects the antenna's physical length as in the example of **Table 2.2** showing the resonant, half-wavelength length for a 20 meter dipole at various electrical heights. Nearby conducting surfaces and materials will also affect resonant length.

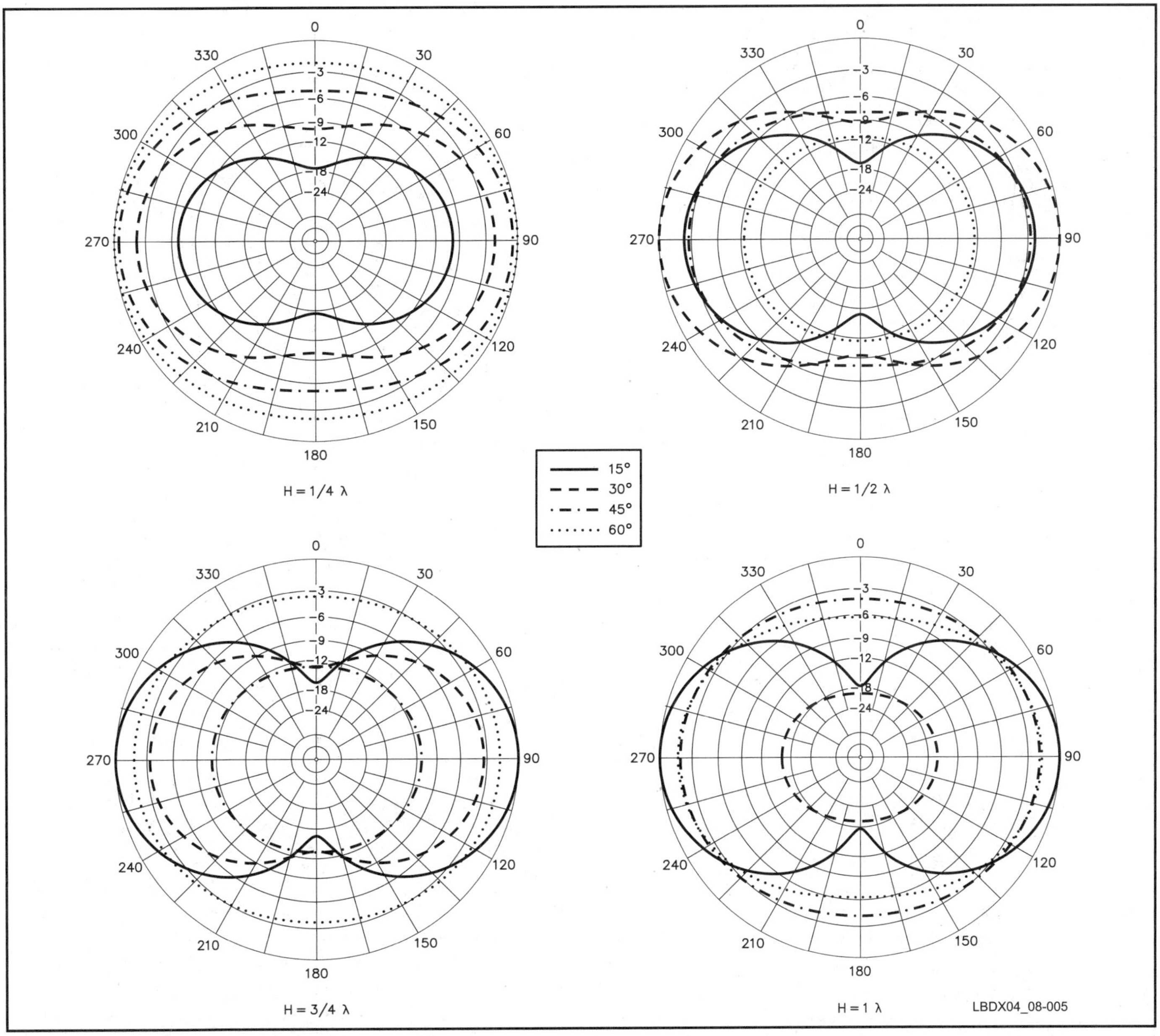

Figure 2.7 — Horizontal radiation pattern for ½-wave horizontal dipole at various heights above ground for wave angles of 15°, 30°, 45° and 60° (modeled over good ground).

Table 2.2
Variation in Dipole Performance with Height

Height in Wavelengths at 14.175 MHz (feet)	*Resonant Length in Feet (L x f)*	*Feed point Impedance in Ω (SWR)*	*Max Gain (dBi) at Angle (Degrees)*
⅛ (8.8)	33.0 (467.8)	31.5 (1.59)	8.3 @ 90
¼ (17.4)	32.9 (466.4)	81.7 (1.63)	6.5 @ 62
½ (34.7)	34.1 (483.4)	69.6 (1.39)	7.9 @ 28
¾ (52.0)	33.4 (473.4)	73.4 (1.47)	7.3 @ 18
1 (69.4)	33.9 (480.5)	71.9 (1.44)	7.7 @ 14
1½ (104.1)	33.8 (479.1)	72.0 (1.44)	7.8 @ 9
2 (138.8)	33.8 (479.1)	72.3 (1.45)	7.9 @ 7

Note: All gain values were calculated using *EZNEC*'s *MININEC* ground. These calculated gains and elevation angles are examples meant to illustrate the large effect that height above ground can have on antenna pattern. They are unlikely to be exactly correct for a real antenna in an actual installation.

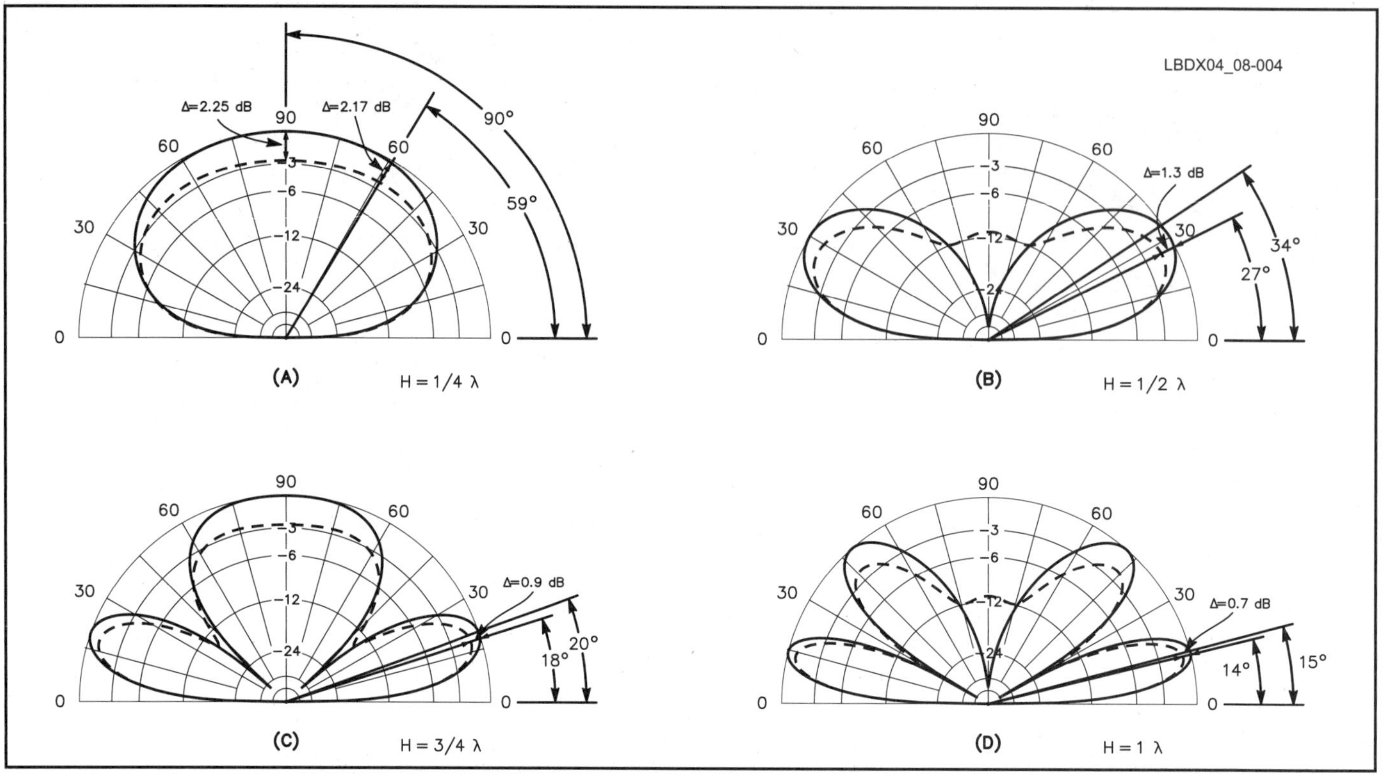

Figure 2.8 — Vertical radiation patterns over two types of ground: saltwater (solid line in each set of plots) and very poor ground (dashed line in each set of plots). The wave angles as well as the gain difference between saltwater and poor ground are given for four antenna heights.

2.1.3 Feed Point Impedance

A feed line is attached directly to the dipole, generally at the center with an insulator separating the antenna's conductor into two sections. Such a dipole is referred to as being *center-fed*. One conductor of the feed line is attached to each section. The point at which the feed line is attached is the dipole's *feed point*. (Off-center-fed dipoles are discussed later in this chapter.)

The dipole's feed point impedance is the ratio of voltage to current at the feed point. Referring back to Figure 2.2A, the feed point impedance of a half-wave dipole will be low at the center (where voltage is minimum and current is maximum) and high on each end (where voltage is maximum and current is minimum).

If a dipole is fed at the center and excited (supplied with power) at the third harmonic, the situation changes to that of Figure 2.2B. The dipole's physical length has not changed but its electrical length at the third harmonic has tripled — it is now three half-wavelengths long. If fed in the center, the same low impedance (low voltage/high current) is presented to the feed line. This situation occurs for all odd harmonics of the dipole's fundamental frequency because the center of the dipole is at a low impedance point and will present a reasonably low SWR to coaxial feed lines.

The situation is reversed if the dipole is excited at an even harmonic. Remove the right-most half-wavelength section in Figure 2.2B as the dipole is now electrically one full wavelength long. At the center of this antenna, voltage is high and current is low so the impedance is high and SWR will be high on any common feed line, coaxial or parallel-conductor. This is the situation at all even harmonics of the dipole's fundamental frequency and is sometimes referred to as *anti-resonance*.

At frequencies in between harmonics, the feed point impedance will take some intermediate value. When fed with parallel-conductor line and a wide-range impedance-matching unit, a dipole can be used on nearly any frequency, including non-resonant frequencies. (An example of such an antenna system is presented in the chapter **General-Purpose MF and HF Antennas**.)

FEED POINT IMPEDANCE IN FREE-SPACE

In free space the theoretical impedance of a half-wavelength antenna made of an infinitely thin conductor is $73 + j42.5\ \Omega$. This antenna exhibits both resistance and reactance. The positive sign in the $+j42.5\ \Omega$ reactive term indicates that the antenna exhibits an inductive reactance at its feed point. The antenna is slightly long electrically, compared to the length necessary for exact resonance, where the reactance is zero.

The feed point impedance of any antenna is affected by the wavelength-to-diameter ratio (λ/dia) of the conductors used. Theoreticians like to specify an "infinitely thin" antenna because it is easier to handle mathematically.

What happens if we keep the physical length of an antenna constant, but change the thickness of the wire used in its construction? Further, what happens if we vary the frequency from well below to well above the half-wave resonance and measure the feed point impedance? **Figure 2.9** graphs the impedance of a 100-foot long, center-fed dipole in free space, made with extremely thin wire — in this case, wire that is only 0.001 inch in diameter. There is nothing particularly significant about the choice here of 100 feet. This is simply a numerical example.

We could never actually build such a thin antenna (and neither could we install it in free space), but we can model how this antenna works using *NEC* antenna modeling software. (See the **Antenna Modeling** chapter for details on antenna modeling.)

The frequency applied to the antenna in Figure 2.9 is varied from 1 to 30 MHz. The x-axis has a logarithmic scale because of the wide range of feed point resistance seen over the frequency range. The y-axis has a linear scale representing the reactive portion of the impedance. Inductive reactance is positive and capacitive reactance is negative on the y-axis. The bold figures centered on the spiraling line show the frequency in MHz.

At 1 MHz, the antenna is very short electrically, with a resistive component of about 2 Ω and a series capacitive reactance of about –5000 Ω. Close to 5 MHz, the line crosses the zero-reactance line, meaning that the antenna goes through half-wave resonance there. Between 9 and 10 MHz the antenna exhibits a peak inductive reactance of about 6000 Ω. It goes through full-wave resonance (again crossing the zero-reactance line) between 9.5 and 9.6 MHz. At about 10 MHz, the reactance peaks at about –6500 Ω. Around 14 MHz, the line again crosses the zero-reactance line, meaning that the antenna has now gone through 3/2-wave resonance.

Between 19 and 20 MHz, the antenna goes through 4/2-wave resonance, which is twice the full-wave resonance or four times the half-wave frequency. If you allow your mind's eye to trace out the curve for frequencies beyond 30 MHz, it eventually spirals down to a resistive component somewhere between 200 and 3000 Ω. Thus, we have another way of looking at an antenna—as a sort of transformer, one that transforms the free-space impedance into the impedance seen at its feed point.

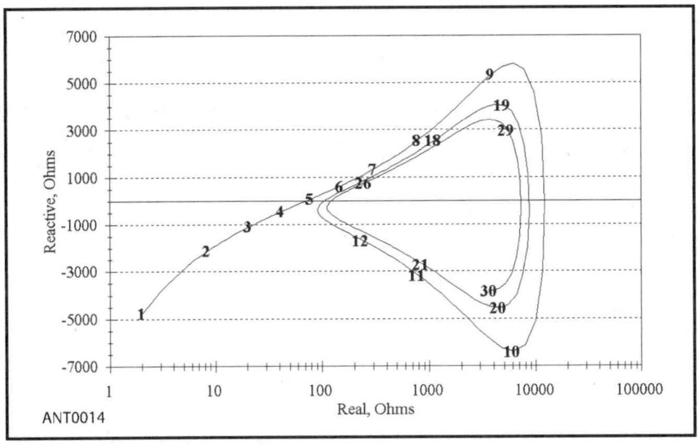

Figure 2.9 — Feed point impedance versus frequency for a theoretical 100-foot long dipole in free space, fed in the center and made of extremely thin 0.001-inch diameter wire. The y-axis is calibrated in positive (inductive) series reactance up from the zero line, and negative (capacitive) series reactance in the downward direction. The range of reactance goes from –6500 Ω to +6000 Ω. Note that the x-axis is logarithmic because of the wide range of the real, resistive component of the feed point impedance, from roughly 2 Ω to 10,000 Ω. The numbers placed along the curve show the frequency in MHz.

Now look at **Figure 2.10**, which shows the same kind of spiral curve, but for a thicker-diameter wire, one that is 0.1 inch in diameter. This diameter is close to #10 AWG wire, a practical size we might actually use to build a real dipole. Note that the y-axis scale in Figure 2.10 is different from that in Figure 2.9. The range is ±3000 Ω in Figure 2.10, while it was ±7000 Ω in Figure 2.9. The reactance for the thicker antenna ranges from +2300 to –2700 Ω over the whole frequency range from 1 to 30 MHz. Compare this with the range

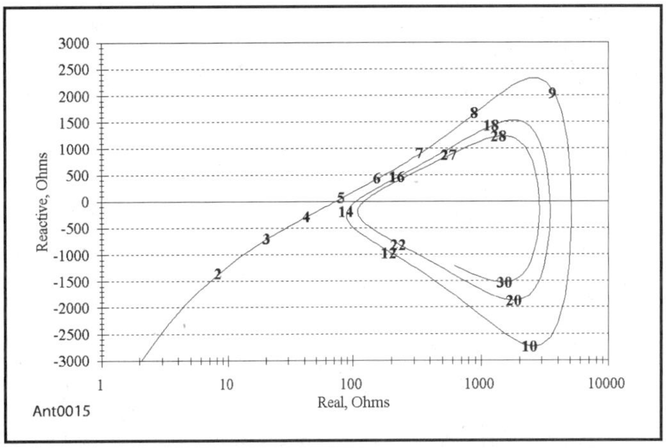

Figure 2.10 — Feed point impedance versus frequency for a theoretical 100-foot long dipole in free space, fed in the center and made of thin 0.1-inch (#10 AWG) diameter wire. Note that the range of change in reactance is less than that shown in Figure 2.9, ranging from –2700 Ω to +2300 Ω. At about 5000 Ω, the maximum resistance is also less than that in Figure 2.9 for the thinner wire, where it is about 10,000 Ω.

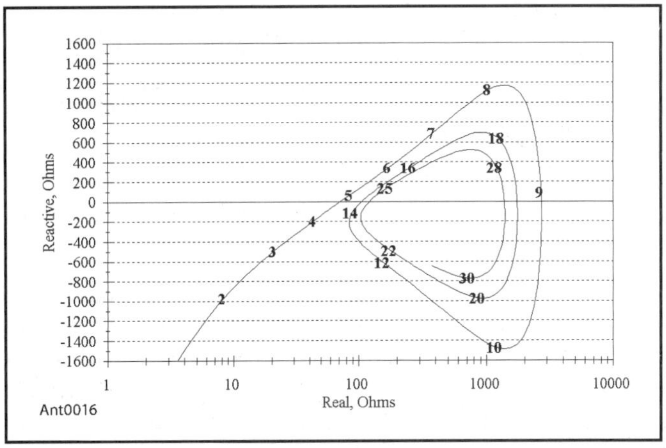

Figure 2.11 — Feed point impedance versus frequency for a theoretical 100-foot long dipole in free space, fed in the center and made of thick 1.0-inch diameter wire. Once again, the excursion in both reactance and resistance over the frequency range is less with this thick wire dipole than with thinner ones.

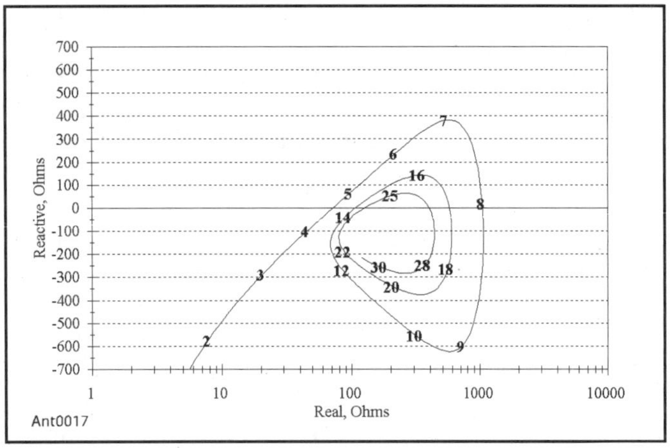

Figure 2.12 — Feed point impedance versus frequency for a theoretical 100-foot long dipole in free space, fed in the center and made of very thick 10.0-inch diameter wire. This ratio of length to diameter is about the same as a typical rod type of dipole element commonly used at 432 MHz. The maximum resistance is now about 1,000 Ω and the peak reactance range is from about –625 Ω to +380 Ω. This performance is also found in "cage" dipoles, where a number of paralleled wires are used to simulate a fat conductor.

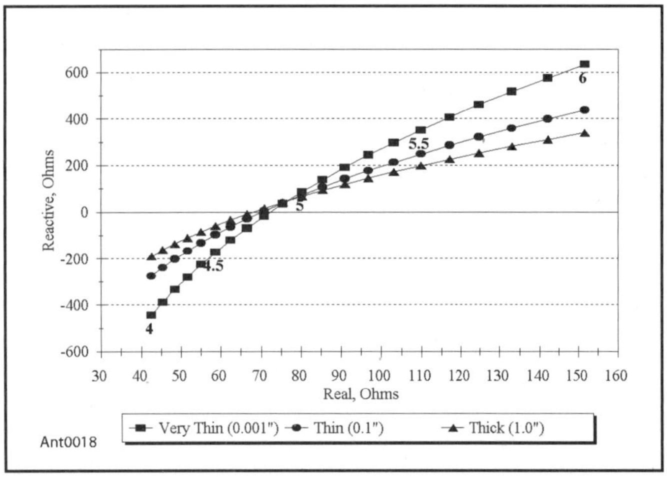

Figure 2.13 — Expansion of frequency range around half-wave resonant point of three center-fed dipoles of three different thicknesses. The frequency is shown along the curves in MHz. The slope of change in series reactance versus series resistance is steeper for the thinner antennas than for the thick 1.0-inch antenna, indicating that the Q of the thinner antennas is higher.

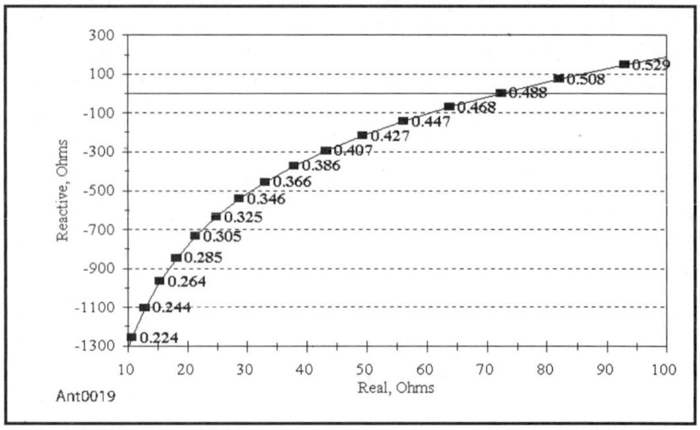

Figure 2.14 — Another way of looking at the data for a 100-foot, center-fed dipole made of #14 AWG wire in free space. The numbers along the curve represent the fractional wavelength, rather than frequency as shown in Figure 2.13. Note that this antenna goes through its half-wave resonance at about 0.488 λ, rather than exactly at a half-wave physical length.

of +5800 to –6400 Ω for the very thin wire in Figure 2.9.

Figure 2.11 shows the impedance for a 100-foot long dipole using really thick, 1.0-inch diameter wire. The reactance varies from +1000 to –1500 Ω, indicating once again that a larger diameter antenna exhibits less of an excursion in the reactive component with frequency. Note that at the half-wave resonance just below 5 MHz, the resistive component of the impedance is still about 70 Ω, just about what it is for a much thinner antenna. Unlike the reactance, the half-wave radiation resistance of an antenna doesn't radically change with wire diameter, although the maximum level of resistance at full-wave resonance is lower for thicker antennas.

Figure 2.12 shows the results for a very thick, 10-inch diameter wire. Here, the excursion in the reactive component is even less: about +400 to –600 Ω. Note that the full-wave resonant frequency is about 8 MHz for this extremely thick antenna, while thinner antennas have full-wave resonances closer to 9 MHz. Note also that the full-wave resistance for this extremely thick antenna is only about 1000 Ω, compared to the 10,000 Ω shown in Figure 2.9. All half-wave resonances shown in Figures 2.9 through 2.12 remain close to 5 MHz, regardless of the diameter of the antenna wire. Once again, the extremely thick, 10-inch diameter antenna has a resistive component at half-wave resonance close to 70 Ω. And once again, the change in reactance near this frequency is very much less for the extremely thick antenna than for thinner ones.

Now, we grant you that a 100-foot long antenna made with 10-inch diameter wire sounds a little odd! A length of 100 feet and a diameter of 10 inches represent a ratio of 120:1 in length to diameter. However, this is about the same length-to-diameter ratio as a 432 MHz half-wave dipole using 0.25-inch diameter elements, where the ratio is 109:1. In other words, the ratio of length-to-diameter for the 10-inch diameter, 100-foot long dipole is not that far removed from what might actually be used at UHF.

Another way of highlighting the changes in reactance and resistance is shown in **Figure 2.13**. This shows an expanded portion of the frequency range around the half-wave resonant frequency, from 4 to 6 MHz. In this region, the shape of each spiral curve is almost a straight line. The slope of the curve for the very thin antenna (0.001-inch diameter) is steeper than that for the thicker antennas (0.1 and 1.0-inch diameters). **Figure 2.14** illustrates another way of looking at the impedance data above and below the half-wave resonance. This is for a 100-foot dipole made of #14 AWG wire. Instead of showing the frequency for each impedance point, the wavelength is shown, making the graph more universal in application.

Just to show that there are lots of ways of looking at the

same data, recall that Figure 2.8 graphs the constant "K" used to multiply the free-space half-wavelength as a function of the ratio between the half-wavelength and the conductor diameter. The curve approaches the value of 1.00 for an infinitely thin conductor, in other words an infinitely large ratio of half-wavelength to diameter.

The behavior of antennas with different λ/diameter ratios corresponds to the behavior of ordinary series-resonant circuits having different values of Q. When the Q of a circuit is low, the reactance is small and changes rather slowly as the applied frequency is varied on either side of resonance. If the Q is high, the converse is true. The response curve of the low-Q circuit is broad; that of the high-Q circuit sharp. So it is with antennas — the impedance of a thick antenna changes slowly over a comparatively wide band of frequencies, while a thin antenna has a faster change in impedance. Antenna Q is:

$$Q = \frac{f_0 \Delta X}{2 R_0 \Delta f} \qquad (5)$$

where f_0 is the center frequency, ΔX is the change in the reactance for a Δf change in frequency, and R_0 is the resistance at f_0. For the "Very Thin," 0.001-inch diameter dipole in Figure 2.9, a change of frequency from 5.0 to 5.5 MHz yields a reactance change from 86 to 351 Ω, with an R_0 of 95 Ω. The Q is thus 14.6. For the 1.0-inch-diameter "Thick" dipole in Figure 2.11, ΔX = 131 Ω and R_0 is still 95 Ω, making Q = 7.2 for the thicker antenna, roughly half that of the thinner antenna.

A lower Q means the antenna's bandwidth is greater. This effect is can be exploited to increase an antenna's bandwidth. On the HF bands, one can make a dipole using multiple conductors in a cage or fan to increase its equivalent diameter d. On the VHF and UHF bands, one can build a dipole with copper pipe. The formulas in Eq 2 are particularly useful for antennas that are short and thick. Again, K is obtained from Figure 2.3 or Eq 3.

Let's review. The dipole can be described as a transducer or as a sort of transformer to a range of free-space impedances. Now, we just compared the antenna to a series-tuned circuit. Near its half-wave resonant frequency, a center-fed λ/2 dipole exhibits much the same characteristics as a conventional series-resonant circuit. Exactly at resonance, the current at the input terminals is in phase with the applied voltage and the feed point impedance is purely resistive. If the frequency is below resonance, the phase of the current leads the voltage; that is, the reactance of the antenna is capacitive. When the frequency is above resonance, the opposite occurs; the current lags the applied voltage and the antenna is inductive. Just like a conventional series-tuned circuit, the antenna's reactance and resistance determines its Q.

FEED POINT IMPEDANCE VERSUS HEIGHT ABOVE GROUND

Radiation directly downward from the antenna will reflect vertically from the ground and, in passing the antenna on the upward journey, induce a current in it. The magnitude and phase of the current depends on the height of the antenna above the reflecting surface and the characteristics of the surface. The total current in the antenna consists of two components: the amplitude of the first is determined by the excitation from the transmitter and the second component is induced in the antenna by the wave reflected from the ground. This second component of current, while considerably smaller than the first at most useful antenna heights, is by no means insignificant. At some heights, the two components will be in phase but at other heights the two components are out of phase. Changing the height of the antenna above ground will change the current amplitude at the feed point (we are assuming that the power input to the antenna is constant). A higher current at the same input power means that the effective resistance of the antenna is lower, and vice versa. In other words, the feed point resistance of the antenna is affected by the height of the antenna above ground because of mutual coupling between the antenna and the ground beneath it.

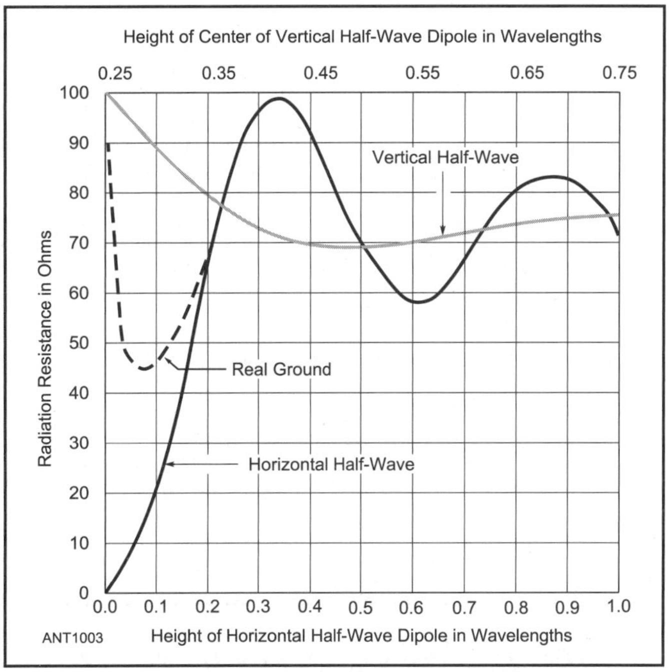

Figure 2.15 — Variation of feed point resistance with height for vertical and horizontal half-wave dipoles. Solid lines show feed point resistance over perfect ground while the dashed line is for real ground (see text).

The electrical characteristics of the ground affect both the amplitude and the phase of reflected signals. For this reason, the electrical characteristics of the ground under the antenna will have some effect on the impedance of that antenna, the reflected wave having been influenced by the ground. Different impedance values may be encountered when an antenna is erected at identical heights but over soils with different characteristics.

Figure 2.15 gives an example of the way the feed point impedance of horizontal and vertical half-wave antennas can vary with height above ground. The height of the vertical half-wave is the distance from the center of the antenna to ground. For horizontally polarized half-wave antennas, the differences between the effects of perfect ground and real earth are negligible if the antenna height is greater than 0.2 λ.

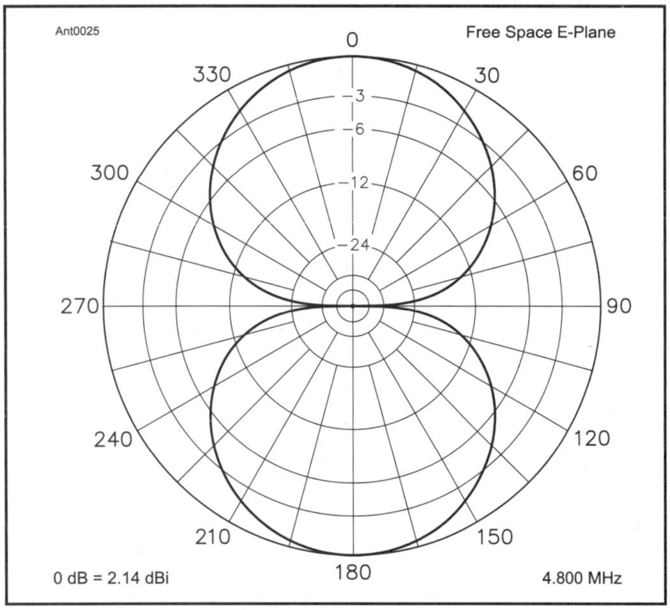

Figure 2.16 — Free-space E-plane radiation pattern for a 100-foot dipole at its half-wave resonant frequency of 4.80 MHz. This antenna has 2.14 dBi of gain. The dipole is located on the line from 90° to 270°.

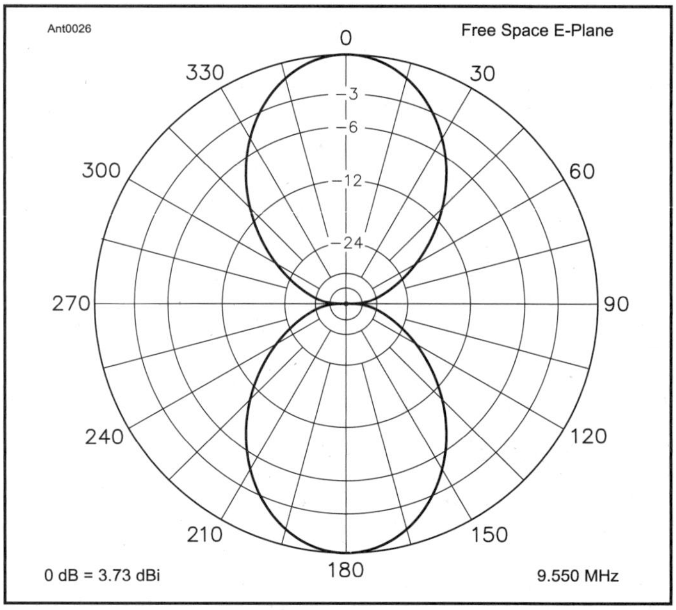

Figure 2.17 — Free-space E-plane radiation pattern for a 100-foot dipole at its full-wave resonant frequency of 9.55 MHz. The gain has increased to 3.73 dBi, because the main lobes have been focused, giving a sharpened pattern compared to Figure 2.16.

At lower heights, the feed point resistance over perfect ground decreases rapidly as the antenna is brought closer to a theoretically perfect ground. However, over real earth, the resistance actually begins increasing again at heights below about 0.08 1 as indicated by the dashed line. The reason for the increasing resistance at very low heights is that the field of the antenna interacts more strongly with the ground increasing ground losses. This increase in loss is reflected in an increased value for the feed point resistance.

Figure 2.15 shows that only for certain electrical heights above ground will the feed point impedance be close to the expected 73.5 Ω free-space value for a half-wave dipole. The feed point impedance will vary from very low when the antenna is close to the ground to a maximum of nearly 100 Ω at 0.34 λ above ground, varying around 73.5 Ω as the antenna is raised farther. The 73.5-Ω value is most likely to be realized in a practical installation when the horizontal dipole is approximately ½, ¾, or 1 λ above ground. This is why few amateur λ/2 dipoles exhibit a center-fed feed point impedance of 73.5 Ω even though they may be resonant.

2.1.4 Effect of Frequency on Radiation Pattern

Earlier, we saw how the feed point impedance of a fixed-length center-fed dipole in free space varies as the frequency is changed. What happens to the radiation pattern of such an antenna as the frequency is changed?

In general, the greater the length of a center-fed antenna, in terms of wavelength, the larger the number of lobes into which the pattern splits. A feature of all such patterns is the fact that the main lobe is always the one that makes the smallest angle with (is closest to) the antenna wire. Furthermore, this angle becomes smaller as the length of the antenna is increased.

Let's examine how the free-space radiation pattern changes for a 100-foot long wire made of #14 AWG wire as the frequency is varied. (Varying the frequency effectively changes the electrical length of a fixed-length wire.) **Figure 2.16** shows the E-plane pattern at the λ/2 resonant frequency of 4.8 MHz. This is a classical dipole pattern, with a gain in free space of 2.14 dBi referenced to an isotropic radiator.

Figure 2.17 shows the free-space E-plane pattern for the same antenna, but now at the full-wave (2 λ/2) resonant frequency of 9.55 MHz. Note how the pattern's two main lobes have become sharper at this frequency, making the gain 3.73 dBi, higher than at the λ/2 frequency.

Figure 2.18 shows the pattern at the 3 λ/2 frequency of 14.6 MHz. More lobes have developed compared to Figure 2.16. This means that the power has been divided into more lobes and consequently the gain decreases a small amount, down to 3.44 dBi. This is still higher than the dipole at its λ/2 frequency, but lower than at its full-wave frequency. **Figure 2.19** shows the E-plane response at 19.45 MHz, the 4 λ/2, or 2 λ, resonant frequency. Now the pattern has reformed itself into only four lobes, and the gain has as a consequence risen to 3.96 dBi.

In **Figure 2.20** the response has become quite complex at the 5 λ/2 resonance point of 24.45 MHz, with ten lobes showing. Despite the presence all these lobes, the main lobes now show a gain of 4.78 dBi. Finally, **Figure 2.21** shows the pattern at the 3λ (6 λ/2) resonance at 29.45 MHz. Despite the fact that there are fewer lobes radiating power than at 24.45 MHz, the peak gain is slightly less at 29.45 MHz, at 4.70 dBi.

The pattern — and hence the gain — of a fixed-length antenna varies considerably as the frequency is changed. Of

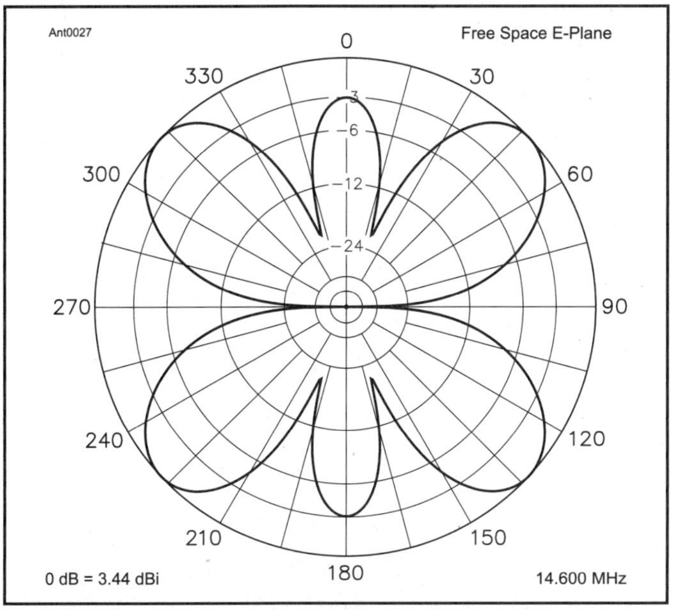

Figure 2.18 — Free-space E-plane radiation pattern for a 100-foot dipole at its 3/2λ resonant frequency of 14.60 MHz. The pattern has broken up into six lobes, and thus the peak gain has dropped to 3.44 dBi.

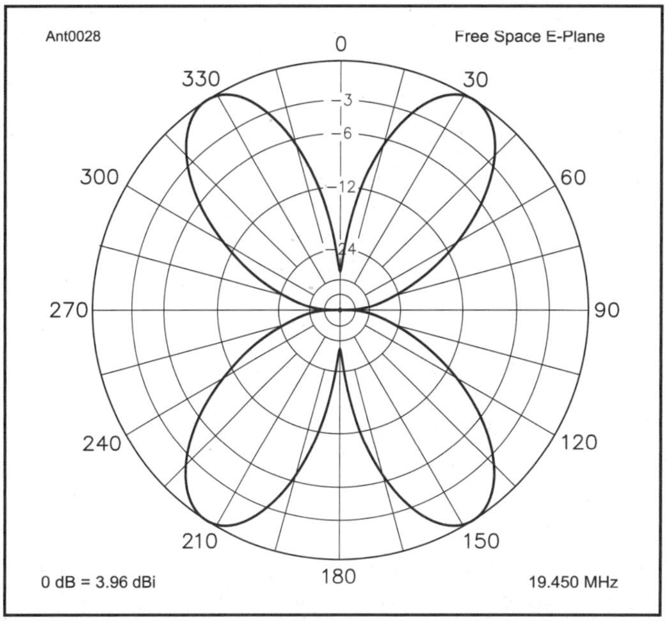

Figure 2.19 — Free-space E-plane radiation pattern for a 100-foot dipole at twice its full-wave resonant frequency of 19.45 MHz. The pattern has been refocused into four lobes, with a peak gain of 3.96 dBi.

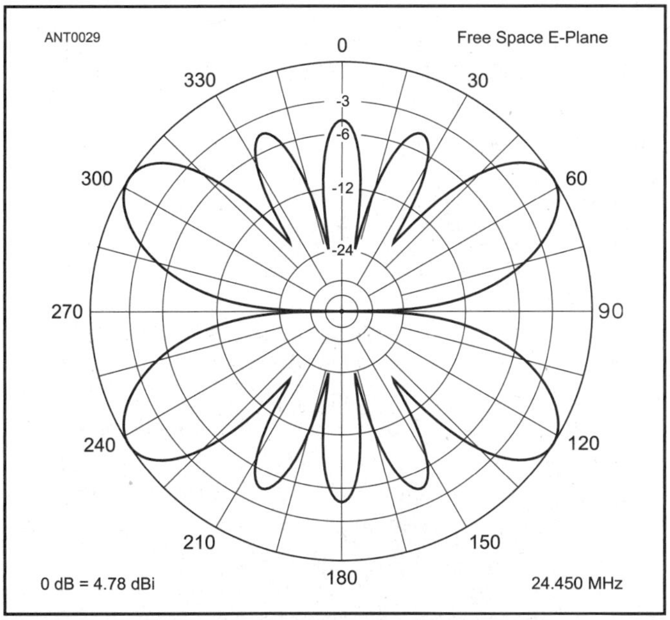

Figure 2.20 — Free-space E-plane radiation pattern for a 100-foot dipole at its 5/2λ resonant frequency of 24.45 MHz. The pattern now has ten lobes, with a peak gain of 4.78 dBi.

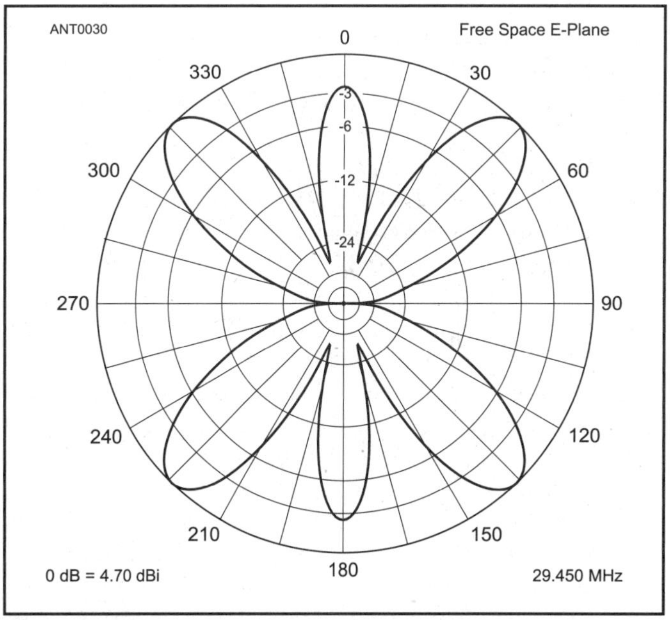

Figure 2.21 — Free-space E-plane radiation pattern for a 100-foot dipole at three times its full-wave resonant frequency of 29.45 MHz. The pattern has returned to six lobes, with a peak gain of 4.70 dBi.

course, the pattern and gain change in the same fashion if the frequency is kept constant and the length of the wire is varied. In either case, the wavelength and distribution of current are is changing. It is also evident that certain lengths reinforce the pattern to provide more peak gain. If an antenna is not rotated in azimuth when the frequency is changed, the peak gain may occur in a different direction than you might like. In other words, the main lobes change direction as the frequency is varied.

2.1.5 Folded Dipoles

Figure 2.22A shows a two-wire *folded dipole* made from parallel 1/2-λ conductors connected together at each end of the antenna. One conductor is continuous from end to end.

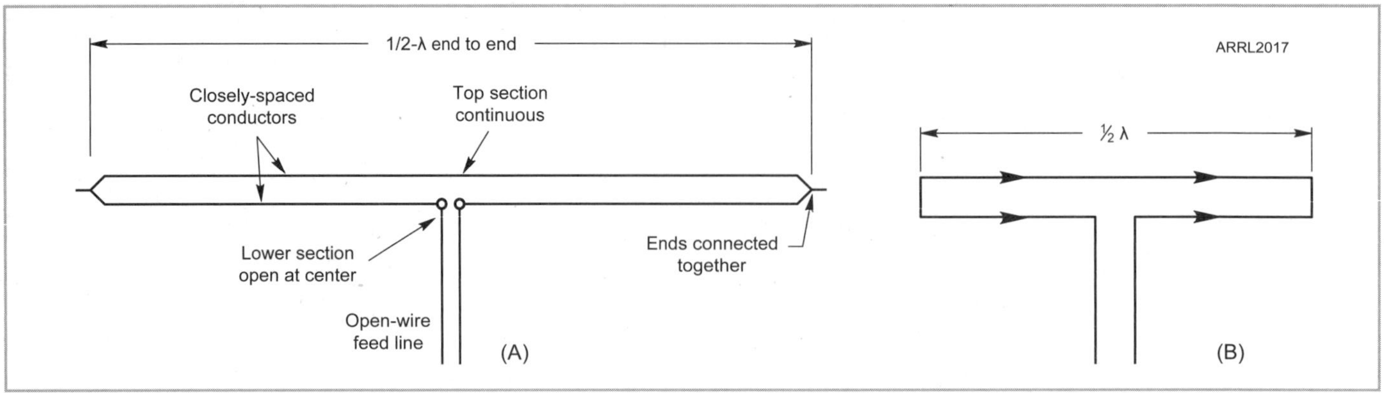

Figure 2.22 — The folded dipole is most often constructed from open-wire transmission line with the ends connected together. The close proximity of the two conductors and the resulting coupling act as an impedance transformer to raise the feed point impedance over that of a single-wire dipole by the square of the number of conductors used.

The remaining conductor is open in the middle where the feed line is attached. Folded dipoles can also have three or even four conductors but the feed line is attached to only one. A folded dipole has exactly the same gain and radiation pattern as a single-wire dipole.

Coupling between the closely spaced conductors causes the currents on each to be equal and in phase, as shown in Figure 2.22B. (This assumes all conductors have identical diameters.) Because the currents are equal, for n conductors the current in each is $1/n^{th}$ of the total current, I_{TOTAL}. This includes the feed point where $I_{FP} = I_{TOTAL} / n$. For a given amount of applied power, P, feed point impedance $Z_{FP} = P / I_{FP}^2$ so $Z_{FP} = P / (I_{TOTAL} / n)^2 = n^2 P / I_{TOTAL}^2$. If n = 1 (a one-wire dipole) then $Z_{FP} = P / I_{TOTAL}^2$ which is the usual dipole feed point impedance.

As more conductors are added, Z_{FP} goes up with the square of n. For the two-wire folded dipole, there are two conductors in the antenna, so the feed point impedance is $2^2 = 4$ times that of a single-wire dipole. A three-wire folded dipole would have $3^2 = 9$ times higher feed point impedance and so forth. If the diameters of the conductors are different, the currents don't divide equally, and the impedance ratio will not be an exact square of the number of conductors.

A common use of the folded dipole is to raise the feed point impedance of the antenna to present a better impedance match to feed lines. For example, if a very long feed line to a dipole is required, open-wire feed line would be preferable because of its lower loss. By raising the dipole's feed point impedance, SWR on the open-wire line is reduced from what it would be if connected to a single-wire dipole. For example, by using a three-wire folded dipole, the 72 Ω feed point impedance of a single-wire dipole is increased to 9 × 72 = 648 Ω which is a good match to 600 Ω open-wire line. A 50 Ω feed point impedance would be increased to 9 × 50 = 450 Ω, a good match for widely available window line.

Another use of the folded dipole is for driven elements of Yagi antennas which often have impedances much lower than 50 Ω. For example, if the feed point impedance of a single-conductor driven element is 15 Ω, SWR on a 50 Ω feed line would be 3.33:1. Changing to a two-conductor folded dipole would raise the feed point impedance to 15 × 4 = 60 Ω, lowering the SWR to 60/50 = 1.2:1.

2.1.6 Vertical Dipoles

A half-wave dipole can also be oriented vertically over ground instead of horizontally, becoming a vertical dipole. The dipole's pattern becomes generally omnidirectional. Practical aspects of vertical dipoles are also discussed in the **General-Purpose MF and HF Antennas** chapter.

In **Figure 2.23A** and B with the bottom of the vertical dipole very close (λ/80) to a saltwater ground plane the vertical dipole can have a gain of 6.1 dBi. Gain drops to about 0 dBi over good soil and lower over poorer soils. As with all vertical antennas, it is mainly the quality of the ground in the antenna's far field (several wavelengths from the antenna and beyond) that determines how good a low-angle radiator the vertical dipole will be as shown in Figure 2.23 and further discussed in the chapter **Effects of Ground**. Raising the λ/2 vertical higher above the ground introduces multiple lobes as shown in Figure 2.23C and D with the antenna's bottom tip λ/8 above ground.

The radiation resistance of the vertical dipole also depends on the height of its lower tip above ground as shown in Figure 2.15. The impedance of vertical and horizontal dipoles vary with height above ground for different reasons: The horizontal dipole receives reflected power from the ground affecting its mutual impedance. The vertical dipole, however, receives less reflected power from the ground and as it is lowered closer to ground, it is increasing its effective height and increasing its gain which has the effect of increasing its radiation resistance. As with the horizontal dipole, the radiation resistance varies above and below the free-space value of 73.5 Ω but not as much as for the horizontal dipole since its feed point is farther above ground. (Effective height of the dipole which is proportional to average current in the dipole divided by its physical length, increases as the antenna becomes lower because average current in the antenna increases. See Zavrel's referenced article on

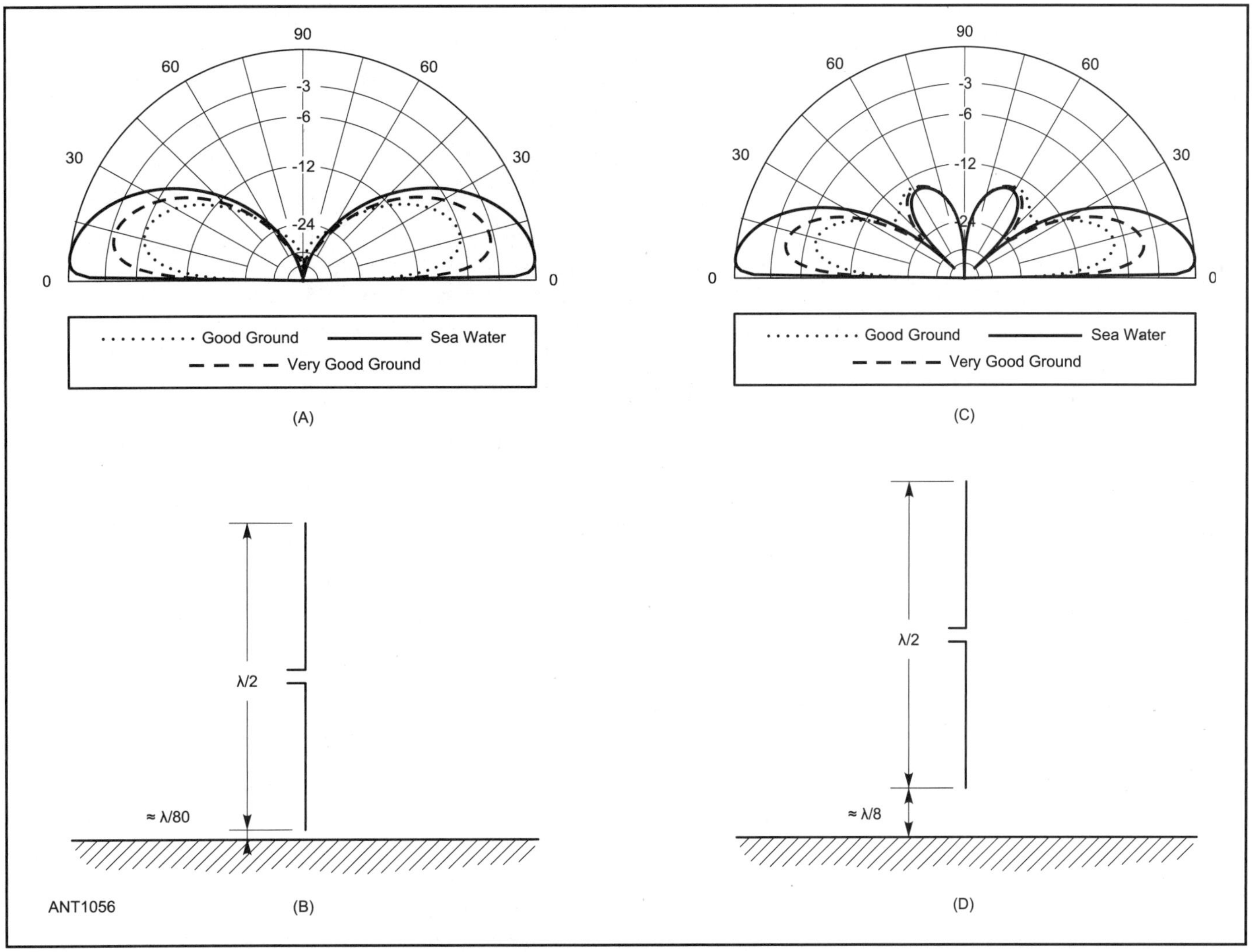

Figure 2.23 — At A and B, vertical radiation patterns over various grounds for a vertical half-wave center-fed dipole with the bottom tip just clearing the ground. The gain is as high as 6.1 dBi over ground and the feed point impedance is 100 Ω. At C and D, the vertical radiation patterns of the half-wave vertical dipole with the bottom tip ⅛ λ off the ground. Note the appearance of lobes in the radiation pattern at high elevation angles.

maximizing radiation resistance.)

In practice, it is not possible to obtain symmetrical currents in the upper and lower halves of the vertical dipole at HF due to the asymmetrical relationship of the two sections to ground. Further, the presence of the feed line introduces a third conductor for common-mode current that can influence the antenna's performance unless decoupled. Thus, the radiation patterns are unlikely to be very close to the ideal patterns shown here.

2.1.7 Off-Center-Fed (OCF) Dipoles

The usual practice is to feed a λ/2 dipole in the center where the feed point impedance is low and makes a suitable match to coaxial cables. The dipole will accept energy from a feed point anywhere along its length, however. (See the section on Feed Point Impedance earlier in this chapter.) A common variation is the *off-center-fed* (*OCF*) dipole where the feed point is offset from the center by some amount and an impedance transformer used to match a coaxial feed line to the higher impedance that is presented away from the center point on several bands. The extreme example of an OCF dipole is the *end-fed half-wave* with the feed point moved all the way to one end.

The OCF dipole feed point and overall length are generally selected for a feed point impedance in the neighborhood of 150 to 300 Ω on several bands. A 4:1 or 6:1 impedance transformer is then used to match the antenna to 50 Ω coaxial cable. The feed point impedance will vary with height above ground and so will the SWR. A choke balun (see the **Transmission Line System Techniques** chapter) should also be used if the impedance transformer does not provide feed line isolation. The transformer creates a convenient point from which to suspend the antenna in an asymmetric inverted-V configuration.

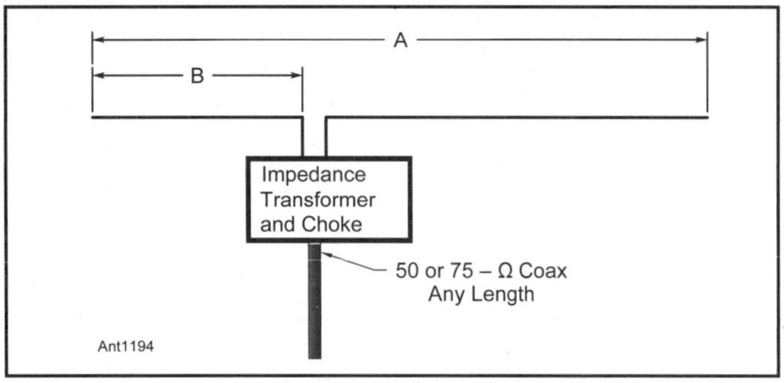

Figure 2.24 — An off-center-fed (OCF) dipole for use on several bands. Length A is generally λ/2 on the lowest frequency of operation. Length B is approximately ⅓ of A. The impedance transformer is usually a 4:1 or 6:1 design. A choke balun may be required for feed line isolation. Feed point impedance and SWR are affected by the antenna's height above ground.

Figure 2.24 shows the basic structure of the OCF dipole. Overall length is λ/2 on the lowest frequency of operation. Various formulas and conventions are used to determine how far from center the feed point is located. (See the Bibliography entry for Richter's in-depth article on the OCF dipole.) The most common location for the feed point is approximately ⅓ of overall length from one end. Specific OCF dipole designs are presented in the chapter on **Multiband HF Antennas**.

The OCF dipole is often mistakenly referred to as a "Windom" or a "coax-fed Windom." The two antennas are not the same, however, since the Windom is essentially a short vertical loaded by an asymmetric flat-top, fed against ground at the base of the vertical section.

Richter also provides a formula for radiation resistance, R_r, at various locations for the feed point on its fundamental and harmonic frequencies. The antenna is assumed to be in free space:

$$R_r = \frac{73.1}{\sin^2\left(m\frac{2X}{\lambda}\times 180\right)} \qquad (6)$$

where m is a multiplying factor (1 at the fundamental, 2 at the second harmonic, 3 at the third harmonic, etc.), λ is the full corrected wavelength, and X is the distance from the end of the antenna. Use the same units, meters or feet, for λ and X. The corrected wavelength includes the effect of the K factor (see the section Effects of Conductor Diameter earlier in this chapter).

Height above ground will affect the resonant length and R_r. The antenna is at different electrical heights on different bands, as well. Thus, it is difficult to make a precise calculation of feed point impedance or choose its location for a specific impedance on more than one band. Antenna modeling is probably the best practical method for designing an OCF dipole for a particular installation.

The OCF dipole is also an asymmetric or unbalanced antenna and there will be coupling to the feed line that affects the feed point impedance and radiation pattern. To minimize coupling use at least one feed line choke to isolate the feed line. One choke should be used at the feed point and others at intervals which prevent feed line resonances on the bands being used. Without decoupling, the antenna will still radiate a signal but the feed point impedance will likely vary a great deal from the design value.

2.2 Monopoles

This section includes updates derived or excerpted from Antenna Physics: An Introduction, Second Edition, *by Robert Zavrel, W7SX. The book goes into much more detail, including derivations of key equations and relationships that are included here. – Ed.*

Another simple form of antenna derived from a dipole is called a *monopole*. The name suggests that this is one half of a dipole, and so it is. The monopole is always used in conjunction with a *ground system* that acts as a sort of electrical mirror. See **Figure 2.25**, where a λ/2 dipole and a λ/4 monopole are compared. Power is applied to the monopole's feed point at the junction of the physical element and the ground system. The image antenna for the monopole is the dotted line beneath the ground system. The image forms the missing half of the antenna, transforming a monopole into the functional equivalent of a dipole.

A *ground-mounted* monopole uses the surface layers of the Earth as its ground surface. This might be through a direct connection such as a ground rod or wires buried in or placed directly on the surface of the ground. This is also referred to as a *driven ground* antenna. Monopoles mounted above but not directly connected to the ground also use the ground as part of the antenna but with less and less effect as the antenna is raised. The complicated interaction between a monopole and the ground is discussed in the chapter **Effects of Ground**.

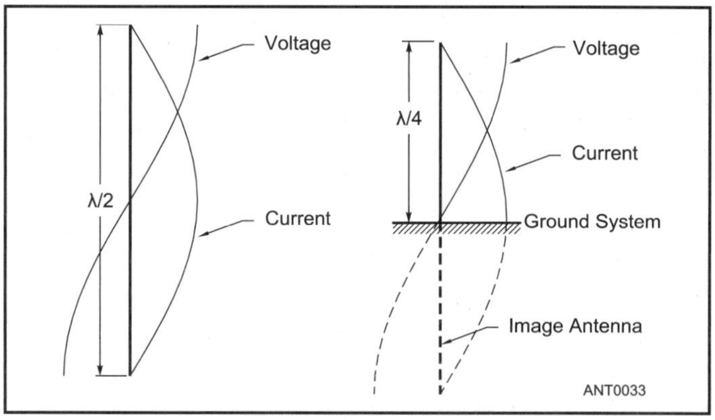

Figure 2.25 — The λ/2 dipole antenna and its λ/4 ground-plane counterpart. The "missing" quarter wavelength is supplied as an image in "perfect" (that is, high-conductivity) ground.

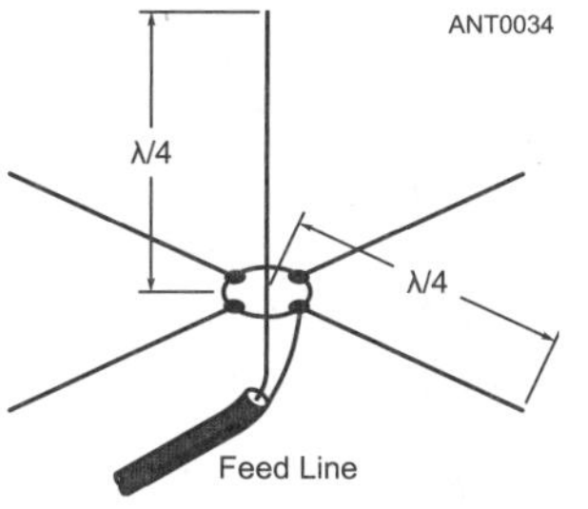

Figure 2.26 — The ground-plane antenna. Power is applied between the base of the vertical radiator and the center of the four ground plane wires.

Monopoles mounted above the ground use a metal surface or a set of conductors to create the ground system as in **Figure 2.26**. The structure is called a *ground plane* and the same name is used for the entire antenna, including the radiating element. Monopoles mounted on the ground with a set of conductors forming the ground system, whether or not they are in actual contact with the ground, are also called ground plane antennas, so the term is somewhat loosely applied.

Since monopoles are usually mounted vertically with respect to the surface of the ground, they are called *vertical monopoles*, or simply *verticals*. A practical ground-mounted vertical's ground system is usually made up of a series of wires extending from the feed point at the base of the antenna, usually symmetrically. An elevated ground plane antenna might use rods or tubing for the same purpose. These are called *radials* since they extend radially around the antenna. Radials in or in close contact with the ground are not resonant but for elevated ground-planes are typically λ/4 long as shown in Figure 2.27. The term *counterpoise* refers to a ground system somewhat above the surface of the ground. See the discussion of counterpoises in the chapter **Effects of Ground**.

The chapter **Effects of Ground** devotes much attention to the requirements for an efficient grounding system for vertical monopole antennas. The chapter **General-Purpose MF and HF Antennas** gives more information on practical ground-plane verticals at HF. Ground-plane antennas at higher frequencies are discussed in the chapters **VHF and UHF Antenna Systems** and **Mobile VHF and UHF Antennas**.

2.2.1 Effective Versus Electrical Height

The word "height" applied to a vertical monopole antenna whose base is on or near the ground has the same meaning as length when applied to λ/2 dipole antennas. Some texts refer to heights in electrical degrees, referenced to a free-space wavelength of 360 degrees, or height may be expressed in terms of the free-space wavelength. There are three different "heights" associated with ground-mounted vertical monopoles. One is the measured *physical height*, h_P, including the image antenna. A second height is the *electrical height*, h_{elec}, which is measured in electrical degrees in terms of the free space wavelength of 360 degrees. It is determined by the distribution of voltage and current along the antenna.

The third height is *effective height*, h_{eff}, which is a function of the physical height. The voltage, V, induced into a monopole antenna with the same polarization as the incident E-field is proportional to the antenna's effective height. i.e., a longer antenna will receive more of the incident E-field energy.

$$h_{eff} = V / E \text{ and } V = h_{eff} E$$

where V is the induced voltage and E is the incident electric field strength in V/m, so h_{eff} is calculated in meters. For example, if E is 20 μV/m and h_{eff} is 6 meters, the induced voltage will be 6 × 20 = 120 μV.

The relationship between h_{eff} and h_P is:

$$h_{eff} = \frac{I_{avg}}{I_{peak}} h_P$$

where I_{avg} is the average current and I_{peak} is the peak value of current along the length of the antenna. Remember that h_P includes the image antenna length.

2.2.2 Radiation Resistance and Feed Point Impedance

MONOPOLE RADIATION RESISTANCE

The aperture of the perfect monopole-plus-ground system is the same as for the dipole but half of the aperture of the full-size λ/2 dipole has been replaced by an electrical image that does not actually exist. This also means that half of the dipole's radiation resistance, R_r, is replaced by the image and so cannot radiate power. The results it that the radiation resistance of a vertical monopole close to λ/4 resonance over a perfect ground system is 36.6 Ω which is one-half of the 73.2 Ω for a resonant λ/2 dipole.

The variation of a monopole's radiation resistance with electrical length or height is shown in **Figure 2.27** from 0 degrees to 270 degrees. Note that the current maximum for monopoles longer than λ/4 will be located above the base of the antenna and have a different value than that at the base. (See the referenced book and articles by Zavrel for a more complete discussion of radiation resistance and feed point impedance.)

MONOPOLE FEED POINT IMPEDANCE

Figure 2.28, which shows the feed point impedance of a vertical antenna made of #14 AWG wire, 50 feet long, located over perfect ground, and fed at the base. Impedance is shown over the whole HF range from 1 to 30 MHz. Again, there is nothing special about the choice of 50 feet for the length of the vertical radiator; it is simply a convenient length for evaluation.

Figure 2.29 shows an expanded portion of the frequency range above and below the λ/4 resonance, but now calibrated in terms of wavelength. Note that this particular antenna goes through λ/4 resonance at a length of 0.244 λ, not at exactly 0.25 λ. The exact length for resonance varies with the diameter of the wire used, just as it does for the λ/2 dipole at its λ/2 resonance. The range shown in Figure 2.29 is from

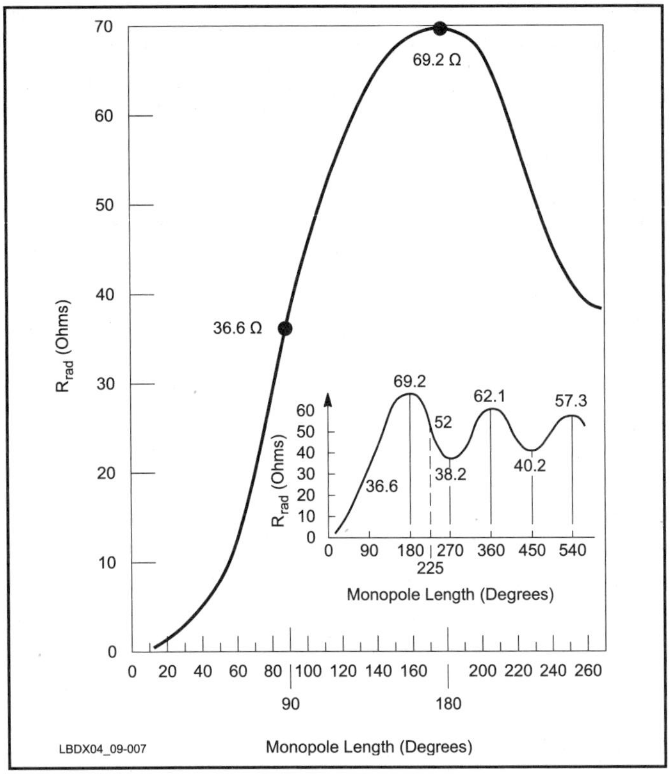

Figure 2.27 — Radiation resistances (at the current maximum) of monopoles with sinusoidal current distribution. The chart can be used for dipoles up to λ/2 in length, corresponding to the λ/4 monopole, but all values must be doubled.

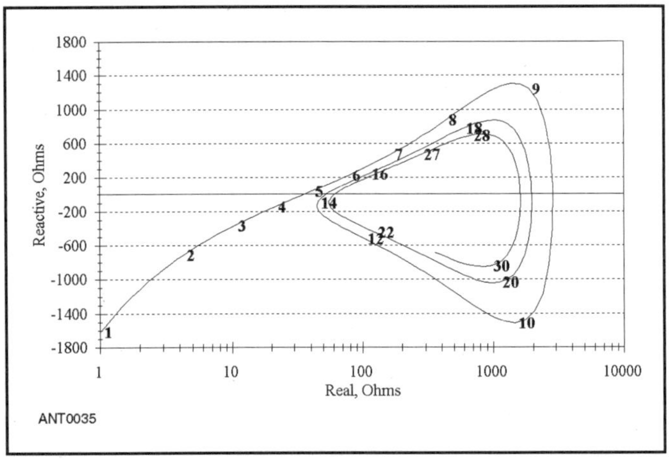

Figure 2.28 — Feed point impedance versus frequency for a theoretical 50-foot-high grounded vertical monopole made of #14 AWG wire. The numbers along the curve show the frequency in MHz. This was computed using "perfect" ground. Real ground losses will add to the feed point impedance shown in an actual antenna system.

0.132 λ to 0.300 λ, corresponding to a frequency range of 2.0 to 5.9 MHz.

The reactive portion of the feed point impedance is highly dependent on the length/diameter ratio of the conductor, as was discussed previously for a horizontal center-fed dipole. The impedance curve in Figures 2.25 and 2.26 is based on a

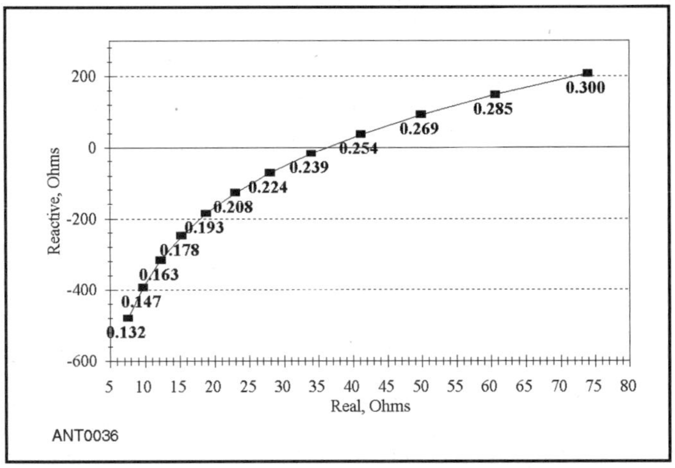

Figure 2.29 — Feed point impedance for the same antennas as in Figure 2.26, but calibrated in wavelength rather than frequency over the range from 0.132 to 0.300 λ, above and below the quarter-wave resonance.

#14 AWG conductor having a length/diameter ratio of about 800 to 1. As usual, thicker antennas can be expected to show less reactance at a given height, and thinner antennas will show more.

2.2.3 Monopole Efficiency

Since an antenna's efficiency is:

$$\eta = \frac{R_r}{R_r + R_{loss}}$$

changing the current distribution along the vertical element can improve efficiency without changing physical height, h_P. This is also discussed in the section on Short Verticals and in the chapter **Mobile and Marine HF Antennas**.

The efficiency of a real vertical antenna over real ground often suffers dramatically compared with that of a λ/2 antenna. Without a fairly elaborate ground system, the efficiency is not likely to exceed 50%, and it may be much less, particularly at physical heights below λ/4. In addition, the gain of a monopole at angles close to the ground plane is highly dependent on the conductivity of the ground plane. Both effects are discussed extensively in the chapter **Effects of Ground**.

There is a simple method to determine R_{loss} for a λ/4 ground-mounted vertical. Adjust the length for resonance (zero reactance but a small amount is acceptable) and measure the feed point impedance resistive component, R_{feed}. Since R_r is very close to 36 Ω:

$$R_{loss} = R_{feed} - 36\ \Omega$$

In this way you can assess the effect of adding radials, changing their lengths, etc. For physical heights, h_P, up to 5/8 wavelengths, you can estimate R_r from Figure 2.28 and use the same equation. **Table 2.3** gives a variety of calculated values for ground-mounted verticals of various lengths from 1/32 to 5/8 wavelengths with no loading.

Table 2.3
R_r for Various Ground-Mounted Vertical Physical Heights With No Loading
Notice that the maximum possible gain for very short verticals converges on the value for an infinitesimal high vertical, or twice the gain of an infinitesimal dipole.

$h_p\lambda$	$h_e\lambda$	dBi	G_{iso}	$A_e\lambda$	$R_r\Omega$
1/32	0.031	4.76	3.00	0.239	0.4
1/16	0.063	4.77	3.00	0.239	1.6
1/8	0.125	4.86	3.06	0.244	6.0
1/4	0.32	5.14	3.27	0.260	36.8
3/8	0.54	5.79	3.79	0.302	92.1
1/2	0.64	6.88	4.88	0.388	99.5
5/8	0.587	8.14	6.52	0.519	62.9

2.2.4 Directivity of a Monopole

The free-space directional characteristics of a λ/4 monopole with its ground plane are very similar to that of a λ/2 antenna in free space. The directivity (*D*) and, thus, gain for the λ/4 monopole over a perfect, infinite ground plane is double (3 dB greater) that of the λ/2 dipole in free space (assuming no losses) because there is no radiation in the hemisphere below the ground plane. Real monopoles over finite ground planes have less gain. Like a λ/2 dipole antenna, the λ/4 monopole has an omnidirectional radiation pattern in the plane perpendicular to the monopole.

Table 2.3 includes the lossless gain values for vertical antennas of various physical lengths. The column labeled "dBi" shows the gain with respect to an isotropic antenna (G_{iso} is just the non-dB numeric equivalent). Note that for verticals λ/8 or shorter, the aperture and gain changes very little.

Ground losses must be taken into account when evaluating the gain of any actual antenna installed over a real ground system as discussed later in this chapter and in the chapter **Effects of Ground**.

DIRECTIVITY FOR HARMONIC OPERATION

Like dipoles, a monopole can also be used on harmonics of its fundamental frequency. The feed point impedance will vary similarly to a dipole's: low for odd harmonics and high on even harmonics. Also like the dipole, the radiation pattern will change dramatically for harmonic operation.

A vertical monopole operated on its harmonics will remain omnidirectional in azimuth, but the elevation pattern will break up into multiple lobes and nulls. Consider Figures 2.15 through 2.21 which show radiation patterns for dipoles at the fundamental, f, through the sixth harmonic, 6f. The elevation pattern for a monopole over a perfect ground plane will be the same as the dipole patterns, turned 90 degrees and divided along the line from 0 to 180 degrees.

As the operating frequency is raised to 3f, a null will develop at about 20 degrees elevation with a wide major lobe just above 45 degrees and a minor lobe at the horizon. Further increases in frequency result in more lobes and nulls, making performance and coverage areas quite different on different bands.

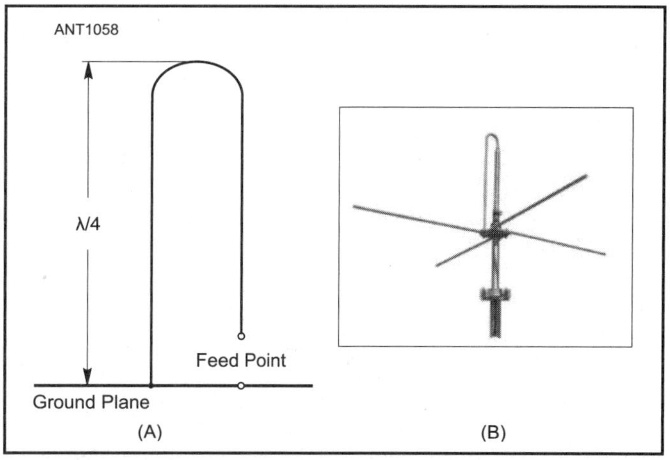

Figure 2.30 — The folded monopole antenna (A) can be understood similarly to the folded dipole with the ground-plane or counterpoise supplying the "missing half" of the antenna with an electrical image. An example of a commercial folded monopole is depicted in (B).

2.2.5 Folded Monopoles

A folded monopole, shown in **Figure 2.30**, can be understood similarly to the folded dipole and the same increase in feed point impedance is achieved (see reference for Cebik). Again, the ground-plane or counterpoise supplies the "missing half" of the antenna with an electrical image. The point opposite the feed point is electrically neutral in the λ/4 folded monopole and so is connected to the ground plane as in Figure 2.30A. An example of a commercial folded monopole is depicted in Figure 2.30B.

The increased feed point impedance of the folded monopole is often misunderstood as reducing ground losses due to the lower current at the feed point. This is incorrect because the radiation resistance and ground losses of both single and multiple-conductor monopoles is the same when correctly normalized to the feed point and so there is no difference when calculating radiation efficiency. For equivalent amounts of power, the same amount of current will flow in the ground system, regardless of the impedance transformation created by folding the conductor.

2.2.6 Loading the Monopole

If the λ/4 monopole antenna is loaded, the current distribution along the antenna will change, as will the average value of current. The current in a λ/4 monopole varies practically sinusoidally (as is the case with a λ/2 dipole) and is highest at the ground system connection. The voltage is highest at the open (top) end and minimum at the ground system as shown in **Figure 2.31A**. If top loading is added as in Figure 2.32B, the current distribution changes, moving the point of maximum current up the antenna. The average value of current in Figure 2.31A is 0.64 I_{peak} and 0.92 I_{peak} in Figure 2.28B. Different amounts of loading will change the average current.

This changes the effective height from $h_{eff} = 0.32\lambda$ for the unloaded antenna to $h_{eff} = 0.46\lambda$ with loading. Radiation

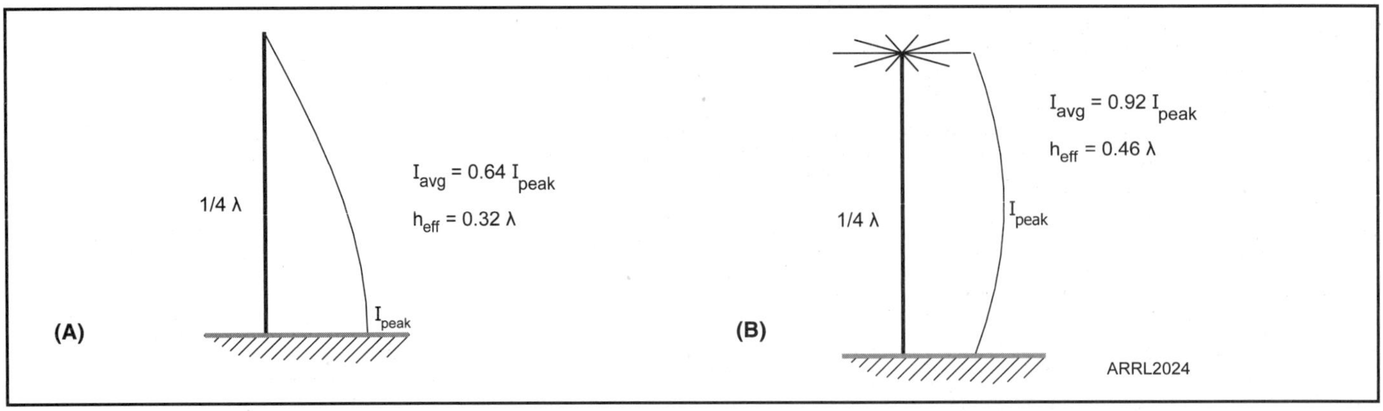

Figure 2.31 — Current distribution along a λ/4 ground-mounted vertical without (A) and with (B) top loading. The unloaded vertical's sinusoidal current distribution waveform starts at 0 (0 degrees) at the top and reaches maximum (90 degrees) at the base feed point.

resistance, R_r, varies with the square of effective height so for the top-loaded vertical, $R_r = 74\ \Omega$.

Figure 2.32 shows how top-loading modifies the current distribution on the vertical. This changes the ratio of I_{avg} and I_{pk}, which modifies h_{eff} and, consequently, R_r, and the feed point resistance, R_{feed}. The usual method is to use a capacity hat of radial wires or tubes. Loading inductors can also be placed in series with the capacity hat to increase the top loading effect. Both the inductor and capacity hat also have resistive losses which must be added to R_{loss} in the efficiency equation. As loading is increased, so will the losses associated with the loading mechanisms. For more information, see the following section on Loading Techniques for Short Verticals. The chapter **Mobile and Marine HF Antennas** discusses the design of popular loading techniques for mobile antennas.

2.2.7 Ground System Effects

The ground-mounted vertical can be broken into three fundamental basic components that directly affect the directivity, efficiency, and gain of the overall antenna system as shown in **Figure 2.33**:

1. The vertical radiator
2. The near-field ground
3. The far-field ground

The height of the vertical radiator and the current distribution along it largely determine the maximum gain (ignoring ground losses for the moment) and R_r of the antenna. The near-field ground is the primary contributor to R_{loss} and with R_r determine the antenna's efficiency. The usual method of reducing near-field losses is to use ground radials that act as a low impedance return path for the antenna's ground currents flowing back to the antenna feed point. Ground characteristics are difficult to characterize accurately.

Limited conductivity of the far-field ground will attenuate the power in the far field, particularly at low elevation angles and further reduces antenna system efficiency. This also shapes the final directivity and gain of the antenna. Because this occurs in the far field, these are considered to be propagation effects. Once a physical height has been chosen, the usual methods of improving antenna system performance available to the average amateur focus on minimizing near-field ground losses through an effective ground system. Far field effects

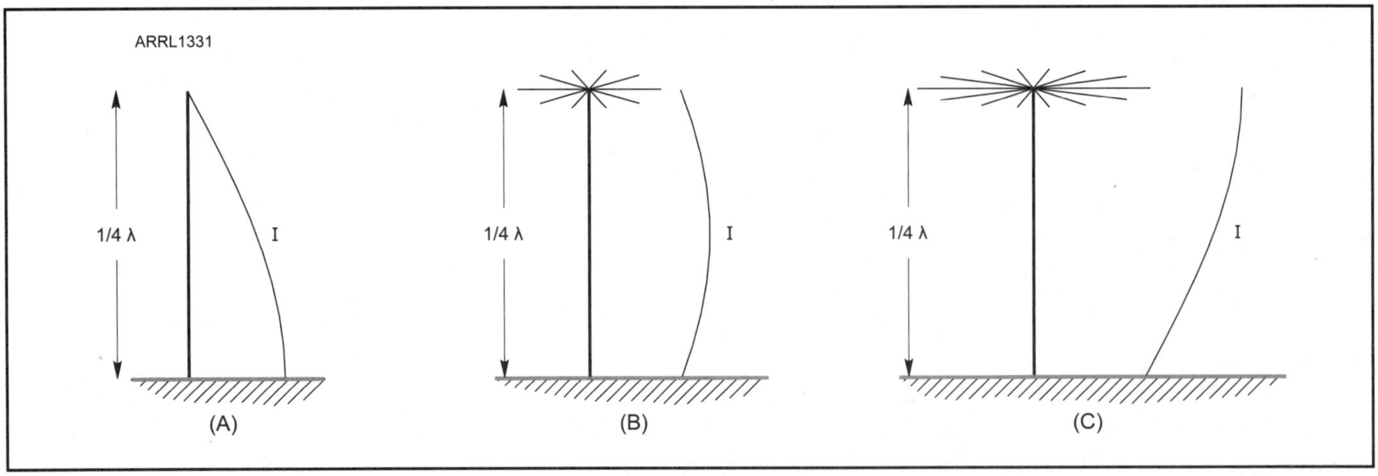

Figure 2.32 — Three configurations for a λ/4 ground-mounted vertical showing the current distribution. (A) shows a typical non-loaded vertical. (B) shows the current distribution that produces maximum R_r. (C) shows the current distribution for maximum R_{feed}.

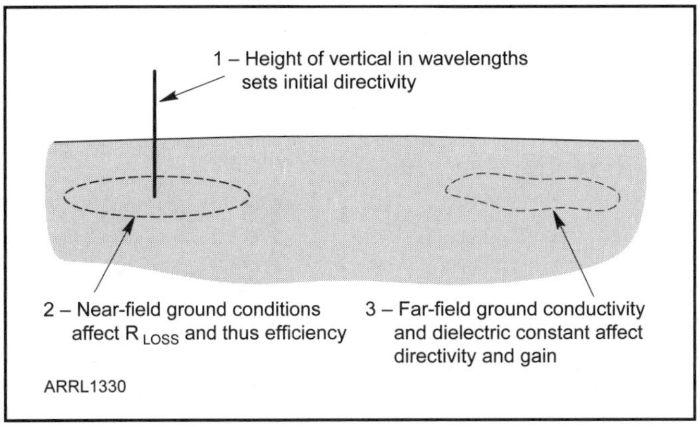

Figure 2.33 — Three factors affecting the performance of a ground-mounted vertical. The height of the vertical radiator determines radiation resistance, R_r, and initial directivity. The near-field ground conditions creates losses that affect the antenna's efficiency and gain. Far-field ground contributes losses that affect directivity at low elevation angles and thus overall system efficiency and gain.

are harder to control but the location of the antenna can optimized. The effects of ground on the antenna system are discussed in more detail in the chapter **Effects of Ground**.

2.2.8 Monopoles Above Ground

Returning to Figure 2.27, we can consider a ground plane antenna in free space. Radiation resistance is approximately 22 Ω. The antenna's gain (ignoring losses) is 1.25 dBi and the radiation pattern is the familiar dipole-like donut. Note that the radials do not direct all of the radiation into the hemisphere above them as a perfect ground plane would do. The purpose of the radials is to create an effective feed system for the antenna and do not affect R_r or gain.

Lowering the antenna toward the Earth, there is a significant effect on the final directivity. At elevations down to about λ/2, R_r remains relatively constant at the free-space 22 Ω value but below that elevation, R_r varies until it returns to about 36 Ω when it reaches the ground. Above λ/2, propagation effects from the ground control the final directivity pattern but the antenna itself is mostly unaffected. As the antenna is lowered to ground level, the image gradually returns and increases antenna gain, h_{eff}, and R_r. In effect, the ground gradually becomes part of the antenna.

There has been a lot of effort made to understand the tradeoffs between ground-mounted verticals and slightly elevated radials. This primarily involves the decreased image in the ground versus an effective decrease in ground losses. Elevated radials are discussed more in the chapter **Effects of Ground**.

2.3 Loading Techniques for Short Verticals

The following section was contributed by Rudy Severns, N6LF, who is well-known for his work with vertical antennas and the effects of ground. The article "The W2FMI Ground-Mounted Short Vertical" by Jerry Sevick, W2FMI, considered one of the best on the subject of short verticals, is included in the online material, as well.

2.3.1 Short Vertical Antennas

A λ/4 vertical can be a simple and efficient antenna but at lower operating frequencies it becomes increasingly difficult to accommodate the full λ/4 height and a set of full-length λ/4 radials. For example, at 3.7 MHz λ/4 ≈ 66 feet, at 1.83 MHz λ/4 ≈ 134 feet, at 475 kHz λ/4 ≈ 518 feet and at 137 kHz λ/4 ≈ 1800 feet — more than ¼ mile! Fortunately, it's not necessary to make the antenna full-size (λ/4). With careful design the size of the antenna can be reduced by half or even much more while retaining reasonable efficiency and radiation pattern. Some form of loading, inductive and/or capacitive, will be needed both for matching and to maximize efficiency. When height falls below 0.1 λ, such as for typical suburban locations, design becomes more difficult but usable antennas are still possible.

The operating bandwidth of a shortened antenna will be reduced because shortened antennas have higher Q. This translates into a more rapid increase of reactance away from resonance. The effect can be mitigated to some extent by using larger-diameter conductors, but bandwidth will still be a problem, particularly on the 3.5 to 4 MHz band which is very wide (13.3%) in proportion to the center frequency of 3.75 MHz.

This section discusses vertical antennas with heights H < λ/4 employing inductive and/or capacitive loading. The focus is on antennas for the current US amateur bands of 80 and 160 meters. (See the **General-Purpose MF and HF Antennas** chapter's section on verticals for 630 and 2200 meters.) The loading techniques discussed here can also be used at higher frequencies and for horizontal antennas. Note that the abbreviation R_r is used for feed point resistance in this section.

One word of advice: very short verticals rely on heavy capacitive top-loading to obtain reasonable efficiency. The details of the loading arrangements will vary at each location due to wide variations in available supports (poles, trees, and so on) or lack thereof. While it is possible to make approximate calculations for each structure to approximate the efficacy of a particular choice, it is much easier to model multiple possibilities using antenna modeling software and then choose the best design. Suitable software is available both free and at modest cost. (See the chapter on **Antenna Modeling**.) The use of modeling software can be of great help in optimizing a design for a particular installation and is strongly recommended.

2.3.2 Efficiency of Short Verticals

The essential problem of short antennas, generally, is that of efficiency. **Table 2.4** illustrates the dramatic effect of reducing an 80-meter vertical's physical height by adding base loading inductance. (Perfect ground and conductors are assumed in this design. The vertical section is 2 inches in diameter.) **Figure 2.34** graphs the feed point resistance (R_r), the magnitude of the capacitive reactance ($|X_C|$) and Q (Q_a) for an ideal vertical as a function of height (H in wavelengths), where $Q_a = |X_C|/R_r$. At resonance $R_r = 36\ \Omega$ and $X_C = 0$ but as H is reduced below resonance R_r falls quickly and $|X_C|$ increases very rapidly. For example, when $H = 0.125\ \lambda$, $R_r \approx 6.5\ \Omega$ and $|X_C| \approx 500\ \Omega$. As H is reduced further, R_r decreases proportional to H^2 and $|X_C|$ increases proportional to $1/H$. As a result, Q_a increases proportional to $1/H^3$! Short antennas have low radiation resistances and high capacitive reactances at the feed point which becomes especially acute on 2200 meters where only very short (in terms of λ) verticals are practical. Short antennas also have very high Q resulting in narrow operating bandwidths.

Table 2.4
Effect of Shortening a Vertical Radiator Below λ/4 Using Inductive Base Loading.
Frequency is 3.525 MHz and for the inductor Q_L=200. Ground and conductor losses are omitted.

Length (feet)	*Length (λ)*	*R_r (Ω)*	*X_C (Ω)*	*R_L (Ω)*	*Efficiency (%)*	*Loss (dB)*
14	0.050	0.96	–761	3.8	20	–7.0
20.9	0.075	2.2	–533	2.7	45	–3.5
27.9	0.100	4.2	–395	2.0	68	–1.7
34.9	0.125	6.8	–298	1.5	82	–0.86
41.9	0.150	10.4	–220	1.1	90	–0.44
48.9	0.175	15.1	–153	0.77	95	–0.22
55.8	0.200	21.4	–92	0.46	98	–0.09
62.8	0.225	29.7	–34	0.17	99	–0.02

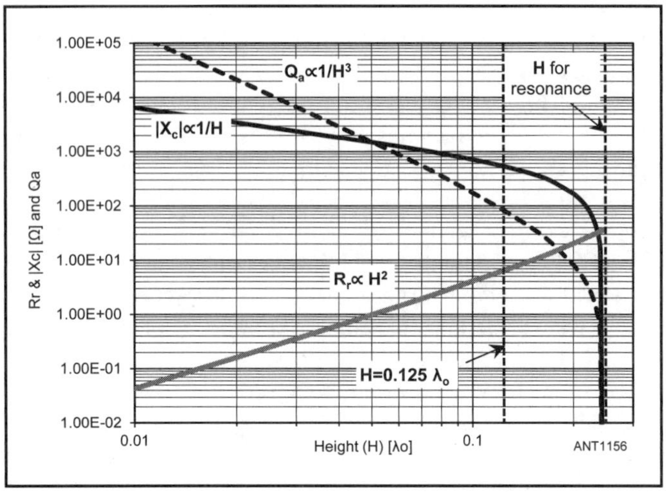

Figure 2.34 — Base impedances and Q of a lossless vertical over perfect ground.

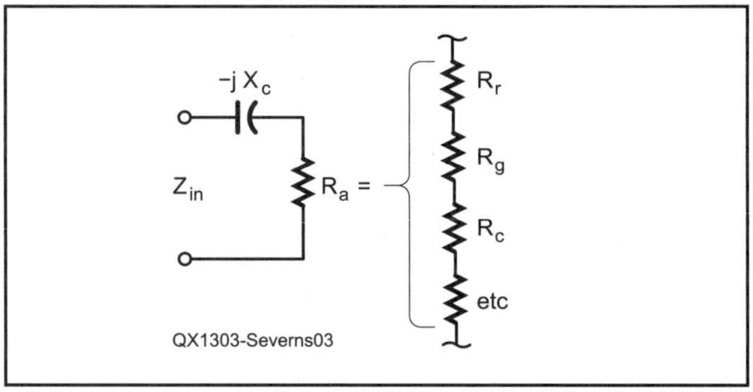

Figure 2.35 — Equivalent circuit for a short vertical including loss elements.

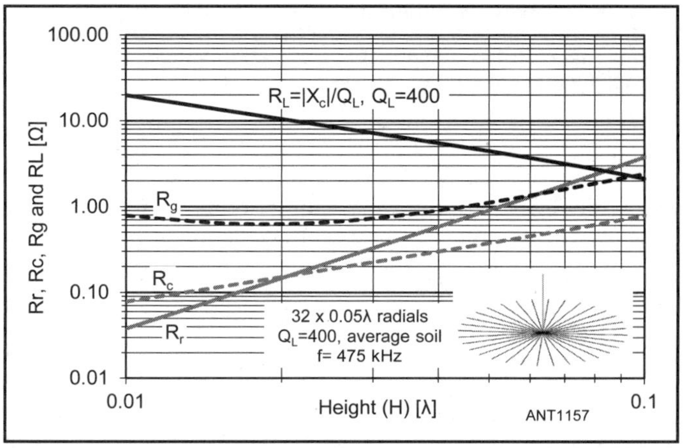

Figure 2.36 — Typical values for R_r, R_c, R_L, and R_g as a function of H.

From Figure 2.35 we can see that a short vertical is essentially a capacitor in series with a resistance as shown by the electrical model in **Figure 2.35**.

In addition to R_r and X_C a real antenna will have several sources of loss:

R_L — Loading coil series resistance where $R_L = X_L/Q_L$, Q_L = inductor Q.

R_g — Equivalent ground loss resistance.

R_c — Conductor resistance.

R_l — Losses due to leakage across the base insulator and insulators at wire ends.

R_{cor} — Corona loss at high voltage points. This can be a problem at higher altitudes.

R_n — Matching network losses.

Figure 2.36 shows an example of typical values of R_r, R_c, R_g and R_L for a short vertical with a radial ground system over average soil. This example is for 630 meters but 160 and 80 meter verticals of the same electrical heights (in λ) will be similar.

Antenna efficiency (η) can be expressed as:

$$\eta = \frac{R}{R_r + (R_L + R_g + R_c + R_l + R_{cor} + R_n)} = \frac{R_r}{R_r + R_{loss}}$$

The design goal is to increase efficiency by increasing R_r and decreasing R_{loss}.

2.3.3 Base Loading

A vertical with $H < \lambda_o/4$ will require some form of loading and matching. The base of the antenna is a convenient point at which to add a loading inductor as shown in **Figure 2.37**.

Ignoring R_l, R_{cor}, and R_n for the present, we can use the preceding equation for efficiency and the loss resistance values in Figure 2.36 to graph the efficiency of that antenna as shown in **Figure 2.38**. H=0.01λ corresponds to a 72-foot vertical at 137 kHz. The efficiency is <0.2%!

The dashed line in Figure 2.39 shows the efficiency when only R_L is included compared to the sum of R_L and R_g (the solid line). This graph has a strong message: *Make the vertical as tall as possible (when H<λ/4) because even small increases in H can substantially improve efficiency.*

This graph also illustrates the low efficiency intrinsic to short base-loaded verticals. The efficiency of very short verticals is dominated by R_L because the resonating inductance value is large and therefore R_L is large ($R_L = X_C/Q_L$). We can increase Q_L to reduce R_L but there are practical limits.

The base of the antenna is a convenient point at which to add a loading inductor, but it's usually not the lowest loss point at which an inductor could be placed. There is an extensive discussion of the optimum location for the loading coil in a short vertical as a function of ground loss and Q_L in the **Mobile and Marine HF Antennas** chapter. This information should be reviewed before using inductive loading. The program *MOBILE.EXE* is available from the online website for this book. It is an excellent tool for designing short, inductively loaded antennas. A loading inductor at the base has no effect on R_r but moving the loading coil from the base to near the middle can increase R_r significantly.

2.3.4 Capacitive Top-Loading

As shown above, inductive loading is not a very efficient way to compensate for reduced antenna height. Capacitive top loading can be much more effective as shown in the example illustrated by **Table 2.5** for a top-loaded 3.525 MHz vertical made of 2-inch tubing. The vertical section is L_1 and the horizontal top-loading section is L_2, also a piece of 2-inch tubing adjusted to make the antenna resonant. As with the previous example, perfect ground and conductors are assumed.

A simple example using a wire suspended between two

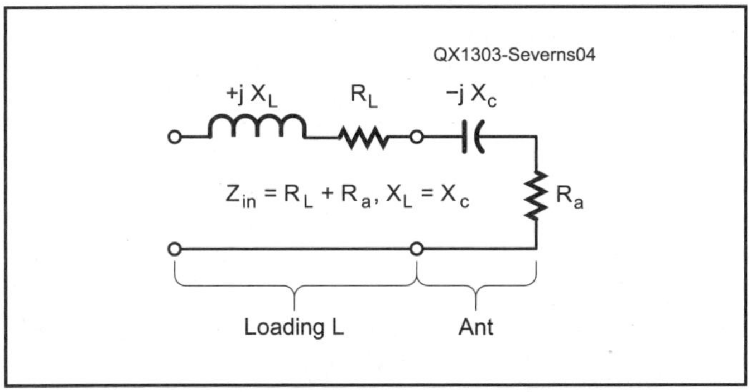

Figure 2.37 — Equivalent circuit for inductive base loading.

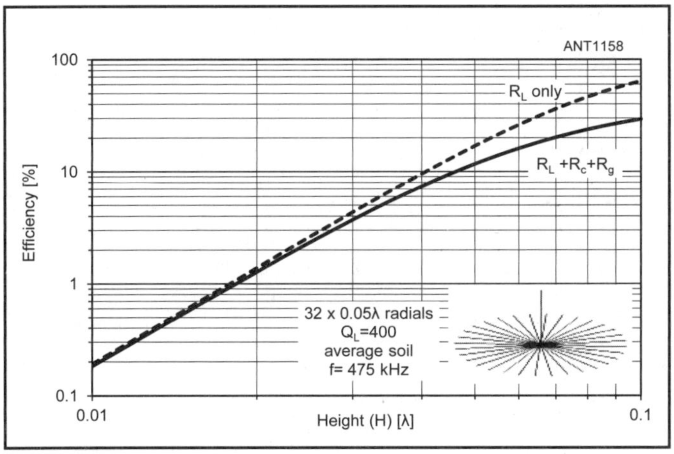

Figure 2.38 — Example of efficiency as a function of height.

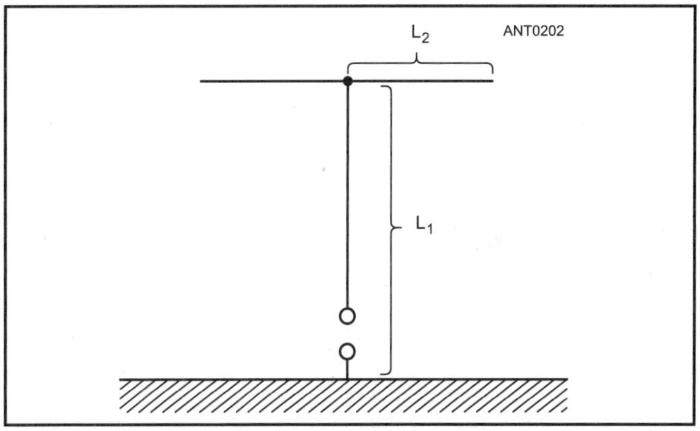

Figure 2.39 — Example of a "T" top-loaded vertical.

Table 2.5
Effect of Shortening a Vertical Using Top Loading

Frequency is 3.525 MHz and vertical is made of 2-inch tubing

L1 (feet)	*L2 (feet)*	*Length (λ)*	*R_r (Ω)*
14.0	48.8	0.050	4.0
20.9	38.6	0.075	8.5
27.9	30.1	0.100	14.0
34.9	22.8	0.125	19.9
41.9	17.3	0.150	25.5
48.9	11.9	0.175	30.4
55.8	7.0	0.200	33.9
62.8	2.4	0.225	35.7

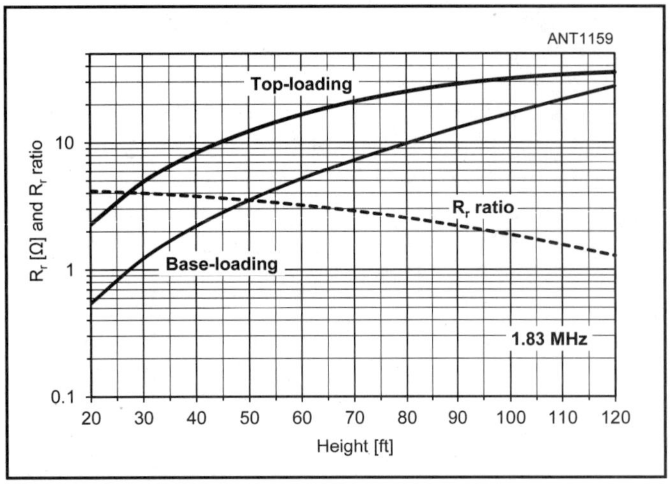

Figure 2.40 — R_r comparison between top and base loading.

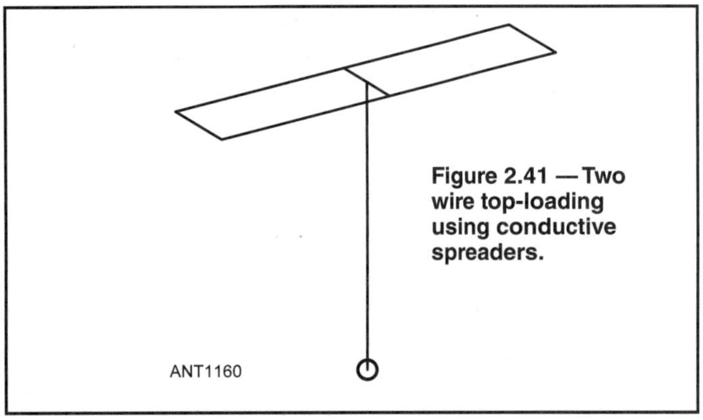

Figure 2.41 — Two wire top-loading using conductive spreaders.

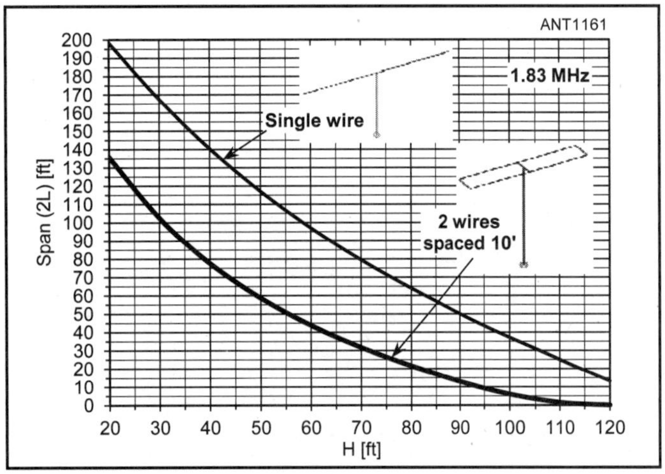

Figure 2.42 — Span (=2L) of the top hat for one or two wires.

supports forming a "T" antenna is shown in **Figure 2.39**. We can model this antenna varying H and adjusting L to resonate the antenna to illustrate the advantages of top-loading. A comparison of R_r between capacitive top-loading and inductive base loading is given in **Figure 2.40**. The dashed line shows the ratio of R_r-top to R_r-base. For a given vertical height, resonating the antenna with top-loading results in much higher radiation resistance R_r.

Compared to inductive base loading, top loading significantly increases R_r while base loading does not. For example, if H=20 feet (≈0.04λ @1.83 MHz) R_r with top loading is over four times that for base loading. Besides increasing R_r, top loading can reduce or eliminate R_L due to reduction in value of the base inductor. The net result can be a dramatic improvement in efficiency! As shown in **Figure 2.41** we can use multiple wires to obtain even more top loading.

Figure 2.42 compares the span (= 2L, see Figure 2.39) of the top-loading wire(s) when using one or two wires. The use of two wires with conductive spreaders such as aluminum tubing significantly reduces the spacing (span) required between the supports for the top-loading wires. The span can be further reduced by using longer spreaders and additional well-spaced parallel wires. The design of this kind of top loading will vary with every installation. Antenna modeling software will be of great assistance for optimizing the design. (L.B. Cebik, W4RNL, also analyzed top-hat loading, comparing it to inductive loading. See the References and Bibliography.)

When no supports other than the vertical itself are available, a top hat using rigid spreaders can be used. **Figure 2.43** is a sketch of this kind of "wagon-wheel" top hat along with the required radius for resonance at 1.83 MHz for a range of vertical heights. A simple way to make a capacitance hat would be to take four to six 8-foot CB mobile whips, arrange them like spokes in a wagon wheel and connect the ends with a peripheral wire. This arrangement will produce a 16-foot diameter hat that is economical and very durable. The whip lengths can be extended further with lengths of aluminum tubing. Another approach for a large hat would be to salvage the hub and spreaders from an old 20-meter quad and use them for the wagon-wheel.

Unfortunately, radii greater than ≈15 feet become increasingly impractical. The rigid supports can be replaced with wires sloping downward as shown in **Figure 2.44** to form

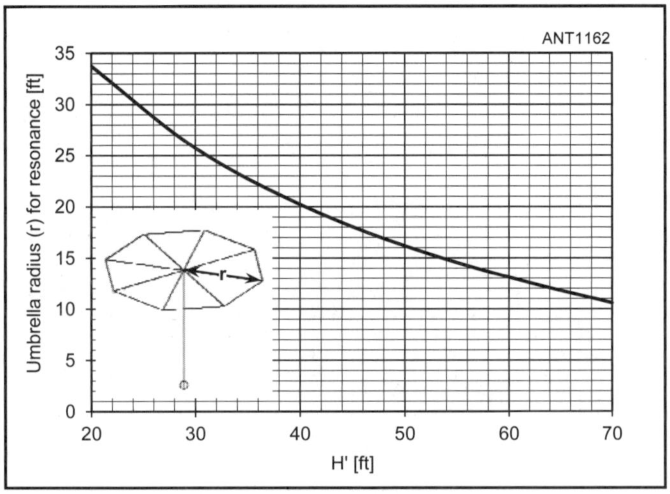

Figure 2.43 — Wagon wheel top-hat example.

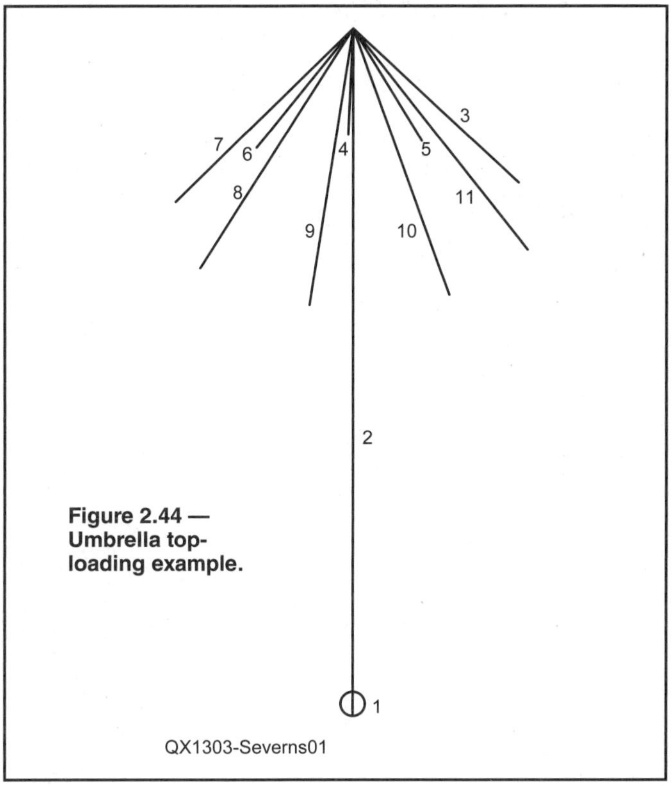

Figure 2.44 — Umbrella top-loading example.

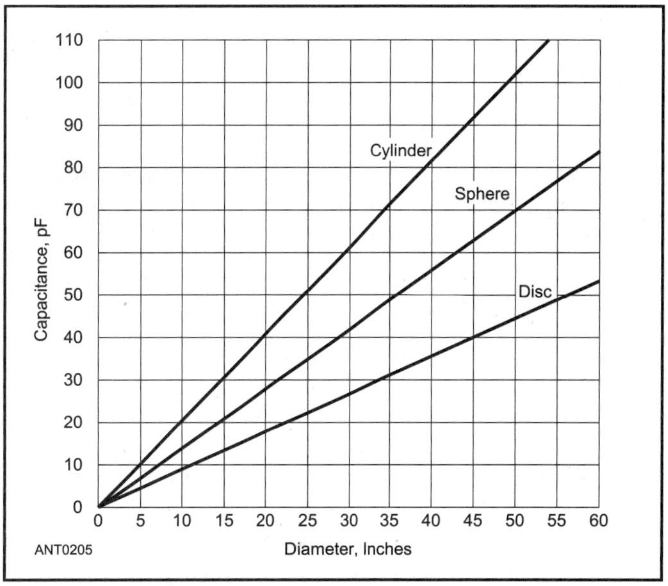

Figure 2.45 — Capacitance of sphere, disc and cylinder as a function of their diameters. The cylinder length is assumed equal to its diameter.

what is called an "umbrella" vertical. To increase the loading effect of the umbrella more wires can be used as well as a skirt wire like that shown in Figure 2.44. The wires can also be made longer but there is a point where the reduction in R_L due to a smaller resonating inductor is offset by the decreasing value of R_r due to canceling currents flowing in the umbrella wires. This is situation where optimization is best done using modeling software.

FINDING CAPACITANCE HAT SIZE

Practically any sufficiently large metallic structure can be used for capacitive loading and the structure doesn't even have to be symmetric, but simple geometric forms such as the sphere, cylinder and disc are preferred because of the relative ease with which their capacitance can be calculated.

While antenna modeling software is very helpful for the design of top loading structures the capacitance of common geometric forms can be estimated from the curves of **Figure 2.45** as a function of size. For the cylinder, the length is specified equal to the diameter. The sphere, disc and cylinder can be constructed from sheet metal, if such construction is feasible, but the capacitance will be almost the same if a "skeleton" type of construction with screen or wire networks or tubing is used.

The required value of the capacitance may be determined using the following procedure. The information in this section is based on a September 1978 *QST* article by Walter Schulz, K3OQF (see References and Bibliography). The physical length of a shortened antenna can be found from:

$$h_{inches} = \frac{11,808}{F_{MHz}} \times \lambda$$

where h is length in inches and λ is the elect\rical length in wavelengths.

Thus, using an example of 7 MHz and a shortened length of 0.167 λ, h = 11808/7 × 0.167 = 282 inches, equivalent to 23.48 feet.

Consider the vertical radiator as an open-ended transmission line, so the impedance and top loading may be determined. The characteristic impedance of a vertical antenna can be found from

$$Z_0 = 60\left[\ln\left(\frac{4h}{d}\right) - 1\right]$$

where

ln = natural logarithm
h = length (height) of vertical radiator in inches (as above)
d = diameter of radiator in inches

The vertical radiator for this example has a diameter of 1 inch. Thus, for this example,

$$Z_0 = 60\left[\ln\left(\frac{4 \times 281}{1}\right) - 1\right] = 361\ \Omega$$

The capacitive reactance required for the amount of top loading can be found from

$$X_c = \frac{Z_0}{\tan\theta}$$

where

X_C = capacitive reactance, ohms
Z_0 = characteristic impedance of antenna
θ = amount of electrical loading, degrees.

This value for a 30° hat is 361/tan 30° = 625 Ω. This capacitive reactance may be converted to capacitance with the following equation,

$$C = \frac{10^6}{2\pi f X_c}$$

where

C = capacitance in pF
f = frequency, MHz
X_C = capacitive reactance, ohms (from above).

For this example, the required C = $10^6/(2\ \pi \times 7 \times 625)$ = 36.4 pF, which may be rounded to 36 pF. A disc capacitor is used in this example. The appropriate diameter for 36 pF of hat capacitance can be found from Figure 2.45. The disc diameter that yields 36 pF of capacitance is 40 inches.

COMBINED LOADING

As an antenna is shortened, the size of the top-loading device will become larger and at some point, it will become impractical to resonate the antenna with top-loading only. In that situation inductive loading, usually placed either at the base or directly between the capacitance "hat" and the top of the antenna, can be added to resonate the antenna. An alternative would be to use linear loading in place of inductive loading.

SHORTENING RADIALS

Very often the space required by full-length radials is not available. Like the vertical portion of the antenna, the radials can also be shortened and loaded in very much the same way. An example of end-loaded radials is given in **Figure 2.46A**. Radials half the usual length can be used with little reduction in efficiency but, as in the case of top loading, the antenna Q will be higher and the bandwidth reduced. As shown in Figure 2.47B**,** inductive loading can also be used. As long as they are not made too short (down to 0.1 λ) loaded radials can be efficient — with careful design.

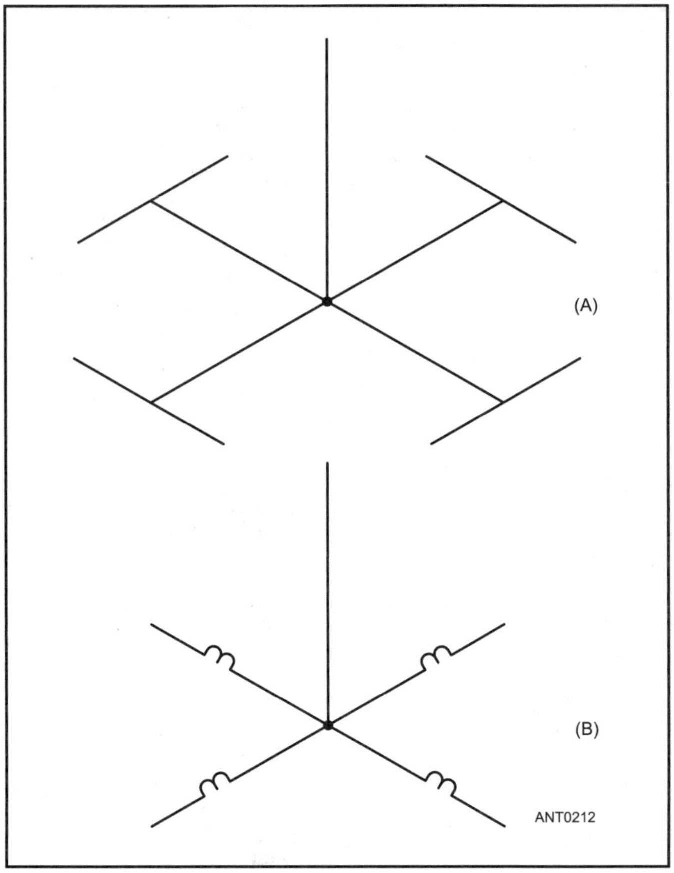

Figure 2.46 — Radials may be shortened by using either capacitive (A) or inductive (B) loading. In extreme cases both may be used but the operating bandwidth will be limited.

2.3.5 General Rules for Loading Vertical Antennas

Sound advice on the design of LF and MF antennas was given many years ago by Woodrow Smith: "The main object in the design of low frequency transmitting antenna systems can be summarized briefly by saying that the general idea is to get as much wire as possible as high in the air as possible and to use excellent insulation and an extensive ground system."

We can codify this advice in order of priority:

- Make the height as great as practical up to the point where H = λ/4.
- Provide as much top loading as possible.
- Make the diameter of the vertical section large. Tubing or a cage of smaller wires will work well.
- If the capacitive loading is insufficient, resonate the antenna with a high-Q inductor placed between the hat and the top of the antenna.
- For buried-ground systems, use as many radials (>0.2 λ) as possible, 32 or more is best.
- If an elevated ground plane is used, use 12 or more radials, 5 or more feet above ground.
- Use of high-quality insulators at the base and wire ends.
- If shortened radials must be used, capacitive loading is preferable to inductive loading.

2.4 References and Bibliography

The dipole and monopole are discussed extensively in the references listed in the References and Bibliography for the **Antenna Fundamentals** chapter.

Cebik, LB, W4RNL, "What Is a Folded Monopole? Skirting the Issue," *Antennex*, 1997.

Devoldere, J., ON4UN, *ON4UN's Low-Band DXing*, Fifth Edition (ARRL, 2010).

Elliott, R., *Antenna Theory and Design* (Wiley, 1981). Reissued by Wiley-IEEE, 2003, pp. 297 – 304.

Hansen, R., "Carter Dipoles and Resonant Dipoles," *Allerton Antenna Applications Symposium*, pp. 282 – 287, Monticello, IL, Sep. 21 – 23, 2010.

Lee, P., *The Amateur Radio Vertical Antenna Handbook*, 2nd Edition (CQ Publishing, 1984).

Richter, J., "Off-Center-Fed Dipole Antennas: Theory and Practice," *HF Dipole Antennas for Amateur Radio* (ARRL, 2019) pp. 105 – 123.

Schelkunoff, S., "Theory of Antennas of Arbitrary Size and Shape," *Proceedings of the IRE*, Sep 1941; corrections Nov 1941 and Jan 1943. Republished in *Proc. IRE*, Sep. 1984.

Silver, W., NØAX, "Hands-On Radio: Experiment #84 — Antenna Height," *QST*, Nov. 2009, pp. 64 – 65.

Silver, W., NØAX, "Hands-On Radio: Experiment #92 — The 468 Factor," *QST*, Sep. 2010, pp. 53 – 54.

Stearns, S., K6OIK, "Antennas: The Story from Physics to Computational Electromagnetics," *Antenna Seminar, ARRL Pacificon*, San Ramon, CA, Oct 19 – 21, 2019.

Stearns, S., K6OIK, "Antenna Modeling for Radio Amateurs," *Antenna Seminar, ARRL Pacificon*, San Ramon, CA, Oct 20 – 22, 2017. Archived at **www.fars.k6ya.org/docs**.

Tai C., and Long, S., "Dipoles and Monopoles," *Antenna Engineering Handbook*, 3rd Edition, R.C. Johnson Ed. (McGraw-Hill, 1993).

Zavrel, R., W7SX, *Antenna Physics: An Introduction, Second Edition* (ARRL, 2020)

Zavrel, R., W7SX, "Maximizing Radiation Resistance in Vertical Antennas," *QEX*, July/Aug. 2009, pp. 28 – 33.

Zavrel, R., W7SX, "Radiation Resistance, Feed Point Impedance and Mythology," *QEX*, Mar./Apr. 2014, pp. 32 – 35.

TABLE OF CONTENTS

Chapter 3 — Online Material

Articles

- "Determination of Soil Electrical Characteristics Using a Low Dipole" by Rudy Severns, N6LF
- "Maximum-Gain Radial Ground Systems for Vertical Antennas" by Al Christman, K3LC
- "Some Thoughts on Vertical Ground Systems over Seawater" by Rudy Severns, N6LF
- "Radiation and Ground Loss Resistances in LF, MF, and HF Verticals: Parts 1 and 2" by Rudy Severns, N6LF
- "Determination of Soil Electrical Characteristics Using a Low Dipole" by Rudy Severns, N6LF
- "The Case of Declining Beverage-on-Ground Performance" by Rudy Severns, N6LF
- FCC Ground Conductivity Map Set

Chapter 3
The Effects of Ground

The **Antenna Fundamentals** chapter dealt mainly with ideal antennas in free space, completely removed from the influence of ground. Real antennas however, are placed over ground and in some cases have components right on or even buried in the ground. The presence of ground can have a profound effect on the behavior of an antenna, including the feed point impedance, the efficiency and the radiation pattern. This chapter is devoted to describing the interactions between antennas and ground and ways to reduce ground losses close to the antenna. For the purposes of this chapter the terms "soil" or "earth" are considered equivalent to "ground". In some cases "ground" may actually be fresh water or seawater.

We will begin by examining the characteristics of typical soils and then proceed to interactions between grounds and antennas. The interaction discussion is divided between two areas around the antenna: the *reactive near field* and the *radiating far field*. The reactive near field only exists very close to the antenna itself, essentially within one wavelength. In this region the antenna acts as though it were a large lumped-constant R-L-C tuned circuit where energy is stored in the fields close to the antenna. Only a portion of this energy is radiated. The RF current in the antenna will induce currents in the ground which in turn will affect the currents in the antenna. These interactions can modify the feed point impedance of an antenna and, due to the currents flowing in the ground, add power losses. This loss represents power supplied to the antenna from the transmitter but not radiated so there is a net reduction in signal for a given power input to the antenna. For vertical antennas located on or near ground, this can be very significant.

In the radiating far field, the presence of ground profoundly influences the radiation pattern of an antenna. (The radiating near field can be neglected as a transition zone between the reactive near field and radiating far field.) The interaction differs depending on the antenna polarization with respect to the ground. For horizontally polarized antennas, the shape of the radiated pattern in elevation plane depends primarily on the antenna's height above ground. For vertically polarized antennas, both the shape and the strength of the radiated pattern in the elevation plane strongly depend on the nature of the ground itself, as well as the height of the antenna above ground.

The material in this chapter assumes a flat ground surface surrounding the antenna. An extensive discussion of how to account for non-flat ground is presented in the chapter **HF Antenna System Design**, including use of the *HFTA* terrain analysis software by Dean Straw, N6BV.

3.1 Effects of Ground in the Reactive Near Field

Sections 3.1 and 3.2 of this chapter were expanded and reworked by the original author, Rudy Severns, N6LF to accommodate the results of work for the 24th edition.

3.1.1 Electrical Characteristics of Ground

One way to investigate the characteristics of a given sample of soil would be to fabricate a simple parallel plate capacitor as shown in **Figure 3.1**. First we might measure the capacitance (C_S) and the shunt resistance (R_S) without any soil between the plates. We would expect to get a very high value for R_s and some modest capacitance proportional to the plate areas and inversely proportional to the plate spacing. If we then fill the space between the plates with the soil we're interested in and repeat the R_s and C_s measurements, the chances are we will see a marked change in both: much lower R_S and higher C_S. What this experiment tells us is that soil acts like a lossy capacitor. When an RF current flows in the soil there will be some loss associated with R_s. The trick is to keep the RF current out of the soil at least near the antenna.

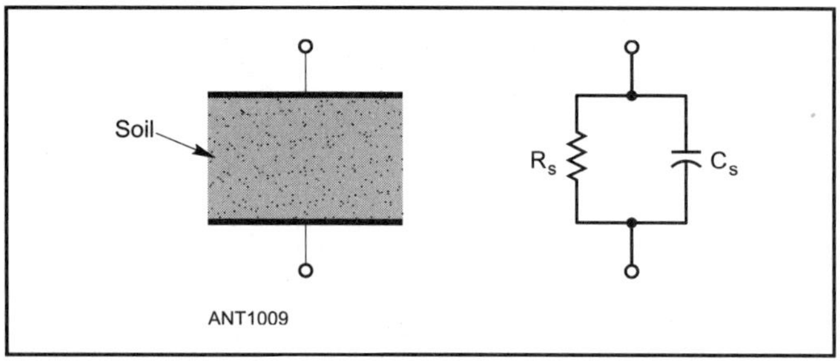

Figure 3.1 — Equivalent circuit for soil characteristics.

R_s is inversely related to the soil conductivity (σ) and C_s is directly related to the relative permittivity (dielectric constant) (ε_r or E_r as represented later in this chapter). We can infer values for σ and ε_r from measurements made on the capacitor with and without the soil between the plates. The unit for σ is siemens per meter (S/m). ε_r is dimensionless. At HF both σ and ε_r are needed for the determination of ground losses or radiation patterns and are an important part of antenna modeling.

A century of measurements on different soils has shown that both σ and ε_r vary over a wide range depending on location, soil composition, stratification of the soil, soil moisture content and many other variables. **Table 3.1** lists typical characteristics for a variety of typical grounds.

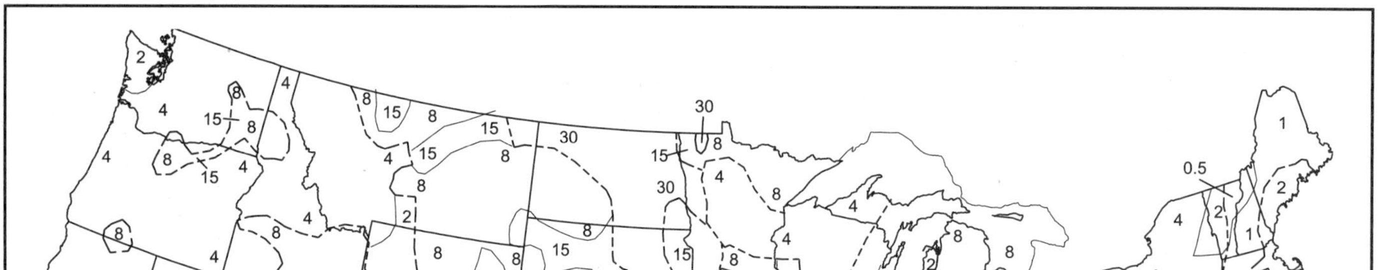

Figure 3.2 — Estimated effective ground conductivity in the United States. FCC map prepared for the Broadcast Service, showing typical conductivity for continental USA. Values are for the band 500 to 1500 kHz. Values are for flat, open spaces and often will not hold for other types of commonly found terrain, such as seashores, river beds, etc.

Table 3.1
Conductivities and Dielectric Constants for Common Types of Earth

Surface Type	*Dielectric Constant*	*Conductivity (S/m)*	*Relative Quality*
Fresh water	80	0.001	
Saline	80	0.5 or more	Salty water but not salt water
Salt water	81	5.0	
Pastoral, low hills, rich soil, typ Dallas, TX, to Lincoln, NE areas	20	0.0303	Very good
Pastoral, low hills, rich soil typ OH and IL	14	0.01	
Flat country, marshy, densely wooded, typ LA near Mississippi River	12	0.0075	
Pastoral, medium hills and forestation, typ MD, PA, NY, (exclusive of mountains and coastline)	13	0.006	
Pastoral, medium hills and forestation, heavy clay soil, typ central VA	13	0.005	Average
Rocky soil, steep hills, typ mountainous	12-14	0.002	Poor
Sandy, dry, flat, coastal	10	0.002	
Cities, industrial areas	5	0.001	Very Poor
Cities, heavy industrial areas, high buildings	3	0.001	Extremely poor

Real soils seldom have these exact pairs of σ and ε_r. For a given value of σ, ε_r can vary widely. Both σ and ε_r tend to increase with soil moisture content so it is normal to have higher ε_r when you have higher σ. However, it is also possible to have moderate values of σ but quite high values for ε_r. Soils with clay particles often have high ε_r. For fresh water at 23° C, $\varepsilon_r = 78$, so you may wonder how soil can have an ε_r higher than water. The higher values are the result of polarization effects that can occur in clay soils. It is quite possible to have $\varepsilon_r > 100$, at least at lower HF frequencies. In general, conductivity will increase with frequency and permittivity will decrease initially at lower HF but level out at higher frequencies.

Much of the data on soil conductivity stems from work at broadcast band frequencies. **Figure 3.2** is a graphic of typical ground conductivity for the United States. While useful for BC (AM broadcast) station planning this graphic is of limited use to amateurs because it averages the conductivity over large areas and the primary concern is ground wave propagation at BC frequencies. Amateurs are usually more concerned with the soil close to their antennas where the conductivity can vary dramatically from the large area average. (A set of detail maps showing ground conductivity is available as the file "FCC-GroundMap.zip" in this chapter's online material.)

Soil characteristics vary not only with location and time of year but also with frequency. **Figures 3.3** and **3.4** show the variation of σ and ε_r with frequency at two locations at a typical amateur QTH (N6LF). See this chapter's section

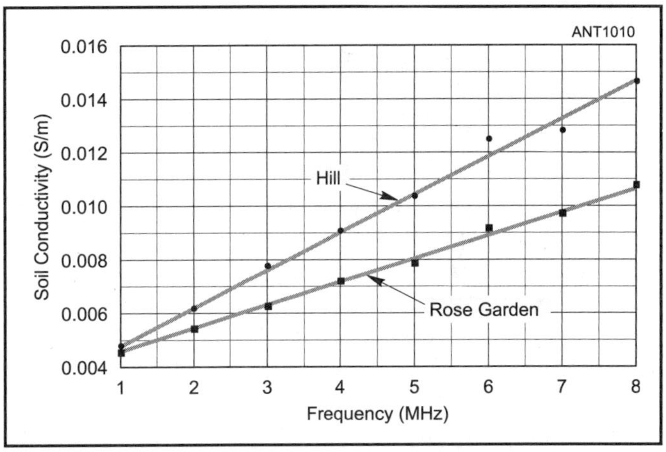

Figure 3.3 — Typical soil conductivity variation with frequency.

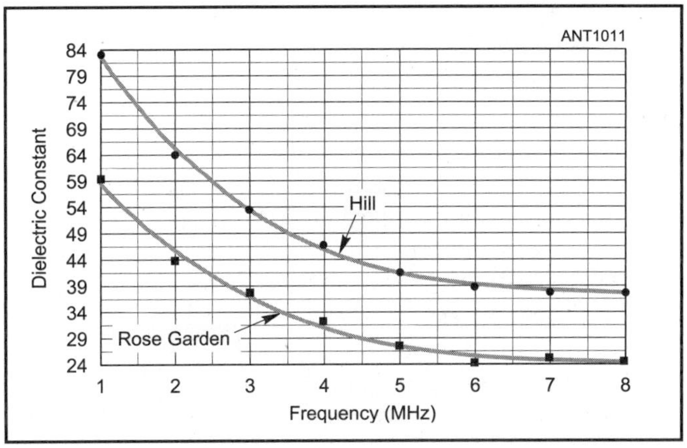

Figure 3.4 — Typical soil permittivity variation with frequency.

Soil Impedance vs Frequency

One of the limitations of presenting maps and charts of soil conductivity and permittivity is the extreme variability of soil. Not only does soil vary electrically from place to place but with frequency as well. In 1975, Longmire and Smith published "A Universal Impedance For Soils" (Nuclear Defense Agency report DNA3788T) that presented a consistent profile for changes in soil impedance with frequency that could be scaled and/or offset to match a specific soil type — the shape of a soil's impedance versus frequency graph was consistent. This report is discussed in Rudy Severn, N6LF's online report "Comments on Longmire and Smith" at **www.antennasbyn6lf.com/design_of_radial_ground_systems**.

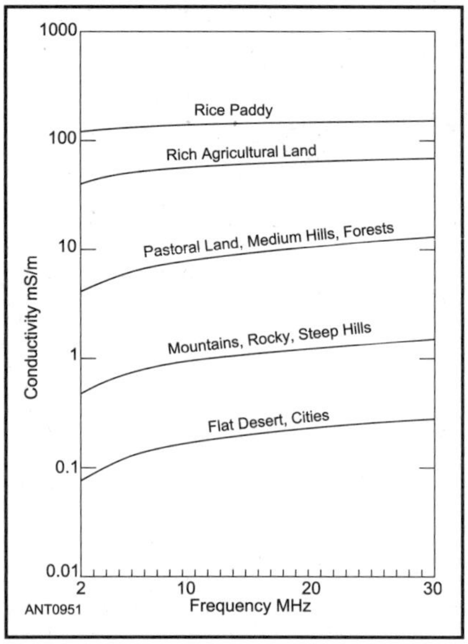

Figure 3.5 — Ground conductivity variation with frequency for different types of soils.

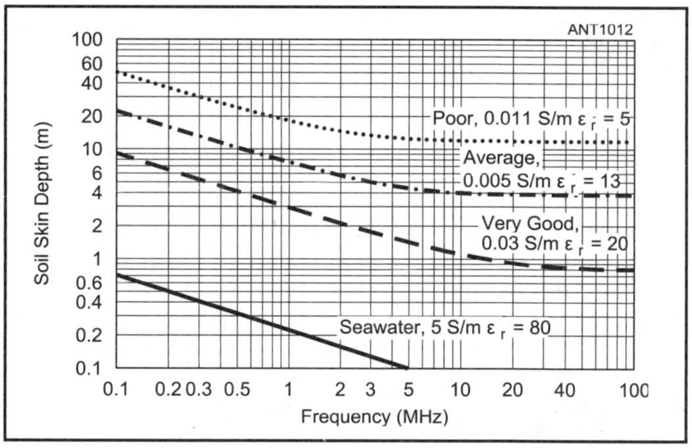

Figure 3.6 — Examples of skin depth variation with frequency for different grounds.

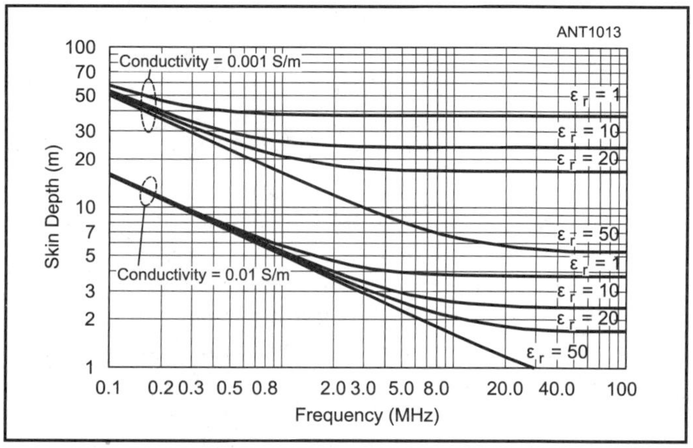

Figure 3.7 — Examples of skin depth as a function of ε_r for two different conductivities.

"Ground Parameters for Antenna Analysis" for methods of measuring ground parameters for antenna modeling and design.

George Hagn and his associates at SRI have made a very large number of ground characteristic measurements at many different places in the world.[1] **Figure 3.5** shows the results of some of this work.

3.1.2 Skin Depth In Soil

It is very probable that the soil at a given location will be stratified (vary with depth) so it will be necessary to take some average value. The question is then "how deep do I have to go to make the average?" This question is best answered by determining the depth to which the fields or the RF currents penetrate the soil. This penetration depth is often expressed in terms of the "skin depth" where the skin depth (δ) is the depth at which the current or the field has been attenuated to 1/e or 37% (e ≈ 2.71828) of its value at the ground surface. Skin depth is also used in the calculation of ground loss.

Knowing σ and ε_r, the skin depth in an arbitrary material can be determined from:

$$\delta = \left(\frac{\sqrt{2}}{\omega\sqrt{\mu\varepsilon}}\right)\left[\sqrt{1+\left(\frac{\sigma}{\omega\varepsilon}\right)^2}-1\right]^{-1/2} \quad (1)$$

where

δ = skin or penetration depth [meters]
ω = 2πf, f = frequency [hertz]
σ = conductivity [siemens/meter, S/m]
$\mu = \mu_r \mu_o$ = permeability
μ_o = permeability of vacuum = $4\pi 10^{-7}$ [henry/meter]
μ_r = relative permeability [dimensionless]
$\varepsilon = \varepsilon_r \varepsilon_o$ = permittivity [farad/meter]
ε_o = permittivity of vacuum = 8.854×10^{-12} [farad/meter]
ε_r = relative permittivity [dimensionless]

A graph of Eq 1 for typical grounds is given in **Figure 3.6**.

Skin depth varies with frequency and soil characteristics. For example, at 1.8 MHz δ varies from about 16 cm in seawater to 15 m in poor soil. As we go up in frequency the skin depth decreases, roughly proportional to $1/\sqrt{f}$, until at some point it flattens out.

The soil types in Figure 3.6 represent the typical values used in antenna modeling. An example of the effect of differing ε_r for soils with σ = 0.001 and 0.01 S/m is shown in **Figure 3.7**. In Figure 3.7, we can see several interesting things. At low frequencies (in the BC band) the values for δ converge and ε_r makes little difference. This is one reason why soil characteristics from BC data seldom include the permittivity. At high frequencies the curves are flat with a value that depends on σ and ε_r.

3.1.3 Wavelength In Soil

Because soil is a complex medium where both σ and ε_r are significantly different from their values in free space, the wavelength in soil (λ) may differ greatly from the wavelength in free space (λ_o). This is important for antennas and radial systems close to or buried in the ground. In general the wavelength in soil will be considerably shorter than the free space

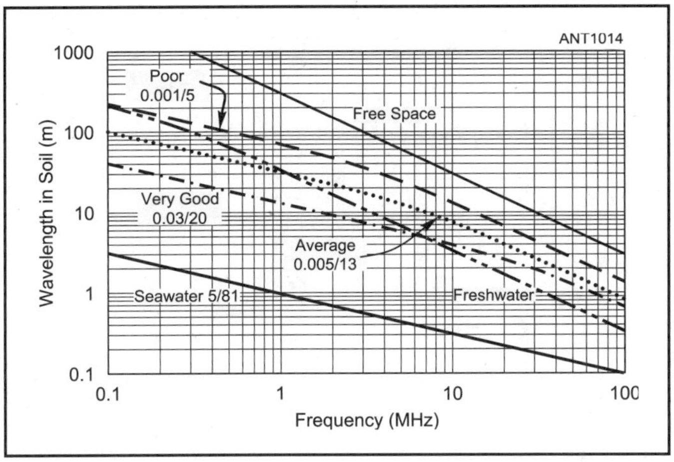

Figure 3.8 — Wavelength in typical soils as a function of frequency.

wavelength and this must be taken into account for wire segmentation during modeling.

The wavelength in free space (λ_o) is:

$$\lambda_o = \frac{299.79}{f(MHz)} \text{ in meters} \quad (2)$$

The wavelength in soil (λ) is:

$$\lambda = \frac{\lambda_o}{\left[\varepsilon_r^2 + \left(\frac{\sigma}{\omega \varepsilon_o}\right)^2\right]^{1/4}} \quad (3)$$

Figure 3.8 gives examples of wavelength as a function of frequency for different soils, salt and fresh water and free space. It can be seen that the wavelength in soil is typically much smaller than in free space.

3.2 Ground Systems for Vertical Monopoles

In this section we look at vertical monopoles which are shorter than $\lambda/2$ and require some sort of ground system in order to make up for the "missing" part of the antenna and, just as importantly, reduce the power dissipated in the near field. (For the purposes of this chapter, the term "vertical" should be understood to represent a vertical monopole antenna mounted on or near the ground.)

Because the losses in the soil near a vertical are a function of the electric and magnetic field intensities close to the antenna we will begin by looking at these fields. The next step will be to show what the actual soil losses are and how that loss is distributed in the soil near the base of the vertical. Finally we'll describe ground systems which can greatly reduce this loss.

3.2.1 Fields Near The Base Of A Vertical

In this section we will be examining the E and H fields at ground level within $\lambda/2$ of the base of typical verticals. (See the **Antenna Fundamentals** chapter for a discussion of E and H fields.) This may seem like an abstract exercise but it's important because it allows us to visualize what's happening in the soil around the base of a vertical, giving us both the amplitude and location of the ground currents and their associated losses. This information will guide us in the design and optimization of ground systems.

A vertical antenna has two field components that induce currents in the ground around the antenna: E_z and H_Φ. **Figure 3.9** shows in a general way the electric-field component (E_z, in V/m) and magnetic-field component (H_Φ, in A/m) in the region near a vertical. Both of these field components will induce currents (I_V and I_H) in the soil. Because the soil near the antenna typically has relatively high resistance this results in power loss in the soil. Power dissipated in the ground is subtracted from the power supplied to the antenna weakening your signal.

As shown in Figure 3.9, the tangential component of the H-field (H_Φ) induces horizontal currents (I_H) flowing radially in the soil. The normal component (perpendicular to the ground surface) of the E-field (E_z) induces vertically flowing currents (I_v) in the soil. These field-induced ground currents will decrease as we go deeper into the soil with the rate of decrease a function of the skin depth in the particular soil.

We can determine E_z and $H\Phi$ from either modeling (near-field calculations with *NEC*-based software, for example) or

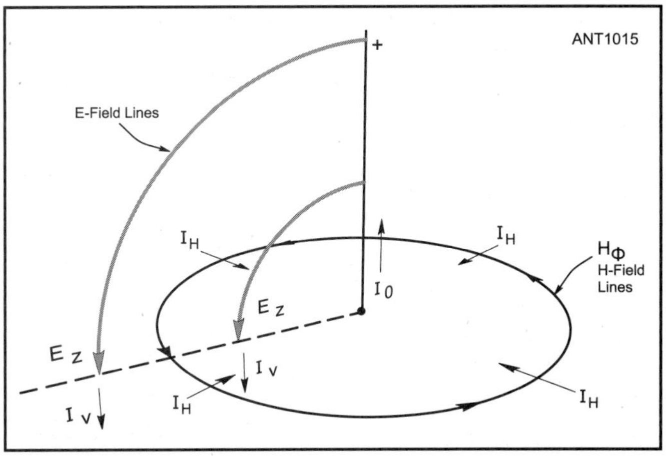

Figure 3.9 — Examples of the fields and currents close to a vertical.

Table 3.2

Base Excitation Currents as a Function of Vertical Height (h) in Wavelengths

The input power is 1500 W.

h (λ)	I_o (A)	R_r (Ω)
0.050	39.7	0.95
0.125	15.1	6.57
0.250	6.45	36.1
0.375	2.53	234

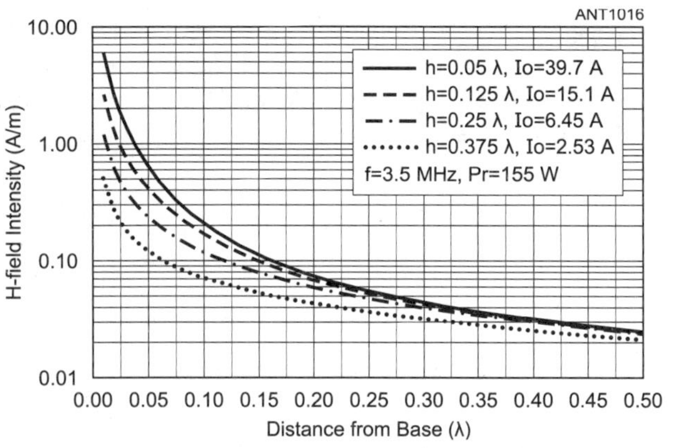

Figure 3.10 — H_Φ intensity as a function of distance from the base in wavelengths. Data is for a vertical at 3.5 MHz.

directly from equations. It turns out that the field intensities close to the base (<λ/2) of a vertical (within λ/2) over real ground are very close to the values for perfect ground. This allows us to use much simpler modeling or equations. The following graphs for field intensities assume perfect ground.

The base currents and the resistive part of the feed point impedance at the base of verticals with different heights (h) are given in **Table 3.2**. These are the values of current which result from an input power of 1.5 kW for an ideal vertical over perfect ground.

Figure 3.10 shows the H-field intensity within λ/2 of the base of the vertical for four different vertical heights (h): 0.05 λ, 0.125 λ, 0.250 λ and 0.375 λ. **Figure 3.11** shows the E-field intensity for the same values of h. Both of these graphs make the same two points:

1) Field intensity increases rapidly as you approach the base, particularly within a radius <λ/8, and

2) The shorter the antenna, the higher the fields for the same power input.

In the case of the E-field, the minimum field occurs for h = 0.25 λ and then increases again as h is increased beyond 0.25 λ. Ground losses are proportional to the square of the field intensity. In other words, if you double the intensity the power loss increases by 4 times! *This tells us that we must pay special attention to the ground system within λ/8 of the base and that short verticals require additional attention to the ground system.*

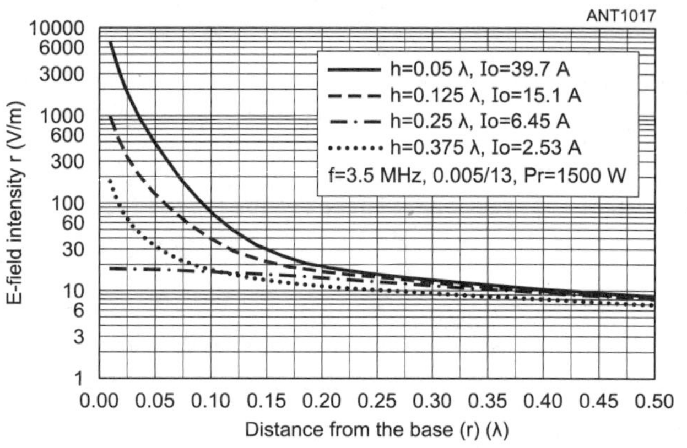

Figure 3.11 — E_Z intensity as a function of distance from the base in wavelengths. Data is for a vertical at 3.5 MHz.

Another point which can be inferred from Figure 3.11 is the very high voltages which can be present on the antenna. The shorter the antenna and the higher the power level, the higher these voltages will be. Verticals taller than λ/4 can also have very high voltages near the base. This is a very real safety hazard! Touching a vertical while transmitting can lead to severe RF burns.

Figures 3.12 and **3.13** show the field intensities for a λ/4 vertical at frequencies from 1.8 to 28 MHz. At a given distance in λ, both E and H fields increase with frequency but, as the dashed line in Figure 3.12 indicates, at a given fixed physical distance the H-field intensity is constant, independent of frequency. However, as the dashed line in Figure 3.13 shows, the E-field at a given physical distance is not constant but increases with frequency.

This behavior may seem a bit strange because it says that the field distributions do not scale linearly with frequency! Keep in mind that the base current at all frequencies was set to 6.45 A (P_r = 1500 W, h = 0.250 λ). As the frequency was changed the height of the vertical was reduced from 135 feet at 1.8 MHz to 8.8 feet at 28 MHz. The high current point on

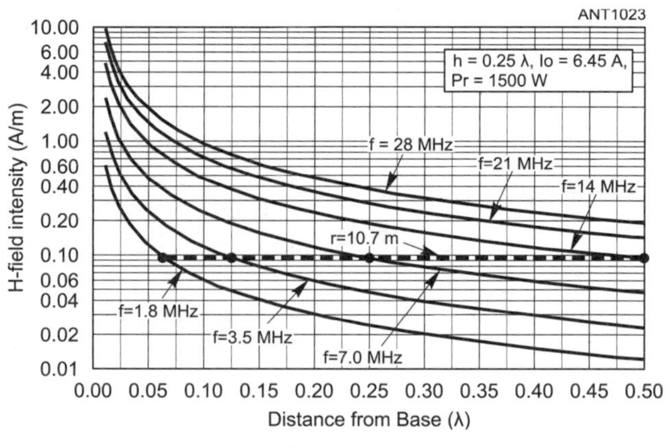

Figure 3.12 — H_Φ intensity as a function of distance from the base in wavelengths. Data is for a λ/4 vertical.

a vertical ($h \leq \lambda/4$) is at the base but the high voltage point is at the top. As we change frequency and alter h, the H-field is primarily influenced by the base current which does not change amplitude or location. However, the E-field is primarily influenced by the high voltage at the top of the vertical which is moving closer to ground as we go up in frequency. Normally we scale the dimensions of the ground system as we go up in frequency. If we elect to use $\lambda/4$ radials they will be approximately 34 feet on 40 meters, 17 feet on 20 meters, etc. The problem is that the fields are not scaling with frequency. At a given distance in λ the fields are higher as we go up in frequency. These observations tell us that for a given size (in λ) ground system *as we go up in frequency the ground loss will increase!*

As shown earlier, soil conductivity typically improves as we go up in frequency but that varies over a wide range and may not help as much as we would like. Better to be conservative and not count on the increase in σ unless you have actually measured your particular soil characteristics.

3.2.2 Radiation Efficiency and Power Losses in the Soil

We can discuss the efficiency of an antenna by using an equivalent circuit model like that shown in **Figure 3.14** for the resistive part of the feed point impedance. We account for the radiated power (P_r) by assuming there is a resistor we call the *radiation resistance* (R_r) through which the antenna base current (I_o) flows. The radiated power (P_r) is then:

$$P_r = R_r I_o^2 \tag{4}$$

Similarly, we can account for the power dissipated in the ground (P_g) by adding a loss resistance (R_g) in series with R_r. The ground loss is:

$$P_g = R_g I_o^2 \tag{5}$$

Additional losses due to conductors, loading coils, etc can also be simulated by adding more series loss resistances to the equivalent circuit but for this discussion we will ignore these additional losses although they can be significant in real antennas. The total input power (P_T) is simply the sum of P_r and P_g

The efficiency (η) of a vertical can be expressed as:

$$\eta = \frac{P_r}{P_r + P_g} = \frac{P_r}{P_T} \tag{6}$$

This can be restated in terms of resistances as:

$$\eta = \frac{R_r}{R_r + R_g} = \frac{1}{1 + \frac{R_g}{R_r}} \tag{7}$$

In essence, efficiency is the ratio of the radiated power to the total input power. Another way of saying this is that efficiency depends on the ratio of ground loss resistance to radiation resistance. The smaller we make R_g the more power will be radiated for a given input power. *Reducing R_g is the purpose of the ground system.*

We can determine P_g near the vertical from the E- and H-fields shown earlier. Given P_g and I_o we can calculate R_g and from that the radiation efficiency. For this discussion we will omit the mathematical details but these can be found in the spreadsheet referenced earlier.

In the following discussion the radiated power is kept constant at 1.5 kW but the total input power may be much greater because it will include the power dissipated in the soil which can become very large for short antennas with limited ground systems. The ground losses shown in **Figures 3.15** and **3.16** are what you would see if the ground system were simply a long stake driven into the soil beneath the antenna. The size of these losses makes it clear why we need to add a radial ground system around the base of a vertical.

Figure 3.15 shows the total ground loss (including both E- and H-field losses) within a radius r (in λ) around the base of verticals with different heights at 3.5 MHz, over average ground (σ = 0.005 S/m and ε_r = 13). For all heights the loss is significant but becomes almost astronomical for very short antennas. For example, the loss associated with the 0.050 λ

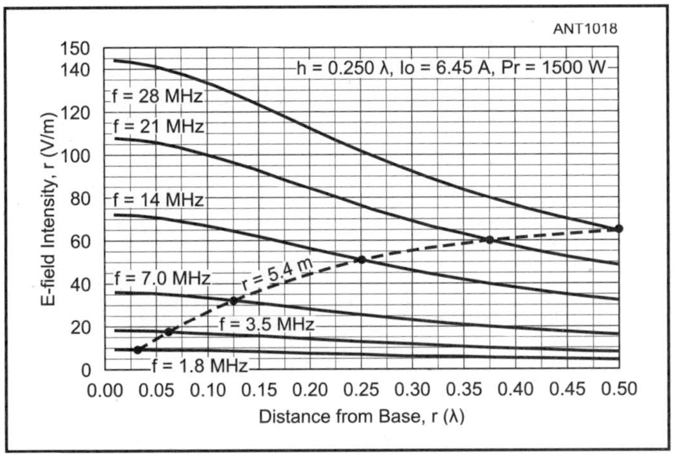

Figure 3.13 — E_z intensity as a function of distance from the base in wavelengths. Data is for a $\lambda/4$ vertical.

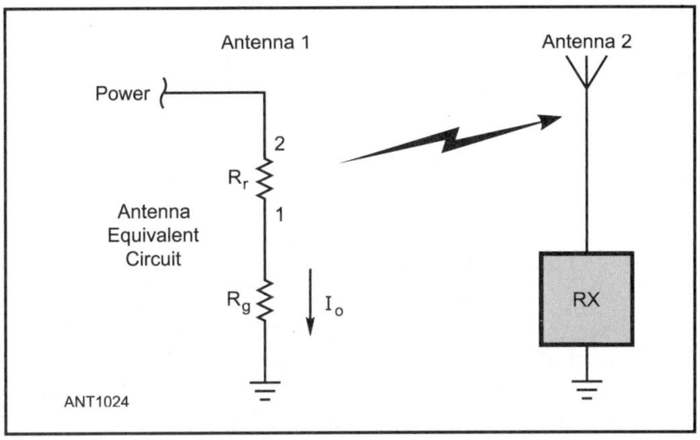

Figure 3.14 — Equivalent circuit for a vertical in terms of R_r and R_g.

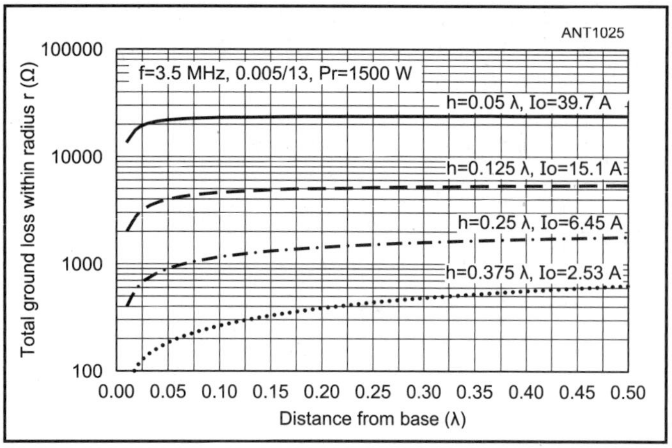

Figure 3.15 — Total ground loss within a fixed radius around verticals of different heights at 3.5 MHz.

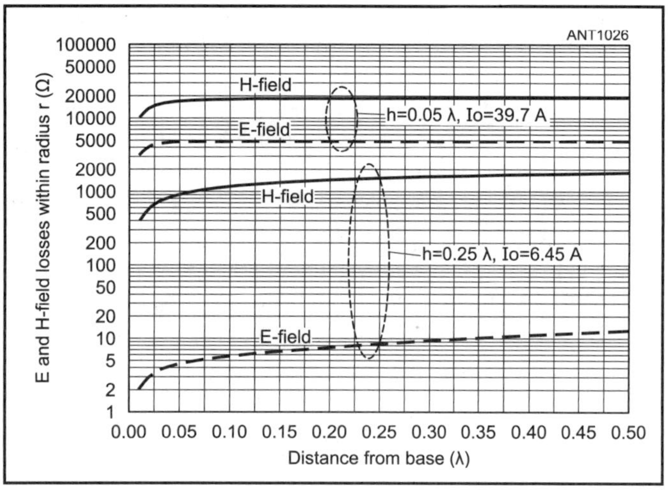

Figure 3.16 — Comparison between E- and H-field losses for two antenna heights.

Table 3.3

Efficiency for Verticals of Different Heights with Ground System Consisting of Only a Ground Stake in Average Soil

Height (h) (λ)	*Efficiency (%)*	*Power Loss (dB)*
0.050	4	–13.8
0.125	21	–6.7
0.250	46	–3.4
0.375	71	–1.5

Saltwater Grounds

Saltwater is often cited or imagined to be the very best ground for an HF vertical antenna. While very good, the issues to be considered are discussed by Rudy Severns, N6LF, in the paper "Some Thoughts on Vertical Ground Systems over Seawater," included with this book's online material.

vertical (about 13 feet for 3.5 MHz) amounts to a signal loss of almost 14 dB; in other words, over 20 kW of power is lost in the ground in order to produce 1.5 kW of radiated power. The efficiency of each antenna (using the values for R_r listed in Table 3.2) is listed in **Table 3.3**.

The efficiencies listed in Table 3.3 make it clear why some additional ground system beyond a simple ground stake is highly desirable in most installations. Keep in mind that these numbers are for one particular ground type (average). Poorer grounds will have even higher losses but better soils will have lower losses. Even a λ/4 vertical will have more than 3 dB of signal loss for a given input power because over half the input power is dissipated in the soil. This shows us that *the shorter the vertical, the more critical the ground system is!*

Figure 3.15 also shows that most of the loss is occurring within λ/8 of the base which correlates with the field intensities shown in Figures 3.12 and 3.13. When designing radial ground systems this ground loss distribution is reflected in the need to increase the number of radials close to the base.

Figure 3.15 shows the total loss in the soil due to both E- and H-fields. However, the relative contribution of each field component to the total varies greatly with the height of the vertical. Figure 3.16 shows a comparison between the E and H losses for h = λ/4 and h = 0.05 λ. For the λ/4 vertical the E-field losses are very small compared to the H-field losses but for the shorter vertical both the E- and H-field losses increase dramatically and the E-field loss is comparable to the H-field loss. In very short verticals the E-field losses can become larger than the H-field losses.

3.2.3 Wire Ground Systems

Figure 3.17 illustrates what we mean by a "radial" ground system. The ground system wires are connected together at

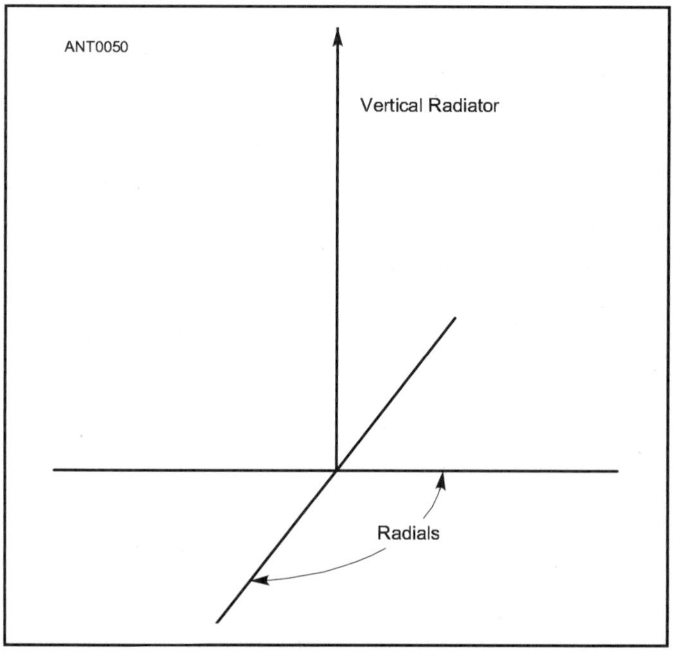

Figure 3.17 — Vertical monopole and radial ground system.

the base of the antenna and arranged radially outward from the base. Why radial wires? Why not wires in circles or some other shape? As shown in Figure 3.9, the H-field lines have the form of circles around the base of the vertical. When the H-field passes over a conductor there will be a current induced in the conductor which flows at right angles to the H-field vector. In a wire ground system the optimum orientation for the wires is at right angles to the H-field (i.e. radially). If the wire were oriented parallel to the H-field (in a circle) there would be no current induced in the wire and the current would simply flow in the soil instead. In some cases where multiple verticals are present (i.e. in an array for example) it may not be practical to use only radial wires. Some form of coarse mesh may be needed.

BURIED OR GROUND SURFACE RADIAL SYSTEMS

There are different ways to install wire ground systems: the wires may be buried in the soil a few inches or lying on the ground surface or elevated several feet above ground or even some combination of these. In addition, in elevated systems there may be interconnections between the radial wires to form what is called a "counterpoise." (See the section "Counterpoise Systems" later in this chapter.) Another possibility is to use a coarse rectangular mesh, either on the ground surface or elevated. We will discuss all these options but for the moment we'll focus on radial systems either buried or lying on the ground surface.

If we know the values for E and H, I_o and the soil electrical characteristics we can determine P_g. We can then determine R_g directly from P_g:

$$R_g = \frac{P_g}{I_o^2} \quad (8)$$

R_g is *not* a resistance unique to a particular ground system that you can measure with some kind of ohmmeter. It is simply the relation between a given excitation current (I_o) and the power dissipated in the soil (P_g) for a given vertical. P_g in turn depends not only on the soil characteristics but on both I_o *and* the details of the vertical itself, i.e. height, loading, etc. For this reason R_g for a given ground system *will change* as we change the vertical even if the soil characteristics and the physical ground system itself are kept constant.

IDEAL GROUND SCREENS

Initially we'll assume that the ground system is ideal: i.e. a high conductivity ground screen that covers the soil from the base out to some radius "r". This ideal ground screen will give us the minimum possible R_g for ground system of a given radius. Later we'll look at R_g for more practical wire ground systems with a limited number of radials to see how they compare. From the ideal ground screen information we will know what the ultimate limits are and can determine when adding more wire might result in only a small improvement. Surprisingly, it does not take a large number of radials to give a good approximation of an ideal ground screen.

Figure 3.18 is an example of R_g as a function of ground

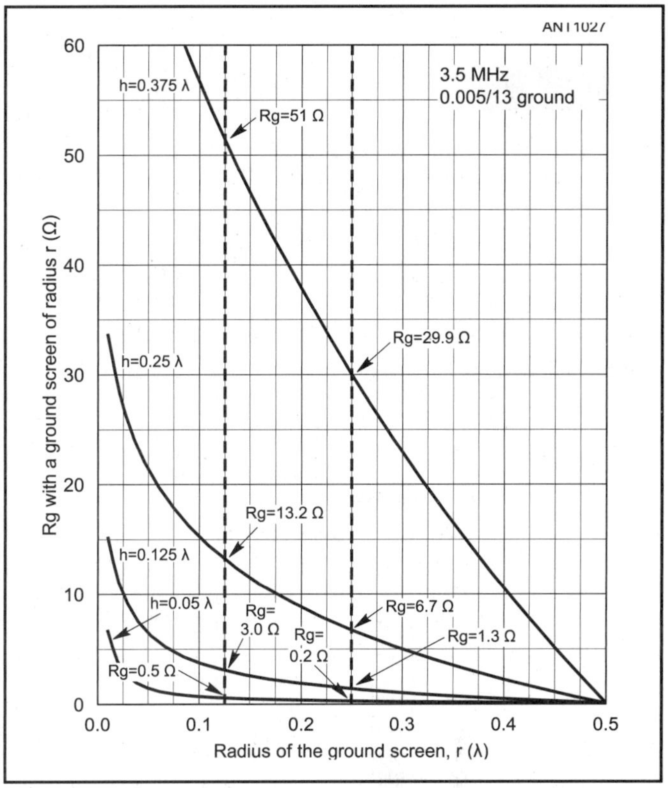

Figure 3.18 — Variation in R_g with ground screen radius. Normalized to include losses out to r = 0.5 λ.

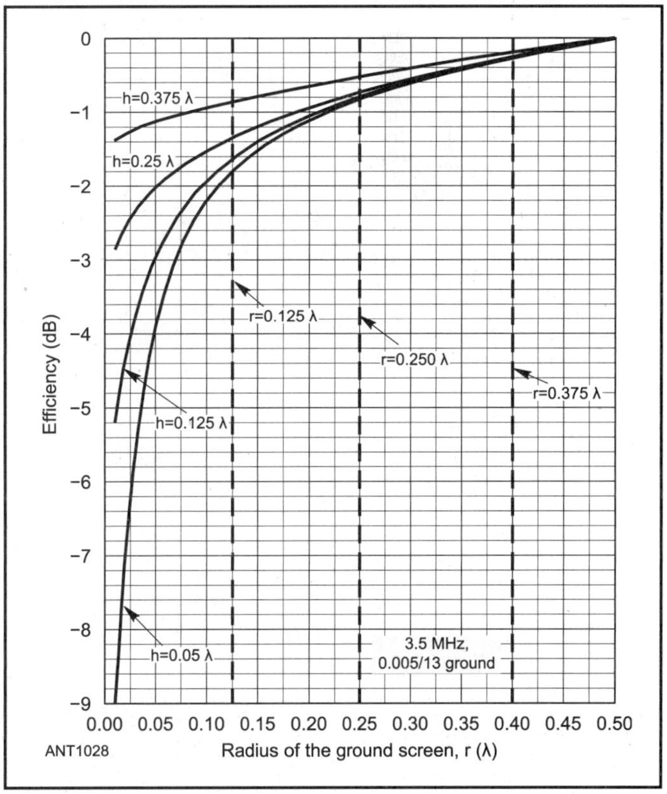

Figure 3.19 — Efficiency in dB as a function of ground screen radius.

screen radius for several antenna heights at 3.5 MHz, over average ground. As we saw in Figure 3.15, near the base of the vertical the total ground loss is large but as we move outward from the base the *additional* ground loss becomes much smaller. This means that the values for R_g fall quickly as r initially increases but as the radius of the screen gets larger, the rate of decrease in R_g slows down.

If we take the values for R_r from Table 3.2 and combine these with the values for R_g in Figure 3.18 and use Eq 8, we can calculate the efficiency as shown in **Figure 3.19**. The efficiency is stated in dB so that this graph tells us directly how much our signal will improve as we expand the radius of the ground screen. For example, for h = 0.25 λ, expanding the screen radius from 0.01 λ to 0.125 λ increases the signal by 1.5 dB. If we further increase the radius to 0.250 λ we pick up another 0.6 dB and if we go to a screen radius of 0.375 λ the gain is an additional 0.4 dB. Clearly there is a substantial advantage to having a screen with a radius of at least λ/8 but as we increase the size, the incremental improvement gets smaller. In general for amateur applications expanding the ground screen radius beyond λ/4 is seldom worth the additional cost and effort at least on the lower bands (160 and 80 meters). But as pointed out earlier, we can make a case for larger ground systems (in terms of λ) at higher frequencies.

Figure 3.18 shows R_g for one frequency and one ground characteristic. **Figures 3.20** and **321** show what happens to R_g as we change frequency or ground characteristics for a given height (λ/4 in this example).

Figure 3.20 is a graph of the changes in R_g with frequency for several different screen radii. This graph is for a λ/4 vertical over average ground. What the figure shows us is that for a given antenna, screen radius (in wavelengths) and ground characteristic, R_g can increase significantly as we go to higher frequencies. For example with r = λ/4, R_g = 7 Ω at 3.5 MHz but at 28 MHz a λ/4 screen has an R_g = 12 Ω. If we increased the screen radius at 28 MHz to 0.375 λ, R_g drops back down to 5 Ω. Expanding the screen radius from λ/4 to 3λ/8 at 28 MHz means extending the radial lengths from 2.7 meters (8.8 feet) to 4 meters (13.2 feet) which is very practical. The message is: *as we go higher in frequency we should consider using a ground screen with a larger electrical radius (in terms of λ) and/or more radials.* Fortunately, as we go up in frequency the wavelength gets shorter so it's easier to add more and/or longer radials for a given total amount of wire.

From Figure 3.21 we see that with lower quality soils R_g is significantly higher and it becomes increasingly important to use a more extensive ground system to maintain efficiency.

REAL WIRE RADIAL SYSTEMS

In practice, ground systems are usually made with wire in the form of a radial fan like that shown in Figure 3.17. How a

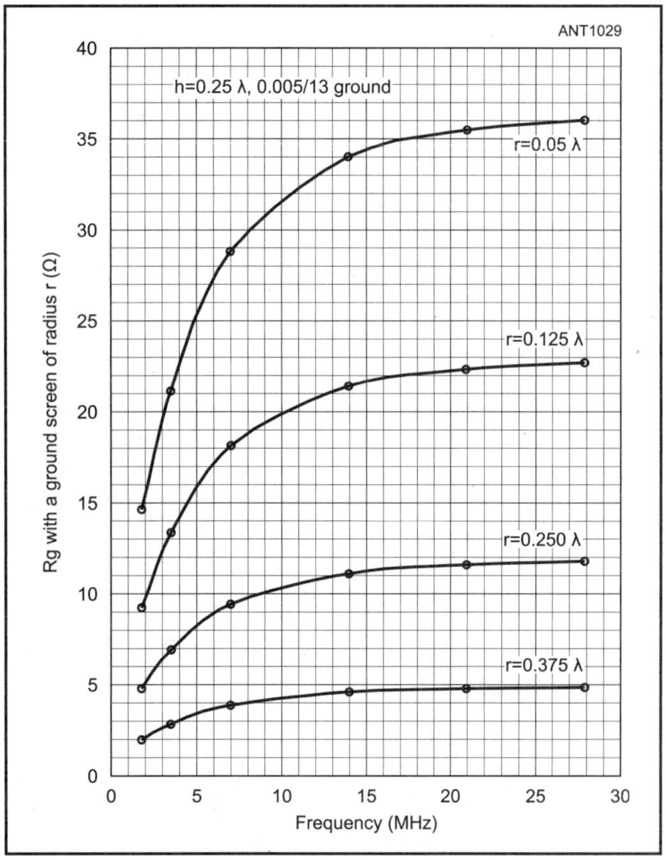

Figure 3.20 — Variation in R_g with frequency for h = λ/4 for several ground screen radii (in wavelengths) over average ground.

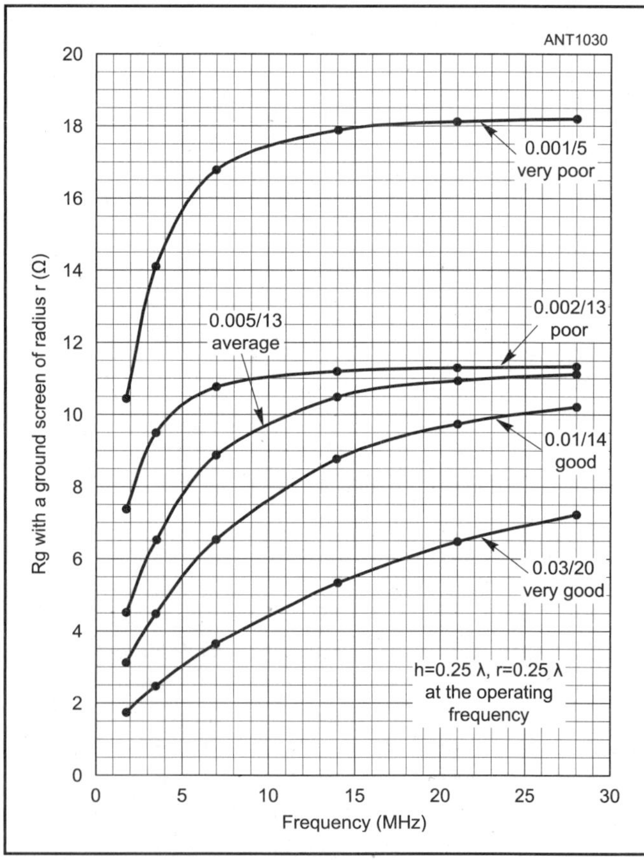

Figure 3.21 — Variation in R_g with frequency for h = r = λ/4 at the operating frequency.

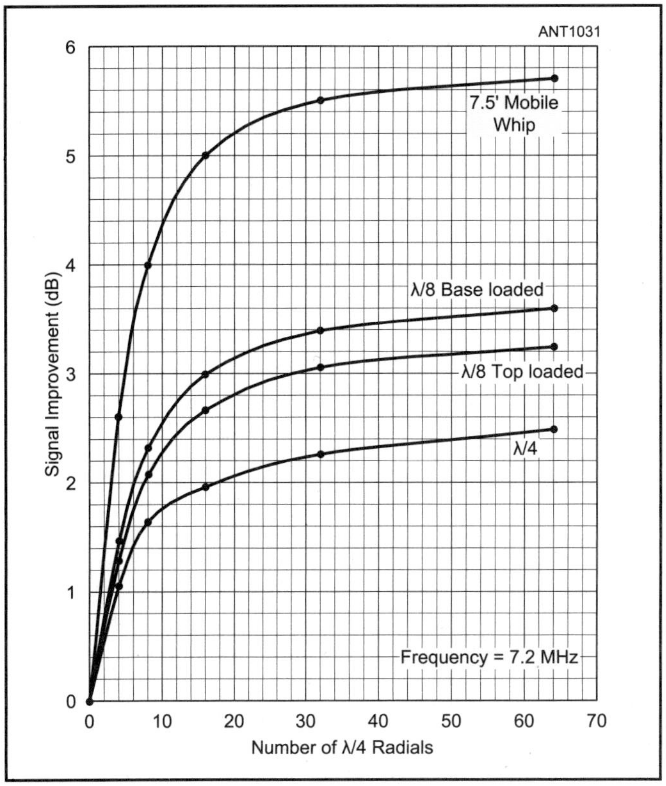

Figure 3.22 — Signal improvement for typical 40-meter verticals as the radial number is increased from 0 to 64.

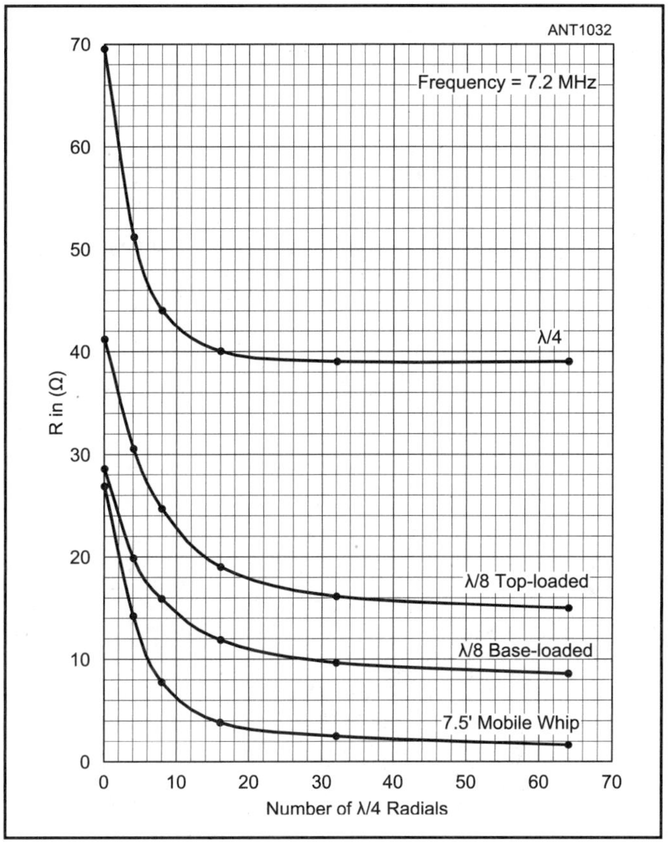

Figure 3.23 — Variation in R_g as a function of the number of λ/4 radials.

particular ground system performs compared to an ideal ground screen can be determined using mathematical analysis or from *NEC* software modeling or from actual measurements on real antennas. All three routes give essentially the same answers but for this discussion we will use actual measurements on real antennas and also some *NEC* modeling results.

Figure 3.22 shows the measured signal improvement as λ/4 (33 feet) radials lying on the ground surface were added to different 40 meter antennas: a λ/4 vertical, a λ/8 vertical with sufficient top-loading to resonate it, a λ/8 vertical resonated with an inductor at the base and a 7.5 foot 40 meter mobile whip.

The measurements began with only a single ground stake, no radials. Figure 3.22 shows the increase in signal strength (for a constant input power) for each antenna as the number of λ/4 radials was increased from 0 to 64. Initially, as radials were added, the signal improved rapidly but by the time there were 16 radials, the rate of increase in signal improvement turned a corner and started to decrease. Going from 32 to 64 radials the improvement was only a fraction of a dB (0.2 dB). What this tells us is that a radial fan with 32 or more radials is a good approximation of an ideal ground screen, at least for λ/4 radials. For short, loaded antennas over poor soils, 64 radials might be justified and should be considered. However, the standard broadcast ground system of 120, 0.4 λ radials would probably be a waste of copper for most amateur installations.

Another important thing we see in Figure 3.22 is that short loaded antennas benefit more from the same ground system. This is because (as shown earlier) the E- and H-field intensities are much higher close to the base of shorter antennas. Note also that in short antennas, moving the loading up the vertical, top-loading for example or placing the loading coil above the base, improves the signal for a given ground system.

The soil over which this experiment was conducted would be rated as very good ($\sigma = 0.015$ S/m, $\varepsilon_r = 30$). Over that soil the improvement going from 0 to 64 radials ranged from 2.5 to 5.7 dB. Over poor or even average soils the improvement would be substantially greater. Figure 3.22 also shows how important is to have at least a simple radial system. Sixteen radials is pretty much the practical minimum, especially over poor soils.

Measuring the signal strength for a given input power to the antenna, as radials are added to the ground system is a very direct way to gauge when adding more radials will give only a small improvement but for most amateurs that's not very practical. There is a simpler way to gauge when there are sufficient radials in the ground system. We can look at the feed point impedance which is a simple, direct measurement. An example of the variation of the resistive part (R_{in}) of the input impedance as radials are added to a ground system is given in **Figure 3.23**. The values are for the same antennas used in Figure 3.22. Note that for the 7.5 foot mobile whip, the series resistance of the loading coil has been subtracted from the measured feed point impedance.

If we assume that $R_{in} = R_r + R_g$ and that R_r is constant as we add radials (a reasonable approximation), then the leveling out of the curves for radial numbers above 16 can be interpreted as meaning that the minimum R_g for that radial length has been reached. Again, we see that 16 radials are pretty much the minimum but by the time we get to 32 radials the rate of change is quite small. Figures 3.22 and 3.23 tell the same tale.

OPTIMIZING RADIAL LENGTHS

In the real world the amount of wire available for a ground system may be limited. How should we use the available wire: a few long radials or a bunch of short ones?[2,3,4] We can use *NEC* modeling to address this question. **Figures 3.24** and **3.25** show the signal improvement as both the number and the length of the radials are changed. Both figures assume f = 1.8 MHz and average ground (σ = 0.005 S/m, ε_r = 13). Figure 3.25 is for h = λ/8 and Figure 3.26 is for h = λ/4. These figures illustrate a number of important points and provide a guide to the optimal use of a given total length of radial wire in the ground system. The 0 dB reference is four λ/8 radials. In both figures we see that when only a few radials are employed (<16) there is very little increase in signal when longer radials are used. In general, from both modeling and experiments, we can say that a few long radials make a poor ground system.[5]

The Broadcast 120-Radial Ground System

The 1937 IRE study by Brown, Lewis, and Epstein is often cited as the origin of the "standard" broadcast ground system with 120 λ/4 radials.[11] This is a misunderstanding of what the report actually describes, which is 120 symmetrically laid-out radials of a length that results in 0.05 λ between the ends of the radials. This length happens to be close to λ/4 and so it was a natural assumption to assume it was specified to make the radial resonant as a λ/4 element. As is stated in this chapter, a radial laying on the ground will have an electrical length quite a bit longer than its physical length, so it is unlikely to act like a wire λ/4 long.

The importance of the ground radial system is to reduce ground losses by providing a low-loss path for current flowing to the feed point that would otherwise flow through soil. Broadcasters, having resources beyond the typical amateur and a strong economic reason to avoid ground losses, opt for the 120-radial system which is well beyond the point of reasonable return for amateurs. As you can see from the numerous graphs in this chapter, nothing special happens when a radial on or below the surface of the ground is λ/4 long. The *NCJ* article by K3LC on optimum radial systems in the References and Bibliography provides excellent guidance on how to get the best results from your radial system, even with far fewer than 120 radials!

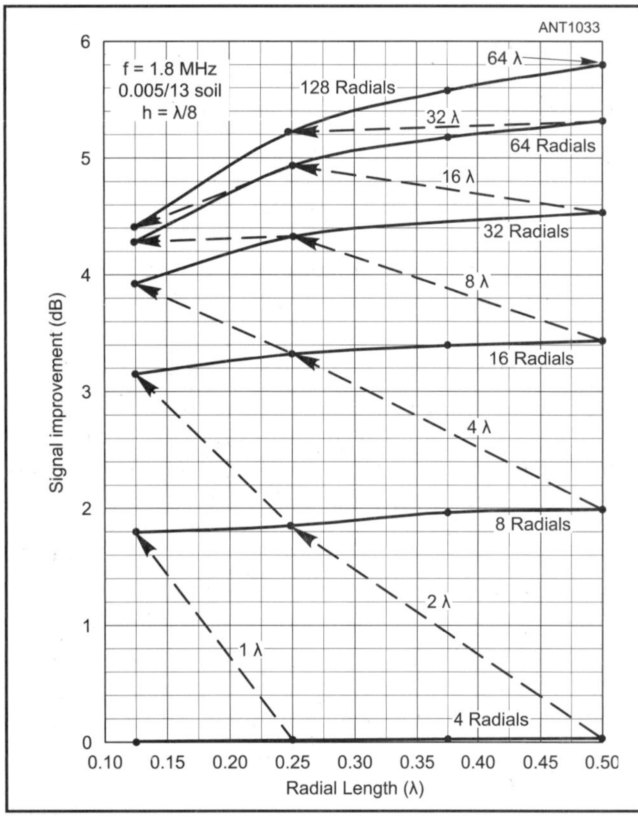

Figure 3.24 — Comparison of radial lengths and number versus signal improvement for a given total amount of radial wire (in λ). In this example h = λ/8.

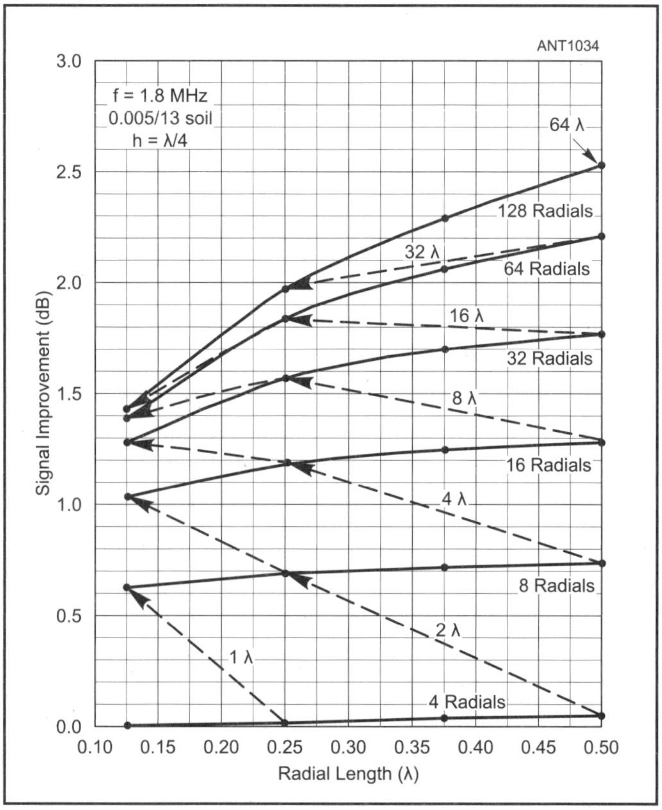

Figure 3.25 — Comparison of radial lengths and number versus signal improvement for a given total amount of radial wire (in λ). In this example h = λ/4.

From the graphs we can see that longer radials become effective only as we increase the number of radials.

These graphs show how to optimize your signal for a given total length of radial wire. The dashed lines connect points which have the same total wire length in the ground system. For example, if we have a total of 2 λ of wire we could make four λ/2 radials or eight λ/4 radials or 16 λ/8 radials. From the graphs we can see that with a total of 2 λ of wire the best radial system would be 16 λ/8 radials. That would be an improvement of over 3 dB for the λ/8 vertical and 1 dB for the λ/4 vertical. Similarly, if 4 λ of wire is available then the optimum use would be 32 λ/8 radials. When we go to 8 λ of wire, things change a bit. For the h = λ/8 antenna either 64 λ/8 or 32 λ/4 radials will work about the same. However, for the h = λ/4 antenna the best use of the wire would be 32 λ/4 radials. The reason that a large number of short radials are effective stems directly from the high field intensities close to the base of a vertical. The first priority is to reduce the losses close to the base. As more wire becomes available and the losses close to the base have been reduced then reducing the losses further out becomes useful. (See the sidebar **Optimum Use of Available Wire**.)

A bit of long held conventional wisdom among amateurs is that "the length of the radials should be similar to the height of the vertical." Figures 3.24 and 3.25 give partial support to that belief. When 8 λ of wire is available, λ/8 radials will work fine for a λ/8 vertical but a smaller number of λ/4 radials work better for the λ/4 vertical. This stems from the much higher field intensities associated with the λ/8 vertical. In short antennas it's very important to use numerous radials close to the base. But in both cases, when sufficient wire is available, there comes a point where fewer but longer radials are a better choice.

Optimum Use Of Available Wire

The table and graph in this section are taken from the March 2004 *National Contest Journal* article "Maximum-Gain Radial Ground Systems for Vertical Antennas," by Al Christman, K3LC. This table and graph contain results for average soil. (The full article is available in the online content for this book.)

Table 3.A lists the "best" number of radials to maximize the gain, for each specific total wire length, on 40, 80, and 160 meters, in average soil. (Results for very poor to very good soil are provided in the original article.) For a given total length of wire, fewer (but longer) radials are needed as we go lower in frequency. Since the wavelength is greater at lower frequencies, the physical height of the quarter-wave vertical-monopole (radiator) also increases as we switch from 40 to 80 to 160 meters. As a result, the displacement currents leaving the vertical element intersect the soil farther from the base of the antenna, and longer radials are needed in order to collect this current.

Figure 3.A1 displays the same data in graphical form. Whenever the computer analysis showed that several different numbers of radials would provide the same peak gain, the average value was calculated and used for the graph. It is reasonable to extrapolate from this graph to 30, 20, and 17 meters for radial systems.

Table 3.A
Optimum Number of Radials versus Total Wire Length for All Three Bands, in Average Soil

Total wire length (ft)	*Optimum number of radials for each band* 40 meters	*80 meters*	*160 meters*
125	9		
250	14	10	
500	21–22	15–16	11–12
1,000	28–31	21–24	17–18
2,000	39–43	29–32	23–28
4,000	62–63	42–46	32–40
8,000	100–104	63–67	45–55
16,000		99–111	76
32,000			99–122

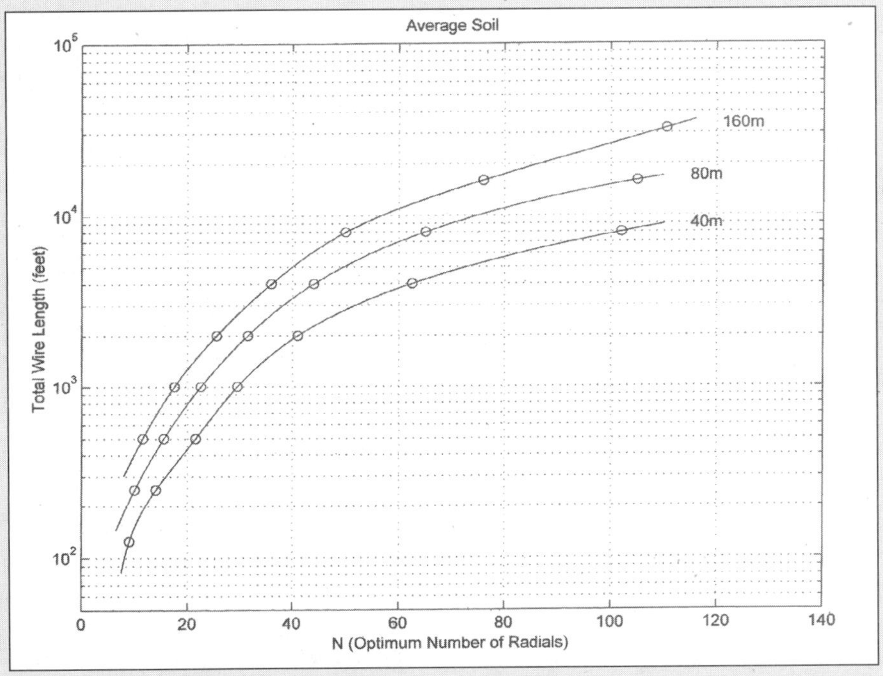

Figure 3.A1 — Optimum number of buried radials for quarter-wave vertical antennas over average soil, as the total wire length is varied, for the three "low bands" 40, 80, and 160 meters.

RADIAL SCREENS WITH MISSING SECTORS

In many installations a vertical may have to be located close to an obstacle such as a building or a driveway and it may not be possible to have a full 360° symmetrical radial field. The lack of radials in a subsector of the radial system will increase ground loss because there aren't any radials in that sector to keep the field out of the soil! In addition there will be pattern distortion. Depending on the size of the missing radial sector and the soil characteristics the signal reduction and pattern distortion can be several dB.[6] This is definitely undesirable but may be unavoidable in some situations. If the obstacle is a building, locating the antenna along one side will result in a 180° missing sector but if the antenna can be moved to a corner of the building the missing sector will only be 90° which will be a significant improvement over the 180° case. From the earlier discussion we know that the ground close to the base is the most critical. If possible, the antenna should be moved away from the structure and a fan of short radials inserted in the missing sector. Of course the building itself may have considerable effect on the antenna. It's generally a good idea to keep antennas as far as possible from structures. *If space is limited in all directions, it may be better to move the antenna away from the structure and accept short radials all the way around.*

3.2.4 Elevated Ground Systems

Ground systems can elevated above ground and electrically isolated from ground. The most common system uses four or more λ/4 radial wires placed a few feet above ground. Another form of elevated system consists of a number of radial wires, which have lengths <λ/4, perhaps with a skirt wire connecting the outer ends of the radial wires as well as interconnecting wires between the radials closer to the base. It is also possible to use an elevated wire mesh. These last two options are often referred to as a "counterpoise" or "capacitive" ground system. A λ/4 vertical with several λ/4 radials (usually four radials) is called a "ground-plane" antenna. Ground plane antennas are discussed in the **Dipoles and Monopoles** chapter.

ELEVATED SYSTEMS WITH SIMPLE RADIAL WIRES

In this section we discuss radial systems made from straight wires of the same length for single band use. Multiband and counterpoise systems are discussed in following sections.

For a number of years there has been much discussion regarding the relative merits of buried or ground surface radials versus elevated radials. Modeling using *NEC* software has consistently indicated that a few elevated radials should perform as well as a large number of radials on the ground. Modeling has also predicted that the signal would improve very quickly with height even for small elevations. To verify these modeling predictions a carefully controlled series of experiments were performed at 7.2 MHz directly comparing the signal from a vertical using either an on-the-ground system with many radials or an elevated system with only a few radials.[7]

The experiment began with the base of the vertical at ground level with four λ/4 radials lying on the ground surface. The signal strength at a remote point was recorded. This was used as the reference level (0 dB). The next step was to elevate the base of the antenna and the four radials in increments from zero to 48 inches. At each point the change in signal from the reference level was recorded. The second part of the experiment left all the radials on the ground surface but starting with four radials incrementally increased the number of radials. **Figure 3.26** shows the results of that experiment. The *NEC* modeling predictions agree well with the observed behavior:

1) Even a small elevation makes a considerable difference in signal and

2) The elevated system is equivalent to the ground system with 32 or more radials.

One additional point should be emphasized regarding this experimental work. For the 4-radial elevated system to work

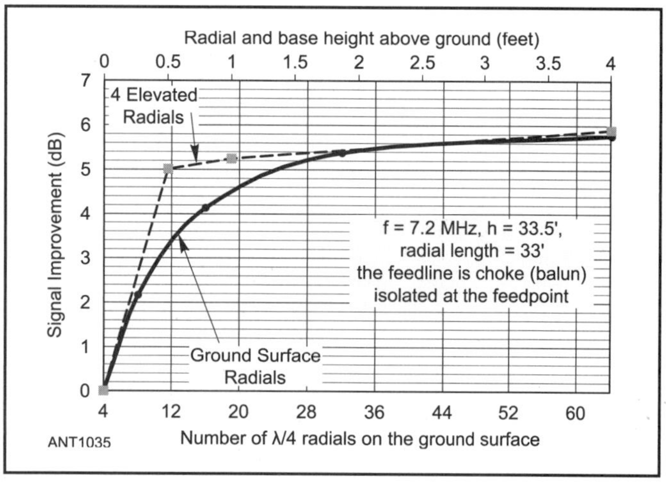

Figure 3.26 — Comparison between four elevated λ/4 radials and λ/4 radials on the ground surface. Note that this graph has two different horizontal axes. The arrows associate the plots with the appropriate horizontal axis.

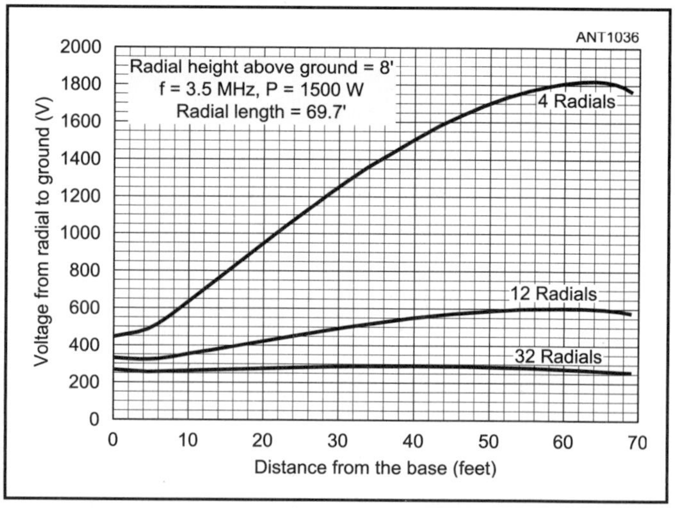

Figure 3.27 — Examples of the voltage from an elevated radial wire to ground with different numbers of radials. The input power to the vertical is 1500 W, the operating frequency is 3.5 MHz and the radial system is elevated 8 feet above ground.

as well as the multiple radial ground based system, during the experiment it was found that very great care had to be taken to assure that the radial geometry was highly symmetric, the radial lengths identical and that the currents in the radials were all equal and in phase with the base current as discussed below.

SAFETY CONSIDERATION WITH ELEVATED RADIALS

Before looking more closely at the assertion that four elevated radials are equivalent to many ground-based radials we need to consider a safety issue. Like a vertical located at ground level, elevated verticals will have high voltages on the vertical but in addition, elevated radial systems can have very high fields and voltages on and near the radial wires. **Figure 3.27** gives examples of the voltage from a radial wire to ground for 4, 12 and 32 $\lambda/4$ radial systems.

Note that the voltages vary from 250 V_{RMS} to nearly 2000 V_{RMS}! These voltages are proportional to $\sqrt{P_{in}}$ so that if you drop the power from 1500 W to 100 W, a factor of 15:1, the voltages only drop by a factor of less than four. Even at 100 W, they are still very high! For safety reasons, elevated radial systems are usually placed well above head-height, 8 feet or more. This is done so that people or animals cannot accidentally run into the wires but also because the high voltages which are present on the radials, particularly at the ends, can cause severe RF burns. This hazard is typical of elevated ground systems.

Given the high potentials at the ends of the radials, high quality insulators should be used at the radial ends. In addition to the high voltages, the E-field intensities are also very high near the radial ends. This means there is a danger of corona discharge which can erode the wire or even damage a plastic insulator. Where the radial wires are attached to the insulators, care should be taken that the wire ends do not form any sharp points which could be sites for corona discharge. Usually a ball of solder is used to cover the wire end. This problem will become more acute as the altitude of the station increases.

SOME ALTERNATIVE ELEVATED SYSTEMS

It is not always practical to have the base of the antenna high above ground. For example, a 20 meter $\lambda/4$ vertical will only be about 17 feet long and made from small diameter aluminum tubing. Elevating this is not a great challenge. But a 160 meter $\lambda/4$ vertical will be about 130 feet high and probably made from tower sections or heavy tubing. It may not be possible to elevate the base of the larger antennas very high. As an alternative, the elevated radials can be arranged in several ways as shown in **Figure 3.28**. The simplest approach when the base is close to ground as shown in (B) is to just slope the radials out at an angle. While this approach can place the radial tips well above head height it still leaves a lot of the radial at low heights. Another approach (C) is to slope the initial portion of the radial upward at an angle of 45° until a height of 8 feet is reached and then to run the rest of the radial out at constant height. These are referred to as "gull wing" radials.[8]

Another problem that often arises (particularly on 80 and 160 meters) is that there may not be enough space for full $\lambda/4$ radials. That's OK because as shown in Figure 3.28D, shorter

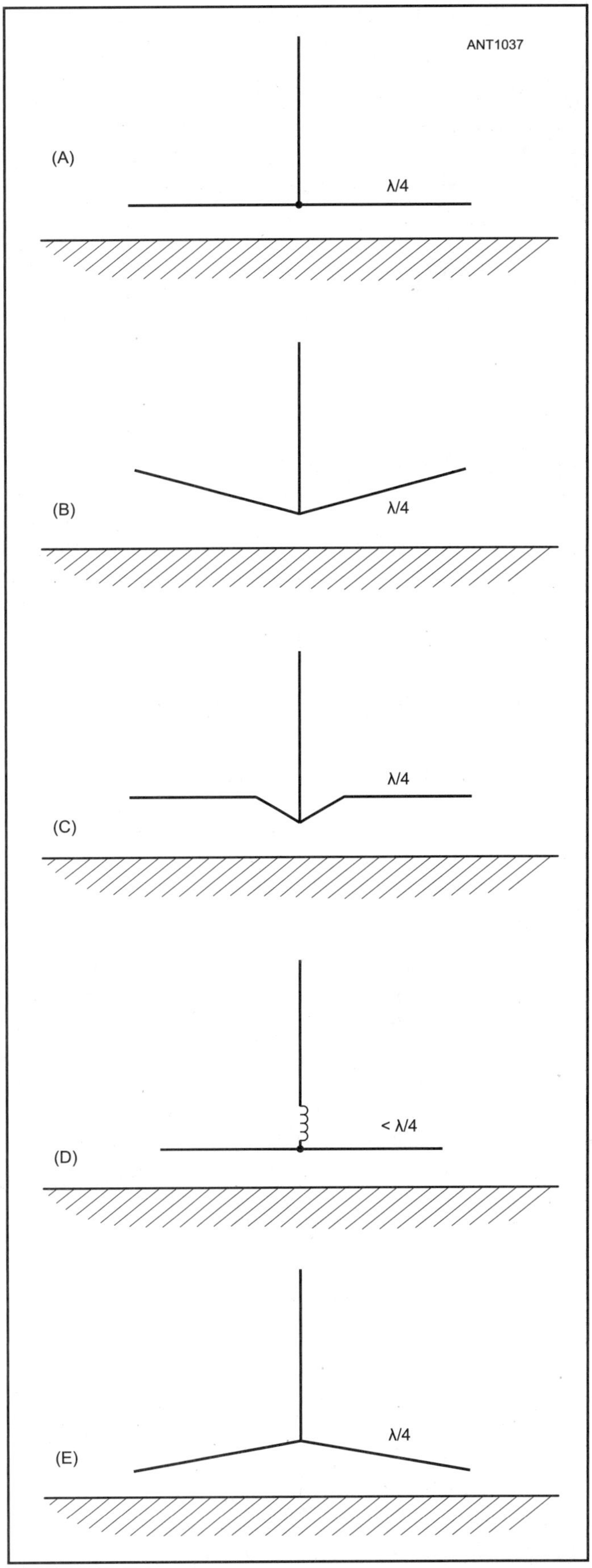

Figure 3.28 — Alternate elevated ground system configurations.

radials can be used and an inductor added at the base of the antenna to resonate it.[9] (Note, it is also possible to place individual inductors in each radial which may be helpful in balancing the current division between the radials.) Another alternative configuration is shown in Figure 3.28E. In this case the base is higher than the ends of the radials. This configuration is often used at 20 meters and above where the radials are self-supporting conductors. When the radials are anchored only at the base some droop to the outer end of the radials would be normal. In some cases the radials are deliberately sloped downward to increase the feed point impedance and provide a lower SWR. It should be pointed out that configurations (B) and (C), where the radials slope upward from the base, will have lower feed point impedances and somewhat reduced SWR bandwidths.[7]

This raises the question: how much is the antenna performance degraded by these alternate schemes? Again this can be addressed either with modeling or experimentally. The earlier experiment comparing elevated and ground surface radials on 40 meters was extended to compare the alternatives given in Figure 3.28. The results are listed in **Table 3.4**. All the alternatives were tested with four radials at 7.2 MHz. The base tuning inductor in option (D) had a Q of 350. The conventional system with all of the radials and the base elevated to the same height is used as the 0 dB reference. Except for (E), there is a small penalty associated with the alternate elevated radial configurations (on the order of –0.5 dB) which may be acceptable in many situations.

PROBLEMS WITH ELEVATED RADIALS

The simplicity of elevated systems with only three or four radials is very attractive because, at least in principle, they can be just as effective as much more extensive ground based systems. However, as pointed out above with regard to the experimental work, elevated systems with small numbers of radials are very sensitive to the mechanical details of the radials: i.e. length, droop, asymmetry in the radial fan, nearby conductors and so forth. The input impedance, current division between radials, radiation pattern, resonant frequency and efficiency of the antenna are all sensitive to even small asymmetries. It has been demonstrated experimentally that radial geometry asymmetry[9] and irregularities in ground characteristics under the radial fan[10] will cause problems. The following discussion explores some of these problems.

Typically a vertical with an elevated ground system will be fabricated with the vertical and the radial lengths calculated from L = 234/f in MHz which is 5% shorter than a free-space λ/4. The common wisdom that 5% shortening should be used is derived from work done in the 1930s but is only an approximation. When the base impedance of an actual antenna is checked, the resonant frequency will often be substantially different from what was expected and dependent on the number of radials as well as their length.

NEC modeling can be used to explore what's happening. The modeling is done in two steps: first model the vertical radiator element over a perfect ground and adjust its length to resonate at the desired frequency (7.2 MHz in this example). This example uses a #12 AWG wire and to be resonant at 7.2 MHz, h = 32.22 feet which is about 5.5% shorter than free space. The next step in the modeling is to add various numbers of horizontal #12 AWG wire radials. Each of the radials has the same length as the vertical (L = 32.22 feet). **Figure 3.29** shows the resonant frequency as a function of the number of radials from 2 to 128.

Table 3.4
Signal Comparison Between Different Elevated Radial Systems

Radial System Configuration	*Relative Signal*
(A) Base and four radials elevated 4 ft	0.00 dB
(B) Base at ground level, radial ends at 4 ft	–0.47 dB
(C) Base at ground level, gull-wing-radials, ends at 4 ft	–0.65 dB
(D) Base and radials at 4 ft, λ/8 radials with L = 2.2 μH	–0.36 dB
(E) Base at 4 ft and radial ends at 3 ft	+0.10 dB

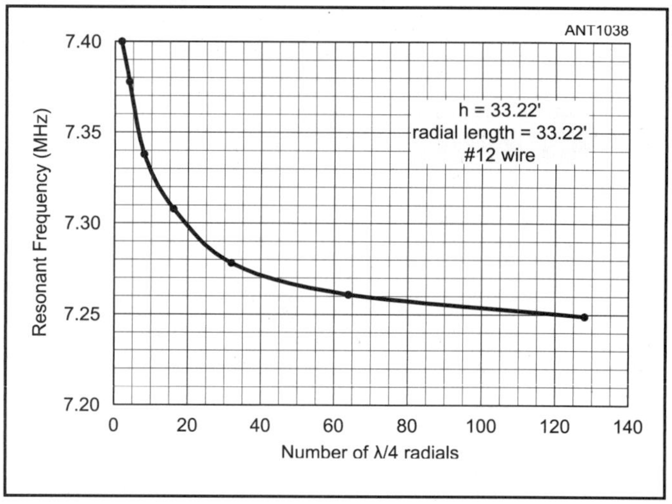

Figure 3.29 — Resonant frequency of a λ/4 vertical for different numbers of elevated radials.

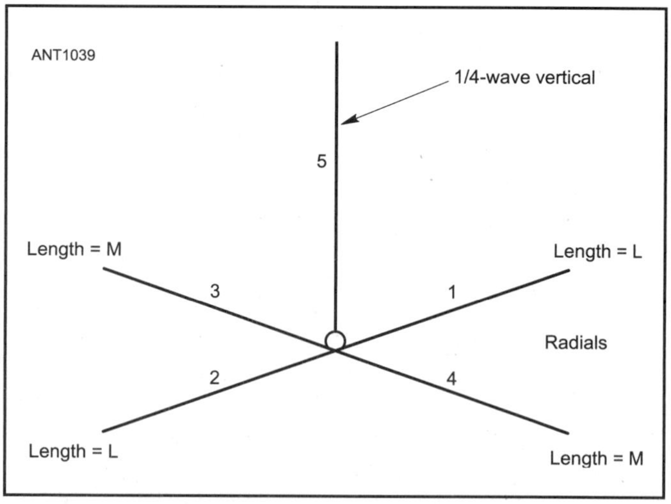

Figure 3.30 — 40-meter λ/4 vertical with four radials. Vertical height is 34.1 feet. The radial lengths are varied.

The resonant frequency of the complete antenna with the radials approaches the desired 7.2 MHz but doesn't quite get there. Even when a large number of λ/4 radials are used, the radial fan is not the same as an infinite ideal ground. In general, elevated systems should start with radials and perhaps the vertical radiator, with a length corresponding to the free space value for λ/4 (L = 246/f in MHz) and then trim the radials to resonate the antenna at the desired frequency. There is another reason for starting with a vertical that is a bit taller. When the radials are trimmed to resonate the antenna, the length of the radials will be somewhat shorter than might be expected. This saves wire and reduces the footprint.

A very common problem in elevated systems is that the radials may not all be exactly the same length. Experimentally this has been shown to cause non-uniform current division between the radials which can have a serious effect on the performance of the antenna.[9] We can model an example to demonstrate how severe this effect can be. Start with a 40 meter λ/4 vertical with four radials as shown in **Figure 3.30**, where the base of the vertical and the radials are placed 8 feet above average ground ($\sigma = 0.005$ S/m, $\varepsilon_r = 13$). Radials 1 and 2 form a pair of opposing radials with a length = L. Radials 3 and 4 are a second opposing pair of radials with length = M. First we model the antenna with all the radials the same length (L = M) and then with radials that differ in length (L ≠ M). The initial length for the vertical and the radials was made 34.1 feet to resonate the antenna at 7.2 MHz.

The modeled feed point impedances (from 7.0 to 7.3 MHz) for three different radial length configurations are compared in **Figure 3.31** which is a graph of R_{in} versus X_{in} ($Z_{in} = R_{in} + jX_{in}$ = feed point impedance) as the frequency is varied from 7.0 to 7.3 MHz. The plot on the left is for the case where all the radials are identical (L = M = 34.1 feet). The looping plot on the right is for case where L = 35.6 feet and M = 33.1 feet, this represents a length error of ±2.9%. The middle plot is for L = 34.6 feet and M = 33.6 feet, which is a length error of ±1.4%. Clearly even small radial length

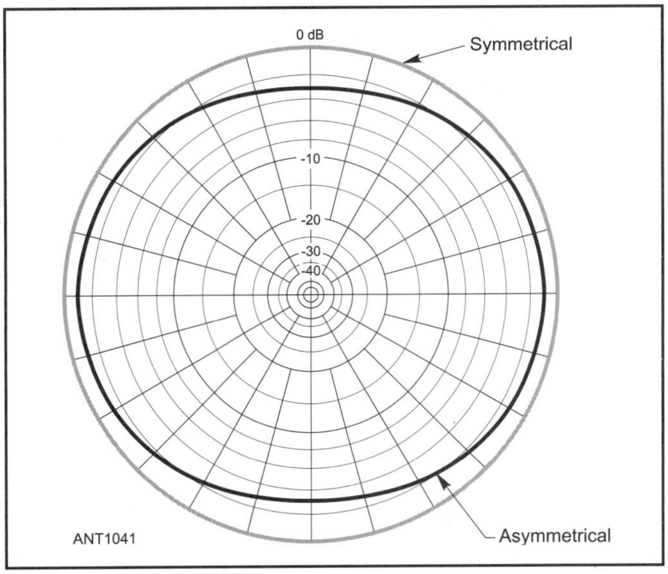

Figure 3.32 — Azimuthal radiation pattern at an elevation angle of 22 degrees, comparing symmetric (L = M = 34.1 feet) and asymmetric (L = 35.1 feet, M = 33.1 feet) radials at 7.25 MHz.

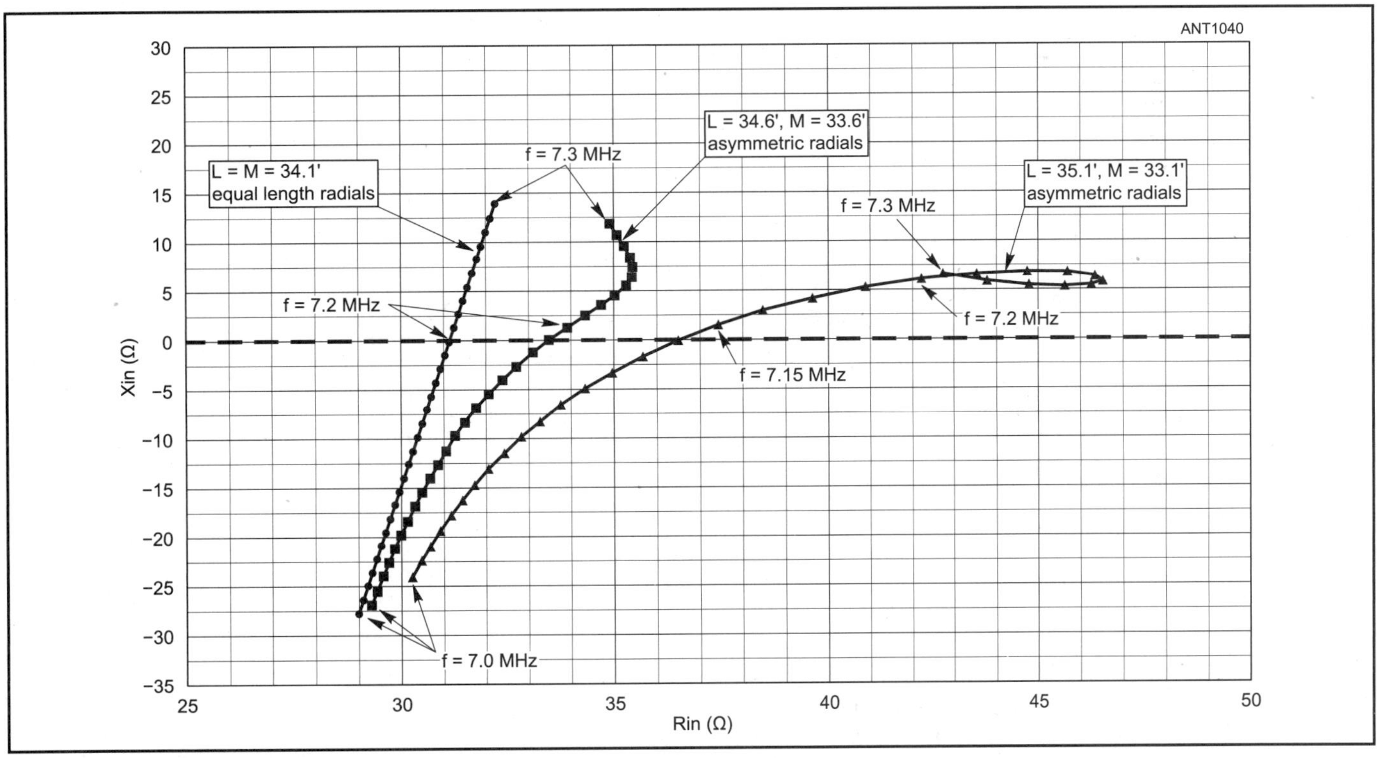

Figure 3.31 — A comparison of the input impedances ($Z_{in} = R_{in} + jX_{in}$) from 7.0 to 7.3 MHz at the feed point of the vertical for symmetric and asymmetric radial lengths.

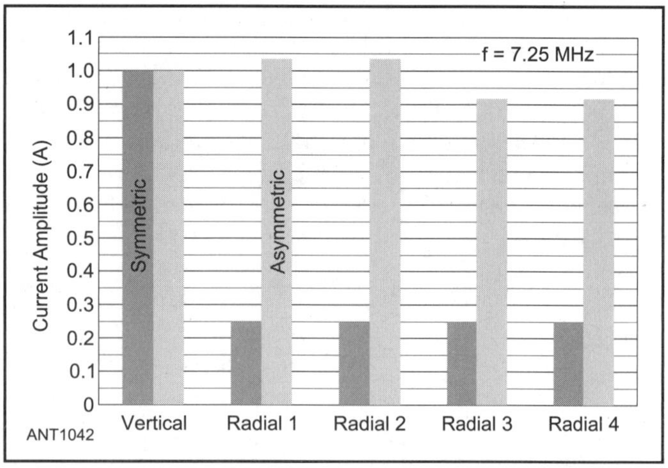

Figure 3.33 — Comparison of radial current at the bases of the vertical and the radials with symmetric (L = M = 34.1 feet) and asymmetric (L = 35.1 feet and M = 33.1 feet) radial lengths. The radials are numbered as shown in Figure 3.30. The current shown is the magnitude.

asymmetry can have a dramatic effect on the feed point impedance and resonant frequency. The resonant frequency is the point at which $X_{in} = 0$.

But feed point impedance is not the only problem created by asymmetric radial lengths. **Figure 3.32** compares modeled radiation patterns between symmetric and asymmetric systems at 7.25 MHz. The amount of pattern distortion varies across the band from a fraction of a dB at 7.0 MHz to 3 dB at 7.25 MHz. Besides the distortion, the gain in all directions is smaller for the asymmetric case. Computing the average gains for the symmetric and asymmetric cases in Figure 3.32, there is about a 1.6 dB difference. What this tells us is that asymmetric radials can lead to significantly higher ground losses!

The pattern distortion and increased ground loss with asymmetric radials occurs because the radial currents with asymmetric radial lengths can be much different from the symmetric case. An example is given in **Figure 3.33**. The graph bars represent the current amplitudes at the base of the vertical and each of the radials immediately adjacent to the base of the vertical. The black bars are for symmetric radial lengths (L = M = 34.1 feet) and the red bars are for asymmetric radials (L = 35.1 feet and M = 33.1 feet). In the symmetric case each of the radials has a current of 0.25 A which sums to 1 A, the current at the base of the vertical. The radial currents are also in phase with the base current.

With asymmetric radials the picture is very different: the current amplitudes are different between radial pairs 1 and 2 and 3 and 4 and the sum of the current amplitudes is *not* 1 A, it is much larger! This would seem to violate Kirchhoff's current law which requires the **vector** sum of the currents at a node to be zero. In this case the radial currents in the two pairs of radials are not in phase with each other or the vertical base current. The current in radials 1 and 2 is shifted by –62° from the base current and the current in radials 3 and 4 is shifted by +89°. The radial currents still sum *vectorially* to 1 A however. These large asymmetric currents go a long way towards explaining the increased loss and pattern distortion.

How can we tell if there is a problem in an existing radial fan? One way is to measure the current amplitudes in the individual radials close to the base of the vertical.[13] If the current amplitudes are significantly different between the radials *and/or* if the sum of the current amplitudes in the radials is greater than the base current then you have a problem. These measurements can be made with a RF ammeter. More accurate measurements which also show the phase can be made using current transformers and an oscilloscope (see the section "Practical Aspects of Phased Array Design" in the **Phased Arrays** chapter) or a vector network analyzer (see the **Antenna and Transmission Line Measurements** chapter).[13]

The sensitivity to asymmetric radial lengths is reduced when a larger number of radials are employed. The primary effect of additional elevated radials (>4) is to reduce the sensitivity to radial asymmetry, nearby conductors, variations in ground conductivity or objects under the radial fan, and, as shown in Figure 3.27, more numerous radials reduce the potentials on the radials. More numerous radials also reduce the E-field intensity below the radial fan. *Whenever possible an elevated ground system should use 10 or more radials.* If you follow that rule you are very likely to get the performance you expect. With small numbers of elevated radials the results can be hit or miss.

Sometimes it's not possible to have a symmetric radial fan of λ/4 radials. This is often the case on 160 or 80 meters. Because each installation will be unique it is difficult to give general advice but certainly the first step should be to model the proposed antenna with different radial options to get a feeling for how well they might work. One option is to keep the radial fan symmetric with a radius smaller than λ/4. You can then resonate the vertical with an inductor as shown in Figure 3.28D, add some top-loading to the vertical, make the vertical taller, or some combination of all three. With short radials it may be helpful to add a skirt wire at the ends of the radials as shown in **Figure 3.34**. Adding a skirt wire to the

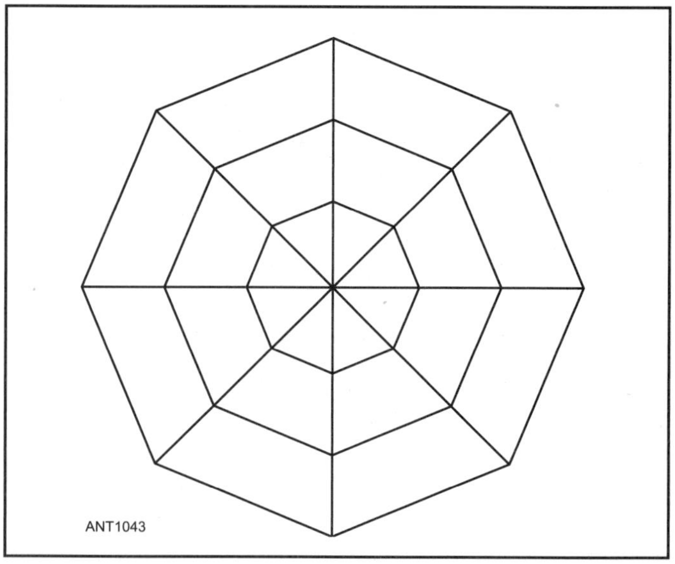

Figure 3.34 — Use of skirt wires to increase the effective size of short radial systems.

radial system will reduce the size of the base loading inductance.

ISOLATION OF ELEVATED GROUND SYSTEMS

In an elevated ground system it is good practice to isolate the feed line with a common mode choke (i.e. a current balun — see the **Transmission Line System Techniques** chapter). Simply attaching a coaxial feed line to the antenna and running it back down to ground can increase ground loss and in some cases have a strong effect on the resonant frequency and behavior of the impedance across the band. The effects of not isolating the feed line vary from slight to severe depending on the details of each installation. Elevated systems with small numbers of radials are particularly sensitive. An additional problem with asymmetric radials is that they can greatly increase the voltage across the balun, leading to larger losses in the balun core.

3.2.5 Notes On Radial Systems

CURRENT DISTRIBUTION ON RADIALS

Ground systems using elevated radials, radials lying on the ground surface, or buried radials can all provide good performance but there are some differences with practical consequences which need to be recognized. As shown in **Figure 3.35**, the current distribution on a λ/4 radial is different for each of those arrangements.

When a radial is placed very close to the ground, the velocity of propagation along the radial is slower so that the radial is effectively electrically longer and the current maximum moves out onto the radial as shown in Figure 3.35B. This has two consequences: First, it can increase the ground loss and, second, it can affect the feed point impedance and resonant frequency of the vertical.[14,15] Note from the current distribution, that the ground surface radial behaves more like an elevated radial than a buried one. Ground surface radials can affect the resonant frequency of the antenna.

An example of this is given in **Figure 3.36**. The experiment from which the data in Figure 3.36 was obtained began with no radials and only a single ground stake. As radials were added, the resonant frequency of the antenna was measured.

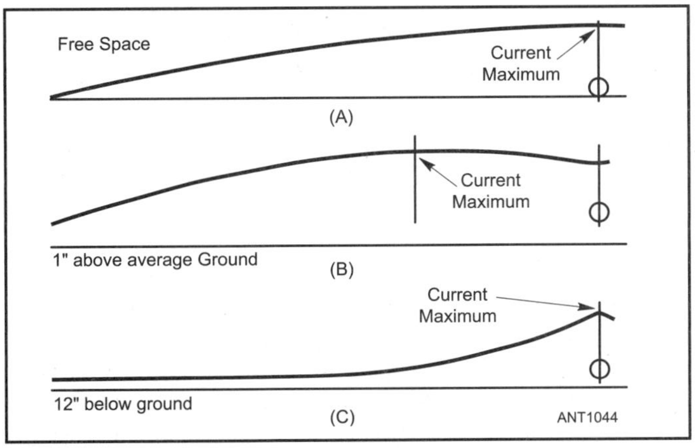

Figure 3.35 — Examples of current distribution on radials in elevated, ground surface and buried systems.

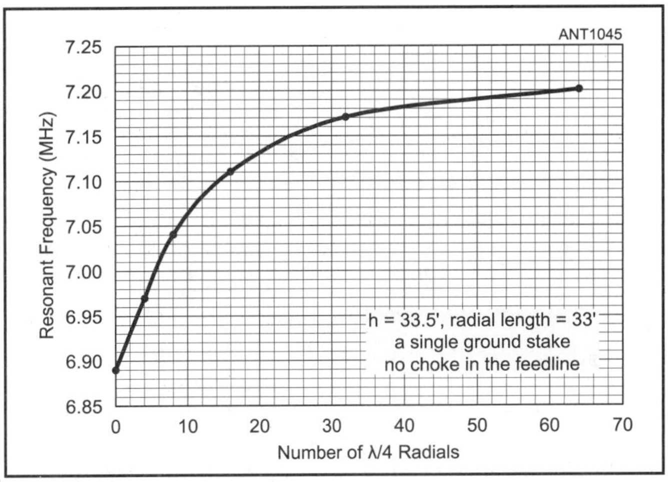

Figure 3.36 — A typical example of the effect on resonant frequency as the number of ground surface radials is varied.

Initially the change in resonant frequency was quite rapid but by 32 radials the rate of change slowed and the resonant frequency stabilized. The additional ground loss when <8 radials are used can be significant, providing another reason for not using a few long radials.[14]

For bare radials, well buried in the soil, the radial current distribution is exponential due to the damping effect of the soil conductivity (Figure 3.35C). In general, changing the radial number in buried systems does not greatly affect resonance except possibly in very low conductivity soils. The change in current distribution as a radial is taken from the surface into the soil is not abrupt. Radials just below the surface will behave much like radials right on the surface. The rate of change depends on the soil characteristics so don't be surprised if you see some shift in resonance as more radials are added with shallow burial.

The goal of the radial system is to give current returning to the feed point a low-loss path through metal rather than soil. It is also more convenient for radials to be on the ground surface or buried. Figure 3.35 indicates that ground surface radials provide a reasonable compromise between efficiency and convenience. The discussion of radial systems in the **General Purpose MF and HF Antennas** chapter shows several practical options, including how to let grass and the animals living in the soil do the job of burying the radials for you.

Advice on Ground Systems

With so many configurations and options available, some advice on ground systems is in order: Buried, ground surface or elevated radial systems can all be very efficient. However, no matter which configuration is chosen, a few long radials are not likely to provide a satisfactory ground system. If you want an efficient ground system don't skimp on the radials! Try to use at least 20 or more radials on the ground and 10 or more in elevated systems.

MULTIBAND RADIAL SYSTEMS

Multiband verticals are very popular which raises the question of what kind of ground system to use with them? In practice, the most common ground system, either on the ground or elevated, has four $\lambda/4$ radials *for each band*. For example if the vertical operates on 7, 14, 21 and 28 MHz there will be a total of 16 radials which is about 280 feet of radial wire. Multiband ground systems have been evaluated experimentally and shown to work very well.[15] Even though the most common system has only four radials on each band, coupling between the radials seems to minimize the elevated system problems discussed earlier. One alternative (for either on-the-ground or elevated systems) for a 40-10 meter vertical would be to use 30 or more 40 meter $\lambda/4$ radials with-out any of the shorter higher band radials. This also works well yielding an improvement of about 1 dB over the standard system. However, 30 radials on 40 meters total about 2100 feet of wire which is almost eight times the total wire in the standard system!

RADIAL WIRE SIZE AND MATERIAL

If the recommended number of radials is used, the wire size used in a radial system usually has only a small effect on the electrical performance. For a given amount of copper it is much better to use many small diameter radials instead of a few large ones. The practical issues are more mechanical than electrical: i.e. is the wire sturdy enough for installation and will it survive burial in the soil or lying on the soil surface for extended periods? In the case of elevated radial systems, is the wire strong enough to be stretched between the supports and, in climates where icing is a problem, is it strong enough for the possible ice load? The wire can be either insulated or bare although in the case of buried radials insulated wire may resist corrosion longer. Wire sizes as small as #22 AWG may be acceptable if there are a large number of radials. Either copper or aluminum wire can be used. Steel wire is very strong and inexpensive but both copper and aluminum wire have much better conductivity. Generally speaking galvanized fence wire should be viewed as an emergency measure. Although aluminum wire is attractive because it's much cheaper than copper, it has much lower corrosion resistance and may not be suitable for buried installations in most soils. Aluminum has the additional problem that it is difficult to solder and usually requires mechanical connections which may not be reliable when exposed to the weather for long periods. Either solid or stranded wire can be used in ground systems although in buried systems solid wire may be more corrosion resistant. For elevated systems where severe ice loading is expected Copperweld or Alumaweld wire can be used. These are steel wires with a thick copper or aluminum cladding. This construction gives both good conductivity and great strength. However, any damage to the cladding will expose bare steel to the elements, resulting in corrosion.

Insulated copper wire will frequently be less expensive than bare wire and in addition, insulated wire of many different kinds is often available inexpensively from surplus sources. It is not necessary to strip the insulation from the wire to use it for radials except to connect it at the base of the antenna or to other radials. In an elevated system loading by the insulation will make the radials electrically 2-3% longer but add little loss. For buried radials the insulation may provide some corrosion protection.

RADIALS FOR DIRECTIONAL ARRAYS

In directional arrays such as four-squares, a full-size radial system for each element is large enough to overlap with the radials from adjacent elements. This is common in AM broadcast antenna systems and the industry's approach is described in the paper "AM Antenna Systems" from Rockwell Media at **rockwellmedia.net/wp-content/uploads/2019/11/AM-Antenna-Systems-White-Paper-1.pdf**.

For directional antennas with multiple towers, the same basic ground system elements (*buried radials – Ed.*) are used, except where radials from different towers intersect, they are terminated into and bonded to a transverse copper strap. For example, in a two-tower array with ¼ wavelength spacing, a transverse copper strap would be installed halfway between the two towers and all the radials that would intersect the other tower's radials would terminate onto the transverse strap.

Amateurs are unlikely to have the budget to use that much copper strap, but a heavy wire will suffice. The job becomes more tedious than difficult, requiring the soldering of many wires, but this provides the best current path and minimizes interactions and variations with weather-related ground conductivity. If insulated radials are used or the radials are buried at different depths, the radial systems can be left overlapping without direct contact, realizing that there may be some interaction between the elements.

Multi-element vertical arrays are often constructed with elevated radial systems. Interaction of the radials can be much stronger than in buried-radial systems. Configuration of elevated radials in these systems was discussed in two *QST* articles by Christman.[20]

3.2.6 Counterpoise Systems

(The following section is largely derived from Rudy Severns, N6LF's paper, LF and MF Antennas for Amateurs, which is included in the online material for the chapter on ***General Purpose HF and MF Antennas****.)*

THE DEFINITION OF COUNTERPOISE

The term "counterpoise" is widely used (and misused) in amateur antenna system design. It is often confused with a set of radial wires that are buried or placed directly on top of the ground. A system of a small number of radials is occasionally referred to as a counterpoise, as well. Finally, it is sometimes applied to a single wire, often laid directly on the ground, attached to a feed point or equipment enclosure. None of these are counterpoises.

Early radio system designers, working at what is today called LF and MF, recognized that an elevated system called a "counterpoise" or "capacitive ground," with dimensions significantly smaller than $\lambda/4$, could be very effective. Summarizing the preceding decades of radio practice, here is the 1952 definition of a counterpoise by Laport (see References and Bibliography):

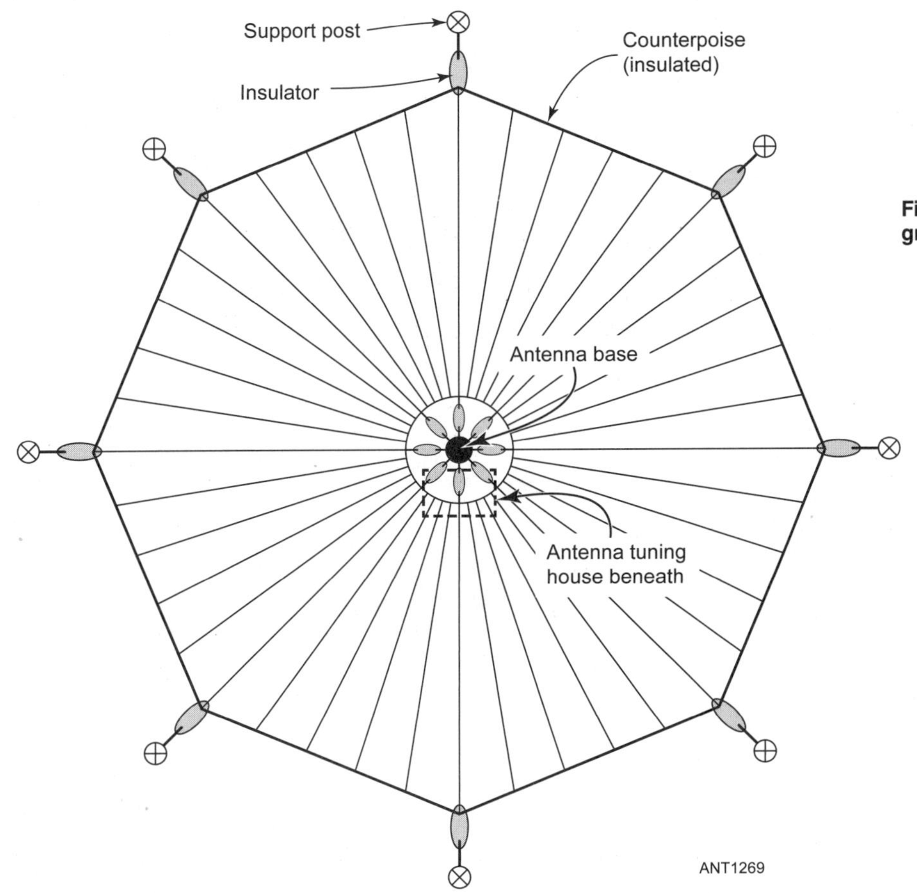

Figure 3.37 — A typical counterpoise ground system. (Figure from Laport)

"The counterpoise is an insulated net of radial wires assembled above the ground to form a large capacitance with the ground. From the earliest days of radio, the merits of the counterpoise as a low-loss ground system have been recognized because of the way in that the current densities in the ground are more or less uniformly distributed over the area of the counterpoise...It is inconvenient structurally to use very extensive counterpoise systems, and this is the principal reason that has limited their application. The size of the counterpoise depends upon the frequency. It should have sufficient capacitance to have a relatively low reactance at the working frequency so as to minimize the counterpoise potentials with respect to ground. The potential existing on the counterpoise may be a physical hazard that may also be objectionable. There should not be any connection to actual ground in the antenna circuit when a counterpoise is used."

The counterpoise definition in amateur publications began to drift in following years, however. LB Cebik, W4RNL, traced the gradual changes in a paper titled "Counterpoise? The Use and Abuse of a Word."[21] He summarizes the types of systems employed with vertical monopole antennas as follows:

"Below the ground, we have buried radial systems with the wire in direct (or nearly direct, if the wires are insulated) contact with the ground. Next, we have the counterpoise alternative to buried radials. The counterpoise size and wire density equaled that of the buried radial system for maximum efficiency. Third, we have elevated radials, where the capacitance between the radials and the ground is too small to be effective in the determination of antenna efficiency. Rather, the radials become part of the antenna structure sufficiently independent of the ground that antenna resonance is a function of the overall antenna size at the operating frequency."

In radial systems, the non-resonant radial wires form a ground plane or ground screen so that return currents in the ground flowing to the feed point do not flow through lossy soil. This reduces ground losses and increases the antenna system's efficiency. In elevated radials, the radial wires and antenna are tuned to create a resonant system above and not connected to the ground. Both cases are discussed in previous sections of this chapter.

Figure 3.37 shows an example of a counterpoise, also from Laport. Rectangular counterpoises, some with a coarse rectangular mesh, were common. Amateurs also used counterpoises. **Figure 3.38** is a sketch of the antenna used for the initial transatlantic tests by amateurs (1BCG) in 1921-22.The operating frequency for the tests was about 1.3 MHz (230 meters). At 1.3 MHz $\lambda/4$ = 189 feet so the 60-foot radius of the counterpoise corresponds to $\approx 0.08\ \lambda$.

Finally, a single wire attached to a feed point merely acts as more of the antenna, even if lying on the ground. It is usually used to lower or stabilize the feed point impedance by effectively moving the feed point away from the end of the

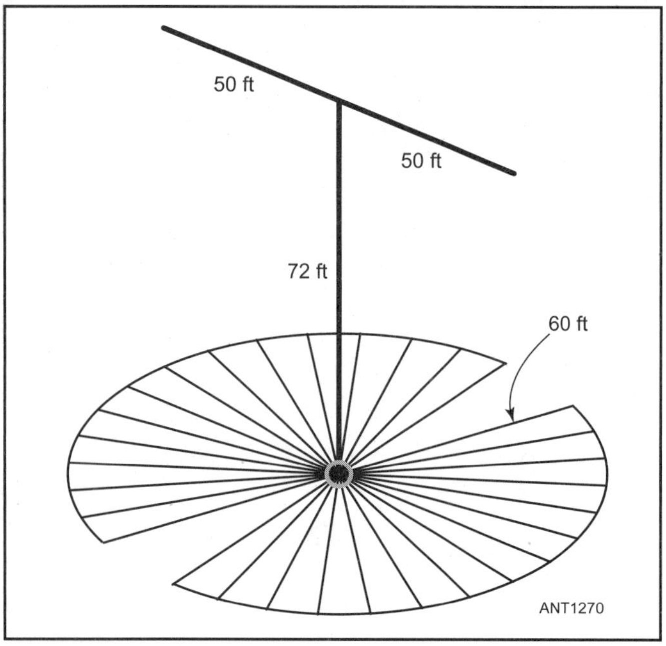

Figure 3.38 — *EZNEC* model of the 1BCG antenna.

antenna where impedance is very high. Quarter-wavelength, resonant wires are sometimes attached to equipment in order to lower RF voltage on the enclosure or attached cables.

Counterpoises vs Buried Radial Systems

It is useful to compare buried radial systems with a counterpoise using a modeled antenna. In many cases the ground under and near the antenna may not be suitable for a buried radial system. For low-frequency antennas, a standard radial system may be impractically large. For this comparison, the antennas modeled are shown in Figure 3.39 – 3.47; a 42-foot vertical with a 16-wire, 25-foot radius umbrella and a counterpoise; and a 50-foot vertical with a 16-wire, 25-foot radius umbrella and a buried radial system. The umbrella, counterpoise, and buried radial system all have skirt wires connecting the end of radial conductors. Note that the tops of the antennas are both at 50 feet, so for the counterpoise vertical, H=42 feet (the vertical conductor length) with the counterpoise 8 feet above the ground, but for the buried radial system vertical, H=50 feet. (For safety reasons the counterpoise is assumed to be installed at least 8 feet above the ground so there is no danger of casual contact with the high potentials on the counterpoise while transmitting. This is a concern for any elevated radial system.)

Figure 3.39 shows a comparison between 64-radial buried radial systems and 16-radial counterpoises as the radius of the ground system is varied from 10' to 50'. In the graphs, solid lines are for the counterpoise and the dashed lines are for the buried radial system.

With 16 radials the counterpoise is superior for radii less than 100 feet. With 64 wires the counterpoise is better out to about 35 feet after which the buried system is better. While the counterpoise looks better electrically, remember that it is a large and very visible elevated structure while the buried or ground surface system is out of sight. In addition, the counterpoise will require posts to support the ends, insulators at the outer periphery, a larger loading inductor, and an isolation choke for the feed line. (Link coupling to a loading coil can also be used for feed line isolation.)

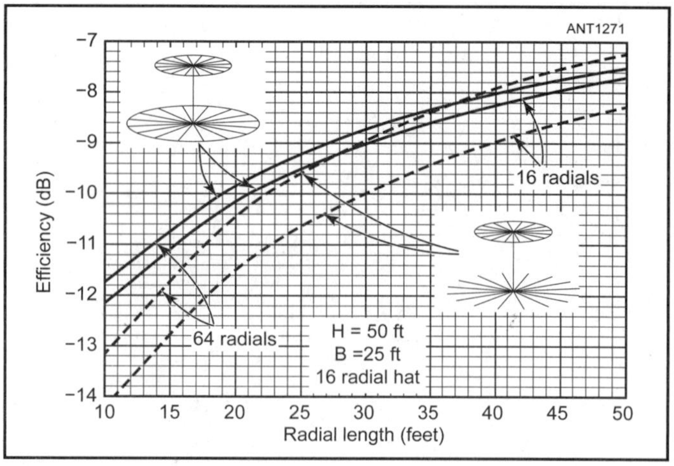

Figure 3.39 — A comparison between buried ground systems and counterpoises.

Electrically short verticals can be analyzed as a resistor in series with a large capacitive reactance (X_C). (See N6LF's paper from which this section is derived for a detailed discussion of how the antenna ground system is modeled.) The series resistance is a combination of all the antenna resistances; radiation resistance (R_g), loss resistance (R_L which includes losses in conductors and the loading coil), and ground losses (R_g).

An important point is that in most cases a loading inductor will be needed to resonate the antenna by canceling X_C. This inductor will have a finite Q and in many cases the inductor loss will be the dominant resistive loss in the system. For each graph the appropriate values for R_L were used assuming a coil Q of 400. (See the paper by N6LF on LF and MF Antennas in the online material for the **General-Purpose MF and HF Antennas** chapter for a thorough analysis of typical short verticals.)

To generalize from this example to other configurations,

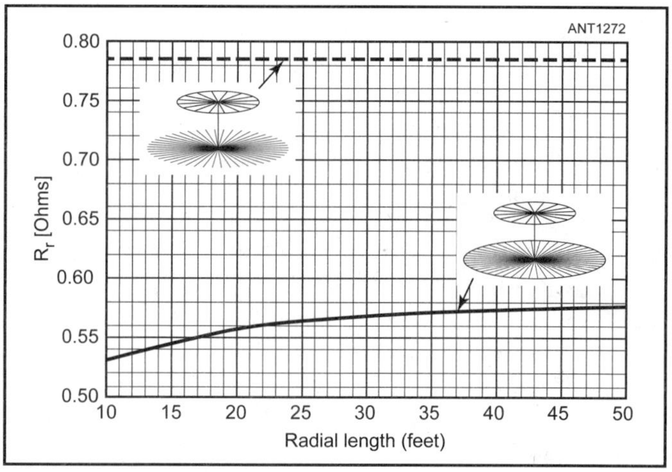

Figure 3.40 — Comparison of R_r between buried and counterpoise ground systems.

it's useful to look in more detail at what's happening to R_r, R_L, and R_g as we vary the radius of the buried radial system or counterpoise.

Figures 3.41 through 3.42 show comparisons as a function of radial length for R_r, R_L, and R_g between a buried system and a counterpoise. As the figures show, there are significant differences between the values and behavior of R_r, R_L, and R_g. In **Figure 3.40**, R_r for the vertical with a counterpoise is significantly lower than for the buried radial system. This is because the length of the vertical is shorter with the counterpoise (42 feet). For short verticals, R_r varies as the square of the length. For example, comparing R_r' at 42 feet to R_r at 50 feet, where R_r=0.79 Ω:

$$R_r' = 0.79(42 / 50)^2 = 0.56\ \Omega$$

This agrees with what we see in Figure 3.40. If we have a vertical taller than 50 feet, the reduction in R_r from shortening its length by 8 feet will be reduced, but if the vertical is shorter than 50 feet the reduction in R_r will be greater.

Figure 3.41 shows how much larger R_L becomes when the counterpoise is employed. This comes from two sources: first, X_C is larger due to the shorter length and second, the counterpoise itself introduces a reactance (X_{CP}) in series with X_C. To resonate the antenna with a loading coil requires inductive reactance of $X_L=X_C+X_{CP}$, and the loading inductor will be larger which increases R_L. (Since the vertical is made shorter by the height of the counterpoise (8 feet) a larger loading inductor is required to resonate the antenna.) Increasing the radius of the counterpoise and the number of radial wires reduces X_{CP} and thus, the required reactance of the loading coil.

The trends in Figures 3.41 and 3.42 would seem to favor the buried ground system but **Figure 3.42** gives the opposite message. It shows the counterpoise is indeed a very efficient ground system which has much lower ground losses as seen in the lower values for R_g. When combine the values in Figure 3.40 and 3.42 are combined using the following equations, you get the result shown in Figure 3.42 where the counterpoise has better efficiency for the smaller radii (<40 feet):

$$R_r \equiv P_r / (I_o^2) \text{ and } R_g \equiv P_g / (I_o^2)$$

In some ways however, the comparison just made between a buried radial system and a counterpoise is a bit misleading because the overall height of the vertical was limited to a fixed value (H=50 feet). If we keep H=50 feet and simply elevate the entire vertical by 8 feet so that the top is now at 58 feet, then the results are different as shown in **Figure 3.43** where the dashed lines are for buried wire ground systems and the solid lines are for a counterpoise with the same number of radials and radius.

What we see in the upper two traces in Figure 3.43 is that simply raising the entire antenna by 8 feet and substituting the counterpoise for the buried wire system makes a large improvement in efficiency and the counterpoise is superior at all radii up to and beyond 50 feet. You could argue that if we

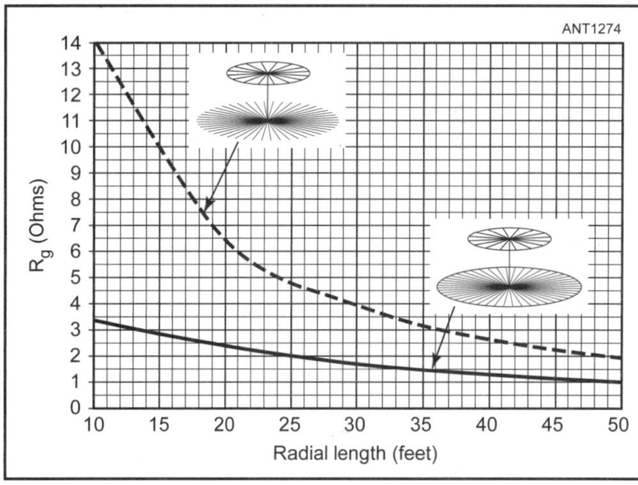

Figure 3.42 — Comparison of R_g between buried and counterpoise ground systems.

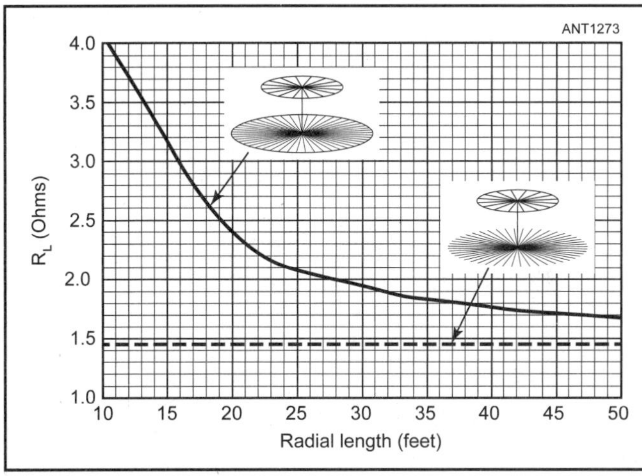

Figure 3.41 — Comparison of R_L between buried and counterpoise ground systems.

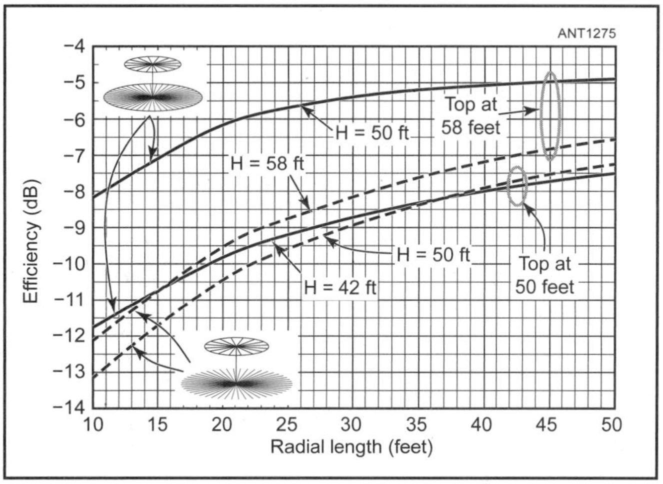

Figure 3.43 — Efficiency comparisons between counterpoise and buried wire ground systems for top heights of 50 feet and 58 feet.

can raise the entire antenna 8 feet we could just as well have simply increased H to 58 feet and retained the buried ground system but as we can see in Figure 3.43, the counterpoise is still better, at least to radii of 50 feet.

The point is that if you have reasonable heights for the vertical and lots of capacitive top-loading but very restricted room for the ground system, then a counterpoise may be the best option. Be careful in drawing general conclusions from the limited examples given here. There are many variables: ground characteristics, height of the top of the vertical, height of the bottom of the vertical, the amount of top-loading, the number of wires in the counterpoise, the radius, etc. The choice between buried wire and counterpoise ground systems is not obvious! The considerable mechanical complexity, vulnerability to ice damage and visual impact of a counterpoise may also work against it. This choice must be made on a case-by-case basis and will probably require modeling with *NEC4* software.

Reiterating safety concerns, just as with an elevated radial system, the counterpoise system will have high voltages present, even when not running high power. Care must be taken to keep these points away from flammable material and from where people might come in contact with them. A true counterpoise, not connected to the ground, cannot serve as part of a lightning protection ground, either. That ground path will have to be provided separately and the effect of that connection and where it is located will have to be taken into account when designing the overall antenna system.

3.3 The Effect of Ground in the Far Field

The properties of the ground in the far field of an antenna are very important, especially for a vertically polarized antenna. Even if the ground-radial system for a vertical has been optimized to reduce ground-return losses in the reactive near field to an insignificant level, the electrical properties of the ground may still diminish far-field performance to lower levels than "perfect-ground" analyses might lead you to expect. The key is that *ground reflections* from horizontally and vertically polarized waves behave very differently.

This section, from earlier editions, uses an alternate convention in which k and ε_r refer to the same quantity and are interchangeable, as are σ and G. Both are in common use in the technical literature.

3.3.1 Reflections In General

First, let us consider the case of flat ground. Over flat ground, either horizontally or vertically polarized down-going waves launched from an antenna into the far field strike the surface and are reflected by a process very similar to that by which light waves are reflected from a mirror. As is the case with light waves, the angle of reflection is the same as the angle of incidence, so a wave striking the surface at an angle of, say, 15° is reflected upward from the surface at 15°.

The reflected waves combine with direct waves (those radiated at angles above the horizon) in various ways. Some of the factors that influence this combining process are the height of the antenna, its length, the electrical characteristics of the ground, and the polarization of the wave. At some elevation angles above the horizon the direct and reflected waves are exactly in phase — that is, the maximum field strengths of both waves are reached at the same time at the same point in space, and the directions of the fields are the same. In such a case, the resultant field strength for that angle is simply the sum of the direct and reflected fields. (This represents a theoretical increase in field strength of 6 dB over the free-space pattern at these angles.)

At other elevation angles the two waves are completely out of phase — that is, the field intensities are equal at the same instant and the directions are opposite. At such angles, the fields cancel each other. At still other angles, the resultant field will have intermediate values. Thus, the effect of the ground is to increase radiation intensity at some elevation angles and to decrease it at others. When you plot the results as an elevation pattern, you will see *lobes* and *nulls*, as described in the **Antenna Fundamentals** chapter.

The concept of an *image antenna* is often useful to show the effect of reflection. As **Figure 3.44** shows, the reflected ray has the same path length (AD equals BD) that it would if it originated at a virtual second antenna with the same char-

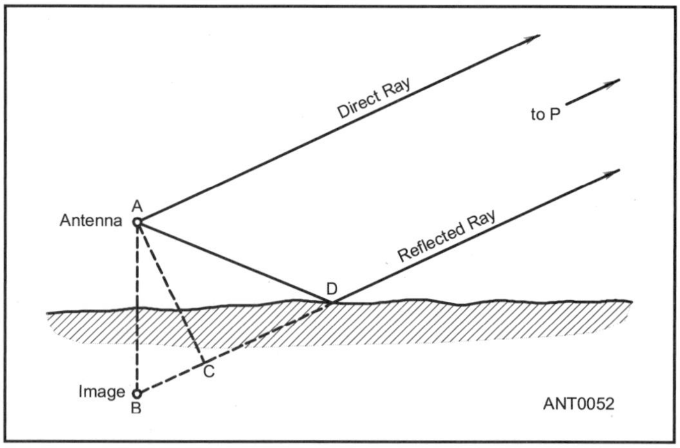

Figure 3.44 — At any distant point, P, the field strength will be the vector sum of the direct ray and the reflected ray. The reflected ray travels farther than the direct ray by the distance BC, where the reflected ray is considered to originate at the image antenna.

acteristics as the real antenna, but situated below the ground just as far as the actual antenna is above it.

Now, if we look at the antenna and its image over perfect ground from a remote point on the surface of the ground, we will see that the currents in a horizontally polarized antenna and its image are flowing in opposite directions, or in other words, are 180° out of phase. But the currents in a vertically polarized antenna and its image are flowing in the *same* direction — they are *in* phase. This 180° phase difference between the vertically and horizontally polarized reflections off ground is what makes the combinations with direct waves behave so very differently.

For horizontally polarized waves, the *Fresnel zone* (*fruh-NEL*) is the region of ground surface from which reflections create the antenna's elevation pattern. The location of the zone changes for different antenna heights, elevation angles, and wavelengths. Point D in Figure 3.44 is in the Fresnel zone for this particular reflection. While the material in this chapter generally assumes that the ground surface is flat, the actual characteristics and topography of the ground in the Fresnel zone are very important in determining the resulting elevation pattern. This is discussed in detail in the chapter **HF Antenna System Design**. That chapter also includes an extensive discussion on the use of *HFTA* terrain analysis and modeling software.

3.3.2 Far-Field Ground Reflections and The Vertical Antenna

A vertical's azimuthal directivity is omnidirectional. A λ/2 vertical over ideal, perfectly conducting earth has the elevation-plane radiation pattern shown by the solid line in **Figure 3.45**. Over real earth, however, the pattern looks more like the shaded one in the same diagram. In this case, the low-angle radiation that might be hoped for because of perfect-ground performance is not realized in the real world.

Now look at **Figure 3.46A**, which compares the computed elevation-angle response for two half-wave dipoles at 14 MHz. One is oriented horizontally over ground at a height of λ/2 and the other is oriented vertically, with its center just over λ/2 high (so that the bottom end of the wire doesn't actually touch the ground). The ground is "average" in dielectric constant (13) and conductivity (0.005 S/m). At a 15° elevation angle, the horizontally polarized dipole has almost 7 dB more

Real-World Ground Surfaces

The material in this chapter deals with the effects of ground assuming that the ground surface around the antenna is flat. This is obviously not the case in the majority of actual installations! Accounting for the effects of non-flat ground is included in the chapter **HF Antenna System Design**, including an extensive discussion on the use of *HFTA* terrain analysis software by Dean Straw, N6BV.

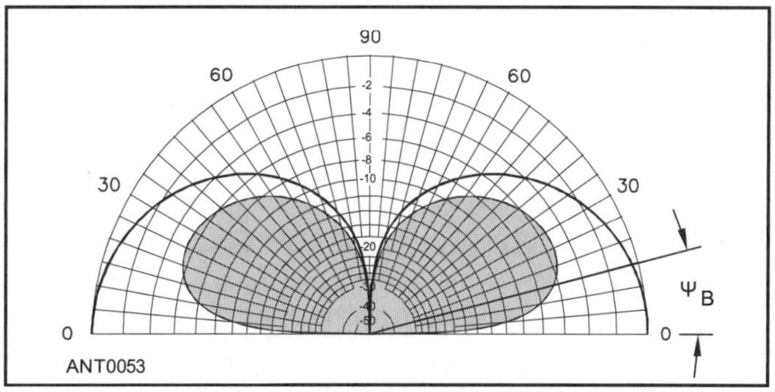

Figure 3.45 — Vertical-plane radiation pattern for a ground-mounted half-wave vertical. The solid line is the pattern for perfect earth. The shaded pattern shows how the response is modified over average earth (k = 13, G = 0.005 S/m) at 14 MHz. ψ is the pseudo-Brewster angle (PBA), in this case 14.8°.

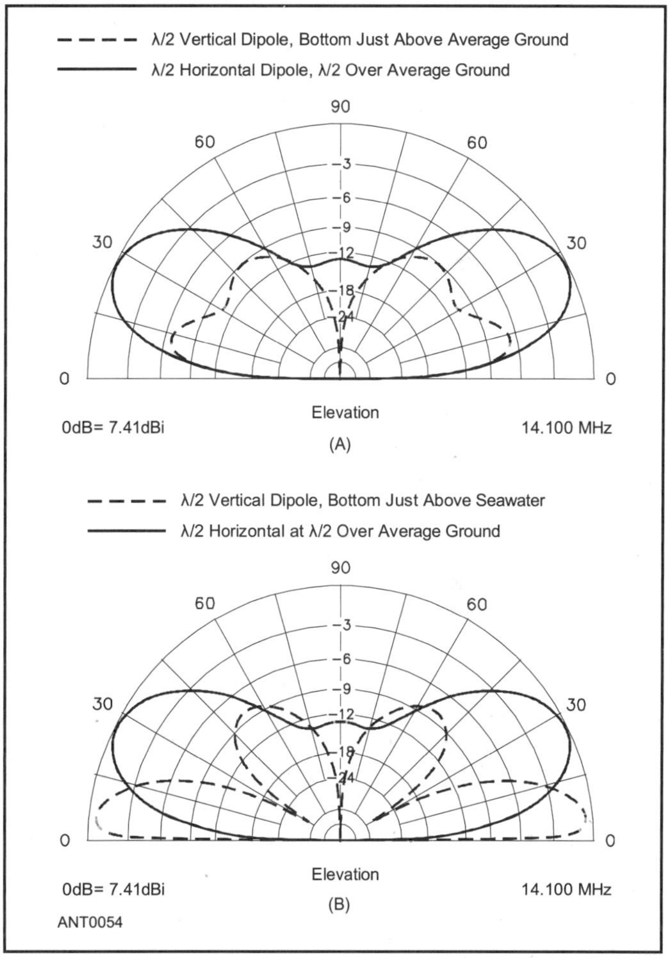

Figure 3.46 — At A, comparison of horizontal and vertical λ/2 dipoles over average ground. Average ground has conductivity of 5 mS/m and dielectric constant of 13. Horizontal dipole is λ/2 high; vertical dipole's bottom wire is just above ground. Horizontal antenna is much less affected by far-field ground losses compared with its vertical counterpart. At B, comparison of 20 meter λ/2 vertical dipole whose bottom wire is just above seawater with λ/2-high horizontal dipole over average ground. Seawater is great for verticals!

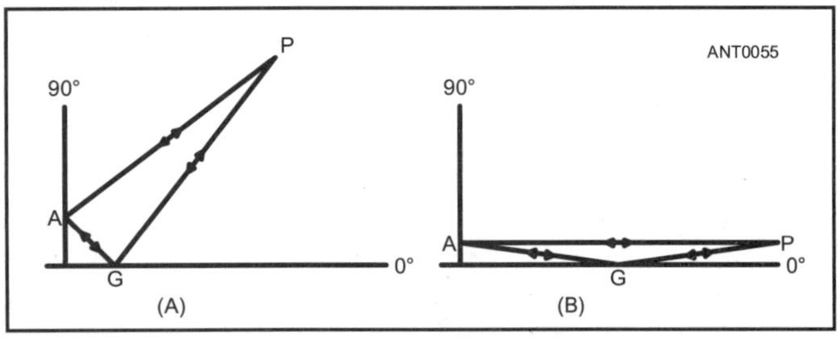

Figure 3.47 — The direct wave and the reflected wave combine at point P to form the pattern (P is very far from the antenna). At A the two paths AP and AGP differ appreciably in length, while at B these two path lengths are nearly equal.

gain than its vertical brother. Contrast Figure 3.46A to the comparison in Figure 3.46B, where the peak gain of a vertically polarized half-wave dipole over seawater, which is virtually perfect for RF reflections, is quite comparable with the horizontal dipole's response at 15°, and exceeds the horizontally polarized antenna dramatically below 15° elevation.

To understand in a qualitative fashion why the desired low-angle radiation from a vertical is not delivered when the ground isn't "perfect," examine **Figure 3.47A**. Radiation from each antenna segment reaches a point P in space by two paths; one directly from the antenna, path AP, and the other by reflection from the earth, path AGP. (Note that P is so far away that the slight difference in angles is insignificant — for practical purposes the waves are parallel to each other at point P.)

If the earth were a perfectly conducting surface, there would be no phase shift of the vertically polarized wave upon reflection at point G. The two waves would add together with some phase difference because of the different path lengths. This difference in path lengths of the two waves is why the free-space radiation pattern differs from the pattern of the same antenna over ground.

Now consider a point P that is close to the horizon, as in Figure 3.47B. The path lengths AP and AGP are almost the same, so the magnitudes of the two waves add together, producing a maximum at zero angle of radiation. The arrows on the waves point both ways since the process works similarly for transmitting and receiving.

With real earth, however, the reflected wave from a vertically polarized antenna undergoes a change in both *amplitude* and *phase* in the reflection process. Indeed, at a low-enough elevation angle, the phase of the reflected wave will actually change by 180° and its magnitude will then subtract from that of the direct wave. At a zero takeoff angle, it will be almost equal in amplitude, but 180° out of phase with the direct wave.

Note that this is very similar to what happens with horizontally polarized reflected and direct waves at low elevation angles. Virtually complete cancellation will result in a deep null, inhibiting any radiation or reception at 0°. For real-world soils, the vertical loses the theoretical advantage it has at low elevation angles over a horizontal antenna, as Figure 3.46A so clearly shows.

The degree that a vertical works better than a horizontal antenna at low elevation angles is largely dependent on the characteristics of the ground around the vertical, as we'll next examine.

3.3.3 The Pseudo-Brewster Angle (PBA) and The Vertical Antenna

Much of the material presented here regarding pseudo-Brewster angle was prepared by Charles J. Michaels, W7XC, and first appeared in July 1987 *QST*, with additional information in *The ARRL Antenna Compendium, Vol 3*[12] and Cebik's article on the PBA.[19]

Most fishermen have noticed that when the sun is low, its light is reflected from the water's surface as glare, obscuring the underwater view. When the sun is high, however, the sunlight penetrates the water and it is possible to see objects below the surface of the water. The angle at which this transition takes place is known as the *Brewster angle*, named for the Scottish physicist, Sir David Brewster (1781-1868).

A similar situation exists in the case of vertically polarized antennas; the RF energy behaves as the sunlight in the optical system, and the earth under the antenna acts as the water. The *pseudo-Brewster angle* (PBA) is the angle at which the reflected wave is 90° out of phase with respect to the direct wave. "Pseudo" is used here because the RF effect is similar to the optical effect from which the term gets its name. Below this angle, the reflected wave is between 90° and 180° out of phase with the direct wave, so some degree of cancellation takes place. The largest amount of cancellation occurs near 0°, and steadily less cancellation occurs as the PBA is approached from below.

The factors that determine the PBA for a particular location *are not related to the antenna itself, but to the ground around it*. The first of these factors is earth conductivity, σ, which is a measure of the ability of the soil to conduct electricity. Conductivity, measured in siemens/meter is the inverse of resistivity. The second factor is the dielectric constant, ε_r, which is a unit-less quantity that corresponds to the capacitive

Table 3.5
Pseudo-Brewster Angle Variation with Frequency, Dielectric Constant, and Conductivity

Frequency (MHz)	*Dielectric Constant*	*Conductivity (S/m)*	*PBA (degrees)*
7	20	0.0303	6.4
	13	0.005	13.3
	13	0.002	15.0
	5	0.001	23.2
	3	0.001	27.8
14	20	0.0303	8.6
	13	0.005	14.8
	13	0.002	15.4
	5	0.001	23.8
	3	0.001	29.5
21	20	0.0303	10.0
	13	0.005	15.2
	13	0.002	15.4
	5	0.001	24.0
	3	0.001	29.8

effect of the earth. (See the section "Electrical Characteristics of Ground" earlier in this chapter for a discussion of both σ and ε.) For both of these quantities, the higher the number, the better the ground (for vertical antenna purposes). The third factor determining the PBA for a given location is the frequency of operation. The PBA increases with increasing frequency, all other conditions being equal.

As the frequency is increased, the role of the dielectric constant in determining the PBA becomes more significant. **Table 3.5** shows how the PBA varies with changes in ground conductivity, dielectric constant and frequency. The table shows trends in PBA dependency on ground constants and frequency.

At angles below the PBA, the reflected vertically polarized wave subtracts from the direct wave, causing the radiation intensity to fall off rapidly. Similarly, above the PBA, the reflected wave adds to the direct wave, and the radiated pattern approaches the perfect-earth pattern. **Figure 3.48** shows the PBA, usually labeled ψ_B.

When plotting vertical-antenna radiation patterns over real earth, the reflected wave from an antenna segment is multiplied by a factor called the *vertical reflection coefficient*, and the product is then added vectorially to the direct wave to get the resultant. The reflection coefficient consists of an attenuation factor, A, and a phase angle, ϕ, and is usually expressed as $A\angle\phi$. (ϕ is always a negative angle, because the earth acts as a lossy capacitor in this situation.) The following equation (see Terman in the "References and Bibliography") can be used to calculate the reflection coefficient for vertically polarized waves, for earth of given conductivity and dielectric constant at any frequency and elevation angle (also called the wave angle in many texts).

$$A_{Vert}\angle\phi = \frac{k'\sin\psi - \sqrt{k' - \cos^2\psi}}{k'\sin\psi + \sqrt{k' - \cos^2\psi}} \qquad (9)$$

where

$A_{Vert}\angle\phi$ = vertical reflection coefficient

ψ = elevation angle

$$k' = k - j\left|\frac{1.8\times10^4\times G}{f}\right|$$

k = dielectric constant of earth (k for air = 1)

G = conductivity of earth in S/m

f = frequency in MHz

j = complex operator ($\sqrt{-1}$)

(Reminder: k and ε_r refer to the same quantity and are interchangeable, as are σ and G. Both are in common use in the technical literature.)

Solving this equation for several points illustrates the effect of earth on vertically polarized signals at a particular location for a given frequency range. Figure 3.48 shows the reflection coefficient as a function of elevation angle at 21 MHz over average earth (G = 0.005 S/m, and k = 13). Note that as indicated by the dashed lines, the phase curve, ϕ, passes through 90° (right-hand axis) as the attenuation curve (A) passes through a minimum at the same wave angle ψ. This is the PBA. At this angle, the reflected wave is not only at a phase angle of 90° with respect to the direct wave, but is so low in amplitude that it does not aid the direct wave by a significant amount. In the case illustrated in Figure 3.48 this elevation angle is about 15°.

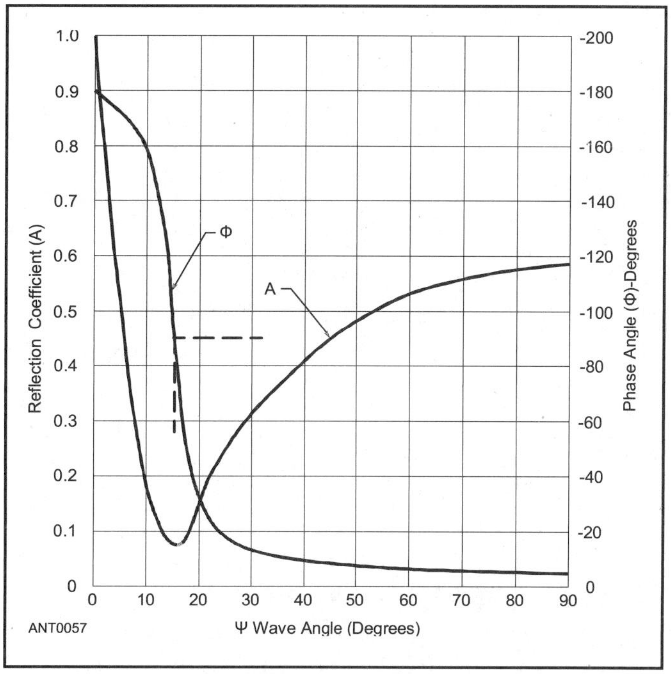

Figure 3.48 — Reflection coefficient for vertically polarized waves. A and ϕ are magnitude and angle for wave angles ψ. This case is for average earth, (k = 13, G = 0.005 S/m), at 21 MHz.

VARIATIONS IN PBA WITH EARTH QUALITY

From Eq 9, it is quite a task to search for either the 90° phase point or the attenuation curve minimum for a wide variety of earth conditions. Instead, the PBA can be calculated directly from the following equation.

$$\psi_B = \arcsin\sqrt{\frac{k-1+\sqrt{\left(x^2+k^2\right)^2(k-1)^2+x^2\left[\left(x^2+k^2\right)^2-1\right]}}{\left(x^2+k^2\right)^2-1}} \qquad (10)$$

where k, G and f are as defined for Eq 9, and

$$x = \frac{1.8\times10^4\times G}{f}$$

Figure 3.49 shows curves calculated using Eq 10 for several different earth conditions, at frequencies between 1.8 and 30 MHz. As expected, poorer earths yield higher PBAs. Unfortunately, at the higher frequencies (where low-angle radiation is most important for DX work), the PBAs are highest. The PBA is the same for both transmitting and receiving.

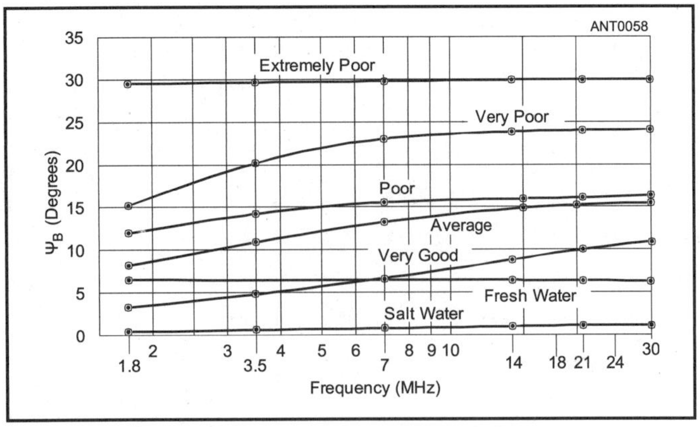

Figure 3.49 — Pseudo-Brewster angle (ψ) for various qualities of earth over the 1.8 to 30-MHz frequency range. Note that the frequency scale is logarithmic. The constants used for each curve are given in Table 3.5.

RELATING PBA TO LOCATION AND FREQUENCY

Table 3.2 presented earlier in this chapter lists the physical descriptions of various kinds of earth with their respective conductivities and dielectric constants, as mentioned earlier. Note that in general, the dielectric constants and conductivities are higher for better earths. This enables the labeling of the earth characteristics as extremely poor, very poor, poor, average, very good, and so on, without the complications that would result from treating the two parameters independently.

Fresh water and salt water are special cases; in spite of high resistivity, the fresh-water PBA is 6.4°, and is nearly independent of frequency below 30 MHz. Salt water, because of its extremely high conductivity, has a PBA that never exceeds 1° in this frequency range. The extremely low conductivity listed for cities (the last case) in Table 3.5 results more from the clutter of surrounding buildings and other obstructions than any actual earth characteristic. The PBA at any location can be found for a given frequency from the curves in Figure 3.49. (The map presented earlier as Figure 3.2 shows approximate conductivity values for different areas in the continental United States.)

3.3.4 Flat-Ground Reflections and Horizontally Polarized Waves

The situation for horizontal antennas is different from that of verticals. **Figure 3.50** shows the reflection coefficient for horizontally polarized waves over average earth at 21 MHz. Note that in this case, the phase-angle departure from 0° never gets very large, and the attenuation factor that causes the most loss for high-angle signals approaches unity for low angles. Attenuation increases with progressively poorer earth types.

In calculating the broadside radiation pattern of a horizontal λ/2 dipole, the perfect-earth image current, equal to the true antenna current but 180° out of phase with it) is multiplied by the horizontal reflection coefficient given by Eq 11 below. The product is then added vectorially to the direct wave to get the resultant at that elevation angle. The reflection coefficient for horizontally polarized waves can be calculated using the following equation.

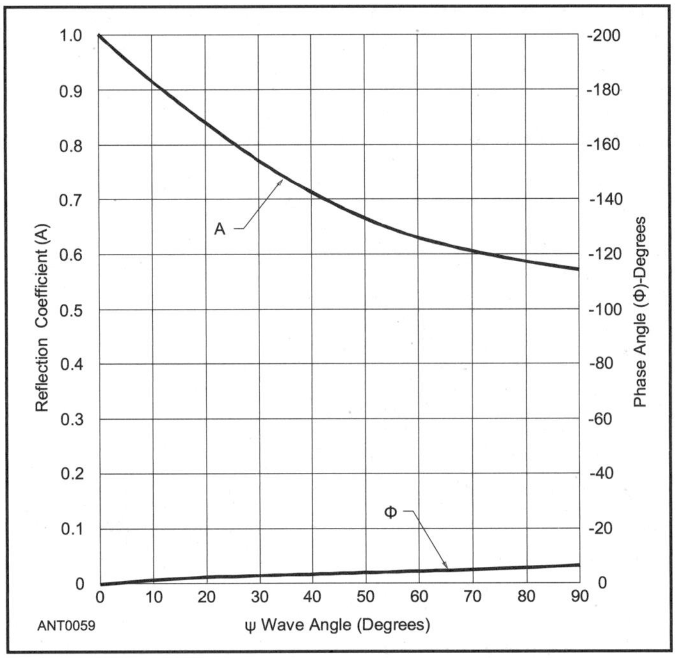

Figure 3.50 — Reflection coefficient for horizontally polarized waves (magnitude A at angle φ), at 21 MHz over average earth (k = 13, G = 0.005 S/m).

$$A_{Horiz}\angle\phi = \frac{\sqrt{k' - \cos^2\psi} - \sin\psi}{\sqrt{k' - \cos^2\psi} + \sin\psi} \qquad (11)$$

where

$A_{Horiz}\angle\phi$ = horizontal reflection coefficient

ψ = elevation angle

$$k' = k - j\left|\frac{1.8\times10^4 \times G}{f}\right|$$

k = dielectric constant of earth

G = conductivity of earth in S/m

f = frequency in MHz

j = complex operator ($\sqrt{-1}$)

For a horizontal antenna near the earth, the resultant pattern is a modification of the free-space pattern of the antenna. (See the "Effects of Local Terrain" section of the **HF Antenna System Design** chapter.) **Figure 3.51** shows how this modification takes place for a horizontal λ/2 antenna over a perfectly conducting flat surface. The patterns at the left show the relative radiation when one views the antenna from the side; those at the right show the radiation pattern looking at the end of the antenna. Changing the height above ground from λ/4 to λ/2 makes a significant difference in the high-angle radiation, moving the main lobe down lower.

Note that for an antenna height of λ/2 (Figure 3.51, bottom), the out-of-phase reflection from a perfectly conducting surface creates a null in the pattern at the zenith (90° elevation angle). Over real earth, however, a *filling in* of this null occurs because of ground losses that prevent perfect reflection of high-angle radiation.

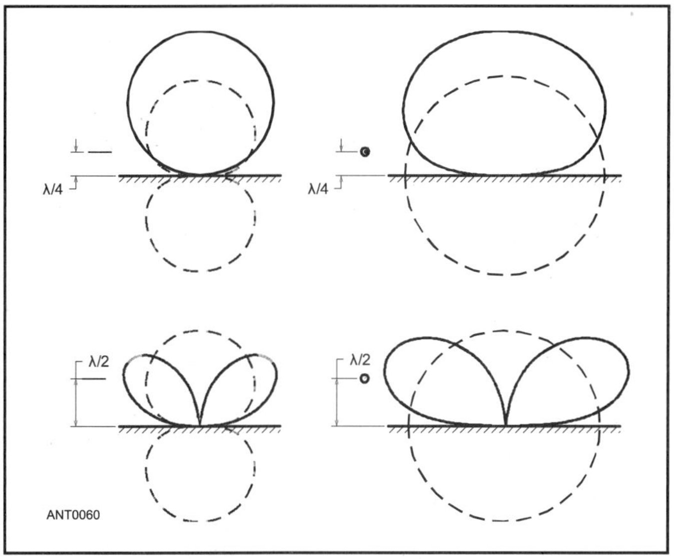

Figure 3.51 — Effect of the ground on the radiation from a horizontal half-wave dipole antenna, for heights of one-fourth and one-half wavelength. Broken lines show what the pattern would be if there were no reflection from the ground (free space).

At a 0° elevation angle, horizontally polarized antennas also demonstrate a null, because out-of-phase reflection cancels the direct wave. As the elevation angle departs from 0°, however, there is a slight filling-in effect so that over other-than-perfect earth, radiation at lower angles is enhanced compared to a vertical. A horizontal antenna will often outperform a vertical for low-angle DX work, particularly over lossy types of earth at the higher frequencies.

Reflection coefficients for vertically and horizontally polarized radiation differ considerably at most angles above ground, as can be seen by comparison of Figures 3.48 and 3.50. (Both sets of curves were plotted for the same ground constants and at the same frequency, so they may be compared directly.) This is because, as mentioned earlier, the image of a horizontally polarized antenna is out-of-phase with the antenna itself, and the image of a vertical antenna is in-phase with the actual radiator.

The result is that the phase shifts and reflection magnitudes vary greatly at different angles for horizontal and vertical polarization. The magnitude of the reflection coefficient for vertically polarized waves is greatest (near unity) at very low angles, and the phase angle is close to 180°. As mentioned earlier, this cancels nearly all radiation at very low angles. For the same range of angles, the magnitude of the reflection coefficient for horizontally polarized waves is also near unity, but the phase angle is near 0° for the specific conditions shown in Figures 3.42 and 3.44. This causes reinforcement of low-angle horizontally polarized waves. At some relatively high angle, the reflection coefficients for horizontally and vertically polarized waves are equal in magnitude and phase. At this angle (approximately 81° for the example case), the effect of ground reflection on vertically and horizontally polarized signals will be the same.

3.3.5 Directive Patterns Over Real Ground

As explained in the **Antenna Fundamentals** chapter, because antenna radiation patterns are three-dimensional, it is helpful in understanding their operation to use a form of representation showing the elevation-plane directional characteristic for different heights. It is possible to show selected elevation-plane patterns oriented in various directions with respect to the antenna axis. In the case of the horizontal half-wave dipole, a plane running in a direction along the axis and another broadside to the antenna will give a good deal of information.

The effect of reflection from the ground can be expressed as a separate *pattern factor*, given in decibels. For any given elevation angle, adding this factor algebraically to the value for that angle from the free-space pattern for that antenna gives the resultant radiation value at that angle. The limiting conditions are those represented by the direct ray and the reflected ray being exactly in-phase and exactly out-of-phase, when both (assuming there are no ground losses) have equal amplitudes. Thus, the resultant field strength at a distant point may be either 6 dB greater than the free-space pattern (twice the field strength), or zero, in the limiting cases.

HORIZONTALLY POLARIZED ANTENNAS

The way in which pattern factors vary with height for horizontal antennas over flat earth is shown graphically in the plots of **Figure 3.52**. The solid-line plots are based on perfectly conducting ground, while the shaded plots are based on typical real-earth conditions. These patterns apply to horizontal antennas of any length. While these graphs are, in fact, radiation patterns of horizontal single-wire antennas (dipoles) as viewed from the axis of the wire, it must be remembered that the plots merely represent pattern factors.

Figure 3.53 shows vertical-plane radiation patterns in the directions off the ends of a horizontal half-wave dipole for various antenna heights. These patterns are scaled so they may be compared directly to those for the appropriate heights in Figure 3.46. Note that the perfect-earth patterns in Figures 3.52A and 3.52B are the same as those in the upper part of Figure 3.51. Note also that the perfect-earth patterns of Figures 3.47B and 3.52D are the same as those in the lower section of Figure 3.51. The reduction in field strength off the ends of the wire at the lower angles, as compared with the broadside field strength, is quite apparent. It is also clear from Figure 3.53 that, at some heights, the high-angle radiation off the ends is nearly as great as the broadside radiation, making the antenna essentially an omnidirectional radiator.

In vertical planes making some intermediate angle between 0° and 90° with the wire axis, the pattern will have a shape intermediate between the broadside and end-on patterns. By visualizing a smooth transition from the end-on pattern to the broadside pattern as the horizontal angle is varied from 0° to 90°, a fairly good mental picture of the actual solid pattern may be formed. An example is shown in **Figure 3.54**. At A, the elevation-plane pattern of a half-wave dipole at a height of λ/2 is shown through a plane 45° away from the favored direction of the antenna. At B and C, the pattern of the same

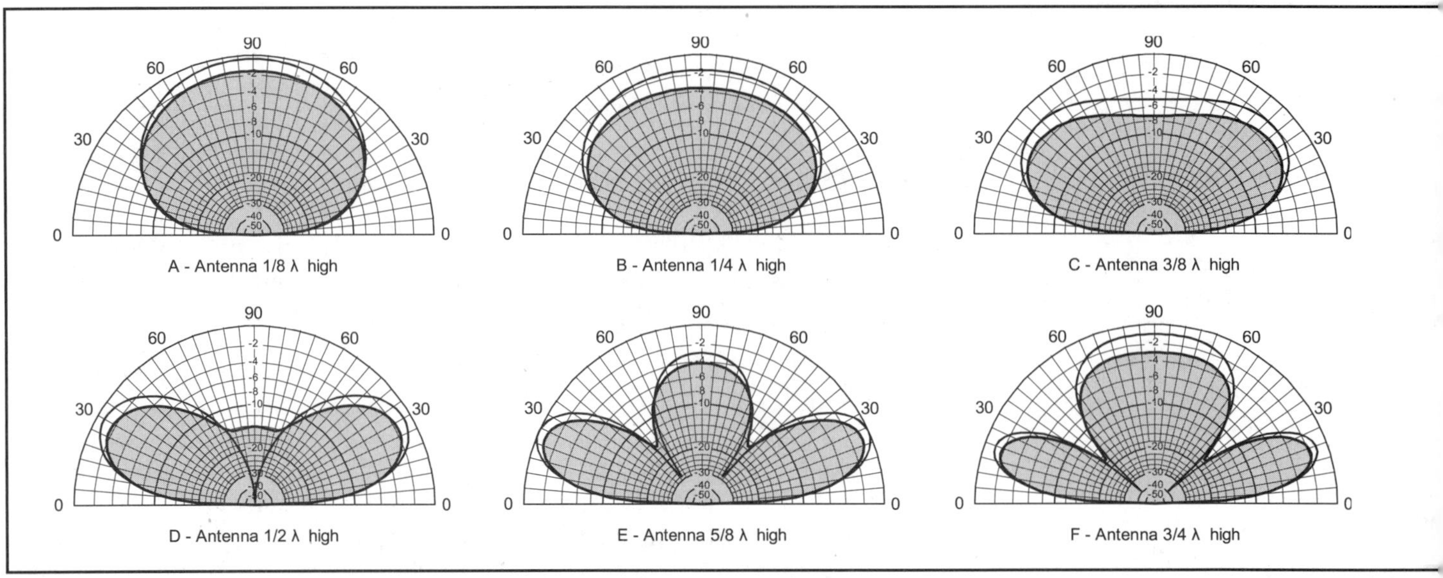

Figure 3.52 — Reflection factors for horizontal dipole antennas at various heights above flat ground. The solid-line curves are the perfect-earth patterns (broadside to the antenna wire); the shaded curves represent the effects of average earth (k = 13, G = 0.005 S/m) at 14 MHz. Add 7 dB to values shown for absolute gain in dBd referenced to dipole in free space, or 9.15 dB for gain in dBi. For example, peak gain over perfect earth at ⅝ λ height is 7 dBd (or 9.15 dBi) at 25° elevation.

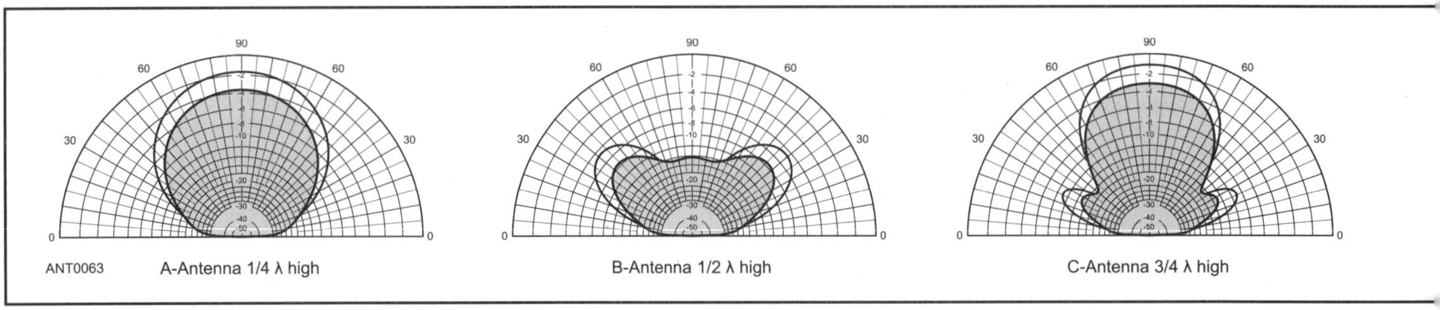

Figure 3.53 — Vertical-plane radiation patterns of horizontal half-wave dipole antennas off the ends of the antenna wire. The solid-line curves are the flat, perfect-earth patterns, and the shaded curves represent the effects of average flat earth (k = 13, G = 0.005 S/m) at 14 MHz. The 0-dB reference in each plot corresponds to the peak of the main lobe in the favored direction of the antenna (the maximum gain). Add 7 dB to values shown for absolute gain in dBd referenced to dipole in free space, or 9.15 dB for gain in dBi.

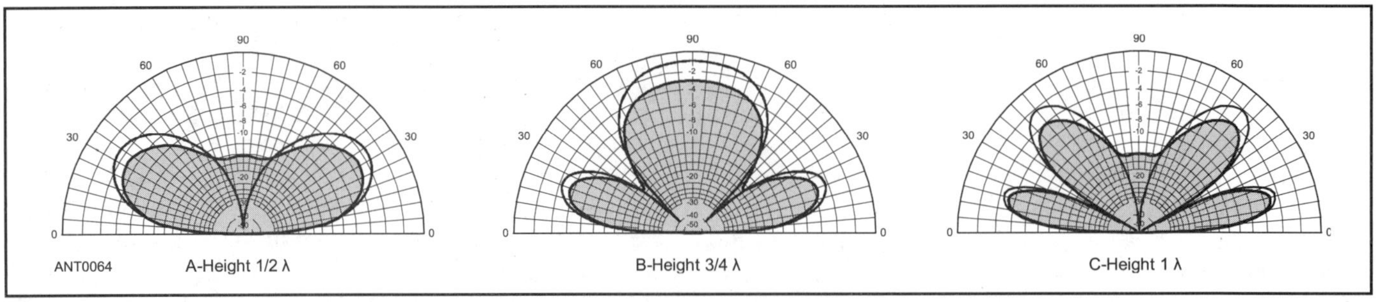

Figure 3.54 — Vertical-plane radiation patterns of half-wave horizontal dipole antennas at 45° from the antenna wire over flat ground. The solid-line and shaded curves represent the same conditions as in Figures 3.52 and 3.53. These patterns are scaled so they may be compared directly with those of Figures 3.52 and 3.53.

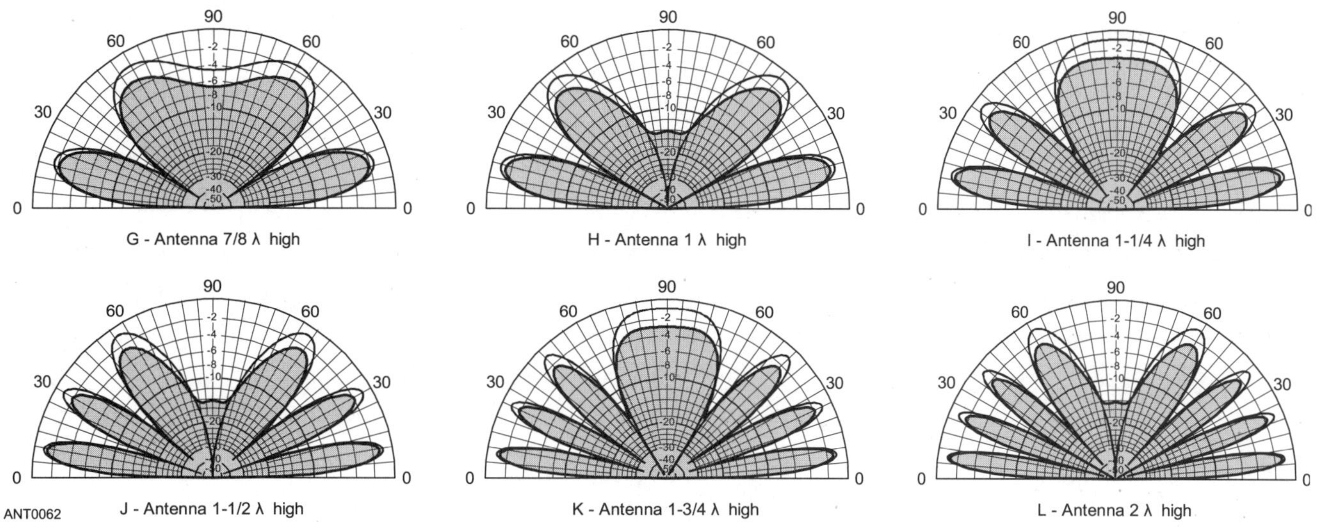

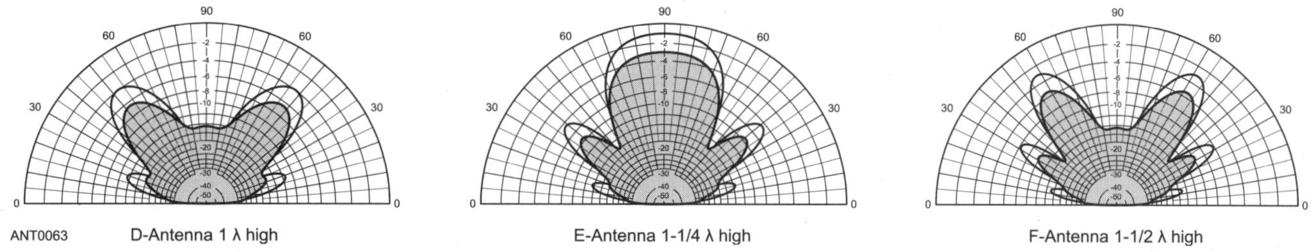

antenna is shown at heights of 3λ/4 and 1λ (through the same 45° off-axis plane). These patterns are scaled so they may be compared directly with the broadside and end-on patterns for the same antenna (at the appropriate heights) in Figures 3.53and 3.54.

The curves presented in **Figure 3.55** are useful for determining heights of horizontal antennas that give either maximum or minimum reinforcement at any desired wave angle. For instance, if you want to place an antenna at a height so that it will have a null at 30°, the antenna should be placed where a broken line crosses the 30° line on the horizontal scale. There are two heights (up to 2 λ) that will yield this null angle: 1 λ and 2 λ.

As a second example, you may want to have the ground reflection give maximum reinforcement of the direct ray from a horizontal antenna at a 20° elevation angle. The antenna height should be 0.75 λ. The same height will give a null at 42° and a second lobe at 90°.

Figure 3.55 is also useful for visualizing the vertical pattern of a horizontal antenna. For example, if an antenna is erected at 1.25 λ, it will have major lobes (solid-line crossings) at 12° and 37°, as well as at 90° (the zenith). The nulls in this pattern (dashed-line crossings) will appear at 24° and 53°.

The Y-axis in Figure 3.55 plots the wave angle versus the height in wavelength above flat ground on the X-axis. Figure 3.55 doesn't show the elevation angles required for actual communications to various target geographic locations of interest. The **Radio Wave Propagation** chapter and this book's online material information give details about the range of angles required for target locations around the world. It is very useful to overlay plots of these angles together with the elevation pattern for horizontally polarized antennas at various heights above flat ground. This will be demonstrated in detail later in the **HF Antenna System Design** chapter.

VERTICALLY POLARIZED ANTENNAS

In the case of a vertical λ/2 dipole or a ground-plane antenna, the horizontal directional pattern is simply a circle at any elevation angle (although the actual field strength will vary, at the different elevation angles, with the height above ground).

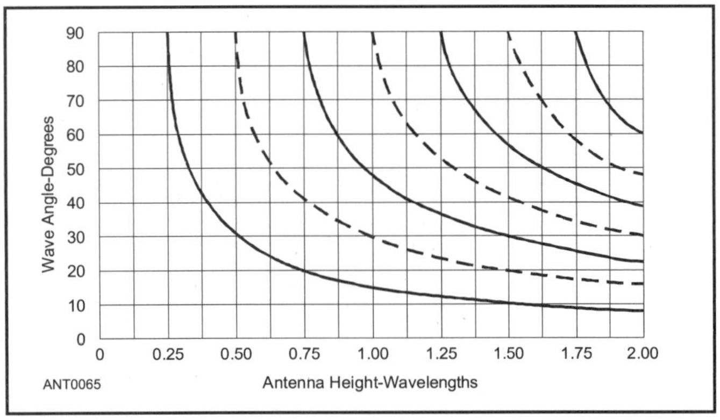

Figure 3.55 — Angles at which nulls and maxima (factor = 6 dB) in the ground-reflection factor appear for antenna heights up to two wavelengths over flat ground. The solid lines are maxima, dashed lines nulls, for all horizontal antennas. See text for examples. Values may also be determined from the trigonometric relationship θ = arc sin (A/4h), where θ is the wave angle and h is the antenna height in wavelengths. For the first maximum, A has a value of 1; for the first null A has a value of 2, for the second maximum 3, for the second null 4, and so on.

Hence, one vertical pattern is sufficient to give complete information (for a given antenna height) about the antenna in any direction with respect to the wire. A series of such patterns for various heights is given in **Figure 3.56**. Rotating the plane pattern about the zenith axis of the graph forms the three-dimensional radiation pattern in each case.

The solid-line curves represent the radiation patterns of the λ/2 vertical dipole at different feed point heights over perfectly conducting ground. The shaded curves in Figure 3.56 show the patterns produced by the same antennas at the same heights over average ground (G = 0.005 S/m, k = 13) at 14 MHz. The PBA in this case is 14.8°.

In short, far-field losses for vertically polarized antennas are highly dependent on the conductivity and dielectric constant of the earth around the antenna, extending far beyond the ends of any radials used to complete the ground return for the near field. Putting more radials out around the antenna may well decrease ground-return losses in the reactive near field for a vertical monopole, but will not increase radiation at low elevation launch angles in the far field, unless the radials can extend perhaps 100 wavelengths in all directions! Aside from moving to the fabled "salt water swamp on a high hill," there is very little that someone can do to change the character of the ground that affects the far-field pattern of a real vertical. Classical texts on verticals often show elevation patterns computed over an "infinitely wide, infinitely conducting ground plane." Real ground, with finite conductivity and less than perfect dielectric constant, can severely curtail the low-angle radiation at which verticals are supposed to excel.

While real verticals over real ground are not a sure-fire method to achieve low-angle radiation, cost versus performance and ease of installation are still attributes that can highly recommend verticals to knowledgeable builders. Practical installations for 160 and 80 meters rarely allow amateurs to put up horizontal antenna high enough to radiate effectively at low elevation angles. After all, a half-wave on 1.8 MHz is 273 feet high, and even at such a lofty height the peak radiation for a horizontal antenna would be at a 30° elevation angle, which is higher than desired for long-distance communication. A simple ground-mounted vertical with a reasonable radial field will almost always give much better results in this case.

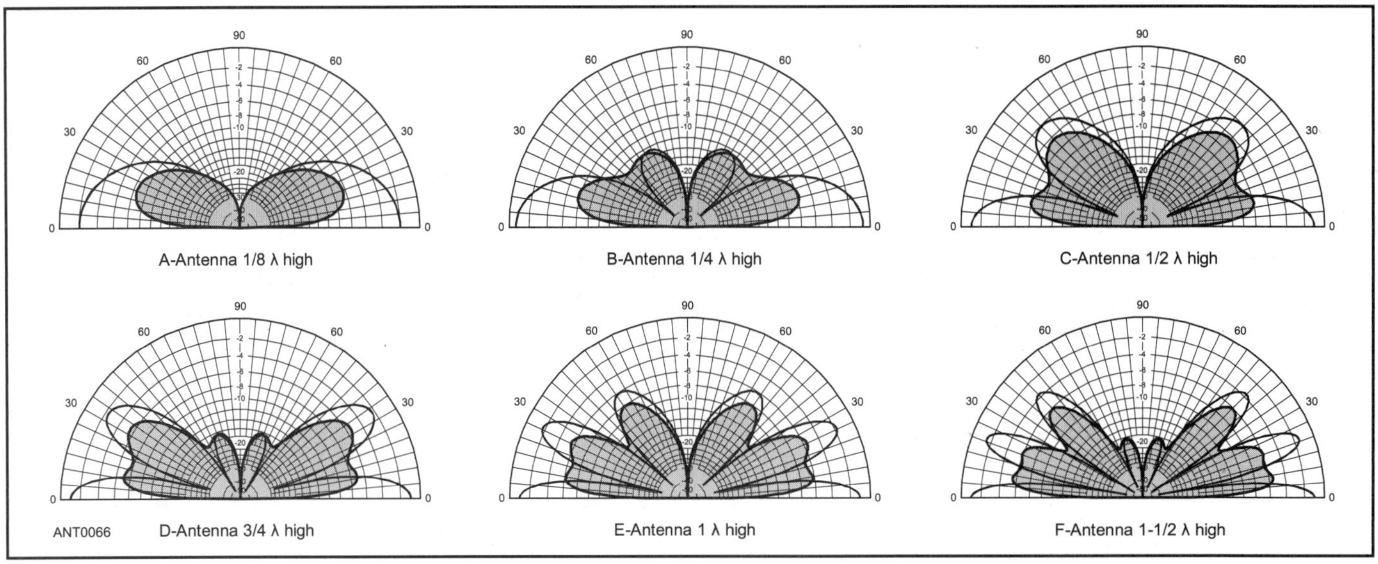

Figure 3.56 — Vertical-plane radiation patterns of a ground-plane antenna above flat ground. The height is that of the ground plane, which consists of four radials in a horizontal plane. Solid lines are perfect-earth patterns; shaded curves show the effects of real earth. The patterns are scaled — that is, they may be directly compared to the solid-line ones for comparison of losses at any wave angle. These patterns were calculated for average ground (k = 13, G = 0.005 S/m) at 14 MHz. The PBA for these conditions is 14.8°. Add 6 dB to values shown for absolute gain in dBd over dipole in free space.

3.4 Ground Parameters for Antenna Analysis

The first part of this section is taken from an article of the same title in The ARRL Antenna Compendium, Vol. 5 by R. P. Haviland, W4MB. The sections on direct and indirect soil measurements have been updated by R. Severns, N6LF.

In the past, amateurs paid very little attention to the characteristics of the earth (ground) associated with their antennas. There are two reasons for this. First, these characteristics were not easy to measure. Second, most hams have to put up with what they have! Further, the ground is not a dominant factor for horizontally polarized antennas such as a tri-band Yagi at 40 feet or higher, or a 2-meter vertical at roof height. For vertically polarized antennas, however, the soil characteristics are very important for the design of ground systems, predictions of efficiency, and elevation radiation patterns. Ground data is useful for antennas mounted at low heights generally, and for such specialized ones as Beverage receiving antennas. The performance of such antennas changes significantly as the ground changes.

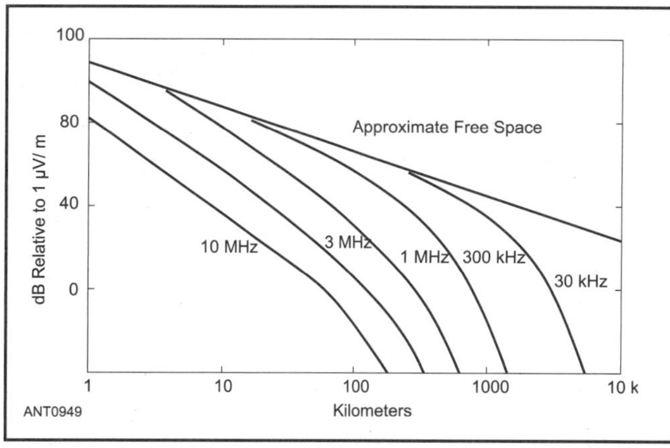

Figure 3.57 — Variation of field strength with distance. Typical field strengths for several frequencies are shown. This is from CCIR data for fairly poor soil, with dielectric constant of 4 and conductivity of 3 mS/m. The curves for good soil are closer to the free-space line, and those for sea water are much closer to the free-space line.

3.4.1 Importance of Ground Parameters

As an example of why ground parameters can be important, we can look at some values for ground wave attenuation with distance. At 10 MHz, *CCIR Recommendation 368* (see Bibliography), gives the distance at which the signal is calculated to drop 10 dB below its free-space level as shown in **Table 3.6**.

Inter-island work in the Caribbean on 40 and 80 meters is easy, whereas 40-meter ground-wave contact is difficult for much of the USA, because of much lower ground conductivity. On the other hand, a Beverage receiving antenna depends on poor ground conductivity to work well.

Figure 3.57 shows a typical set of expected propagation curves for vertically polarized signals over a range of frequencies. This data is also from *CCIR Recommendation 368* for relatively poor ground, with a dielectric constant of 4 and a conductivity of 3 mS/m (one milliSiemens/meter is 0.001 mho/meter). The same data is available in the *Radio Propagation Handbook*. There are equivalent FCC curves, found in the book *Reference Data for Radio Engineers*, but only the ones near 160 meters are useful. In Florida the author has difficulty hearing stations across town on ground wave, an indication of the poor soil conditions — reflected sky-wave signals are often stronger.

Table 3.6
Distance at which a 10 MHz Signal Attenuates 10 dB Over Various Ground Types

Ground type	*Conductivity (mS/meter)*	*Distance for 10 dB Drop (km)*
Seawater	5000	100
Rich soil	30	15
Moderate	3	0.3

3.4.2 Securing Ground Data

There are two basic ways to approach this matter of ground data. One is to use generic ground data typical to the area. The second is to make direct measurements, which, with the introduction of moderately priced vector network analyzers (VNA), has become much easier. The easiest approach may be a combination of techniques — make some simple measurements using the low frequency conductivity procedure outlined below and then combine this with the generic data to make a better estimate. For 2,200 and 630 meters the 50/60 Hz conductivity is usually adequate. For 160 meters and higher amateur bands, a more sophisticated measurement yielding both conductivity (σ) and dielectric constant (ε_r) (also referred to as the relative permittivity) can be helpful. The simple approach of only measuring the LF conductivity may not be highly accurate for HF antennas but will still be much better than simply inserting some arbitrary preset values into an analysis program. Having a good set of values to plug into an analysis can be of great help in evaluating the true worth of a new antenna project, especially if the antenna is horizontally polarized.

GENERIC DATA

In connection with its licensing procedure for broadcast stations, the FCC has published generic data for the entire country. This map was presented earlier as Figure 3.2, showing the "estimated effective ground conductivity in the United States." A range of 30:1 is shown, from 1 to 30 mS/m. An equivalent chart for Canada has been prepared, originally by DOT, now DOC.

Of course, some judgment is needed when trying to use this data for your location. Broadcast stations are likely to be

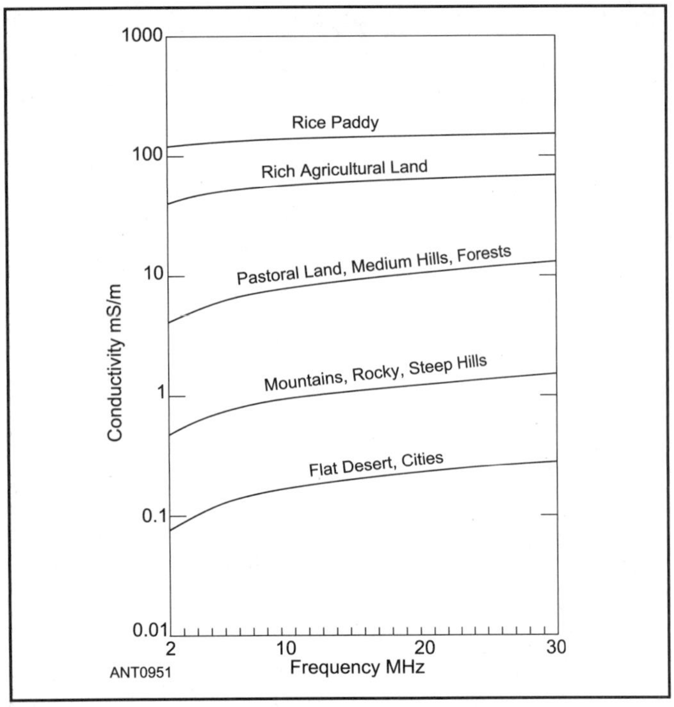

Figure 3.58 — Typical terrain conductivities versus frequency for 5 types of soils. This was measured by SRI. Units are mS/m. Conductivity of seawater is usually taken as 5000 mS/m. Conductivity of fresh water depends on the impurities present, and may be very low. To extrapolate conductivity values (for 500 to 1500 kHz) shown in Figure 3.2 for a particular geographic area to a different frequency, move from the conductivity at the left edge of this figure to the desired frequency. For example, in rocky New Hampshire, with a conductivity of 1 mS/m at BC frequencies, the effective conductivity at 14 MHz would be approximately 4 mS/m.

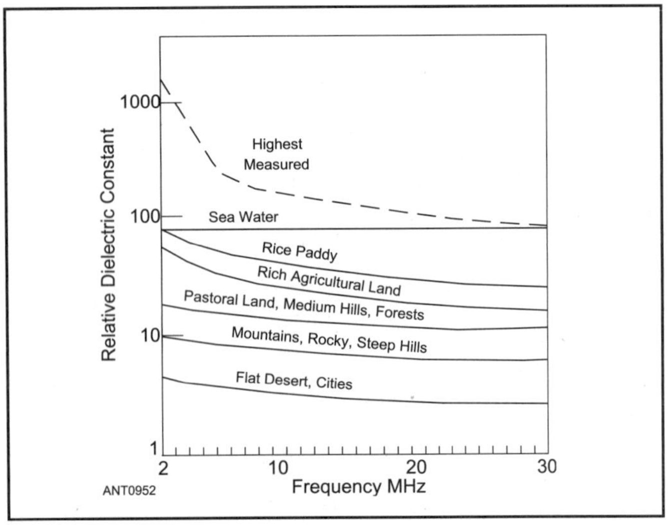

Figure 3.59 — Typical terrain relative dielectric constant for the 5 soil types of Figure 3.58, plus sea water. The dashed curve shows the highest measured values reported, and usually indicates mineralization.

in open areas, so the data should not be assumed to apply to the center of a city. And a low site near the sea is likely to have better conductivity than the generic chart. Other than such factors, this chart gives a good first value and a useful cross-check if some other method is used.

Still another FCC-induced data source is the license application of your local broadcast station. This includes calculated and measured coverage data. This may include specific ground data, or comparison of the coverage curves with the CCIR or FCC data to give the estimated ground conductivity. Another set of curves for ground conditions are those prepared by SRI (see References). These give the conductivity and dielectric constant versus frequency for typical terrain conditions. These are reproduced as **Figures 3.58** and **3.59**. By inspecting your own site, you may select the curve most appropriate to your terrain. The curves are based on measurements at a number of sites across the USA and are averages of the measured values.

Figures 3.60 through **3.62** are data derived from these measurements. Figure 3.60 gives the ground-dissipation factor. Sea water has low loss (a high dissipation factor), while soil in the desert or in the city is very lossy, with a low dissipation factor. Figure 3.61 gives the skin depth, the distance for the signal to decease to 63% of its value at the surface. Penetration is low in high-conductivity areas and deep in low-conductivity soil. Finally, Figure 3.62 shows the wavelength in the earth. For example, at 10 meters (30 MHz), the wavelength in sea water is less than 0.3 meters. Even in the desert, the wavelength has been reduced to about 6 meters at this frequency. This is one reason why buried antennas have peculiar properties. Lacking other data, it is suggested that the values of Figure 3.58 and 3.59 be used in computer antenna modeling programs.

MEASURING GROUND CONDITIONS

M.C. Waltz, W2FNQ developed a simple technique to measure low-frequency soil conductivity, which has been used by Jerry Sevick, W2FMI. The test setup is drawn in **Figure 3.63** and uses a very old technique of 4-terminal resistivity measurements. For probes of 9/16-inch diameter (a standard grounding rod size), spaced 18 inches and penetrating 12 inches into the earth, the conductivity is:

$$G = 21\ V_1/V_2\ \text{mS/m} \qquad (12)$$

The voltages are conveniently measured by a digital voltmeter, to an accuracy of about 2%. In soil suitable for farming, the probes can be copper or aluminum. The strength of iron or Copperweld may be needed in hard soils. A piece of 2 × 6 inch or 4 × 4 inch lumber with guide holes drilled through it will help maintain proper spacing and vertical alignment of the probes, greatly speeding up the measurement process. Use care when measuring — there is a potential shock hazard! An isolating transformer with a 24 V secondary should be used instead of 120 V directly to reduce the danger. Ground conditions vary quite widely over even small areas. It is best to make a number of measurements around the area of the antenna and average the measured values.

While this measurement gives only the low-frequency con-

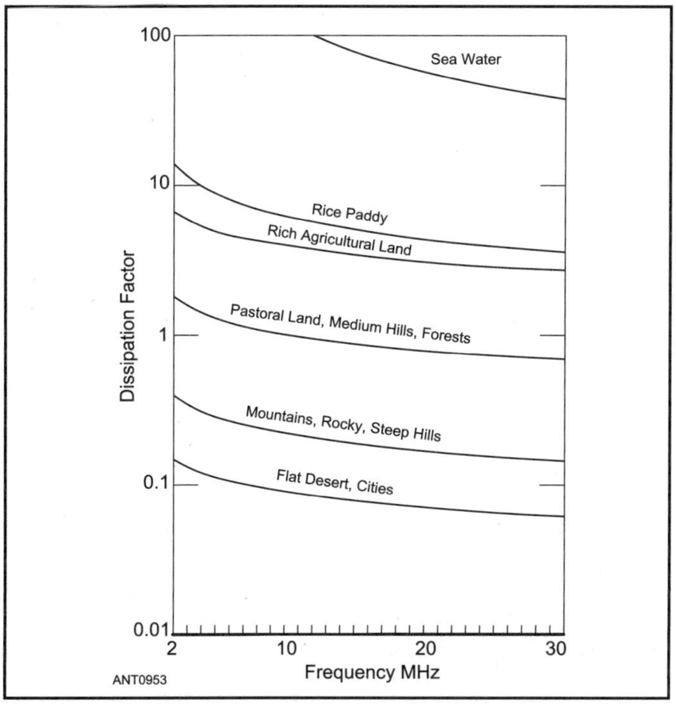

Figure 3.60 — Typical values of dissipation factor. The soil behaves as a leaky dielectric. These curves showing the dimensionless dissipation factor versus frequency for various types of soils and for sea water. The dissipation factor is inversely related to soil conductivity. Among other things, a high dissipation factor indicates that a signal penetrating the soil or water will decrease in strength rapidly with depth.

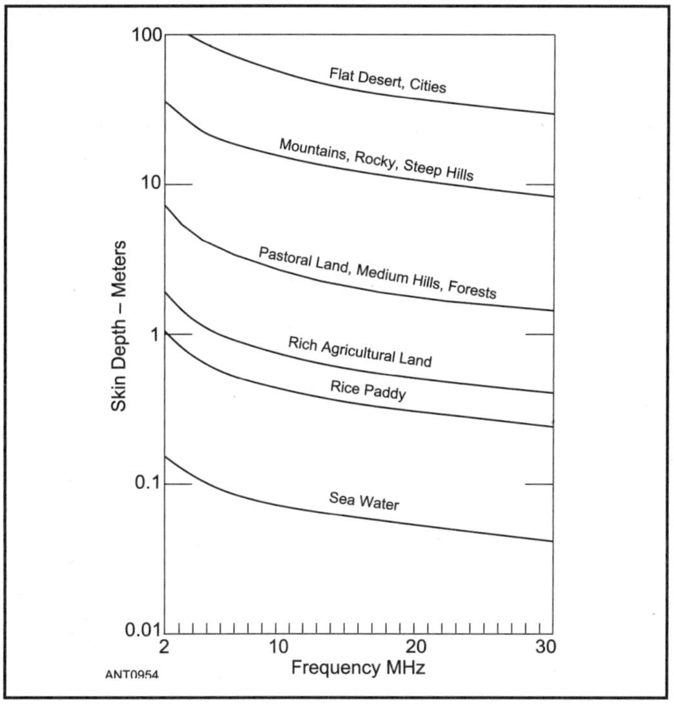

Figure 3.61 — Typical values of skin depth. The skin depth is the depth at which a signal will have decreased to 1/e of its value at the surface (to about 30%). The effective height above ground is essentially the same as the physical height for sea water, but may be much greater for the desert. For practical antennas, this may increase low-angle radiation, but at the same time will increase ground losses.

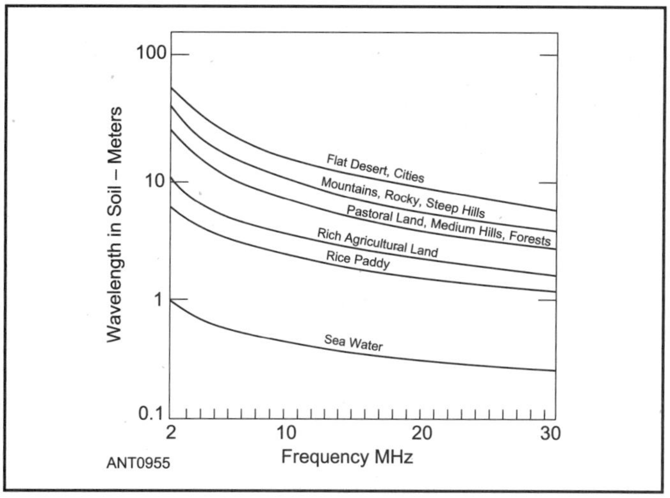

Figure 3.62 — Typical values of wavelength in soil. Because of its dielectric constant, the wavelength in soils and water will be shorter than that for a wave traveling in air. This can be important, since in a Method of Moment the accuracy is affected by the number of analysis segments per wavelength. Depending on the program being used, adjust the number of segments for antennas wholly or partly in the earth, for ground rods, and for antennas very close to earth.

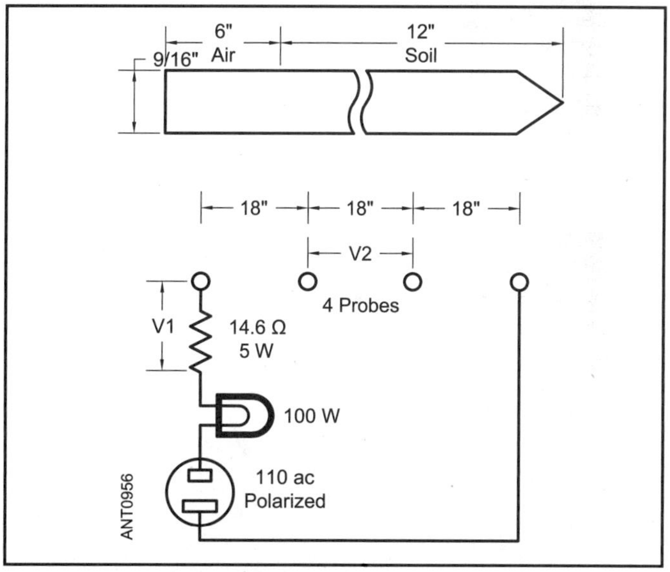

Figure 3.63 — Low-frequency conductivity measurement system. A 60-Hz measuring system devised by W2FNQ and used by W2FMI. The basic system is widely used in geophysics. Use care to be certain that the plug connection is correct. A better system would use a lower voltage and an isolation transformer. Measure the value of V_2 with no power applied — there may be stray ground currents present, especially if there is a power station or an electric railway close.

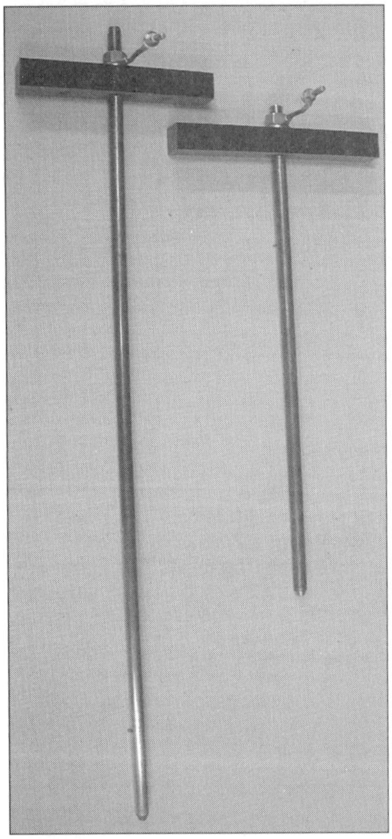

Figure 3.64 —Typical monopole probes, 12-inch and 19-inch examples.

ductivity, it can be used to select curves in Figure 3.64 to give an estimate of the conductivity for the common ham bands. Assume that the 50/60 Hz value is valid at 2 MHz and find the correct value on the left axis. Move parallel to the curves on the figure to develop the estimated curve for other soil conditions. This will give a value for conductivity which is a bit low but still very helpful.

A small additional refinement is possible. If the dielectric constant from Figure 3.65 is plotted against the conductivity from Figure 3.64 for a given frequency, a scatter plot develops, showing a trend to higher dielectric constant as conductivity increases which is mostly due to variations in moisture content. As the moisture content increases both conductivity and relative dielectric constant increase. At 14 MHz, the relation is:

$$k = \sqrt{1000 / G} \tag{13}$$

where k is the dielectric constant and G is the measured conductivity. Using these values in *MININEC* or *NEC* calculations should give better estimates than countrywide average values.

DIRECT MEASUREMENT OF GROUND PROPERTIES

More reliable values for both the conductivity and ε_r can be obtained from a direct measurement at the operating frequency using a soil probe and a vector network analyzer (VNA). This is particularly important at HF where the value

Figure 3.66 — Monopole probe inserted through the mesh sheet.

Figure 3.65 — Two open-wire line (OWL) probes (top) and a detail of probe construction (bottom).

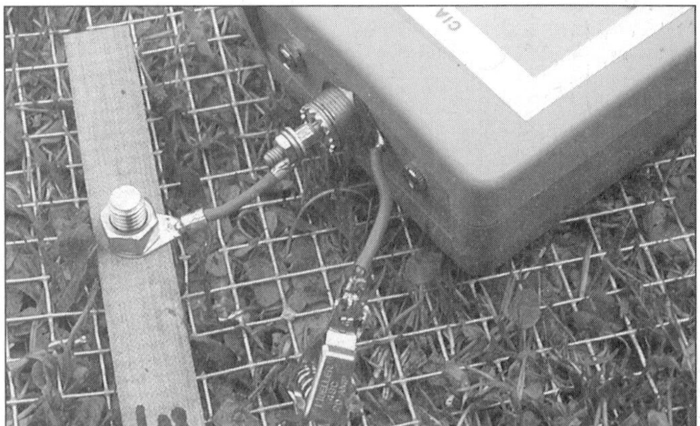

Figure 3.67 — Monopole probe impedance measurement.

for ε_r is as important as the conductivity. A probe is inserted into the soil and the impedance (Z=R+*j*X) measured at the probe terminals. Values for conductivity and ε_r can then be calculated using the measured impedances.

Examples of amateur-made probes are shown **in Figure 3.64** and **3.65.** The probe rods can be aluminum, brass or even steel, the choice makes little difference in the measurement. A short length of clothesline can be placed around the horizontal handle before inserting the probe into the soil, making it much easier to extract the probe when measurements are complete.

The monoprobe shown in Figure 3.58 is inserted into the soil through a hole in a conducting sheet which can be a 36 × 36 inch galvanized mesh as shown in **Figure 3.66**. Z is measured between the probe and the mesh as shown in **Figure 3.67**.

The open-wire line (OWL) probe and technique are described in George Hagn's article and Severn's Nov/Dec 2006 *QEX* article).[1, 22] This was the technique used by Hagn to secure the data for Figures 3.58 through 3.62. Examples of homemade OWL probes are shown in Figure 3.65A and B. **Figure 3.68** shows the OWL impedance measurement setup. A coaxial common-mode choke can be used to isolate the balanced probe terminals from an unbalanced measuring instrument. (See the **Transmission Line System Techniques** chapter for examples of chokes.)

Some care must be taken while making OWL probe impedance measurements because the measured values for Z are very sensitive to stray capacitance between the VNA-plus-connecting cable to ground. An example of isolation using a choke is shown in Figure 3.68. Murata series 5000 common-mode chokes are a good choice for a choke and inexpensive. Another option is to use one of the surface mount RF signal

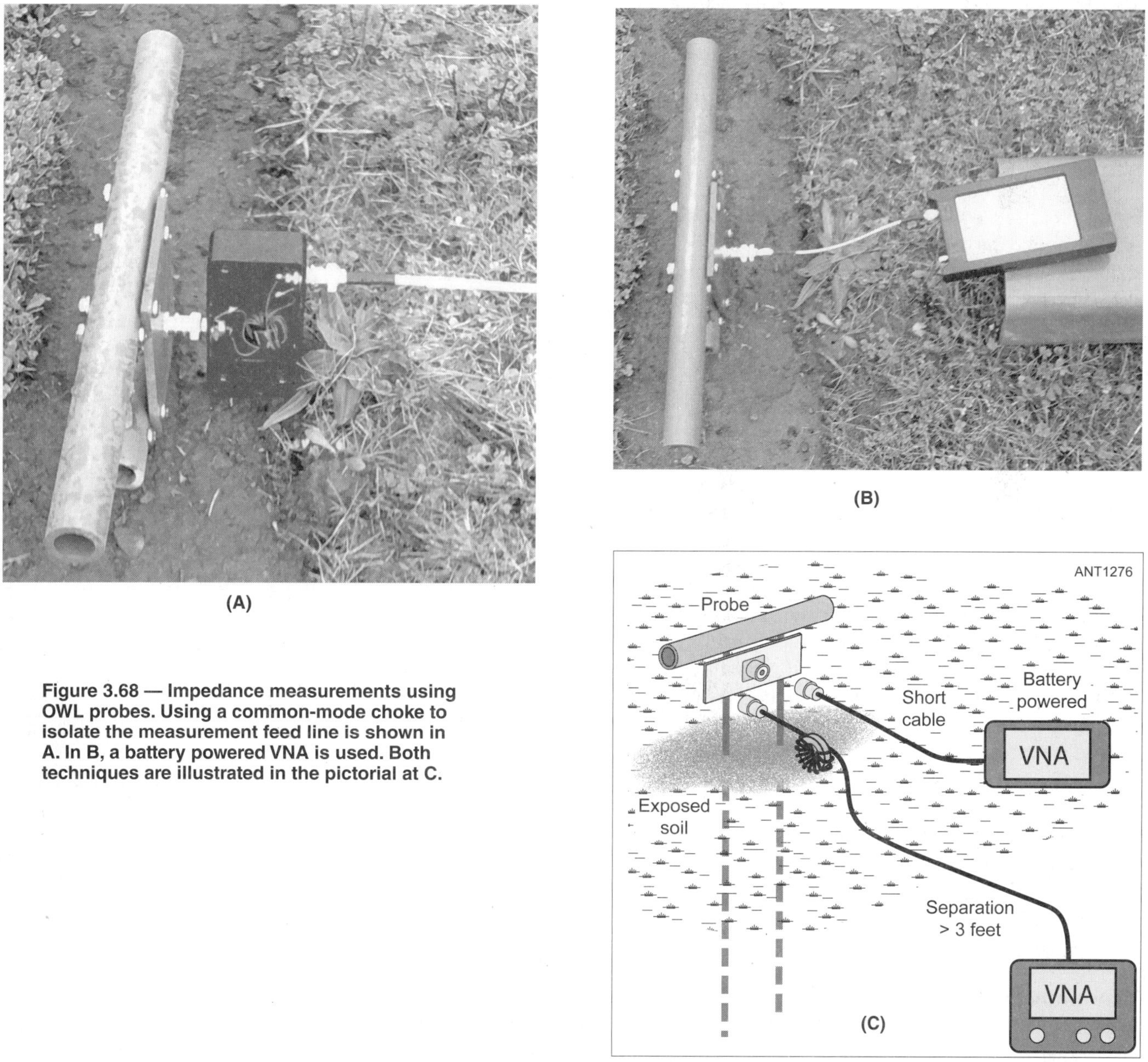

Figure 3.68 — Impedance measurements using OWL probes. Using a common-mode choke to isolate the measurement feed line is shown in A. In B, a battery powered VNA is used. Both techniques are illustrated in the pictorial at C.

transformers like the Coilcraft SWB series. Another alternative, also shown in Figure 3.68, is to use an isolated battery powered VNA supported above the soil surface.

As shown in Figures 3.8 and 3.62, the wavelength in soil at a given frequency will be much shorter than in air. This means the self-resonant frequency of the probe will be much lower in frequency than it would be in air. Probe self-resonance greatly complicates the derivation of soil characteristics from impedance measurements. At frequencies up to about 4 MHz however, a 12" – 24" probe will be electrically short enough to allowing us to model the probe as simple capacitors and use simple algebra to convert Z to soil parameters with sufficient accuracy. At a given frequency (f_{MHz}) where $Z = R + jX$ and the capacitance of the probe in air is C_o pF, σ and ε_r can be calculated from:

$$\sigma = \frac{8.84}{C_o}\left[\frac{R}{R^2+X^2}\right]$$

$$\varepsilon_r = \frac{10^6}{2\pi f_{MHZ} C_o}\left[\frac{X}{R^2+X^2}\right]$$

Probe impedances in air will be very high with very little resistive loss, so it is only necessary to measure the capacitance (C_o) not the full complex impedance. This can be done with a digital capacitance meter or if one is not available it is possible to estimate the capacitance from the probe dimensions. It should be kept in mind that we are not trying to make 1% measurements.

For an OWL probe:

$$C_o \approx \frac{3.677}{\log(\frac{2D}{d})} \text{pF / foot}$$

For a monoprobe:

$$C_o \approx \frac{7.36h}{\log(\frac{2h}{d}) - 0.4} \text{pF}$$

Where D is the center-to-center spacing between probe rods, d is the rod diameter and h is the rod length. D, d, and h have the dimensions of feet. In the case of an OWL probe, D/d is independent of units as long as both are the same.

At higher frequencies we cannot assume the probe is a simple capacitor, it is a length of transmission line buried in a lossy dielectric and much more complex math is required to extract the soil parameters. That math can be challenging but fortunately programs for converting the impedance data from OWL probes to soil parameters is readily available online at **ham-radio.com/k6sti** and **dl1glh.de/groundconductivity.html**.

Keep in mind that soil parameters will vary widely with season (rain versus dry) and at different places around the site. Measured values for σ and ε_r can vary as much as ±20%, reflecting seasonal variations, measurement errors and approximations in the calculations. However, knowing σ and ε_r within 20% is still a vast improvement over a value from a large-scale regional map!

INDIRECT MEASUREMENT

The following material is a condensed version of the article "Determination of Soil Electrical Characteristics Using a Low Dipole" by R. Severns, N6LF, which is included with this book's online information.

The direct measurements discussed in the previous section are effective for determining reasonable values for soil characteristics, but the probe technique characterizes the soil in only a relatively small volume at one point in the site. You can get a more general characterization by repeating the measurements at several points spread over the site. This works, but there is another way to get average values over a large area. The terminal impedance and resonant frequency of an antenna will vary with both the height above ground and the electrical characteristics of the soil, it is possible to measure the feed point impedance of a low dipole and, from modeling of the antenna, determine the effective or average characteristics of the soil under the antenna.

The first question is "what height (z) should we use." Using a dipole with its length adjusted to maintain resonance at 3.7 MHz as the height and ground characteristics are varied, **Figure 3.69** illustrates how the feed point resistance, R, varies with height for a range of ground parameter pairs.

From Figure 3.69 it would appear that any height from 1 foot to 10 feet should give good resolution for determining the appropriate values of σ and ε_r. It should be noted that it is not necessary for the dipole to be resonant but if the dipole is close to resonance the values for R and X will be within the range of 10 to a few hundred ohms. That range is compatible with typical moderately priced impedance measurement instruments.

Erecting the test dipole can be greatly simplified if standard electric fence wire and insulating hardware are used. **Figures 3.70** and **3.71** are photos of a typical test antenna using standard electric fence hardware widely available in hardware and farm stores. #17 AWG aluminum electric fence wire is suspended at 36 inches from 5-foot fiberglass wands driven ≈1 foot deep, with plastic wire clips which slide up/down the wands for height adjustment. The wands were spaced 10 to 20 feet apart and the wire anchored at the ends to steel fence posts 6 to 10 feet away from the ends of the wire. Multiple support points and significant wire tension can keep the droop to less than 0.25 inch. High quality insulators and nonconducting Dacron line were used at the wire ends. Figure 3.64 shows the Budwig center connecter and the common-mode choke or choke balun at the feed point as described in the **Transmission Line System Techniques** chapter.

As an example of the procedure, we'll assume a horizontal center-fed dipole made with #17 AWG aluminum wire at a convenient working height above ground (z) of 36 inches and the availability of antenna modeling software capable of properly calculating the effect of real ground, *NEC-4* for example.

After tuning to resonance at 3.5 MHz the length (L) is 131.11 feet. The measured feed point impedance (Z) at 3.5 MHz is $80.26 + j0\ \Omega$. With this information, we determine the values for soil conductivity (σ [S/m]) and relative dielectric constant (ε_r) at 3.5 MHz. The first step is to create a *NEC-4* model with #17 Al wire, L = 131.11 feet and z = 36 inches.

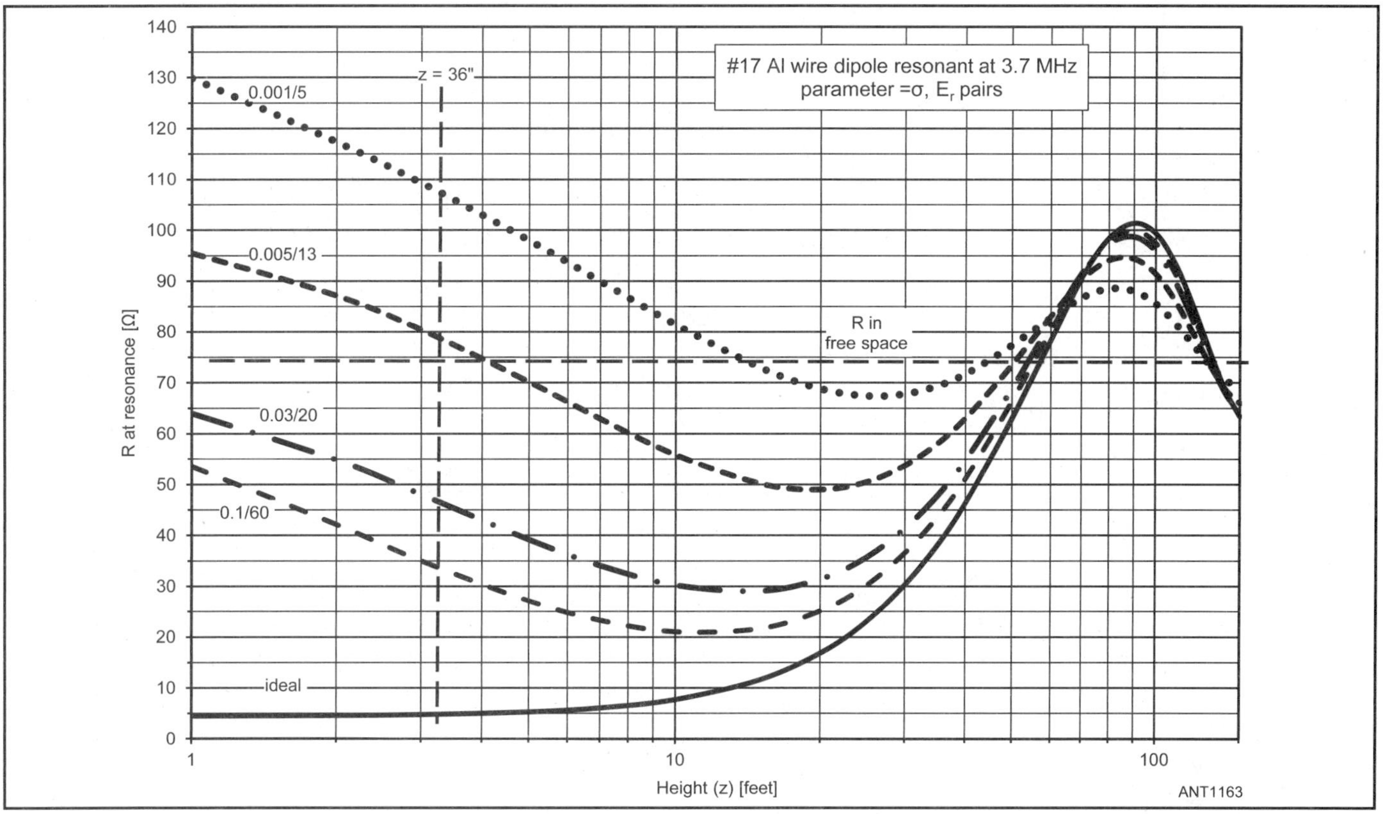

Figure 3.69 — Variation of feed point resistance, R, with height.

Figure 3.70 — Center connector and feed point support. Note use of a common-mode choke at the feed point to isolate the feed line shield.

Since we do not know the values for σ or ε_r we'll have to run the model repeatedly with a range of possible values for σ and ε_r. If we're too far off in our choice of values the process will show us and point the way to go! In this case the trial values will be $0.001 < \sigma < 0.01$ [S/m] and $1 < \varepsilon_r < 50$ which covers a wide range of typical soils. Running the model repeatedly we can determine Z for a matrix of σ and ε_r values. A spreadsheet is a good way to keep track as shown in **Table 3.7**.

A quick scan of the table shows that for $\varepsilon_r > 20$ and $\sigma = 0.009$ there are no resonances (i.e., Xi transitions from + to

Figure 3.71 — Test antenna supported with fiberglass wands.

Table 3.7
Calculated Values for R and X (in Ω)

	sigma (σ) = 0.001		*0.002*		*0.003*		*0.004*		*0.005*	
ε_r	R_i	X_i	R_i	*Xi*	*Ri*	*Xi*	R_i	X_i	R_i	X_i
1	130.09	19.13	101.36	26.50	87.12	21.85	78.76	16.71	73.05	11.68
5	111.22	6.37	100.86	14.26	90.05	14.03	82.13	11.41	76.23	7.80
10	104.72	2.61	98.61	5.33	91.25	6.14	84.60	5.07	79.06	2.86
15	102.87	−2.21	97.34	−0.71	91.51	0.05	85.88	−0.54	80.89	−1.84
20	101.52	−7.01	96.45	−5.80	91.38	−5.29	86.51	−5.45	82.05	−6.16
30	98.60	−16.00	94.56	−14.75	90.58	−13.97	86.74	−13.62	83.09	−13.65
40	95.71	−23.00	92.50	−21.72	89.30	−20.79	86.17	−20.18	83.14	−19.86
50	93.00	−28.50	90.40	−27.27	87.77	−26.30	85.17	−25.60	82.62	−25.14

	sigma (σ) = 0.001		*0.007*		*0.008*		*0.009*		*0.01*	
ε_r	R_i	X_i	R_i	X_i	R_i	X_i	R_i	X_i	R_i	X_i
1	68.88	7.27	65.65	3.47	63.06	0.18	60.90	−2.72	59.06	−5.29
5	71.75	4.30	68.23	1.11	65.36	−1.76	62.97	−4.35	60.94	−6.69
10	74.57	0.39	70.90	−2.06	67.84	−4.39	65.27	−6.58	63.06	−8.62
15	76.62	−3.51	72.99	−5.32	69.89	−7.15	67.22	−8.94	64.90	−10.69
20	78.07	−7.24	74.58	−8.53	71.52	−9.93	68.84	−11.36	66.47	−12.85
30	79.71	−13.99	76.60	−14.56	73.77	−15.30	71.20	−16.21	68.88	−17.21
40	80.26	−19.79	77.54	−19.96	75.00	−20.33	72.63	−20.84	70.45	−21.44
50	80.15	−24.88	77.78	−24.81	75.52	−24.88	73.39	−25.09	71.38	−25.39

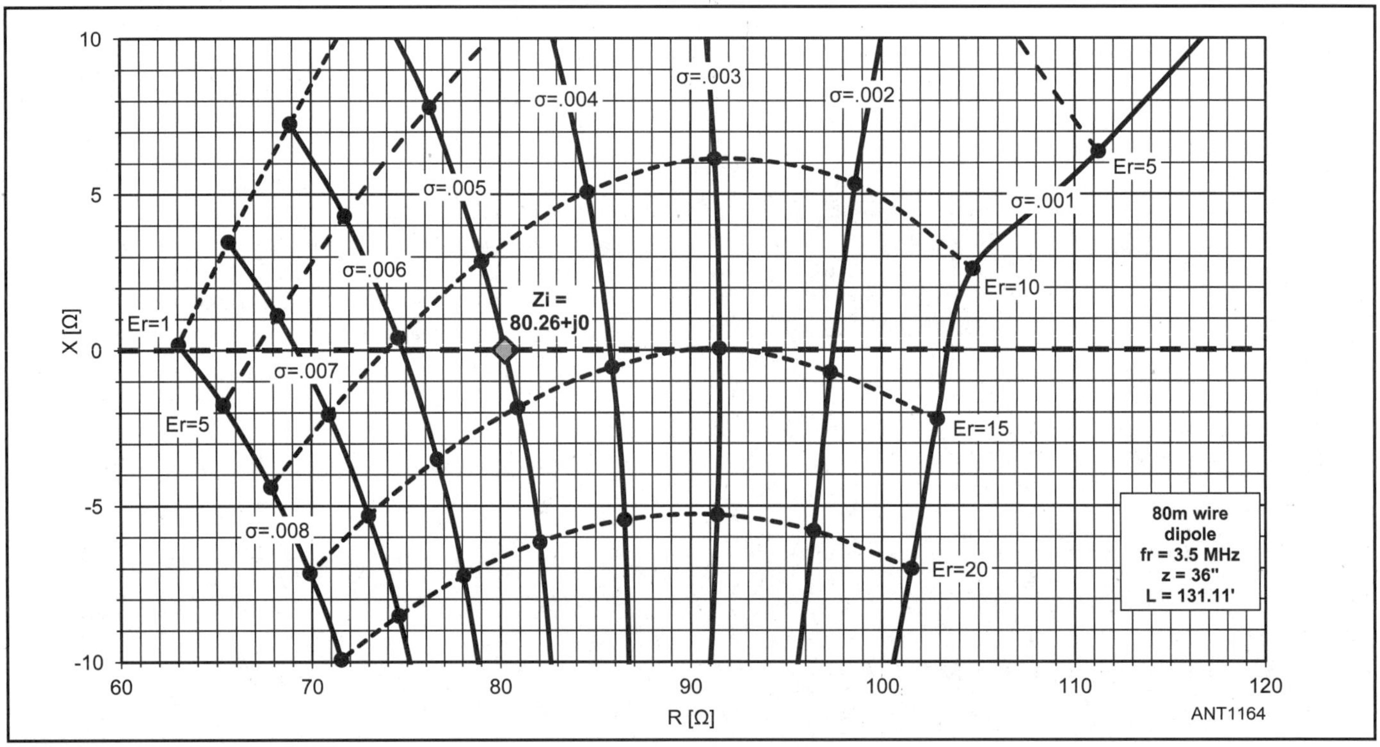

Figure 3.72 — Graph of Z = R + *j*X for a range of σ and ε_r values at 3.5 MHz.

–) so we don't need to graph all the values. Using the spreadsheet we can graph a more restricted data as set shown in **Figure 3.72**, a graph of R versus X for the feed point impedance (Z = R + *j*X) with constant σ and ε_r contours. The solid lines are constant values of σ and the dashed lines constant values of ε_r.

The measured value of Z for the antenna at 3.5 MHz is 80.26 + *j*0 Ω. A dot with a label has been placed at that value on the graph. What we see is that our matrix of values has bracketed this value. The σ = 0.005 S/m line passes right through Z. We can also see that Z lies between ε_r = 10 and ε_r = 15 lines, with a bias towards ε_r = 15. Interpolation gives a value for $\varepsilon_r \approx 13$. At this point we could repeat the process for multiple values of ε_r around ε_r = 13 to refine the answer further but from a practical point of view we're close enough! σ = 0.005 S/m and ε_r = 13, which is average soil.

3.5 References and Bibliography

References

1. Hagn, G., SRI, "HF Ground Measurements at the Lawrence Livermore National Laboratory (LLNL) Field Site," *Applied Computational Electromagnetics Society Journal and Newsletter*, Vol 3, Number 2, Fall 1988.
2. Christman, Al, K3LC, "Ground System Configurations for Vertical Antennas," *QEX*, July/Aug. 2005, pp. 28 – 37.
3. Christman, Al, K3LC, "Maximum-Gain Radial Ground Systems for Vertical Antennas," *National Contest Journal*, Mar./Apr. 2004, pp. 5 – 10.
4. Stanley, John, K4ERO, "Optimum Ground Systems for Vertical Antennas," *QST*, Dec. 1976, pp. 13 – 15.
5. Severns, R., N6LF, "An Experimental Look at Ground Systems for HF Verticals," *QST*, Mar. 2010, pp. 30 – 33.
6. Severns, R., N6LF, "Experimental Determination of Ground System Performance for HF Verticals, Part 7, Ground Systems with Missing Sectors," *QEX*, Jan./Feb. 2010, pp. 18 – 19.
7. Severns, R., N6LF, "Experimental Determination of Ground System Performance for HF Verticals, Part 3, Comparisons Between Ground Surface and Elevated Radials," *QEX*, Mar./Apr. 2009, pp. 29 – 32.
8. Christman, Al, K3LC, "Gull-Wing Vertical Antennas," *National Contest Journal*, Nov./Dec. 2000, pp. 14 – 18.
9. Weber, D., K5IU, "Optimum Elevated Radial Vertical Antennas," *Communications Quarterly*, Spring 1997, pp. 9 – 27.
10. Doty, Frey and Mills, "Efficient Ground Systems for Vertical Antennas," *QST*, Feb. 1983, pp. 20 – 25.
11. Brown, Lewis, and Epstein, "Ground Systems as a Factor in Antenna Efficiency," *Proc. IRE*, June 1937.
12. Michaels, C., "Horizontal Antennas and the Compound Reflection Coefficient," *The ARRL Antenna Compendium*, Vol. 3 (ARRL, 1992).
13. Severns, R., N6LF, "Experimental Determination of Ground System Performance for HF Verticals, Part 1, Test Setup and Instrumentation", *QEX*, Jan./Feb. 2009, pp. 21 – 25.
14. Severns, R., N6LF, "Experimental Determination of Ground System Performance for HF Verticals, Part 2, Test Setup and Instrumentation," *QEX*, Jan./Feb. 2009, pp. 48 – 52.
15. Severns, R., N6LF, "Experimental Determination of Ground System Performance for HF Verticals, Part 6, Ground Systems for Multi-band Verticals," *QEX*, Nov./Dec. 2009, pp. 19 – 24
16. Severns, R., N6LF, "Measurement of Soil Electrical Parameters at HF," *QEX*, Nov./Dec. 2006, pp 3-8.
17. Maxwell, M., *Reflections III — Transmission Lines and Antennas*, 3rd Edition (CQ Communications, 2010).
18. Lindell, I., Alanen, E., Mannerslo, K., "Exact Image Method for Impedance Computation of Antennas Above the Ground," *IEEE Trans. On Antennas and Propagation*, AP-33, Sep. 1985, pp. 937 – 945.
19. Cebik, L., W4RNL, "The Pseudo-Brewster Angle Revisited," available online from various sources.
20. Christman, A., K3LC, "Radial Configurations for Elevated Four-Square Arrays" and "More Radial Configurations for Elevated Four-Square Arrays," *QST*, Apr. 2022, pp. 30 – 33 and Jan. 2023, pp. 30 – 34.
21. Cebik, L., "Counterpoise? The Use and Abuse of a Word," available on-line from various sources.
22. Severns, Rudy, N6LF, "Measurement of Soil Electrical Parameters at HF," *QEX*, Nov. 2006. pp. 3-9.

Bibliography

Source material and more extended discussion of the topics covered in this chapter can be found in the references listed below and in the textbooks listed at the end of the **Antenna Fundamentals** chapter.

Boothe, B., "The Minooka Special," *QST*, Dec. 1974, pp. 15 – 19, 28.

Brown, G., "The Phase And Magnitude Of Earth Currents Near Radio Transmitting Antennas," *Proc. IRE*, Feb. 1935, pp. 168 – 182.

CCIR Recommendation 368, Documents of the CCIR XII Plenary Assembly, ITU, Geneva, 1967.

Collin R., and Zucker, F., *Antenna Theory*, Chap 23 by J. Wait, Inter-University Electronics Series (McGraw-Hill, 1969), Vol 7, pp. 414 – 424.

Hulick, T., "A Two-Element Vertical Parasitic Array For 75 Meters," *QST*, Dec. 1995, pp. 38 – 41.

Jones, R., "A 7-MHz Vertical Parasitic Array," *QST*, Nov. 1973, pp. 39 – 43, 52.

Laport, E., *Radio Antenna Engineering* (McGraw-Hill, 1952).

Larsen, T., "The E-Field and H-Field Losses Around Antennas With a Radial Ground Wire System," *Journal of Research of the National Bureau of Standards*, D. Radio Propagation, Vol. 66D, No. 2, Mar.-Apr. 1962, pp. 189 – 204.

McNamara, D., Pistorius, C., Malherbe, J., *Introduction to the Geometrical Theory of Diffraction* (Artech House, 1994).

Michaels, C., "Some Reflections on Vertical Antennas," *QST*, July 1987, pp. 15 – 19.

Radio Broadcast Ground Systems, available from Smith Electronics, Inc, 8200 Snowville Rd, Cleveland, OH 44141.

Reference Data for Radio Engineers, 5th Edition (Howard W. Sams, 1968), Chapter 28.

Severns, R., "Verticals, Ground Systems and Some History," *QST*, July 2000, pp 38-44.

Severns, R., "Comments on Longmire and Smith," Dec 2016. Available from **www.antennasbyn6lf.com/measurement_of_soil_characteristics/**.

Sevick, J., "The Ground-Image Vertical Antenna," *QST*,

July 1971, pp. 16 – 19, 22
Sevick, J., "The W2FMI 20-Meter Vertical Beam," *QST*, June 1972, pp. 14 – 18.
Sevick, J., "The W2FMI Ground-Mounted Short Vertical," *QST*, Mar. 1973, pp. 13 – 18, 41.
Sevick, J., "A High Performance 20-, 40- and 80-Meter Vertical System," *QST*, Dec. 1973, pp. 30 – 33.
Sevick, J., "The Constant-Impedance Trap Vertical," *QST*, Mar. 1974, pp. 29 – 34.
Sevick, J., "Short Ground-Radial Systems for Short Verticals," *QST*, Apr. 1978, pp. 30 – 33.
Sevick, J., "Measuring Soil Conductivity," *QST*, Mar. 1981, pp. 38 – 39.
Stanley, J., "Optimum Ground Systems for Vertical Antennas," *QST*, Dec. 1976, pp. 13 – 15.
Terman, F., *Radio Engineers' Handbook*, 1st Edition (McGraw-Hill, 1943).

TABLE OF CONTENTS

Chapter 4 — Online Material

Articles
- Antenna Book Table 4.3 expanded for other locations
- "Build a Homebrew Radio Telescope" by Mark Spencer, WA8SME
- "Frequency Selection for HF Operation"
- Propagation Prediction Tables
- "The Reverse Beacon Network" by Pete Smith, N4ZR and Ward Silver NØAX
- "Using Propagation Predictions for HF DXing" by Dean Straw, N6BV

Chapter 4
Radio Wave Propagation

The function of an antenna is to radiate or receive radio waves which travel or *propagate* through the Earth's atmosphere or space. Radio communication is carried out by means of these propagating waves which are key to amateur radio. To communicate effectively by radio, then, it is important to understand the basics of propagation. This allows you to select or design-and-build effective antennas and antenna systems that are well-suited for the types of propagation expected to occur.

This content has been comprehensively updated from previous editions by Carl Luetzelschwab, K9LA, with additional input from Frank Donovan, W3LPL, Ward Silver, NØAX, Ethan Miller, K8GU, and Hermann Schumacher, DF2DR. The section on VHF/UHF mobile propagation was originally contributed by Alan Bloom, N1AL. More detailed discussions and the underlying mathematics of radio propagation physics can be found in the references listed at the end of this chapter.

The material in this chapter is adapted from ARRL's propagation book, *Here to There: Radio Wave Propagation*. This chapter focuses on the fundamentals of radio waves and the types of propagation that pertain to antenna selection and design. You can find more about solar behavior and space weather in the *ARRL Handbook* as well.

4.1 Fundamentals of Radio Wave Propagation

Radio belongs to a family of electromagnetic radiation that includes infrared (radiation heat), visible light, ultraviolet, X-rays, and the even shorter-wavelength gamma and cosmic rays. Radio has the longest wavelength of this group and thus the lowest frequency. See **Table 4.1** (See the **Antenna Fundamentals** chapter for more about electromagnetic waves.)

Electromagnetic radio waves are composed of an interrelated electric and magnetic field. The electric and magnetic components are oriented at right angles to each other and are also perpendicular to the direction of travel. The polarization of a radio wave is usually designated the same as the orientation of its electric field. This relationship can be visualized in **Figure 4.1**. Unlike sound waves or ocean waves, electromagnetic waves need no propagating medium, such as air or water. This property enables electromagnetic waves to travel through the vacuum of space.

A radio wave far enough from its source to appear flat is called a *plane wave*. From here on, we will be discussing plane

Table 4.1
The Electromagnetic Spectrum

Radiation	*Frequency*	*Wavelength*
Radio	10 kHz – 300 GHz	30 km – 1 mm
Infrared	300 GHz – 428.6 THz	1 mm – 700 nm
Visible light	428.6 THz – 750 THz	700 nm – 400 nm
Ultraviolet	750 THz – 3 × 10³ THz	400 nm – 100 nm
Extreme Ultraviolet	3 × 10³ THz – 3 × 10⁴ THz	100 nm – 10 nm
"Soft" X-ray	3 × 10⁴ THz – 3 × 10⁵ THz	10 nm – 1 nm
"Hard" X-ray	3 × 10⁵ THz – 3 × 10⁶ THz	1 nm – 0.1 nm

Optics and Radio Propagation

The four terms *reflection*, *refraction*, *scatter*, and *diffraction* were terms used in optics long before the radio age began. Radio propagation is nearly always a mix of these phenomena, and it may not be easy to identify or separate them while they are happening when we are on the air. This book tends to rely on the words *bending* (refraction) and *scattering* in its discussions, with appropriate modifiers as needed. The important thing to remember is that any alteration of the path taken by energy radiated from an antenna is almost certain to affect on-the-air results

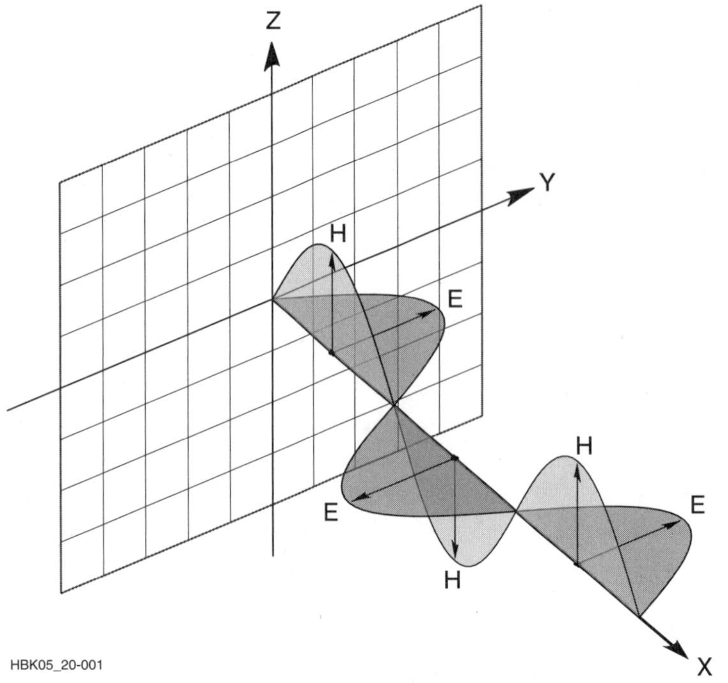

Figure 4.1 — Electric and magnetic field components of the electromagnetic wave. The polarization of a radio wave is the same direction as the plane of its electric field.

waves. The path taken by a plane wave is called a *ray*. Traced from its source to any point on a spherical surface, such as the Earth, the ray will follow a straight line on an *azimuth-equidistant* map projection with the source at its center. (See the sidebar Azimuth-Equidistant Maps.)

4.1.1 Free-Space Attenuation

Free-space attenuation results from the spherical spreading of radio energy as it travels from its source. See **Figure 4.2**. Attenuation grows rapidly with distance because signal strength decreases with the square of the distance from the source. (The signal's field strength in V/m decreases linearly with distance and its power density in W/m^2 decreases with the square of the distance.) In free space, if the distance between transmitter and receiver is increased from 1 to 10 kilometers (0.6 to 6 miles), the signal power will be reduced by a factor of 100 (20 dB). Attenuation increases with frequency as well (for example, doubling the frequency results in 6 dB more free-space path loss). Free-space attenuation (path loss) is

$$L_{fs} = 32.45 + 20 \log d + 20 \log f$$

where

L_{fs} = free pace path loss in dB
d = distance in km
f = frequency in MHz

Enhancements (especially at VHF and above) can increase signal strength at the receiver due to multi-path and scattering. Thus, in the real world, the L_{fs} equation above may not be valid.

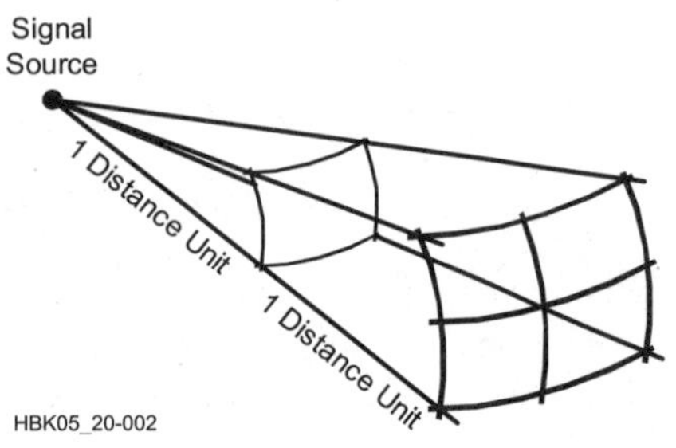

Figure 4.2 — Radio energy disperses as the square of the distance from its source. For the change of one distance unit shown the signal's power per unit of area is only one quarter as strong. Each spherical section has the same surface area.

4.1.2 Refraction

Electromagnetic waves travel in straight lines until they are deflected by something. Radio waves are *refracted*, or bent, slightly when traveling from one medium to another. *Refraction* is the bending of a ray as it passes from one medium to another at an angle. The apparent bending of a pencil partially immersed in a glass of water demonstrates this principle quite dramatically. The mediums should extend much more than one wavelength around the region in which refraction takes place. Radio waves behave no differently from other familiar forms of electromagnetic radiation in this regard.

Refraction is caused by a change in the velocity of a wave when it crosses the boundary between one propagating medium and another. If this transition is made at an angle, one portion of the wavefront slows down (or speeds up) before

Azimuth-Equidistant Maps

We can better understand several aspects of long-distance propagation if you become accustomed to thinking of the Earth as a ball. This is easy if you use a globe frequently. A flat map of the world, of the *azimuthal-equidistant* projection type, is a useful substitute.

Also known as *great circle maps*, these are a projection of the Earth's surface that make all great circle paths straight lines from a given location, which is at the center of the map. The outer perimeter of the map is usually 20,000 kilometers (12,427 miles or halfway around the world), but some mapping programs allow other distances.

An azimuth-equidistant map centered on Newington, Connecticut, is shown in **Figure 4.A**. NS6T provides an online service at **ns6t.net/azimuth/azimuth.html** to generate these maps centered on any location. The ARRL World Map is an azimuth-equidistant map centered on Wichita, Kansas. The program *DX Atlas* by VE3NEA **dxatlas.com**) also presents azimuth-equidistant maps, and it offers the option of adding the terminator (the gray line), the auroral oval and ionospheric parameters.

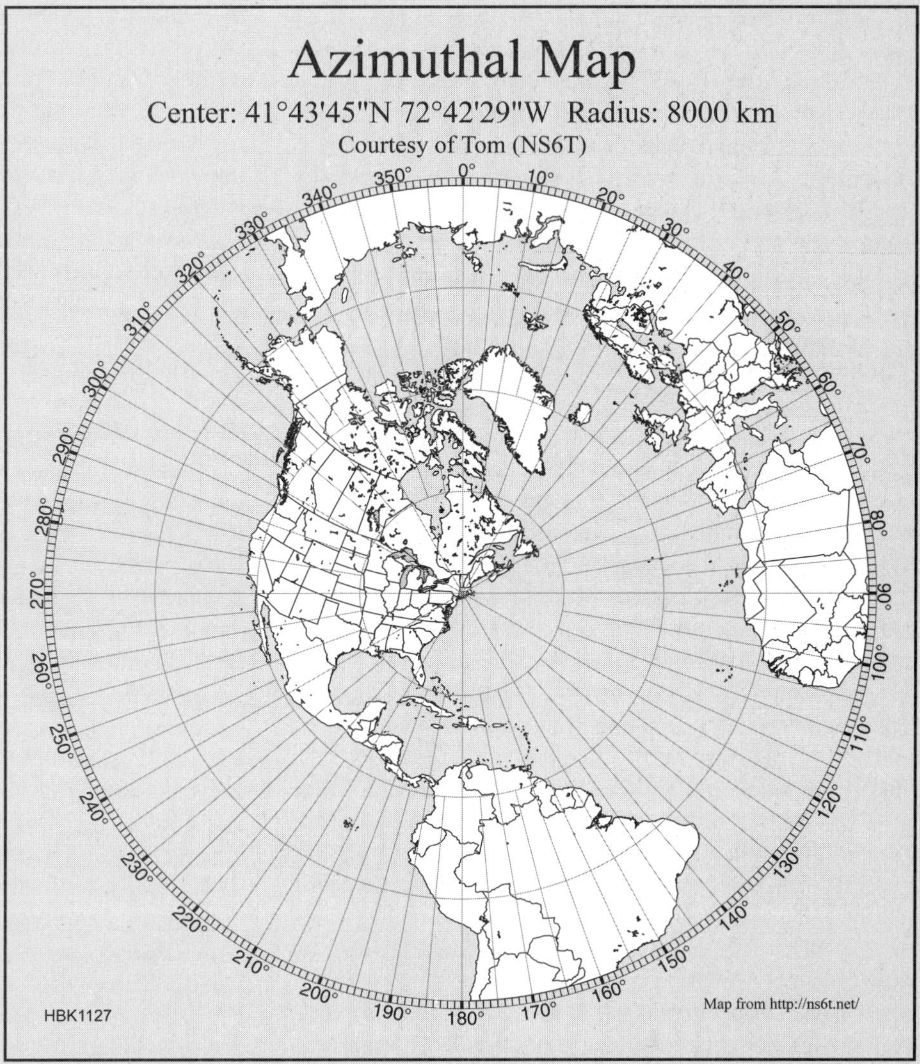

Figure 4.A — Azimuth-equidistant map centered on W1AW out to 8,000 kilometers. (Map provided by ns6t.net/azimuth/azimuth.html)

the other, thus bending the wave slightly. This is shown schematically in **Figure 4.3**.

The amount of bending increases with the ratio of the *refractive indexes* of the two media. The refractive index, n, is simply the velocity of a radio wave in free space divided by its velocity in the medium.

Radio waves are commonly refracted when they travel through different regions of the atmosphere, whether the highly charged ionospheric regions roughly 100 kilometers (60 miles) above the ground and higher, or the weather-sensitive area near the Earth's surface. There is slight atmospheric bending in our HF bands. It becomes noticeable at 28 MHz, more so at 50 MHz, and it is much more of a factor in the higher VHF range and in UHF and microwave propagation.

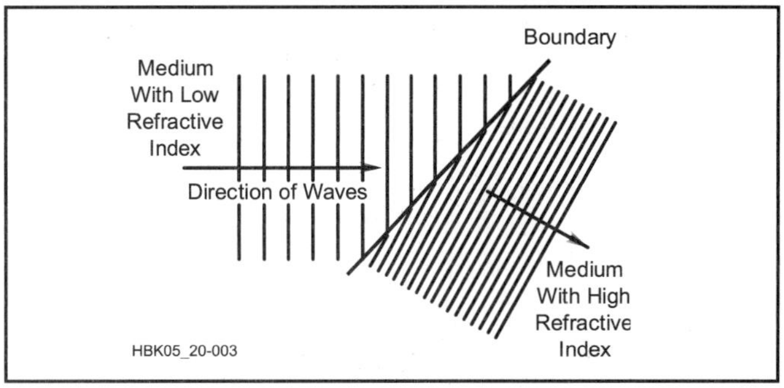

Figure 4.3 — Radio waves are refracted as they pass at an angle between dissimilar media. The lines represent the crests of a moving wave front and the distance between them is the wavelength. The direction of the wave changes because one end of the wave slows down before the other as it crosses the boundary between the two media. The wavelength is simultaneously shortened, but the wave's frequency (number of crests that pass a certain point in a given unit of time) remains constant.

4.1.3 Reflection

Reflection occurs at any boundary between materials of differing dielectric constant when the extent of the materials is on the order of at least one wavelength at the frequency considered. (The surface should extend at least one wavelength from the point at which reflection takes place.) Familiar examples with light are reflections from water surfaces and windowpanes. Both water and glass are transparent for light, but their dielectric constants are very different from that of air. Light waves, being very short, seem to bounce off both surfaces. Radio waves, being much longer, are practically unaffected by glass, but will be at least partially reflected by a large body of conductive material such as soil or water.

The degree of bending of radio waves at boundaries between air masses which have different indexes of refraction increases with the radio frequency. When the ratio of the refractive indexes of two media is great enough, radio waves can be reflected, just like light waves striking a mirror.

At amateur frequencies above 30 MHz, reflections from a variety of large objects, such as water towers, buildings, airplanes, mountains, and the like, can provide a useful means of extending over-the-horizon paths by several hundred kilometers. Two stations need only beam toward a common reflector, whether stationary or moving. A metal surface works well as a reflector if it is several wavelengths in diameter.

Maximum range is limited by the radio line-of-sight distance of both stations to the reflector and by reflector size and shape. The reflectors must be many wavelengths in size and ideally have flat surfaces. Large airplanes make fair reflectors and may provide the best opportunity for long-distance contacts. The calculated limit for airplane reflections is 900 kilometers (560 miles), assuming the largest jets fly no higher than 12,000 meters (40,000 feet), but actual airplane reflection contacts are likely to be considerably shorter.

Reflections from the ionosphere can occur when the plasma frequency (which is proportional to the square root of electron density) is much, much greater than the operating frequency. A good example is sporadic E propagation since it occurs from a very thin layer (a couple kilometers) that isn't thick enough to return signals to Earth via refraction.

In material of a given electrical conductivity, long waves penetrate deeper than short ones, and so require a thicker conductive region for good reflection. Thin metal, which is highly conductive, is a good reflector of even long-wavelength radio waves. The Earth is a rather lossy reflector of radio signals. With poorer conductors, such as the Earth's crust, long waves may penetrate quite a few feet below the surface. Soil conductivity ranging from poor (sand and rock) to high (wet, rich soil) affects reflection. Water will also reflect radio waves depending on its conductivity. Distilled water is an insulator but salt- and brackish water have high conductivity. The polarization of the reflected wave also affects reflection losses.

4.1.4 Scattering

The direction of radio waves can also be altered through *scattering*. Scatter is the redirection of an electromagnetic wave in many directions from a medium when the volume of material from which the scattering takes place is much smaller than one wavelength at the frequency being considered. Scattering results in additional loss.

The effect seen by a beam of light attempting to penetrate fog is a good example of light-wave scattering. Even on a clear night, a highly directional searchlight is visible due to a small amount of atmospheric scattering perpendicular to the beam. Radio waves are similarly scattered when they encounter randomly arranged objects of wavelength size or smaller, such as masses of electrons or water droplets. When the density of scattering objects becomes great enough, they behave more like a propagating medium with a characteristic refractive index.

4.1.5 Knife-Edge Diffraction

Radio waves can also pass behind solid objects with sharp upper edges, such as a mountain range, by *knife-edge diffraction*. This is a common natural phenomenon that affects light, sound, radio and other coherent waves, but it is difficult to comprehend. **Figure 4.4** depicts radio signals approaching an idealized knife-edge. The portion of the radio waves that strike the base of the knife-edge is entirely blocked, while that portion passing several wavelengths above the edge travel on relatively unaffected. It might seem at first glance that a knife-edge as large as a mountain, for example, would completely prevent radio signals from appearing on the other side but that is not quite true. Something quite unexpected happens to radio signals that pass just over a knife-edge.

Normally, radio signals along a wave front interfere with each other continuously as they propagate through unobstructed space, but the overall result is a uniformly expanding wave. When a portion of the wave front is blocked by a knife-edge, the resulting *interference pattern* is no longer uniform. This can be understood by visualizing the radio signals right at the knife-edge as if they constituted a new and separate transmitting point, but in-phase with the source wave at that point. The signals adjacent to the knife-edge still interact with signals passing above the edge, but they cannot interact with signals that have been obstructed below the edge. The resulting interference pattern no longer creates a uniformly expanding wave front, but rather appears as a pattern of alternating strong and weak bands of waves that spread in a nearly 180° arc behind the knife-edge.

The crest of a range of hills or mountains 50 to 100 wavelengths long can produce knife-edge diffraction at UHF and microwave frequencies. Hillcrests that are clearly defined and free of trees, buildings and other clutter make the best knife-edges, but even rounded hills may serve as a diffracting edge. Alternating bands of strong and weak signals, corresponding to the interference pattern, will appear on the surface of the Earth behind the mountain, known as the *shadow zone*. The phenomenon is generally reciprocal, so that two-way communication can be established under optimal conditions. Knife-edge diffraction can make it possible to complete paths of 100 kilometers or more that might otherwise be entirely obstructed by mountains or seemingly impossible terrain.

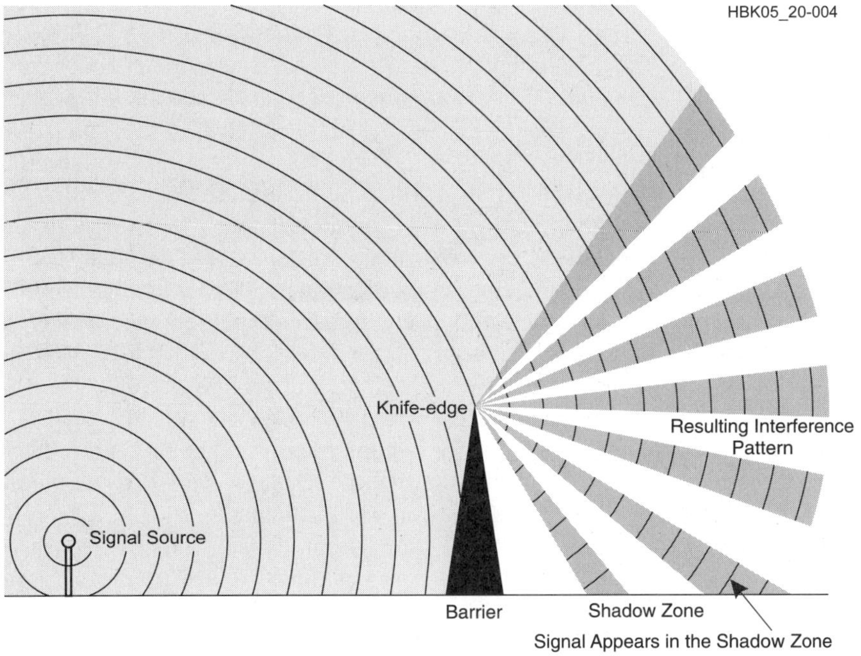

Figure 4.4 — Knife-edge diffraction occurs when VHF and UHF radio waves, light and other waves are diffracted around the sharp edge of a solid object that is large in terms of wavelengths. Diffraction results from interference between waves right at the knife-edge and those that are passing above it. Some signals appear behind the knife-edge as a consequence of the interference pattern. Hills or mountains can serve as natural knife-edges at radio frequencies.

4.1.6 Surface Wave

In general, an electromagnetic *surface wave* travels along an interface between two media with different dielectric constants or refractive indexes. At HF and lower frequencies, the most common type of surface wave travels along the surface of the Earth and is one aspect of what amateurs refer to as "ground wave" as discussed in the next section.

Surface waves can also travel along an interface between media with two different dielectric constants such as layers of air with different temperatures and humidity or insulating layers with different dielectric constants. At microwave frequencies, surface waves travel along the interface between a dielectric (such as air or plastic) and a conducting surface, usually metal.

A radio wave could be traveling in contact with the ground as a surface wave. As the frequency is raised, the distance over which surface waves can travel without excessive energy loss becomes smaller and smaller. The surface wave can provide coverage up to about 100 miles in the standard AM broadcast band during the daytime, but attenuation is high.

Power from a surface wave is typically not transferred to either medium. The wave's E and H fields become weaker exponentially with perpendicular distance from the interface. This property is called *evanescence*. If one or both of the mediums are lossy or leaky, some power can be transferred from the surface wave, and it will become weaker with increasing distance traveled along the interface.

4.1.7 Ground Wave

Ground wave propagation includes two types of radio wave propagation. One type is surface waves (as discussed above) that travel along the Earth's surface. The second is diffracted waves that are bent around the curvature of the Earth by a special form of diffraction that primarily affects longer-wavelength, vertically polarized radio waves.

Ground wave propagation is most apparent in the 3.5 MHz (80 meter), 1.8 MHz (160 meter), 475 kHz (630 meter), and 137 kHz (2200-meter) amateur bands, where practical ground wave distances may extend beyond 200 kilometers (120 miles). It is also the primary mechanism used by AM broadcast stations in the medium-wave bands. The term ground wave is often mistakenly applied to any short-distance communication, but the actual mechanism is unique to the longer-wave bands.

Ground wave is most useful during the day at 1.8 and 3.5 MHz, when D region absorption makes skywave propagation more difficult. Vertically polarized antennas with excellent ground systems provide the best results. Ground wave losses are reduced considerably over saltwater and are highest over dry and rocky land. As can be seen from **Figure 4.5**, the range

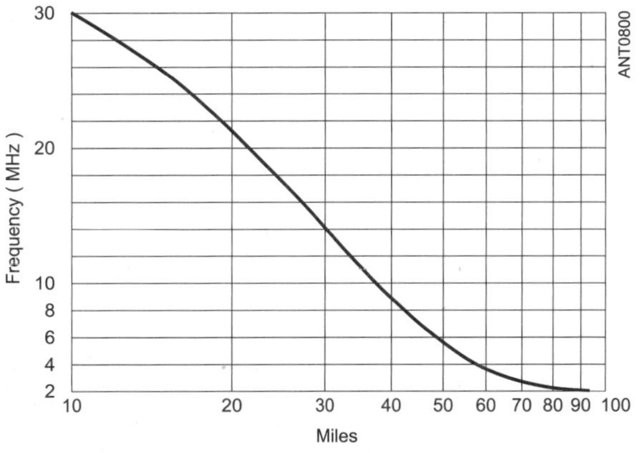

Figure 4.5 — Typical HF ground wave range as a function of frequency.

increases with decreasing frequency. The results in this figure are for a typical 100-watt station using a vertical over average ground in a rural noise environment. Your results will vary depending on your power level, antenna, ground conditions, and local noise level.

Radio waves are bent slightly as they pass over a sharp edge, but the effect extends to edges that are considerably rounded. At medium and long wavelengths, the curvature of the Earth looks like a rounded edge. Bending results when the lower part of the wave front loses energy due to currents induced in the ground. This slows down the lower part of the wave, causing the entire wave to tilt forward slightly. This tilting follows the curvature of the Earth, thus allowing low- and medium wave radio signals to propagate over distances well beyond line of sight.

The better the station in terms of external noise level, antenna gain, and transmit output power, the greater the surface wave range. Vertically polarized antennas are preferred, which tends to limit amateur surface-wave communication to the bands and installations for which large vertical antennas can be erected.

4.1.8 The Space or Direct Wave

Propagation between two antennas situated within line of sight of each other is shown in **Figure 4.6**. Energy traveling directly between the antennas is attenuated to about the same degree as in free space. Unless the antennas are very high or quite close together, an appreciable portion of the energy is reflected from the ground. This reflected wave combines with direct radiation to affect the actual signal received.

In most communication between two stations on the ground, the angle at which the wave strikes the ground will be small. For a horizontally polarized signal, such a reflection reverses the phase of the wave. If the distances traveled by both parts of the wave were the same, the two parts would arrive out of phase, and would therefore cancel each other. The ground-reflected ray in Figure 4.6 must travel a little farther, so the phase difference between the two depends on the lengths of the paths, measured in wavelengths. The wavelength in use is important in determining the useful signal strength in this type of communication. Let's analyze an example of why the wavelength in use is important.

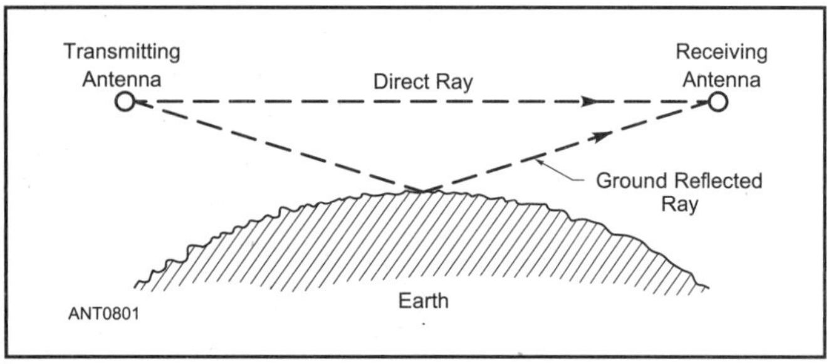

Figure 4.6 — The ray traveling directly from the transmitting antenna to the receiving antenna (direct wave) combines with a ray reflected from the ground (ground reflected ray) to form the space wave. For a horizontally polarized signal, a reflection as shown here reverses the phase of the ground-reflected ray.

If the difference in path length is 3 meters, the phase difference with 160 meter waves would be only $360° \times 3/160 = 6.8°$. This is a negligible difference from the 180° shift caused by the reflection, so the effective signal strength over the path would still be very small because of cancellation of the two waves. But with 6-meter radio waves the phase length would be $360° \times 3/6 = 180°$. With the additional 180° shift on reflection, the two rays would add. Thus, the space wave is a negligible factor at low frequencies, but it can be increasingly useful as the frequency is raised. It is a dominant factor in local amateur communication at 50 MHz and higher.

Interaction between the direct and reflected waves is the principal cause of *mobile flutter* observed in local VHF communication between fixed and mobile stations. The flutter effect decreases once the stations are separated enough so that the reflected ray becomes inconsequential. The reflected energy can also confuse the results of field-strength measurements during tests on VHF antennas.

As with most propagation explanations, the space-wave picture presented here is simplified, and practical considerations dictate modifications. There is always some energy loss when the wave is reflected from the ground. Further, the phase of the ground-reflected wave is not shifted exactly 180°, so the waves never cancel completely. At UHF, ground-reflection losses can be greatly reduced or eliminated by using highly directive antennas. By confining the antenna pattern to something approaching a flashlight beam, nearly all the energy is in the direct wave. The resulting energy loss is low enough that microwave relays, for example, can operate with moderate power levels over many miles. Thus, we see that, while the space wave is inconsequential below about 20 MHz, it can be a prime asset in the VHF realm and higher.

4.1.9 Polarization

ANTENNA POLARIZATION SELECTION

If effective communication over long distances were the only consideration, we might be concerned mainly with radiation of energy at the lowest possible angle above the horizon. As an example, our 1.8 and 3.5 MHz bands are well-suited for short-distance communication because they serve that purpose with antennas that are not difficult or expensive to put up. Out to a few hundred miles, simple wire antennas for these bands do well, even though their radiation is mostly at high angles above the horizon. (See the section on NVIS Communication.) Vertical antenna systems are generally better for long-distance use on these frequencies although they require extensive ground systems for good performance.

Horizontal antennas that radiate well at low angles are most easily erected for 7 MHz and higher frequencies — horizontal wires and arrays are almost standard practice for work on 7 through 29.7 MHz. Vertical antennas, such as a single omnidirectional antenna of multiband design, are also used in this frequency range. An antenna of this type may be a good solution to the space problem for a city dweller on a small lot, or even for the resident of an apartment building.

High-gain antennas are almost always used at 50 MHz and higher frequencies, and most of them are horizontal. The principal exception is mobile communication with FM through repeaters. The height question is answered easily for VHF enthusiasts — the higher the better.

The theoretical and practical effects of height above ground at HF are treated in detail in the chapter **Effects of Ground** in this book. Note that it is the electrical height in *wavelengths* that is important — a good reason to think in the metric system, rather than in feet and inches, since our bands are generally referred to in terms of meters.

In working locally on any amateur frequency band, best results will be obtained with the same polarization at both stations, except on rare occasions when polarization rotation is caused by terrain obstructions or reflections from buildings. Where such a shift is observed, mostly above 100 MHz or so, horizontal polarization tends to work better than vertical. This condition is found primarily on short paths, so it is not too important.

THE ELECTRON GYRO-FREQUENCY

On 1.8 MHz two interesting effects occur because the operating frequency is close to the ionosphere's electron gyro-frequency. (The *gyro-frequency* is the frequency at which an electron will spiral around a particular magnetic field line.) Waves entering the ionosphere split into an *extraordinary* (X) and *ordinary* (O) wave which propagate somewhat differently as discussed in the section Propagation Below 1.8 MHz. First, the extraordinary wave suffers significantly higher absorption than the ordinary wave, so for all intents and purposes only one characteristic wave propagates on 160 meters. Second, the ordinary wave is highly elliptical, approaching linear polarization. For stations at mid to high northern latitudes, vertical polarization couples the most energy into the ordinary wave — thus vertical polarization is generally the best way to go on 160 meters. But other effects, like disturbances to propagation or high angle modes, sometimes dictate horizontal polarization. This is the origin of the oft-repeated statement on 160 meters that "you can't have enough antennas on Top Band."

4.2 Skywave or Ionospheric Propagation

The Earth's atmosphere, which reaches to more than 600 kilometers (370 miles) altitude, is usually divided into a number of regions based on a transitioning characteristic of the atmosphere — like temperature. For propagation purposes, the important regions are shown in **Figure 4.7**. The weather-producing *troposphere* lies between the surface and an average altitude of 10 kilometers (6 miles). Between 10 and 50 kilometers (6 and 30 miles) are the *stratosphere* and the embedded *ozonosphere*, where ultraviolet-absorbing ozone reaches its highest concentrations. About 99% of atmospheric gases are contained within these two lowest regions.

Above 50 kilometers to about 600 kilometers (370 miles) is the *ionosphere*, notable for its effects on radio propagation. At these altitudes, atomic oxygen, molecular oxygen, molecular nitrogen, and nitric oxide predominate under very low pressure and are the important species to consider for propagation. High-energy solar EUV and X-ray radiation ionize these constituents, creating a broad region where ions are created in relative abundance. The ionosphere is subdivided into distinctive D, E, and F regions.

4.2.1 Structure of the Geomagnetic Field

The north and south *magnetic poles* are the points on the surface of Earth at which the planet's magnetic field points vertically downward for the north pole and vertically upward for the south pole. In other words, if a magnetic compass needle is allowed to rotate in three dimensions, it will point straight down at the north magnetic pole and straight up at the south magnetic pole. There is only one location in each hemisphere where this occurs, near (but distinct from) the *geographic* north and south poles.

Related to the magnetic poles are the *geomagnetic* poles. These are the poles of an ideal dipole model of the Earth's magnetic field that most closely fits the Earth's actual magnetic field. The geomagnetic poles are the centers of the auroral ovals. **Figure 4.8** shows the location of the north magnetic pole and how it has drifted much more than the geomagnetic north pole over the years. The figure also shows how the location of the north geomagnetic pole has drifted very little during the same period.

GEOMAGNETIC EQUATOR

The geomagnetic equator is the great circle of the Earth whose plane is perpendicular to the axis of the geomagnetic field. **Figure 4.9** shows the geomagnetic equator (the dashed line) on a Mercator projection of Earth as of 1990. The geomagnetic equator is important for *trans-equatorial propagation* (TEP) and will be discussed later. This is also called the *magnetic dip equator.*

4.2.2 The Ionosphere

The ionosphere plays a basic role in long-distance communications in all the amateur bands from 1.8 MHz to 50 MHz. The effects of the ionosphere are less apparent at the very high frequencies (30 – 300 MHz), but they persist for terrestrial communications at least through 432 MHz. The basic physics of ionospheric propagation was largely worked out by the 1930s, yet both amateur and professional experimenters made further discoveries throughout the 1930s, 1940s, and 1950s, and continue to make important new discoveries to this day. Sporadic E, aurora, trans-equatorial, meteor scatter, and several types of field-aligned scattering were among many additional ionospheric phenomena that required explanation and are still not fully understood.

REGION VERSUS LAYER

The commonly used term "layer" could lead to the erroneous assumption that the ionosphere consists of distinct thin sheets separated by emptiness in between. This is not so. The ionosphere is a continuous electronic density versus altitude, with definite peaks and inflection points that define the D, E, F_1, and F_2 regions. (F_1 and F_2 are often labeled F1 and F2 — this chapter will use the subscripted version. These notations are equivalent and refer to the same region.) Studies have shown that the height of the various regions may also vary by latitude. On-going ionospheric physics research seeks to better understand the geographic and day-to-day variability of the ionosphere.

4.2.3 Ionospheric Refraction

The degree of bending (refraction) of a wave path in an ionized region depends on the density of the ionization and the length of the wave (inversely related to the square of its frequency). The bending at any given frequency or wavelength will increase with increased ionization density and will bend away from the region of most-intense ionization. For a given ionization density, bending increases with longer wavelengths (that is, bending decreases with higher frequencies).

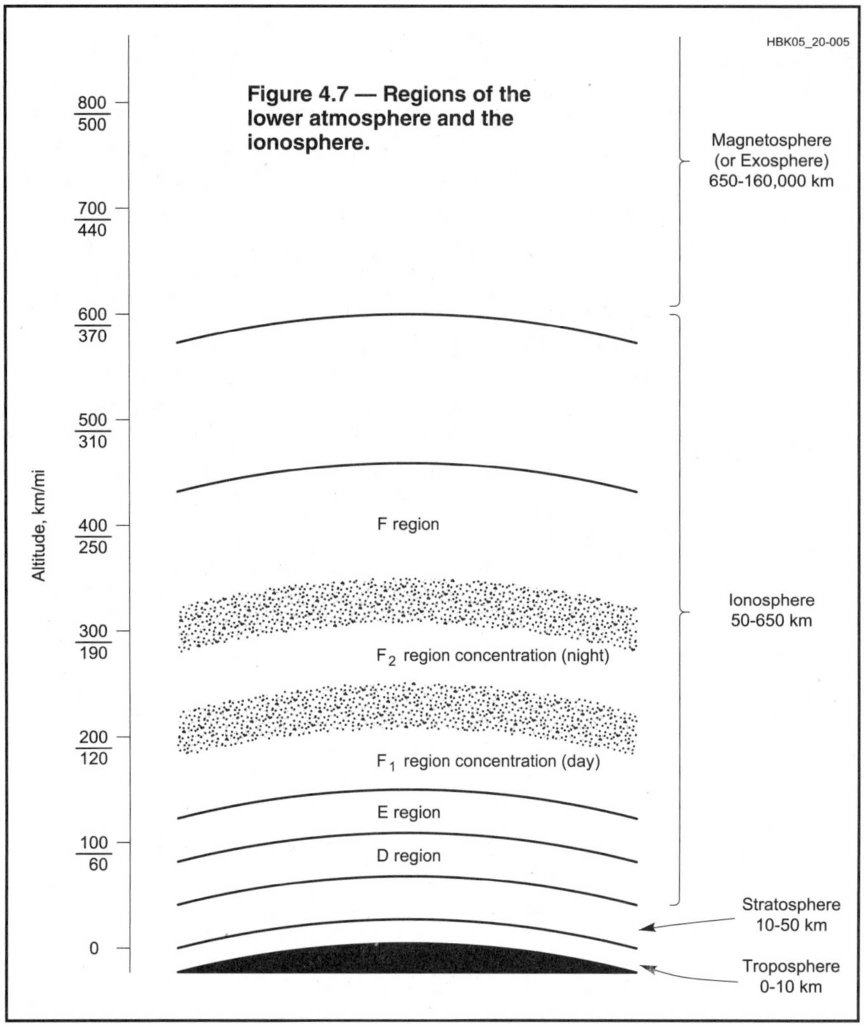

Figure 4.7 — Regions of the lower atmosphere and the ionosphere.

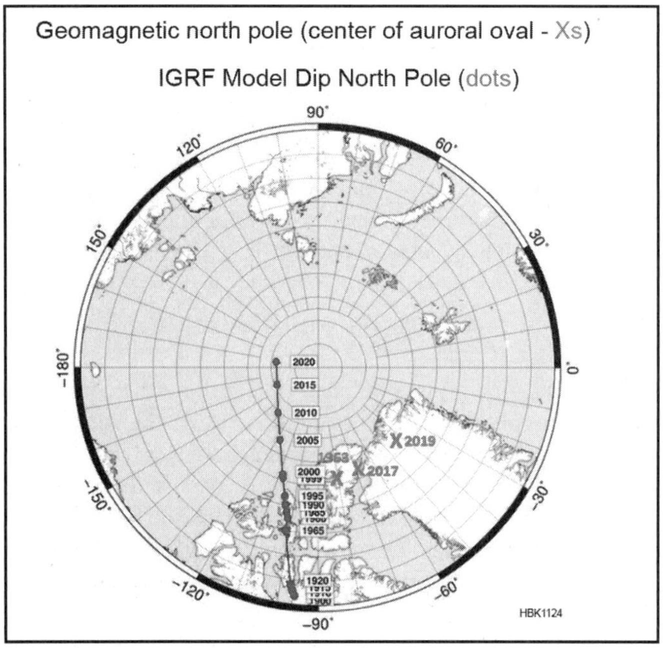

Figure 4.8 — Geomagnetic poles. The location of the north magnetic pole is shown as dots. The geomagnetic north pole locations are labeled with an X.

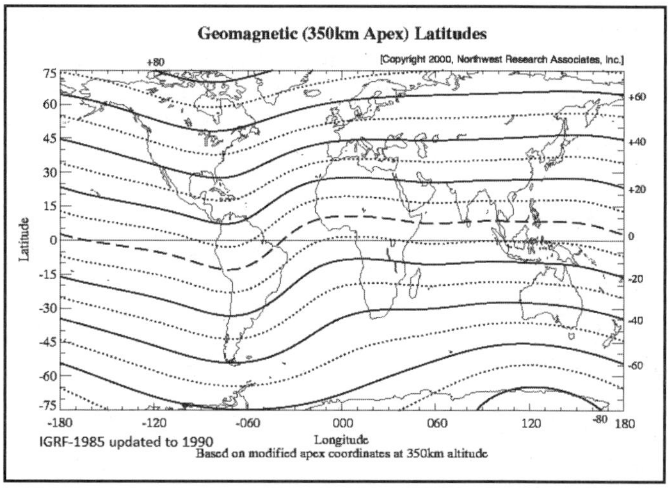

Figure 4.9 — Geomagnetic equator (the dashed line) on a Mercator projection of Earth as of 1990 (thanks Northwest Research Associates).

VIRTUAL HEIGHT

Although refraction is the primary mechanism of ionospheric propagation, it is usually more convenient to think of the process as a reflection. The *virtual height* of an ionospheric region is the equivalent altitude of a reflection that would produce the same effect as the actual refraction. The virtual height of any ionospheric region can be determined using an ionospheric sounder, or *ionosonde*, a sort of vertically oriented radar. The ionosonde sends pulses that sweep over a wide frequency range, generally from 2 MHz to 20 MHz or higher, straight up into the ionosphere. The frequencies of any echoes are recorded against time and then plotted as distance on an *ionogram*. **Figure 4.10** depicts a simple vertical incidence ionogram. (Ionospheric sounding is described in the following sections.)

CRITICAL FREQUENCY

The highest frequency that returns signals from the E and F regions at vertical incidence is known as the *vertical incidence* or *critical frequency*. When the frequency of a vertically incident signal is raised above the critical frequency of an ionospheric region, that portion of the ionosphere is unable to refract the signal back to Earth.

The critical frequency is a function of ion density. The higher the ionization at a particular altitude, the higher becomes the critical frequency. Strictly speaking, the critical frequency is the term applicable to the peak electron density of a region. Physicists relate any electron density in any part of the ionosphere to a plasma frequency, because technically gases in the ionosphere are in a plasma, or partially ionized state. F region critical frequencies commonly range from about 1 MHz and occasionally reach as high as 15 MHz during periods of intense ionization.

4.2.4 Faraday Rotation

Magnetic and electrical forces rotate the polarization of radio waves passing through the ionosphere. For example, signals that leave the Earth as horizontally polarized and return after a reflection from the Moon rarely arrive with the same polarization. Additional path loss can occur when polarization is shifted by 90 degrees, resulting in fading of the received signal.

Faraday rotation is difficult to predict and its effects change over time and with operating frequency. At 144 MHz, the polarization of space waves may shift back into alignment with the antenna within a few minutes. Often, just waiting can solve the Faraday rotation problem. At 432 MHz, it may take half an hour or longer for the polarization to become realigned. Use of circular polarization completely eliminates this problem but creates a new problem for EME paths. The sense of circularly polarized signals is reversed with reflection, so two complete antenna systems are normally required, one with left-hand polarization and one with right-hand polarization. As a result, circularly polarized antennas are not commonly used for amateur EME communications.

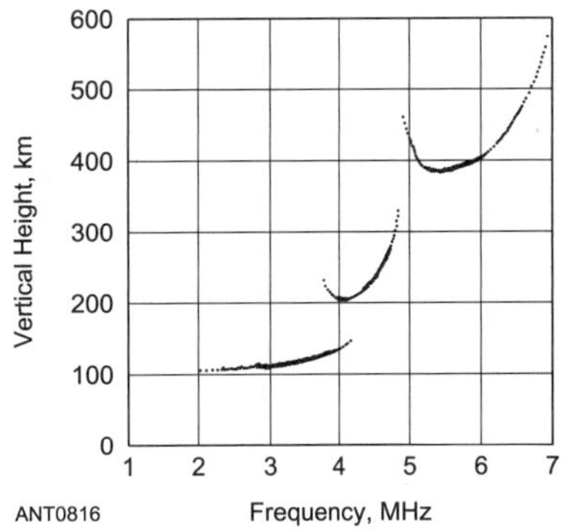

Figure 4.10 — Very simplified ionogram from a vertical-incidence sounder. The lowest trace is for the E region; the middle trace for the F_1 region and the upper trace for the F_2 region.

4.2.5 Maximum and Lowest Usable Frequencies

A signal entering an ionospheric region above its critical frequency will not be reflected and may be lost to space. However, a signal above the critical frequency may be returned to Earth if it enters the region at an *oblique angle*, rather than at vertical incidence. This is fortunate because it permits two widely separated stations to communicate on significantly higher frequencies than the critical frequency. See **Figure 4.11**. This creates *skip propagation* in which the signal travels from the Earth, into the ionosphere, and back to the Earth where it can be received.

Lower frequency signals can be returned to Earth at higher incident angles whereas a higher frequency signal at the same angle will not. The higher frequency signal will need to enter the ionosphere at a lower angle if it is to be returned to Earth. The lower angle also means it will be returned to Earth at a more distant point. The combination of ionization, angle of incidence, and fixed station locations creates ranges of frequencies for which communication is possible. This is discussed in the following sections.

MAXIMUM USABLE FREQUENCY (MUF)

The highest frequency supported by the ionosphere for reliable communications between two stations is the *maximum usable frequency* (MUF) for that path. If the separation between the stations is increased, a still higher frequency can be supported at lower launch angles. The MUF for this longer path is higher than the MUF for the shorter path because more refraction can occur for radio waves that encounter the ionosphere at more oblique angles. When the distance is increased to the maximum one-hop distance, the launch angle of the signals between the two stations is zero (that is, the ray path is tangential to the Earth at the two stations) and the MUF for this path is the highest that can be supported by that region of the ionosphere at that location, although antennas producing adequate radiation at very low angles may not be feasible at many locations. This

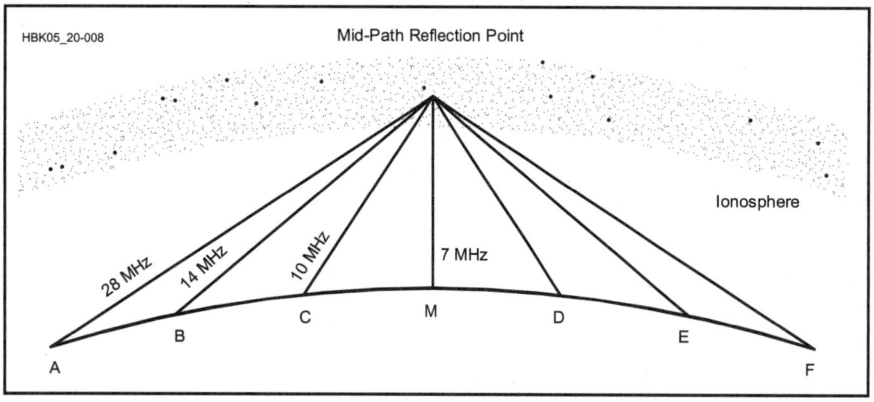

Figure 4.11 — The relationships between critical frequency, maximum usable frequency (MUF) and skip zone can be visualized in this simplified, hypothetical case. The critical frequency is 7 MHz, allowing lower frequencies to be used for short-distance ionospheric communication by stations in the vicinity of point M. These stations cannot communicate by the ionosphere at 14 MHz. Stations at points B and E (and beyond) can communicate because signals at this frequency are refracted back to Earth because they encounter the ionosphere at an oblique angle of incidence. At greater distances, higher frequencies can be used because the MUF is higher at the larger angles of incidence (low launch angles). In this figure, the MUF for the path between points A and F, with a small launch angle, is shown to be 28 MHz. Each pair of stations can communicate at frequencies at or below the MUF of the path between them, but not below the LUF — see text.

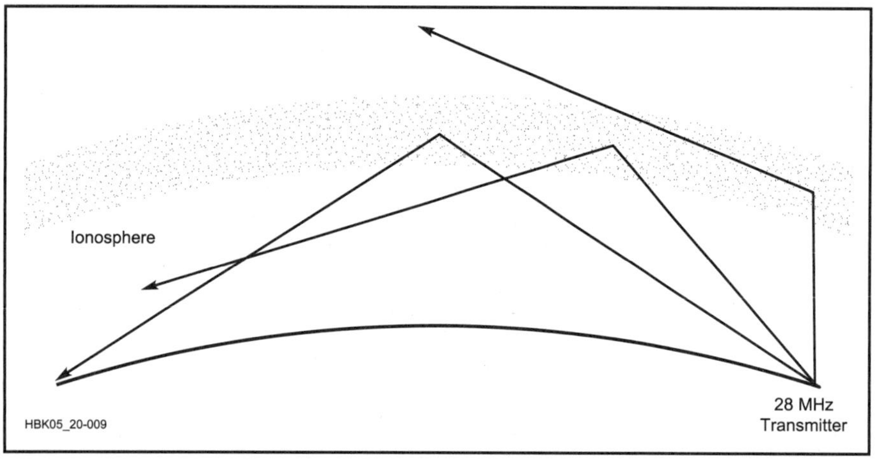

Figure 4.12 — Signals at the MUF propagated at a low angle to the horizon provide the longest possible one-hop distances. In this example, 28 MHz signals entering the ionosphere at higher angles are not refracted enough to bring them back to Earth.

maximum distance is about 4,000 kilometers (2,500 miles) for the F_2 region and about 2,000 kilometers (1,250 miles) for the E region. See **Figure 4.12**. This is illustrated more completely in **Figure 4.13** which gives maximum single-hop distances for E and F region propagation at various wave angles.

The MUF is a function of path, time of day, season, location, solar UV and X-ray radiation levels and ionospheric disturbances. For vertically incident waves, the MUF is the same as the critical frequency. The seasonal variation of the MUF is due to the ratio of atomic oxygen to molecular nitrogen (O/N_2) in the atmosphere. Atomic oxygen is critical for electron production, whereas molecular nitrogen is critical for the recombination of electrons and ions. During the winter months in the northern hemisphere, the ratio of O/N_2 is higher, resulting in higher daytime MUFs.

Precisely speaking, a maximum usable frequency or MUF is defined for communication between two specific points on the Earth's surface, for the conditions existing at the time, including the minimum elevation angle that the station can launch at the frequency in use. (This practical form of MUF is sometimes called the *operational MUF*). At the same time and for the same conditions, the MUF from either of these two points to a third point may be different.

Therefore, the MUF cannot be expressed broadly as a single frequency, even for any given location at a particular time. The ionosphere is never uniform, and in fact at a given time and for a fixed distance, the MUF changes significantly with changes in compass direction for almost any point on the Earth. Under usual conditions, the MUF will always be highest in the direction toward the Sun — to the east in the morning, to the south at noon (from northern latitudes), and to the west in the afternoon and evening.

While vertical-incidence critical frequencies from ionosondes are interesting from a scientific point of view, hams are far more concerned about how we can exploit propagation conditions to communicate, preferably at long distances. The MUF for a 4,000-kilometer (2,500 mile) distance using the F_2 region is about 3.5 times the vertical-incidence critical f_oF_2 frequency existing at the path midpoint. For one-hop signals, if a uniform ionosphere is assumed, the MUF decreases with shorter distances along the path. This is true because the higher-frequency waves must be launched at higher elevation angles for shorter ranges, and at these launch angles they are not bent sufficiently to reach the Earth. Thus, a lower frequency (where more bending occurs) must be used.

M FACTOR

It is important to realize that the takeoff angle from an antenna is not the same as the angle-of-incidence to the refracting region of the ionosphere. (See **Figure 4.14**) The angle of incidence on the ionosphere of a wave launched from the ground depends on the launch angle and the height of the ionospheric region. Due to the spherical geometry of the Earth-ionosphere system, the angle of incidence on the ionosphere does not approach zero as the antenna launch angle approaches

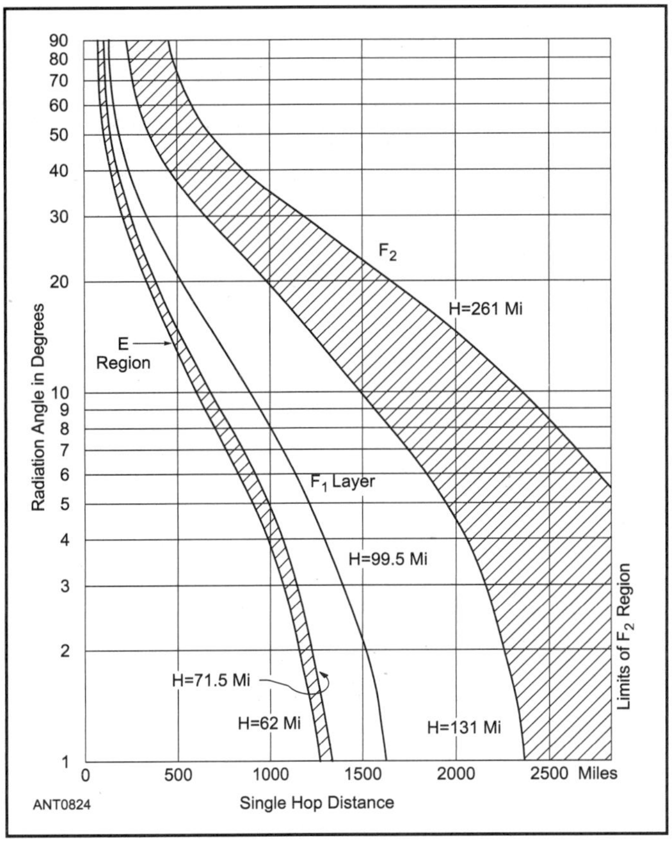

Figure 4.13 — Distance plotted against wave angle (one-hop transmission) for the nominal range of heights for the E, F_1, and F_2 regions.

zero. As a result, for path lengths at the limit of one-hop propagation, the MUF can be several times the critical frequency.

The ratio between the MUF and the critical frequency is known as the *M-factor*. The M-factor can be estimated using simple geometry in a spherical model of the Earth-ionosphere system. As the launch angle approaches zero, the angle of incidence to the ionosphere is limited to approximately 19° and 11° for the F_2 and E regions, respectively. This limits the M-factor to approximately 3 and 5 for these two regions (from the equation MUF = 1 / the sine of the angle of incidence on the ionosphere). See **Table 4.2** for typical M-factors of the various regions.

FREQUENCY OF OPTIMUM TRAFFIC (FOT)

Our understanding of the variation of the ionosphere is statistical in nature. Our predictions of the MUF are monthly median values, which imply a 50% probability. To achieve a higher probability, you should move down in frequency until the probability of the path is greater than 50%. Moving down to a frequency that is 0.85 times the monthly median MUF gives a probability of 90%. This lower frequency is the FOT — *frequency of optimum traffic*. This is illustrated in **Figure 4.15**.

LOWEST USEABLE FREQUENCY (LUF)

The MUF at any time on a particular path is just that — the *maximum* usable frequency. Frequencies below the MUF will also propagate along the path, but ionospheric absorption and noise at the receiving location (due to man-made noise and/or noise from local or distant thunderstorms) may make the received signal-to-noise ratio too low to be usable. This happens because signal absorption increases proportionally to the square of the decrease in frequency. In this case, the frequency is said to be below the *lowest usable frequency* (LUF). The frequency nearest the point where reception became unusable would be the LUF. This occurs most frequently

Table 4.2
Maximum Usable Frequency Factors (M-factors) for 2000 km E Hops and 3000 km F Hops

Region	*Maximum Critical Frequency (MHz)*	*M-factor*	*Useful Operating Frequencies (MHz)*
F_2	15.0	3.3-4.0	1-60
F_1*	5.5	4.0	10-20
E*	4.0	4.8	5-20
Es	30.0	5.3	20-160

*Daylight only

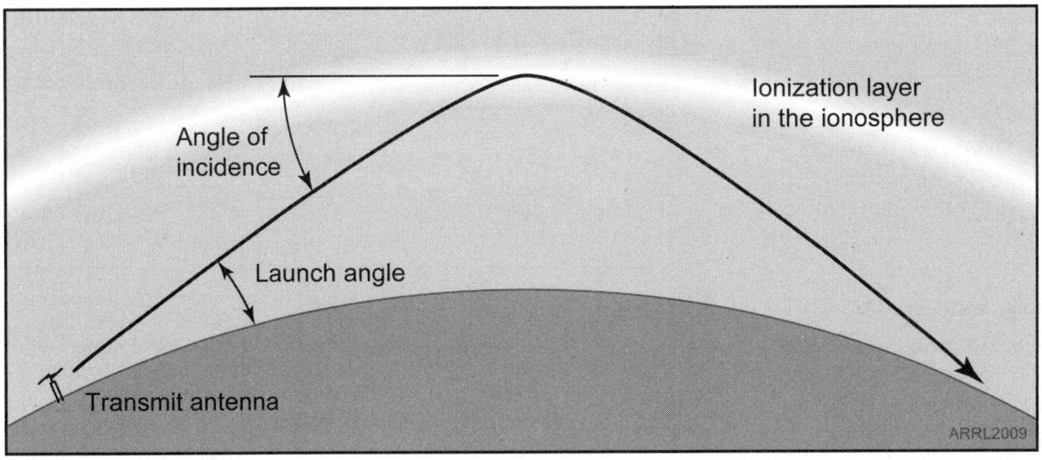

Figure 4.14 — A wave's launch angle at the transmitting antenna is not the angle of incidence for the wave at the ionosphere because of the spherical nature of the Earth-Ionosphere geometry.

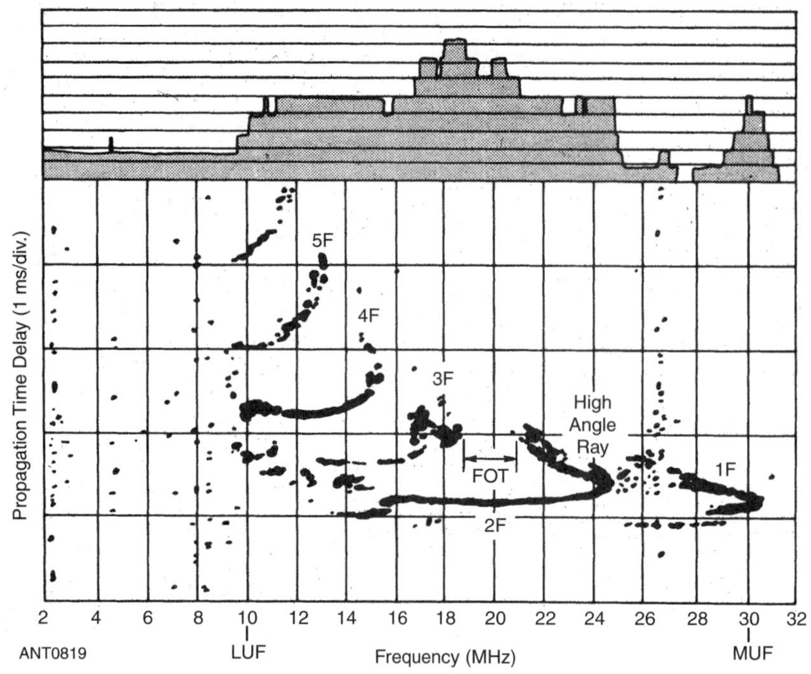

Figure 4.15 — HF oblique-sounder ionogram. This is a typical oblique ionosonde measurement on a 2,500-mile path from Hawaii to southern California during midmorning in March at a medium level of solar activity. Six possible modes (hops) are shown. The "FOT" is the frequency of optimum traffic, considered most reliable for this path/time.

below 10 MHz, where atmospheric and man-made noises are most troublesome.

The LUF can be lowered somewhat by the use of high power and directive antennas, or through the use of communication modes that permit reduced receiver bandwidth or are less demanding of SNR — CW or PSK31 instead of SSB, for example. Digital modes such as JT65 and FT8 that use coding techniques to reject noise can extend the LUF below what is available to analog modes, as well. This is not true of the MUF, which is limited by the physics of ionospheric refraction, no matter how high your transmitter power or how narrow your receiver bandwidth.

The LUF can be higher than the MUF. When the LUF is higher than the MUF, there is no frequency that supports communication on the particular path at that time.

For example, when solar activity is very high at the peak of a solar cycle, the LUF often rises higher than 14 MHz on the morning Eastern US-to-Europe path on 20 meters. Just before sunrise in the US, the 20-meter band will be first to open to Europe, followed shortly by 15 meters, and then 10 meters as the Sun rises further. By mid-morning, however, when 10 and 15 meters are both wide open, 20 meters will become very marginal to Europe, even when both sides are running maximum legal power levels. By contrast, stations on 10 meters can be worked readily with a transmitter power of only 1 or 2 W, indicating the wide range between the LUF and the MUF.

4.2.6 Skip Zone and Skip Distance

Figure 4.16 shows we can communicate with the point on the Earth labeled "A" (where Wave #3 arrives), but not any closer to our transmitter site. When the critical angle is less than 90° (that is, directly overhead) there will always be a region around the transmitting site where an ionospherically propagated signal cannot be heard or is heard weakly.

The area between the outer limit of the ground wave range and the inner edge of signals returning from the ionosphere is the *skip zone*. The outer edge of the skip zone is the distance between the originating site and the beginning of the ionospheric return and is called the *skip distance*. This terminology should not be confused with ham jargon such as "skip is in," referring to the fact that a band is open for skywave propagation. Since both skip distance and maximum ground wave distance change with frequency, the size of the skip zone also changes with frequency.

The size of the skip zone is closely related to MUF. When two stations are unable to communicate with each other on a

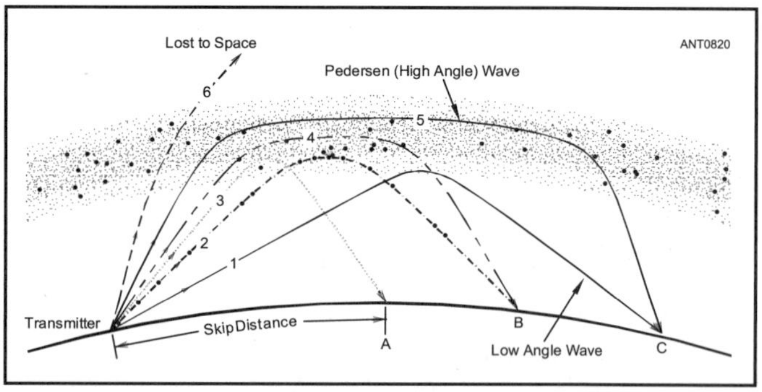

Figure 4.16 — Very simplified smooth-Earth/ionosphere diagram showing how the ground range from transmitter to receiver can vary as the elevation angle is gradually raised. The Pedersen wave, launched at a relatively high angle, has the same ground range as the low-angle wave #1. It may be slightly weaker and less stable, having traveled for a long distance in the ionosphere.

particular frequency because the ionosphere is unable to refract the signal enough from one to the other through the required angle — that is, the operating frequency is above the MUF — the stations are in the skip zone for that frequency. Stations within the skip zone may be able to work each other on a lower frequency, by ground wave, or by other mechanisms if they are close enough. There is no skip zone at frequencies below the MUF for that path. The signal may often be heard to some extent within the skip zone through various forms of scattering (discussed in detail later), but it will ordinarily be marginal in strength. When the skip distance is short, both ground wave and skywave signals may be received near the transmitter. In such instances the skywave frequently is stronger than the ground wave, even as close as a few miles from the transmitter. The ionosphere is an efficient communication medium under favorable conditions.

Comparatively, the ground wave is not efficient. If the radio wave leaves the Earth at an angle of zero degrees, just at the horizon, the maximum distance that may be reached under usual ionospheric conditions in the F_2 region is about 4,000 kilometers (2,500 miles). This 4,000-kilometer value is for the higher HF frequencies. Lower frequencies result in shorter hops.

For signals below the critical frequency, there is no skip zone, making these frequencies very useful for regional communication. Deliberately launching a signal at high elevation angles in order to cover a region around the transmitting antenna is very useful for emergency and disaster relief communications, for example. This type of propagation is referred to as *near vertical incidence skywave* (NVIS) and is described in a subsequent section.

4.2.7 Multi-Hop Propagation

As mentioned previously in the discussion about Figure 4.16, the Earth itself can act as a reflector for radio waves, resulting in multiple hops. Thus, a radio signal can be reflected from the reception point on the Earth back into the ionosphere, reaching the Earth a second time at a still more-distant point. This effect is illustrated in **Figure 4.17**, where a single ionospheric region is depicted, although this time we show both the region and the Earth beneath it as curved rather than flat. The wave identified as "Critical Angle" travels from the transmitter via the ionosphere to point A, in the center of the drawing, where it is reflected upwards and travels through the ionosphere to point B, at the right. This shows a two-hop signal path.

As in the simplified case in Figure 4.16, the distance at which a ray eventually reaches the Earth depends on the launch elevation angle at which it left the transmitting antenna.

The information in Figure 4.17 is greatly simplified. On actual communication paths the picture is complicated by many factors. One is that the transmitted energy spreads over a considerable area after it leaves the antenna. Even with an antenna array having the sharpest practical beam pattern, there is what might be described as a *cone of radiation* centered on the wave lines (rays) shown in the drawing. The reflection/refraction in the ionosphere is also highly variable and is the cause of considerable spreading and scattering.

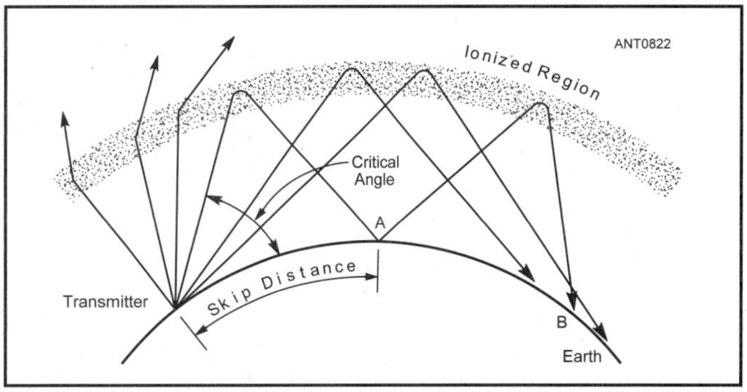

Figure 4.17 — Behavior of waves encountering a simple curved ionospheric region over a curved Earth. Rays entering the ionized region at angles above the critical angle are not bent enough to be returned to Earth and are lost to space. Waves entering at angles below the critical angle reach the Earth at increasingly greater distances as the launch angle approaches the horizontal. The maximum distance that may normally be covered in a single hop is 4,000 kilometers for low elevation angle signals at the upper end of HF. Greater distances are covered with multiple hops. Lower frequencies have shorter maximum hop distances due to more bending in the ionosphere.

Under some conditions it is possible for as many as four or five signal hops to occur over a radio path, as illustrated by the oblique ionogram in Figure 4.15. But no more than two or three hops is the norm. In this way, HF communication can be conducted over many thousands of miles.

An important point should be recognized with regard to signal hopping. A significant loss of signal occurs with each hop — especially at lower HF frequencies. The D and E regions of the ionosphere absorb energy from signals as they pass through, and the ionosphere tends to spread and scatter the radio energy in various directions, rather than confining it in a tight bundle. The roughness of the Earth's surface also scatters the energy at the Earth reflection point.

Assuming that both waves do reach point B in Figure 4.17, the low-angle wave will contain more energy at point B. This wave passes through the lower regions just twice, compared to the higher-angle route, which must pass through these regions four times, plus encountering an Earth reflection. Measurements indicate that although there can be great variation in the relative strengths of the two signals — the one-hop signal will generally be from 7 to 10 dB stronger, although many amateurs may not be capable of transmitting at these low angles. The nature of the terrain at the mid-path reflection point for the two-hop wave, the angle at which the wave is reflected from the Earth, and the condition of the ionosphere in the vicinity of all the refraction points are the primary factors in determining the signal-strength ratio.

The loss per hop becomes significant at greater distances. It is because of these losses that no more than four or five propagation hops are useful; the received signal becomes too weak to be usable over more hops. Although modes other than signal hopping also account for the propagation of radio waves over thousands of miles, backscatter studies of actual radio propagation have displayed signals with as many as five hops. So, the hopping mode is arguably the most prevalent method for long-distance communication.

Figure 4.18 shows another way of looking at propagation — at a *geographic area*. Figure 4.18 shows 15-meter signal levels across the US as they propagate from a transmitting station in San Francisco. This simulation of propagation conditions is for the month of November, with a medium level of solar activity (SSN = 50) at 22 UTC. Figure 4.18 was created using the *VOAAREA* software program, part of the *VOACAP* software suite. Transmitter power is assumed to be 1,500 W, with 3-element Yagis, 55 feet high, at the transmitter and at each receiving location.

From the transmitter out to about 50 miles, signals are moderate, at about S-5 on an S meter. Beyond that coverage area to almost ⅓ of the way across the country (to Colorado), there is a large and distinctive skip zone, where only very weak signals return to Earth (S-1 or less). Beyond Colorado, signals rapidly build up to S-9+10 dB across the middle of the US, falling to S-9 and then to S-7 in the vicinity of Chicago, Illinois. Beyond Chicago, the signals drop to S-5 in a swath from Michigan and part of Ohio down to Alabama. All along the US East Coast, signals come back strong at S-9.

The reason why the signals in Figure 4.18 drop down to S-5 in the Midwest is that the necessary elevation angles to cover this region in a single F_2 hop are extremely low even at a moderate level of solar activity. To achieve launch angles as low as 1 degree requires either very high antenna heights or a high mountaintop location. Beyond the Midwest, out to the US East Coast, two F_2 hops are required, with higher elevation angles and hence greater antenna gain for moderate antenna heights.

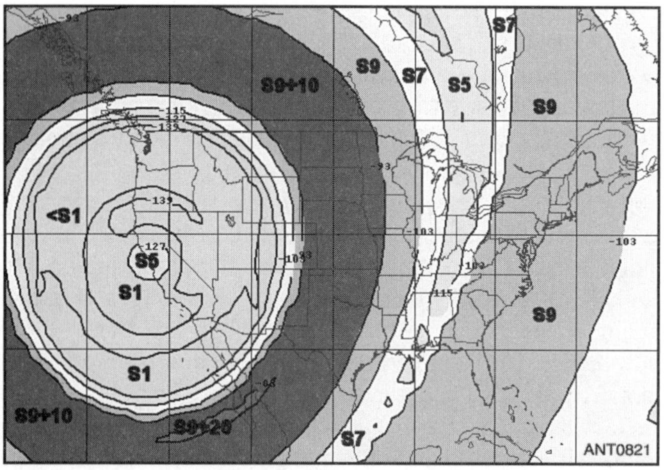

Figure 4.18 — Modified *VOAAREA* plot for 21.2 MHz from San Francisco to the rest of the US, annotated with signal levels in S units, as well as signal contours in dBW (dB below a watt). Antennas are assumed to be 3-element Yagis at 55 feet above flat ground; the transmitter power is 1,500 W; the month is November with SSN=50, a moderate level of solar activity, at 22 UTC. The most obvious feature is the large "skip zone" centered on the transmitter in San Francisco, extending almost one third of the distance across the US.

4.2.8 NVIS Propagation

The purpose of *near vertical incidence skywave* (*NVIS*) propagation is to bridge the gap between where ground wave is too weak and where the skip zone ends. Because NVIS communication operates with high angles of incidence to the ionosphere, signals return to Earth relatively close to the transmitter. By using frequencies close to and below the critical frequency and antennas that produce radiation at high elevation angles, NVIS operation exhibits no skip zone and communications can be maintained over these relatively short distances. (See the previous section on Ionospheric Refraction for an explanation of critical frequency.)

Frequency selection is important since absorption in the ionosphere increases as frequency decreases. By operating close to the critical frequency, absorption is minimized. This requires knowing the critical frequency which is continually changing from day to day and hour to hour. Propagation modeling software can be used to predict the critical frequency for propagation between two locations. You can also use real-time ionosonde data as discussed previously if the data is representative of your location and the desired communication path. You should expect to change bands as conditions change to remain close to the critical frequency.

A desirable NVIS antenna radiation pattern is one in which most of the radiation is upward at elevation angles greater than 45 degrees. This is referred to as *NVIS directivity*, defined as the average directivity for elevation angles between 70° and 90°. To maximize NVIS directivity, the most common antenna is a $\lambda/2$ dipole mounted approximately 0.2 λ above the ground. Lower heights incur greater ground losses and larger heights result in more low-angle radiation such that NVIS performance begins to suffer.

Figure 4.19 summarizes the relationship of antenna height, NVIS directivity, and ground loss at 5.4 MHz. NVIS directivity varies only slowly with height with an optimum at 0.09 λ, whereas the antenna gain has a distinct optimum at 0.19 λ, sharply decreasing at low heights due to excessive ground loss. The same general relationship exists for the other amateur bands below 10 MHz, leading to the overall recommendation for antenna heights of approximately 0.2 λ for best NVIS performance. (This paragraph and figure are from the referenced 2015 article by Witvliet and others — see below.) In the December 2005 issue of *QST*, Dean Straw, N6BV, used the *VOACAP* propagation prediction program to analyze a variety of NVIS paths centered on San Francisco. His analysis showed area coverage maps (signal strength contours versus distance from the transmitter) and elevation patterns of antennas at various heights. His analysis allowed him to formulate a very nice summary: *As a rule-of-thumb, for ham band NVIS, I would recommend that 40 meters be used during the day, 80 meters during the night.* Additionally, during a winter night near solar minimum, changing to 160 meters may be necessary.

Another way to summarize this would be to say that 40 meters (7 MHz) and 60 meters (5 MHz) are good for NVIS propagation out to a couple hundred miles during the day, with 40meters better at solar maximum and 60 meters better

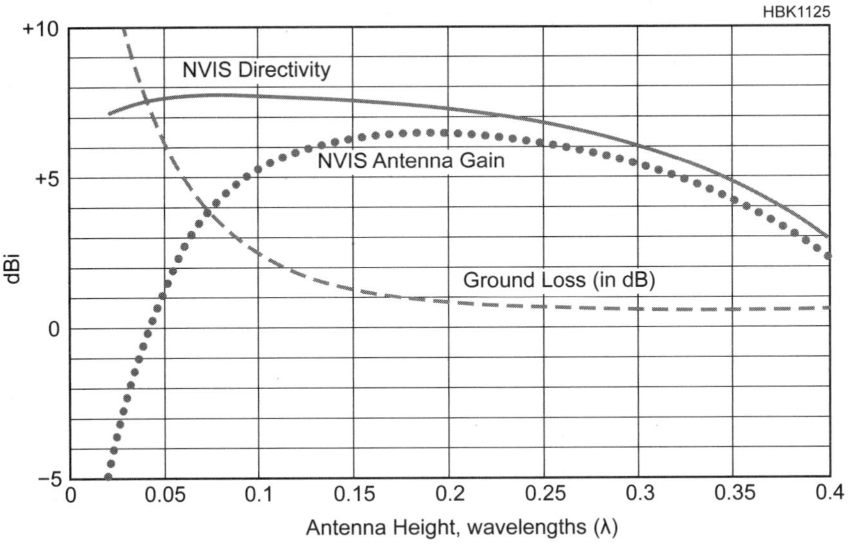

Figure 4.19 — The relationship between antenna height, ground loss (farmland soil), and NVIS directivity at 5.4 MHz. The same general relationship exists on the lower HF bands and for different types of soil. (Graph from Witvliet et al.)

at solar minimum. Likewise, 80 meters (4 MHz, aka 75 meters) and 160 meters (1.9 MHz) are good for NVIS propagation out to several hundred miles during the night, with 80 meters better at solar maximum and 160 meters better at solar minimum (especially during the winter months). This summary takes into account both the MUF and ionospheric absorption.

Along with N6BV's *QST* article, there are several excellent articles that apply to the amateur radio use of NVIS. Ed Farmer, AA6ZM, contributed an overview of NVIS techniques, including an analysis of antenna type and pattern with height, in January 1995 *QST* which is also listed in the References section. Perhaps the most complete current article on NVIS is from 2015: "Near Vertical Incidence Skywave Propagation: Elevation Angles and Optimum Antenna Height for Horizontal Dipole Antennas" by Witvliet and others in the *IEEE Antennas and Propagation Magazine*. It presents a comprehensive discussion of propagation, coverage, and antenna systems. Tom Kamp, DF5JL, also studied NVIS on the 60 meter band in Germany. See the References entry for his article in *CQ DL*. NVIS antenna systems are also discussed in this book's chapter on **HF Antenna System Design**.

4.2.9 E Region Propagation

The lowest portion of the ionosphere useful for long-distance communication by amateurs, the *E region* lies between 90 and 150 kilometers (60 and 90 miles) altitude, but a narrower region centered at 95 to 120 kilometers (60 to 70 miles) is more important for radio propagation. In the E region, nitric oxide and molecular oxygen are ionized by short-wavelength UV and long-wavelength X-ray radiation (so-called soft X-rays). Normally, the E region exists primarily during daylight hours, because, like the D region, it requires a constant source of ionizing radiation. Recombination is not as fast as in the denser D region and absorption is much less. The E region has a daytime critical frequency that varies between 3 and 4 MHz with the solar cycle. At night, the normal E region decays to a minimum critical frequency of 0.3 – 0.5 MHz, which is still enough to refract low elevation angle 1.8 MHz signals.

DAYTIME AND NIGHTTIME E REGION

In the E region, at intermediate atmospheric density, ionization varies with the Sun angle above the horizon (the solar zenith angle), but solar EUV radiation is not the sole ionizing agent. Solar X-rays, meteors, and meteoroids entering this portion of the Earth's atmosphere also play a part. Ionization increases rapidly after sunrise, reaches maximum around noon local time, and drops off quickly after sundown. The minimum is after midnight, local time. As with the D region, the E region absorbs wave energy in the lower-frequency amateur bands when the Sun angle is high,

The E region plays a small role in propagating HF signals but can be a major factor limiting propagation during daytime hours. Its usual critical frequency of 3 to 4 MHz, with an M-factor of about 5, suggests that single-hop E region skip might be useful between 5 and 20 MHz at distances up to 2,300 kilometers (1,400 miles). In practice this is not the case, because the potential for E region skip is severely limited by D region absorption. Signals radiated at low angles at 7 and 10 MHz, which might be useful for the longest-distance contacts, are largely absorbed by the D region. Only high-angle signals pass through the D region at these frequencies, but high-angle E region skip is typically limited to 1,200 kilometers (750 miles) or so. Signals at 14 MHz penetrate the D region at lower angles at the cost of some absorption, but the casual operator may not be able to distinguish between signals propagated by the E region or higher-angle F region propagation.

E REGION BLANKETING

On 20 meters during the summer, on 40 and 30 meters around mid-day and on 160 meters mostly during the evening hours or well into the night in summer, the E region can block a signal from getting to the higher F region that offers longer hops. This is known as *E region blanketing*, in reference to the E region acting as a blanket that signals cannot get through. This is often mistakenly considered to be ionospheric absorption from the D region.

E region blanketing is the primary reason why long-distance propagation is uncommon during midday on the 40- and 30-meter bands and during the summer on 20 meters. D region absorption — to a much lesser degree — also degrades midday long-distance propagation on 40 and 30 meters, mainly by absorbing most low angle, multi-hop, E region propagation.

AURORA

If the orientation of the interplanetary magnetic field from a powerful CME is aligned southward (opposite to that of the Earth's geomagnetic field, indicated by negative reported values of Bz) and if the magnetic field strength is greater than about 5 nanoteslas and *if both conditions* persist for about two hours or more, the geomagnetic field can partially collapse and the particles normally trapped there can be deposited into the Earth's upper atmosphere. This produces a visible or radio *aurora*. An aurora is visible if the time of entry is after dark.

The visible aurora is an optical signature of the plasma curtain capable of refracting radio waves in the range of frequencies above about 20 MHz. D-region absorption increases between midnight and sunrise on lower frequencies during auroras. The exact frequency ranges depend on many factors: time, season, position with relation to the Earth's auroral regions, and the level of solar activity at the time, to name a few.

Radar signals as high as 3 GHz have been scattered by the *aurora borealis* or northern lights (*aurora australis* in the Southern Hemisphere), but amateur aurora contacts are common only from 28 through 432 MHz. By pointing directional antennas generally northward toward the center of aurora activity, oblique paths between stations up to 2,300 kilometers (1,400 miles) apart can be completed. See **Figure 4.20**. High power and large antennas are not necessary. Stations with small Yagis and as little as 10 W output have used auroras on frequencies as high as 432 MHz, but contacts at 902 MHz and higher are exceedingly rare. Auroral propagation works just as well in the Southern Hemisphere, in which case antennas must be pointed generally southward.

AURORAL PROPAGATION AND ABSORPTION

The auroral effect on VHF waves is another amateur discovery, this one dating back to the 1930s. The discovery came coincidentally with improved transmitting and receiving techniques then. The returning signal is diffused in frequency by the diversity of the auroral curtain as a refracting (scattering) medium. The result is a modulation of a CW signal, from just a slight burbling sound to what is best described as a "keyed roar." Before SSB took over in VHF work, voice was all but useless for auroral paths. A sideband signal suffers, too, but its narrower bandwidth helps to retain some degree of understandability. Distortion induced by a given set of auroral conditions increases with the frequency in use. 50-MHz signals are much more intelligible than those on 144 MHz on the same path at the same time. On 144 MHz, CW is almost mandatory for effective auroral communication.

It is not necessary to see an aurora to make auroral contacts. Useful auroras may be 500 – 1,000 kilometers (310 – 620 miles) away and below the visual horizon. Antennas should be pointed generally north and then moved east and west to peak signals, because auroral ionization is field aligned. This means that for any pair of stations, there is an optimal direction for aurora scatter. Offsets from north are usually greatest when the aurora is closest and often provide the longest contacts.

There may be some advantage to antennas that can be elevated, especially when auroras are high in the sky. The antenna requirements for auroral work are mixed. High gain helps, but the area of the aurora yielding the best returns sometimes varies rapidly, so sharp directivity can be a disadvantage. So could a very low radiation angle, or a beam pattern very sharp in the vertical plane. Experience indicates that few amateur antennas are sharp enough in either plane to present a real handicap. The beam heading for maximum signal can change, however, so a bit of scanning in azimuth may turn up some interesting results. A very large array, such as is commonly used for moonbounce (with azimuth-elevation control), should be worthwhile.

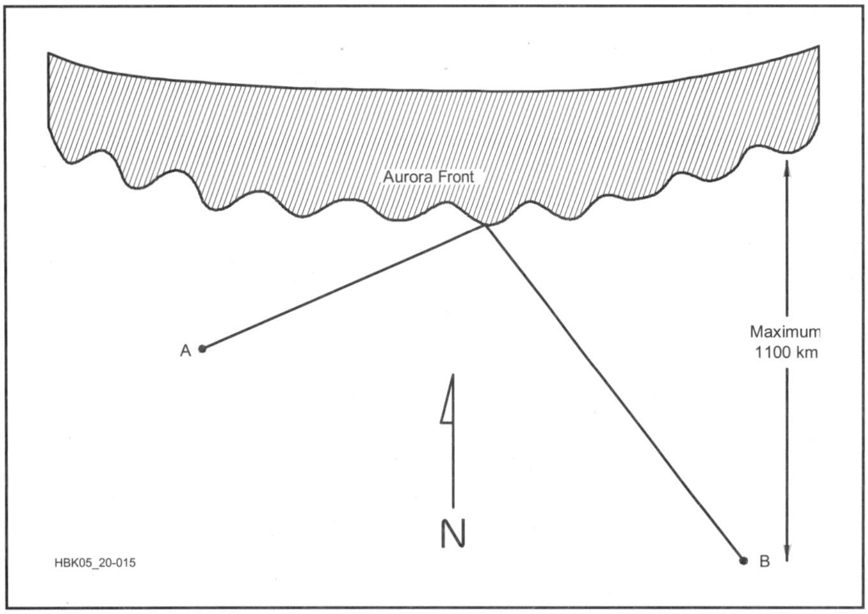

Figure 4.20 — Point antennas generally north to make oblique long-distance contacts on 28 through 432 MHz via aurora scattering. Optimal antenna headings may shift considerably to the east or west depending on the location of the aurora. This necessitates moving your antenna azimuth direction for best propagation and continuing to do so as the aurora progresses.

4.2.10 F Region Propagation

Most of our long-distance communication capability stems from the tenuous outer reaches of the Earth's atmosphere known as the *F region*. At heights above 100 miles, ions and electrons recombine more slowly, allowing the region to hold its ability to reflect wave energy back to Earth well into the night.

The F region, from 150 kilometers (90 miles) to over 400 kilometers (250 miles) altitude, is by far the most important for long-distance HF communications. F-region oxygen atoms are ionized primarily by ultraviolet radiation. During the day, ionization reaches maxima in two distinct regions, called F_1 and F_2. The F_1 region forms between 150 and 250 kilometers (90 and 160 miles), is most prevalent in the summer months, and disappears at night. The F_2 region extends above 250 kilometers (160 miles), with a peak of ionization around 300 kilometers (190 miles). At night, F-region ionization collapses into one broad region at 300 – 400 kilometers (190 – 250 miles) altitude. Ions recombine very slowly at these altitudes because atmospheric density is relatively low. Maximum ionization levels change significantly with time of day, season, and year of the solar cycle.

The region's height may be from 160 to more than 500 kilometers (100 to over 310 miles), depending on the season of the year, the latitudes, the time of day, and, most capricious of all, what the Sun has been doing in the last few minutes and in perhaps the last three days before the attempt is made. The MUF between Eastern US and Europe, for example, has been anything from 7 to 70 MHz, depending on the conditions mentioned above, plus the point in the long-term solar-activity cycle at which the check is made. The MUF between North America and Europe frequently drops below 7 MHz during the fall and winter during solar minimum years, mainly because of the electron depletion effects of the mid latitude trough which is discussed later.

Propagation information tailored to amateur needs is published by the ARRL via propagation bulletins that are issued every Friday — they are archived at **arrl.org/w1aw-bulletins-archive-propagation**. Finally, solar and geomagnetic field data, transmitted hourly and updated eight times daily, are given in brief bulletins carried by the US Time Standard stations, WWV (18 minutes past the hour) and WWVH (45 minutes past the hour), and also on internet websites. More information on these services is provided later.

F_1 REGION

The daytime F_1 region is not very important to HF communication. In reality it is an inflection point in the electron density profile, not a peak. It exists only during daylight hours and is largely absent in winter. Radio signals below 10 MHz are not likely to reach the F_1 region, because they are either absorbed by the D region or refracted by the E region. Signals higher than 20 MHz that pass through both of the lower ionospheric regions are likely to pass through the F_1 region as well, because the F_1 MUF rarely rises above 20 MHz. Absorption diminishes the strength of any signals that continue through to the F_2 region during the day. Some useful F_1 region refraction may take place between 10 and 20 MHz during summer days, yielding paths as long as 3,000 kilometers (1,900 miles), but these would be practically indistinguishable from F_2 skip. The F_1 region often blankets F_2 region daytime 20-meter propagation during the summer, preventing most mid-day long distance propagation. Hams often mistakenly refer to this as "absorption."

F_2 REGION

The F_2 region forms between 250 and 400 kilometers (160 and 250 miles) during the daytime and persists throughout the night as a single consolidated F region 50 kilometers (30 miles) higher in altitude. Typical ion densities are the highest of any ionospheric region, with the possible exception of some unusual E region phenomena. In contrast to the other ionospheric regions, F_2 ionization varies considerably with time of day, season, and position in the solar cycle, but it is never altogether absent. These two characteristics make the F_2 region the most important for long-distance HF communications.

The F_2 region MUF is nearly a direct function of ultraviolet (UV) solar radiation, which in turn closely follows the solar cycle. During the lowest years of the cycle, the daytime MUF may climb above 14 MHz for only a few hours a day. In contrast, the MUF may rise beyond 50 MHz during peak years of strong solar cycles and stay above 14 MHz throughout the night. The virtual height of the F_2 region averages 330 kilometers (210 miles) but varies between 200 and 400 kilometers (120 and 250 miles). Maximum one-hop distance is about 4,000 kilometers (2,500 miles). Near vertical incidence skywave propagation just below the critical frequency provides reliable coverage out to 200 – 300 kilometers (120 – 190 miles) with no skip zone. It is most often observed on 7 MHz during the day.

At any given location on Earth, in general both F_2 region ionization and MUF at that point build rapidly at sunrise, usually reaching a maximum after local noon, and then decrease to a minimum at night prior to the next sunrise. Depending on the season, the MUF is generally highest on paths directed within 20 degrees of the Equator and lower toward the poles. For this reason, transequatorial paths may be open at a particular frequency when all other paths are closed, especially when TEP propagation is available during late afternoon and evening.

MULTI-HOP F REGION PROPAGATION

Most HF communication beyond 4,000 kilometers (2,500 miles) takes place via multiple ionospheric hops. Radio signals are reflected from the Earth back toward the ionosphere for additional ionospheric refractions. A series of ionospheric refractions and terrestrial reflections commonly create paths half-way around the Earth. Each hop involves additional attenuation and absorption, so the longest-distance signals tend to be the weakest.

Multi-hop paths can take on many different configurations, as shown in the examples of Figure 4.20. E region (especially sporadic E) and F region hops may be mixed. In practice, multi-hop signals arrive via many different paths, which often increases the problems of fading. Analyzing multi-hop paths is complicated by the effects of D- and E-region absorption, possible reflections from the tops of sporadic E regions, disruptions in the auroral zone, and other phenomena.

In general, when a band is opening and closing as the MUF changes, extremely low elevation angles are dictated (less than 5 degrees). During the main part of the opening, the elevation angle is higher — generally in the range of 5 to 20 degrees. This is why stations with extremely high antennas (which have higher gain at lower takeoff angles) perform better at band openings and closings.

It is possible for signals to be propagated completely around the world and arrive back at their originating point. Multiple reflections within the F region may at times bypass ground reflections altogether, creating what are known as *chordal hops*, with lower total attenuation. It takes a radio signal about 0.14 seconds to make an around-the-world trip.

F-REGION LONG PATH

Most HF communication takes place along the shortest great-circle path between two stations. Short-path propagation is always less than 20,000 kilometers (12,000 miles) — halfway around the Earth. Nevertheless, it may be possible at times to make the same contact in exactly the opposite direction via the *long path*. The long-path distance will be 40,000 kilometers (25,000 miles) minus the short-path length. Signal strength via the long path is usually considerably less than the more direct short path. When both paths are open simultaneously, there may be a distinctive sort of echo on received signals. The time interval of the echo represents the difference between the short-path and long-path distances.

Sunlight is a required element in long-haul communication via the F region above about 10 MHz, but due to recombination being a slow process after sunset, propagation may be supported in the dark ionosphere for many hours after sunset. This fact tends to define long-path timing and antenna aiming. Both are essentially the reverse of the "normal" for a given circuit. We know also that salt-water paths work better than overland ones. This can be significant in long-path work.

F-REGION SCATTER MODES

Special forms of F-region scattering can create unusual paths within the skip zone. *Backscatter* and *sidescatter* signals are usually observed just below the MUF for the direct path and allow communications not normally possible by other means.

Two stations communicating over a single ionospheric hop can often be heard to some degree by other stations at almost any point along the way. Some of the wave energy is scattered in all directions, including back to the starting point and farther. Backscatter and sidescatter are closely related and the terminology does not precisely distinguish between the two. Backscatter usually refers to single-hop signals that have been scattered by the Earth or the ocean at some distant point back toward the transmitting station.

Stations using backscatter point their antennas toward a common scattering region at the one-hop distance, rather than toward each other. Backscattered signals are generally weak and have the characteristic "hollow" sound of signals combining from many paths. Useful communication distances range from 100 kilometers (60 miles) to the normal one-hop distance of 4,000 kilometers (2,500 miles).

Backscatter functions like a sort of HF ionospheric radar. **Figure 4.21** shows a schematic for a simple backscatter path. The signal launched from point A travels through the ionosphere back to earth at Point S, the scattering point. Here, the rough terrain of the land or ocean waves scatters signals in many directions, one of which propagates a weak signal back through the ionosphere to land at point B. Point B would normally be in the no-signal skip zone between A and S. Two stations spaced a few hundred kilometers apart can often communicate via a backscatter path near the MUF. See **Figure 4.22**.

Because backscatter involves mainly scattering from the Earth at the point where the strong ionospherically propagated signal comes down, it is a part of HF over-the-horizon radar techniques. Amateurs using sounding techniques have shown that you can tell to what part of the world a band is usable (single-hop F) by probing the backscatter with a directive antenna and high transmitter power, especially when the Earth contact point is over the open ocean. In fact, that's where the mode is at its best, because ocean waves can be efficient backscatter reflectors. Sidescatter is similar to backscatter, except the ground scatter zone is off the direct line between participants. A typical example, often observed during the lowest years of the solar cycle, is communication on 28 MHz between the eastern US (and adjacent areas of Canada) and much of the European continent. Often, this may start as "backscatter chatter" between Europeans whose antennas are turned toward the Azores. Then suddenly the North Americans join the fun, perhaps for only a few minutes, but sometimes much longer, with beams also pointed

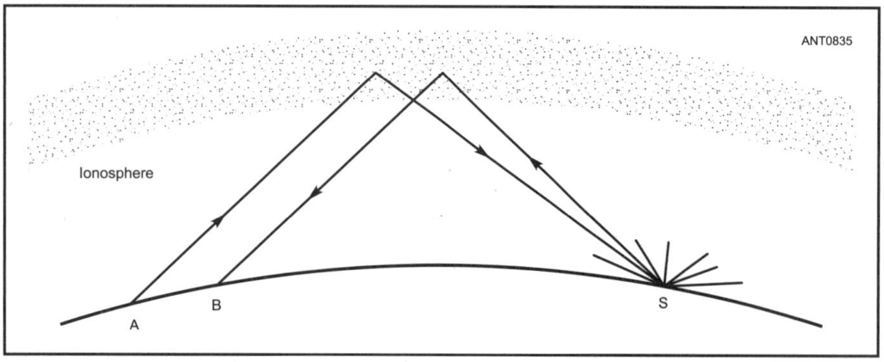

Figure 4.21 — Schematic of a simple backscatter path. Stations A and B are too close to make contact via normal F-region ionospheric refraction. Signals scattered back from a distant point on the Earth's surface (S), often the ocean, may be accessible to both A and B to create a backscatter circuit.

Propagation Summary, By Band

LOW FREQUENCY (LF) BANDS AND MEDIUM FREQUENCY (MF) BANDS

135.7 – 137.8 kHz (2200 meters) and 472-479 kHz (630 meters)

See the section Propagation Below 1.8 MHz.

1.8 – 2.0 MHz (160 meters)

160 meters suffers from daytime D region absorption. Daytime communication is limited to ground wave coverage and a single E hop out to about 1,500 kilometers for well-equipped stations (running the full legal limit, quarter-wave verticals with a good ground system, and a low noise receiving environment). At night, the D region quickly disappears and worldwide 160 meter communication becomes possible via F_2 region propagation. Atmospheric and man-made noise limits propagation. Tropical and mid latitude thunderstorms cause high levels of static in summer, making winter evenings the best time to work DX at 1.8 MHz. A proper choice of receiving antenna (Beverage, 4-square, small loop) can often significantly reduce the amount of received noise to improve the signal-to-noise ratio.

HIGH FREQUENCY (HF) BANDS (3 – 30 MHz)

A wide variety of propagation modes are useful on the HF bands. The 80 meter band shares many daytime characteristics with 160 meters. The 60 meter band is better for daytime communications than the 160- and 80 meter bands and is much more similar to the 40 meter band. The transition between bands primarily useful at night or during the day appears around 10 MHz. Most long-distance contacts are made via F_2 region skip. Above 21 MHz, more exotic propagation, including TE, sporadic E, aurora, and meteor scatter, begins to be practical.

3.5 – 4.0 MHz (80 meters for the lower end, 75 meters for the higher end)

The lowest HF band is similar to 160 meters in many respects. Daytime absorption is significant, but not quite as extreme as at 1.8 MHz. At night, signals are often propagated halfway around the world. As at 1.8 MHz, atmospheric noise is a nuisance, especially during the summer, making winter the most attractive season for the 80/75 meter DXer.

5.3 – 5.4 MHz (60 meters)

The distance covered during daytime propagation is very similar to the 40 meter band. At night, worldwide propagation is possible in spite of the relatively low power limit. Signal strengths will typically be higher than on 80 meters but not as high as on 40 meters.

7.0 – 7.3 MHz (40 meters)

The popular 40 meter band supports daytime NVIS propagation and E region propagation out to about 500 miles during most daylight hours. D region absorption is much less severe than on the 160- and 80 meter bands except at low elevation angles, so short-distance skip via the E and F regions is possible. During the day, a typical station can cover a radius of approximately 800 kilometers (500 miles). At night, reliable worldwide communication via F_2 is common on the 40 meter band.

Atmospheric noise is much less troublesome than on 160 and 80 meters, and 40 meter DX signals are often of sufficient strength to override even high-level summer static. For these reasons, 40 meters is the lowest-frequency amateur band considered reliable for DX communication in all seasons. Even during the lowest point in the solar cycle, 40 meters is usually open for worldwide DX throughout the night.

10.1 – 10.15 MHz (30 meters)

The 30 meter band is unique because it shares characteristics of both daytime and nighttime bands. D region absorption is not a significant factor. Communication up to 3,000 kilometers (1,900 miles) is typical during the daytime, and this extends halfway around the world via all-darkness paths. The band is generally open via F_2 on a 24-hour basis, but during a solar minimum, the MUF on many DX paths often drops below 10 MHz shortly after sunset. DX paths may drop below 10 MHz at night. Under these conditions, 30 meters adopts the characteristics of the daytime bands at 14 MHz and higher.

14.0 – 14.35 MHz (20 meters)

The 20 meter band is traditionally regarded as the amateurs' primary long-haul DX favorite. Regardless of the 11-year solar cycle, 20 meters can be depended on for at least a few hours of worldwide F_2 propagation during the day. During solar-maximum periods, 20 meters will often stay open to distant locations throughout the night. Skip distance is usually appreciable and is always present to some degree. Daytime E region and sporadic E propagation from sunrise through midnight propagation may be detected along very short paths. Atmospheric noise is not a serious consideration, even in the summer. Because of its popularity, 20 meters tends to be very congested during the daylight hours.

18.068 – 18.168 MHz (17 meters)

The 17 meter band is similar to the 20 meter band in many respects, but the effects of fluctuating solar activity on F_2 propagation are more pronounced. During the years of high solar activity, 17 meters is reliable for daytime and early-evening long-range communication, often lasting well after sunset. During moderate years, the band may open only during sunlight hours and close shortly after sunset. At solar minimum, 17 meters will open to middle and equatorial latitudes, but only for short periods during midday on north-south paths.

21.0 – 21.45 MHz (15 meters)

The 15 meter band has long been considered a prime DX band during solar cycle maxima, but it is sensitive to changing solar activity. During peak years, 15 meters

(Continued on next page.)

is reliable for daytime F_2 region DXing and will often stay open well into the night. During periods of moderate solar activity, 15 meters is basically a daytime-only band, closing shortly after sunset. During solar minimum periods, 15 meters may not open at all except for infrequent north-south transequatorial circuits. Sporadic E is observed occasionally in early summer and mid-winter, although the effects are not as pronounced as on the higher frequencies.

24.89 – 24.99 MHz (12 meters)

This band offers propagation that combines the best of the 10- and 15 meter bands. Although 12 meters is primarily a daytime band during low and moderate sunspot years, it may stay open well after sunset during the solar maximum. During years of moderate solar activity, 12 meters opens to the low and middle latitudes during the daytime hours, but it seldom remains open after sunset. Periods of low solar activity seldom cause this band to go completely dead, except at higher latitudes. Occasional daytime openings, especially in the lower latitudes, are likely over north-south paths. The main sporadic E season on 24 MHz lasts from late spring through summer and short openings may be observed in mid-winter.

28.0 – 29.7 MHz (10 meters)

The 10 meter band is well known for extreme variations in characteristics and a variety of propagation modes. During solar maxima, long-distance F_2 propagation is so efficient that very low power can produce strong signals halfway around the globe. DX is abundant with modest equipment. Under these conditions, the band is usually open from sunrise to a few hours past sunset. During periods of moderate solar activity, 10 meters usually opens only to low and transequatorial latitudes around noon. During the solar minimum, there may be no F_2 propagation at any time during the day or night.

Sporadic E is fairly common on 10 meters, especially May through August, although it may appear at any time. Short skip, as sporadic E is sometimes called on the HF bands, has little relation to the solar cycle and occurs regardless of F region conditions. It provides single-hop communication from 300 to 2,300 kilometers (190 to 1,400 miles) and multiple-hop opportunities of 4,500 kilometers (2,800 miles) and farther.

Ten meters is a transitional band in that it also shares some of the propagation modes more characteristic of VHF. Meteor scatter, aurora, and auroral E provide the means of making contacts out to 2,300 kilometers (1,400 miles) and TEP propagation often supports propagation to 5,000 miles and farther, but these modes often go unnoticed at 28 MHz. Techniques similar to those used at VHF can be very effective on 10 meters, as signals are usually stronger and more persistent. These exotic modes can be more fully exploited, especially during the solar minimum when F_2 DXing has waned.

VERY HIGH FREQUENCY (VHF) BANDS (30 – 300 MHz)

A wide variety of propagation modes are useful in the VHF range. F-region skip appears on 50 MHz during solar cycle peaks. Sporadic E and several other E-region phenomena are most effective in the VHF range. Still other forms of VHF ionospheric propagation, such as field-aligned irregularities (FAI) and transequatorial propagation (TEP), are rarely observed at VHF. Tropospheric propagation, which is not a factor at HF, becomes increasingly important above 50 MHz.

50 – 54 MHz (6 meters)

The lowest amateur VHF band shares many of the characteristics of both lower and higher frequencies. In the absence of any favorable ionospheric propagation conditions, well-equipped 50 MHz stations work regularly over a radius of 300 kilometers (190 miles) via tropospheric scatter, depending on terrain, power, receiver capabilities, and antenna. Weak-signal troposcatter allows the best stations to make 500 kilometers (310 mile) contacts nearly any time. Weather effects may extend the normal range by a few hundred kilometers, especially during the summer months, but true tropospheric ducting is rare.

During the peak of the 11-year solar cycle (especially during the winter months), worldwide 50 MHz DX is possible via the F_2 region during daylight hours. F_2 backscatter provides an additional propagation mode for contacts as far as 4,000 kilometers (2,500 miles) when the MUF is just below 50 MHz. TEP paths as long as 8,000 kilometers (5,000 miles) across the magnetic equator are common around the spring and fall equinoxes of peak solar cycle years.

Sporadic E is probably the most common and certainly the most popular form of propagation on the 6-meter band. Single-hop E-skip openings may last many hours for contacts from 600 to 2,300 kilometers (370 to 1,400 miles), primarily during the spring and early summer. Multiple-hop E_s provides transcontinental contacts mostly during June and July in the northern hemisphere at distances of 10,000 kilometers or more and contacts between the US and South America, Europe and Japan via multiple-hop E-skip occur nearly every summer.

Other types of E region ionospheric propagation make 6 meters an exciting band. Maximum distances of about 2,300 kilometers (1,400 miles) are typical for all types of E-region modes. TEP propagation is frequent for stations in within range of the geomagnetic equator and for stations farther away when sporadic E propagation couples to TEP. Propagation via FAI often provides additional hours of contacts immediately following sporadic E events. Auroral propagation often makes its appearance in late afternoon when the geomagnetic field is disturbed. Closely related auroral E propagation may extend the 6 meter range to 4,000 kilometers (2,500 miles) and sometimes farther across the northern states and Canada and Alaska and occasionally to Scandinavia, usually after midnight. Meteor scatter provides reliable contacts almost every day around sunrise and especially during one of the dozen or so prominent annual meteor showers.

144 – 148 MHz (2 meters)

Ionospheric effects are significantly reduced at 144 MHz, but they are far from absent. F-region propagation is unknown except for TEP, which is responsible for the current 144 MHz terrestrial DX record of nearly 8,000 kilometers (5,000 miles). Sporadic E occurs as high as 144 MHz less than a tenth as often as at 50 MHz, but the usual maximum single-hop distance is the same, about 2,300 kilometers (1,400 miles). Multiple-hop sporadic E contacts greater than 3,000 kilometers (1,900 miles) have occurred from time to time across the continental US, as well as across Southern Europe.

Auroral propagation is quite similar to that found at 50 MHz, except that signals are weaker and more Doppler-distorted. Auroral E contacts are rare. Meteor-scatter contacts are limited primarily to the periods of the great annual meteor showers and require much patience and operating skill. Contacts have been made via FAI on 144 MHz, but its potential has not been fully explored.

Tropospheric effects improve with increasing frequency, and 144 MHz is the lowest VHF band at which terrestrial weather plays an important propagation role. Weather-induced enhancements may extend the normal 300- to 600-kilometer (190- to 370-mile) range of well-equipped stations to 800 kilometers (500 miles) and more, especially during the summer and early fall. Tropospheric ducting extends this range to 2,000 kilometers (1,200 miles) and farther over the continent and at least to 4,000 kilometers (2,500 miles) over some well-known all-water paths, such as that between California and Hawaii.

222 – 225 MHz (135 cm)

The 135 cm band shares many characteristics with the 2 meter band. The normal working range of 222 MHz stations is nearly as far as comparably equipped 144 MHz stations. The 135 cm band is slightly more sensitive to tropospheric effects, but ionospheric modes are more difficult to use. Auroral and meteor-scatter signals are somewhat weaker than at 144 MHz, and sporadic E contacts on 222 MHz are extremely rare. FAI and TEP may also be well within the possibilities of 222 MHz, but reports of these modes on the 135 cm band are uncommon. Increased activity on 222 MHz will eventually reveal the extent of the propagation modes on the highest of the amateur VHF bands.

ULTRA-HIGH FREQUENCY (UHF) BANDS (300 – 3000 MHz) AND HIGHER

Tropospheric propagation dominates the bands at UHF and higher, although some forms of E region propagation are still useful at 432 MHz. Above 10 GHz, atmospheric attenuation increasingly becomes the limiting factor over long-distance paths. Reflections from airplanes, mountains and other stationary objects may be useful adjuncts to propagation at 432 MHz and higher.

420 – 450 MHz (70 cm)

The lowest amateur UHF band marks the highest frequency on which ionospheric propagation is commonly observed. Auroral signals are weaker and more Doppler distorted; the range is usually less than at 144 or 222 MHz. Meteor scatter is much more difficult than on the lower bands, because bursts are significantly weaker and of much shorter duration. Although sporadic E and FAI are unknown as high as 432 MHz and probably impossible, TEP may be possible.

Well-equipped 432 MHz stations can expect to work over a radius of at least 300 kilometers (190 miles) in the absence of any propagation enhancement. Tropospheric refraction is more pronounced at 432 MHz and provides the most frequent and useful means of extended-range contacts. Tropospheric ducting supports contacts of 1,500 kilometers (930 miles) and farther over land. The current 432 MHz terrestrial DX record of more than 4,000 kilometers (2,500 miles) was accomplished by ducting over water.

902 – 928 MHz (33 cm) and Higher

Ionospheric modes of propagation are nearly unknown in the bands above 902 MHz. Auroral scatter may be just within amateur capabilities at 902 MHz, but signal levels will be well below those at 432 MHz. Doppler shift and distortion will be considerable, and the signal bandwidth may be quite wide. No other ionospheric propagation modes are likely, although high-powered research radars have received echoes from auroras and meteors as high as 3 GHz.

Almost all extended-distance work in the UHF and microwave bands is accomplished with the aid of tropospheric enhancement. The frequencies above 902 MHz are very sensitive to changes in the weather. Tropospheric ducting occurs more frequently than in the VHF bands and the potential range is similar. At 1296 MHz, 2,000-kilometer (1,200-mile) continental paths and 4,000-kilometer (2,500-mile) paths between California and Hawaii have been spanned many times. Contacts of 1,000 kilometers (620 miles) have been made on all bands through 10 GHz in the US and over 1,600 kilometers (1,000 miles) across the Mediterranean Sea. Well-equipped 903 and 1296 MHz stations can work reliably up to 300 kilometers (190 miles), but normal working ranges generally shorten with increasing frequency.

Other tropospheric effects become evident in the GHz bands. Evaporation inversions, which form over very warm bodies of water, are usable at 3.3 GHz and higher. It is also possible to complete paths by scattering from rain, snow and hail in the lower GHz bands. Above 10 GHz, attenuation caused by atmospheric water vapor and oxygen become the most significant limiting factors in long-distance communication.

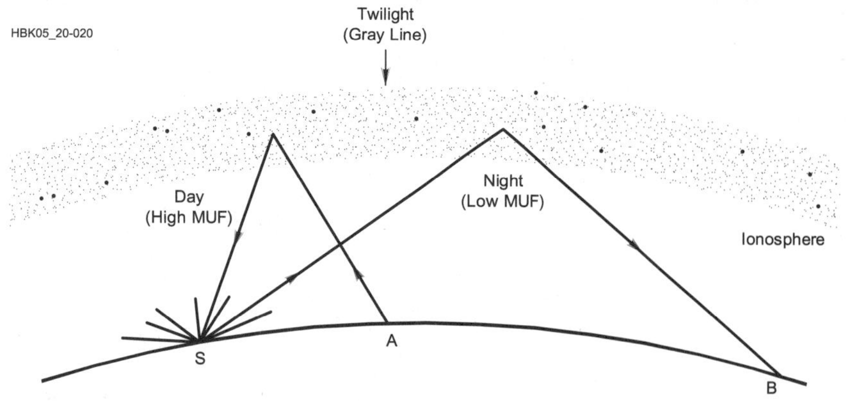

Figure 4.22 — Backscatter path across the gray line. Stations A and B are too close to make contact via normal ionospheric refraction, but may hear each other's signals scattered from point S. Station A makes use of a high-angle refraction on the day side of the gray line, where the MUF is high. Station B makes use of a night-time refraction, with a lower MUF and lower angle of propagation. Note that station A points away from B to complete the circuit.

toward the Azores. Duration of the game can be extended, at times, by careful reorientation of antennas at both ends, as with backscatter. The secret, of course, is to keep hitting the highest-MUF area of the ionosphere and the most favorable ground-reflection points.

The favorable route is usually, but not always, south of the great-circle heading (for stations in the Northern Hemisphere). There can also be sidescatter from the auroral regions. Sidescatter signals are stronger than backscatter signals using the same general area of ground scattering.

Sidescatter signals have been observed frequently on the 14-MHz band, and can take place on any band where there is a large window between the MUF and the LUF. For sidescatter communications to occur, the thing to look for is a common area to which the band is open from both ends of the path (the Azores, in the above example) when there is no direct-path opening. It helps if the common area is in the open ocean, where there is less scattering loss than over land.

Forward scatter is similar to the other two scatter modes. In this case, the MUF may not be high enough for pure refraction to a distant point, but some scatter energy may go forward to complete the path.

4.3 VHF/UHF Non-Ionospheric Propagation

4.3.1 Line of Sight

At one time it was thought that communications in the VHF range and higher would be restricted to line-of-sight paths. Although this has not proven to be the case even in the microwave region, the concept of line of sight is still useful in understanding tropospheric propagation. In the vacuum of space or in a completely homogeneous medium, radio waves do travel essentially in straight lines, but these conditions are almost never met in terrestrial propagation.

Radio waves traveling through the troposphere are ordinarily refracted slightly earthward. The normal drop in temperature, pressure, and water-vapor content with increasing altitude changes the index of refraction of the atmosphere enough to cause refraction. Under average conditions, radio waves are refracted toward Earth enough to make the horizon appear 15 percent farther away than the visual horizon. Under unusual conditions, tropospheric refraction may extend this range significantly.

A simple formula can be used to estimate the distance to the radio horizon under average conditions:

$$d = \sqrt{2h}$$

where

d = distance to the radio horizon, miles

h = height above average terrain, feet.

and

$$d = \sqrt{17h}$$

where

d = distance to the radio horizon, kilometers

h = height above average terrain, meters.

The distance to the radio horizon for an antenna 30 meters (98 feet) above average terrain is thus 22.6 kilometers (14 miles). A station on top of a 1000-meter (3,280 foot) mountain has a radio horizon of 130 kilometers (80 miles).

4.3.2 Beyond Line of Sight

From Figure 4.7 it appears that use of the space wave depends on direct line of sight between the antennas of the communicating stations. This is not literally true, although that belief was common in the early days of amateur communication on frequencies above 30 MHz. When equipment became available that operated more efficiently and after antenna techniques were improved, it soon became clear that VHF waves were actually being bent or scattered in several ways, permitting reliable communication beyond visual distances between the two stations. This was found true even with low power and simple antennas. The average communication range can be approximated by assuming the waves travel in straight lines, but with the Earth's radius increased by one-third. The distance to the *radio horizon* is then given as

$$D_{miles} = 1.415\sqrt{H_{feet}}$$

or

$$D_{km} = 4.124\sqrt{H_{meters}}$$

where H is the height of the transmitting antenna, as shown in **Figure 4.23**.

The formula assumes that the Earth is smooth out to the horizon, so any obstructions along the path must be taken into consideration. For an elevated antenna, the communication distance is equal to D + D1, that is, the sum of the distances to the horizon of both antennas. Radio horizon distances are given in graphic form in **Figure 4.24**. Two stations on a flat plain, one with its antenna 60 feet above ground and the other 40 feet, could be up to about 20 miles apart for strong-signal line-of-sight communication (11 + 9 miles). The terrain is almost never completely flat, however, and variations along the way may add to or subtract from the distance for reliable communication. Remember that energy is absorbed, reflected, or scattered in many ways in nearly all communication situations. The formula or the chart will be a good guide for estimating the potential radius of coverage for a VHF FM repeater, assuming the users are mobile or portable with simple, omnidirectional antennas. Coverage with optimum home-station equipment, high-gain directional arrays, and SSB or CW is quite a different matter. A much more detailed method for estimating coverage on frequencies above 50 MHz is given later in this chapter.

For maximum use of the ordinary space wave, it is important to have the antenna as high as possible above nearby buildings, trees, wires and surrounding terrain. A hill that rises above the rest of the countryside is a good location for an amateur station of any kind, and particularly so for extensive coverage on the frequencies above 50 MHz. That highest point is not necessarily the best location for the antenna, though. In the example shown in **Figure 4.25**, the hilltop would be a good site in all directions. But if maximum performance to the right is the objective, a downslope in terrain results in more energy at lower elevation angles. This would involve a trade-off with reduced coverage in the opposite direction. Conversely, an antenna situated on the left side, lower down the hill, might do well to the left, but almost certainly would be inferior in performance to the right. For more information about the effect of terrain on ground reflections and elevation patterns, see the discussion of the Fresnel zone in this book's chapter on **HF Antenna System Design**.

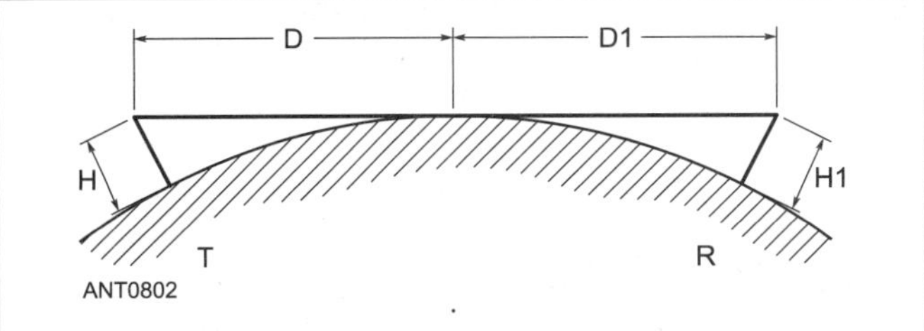

Figure 4.23 — The distance D to the horizon from an antenna of height H is given by equations in the text. The maximum line-of-sight distance between two elevated antennas is equal to the sum of their distances to the horizon as indicated here.

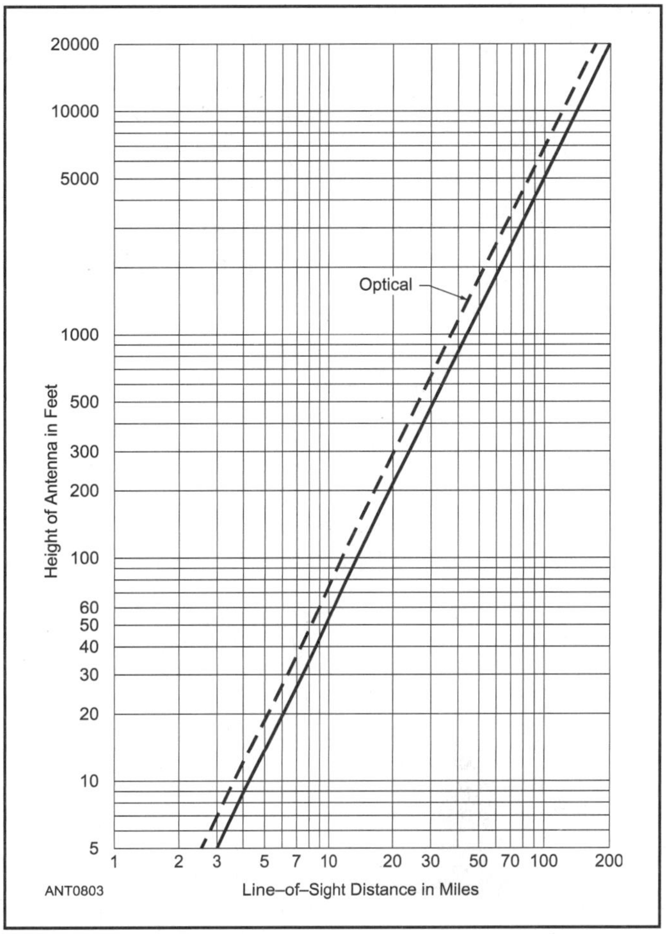

Figure 4.24 — Distance to the horizon from an antenna of given height. The solid curve includes the effect of atmospheric refraction. The optical line-of-sight distance is given by the dashed line.

POLARIZATION FACTORS ABOVE 50 MHZ

In most VHF communication over short distances, the polarization of the space wave tends to remain constant. *Polarization discrimination* is the difference in an antenna's response to signals having the same and opposite polarizations to that of the antenna. The difference can be significant, in excess of

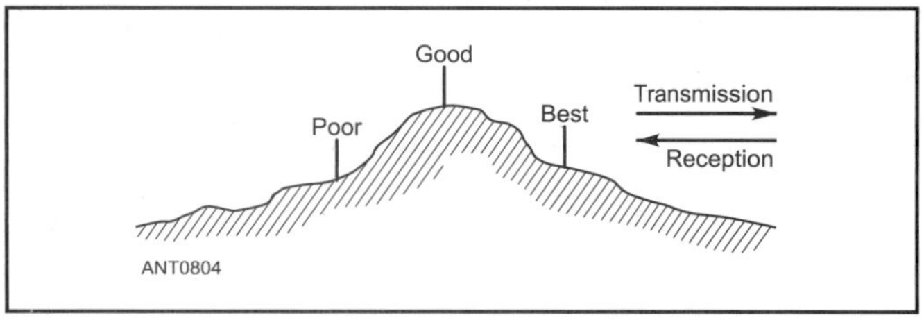

Figure 4.25 — Propagation conditions are generally best when the antenna is located near the top of a downslope facing the distant station. Communication is poor when there is a sharp rise immediately in front of the antenna in the direction of communication.

20 dB, so the same polarization should be used at both ends of the circuit. Horizontal polarization is the standard for terrestrial communications, except for mobiles which use vertical polarization. Circular polarization is standard for most satellite communications.

Horizontal systems are popular, in part because they tend to reject man-made noise, much of which is vertically polarized. There is some evidence that vertical polarization shifts to horizontal in hilly terrain, more readily than horizontal shifts to vertical. With large arrays, horizontal systems may be easier to erect, and they tend to give higher signal strengths over irregular terrain, if any difference is observed.

Practically all work with VHF mobiles is now handled with vertical systems. For use in a VHF repeater system, the vertical antenna can be designed to have gain without losing the desired omnidirectional quality. In the mobile station, a small vertical whip has obvious aesthetic advantages. Often, a whip used for broadcast reception can be pressed into service for 144 MHz FM. A car-top mount is preferable, but the broadcast whip is a practical compromise. Tests with at least one experimental repeater have shown that horizontal polarization can give a slightly larger service area, but mechanical advantages of vertical systems have made them the almost unanimous choice in VHF FM communication.

4.3.3 Reliable VHF Coverage

In the preceding sections we discussed means by which amateur bands above 50 MHz may be used intermittently for communication far beyond the visual horizon. In emphasizing distance, we should not neglect a prime asset of the VHF band: reliable communication over relatively short distances. The VHF region is far less subject to disruption of local communication than are frequencies below 30 MHz. Since much amateur communication is essentially local in nature, our VHF assignments can carry a great load, and such use of the VHF bands helps solve interference problems on lower frequencies.

Because of age-old ideas, misconceptions about the coverage obtainable in our VHF bands persist. This reflects the thoughts that VHF waves travel only in straight lines, except when the DX modes described above happen to be present. However, let us survey the picture in the light of modern wave-propagation knowledge and see what the bands above 50 MHz are good for on a day-to-day basis, ignoring the anomalies that may result in extensions of normal coverage.

It is possible to predict with fair accuracy how far you should be able to work consistently on any VHF or UHF band, provided a few simple facts are known. The factors affecting operating range can be reduced to graph form, as described in this section. The information was originally published in the November 1961 *QST* by D. W. Bray, K2LMG, (see the Bibliography at the end of this chapter).

VHF/UHF STATION GAIN

To estimate your station's capabilities, two basic numbers must be determined: *path loss* and *station gain*. Path loss is readily determined from the nomogram in **Figure 4.26**. This gives path loss over smooth Earth, for 99% reliability.

For 50 MHz, lay a straightedge from the distance between stations (left side) to the appropriate distance at the right side. For 1296 MHz, use the full scale, right center. For 144, 222 and 432, use the dot in the circle, square or triangle, respectively. Example: At 300 miles the path loss for 144 MHz is 214 dB.

To be meaningful, the losses determined from this nomogram are necessarily greater than simple free-space path losses. As described in an earlier section, communication beyond line-of-sight distances involves propagation modes that increase the path attenuation with distance.

Station gain is made up of seven factors: receiver sensitivity, transmitted power, receiving antenna gain, receiving antenna height gain, transmitting antenna gain, transmitting antenna height gain, and required signal-to-noise ratio. This looks complicated, but it really boils down to an easily made evaluation of receiver, transmitter, and antenna performance.

The information in this section is based on CW operation, so the additional signal needed for other modes must be subtracted. Use a figure of 3 dB for SSB. Fading losses must be accounted for also. It has been shown that for distances beyond 100 miles, the signal will vary plus or minus about 7 dB from the average level, so 7 dB must be subtracted from the station gain for high reliability. For distances under 100 miles, fading diminishes almost linearly with distance. For 50 miles, use –3.5 dB for fading.

The largest of the eight factors involved in station design is receiver sensitivity. This is obtainable from **Figure 4.27**, if you know the approximate receiver noise figure and transmission-line loss. If you can't measure noise figure, assume 3 dB for 50 MHz, 5 for 144 or 222, 8 for 432 and 10 for 1296 MHz, if you know your equipment is working moderately well. These noise figures are well on the conservative side for modern solid-state receivers. Line loss can be taken from tables of transmission line characteristics, assuming no additional loss due to impedance mismatches.

Next, lay a straightedge between the appropriate points at either side of Figure 4.29, to find effective receiver sensitivity in decibels with respect to 1 watt (dBW). Use the narrowest bandwidth that is practical for the emission intended, with the receiver you will be using. For CW, an average value for effective operation is about 500 Hz. Phone bandwidth can be taken

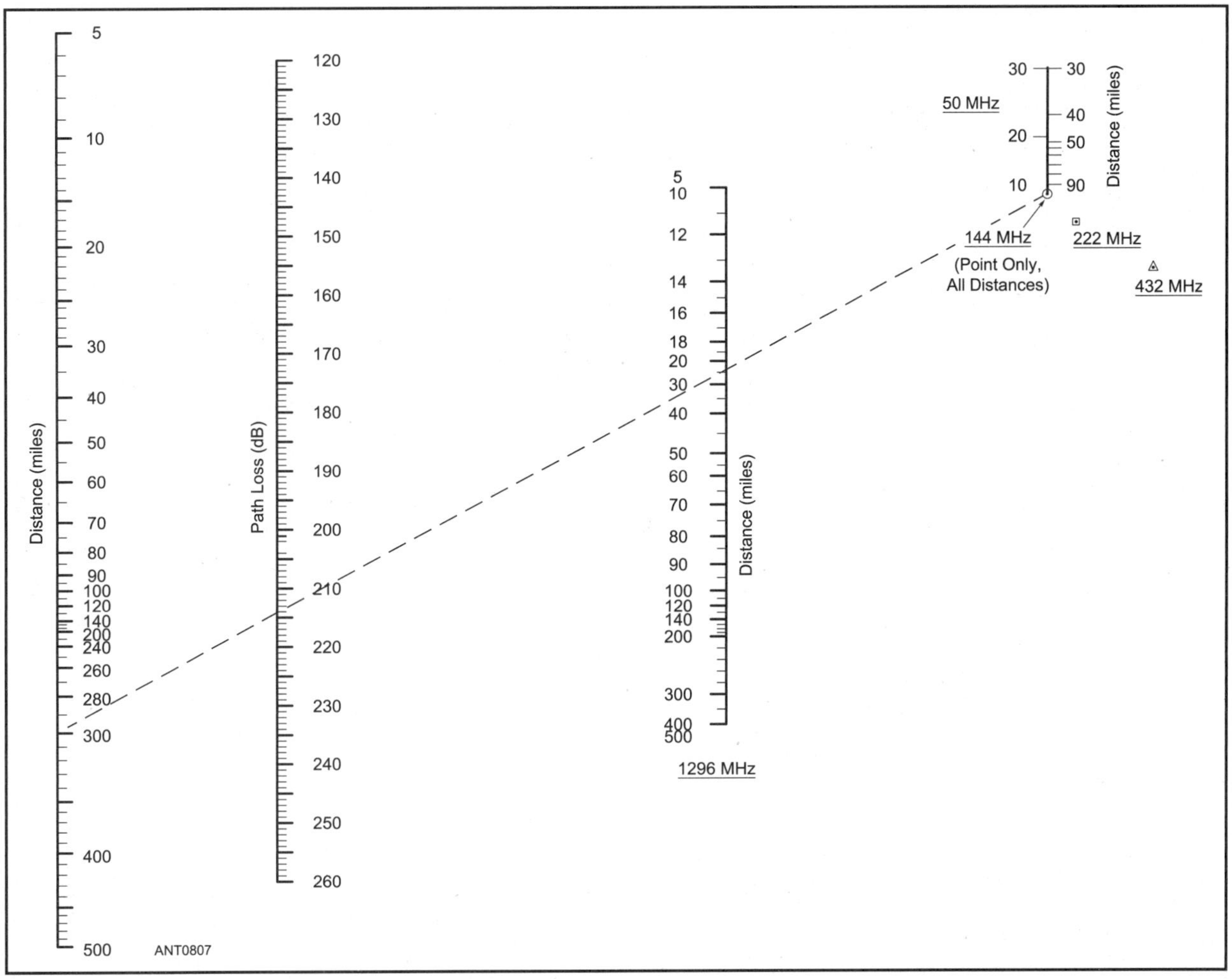

Figure 4.26 — Nomogram for finding the capabilities of stations on amateur bands from 50 to 1300 MHz. Either the path loss for a given distance or vice versa may be found if one of the two factors is known. See the text for instructions for use and an example calculation.

from the receiver instruction manual, but it usually falls between 2.1 to 2.7 kHz. For example, if using phone with a bandwidth of 2.4 kHz and a receiver with a noise figure of 6 dB, the line between those two points will give a receiver sensitivity of –163 dBW. If line loss is 5 dB, receiver sensitivity changes to –163 dBW + 5 dB = –158 dBW.

Antenna gain is next in importance. Gains of amateur antennas are often exaggerated. For well-designed Yagis the gain (over isotropic) runs close to 10 times the boom length in wavelengths. (Example: A 24-foot Yagi on 144 MHz is 3.6 wavelengths long; 3.6 × 10 = 36, and $10 \log_{10} 36 = 15.5$ dBi in free space.) Add 3 dB for stacking, where used properly. Add 4 dB more for ground reflection gain. Ground reflection gain varies in amateur work but averages out near this figure.

We have one more plus factor — antenna height gain, obtained from **Figure 4.28**. Note that this is greatest for short distances. The left edge of the horizontal center scale is for 0 to 10 miles, the right edge for 100 to 500 miles. Height gain for 10 to 30 feet is assumed to be zero. For 50 feet the height gain is 4 dB at 10 miles, 3 dB at 50 miles, and 2 dB at 100 miles. At 80 feet the height gains are roughly 8, 6 and 4 dB for these distances. Beyond 100 miles the height gain is nearly uniform for a given height, regardless of distance. An example is provided in the figure for an antenna height of 50 feet and a distance of 50 miles, giving a gain of 3.5 dB.

Transmitter power output must be stated in dBW (decibels with respect to 1 watt). If you have 500 W output, add 10 log (500/1), or 27 dB, to your station gain. The transmission-line loss must be subtracted from the station gain. So must the required signal-to-noise ratio.

Add all the plus and minus terms (they are added and subtracted since they are in terms of dB) to get the station gain. This 'station gain' value is equivalent to the amount of loss that your station can tolerate to meet the required signal-to-ratio goal that you specified.

Enter this value as the path loss in Figure 4.28. Or work it the other way around: Find the path loss for the distance you want to cover from the nomogram and then figure out what station changes will be needed to overcome it.

For example, over a 50-mile path on 144 MHz, with a receiver sensitivity of –163 dBW, transmitted power of

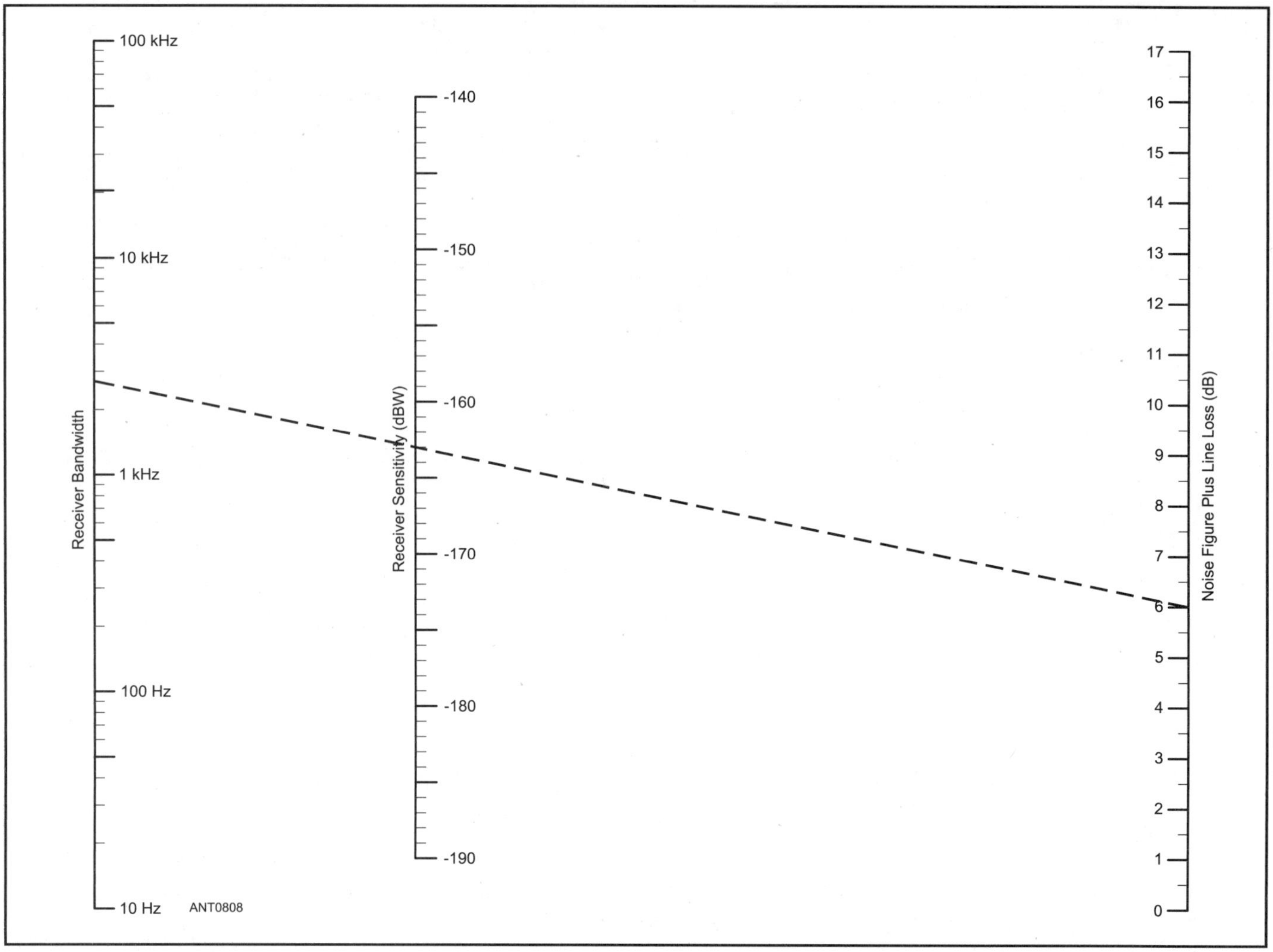

Figure 4.27 — Nomogram for finding effective receiver sensitivity.

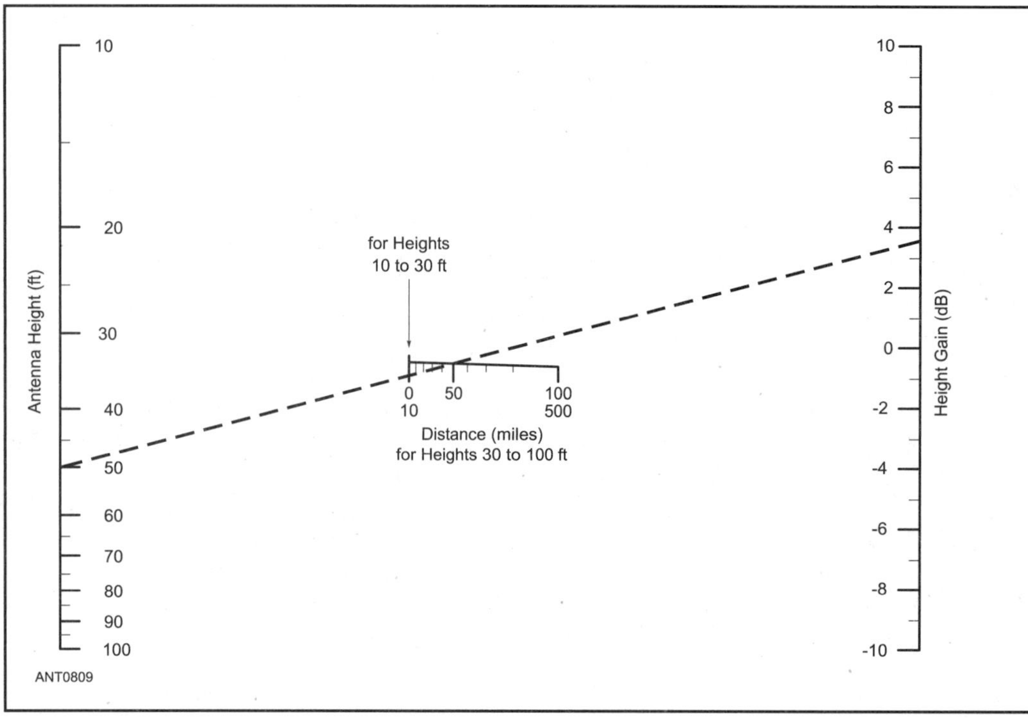

Figure 4.28 — Nomogram for determining antenna-height gain.

20 dBW (100 watts) less 5 dB of line loss for 15 dBW, receiving antenna gain of 15 dB plus height gain of 3.5 dB, transmitting antenna gain of 12 dB plus height gain of 4.5 dB, station gain would be:

163 dBW + 15 dBW (transmitted power with 5 dB line loss) + 15 dB (receive antenna gain) + 3.5 dB (receive antenna height gain) + 12 dB (transmitted antenna gain) + 4.5 dB (transmitted antenna height gain) – 3.5 dB (fading) = 209.5 dB. That's the allowable loss for a 0 dB SNR for CW in a 500 Hz, which can be translated to distance via Figure 4.28. Note that receiver sensitivity is treated as a positive value since this is a calculation of allowable loss and not an absolute power.

EVALUATING COVERAGE

The significance of all this becomes more obvious when we see path loss plotted against frequency for the various bands, as in **Figure 4.29**. At the left this is done for 50 percent reliability. At the right is the same information for 99% reliability. For near-perfect reliability, a path loss of 195 dB (easily encountered at 50 or 144 MHz) is involved in 100-mile communication. But look at the 50% reliability curve: The same path loss takes us out to well over 250 miles. Few amateurs demand near-perfect reliability. By choosing our times, and by accepting the necessity for some repeats or occasional loss of signal, we can maintain communication out to distances far beyond those usually covered by VHF stations.

Working out a few typical amateur VHF station setups with these curves will show why an understanding of these factors is important to any user of the VHF spectrum. Note that path loss rises very steeply in the first 100 miles or so. This is no news to VHF operators; locals are very strong, but stations 50 or 75 miles away are much weaker. What happens beyond 100 miles is not so well known to many of us.

From the curves of Figure 4.29, we see that path loss levels off markedly at what is the approximate limit of working range for average VHF stations using wideband modulation modes. Work out the station gain for a 50-W station with an average receiver and antenna, and you'll find that it comes out around 180 dB. This means you'd have about a 100-mile working radius in average terrain, for good but not perfect reliability. Another 10 dB may extend the range to as much as 250 miles. Changing from wideband modes such as FM or AM phone to SSB and CW makes a major improvement in daily coverage on the VHF bands.

A bigger antenna, a higher one if your present beam is not at least 50 feet up, an increase in power to 500 W from 50 W, an improvement in receiver noise figure if it is presently poor — any of these things can make a big improvement in reliable coverage. Achieve all of them, and you will have very likely tripled your operating range, thanks to that hump in the path-loss curves. This goes a long way toward explaining why using a 10-W packaged station with a small antenna, fun though it may be, does not begin to show what the VHF bands are really good for.

EFFECTS OF TERRAIN AT VHF/UHF

The coverage figures derived from the above procedure are for smooth terrain. What of stations in mountainous country? Although an open horizon is generally desirable for the VHF station site, mountain country should not be considered hopeless. Help for the valley dweller often lies in the optical phenomenon known as *knife-edge diffraction*. A flashlight beam pointed at the edge of a partition does not cut off sharply at the partition edge, but is diffracted around it, partially illuminating the shadow area. A similar effect is observed with VHF waves passing over ridges; there is a shadow effect, but not a complete blackout. If the signal is strong where it strikes the mountain range, it will be heard well at the bottom of a valley on the far side. (See the prior section on "Fundamentals of Radio Wave Propagation" for more information on diffraction.)

This is familiar to all users of VHF communications equipment who operate in hilly terrain. Where only one ridge lies in the way, signals on the far side may be almost as good as on the near side. Under ideal conditions (a very high and

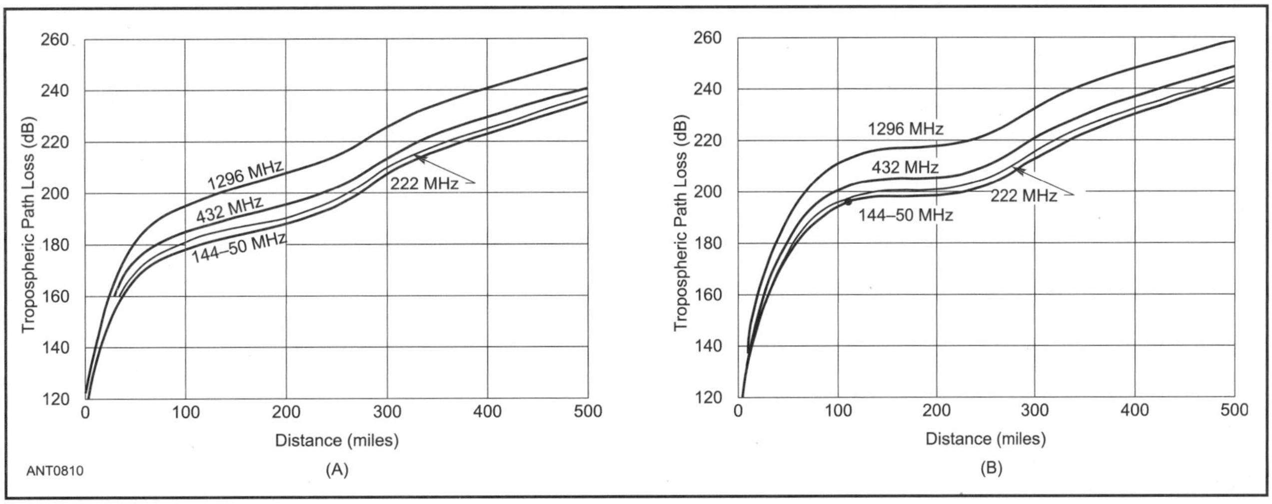

Figure 4.29 — Path loss versus distance for amateur frequencies above 50 MHz. At A are curves for 50% of the time; at B, for 99% of the time. The curves at A are more representative of amateur radio requirements.

sharp-edged obstruction near the midpoint of a long-enough path so that signals would be weak over average terrain), knife-edge diffraction may yield signals even stronger than would be possible with an open path.

The obstruction must project into the radiated beam of the antennas used. Often mountains that look formidable to the viewer are not high enough to have an appreciable effect, one way or the other. Since the normal radiation pattern from a VHF array is several degrees above the horizontal, mountains that are less than about three degrees above the horizon, as seen from the antenna, are missed by the radiation from the array. Moving the mountains out of the way would have substantially no effect on VHF signal strength in such cases.

Rolling terrain, where obstructions are not sharp enough to produce knife-edge diffraction, still does not exhibit a complete shadow effect. There is no complete barrier to VHF propagation — only attenuation, which varies widely as the result of many factors. Thus, even valley locations are usable for VHF communication. Good antenna systems, preferably as high as possible, the best available equipment, and above all, the willingness and ability to work with weak signals may make outstanding VHF work possible, even in sites that show little promise to casual inspection.

The availability of detailed terrain data for much of the world enables software to make good predictions about coverage of VHF, UHF, and microwave signals. The software is freely available to amateurs and can run online in a web browser without having to download and install software or large databases on a PC.

Radio Mobile Online is the online version of the popular RF propagation tool *Radio Mobile* by Roger Coudé, VE2DBE. (**ve2dbe.com/rmonline.html**) It uses digital terrain information and a mathematical model to simulate coverage between two fixed sites (such as a radio link) or between a fixed site and a mobile (such as for repeater coverage).

The digital terrain information comprises two databases totaling 198 gigabytes: elevation and land cover. Its online model works with amateur bands from 30 MHz to 25 GHz. The information for both stations (whether fixed/fixed or fixed/mobile) is comprehensive. The model is based on the public-domain ITM model augmented with the author's decades of experience in the mobile communications field. To use the service, you must first create a personal account (free) and your calculations will be saved on the system's server for online access.

A tutorial on *Radio Mobile Online* is available from KØLWC — search YouTube.com for "Radio Mobile Tutorial — How To Predict Ham Radio Coverage." Two sample outputs from *Radio Mobile Online* are shown in **Figure 4.30**: a coverage signal-strength contour map (A) and a point-to-point link map (B). Both figures are from **ve2dbe.com/rme.html**.

The AREDN network (**arednmesh.org**) operates Ubiquiti microwave wireless networking equipment with customized firmware to support amateur radio operation. Nodes in the network are situated to offer coverage to individual stations, relay stations, and stations deployed in disaster response. To support network node setup and site selection, the Ubiquiti *AirLink* planning tool software is used. (**link.ubnt.com**)

After creating a free account, the simplest way to get started is to enter your own address, zoom out on the map, and select a nearby location. The link profile will automatically be generated as shown in **Figure 4.31** which shows

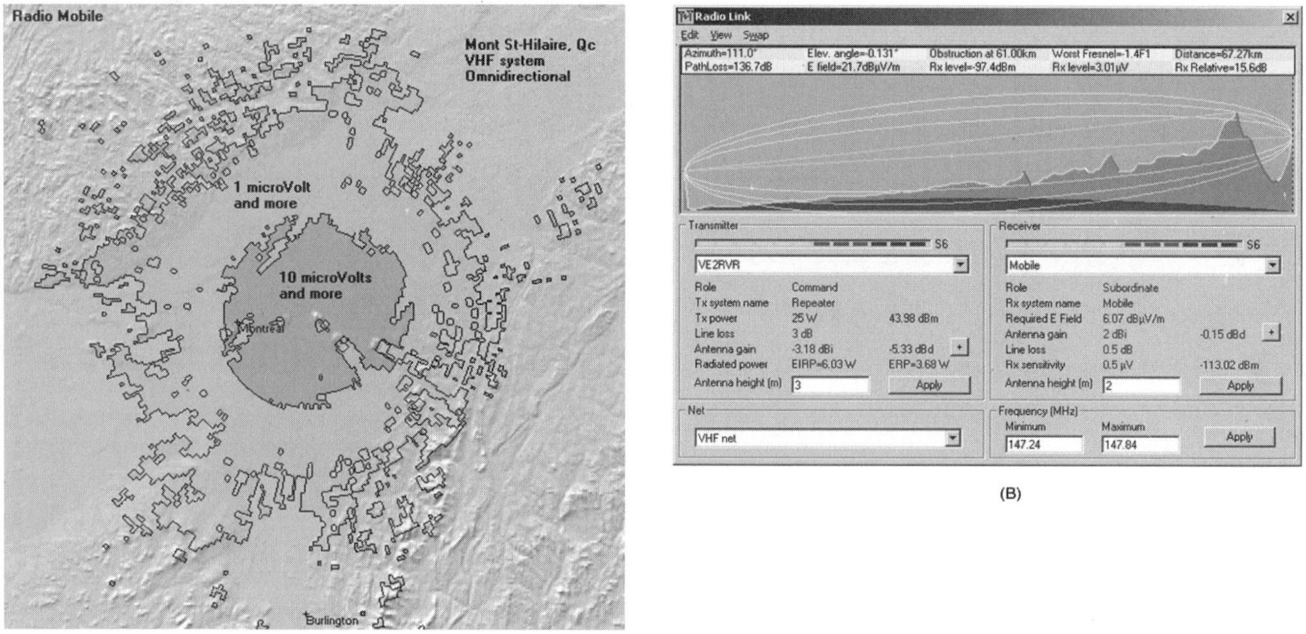

HBK1128

(A)

(B)

Figure 4.30 — *Radio Mobile* signal-strength contour map (A) and a point-to-point link map (B). Both figures are from ve2dbe.com/rme.html.

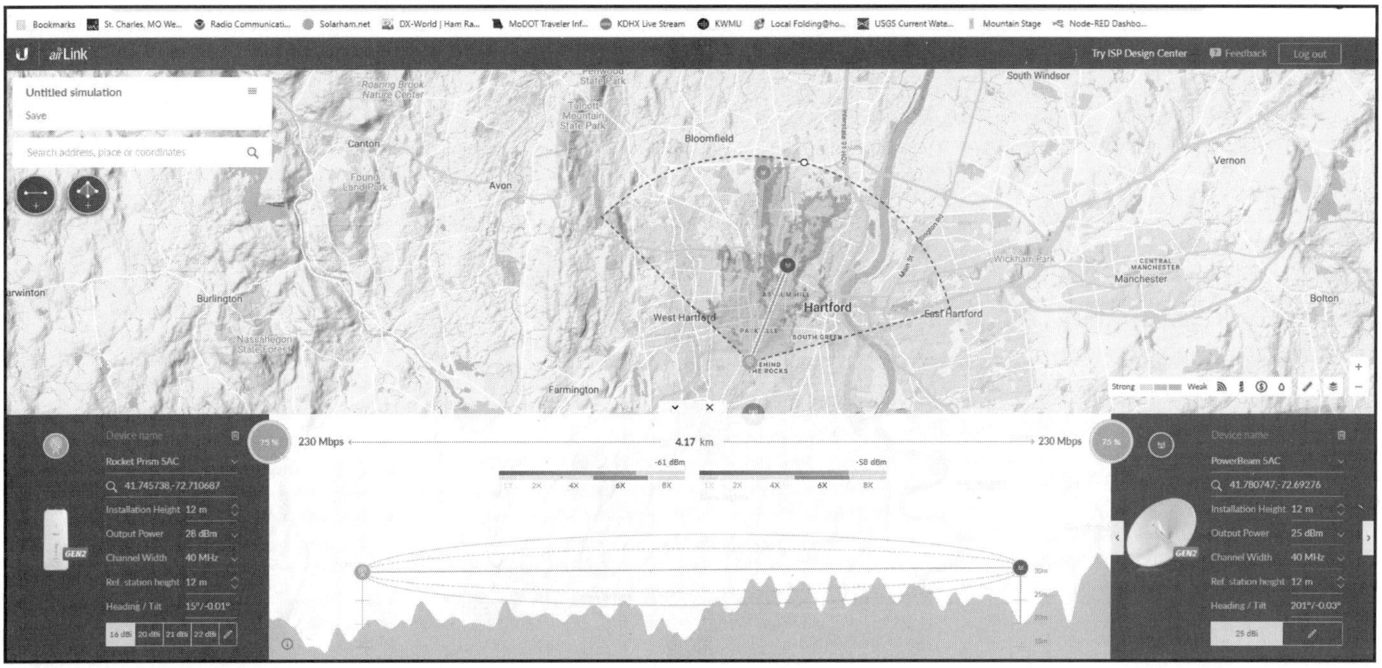

Figure 4.31 — A coverage map for 5 GHz wireless links from the Ubiquiti AirLink coverage and site evaluation tool at link.ui.com. The map and link profile show coverage between ARRL headquarters in Newington, CT and nearby Hartford, CT.

coverage between ARRL headquarters in Newington and nearby Hartford, CT over a 5 GHz link. A step-by-step tutorial is available at **beetledigital.com/how-to-use-the-ubiquiti-airlink-tool-to-design-and-plan-wireless-network-links**.

These are two popular and low-cost tools available to amateurs for use in the VHF, UHF, and microwave range. The AREDN group has developed a very useful guide to wireless network design at **arednmesh.readthedocs.io/en/stable/arednNetworkDesign/network_modeling.html**. It discusses other packages, services, and techniques that you may find valuable. This is an active group and regularly updates its website, so this is a good source for coverage and propagation planning and site evaluation.

4.3.4 Tropospheric Scatter

Contacts beyond the radio horizon out to a working distance of 100 to 500 kilometers (60 – 310 miles), depending on frequency, equipment, and local geography, are made every day without the aid of obvious propagation enhancement. At 1.8 and 3.5 MHz, local communication is due mostly to ground wave. At higher frequencies, especially in the VHF range and above, the primary mechanism is scattering in the troposphere, *tropospheric scatter* or *troposcatter*.

Most amateurs are unaware that they use troposcatter even though it plays an essential role in most local communication. Radio signals through the VHF range are scattered primarily by wavelength-sized gradients in the index of refraction of the lower atmosphere due to turbulence, along with changes in temperature. Radio signals in the microwave region can also be scattered by rain, snow, fog, clouds and dust. That tiny part that is scattered forward and toward the Earth creates the over-the-horizon paths. Troposcatter path losses are considerable and increase with frequency.

Tropospheric scatter is always with us. Its effects are often hidden, masked by more effective propagation modes on the lower frequencies. But beginning in the VHF range, scatter from the lower atmosphere extends the reliable range markedly if we make use of it. Troposcatter is what produces that nearly flat portion of the curves that were used in the section on calculating reliable VHF coverage range). With a decent station, you can consistently make troposcatter contacts out to 300 miles on the VHF and even UHF bands, especially if you don't mind weak signals and something less than 99% reliability.

Ionospheric scatter works much the same as the tropospheric version, except that the scattering medium is higher up, mainly the E region of the ionosphere, but with some help from the D and F regions, too. Ionospheric scatter is useful mainly above the MUF, so its useful frequency range depends on geography, time of day, season, and the state of the Sun.

With near maximum legal power, good antennas and quiet locations, scatter from material 50 to 150 miles or so above the Earth can also fill in the skip zone with marginally readable signals scattered from ionized trails of meteors, from small areas of random ionization, from cosmic dust, from satellites, and from whatever else may come into the antenna's radiation pattern.

The maximum distance that can be linked via troposcatter is limited by the height of a scattering volume common to two stations, shown schematically in **Figure 4.32**. The highest altitude for which scattering is efficient at amateur power levels is about 10 kilometers (6 miles). An application of the distance-to-the-horizon formula yields 800 kilometers (500 miles) as the limit for troposcatter paths, but typical maxima

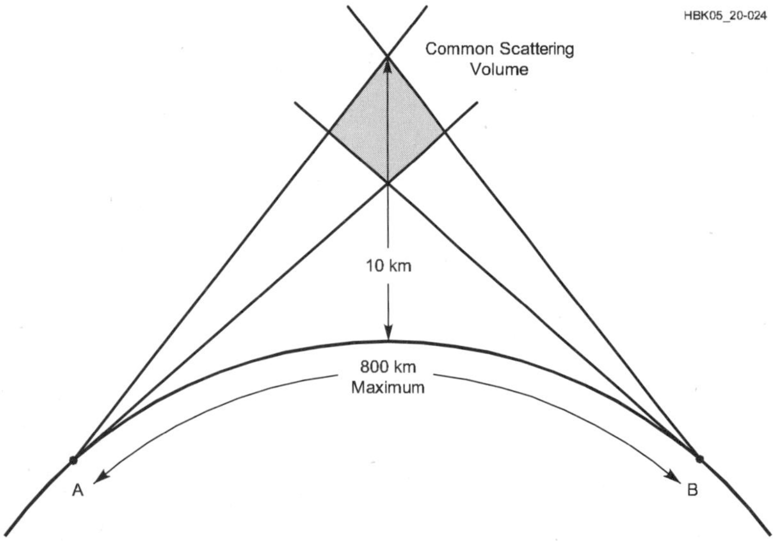

Figure 4.32 — Tropospheric-scatter path geometry. The lower boundary of the common scattering volume is limited by the take-off angle of both stations. The upper boundary of 10 kilometers (6 miles) altitude is the limit of efficient scattering in the troposphere. Signal strength increases with the scattering volume.

are about half that. Tropospheric scatter varies little with season or time of day, but it is difficult to assess the effect of weather on troposcatter alone. Variations in tropospheric refraction, which is very sensitive to the weather, probably account for most of the observed day-to-day differences in troposcatter signal strength.

Troposcatter does not require special operating techniques or equipment, as it is used unwittingly all the time. In the absence of all other forms of propagation, especially at VHF and above, the usual working range is essentially the maximum troposcatter distance. Ordinary working range increases most dramatically with antenna height, because that lowers the take-off angle to the horizon. Working range increases less quickly with antenna gain and transmitter power. For this reason, a mountaintop is the choice location for extending ordinary troposcatter working distances.

PRECIPITATION SCATTER

Scatter from raindrops is a special case of troposcatter practical in the 3.3 to 24 GHz range. Stations simply point their antennas toward a common area of rain. A certain portion of radio energy is scattered by the raindrops, making possible over-the-horizon or obstructed-path contacts, even with low power. The theoretical range for rain scatter is as great as 600 kilometers (370 miles), but the experience of amateurs in the microwave bands suggests that expected distances are less than 200 kilometers (120 miles). Snow and hail make less efficient scattering media unless the ice particles are partially melted. Smoke and dust particles are too small for extraordinary scattering, even in the microwave bands.

4.3.5 Refraction and Ducting in the Troposphere

Radio waves are refracted by natural gradients in the index of refraction of air with altitude, due to changes in temperature, humidity and pressure. Refraction under standard atmospheric conditions extends the radio horizon somewhat beyond the visual line of sight. Favorable weather conditions further enhance normal tropospheric refraction, lengthening the useful VHF and UHF range by several hundred kilometers and increasing signal strength. Higher frequencies are more sensitive to refraction, so its effects may be observed in the microwave bands before they are apparent at lower frequencies.

Ducting takes place when refraction is so great that radio waves are bent back to the surface of the Earth. When tropospheric ducting conditions exist over a wide geographic area, signals may remain very strong over distances of 1,500 kilometers (930 miles) or more. Ducting results from the gradient created by a sharp increase in temperature with altitude, quite the opposite of normal atmospheric conditions. A simultaneous drop in humidity contributes to increased refractivity.

Normally the temperature steadily decreases with altitude, but at times there is a small portion of the troposphere in which the temperature increases and then again begins decreasing normally. This is called an *inversion*. Useful temperature inversions form between 250 and 2,000 meters (800-6,500 feet) above ground. The elevated inversion and the Earth's surface act something like the boundaries of a natural open-ended waveguide. Radio waves of the right frequency range caught inside the duct will be propagated for long distances with relatively low losses. Several common weather conditions can create temperature inversions.

To support radio wave propagation via tropospheric ducting, a temperature inversion is required to be a minimum depth depending on frequency. The inversion's depth is the difference in altitude from when the temperature begins increasing and when it again starts decreasing. Figure 12 in Chapter 3 of the *ARRL UHF/Microwave Experimenter's Manual* (1990) gives the required inversion depth at various frequencies versus the index of refraction (N) lapse rate per meter. Pulling data from that figure at a typical lapse rate of $\Delta N/m = 0.157$ gives **Figure 4.33** which shows required inversion depths for ducting to occur at different frequencies. The original data only went down to 432 MHz (70 cm), so the line was extrapolated down to 50 MHz (6 meters). This plot tells us that the inversion depth under typical conditions for tropospheric ducting on 2 meters needs to be around 550 meters and it needs to be around 1,000 meters on 6 meters. If the lapse rate is greater than 0.157, less inversion depth is required.

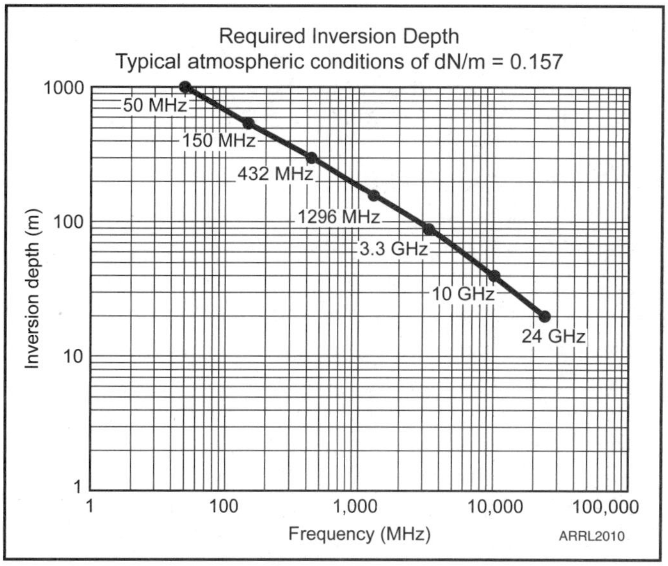

Figure 4.33 — Required temperature inversion depth to form a duct at different frequencies.

4.4 Propagation Predictions for HF Operation

4.4.1 The Big Picture Overhead

There are about 150 vertical-incidence ionosondes around the world. Ionosondes are located on land, even on a number of islands. There are gaps in sounder coverage, however, mainly over large expanses of open ocean. The compilation of all available vertical-incidence data from the worldwide network of ionospheric sounders results in global f_oF_2 maps, such as the map shown in **Figure 4.34**, a simulation from the highly sophisticated *PropLab Pro* computer program. (**spacew.com** — listed under Geophysical Software)

This simulation is for 1900 UTC, about four hours before East Coast sunset on September 15, 1998, with a high level of solar activity (SSN of 85) and a planetary A_p index of 5, indicating quiet geomagnetic conditions. (These conditions are typical of what is expected as we move into the peak years of solar Cycle 25.) The contours of f_oF_2 peak over the South American longitudes at around 12 MHz. If you look carefully, you'll see two peaks on either side of the geomagnetic equator, at about –12 degrees geographic latitude in the South American sector.

These two peaks in f_oF_2 form what is known as the *equatorial anomaly* and are caused by upwelling "fountains" of high electron concentration located in daylight areas about ±20 degrees from the geomagnetic equator. The equatorial anomaly is important in transequatorial propagation. Those LU stations in Argentina that you can hear on 28 MHz from the US in the late afternoon, even during low portions of the solar cycle when other stations to the south are not coming through, are benefiting from transequatorial propagation, a form of *chordal hop* propagation, because signals going through this area remain in the ionosphere without lossy intermediate hops to the ground.

From records of f_oF_2 profiles, the underlying electron densities along a path can be computed. And from the electron density profiles computerized "ray tracing" may be done throughout the ionosphere to determine how a wave propagates

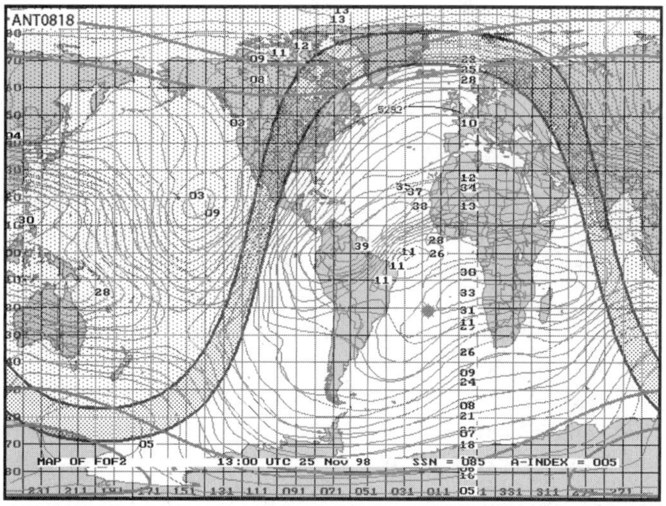

Figure 4.34 — Computer simulation of the f_oF_2 contours for 15 September 1998 at 1900 UTC, for an SSN of 85 and a quiet planetary Ap index of 5. Note the two regions of high f_oF_2 values in the South American sector (they can be seen better in a plot of MUF contours). These are the "equatorial anomalies," regions of high electronic density in the F_2 region that often allow chordal-hop north-south propagation. See also Figure 6. (*PropLab Pro V3* simulation courtesy of K9LA)

from a transmitter to a particular receiver location. *PropLab Pro* can do complex ray tracings that explicitly include the effect of the Earth's magnetic field, even taking into effect ionospheric stormy conditions.

4.4.2 MUF Prediction

F region MUF prediction is key to forecasting HF communications paths at particular frequencies, dates, and times, but forecasting is complicated by several variables. Solar radiation varies over the course of the day, season, year, and solar cycle. Additionally, the ionization at any given point in the world depends on geomagnetic field activity and events in the lower atmosphere coupling up to the ionosphere. These regular intervals provide the main basis for prediction, yet recurrence is far from reliable. In addition, forecasts are predicated on a quiet geomagnetic field, but the condition of the Earth's magnetic field is most difficult to predict weeks or months ahead. For professional users of HF communications, uncertainty is a nuisance for maintaining reliable communications paths, while for many amateurs it provides an aura of mystery and chance that adds to the fun of DXing. Nevertheless, many amateurs want to know what to expect on the HF bands to make best use of available on-the-air time, plan contest strategy, ensure successful net operations, or engage in other activities.

An online supplement, "Frequency Selection for HF Operation" will help you make the best use of your operating time. You may also find an article that appeared in the Summer 2008 issue of *CQ VHF* to be helpful with your 6-meter endeavors: "Predicting 6-meter F2 Propagation," by Carl Luetzelschwab, K9LA.

4.4.3 Solar and Geophysical Data

An amateur radio operator interested in long-distance QSOs on any frequency benefits from being aware of where we are in a solar cycle (solar data), the current state of the Sun's activity with respect to disturbances (geophysical data), and the fundamentals of propagation.

For solar and geophysical data, the internet can be very helpful. For example, the Space Weather Prediction Center (SWPC) website at **swpc.noaa.gov** can supply pretty much everything you need to know either directly or through links. Sites providing information on more specific propagation topics are referenced in the appropriate sections. Another source of useful information about solar disturbances is **spaceweather.com**.

The SWPC home page includes indexes for radio blackouts, solar radiation storm impacts, and geomagnetic storm impacts. Three videos are provided of recent X-ray activity (for assessment of solar radiation storms and radio blackouts), recent coronal mass ejections (for assessment of geomagnetic storms), and predicted visible aurora (which gives a general indication of radio aurora conditions). Plots of X-ray flux (for assessment of solar radiation storms and radio blackouts), proton flux (for assessment of solar radiation storms), and recent K-index activity (for assessment of geomagnetic storms) are also provided.

Under these six images are numerous links in three major categories of additional data: ABOUT SPACE WEATHER, PRODUCTS AND DATA, and DASHBOARDS for specific interests. The RADIO dashboard is most suited to amateur radio.

There are also links to other sources of space weather information not on the SWPC website. For example, entering a specific topic (such as STEREO) in the search area in the upper right-hand corner of the home page brings up many links to NASA's STEREO mission (which allows us to look at the backside of the Sun).

Some other useful websites are: **dx.qsl.net/propagation**, **solen.info/solar**, **solarham.net**, and **hfradio.org/propagation.html**. You may also access propagation information via your preferred spotting network server. Use the command SH/WWV/*n*, where *n* is the number of spots you wish to see (five is the default).

Another excellent method for obtaining an "equivalent sunspot number" (SSN_e) is to go to the Space Weather site of NorthWest Research Associates at **spawx.nwra.com/spawx/ssne24.html**. NWRA compares real-time ionospheric sounder data around the world with predictions using various levels of SSN looking for the best match. They thus "back into" the actual effective sunspot number. Note that this is necessarily a best fit of ionospheric sounder data to an equivalent sunspot number — it's not a perfect fit for all the data due to the dynamic hour-to-hour variability of the worldwide F_2 region. **Figure 4.35** is a typical NWRA graph, which covers the week ending October 6, 2002. This graph was selected as it clearly shows the effects of a typical space weather event: In this case, a sudden decrease in SSN_e after a geomagnetic storm depressed SSN_e by more than 50 percent.

GEOPHYSICAL DATA FROM WWV AND WWVH

The standard time station WWV (Fort Collins, Colorado) transmits on 2.5, 5, 10, 15 and 20 MHz, and WWVH (Kauai, Hawaii) transmits on 2.5, 5, 10, and 15 MHz. Both of these

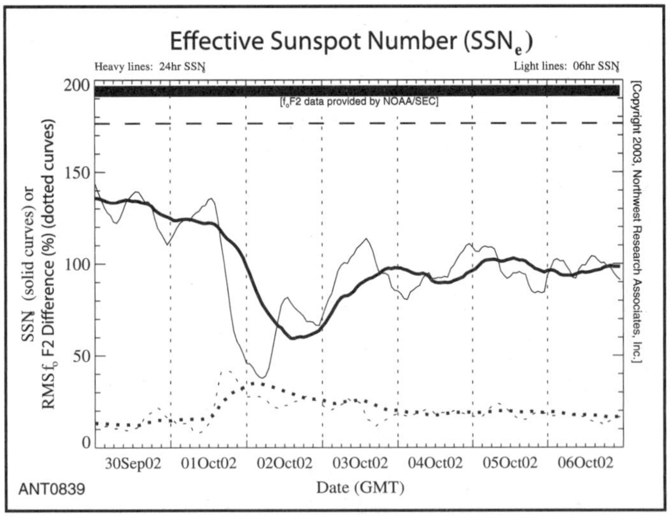

Figure 4.35 — Effective Sunspot Number (SSN_e) produced by NWRA. Note the large drop in effective SSN due to a geomagnetic storm commencing Oct 1, 2002.

stations are popular for propagation monitoring. They transmit 24 hours a day. Daily monitoring of these stations for signal strength and quality can quickly provide a good basic indication of propagation conditions. In addition, each hour they broadcast the geomagnetic A- and K-indexes, the 2800 MHz (10.7 cm) solar flux, and a short forecast of conditions for the next day. These are heard on WWV at 18 minutes past each hour and on WWVH at 45 minutes after the hour. The same information, along with a lot more space weather information, is available at various websites, such as **swpc.noaa.gov** or **spaceweather.com**. The K-index is updated every three hours, while the A-index and solar flux are updated after 2100 UTC. These data are useful for making predictions on home computers, especially when averaged over several days of solar flux observations. For more details about the broadcasts, visit **nist.gov/pml/time-and-frequency-division/radio-stations/wwv/wwv-and-wwvh-digital-time-code-and-broadcast**.

A word of caution — it's easy to be overwhelmed with all this data. In fact, much of it has little direct relevance to radio propagation — but it certainly is colorful and interesting to look at! However, knowing the level and trends of the 10.7 cm solar flux and sunspot number, along with the level and trend of the K-index, will give a good picture of HF propagation. The more esoteric parameters may serve to aid in the analysis of a particular propagation observation.

4.4.4 MUF Forecasts

LONG-RANGE

Long-range forecasts several months ahead provide only the most general form of prediction. For many years, ARRL published charts similar to **Figure 4.36** to forecast average propagation for a one-month period over specific paths. These charts are no longer published, but customizable charts are available online from **voacap.com/hf**. Charts like the one in Figure 4.36 assumed a single average solar flux value for the entire month and they assumed that the geomagnetic field is undisturbed.

The uppermost curve in Figure 4.36 shows the highest frequency that will be propagated on at least 10% of the days in the month. The given values might be exceeded considerably on a few rare days. On at least half the days, propagation should be possible on frequencies as high as the middle curve. Propagation will exceed the lowest curve on at least 90% of the days. The exact MUF on any particular day cannot be determined from these statistical charts, but you can determine when you should start monitoring a band to see if propagation actually does occur that day — particularly at frequencies above 14 MHz.

SHORT-RANGE

Short-range forecasts of a few days ahead are marginally more reliable than long-range forecasts because underlying solar indexes and geomagnetic conditions can be anticipated with greater confidence. The tendency for disturbances caused by solar phenomena to recur at 27-day intervals may enhance short-term forecasts. (See the section 27-Day Recurrence)

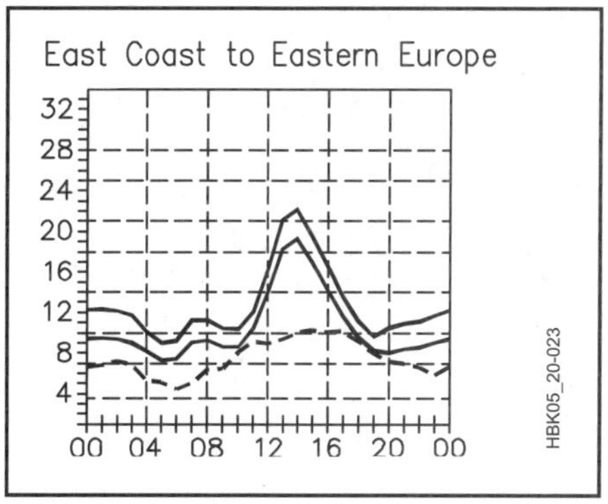

Figure 4.36 — For many years, ARRL published propagation prediction charts like this one for East Coast to Eastern Europe for April 2001. An average 2800 MHz (10.7 cm) solar flux of 159 was assumed for the month. On 10% of the days, the highest frequency propagated is predicted to be at least as high as the uppermost curve (the Highest Possible Frequency, or HPF, approximately 33 MHz), and for 50% of the days as high as the middle curve, the MUF. The lowest curve shows the Lowest Usable Frequency (LUF) for a 1,500-W CW transmitter.

Daily forecasts may not be any better, as the ionosphere does not instantly react to small changes in solar flux (or sunspot number) and geomagnetic field indexes. Regardless of these limitations, it is always good to know the current solar and geophysical data, as well as understanding warnings provided by observations of the Sun in the visual to X-ray range.

4.4.5 HF Propagation Prediction Software

Like predicting the weather, predicting HF propagation — even with the best computer software available — is not an exact science. The processes that cause a signal to propagate from one point on the Earth to another are enormously complicated and subject to an incredible number of variables. Experience and knowledge of propagation conditions (as related to solar activity, especially unusual solar activity, such as flares or coronal mass ejections) are needed when you actually get on the air to check out the bands. Keep in mind, too, that prediction software is written mainly to calculate propagation for great-circle paths via the F region. Scatter, skew-path, auroral, and other such propagation modes may provide contacts when computer predictions indicate no contacts are possible.

Modern programs are designed for quick-and-easy predictions of propagation parameters. See **Table 4.3** for a listing of a number of popular PC programs. (An online description of propagation prediction software is also available at **astrosurf.com/luxorion/qsl-review-propagation-software.htm**.)

The basic input information required is the smoothed sunspot number (R_{12}) or smoothed solar flux (F_{12}), the date (month and day), and the latitudes and longitudes at the two ends of the radio path. The latitude and longitude, of course, are used to determine the great-circle radio path. Most commercial programs tailored for ham use allow you to specify locations by the call sign. The date is used to determine the latitude of the Sun, and this, with the smoothed sunspot number (or

Table 4.3
Features and Attributes of Several Currently Available Propagation Prediction Programs

	ASAPS V. 7	*VOACAP Windows*	*W6ELProp V. 2.70*	*PropLab Pro 3.1*
User Friendliness	Good	Good	Good	Good
Operating System	*Windows*	*Windows*	*Windows*	*Windows*
Uses K-index	No	No	Yes	Yes
User library of QTHs	Yes/Map	Yes	Yes	No
Bearings, distances	Yes	Yes	Yes	Yes
MUF calculation	Yes	Yes	Yes	Yes
LUF calculation	Yes	Yes	No	Yes
Wave angle calculation	Yes	Yes	Yes	Yes
Vary minimum wave angle	Yes	Yes	Yes	Yes
Path regions and hops	Yes	Yes	Yes	Yes
Multipath effects	Yes	Yes	No	Yes
Path probability	Yes	Yes	Yes	Yes
Signal strengths	Yes	Yes	Yes	Yes
S/N ratios	Yes	Yes	Yes	Yes
Long path calculation	Yes	Yes	Yes	No
Antenna selection	Yes	Yes	Indirectly	Yes
Vary antenna height	Yes	Yes	Indirectly	Yes
Vary ground characteristics	Yes	Yes	No	No
Vary transmit power	Yes	Yes	Indirectly	Yes
Graphic displays	Yes	Yes	Yes	2D/3D
UT-day graphs	Yes	Yes	Yes	Yes
Area Mapping	Yes	Yes	Yes	Yes
Documentation	Yes	Online	Yes	Yes
Price class	$AUD385[1]	free[2]	free[3]	$240[4]

Prices are for early 2017 and are subject to change.

[1]See **www.sws.bom.gov.au/Products_and_Services/1/2**. Be advised that the price will depend on the exchange rate.
[2]*VOACAP* available at **its.ntia.gov/research-topics/radio-propagation-software/high-frequency/voacap-propagation-model**
[3]*W6ELProp, see* **www.qsl.net/w6elprop**
[4]*PropLab Pro V3, see* **www.spacew.com/proplab**

smoothed 10.7 cm solar flux converted to a smoothed sunspot number), is used to determine the properties of the ionosphere at critical points on the path.

A very powerful package, *VOACAP* is based on *IONCAP*, short for Ionospheric Communications Analysis and Prediction. It was written by an agency of the US government and has been under development for several decades. The *IONCAP* program has a well-deserved reputation for being difficult to use, since it came from the world of FORTRAN punch cards and mainframe computers.

VOACAP is a version of *IONCAP* adapted to Voice of America predictions, but this one includes a sophisticated Windows interface. The Voice of America (VOA) started work on *VOACAP* in the early 1990s and continued for several years before funding ran out. The program was maintained by a single, dedicated computer scientist, Greg Hand, at NTIA/ITS (Institute for Telecommunication Sciences), an agency of the US Department of Commerce in Boulder, Colorado. More information about VOACAP and an online app tailored for amateur radio are available from **voacap.com**.

ONLINE PREDICTION SERVICES

The website **voacap.com** by Jari Perkiömäki, OH6BG/OG6G uses *VOACAP*. As mentioned earlier, *VOACAP* includes the above-the-MUF mode mentioned earlier in the section on HF Scatter modes. If a path can withstand a bit more loss, the QSO may be completed although the MUF is *below* the operating frequency. And with FT8 having the ability to decode signals farther down into the noise than CW, it essentially says the MUF can be even lower than for CW. FT8 will certainly help in the summer sporadic E season, as well.

Another website, **soundbytes.asia/proppy**, is produced in collaboration with the RSGB's Propagation Studies Committee. Both websites have instructions on how to input your data to generate the desired prediction.

PC PREDICTION SOFTWARE

Table 4.3 contains information about four available software packages for the home PC. These programs generally allow predictions from 3 to 30 MHz. Unfortunately, prediction programs are not available on 160 meters (because of an incomplete understanding of the lower ionosphere and ducting mechanisms that contribute to propagation on that band) and on 6 meters (because of an incomplete understanding of openings when the MUF is not predicted to be high enough). As a general guideline, you should look for 160-meter openings when the path to your target station is in darkness and around sunrise/sunset, and you should look for 6 meter F_2 openings in the daytime during winter months near solar maximum.

PROPAGATION PREDICTION TABLES

The supplemental information contains detailed propagation-prediction tables from more than 150 locations around the world for six levels of solar activity, for the 12 months of the year. Again, keep in mind that these long-range forecasts assume quiet geomagnetic conditions.

4.4.6 Beacons

Automated *beacons* in the higher amateur bands can also be useful adjuncts to propagation watching. Beacons are ideal for this purpose because most are designed to transmit 24 hours a day.

One of the best organized beacon systems is the International Beacon Project, sponsored by the Northern California DX Foundation (NCDXF) and International Amateur Radio Union (IARU). The beacons operate 24 hours a day at 14.100, 18.110, 21.150, 24.930 and 28.200 MHz. Eighteen beacons on five continents transmit in successive 10-second intervals (each beacon transmits once every 3 minutes). More on this system can be found at the Northern California DX Foundation website **ncdxf.org** (click on the *Beacons* link on the left).

A network of automated SDR receivers called the Reverse Beacon Network (**reversebeacon.net**) has been created to monitor the HF CW sub-bands for signals that are automatically decoded by *CW Skimmer* and *RTTY Skimmer* software by VE3NEA (**dxatlas.com**). The network logs signal strength and location for the received signals, providing real-time information on HF propagation.

A list of many 28 MHz beacons can be found on the 10-10 International website, **ten-ten.org** (look for BEACONS under the RESOURCES link). Beacons often include location as part of their automated message, and many can be located from their call sign. Thus, even casual scanning of beacon sub-bands can be useful. **Table 4.4** provides the frequencies where beacons useful to HF propagation are most commonly placed.

There are also many beacons on VHF and higher bands. "G3USF's Worldwide List of 50 MHz Beacons" may be found at **keele.ac.uk/depts/por/50.htm**. Information on North American beacons on 144 MHz and up is maintained by Ron Klimas, WZ1V, at **newsvhf.com/beacons2.html**.

4.4.7 Other Methods for Real-Time Assessment of the Ionosphere

Since our propagation predictions are statistical in nature over a month's time frame, determining what the ionosphere is doing in real time can be important. In addition to using the effective sunspot number, making direct observations of our bands, monitoring WWV/WWVH on their five frequencies, and monitoring the IARU/NCDXF and other beacons as discussed previously, there are other ways to get a real-time picture of the ionosphere.

An emerging method is to use the Reverse Beacon Network (RBN — briefly discussed previously) to construct maps of F_2 region critical frequencies (foF2). Applying some simple rules will allow MUFs (maximum useable frequencies) to be easily estimated. See the RBN article in the online material for general information about the RBN. Also expect to see a technical article about foF2 maps generated from RBN data in the future.

Watching the worldwide spotting network is another way to assess real-time propagation. An extension of this is to plot the paths from the spots on a worldwide map. One such presentation is at **dxmaps.com/spots/map.php**.

Time-and-frequency stations are also excellent indicators since they operate around the clock. See **Table 4.5**. WWV and WWVH are also reliable sources of propagation data, hourly, as discussed in more detail elsewhere in this chapter. And the NCDXF/IARU beacon system (**ncdxf.org/pages/beacons.html**) can give you a real-time picture of worldwide propagation on our 20-, 17-, 15-, 12-, and 10 meter bands.

Table 4.5
Time and Frequency Stations Useful for Propagation Monitoring

Call	*Frequency (MHz)*	*Location*
WWV	2.5, 5, 10, 15, 20	Ft Collins, Colorado
WWVH	Same as WWV except 20	Kekaha, Kauai, Hawaii
CHU	3.330, 7.850, 14.670	Ottawa, Ontario, Canada
RID*	5.004, 10.004, 15.004	Irkutsk, Russia
RMW	4.996, 9.996, 14.996	Novosibirsk, Russia
VNG	2.5, 5, 8.634, 12.984, 16	Lyndhurst, Australia
BPM	2.5, 5, 10, 15	Xiang, China
BSF	5, 15	Taoyuan, Taiwan
LOL	5, 10, 15	Buenos Aires, Argentina

*The call sign, taken from an international table, may not be the one used during actual transmission. Locations and frequencies appear to be accurate as provided.

Table 4.4
Popular Beacon Frequencies

MHz	*Comments*
14.100, 18.110, 21.150, 24.930, 28.200	Northern California DX Foundation beacons
28.2-28.3	Several dozen beacons worldwide
50.0-50.1	Most US beacons are within 50.06-50.08 MHz
70.03-70.13	Beacons in England, Ireland, Gibraltar and Cyprus

PROPAGATION REPORTING NETWORKS

Other real-time methods to assess propagation include the world-wide DX spotting networks and the Reverse Beacon Network. A good example of a spotting network website is DX Summit sponsored by Radio Arcala in Finland. The website DX Maps also displays contacts by band. Visit **dxsummit.fi** or **dxmaps.com** to see who is working what. You can filter spots by band, mode, and geographical area to mitigate being inundated with spots that likely don't help.

The Reverse Beacon Network (RBN: **reversebeacon.net**) consists of many software defined radios (SDRs) that monitor multiple bands at once and decode CW messages. By transmitting a properly formatted short message using your call sign (typically CQ CQ CALL CALL) from these monitoring stations will be published and logged, including your signal-to-noise ratio (SNR) at each location. This allows you to assess propagation to the various monitors, and can also be used for A/B antenna comparisons. The RBN is described in an October 2016 *QST* article by Pete Smith, N4ZR, and Ward Silver, NØAX (see the Bibliography).

Two other propagation reporting networks are PSKReporter (**pskreporter.info/pskmap.html**) and WSPRnet (**wsprnet.org**). PSKReporter accepts reception reports for digital signals from stations around the world. The contacts are displayed on a world map which can be configured to show contacts by band and by one of several popular digital modes. Most logging software can be set up to automatically relay reception data to PSKReporter.

WSPR is a special mode that is part of the *WSJT-X* software suite (**physics.princeton.edu/pulsar/k1jt/wsjtx.html**). It is intended to act as a propagation assessment tool, using very low power. Stations receiving WSPR transmissions send reception reports to the WSPRnet server where they are stored and displayed on a map, sortable by band.

4.5 VHF/UHF Mobile Propagation

Most amateurs are aware that radio signals in free space obey the inverse-square law: the received signal power is inversely proportional to the square of the distance between the transmitting and receiving antennas. That law applies *only* if the transmitting and receiving antennas have an obstruction-free radio path between them.

Imagine two operators using hand-held 144 MHz radios, each standing on a mountaintop so that they have a direct line of sight. How far apart can they be and still maintain reliable communications? Assume 5 W transmitter power, 0 dBi antenna gain (2.15 dB worse than a dipole), 5 dB receiver noise figure, 10 dB S/N ratio and 12 kHz receiver bandwidth.

Ask experienced VHF mobile operators that question, and you'll generally get guesses in the range of 20 to 30 miles because their experience tells them that's about the most you can expect when not communicating through a repeater. The correct answer, however, is over 9,500 kilometers (5,938 miles) based on the parameters given in the previous paragraph! This also explains how some operators have been able to work the amateur radio station on the International Space Station using only a handheld transceiver.

The discrepancy is explained by the fact that the line-of-sight scenario is not realistic for a mobile station located close to ground level. At distances greater than a few miles, there usually is no line of sight — the signal is reflected at least once on its journey from transmitter to receiver. As a result, path loss is typically proportional to distance to the third or fourth power, not the second power as the inverse-square law implies.

4.5.1 Rayleigh Fading

Not only is the signal reflected, but it usually arrives at the receiver by several different paths simultaneously. See **Figure 4.37**. Because the length of each path is different, the signals are not in phase. If the signals on two paths happen to be 180 degrees out of phase and at the same amplitude, they will cancel. If they are in phase, then their amplitudes will add. As the mobile station moves about, the phases of the various paths vary in a random fashion. However, they tend to be uncorrelated over distances greater than $\lambda/4$ or so, which is about 20 inches on the 2 meter band. That is why if the repeater you are listening to drops out when you are stopped at a traffic light, you can often get it back again by moving forward a few inches.

Because there are typically dozens of paths, it is rare for their amplitudes and phases to be such that they all cancel perfectly. Fades of 20 to 30 dB or more are common. The range of signal strengths has a *Rayleigh* distribution, named after the physicist/mathematician who first derived the mathematical formula. That is why the phenomenon is called

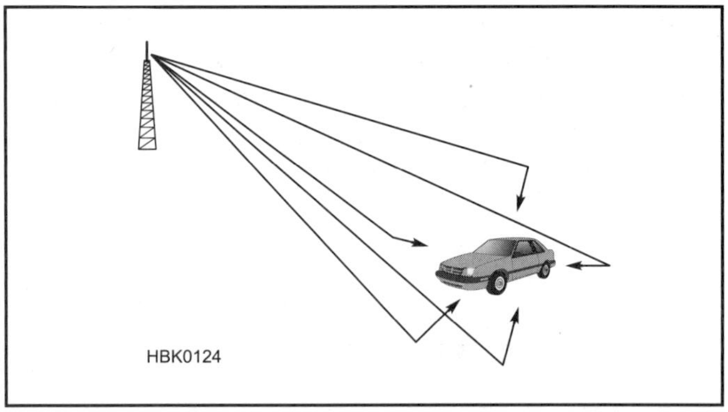

Figure 4.37 — At distances greater than a few miles, there normally is no line-of-sight path to the mobile station. Each path experiences at least one reflection. (Because the radio paths are reciprocal, propagation is the same for both receiving and transmitting.)

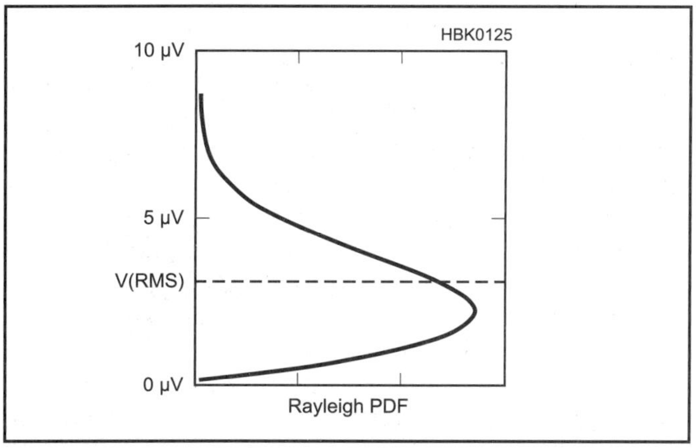

Figure 4.38 — The horizontal axis is the Rayleigh probability density function, which is the relative probability of the different signal levels shown on the vertical axis. For this graph, an RMS value of 3.16 μV has been selected to represent a typical average signal level in a marginal-coverage area.

Rayleigh fading. **Figure 4.38** shows the relative probability of various signal strengths. **Figure 4.39** is the same graph plotted on a logarithmic (dB) scale, along with a typical plot of signal strength versus time as the mobile station moves down the road.

The closer a reflecting object is to the antenna, the smaller is the path loss. A reflector that is close to the transmitter or receiver antenna gives a much stronger signal than one located halfway between. Even a weak reflector, such as a tree branch or telephone pole, is significant if it is close to the mobile station. Because there are many such close-in reflectors, many rays arrive from all directions.

Rays arriving from in front of a forward-moving vehicle experience a positive Doppler frequency shift and rays from the rear have a negative Doppler shift. Those from the sides are somewhere in-between, proportional to the cosine of the angle of arrival. The received signal is the sum of all those rays, which results in *Doppler spreading* of the signal as illustrated in **Figure 4.40**. At normal vehicle speeds on the 2-meter band, the Doppler spread is only plus and minus 10 or 15 Hz, calculated from:

$$\text{Doppler frequency} = F_c \, v/c$$

where

F_c = the carrier frequency
v = the vehicle speed
c = the speed of light

Use the same units for v and c. On an FM voice signal the only effect is a slight distortion of the audio, but Doppler spreading can severely affect digital signals, as will be discussed later.

4.5.2 Multipath Propagation

In addition to scattering by local reflectors, it is not uncommon also to have more than one main radio path caused by strong reflectors, such as large metal buildings located some distance away. See **Figure 4.41**. Each main path typically does not reach the mobile station directly but is separately Rayleigh-faded by the local reflectors.

As the mobile station moves around, the shadowing of various paths by intervening hills, buildings and other objects causes the average signal level to fade in and out, but at a much slower rate than Rayleigh fading. This is called *shadowing* or *slow fading*. It is also called *log-normal fading* because the distribution of average signal levels tends to follow a log-normal curve. That means that the *logarithm* of the signal level (on a dB scale, if you will) has a *normal* distribution (the famous bell-shaped curve). This effect typically causes the average signal level on each path to vary plus and minus 10 – 20 dB (at two standard deviations) from the mean value. This is in addition to the signal variation due to Rayleigh fading.

4.5.3 Effect on the Receiver

Radio direction finding (RDF) enthusiasts have noticed that their RDF receivers usually do not give stable or accurate indications unless located on a hilltop or other location with a clear line of sight to the transmitter because of fading and reflections. The best technique is to record the bearing to the hidden transmitter from a high location clear of nearby reflecting objects, then drive in that direction and take the next

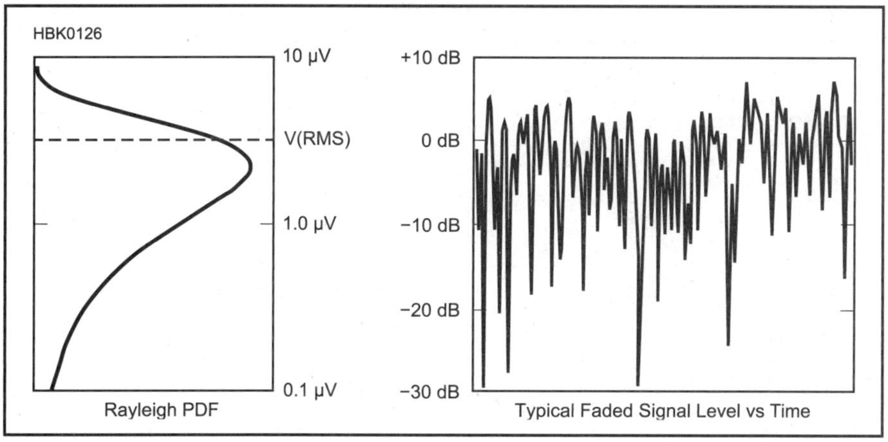

Figure 4.39 — The graph at the left is the same as Figure 4.38 except that the vertical axis is logarithmic (proportional to dB). At the right is a typical Rayleigh-faded signal of the same RMS voltage level plotted using the same vertical scale.

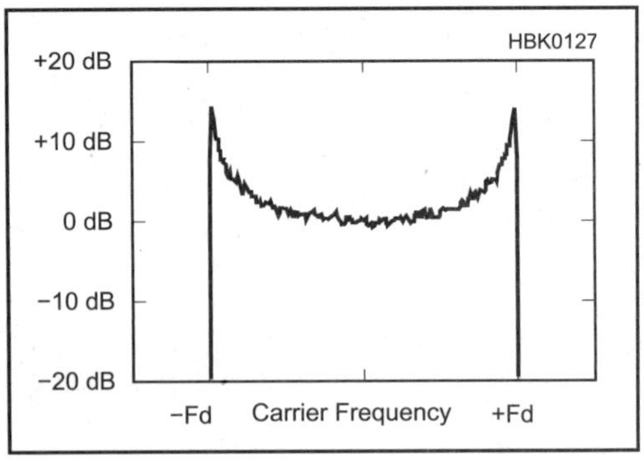

Figure 4.40 — Rayleigh fading spectrum of a CW (unmodulated) signal. The maximum deviation (F_d) from the center frequency is typically less than 10 – 15 Hz on the 144 MHz band.

reading from another (hopefully closer) hilltop. Only when close to the transmitter will the RDF equipment typically give good readings while the vehicle is in motion.

Under weak-signal conditions Rayleigh fading causes the signal to drop out periodically, even though the average signal level would be high enough to maintain reliable communications if there were no fading. *Picket fencing*, as such rapid periodic dropout is called, is a common occurrence when traveling in a weak-signal area at highway speeds. With analog modulation, normally the only solution is to stop at a location where the signal is strong. Moving the vehicle forward or backward a few inches is often enough to change an unreadable signal to solid copy.

Another possible solution is to employ *diversity reception*. If two (or more) mobile receiver antennas are spaced a half-wavelength or more apart, their Rayleigh fading will be almost entirely uncorrelated. That means it is relatively rare for both to experience a deep fade at the same time. The receiving system must have circuitry to determine which of the antennas has a stronger signal at any given time and to automatically combine the signals using some scheme that minimizes the probability of signal dropout. (One engineering text that has a fairly readable discussion of fading is *Cellular Radio Performance Engineering* by Asha Mehrotra; see the "References and Bibliography" section at the end of this chapter.)

One low-tech scheme is to use stereo headphones with each channel connected to a separate receiver and antenna. That method works better with linear modulation (AM, SSB, CW) than with FM because of the noise burst that occurs when the FM signal drops out.

Diversity antennas can also be used at a repeater site. The conditions are different because most repeater antennas are located in the clear with few local reflectors. The diversity antennas must be located much farther apart, typically on the order of 10 to 20 wavelengths, for the fading to be uncorrelated.

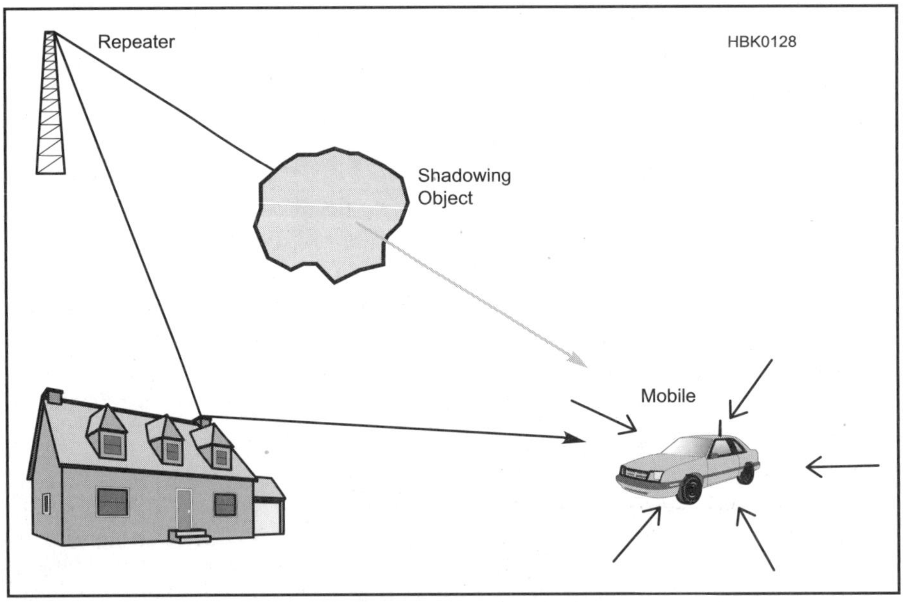

Figure 4.41 — When multipath propagation occurs, each main path is typically Rayleigh-faded by multiple reflectors close to the mobile station.

4.6 Special Propagation Modes and Topics

4.6.1 *WSJT-X* and WSPR

WSJT-X (Weak Signal Communication, see **physics.princeton.edu/pulsar/k1jt/wsjtx.html**) is a suite of digital protocols optimized for weak signal communications on HF and VHF/UHF. See the **Digital Protocols and Modes** chapter's section on Structured Digital Modes and this chapter's section on E Region Propagation. This software provides a wide variety of modes tailored for specific types of propagation. For example, FT8 is optimized for use at HF and 50 MHz; JT65 for EME on the VHF and UHF bands; JT4 for microwave bands; and JT9 especially for LF and MF. Any of these modes can be used on the HF bands to make worldwide contacts with a few watts and compromise antennas.

The WSPR mode implements a protocol that is specially designed for probing potential propagation paths with low power transmissions. WSPR transmissions carry a station's call sign, Maidenhead grid locator, and transmitter power in dBm. The program can decode signals with a signal-to-noise ratio as low as –28 dB in a 2500 Hz bandwidth. For more details on the WSPR mode and its capabilities, visit **wsprnet.org** and **physics.princeton.edu/pulsar/k1jt/wspr.html**.

4.6.2 Sporadic E

Short skip, long familiar on the 10-meter band during the summer months, affects the VHF bands as high as 222 MHz. *Sporadic E*, as this phenomenon is properly called, commonly propagates 28, 50, and, less frequently, 144 MHz radio signals between 500 and 2,300 kilometers (300 and 1,400 miles). Signals are apt to be exceedingly strong, allowing even modest stations to make E_s contacts. At 21 MHz, the skip distance may only be a few hundred kilometers. During the most intense E_s events, skip may shorten to less than 200 kilometers (120 miles) on the 10 meter band, with no skip zone at all on 15 meters. The first confirmed 220 MHz E_s contact was made in June 1987, but such contacts are likely to remain very rare. Much of what is known about E_s has come as the result of amateur pioneering in the VHF range.

Note that sporadic E can also be written as *E-subscript-s* (E_s) but is often written simply as Es (pronounced E ess). Sporadic E results from ionization at the lower boundary of the E-region, but of different origin and communication potential from the E region that affects mainly our lower amateur frequencies. *[The amateur and professional literature is not strict about the name, with sporadic E and Es used regularly — Ed.]*

4.6.3 Meteor Scatter

Contacts between 800 and 2,300 kilometers (500 and 1,400 miles) can be made at 28 through 432 MHz via reflections from the ionized trails left by meteors as they travel through the atmosphere. The kinetic energy of meteors no larger than grains of rice are sufficient to ionize a column of air 20 kilometers (12 miles) long in the E region. The particle itself evaporates and never reaches the ground, but the ionized column may persist for a few seconds to a minute or more before it dissipates. This is enough time to make very brief contacts by reflections from the ionized trails. Millions of meteors enter the Earth's atmosphere every day, but few have the required size, speed, and orientation to the Earth to make them useful for meteor-scatter propagation.

Radio signals in the 30 to 100 MHz range are reflected best by meteor trails, making the 50 MHz band prime for meteor-scatter work. The early morning hours around dawn are usually the most productive, because the morning side of the Earth faces in the direction of the planet's orbit around the Sun. The relative velocity of meteors that head toward the Earth's morning side are thus increased by up to 30 km/sec, which is the average rotational speed of the Earth in orbit. See **Figure 4.42**. The maximum velocity of meteors in orbit around the Sun is 42 km/sec. Thus when the relative velocity of the Earth is considered, most meteors must enter the Earth's atmosphere somewhere between 12 and 72 km/sec.

4.6.4 Propagation for Space Communications

Communication of all sorts into space has become increasingly important. Amateurs confront extraterrestrial propagation when accessing satellite repeaters and transponders or when using the Moon as a reflector. (More information on operating modes such as Earth-Moon-Earth and Satellites may be found in this book's chapter on **Antennas for Space Communications** and the supplemental chapter **Space Communications** of the *ARRL Handbook*.) Special propagation problems arise from signals that travel from the Earth through the ionosphere (or a substantial portion of it) and back again. Tropospheric and ionospheric phenomena, so useful for terrestrial paths, are unwanted and serve only as a nuisance for space communication. A phenomenon known as *Faraday rotation* may change the polarization of radio waves traveling through the ionosphere, presenting special problems to receiving weak signals. Cosmic noise also becomes an important factor when antennas are intentionally pointed into space.

SCINTILLATION

Extraterrestrial signals experience scintillation fading as they travel through the lower atmosphere, as described previously in the Tropospheric Fading section. However, they also

Aircraft and Spacecraft Scatter

If you run appropriate power, have a low noise receive system, and a decent antenna system, you may be able to complete a QSO using aircraft or satellite scatter. Your best chance of doing this is with the *WSJT-X* mode MSK144. For aircraft scatter observations, visit **w3sz.com/AircraftScatter.htm**.

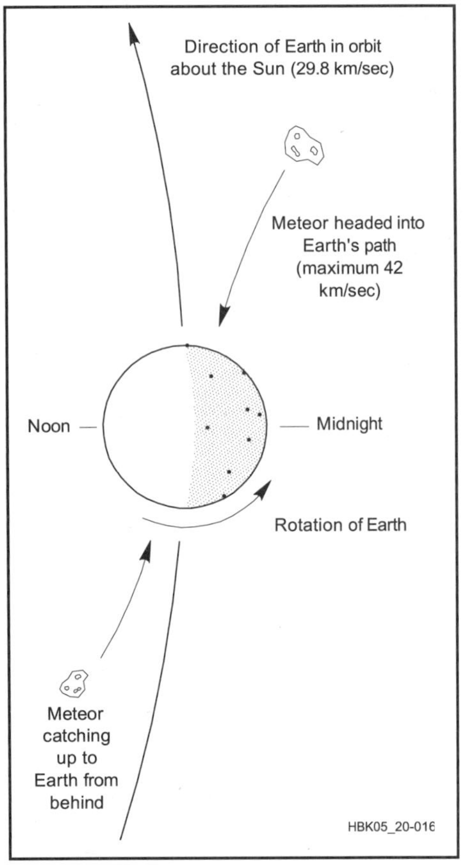

Figure 4.42 — The relative velocity of meteors that meet the Earth head-on is increased by the rotational velocity of the Earth in orbit. Fast meteors strike the morning side of the Earth because their velocity adds to the Earth's rotational velocity, while the relative velocity of meteors that "catch up from behind" is reduced.

experience scintillation fading as they traverse the ionosphere. The rule of thumb is that the ionosphere dominates below about 2 GHz and the atmosphere is usually more significant above 2 GHz. Scintillation in the ionosphere is more complex than in the troposphere because, unlike the troposphere, the ionosphere is highly anisotropic and irregularities tend to be aligned with the Earth's magnetic field lines.

Ionospheric scintillation varies with geographical location, time of day, the 11-year solar cycle, and the presence of geomagnetic storms. **Figure 4.43** shows the worldwide areas where scintillation occurs. The most intense and frequent disturbances are located within about 20 degrees of the Earth's magnetic equator. Fading depth in this region can be as much as 20 to 30 dB. Scintillation fading is also common within about 30 degrees of the magnetic poles, although fade depths are generally no more than 10 dB or so. At mid-latitudes ionospheric scintillation is rarely an issue. Longitude is important as well. In South America, Africa and India, the frequency of occurrence is significantly lower in the southern hemisphere winter (May-August) than during the summer, while the pattern is exactly opposite in the central Pacific.

In the polar regions disturbances occur at any time of day, but in the equatorial region they tend to start an hour or so after sunset and last a few hours, from perhaps 1900-2400 local time at the equinox. Fading events are sometimes intermittent, starting abruptly and ending abruptly some minutes later. The signal paths to slow-moving sources, such as the moon or geostationary satellites, tend to fade about once or twice a second. While not much work has been done on characterizing ionospheric scintillation fading on low Earth orbit (LEO) satellites, the fade rate should be more than an order of magnitude faster since the orbital period is typically on the order of 1.5 hours, much less than the Earth's 24-hour rotational period.

As might be expected, sunspots also are important. Ionospheric scintillation is much more severe and frequent near the peak of the 11-year solar cycle. Geomagnetic storms also have a major effect, principally in the polar regions.

EARTH-MOON-EARTH

Amateurs have used the Moon as a reflector on the VHF and UHF bands since 1960. Maximum allowable power and large antennas, along with the best receivers, are normally required to overcome the extreme free-space and reflection

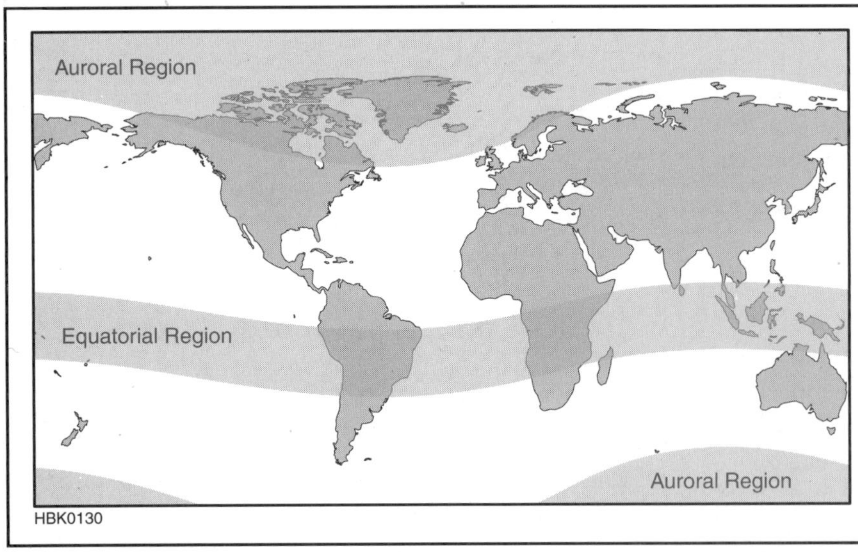

Figure 4.43 — Regions of the world (shaded) where scintillation fading effects can be especially severe.

losses involved in Earth-Moon-Earth (EME) paths, also known as *moonbounce*. More modest stations make EME contacts by scheduling operating times when the moon is at perigee (on the horizon). The Moon, which presents a target only one-half degree wide, reflects only 7% of the radio signals that reach it. Techniques must cope with Faraday rotation, cosmic noise, Doppler shift (due to the Moon's movements), and other difficulties. Despite the problems involved, thousands of amateur stations have made contacts via the Moon on all bands from 50 MHz to 10 GHz.

In communication over the Earth-Moon-Earth (EME) route, the polarization picture is blurred, as might be expected with such a multi-layered medium. If the Moon were a flat target, we could expect a 180° phase shift from the Moon reflection process, but it is not flat. This plus the Moon's *libration* (its slow oscillation, as viewed from the Earth), and the fact that waves must travel both ways through the Earth's entire atmosphere and magnetic field, provide other variables that confuse the phase and polarization issue. Most EME enthusiasts choose to take their chances with phase and polarization problems rather than attempt large tracking arrays.

SATELLITES

Accessing amateur satellites generally does not involve huge investments in antennas and equipment, yet station design does have to account for the special challenges of space propagation. Free-space loss is a primary consideration, but it is manageable when satellites are only a few hundred kilometers distant. Free-space path losses to satellites in high-Earth orbits are considerably greater, and appropriately larger antennas and higher powers are needed.

Satellite frequencies below 30 MHz can be troublesome. Ionospheric absorption and refraction may prevent signals from reaching space, especially to satellites at very low elevations. In addition, man-made and natural sources of noise are high. VHF and especially UHF are largely immune from these effects, but free-space path losses are greater. Problems related to polarization, including Faraday rotation, intentional or accidental satellite tumbling and the orientation of a satellite's antenna in relation to terrestrial antennas, are largely overcome by using circularly polarized antennas.

4.6.5 Propagation Below 1.8 MHz

Two bands in the LF/MF spectrum have been allocated for access by amateurs: 135.7 – 137.8 kHz and 472 – 479 kHz. It would be beneficial to review some fundamental issues with respect to propagation in these bands.

Three important parameters are involved in determining the likelihood of a signal getting from Point A to Point B. These parameters are refraction, absorption, and polarization. Although propagation is certainly different on 160 meters and 6 meters, an electromagnetic wave on both bands follows the same laws of physics. Thus, understanding how these three parameters change versus frequency will give us insight into propagation on our potentially new low frequency bands.

The amount of refraction incurred by an electromagnetic wave at a given frequency with a given electron density profile is inversely proportional to the square of the frequency. The second concept is that the amount of absorption incurred by an electromagnetic wave at a given frequency with a given electron density profile is also inversely proportional to the square of the frequency.

When undergoing refraction, as frequency decreases, the wave is bent more. Thus, waves at MF and LF don't get as high into the ionosphere as an HF wave. This results in shorter hops. For example, a 28 MHz wave easily gets up to 300 kilometers (186 miles) and gives maximum hops around and above 4000 kilometers (2,485 miles). A wave at 1.8 MHz only gets up to something like 200 kilometers (124 miles) and gives maximum hops around 2,000 kilometers (1,243 miles). And a wave at 150 kHz only gets up to around 80 kilometers (50 miles), with maximum hops of around 1,400 kilometers (870 miles). Thus, propagation on 2200 meters and 630 meters is via shorter hops than at HF.

Regarding absorption, as frequency decreases, the wave incurs more absorption. This is why a long-distance signal on 10 meters can be well over S-9 while a long-distance signal on 160 meters is normally much nearer the noise level. But an interesting phenomenon occurs below 160 meters, which is just above the electron gyro-frequency — the frequency at which an electron spirals around a magnetic field line. Absorption actually decreases as the frequency goes below about 1.5 MHz. Also, as stated above, 150 kHz waves don't get too high into the absorbing region at night (the lower E region). For more details, read the article titled "Physics of Propagation — Refraction Absorption Polarization" in the *Basic Concepts* link at the website **k9la.us**. In theory, highly elliptical polarization will prevail for these low frequencies. It is likely that most antennas for these bands will be vertically polarized, which should be a good match to couple the most energy into the ordinary wave.

In summary, skywave propagation on 135.7 – 137.8 kHz and 472 – 479 kHz will be via short hops between the ionosphere and the D region, but with lower absorption than on 160 meters. Tempering this lower absorption are the lower antenna efficiencies and higher received noise levels. Thus, these bands will be tough, but the amateur radio spirit will make the most of them.

This is a cursory look at propagation on our new bands. For a more detailed discussion, the December 2018 article titled "Propagation on 630m and 2200m" in the Monthly Feature link at **k9la.us** looks at both the ordinary wave and the extraordinary wave at these frequencies.

4.7 References and Bibliography

Barter, A., G8ATD, Ed., *VHF/UHF Handbook*, 2nd edition (RSGB, 2008). Includes a chapter on VHF/UHF propagation.

Donovan, F., W3LPL, "What to Expect During the Rising Years of Solar Cycle 25," *QST*, May 2021, pp. 57 – 59.

Farmer, E., AA6ZM, "A Look at NVIS Techniques," *QST*, Jan. 1995, pp. 39 – 42.

Jacobs, G., Cohen, T., Rose, R., *The New Shortwave Propagation Handbook* (CQ Communications, Inc., 1995).

Kamp, T., DF5JL, "60-m-Band: Chancen für NVIS," *CQ DL*, Feb. 2017, pp. 48 – 50.

Luetzelschwab, C., K9LA, **k9la.us**. Many articles about the ionosphere and propagation as they relate to 160 meters, HF, VHF, contesting, and more.

Maanen, E., PA3DES, Witvliet, B., PA5BW, Visser, G. PAØSIR, Westenberg, A., PAØA, "Elevation Angle Measurements During a Local Contest," *QST*, Jan. 2006, pp. 28 – 30.

McNamara, L., *Radio Amateur's Guide to the Ionosphere* (Krieger Publishing Company, 1994). Excellent, quite-readable text on HF propagation.

Muldrew, D., and Maliphant, R., "Long Distance One-Hop Ionospheric Radio Wave Propagation," *Journal of Geophysical Research*, May 1962.

Silver, W., NØAX, and Smith, P., N4ZR, "The Reverse Beacon Network," *QST*, Oct. 2016.

Straw, D., N6BV, "What's the Deal About 'NVIS'?" *QST*, Dec. 2005, p. 38.

TABLE OF CONTENTS

Chapter 5 — Online Material

Articles

- "How to Start Modeling Antennas Using EZNEC" by Greg Ordy, W8WWV
- "Modeling Insulated Wire" by Steve Stearns, K6OIK
- "Understanding the Changes to the FCC RF Exposure Rules" by Ed Hare, W1RFI

Chapter 5

Antenna Modeling and RF Exposure

In this chapter's first set of sections, we'll look at modeling antennas on a PC. We'll evaluate some typical antennas over flat ground and in free space. Once characterized — or even optimized for certain characteristics — these antennas can then be analyzed over real terrain using *HFTA*, a Windows program that performs *high-frequency terrain assessment*, and the other tools discussed in the **HF Antenna System Design** chapter. Steve Stearns, K6OIK, contributed the new sections on optimization, "thin-wire antennas," and electromagnetic modeling software. (Also see K6OIK's 2017 Pacificon presentation on antenna modeling, "Antenna Modeling for Radio Amateurs" at **fars.k6ya.org/docs/k6oik**.)

Previous editions of this book have included a special set of *EZNEC-ARRL* antenna model files that ran with demo versions of *EZNEC*. (**eznec.com**) With the retirement of its author, W7EL, the program is now free as *EZNEC Pro+ v. 7.0* and it will run all *EZNEC-ARRL* models without limitations. Model files referenced in this chapter are included with this book's online material.

5.1 Overview: Antenna Analysis By Computer

As pointed out in **The Effects of Ground** chapter, irregular local terrain can have a profound effect on the launch of HF signals into the ionosphere. A *system approach* as described in the **HF Antenna System Design** chapter is needed to create a scientifically planned station. Antenna modeling programs do not generally take into account the effects of irregular terrain and by "irregular" we mean any sort of ground that is not flat. Most modeling programs based on *NEC-2* or *MININEC* do model reflections, but they do not model diffractions.

On the other hand, while a ray-tracing program like *HFTA* (HF Terrain Assessment by Dean Straw, N6BV — described in the **HF Antenna System Design** chapter) does take into account diffraction, it doesn't explicitly factor in the mutual impedance between an antenna and the ground. Instead, *HFTA* makes the basic assumption that the antenna is mounted sufficiently high above ground so that the mutual impedance between an antenna and the ground is minimal.

In this chapter we'll look at modeling the antennas themselves on the PC. We'll evaluate some typical antennas over flat ground and also in free space. Once characterized — or even optimized for certain characteristics — these antennas can then be analyzed over real terrain using *HFTA* and the other tools discussed in the **HF Antenna System Design** chapter.

Previous editions of this book have included *EZNEC-ARRL*, a version of *EZNEC* antenna modeling software that worked with a special set of model files. Effective with this edition *EZNEC 7.0* (now free from **eznec.com**) will run all *EZNEC-ARRL* models from previous editions, without limitations. Model files including those referenced in this chapter are included with this book's downloadable supplemental information. Models created for earlier editions of this chapter have filenames beginning with "Ch8".

5.1.1 A Short History of Antenna Modeling

With the proliferation of personal computers since the early 1980s, amateurs and professionals alike have made significant strides in computerized antenna system analysis. It is now possible for the amateur with a relatively inexpensive computer to evaluate even complicated antenna systems. Amateurs can obtain a keener grasp of the operation of antenna systems — a subject that has been a great mystery to many in the past. We might add that modern computing tools allow hams to debunk overblown claims made about certain antennas.

The most commonly encountered programs for antenna analysis are those derived from a program developed at US government laboratories called *NEC*, short for "Numerical Electromagnetics Code." *NEC* uses a so-called *Method of Moments (MoM)* algorithm. (The name derives from a numerical method of dealing with accumulated errors in fields generated by current distributed along an antenna.) If you want to delve into details about the Method of Moments, see the excellent chapter in *Antennas*, 2nd edition, by John Kraus, W8JK. See also the article "Programs for Antenna Analysis by the Method of Moments," by Bob Haviland, W4MB, in *The ARRL Antenna Compendium*, Vol 4.

The mathematics behind the MoM algorithm are pretty formidable, but the basic principle is simple. An antenna is broken down into a number of straight-line wire segments, and the field resulting from the RF current in each segment is evaluated by itself and also with respect to other mutually coupled segments. Finally, the field from each contributing segment is vector-summed to yield the total field, which can be computed for any elevation or azimuth angle desired. The effects of flat-earth ground reflections, including the effect of ground conductivity and dielectric constant, may be evaluated as well.

In the early 1980s, *MININEC* was written in *BASIC* for use on personal computers. Because of limitations in memory and speed typical of personal computers of the time, several simplifying assumptions were necessary in *MININEC*, limiting potential accuracy. Perhaps the most significant limitation was that perfect ground was assumed to be directly under the antenna, even though the radiation pattern in the far field did take into account real ground parameters. This meant that antennas modeled closer to ground than approximately 0.2 λ sometimes gave erroneous impedances and inflated gains, especially for horizontal polarization. Despite some limitations, *MININEC* represented a remarkable leap forward in analytical capability. See Roy Lewallen's (W7EL) "*MININEC* — the Other Edge of the Sword" in Feb 1991 *QST* for an excellent treatment on pitfalls when using *MININEC*.

Because source code was made available when *MININEC* was released to the public, a number of programmers produced some very capable commercial versions for the amateur market, many incorporating exciting graphics showing antenna patterns in 2D or 3D. These programs also simplify the creation of models for popular antenna types, and several come with libraries of sample antennas.

By the end of the 1980s, the speed and capabilities of personal computers had advanced to the point where PC versions of *NEC* became practical, and several versions are now available to amateurs. The most recent public-domain version is *NEC-2* and this is the computational core that we'll use as an example throughout this chapter.

Like *MININEC*, *NEC-2* is a general-purpose modeling package and it can be difficult to use and relatively slow in operation for certain specialized antenna forms. Thus, custom commercial software has been created for more user-friendly and speedier analysis of specific antenna varieties, mainly Yagi arrays described in the chapter on **HF Beam Antennas**. Also see the sidebar, "Commercial Implementations of *MININEC* and *NEC-2* Programs."

The following material on antenna modeling is by necessity a summary since entire books have been written on this subject. Additional antenna modeling resources may be found on the ARRL website at **www.arrl.org/antenna-modeling**. We also strongly recommend that you read the **HELP** files available with the demo version of *EZNEC*. It contains a wealth of practical information on the finer points of antenna modeling.

5.1.2 Comparing *NEC-2* To *NEC-4*

The following section was contributed by Greg Ordy, W8WWV, for the 23rd edition. Another article by Greg, "How to Start Modeling Antennas Using EZNEC," is included with this book's downloadable supplemental information.

Two popular antenna modeling programs are *EZNEC* and *4nec2*. They are GUI (graphical user interface) shells or wrappers that use an *NEC* modeling engine to perform the antenna simulation. (See the Reference section of this chapter for website URLs for these and other programs referenced in this sidebar.)

NEC versions 2 and 4 are the engines available to both modeling programs. I know of no antenna model where *NEC-2* would be preferred to *NEC-4*. *NEC-2* survives because it is free software in the public domain and *NEC-4* is licensed software that costs at least several hundred dollars. Programs such as *EZNEC* and *4nec2* have augmented *NEC-2* with extensions that address the most serious of the *NEC-2* deficiencies in many models.

For a wide range of models, the choice between the *NEC-2* and *NEC-4* engines makes little difference. There are models, however, where the choice of engine becomes very important if the highest accuracy is desired.

This section highlights the differences between *NEC-2* and *NEC-4*. Most of the information comes from the references included at the end of this chapter.

HISTORY

The acronym "NEC" (pronounced *neck*) stands for Numerical Electromagnetics Code. It is not related to the other popular use of "NEC," which is used to refer to the National Electrical Code.

•*NEC-1* was developed at the Lawrence Livermore National Lab (LLNL) in 1977. *NEC-1* built upon versions 1 and 2 of the *Antenna Modeling Program* (*AMP*). They date back to 1974.

•*NEC-2* was released in 1980, with its user manual dated January, 1981.

•*NEC-3* was released in 1983. It addressed one of the two major problems with *NEC-2*. It accepts buried wires and wires that penetrate a lossy media (such as ground).

•*NEC-4* was released in 1992. It extends and improves *NEC-3*. In particular, it corrects the problem that *NEC-2* and *NEC-3* have with stepped-radius wires and junctions of tightly coupled wires.

•*NEC-5* was released in 2019. It improves on *NEC-4* modeling of surface fields and the use of *MININEC* ground. It lacks several important features of *NEC-3* and *NEC-4* as described in its summary below.

NEC-2 is the most recent version that was released in source form without restriction into the public domain. It has been used as part of many commercial and freeware software packages. It's safe to say that more models have been run using *NEC-2* than any other modeling engine.

NEC-4 and *NEC-5* are both available as discussed at **ipo.llnl.gov/technologies/software/nec-v50-numerical-electromagnetic-code**. To use *NEC-4* or *NEC-5* legally, it is necessary to obtain a license from the Lawrence Livermore National Laboratory (LLNL). The one-time license fee is a function of the intended use of the software (personal/commercial/foreign). The cost of US non-commercial individual licenses for *NEC-4* is $300 and $110 for *NEC-5* as of 2023. The license is non-transferable.

It is worth mentioning the *MININEC* modeling engine. *MININEC* was developed as a personal computer antenna modeling engine in 1982. It is written in the *BASIC* programming language as opposed to *FORTRAN*. Despite the name, it is not a cut down version of *NEC*. While it has its own set of issues and concerns, in some areas it is considered to be superior to *NEC-2* and potentially even *NEC-4*. In particular, *MININEC* models stepped-radius wires accurately. *MININEC* has its place in the world of antenna modeling. It is available in packages such as *MMANA-GAL*.

In recent years, the specific version of *NEC-4* that has been in common use is version 4.1. In 2011, version 4.2 was released, and is included along with version 4.1 as part of the licensed package from LLNL. Version 4.2 adds a new ground option. In some cases, it can be more accurate than the 4.1 version. Using it, however, does slow down running the model. References to *NEC-4* in this chapter refer to version 4.2.

The modeling engine software dates back to the time before the explosive expansion of the power and capacity of the personal computer (PC). To provide the highest performance at the time, programmers included options such as using single-precision (32-bit) floating point computations as opposed to double-precision (64-bit) computations. After almost 30 years of PC evolution, there are very few cases where using the single-precision modeling engine makes sense. There are cases, however, where the added precision of the double-precision engine improves accuracy.

NEC-4 is more "idiot proof" than *NEC-2*. It has fewer idiosyncrasies and modeling guidelines to follow. To use *NEC-2* effectively, you have to be smarter about how to use it, and even when to use it.

It has always been possible to directly use the *NEC* engines and skip the wrapper or shell programs. Be aware, however, that the original interface to *NEC* is a deck of punch cards on the input and a long plain-text output file intended for a line printer.

AVERAGE GAIN TEST

Regardless of the engine in use, it's always a good idea to use the Average Gain Test to evaluate and increase the confidence in a model. When a model is inappropriate due to guideline violations or using an engine incorrectly, the Average Gain Test can highlight the problem.

A lossless antenna in a lossless environment radiates all incoming power in some direction. To perform the average gain test all loss is removed from the model. When run, all of the power supplied to the antenna should be captured in the pattern. If not, then the lossless antenna is acting like an amplifier or attenuator, and since it is neither, something is wrong with the results.

Details on running the Average Gain Test can be found in the program documentation. Often times the model can be adjusted to pass the test. If that is not possible, then the numeric results of the test can be used to correct the reported gain. If your model results are way out of line with common sense expectations, then it is a good time to run the Average Gain Test.

DIFFERENCES BETWEEN *NEC-2* AND *NEC-4*

Stepped Wire Diameters

Perhaps the biggest failing with *NEC-2* is its handling of stepped-diameter wire connections. If this sounds esoteric, please remember that telescoping aluminum tubing is used to construct many of our antennas — including the Yagi and ground-mounted vertical. When an aluminum tube slides into the next size tube, a stepped diameter situation is created. This problem was finally addressed in *NEC-4*, and *MININEC* does a good job of handling it appropriately too. If not for the ability to compensate or correct for this problem, I think it's fair to say that *NEC-2* would have been abandoned — certainly for this class of antenna.

The correction approach used was developed by Dr David Leeson, W6NL. It is detailed in Chapter 8 of his book *Physical Design of Yagi Antennas*. Although more than 20 years old, it is still a valuable reference for anyone who is serious about the mechanical or electrical aspects of a Yagi.

The stepped-diameter correction algorithm converts a set of coaxial and stepped wires into a single wire with a uniform diameter. The before and after elements are considered to be equivalent. The shell programs such as *EZNEC* and *4nec2* identify correction opportunities and apply the algorithm automatically. The corrected model is sent to the engine. Although this tight integration makes it painless to use the correction algorithm, there are a set of conditions that have to be true in order for the correction to be used. Needless to say, if you believe that the correction is being used, but it isn't, the results will contain more error than expected.

The *EZNEC* Help documentation contains a complete list of the constraints on the use of the stepped diameter correction algorithm. The program also indicates when the correction is in use, but in the heat of modeling the antenna it's possible to lose track of the status information.

Over-simplified, the correction can only be applied to a set of two or more wires that are all collinear with more than one diameter. In addition, if there the model includes a Source or Load or Transmission Line, they must be connected in the middle of the element. These constraints allow for many antenna designs, but not all.

Situations that would disable the correction include:

1. Phasing lines running down and connecting to a set of elements.

2. Loading coils or traps not located at the center of an element.

3. Wires located near the feed point used in matching networks such as the gamma or hairpin (beta) match.

4. Non-collinear wires, such as squares or rectangles in antennas such as quads or the Moxon rectangle.

5. Wires groups that are not within 15% of the half-wave resonant length (or quarter-wave resonant length for wire groups with one end grounded).

If a particular model cannot use stepped diameter correction where needed, and the highest accuracy is desired, then moving to *NEC-4* (or possibly *MININEC*) is the solution.

While on the topic of modeling Yagis, two more correction or compensation situations warrant mention. They are compensating for the metal boom and the metal element clamps. The metal present in the boom and clamps changes the electrical length of the element.

Boom compensation for the typical HF Yagi is on the order of ⅛ inch per half-element. This might be smaller than the construction accuracy and is often ignored. On VHF and UHF antennas, however, where the ratio between element length and boom diameter is much lower, boom compensation is an important consideration.

The element clamping scheme also creates a need for compensation, even at HF. There are a number of clamp styles and sizes in use. Chapter 9 of Leeson's book tackles the problem of clamp compensation. The approach converts the dimensions of a set of clamp styles into a wire length and diameter that can be located at the center of the element in the model. This is a clever way to incorporate the effect of the clamp by turning it into something that can be modeled — a fat and short wire located at the middle of the element.

The *AutoEZ* program from Dan Maguire, AC6LA, is a very powerful antenna modeling tool that uses *EZNEC* as its engine. *AutoEZ* incorporates the Leeson clamp models. By selecting a clamp style and then entering the dimensions, *AutoEZ* will compute the equivalent wire and add it to the model.

When you come across a model for a Yagi, especially an existing commercial product, and if you find a relatively short and fat wire at the center of an element it's safe to assume that it is a proxy for the element to boom mounting clamp. This is most certainly true if the wire is not part of the mechanical specification. It's intended to represent the effect of the actual metal clamp in the model.

The errors introduced by not using stepped diameter correction (*NEC-2* only) or boom compensation, or clamp compensation, all shift the model performance in the same direction. If they are not used the model results will be shifted downward in frequency. The natural response to this result is to scale the design up in frequency, usually by shortening the elements. Now, the model results line up with the target frequency. Unfortunately, if you build the antenna using those dimensions and then measure characteristics such as SWR, you will find that the antenna performance characteristics have been shifted above the target frequency. This leads to the sad realization that your elements are too short.

The impact on the actual antenna is on the order of ½ inch per each element end on 15 meters, a little bit more on 20 meters and a little bit less on 10 meters. This is for antennas modeled using *NEC-4*, but without clamp or boom compensation added to the model. The size and shape of the clamps will influence the amount of compensation needed.

If the short and fat wire representing the clamp contains a *NEC* Source, representing the feed point, then even with *NEC-4* there can be additional error in the results depending upon the segment length to diameter ratio. One approach is to use the average gain as a correction factor, or, use the stepped diameter correction algorithm even with the *NEC-4* engine.

Wires Below Ground

Since before the classic paper "Ground Systems as a Factor in Antenna Efficiency" was published by Brown, Lewis and Epstein in 1937 (see the Bibliography for the chapter **Effects of Ground**), the topic of ground wires and radials for ground mounted vertical antennas has been discussed. If you wish to explore what modeling predicts about wires very near to the surface of the ground or buried below the surface, you must use *NEC-4*.

Rudy Severns, N6LF, has written extensively on the topic of radial systems, using both *NEC-4* models and field measurements. His work is an excellent place to start any investigation of this topic, including the material in this book's chapter, **Effects of Ground**.

If it's not possible to model ground radial systems with *NEC-2* then that should surely imply that vertical antennas with ground radials can only be modeled with *NEC-4*. Fortunately, that is not the case. *NEC-2* has been modified to include a ground type that was first used with the *MININEC* engine. This is called the *MININEC*-type ground. In this case, the approach is to directly connect the vertical to *MININEC* ground, and then add a resistive Load at the base with a value chosen to simulate the expected ground resistance for the presumed radial system. Many sources such as the *ARRL Antenna Book* and ON4UN's *Low-Band DXing* explore the topic of ground resistance for radial fields.

The use of the *MININEC*-type ground turns out to be a good solution because the truth is that in most cases precise modeling of radial fields is work. The *MININEC* solution works well for investigating vertical antennas and arrays of vertical antennas.

If you are investigating buried wires with *NEC-4*, but sure to use the real/high accuracy ground model. There is a general admonition against using the *MININEC* ground with any horizontal wire less than 0.2 wavelength above ground. Elevated radials can be modeled with either *NEC-2* or *NEC-4*.

Other Differences

Here are two other differences between *NEC-2* and *NEC-4* that should be kept in mind while modeling. Quoting from the *EZNEC* Help documentation, "*NEC* has some difficulty in accurately modeling multiple wires joining at a very acute angle, such as with a "fan" antenna, the difficulty being greater with *NEC-2* than with *NEC-4*."

Similarly, quoting from the *NEC-4.2* User's Manual, "The size of the segments determines the resolution in solving for the current on the model, since the current is computed at the center of each segment. Earlier versions of *NEC* suffered a loss of precision or complete failure of the solution when very short segments were used, but this problem has been corrected in *NEC-4*. The extremely short segments can be used with *NEC-4*, subject to limitations related to the wire radius as discussed below."

Geometry and Segmentation Checks

NEC models are built out of wires. Wires are divided into segments. The number of segments on a wire is specified by the modeler. The number of segments on a wire is an important factor in determining the accuracy of the results as well as the time it takes to run the model.

Segments are like the story of Goldilocks and the Three Bears — you do not want to use too few or too many segments. There are many trade-offs in determining the best number of segments to use.

1. Reducing the number of segments speeds up the *NEC* engine. Past some point, accuracy suffers because the model becomes too coarse.

2. Increasing the number of segments slows down the *NEC* engine. The maximum number of allowed segments may be limited by the modeling package. At some point, too many segments can reduce the accuracy of the results. This is especially true with *NEC-2*. Blindly using more segments is not a solution.

3. There are situations where it's desirable to align segments between closely-spaced wires.

Fortunately the programs that drive the *NEC* engines include a number of checks that follow the *NEC* segmentation guidelines. You should observe their warnings related to the segment count. It is also possible to have the program automatically segment the wires in a model.

If there is any concern about the number of segments being used, you should perform segment convergence testing. The idea is to run a model, note the results (SWR, impedance, gain, F/B, and so on), change the number of segments, and run it again. As you move from too few toward too many segments, the results should converge. Once they do, there is little point in increasing the number of segments. If you notice that the results are very sensitive to segmentation, then it may signal a model that challenges the *NEC* engines, and might have less trustworthy results. *4nec2* has a convergence test option that automates much of the process. Getting segmentation correct is more important with *NEC-2* than *NEC-4*.

5.1.3 Thin-Wire Modeling Software

The earliest programs for antenna and electromagnetic scattering analysis were "thin-wire" programs. Thin-wire programs use mathematical simplifying approximations to model antennas made of one or more straight wires, rods, and tubes, whose diameters are constant and small. Current flows longitudinally on the surface of the wire due to skin effect. Current is assumed to have circular symmetry, i.e. uniform distribution around a wire's axis. Thin-wire codes cannot model proximity effect, when current is not equal and symmetric around a wire. *NEC* and *MiniNEC* are well known examples of early thin-wire programs. *EZNEC*, *4nec2*, *Nec2Go*, and *MMANA-GAL* provide graphical user interfaces (GUIs) and improvements to these programs. However, other thin-wire programs, developed later, are more accurate. Some alternative, low-cost thin-wire codes, *AWAS*, *MBC*, *AnSim*, *AN-SOF*, and the recent *NEC-5*, are introduced here.

AWAS

AWAS is an acronym for Analysis of Wire Antennas and Scatterers. The program was developed by A.R. Djordjevi⍰ at University of Belgrade, T.K. Sarkar at Syracuse University, and others. It built upon the earlier program WireZeus, and the same equations were used in the later programs *WIPL-D*, *Tides*, and *HOBBIES*. These programs are referred to as the Popović family of computational electromagnetics programs due to their common origin and relations. Current in wires is represented by using long segments and polynomial basis functions. The advantage of polynomial basis functions is that segments can be fewer and longer. Wires are not restricted to being straight or constant diameter. Curved wires are defined by parametric equation. A helix or coil can be defined as a single wire. Diameter can vary linearly along the length of a wire because wires are treated as right circular cones or circular frustums instead of cylinders. *AWAS* has a very accurate real ground model. *AWAS* has a Sommerfeld ground subroutine developed by T.K. Sarkar and A.R. Djordjević. It is very accurate, possibly more accurate than the one in the *NEC* programs. This ground code wasn't migrated into subsequent programs in the Popović family. Modern surface codes, such as *FEKO*, *HOBBIES*, and *WIPL-D*, treat real ground directly as a dielectric surface of large but finite extent. The surface need not be flat.

The thin-wire codes in *AWAS*, *WIPL-D*, and *HOBBIES* were written by different programmers from the same equations. Some differences are in *AWAS* the user sets segmentation, polynomial degree and computation precision, whereas in *HOBBIES* segmentation and polynomial degree are automated, and computation is double precision by default. Unlike *NEC*, *AWAS* accuracy is not sensitive to acute angle junctions between wires. The program is numerically stable if wires are near other wires although, being a thin-wire code, proximity effect is not modeled. *AWAS* is a good alternative to *NEC* for thin-wire antennas. *AWAS* 2 runs under Windows and has an intuitive GUI or user interface but lacks an optimizer. The program is no longer sold or supported, but a free "demo" version of *AWAS* 2 can be downloaded from the University of Belgrade. See the software list elsewhere in this chapter.

ANSIM AND *MBC*

MBC is the multi-radius bridge current program developed by M. Tilston under K.G. Baslmain at University of Toronto. The equations are in the published literature and have long been considered the gold standard for thin-wire antenna analysis. The *MBC* program is not available commercially. It can be obtained from University of Toronto but is replaced by a newer program *AnSim*. *AnSim* is a thin-wire moment method electromagnetic modeling program based on the high-accuracy *MBC* equations. As with *AWAS*, a major advantage of *AnSim* over *NEC* and *MiniNEC* is better accuracy with fewer restrictions or limitations on how wires are placed or segmented. *AnSim* can model electrically small and medium size structures made of wires, plates, and dielectric blocks.

AnSim has limited ground models. Infinite flat perfect electrical conducting (PEC) ground is simulated by the method of images, i.e., by putting a reflected copy of an antenna symmetrically below a hypothetical ground interface. *AnSim* also allows ground to be modeled as a dielectric surface of large but finite size meshed into patches. The same approach is found in *FEKO*, *HOBBIES* and *WIPL-D*. The advantage of this approach is ground can be either flat or irregular, like real terrain. *AnSim* lacks an optimizer. The program has a script editor for entering commands. A GUI user interface was being developed and may be available. *AnSim* is sold and supported by Phoenix Antenna Systems in Perth, Ontario, Canada. An evaluation version is free, and the full program is affordably priced. See the software list elsewhere in this chapter.

AN-SOF

AN-SOF is a descendant of the *EMMCAP* electromagnetic program at the University of Buenos Aires, Argentina developed by T. Golden and L.A. Dorado. *AN-SOF* uses of an exact EFIE (electric field integral equation) and exact kernel, which give it high accuracy under geometric conditions which programs that use approximate kernels cannot model accurately, such as close-spaced wires and acute angle junctions. *AN-SOF* uses triangle basis functions like *MiniNEC*. Wires can be curved, not just straight. A curve is defined parametrically as in *HOBBIES* and *WIPL-D*. Wires are treated as circular cylinders of constant diameter. *AN-SOF* follows the *NEC*-2 and *NEC*-4.2 convention of putting sources at the middle of segments, whereas *AWAS*, *WIPL-D*, *HOBBIES*, and *NEC-5* put sources at wire ends.

AN-SOF uses a two-part approach to modeling real ground that's somewhat like *MiniNEC*: using one method to calculate antenna currents and switching to another for calculating far field patterns. Given ground dielectric properties, antenna currents are calculated assuming not one but many images. The method is based on a generalization of a reflection coefficient ground for dielectric surfaces that results in an infinite series of images of different strengths and depths. The generalized image method was developed by J.R. Wait and is an asymptotic approximation to Sommerfeld ground. The method of moments is used to calculate the currents in the antenna and its images. Having found the currents, far field radiation and antenna pattern are calculated by using conventional single-image complex Fresnel reflection coefficient ground as done in other programs. Although *AN-SOF* lacks an optimizer, it does have a good, intuitive user interface, and it allows ARRL scaling of polar plots. *AN-SOF* is sold and supported by Golden Engineering, in Montevideo, Uruguay. An evaluation version is free, and the full program is affordably priced. See the software list elsewhere in this chapter.

NEC-5

NEC-5 was developed by G.J. Burke and released in late 2019. It is the final iteration of the *NEC* family. Program development stopped with Burke's death in 2021. *NEC-5* uses different mathematics from earlier *NEC*s and is more stable numerically. It uses a "two-potential" MoM formulation like the one used in the Popović family of modeling codes. The two-potential integral formulation leads to exact an EFIE and exact kernel. *NEC-5* still uses straight line wires divided into segments and three-term (sine, cosine, and constant) basis functions. Although *NEC-5* is probably the most accurate member of the *NEC* family, the software is the least developed. The software lacks important features that were in *NEC*-3 and *NEC*-4, such as an insulated wire option. Sources are handled differently too: sources are placed at junctions as in the Popović family (*AWAS*, *WIPL-D*, and *HOBBIES*), instead of mid-segment as in earlier *NEC*s. *NEC-5* lacks an optimizer. It runs by using a command script but also includes a simple user interface GUI. LLNL provides GUI source code with *NEC-5*. Consequently, amateur software developers can continue to add features. *EZNEC*'s final version can run *NEC-5* if installed. Given that *EZNEC* and *AutoEZ* can run *NEC-5*, one can use them to operate *NEC-5*. The trio of *NEC-5*, *EZNEC*, and *AutoEZ* are powerful and affordably priced. *NEC-5* is available from Lawrence Livermore National Laboratory. See the software list elsewhere in this chapter.

COMPARISONS OF THIN-WIRE CODES

Amateurs have studied and compared the thin-wire codes mentioned above. AC6LA examined numerical results for differences. GØKSC compared numerical results to experimental measurements. K6OIK examined the mathematical equations of the algorithms in the various programs. It was found that *AN-SOF* and *NEC*-5 agree with each other, and *NEC*-2 and *NEC-4.2* agree with each other. However, the pairs disagree. Thus, *NEC-5* and *AN-SOF* are of comparable accuracy or inaccuracy. In side-by-side comparisons *AN-SOF* is roughly equivalent to *NEC*-5 and appears to be more accurate than NEC-4.2. The reason is thought to be because *AN-SOF* and *NEC-5* both use a two-potential EFIE and exact kernel, similar to that in *AWAS*, *WIPL-D*, and *HOBBIES*, whereas *NEC*-2 and *NEC*-4.2 use different integral equations with approximate kernels in their MoM calculations. Other known differences are each program handles sources and ground differently.

Delta-gap sources are known to be a source of error in MoM codes. K6OIK calculated the broadband impedance function of a dipole in free space by six methods (Induced EMF, *NEC*-2, *NEC*-4, *FEKO*, *WIPL-D*, and *HOBBIES*). It was found that differences in calculated impedance can be explained as differences in shunt capacitance at the source. Differences among MoM methods can be attributed to different shunt capacitances inherent to their delta-gap sources: *FEKO*'s delta-gap source has the smallest shunt capacitance; *HOBBIES* and *WIPL-D*'s delta-gap sources have the greatest shunt capacitance; and *NEC*-2 and *NEC*-4's delta-gap sources have intermediate shunt capacitance. This finding explains why MoM programs have difficulty determining resonant frequency to great accuracy. Modern programs generally have better formulations than earlier programs.

Apart from accuracy, an important difference for hams is that the thin-wire codes mentioned above as well as *WIPL-D* and *HOBBIES*, do not suffer restrictions on wire proximity to other wires, nor segment length-to-diameter ratio. However thin-wire codes cannot not model proximity effect, even though they may be numerically stable if wires are near to each other. Thus *NEC-5* is numerically stable if wires are in proximity of each other, but *NEC-5* does not model proximity effect, nor do the other thin-wire codes. Surface-meshing codes like *FEKO*, *WIPL*, *WIPL-D*, or *HOBBIES* can model proximity effect if wires are treated as surfaces instead of wires. Applications where proximity effect is important are the electromagnetic modeling of low-Z open wire lines or computing the effect of nearby trees on antenna performance.

Numerical comparisons merely show that different programs produce different results. The question remains which program is the most accurate. This question is best answered by comparing against standard test cases for which exact solution is possible. The case of small loops over ground can be analyzed exactly. The solution is in papers by J.R. Wait. A validation test would reveal which program gives the most accurate answer.

5.1.4 Antenna and Electromagnetic Modeling Software

Originally developed at Lawrence Livermore National Laboratory (LLNL) in the 1970s, the program *Numerical Electromagnetics Code* or *NEC* is publicly available for general use and is available for personal computers running Windows, Linux, and macOS. The public version is *NEC-2* (**nec2.org**). An updated version, *NEC-4*, is only available as a compiled binary program under license agreement with LLNL and is best-known to amateurs in the *EZNEC-Pro* software. More information about *NEC* is available from **en.wikipedia.org/wiki/Numerical_Electromagnetics_Code**.

The *NEC* program family (*NEC*, *NEC-BSC*, *NEC-2*, *NEC-3*, *NEC-4*, *NEC-5*) and the alternative implementation *MININEC* are examples of "thin wire" codes, which are limited to modeling antennas made of wires, rods, and tubes, with materials limited to non-magnetic metals.

Other programs have appeared since *NEC* was developed. For example, a series of programs started from B.D. Popović's modeling program *WireZeus*, including *AWAS*, *WIPL*, *WIPL-D*, and *HOBBIES*. *WireZeus* and *AWAS* are thin-wire programs. *WIPL* and *WIPL-D* are professional modeling programs. *HOBBIES* was released in 2012 and is the newest member of the family. Most of these programs are characterized as electromagnetic "solvers" or simulators that use advanced numerical algorithms to handle general geometries and arbitrary materials.

Programs such as *WIPL-D*, *FEKO*, and *HOBBIES* have, in addition to a thin-wire code, the ability to model surfaces. Just as a thin-wire code divides a wire into segments, a surface code divides a surface into elementary patches by using a process called meshing. Surface meshing allows one to model irregularly shaped antennas on irregularly shaped platforms such as automobiles, airplanes, ships, and even irregular terrain. Consequently, models can include structural features such as buildings, trees, hills, and valleys, which a thin-wire code cannot model.

Figure 5.1 shows a mesh model of a car's surface modeled as metal skin over ground. A 2 meter whip is mounted on the car's roof. The surface has been meshed into small triangular patches by the modeling program *FEKO*. A source placed at the base of the whip feeds it against the roof. *FEKO* computes RF current in each triangular patch. **Figure 5.2** shows the computed currents. Magnitude is shown as shades of grey and current direction is depicted by small arrows. Note how current concentrates around the edges of the windows. (See KI6BDR's October 2016 *QST* article for more about this type of modeling.)

HOBBIES is an acronym for "Higher Order Basis Based Integral Equation Solver." It runs on an ordinary Windows PC, is feature rich, and is affordable for amateurs. However, it does require learning modern CAD geometry modeling and uses *GiD* version 10 for specifying model geometries.

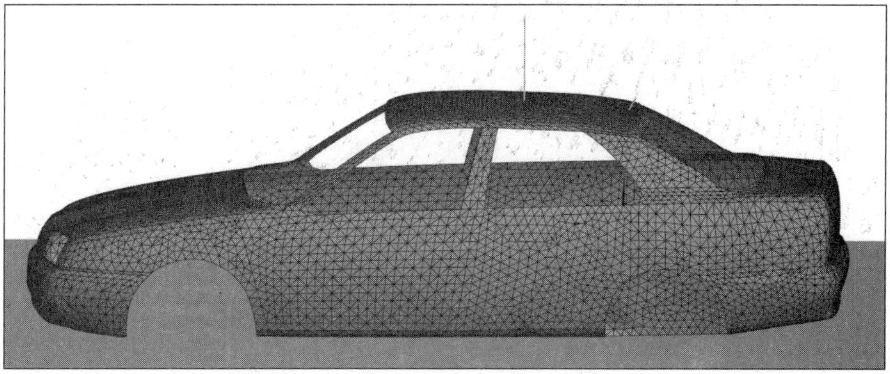

Figure 5.1 — *FEKO* model of a 2 meter whip on the roof of a car.

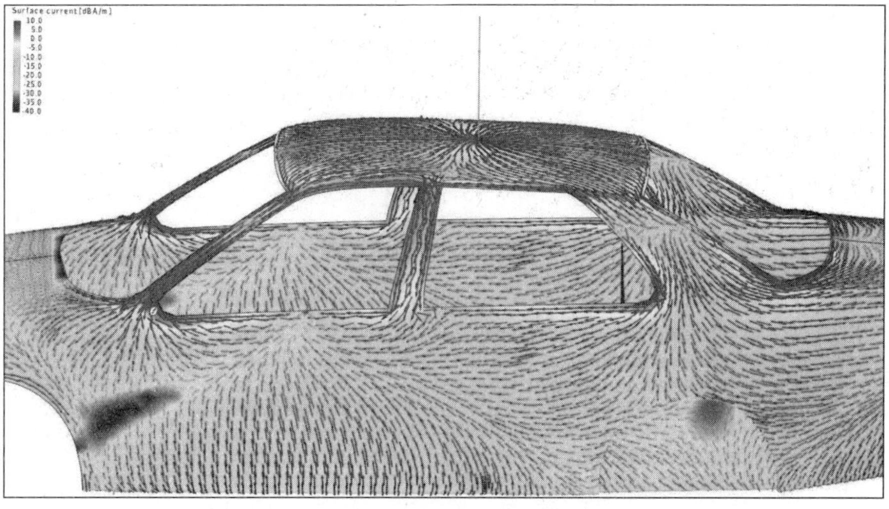

Figure 5.2 — Skin currents in car are concentrated around the windows.

Table 5.1
Antenna and Electromagnetic Modeling Software

Program	*Type*	*Optimizer*	*Price*	*Source*	*Notes*
AN-SOF	Thin Wire	AutoEZ	$450	**antennasimulator.com**	
ANSim Pro	Thin Wire	No	$325	**em-bench.myshopify.com**	Command script program
AutoEZ	GUI shell	Yes	$80	**ac6la.com/autoez.html**	Free demo version available
AWAS 2.0 Demo version	Thin Wire	No	Free	**oet.etf.bg.ac.rs/software.htm**	Contact K6OIK for his instruction manual and example files.
AWR Analyst	FEM	Yes	Expensive	**cadence.com/en_US/home/tools/system-analysis/rf-microwave-design/awr-analyst-software.html**	Free trial
AWR Axiem	3D Planar MoM	Yes	Expensive	**cadence.com/en_US/home/tools/system-analysis/rf-microwave-design/awr-axiem-analysis.html**	Free trial
CocoaNEC	GUI shell			**w7ay.net/site/Applications/cocoaNEC**	Mac OS X version of NEC-2
COMSOL Multiphysics	Multiple	Yes	Expensive	**comsol.com**	Free trial
CST Studio Suite	Multiple	Yes	Expensive	**3 ds.com/products-services/simulia/products/cst-studio-suite**	Free trial
EZNEC	GUI shell	AutoEZ	Free	**eznec.com**	GUI for NEC-2, -4, and -5
FEKO	Multiple	Yes	Expensive	**altair.com/feko-applications** Free trial	
4nec2	GUI shell	Yes	Free	**qsl.net/4nec2**	GUI for NEC-2
GprMax	FDTD	unknown	Free	**gprmax.com/about.shtml**	
HFSS	FEM	Yes	Expensive	**ansys.com/products/electronics/ansys-hfss**	Free trial
HOBBIES Professional	Wires & Surfaces	Yes	$5,000	Currently unavailable	Contact K6OIK for status.
HOBBIES Academic	Wires & Surfaces	Yes	Free	Currently unavailable	Contact K6OIK for status.
Meep	FDTD	unknown	Free	**meep.readthedocs.io/en/latest**	
MiniNEC Pro	Thin Wire	No	$30	**blackcatsystems.com/software/mininec-antenna-analysis-modeling-software.html**	
MMANA-GAL	GUI shell	Yes	Free	**gal-ana.de/promm**	GUI for MiniNEC
Nec2Go	GUI shell	No	$40	**nec2go.com**	GUI for NEC-2
NEC-2	Thin Wire	AutoEZ	Free	**nec2.org**	Command script program
NEC-4.2	Thin Wire+	AutoEZ	$300	**ipo.llnl.gov/technologies/software/nec-v50-numerical-electromagnetic-code**	Command script program
NEC-5	Thin Wire+	AutoEZ	$110	**ipo.llnl.gov/technologies/software/nec-v50-numerical-electromagnetic-code**	
openEMS	FDTD	unknown	Free	**openems.de**	
Quick Yagi 4	Yagi-Uda	No	Free	**hamuniverse.com/quickyagiqy4.html**	WA7RAI calculator for Yagi-Uda design
SEMCAD X	FDTD	Yes	Expensive	**speag.swiss/products/semcad/solutions/antenna-solution**	Free trial
SEMCAD X Student Version	FDTD	Yes	Free	**speag.swiss/products/semcad/academic**	

See also **en.wikipedia.org/wiki/Comparison_of_EM_simulation_software.**

Table 5.2
CAD Software Tools for Working with Geometries and Meshes for CEM Programs and 3D Printing

Program	*Type*	*Price*	*Source*	*Notes*
AutoCAD	Student version	Free	**autodesk.com/education/free-software/autocad**	
Autodesk	Student version	Free	**autodesk.com/education/free-software/featured**	
Autodesk	Professional	Expensive	**autodesk.com**	Free trial
BlocksCAD		Free	**blockscad3d.com**	
FreeCAD		Free	**freecadweb.org**	
Meshmixer		Free	**meshmixer.com**	Triangle mesh editor
Microsoft 3D Builder		Free	**apps.microsoft.com/store/detail/3d-builder/9WZDNCRFJ3T6**	Bundled with Window
Microsoft 3D Viewer	Viewer	Free	**apps.microsoft.com/store/detail/3d-viewer/9NBLGGH42THS**	Bundled with Window
OpenSCAD		Free	**openscad.org**	
Rhinoceros	Professional	Expensive	**rhino3d.com**	Free trial
ShareCAD viewer	Viewer	Free	**beta.sharecad.org**	Web-based CAD viewer
Solidworks	Professional	Expensive	**solidworks.com**	Free trial

These programs are useful for viewing, making, and manipulating geometries and meshes; as well as converting among file types.

Modern programs for computational electromagnetics do a better job of telling us what is going on with antennas and structures. Such programs not only allow more complex models, but their graphical output can improve our understanding of electromagnetic physics, fields, and wave propagation. See K6OIK's 2017 Pacificon presentation on antenna modeling, "Antenna Modeling for Radio Amateurs," at **fars.k6ya.org/docs/k6oik**. **Tables 5.1** and **5.2** list antenna and electromagnetic modeling software as well as CAD software for working with geometries and meshes.

5.2 The Basics of Antenna Modeling

This chapter will discuss the following antenna-modeling topics for *NEC-2*-based modeling software, using *EZNEC* as an example:

- Program outputs
- Wire geometry
- Segmentation, warnings and limitations
- Source (feed point) placement
- Environment, including ground types and frequency
- Loads and transmission lines
- Testing the adequacy of a model

While *EZNEC* is very popular among amateurs, two other programs also based on *NEC-2* are available and widely used. *4nec2* is a free application by Arie Voors that offers a graphical interface for design and a 3-D color radiation pattern display many find useful. For Apple Macintosh users, Kok Chen, W7AY, has published *cocoaNEC 2.0*. Both can use the *NEC-4* engine, assuming licensing permission has been obtained. See the Reference section of this chapter for more modeling software and links to the websites for these and other programs. Actively supported programs such as *4nec2* also offer tutorials via the website and user groups may also be available online to help with questions and share model files.

Antenna Modeling Tutorial – W8WWV

In addition to the material here, an additional tutorial on antenna modeling using *EZNEC* has been contributed by Greg Ordy, W8WWV, and is included with this book's online content. It features alternate perspectives on topics in this chapter and covers additional material in depth. The tutorial was originally presented in support of a presentation at Contest University in 2011.

5.2.1 Program Outputs

Instruction manuals for software programs traditionally start out describing in detail the input data needed by the program. They then demonstrate the output data the program can generate. We feel it is instructive, however, to turn things around and start out with a brief overview of the output from a typical antenna-modeling program.

We'll look at the output from public-domain *NEC-2*. Next, we'll look at the output information available from commercial adaptations of *NEC-2*, using *EZNEC* as an example. (From this point on in the chapter, *EZNEC* will refer to *EZNEC Pro+ 7.0*, the now-free version.) After this brief overview of the output data, we'll look in detail at the input data needed to make a modeling program work. In the following discussions it will be very instructive if you to bring up *EZNEC* on your computer and open the specific modeling files used in each example.

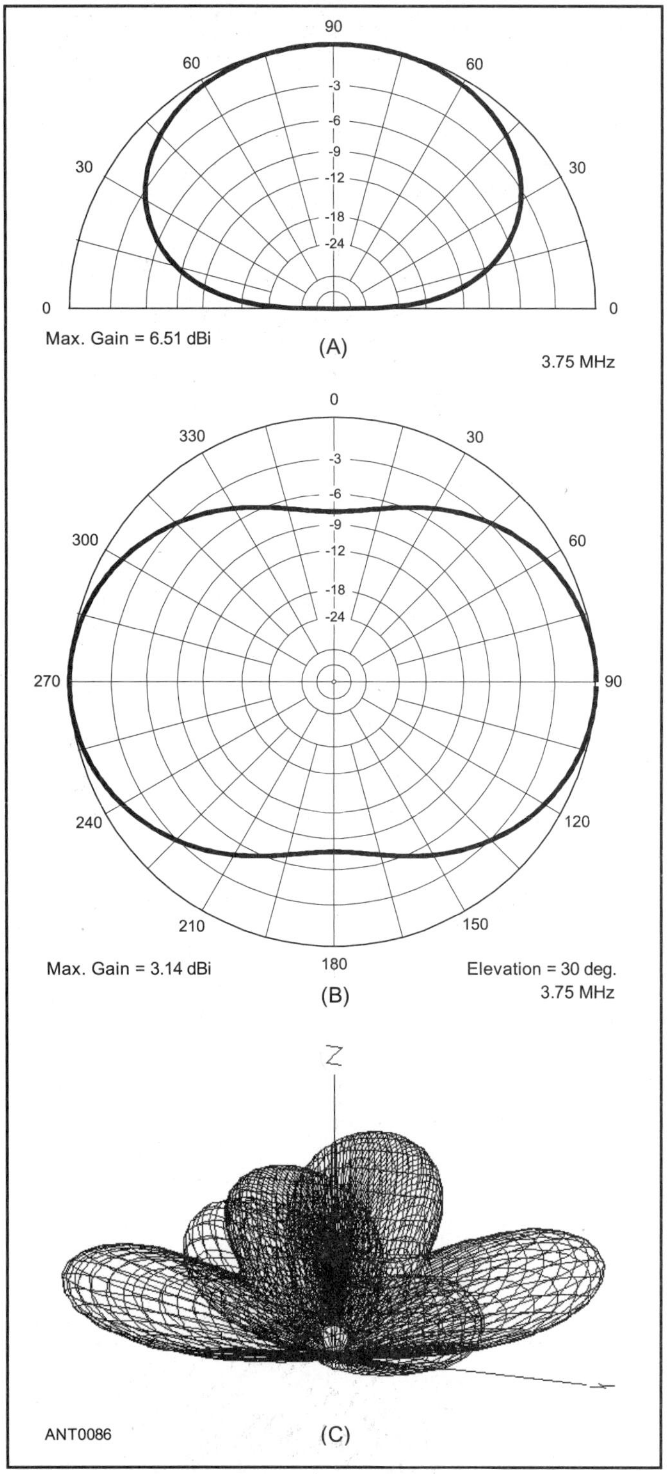

Figure 5.3 — At A, far-field elevation-plane pattern for a 135- foot-long horizontal dipole, 50 feet above flat ground, at 3.5 MHz. At B, the far-field azimuth-plane pattern at an elevation angle of 30°.

NATIVE *NEC-2*

Native *NEC-2* was written in the Fortran language, which stands for *Formula Translation*. The original program used Hollerith punch-cards to enter the program and input data. The output of the program was raw numeric data printed on many sheets of paper. Commercial software that uses the *NEC-2* computational core algorithms shields provides much easier methods of entering antenna design information and generates graphic output that is much easier to understand. Numerical tables are provided where they are useful, such as for source impedance and SWR at a single frequency, or the characteristics of a load or a transmission line. *EZNEC* produces the following types of graphs:

- Polar (linear-dB or ARRL-style) graphs of the far-field elevation and azimuth responses.
- 3-D wire-frame graph of the total far-field response.
- Graph of the SWR across a frequency band.
- Graphical display of the RF currents on various conductors in a model.
- Rotatable, zoom-able 3-D views of the wires used to make a model.
- Output to programs capable of generating Smith charts and performing other analysis

Figure 5.3 shows the computed far-field 2-D elevation and azimuth patterns for a 135-foot long horizontal dipole, mounted in a flattop configuration 50 feet above flat ground. These figures were generated using *EZNEC* at 3.75 MHz. Figure 5.3C shows a 3-D wire-frame picture of the far-field response, but this time at 14.2 MHz.

Figure 5.4 shows the computed SWR curve over the frequency range 3.0 to 4.0 MHz for this dipole, fed with lossless 50-Ω transmission line. *EZNEC* generated this plot using the "SWR" button. Figures 5.3 and 5.4 are typical of the kind of graphical outputs that commercial implementations of the *NEC-2* computing core can produce — a vast improvement over tables of numbers from a mainframe computer's line

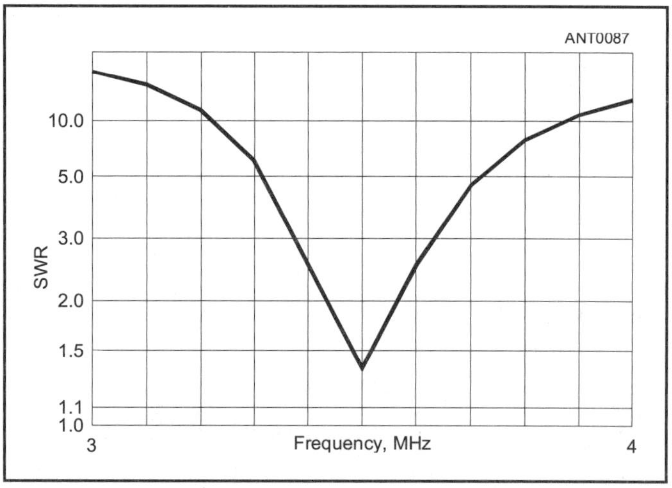

Figure 5.4 — SWR curve for 135-foot flattop dipole over the frequency range 3.0 to 4.0 MHz for a 50-Ω feed line. This antenna is an example and is not optimized for the amateur band.

printer! Now, let's get into the details of what kind of input data is required to run a typical method-of-moments antenna-modeling program.

5.2.2 Program Inputs: Wire Geometry

COORDINATES IN AN X, Y AND Z WORLD

The most difficult part of using a *NEC*-type of modeling program is setting up the antenna's geometry — you must condition yourself to think in three-dimensional, Cartesian coordinates. Each end point of a wire is represented by three numbers: an x, y and z coordinate. These coordinates represent the distance from the origin (x-axis), the width of an antenna (y-axis), and the height (z-axis).

An example should help sort things out. **Figure 5.5** shows a simple model of a 135-foot center-fed dipole, made of #14 copper wire placed 50 feet above flat ground. The common term for this antenna is "flattop dipole." For convenience, the ground is located at the origin of the coordinate system, at (0, 0, 0) feet, directly under the center of the dipole. **Figure 5.6** shows the *EZNEC* spreadsheet-like input data for this antenna. (Use model file: **Ch8-Flattop Dipole.EZ.**) *EZNEC* allows you to specify the type of conductor material from its main window, using the **WIRE LOSS** button to open a new window. We will click on the **COPPER** button for this dipole.

Above the origin, at a height of 50 feet on the z-axis, is the dipole's *feed point*, called a *source* in *NEC* terminology. The width of the dipole goes toward the left (that is, in the "negative-y" direction) one-half the overall length of 135 feet, or –67.5 feet. Toward the right, our dipole's other end is at +67.5 feet. The x-axis dimension of our dipole is zero, meaning that the dipole wire is parallel to and directly above the x-axis. The dipole's ends are thus represented by two points, whose coordinates are (0, –67.5, 50) and (0, 67.5, 50) feet. The use of parentheses with a sequential listing of (x, y, z) coordinates is a common practice among antenna modelers to describe a wire end point.

Figure 5.5B includes some other useful information about this antenna beyond the wire geometry. Figure 8.3B overlays the wire geometry, the current distribution along the wire and the far-field azimuth response, in this case at an elevation angle of 30°.

Although not shown specifically in Figure 5.5, the thickness of the antenna is the diameter of the wire, #14 AWG. Note that native *NEC* programs specify the *radius* of the wire, rather than the diameter, but programs like *EZNEC* use the more intuitive diameter of a wire rather than the radius. *EZNEC* (and other commercial programs) also allows the user to specify the wire as an AWG gauge, such as #14 or #22, for example.

We've represented our simple dipole in Figure 5.5 using a single, straight wire. In fact, all antenna models created for method-of-moments programs are made of combinations of straight wires. This includes even complex antennas, such as helical antennas or round loops. (The mathematical basis for modeling complex antennas is that they can be simulated using straight-wire polygons. A circular loop, for example, can be modeled using an octagon.)

SEGMENTATION AND SPECIFYING A SOURCE SEGMENT

We've specified the physical geometry of this simple one-wire dipole. Now several more modeling details surface — you must specify the number of *segments* into which the dipole is divided for the method-of-moments analysis and you must somehow feed the antenna. The *NEC-2* guideline for setting

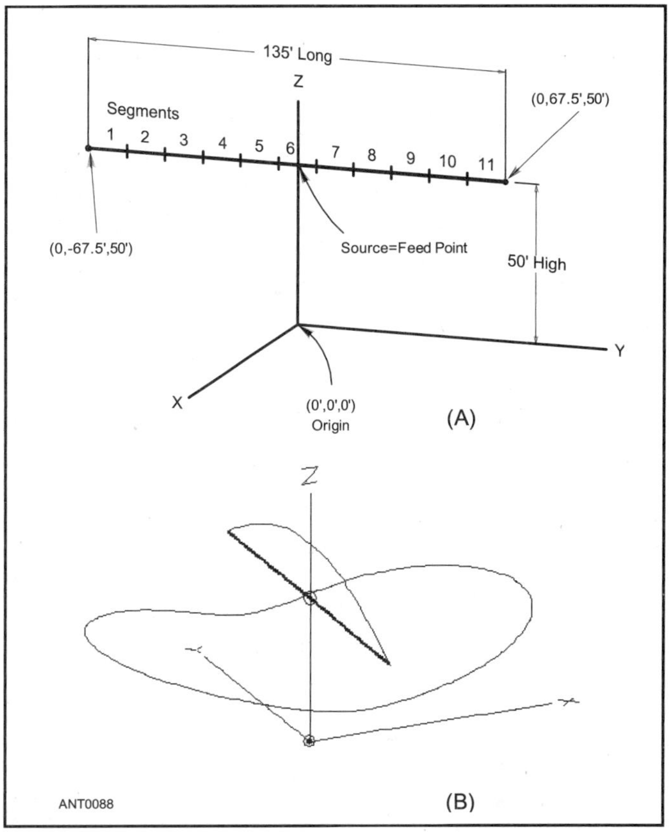

Figure 5.5 — At A, simple model for a 135-foot long horizontal dipole, 50 feet above the ground. The dipole is over the y-axis. The wire has been segmented into 11 segments, with the center of segment number 6 as the feed point. The left-hand end of the antenna is –67.5 feet from the center feed point and that the right-hand end is at 67.5 feet from the center. At B, *EZNEC* "View Antenna" screen, showing geometry of wire and the x, y and z axes. Overlaid on the wire geometry drawing are the current distribution along the wire and the far-field azimuthal response at an elevation angle of 30°.

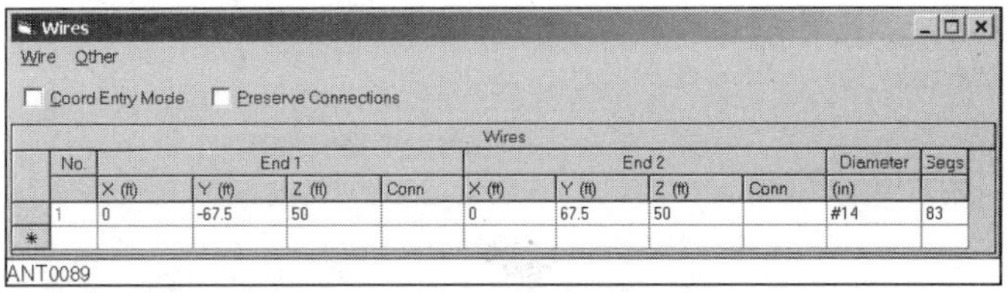

No.	End 1 X (ft)	End 1 Y (ft)	End 1 Z (ft)	End 1 Conn	End 2 X (ft)	End 2 Y (ft)	End 2 Z (ft)	End 2 Conn	Diameter (in)	Segs
1	0	-67.5	50		0	67.5	50		#14	83
*										

Figure 5.6 — *EZNEC* "View Wires" data entry screen for simple flattop dipole in Figure 5.5. The numbers shown are in feet, except for the wire diameter, which *EZNEC* allows you to specify as an AWG gauge, in this case #14. Note that 83 segments have been specified for this antenna for analysis over the range from 3.5 to 29.7 MHz.

the number of segments is to use at least 10 segments per half-wavelength. This is a general rule of thumb, however, and in many models more dense segmentation is mandatory for good accuracy.

In Figure 5.5, we've specified that the dipole be divided into 11 segments for operation on the 80 meter band. This follows the rule of thumb above, since the 135-foot dipole is about one half-wavelength long at 3.5 MHz.

Setting the Source Segment

The use of 11 segments, an odd rather than an even number such as 10, places the dipole's feed point (a feed point is referred to as a *source* in *NEC*-speak, a word choice that can befuddle beginners) right at the antenna's center, at the center of segment number six. In concert with the "EZ" in its name, *EZNEC* makes choosing the source segment easy by allowing the user to specify a percentage along the wire, in this case 50% centers the source in the middle of the segment.

At this point you may very well be wondering why no center insulator is shown in the middle of our center-fed dipole. After all, a real dipole would have a center insulator. However, method-of-moments programs assume that a source generator is placed across an infinitely small gap in the antenna wire. While this is convenient from a mathematical point of view, the unstated use of such an infinitely small gap often confuses newcomers to the world of antenna modeling. We'll get into more details, caveats and limitations in source placement later in this chapter. For now, just trust that the model we've just described with 11 segments, fed at segment 6, will work well over the entire amateur band from 3.5 to 4.0 MHz.

Now, let's consider what would happen if we want to use our 135-foot long dipole on all HF amateur bands from 3.5 to 29.7 MHz, rather than just from 3.5 to 4.0 MHz. Instead of feeding such an antenna with coax cable, we would feed it with open-wire line and use an antenna tuner in the station to create a 50-Ω load for the transmitter. To comply with the segmentation rule above, the number of segments used in the model should vary with frequency — or at least be segmented at or above the minimum recommended level at the highest frequency used. This is because a half-wavelength at 29.7 MHz is 16.6 feet, while a half-wavelength at 3.5 MHz is 140.6 feet. So the number of segments for proper operation on 29.7 MHz should be 10 × 135/16.6 = 81. We'll be a little more conservative than the minimum requirement and specify 83 segments. Figure 5.6 shows the *EZNEC* input spreadsheet for this model. (Use model file: **Ch8-Multiband Dipole.EZ.**)

The penalty for using more segments in a program like *NEC* is that the program slows down roughly as the square of the segments — double the number of segments and the speed drops by a factor of four (two squared). Using too few segments will result in inaccuracies, particularly in computing the feed point impedance. We'll delve into the area of segmentation density in more detail later when we discuss testing the adequacy of a model.

Segment Length-to-Wire-Diameter Ratio

Even if you're willing to live with the slowdown in computing speed for situations involving a large number of wire segments, you should make sure the ratio between the segment length and the diameter of any wire is greater than 1:1. This is to say that the length of each segment should be longer than the diameter of the wire to avoid internal limitations in the *NEC* program.

For the #14 wire specified in this simple 135-foot long dipole, it's pretty unlikely that you'll bump up against this limitation for any reasonable level of segmentation. After all, #14 wire has a diameter of 0.064 inch and 135 feet is 1620 inches. To stay above a segment length of 0.064 inch, the maximum number of segments is 1620/0.064 = 25,312. This is a very large number of segments and it would take a very long time to compute, assuming that your program can handle that many segments.

Staying above a 1:1 ratio in segment length to wire diameter can be more challenging at VHF/UHF frequencies, however. This is particularly true for fairly large "wires" made of aluminum tubing. Incidentally, this is another point where newcomers to antenna modeling can be led astray by the terminology. In a *NEC*-type program, all conductors in a model are considered to be wires, even if they consist of hollow aluminum or copper tubes. The skin effect keeps the RF current in any conductor confined to the outer surface of that conductor, and thus it doesn't matter whether the conductor is hollow or solid, or even a number of wire strands twisted together.

Let's look at a half-wave dipole at 420 MHz. This would be about 14.1 inches long. If you use ¼-inch diameter tubing for this dipole, the maximum segment length meeting the 1:1 diameter-to-length ratio requirement is also ¼ inch long. The maximum number of segments then would be 14.1/0.25 = 56.4, rounded down to 56. From this discussion you should now understand why method-of-moments programs are known for using a "thin-wire approximation." Really fat conductors can get you into trouble, particularly at VHF/UHF.

SOME CAVEATS AND LIMITATIONS CONCERNING GEOMETRY

Example: Inverted-V Dipole

Now, let's get a little more complicated and specify another 135-foot-long dipole, but this time configured as an inverted V. As shown in **Figure 5.7**, you must now specify two wires. The two wires join at the top, at (0, 0, 50) feet. (Again, the program doesn't use a center insulator in the model.)

If you are using a native version of *NEC*, you may have to go back to your high-school trigonometry book to figure out how to specify the end points of our "droopy" dipole, with its 120° included angle. Figure 5.7 shows the details, along with the trigonometric equations needed. *EZNEC* is indeed more "easy" here since it allows you to tilt the ends of each wire downward an appropriate number of degrees (in this case –30° at each end of the dipole) to automatically create an inverted-V configuration. **Figure 5.8A** shows the *EZNEC* spreadsheet describing this inverted-V dipole with a 120° included angle between the two wires.

See the *EZNEC* **HELP** section under "Wire Coordinate

Shortcuts" for specific instructions on how to use the "elevation rotate end" shortcut "RE-30" to create the sloping wires easily by rotating the end of the wire down 30°. Now the specification of the source becomes a bit more complicated. The easiest way is to specify two sources, one on each end segment at the junction of the two wires. *EZNEC* does this automatically if you specify a so-called *split-source* feed. Figure 5.8B shows the two sources as two open circles at the top ends of the two wires making up the inverted-V dipole. What *EZNEC* is doing is creating two sources, one in each of the segments immediately on either side of the junction of the two wires. *EZNEC* sums up the two source impedances to provide a single result.

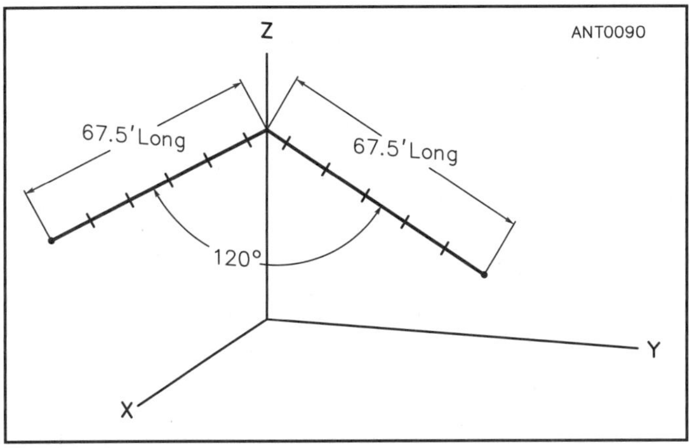

Figure 5.7 — Model for an inverted-V dipole with an included angle between the two legs of 120° apex at 50' high. Sine and cosine functions are used to describe the heights of the end points for the sloping arms of the antenna.

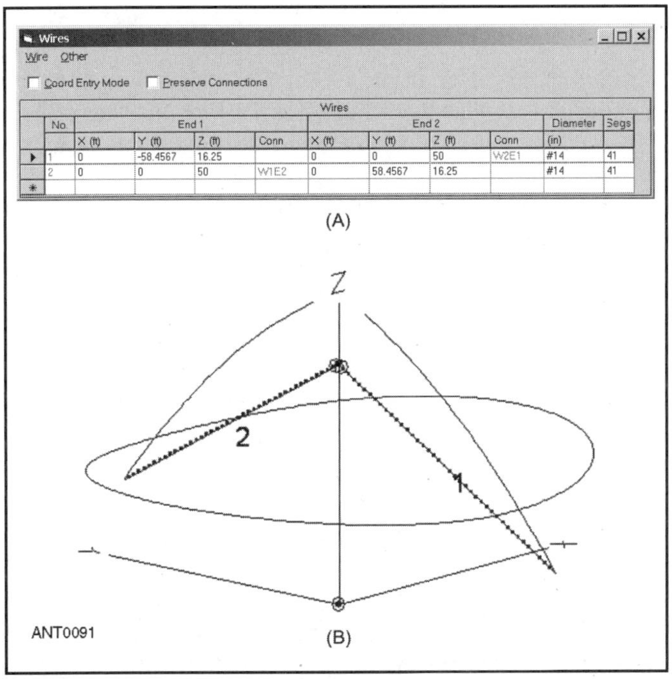

No	End 1 X (ft)	End 1 Y (ft)	End 1 Z (ft)	End 1 Conn	End 2 X (ft)	End 2 Y (ft)	End 2 Z (ft)	End 2 Conn	Diameter (in)	Segs
1	0	-58.4567	16.25		0	0	50	W2E1	#14	41
2	0	0	50	W1E2	0	58.4567	16.25		#14	41

Figure 5.8 — At A, *EZNEC* "View Wires" data entry screen for inverted-V dipole in Figure 5.7. Now the ends of the inverted-V dipole are 16.25 feet above ground, instead of 50 feet for the flattop dipole. At B, *EZNEC* "View Antenna" screen, with overlay of geometry, current distribution and azimuth plot.

Navigating in the View Antenna Window

At this point it's worthwhile to explore some of the ways you can look at the antenna you've designed using the *EZNEC* **VIEW ANT** button on the main window. Bring up the file **Ch8-Inverted V Dipole.EZ** in *EZNEC*, and click on the **VIEW ANT** button. You will see a small inverted-V dipole raised over the (0, 0, 0) origin on the ground directly under the feed point of the inverted-V dipole. First, "rotate" the dipole by holding down the left-mouse button and moving the mouse. You can orient the picture any way you wish.

Let's take a closer look at the junction of the two wires at the feed point. Click the **CENTER ANT IMAGE** checkbox toward the bottom of the window to anchor the center of the image at the center of the window, then move the *Zoom* slider upward to zoom in on the image. At some point the junction of the two slanted wires will move up beyond the edge of the window, so you will need to click on the left-hand side of the **Z MOVE IMAGE** slider to bring the junction back into view. You should be able to see a zoomed view of the junction along with the two open circles that represent the location of the split sources in the middle of the segments adjacent to the wire junction.

Place the mouse cursor over one of the slanted wires and double click the left-mouse button. *EZNEC* will now identify that wire and show its length, as well as the length of each segment on that wire.

Short, Fat Wires and the Acute-Angle Junction

Another possible complication can arise for wires with short, fat segments, particularly ones that have only a small included angle between them. These wire segments can end up inter-penetrating within each other's volumes, leading to problems in a model. Once you think of each wire segment as a thick cylinder, you can appreciate the difficulty in con-

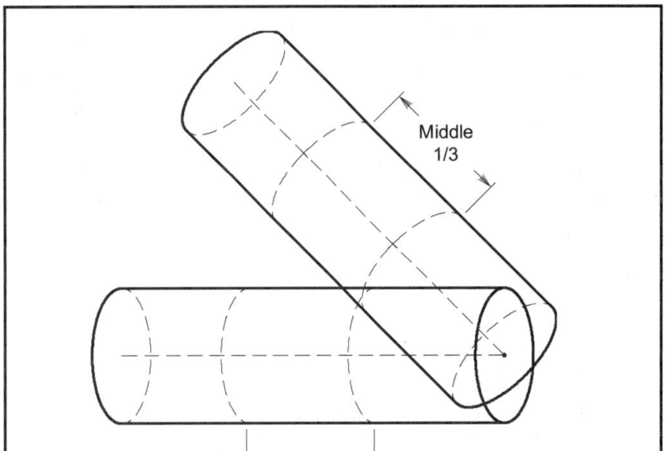

Figure 5.9 — A junction of two short, fat wire segments at an acute angle. This results in inter-penetration of the two wire volumes beyond the middle-⅓ recommended limit.

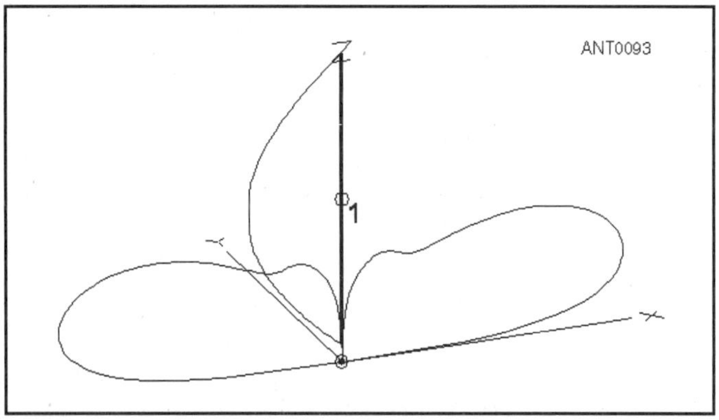

Figure 5.10 — A vertical half-wave dipole, created by turning the dipole in Figure 5.5 on its end, with a minimum height at the lower end of 8 feet to keep the antenna away from people and animals. The current distribution and the elevation pattern for this antenna are also shown overlaid on the wire geometry.

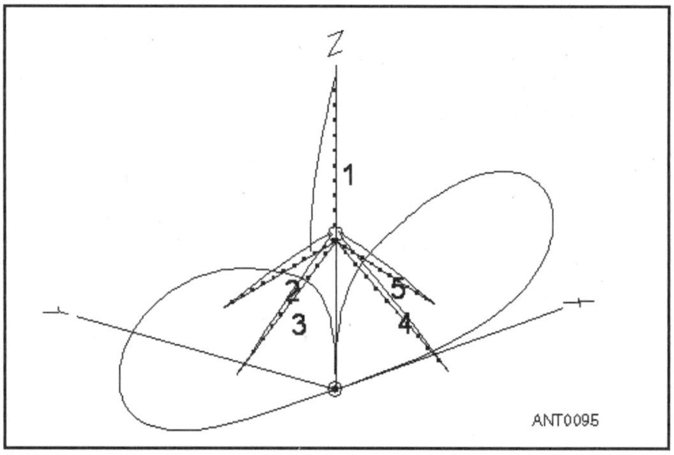

Figure 5.12 — *EZNEC* "View Antenna" screen for the ground-plane antenna with its four radials tilted downward by 40° to improve the SWR at the feed point.

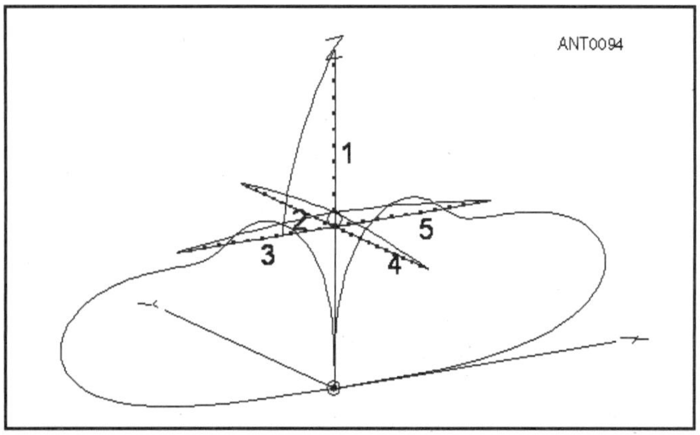

Figure 5.11 — A vertical ground-plane antenna. The radials and the bottom of the vertical radiator are located 15 feet off the ground in this model. The current distribution along each wire and the far-field elevation-plane pattern are overlaid on the antenna geometry.

necting two wires together at their ends. The two wires always inter-penetrate each other's volume to some extent. **Figure 5.9** depicts this problem graphically for two short, fat wires joined at their ends at an acute angle. A rule of thumb is to avoid creating junctions where more than ⅓ of the wire volumes inter-penetrate. You can achieve this by using longer segment lengths or thinner wire diameters.

SOME OTHER PRACTICAL ANTENNA GEOMETRIES

A Vertical Half-Wave Dipole

If you turn the 135-foot-long horizontal dipole in Figure 5.3 on its end you will create a vertical half-wave dipole that is above the origin of the x, y and z axes. See **Figure 5.10** where the bottom end of the dipole is placed 8 feet off the ground to keep it away from humans and animals at (0, 0, 8) feet. The top end is thus at 8 + 135 = 143 feet off the ground at (0, 0, 143). Figure 5.10 also shows the current distribution and the elevation pattern for this antenna. (Use *EZNEC* model file: **Ch8-Vertical Dipole.EZ.**)

A Ground-Plane Antenna

The ground-plane model is more complicated than previous ones because a total of five wires are now needed: one for the vertical radiator and four for the radials. **Figure 5.11** shows the *EZNEC* view for a 20 meter ground plane mounted 15 feet off the ground (perhaps on a garage roof), with the overlay of both the current distribution and the elevation-plane plot. (Use *EZNEC* model file: **Ch8-GP.EZ.**) Note that the source has been placed at the bottom segment of the vertical radiator. Once again, the program needs no bottom insulator since all five wires are connected together at a common point. *EZNEC* reports that this antenna has a resonant feed-point impedance of about 22 Ω, which would show an SWR of 2.3:1 for a 50-Ω coax feed line if no matching system is used, such as a gamma or hairpin match.

Figure 5.12 shows the same antenna, except that the radials have now been tilted downward by 35° to raise the feed point impedance to present an almost perfect 50-Ω match (SWR = 1.08:1). In addition, the length of the radiator in this model was shortened by 6 inches to re-resonate the antenna. (Use *EZNEC* model file: **Ch8-Modified GP.EZ.**) The trick of tilting the radials downward for a ground-plane antenna is an old one, and the modeling programs validates what hams have been doing for years.

A 5-Element Horizontal Yagi

This is a little more challenging modeling exercise. Let's use a 5-element design on a 40-foot boom, but rather than using telescoping aluminum tubing for the elements, we'll use #14 wire. The *SCALE* program (available in the online material) converted the aluminum-tubing 520-40.YW to a design using #14 AWG copper wire. **Table 5.3** shows the

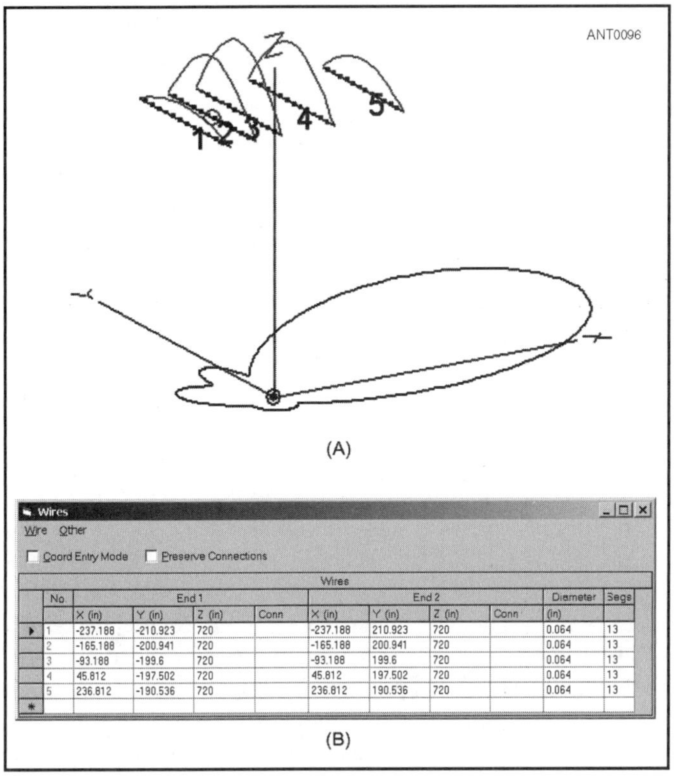

	Wires									
No.	End 1				End 2				Diameter	Segs
	X (in)	Y (in)	Z (in)	Conn	X (in)	Y (in)	Z (in)	Conn	(in)	
1	-237.188	-210.923	720		-237.188	210.923	720		0.064	13
2	-165.188	-200.941	720		-165.188	200.941	720		0.064	13
3	-93.188	-199.6	720		-93.188	199.6	720		0.064	13
4	45.812	-197.502	720		45.812	197.502	720		0.064	13
5	236.812	-190.536	720		236.812	190.536	720		0.064	13
*										

Figure 5.13 — At A, geometry for 5-element Yagi on a 40-foot boom, mounted 720 inches (60 feet) above flat ground, with an overlay of current and the azimuth pattern. At B, *EZNEC* "View Wires" screen for this antenna. This design uses #14 wire for simplicity.

element lineup for this antenna. (Later in this chapter we'll see what happens when telescoping aluminum tubing is used in a real-world Yagi design.)

Some explanations of what Table 5.3 means are in order. First, only one half of each element is shown. The *YW* program (*Yagi for Windows*), included with this book's downloadable supplemental information, computes the other half of the Yagi automatically, essentially mirroring the other half on the opposite side of the boom. Having to enter the dimensions for only half of a real-world Yagi element that uses telescoping aluminum tubing is much easier this way.

Second, the placement of the elements along the boom starts at 0.0 inches for the reflector. The distance between adjacent elements defined in this particular file is the spacing between the element itself and the element just before it. For example, the spacing between the driven element and the reflector is 72 inches and the spacing between the first director and the driven element is also 72 inches. The spacing between the second director and the first director is 139 inches.

Figure 5.13A shows the wire geometry for this Yagi array when it is mounted 720 inches (60 feet) above flat ground and Figure 5.13B shows the *EZNEC* Wires spreadsheet that describes the coordinates. (Use *EZNEC* model file: **Ch8-520-40W.EZ.**) You can see that the x-axis coordinates for the elements have been automatically moved by the *SCALE* program so that the center of the boom is located directly above the origin. This makes it easier to evaluate the effects of stacking different monoband Yagis on a rotating mast in a typical "Christmas Tree" arrangement, such as 20, 15 and 10 meter monobanders on a single rotating mast sticking out of the top of the tower.

Table 5.3
520-40W.YW, using #14 AWG wire 0.064-inch diameter from 520-40H.YW
14.000 14.174 14.350 MHz

Element Spacing (inches)	*Half-element Length (inches)*
0.000	210.923
72.000	200.941
72.000	199.600
139.000	197.502
191.000	190.536

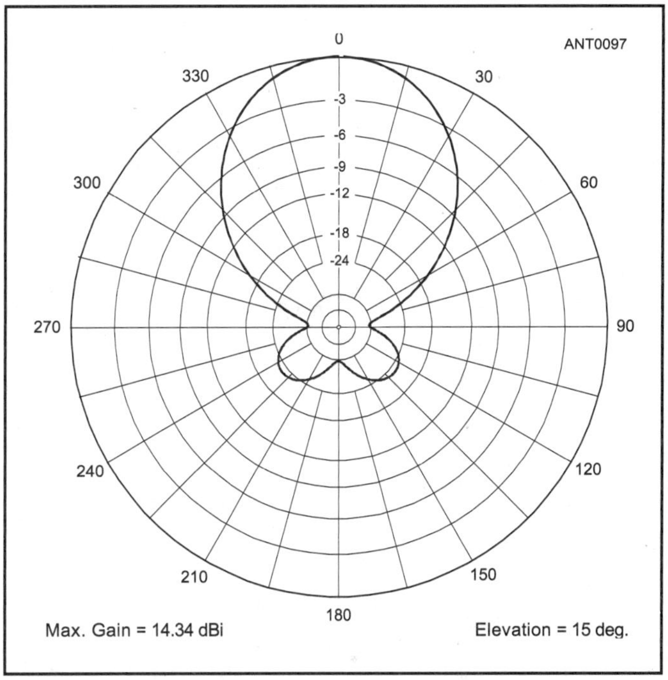

Figure 5.14 — *EZNEC* azimuth-plane pattern at an elevation angle of 15° for #14 AWG wire Yagi described in Figure 5.13.

Figure 5.14 shows the computed azimuth pattern for this Yagi at 14.175 MHz, at an elevation angle of 15°, the angle at which the peak of the forward lobe occurs for this height above flat ground. The antenna exhibits excellent gain at 13.1 dBi, as well as a clean pattern behind the main lobe. The worst-case front-to-rear ratio at any point from 90° to 270° in azimuth is better than 23 dB. *EZNEC* says the feed point impedance is 25 – *j* 23 Ω, just the right impedance suited for a simple hairpin or gamma match.

A Monoband 2-Element Quad

Unlike a Yagi, with its elements existing only in the x-y plane, a quad type of beam is a three-dimensional antenna. A quad loop has height in the z-axis, as well as width and length in the x-y plane. Each individual loop for a monoband quad consists of four wires, joined together at the corners. **Fig-**

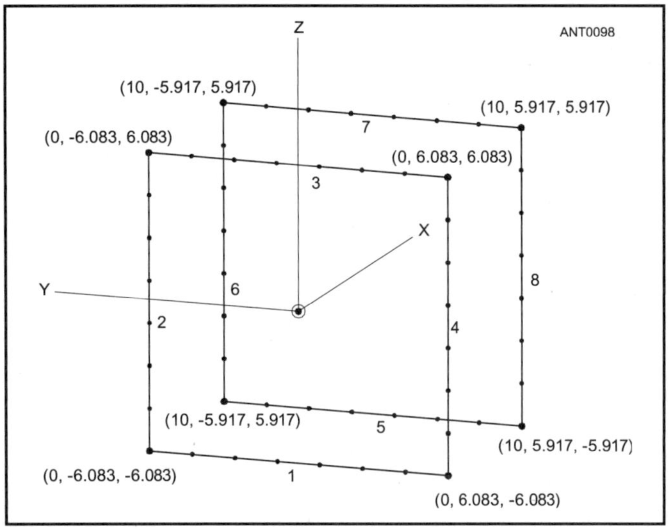

Figure 5.15 — Wire geometry for a 2-element quad, with a reflector and driven element. The x-axis is the axis of symmetry for this free-space model.

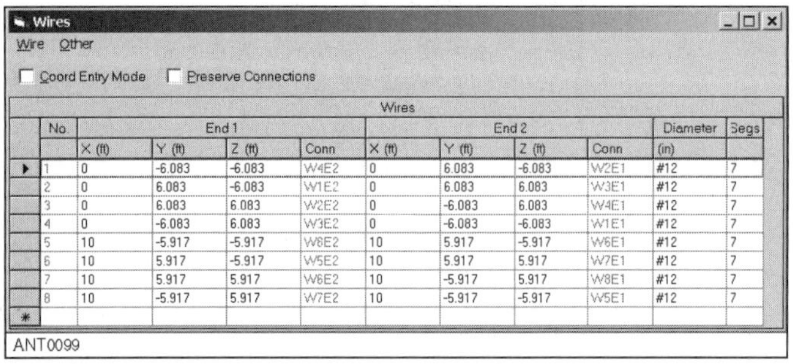

Wires

Wire Other

Coord Entry Mode Preserve Connections

No.	End 1 X (ft)	End 1 Y (ft)	End 1 Z (ft)	End 1 Conn	End 2 X (ft)	End 2 Y (ft)	End 2 Z (ft)	End 2 Conn	Diameter (in)	Segs
1	0	-6.083	-6.083	W4E2	0	6.083	-6.083	W2E1	#12	7
2	0	6.083	-6.083	W1E2	0	6.083	6.083	W3E1	#12	7
3	0	6.083	6.083	W2E2	0	-6.083	6.083	W4E1	#12	7
4	0	-6.083	6.083	W3E2	0	-6.083	-6.083	W1E1	#12	7
5	10	-5.917	-5.917	W8E2	10	5.917	-5.917	W6E1	#12	7
6	10	5.917	-5.917	W5E2	10	5.917	5.917	W7E1	#12	7
7	10	5.917	5.917	W6E2	10	-5.917	5.917	W8E1	#12	7
8	10	-5.917	5.917	W7E2	10	-5.917	-5.917	W5E1	#12	7

ANT0099

Figure 5.16 — *EZNEC* "View Wires" screen showing the coordinates used for the quad in Figure 5.15. Note how the x-axis describes the position of an element on the 10-foot boom and also is the axis of symmetry for each element. The values for the z-axis and y-axis vary above and below the axis of symmetry.

ure 5.15 shows the coordinates for a 2-element 15-meter quad, consisting of a reflector and a driven element on a 10-foot boom.

You can see that the axis of symmetry, the x-axis, runs down the center of this model, meaning that the origin of this particular x, y and z-coordinate scheme is in the center of the reflector. The (0, 0, 0) origin is placed this way for convenience in assigning corner coordinates for each element. For actual placement of the antenna at a particular height above real ground, the heights of all z-axis coordinates are changed accordingly. *EZNEC* has a convenient built-in function to change the height of all wires at a single stroke.

Figure 5.16 shows the input *EZNEC* spreadsheet for this quad in free space, clearly showing the symmetrical nature of the corner coordinates. (Use *EZNEC* model file: **Ch8-Quad. EZ.**) This is a good place to emphasize that you should enter the wire coordinates in a logical sequence. The most obvious example in this particular model is that you should group all the wires associated with a particular element together — for example, the four wires associated with the reflector should be in one place. In Figure 5.16 you can see that all four wires with an x-coordinate of zero represent the reflector.

It's best to follow a convention in entering wires in a loop structure in a logical fashion. The easiest way is to connect the end point of one wire to the starting point of the next wire. For example, in Figure 5.15 you can see that the left-hand end of Wire 1 is connected to the bottom of Wire 2, and that the top of Wire 2 connects to the left-hand end of Wire 3. In turn, Wire 3 connects to the top of Wire 4, whose bottom end connects to the right-hand end of Wire 1. The pattern is known as "going around the horn" meaning that the connections proceed smoothly in one direction, in this case in a clockwise direction.

You can see that the entry for the wires making up the elements in the 5-element Yagi in Figure 5.13B also proceeded in an orderly fashion by starting with the reflector, then the driven element, then director 1, then director 2 and finally director 3. This doesn't mean that you couldn't mix things up, say by specifying the driven element first, followed by director 3, and then the reflector, or whatever. But it's a pretty good bet that doing so in this quasi-random fashion will result in some confusion later on when you revisit a model, or when you let another person use or review your model.

5.2.3 The Modeling Environment

THE GROUND

Above, when considering the 135-foot dipole mounted 50 feet above flat earth, we briefly mentioned the most important environmental item in an antenna model — the ground beneath it. Let's examine some of the options available in the *NEC-2* environment in *EZNEC*:

- Free space
- Perfect ground
- *MININEC* type ground
- "Fast" type ground
- Sommerfeld-Norton ground.

The free space environment option is pretty self-explanatory — the antenna model is placed in free space away from the influence of any type of ground. This option is useful when you wish to optimize certain characteristics of a particular antenna design. For example, you might wish to optimize the front-to-rear ratio of a Yagi over an entire amateur band and this might entail many calculation runs. The free-space option will run the fastest because there is no ground interaction to compute.

Perfect ground is useful as a reference case, especially for vertically polarized antennas over real ground. Antenna evaluations over perfect ground are shown in most classical antenna textbooks, so it is useful to compare models for simple antennas over perfect ground to those textbook cases.

MININEC type ground is useful when modeling vertical wires, or horizontal wires that are higher than 0.2 λ above ground. A *MININEC* type ground will compute faster than either a "Fast" ground or a Sommerfeld-Norton type of ground because it assumes that the ground under the antenna is perfect, while still taking into account the far-field reflections for ground using user-specified values of ground conductivity and dielectric constant. The fact that the ground under the antenna is treated as perfect allows the *NEC-2* user of a *MININEC* type ground to specify wires that touch (but don't go below) the ground surface, something that only users of the advanced *NEC-4* program can do with the more accurate Sommerfeld-Norton type of ground described below. (*NEC-4* is presently not in the public domain. Software based on *NEC-4*, such as *EZNEC-PRO*, requires an additional license from the copyright holder — the US government. See the section Comparing *NEC-2* to *NEC-4* at the beginning of this chapter.) The ability to model grounded wires is useful with vertical antennas. The modeler must be wary of the feed-point source impedances reported for either horizontally or vertically polarized wires because of the perfect-ground assumption inherent in a *MININEC*-type ground.

The "Fast" type of ground is a hybrid type of ground that makes certain simplifying assumptions to speed up calculations, provided that horizontal wires are higher than about 0.1 λ above ground. With today's high-speed computers, the simplifications are no longer required and the Sommerfeld-Norton model is preferred.

The Sommerfeld-Norton ground (referred to in *EZNEC* as "High Accuracy" ground) is preferable to the other ground models because it has essentially no practical limitations for wire height. It has the disadvantage that it runs about four times slower than a *MININEC* type of ground but today's fast computers make that almost a non-issue. Again, *NEC-2*-based programs cannot model wires that penetrate into the ground (although there are workarounds described below).

As mentioned above, for any type of ground other than perfect ground or free space the user must specify the conductivity and dielectric constant of the soil. (See the section "Ground Parameters for Antenna Analysis" in the chapter **Effects of Ground**.) *EZNEC* allows selection of several user-friendly categories, where σ is conductivity in siemens/meter and ε is dielectric constant:

- Extremely poor: cities, high buildings (σ = 0.001, ε = 3)
- Very Poor: cities, industrial (σ = 0.001, ε = 5)
- Sandy, dry (σ = 0.002, ε = 10)
- Poor: rocky, mountainous (σ = 0.002, ε = 13)
- Average: pastoral, heavy clay (σ = 0.005, ε = 13)
- Pastoral: medium hills and forestation (σ = 0.006, ε = 13)
- Flat, marshy, densely wooded (σ = 0.0075, ε = 12)
- Pastoral, rich soil, US Midwest (σ = 0.010, ε = 14)
- Very Good: pastoral, rich, central US (σ = 0.0303, ε = 20)
- Fresh water (σ = 0.001, ε = 80)
- Saltwater (σ = 5, ε = 80)

Let's use *EZNEC*'s ability to overlay one or more plots together on one graph to compare the response of the vertical ground plane antenna in Figure 5.11 for two different types of ground: Saltwater and Poor. Open the **Ch8-GP.EZ** file in *EZNEC*. Click the **GROUND DESCRIP** button and then right-click anywhere in the Media window that opens up. Choose first the "Poor: rocky, mountainous" option button, click **OK** and then **FF PLOT**. When the elevation plot appears, click the File menu at the top of the main window, and then **SAVE AS.** Choose an appropriate name for the trace, perhaps "Poor Gnd.PF."

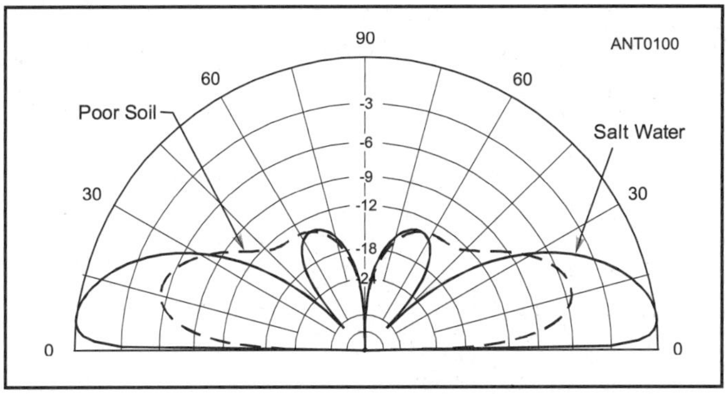

Figure 5.17 — A comparison of the elevation response for the vertical ground plane in Figure 5.11 over saltwater and over "poor: rocky, mountainous" soil. Saltwater works wonders for verticals, providing excellent low-angle signals.

Go back and select saltwater using **GROUND DESCRIP** and follow the same procedure to compute the far-field plot for saltwater ground. Now, add the Poor Gnd.PF trace, by clicking menu selection **FILE, ADD TRACE**. **Figure 5.17** shows this comparison, which greatly favors the saltwater environment, particularly at low elevation angles. At 5° the ground plane mounted over saltwater has about a 10 dB advantage compared to its landlocked cousin.

You might be wondering what happens if we move the ground-plane antenna down closer to the ground. The lower limit to how closely radials may approach lossy ground is 0.001 λ or twice the diameter of the radial wire. A distance of 0.001 λ is about 6 inches at 1.8 MHz and 0.4 inch at 30 MHz. While *NEC-2*-based programs cannot model wires that penetrate the ground, radial systems just above the ground with more than about eight radial wires can provide a work-around to simulate a direct-ground connection.

MODELING ENVIRONMENT: FREQUENCY

It's always a good idea to evaluate an antenna over a range of frequencies, rather than simply at a single spot frequency. Trends that become quite apparent on a frequency sweep are often lost when looking simply at a single frequency. Native *NEC-2* has built-in frequency sweep capabilities but once again the commercial programs make the process easier to use and understand. You saw in the SWR curve of Figure 8.2 the result of one such frequency sweep using *EZNEC*. **Figure 5.18** shows a frequency sweep of the azimuth response for the 5-element Yagi in Figure 5.13 across the 20 meter band, using steps of 117 kHz so there are four evaluation frequencies. At 14.0 MHz this Yagi's gain is down a small amount compared to the gain at 14.351 MHz but the rearward pattern is noticeably degraded, dropping to a front-to-back ratio of just under 20 dB.

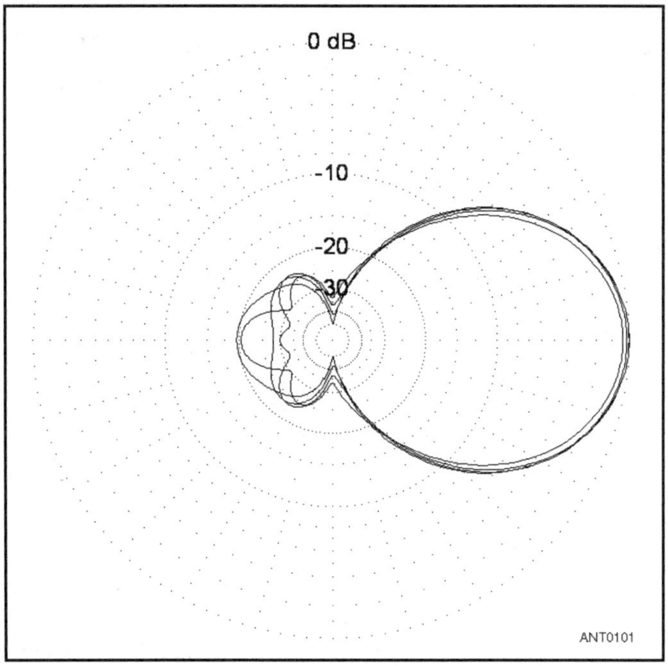

Figure 5.18 — Frequency sweep of 5-element Yagi described in Figure 5.13, showing how the azimuth pattern changes with frequency.

EZNEC can save frequency sweeps of elevation (or azimuth) patterns to a series of output plot files. In essence, the program automates the process described above for saving a plot to disk and then overlaying it on another plot. *EZNEC* can save the following parameters to a text file for later analysis (or perhaps importation into a spreadsheet) chosen by the user:

- Source data
- Load data
- Pattern data
- Current data
- *MicroSmith* numeric data
- Pattern analysis summary.

FREQUENCY SCALING

EZNEC has a very useful feature that allows you to create new models scaled to a new frequency. You invoke the algorithm used to scale a model from one frequency to another by checking the **RESCALE** box after you've clicked the **FREQUENCY** button. *EZNEC* will scale all model dimensions (wire length, height and diameter) except for one specific situation — if you originally specified wire size by AWG gauge, the wire diameter will stay the same at the new frequency. For example, #14 copper wire for a half-wave 80-meter dipole will stay #14 copper wire when the antenna is scaled to become a 20-meter half-wave dipole. If, however, you specified diameter as a floating point numeric value originally (such as 0.064 inch), the diameter will be scaled by the ratio of new to old frequency, along with wire length and height.

Start up *EZNEC* and open up the file **Ch8-520-40W.EZ** for the 5-element 20-meter Yagi on a 40-foot boom. Click the **FREQUENCY** box and then check the **RESCALE** check box. Now, type in the frequency of 28.4 MHz and click **OK**. You have quickly and easily created a new 5-element 10-meter Yagi, that is mounted 29.9949 feet high, the exact ratio of 28.4 MHz to 14.1739 MHz, the original design frequency on 20 meters. Click the **FF PLOT** button to plot the azimuth pattern for this new Yagi. You will see that it closely duplicates the performance of its 20 meter sibling. Click **SRC DAT** to see that the source impedance is 25.38 – j 22.19 Ω, again very close to the source data for the 20-meter version.

5.2.4 Revisiting Source Specification

SENSITIVITY TO SOURCE PLACEMENT

Earlier, we briefly described how to specify a source on a particular segment using *EZNEC*. The sources for the relatively simple dipole, Yagi and quad models investigated so far have been in the center of an easy-to-visualize wire. The placement for the source on the vertical ground plane was at the bottom of the vertical radiator, an eminently logical place. In the other cases we specified the position of the source at 50% of the distance along a wire, given that the wire being fed had an odd number of segments. Please note that in each case so far, the feed point (source) has been placed at a relatively low-impedance point, where the current changes relatively slowly from segment to segment.

Now we're going to examine some subtler source-placement problems. *NEC-2* is well-known as being very sensitive to source placement. Significant errors can result from a haphazard choice of the source segment and the segments surrounding it.

Let's return to the inverted-V dipole in Figure 5.7. The first time we evaluated this antenna (**Ch8-Inverted V Dipole.EZ**) we specified a split source in *EZNEC*. This function uses two sources, one on each of the segments immediately adjacent to the junction of the two downward slanting wires.

Another common method to create a source at the junction of two wires that meet at an angle is to separate these two slanted wires by a short distance and bridge that gap with a short straight wire which is fed at its center. **Figure 5.19** shows a close-up of this scheme in which the length of the segments surrounding the short middle wire are purposely made equal to the length of the middle wire bridging the gap between them. The segmentation for the short middle wire is set to one. **Table 5.4** lists the source impedance and the maximum gain the *EZNEC* computes for four different models:

1. **Ch8-Inverted V Dipole.EZ** (the original model)
2. **Ch8-Inverted V Dipole Triple Segmentation.EZ**
3. **Ch8-Modified Inverted V Dipole.EZ** (as shown in

Table 5.4
135-Foot Inverted-V Dipole at 3.75 MHz

Case	*Segments*	*Source Impedance (Ω)*	*Max. Gain (dBi)*
1	82	72.64 + j 128.2	4.82
2	246	73.19 + j 128.9	4.82
3	67	73.06 + j 129.1	4.85
4	401	76.21 + j 135.2	4.67

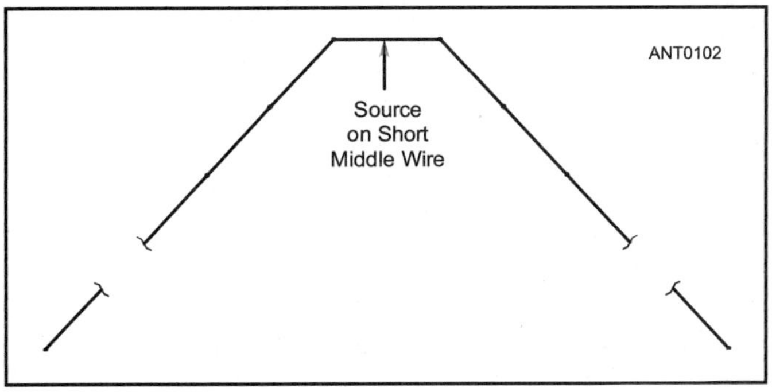

Figure 5.19 — Model of inverted-V dipole using a short center wire on which the source is placed.

Figure 5.19, for the middle wire set to be 2 feet long)

4. **Ch8-Mod Inverted V Poor Segmentation.EZ** (where the number of segments on the two slanted wires have been increased to 200)

Case 2 shows the effect of tripling the number of segments in Case 1. This is a check on the segmentation, to see that the results are stable at a lower level compared to a higher level of segmentation (which theoretically is better although slower in computation). We purposely set up Case 4 so that the lengths of the segments on either side of the single-segment middle wire are significantly different (0.33 feet) compared to the 2-foot length of the middle wire.

The feed point and gain figures for the first three models are close to each other. But you can see that the figures for the fourth model are beginning to diverge from the first three, with about a 5% overall change in the reactance and resistance compared to the average values, and about a 3% change in the maximum gain. This illustrates that it is best to keep the segments surrounding the source equal or at least close to equal in length. We'll soon examine a figure of merit called the *Average Gain* test, but it bears mentioning here that the average gain test is very close for the first three models and begins to diverge for the fourth model.

Things get more interesting if the source is placed at a high-impedance point on an antenna — for example, in the center of a full-wave dipole — the value computed for the source impedance will be high and the results will be quite sensitive to the segment lengths. We'll repeat the computations for the same inverted-V models, but this time at twice the operating frequency, at 7.5 MHz.

Table 5.5 summarizes the results. The source impedance is high, as expected. Note that the resistance term varies quite a bit for all four models with a range of about 23% around the average value. Interestingly, the poorly segmented model's resistance falls in between the other three. The reactive terms are closer for all four models but still cover a range of 4% around the average value. Maximum gain shows the same tendency to be somewhat lower in the fourth model compared to the first three and thus looks as potentially untrustworthy at 7.5 MHz as it does at 3.75 MHz.

This is, of course, a small sampling of segmentation schemes and caution dictates that you shouldn't take these results as being representative of all possibilities. Nevertheless, the lesson to be learned here is that the feed-point (source) impedance can vary significantly at a point where the current is changing rapidly, as it does at a high impedance point on the antenna. Another general conclusion that can be drawn from Tables 5.4 and 5.5 is that more segments, particularly if they surround the source segment improperly, is not necessarily better.

Table 5.5
135-Foot Inverted-V Dipole at 7.5 MHz

Case	*Segments*	*Source Impedance (Ω)*	*Max. Gain (dBi)*
1	82	2297 – *j* 2668	5.67
2	246	1822 – *j* 2553	5.66
3	67	1960 – *j* 2583	5.66
4	401	2031 – *j* 2688	5.48

VOLTAGE AND CURRENT SOURCES

Before we leave the topic of sources, you should be aware that programs like *EZNEC* and others have the ability to simulate both voltage sources and current sources. Although native *NEC-2* has several source types, voltage sources are the most commonly used by amateurs. Native *NEC-2* doesn't have a current source but a current source is nothing more than a voltage source delivering current through a high impedance. Basic network theory says that every Thevenin voltage source has a Norton current source equivalent.

Various commercial implementations of *NEC-2* approach the creation of a current source in slightly different fashions. Some use a high value of inductive reactance as a series impedance, while others use a high value of series resistance. Why would we want to use a current source instead of a voltage source in a model? The general-purpose answer is that models containing a single source at a single feed point can use a voltage source with no problems. Models that employ multiple sources, usually with different amplitudes and different phase shifts, do best with current sources.

For example, *driven arrays* feed RF currents at different amplitudes and phase shifts into two or more elements. The impedances seen at each element may be very different — some impedances might even have negative values of resistance, indicating that power is flowing out of that element into the feed system due to mutual coupling to other elements. Having the ability to specify the amplitude and phase of the current rather than a feed voltage at a feed point in a program like *EZNEC* is a valuable tool.

Next, we examine one more important aspect of building a model — setting up loads. After that, we'll look into two tests for the potential accuracy of a model. These tests can help identify source placement, as well as other problems.

5.2.5 Loads

Many ham antennas, in particular electrically short ones, employ some sort of *loading* to resonate the system. Sometimes loading takes the form of *capacitance hats*, but these can and should be modeled as wires connected to the top of a vertical radiator. A capacitance hat is not the type of loading we'll explore in this section.

Here, the term *loads* refers to discrete inductances, capacitances and resistances that are placed at some point (or points) in an antenna system to achieve certain effects. One fairly common form of a load is a *loading coil* used to resonate an electrically short antenna. Another form of load often seen in ham antennas is a *trap*. *EZNEC* has a special built-in function to evaluate parallel-resonant traps, even at different frequencies beyond their main parallel resonance.

Just for reference a more subtle type of load is a *distributed* material load. We encountered just such a load in our first model antenna, the 135-foot long flattop dipole — although we didn't identify it specifically as a load at that time. Instead, it was identified as a "wire loss" associated with copper.

The *NEC-2* core program has the capability of simulating a number of built-in loads, including distributed material and discrete loads. *EZNEC* implements the following discrete loads:

- Series R ± *j* X loads.
- Series R-L-C loads, specified in Ω of resistance, μH of inductance and pF of capacitance.
- Parallel R-L-C loads, specified in Ω of resistance, μH of inductance and pF of capacitance.
- Trap loads, specified in Ω of resistance in series with μH of inductance, shunted by pF of capacitance, at a specific frequency.
- Laplace loads, specified as mathematical Laplace coefficients (sometimes used in older modeling programs and left in *EZNEC* for backward compatibility).

It is important to recognize that the discrete loads in an antenna modeling program *do not radiate* and they have zero size which is why *NEC-2* discrete loads are described as being *mathematical loads*. The fact that *NEC-2* loads do not radiate means that the popular mobile antennas that use helical loading coils wound over a length of fiberglass whip cannot be modeled accurately with *NEC-2* because such coils do radiate.

Let's say that we want to put an air-wound loading coil with an unloaded Q of 400 at the center of a 40-foot long, 50-foot high, flattop dipole so that it is resonant at 7.1 MHz. The schematic of this antenna is shown in **Figure 5.20**. Examine the modeling file **Ch8-Loaded Dipole.EZ** to see how a discrete series-RL load is used to resonate this short dipole at 7.1 MHz, with a feed-point (source) impedance of 25.3 Ω. This requires a series resistance of 1.854 Ω and an inductive reactance of +741.5 Ω. Note that we again used a single wire to model this antenna, and that we placed the load at a point 50% along the length of the wire.

Specifying this value of reactance represents a 16.62 μH coil with an unloaded Q of 741.5/1.854 = 400 which is just what we wanted. Let's assume for now that we use a perfect transformer to transform the 25.3-Ω source impedance to 50 Ω. If we now attempt to run a frequency sweep over the whole 40 meter band from 7.0 to 7.3 MHz, the load reactance and resistance will not change, since we specified fixed values for reactance and resistance. Hence, the source impedance will be correct only at the frequency where the reactance and resistance are specified since the reactance of an actual coil changes with frequency.

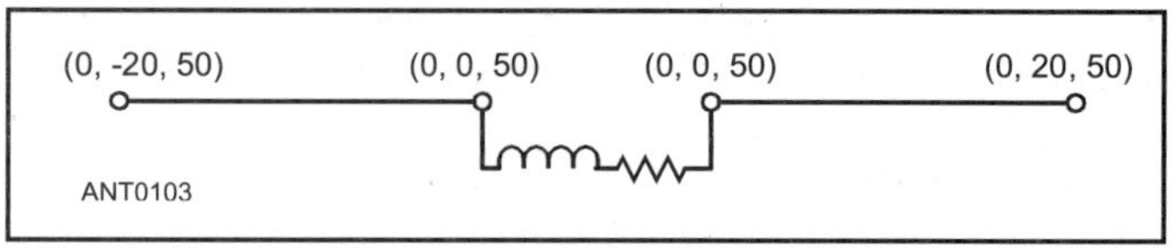

Figure 5.20 — Schematic diagram of a 40-foot long flattop dipole with a loading coil placed at the center. This coil has an unloaded Q of 400 at 7.1 MHz.

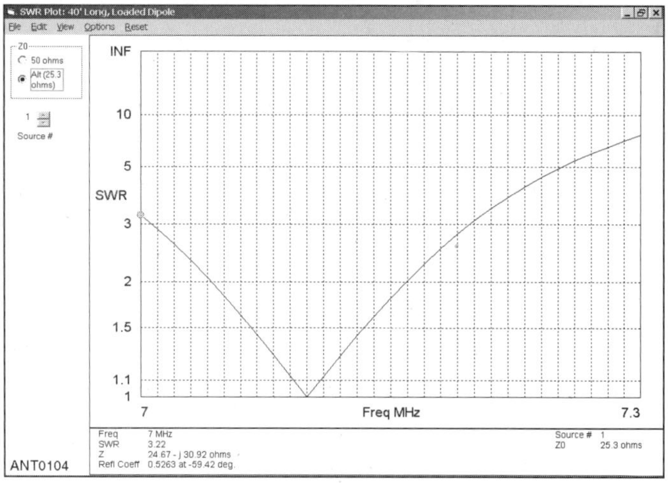

Figure 5.21 — SWR graph of the loaded 40-foot long flattop dipole shown in Figure 5.20.

Let's use a different type of load, a 16.62 μH coil with a series 1.854-Ω resistance at 7.1 MHz. We'll let *EZNEC* take care of the details of computing both the reactance and the changing series resistance at various frequencies. The degree that both reactance and series loss resistance of the coil change with frequency may be viewed using the **LOAD DAT** button from the main *EZNEC* window. By specifying inductance or capacitance, the model's reactance will change with frequency as expected.

Figure 5.21 shows the computed SWR curve for a 25.3-Ω **ALT SWR Z0** reference resistance. The 2:1 SWR bandwidth is about 120 kHz. As could be expected, the antenna has a rather narrow bandwidth because it is electrically short.

5.2.6 Accuracy Tests

There are two tests that can help identify accuracy problems in a model:

- The Convergence test.
- The Average Gain test.

CONVERGENCE TEST

The idea behind the Convergence test is simple: If you increase the segmentation in a particular model and the results change more than you'd like, increase the segmentation until

Table 5.6
135-Foot Flattop Dipole at 3.75 MHz

Segments	*Source Impedance (Ω)*	*Max. Gain (dBi)*
3	85.9 + *j* 128.0	6.34
5	86.3 + *j* 128.3	6.45
7	86.8 + *j* 128.8	6.48
11	87.9 + *j* 129.5	6.50
23	88.5 + *j* 130.3	6.51
45	89.0 + *j* 130.8	6.51
101	89.4 + *j* 131.1	6.51

Table 5.7
5-element Wire Yagi at 14.1739 MHz

Segments	*Source Impedance (Ω)*	*Max. Gain (dBi)*	*180° F/B (dB)*	*F/R (dB)*
3	28.5 – *j* 30.6	12.79	23.2	22.4
5	26.3 – *j* 25.6	13.02	30.5	23.1
7	25.6 – *j* 24.0	13.07	34.8	23.1
11	25.1 – *j* 22.9	13.09	39.9	23.1
25	24.9 – *j* 22.0	13.10	43.7	23.1
99	24.7 – *j* 21.5	13.10	44.2	23.1

the computations converge to a consistent answer. This process has the potential for being subjective but simple antenna models do converge quickly. In this section, we'll review several more of the antennas discussed previously to see how they converge.

Let's go back to the simple dipole in Figure 5.5. The original segmentation was 11 segments, but we'll start with a very low value of segmentation of three, well below the minimum recommended level. **Table 5.6** shows how the source impedance and gain change with increase in segmentation at 3.75 MHz. For this simple antenna, the gain levels off at 6.50 dBi when segmentation has reached 11 segments. Going to ten times the minimum-recommended level (to 111 segments) results in an increase of only 0.01 dBi in the gain.

Arguably, the impedance has also stabilized by the time we reach a segmentation level of 11 segments, although purists may opt for 23 segments. The tradeoff is a slowdown in computational speed.

Let's see how the 5-element Yagi model converges with changes in segmentation level. **Table 5.7** shows how the source impedance, gain, 180° front-to-back ratio and worst-case front-to-rear ratio change with segmentation density. By the time the segmentation has reached 11 segments per wire, the impedance and gain have stabilized quite nicely, as has the F/R. The 180° F/B is still increasing with segmentation level until about 25 segments, but a relatively small shift in frequency will change the maximum F/B level greatly. For example, with 11 segments per wire, shifting the frequency to 14.1 MHz — a shift of only 0.5% — will change the maximum 180° F/B from almost 50 dB down to 27 dB. For this reason the F/R is considered a more reliable indicator of the adequacy of the segmentation level than is F/B.

AVERAGE GAIN TEST

The theory behind the Average Gain test is a little more involved. Basically, if you remove all intentional losses in a model and if you place the antenna either in free space or over perfect ground, then all the power fed to the antenna should be radiated by it. Internally, the program runs a full 3-D analysis, adding up the radiated power in all directions and dividing that sum by the total power fed to the antenna. Ideally, the ratio of input power to radiated power should be unity. Since *NEC-2* is very sensitive to source placement, as mentioned before, the Average Gain test is a good indicator if something is wrong with the specification of the source.

Various commercial versions of *NEC-2* handle the Average Gain test in different ways. *EZNEC* requires the operator to turn off all distributed losses in wires or set to zero any discrete resistive losses in loads. Next set the ground environment to free space (or perfect ground) and request a 3-D pattern plot. *EZNEC* will then report the average gain, which will be 1.000 if the model has no problems. Average gain can be lower or higher than 1.000 but if it falls within the range 0.95 to 1.05 it is usually considered adequate.

As L. B. Cebik, W4RNL has stated: "Like the convergence test, the average gain test is a necessary but not a sufficient condition of model reliability." Pass both tests, however, and you can be pretty well sure that your model represents reality. Pass only one test, and you have reason to worry about how well your model represents reality.

Once again, open the model file **Ch8-Mod Inverted V Poor Segmentation.EZ** and set **WIRE LOSS** to zero, **GROUND TYPE** to **FREE SPACE** and **PLOT TYPE** to **3-DIMENSIONAL**. Click on the **FF PLOT** button. *EZNEC* will report that the Average Gain is 0.955 = –0.2 dB. This is very close to the lower limit of 0.95 considered valid for excellent accuracy. This is a direct result of forcing the segment lengths adjacent to the source segment to be considerably shorter than the source segment's length. The gain reported using this test would be approximately –0.2 dB from what it should be — just what Table 5.5 alludes to also.

Now, let's revisit the basic model **Ch8-Inverted V Dipole. EZ** and look at Case 2 in Table 5.5. Case 2 amounts to a Convergence test for the basic inverted-V model. Since the impedance and gain changes were small comparing the basic model to the one using three times the number of segments, the model passed the Convergence test. The Average Gain test for the basic model yields a value of 0.991, well within the limits for good accuracy. This model has thus passed both tests and can be considered accurate.

Running the Average Gain test for the 5-element Yagi (using 11 segments per wire and whose convergence we examined in Table 5.7) yields a value of 0.996, again well within the bounds indicating a good model. The simple flattop dipole with 11 segments at 3.75 MHz yields an Average Gain result of 0.997, again indicating a very accurate model.

5.2.7 Other Possible Model Limitations

Programs based on the *NEC-2* core computational code have several well-documented limitations that you should know about. Some limitations have been removed in the restricted-access *NEC-4* core (which is not generally available to users), but other limitations still exist, even in *NEC-4*.

CLOSELY SPACED WIRES

If wires are spaced too close to each other, the *NEC-2* core can run into problems. If the segments are not carefully aligned, there also can be problems with accuracy. The worst-case situation is where two wires are so close together that their volumes actually merge into each other as we discussed earlier for wire junctions. This can happen where wires are thick, parallel to each other and close together. You should keep parallel wires separated by at least several wire diameters.

For example, #14 AWG wire is 0.064 inch in diameter. The rule then is to keep parallel #14 wires separated by more than $2 \times 0.064 = 0.128$ inch. And you should run the Convergence test to assure yourself that the solution is indeed converging when you have closely spaced wires, especially if the two wires have different diameters. To model antennas containing closely spaced wires, very often you will need many more segments than usual and you must also carefully ensure that the segments align with each other.

Things can get a little more tricky when wires cross over or under each other, simply because such crossings are sometimes difficult to visualize. Again, the rule is to keep crossing wires separated by more than two diameters from each other and if you intend to join two wires together, make sure you do so at the ends of the two wires, using identical end coordinates. When any of these rules are violated, the Convergence and Average Gain tests will usually warn you of potential inaccuracies.

Parallel-Wire Transmission Lines and LPDAs

A common example of problems with closely spaced wires is when someone attempts to model a parallel-wire transmission line. *NEC-2*-based programs usually do not work as well in such situations as do *MININEC*-based programs. The problems are compounded if the diameters are different for the two wires simulating a parallel-wire transmission line. In *NEC-2* programs, it is usually better to use the built-in "perfect transmission line" function than to try to model closely spaced parallel wires as a transmission line.

For example, a Log Periodic Dipole Array (LPDA) is composed of a series of elements fed using a transmission line that reverses the phase 180° at each element. (See the **Frequency Independent Antennas** chapter.) In other words, the elements are connected to a transmission line that reverses connections left-to-right at each element. It is cumbersome to do so, but you could model such a transmission line using separate wires in *EZNEC* but it is a potentially confusing and a definitely painstaking process. Further, the accuracy of the resulting model is usually suspect, as shown by the Average Gain test.

It is far easier to use the **TRANS LINES** function from the *EZNEC* main window to accurately model an LPDA.

FAT WIRES CONNECTED TO SKINNY WIRES

Another inherent limitation in the *NEC-2* computational core shows up when modeling many Yagis and some quads: popular amateur antennas.

Tapered Elements

As mentioned before, many Yagis are built using telescoping aluminum tubing. This technique saves weight and makes for a more flexible and usually stronger element, one that can survive wind and ice loading better than a single-diameter "monotaper" element design. Many vertical antennas are also constructed using telescoping aluminum tubing.

Unfortunately, native *NEC-2* doesn't model accurately such *tapered elements*, as they are commonly called. There is, however, a sophisticated and accurate workaround for such elements, called the *Leeson corrections*. Derived by Dave Leeson, W6NL, from pioneering work by Schelkunoff at Bell Labs, these corrections compute the diameter and length of an element that is electrically equivalent to a tapered element. This monotaper element is much easier to use in a program like *NEC-2*. (See the **HF Beam Antennas** chapter for more information on tapered elements.)

EZNEC and other *NEC-2* programs can automatically invoke the Leeson corrections, providing that some basic conditions are met — and happily, these conditions are true for the telescoping aluminum-tubing elements commonly used as Yagi elements. *EZNEC* gives you the ability to disable or enable Leeson corrections, under the **OPTION** menu, under **STEPPED DIAMETER CORRECTION**, *EZNEC*'s name for the

Table 5.8
5-element Yagi at 14.1739 MHz with Telescoping Aluminum Elements

	With Leeson Corrections			*Without Leeson Corrections*		
Freq (MHz)	*Source Impedance (Ω)*	*Gain (dBi)*	*F/R (dB)*	*Source Impedance (Ω)*	*Gain (dBi)*	*F/R (dB)*
14.0	23.2 – *j* 26.5	14.82	23.3	22.4 – *j* 12.7	14.92	23.1
14.1	22.7 – *j* 20.5	14.87	22.8	18.6 – *j* 12.5	14.70	21.6
14.2	22.8 – *j* 14.8	14.87	22.7	6.6 – *j* 4.6	14.01	16.2
14.3	22.5 – *j* 11.9	14.76	21.5	1.9 + *j* 10.6	10.61	3.1
14.4	14.5 – *j* 10.5	14.45	19.9	1.6 + *j* 23.7	11.15	–11.4

Leeson corrections. Open the modeling file **520-40H.EZ**, which contains tapered aluminum tubing elements and compare the results using and without using the Leeson corrections.

Table 5.8 lists the differences over the 20-meter band, with the 5-element Yagi at a height of 70 feet above flat ground. You can see that the non-Leeson corrected figures are very different from the corrected ones. At 14.3 MHz, the pattern for the non-corrected Yagi has degenerated to a F/R of 3.1 dB, while at 14.4 MHz, just outside the top of the amateur band, the pattern for the non-corrected antenna actually has reversed. Even at 14.2 MHz, the non-corrected antenna shows a low source impedance, while the corrected version exhibits smooth variations in gain, F/R and impedance across the whole band, just as the actual antenna exhibits.

Some Quads

Some types of cubical quads are made using a combination of aluminum tubing and wire elements, particularly in Europe where the "Swiss" quad has a wide following. Again, *NEC-2*-based programs don't handle such tubing/wire elements well. It is best to avoid modeling this type of antenna, although there are some ways to attempt to get around the limitations, ways that are beyond the scope of this chapter.

5.2.8 Near-Field Outputs

FCC regulations set limits on the maximum permissible exposure (MPE) allowed from the operation of radio transmitters. These limits are expressed in terms of the electric (V/m) and magnetic fields (A/m) close to an antenna. *NEC-2*-based programs can compute the electric and magnetic near fields and the FCC accepts such computations to demonstrate that an installation meets their regulatory requirements. See the section on "RF Safety" in the **Safe Practice**s chapter. We'll continue to use the 5-element Yagi at 70 feet to demonstrate a near-field computation. Open **Ch8-520-40H.EZ** in *EZNEC* and choose **SETUPS** and then **NEAR FIELD** from the menu at the top of the main window. Let's calculate the E-field and H-field intensity for a power level of 1500 W (chosen using the **OPTIONS, POWER LEVEL** choices from the main menu) in the main beam at a fixed distance, say 50 feet, from the tower base. We'll do this at various heights, using 10-foot increments of height, in order to see the lobe structure of the Yagi at 70 feet height.

Table 5.9 summarizes the total H- and E-field intensities as a function of height. As you might expect, the fields are strongest directly in line with the antenna at a height of 70 feet. At ground level, the total fields are well within the FCC limits for RF exposure for both fields. In fact, the fields are within the FCC limits if someone were to stand at the tower base, directly under the antenna.

Table 5.9
E- and H-Field Intensities for 1500 W into 5-Element Yagi at 70 Feet on 14.2 MHz

Height (Feet)	*H-Field (A/m)*	*E-Field (V/m)*
0	0.04	4.1
10	0.03	13.8
20	0.04	20.6
30	0.06	22.6
40	0.08	25.8
50	0.10	33.8
60	0.12	41.5
70	0.12	44.3

5.3 Optimizers in Antenna Modeling

Optimizers have become an indispensable tool of design engineering in technical fields. Optimizers are found in professional software tools for computer aided engineering (CAE), computer aided design (CAD), computational electromagnetics (CEM), electronic design automation (EDA), circuit analysis, electrical engineering, mechanical engineering, and physics. Some inexpensive antenna modeling software tools have optimizers too.

An optimizer relieves a designer of the drudgery of finding the best design by trial and error. A computer can analyze many trial designs in far less time than it takes to build and test prototypes. The fastest and most efficient optimizers have a search strategy or method like artificial intelligence (AI). Optimization is a powerful tool in the hands of a skilled designer. Optimization is best used in the later stages of a design process to fine-tune key elements. However, using an optimizer blindly or expecting too much from it can disappoint. An optimizer is no substitute for human skill.

5.3.1 Variables, Goals, and Methods

To perform an optimization, a modeling program needs three things to be specified: variables, goals, and an optimization method. Each is described below.

VARIABLES

The variables in a model are the parameters that are to be varied in search of better performance. Examples of variables are antenna dimensions, height above ground, orientation in space, the values of lumped elements in loads and traps, and the lengths and impedances of transmission line segments. Electromagnetic simulations take more time than circuit simulations to run. To avoid excessive run times, the number of variables should be kept small, to no more than about 20 variables.

Each variable represents a number that lies in an interval. For each variable, an initial value, and a range of allowed values are specified. The endpoints of the interval (minimum and maximum values) must be specified for each variable. For example, the height of a horizontal dipole might be specified as a variable h that lies between 10 feet and 100 feet. Dipole heights are positive numbers, whereas the height of radials can be positive (elevated) or negative (buried).

For successful optimizations, initial values should be close to the optimal values being sought. The range of values should be no larger than necessary. Given good initial values, an optimizer can more quickly converge or "zero in" on optimum values for all variables.

GOALS

A goal is a specification of what the optimizer should try to achieve and when to stop. Examples of optimization goals are:

•Gain in a specified direction > 6 dBi

•Maximum gain in a specified sector < –10 dBi

•SWR < 2 at all frequencies in a specified band of frequencies

•Front-to-rear ratio > 8 dB at a specified frequency

More generally, a goal is a statement of a mathematical inequality between an antenna or circuit response and a threshold number.

Is *more* better or is *less* better? That is the question. If more is better, the response is called a "figure of merit," a "fitness" function, or an "objective" function. If less is better, the response is called a "cost" or "penalty" function. The distinction is unimportant as either can be converted into the other by adding a minus sign. A cost function is a mapping from values of the variables to a single real number.

A goal can be defined for a very specific condition such as a specific frequency and direction. A goal can also be defined over a range of frequencies or directions. In this case, the cost function is defined to be the average, maximum, minimum, or root-mean-square of the values at the individual frequencies or directions. Multiple goals can be combined by making a compound cost function, a weighted sum of the errors associated with individual goals. The weights can be set to emphasize one goal relative to the other goals. For example, this allows one to specify whether high gain or low SWR is more important.

OPTIMIZATION METHODS

An optimization method is a strategy that is used to achieve the goal. One can avoid optimization altogether by evaluating the cost function at all points on a grid.

•*Compute All Points.* A brute force computation in which the cost function is computed at all points on a uniform search grid in the variable space. Values are compared after all are computed. This method is useful when the number of variables is one or two but impossibly slow when there are many variables.

Guided search methods, by contrast, determine where to search next based on previous calculations. Derivative-based methods use concepts from calculus, such as derivatives (slopes and curvature) of a cost function. These methods find local minima but are easily trapped and don't always find the global (overall) minimum of the cost function.

•*Gradient Descent.* A search that follows the trajectory of the steepest slope of a cost function.

•*Conjugate Gradient.* A variant of gradient descent that uses (or estimates) second derivatives, i.e., curvature in addition to slope.

Nonderivative-based methods use heuristics (rules) or "metaheuristics" instead of calculus concepts to perform "intelligent" search. Such methods use the cost values at previous points to guide the search. They work when a cost function is not continuous (has jumps) or when variables are discrete instead of continuous. Such methods are not so easily trapped in local minima and are more likely to find the global minimum of a cost function.

•*Random Search.* A search in which a random number generator generates random values for variables for which the cost function is computed, sometimes using an estimated probability distribution to bias the search for the minimum. This is also called a "Monte Carlo" search.

•*One Variable at a Time.* A simple pattern search over a single variable at a time, holding the rest constant. Search cycles through all variables and repeats.

•*Hooke-Jeeves method.* A pattern search method that searches along one direction one at a time to find the best value. Search chooses many different directions to explore.

•*Powell's method.* A pattern search method that's more sophisticated than Hooke-Jeeves.

•*Nelder-Mead* (also called the Amoeba, Nonlinear Simplex, or Downhill Simplex method). A method in which an N+1-legged spider crawls through a search space, keeping N legs planted and testing locations with its extra leg according to rules of reflection and extension. When a better (lower elevation) point is found, the spider plants its foot and lifts whichever foot is at highest elevation. The leg in the worst (highest) location is moved in each round.

•*Simulated Annealing (SA)* method. A statistical search method based on principles of heating and cooling in metallurgy. This method is included in the circuit analysis program *QucsStudio*.

•*Alternating Projection.* A search method based on mathematical projections onto convex sets.

Metaheuristic optimization methods are non-derivative methods inspired by nature and include methods based on evolution or social group intelligence.

•*Genetic Algorithm (GA).* A statistical random search based on the Darwinian evolutionary processes of adaptation, selection, survivability, and mutation.

•*Particle Swarm Optimization (PSO).* A statistical random search based on the movement and intelligence of swarms of bees looking for food. This method is included in the antenna modeling program *HOBBIES*.

•*Grey Wolf Optimization (GWO).* A statistical random search based on the hunting process of packs of grey wolves.

•*Brain Storm Optimization (BSO).* A statistical random search based on group intelligence, i.e., the collective behavior of human beings in solving problems.

Finally, "Cocktail" methods combine several different

methods. A cocktail method consists of running several iterations of one method followed by iterations of a second method, then a third method. This approach can help escape being from being trapped in a local minimum. One might run Nelder-Mead for a thousand iterations, interrupted every 200 iterations for a hundred iterations of Random Search. Once it becomes clear that the search is in the vicinity of the global minimum, one can switch to Gradient Descent for rapid convergence.

5.3.2 Software

Although optimizers are a standard feature of professional programs, not all amateur antenna modeling programs have optimizers. However, some low-cost programs do have optimizers, such as the GUI shell program *AutoEZ* and the electromagnetic program *HOBBIES*, as well as several low-cost circuit simulation programs. The programs *ARRL Radio Designer (ARD)* and Ansoft *Serenade SV*, and the free circuit simulator *QucsStudio* have good optimizers. **Table 5.10** summarizes the optimizers in software programs for antenna modeling that amateurs are likely to use, while **Table 5.11** does the same for commercial "computational electromagnetic" (CEM) software, and **Table 5.12** does the same for circuit analysis programs.

The optimizer setup process in *AutoEZ* is straightforward. *AutoEZ* allows variables to be defined by numbers or by equations. Variables defined by numbers can be optimized, but variables defined by equations cannot be optimized. The optimizer Setup screen is shown in **Figure 5.22**. The Setup screen is where a user selects the variables to optimize.

AutoEZ's Optimize sheet has two parts. The top of the sheet has an input block shown in **Figure 5.23**. A user specifies a compound goal as a weighted combination of six different antenna performance metrics. The user also specifies a range of frequencies and directions. Clicking the Start button launches the optimization. Some patience is required as the calculations can take a while to complete.

Figure 5.22 — *AutoEZ* optimizer setup screen. (by permission of AC6LA)

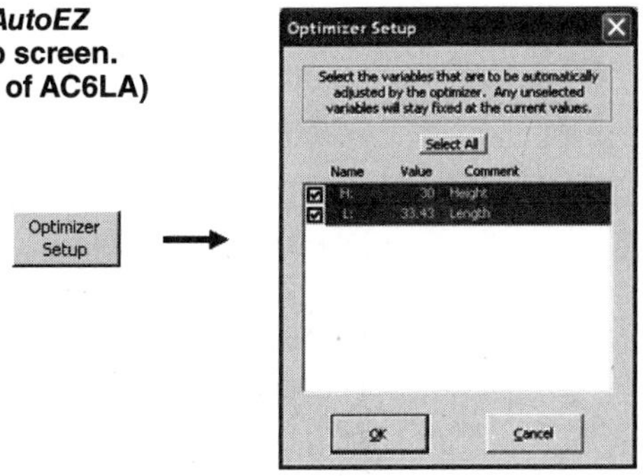

Table 5.10
Optimizers in Wire Antenna Modeling Software.

Program	*Type*	*Optimizer*	*Methods*	*Number of Variables*
AN-SOF	Antennas	No	None. Use *AutoEZ*	0
ANSim	Antennas	No	None	0
AutoEZ	GUI shell	Yes	Nelder-Mead, limited goal types	No limit
AWAS	Antennas	No	None	0
EZNEC	GUI shell	No	None. Use *AutoEZ*	0
4nec2	GUI shell	Yes	Gradient Descent, Genetic Algorithm	No limit
HOBBIES Acad.	General EM	Yes	Powell, Nelder-Mead, Particle Swarm	No limit
MiniNEC	Antennas	No	None	
MMANA-GAL	Antennas	Yes	Simple search, limited goal types	128
Nec2Go	Antennas	No	None	0
NEC-2	Antennas	No	None. Use *AutoEZ*	0
NEC-4.2	Antennas	No	None. Use *AutoEZ*	0
NEC-5	Antennas	No	None. Use *AutoEZ*	0
WIPL-D Lite	Antennas	Yes	Method not stated, limited goal types	2

Table 5.11
Optimizers in Commercial CEM Software.

Program	*Type*	*Optimizer*	*Methods*	*Number of Variables*
Altair FEKO	General EM	Yes	Multiple methods	No limit
ANSYS HFSS	General EM	Yes	Multiple methods	No limit
COMSOL	General EM	Yes	Multiple methods	No limit
Dassault/CST Microwave Studio	General EM	Yes	Multiple methods	No limit
Cadence/AWR Microwave Office	General EM	Yes	Multiple methods	No limit
HOBBIES Pro	General EM	Yes	Powell, Nelder-Mead, Particle Swarm	No limit
Remcon XFdtd	General EM	Yes	Multiple methods	No limit
Sonnet Suites	General EM	Yes	Multiple methods	No limit
Speag SEMCAD	General EM	Yes	Multiple methods	No limit
WIPL-D Pro	General EM	Yes	Multiple methods	No limit

Table 5.12
Optimizers in Circuit Analysis Software.

Program	*Type*	*Optimizer*	*Methods*	*Number of Variables*
Keysight/Agilent ADS	Circuits	Yes	Many methods	Unlimited
Ansoft Serenade SV	Circuits	Yes	Random, Gradient Descent, MiniMax, Levenberg-Marquardt	4
ARRL Radio Designer	Circuits	Yes	Random, Gradient Descent	10
Cadence/AWR Microwave Office	Circuits	Yes	Many methods	No limit
Puff	Circuits	No	None	0
QucsStudio	Circuits	Yes	Grid Search, Random, Coordinate Descent, Hooke-Jeeves, Nelder-Mead, Simulated Annealing, Modified Controlled Random Search, Differential Evolution, DEGL/SAW	No limit
RFSim99	Circuits	No	None	0

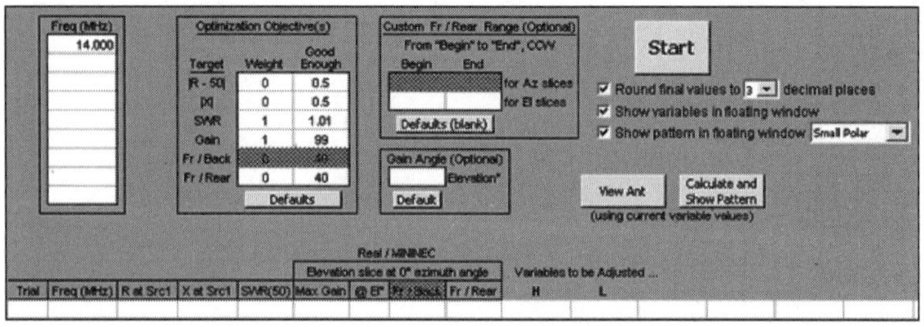

Figure 5.23 — *AutoEZ*'s Optimize sheet input block. (by permission of AC6LA)

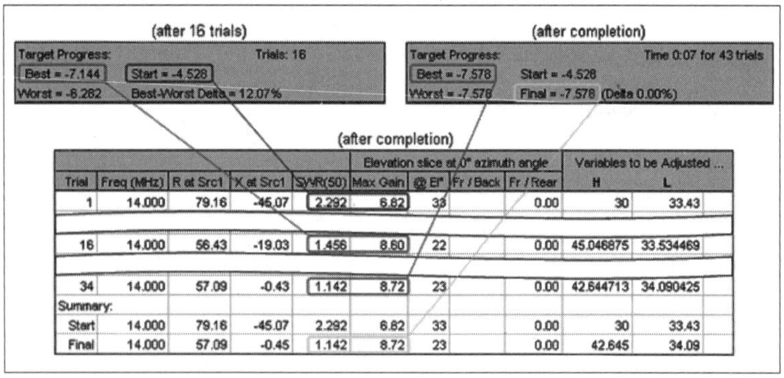

Figure 5.24 — *AutoEZ*'s optimization trial results and summary. (by permission of AC6LA)

The lower part of the *AutoEZ*'s Optimize sheet under the input block is where the optimization results appear. **Figure 5.24** shows the format. The block shows the results of trials followed by a Summary section that shows the first and last trials.

5.3.3 Example: Finding the Best Shape of a Wire Antenna

Although varying the inductance and position of a loading coil would make for a simple example of optimization, the following more complex problem shows the power of optimization. Dipoles and loops are made by a wire that is either straight or curved into a circle. A reasonable question to ask is whether these are the only good shapes or is there, perhaps, a different shape that is better — a shape that is neither straight nor a circle? For a given frequency and a given length of wire, what shape is best in the sense of having maximum gain in a given direction? For this example, we consider the limited class of shapes that are planar (so the wire lies in a plane) and that have left/right mirror symmetry (so the wire must be center fed). Dipoles and loops have these geometric properties. We want to know whether any better antenna having these geometric properties exists. Optimizers can answer these questions by computation and search.

The first step is to model an antenna whose shape depends on variables. There are two ways to specify a shape. If the approximate shape is known in advance, a parametric equation can represent the shape, and the variables are simply the coefficients of the equation. Since we know nothing about the shape, a parametric equation cannot be specified, and a different approach is used that imposes less restriction on allowable shapes. We consider a curved wire made of many short, straight wires in which the direction of each varies. Hence the wire can curve smoothly or zig-zag. The constraints will be that all wires lie in a common plane and the antenna has two symmetric arms. The arms may be straight or may bend forming a circle; so, dipoles and loops are included as allowed geometries. As most wire antenna programs assume wires are straight between two endpoints, we define each wire's endpoints by equation as shown in **Table 5.13**.

The variables x_{ij} and z_{ij} are coordinates of the ends of the wires. The subscript "*i*" is the wire number, and the subscript "*j*" is the end number. The range of the subscripts is $i = 1, \ldots, N$, and $j = 1$ or 2. h is the length of each wire, and angles $\theta_1, \ldots, \theta_N$ are variables defined by recursive equations as follows.

Table 5.13
Wire Table

	End 1			End 2		
Wire no.	*x1*	*y1*	*z1*	*x2*	*y2*	*z2*
1	0	0	0	x11 + h sin θ 1	0	x11 + h cos θ 1
2	x12	0	z12	x21 + h sin θ 2	0	z21 + h cos θ 2
			•			
			•			
			•			
N	xN-1,2	0	z N-1,2	xN,1 + h sin θ N	0	zN,1 + h cos θ N

$$\theta_1 = \delta_1$$
$$\theta_2 = \delta_1 + \delta_2$$
$$\theta_3 = \delta_1 + \delta_2 + \delta_3$$
$$\vdots$$
$$\theta_n = \theta_{n-1} + \delta_n$$
$$\vdots$$
$$\theta_N = \theta_{N-1} + \delta_N$$

Variables $\theta_1, \cdots, \theta_N$ represent the directions of each wire. Variables $\delta_1, \cdots, \delta_N$ represent the incremental change in direction from one wire to the next. These latter variables are the ones that will be optimized. Each is given the initial value of zero and a range of –90 to 90 degrees. The range is restricted so that arms of the antenna curve gradually through space; wire junctions cannot form acute angles. The variables to be optimized are the change in direction or angle from a wire to its successor.

A second arm consisting of wires N+1 through $2N$ is defined by symmetry: x values are negated while y and z values are the same. The total length L of the antenna is the sum of the lengths of the wires in both arms, $L = 2Nh$. The second arm introduces no additional variables. Thus, there are N variables to be optimized, $\delta_1, \cdots, \delta_N$, which are the changes in direction from a wire to its successor.

The second step is to specify a goal. We consider a simple goal: obtain maximum gain in a given direction. We seek the wire shape that, for a given length, achieves maximum gain in a specified direction at a specified frequency. Small dipoles and loops have the property that there is a plane that contains both the wire and the direction of maximum gain. In the case of a loop, there is one such plane. In the case of a dipole, there is an infinite number of such planes. To allow the possibility that either a dipole or a loop actually is the optimal shape, we must not exclude either of their shapes beforehand. Accordingly, the goal will be to achieve maximum possible gain in a direction that lies in a plane that contains the wire. Because the wire lies in the x-z plane, it is convenient to choose either the x or y axis as the direction in which gain is to be maximized. The choice being arbitrary, the z axis is chosen. The plane and direction in the plane give the direction in standard spherical coordinates as $\theta = 0$ and $\phi = 0$, where θ is polar angle measured from the z axis, and ϕ is azimuth measured from the x axis. (Don't confuse these angles with the subscripted angles that define wire directions.) Thus, the goal is to maximize the gain in the direction specified by $\theta = \phi = 0$.

The third part in setting up the optimization is to specify the method of search. Nelder-Mead is chosen in part because most of the optimizers that amateurs encounter have it. The Nelder-Mead method is tried-and-true. It is not easily fooled or trapped in local minima and usually finds the global optimal solution for the variables if the number of local minima is not excessively great.

The antenna modeling program *HOBBIES* was used to implement the model above for 27 different electrical lengths of wire, ranging from ½ to 6 wavelengths. Each wire has a single segment. The number of wires $2N$ was increased as total length increased to keep each wire a fixed fraction of a wavelength. N was always made large enough to accurately represent a smooth curve. Wires were modeled as lossless perfect electrical conductor (PEC). Wire diameter was fixed at 1/22,000th wavelength, which corresponds to AWG #10 wire for a 5-MHz half-wavelength dipole. The number of variables increased with wire length, and optimizer runs took progressively longer to complete. The points in the graph were computed from left to right, from small N to large. With each increment in N, the previous variable values were used to initialize the next case. The half-wavelength case (12 variables, 800 iterations) ran in tens of minutes. The three-wavelength case (20 variables, 2,000 iterations) ran in several hours. The six-wavelength case ran overnight.

The optimizer tells us how well the goal is met and the best values of the variables. **Figure 5.25** shows the results of 27 runs. Shown for comparison is the gain of a straight dipole. It is evident that an optimally shaped wire outperforms a

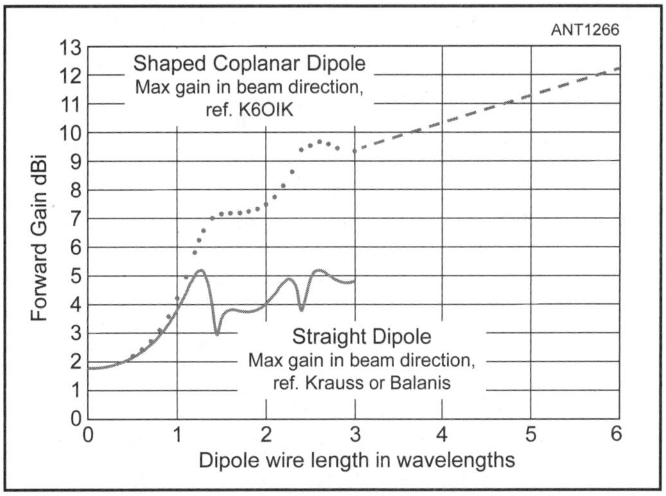

Figure 5.25 — Gain of shaped planar wire antenna found by optimizer.

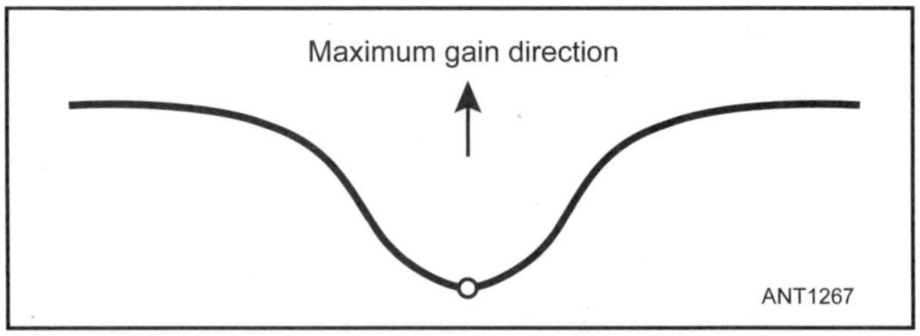

Figure 5.26 — A 1.4-wavelength shaped wire antenna optimized for maximum directive gain.

straight dipole at all wire lengths. Not shown in the figure, the optimally shaped wire also outperforms a circular loop at all wire lengths. Unlike the dipole and loop patterns, which are omnidirectional, the shaped wire's pattern is directional and has a pronounced main lobe.

The values of the variables $\delta_1, \cdots, \delta_N$ found by the optimizer give the best shape of wire for a given total wire length $L = 2Nh$. Figure 5.25 shows the gain of the best-shaped wire at each length. **Figure 5.26** shows one case: the best shape for a wire of 1.4 wavelengths. The shape has been likened to a cowboy hat or a long-horned steer viewed in profile. As the wire length increases from zero to many wavelengths, the shape changes progressively from a straight dipole to a cowboy hat, and eventually converges to a Vee dipole with wiggles and arms that open to an optimal angle. Thus, a dipole morphs into a traveling-wave Vee antenna as it gets longer provided the curve is chosen for maximum gain for each length.

The shape shown in Figure 5.26 was first reported by H. H. Meinke and F. M. Landstorfer in a 1975 German patent. It is commonly referred to as a Landstorfer antenna and became the subject of many papers in the professional antenna literature and articles in amateur radio magazines, cf. *CQ DL* (Apr.1982), RadCom (Dec. 1982), and *Ham Radio* (Nov. 1983 and Apr.1986). For more on this antenna see the Bibliography.

The example above illustrates the simple problem of optimizing the geometric shape of a wire to achieve maximum gain at a single frequency. The example only scratches the surface of what optimizers can do. There are many ways to vary the problem. Some are obvious, such as expanding frequencies or adding goals to limit SWR. Some are less obvious, such as changing the direction of maximum gain to be perpendicular to the plane of the wire antenna instead of lying in the plane, or even allowing the wire to be a general nonplanar 3D curve. Or the length L of wire might be cut into sections, and the sections arranged but not connected (as in a multi-element Yagi-Uda beam), or else connected in multi-wire junctions (as in log-periodic, T, lazy H, and fan dipole antennas). Traps, loads, transmission lines, and match networks can be included and jointly optimized within an antenna model. Optimizers are a powerful tool in the study and design of interesting antennas that push the boundary of design.

5.4 Advanced Modeling Techniques

This section describes some useful advanced techniques for modeling with thin-wire electromagnetic programs. Three techniques are described: how to model insulated wires, how to use wire grids to model metallic surfaces, and how to use generalized wire grids to model dielectric surfaces and objects.

5.4.1 Insulated Wire

Amateur antennas often incorporate insulated wire, i.e., wire that is covered with a dielectric or magnetic material. Some programs, notably *NEC-3* and later versions, feature insulated wire capability in their calculating engine. Most thin-wire calculating engines do not have the capability to model insulated wire. In some cases, an aftermarket user interface GUI program provides insulated wire capability that the main calculating engine lacks. It is therefore important to know how to model insulated wire both for using programs that lack the capability and for getting an independent check on those programs that do. The simplest approach to modeling insulated wires is to replace them with *equivalent* uninsulated wires.

"Modeling Insulated Wire," a detailed paper by K6OIK describing three different methods for specifying equivalent wires is provided in the online material. The two methods most useful to amateurs are summarized here.

W4RNL Method: This method is due to L.B. Cebik, W4RNL, and is used in *4nec2* with the *NEC-2* engine. Given an insulated wire having conductor radius a and insulation outer radius b, the equivalent wire is an uninsulated wire having the same radius a plus distributed inductive loading in the amount

$$L = \frac{\mu_0}{2\pi}\left(\frac{b}{a}\varepsilon_r\right)^{\frac{1}{12}}\left(1-\frac{1}{\varepsilon_r}\right)\ln\left(\frac{b}{a}\right) \text{ henries per meter}$$

where

L is the distributed inductance in henries per meter

ε_r is the dielectric constant of the insulation

μ_0 is the magnetic permeability of vacuum in henries per meter

b is the outer radius of the dielectric insulation

a is the outer radius of the metal wire

ln(·) is the natural (base e) logarithm function

The W4RNL method defines equivalent wires by adding distributed inductance alone. Conductor diameter and metal conductivity are not changed. This method is not based on electromagnetic theory but comes from an ad hoc curve fit to the insulated wire correction inside *NEC-4*. Absolute accuracy has not been established.

K6OIK Method: This method is due to S.D. Stearns, K6OIK. According to this method, the equivalent wire is an uninsulated wire having an intermediate outer radius a' plus distributed inductive loading and modified conductivity. The radius and loading are given by

$$a' = a\left(\frac{b}{a}\right)^{\left(1-\frac{1}{\varepsilon_r}\right)}$$

$$L = \frac{\mu_0}{2\pi}\left(1-\frac{1}{\varepsilon_r}\right)\ln\left(\frac{b}{a}\right) \text{ henries per meter}$$

Skin effect loss is accounted for by decreasing the metal conductivity to

$$\sigma'_{effective} = \sigma\left(\frac{a}{b}\right)^{2\left(1-\frac{1}{\varepsilon_r}\right)} \text{ siemens per meter}$$

The radius lies between those of the other methods, i.e., $a < a' < b$. So, the diameter of the equivalent wire lies between the diameter of the conductor and the diameter of the insulation of the insulated wire. Distributed inductance and metal conductivity are less than in the other methods. The adjustments to wire diameter, loading and conductivity depend only on geometry and are frequency independent. The K6OIK adjustments are exact at frequencies for which the diameter is very small compared to a wavelength assuming the wire is in vacuum or air.

5.4.2 Wire-Grid Modeling

Thin-wire programs for electromagnetic analysis historically preceded surface codes. As a result, wire grids were used for modeling surfaces before surface programs came into prominence. A *wire grid* is a system of wires each of which is connected at one or both ends.

GUIs for thin-wire antenna programs often have commands for constructing simple wire grids. *4nec2*, *AN-SOF*, *EZNEC*, and *MMANA-GAL* have tools for making wire grids and radial systems based on copy/paste, shifts, and symmetries. *4nec2* includes a library of examples, which includes 14 wire grid models of a horn antenna, airplane, ship, tank, and several vehicles. The models run under *NEC-2*.

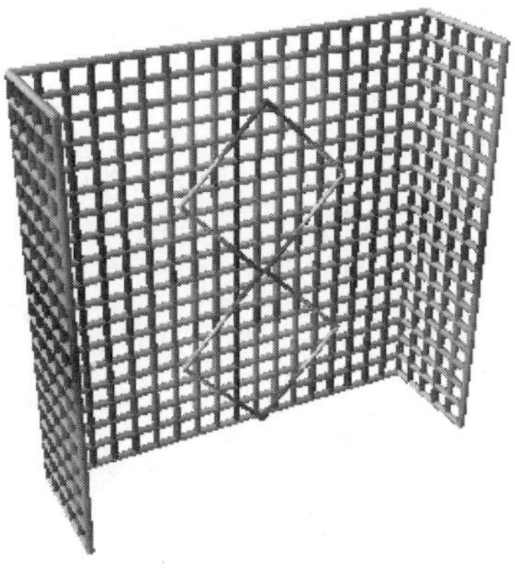

Figure 5.27 — Biquad with wire grid reflector modeled in *4nec2*.

Figure 5.27 shows an example, a biquad antenna with wire-grid reflector created by W6ZTM, also known as a *twin-quad* or *double-diamond*. It consists of two bi-squares, unfolded to create two W's which are joined at their tips and fed in opposition at their midpoints. Without a reflector, the biquad has a dipole-type doughnut pattern and horizontal polarization when oriented as shown.

The reflector is a wire-grid that has 551 vertices, 1054 edges, and 504 square faces. The unit cell is a square patch 1⁄20 of a wavelength on a side. With the reflector, gain is approximately 11 dBi. This antenna is popular for Wi-Fi and DTV applications and the bi-squares can be replaced by double-circles.

For more complicated shapes, *computer-aided design* (CAD) programs allow one to create and manipulate 3D objects in a computer. Such programs allow one to create and manipulate the geometries of objects needed for antenna modeling, 3D printing, and mechanical design. Many programs can create a mesh from a geometry in a process called *meshing*. The Bibliography section lists CAD programs that are free or have free trials. **Table 5.14** is a list of standard file formats used by CAD programs to store geometry and mesh data.

Table 5.14. CAD Geometry and Mesh File Formats

Geometry	*Formats*	*Mesh Formats*
IGES	NASTRAN	mesh
DXF	STL	mesh
Parasolid	VRML	mesh
ACIS	3DStudio	mesh
VDA	CGNS	mesh
Rhinoceros	HOBBIES	mesh
Shapefile	Surface	mesh
XYZ points	Ply	mesh
KML	VTK	voxels
	XYZ	nodes

MESHES AND GRIDS

Knowledge of basic ideas from computational geometry or computer graphics can aid our understanding. A *mesh* divides a surface or volume into smaller surface patches or volumes that fill the original surface or volume without overlap. Surface meshing divides an area into *polygons*. Volume meshing divides a volume into *polyhedra*.

A mesh is an example of a *tessellation* in mathematics. A surface mesh consists of faces (F), edges (E), and vertices (V). In a surface mesh, every edge is connected at both ends to other edges at vertices. Such meshes satisfy Euler's formula $V - E + F = 2$.

A wire grid is similar to a *graph* in mathematics. A wire grid has wires and nodes just as a graph has edges and vertices. The difference is a wire grid has both geometric (size) and topological (shape) properties, whereas graphs are purely topological objects.

Care is needed to avoid confusion and to distinguish between geometric and topological ideas. In the language of graph theory, a *vertex* is a junction of edges but carries no concept of spatial location. *Nodes*, by contrast, are points in space where wires meet. Nodes are geometric entities that define the ends of wires, whereas vertices are topological entities that define the connections of edges.

Nodes and their associated edges are referred to collectively as a *wireframe*. A wireframe is a geometric entity that can be converted into a wire grid by adding attributes such as diameter and conductivity. However not all wire grids have wires that are connected at both ends. Some wire grids have wires connected at only one end. Examples include dipole and vertical antennas and radial systems. Hence, while the wireframe or edges of a mesh can be used to define a wire grid, not all wire grids correspond to meshes. To learn more about computational geometry and computer graphics see the websites given in the Bibliography.

Surface EM (electromagnetic) codes, such as *FEKO*, *HOBBIES*, or *WIPL-D*, compute currents in the *faces*. Alternatively, the *edges* of a mesh define a wireframe, which if converted to a wire grid or wire table can be used by a thin-wire code, such as *AN-SOF*, *ANSim*, *AWAS*, *MiniNEC*, or *NEC*. At present, most popular thin-wire GUIs (graphical user interfaces), such as *AutoEZ*, *EZNEC*, *4nec2*, *MMana-GAL*, and *Nec2Go*, do not have the capability to import mesh files and convert them to a wire grid or wire table, but professional *computational electromagnetics* (CEM) programs do.

Today's CAD programs can be used to define antenna geometries and create meshes. A feature that would import mesh files and create wire grids automatically from them would be a powerful addition to the popular GUI programs that amateurs use for antenna modeling.

CONSTRUCTING GRIDS

When making wire grids, the important variables are wire length, segment length, wire diameter, and wire conductivity. Numerical simulation results are sensitive to wire radii. The *equal area rule* (*EAR*) is a rule of thumb for setting the wire diameter in wire-grid modeling. The EAR rule states that the surface area of the wires should be made equal to the surface area of the solid surface being modeled. The rule applies to wires in the direction of the electric field. Near fields are seldom linearly polarized. Elliptical polarization is the norm. Consequently, the EAR rule should be satisfied by grid wires in all directions. For a rectangular grid, in the X-Y plane for example, the wires in the X direction and Y direction should independently satisfy the rule. Wire circumference should equal center-to-center grid spacing in the direction perpendicular to wires. Consequently, the diameter of the wires in the X direction should be $1/\pi$ times the length of the wires in the Y direction, and conversely the diameter of the wires in the Y direction should be $1/\pi$ times the length of the wires in the X direction.

Because *NEC* prefers the ratio of segment length to diameter to be greater than unity, it follows that there should be no more than one or two segments per wire when making wire grid models for *NEC*. Smaller wires and closer grid spacing leads to greater accuracy at the expense of more computation.

When the equal area rule is satisfied, the surface impedance of the grid is approximately equal to the surface impedance of the area being gridded. More importantly, the difference between near fields of the grid and the continuous surface are smallest, and the wire grid becomes an accurate representation of the continuous surface it models. For more information, tips, and rules of thumb for making rectangular and non-rectangular grids see the references in the Bibliography.

5.4.3 Modeling Surfaces

When modeling a surface, a wire-grid model is sometimes better than a surface patch model. For example, when *NEC* is used to model open surfaces, a wire grid must be used because the magnetic field integral equation (MFIE) in *NEC's* patch model is only appropriate for closed bodies. So, although the reflector in Figure 5.27 could be modeled using *NEC's* patch commands, the wire grid is preferable. Other surface programs don't have this restriction.

Figure 5.28 shows a sphere meshed with quadrilaterals. The mesh was created by *HOBBIES*. Current on the surface of the sphere can be represented either as currents in the faces of the quadrilaterals or as currents in the wires of the wire frame. The mesh need not be made of quadrilaterals. A quadrilateral mesh can be converted to a triangular mesh by dividing each quadrilateral into two triangles by adding an edge that connects a vertex to its opposite. The number of new faces created equals the number of new edges, so Euler's formula remains correct. The near fields, whether computed from currents in the faces or wires of a grid, will be nearly equal provided the EAR rule is satisfied (as well as other wire restrictions that a particular thin-wire program may require).

Figure 5.29 shows KI6BDR's triangular mesh of a car that was featured in *QST*, October 2016. The car has a vertical whip on the roof for 2-meter operation. The mesh is finer than

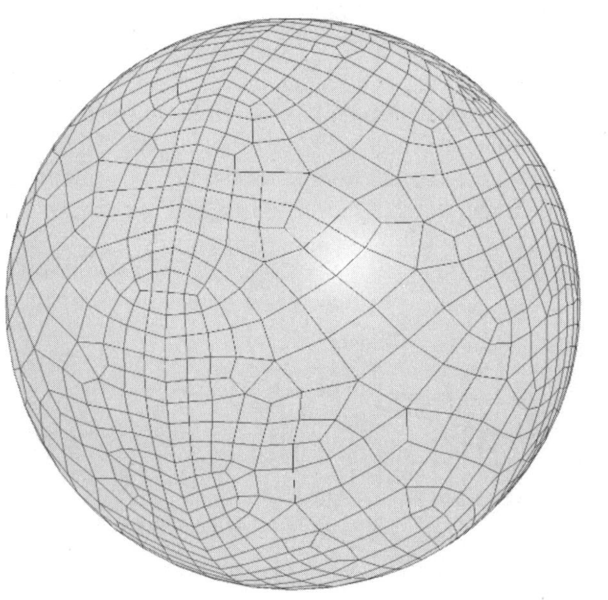

Figure 5.28 — A sphere meshed with quadrilaterals. [K6OIK, 2018]

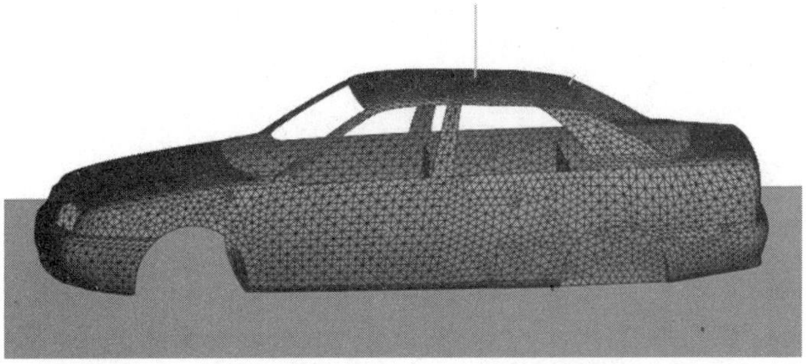

Figure 5.29 — A car meshed with triangles. [KI6BDR, *QST*, October 2916]

needed. For 2-meter band modeling, triangles having 8-inch sides are sufficient. The mesh was created by *FEKO* which computed currents in the triangular patches. However, if the wireframe of the mesh had been used to make a wire grid model of the car, the computed fields would be the same provided the wire diameter satisfies the EAR rule.

Not surprisingly, wire grids can be used to model objects that have metal conducting surfaces, such as individual antennas or antennas on automobiles, airplanes, and ships for example. **Figures 5.30** and **31** show wire grid models of a horn antenna for 8 GHz and a US Navy destroyer. The ship model is for 20 MHz and has 8,031 wires and 16 antennas onboard. **Figure 5.32** shows a triangle mesh of a state-of-the-art UAV aircraft. The program *FEKO* created the mesh for surface patch computations. The mesh's wireframe could be converted to a wire grid for computation by a thin-wire program.

MODELING DIELECTRIC SURFACES

Wire grids can model dielectric objects, too. The approximation is best for lossy dielectric objects – those for which a wave entering attenuates before reaching the other side, similar to the skin effect for RF current in conductors. The method involves making the grid appear to have the same surface impedance as the dielectric object. To do this, the grid should satisfy the same Leontovich boundary condition as the object's surface. This can be done by adding distributed loading to the wires in a wire grid in a manner analogous to the K6OIK insulated wire method above. The wire diameter is set to satisfy the EAR rule. Then distributed loading is added to the wires to achieve a surface impedance equal to that of the object being modeled.

$$\eta = \sqrt{\frac{j\,2\pi f \mu_r \mu_0}{\sigma + j\,2\pi f \varepsilon_r \varepsilon_0}} \text{ ohms per square meter}$$

where

η is surface impedance in ohms per square meter
j is $\sqrt{-1}$
f is frequency in hertz
μ_r is relative permeability of material compared to vacuum
μ_0 is magnetic permeability of vacuum in henries per meter
ε_r is dielectric constant (relative permittivity of material compared to vacuum)
ε_0 is electric permittivity of vacuum in farads per meter
σ is conductivity of material in siemens per meter

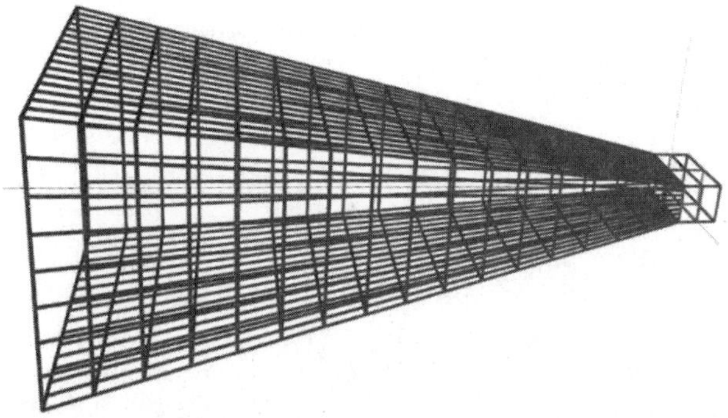

Figure 5.30 — Horn antenna wire grid. [Hasan, 2022]

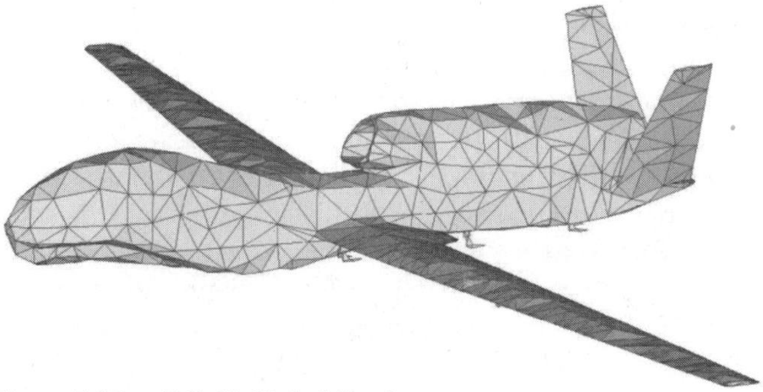

Figure 5.32 — RQ-4A Global Hawk triangle mesh. [K6OIK, 2008]

Figure 5.31 — USS Mitscher DDG-57 Destroyer wire grid. [Lysiak, 2008]

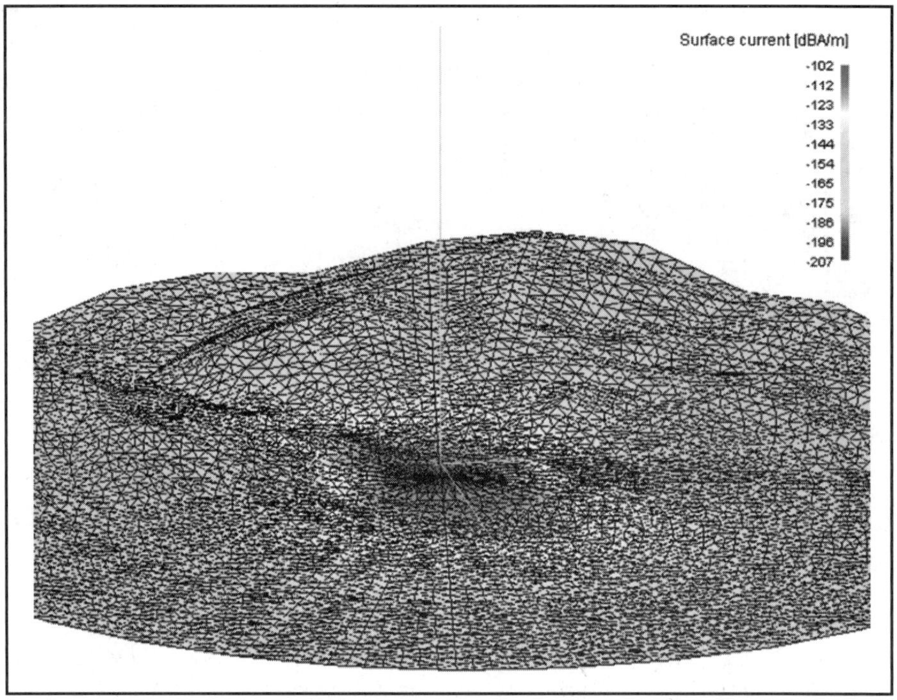

Figure 5.34 — Ground terrain modeling by surface mesh and MoM. [K6OIK, 2006]

Figure 5.33 — Wire-grid model of a tree trunk.

A tree trunk is an example of a lossy dielectric object that can be modeled by a wire grid. KE4PT and W4RQ modeled a tree trunk as a lossy dielectric cylinder [*QST*, February 2018]. Their model consists of a single wire and distributed loading chosen to satisfy an impedance boundary condition. Current in a wire is assumed to be equally distributed around the wire's circumference. The single-wire tree model cannot model azimuthal currents nor proximity effect for which the current is unequally distributed around the tree's circumference.

The tree model can be improved by using more wires to form a wire grid of the tree's surface as shown in **Figure 5.33**. Six vertical wires, each centered on a tree trunk's perimeter are sufficient to capture proximity effect. The spacing of the wires must satisfy the usual rules for wires. The diameters of the wires must satisfy EAR. If six vertical wires are used, the sum of the circumferences of the vertical wires must equal the circumference of the tree. With six vertical wires, the wire diameter should be 1/6th of the diameter of the tree. The horizontal wires form hexagons at different vertical heights. The hexagons should be separated vertically by a tenth of a wavelength or less. Horizontal wire diameter is 1/π times the vertical separation. The radial end caps have little effect and may be dropped.

Finally, all wires are assigned distributed loading that satisfies the surface impedance boundary condition according to the loading method described by Tsunekawa and Ando, and material parameters given by KE4PT and W4RQ.

The theory of surface impedance boundary conditions originated from M.A. Leontovich and is explained by T.B.A. Senior and others. Subsequent research extended Leontovich's original work. The subject is having a resurgence due to interest in the unique electromagnetic properties of meta-surfaces. More information can be found in the references in the Bibliography.

MODELING GROUND WITH SURFACE MESHING CODES

Moment method surface programs like *FEKO*, *WIPL-D*, and *HOBBIES*, give a new way to model ground — as a dielectric surface. This method is fundamentally different from the methods such as a Sommerfeld ground or a Fresnel complex reflection coefficient ground. With a simple reflection coefficient ground, an antenna must be well above the ground surface for antenna currents to be computed accurately.

Surface meshing programs treat the ground as a dielectric surface. The surface is meshed like any surface. The advantage of modeling the ground using a dielectric surface is that the surface can have any shape. The surface does not need to be flat, infinite, or smooth. It can be irregular. One can import topographic terrain data and convert it to a mesh. By making the ground terrain and trees part of an antenna model, terrain and trees are accounted for. Separate terrain analysis by a ray-tracing program like *HFTA* is unnecessary because the terrain is included in the antenna model. (See the **HF Antenna System Design** chapter for more information on the *HFTA* program.)

The surface used for ground does not need to enclose a

volume but must be electrically large to minimize diffraction at its edges. A reasonably accurate ground model is one that extends one wavelength in all compass directions and 5 to 10 wavelengths in the directions where the elevation pattern is desired. Edge diffraction leads to radiation into the lower half space below the ground surface. Normally this radiation is many dB down and can be ignored. One can define graph limits to simply not show radiation into the lower half space. The far field pattern at low elevation angles converges to the same pattern as a Sommerfeld ground would predict in the limit as the ground plane size increases to infinity. Of course, real ground is neither flat nor infinite nor smooth, which is precisely the reason for not using Sommerfeld or reflection coefficient ground models.

Figure 5.34 shows an example of terrain modeled as an irregular dielectric surface at HF. Terrain data was imported and meshed out to 4 miles in all directions from the QTH at the center. Average soil parameters were assigned. The program *FEKO* was used to compute antenna patterns for antennas at several heights and frequencies in the 160-meter, 75-meter, and 40-meter bands. The 160-meter cases required 4,000 triangles in the mesh, used 530 Mb of memory, and ran in under 8 minutes. However, the 40-meter cases required 39,000 triangles in the mesh, used nearly 10 Gb of memory, and took 52 hours to run. The computer had a compute speed of 53 Gflops/s. *HFTA* does terrain analysis by ray tracing in less time, but the method of moments provides elevation pattern information that is more accurate. One can analyze smaller models on a fast desktop computer. All of the pieces needed to carry out such an analysis are available to an industrious amateur.

In summary, a wire grid can model a dielectric surface, and the surface can be irregular. The wires must be sized to satisfy EAR and must be loaded to model dielectric per the Leontovich boundary condition. A properly loaded wire grid can model irregular ground. However, there are two limitations to consider before using a wire grid as a ground model. First, the antenna cannot penetrate the ground. The wire-grid ground method does not permit the antenna to touch the ground nor to be buried. Second, many segments are needed, which may exceed the number of segments that a thin-wire program allows. Computation time can be enormous. Surface programs, such as *FEKO*, *WIPL-D*, and *HOBBIES*, overcome these limitations. They are based on the surface equivalence principle, and an antenna can be in more than one domain, e.g., partially in the air and below ground. These programs use exact surface integral equations instead of thin-wire approximations. *WIPL-D* and *HOBBIES* use higher-order basis functions which reduces the number of variables by allowing larger surface patches up to 100 times larger than the grid spacing required by a wire grid for *NEC*.

5.5 References and Bibliography

5.5.1 Introduction to Antenna Modeling

Harrington, R., "Matrix Methods for Field Problems," *Proceedings of the IEEE*, Vol. 55, No. 2, Feb. 1967, pp. 136 – 149.

Harrington, R., *Field Computation by Moment Methods*, Macmillan, 1968. Republished by Krieger, 1981; IEEE-Wiley, 1993.

Haviland, R., W4MB, "Programs for Antenna Analysis by the Method of Moments," *ARRL Antenna Compendium, Vol.* 4, ARRL, 1995, pp. 69 – 73.

Rawle, W., "The Method of Moments: A Numerical Technique for Wire Antenna Design," *High Frequency Electronics*, Feb. 2006, pp. 42 – 47. Available from **highfrequencyelectronics.com/Feb06/HFE0206_Rawle.pdf**

Stamm. J., and Breakall, J., "Comparison of Results for the NEC4, WIPL-D, and EIGER Antenna Modeling Programs," *17th Annual Review of Progress in Applied Computational Electromagnetics*, Monterey, CA, Mar. 19 – 23, 2001, pp. 269 – 276.

Stearns, S., K6OIK, *Antenna Modeling for Radio Amateurs*, ARRL Pacificon Antenna Seminar, San Ramon, CA, Oct. 20 – 22, 2017. Available from **fars.k6ya.org/docs/k6oik**

5.5.2 Thin Wire, Triangle Function Basis Method (MiniNEC, MMANA-GAL)

Belrose, J., VE2CV, "Modeling HF Antennas with MININEC: Guidelines and Tips from a Code-Users Notebook," *ARRL Antenna Compendium, Vol.* 3, ARRL, 1992, pp. 156 – 164.

Cebik, L., W4RNL, "NEC and MININEC Antenna Modeling Programs: A Guide to Further Information," *QEX*, Mar./Apr.1998, pp. 47 – 49.

Cebik, L., W4RNL, "Notes on Modeling LPDAs in MININEC," *QST*, June 2004, pp. 72 – 73.

Julian, A., Logan, J., and Rockway, J., *MININEC: A Mini-Numerical Electromagnetics Code*, Tech Doc 516, DTIC ADA121535, Naval Ocean Systems Center, Sep. 6, 1982. Available from **dtic.mil/dtic/tr/fulltext/u2/a121535**

Lewallen, R., W7EL, "MININEC: The Other Side of the Sword," *QST*, Feb. 1991, pp. 18 – 22.

MiniNEC Pro software from Black Cat Systems: **blackcatsystems.com/software/mininec-antenna-analysis-modeling-software.html**. Latest version of MiniNEC. Low cost.

MMANA-GAL software: **gal-ana.de/basicmm/en**. Basic version is free. A user interface for MiniNEC.

Nichols, S., GØKYA, *An Introduction to Antenna Modeling*, RSGB, 2014.

Logan J., and Rockway, J., *The New MININEC (Version 3): A Mini-Numerical Electromagnetic Code*, Tech. Doc. 938, DTIC ADA181682, Naval Ocean Systems Center, Sep. 1986. Available from **dtic.mil/dtic/tr/fulltext/u2/a181682**

Rockway, J., Logan, J., Li, S., and Tam, D., *The MININEC System: Microcomputer Analysis of Wire Antennas*, Artech House, 1988.

Rockway J., and Logan, J., N6BRF, "Wire Modeling Limitations of NEC and MININEC for Windows," *QEX*, May/June 1998, pp. 17 – 21.

Wait, J., "Complex Image Theory Revisited," IEEE Antennas and Propagation Magazine, vol. 33, no. 4, pp. 27 – 29, Aug. 1991.

Wait, J., "Radiation Resistance of a Small Circular Loop in the Presence of a Conducting Ground," Journal of Applied Physics, vo. 24, no. 5, pp. 646 – 649, May 1953.

Wait, J., "Mutual Electromagnetic Coupling of Loops Over a Homogeneous Ground," Geophysics, vol. 20, no. 3, pp. 630 – 637, July 1955. "Additional Note," vol. 21, no. 2, pp. 479 – 484, 1956.

5.5.3 Thin Wire, 3-Term Sinusoid Basis Method (NEC-*n*, EZNEC, 4nec2)

AutoEZ software: **ac6la.com**. Supplemental software that adds useful features to EZNEC.

Burke, G., and Poggio, A., *Numerical Electromagnetics Code (NEC)-Method of Moments*, Part I: Program Description-Theory; Part II: Program Description-Code; Part III, Users Guide, Tech. Rept. UCID-18834, DTIC ADA956129, Lawrence Livermore Laboratory, Jan. 1981. Available from **dtic.mil/dtic/tr/fulltext/u2/a956129**

Burke, G., *Numerical Electromagnetics Code – NEC-4, Method of Moments*, Part I: User's Manual (NEC 4.1); Part II: Program Description-Theory, Tech. Report UCRL-MA-109338, Jan. 1992. Available from **puhep1.princeton.edu/~mcdonald/examples/NEC_Manuals**

Burke, G., E. K. Miller, and A. I. Poggio, "The Numerical Electromagnetics Code (NEC) — A Brief History," *IEEE International Symposium on Antennas and Propagation*, Monterey, CA, June 20 – 26, 2004, pp. 2871 – 2874. Available from **e-reports-ext.llnl.gov/pdf/304278**

Burke, G., *NEC-5 Validation Manual*, Tech. Report LLNL-SM-791163, Lawrence Livermore National Laboratory, Sep. 24, 2019. Available from **ipo.llnl.gov/technologies/software/nec-v50-numerical-electromagnetic-code**. NEC-5 uses a "two-potential" method (pioneered in WIPL and HOBBIES) and triangle basis functions (as in MiniNEC).

Cebik, L., W4RNL, "A Beginner's Guide to Using Computer Antenna Modeling Programs," *ARRL Antenna Compendium, Vol.* 3, ARRL, 1992, pp. 148 – 155.

Cebik, L., W4RNL, "NEC and MININEC Antenna Modeling Programs: A Guide to Further Information," *QEX*, Mar./Apr.1998, pp. 47 – 49.

Cebik, L., W4RNL, "A Beginners Guide to Modeling with *NEC*," a four-part series in *QST*: Part 1, Nov. 2000, pp. 34 – 38; Part 2, Dec. 2000, pp. 40 – 44; Part 3, Jan. 2001, pp. 44 – 48; Part 4, Feb. 2001, pp. 31 – 35.

Cebik, L., W4RNL, *ARRL Antenna Modeling Course*, ARRL, 2003.

Cebik, L., W4RNL, "Antenna Options: Modeling Software," Part 1, *QEX*, Sep./Oct. 2005, pp. 54 – 59; Part 2, *QEX*, Nov./Dec. 2005, pp. 50 – 56.

EZNEC software: **eznec.com**. Free, except Pro/4 version discontinued. A well developed user interface.

4nec2 software: **qsl.net/4nec2**. Free user interface and NEC-2.

NEC-4.2 and *NEC-5* software: **ipo.llnl.gov/technologies/software/nec-v50-numerical-electromagnetic-code**. Individual licenses are inexpensive.

Rockway J., and Logan, J., N6BRF, "Wire Modeling Limitations of NEC and MININEC for Windows," *QEX*, May/June 1998, pp. 17-21.

Silver, W., NØAX, *Antenna Modeling for Beginners: An Introductory Guide to Using Antenna Modeling Software*, ARRL, 2012.

Tracy, M., KC1SX, "EZNEC 3.0 for Windows," *QST*, Sep. 2000, p 66.

5.5.4 Thin Wire, Multi-Radius Bridge Current Method (MBC, ANSim)

ANSim software: **em-bench.com**. Free demo version offered. GUI requires Python is installed.

MBC software: Contact Professor Robert Paknys, VE2JBP, emeritus at Concordia University.

Tilston, M., *Thin-Wire Reciprocal Multiradius Implementation of the Electromagnetic Moment Method*, Ph.D. dissertation, University of Toronto, 1989.

Tilston M., and Balmain, K., "On the Suppression of Asymmetric Artifacts Arising in an Implementation of the Thin-Wire Method of Moments," *IEEE Transactions on Antennas and Propagation*, Vol. 38, No. 2, Feb. 1990, pp. 281 – 285.

Tilston M., and Balmain, K., "A Multiradius, Reciprocal Implementation of the Thin-Wire Moment Method," *IEEE Transactions on Antennas and Propagation*, Vol. 38, No. 10, Oct. 1990, pp. 1636 – 1644.

5.5.5 Surface Methods (FEKO, WIPL-D, HOBBIES)

Elsherbeni, A., Nayeri, P., and Reddy, C., Antenna Analysis and Design using FEKO Electromagnetic Simulation Software, SciTech Publishing, 2014.

FEKO software: **altair.com/feko-applications**. Free trial.

FEKO Student Edition software: **altairuniversity.com/feko-student-edition**. Low cost.

García-Doñoro, D., Zhang, Y., Zhao, W., Sarkar, T., García-Castillo, L., and Salazar-Palma, M., "HOBBIES: A New Electromagnetic Simulator," IEEE International Symposium on Antennas and Propagation (APS-URSI), July 2010.

HOBBIES software: **hobbies-em.com**. Free Academic version with book purchase.

Kolundzija, B., Ognjanovic, J., and Sarkar, T., WIPL-D Microwave: Circuit and 3D EM Simulation for RF & Microwave Applications — Software and User's Manual, Artech House, 2006.

Rao, S., Wilton, D., and Glisson, A., "Electromagnetic Scattering by Surfaces of Arbitrary Shape," IEEE Transactions on Antennas and Propagation, Vol. 30, No. 3, May 1982, pp. 409 – 418.

Snyder, K., KI6BDR, "Near Fields of a Mobile Mounted 2 Meter Antenna," *QST*, Oct. 2016, pp. 33 – 35.

Stearns, S., K6OIK, "HOBBIES Software for Computational Electromagnetics," QEX, pp. 17 – 21, Nov./Dec. 2020, pp. 17 – 21.

WIPL-D software: **wipl-d.com**. Free trial.

Zhang, Y., Sarkar, T., Zhao, X., Garcia-Doñoro, D., Zhao, W., Salazar-Palma, M., and Ting, S., Higher Order Basis Based Integral Equation Solver (HOBBIES), Wiley, 2012.

5.5.6 Higher Order Basis Methods (WireZeus, AWAS, WIPL-D, HOBBIES)

AWAS (Analysis of Wire Antennas and Scatterers) software, version 2.0. Free demo version at University of Belgrade: **oet.etf.bg.ac.rs/software.htm**. AWAS is the thin wire predecessor of WIPL-D and HOBBIES. It has a highly accurate Sommerfeld ground model. Contact Steve Stearns, K6OIK, for instructions and examples files.

Djordjević, A., Bazdar, M., Petrovic, V., Olcan, D., Sarkar, T., and Harrington, R., *AWAS for Windows Version 2.0, Analysis of Wires and Scatterers – Software and User's Manual*, Artech House, 2002.

Kolundžija B., and Djordjević, A., *Electromagnetic Modeling of Composite Metallic and Dielectric Structures*, Artech House, 2002.

Kolundžija, B., Ognjanović, J., and Sarkar, T., *WIPL-D Microwave: Circuit and 3D EM Simulation for RF & Microwave Applications – Software and User's Manual*, Artech House, 2005.

Popović, B., "Polynomial Approximation of Current along Thin Symmetrical Cylindrical Dipoles," *Proceedings of the IEE* (British), Vol. 117, No. 5, May 1970, pp. 873 – 878.

Popović, B., "On Polynomial Approximation of Current Along Thin Asymmetrical Cylindrical Dipoles," *IEEE Transactions on Antennas and Propagation*, Vol. 19, No. 1, Jan. 1971, pp. 117 – 120.

Popović, B., Dragović, M., and Djordjević, A., *Analysis and Synthesis of Wire Antennas*, Research Studies Press, Division of Wiley, 1982.

Popović, B., *CAD of Wire Antennas and Related Radiating Structures*, Research Studies Press, Division of Wiley, 1991.

B. D. Popović and B. M. Kolundžija, *Analysis of Metallic Antennas and Scatterers*, IEE Press, 1994.

Zhang Y., and Sarkar, T., *Parallel Solution of Integral Equation-Based EM Problems in the Frequency Domain*, Wiley, 2009.

Zhang, Y., Sarkar, T., Zhao, X., Garcia-Doñoro, D., Zhao, W., Salazar-Palma, M., and Ting, S., *Higher Order Basis Based Integral Equation Solver (HOBBIES)*, Wiley, 2012.

5.5.7 Finite Difference Time Domain (FDTD) Method

Kunz, K., Luebbers, R., and Hunsberger, F, "A Thin Dipole Antenna Demonstration of the Antenna Modeling Capabilities of the Finite Difference Time Domain Technique," *Applied Computational Electromagnetics Society Journal*, Vol. 5, No. 1, Mar. 1990, pp. 2 – 18. Available from **aces-society.org/journal.php**

Meep software: **meep.readthedocs.io/en/latest**. Open source Finite Difference Time Domain (FDTD) modeling code that uses Python or Scheme. Free.

OpenEMS software: **openems.de/start**. Finite Difference Time Domain (FDTD modeling code that uses Matlab or Octave. Free.

Taflove, A., "Application of the Finite-Difference Time-Domain Method to Sinusoidal Steady-State Electromagnetic-Penetration Problems," *IEEE Transactions on Electromagnetic Compatibility*, Vol. 22, No. 3, Aug. 1980, pp. 191 – 202.

Yee, K., "Numerical Solution of Initial Boundary Value Problems Involving Maxwell's Equations in Isotropic Media," *IEEE Transactions on Antennas and Propagation*, Vol. 14, No. 3, May 1966, pp. 302 – 307.

5.5.8 Modeling Ground

Burke, G., Ground Model Options in the NEC-4.2 Antenna Code, Tech. Report LLNL-TR-490316, Lawrence Livermore National Laboratory, July 6, 2011. Available from **e-reports-ext.llnl.gov/pdf/498709.pdf**

Kolundžija, B., Đordević, A., and Petrović, V., "A New Approach for Accurate Analysis of Antennas Above Real Ground," *IEEE International Symposium on Antennas and Propagation*, Orlando, FL, July 11 – 16, 1999.

Haviland, R., W4MB, "Ground parameters for Antenna Analysis," *ARRL Antenna Compendium, Vol.* 5, ARRL, 1996, pp. 96 – 100.

Sarkar, T., Dyab, W., Abdallah, M.., Salazar-Palma, M., Prasad, M., and Ting, S-W., "Application of the Schelkunoff Formulation to the Sommerfeld Problem of a Vertical Electric Dipole Radiating Over an Imperfect Ground," *IEEE Transactions on Antennas and Propagation*, Vol. 62, No. 8, Aug. 2014, pp. 4162 – 4170.

Stearns, S., K6OIK, "Terrain Modeling," in *Antenna Modeling for Radio Amateurs*, ARRL Pacificon Antenna Seminar, San Ramon, CA, Oct. 20 – 22, 2017, pp. 95 – 100. Available from **fars.k6ya.org/docs/k6oik**

5.5.9 Modeling Insulated Wire

Cebik, L., "Insulated Wires – The NEC-2 Way," Note 83. (**on5au.be/content/amod/amod83.html**)

Lee J., and Balmain, K., "Wire Antennas Coated with Magnetically and Electrically Lossy Material," *Radio Science*, Vol. 14, No. 3, May – June 1979, pp. 437 – 445. Available from **onlinelibrary.wiley.com/journal/10.1002/(ISSN)1944-799X/issues**

Popović, B., Djordjević A., and Kirćanski, N., "Simple Method for Analysis of Dielectric-Coated Wire Antennas," IERE, *The Radio and Electronic Engineer*, Vol. 51, No. 3, Mar. 1981, pp. 141 – 145.

Popović B., and Nešic, A., "Generalisation of the Concept of Equivalent Radius of Thin Cylindrical Antennas," *IEE Proceedings*, Part H, Vol. 131, No. 3, June 1984, pp. 153 – 158.

Richmond J., and Newman, E., "Dielectric Coated Wire Antennas," *Radio Science*, Vol. 11, No. 1, Jan. 1976, pp. 13 – 20.

Stearns, S., K6OIK, "Antenna Modeling for the Radio Amateur," ARRL Pacificon Antenna Seminar, San Ramon, CA, Oct. 17 – 19, 2008. **fars.k6ya.org/docs/k6oik**

Stearns, S., K6OIK, "Insulated Wire," in *Antenna Modeling for Radio Amateurs*, ARRL Pacificon Antenna Seminar, San Ramon, CA, Oct. 20 – 22, 2017, pp. 58 – 62. Available from **fars.k6ya.org/docs/k6oik**

Yurkov, A., RA9MB, "Simulation of Wire Antennas Coated by Dielectric by Means of NEC-2 Code," (translated from Russian) Radio Communication Technology, vol. 24, no. 1, pp. 27 – 32, 2015.

5.5.10 Modeling Trees and Forests

Olcan D., and Kolundžija, B., "Precise and Efficient EM Modeling of Trees with WIPL-D Code," *20th Annual Review of Progress in Applied Computational Electromagnetics*, Syracuse NY, Apr.19 – 23, 2004.

Olcan D., and Kolundžija, B., "On the Simulation of RCS from Trees and Forests above Real Finite Ground," *European Conference on Antennas and Propagation* (EuCAP), Nice, France, Nov. 6 – 10, 2006.

Siwiak, K., KE4PT, and R. Quick, W4RQ, "Live Trees Affect Antenna Performance," *QST*, Feb. 2018, pp. 33 – 37.

Stearns, S., K6OIK, "Antenna Performance in a Forest of Trees," *QST*, Aug. 2020, pp. 34 – 37.

5.5.11 Terrain Analysis by Method of Moments

Snyder, K., KI6BDR, Snyder, "Meshing Silicon Valley," Applied Computational Electromagnetics Society Conference, Miami, FL, Mar. 2006.

Stearns, S., K6OIK, "Novel and Strange Ideas in Antennas and Impedance Matching," ARRL Pacificon Antenna Seminar, San Ramon, CA, Oct. 13 – 16, 2006. Also in "Antenna Modeling for Radio Amateurs," ARRL Pacificon, Oct. 20 – 22, 2017. **fars.k6ya.org/docs/k6oik**

5.5.12 Wire Grid Modeling of Conductive and Dielectric Surfaces

Austin B., and Najm, R., "Wire-Grid Modelling of Vehicles with Flush-Mounted Window Antennas," *7th International Conference on Antennas and Propagation*, pp. 950 – 953, York, UK, April 15 – 18, 1991.

Caloz C., and Achouri, K., *Electromagnetic Metasurfaces: Theory and Applications*, IEEE Press, Wiley, 2019.

Golden, T., *Equivalent Wire-Grids for the Electromagnetic Modeling of Conducting Surfaces*, R&D Report, Golden Engineering, July 10, 2020. **antennasimulator.com/index.php/resources**

Hasan, A., Klyukin, D., Kvasnikov, A., Komnatnov, M., and Kuksenko, S., "On Wire-Grid Representation for Modeling Symmetrical Antenna Elements," *Symmetry*, Vol. 14, no. 7, paper 1354, June 2022. **doi.org/10.3390/sym14071354**

Hubing T., and Kauffman, J., "Modeling Electrically Small, Thin Surfaces with Wire Grids," *Applied Computational Electromagnetics Society Journal*, vol. 5, no. 1, pp. 19 – 24, Mar. 1990. **aces-society.org/journal.php**

Hoppe D., and Rahmat-Samii, Y., *Impedance Boundary Conditions in Electromagnetics*, CRC Press, Taylor & Francis, 1995.

Abou-Jaoude R., and Walton, E., "Numerical Modeling of On-Glass Conformal Automobile Antennas," *IEEE Transactions on Antennas and Propagation*, vol. 46, no. 6, pp. 845 – 852, June 1998.

Kolundžija, B., Miodrag, T., and Djordjević, A., "Optimal Wire-Grid Modeling Based on Conversion of Solid Surface Model," *IEEE International Symposium on Antennas and Propagation*5, Boston, MA, July 8 – 13, 2001, pp. 592 – 59.

Larsen, T., "A Survey of the Theory of Wire Grids," *IRE Transactions on Microwave Theory and Techniques*, Vol. 10, No. 3, May 1962, pp. 191 – 201.

Leontovich, M., Investigations *on Radiowave Propagation, Part II*, USSR Academy of Sciences, Moscow, 1948.

Ludwig, A., "Wire Grid Modeling of Surfaces," *IEEE Transactions on Antennas and Propagation*, Vol. 35, No. 9, Sep. 1987, pp. 1045 – 1048.

Lysiak, K., "Comparison of Three Major MOM Codes for a Large Wire-Grid Ship Model," *24th Annual Review of Progress in Applied Computational Electromagnetics*, Niagara Falls, Canada, Mar. 30 – Apr.4, 2008, pp. 341 – 347.

Oyekanmi L., and Watkins, J., "Selecting Wire Radius for Grid/Mesh Models," *Applied Computational Electromagnetics Society Journal*, vol. 5, no. 2, pp. 43 – 57, July 1990. **aces-society.org/journal.php**

Paknys R., and Raschkowan, L., "Moment Method Surface Patch and Wire Grid Accuracy in the Computation of Near Fields," *Applied Computational Electromagnetics Society Journal*, vol. 12, no. 3, pp. 16 – 25, Nov. 1997. **aces-society.org/journal.php**

Pelosi, G., and Ufimtsev, P., "The Impedance-Boundary Condition," *IEEE Antennas and Propagation Magazine*, vol. 38, no. 1, pp. 31 – 35, February 1996.

Richmond, J., "A Wire-Grid Model for Scattering by Conducting Bodies," *IEEE Transactions on Antennas and Propagation*, Vol. 14, No. 6, Nov. 1966, pp. 782 – 786.

Rubinstein, A., Rachidi, F., and Rubinstein, M., "On Wire-Grid Representation of Solid Metallic Surfaces," *IEEE Transactions on Electromagnetic Compatibility*, Vol. 47, No. 1, Feb. 2005, pp. 192 – 195.

Rubinstein, A., Rostamzadeh, C., Rubinstein, M., and Rachidi, F., "On the Use of the Equal Area Rule for the Wire-Grid Representation of Metallic Surfaces," *17th International Zurich Symposium on Electromagnetic Compatibility*, Singapore, Feb. 27 – Mar. 3, 2006, pp. 212 – 215.

Rubinstein, A., Rubinstein, M., and Rachidi, F., "A Physical Interpretation of the Equal Area Rule," *IEEE Transactions on Electromagnetic Compatibility*, vol. 48, no. 2, pp. 258 – 263, May 2006.

Sarolic, A., Modlic, B., and Poljak, D., "Measurement Validation of Ship Wiregrid Models of Different Complexity," IEEE International Symposium on Electromagnetic Compatibility, pp. 147 – 150, Montreal, QC, Canada, Aug.. 13 – 17, 2001.T. B. A. Senior, "Impedance Boundary Conditions for Imperfectly Conducting Surfaces," *Applied Scientific Research*, Section B, Vol. 8, No. 1, pp. 418 – 436, Dec. 1960, pp. 418 – 436.

Senior, T., Impedance Boundary Conditions for Imperfectly Conducting Surfaces, *Applied Scientific Research*, Section B, vol. 8, no. 1, paper 418, pp. 418 – 436, Dec. 1960.

Senior, T., "Approximate Boundary Conditions," *IEEE Transactions on Antennas and Propagation*, vol. 29, no. 5, pp. 826 – 829, Sep. 1981.

Stearns, S., K6OIK, "Wire Grid Modeling," in *Antenna Modeling for Radio Amateurs*, ARRL Pacificon Antenna Seminar, San Ramon, CA, Oct. 20 – 22, 2017, pp. 69 – 72. **fars.k6ya.org/docs/k6oik**

Trueman C., and Kubina, S., "Verifying Wire-Grid Model Integrity with Program *CHECK*," *Applied Computational Electromagnetics Society Journal*, vol. 5, no. 2, pp. 17 – 42, July 1990. **aces-society.org/journal.php**

Trueman, C., and Kubina, S., "Fields of Complex Surfaces using Wire Grid Modelling," *IEEE Transactions on Magnetics*, vol. 27, no. 5, pp. 4262 – 4267, Sep. 1991.

Tsunekawa, K., and Ando, A., "Advanced Wire Grid Method for Solving the Scattered Field of a Lossy Dielectric Object," *IEEE International Symposium on Antennas and Propagation*, Chicago, IL, June 18 – 25, 1992, pp. 797 – 800.

Wait, J., "On the Theory of Scattering from a Periodically Loaded Wire Grid," *IEEE Transactions on Antennas and* Propagation, vol. 25, no. 3, pp. 409 – 413, May 1977.

Wang, D-S., "Limits and Validity of the Impedance Boundary Condition on Penetrable Surfaces," *IEEE Transactions on Antennas and Propagation*, vol. 35, no. 4, pp. 453 – 457, April 1987.

Yuferev S., and Ida, N., *Surface Impedance Boundary Conditions: A Comprehensive Approach*, CRC Press, Taylor & Francis, 2010.

5.5.13 Yagi Antenna Optimization

S.Baskar, A. Alphones, P. N. Suganthan, and J. J. Liang, "Design of Yagi-Uda Antennas Using Comprehensive Learning Particle Swarm Optimization," *IEE Proceedings on Microwaves, Antennas and Propagation*, Vol. 152, No. 5, May 2005, pp. 340 – 346.

Z. Bayraktar, P. L. Werner, and D. H. Werner, "The Design of Miniature Three-Element Stochastic Yagi-Uda Arrays Using Particle Swarm Optimization," *IEEE Antennas and Wireless Propagation Letters*, Vol. 5, Mar. 2006, pp. 22 – 26.

Cebik, L., W4RNL, "Notes on the OWA Yagi," *QEX*, July/Aug. 2002, pp. 22 – 34.

Cheng, D., "Gain Optimization for Yagi-Uda Arrays," *IEEE Antennas and Propagation Magazine*, Vol. 33, No. 3, June 1991, pp. 42 – 45.

Donnelly, D., K2SS, "Optimizing Gain on Yagi Antennas," *Ham Radio*, Mar. 1988, pp. 21 – 24.

Hopkins, D., VK4ZF, "Computer-Aided Design of Long VHF Yagi Antennas," *Ham Radio*, May 1986, pp. 28 – 33.

Jones E., and Joines, W., "Design of Yagi-Uda Antennas Using Genetic Algorithms," *IEEE Transactions on Antennas and Propagation*, Vol. 45, No. 9, Sep. 1997, pp. 1386 – 1392.

Khodier M., and Al-Aqil, M., "Design and Optimisation of Yagi-Uda Antenna Arays," *IET Proceedings on Microwaves, Antennas and Propagation*, Vol. 4, No. 4, 2010, pp. 426 – 436.

Lawson, J., W2PV, "Yagi Antenna Design, Part 2: Experiments Confirm Computer Analysis," *Ham Radio*, Feb. 1980, pp. 19 – 27.

Leeson, D., W6QHS, *Physical Design of Yagi Antennas*, ARRL, 1992.

Rattan, M., Patterh, M., and Sohi, S., "Optimization of Yagi-Uda Antenna Using Simulated Annealing," *Journal of Electromagnetic Waves and Applications*, Vol. 22, No. 2/3, Feb. 2008, pp. 291 – 299.

Venkatarayalu, N., and Ray, T., "Optimum Design of Yagi-Uda Antennas Using Computational Intelligence," *IEEE Transactions on Antennas and Propagation*, Vol. 52, No. 7, July 2004, pp. 1811 – 1918.

YagiCAD software: **yagicad.com/yagicad/YagiCAD.htm**. Developed by P. McMahon, VK3DIP. Free.

YW (Yagi for Windows) software, version 2.0. Developed by R. D. Straw, N6BV, and included in ARRL Antenna Book Software.

5.5.14 CAD Tools for Working with Geometries and Meshes

AutoCAD software: **autodesk.com/education/free-software/autocad**. Free to students.

Autodesk software: **autodesk.com/education/free-software/featured**. Free to students.

Autodesk software: **autodesk.com**. Professional. Free trial.

BlocksCAD software: **blockscad3d.com**. Free.

FreeCAD software: **freecadweb.org**. Free.

Meshmixer software: **meshmixer.com**. Triangle mesh editor. Free.

Microsoft 3D Builder software. Bundled in Windows. Free.

Microsoft 3D Viewer software. Viewing only. Bundled in Windows. Free.

OpenSCAD software: **openscad.org**. Free.

Rhinoceros software: **rhino3d.com**. Professional. Free trial.

ShareCAD viewer: **beta.sharecad.org**. Web-based CAD viewer.

Solidworks software: **solidworks.com**. Professional. Free trial.

Tinkercad software: **tinkercad.com**. Free.

5.5.15 Computational Geometry for Computer-Aided Design (CAD)

Mesh generation: **en.wikipedia.org/wiki/Mesh_generation**

Point-set triangulation: **en.wikipedia.org/wiki/Point-set_triangulation**

Polygon mesh: **en.wikipedia.org/wiki/Polygon_mesh**

Regular grid: **en.wikipedia.org/wiki/Regular_grid**

Tessellation: **en.wikipedia.org/wiki/Tessellation_(computer_graphics)**

Types of mesh: **en.wikipedia.org/wiki/Types_of_mesh**

Unstructured grid: **en.wikipedia.org/wiki/Unstructured_grid**

Wire-frame model: **en.wikipedia.org/wiki/Wire-frame_model**

5.5.16 Optimization and Optimizers

Hooke R., and Jeeves, T., "'Direct Search' Solution of Numerical and Statistical Problems," *Journal of the ACM*, Vol. 8, No. 2, pp. 212 – 219, April 1961.

Nelder J., and Mead, R., "A Simplex Method for Function Minimization," *The Computer Journal*, vol. 7, no. 4, pp. 308 – 313, January 1965; correction vol. 8, no. 1, p. 27, April 1965.

Gucker, G., *Stochastic Gradient Algorithms* for *Searching Multidimensional Multimodal Surfaces*, Tech. Rept. 6778 – 7, AD0738965, Stanford University, Oct. 1969.

Fletcher, R., "A New Approach to Variable Metric Algorithms," *The Computer Journal*, vol. 13, no. 3, pp. 317 – 322, Aug. 1970.

Kirkpatrick, S., Gelatt, Jr., C., and Vecchi, M., "Optimization by Simulated Annealing," *Science*, vol. 220, no. 4598, pp. 671 – 780, May 13, 1983.

Bandler, J., Kellermann, W., and Madsen, K., "A Superlinearly Convergent Minimax Algorithm for Microwave Circuit Design," IEEE *Transactions on Microwave Theory and Techniques*, vol. 33, no. 12, pp. 1519 – 1530, Dec. 1985.

Křivý, I., and Tvrdík, J., "The Controlled Random Search Algorithm in Optimizing Regression Models," *Computational Statistics & Data Analysis*, vol. 20, no. 2, pp. 229 – 234, Aug. 1995.

Kennedy, J., and Eberhart, R., "Particle Swarm Optimization," *Int. Conf. on Neural Networks*, pp. 1942 – 1948, Perth, Australia, Nov. 27 – Dec. 1, 1995.

Storn, R., and Price, K., "Differential Evolution – A Simple and Efficient Heuristic for Global Optimization over Continuous Spaces," *Journal of Global Optimization*, vol. 11, no. 4, pp. 341 – 359, Dec. 1997.

Robinson, J., and Rahmat-Samii, Y., "Particle Swarm Optimization in Electromagnetics," *IEEE Transactions on Antennas and Propagation*, vol. 52, no. 2, pp. 397 – 407, February 2004.

Powell, M., "The BYOBYQA Algorithm for Bound Constrained Optimization Without Derivatives," Tech. Rept. DAMTP 2009/NA06, Department of Applied Mathematics and Theoretical Physics, Cambridge University, UK, 2009.

Das, S., et al., "Differential Evolution Using a Neighborhood-Based Mutation Operator," *IEEE Transactions on Evolutionary Computation*, vol. 13, no. 3, pp. 526 – 553, Junee 2009.

Zhan, Z-H., et al., "Adaptive Particle Swarm Optimization," *IEEE Transactions on Systems, Man, and* Cybernetics, vol. 39, no. 6, pp. 1362 – 1381, Dec. 2009.

Li, X., and Guo, Y., "The Grey Wolf Optimizer for Antenna Optimization Designs," *IEEE Antennas & Propagation Magazine,* vol. 64, no. 6, pp. 29 – 40, Dec. 2022.

Alternating Projection: **en.wikipedia.org/wiki/Projections_onto_convex_sets**

Conjugate Gradient method: **en.wikipedia.org/wiki/Conjugate_gradient_method**

Gradient Descent method: **en.wikipedia.org/wiki/Gradient_descent**

Design Optimization: **en.wikipedia.org/wiki/Multidisciplinary_design_optimization**

Metaheuristic methods: **en.wikipedia.org/wiki/Table_of_metaheuristics**

Nelder-Mead method: **en.wikipedia.org/wiki/Nelder-Mead_method**

Pattern Search methods: **en.wikipedia.org/wiki/Pattern_search_(optimization)**

Powell method: **en.wikipedia.org/wiki/Powell%27s_method**

Simulated Annealing method: **en.wikipedia.org/wiki/Simulated_annealing**

AutoEZ optimizer: **ac6la.com/aeoptimize.html**

De Canck, M., ON5AU, *Advanced Antenna Modeling* (2019). **on5au.be/advanced_modeling_book.html**

5.5.17 Optimum Shaped Wire Antennas

Meinke, H., Landstorfer, F., Flachenecker, G., Lindenmeier, H., and Enders, N., *Antenne mit einem Dipol, dessen Leiter einen trappenförmigen Verlauf haben*, German patent DE 25 52 043, applied Nov. 20, 1975, released Dec. 17, 1981.

Landstorfer, F., "Zur optimalen Form von Linearantennen," *Frequenz*, vol. 30, no. 12, pp. 344 – 349, Jan. 1976.

Landstorfer, F., "A New Type of Directional Antenna," *IEEE International Symposium on Antennas and Propagation*, pp. 169 – 172, Amherst, MA, Oct. 11 – 15, 1976.

Thiele, G., and Ekelman, E., "Design Formulas for Vee Dipoles," *IEEE Transactions on Antennas and Propagation*, vol. 28, no. 4, pp. 588 – 590, July 1980.

Cheng, D., and Liang, C., "Shaped Wire Antennas with Maximum Directivity," *Electronics Letters*, vol. 18, no. 19, pp. 816 – 818, Sep. 1982.

Landstorfer, F., and Sacher, R., *Optimisation of Wire Antennas*, Research Studies Press, John Wiley & Sons, 1985, ISBN 0863800254.

Rabelo, R., and Terada, M., "Analysis and Optimization of Wire Antennas over the Internet," *IEEE Antennas and Propagation Magazine*, vol. 52, no. 1, pp. 188 – 194, Feb. 2010.

Stearns, S., K6OIK, "The Shape of Antennas Yet to Come," presented to Foothills Amateur Radio Society, Dec. 17, 2021. **fars.k6ya.org/docs/k6oik**

5.5.18 Method of Moments (MoM) Antenna Software

AN-SOF by Golden Engineering: **antennasimulator.com**
ANSim by Phoenix Antenna Systems: **em-bench.myshopify.com**
AutoEZ by AC6LA: **ac6la.com/autoez.html**
AWAS 2.0, University of Belgrade: **oet.etf.bg.ac.rs/software.htm**. Contact K6OIK for manual.
EZNEC by W7EL: **eznec.com**
FEKO by Altair: **altair.com/feko-applications**
4nec2 by Arie Voors: **qsl.net/4nec2**
HOBBIES: No longer supported. Contact K6OIK.
MiniNEC Pro: **blackcatsystems.com/software/mininec-antenna-analysis-modeling-software.html**
MMANA-GAL: **gal-ana.de/promm**
NEC-2: **nec2.org**
NEC-5 and *NEC-4.2*: **ipo.llnl.gov/technologies/software/nec-v50-numerical-electromagnetic-code**
WIPL-D Lite: **wipl-d.com/products/wipl-d-starter**
WIPL-D Pro: **wipl-d.com**

5.5.19 Finite Difference Time Domain (FDTD) Antenna Software

GprMax: **gprmax.com/about.shtml**
Meep: **meep.readthedocs.io/en/latest**
openEMS: **openems.de**

5.5.20 Circuit Simulation Software

QucsStudio: **qucsstudio.de**

Notes

Notes

Notes

Notes

Notes

The ARRL
Antenna Book

FOR RADIO COMMUNICATIONS

ARRL The National Association for Amateur Radio®

Twenty-Fifth Edition

- **Volume 1: Antenna Fundamentals — Chapters 1 – 5**
- **Volume 2: Basic Antenna Types and MF and HF Antennas — Chapters 6 – 14**
- **Volume 3: VHF, UHF, and Microwave Antennas and Specialty Applications — Chapters 15 – 24**
- **Volume 4: Building and Maintaining Antenna Systems and Index — Chapters 25 – 29**

Editor
Ward Silver, NØAX

Production Editor
Mark Derks, KC1RVQ

Production
Chris Dunham
Jodi Morin, KA1JPA
David F. Pingree, N1NAS

Editorial Assistant
Maty Weinberg, KB1EIB

Contributors to the Twenty-Fifth Edition
Alan Applegate, KØBG
Jim Breakall, WA3FET
Jim Brown, K9YC
Wyatt Dirks, ACØRA
Frank Donovan, W3LPL
James Duffey, KK6MC
Peter East
Doug Grant, K1DG
Rus Healy, K2UA
Matt Heere, N3NWV
Wolfgang Herrmann
Chuck Higgins
Martin Huml, OK1FUA
Dale Hunt, WB6BYU
Justin Johnson, GØKSC
Al Katz, K2UYH
Hal Kennedy, N4GG
Pete Kobak, KØBAK
Steve Kostro, N2CEI
Sean Kutzko, KX9X
Greg Lapin, N9GL
Dan Layne, ADØCY
Roy Lewellan, W7EL
Fred Maas, KT5X
Richard Marsden
Eduard Mol
Dave Mueller, N2NL
Mario Natali, IØNAA
Ron Ochu, KOØZ
Dave Olean, K1WHS
Steve Olney, VK2XV
Charles Osborne, K4CSO
Jonathan Pettingale
Alex Pettit, KK4VB
John Portune, W6NBC
Bruce Randall, NT4RT
Whitman Reeve
Rick Rosen, K1DS
Rich Russell, ACØUB
Charles Scharlau, NZØI
Tom Schiller, N6BT
Rudy Severns, N6LF
Lowell Silver, WW7SAR
Andrea Slack, K2EZ
Jim Smith, KKØU
John Stanley, K4ERO
Steve Stearns, K6OIK
Patrick Stoddard, WD9EWK
Chip Sufitchi, N2YO
Dave Typinkski, AJ4CO
Ray Uberecken, AAØL
Paul Wade, W1GHZ
Bill Wortman, N6MW
Ken Wyatt, KX7KW

Printed in the USA

ISBN: 978-1-62595-175-5 Hardcover
978-1-62595-176-2 Softcover
978-1-62595-183-0 Four-Volume Set

Kindle eBook Editions
ISBN: 978-1-62595-177-9 — Volume 1
ISBN: 978-1-62595-179-3 — Volume 2
ISBN: 978-1-62595-180-9 — Volume 3
ISBN: 978-1-62595-182-3 — Volume 4

Twenty-Fifth Edition, Second Printing

We strive to produce books without errors. Sometimes mistakes do occur, however. When we become aware of problems in our books (other than obvious typographical errors), we post corrections on the ARRL website. If you think you have found an error, please check **www.arrl.org/arrl-antenna-book-reference** for corrections. If you don't find a correction there, please let us know by sending an email to **pubsfdbk@arrl.org**.

Table of Contents

A detailed Table of Contents is included at the beginning of each chapter.
See the **Project Index** for a complete list of projects.

VOLUME 2

Basic Antenna Types and MF and HF Antennas

VOLUME 3

VHF, UHF, and Microwave Antennas and Specialty Applications

VOLUME 4

Building and Maintaining Antenna Systems and Index

TABLE OF CONTENTS

Chapter 6 — Online Material

Articles
- "A Four Wire Steerable V Beam for 10 through 40 Meters" by Sam Moore, NX5Z

Chapter 6

Long-Wire and Traveling-Wave Antennas

The power gain and directive characteristics of electrically long wires (that is, wires that are long in terms of wavelength) make them useful for long-distance transmission and reception on the higher frequencies. Long wires can be combined to form antennas of various shapes that increase the gain and directivity over a single wire. The term *long wire*, as used in this chapter, means any such configuration, not just a straight-wire antenna. The material in this chapter applies to antennas used at MF and HF. A related antenna used for receiving, the Beverage, is discussed in the **Receiving Antennas** chapter.

6.1 Overview

6.1.1 Long Wires Versus Multielement Arrays

In general, the gain obtained with long-wire antennas is not as great, when the space available for the antenna is limited, as you can obtain from driven dipole arrays or from a parasitic array such as a Yagi or quad. However, the long-wire antenna has advantages of its own that tend to compensate for this deficiency. The construction of long-wire antennas is simple both electrically and mechanically, and there are no especially critical dimensions or adjustments. The long-wire antenna will work well and give satisfactory gain and directivity over a 2-to-1 frequency range. In addition, it will accept power and radiate well on any frequency for which its overall length is not less than about a half wavelength. Since a wire is not electrically long, even at 28 MHz, unless its physical length is equal to at least a half wavelength on 3.5 MHz, any long-wire can be used on all amateur bands that are useful for long-distance communication.

Between two directive antennas having the same theoretical gain, one a multielement array and the other a long-wire antenna, many amateurs have found that the long-wire antenna seems more effective in reception. One possible explanation is that there is a *diversity effect* with a long-wire antenna because it is spread out over a large distance, rather than being concentrated in a small space, as would be the case with a Yagi, for example. This may raise the average level of received energy for ionospheric-propagated signals. Another factor is that long-wire antennas have directive patterns that can be extremely sharp in the horizontal (azimuthal) plane. This is an advantage that other types of multielement arrays do not have, but it can be a double-edged sword too. We'll discuss this aspect in some detail in this chapter.

6.1.2 General Characteristics of Long-Wire Antennas

Whether the long-wire antenna is a single wire running in one direction or is formed into a V-beam, rhombic, or some other configuration, there are certain general principles that apply and some performance features that are common to all types. The first of these is that the power gain of a long-wire antenna as compared with a half-wave dipole is not considerable until the antenna is really long (its length measured in wavelengths rather than in a specific number of feet). The reason for this is that the fields radiated by elementary lengths

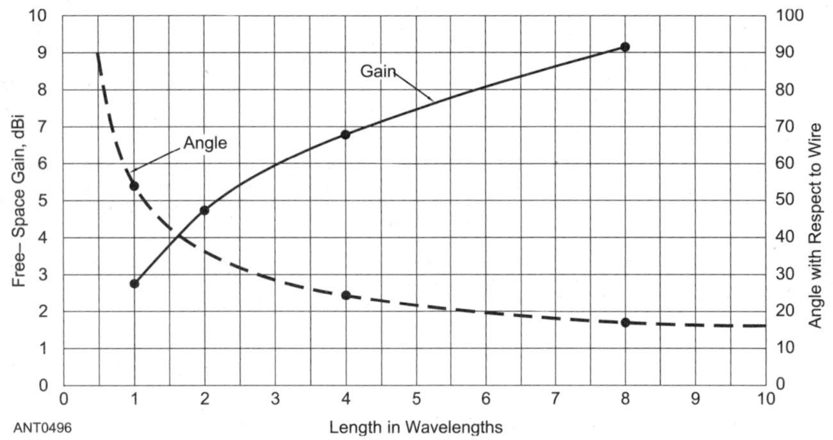

Figure 6.1 — Theoretical gain of a long-wire antenna, in dBi, as a function of wire length. The angle, with respect to the wire, at which the radiation intensity is maximum also is shown.

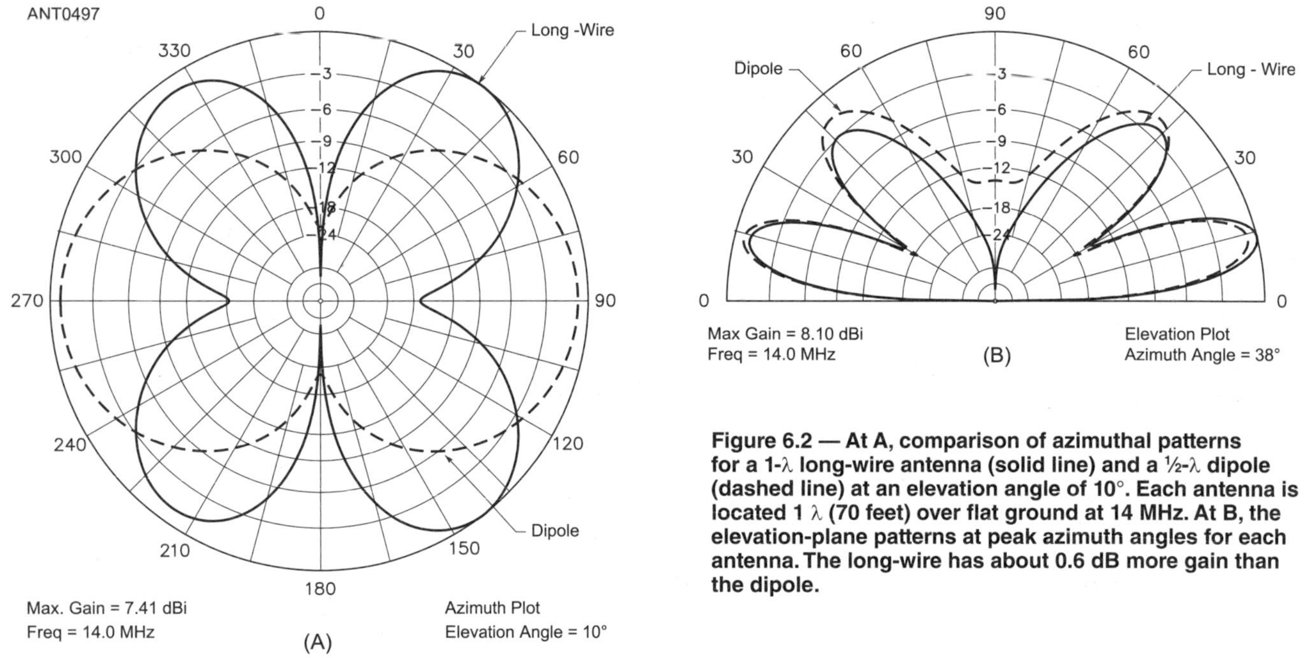

Figure 6.2 — At A, comparison of azimuthal patterns for a 1-λ long-wire antenna (solid line) and a ½-λ dipole (dashed line) at an elevation angle of 10°. Each antenna is located 1 λ (70 feet) over flat ground at 14 MHz. At B, the elevation-plane patterns at peak azimuth angles for each antenna. The long-wire has about 0.6 dB more gain than the dipole.

of wire along the antenna do not combine, at a distance, in as simple a fashion as the fields from half-wave dipoles used in other types of directive arrays.

There is no point in space, for example, where the distant fields from all points along the wire are exactly in phase (as they are, in the optimum direction, in the case of two or more collinear or broadside dipoles when fed with in-phase currents). Consequently, the field strength at a distance is always less than would be obtained if the same length of wire were cut up into properly phased and separately driven dipoles. As the wire is made longer, the fields combine to form increasingly intense main lobes, but these lobes do not develop appreciably until the wire is several wavelengths long. See **Figure 6.1**.

The longer the antenna, the sharper the lobes become, and since it is really a hollow cone of radiation about the wire in free space, it becomes sharper in both planes. Also, the greater the length, the smaller the angle with the wire at which the maximum radiation lobes occur. There are four main lobes to the directive patterns of long-wire antennas; each makes the same angle with respect to the wire.

Figure 6.2A shows the azimuthal radiation pattern of a 1-λ long-wire antenna, compared with a ½-λ dipole. Both antennas are mounted at the same height of 1 λ above flat ground (70 feet high at 14 MHz, with a wire length of 70 feet) and both patterns are for an elevation angle of 10°, an angle suitable for long-distance communication on 20 meters. The long-wire in Figure 6.2A is oriented in the 270° to 90° direction, while the dipole is aligned at right angles so that its characteristic figure-8 pattern goes left-to-right. The 1-λ long-wire has about 0.6 dB more gain than the dipole, with four main lobes as compared to the two lobes from the dipole.

You can see that the two lobes on the left side of Figure 6.2A are about 1 dB down compared to the two lobes on the right side. This is because the long-wire here is fed at the left-hand end in the computer model. Energy is radiated as a wave travels down the wire and some energy is also lost to ohmic resistance in the wire and the ground. The forward-going wave then reflects from the open-circuit at the right-hand end of the wire and reverses direction, traveling toward the left end, still radiating as it travels. An antenna operating in this way has much the same characteristics as a transmission line that is terminated in an open circuit — that is, it has standing waves on it. Unterminated long-wire antennas are often referred to as standing wave antennas. As the length of a long-wire antenna is increased, a moderate front-to-back ratio results, about 3 dB for very long antennas.

Figure 6.2B shows the elevation-plane pattern for the long-wire and for the dipole. In each case the elevation pattern is at the azimuth of maximum gain — at an angle of 38° with respect to the wire-axis for the long-wire and at 90° for the dipole. The peak elevation for the long-wire is very slightly lower than that for the dipole at the same height above ground, but not by much. In other words, the height above ground is the main determining factor for the shape of the main lobe of a long-wire's elevation pattern, as it is for most horizontally polarized antennas.

The shape of the azimuth and elevation patterns in Figure 6.2 might lead you to believe that the radiation pattern is simple. **Figure 6.3** is a 3-D representation of the pattern from a 1-λ long-wire that is 1 λ high over flat ground. Besides the main low-angle lobes, there are strong lobes at higher angles. Things get even more complicated when the length of the long-wire increases.

DIRECTIVITY

Because many points along a long wire are carrying currents in different phases (with different current amplitudes as well), the field pattern at a distance becomes more complex as the wire is made longer. This complexity is manifested in a series of minor lobes, the number of which increases with the wire length. The intensity of radiation from the minor lobes is frequently as great as, and sometimes greater than, the radiation from a half-wave dipole. The energy radiated in the minor lobes is not available to improve the gain in the major lobes, which is another reason why a long-wire antenna must be long to give appreciable gain in the desired directions.

Figure 6.4 shows an azimuthal-plane comparison between a 3-λ (209 feet long) long-wire and the comparison ½-λ dipole. The long-wire now has eight minor lobes besides the four main lobes. Note that the angle the main lobes make with respect to the axis of the long-wire (also left-to-right in Figure 6.4) becomes smaller as the length of the long-wire increases. For the 3-λ long-wire, the main lobes occur 28° off the axis of the wire itself.

Other types of simple driven and parasitic arrays do not have minor lobes of any great consequence. For that reason they frequently seem to have much better directivity than long-wire antennas, because their responses in undesired directions are well down from their response in the desired direction. This is the case even if a multielement array and a long-wire antenna have the same peak gain in the favored direction. **Figure 6.5** compares the same 3-λ long-wire with a 4-element Yagi and a ½-λ dipole, again both at the same height as the long-wire. Note that the Yagi has only a single rear lobe, down about 21 dB from its broad main lobe, which has a 3-dB beamwidth of 63°. The 3-dB beamwidth of the long-wire's main lobes (at a 28° angle from the wire axis) is far more narrow, at only 23°.

For amateur work, particularly with directive antennas that

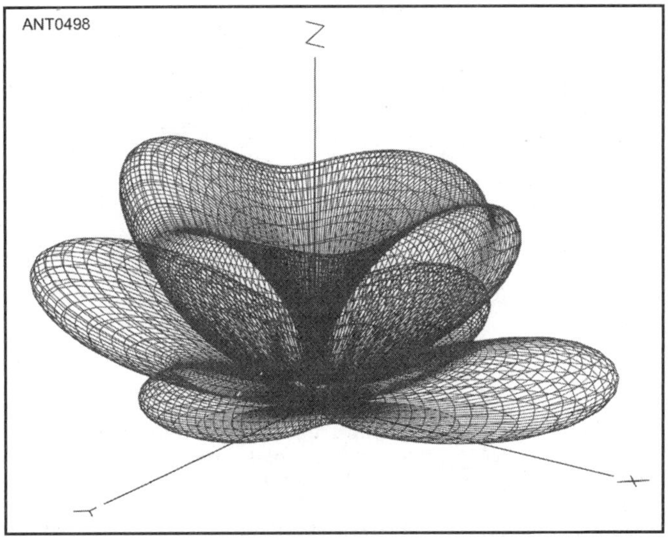

Figure 6.3 — A 3-D representation of the radiation pattern for the 1-λ long-wire shown in Figure 6.2. The pattern is obviously rather complex. It gets even more complicated for wires longer than 1 λ.

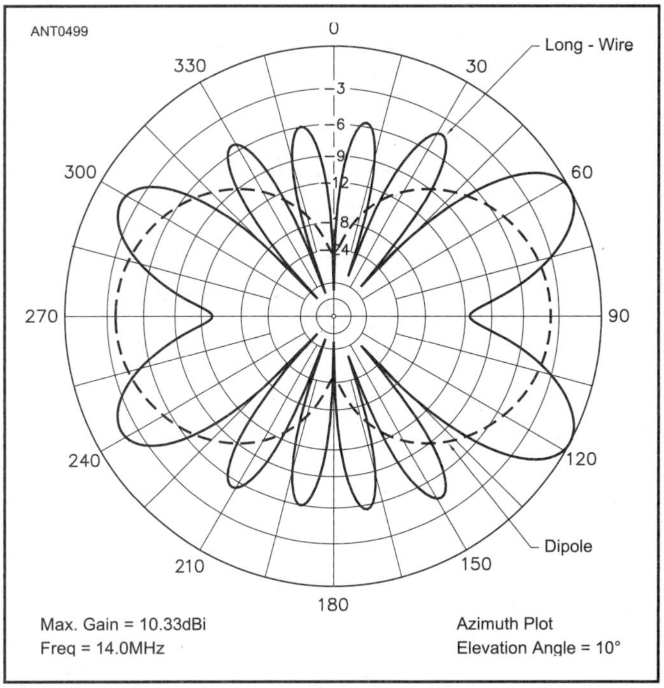

Figure 6.4 — An azimuthal-plane comparison between a 3-λ (209 feet long) long-wire (solid line) and the comparison ½-λ dipole (dashed line) at 70 feet high (1 λ) at 14 MHz.

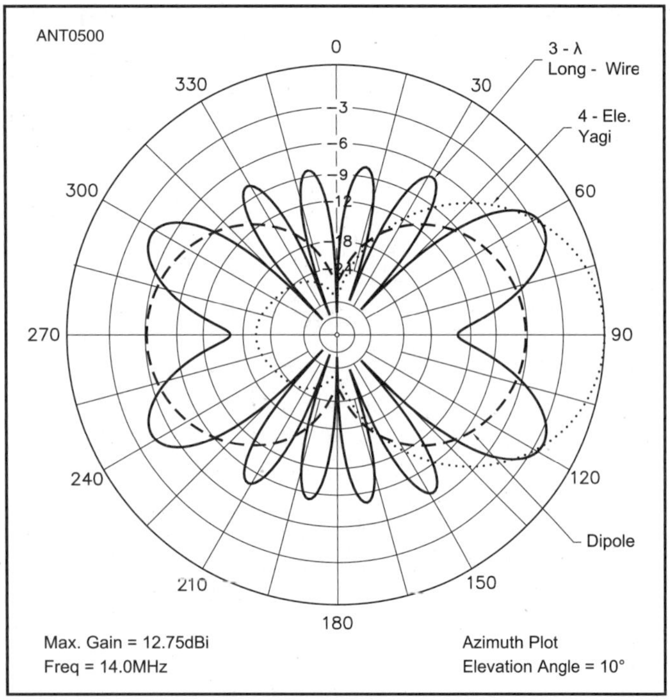

Figure 6.5 — A comparison between the 3-λ long-wire (solid line) in Figure 6.4, a 4-element 20-meter Yagi on a 26-foot boom (dotted line), and a ½-λ dipole (dashed line), again at a height of 70 feet. The main lobes of the long-wire are very narrow compared to the wide frontal lobe of the Yagi. The long-wire exhibits an azimuthal pattern that is more omnidirectional in nature than a Yagi, particularly when the narrow, deep nulls in the long-wire's pattern are filled-in due to irregularities in the terrain under its long span of wire.

cannot be rotated, the minor lobes of a long-wire antenna have some advantages. Although the nulls in the computer model in Figure 6.5 are deeper than 30 dB, they are not so dramatic in actual practice. This is due to irregularities in the terrain that inevitably occur under the span of a long wire. In most directions the long-wire antenna will be as good as a half-wave dipole, and in addition will give high gain in the most favored directions, even though that is over narrow azimuths.

Figure 6.6A compares the azimuth responses for a 5-λ long-wire (350 feet long at 14 MHz) to the same 4-element Yagi and dipole. The long-wire now exhibits 16 minor lobes in addition to its four main lobes. The peaks of these sidelobes are down about 8 dB from the main lobes and they are stronger than the dipole, making this long-wire antenna effectively omnidirectional. Figure 6.6B shows the elevation pattern of the 5-λ long-wire at its most effective azimuth compared to a dipole. Again, the shape of the main lobe is mainly determined by the long-wire's height above ground, since the peak angle is only just a bit lower than the peak angle for the dipole. The long-wire's elevation response breaks up into numerous lobes above the main lobes, just as it does in the azimuth plane.

For the really ambitious, **Figure 6.7** compares the performance for an 8-λ (571 feet) long-wire antenna with a 4-element Yagi and the ½-λ dipole. Again, in actual practice, the nulls would tend to be filled in by terrain irregularities, so a very long antenna like this would be a pretty potent performer.

CALCULATING LENGTH

In this chapter, lengths are discussed in terms of wavelengths. Throughout the preceding discussion the frequency in the models was held at 14 MHz. Remember that a long-wire that is 4 λ long at 14 MHz is 8 λ long at 28 MHz.

There is nothing very critical about wire lengths in an antenna system that will work over a frequency range includ-

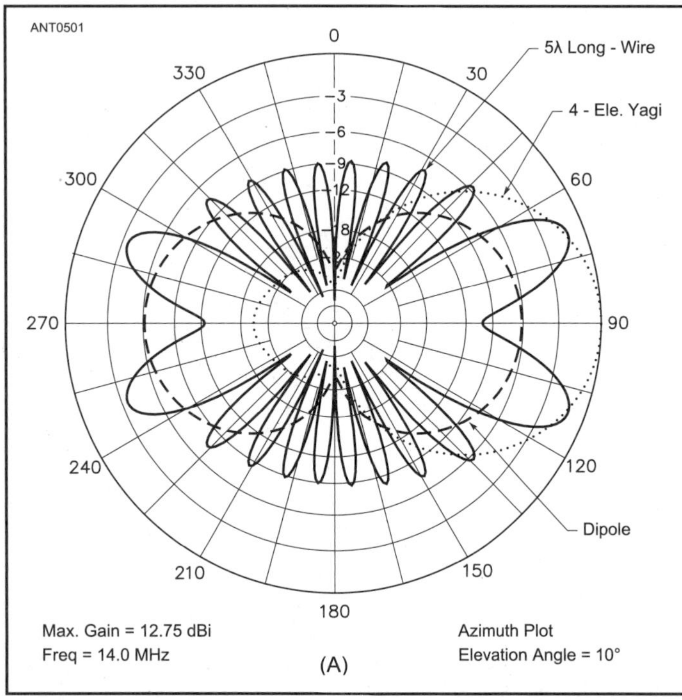

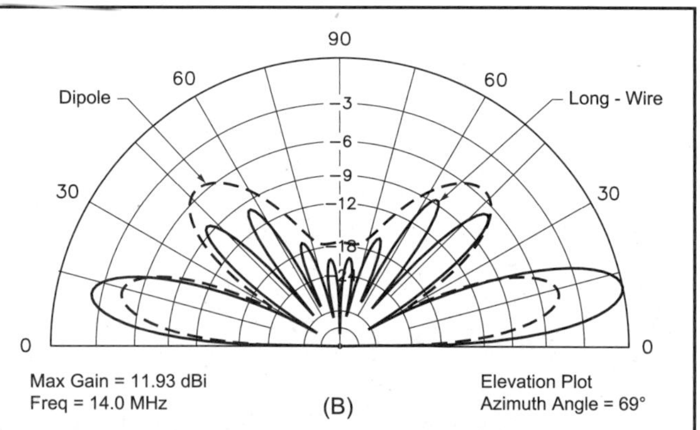

Figure 6.6 — At A, the azimuth responses for a 5-λ long-wire (350 feet long at 14 MHz — solid line) to the same 4-element Yagi (dotted line) and dipole (dashed line) as in Figure 6.5. At B, the elevation-plane responses for the long-wire (solid line) and the dipole (dashed line) by themselves. Note that the elevation angle giving peak gain for each antenna is just about the same. The long-wire achieves gain by compressing mainly the azimuthal response, squeezing the gain into narrow lobes; not so much by squeezing the elevation pattern for gain.

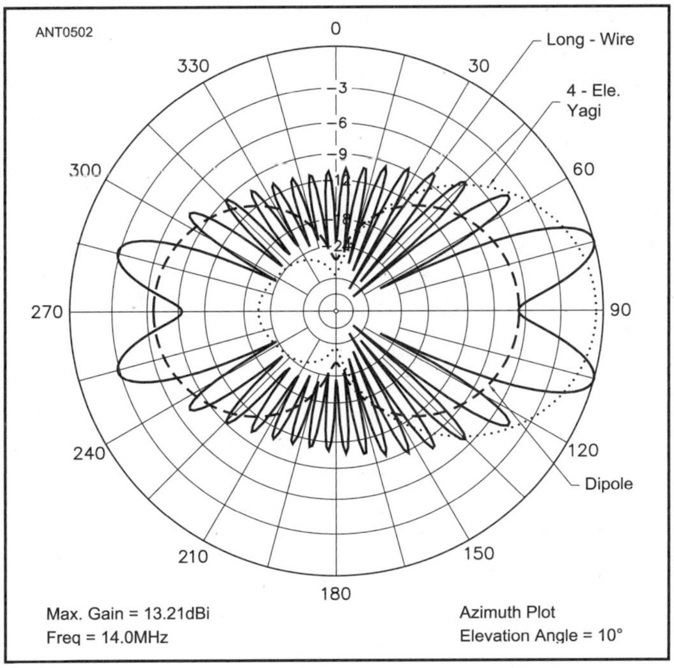

Figure 6.7 — The azimuthal-plane performance for an 8-λ (571 feet) long-wire antenna (solid line), compared with a 4-element Yagi (dotted line) and a ½-λ dipole (dashed line).

ing several amateur bands. The antenna characteristics change very slowly with length, except when the wires are short (around one wavelength, for instance). There is no need to try to establish exact resonance at a particular frequency for proper antenna operation.

The formula for determining the lengths for harmonic wires is:

$$\text{Length (feet)} = \frac{984\,(N - 0.025)}{f\,(MHz)}$$

where N is the antenna length in wavelengths. In cases where precise resonance is desired for some reason (for obtaining a resistive load for a transmission line at a particular frequency, for example) it is best established by trimming the wire length until the standing-wave ratio on the line is minimum.

TILTED WIRES

In theory, it is possible to maximize gain from a long-wire antenna by tilting it to favor a desired elevation take-off angle. Unfortunately, the effect of real ground under the antenna negates the possible advantages of tilting, just as it does when a Yagi or other type of parasitic array is tilted from horizontal. You would do better keeping a long-wire antenna horizontal, but raising it higher above ground, to achieve more gain at low takeoff angles.

6.1.3 Feeding Long Wires

A long-wire antenna can be fed at the end using an impedance transformer and coaxial cable, similarly to an end-fed half-wave (EFHW) as described in the **Multiband HF Antennas** chapter. On bands for which its length is an integer number of half-wavelengths, the feed point impedance will be highest. This is classic long-wire multiband operation. On other bands, the impedance at the wire's end will be high enough for the fixed ratio of the impedance transformer to present an acceptable impedance to the coax.

The antenna can also be fed directly by coaxial cable at a current maximum where feed point impedance is relatively low. The current maximum position will change with frequency, but coax feed can be used for bands at which the current maximums coincide.

It is important to note that for antennas with more than one current maximum (typically 1 λ or longer), the position of the feed point will alter the current distribution on the antenna and that will affect the antenna's radiation pattern. In such cases, modeling will help determine the best place to feed the antenna.

Another common method of feeding a long-wire is to use open-wire line connected at the end of the antenna. Any convenient line length can be used if you match the transmitter to the line's input impedance using an antenna tuner. Using coaxial cable to feed long wires directly can lead to excessive losses if the SWR is high and so open-wire line is the usual choice. If the feed line is resonant (an integer multiple of one-half wavelength on the lowest band) This system will work on all harmonically related bands (such as 80, 40, 20, 15, and 10 meters) down to the band, if any, at which the antenna is only a half-wave long.

The unbalanced open-wire end-fed arrangement has the disadvantage that there is likely to be a considerable common-mode current on the line. In addition, the antenna reactance changes rapidly with frequency. Consequently, when the wire is several wavelengths long, a relatively small change in frequency — a fraction of the width of a band — may require major changes in the adjustment of the antenna tuner. The problem can be overcome by using multiple long wires in a V or rhombic shape, as described below.

Two arrangements for using nonresonant lines are given in **Figure 6.8**. The one at A is useful for one band only since the matching section must be a quarter-wave long, approxi-

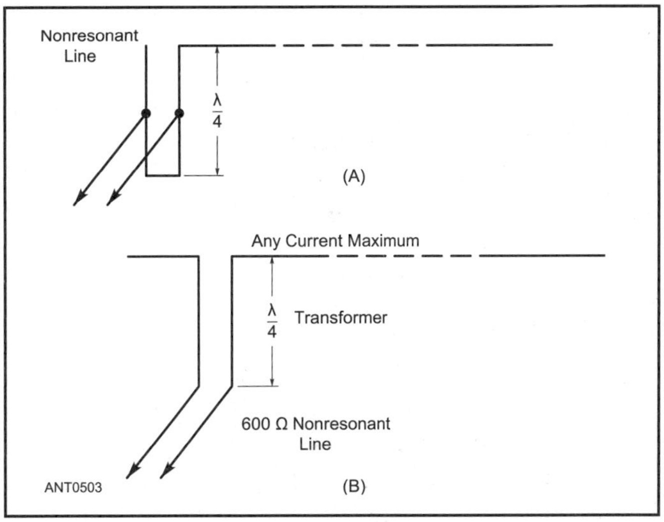

Figure 6.8 — Methods for feeding long single-wire antennas.

mately, unless a different matching section is used for each band. In B, the λ/4 transformer (Q-section) impedance can be designed to match the antenna to the line. You can determine the value of radiation resistance using a modern modeling program or you can measure the feed point impedance. Although it will work as designed on only one band, the antenna can be used on other bands by treating the line and matching transformer as a resonant line.

6.2 The V-Beam Antenna

The directivity and gain of long wires may be increased by using two wires placed in relation to each other such that the fields from both combine to produce the greatest possible field strength at a distant point. The principle is similar to that used in designing multielement arrays.

One possible method of using two (or more) long wires is to place them in parallel, with a spacing of ½ λ or so, and feed the two in phase. In the direction of the wires the fields will add in phase. However, the takeoff angle is high directly in the orientation of the wire, and this method will result in rather high-angle radiation even if the wires are several wavelengths long. With a parallel arrangement of this sort the gain should be about 3 dB over a single wire of the same length, at spacings in the vicinity of ½ wavelength.

Instead of using two long wires parallel to each other, they may be placed in the form of a horizontal V, with the included angle between the wires equal to twice the angle made by the main lobes referenced to the wire axis for a single wire of the same physical length. For example, for a leg length of 5 λ, the angle between the legs of a V should be about 42°, twice the angle of 21° of the main lobe referenced to the long-wire's axis. See Figure 6.6A.

The plane directive patterns of the individual wires combine along a line in the plane of the antenna and bisecting the V, where the fields from the individual wires reinforce each other. The sidelobes in the azimuthal pattern are suppressed by about 10 dB, so the pattern becomes essentially bidirectional. See **Figure 6.9**.

The included angle between the legs is not particularly critical. This is fortunate, especially if the same antenna is used on multiple bands, where the electrical length varies directly with frequency. This would normally require different included angles for each band. For multiband V-antennas, a compromise angle is usually chosen to equalize performance.

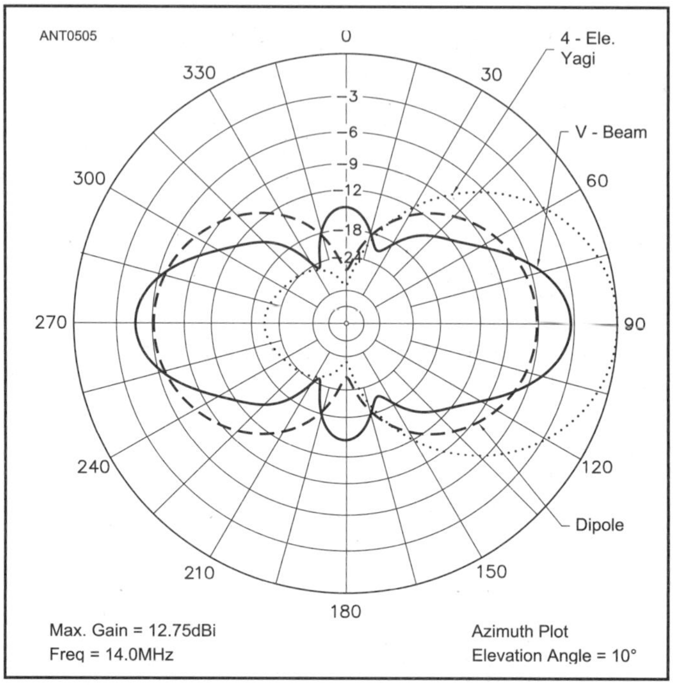

Figure 6.10 — Azimuthal-plane pattern at 10° elevation angle for a 14-MHz V-beam (solid line) with 1-λ legs (68.5 feet long), using an included angle of 75° between the legs. The V-beam is mounted 1 λ above flat ground, and is compared with a ½-λ dipole (dashed line) and a 4-element 20-meter Yagi on a 26-foot boom (dotted line).

Figure 6.10 shows the azimuthal pattern for a V-beam with 1-λ legs, with an included angle of 75° between the legs, mounted 1 λ above flat ground. This is for a 10° elevation angle. At 14 MHz the antenna has two 70-foot high, 68.5-foot long legs, separated at their far ends by 83.4 feet.

Figure 6.9 — Two long wires and their respective patterns are shown at the left. If these two wires are combined to form a V with an angle that is twice that of the major lobes of the wires and with the wires excited out of phase, the radiation along the bisector of the V adds and the radiation in the other directions tends to cancel.

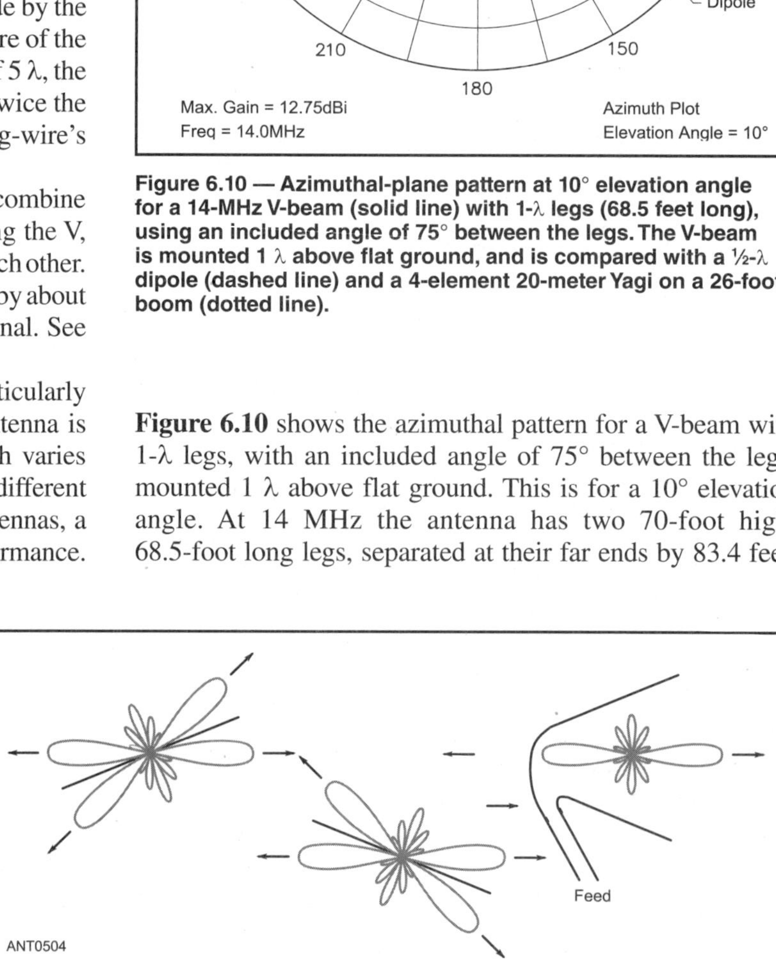

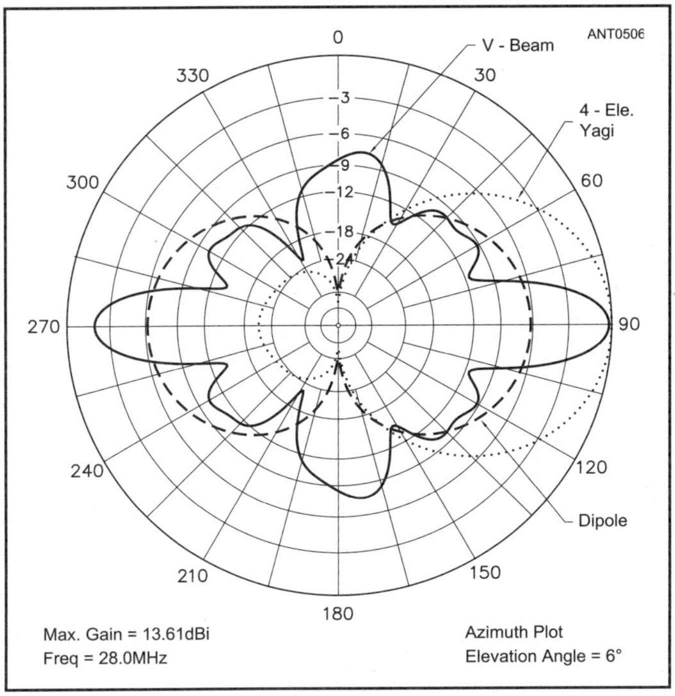

Figure 6.11 — The same V-beam as in Figure 6.10 at 28 MHz (solid line), at an elevation angle of 6°, compared to a 4-element Yagi (dotted line) and a dipole (dashed line). The V-beam's pattern is very narrow, at 18.8° at the 3-dB points, requiring accurate placement of the supports poles to aim the antenna at the desired geographic target.

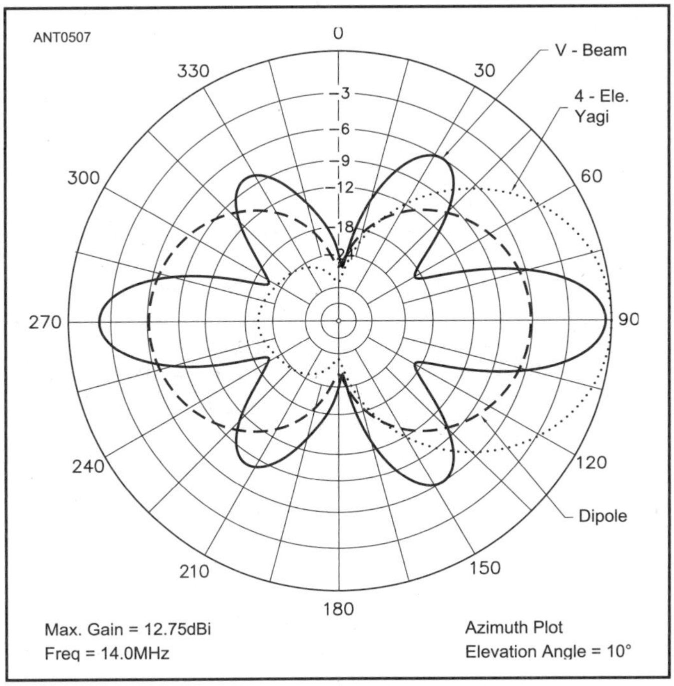

Figure 6.12 — Azimuthal pattern for a V-beam (solid line) with 2-λ legs (137 feet at 14 MHz), with an included angle of 60° between them. The height is 70 feet, or 1 λ, over flat ground. For comparison, the response for a 4-element Yagi (dotted line) and a dipole (dashed line) are shown. The 3-dB beamwidth has decreased to 23.0°.

For comparison, the azimuthal patterns for the same 4-element Yagi and ½-λ dipole used previously for the long-wires are overlaid on the same plot. The V has about 2 dB more gain than the dipole but is down some 4 dB compared to the Yagi, as expected for relatively short legs.

Figure 6.11 shows the azimuthal pattern for the same antenna in Figure 6.10, but at 28 MHz and at an elevation angle of 6°. Because the legs are twice as long electrically at 28 MHz, the V-beam has compressed the main lobe into a narrow beam that now has a peak gain equal to the Yagi, but with a 3-dB beamwidth of only 18.8°. Note that you could obtain about 0.7 dB more gain at 14 MHz, with a 1.7-dB degradation of gain at 28 MHz, if you increase the included angle to 90° rather than 75°.

Figure 6.12 shows the azimuthal pattern for a V-beam with 2-λ legs (137 feet at 14 MHz), with an included angle of 60° between them. As usual, the assumed height is 70 feet, or 1 λ at 14 MHz. The peak gain for the V-beam is just about equal to that of the 4-element Yagi, although the 3-dB nose beamwidth is narrow, at 23°. This makes setting up the geometry critical if you want to maximize gain into a particular geographic area. While you might be able to get away with using convenient trees to support such an antenna, it's far more likely that you'll have to use carefully located towers to make sure the beam is aimed where you expect it to be pointed.

For example, in order to cover all of Europe from San Francisco, an antenna must cover from about 11° (to Moscow) to about 46° (to Portugal). This is a range of 35° and signals from the V-beam in Figure 6.12 would be down some 7 dB over this range of angles, assuming the center of the beam is pointed exactly at a heading of 28.5°. The 4-element Yagi on the other hand would cover this range of azimuths more consistently, since its 3-dB beamwidth is 63°.

Figure 6.13 shows the same V-beam as in Figure 6.12, but this time at 28 MHz. The peak gain of the main lobe is now about 1 dB stronger than the 4-element Yagi used as a reference, and the main lobe has two nearby sidelobes that tend to broaden out the azimuthal response. At this frequency the V-beam would cover all of Europe better from San Francisco.

Figure 6.14 shows a V-beam with 3-λ (209 feet at 14 MHz) legs with an included angle of 50° between them. The peak gain is now greater than that of a 4-element Yagi, but the 3-dB beamwidth has been reduced to 17.8°, making aiming the antenna even more critical. **Figure 6.15** shows the same V-beam at 28 MHz. Here again, the main lobe has nearby sidelobes that broaden the effective azimuth to cover a wider area.

Figure 6.16 shows the elevation-plane response for the same 209-foot leg V-beam at 28 MHz (3-λ at 14 MHz), compared to a dipole at the same height of 70 feet. The higher-gain V-beam suppresses higher-angle lobes, essentially stealing energy from them and concentrating it in the main beam at 6° elevation.

The same antenna can be used at 3.5 and 7 MHz. The gain will not be large, however, because the legs are not very long at these frequencies. **Figure 6.17** compares the V-beam versus a horizontal ½-λ 40-meter dipole at 70 feet. At low elevation angles there is about 2 dB of advantage on 40 meters.

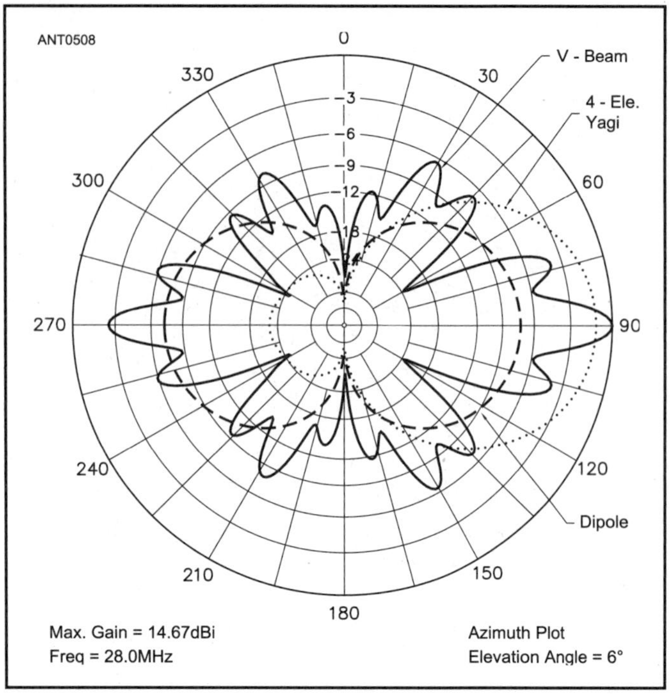

Figure 6.13 — The same 2-λ-per-leg V-beam (solid line) as in Figure 6.12, but at 28 MHz and at a 6° takeoff elevation angle. Two sidelobes have appeared flanking the main lobe, making the effective azimuthal pattern wider at this frequency.

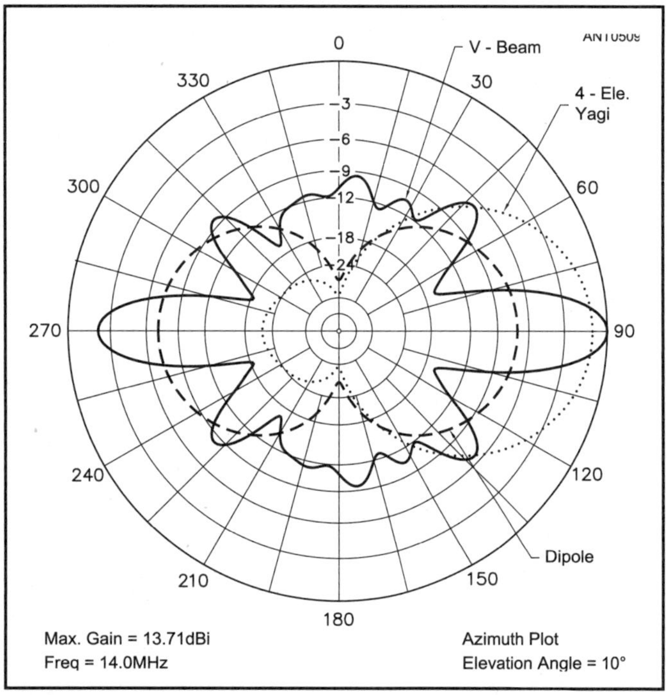

Figure 6.14 — A V-beam (solid line) with 3-λ (209 feet at 14 MHz) legs using an included angle of 50° between them, compared to a 4-element Yagi (dotted line) and a dipole (dashed line). The 3-dB beamwidth has now decreased to 17.8°.

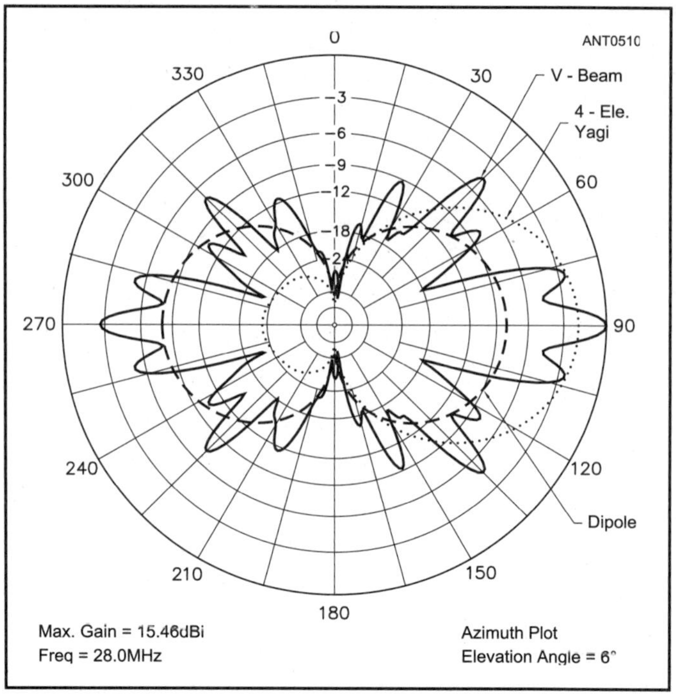

Figure 6.15 — The same 209-foot-per-leg V-beam as Figure 6.14, but at 28 MHz. Again, the two close-in sidelobes tend to spread out the azimuthal response some at 28 MHz.

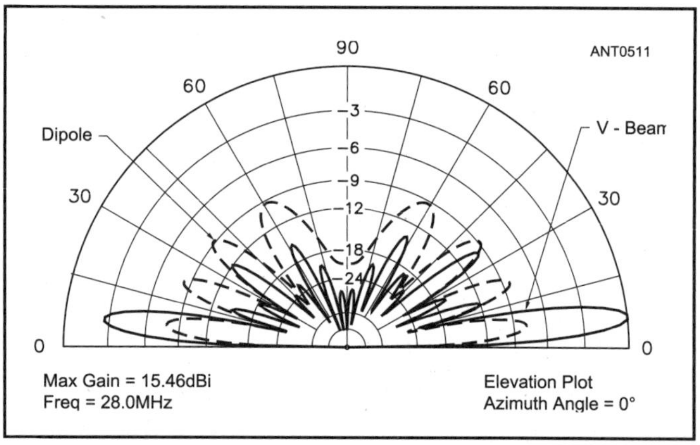

Figure 6.16 — The elevation-plane of the 209-foot-per-leg V-beam (solid line) compared to the dipole (dashed line). Again, the elevation angle for peak gain corresponds well to that of the simple dipole at the same height.

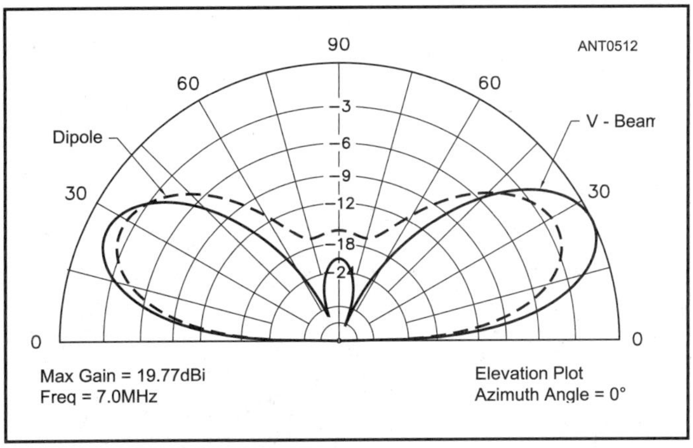

Figure 6.17 — Elevation pattern for the same 209-foot-per-leg V-beam (solid line), at 7 MHz, compared to a 40 meter dipole (dashed line) at the same height of 70 feet.

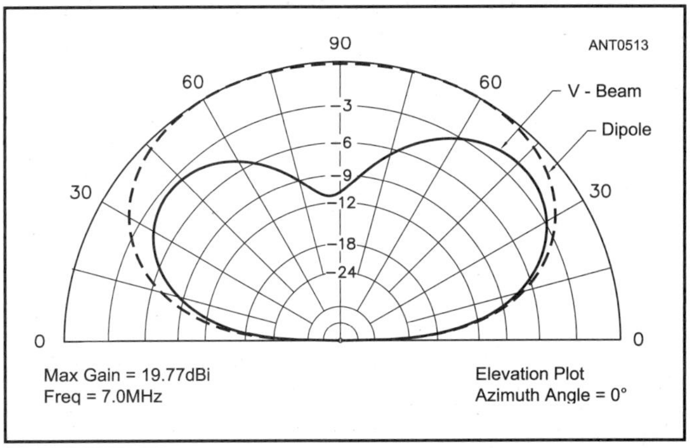

Figure 6.18 — Elevation pattern for the same 209-foot-per-leg V-beam (solid line), at 3.5 MHz, compared to an 80 meter dipole at 70 feet (dashed line).

Figure 6.18 shows the same type of comparison for 80 meters, where the 80 meter dipole is superior at all angles.

OTHER V COMBINATIONS

A gain increase of about 3 dB can be had by stacking two V-beams one above the other, a half wavelength apart, and feeding them with in-phase currents. This will result in a lowered angle of radiation. The bottom V should be at least a quarter wavelength above the ground, and preferably a half wavelength. This arrangement will narrow the elevation pattern and it will also have a narrow azimuthal pattern.

The V antenna can be made unidirectional by using a second V placed an odd multiple of a quarter wavelength in back of the first and exciting the two with a phase difference of 90°. The system will be unidirectional in the direction of the antenna with the lagging current. However, the V reflector is not normally employed by amateurs at low frequencies because it restricts the use to one band and requires a fairly elaborate supporting structure. Stacked Vs with driven reflectors could, however, be built for the 200- to 500-MHz region without much difficulty.

FEEDING THE V-BEAM

The V-beam antenna is most conveniently fed with tuned open-wire feeders with an antenna tuner, since this permits multiband operation. Although the length of the wires in a V-beam is not at all critical, it is important that both wires be the same electrical length. If a single band matching solution is desired, probably the most appropriate matching system is that using a stub or quarter-wave matching section.

6.3 The Resonant Rhombic Antenna

The diamond-shaped or rhombic antenna shown in **Figure 6.19** can be looked upon as two acute-angle V-beams placed end-to-end. This arrangement is called a resonant rhombic. The leg lengths of the resonant rhombic must be an integral number of half wavelengths to avoid reactance at its feed point.

The resonant rhombic has two advantages over the simple V-beam. For the same total wire length it gives somewhat greater gain than the V-beam. A rhombic with 3 λ on a leg, for example, has about 1 dB gain over a V antenna with 6 wavelengths on a leg. **Figure 6.20** compares the azimuthal pattern at a 10° elevation for a resonant rhombic with 3 λ legs on 14 MHz, compared to a V-beam with 6 λ legs at the same height of 70 feet. The 3-dB main lobe beam-width of the resonant rhombic is only 12.4° wide, but the gain is very high at 16.26 dBi.

The directional pattern of the rhombic is less frequency sensitive than the V when the antenna is used over a wide frequency range. This is because a change in frequency causes the major lobe from one leg to shift in one direction while the lobe from the opposite leg shifts the other way. This automatic compensation keeps the direction the same over a considerable frequency range. The disadvantage of the rhombic as compared with the V-beam is that an additional support is required. Some authors also report success with "half-rhombics" oriented vertically over ground with and without a counterpoise. (See References and Bibliography entry for Orr.)

The same factors that govern the design of the V-beam apply in the case of the resonant rhombic. The optimal apex angle A in Figure 6.19 is the same as that for a V having an equal leg length. The diamond-shaped antenna also can be

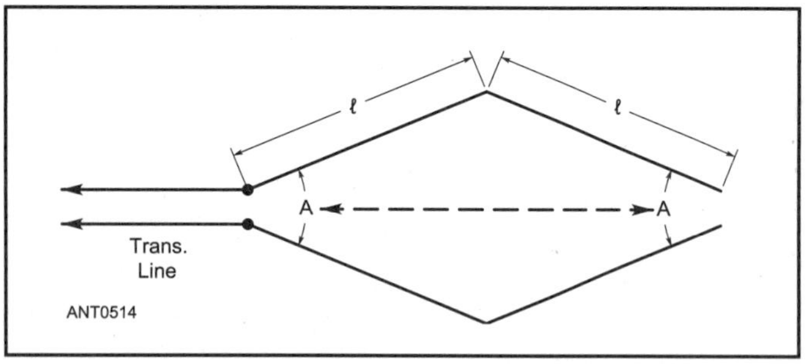

Figure 6.19 — The resonant rhombic or diamond-shaped antenna. All legs are the same length, and opposite angles of the diamond are equal. Length ℓ is an integral number of half wavelengths for resonance.

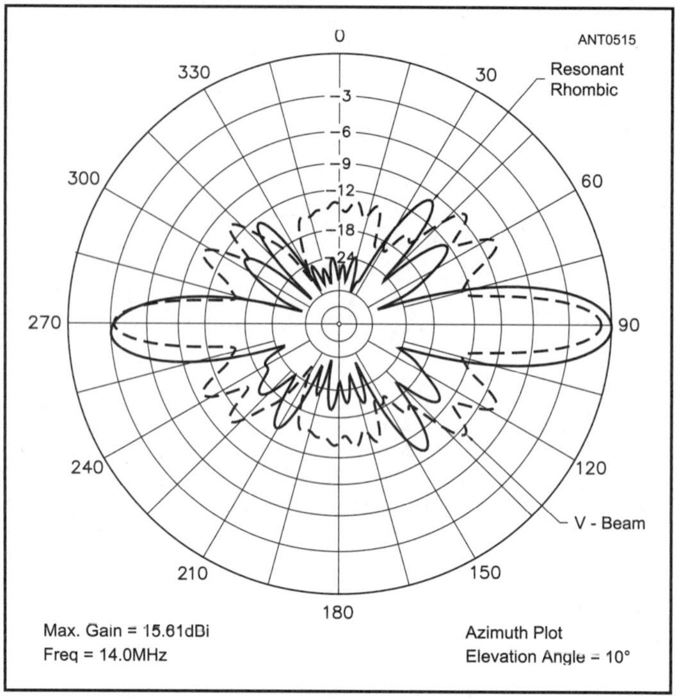

Figure 6.20 — Azimuthal-plane pattern of resonant (unterminated) rhombic (solid line) with 3-λ legs on 14 MHz, at a height of 70 feet above flat ground, compared with a 6-λ per leg V-beam (dashed line) at the same height. Both azimuthal patterns are at a takeoff angle of 10° . The sidelobes for the resonant rhombic are suppressed to a greater degree than those for the V-beam

operated as a terminated antenna, as described later in this chapter, and much of the discussion in that section applies to the resonant rhombic as well.

The resonant rhombic has a bidirectional pattern, with minor lobes in other directions, their number and intensity depending on the leg length. In general, these sidelobes are suppressed better with a resonant rhombic than with a V-beam. When used at frequencies below the VHF region, the rhombic antenna is always mounted with the plane containing the wires horizontal. The polarization in this plane, and also in the perpendicular plane that bisects the rhombic, is horizontal. At 144 MHz and above, the dimensions are such that the antenna can be mounted with the plane containing the wires vertical if vertical polarization is desired.

When the rhombic antenna is to be used on several HF amateur bands, it is advisable to choose the apex angle, A, on the basis of the leg length in wavelengths at 14 MHz. Although the gain on higher frequency bands will not be quite as favorable as if the antenna had been designed for the higher frequencies, the system will still work well at the low angles that are necessary at such frequencies.

The resonant rhombic has lots of gain, but you must not forget that this gain comes from a radiation pattern that is very narrow. This requires careful placement of the supports for the resonant rhombic to cover desired geographic areas. This is definitely not an antenna that allows you to use just any convenient trees as supports!

Even if you cannot place its corners exactly, the rhombic can still give good performance. (See the References and Bibliography entry for Hallas.) The main lobe broadens and peak gain is lower but the author found it to be a very effective antenna.

The resonant rhombic antenna can be fed in the same way as the V-beam. Resonant feeders are necessary if the antenna is to be used in several amateur bands.

6.4 Terminated Long-Wire Antennas

All the antenna systems considered so far in this chapter have been based on operation with standing waves of current and voltage along the wire. Although most hams use antenna designs based on using resonant wires, resonance is by no means a necessary condition for the wire to radiate and intercept electromagnetic waves efficiently, as discussed in the **Antenna Fundamentals** chapter. The result of using nonresonant wires is reactance at the feed point, unless the antenna is terminated with a resistive load.

In **Figure 6.21**, suppose that the wire is parallel with the ground (horizontal) and is terminated by a load Z equal to its characteristic impedance, Z_{ANT}. The wire and its image in the ground create a transmission line. The load Z can represent a receiver matched to the line. The *terminating resistor* R is also equal to the Z_{ANT} of the wire. A wave coming from direction X will strike the wire first at its far end and sweep across the wire at some angle until it reaches the end at which Z is connected. In so doing, it will induce voltages in the antenna, and currents will flow as a result. The current flowing toward Z is the useful output of the antenna, while the current flowing backwards toward R will be absorbed in R. The same thing is true of a wave coming from the direction X'. In such an antenna there are no standing waves, because all received power is absorbed at either end.

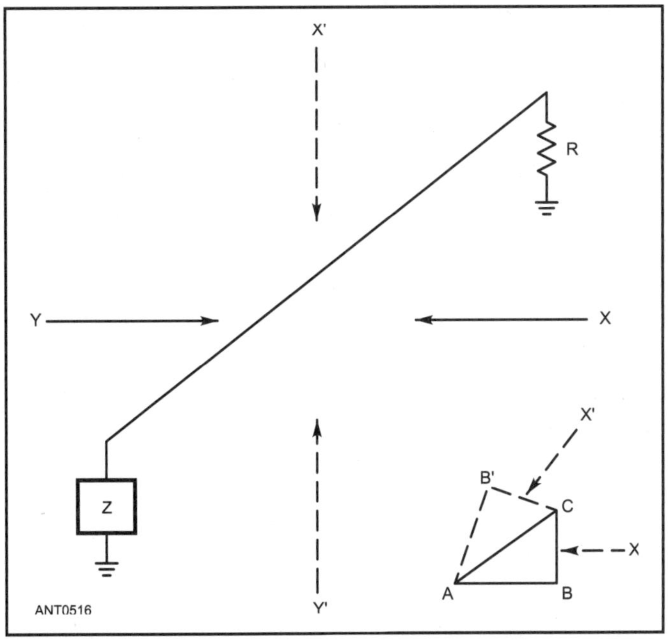

Figure 6.21 — Layout for a terminated long-wire antenna.

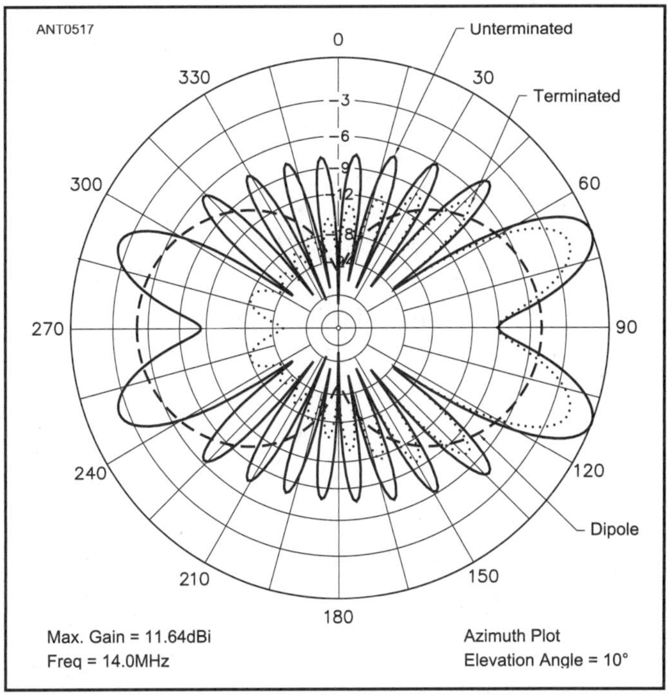

Figure 6.22 — Azimuthal-plane pattern for 5-λ long-wire antenna at 14 MHz and 70 feet above flat ground. The dotted line shows the long-wire terminated with 600-Ω to ground, while the solid line is for the same antenna unterminated. For comparison, the response for a ½-λ dipole (dashed line) is overlaid with the two other patterns. You can see that the terminated long-wire has a good front-to-back pattern, but it loses about 2 dB in forward gain compared to the unterminated long-wire.

The greatest possible power will be delivered to the load Z when the individual currents induced as the wave sweeps across the wire all combine properly on reaching the load. The currents will reach Z in optimum phase when the time required for a current to flow from the far end of the antenna to Z is exactly one-half cycle longer than the time taken by the wave to sweep over the antenna. A half cycle is equivalent to a half wavelength greater than the distance traversed by the wave from the instant it strikes the far end of the antenna to the instant that it reaches the near end. This is shown by the small drawing, where AC represents the antenna, BC is a line perpendicular to the wave direction, and AB is the distance traveled by the wave in sweeping past AC. AB must be one-half wavelength shorter than AC. Similarly, AB' must be the same length as AB for a wave arriving from X'.

A wave arriving at the antenna from the opposite direction Y (or Y'), will similarly result in the largest possible current at the far end. However, since the far end is terminated in R, which is equal to Z, all the power delivered to R by the wave arriving from Y will be absorbed in R. The current traveling to Z will produce a signal in Z in proportion to its amplitude. If the antenna length is such that all the individual currents arrive at Z in such phase as to add up to zero, there will be no current through Z. At other lengths the resultant current may reach appreciable values. The lengths that give zero amplitude are those which are odd multiples of ¼ λ, beginning at ¾ λ. The response from the Y direction is greatest when the antenna is any even multiple of ½ λ long; the higher the multiple, the smaller the response.

DIRECTIONAL CHARACTERISTICS

Figure 6.22 compares the azimuthal pattern for a 5-λ long 14-MHz long-wire antenna, 70 feet high over flat ground, when it is terminated and when it is unterminated. The rearward pattern when the wire is terminated with a 600 Ω resistor is reduced about 15 dB, with a reduction in gain in the forward direction of about 2 dB.

For a shorter leg length in a terminated long-wire antenna, the reduction in forward gain is larger — more energy is radiated by a longer wire before the forward wave is absorbed in the terminating resistor. The azimuthal patterns for terminated and unterminated V-beams with 2-λ legs are overlaid for comparison in **Figure 6.23**. With these relatively short legs the reduction in forward gain is about 3.5 dB due to the terminations, although the front-to-rear ratio approaches 20 dB for the terminated V-beam. Each leg of this terminated V-beam use a 600-Ω non-inductive resistor to ground. Each resistor would have to dissipate about one-quarter of the transmitter power. For average conductor diameters and heights above ground, the Z_{ANT} of the antenna is of the order of 500 to 600 Ω.

6.4.1 The Terminated Rhombic Antenna

The highest development of the long-wire antenna is the *terminated rhombic*, shown schematically in **Figure 6.24**. It consists of four conductors joined to form a diamond, or *rhombus*. All sides of the antenna have the same length and the opposite corner angles are equal. The antenna can be considered as being made up of two V antennas placed end to end and terminated by a noninductive resistor to produce

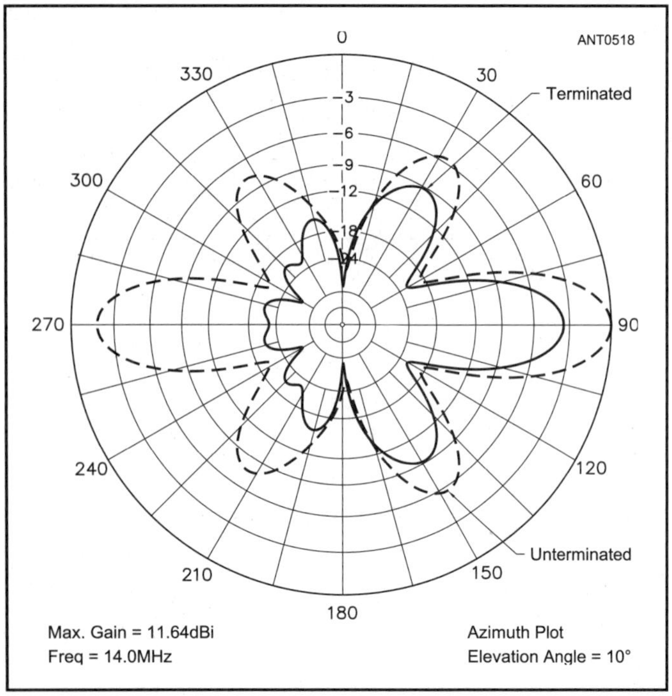

Figure 6.23 — The azimuthal patterns for a shorter-leg V-beam (2-λ legs) when it is terminated (solid line) and unterminated (dashed line). With shorter legs, the terminated V-beam loses about 3.5 dB in forward gain compared to the unterminated version, while suppressing the rearward lobes as much as 20 dB.

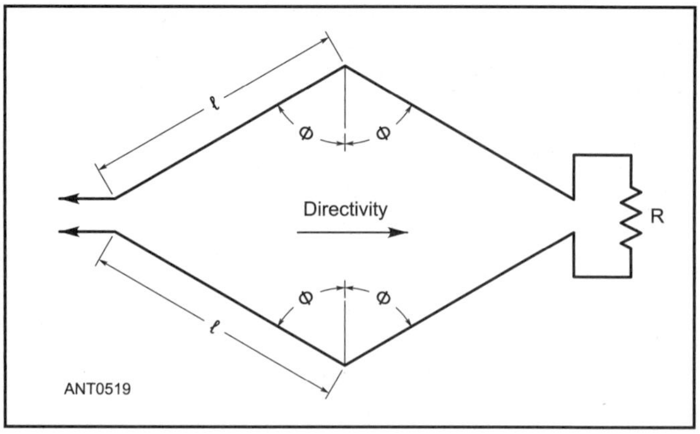

Figure 6.24 — The layout for a terminated rhombic antenna.

a unidirectional pattern. The terminating resistor is connected between the far ends of the two sides, and is made approximately equal to the characteristic impedance of the antenna as a unit. The rhombic may be constructed either horizontally or vertically, but is practically always constructed horizontally at frequencies below 54 MHz, since the pole height required is considerably less. Also, horizontal polarization is equally, if not more, satisfactory at these frequencies over most types of soil.

The basic principle of combining lobes of maximum radiation from the four individual wires constituting the rhombus or diamond is the same in either the terminated type or the resonant type described earlier in this chapter.

TILT ANGLE

In dealing with the terminated rhombic, it is a matter of custom to talk about the tilt angle (ϕ in Figure 6.24), rather than the angle of maximum radiation with respect to an individual wire. **Figure 6.25** shows the tilt angle as a function of the antenna leg length. The curve marked "0°" is used for a takeoff elevation angle of 0°; that is, maximum radiation in the plane of the antenna. The other curves show the proper tilt angles to use when aligning the major lobe with a desired takeoff angle. For a 5° takeoff angle, the difference in tilt angle is less than 1° for the range of lengths shown.

The broken curve marked "optimum length" shows the leg length at which maximum gain is obtained at any given takeoff angle. Increasing the leg length beyond the optimum will result in less gain, and for

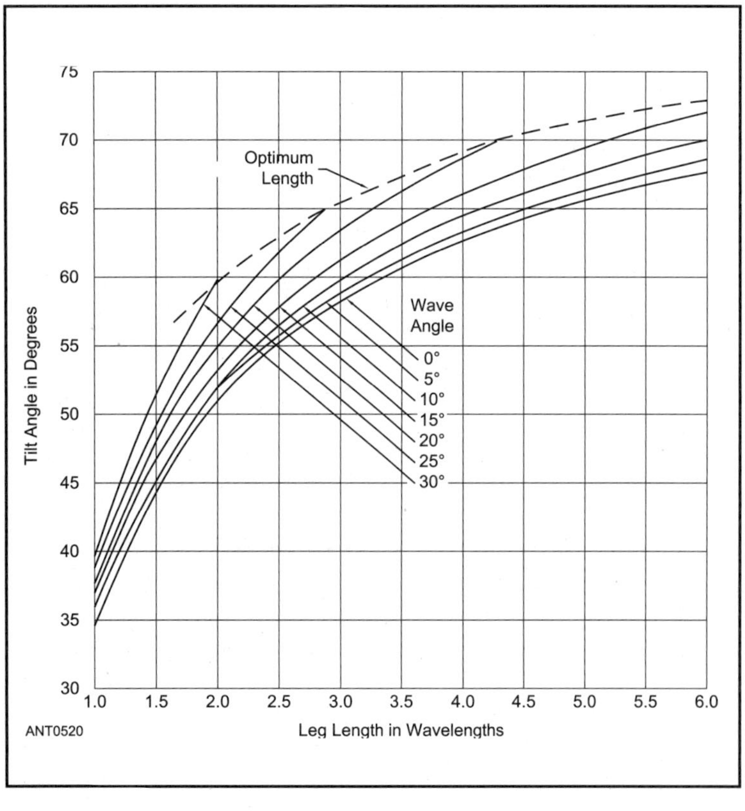

Figure 6.25 — Rhombic-antenna design chart. For any given leg length, the curves show the proper tilt angle to give maximum radiation at the selected takeoff angle. The broken curve marked "optimum length" shows the leg length that gives the maximum possible output at the selected takeoff angle. The optimum length as given by the curves should be multiplied by 0.74 to obtain the leg length for which the takeoff angle and main lobe are aligned.

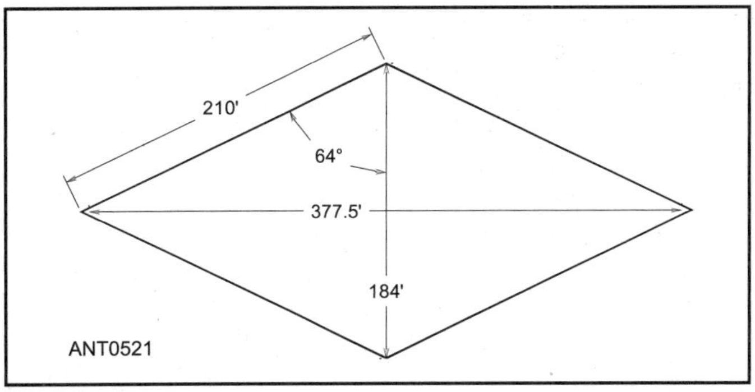

Figure 6.26 — Rhombic antenna dimensions for a compromise design between 14- and 28-MHz requirements, as discussed in the text. The leg length is 6 λ at 28 MHz, 3 λ at 14 MHz.

that reason the curves do not extend beyond the optimum length. Note that the optimum length becomes greater as the desired takeoff angle decreases. Leg lengths over 6 λ are not recommended because the directive pattern becomes so sharp that the antenna performance is highly variable with small changes in the angle, both horizontal and vertical, at which an incoming wave reaches the antenna. Since these angles vary to some extent in ionospheric propagation, it does not pay to attempt to try for too great a degree of directivity.

MULTIBAND DESIGN

When a rhombic antenna is to be used over a considerable frequency range, a compromise must be made in the tilt angle. **Figure 6.26** gives the design dimensions of a suitable compromise for a rhombic that covers the 14 to 30 MHz range well. **Figure 6.27** shows the azimuth and elevation patterns for this antenna at 14 MHz, at a height of 70 feet over flat ground. The comparison antenna in this case is a 4-element Yagi on a 26-foot boom, also 70 feet above flat ground. The rhombic has about 2.2 dB more gain, but its azimuthal pattern is 17.2° wide at the 3 dB points, and only 26° at the –20 dB points! On the other hand, the Yagi has a 3-dB beamwidth of 63°, making it far easier to aim at a distant geographic location. Figure 6.27B shows the elevation-plane patterns for the same antennas above. As usual, the peak angle for either horizontally polarized antenna is determined mainly by the height above ground.

The peak gain of a terminated rhombic is less than that of an unterminated resonant rhombic. For the rhombic of Figure 6.26, the reduction in peak gain is about 1.5 dB. **Figure 6.28** compares the azimuthal patterns for this rhombic with and without an 800-Ω termination.

Figure 6.29 shows the azimuth and elevation patterns for the terminated rhombic of Figure 6.26 when it is operated at 28 MHz. The main lobe becomes very narrow, at 6.9° at the 3-dB points. However, this is partially compensated for by the appearance of two sidelobes each side of the main beam. These tend to spread out the main pattern some. Again, a 4-element Yagi at the same height is used for comparison.

TERMINATION

Although the difference in the gain is relatively small with terminated or unterminated rhombics of comparable design, the terminated antenna has the advantage that over a wide frequency range it presents an essentially resistive and constant load to the transmitter. In a sense, the power dissipated in the terminating resistor can be considered power that would have been radiated in the other direction had the resistor not been there. Therefore, the fact that some of the power (about one-

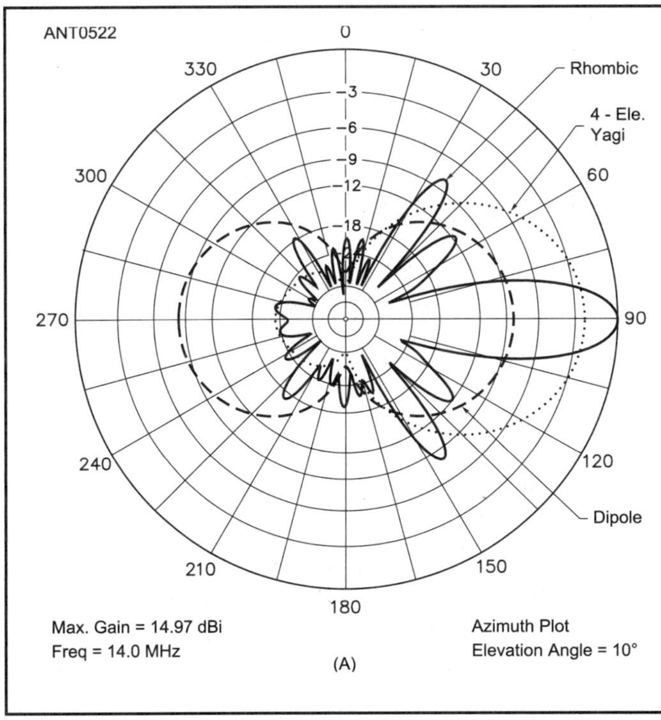

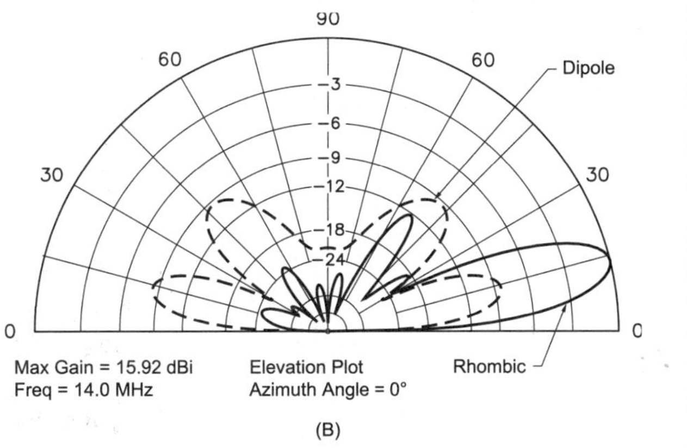

Figure 6.27 — At left, azimuthal pattern for 3-λ (at 14 MHz) terminated rhombic (solid line) shown in Figure 6.26, compared with 4-element 20-meter Yagi (dotted line) on a 26-foot boom and a 20-meter dipole (dashed line). All antennas are mounted 70 feet (1 λ) above flat ground. The rearward pattern of the terminated rhombic is good and the forward gain exceeds that of the Yagi, but the frontal lobe is very narrow. Above, elevation-plane pattern of terminated rhombic compared to that of a simple dipole at the same height.

third) is used up in heating the resistor does not mean that much actual loss in the desired direction.

The characteristic impedance of an ordinary rhombic antenna, looking into the input end, is in the order of 700 to 800 Ω when properly terminated in a resistance at the far end.

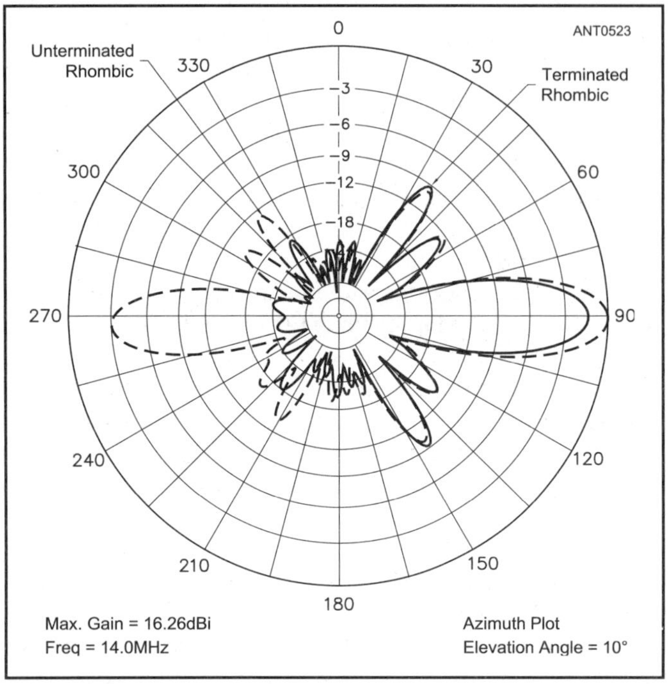

Figure 6.28 — Comparison of azimuthal patterns for terminated (solid line) and unterminated (dashed line) rhombic antennas, using same dimensions as Figure 6.26 at a frequency of 14 MHz. The gain tradeoff is about 1.5 dB in return for the superior rearward pattern of the terminated antenna.

The terminating resistance required to bring about the matching condition usually is slightly higher than the input impedance because of the loss of energy through radiation by the time the far end is reached. The correct value usually will be found to be of the order of 800 Ω, and should be determined experimentally if the flattest possible antenna is desired. However, for average work a noninductive resistance of 800 Ω can be used with the assurance that the operation will not be far from optimum.

The terminating resistor must be practically a pure resistance at the operating frequencies; that is, its inductance and capacitance should be negligible. Ordinary wire-wound resistors are not suitable because they have far too much inductance and distributed capacitance. Small carbon resistors have satisfactory electrical characteristics but will not dissipate more than a few watts and so cannot be used, except when the transmitter power does not exceed 10 or 20 watts or when the antenna is to be used for reception only. The special resistors designed either for use as dummy antennas or for terminating rhombic antennas should be used in other cases. To allow a factor of safety, the total rated power dissipation of the resistor or resistors should be equal to half the power output of the transmitter.

To reduce the effects of stray capacitance it is desirable to use several units, say three, in series even when one alone will safely dissipate the power. The two end units should be identical and each should have one fourth to one third the total resistance, with the center unit making up the difference. The units should be installed in a weatherproof housing at the end of the antenna to protect them and to permit mounting without mechanical strain. The connecting leads should be short so that little extraneous inductance is introduced.

Alternatively, the terminating resistance may be placed at

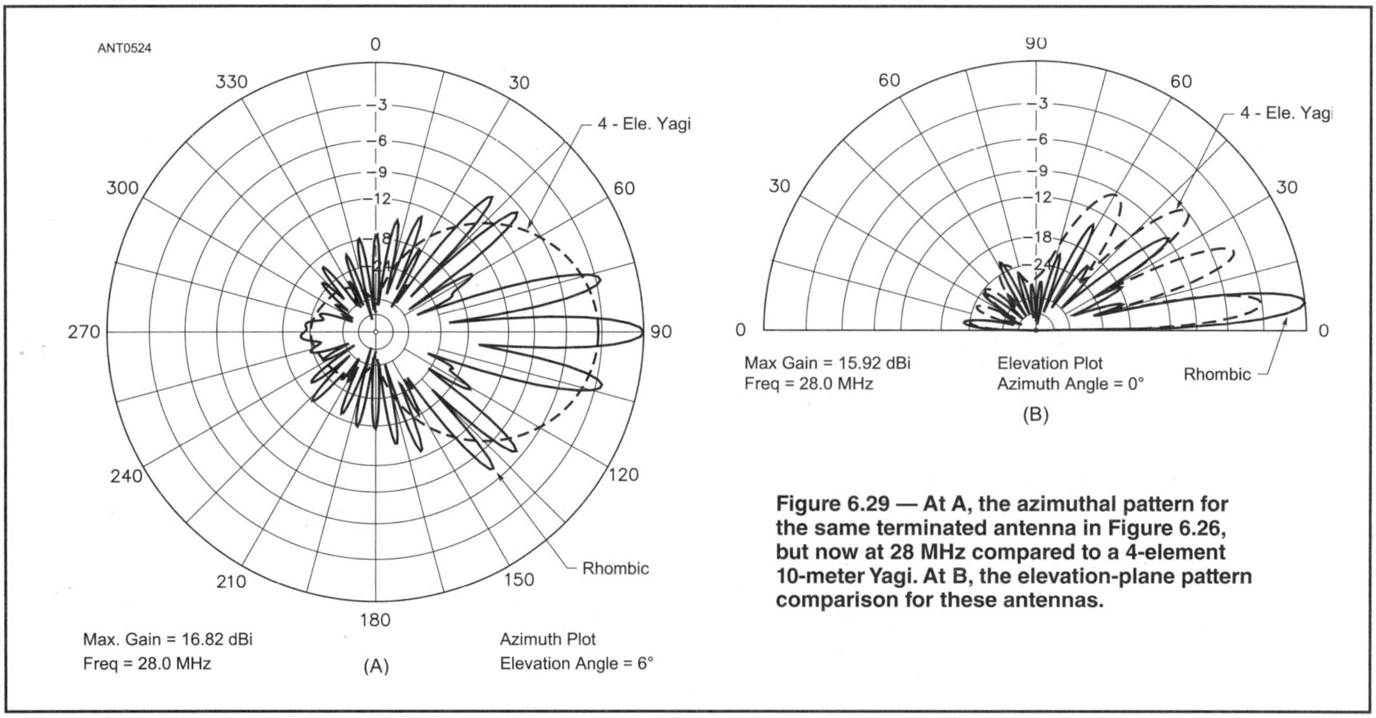

Figure 6.29 — At A, the azimuthal pattern for the same terminated antenna in Figure 6.26, but now at 28 MHz compared to a 4-element 10-meter Yagi. At B, the elevation-plane pattern comparison for these antennas.

the end of an 800-Ω line connected to the end of the antenna. This will permit placing the resistors and their housing at a point convenient for adjustment rather than at the top of the pole. Resistance wire may be used for this line, so that a portion of the power will be dissipated before it reaches the resistive termination, thus permitting the use of lower wattage lumped resistors.

If the rhombic is to be used on a single-band, Hallas (see the References and Bibliography) presents an interesting method of using an antenna tuner and a more common 50-Ω dummy load to create a "tunable load" that can be adjusted for the best performance.

MULTI-WIRE RHOMBICS

The input impedance of a rhombic antenna constructed as in Figure 6.26 is not quite constant as the frequency is varied. This is because the varying separation between the wires causes the characteristic impedance of the antenna to vary along its length. The variation in Z_{ANT} can be minimized by a conductor arrangement that increases the capacitance per unit length in proportion to the separation between the wires.

The method of accomplishing this is shown in **Figure 6.30**. Three conductors are used, joined together at the ends but with increasing separation as the junction between legs is approached. For HF work the spacing between the wires at the center is 3 to 4 feet, which is similar to that used in commercial installations using legs several wavelengths long. Since all three wires should have the same length, the top and bottom wires should be slightly farther from the support than the middle wire. Using three wires in this way reduces the Z_{ANT} of the antenna to approximately 600 Ω, thus providing a better match for practical open-wire line, in addition to smoothing out the impedance variation over the frequency range.

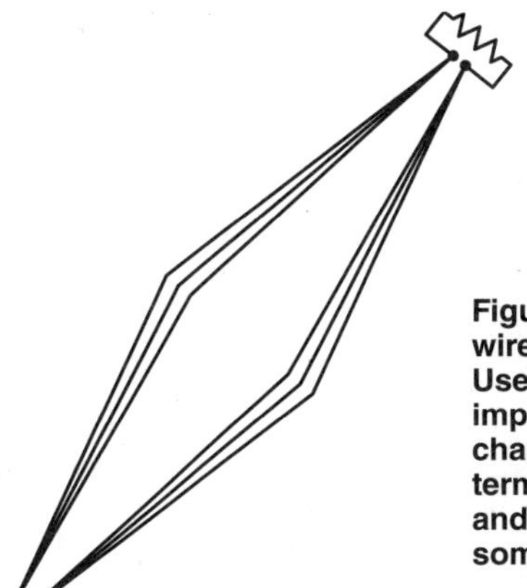

Figure 6.30 — Three-wire rhombic antenna. Use of multiple wires improves the impedance characteristic of a terminated rhombic and increases the gain somewhat.

A similar effect (although not quite as favorable) is obtained by using two wires instead of three. The 3-wire system has been found to increase the gain of the antenna by about 1 dB over that of a single-conductor version.

FRONT-TO-BACK RATIO

It is theoretically possible to obtain an infinite front-to-back ratio with a terminated rhombic antenna, and in practice very large values can be had. However, when the antenna is terminated in its characteristic impedance, the infinite front-to-back ratio can be obtained only at frequencies for which the leg length is an odd multiple of a quarter wavelength. The front-to-back ratio is smallest at frequencies for which the leg length is a multiple of a half wavelength.

When the leg length is not an odd multiple of a quarter-wave at the frequency under consideration, the front-to-back ratio can be made very high by decreasing the value of terminating resistance slightly. This permits a small reflection from the far end of the antenna, which cancels out the residual response at the input end. With large antennas, the front-to-back ratio may be made very large over the whole frequency range by experimental adjustment of the terminating resistance. Modification of the terminating resistance can result in a splitting of the back null into two nulls, one on either side of a small lobe in the back direction. Changes in the value of terminating resistance thus permit steering the back null over a small horizontal range so that signals coming from a particular spot not exactly to the rear of the antenna may be minimized.

METHODS OF FEED

If the broad frequency characteristic of the terminated rhombic antenna is to be utilized fully, the feeder system must be similarly broadbanded. Open-wire transmission line of the same characteristic impedance as that shown at the antenna input terminals (approximately 700 to 800 Ω) may be used. Data for the construction of such lines is given in the chapter on **Transmission Lines**. While the usual matching stub can be used to provide an impedance transformation to more satisfactory line impedances, this limits the operation of the antenna to a comparatively narrow range of frequencies centering about that for which the stub is adjusted. Probably a more satisfactory arrangement would be to use a coaxial transmission line and a broadband transformer balun at the antenna feed point.

6.5 Project: Four-Wire Steerable V-Beam for 10 through 40 Meters

A simple arrangement of four wires can be used to work multiple bands and have antenna gain in different directions without using a rotator. A version of this antenna was described in *QST* (see References and Bibliography entry for Colvin) and is included in ARRL's *Wire Antenna Classics*. That version had wires 584 feet long. In this version, built by Sam Moore, NX5Z, each wire is only 106 feet long. Many DX stations have had great success with this type of antenna. (The full *QST* article by NX5Z is included in the online material.)

ANTENNA CHARACTERISTICS

An unterminated V-beam gain pattern is bidirectional with two main gain lobes 180° apart if the leg lengths are at least a wavelength long. In **Figure 6.31**, a long wire antenna at the left is shown to have a gain pattern of four major lobes. Another long wire antenna positioned 45° from the first is also shown. If these are combined to form a V, it has the gain pattern as shown to the right in Figure 6.31.

In this design, four 106-foot wires are spaced at 45°. The length of the wire is not as important as that they all be the same length. The author installed his V-beam with the apex and relay control box at a height of 40 feet with the wire ends 10 feet off the ground in a sloping V configuration. This V-beam's gain approximates that of a three element Yagi on 10, 12, 15 and 17 meters and is within a few dB on 20 meters. The antenna provides useful operation on 30 and 40 meters, with essentially an omnidirectional pattern on 40. The beam direction is controlled by simply switching two switches in the station.

This antenna may also be built with wire lengths as short as 60 feet to more easily fit on a city lot. There will be a small decrease in gain. The V-beam gain increases with the length of the wires. The longer the wires, the greater the gain. As the wire lengthens, however, the beamwidth narrows. The gains and beamwidths of 106 and 60 foot versions are shown in **Table 6.1**, based on *EZNEC* analysis. (*EZNEC* modeling software is discussed in the **Antenna Modeling** chapter.) As a reference, the typical two-element Yagi has 6 to 7 dBi gain while a three element Yagi can be expected to have a 7.5 to 8.1 dBi gain, depending on design, especially boom length.

The azimuth pattern looking down on a V-beam is shown in **Figure 6.32**. If the height of the V-beam is less than ½ wavelength, the gain pattern will distort and make the antenna more omnidirectional.

To reduce the gain lobe to the rear of the V-beam you can terminate the wire ends with a resistor. An unterminated version has gain in both directions. If terminated, the antenna would need eight wires instead of four to have gain in all directions.

Since this antenna may be used for multiband operation, the gain waveform changes somewhat depending on the frequency of operation. The higher the frequency, the greater the gain, since the frequency to wire length ratio changes. For example, if your V-beam is 1 wavelength long at 20 meters, it is 2 wavelengths long at 10 meters, thus causing greater gain and narrower beamwidth as shown in Table 6.1. While essentially bidirectional on the upper bands, there is a 1 to 2 dB front to back ratio, with the maximum signal to the open end of the V. The beamwidth shown in Table 6.1 is of the front beam, with the rear beam generally somewhat narrower. A horizontal, rather than sloping, V-beam will be more symmetrical.

Table 6.1
Gain and Beamwidth of the V-Beam on Each Band

Frequency (MHz)	*Gain at 106' (dBi)*	*3 dB Beamwidth at 106' (°)*	*Gain at 60' (dBi)*	*3 dB Beamwidth at 60' (°)*
7.15	1.9*	Omnidirectional	2.4*	Omnidirectional
10.12	3.6	133	3.7*	Omnidirectional
14.15	6.7	71	4.1	137
18.11	8.5	42	4.1	136
21.2	9.1	33	6.0	63
24.93	9.7	28	6.1	61
28.3	10.7	23	7.3	40

*Essentially omnidirectional with maximum gain nearly perpendicular to the wire bisector.

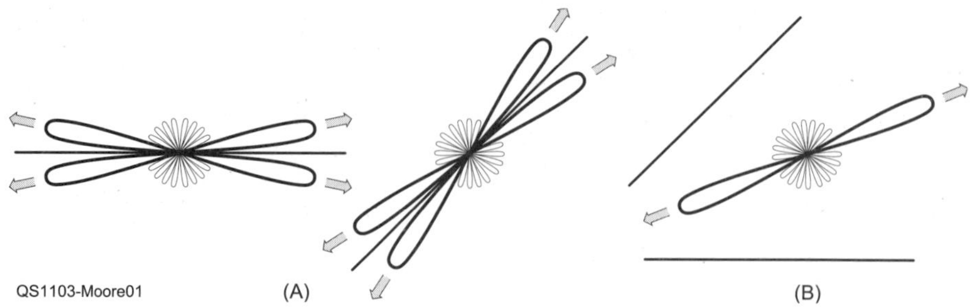

Figure 6.31 — The azimuth patterns of two long wire antennas are shown at (A). If the two are combined in phase to form a V, the resulting pattern is shown at (B).

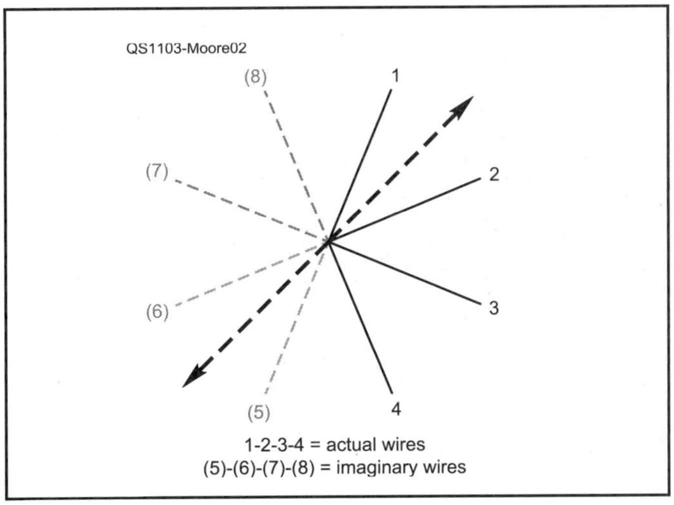

Figure 6.32 — The selectable azimuth looking down on the V-beam. The arrow shows directions of maximum radiation with wires 1 and 2 connected.

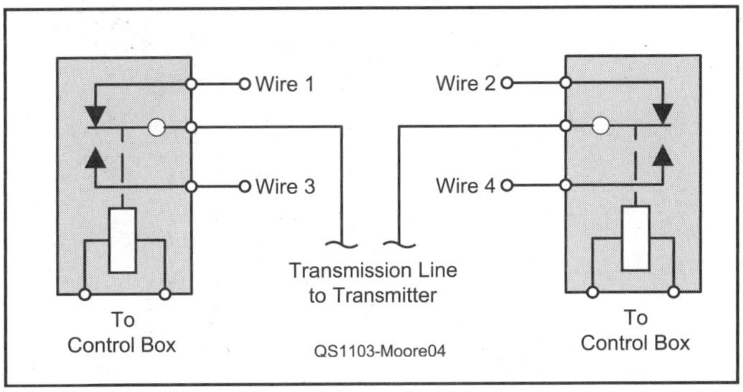

Figure 6.34 — Schematic diagram of the relay box used to remotely select the V-beam wires.

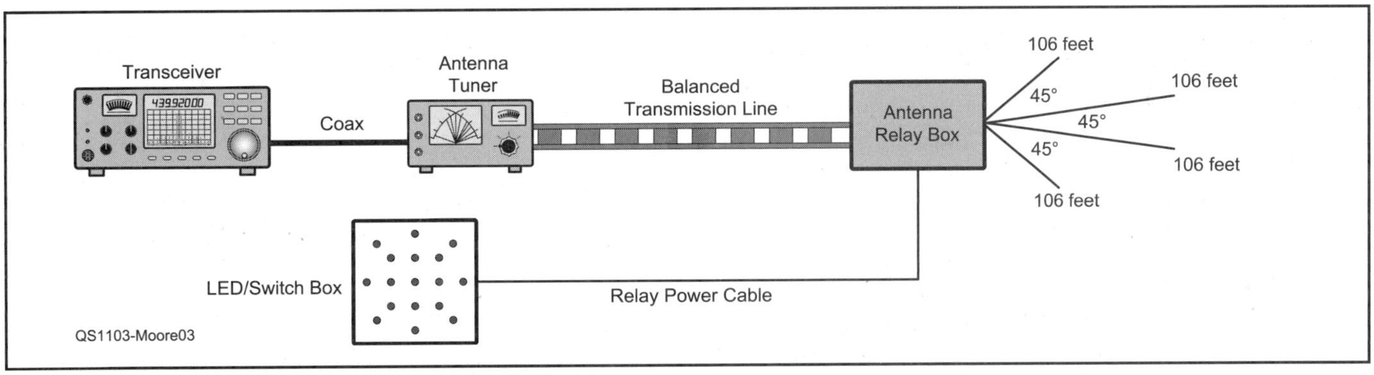

Figure 6.33 — The block diagram of the V-beam system. The antenna tuner must be able to accept balanced transmission line and a built-in or external 4:1 balun is necessary.

The block diagram of the V-beam system is shown in **Figure 6.33**. The antenna tuner must be able to accept balanced transmission line and a built in or external 4:1 balun is necessary. The author made a homebrew air core external 4:1 balun using 1 inch PVC pipe and used a small automatic antenna tuner.

CONTROLS AND INDICATORS

The LED switch box supplies power to the relays in the antenna relay box at the center of the V-beam via a three wire cable such as three wire electrical zip cord. Smaller wires would work.

The relay box schematic is shown in **Figure 6.34**. Only two switches are needed to power relays 1 and 2. Relay 1 switches between wire 1 and 3 and relay 2 switches between wire 2 and 4. Note that wire 4 is used in combination with wire 1 instead of (imaginary) wire 5. This obtuse angle yields the about same gain and waveform as wire 4 to 5 would have offered, without having to string another wire. **Figure 6.35** shows an assembled relay box in a power entry PVC cover.

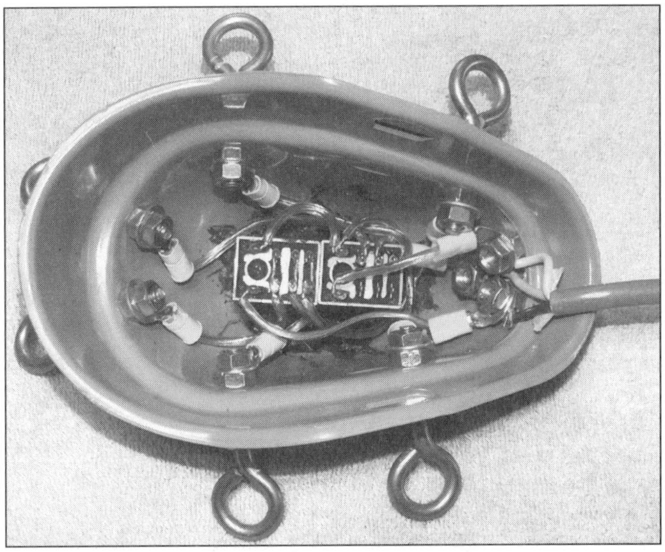

Figure 6.35 — Relay box assembled in a power entry PVC cover.

Figure 6.36 — The V-beams are supported by a mast and a standard PVC power-entry cover. Relays are used to switch between the beams and are controlled through a separate control cable.

The assembled control head is mounted as shown in **Figure 6.36**. Total cost is around $50, not counting the balun and balanced transmission line. For the four wires, the author used electric fence wire, which accepted solder surprisingly well. You can buy a ¼ mile roll of electric fence wire inexpensively at agricultural supply stores. You may also have a few necessary parts in your junk box.

6.6 References and Bibliography

Source material and more extended discussion of topics covered in this chapter can be found in the references given below.

Bruce, E., "Developments in Short-Wave Directive Antennas," *Proc IRE*, Aug. 1931.

Bruce, E., Beck, A., and Lowry, L., "Horizontal Rhombic Antennas," *Proc IRE*, Jan. 1935.

Carter, P., Hansel C., and Lindenblad, N., "Development of Directive Transmitting Antennas by R.C.A. Communications," *Proc IRE*, Oct. 1931.

Colvin, L., DL4ZC (W6KG), "Multiple V Beams," *QST*, Aug. 1956.

Devoldere, J., *ON4UN's Low-Band DXing*, 5th Edition (ARRL, 2010).

Hallas, J., W1ZR, "Achieving Near Perfection with the Imperfect Rhombic," *QST*, Nov. 2004, pp. 28 – 32.

Harper, A., *Rhombic Antenna Design* (D. Van Nostrand Co., Inc., 1941).

Laport, E., "Design Data for Horizontal Rhombic Antennas," *RCA Review*, Mar. 1952.

Miller, G., *Modern Electronic Communication* (Prentice Hall, 1983).

Moore, S., NX5Z, "A Four Wire Steerable V Beam for 10 through 40 Meters," *QST*, Mar. 2011, pp. 30 – 33.

Mullaney, J., Capt., W4HGU, "The Half-Rhombic Antenna," *QST*, Jan. 1946, pp. 28 – 31.

Orr, M., AA2PE, "The Tilted Half-Rhombic Antenna," *Antenna Compendium*, Vol. 4 (ARRL, 1999), pp. 5-10 – 5-13.

Terman, F., *Radio Engineering*, 2nd Edition (McGraw-Hill, 1937).

TABLE OF CONTENTS

Chapter 7 — Online Material

Articles

- "A Four-Square with Eight Directions of Fire," by Al Christman, K3LC
- "A Wire Eight-Circle Array (for 7 MHz)" by Tony Preedy, G3LNP
- "A Study of Tall Verticals" by Al Christman, K3LC
- "Compact Four-Squares," by Al Christman, K3LC
- Manual Calculations for Arrays
- "Simplifying the Four-Square" by Tony Preedy, G3LNP
- "Tall Vertical Arrays" by Al Christman, K3LC
- "The Simplest Phased Array Feed System—That Works" by Roy Lewellan, W7EL

Support Files

- *EZNEC* modeling files for examples in this chapter
- *Arrayfeed1* software for calculating array feed system values

Chapter 7
Phased Arrays

From the chapter "Arrays of Dipoles And of Apertures" in *Antennas, 2nd Edition* by Kraus, "the term *phased array* has come to mean an array of many elements with the phase (and also…the amplitude) of each element being a variable, providing *control of the beam direction* (this is known as *beam steering*) and pattern shape, including sidelobes." For amateurs, the number of elements is usually two to four, although arrays with as many as nine elements are in use. "An objective of a phased array is to accomplish beam steering without (physically) rotating the entire array."

Kraus also describes beam steering by mechanical switching of different lengths of feed line to each of the array elements. This is not uncommon for simple amateur arrays, being very cost-effective and relying on using the right lengths of feed line as a *phase shifter*. For example, the popular Four-Square array can be switched in eight directions by this technique. LC circuits can also be used as phase shifters. Given that fully rotatable beam antennas for 80 and 160 meters are impractical for all but a very few stations, the ground-mounted array is a very satisfactory solution.

This chapter begins with an overview of the principles that govern the design and function of phased arrays. This includes careful definition of terms and the techniques that are used to create currents in each element with the necessary phase relationships. The second section of the chapter shows how these techniques are used for practical arrays of monopoles (verticals) with useful pattern control for typical amateur operation. (Even though the array may have any physical orientation, the term "vertical" will be used instead of monopole because of its familiarity to amateurs.) The interested amateur may also want to read the comprehensive professional review of phased array patterns by Delos and others in *Microwaves & RF* or the *Phased Array Antenna Handbook* by Mailloux (see the References and Bibliography section).

This chapter has been comprehensively updated for the 25th edition by Roy Lewallen, W7EL, author of the now-free *EZNEC* antenna modeling software (**eznec.com**). As described in a later sidebar, *EZNEC* models are used to illustrate key ideas and a package of models is available to the reader in the online material. Reading the **Antenna Modeling** chapter, then following along in the examples given here is an excellent way to get started with antenna modeling.

7.1 Creating Gain and Directivity

The gain and directivity offered by an array of elements represents a worthwhile improvement both in transmitting and receiving. Power gain in an antenna is the same as an equivalent increase in the transmitter power. But unlike increasing the power of your own transmitter, antenna gain works equally well on signals received from the favored direction. In addition, the directivity reduces the strength of signals coming from the directions not favored, and so helps discriminate against interference.

One common method of obtaining gain and directivity is to combine the radiation from a group of $\lambda/2$ dipoles to concentrate it in a desired direction. A few words of explanation may help make it clear how power gain is obtained.

In **Figure 7.1**, imagine that the four circles, A, B, C and D, represent four dipoles so far separated from each other that the coupling between them is negligible. Also imagine that point P is so far away from the dipoles that the distance from P to each one is exactly the same (obviously P would have to be much farther away than it is shown in this drawing). Under these conditions the fields from all the dipoles will add up at

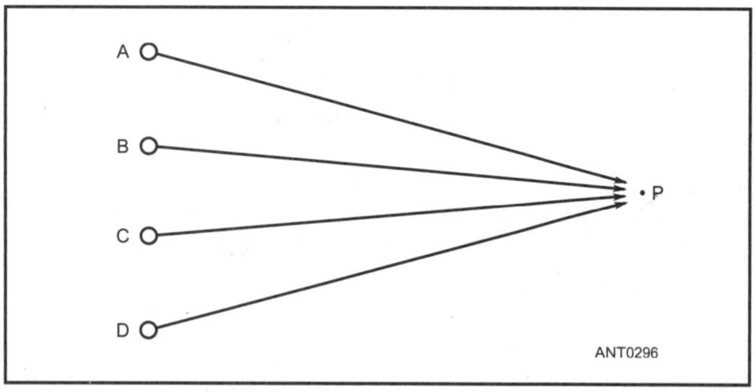

Figure 7.1 — Fields from separate antennas combine at a distant point, P, to produce a field strength that exceeds the field produced by the same power in a single antenna.

Table 7.1
Comparison of Dipoles with Negligible Coupling (See Figure 7.1)

Dipoles	*Relative Output Power*	*Relative Input Power*	*Power Gain*	*Gain in dB*
A only	1	1	1	0
A and B	4	2	2	3
A, B and C	9	3	3	4.8
A, B, C and D	16	4	4	6

P if all four are fed RF currents in the same phase.

Let us say that a certain current, I, in dipole A will produce a certain value of field strength, E, at the distant point P. The same current in any of the other dipoles will produce the same field at P. Thus, if only dipoles A and B are operating, each with a current I, the field at P will be 2E. With A, B and C operating, the field will be 3E, and with all four operating with the same I, the field will be 4E. Since the power received at P is proportional to the square of the field strength, the relative power received at P is 1, 4, 9, or 16, depending on whether one, two, three or four dipoles are operating.

Now, since all four dipoles are alike and there is no coupling between them, the same power must be put into each in order to cause the current I to flow. For two dipoles the relative power input is 2, for three dipoles it is 3, for four dipoles 4, and so on. The actual gain in each case is the relative received (or output) power divided by the relative input power. Thus, we have the results shown in **Table 7.1**. The power ratio is directly proportional to the number of elements used.

It is well to have clearly in mind the conditions under which this relationship is true:

1) The fields from the separate antenna elements must be in-phase at the receiving point.

2) The elements are identical, with equal currents in all elements.

3) The elements must be separated in such a way that the current induced in one by another is negligible; that is, the radiation resistance of each element must be the same as it would be if the other elements were not there.

Very few antenna arrays meet all these conditions exactly. However, the power gain of a directive array using dipole elements with optimum values of element spacing is approximately proportional to the number of elements. Another way to say this is that a gain of approximately 3 dB will be obtained each time the number of elements is doubled, assuming the proper element spacing is maintained. It is possible, though, for an estimate based on this rule to be in error by a ratio factor of two or more (gain error of 3 dB or more), especially if mutual coupling is not negligible.

7.1.1 Definitions

Some definitions are repeated here from the **Antenna Fundamentals** chapter for the convenience of the reader. Refer to that chapter for more extensive discussion of basic concepts.

An *element* in a phased array is usually a $\lambda/2$ radiator (dipole) or a $\lambda/4$ vertical (monopole) element above ground. However, other lengths and other shapes, for example loops or inverted vees, can be used. For simplicity, antenna elements in multielement arrays of the type considered in this chapter are always either *parallel*, as in **Figure 7.2A**, or *collinear* (end-to-end), as in Figure 7.2B. Other relative orientations are possible, but analysis requires modeling software. Figure 7.2C shows an array combining both parallel and collinear elements. The elements can be either horizontal or vertical, depending on whether horizontal or vertical polarization is desired. Except for space communications, there is seldom any reason for mixing polarization, so arrays are customarily constructed with all elements similarly polarized.

A *driven element* is one supplied power from the transmitter, usually through a transmission line. A *parasitic element* is one that obtains power solely through coupling to another element in the array because of its proximity to such an element. Parasitic elements are usually a few percent shorter or longer than resonant to create the phase shift of current necessary for creating the desired pattern.

A *driven array* is one in which all the elements are driven elements. A *parasitic array* is one in which one or more of the elements are parasitic elements. At least one element must be a driven element since you must somehow introduce power into the array.

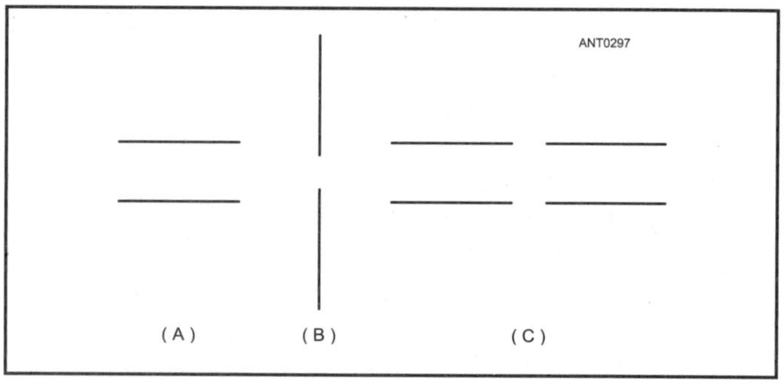

Figure 7.2 — At A, parallel and at B, collinear antenna elements. The array shown at C combines both parallel and collinear elements.

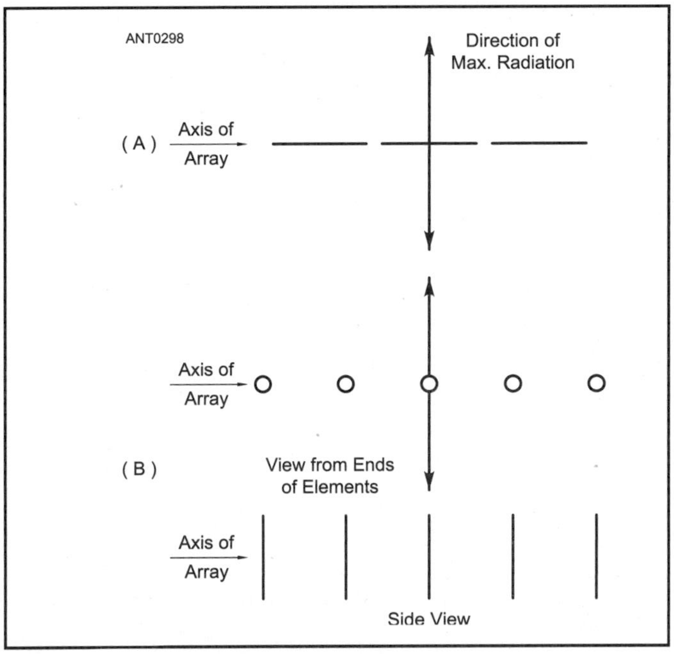

Figure 7.3 — Representative broadside arrays. At A, collinear elements, with parallel elements at B.

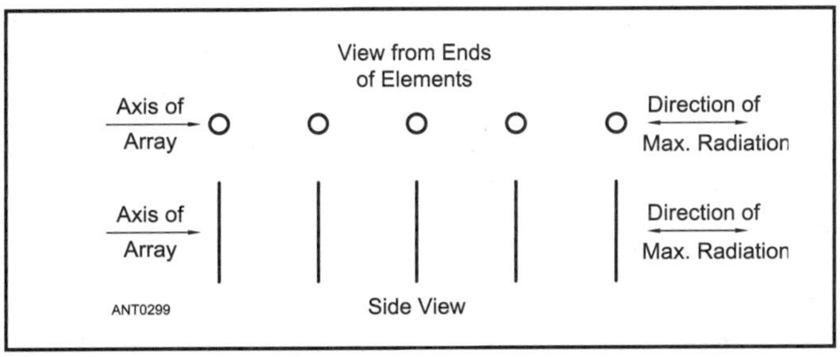

Figure 7.4 — An end-fire array. Practical arrays may combine both broadside directivity (Figure 7.3) and end-fire directivity, including both parallel and collinear elements.

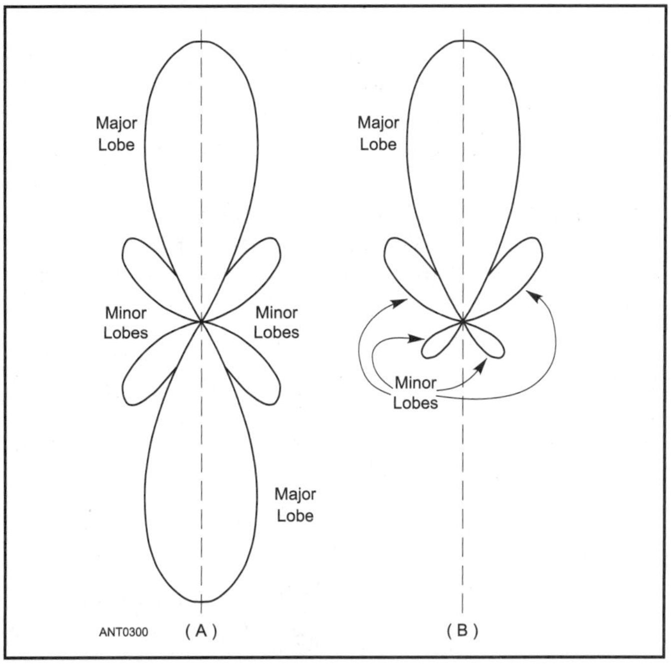

Figure 7.5 — At A, typical bidirectional pattern and at B, unidirectional directive pattern. These drawings also illustrate the application of the terms *major* and *minor* to the pattern lobes.

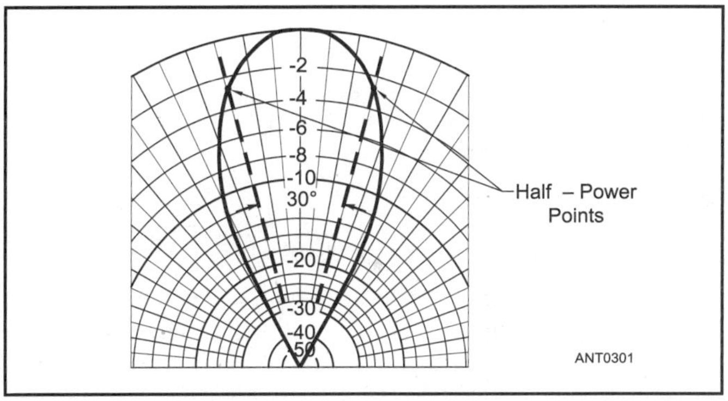

Figure 7.6 — The width of a beam is the angular distance between the directions at which the received or transmitted power is half the maximum power (–3 dB). Each angular division of the pattern grid is 5°.

A *broadside array* is one in which the principal direction of radiation is perpendicular to the axis of the array and to the plane containing the elements, as shown in **Figure 7.3**. The elements of a broadside array may be collinear, as in Figure 7.3A, or parallel (two views in Figure 7.3B).

An *end-fire array* is one in which the principal direction of radiation coincides with the direction of the array axis. This definition is illustrated in **Figure 7.4**. An end-fire array usually consists of parallel elements. They cannot be collinear, as linear elements do not radiate straight off their ends. A Yagi is a familiar form of an end-fire array.

A *bidirectional array* is one that radiates equally well in either direction along the line of maximum radiation. A bidirectional pattern is shown in **Figure 7.5A**. A *unidirectional array* is one that has only one principal direction of radiation, as the pattern in Figure 7.5B shows.

The *major lobes* of the directive pattern are those in which the radiation is a maximum. Lobes of lesser radiation intensity are called *side* or *minor* lobes. The *beamwidth* of a directive antenna is the width, in degrees, of the major lobe between the two directions at which the relative radiated power is equal to one half its value at the peak of the lobe. At these *half-power points* the field intensity is equal to 0.707 times its maximum value, or in other words, is down 3 dB from the maximum. **Figure 7.6** shows a lobe having a beamwidth of 30°.

Unless specified otherwise, the term *gain* as used in this section is the power gain over an isotropic radiator in free space. The gain can also be compared with a $\lambda/2$ dipole of the same orientation and height as the array under discussion and having the same power input.

Gain may either be measured experimentally or determined by calculation. Experimental measurement is difficult and often subject to considerable error, for two reasons. First, errors normally occur in measurement because the accuracy of simple RF measuring equipment is relatively poor — even high-quality instruments suffer in accuracy compared with their low-frequency and dc counterparts. And second, the accuracy depends considerably on conditions — the antenna site, including height, terrain characteristics, and surroundings — under which the measurements are made. (At UHF and higher frequencies, an anechoic chamber can measure an antenna as if it was in a free space environment.)

Calculations are frequently based on the measured or theoretical directive patterns of the antenna. The theoretical *free space gain* of a unidirectional array may be determined approximately from:

$$G \approx \frac{41,253}{H_{3dB} \times E_{3dB}}$$

where

G = decibel gain over a dipole in its favored direction

H_{3dB} = half-power beamwidth in degrees of the H-field pattern

E_{3dB} = vertical half-power beamwidth in degrees of the E-field pattern.

This equation, strictly speaking, applies only to lossless antennas having approximately equal and narrow E- and H-plane beam widths — up to about 20° — and no large minor lobes. The error may be considerable when the formula is applied to simple directive antennas having relatively large beam widths. The error is in the direction of making the calculated gain larger than the actual gain.

PHASE, POLARITY, AND DIRECTION

The term *phase* has the same meaning when used in connection with the currents flowing in antenna elements as it does in ordinary circuit work. Reiterating from the **Antenna Fundamentals** chapter, it is important to distinguish between phase and *polarity*. Polarity is simply a convention that assigns a positive and negative direction or convention, while phase refers to the part of the cycle of an ac voltage or current. But it's essential to realize that both are relative values and are meaningless without some reference — and the reference is largely arbitrary! For example, suppose an RF voltage is applied to a transmission line that delays the phase of the voltage by 1/8 of a cycle. Relative to the input end, the phase at the far end of the line will be –45°. But if we consider the phase of the voltage at the far end of the line to be zero, then the phase of the source at the input is +45°. So, the voltage at neither end of the line has an absolute value of phase — but the relative phase between ends is constant.

What happens when we reverse the source terminals, which reverses the polarity of the applied signal? If the source was and is still considered to be producing a voltage with zero phase at its positive terminal relative to its negative terminal, then the phase of the voltage applied to the transmission line has reversed, becoming 180° (or –180° which is the same thing), and the voltage at the other end of the line has also

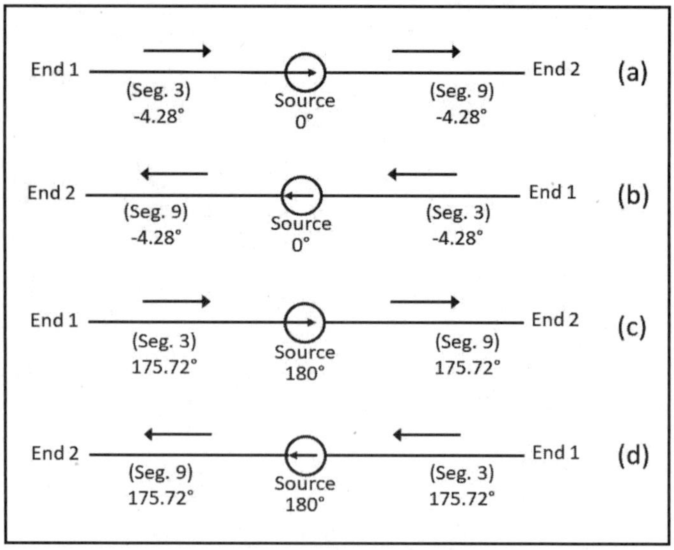

Figure 7.7 — Four equivalent ways of representing the phase and direction of the same current in a model of a dipole (see text).

changed phase by 180° to become +135° or –225°. But we could use the voltage at the line input to be the zero-degree reference, in which case we're back to –45° at the line output.

In dealing with currents on wires, we also have the concept of current *direction*. Alas, this is also arbitrary, and causes a great deal of confusion when modeling antennas. If a current can be said to be flowing in one direction in a wire with a phase angle of –45°, it can be said equally correctly to be flowing in the other direction with a phase angle of –225° or +135. In phased arrays, we're interested in the physical direction or phase of one current relative to that in other elements. **Figure 7.7A** shows the phases of the currents at the source and approximately the midpoint of each wire half when modeling the *Dipole1.EZ* example antenna with *EZNEC*. (Filenames with the .EZ extension refer to *EZNEC* models that are included in the online material.) In Figure 7.7B, the wire ends have been reversed. The program defines the direction of positive current flow from a source to be in the wire end 1 — end 2 direction, so physically reversing the wire ends also physically reverses the source, keeping its orientation the same relative to the wire ends. The phase of the currents in the wires, *relative to the phase of the source*, hasn't changed. However, the *physical* direction of the current has changed, and another element will see this as a 180° change in this element's current — it was at an angle of –4.28° to the right and now it's –4.28° to the left.

The same physical phase reversal can be created by leaving the ends as they were and changing the source phase to 180° as in Fig. 7.7C. This is equivalent to turning the source around or reversing a feed-line connection. The "direction" of current is still defined as from end 1 to end 2 (left to right) but its phase is reversed — so the currents are exactly the same as in B and 175.72° in one direction is exactly the same as –4.28° in the other. Finally, both reversing the wire and the phase of the source as in Fig. 7.7D results in exactly the same currents as in A.

The phasing of driven elements depends on the direction of the element, the phase of the applied currents, and the point at which the current is applied. In many arrays used by amateurs, the currents applied to the elements are exactly in- or exactly out-of-phase with each other. Also, the axes of the elements are nearly always in the same direction, since parallel or collinear elements are invariably used. The currents in driven elements in such systems therefore are usually either exactly in or exactly out-of-phase with each other.

It is possible to use phase differences of less than 180° in driven arrays. One important case is where the current in one set of elements differs by 90° from the current in another set (said to be *in quadrature*). However, making provision for proper phasing in such systems is considerably more complex than in the case of simple 0° or 180° phasing, as described in a later section of this chapter.

In parasitic arrays the phase of the currents in the parasitic elements depends on the spacing and tuning as described later and can take on various values of relative amplitude and phasing.

GROUND EFFECTS

The effect of the ground is the same with a directive antenna as it is with a simple dipole antenna. The reflection factors discussed in the chapter **Effects of Ground** may therefore be applied to the vertical pattern of an array, subject to the same modifications mentioned in that chapter. In cases where the array elements are not all at the same height, the reflection factor for the mean height of the array may be used for a close approximation. The mean height is the average of the heights measured from the ground to the centers of the lowest and highest elements.

7.1.2 Mutual Coupling

Consider two λ/2 elements that are fairly close to each other. Assume that power is applied to only one element ("element 1"), causing current to flow. This creates an electromagnetic field which induces a voltage in the second element and causes current to flow in it as well. The current flowing in element 2 will in turn induce a voltage in element 1, causing a change in the current flowing there. The total current in 1 is then the sum (taking phase into account) of the original current and the induced current. This effect is called *mutual coupling*. The ratio of voltage induced in the second element to the current in the first is called *mutual impedance*. It's determined only by the array geometry and used in calculating the effects of mutual coupling. It's not the same as and shouldn't be confused with *coupled impedance*, described next.

The element 1 current will be different than if element 2 were not there. This indicates that, if element 1 is driven, the presence of the second element has changed the feed-point impedance of the first. This change is called *coupled impedance* which has both resistive and reactive components. The actual impedance of an antenna element is the sum of its self-impedance (the impedance with no other antennas present) and its coupled impedances from all other elements in the array (or, often unintentionally, other antennas in the vicinity). The induced current and coupled impedance in an element depends on the spacing between it and other elements, and their lengths, diameters, orientations, and currents. The coupled impedance (feed-point impedance change) of a driven element often has a major impact on the current being delivered to it by the feed system. Consequently, the feed system must be designed to deliver the desired current — magnitude and phase — to the feed-point impedance which includes the coupled impedance(s).

It's important to realize that the altered current in the first element determines the coupled impedance in the second element and vice versa, and in many cases the resulting coupled impedance is different for each element in an array. Because all elements in an array interact with all others, analysis of all element interactions must be done simultaneously. Practically, this can only be done with a computer except for some simple cases.

In the discussion of the several preceding paragraphs, power is applied to only one of the two elements. Do not interpret this to mean that mutual coupling exists only in driven or parasitic arrays! It is important to remember that mutual coupling exists between any two conductors that are located near one another — regardless of whether they're part of an intentional array.

Because these phenomena affect nearby antennas in the same manner as elements in intentional arrays, "antenna" and "element" are used interchangeably in the following discussion.

AMPLITUDE OF INDUCED CURRENT

The induced current will be largest when two antennas are close together and are parallel. Under these conditions the voltage induced in the second antenna by the first, and in the first by the second, has its greatest value and causes the greatest induced current. The coupling decreases as the parallel antennas are moved farther apart.

The coupling between collinear antennas is comparatively small, and so the mutual impedance between such antennas is likewise small. It is not negligible, however.

PHASE RELATIONSHIPS

When the separation between two antennas is an appreciable fraction of a wavelength, a measurable amount of time elapses before the field from antenna 1 reaches antenna 2. There is a similar time lapse before the field set up by the current in number 2 gets back to induce a current in number 1. Hence the current induced in antenna 1 by antenna 2 will have a phase relationship with the original current in antenna 1 that depends on the spacing between the two antennas as well as other factors.

The induced current can range all the way from being completely in-phase with the original current to being completely out-of-phase with it. If the currents are in-phase, the total current is larger than the original current and the antenna feed-point impedance is reduced. If the currents are out-of-phase, the total current is smaller, and the impedance is increased.

Except in the special cases when the induced current is exactly in or out-of-phase with the original current, the induced current causes the phase of the total current to shift with respect to the applied voltage. Consequently, the presence of a second antenna nearby may cause the impedance of an otherwise resonant antenna to be reactive — that is, the antenna will be detuned from resonance — even though its self-impedance is entirely resistive. The amount of detuning depends on the magnitude and phase of the induced current.

TUNING CONDITIONS

The length and diameter of an antenna and its spacing from other antennas affect the magnitude and phase of the current induced into it, and therefore the magnitude and phase of currents it induces into other antennas. If it is a parasitic element, changing one or more of those physical properties is the only way to alter these currents.

7.1.3 Mutual Coupling and Gain

The mutual coupling between antennas or antenna elements can have a significant effect on the amount of current that will flow for a given amount of power supplied, which then determines the radiated field strength from the antenna in various directions. *But the total amount of power radiated from any combination of antennas and elements, excluding loss, must always equal the total amount of power applied.* So, any increases in field strength in some directions must result in corresponding decreases in others.

Loss can usually be analyzed separately, but sometimes mutual coupling can result in increased or decreased losses due to its effect on current amplitudes. One example is the W8JK antenna (see the **Broadside and End-Fire Dipole Arrays** chapter). In this antenna consisting of two parallel, closely spaced, out-of-phase dipole elements, mutual coupling greatly reduces the feed-point impedances and increases the currents for a given power input. The result is greater wire I^2R loss which converts more of the applied power to heat and consequently reduces the gain below what would be expected without loss. At very close spacings and/or small wire sizes, the amount of loss can equal or exceed the gain in the best directions, resulting in zero or negative gain in all directions.

Other things being equal, if the mutual coupling between two antennas is such that the currents are greater for the same total power than would be the case if there was no coupling, the power gain in the most favored direction will be greater than shown in Table 7.1. If mutual coupling results in a reduction of current, the gain will be less than if there was no coupling. But the same increase in current caused by mutual coupling that increases gain also increases loss, placing a practical limit to how much overall performance can be gained as in the example above.

The calculation of coupled impedance between antennas is a complex problem. Data for two simple but important cases are graphed in **Figures 7.8** and **7.9**. These graphs do not show the coupled impedance, but instead show a more useful quantity — the feed-point resistance measured at the center of an antenna as it is affected by the spacing between two antennas.

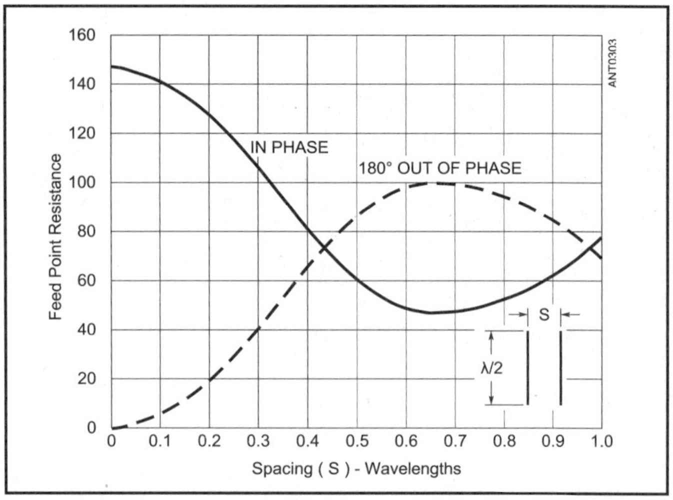

Figure 7.8 — Feed-point resistance measured at the center of one element as a function of the spacing between two parallel ½-λ self-resonant antenna elements. For ground-mounted ¼ λ vertical elements, divide these resistances by two.

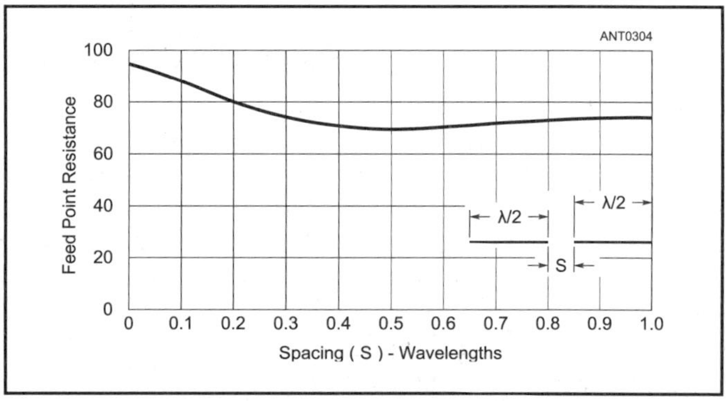

Figure 7.9 — Feed-point resistance measured at the center of one element as a function of the spacing between the ends of two collinear self-resonant ½-λ antenna elements operated in phase.

As shown by the solid curve in Figure 7.8, the feed-point resistance at the center of either antenna, when the two are *self-resonant*, parallel, and operated in-phase, decreases as the spacing between them is increased until the spacing is about 0.7 λ. This is a broadside array since the fields from the two elements arrive in-phase at a distant point placed on a line perpendicular to the line joining the two antennas. (Self-resonance means the antenna is resonant in the absence of mutual coupling with any other antenna.) The maximum gain is achieved from a pair of such elements when the spacing is in this region because the current is larger for the same power.

The dashed line in Figure 7.8, representing two antennas operated 180° out-of-phase (this is an end-fire array in which maximum radiation is in the plane of the elements), cannot be interpreted quite so simply. The feed-point resistance

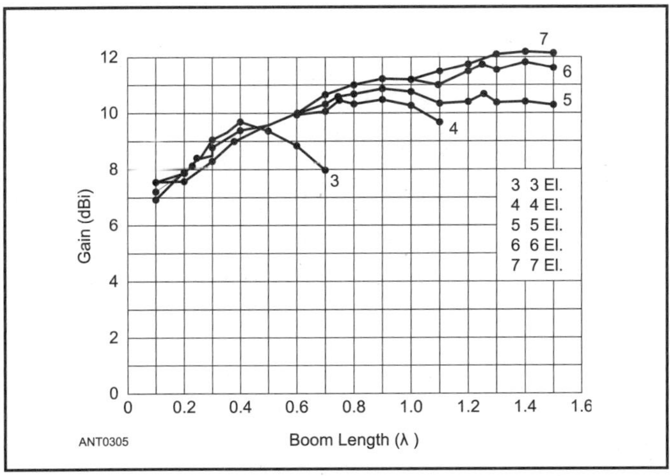

Figure 7.10 — Yagi gain for 3, 4, 5, 6 and 7-element beams as a function of boom length. (From *Yagi Antenna Design*, J. Lawson, W2PV.)

decreases with spacing down to less than about 0.6 λ. However, for the range of spacings considered, only when the spacing is 0.5 λ do the fields from the two antennas add up exactly in phase at a distant point in the favored direction. At smaller spacings the fields become increasingly out-of-phase, so the total field is less than the simple sum of the two. Smaller spacings thus decrease the gain while the reduction in feed-point resistance is increasing it. For a lossless antenna, the gain goes through a maximum when the spacing is in the region of 1/8 λ. However, when loss is considered, the low resistance and consequent high current for a given power input could significantly reduce the actual gain that is achieved.

The feed-point resistance curve for two collinear elements in-phase, Figure 7.9, shows that the feed-point resistance decreases and goes through a broad minimum in the region of 0.4 to 0.6 λ spacing between the adjacent ends of the antennas. As the minimum is not significantly less than the feed-point resistance of an isolated antenna, the gain will not exceed the gain calculated based on uncoupled antennas. That is, the best that two collinear elements will give, even with optimum spacing, is a power gain of about 2 (3 dB). When the separation between the ends is very small — the usual method of operation — the gain is reduced.

7.1.4 Gain and Array Dimensions

The gain of an array is principally determined by the dimensions of the array, so long as there are a minimum number of elements. A good example of this is the relationship between boom length, gain and number of elements for an array such as a Yagi. **Figure 7.10** compares the gain versus boom length for Yagis with different numbers of elements. For a given number of elements, notice that the gain increases as the boom length increases, up to a maximum. Beyond this point, longer boom lengths result in less gain for a given number of elements. This observation does not mean that it is always desirable to use only the minimum number of elements. Other considerations of array performance, such as front-to-back ratio, minor lobe amplitudes or operating bandwidth, may make it advantageous to use more than the minimum number of elements for a given array length. A specific example of this is illustrated in the discussion of the Half-square, Bobtail curtain, and Bruce array in the chapter **Broadside and End-Fire Dipole Arrays**.

In a broadside array the gain is a function of both the length and width of the array. The gain can be increased by adding more elements (with additional spacing) or by using longer elements (>λ/2), although the use of longer elements requires proper attention to current distribution in the elements. In general, in a broadside array the element spacing that gives maximum gain for a minimum number of elements is in the range of 0.5 to 0.7 λ. Broadside arrays with elements spaced for maximum gain will frequently have significant side lobes and associated narrowing of the main lobe beamwidth. Side lobes can be reduced by using more than the minimum number of elements, spaced closer than the maximum gain distance.

Additional gain can be obtained by expanding the array into a third dimension. An example of this is the stacking of end-fire arrays in a broadside configuration. In the case of stacked short end-fire arrays, maximum gain occurs with spacings in the region of 0.5 to 0.7 λ. However, for longer higher-gain end-fire arrays, larger spacing is required to achieve maximum gain. This is important in VHF and UHF arrays, which often use long-boom Yagis. (See the discussions of stacking Yagi antennas in the **HF Antenna System Design** and **VHF, UHF, and Microwave Antenna Systems** chapters.)

7.2 Driven Arrays

Phenomena described in the preceding section apply to multielement arrays of both types, driven and parasitic, since mutual coupling occurs regardless. However, there are special considerations for driven arrays that do not necessarily apply to parasitic arrays, and vice versa. Such considerations for Yagi and quad parasitic arrays are presented in the **HF Beam Antennas** chapter. The remainder of this chapter is devoted to driven arrays.

Driven arrays in general are either broadside or end-fire, and may consist of collinear elements, parallel elements, or a combination of both. From a practical standpoint, the maximum number of usable elements depends on the frequency and the space available for the antenna. Fairly elaborate arrays, using as many as 16 or even 32 elements, can be installed in a rather small space when the operating frequency is in the VHF range and more at UHF. At lower frequencies the construction of antennas with a large number of elements is impractical for most amateurs.

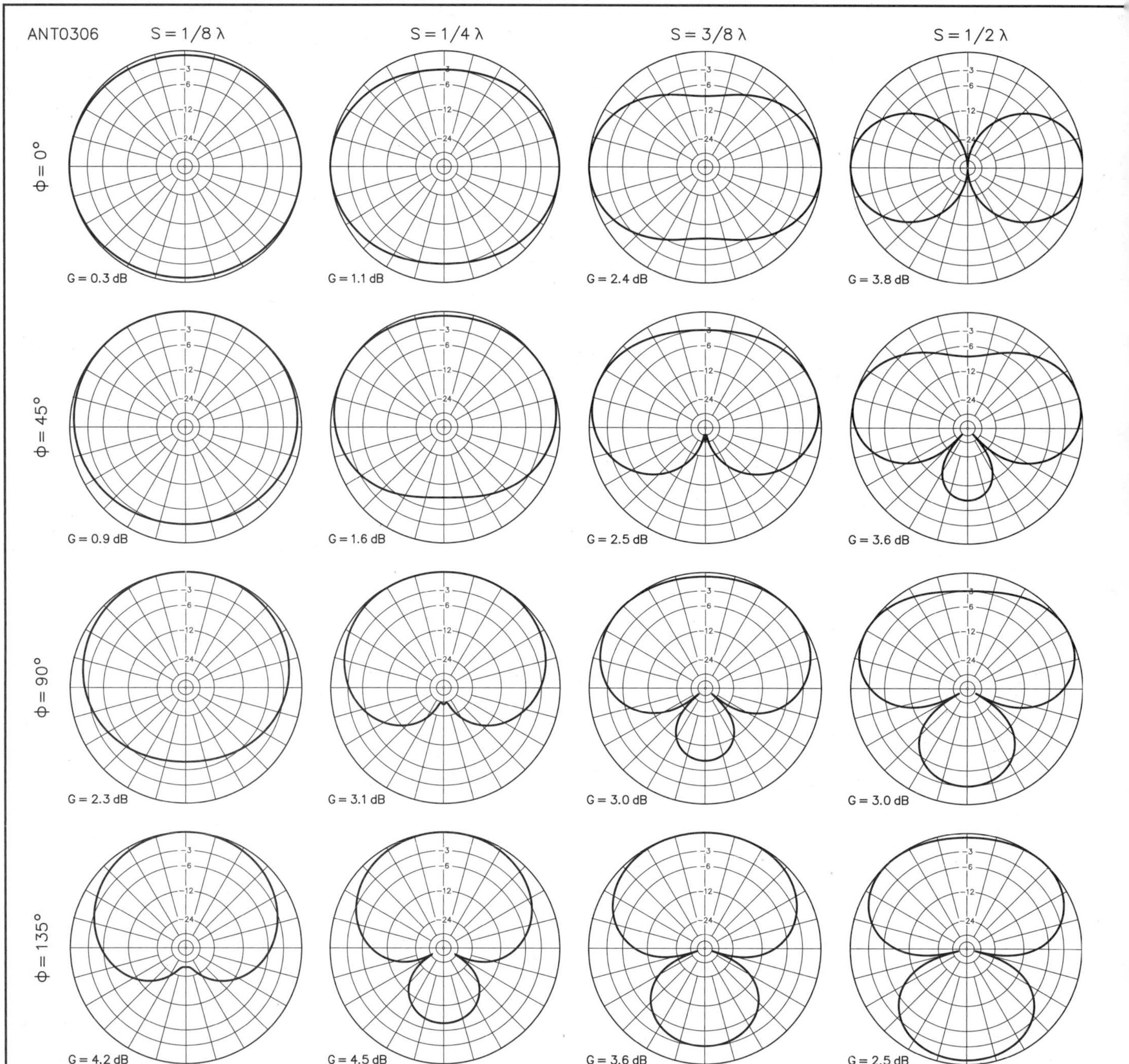

Figure 7.11 — H plane patterns of two identical parallel driven elements, spaced and phased as indicated (S = spacing, ϕ = phasing). The elements are aligned with the vertical (0°-180°) axis, and the element nearer the 0° direction (top of page) is of lagging phase at angles other than 0°. The two elements are assumed to be thin and self-resonant, with equal-amplitude currents flowing at the feed point. See text regarding current distributions. The gain figure associated with each pattern indicates that of the array over a single

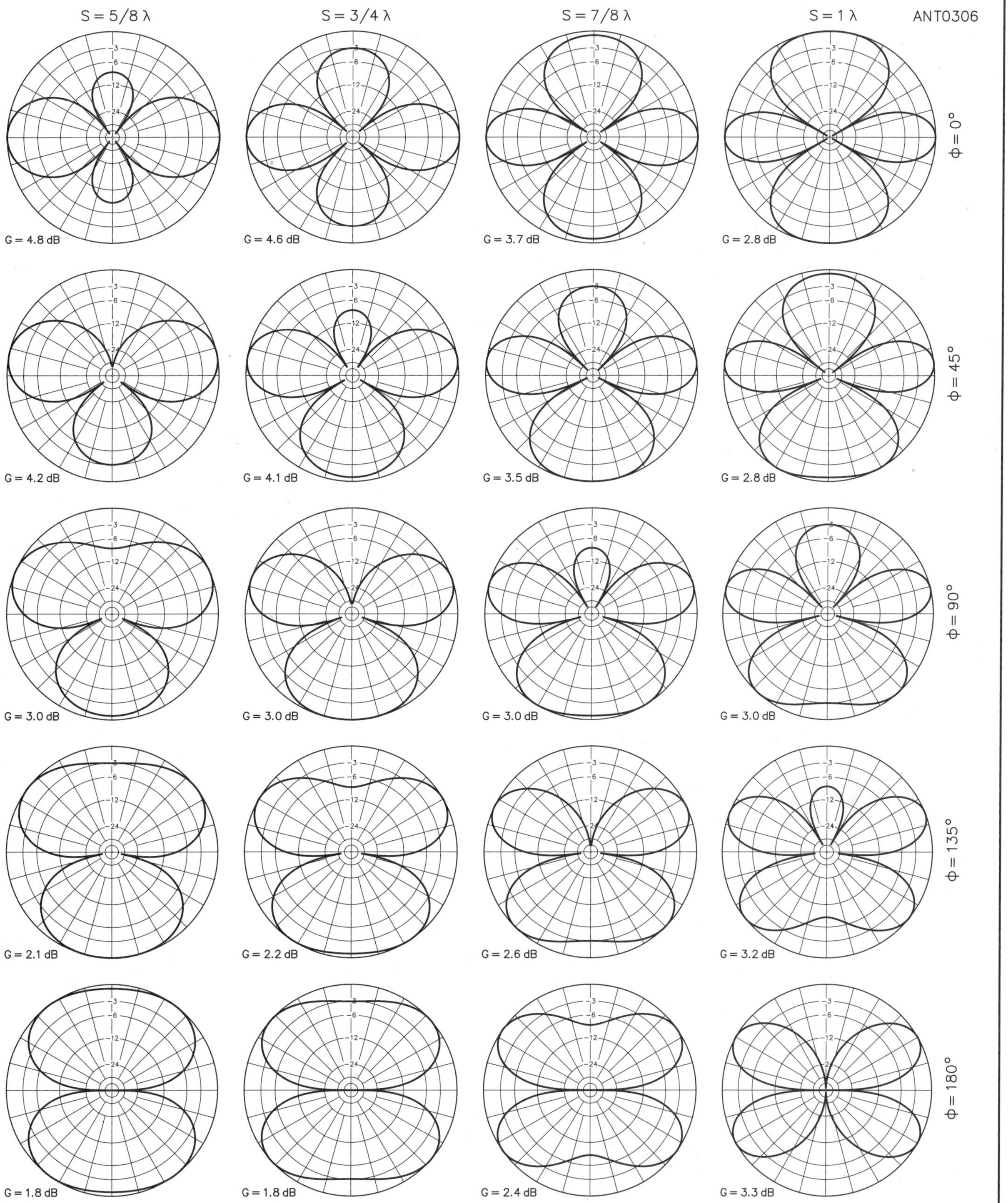

element. The plots represent the horizontal or azimuth pattern at a 0° elevation angle of two 1/4-λ vertical elements over a perfect conductor, or the free-space vertical or elevation pattern of two horizontal 1/2-λ elements when viewed on end, with one element above the other. (Patterns computed with *ELNEC* — the predecessor to *EZNEC*.)

Of course, the simplest of driven arrays is one with just two elements. If the elements are collinear, they are always fed in-phase. The effects of mutual coupling are not great, as illustrated in Figure 7.9. Therefore, feeding power to each element in the presence of the other presents no significant problems. But this may not be the case when the elements are parallel to each other. However, because the combination of spacing and phasing arrangements for parallel elements is infinite, the number of possible radiation patterns is endless.

This is illustrated in **Figure 7.11** for arrays of just two elements. When the elements are fed in-phase, a broadside pattern always results. At spacings of less than 5λ/8 with the elements fed 180° out-of-phase, an end-fire pattern always results. With intermediate amounts of phase difference, the results cannot be so simply stated. Patterns evolve that are not symmetrical in all four quadrants.

Because of the effects of mutual coupling between the two driven elements, for a given power input greater or lesser currents will flow in each element with changes in spacing and phasing as described earlier. This, in turn, affects the gain of the array in a way that cannot be shown merely by plotting the *shapes* of the patterns, as has been done in Figure 7.11. Therefore, supplemental gain information is also shown in Figure 7.11, adjacent to the pattern plot for each combination of spacing and phasing. The gain figures shown are referenced to a single element. For example, a pair of elements fed 90° out of phase with a spacing of λ/4 will have a gain in the direction of maximum radiation of 3.1 dB over a single element.

It is characteristic of broadside arrays that the power gain is proportional to the length of the array but is substantially independent of the number of elements used, *provided the optimum element spacing is not exceeded.* This means, for example, that a five-element array and a six-element array will have nearly the same gain, provided the elements in both are spaced so the overall array length is the same. Although this principle is seldom used for the purpose of reducing the number of elements because of complications introduced in feeding power to each element in the proper phase, it does illustrate the fact that there is nothing to be gained, in terms of more gain, by increasing the number of elements if the space occupied by the antenna is not increased proportionally.

The maximum gain in the smallest linear dimensions will result when the antenna combines both broadside and end-fire directivity and uses both parallel and collinear elements. In this way the antenna is spread over a greater volume of space, which has the same effect as extending its length to a much greater extent in one linear direction.

7.2.1 Current Distribution in Phased Arrays

In the plots of Figure 7.11, the two elements are assumed to be identical and self-resonant. In addition, currents of equal amplitude are assumed to be flowing at the feed point of each element, a condition that most often will not exist in practice without devoting special consideration to the feeder system. Such considerations are discussed in the next section of this chapter.

Most literature for radio amateurs concerning phased arrays assumes that if all elements in the array are identical, the *current distribution* (the way the current changes along the element) in all the elements will be identical. This distribution is presumed to be that of a single, isolated element, or nearly sinusoidal. However, information published in the professional literature as early as the 1940s indicates the existence of dissimilar current distributions among the elements of phased arrays. (See Harrison and King entries in the References and Bibliography.) Lewallen, in the July 1990 *QST*, pointed out the causes and effects of dissimilar current distributions.

In essence, even though the two elements in a phased array may be identical and have exactly equal currents of the desired phase flowing *at the feed point*, the amplitude and phase relationships degenerate with departure from the feed point. This happens any time the phase relationship is not 0° or 180°. Thus, the field strengths produced at a distant point by the individual elements may differ. This is because the field from each element is determined by the *distribution* of the current, as well as its magnitude and phase. (Note that the current itself flows in the familiar sine wave. Distribution is the magnitude of that sine wave along the element.)

The effects are minimal with shortened elements — verticals less than λ/4 or dipoles less than λ/2 long. The effects on radiation patterns begin to show at lengths greater than resonance and become profound with longer elements — λ/2 or longer verticals and 1 λ or longer center-fed elements. These effects are less pronounced with thin elements. The amplitude and phase degeneration takes place because the shape of the current distribution of the currents in the array elements becomes farther from being sinusoidal. This is true even in two-element arrays with phasing of 0° or 180°, although in these two special cases they do remain identical so don't affect the pattern.

The pattern plots of Figure 7.11 take element current distributions into account. The visible results of dissimilar distributions are incomplete nulls in some patterns and the development of very small minor lobes in others. For example, the pattern for a phased array with 90° spacing and 90° phasing has traditionally been published in amateur literature as a cardioid with a perfect null in the rear direction. Figure 7.11, calculated for 7.15 MHz self-resonant dipoles of #12 AWG wire in free space, shows a minor lobe at the rear and only a 33 dB front-to-back ratio due to the current distribution difference between the elements that results from mutual coupling. Element current distribution and its effects are discussed further in the section "Feeding Phased Arrays" later in this chapter.

7.3 Phased Array Techniques

Phased arrays of vertical elements have become increasingly popular for amateur use, particularly on the lower frequency bands where they provide one of the few practical methods to obtain substantial gain and directivity.

The operation and limitations of phased arrays, how to design feed systems to make them work properly and how to make necessary tests and adjustments are discussed in the pages that follow. The examples deal primarily with vertical HF arrays, but the principles apply to VHF and UHF arrays and arrays made from other element types as well. Unless otherwise noted, in the rest of this chapter, the term "phased array" will refer to arrays of vertical ground-plane elements. Step-by-step instructions for some examples are given in the sidebar *EZNEC* Examples.

7.3.1 Overview

Much of this chapter is devoted to techniques for feeding phased arrays. This is not a simple problem only of connecting array elements through "phasing lines" — transmission lines of the desired electrical lengths. Except for a very few special cases, this approach won't achieve the desired array pattern. Other proposed universal solutions, such as hybrid couplers or Wilkinson or other power dividers, also usually fail to achieve the necessary phasing in almost all cases.

Very briefly, the reason why the simple phasing line approach fails is that the delay of current or voltage in a transmission line equals the line's electrical length only if the line is terminated in its characteristic impedance. And in phased arrays, element feed-point impedances are profoundly affected by mutual coupling.

Consequently, even if each element has the correct impedance when isolated, it won't when all elements are excited. Furthermore, transmission lines that are not terminated in their characteristic impedance will transform both the voltage and current magnitude. The net result is that the array elements will have neither the correct magnitudes nor phases of current necessary for proper operation except in a few special cases. This isn't a minor effect of concern only to perfectionists, but often a major one that causes significant pattern distortion and poor or mis-located nulls.

7.3.2 Fundamentals of Phased Vertical Arrays

The performance of a phased array is determined by several factors. Most significant among these are the characteristics of a single element, reinforcement or cancellation of the fields from the elements and the effects of mutual coupling. To understand the operation of phased arrays, it is first necessary to understand the operation of a single antenna element.

Of primary importance is the strength of the field produced by the element, assumed to be a vertical ground-plane. The field radiated from a linear (straight) element, such as a dipole or vertical monopole, is proportional to the sum of the elementary currents flowing in each part of the antenna element. For this discussion it is important to understand what determines the current in a single element.

The amount of current flowing at the base of a ground mounted vertical or ground-plane antenna is given by the familiar formula

$$I = \sqrt{\frac{P}{R}}$$

where

P is the power supplied to the antenna

R is the feed-point resistance.

R consists of two parts, the loss resistance, and the radiation resistance. The loss resistance, R_L, includes losses in the conductor, in the matching and loading components and dominantly (in the case of ground-mounted verticals) in ground losses. The power consumed by the radiation resistance, R_r, is actually the power that is radiated, so maximizing the power consumed by the radiation resistance is desirable. However, the power consumed by the loss resistance truly is dissipated and lost as heat, so resistive losses should be made as small as possible.

The radiation resistance of an element can be derived from electromagnetic field theory, being a function of antenna length, diameter, and geometry. Graphs of radiation resistance versus antenna length are given in the **Dipoles and Monopoles** chapter. The radiation resistance of a thin resonant λ/4 ground-mounted vertical is about 36 Ω. A resonant λ/2 dipole in free space has a radiation resistance of twice this amount, about 73 Ω. Reducing the antenna lengths by one-half drops the radiation resistances to approximately 7 and 14 Ω, respectively.

The radiation resistance of a large variety of antennas can easily be determined by using *EZNEC*. The radiation resistance is simply the feed-point resistance (the resistive part of the feed-point impedance) when all losses have been set to zero. For ground mounted verticals use Perfect or MININEC-type ground to assure zero ground loss. This is one of the powerful benefits of modeling. Because it's not possible to set all the loss resistances of a real antenna to zero, the radiation resistance of an antenna can't directly be measured unless the losses are known to be negligible. Accurate measurements can be difficult to make in any case.

RADIATION EFFICIENCY

To generate a stronger field from a given radiator, it is necessary to increase the power P (the brute-force solution), decrease the loss resistance R_L (by putting in a more elaborate ground system for a vertical, for instance) or to somehow decrease the radiation resistance R_r so more current will flow with a given power input. This can be seen by expanding the formula for feed-point current as

$$I = \sqrt{\frac{P}{R_r + R_L}}$$

EZNEC Examples

This appendix contains step-by-step procedures using *EZNEC* to illustrate various topics discussed in this chapter. Instructions for getting and installing *EZNEC* and the various modeling files described in the text can be found in the previous sidebar "Using *EZNEC*."

EZNEC Example — Mutual Coupling and Feed-Point Impedance

This example illustrates the effect of mutual coupling on feed-point impedance. Open the *AntBk_Cardioid.EZ* file, which is a 2-element vertical array mounted over "perfect" ground and fed with ideal current sources that are equal in amplitude and 90° different in phase. Click the **VIEW ANT** button to see a diagram of the antenna, a 2-element array of vertical elements. Click on the **WIRES** line in the main window to open the **WIRES** Window. Click the button at the left of the **WIRE 2** line, and then press the **DELETE** key on your keyboard to delete wire #2. After clicking **OK**, note that one of the verticals (and the source at its base) has disappeared from the **VIEW ANTENNA** display, leaving a single element. Click **SRC DAT** and note that the feed-point impedance of this single vertical is about $36 + j\,0\ \Omega$ — it's very nearly resonant.

Next, in the **WIRES** Window, open the **EDIT** menu at the top and click **UNDO DELETE WIRE(S)** to restore the second element. Click **SRC DAT** again and notice that the feed-point impedance of wire #1 is now about $20 - j\,19\ \Omega$. The feed-point impedance of the second element, which is identical to the first, is about $51 + j\,19\ \Omega$. This difference between elements and the change from the self-impedance of $36 + j\,0\ \Omega$ is due to mutual coupling. As you see, it's not at all a minor effect.

As an additional exercise, change the magnitude or phase angle of the source at the base of wire #2 (click on the **SOURCES** line in the main window to open the **SOURCES** Window), then click **SRC DAT** after making changes to see how this changes the feed-point impedances of both elements.

EZNEC Example — Nulls

This example illustrates the effect of current magnitude on nulls and gain. Again, open the *AntBk_Cardioid.EZ* file. Click the **FF PLOT** button to generate the azimuth pattern of an ideal array. Save the plot for future reference as follows: In the plot window, open the **FILE** menu and select **SAVE TRACE AS**. Enter the name **CARDIOID** and click **SAVE**. Now, in the main window click on the **SOURCES** line to open the **SOURCES** Window. Change the amplitude of source 1 from 1 to 1.1, and of source 2 from 1 to 0.9 and press **ENTER** on your keyboard so that *EZNEC* will accept the changes.

Click **FF PLOT** to generate a pattern with the new currents. In the plot window, open the File menu and select **ADD TRACE**. Enter the name **CARDIOID** and click **OPEN**. You should now see the original plot and new plot overlaid. Notice that the null is much less deep with the altered currents, but the forward patterns are nearly identical. By clicking on the names of the traces in the upper left, **PRIMARY** and **CARDIOID**, you can see in turn the gain and front-to-back ratio of each of the traces. The original, **CARDIOID**, has a front-to-back ratio of about 32 dB, while Primary, the new plot, has a ratio of about 22 dB. The forward gain, however, differs by only 0.03 dB, a completely insignificant amount.

EZNEC Example — "Phasing-Line" Feed

This example illustrates the effect of using a "phasing-line" feed with no regard for the feed-point impedance changes caused by mutual coupling. Open the file *AntBk_Cardioid_TL_Deg_Example.EZ*. This is a model of an array fed with transmission lines whose lengths were designed in a previous section using the *Arrayfeed1* program to account for the actual load impedances of elements in a phased array (See Figure 7.21). In that section, the model transmission line lengths were specified in feet but in this model, they are specified in electrical degrees for convenience in illustrating the point of this example. Unlike real lines, degree-specified model lines will change physical length with frequency, so this model shouldn't be used to evaluate array performance at different frequencies.

In the main window, click on the **TRANS LINES** line to open the Transmission Lines window. In it you can see that the lengths of the feed lines, both of which are connected to the same source via Virtual Segment V1, are about 69° and 156°. This is a difference of 87° which in this case is close to the desired 90° current phase angle difference, but that's coincidental and in many cases the length difference is not close. (For example, the lines in the *EZNEC* example file *CardTL.EZ* have a length difference of about 74° for 90° phasing.) As we'll see, just any two lengths differing by about 90° won't produce the pattern we're expecting.

In the main window, click the **CURRENTS** button and look at the current shown for segment 1 of wires 1 and 2. These are the currents at the element feed points. The element 2: element 1 ratio of the magnitude of currents is 0.56316/0.56314 = 1.0000, and the phase difference is –146.58°– (–56.57°) = –90.01°. The current ratio is almost exactly correct.

Let's try a couple of other length of "phasing lines" differing in length by 90°. First, click the **FF PLOT** button to generate the azimuth pattern of the original model. Save the plot for future reference as follows: In the 2D Plot window, open the **FILE** menu and select **SAVE TRACE AS**. Enter the name **CARDTL** and click **SAVE**. Now in the **TRANSMISSION LINES** Window, change the length of line number 1 from 68.56° to 90°. *Important:* In the line 1 Length box, enter **90D** to make the line 90° long. If you omit the "d," it will become 90 feet long! Similarly, change the length of line 2 to 180° by entering **180D** in the line 2 Length box, then press **ENTER** on your keyboard so that *EZNEC* will accept the last change.

Click **FF PLOT** to generate a pattern with the new line lengths. In the plot window, open the **FILE** menu and select **ADD TRACE**. Enter the name **CARDTL** and click **OPEN**. You should now see the original plot and new plot overlaid together. Notice that the gain of the modified model is about 0.3 dB greater than the original, but the front-to-back ratio has degraded by 18 dB.

Experiment with different combinations of line lengths that differ by 90° — for example, 45° and 135° (don't forget the 'd'!), or change the impedance of one or both lines and you'll see that you can get a wide variety of patterns. None, however, are likely to be as close to the ideal cardioid pattern as the original, and the front-to-back ratio suffers the worst from an incorrect current ratio.

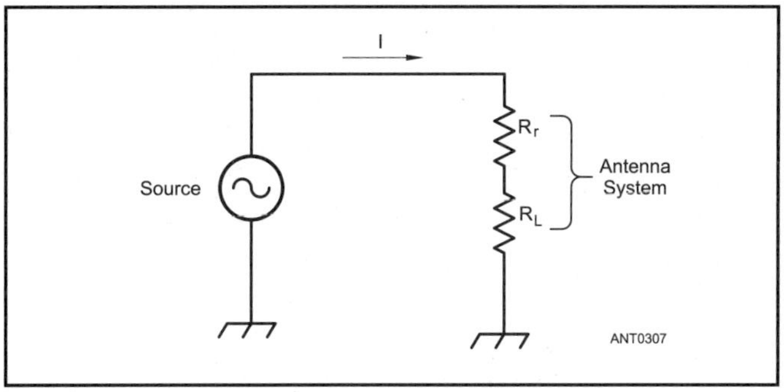

Figure 7.12 — Simplified equivalent circuit for a single-element resonant antenna. Rr represents the radiation resistance, and R_L the ohmic losses in the total antenna system.

Splitting the feed-point resistance into components R_r and R_L easily leads to an understanding of element efficiency. The efficiency of an element is the proportion of the total applied power that is actually radiated and not converted to heat by losses. The roles of R_r and R_L in determining efficiency can be seen by analyzing a simple equivalent circuit, shown in **Figure 7.12**.

The power consumed by R_r (the radiated power) equals I^2R_r. The total power supplied to the antenna system is

$$P = I^2 (R_r + R_L)$$

so the efficiency (the fraction of supplied power that is radiated) is

$$\text{Eff}(\eta) = \frac{I^2 R_r}{I^2 (R_r + R_L)} = \frac{R_r}{R_r + R_L}$$

Efficiency is frequently expressed in percent but expressing it in decibels relative to a 100%-efficient radiator gives a better idea of what to expect in the way of signal strength. The field strength of an element relative to a lossless but otherwise identical element is

$$\text{Relative field strength} = 10 \log \frac{R_r}{R_r + R_L} \text{ dB}$$

This will, of course, always be a negative value if R_r isn't zero.

For example, information presented by Sevick in March 1973 *QST* shows that a λ/4 ground-mounted vertical antenna with four 0.2 λ radials has a feed-point resistance of about 65 Ω (see the References and Bibliography section). The efficiency of such a system is 36/65 = 55.4%. It is rather disheartening to think that, of 100 W fed to the antenna, only 55 W is being radiated, with the remainder literally warming up the ground. Yet the signal will be only 10 log (36/65) = –2.57 dB relative to the same vertical with a perfect ground system. In view of this information, trading a small reduction in signal strength for lower cost and greater simplicity may become an attractive consideration.

So far, only the current at the base of a resonant antenna has been discussed, but the field is proportional to the sum of currents in each tiny part of the antenna. The field is a function of not only the magnitude of current flowing at the feed point, but also the distribution of current along the radiator and the length of the radiator. Nothing can be done at the feed point to change the current distribution, so for a given element the field strength is proportional to the feed-point current. However, changing the radiator length or loading it at some point other than the feed point will change the current distribution.

More information on shortened or loaded radiators may be found in the **Dipoles and Monopoles** and **General-Purpose MF and HF Antennas** chapters and in the References and Bibliography section. The current distribution is also changed by mutual coupling to other array elements, although for most arrays this has only a minor effect on the pattern. This is discussed later in more detail. A few other important facts follow.

1) If there is no loss, the field from even an infinitesimally short radiator is less than ½ dB weaker than the field from a half-wave dipole or quarter-wave vertical. Without loss, all the supplied power is radiated regardless of the antenna length, so the only factor influencing gain is the slight difference in the patterns of very short and λ/2 antennas which results from different current distributions. A short antenna has a very low radiation resistance, resulting in a heavy current flow over its short length. In the absence of loss, this generates a field strength comparable to that of a longer antenna with less current. Where loss is present — that is, in practical antennas — shorter radiators usually don't do so well, since the low radiation resistance leads to lower efficiency for a given loss resistance. Nevertheless, reasonably short antennas can achieve good efficiency if care is taken to minimize loss.

2) Caution must be used in calculating the efficiency of folded antennas. Folding transforms both the radiation resistance and loss resistance by the same factor, so their ratio and therefore the efficiency remains the same. It's easy to show that in a ground-mounted vertical antenna, folding reduces the current flowing from the feed line to the ground system by a factor of two due to the impedance transformation. However, the folded antenna has an additional connection to ground, which also carries half the original ground current. The result is that the same amount of current flows into the ground system whether unfolded or folded, resulting in the same ground system loss. Analyses purporting to show otherwise invariably transform the radiation resistance but neglect to also transform the loss resistance and therefore reach an incorrect conclusion.

3) The current flowing in an element with a given power input can be increased or decreased by mutual coupling to other elements. The effect is equivalent to changing the element radiation resistance. Mutual coupling is sometimes thought of as a minor effect, but often it is not minor!

FIELD REINFORCEMENT AND CANCELLATION

The mechanism by which phased arrays produce gain, and the role of mutual coupling in determining gain, were covered earlier in this chapter. One important point that can't be emphasized enough is that all antennas must abide by the law of *conservation of energy*. No antenna can radiate more power than supplied to it, and in fact always radiates less. The total amount of power it radiates is the amount it's supplied less the amount lost as heat. This is true of all antennas, from the smallest "rubber duck" to the most gigantic array.

GAIN

Gain is strictly a relative measure, so the term is completely meaningless unless accompanied by knowledge of just what it is relative to. One useful measure for phased array gain is *gain relative to a single similar element.* This is the increase in signal strength that would be obtained by replacing a single element by an array made from elements just like it. In some instances, such as investigating what happens to array performance when all elements become more lossy, it's useful to state gain relative to a more absolute, although unattainable standard: a lossless element.

And the most universal reference for gain is another unattainable standard, the isotropic radiator. This fictional antenna radiates absolutely equally in all directions. It's very useful because the field strength resulting from any power input is readily calculated, so if the gain relative to this standard is known, the field strength is also known for any radiated power. Gain relative to this reference is referred to as dBi, and it's the standard used by most modeling programs including *EZNEC.* To find the gain of an array relative to a single element or other reference antenna such as a dipole, model both the array and the single element or other reference antenna in the same environment and subtract their dBi gains. Don't rely on some assumption about the gain of a single element — many people assume values that can be very wrong.

NULLS

Pattern nulls are very often more important to users of phased arrays than gain because of their importance in reducing noise and interference when receiving. Consequently, a good deal of emphasis is, and should be, placed on achieving good pattern nulls. Unfortunately, good nulls are much more difficult to achieve than positive gain and they are much more sensitive to array and feed-system imperfections.

As an illustration, consider two elements that each produce a field strength of exactly 1 millivolt per meter (mV/m) at some distance many wavelengths from the array. In the direction in which the fields from the elements are in-phase, a total field of 2 mV/m results. In the direction in which they're out-of-phase, zero field results. The ratio of maximum to minimum field strength of this array is 2/0, or infinity.

Now suppose, instead, that one field is 10% high and the other 10% low — 1.1 and 0.9 mV/m, respectively. In the forward direction, the field strength is still 2 mV/m, but in the canceling direction, the field will be 0.2 mV/m. The front-to-back ratio of field strengths has dropped from infinity to 2/0.2, or 20 dB. (Actually, slightly more power is required to redistribute the field strengths this way, so the forward gain is reduced — but by only a small amount, less than 0.1 dB.) For most arrays, unequal fields from the elements have a minor effect on forward gain but a major effect on pattern nulls. This is illustrated by ***EZNEC* Example — Nulls** in the sidebar

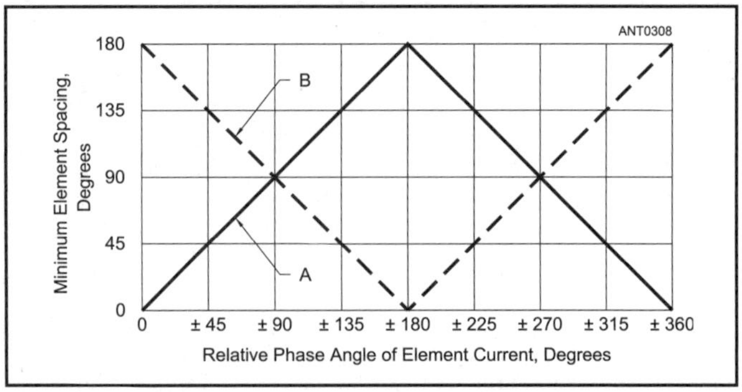

Figure 7.13 — Minimum element spacing required for total field reinforcement, curve A, or total field cancellation, curve B. Total cancellation results in pattern nulls in one or more directions. Total reinforcement does not necessarily mean there is gain over a single element, as the effects of loss and mutual coupling must also be considered.

"*EZNEC* Examples". (Refer to this sidebar for other ***EZNEC* Example** exercises.)

But deep nulls aren't assured even with perfect current balance. **Figure 7.13** shows the minimum spacing required for total field reinforcement or cancellation. If the element spacing isn't adequate, there may be no direction in which the fields are completely out-of-phase (see curve B of Figure 7.13). Slight physical and environmental differences between elements will invariably affect null depths, and null depths will also vary with elevation angle.

However, a properly designed and fed array can produce very impressive nulls. The key to producing good nulls, like producing gain, is controlling the strengths and phases of the fields from the elements. Just how to accomplish that is the subject of most of the remainder of this section. But be sure to keep in mind that producing good nulls is generally a much more difficult task than producing close to the predicted gain.

MUTUAL COUPLING

Mutual coupling was discussed briefly earlier in this chapter. Because it has an important and profound effect on both the performance and feed system design of phased arrays, it will be covered in greater depth here.

Mutual coupling refers to the effects which the elements in an array have on each other. Mutual coupling can occur intentionally or entirely unintentionally. People with multiple antennas on a small lot (or car top) often discover that a better description of their system is a single antenna with multiple feed points. Current is induced in conductors in various antennas by mutual coupling, causing them to act like parasitic elements which re-radiate and distort the antenna's pattern. Mutual coupling and its effects are present whether or not the elements are driven.

Suppose that two driven elements are many wavelengths from each other. Each has some voltage and current at its feed point. For each element, the ratio of this voltage to current is the element self-impedance. If the elements are brought close to each other, the current in each element will change in magnitude and phase because of coupling with the field from the other element. The field from the first element changes the current in the second. This changes the field from the second which alters the current in the first, and so forth creating an equilibrium condition in which the currents in all elements (hence, their fields) are totally interdependent.

The feed-point impedances of all elements also are changed from their values when far apart, and all are dependent on each other. In a driven array, the changes in feed-point impedances can cause additional changes in element currents because the operation of many feed systems depends on the element feed-point impedances. Significant mutual coupling occurs at spacings as great as a wavelength or more.

Connecting the elements to a feed system to form a driven array does not eliminate the effects of mutual coupling. In fact, in many driven arrays the mutual coupling has a greater effect on antenna operation than the feed system does. All feed-system designs must account for the impedance changes caused by mutual coupling if the desired current balance and phasing are to be achieved.

Several general statements can be made regarding the effects of mutual coupling on phased-array systems (unless loss is high enough to swamp mutual coupling effects as discussed in the next section).

1) The resistances and reactances of all elements of an array generally will be substantially different from the values when the elements are isolated (that is, very distant from other elements).

2) If the elements of a two-element array are identical and have equal currents that are in-phase or 180° out-of-phase, the feed-point impedances of the two elements will be equal. But they will be different than for an isolated element. If the two elements are part of a larger array, their impedances can be very different from each other.

3) If the elements of a two-element array have currents that are neither in-phase (0°) nor out-of-phase (180°), their feed-point impedances will not be equal. The difference will be considerable in typical amateur arrays.

4) The feed-point resistances of the elements in a closely spaced, 180° out-of-phase array will be very low, resulting in poor efficiency due to ohmic losses unless care is taken to minimize loss. This is also true for any other closely spaced array with significant predicted gain.

It's essential to realize that this is not a minor effect and one that can be overlooked or ignored. See ***EZNEC* Example — Mutual Coupling and Feed-Point Impedance** for an illustration of these phenomena.

LOSS RESISTANCE, MUTUAL COUPLING AND ANTENNA GAIN

Loss reduces the effects of mutual coupling because the feed-point impedance change (coupled impedance) resulting from mutual coupling is effectively in series with loss resistance. If the loss is great enough, two important results occur. First, the feed-point impedance becomes relatively independent of the presence of nearby current-carrying elements. This greatly simplifies feed system design — the simple "phasing-line" or hybrid-coupler feed system described below is adequate *provided that all elements are physically identical, and the feed point of each element is matched to the Z_0 of the feed line and, if used, the hybrid coupler.*

The impedance matching restrictions are necessary to ensure that the phasing line or hybrid coupler performs as expected. Identical elements are needed so that equal element currents will result in equal fields from the elements.

In the absence of mutual coupling effects, the maximum gain of an array of identical elements relative to a single (similarly lossy) element is simply 10 log(N), where N is the number of elements — providing that spacing is adequate for the fields to fully reinforce in some direction. If spacing is less, maximum gain will also be less. Of course, the array gain relative to a single *lossless* element will be very low, most likely a sizeable negative number when expressed in dB. So, intentionally introducing loss isn't a wise idea for a transmitting array. It is sometimes an advantageous thing to do for an HF receiving array, however, as explained in the following section.

Close spaced arrays with 180° out-of-phase elements such as the W8JK phased array (see the **Broadside and End-Fire Dipole Arrays** chapter) depend heavily on mutual coupling to achieve their gain because there are no directions in which the fields completely reinforce. Introduction of any loss to these arrays, which reduces the mutual coupling effects, has a profound effect on the gain. This is also true of many parasitic arrays. Consequently, parasitic or close-spaced driven arrays often produce disappointing results when made from grounded vertical elements unless each has a fairly elaborate (and therefore very low-loss) ground system.

If you place two low-loss elements very close together and feed them in-phase, mutual coupling reduces the array gain to essentially that of a single element, so there's no advantage to this configuration over a single element. However, if you have a single lossy element, for example a short vertical having a relatively poor ground system, you can improve the gain by up to almost 3 dB by adding a second, close spaced, element and ground system and feeding the two in-phase. Another way to look at this technique is that you're putting two equal ground system resistances in parallel, which effectively cuts the loss in half. The gain you can realize in practice depends on such things as the ground system overlap, but it might be a practical way to improve transmitting array performance in some situations.

7.3.3 Feeding Phased Arrays

The previous section explains why the fields from the elements must be very close to the ratios required by the array design. Since the field strengths are proportional to the currents in the elements, controlling the fields requires controlling the element currents. Since the desired current magnitude ratio is 1:1 for virtually all two-element and for most (but not all) larger amateur arrays, special attention is paid to methods of assuring equal element currents. But we will also examine other current ratios.

THE ROLE OF ELEMENT CURRENTS

The field from a conductor is proportional to the current flowing on it. So, if we're to control the relative strengths and phases of the fields from the elements, we have to control their currents. We usually do this by controlling the currents at the element feed points. But because the field from an element depends on the current everywhere along the element, elements having identical feed-point currents will produce different fields if they have different current distributions — that is, if the way the current varies along the lengths of the elements is different.

The previous section "Current Distribution in Phased Arrays" explained that mutual coupling alters the current distribution, so in many arrays the current distributions will be different on the elements and consequently the relationship between the elements' total fields won't be the same as that between the feed-point currents. Fortunately, this effect is relatively minor in thin, λ/4 vertical or λ/2 dipole elements. The most common arrays are made from elements in this category, so we can generally get very nearly the desired ratio of fields by creating the same ratio of feed-point currents. Exceptions are detailed in the following section.

FEED-POINT VS ELEMENT CURRENT

For most antennas, environmental factors are likely to cause greater performance anomalies than current distribution differences, and both can be corrected with minor feed system adjustments. The difference between field and feed-point current ratios can become very significant, however, if the elements are very fat and/or close to λ/2 (monopole) or 1 λ (dipole) long. In those cases, most of the feed systems described here won't produce the desired field ratios without major adjustment or modification, except in the special cases of 2-element arrays with identical elements having feed-point currents in-phase or 180° out-of-phase. In those special cases, the element current distributions are the same for the same reason the feed-point impedances are equal — each element "sees" exactly the same amplitude and phase field from the other element. This is explained later in the feed system sections.

To get an idea of just how large an element must be for the current distribution difference to disturb the pattern of an array with correct feed-point currents, a two-element cardioid array of quarter wave vertical elements was modeled at 10 MHz. With thin, 0.1-inch diameter elements the front-to-back ratio was 35 dB, the very small reverse lobe caused by slightly unequal element current distributions. Increasing the element diameter to 20 inches decreased the front-to-back ratio to 20 dB. Returning the front-to-back ratio of the array of 20-inch elements to >35 dB required changing the feed-point current ratio from the nominal value of 1.0 at an angle of 90° to 0.88 at 83°.

The same array was again modeled with 0.1-inch diameter elements where it has a front-to-back ratio of 35 dB, then the elements were lengthened. The front-to-back ratio dropped to 20 dB at an element length of 36 feet, or about 0.37λ. In that case, adjustment of the feed-point current ratio to about 0.9 at about 83° restored a good front-to-back ratio.

In the discussion and development which follow, the assumption is made that the fields will be very nearly proportional to the feed-point currents. If the elements are fat or long enough to make this assumption untrue, some adjustment of feed-point current ratio will be necessary to achieve the desired pattern, particularly nulls. Most feed systems can be designed for any current ratio. Modeling will reveal the ratio required for the desired pattern, and then the feed system can be designed accordingly.

7.3.4 Common Phased Array Feed Systems

This section will first describe several popular approaches to feeding phased arrays that often don't produce the desired results and why they don't work as well as hoped. It also briefly discusses systems that could be used but often aren't appropriate or optimum for amateur arrays.

This will be followed in the next section by detailed descriptions of array feed systems that do produce the predicted element current ratios and array patterns.

THE "PHASING-LINE" APPROACH

For an array to produce the desired pattern, the element currents must have the required magnitude and the required phase relationship. As explained above, this can generally be achieved well enough by causing the feed-point currents to have that same relationship.

On the surface, this sounds easy — just make sure that the difference in electrical lengths of the feed lines to the elements equals the desired phase angle. Unfortunately, this approach doesn't usually achieve the desired result. The first problem is that the phase shift through the line is not necessarily equal to its electrical length but in general depends on the impedance terminating the line. The current (or, for that matter, voltage) delay in a transmission line is equal to its electrical length in only a few special cases — cases which don't exist in most amateur arrays. The impedance of an element in an array is frequently very different from the impedance of an isolated element and the impedances of all the elements in an array can be different from each other.

See ***EZNEC* Example — Mutual Coupling and Feed-Point Impedance** for a graphic illustration of the effect of mutual coupling on feed-point impedance. Also look at the Four-Square array example in the "Phased Array Design Examples" section. That array has one element with a *negative* feed-point resistance if ground loss is low. This means that the element is delivering power to, rather than absorbing power from, the feed system. It gets this power from the other elements via mutual coupling. Without mutual coupling, the feed-point resistance of that same element would be about 36 Ω plus ground loss.

So, because of mutual coupling, the elements seldom provide a matched load for the element feed lines. The effect of mismatch on phase shift can be seen in **Figure 7.14**. Observe what happens to the phase of the current and voltage on a line terminated by a purely resistive impedance that is lower than the characteristic impedance of the line (Figure 7.14A). At a point 45° from the load the current has advanced less than 45° and the voltage more than 45°. At 90° from the load both are advanced 90°. At 135° the current has advanced more and the voltage less than 135°. This apparent slowing down and

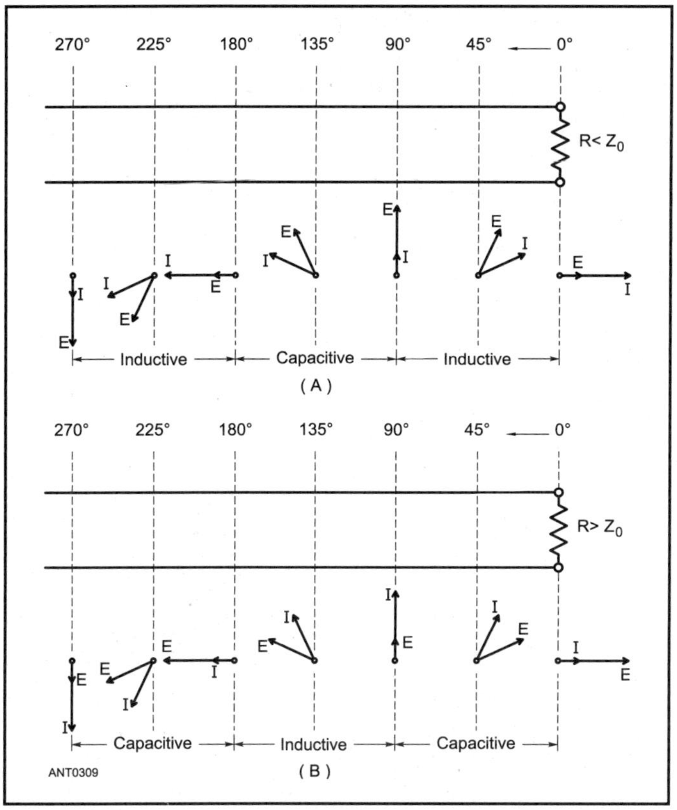

Figure 7.14 — Resultant voltages and currents along a mismatched line. At A, R less than Z0, and at B, R greater than Z0.

speeding up of the current and voltage waves is caused by interference between the forward and reflected waves. It occurs on any line that is not terminated with an impedance equal to its characteristic impedance. (Note that Z_0 of a lossless transmission line is purely resistive while Z_0 of a line with loss includes a reactive component. This reactive component is usually negligible unless loss is unusually high.) If the load resistance is greater than the characteristic impedance of the line as shown in Figure 7.14B, the voltage and current exchange angles. Adding reactance to the load causes additional phase shift. *The only cases in which the current (or voltage) delay is equal to the electrical length of the line are*:

1) When the line is flat; that is, terminated in a load equal to its characteristic impedance;

2) When the line length is an integral number of half wavelengths;

3) When the line length is an odd number of quarter wavelengths, and the load is purely resistive; and

4) When other specific lengths are used for specific load impedances.

These apply to most cases of the arrays described in the **Broadside and End-Fire Dipole Arrays** chapter.

Just how much phase error can be expected if two feed lines are simply hooked up to form an array? There is no simple answer. Some casually designed feed systems might deliver satisfactory results, but most will not. See the ***EZNEC* Example — "Phasing-Line" Feed** for the typical consequences of using this sort of feed system.

A second problem with simply connecting feed lines of different lengths to the elements is that the lines will change the *magnitudes* of the currents. The magnitude of the current (or voltage) out of a line does not equal the magnitude into that line, except in cases 1, 2, and 4 above. The feed systems presented later in this chapter assure currents that are correct in both magnitude and phase.

The elementary phasing-line approach *will* work in three very special but common situations. If the array consists of only two identical elements and those elements are fed in-phase, mutual coupling will modify the element impedances, but both will be modified by the same amount. Consequently, if the two elements are fed through equal-length transmission lines, the lines will transform and delay the currents equally and result in equal, in-phase currents at the element feed points.

Similarly, an array of two identical elements fed 180° out of phase will have the same feed-point impedances and can be fed with two lines of any length so long as one line is an electrical half wavelength longer than the other or one twin-lead type line is reversed. But this can't be extended to any two elements in a larger array, since mutual coupling to the other elements can result in different feed-point impedances. Methods will be described later which do assure a correct current ratio in this situation.

The third application in which the phasing-line approach works is in receiving arrays where the elements are very short in terms of wavelength and/or very lossy. In either of these cases, mutual coupling between elements is much less than an element's self-impedance. This allows the elements to be individually matched to the feed lines, with no significant change taking place when the elements are formed into an array. Under those conditions, the transmission lines can be matched, and the lines used as simple delay lines with easily predictable phase shift and with no transformation of current or voltage magnitude other than cable loss. This is discussed in the later section on receiving antennas.

ON4UN's Low-Band DXing (see References and Bibliography) describes a modified phasing-line feed system method which works in some special cases where the feed-point resistances are favorable to the approach. First, a quarter-wave transmission line is connected to each element. Then a shunt inductor or capacitor is added at the input of the line to the lagging element to make the impedance at that point purely resistive. If the resulting resistance is close to the characteristic impedance of an available transmission line (e.g., 50 or 75 Ω), a simple delay line can be used to feed that element. See the *EZNEC* model *AntBk_Cardioid_Modified_Phasing_Line_Feed.EZ* and its accompanying Antenna Notes file for an example. When impedance values allow this feed method, it saves only one component compared to the L network feed method and isn't fully adjustable. And it has one component more than the "simplest" method to be described. The bandwidth of arrays fed using this method isn't significantly different from the other feed systems, so there's no clear advantage to using it. However, it might be a viable approach under some circumstances. More design information is available in *Low-Band DXing*.

The interested reader is encouraged to consult *Low-Band DXing* as it contains examples of additional feed system design approaches that produce a desired element magnitude and phase ratio.

Many arrays *can* be correctly fed with a feed system consisting of only transmission lines, but the technique requires knowledge of the element feed-point impedances in a correctly fed array. Line lengths can then be computed that provide the correct ratio of currents into those particular load impedances. The line lengths generally differ by an amount that's considerably different from the element phase angle difference, and appropriate line lengths can't always be found for all arrays. This technique is described more fully in the section "The Simplest Phased Array Feed System…That Works" later in this chapter and illustrated in the examples in "Phased Array Design Examples."

THE WILKINSON DIVIDER

The *Wilkinson divider*, sometimes called the *Wilkinson power divider*, was once heavily promoted as a means to distribute power among the elements of a phased array. While it's a very useful device for other purposes, it won't produce the desired current ratios in antenna elements. In most phased arrays, element feed-point resistances are different and therefore require different amounts of power to achieve the desired equal magnitude currents. (See the section on mutual coupling above.) A Wilkinson divider is intended to deliver equal powers, not currents, to multiple loads. And it won't even do that when the load impedances are different. It might be useful in combining element outputs in receiving arrays in which element losses are high enough to swamp mutual coupling effects and to affect impedance matches.

THE HYBRID COUPLER

Hybrid couplers are promoted as solving the problem of achieving equal magnitude currents with a 90° phase difference between elements. Unfortunately, standard 90° hybrids provide equal magnitude, quadrature (90° phased) currents only when the load impedances are equal and correct. And this simply isn't true of arrays with quadrature-fed elements, except for arrays consisting of short and/or lossy elements, usually suitable only for receiving. In those arrays, the hybrid coupler can be useful for the same reasons as the phasing-line approach, discussed in an earlier section.

Hybrid couplers can, however, be useful for feeding transmitting or low-loss phased arrays if suitably modified to function when terminated with the particular impedances presented by an array's elements. In *Low-Band DXing*, methods are described to modify the standard 90° hybrid design to provide approximate hybrid functionality when terminated in typical phased array impedance values. Methods are quite involved, as evidenced by more than 20 pages devoted to the topic. It is important to realize that no passive network, including the hybrid coupler, can provide equal magnitude 90° phased currents in loads with arbitrary impedances. See the section "The Magic Bullet" below for more information.

"CROSSFIRE" FEED METHOD

Tom Rauch, W8JI, has described a "crossfire" feed system which can produce a deep null in one direction over an exceptionally wide bandwidth. This method, generally suitable only for lossy receiving arrays, is described in more detail in this chapter's section "Receiving Arrays and Broadbanding."

LARGE ARRAY FEED SYSTEMS

The author once worked on a radar system where the transmit array consisted of over 5,000 separate dipole elements and the receive array over 4,000 pairs of crossed dipoles, all over a metal reflecting plane which was the sloping side of a 140-foot-high building. In such large arrays, each element is in essentially the same environment as every other element except near the array edges, so almost all elements have nearly the same feed-point impedance. While producing the phase shifts and magnitude tapers is a considerable mathematical challenge, the problem of unequal element feed-point impedances can largely be ignored. Consequently, feed methods for these large arrays are generally not suitable for typical amateur arrays of a few elements. (See **microwaves101.com/encyclopedias/butler-matrix** for an example of how such large arrays are designed and operate.)

THE BROADCAST APPROACH

Networks can be designed to transform the element base impedances from their values in an excited array to, say, 50 Ω resistive. Then another network can be inserted at the junction of the feed lines to properly divide the power among the elements (not necessarily equally!). And finally, additional networks must be added to correct for the phase shifts and magnitude transformations of the other networks. This general approach is used by the broadcast industry, in installations that are typically adjusted only once for a particular frequency and pattern.

Although this technique can be used to correctly feed any type of array, design is difficult, and adjustment is tedious since all adjustments interact. When the relative currents and phasing are adjusted, the feed-point impedances change, which in turn affect the element currents and phasing, and so on. A further disadvantage of using this method is that switching the array direction is generally impossible. Information on applying this technique to amateur arrays can be found in Paul Lee's book on verticals, listed in the References and Bibliography.

THE "MAGIC BULLET"

Long ago, this chapter proposed a theoretical passive circuit that would provide equal-magnitude, 90° phased currents into two loads without respect to the load impedances. This would be a circuit that would guarantee exactly the correct currents in any two elements. In 1996, Kevin Schmidt, W9CF, formulated a mathematical proof that such a circuit — in fact, one resulting in any relative phase other than 0° or 180° — cannot exist if restricted to reciprocal elements. (That is, it can't exist unless directional components such as ferrite circulators are used.) Thus, to design a network to feed elements with currents of any other phase angle other than 0 or 180°, we must know the impedance of at least one element, and correct feed system operation depends on that impedance. There's no way around this requirement. The proof is available at the time of this writing at **w9cf.github.io/articles/magic/index.html**.

7.3.5 Recommended Feed Methods For Amateur Arrays

The following feed methods can produce element feed-point currents having a desired magnitude and phase relationship, resulting in desirable and predictable patterns. Most methods require knowing the feed-point impedance of all, or all but one, array elements *when the array element currents are the correct values*. This isn't possible to measure directly, because if the element currents were correct, the feed system would already be working properly, and no further design would be necessary.

Usually, the only practical way and by far the easiest way to get this information is by computer modeling. Modeling programs such as *EZNEC* allow you to construct an ideal array with perfect element currents then look at the resulting feed-point impedances. The alternative requires high accuracy measurements just to determine the self and mutual impedances. The coupled impedance is then calculated for each element from its self-impedance and the mutual impedances between it and all other elements. Finally, this is added to the self-impedance to get the feed-point impedance of that element and the process repeated for each element. The file "Manual Calculations for Arrays" in the online material contains equations and manual techniques from previous editions of *The ARRL Antenna Book* for those who are interested. You can also find a great deal of additional information in many of the texts listed in the References and Bibliography, particularly Jasik and Johnson.

Because of its simplicity and versatility, the computer modeling approach is highly recommended and it's the one used for the array design examples in this chapter.

The program *Arrayfeed1* does the calculations to simplify designing feed systems using this method and is used in the examples that follow. It's available in the online material and from **eznec.com/misc/Arrayfeed**.

CURRENT FORCING WITH λ/4 LINES — ELEMENTS IN-PHASE OR 180° OUT-OF-PHASE

The feed method introduced here has been used in its simplest form to feed television receiving antennas and other arrays, as presented by Jasik, pages 2-12 and 24-10 or Johnson, on his page 2-14. However, until first presented in the *ARRL Antenna Book*, this feed method was not widely applied to amateur arrays.

The method takes advantage of an interesting property of λ/4 transmission lines. (All references to lengths of lines are electrical length and lines are assumed to have negligible loss.) See **Figure 7.15**. The magnitude of the *current out* of a λ/4 transmission line is equal to the *input voltage* divided by the characteristic impedance of the line. This is independent of the load impedance. (Note that the power into and out of the line is the same, less losses in the line.) In addition, the phase

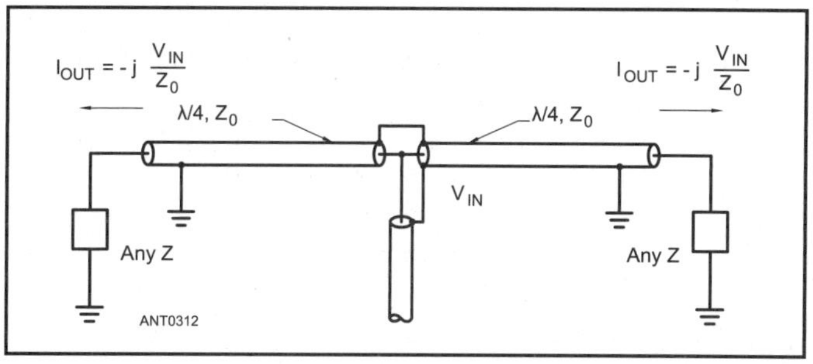

Figure 7.15 — A useful property of λ/4 transmission lines; see text. This property is utilized in the "current-forcing" method of feeding an array of coupled elements.

of the output current lags the phase of the input voltage by 90°, also independent of the load impedance. These properties can be used to advantage in feeding arrays with certain phase angles between elements.

If any number of loads are connected to a common driving point through λ/4 lines of equal impedance, the currents in the loads will be *forced* to be equal and in-phase, regardless of the load impedances. So, any number of in-phase elements can be correctly fed using this method, regardless of how their impedances might have been changed by mutual coupling. Arrays that require unequal currents can be fed through λ/4 lines of unequal impedances to achieve other current ratios.

The properties of λ/2 lines also are useful. Since the current out of a λ/2 line equals the input current shifted 180° regardless of the load impedance, any number of half wavelengths of line may be added to the basic λ/4, and the current and phase forcing property will be preserved. For example, if one element is fed through a λ/4 line and another element is fed from the same point through a 3λ/4 line of the same characteristic impedance, the currents in the two elements will be forced to be equal in magnitude and 180° out-of-phase, regardless of the feed-point impedances of the elements.

If an array of two, and only two, identical elements is fed in-phase or 180° out-of-phase with equal magnitude currents, both elements have the same feed-point impedance. The reason is that each element "sees" exactly the same thing when looking at the other. In an in-phase array, each sees another element with an identical current; in an out-of-phase array, each sees another element with an equal magnitude current that's 180° out-of-phase, the same distance away in both cases. This isn't true in something like a 90° fed array, where one element sees another with a current leading its current by 90° while the other sees another element with a lagging current.

So, with arrays of two identical in-phase or 180° out-of-phase elements, feeding the elements through feed lines of equal length and Z_0 (in-phase) or lengths differing by 180° (out-of-phase) will lead to the correct current magnitude ratio and phase difference, regardless of the line length and regardless of how much the element feed-point impedances depart from the lines' Z_0.

As described above, unless the feed-point impedances equal the line Z_0 or the lines are an integral number of half wavelengths long, the magnitudes of the currents out of the lines will not be equal to the input magnitudes, and the phase will not be shifted an amount equal to the electrical lengths of the lines. But both lines will produce the same transformation and phase shift because their load impedances are equal, resulting in a properly fed array. In practice, however, feed-point impedances of elements frequently are different even in these arrays because of such things as different ground systems (for ground mounted vertical elements), proximity to buildings or other antennas, or different heights above ground (for horizontal or elevated vertical elements).

And in many larger arrays, two or more elements must be fed either in-phase or out-of-phase with equal currents but coupling to other elements can cause their impedances to change unequally — sometimes extremely so. Using the current-forcing method allows the feed system designer to ignore all these effects, while guaranteeing equal and correctly phased currents in any combination and number of 0° and 180° fed elements, regardless of their feed-point impedances.

This method is used to develop feed systems for the Four-Square and 4-element rectangular arrays in the "Practical Array Design" section. The front and rear elements of a Four-Square antenna provide a good example of elements having very different feed-point impedances that are forced to have equal out-of-phase currents.

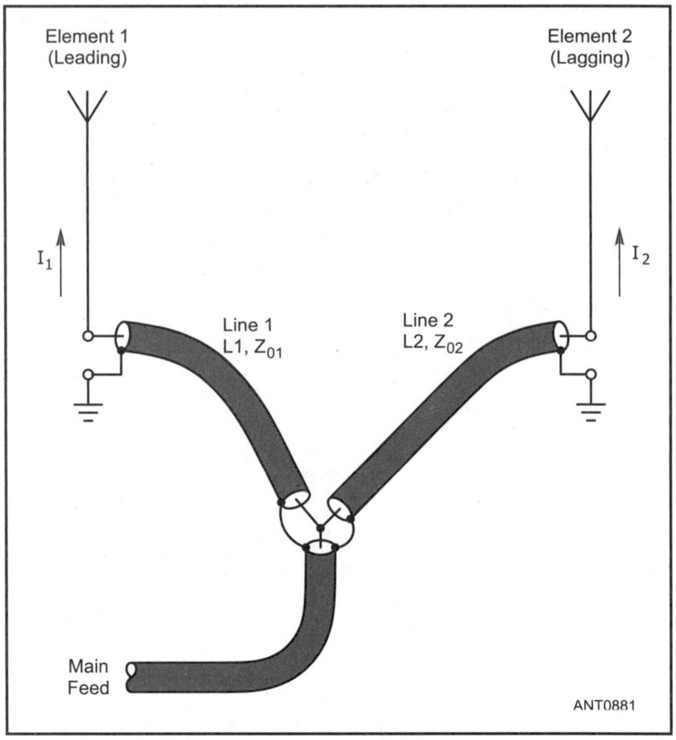

Figure 7.16 — "Simplest" feed system for 2-element array. No matching or phasing network is used here, only transmission lines.

"THE SIMPLEST PHASED ARRAY FEED SYSTEM...THAT WORKS"

This is the title of an article in *The ARRL Antenna Compendium, Vol 2*, which describes how arrays can be fed with a feed system consisting of only transmission lines. (The article is available for viewing at **eznec.com/Amateur/Articles/Simpfeed.pdf**.) Program *Arrayfeed1* does the calculations to simplify designing feed systems using this method. It's available in the online material and from **eznec.com/misc/Arrayfeed/**.

As explained earlier in "The 'Phasing-Line' Approach" section above, this method requires knowing what the element feed-point impedances will be in a correctly fed array *with the correct element currents present*. Feed-line lengths can then be computed for most but not all arrays. These lengths will produce the desired current ratio (magnitude and phase) in array elements that do present those feed-point impedances. If you know the load impedances connected to transmission lines whose inputs are connected to a common source, it's relatively simple to calculate the resulting load currents for any transmission line lengths. However, the reverse problem is much more difficult; that is, given the load impedances and desired currents to calculate the required cable lengths.

One way to solve the problem is to choose some feed-line lengths, solve for the currents, examine the answer, adjust the feed-line lengths, and try again until the desired currents are obtained. The author used this iterative approach, using first a programmable calculator and later a computer, for some time before developing a direct way of solving for the transmission-line lengths. The direct solution method is described briefly in the *Compendium* article and is used by the *Arrayfeed1* program.

Figure 7.16 shows the basic "simplest" system applied to a two-element array. Although it resembles an elementary phasing-line system as described earlier, the critical difference is that the lengths of Lines 1 and 2 are calculated to provide the correct current relative magnitude transformation and phase shift when terminated with the actual feed-point impedances.

The advantage to using this "simplest" feed system is indeed its simplicity. It's no more complicated than the elementary phasing-line approach but it works as planned. The disadvantage over some other methods is that there's no convenient adjustment to compensate for environmental factors, array imperfections, or inaccurate feed-point impedances.

Also, while unusual, it's possible that no suitable feed-line lengths can be found for some arrays, or at least none with practical feed-line characteristic impedances. The difference in electrical feed-line lengths almost never equals the difference in phase angles between element currents. This is because of the different line delays resulting from different feed-point impedances.

Arrayfeed1 can do the calculations for any two elements (alone or in a larger array), a Four-Square array, or a rectangular array in which two in-phase elements are driven at any current magnitude and phase relative to the other two in-phase elements. These possibilities cover many common arrays.

Arrayfeed1 can also be applied to suitable groups of elements in other types of arrangements. The required knowledge of element feed-point impedances in a correctly fed array can be obtained using *EZNEC*. Examples of the design of a "simplest" feed system for several different arrays using *EZNEC* and *Arrayfeed1* can be found in the "Phased Array Design Examples" section.

When a solution is possible for a given choice of line characteristic impedances, a second solution with different lengths is always available. See the comments in the introductory part of the "Phased Array Design Examples" section about choosing which solution might be most suitable.

AN ADJUSTABLE L-NETWORK FEED SYSTEM

Adjustment of the current ratio of any two elements requires varying two independent quantities: for example, the magnitude and phase of the current ratio. Two degrees of freedom — adjustments that are at least partially independent — are required. The "simplest" all-transmission line feed system described earlier adjusts the lengths of the two transmission lines to achieve the correct ratio.

But if the antenna characteristics aren't well known — for example, if the ground resistance isn't known even approximately — then the initial "simplest" design won't be optimum, and adjustment can be difficult and tedious. The current-forcing method produces correct currents independently of the element characteristics, so it doesn't require adjustment when the elements are identical. But it's suitable only for feeding elements in-phase or 180° out-of-phase and a few fixed current-magnitude ratios.

The addition of a simple network as shown in **Figure 7.17** allows you to easily adjust feeding of element pairs at other relative phase angles and/or magnitude ratios. Any desired current ratio (magnitude and phase) can be obtained with two elements fed with any lengths of wire, equal or unequal, by adding a network.

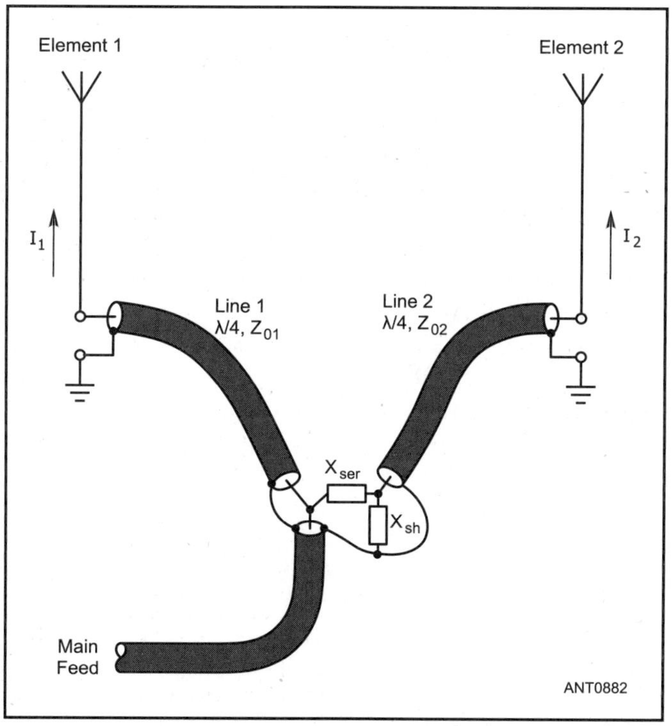

Figure 7.17 — The addition of a simple L network to Figure 7.16 allows you to easily adjust feeding of element pairs at other relative phase angles and/or magnitude ratios.

However, calculations for the general case are complex. The problem becomes much simpler if the transmission lines are restricted to lengths of odd multiples of λ/4, forming a modified "forcing" system that includes an added network. There are at least three additional advantages to this scheme. First is that a λ/4 line is easy to measure, even if the velocity factor isn't known. This procedure is described in the **Antenna and Transmission Line Measurements** chapter.

Second is that the feed system becomes completely insensitive to the feed-point impedance of one of the two elements. And the third is that the transmission lines of "forcing" systems feeding groups of elements in larger arrays can be used in place of the normal λ/4 lines. This greatly simplifies both the design of feed systems for larger arrays and the feed systems themselves. Note that both lines can be changed to 3λ/4 if necessary to span the physical distance between elements, but both lines must be the same 3λ/4 length and doing so might decrease the bandwidth.

This basic feed method can be used for any pair of elements, or for two groups of elements having forced equal currents. (See Feeding Four Element and Larger Arrays below) Many networks can accomplish the desired function, but a simple L network is adequate for most feed systems. The network can be designed to produce a phase lead or phase lag. The basic two-element L network feed system is shown in Figure 7.17. Many variations of this general method can be used, but the equations, program, and method to be discussed here apply only to the feed system shown.

If the phase angle of I_2/I_1 is negative (element 2 is lagging element 1), the L network will usually resemble a low-pass network (X_{ser} is an inductor and X_{sh} is a capacitor). But if the phase angle is positive (element 2 leading element 1), the L network will resemble a high-pass network (X_{ser} is a capacitor and X_{sh} is an inductor). Some current ratios and feed-point impedances result in both components being inductors or both being capacitors.

If it's desired to maintain symmetry in the feed system, X_{ser} can be divided into two components, each being inserted in series with a transmission line conductor. If X_{ser} is an inductor, the new components will each have half the value of the original X_{ser}, as shown in **Figure 7.18**. If X_{ser} is a capacitor, each of the new components will be twice the original value of X_{ser}.

Because of the current-forcing properties of λ/4 lines, we need to make the ratio of voltages at the inputs of the lines equal to the desired ratio of currents at the output ends of the lines; that is, at the element feed points. The job of the L network is to provide the desired voltage magnitude and phase transformation. If the output-to-input voltage ratio of the network is, say, 2.0 at an angle of –60°, then the ratio of element currents (I_2/I_1) will be 2.0 at an angle of –60°. The voltage transformation of the network is affected by the impedance of element 2, but not by the impedance of element 1. So only the impedance of element 2 must be known to design the feed system.

Equations for designing the L network are given in Manual

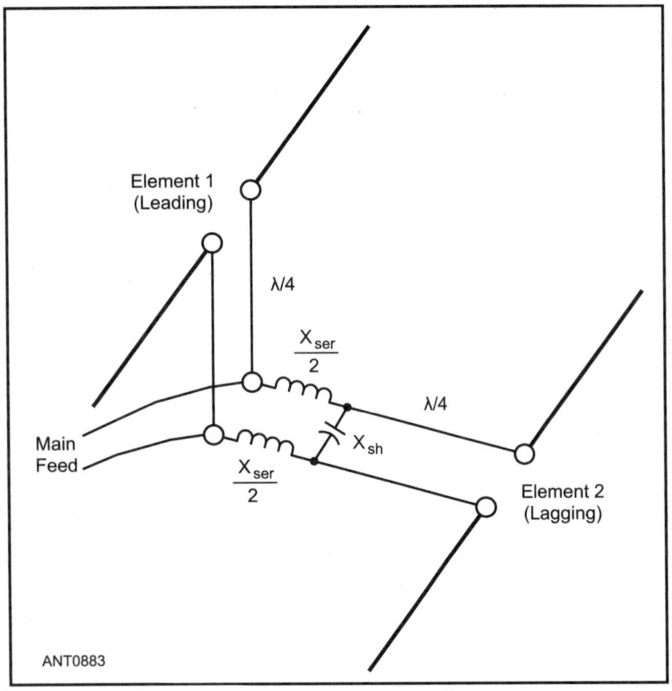

Figure 7.18 — Symmetrical feed system similar to Figure 7.17, in which the feed network is split into two symmetrical parts.

Calculations for Arrays in the online material, but the program *Arrayfeed1* makes it unnecessary to solve them. Only the feed-point impedance of the element or group of elements fed through the network must be known to design the network. This can best be determined by modeling the array with *EZNEC*. The impedance can be manually calculated for some simple element- and array-types by using the equations in Appendix B, but those types of elements and arrays are simple to model.

Examples of the design of L network feed systems for several different arrays using *EZNEC* and *Arrayfeed1* can be found in the "Phased Array Design Examples" section. A similar application of this feed system and a spreadsheet program for calculation was developed by Robye Lahlum, W1MK, and described in *Low-Band DXing* (see the References and Bibliography). *Arrayfeed1* can be used for the applications of the feed system described in that book if desired.

ADDITIONAL CONSIDERATIONS

Feeding 4-Element and Larger Arrays

Both the "simplest" and L network feed systems described above can be extended to feeding larger arrays having two groups of elements in which all the elements in a group are in-phase or 180° out-of-phase with each other — basically, any group that can be fed with the current-forcing method. The elements in each group are connected to a common point with λ/4 or 3λ/4 lines to force the currents to be in the correct ratio within the group. Then the "simplest" or L network feed system can be used to produce the correct phasing between the two groups, just as it does between two individual elements.

Two common arrays that fit this description are the Four-Square and the 4-element rectangular array. But more elaborate arrays could be constructed and fed using this method, such as a pair of binomial arrays. (A single binomial array is described in the "Phased Array Design Example" section below.) The *Arrayfeed1* program incorporates additional calculations necessary for designing Four-Square and 4-element rectangular arrays.

What If the Elements Aren't Identical?

Getting the desired pattern requires getting the correct relative magnitude and phase of the fields from the elements. If the elements are identical, which we've generally assumed up to this point, then producing currents of the desired magnitude and phase will create the desired *fields* (neglecting mutual coupling current distribution effects, discussed elsewhere).

But what if the elements aren't identical? Fortunately, the feed systems described here can still be used for any 2-element array and some more complex arrays, provided that the system can be accurately modeled. But a slightly different approach is required than for identical elements.

The first step is to model the array with a current source at the feed point of each element. Next, the magnitudes and phases of the model source currents are varied until the desired pattern is achieved. Then the ratio of feed-point (source) currents is calculated. This value and the feed-point impedances reported by the model are used for the feed system design. The feed system will produce the same ratio of currents as the model, resulting in the same pattern.

In general, this approach won't work with shunt fed towers or gamma-fed elements because of the difficulty of accurately modeling those structures as described in the following section.

Shunt- and Gamma-Fed Towers and Elements

In a shunt-, gamma-, or similarly fed tower or element, the feed-point current isn't the same as the main current flowing in the element. The ratio between the feed-point current and element current isn't a constant but depends on several factors. The ratio of currents in shunt- or gamma-fed elements is typically different — often vastly different — from the currents at the feed points. This complicates the design of feed systems for arrays of these elements.

An even more limiting problem is that the feed-point impedances are difficult to determine. The feed-point impedances of one or more elements in a properly fed array must be known to design a feed system for anything but 2-element in-phase or 180° out-of-phase arrays.

The only practical way to get this information for a shunt or gamma fed array is by modeling an array having the desired element currents. But Cebik has pointed out ("Two Limitations of NEC-4" — see the References and Bibliography) that common antenna analysis programs, including *EZNEC*, have difficulty accurately modeling folded dipoles with unequal diameter wires. (This applies to MININEC and NEC-2, -4.1, -4.2, and -5 programs and programs such as *EZNEC* which use them for calculations.) The same problem applies to shunt and gamma-fed elements when the element diameter is significantly different from the diameter of the shunt or gamma feed wire. Without accurate feed-point impedances, feed systems can't be designed to work without adjustment.

Loading, Matching, and Other Networks

Adding a component such as a loading inductor in series with an element or element feed point won't change the ratio of element current to feed-point current. As a result, a feed system designed to produce a particular ratio of element currents will still function properly if the elements contain series components. The extra feed-point impedance introduced by the loading component(s) must be considered when designing the feed system, however. Similarly, end or top loading won't alter the relationship between feed point and element current, provided that the current distribution in the elements is essentially the same. (See "Feed-Point vs Element Current" previously.)

However, insertion of any *shunt* component or network containing a shunt component *will* alter the relationship between feed point and element current because it will divert part of the feed-line current that would otherwise flow into the antenna, and that part depends on the feed-point impedance. As a result, a feed system designed to deliver correct currents at the feed points will produce incorrect element currents and therefore an incorrect pattern. So, any components or networks other than a series loading component should be avoided at any place in the feed system on the antenna side of the point at which the feed system splits to go to the various elements.

There are a few exceptions to this rule. If the feed-point impedances of the elements when in the excited array are equal, then identical networks with or without shunt components can be put at the feed points of the elements and the proper element current ratio maintained — so long as the feed system is designed to deliver the proper feed-point current ratio with the networks in place. Equal element impedances occur in arrays having only two identical elements fed in-phase or 180° out-of-phase, or arrays of any number of elements where the elements are electrically short and/or very lossy.

Baluns in Phased Arrays

For purposes of achieving the correct array pattern, baluns aren't usually required when feeding grounded vertical elements with coaxial cable feed lines. However, a balun (more specifically, a *current balun* or *choke balun*) might be desirable if current induced onto the outside of the feed line by mutual coupling to the elements is causing RF in the station. And with arrays of dipole or other elevated elements, baluns can be important to achieve the proper element current ratio, as explained below. See the **Transmission Line System Techniques** chapter for information on baluns and feed-line chokes.

First, however, the general rules for using baluns in phased arrays will be stated. Here, "main feed line" means the feed line going from the transmitter or receiver to the common point where the system splits to feed the various elements. "Phasing-system lines" means any transmission lines between that common point and any element. The rules are:

Rule 1: A balun or baluns should be used as necessary to suppress unbalanced (common-mode) current on the main feed line. Unbalanced current can occur on either coax or parallel-conductor lines. A balun usually isn't required when feeding grounded elements with coaxial feed line from an unbalanced transceiver or tuner. If current on the outside of the shield of a coaxial feed line to a grounded element is a problem, the feed line can be buried or run in buried conduit to reduce coupling to the antenna or to the radial system. Additional chokes can be added anywhere along the feed line.

"Baluns: What They Do and How They Do It", listed in the References and Bibliography, describes conducted-imbalance (common-mode) currents. Imbalance can also be caused by mutual coupling to the array elements. Common-mode currents have at least two undesirable effects on array performance. First, the imbalance current can flow from the main feed line to the phasing system lines, not necessarily splitting in the right proportion to maintain the correct element current ratio. This can affect the array pattern. In practice, however, this effect is likely to be small unless the common-mode current is unusually large. Even a small common-mode current, however, results in main feed-line radiation, and even a small amount of radiation can significantly degrade array pattern nulls. Any type of current balun can be used on the main feed line, at any place along the line, without any effect on the array pattern except to the extent that it reduces common mode current.

Rule 2: No balun or any other component or network should be inserted in any phasing system line that will alter the line length or characteristic impedance. This means that baluns in phasing system lines must be of a type made from the phasing line itself. Options include placing ferrite cores along the outside of the feed line (the W2DU "bead" balun is also an option); an air-core wound-coax choke made by winding part of the line into an approximately self-resonant or otherwise high-impedance coil; or winding part of the line onto a ferrite core or rod to make a several-turn winding. When coaxial cable is used, the feed system characteristics are dictated by the conditions *inside* the cable. Any cores placed on or winding of the *outside* prevents common-mode current on the outside but otherwise have no effect on the phasing performance. This rule applies equally to parallel-wire line, where the balun affects only common-mode current (equivalent to current on the outside of coax) while the phasing performance depends on differential mode current (equivalent to the current on the inside of coax).

Baluns are important when feeding dipoles or other elevated arrays, unless a fully balanced tuner is used. This is because common-mode current represents a diversion of some of the current that should be going to the array elements. The presence of common-mode current means that the element currents are being altered from the desired ratio and therefore the pattern won't be as intended. A balun should be placed wherever a path for current exists other than equally along a parallel-line conductor or on the inside of a coaxial line. Such a path exists, for example, where a coaxial cable connects to a dipole as shown in Figure 1 of the balun article referenced above. The

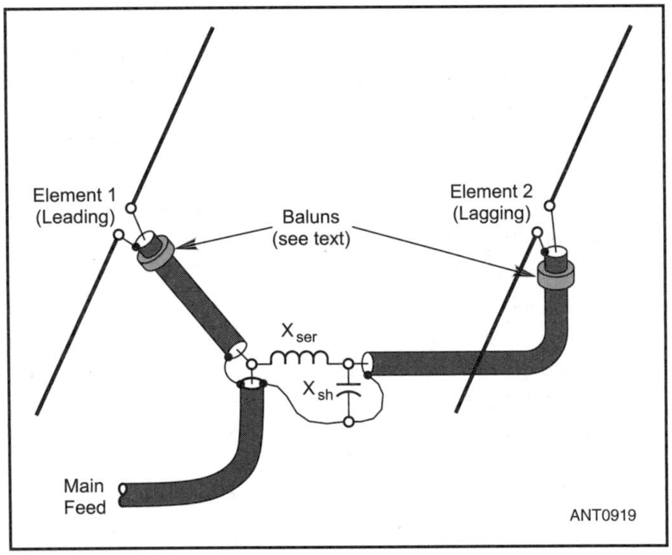

Figure 7.19 — Adding choke baluns to a two-dipole feed system to get rid of common-mode currents induced onto the coax shields.

path causes current on the outside of the coax and unbalanced current in the dipole halves. Or a path can exist where a parallel-conductor transmission line connects to an unbalanced tuner or to a coaxial line (shown in Figure 2 of the above balun article) and causes current on the outside of the coax or tuner and unbalanced current in the parallel conductors. A balun presents a high impedance to these currents, thereby reducing their magnitudes, but remember that all baluns must conform to the rules above. **Figure 7.19** shows recommended balun locations for a coax-fed dipole array using an L network feed system.

Receiving Arrays and Broadbanding

While it might not be entirely intuitive, an array designed for a particular gain and pattern for transmitting that considers mutual coupling, element currents, field reinforcement and cancellation, and so forth, will perform exactly the same when receiving. So, a receiving array can be designed by approaching the problem as though the array were to be used for transmitting or vice-versa.

However, at HF and below, the system requirements for transmitting and receiving antennas are different, so receiving-only arrays can be designed that aren't suitable for transmitting but are perfectly adequate for receiving in that frequency range. The reason is that at HF and below, atmospheric noise is typically much greater than a receiver's internally generated noise. Lowering a receiving antenna's gain and efficiency reduces the signal and atmospheric noise both by the same factor. Because the overall noise is for practical purposes all atmospheric noise, the signal/noise ratio isn't affected by antenna efficiency.

Of course, a point can be reached where the atmospheric noise is so reduced by inefficiency that the receiver itself becomes the dominant source of noise, but this typically doesn't happen until the antenna is extremely inefficient. When transmitting, reduced efficiency lowers the transmitted signal and therefore the strength of the received signal but it has no effect on the receiving station's noise. So reduced efficiency of a transmitting antenna results in a reduced signal/noise ratio at the receiving end, and consequently should be avoided.

Mutual coupling effects can be minimized by increasing the loss (and therefore reducing the efficiency) of the elements or by reducing the element sizes to a small fraction of a wavelength. Doing the second without the first isn't usually a good idea because the feed-point impedance tends to change rapidly with frequency for very small elements, making an antenna that works well over only a narrow bandwidth. But increasing loss broadens the bandwidth, even for small elements, as well as reducing mutual coupling effects. So, this approach is often taken for designing a receiving-only array. With mutual coupling effects minimized because of loss, feed-system design becomes relatively simple, provided a few simple rules are followed. (See Loss Resistance, Mutual Coupling, and Antenna Gain above.)

Readers may also be interested in receiving arrays of active vertical elements. An active vertical consists of an electrically short whip connected to a very high impedance amplifier (FET or op amp) so the impedance of the antenna is not a factor. Active short verticals can have extremely wide bandwidth from 100 kHz to 100 MHz and are often used in directional receive arrays. Active antennas are discussed in the **Receiving Antennas** chapter.

The "crossfire" feed method described by Tom Rauch, W8JI (**w8ji.com/crossfire_phasing.htm**) discussed in more detail in the design examples, delivers an array with very wide pattern bandwidth. That is, the pattern shape, particularly null direction and depth, stay nearly constant over a very large range of frequencies. The method requires elements whose feed-point impedances stay nearly constant over the frequency range, which for a wide range requires high loss and low efficiency. As explained earlier, however, this is acceptable for HF and MF receiving arrays, and receiving arrays are usually where deep and predictable pattern nulls are the most important. The basic idea for two elements is to use a delay line between the two whose delay in electrical degrees equals the element spacing in degrees then add a frequency-independent phase inversion to the line. A distant signal arriving at the first element creates a wave in the delay line connected to its feed point. That same signal arrives at the second element at the same time the wave from the first element reaches the end of the delay line. The signal from the second element is added to the inverted wave from the first element and, if the two are the same amplitude, complete cancellation occurs. This is independent of the frequency and of the element spacing, provided that the delay line's electrical length is the same as the element spacing and that it's terminated in its characteristic impedance. See the design examples for additional information.

For ungrounded elements like dipoles, the inversion can easily be created simply by reversing the feed-line connections to one element by giving the phasing line a half twist. Or a broadband inverting transformer can be used for either grounded or ungrounded elements. Proper termination of the line is usually accomplished with a resistor whose impedance is much less than that of the short elements.

Feed-Line Shield Connections

It is common for antenna system designs to assume that the shields of coaxial feed lines and phasing lines are isolated from the antennas. (See the previous discussion of "Baluns in Phased Arrays.") It is also often assumed that the shield connections are so short that the connection length can be ignored. However, an unintended or uncontrolled current path between the feed-line shields may upset the array's phasing and current levels. See this chapter's "Practical Aspects of Phased Array Design" subsection on "Improving Array Switching Systems."

If the shields of the feed lines are switched, the feed lines should be run directly to the relays or switches and connected there. If several shields are to be connected together, use a single terminal or an electrically short copper strip or wide braid as a common connecting point. Keep shield connections as short as possible. If the antenna switching and control circuits are placed in metal enclosures, use caution when using bulkhead or feedthrough connectors since they will create an uncontrolled connection between the feed-line shields. Do not use bulkhead connectors in a metal enclosure if the shields are also switched, since the enclosure connection will bypass the switched connection.

If a plastic or fiberglass enclosure is used, bulkhead connectors may be used. Remember that the entire length of the feed line includes the connectors and any internal lengths of cable that must be accounted for.

Lightning protection is another source of unwanted connections between feed lines. For example, the ground connection of gas-discharge tube lightning arrestors in a system of multiple feed lines can create an undesired additional connection between the shields. For ground-mounted vertical arrays, such as a Four-Square, a lightning arrestor should be located at each element's feed point where the feed-line shield connects to the ground system. Otherwise, the arrestor should be located on the array's main feed line from the station.

7.4 Phased Array Design Examples

This section, also written by Roy Lewallen, W7EL, presents examples of feed-system design for several kinds of array using the design principles given in previous sections. All but the dipole example array are assumed to be made of $\lambda/4$ vertical elements. The dipole example illustrates that the same method can be used for arrays of any shape of elements, including dipole, square (quad) and triangular. Likewise, the methods shown here apply equally well to VHF and UHF arrays. The first example includes more detail than the remaining ones, so you should read it before the others. Following the array design examples using the "simplest" and L network feed systems is an example of a receiving array using two different configurations of "crossfire" feed.

All modeling examples use *EZNEC Pro/2+* v. 7.0, available at no cost, and various example files as described in the sidebar "Using *EZNEC*." Users with the optional NEC-4.2 or NEC-5 calculating engines can directly model buried radials using REAL, HIGH- or EXTENDED-ACCURACY ground in place of PERFECT or MININEC type ground with added resistance for ground loss. Expect small differences in calculated results when using different engines.

EZNEC has the capability of incorporating L networks in the model, so the L network fed array can also be analyzed for accuracy, frequency changes, and the effects of component and ground loss variation. *AntBk_Cardioid_L_Network_Example.EZ* models the cardioid array with L network feed system.

EZNEC also has provisions for including transmission line loss. It can be instructive to add various amounts of loss to see the effect on pattern and bandwidth. The effect of loss will generally be most apparent on lines running with a high SWR, i.e., when terminated in an impedance very different from the characteristic impedance.

In the following sections, text in SMALL CAPS denotes a menu or function button label or an input to the *EZNEC* software used in the creation and modeling of these examples.

In addition to the set of examples, there is a great deal of additional information and numerous other array designs in Chapter 11, "Phased Arrays" of *Low-Band DXing* by John Devoldere, ON4UN (see the References and Bibliography section). Al Christman, K3LC, has also written excellent articles (included with the online material) examining the

Using *EZNEC*

Antenna modeling program *EZNEC* is used extensively in this chapter to illustrate phased array operation and aid in feed-system design. You should download and install the free *EZNEC* Pro/2+ v. 7.0 from **eznec.com** to gain the most from this chapter. The example model files which are used are available for download from **eznec.com/ARRL_Antenna_Book.htm**. The names of the model (description) files used in this chapter begin with AntBk_ and have the extension .EZ. After unzipping the downloaded file, copy the files to your *EZNEC* description file directory which is C:\EZNEC 7.0\Docs\Ant\ unless you've changed it during installation. (You can put them in a separate folder at that location if you'd prefer.) You should then see the files or new folder when you select File/Open or click the Open button in *EZNEC*. In addition, supporting files such as *Arrayfeed1* and other models for antennas in the *Antenna Book* are available here.

effect of lengthening λ/4 verticals to 5λ/8, both for single elements and for the 2-element cardioid pattern and 4-square array on 80 meters. His study of phased vertical arrays also provides useful information to the antenna system builder (see the References and Bibliography).

7.4.1 General Array Design Considerations

If either the "simplest" feed system (Figure 7.16) or L network feed system (Figure 7.17) is used, the feed-point impedance of one or more elements — *when the array elements all have the correct currents* — must be known. By far the best way to determine this is by modeling. If accurate modeling isn't practical for some reason, an estimate should be made from an approximate model, and you should expect to have to adjust the feed system after building and installing it. The L network system is more practical in that case because of its simpler adjustability.

Manual calculation methods for some simple configurations are given in Manual Calculations for Arrays in the online material, but calculation is tedious and the configurations for which this method works are the very ones which are easiest to model. *EZNEC* is used in the following examples to determine feed-point impedance. Space doesn't permit detailed instructions here on creating the models, so they are included in complete form in the online content. They should provide a convenient starting point for any variations you might like to try. See the *EZNEC* manual (accessed by clicking **HELP/CONTENTS** in the main *EZNEC* window) for help in using this program.

In the following examples, vertical elements are close to λ/4 high and dipole elements close to λ/2, and their lengths have been adjusted for resonance (zero or near zero feed-point reactance) when all other elements are absent or open circuited. There's no need in practice to make the elements self-resonant — it's simply used as a convenient reference point for these examples. You'll also find it interesting to see how much reactance is present at the feed points of the elements when in the arrays with the desired currents, knowing that it's very nearly zero when only one element is present.

In any real grounded vertical array, there is ground loss associated with each element. The amount of loss depends on the length and number of ground radials and on the type and wetness of the ground under and around the antenna. This resistance becomes part of the feed-point resistance, so it must be explicitly included in the model used to determine feed-point impedance unless using an NEC-4.2 or -5 calculating engine to directly model the buried wires. The 90° Fed, 90° Spaced Array example below discusses how this is done. **Figure 7.20** gives resistance values for typical ground systems, based on measurements by Sevick (July 1971 and March 1973 *QST*). The values of feed system components based on Figure 7.20 will be reasonably close to correct, even if the ground characteristics are somewhat different than Sevick's.

Feed systems for the design example arrays to follow are based on the resistance values given in **Table 7.2**.

Elevated radial systems also have some ground loss, although it can be considerably less than a system with the same number of buried radials. This loss will be automatically included in the feed-point impedance of a model which includes elevated radials, so no further estimation is required. Be sure to use Real, High-Accuracy ground type when modeling an elevated radial system with *EZNEC*. In other *NEC* based programs, this might be referred to as Sommerfeld type ground. More information can be found in the *EZNEC* manual. See also the chapter **Effects of Ground** for a discussion of elevated radials.

The matter of matching the array for the best SWR on the feed line to the station is not dealt with here, since it's a separate problem from that of the main topic, which is designing feed systems to produce a desired pattern. No change made in the single feed line running to the station, including insertion or adjustment of a matching network, will have any impact on the shape of the array pattern. Some of the simpler arrays provide a match that is close to 50 or 75 Ω, so no further matching is required. However, as shown by the program *Arrayfeed1*, many larger arrays present a less favorable impedance for direct connection and will require matching if a low SWR on the main feed line is required. If matching is necessary, the appropriate network should be placed in the single feed line running to the station. Attempts to improve the match by adjustment of the phasing L network, individual element

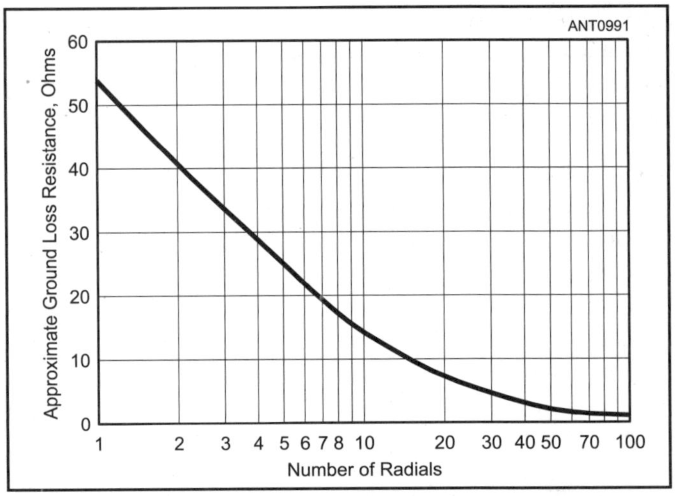

Figure 7.20 — Approximate ground system loss resistance of a resonant λ/4 ground-mounted vertical element versus the number of radials, based on measurements by Jerry Sevick, W2FMI. Moderate length radials (0.2 to 0.4 λ) were used for the measurements. The exact resistance, especially for only a few radials, will depend on the nature of the soil under the antenna. Add 36 Ω for the approximate feed-point resistance of a thin resonant λ/4 vertical.

Table 7.2
Feed-Line Resistance Values for Design Examples

Number of Radials	*Loss Resistance, Ω*
4	29
8	18
16	9
Infinite	0

lengths, matching at the element feed points or individual element feeder lengths will usually ruin the current balance and therefore the pattern of the array. Program *TLW*, included with the online material, can be used for designing an appropriate matching network. The network can be included in the *EZNEC* model to verify its accuracy and to evaluate the change in SWR as frequency is changed. Additional information on impedance matching may be found in the **Transmission Line System Techniques** chapter.

CHOOSING *ARRAYFEED1* SOLUTIONS

When designing a feed system for a two-element array, the *Arrayfeed1* program allows you to choose the characteristic impedances of the two transmission lines going to the elements, which don't have to be equal, so you have your choice of more than one solution. However, directional array switching is much more difficult if the lines have different impedances, so in general you should use the same characteristic impedances.

For larger arrays, *Arrayfeed1* requires the feed lines to all elements to have the same impedance. Usually, you can simply choose convenient impedances. But in general, you should avoid L network solutions where component reactance (X) values are vastly different (say, more than three times or less than one third as large) as the line characteristic impedances. Such networks will become more critical to adjust, and both the impedance and pattern will change more rapidly with changes in frequency. You can usually avoid this situation by choosing feed-line impedances that are in the same ballpark as the element feed-point impedances. The last example in the "Practical Array Design" section illustrates this problem and its solution.

When designing a "simplest" feed system, the most broadbanded and least critical system is usually one where the difference in electrical feed-line lengths is closest to the relative element phase angle. Here, "broadbanded" means that the pattern changes less with frequency, not necessarily that the SWR changes less. However, an array that's broadbanded in the pattern sense is usually also relatively broadbanded with respect to SWR.

Arrayfeed1 reports the impedance seen at the main array feed point. While it might be tempting to choose the solution producing the lowest SWR on the main feed line, you'll end up with a less critical and more broadbanded system if you base your choice on the criteria given above and provide separate impedance matching at the array's main feed point when necessary.

7.4.2 90° Fed, 90° Spaced Vertical Array

This example illustrates the design of both "simplest" and L network feed systems for a 2-element, 90° spaced and fed vertical array. The first task when using either feed system is to determine the feed-point impedances of the elements when placed in an array having the desired element currents. The "simplest" feed system method requires knowledge of both element impedances, while the L network system requires you to know only one. It's equally easy to determine both as it is to find just one, using *EZNEC*. (Manual Calculations for Arrays in the online material contains equations for those interested in manual methods or for more insight as to how the impedances come about.) The first step is to specify the antenna we want. For this example, we'll specify:

- Frequency: 7.15 MHz.
- Two identical, one inch (2.54 cm) diameter, 33 feet (10.06 meter) long elements spaced 90 electrical degrees, with element currents equal in magnitude and 90° different in phase.
- 8 buried radial wires, 0.3 λ long, under each element.

A model of this antenna has been created and furnished with *EZNEC*. So, the next step is to start *EZNEC*, click the **OPEN** button and locate and double-click example file *AntBk_Cardioid_Example.EZ* in the list to open it.

This *EZNEC* example model uses a *MININEC*-type ground which is the same as perfect ground when calculating antenna currents and impedances. A real antenna would have some additional resistive loss due to the finite conductivity of the ground system. The only way to model a buried radial ground system with an *NEC-2* calculating engine like the one built into *EZNEC* Pro/2+ is to create radial wires just above the ground (using the Real, High-Accuracy ground type), because *NEC-2* can't handle buried conductors. This provides only a moderate approximation of a buried system. (Buried conductors can be directly modeled with the optional *EZNEC* NEC-4.2 or -5 calculating engines but won't be described here.)

Another way to estimate ground-system resistance is to measure the feed-point resistance of a single element, then subtract from it the resistance reported for a model of that element over perfect (or *MININEC*-type) ground. For most uses, however, an adequate approximation can be made by simply referring to the graph of Figure 7.20. As stated previously, the feed system design depends on the feed-point impedances of the elements, which in turn depend on the ground system resistance. So, the ground system resistance must be known, or the ground system modeled, approximately anyway, before designing the feed system. At the end of this example, we'll investigate the effect on the pattern of changes in the ground system or errors in estimating the resistance.

For 8 radials, Figure 7.20 shows the ground system resistance to be about 18 Ω. This is included in the example model as a simple series resistive load at the feed point of each element. Click the **SRC DAT** button to see the feed-point impedances of the two elements. In this model, Source 1 is at the base of Wire 1 (element 1), and Source 2 is at the base of Wire 2 (element 2). Notice in the **SOURCE DATA** display that the Source 1 current has been specified at 1 amp at 0°, and Source 2 is 1 amp at –90°. So, element 2 is the lagging element. You should see impedances of $37.67 - j19.09\ \Omega$ for element 1 and $68.97 + j18.68\ \Omega$ for element 2. These are the feed-point impedances resulting when the array is ideally fed, with equal magnitude and 90° phased currents. Record these values for use in *Arrayfeed1*.

Click the **FF PLOT** button to generate a plot of the azimuth pattern at an elevation angle of 10°. In the 2D Plot Window, open the **FILE** menu and select **SAVE TRACE AS**. Enter **CARDIOID — IDEAL FEED** in the **FILE NAME** box, then click **SAVE**. This saves the cardioid pattern plot so you can compare it later to the pattern you get with the transmission line feed system.

Now it's time to design the feed system. Refer to the appropriate subheading below for the design of each of the two kinds of feed systems. Both systems use the *Arrayfeed1* program.

"SIMPLEST" (TRANSMISSION LINE ONLY) FEED SYSTEM

Start *Arrayfeed1*. In the **ARRAY TYPE** frame, select **TWO ELEMENT**. In **FEED SYSTEM TYPE**, select **SIMPLEST**. In the **INPUTS** frame, enter the following values:

Frequency MHz = 7.15; Feed-point impedances – Leading Element: R ohms = 37.67, X ohms = -19.09; Lagging Element: R ohms = 68.97, X ohms = 18.68 (these are the element R and X values from *EZNEC*). We'll be discussing the array input impedance, so check the **CALC ZIN** box near the lower left corner of the main window if it's not already checked.

We're free to choose any transmission line characteristic impedances we want, so long as we can get cables with those impedances. And the two cables don't have to have the same characteristic impedances. Each choice will lead to a different set of solutions. But sometimes a solution isn't possible, which then requires choosing different line impedances. Let's try 50 Ω for both lines. Enter **50** in both **Z0** boxes.

Finally, enter **1** for the **LAGGING:LEADING I MAG**, and **–90** for the **PHASE**. Click **FIND SOLUTIONS**. The result is no solution! So, enter 75 into both the line impedance boxes and click **FIND SOLUTIONS** again. You should now see two sets of results in the Solutions frame, electrical lengths of 68.65° and 155.80° for the first solution and 131.67° and 184.92° for the second. (Notice that the difference in length between the two lines isn't 90° for either solution, although the first solution is quite close. It's normal for the feed-line length difference to not be the same as the phase difference, due to the unequal element feed-point impedances caused by mutual coupling.)

The solution with a line length difference closest to the element phase difference is usually preferable. Also, all else being equal, the solution with shortest lines is better, providing that the lines will physically reach the elements. This is because the current magnitude and phase will change less with frequency than for a longer-length solution. However, there might be some cases where the change with frequency luckily compensates for the changing electrical distance between elements, so it's not a bad idea to model both solutions unless you plan on using the antenna over only a narrow frequency range.

In this case, the first solution looks best in all respects. The sum of the two lines in the first solution is about 224 electrical degrees. Assuming the lines have a velocity factor of 0.66, the total length of the lines will be more than 148 physical degrees. Since our two elements are spaced 90 physical degrees apart, the lines will comfortably reach a common point. If they didn't, we could either use the second solution's lengths, use cable with a higher velocity factor, or add a half electrical wavelength to both the lines in the first solution.

The impedance **ZIN** shown by *Arrayfeed1* is the impedance

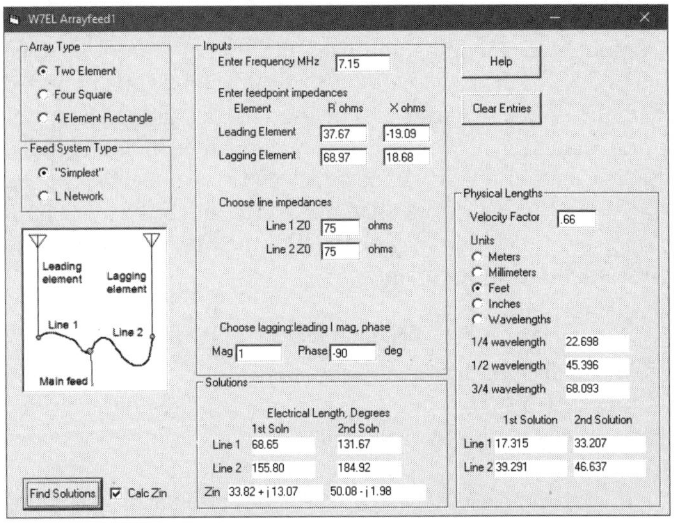

Figure 7.21 — Screen capture from *Arrayfeed1* program for "Simplest" 2-element phased array shown in Figure 7.16 and whose feed-point impedances are modeled by *EZNEC*.

at the input to the feed system, so it's the impedance that will be seen by the main feed line. The second solution provides nearly a perfect match for a 50 Ω transmission line. But the first solution is otherwise good for nearly all applications. Also, a 50 Ω line connected to the first solution's feed system would have an SWR of only 1.65:1, which wouldn't require any matching under most circumstances. Normal line loss would reduce the SWR even more at the transmitter end of the feed coax.

You might need to measure the transmission line Z_0 and use it in *Arrayfeed1* for a more accurate result. Methods are described in the **Antenna and Transmission Line Measurements** chapter including a step-by-step process for this example array. If the transmission lines don't have solid dielectric, you'll need to measure the velocity factor (VF) for use in *Arrayfeed1*.

Enter the cable velocity factor and make your choice of units in the **PHYSICAL LENGTHS** frame. The design is now complete; all you must do is cut two lines to the specified lengths and connect one from a common feed point to each element as shown in Figure 7.16, or schematically in the **Figure 7.21** *Arrayfeed1* screen capture.

Next, we'll design an L network feed system for the same array.

L NETWORK FEED SYSTEM

In *Arrayfeed1*, select **L NETWORK** in the **FEED SYSTEM TYPE** frame. The program doesn't need to know the leading element impedance to calculate the L network values, but it does need it to calculate the array input impedance. If you want to know the input impedance, check the **ZIN** box at the lower left corner of the main window, otherwise you can uncheck it and the input box for the leading element Z will disappear. The values from the "Simplest" analysis should still be present in the appropriate boxes; if not, refer to the "Simplest" feed system

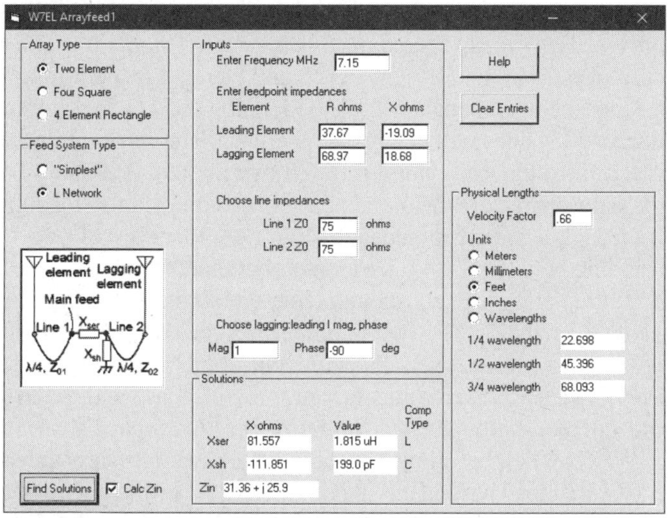

Figure 7.22 — Screen capture from *Arrayfeed1* program for L network feed system using "current-forcing" properties of λ/4 feed lines.

design above and re-enter the values. Again, we'll use 75 Ω for the line impedances that gave us a solution for the "Simplest" feed system. This feed system is more versatile, though, so we could use 50 Ω lines if desired.

Click **FIND SOLUTION** and see the results in the **SOLUTION** frame and the screen capture in **Figure 7.22**.

With 75-Ω lines, the L network consists of a series inductor of 1.815 μH and a shunt capacitor of 199.0 pF, connected as shown in the diagram in the left part of the program window. To find the physical length of the λ/4 lines, enter the velocity factor and choice of units in the **PHYSICAL LENGTHS** frame.

The main feed-point impedance of 31.36 + *j*25.9 Ω would result in about a 2.2:1 SWR on a 50-Ω feed line, which would be acceptable for many applications. It could easily be reduced to 1.6:1 by the simple addition of a series capacitor of 25.9 Ω reactance (859 pF) at the main feed point or, of course, reduced to 1:1 with a simple L network or other matching system designed with the *TLW* program.

These L and C values aren't standard values. If made from discrete components, series and/or parallel combinations would have to be used. But you can quickly see the impact of just using the nearest standard values by substituting them into the *EZNEC* model. And the model can also tell you what voltages and currents will be present at the components for a given applied power — see "Power Level selection" in the *EZNEC* manual.

PATTERN VERIFICATION AND EFFECT OF LOSS RESISTANCE

L network *EZNEC* model *AntBk_Cardioid_TL_Example.EZ* contains the "simplest" feed system just designed. Open it with *EZNEC*. In the **VIEW ANTENNA DISPLAY** you can see the boxes representing the transmission lines placed at the element bases. The other ends of the transmission lines are shown going to "virtual segment" (connection point) V1. (See "Virtual Segment" in the *EZNEC* manual index for information.) The source is also connected to V1. In *EZNEC*, the model locations of the ends of transmission line models don't have to be the same as the physical locations, so the view isn't a precise representation of what the actual setup would look like.

Click **FF PLOT** to generate a 2D pattern of the antenna. In the 2D Plot Window, open the **FILE** menu and select **ADDTRACE**. Select **CARDIOID — IDEAL FEED** (which you saved earlier) and click **OPEN**. The added plot overlays perfectly, indicating that the pattern using this feed system is identical to the pattern we got with perfect current sources at each feed point.

To check the feed-point currents, click the **CURRENTS** button. In the resulting table, you can see that **WIRE 1 SEGMENT 1** (feed point) current is 0.56457 A at a phase of –56.70° and **WIRE 2 SEGMENT 1** (feed point) current is 0.56451 A at –146.69°. (If you see the correct phase angles but wrong magnitudes, open the main window **OPTIONS** menu, select **POWER LEVEL**, and make sure the **ABSOLUTE V, I SOURCES** box is checked.) The ratio is 1.0001 at an angle of –89.99°, which is well within normal error bounds for the desired 1 at –90°.

As a check on *Arrayfeed1*, click the **SRC DAT** button to find the impedance seen by the source. This would be the impedance at the main feed-line connection in the real array. *EZNEC* reports 33.94 + *j*13.13 Ω, very close to the 33.82 + *j*13.07 Ω given by *Arrayfeed1* in Figure 7.21. Small differences of this order are normal. This provides a further check that the *EZNEC* model is correctly analyzing the *Arrayfeed1* feed system.

This *EZNEC* model uses lossless transmission lines of a fixed physical length rather than a fixed electrical length (number of degrees), so they'll behave like real lines as the frequency is changed. By changing the *EZNEC* frequency and re-running the 2D plot, you see that the front-to-back ratio degrades at 7.0 and 7.3 MHz. A slight adjustment of one-or-more line lengths, or a new *Arrayfeed1* solution at a slightly different frequency might produce a better compromise for some uses.

You could also evaluate the second *Arrayfeed1* solution or use different line impedances. Note that two identical transmission lines can be connected in parallel to simulate a line of one-half the line Z_0 as done in the next section. But keep the two line impedances equal if you anticipate doing array direction switching. The effect of varying ground system resistance can also be evaluated by clicking the **LOADS** line in the main window and changing the load resistance values. For example, if the ground system resistance were 9 Ω instead of the 18 Ω we have assumed and designed for, the front-to-back ratio would drop from about 32 to about 21 dB. Note that changing the *EZNEC* ground conductivity in this model has no effect on the feed-point current ratio, whereas in a real antenna it will because it changes the ground system loss resistance. With a *MININEC* type ground, it's used only for pattern calculation — the ground is assumed perfect during impedance and current calculations, and the only ground system loss resistance in the model is what we've specifically put in as loads.

Not surprisingly, the forward gain is affected very little by changes in frequency or ground system loss. To find the gain relative to a single element, compare the reported dBi gain of **AntBk_Cardioid_Example** with the same model with one of the elements deleted. You'll find it's very close to 3.0 dB. The 90° fed, 90° spaced array is a special case of array where the effects of mutual coupling on the two elements are opposite and cancel, resulting in the same gain as if mutual coupling didn't exist. But mutual coupling most certainly does exist!

The second solution presents a more favorable main feed-point impedance, so it would be tempting to use it instead of the first. Modeling the second solution by changing the feed-line lengths to the second solution lengths shows that the front-to-back ratio deteriorates more at the band edges. This might be tolerable if restricted frequency use is anticipated. But it does illustrate that the solution with shorter lines is generally more broadbanded and that the choice of solution shouldn't necessarily be based solely on the one giving the most favorable impedance.

7.4.3 A Three-Element Binomial Broadside Array

An array of three in-line elements spaced λ/2 apart and fed in-phase gives a pattern that is generally bidirectional. If the element currents are equal, the resulting pattern has a forward gain of 5.7 dB (for lossless elements) compared to a single element, but it has substantial side lobes. If the currents are tapered in a binomial coefficient 1:2:1 ratio (twice the current in the center element as in the two end elements), the gain drops slightly to just under 5.3 dB but the main lobes widen and the side lobes disappear.

The array is shown in **Figure 7.23**, and an *EZNEC* model of the antenna over perfect ground to show the ideal pattern is provided as *AntBk_Binomial_Example.EZ*. To obtain a 1:2:1 current ratio in the elements, each end element is fed through a 3λ/4 line of impedance Z_0. Line lengths of 3λ/4 are chosen because λ/4 lines will not physically reach. The center element is fed from the same point through two parallel 3λ/4 lines of the same characteristic impedance. This is equivalent to feeding it through a line of impedance $Z_0/2$. (Another way of understanding how this works is to remember that the current out of a quarter electrical wavelength line equals its input voltage divided by its Z_0. The two parallel lines have the same input voltage, so each delivers the same current as a normal single line.) The currents are thus forced to be in-phase and to have the correct ratio. *AntBk_Binomial_TL_Example.EZ* is an *EZNEC* model that shows this feed system with lossless transmission lines. The reader is encouraged to experiment with this model to see the effect of changes in frequency, the addition of loss resistance (as resistive loads at the element feed points), transmission line loss, and other alterations on the array pattern and gain. You should also replace the perfect ground with *MININEC* type of ground to show how radiation patterns over real ground differ from the theoretical perfect-ground pattern.

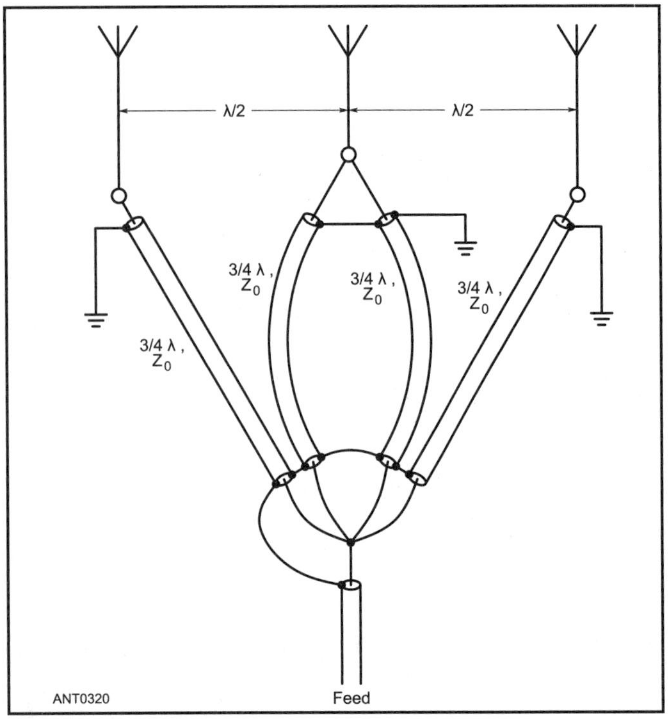

Figure 7.23 — Feed system for the three element 1:2:1 binomial array. All feed lines are 3/4 electrical wavelength long and have the same characteristic impedance.

7.4.4 A Four-Square Array

Several types of feed system are used for feeding this popular array, and most share a common problem — they don't provide the correct element current ratio — although some produce a workable approximation. The feed systems described here can produce exactly the correct current ratio. The only significant variable is the element feed-point impedances, so the quality of the result depends on your ability to model the feed-point impedances of a correctly fed array. As in the examples above, *EZNEC* will be used for that purpose and *Arrayfeed1* for the design of the feed system itself.

The most extensive collection of practical design

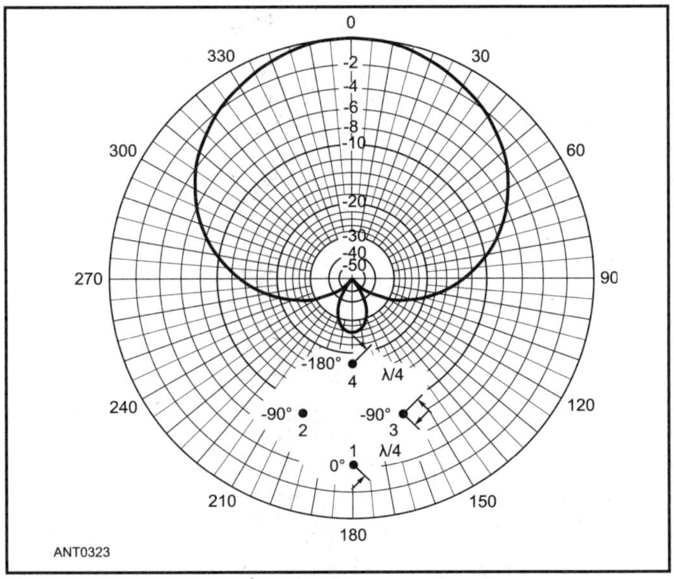

Figure 7.24 — Pattern and layout of the four-element Four-Square array. Gain is referenced to a single similar element; add 5.5 dB to the scale values shown.

information for the Four-Square (and other arrays) is in *Low-Band DXing, 5th Edition*, by John DeVoldevere, ON4UN. The chapter on Phased Arrays contains detailed construction and tuning guidelines for a variety of arrays. Additional articles about these popular antennas are available in the online material, including reduced-space approaches. An excellent online resource with several array design examples and references can be found on the AC6LA website: **ac6la.com/aecollection8.html**.

In this array (see **Figure 7.24**), four elements are placed in a square with λ/4 sides. (A variation of the Four-Square uses wider spacing.) The rear and front elements (1 and 4) are 180° out-of-phase with each other. The side elements (2 and 3) are in phase with each other and 90° delayed from the front element. The magnitudes of the currents in all four elements are equal. The front and rear elements can be forced to be 180° out-of-phase and have equal currents by using the current-forcing method described earlier. One element is connected to a line that is either λ/4 or 3λ/4 long, the other to a line that is λ/2 longer, and the two lines to a common point.

Likewise, the two side element currents are forced to be equal by connecting them to a common point via λ/4 or 3λ/4 lines. **Figure 7.25** shows the basic current-forcing system.

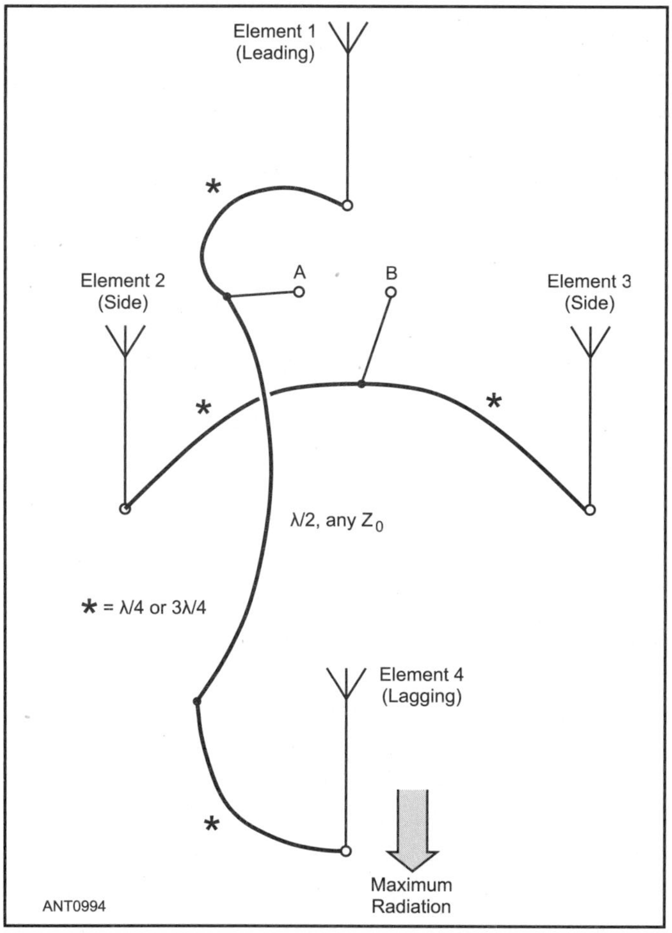

Figure 7.25 — "Simplest" feed system for the Four-Square array in Figure 7.24. Grounds and cable shields have been omitted for clarity.

If the pattern is to be electrically rotated, it is necessary to bring lines from all four elements to a common location. If solid-polyethylene dielectric coaxial cable, which has a dielectric constant of 0.66, is used, λ/4 lines won't reach the center of the array. So, 3λ/4 lines must be used. Alternatively, you can use λ/4 lines with foam or other dielectric having a velocity factor of more than about 0.71 (plus a little extra margin). Be careful, though — foam dielectric cables frequently have velocity factors which are lower than specified and should always be measured as described in the **Antenna and Transmission Line Measurements** chapter. Whatever your choice, three of the lines must be the same length (λ/4 or 3λ/4) and the fourth must be λ/2 longer.

In this array, the side elements (2 and 3) have equal impedances, but the rear and front (1 and 4) are different from each other and from the side elements. We must know the feed-point impedances of the front, rear, and side elements in order to design the "simplest" feed system, but only the side element impedances are needed to design the L network system. Knowledge of all feed-point impedances is necessary if the array main feed-point impedance **ZIN** is to be calculated. *EZNEC* model *AntBk_4Square_Example.EZ* shows a 40-meter Four-Square array with 18 Ω of loss resistance at each element, to approximate an 8-radial-per-element ground system. (See the cardioid array example above for more information about modeling ground system loss.) Opening the file in *EZNEC* and clicking the **SRC DAT** button gives the following impedances:

Source 1: 16.4 – *j*15.85 Ω
Sources 2 and 3: 57.47 – *j*19.45 Ω
Source 4: 77.81 + *j*54.8 Ω

It's interesting to note that the resistive part of source 1 is less than the 18 Ω of loss resistance we intentionally added to simulate ground system loss. That means the element 1 feed-point resistance would be *negative* if the ground resistance were less than about 1.5 Ω. This isn't uncommon in phased arrays and simply means that the element is feeding power into the feed system. This power comes via mutual coupling from the other elements.

"SIMPLEST" (TRANSMISSION LINE ONLY) FEED SYSTEM

To design a "simplest" feed system, start *Arrayfeed1*. In the **ARRAY TYPE** frame, select **4 SQUARE**, and select **SIMPLEST** in the **FEED SYSTEM TYPE** frame. In the **INPUTS** frame, enter the frequency and the impedances from *EZNEC*:

Frequency = 7.15 MHz
Leading Element: R = 16.4, X = –15.85
Side elements: R = 57.47, X = –19.45
Lagging Element: R = 77.81, X = 54.8

We'll try using 50 Ω for all lines, so enter **50** into the next three boxes.

Enter **1** for the lagging:leading I magnitude and –90 for the phase.

Click **FIND SOLUTIONS**.

The result is shown in the **SOLUTIONS** frame, shown in

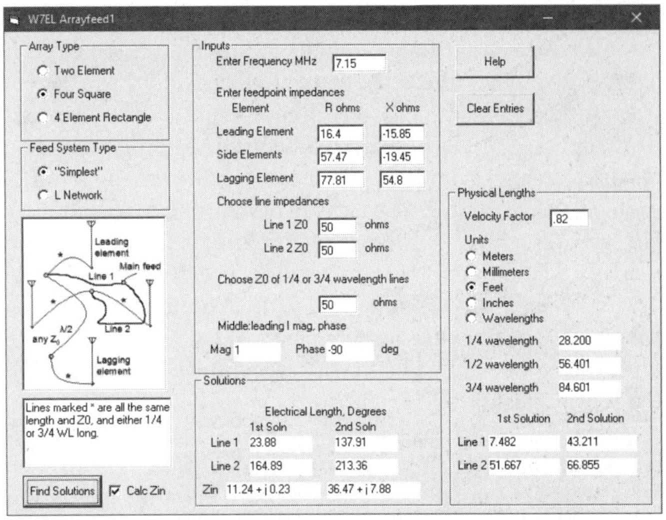

Figure 7.26 — Screen capture from *Arrayfeed1* for "Simplest" feed system for Four-Square feed system shown in Figure 7.25.

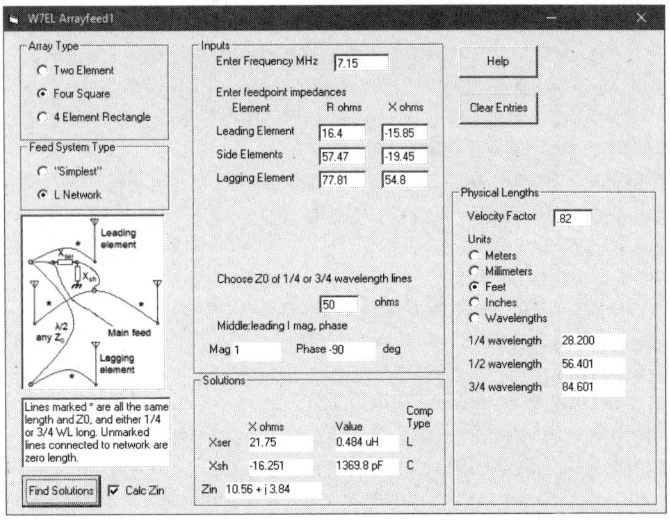

Figure 7.27 — L network setup for Four-Square array in Figure 7.25, fed with λ/4 (or 3λ/4) current-forcing feed system.

Figure 7.26. As always when any solution exists, there are two to choose from. The one with the shortest lines is generally preferable, so we'll choose it. For this example, we'll use λ/4 lines with velocity factor of 0.82. So, enter **0.82** in the **VELOCITY FACTOR** box in the **PHYSICAL LENGTHS** frame, and read the physical lengths from that frame. The ¼ wavelength lines (marked in the *Arrayfeed1* diagram with an asterisk) are 28.2 feet, line 1 is 7.482 feet and line 2 is 51.667 feet. The "simplest" feed system solution is shown in Figure 7.26, and the complete feed system consists of this connected to the array of Figure 7.25.

Again, though, be sure to measure your transmission line's velocity factor before finalizing the design — it might not be the 0.82 you thought. Velocity factor in specifications is a nominal value and can vary from batch to batch of the same cable from the same manufacturer and between manufacturers.

EZNEC model *AntBk_4Square_TL_Example.EZ* simulates the array fed with this system. Comparison of the pattern plot to one from ideal-current model *AntBk_4Square_Example.EZ* and examination of the element currents verify that the feed system is producing the desired pattern and element currents. You can use *AntBk_4Square_Example.EZ* to investigate the effect of frequency change, ground loss, and other changes on the array gain and pattern.

L NETWORK FEED SYSTEM

To design the L network feed system, simply change the **FEED SYSTEM TYPE** to **L NETWORK** and click **FIND SOLUTIONS**. The results you should see are a 0.484 μH inductor for the series component X_{ser}, and a 1369.8 pF capacitor for the shunt component X_{sh}. The L network feed system is shown in **Figure 7.27**, and the complete feed system consists of this L network connected to the array of Figure 7.25.

EZNEC model *AntBk_4Square_L_Network_Example.EZ* simulates the array fed with this system. You can compare it with the idealized feed system array and use it to see the effects of various parameter changes as you did with the "simplest" feed system model. Arrays have also been built using this feed system and the element currents measured, with exactly the expected results.

This array, particularly the null depth, is more sensitive to adjustment than the 2-element 90° fed, 90° spaced array. Adjustment procedures and a method of remotely switching the pattern direction are described in the "Practical Aspects of Phased Array Design" section below.

7.4.5 A 4-Element Rectangular Array

The 4-element rectangular array shown with its pattern in **Figure 7.28** has appeared numerous times in amateur publications. However, many of the accompanying feed systems fail

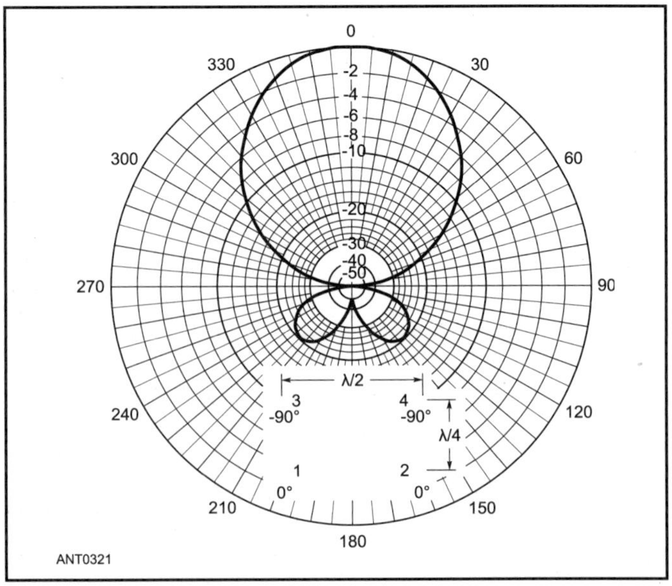

Figure 7.28 — Pattern and layout of the four-element rectangular array. Gain is referenced to a single similar element; add 6.8 dB to the scale values shown.

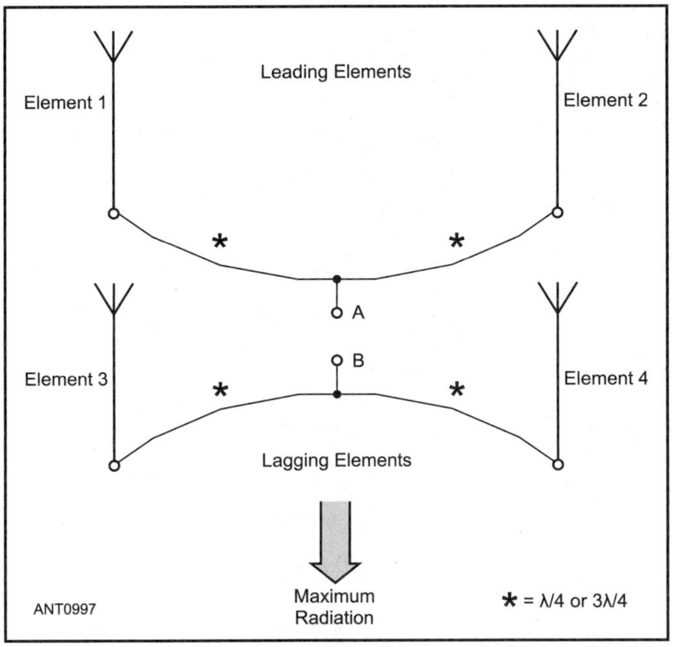

Figure 7.29 — "Simplest" feed system for four-element rectangular array, using four equal-length λ/4 (or 3λ/4) cables.

to deliver currents in the proper amounts and phases to the various elements. The array can be correctly fed using the principles discussed in this chapter and the design methods that follow.

Elements 1 and 2 can be forced to be in-phase and to have equal currents by feeding them through 3λ/4 lines. (As in the binomial and Four-Square array examples, 3λ/4 lines are chosen because λ/4 lines won't physically reach.) The currents in elements 3 and 4 can similarly be forced to be equal and in-phase. **Figure 7.29** shows the "current-forcing" feed system. Elements 3 and 4 are made to have currents of equal magnitude but of 90° phase difference from elements 1 and 2 by use of either a "simplest" all-transmission line feed system or an L network feed system. Both will be designed in this example.

For this array, we have to know the feed-point impedances of two elements (one of each pair) in order to design either type of feed system. *EZNEC* model *AntBk_Rectangular_Example.EZ* shows a 20-meter rectangular array with 18 Ω of loss resistance at each element, again to approximate an 8-radial per element ground system. (See the cardioid array example above for more information about modeling ground system loss.) Open the file in *EZNEC* and click the **SRC DAT** button to find the following feed-point impedances:

Sources 1 and 2: 21.44 – *j* 21.15 Ω

Sources 3 and 4: 70.85 – *j* 5.085 Ω

"SIMPLEST" (TRANSMISSION-LINE ONLY) FEED SYSTEM

To design a "simplest" feed system, start program *Arrayfeed1*. In the **ARRAY TYPE** frame, select **4 ELEMENT RECTANGLE**, and select **SIMPLEST** in the **FEED SYSTEM TYPE** frame. In the **INPUTS** frame, enter the frequency and the impedances from *EZNEC*:

Frequency = 14.15 MHz

Leading Elements R = 21.44, X = –21.15

Lagging Elements R = 70.85, X = –5.085

We'll use 50 Ω for all lines, so enter **50** into the next three boxes.

Enter **1** for the **LAGGING:LEADING I MAGNITUDE** and **–90** for the phase.

Click **FIND SOLUTIONS**.

The result will be "No Solution" — indicating this combination of line impedances can't be used. Several other combinations also produce this result but making lines 1 and 2 each 75 Ω and the 3λ/4 lines 50 Ω does produce a solution. Enter **75** into the **LINE 1 Z0** and **LINE 2 Z0** boxes and leave **50** in the **CHOOSE Z0 OF 1/4 OR 3/4 WAVELENGTH LINES** box, then click the **FIND SOLUTIONS** button. There won't be any problem making lines 1 and 2 reach, so we'll choose the first solution because the lines are shorter. The physical lengths of all the lines are shown in the **PHYSICAL LENGTHS** frame when the velocity factor is entered in the appropriate box. Assuming we use coax with a velocity factor of 0.66 (and the example frequency of 14.15 MHz), the lengths are:

Line 1: 5.012 feet

Line 2: 20.162 feet

3λ/4 lines (marked with an asterisk in the *Arrayfeed1* diagram): 34.408 feet

The lines are connected following the diagram in the upper left part of the *Arrayfeed1* window. This completes the "simplest" feed system design. *EZNEC* model *AntBk_Rectangular_TL_Example.EZ* simulates an array fed with this system.

Comparison of the pattern plot to one from ideal-current *AntBk_Rectangular_Example.EZ*, and examination of the element currents verify that the feed system is producing the desired pattern and element currents.

L-NETWORK FEED SYSTEM

To design the L network feed system using *Arrayfeed1*, change the **FEED SYSTEM TYPE** to **L NETWORK** and click **FIND SOLUTIONS**. The resulting L network values are a 0.198 μH inductor for the series component X_{ser} and a 683.3 pF capacitor for the shunt component X_{sh}. *EZNEC* model *AntBk_Rectangular_L_Network_Example.EZ* simulates an array fed with this system.

7.4.6 120° Fed, 60° Spaced Dipole Array

This example shows the design of "simplest" and L network feed systems for a 2-element 20-meter dipole array, rather than a vertical array. No special accommodation is required for the array made from dipoles rather than vertical elements — the same methods can be used regardless of element shape. This example also shows that both the "simplest" and L network feed systems can readily be applied to elements that use phase angles other than 90°.

Any 2-element array made with identical elements spaced λ/2 or closer and having equal magnitude currents with a relative phase angle of 180° minus the spacing will produce a unidirectional pattern with a good null to the rear. In practice, very close spacings lead to very low feed-point resistances, with consequent losses and very narrowband characteristics. But this 60° spaced array is well within the range of practical realization. File *AntBk_Dipole_Array_Example.EZ* is a model created for this array, with ideal element currents. Open this file in *EZNEC* and click **FF PLOT** to show the pattern at an elevation angle of 10°. You can save this pattern for later comparison to the pattern with a "simplest" feed system by opening the **FILE** menu in the **2D PLOT** window, selecting **SAVE TRACE AS**, entering a name for the trace file and clicking **SAVE**.

Following the same procedure as in the previous examples, we begin the array design by finding the element feed-point impedances in the ideally fed array using *EZNEC* numbers. Having already opened *AntBk_Dipole_Array_Example.EZ*, all that's needed is to click **SRC DAT**. The results are:

Leading element (source 1): 36.06 – j 46.33 Ω

Lagging element (source 2): 49.59 + j 51.09 Ω

"SIMPLEST" (TRANSMISSION LINE ONLY) FEED SYSTEM

Select **TWO ELEMENT** for the **ARRAY TYPE** in *Arrayfeed1* and "**SIMPLEST**" for the **FEED SYSTEM TYPE.** Enter the frequency of 14.15 MHz and enter the element feed-point impedances from *EZNEC* into the appropriate boxes in the **INPUTS** frame as in the previous examples. For line impedances, the section describing the "simplest" feed system recommends against choosing one which is very different from the element feed-point impedances, but as an experiment let's try 300 Ω for the two lines and see what happens. Enter **300** in the **LINE 1 Z0** and **LINE 2 Z0** boxes. Finally, enter the **LAGGING:LEADING I MAG, PHASE** of 1 for **MAG** and **–120** for **PHASE.**

Click **FIND SOLUTIONS.** For this example, we'll assume that TV-type twin-lead with a velocity factor of 0.8 is being used. So, enter **0.8** for the **VELOCITY FACTOR** and read the physical line lengths in the **PHYSICAL LENGTHS** frame. A model of the array using the first solution has been created as *AntBk_Dipole_Array_TL_Example.EZ.* Open this file in *EZNEC* and click **FF PLOT.** You should see that the plot is virtually identical to the one saved earlier from the ideal-current model. Note the gain and front-to-back ratio of 8.8 dBi and 30.92 dB respectively reported in the data box below the 2D plot.

Don't subtract 2.15 dB to find the gain relative to a single element! This isn't a free-space model, and the gain of a single dipole over ground is much greater than 2.15 dBi. Instead, delete one of the elements in the model to find the gain of a single element (4.88 dBi) and subtract that value from the reported dBi array gain. This shows that the gain of the array is about 3.9 dB relative to a single dipole at the same height. You can use the undo feature or re-open the file to restore the array.

Now, change Frequency to **14.0** MHz. Click **FF PLOT** again. The gain decreases a little, to 8.54 dBi and the front-to-back ratio also decreases, to 21.82 dB. At 14.3 MHz, the gain is slightly higher, 9.04 dBi, but the front-to-back is again worse, down to 18.59 dB. But this isn't bad overall.

Let's look at the second solution. Click the **TRANS LINES** line in the main *EZNEC* window to open the **TRANSMISSION LINES** window. Change the length of the first line to **28.865** feet, the second to **28.365** feet, and press the Enter key to complete the change. Change the **FREQUENCY** back to 14.15 MHz and click **FF PLOT.** You should see the same pattern as for both the first solution and for the ideal current model. But now change the **FREQUENCY** to **14.0** MHz, click **FF PLOT**, and look at the pattern.

What happened? The gain drops to 5.95 dBi and the front-to-back to only 3.1 dB. The array is now nearly bi-directional! It's almost as bad at 14.3 MHz. So, we've created a system that is terribly sensitive to small changes. The chance of it working correctly even at the design frequency is slim, because there are inevitably some differences between the model and real antenna.

We did have a clue this might happen. As stated in the section describing the "simplest" feed system, the best choices for line Z_0 and for the resulting solution give a difference in electrical line lengths about equal to the desired phase delay of the current. The difference in electrical line lengths for the first solution was about 152° — not as close to the 120° current phase difference as we'd like (indicating reasonably matched feed lines), but much better than the mere 9.7° difference of the lines for the second solution. While the 300-Ω line Z_0 is quite different from the element feed-point impedances, the first solution result is quite good. If desired, you can try other line impedance values in *Arrayfeed1* and evaluate the results with *EZNEC.*

Please see the information about baluns in the "Baluns in Phased Arrays" section. Baluns are placed the same as in Figure 7.19, which shows the L network feed system.

L-NETWORK FEED SYSTEM

To design an L network feed system, change the *Arrayfeed1* **FEED SYSTEM TYPE** to **L NETWORK** and click **FIND SOLUTIONS.** The results aren't good ones to use. The component reactance magnitudes of about 1,571 and 2,586 Ω are more than five times the 300 Ω Z_0 of the feed lines. As explained in the section describing the L network feed system, it's undesirable to have such a large ratio of component reactance to line Z_0. Among other problems, the inductor and capacitor values are quite extreme, and capacitor stray inductance and inductor capacitance would have a significant impact on performance.

The problem occurs because the feed-line impedance we chose is much larger than the element feed-point impedances, so the λ/4 lines transform the feed-point impedances to much higher values at the L network and main feed point. This feed system would be extremely critical, narrowbanded and difficult to adjust. We can do better by choosing feed-line impedances that aren't too drastically different than the element feed-point impedances. In this case, 50 or 75 Ω would be a much better choice than 300. Let's try 75.

In *Arrayfeed1*, change the **LINE 1 Z0** and **LINE 2 Z0** impedances from **300** to **75** and click **FIND SOLUTIONS.** L network component reactance magnitudes are now about 98 and 162 Ω, much better than before. This will be a relatively uncritical and broadbanded feed system.

Again, be sure to read the information about baluns in the "Baluns in Phased Arrays" section. Figure 7.19 shows the completed feed system including baluns. *EZNEC* example file *AntBk_Dipole_Array_L_Network_Example.EZ* is a model of the array with L network feed. It does not include baluns since the transmission line models support only differential mode currents and therefore have the implicit effect of including ideal baluns.

7.4.7 Crossfire Receiving Array

While any transmitting array can be used for receiving with the same gain and directivity, inefficient (lossy) arrays do well for HF and MF receiving but not for transmitting. High loss brings the potential for exceptionally wide bandwidth, simplified feed systems and compact size, so receive-only arrays are worth considering for many installations. The following example is for a simple 2-element array using the "crossfire" phasing principle discussed earlier. The same methods can be used for more complex receiving arrays.

The general principle of "crossfire" phasing is to connect the elements together with a delay line having an electrical length equal to the distance between the elements. A frequency-independent phase inversion (such as a wide-band transformer or physical connection reversal of one of the transmission lines) is added somewhere in the feed system path to one but not both elements, causing frequency-independent cancellation of the signals from the two elements when the signal is coming from one end-fire direction. The result is a potentially deep pattern null in one direction over a wide range of frequencies. The pattern can be reversed using methods described later and more elaborate arrays such as the Four-Square can be designed to allow additional null directions by directional switching. Since transmission lines invariably have velocity factors less than one, a single delay line of the proper electrical length is too short in practice to reach between the elements. The method works equally well, however, using a line to each element from a common point, the only requirement being that the difference in their electrical lengths equals the correct delay length. That is how these example designs were created.

There are several ways to create a time delay but about the only practical way of achieving a constant time delay over a wide frequency range is to use a transmission line terminated in its characteristic impedance. The termination must, of course, maintain its impedance over the wide frequency range. The straightforward method of designing a receiving antenna therefore requires the voltages across the transmission lines from the elements to somehow be added together but with each line properly terminated. This could be done with active circuits, for example, by terminating each line with a terminating resistor connected to the high impedance input of an amplifier or buffer circuit whose outputs can then be added (or subtracted, since an inversion is required somewhere) without affecting the transmission line termination. Passive methods include terminating each transmission line with a matched attenuator, then resistively combining the attenuator outputs. This effectively isolates the terminating impedance from the summing circuitry. Another passive method is to use a hybrid combiner (see **Figure 7.30**). The potential advantage of this method is its greater relative efficiency, resulting in higher signal (and noise) level output than the attenuator method. This is important only if the received signal level is otherwise small enough that receiver noise becomes apparent.

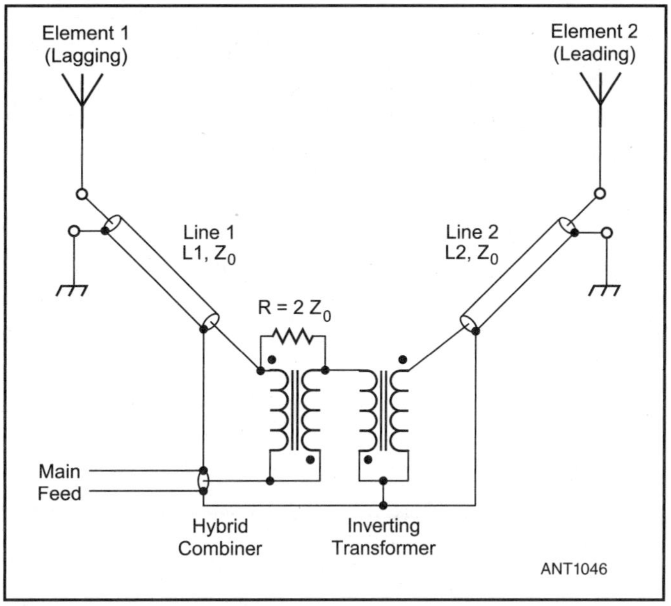

Figure 7.30 — "Crossfire" array with hybrid termination.

Yet another approach to the "crossfire" system is to design the array as though it were to be used for transmitting, even though its low efficiency would make it impractical for that purpose. The difference in this approach is that the transmission line termination is done at the element feed points rather than at the summing point (see **Figure 7.31**). Reciprocity assures that the same directional properties will exist when using the array for receiving, even though the transmission

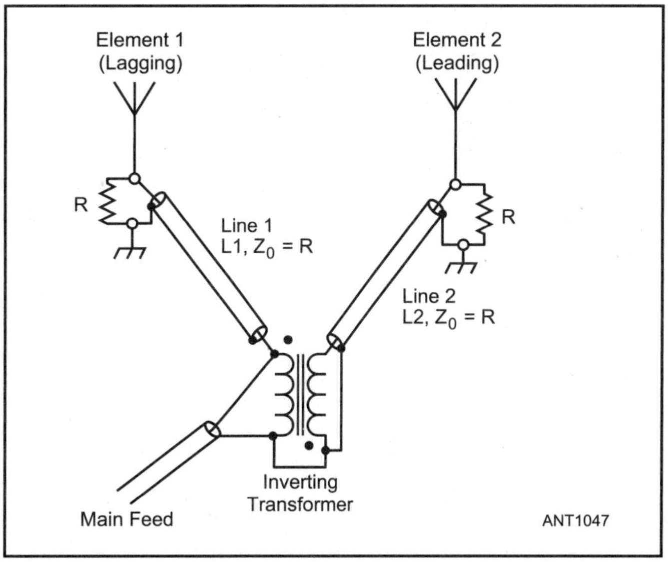

Figure 7.31 — "Crossfire" array with transmitting-type design.

lines are properly terminated at their source rather than load ends. The following examples show this approach and the hybrid-terminated receiving array approach for comparison and to illustrate that the reciprocal principle applies. Both arrays have the same pattern when receiving as they do for the modeled transmitting case.

TRANSMITTING ARRAY TYPE DESIGN

A two-element "crossfire" array with transmitting-type design is included as *EZNEC* example file *AntBk_Crossfire1_Example.EZ.* The array consists of two 30-foot-high vertical elements spaced 60 feet apart. These correspond to approximately 1⁄16 wavelength and 1⁄8 wavelength respectively at 1.85 MHz. Two transmission lines are used. This is being called a "transmitting array type design" because the transmission lines are terminated at the load end when driven by a transmitter but are terminated at the load end in some other impedance when receiving.

Because the elements are electrically short, they have high feed-point impedances, so parallel 50 Ω resistors at each feed point provide stable, near 50 Ω terminations for the transmission lines regardless of mutual coupling effects. The difference between the electrical lengths of the lines is 60 feet, the same as the line spacing. An ideal transformer is used in the model to create the required phase inversion to one of the elements.

The model could be simplified by deleting the transformer and specifying a reverse connection of one of the transmission lines, and the result would be the same. However, this model more closely represents how the antenna would be implemented. It shows a front-to-back ratio of better than 30 dB at 1 and 1.85 MHz, dropping to just over 21 dB at 4 MHz. Degradation at the higher frequency is due to the lower feed-point impedance as the element becomes electrically longer. This dilutes the effectiveness of the 50 Ω feed-point swamping resistors. Modeling with *EZNEC*'s plane wave excitation confirms that the front-to-back ratio is the same when receiving as when transmitting.

An 18 Ω resistor is included at each element base to simulate the resistance of a moderately good ground system. However, because its value is small compared to the large impedance of a short element, it has no significant effect on the array performance. This indicates that the array will work as well without an elaborate ground system.

RECEIVING ARRAY TYPE DESIGN — HYBRID TERMINATION

EZNEC example *AntBk_Crossfire_Hybrid_Feed_Example.EZ* uses the opposite design approach from the previous example. Instead of being terminated at the elements, the transmission lines are terminated at the ends where a receiver would be connected. Termination and signal addition are done with a hybrid combiner circuit consisting of a transformer and resistor. The inverting transformer is replaced in this model with a connection reversal of one of the transmission lines, for simplicity, as explained in the previous example. This model shows an improved front-to-back ratio of about 29 dB at 4 MHz compared to about 22 dB for the transmitting-type design, due to reduced sensitivity to mutual coupling and element impedance effects. Front-to-back ratio of the two examples is about the same at 1 and 1.85 MHz. This system is more efficient than the previous design, with signal (and noise) levels of about 5 dB greater at 4 MHz increasing to 14 dB greater at 1 MHz. But this won't improve system's signal-to-noise ratio unless the level of atmospheric noise is so low that receiver noise is audible.

Note: Modeling of the hybrid circuit and similar structures is difficult, requiring some experimenting and compromising to satisfy the requirements of the *NEC* calculating engines, which aren't designed for analysis of loaded, electrically small structures. Often, a suitable compromise isn't possible. This design is included for illustration only, not as encouragement to attempt to construct similar models.

7.5 Practical Aspects of Phased Array Design

With almost any type of antenna system, there is much that can be learned from experimenting with, testing, and using various array configurations. In this section, W7EL shares the benefit of years of his experience from building, adjusting, and using phased arrays. There is much more work to be done in most of the areas covered here, and the reader is encouraged to build on this work.

7.5.1 Theory Vs. Practice

All the following sections assume that the physical and/or electrical characteristics of an array and the transmission lines feeding it are known. While it's impossible to know these exactly, the results and methods described will usually produce very close to the predicted gains and forward patterns if care is taken in measurements and modeling. Only if very deep nulls are required will high precision or adjustment usually be necessary.

Except in a few special cases, successful feed system design requires knowledge of the feed-point impedance of some or all elements when the currents in it and all the other elements are those required for the desired array pattern. There's no practical way to measure this, so emphasis will be placed on modeling. The impedance of an isolated element can either be modeled or measured, but there's no benefit in manually calculating the resulting impedances in the active array — a modeling program does the same calculations, and in addition can do them for geometries which aren't possible to calculate manually.

Currently available modeling programs such as the free *EZNEC Pro/2+* v. 7.0 are very powerful and strikingly accurate since they employ the same mathematical rules that govern operation of real antennas. But the results are only as good as the description they're given of the physical antenna. Invariably there are some factors not known or taken into account in the model which will affect results. In other words, the results will be a strikingly accurate description of the performance of the *model*, not necessarily of the (imperfectly) modeled antenna. Again, though, the results will be very good in most cases, particularly for simple geometries, if care is taken — except when needing the best possible pattern nulls. A very powerful feature of system modeling programs such as *EZNEC* is that they allow you to create a complete model of an array including the feed system, then change various parameters such as ground loss resistance and transmission line characteristics and see the effect on the pattern caused by measurement or estimation errors. It also allows you to see how the array will perform at frequencies other than the specific frequency used for the feed system design.

Feed lines used in the previously described systems must be very close to the correct length to produce the expected patterns, and for all but the current "forcing" method the transmission line characteristic impedance Z_0 also must be closely known. Even small errors in either quantity can significantly degrade the front-to-back ratio, but much larger errors are usually required to have a significant impact on the shape and gain of the main lobe or lobes.

Errors in element feed-point impedance have similar effects on the main lobe(s) and front-to-back ratio, and these often involve some estimation and guesswork. Consequently, achieving very deep nulls invariably requires some adjustment of the feed system as described below in the section "Adjusting Phased Array Feed Systems." However, a useful front-to-back ratio can often be achieved without additional adjustment, and the need to adjust for the best front-to-back ratio can be minimized by using carefully measured transmission lines for the feed system.

Even seemingly simple measurements of transmission line electrical length and characteristic impedance (Z_0) can be surprisingly difficult to do with good accuracy. Practical methods using commonly available equipment are described in detail in the **Antenna and Transmission Line Measurements** chapter.

The feed system techniques in this chapter all assume that transmission lines have no loss. This is obviously not true, but the effects of loss are generally negligible. Transmission line loss can, however, be included in a model to see what its effect will be in a specific array, and adjustments can be made to compensate if needed. Methods for measuring feed-line loss are also described in the **Antenna and Transmission Line Measurements** chapter.

Sensitivity of an array to changes in frequency and in calculation or estimation errors of transmission line loss, Z_0, velocity factor, or other parameters is increased when lines have a high SWR, that is, when the impedances seen by the lines are very different from their Z_0. This isn't always possible to avoid, but some methods allow more than one option so this should be kept in mind when choosing. Although all options may seem to produce the same result, that will be true only if all parameters are exactly as calculated or modeled. Modeling the complete array, including feed system, allows you to intentionally introduce errors to quickly evaluate their effects.

7.5.2 Adjusting Phased Array Feed Systems

If a phased array is constructed only to achieve forward gain and care is taken in designing and implementing the feed system, adjusting it is seldom worthwhile. This is because the forward gain of most arrays is relatively insensitive to either the magnitude or phase of the relative currents flowing in the elements. If, however, good rejection of unwanted signals is desired, adjustment may be required. And achieving very deep nulls will almost surely require some adjustment. The effect of errors in transmission line lengths and Z_0, and in determining the element feed-point impedances, can be seen by modifying the appropriate values in a model which includes the feed system and comparing the patterns to those of the unmodified models.

The in-phase and 180° out-of-phase current-forcing method supplies very well-balanced and well-phased currents to elements without adjustment and requires only that the transmission lines be the same and very close to a quarter wavelength (or odd multiple). If the pattern of an array fed using this method is unsatisfactory, it's generally the result of environmental differences — where the elements, even though furnished with correct currents, aren't generating the necessary fields. Such an array can be optimized in a single direction, but a more general approach than the current-forcing method must be taken. Some possibilities are described by Paul Lee and Forrest Gehrke (see References and Bibliography).

Unlike the current-forcing method, the "simplest" and L network feed systems described earlier in this chapter are dependent on the self- and mutual-impedance of one or more elements. The required transmission-line lengths or L network component values can be computed to a high level of precision, but the results are only as good as the knowledge of the relevant feed-point impedances and transmission line characteristics.

While the "simplest" feed system doesn't readily lend itself to adjustment, the components of an L network can easily be made adjustable or can be experimentally changed in increments. A practical approach is to model the array as accurately as possible, design and build the feed system based on the model results, and then adjust the network for the best performance.

Simple arrays such as the two-element 90°-fed-and-spaced array can be adjusted as follows. Place a low-power signal source at a distance from the array (preferably several wavelengths), in the direction a null should be. While listening to the signal on a receiver connected to the array, alternately adjust the two L network components for the best rejection of the signal.

This has proved to be a very good way to adjust 2-element arrays. However, variable results were obtained when a Four-Square array was adjusted using this technique. The probable

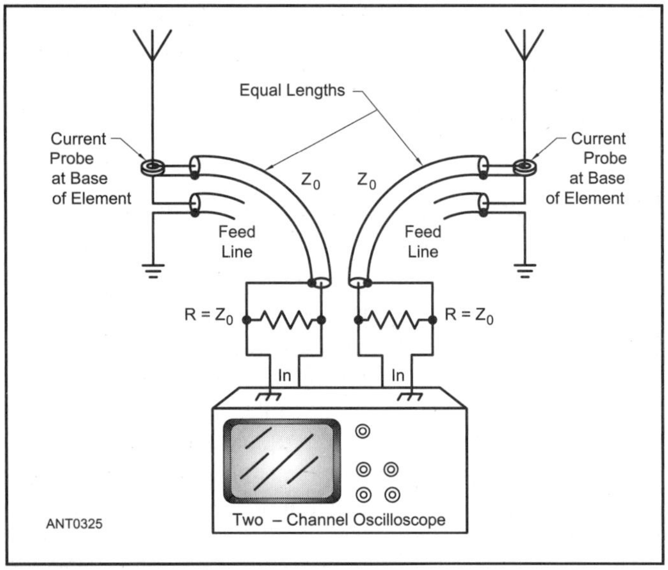

Figure 7.32 — One method of measuring element currents in a phased array. Details of the current probe are given in Figure 7.33. Caution: Do not run high power to the antenna system for this measurement, or damage to the test equipment may result.

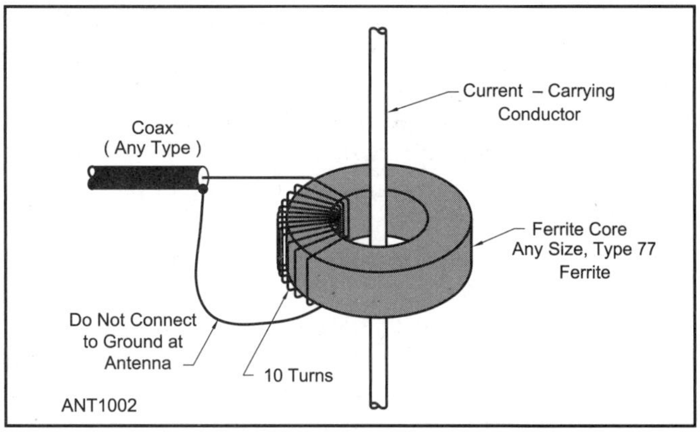

Figure 7.33 — The current probe for use in the test setup of Figure 7.32. The ferrite core is of type 77 ferrite or type 2 powdered iron and may be any size. The coax line must be terminated at the opposite end with a resistor equal to its characteristic impedance. You should build this probe in a plastic or metal box to provide mechanical ruggedness.

reason is that more than one combination of current balance and phasing can produce a null in a given direction, but each produces a different overall pattern, some with undesirable sidelobes. So, a different method must be used for adjusting more complex arrays. This involves measuring the element currents in some way and adjusting the network until the currents are correct. After adjusting the currents, small adjustments can be made to deepen the null(s) if desired.

MEASURING ELEMENT CURRENTS

You can measure the element currents in several ways. One way is to measure them directly at the element feed points, as shown in **Figure 7.32**. A dual-channel oscilloscope is required to monitor the currents. This method is the most accurate and it provides a direct indication of the actual relative magnitudes and phases of the element currents. The current probe is shown in **Figure 7.33**. Another current probe design by W8JI is presented at **w8ji.com/building_a_current_meter.htm** and shown in the **Antenna and Transmission Line Measurements** chapter.

Instead of measuring the element currents directly, you could measure them indirectly by measuring the voltages on the feed lines an electrical λ/4 or 3λ/4 distance from the array. The voltages at these points are directly proportional to the element currents. This introduces additional variables that can reduce the accuracy of the result, but in general it's accuracy is adequate. The 2-element arrays fed with the L network system and all the four-element arrays presented earlier have λ/4 or 3λ/4 lines from all elements to a common location, making this second measurement method convenient. The voltages can be observed with a dual-channel oscilloscope, or, to adjust for equal-magnitude currents and 90° phasing, you can use the test circuit shown in **Figure 7.34**.

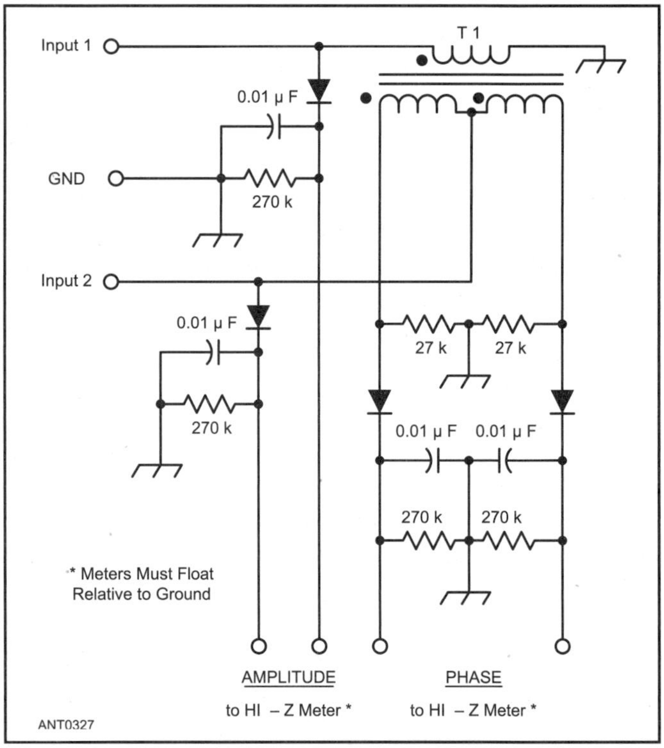

Figure 7.34 — Quadrature test circuit. All diodes are germanium, such as 1N34A, 1N270, or equiv. Hot carrier or silicon diodes can be used at higher power levels. All resistors are ¼ or ½ W, 5% tolerance. Capacitors are ceramic. Alligator clips are convenient for making the input and ground connections to the array.
T1 — 7 trifilar turns on an Amidon FT-37-43, -75, -77, or equivalent ferrite toroid core.

The test circuit is connected to the feed lines of two elements that are to be adjusted for 90° phasing (such as elements 1 and 2, or 2 and 4 of the Four-Square arrays of Figures 7.24 and 25). Adjust the L network components alternately until both meters read zero. You can verify proper operation of the test circuit by disconnecting one of the inputs. The phase output should remain close to zero. If not, there is an undesirable imbalance in the circuit, which must be corrected. Another way to verify is to first adjust the L network, so the tester indicates correct phasing (zero volts at the phase output). Then reverse the tester input connections to the elements. The phase output should remain close to zero.

7.5.3 Directional Switching of Arrays

One ideal directional-switching method would take the entire feed system including the lines to the elements and physically rotate it. The smallest possible increment of rotation would depend on the symmetry of the array — the feed system would need to rotate until the array again looks the same to it. For example, any 2-element array can be rotated 180° (although that wouldn't accomplish anything if the array is bidirectional to begin with). The 4-element rectangular array of Figures 7.28 and 29 can also be reversed, and the Four-Square array of Figures 7.24 and 25 can be switched in 90° increments.

Smaller switching increments can be accomplished only by reconfiguring the feed system, including any network if used, effectively creating a different kind of array. Switching in smaller increments than dictated by symmetry will create a different pattern in some directions than in others and must be thoughtfully done to maintain equal and properly phased element currents. The methods illustrated here will deal only with switching in increments related to the array symmetry, except for one: a 2-element broadside/end-fire array.

In all arrays, the success of directional switching depends on the elements and ground systems being identical so that equal element currents result in equal fields. It's even more important in arrays fed with any method other than current forcing because the effectiveness of those methods depends on the element feed-point impedances. Few of us can afford the luxury of having an array many wavelengths away from all other conductors, so an array will nearly always perform somewhat differently in each direction. Therefore, the array should be adjusted when steered in the direction requiring the most signal rejection in the nulls. Forward gain will, for all practical purposes, be equal in all the switched directions since gain is much more tolerant of error than are nulls.

BASIC SWITCHING METHODS

Following is a discussion of basic switching methods, how to power relays through the main feed line, and other practical considerations. In diagrams, grounds are frequently omitted to aid clarity, but connections of the ground conductors must be carefully made. In fact, it is recommended that the ground conductors be switched just as the center conductors are, as explained in more detail in "Improving Array Switching Systems" below. In all cases, interconnecting lines must be very short.

A pair of elements spaced λ/2 apart can readily be switched between broadside and end-fire bidirectional patterns, using the current-forcing properties of λ/4 lines. The method is shown in **Figure 7.35**. The switching device can be a relay powered via a separate cable or by dc sent along the main feed line as shown in Figure 7.40.

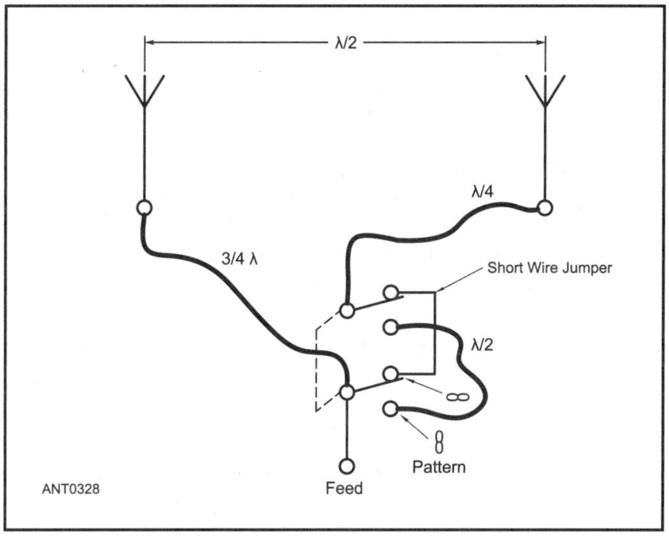

Figure 7.35 — Two-element broadside/end-fire switching. All lines must have the same characteristic impedance. Grounds and cable shields have been omitted for clarity.

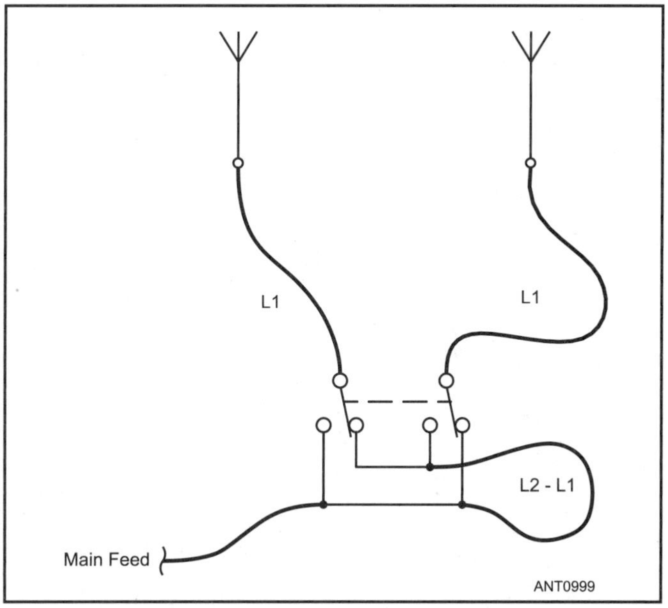

Figure 7.36 — Directional switching for 90° fed, 90° spaced 2-element array fed with a "simplest" feed system. The switch adds line to create a total length of L2 to the right-hand element.

Figure 7.36 shows directional switching of a 90° fed, 90° spaced array fed with a "simplest" feed system, where L1 and L2 are the required lengths of the two feed lines from *Arrayfeed1*. **Figure 7.37** shows how to switch the same array when fed with an L network, current-forcing system.

The rectangular array of Figure 7.28 can be switched in a similar manner, as shown in **Figure 7.38**. To switch a "simplest" fed rectangular array, use the switching circuit of Figure 7.36, but connect the two equal length lines to points A and B of Figure 7.29 in place of the two elements shown in Figure 7.36.

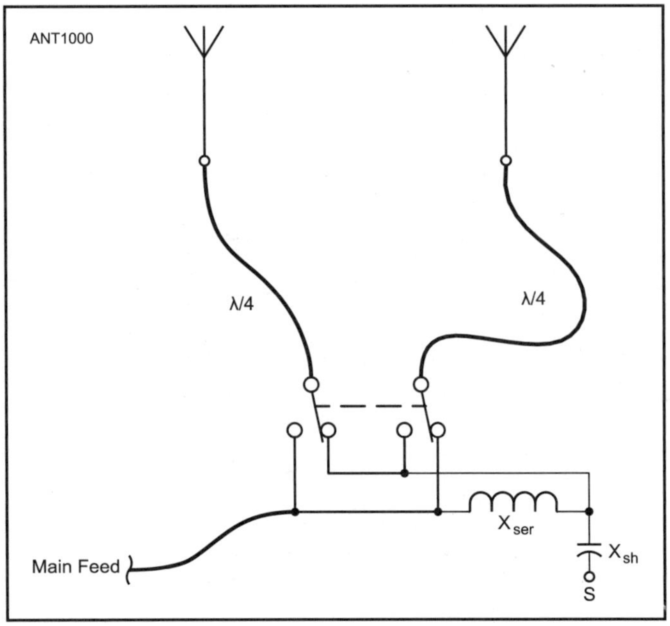

Figure 7.37 — Directional switching for 90°, 90° spaced 2-element array fed with an L network, current-forcing feed system.

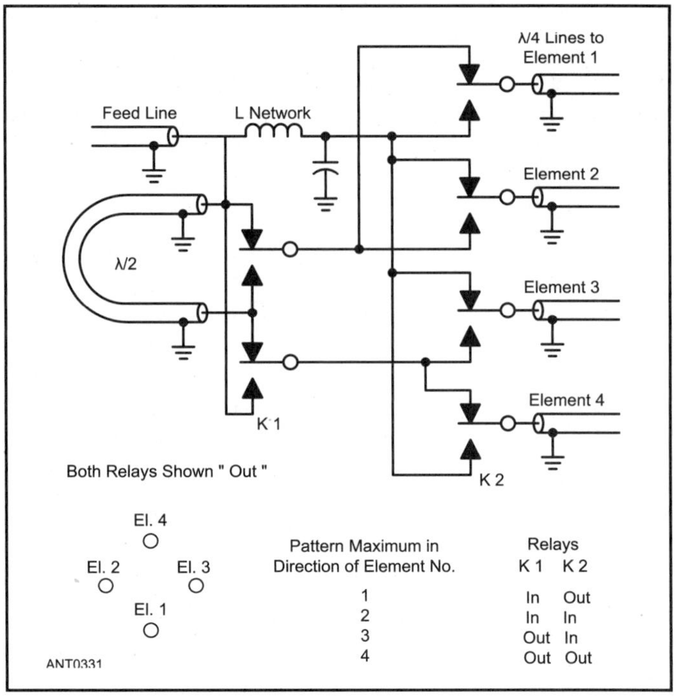

Figure 7.39 — Directional switching of the Four-Square array. All interconnections must be very short.

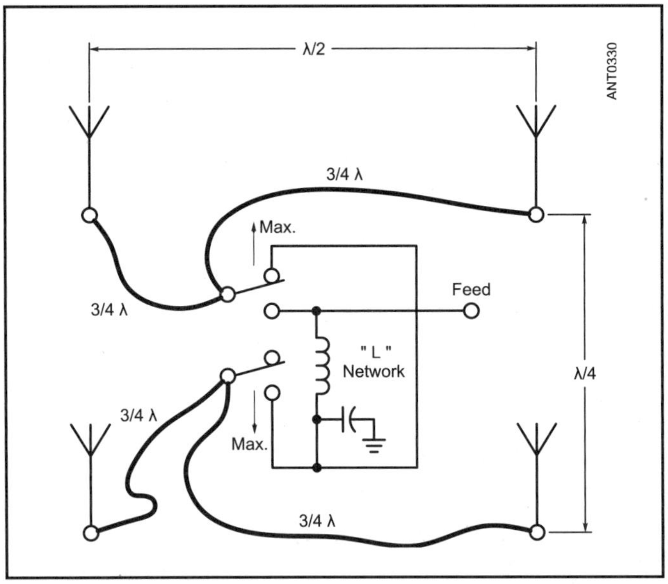

Figure 7.38 — Directional switching of a four-element rectangular array. All interconnections must be very short. As usual, grounds and cable shields have been omitted for clarity.

Switching the direction of an array in increments of 90°, when permitted by symmetry, requires at least two relays. A method of 90° switching of the Four-Square array with L network feed is shown in **Figure 7.39**.

POWERING RELAYS THROUGH FEED LINES

All the above switching methods can be implemented without additional wires to the switch box. A single-relay system is shown in **Figure 7.40A**, and a two-relay system in Figure 7.40B. Small 12- or 24-V dc power relays can be used in either system at power levels up to at least a few hundred watts. Do not attempt to change directions while transmitting, however. Blocking capacitors C1 and C2 should be good quality ceramic or transmitting mica units of 0.01 to 0.1 μF. No problems have been encountered using 0.1 μF, 300-V monolithic ceramic units at RF output levels up to 300 W, but an *EZNEC* model can be used to find the voltages which will be present — just remember that *EZNEC* reports RMS values, while the capacitor should be rated for well above the peak voltage. C2 may be omitted if the antenna system is an open circuit at dc. C3 and C4 should be ceramic, 0.001 μF or larger.

In Figure 7.40B, capacitors C5 through C8 should be selected with the ratings of their counterparts in Figure 7.40A, as given above. Electrolytic capacitors across the relay coils, C9 and C10 in Figure 7.40B, should be large enough to prevent the relays from buzzing, but not so large as to make relay operation too slow. Final values for most relays will be in the range from 10 to 100 μF. They should have a voltage rating of at least double the relay coil voltage. Some relays do not require this capacitor. All diodes are 1N4001 or similar. A rotary switch may be used in place of the two toggle switches in the two-relay system to switch the relays in the desired sequence.

IMPROVING ARRAY-SWITCHING SYSTEMS

The extra circuitry involved in switching arrays can degrade array performance by altering the relative currents fed to each element. One common cause is current sharing in common ground conductors, even when connections are kept very short. The author has seen a 30° phase shift in voltage along a 4-inch piece of #12 AWG wire in a 40-meter array feed system.

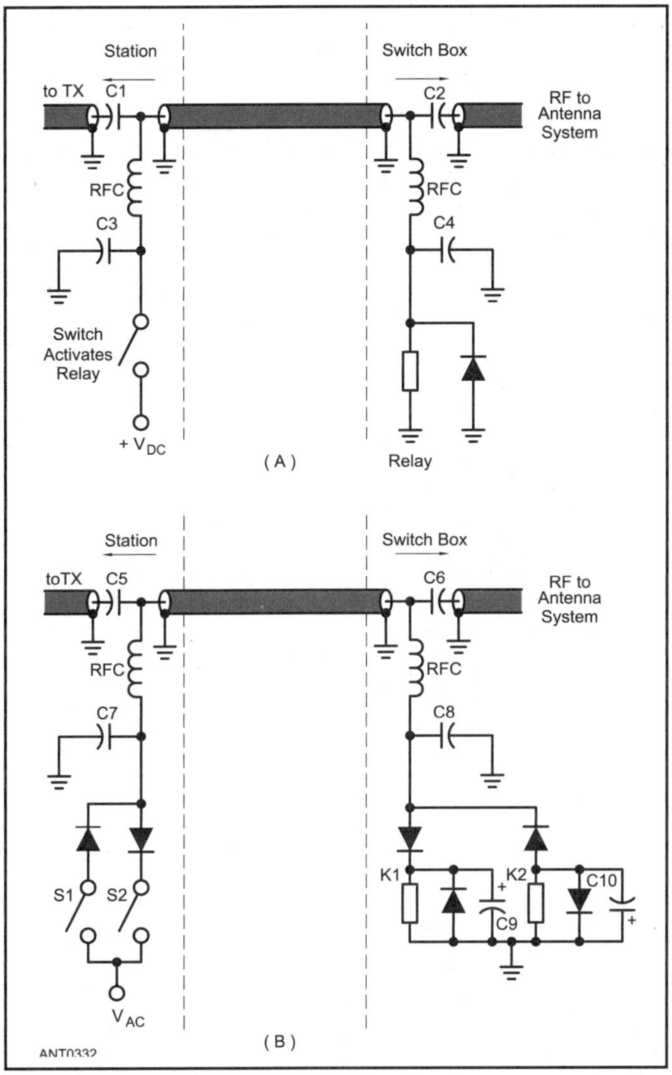

Figure 7.40 — Remote switching of relays. See text for component information. A one-relay system is shown at A, and a two-relay system at B. In B, S1 activates K1, and S2 activates K2. In addition, it is recommended to bypass each diode with a 0.001 – 0.01 μF disc ceramic capacitor to avoid generating harmonics and mixing products when transmitting.

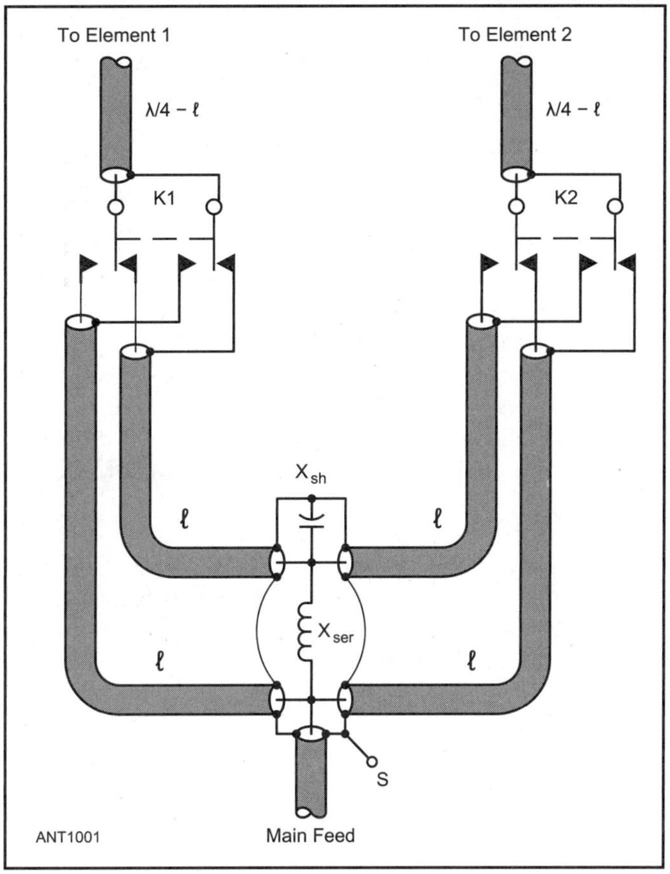

Figure 7.41 — A carefully designed L network, current-forcing switching system that switches both hot and shield conductors in feed coaxes.

When the two conductors of a feed line are physically separated from each other the characteristic impedance increases. This is especially true when the main lines are coaxial cables. If currents from two elements share the ground conductor of a split line, a relatively large voltage drop results. Voltage changes λ/4 from the elements translate to current changes at the elements. Although keeping all leads extremely short is sometimes adequate, the best way to reduce current sharing problems is to keep the two conductors of each transmission line as close together as possible, and switch both conductors on each line rather than just a single or "hot" conductor.

An example of a carefully designed switching system is shown in **Figure 7.41**. It avoids the problem of shared ground conductor currents, as well as another common problem, namely that effective line lengths are often different along different switching paths. Notice how the path from the main feed point travels through a single line to each element with no common ground connections to other lines except at the main feed point. Notice also that the distance doesn't change as the direction is switched. The λ/4 lines going to the two elements must be shortened by the length ℓ of the lines on the feed side of the relays so that the total line length from the main feed point to each element is λ/4 (or 3λ/4).

You can see that in either relay position there's an open-ended stub of length ℓ connected at the main feed point and another at the output end of the L network. If considerably shorter than λ/4, these will add capacitance at those points. Extra capacitance at the main feed point will alter the overall impedance seen by a transmitter but won't otherwise have any effect on the array or its performance. The one at the output of the L network will, however, change the transformation and phase shift properties of the network. But it's easy to compensate — the value of the shunt capacitor element is simply reduced by the amount of capacitance added by the stub. The amount of capacitance for any kind of short transmission line can be calculated from:

$$C\,(\text{pF}/\text{ft}) = \frac{1017}{Z_0\,\text{VF}}$$

or

$$C\,(\text{pF}/\text{m}) = \frac{3336}{Z_0\,\text{VF}}$$

where Z_0 = the characteristic impedance of the line and VF = the velocity factor. This works out to 31 pF/foot or 101 pF/meter for 50 Ω solid polyethylene insulated coax which has a velocity factor of 0.66. For more accuracy or longer lines, the input impedance of an open-circuited stub must be calculated or determined with *TLW* or a similar program and added in parallel with the connection point. The stubs can be added to the complete *EZNEC* model for verification or adjustment.

The general principles illustrated in Figure 7.41 can be extended to other switching systems. If switching the ground conductors as described above isn't practical, use of a metal box for the switching circuitry is recommended so that the relatively large surface area of the box can be used for the common ground conductors, minimizing their inductance. Always keep leads extremely short.

7.5.4 Measuring Element Self- and Mutual-Impedances

The need for measuring element self- and mutual-impedances has been made largely unnecessary with the ready availability of modeling software. Few amateurs appreciate the considerable difficulty of making accurate impedance measurements, and accurate mutual impedance measurements are very difficult even with professional test equipment and skills because they require accurate evaluation of small impedance changes. Despite the limitations of computer modeling, results very often are better than measured values because of the multiple factors affecting measurement accuracy.

Those who are interested in measuring self and mutual impedances can find more detailed information about doing so in "Mutual Calculations for Arrays" in the online material. The information there is from earlier editions of the *Antenna Book*.

7.6 References and Bibliography

Source material and more extended discussion of topics covered in this chapter can be found in the references given below and in the textbooks listed at the end of the **Antenna Fundamentals** chapter.

Atchley, D., Stinehelfer, H., and White, J., "360°-Steerable Vertical Phased Arrays," *QST*, Apr. 1976, pp. 27 – 30.

Brown, G., "Directional Antennas," *Proc. IRE*, Vol. 25, No. 1, Jan. 1937, pp. 78 – 145.

Brown, G., Lewis, R., and Epstein, J., "Ground Systems as a Factor in Antenna Efficiency," *Proc. IRE*, June 1937, pp. 753 – 787.

Brown G., and Woodward, O., "Experimentally Determined Impedance Characteristics of Cylindrical Antennas," *Proc. IRE*, Apr. 1945.

Brown, J., "Transmitting Chokes for the 160-10M Bands," **k9yc.com/publish.htm**.

Cebik, L., "Two Limitations of NEC-4", various online sources.

Christman, A., "Feeding Phased Arrays: An Alternate Method," *Ham Radio*, May 1985, pp. 58 – 59, 61 – 64.

Christman, A., "Tall Vertical Arrays," *QEX*, Nov./Dec. 2012, pp. 24 – 35.

Christman, A., "A Study of Tall Verticals," *QEX*, May/June 2011, pp. 26 – 31.

Christman, A., "A Study of Tall Verticals (Letters to the Editor)," *QEX*, Sep./Oct. 2011, pp. 38 – 39.

Christman, A., "A Study of Phased Vertical Arrays," *QEX*, Mar./Apr. 2005, pp. 46 – 51.

Delos, P., Broughton, B., Kraft, J., "Phased-Array Antenna Patterns," *Microwaves & RF*, Parts 1 – 5, May/June, July/Aug., Sep./Oct., Nov./Dec., 2020, and Jan./Feb. 2021.

Devoldere, J., *ON4UN's Low-Band DXing*, 5th edition. (ARRL, 2010).

EZNEC is an antenna-modeling computer program for Microsoft *Windows*. See **eznec.com** for full information.

Gehrke, F., "Vertical Phased Arrays," in six parts, *Ham Radio*, May, June, July, Oct., and Dec. 1983, and May 1984.

Harrison, C., and King, R., "Theory of Coupled Folded Antennas," *IRE Trans on Antennas and Propagation*, Mar. 1960, pp. 131 – 135.

Hayward W., and DeMaw, D., *Solid State Design for the Radio Amateur* (ARRL, 1977).

Hayward, W., *Radio Frequency Design* (ARRL, 1994).

Jasik, H., *Antenna Engineering Handbook*, 1st ed. (McGraw-Hill, 1961). Later editions are edited by Richard C. Johnson.

King R., and Harrison, C., "Mutual and Self-Impedance for Coupled Antennas," *Journal of Applied Physics*, Vol. 15, June 1944, pp. 481 – 495.

King, R., "Self- and Mutual Impedances of Parallel Identical Antennas," *Proc. IRE*, Aug. 1952, pp. 981 – 988.

King, R., *Theory of Linear Antennas* (Harvard University Press, 1956), p. 275ff.

Kohler, H., "Antenna Design for Field-Strength Gain," *Proc. IRE*, Oct. 1944, pp. 611 – 616.

Kraus, J., "Antenna Arrays with Closely Spaced Elements," *Proc. IRE*, Feb., 1940, pp. 76 – 84.

Kraus, J., *Antennas*, 2nd ed. (McGraw-Hill, 1988).
Johnson, R., *Antenna Engineering Handbook*, 3rd ed. (McGraw-Hill, 1993). This is a later edition of the volume by the same name edited by H. Jasik.
Laport, E., *Radio Antenna Engineering* (McGraw-Hill, 1952).
Lawson, J., "Simple Arrays of Vertical Antenna Elements," *QST*, May 1971, pp. 22 – 27.
Lee, P., *The Amateur Radio Vertical Antenna Handbook*, 2nd ed. (CQ Publishing, Inc., 1984).
Lewallen, R., "Baluns: What They Do and How They Do It," *The ARRL Antenna Compendium, Vol. 1* (ARRL, 1985). Also available for viewing at **eznec.com/Amateur/Articles/Baluns.pdf**.
Lewallen, R., "The Impact of Current Distribution on Array Patterns," Technical Correspondence, *QST*, July 1990, pp. 39 – 40. Also available for viewing at **eznec.com/Amateur/Articles/Current_Dist.pdf**
Mailloux, R., *Phased Array Antenna Handbook*, 3rd Edition (Artech House, 2018).
Maxwell, M., "Some Aspects of the Balun Problem," *QST*, Mar. 1983, pp. 38 – 40.
Sevick, J., "The Ground-Image Vertical Antenna," *QST*, July 1971, pp. 16 – 19, 22.
Sevick, J., "The W2FMI Ground-Mounted Short Vertical," *QST*, Mar. 1973, pp. 13 – 28,41.
Wilkinson, E., "An N-Way Hybrid Power Divider," *IRE Transactions on Microwave Theory and Techniques*, Jan. 1960.

TABLE OF CONTENTS

Chapter 8 — Online Material

Articles

- “A Dipole Curtain for 15 and 10 Meters” by Mike Loukides, W1JQ
- “Bob Zepp: A Low Band, Low Cost, High Performance Antenna – Parts 1 and 2” by Robert Zavrel, W7SX
- “Curtains for You” by Jim Cain, K1TN (and feedback)
- “Hands-On Radio Experiment #133 – Extended Double Zepp Antenna” by Ward Silver NØAX
- “Some Notes on Two-Element Horizontal Phased Arrays – Parts 1 – 4,” by L. B. Cebik W4RNL
- “The Extended Double Zepp Revisited” by Jerry Haigwood, W5JH
- “The Extended Lazy H Antenna” by Walter Salmon, VK2SA
- “The Multiband Extended Double Zepp and Derivative Designs” by Robert Zavrel, W7SX
- “The N4GG Array” by Hal Kennedy, N4GG
- “The W8JK Antenna: Recap and Update” by John Kraus, W8JK
- “Try the FD Special” by Roy Lewallen, W7EL

Chapter 8

Broadside and End-Fire Dipole Arrays

Dipoles — referring to elements which are usually λ/2 long — can be arranged to produce arrays, creating antennas with evocative names like the Lazy-H, ZL Special, Sterba Curtain, Half Square, and so forth. In these arrays, the elements are driven through specific lengths of feed line and by connections to other elements. Clever arrangement of the elements produces the necessary phase relationships of currents in each element to create the desired pattern.

This chapter was originally contributed by Rudy Severns, N6LF, and is written from the familiar amateur perspective of using these antennas at HF with wire elements. Much of the material translates easily to VHF and higher frequencies where the elements are made of tubing or rod. At microwave frequencies, the elements are often slot antennas. The reader will find several projects for designing and constructing these antennas in the "References and Bibliography" and with this book's online material.

Several *EZNEC* model files for antennas in this chapter are available in a zipped file in the online material. After unzipping the file, copy the files to your *EZNEC* description file directory. You should then see the files or new folder in *EZNEC*.

PHASE AND POLARITY

The term phase has the same meaning when used in connection with the currents flowing in antenna elements as it does in ordinary circuit work. Reiterating from the **Antenna Fundamentals** chapter, it is important to distinguish between phase and polarity. *Polarity* is simply a convention that assigns a positive and negative direction or convention, while *phase* refers to the part of the cycle of an ac voltage or current. See the in-depth discussion of phase, polarity, and direction in the **Phased Arrays** chapter.

ARRAY DEFINITIONS

These definitions were introduced in the **Phased Arrays** chapter but are repeated here for reference.

A *driven array* is one in which all the elements are driven elements with power supplied to them through feed lines or direct connections to other elements. A *parasitic array* is one in which one or more of the elements are parasitic elements. The antennas described in this chapter are driven arrays.

A *broadside array* is one in which the principal direction of radiation is perpendicular to the axis of the array and to the plane containing the elements. The elements of a broadside array may be *collinear* or *parallel.*

An *end-fire* array is one in which the principal direction of radiation coincides with the direction of the array axis. An end-fire array usually consists of parallel elements. They cannot be collinear, as linear elements do not radiate straight off their ends.

A *bidirectional* array is one that radiates equally well in either direction along the line of maximum radiation. A *unidirectional* array is one that has only one principal direction of radiation.

PHASING IN ARRAY ELEMENTS

In this chapter's antenna diagrams, the relative direction of current flow in the various elements and connecting lines is shown by arrows. Phasing in the different elements may be determined either based on current direction or polarity of voltages. There are two rules to remember:

1) Starting from an element's open end, the current directions reverse in each successive λ/2 element. In terms of voltage, the polarity reverses at each λ/2 point, starting from an open end.

2) Currents in transmission lines flow in opposite directions in adjacent wires. In terms of voltage, polarities always must be opposite.

Examples of the use of current direction and voltage polarity are given at A and B, respectively, in **Figure 8.1**. The λ/2 points in the system are marked by small circles. When current in one section flows toward a circle, the current in the next section must also flow toward it, and vice versa.

In the 4-element antenna shown at A, the current in the upper right-hand element cannot flow toward the phasing line

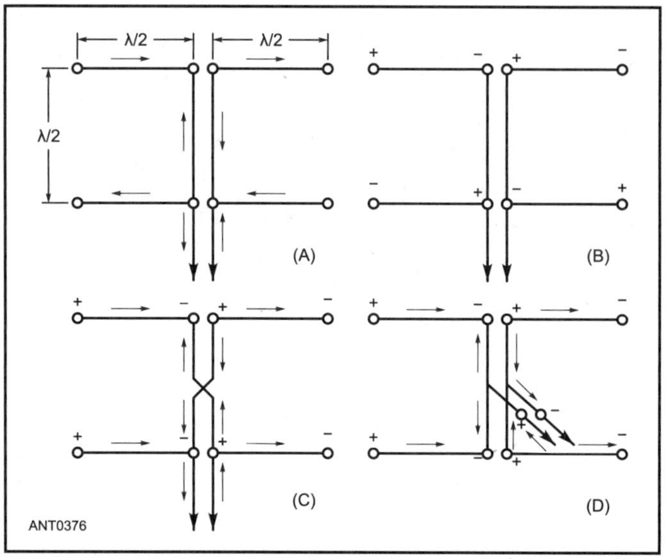

Figure 8.1 — Methods of showing and checking the phase of currents in elements and phasing lines.

because then the current in the right-hand section of the phasing line would have to flow upward and thus would be flowing in the same direction as the current in the left-hand wire. The phasing line would simply act like two wires in parallel in such a case. Of course, all arrows in the drawing could be reversed, and the net effect would be unchanged. (A phasing line is a transmission line connected between elements that controls the phase of the elements.)

C shows the effect of transposing the phasing line. This transposition reverses the direction of current flow in the lower pair of elements, as compared with A, and thus changes the array from a combination collinear and end-fire arrangement into a collinear-broadside array.

The drawing at D shows what happens when the transmission line is connected at the center of a section of phasing line. Viewed from the main transmission line, the two parts of the phasing line are simply in parallel, so the half wavelength is measured from the antenna element along the upper section of phasing line and thence along the transmission line. The distance from the lower elements is measured in the same way. Obviously, the two sections of phasing line should be the same length. If they are not, the current distribution becomes quite complicated; the element currents are neither in-phase nor 180° out-of-phase, and the elements at opposite ends of the lines do not receive the same current. To change the element current phasing at D into the phasing at A, simply transpose the wires in one section of the phasing line. This reverses the direction of current flow in the antenna elements connected to that section of phasing line.

8.1 Broadside Arrays

Broadside arrays can be made up of collinear or parallel elements or combinations of the two. They can provide performance comparable to rotatable beams at very low cost if the amateur has the necessary supports.

8.1.1 Collinear Arrays

Collinear arrays are always operated with the elements in-phase. (If alternate elements in such an array are out-of-phase, the system simply becomes a harmonic type of antenna.) A collinear array is a broadside radiator, the direction of maximum radiation being at right angles to the line of the antenna.

POWER GAIN

Because of the nature of the mutual impedance between collinear elements, the feed point resistance (compared to a single element, which is ≈73 Ω) is increased as shown in the **Phased Arrays** chapter. For this reason, the power gain does not increase in direct proportion to the number of elements. The gain with two elements, as the spacing between them is varied, is shown by **Figure 8.2**. Although the gain is greatest when the end-to-end spacing is in the region of 0.4 to 0.6 λ, the use of spacings of this order is inconvenient to build and introduces problems in feeding the two elements. As a result, collinear elements are almost always operated with their ends quite close together — in wire antennas, usually with just a strain insulator between.

With very small spacing between the ends of adjacent elements the theoretical power gain of collinear arrays, assuming the use of #12 AWG copper wire, is approximately as follows over a dipole in free space:

2 collinear elements — 1.6 dB
3 collinear elements — 3.1 dB
4 collinear elements — 3.9 dB

More than four elements are rarely used.

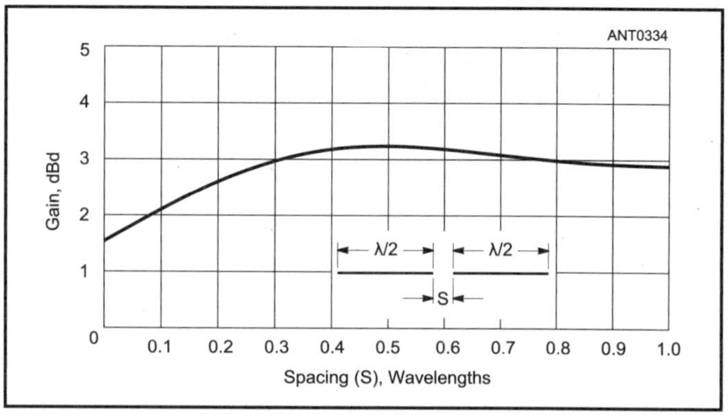

Figure 8.2 — Gain of two collinear λ/2 elements as a function of spacing between the adjacent ends.

DIRECTIVITY

The directivity of a collinear array, in a plane containing the axis of the array, increases with its length. Small secondary lobes appear in the pattern when more than two elements are used, but the amplitudes of these lobes are low enough so that they are usually not important. In a plane at right angles to the array the directive diagram is a circle, no matter what the number of elements. Collinear operation, therefore, affects only E-plane directivity, the plane containing the antenna.

When a collinear array is mounted with the elements vertical, the antenna radiates equally well in all geographical directions. An array of such stacked collinear elements tends to confine the radiation to low vertical angles. This configuration is common in base station antennas for VHF and UHF and is discussed in the **VHF and UHF Antenna Systems** chapter.

If a collinear array is mounted horizontally, the directive pattern in the vertical plane at right angles to the array is the same as the vertical pattern of a simple λ/2 antenna at the same height as discussed in the chapter **Effects of Ground**.

8.1.2 Two-Element Arrays

The simplest and most popular collinear array is one using two elements, as shown in **Figure 8.3**. This system is commonly known as two half-waves in phase. The directive pattern in a plane containing the wire axis is shown in **Figure 8.4**, which shows superimposed patterns for a dipole and 2, 3 and 4-element collinear arrays. Depending on the conductor size, height, and similar factors, the impedance at the feed point can be expected to be in the range of 4 to 6 kΩ, for wire antennas. If the elements are made of tubing having a low λ/diameter (wavelength to diameter) ratio, values as low as 1 kΩ are representative. The system can be fed through an open-wire tuned line with negligible loss for ordinary line lengths, or a matching section may be used if desired.

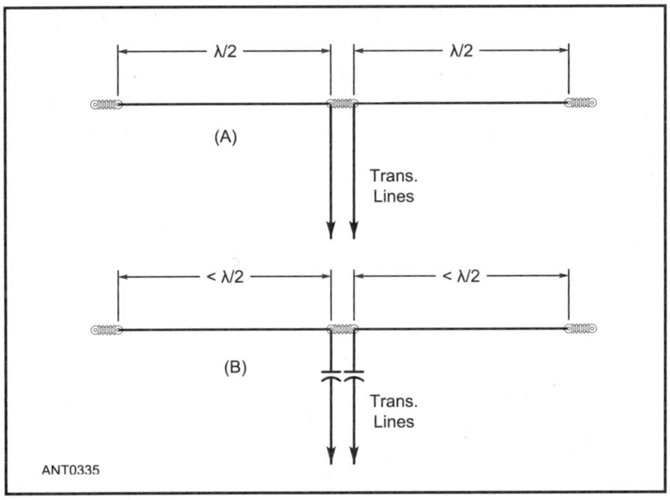

Figure 8.3 — At A, two-element collinear array (two half-waves in phase). The transmission line shown would operate as a tuned line. A matching section can be substituted and a nonresonant line used if desired, as shown at B, where the matching section is two series capacitors.

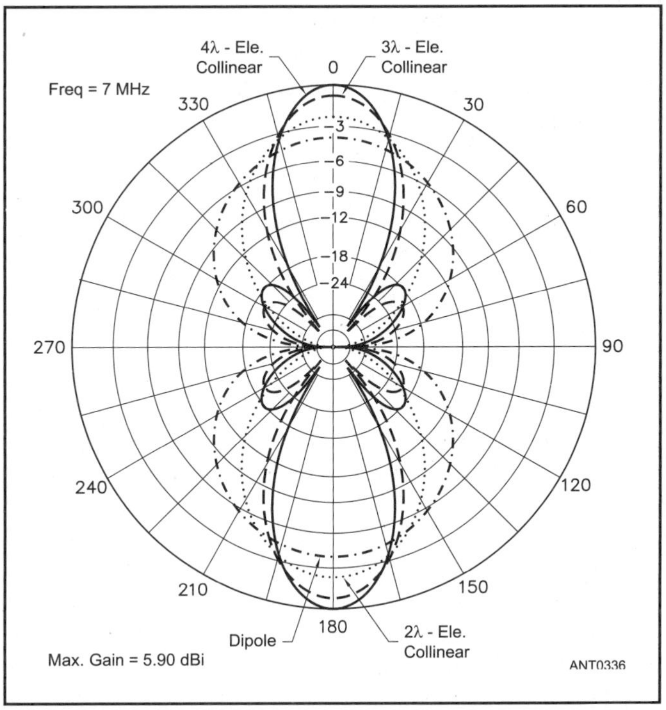

Figure 8.4 — Free-space E-plane directive diagram for dipole, 2, 3 and 4-element collinear arrays. The solid line is a 4-element collinear; the dashed line is for a 3-element collinear; the dotted line is for a 2-element collinear and the dashed-dotted line is for a λ/2 dipole.

Several arrangements for matching the feed line to this antenna are described in the chapter **Transmission Line System Techniques**. If elements somewhat shorter than λ/2 are used, then additional matching schemes can be employed at the expense of a slight reduction in gain. When the elements are shortened two things happen — the impedance at the feed point drops and the impedance has inductive reactance that can be tuned out with simple series capacitors, as shown in Figure 8.3B.

Note that these capacitors must be suitable for the power level. Small doorknob capacitors, such as those frequently used in power amplifiers, are suitable. By way of an example, if each side of a 40 meter 2-element array is shortened from 67 to 58 feet, the feed point impedance drops from nearly 6000 Ω to about 1012 Ω with an inductive reactance 1800 Ω. The reactance can be tuned out by inserting 25 pF capacitors at the feed point. The 1012 Ω resistance can be transformed to 200 Ω using a λ/4 matching section made of 450-Ω ladder line and then transformed to 50 Ω with a 4:1 balun. Shortening the array as suggested reduces the gain by about 0.5 dB.

Another scheme that preserves the gain is to use a 450-Ω λ/4 matching section and shorten the antenna only slightly to have a resistance of 4 kΩ. The impedance at the input of the matching section is then near 50 Ω and a simple 1:1 balun can be used. Many other schemes are possible. The free-space E-plane response for a 2-element collinear array is shown in Figure 8.4, compared with the responses for more elaborate collinear arrays described below.

8.1.3 Three- And Four-Element Arrays

In a long wire the direction of current flow reverses in each λ/2 section. Consequently, collinear elements cannot simply be connected end to end; there must be some means for making the current flow in the same direction in all elements. When more than two collinear elements are used it is necessary to connect phasing stubs between adjacent elements to bring the currents in all elements in-phase. In **Figure 8.5A** the direction of current flow is correct in the two left-hand elements because the shorted λ/4 transmission line (stub) is connected between them. This stub may be looked upon simply as the alternate λ/2 section of a long-wire antenna folded back on itself to cancel its radiation. In Figure 8.5A the part to the right of the transmission line has a total length of three half wavelengths, the center half wave being folded back to form a λ/4 phase-reversing stub. No data are available on the impedance at the feed point in this arrangement, but various considerations indicate that it should be over 1 kΩ.

An alternative method of feeding three collinear elements is shown in Figure 8.5B. In this case power is applied at the center of the middle element and phase-reversing stubs are used between this element and both of the outer elements. The impedance at the feed point in this case is somewhat over 300 Ω and provides a close match to 300 Ω line. The SWR will be less than 2:1 when 600-Ω line is used. Center feed of this type is somewhat preferable to the arrangement in Figure 8.5A because the system is balanced as a whole. This assures more uniform power distribution among the elements. In Figure 8.5A, the right-hand element is likely to receive somewhat less power than the other two because a portion of the input power is radiated by the middle element before it can reach the element located at the extreme right.

A four-element array is shown in Figure 8.5C. The system is symmetrical when fed between the two center elements as shown. As in the three-element case, no data are available on the impedance at the feed point. However, the SWR with a 600 Ω line should not be much over 2:1.

Figure 8.4 compares the directive patterns of 2, 3 and 4-element arrays. Collinear arrays can be extended to more than four elements. However, the simple 2-element collinear array is the type most frequently used, as it lends itself well to multiband operation. More than two collinear elements are

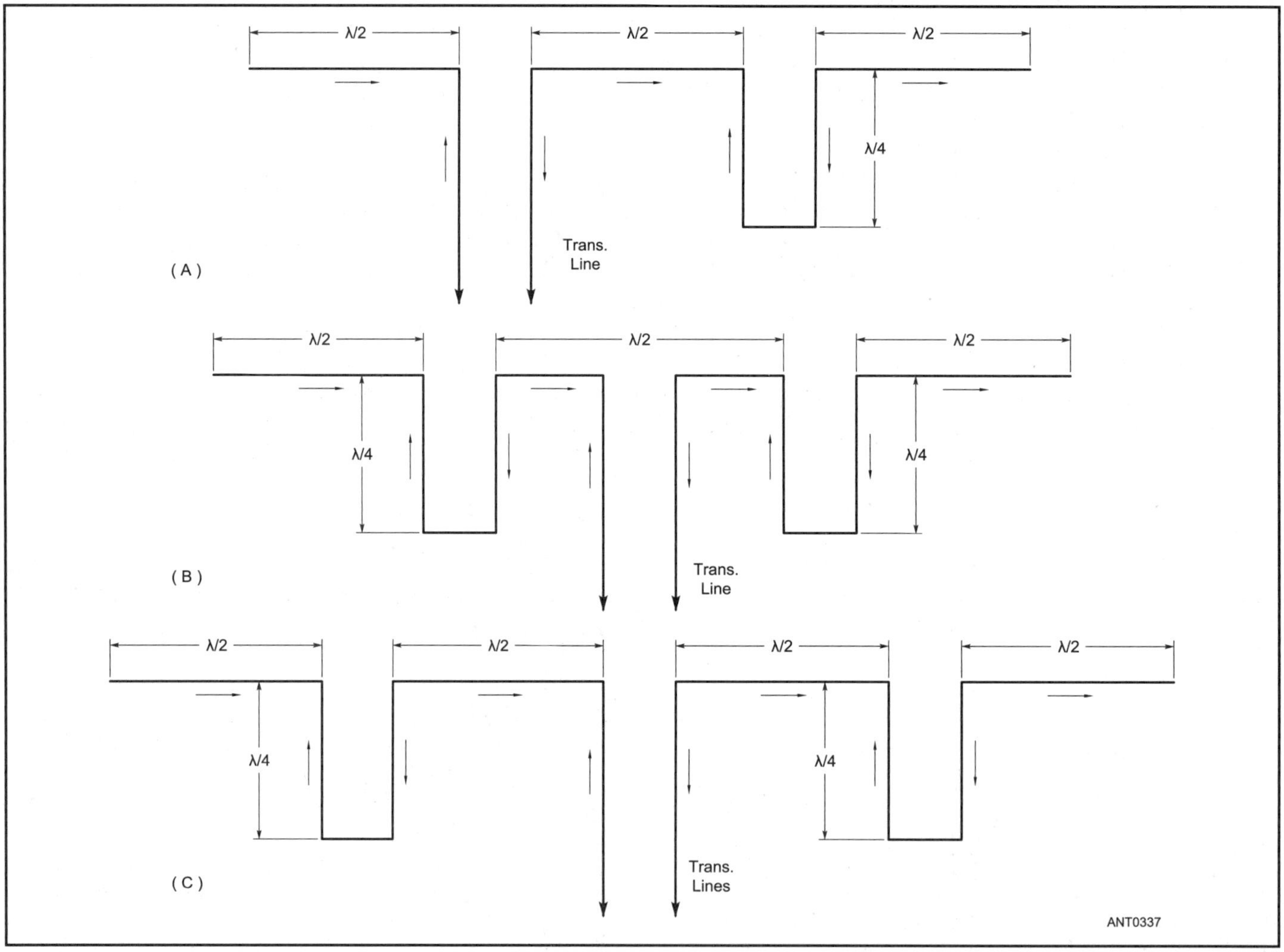

Figure 8.5 — Layouts for 3- and 4-element collinear arrays. Alternative methods of feeding a 3-element array are shown at A and B. These drawings also show the current distribution on the antenna elements and phasing stubs. A matched transmission line can be substituted for the tuned line by using a suitable matching section.

seldom used because more gain can be obtained from other types of arrays.

8.1.4 Collinear Array Adjustment

In any of the collinear systems described, the lengths of the radiating elements are the same as for λ/2 dipoles. The lengths of the phasing stubs can be found from the equations given in the chapter **Transmission Line System Techniques** for the type of line used. If the stub is open-wire line (500 to 600 Ω impedance) you may assume a velocity factor of 0.975 in the formula for a λ/4 line. On-site adjustment is, in general, an unnecessary refinement. If desired, however, the following procedure may be used when the system has more than two elements.

Disconnect all stubs and all elements except those directly connected to the transmission line (in the case of a feed such as is shown in Figure 8.5B leave only the center element connected to the line). Adjust the elements to resonance, using the still-connected element. When the proper length is determined, cut all other elements to the same length. Make the phasing stubs slightly long and use a shorting bar to adjust their length. Connect the elements to the stubs and adjust the stubs to resonance, as indicated by maximum current in the shorting bars or by the SWR on the transmission line. If more than three or four elements are used it is best to add elements two at a time (one at each end of the array), resonating the system each time before a new pair is added.

8.1.5 The Extended Double Zepp

One method to obtain higher gain that goes with wider spacing in a simple system of two collinear elements is to make the elements somewhat longer than λ/2. As shown in **Figure 8.6**, this increases the spacing between the two in-phase λ/2 sections at the ends of the wires. The section in the center carries a current of opposite phase, but if this section is short the current will be small; it represents only the outer ends of a λ/2 antenna section. Because of the small current and short length, the radiation from the center is small. The optimum length for each element is 0.64 λ. At greater lengths the system tends to act as a long-wire antenna, and the gain decreases.

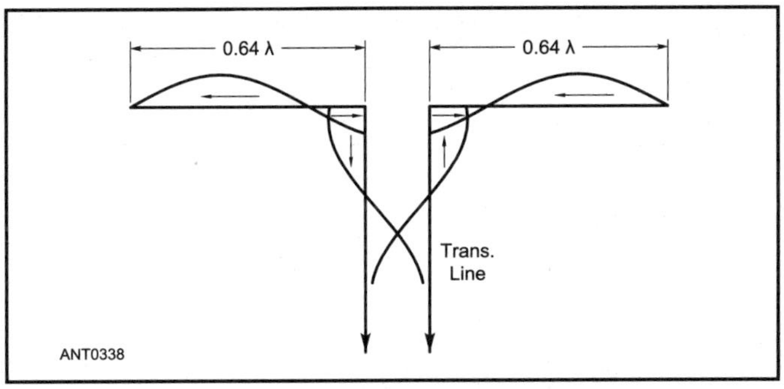

Figure 8.6 — The extended double Zepp. This system gives somewhat more gain than two λ-sized collinear elements.

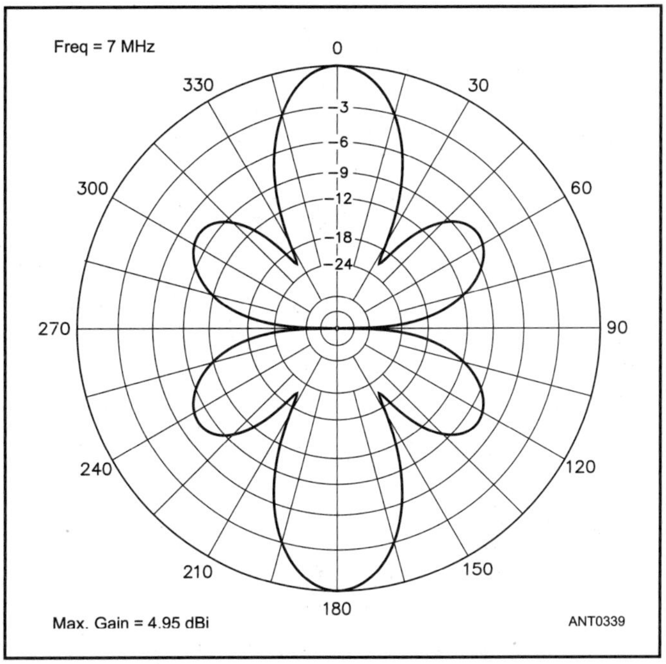

Figure 8.7 — E-plane pattern for the extended double Zepp of Figure 8.6. This is also the horizontal directional pattern when the elements are horizontal. The axis of the elements lies along the 90°-270° line. The free-space array gain is approximately 4.95 dBi.

This system is known as the *extended double Zepp* or *EDZ*, first described in *QST* in 1938 by Hugo Romander, W2NB (see the Bibliography). The gain over a λ/2 dipole is approximately 3 dB, as compared with about 1.6 dB for two collinear λ/2 dipoles. The directional pattern in the plane containing the axis of the antenna is shown in **Figure 8.7**. As in the case of all other collinear arrays, the free-space pattern in the plane at right angles to the antenna elements is the same as that of a λ/2 antenna — circular. The article "The Extended Double Zepp Revisited" by Jerry Haigwood, W5JH from the September 2006 *QST* provides dimensions for the EDZ on 40 through 10 meters, along with building tips. (The article is also included with this book's online material. An analysis of the EDZ and related designs by Zavrel is listed in the Bibliography.)

This antenna is not resonant at the operating frequency so that the feed point impedance is complex (R ± *j*X). A typical example of the variation of the feed point impedance over the band for a 40 meter double-extended Zepp is shown in **Figure 8.8**. This antenna is normally fed with open-wire transmission line to an antenna tuner. This allows the antenna to be used on multiple bands, although SWR may be high on some bands. Impedance at the antenna tuner depends on the antenna's feed point impedance and the length of the feed line.

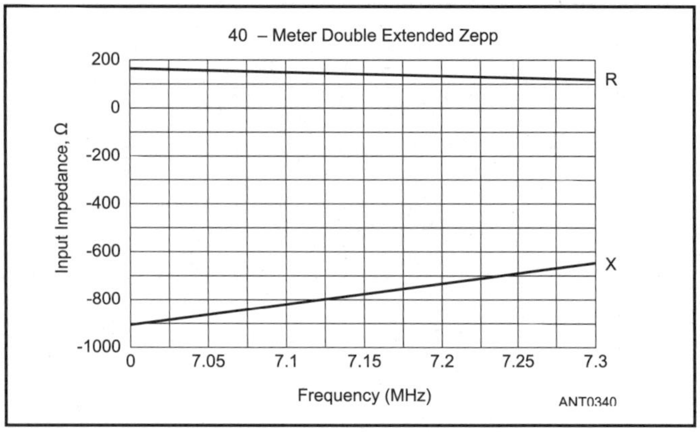

Figure 8.8 — Resistive and reactive feed point impedance of a 40 meter extended double Zepp in free space.

VARIATIONS ON THE EXTENDED DOUBLE ZEPP

By selecting the right length of feed line, the feed point impedance of the EDZ can be brought close to 50 Ω on the desired band. At that point, a choke balun can be used to create a transition to coaxial feed line. (See the **Transmission Line System Techniques** chapter regarding baluns.) The general design is shown in **Figure 8.9**.

Table 8.1 shows EDZ designs by W5JH including the length of the antenna (L_d) and 450 Ω window line (L_f) that result in approximately 50 Ω impedance at the end of the window line. (Note that the design also includes height above average ground which also affects feed point impedance.) The 50 Ω point is created only on the band shown in the table — SWR will be greatly different on other bands. (The original article by W5JH is provided with the online material.) The exact length of antenna which reproduces these impedances will depend on height above ground and type of ground. Be prepared to adjust antenna length and window line length.

If it is desired to place the coax transition point farther from the antenna, add an integer number of λ/2 of feed line which will create a 50 Ω impedance at which the transition can be made. The transmission line software *TLW* can be used to calculate physical lengths for a variety of different types of parallel conductor feed lines.

Figure 8.10 is a 20/15 meter variation of the EDZ with low SWR on both bands through careful selection of the length of

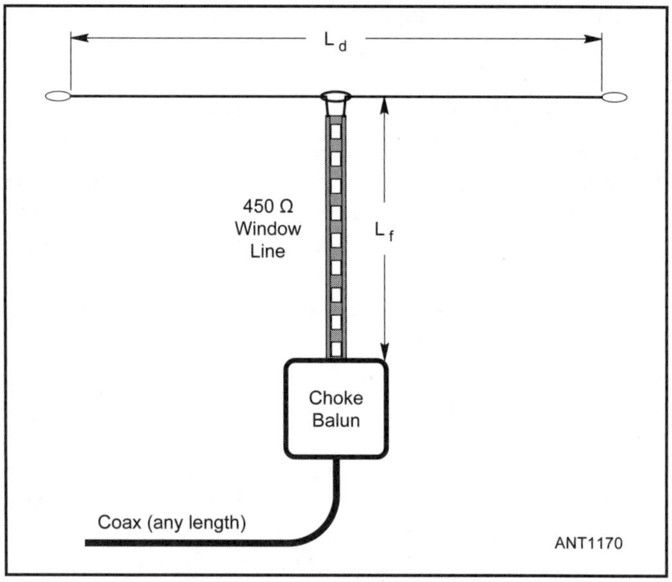

Figure 8.9 — A general design for Extended Double Zepp antennas by W5JH. Values for L_d (doublet length) and L_f (feed line length) are given in Table 8.1.

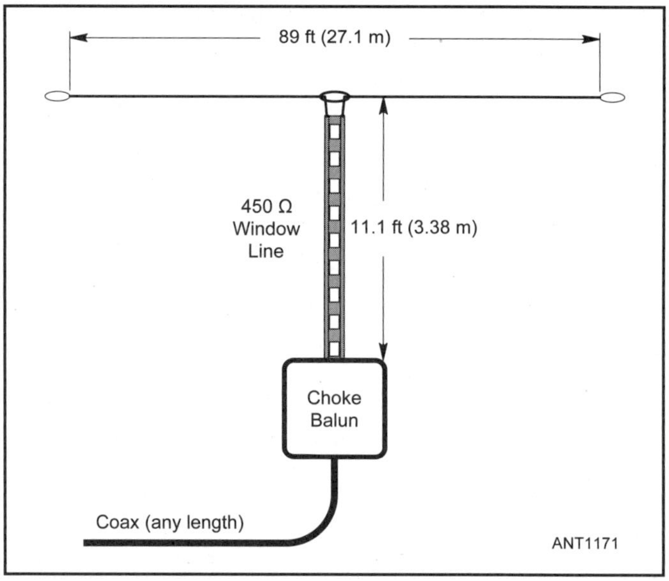

Figure 8.10 — The 20/15 meter EDZ with feed line length specified for low SWR on 20 and 15 meters.

Table 8.1
EDZ Antenna Dimensions for Different Bands

Freq (MHz)	L_d *(ft)*	*Min Height (ft)*	*Antenna Z (Ω)*	*Feed point Z (Ω)*	L_f *(ft)*	*SWR*
7.075	175.2	66	170.5 –*j*976.1	47.10 +*j*0.25	21.65	1.062:1
10.11	122.6	47	163.0 –*j*934.1	47.80 –*j*0.07	14.85	1.046:1
14.175	87.5	34	155.3 –*j*889.6	48.62 –*j*0.36	10.35	1.029:1
18.1	68.5	30	133.4 –*j*848.9	44.63 +*j*0.38	7.92	1.121:1
21.2	58.5	30	132.1 –*j*799.9	47.65 –*j*0.28	6.56	1.050:1
24.9	49.8	30	156.7 –*j*772.3	58.60 –*j*0.10	5.51	1.172:1
28.2	44	30	169.8 –*j*772.4	63.24 +*j*0.26	4.88	1.265:1

L_d is the antenna length, L_f is the length of the matching feed line the 450 Ω feed line section. The pattern for 20 meters shown in **Figure 8.11** is similar to the classic EDZ of Figure 8.7 but because of the extra length on 15 meters, the pattern takes on a clover-leaf shape as shown in **Figure 8.12**. A 74.7-foot section of 450 Ω window line (two λ/2 on 20 meters and 3λ/2 on 15 meters) brings the coax transition point to ground level and SWR is less than 1.5:1 on both bands. (The original article, "Hands-On Radio, Experiment #133," is provided with the online material.)

A versatile antenna for 40, 80, and 160 meters, the "Bob Zepp" by Robert Zavrel, W7SX, is shown in **Figure 8.13**. This design uses traps to isolate sections of the antenna at

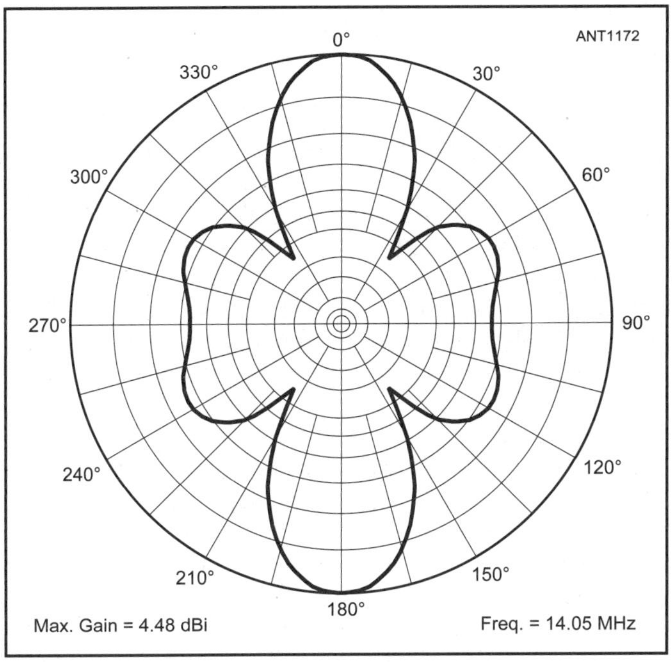

Figure 8.11 — The pattern of the 20/15 meter EDZ on 20 meters.

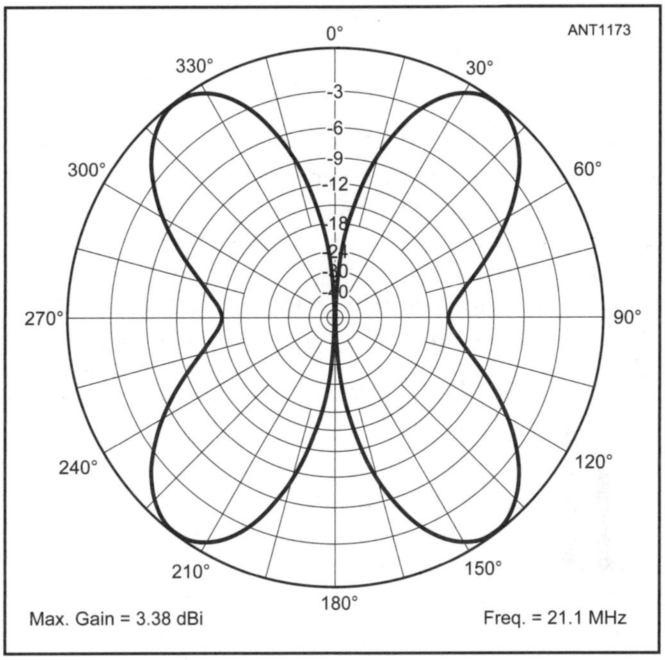

Figure 8.12 — The pattern of the 20/15 meter EDZ on 15 meters.

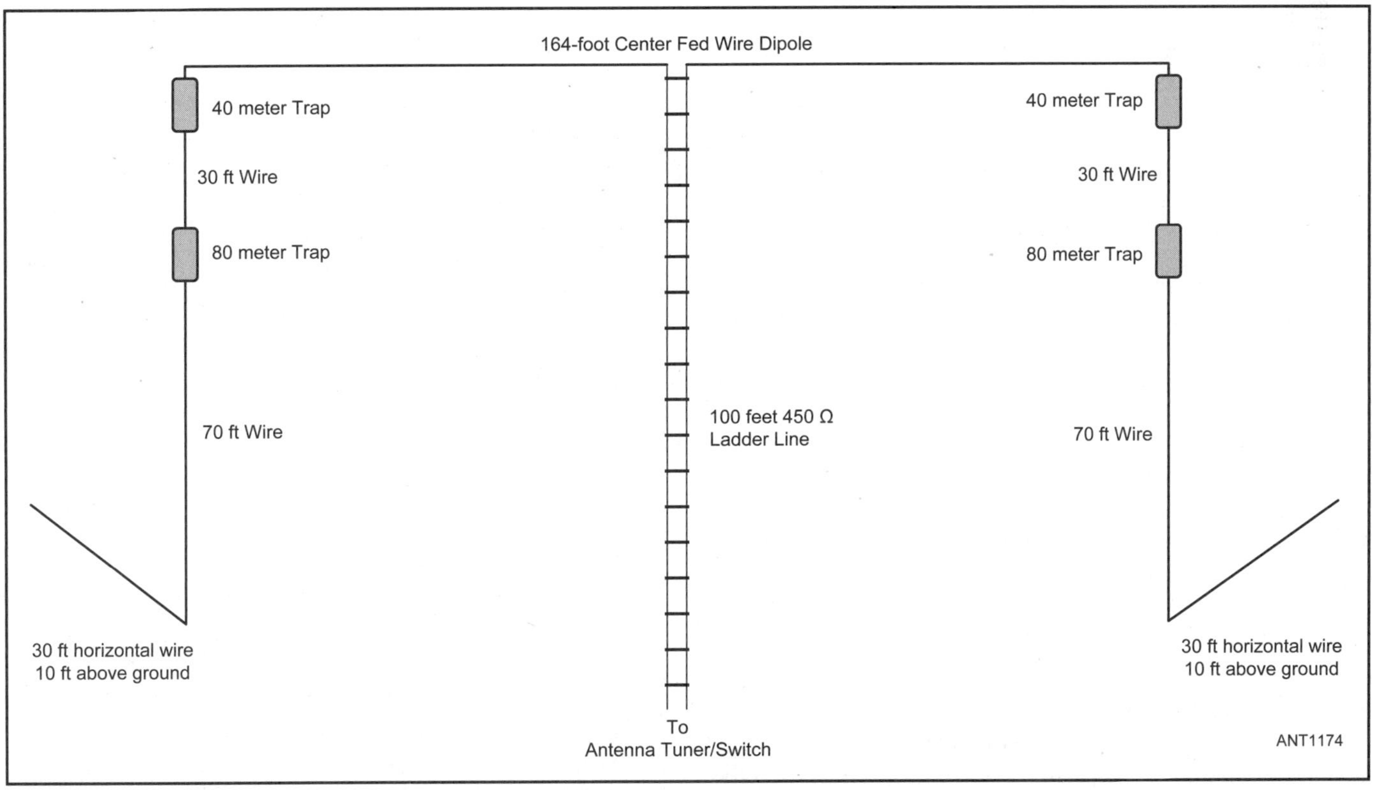

Figure 8.13 — The Bob-Zepp is an EDZ on 40 meters, an extended dipole on 80 meters, and a bi-directional array on 160 meters.

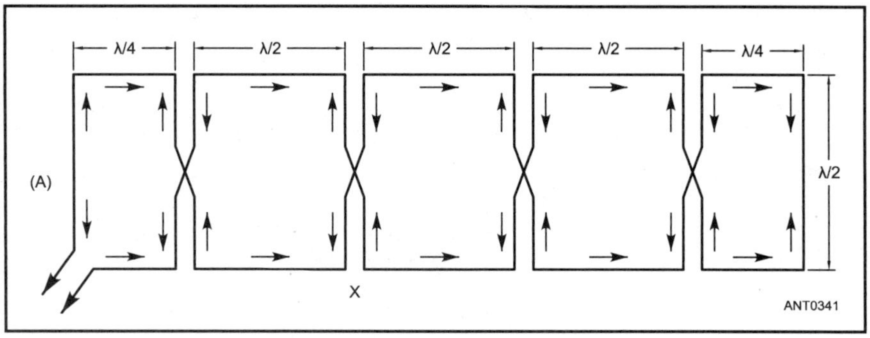

Figure 8.14 — Typical Sterba array, an 8-element version.

40 meters where it acts as a true EDZ. The antenna becomes an extended dipole on 80 meters and a two-vertical bidirectional, top-fed array on 160 meters. The antenna was designed for a height of around 110 feet. Significantly higher installation will affect the antenna's feed point impedance. The antenna can be installed at lower heights, but the lower vertical sections and horizontal wires will have to be altered. See the original article, provided with the online material, for the complete design, including a full-power tuning unit and numerous construction drawings.

8.1.6 The Sterba Curtain

Two collinear arrays can be combined to form the *Sterba array*, often called the *Sterba curtain*. An 8-element example of a Sterba array is shown in **Figure 8.14**. The four λ/4 elements joined on the ends are equivalent to two λ/2 elements. The two collinear arrays are spaced λ/2 and the λ/4 phasing lines connected together to provide λ/2 phasing lines. This arrangement has the advantage of increasing the gain for a given length and increasing the E-plane directivity, which is no longer circular. An additional advantage of this array is that the wire forms a closed loop. For installations where icing is a problem a low voltage dc or low frequency (50 or 60 Hz) ac current can be passed through the wire to heat it for deicing. The heating current is isolated from RF by decoupling chokes. This is standard practice in commercial installations.

The number of sections in a Sterba array can be extended as far as desired but more than four or five are rarely used because of the slow increase in gain with extra elements, the narrow H-plane directivity, and the appearance of multiple sidelobes. When fed at the point indicated the impedance is about 600 Ω. The antenna can also be fed at the point marked X. The impedance at this point will be about 1 kΩ. The gain of the 8-element array in Figure 8.14 will be between 7 to 8 dB over a single element. A 10 meter Sterba curtain is described in the article, "Curtains for You," by Jim Cain, K1TN, that is included with the online material.

8.2 Parallel Broadside Arrays

To obtain broadside directivity with parallel elements the currents in the elements must all be in-phase. At a distant point lying on a line perpendicular to the axis of the array and also perpendicular to the plane containing the elements, the fields from all elements add up in phase. The situation is similar to four parallel λ/2 dipoles fed together as a broadside array.

Broadside arrays of this type theoretically can have any number of elements. However, practical limitations of construction and available space usually limit the number of broadside parallel elements. These practical aspects of building a dipole curtain are illustrated in the article "A Dipole Curtain for 15 and 10 Meters" by Mike Loukides, W1JQ, in the Aug 2003 *QST* article with the online material.

8.2.1 Power Gain

The power gain of a parallel-element broadside array depends on the spacing between elements as well as on the number of elements. The way in which the gain of a two-element array varies with spacing is shown in **Figure 8.15**. The greatest gain is obtained when the spacing is in the vicinity of 0.67 λ.

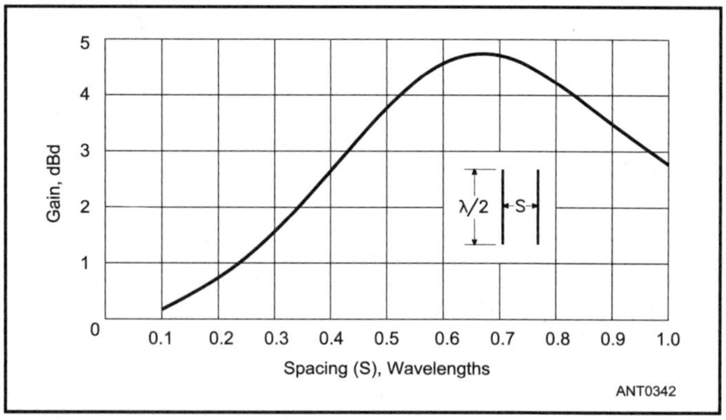

Figure 8.15 — Gain as a function of the spacing between two parallel elements operated in-phase (broadside).

The theoretical gains of broadside arrays having more than two elements are approximately as follows:

No. of Parallel Elements	*dB Gain with λ/2 Spacing*	*dB Gain with 3λ/4 Spacing*
3	5.7	7.2
4	7.1	8.5
5	8.1	9.4
6	8.9	10.4

The elements must, of course, all lie in the same plane, and all must be fed in-phase.

8.2.2 Directivity

The sharpness of the directive pattern depends on spacing between elements and number of elements. Larger element spacing will sharpen the main lobe, for a given number of elements, up to a point as was shown in Figure 8.2. The two-element array has no minor lobes when the spacing is λ/2, but small minor lobes appear at greater spacings. When three or more elements are used the pattern always has minor lobes.

8.3 Other Forms of Broadside Arrays

For those who have the available room, multielement arrays based on the broadside concept have something to offer. The antennas are large but of simple design and non-critical dimensions; they are also very economical in terms of gain per unit of cost.

Large arrays can often be fed at several different points. However, the pattern symmetry may be sensitive to the choice of feed point within the array. Nonsymmetrical feed points will result in small asymmetries in the pattern, but these are not usually of great concern.

Arrays of three and four elements are shown in **Figure 8.16**. In the 3-element array with λ/2 spacing at A, the array is fed at the center. This is the most desirable point in that it tends to keep the power distribution among the elements uniform. However, the transmission line could alternatively be connected at either point B or C of Figure 8.16A, with only slight skewing of the radiation pattern.

When the spacing is greater than λ/2, the phasing lines must be 1 λ long and are not transposed between elements. This is shown Figure 8.16B. With this arrangement, any element spacing up to 1 λ can be used, if the phasing lines can be folded as suggested in the drawing.

The 3-element array at C is fed at the center of the system to make the power distribution among elements as uniform as possible. However, the transmission line could be connected at either point B, C, D or E. In this case the section of phasing line between B and D must be transposed to make the currents flow in the same direction in all elements. The 4-element array at C and the 3-element array at B have approximately the same gain when the element spacing in the array at B is 3λ/4.

An alternative feeding method is shown in Figure 8.16D. This system can also be applied to the 3-element arrays and will result in better symmetry in any case. It is necessary only to move the phasing line to the center of each element, making connection to both sides of the line instead of one only.

The free-space pattern for a 4-element array with λ/2 spacing is shown in **Figure 8.17**. This is also approximately the pattern for a 3-element array with 3λ/4 spacing.

Larger arrays can be designed and constructed by following the phasing principles shown in the drawings. No accurate

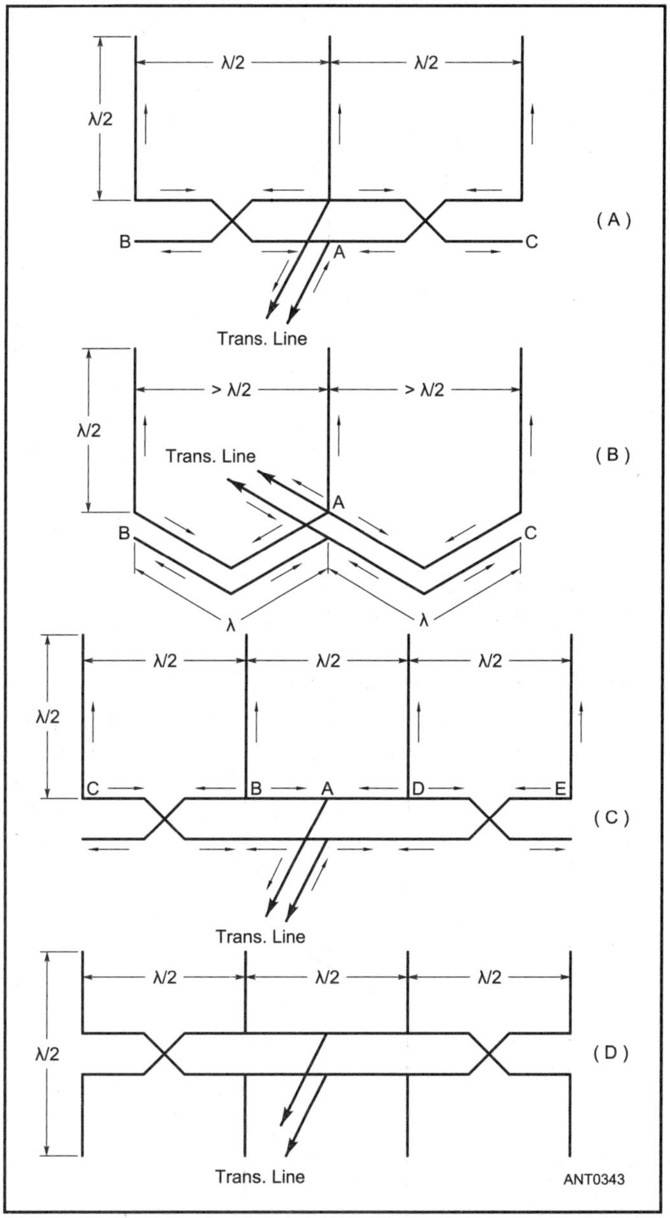

Figure 8.16 — Methods of feeding 3- and 4-element broadside arrays with parallel elements.

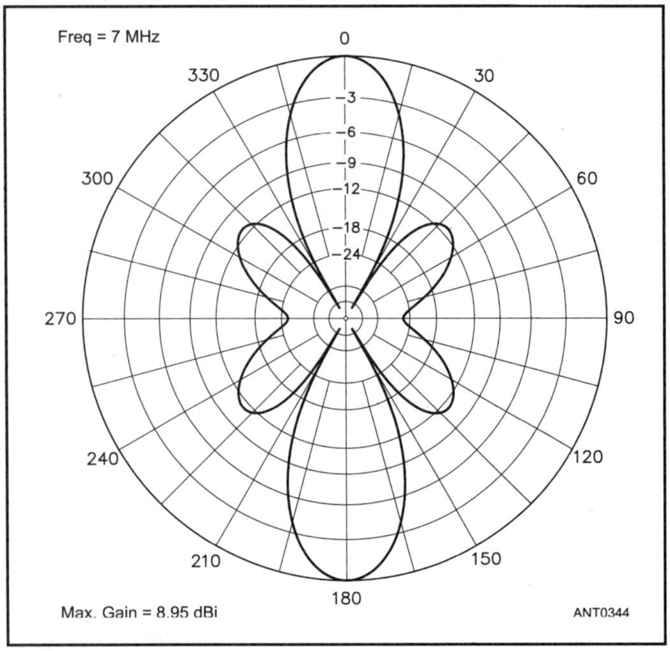

Figure 8.17 — Free-space E-plane pattern of a 4-element broadside array using parallel elements (Figure 8.16). This corresponds to the horizontal directive pattern at low wave angles for a vertically polarized array over ground. The axis of the elements lies along the 90° – 270° line.

figures are available for the impedances at the various feed points indicated in Figure 8.16. You can estimate it to be in the vicinity of 1 kΩ when the feed point is at a junction between the phasing line and a λ/2 element, becoming smaller as the number of elements in the array is increased. When the feed point is midway between end-fed elements as in Figure 8.16C, the feed point impedance of a 4-element array is in the vicinity of 200 to 300 Ω, with 600 Ω open-wire phasing lines. The impedance at the feed point with the antenna shown at D should be about 1.5 kΩ.

8.3.1 Nonuniform Element Currents

The pattern for a 4-element broadside array shown in Figure 8.16 has substantial sidelobes. This is typical for arrays more than λ/2 wide when equal currents flow in each element. Sidelobe amplitude can be reduced by using nonuniform current distribution among the elements. Many possible current amplitude distributions have been suggested. All of them have reduced current in the outer elements and greater current in the inner elements. This reduces the gain somewhat but can produce a more desirable pattern. One of the common current distributions is called *binomial current grading*. In this scheme the ratio of element currents is set equal to the coefficients of a polynomial. For example:

$$1\,x + 1, \Rightarrow 1, 1$$

$$(x+1)^2 = 1x^2 + 2x + 1, \;\Rightarrow 1, 2, 1$$

$$(x+1)^3 = 1x^3 + 3x^2 + 3x + 1, \;\Rightarrow 1, 3, 3, 1$$

$$(x+1)^4 = 1x^4 + 4x^3 + 6x^2 + 6x + 1, \;\Rightarrow 1, 4, 6, 4, 1$$

In a 2-element array the currents are equal, in a 3-element array the current in the center element is twice that in the outer elements, and so on.

8.3.2 Half-Square Antenna

On the low-frequency bands (40, 80 and 160 meters) it becomes increasingly difficult to use λ/2 elements because of their size. The half-square antenna is a 2-element broadside array with λ/4-high vertical elements and λ/2 horizontal spacing. See **Figure 8.18**. The free-space H-plane pattern for this array is shown in **Figure 8.19**. The antenna gives modest (4.2 dBi) but useful gain and has the advantage of only λ/4 height. Like all vertically polarized antennas, real-world performance depends directly on the characteristics of the ground surrounding it.

The half-square can be fed either at the point indicated or at the bottom end of one of the vertical elements using a voltage-feed scheme, such as for the Bobtail curtain described below. The feed point impedance is in the region of 50 Ω when

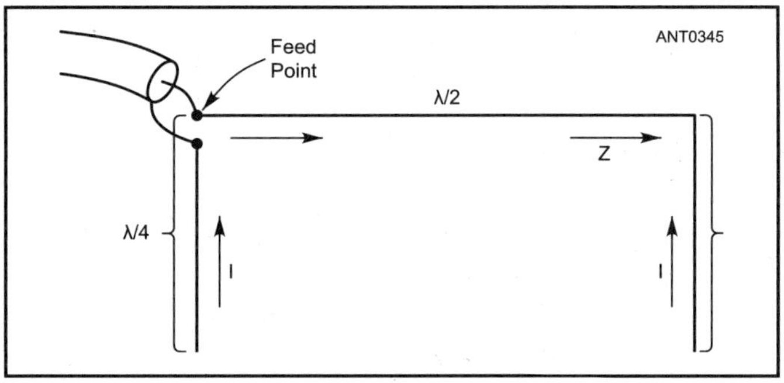

Figure 8.18 — Layout for the half-square antenna.

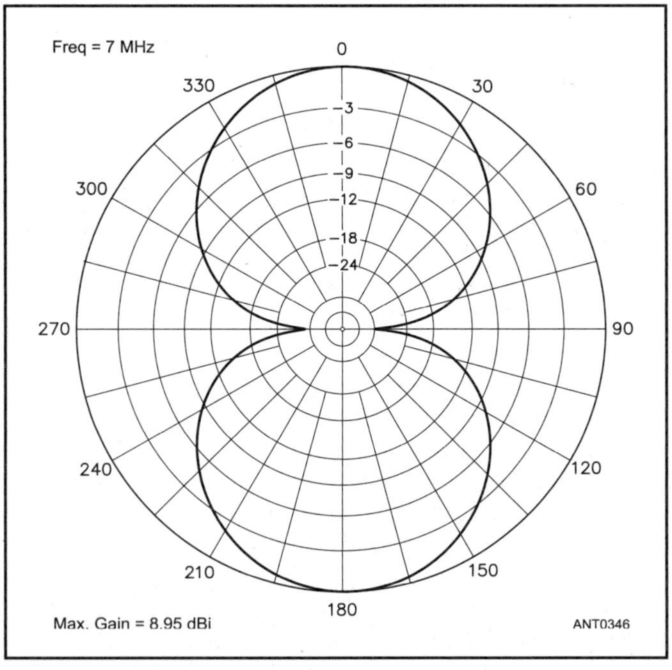

Figure 8.19 — Free-space E-plane directive pattern for the half-square antenna.

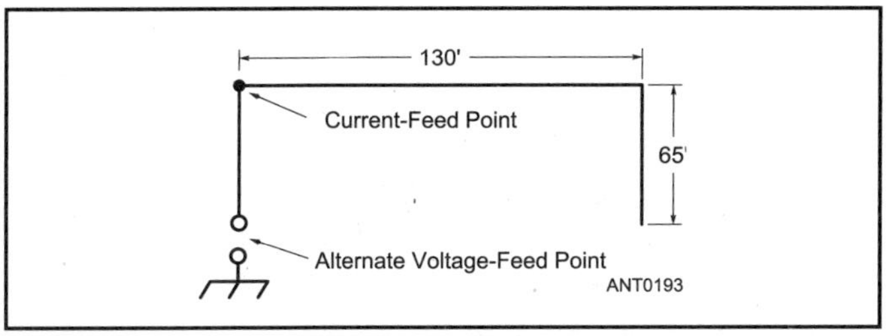

Figure 8.20 — Typical 80 meter half-square, with λ/4-high vertical legs and a λ/2-long horizontal leg. The antenna may be fed at the bottom or at a corner. When fed at a corner, the feed point is a low-impedance, current-feed. When fed at the bottom of one of the wires against a small ground counterpoise, the feed point is a high-impedance, voltage-feed.

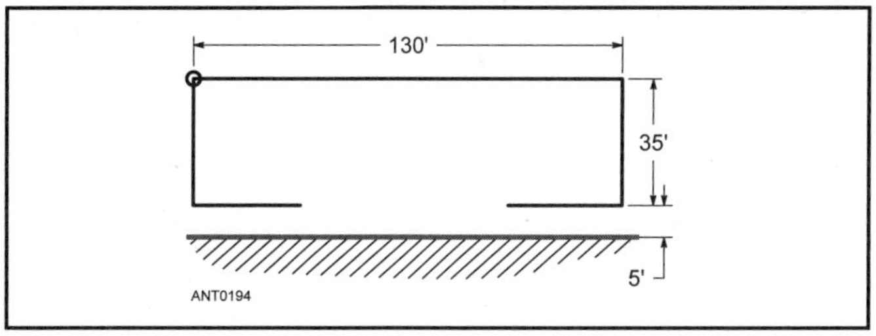

Figure 8.21 — An 80-meter half-square configured for 40-foot-high supports. The ends have been bent inward to re-resonate the antenna. The performance is compromised surprisingly little.

fed at a corner as shown in Figure 8.18. The SWR bandwidth is typically quite narrow as shown in the following design examples.

VARIATIONS ON THE HALF-SQUARE ANTENNA

The following section was originally presented in *The ARRL Antenna Compendium Vol 5*, by Rudy Severns, N6LF.

A simple modification to a standard dipole is to add two λ/4 vertical wires, one at each end, as shown in **Figure 8.20**. This makes a *half-square antenna*. The antenna can be fed at one corner (low-impedance, current fed) or at the lower end of one of the vertical wires (high-impedance, voltage fed). Other feed arrangements are also possible.

The "classical" dimensions for this antenna are λ/2 (131 feet at 3.75 MHz) for the top wire and λ/4 (65.5 feet) for the vertical wires. However, there is nothing sacred about these dimensions! They can vary over a wide range and still obtain nearly the same performance.

This antenna is two λ/4 verticals, spaced λ/2, fed in-phase by the top wire. The current maximums are at the top corners. The theoretical gain over a single vertical is 3.8 dB. An important advantage of this antenna is that it does not require the extensive ground system and feed arrangements that a conventional pair of phased λ/4 verticals would.

COMPARISON TO A DIPOLE

In the past, one of the things that has turned off potential users of the half-square on 80 and 160 meters is the perceived need for λ/4 vertical sections. This forces the height to be >65 feet on 80 meters and >130 feet on 160 meters. That's not really a problem. If you don't have supports of the necessary height, there are several things you can do. For example, just fold the ends in, as shown in **Figure 8.21**. This compromises the performance surprisingly little.

It is helpful to compare the examples given in Figures 8.19 and 8.20 to dipoles at the same height. Two heights, 40 and 80 feet, and average, very good and sea water grounds, were used for this comparison. It is also assumed that the lower end of the vertical wires had to be a minimum of 5 feet above ground.

At 40 feet the half-square has been folded quite a bit, with

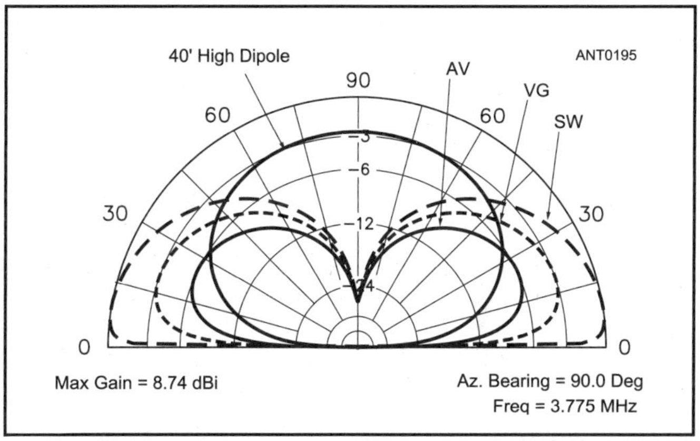

Figure 8.22 — Comparison of 80-meter elevation response of 40-foot high, horizontally polarized dipole over average ground and a 40-foot high, vertically polarized half-square, over three types of ground: average (conductivity σ = 5 mS/m, dielectric constant ε = 13), very good (σ = 30 mS/m, ε = 20) and salt water (σ = 5000 mS/m, ε = 80). The quality of the ground clearly has a profound effect on the low-angle performance of the half-square. Even over average ground, the half-square outperforms the low dipole below about 32°.

only 35-foot long (≈ λ/8) vertical sections. The elevation-plane comparison between this antenna and a dipole of the same height is shown in **Figure 8.22**. Over average ground the half-square is superior below 32° and at 15° is almost 5 dB better. That is a worthwhile improvement. If you have very good soil conductivity, like parts of the lower Midwest and South, then the half-square will be superior below 38° and at 15° will be nearly 8 dB better. For those fortunate few with saltwater frontal property the advantage at 15° is 11 dB! Notice also that above 35°, the response drops off rapidly. This is great for DX but is not good for local work.

Figure 8.23 shows the azimuthal-plane pattern for the 80-meter half-square antenna in Figure 8.21, but this time compared with the response of a flattop horizontal dipole that is 100 feet high. These comparisons are for average ground and are for an elevation angle of 5°. The message here is that the lower your dipole and the better your ground, the more you have to gain by switching from a dipole to a half-square. The half-square antenna looks like a good bet for DXing.

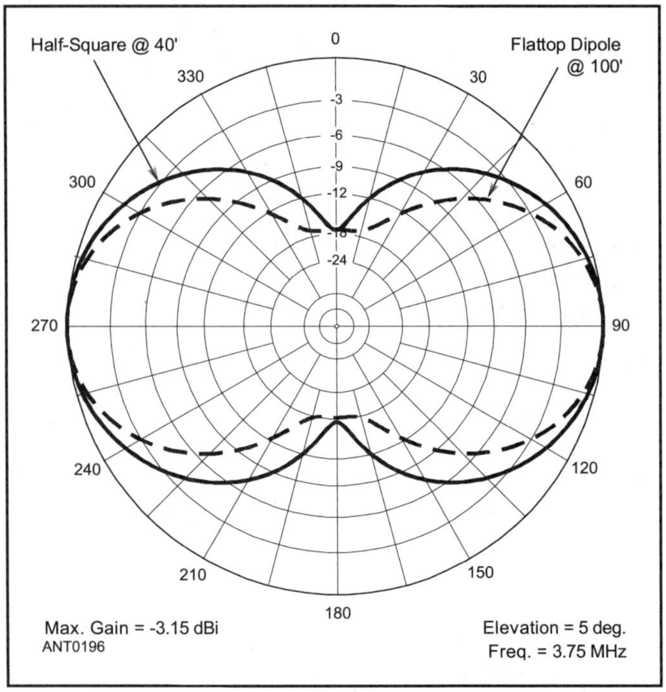

Figure 8.23 — 80-meter azimuth patterns for shortened half-square antenna (solid line) compared with flattop dipole (dashed line) at 100 feet height. Average ground is assumed for these cases.

CHANGING THE SHAPE OF THE HALF SQUARE

Just how flexible is the shape? There are several common distortions of practical importance. Some have very little effect, but a few are fatal to the gain. Suppose you have either more height and less width than called for in the standard version or more width and less height, as shown in **Figure 8.24A**.

The effect on gain from this type of dimensional variation is given in **Table 8.2**. For a top length (LT) varying between 110 and 150 feet, where the vertical wire lengths (L_V) readjusted to resonate the antenna, the gain changes only by 0.6 dB. For a 1-dB change the range of L_t is 100 to 155 feet, a pretty wide range.

Another variation results if we vary the length of the horizontal top wire and readjust the vertical wires for resonance, while keeping the top at a constant height. See Figure 8.24B. Table 8.2 shows the effect of this variation on the peak gain. For a range of L_t = 110 to 145 feet, the gain changes only 0.65 dB.

Table 8.2

Variation in Gain with Change in Horizontal Length

Vertical Height Readjusted for Resonance (Figure 8.24A)

L_t (feet)	*L_V (feet)*	*Gain (dBi)*
100	85.4	2.65
110	79.5	3.15
120	73.7	3.55
130	67.8	3.75
140	61.8	3.65
150	56	3.05
155	53	2.65

Vertical Length Readjusted for Resonance, but Horizontal Wire Kept at Constant Height (Figure 8.24B)

L_t (feet)	*L_V (feet)*	*Gain (dBi)*
110	78.7	3.15
120	73.9	3.55
130	68	3.75
140	63	3.35
145	60.7	3.05

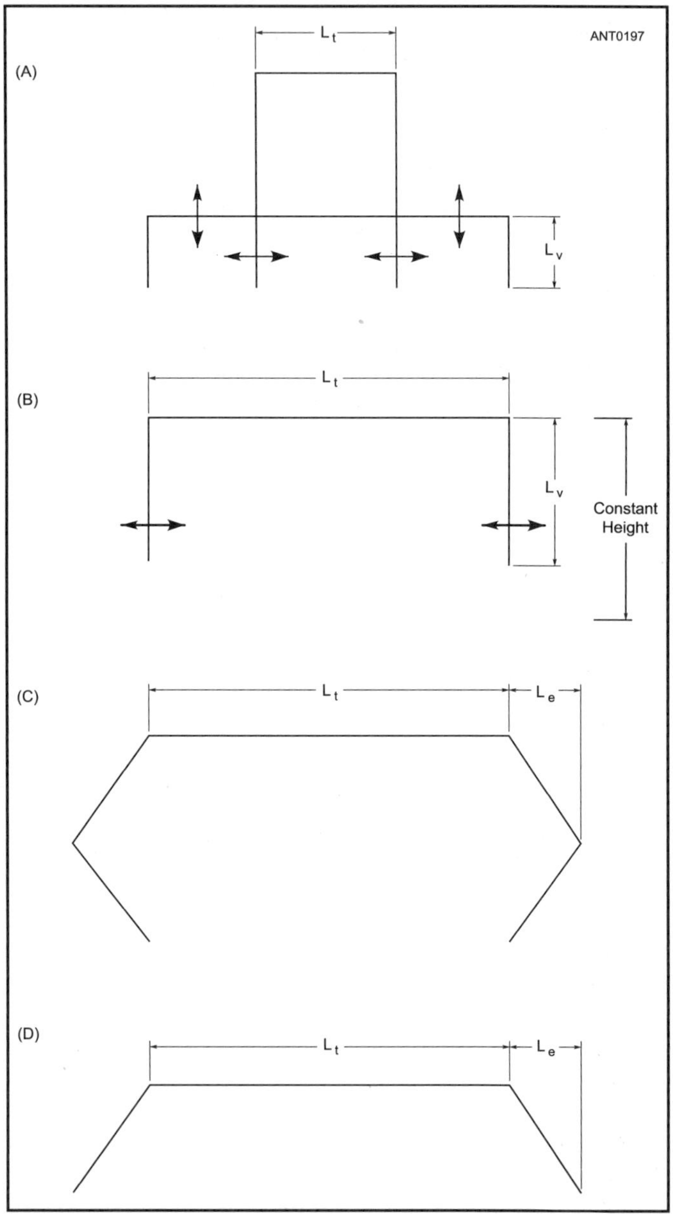

Figure 8.24 — Varying the horizontal and vertical lengths of a half-square. At A, both the horizontal and vertical legs are varied, while keeping the antenna resonant. At B, the height of the horizontal wire is kept constant, while its length and that of the vertical legs is varied to keep the antenna resonant. At C, the length of the horizontal wire is varied and the legs are bent inwards in the shape of "vees." At D, the ends are sloped outward and the length of the flattop portion is varied. All these symmetrical forms of distortion of the basic half-square shape result in small performance losses.

The effect of bending the ends into a V shape, as shown in Figure 8.24C, is given in **Table 8.3**. The bottom of the antenna is kept at a height of 5 feet and the top height (H) is either 40 or 60 feet. Even this gross deformation has only a relatively small effect on the gain. Sloping the ends outward as shown in Figure 8.24D and varying the top length also has only a small effect on the gain. While this is good news because it allows you dimension the antenna to fit different QTHs, not all distortions are so benign.

Suppose the two ends are not of the same height, as illustrated in **Figure 8.25**, where one end of the half-square is 20 feet higher than the other. The elevation-plane radiation pattern for this antenna is shown in **Figure 8.26** compared to a dipole at 50 feet. This type of distortion does affect the pattern. The gain drops somewhat, and the zenith null goes away. The nulls off the end of the antenna also go away, so that there is some end-fire radiation. In this example the difference in height is fairly extreme at 20 feet. Small differences of 1 to 5 feet do not affect the pattern seriously.

If the top height is the same at both ends but the length of the vertical wires is not the same, then a similar pattern distortion can occur. The antenna is very tolerant of symmetrical distortions, but it is much less accepting of asymmetrical distortion.

Table 8.3
Gain for Half-Square Antenna, Where Ends are Bent Into V-Shape (see Figure 8.24C)

Height ⇒ *L_t* *(feet)*	*H=40 feet* *L_v* *(feet)*	*H=40 feet* *Gain* *(dBi)*	*H=60 feet* *L_e* *(feet)*	*H=60 feet* *Gain* *(dBi)*
40	57.6	3.25	52.0	2.75
60	51.4	3.75	45.4	3.35
80	45.2	3.95	76.4	3.65
100	38.6	3.75	61.4	3.85
120	31.7	3.05	44.4	3.65
140	—	—	23	3.05

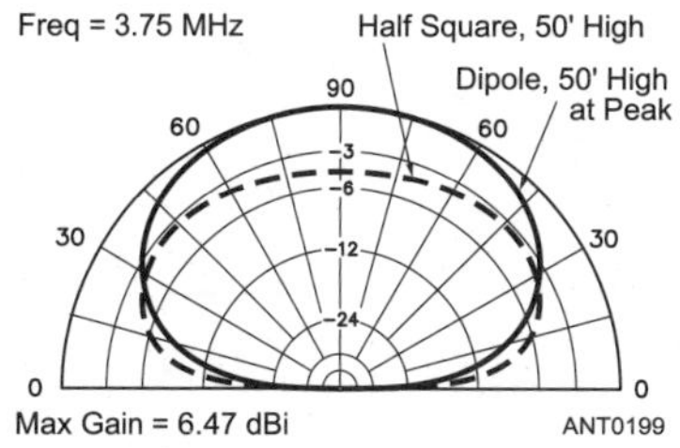

Figure 8.26 — Elevation pattern for the asymmetrical half-square compared with pattern for a 50-foot-high dipole. This is over average ground, with a conductivity of 5 mS/m and a dielectric constant of 13. Note that the zenith-angle null has filled in and the peak gain is lower compared to conventional half-square over the same kind of ground.

What if the length of the wires is such that the antenna is not resonant? Depending on the feed arrangement, that may or may not matter. We will look at that issue later, in the section on patterns versus frequency. The half-square antenna, like the dipole, is very flexible in its proportions.

HALF-SQUARE FEED POINT IMPEDANCE

There are many different ways to feed the half-square. Traditionally the antenna has been fed either at the end of one of the vertical sections, against ground, or at one of the upper corners as shown in Figure 8.20.

For voltage feed at the bottom against ground, the impedance is very high, on the order of several thousand ohms. For current feed at a corner, the impedance is much lower and is usually close to 50 Ω. This is very convenient for direct feed with coax.

The half-square is a relatively high-Q antenna (Q ≈ 17). **Figure 8.27** shows the SWR variation with frequency for this feed arrangement. An 80 meter dipole is not particularly wideband either, but a dipole will have less extreme variation in SWR than the half-square.

Figure 8.25 — An asymmetrical distortion of the half-square antenna, where the bottom of one leg is purposely made 20 feet higher than the other. This type of distortion does affect the pattern!

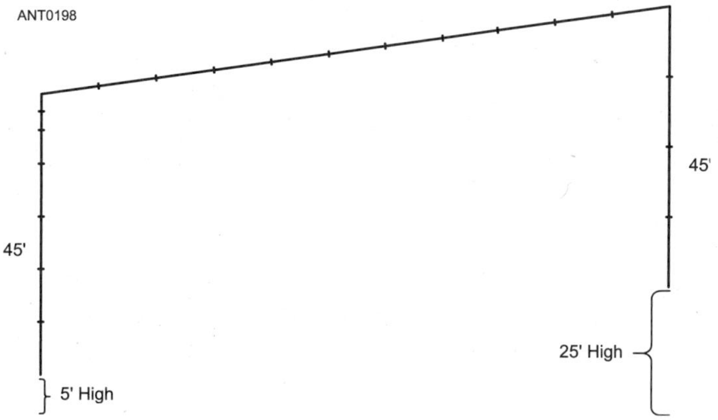

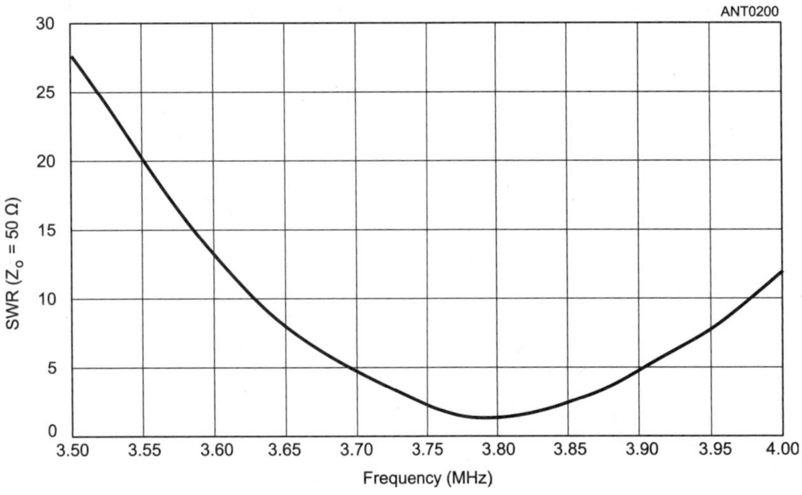

Figure 8.27 — Variation of SWR with frequency for current-fed half-square antenna. The SWR bandwidth is quite narrow.

PATTERNS VERSUS FREQUENCY

Impedance is not the only issue when defining the bandwidth of an antenna. The effect on the radiation pattern of changing frequency is also a concern. For a voltage-fed half-square, the current distribution changes with frequency. For an antenna resonant near 3.75 MHz, the current distribution is nearly symmetrical. However, above and below resonance the current distribution increasingly becomes asymmetrical. In effect, the open end of the antenna is constrained to be a voltage maximum, but the feed point can behave less as a voltage point and more like a current maximum. This allows the current distribution to become asymmetrical.

The effect is to reduce the gain by –0.4 dB at 3.5 MHz and by –0.6 dB at 4 MHz. The depth of the zenith null is reduced from –20 dB to –10 dB. The side nulls are also reduced. Note that this is exactly what happened when the antenna was made physically asymmetrical. Whether the asymmetry is due to current distribution or mechanical arrangements, the antenna pattern will suffer.

When current feed at a corner is used, the asymmetry introduced by off-resonance operation is much less, since both ends of the antenna are open circuits and constrained to be voltage maximums. The resulting gain reduction is only – 0.1 dB. It is interesting that the sensitivity of the pattern to changing frequency depends on the feed scheme used.

Of more concern for corner feed is the effect of the transmission line. The usual instruction is to simply feed the antenna using coax, with the shield connected to vertical wire and the center conductor to the top wire. Since the shield of the coax is a conductor, more or less parallel with the radiator, and is in the immediate field of the antenna, you might expect the pattern to be seriously distorted by this practice. This arrangement seems to have very little effect on the pattern. The greatest effect is when the feed line length was near a multiple of λ/2. Such lengths should be avoided.

Of course, you may use a choke balun at the feed point if you desire. This might reduce the coupling to the feed line even further, but it doesn't appear to be worth the trouble. In fact, if you use an antenna tuner in the shack to operate away from resonance with a very high SWR on the transmission line, a balun at the feed point would be stressed by high common-mode voltages. (See the **Transmission Line System Techniques** chapter.)

VOLTAGE-FEED AT ONE END OF ANTENNA: MATCHING SCHEMES

Several straightforward means are available for narrowband matching. However, broadband matching over the full 80 meter band is much more challenging. Voltage feed with a parallel-resonant circuit and a modest local ground, as shown in **Figure 8.28**, is the traditional matching scheme for this antenna. Matching is achieved by resonating the circuit at the desired frequency and tapping down on the inductor in Figure 8.28A or using a capacitive divider (Figure 8.28B). It is also possible to use a λ/4 transmission-line matching scheme, as shown in Figure 8.28C.

If the matching network shown in Figure 8.28B is used, typical values for the components would be: L = 15 µH, C1 = 125 pF and C2 = 855 pF. At any single point the SWR can be made very close to 1:1 but the bandwidth for SWR < 2:1 will be very narrow at <100 kHz. Altering the L-C ratio doesn't make very much difference. The half-square antenna has a well-earned reputation for being narrowband.

8.3.3 Bobtail Curtain

The antenna system in **Figure 8.29**, called a Bobtail curtain, was originally described by Woodrow Smith, W6BCX, in 1948 (see Bibliography for this and other articles on the Bobtail.) It uses the principles of co-phased verticals to produce a broadside, bidirectional pattern providing approximately 5.1 dB of gain over a single λ/4 element. The antenna performs as three in-phase, top-fed vertical radiators approximately λ/4 in height and spaced approximately λ/2. It is most effective for low-angle signals and makes an excellent long-distance antenna for 1.8, 3.5, or 7 MHz.

The three vertical sections are the actual radiating

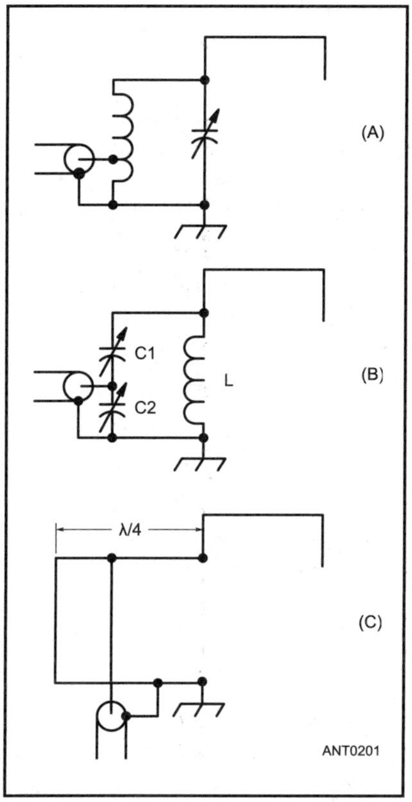

Figure 8.28 — Typical matching networks used for voltage-feeding a half-square antenna. See Figure 8.24 for feeding the antenna at a corner where the lower impedance is suitable for a direct connection to coaxial cable.

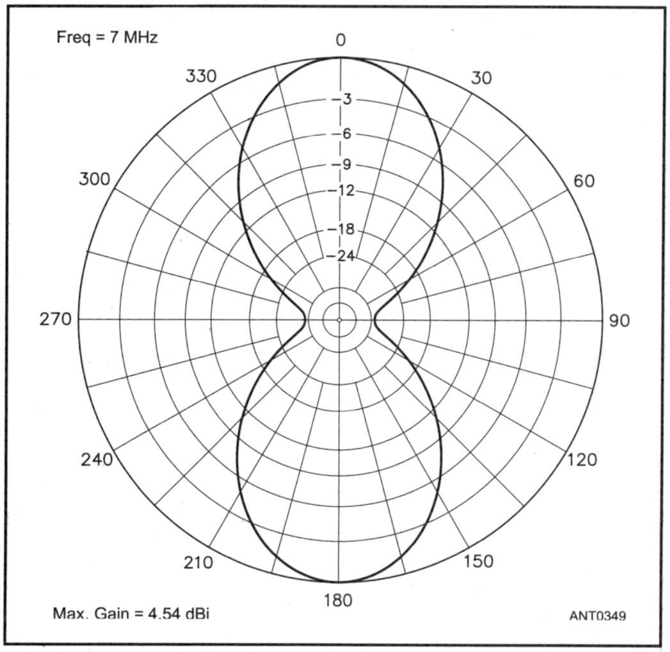

Figure 8.30 — Calculated free-space E-plane directive diagram of the Bobtail curtain shown in Figure 8.29. The array lies along the 90° – 270° axis.

components, but only the center element is fed directly. The two horizontal parts, A, act as phasing lines and contribute very little to the radiation pattern. Because the current in the center element must be divided between the end sections, the current distribution approaches a binomial 1:2:1 ratio. The radiation pattern is shown in **Figure 8.30**.

The vertical elements should be as vertical as possible. The height for the horizontal portion should be slightly greater than B, as shown in Figure 8.29. The tuning network is resonant at the operating frequency. The L/C ratio should be fairly low to provide good loading characteristics. As a starting point, a maximum capacitor value of 75 to 150 pF is recommended, and the inductor value is determined by C and the operating frequency. The network is first tuned to resonance and then the tap point is adjusted for the best match. A slight readjustment of C may be necessary. A link coil consisting of a few turns can also be used to feed the antenna.

A feeling for the matching bandwidth of this antenna can be obtained by looking at a feed point located at the top end of the center element. The impedance at this point will be approximately 32 Ω. An SWR plot (for Z0 = 32 Ω) for an 80 meter Bobtail curtain at this feed point is shown in **Figure 8.31**. However, it is not advisable to actually connect a feed line at

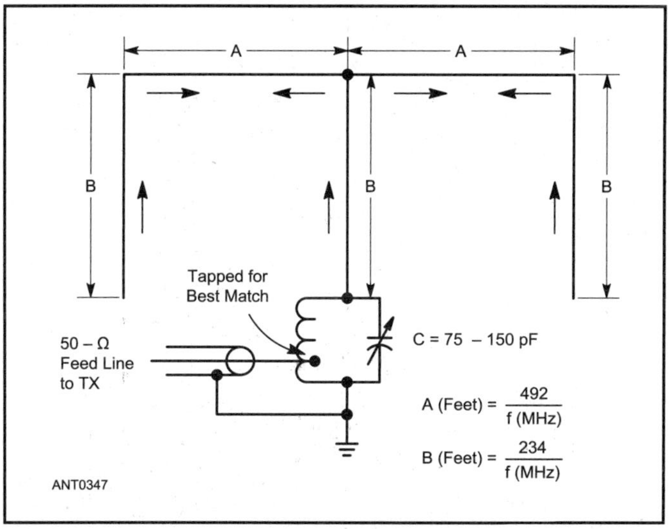

Figure 8.29 — The Bobtail curtain is an excellent low-angle radiator having broadside bidirectional characteristics. Current distribution is represented by the arrows. Dimensions A and B (in feet, for wire antennas) can be determined from the equations.

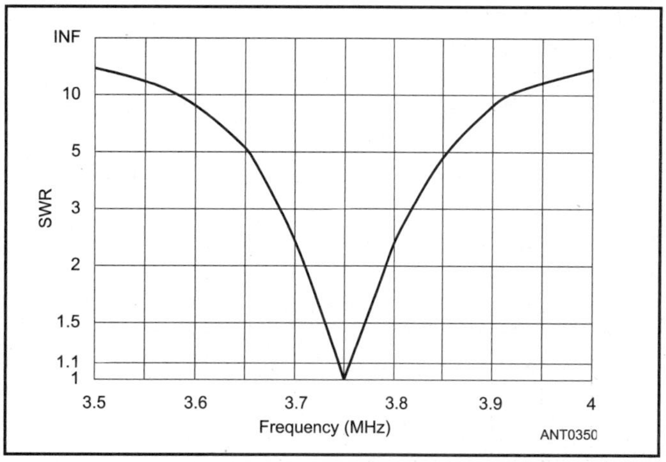

Figure 8.31 — Typical SWR plot for an 80-meter Bobtail curtain in free space. This is a narrow-band antenna.

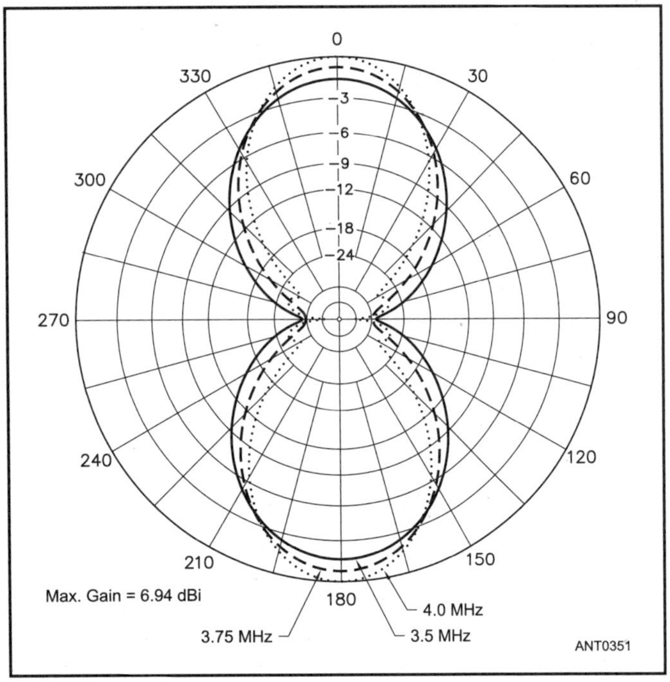

Figure 8.32 — 80-meter Bobtail curtain's free-space E-plane pattern variation over the 80-meter band.

this point since it would detune the array and alter the pattern. This antenna is relatively narrow band. When fed at the bottom of the center element as shown in Figure 8.29, the SWR can be adjusted to be 1:1 at one frequency but the operating bandwidth for SWR less than 2:1 may be even narrower than Figure 8.31 shows. For 80 meters, where operation is often desired in the CW DX portion (3.510 MHz) and in the phone DX portion (3.790 MHz), it will be necessary to retune the matching network as you change frequency. This can be done by switching a capacitor in or out, manually or remotely with a relay.

While the match bandwidth is quite narrow, the radiation pattern changes more slowly with frequency. **Figure 8.32** shows the variation in the pattern over the entire band (3.5 to 4.0 MHz). As would be expected, the gain increases with frequency because the antenna is larger in terms of wavelengths. The general shape of the pattern, however, is quite stable.

The Bobtail can also be fed at one corner using coax as shown for the Half-Square antenna in Figure 8.22. As with the Half-Square, the coax shield should be connected to the vertical element and λ/2 lengths of coax should be avoided. The same caution regarding the use of a choke balun apply as with the Half-Square.

A variation of the Bobtail for multiple bands, the N4GG Array, is described in a July 2002 *QST* article that is included with the online material. The antenna covers multiple bands using parallel wires similarly to a fan dipole and with vertical wires acting as they do in the Bobtail curtain.

8.3.4 The Bruce Array

Four variations of the Bruce array are shown in **Figure 8.33**. The Bruce is simply a wire folded so that the vertical sections carry large in-phase currents, while the horizontal sections carry small currents flowing in opposite directions with respect to the center of a section (indicated by dots). The radiation is vertically polarized. The gain is proportional to the length of the array but is somewhat smaller than you can obtain from a broadside array of λ/2 elements of the same length. This is because the radiating portion of the elements is only λ/4.

The Bruce array has several advantages over the Half-Square and Bobtail curtain:

1) The array is only λ/4 high. This is especially helpful on 80 and 160 meters, where the height of λ/2 supports becomes impractical for most amateurs.

2) The array is very simple. It is just a single piece of wire folded to form the array.

3) The dimensions of the array are very flexible. Depending on the available distance between supports, any number of elements can be used. The longer the array, the greater the gain.

4) The shape of the array does not have to be exactly 1.05 λ/4 squares. If the available height is short but the array can be made longer, then shorter vertical sections and longer horizontal sections can be used to maintain gain and resonance. Conversely, if more height is available but width is restricted then longer vertical sections can be used with shorter horizontal sections.

5) The array can be fed at other points more convenient for a particular installation.

6) The antenna is relatively low Q, so that the feed point impedance changes slowly with frequency. This is very helpful on 80 meters, for example, where the antenna can be relatively broadband.

7) The radiation pattern and gain is stable over the width of an amateur band.

Note that the nominal dimensions of the array in Figure 8.33 call for section lengths = 1.05 λ/4. The need to use slightly longer elements to achieve resonance is common in large wire arrays. A quad loop behaves in the same manner. This is quite different from wire dipoles, which are typically shortened by 2 – 5% to achieve resonance.

Figure 8.34 shows the variations in gain and pattern for 2 to 5-element 80-meter Bruce arrays. **Table 8.4** lists the gain over a vertical λ/2 dipole, a 4-radial ground-plane vertical and the size of the array. The gain and impedance parameters listed are for free space. Over real ground the patterns and gain will depend on the height above ground and the ground characteristics. Copper loss using #12 AWG conductors is included.

Worthwhile gain can be obtained from these arrays, especially on 80 and 160 meters, where any gain is hard to come by. The feed point impedance is for the center of a vertical section. From the patterns in Figure 8.34 you can see that sidelobes start to appear as the length of the array is increased beyond 3λ/4. This is typical for arrays using equal currents in the elements.

It is interesting to compare the Bobtail curtain (Figure 8.29) with a 4-element Bruce array. **Figure 8.35** compares the radiation patterns for these two antennas. Even though the

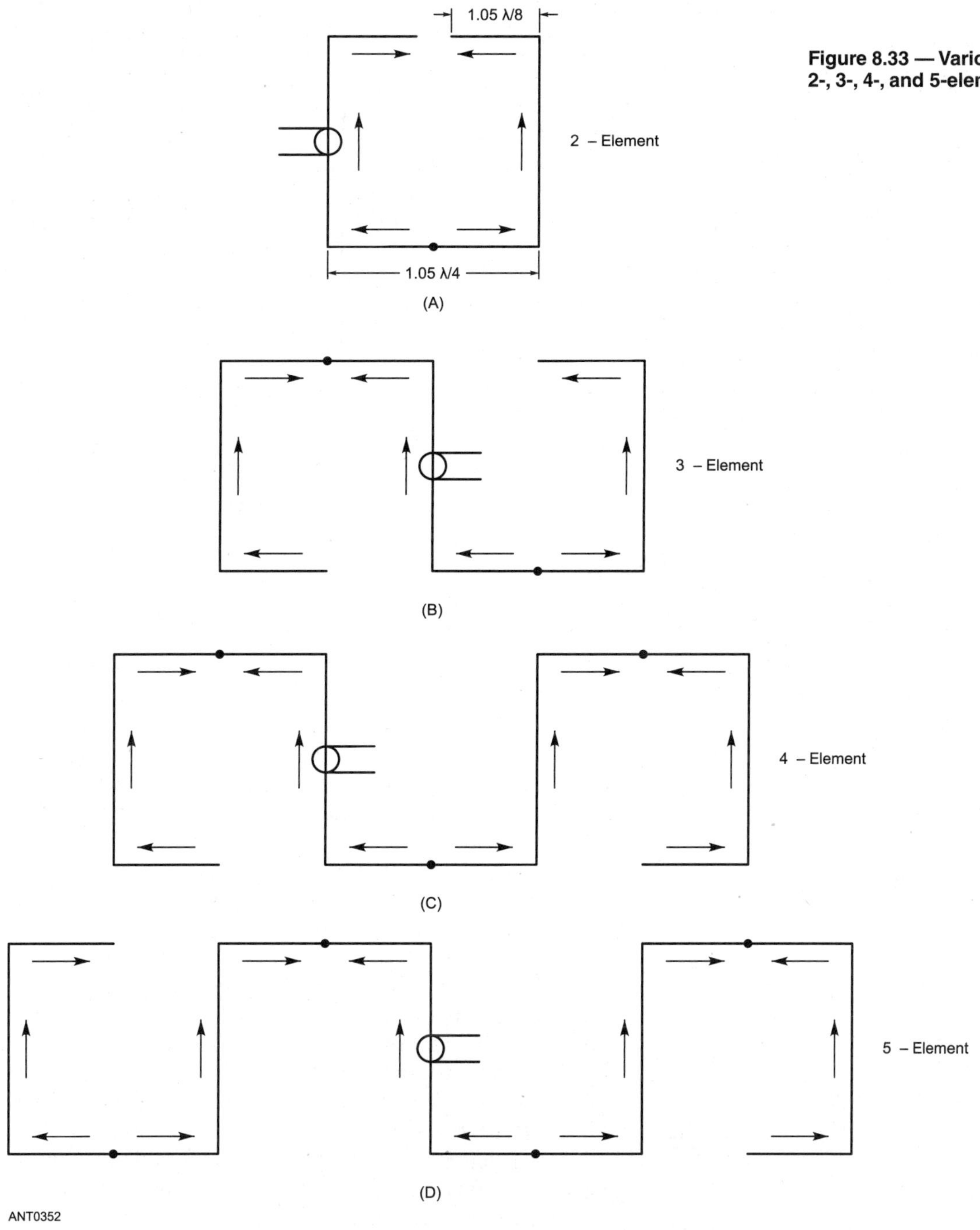

Figure 8.33 — Various Bruce arrays: 2-, 3-, 4-, and 5-element versions.

Bruce is shorter (3λ/4) than the Bobtail (1 λ), it has slightly more gain. The matching bandwidth is illustrated by the SWR curve in **Figure 8.36**. The 4-element Bruce has over twice the match bandwidth (200 kHz) than does the Bobtail (75 kHz in Figure 8.31). Part of the gain difference is due to the binomial current distribution — the center element has twice the current as the outer elements in the Bobtail. This reduces the gain slightly so that the 4-element Bruce becomes competitive. This is a good example of using more than the minimum number of elements to improve performance or to reduce size. On 160 meters the 4-element Bruce will be 140 feet shorter than the Bobtail, a significant reduction. If additional space is available for the Bobtail (1 λ) then a 5-element Bruce could be used, with a small increase in gain but also introducing some sidelobes.

The 2-element Bruce and the half-square antennas are both 2-element arrays. However, since the spacing between radiators is greater in the half-square (λ/2) the gain of the half-square is about 1 dB greater. If space is available, the half-square would be a better choice. If there is not room for a half-square then the Bruce, which is only half as long (λ/4), may be a good alternative. The 3-element Bruce, which has the same length (λ/2) as the half-square, has about 0.6 dB more gain than the half-square and will have a wider match bandwidth.

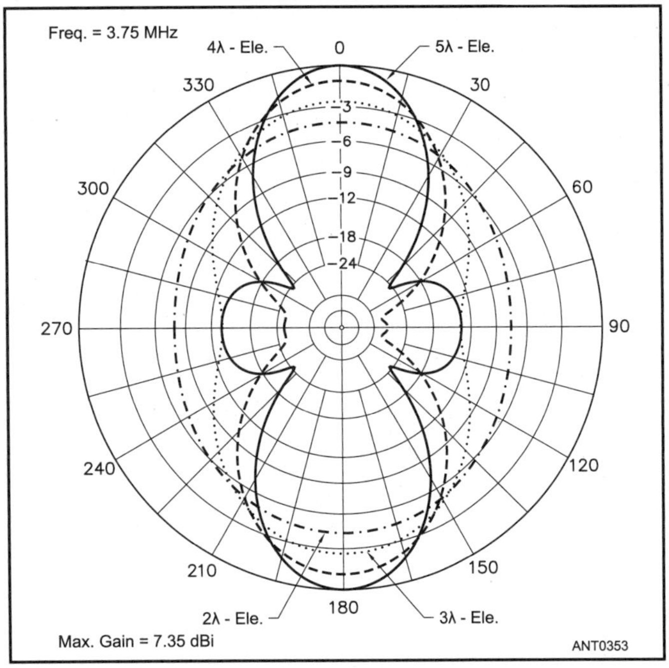

Figure 8.34 — 80-meter free-space E-plane directive patterns for the Bruce arrays shown in Figure 8.33. The 5-element's pattern is a solid line; the 4-element is a dashed line; the 3-element is a dotted line, and the 2-element version is a dashed-dotted line.

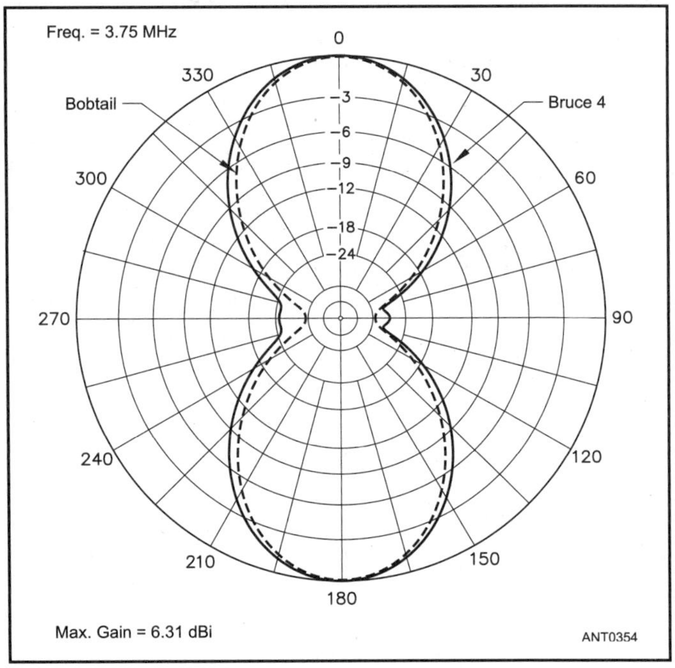

Figure 8.35 — Comparison of free space patterns of a 4-element Bruce array (solid line) and a 3-element Bobtail curtain (dashed line).

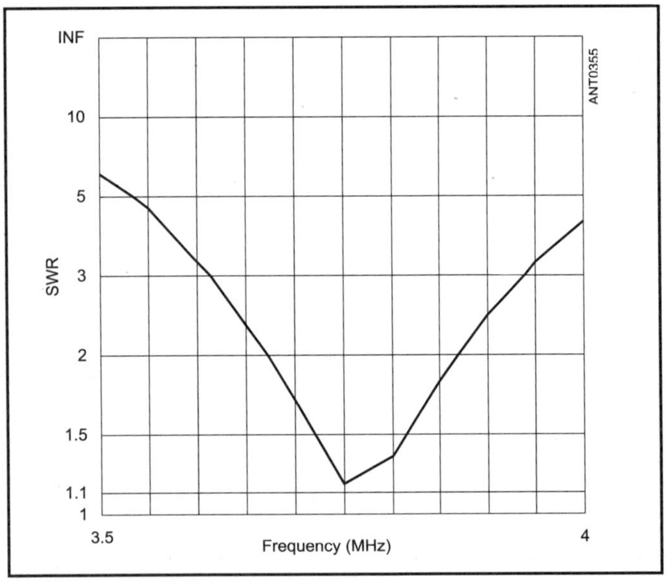

Figure 8.36 — Typical SWR curve for a 4-element 80-meter Bruce array.

The Bruce antenna can be fed at many different points and in different ways. In addition to the feed points indicated in Figure 8.33, you may connect the feed line at the center of any of the vertical sections. In longer Bruce arrays, feeding at one end will result in some current imbalance among the elements but the resulting pattern distortion is small. Actually, the feed point can be anywhere along a vertical section. One very convenient point is at an outside corner. The feed point impedance will be higher (about 600 Ω). A good match for 450 Ω ladder-line can usually be found somewhere on the vertical section. It is important to recognize that feeding the antenna at a voltage node (dots in Figure 8.33) by breaking the wire and inserting an insulator, completely changes the current distribution. This will be discussed in the section on end-fire arrays.

A Bruce can be fed unbalanced against ground or against a counterpoise as shown in **Figure 8.37**. Because it is a vertically polarized antenna, the better the ground system, the better the performance. As few as two elevated radials can be used as shown in Figure 8.37B, but more radials can also be used to improve the performance, depending on local ground constants. The original development of the Bruce array in the late 1920s used this feed arrangement.

Table 8.4

Bruce Array Length, Impedance and Gain as a Function of Number of Elements

Number Elements	*Gain Over λ/2 Vertical Dipole*	*Gain over λ/4 Ground-Plane*	*Array Length Wavelengths*	*Approx. Feed Z, Ω*
2	1.2 dB	1.9 dB	¼	130
3	2.8 dB	3.6 dB	½	200
4	4.3 dB	5.1 dB	¾	250
5	5.3 dB	6.1 dB	1	300

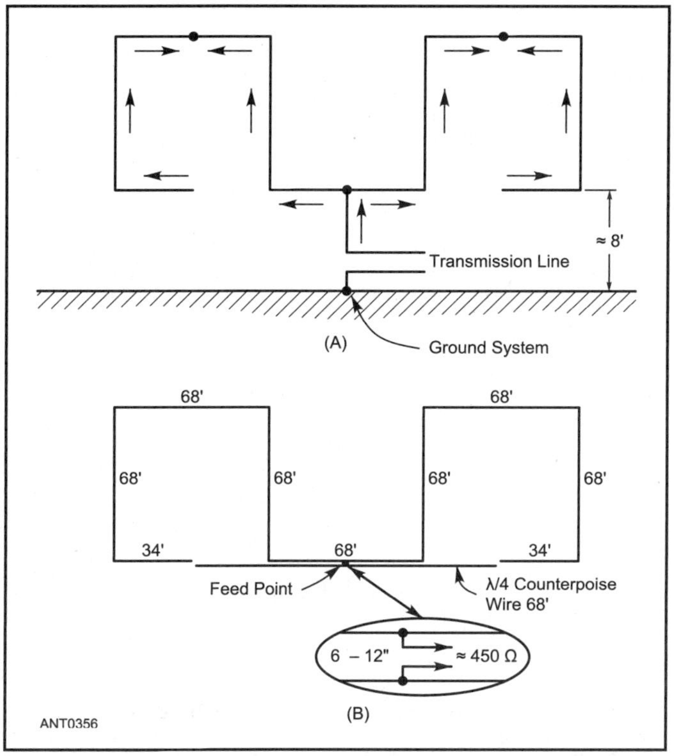

Figure 8.37 — Alternate feed arrangements for the Bruce array. At A, the antenna is driven against a ground system and at B, it uses a two-wire counterpoise.

8.3.5 The Lazy-H Array

The 4-element array shown in **Figure 8.38** is commonly known as the Lazy H. It consists of four λ/2 elements arranged in two parallel collinear arrays, all operated in-phase to give broadside directivity. The gain and directivity will depend on the spacing, as in the case of a simple parallel-element broadside array. The spacing may be chosen between the limits shown on the drawing, but spacings below 3λ/8 are not worthwhile because the gain is small. Estimated gains compared to a single element are:

3λ/8 spacing — 4.2 dB
λ/2 spacing — 5.8 dB
5λ/8 spacing — 6.7 dB
3λ/4 spacing — 6.3 dB

Half-wave spacing is generally used. Directive patterns for this spacing are given in **Figures 8.39** and **8.40**. With

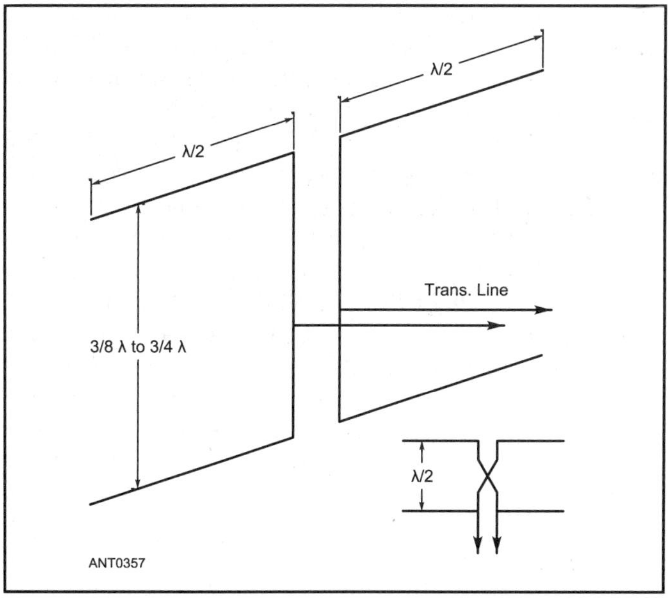

Figure 8.38 — Four-element broadside array ("lazy H") using collinear and parallel elements.

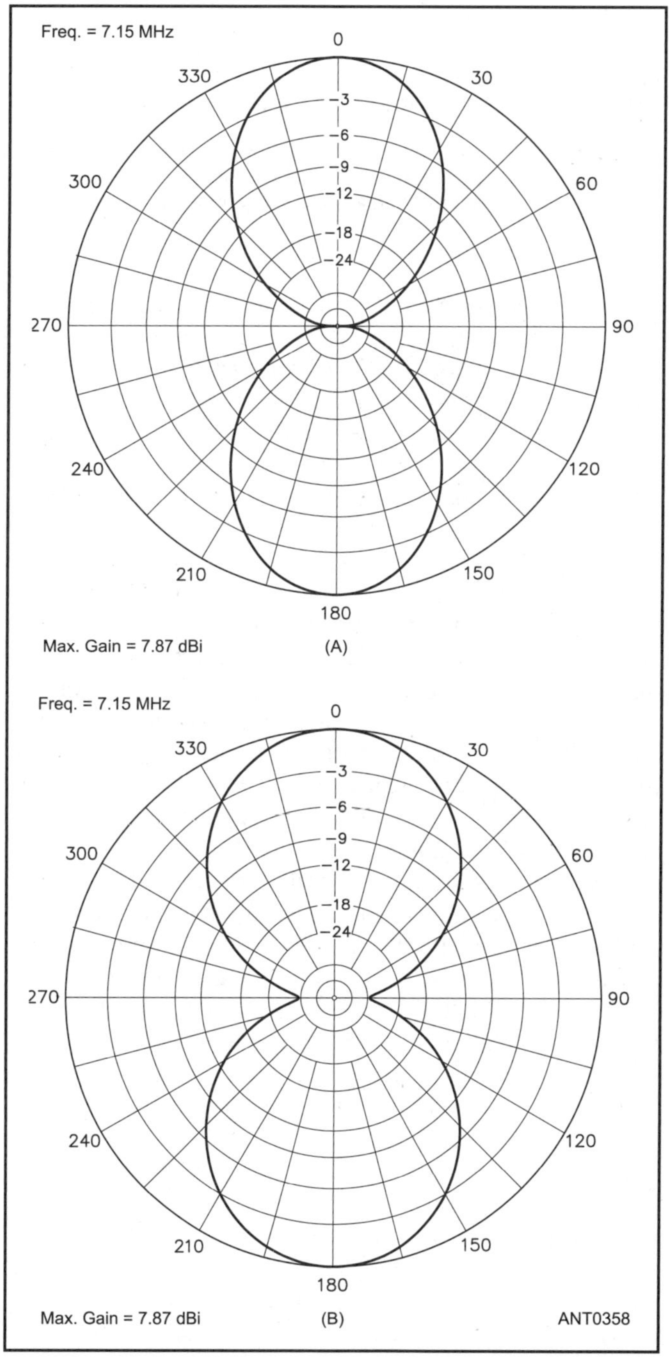

Figure 8.39 — Free-space directive diagrams of the 4-element antenna shown in Figure 8.38. At A is the E-plane pattern. The axis of the elements lies along the 90° – 270° line. At B is the free-space H-plane pattern, viewed as if one set of elements is above the other from the ends of the elements.

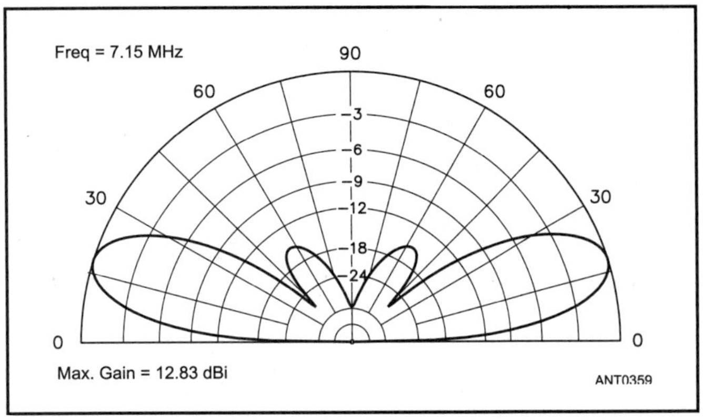

Figure 8.40 — Vertical pattern of the 4-element broadside antenna of Figure 8.38, when mounted with the elements horizontal and the lower set λ/4 above flat ground. Stacked arrays of this type give best results when the lowest elements are at least λ/2 high. The gain is reduced and the wave angle raised if the lowest elements are too close to ground.

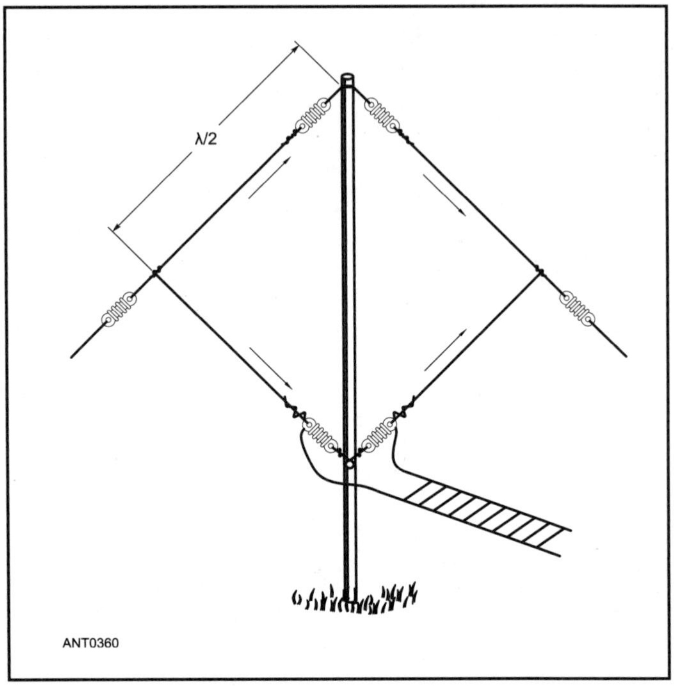

Figure 8.41 — The bi-square array. It has the appearance of a loop but is not a true loop because the conductor is open at the top. The length of each side, in feet, is 480/f (MHz).

λ/2 spacing between parallel elements, the impedance at the junction of the phasing line and transmission line is resistive and in the vicinity of 100 Ω. With larger or smaller spacing the impedance at this junction will be reactive as well as resistive. Matching stubs are recommended in cases where a non-resonant line is to be used. They may be calculated and adjusted as described in the **Transmission Line System Techniques** chapter.

The system shown in Figure 8.38 may be used on two bands having a 2-to-1 frequency relationship. It should be designed for the higher of the two frequencies, using 3λ/4 spacing between parallel elements. It will then operate on the lower frequency as a simple broadside array with 3λ/8 spacing.

An alternative method of feeding is shown in the small diagram in Figure 8.38. In this case the elements and the phasing line must be adjusted exactly to an electrical half wavelength. The impedance at the feed point will be resistive and on the order of 2 kΩ.

A variation of this antenna called the "extended Lazy H" makes an effective broadside antenna on its fundamental and several higher bands. It is a good Field Day antenna if the supports are available for the lower elements to be at least λ/4 above the ground. It can be used with a tuner on all HF bands and as a top-loaded vertical by connecting the feed line conductors together and driving it against a ground system. A version for 7, 14, and 21 MHz is described in "The Extended Lazy H Antenna," by Walter Salmon, VK2SA (in the October 1955 *QST* and included with this book's online material.

8.3.6 The Bi-Square Array

A development of the Lazy H, known as the Bi-Square, is shown in **Figure 8.41**. The gain of the Bi-Square is somewhat less than that of the Lazy-H, but this array is attractive because it can be supported from a single pole. It has a circumference of 2 λ at the operating frequency and is horizontally polarized.

The Bi-Square antenna consists of two 1 λ radiators, fed 180° out-of-phase at the bottom of the array. The radiation resistance is 300 Ω, so it can be fed with either 300- or 600-Ω line. The free space gain of the antenna is about 5.8 dBi, which is 3.7 dB more than a single dipole element. Gain may be increased by adding a parasitic reflector or director. Two Bi-Square arrays can be mounted at right angles and switched to provide omnidirectional coverage. In this way, the antenna wires may be used as part of the guying system for the pole.

Although it resembles a loop antenna, the Bi-Square is not a true loop because the ends opposite the feed point are open. However, identical construction techniques can be used for the two antenna types. Indeed, with a means of remotely closing the connection at the top for lower frequency operation, the antenna can be operated on two harmonically related bands. As an example, an array with 17 feet per side can be operated as a Bi-Square at 28 MHz and as a full-wave loop at 14 MHz. For two-band operation in this manner, the side length should favor the higher frequency. The length of a closed loop is not as critical.

8.4 End-Fire Arrays

The term end-fire covers several different methods of operation, all having in common the fact that the maximum radiation takes place along the array axis, and that the array consists of a number of parallel elements in one plane. End-fire arrays can be either bidirectional or unidirectional. In the bidirectional type commonly used by amateurs there are only two elements, and these are operated with currents 180° out-of-phase. Even though adjustment tends to be complicated, unidirectional end-fire driven arrays have also seen amateur use, primarily as a pair of phased, ground-mounted λ/4 vertical elements. Extensive discussion of this array is contained in the **Phased Arrays** chapter.

Horizontally polarized unidirectional end-fire arrays see little amateur use except in log-periodic arrays (described in the **Frequency-Independent Antennas** chapter). Instead, horizontally polarized unidirectional arrays usually have parasitic elements (described in the **HF Beam Antennas** chapter) and are called Yagis.

8.4.1 Two-Element End-Fire Array

In a 2-element array with equal currents out-of-phase, the gain varies with the spacing between elements as shown in **Figure 8.42**. The maximum gain occurs at about 0.1 λ spacing. Below that the gain drops rapidly due to conductor loss resistance.

The feed point resistance for either element is very low at the spacings giving greatest gain, as shown in the **Phased Arrays** chapter. The spacings most frequently used are λ/8 and λ/4, at which the resistances of center-fed λ/2 elements are about 9 and 32 Ω, respectively.

The effect of conductor resistance on gain for various spacings is shown in **Figure 8.43**. Because current along the element is not constant (it is approximately sinusoidal), the resistance shown is the equivalent resistance (R_{eq}) inserted at the center of the element to account for the loss distributed along the element.

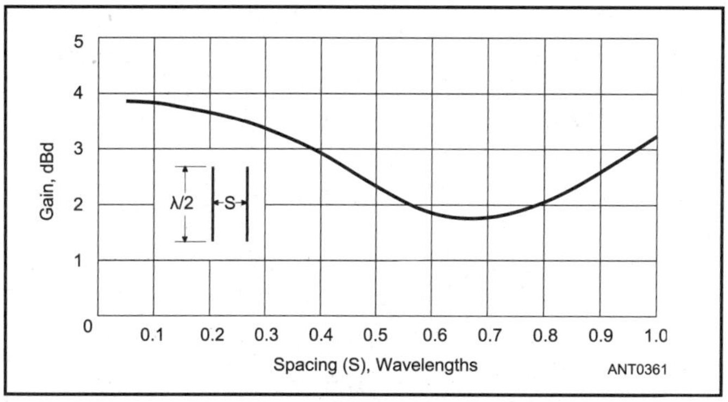

Figure 8.42 — Gain of an end-fire array consisting of two elements fed 180° out-of-phase, as a function of the spacing between elements. Maximum radiation is in the plane of the elements and at right angles to them at spacings up to λ/2, but the direction changes at greater spacings.

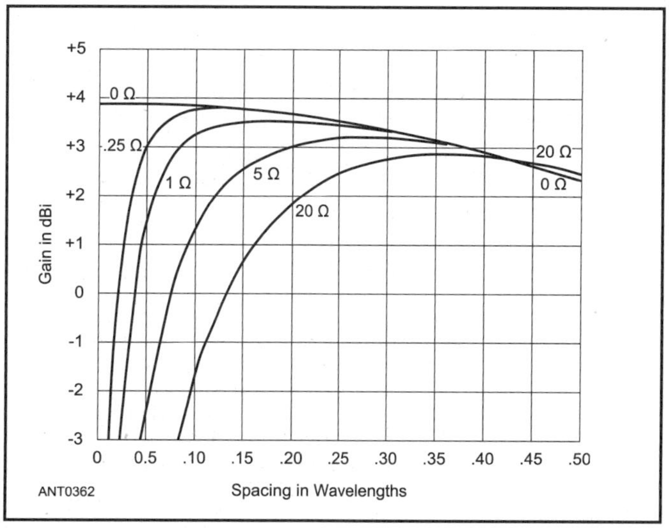

Figure 8.43 — Gain over a single element of two out-of-phase elements in free space as a function of spacing for various loss resistances.

The equivalent resistance of a λ/2 element is one half the ac resistance (R_{ac}) of the complete element. R_{ac} is usually >> R_{dc} due to skin effect. For example, a 1.84 MHz dipole using #12 AWG copper wire will have the following Req:

Wire length = 267 feet

$R_{dc} = 0.00159$ [Ω/foot] × 267 [feet] = 0.42 Ω

$F_r = R_{ac}/R_{dc} = 10.8$

$R_{eq} = (R_{dc}/2) \times F_r = 2.29\ \Omega$

For a 3.75 MHz dipole made with #12 AWG wire, R_{eq} = 1.59 Ω. In Figure 8.43, it is clear that end-fire antennas made with #12 AWG or smaller wire will limit the attainable gain because of losses. There is no point in using spacings much less than λ/4 if you use wire elements. If instead you use elements made of aluminum tubing, then smaller spacings can be used to increase gain. However, as the spacing is reduced below λ/4 the increase in gain is quite small even with good conductors. Closer spacings give little gain increase but can drastically reduce the operating bandwidth due to the rapidly increasing Q of the array.

UNIDIRECTIONAL END-FIRE ARRAYS

Two parallel elements spaced λ/4 apart and fed equal currents 90° out-of-phase will have a directional pattern in the plane at right angles to the plane of the array. See **Figure 8.44**. The maximum radiation is in the direction of the element in which the current lags. In the opposite direction the fields from the two elements cancel.

When the currents in the elements are neither in-phase nor 180° out-of-phase, the feed point resistances of the elements

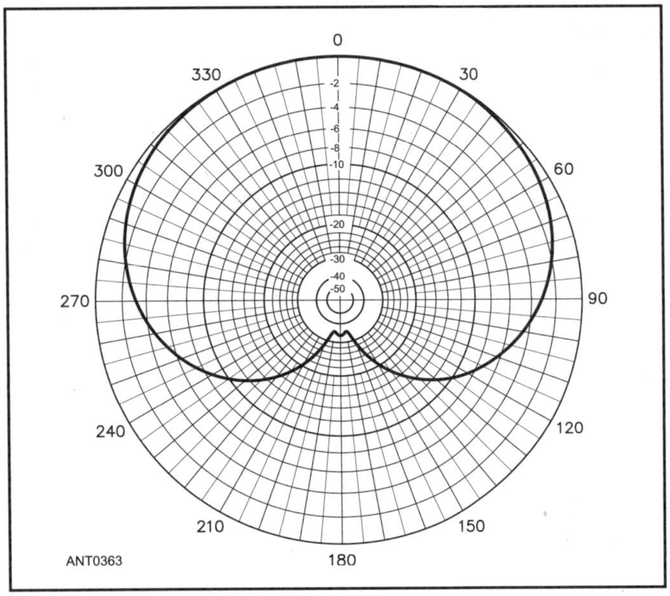

Figure 8.44 — Representative H-plane pattern for a 2-element end-fire array with 90° spacing and phasing. The elements lie along the vertical axis, with the uppermost element the one of lagging phase. Dissimilar current distributions are taken into account.

are not equal. This complicates the problem of feeding equal currents to the elements, as discussed in the **Phased Arrays** chapter and in the section on the "The ZL Special and HB9CV Arrays" later in this chapter.

More than two elements can be used in a unidirectional end-fire array. The requirement for unidirectivity is that there must be a progressive phase shift in the element currents equal to the spacing, in electrical degrees, between the elements.

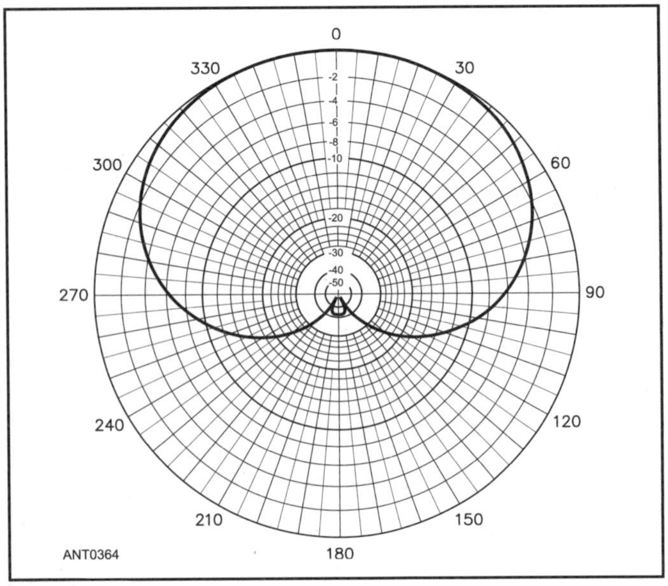

Figure 8.45 — H-plane pattern for a 3-element end-fire array with binomial current distribution (the current in the center element is twice that in each end element). The elements are spaced λ/4 apart along the 0° – 180° axis. The center element lags the lower element by 90°, while the upper element lags the lower element by 180° in phase. Dissimilar current distributions are taken into account.

The amplitudes of the currents in the various elements also must be properly related. This requires binomial current distribution. In the case of three elements, this requires that the current in the center element be twice that in the two outside elements, for 90° (λ/4) spacing and element current phasing. This antenna has an overall length of λ/2. The directive diagram is shown in **Figure 8.45**. The pattern is similar to that of Figure 8.44, but the 3-element binomial array has greater directivity, evidenced by the narrower half-power beamwidth (146° versus 176°). Its gain is 1.0 dB greater.

8.4.2 The W8JK Array

John Kraus, W8JK, described his bidirectional flat-top W8JK beam antenna in 1940. See **Figure 8.46**. (His June 1982 *QST* article "The W8JK Recap and Update" is included with the online material.) Two λ/2 elements are spaced λ/8 to λ/4 and driven 180° out-of-phase. The free-space radiation pattern for this antenna, using #12 AWG copper wire, is given in **Figure 8.47**. The pattern is representative of spacings between λ/8 and λ/4 where the gain varies less than 0.5 dB. The gain over a dipole is about 3.3 dB (5.4 dBi referenced to an isotropic radiator), a worthwhile improvement. The feed point impedance (including wire resistance) of each element is about 11 Ω for λ/8 spacing and 33 Ω for λ/4 spacing. The feed point impedance at the center connection will depend on the length and Z_0 of the connecting transmission line.

Kraus gave several other variations for end-fire arrays, some of which are shown in **Figure 8.48**. The ones fed at the center (A, C, and E) are usually horizontally polarized flat-top beams. The end-fed versions (B, D, and F) are usually vertically polarized, where the feed point can be conveniently near ground. The direction of maximum radiation is in the plane of the antennas.

A practical variation of Figure 8.48B is given in **Figure 8.49**. In this example, the height is limited to λ/4 so the ends can be bent over as shown, producing a 2-element Bruce array. This reduces the gain somewhat but allows much shorter supports, an important consideration on the low bands. If additional height is available, then you can achieve some additional gain. The upper ends can be bent over to fit the available height. The feed point impedance will greater than 1 kΩ.

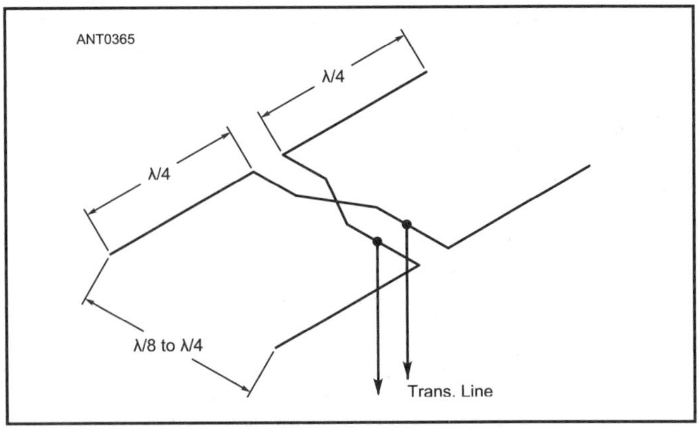

Figure 8.46 — A 2-element W8JK array.

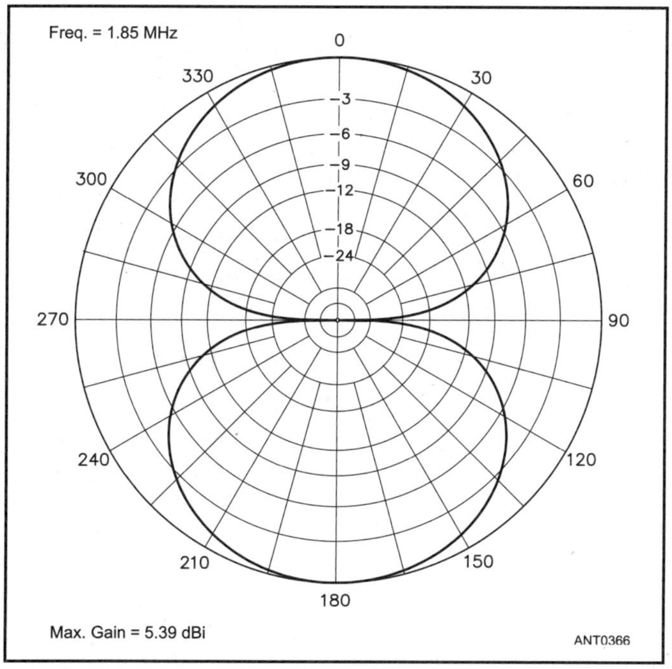

Figure 8.47 — Free-space E-plane pattern for the 2-element W8JK array.

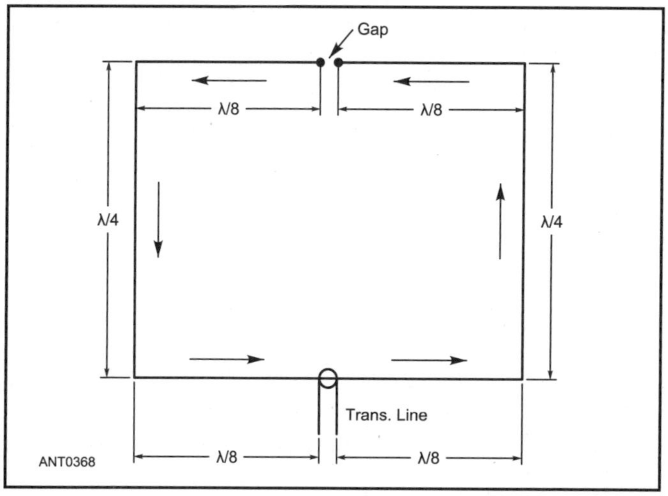

Figure 8.49 — A 2-element end-fire array with reduced height.

The article "Building the W8JK" by Suggs (see Bibliography) shows how to build a W8JK beam that covers 20 through 6 meters.

8.4.3 Four-Element End-Fire and Collinear Arrays

The array shown in **Figure 8.50** combines collinear in-phase elements with parallel out-of-phase elements to give both broadside and end-fire directivity. It is a two-section W8JK. The approximate free-space gain using #12 AWG copper wire is 4.9 dBi with λ/8 spacing and 5.4 dBi with λ/4 spacing. Directive patterns are given in **Figure 8.51** for free space, and in **Figure 8.52** for heights of 1 λ and λ/2 above flat ground.

The impedance between elements at the point where the phasing line is connected is of the order of several thousand ohms. The SWR with an unmatched line consequently is quite high, and this system should be constructed with open-wire line (500 or 600 Ω) if the line is to be resonant. With λ/4 element spacing the SWR on a 600 Ω line is estimated to be in the vicinity of 3 or 4:1.

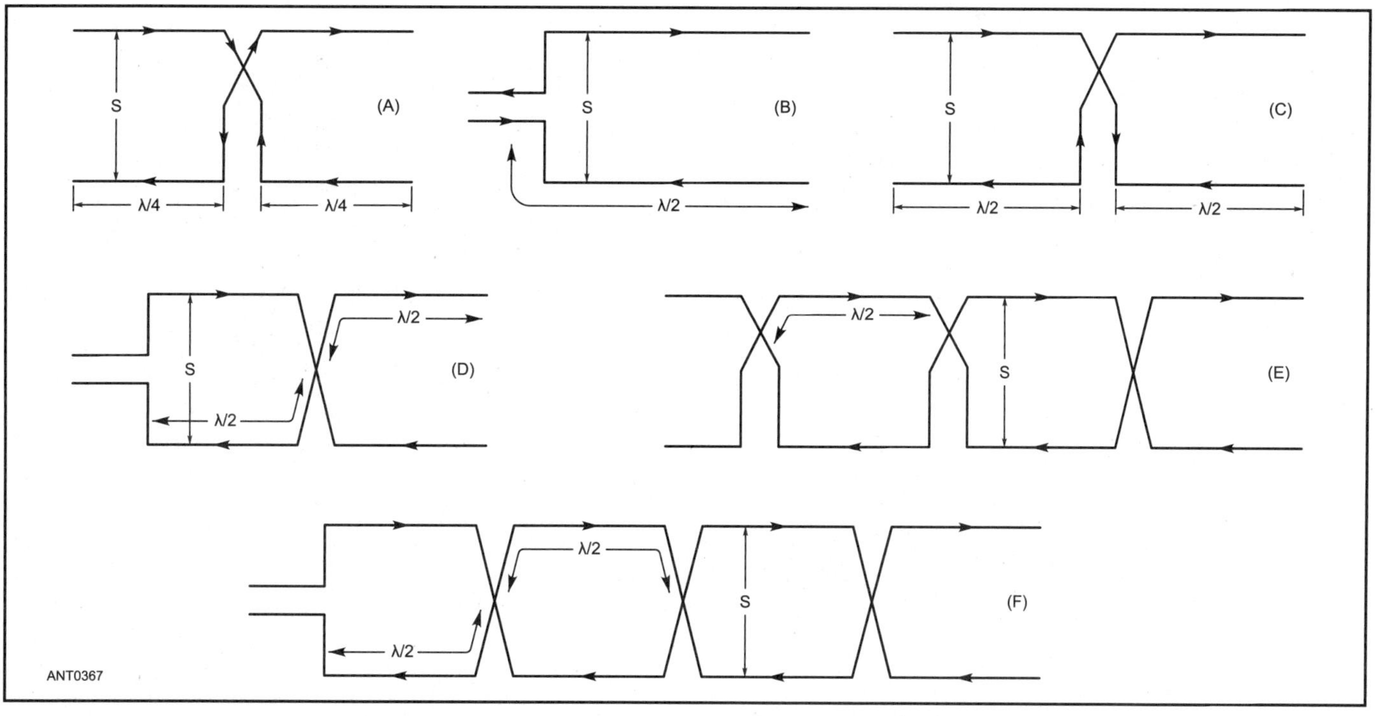

Figure 8.48 — Six other variations of W8JK "flat-top beam" antennas.

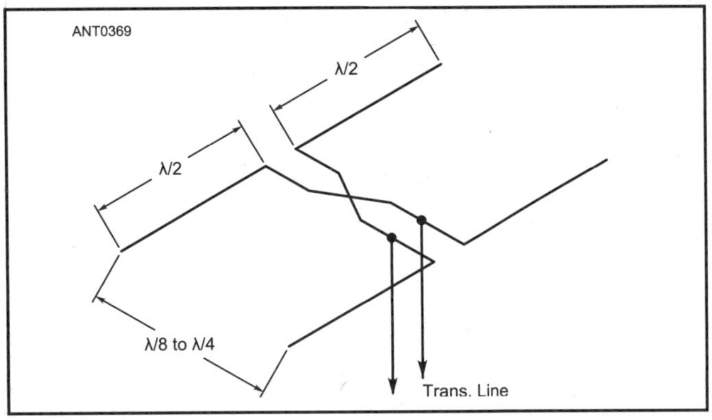

Figure 8.50 — A 4-element array combining collinear broadside elements and parallel end-fire elements, popularly known as a two-section W8JK array.

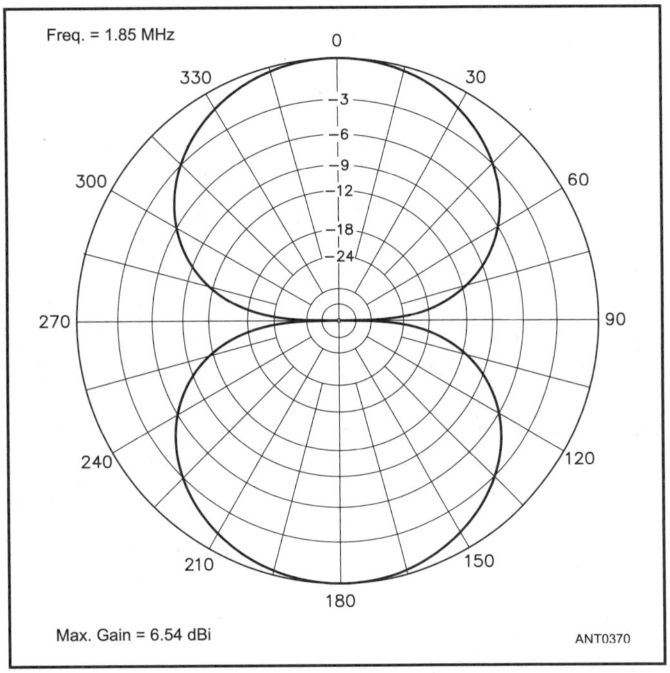

Figure 8.51 — Free-space E-plane pattern for the antenna shown in Figure 8.50, with λ/8 spacing. The elements are parallel to the 90° – 270° line in this diagram. Less than a 1° change in half-power beamwidth results when the spacing is changed from λ/8 to λ/4.

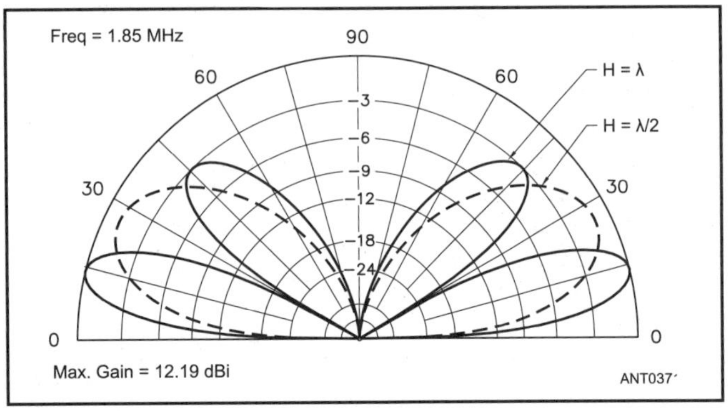

Figure 8.52 — Elevation-plane pattern for the 4-element antenna of Figure 8.50 when mounted horizontally at two heights over flat ground. Solid line = 1 λ high; dashed line = λ/2 high.

To use a matched line, you could connect a closed stub 3λ/16 long at the transmission-line junction shown in Figure 8.50. The transmission line itself can then be tapped on this matching section at the point resulting in the lowest line SWR. This point can be determined by trial.

This type of antenna can be operated on two bands having a frequency ratio of 2 to 1, if a resonant feed line is used. For example, if you design for 28 MHz with λ/4 spacing between elements, you can also operate on 14 MHz as a simple 2-element end-fire array having λ/8 spacing.

COMBINATION DRIVEN ARRAYS

You can readily combine broadside, end-fire, and collinear elements to increase gain and directivity, and this is in fact usually done when more than two elements are used in an array. Combinations of this type give more gain, in a given amount of space, than plain arrays of the types just described. Since the combinations that can be worked out are almost endless, this section describes only a few of the simpler types.

The accurate calculation of the power gain of a multi-element array requires a knowledge of the mutual impedances between all elements, as discussed in earlier sections. For approximate purposes it is sufficient to assume that each set (collinear, broadside, end-fire) will have the gains as given earlier, and then simply add up the gains for the combination. This neglects the effects of cross-coupling between sets of elements. However, the array configurations are such that the mutual impedances from cross-coupling should be relatively small, particularly when the spacings are λ/4 or more, so the estimated gain should be reasonably close to the actual gain. Alternatively, an antenna modeling program, such as *EZNEC*, can give good estimates of all parameters for a real-world antenna, providing that you take care to model all applicable parameters.

8.4.4 The ZL Special and HB9CV Arrays

The following material is extracted from the four-part series on two-element horizontal arrays by L.B. Cebik, W4RNL in the Nov 2001 through May 2002 issues of the *National Contest Journal* and "Try the FD Special" by Roy Lewallen, W7EL, in the June 1984 issue of *QST*. All of these articles are included in the online material.

THE ZL SPECIAL

Figure 8.53 shows the simple elements of a ZL-Special. The two elements bear "forward" and "rear" element labels, where the main lobe is in the direction of the forward element. The customary standard ZL Special feed point is at the junction of the phase line and the forward element. The antenna is designed for use on a single band and can be fixed or rotatable.

Generally, the ZL Special consists of two dipole elements spaced by ⅛ wavelength and fed with a phase difference of 135–160 degrees. (A phase difference of 180 degrees creates the W8JK array.) There are a number of variations in how the array is constructed, fed, and matched with forward

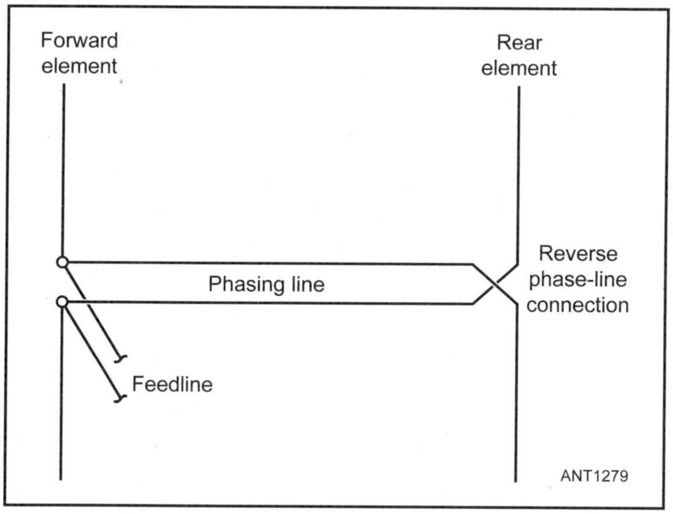

Figure 8.53 — The basic elements of a ZL Special 2-element horizontal array.

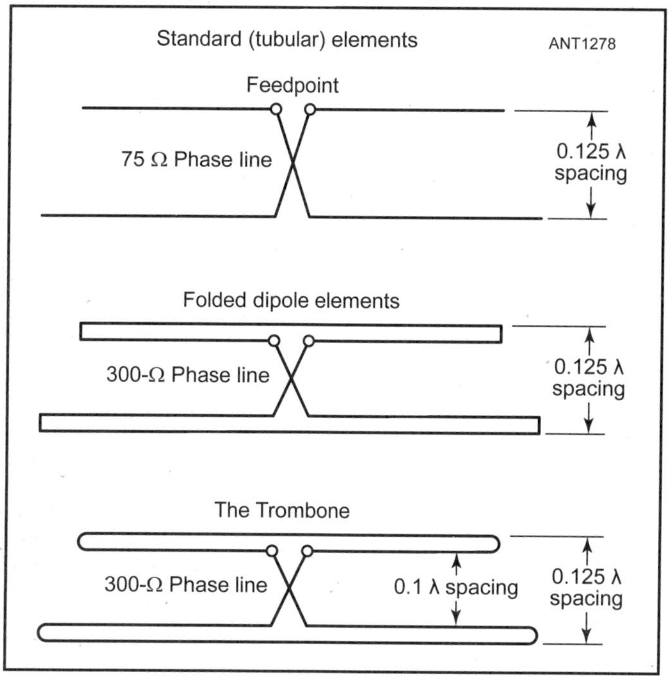

Figure 8.54 — Outline sketches of several classic ZL Special designs.

gain in the range of 6 – 6.5 dBi (free space) and F/B ratio of 18 – 25 dB. See **Figure 8.54**.

This sounds simple but the challenge is to create equal currents in the elements. Mutual coupling as discussed in the **Phased Array** chapter has a big effect on element impedances. If equal currents do flow, the element impedances vary quite a bit from 75 + *j*0 Ω of a single resonant dipole. For example, at 135 degrees difference, the leading element has a feed point impedance of 28 – *j*46 Ω with the lagging element at 28 + *j*46 Ω. As spacing changes so will the feed point impedances. The lower impedance increases current in the elements, helping to overcome the partial cancellation caused by the closely spaced elements.

The other complicating factor is that with few exceptions, the phase delay of current in a transmission line not terminated in its characteristic impedance doesn't equal the electrical length of the line. The difference can be significant!

Many possible ZL Special designs prove to be unfeasible. The requisite characteristic impedance of the phasing line may not exist and cannot be constructed. The required line length may be shorter than the distance between the elements, or it may be excessively long.

The key to successful ZL Special design is to find a set of element lengths and a spacing that meets two conditions. First, the relative current magnitude and phase angle on the individual elements must provide a satisfactory pattern in terms of gain and front-to-back ratio. Second, the impedances of the elements under the first condition must permit the design of a phasing line (or pair of lines) that employs an available or achievable characteristic impedance and that allows the requisite current division and transformation.

The FD Special version of the antenna designed by W7EL for 20 meters is shown in **Figure 8.55**. It uses a pair of folded dipoles with a higher feed point impedance than for single-conductor elements. Spacing is 1/8th wavelength and the phasing line is 300 Ω twinlead with one-half twist. The resulting feed point impedance is close to 50 Ω but slightly inductive which is canceled with two 330 pF capacitors in series with the feed line. A choke balun isolates the feed line from the balanced element.

Element phasing varies from 135 degrees at 14.0 MHz to 148 degrees at 14.35 MHz, with current ratios from 1.04 to 1.13:1. Gain is fairly constant at about 4.5 dBd (free space) across the band.

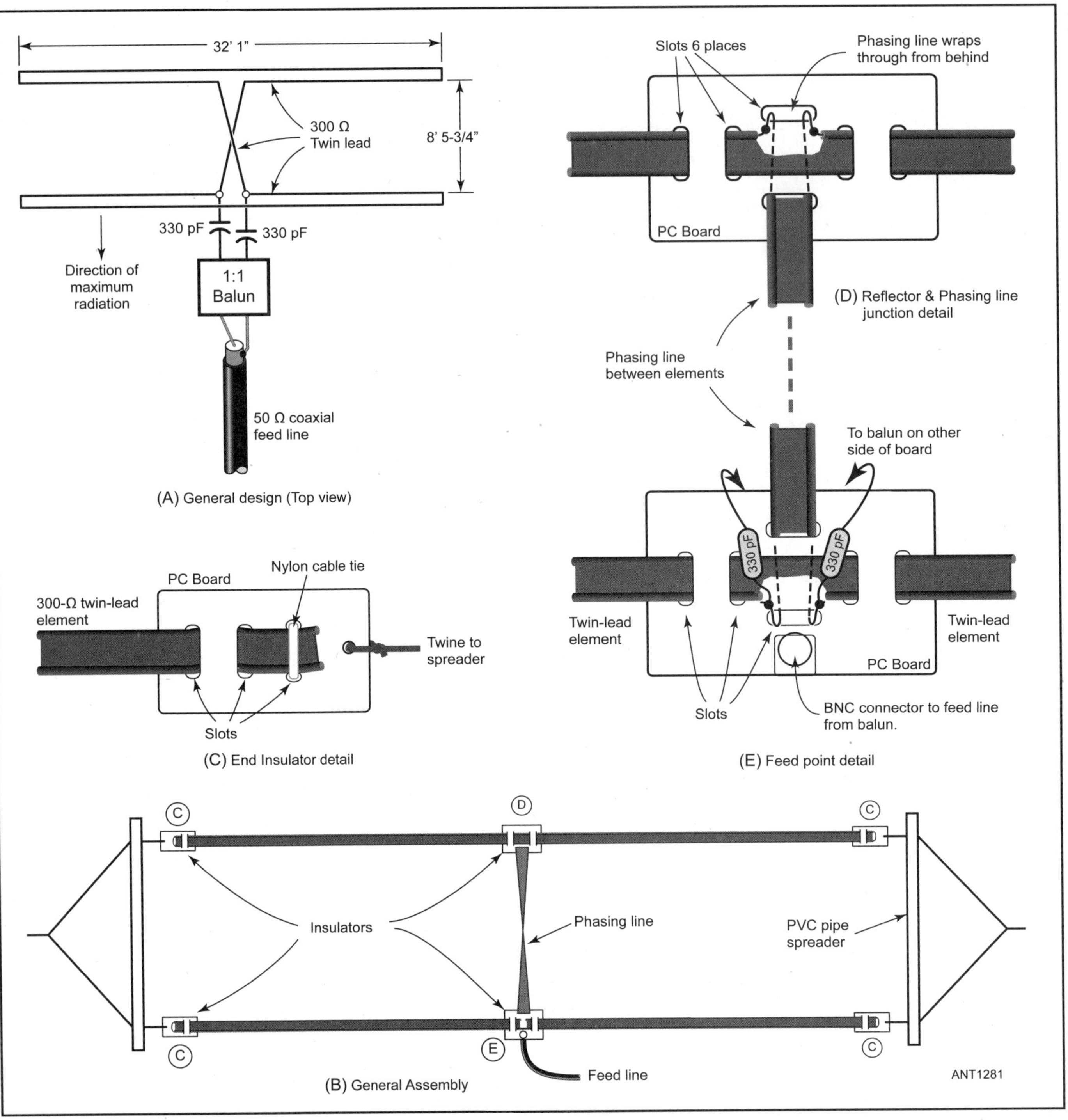

Figure 8.55 — Electrical dimensions for the 20-meter W7EL version of the ZL Special (A). Part B shows how the antenna is assembled on spreaders of PVC pipe.

THE HB9CV ARRAY

The original HB9CV design, shown at the top of **Figure 8.56**, attempted to permit the use of 300 Ω (or other parallel line available in the 1950s) with single tubular elements by the use of T or double gamma-match sections. A later version, shown in the lower part of the figure varied the feed line system for use with 75 Ω coaxial cable. By setting the gamma match in opposite directions on the two elements, the coax shield could connect to the element centers and to the boom.

Later versions of the HB9CV do, in fact, use the boom as one of a pair of lines, with a small diameter line forming the partner. Since the smaller wire in a parallel line with different diameter wires generally determines the line impedance, a single line could run from the rear element connection to the forward element connection and serve both as half the phase line and as the gamma section. 10, 6, and 2 meter versions of the HB9CV using this approach can be found in the **VHF, UHF, and Microwave Antennas** chapter, designed by DK7ZB.

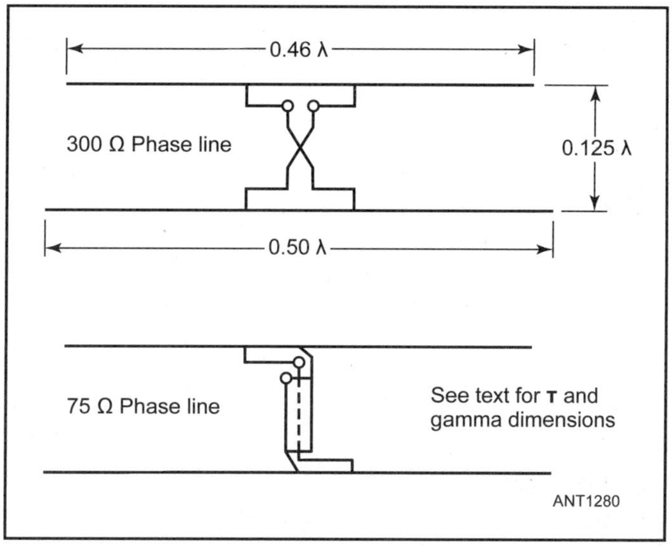

Figure 8.56 — General outlines of two versions of the HB9CV array.

8.5 References and Bibliography

Cain, J., "Curtains for You," *QST*, Oct 1991, pp 26 – 30. See also Feedback in Dec 1991 *QST*, p. 73. This article also appears in *ARRL's Wire Antenna Classics*.

Cebik, L., "Some Notes on Two-Element Horizontal Phased Arrays," in four parts, *NCJ*, Nov./Dec. 2001, pp. 4 – 10; Jan./Feb. 2002, pp. 4 – 9; Mar./Apr. 2002, pp. 3 – 8; and May/June 2002, pp. 3 – 8.

Cooper, D., "The Bi-Square Array," *Ham Radio*, May 1990, pp. 42 – 44.

Haigwood, J., "The Extended Double Zepp Revisited," *QST*, Sep. 2006, pp. 35 – 36.

Jasik, H., *Antenna Engineering Handbook*, 1st ed. (McGraw-Hill, 1961). Later editions are edited by Richard C. Johnson.

Jordan, G., "Understanding the ZL Special Antenna," *Ham Radio*, May 1976, pp. 38 – 40.

Kennedy, H., "The N4GG Array", *QST*, July 2002, pp. 35 – 39.

Kraus, J., *Antennas*, 2nd edition (McGraw-Hill, 1988).

Kraus, J., "The W8JK Antenna Recap and Update," *QST*, June 1982, pp. 11 – 14.

Johnson, R., *Antenna Engineering Handbook*, 3rd edition (McGraw-Hill, 1993). This is a later edition of the volume by the same name edited by H. Jasik.

Laport, E., *Radio Antenna Engineering* (McGraw-Hill, 1952).

Lewallen, R., "Try the FD Special," *QST*, June 1984, pp. 21 – 24.

Loukides, M., "A Dipole Curtain for 15 and 10 Meters," *QST*, Aug. 2003, pp. 34 – 38.

Romander, H., "The Extended Double-Zepp Antenna," *QST*, June 1938, pp. 12 – 16. This article also appears in *More Wire Antenna Classics*, published by ARRL.

Salmon, W., "The Extended Lazy H Antenna," *QST*, Oct. 1955, p. 20.

Smith, W., "Bet My Money on a Bobtail Beam," *CQ*, Mar. 1948, pp. 21 – 23.

Smith, W., "The Bobtail Curtain and Inverted Ground Plane, Part One," *Ham Radio*, Feb. 1983, pp. 82 – 86.

Smith, W., "Bobtail Curtain Follow-Up: Practical DX Signal Gain" *Ham Radio*, Mar. 1983, pp. 28 – 30.

Suggs, D., "Building the W8JK Beam," *QST*, Sep. 2005, pp. 31 – 35.

Zavrel, R., "The Multiband Extended Double Zepp and Derivative Designs," *QEX*, July/Aug. 1999, pp. 34 – 40.

Zimmerman, R., "A Simple 50-Ohm Feed for W8JK Beams," *QST*, June 1999, pp. 41 – 42, 47. See also Feedback in July 1999 *QST*, p. 63.

TABLE OF CONTENTS

Chapter 9 — Online Material

Articles

- "An Update on Compact Transmitting Loops" by John Belrose, VE2CV
- "A Closer Look at Horizontal Loop Antennas" by Doug Demaw, W1FB
- "The Horizontal Loop — An Effective Multipurpose Antenna" by Scott Harwood, K4VWK
- "Small Gap-resonated HF Loop Antenna Fed by a Secondary Loop" by Kai Siwiak, KE4PT and R. Quick, W4RQ
- "Active Loop Aerials for HF Reception Part 1: Practical Loop Aerial Design, and Part 2: High Dynamic Range Aerial Amplifier Design," by Chris Trask, N7ZWY
- "RF Exposure Compliance Distances for Transmitting Loops, and Transmitting Loop Current," Kai Siwiak, KE4PT
- "Small Gap-resonated HF Loop Antenna Fed by a Secondary Loop," Kai Siwiak, KE4PT, and Richard Quick, W4RQ
- "A Delta Loop for 160 Meters," by Ted Algren, KA6W

Chapter 9
Loop Antennas

A loop antenna is a closed-circuit antenna independent of a ground system or counterpoise — that is, one in which a conductor is formed into one or more turns so its two ends are close together. Loops can be divided into two general classes — large loops in which both the conductor length and the loop dimensions are comparable with the wavelength and small loops in which both the total conductor length and the maximum linear dimension of a turn are very small compared with the wavelength.

Material on quad and delta loops is adapted from Chapter 10 of *Low-Band DXing,* 5th edition by John Devoldere, ON4UN. Material on small HF gap-resonated loops was written by Kai Siwiak, KE4PT, based on the *QEX* and *QST* loop articles with coauthor Rich Quick, W4RQ, and includes material from earlier editions that was contributed by Domenic Mallozzi, N1DM. Additional discussion of loop antennas can be found in these chapters: **General Purpose MF and HF Antennas, Multiband HF Antennas,** and **Receiving Antennas**. *Wave angle*, when used in this chapter, is synonymous with elevation angle.

9.1 Large Loops

Resonant loop antennas have a circumference of 1 λ. The exact shape of the loop is not particularly important. In free space, the loop with the highest gain, however, is the loop with the shape that encloses the largest area for a given circumference. This is a circular loop, which is difficult to construct. Second best is the square loop (quad), and in third place comes the equilateral triangle (delta) loop (see the reference for Dietrich and for Stanley, K4ERO).

The maximum gain of a 1 λ loop over a λ/2 dipole in free space is approximately 1.35 dB. Delta loops are used extensively on the low bands at apex heights of λ/4 to 3λ/8 above ground. At such heights the vertically polarized loops far outperform dipoles or inverted-V dipoles for low-angle DXing, assuming good ground conductivity.

Loops are generally erected with the plane of the loop perpendicular to the ground. Whether or not the loop produces a vertically or a horizontally polarized signal (or a combination of both) depends only on how (or on which side) the loop is being fed.

Another variety of large loop antenna comprises the horizontally mounted loops, which have the plane of the loop parallel to the ground. These antennas produce horizontal radiation with takeoff angles determined, as usual, by the height of the horizontal loop over ground.

9.1.1 The Square or Quad Loop

Belcher, WA4JVE; Casper, K4HKX; and Dietrich, WAØRDX, have published studies comparing the horizontally polarized vertical quad loop with a dipole. (See the References and Bibliography section.) A horizontally polarized quad loop antenna (**Figure 9.1A**) can be seen as two short, end-loaded dipoles stacked λ/4 apart, with the top antenna at λ/4 and the bottom one just above ground level. The total length for a resonant loop is approximately 5 to 6% longer than the free-space wavelength.

There is no broadside radiation from the vertical wires of the quad because of the current opposition in the vertical members. In a similar manner, the vertically polarized quad loop consists of two top-loaded, λ/4 vertical dipoles, spaced λ/4 apart. Figure 9.1 shows how the current distribution along the elements produces cancellation of radiation from certain parts of the antenna, while radiation from other parts (the horizontally or vertically stacked short dipoles) is reinforced.

The square quad can be fed for either horizontal or vertical polarization merely by placing the feed point at the center of a horizontal arm or at the center of a vertical arm. At the higher frequencies in the HF range, where the quads are typically half to several wavelengths high, quad loops are usually fed to produce horizontal polarization, although there is no specific reason for this except maybe from a mechanical standpoint. Polarization by itself is of little importance at HF because of random rotation in the ionosphere.

QUAD LOOP IMPEDANCE

The radiation resistance of an equilateral quad loop in free space is approximately 120 Ω. The radiation resistance for a quad loop as a function of its height above ground is given in **Figure 9.2**. The impedance data were obtained by modeling

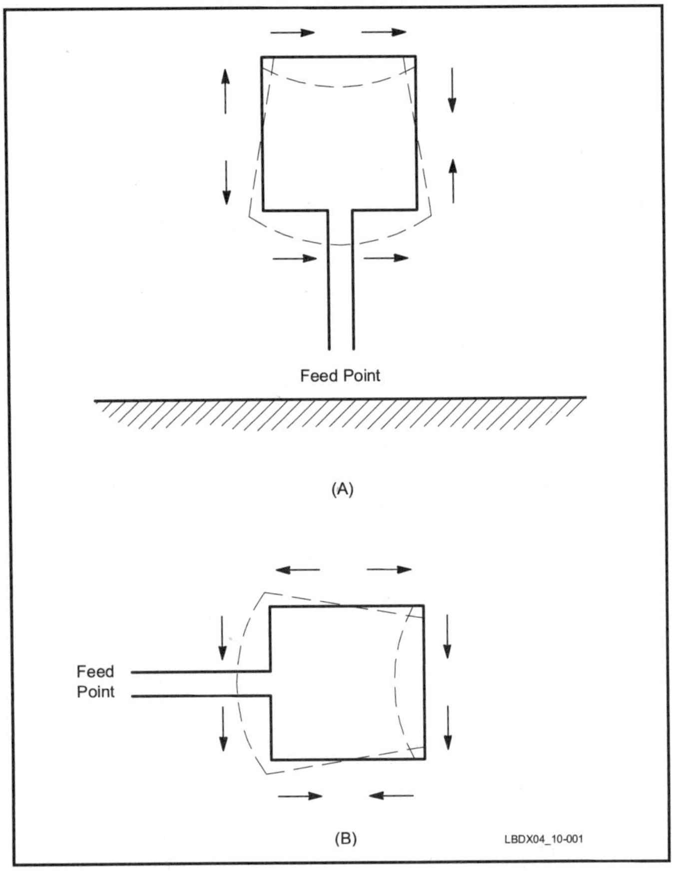

Figure 9.1 — Quad loops with a 1-λ circumference. The current distribution is shown for (A) horizontal and (B) vertical polarization. Note how two opposite legs have opposing currents. This results in cancellation of radiation in directions along the axis of those legs in the plane of the loop. Currents in the other legs are in-phase and reinforce each other in the broadside direction (perpendicular to the plane of the antenna).

an equilateral quad loop over three types of ground (very good, average, and very poor ground) using *NEC*.

The reactance data can assist you in evaluating the influence of the antenna height on the resonant frequency. The loop antenna was first modeled in free space to be resonant at 3.75 MHz and the reactance data were obtained with those free-space resonant-loop dimensions.

For the vertically polarized quad loop, the resistive part of the impedance changes very little with the type of ground under the antenna. The feed point reactance is influenced by the ground quality, especially at lower heights. For the horizontally polarized loop, the radiation resistance is noticeably influenced by the ground quality, especially at low heights. The same is true for the reactance.

QUAD LOOP PATTERNS — VERTICAL POLARIZATION

The vertically polarized quad loop in Figure 9.1B can be considered as two shortened top-loaded vertical dipoles, spaced λ/4 apart. Broadside radiation from the horizontal elements of the quad is canceled, because of the opposition of currents in the vertical legs. The wave (elevation) angle in the broadside direction will be essentially the same as for either of the vertical members. The resulting radiation angle will depend on the quality of the ground up to several wavelengths away from the antenna, as is the case with all vertically polarized antennas.

The quality of the reflecting ground will also influence the gain of the vertically polarized loop to a great extent. The quality of the ground is as important as it is for any other vertical antenna, meaning that vertically polarized loops close to the ground will not work well over poor soil.

Figure 9.3 shows both the azimuth and elevation radiation

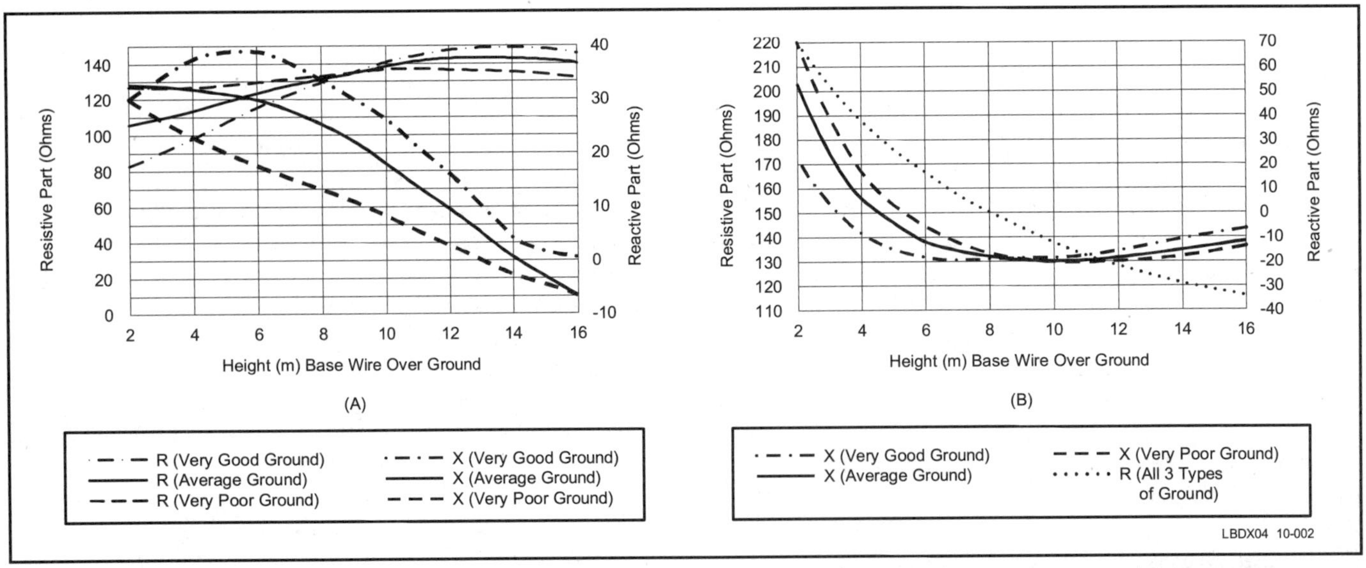

Figure 9.2 — Radiation resistance and feed point resistance for square loops at different heights above real ground. The loop was first dimensioned to be resonant in free space (reactance equal to zero), and those dimensions were used for calculating the impedance over ground. At A, for horizontal polarization, and at B, for vertical polarization. Analysis was with *NEC* at 3.75 MHz.

patterns of a vertically polarized quad loop with a top height of 0.3 λ (bottom wire at approximately 0.04 λ). This is a very realistic situation, especially on 80 meters. The loop radiates an excellent low-angle wave (lobe peak at approximately 21°) when operated over average ground. Over poorer ground, the wave angle would be closer to 30°. The horizontal directivity, Figure 9.3C, is rather poor, and amounts to approximately 3.3 dB of side rejection at any wave angle.

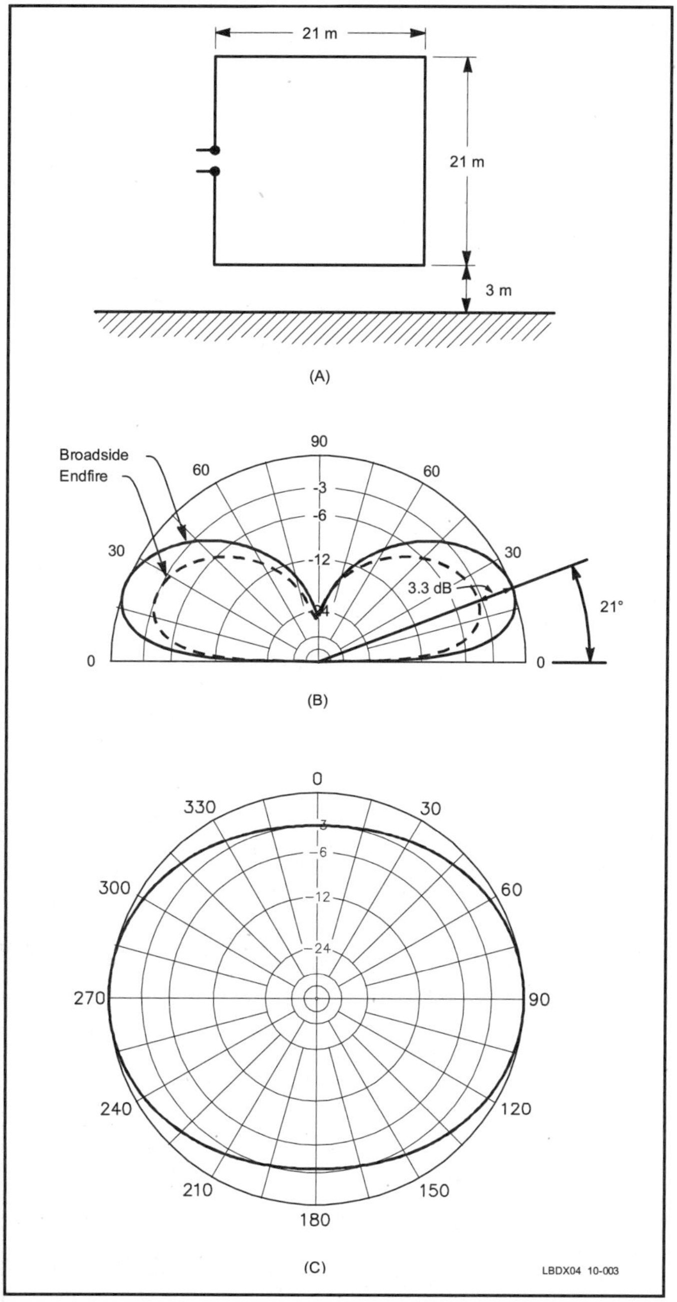

Figure 9.3 — Shown at A is a square loop, with its elevation-plane pattern at B and azimuth pattern at C. The patterns are generated for good ground. The bottom wire is 0.0375 λ above ground (3 meters or 10 feet on 80 meters). At C, the pattern is for a wave angle of 21°.

QUAD LOOP PATTERNS — HORIZONTAL POLARIZATION

A horizontally polarized quad-loop antenna (two stacked short dipoles) produces a wave angle that is dependent on the height of the loop. The low horizontally polarized quad (top at 0.3 λ) radiates most of its energy right at or near zenith angle (straight up).

Figure 9.4 shows directivity patterns for a horizontally

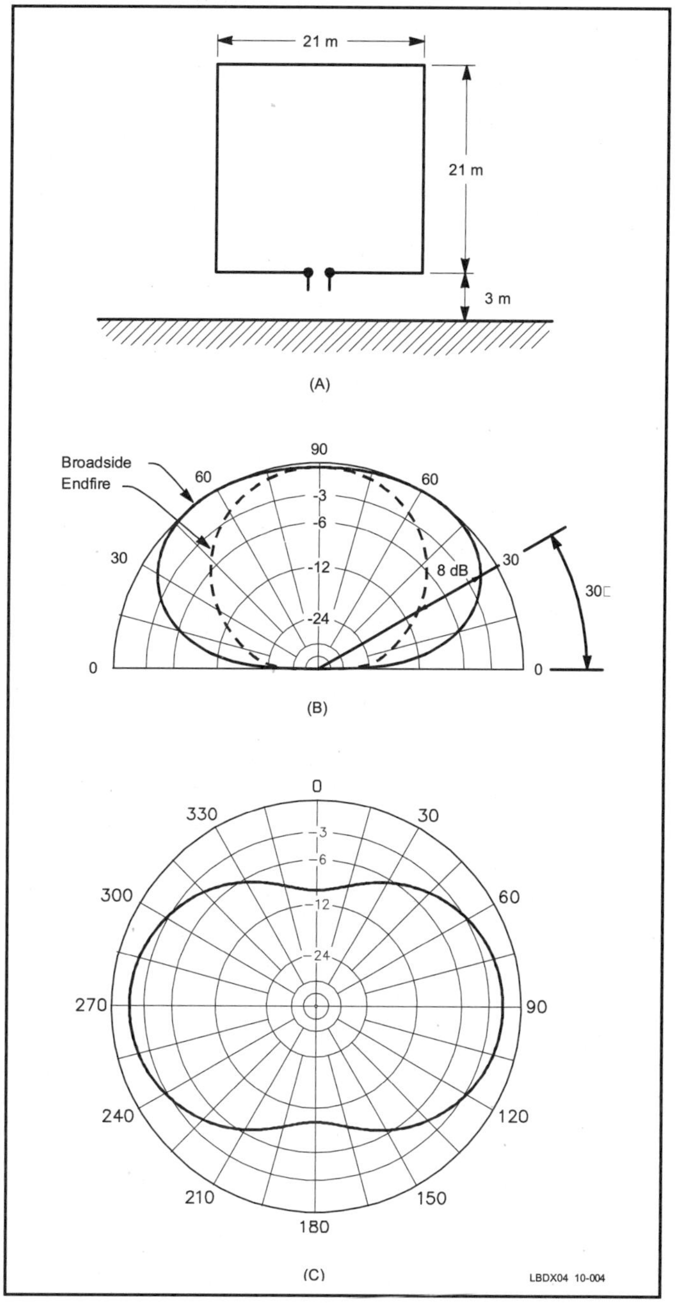

Figure 9.4 — Azimuth and elevation patterns of the horizontally polarized quad loop at low height (bottom wire 0.0375 λ above ground). At an elevation angle of 30°, the loop has a front-to-side ratio of approximately 8 dB.

polarized loop. The horizontal pattern, Figure 9.4C, is plotted for a wave angle of 30°. At low wave angles (20° to 45°), the horizontally polarized loop shows more front-to-side ratio (5 to 10 dB) than the vertically polarized rectangular loop.

VERTICAL VERSUS HORIZONTAL POLARIZATION — QUAD LOOPS

Vertically polarized loops should be used only where very good ground conductivity is available. From **Figure 9.5A** we see that the gain of the vertically polarized quad loop, as well as the wave angle, does not change very much as a function of the antenna height. This makes sense, since the vertically polarized loop is in the first place two phased verticals, each with its own quarter-wavelength radial.

However, the gain is drastically influenced by the quality of the ground. At low heights, the gain difference between very poor ground and very good ground is a solid 5 dB! The wave angle for the vertically polarized quad loop at a low height (bottom wire at 0.03 λ) varies from 25° over very poor ground to 17° over very good ground. Vertically polarized delta loops at low height always require a good ground screen underneath the antenna (unless they are over excellent or perfect ground), exactly in the same way that a vertical with only one or two radials requires a good ground underneath the radials.

With a horizontally polarized quad loop the wave angle is very dependent on the antenna height, but not so much on the quality of the ground. At very low heights, the main wave angle varies between 50° and 60° (but is rather constant all the way up to 90°). As far as gain is concerned, there is a 2.5-dB gain difference between very good and very poor ground, which is only half the difference we found with the vertically polarized loop. Comparing the gain to the gain of the vertically polarized loop, we see that at very low antenna heights the gain is about 3 dB better than for the vertically

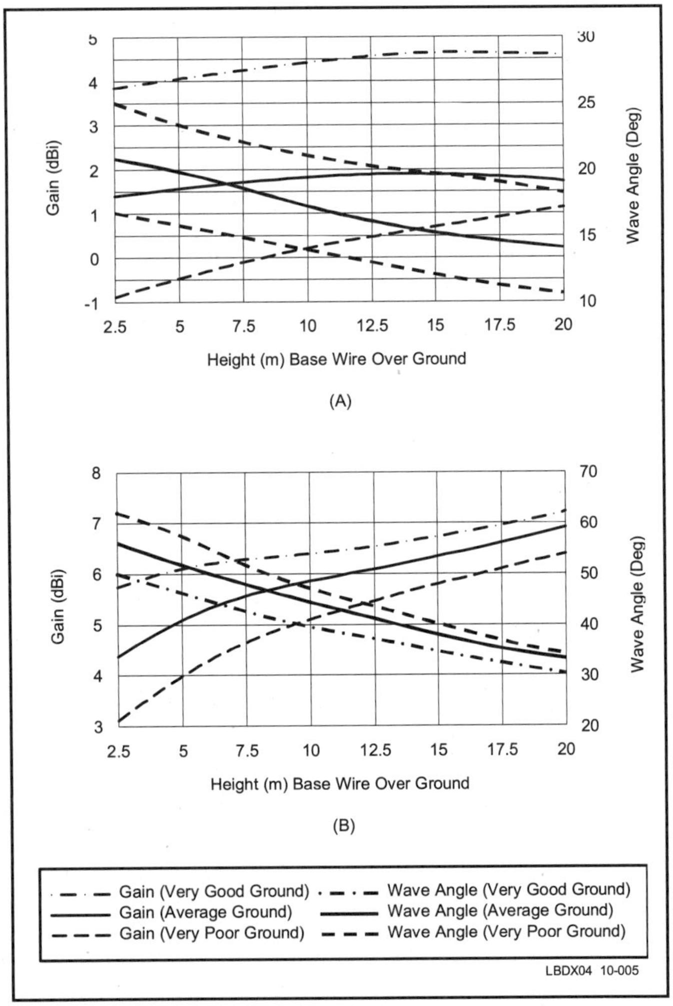

Figure 9.5 — Radiation angle and gain of the horizontally and the vertically polarized square loops at different heights over good ground. At A, for vertical polarization, and at B, for horizontal polarization. Note that the gain of the vertically polarized loop never exceeds 4.6 dBi, but its wave angle is low for any height (14 to 20°). The horizontally polarized loop can exhibit a much higher gain provided the loop is very high. Modeling was done over average ground for a frequency of 3.75 MHz, using *NEC*.

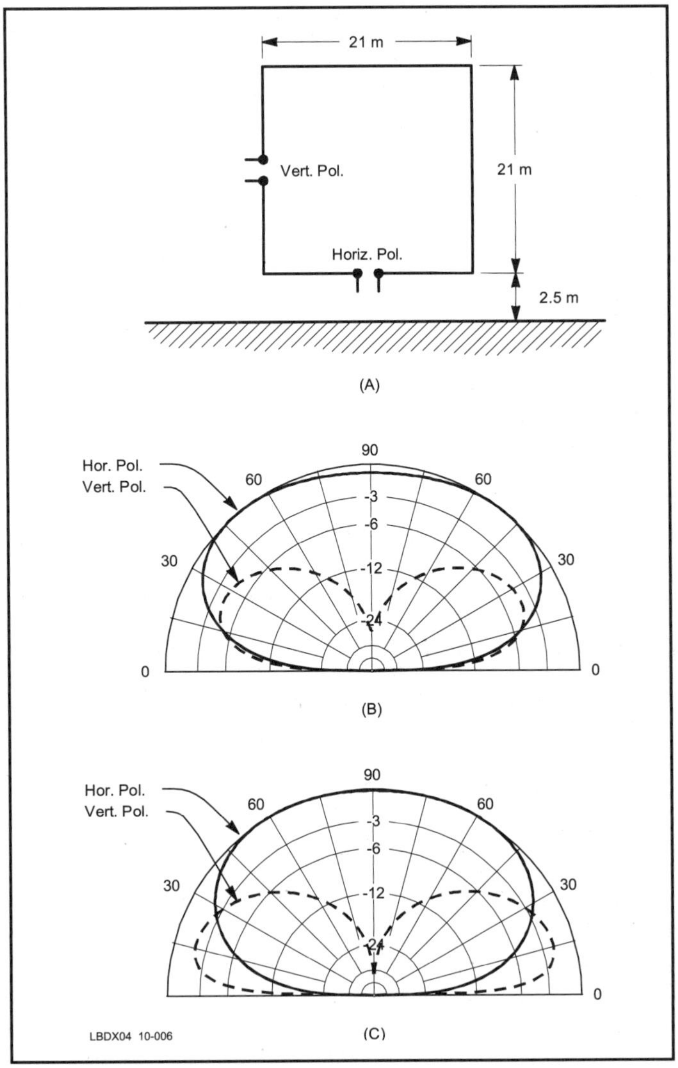

Figure 9.6 — Superimposed patterns for horizontally and vertically polarized square quad loops (shown at A) over very poor ground (B) and very good ground (C). In the vertical polarization mode the ground quality is of utmost importance, as it is with all verticals.

polarized loop. But this gain exists at a high wave angle (50° to 90°), while the vertically polarized loop at very low heights radiates at 17° to 25°.

Figure 9.6 shows the vertical-plane radiation patterns for both types of quad loops over very poor ground and over very good ground on the same dB scale.

RECTANGULAR QUAD LOOPS

A rectangular quad loop with unequal side dimensions can be used with very good results on the low bands. The vertical and the horizontal radiation patterns for this quad loop over good ground are shown in **Figure 9.7**. The horizontal directivity is approximately 6 dB (front-to-side ratio).

Even in free space, the feed point impedance of the two configurations of this rectangular loop is not the same. When fed in the center of a short side, the radiation resistance of the antenna in Figure 9.7 at resonance is 44 Ω. When fed in the center of one of the long sides, the resistance is 215 Ω. Over real ground the feed point impedance is different in both configurations as well; depending on the quality of the ground, the impedance can vary by 40 to 90 Ω.

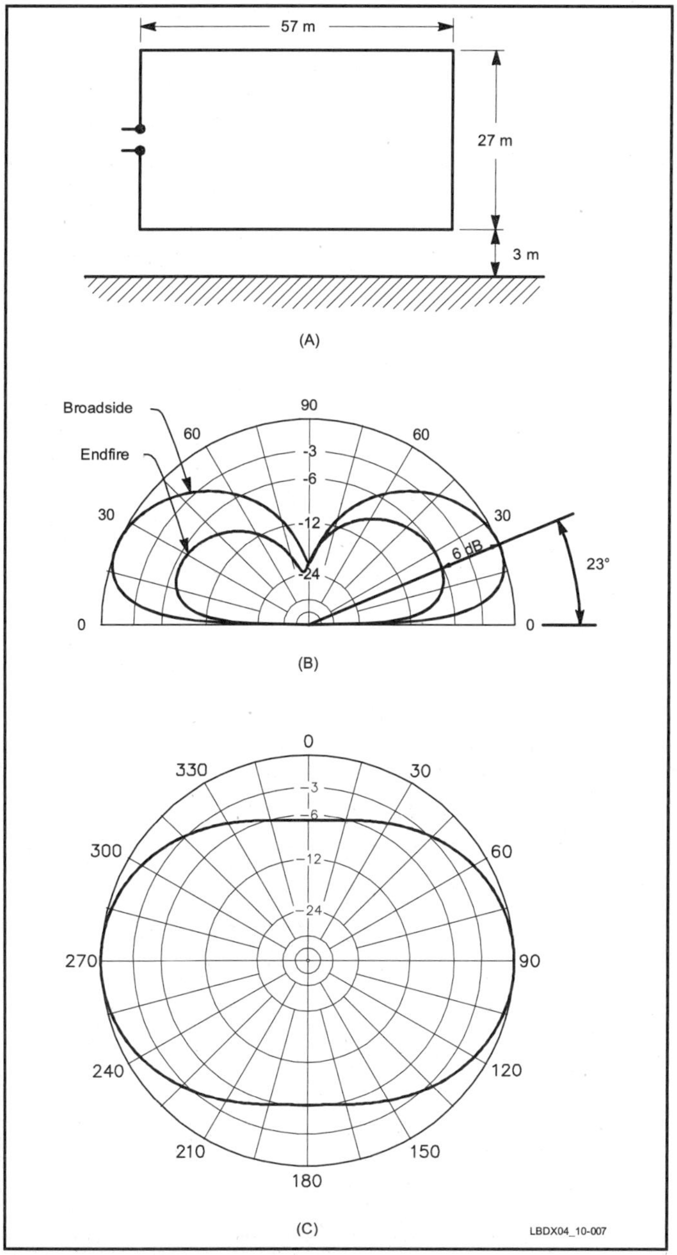

Figure 9.7 — At A, a rectangular loop with its baseline approximately twice as long as the vertical height. At B and C, the vertical and horizontal radiation patterns, generated over good ground. The loop was dimensioned to be resonant at 1.83 MHz. The azimuth pattern at C is taken at a 23° elevation angle.

FEEDING THE QUAD LOOP

The quad loop feed point should be in the middle of the vertical or the horizontal wire. A balun should be used as described in the **Transmission Line System Techniques** chapter. Alternatively, you could use open-wire feeders (for example, 450 Ω line). The open-wire-feeder alternative has the advantage of being a lightweight solution. With a tuner you will be able to cover a wide frequency range with no compromises. (See Feeding Large Loops following the section on Delta Loops.)

9.1.2. Triangular or Delta Loops

Because of its shape, the delta loop with the apex on top is a very popular antenna as it needs only one support. As for the quad configuration, the length of the resonant delta loop is approximately 1.05 to 1.06 λ.

In free space the equilateral triangle produces the highest gain and the highest radiation resistance for a three-sided loop configuration. As we deviate from an equilateral triangle toward a triangle with a long baseline, the effective gain and the radiation resistance of the loop will decrease for a bottom corner-fed delta loop. In the extreme case (where the height of the triangle is reduced to zero), the loop has become a half-wavelength-long transmission line that is shorted at the end, which shows a zero-Ω input impedance (radiation resistance), and thus zero radiation.

Just as with the quad loop, we can switch from horizontal to vertical polarization by changing the position of the feed point on the loop. For horizontal polarization the loop is fed either at the center of the baseline or at the top of the loop. For vertical polarization the loop should be fed on one of the sloping sides, at λ/4 from the apex of the delta. **Figure 9.8** shows the current distribution in both cases.

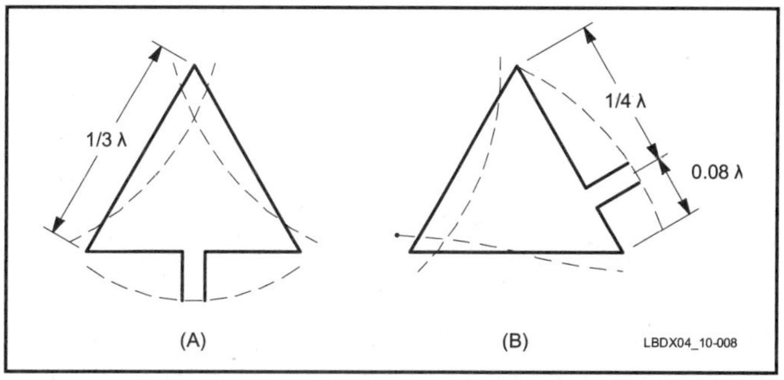

Figure 9.8 — Current distribution for equilateral delta loops fed for (A) horizontal and (B) vertical polarization.

DELTA LOOP PATTERNS — VERTICAL POLARIZATION

As shown in **Figure 9.9**, in the vertical-polarization mode the delta loop can be seen as two sloping quarter-wave verticals (their apexes touch at the top of the support), while the baseline (and the part of the sloping section under the feed point) takes care of feeding the "other" sloping section with the correct phase. The top connection of the sloping verticals can be left open without changing anything about the operation of the delta loop. The same is true for the baseline, where the middle of the baseline could be opened without changing anything. These two points are the high-impedance points of the antenna. Either the apex or the center of the baseline must be shorted, however, in order to provide feed voltage to the other half of the antenna. Normally, of course, we use a fully closed loop in the standard delta loop, although for single-band operation this is not strictly necessary.

Assume we construct the antenna with the center of the horizontal bottom wire open. Now we can see the two half baselines as two λ/4 radials, one of which provides the necessary low-impedance point for connecting the shield of the coax. The other radial is connected to the bottom of the second sloping vertical, which is the other sloping wire of the delta loop. This is similar to a λ/4 vertical using a single elevated radial. The current distribution in the two quarter-wave radials is such that all radiation from these radials is effectively canceled.

The vertically polarized delta loop is really an array of two λ/4 verticals, with the high-current points spaced 0.25 λ to 0.3 λ, and operating in phase. The fact that the tops of the verticals are close together does not influence the performance to a large degree. The reason is that the current near the apex of the delta is at a minimum (it is current that creates radiation!). You can open the apex and move the vertical wires apart if you have a very tall support, in which case you will increase the gain of the antenna somewhat.

Considering a pair of phased verticals, we know from the chapter **Effects of Ground** that the quality of the ground will be very important as to the efficient operation of the antenna. This does not mean that the delta loop requires radials. It has two elevated radials that are an integral part of the loop and take care of the return currents. The presence of the (lossy) ground under the antenna is responsible for near-field losses, unless we can shield it from the antenna by using a ground screen or a radial system, which should not be connected to the antenna.

As with all vertically polarized antennas, the quality of the ground within a radius of several wavelengths will determine the low-angle radiation of the loop antenna.

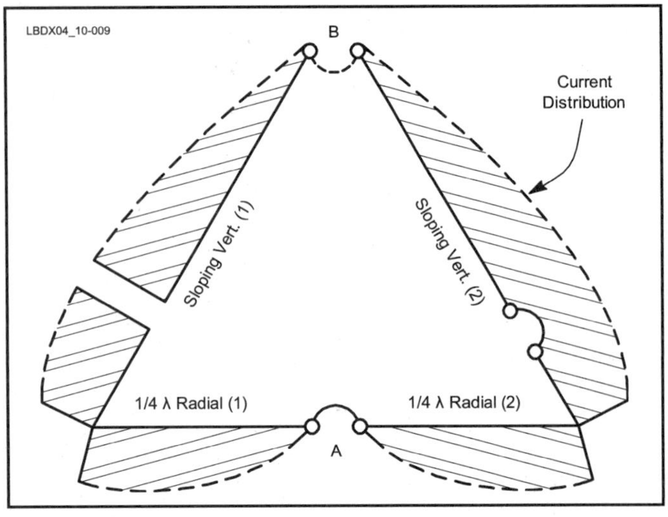

Figure 9.9 — The delta loop can be seen as two λ/4 sloping verticals, each using one radial. Because of the current distribution in the radials, the radiation from the radials is effectively canceled.

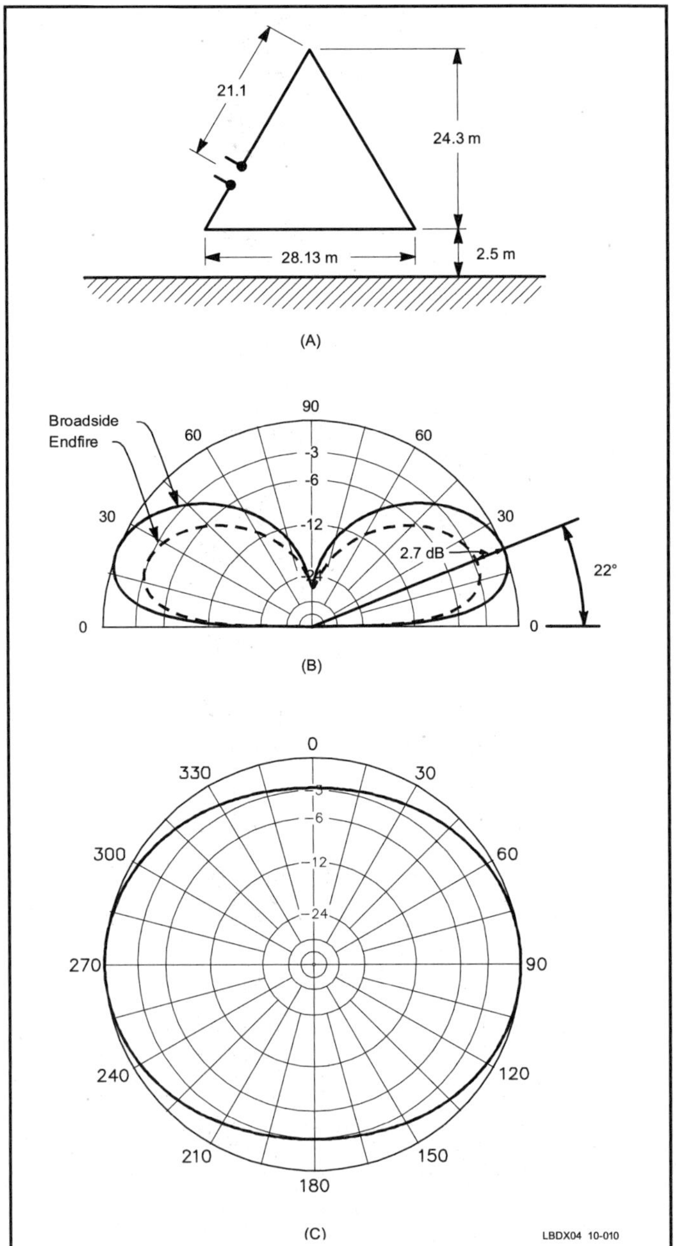

Figure 9.10 — Configuration and radiation patterns for a vertically polarized equilateral delta loop antenna. The model was calculated over good ground, for a frequency of 3.8 MHz. The elevation angle for the azimuth pattern at C is 22°.

The Equilateral Triangle

Figure 9.10 shows the configuration as well as both the broadside and the end-fire vertical radiation patterns of the vertically polarized equilateral-triangle delta loop antenna. The model was constructed for a frequency of 3.75 MHz. The baseline is 2.5 meters above ground, which puts the apex at 26.83 meters. The model was made over good ground. The delta loop shows nearly 3 dB front-to-side ratio at the main wave angle of 22°. With average ground the gain is 1.3 dBi.

The Compressed Delta Loop

Figure 9.11 shows an 80 meter delta loop with the apex at 24 meters and the baseline at 3 meters. This delta loop has a long baseline of 30.4 meters. The feed point is again located λ/4 from the apex.

The front-to-side ratio is 3.8 dB. The gain with average ground is 1.6 dBi. In free space the equilateral triangle gives a higher gain than the "flat" delta. Over real ground and in the vertically polarized mode, the gain of the flat delta loop is 0.3 dB better than the equilateral delta, however. This must be explained by the fact that the longer baseline yields a wider separation of the two "sloping" verticals, yielding a slightly higher gain.

For a 100-kHz bandwidth (on 80 meters) the SWR rises to 1.4:1 at the edges. The 2:1 SWR bandwidth is approximately 175 kHz.

The Bottom-Corner-Fed Delta Loop

Figure 9.12 shows the layout of the delta loop being fed at one of the two bottom corners. The antenna has the same apex and baseline height as the compressed delta loop. Because of the "incorrect" location of the feed point, cancellation of radiation from the base wire (the two "radials") is not 100% effective, resulting in a significant horizontally polarized radiation component. The total field has a uniform gain coverage (within 1 dB) from 25° to 90°. This may be a disadvantage for the rejection of high-angle signals when working DX at low wave angles.

Due to the "incorrect" feed point location, the end-fire radiation (radiation in line with the loop) has become asymmetrical. The horizontal radiation pattern shown in Figure 9.12D is for a wave angle of 29°. Note the deep side null (nearly 12 dB) at that wave angle. The loop actually radiates its maximum signal about 18° off the broadside direction. This feed point configuration (in the corner of the compressed loop) is to be avoided, as it really degrades the performance of the antenna.

DELTA LOOP PATTERNS — HORIZONTAL POLARIZATION

In the horizontal polarization mode, the delta loop can be seen as an inverted-V dipole on top of a very low dipole with its ends bent upward to connect to the tips of the inverted V. The loop will act as any horizontally polarized antenna over real ground; its wave angle will depend on the height of the antenna over the ground.

Figure 9.13 shows the vertical and the horizontal radiation patterns for an equilateral-triangle delta loop, fed at the center of the bottom wire. As anticipated, the radiation is maximum at the zenith. The front-to-side ratio is around 3 dB for a 15 to 45° wave angle. Over average ground the gain is 2.5 dBi. So far we have only spoken about relative patterns. What about real gain figures from the vertically and the horizontally polarized delta loops?

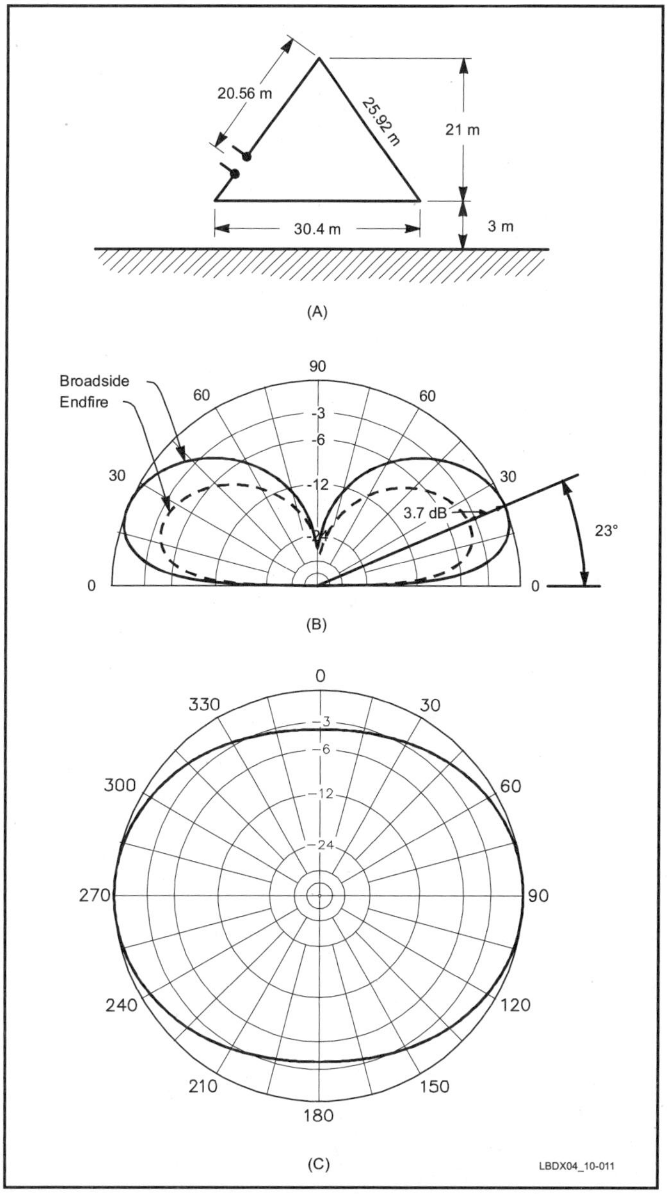

Figure 9.11 — Configuration and radiation patterns for the "compressed" delta loop, which has a baseline slightly longer than the sloping wires. The model was dimensioned for 3.8 MHz to have an apex height of 24 meters and a bottom wire height of 3 meters. Calculations are done over good ground at a frequency of 3.8 MHz. The azimuth pattern at C is for an elevation angle of 23°. Note that the correct feed point remains at λ/4 from the apex of the loop.

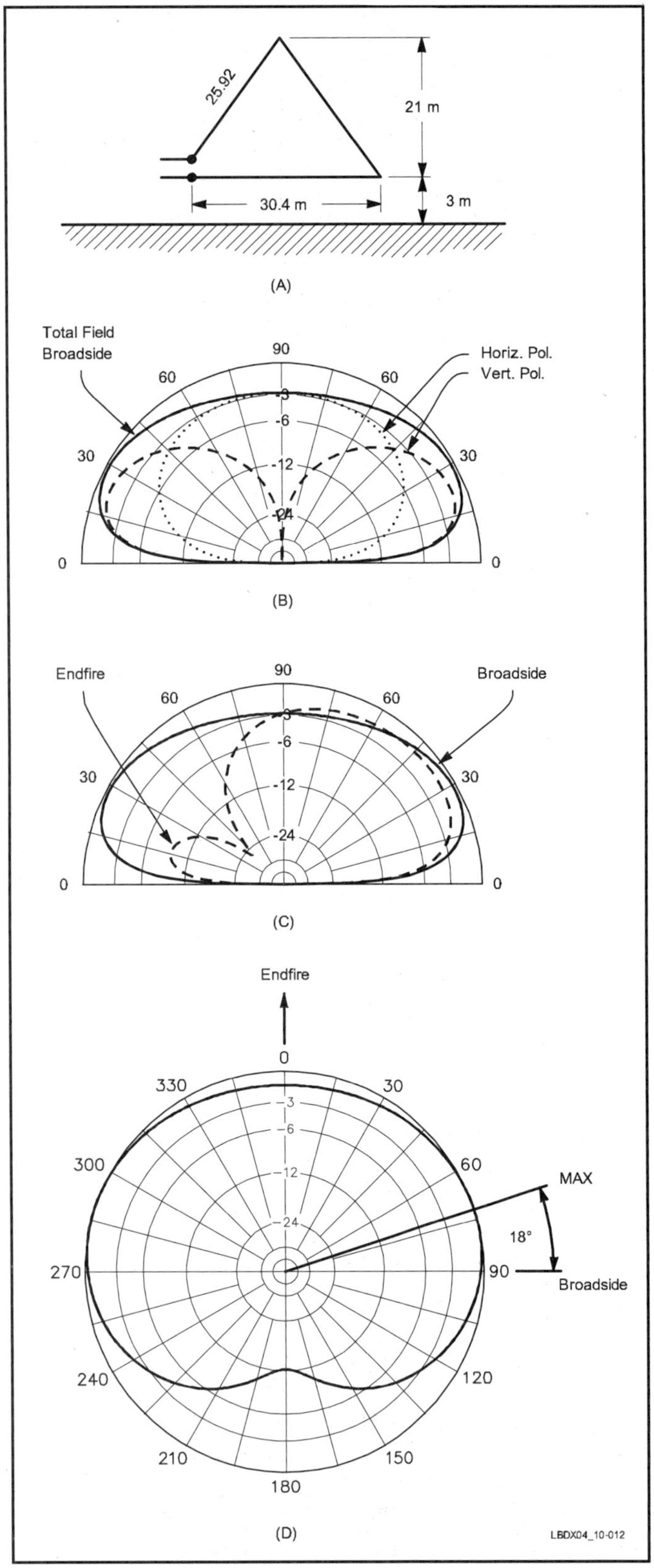

Figure 9.12 — Configuration and radiation patterns for the compressed delta loop of Figure 9.11 when fed in one of the bottom corners at a frequency of 3.75 MHz. Improper cancellation of radiation from the horizontal wire produces a strong high-angle horizontally polarized component. The delta loop now also shows a strange horizontal directivity pattern (at D), the shape of which is very sensitive to slight frequency deviations. This pattern is for an elevation angle of 29°.

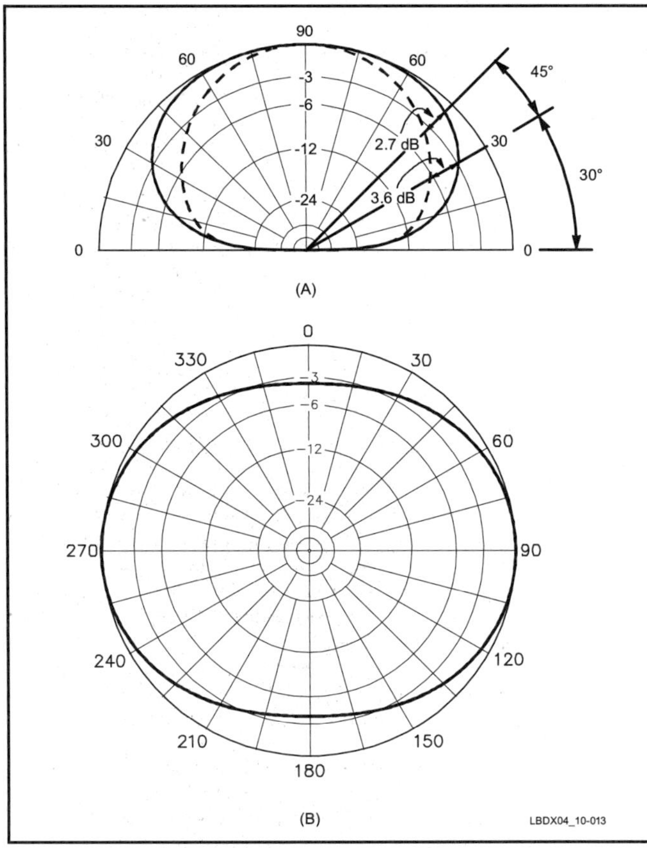

Figure 9.13 — Vertical and horizontal radiation patterns for an 80 meter equilateral delta loop fed for horizontal polarization, with the bottom wire at 3 meters. The radiation is essentially at very high angles, comparable to what can be obtained from a dipole or inverted-V dipole at the same (apex) height.

VERTICAL VERSUS HORIZONTAL POLARIZATION — DELTA LOOPS

Figure 9.14 shows the superimposed elevation patterns for vertically and horizontally polarized low-height equilateral triangle delta loops over two different types of ground (same dB scale).

Over very poor ground, the horizontally polarized delta loop is better than the vertically polarized loop for all wave angles above 35°. Below 35° the vertically polarized loop takes over, but quite marginally. The maximum gain of the vertically and the horizontally polarized loops differs by only 2 dB, but the big difference is that for the horizontally polarized loop, the gain occurs at almost 90°, while for the vertically polarized loop it occurs at 25°.

One might argue that for a 30° elevation angle, the horizontally polarized loop is as good as the vertically polarized loop. It is clear, however, that the vertically polarized antenna gives good high-angle rejection (rejection against local signals), while the horizontally polarized loop will not.

Over very good ground, the same thing that happens with any vertical happens with a vertically polarized delta: The performance at low angles is greatly improved with good ground. The vertically polarized loop is still better at any wave angle under 30° than when horizontally polarized.

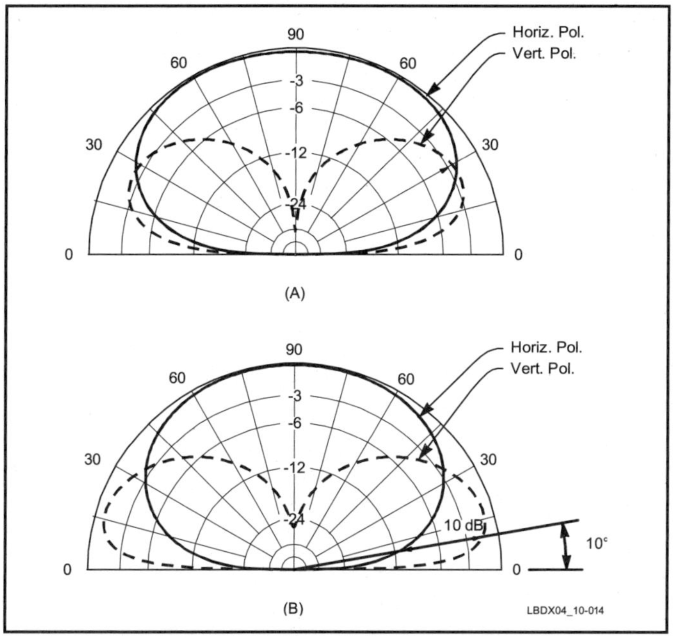

Figure 9.14 — Radiation patterns of vertically and horizontally polarized delta loops on the same dB scale. At A, over very poor ground, and at B, over very good ground. These patterns illustrate the tremendous importance of ground conductivity with vertically polarized antennas. Over better ground, the vertically polarized loop performs much better at low radiation angles, while over both good and poor ground the vertically polarized loop gives good discrimination against high-angle radiation. This is not the case for the horizontally polarized loop.

At a 10° radiation angle, the difference is as high as 10 dB.

In conclusion, over very poor ground, vertically polarized loops do not provide much better low-angle radiation when compared to the horizontally polarized loops. They have the advantage of giving substantial rejection at high angles, however. Over good ground, Figure 9.14 shows that the vertically polarized loop will give up to 10 dB and more gain at low radiation angles as compared to the horizontally polarized loop, in addition to its high-angle rejection.

FEEDING THE DELTA LOOP

The feed point of the delta loop in free space is symmetrical. At high heights above ground the loop feed point is to be considered as symmetrical, especially when we feed the loop in the center of the bottom line (or at the apex), because of its full symmetry with respect to the ground.

Figure 9.15 shows the radiation resistance and reactance for both the horizontally and the vertically polarized equilateral delta loops as a function of height above ground. At low heights, when fed for vertical polarization, the feed point is to be considered as asymmetric, whereby the "cold" point is the point to which the "radials" are connected. The center conductor of a coax feed line goes to the sloping vertical section. Many users have, however, used (symmetric) open-wire line to feed the vertically polarized loop (for example, 450 Ω line).

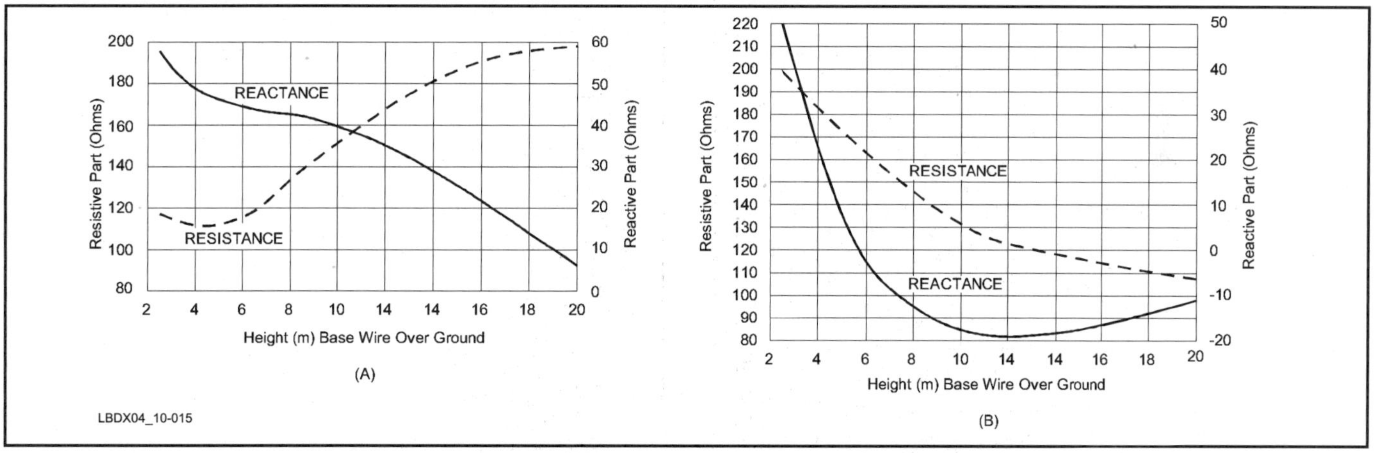

Figure 9.15 — Radiation resistance of (A) horizontally and (B) vertically polarized equilateral delta loops as a function of height above average ground. The delta loop was first dimensioned to be resonant in free space (reactance equals zero). Those dimensions were then used for calculating the impedance over real ground. Modeling was done at 3.75 MHz over good ground, using *NEC*.

Table 9.1
Loop Antennas for 160 Meters

Description	*Feeding Method*	*Gain (dBi)*	*Elevation Angle (degrees)*
Diamond loop, bottom 2.5 meters high	Fed in side corner	2.15	18.0
Square loop, bottom 2.5 meters high	Fed in center of one vertical wire	2.06	20.5
Inverted equilateral delta loop (flat wire on top)	Fed λ/4 from bottom	1.91	20.9
Regular equilateral delta loop	Fed λ/4 from top	1.90	18.1

DELTA LOOP GAIN AND RADIATION ANGLE

Figure 9.16 shows the gain and the main-lobe radiation angle for the equilateral delta loop at different heights. The values were obtained by modeling a 3.8-MHz loop over average ground using *NEC*.

Earl Cunningham, K6SE, investigated different configurations of single element loops for 160 meters, and came up with the results listed in **Table 9.1** (modeling done with *EZNEC* over good ground). These data correspond surprisingly well with those shown in Figure 9.16 (where the ground was average), which explains the slight difference in gain.

9.1.3 Feeding Large Loops

Most practical 1 λ loops present a feed point impedance between 50 and 150 Ω, depending on the exact geometry and coupling to other antennas. Delta loops tend to have impedances at the low end of the range and quads somewhat higher. The shape of the loop and proximity to the ground will affect the feed point impedance.

Measure the feed point impedance using a noise bridge or antenna analyzer connected directly to the antenna terminals. A section of feed line that is λ/2 long (or an integer multiple of λ/2) connected to the feed point will also allow direct measurement of feed point impedance if the feed point can't be reached directly.

If the impedance exceeds 50 to 70 Ω, a λ/4 feed line transformer can be used to reduce the feed point impedance to a more acceptable value (see the **Transmission Line System Techniques** chapter). If the impedance is much higher than 150 Ω, feeding via 450 Ω open-wire feeders may be warranted. Alternatively, you could use an unun (unbalanced-to-unbalanced) transformer, which can be made to cover a very wide range of impedance ratios.

To keep RF current from flowing on the outside of the coaxial feed line, use a balun or current choke at the loop feed point. RF current flowing on the feed point can distort the pattern of the loop and result in unnecessary noise pickup. For details on ununs, baluns, and common-mode chokes see the **Transmission Line System Techniques** chapter.

It is important to consider that the loop's impedance will likely vary quite a bit from band to band. An impedance matching scheme such as a fixed-ratio transformer or tuned feed line sections is not likely to work well on multiple bands. To use a large loop on multiple bands, it is more practical to use a low-loss feed line (window or ladder line are recommended) connected to an adjustable impedance matching unit at a convenient location as described in the following section.

9.1.4 Horizontal Loops

A large loop, installed horizontally over ground is an excellent multiband antenna. A 1 λ circumference loop installed at a height of λ/2 or lower has a radiation pattern similar to a λ/2 dipole at the same height — omnidirectional, high-angle radiation. As the frequency of operation increases the current distribution around the loop and the radiation pattern become more complex. The radiation lobes peak at lower elevation angles, generally approximating the angle of peak radiation for a dipole at the same electrical height.

The exact performance of the loop depends on shape, height, and frequency of use. Radiation patterns for a square, horizontal loop cut for 3.8 MHz at a height of 30 feet (about λ/8) over average ground are shown in **Figure 9.17**. On the fundamental, the pattern is essentially omnidirectional at high-angles, making it an excellent choice for regional communications using NVIS propagation. On the higher bands (patterns are shown for 14.2 and 28.3 MHz) the patterns break up into multiple lobes at lower elevation angles. In practice, the deep nulls are filled in to some degree by reflections from terrain and scatter propagation.

Feed point impedance on the 1 λ resonant frequency is approximately 100 Ω and rises to a few hundred ohms or higher at higher frequencies. Because of the varying feed point impedance, it is recommended that the antenna be fed with parallel conductor transmission line to reduce feed line loss. An impedance matching unit can then be used to convert whatever impedance is presented by the feed line to 50 Ω for coaxial cable.

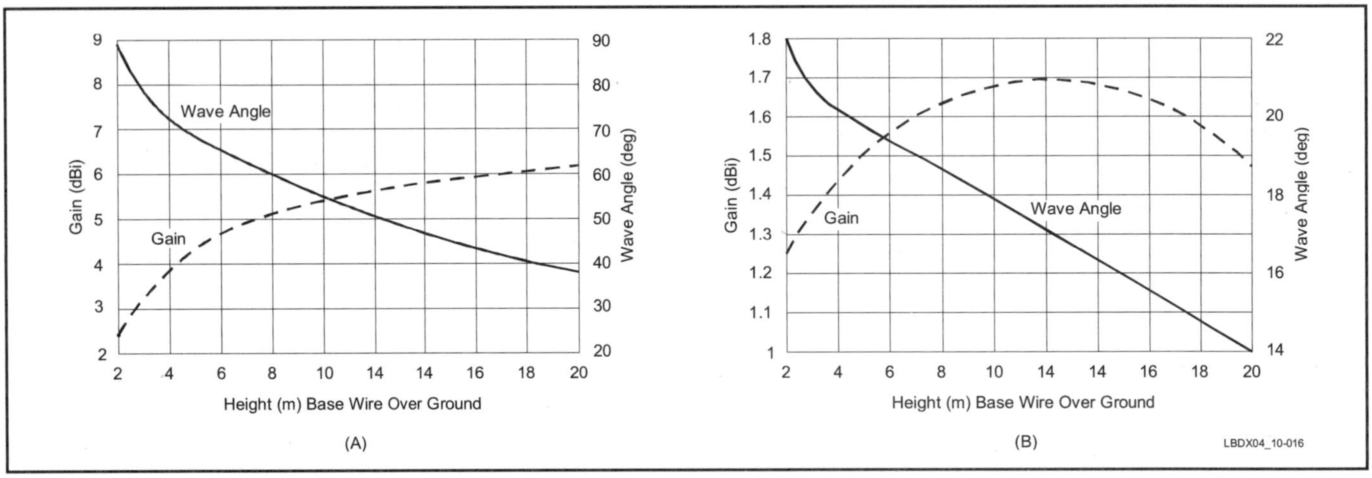

Figure 9.16 — Gain and radiation angle of (A) horizontally and (B) vertically polarized equilateral delta loops as a function of the height above ground. Modeling was done at 3.75 MHz over average ground, using *NEC*.

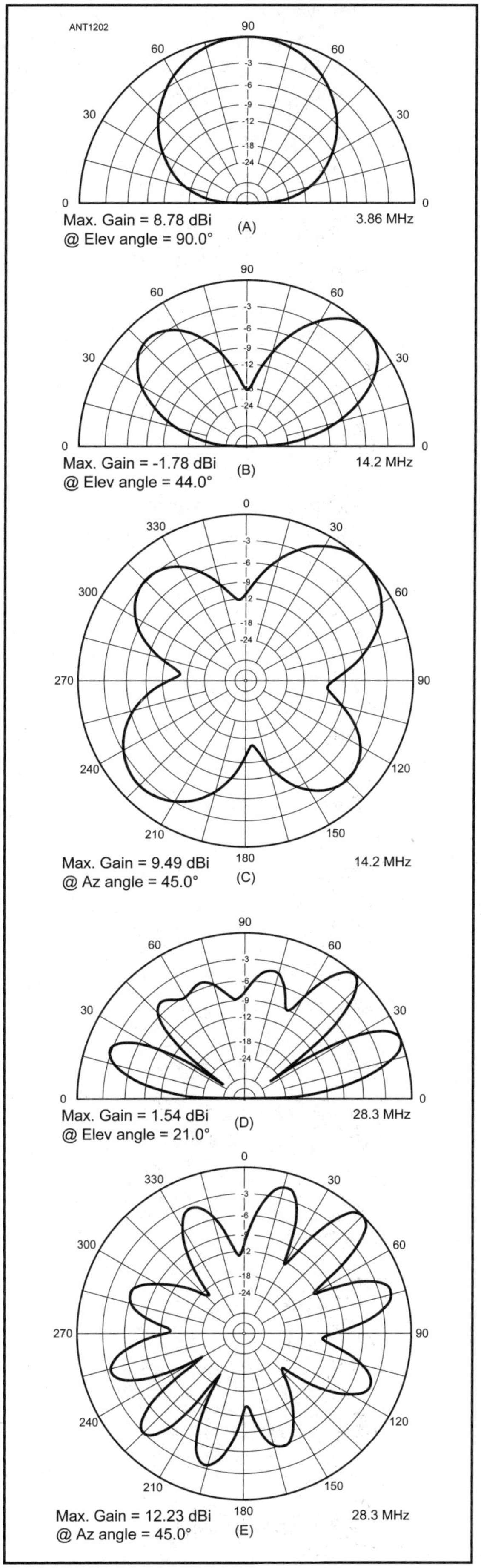

Figure 9.17 — Radiation patterns for a 1 λ square, horizontal loop cut for 3.8 MHz and installed at 30 feet above average ground. On the fundamental (A), the loop has an omnidirectional, high-angle pattern. At higher frequencies (B – 14.2 MHz, C – 28.3 MHz) the pattern breaks up into multiple lobes at lower elevation angles.

9.1.5 Half-Wave Loops

The smallest size of "large" loop generally used is one having a conductor length of ½ λ. The conductor is usually formed into a square, as shown in **Figure 9.18**, making each side ⅛ λ long. When fed at the center of one side, the current flows in a closed loop as shown in Figure 9.18A. The current distribution is approximately the same as on a ½ λ wire, and so is a maximum at the center of the side opposite the terminals X-Y, and a minimum at the feed point. This current distribution causes the field strength to be a maximum in the plane of the loop and in the direction looking from the low-current side to the high-current side. (See the referenced article by Cebik for additional discussion of this configuration.)

If the side opposite the feed point is opened at the center as shown in Figure 9.18B (strictly speaking, it is then no longer a loop because it is no longer a closed circuit), the direction of current flow remains unchanged but the maximum current flow and lowest impedance occurs at the feed point. This reverses the direction of maximum radiation.

The radiation resistance at a current maximum (which is also the resistance at X-Y in Figure 9.18B) is on the order of 50 Ω. The impedance at the feed point in Figure 9.18A is a few thousand ohms. This can be reduced by using two identical loops side by side with a few inches spacing between them and applying power between terminal X on one loop and terminal Y on the other.

Unlike a ½ λ dipole or a small loop, there is no direction in which the radiation from a loop of the type shown in Figure 9.18 is zero. There is appreciable radiation in the direction perpendicular to the plane of the loop, as well as to the "rear" — the opposite direction to the arrows shown. The front-to-back (F/B) ratio is approximately 4 to 6 dB. The small size and the shape of the directive pattern result in a loss of about 1 dB when the field strength in the optimum direction from such a loop is compared with the field from a ½ λ dipole in its optimum direction.

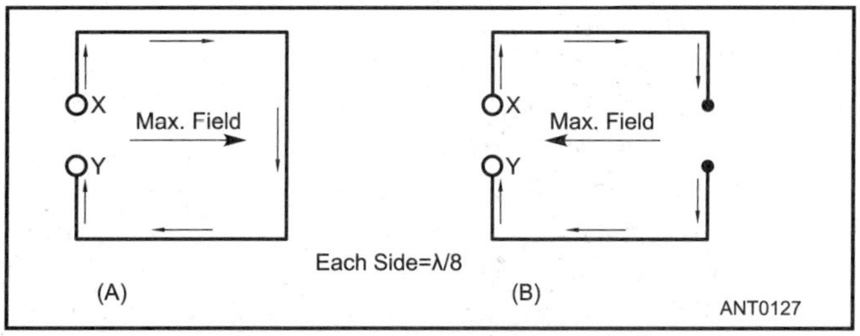

Figure 9.18 — Half-wave loops, consisting of a single turn having a total length of ½ λ.

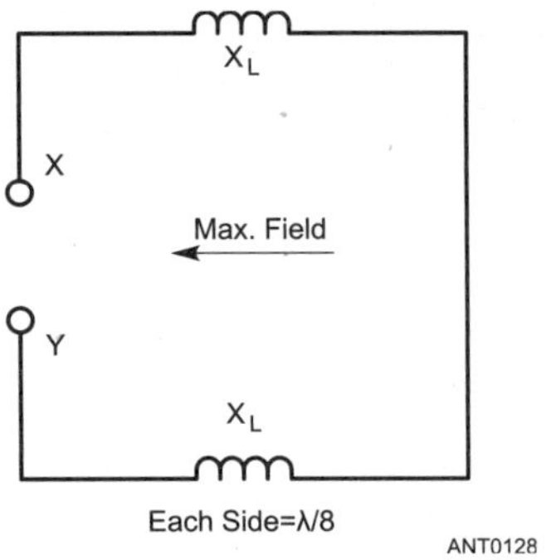

Figure 9.19 — Inductive loading in the sides of a ½-λ loop to increase the directivity and gain. Maximum radiation or response is in the plane of the loop, in the direction shown by the arrow.

The ratio of the forward radiation to the backward radiation can be increased, and the field strength likewise increased at the same time to give a gain of about 1 dB over a dipole, by using inductive reactances to "load" the sides joining the front and back of the loop. This is shown in **Figure 9.19**. The reactances, which should have a value of approximately 360 Ω, decrease the current in the sides in which they are inserted and increase it in the side with the feed point. This increases the directivity and thus increases the efficiency of the loop as a radiator. Lossy coils can reduce this advantage greatly.

9.2 Small Receiving Loops

A "small" loop can be considered to be simply a rather large coil, and the current distribution in such a loop is the same as in a coil. That is, the current has the same phase and the same amplitude in every part of the loop. To meet this condition, the total length of conductor in the loop must not exceed about 0.085 λ. Loops with circumference up to 0.3 λ and intended for transmitting are described in the Small Transmitting Loops section.

The electrically small loop antenna has existed in various forms for many years. Amateur applications of the small loop include direction finding and low-noise directional receiving antennas for 3.5 MHz and below. Because the design of transmitting and receiving loops requires some different considerations, the two situations are examined separately, beginning with receiving loops in this section.

Applications of small loops are presented in the **Receiving Antennas** and the **Stealth and Limited-Space Antennas** chapters. Ferrite-core receiving loop antennas, such as the ferrite *loopstick* found in portable AM broadcast-band receivers, are discussed in the chapter on **Receiving Antennas**. Numerous supporting papers and articles are listed in the References and Bibliography section at the end of this chapter.

9.2.1 The Basic Small Loop

A loop is considered electrically small when the current around the perimeter of the loop is in phase (i.e. assumed to be constant). This requires the total conductor length to be less than 0.085 λ. This constraint results in a very predictable figure-eight radiation pattern, shown in **Figure 9.20**. For greater circumferences, the current can no longer be treated as having a constant phase.

The simplest loop is a 1-turn untuned loop with a load connected to a pair of terminals located in the center of one of the sides, as shown in **Figure 9.21**. How its pattern is developed is easily pictured if we look at some "snapshots" of the antenna relative to a signal source. **Figure 9.22** shows a loop from above and the instantaneous received voltage wave. Note that points A and B of the loop are receiving the same instantaneous voltage. This means that no current will flow through the loop, because there is no current flowbetween points of equal potential. A similar analysis of **Figure 9.23**, with the loop turned 90° from the position represented in Figure 9.22, shows that this position of the loop provides maximum response. Figure 9.20 shows the ideal radiation pattern for a small loop.

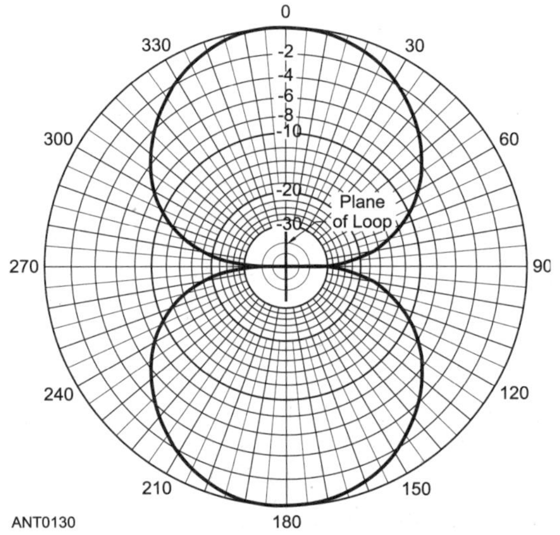

Figure 9.20 — Calculated small loop antenna radiation pattern.

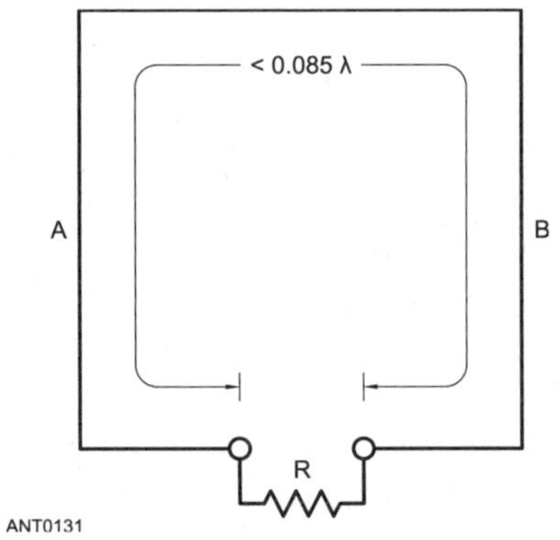

Figure 9.21 — Simple untuned small loop antenna.

The small loop is insensitive to the electric field. The voltage across the loop terminals is given by

$$V = \frac{2\pi ANH\eta_0 \cos(\theta)}{\lambda} \qquad (1)$$

where

V = voltage across the loop terminals
A = area of loop in square meters
N = number of turns in the loop
H = RF magnetic field strength in amperes per meter
η_0 = 376.73 Ω, the intrinsic impedance of free space
θ = angle between the plane of the loop and the signal source (transmitting station)
λ = wavelength of operation in meters

This equation comes from a term called *effective height*. The effective height refers to the height (length) of a vertical piece of wire above ground that would deliver the same voltage to the receiver. The equation for effective height is

$$h_{eff} = \frac{2\pi NA}{\lambda} \qquad (2)$$

where h_{eff} is in meters and the other terms are as for Eq 1. (Effective height is explained in more detail in the **Dipoles and Monopoles** chapter.)

A few minutes with a calculator will show that, with the constraints previously stated, the loop antenna will have a very small effective height. This means it will deliver a relatively small voltage to the receiver, even with a large transmitted signal.

INDUCTANCE OF THE SMALL LOOP

The loop forms an inductor having a very small ratio of winding length to diameter. The equations for calculating inductance given in most radio handbooks assume that the inductor coil is longer than its diameter. However, F. W. Grover of the US National Bureau of Standards has provided equations for inductors of common cross-sectional shapes and small length-to-diameter ratios. (See the References and Bibliography section at the end of this chapter.) Grover's equations are shown in **Table 9.2**. Their use will yield relatively accurate results which are easily worked out with a scientific calculator or PC.

9.2.2 Tuning the Small Loop

We can tune the loop by placing a capacitor across the

Why Is It Called a Magnetic Loop?

Loops smaller than approximately 0.05 λ in circumference are called "magnetic" loops because the voltage they develop is primarily created by the changing magnetic flux (H field) through the loop. This is as opposed to an "electric" dipole in which the voltage is created by the changing electric potential (E field) along the length of the antenna. See Siwiak's referenced Technical Note in Mar./Apr. 2020 *QEX* for an explanation of the 0.05 λ size limit.

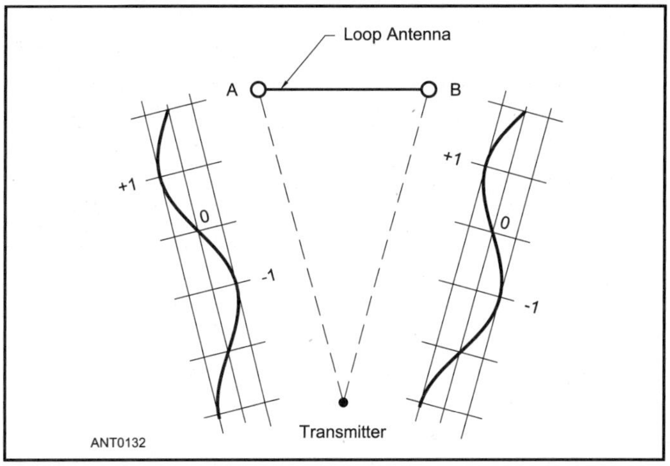

Figure 9.22 — Example of orientation of loop antenna that does not respond to a signal source (null in pattern).

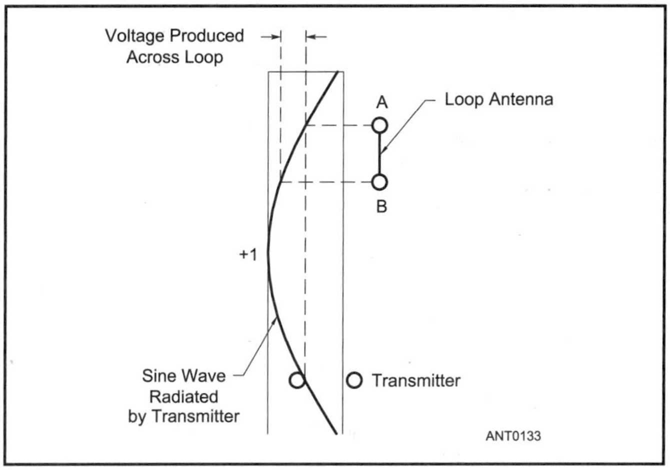

Figure 9.23 — Example of orientation of loop antenna for maximum response.

Table 9.2
Inductance Equations for Short Coils (Loop Antennas)

Triangle:

$$L\,(\mu H) = 0.006\,N^2\,s\left[\ln\left(\frac{1.1547\,s\,N}{(N+1)\,\ell}\right) + 0.65533 + \frac{0.1348\,(N+1)\,\ell}{s\,N}\right]$$

Square:

$$L\,(\mu H) = 0.008\,N^2\,s\left[\ln\left(\frac{1.4142\,s\,N}{(N+1)\,\ell}\right) + 0.37942 + \frac{0.3333\,(N+1)\,\ell}{s\,N}\right]$$

Hexagon:

$$L\,(\mu H) = 0.012\,N^2\,s\left[\ln\left(\frac{2\,s\,N}{(N+1)\,\ell}\right) + 0.65533 + \frac{0.1348\,(N+1)\,\ell}{s\,N}\right]$$

Octagon

$$L\,(\mu H) = 0.016\,N^2\,s\left[\ln\left(\frac{2.613\,s\,N}{(N+1)\,\ell}\right) + 0.75143 + \frac{0.07153\,(N+1)\,\ell}{s\,N}\right]$$

where

N = number of turns
s = side length in cm
ℓ = coil length in cm

antenna terminals. This causes a larger voltage to appear across the loop terminals because of the Q of the parallel resonant circuit that is formed with the loop inductance. The resulting voltage is that given by Eq 1 multiplied by the loaded Q of the resonant circuit. Loaded Q values of 100 or greater are easy to obtain with careful loop construction.

The value of a tuning capacitor to resonate a loop is easy to calculate from the standard resonance equations. The total tuning capacitance, however, must also include the value of distributed capacitance of the loop winding. This capacitance is caused by the slight voltage difference between adjacent turns of the coil. The distributed capacitance appears as a capacitance across the loop terminals in parallel with and adding to the tuning capacitor's value. Therefore, when determining the value of the tuning capacitor, the distributed capacitance must be subtracted from the total capacitance required to resonate the loop. The distributed capacitance also determines the highest frequency at which a particular loop can be used, because it is the minimum capacitance obtainable.

Table 9.3
Values of the Constant H for Distributed Capacitance

Length to Diameter Ratio	*H*
0.10	0.96
0.15	0.79
0.20	0.78
0.25	0.64
0.30	0.60
0.35	0.57
0.40	0.54
0.50	0.50
1.00	0.46

As with all other capacitances, the value of the distributed capacitance is based on the physical dimensions of the coil. An exact mathematical analysis of its value is a complex problem. A simple approximation is given by Medhurst (see Bibliography) as:

$$C = HD \qquad (3)$$

where

C = distributed capacitance in pF

H = a constant related to the length-to-diameter ratio of the coil (**Table 9.3** gives H values for length-to-diameter ratios typical of receiving loops.)

D = diameter of the winding in cm

Medhurst's work was with coils of round cross section. For loops of square cross section the distributed capacitance is given by Bramslev (see Bibliography) as

$$C = 60S \qquad (4)$$

where

C = the distributed capacitance in pF

S = the length of the side in meters

If you convert the length in this equation to centimeters, you will find Bramslev's equation gives results in the same order of magnitude as Medhurst's equation.

9.2.3 Electrostatically Shielded Loops

Many receiving loop antennas include an electrostatic shield. This shield generally takes the form of a tube around the winding, made of a conductive but nonmagnetic material (such as copper or aluminum). Its purpose is to maintain loop balance with respect to ground, by forcing the capacitance between all portions of the loop and ground to be identical. This is illustrated in **Figure 9.24**. It is necessary to maintain electrical loop balance to eliminate what is referred to as the *antenna effect*. When the antenna becomes unbalanced it appears to act partially as a small vertical antenna. This vertical pattern gets superimposed on the ideal figure-eight pattern, distorting the pattern and filling in the nulls. The type of pattern that results is shown in **Figure 9.25**. Since the null is used to reject noise or interference, it is important to preserve balance in the loop's construction for best performance.

Adding the shield has the effect of reducing the loop's terminal voltage somewhat, but this loss is generally offset by the increase in null depth of the loops. Proper balance of the loop antenna requires that the load on the loop also be balanced. This is usually accomplished by use of a balun or a balanced input preamplifier.

One important point regarding the shield is that it cannot form a continuous electrical path around the loop perimeter, or it will appear as a shorted coil turn. Usually the insulated break is located opposite the feed point to maintain symmetry. (The loop conductor itself is not broken.) Another point to be

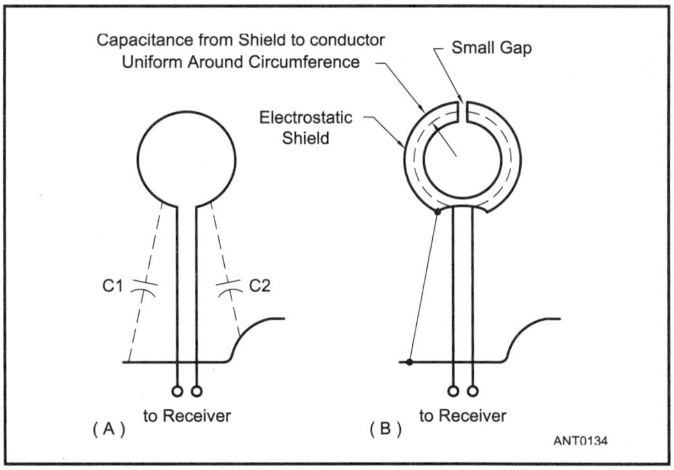

Figure 9.24 — At A, the loop is unbalanced by capacitance to its surroundings. At B, the use of an electrostatic shield overcomes this effect.

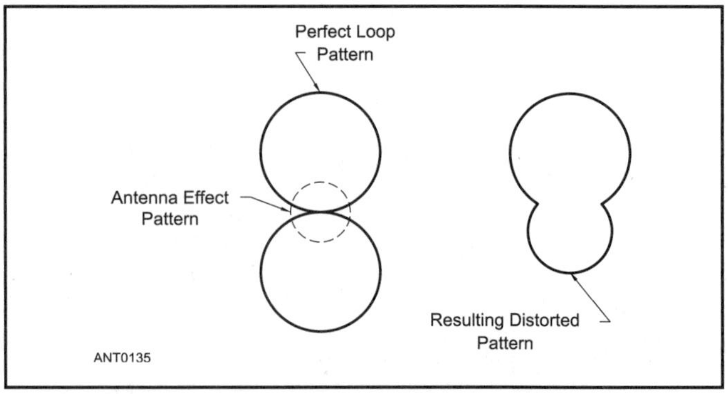

Figure 9.25 — Distortion in loop pattern resulting from antenna effect.

considered is that the shield should have a much larger cross-sectional diameter than the loop conductor, or it will lower the Q of the loop.

Various construction techniques have been used in making shielded loops. Genaille located his loop winding inside aluminum conduit, while True constructed an aluminum shield can around his winding. Others have used pieces of semi-flexible hard line to form a loop, using the outer conductor as a shield. DeMaw used flexible coax with the shield broken at the center of the loop conductor in a multi-turn loop for 1.8 MHz. Goldman uses another shielding method for broadcast receiver loops. His shield is in the form of a cylindrical "barrel" made of hardware cloth, with the loop in its center. (See Bibliography for articles by DeMaw and Goldman.) All these methods provide sufficient shielding to maintain the balance.

It is important to consider the effect of the shield configuration on antenna Q. A short letter by N1DM in *QEX* (July/Aug. 1998, see Bibliography) discusses the Q of a loop antenna with a U-type shield versus a full box shield. His data shows between 54% and 89% degradation of Q for the full box case on an otherwise identical antenna configuration.

The extra capacitance of coaxial cable (flexible or hardline) used to construct a shielded loop will limit the loop's higher frequency tuning range. Those designing a loop of this type should consider this when selecting the loop inductance. This parasitic capacitance must be taken into account to obtain the desired higher frequency tuning point.

It is possible, as Nelson shows, to construct an unshielded loop with good nulls (60 dB or better) but only if the loop circumference is very small (less than 0.0005λ) and by paying great attention to symmetry.

9.2.4 Small Loop Loss Factors

Loop losses include several factors: (1) the Q of the loop conductor by itself, (2) the effect of the load, (3) the effect of the electrostatic shield, and (4) the Q of the tuning capacitor. The major loss factor is the resistance of the loop conductor. The ac resistance of the conductor caused by skin effect is the major consideration. For a copper conductor, ac resistance is:

$$R_{ac} = \frac{0.996 \times 10^{-6} \sqrt{f}}{d} \qquad (5)$$

where

R_{ac} = ac resistance in ohms per foot
f = frequency in Hz
d = conductor diameter in inches

Inductor Q is then the reactance of the inductor ($X_L = 2\pi fL$) divided by the ac resistance. Ac resistance for other metals can be obtained by multiplying R_{ac} by the resistance of the metal relative to that of copper. (Multi-turn loops also experience loss from the proximity effect described later in this section.)

Improvement in Q can be obtained in some cases by the use of Litz wire (short for Litzendraht). Litz wire consists of strands of individual insulated wires that are woven into bundles in such a manner that each conductor occupies each location in the bundle with equal frequency. Litz wire has a reduced ac resistance when compared to an equivalent cross section solid or stranded wire, taking into account the skin depth of conductors as frequency increases.

Litz wire's improvement in ac resistance is due to the fact that the insulated individual strands result in more area of the total cross section of the conductor being in the skin depth region than for an equivalent diameter solid or stranded wire. (Stranded wire at ac acts the same as a solid wire of the same outside diameter.) Over 60% of the ac current is in this skin depth region. Therefore, skin depth is more important to the calculation of ac resistance than the total conductor diameter.

Figure 9.26 shows an example of the skin depth of a solid conductor with radius R and a piece of Litz wire with an equivalent radius R. You can see that the cross-sectional area of the current carrying skin effect region is double that of the solid wire. Litz wire is available in many configurations and the determining factor for the selection of a particular Litz wire starts with determining the optimum diameter of the individual insulated wire strands used in the construction of the particular cable. **Table 9.4** gives the values of optimum wire size based on frequency of use.

Properly-sized Litz wire results in improved Q over solid or stranded wire of equivalent size, up to about 2.8 MHz. Above 2.8 MHz other effects quickly reduce the advantage of Litz

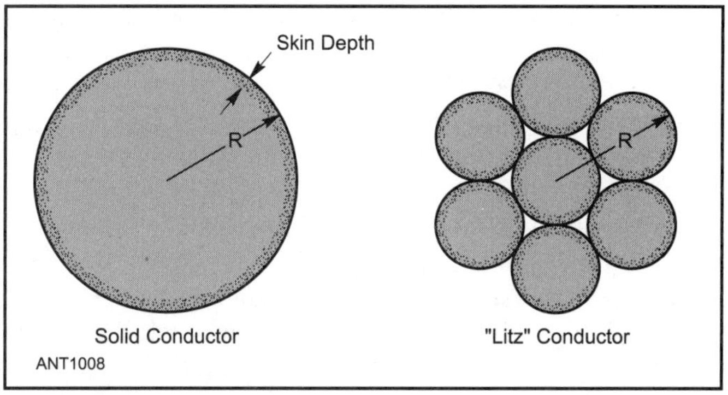

Figure 9.26 — Comparison of skin depth between conventional (A) and Litz wire (B).

Table 9.4
Optimum Sizes of Individual Wires Used in Litz Wire

Frequency Range	*Optimum AWG*
60 Hz to 1 kHz	28
1 kHz to 10 kHz	30
10 kHz to 20 kHz	33
20 kHz to 50 kHz	36
50 kHz to 100 kHz	38
100 kHz to 200 kHz	40
200 kHz to 350 kHz	42
350 kHz to 850 kHz	44
850 kHz to 1.4 MHz	46
1.4 MHz to 2.8 MHz	48

(After Table 2 from New England Wire Technologies "Litz Wire Technical Information")

wire. When using Litz wire it is important to realize that the ends of the Litz wire must be properly prepared so that all the strands of the wire are soldered to the connections of the capacitor and output connector. For those interested in the use of Litz wire the most common modern application is in high-efficiency transformers and inductors in the kHz and low MHz range. Technical journals on transformer and magnetic design still present articles on the use of Litz wire with some regularity.

The Q of the tuned circuit of the loop antenna is also determined by the Q of the capacitors used to resonate it. In the case of air variables or dipped micas typically used this is not usually a problem. But if variable-capacitance diodes are used to remotely tune the loop, pay particular attention to the manufacturer's specification for Q of the diode at the frequency of operation. The tuning diodes can have a significant effect on circuit Q.

Now we consider the effect of load impedance on loop Q. In the case of a directly coupled loop (as in Figure 9.21), the load is connected directly across the loop terminals, causing it to be treated as a parallel resistance in a parallel-tuned RLC circuit. Obviously, if the load is of a low value, the Q of the loop will be low. A simple way to correct this is to use a transformer to step up the load impedance that appears across the loop terminals. In fact, if we make this transformer a balun, it also allows us to use our unbalanced receivers with the loop and maintain loop symmetry. Another solution is to use what is referred to as an inductively coupled loop, such as DeMaw's four-turn electrostatically shielded loop. A one-turn link is connected to the receiver. This turn is wound with the four-turn loop. In effect, this builds the transformer into the antenna.

Another solution to the problem of load impedance on loop Q is to use an active preamplifier with balanced input and unbalanced output. This method also has the advantage of amplifying the low-level output voltage of the loop to where it can be used with a receiver of even mediocre sensitivity.

In recent years, there has been a significant amount of technical interest in this area driven by low-band DXers and AM band DXers. They have discovered that one of the critical issues to maximize performance of a loop/preamp combination is the dynamic range of the preamp. A poorly designed preamp may overload from local broadcast stations or have a poor noise figure itself which limits the ultimate performance observed with the loop antenna. See the **Receiving Antennas** chapter for a discussion of preamps, including a two-part *QEX* article by Chris Trask, N7ZWY, in the online material covering preamp requirements and design.

The Q of a loop when used with a balanced preamplifier having high input impedance may be so high as to be unusable in certain applications. An example of this situation would occur where a loop is being used to receive a 5 kHz wide AM signal at a frequency where the bandwidth of the loop is only 1.5 kHz. In this case the detected audio might be very distorted. The solution to this is to connect a loading resistor across the loop terminals to reduce Q and match the antennas bandwidth to the signal. The chapter **Receiving Antennas** also contains information about preamplifiers for use with loop antennas.

PROXIMITY LOSS

In the case of multi-turn loops there is an additional loss related to a term called *proximity effect*. The proximity effect occurs in cases where the turns are closely spaced (such as being spaced one wire diameter apart). As these current-carrying conductors are brought close to each other, the current density around the circumference of each conductor is redistributed. The result is that more current per square meter is flowing at the surfaces adjacent to other conductors. This means that the loss is higher than a simple skin-effect analysis would indicate, because the current is bunched so it flows through a smaller cross section of the conductor than if the other turns were not present.

As the efficiency of a loop approaches 90%, the proximity effect is less serious. But unfortunately, the less efficient the loop, the worse the effect. For example, an 8-turn transmitting loop with an efficiency of 10% (calculated by the skin-effect method) actually only has an efficiency of 3% because of the additional losses introduced by the proximity effect. It is for this reason that the transmitting loop is typically just one turn. The higher inductance of multi-turn loops also limits the higher frequency range unless the circumference is very small. This generally limits the use of multi-turn loops to receiving applications.

If you are contemplating construction of a multi-turn transmitting loop, you might want to consider spreading the conductors apart to reduce this effect. G. S. Smith includes graphs that detail this effect in his 1972 IEEE paper and Trask examined the details of this loss for receiving antennas and recommends spacing turns at least five wire diameters to reduce this effect. His recommendation also applies to the transmitting variety of the loop antenna.

9.2.5 Using Small Tuned Loops

Most amateur receiving loops are of the tuned variety, consisting of one or more turns. You can use a small tuned loop antenna to improve reception under certain conditions, especially at the lower amateur frequencies. This is particularly true when high levels of man-made noise are prevalent, when harmonic energy from a nearby AM broadcast station falls in the band, or when interference is created by a strong amateur signal in the immediate area. A properly constructed and tuned small loop will exhibit approximately 30 dB of front-to-side response, the minimum response being at right angles to the plane of the loop. Therefore, noise and interference can be reduced significantly or completely nulled out, by rotating the loop so that it is sideways to the noise or interference source.

Generally speaking, small balanced loops are far less responsive to man-made noise than are the larger antennas used for transmitting and receiving. But a trade-off in performance must be accepted when using the loop, for the strength of received signals will be 10 or 15 dB less than when using a full-size resonant antenna. This condition is not a handicap on 1.8 or 3.5 MHz, provided the station receiver has normal sensitivity and overall gain. Adding a preamp may increase signal level but also re-introduces the possibility of overload and intermodulation.

LOOP SELECTIVITY

The loop's narrow frequency response can also become an advantage in rejecting unwanted signals. Not only are out-of-band signals rejected, but unwanted signals within an amateur band can also be attenuated by the loop's narrow bandwidth.

Consider a situation where the inherent selectivity due to the loop's frequency response is helpful. Assume we have a loop with a loaded Q of 100 at 1.805 MHz. While listening to a DX station on 1.805 MHz we are suffering strong interference from a local station 10 kHz away. Switching from a dipole to a small loop will reduce the strength of the off-frequency signal by 6 dB (approximately one S unit) compared to the DX station. This, in effect, increases the receiver's dynamic range. In fact, that farther off-frequency the interfering station is, the greater the attenuation.

INTERFERING SIGNAL REJECTION

Another use of the tuned loop is by using the nulls in its pattern to reject on-frequency (or slightly off-frequency) interference. For example, say we are working a DX station to the north, and another local station to our west is engaged in a contact is just 1 kHz away. We can simply rotate our loop to aim its null to the west, and now the DX station should be readable while the local will be knocked down by several tens of dB. This is quite a noticeable difference.

Of course, this method of nulling will be effective only if the interfering station and the desired station are not in the same direction (or in exact opposite directions) from our location. If the two stations are in the same direction, both stations would be in the null. Luckily the nulls are very sharp, so as long as the station directions are different by at least 10°, using the loop null will be helpful in reducing interference.

REDUCING NOISE

A similar use of the nulling capability is to eliminate local noise interference, such as that from a neighbor's house or other nearby noise source. By aiming the null at the noise source, the noise should be greatly reduced. It is important to realize that the loop will only reject noise or interference in one direction at a time. Noise from multiple directions, as is common from power lines and in urban or suburban locations, will only be partially rejected.

The noise from approaching or regional storms (with attendant atmospheric noise) could be reduced considerably by rotating the loop nulls away from the location of the storm. If the storm is sufficiently distant, the noise may be received as sky-wave signals as discussed in the next section.

PROPAGATION EFFECTS ON NULL DEPTH

After building a balanced loop you may find it does not approach the theoretical performance in the null depth. Small loops oriented vertically do not exhibit meaningful directivity when receiving sky-wave signals.

When mounted vertically, the loop's directivity relates primarily to ground-wave signals since the nulls are in the horizontal plane. This is a feature in disguise, for when nulling out local noise or interference, one is still able to copy sky-wave

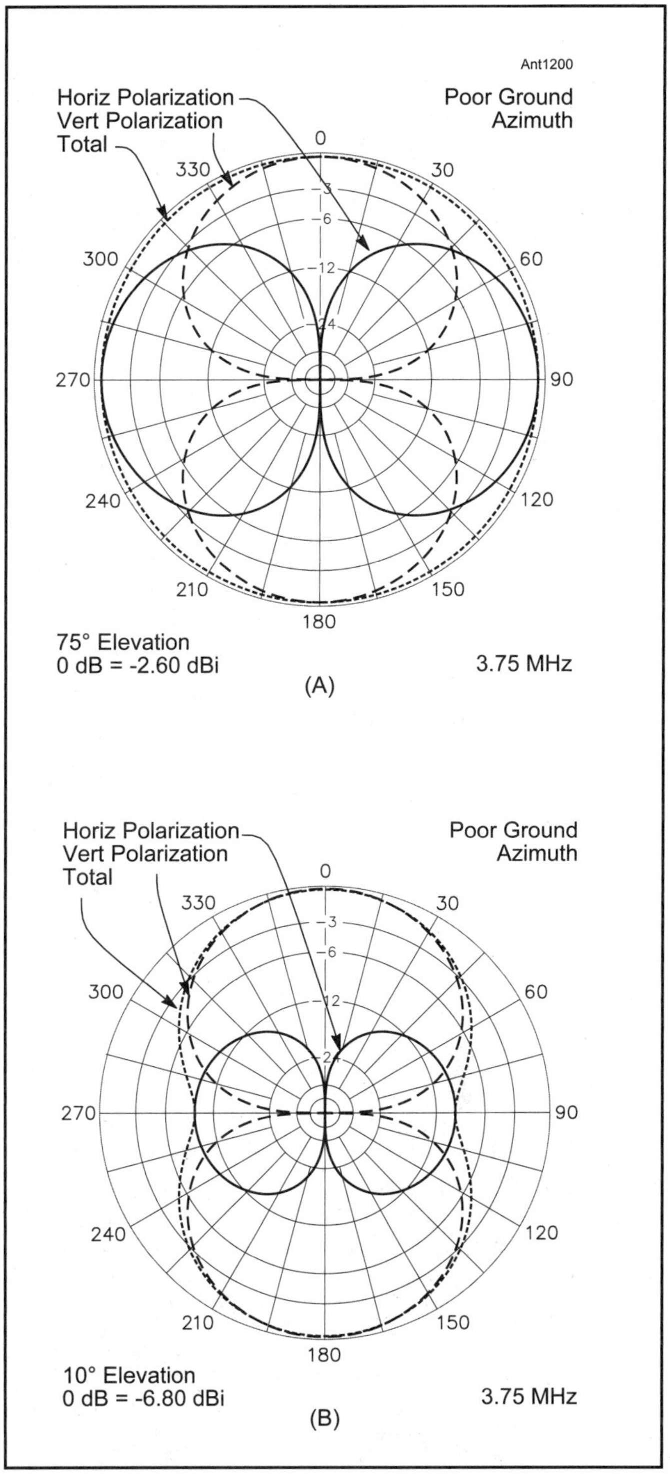

Figure 9.27 — The total azimuthal pattern for a vertically-oriented (along 0° and 180° axis) 3.4 meter diameter loop mounted with its center 2.5 meters over poor ground (3 mS/m conductivity and dielectric constant of 13). The total pattern for an elevation angle of 75° (A) and 10° (B). See the Bibliography entry for Belrose from the November 1993 *QST* for the full article.

signals from any direction.

Nulls have their full effect only when the signal is arriving perpendicular to the axis of rotation of the loop. At incidence angles other than perpendicular, the position and depth of the nulls deteriorate. Bond explained this issue in his book on direction finding in 1944 along with the math to calculate the performance (see the Bibliography). The problem can be even further influenced by the fact that if the loop is situated over less than perfectly conductive ground, the wave front will appear to tilt or bend.

Tilting the loop away from vertical may improve performance under some propagation conditions, to account for the vertical angle of arrival. **Figure 9.27** illustrates the difference in total antenna response at low and high elevation angles for a small transmitting loop over real ground, typical of small receiving loops as well. (The 1993 *QST* article by Belrose listed in the Bibliography is included in the downloadable set of supplemental articles.) The high-angle response is nearly omnidirectional, while the low-angle response shows the pattern nulls at right angles to the plane of the loop.

Another cause of apparent poor performance in the null depth can be from polarization error. If the polarization of the signal is not completely linear, the nulls will not be sharp. In fact, for circularly polarized signals, the loop might appear to have almost no nulls. Propagation effects are discussed further in the **Receiving Antennas** chapter.

SITING EFFECTS ON THE LOOP

The location of the loop has an influence on its performance that at times may become quite noticeable. For ideal performance the loop should be located outdoors and clear of any large conductors, such as metallic downspouts and towers. A VLF loop, when mounted this way, will show good sharp nulls spaced 180° apart if the loop is well balanced.

Most hams locate their loop antennas near their operating position. If you choose to locate a small loop indoors, its performance may show nulls of less than the expected depth, and some skewing of the pattern. For precision direction finding there may be some errors associated with wiring, plumbing, and other metallic construction members in the building. Also, a strong local signal may be reradiated from the surrounding conductors so that it cannot be nulled with any positioning of the loop. There appears to be no known method of curing this type of problem. All this should not discourage you from locating a loop indoors; this information is presented here only to give you an idea of some pitfalls. Many hams have reported excellent results with indoor mounted loops, in spite of some of the problems.

Locating a receiving loop in the field of a transmitting antenna may cause a large voltage to appear at the receiver antenna terminals. This may be sufficient to destroy sensitive RF amplifier transistors or front-end protection diodes. This can be solved by disconnecting your loop from the receiver or shorting it during transmit periods. This can obviously be done automatically with a relay that operates when the transmitter is activated, controlled by a PTT or amplifier relay signal.

9.3 Small Transmitting Loops

This section addresses small loops with a physical circumference of less than 0.3 λ that are used for both transmitting and receiving. As a consequence of their relatively large size (compared to a very small receiving loop), transmitting loops have a non-uniform current distribution along their circumference. This leads to some performance changes from the small receiving loops discussed in the preceding section.

While there are several practical loop designs, this section is primarily concerned with the popular gap-resonated loop designs that are fed with a small coupling loop inside the main loop. The loop is generally tuned with a variable capacitor across a gap in the loop conductor.

The analysis presents expressions for the loop impedance and the effects of the secondary feeding loop. Formulas for the loop current and loop impedance are given which lead to an accurate determination of close-near-fields and far field null depths for loops with circumference up to 0.3 λ. Details are also provided about the loop near fields and far-field null filling that are a direct result of including the non-uniform loop current. The material also covers how loop currents can couple to the coaxial feed line shield, and also to the ground. Additional recent articles are included in the References and Bibliography section, as well.

EVOLUTION OF SMALL TRANSMITTING LOOPS

The small gap-resonated circular loop antenna with a circumference of less than 0.3 λ has received much attention for use on HF since John H. Dunlavy, Jr. (see the Bibliography) patented his efficient small loop that can be tuned over wide bandwidths. The now-expired patent spawned a multitude of homebrew loops and several commercial products aimed at hams.

Loop studies and analyses date back more than a century to the earliest days of radio. A loop antennas theme issue of *QEX* (see the Bibliography) features articles from several authors who investigate the patterns, efficiency, matching, coupling to ground, and other aspects of small HF gap-resonated loops.

In the March 1968 *QST*, Lew McCoy, W1ICP, introduced the so-called "Army Loop" to radio amateurs. This was an amateur version of a loop designed for portable use by Patterson of the US Army and described in 1967. The Army Loop is diagrammed in **Figure 9.28A**, showing that this is a parallel tuned circuit fed by a tapped-capacitance impedance-matching network. As shown in the figure, the loop does not illustrate the best low-loss construction practices and is

Minimum Safe Distances for Small Transmitting Loops

Amateurs using small transmitting loops should also be cautious about RF exposure from high field strengths near the antenna. The antenna's high Q means that a lot of energy is present near the antenna, even at low power. DeNeef (see the Bibliography) used *NEC* to model and analyze fields from a small loop to determine the minimum compliance distances for a 1.0-meter loop at 7, 14, 18, 21, and 28 MHz using power levels of 10, 100, and 400 watts. The ICNIRP distances calculated in **Table 9A** are more conservative than the IEEE standards used by the FCC, but not dramatically. (ICNIRP, or International Commission on Non-Ionizing Radiation Protection, standards use the terms "occupational" and "general public" where the IEEE uses "controlled" and "uncontrolled," respectively.) Because these loops are often used in portable operation and close to living areas, careful attention should be paid to ensuring RF exposure limits are not exceeded and that people cannot inadvertently get too close to the antenna.

Table 9A
Minimum Compliance Distances from Small (1.0 m diameter) Transmitting Loops

Freq (MHz)	*Controlled Distance (m)*	*Uncontrolled Distance (m)*
Power = 10 watts		
7	2.1	2.9
14	2.1	2.9
18	1.9	2.7
21	1.8	2.5
28	1.5	2.1
Power = 100 watts		
7	3.3	4.3
14	3.4	4.7
18	3.1	4.4
21	2.9	4.2
28	2.5	3.9*
Power = 400 watts		
7	4.2	5.7
14	4.5	6.5
18	4.2	6.9*
21	4.0	7.1*
28	3.7*	7.3*

** - Electric field limit*

UK ICNIRP compliance distances from "Safety distances for small HF loop antennas," AE7PD, Radcom, June 2017, pages 46 and 47.

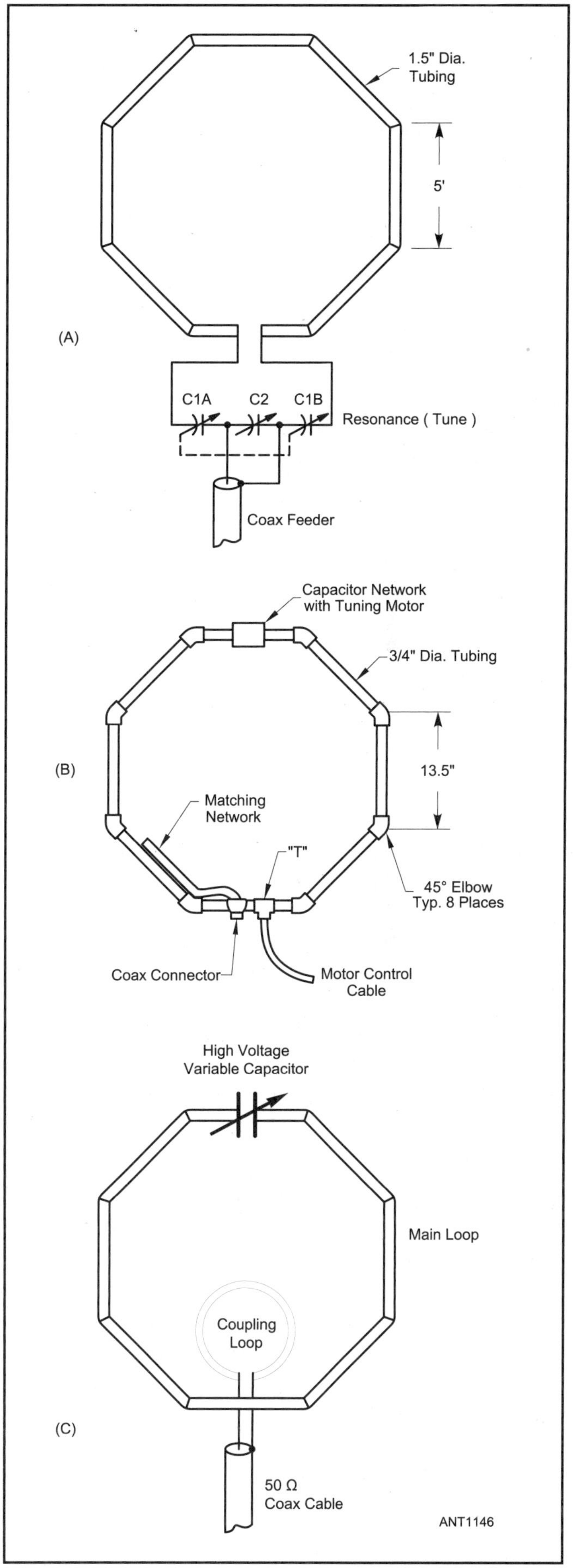

Figure 9.28 — At A, a simplified diagram of the Army Loop. At B, the W5QJR loop. At C, using a smaller loop connected to the transmission line to couple into the larger transmitting loop.

included here only in reference to the loop antenna's history with amateurs.

The Hart "high-efficiency" loop was introduced in the June 1986 *QST* by Ted Hart, W5QJR. It is shown schematically in Figure 9.28B and has the series-tuning capacitor separate from the matching network. The Hart matching network is basically a form of gamma match.

The most popular design today is a smaller loop connected to the transmission line to couple into the larger transmitting loop as shown in Figure 9.28C. In addition, Steve Yates, AA5TB, has published a website (**aa5tb.com/loop.html**) with a great deal of information on these small antennas and links to many designs.

9.3.1 Loop Current

Figure 9.29 shows the circular loop geometry rendered in *4nec2* software. In their *QEX* article, K. Siwiak, KE4PT and R. Quick, W4RQ studied the effect of current variation on loops up to 0.3 λ in circumference. (See the Bibliography and the article is included in the downloadable supplemental material for this book.) In their article, the primary loop radius is b = 0.4534 m (a loop diameter of 3 ft), the loop conductor radius is a = 0.00406 m, the angular extent along the loop circumferences is ϕ, with the loop gap located at ϕ = 0. The resonating capacitor, with a $Q_c = 2400$ in the model, connects across the gap at the bottom of the primary loop. The secondary feeding loop radius is $b_2 = 0.077$ m (diameter of 6 inches) and the conductor radius is $a_2 = 0.002$ m. A coaxial cable feed line connects across a gap at the bottom of the smaller secondary loop. The loop centers are displaced by 0.343 m. The length of the coax feed line can be varied to study the effect on the common mode current coupling to the shield of the feeding coax cable.

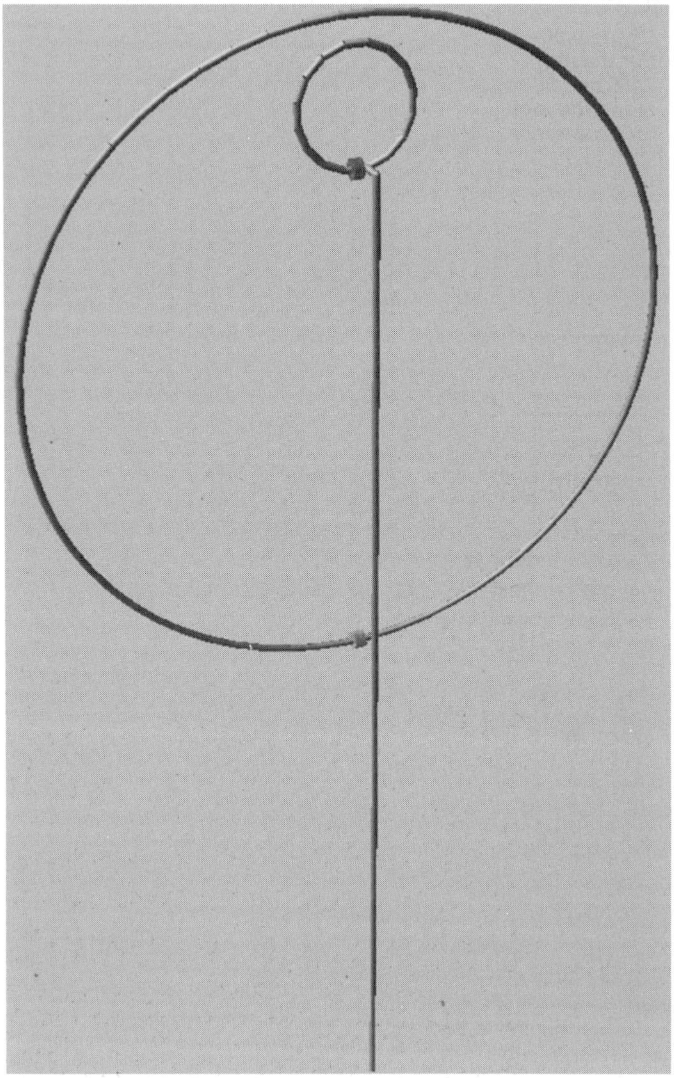

Figure 9.29 — The electrically small HF loop includes a primary loop and a secondary feeding loop in slightly displaced parallel planes, so that the feeding cable does not touch the bottom of the primary loop. A resonating capacitor connects across a gap at the bottom of the primary loop.

THE DUNLAVY LOOP PATENT

John Dunlavy revealed in his US patent 3,588,905 that a one-turn primary loop antenna having a circumference of less than three-eighths of a wavelength and interrupted along its length by a gap, with a tuning capacitor connected across the gap, can be tuned over a wide tuning range. A single-turn secondary loop, much smaller than the primary loop, is inductively coupled to the primary loop. Both loops are in the same or parallel planes. The secondary loop diameter is selected to bear an optimum relationship to the diameter of the primary loop so that variation in feed impedance is minimized over the band of operation. A low-impedance transmission line (50 Ω) connects to the terminals of the secondary loop.

LOOP CURRENT DENSITY

Unlike the small receiving loops discussed previously, the current around the loop is not a constant (I_0), but varies around the loop,

$$I(\phi) = I_o \left\{1 - 2C_\lambda^2 \cos(\phi)\right\} \quad (6)$$

where C_λ is the loop circumference in wavelengths and ϕ is the angular position around the loop. The full description of loop current is a Fourier series. Eq 6 is an abbreviated result, and the detailed equations are in the original *QEX* article by Siwiak and Quick (see Bibliography).

The current's amplitude variation depends solely on the loop circumference in wavelengths. The bigger the loop, the more current varies around the loop. **Figure 9.30** shows the loop current for our loop at 7, 14, and 30 MHz, where C_λ is 0.067, 0.133, and 0.285 λ respectively. The loop current Eq 6 is valid for $C_\lambda < 0.3$, and is used to solve for the loop fields in classical fashion.

9.3.2 Loop Impedance and Q

The transmitting loop is treated as a parallel-tuned circuit with a large inductor acting as the radiator. Calculation of the loop inductance may be carried out with the approximate equations in Table 9.2. Avoid equations for long solenoids

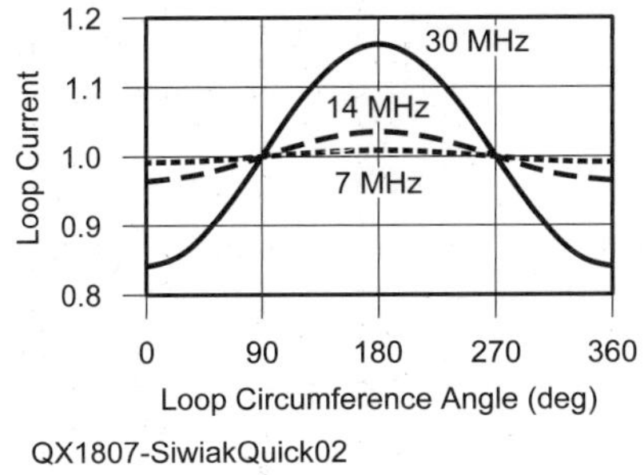

Figure 9.30 — Loop currents at 7, 14, and 30 MHz along the loop circumference.

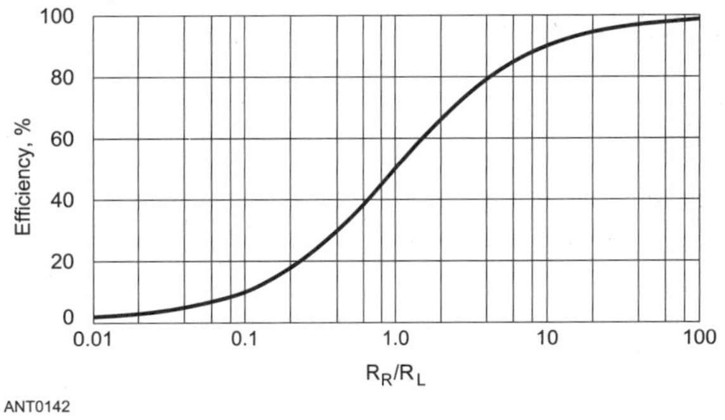

Figure 9.31 — Effect of ratio of R_R/R_L on loop efficiency.

Table 9.5
Transmitting Loop Equations

$$X_L = 2\pi fL$$

$$Q = \frac{f}{\Delta f} = \frac{X_L}{2(R_R + R_L)}$$

$$R_R = 3.12 \times 10^4 \left(\frac{NA}{\lambda^2}\right)^2 = 197 C_\lambda^4 N^2$$

$$V_C = \sqrt{2PX_LQ}$$

$$I_L = \sqrt{\frac{PQ}{X_L}}$$

where
- X_L = inductive reactance in ohms
- f = frequency in Hz
- Df = bandwidth in Hz
- R_R = radiation resistance in ohms
- R_L = loss resistance in ohms (see text)
- C_λ = loop circumference in wavelengths
- N = number of turns
- A = area enclosed by loop in square meters
- λ = wavelength at operating frequency in meters
- V_C = peak voltage across capacitor
- P = power in watts
- I_L = resonant circulating RMS current in the loop

found in most texts. Other fundamental equations for transmitting loops are given in **Table 9.5**.

LOOP RADIATION RESISTANCE

The approximate radiation resistance of a loop is given in Table 9.6. The radiation resistance of a small transmitting loop is usually very small. For example, a 1-meter diameter, single-turn circular loop has a radius of 0.5 meters and an enclosed area of $\pi \times 0.5^2 = 0.785$ m^2. Operated at 14.0 MHz, the free-space wavelength is 21.4 meters and this leads to a computed radiation resistance of only $3.12 \times 10^{-4} (0.785/21.4^2)^2 = 0.092\ \Omega$. (The *QEX* article by Siwiak and Quick article provides a more accurate treatment of radiation resistance.)

The loop also has losses, both ohmic and from skin effect. By using this information, the radiation efficiency of a loop can be calculated from the simple formula:

$$\eta = \frac{R_R}{R_R + R_L} \quad (7)$$

where
- η = antenna efficiency
- R_R = radiation resistance in Ω
- R_L = loss resistance in Ω, which includes the loop's conductor loss plus the loss in the tuning capacitor and any current-carrying mechanical connections.

A simple ratio of R_R to R_L shows the effects on the efficiency, as can be seen from **Figure 9.31**. This graph is presented as a general illustration of how losses affect loop efficiency and is not a representation of any specific design. See the discussion on loop efficiency at the end of this section which compares a simulation, calculation, and measurement of efficiency.

The loss resistance is primarily the ac resistance of the conductor (Eq 5). A transmitting loop generally requires the use of copper conductors of at least ½ inch in diameter in order to obtain reasonable efficiency. Tubing is as useful as a solid conductor because high-frequency currents flow only along a very small depth of the surface of the conductor; the center of the conductor has almost no effect on current flow. Note that the R_L term above must also include the effect of the loss in the tuning capacitor and mechanical connections. Practical details for curbing losses are covered later in this chapter in the section on loop construction.

LOOP IMPEDANCE AND Q

The loop impedance with just radiation loss is

$$Z_{loop} = [\text{Radiation resistance}] + j[\text{Loop reactance}] \quad (8)$$

The loaded radiation Q of the antenna is,

$$Q_{rad} = \frac{1}{2} \frac{\text{Loop reactance}}{\text{Radiation resistance}} \tag{9}$$

The primary loop loss resistance is due to the main loop conductor losses represented by R_{loss}. The Q_c of the resonating capacitor also contributes to losses in the form of parallel resistance across the capacitor, so that the net loaded Q_L of the antenna is

$$Q_L = \frac{0.5}{\frac{1}{Q_C} + \frac{[\text{Radiation resistance}] + R_{loss}}{[\text{Loop reactance}]}} \tag{10}$$

from which we can determine the loop radiation efficiency in dB as,

$$\text{eff}_{dB} = 10 \log\left[\frac{Q_L}{Q_{rad}}\right] \tag{11}$$

The value of Q_L can be obtained from Eq 10 or from direct measurements. See the referenced Siwiak and Quick article for a complete discussion of the loop's radiation resistance.

LOOP BANDWIDTH

It is important to note that small transmitting loops (like their receiving cousins) have very narrow 2:1 SWR bandwidths. If you wish to operate across a wide range of frequencies, expect to do some tuning. In commercial loops this is usually done remotely using a small stepper or gear motor to adjust the tuning capacitor. This type of antenna may require retuning for frequency changes as small as 5 kHz. If you are using any wide-band mode such as AM or a high-speed digital mode, you might wish to sacrifice a little efficiency to obtain the required bandwidth.

9.3.3 The Secondary Feeding Loop

For the given example loop dimensions, the mutual coupling inductance M_{12} between the main and feeding loops is 57.3 nH, and the secondary feed loop inductance L_{feed} is 0.361 μH, obtained using the Jordan and Balmain high frequency extension to the Neumann formula, (see References and Bibliography). The main loop impedance, including the resonating capacitor, is transformed using a circuit model developed by Milton Cram, W8NUE, (see Bibliography) to the feed point impedance by,

$$R_{feed} = j\omega L_{feed} + \frac{(\omega M_{12})^2}{R_{total} + j\Delta} \tag{12}$$

R_{total} is the sum of the radiation resistance and conductor and capacitor series equivalent losses and $\omega = 2\pi f$. Interestingly, the feed point reactance is not zero when the main loop is resonant because of the feed-loop inductance L_{feed} term in Eq 12. The $j\Delta$ term simulates the tuning of the main loop to bring R_{feed} close to 50 Ω. Apply Eq 12 and choose Δ at each frequency for a best match to 50 Ω. In practice, this is accomplished by adjusting the main loop resonating capacitor.

Figure 9.32 shows that the load resistance presented to the coax feed line stays close to 50 Ω (SWR is under 1.3:1) across the 7 to 29.7 MHz operating range of this loop. This illustrates a key characteristic taught by Dunlavy in his 1971 patent — the feeding loop diameter and location bear an optimum relationship to the diameter of the primary loop so that variation in feed impedance is minimized over the band of operation.

9.3.4 Fields at the Loop Center and in the Far Field Null

Since the loop current term varies with ϕ an electric field E_{center} and a magnetic field H_{center} can be found at the loop center, and the magnetic field can be approximated from the single turn solenoid equation, -both equations are in the *QEX* article. The resulting wave impedance Z_W at the loop center is a measure of how well the loop discriminates between the electric and magnetic fields and is,

$$Z_W = \frac{E_{center}}{H_{center}} = -j\eta_0 C_\lambda \tag{13}$$

clearly revealing the dependence of Z_W on the loop circumference. Without the ϕ dependent current variation term E_{center} and Z_W would be erroneously reported as zero!

Likewise, the far-field peak-to-null ratio depends on the current variation term in a simple manner for $C_\lambda < 0.3$,

$$N_{dB} = -20\log(2C\lambda) \tag{14}$$

Figure 9.33 shows the null depth across 7 to 30 MHz. The

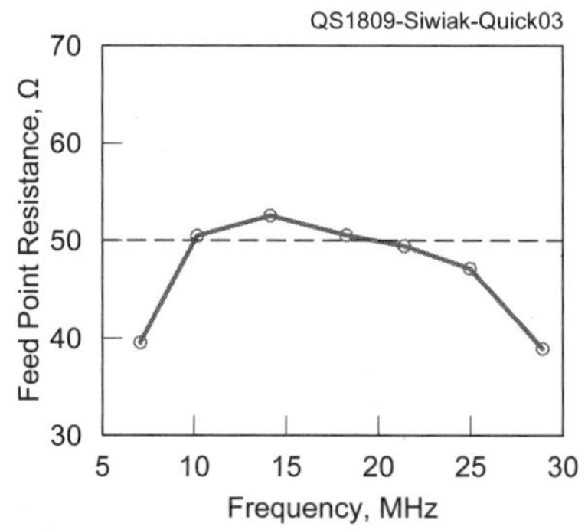

Figure 9.32 — The approximate impedance presented to the coax feed line is not much different from 50 Ω across the entire 7 to 29.7 MHz tuning and operating range of this loop.

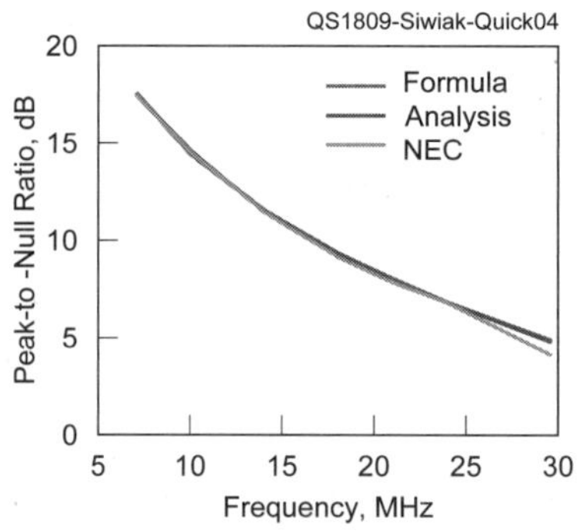

Figure 9.33 — The depth of the null steadily decreases across the 7 to 30 MHz operating band. Omitting the current variation term results in an erroneous infinite null depth.

null depth using Eq 14 is compared to the null depth found with a detailed loop near-field analysis, and the null simulated by a *4nec2* model. The polarization at the horizon of the vertically oriented loop is vertical. The far-field null and E_{center}, however, are horizontally polarized — same as the orientation of the electric field across the main loop gap for the geometry of Figure 9.29. The null becomes shallower as the frequency increases for a fixed-size loop, and at $C_\lambda = 1$ — the classic one wavelength circumference loop — becomes the peak gain (in horizontal polarization) of about +4 dBi in the broadside direction.

9.3.5 Coupling to a Coaxial Feed Line

A coaxial cable feeds the secondary loop directly as shown in **Figure 9.34**, so common-mode currents (CMC) can be picked up on the coax feed line shield. Vary the length of the coax and search for the maximum current on the coax cable shield, just like on a previous *QST* study (see Bibliography) by Quick and Siwiak involving common-mode currents on the feed line to a dipole. They found that the maximum CMC occurs for a coaxial feed line length of 0.45 λ and recommend that common-mode chokes (ferrites) should be installed on the feed line coax at least a loop diameter away from the loop, and at intervals of less than about 0.3 λ (at the highest frequency).

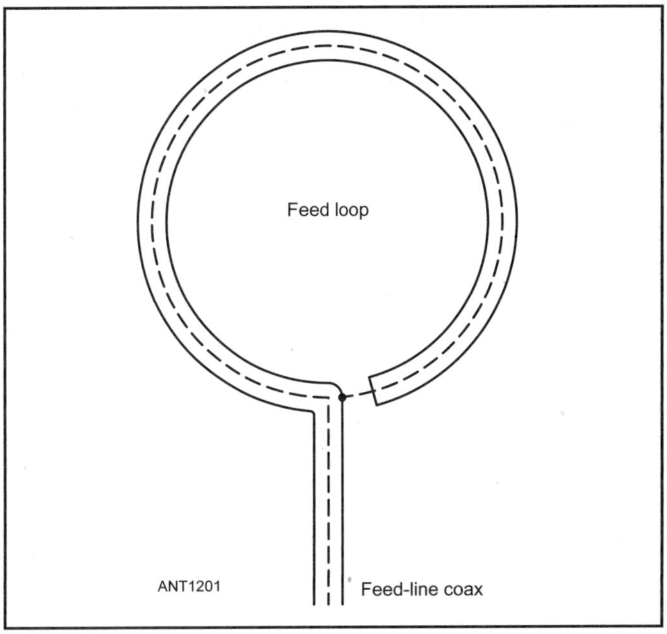

Figure 9.34 — Detail of the secondary feeding loop.

9.3.6 Vertical Loop Coupling to the Ground

Siwiak and Quick estimated the loop coupling to the ground by calculating the mutual inductance between the primary loop and its image in the ground, normalized to the loop self-inductance and expressed in percent. That coupling affects the impedance of the loop antenna system. They calculated the coupling for a perfect electric conductor (PEC) ground, and then estimated the effects over an "average" ground, and a "poor" ground, by reducing the PEC ground coupling by the magnitude of the refection coefficient for normal incidence on the ground. The reflection coefficient magnitude is 1.0 for the PEC, 0.59 for the "average" ground, and 0.39 for "poor" ground at 14.1 MHz. See **Figure 9.35**.

Coupling of the vertical loop to a perfect ground is small, and decreases for realistic ground parameters. Even for a worst-case PEC ground, the coupling is less than 0.5% for a loop with its center more than one loop diameter above ground. Additional discussion of loop-ground coupling is available in the article by DeNeef listed in the Bibliography.

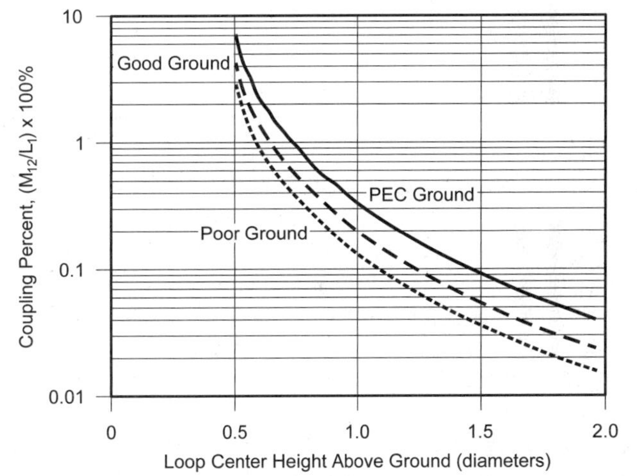

QX1807-SiwiakQuick04

Figure 9.35 — The percent coupling to ground for this small HF loop is insignificant when the loop center is at least one loop diameter above the ground. Coupling is strongest for a PEC ground, and decreases significantly for realistic ground parameters.

9.3.7 Efficiency of the Small Loop

Three methods are compared to estimate the radiation efficiency of the small loop. In one calculated method the total loaded Q_L is found using Eq 10 and compared that to the loaded Q_{rad} of Eq 9, and then Eq 11 is applied for the efficiency.

In a second method measure *loaded* Q_L using a matched transmitter, and apply the classic bandwidth formula,

$$Q_L = \frac{\sqrt{F_H F_L}}{F_H - F_L} = \frac{\text{Frequency}}{\text{Bandwidth}} \qquad (15)$$

where F_H and F_L correspond to the 2.236:1 SWR points on each side of the center frequency.

Finally, the completely independent NEC model is used to simulate efficiency. **Figure 9.36** shows that all three methods of determining efficiency are within 0.5 dB of each other across 7 to 29.7 MHz, inspiring confidence in all three methods.

Austin, Boswell, and Perks published a short paper discussing loss mechanisms for small loops (see the Bibliography) recognizing the extremely low radiation resistance made any

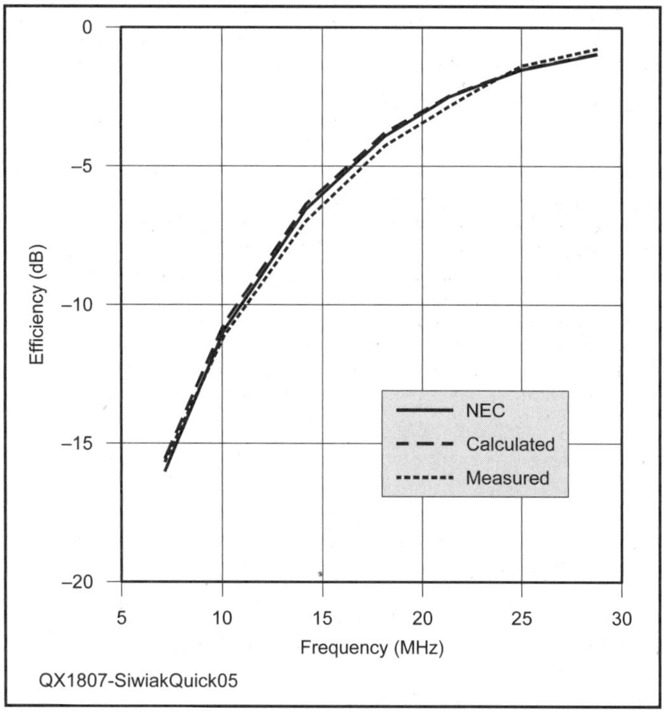

Figure 9.36 — Small loop efficiency simulated in *NEC* (solid), calculated from loop equations (long dashes), and measured using the Q method (short dashes). The close agreement among three methods validates the analysis, the Q measurements, and the completely independent *NEC* model.

resistive losses a major factor in antenna efficiency. Losses in the tuning capacitor contacts were found to be a major source of resistive loss. Ground losses were found to decrease rapidly as the loop was raised above ground and inconsequential above 0.1 λ. Those authors also used a toroidal coupling transformer to couple to the loop.

In addition, Boswell, Tyler, and White published an article in a professional journal which considers the efficiency of a simple 1-meter diameter loop made from 22 mm diameter copper tube from 3–10 MHz (see the Bibliography). Their results were 0.25 percent on 80 meters and 18 percent on 30 meters. The low numbers should not preclude considering the loop as in many cases this is the only reasonable solution to transmitting for the amateurs in a restricted space antenna situation.

9.4 Loop Antenna Projects

9.4.1 Small Transmitting Loop Construction

This section is based on material contributed by Rich Quick, W4RQ. There is an optimum balance between the size of the HF loop, the thickness of the conductor and the relative size of the feeding loop that results in an antenna that has the widest tuning range with close to 50 Ω feeding impedance. Small HF loops are typically between 0.08 and 0.4 λ in circumference over the operating range. That works out to approximately 3 m in circumference for an HF loop the covers 7 to 30 MHz. All else equal, a larger loop will be more efficient than a smaller loop.

It is important to note that the usual construction techniques for antennas can create excessive losses in small loops. Loop radiation resistance is on the order of fraction of an ohm. To minimize resistive losses, solder or weld all connections that carry antenna current. Sliding contacts, clamped joints and connections, and undersized conductors can hurt performance significantly. Bolted connections should be avoided because at RF these joints generally have high resistance, especially after being subjected to weathering.

A circular shape provides the best radiation resistance to conductor loss resistance ratios, hence is most efficient. Loop conductor diameter should be between 1 and 5 cm. Larger conductors help reduce losses. For a given cross-section, tubular conductors will be less lossy than flat conductors because of current bunching at flat conductor edges.

The lowest loss material practical for HF antennas is pure electrical grade copper. T60 aluminum has 66% the conductivity of electrical grade copper. Avoid the high loss of construction grade aluminum and soft copper tubing. Trask has also noted that some flexible copper tubing is made from a lead/copper alloy. This should be avoided due to increased resistivity. Rigid copper tubing, refrigeration tubing or large copper wire should be used. Often it is convenient to fashion loops from RG-213, LMR-400, LMR-500, or LMR-600 coax cable. These work well, especially for portable loop antennas.

Components in a resonant transmitting loop are subject to both high currents and voltages as a result of the large circulating currents found in the high-Q tuned circuit formed by the antenna. Loop circulating current can range from about 8 A at 10 W to more than 25 A at 100 W RF input power. This makes it important that any fixed capacitors have a high RF current rating, such as transmitting micas or the Centralab 850 series. Be aware that even a 100-W transmitter can develop currents in the tens of amperes, and peak voltages across the tuning capacitor in excess of 5000 V. This consideration also applies to any conductors used to connect the loop to the capacitors. A #14 AWG wire used as a jumper and carrying a high current may create more loss than the rest of the loop conductor! It is therefore best to use copper strips or large diameter wire to make any connections. Make the best electrical connection possible, using soldered or welded joints.

The tuning capacitor should be very high Q — greater than 1000 to 4000. For example, a very high-quality tuning capacitor with no mechanical wiping contacts, such as a vacuum-variable or a transmitting butterfly capacitor, might have an unloaded Q of about 5000. This implies a series loss resistance of less than about 0.02 Ω for a capacitive reactance of 100 Ω. This relatively tiny loss resistance can become significant, however, when the radiation resistance of the loop is only on the order of 0.1 Ω!

The highest Q capacitors are vacuum variables, followed by butterfly air dielectric capacitors. Both are hard to find and can be expensive. Often an inexpensive but lower-Q dual-stator capacitor is used. The key is to find a variable range that can resonate with the loop inductance over the desired tuning range. Avoid capacitors that have mechanical 'wiper' connections in the circuit path. They are lossy. Ensure that the capacitor peak voltage rating is sufficient for the loop transmitter power. Use heavy gauge wire to connect the capacitor to the loop.

Because of the sensitivity to loss in the tuning capacitor, there is a lot of interest in designs that avoid sliding or wiping contacts. For example, a "trombone" tuning capacitor design uses concentric copper tubing to construct the linear version of a butterfly capacitor, controlled by a dc motor similarly to a screwdriver mobile antenna.

An enclosure surrounding the capacitor can affect the Q, and therefore losses, dramatically. Use a polymeric enclosure that does not absorb RF energy. Wooden enclosures might be inexpensive and look good, but some types of wood can absorb RF energy.

If you are not sure about an enclosure's effect on RF, place a piece of the material in a microwave oven together with a glass of water. (The material must *not* have any metal fasteners or fittings.) The water provides a load for the oven in the case of a low-loss material. Run the oven on high for about 1 minute. If the material heats up, it absorbs RF, don't use it. Be careful, a piece of lossy material can become hot enough to burn you!

For portable loops that use coax and coaxial connectors for portable assembly, use gold plated connectors since gold will not corrode or oxidize.

A secondary feeding loop circumference should be about 20% of the main loop circumference, and located opposite the capacitor resonated gap. A gamma-type feeding arrangement is also possible for single band operation, but the gamma match can distort the radiation pattern of the loop. A toroidal transformer match might also be suitable for single-band operation.

9.4.2 A Vertically Polarized Quad Loop for 7 MHz

This design is an effective but simple vertical 7 MHz antenna that has a theoretical gain of approximately 1 dB over a dipole. In feet, the overall length of a 1-λ loop is determined from the formula 1005/f (MHz). Hence, for operation at 7.125 MHz the overall wire length will be 141 feet.

Such a loop need not be square, as illustrated in **Figure 9.37**. It can be trapezoidal, rectangular, circular, or some distorted configuration in between those shapes. For best results, however, you should attempt to make the loop as square as possible. The more elongated and narrow the shape, the greater the cancellation of energy in the system, and the less effective

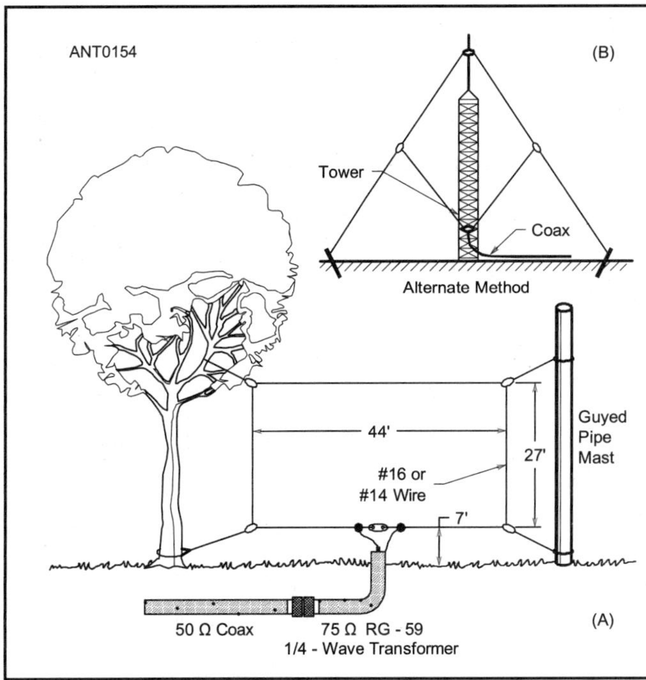

Figure 9.37 — At A, details of the rectangular full-wave loop. The dimensions given are for operation at 7.05 MHz. The height above ground was 7 feet in this instance, although improved performance should result if the builder can install the loop higher above ground without sacrificing length on the vertical sides. At B, illustration of how a single supporting structure can be used to hold the loop in a diamond-shaped configuration. Feeding the diamond at the lower tip provides radiation in the horizontal plane. Feeding the system at either side will result in vertical polarization of the radiated signal.

it will be. In the extreme case, the antenna loses its identity as a loop and becomes a folded dipole.

You can feed the loop in the center of one of the vertical sides if you want vertical polarization. For horizontal polarization, you feed either of the horizontal sides at the center. Since optimum directivity occurs at right angles to the plane of the loop (or in more simple terms, broadside to the loop), you should hang the loop to radiate the maximum amount in some favored direction.

Figure 9.38A shows the azimuthal response at a take-off angle of 15°, a typical angle for 40-meter DX, for vertical and horizontal feed systems over ground with "average" conductivity and dielectric constant. Figure 10.47A includes, for reference, the response of a flattop dipole 50 feet high. For the low elevation angles that favor DX work, the optimal feed point is at the center of one of the vertical wires. Feeding the loop at one of the corners at the bottom gives a compromise result for both local and DX work. The actual impedance is roughly the same at each point: bottom horizontal center, corner or vertical side center.

This same loop antenna in Figure 10.46A fed vertically may be used on the 14 and 21 MHz bands, although its pattern will not be as good as that on its fundamental frequency, and you will have to use an open-wire transmission line to feed the loop for multiband use. **Figure 9.39** shows the peak lobe

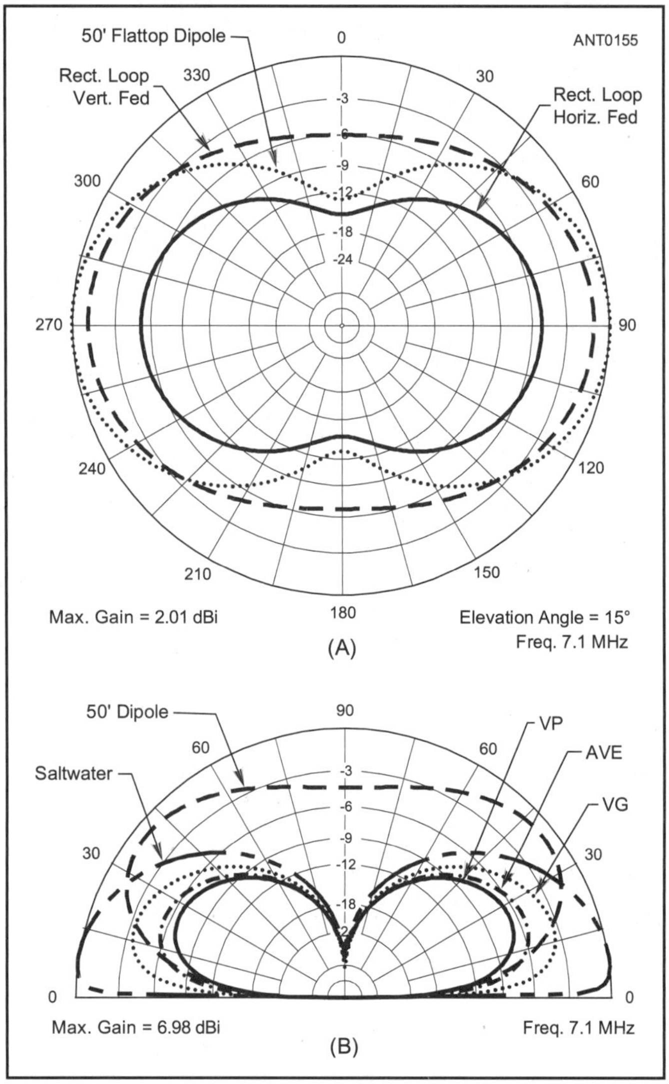

Figure 9.38 — At A, azimuthal plane responses for the vertically and horizontally polarized 7-MHz loop, compared to a flattop 50-foot-high dipole, all at a takeoff angle of 15° for DX work. The solid line is for feeding the loop horizontally at the bottom; the dashed line is for feeding the loop vertically at a side, and the dotted line is for a simple flattop horizontal dipole at 50 feet in height. For DX work, the vertically polarized loop is an excellent performer.

of the loop 14 MHz, at a 45° angle to the plane of the loop, compared to the peak response for a simple half-wave 20-meter dipole, 30 feet high. The gain from a simple flattop dipole, mounted at 30 feet, will be superior to the loop operated on a harmonic frequency.

Just how you erect such a loop will depend on what is available in your backyard. Trees are always handy for supporting loop antennas. A disadvantage to the rectangular loop shown in Figure 9.37A is that two 34-foot-high supports are needed, although in many instances a house may be high enough to serve as one of these supports. If you have a tower higher than about 50 feet, Figure 10.46B demonstrates how you can use it to support a diamond-shaped loop for 40 meters. The elevation and azimuthal responses are almost the same

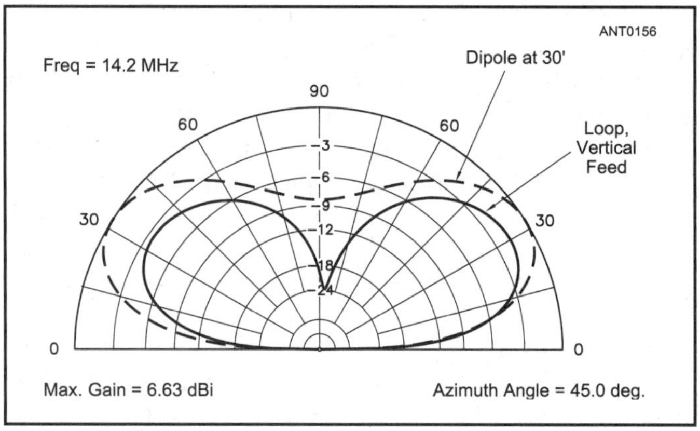

Figure 9.39 — Elevation-plane response of 7-MHz loop used on 14.2 MHz. This is for a feed point at the center of one of the two vertical wires. The dashed line is the response of a flattop 20-meter dipole at 30 feet in height for comparison.

for either loop configuration, rectangular- or diamond-shaped.

On 40 meters, the length in feet of the matching transformer section, an electrical 1/4 λ of 75-Ω coax cable, is 246 by the operating frequency in MHz, then multiplying that number by the velocity factor of the cable being used. Thus, for operation at 7.125 MHz, 246/7.125 MHz = 34.53 feet. If coax with solid polyethylene insulation is used, a velocity factor of 0.66 must be employed. Foam-polyethylene (FPE) coax has a velocity factor of 0.80. Assuming RG-59 is used, the length of the matching transformer becomes 34.53 (feet) × 0.66 = 22.79 feet, or 22 feet, 9½ inches.

9.4.3 A Vertically Polarized Delta Loop for 14 MHz

Two common methods of building a delta loop for the 14 MHz band are shown in **Figure 9.40**. (The design is from *Practical Wire Antennas* published by the RSGB.) Both radiate vertically polarized signals and so ground quality will have an effect on the antenna's efficiency. The total length of wire in the loop should be approximately 1005/f (MHz) = 71 feet for a resonant frequency of 14.15 MHz. For the optimum pattern, the loop should be equilateral with all three sides about the same length.

The antenna in Figure 9.40A has an effective height of about λ/2. The placement of the feed point, at one of the upper corners, configures the antenna to provide low-angle radiation for DX operation. The feed line should be suspended so that it runs directly away from the corner of the loop.

Figure 9.40B inverts the delta to use a single supporting mast or the antenna can be suspended from a tree. The effective height of this antenna is much lower than the "flat-top" version in Figure 9.40A and so the elevation angle of maximum radiation will be higher. Nevertheless, the convenience of this configuration makes it a good choice for Field Day and portable operation. The orientation of the feed line is much less important in this configuration — it can simply drop vertically to the ground.

A commercial dipole center insulator with the built-in sup-

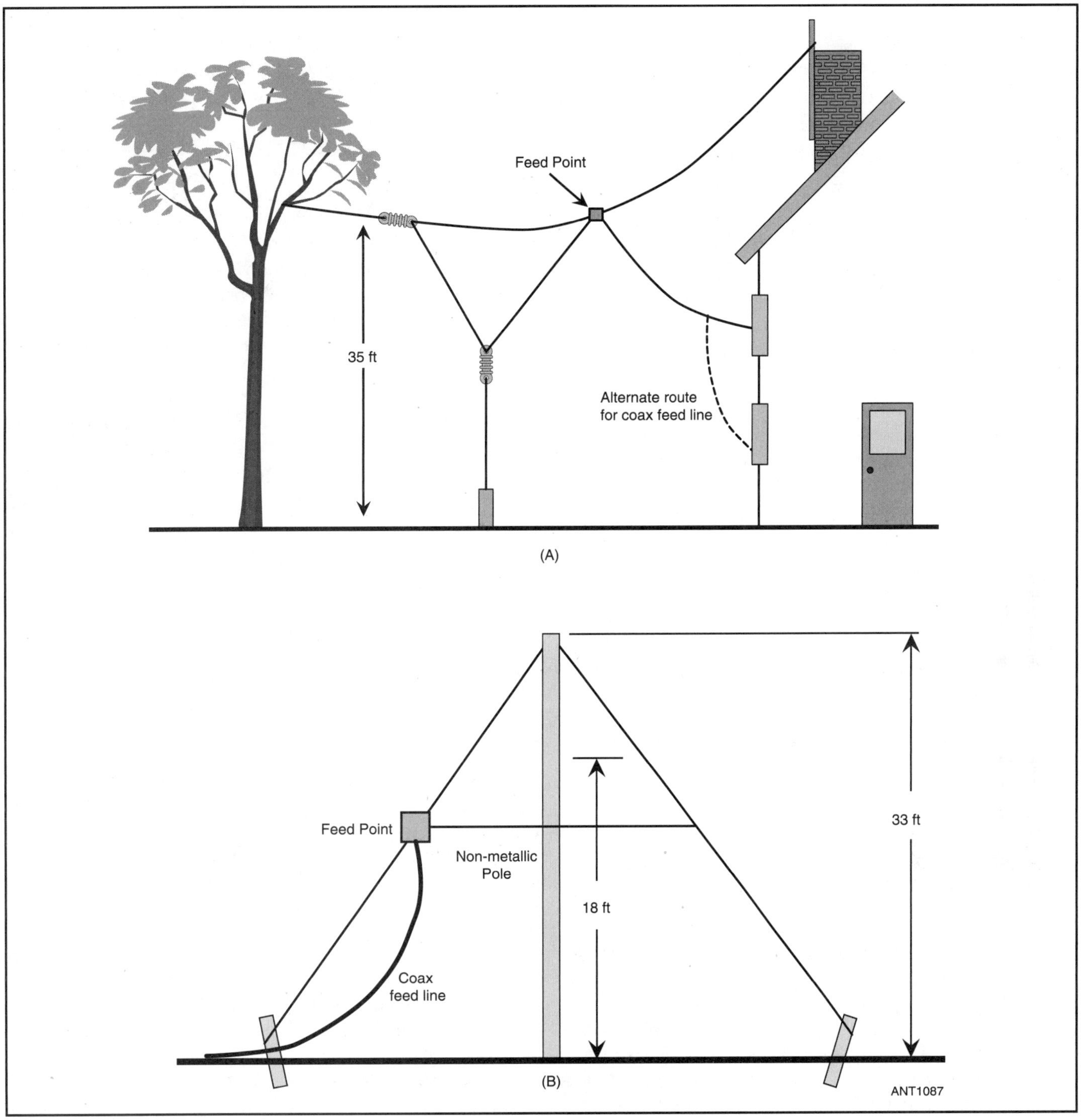

Figure 9.40 — A "flat top" delta loop for 14 MHz is shown at A. Suspended from two supports, its elevation angle of maximum radiation is approximately 20 degrees. The version at B uses only a single support but its effective height is lower than for A, raising the angle of peak signal.

port point will also work well at the feed point. The antenna can be supported with lightweight fishing line or nylon cord. It is recommended that the lower corner or corners be secured as well to prevent the antenna from moving too much in the wind. Insulators can be installed as in Figure 9.44.

The feed point impedance of both configurations will be 100 – 150 Ω and a quarter-wave matching section of feed line can be used to match the loop to 50 or 75 Ω feed line. A choke balun should be used at the feed point. (See the **Transmission Line System Techniques** chapter for more information on quarter-wave matching sections and choke baluns.)

9.4.4 Two-Band Loop for 30 and 40 Meters

The following antenna design is from a *QST* "Hints and Kinks" entry by James Brenner, NT4B, in the May 1989 issue. The version shown in **Figure 9.41** is fed at the apex of a delta

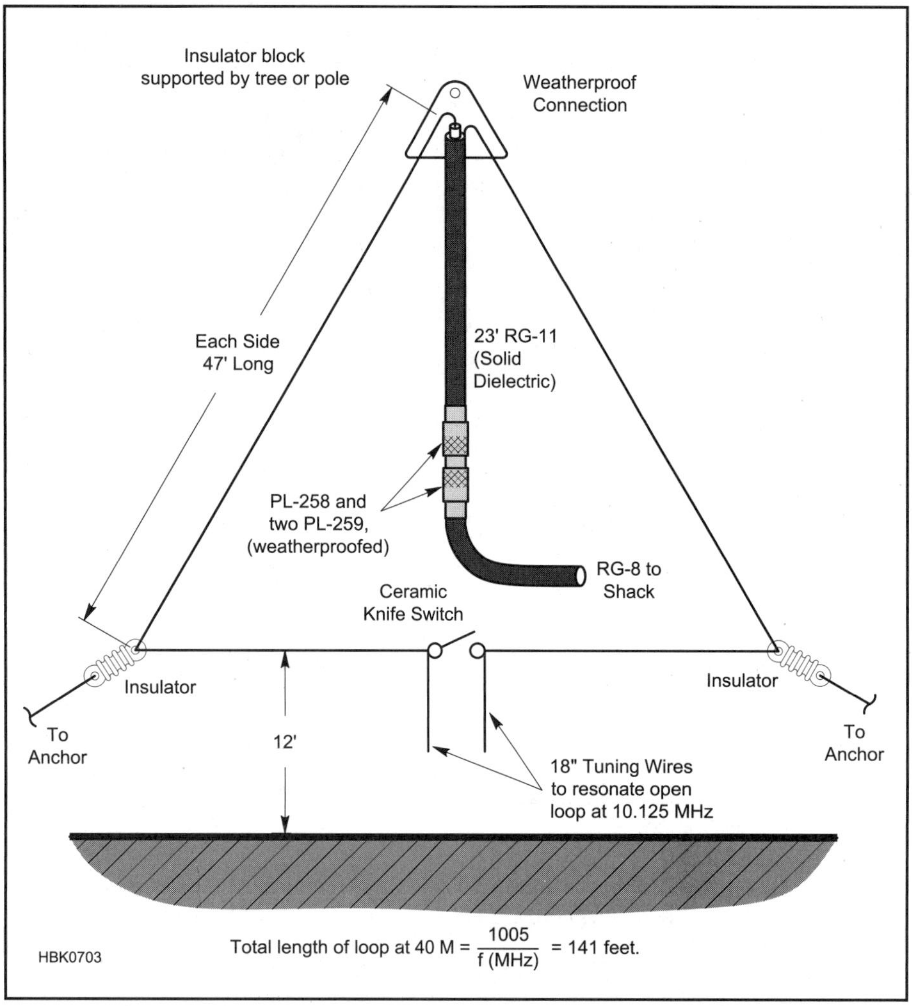

Figure 9.41 — NT4B's 30- and 40-meter loop is fed at the top via a quarter-wave 40 meter matching transformer made of 75 Ω coax. Note the 18-inch tuning wires used to lower the antenna's 30-meter resonance from 10.5 to 10.1 MHz. Adjust the length of these wires to set the 30-meter resonant frequency.

loop but can be adapted to a square or quad loop shape.

The original design was derived from "The Mini X-Q Loop" in *All About Cubical Quad Antennas* by Bill Orr, W6SAI (now out of print) which is 1.5 λ in circumference, with an open circuit opposite the feed point. That antenna has approximately 1 dB of additional gain over a 1 λ loop. Since 30 and 40 meters are close to the same 1:1.5 λ ratio, one loop can be converted between 1 λ on 40 meters and 1.5 λ on 30 meters with a switch.

A large, ceramic SPST knife switch is installed in the center of the delta loop's bottom leg as shown in Figure 9.41. With the switch open, the loop acts as 1½ λ loop at 10.5 MHz, so 18-inch wires were added to the loop on either side of the switch to lengthen the antenna and lower the resonant frequency to 10.1 MHz. Closing the switch shorts out the wires and the loop becomes a regular 1 λ continuous loop for 40 meters.

Note that there is fairly high voltage present at the switch when transmitting on 30 meters. If a relay is used, be sure the contact spacing is sufficient to avoid arcing or use additional pairs of contacts to increase the overall spacing. (See the discussion of broad-banding 160-meter antennas in the **General Purpose MF and HF Antennas** chapter for suggestions on modifying relays for this application.)

The antenna is fed through a quarter-wave transformer (see the **Transmission Line System Techniques** chapter) of 75-Ω RG-11 coax, approximately 23 feet long. The 40-meter configuration of the loop is reported to have a satisfactory SWR of less than 2:1 on 15 meters. In addition, the 30 meter configuration can be used successfully on 80 meters with the use of an antenna tuner.

9.4.5 Nested Loop Antennas for Multiple Bands

The following nested loop array for the 20-, 17-, 15-, 12-, and 10-meter bands can be used as a permanent station antenna or for portable operation. The design was originally described in the *QST* article "Nested Loop Antenna" by Scott Davis, N3FJP, which is included with this book's downloadable supplemental information.

The square loops are constructed as in **Figure 9.42** according to the dimensions in **Table 9.5**. The loops hang in the vertical plane in a diamond shape and are fed at the bottom corner to radiate with horizontal polarization. The perimeter, P, of each loop is calculated by dividing 1005 feet by the frequency in MHz.

The PVC pipe horizontal cross support spans between opposite loop corners, and is 1.41 times the largest loop side length, S. Brass screws are used to hold the PVC together. If the horizontal PVC pipe sags excessively, reinforce it with lengths of 1 × 2-inch pressure-treated wood taped to the pipe.

To construct the antenna, start with the largest loop (20 meters). Cut a 70.9-foot piece of wire for the loop perimeter and divide it by 4 to determine the length, S, of each side. Arrange the wire on the PVC structure, temporarily taping it to find where the wire should pass through the PVC pipe. Drill holes through the pipe for the loop wire. After you run the wire through the holes, wrap a bit of electrical tape on each side of the wire next to the pipe to keep the wire from sliding and to give the pipe additional support. Repeat for each band in sequence from large to small loop perimeter.

At the feed point, rather than sliding the wires through the PVC pipe, use brass wood screws into the PVC to secure the ends of the wires. An SO-239 can be installed at this point or pigtails on the feed line can be attached to the loop wire. Be sure to waterproof this connection.

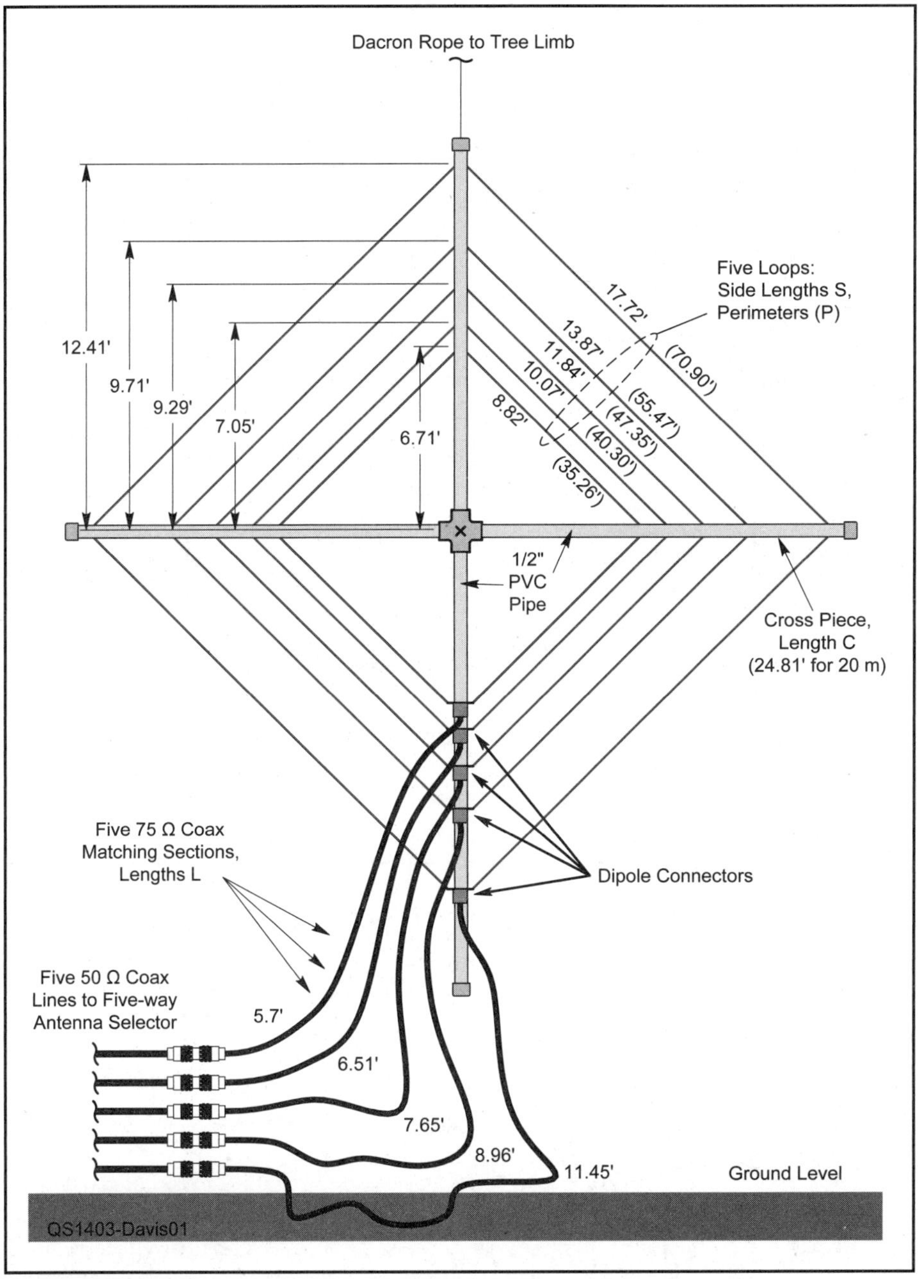

Figure 9.42 — The five-band version of the nested loops with construction notes. See the supplemental article for more construction details and a parts list.

Table 9.5
Nested Loop Antenna Dimensions
75 Ω Matching Section Cable with 0.66 Velocity Factor (VF)

Band	*Freq (MHz)*	*Perimeter, P (ft)*	*Side, S (ft)*	*Cross Support (ft)*	*Distance From Center, D (ft)*	*Length, L of 75 Ω cable (ft)*
20 m	14.175	70.9	17.8	24.8	12.4	11.5
17 m	18.118	55.5	13.9	19.4	9.7	9.0
15 m	21.225	47.4	11.8	16.6	9.3	7.7
12 m	24.940	40.3	10.1	14.1	7.1	6.5
10 m	28.500	35.3	8.8	12.3	6.2	5.7

The loop feed point impedance is about 100 Ω, so individual λ/4 matching sections of 75 Ω coax (RG-59 or RG-11) can be used convert the impedance to 50 Ω. Table 9.5 gives the matching section lengths, taking into account velocity factor VF = 0.66 for solid polyethylene coaxial cable. The formula for the matching sections is L = (246 × VF) / f (MHz). A multi-position remote coax switch can be used to use a single feed line to the antenna with the matching sections connected between the loops and the switch.

9.4.6 Horizontal Loop "Skywire"

A horizontal full-wavelength loop is a very effective omnidirectional antenna for regional communications on its fundamental frequency where its radiation is a maximum at high angles and makes good use of NVIS propagation. The loop is also useful on higher bands where the pattern begins to divide into multiple lobes at lower elevation angles.

While the feed point impedance might be reasonably low on some bands, using a coax feed line will result in significant losses on others. The best way to feed this versatile antenna is with parallel-wire window or ladder line using an antenna tuner in the station.

The Loop Skywire is shown in **Figure 9.43**. The antenna has one wavelength of wire in its perimeter at the design or fundamental frequency. If you choose to calculate L_{total} in feet, L_{total} = 1005 / f (MHz).

Loop shapes other than a square are possible, but the larger the area enclosed by the loop, the better its performance will be. (A circle encloses the maximum area but this is rarely practical.) The Loop Skywire can also be operated as a vertical antenna with top-hat loading by tying both feed line conductors together at the antenna tuner. This method requires good ground system as described in the previous section on Random-Wire Antennas.

Although the loop can be made for any band or frequency of operation, the following Loop Skywires are good performers.

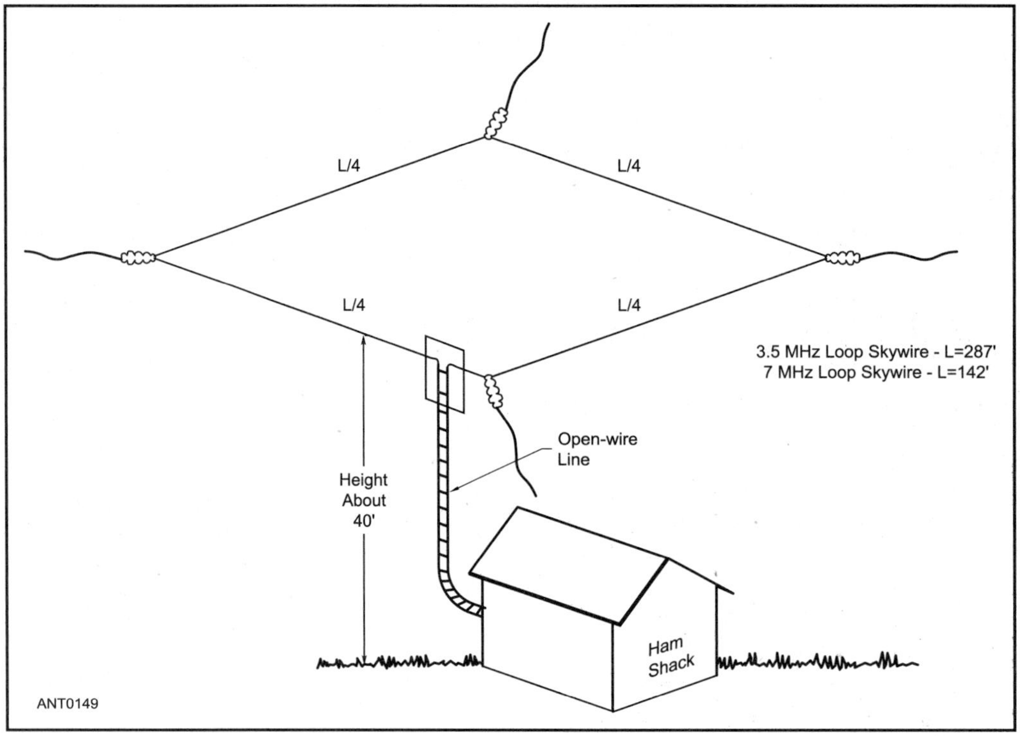

Figure 9.43 — A complete view of the Loop Skywire. The square loop is erected horizontally.

1.8 MHz Loop Skywire
(1.8-28 MHz loop)
Total loop perimeter: 532 feet
Square side length: 133 feet

3.5 MHz Loop Skywire
(3.5-28 MHz loop and 1.8-MHz vertical)
Total loop perimeter: 272 feet
Square side length: 68 feet

7 MHz Loop Skywire
(7-28 MHz loop and 3.5-MHz vertical)
Total loop perimeter: 142 feet
Square side length: 35.5 feet

The actual total length can vary from the above by a few feet, as the length is not at all critical. Do not worry about tuning and pruning the loop to resonance as it will not make a significant difference in performance.

Bare #14 AWG wire is used in the loop. Copper-clad steel wire is recommended for the 3.5 MHz version. **Figure 9.44** shows the placement of the insulators at the loop corners. Two common methods are used to attach the insulators. Either lock or tie the insulator in place with a loop wire tie, as shown in Figure 9.44A, or leave the insulator free to "float" or slide along the wire, Figure 9.44B. Most loop users float at least two insulators. This allows pulling the slack out of the loop once it is in the air, and eliminates the need to have all the supports exactly placed for equal tension in each leg. Floating two opposite corners is recommended. **Figure 9.45** shows a corner insulator and feed point insulator for parallel-conductor lines made from PVC pipe. Other methods of attaching

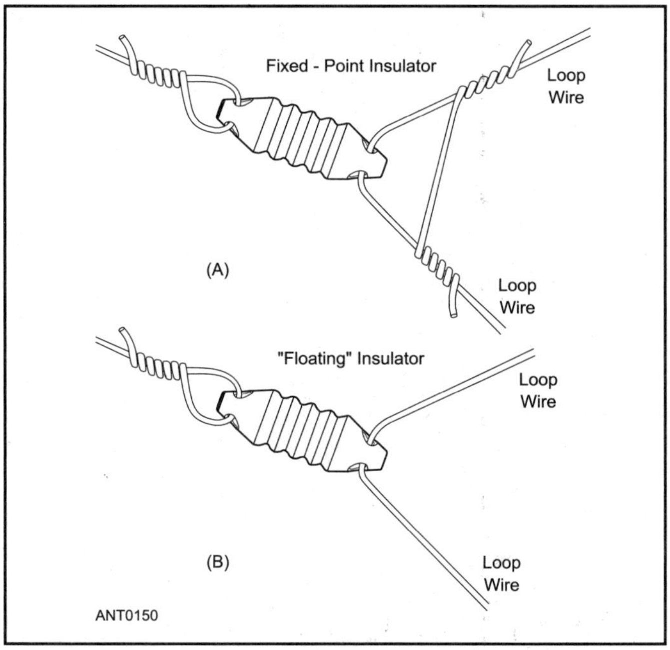

Figure 9.44 — Two methods of installing the insulators at the loop corners.

feed lines can be found in the **Antenna Materials and Construction** chapter.

The feed point can be positioned anywhere along the loop that you wish. However, most users feed the Skywire at a corner. Placing the feed point a foot or so from one corner allows the feed line to exit more freely and keeps the feed line

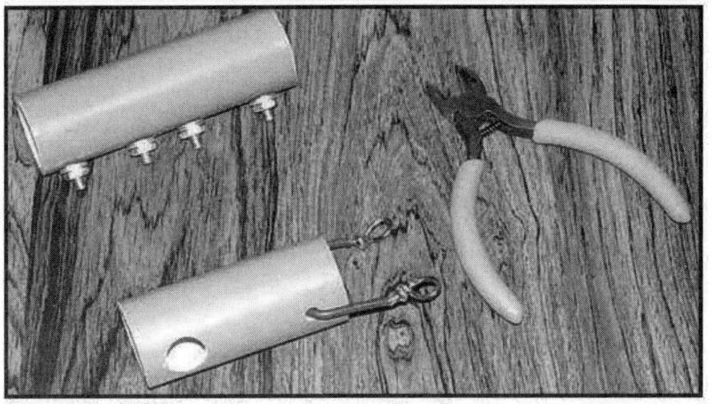

Figure 9.45 — Insulators can also be made from inexpensive PVC pipe. At the top is a feed point insulator for parallel-conductor feed lines. The feed line is attached to the inner terminals and the loop to the output terminals which are jumpered together (jumpers not shown). At the bottom is a corner insulator. The support line is tied through the larger holes and the loop conductors attached to the eyes of the jumper.

free of the loop support.

Generally, a minimum of four supports is required. If trees are used for supports, then at least two of the ropes or guys used to support the insulators should be counterweighted and allowed to move freely. The feed line corner is almost always tied down, however. Very little tension is needed to support the loop (far less than that for a dipole). Thus, counterweights are light. Several such loops have been constructed with bungee cords tied to three of the four insulators. This eliminates the need for counterweights.

Figure 9.46A shows the azimuth performance on 7.2 MHz of a 142-foot long, 7 MHz Loop Skywire, 40 feet high at an elevation angle of 10°, compared to a regular flattop ½ λ dipole at a height of 30 feet. The loop comes into its own at higher frequencies. Figure 10.22B shows the response at 14.2 MHz, compared again to a ½ λ, 14.2 MHz dipole at a height of 30 feet. Now the loop has several lobes that are stronger than the dipole. Figure 9.46C shows the response at 21.2 MHz, compared to a dipole. Now the loop has superior gain compared to the 1/2 λ dipole at almost any azimuth. In its favored direction on 21.2 MHz, the loop is 8 dB stronger than the dipole.

Recommended height for the antenna is 40 feet or more. Higher is better, especially if you wish to use the loop in the vertical mode. However, successful local and DX operation have been reported in several cases with the antenna at 20 feet.

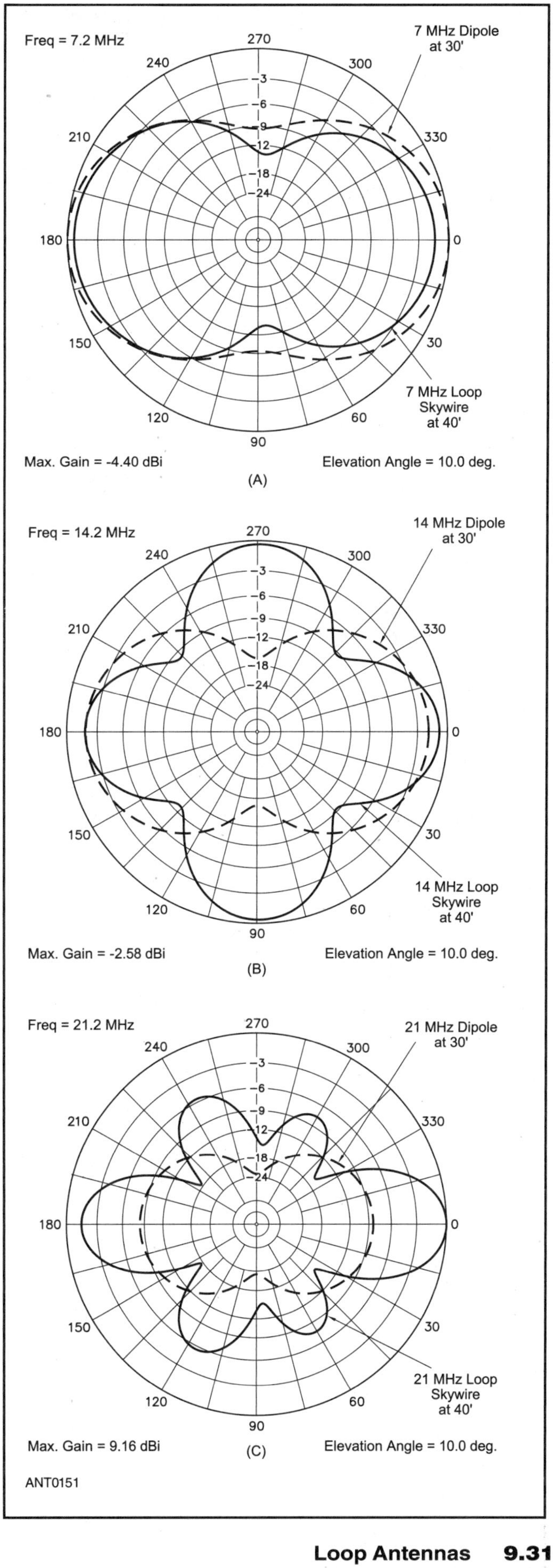

Figure 9.46 — At A, azimuth response of 142-foot long, 7-MHz Loop Skywire, 40 feet in the air at 7.2 MHz, compared with a ½-λ dipole 30 feet in the air. At B, response of same Loop Skywire at 14.2 MHz, compared with a 1/2-λ 14.2-MHz dipole 30 feet in the air. Now the loop has some advantages in certain directions. At C, response of the same Loop Skywire at 21.2 MHz compared to a 21.2-MHz dipole at 30 feet. Here, the Loop Skywire has more gain in almost all directions than the simple dipole. All azimuth patterns were made at 10° elevation.

9.5 References and Bibliography

Source material and more extended discussion of topics covered in this chapter can be found in the references given below and the textbooks listed in the "References and Bibliography" section of the **Antenna Fundamentals** chapter.

Anderson, C., "A Crossed-Loop/Goniometer DF Antenna for 160 Meters," *The ARRL Antenna Compendium, Vol 1* (ARRL, 1985), pp. 127 – 132.

Austin, B., et al, "Loss Mechanisms in the Electrically Small Loop Antenna," *IEEE Antennas & Propagation Magazine*, Aug. 2014, pp. 142 – 147.

Belcher, D., "Loops vs Dipole Analysis and Discussion," *QST*, Aug. 1976, pp. 34 – 37.

Belrose, J., "Ferromagnetic Loop Aerials," *Wireless Engineer*, Feb. 1955, pp. 41 – 46.

Belrose, J., "An Update on Compact Transmitting Loops," *QST*, Nov. 1993, pp. 37 – 40.

Bond, D., *Radio Direction Finders*, 1st edition (McGraw-Hill).

Boswell, A., Tyler, A., and White, A., "Performance of a Small Loop Antenna in the 3 – 10 MHz Band," *IEEE Antennas and Propagation Magazine*, Apr. 2005, pp. 51 – 56.

Bowers B., and Bryant, J., "Very Large Ferrite Loops," *Fine Tuning Proceeding 1994 – 1995* (John H. Bryant, 1994), pp. 18-1 to 18-6.

Cebik, L., "A Comparison of Closed and Interrupted Loop Antennas for 40 Meters," 15 Jan. 2006.

Collins, B., "A New High Performance HF Receiving Antenna," *Proceedings of the International Conference on Antennas and Propagation*, 28 – 30 Nov. 1978, (Institution of Electrical Engineers), pp. 80 – 81.

Cram, M., W8NUE, "Small Transmitting Loop Antennas: a Different Perspective on Determining Q and Efficiency", *QEX*, July/Aug., 2018, pp. 3 – 8.

DeNeef, P., AE7PD, "Safety distances for small HF loop antennas," *Radcom*, June 2017, pp. 46 – 47.

DeMaw, D., "Beat the Noise with a Scoop Loop," *QST*, July 1977, pp. 30 – 34.

DeMaw D., and Aurick, L., "The Full-Wave Delta Loop at Low Height," *QST*, Oct. 1984, pp. 24 – 26.

DeNeef, P., "Effects Due to Ground for Small Transmitting Loop Antennas", *QEX*, July/Aug. 2018, pp. 9 – 11.

Dietrich, J., "Loops and Dipoles: A Comparative Analysis," *QST*, Sep. 1985, pp. 24 – 26.

Dunlavy, J., "Wide Range Tunable Transmitting Loop Antenna", US Patent 3,588,905, issued June 28, 1971.

Fischer, D., "The Loop Skywire," *QST*, Nov 1985, pp. 20 – 22. Also "Feedback," *QST*, Dec. 1985, p 53.

Genaille, R., "V.L.F. Loop Antenna," *Electronics World*, Jan. 1963, pp. 49 – 52.

Glasgow, R., *Principles of Radio Engineering* (McGraw-Hill, 1936).

Goldman, S., "A Shielded Loop for Low Noise Broadcast Reception," *Electronics*, Oct. 1938, pp. 20 – 22.

Grover, F., *Inductance Calculation-Working Formulas and Tables* (D. VanNostrand Co, Inc, 1946).

Hagan, J., "A Large Aperture Ferrite Core Loop Antenna for Long and Medium Wave Reception," *Loop Antennas Design and Theory*, M.G. Knitter, Ed. (National Radio Club, 1983), pp. 37 – 49.

Harwood, S., "The Horizontal Loop — An Effective Multipurpose Antenna," *QST*, Nov. 2006, pp. 42 – 44.

Hawkins, H., "A Low Budget, Rotatable 17 Meter Loop," *QST*, Nov. 1997, p 35.

Jones, B., "A Home-Brew Loop Tuning Capacitor," *QST*, Nov. 1994, pp. 30 – 32.

Jordan E., and Balmain, K., Chapter 10, *Electromagnetic Waves and Radiating Systems*, Second Edition, (Prentice–Hall, 1968).

Lambert, J., "A Directional Active Loop Receiving Antenna System," *Radio Communication*, Nov. 1982, pp. 944 – 949.

Langford-Smith F., (ed.), *Radiotron Designers Handbook*, 4th ed, (Wireless Press, 1953), p. 466

Lankford, D., "Loop Antennas, Theory and Practice," *Loop Antennas Design and Theory*, M. G. Knitter, Ed. (National Radio Club, 1983), pp. 10 – 22.

Loop Antenna Issue, *QEX*, Jul/Aug. 2018.

Mallozzi, D., "Q of Shielded Loop Antennas," *QEX*, July/Aug. 1998, pp. 59 – 60.

Marris, R., "The Optima 160/80-Meter Receive Antenna," *The ARRL Antenna Compendium Vol 6*, (ARRL), pp. 45 – 48.

New England Wire Technologies, Litz Wire Technical Information (**newenglandwire.com**)

Polydoroff, W., High *Frequency Magnetic Materials — Their Characteristics and Principal Applications* (Wiley, 1960).

Quick, R., W4RQ, and K. Siwiak, KE4PT, "Does Your Antenna Need a Choke or a Balun?," *QST*, Mar., 2017, pp. 30 – 33.

Reference Data for Radio Engineers, 6th ed. (Howard W. Sams & Co, 1977).

Simpson, T., and Cahill, J., "The Electrically Small Elliptical Loop with an Oblate Spheroidal Core," *IEEE Antennas and Propagation Magazine*, Vol. 49, No. 5, Oct. 2007, pp. 83 – 92.

Simpson T., and Zhu, Y., "The Electrically Small Multi-Turn Loop with a Spheroidal Core," *IEEE Antennas and Propagation Magazine*, Vol. 48, No. 5, Oct. 2006, pp. 54 – 65.

Siwiak, K., KE4PT, and R. Quick, W4RQ, "Small Gap-resonated HF Loop Antenna Fed by a Secondary Loop", *QEX*, July/Aug., 2018, pp. 12 – 17.

Siwiak, L., KE4PT, and R. Quick, W4RQ, "Small Gap-resonated HF Loop Antennas", *QST*, Sep. 2018, pp. 30 – 33.

Smith, G., "Radiation Efficiency of Electrically Small Multiturn Loop Antennas," *IEEE Trans on Ant and Prop*, Sep. 1972, pp. 656 – 657.

Stanley, J., K4ERO, "Some Additional Geometries for Loop Antennas," *QEX*, July/Aug., 2018, pp. 31 – 36.

Sullivan, C., "Optimal Choice for the Number of Strands in a Litz-Wire Transformer Winding," *IEEE Transactions on Power Electronics*, Mar 1999, Vol. 14, No. 2, pp. 283 – 291 (available from **engineering.dartmouth.edu/inductor/papers/litzj.pdf**)

Trask, C., "Active Loop Aerials for HF Reception Part 1: Practical Loop Aerial Design," *QEX*, July/Aug. 2003, pp. 35 – 42.

Trask, C., "Active Loop Aerials for HF Reception Part 2: High Dynamic Range Aerial Amplifier Design," *QEX*, Sep./Oct. 2003, pp. 44 – 49.

True, J., "Low-Frequency Loop Antennas," *Ham Radio*, Dec. 1976, pp. 18 – 24.

TABLE OF CONTENTS

Chapter 10 — Online Material

Articles
- 5-Band LPDA Construction Project and Telerana Construction Project
- "An Updated 2 Meter LPDA" by Andrzej Przedpelsi, KØABP
- Log Periodic-Yagi Arrays
- "Practical High-Performance HF Log Periodic Antennas" by Bill Jones, K8CU
- "Six Band, 20 through 6 Meter LPDA" by Ralph Crumrine, NØKC
- "The Log Periodic Dipole Array" by Peter Rhodes, K4EWG
- "Using LPDA TV Antennas for the VHF Ham Bands" by John Stanley, K4ERO
- "Vee Shaped Elements vs Straight Elements" by John Stanley, K4ERO
- "An LPDA for 2 Meter Plus," by L.B. Cebik, W4RNL
- "Wideband 10 Meter LPDA," by Justin Johnson, GØKSC
- "A VHF/UHF Discone Antenna" by Bob Patterson, K5DZE
- "Three-Band Log-Periodic Antenna" by Robert Heslin, K7RTY/2
- "Using LPDA TV Antennas for the VHF Ham Bands" by John Stanley, K4ERO
- "HF Discone Antennas"
- "HF Discone Antenna Projects" by W8NWF
- "The HF Discone Antenna" by John Belrose, VE2CV
- "Classical LPDA Design Method – Antenna Book 24th Edition"

Support Files
- 2element 10 Meter LPDA - G0KSC.ez
- 10M LPDA - Patterns and SWR Sweep.zip
- 10M LPDA Assembly Drawings.pdf
- LPDA1 – Design Example, both .EZ and .NEC files
- LPDA Design Figures.zip

Chapter 10

Frequency-Independent Antennas

True *frequency-independent antennas* are those with feed point impedances and radiation patterns (including gain, polarization, and shape) that don't change with frequency. This is different from a "wide-band" design that has an operating bandwidth larger than that of "tuned" antennas but cannot be extended to all frequencies.

To behave independently of frequency, the antenna must comply with *Rumsey's principle* that requires the antenna shape to be specified only in terms of angles, not fixed lengths. (V. H. Rumsey developed this principle and other important ideas about frequency independent antennas during the 1950s and 1960s. See the "References and Bibliography" section for his 1966 book on the subject.)

Design only in terms of angles means that the structure must extend infinitely, both at large and small scales. Practical antennas have edges that truncate and are designed so that currents flowing on them are gradually attenuated until they become negligible at the point of truncation, whether that is a surface, edge, or point.

There are an infinite number of shapes defined solely by angles and there are many different antennas based on those shapes. However, amateurs are most familiar with a few that are covered in this chapter: the *log-periodic dipole array* (*LPDA*), and the *discone*. The *logarithmic spiral* is also encountered occasionally, usually for EMC testing at UHF and microwave frequencies. There are many other possibilities and with the advanced fabrication capabilities available to amateurs such as 3D printing and CNC machining. Experimentation is encouraged!

10.1 Classical LPDA Design

The *log-periodic dipole array* (*LPDA*), usually referred to as a "log periodic" is one family of frequency-independent antennas. The LPDA forms a directional antenna with relatively constant characteristics across a wide frequency range. Log periodics are a good compromise when multiple antennas aren't convenient, but the tradeoff is reduced gain. Some designs have little more gain than a dipole at each frequency even though they have many elements for wide frequency coverage.

The log-periodic approach may also be used with parasitic elements to achieve specific characteristics within a narrow frequency range. Common names for such hybrid arrays are the log-cell Yagi or the Log-Yagi. (Information on the log-cell Yagi is included with this book's online material.) This chapter was contributed originally by L. B. Cebik, W4RNL, with additional contributions from John Stanley, K4ERO.

The LPDA is the most popular form of log-periodic systems which also include zigzag, planar, trapezoidal, slot, and V forms. The appeal of the LPDA version of the log periodic antenna owes much to its structural similarity to the Yagi-Uda parasitic array. This permits the construction of directional LPDAs that can be rotated — at least within the upper HF and higher frequency ranges. Nevertheless, the LPDA has special structural as well as design considerations that distinguish it from the Yagi. Different construction techniques for both wire and tubular elements are illustrated later in this chapter.

The LPDA in its present form derives from the pioneering work of D. E. Isbell at the University of Illinois in the late 1950s. Although you may design LPDAs for large frequency ranges — for example, from 3 to 30 MHz or a little over 3 octaves — the most common LPDA designs that radio amateurs use are limited to a one-octave range, usually from 14 to 30 MHz. Amateur designs for this range tend to consist of linear elements. However, experimental designs for lower frequencies have used elements shaped like inverted Vs and some versions use vertically oriented ¼-λ elements over a ground system.

Figure 10.1 shows the parts of a typical LPDA. The structure consists of a number of linear elements, the longest of which is approximately ½ λ long at the lowest design frequency. The shortest element is usually about ½ λ long at a frequency well above the highest operating frequency. The

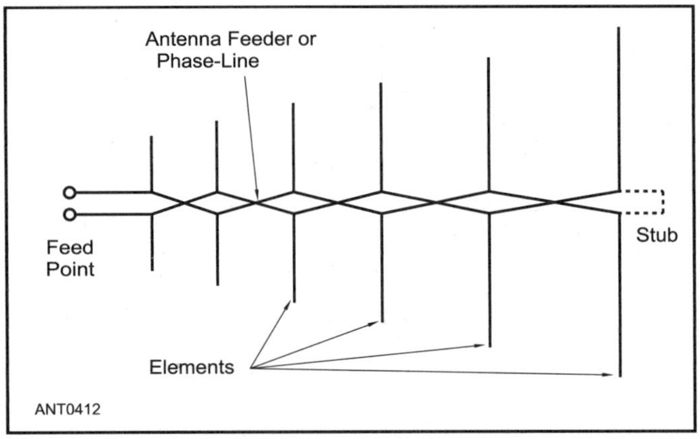

Figure 10.1 — The basic components of a log periodic dipole array (LPDA). The forward direction is to the left in this sketch. Many variations of the basic design are possible.

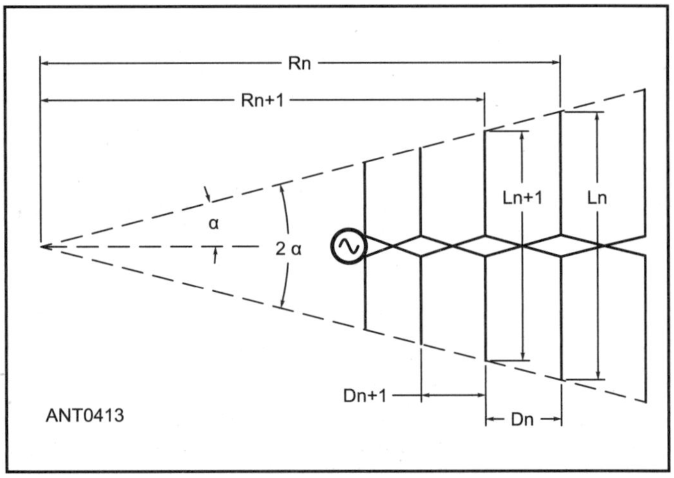

Figure 10.2 — Some fundamental relationships that define an array as an LPDA. See the text for the defining equations.

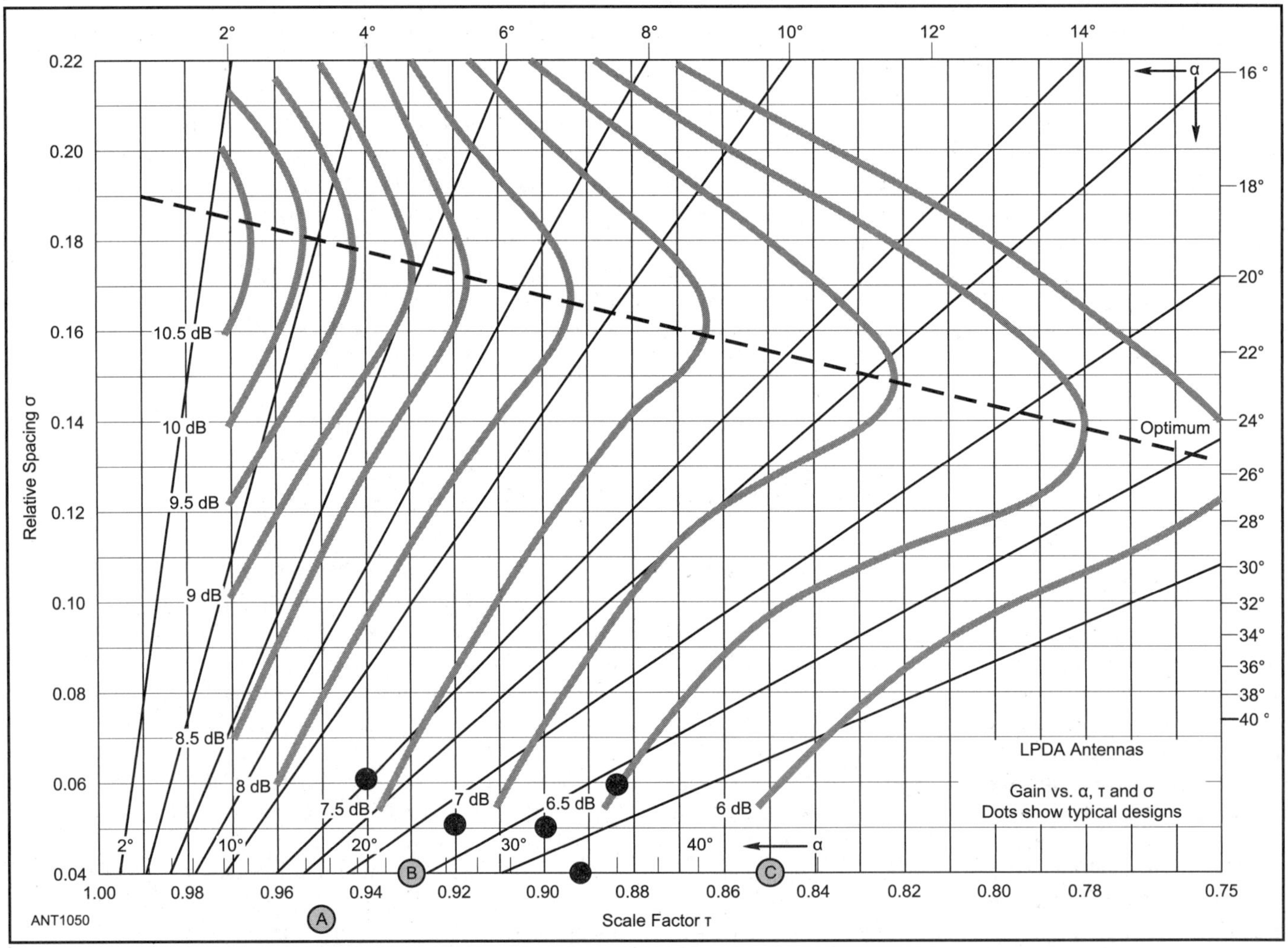

Figure 10.3 — The optimum value for σ lies on the straight line that intersects the constant-gain curves at different values of τ. Using the optimum value of σ to design an LPDA usually results in antenna that is impractically large at HF (see text). Points labeled "A," "B," and "C" represent values of σ and τ that result in practical antenna sizes as in the design examples in this section; "9302," "8904," and "8504," respectively. (Graphic after Carrel, et al; see Bibliography.)

antenna feeder, also informally called the *phase-line*, connects the center points of each element in the series, with a phase reversal or *crossover* between each element. A tuning stub or "hairpin" consisting of a shorted length of parallel-wire feed line is often added at the back of an LPDA.

The arrangement of elements and the method of feed yield an array with relatively constant gain and front-to-back ratio across the designed operating range. In addition, the LPDA exhibits a relatively constant feed point impedance, simplifying matching to a transmission line.

CLASSICAL DESIGN PROCEDURE

For the amateur designer, the most fundamental facets of the LPDA revolve around three interrelated design variables: α (alpha), τ (tau), and σ (sigma). Any one of the three variables may be defined by reference to the other two.

Figure 10.2 shows the basic components of an LPDA. The angle α defines the outline of an LPDA and permits every dimension to be treated as a radius or the consequence of a radius (R) of a circle. The most basic structural dimensions are the element lengths (L), the distance (R) of each element from the apex of angle α, and the distance between elements (D). A single design constant, τ, defines all of these relationships in the following manner:

$$\tau = \frac{R_{n+1}}{R_n} = \frac{D_{n+1}}{D_n} = \frac{L_{n+1}}{L_n} \qquad (1)$$

where element n and n+1 are successive elements in the array working toward the apex of angle α. The value of τ is always less than 1.0 although effective LPDA design requires values as close to 1.0 as may be feasible.

The variable τ defines the relationship between successive element spacings but it does not itself determine the initial spacing between the longest and next longest elements upon which to apply τ successively. The initial spacing also defines the angle α for the array. Hence, we have two ways to determine the value of σ, the relative spacing constant:

$$\sigma = \frac{1-\tau}{4\tan\alpha} = \frac{D_n}{2L_n} \qquad (2)$$

where D_n is the distance between any two elements of the array and L_n is the length of the longer of the two elements. From the first of the two methods of determining the value of σ, we may also find a means of determining α when we know both τ and σ.

For any value of τ, we may determine the optimal value of σ:

$$\sigma_{opt} = 0.243\tau - 0.051 \qquad (3)$$

The combination of a value for τ and its corresponding optimal value of σ yields the highest performance of which an LPDA is capable. For values of τ from 0.80 through 0.98, the value of optimal σ varies from 0.143 to 0.187, in increments of 0.00243 for each 0.01 change in τ. This is illustrated by the graph in **Figure 10.3**, originally published by Carrel and updated by Butson and Thompson (see the Bibliography).

Practically however, using the optimal value of σ usually yields a total array length that is beyond amateur construction or tower/mast support capabilities. A design procedure more likely to result in a useful design at HF is to reduce σ until maximum gain begins to fall significantly. Consequently, amateur LPDAs usually employ compromise values of τ and σ that yield lesser but acceptable performance. The sidebar "Determining LPDA Design Parameters Quickly" shows how to arrive at acceptable values of τ and σ for use in *LPCAD* and similar software.

For a given frequency range, increasing the value of τ increases both the gain and the number of required elements. Increasing the value of σ increases both the gain and the overall boom length. A τ of 0.96 — which approaches the upper maximum recommended value for τ — yields an optimal σ of about 0.18, and the resulting array grows to over 100 feet long for the 14 to 30 MHz range. The maximum free space gain is about 11 dBi, with a front-to-back ratio that approaches 40 dB. Normal amateur practice, however, uses values of τ from about 0.88 to 0.95 and values of σ from about 0.03 to 0.06.

Standard design procedures usually assign to the rear element a resonant frequency about 7% lower than the lowest design frequency with a physical length 5% lower than a free-space half wavelength. The upper frequency limit of the design is ordinarily set at about 1.3 times the highest design frequency. Since τ and σ set the increment between successive element lengths, the number of elements becomes a function of when the shortest element reaches the dipole length for the adjusted upper frequency.

The adjusted upper frequency limit results from the

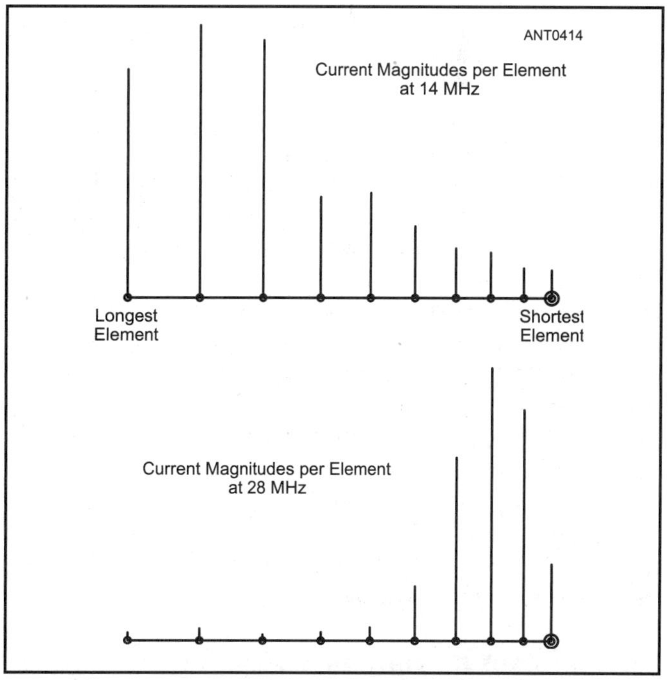

Figure 10.4 — The relative current magnitude on the elements of an LPDA at the lowest and highest operating frequencies for a given design. Compare the number of "active" elements, that is, those with current levels at least 1⁄10 of the highest level.

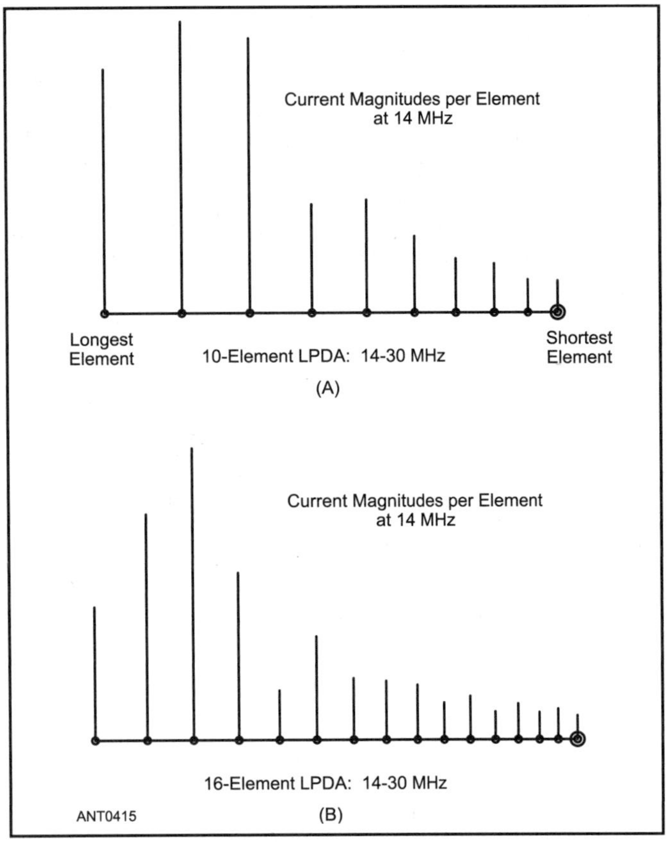

Figure 10.5 — Patterns of current magnitude at the lowest operating frequency of two different LPDA designs: a 10-element low-τ design and a 16-element higher-τ design.

behavior of LPDAs with respect to the number of active elements. See **Figure 10.4**, which shows an edge view of a 10-element LPDA for 20 through 10 meters. The vertical lines represent the peak relative current magnitude for each element at the specified frequency. At 14 MHz, virtually every element of the array shows a significant current magnitude. However, at 28 MHz, only the forward 5 elements carry significant current. Without the extended design range to nearly 40 MHz, the number of elements with significant current levels would be severely reduced, along with upper frequency performance.

The need to extend the design equations below the lowest proposed operating frequency varies with the value of τ. In **Figure 10.5**, we can compare the current on the rear elements of two LPDAs, both with a σ value of 0.04. The upper design uses a τ of 0.89, while the lower design uses a value of 0.93. The most significant current-bearing element moves forward with increases in τ, reducing (but not wholly eliminating) the need for elements whose lengths are longer than a dipole for the lowest operating frequency.

10.1.1 LDPA Design Software

Originally, LPDA design proceeded through a series of design equations intended to yield the complete specifications for an array. More recent techniques available to radio amateurs include basic LPDA design software and antenna modeling

Determining LPDA Design Parameters Quickly

LPCAD provides a very efficient method of arriving at a preliminary design for an LPDA. When using the program, it is much faster to use the boom length and number of elements input method rather than the τ and σ data entry method. Those parameters will be calculated along with all dimensions. For those wishing to better understand the trade-offs or who wish to use the formulas in this chapter to calculate all dimensions, the following procedure will help you to arrive at starting values of τ and σ that will make your progress towards a final design much more rapid.

The first consideration is the frequency range to be covered. Many ham LPDA designs cover about a 2:1 frequency range, for example, 14 to 29 MHz. Extending the high end of the frequency range adds little to the size and cost and can smooth out operation at the higher frequencies of interest. An LPDA with an 8:1 range will be only 1.8 times longer than an antenna with a 2:1 range, while the frequency range is 4 times more. However, use of a very wide-band LPDA only at its high frequency end means most of the size and weight of the antenna is wasted since the largest parts of the antenna are inactive. Covering only the desired frequencies reduces boom length. **Figure 10.A** illustrates the different sizes of arrays having the same gain but different frequency coverage.

The lowest frequency determines the longest element length, which will be somewhat more than ½ wavelength at that frequency. An approximate length in feet can be gotten by dividing 500 by the frequency in MHz. This dimension is pretty much fixed, although in some large designs it is reduced somewhat by inductive or capacitive loading of the lowest frequency elements.

The boom length is the next logical parameter to be determined and will be based on what you can afford to build, raise, support and rotate versus the performance to be expected for various boom options. For a given frequency ratio — the longer the boom, the higher the gain. **Figure 10.B** will give some guidance as to the gain to be expected for a given boom length. The boom length factor (BLF) represents the boom length compared to the length of the longest element. As a rough

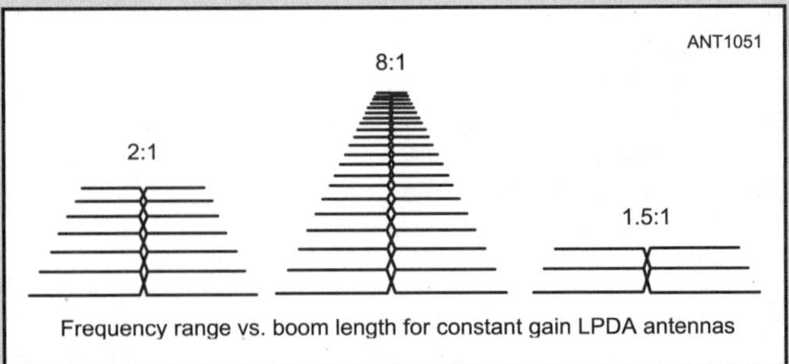

Figure 10.A — Comparison of boom length for several LPDA designs with the same gain but different ratios of maximum to minimum frequency coverage.

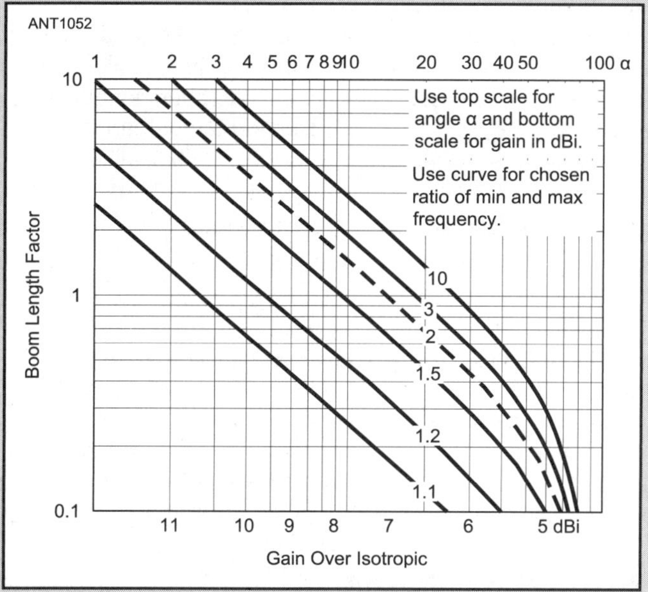

Figure 10.B — A chart relating gain, boom length factor, and α.

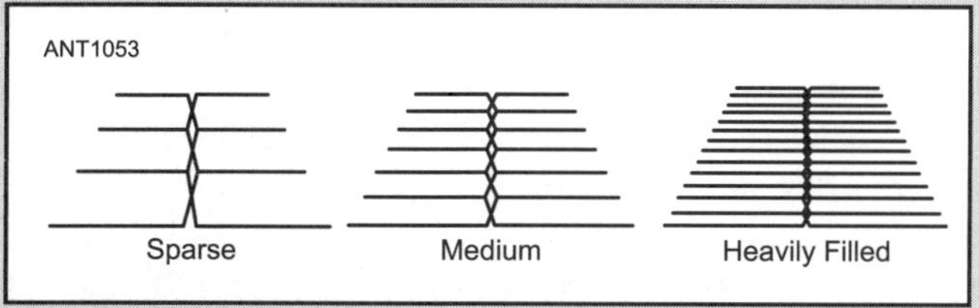

Figure 10.C — Illustration of sparse to heavily filled arrays.

rule of thumb, each doubling of the boom length will add about 1 dB of gain. Beyond a certain point, doubling the boom length to add 1 dB will be uneconomical. Keeping the boom to a reasonable length means accepting either a reduced frequency range or a lower gain.

Once a boom length is chosen, based on a trade-off between mechanical limits and the desired gain value and frequency range ratio, we can also read off the angle using this same graph, since gain is closely associated with α.

Once α is determined along with the frequency coverage, the shape and size of the antenna is defined. We must now decide how to fill up this outlined shape with elements. In other words, determine the number, length and spacing of the elements. The chart of Figure 10.3 is useful in this. This chart plots gain as a function of τ, σ and α. The α values are represented by the slanted lines between the marked degree values on the chart. Since we have determined the desired value of α, we can use that value to choose values of τ and σ.

For a given α value (slanted lines), follow the corresponding line diagonally along and you will see the lower part of the line lies more or less parallel to a constant gain curve. The portion of a constant-α line on the bottom left represents a heavily filled array (more elements, closely spaced) while upper right portion of the constant, a line represents a sparsely filled array (fewer elements). (**see Figure 10.C**)

Notice that as the number of elements decreases (moving right and upwards on the constant-α gain may fluctuate a bit but will reach a point where it quickly falls off as you cross the "optimum σ" line. Values above this line represents a design that has too few elements to realize good performance. The interpretation of this "optimum σ" value is not obvious and perhaps its name was poorly chosen. It may be intended to give the value of σ for maximum gain for a given value of τ. However, this approach disregards the boom length factor. Designs that first choose τ and then use the optimum σ will be outrageously long — of course they will have high gain!

For a given antenna outline (length vs. width or constant-α) "optimum σ" is optimum only in that it indicates the point where further reduction in the number of elements will cause the gain to drop off. Thus, it gives a design with the least number of elements for a given gain. However, this design will not have the smoothest gain and SWR vs. frequency and the F/B ratio may be inadequate. For this reason, very few practical designs fall near the "optimum σ" line. All of the designs in this chapter fall well below the "optimum σ" as shown by the dots on the chart.

For designs with a higher α value, the gain should increase a bit with the more heavily filled arrays. Very narrow (long boom) arrays may have less gain if too heavily filled. It is not wise to approach the "optimum σ" line too closely in an attempt to reduce the element number. On the other hand, designs too far down the sloping a curve will be more expensive to build and will have greater wind load due to having more elements than necessary.

Having selected from the chart some τ and σ combinations that give us the design we want, we can now use software such as *LPCAD* to generate a detailed design using those values of τ and σ. This should result in a boom length fairly close to that desired. A second pass through the program may use the exact boom length desired and the number of elements calculated in the first pass, plus or minus one element. The program will then give us all of the mechanical dimensions, and also generate files for *NEC* analysis. Alternatively, we can proceed with the manual method given elsewhere in this chapter. If using that method we should come out close to the desired design on the first try and avoid multiple guesses as to where to start.

Several different designs that are close to that desired should be prepared as *NEC* files and an analysis done using *NEC*. This may show that a sparsely filled array may have excessive variation of gain and SWR vs. frequency or an inadequate F/B ratio or have other problems such as a weakness on an important frequency.
— *John Stanley, K4ERO*

software. One good example of LPDA design software is *LPCAD* by Roger Cox, WBØDGF, available for downloading from **w8io.com/LPCAD.htm**. The user begins by specifying the lowest and highest frequencies in the design. The user then selects values for τ and σ or choices for the number of elements and the total length of the array. With this and other input data, the program provides a table of element lengths and spacings, using the adjusted upper and lower frequency limits described earlier. (A spreadsheet to assist with performing the calculations in this chapter was contributed by Dennis Miller, KM9O, and is available for downloading from the online material.)

The program also requests the diameters of the longest and shortest elements in the array, as well as the diameter of average element. From this data, the program calculates a recommended value for the characteristic impedance of the phase-line connecting the elements and the approximate resistive value of the input impedance. Among the additional data that *LPCAD* makes available is the spacing of conductors to achieve the desired characteristic impedance of the phase-line. These conductors may be round — as we would use for a wire phase-line — or square — as we might use for double-boom construction.

An additional vital output from *LPCAD* is the conversion of the design into antenna modeling input files of several formats, including versions for *AO* and *NEC4WIN* (both *MININEC*-based programs), and a version in the standard *.NEC format usable by many implementations of *NEC-2* and *NEC-4*, including *NECWin Plus*, *GNEC*, and *EZNEC Pro*. Every proposed LPDA design should be verified and optimized by means of antenna modeling, since basic design calculations rarely provide arrays that require no further work before construction. Moreover, some of the design equations are based upon approximations and do not completely predict LPDA behavior. Despite these limitations, most of the sample LPDA designs shown later in this chapter are based directly upon the fundamental calculations.

Modeling LPDA designs is most easily done on a version of *NEC*. The transmission line (TL) facility built into *NEC-2* and *NEC-4* alleviates the problem of modeling the phase-line as a set of physical wires, each section of which has a set of constraints in *MININEC* at the right-angle junctions with the elements. Although the *NEC* TL facility does not account for losses in the lines, the losses are ordinarily low enough to neglect.

NEC models do require some careful construction to obtain the most accurate results. Foremost among the cautions is the need for careful segmentation, since each element has a different length. The shortest element should have about 9 or 11 segments, so that it has sufficient segments at the highest modeling frequency for the design. Each element behind the shortest one should have a greater number of segments than the preceding element by the inverse of the value of τ. However, there is a further limitation. Since the transmission line is at the center of each element, *NEC* elements should have an odd number of segments to hold the phase-line centered. Hence, each segmentation value calculated from the inverse of τ must be rounded up to the nearest odd integer.

Initial modeling of LPDAs in *NEC-2* should be done with uniform-diameter elements, with any provision for stepped-diameter element correction turned off. Since these correction factors apply only to elements within about 15% of dipole resonance at the test frequency, models with stepped-diameter elements will correct for only a few elements at any test frequency. The resulting combination of corrected and uncorrected elements will not yield a model with assured reliability.

Once one has achieved a satisfactory model with uniform-diameter elements, the modeling program can be used to calculate stepped-diameter substitutes. Each uniform-diameter element, when extracted from the larger array, will have a resonant frequency. Once this frequency is determined, the stepped-diameter element to be used in final construction can be resonated to the same frequency. Although *NEC-4* handles stepped diameter elements with much greater accuracy than *NEC-2*, the process just described is also applicable to *NEC-4* models for the greatest precision.

10.1.2 LDPA Behavior

Although LPDA behavior is remarkably uniform over a wide frequency range compared to narrow-band designs, such as the Yagi-Uda array, it nevertheless exhibits very significant variations within the design range. **Figure 10.6** shows several facets of these behaviors. Figure 10.6 shows the free-space gain for three LPDA designs using 0.5-inch diameter aluminum elements. The designations for each model list the values of τ (0.93, 0.89, and 0.85) and of σ (0.02, 0.04, and 0.06) used to design each array. The resultant array lengths are listed with each designator. The total number of elements varies from 16 for "9302" to 10 for "8904" to 7 for "8504."

First, the gain is never uniform across the entire frequency span. The gain tapers off at both the low and high ends of the design spectrum. Moreover, the amount of gain undulates across the spectrum, with the number of peaks dependent upon

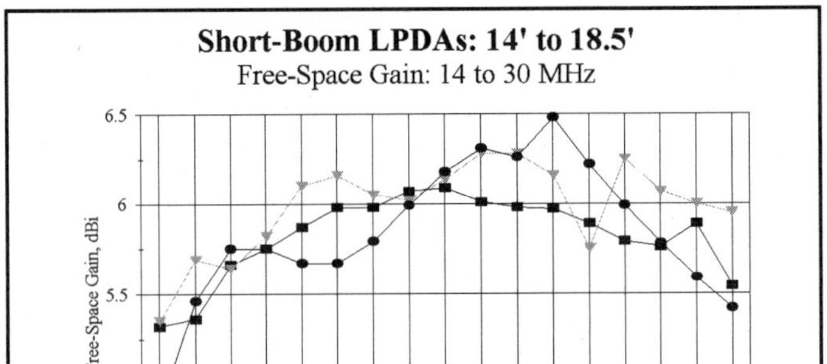

Figure 10.6 — The modeled free-space gain of three relatively small LPDAs of different design. Note the relationship of the values of τ and of σ for these arrays with quite similar performance across the 14-30 MHz span.

the selected value of τ and the resultant number of elements. The front-to-back ratio tends to follow the gain level. In general, it ranges from less than 10 dB when the free-space gain is below 5 dBi to over 20 dB as the gain approaches 7 dBi. The front-to-back ratio may reach the high 30 dBs when the free-space array gain exceeds 8.5 dBi. Well-designed arrays, especially those with high values of τ and σ, tended to have well-controlled rear patterns that result in only small differences between the 180° front-to-back ratio and the averaged front-to-rear ratio.

Since array gain is a mutual function of both τ and σ, average gain becomes a function of array length for any given frequency range. Although the gain curves in Figure 10.6 interweave, there is little to choose among them in terms of average gain for the 14 to 18-foot range of array lengths. Well-designed 20 to 10 meter arrays in the 30-foot array length region are capable of about 7 dBi free-space gain, while 40-foot arrays for the same frequency range can achieve about 8 dBi free-space gain.

Exceeding an average gain of 8.5 dBi requires at least a 50-foot array length for this frequency range. Long arrays with high values of τ and σ also tend to show smaller excursions of gain and of front-to-back ratio in the overall curves. In addition, high-τ designs tend to show higher gain at the low frequency end of the design spectrum.

The frequency sweeps shown in Figure 10.6 are widely spaced at 1 MHz intervals. The evaluation of a specific design for the 14 to 30-MHz range should decrease the interval between check points to no greater than 0.25 MHz in order to detect frequencies at which the array may show a *performance weakness*. Weaknesses are frequency regions in the overall design spectrum at which the array shows unexpectedly lower values of gain and front-to-back ratio. In Figure 10.6 note the unexpected decrease in gain of model "8904" at 26 MHz. The other designs also have weak points, but they fall between the frequencies sampled.

In large arrays, these regions may be quite small and may occur in more than one frequency region. The weakness results from the harmonic operation of longer elements to the rear of those expected to have high current levels. Consider a 7-element LPDA with a boom about 12.25-feet long for 14 to 30 MHz using 0.5-inch aluminum elements. At 28 MHz, the rear elements operate in a harmonic mode as shown by the high relative current magnitude curves in **Figure 10.7**. The result is a radical decrease in gain, as shown in the "No Stub" curve of **Figure 10.8**. The front-to-back ratio also drops as a result of strong radiation from the long elements to the rear of the array.

Early designs of LPDAs called for terminating transmission-line stubs as standard practice to help eliminate such weak spots in frequency coverage. In contemporary designs, their use tends to be more specific for eliminating or moving frequencies that show gain and front-to-back weakness. (Stubs have the added

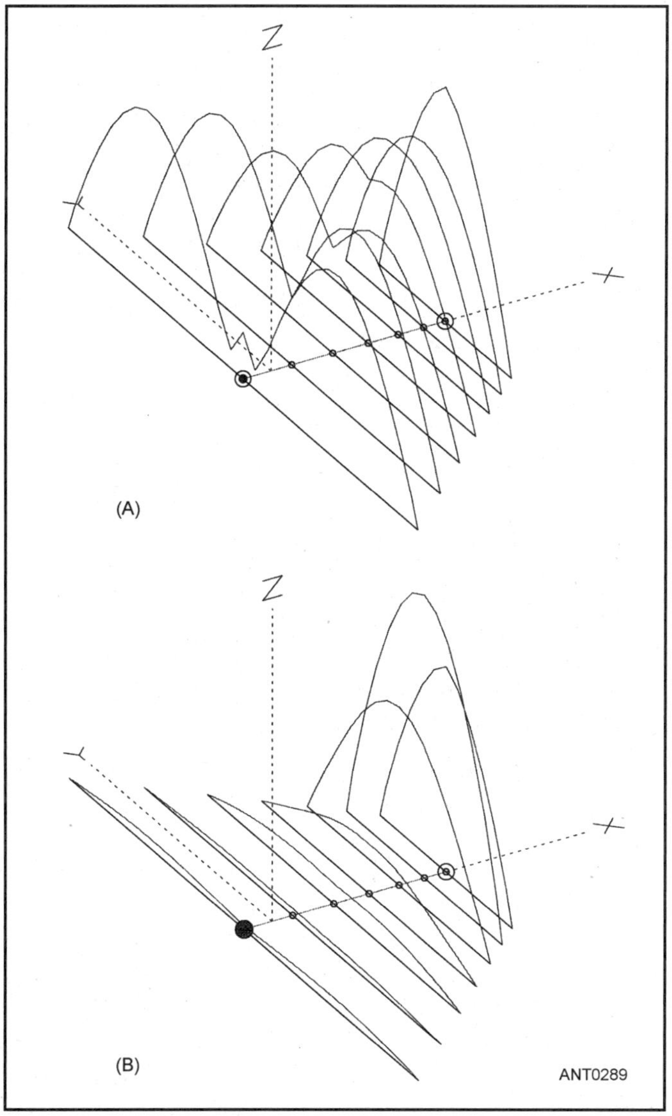

Figure 10.7 — The relative current magnitude on the elements of model "8504" at 28 MHz without and with a stub. Note the harmonic operation of the rear elements before a stub is added to suppress such operation.

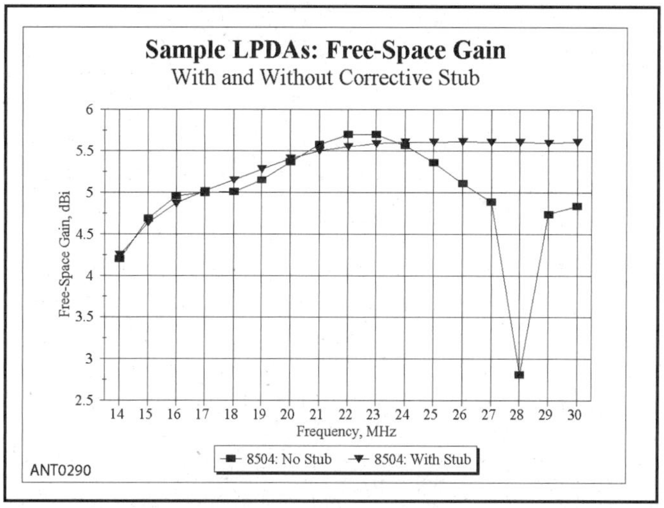

Figure 10.8 — A graph of the gain of model "8504" showing the frequency region in which a "weakness" occurs and its absence once a suitable stub is added to the array.

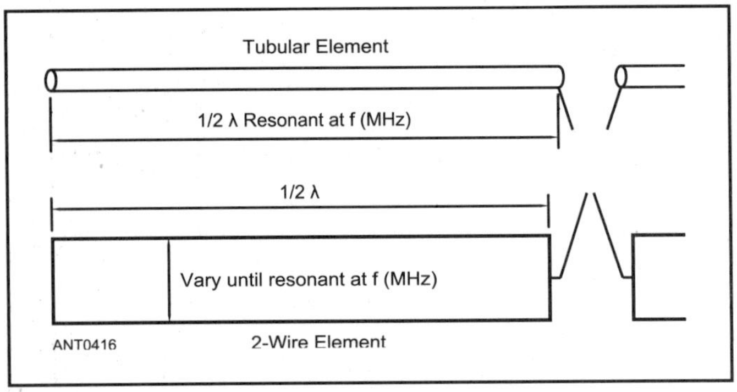

Figure 10.9 — A substitute for a large-diameter tubular element composed of two wires shorted at both the outer ends and at the center junction with the phase-line.

function of keeping both sides of each element at the same dc level of static charge or discharge.) The model dubbed "8504" was fitted (by trial and error) with an 18-inch shorted stub of 600-Ω transmission line. As Figure 10.7B shows, the harmonic operation of the rear elements is attenuated. The "stub" curve of Figure 10.8 shows the smoothing of the gain curve for the array throughout the upper half of its design spectrum. In some arrays showing multiple weaknesses, a single stub may not eliminate all of them. However, it may move the weaknesses to unused frequency regions. Where full-spectrum operation of an LPDA is necessary, additional stubs located at specific elements may be needed.

Most LPDA designs benefit (with respect to gain and front-to-back ratio) from the use of larger-diameter elements. Elements with an average diameter of at least 0.5-inch are desirable in the 14 to 30 MHz range. However, standard designs usually presume a constant element length-to-diameter ratio. In the case of *LPCAD*, this ratio is about 125:1, which assumes an even larger diameter. To achieve a relatively constant length-to-diameter ratio in the computer models, you can set the diameter of the shortest element in a given array design and then increase the element diameter by the inverse of τ for each succeeding longer element. This procedure is often likely to result in unreasonably large element diameters for the longest elements, relative to standard amateur construction practices.

Since most amateur designs using aluminum tubing for elements employ stepped-diameter (tapered) elements, roughly uniform element diameters will result unless the LPDA mechanical design tries to lighten the elements at the forward end of the array. This practice may not be advisable, however. Larger elements at the high end of the design spectrum often counteract (at least partially) the natural decrease in high-frequency gain and show improved performance compared to smaller diameter elements.

An alternative construction method for LPDAs uses wire throughout. At every frequency, single-wire elements reduce gain relative to larger-diameter tubular elements. An alternative to tubular elements appears in **Figure 10.9**. For each element of a tubular design, there is a roughly equivalent two-wire element that may be substituted. The spacing between the wires is determined by taking one of the modeled tubular elements and finding its resonant frequency. A two-wire element of the same length is then constructed with shorts at the far ends and at the junctions with the phase-line. The separation of the two wires is adjusted until the wire element is resonant at the same frequency as the original tubular element. The required separation will vary with the wire chosen for the element. Models used to develop these substitutes must pay close attention to segmentation rules for *NEC* due to the short length of segments in the end and center shorts, and to the need to keep segment junctions as exactly parallel as possible with close-spaced wires.

10.1.3 Feeding and Constructing the LPDA

Original design procedures for LPDAs used a single, ordinarily fairly high, characteristic impedance for the phase-line (antenna feeder). Over time, designers realized that other values of impedance for the phase-line offered both mechanical and performance advantages for LPDA performance. Consequently, for the contemporary designer, phase-line choice and construction techniques are almost inseparable considerations.

High-impedance phase-lines (roughly 200 Ω and higher) are amenable to wire construction similar to that used with ordinary parallel-wire transmission lines. They require careful placement relative to a metal boom used to support individual elements (which themselves must be insulated from the support boom). Connections also require care. If the phase-line is given a half-twist between each element, the construction of the line must ensure constant spacing and relative isolation from metal supports to maintain a constant impedance and to prevent shorts.

Along with the standard parallel-wire line, shown in **Figure 10.10A**, there are a number of possible LPDA structures using booms to support the elements and to create relatively low-impedance (under 200 Ω) phase-lines. Figure 10.10B shows the basics of a twin circular tubing boom with the elements cross-supported by insulated rods. Figure 10.10C shows the use of square tubing with the elements attached directly to each tube by through-bolts. Figure 10.10D illustrates the use of L-stock, which may be practical at VHF frequencies. Each of these sketches is incomplete, however, since it omits the necessary stress analyses that determine the mechanical feasibility of a structure for a given LPDA project.

The use of square boom material requires some adjustment when calculating the characteristic impedance of the phase-line. For conductors with a circular cross-section,

$$Z_0 = 120 \cosh^{-1} \frac{D}{d} \qquad (4)$$

where D is the center-to-center spacing of the conductors and d is the outside diameter of each conductor, both expressed in the same units of measurement. Since we are dealing with closely spaced conductors, relative to their diameters, the use of this version of the equation for calculating the characteristic impedance (Z_0) is recommended. For a square conductor,

$$d \approx 1.18\ w \qquad (5)$$

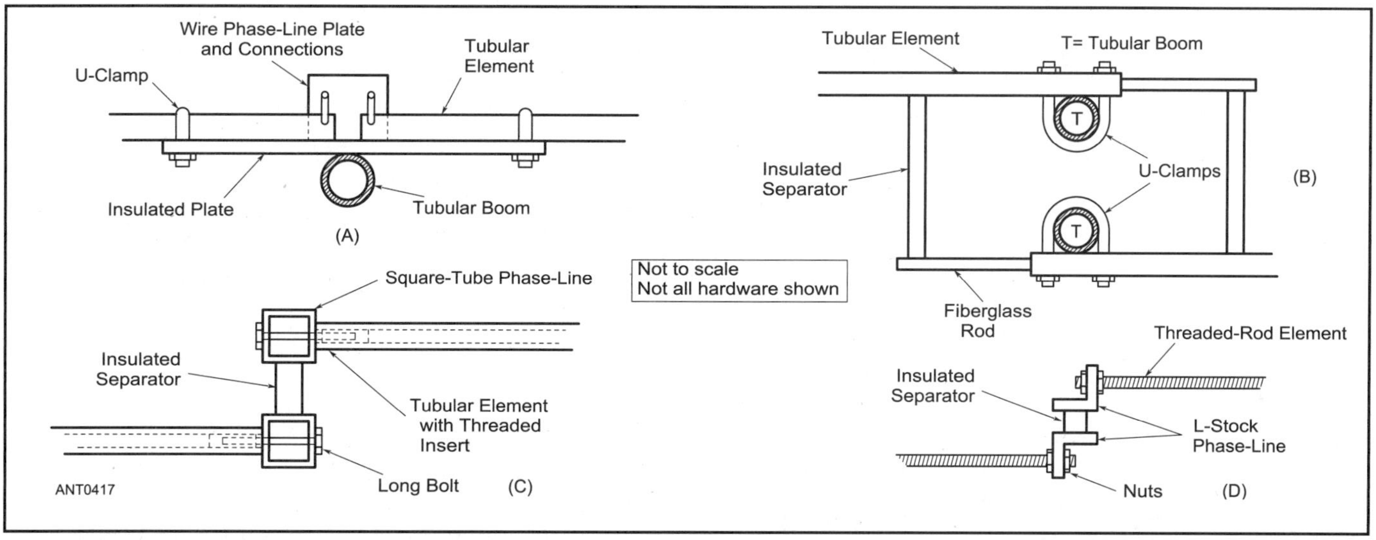

Figure 10.10 — Four (of many) possible construction techniques, shown from the array end. In A, an insulated plate supports and separates the wires of the phase-line, suitable with wire or tubular elements. A dual circular boom phase-line also supports the elements, which are cross-supported for boom stability. Square tubing is used in C, with the elements joined to the boom/phase-line with through-bolts and an insert in each half element. The L-stock shown in D is useful for lighter VHF and UHF arrays.

where d is the approximate equivalent diameter of the square tubing and w is the width of the tubing across one side. Thus, for a given spacing, a square tube permits you to achieve a lower characteristic impedance than round conductors. However, square tubing requires special attention to matters of strength, relative to comparable round tubing.

Electrically, the characteristic impedance of the LPDA phase-line tends to influence other performance parameters of the array. Decreasing the phase-line Z_0 also decreases the feed point impedance of the array. For small designs with few elements, the decrease is not fully matched by a decrease in the excursions of reactance. Consequently, using a low impedance phase-line may make it more difficult to achieve a 2:1 or less SWR for the entire frequency range. However, higher-impedance phase-lines may result in a feed point impedance that requires the use of an impedance-matching balun.

Decreasing the phase-line Z_0 also tends to increase LPDA gain and front-to-back ratio. There is a price to be paid for this performance improvement — weaknesses at specific frequency regions become much more pronounced with reductions in the phase-line Z_0. For a specific array you must weigh carefully the gains and losses, while employing one or more transmission line stubs to get around performance weaknesses at specific frequencies.

Depending upon the specific values of τ and σ selected for a design, you can sometimes select a phase-line Z_0 that provides either a 50-Ω or a 75-Ω feed point impedance, holding the SWR under 2:1 for the entire design range of the LPDA. The higher the values of τ and σ for the design, the lower the reactance and resistance excursions around a central value. Designs using optimal values of σ with high values of τ show a very slight capacitive reactance throughout the frequency range. Lower design values obscure this phenomenon due to the wide range of values taken by both resistance and reactance as the frequency is changed.

At the upper end of the frequency range, the source resistance value decreases more rapidly than elsewhere in the design spectrum. In larger arrays, this can be overcome by using a variable Z_0 phase-line for approximately the first 20% of the array length. This technique is, however, difficult to implement with anything other than wire phase-lines. Begin with a line impedance about half of the final value and increase the wire spacing evenly until it reaches its final and fixed spacing. This technique can sometimes produce smoother impedance performance across the entire frequency span and improved high-frequency SWR performance.

Designing an LPDA requires as much attention to designing the phase-line as to element design. It is always useful to run models of the proposed design through several iterations of possible phase-line Z_0 values before freezing the structure for construction.

10.1.4 Special Design Corrections

The curve for the sample 8504 LPDA in Figure 10.8 revealed several deficiencies in standard LPDA designs. The weakness in the overall curve was corrected by the use of a stub to eliminate or move the frequency at which rearward elements operated in a harmonic mode. In the course of describing the characteristics of the array, we have noted several other means to improve performance. Fattening elements (either uniformly or by increasing their diameter in step with τ) and reducing the characteristic impedance of the phase-line are capable of small improvements in performance. However, they cannot wholly correct the tendency of the array gain and front-to-back ratio to fall off at the upper and lower limits of the LPDA frequency range.

One technique sometimes used to improve performances

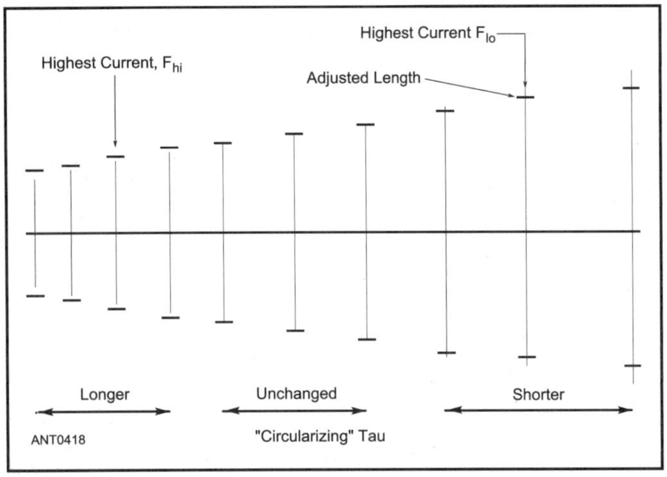

Figure 10.11 — A before and after sketch of an LPDA, showing the original lengths of the elements and their adjustments from diminishing the value of τ at both ends of the array. See the text for the amount of change applicable to each element.

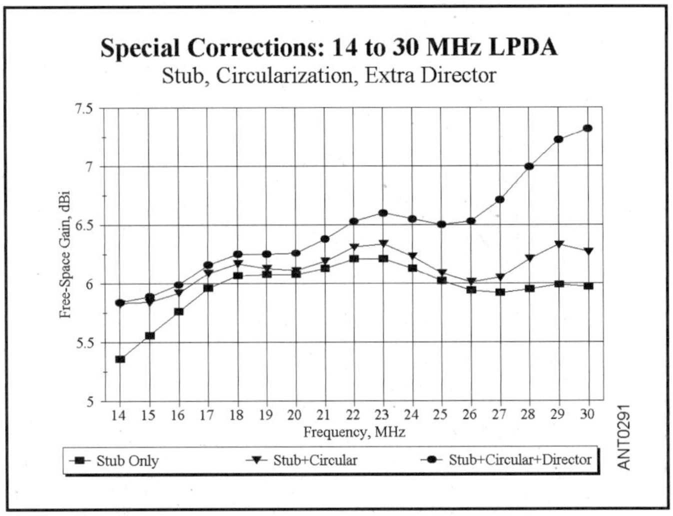

Figure 10.12 — The modeled free space gain from 14 to 30 MHz of an LPDA with τ of 0.89 and σ of 0.04. Squares: just a stub to eliminate a weakness; Triangles: with a stub and circularized elements, and Circles: with a stub, circularized elements and a parasitic director.

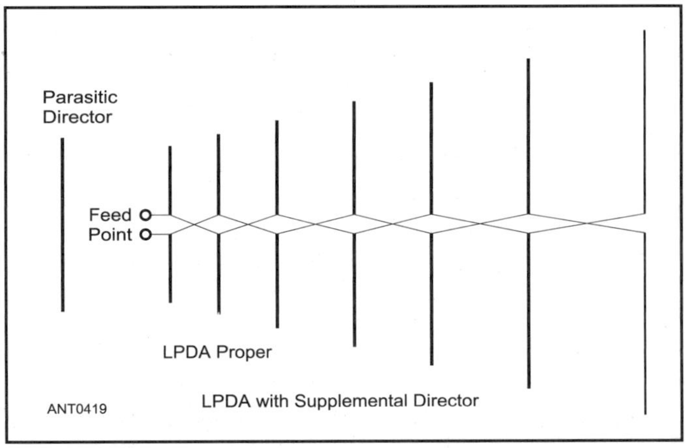

Figure 10.13 — A generalized sketch of an LPDA with the addition of a parasitic director to improve performance at the higher frequencies within the design range.

near the frequency limits is to design the LPDA for upper and lower frequency limits much higher and lower than the frequencies of use. This technique unnecessarily increases the overall size of the array and does not eliminate the downward performance curves. Increasing the values of τ and σ will usually improve performance at no greater cost in size than extending the frequency range. Increasing the value of τ is especially effective in improving the low frequency performance of an LPDA.

Working within the overall size limits of a standard design, one may employ a technique of *circularizing* the value of τ for the rear-most and forward-most elements. See **Figure 10.11**, which is not to-scale relative to overall array length and width. Locate (using an antenna-modeling program) the element with the highest current at the lowest operating frequency, and the element with the highest current at the highest operating frequency. The adjustments to element lengths may begin with these elements or — at most — one element further toward the array center. For the first element (counting from the center) to be modified, reduce the value of τ by about 0.5%. For a rearward element, use the inverse of the adjusted value of τ to calculate the new length of the element relative to the unchanged element just forward of the change. For a forward element, use the new value of τ to calculate the new length of the element relative to the unchanged element immediately to the rear of it.

For succeeding elements outward, calculate new values of τ from the adjusted values, increasing the increment of decrease with each step. Second adjusted elements may use values of τ about 0.75% to 1.0% lower than the values just calculated. Third adjusted elements may use an increment of 1.0% to 1.5% relative to the preceding value.

Not all designs require extensive treatment. As the values of τ and σ increase, fewer elements may require adjustment to obtain the highest possible gain at the frequency limits, and these will always be the most outward elements in the array. A second caution is to check the feed point impedance of the array after each change to ensure that it remains within design limits.

Figure 10.12 shows the free-space gain curves from 14 to 30 MHz for a 10-element LPDA with an initial τ of 0.89 and a σ of 0.04. The design uses a 200-Ω phase-line, 0.5-inch aluminum elements, and a 3-inch 600-Ω stub. The lowest curve shows the modeled performance across the design frequency range with only the stub. Performance at the frequency limits is visibly lower than within the peak performance region. The middle curve shows the effects of circularizing τ. Average performance levels have improved noticeably at both ends of the spectrum.

In lieu of, or in addition to, the adjustment of element lengths, you may also add a parasitic director to an LPDA, as shown in **Figure 10.13**. The director is cut roughly for the highest operating frequency. It may be spaced between 0.1 λ and 0.15 λ from the forward-most element of the LPDA. The exact length and spacing should be determined experimentally (or from models) with two factors in mind. First, the element should not adversely affect the feed point impedance at the highest operating frequencies. Close spacing of the director has the greatest effect on this impedance. Second, the exact spacing and element length should be set to have the most desired effect on the overall performance curve of the array. The mechanical impact of adding a director is to increase overall array length by the spacing selected for the element.

The upper curve in Figure 10.12 shows the effect of adding a director to the circularized array already equipped with a stub. The effect of the director is cumulative, increasing the upper range gain still further. Note that the added parasitic director is not just effective at the highest frequencies within the LPDA design range. It has a perceptible effect almost all the way across the frequency span of the array, although the effect is smallest at the low-frequency end of the range.

The addition of a director can be used to enhance upper frequency performance of an LPDA, as in the illustration, or simply to equalize upper frequency performance with mid-range performance. High-τ designs, with good low-frequency performance, may need only a director to compensate for high-frequency gain decrease. One potential challenge to adding a director to an LPDA is sustaining a high front-to-back ratio at the upper frequency range. (See this book's online material for more information on such log-Yagi arrays.)

Throughout the discussion of LPDAs, the performance curves of sample designs have been treated at all frequencies alike, seeking maximum performance across the entire design frequency span. Special compensations are also possible for ham-band-only LPDA designs. They include the insertion of parasitic elements within the array as well as outside the initial design boundaries. In addition, stubs may be employed not so much to eliminate weaknesses, but only to move them to frequencies outside the range of amateur interests.

10.1.5 Band-Optimized LPDA (BOLPA)

The traditional approach to designing LPDAs assumes that the antenna parameters for gain, pattern ratios, feed point impedance, and so on are maintained across the entire range of frequencies. For example, the usual amateur HF LPDA covers 14 to 30 MHz in one continuous range. This approach allows the antenna manufacturer to provide the same antenna to all services within that range — a very reasonable thing to do from a business standpoint. The amateur service, however, only makes use of a few bands within that range: 20, 17, 15, 12, and 10 meters. Relaxing the continuous-coverage requirement to only cover the amateur bands allows the antenna designer some flexibility. The Band-Optimized LPDA or BOLPA is best explained in terms of the process by which the design was created by Justin Johnson, GØKSC (**g0ksc.co.uk**).

The wide bandwidth of log-periodics was adapted to the design of single-band Yagi antennas by replacing the single driven element with a two- or three-element array called a "log-cell." (See the References and Bibliography entry for Cebik.) The log-cell acted as a single compound element with the active dipole moving back and forth within the log-cell

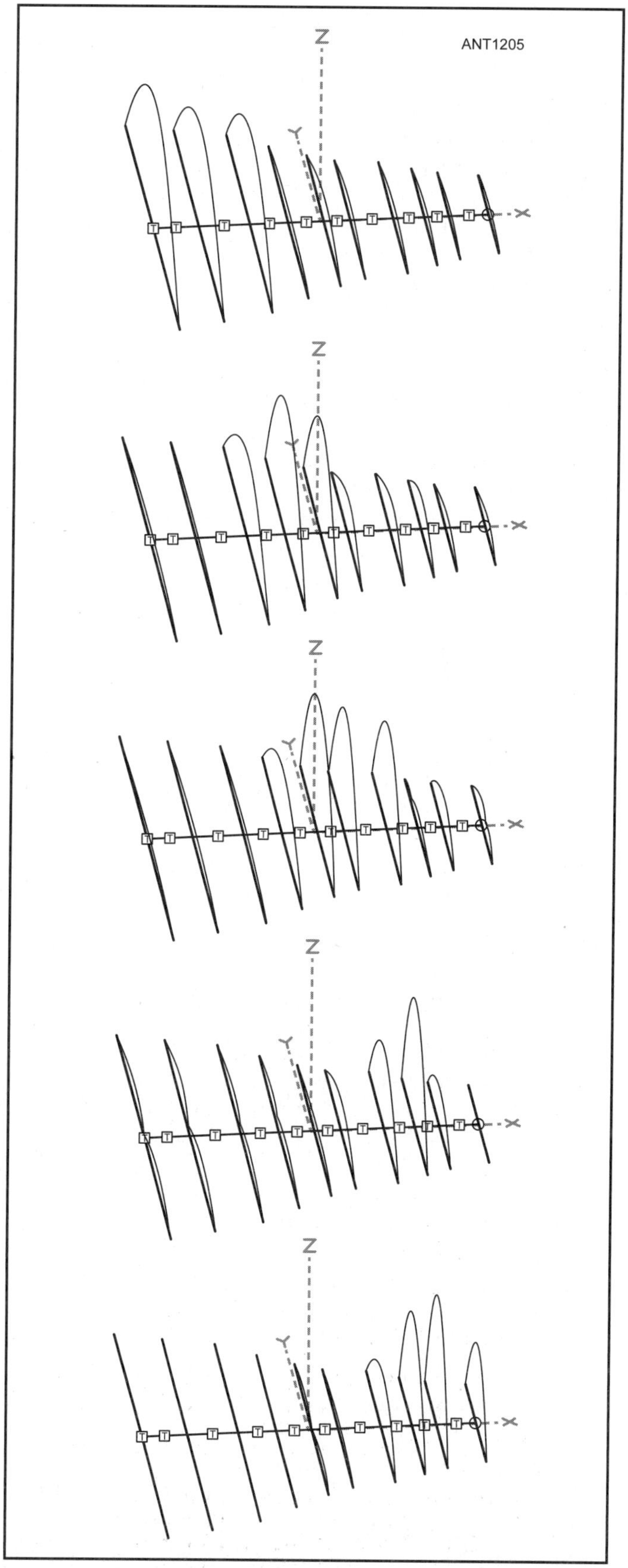

Figure 10.14 — Individual single-band LPDA cells are arranged along a single boom producing consistent performance across the amateur bands in one antenna. These *EZNEC Pro/4* models show current amplitudes on each element from 20 meters (A) through 10 meters (E).

from one end of the band to the other. This technique was mostly used on 20 meters to give better performance from 14.0 to 14.35 MHz. The movement of the active region back and forth along the axis of the antenna with frequency made it difficult to optimize the pattern ratios and gain across the band, since the reflector and director elements were fixed in place.

The log-cell Yagi was replaced by the Optimized Wideband Array (OWA) design, created by Jim Breakall, WA3FET. (See the **HF Beam Antennas** chapter.) Using the "open sleeve" method of very closely-spaced elements, the OWA Yagi moved the first director very close to the driven element, creating a compound driven element. This enabled a 50-Ω feed point impedance and stabilized gain and other pattern ratios across a wider bandwidth. The open-sleeve technique is discussed for dipoles in the **General Purpose MF and HF Antennas** chapter. It is a popular technique on the VHF and UHF bands, as well. (See the Bibliography entry for Cebik.)

The BOLPA revisits the log-cell concept by creating individual monoband LPDA cells along the antenna axis. Each cell is a short-boom three-element LPDA for a single amateur band. The cells are designed for 50-Ω direct feed, optimized across the band, and spaced for minimal interaction. Additional elements are inserted between the cells to provide coverage of intermediate bands. A 50-Ω feed point impedance also eliminates the LPDA's usual 4:1 impedance transformer.

The initial design consisted of monoband log cells for 20, 15, and 10 meters along a single common feed line that serves as the boom. Each cell has three elements and was optimized for performance. Then the spacing between the cells was optimized along the boom and the individual cells re-optimized. This produced good performance on the three contest bands but 17 and 12 meters were left out.

The next step was to add a single 17 meter element between the 20 meter cell and the 15 meter cell. (The 17 meter element is the fourth from the left with the highest current in **Figure 10.14A**.) With the 17 meter element present, the original three cells were shifted on the boom to maintain performance without adjusting any element lengths except that of the 17 meter element. The additional element acted as a fourth element on 15 meters with some improvement in gain. The process is continued to add 12 meters with another element between the 15 and 10 meter cells. Final boom length is 26 feet and free-space gain is approximately 7 dBi with front-to-back ratios of 20 dB or more.

The sequence of current models in Figures 10.14A-E shows how the active area of the antenna moves forward along the boom with increasing frequency—typical of LPDA antennas. Depending on the designer's requirements and resources a BOLPA covering 40-30-20 meters is possible or the BOLPA described here could be extended to 6 meters with another cell in front of the 10 meter cell. Start with a basic log cell and treat it as a fixed component as the additional bands are added. Automatic optimizer software (see the **Antenna Modeling** chapter) makes this job a lot easier.

10.2 Updated LPDA Design

The classical design procedure for the LPDA in the previous section relies mostly on a series of calculations involving the various user-selected scale factors (τ, σ, and α), the dipole element lengths, the spacings between elements, and the resulting boom length. One of the main problems with this method is that the boom length can become unbuildably long and the procedure must be repeated over and over until a reasonable boom length is obtained.

The radiation pattern performance is mediocre for most designs using the classical procedure, as well, particularly over real ground. The peak gain to peak sidelobe level ratio (front-to-back or front-to-rear ratios) for these antennas varies from poor to average (10 – 15 dB). Yagis with 3 or more elements that have been optimized for minimum backlobes can easily have better performance of greater than 20 dB.

For the purpose of improving the design procedure of log-periodic antennas, an updated procedure was presented in a paper by Jim Breakall, WA3FET, in 1996 (see the References and Bibliography entry for Breakall and Solis). The new procedure requires as inputs the frequency range of operation, the boom length, L, and either the scale factor, τ, or the number of dipole elements, N. The resulting design can then be simulated numerically in antenna modeling software to verify the gain, and peak gain to peak sidelobe level ratio. This procedure makes much more sense in that the constraints of boom length and the number of elements are selected at the outset for what a user can physically construct.

10.2.1 Updated LPDA Design Procedure

The updated procedure starts with the given boom length, L, the frequency range of operation, $f_L < f < f_H$, and either the scale factor, τ, or the number of dipole elements, N, as user input parameters.

1. First, pick f_L the lowest frequency of operation. Then, $f_1 = f_L$.
2. Pick f_H = the highest frequency of operation.
3. Pick $f_N = 1.5 \times f_H$ to insure a smooth transition at the upper cutoff frequency. This has been found experimentally to insure a good SWR at the highest frequency, f_H, by adding a few extra shorter elements at the feed point.
4. The user next picks the boom length, L.

5. The user then has a choice to either pick the scale factor τ (scaling factor between lengths and spacings of adjacent elements) or the total number of dipole elements, N.

If the user chooses τ, then the number of dipole elements will be given by the following equation:

$$N = \left\{ \frac{[\log(\frac{f_1}{f_N})]}{\log(\tau)} \right\} + 1 \qquad (6)$$

where N is the nearest integer of the value found.

If the user chooses N, then the τ factor is given by the following formula:

$$\tau = 10^{\left(\frac{\log(\frac{f_1}{f_N})}{N-1} \right)} \qquad (7)$$

The length of longest element $l_1 = \lambda_1 / 2$, where $\lambda_1 = c/f_1$. Thus, $l_1 = 492/f_1$ in ft when f_1 is in MHz.

The distance between the first and the second dipole elements is then given by: $d_{12} = L\left[(\tau - 1)/\left(\tau^{(N-1)} - 1\right)\right]$ (8)

One then calculates the second longest element length as $l_2 = \tau\, l_1$, and then $l_3 = \tau\, l_2$, etc. Similarly, the distance $d_{23} = \tau\, d_{12}$, $d_{34} = \tau\, d_{23}$, etc.

It has been found that the diameters of the dipole elements can be chosen and scaled with the τ factor or kept all the same diameter for all elements with little effect on the final SWR and performance. It has also been found experimentally that the characteristic impedance of the transmission line can be chosen and will result in an input impedance about half of the characteristic impedance that was picked.

Furthermore, it has also been found that one can pick either τ or N and construct a model and find the front-to-back (rear) and SWR performance and then choose another τ or N and construct another model and determine if the front-to-back (rear) and SWR is better or worse. By iterating on the value of τ or N, and continuing to model, one can then find an optimum design.

As an example, select the frequency range of 13 to 30 MHz. Then, f_1 = 13 MHz and f_N = 1.5 × 30 MHz = 45 MHz. The antenna is placed at a height of 55 ft over average ground (relative dielectric constant = 13 and conductivity = .005 S/m). The boom length, L, is chosen to be 48 ft. (It is useful to compare the results to those of the same design using the classical procedure in the online material.)

Many τ factors were tried for this design and an optimum τ was found to be equal to 0.93. This design is shown here. Then, the number of dipole elements for this design is:

$$N = \left\{ \frac{[\log(\frac{13}{45})]}{\log(0.93)} \right\} + 1 = 18 \text{ elements}$$

The lengths of the dipole elements are:

$l_1 = 492/13 = 37.8$ feet

$l_2 = \tau\, l_1 = 0.93\,(37.8) = 35.2$ feet

$L_N = \tau^{N-1}\, l_1 = 11.0$ feet

The spacings between dipole elements are:

$d_{12} = L[(\tau - 1)/(\tau^{(N-1)} - 1] = 4.74$ feet

$d_{23} = \tau\, d_{12} = 4.41$ feet

$d_{N-1,N} = \tau^{N-2}\, d_{12} = 1.48$ feet

This design is then modeled in EZNEC with a diameter of every dipole element equal to 1 inch. (The model file is included in the online material as LPDA1-Design Example as both EZNEC and NEC files.)

The SWR is shown for the log periodic design in free space first in **Figure 10.15**. (The reference impedance for calculating SWR is 100 Ω, shown as "SWR(100).") The gain, front-to-back ratio, and front-to-rear ratio are shown in **Figure 10.16** for the log periodic in free space. The free-space azimuth pattern at the mid-band frequency of 21.5 MHz is shown in **Figure 10.17**. (Azimuth patterns for frequencies from

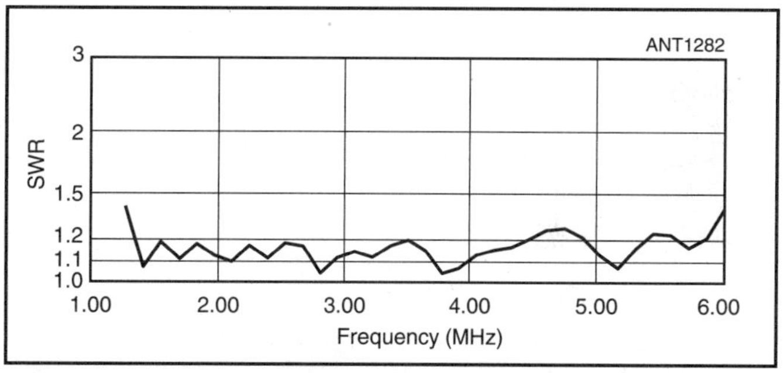

Figure 10.15 — SWR in free space. Performance at 55 feet above ground is nearly indentical.

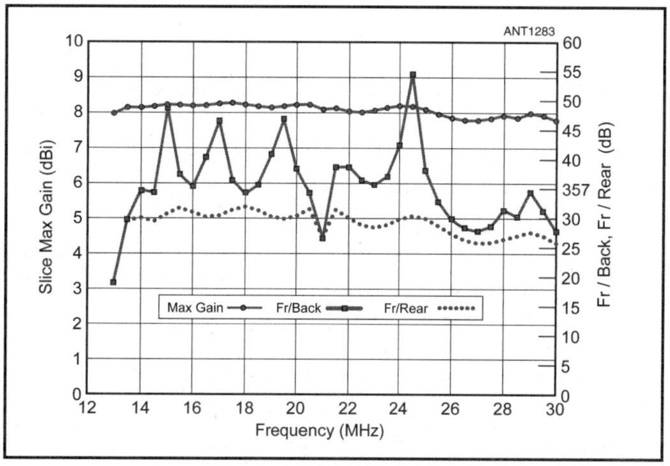

Figure 10.16 — Gain, front-to-back, and front-to-rear in free space.

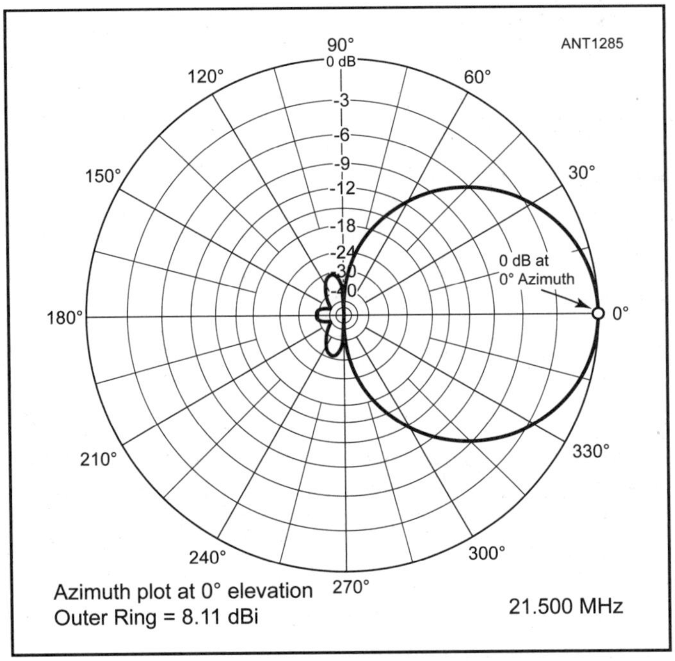

Figure 10.17 — Free-space azimuth pattern at 21.5 MHz.

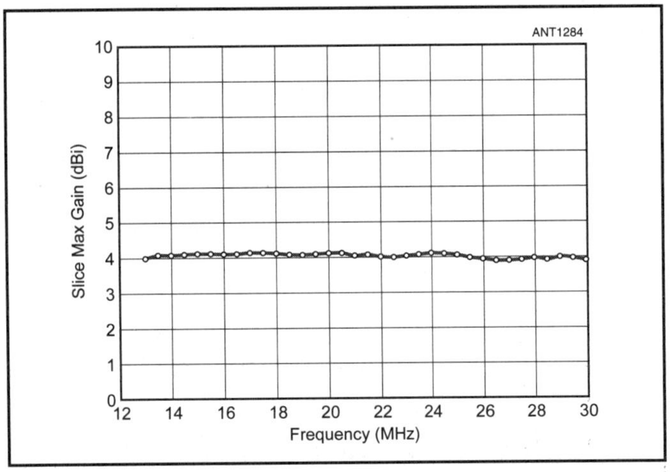

Figure 10.18 — Gain at peak elevation angle at 55 feet over average ground.

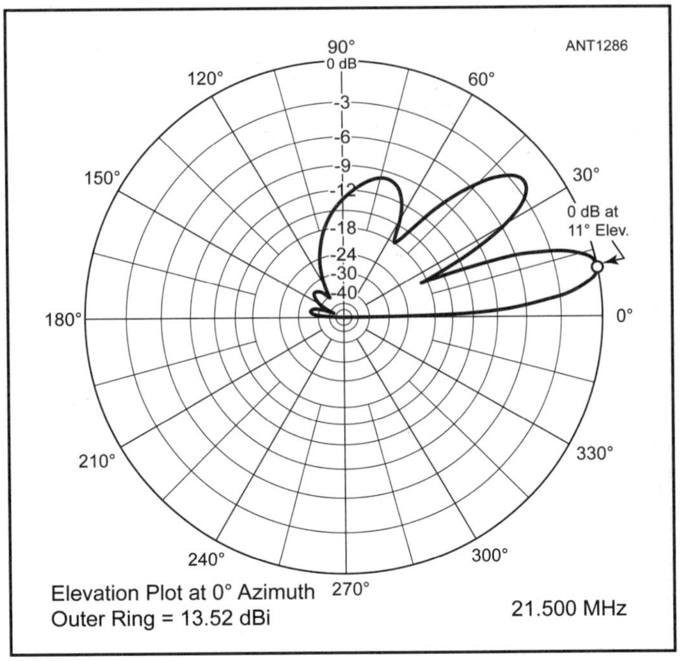

Figure 10.19 — Elevation pattern 55 feet over average ground at 21.5 MHz.

14 through 29 MHz are included in the online material as Figures 10.17A – E.)

The SWR is shown for the log periodic at a height of 55 ft in Figure 10.15B with almost the same SWR performance as in free space. The gain at the peak elevation angle at 55 ft over average ground is shown in **Figure 10.18**. Elevation patterns at the mid-band frequency of 21.5 MHz is shown in **Figure 10.19**. (Elevation patterns at 55 feet over average ground for frequencies from 14 through 29 MHz are included in the online material as Figure 10.19A – E.)

10.3 Discone Antenna Design

The name "discone" is a combination of the words, "disc" and "cone." Although people often describe a discone by its design-center frequency (for example, a "20-meter discone"), discones work very well over a wide frequency range, as much as several octaves. The discone is very popular as a low-power or receive-only antenna for VHF/UHF use, as well, with several commercial versions available along with numerous buildable designs. Its design is related to the biconical dipole.

The antenna produces a vertically polarized signal at a low-elevation angle, and it presents a good match for 50 Ω coax over its operating range. Two excellent articles by Belrose and Stearns (see References and Bibliography) describe the discone's operation and design. This chapter's online material information includes additional information on the discone and a pair of designs for HF use.

The dimensions of a discone, shown in **Figure 10.20**, are determined by the range of frequencies over which it is to be used. The cone angle, ϕ, is typically around 60 degrees and the height of the discone, L_V, is approximately $\lambda/4$ at the lowest frequency of use. The disc diameter, D, is approximately 0.7 of the cone's maximum width at the bottom. The 3 – 30 MHz discone photographed in **Figure 10.21** gives an

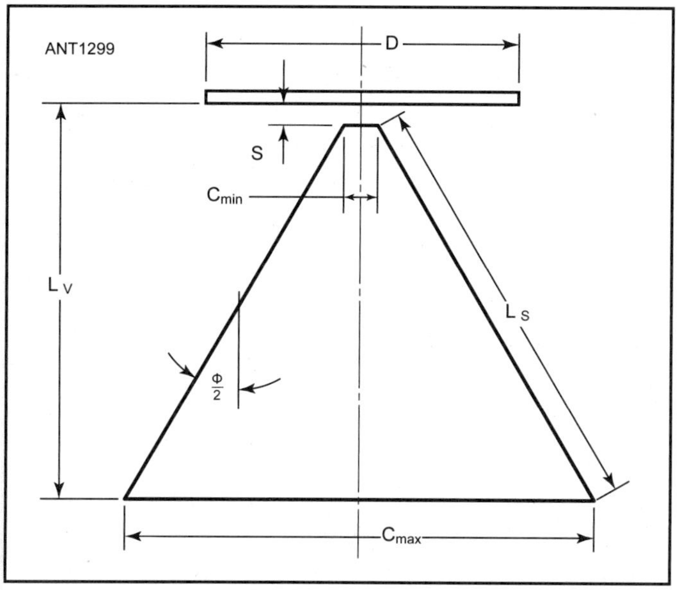

Figure 10.20 — Basic discone dimensions from Belrose's article in the online material. C_{min} and C_{max} are the minimum and maximum cone diameter, respectively.

Figure 10.21 — This 80-foot, 3-30 MHz discone is located at the Titan Missile Museum south of Tucson, Arizona.

idea of the size of a discone covering all of the HF bands. (See the article "Discone DXpedition" by WB5DYG in the February 2015 *QST* to learn about this antenna and how it is made available for visiting amateurs to use.) Smaller versions that cover the higher HF bands, such as 20 – 10 meters, are presented in the article "HF Discone Projects," in the online material.

Along with its wide frequency coverage, another advantage of the discone is that its maximum current region is near the top of the antenna, where radiation far above the ground reduces losses in the soil and from nearby obstructions. The cone-like skirt of the discone radiates the signal — radiation from the disc on top is minimal. This is because the currents flowing in the sloping skirt wires all have a common vertical component, while radiation from the symmetrical horizontal currents in the disc elements oppose each other and cancel out.

10.4 HF LPDA Projects

This section presents a pair of LPDA designs — a fixed array of wire dipoles for the 3.5 and 7 MHz bands and a rotatable array for the five amateur bands from 14 through 30 MHz. In addition, the *QST* article "Practical High Performance HF Log Periodic Antennas" by Bill Jones, K8CU is provided with this book's online material for additional design information.

10.4.1 LPDA for 3.5 or 7 MHz

These wire log-periodic dipole arrays for the lower HF bands are simple in design and easy to build. They are designed to have reasonable gain, be inexpensive and lightweight, and may be assembled with stock items found in large hardware stores. These antennas were first described by John J. Uhl, KV5E, in *QST* for August 1986. **Figure 10.22** shows one method of installation. You can use the information here as a guide and point of reference for building similar LPDAs.

If space is available, the antennas can be rotated or repositioned in azimuth after they are completed. A 75-foot tower and a clear turning radius of 120 feet around the base of the tower are needed. The task is simplified if you use only three anchor points, instead of the five shown in Figure 10.22. Omit the two anchor points on the forward element, and extend the two nylon strings used for element stays all the way to the forward stay line.

DESIGN OF THE LOG-PERIODIC DIPOLE ARRAYS

Design constants for the two arrays are listed in **Tables 10.1** and **10.2.** The classical design procedure for arriving at the dimensions and other parameters of these arrays was

Table 10.1
Design Parameters for the 3.5-MHz Single-Band LPDA

$f1 = 3.3$ MHz	Element lengths:
$f_n = 4.1$ MHz	$\ell 1 = 149.091$ feet
$B = 1.2424$	$\ell 2 = 125.982$ feet
$\tau = 0.845$	$\ell 3 = 106.455$ feet
$\sigma = 0.06$	$\ell 4 = 89.954$ feet
Gain = 5.9 dBi = 3.8 dBd	Element spacings:
$\cot \alpha = 1.5484$	$d_{12}=17.891$ feet
$B_{ar} = 1.3864$	$d_{23} = 15.118$ feet
$B_s = 1.7225$	$d_{34} = 12.775$ feet
L = 48.42 feet	Element diameters
N = 4.23 elements (decrease to 4)	All = 0.0641 inches
Z_t = 6-inch short jumper	ℓ/diameter ratios:
$R_0 = 208\ \Omega$	$\ell/diam_4 = 16840$
$Z_{AV} = 897.8\ \Omega$	$\ell/diam_3 = 19929$
$\sigma' = 0.06527$	$\ell/diam_2 = 23585$
$Z_0 = 319.8\ \Omega$	$\ell/diam_1 = 27911$

Antenna feeder: #12 AWG wire spaced 0.58 inches
Balun: 4:1
Feed line: 52-Ω coax

Table 10.2
Design Parameters for the 7-MHz Single-Band LPDA

$f1 = 6.9$ MHz	Element lengths:
$f_n = 7.5$ MHz	$\ell 1 = 71.304$ feet
$B = 1.0870$	$\ell 2 = 60.252$ feet
$\tau = 0.845$	$\ell 3 = 50.913$ feet
$\sigma = 0.06$	$\ell 4 = 43.022$ feet
Gain = 5.9 dBi = 3.8 dBd	Element spacings:
$\cot \alpha = 1.5484$	$d_{12} = 8.557$ feet
$B_{ar} = 1.3864$	$d_{23} = 7.230$ feet
$B_s = 1.5070$	$d_{34} = 6.110$ feet
L = 18.57 feet	Element diameters:
N = 3.44 elements (increase to 4)	All = 0.0641 inches
Z_t = 6-inch short jumper	ℓ/diameter ratios:
$R_0 = 208\ \Omega$	$\ell 4/diam_4 = 8054$
$Z_{AV} = 809.3\ \Omega$	$\ell 3/diam_3 = 9531$
$\sigma' = 0.06527$	$\ell 2/diam_2 = 11280$
$Z_0 = 334.2\ \Omega$	$\ell 1/diam_1 = 13349$

Antenna feeder: #12 AWG wire spaced 0.66 inches
Balun: 4:1
Feed line: 52-Ω coax

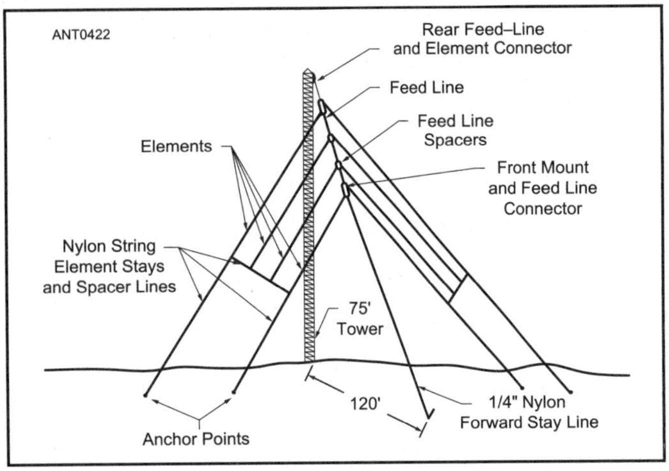

Figure 10.22 — Typical lower-HF wire 4-element log periodic dipole aray erected on a tower.

presented earlier in this chapter. The primary differences between these designs and one-octave upper HF arrays are the narrower frequency ranges and the use of wire, rather than tubing, for the elements. As design examples for the LPDA, you may wish to work through the step-by-step procedure and check your results against the values in Tables 10.1 and 10.2. You may also wish to compare these results with the output of an LPDA design software package such as *LPCAD*.

From the design procedure, the feeder wire spacings for the two arrays are slightly different, 0.58 inch for the 3.5-MHz array and 0.66 inch for the 7-MHz version. As a compromise toward the use of common spacers for both bands, a spacing of ⅝ inch is quite satisfactory. Surprisingly, the feeder spacing is not at all critical here from a matching standpoint, as may be verified from the equations presented earlier in this chapter. Increasing the spacing to as much as ¾ inch results in an R_0 SWR of less than 1.1:1 on both bands.

CONSTRUCTING THE ARRAYS

Construction techniques are the same for both the 3.5 and the 7-MHz versions of the array. Once the designs are completed, the next step is to fabricate the fittings; see **Figure 10.23** for details. Cut the wire elements and feed lines to the proper sizes and mark them for identification. After the wires are cut and placed aside, it will be difficult to remember which is which unless they are marked. When you have finished fabricating the connectors and cutting all of the wires, the antenna can be assembled. Use your ingenuity when building one of these antennas; it isn't necessary to duplicate these LPDAs precisely.

The elements are made of standard #14 AWG stranded copper wire. The two parallel-wire feed lines are made of #12 AWG solid copper-clad steel wire, such as Copperweld. Copperweld will not stretch when placed under tension. The front and rear connectors are cut from ½-inch thick polycarbonate sheeting, and the feed line spacers from ¼-inch acrylic sheeting.

Study the drawings carefully and be familiar with the way the wire elements are connected to the two feed lines, through the front, rear and spacer connectors. Details are sketched in **Figures 10.24** and **10.25**. Connections made in the way shown in the drawings prevent the wire from breaking. All of the rope, string, and connectors must be made of materials that can withstand the effects of tension and weathering. Use nylon rope and strings, the type that yachtsmen use. Figure 10.22 shows the front stay rope coming down to ground level at a point 120 feet from the base of a 75-foot tower. Space may not be available for this arrangement in all cases. An alternative installation technique is to put a pulley 40 feet up in a tree and run the front stay rope through the pulley and down to ground level at the base of the tree. The front stay rope will have to be tightened with a block and tackle at ground level.

Putting an LPDA together is not difficult if it is assembled

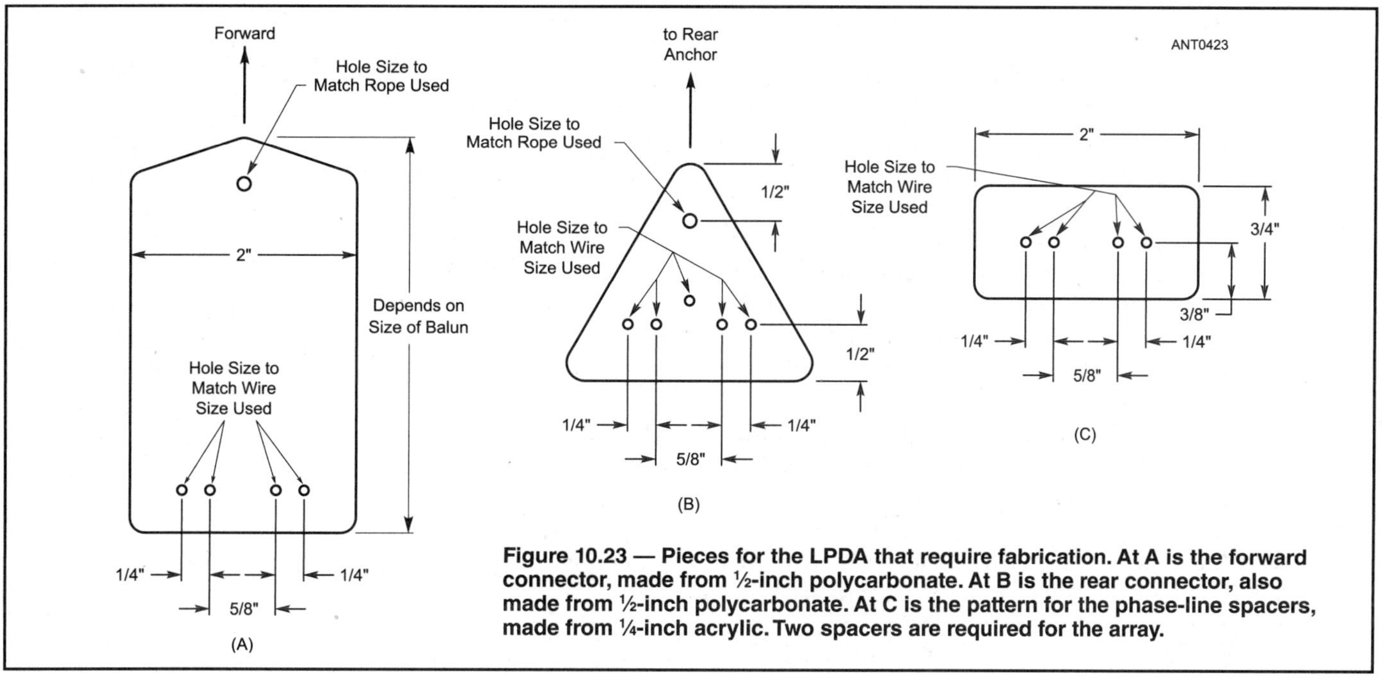

Figure 10.23 — Pieces for the LPDA that require fabrication. At A is the forward connector, made from ½-inch polycarbonate. At B is the rear connector, also made from ½-inch polycarbonate. At C is the pattern for the phase-line spacers, made from ¼-inch acrylic. Two spacers are required for the array.

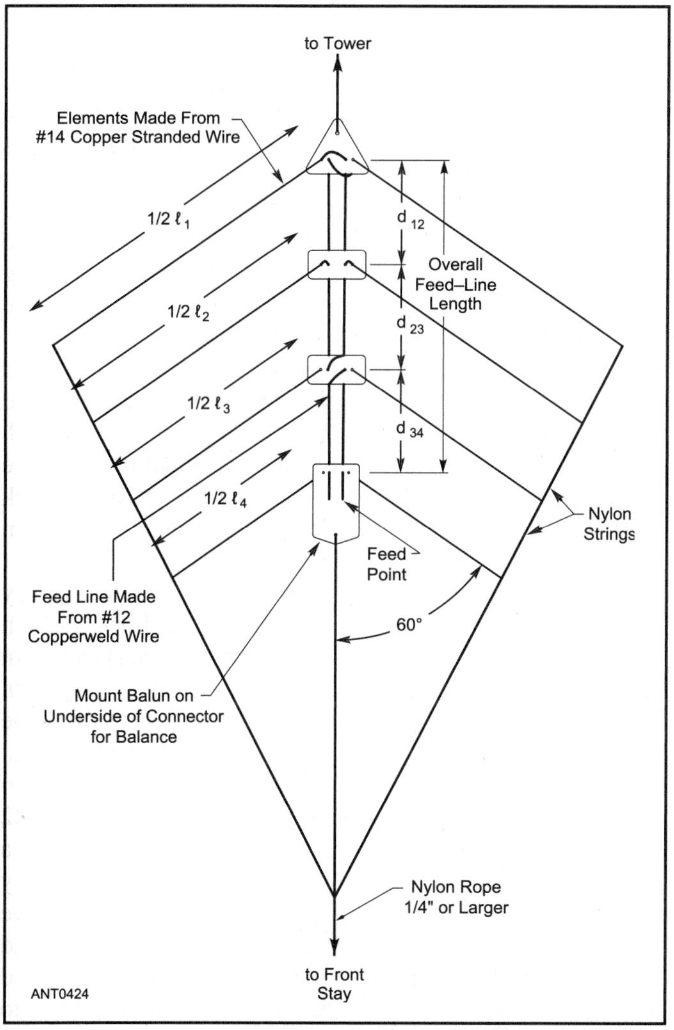

Figure 10.24 — The generic layout for the lower HF wire LPDA. Use a 4:1 balun on the forward connector. See Tables 10.1 and 10.2 for dimensions.

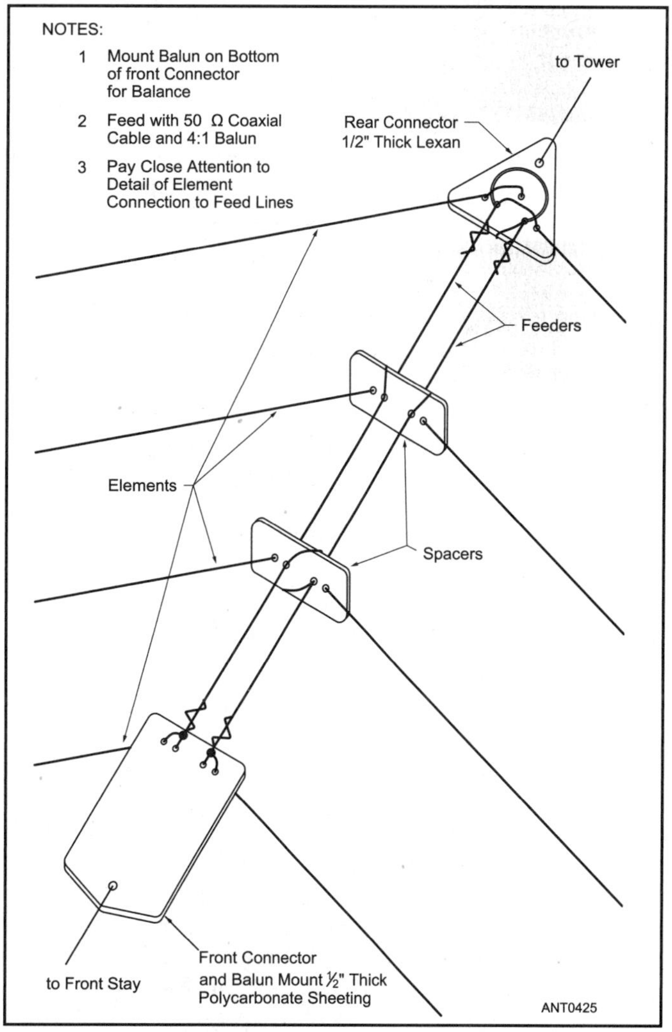

Figure 10.25 — Details of the electrical and mechanical connections of the elements to the phase-line. Knots in the nylon rope stay line are not shown.

in an orderly manner. It is easier to connect the elements to the feeder lines when the feed line assembly is stretched between two points. Use the tower and a block and tackle. Attaching the rear connector to the tower and assembling the LPDA at the base of the tower makes raising the antenna into place a much simpler task. Tie the rear connector securely to the base of the tower and attach the two feeder lines to it. Then thread the two feed line spacers onto the feed line. The spacers will be loose at this time, but will be positioned properly when the elements are connected. Now connect the front connector to the feed lines. A word of caution: Measure accurately and carefully! Double-check all measurements before you make permanent connections.

Connect the elements to the feeder lines through their respective plastic connectors, beginning with element 1, then element 2, and so on. Keep all of the element wires securely coiled. If they unravel, you will have a tangled mess of kinked wire. Recheck the element-to-feeder connections to ensure proper and secure junctions (see Figures 10.17 and 10.18). Once you have completed all of the element connections, attach the 4:1 balun to the underside of the front connector. Connect the feeder lines and the coaxial cable to the balun.

You will need a separate piece of rope and a pulley to raise the completed LPDA into position. First secure the eight element ends with nylon string, referring to Figures 10.22 and 10.24. The string must be long enough to reach the tie-down points. Connect the front stay rope to the front connector, and the completed LPDA is now ready to be raised into position. While raising the antenna, uncoil the element wires to prevent their getting away and tangling up into a mess. Use care! Raise the rear connector to the proper height and attach it securely to the tower, then pull the front stay rope tight and secure it. Move the elements so they form a 60° angle with the feed lines, in the direction of the front, and space them properly relative to one another. By adjusting the end positions of the elements as you walk back and forth, you will be able to align all the elements properly. Now it is time to hook your radio to the system and make some contacts.

PERFORMANCE

The reports received from these LPDAs were compared with an inverted-V dipole. All of the antennas are fixed; the LPDAs radiate to the northeast, and the dipole to the northeast and southwest. The apex of the dipole is at 70 feet, and the 40- and 80-meter LPDAs are at 60 and 50 feet, respectively. Basic array gain was apparent from many of the reports received. During pileups, it was possible to break in with a few tries on the LPDAs, yet it was impossible to break the same pileups using the dipole. The gain of the LPDAs is several dB over the dipole. For additional gain, experimenters may wish to try a parasitic director about ⅛ λ ahead of the array. Director length and spacing from the forward LPDA element should be field-adjusted for maximum performance while maintaining the impedance match across each of the bands.

Wire LPDA systems offer many possibilities. They are easy to design and to construct: real advantages in countries where commercially built antennas and parts are not available at reasonable cost. The wire needed can be obtained in all parts of the world, and cost of construction is low. If damaged, the LPDAs can be repaired easily with pliers and solder. For those who travel on DXpeditions where space and weight are large considerations, LPDAs are lightweight but sturdy, and they perform well.

10.4.2 LPDA for 13 to 30 MHz

A rotatable log periodic array designed to cover the frequency range from 13 to 30 MHz is pictured in **Figure 10.26**. This is a large array having a free-space gain that varies from 6.6 to over 6.9 dBi, depending upon the operating portion of the design spectrum. This antenna system was originally described by Peter D. Rhodes, K4EWG, in the November 1973 issue of *QST*.

The characteristics of this array are:

1) Half-power beamwidth, 43° (14 MHz)
2) Design parameter $\tau = 0.9$
3) Relative element spacing constant $\sigma = 0.05$
4) Boom length, L = 26 feet
5) Longest element $\lambda 1$ = 37 feet 10 inches.
6) Total weight, 116 pounds
7) Wind-load area, 10.7 square feet
8) Required input impedance (mean resistance), $R_0 = 72\ \Omega$, Z_t = 6-inch jumper #18 AWG wire
9) Average characteristic dipole impedance, Z_{AV}: 337.8 Ω
10) Impedance of the feeder, Z_0: 117.1 Ω
11) Feeder: #12 AWG wire, close spaced
12) With a 1:1 toroid balun at the input terminals and a 72-Ω coax feed line, the maximum SWR is 1.4:1.

The mechanical assembly uses materials readily available from most local hardware stores or aluminum supply houses. A complete set of tables and assembly drawings are included in the original article included with this book's online material.

Experimenters may wish to improve the performance of the array at both the upper and lower frequency ends of the

Figure 10.26 — The 13 – 30 MHz log periodic dipole array.

design spectrum so that it more closely approaches the performance in the middle of the design frequency range. The most apt general technique for raising both the gain and the front-to-back ratio at the frequency extremes would be to circularize τ as described earlier in this chapter. However, other techniques may also be applied.

10.4.3 Wideband 10 Meter LPDA

A new design by Justin Johnson, GØKSC, of Innovantennas (**innovantennas.com/en**) presents an easy-to-build, 2-element, 10 meter LPDA (BOLPA-2 – Band Optimised Log Periodic Array) that covers the entire 10-meter band. Despite having only two elements, the antenna has a very consistent pattern and low SWR. **Figure 10.27** shows the antenna's free-space radiation pattern at 28.5 MHz with 6.15 dBi gain (4 dBd) and 21 dB F/B. The SWR sweep in **Figure 10.28** shows 1.5:1 or less from 28 to 30 MHz. Additional radiation patterns at 28.0, 29.0, 29.5, and 30 MHz are provided in the online material.

How is this a Log Periodic with only 2 elements? It follows the concept of the log-cell with both elements driven (electrically connected to the feed line) and out of phase with one another. The commercially available BOLPA-3 and BOLPA-4 follow on from this with 3 and 4 out-of-phase driven elements.

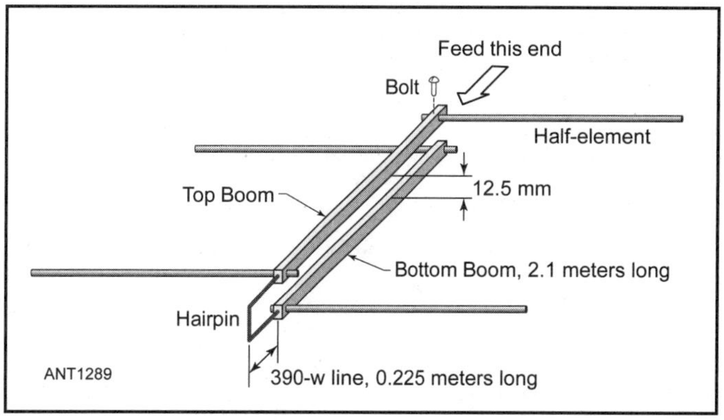

Figure 10.29 — Basic assembly for the 2-element 10-meter BOPLA-2 LPDA antenna designed by GØKSC.

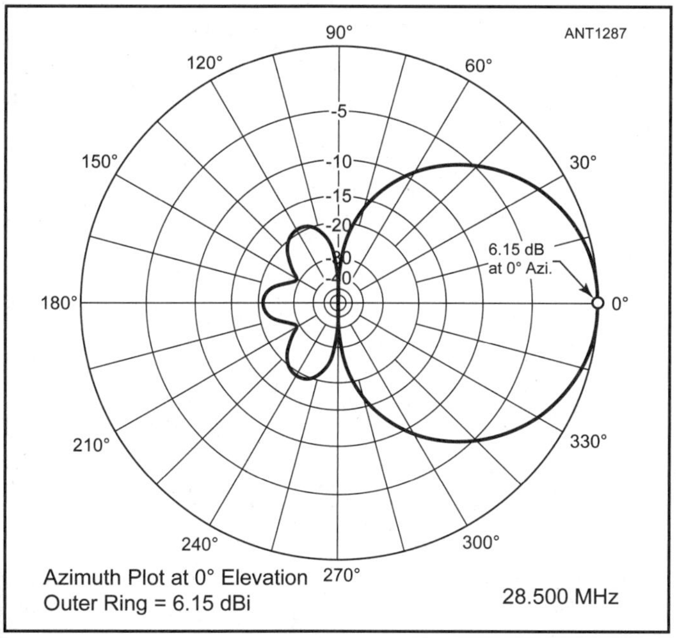

Figure 10.27 — 2-element 10 Meter LPDA free-space radiation pattern at 28.5 MHz.

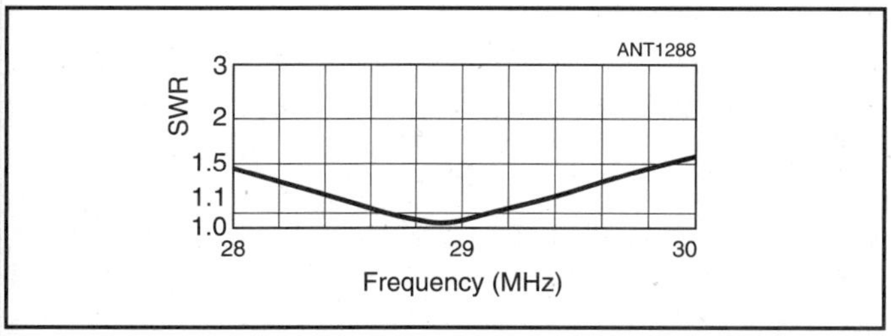

Figure 10.28 — SWR sweep from 28 to 30 MHz of the antenna in Figure 10.27.

The simplicity of the design makes this an attractive project for anyone with a basic workshop and materials. As with any log periodic, a parallel tuned feed line between the antennas is used to invert the phase of each element in the array. In this instance, the twin booms form the tuned (balanced) feed line, the impedance of which is controlled by the spacing between them. The phase switching is accomplished by the element halves alternating between the top and the bottom booms as shown in the assembly sketch, **Figure 10.29**. The booms are constructed from 1-inch square aluminum, 2.1 meters long, with a spacing between boom faces of 12.5 mm (½ inch).

Each element in an LPDA is fed in the center with the alternate elements out-of-phase. In this example, the first element half on the left is connected at its end to the top square boom by being fitted into and through the boom, secured by a single stainless-steel bolt. The right side of this element must then exit the bottom boom in the opposite direction to the other half of the dipole. The second element at the opposite end of the boom has its element halves connected to the opposite booms from the first element.

HAIRPIN MATCH

While it is not always used, a hairpin match (a single-turn inductor between the top and bottom booms) on this antenna enhances front-to-back ratio and forward gain. As boom impedance impacts performance and feed point impedance, the associated impedance of the hairpin is important and will impact performance and feed point impedance.

While correct boom separation ensures the feed-line impedance to each element is correct, with the hairpin separation is a bit more variable, which allows for adjustment to optimize the antenna. Should the antenna be used for portable or Field Day operating, then a piece of wire "looped" between the two booms at the rear and trimmed for lowest SWR will be fine. If a more permanent installation is required, use something

Figure 10.30 — Boom-to-mast mount (see text). Four 48.5 mm Stauff PA (polycarbonate) clamps hold a plate of clear UV-resistant Perspex (Plexiglass) to the round 2-inch mast with the square booms mounted on the two center clamps. A large stainless steel washer is used under all bolts as a "load spreader."

Table 10.3
10 Meter 2-element LPDA Tubing Sizes

EZNEC diameter (mm)	*Tubing size (inches)*
15.88	0.625 (5⁄8)
12.7	0.500 (1⁄2)
9.525	0.375 (3⁄8)

more substantial, such as flat, bar aluminum pieces formed into a U. Approximate impedance of the hairpin should be 390 Ω.

BOOM TO MAST PLATE

The boom-to-mast arrangement is a little more complex than it would be with a Yagi. Because both booms are "hot" and form a part of the antenna, feed lines between elements, some form of insulator needs to be placed as shown in **Figure 10.30**. The mast is shown sideways in the photograph, so the clamps can be seen more clearly. The booms are actually mounted one above the other and not side-by-side.

A Stauff clamp bolted through each boom and then directly to the mast works well. 48.5 mm clamps were a little small for a 2-inch mast (50.8 mm) but deformed around the mast as tightened and ensured a secure fit. It may be that this option would loosen after repeated installations such as for Parks On The Air or other temporary stations. However, for a permanent installation this method has proved satisfactory.

ANTENNA MODELING AND TUBING SIZES

The antenna's *EZNEC* design file is also included in the online material. The elements are specified with metric (mm) measurements, but the diameters are chosen to match the inch-based telescoping tubing sizes available in North America. (See the **Antenna Materials and Construction** chapter.) Use 0.049-inch or 0.058-inch wall thickness tubing as specified in **Table 10.3**. Lengths are given in the model's Wires table in millimeters (0.001-meter) increments. This represents an excellent compromise between accuracy of the model and specifying available materials for North American builders.

10.5 VHF/UHF LPDA and Discone Projects

Operation across the VHF and UHF bands with a single antenna makes the log-periodic and discone antennas a popular choice for the amateur with limited antenna options.

10.5.1 VHF/UHF LPDA Projects

The Tennadyne T-28 shown in **Figure 10.31** is an example of an LPDA that covers 50 to 1300 MHz with a boom length of only 12 feet. In addition, the antenna looks very much like a TV-receive antenna, attracting much less attention than a stack of monoband Yagis for the same frequency range!

Two VHF log-periodic designs are presented by reprints of *QST* articles included in the online material. The first is a 4-element, single-band design covering 2 meters and adjacent

Yes, That's a TV Antenna!

If you have noticed the similarities between TV-receive antennas and the log-periodic antennas that cover the amateur VHF and lower UHF bands, you are not alone! In an article included in the online material, John Stanley, K4ERO, shows how to modify a mid-sized log-periodic originally designed for receiving TV broadcasts into a stealthy, yet effective ham antenna covering 50 through 222 MHz. For hams limited to TV-antennas only, this might be a good solution to getting on at least a few of the ham bands and you can answer truthfully when asked about your new "TV antenna"!

Figure 10.31—The Tennadyne T-28 covers 50 – 1300 MHz with a boom length of 12 feet.

Table 10.4
Element Dimensions (see Figure 10.31)

	Element Spacing (inches)	
D_{12}	1 to 2	$4\tfrac{3}{8}$
D_{23}	2 to 3	$4\tfrac{1}{32}$
D_{34}	3 to 4	$3\tfrac{11}{16}$
	Element Lengths (inches)	
L11, L12	$20\tfrac{7}{32}$	
L21, L22	$19\tfrac{1}{32}$	
L31, L32	$17\tfrac{1}{32}$	
L41, L42	$16\tfrac{1}{8}$	

L11, L12, L31, and L32 are mounted on the outside of the rails. An alternate set of lengths is provided in the complete article.

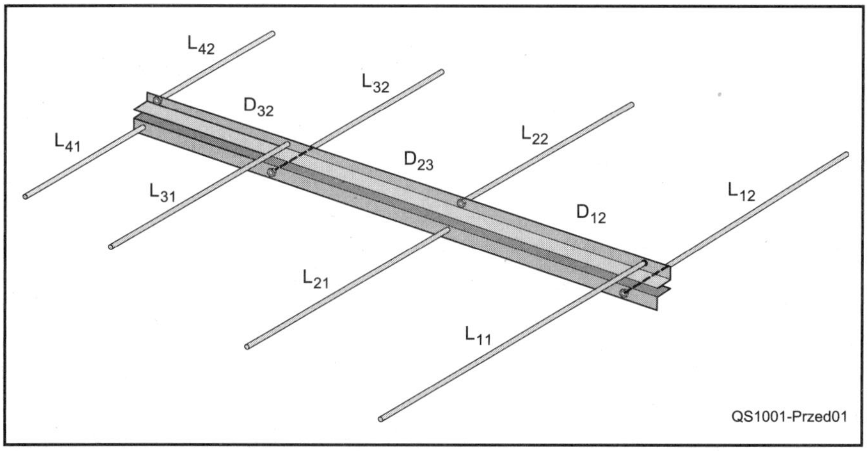

Figure 10.32 — Layout of the 2-meter LPDA as seen from the top. Alternate elements are attached to parallel sub booms that are insulated from each other. Dimensions for each designated element are given in Table 10.4.

frequencies, "An Updated 2 Meter Log Periodic Dipole Array," by Andy Przedpelski, KØABP. The antenna, described in **Figure 10.32** and **Table 10.4** covers 142 – 150 MHz and makes a great upgrade from a J-pole or other vertical antenna optimized for repeater coverage. The antenna described in "An LPDA for 2 Meter Plus," by L.B. Cebik, W4RNL, covers 130–170 MHz and is also included in the online material.

The second design is a three-band log-periodic covering the 144, 222, and 432 MHz bands, "A Three-Band Log-Periodic Antenna," by Robert Heslin, K7RTY, from the June 1963 *QST*. This antenna is shown in **Figure 10.33**. The design is based on the same principles used today and the antenna is similar to commercial models covering the same frequency range.

For higher frequencies, it is practical to etch antennas directly on printed circuit board material. **Figure 10.34** shows three LPDA examples designed by Kent Brittain, WA5VJB, whose "Cheap Yagi" designs appear in the **VHF, UHF, and Microwave Antennas** chapter. Coaxial feed line attaches to an SMA connector at the front of the antenna with the cable run along the "boom" in the center of the antenna. **Figure 10.35** shows how Ken Wyatt, KX7KW, uses the antenna for table-top EMC troubleshooting as described at **edn.com/pc-board-log-periodic-antennas.** An article by WA5VJB about designing these antennas and the effects of PCB material is available at **wa5vjb.com/references/DesigningPCB-LPs.pdf.** A wide-band Vivaldi antenna fabricated on PCB material is shown in the **VHF, UHF, and Microwave Antennas** chapter, as well.

10.5.2 VHF/UHF Discone Project

The wideband discone antenna is a very popular omnidirectional antenna for scanner use at VHF and above with numerous commercial models available. Discones for VHF and UHF coverage are simple to build, such as the design shown in **Figure 10.36** from the May 2003 *QST* article, "A VHF/UHF Discone Antenna," by Bob Patterson, K5DZE. The antenna is constructed from wire-mesh hardware cloth, but sheet metal or heavy screen can be used. As the author mentions in the article, even aluminum foil on cardboard worked fine as an indoor receive antenna!

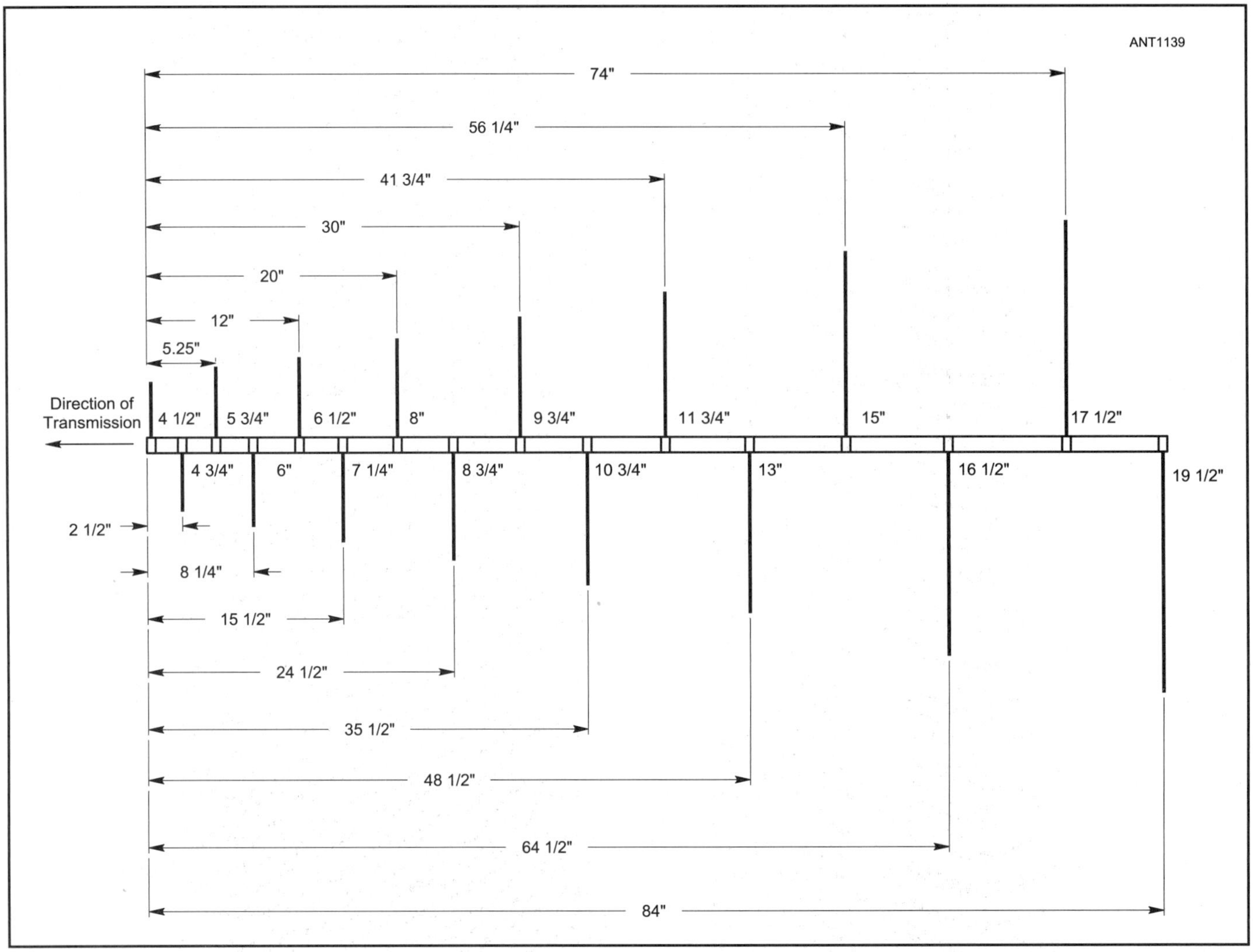

Figure 10.33 — Outline and sketch of the dimensions for the log-periodic antenna covering the 144, 222, and 432 MHz Bands.

Figure 10.34 — These three PCB log-periodic antennas cover 400 MHz to 11 GHz in three bands: 400-1000 MHz, 850-6500 MHz, and 2-11 GHz. Each antenna has a gain of approximately 6 dBi. A 6-inch rule at the bottom of the photo provides scale. These antennas are designed by Kent Brittain, WA5VJB, and sold by Kent Electronics (wa5vjb.com). (Photo courtesy Ken Wyatt, KX7KW)

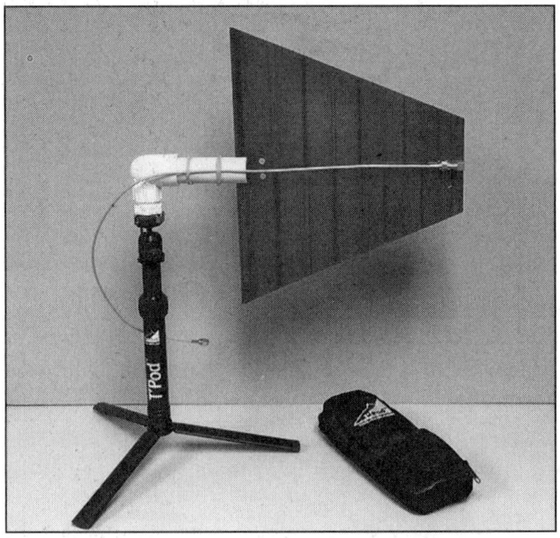

Figure 10.35 — Mounted on a tripod using PVC plumbing adapters, this small LPDA is used for table-top EMC troubleshooting. Note the placement of the coaxial cable and that it secured to stay in place along the center of the antenna. (Photo courtesy Ken Wyatt, KX7KW)

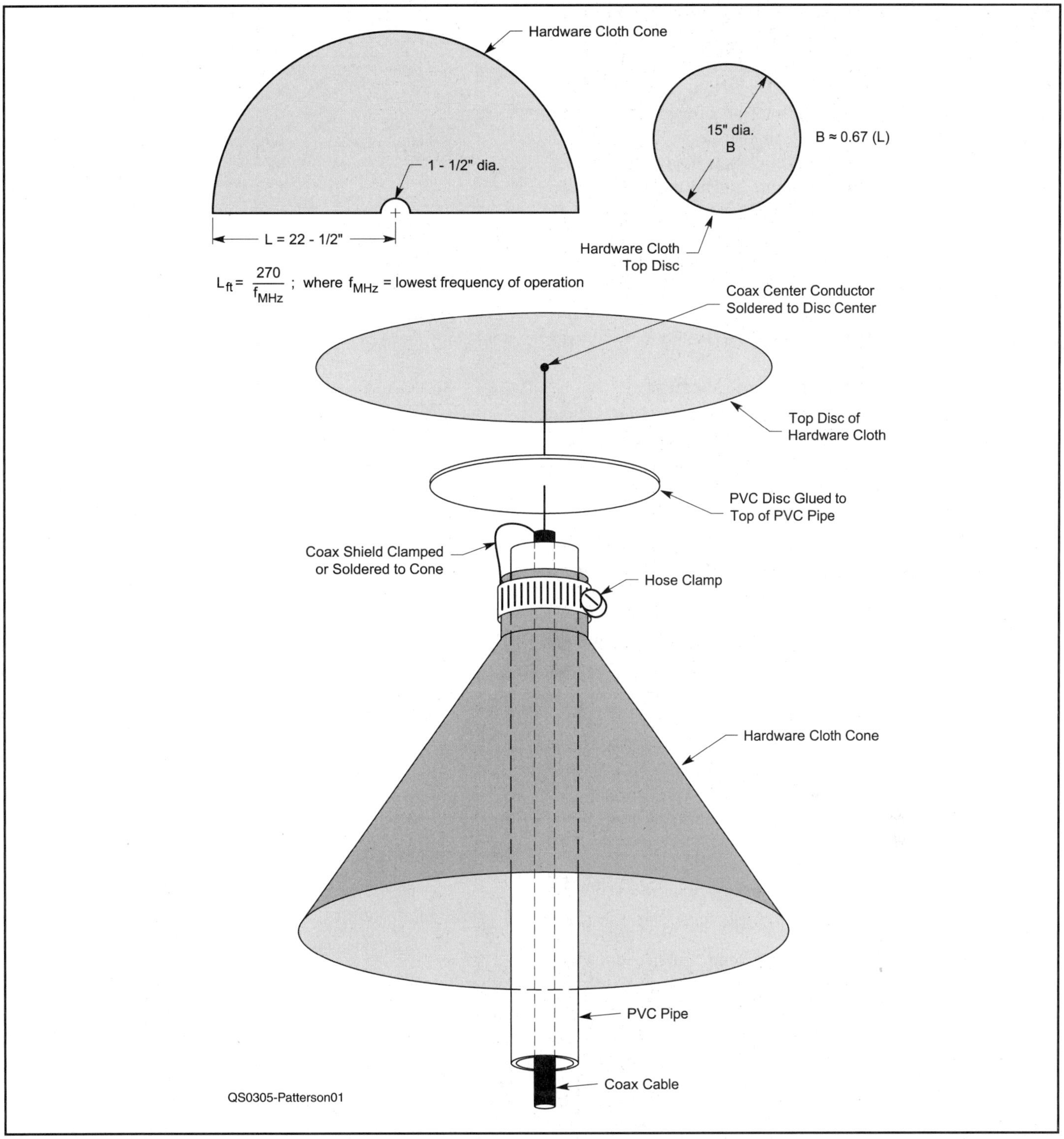

Figure 10.36 — Construction details for a VHF/UHF discone antenna. The largest dimension of the discone is determined by the lowest frequency of use.

10.6 Miscellaneous Frequency Independent Antennas

As mentioned at the beginning of this chapter, there are an infinite number of patterns, both 2D and 3D, that are defined completely in terms of angles. If a means of feeding power to them can be devised, it is likely that someone has made an antenna in that pattern! This section touches on several types of designs that are sometimes encountered by amateurs. These are generally used at and above VHF for EMC testing, satellite and terrestrial microwave communication, various measurement activities, and so forth. Once you become interested in this type of antenna, there is a lot to learn! (Material in this section is based on *Antennas, 2nd edition* by Kraus from the chapter on Broadband and Frequency-Independent Antennas.)

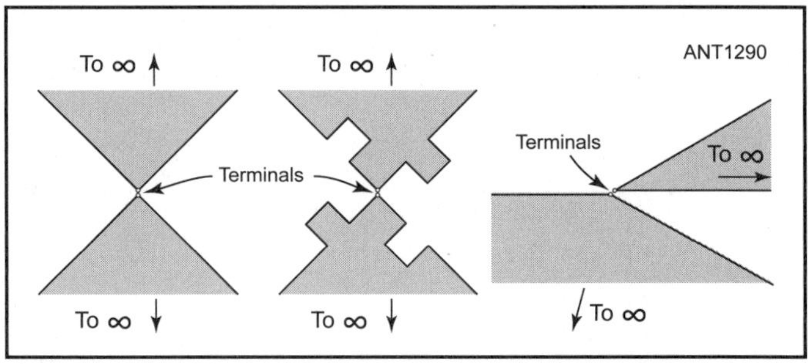

Figure 10.37 — Three self-complementary planar antennas. All of the figures extend to infinity as indicated. They all have a feed point impedance of 188 Ω. (after Kraus, 2nd edition, Figure 15-3)

10.6.1 Self-Complementary Antennas

Prior to the development of true frequency independent antennas, it was observed by Mushiake and Uda in 1949 that self-complementary planar antennas have a constant impedance of ½ the impedance of free space or 188 Ω, no matter what the exact shape is. A self-complementary planar antenna has a conductive (metal) area and an open area that are congruent (geometrically identical) and that can be made to overlay each other exactly by sliding or rotating the shapes without changing them. Three examples are shown in **Figure 10.37**. A casual online search on the subject will turn up hundreds of example antennas.

10.6.2 Logarithmic Spirals

A *logarithmic spiral* or *log spiral* as shown in **Figure 10.38** is defined by the equation:

$$r = a^{\theta} \ or \ \ln(r) = \theta \ln a$$

where

r = distance from the center to a point, P, on the spiral
θ = angle with respect to the x axis
a = a constant
ln = natural logarithm, sometimes abbreviated $\log_e$

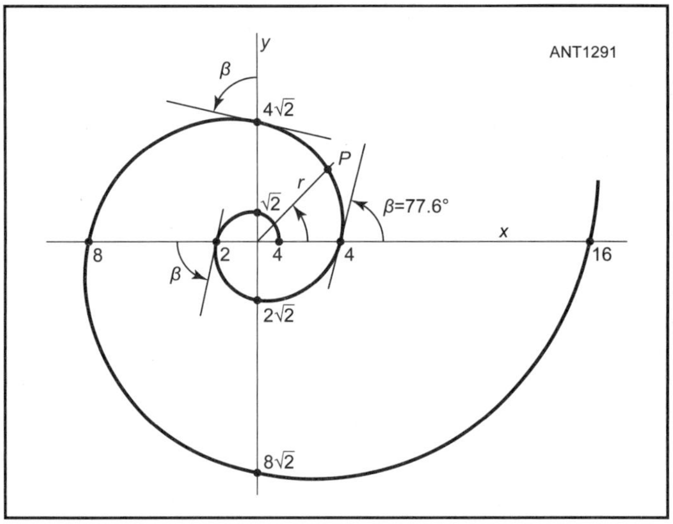

Figure 10.38 — A logarithmic or log spiral. (after Kraus, 2nd edition, Figure 15-4)

The angle between the spiral and the line from the center to P is also determined by quantity ln a.

If a second, third, and fourth identical log spirals are created but rotated to fit exactly between each other, an antenna can be created by filling the space between alternating pairs of spirals with metal. This creates two spiral conductors wrapped around each other as in **Figure 10.39**. The RF energy travels along the spiral (in this case counter-clockwise), creating radiation that is right-circularly polarized (RCP) out of the page and left-circularly polarized behind the page.

The high-frequency limit of operation is determined by the spacing of the spirals at the feed point terminals and the low-frequency limit by the overall diameter. A typical ratio of high-to-low frequencies might be as much as 20 or 25:1. As shown in the figure, it is common practice to cut spiral into a metal ground plane or etch the pattern into PCB material. Coaxial feed line is laid on and bonded to one arm and the

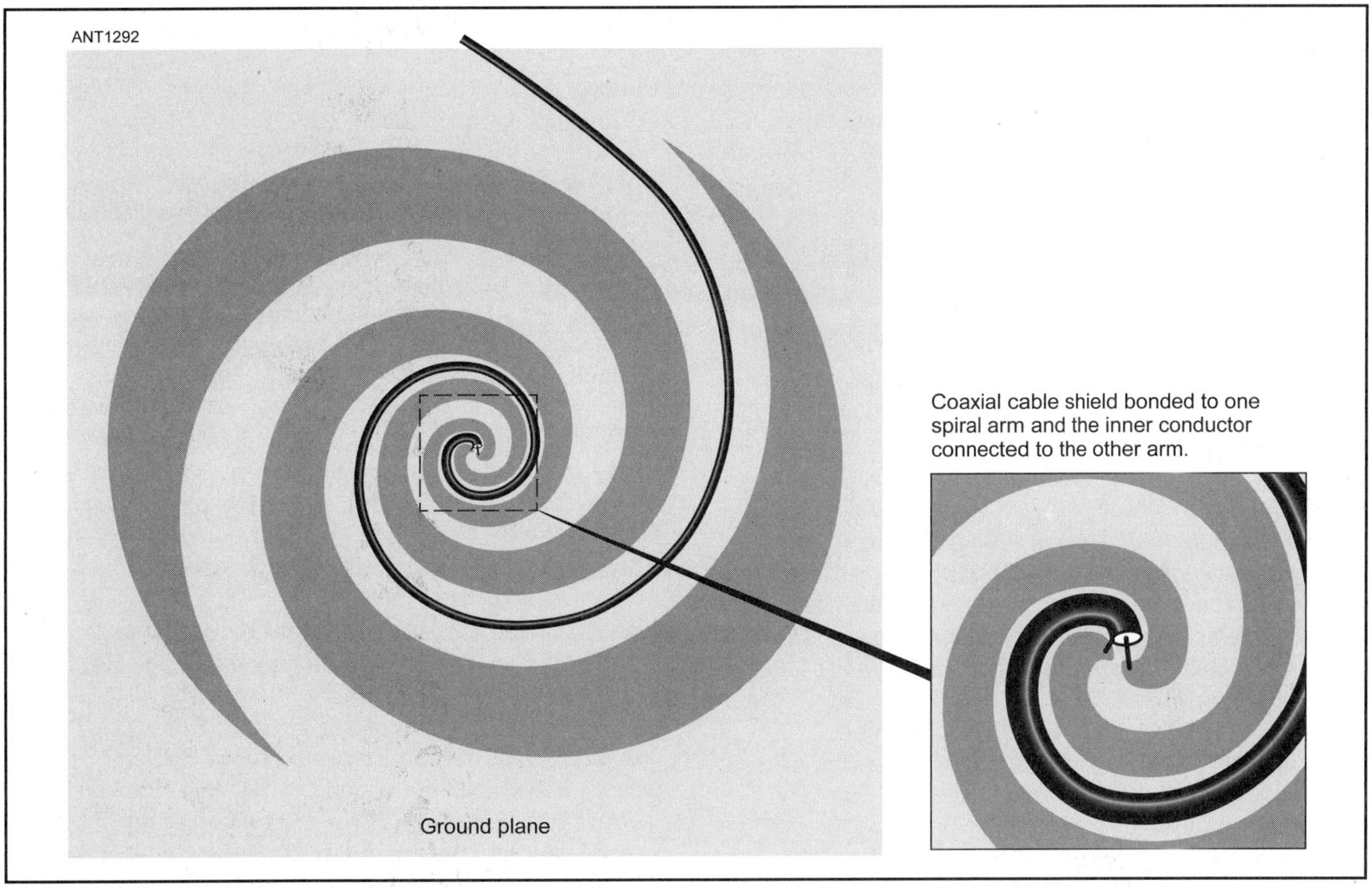

Figure 10.39 — A frequency independent planar spiral antenna cut from a large ground plane or etched on PCB material. The antenna generates circularly polarized radiation broadside to the page — see text. (after Kraus, 2nd edition, Figure 15-6)

center conductor attached to the other arm. Practical antennas of this type have feed point impedances in the range of 50 to 100 Ω.

10.6.3 Conical Logarithmic Spirals

The multiple interlaced spirals of the planar log spiral can be wrapped around a conical surface and made uni-directional. The spiral can be made from strips of metal or even from coaxial cable. The feed point is at the apex of the cone and the radiation is circularly polarized in the direction of the apex.

The input impedance is in the range of 100 and 150 Ω for a pitch angle (the angle of the strip along the cone) of 17 degrees and cone body angles of 20 to 60 degrees. Smaller (narrower) cones have higher F/B ratios. The bandwidth is determined by the ratio of the base diameter (approximately $\lambda/2$ at the lowest frequency) to the truncated apex diameter (approximately $\lambda/4$ at the highest frequency).

Because of their well-controlled radiation patterns and extremely wide bandwidth, compact conical spiral antennas of this type are well-suited for testing and measurement purposes. They are often found in EMC labs for troubleshooting and certifying equipment. Examples of these antennas by the ETS-Lindgren company are presented along with performance data at **ets-lindgren.com/get-manuals/3102(1).pdf**.

10.7 References and Bibliography

Source material and more extended discussion of the topics covered in this chapter can be found in the references listed below and in the textbooks listed at the end of the **Antenna Fundamentals** chapter.

Allen, D., "The Log Periodic Loop Array (LPLA) Antenna," *The ARRL Antenna Compendium, Vol. 3*, pp. 115 – 117.

Balanis, C., *Antenna Theory, Analysis and Design*, 2nd Ed. (New York: John Wiley & Sons, 1997) Chapter 9.

Belrose, J., "The HF Discone Antenna," *QST*, July 1975, pp. 11 – 14, 56.

Breakall J., and Rodriguez Solis, R., "A New Design Method for Low Sidelobe Level Log-Periodic Dipole Antennas." *Applied Computational Electromagnetics Society Journal* 11(2): July, 1996, pp. 100 – 107.

Butson, P., and Thompson, G., "A Note on the Calculation of the Gain of Log-Periodic Dipole Antennas," *IEEE Trans on Antennas and Propagation*, Vol. AP-24, No. 1, Jan. 1976, pp. 105 – 106.

Carrel, R., "The Design of Log-Periodic Dipole Antennas," *1961 IRE International Convention Record.*

Cebik, L., "Notes on HF Discone Antennas," widely available online.

Cebik, L., "Notes on Standard Design LPDAs for 3-30 MHz," *QEX*, Part 1, May/June 2000, pp. 23 – 38; Part 2, July/Aug. 2000, pp. 17 – 31.

Cebik, L., "The Monoband Log-Cell Yagi Revisited, Part 1: An Introduction to the Log-Cell Yagi and Some Standards of Comparison and Part 2: Element Phasing and Log-Cell Design," various sources online.

Cebik, L., "Notes on the OWA Yagi," *QEX*, July/Aug. 2002, pp. 22 – 34.

DuHamel R., and Isbell, D., "Broadband Logarithmically Periodic Antenna Structures," *1957 IRE National Convention Record*, Part 1.

Fisher, J., "Development of the W8JF Waveram: A Planar Log-Periodic Quad Array," *The ARRL Antenna Compendium, Vol. 1*, pp. 50 – 54.

Heitner, K., "A Wide-Band, Low-Z Antenna — New Thoughts on Small Antennas," *The ARRL Antenna Compendium, Vol. 1*, pp. 48 – 49.

Isbell, D., "Log-Periodic Dipole Arrays," *IRE Transactions on Antennas and Propagation*, Vol. AP-8, No. 3, May 1960.

Johnson, J., "Enhancing performance of stacked Yagi arrays and BOLPA Development," Dayton Hamvention Antenna Forum 2017, **kkn.net/dayton2017/Yagi-Stack-Enhance-BOLPA-Dayton-2017.pdf**.

Kraus, J., *Antennas*, 2nd Ed. (New York: McGraw-Hill, 1988), Chapter 15.

Johnson, R., Ed., *Antenna Engineering Handbook*, 3rd Ed. (New York: McGraw-Hill, 1993), Chapters 14 and 26.

Luetzelschwab, C., K9LA, "Log Periodic Dipole Array Improvements," *The ARRL Antenna Compendium, Vol. 6*, pp. 74 – 76.

Luetzelschwab, C., K9LA, "More Improvements to an LPDA," *The ARRL Antenna Compendium, Vol. 7*, pp. 121 – 122.

Mayes P., and Carrel, R., "Log Periodic Resonant-V Arrays," *IRE Wescon Convention Record*, Part 1, 1961.

Mayes, P., Deschamps, G., and Patton, W., "Backward Wave Radiation from Periodic Structures and Application to the Design of Frequency Independent Antennas," *Proc. IRE*.

Milner, C., "Log Periodic Antennas," *QST*, Nov. 1959, pp. 11 – 14.

Orr W., and Cowan, S., *Beam Antenna Handbook*, pp. 251 – 253.

Rumsey, V., *Frequency Independent Antennas* (Academic Press, 1966).

Stearns, S., "All About the Discone Antenna: Antenna of Mysterious Origin and Superb Broadband Performance," *QEX*, Jan./Feb. 2007, pp. 37 – 44.

Stutzman W., and Thiele, G., *Antenna Theory and Design*, 2nd Edition (Wiley, 1998), Chapter 6.

Wyatt, K., "PC board log periodic antennas," *EDN*, 15 Dec. 2012, **edn.com/pc-board-log-periodic-antennas**

TABLE OF CONTENTS

Chapter 11 — Online Material

Articles

- "Designing a Shortened Antenna" by Luiz Duarte Lopes, CT1EOJ
- "A 6-Foot-High 7-MHz Vertical" by Jerry Sevick, W2FMI
- "A Horizontal Loop for 80-Meter DX" by John Belrose, VE2CV
- "A Gain Antenna for 28 MHz" by Brian Beezley, K6STI
- "A Low-Budget, Rotatable 17 Meter Loop" by Howard Hawkins, WB8IGU
- "A Simple Broadband Dipole for 80 Meters" by Frank Witt, AI1H
- "A Wideband Dipole for 75 and 80 Meters" by Ted Armstrong, WA6RNC
- "A Wideband 80 Meter Dipole" by Rudy Severns, N6LF
- "Broad-Band 80-Meter Antenna" by Allen Harbach, WA4DRU
- "Broad-banding a 160 m Vertical Antenna" by Grant Saviers, KZ1W
- "Inductively Loaded Dipoles"
- "Off-Center Loaded Antennas" by Jerry Hall, K1PLP
- "The 3/8-Wavelength Vertical — A Hidden Gem" by Joe Reisert, W1JR
- "The 160-Meter Sloper System at K3LR" by Al Christman, KB8I, Tim Duffy, K3LR and Jim Breakall, WA3FET
- "The 'C-Pole' — A Ground Independent Vertical Antenna" by Brian Cake, KF2YN
- "The Compact Vertical Dipole"
- "The Half-Delta Loop — A Critical Analysis and Practical Deployment" by John Belrose, VE2CV and Doug DeMaw, W1FB
- "The K1WA 7-MHz Sloper System"
- "The K4VX Linear-Loaded Dipole for 7 MHz" by Lew Gordon, K4VX
- "The Story of the Broadband Dipole" by Dave Leeson, W6NL
- "The W2FMI Ground-Mounted Short Vertical" by Jerry Sevick, W2FMI
- "Use Your Tower as a Dual-Band, Low-Band DX Antenna" by Ted Rappaport, N9NB, and Jim Parnell, W5JAW
- "Shunt Feed Towers for Operation On the Lower Amateur Frequencies" by Earl Cunningham, K6SE
- "Shunt Feed Your Tower on 160 Meters" by Mike Taylor, AC9C

Chapter 11

General Purpose MF and HF Antennas

The antennas in this chapter are based on the dipole and the ground-plane monopole or "vertical" described in the **Dipoles and Monopoles** chapter. These simple antennas can be combined into arrays for additional directivity as described in the **Phased Arrays** and **Broadside and End-Fire Dipole Arrays** chapters. Practical information on building these antennas is covered in the **Antenna Materials and Construction** chapter.

This chapter presents practical designs most often used as resonant antennas on the MF and HF amateur bands. This is not to say that the antennas can only be used on a single band or below 30 MHz — many can be used on several bands as discussed in the **Multiband HF Antennas** chapter and the same principles can be used to create VHF and UHF antennas. Nevertheless, in these examples the discussion will be mainly concerned with use below 30 MHz.

For most amateurs, the choice of what type of antenna to install and whether it is installed horizontally or vertically is one of necessity and is driven by constraints such as whether trees or a tower are available, restrictions on external antennas, and aesthetic concerns. The goal of this chapter is to present a variety of options so that given the circumstances, the best choice or choices for the desired purpose can be made.

On the lower HF bands, a horizontally polarized antenna at low heights (in terms of wavelengths) provides good regional coverage via NVIS propagation. As a result, horizontal antennas like dipoles are very popular on these bands for short range and regional communications, nets, and rag chewing. Another advantage of horizontal antennas is that they do not require extensive ground systems to be efficient.

For a horizontally polarized antenna to be effective for typical DX communications, heights of $\lambda/2$ to 1 λ are considered to be a minimum. As we go down in frequency these heights become harder to realize. For example, a 160-meter dipole at 70 feet is only 0.14 λ high, the equivalent of a 20-meter dipole only 9 feet off the ground! This antenna will be very effective for local and short distance QSOs but not very good for DXing.

As a result, on our MF bands (630 and 160 meters) and the lower HF bands, vertical antennas become increasingly attractive — especially for making DX contacts — because they provide a means for lowering the radiation angle. This is especially true where practical heights for horizontally polarized antennas are too low. On our LF band with a wavelength of 2200 meters, vertical antennas are the only practical option, requiring special loading techniques to resonate at that low frequency.

Performance of a vertically polarized antenna is determined by several factors:

- Electrical height of the vertical portion of the radiator
- The ground or counterpoise system efficiency if one is used
- Ground characteristics in the near- and far-field regions
- The efficiency of loading elements and matching networks

Determining whether a horizontal or vertical antenna is appropriate depends on the intended use of the antenna. The chapter **HF Antenna System Design** will extend the discussion beyond individual antennas to the selection of antennas for a desired purpose, such as DX versus local or continental coverage.

11.1 Dipole Antennas

11.1.1 Dipole Basics

Half-wave dipoles and variations of these can be a very good choice for an HF antenna. Where only single-band operation is desired, the λ/2 antenna fed with 50- or 75-Ω coaxial cable is a popular and inexpensive antenna. It can also be used on the third harmonic with some adjustment as explained in the project at the end of this section. The basic and most common construction for MF and HF dipoles is shown in **Figure 11.1**. For more information about constructing wire antennas in general, see the **Antenna Materials and Construction** chapter.

The length of the λ/2 dipole in feet is often stated as ℓ = 468/f (MHz) although this rarely results in an antenna resonant at the desired frequency as discussed in the chapter **Dipoles and Monopoles**. It is more practical to begin with a length of 485/f or 490/f (**Table 11.1** gives lengths for each of the ham bands from 1.8 through 50 MHz) and then adjust the antenna according to the following procedure:

1) Assemble the antenna with length l_1 for a desired frequency of f_1 but do not make the attachments to the end insulators permanent. Twisting or clamping the antenna wire at the

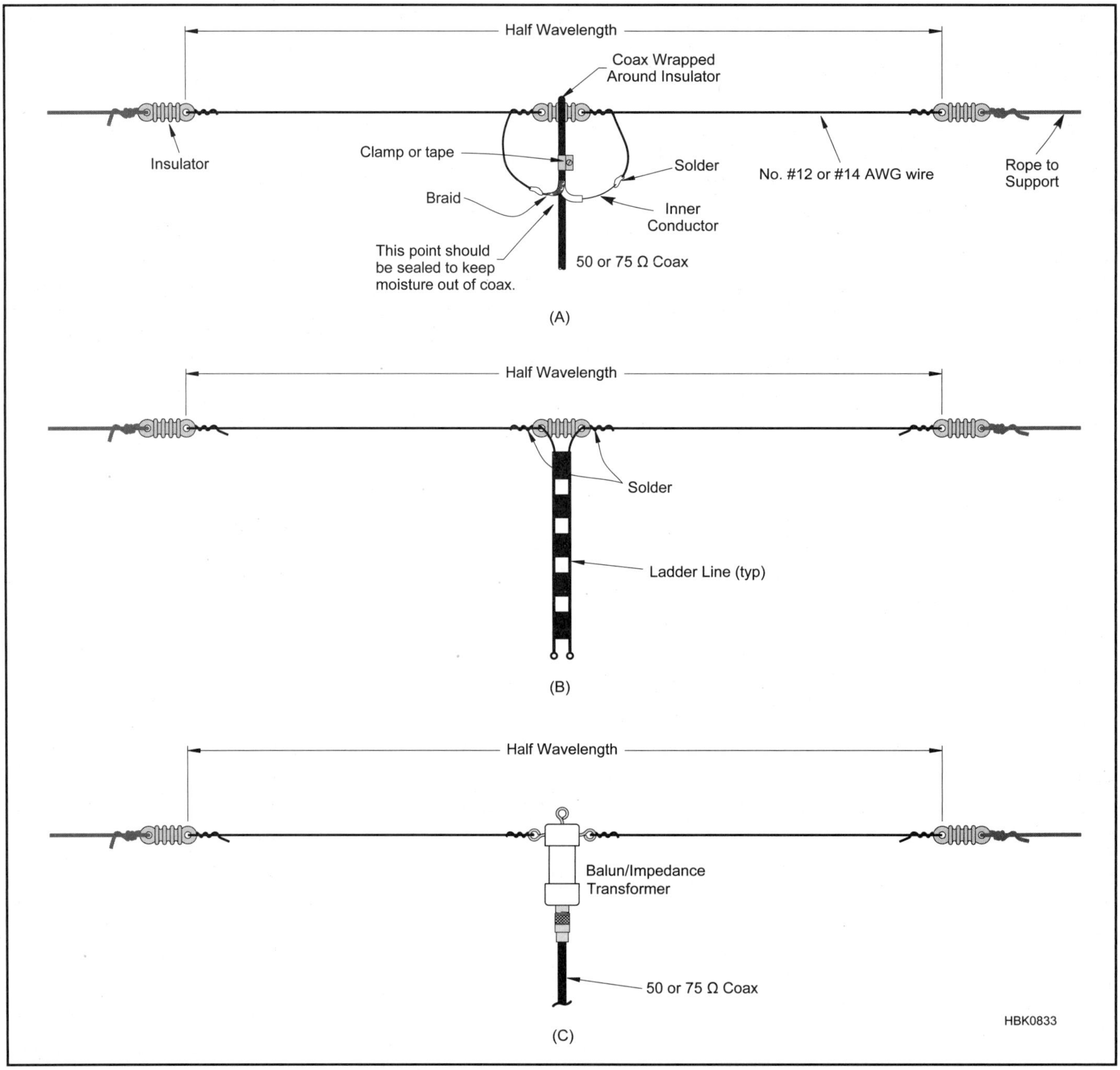

Figure 11.1 — Details of center-fed coax-fed dipole construction. The dipole at B can also be fed with open-wire or ladder-line. Note that the electrical length of the dipole extends to the tips of the loops of wire attached to the insulators.

Dipoles and NVIS Operation

Dipoles are by far the most popular antenna used for NVIS communication, mostly for public service operations. At a height of 0.2 wavelengths, the dipole is efficient and has good gain at high elevation angles. Lower heights increase ground losses and reflector wires on the ground do not provide significant improvements in signal strength. See the NVIS sections of the chapters **Propagation of Radio Waves** and **HF Antenna System Design** for more information.

insulators will suffice during adjustment.

2) Raise the antenna to its desired position and determine the frequency of lowest SWR, f_2.

3) Assuming that f_2 is too low (the antenna is too long), calculate the desired length $l_2 = l_1 \times f_2 / f_1$. Trim the antenna to the desired length by removing equal amounts of wire on each end to maintain electrical balance at the feed point.

Example: A dipole intended to be used at 14.250 MHz is initially built with a physical length of 490 / 14.250 = 34.4 feet (34 feet 5 in). Once in place, f_2 is determined to be 13.795 MHz. Using step 3, the desired length should be 34.4 × 13.795 / 14.250 = 33.3 feet and the antenna is 34.4 – 33.3 = 1.1 feet (1 foot 1 inch) too long. Remove 6.5 inches from each end of the antenna.

Using insulated wire, such as THHN electrical service wire, will shorten the antenna by a few percent, depending on the type of insulation. If the insulation has a clear sheath, it will degrade from solar UV and eventually shed from the wire but will have little effect on antenna performance.

When shortening the antenna to resonance on the desired frequency, simply pull the wire through the insulators and wrap it around the main antenna wire. It is not necessary to cut off surplus wire as you tune the antenna or solder it to the main wire. After the antenna has been tuned, then you can trim the wire and make the insulator attachment permanent.

Coaxial lines present support problems as a concentrated weight at the center of the antenna, tending to pull the center of the antenna down, so care must be taken to make the feed point connections strong and provide support for the cable. If a center support or conveniently located tree is available, insulators with a rope attachment point can be used to support the weight.

The feed line should come away from the antenna at right angles for the longest practical distance so as to preserve electrical balance and minimize coupling of the antenna to the feed line shield's outer surface. Adding a choke balun at the feed point helps to electrically isolate the shield surface and prevent common-mode current from flowing on the feed line. (See the **Transmission Line System Techniques** chapter for a discussion of the use of choke baluns.)

BENDING A DIPOLE

If you do not have sufficient length between the supports, simply hang as much of the center of the antenna as possible between the supports and let the ends hang down as in **Figure 11.2**. The ends can be straight down or may be at an angle as indicated but in either case should be secured so that they do not move in the wind. As long as the center portion between the supports is at least λ/4, the radiation pattern will be very nearly the same as a full-length dipole.

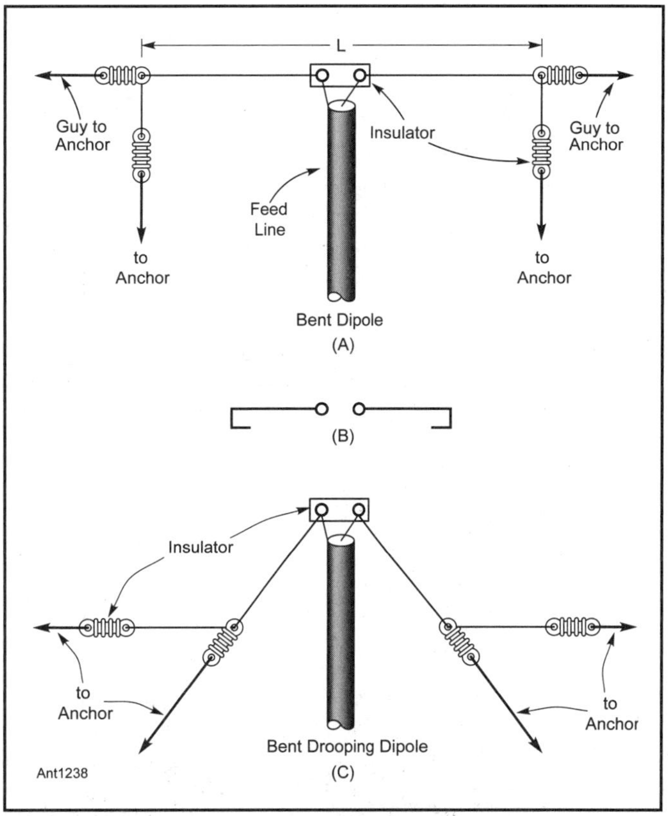

Figure 11.2 — When space is limited, the ends may be bent downward as shown at A, or back on the radiator as shown at B. The bent dipole ends may come straight down or be led off at an angle away from the center of the antenna. An inverted-V as shown in C can be erected with the ends bent parallel to the ground when the support structure is not high enough.

Table 11.1
Starting Lengths for Amateur Band Dipoles

Freq	*Length in feet*		
(MHz)	*468/f*	*485/f*	*490/f*
1.85	253.0	262.2	264.9
3.6	130.0	134.7	136.1
3.9	120.0	124.4	125.6
5.3	88.3	91.5	92.5
7.1	65.9	68.3	69.0
10.1	46.3	48.0	48.5
14.15	33.1	34.3	34.6
18.1	25.9	26.8	27.1
21.2	22.1	22.9	23.1
24.9	18.8	19.5	19.7
28.2	16.6	17.2	17.4
29	16.1	16.7	16.9
50.1	9.3	9.7	9.8

The resonant length of the wire will be somewhat shorter than a full-length dipole and can best be determined by experimentally adjusting the length of ends, which may be conveniently near ground. Keep in mind that there can be very high voltages at the ends of the wires and for safety the ends should be kept out of reach.

Folding the ends back on the antenna is a type of *linear loading*. Both types of loading are discussed later in this chapter. While both techniques are efficient, it will also reduce the matching bandwidth — as does any form of loading.

11.1.2 Inverted-V Dipole

If only a single support is available, the halves of a dipole may be sloped to form an inverted-V dipole, as shown in **Figure 11.3**. This also reduces the horizontal space required for the antenna. L.B. Cebik, W4RNL, wrote several articles on the inverted-V, including "Antennas From the Ground Up" (available from several online sources) that shows the antenna's evolution from the dipole. Cebik discusses the effects of moving the arms on pattern and impedance along with other variations in antenna geometry.

There will be some difference in performance between a horizontal dipole and the inverted-V as shown by the radiation patterns in **Figure 11.4**. There is small loss in peak gain and the pattern is less directional.

Bringing a dipole's wires toward each other results in a decrease of the resonant frequency and a decrease in feed point impedance and bandwidth. (This is true whether the dipole is constructed as an inverted-V or not.) Thus, to maintain the same resonant frequency, the length of the dipole must be decreased somewhat over that of the horizontal configuration.

The amount of shortening required varies with the circumstances of the installation but a reasonable rule of thumb would be 5% for every 45 degrees that the legs of the dipole are lowered from horizontal. Start with an initial length for a horizontal dipole and then trim it in the inverted-V configuration according to the procedure given for horizontal dipoles.

The angle at the apex is not critical, although angles smaller than 90° begin to compromise performance significantly. Because of the lower feed point impedance, a 50-Ω feed line should be used.

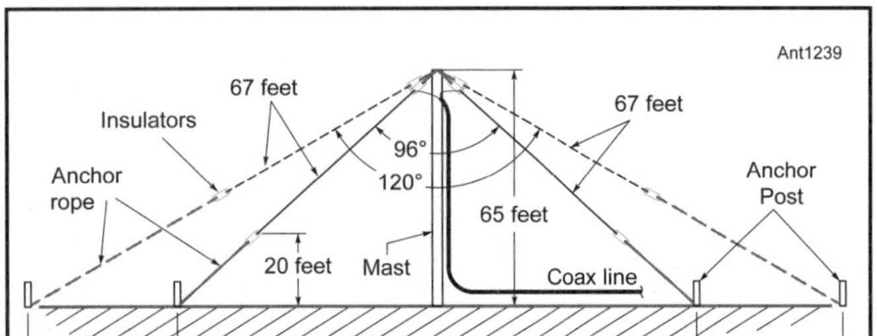

Figure 11.3 — The inverted-V dipole. Two different configurations for 80 meter inverted-V dipoles are shown — one for a 120° apex angle and one for a 96° apex angle.

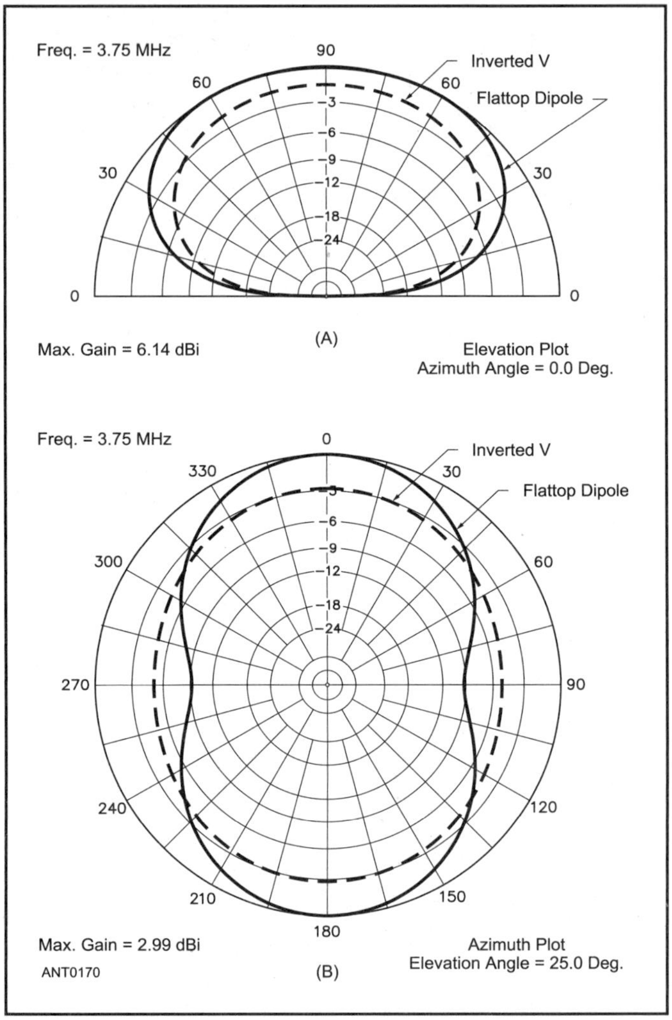

Figure 11.4 — At A, elevation and at B, azimuthal radiation patterns comparing a normal 80-meter dipole and an inverted-V dipole. The center of both dipoles is at 65 feet and the ends of the inverted-V are at 20 feet. The frequency is 3.750 MHz.

If a close match to the feed line impedance is desired, the usual procedure is to adjust the angle for lowest SWR while keeping the dipole resonant by adjusting its length. Bandwidth may be increased by using multiconductor elements, such as a cage or fan configuration as discussed below.

11.1.3 Sloping Dipoles

Another variation of the single-support configuration is the *λ/2 sloping dipole* shown in **Figure 11.5A**. This antenna is also known as a *sloper* or *half-wave sloper* to distinguish it from the *half sloper* described in the section on vertically-polarized antennas. The feed point impedance depends on the height of the antenna above ground, the characteristics of the ground, and the angle the antenna makes with the ground. In most cases, an acceptable SWR for coaxial cable can be achieved by altering the direction and height. Losses increase as the antenna ends approach the support or the ground, so the same cautions about the height of the antenna ends apply as for the inverted-V antenna.

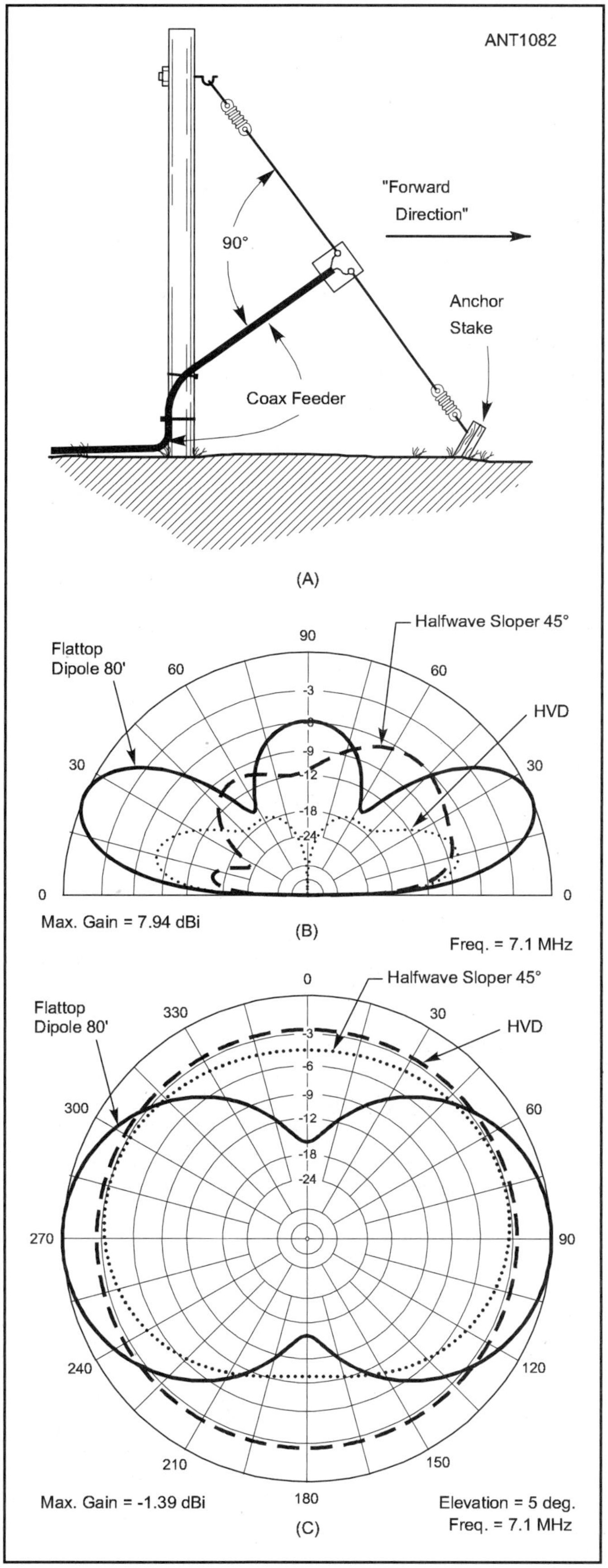

Figure 11.5 — Example of a sloping λ/2 dipole, or *full sloper*. On the lower HF bands, maximum radiation over poor to average ground is off the sides and in the *forward direction* indicated if a non-conductive support is used. A metal support will alter this pattern by acting as a parasitic element. B shows the elevation patterns for the same antennas. Note that the sloping half wave dipole has more energy at higher elevation angles than either the flattop dipole or HVD. C compares the 40-meter azimuthal patterns at a DX takeoff angle of 5° for three configurations: a flattop dipole, a dipole tilted down 45° and an HVD (half wave vertical dipole).

The amount of slope from horizontal can vary from 0°, where the dipole is in a flattop configuration, all the way to 90°, where the dipole becomes fully vertical. The latter configuration is sometimes called a *Halfwave Vertical Dipole* (*HVD*) and is discussed in the section on vertically polarized antennas.

This antenna slightly favors the direction of the antenna's slope as shown in Figure 11.6B. With a non-conducting support and average to poor ground, signals off the back are weaker than those off the front. With a non-conducting mast and good ground, the response is approximately omnidirectional with no gain in any direction.

A conductive support such as a tower acts as a parasitic element. The parasitic effects vary with ground quality, support height, and other conductors on the support (such as a beam at the top or other wire antennas). With such variables, performance is very difficult to predict but that is no reason not to put up the antenna and experiment with it. Many hams report good results with a sloper. To prevent coupling to the feed line, route the coax away from the feed point at 90° from the antenna as far as possible and use a choke balun at the feed point.

An intensive modeling study on feeding the closely-related HVD was done for the book *Simple and Fun Antennas for Hams* (see References and Bibliography). This study indicated that directing the feed line at an angle down to the ground of as little as 30° from the antenna can work with only minor interaction, provided that feed line chokes were employed at the feed point and a quarter-wavelength down the line from the feed point. (See the **Transmission Line System Techniques** chapter.)

The sloping half-wave dipole in Figure 11.6 exhibits about 5 dB of front-to-back ratio, although even at its most favored direction it doesn't quite have the same maximum gain as the HVD or the flattop dipole.

Two systems of multiple sloping dipoles are presented in articles included with this book's online material. A system designed for 7 MHz by K1WA and another for 1.8 MHz by K3LR give the builder some directivity while only requiring a single support. These systems can also be adapted to other bands. See also the References and Bibliography entry for *Sloper Antennas* by Juergen Weigl, OE5CWL.

11.1.4 Zepp Antenna

Other than to obtain a convenient feed point impedance and to be somewhat balanced, there is no reason why a dipole has to be fed exactly at the center. In the early days, the λ/2 dipole was often fed at one end where it was called an

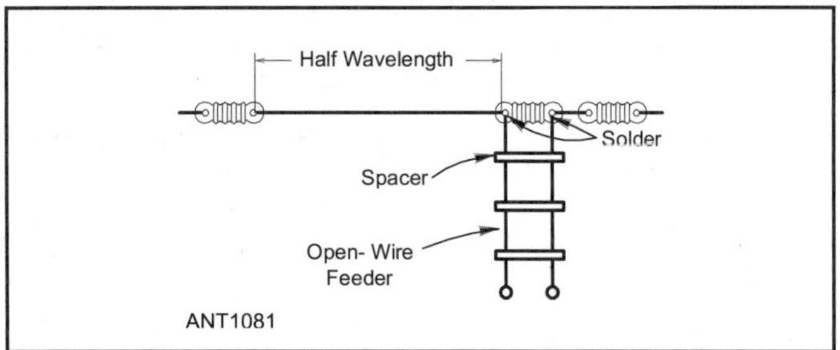

Figure 11.6 — An end-fed Zepp with a parallel-wire feed line connected at one end. Tuned feeders can be used to create a low impedance point for connecting coax as described in the text.

"End-fed Zepp" after the Zeppelin airships from which it was first deployed.

Figure 11.6 shows a typical end-fed Zepp with a parallel-wire feed line. Since the feed line is connected at a low-current/high-voltage point on the antenna, the feed point impedance is quite high and often in the neighborhood of 3000-5000 Ω. This is too high to present a match to even the widest-spaced parallel-wire lines, so *tuned feeders* are often employed in which the feed line is an odd number of quarter-wavelengths long. Such a feed line transforms a high impedance into a low impedance as described in the **Transmission Lines** chapter, allowing low-impedance feed lines such as coax to be connected at a point with a more manageable SWR.

Coaxial feed lines can connect directly to an end-fed Zepp or end-fed half-wave (EFHW) by using an impedance transformer at the antenna's feed point. Because of the high impedance, a turns ratio of 7:1 or 8:1 (impedance ratio of 49:1 or 64:1) is required. See the EFHW section of the **Multiband HF Antennas** chapter for transformer design information. Owen Duffy, VK1OD, presents an 8:1 autotransformer in his article "Small efficient matching transformer for an EFHW" at **owenduffy.net**.

Feed points that are not in the center of the antenna (i.e. – asymmetric) are intrinsically unbalanced and significant amounts of common-mode current will flow on the feed line, whether parallel-conductor or coaxial, unless blocked by choke baluns as described in the **Transmission Line System Techniques** chapter. End-fed antennas include the feed line's common-mode path as part of the antenna system. Off-center-fed antennas may or may not include the common-mode path depending on the antenna design.

11.1.5 Folded Dipoles

Figure 11.7 shows a two-wire *folded dipole* made from parallel ½-wavelength conductors connected together at each end of the antenna. Folded dipoles can also have three or even four conductors. One conductor is open in the middle where the feed line is attached. The remaining conductors are continuous from end to end. A folded dipole has the same gain and radiation pattern as a single-wire dipole.

As more conductors are added, feed point impedance goes up with the square of the number of conductors. For the two-wire folded dipole, the feed point impedance is $2^2 = 4$ times that of a single-wire dipole. A three-wire folded dipole would have $3^2 = 9$ times higher feed point impedance and so forth. (See the **Dipoles and Monopoles** chapter for more about why folded dipoles work this way.)

A common use of the folded dipole is to raise the feed point impedance of the antenna to present a better impedance match to feed lines. For example, if a very long feed line to a dipole is required, open-wire feed line would be preferable because of its lower loss. By raising the dipole's feed point impedance, SWR on the open-wire line is reduced from what it would be if connected to a single-wire dipole. For example, by using a three-wire folded dipole, the 72 Ω feed point impedance of a single-wire dipole is increased to 9 × 72 = 648 Ω which is a good match to 600 Ω open-wire line. A 50 Ω feed point impedance would be increased to 9 × 50 = 450 Ω, the same as widely available window line.

11.1.6 Broadband Dipoles

Producing a dipole with an SWR bandwidth covering an entire amateur band is difficult for the 160 meter and 80 meter bands due to their relative spans: approximately 10.5% for the 160 meter band and 13.4% for the 80 meter band from the lowest to the highest frequency of the allocation. Most single-wire dipoles have an SWR bandwidth of a few percent in comparison, making it difficult to cover these widest of our bands with just one antenna. Given the importance of 80 meters to a wide variety of operating activities, that band has received the most attention. The higher HF bands are much narrower

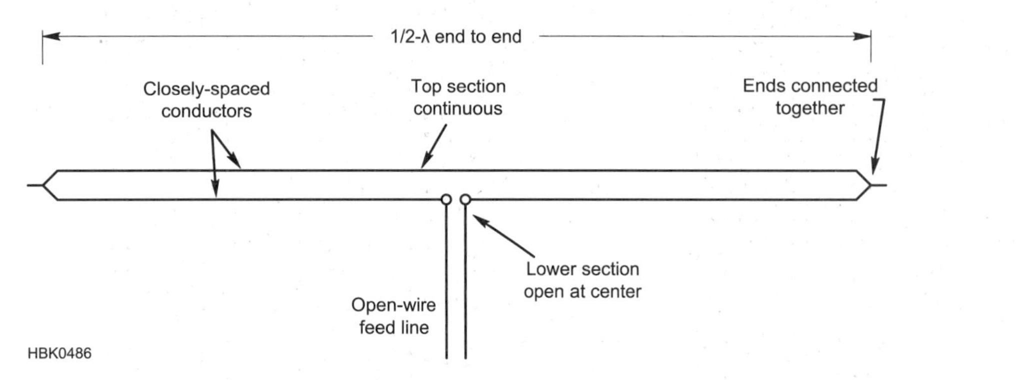

Figure 11.7 — The folded dipole is constructed from open-wire transmission line with the ends connected together. The close proximity of the two conductors and the resulting coupling act as an impedance transformer to raise the feed point impedance over that of a single-wire dipole by the square of the number of conductors used.

in comparison and generally can be covered by a single-wire dipole.

As Dave Leeson, W6NL, notes in his *QEX* article "The Story of the Broadband Dipole," the search for a broadband 80 meter dipole antenna has a long history, culminating in a series of articles by Frank Witt, AI1H. Beginning with the antennas seen in early radio, a wide range of broadbanding concepts has been explored. Here are some major categories:

- Cage, parallel wire, fan or bow-tie dipoles that have a large equivalent diameter or that approximate conical dipoles
- Multiple dipoles with staggered resonances connected in parallel
- Dipoles with a coupled resonator "open sleeve" element
- "Bazooka" dipoles with coaxial radiating elements
- Dipoles with lumped reactive matching networks
- Dipoles with coaxial radiating and matching elements
- Dipoles with resonant feed line matching

There are mechanical and reliability problems associated with many of the broadbanding schemes. Cage and multiple dipole configurations have a tendency to become twisted in the wind, are difficult to construct and install, take up more space, and have greater visual impact than a simple dipole. Coaxial radiating elements require sealing from the weather, are heavy to support and are not necessarily strong enough to avoid stretching under load. In addition, some of the published designs don't have sufficient broadband response, while others are broadband mainly because of losses in the matching network.

The remainder of this section will present examples of each, focusing on the first and last categories as giving the best performance. See Leeson's article for a thorough list of references. All of these methods are discussed in the article "Broadband Antenna Matching" by Frank Witt, AI1H, which is included with the **Transmission Line System Technique** chapter's online material.

INCREASING ANTENNA DIAMETER

The simplest way to increase the SWR bandwidth of a single-wire dipole is to increase the thickness of the wire (the length-to-diameter ratio) as discussed in the **Antenna Fundamentals** and **Dipoles and Ground-Planes** chapters. This reduces the change in reactance with frequency, causing the SWR to vary more gradually with frequency, as well. Since the range of available wire sizes is quite limited in the potential effect on bandwidth at MF and HF, the technique of employing multiple wires is used to create a larger-diameter conductor.

There are three common methods of using multiple wires in this way: the cage, the fan and the open-sleeve. The cage shown in **Figure 11.8** is a very old design, having been employed during the early days of "wireless" to increase bandwidth of antennas used for spark signals with their very wide bandwidths. The cage consists of several wires (three or more) held apart by spreaders (insulating or non-insulating) and connected together at the ends and at the feed point. A project describing the cage dipole in use at W1AW on 80 meters is included at the end of this section. The "flat-top" antenna of several wires in parallel instead of in a cage was also used in this way.

In fact, to increase bandwidth it is not necessary to increase the antenna's diameter over the entire length, just near the ends. Thus, a simplified variation on the cage is to create a "bow-tie" with just two wires in each leg of the dipole. The wires are tied together at the feed point and spread apart up to 10 feet at the ends of the dipole where they may be connected together or left separate. The bow-tie or "skeleton biconical dipole" was discussed by Hallas in the May 2005 *QST* (see the References and Bibliography).

In both cases, extra tethers are usually required at the ends of the cage or fan to keep the antenna from twisting in the wind. This is less of a problem with the cage design which uses multiple spreaders to keep the wires apart. Such antennas provide excellent electrical performance at the cost of some mechanical complexity and extra weight. They may not be suitable in areas where heavy icing or high wind speed is common.

A second method creates a fan of two or more dipoles with close but not identical resonant frequencies. This is illustrated in **Figure 11.9** in which three dipoles are cut for the bottom, middle and top frequencies in the 80 meter band (3.5, 3.75, and 4 MHz) and fed in parallel at the feed point. This is similar to the bow-tie mentioned in the previous paragraph

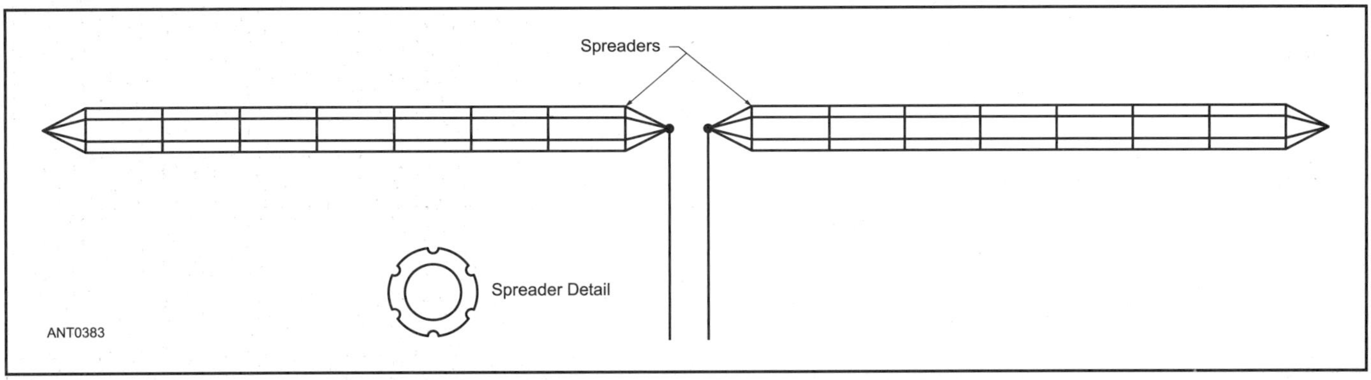

Figure 11.8 — Construction of a cage dipole. The spreaders need not be of conductive material and should be lightweight. Between adjacent conductors, the spacing should be 2% of overall length or less. The number of spreaders and their spacing along the dipole should be sufficient to maintain a relatively constant separation of the radiator wires. The spreaders can be round as shown in the detail or any suitable cross arrangement.

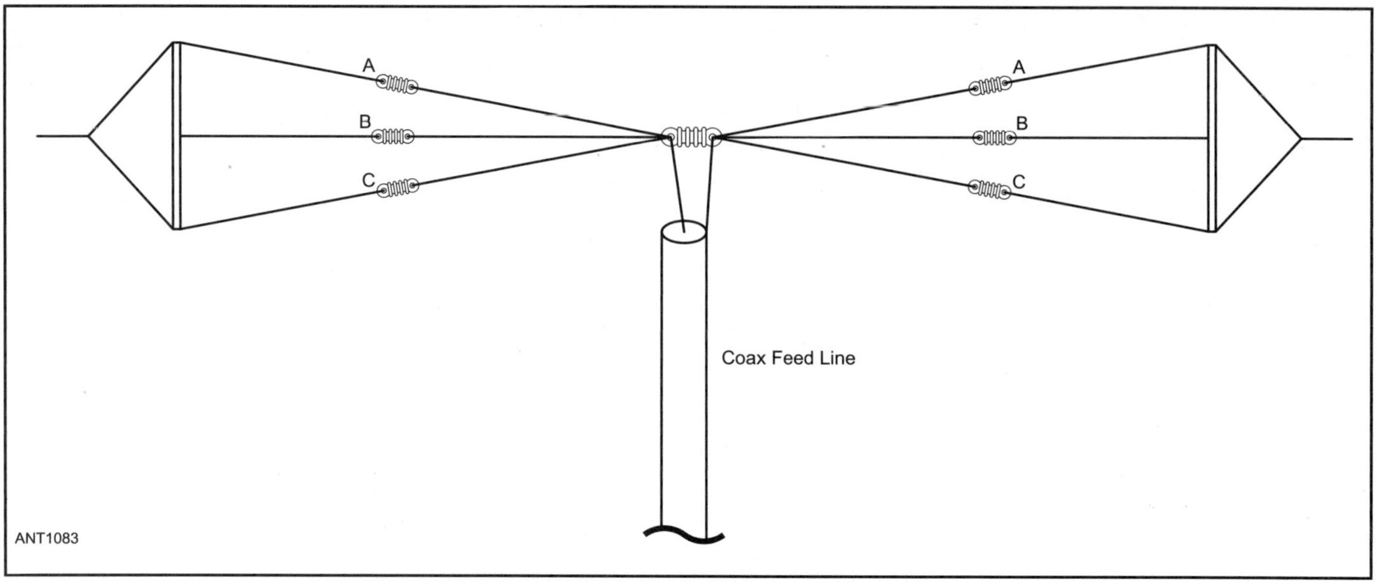

Figure 11.9 — A broad-banded "fan" dipole. The three dipoles a, b and c are cut to resonate at the band edges and center band frequency. This creates a single antenna that can be used over the entire 3.5 MHz band. On 80 meters, the dipole cut for 3.5 MHz will be approximately 7 feet longer than the one cut for 4 MHz. (Figure 11.10 from *Practical Wire Antennas*, courtesy RSGB — see References and Bibliography.)

but the ends of the dipoles are not connected together. A nonconducting spreader is used to hold the wires apart.

The dipole impedances interact to some degree depending on how different the resonant frequencies are. Modeling is recommended at the expected height above ground but may not give completely accurate results due to the very shallow angle at which the wires join at the feed point. Expect some adjustments as the three dipoles are adjusted to give the desired SWR curve across the band. Two dipoles can cover approximately two-thirds of the band.

PROJECT: COUPLED-RESONATOR 75/80 METER DIPOLE

The third method is to place a parasitic dipole extremely close to the driven dipole so that it couples to the driven dipole and essentially operates in parallel with it. This technique has been refined in several recent articles. **Figure 11.10** shows

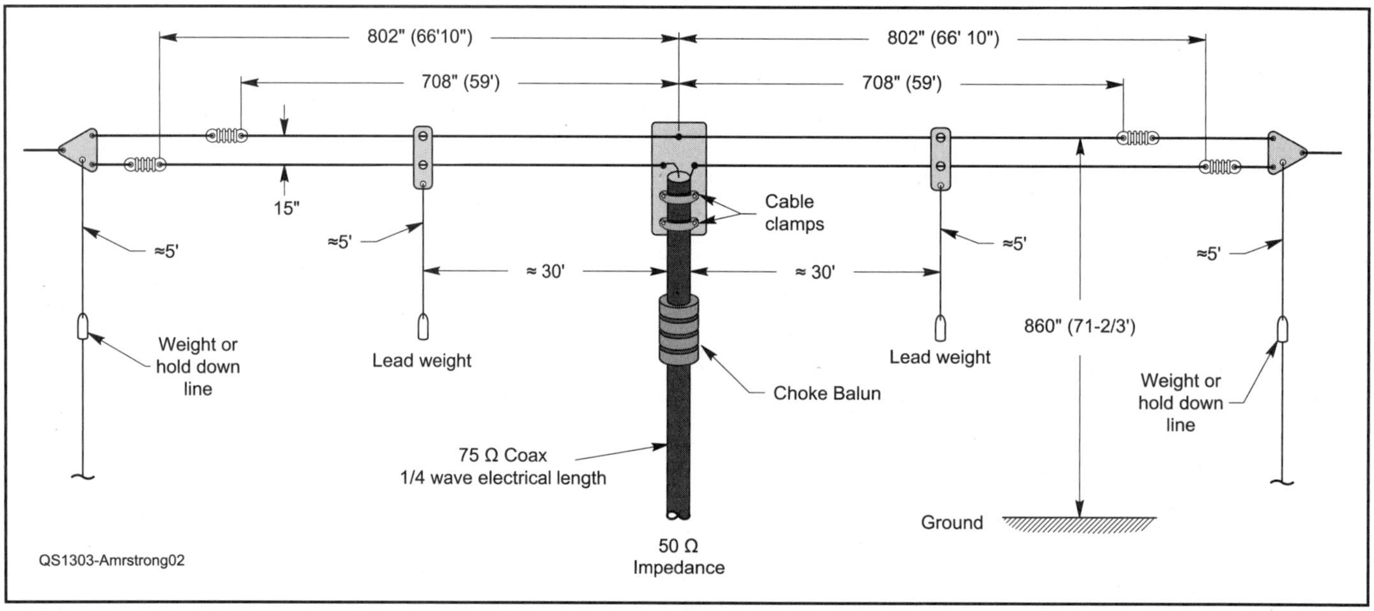

Figure 11.10 — Construction details and dimensions of the coupled resonator 75/80-meter antenna including a λ/4 impedance transformer of 75 Ω coax. Based on a coax velocity factor of 0.66 (66%) for solid polyethylene dielectric coax, the λ/4 transformer should be 43.3 feet long at 3.75 MHz. Measure and build the λ/4 transformer as described in the Transmission Line System Techniques chapter. The original article recommended type #43 ferrite cores but type #31 will provide more choking impedance at these frequencies.

an implementation by Ted Armstrong, WA6RNC, from the March 2013 issue of *QST*. (This article is included with this book's online material along with a previous design from Rudy Severns, N6LF. Check *The Antenna Book's* website at **arrl.org/arrl-antenna-book-reference** for more information about broadband designs for antennas on 160 through 40 meters.)

An isolated wire is placed next to a dipole cut for the low end of the band. The shorter wire has a higher resonant frequency and becomes the radiator as energy shifts to it higher in the band. This antenna's SWR was less than 1.7:1 across the entire 75/80 meter band and lower at most frequencies.

PROJECT: W1AW 80 METER CAGE DIPOLE

The 80 meter cage antenna used at W1AW is based loosely on a design that appeared in a December 1980 *QST* article by Allen Harbach, WA4DRU. (See the References and Bibliography and this book's online material.) The antenna is used primarily for W1AW's scheduled transmissions. It is also used for regular visitor operations as well. The resonant frequency of the antenna is 3627 kHz but the overall SWR is less than 2:1 from 3580 to 3995 kHz.

The W1AW cage antenna differs from the original article in that it's meant to be in place for a long period of time. So, most parts of the antenna are designed more ruggedly than in the Harbach design.

Each leg of the dipole is a cage made of four 80 meter dipole antennas of #14 AWG stranded copper wire tied together both at the ends and at the feed point as shown in **Figure 11.11**. Although Copperweld or an equivalent heftier wire could have been used, this size wire was easy to work with. The four wires forming each leg of the dipole are separated using a similar crosspiece. There is a crosspiece near the feed point and the ends. The spacing between the wires is three feet.

Figure 11.11 — The center insulator is constructed from a PVC pipe tee with end caps covering each end. A similar crosspiece separates the wires at the ends of the antenna. Stainless steel eyebolts are used to hold the ends of the cage and solder lugs with jumpers connect to the cage wires. Inside the tee, jumpers connect the eyebolts to the SO-239 on the third pipe cap. A piece of PVC pipe is U-bolted to the inner crosspieces of both cage sections. The entire assembly is supported by a sidearm from one of the W1AW towers. The antenna was constructed by W1AW Chief Operator Joe Carcia, NJ1Q.

Each cage wire passes through one leg of each crosspiece. A keeper wire is soldered across around the end of the PVC tube to the antenna wire on either side. This keeps the crosspiece from moving up and down the antenna. Exterior silicone caulk is applied to the hole in the tubing to seal it from moisture. Inside the crosspieces are oak dowels and the ends of the crosspieces are also capped. This adds rigidity to the crosspiece.

The feed point assembly is a homebrew PVC center insulator consisting of a pair of 2-inch end caps attached to both ends of a 6-inch long, 2-inch PVC pipe tee. A stainless steel eyebolt with two solder lugs is mounted in the middle of each cap. One solder lug is on the outside of the cap for a connection to the antenna and the other is inside for connection to the SO-239 coax connector. The SO-239 is mounted on a third cap attached to the middle tee section.

An 8-turn coaxial choke is connected to the antenna at the center insulator. The choke is made from RG-213 coax using designs included in the **Transmission Line System Techniques** chapter.

The center insulator assembly is bolted to a 4-foot length piece of 1-inch PVC pipe. The inner crosspieces are also bolted to this section of pipe as shown in the figure. This provides added support to the antenna. The center insulator and length of PVC are secured to the tower using a side-arm.

At the feed point, the four wires of each leg are brought together and looped through the eye-bolt. They are then twisted and soldered together and a short jumper of wire connects the twisted wires to a solder lug on the eyebolt. Inside each end cap, a jumper wire connects a second solder lug on the eyebolt to the SO-239 on the remaining cap.

At the outside ends of the cage, all four wires are brought to a common point, twisted together, and then attached to a strain insulator. The strain insulator and two of the crosspiece arms are tied off to the supports. This keeps the antenna legs from twisting in the breeze.

Tuning the antenna can be a bit tricky since each leg (wire) must be trimmed the same amount. It is best to start off with wire lengths calculated using the lowest operating frequency (for example, 3500 kHz). After trimming, the overall length of the antenna will be slightly smaller than that of a single-wire 80 meter dipole. This is because the radiating element is three feet in diameter — much thicker than a single-wire dipole.

While construction of this antenna is a bit more involved than that of a regular dipole, the result is a broadband antenna that doesn't require a tuner. The design specifications can also be recalculated to fit other amateur bands.

"BAZOOKA" DIPOLES

The initial design for the "Double Bazooka" antenna was published in 1968 by W8TV (see the References and Bibliography). It uses twin matching stubs of RG-58 coax connected in series with each leg of the dipole at the feed point (see **Figure 11.12A**) A variation called the "Crossed Bazooka" (Figure 11.12B) connects the stubs in parallel across the feed point. Each stub is electrically $\lambda/4$ inside the coax, shorted at the outside end. The outer surface of the coax shield and the

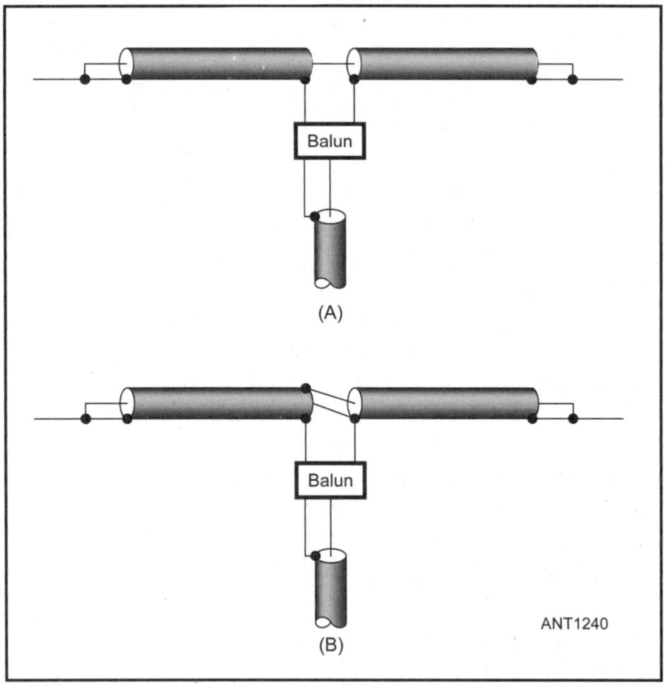

Figure 11.12 — The construction of a double bazooka dipole (A) showing the construction of the stubs and connection to the feed line. A variation is the crossed bazooka antenna at B.

extension wires from the shorted end of the coax form the radiating surface of the dipole. The original article suggested parallel-conductor extensions but most builders use a single wire today.

The theory of the double bazooka is as follows: The shorted stubs act as open circuits at their resonant frequency. As the frequency is increased, the stubs are electrically longer and become capacitive above resonance. The feed point impedance of the antenna, however, becomes inductive. So, the reactance of the stubs acts to cancel some of the antenna's reactance, increasing the SWR bandwidth of the antenna.

Walt Maxwell, W2DU, addressed the issues of stub losses, reactance cancellation, and other issues in a *QST* "Technical Correspondence" column and a *Ham Radio* article (see References and Bibliography), attributing the reported larger bandwidth primarily to feed line losses and increased conductor diameter. A similar analysis by Owen Duffy, VK1OD (**owenduffy.net/antenna/DoubleBazooka/index.htm**) yields the same conclusions. Although the antenna is popular and gives adequate service, a regular dipole will work just as well and is simpler to construct.

LC MATCHING NETWORK

This matching technique in **Figure 11.13** uses a parallel LC circuit for reactance compensation instead of stubs. It also acts as a step-up transformer voltage balun — see the **Transmission Line System Techniques** section on Three-Winding Voltage Baluns. The network is connected across the feed point of a single-wire dipole and adjusted for best performance.

The 3:1 SWR points can be kept entirely within the 80-meter band, rising to near 3:1 at 3.5 and 4 MHz. The version known

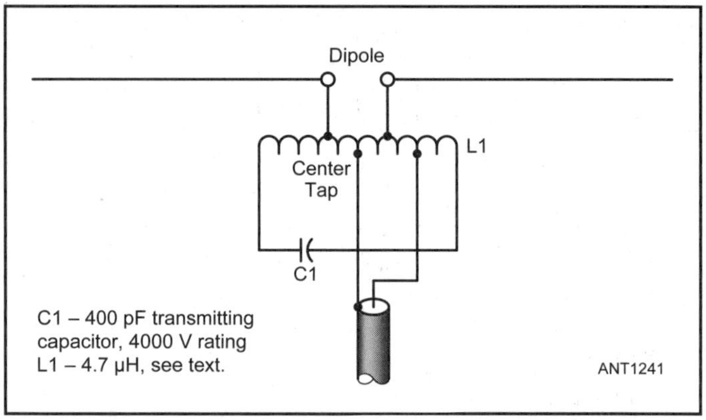

Figure 11.13 — The matching network for the DXer's Delight.

as the "DXer's Delight" sacrifices low SWR at the high end of the band for low SWR near 3.5 and 3.8 MHz — both prime frequencies for DXing and contesting. The DXer's Delight version uses a 400 pF / 4 kV transmitting capacitor for C1 and a 4.7 µH inductor made from 8½ turns of B&W type 3029 coil stock (6 turns per inch, 2½ inch diameter, #12 AWG wire). Regardless of which version you prefer, use a variable capacitor and tapped coil to perform adjustments at low power levels and then replace the components with high-power equivalents.

PROJECT: TRANSMISSION LINE RESONATOR (TLR) WIDEBAND 80 METER DIPOLE

The TLR system was introduced to amateurs by Frank Witt, AI1H, in the article "A Simple Broadband Dipole for 80 Meters." (The article is excerpted here and is included in this chapter's online material along with a more complete treatment of broadband matching techniques in the article "Broadband Antenna Matching" in the online material for the **Transmission Line System Techniques** chapter.) In this system-level approach, the single-wire dipole antenna itself is not modified. The system uses a broadband match as shown in **Figure 11.14A**. The key broadbanding element is the *transmission*

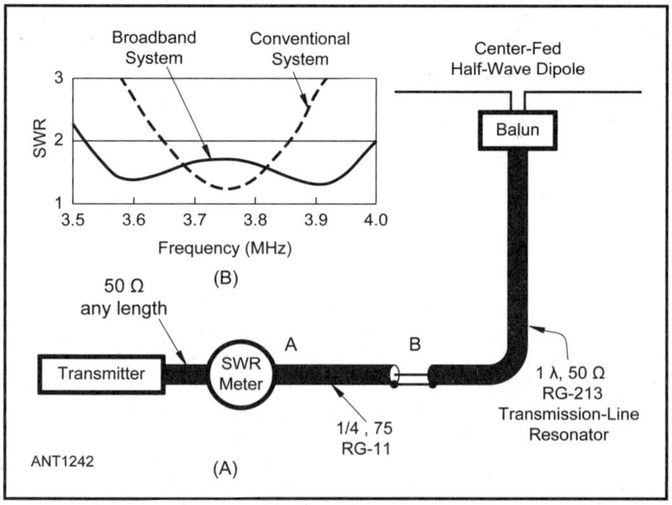

Figure 11.14 — Broadband TLR-based feed line match (see text).

line resonator (TLR). Part of the transmission line compensates for the reactance presented by the dipole away from its resonant frequency. This part of the line is a multiple of an electrical half-wavelength. The quarter-wavelength segment acts as a synchronous transformer to present an appropriate source impedance to the TLR. (See the **Transmission Line System Techniques** chapter.) See the References and Bibliography entry for Cebik's more detailed analysis and modeling of the TLR match.

The antenna is a center-fed half-wavelength dipole trimmed for resonance at 3.75 MHz. The transmission line is segmented into one electrical wavelength of 50-Ω coax and an electrical quarter-wavelength of 75-Ω coax. The values in **Table 11.2** were calculated using standard formulas for electrical length of feed lines (see the **Transmission Lines** chapter) with a velocity factor (VF) of 0.66, corresponding to coax with a solid polyethylene dielectric, such as RG-11 and RG-213. The actual values resulted from tuning the antenna, manufacturing variations in VF from published values, and stretching of the coax. The antenna was installed as an inverted-V with an included angle of 140° and a height of 60 feet. Dipole wire size was #14 AWG but that is not critical.

The system's SWR at the transmitter is shown in Figure 11.14B. The SWR curve labeled "Conventional System" is the same dipole fed with about ¾-wavelength of RG-213. The broadband system's SWR bandwidth is 2.2 times that of the conventional system. Additional mismatch loss caused by SWR in the matching sections is less than 0.5 dB anywhere in the band. Witt's 1993 article shows loss for several system configurations, including an unmatched system.

The feed line length may be extended by adding any required length of 50-Ω coax between the transmitter and the quarter-wave segment (point A in Figure 11.14A). A choke balun should be installed at the antenna's feed point. If the feed line is formed into a coiled-coax choke, that length of line must be included in the one-wavelength segment.

Tuning is accomplished by adjusting the dipole and resonator lengths. You can tune the antenna without connectors at the junction of the resonator and ¼-wavelength section by tack-soldering them together and testing with an antenna analyzer or very low power. Start with the dipole legs each 2 feet longer than calculated and the extra wire wrapped back around the antenna — it can be trimmed after adjustment is complete. (Existing 80 meter dipoles can be converted to the TLR system by trimming them to resonance at a mid-band frequency and adding the necessary resonator segments.

Begin by adjusting the dipole legs, each by the same amount, so that the SWR curve measured at the transmitter is symmetrical about the center frequency. To offset the frequency of minimum SWR, adjust the length of the one-wavelength resonator using the formula $L_{NEW} = L_{OLD} (f_0 - \Delta f) / f_0$, where f_0 is the design frequency of the antenna system and Δf is the required offset. The length of the quarter-wavelength segment does not need to be changed to shift the SWR curve.

Table 11.2
Broadband 80 Meter Dipole Calculated and Actual Lengths

	Calculated	*Actual*
¼-λ Coax	43.3 ft	43.3 ft
1-λ Coax	173.1 ft	170.5 ft*
Dipole	124.5 ft	122.7 ft

* includes feed line used for choke balun

11.1.7 Linear Loaded Dipoles

This method of shortening radiators can be applied to almost any antenna configuration — including parasitic arrays. *Linear loading* basically consists of folding the antenna to reduce its physical length while attempting to preserve overall signal radiation. This folding can significantly reduce the required length for resonant antennas. For example, it is easy to make a resonant antenna that is as much as 30 to 40 percent shorter than an ordinary dipole for a given band.

Depending on what section of the antenna is folded and the spacing of the conductors, radiation from the folded section is relatively low because of field cancellation but because current is not constant over the length of the folded section, cancellation is not total and there is some radiation. Careful selection of what sections to fold and the use of modeling can result in an effective antenna.

Linear loading introduces relatively little loss, does not degrade directivity patterns, and has low enough Q to allow reasonably good bandwidth. Note that linear loading increases the mechanical complexity of the antenna which can cause reliability problems, especially when it is used on large antennas exposed to the wind and icing.

When experimenting with linear loading, try a length of conductor 10% to 20% longer than the difference between the shortened antenna and the full-size dimension of the folded sections. Then use the "cut-and-try" method, varying both the spacing and length of the folded section to optimize the match. A hairpin match at the feed point can be useful in achieving a 1:1 SWR at resonance.

Linear-loaded dipole antennas can be mounted either horizontally or vertically. Using different diameter conductors in linear-loaded antenna configurations yields different results, depending on whether the larger or small diameter conductor is fed.

PROJECT: LINEAR LOADED DIPOLE FOR 7 MHZ

Lew Gordon, K4VX, designed a simple 40-meter dipole using linear loading with 450 Ω window line as the folded section. Complete details are available in his article "The K4VX Linear-Loaded Dipole for 7 MHz" that is included in the online material.

Linear loading is added at the center of the dipole to simplify the mechanical design and avoid extra insulators along the antenna if the loading was added midway along the antenna. Modeling showed little difference between these two approaches. **Figure 11.15** shows the overall approach and a construction detail at the center.

Rather than place the window line below the main single-wire span, the single-wire span is threaded through holes in

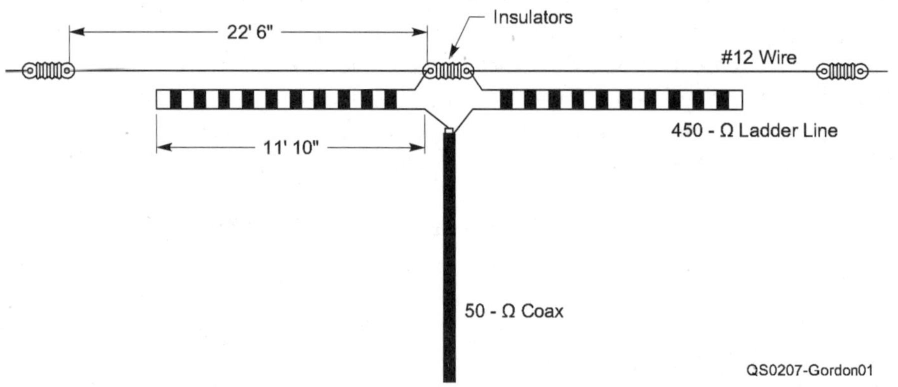

Figure 11.15 — Layout of the K4VX linear-loaded dipole (A). The #12 AWG wire is threaded through the window line but is shown separately for clarity. (See text.) At B are details of the feed point. Note that the coax feed line requires mechanical support to make the connections to the window line, which must be carefully waterproofed.

the window line spaced every 6 inches or so. This keeps the single wire centered in the window line. Concerns that the three conductors are spaced too closely were not realized on the air.

11.1.8 Inductively Loaded Dipoles

Dipoles can also be shortened by inserting loading coils in the antenna, similarly to vertical whips. Two identical coils are used, one on each side of the feed point, placed equal distances from the feed point. To accomplish a specific amount of shortening, the farther from the feed point the coils are placed, the more reactance they must have.

The most serious drawback associated with inductive loading is loss in the coils themselves. It is important that you use "high-Q" inductors made from reasonably large wire or tubing to minimize this problem. Close winding of turns should also be avoided if possible. A good compromise is to use some off-center inductive loading in combination with capacitive end loading, keeping the inductor losses small and the efficiency as high as possible.

Some examples of off-center coil loading and capacitive-end loading are shown in **Figure 11.16**. This technique was described by Jerry Hall, K1TD in September 1974 *QST* and by Luiz Duarte Lopes, CT1EOJ in October 2003 *QST*. (These articles are included with this book's online material and are listed in the References and Bibliography.) Approximate inductive reactances for single-band resonance (for the antenna in Figure 11.16A only) may be determined with the aid of **Figure 11.17**. The final values will depend on the proximity of surrounding objects in individual installations and must be determined experimentally.

The use of high-Q low-loss coils is important for maximum efficiency. This is particularly important if high power is to be used. Several calculators and online articles are available to guide the antenna builder. Serge Stroobandt, ON4AA, has made available a sophisticated inductor design calculator web page at **hamwaves.com/antennas/inductance.html** which takes into account a number of important effects that affected the accuracy of earlier calculators. Tom Rauch, W8JI, has published a great deal of information about loading coils at **w8ji.com/loading_inductors.htm**. And as an example of building high-performance loaded antennas, Steve Babcock, VE6WZ, shows construction methods for some rugged 80-meter Yagi loading coils at **qsl.net/ve6wz/coil.htm**. One caveat for the coil winder — if you use copper tubing instead of wire, it is specified by its inside diameter, not the outside diameter, as is done for wire.

An antenna analyzer is recommended for use during adjustment of the system. Note that the minimum inductance required is for a center-loaded dipole where the coil is at the feed point (B=0). If the inductive reactance is read from Figure 11.17 for a dimension B of zero, one coil having approximately twice this reactance can be used near the center of the dipole.

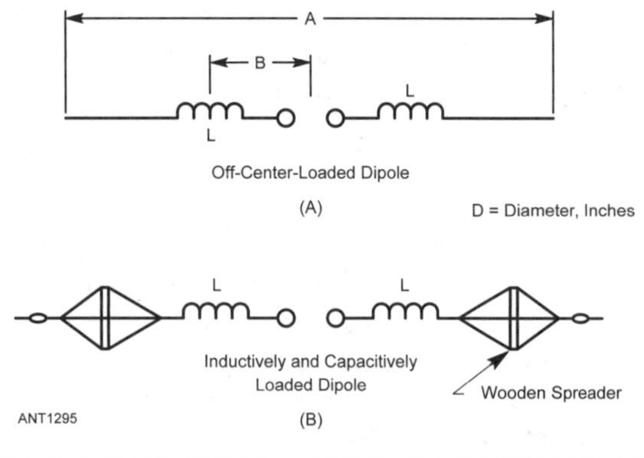

Figure 11.16 — At A is a dipole antenna lengthened electrically with off-center loading coils. For a fixed dimension A, greater efficiency will be realized with greater distance B, but as B is increase, L must be larger in value to maintain resonance. At B, capacitive loading of the ends will reduce the required inductance of the coils.

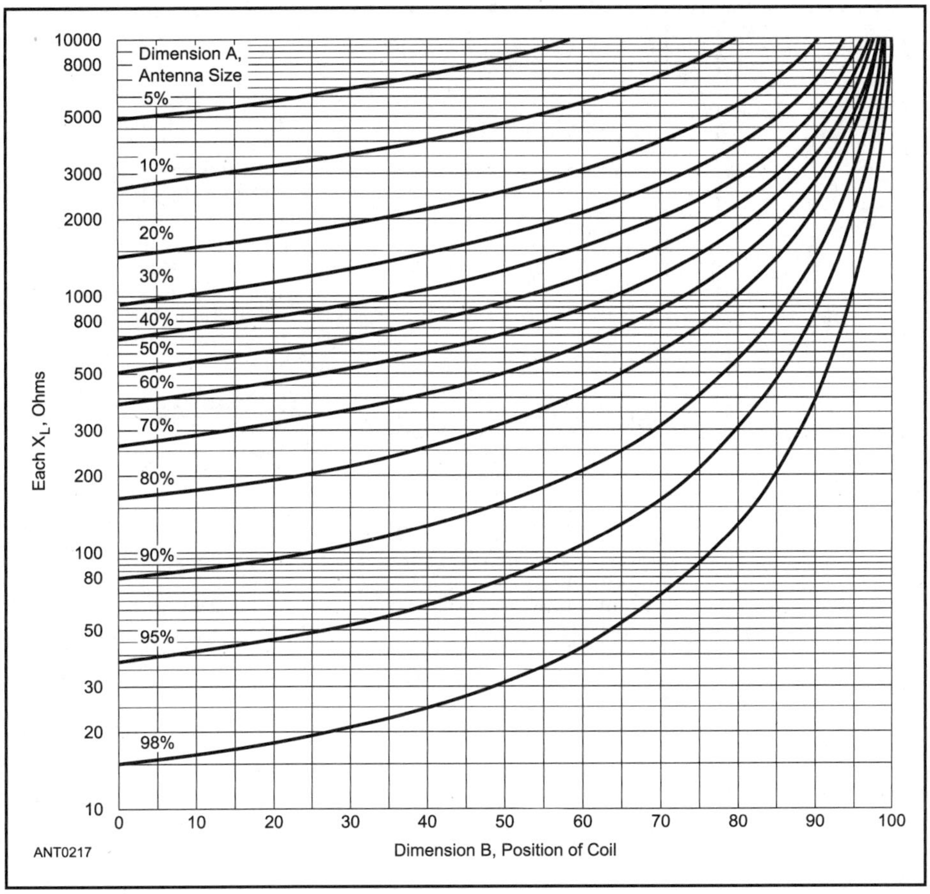

Figure 11.17 — Chart for determining approximate inductance values for off-center-loaded dipoles shown in Figure 11.16A. At the intersection of the appropriate curve from the body of the chart for dimension A and proper value for the coil position from the horizontal scale at the bottom of the chart, read the required inductive reactance for resonance from the scale at the left. Dimension A is expressed as percent length of the shortened antenna with respect to the length of a half-wave dipole of the same conductor material (that is, how much shorter than a full-size λ/2 dipole). Dimension B is expressed as the percentage of coil distance from the feed point to the end of the antenna. For example, a shortened antenna, which is 50% or half the size of a half-wave dipole (λ/4 overall) with loading coils positioned midway between the feed point and each end (50% out), would require coils having an inductive reactance of approximately 950 Ω at the operating frequency for antenna resonance.

11.2 Vertical Antennas

11.2.1 The Half-Wave Vertical Dipole (HVD)

The simplest form of vertical is that of a half-wave vertical dipole, an HVD. This is a horizontal dipole turned 90° so that it is perpendicular to the ground under it. Of course, the top end of such an antenna must be at least a half wave above the ground or else it would be touching the ground. This poses quite a construction challenge if the builder wants a free-standing low-frequency antenna. Hams fortunate enough to have tall trees on their property can suspend wire HVDs from these trees. Similarly, hams with two tall towers can run rope catenaries between them to hold up an HVD.

A vertical half-wave dipole has some operational advantages compared to a more-commonly used vertical configuration — the quarter-wave vertical used with some sort of above-ground counterpoise or an on-ground radial system. See **Figure 11.18**, which shows the two configurations discussed here. In each case, the lowest part of each antenna is 8 feet above ground, to prevent passersby from being able to touch any live wire. Each antenna is assumed to be made of #14 AWG wire resonant on 80 meters.

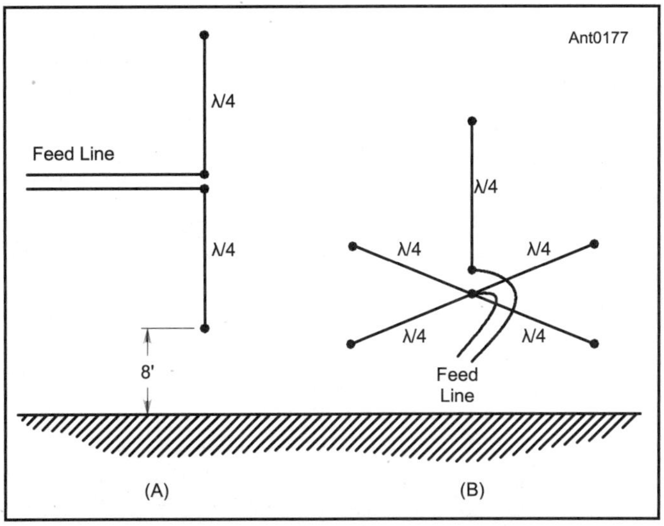

Figure 11.18 — At A, an 80-meter half-wave vertical dipole elevated 8 feet above the ground. The feed line is run perpendicularly away from the dipole. At B, a "ground plane" type of quarter-wave vertical, with four elevated resonant radials. Both antennas are mounted 8 feet above the ground to keep them away from passersby.

Figure 11.19 compares elevation patterns for the two

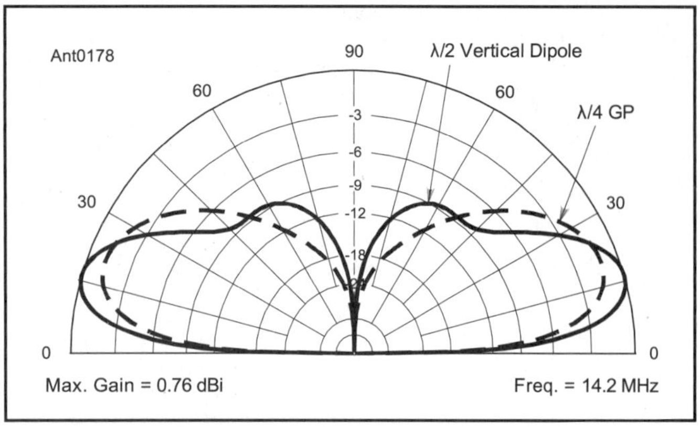

Figure 11.19 — A comparison of the elevation patterns for the two antennas in Figure 11.16. The peak gain of the HVD is about 1.5 dB higher than that for the quarter-wave ground plane radiator with radials.

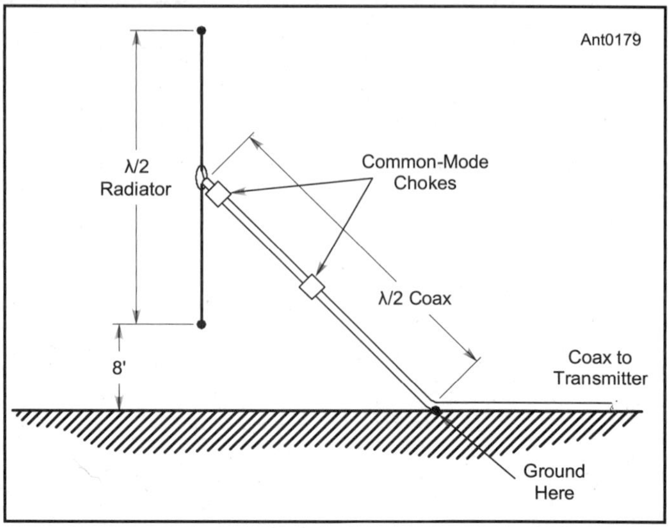

Figure 11.20 — A 20-meter HVD whose bottom is 8 feet above ground. This is fed with a λ/2 of RG-213 coax. This system uses a common-mode choke at the feed point and another λ/4 down the line. The resulting azimuthal radiation pattern is within 0.4 dB of being perfectly circular. The "wingspan" of this antenna system is 27 feet from the radiator to the point where the coax comes to ground level.

antennas for "average ground." You can see that the half-wave vertical dipole has about 1.5 dB higher peak gain, since it compresses the vertical elevation pattern down somewhat closer to the horizon than does the quarter-wave ground plane. Another advantage to using a half-wave radiator besides higher gain is that less horizontal "real estate" is needed compared to a quarter-wave vertical with its horizontal radials.

The obvious disadvantage to an HVD is that it is taller than a quarter-wave ground plane. This requires a higher support (such as a taller tree) if you make it from wire, or a longer element if you make it from telescoping aluminum tubing.

CENTER-FEEDING A VERTICAL DIPOLE

To minimize coupling between the feed line and the antenna, you must arrange the feed line so that it is perpendicular to the half-wave radiator. This means you must support the coax feed line above ground for some distance before bringing the coax down to ground level. A question immediately arises: How far must you go out horizontally with the feed line before going to ground level to eliminate common-mode currents that are radiated onto the coax shield? Such common-mode currents will affect the feed point impedance as well as the radiation pattern for the antenna system. Quite a bit of distortion in the azimuthal pattern can be created if common-mode currents aren't suppressed, usually by using a common-mode choke balun.

Constructing such a choke is very simple: ferrite beads of type #31 or #43 mix are slipped over the coax (before the connectors are soldered on or else they won't fit!) and taped in place. The only problem with this scheme is that an additional support (some sort of "skyhook") is required to support the coax horizontally. Let's try to simplify the installation, by slanting the feed line coax down to ground from the feed point at a fairly steep angle of about 30° from vertical. See **Figure 11.20**.

Note that the bottom end of the coax in Figure 11.20 is grounded to a ground rod. This serves as a mechanical connection to hold the coax in place and it provides some protection against lightning strikes. Now, as a purely practical matter, just how picky are we being here? What if we skip the second common-mode choke and just use one at the feed point? The computer model predicts that there will be some distortion in the azimuthal pattern — about 1.1 dB worth. Whether this is serious is up to you. However, you may find other problems with common-mode currents on the coax shield — problems such as "RF in the shack" or variable SWR readings depending on the way coax is routed in the station. The addition of three extra ferrite beads to suppress the common-mode currents is cheap insurance.

A variation on the HVD that is shortened through the use of capacitive loading is the *Compact Vertical Dipole (CVD)*. An article describing the CVD is included with this book's online material.

PROJECT: END-FED VERTICAL DIPOLE

The problem of coupling between the feed line and the antenna is minimized by making the feed line part of the antenna as shown in **Figure 11.21**. In this design by Jim Brown, K9YC, and Glen Brown, W6GJB, a quarter-wavelength of coaxial cable forms the bottom leg of the dipole and is terminated by a high-impedance ferrite choke. The upper leg is a conventional quarter-wave wire. This design requires just the single support and has none of the issues with trying to run the feed line at right angles to the antenna. The full presentation by K9YC is available at **k9yc.com/VerticalDipole.pdf**.

The challenge with this design is to ensure that the ferrite choke at the bottom of the antenna has enough choking impedance at the frequency of use to act like an insulator and to not overheat from the high voltage across it. The choke can be made of type #31 or #43 ferrite and should be resonant (for

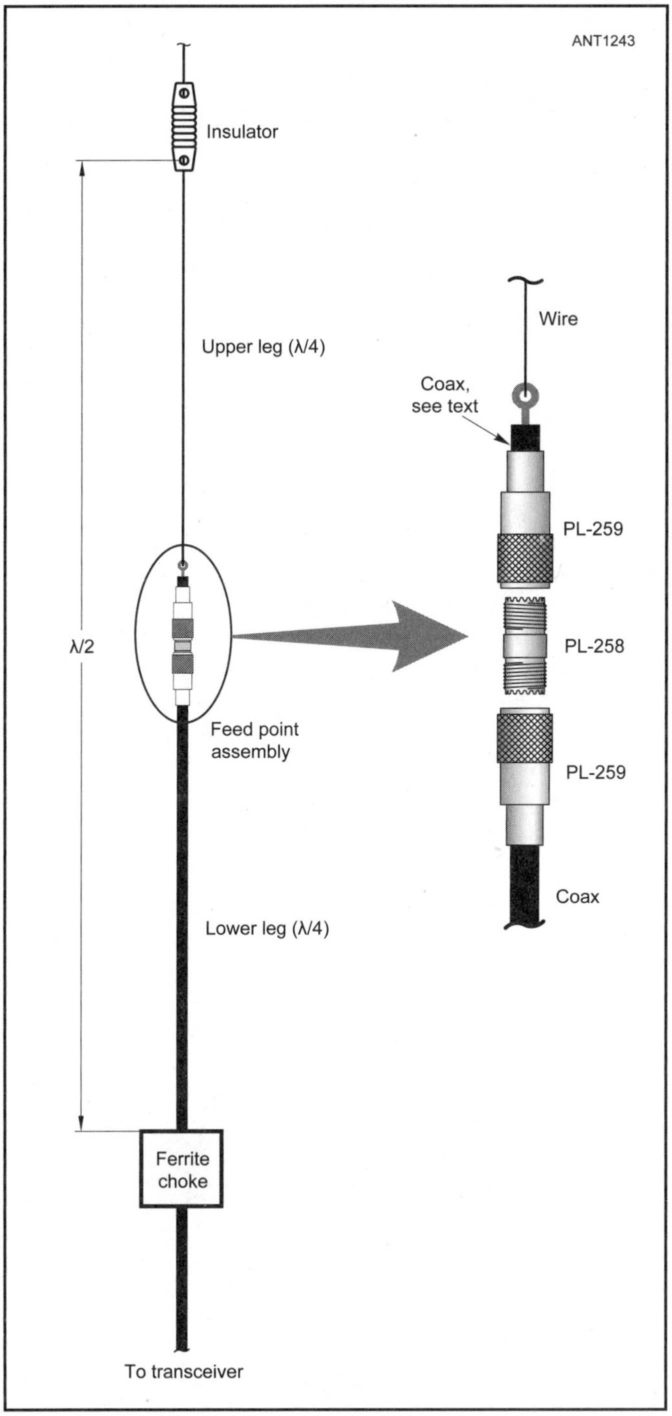

Figure 11.21 — The vertical dipole shown here uses coaxial cable as the lower leg. The upper and lower leg are connected together by using a PL-259 on each and a PL-258 to join them. The feed point assembly must be carefully weatherproofed. A ferrite choke acts as an insulator for the bottom leg (see text).

maximum impedance) near the operating frequency. If the choking impedance is not high enough, the choke will overheat. For 1500 W operation, at least 10 kΩ is needed; for 500 W, 5 kΩ; and for 100 W or less, 1–2 kΩ is sufficient. (See the sections on ferrite chokes in the **Transmission Line System Techniques** chapter.)

Use coaxial cable with a copper braid shield for power levels above QRP. Do not use cable with only a foil or thin braid shield. The dipole's impedance will be around 70 Ω so either 75-Ω or 50-Ω coax will work.

The center connection is shown as a detail in Figure 11.19 and consists of two PL-259 plugs connected by a PL-258 "barrel" connector. The upper PL-259 should have a short piece of RG-8 or RG-213 inserted with only the center conductor soldered to the connector pin. The coax should extend out of the body of the connector for at least ¼ inch so that the center conductor is insulated from the body by the coax jacket. Extend the center conductor for an inch or so and form a small loop. The upper leg wire is then soldered to this loop. The lower PL-259 is fitted to the coax feed line as usual.

11.2.2 Monopole Verticals With Ground-Plane Radials

For best performance the vertical portion of a ground-plane type of antenna should be λ/4 or more, but this is not an absolute requirement. With proper design, antennas as short as 0.1 λ or even less can be efficient and effective. Antennas shorter than λ/4 will be reactive and some form of loading and perhaps a matching network will be required.

If the radiator is made of wire supported by nonconducting material, the approximate length for λ/4 resonance can be found from:

$$\ell_{feet} = \frac{234}{f_{MHz}} \quad (1)$$

The same cautions about the effects of ground and wire or tubing diameter apply to this equation for verticals. For a tower, the resonant length will be shorter still. It is recommended that the builder start a few percent long and trim the antenna to length based on measurements taken with the antenna in place. (See the **Dipoles and Monopoles** chapter.)

The effect of ground characteristics on losses and elevation pattern is discussed in detail in the chapter **Effects of Ground**. The most important points made in that discussion are the effect of ground characteristics on the radiation pattern and the means for achieving low ground-loss resistance in a buried ground system. As ground conductivity increases, low-angle radiation improves. This makes a vertical very attractive to those who live in areas with good ground conductivity. If your QTH is on a saltwater beach, then a vertical would be very effective, even when compared to horizontal antennas at great height.

When a radial ground system is used, the efficiency of the antenna will be limited by the loss resistance of the ground system. The ground can be a number of radial wires extending out from the base of the antenna for about λ/4. Note that radial wires on or in the ground are not tuned or resonant in the conventional sense because they couple strongly to the soil. That they are often specified to be λ/4 long stems from a study of broadcast antenna systems in the 1930s that resulted in the current standard of 120 radials, each long enough such that their ends are approximately 0.05-λ apart. That system was designed to minimize ground losses and the radials are approximately ¼ λ long but not resonant or tuned.

If the radials are to be buried, only enough depth to keep them in the ground throughout the year is required. Burying radials more than a few inches deep reduces their effectiveness, particular on the higher HF bands. Radial wires should be close to or on the surface for maximum effectiveness. Radials can be effectively hidden by mowing grass or cutting vegetation as short as practical, then laying down the radials and holding them against the ground with lawn staples or short lengths of iron rebar-tie wire bent double. Grass will quickly grow over the radials, effectively hiding them while keeping them close to the soil surface. The action of plant growth, insects, and worms will eventually bury the radials in soil.

Driven ground rods, while needed for electrical safety and for lightning protection, are of little value as an RF ground for a vertical antenna, except perhaps in marshy or beach areas. As pointed out, many long radials are desirable. In general, however, a large number of short radials are preferable to only a few long radials, although the best system would have 60 or more radials longer than λ/4. An elevated system of radials or a ground screen (*counterpoise*) may be used instead of buried radials, and can result in an efficient antenna. **Figure 11.22** illustrates the difference between buried and elevated radial systems. The reader is directed to the chapter **Effects of Ground** for a discussion of ground plane radial systems for vertical monopole antennas.

11.2.3 Ground-Plane Antennas

The ground-plane antenna is a λ/4 vertical with four radials, as shown in **Figure 11.23**. The entire antenna is elevated above ground. A practical example of a 7-MHz ground-plane antenna is given in **Figure 11.24**. As explained earlier, elevating the antenna reduces the ground loss and lowers the radiation angle somewhat. The radials are sloped downward to make the feed point impedance closer to 50 Ω. (Also see the discussion of elevated radial systems in the chapter **Effects of Ground**.)

The feed point impedance of the antenna varies with the height above ground, and to a lesser extent varies with the ground characteristics. **Figure 11.25** is a graph of radiation resistance (R_R) for a ground-plane antenna with the radials parallel to the ground. R_R is plotted as a function of height above ground. Notice that the difference between perfect ground and average ground ($\varepsilon = 13$ and $\sigma = 0.005$ S/m) is small, except when quite close to ground. Near ground R_R is between 36 and 40 Ω. This is a reasonable match for 50-Ω feed line but as the antenna is raised above ground R_R drops to approximately 22 Ω, which is not a very good match. The feed point resistance can be increased by sloping the radials

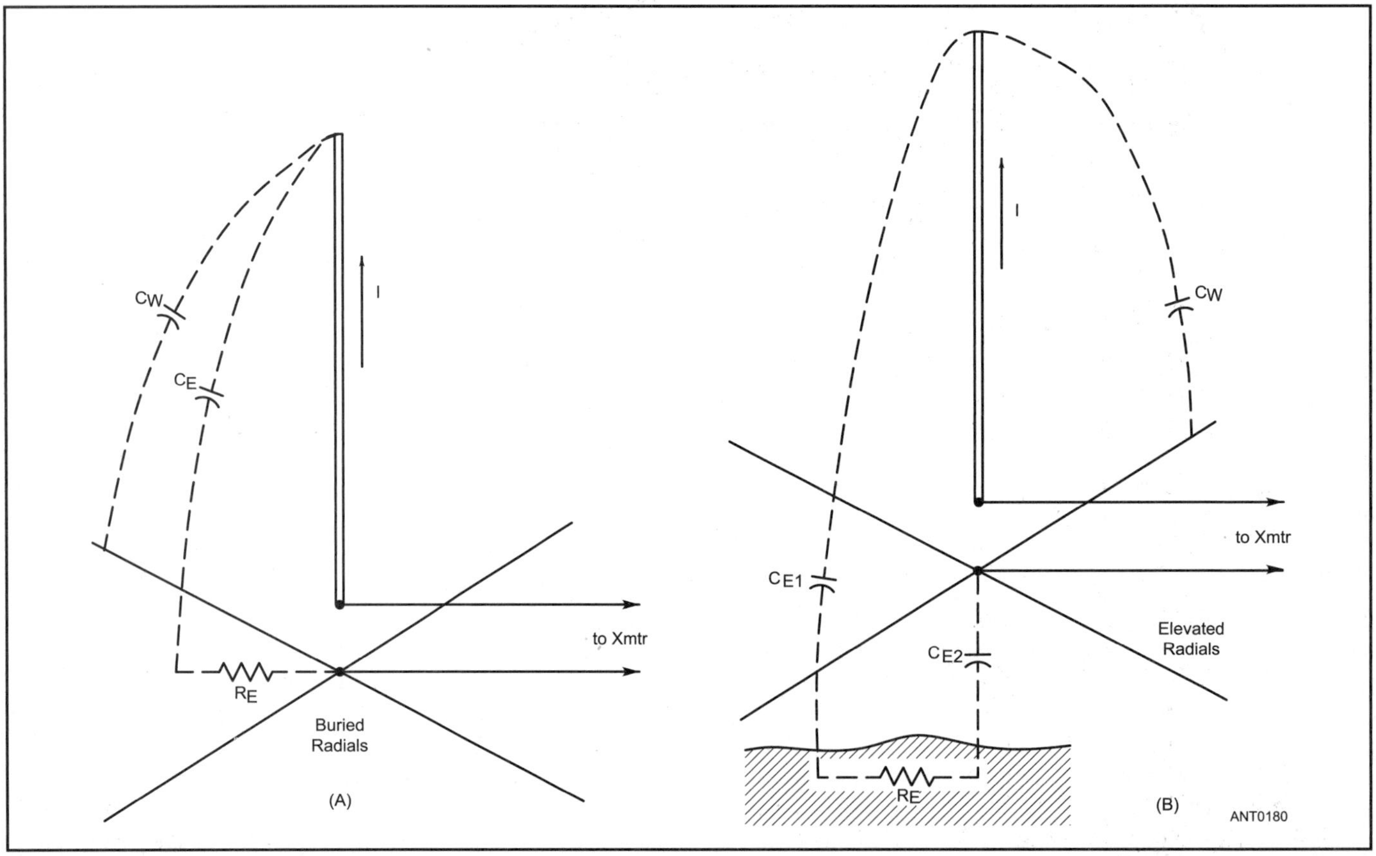

Figure 11.22 — How earth currents affect the losses in a short vertical antenna system. At A, the current through the combination of C_E and R_E may be appreciable if C_E is much greater than C_W, the capacitance of the vertical to the ground wires. This ratio can be improved (up to a point) by using more radials. By raising the entire antenna system off the ground, C_E (which consists of the series combination of C_{E1} and C_{E2}) is decreased while C_W stays the same. The radial system shown at B is sometimes called a counterpoise.

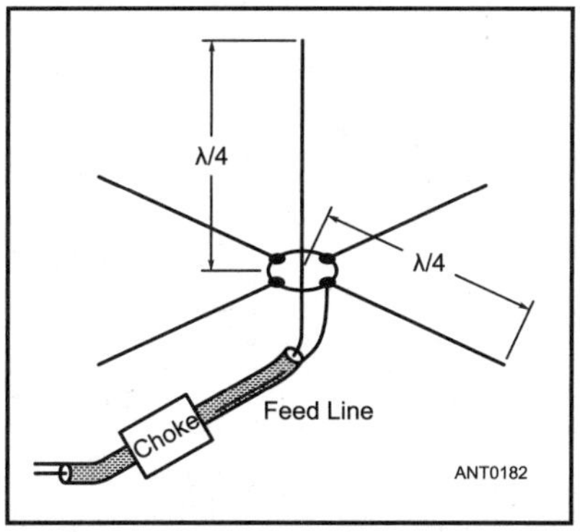

Figure 11.23 — The ground-plane antenna. Power is applied between the base of the vertical radiator and the center of the ground plane, as indicated in the drawing. Decoupling from the transmission line and any conductive support structure is highly desirable.

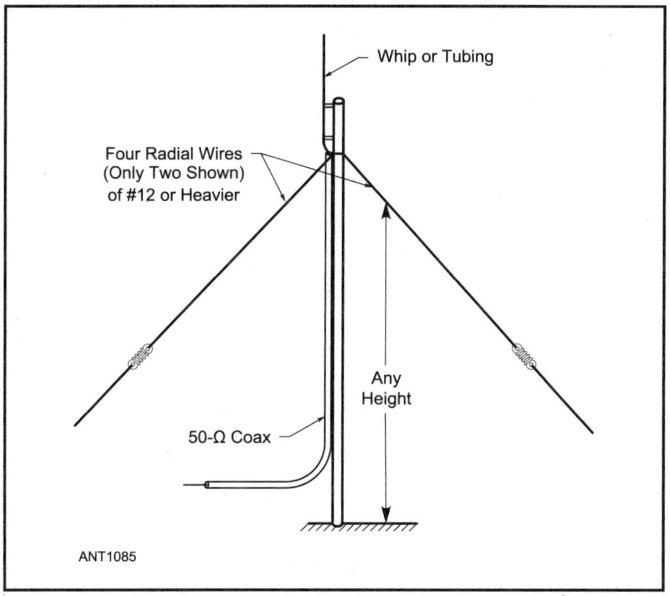

Figure 11.24 — A ground-plane antenna is effective for DX work on 7 MHz. Although its base can be any height above ground, losses in the ground underneath will be reduced by keeping the bottom of the antenna and the ground plane as high above ground as possible. Feeding the antenna directly with 50-Ω coaxial cable will result in a low SWR. The vertical radiator and the radials are all λ/4 long electrically. The radial's physical length will depend on their length-to-diameter ratios, the height over ground and the length of the vertical radiator, as discussed in text.

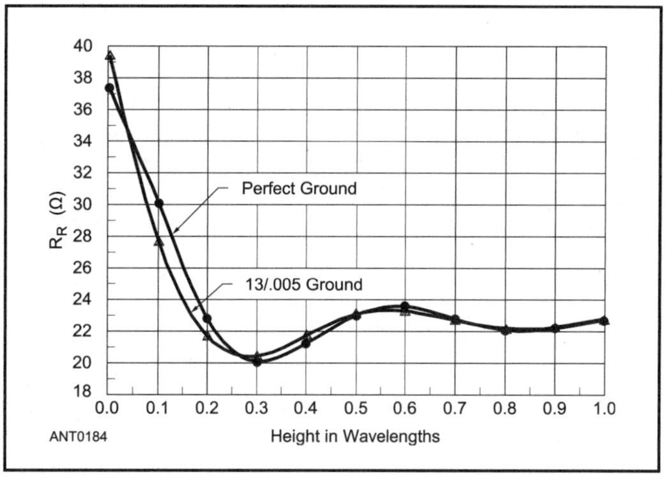

Figure 11.25 — Radiation resistance of a 4-radial groundplane antenna as a function of height over ground. Perfect and average ground are shown. Frequency is 3.525 MHz. Radial angle (θ) is 0°.

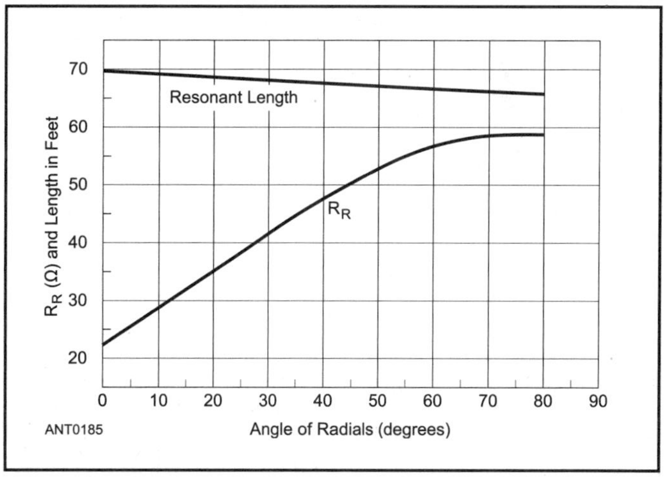

Figure 11.26 — Radiation resistance and resonant length for a 4-radial ground-plane antenna > 0.3 l above ground as a function of radial droop angle (θ).

downward, away from the vertical section.

The effect of sloping the radials is shown in **Figure 11.26**. The graph is for an antenna well above ground (> 0.3 λ). Notice that R_R = 50 Ω when the radials are sloped downward at an angle of 45°, a convenient value. The resonant length of the antenna will vary slightly with the angle. In addition, the resonant length will vary a small amount with height above the ground. It is for these reasons, as well as the effect of conductor diameter, that some adjustment of the radial lengths is usually required. When the ground-plane antenna is used on the higher HF bands and at VHF, the height above ground is usually such that a radial sloping angle of 45° will give a good match to 50-Ω feed line.

The effect of height on R_R with a radial angle of 45° is shown in **Figure 11.27**. Below 7 MHz, it is seldom possible to elevate the antenna a significant portion of a wavelength and the radial angle required to match to 50-Ω line is usually of the order of 10° to 20°. To make the vertical portion of the antenna as long as possible, it may be better to accept a slightly poorer match and keep the radials parallel to ground.

The principles of the folded dipole discussed earlier can also be applied to the ground-plane antenna, as shown in **Figure 11.28**. This is the *folded monopole* antenna. The feed point resistance can be controlled by the number of parallel vertical conductors and the ratios of their diameters.

As mentioned earlier, it is important in most installations to isolate the antenna from the feed line and any conductive

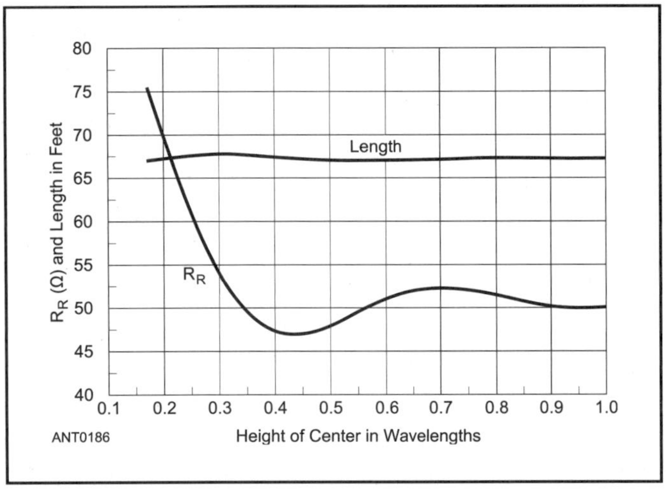

Figure 11.27 — Radiation resistance and resonant length for a 4-radial ground-plane antenna for various heights above average ground for radial droop angle θ = 45°.

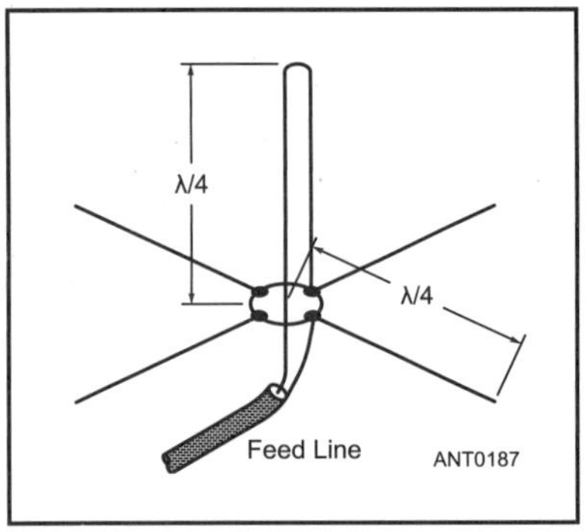

Figure 11.28 — The folded monopole antenna. Shown here is a ground plane of four λ/4 radials. The folded element may be operated over an extensive counterpoise system or mounted on the ground and worked against buried radials and the earth. As with the folded dipole antenna, the feed point impedance depends on the ratios of the radiator conductor sizes and their spacing.

supporting structure. This is done to minimize the return current conducted through the ground. A return current on the feed line itself or the support structure can drastically alter the radiation pattern, usually for the worse. For these reasons, a choke balun (see the chapter **Transmission Line System Techniques**) or other isolation scheme must be used. 1:1 baluns are effective for the higher bands but at 3.5 and 1.8 MHz commercial baluns often have insufficient impedance to provide adequate isolation. It is very easy to recognize when the isolation is inadequate. When the antenna is being adjusted while watching an isolated impedance or SWR meter, adjustments may be sensitive to your touching the instrument. After adjustment and after the feed line is attached, the SWR may be drastically different. When the feed line is inadequately isolated, the apparent resonant frequency or the length of the radials required for resonance may also be significantly different from what you expect because the feed line's outer surface has become part of the antenna.

A choke balun is used to isolate the feed line shield as discussed in the **Transmission Line System Techniques** chapter. An impedance of 500 Ω is usually sufficient but in difficult cases, up to 5000 Ω may be required. A ferrite choke with resistive material is preferred but the wound-coax choke shown in **Figure 11.29** will do the job on a single-band antenna. For 1.8 MHz, 30 turns of RG-213 wound on a 14-inch length of 8-inch diameter PVC pipe will make a very good choke balun that can handle full legal power continuously. A smaller choke could be wound on 4-inch diameter plastic drain pipe using RG-8X or a Teflon insulated cable. The important point here is to isolate or decouple the antenna from the feed line and support structure.

11.2.4 PROJECT: The ⅜ λ Vertical

The ⅜-wavelength vertical shown in **Figure 11.30** is often overlooked. It has several advantages over the ¼ λ vertical while adding just 50% to the height. When ground-mounted, its vertical takeoff angle is a few degrees lower than the ¼-λ which is important for DX contacts. The current maximum is above the ground by ¼ λ which keeps the maximum radiation point clear of ground clutter. It is easily matched, has a wide operating bandwidth, and a high radiation resistance. ⅜-λ and ⅝-λ verticals are discussed in the Vertical Antennas chapter of *ON4UN's Low-Band DXing*.

For 14 MHz and higher-frequency operation, the antenna can be constructed from copper wire or aluminum tubing as shown in **Figure 11.31A**. This is an easy antenna to construct from the pieces of old antennas on the upper HF bands. A complete construction article by Joe Reisert, W1JR, is included in this chapter's online material.

The antenna has a series impedance of approximately 200 Ω of resistance in series with an inductive reactance of 300 to 700 Ω. A series capacitor will cancel the inductive reactance and a 4:1 impedance transformer will convert the

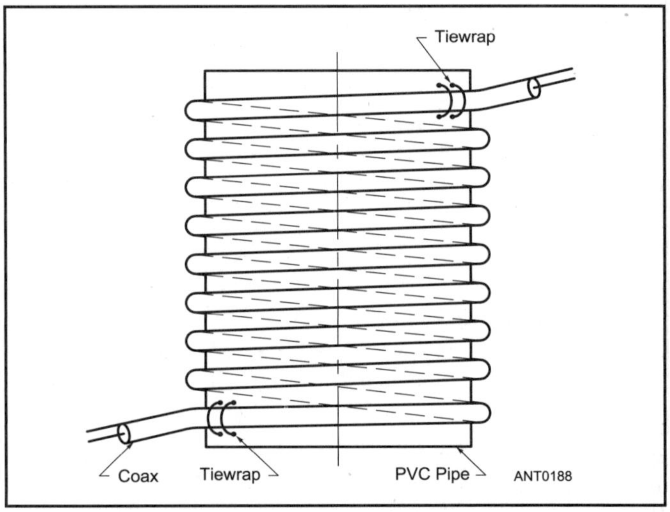

Figure 11.29 — A wound-coax choke balun with sufficient impedance to isolate the antenna properly can be made by winding coaxial cable around a section of plastic pipe. Suitable dimensions are given in the text.

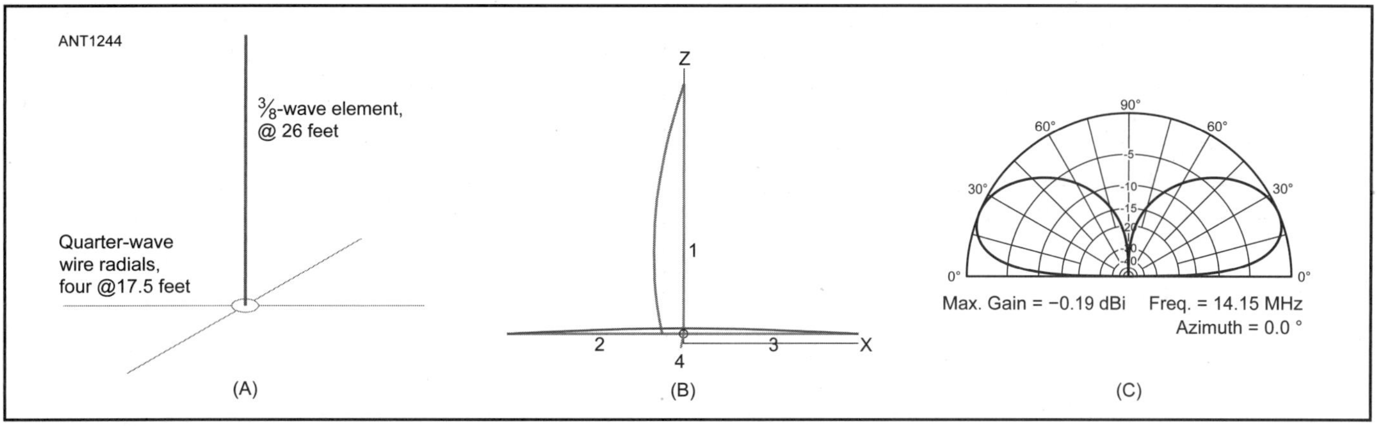

Figure 11.30 — The ⅜ λ vertical (A) is 50% taller than the ¼ λ vertical. Note that the current distribution on the vertical (B) shows the current maximum to be about ¼ λ above the base. The elevation pattern of the ground mounted ⅜ λ wavelength vertical is shown at C.

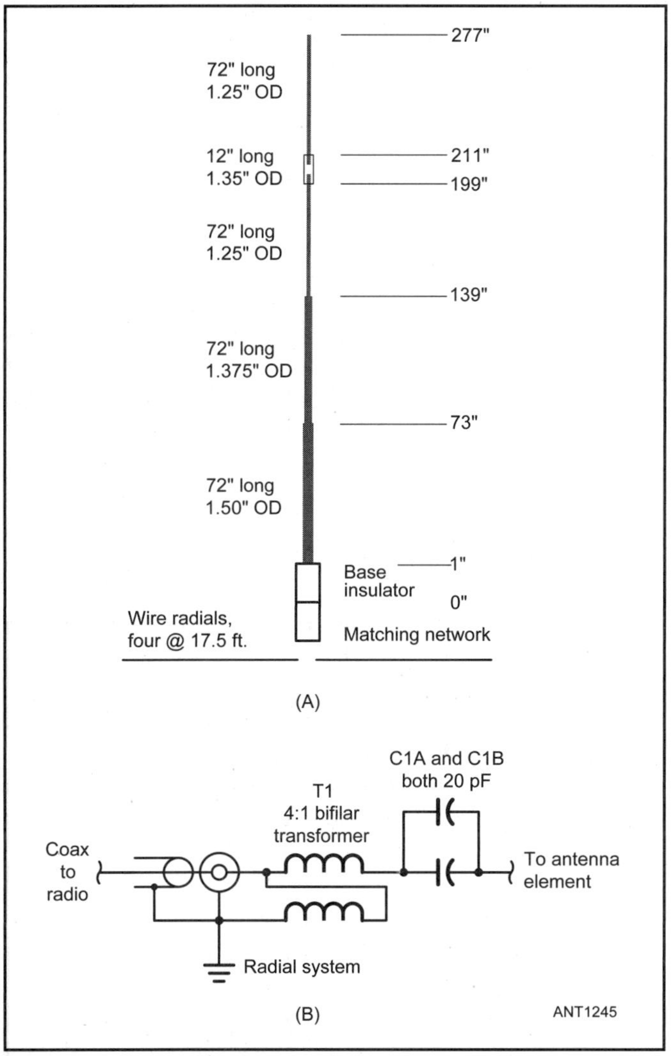

Figure 11.31 — Tubing construction for the ⅜ λ vertical (A) can use clamps, rivets, or screws to hold the sections together. The 12-inch section near the top is a splice to connect the top section. The impedance matching network is shown at B — see text for component values and rating.

remaining resistance to approximately 50 Ω for attaching a coaxial feed line. At 14 MHz, the typical series capacitance required is 40 to 50 pF and is not critical. See the **Transmission Line System Techniques** chapter for 4:1 transformer designs.

Ground radials can be made of wire of any gauge or material. PVC-insulated #14 AWG is recommended so that the ground characteristics will have little effect on long-term performance. Only four radials are needed, approximately 15–17 feet long at 14 MHz, and length is not critical.

11.2.5 Shunt-Feeding Towers

A tower can be used as a vertical ground plane antenna, provided that a good ground system is in place. Most towers are grounded at the base, so they aren't fed directly. The most common method of feeding a grounded tower as a vertical antenna is *shunt feeding*. Shunt feeding is basically a gamma match applied to a tower. (More detail about the gamma match is available in the "T and Gamma Matching" section of the **Transmission Line System Techniques** chapter.) Other matching networks, such as the L and omega networks, are also described in the same chapter.

The shunt-fed tower is at its best on 3.5 and 1.8 MHz when a separate or full-size λ/4 vertical antenna can't be constructed. Shunt feeding is also practical on 7 MHz for short towers. Beams on a mast at and above the top of the tower provide additional top loading to increase its electrical length. If a tower is too short for the gamma match approach, an omega match may be required to match inductive feed point impedances below 50 Ω.

There is no single formula that will generate the dimensions and component values for shunt feeding. Building your system will be a process of measuring and adjusting for the best match and bandwidth. You may want to ask others with successful shunt-fed towers about their systems and how they arrived at the basic dimensions. There are a lot of online resources showing various shunt-fed towers.

A detailed discussion of using towers as vertical antennas for low-band operation can be found in the Vertical Antennas chapter of *ON4UN's Low-Band DXing*. A 30-minute video on shunt-feeding towers by Steve Babcock, VE6WZ, is available online at **youtube.com**: search for "How To Shunt Feed A Tower VE6WZ". He has also made a short video on omega matching: search for "Omega Match-Shunt Fed Tower VE6WZ" to find it. Tom Rauch, W8JI, also discusses gamma and omega matching at **w8ji.com/omega_and_gama_matching.htm**.

PREPARING THE TOWER

Usually, some work must be done before shunt-feeding is tried. If present, metallic guys should be broken up with insulators. They can be made to provide top loading, if needed, by judicious placement of the first insulators. Don't overdo it; there is no need to "tune the radiator to resonance" in this way since a shunt feed is employed. Resonant guy wire lengths for the band on which the tower will radiate should be avoided.

If the tower is attached above its base to a building, insulate the tower from the supporting material. If you are using a standard house bracket insert a PVC or Teflon sleeve between the U-bolt and tower leg. Otherwise, a high-strength plastic sheet, such as Delrin or HDPE, can be used as an insulator.

All feed lines should be taped to the tower, on the inside, and run all the way to ground level where voltage on the tower is very low. It is not necessary to bond shielded cables to the tower electrically except for lightning protection, but there should be no exceptions to the down-to-the-ground rule. It is also not necessary to bond the nested sections of a crank-up tower together if the tower is all-metal construction.

A good ground system for RF transmitting, well-bonded to the tower base, is required as described in the **Effects of Ground** chapter. Losses in the ground system affect the shunt feed adjustments and will reduce radiated signal strength. Lightning ground systems consisting solely of ground rods or ground rods with a single buried radial wire are insufficient. The feed point network must be well-bonded to the ground system, as well.

BASIC SHUNT-FEED STRUCTURE

The basic shunt-fed tower is shown in **Figure 11.32**. The tower is assumed to be well-grounded as described in the preceding section and of all-metal construction. The tower diameter will affect the behavior of the shunt-feed system and the frequency of the impedance match as described below.

The shunt-feed gamma consists of the *shunt arm* installed up the tower at the *tap point*, the *gamma wire*, and the *gamma capacitor*. The simplest description of the system is that the shunt arm is installed at a point where the gamma wire transforms the impedance at the tap point to $50 + jX\ \Omega$ and X is some amount of inductive reactance. The gamma capacitor is then selected to cancel the inductive reactance, leaving a feed point impedance of $50 + j0\ \Omega$. The diameter of the gamma wire, its spacing from the tower, and the placement of the shunt arm on the tower all affect the frequency at which the tower can be matched.

The shunt arm can be made of any conducting metal bar, rod, tubing, or angle that is strong enough to hold the gamma wire or wires in the expected weather conditions. If the tower will be climbed regularly, plan for the arm to be stepped on or gripped as the climber goes up and down. The gamma wire should be at least 12 inches from the tower and be #12 AWG or heavier wire or tubing. Two or three gamma wires in parallel will increase matching bandwidth.

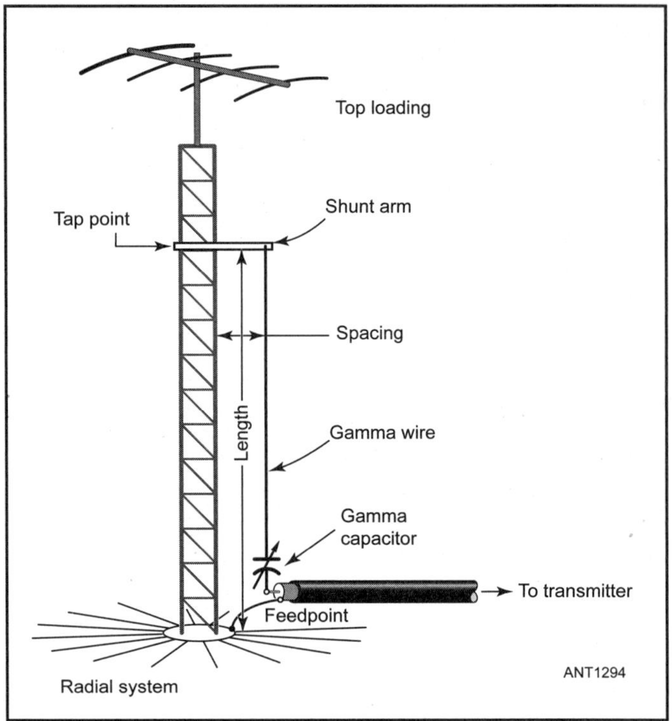

Figure 11.32 — The basic structure for shunt feeding a tower using a gamma match system.

Top Loading

In a typical grounded-tower-and-Yagi installation, a top-mounted Yagi acts as a large capacitance hat, top loading the tower. Fortunately, top loading is the most efficient means of loading a vertical antenna. The examples in **Table 11.3** give an idea of how much top loading might be expected from typical amateur antennas. The values listed in the *Equivalent Loading* column approximate the vertical height replaced by the antennas listed in a top-loaded vertical. Note that for all but the 10 meter antennas, the equivalent loading equals or exceeds a quarter wavelength on 40 meters. See the section Loading Techniques for Short Antennas in the **Dipoles and Monopoles** chapter for more information about top-loading.

Capacitor Ratings

The gamma capacitor must be capable of carrying the current at whatever power level the station uses. RMS current through the capacitor is $I = \sqrt{(P / Z_0)}$ where Z_0 is the feed line characteristic impedance, assumed to be 50 Ω in this section. At full power, current in a 50 Ω system will be $\sqrt{(1500 / 50)} = 5.5\ A_{RMS}$.

The capacitor must also withstand the peak voltage created by that current: $V_{PK} = I_{PK} \times X_C = I_{RMS} \times 1.4 \times X_C$. This can

Table 11.3
Effective Loading of Common Yagi Antennas

Antenna	*Boom Length (feet)*	*Area (sq ft)*	*Equivalent Loading (feet)*
3L 20	24	768	39
5L 15	26	624	35
4L 15	20	480	31
3L15	16	384	28
5L 10	24	384	28
4L 10	18	288	24
3L 10	12	192	20
TH7	24	-	40 (estimated)
TH3	14	-	27 (estimated)

be several kV so a high-voltage transmitting capacitor will be required. For example, if SWR is 2:1, capacitor current would be 35 percent higher and peak voltage would double over that for an SWR of 1:1.

Safety factors for the current rating should be at least 50 percent at the maximum expected SWR. Avoid variable capacitors with small current-carrying bearings or sliding contacts that can heat up. Voltage safety rating should be enough to handle SWR spikes and environmental stresses like moisture or dust so at least double the peak expected voltage. Remember that using the antenna on frequencies where SWR is high will also raise the voltage across the capacitor.

Necessary Test Equipment

To adjust the shunt-feed system, you should have an antenna analyzer capable of measuring R and X in the band to be used. Wider frequency ranges are useful, particularly at the beginning of the process when the feed point impedance at the design frequency is likely to be quite different than $50 + j0\ \Omega$! Measurements should be made in the series circuit form: R_S and X_S. This is the default on most analyzers.

If you are using a vector analyzer, you'll get the best results if you perform a calibration that includes any connecting cable between the analyzer and feed point. See the **Antenna and Transmission Line Measurements** chapter for more about analyzer calibration.

SHUNT ARM PLACEMENT

Deciding where to place the shunt arm and how long to make it are often daunting for first-time shunt-feed builders. Three methods of obtaining a reasonable initial position and adjusting the position are presented here: modeling and a gamma calculator, using charts from *Low-Band DXing*, and starting at the top and estimating a lower position. To account for any top loading from beams, use Table 11.3 to add additional length to the tower. Lengths of mast or whips above one or more beams will increase electrical length but not equivalent to their full physical length. An initial shunt arm length of 18 to 24 inches is a good starting point.

Modeling

If you can model the tower (including any top loading), find an approximate base impedance and use the *GAMMAMW9a* spreadsheet at **n6mw.jimdofree.com/antenna-matching** to calculate where the shunt arm should be placed. For a standard triangular lattice tower, the effective diameter is about ¾ of the face width. (A 12-inch tower face results in a 9-inch effective diameter.)

The design example in the T and Gamma Matching section referenced earlier starts with a 50-foot tower with a beam on the top being shunt-fed on 80 meters. Modeling calculates a base impedance of $20 + j100\ \Omega$. After the first calculation results in a tap point above the tower top, gamma wire spacing is changed to 24 inches, resulting in a gamma wire length of 44 feet (this is where the shunt arm is placed), and a series capacitor of 51 pF. This is a typical process: model the base impedance, choose an initial gamma wire diameter and spacing, and if the results are not suitable, adjust the choices until the resulting dimensions are practical.

Charts from ON4UN's Low-Band DXing

Another option is to use the chart of estimated electrical lengths from *Low-Band DXing* in **Figure 11.33**. The chart assumes a frequency of 1.825 MHz and an effective tower diameter of 30 cm (14 inches) which is close to Rohn 25. Start at your tower's height on the horizontal axis, not including any top-loading antennas. Pick the type of the antenna closest to the top of the tower and travel upward from the horizontal axis until you reach the line representing your chosen antenna type. Travel horizontally to the vertical axis to read electrical length in degrees. If the lowest antenna is the largest — a typical installation — it will have the greatest effect so it will do the most to determine electrical length. For example, at 1.825 kHz,

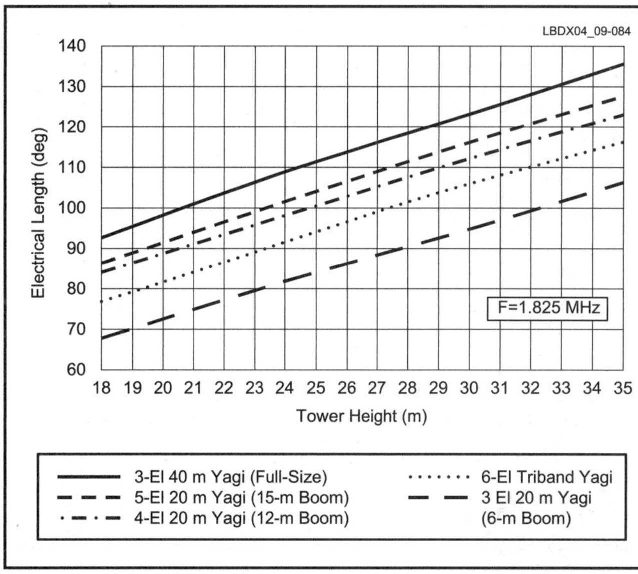

Figure 11.33 — Electrical length of a tower loaded with a Yagi antenna. The chart is valid for 160 meters (1.825 MHz) and a tower with an equivalent diameter of 30 cm (12 inches). For a larger tower diameter, the electrical length will be shorter (4 to 7 degrees for a 60 cm (24 inch) tower).

a 25-meter tower with a large (6-element) tribander on top is estimated at 99 degrees of electrical length. (The tower will be approximately twice as long, electrically, on 80 meters.)

If you have more antennas above the largest, use the next largest antenna type when picking electrical length. In tower configurations with multiple antennas of various sizes, at various heights, the charts from *Low-Band DXing* are a good starting point using the largest antenna. To one degree or another however, all the antennas will affect the apparent electrical length of the combination, meaning you'll arrive at the final match through repeated adjust-and-measure cycles.

You can also create a feed point in the wire (see the VE6WZ video) and use an analyzer to find the lowest resonant frequency. This is the ¼-wavelength resonant frequency, $f_{1/4\text{-wave}}$. Calculate the electrical length at your design frequency as: L (in degrees) = 90 × (f_{design} / $f_{1/4\text{-wave}}$). For example, if your tower is ¼-wavelength long at 1.7 MHz, its electrical length at 1.85 MHz is 90 × (1.85 / 1.7) = 97 degrees. At 3.8 MHz, it will be 90 × (3.8 / 1.7) = 163 degrees.

Now use the chart in **Figure 11.34** to pick an initial tap height and series capacitance. For the 99-degree tower on 160 meters and a gamma wire spacing of 1 meter, the chart indicates a tap position of 14 meters. The series capacitor is approximately 450 pF.

Start At the Top

Another more empirical method is to attach the shunt arm at the top of the tower and use an antenna analyzer at the feed point to measure R and X. Resistance should be greater than 50 Ω and reactance should be positive (inductive). If these conditions are met, proceed to the next paragraph. If resistance is less than 50 Ω, try increasing the gamma wire spacing. If reactance is negative (capacitive), then the tower is too short for the single series capacitor method.

Assuming R > 50 Ω, lower the tap position. Start by calculating R / tower height to get a rough estimate of ohms / foot of height. Subtract 50 from R and divide the result by the ohms / foot of height. Move the shunt arm down by that much and remeasure the feed point R and X. If the initial R is 90 Ω on a 100-foot tower, that is 90 / 100 = 0.9 ohms / foot. To reduce R by 90 – 50 = 40 Ω, lower the tap position by 40 × 0.9 = 36 feet. (This height is likely to be too low because of the over-simplified calculation.) Remeasure feed point R and X, then move the shunt arm until R is close to 50 Ω. If you are unable to find a shunt arm height that produces R = 50 Ω, increase the gamma wire spacing from the tower and repeat the procedure.

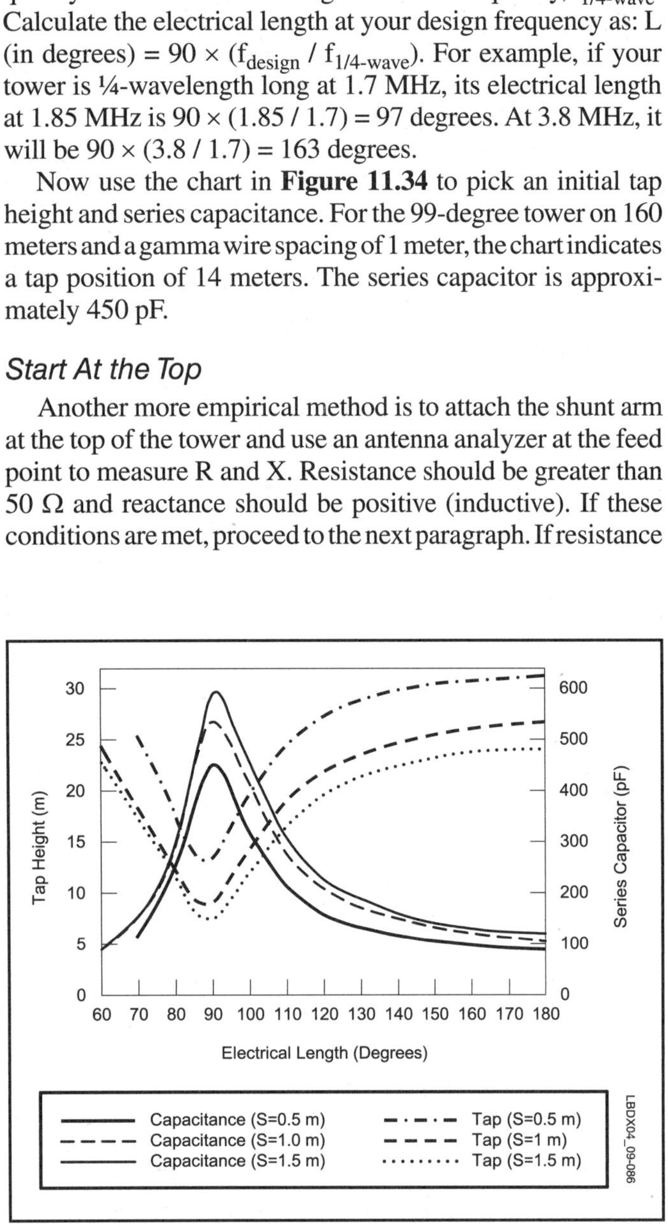

Figure 11.34 — Tap height and values of the gamma series capacitor for a shunt-fed tower at 1.835 MHz. The tower diameter is 25 cm (10 inches) and the gamma wire has a diameter of 10 mm (0.4 inches). Three sets of curves are shown for three spacings (S). The spacing is the distance from the wire to the tower center.

TUNING THE SHUNT-FEED MATCH

There are two basic rules for adjusting the shunt arm and gamma wire spacing:

•Moving the shunt arm up increases R

•Increasing the gamma wire spacing increases R and decreases X

Follow these rules until you have an acceptable value of R and the value of X is inductive at the feed point. A good rule of thumb for field adjustments with minimal instrumentation is that if moving the base of the gamma wire (raising feed point resistance) away from the tower lowers SWR, the tap point should be higher on the tower, resulting in a higher feed point resistance.

Now calculate the necessary capacitance in pF = 10^{12} / 2 π f X_L. If X_L = 100 Ω and the operating frequency is 1.85 MHz, the necessary capacitance in pF is 10^{12} / 6.28 × 1.85 × 10^6 × 100 = 860 pF. For 1500 W operation, the current of 5.5 A_{RMS} will produce 5.5 A × 1.4 × 100 Ω = 770 V_{PK}. A 2 kV rating will do the job. Smaller capacitors with higher reactance will experience higher voltages.

Fine-Tuning Resistance

The feed point resistance can be fine-tuned simply by moving the feed point end of the gamma wire closer to or away from the tower. The same rule for gamma wire spacing applies. This may be enough adjustment that you do not have to adjust the position of the shunt arm. This will also affect the reactance value, requiring retuning of the series capacitor.

Increasing SWR Bandwidth

Increasing the diameter of the gamma wire will also increase the SWR bandwidth of the matched tower. If you use a single wire, use the heaviest gauge you have or tubing. Consider using multiple wires at least one-half the tower-to-gamma-wire spacing apart. This will create an effective diameter much greater than a single wire. The result is a slower change in feed point impedance with frequency, increasing SWR bandwidth. The video by VE6WZ shows one method of supporting double gamma wires. Another method is a single shunt arm with a crossarm at the end.

Determining Capacitor Value

During the initial phase of low-power testing, use a low-voltage variable capacitor to tune the match for minimum SWR. You can then measure the capacitor's value and replace it with a fixed-value transmitting capacitor or several in parallel to create the necessary value. This is much less expensive than a high-voltage variable capacitor capable of full-power operation. Remember that current through the capacitors divides according to value with the highest current through the largest capacitance.

If you change the frequency of the match as described in the next paragraph, the required value of capacitance will also change. You may wish to adjust the frequency of the match slightly to use a standard capacitor value.

Changing Frequency of the Match

It's common to achieve a good match but then need to change the frequency of the match. For example, you might obtain a feed point impedance very close to 50 Ω but too close to the band edge or too high in the band. To raise (or lower) the frequency at which R = 50 Ω, lower (or raise) the tap position by a few feet and measure how much the frequency for R = 50 Ω has changed. This will show you in kHz / foot how much to move the tap to get to the desired frequency. You can also move the gamma wire as described above and change the capacitor value. By carefully adjusting the match frequency, you can place the useable SWR bandwidth where you want to operate.

L and Omega Matching Networks

Figure 11.35 shows four variations of the shunt-feed matching networks beginning with the single-series capacitor at A. A standard L-C L network is shown at B for load impedances greater than 50 Ω. Two forms of the omega network (an L network of two capacitors) are shown at C and D with C being the most common configuration.

There are many L-C network calculators online, such as the multi-configuration version by John Wetherell that computes component values for 16 different types of networks. A negative value of the component means that the network configuration will not match the specified load. (**home.sandiego.edu/~ekim/e194rfs01/jwmatcher/matcher2.html**)

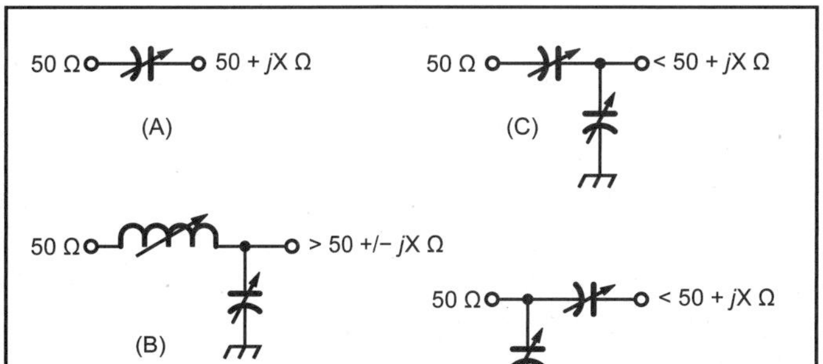

Figure 11.35 — Four networks used for shunt-feed matching (see text).

The omega network is used when the resistive impedance at the end of the gamma wire is less than 50 Ω. Another situation would be for short towers with lots of top loading, resulting in a calculated tap position above the top of the tower. Lowering the tap to a position on the tower will also lower the resulting value of R below 50 Ω.

Values for the omega capacitors can be calculated using the online L network calculator or you can temporarily connect two variable capacitors in the omega configuration and adjust them for the best match as with an antenna tuner. There is a fair amount of interaction between the two capacitors when adjusted so a graphical analyzer that can show the results of a swept frequency measurement is particularly useful here. A very sharp match may indicate that the network is not delivering power to the antenna — see the section on T matching networks in the **Transmission Line System Techniques** chapter for more information about this problem.

Note that both capacitors will need to be rated for full voltage. The omega match is a high-Q, low bandwidth network and will result in a narrow SWR bandwidth; 20-30 kHz is normal. You may get better bandwidth by using a regular L-C network (Figure 11.35B) and winding your own inductor from heavy wire or small-diameter copper tubing. The closer the feed point impedance is to 50 Ω, the lower the Q of the L-C network (and the wider the bandwidth) will be.

Slant-Wire Feed

A variation on moving the gamma wire away from the tower base to increase the feed point R, *slant-wire feed* is used in the broadcast field. It consists of attaching a single wire to the tap point with the other end some distance away. As with the gamma matching system, the attachment point is varied until the feed point resistance is 50 Ω. A series reactance (usually a capacitor) cancels whatever reactance is present. A fair amount of experimentation is required, and the feed point must be well-grounded to the same system used by the tower. Because the wire is slanted, there will be some horizontally polarized radiation, as well. A short *NCJ* article by Mike Taylor, AC9C, describing his 160 meter slant-wire system is included in the online material.

PROJECT: 1.8 and 3.5 MHz Shunt Fed Tower

A shunt-fed tower for 1.8 and 3.5 MHz designed by Earl Cunningham, K6SE, is shown in **Figure 11.36**. Two separate shunt-feed structures are used: a full-size gamma match for 160 meters and a shortened omega match for 80/75 meters. The tower is too high electrically to be gamma matched on 80/75 meters, so a very low tap position is used that gives a feed point resistance less than 50 Ω that is suitable for the omega match. The article "Shunt Feed Towers for Operation on the Lower Amateur Frequencies" is included in the online material.

Ted Rappaport, N9NB, and Jim Parnell, W5JAW, have developed a system for using a 65-foot tower on both 3.5 and 1.8 MHz with a single feed line. Their approach is detailed in the article "Use Your Tower as a Dual-Band, Low-Band DX Antenna" which is included in the online material. **Figure 11.37** shows the basic approach.

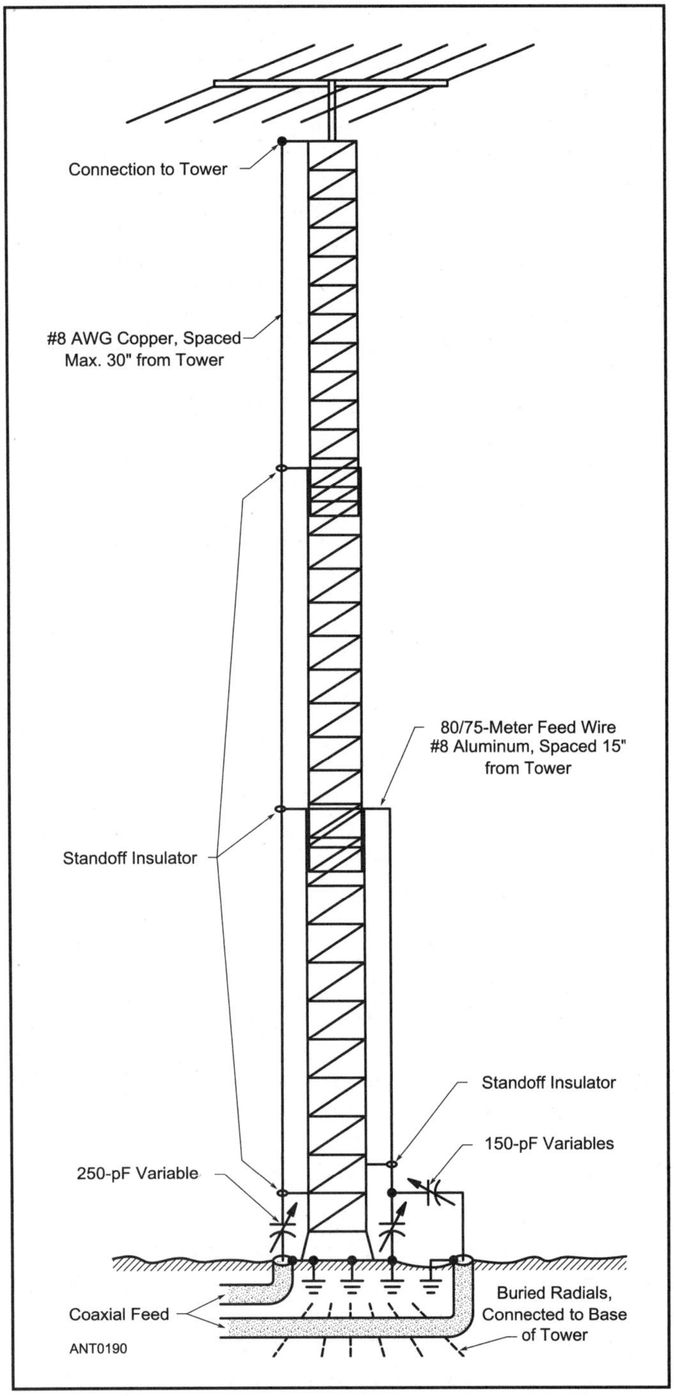

Figure 11.36 — Principal details of the shunt-fed tower at K6SE. The 1.8-MHz feed, left side, connects to the top of the tower through a horizontal arm of 1-inch diameter aluminum tubing. The other arms have standoff insulators at their outer ends, made of 1-foot lengths of plastic water pipe. The connection for 3.5 to 4 MHz, right, is made similarly, at 28 feet, but two variable capacitors are used to permit adjustment of matching with large changes in frequency.

11.2.6 Elevated Ground-Plane Antennas

This section describes a simple and effective means of using a grounded tower, with or without top-mounted antennas, as an elevated ground-plane antenna for 80 and 160 meters. It first appeared in a June 1994 *QST* article by Thomas Russell, N4KG. Elevated radial systems for vertical ground-plane antennas are discussed in the **Effects of Ground** chapter by Rudy Severns, N6LF.

FROM SLOPER TO VERTICAL

Recall the quarter-wavelength sloper, also known as the *half sloper*. (The half sloper is covered later in this chapter in more detail.) It consists of an isolated quarter wavelength of wire, sloping from an elevated feed point on a grounded tower. Best results are usually obtained when the feed point is somewhere below a top-mounted Yagi antenna. You feed a sloper by attaching the center conductor of a coaxial cable to the wire and the braid of the cable to the tower leg. Now, imagine four (or more) slopers, but instead of feeding each individually, connect them together to the center conductor of a single feed line. Voilà! Instant elevated ground plane.

Now, all you need to do is determine how to tune the antenna to resonance. With no antennas on the top of the tower, the tower can be thought of as a fat conductor and should be approximately 4% shorter than a quarter wavelength in free

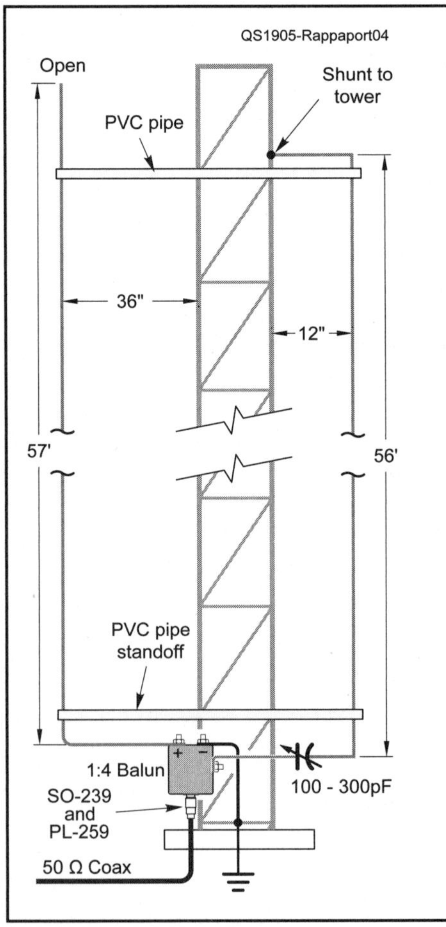

Figure 11.37 — Feed system design of the dual-band 80/160-meter vertical antenna using an existing tower and single coax feed line.

space. Calculate this length and attach four insulated quarter-wavelength radials at this distance from the top of the tower. For 80 meters, a feed point 65 feet below the top of an unloaded tower is called for. The tower guys must be broken up with insulators for all such installations. For 160 meters, 130 feet of tower above the feed point is needed.

CONSTRUCTION EXAMPLES

Consider this example: A TH7 triband Yagi mounted on a 40-foot tower. The TH7 has approximately the same overall dimensions as a full-sized 3-element 20 meter beam, but has more interlaced elements. Its equivalent loading is estimated to be 40 feet. At 3.6 MHz, 65 feet of tower is needed without loading. Subtracting 40 feet of equivalent loading, the feed point should be 25 feet below the TH7 antenna.

Ten λ/4 (65-foot) radials were run from a nylon rope tied between tower legs at the 15-foot level, to various supports 10 feet high. Nylon cord was tied to the insulated, stranded, #18 AWG wire, without using insulators. The radials are all connected together and to the center of an exact half wavelength (at 3.6 MHz) of RG-213 coax, which will repeat the antenna feed impedance at the other end. **Figure 11.38** is a drawing of the installation. The author used a Hewlett-Packard low-frequency impedance analyzer to measure the input impedance across the 80 meter band. An exact resonance (zero reactance) was seen at 3.6 MHz, just as predicted. The radiation resistance was found to be 17 Ω. The next question is, how to feed and match the antenna.

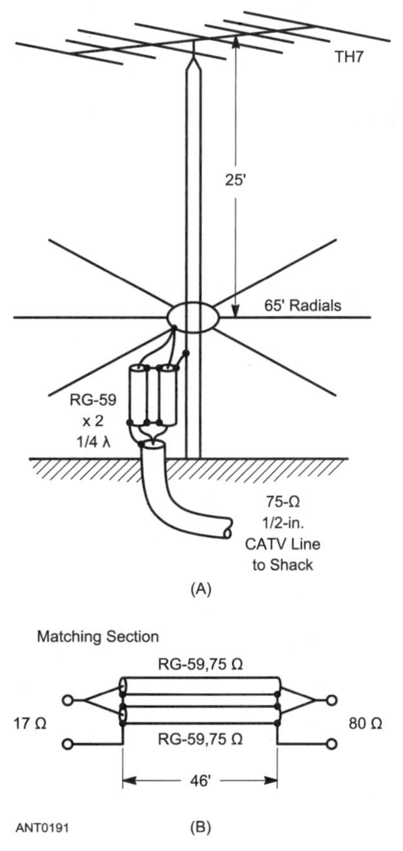

Figure 11.38 — At A, an 80-meter top-loaded, reverse-fed elevated ground plane, using a 40-foot tower carrying a TH7 triband Yagi antenna. At B, dimensions of the 3.6-MHz matching network, made from RG-59.

One good approach to 80 meter antennas is to tune them to the low end of the band, use a low-loss transmission line, and switch an antenna tuner in line for operation in the higher portions of the band. With a 50-Ω line, the 17-Ω radiation resistance represents a 3:1 SWR, meaning that an antenna tuner should be in-line for all frequencies. For short runs, it would be permissible to use RG-8 or RG-213 directly to the tuner. If you have a plentiful supply of low-loss 75-Ω CATV rigid coax, you can take another approach.

Make a quarter-wave (70 feet × 0.66 velocity factor = 46 feet) 37-Ω matching line by paralleling two pieces of RG-59 and connecting them between the feed point and a run of the rigid coax to the transmitter. The magic of quarter-wave matching transformers is that the input impedance (R_i) and output impedance (R_o) are related by:

$$Z_0^2 = R_i \times R_o \qquad (2)$$

For $R_i = 17\ \Omega$ and $Z_0 = 37\ \Omega$, $R_o = 80\ \Omega$, an almost perfect match for the matching section made from 75-Ω CATV coax. The resulting 1.6:1 SWR at the transmitter is good enough for CW operation without a tuner.

160 METER OPERATION

On the 160 meter band, a resonant quarter-wavelength requires 130 feet of tower above the radials. That's a pretty tall order. Subtracting 40 feet of top loading for a 3-element 20 meter or TH7 antenna brings us to a more reasonable 90 feet above the radials. Additional top loading in the form of more antennas will reduce that even more.

Another installation, using stacked TH6s on a 75-foot tower, is shown in **Figure 11.39**. The radials are 10 feet off the ground.

11.2.7 Broadbanding 1.8 and 3.5 MHz Verticals

As with dipoles, the SWR bandwidth of a vertical antenna can be increased by making the antenna thicker. For example, a wire about a foot from and in parallel with a tower used as a vertical will increase the SWR bandwidth noticeably. Up to three wires will have a similar effect. Even so, it is unlikely for the antenna to cover the entire band without some form of retuning.

In the article "Broad-banding a 160 m Vertical Antenna," Grant Saviers, KZ1W, shows a method of switching in different values of series capacitance using relays. Series capacitance is a common matching technique for inverted-L and top-loaded antennas. Rather than use a continuously adjustable, motor-driven capacitor, KZ1W switches between several band segments. He also discusses the use of inexpensive relays to perform the switching. (The article is included in the online material.)

In an email on the AntennaWare reflector from September 2018, Guy Olinger, K2AV, discusses modifying a DPDT relay to become a *shorting-bar relay*. These relays have two sets of

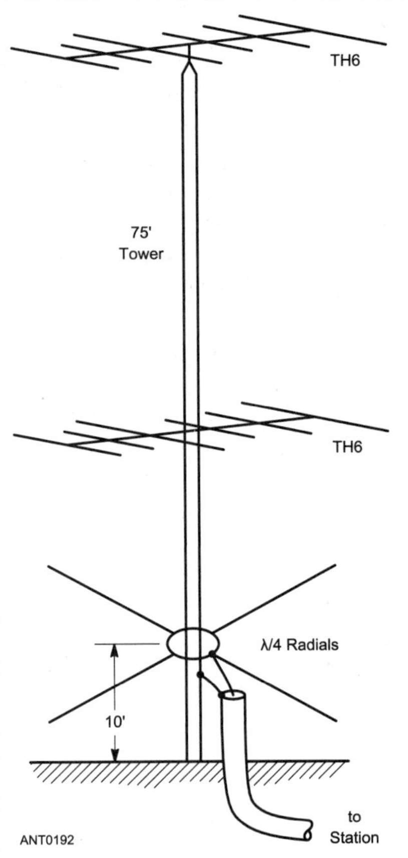

Figure 11.39 — A 160-meter antenna using a 75-foot tower carrying stacked triband Yagis.

contacts that are shorted together by a direct, low-impedance conductor when the relay is activated. This is ideal for selecting series capacitors or inductor taps. Not all DPDT relays are suitable for modification in this way. A graphic is available at **qsl.net/ei7ba/remote.htm** and more discussion is available in the original post which is available in the searchable archives for the reflector during September 2018 at **lists.contesting.com/_antennaware**. You can read more about switching tuning networks with high voltage or current present in the Folded Counterpoise (FCP) web pages at **k2av.com**.

The Effect of Trees

Wire verticals and vertical dipoles are often used close to trees which make great supports for them. While the absorption of VHF and UHF signals by foliage is well known, the effect at HF is less pronounced. In a QST article (see References and Bibliography), Kai Siwiak, KE4PT, and Richard Quick, W4RQ, studied the effects of placing HF antennas close to live trees. The tree acts as a parasitic element if the antenna is closer than 0.2 λ to the trunk of the tree with the effect diminishing quickly at larger distances. Loss through a group of trees depends on foliage and tree density so there is a summer/winter variation. Trees also affect propagation as the top of the forest canopy may act to guide waves along the foliage "surface."

11.3 Inverted-L and T Antennas

11.3.1 The Inverted-L

The antenna shown in **Figure 11.40** is called an *inverted-L.* Simple and easy to construct, it is a form of top-loaded vertical, with asymmetrical top loading. Like all ground-plane verticals, a good ground system or elevated radials are required. It can also be imagined as a "bent vertical," just less than 3/8λ long. A yardarm attached to a tower, or a tree limb can be used to support the vertical section. As with any vertical, for best results the vertical section should be as long as possible.

The shape results in both vertical and horizontal polarization because the currents in the top wire do not cancel like they do in a symmetrical-T vertical which is described in the next section. This is not necessarily a bad thing because it eliminates the overhead null present of a true vertical. The azimuthal radiation pattern is slightly asymmetrical with an approximate 1 to 2 dB increase in the direction opposite to the horizontal wire. This allows for good communication at short ranges as well as for DX and makes the inverted-L a good general-purpose antenna. (The article "The L Antenna" by L.B. Cebik, W4RNL provides a great deal of additional information and is available online.)

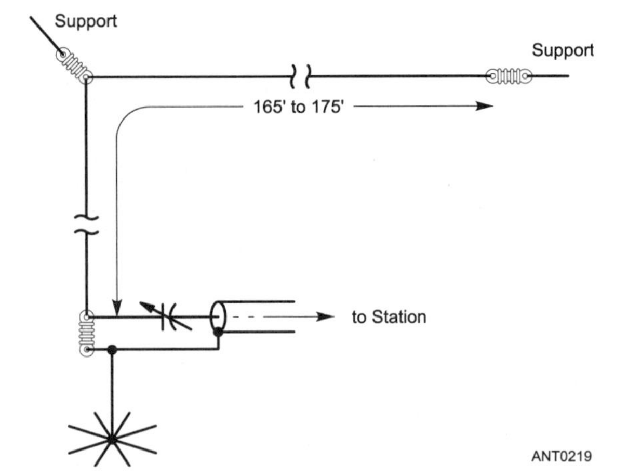

Figure 11.40 — The 1.8-MHz inverted-L. Overall wire length is 165 to 175 feet. The variable capacitor has a capacitance range from 100 to 800 pF, at 3 kV or more. Adjust antenna length and variable capacitor for lowest SWR.

INVERTED-L DESIGN BASICS

A resonant λ/4 vertical monopole has a feed point impedance of approximately 36 Ω. At an electrical length of 102° a vertical wire antenna will have a feed point impedance of approximately 50 + *j*100 Ω over good ground. Bending part of the vertical so that it is parallel to the ground or sloping will change the feed point impedance and lower the resistive component. As with the vertical antenna, the length is increased to raise the feed point impedance.

As a design example on 160 meters, with a 20-meter (65.6-foot) vertical section and the remaining horizontal wire parallel to the ground, the resonant λ/4 antenna (now 41.25 meters long overall) has a feed point impedance of only 20 Ω. With the horizontal section extended to 39.7 meters (130.25 feet) the feed point impedance becomes 50 + *j* 517 Ω. A series capacitor of 169 pF then cancels the inductive reactance and reduce SWR to 1:1. Various other combinations of vertical and horizontal lengths behave similarly as shown in **Table 11.4**. (The dimensions can be scaled to 80 meters with approximately half the series capacitance required.) SWR bandwidth should be about 50 kHz on 160 meters and 100 kHz on 80 meters. The inverted-L is an excellent antenna on which to practice modeling and matching skills. **Table 11.5** suggests some overall lengths for inverted-L antennas on other bands. Table 11.5 gives some idea of how the feed point impedance will change as the antenna's vertical and horizontal proportions change.

As for other wire antennas, the bandwidth of the antenna can be improved by increasing its effective diameter. For example, making the vertical radiator out of window or ladder line with both conductors tied together at the top and bottom will show some increase in operating bandwidth, the horizontal section remaining a single wire. Jim Brown, K9YC, reports using a pair of #10 AWG wires spaced 10 inches apart increased the 1.5:1 SWR bandwidth of his 160-meter L to 100 kHz. A tower can also be used as the vertical section as described below.

Table 11.4
Inverted L Dimensions for 50 Ω Feed-Point Resistance at 1.82 MHz

Vertical Height (m)	*Horizontal Length (m)*	*Inductive Reactance (Ω)*	*Capacitance Required (pF)*
10	59.9	1238	71
20	39.7	517	169
30	21	235	372

Results modeled by *EZNEC* 5.0 over real ground. Capacitance specified as series value to cancel inductive reactance.

Table 11.5
Inverted-L Overall Length for 80–30 Meters

Frequency (kHz)	*Feet*	*Inches*
3550	65.9	791
3750	62.4	749
3960	59.2	711
5370	43.6	523
7025	33.3	400
7125	32.8	394
7225	32.4	389
10125	23.1	277

INVERTED-L VARIATIONS

If you don't have the space for the inverted-L shown in Figure 11.40 (with its 115-foot horizontal section) and if you don't have a second tall supporting structure to make the top wire horizontal, consider sloping the top wire down toward ground. **Figure 11.41** illustrates such a setup on 160 meters, with a 60-foot high vertical section and a 79-foot sloping wire. As always, you will have to adjust the length of the sloping wire to fine-tune the resonant frequency. For a good ground radial system, the feed point impedance is about 12 Ω, which may be transformed to 50 Ω with a 25-Ω quarter-wave transformer consisting of two paralleled 50-Ω quarter-wave coaxes. The peak gain will decrease about 1 dB compared to the inverted-L shown in Figure 9.52. **Figure 11.42** overlays the elevation responses for average ground conditions. The

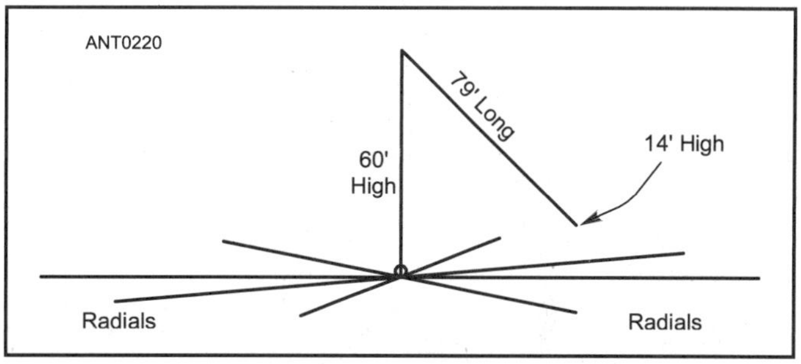

Figure 11.41 — Sketch showing a modified 160 meter inverted-L, with a single supporting 60-foot-high tower and a 79-foot-long slanted top-loading wire. The feed point impedance is about 12 Ω in this system, requiring a quarter-wave matching transformer made of paralleled 50-Ω coaxes.

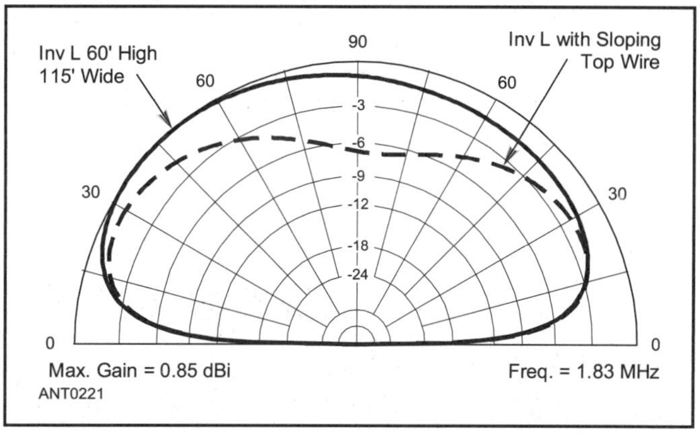

Figure 11.42 — Overlay of the elevation responses for the inverted-L antennas in Figure 11.40 (solid line) and Figure 11.41 (dashed line). The gains are very close for these two setups, provided that the ground radial system for the antenna in Figure 11.41 is extensive enough to keep ground losses low.

2:1 SWR bandwidth will be about 30 kHz, narrower than the larger system in Figure 11.40.

If a large ground system is not practical, such as for a temporary installation, you can use a single elevated radial as shown in **Figure 11.43**. For the dimensions shown in the figure, $Z_i = 50 + j\,498\ \Omega$ at 1.83 MHz, requiring a 175-pF series resonating capacitor. The azimuthal radiation pattern is shown in **Figure 11.44** compared to the inverted-L with a full ground system in Figure 11.52. Note that the 1 to 2 dB asymmetry is now in the direction of the horizontal wires, just the opposite of that for a symmetrical ground system. The 2:1 SWR bandwidth is about 40 kHz, with the series capacitor adjusted for minimum SWR.

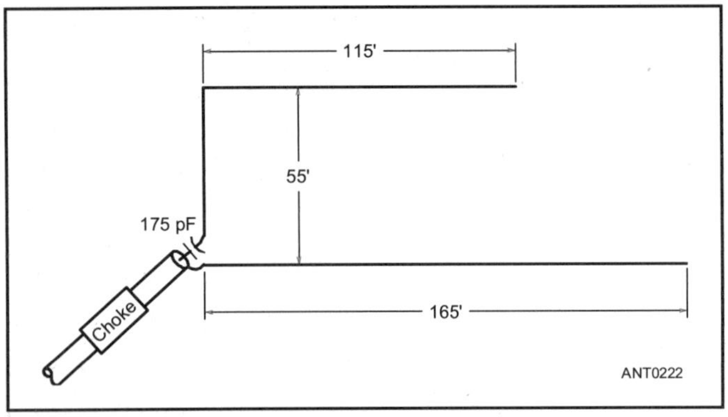

Figure 11.43 — A single elevated radial can be used for the inverted-L. This changes the directivity slightly. The series tuning capacitor is approximately 175 pF for this system.

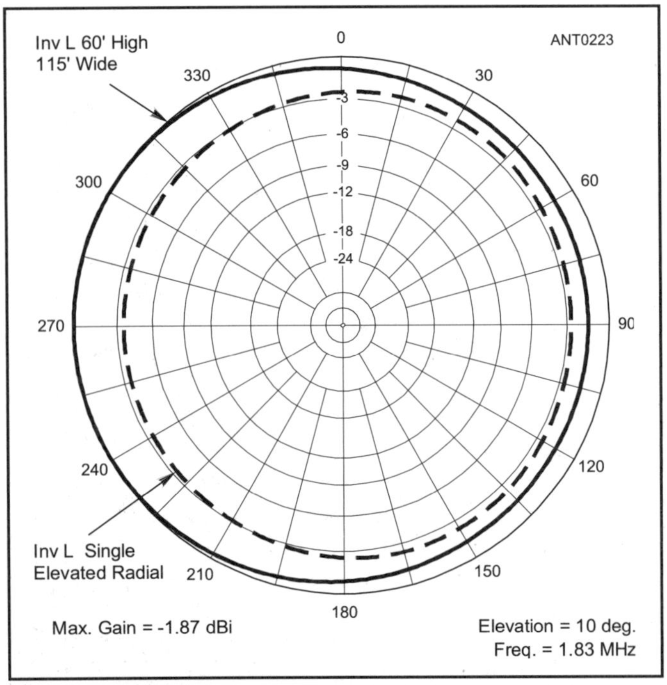

Figure 11.44 — Azimuthal pattern comparison for inverted-L antennas shown in Figure 11.40 (solid line) and the compromise, single-radial system in Figure 11.43 (dashed line). This is for a takeoff angle of 10°.

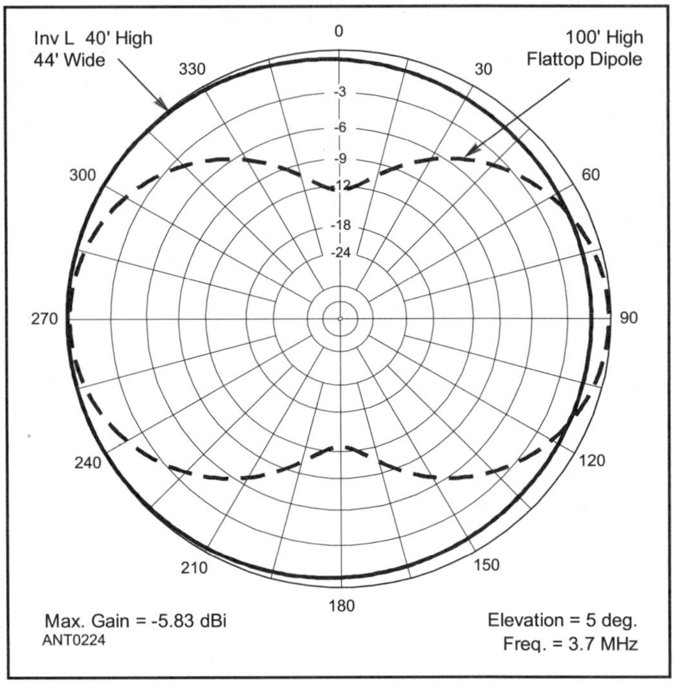

Figure 11.45 — Azimuthal pattern at a takeoff angle of 5° for an 80-meter version of the inverted-L (solid line) in Figure 11.40, compared to the response for a 100-foot-high flattop dipole (dashed line).

Figure 11.45 shows the azimuthal response at a 5° elevation angle for an 80-meter version of the inverted-L in Figure 11.52. The peak response occurs at an azimuth directly behind the direction in which the horizontal portion of the inverted-L points. For comparison, the response for a 100-foot-high flattop dipole is also shown. The top wire of this antenna is only 40 feet high and the 2:1 SWR bandwidth is about 150 kHz wide with a good, low-loss ground-radial system.

Figure 11.45 illustrates that the azimuth response of an inverted-L is nearly omnidirectional. This gives such an antenna an advantage in certain directions compared to a flattop dipole, which is constrained by its supporting mounts (such as trees or towers) to favor fixed directions. For example, the flattop dipole in Figure 11.45 is at its weakest at azimuths of 0 and 180 degrees, where it is down about 12 dB compared to the inverted-L.

PROJECT: TOWER-BASED INVERTED-L

Figure 11.46 shows the method used by Doug DeMaw, W1FB, to gamma match (shunt-feed) his self-supporting 50-foot tower operating as an inverted-L. A wire cage acts as the gamma wire. The tuning capacitor is fashioned from telescoping sections of 1¼ and 1½-inch aluminum tubing with polyethylene tubing serving as the dielectric. This capacitor is more than adequate for power levels of 100 W. The horizontal wire connected to the top of the tower provides the additional top loading. (See the previous section on Shunt-Feeding Towers for more information on this approach.)

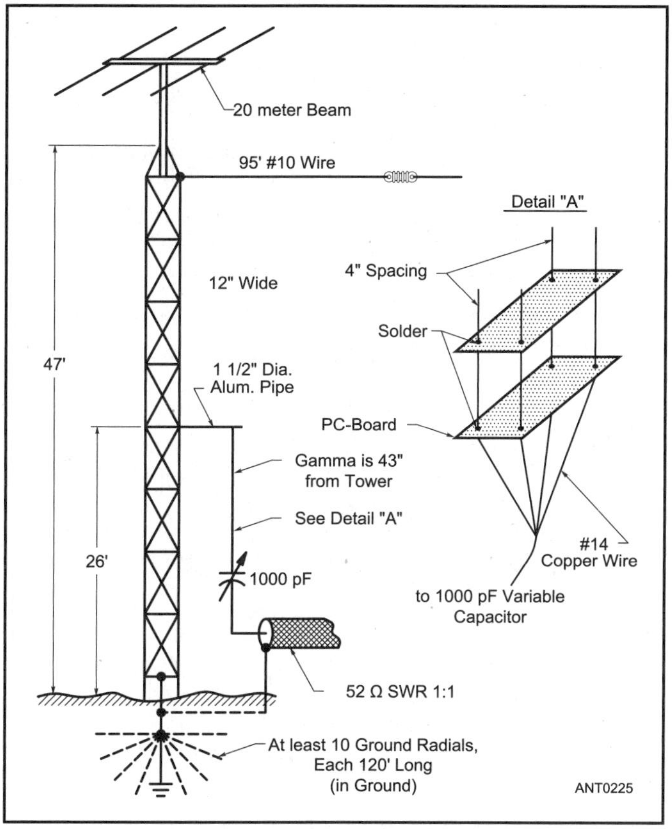

Figure 11.46 — Details and dimensions for gamma-match feeding a 50-foot tower as a 1.8-MHz vertical antenna. The rotator cable and coaxial feed line for the 14-MHz beam is taped to the tower legs and run into the station from ground level. No decoupling networks are necessary.

11.3.2 T Antennas

The T is basically a shortened ¼ λ vertical with the flat-top T section acting as capacitive top-loading to lengthen the antenna electrically. **Figure 11.47** shows a flat-top T vertical. As with the inverted-L, the vertical section (H) should be as large as possible (up to λ/4) for best results. Length of the horizontal T section is then adjusted to achieve resonance.

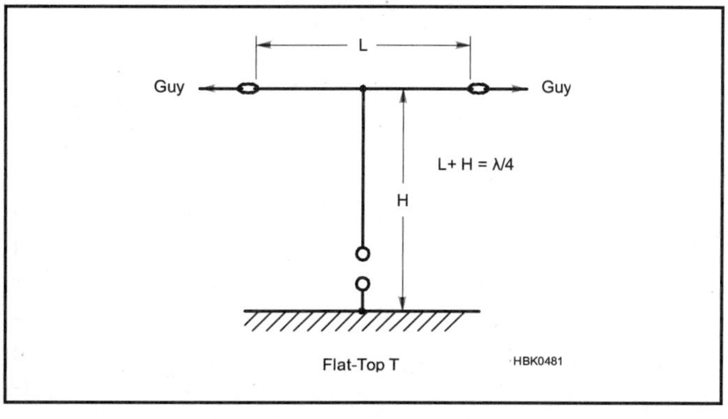

Figure 11.47 — The T antenna is basically a shortened ¼ λ vertical with the flat-top T section acting as capacitive loading to lengthen the antenna electrically.

Maximum radiation is polarized vertically despite the horizontal top-loading wire because current in each symmetrical half creates out-of-phase radiation that cancels. A good ground system is required just as for any other ground-plane antenna.

A sidearm or a length of line attached to a tower can be used to support the vertical section of the T antenna. (Keep the vertical section of the antennas as far from the tower as is practical. Certain combinations of tower height and top loading can create a resonance that interacts severely with the antennas — a 70-foot tower and a 5-element Yagi are approximately resonant on 160 meters, for example.) If the tower has an insulated base or is shunt-fed, it can be used as a T antenna by attaching horizontal or sloping wires at the top of the tower.

An ordinary dipole can be used as a T antenna by shorting the feed-line terminals together and driving the single resulting wire against the station ground system as shown in **Figure 11.48A**. The feed line becomes the vertical section of the antenna although it does not need to be completely vertical for this configuration. Radials can be attached temporarily to create the ground system. Because the feed-line shield becomes part of the antenna, a choke at the shorting point will likely be required to block common-mode RF current from the station.

A photo of a typical feed-through panel configured for using a dipole fed with window line in this way is shown in Figure 11.60B. Coax to the antenna tuner is reconnected with one conductor connected to the shorted external feed line and the other conductor is connected to the station ground system outside. Dipoles fed with coax can also be used by shorting the feed line at the feed-through panel and treating both conductors as a single wire.

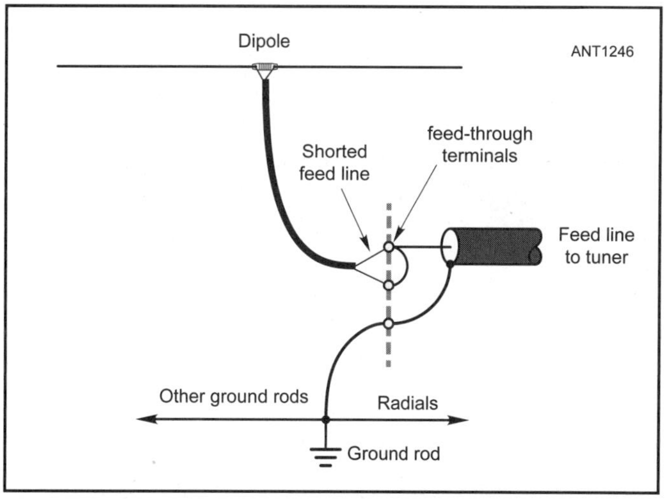

Figure 11.48 — Converting a dipole to a T antenna (A) can be done by shorting the feed line terminals and driving both conductors against the station ground. B shows a typical feed-through panel with a shorting jumper attached to the feed-through terminals. One conductor of the feed line to a balanced tuner is attached to the shorted terminals and the other to the station ground terminal. Powerpole connectors are used to allow quick feed line configuration changes.

11.4 Half-Sloper Antennas

Sloping dipoles and λ/2 dipoles can be very useful antennas on the low bands. These antennas can have one end attached to a tower, tree or other structure and the other end near ground level, elevated high enough so that passersby can't contact them, of course. The following section gives a number of examples of these types of antennas. The book *Sloper Antennas* by Juergen Weigl, OE5CWL, contains analysis and many designs of half-sloper antennas (see the References and Bibliography).

Perhaps one of the easiest antennas to install is the λ/4 sloper shown in **Figure 11.49**. As pointed out above, a sloping λ/2 dipole is known among radio amateurs as a *sloper* or sometimes as a *full sloper*. If only one half of it is used, it becomes a *half sloper*. The performance of the two types of sloping antennas is similar — they exhibit some directivity in the direction of the slope and radiate vertically polarized energy at low angles respective to the horizon. The amount of directivity will range from 3 to 6 dB, depending upon the individual installation, and will be observed in the slope direction.

The main advantage of the half sloper over the full half wave-long sloping dipole is that its supporting tower needn't be as high. Both the half sloper and the full sloper place the feed point (the point of maximum current) high above lossy ground. But the half-sloper only needs half as much wire to build the antenna for a given amateur band. The disadvantage of the half sloper is that it is sometimes difficult or even impossible to obtain a low SWR when using coaxial-cable feed, especially without a good isolating choke balun. (See the section above on isolating ground-plane antennas.)

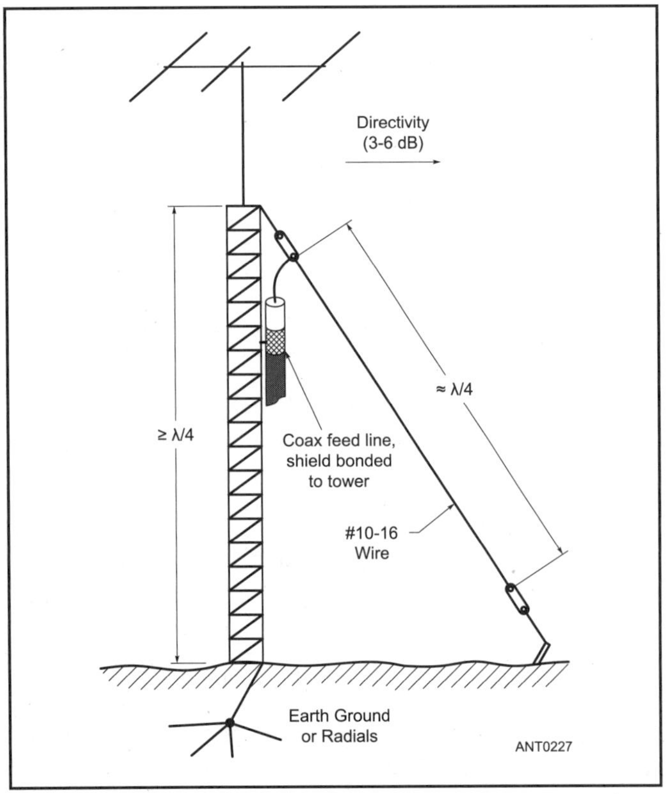

Figure 11.49 — The λ/4 "half sloper" antenna.

Other factors that affect the feed-impedance are tower height, height of the attachment point, enclosed angle between the sloper and the tower, and what is mounted atop the tower (HF or VHF beams). Further, the quality of the ground under the tower (ground conductivity, radials, etc) has a marked effect on the antenna performance. The final SWR can vary (after optimization) from 1:1 to as high as 6:1. Generally speaking, the closer the low end of the slope wire is to ground, the more difficult it will be to obtain a good match.

The half sloper can be an excellent DX antenna. Hams usually install them on a metal supporting structure such as a mast or tower. Assuming coax feed line, the center conductor is connected to the sloping wire and the shield to the support. The support needs to be grounded at the lower end, preferably to a buried or on-ground radial system. If a nonconductive support is used, the outside of the coax braid becomes the return circuit and should be grounded at the base of the support.

As a starting point you can attach the sloper so the feed point is approximately λ/4 above ground. If the tower is not high enough to permit this, the antenna should be fastened as high on the supporting structure as possible. Start with an enclosed angle of approximately 45°, as indicated in Figure 11.61. Cut the wire to the length determined from

$$\ell = \frac{260}{f_{MHz}}$$

This will allow sufficient extra length for pruning the wire for the lowest SWR. A metal tower or mast becomes an operating part of the half sloper system. In effect, it and the slope wire function somewhat like an inverted-V dipole antenna. In other words, the tower operates as the missing half of the dipole. Hence its height and the top loading (beams) play a significant role.

Detailed modeling indicates that a sufficiently large mass of metal (that is, a large, all-metal Yagi) connected to the top of the tower acts like enough of a "top counterpoise" that the tower may be removed from the model with little change in the essential characteristics of the half-sloper system. Consider an installation using a freestanding 50-foot tower with a large 5-element 20 meter Yagi on top. This Yagi is assumed to have a 40-foot boom oriented east-west, 90° from the north-facing slanted 80 meter half-sloper. The best SWR that could be reached by changing the length and slant angle for this sloper is 1.67:1, representing a feed point impedance of 30.1 – j 2.7 Ω. The peak gain at 3.8 MHz is 0.97 dBi at an elevation angle of 70°. **Figure 11.50** shows the azimuth-plane pattern for this half sloper, compared to a 100-foot high flattop dipole for reference, at an elevation angle of 5°.

Removing the tower from the model resulted in a feed point impedance of 30.1 – j 1.5 Ω and a peak gain of 1.17 dBi. The tower is obviously not contributing much in this setup, since

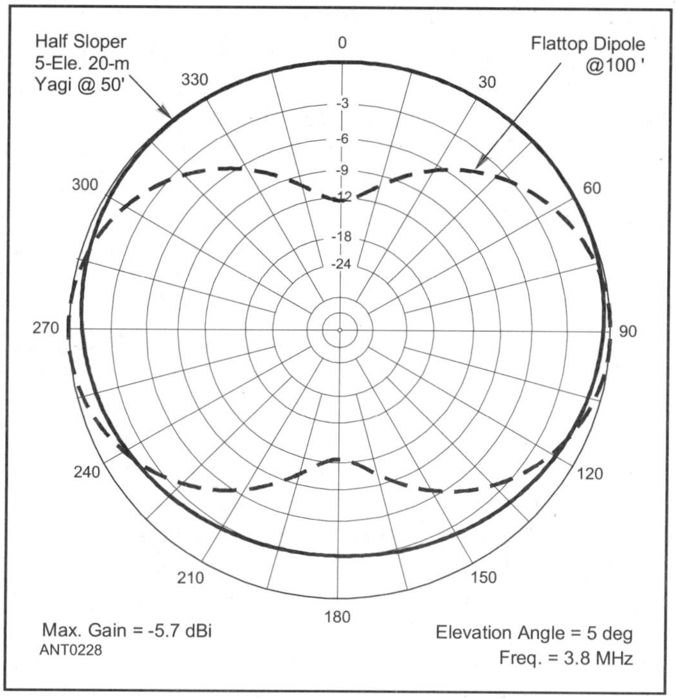

Figure 11.50 — Radiation pattern for a typical half sloper (solid line) mounted on a 50-foot-high tower with a large 5-element 20 meter beam on the top compared to that for a flattop dipole (dashed line) at 100 feet. At a 5° takeoff angle typical for DX work on 80 meters, the two antennas are pretty comparable in the directions favored by the high dipole. In other directions, the half sloper has an advantage of more than 10 dB.

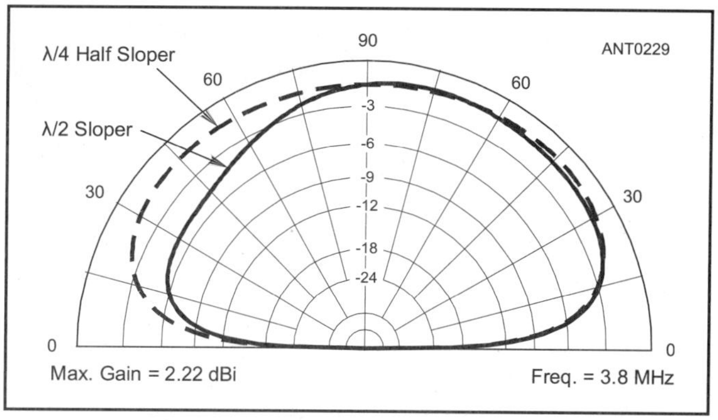

Figure 11.51 — Comparison of elevation patterns for a full-sized half wave sloper (solid line) on a 100-foot tower and a half sloper (dashed line) on a 50-foot tower with a 5-element 20-meter Yagi acting as a top counterpoise. The performance is quite comparable for these two systems.

the mass of the large 20 meter Yagi is acting like an elevated counterpoise all by itself. It's interesting to rotate the boom of the model Yagi and observe the change in SWR that occurs on the half-sloper antenna. With the boom turned 90°, the SWR falls to 1.38:1. This level of SWR change could be measured with amateur-type instrumentation.

On the other hand, substituting a smaller 3-element 20 meter Yagi with an 18-foot boom in the model does result in significant change in feed point impedance and gain when the tower is removed from the model, indicating that the "counterpoise effect" of the smaller beam is insufficient by itself. Interestingly enough, the best SWR for the half sloper/tower and the 3-element Yagi (with its boom in line with the half sloper is 1.33:1), changing to 1.27:1 with the boom turned 90°. Such a small change in SWR would be difficult to measure using typical amateur instrumentation.

In any case, the 50-Ω transmission line feeding a half sloper should be taped to the tower leg at frequent intervals to make it secure. The best method is to bring it to earth level, then route it to the operating position along the surface of the ground if it can't be buried. This will ensure adequate RF decoupling, which will help prevent RF energy from affecting the equipment in the station. Rotator cable and other feed lines on the tower or mast should be treated in a similar manner.

Adjustment of the half sloper is done while measuring SWR. A compromise can usually be found between the enclosed angle and wire length, providing the lowest SWR attainable in the center of the chosen part of an amateur band. If the SWR "bottoms out" at 2:1 or lower, the system will work fine without using an antenna tuner, provided the transmitter can work into the load. Typical optimum values of SWR for 3.5 or 7-MHz half slopers are between 1.3:1 and 2:1. A 100-kHz bandwidth is normal on 3.5 MHz, with 200 kHz being typical at 7 MHz.

If the lowest SWR possible is greater than 2:1, the attachment point can be raised or lowered to improve the match. Readjustment of the wire length and enclosed angle may be necessary when the feed point height is changed. If the tower is guyed, the guy wires will need to be insulated from the tower and broken up with additional insulators to prevent resonance.

At this point you may be curious about which antenna is better — a full sloper or a half sloper. The peak gain for each antenna is very nearly identical. **Figure 11.51** overlays the elevation-plane pattern for the full-sized half wave sloper on a 100-foot tower and for the half sloper shown in Figure 11.61 on a 50-foot tower with a 5-element 20 meter Yagi on top. The full-sized half wave sloper has more front-to-back ratio, but it is only a few dB more than the half sloper. **Figure 11.52** compares the azimuthal patterns at a 5° takeoff angle for a 100-foot high flattop dipole and a half-sloper system on a 50-foot tower with a 3-element 20 meter Yagi on top.

Despite the frustration some have experienced trying to achieve a low SWR with some half-sloper installations, many operators have found the half sloper to be an effective and low-cost antenna for DX work.

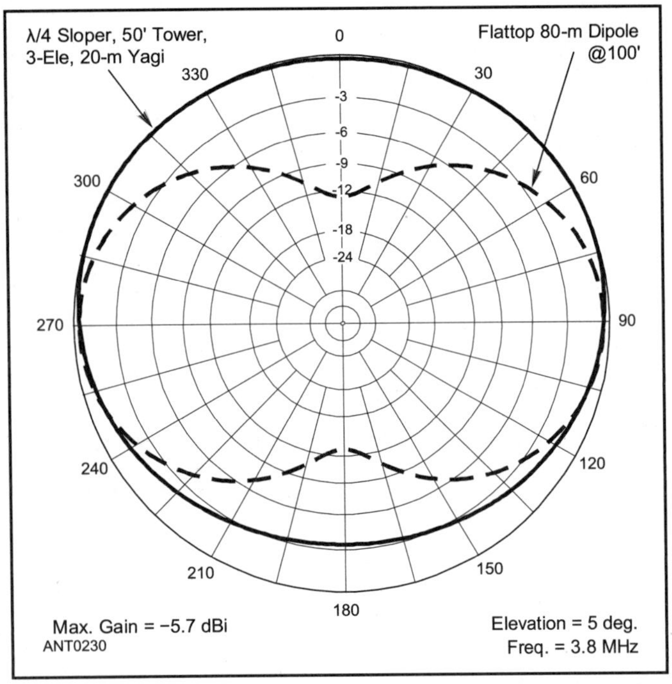

Figure 11.52 — Comparing the azimuthal response of a half-sloper (solid line) on a 50-foot tower with a 3-element 20-meter Yagi on top to that of a flattop dipole (dashed line) at 100 feet. The two are again quite comparable at a 5° takeoff angle.

11.4.1 1.8-MHz Half-Slopers Using Towers

The material on Monopoles in the **Dipoles and Monopoles** chapter covers loading, losses, and efficiency of electrically short verticals. Prominent 1.8-MHz operators who have had success with the half sloper antenna suggest a minimum tower height of 50 feet. Dana Atchley, W1CF, used the configuration sketched in **Figure 11.53**. He reported that the uninsulated guy wires act as an effective counterpoise for the sloping wire. In **Figure 11.54** is the feed system used by Doug DeMaw, W1FB, on a 50-foot self-supporting tower. The ground for the W1FB system is provided by buried radials connected to the tower base. Jack Belrose, VE2CV and DeMaw also described an interesting method of using a sloping wire to create a "half delta" usable on the lower HF bands. The system is described in the Sep 1982 *QST* article, "The Half-Delta Loop: A Critical Analysis and Practical Deployment" which is also included with this book's donline material.

Shunt feeding a grounded tower is often used on 80and 160 meters as described in the section on "Shunt Feeding Towers" earlier in this chapter. The shunt-fed tower is at its best on 1.8 MHz, where a full λ/4 vertical antenna is rarely possible. Almost any tower height can be used. An HF beam at the top provides some top loading.

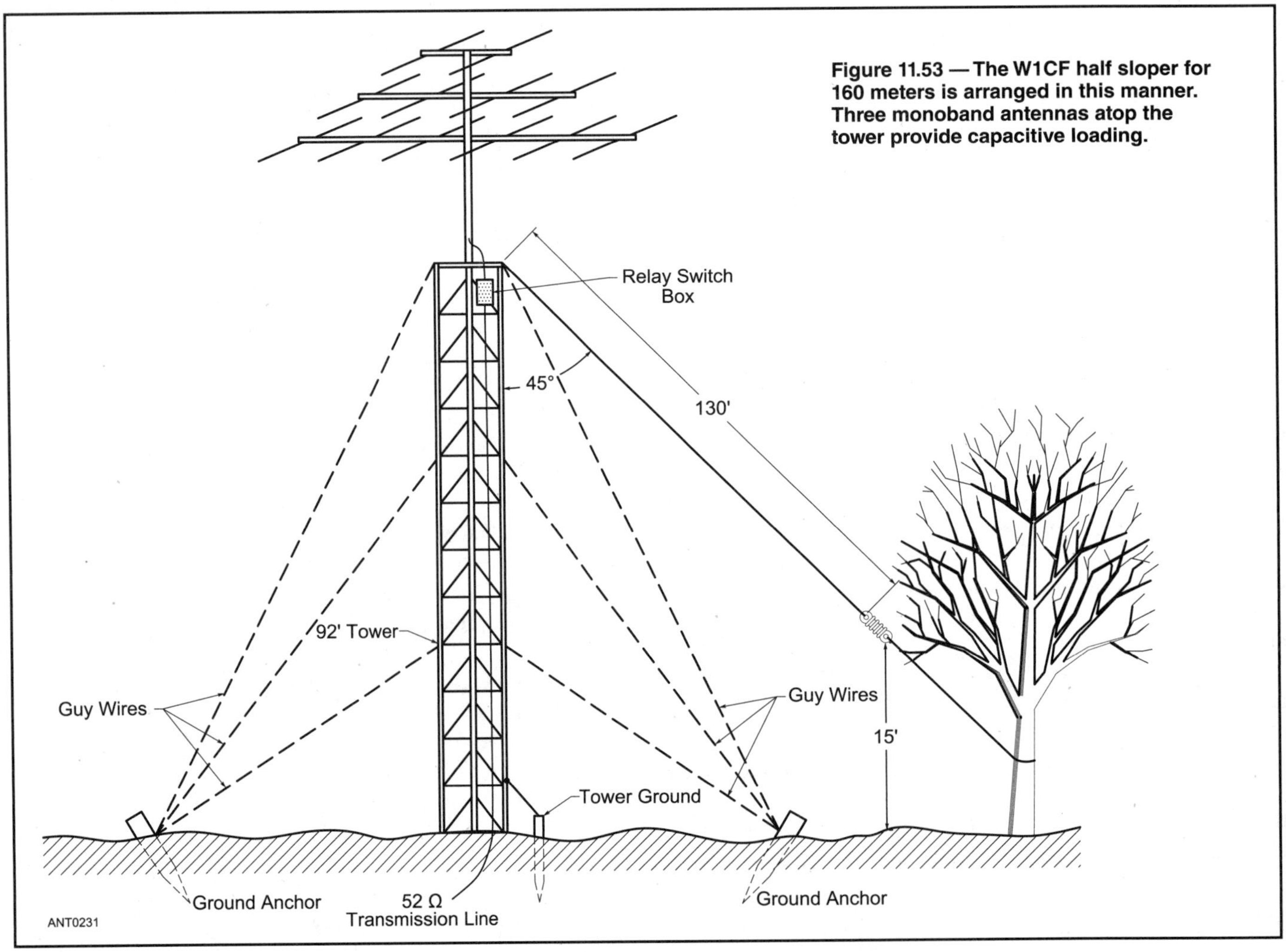

Figure 11.53 — The W1CF half sloper for 160 meters is arranged in this manner. Three monoband antennas atop the tower provide capacitive loading.

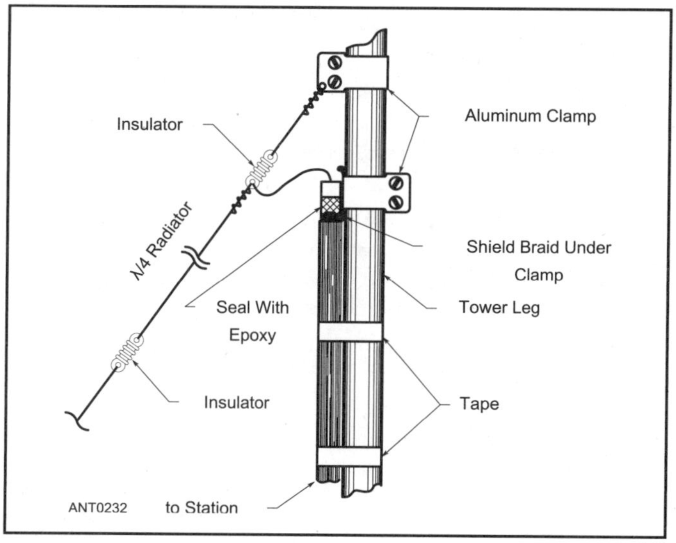

Figure 11.54 — Feed system used by W1FB for 1.8 MHz half-sloper on a 50-foot self-supporting tower.

11.5 LF and MF Antennas

This material is excerpted and summarized from "LF and MF Antennas for Amateurs" by Rudy Severns, N6LF. Rudy participated as experimental licensee WD2XSH/20 during the evaluation period for the new bands and gained much valuable experience. This section is a collection of theory and practical recommendations for transmitting antennas at the long wavelengths below the AM broadcast band. The full document is downloadable from ***antennasbyn6lf.com*** *and the interested reader is urged to read the full presentation.*

A hundred years ago, amateurs were restricted to wavelengths below 200 meters (f >1.5 MHz). We now have allocations at 2200 meters (135.7-137.8 kHz, LF) and 630 meters (472-479 kHz, MF). However, amateurs have very little experience at these frequencies and it turns out that design and construction of antennas for the new bands is substantially different from HF. The primary purpose of these notes is practical advice on LF/MF transmitting antennas. There is a perception that substantial acreage is required for the antennas on these bands. That is not the case! Those with small properties can be successful,but we have to know how!

There are differences between LF/MF and HF which impact antenna design:

1) Wavelengths are much longer so that any practical antenna will be electrically small.

2) Soil electrical characteristics change substantially going from HF down to LF/MF.

3) Power limitations are in terms of power radiated from the antenna rather than maximum transmitter output power although there are also limits on transmitter power.

LONG WAVELENGTHS

At 1.9 MHz, the wavelength (λ) ≈ 518 feet, so a λ/4 vertical will be ≈ 130 feet high. If you divide 1.9 MHz by 4 you get 475 kHz, right in the middle of the new 630 meter band. λ/4 on 160 meters is only ≈ λ/16 at 475 kHz. The 2200-meter band is another factor of 3.5 lower in frequency, so a λ/4 vertical on 160 meters is only ≈ 0.018 λ on 137 kHz. At 475 kHz, λ ≈ 2071 feet, so a λ/4 vertical would be ≈ 500 feet high. At 137 kHz λ/4 ≈ 1800 feet! In any case, the FCC has limited the maximum height to 197 feet (60 meter), which is still only 0.095 λ at 137 kHz.

The focus of this discussion is on antennas with heights (H) practical for amateurs, i.e. H = 20–100 feet (H ≈ 0.01 – 0.05 λ at 475 kHz and H ≈ 0.003 – 0.015 λ at 137 kHz). In terms of electrical height these are certainly "short" antennas, with very low radiation resistance (R_r), narrow matched SWR bandwidth and low efficiency. A major part of the design effort for LF/MF antennas is directed towards obtaining adequate efficiency.

SOIL CHARACTERISTICS

Because ground electrical characteristics have a profound effect, some basic information on soil electrical characteristics will be needed. At 100 Hz soil conductivity σ ≈ 0.09 S/m and is relatively constant up to ≈1 MHz but beyond that point σ increases rapidly with frequency. Relative permittivity of soil, ε_r behaves just the opposite, with a value of approximately 200 at 137 kHz that decreases with frequency, reaching approximately 70 at 475 kHz. Permittivity continues to decrease with an approximate value of 50 at 1 MHz and 23 at 10 MHz. At a given QTH, with the same soil, the electrical characteristics will be very different between HF and LF/MF.

EIRP AND RADIATED POWER

Power limits on 2200 and 630 meters are stated in terms of *effective isotropic radiated power* (EIRP). The "isotropic" in EIRP refers to an idealized antenna in free space which radiates power uniformly in all directions. On 630 meters 5 W EIRP is allowed and on 2200 meters the allowed EIRP is 1 W, which means the total radiated power, P_r, is about 1.7 W on 630 meters and 0.33 W on 2200 meters.

At HF, antenna efficiencies are typically >90% and the focus is on antenna gain. On LF/MF our goal is to achieve sufficient efficiency that we can radiated the allowed power with the available transmitter power. This is a fundamentally different mindset! We have the choice of a large efficient antenna with small input power (P_i) or a small inefficient antenna with a large input power. Most installations will be a balance between the two extremes.

A transmitter output power of 100 W is generally pretty easy to obtain, and 100 W is frequently assumed in later chapters unless stated otherwise. In addition to the EIRP power limit, the FCC has also limited the input power to the antenna to 500 W PEP on 630 meters and 1.5 kW PEP on 2200 meters. However, given limitations due to the high voltages associated with these power levels, from a practical point of view these limits are moot. (P_r can be measured directly as described in Severn's material.)

A major part of the design effort for LF/MF antennas is directed at obtaining adequate efficiency. Given practical height limitations, most LF/MF antennas will require loading inductors for resonance and matching. In many cases the losses in this inductor will determine the efficiency of the antenna. Much of the design effort is directed towards first minimizing the required inductance (L) with height and top-loading and then maximizing inductor "Q" (Q_L)

FUNDAMENTAL ADVICE

Repeating Woodrow Smith's summary of LF/MF antenna design given earlier in the discussion on short verticals:

"The main object in the design of low frequency transmitting antenna systems can be summarized briefly by saying that the general idea is to get as much wire as possible as high in the air as possible and to use excellent insulation and an extensive ground system."

This simple advice should be taken literally! This advice can be organized in order of priority:

1) Make the vertical as tall as you can.

2) Use as much capacitive top-loading as practical.

3) Use carefully placed high-Q loading coils.

4) Put substantial effort into the ground system, with the radial density high near the base of the vertical and under the top-loading hat.

5) Minimize conductor losses by using multiple wires and/or large diameter conductors.

6) Use high quality insulators, at the base and at wire ends.

11.5.1 Efficiency of Very Short Verticals

The material on Monopoles in the **Dipoles and Monopoles** chapter covers loading, losses, and efficiency of electrically short verticals." Refer to Chapter 2 of N6LF's notes for detailed discussion of the material in this section.

Efficiency of a short vertical antenna depends on the radiation resistance (R_r), ground resistance (R_g) and the loss resistance in the antenna itself (R_{loss}):

$$\eta = \frac{R_r}{R_r + R_g + R_{loss}} \qquad (8)$$

Figure 11.55 graphs R_r for a lossless #12 AWG wire vertical for H = 20 to 100 feet at 137 and 475 kHz. We can see that R_r is very small even for heights of 100 feet. A λ/4 vertical would have $R_r \approx 36\ \Omega$ but in LF/MF antennas R_r is typically smaller by a factor of 100 to 1000!

Conductors larger than #12 wire are often employed by using multiple parallel wires (up to a few inches spacing) or a cage of vertical wires is created with all of the wires connected together at the top. For diameters up to a few feet, eight wires are more than adequate but for very large diameters, say 10 to 40 feet, adding more wires to the cage may be worth doing. Using a larger diameter conductor or more wires has the immediate benefit of reducing conductor loss (R_c) which is a component of R_{loss}. The bottom ends of the vertical wires are typically connected together with a skirt wire like similarly to the top of the cage. The bottom skirt wire is then driven against ground or inductors are placed in each downlead and only one or two are driven. **Figure 11.56** shows the variation in R_r at 475 and 137 kHz as the conductor diameter is varied from #12 wire to 40 feet over heights from 20 to 100 feet. For a thin wire vertical (no inductor or capacitive loading) with a triangular current distribution:

$R_r \approx 0.003 G_v^2\ \Omega$

where G_v is the electrical length of the wire in degrees.

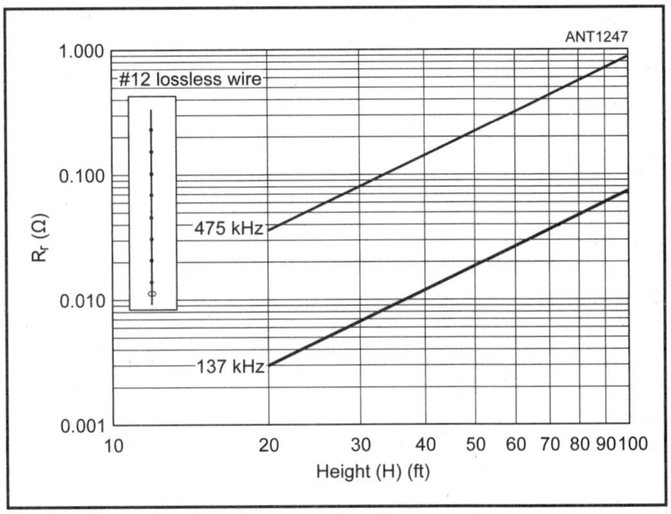

Figure 11.55 — Example of R_r variation with vertical height (H).

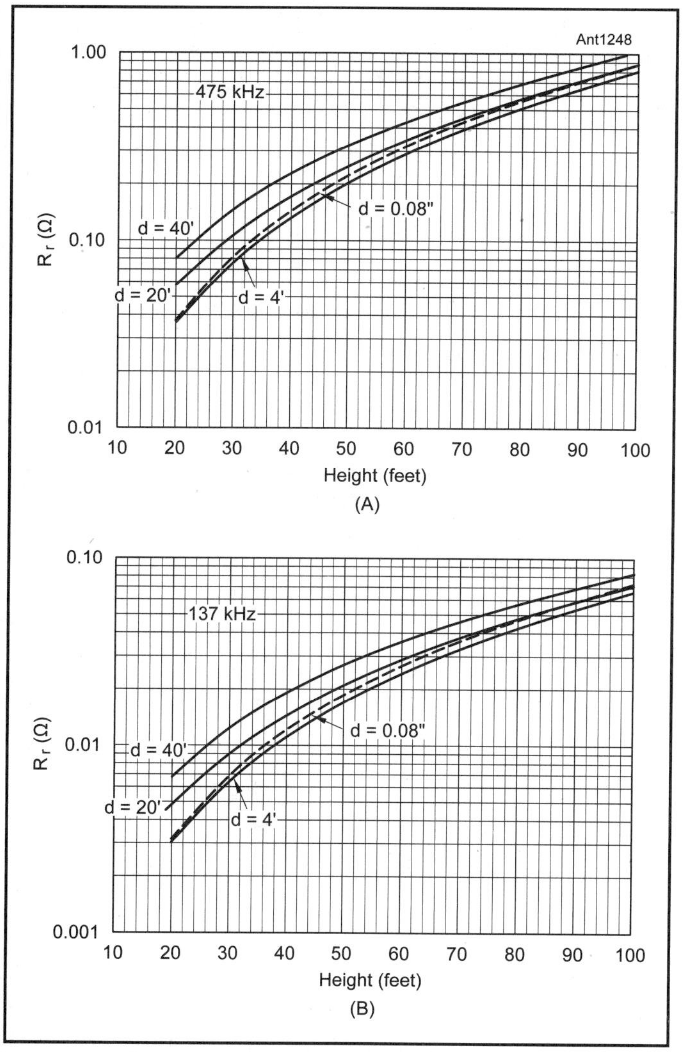

Figure 11.56 — Effects of conductor diameter on R_r at 475 kHz (A) and 137 kHz (B).

LOADING INDUCTORS

Loaded Q (Q_L) can range from 100 to >1000. In general, for a given inductor, Q_L at 137 kHz will be ≈0.54 Q_L at 475 kHz or a little less if the Q_L at 475 kHz is near a peak value due to the effects of self-resonance. While very high Q_L inductors are possible most of this discussion will assume Q_L=200 at 137 kHz and 400 at 475 kHz because these values are practical with modest effort but keep in mind that higher values are possible.

We can get a good feeling for the effect of loading inductor losses (R_L) on efficiency by assuming input $R = R_L + R_r$ (i.e. ignoring other losses) and calculate the efficiency. Q_L = 200 at 137 kHz and 400 at 475 kHz are assumed. With H = 20 feet, at 137 kHz η = 0.0024% and at 475 kHz η = 0.20% even without accounting for any other losses! Increasing H to 100 feet makes a great difference. At 137 kHz η = 0.24% which is still very low but a factor of 100 improvement. With 100 W output from the transmitter, to radiate the maximum allowed power the antenna will have to have η > 2% at 475 kHz and η >0.33% at 137 kHz.

To radiate the maximum allowed power, a minimum height of 45 feet on 630 meters and >100 feet on 2200 meters is needed. A small change in height means a large change in efficiency! Maximizing height is a vital for improving efficiency.

R_L is the dominant loss throughout this range of H, especially as we go lower in frequency. This observation is important because it tells us what our design priorities must be. The value of R_L is tied directly to the value of X_L ($|X_L| \approx |X_C|$) through Q_L. Again, the message is very clear: To reduce R_L we must reduce X_C! Once height has been maximized, top-loading becomes the primary tool for reducing X_C.

Chapter 6 of the online notes, "Design and Fabrication of High-Q Tuning Inductors for LF/MF Antennas" goes into detail about the construction of high-Q inductors for transmitting use at LF/MF. The information in the chapter can also be applied to inductors for the lower HF bands, as well. The chapter also covers the design and construction of variable inductors and variometers.

FEED POINT VOLTAGES AND CURRENTS

Unfortunately, low efficiency is not the only challenge. Base currents (I_O) and feed point voltages (V_O) can be very high. The following discussion is for a simple vertical without the top-loading that significantly reduces I_O and V_O.

Figure 11.57A shows base current (I_O in A_{RMS}) as a function of H. These are the RMS currents required to produce the allowed radiated power on each band. Figure 11.57B shows the input power, P_i, required to produce the allowed radiated power, P_r, on each band for a given loading inductor Q. If you wish to use a simple 20 feet vertical on 137 kHz radiating the maximum allowed power you'll have to provide $P_i \approx 9$ kW!

The input current (I_O) flows through the short-vertical series circuit of the sum of radiation resistance and all losses (R_a), the antenna's inductive reactance ($+X_a$) and capacitive reactance ($-X_c$). In short antennas R_a and X_a are very small compared to X_C. The input reactance is thus $X_i \approx X_C$ and X_C will be very large. The voltage across the feed point ($V_O = I_O \times X_i$) will be very high as indicated in **Figure 11.58A**. A 20-foot vertical at 137 kHz with $P_i \approx 9$ kW and $P_r = 0.33$ W will have $V_O \approx 300$kV! Given the very modest radiating power allowed, these voltage levels can come as an unpleasant surprise when an increase in transmitter power unexpectedly causes the loading coil to go up in flames or there is arcing across the base insulator or within tuning network components! For most amateurs $P_i \leq 100$ W is more realistic but even at this greatly reduced power V_O can still be many kV as shown in Figure 11.58B. This further reinforces the advice to minimize X_C. We must be very respectful of the voltages present on these antennas even at seemingly low power levels. *Be careful*!

11.5.2 Capacitive Top-Loading

A critical part of achieving higher efficiency is the reduction of capacitive reactance at the feed point because this reduces the size of the loading inductor and its associated R_L. In addition, steps to reduce inductance can also increase R_r at the same time. Other top-loading schemes which reduce X_i also

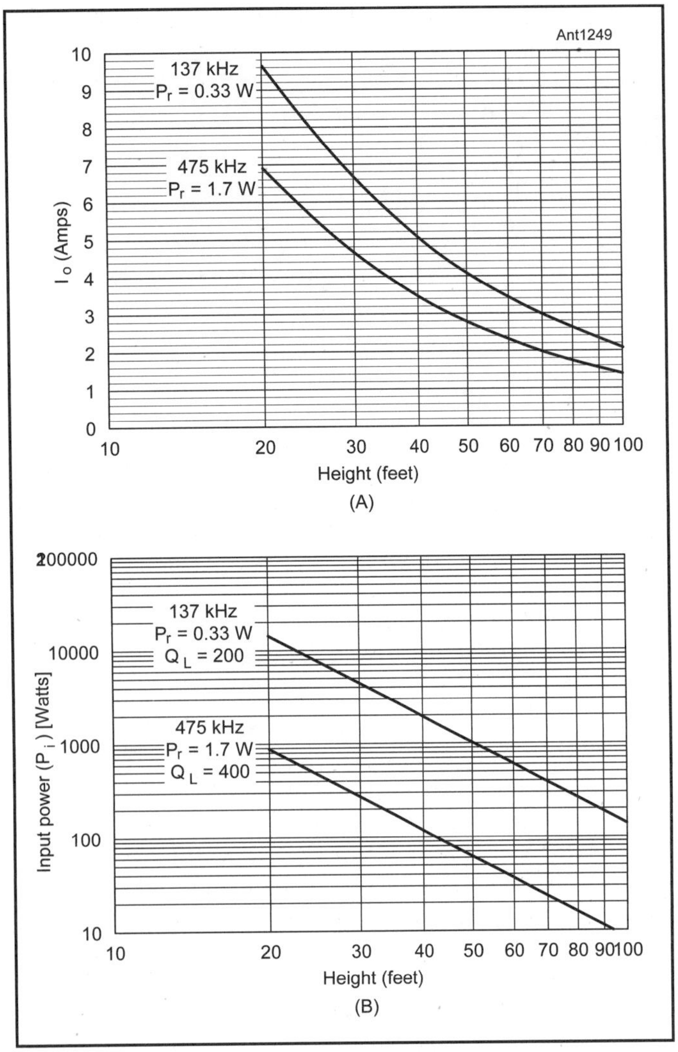

Figure 11.57 — Base current (I_O) and voltage (V_O) versus antenna height (H) at A. B shows the required input power (P_i) to radiate the maximum allowed power, (P_r) versus height with the same loading inductors as in the previous section.

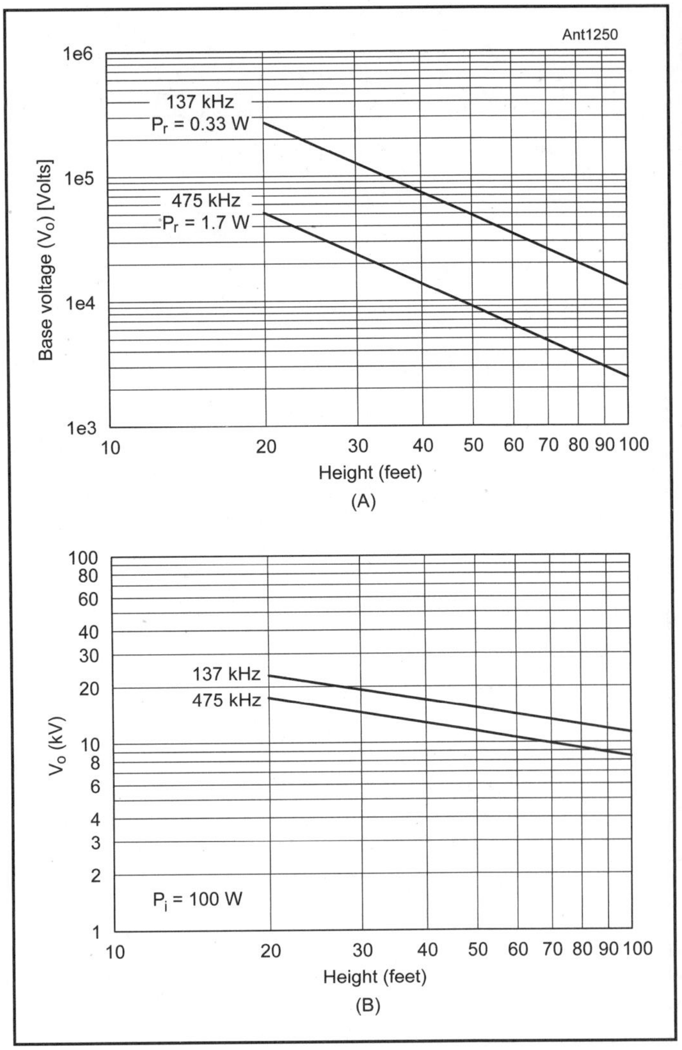

Figure 11.58 — Base voltage when radiating allowed P_r (A) and with P_i = 100 W (B).

reduce R_r rather than increase it and at some point the efficiency may actually start to fall even though we continue to reduce X_i. This happens in top-loading schemes with sloping wires with currents opposing the current in the main vertical conductor.

For amateurs new to LF/MF the first antenna needs to be as simple as possible to get on the air. We'll start by keeping it simple and assuming only one or two supports, a roll of wire and some insulators. We'll begin with an example showing R_r, X_i and efficiency and then go on to explain why top-loading is so effective in reducing X_i and increasing R_r.

It is assumed that height (H) has been made as tall as practical and we are now turning to capacitive top-loading to improve efficiency. The same tuning inductor Q_L = 400 at 475 kHz and 200 at 137 kHz is assumed. Many different variables affect the capacitance introduced by the top-loading structure:

- The number and/or length of umbrella wires
- Whether or not there is a skirt tying the ends of the wires together
- The location of the tuning inductor along the vertical conductor
- Conductor size

Keep in mind that efficiency determined from only R_L and R_r is an upper limit, i.e. the best we can do. Adding more losses only reduces efficiency. Once we've reduced R_L as much as possible we can deal with other losses. Reducing X_i has the further benefit of reducing the voltages and currents at the base. For the most part, if you make some change in your antenna which reduces the inductance required to resonate the antenna that change is likely to improve your efficiency. This is a useful guide when experimenting. One important point, most examples have symmetric wire arrangements because it's easier to model but symmetry is not required!

EFFICIENCY WITH TOP-LOADING

The T antenna shown in **Figure 11.59A** and B illustrates how effective capacitive top-loading is. Efficiency is shown in percent (%) as a function of the length of the top wire (L)

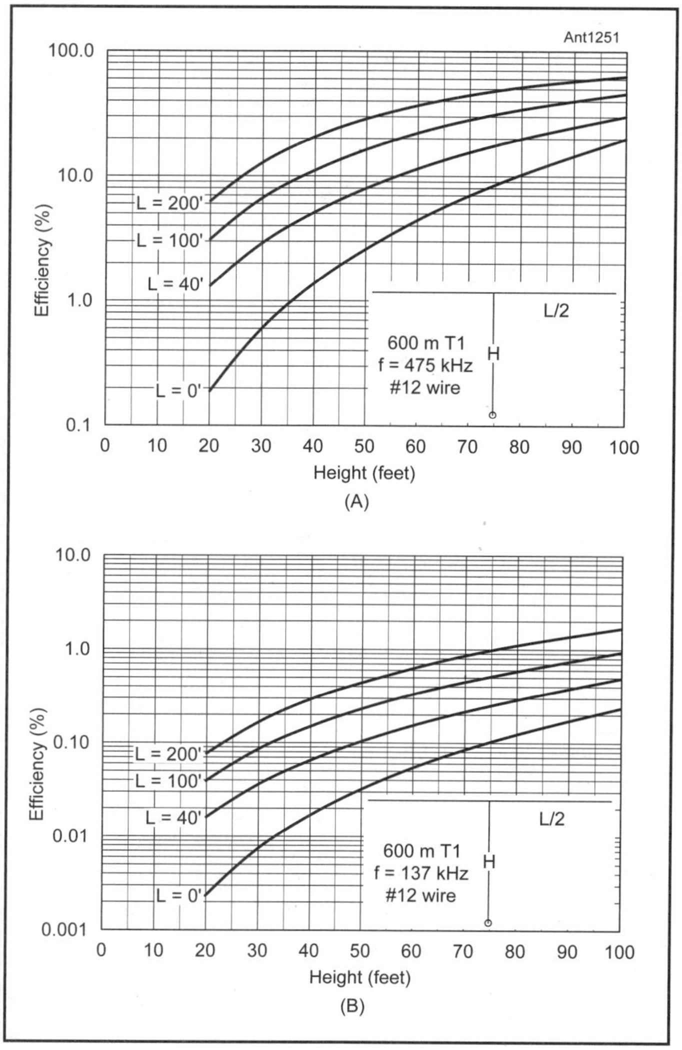

Figure 11.59 — Efficiency at 475 kHz (A) and 137 kHz (B) with base tuning.

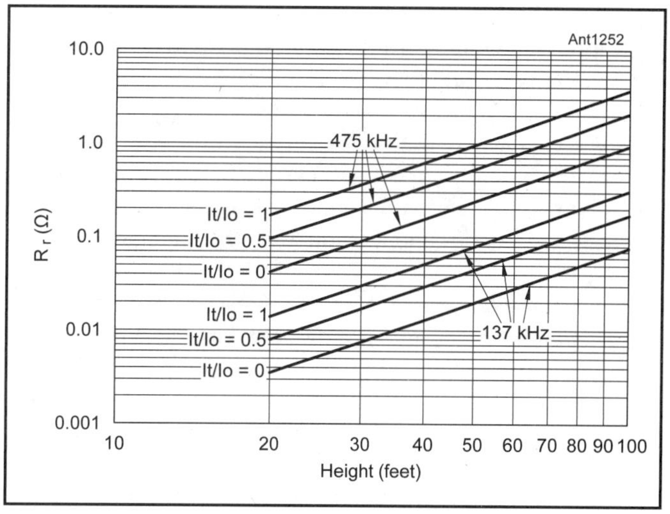

Figure 11.60 — R_r as a function of height in feet.

at 475 and 137 kHz. The L = 0 trace represents the case with no top-loading, just the bare vertical with a loading coil at the base. In these models the value of the tuning inductor was adjusted to maintain resonance as L and H were changed. #12 wire conductors are assumed. Even a small amount of top-loading increases efficiency. As an example, for L = 0 and H = 20 feet, $\eta = 0.19\%$ at 475 kHz. Keeping H = 20 feet but adding a 40 foot top-wire, $\eta = 1.3\%$, a factor of 6.8! Taking L to 100 feet increases the efficiency by a factor of 18! Height and capacitive top-loading are keys to improving efficiency!

R_r WITH TOP-LOADING

Top-loading can also improve R_r. R_r increases substantially as we add more top wire. **Figure 11.60** shows a graph for R_r as a function of antenna height with the current ratio I_t/I_O as a parameter (I_t is the current at the top of the antenna). The I_t/I_o ratio is varied from 0 (no top-loading) to 1, which corresponds to heavy top-loading and constant current on the vertical radiator (i.e. $I_t/I_O = 1$). As I_t/I_O goes from zero to 1, R_r increases by a factor of 4.

What happens to V_O and I_O when top-loading is present?

$$I_O = \sqrt{\frac{P_r}{R_r}} \text{ and}$$

$$V_O = X_i I_O = X_i \sqrt{\frac{P_r}{R_r}}$$

Even a small amount of top-loading significantly reduces I_O and greatly reduces V_O! This is important because despite the low radiated powers V_O can easily approach 1 kV on 630 meters and be even higher on 2200 meters. This must be kept in mind when selecting a base insulator. It's clear that substantial height (H) and top-loading are required on 2200 meters if V_O is to be kept below 10 kV. Given that we are trying to radiate only 330 mW that may come as a shock!

TOP-LOADING AND SYMMETRY

In the real world, the T antenna will not be perfectly symmetric and there will be some sag in the support wire, as well. With 5 feet of sag in the 475 kHz antenna the efficiency drops by ≈1.5%. At 137 kHz the efficiency drops from 0.23% to 0.20%.

In some installations it may be more convenient to attach the vertical portion of the antenna at a point other than the center of the top-wire. We can attach the downlead anywhere along the wire and we are also free to place the ground end of the downlead pretty much where we want with little effect on efficiency.

Using supports already on hand (trees, poles, structures), the two ends of the top-wire are likely to be at different heights. For example, the top-wire might slope downward from one end to the other. This is not necessarily a bad thing. For a given center height of 50 feet (see the graphs in the online notes), when we raise one end of the top wire the efficiency goes up even though we've lowered the other end.

Sometimes only one support will be available and the

top-wires will have to slope downward from the center support. If the height of the vertical is assumed to be 50 feet and the two top-wires are 50 feet long, efficiency is a strong function of the height of the top-wire ends. The higher the better! This is a case where the current in the top-wires has a component that partially cancels the current in the vertical, reducing R_r. The ends of the top-loading wires can also be bent toward the ground without significant effects on efficiency.

USING A TOWER AS A SUPPORT

What is the effect of coupling between a grounded tower and the vertical downlead? The simple answer is that it will reduce the efficiency somewhat but usually only a few percent because the tower, even with multiple Yagis for loading, is unlikely to be resonant anywhere near 475 kHz, not to mention 137 kHz. The coupling can be minimized by spacing the top anchor point as far out from the tower as possible; several feet would be helpful. Pulling the bottom and/or the top of the downlead away from the tower as shown in **Figure 11.61** can also help. The effect on a specific installation is best explored with modeling.

11.5.3 More Complicated Top-Loading

While a vertical with a single top-wire is attractive, we're not limited to a single wire for top-loading. More complex top-loading can substantially improve efficiency. For example, we can add spreaders and use two or more wires in a rectangular arrangement.

One note, the T models in this section use only a single conductor down-lead. When multiple top wires are used, the down-lead can also have multiple wires for at least part of its length which can make a small improvement in efficiency by reducing conductor loss, R_C. One of the problems with spreaders is that they tend to rotate and twist. Extra downleads can act as stabilizers. Light non-conducting lines can also be used to stabilize the spreaders.

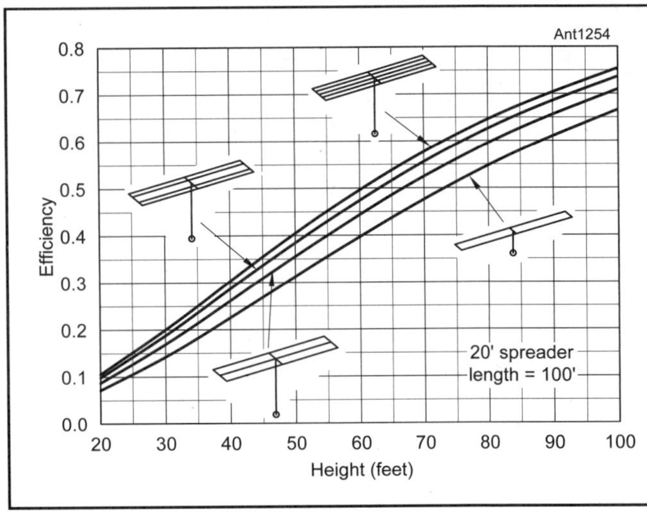

Figure 11.62 — Comparison of top-loaded verticals using a single wire or multiple horizontal wires with 20-foot spreaders.

Adding more wire to the hat reduces X_C and leads to higher efficiency but as can be seen in the test cases shown in **Figure 11.62**, for a given spreader length the rate of improvement falls pretty quickly and in this case using more than five wires gains very little. Adding more wire to the top-hat helps to reduce X_C but there's an important drawback to more wire up in the air: if you live in a area with ice storms, the antenna becomes much more vulnerable.

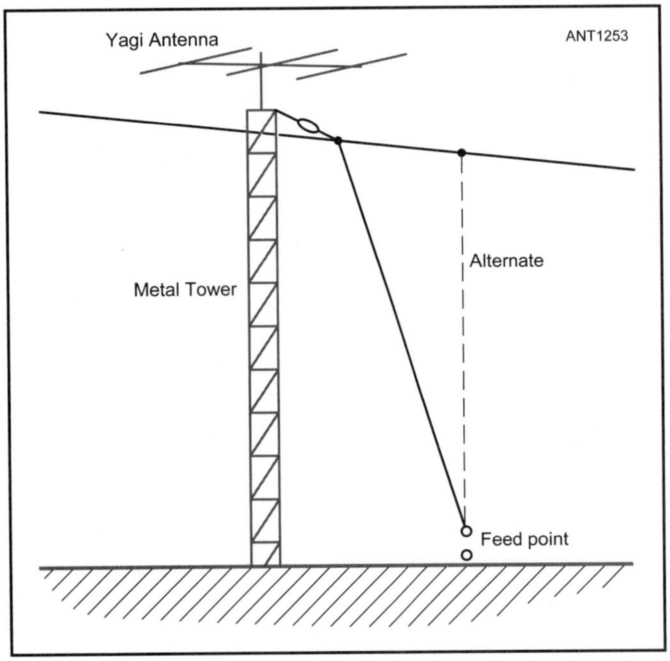

Figure 11.61 — Using a grounded tower as a center support.

UMBRELLA TOP-LOADING

Top-loading structures like umbrellas are often used when only one support (the vertical itself for example) is available. There are many ways to construct an umbrella top-hat:

- Use several rigid radial supports like a wagon wheel. For example, the supports can be aluminum tubing or fiberglass poles supporting wires or some combination of the two. The practical limit for a self-supporting structure is probably a radius of 20 feet or so. Using the hub(s) and spreaders from an old HF quad can be a very simple way to fabricate an umbrella.
- Set up a circle of poles (3 to 8) at some distance from the base of the vertical to support the far ends of the umbrella wires. The radial dimension of the umbrella could be quite large but the length of the poles, which establishes the height of the outer rim of the umbrella, is a limiting factor.
- Attach a number of wires to the top of the vertical, sloping them downward at an angle towards anchor points located at some distance from the base of the vertical.
- You can add sloping wires to the outer rim of a horizontal umbrella to increase the loading.
- You do not have to connect the outer ends of the umbrella radial wires together with a "skirt" wire but a skirt-wire

significantly increases the loading effect of an umbrella of a given radius.

• While most of this discussion shows symmetric umbrellas, symmetry is not required. Supports at different distances with different heights can also be used to good effect. Take advantage of what's on site!

As few as three wires (with a skirt!) can make a useful umbrella but the performance improves substantially as you go from three to four and then eight wires. While the jump from four to eight wires gives a useful improvement the law of diminishing returns starts to set in and sixteen wires is about the useful limit. The online notes go into some detail about umbrellas that slope downward from the center hub.

11.5.4 Conductor Loss, R_C

Copper or aluminum wire and aluminum tubing are typical conductors. The wire may be bare or insulated. The choice of conductor is usually a matter of what's on hand and/or what's economical. Insulated wire intended for home wiring is of often the most economic choice for copper wire. Aluminum electric fence wire, available in #14 or #17 AWG, is a less expensive choice but aluminum has greater resistance than copper and because soldering aluminum is often not satisfactory, joints require special attention. We can work around the resistance issue by using multiple, well-spaced, wires in parallel.

Given the very low efficiencies compared to typical antennas at HF and above, some attention must be made to the loss created skin effect, δ, which forces ac current to flow in a thin layer on the outside of a conductor. At 137 kHz δ = 7.03 mils (0.00703 inches) and at 475 kHz δ = 3.78 mils. (1 mil = 0.001 inch) At RF, ac resistance can be calculated as $R_{ac} = R_{dc} \times K_s$, where K_s is a factor attributed to skin effect that depends on the ratio of conductor thickness, d, to skin depth, δ. **Table 11.6** shows K_s on both bands for several wire sizes.

11.5.5 Loading Inductor Placement

Top-loading is not the only means for increasing R_r. We can move the tuning inductor or even only a portion of it, from the base up into the vertical. We can also move the feed point higher in the vertical. Multiple inductors can be useful in what referred to as "multiple-tuning," a technique using multiple inductors in multiple downleads to manipulate the feed point impedance and distribution of current between parallel wires.

In HF mobile verticals it has long been standard practice to move the loading inductor from the base up into the vertical to increase R_r. We can do the same for LF/MF verticals. Traditionally the entire loading inductance is moved up. However, there are advantages to moving only a portion of the loading inductance up into the antenna and retaining the remainder (L_{base}) at the base. For a given Q_L, R_L will increase as the inductor is moved up. Despite this increase in R_L moving the inductor up generally improves efficiency. The peak efficiency occurs for heights of 40 to 50% of H. How much does this increase our signal? For $L_{base} = 0$, i.e. we move all the inductance up, we can get about 0.74 dB of improvement at a height of ≈35%. By making L_{base}=1500 Ω of the total reactance we can pick up another 0.25 dB for a total improvement of almost 1 dB which is probably worth doing.

Even if a modest increase in signal is not compelling there are other reasons for using two inductors. Even when resonated it will still be necessary to match the feed point impedance to the feed line which can be done very simply by tapping the base inductance as shown in **Figure 11.63**.

A base inductor is also a convenient point to retune the antenna when necessary. Over the course of the seasons as the soil characteristics change, the tuning often shifts, primarily due to variations in effective loading capacitance as the soil conductivity changes with moisture content. Small heavily loaded verticals typically have very narrow bandwidths. In most cases some arrangement for adjusting the inductance will be needed. This can be readily done by using a variometer (Figure 11.63B) or a separate small roller inductor in series. One additional advantage of not putting all the inductance up high is the reduced weight of the elevated inductor.

TOP-LOADING COMBINED WITH INDUCTOR TOP-LOADING

Moving the inductor higher into a heavily top-loaded vertical has less effect on the current distribution. As a result efficiency improvements are much smaller. For the example of a T antenna with H = 50 feet and a single 100-foot top-wire,

Table 11.6
K_s for Typical Wire Sizes

Wire #	*137 kHz Ks*	*475 kHz Ks*
8	4.82	8.76
10	3.87	7.00
12	3.10	5.55
14	2.53	4.49
16	2.06	3.61
18	1.15	1.93

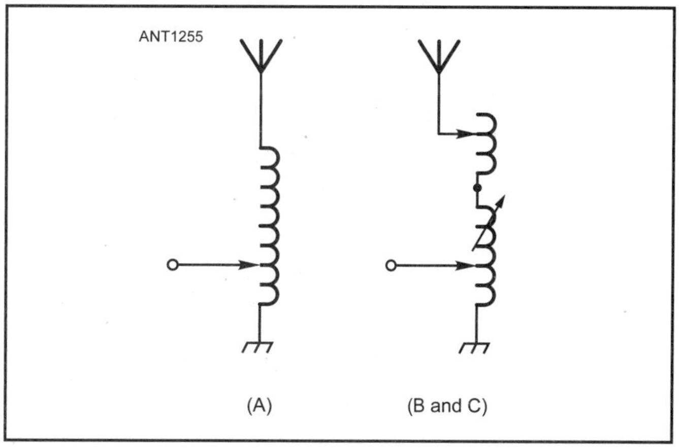

Figure 11.63 — Feed point impedance matching with the base inductor.

as the inductor is moved up there is some improvement in the signal but not a lot, only 0.4 dB even when two inductors are used.

When there is a much larger top-hat (for example, three wires 100 feet long by 20 feet wide) the improvement from elevating the inductor is even smaller, <0.15 dB This small an improvement is not worth the hassle of mounting an inductor high in the antenna.

In heavily top-loaded verticals there appears to be little improvement in efficiency from elevating the loading inductor. On the other hand if the top-loading is less, I_t/I_O ratio <0.4 – 0.5, and more top-loading is not practical then moving the coil up may help. This has to be evaluated on a case-by-case basis using modeling.

11.5.6 Grounded Tower Verticals for LF/MF

A grounded tower with attached HF antennas and associated cabling is sometimes available. For an LF/MF antenna the tower may be simply a support but it can also be a radiator. One way we might do this is shown in **Figure 11.64** where the loading inductor and the feed point have been moved to the top of the tower. The top-loading wires are insulated from the tower and connected to one end of the loading inductor. The other end of the inductor is connected to the top of the tower. A coaxial feed line runs up the tower with the shield connected to the top of the tower. The coax center conductor is connected to a tap on the loading inductor to provide a match. Although not shown, it is possible to have a mast with HF Yagis extending above the top of the tower which will add some additional capacitive loading. The downside of this scheme is that all the adjustments must be made at the top of the tower.

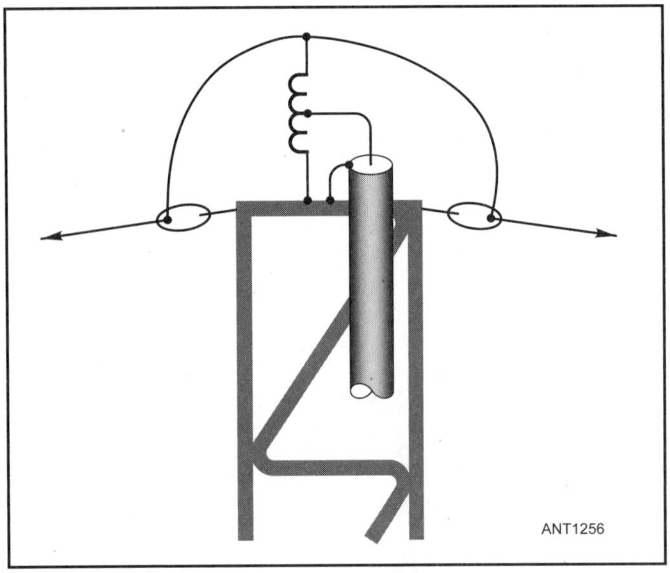

Figure 11.64 — A grounded tower radiator with the feed point and loading inductor at the top with top-loading.

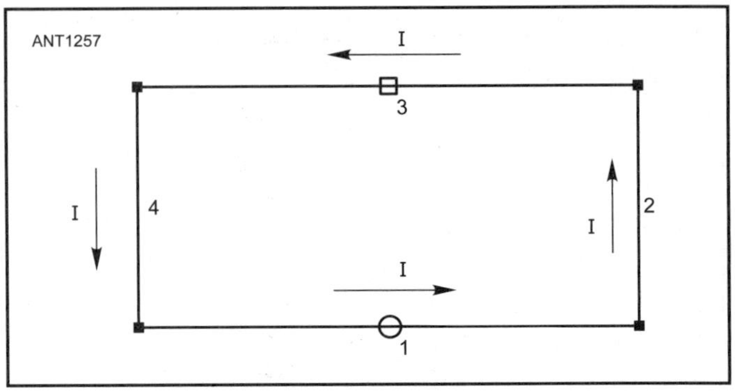

Figure 11.65 — Loop antenna for LF/MF operation.

Shunt feeding a grounded tower is often used on 80 and 160 meters as described in the section on Shunt Feeding Towers earlier in this chapter. Unfortunately, this scheme works only if H >0.7 λ/4 at the operating frequency. At 475 kHz that would mean H >350 feet, much taller than most amateur towers. See the online notes for an extended discussion of this situation including creating and tuning a folded monopole.

11.5.7 Loop Antennas at MF

Ground systems are a nuisance and sometimes impractical. We might consider using a transmitting loop like that shown in **Figure 11.65**. In this example the horizontal wires are 100 feet long and the vertical wires 50 feet, all #12 AWG copper wire. The bottom wire is 8 feet above average ground (0.005/13) and f = 475 kHz. The antenna is resonated with a capacitive load at the center of the upper wire (point 3) where X_C = 537 Ω. As is typical for small loops the current amplitude around the loop varies only ±5% with very little phase difference and for the values given, the radiation efficiency will be ≈1.8%. John Andrews, W1TAG and WE2XGR/3, has used a similar loop made with RG-8 coax (diameter ≈ 0.3 inch). This increases the efficiency to ≈ 3.1%. Not great but if 100 W of input power is available then the maximum EIRP can be reached. Even if we used superconducting wire the efficiency would still be limited to ≈ 4.2% due to near-field ground losses, a factor often overlooked in transmitting loops. Poorer soil would mean even lower efficiency. These efficiencies are not very encouraging but then transmitting loop antennas are not known for their efficiency! Inductive and capacitive loading can also be used, increasing efficiency to 9 to 12%.

11.5.8 LF/MF Ground Systems

The discussion of efficiency focused on loss introduced by the tuning inductor which is reasonable given that R_L often represents a major loss in short antennas. While much can done to reduce X_i and increase R_r, there are practical limits and at some point we have to start thinking about reducing other losses. A substantial portion of the power supplied to the antenna may be absorbed in the soil near the base. To reduce this loss we install a ground system:

- what type of ground system?
- how many radials?
- radial lengths?
- what performance can we expect?
- optimum use of a fixed amount of wire?

A range of examples have been chosen to provide general guidance but none should be taken as exact numerical descriptions for all cases. You will have to do some measurements, modeling and/or calculations to arrive at the best solution for your unique situation.

CHOICES FOR GROUND SYSTEMS

Ground systems can take several forms:

1. A radial wire fan lying on the ground surface or buried a few inches.
2. A rectangular grid of wires
3. Single or multiple ground rods.
4. Elevated wires in the form of a counterpoise or "capacitive" ground.
5. Combinations of the above. Limited only by imagination!

The choice of ground system will be dictated by the operating wavelength, available space, soil mechanical characteristics (i.e. sandy loam or tree stumps and boulders), available resources, etc. Because of the much longer wavelengths at LF/MF and significant differences in soil electrical characteristics between LF/MF and HF, the ground systems may be significantly different from what we are accustomed to at HF.

The instructions for an adequate ground system can be stated as:

1) Use at least 50 radials. Most backyards will only have room for 30 to 40-foot radials. Where possible the radials should be somewhat longer than the height of the vertical

2) In the case of a very large top-hat, the radials should extend out to 1.25 times the top-hat radius if possible.

3) When a large number of radials are used the wire size is not important. The wire needs to be strong enough to be installed and survive in its environment.

4) Almost any metal can be used for the radials but the usual choice is insulated copper house wiring because it is usually cheaper than the same wire bare. For an elevated system #17 AWG aluminum electric fence wire can be used. However, lying on the surface or buried, aluminum wire may degrade quickly

5) If the radials are lying on the surface, use lots of staples to keep them close to the ground so mowing or other traffic will not damage them.

6) Use at least one ground stake at the base for safety.

Consider three ground systems with an inductor Q_L of 400: a single 8-foot × ⅝-inch ground rod or stake, 32 10-foot radials and 32 25-foot radials. With the single ground rod the ground loss (R_g) is so large that R_L doesn't matter very much. As soon as we add even the small radial system (32 × 10 feet) the efficiency increases by almost an order of magnitude and expanding the radial lengths to 25 feet yields another factor of ≈3× in efficiency. This is due to reductions in R_g with longer radials.

Increasing the number of radials is very helpful. In fact, we could have gone to 64 radials and still have had some useful improvement over 32. Second, the efficiency improves rapidly up to lengths of 60 to 70 feet before the point of decreasing returns is reached. By 150 feet there isn't much point in making the radials longer. Independent of radial number, that point in this example corresponds to a radial length of approximately 65 to 70 feet which is a bit more than H (50 feet).

Once you have greatly reduced the losses near the base of the antenna, adding more close-in copper doesn't buy much. At some point it's more useful to put the copper further out and reduce more distant losses, which may be smaller but still significant. The difference in break point (in terms of radial number) stems from differences in the field intensities around the two antennas. For the same power, the fields near the base of a short vertical will be much higher than those for the ¼-wave vertical so we need to put more effort into reducing the close-in power losses.

ELEVATED GROUND SYSTEMS

In many cases the ground under and near the antenna may not be suitable for a buried radial system. Systems with elevated wires are well known at HF, i.e. ground-plane verticals, but these systems typically use radials with lengths close to $\lambda/4$. For amateur installations this will not be possible but all is not lost. In the early days of radio it was recognized that an elevated system called a "counterpoise" or "capacitive ground," with dimensions significantly smaller than $\lambda/4$, could be very effective.

If you have reasonable heights for the vertical and lots of capacitive top-loading but very restricted room for the ground system, then a counterpoise may be the best option. But we have to be careful in drawing general conclusions. There are many variables: ground characteristics, height of the top of the vertical, height of the bottom of the vertical, the amount of top-loading, the number of wires in the counterpoise, the radius, etc. The choice between buried wire and counterpoise ground systems is not obvious! The considerable mechanical complexity, vulnerability to ice damage and visual impact of a counterpoise may also militate against it. This choice has to be made on a case-by-case basis and will probably require modeling with *NEC4* software.

11.6 References and Bibliography

Source material and more extended discussion of topics covered in this chapter can be found in the references given below.

ARRL's Wire Antenna Classics (ARRL, 1999).

Atchley, D., Jr., "Putting the Quarter-Wave Sloper to Work on 160," *QST*, July 1979, pp. 19 – 20.

Beezley, B., "A Gain Antenna for 28 MHz," *QST*, July 1994, pp. 70.

Belrose, J., "Transmission Line Low Profile Antennas," *QST*, Dec. 1975, pp. 19 – 25.

Belrose, J., "Terminated Folded Dipole," Technical Correspondence, *QST*, May 1994, p. 88.

Belrose, J., "A Horizontal Loop for 80-Meter DX," *QST*, Aug. 2002, pp. 30 – 35.

Belrose J., and DeMaw, D., "The Half-Delta Loop: A Critical Analysis and Practical Deployment," *QST*, Sep. 1982, pp. 28 – 32.

Braskamp, L., AA6GL, "MOBILE, a Computer Program for Short HF Verticals," *The ARRL Antenna Compendium* Vol. 4 (ARRL, 1995), pp. 92 – 96.

Brown, G., "The Phase and Magnitude of Earth Currents Near Radio Transmitting Antennas," *Proc. IRE*, Vol. 23, No. 2, Feb. 1935, pp. 168 – 182.

Brown, G., Lewis R., and Epstein, J., "Ground Systems as a Factor in Antenna Efficiency," *Proc. IRE*, Vol. 25, No. 6, June 1937, pp. 753 – 787.

Cake, B., *The Antenna Designer's Notebook* (ARRL, 2009).

Cake, B., "The 'C Pole' — A Ground Independent Vertical Antenna," *QST*, Apr. 2004, pp. 37 – 39.

Carr, P., "A Two Band Half-Square Antenna with Coaxial Feed," *CQ*, Sep. 1992, pp. 40 – 45.

Carr, P., "A DX Antenna for 40 Meters," *CQ*, Sep 1994, pp. 40 – 43.

Carr, P., "The N4PC Loop," *CQ*, Dec. 1990, pp. 11 – 15.

Cebik, L., "Best way (of) Loading the Half-Length 80M Vertical Monopoles – V," various online sources.

Cebik, L., "Notes on a Wide-Band 50/75-Ohm Coax Feed System for Low HF Band Dipoles and Vees," 2007, various online sources.

Christman, "Elevated Vertical Antennas for the Low Bands," *The ARRL Antenna Compendium* Vol. 5 (ARRL, 1996), pp. 11 – 18.

DeMaw, D., "Additional Notes on the Half Sloper," *QST*, July 1979, pp. 20 – 21.

DeMaw, D., Aurick, L., "The Full-Wave Delta Loop at Low Height," *QST*, Oct. 1984, pp. 24 – 26.

Devoldere, J., *ON4UN's Low Band DXing,* 5th Edition (ARRL, 2010).

Dodd, P., *The LF Experimenter's Source Book* (RSGB, 1996).

Doty, A., Frey J., and Mills, H., "Efficient Ground Systems for Vertical Antennas," *QST*, Feb. 1983, pp. 9 –12.

Ford, S., "Lowfing on 1750 Meters," *QST*, Oct. 1993, pp. 67 – 68.

Fosberg, R., "Some Notes on Ground Systems for 160 Meters," *QST*, Apr. 1965, pp. 65 – 67.

Gordon, L., K4VX, "The K4VX Linearly Loaded Dipole for 7 MHz," *QST*, July 2002, pp. 40 – 42.

Hall, J., "Off-Center-Loaded Dipole Antennas," *QST*, Sep. 1974, pp. 28 – 34.

Hallas, J., "The Fan Dipole as a Wideband and Multiband Antenna Element," *QST*, May 2005, pp. 33 – 35.

Hallas, J., "A Close Look at the Terminated Folded Dipole Antenna," Getting on the Air, *QST*, Sep. 2010, pp. 51 – 52.

Harbach, A., "Broad-Band 80-Meter Antenna," *QST*, Dec. 1980, pp. 36 – 37.

Hawkins, H., "A Low-Budget, Rotatable 17 Meter Loop," *QST*, Nov. 1997, p. 35.

Heys, J., *Practical Wire Antennas* (RSGB, 1989).

Hubbell, G., "Feeding Grounded Towers as Radiators," *QST*, June 1960, pp. 32 – 33, 140, 142.

Hutchinson C., and Straw, R., *Simple and Fun Antennas for Hams* (ARRL, 2002).

Lee, P., *The Amateur Radio Vertical Antenna Handbook*, 1st Edition (Cowan Publishing Corp, 1974).

Leeson, D., "The Story of the Broadband Dipole," *QEX*, Nov./Dec. 2018, pp. 18 –22.

Lopes, L., "Designing a Shortened Antenna" *QST*, Oct. 2003, pp. 28 – 32.

Lopez, A., "Double-Tuned Impedance Matching," *IEEE Antennas and Propagation Magazine*, Vol. 54, No. 2, Apr. 2012, pp. 109 – 115.

Maxwell, W., "Technical Correspondence: The Double Bazooka Antenna — How Broad Is It?" *QST*, Sep. 1976, pp. 29 – 30.

Maxwell, W., "A Revealing Analysis of the Coaxial Antenna," *Ham Radio*, Aug. 1976, p. 46.

Michaels, C., "Some Reflections on Vertical Antennas," *QST*, July 1987, pp. 15 – 19; feedback *QST*, Aug. 1987, p. 39.

More Wire Antenna Classics (ARRL, 1999).

Moxon, L., *HF Antennas for All Locations*, 2nd Edition (RSGB, 1993).

Poole, I., *Practical Wire Antennas 2* (RSGB, 2005).

"Prediction of Sky-wave Field Strength at Frequencies Between About 150 and 1700 kHz," ITU Doc 3/14, Radiocommunication Study Groups, Feb 1995.

Rappaport T., and Parnell, J., "Use Your Tower as a Dual-Band, Low-Band DX Antenna," *QST*, May 2019, pp. 41 – 45.

Rashed and Tai, "A New Class of Wire Antennas," 1982 International Symposium Digest, *Antennas and Propagation*, Vol 2, published by IEEE.

Russell, T., "Simple, Effective, Elevated Ground-Plane Antennas," *QST*, June 1994, pp. 45 – 46.
Saviers, G., "Broad-banding a 160 m Vertical Antenna," *QEX*, May/June 2019, pp. 9 – 12.
Schnell, F., "The Flagpole Deluxe," *QST*, Mar. 1978, pp. 29 – 32.
Schultz, W., "Designing a Vertical Antenna," *QST*, Sep. 1978, p. 19.
Severns, R., "A Wideband 80-Meter Dipole," *QST*, July 1995, pp. 27 – 29.
Severns, R., "A Closer Look at Vertical Antennas With Elevated Ground Systems," *QEX*, Mar./Apr. 2012, pp. 32 – 44 and May/June 2012, pp. 24 – 33.
Severns, R., "Some Ideas for Short 160 Meter Verticals," *QEX*, May/June 2013, pp. 35 – 46
Sevick, J., "The Ground-Image Vertical Antenna," *QST*, July 1971, pp. 16 – 19, 22.
Sevick, J., "The W2FMI Ground-Mounted Short Vertical," *QST*, Mar. 1973, pp. 13 – 18, 41.
Sevick, J., "The Constant-Impedance Trap Vertical," *QST*, Mar. 1974, pp. 29 – 34.
Sevick, J., "Short Ground-Radial Systems for Short Verticals," *QST*, Apr. 1978, pp. 30 – 33.
Sevick, J., *Transmission Line Transformers* (Noble Publishing, 1996).
Siwiak K., and Quick, R., "Live Trees Affect Antenna Performance," *QST*, Feb. 2018, pp. 33 – 37.
Stanford, J., "Linear-Loaded Short Wire Antennas," *The ARRL Antenna Compendium* Vol. 5 (ARRL, 1996), pp. 105 – 107.
Vertical Antenna Classics (Newington: ARRL, 1995)
Watt, A., *VLF Radio Engineering* (Pergamon Press, 1967) (out of print).
Weigl, J., *Sloper Antennas* (CQ Communications, 2009).
Witt, F., AI1H, "A Simple Broadband Dipole for 80 Meters," *QST*, Sep. 1993, pp. 27 – 30, 76.

TABLE OF CONTENTS

Chapter 12 — Online Material

Articles

- "A Compact Multiband Dipole" by Zack Lau, W1VT
- "A No Compromise Off-Center Fed Dipole for Four Bands" by Rick Littlefield, K1BQT
- "A Triband Dipole for 30, 17, and 12 Meters" by Zack Lau, W1VT
- "An Effective Multi-Band Aerial of Simple Construction" by Louis Varney, G5RV (Original G5RV article)
- "An Experimental All-Band Non-directional Transmitting Antenna," by G.L. Countryman, W3HH
- "An Improved Multiband Trap Dipole Antenna" by Al Buxton, W8NX
- "Broadband Transmitting Wire Antennas for 160 through 10 Meters" by Floyd Koontz, WA2WVL
- "Cat Whiskers — The Broadband Multi-Loop Antenna" by Jacek Pawlowski, SP3L
- "End-Fed Antennas" by Ward Silver, NØAX
- "HF Discone Antennas"
- "HF Discone Antenna Projects" by W8NWF
- "Nested Loop Antennas" by Scott Davis, N3FJP
- "Revisiting the Double-L" by Don Toman, K2KQ
- "Six Band Loaded Dipole Antenna" by Al Buxton, W8NX
- "The HF Discone Antenna" by John Belrose, VE2CV
- "The J78 Antenna: An Eight-band Off-Center-Fed HF Dipole" by Brian Machesney, K1LI/J75Y
- "The Multimatch Antenna System" by Chester Buchanan, W3DZZ
- "The Open Sleeve Antenna" by Roger Cox, WBØDGF
- "The Open-Sleeve Antenna" from previous editions
- "Two New Multiband Trap Dipoles" by Al Buxton, W8NX
- "Wideband 80 Meter Dipole" by Rudy Severns, N6LF
- "The Coupled-Resonator Principle: A Flexible Method for Multiband Antennas" by Gary Breed, K9AY
- "An Off Center End Fed Dipole for Portable Operation on 40 to 6 Meters" by Kai Siwiak, KE4PT

Chapter 12
Multiband HF Antennas

For operation on a number of bands below 30 MHz, it would be impractical for most amateurs to put up a separate antenna for each band. But this is not necessary — for example, a dipole one half-wavelength long on the lowest frequency band to be used can be operated readily on higher frequencies. In fact, most common antennas can be used on multiple bands through the use of antenna tuners and other techniques. What is usually referred to as a "multiband antenna," however, is one for which a method has been devised that allows the antenna to operate on a number of bands while still offering a good match to a transmission line, usually coaxial cable.

When a single physical antenna is used on different bands, one must be aware that the changing electrical heights and lengths lead to changes in the feed point impedance and the azimuth and the elevation patterns of the antenna as described in the chapters **Antenna Fundamentals** and **Dipoles and Monopoles**. For example, a horizontal wire antenna at an electrical height of $\lambda/2$ on 20 meters is $2\lambda/3$ high on 15 meters and $\lambda/4$ on 40 meters, leading to very different elevation patterns than if the antenna were at the same electrical height on all bands. Similarly, the elevation pattern and feed point impedance of a single vertical antenna will also change dramatically on different bands.

In fact, it is usually more effective to consider the installation as a "multiband antenna system" in which the antenna, feed line, and any impedance matching devices are considered together — as a package. By thinking about the performance of the antenna on different bands you can select a combination of system elements that result in good performance on all bands and not just one.

This chapter describes a number of antennas and antenna systems that are designed to be used on two or more of the HF bands. Separate chapters cover nonresonant **Long-Wire and Traveling Wave Antennas** as well as the popular **HF Beam Antennas**. See the **Transmission Line System Techniques** chapter for more information on using feed lines and impedance matching circuits.

HARMONIC RADIATION FROM MULTIBAND ANTENNAS

Since a multiband antenna is intentionally designed for operation on a number of different frequencies, any harmonics or spurious emissions with frequencies that happen to coincide with one of the antenna resonant frequencies will be radiated with very little, if any, attenuation. Particular care should be exercised, therefore, to prevent such harmonics from reaching the antenna. While output signals from properly operating commercial equipment are typically quite "clean," they can be misadjusted or defective, or another piece of equipment in the antenna system could be acting to create unwanted signals. As a consequence, amateurs need to remain vigilant about signal quality and harmonic content.

If there is any concern about whether harmonics or spurious emissions are being generated, it is advisable to conduct tests with other nearby amateur stations. If harmonics of the transmitting frequency or spurious emissions can be heard at a distance of, say, a mile or so, more selectivity should be added to the system since a harmonic that is heard locally, even if weak, may be quite strong at a distance because of propagation conditions. In addition, the source of spurious emissions should be identified and the unwanted signals removed.

12.1 Simple Wire Antennas

12.1.1 End-Fed Random Wire Antennas

The simplest multiband antenna is a random length of wire, attached directly to the output of a transmitter or antenna tuner. Power can be fed to the wire on practically any frequency using one or the other of the methods shown in **Figure 12.1**. If the wire is approximately 67 or 137 feet long ($\lambda/4$ or $\lambda/2$ on 80 meters) the end impedance will be high on the bands that are harmonics of 80 meters and it can be fed through a tuned circuit, as in **Figure 12.2**. Many antenna tuners have the option to feed an end-fed random wire in this way. Use an SWR meter between the transmitter and the matching network to adjust for minimum SWR. (This is a variation of the End-Fed Half-Wave (EFHW) antenna popular for portable use and discussed later in this chapter and in the **Portable Antennas** chapter.)

It is also possible to use an antenna's feed line as part of the antenna (also see the Inverted-L and T Antennas section in the **General Purpose MF and HF Antennas** chapter). Short the feed line conductors together and feed them directly as a random wire as shown in Figure 10.1. This technique may be

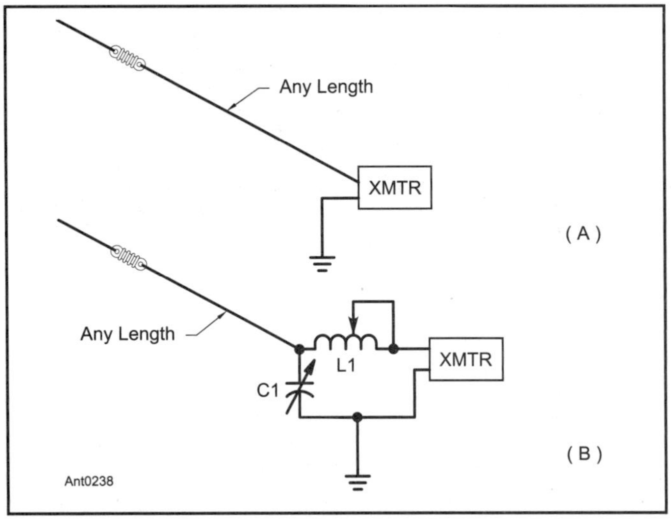

Figure 12.1 — At A, a random-length wire can be driven directly from the pi-network output of a vacuum-tube transmitter. At B, an L network (or antenna tuner) can be used with solid state transmitters that do not have tunable output networks. C1 should have plate spacing sufficient for at least several hundred volts; a maximum capacitance of 100 pF is sufficient if L1 is 20 to 25 µH. A suitable coil would consist of 30 turns of #12 AWG wire, 2½ inches diameter, 6 turns per inch. Bare wire should be used so the tap can be placed as required for loading the transmitter.

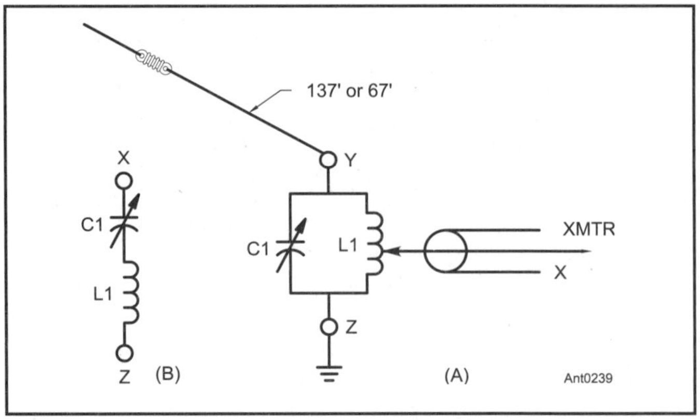

Figure 12.2 — If the antenna length is 135 to 140 feet, a parallel-tuned coupling circuit (A) can be used on each amateur band from 3.5 through 30 MHz, with the possible exception of the 10-, 18- and 24-MHz bands. C1 should be from 500-1000 pF with plate spacing capable of withstanding several hundred volts. L1 should be chosen to resonate with 20-80% of C1's maximum value. If the wire is 67 feet long, series tuning can be used on 3.5 MHz as shown at the left; parallel tuning will be required on 7 MHz and higher frequency bands. The L network shown in Figure 12.1B is also suitable for these antenna lengths.

useful in situations requiring improvisation, such as after a natural disaster when antenna systems have been damaged.

The primary disadvantage of all such directly-fed systems is that the antenna system is composed of the random wire plus all of the station equipment enclosures and the station ground system. The point at which the antenna is connected can be thought of as a randomly chosen feed point in an antenna that has one end tied to ground. As such, there is a good chance that you will have "RF hot spots" in your station because of the RF current in the antenna system. RF voltages and current from such an antenna system can also create high RF fields in the station as discussed in the RF Safety section of the **Safe Practices** chapter.

RF voltages within the station can often be minimized by choosing an antenna and ground wire length so that the low feed point impedance at a current maximum occurs at or near the transmitter. A short connection (several feet or less) with heavy wire or strap to a ground rod or metallic water pipe that runs through ground may be sufficient on the lower bands but most ground connections are not short enough to minimize RF voltage by themselves. Regardless of how you address this issue, begin by connecting all equipment enclosures together (bonding) to prevent significant voltages from existing between pieces of equipment. See the *ARRL Handbook's* chapter on **Assembling a Station** for more information on managing the RF in your station.

Using an antenna wire length close to $\lambda/4$ (65 feet at 3.6 MHz, 33 feet at 7.1 MHz), or an odd multiple of $\lambda/4$ ($\frac{3}{4}\lambda$ is 195 feet at 3.6 MHz, 100 feet at 7.1 MHz, 50 feet at 14 MHz, etc) may be helpful. The goal is to place the antenna system's connection to the transmitter or antenna tuner at a point of low voltage. Obviously, this can be done for only one band even in the case of harmonically related bands, since the wire length that presents a current maximum at the transmitter will present a voltage maximum at two (or four) times that frequency.

Another possibility is to attach a counterpoise wire to the transmitter or antenna tuner enclosure. The counterpoise length is adjusted so that RF voltage on the station equipment is minimized. The length may or may not be $\lambda/4$ at the operating frequency since the impedance at the end of the antenna wire is unknown. Be prepared to experiment with different lengths. Different wires can be attached at different frequencies.

Another option is to use an "artificial ground" such as the MFJ-931 (**mfjenterprises.com**) as in **Figure 12.3** that tunes the counterpoise on different frequencies. It is also possible in many cases to use an ordinary 100-W antenna tuner to accomplish the same thing — tuning the random-length coun-

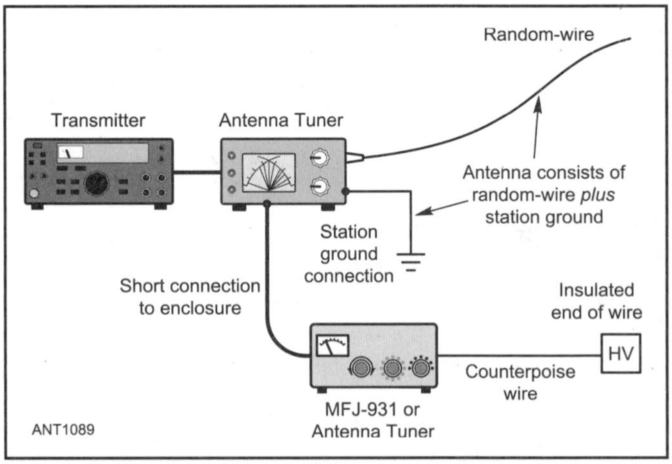

Figure 12.3 — An "artificial ground" can be used to tune a random length of wire to minimize RF voltage on station equipment enclosures.

terpoise to present a low impedance at the transmitter or antenna tuner enclosure.

If you do use a counterpoise, be sure to insulate the unattached end because like all unconnected ends of antennas, there will likely be enough RF voltage to cause an RF burn, particularly at 100 W or higher.

12.1.2 End-Fed Half-Wave Antennas

FEED LINE RADIATION

Feed lines pick up the antenna's radiated signal when they are not symmetrically oriented with respect to the antenna and its radiated field. For example, a feed line that approaches a center-fed dipole at an angle other than 90° will couple more strongly to the closer leg of the antenna. The closer the feed line is to one leg, the more energy it will pick up. Feed lines to an end-fed antenna almost always carry common-mode current because they are connected at one end and not the middle. Common-mode feed line current and techniques for minimizing it are addressed in the chapter **Transmission Line System Techniques**.

It should be emphasized that any radiation from a feed line is not "lost" energy and is not necessarily harmful. Whether or not feed line radiation is important depends entirely on the antenna system being used. In the case of a multiband antenna where general coverage is desired, if the feed line happens to radiate, such energy could actually have a desirable effect. From a practical standpoint where you are not concerned with a directive pattern, much time and labor can be saved by ignoring possible feed line radiation.

ZEPP ANTENNA

The *Zepp* antenna is shown in **Figure 12.4**. The antenna length is λ/2 long at the operating frequency. (This name came about because the antenna's first documented use was on *Zeppelin* airships. The antenna was hung from one end and trailed below the airship.) The Zepp was originally designed to be used on one frequency and often employed *tuned feeders* as described in the "Quarter-Wave Transformer" section of the **Transmission Line System Techniques** chapter.

An antenna tuner with a balanced output can provide multiband coverage with a Zepp antenna using any length of open-wire feed line, as shown in Figure 12.4. This may or may not result in a tunable impedance. When the tuner has difficulty obtaining a match, try inserting or removing a λ/8 section of feed line, to change the impedance at the end of the feed line.

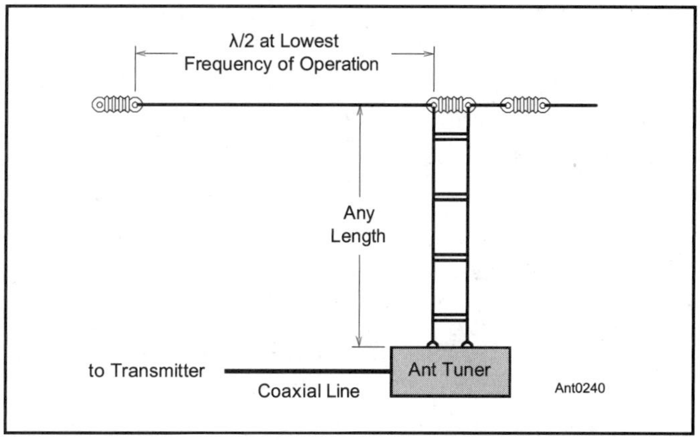

Figure 12.4 — An end-fed Zepp antenna for multiband use.

END-FED HALF-WAVE (EFHW)

Figure 12.5 shows the EFHW antenna, a half-wavelength dipole fed at one end. The open end is supported by a tree, mast, or other convenient support and the feed point can be close to the ground or elevated. The EFHW is popular for portable and temporary operation in the sloping configuration shown. It can also be installed horizontally with the feed point elevated. Both ends of the antenna will be high voltage points and should be kept clear of tree branches and leaves, and away from where they can be touched accidentally.

The EFHW can be used on any band for which it is an integer number of half-wavelengths long. It may also present an acceptable impedance on non-resonant bands. For example, an EFHW for 80 meters will also be useable on 40, 20, 15, and 10 meters, as well as 30, 17, and 12 meters with a somewhat higher SWR.

The feed point's high impedance must be transformed to a lower value so that coaxial feed line can be used. A 49:1 impedance transformer (7:1 turns ratio) as shown in

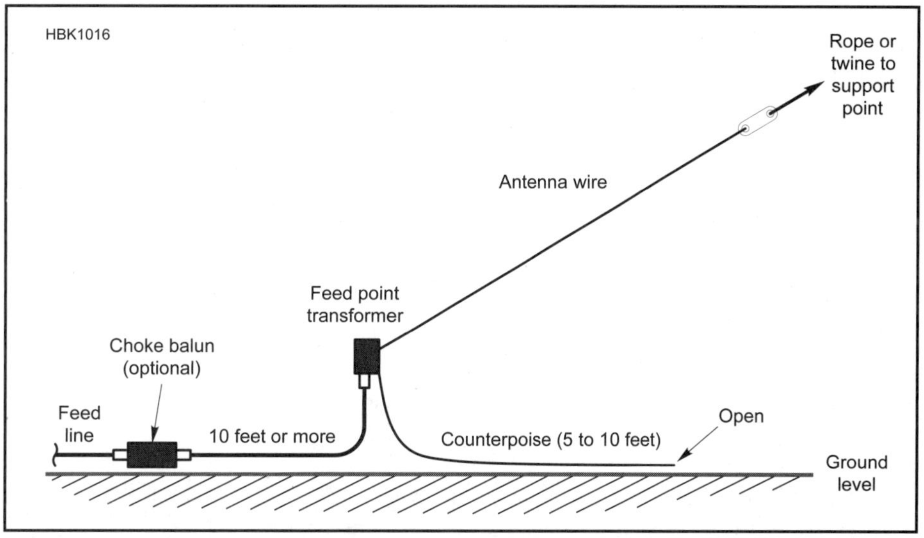

Figure 12.5 — The end-fed half-wave (EFHW) antenna. For permanent use, it can be installed horizontally with both ends well above ground. For portable or temporary use, it is common to keep the feed point near ground level and support the other end with a mast or with a rope over a tree branch. The counterpoise is connected to the feed line shield.

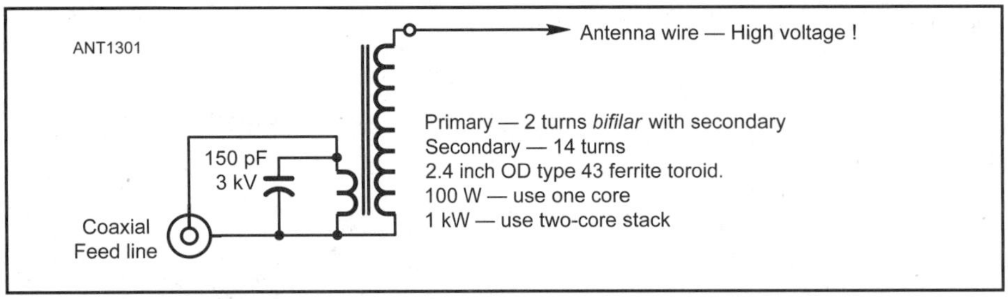

Figure 12.6 — A common 49:1 impedance transformer design. The transformer consists of a 2-turn primary and 14-turn secondary. The primary and first two secondary turns are bifilar. The 150 pF capacitor compensates for transformer inductance above 20 MHz. A low-loss 3 kV ceramic capacitor is recommended.

Figure 12.6 is a popular choice, although ratios from 9:1 to 64:1 may work acceptably. Placing the feed point a few feet from one end will also lower the impedance somewhat.

The EFHW requires a counterpoise to stabilize the feed point impedance. The outer surface of the coaxial feed line often serves as the counterpoise, so do not use a feed point choke balun without a counterpoise wire. If it is necessary to reduce common-mode RF current on the feed line, a choke balun or "line isolator" can be added 10 feet or more from the feed point, or where the feed line reaches ground level if the feed point is elevated. In the common configuration with the feed point near the ground, a few feet of wire connected to the feed line shield will supplement the feed line shield in stabilizing the feed point impedance. If installed with the feed point elevated, the feed line shield serves as a counterpoise and a separate wire is not used. A ground rod is not necessary for temporary installations.

The EFHW is also λ/4 long on the band that is one-half its lowest design frequency, i.e., an 80 meter EFHW is also λ/4 long on 160 meters. This creates a "bonus" use of the EFHW as a ground-plane antenna without reconfiguring the wire. Disconnect the antenna from the feed point transformer, add a ground system such as radial wires or rolls of wire mesh, and drive the antenna directly from the feed line. An antenna tuner or matching network at the antenna can be used to remove any impedance mismatch.

The EFHW antenna is explained in more detail in a presentation from Steve Dick, K1RF, at **gnarc.org/wp-content/uploads/The-End-Fed-Half-Wave-Antenna.pdf** and on the website of Steve Yates, AA5TB, at **aa5tb.com/efha.html**.

12.1.3 Center-Fed Antennas

A center-fed single-wire antenna can be made to accept power and radiate it with high efficiency on any frequency higher than its fundamental resonant frequency and, with a reduction in efficiency and bandwidth, on frequencies as low as one-half the fundamental.

In fact, it is not necessary for an antenna to be a full half-wavelength long at the lowest frequency. An antenna can be considerably shorter than ½ λ, even as short as ¼ λ, and still be a very efficient radiator. The use of electrically short antennas creates high feed point impedances that can result in high voltages and currents in antenna tuners and feed lines. Losses may also be higher as discussed in the **Transmission Line System Techniques** chapter.

The simplest and most flexible (and also least expensive) all-band antennas are those using parallel-wire feed lines to the center of the antenna, as in **Figure 12.7**. Because each half of the flattop is the same length, the feed line currents will be balanced at all frequencies unless, of course, imbalance is introduced by one half of the antenna being closer to ground (or a grounded object) than the other. To maintain balance of the current in each antenna leg and minimize common-mode current on the feed line, the feed line should be run at right angles to the antenna, preferably for a distance of at least λ/4 from the feed point.

Center-feed creates a better feed line balance and generally results in a lower SWR on parallel-wire line having a characteristic impedance of 300 to 600 Ω. Ladder or window line is satisfactory for all but possibly high power installations (over 500 W), where heavier wire and wider spacing are desirable to handle the larger currents and voltages that may be present at high SWR.

The best type of antenna tuner to use in such an installation is a balanced type designed for coaxial feed line on the input and parallel-wire line on the output. An unbalanced tuner can also be used but because one wire of the output feed line is connected to the enclosure, RF current and voltages are more likely to be present in the station.

Use of a balun at the output of an unbalanced antenna tuner to allow the use of balanced feed line is a common technique. The balun is usually an impedance transformer (voltage balun) with a 4:1 ratio in anticipation of higher impedances on balanced feed lines. This is not always the case as discussed in the chapters on **Transmission Lines** and **Transmission Line System Techniques**. The electrical length of the feed line and SWR at the antenna determine what impedance appears at the transmitter end of the feed line. Expect a wide range of imped-

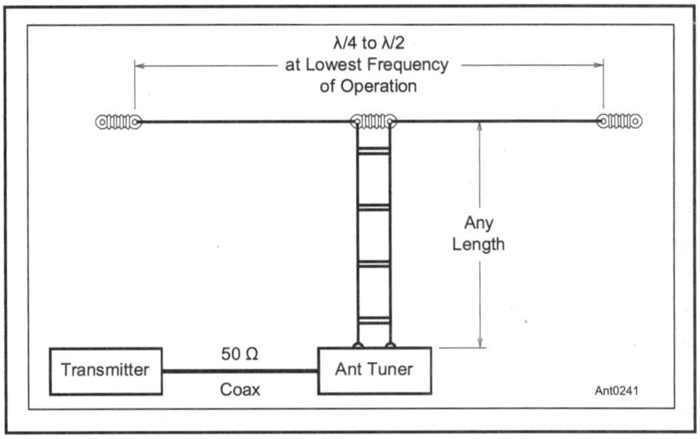

Figure 12.7 — A center-fed antenna system for multiband use.

ance values and be careful not to overstress the balun.

It is not recommended that coaxial feed line be used between the antenna tuner and antenna. At frequencies where the SWR is high, feed line loss in coaxial cable runs of more than 50 feet at HF can quickly become very high. (See the **Transmission Line System Techniques** chapter).

The length of the antenna is not critical, nor is the length of the line. As mentioned earlier, the length of the antenna can be considerably less than λ/2 and still be very effective. If the overall length is at least λ/4 at the lowest frequency, a quite usable system will result. Some experimentation will likely be necessary to find the length that works best at a specific location on the bands required.

Several nonresonant lengths of center-fed doublets have become popular in recent years for their consistent radiation patterns on several bands. For example, L.B. Cebik, W4RNL recommended the following lengths because the pattern remains at peak gain broadside to the antennas:

- 44 feet covers 10, 12, 15, 17, 20, 30, 40 meters
- 66 feet covers 15, 17, 20, 30, 40, 60 meters
- 88 feet covers 20, 30, 40, 60, 80 meters

Cebik's article "My Top Five Backyard Multi-Band Wire HF Antennas" explains why these simple antennas are such good performers (see the References and Bibliography).

The impedance of the antenna varies broadly from band to band, requiring the use of a wide-range tuner, but that is an acceptable tradeoff. Gain drops with frequency but not a great deal below that of a λ/2 dipole. Height above ground also influences the elevation pattern of the antenna as it does for a dipole, so install the antenna as high as you can for best performance at long distances.

PROJECT: 135-FOOT, 80 TO 10 METER DIPOLE

As mentioned previously, one of the most versatile antennas around is a simple 80 meter dipole, center-fed with open-wire transmission line and used with an antenna tuner in the station. A 135-foot long dipole hung horizontally between two trees or towers at a height of 50 feet or higher works very well on 80 through 10 meters. Such an antenna system has significant gain at the higher frequencies. The antenna can also be used on 1.8 MHz as a λ/4 antenna with some reduction in efficiency. (See the previous section for non-resonant dipole lengths that work well.)

Flattop or Inverted V Configuration?

There is no denying that the inverted V mounting configuration is very convenient, since it requires only a single support. The flattop configuration, however, where the dipole is mounted horizontally, gives more gain at the higher frequencies. **Figure 12.8** shows the 80 meter azimuth and elevation patterns for two 135-foot long dipoles. The first is mounted as a flattop at a height of 50 feet over flat ground with a conductivity of 5 mS/m and a dielectric constant of 13, typical for average soil. The second dipole uses the same length of wire, with the center apex at 50 feet and the ends drooped

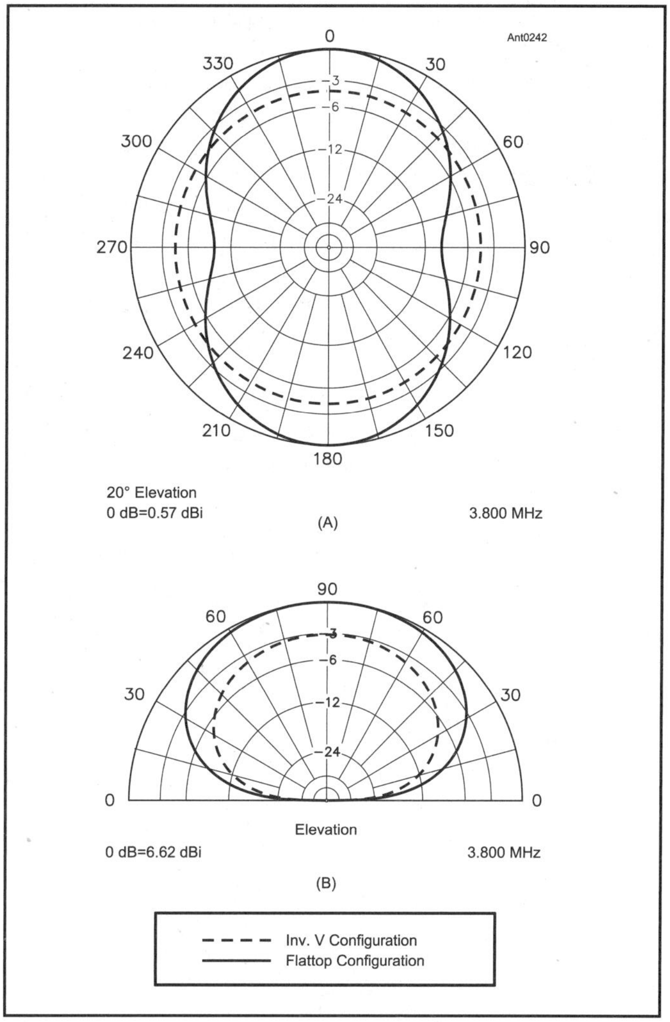

Figure 12.8 — Patterns on 80 meters for 135-foot, center-fed dipole erected as a horizontal flattop dipole at 50 feet, compared with the same dipole installed as an inverted V with the apex at 50 feet and the ends at 10 feet. The azimuth pattern is shown at A, where the dipole wire lies in the 90° to 270° plane. At B, the elevation pattern, the dipole wire comes out of the paper at a right angle. On 80 meters, the patterns are not markedly different for either flattop or inverted V configuration.

down to be suspended 10 feet off the ground. This height is sufficient so that there is no danger to passersby from RF burns.

At 3.8 MHz, the flattop dipole has about 4 dB more peak gain than its drooping cousin. On the other hand, the inverted V configuration gives a pattern that is more omnidirectional than the flattop dipole, which has nulls off the ends of the wire. Omnidirectional coverage may be more important to net operators, for example, than maximum gain.

Figure 12.9 shows the azimuth and elevation patterns for the same two antenna configurations, but this time at 14.2 MHz. The flattop dipole has developed four distinct lobes at a 10° elevation angle, an angle typical for 20 meter skywave communication. The peak elevation angle gain of 9.4 dBi occurs at about 17° for a height of 50 feet above flat ground for the flattop dipole. The inverted V configuration is again nominally more omnidirectional, but the peak gain is down some 6 dB from the flattop.

The antenna's peak is reduced even more at 28.4 MHz for the inverted V configuration. Here the peak gain is down about 8 dB from that produced by the flattop dipole, which exhibits eight lobes at this frequency with a maximum gain of 12.5 dBi at about 7° elevation. See the comparisons in **Figure 12.10**.

Whatever configuration you choose to mount the 135-foot dipole, you will want to feed it with some sort of low-loss open-wire feed line. For example, 450-Ω window line is popular for this application. Be sure to twist the line once or twice per foot to keep it from twisting excessively in the wind. (Do not twist it so much that the wire spacing is reduced.) Make sure also that you provide some mechanical support for the line at the junction with the dipole wires. This will prevent flexing of the transmission-line wire, since excessive flexing will result in breakage. (See the **Antenna Materials and Construction** chapter)

12.1.4 Dual-Band Dipoles

Dipoles have harmonic resonances near odd multiples of their fundamental resonances. Because 21 MHz is the third harmonic of 7 MHz, 7-MHz dipoles are harmonically resonant in the popular ham band at 21 MHz. This is attractive because it allows you to install a 40 meter dipole, feed it with coax, and use it without an antenna tuner on both 40 and 15 meters. This scheme can also be used to build a dipole that can be used on 80 and 30 meters and on 75 and 10 meters.

But there's a catch: The third harmonic resonance is actually higher than three times the fundamental resonant frequency. This is because there is no end effect in the center portion of the antenna where there are no insulators. The online article "End Effect and Harmonic Antenna Design" by Bob

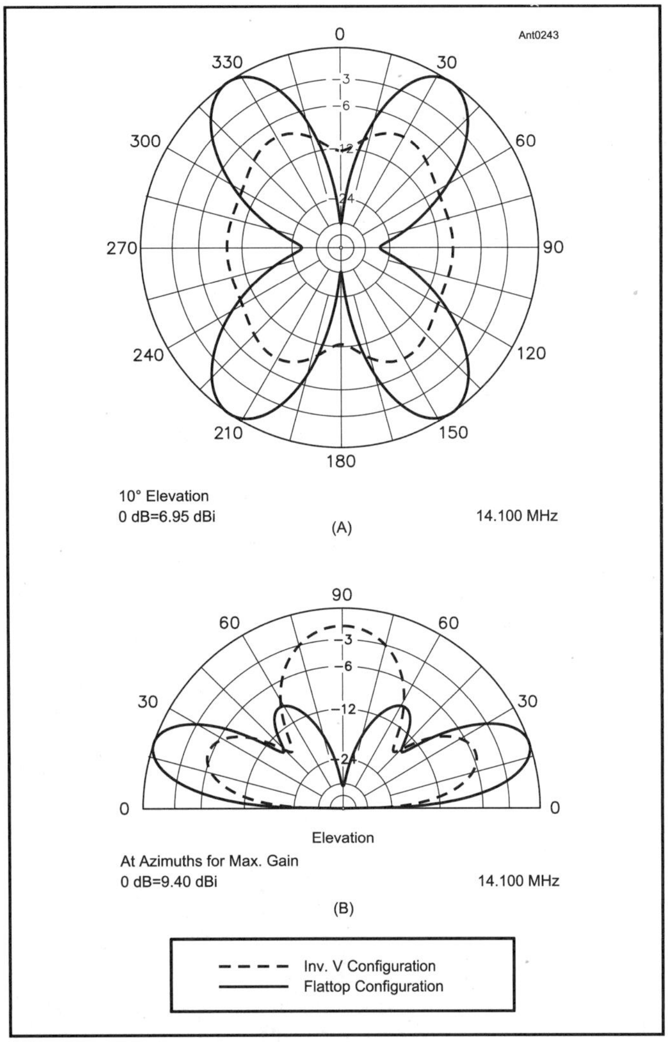

Figure 12.9 — Patterns on 20 meters for two 135-foot dipoles. One is mounted horizontally as a flattop and the other as an inverted V with 120° included angle between the legs. The azimuth pattern is shown in A and the elevation pattern is shown in B. The inverted V has about 6 dB less gain at the peak azimuths, but has a more uniform, almost omnidirectional, azimuthal pattern. In the elevation plane, the inverted V has a large high-angle lobe, making it a somewhat better antenna for local communication, but not quite so good for DX contacts at low elevation angles.

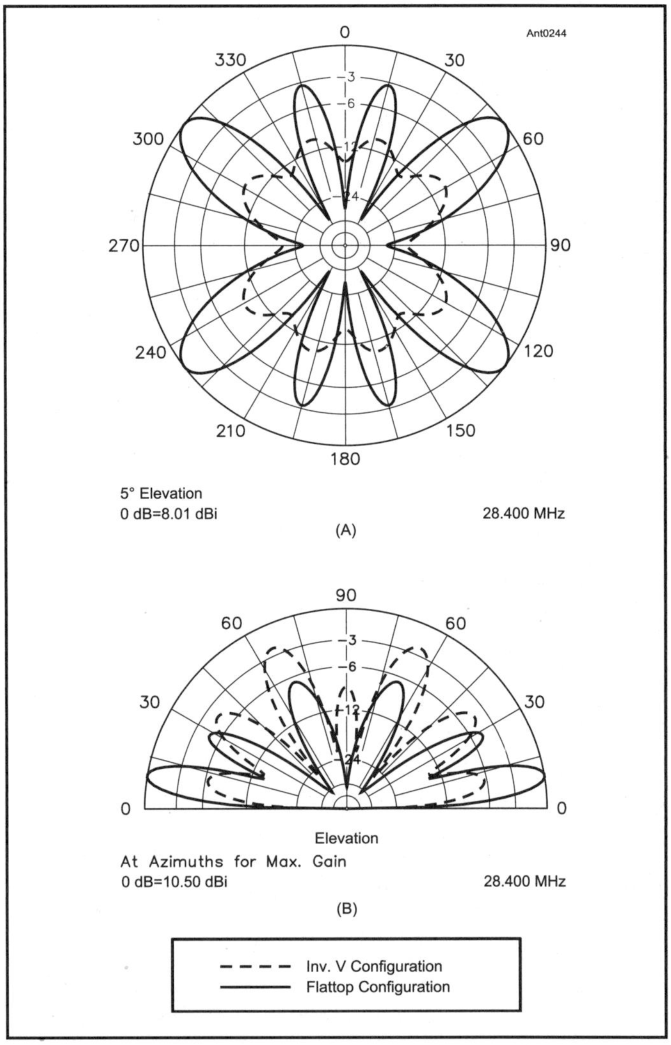

Figure 12.10— Patterns on 10 meters for same antenna configurations as in Figs 10.6 and 10.7. Once again, the inverted V configuration yields a more omnidirectional pattern, but at the expense of almost 8 dB less gain than the flattop configuration at its strongest lobes.

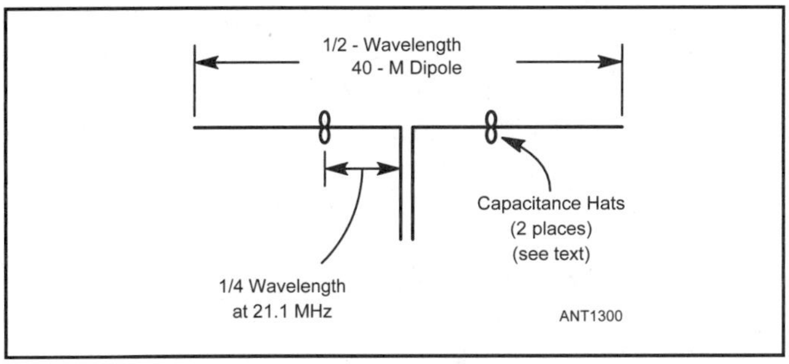

Figure 12.11 — Figure-8-shaped capacitance hats, made and placed as described in the text, can resonate a 40 meter dipole anywhere in the 15 meter band. The same technique can be used for 80/30 meter and 75/10 meter dipoles.

Glorioso, W1IS, and Bob Rose, KC1DSQ analyzes and models the 40/15-meter dipole, including a discussion of end effect, at **batteryeliminatorstore.com/blogs/ocf-masters-articles/end-effect-and-harmonic-antenna-design**.

An easy fix for this, as shown in **Figure 12.11**, is to add capacitive loading to the antenna about $\lambda/4$ wavelength (at the third harmonic) away from the feed point in both legs of the dipole. These *capacitance hats* (see the **Dipoles and Monopoles** chapter) lower the antenna's resonant frequency on the third harmonic without substantially affecting resonance on the fundamental.

Measure, cut, and adjust the dipole to resonance at the desired fundamental frequency. Then, cut two 2-foot-long pieces of solid #12 or #14 AWG wire and solder the ends of each one together to form loops. Twist the loops in the middle to form figure eights, and strip and solder the wires where they cross. Move the hats back and forth along the antenna until the desired SWR is achieved on the third harmonic (about a third of the way out from the feed point — placement isn't critical), then strip the antenna wire and solder or clamp the hats to the antenna. Adjusting the shape of the loops, making their area larger or smaller will also change the SWR.

12.1.5 The G5RV and Related Multiband Antennas

A variation on the center-fed antenna that does not require a lot of space, is simple to construct and low in cost is the G5RV. Designed in England by Louis Varney, G5RV, some years ago, it has become quite popular in the US. (The original article by G5RV in the *RSGB Bulletin* is included with this book's online material.) The G5RV design is shown in **Figure 12.12**. The antenna may be used from 3.5 through 30 MHz, although the use of an antenna tuner should be expected on any band except 14 MHz as Varney himself recommended. Low SWR with coax feed and no matching network on bands other than 14 MHz probably indicates excessive losses in the coax. In fact, an analysis of the G5RV feed point impedance shows there is no length of balanced line of any characteristic impedance that will transform the terminal impedance to the 50 to 75-Ω range on all bands.

Compared to a standard $\lambda/2$ 20 meter dipole at 50 feet and a 132-foot long center-fed doublet at 50 feet (also discussed in the next section on Windom-style antennas), on 20 meters the G5RV is within 1 dB of peak gain of either antenna. The G5RV has a four-lobed pattern that is somewhat more omnidirectional than either dipole or doublet. This is somewhat of an advantage for a wire antenna which cannot be rotated. The G5RV patterns for other frequencies are similar to those shown for the 135-foot dipole in the previous section.

The portion of the G5RV antenna shown as horizontal in Figure 12.9 may also be installed in an inverted V dipole arrangement, subject to the same loss of peak gain mentioned above for the 135-foot dipole. Or instead, up to ⅙ of the total length of the antenna at each end may be dropped vertically or semi-vertically, or bent at a convenient angle to the main axis of the antenna, to cut down on the requirements for real estate.

A useful variation on the G5RV theme was designed by Brian Austin, ZS6BKW (now GØGSF — see the References and Bibliography). It is shown in **Figure 12.13** and has a very similar radiation pattern to the G5RV. It is almost 10 feet

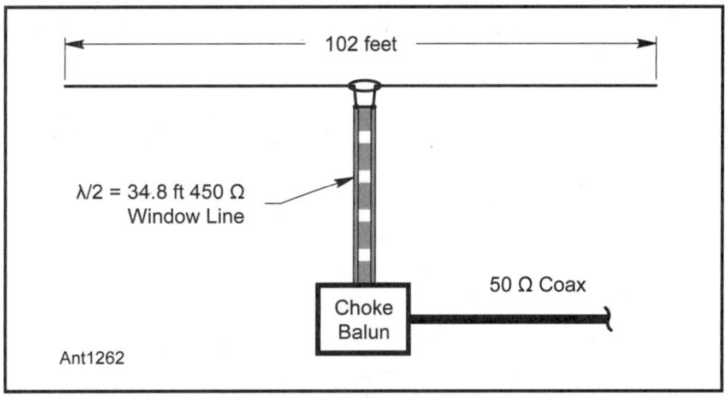

Figure 12.12 — The G5RV multiband antenna covers 3.5 through 30 MHz. It may be fed directly on 14 MHz but its designer recommends a matching network or antenna tuner on other bands.

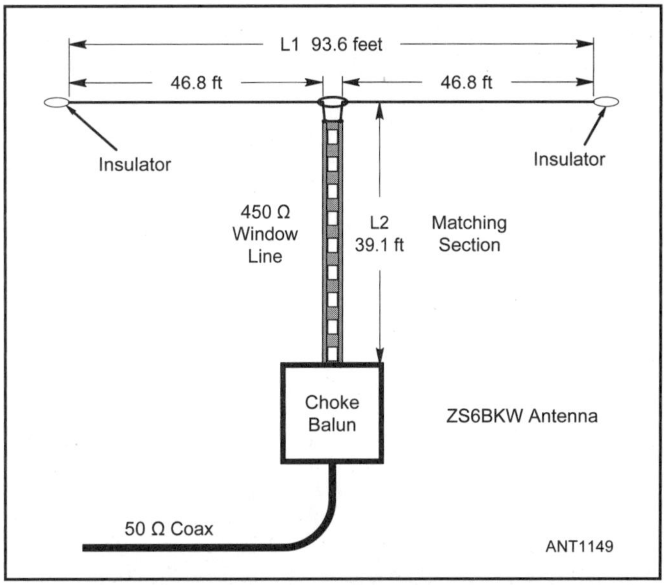

Figure 12.13 — The ZS6BKW multiband antenna is a development of the G5RV. The length of the matching section of feed line is based on a velocity factor of 0.91 (91%). (See text.)

shorter than the G5RV and uses a slightly longer length of 450-Ω window line to create the 50-Ω impedance point. The antenna is usable without a tuner in some portion of the 7, 14, 18, 24, and 28 MHz bands. The SWR is high on 3.5 MHz but within range of a tuner.

All of the G5RV/ZS6BKW and other variations that use a section of balanced transmission line to create a 50-Ω point require the use of a 1:1 choke balun as described in the chapter on **Transmission Line System Techniques**. Without the isolation between the balanced and unbalanced feed lines, the coaxial feed line will pick up significant common-mode RF current and allow noise picked up on the coax to enter the feed line.

The weight of the balun should be supported by sturdy construction and strain relief for the parallel conductor feed line. For this antenna, window line using stranded conductors is preferable to copper-plated steel since the repeated flexing will eventually cause the conductors to break. The window line matching section should be oriented at right angles to the top section (or symmetrically in an inverted V configuration) to avoid unbalancing the antenna system which would exacerbate common-mode current problems and distort the pattern unpredictably.

PROJECT: TRIBAND DIPOLE FOR 30, 17, AND 12 METERS

(This antenna was originally described in the article "A Triband Dipole for 30, 17, and 12 Meters" by Zack Lau, W1VT in the Mar/Apr 2015 issue of *QEX*. A similar design by W1VT for 10, 20, and 40 meters described in the article "A Compact Multiband Dipole" from the March 2016 issue of *QST*. Both articles are available in this chapter's online material.)

The triband dipole in **Figure 12.14** is a 58-foot doublet made out of #14 THHN solid house wire fed with a 35.5-foot matching section of 600-Ω ladder line having a velocity factor of 0.91. It has modest gain over a dipole on all three of the design bands and SWR at the balun input is approximately 1.5 on 17 meters and 2.4 on 12 meters and 30 meters. These values of SWR are all easily matched to 1:1 with the auto-tuner built into many transceivers or with an external impedance matching unit.

If the antenna will be installed at a significantly different height, the length of the doublet and the open-wire feed line may have to be adjusted to obtain resonances in all three bands. The length of the 50 Ω cable feed line isn't critical.

The #14 THHN house wire has two layers of insulation: 15 mils of PVC and another 4 mils of nylon. While the nylon typically flakes off in less than a year, the antenna was modeled in *EZNEC* using an insulation thickness of 19 mils and a dielectric constant of 3.5. Changing the insulation thickness to 15 mils doesn't appreciably change the resonance points.

The somewhat low velocity factor of the open-wire line assumes it is constructed using the same insulated wire as the doublet. A spacing of 3 inches is recommended for constructing the 600 Ω open-wire line. If other wire types are used to construct the open-wire line, an accurate approximation for the impedance of air-insulated open-wire line is $Z_0 = 276 \log_{10} (2S/d)$ where S is the center-to-center distance between the conductors and d is the diameter of the conductors in the same units as S. (Z_0 for open-wire line is discussed in detail in the **Transmission Lines** chapter.)

A 1:1 choke balun is required between the open-wire feed line and the coax to prevent the outside of the coax shield from becoming a radiating antenna element and affecting feed point impedance. W1VT recommends 11 turns of solid dielectric RG-58 on an FT-140-43 core as working well from 10 to 30 MHz. Other ferrite-core choke designs can be found in the **Transmission Line System Techniques** chapter.

Assuming a height of 42 feet, on 17 meters the antenna has a clean, bidirectional pattern, with maximum gain of 8.9 dBi broadside to the wires. (All antenna patterns are included in the full article.) On 30 meters, the antenna also has gain broadside to the wires of 6.9 dBi. On 12 meters the antenna has an azimuthal radiation pattern with four main lobes having 8.2 dBi gain, 50° off broadside. The broadside lobes are 3.5 dB weaker than the main lobes.

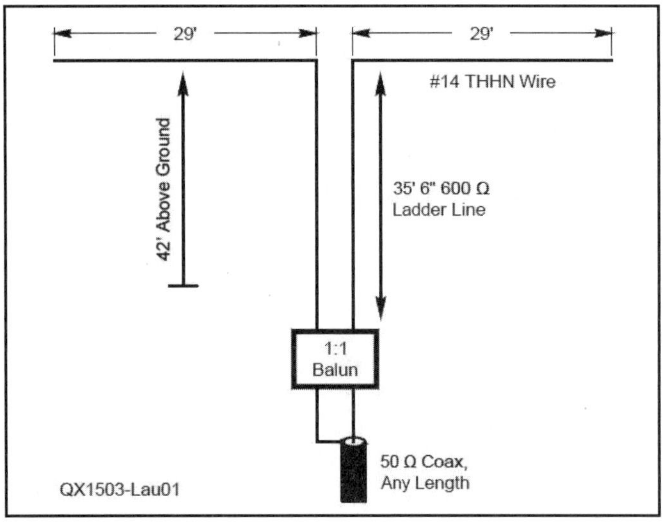

Figure 12.14 — Basic configuration of the triband dipole, consisting of a 58-foot doublet fed with 35.5 feet of 600-Ω open-wire feed line. SWR at the balun input is approximately 1.5:1 on 17 meters and 2.4:1 on 12 meters and 30 meters.

12.1.6 The Windom and Carolina Windom

An antenna that enjoyed popularity in the 1930s and into the 1940s was what we now call the Windom. It was known at the time as a "single-feeder Hertz" antenna, after being described in the September 1929 *QST* by Loren G. Windom, W8GZ (see References and Bibliography).

The Windom antenna, shown in **Figure 12.15**, is fed with a single wire, attached approximately 14% off center. In theory, this location provides a match for the single-wire transmission line, which is driven against an earth ground. Because the single-wire feed line is not inherently well balanced and because it is brought to the operating position, "RF in the shack" is a likely result of using this antenna. For that reason, the true single-feed-wire Windom antenna is rarely used

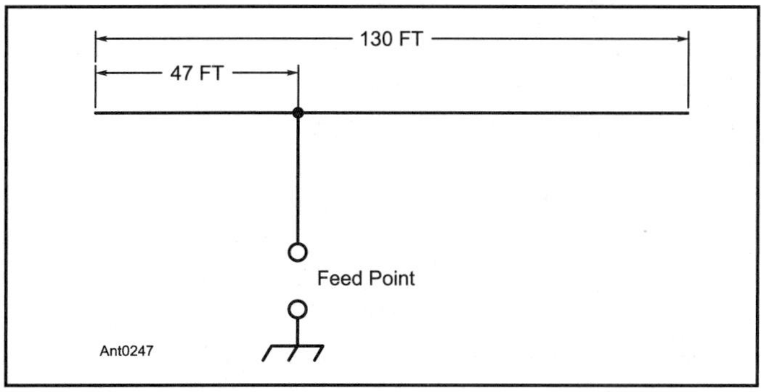

Figure 12.15 — The Windom antenna, cut for a fundamental frequency of 3.75 MHz. The single-wire feed line, connected 14% off center, is brought into the station and the system is fed against ground. The antenna is also effective on its harmonics.

although the name is often given to wires with non-centered feed points as described in the next section.

A recent variation is called the "Carolina Windom," apparently because two of the designers, Edgar Lambert, WA4LVB, and Joe Wright, W4UEB, lived in coastal North Carolina (the third, Jim Wilkie, WY4R, lived in nearby Norfolk, Virginia). One of the interesting parts about the Carolina Windom is that it turns a potential disadvantage — feed line radiation — into a potential advantage.

Figure 12.16 is a diagram of a flattop Carolina Windom, which uses a 50-foot wire joined with an 83-foot wire at the feed point insulator. This resembles the layout shown in Figure 12.12 for the original W8GZ Windom. The "Vertical Radiator" for the Carolina Windom is a 22-foot piece of RG-8X coax, with a "line isolator" (feed line choke) at the bottom end and a 4:1 "matching unit" (impedance transformer) at the top. The system takes advantage of the asymmetry of the horizontal wires to purposely induce current onto the outer shield surface of the vertical coax section. Note that the matching unit is a voltage balun transformer, which purposely does

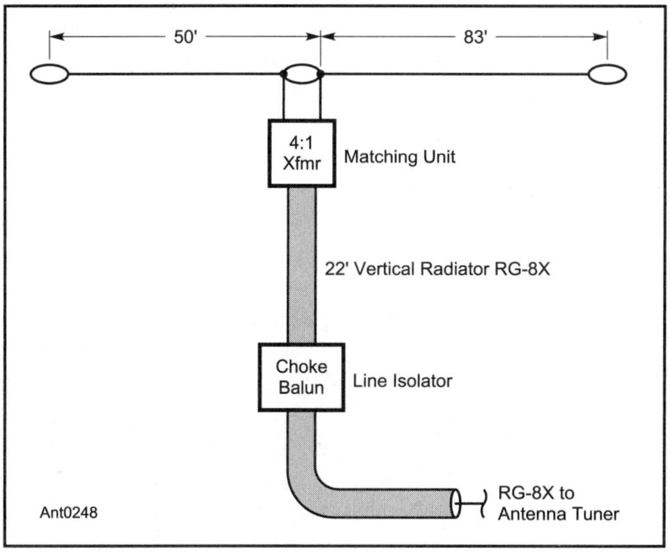

Figure 12.16 — Layout for flattop "Carolina Windom" antenna.

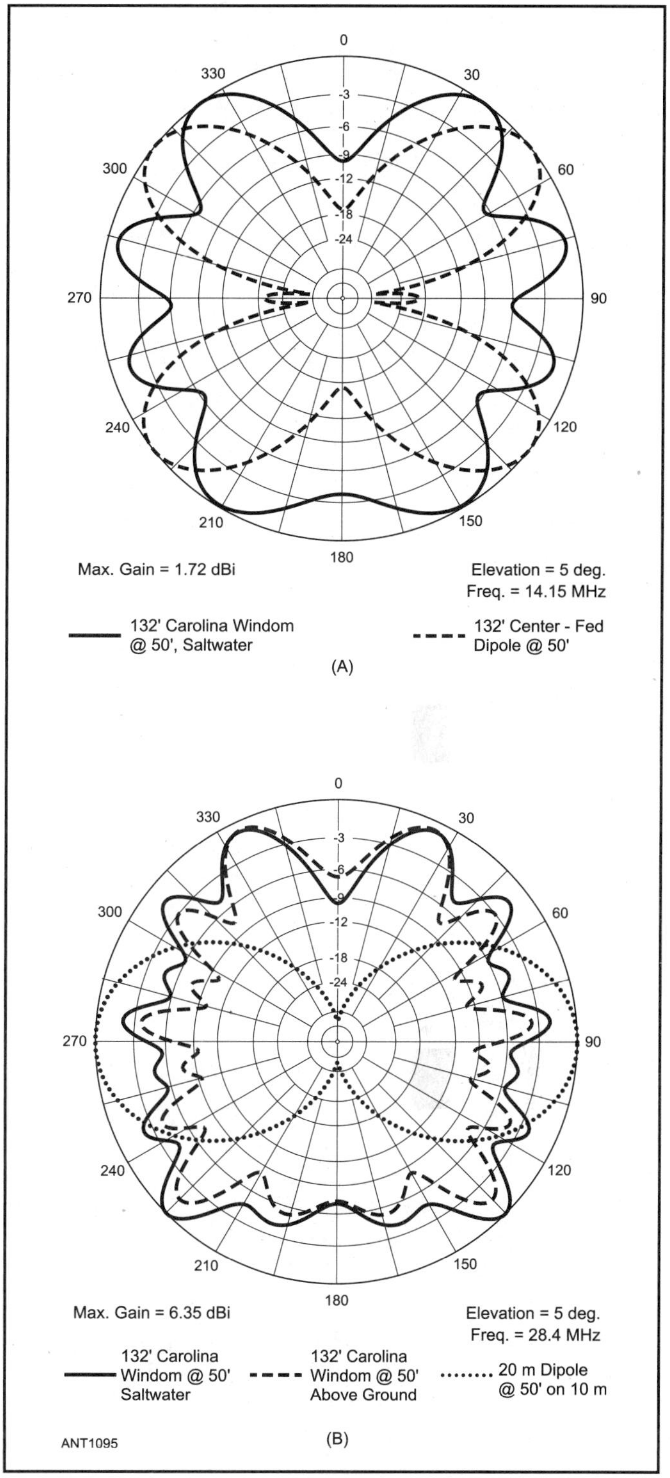

Figure 12.17 — At A, 20 meter azimuth patterns for a 132-foot long off-center fed Carolina Windom and a 132-foot long center-fed flattop dipole on 20 meters, both at a height of 50 feet above saltwater. The response for the Carolina Windom is more omnidirectional because the vertically polarized radiation from the 22-foot long vertical RG-8X coax fills in the deep nulls. At B, 10 meter azimuthal responses for a 132-foot long, 50-foot high Carolina Windom over saltwater (solid line) and over average ground (dashed line), compared to that for a 20 meter half-wave dipole at 50 feet (dotted line).

not act like a common-mode choke balun. You must use an antenna tuner with this system to present a 1:1 SWR to the transmitter on the amateur bands from 80 through 10 meters.

The radiation resulting from current induced onto the 22-foot vertical coax section tends to fill in the deep nulls that would be present if the 132 feet of horizontal wire were center fed. Over saltwater, the vertical radiator can give significant gain at the low elevation angles needed for DX work. Indeed, field reports for the Carolina Windom are most impressive for stations located near or on saltwater. Over average soil the advantage of the additional vertically polarized component is not quite so evident. **Figure 12.17A** compares a 50-foot high Carolina Windom on 14 MHz over saltwater to a 50-foot high, 132-foot long, flattop center-fed dipole. The Carolina Windom has a more omnidirectional azimuthal pattern, a desirable characteristic in a 132-foot long wire antenna that is not normally rotated to favor different directions.

Another advantage of the Carolina Windom over a traditional Windom is that the coax feed line between the transmitter and common-mode choke balun does not radiate. Since the feed line is not always operating at a low SWR on various ham bands, use the minimum length of feed coax possible to reduce losses in the coax.

Figure 12.14B shows the azimuth responses for a 50-foot high flattop Carolina Windom on 28.4 MHz over saltwater and over average soil. The pattern for a 50-foot high, flattop 20 meter dipole operated on 28.4 MHz is also shown, since this 20 meter dipole can also be used as a multiband antenna, when fed with open-wire transmission line rather than with coax. Again, the Carolina Windom exhibits a more omnidirectional pattern, even if the pattern is somewhat lopsided at the bottom.

12.1.7 Off-Center-Fed (OCF) Dipoles

The usual practice is to feed a λ/2 dipole in the center where the feed point impedance is low and makes a suitable match to coaxial cables. The dipole will accept energy from a feed point anywhere along its length, however, assuming that the source is matched to the higher impedance that is presented away from the center point. (As discussed in the **Dipoles and Monopoles** chapter, if the feed point is moved away from the center of the dipole, the impedance rises because current is dropping while voltage is rising.)

The *off-center-fed* dipole takes advantage of placing the feed point in a location along the dipole at which the impedance is similar on more than one band, generally in the neighborhood of 150 – 300 Ω. A suitable impedance transformer is then used to reduce the feed point impedance to something closer to 50 Ω. As with all horizontal antennas above ground, changes in height will affect feed point impedance and SWR. The article "Off-Center-Fed Dipoles — Theory and Practice" by Jackson R. Richter, WBØUSA, analyzes the OCF in detail (see the References and Bibliography).

Figure 12.18 shows an off-center-fed or *OCF* dipole. Because it is similar in appearance to the Windom of Figure 12.15, this antenna is often mistakenly called a "Windom," or sometimes a "coax-fed Windom." The two antennas are not the same, since the Windom is driven against an earth ground, while the OCF dipole is fed like a regular dipole — just not at its center. The extreme case of an OCF is the end-fed Zepp. where the feed point is moved all the way to the end of the antenna. (See the previous sections.)

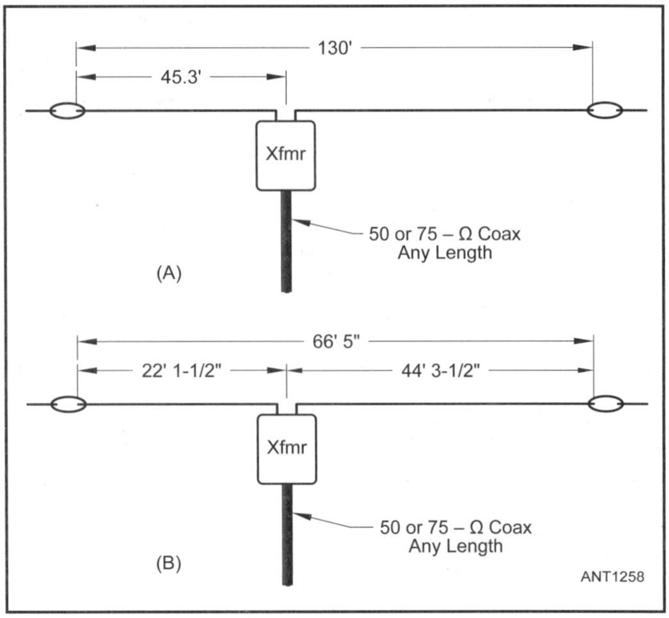

Figure 12.18 — The off-center-fed (OCF) dipole for 3.5, 7, and 14 MHz (A) and for 7, 14, 28, and 50 MHz (B) A 1:4 or 1:6 impedance transformer and choke balun are used at the feed point.

The OCF dipole of Figure 12.18, fed ⅓ of its length from one end, may be used on its fundamental and even harmonics. Its free-space antenna-terminal impedance at 3.5, 7, and 14 MHz is on the order of 150 to 200 Ω. At the 6th harmonic, 21 MHz, the antenna is three wavelengths long and fed at a voltage maximum instead of a current maximum. The feed point impedance at this frequency is high, a few thousand ohms, so the antenna is unsuitable for use on this band.

An alternate design that works on 40, 20, 10, and 6 meters is also described in the article "A No Compromise Off-Center Fed Dipole for Four Bands" by Rick Littlefield, K1BQT. An 8-band model that operates on the HF bands from 80 through 10 meters (except 60 meters) is described in the paper "The J78 Antenna: An Eight-band Off-Center-Fed HF Dipole" by Brian Machesney, K1LI/J75Y. Both articles are included in the online material for this chapter.

All OCF dipole antennas require an impedance transformer at the feed point. A 4:1 or 6:1 ratio typically provides good results. A choke balun on the feed line is also required to reduce coupling between the antenna and feed line, which can affect feed point impedance. Height above ground also affects feed point impedance — be prepared to adjust the dipole length to achieve an acceptable match on the bands to be used.

COMMON-MODE CURRENT

Because the OCF dipole is not fed at the center of the radiator, the feed line is not placed symmetrically with respect to the antenna's radiated field. As a result, common-mode current will flow on the feed line, usually a coaxial cable. How much current flows depends on the impedance of the coaxial cable's outer surface which, in turn, depends on the orientation of the cable, how long it is, height above ground, and so forth. (Some of the common-mode current results from the slightly unequal impedances presented by the OCF legs but most of the shield current is induced by the asymmetric location in the antenna's field.)

Regardless of how the common-mode current is caused to flow on the feed line, it is generally viewed as undesirable and a choke balun is used to increase the impedance of coaxial cable's outer surface. Radiation from the feed line may not be a problem in your installation and may even improve the antenna's radiation pattern by filling in nulls. (See "Feed Line Radiation" above.) In that case, no balun is required. (Choke baluns are discussed in the chapter **Transmission Line System Techniques**.)

PROJECT: OFF-CENTER END FED (OCEF) DIPOLE FOR 40 TO 6 METERS

In this project's design, Kai Siwiak, KE4PT, describes a dipole antenna that is physically connected to the feed line at one end of the main antenna wire. Electrically, though, it is off-center fed including the desired section of the feed line's shield and an additional "droop" wire, becoming the "OCEF" dipole. The antenna is designed to roll up into a compact and convenient package for travel. (The full article is available with this book's online material.)

As shown in **Figure 12.19**, the OCEF consists of two dipole legs and an optional droop wire. For the wire-only portion, use 30 feet of Teflon insulated #20 AWG stranded copper wire, beginning at the egg insulator at A. The wire is then soldered to the center conductor of miniature RG-174 coax, forming the physical feed point at B. The second radiating section is the outer shield of the 11.5-foot-long section of RG-174 coax extending from the electrical feed point to the ferrite chokes at C.

A 2-foot droop wire can be attached at C to extend the lower frequency range of the antenna and lower the impedance of the antenna near the chokes. This combination of lengths results in impedances that are relatively easy for the antenna tuning unit to match even though the antenna is not self-resonant in any of the ham bands.

The OCEF can be held away from a vertical structure with a nonconductive fishing pole or other support in either a horizontal or vertical orientation. Alternately, the antenna can be tied to a support structure such as a tree and stretched out back toward the operating position. In either arrangement, the antenna tuning unit should be able to create a 50 Ω match on any band from 40 to 6 meters. If 80 meter operation is desired

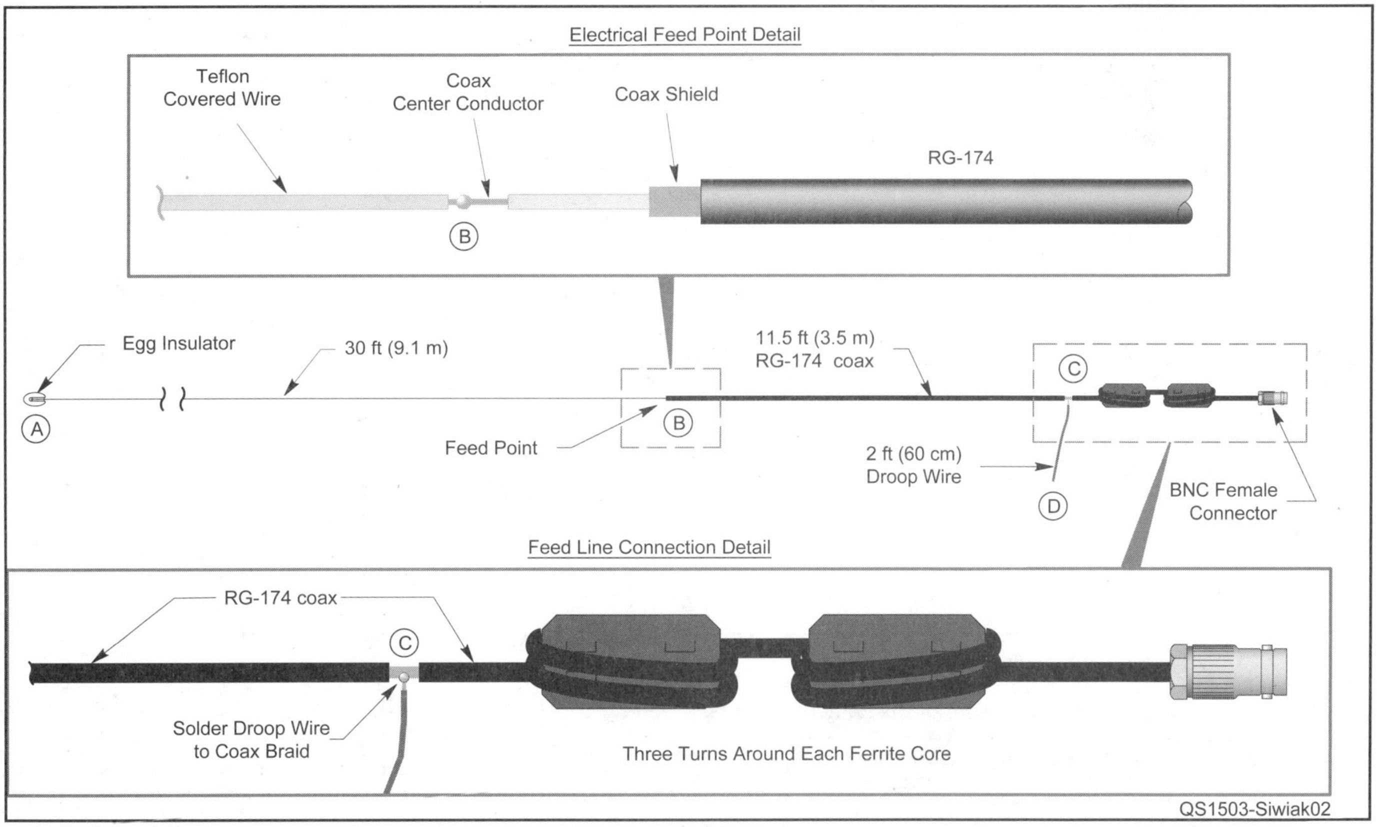

Figure 12.19 — The OCEF dipole detail shows the radiating portions A-C, details of the electrical feed point, and the common-mode chokes at the feed line connection. See the full article in the online material for construction details.

and an antenna tuner will not create the required match, additional lengths of wire can be clipped onto the droop wire until a match can be obtained. This extra piece of wire must be removed to operate on the higher bands, however.

The OCEF dipole allows all sorts of portable operation but pay attention to RF exposure — with 50 W of RF input power, the minimum safe distance rises from about 1 foot on 80 meters to 10 feet on 10 and 6 meters.

12.1.8 Fan Dipole Antennas

The antenna system shown in **Figure 12.20A** consists of a group of center-fed dipoles, all connected in parallel at the point where the transmission line joins them. Each of the dipole elements is individually constructed to be an electrical λ/2 at different frequencies. This is often referred to as a "fan dipole," although that term is also applied to a dipole constructed as a bow-tie to increase operating bandwidth. (See the section "Broadband Dipoles" in the chapter **General Purpose MF and HF Antennas**.) The general idea is that the feed point impedance of the dipoles far from resonance will be high enough that nearly all of the signal power is radiated by the resonant dipole and not by the nonresonant dipoles.

In theory, the 4-wire antenna of Figure 12.20A can be used with a coaxial feed line on five bands. The four wires are prepared as parallel-fed dipoles for 3.5, 7, 14, and 28 MHz. The 7-MHz dipole is intended to be used on its 3rd harmonic for 21-MHz operation to cover a fifth band. However, in practice it has been found difficult to get a good match to coaxial line on all bands.

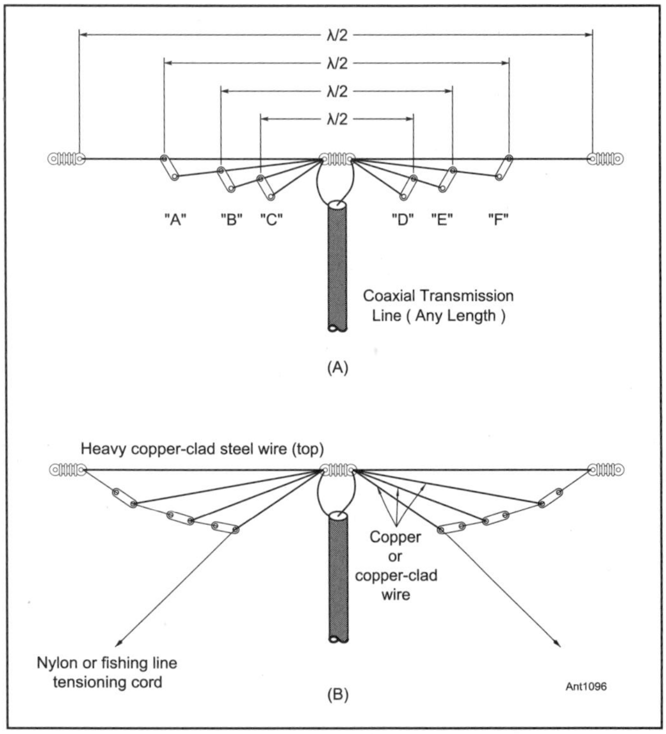

Figure 12.20 — At A, multiband antenna using paralleled dipoles all connected to a common low-impedance transmission line. The half-wave dimensions may be either for the centers of the various bands or selected to fit favorite frequencies in each band. Because of interaction among the various dipoles, the builder should expect to adjust lengths for resonance on each band. B shows a method of constructing the dipole that offers less interaction between the dipoles, making it easier to tune.

The λ/2 resonant length of any one dipole in the presence of the others is not the same as for a dipole by itself due to interaction and attempts to optimize all four lengths can become a frustrating procedure. The problem is compounded because the optimum tuning changes in a different antenna environment, so what works for one amateur may not work for another. The builder should start with a single dipole longer than resonance as discussed in the **Dipoles and Monopoles** chapter and be prepared to make repeated adjustments to the dipole lengths as more dipoles are added to the antenna.

Even if a perfect match cannot be obtained on all bands, many amateurs with limited antenna space are willing to accept the mismatch on some bands just so they can operate on those frequencies using a single coax feed line. The fewer dipoles that are used in parallel, the easier it will be to adjust them for the desired performance.

If an attempt is made to model the multi-wire dipole, take extra care to define the feed point construction carefully. As noted in the **Antenna Modeling** chapter, wires that are very close to each other or that join at small angles are hard to model so that the results reflect actual performance.

The multiple-dipole antenna can be fed with parallel-wire feed line and an antenna tuner but that negates the intended advantage of the design over a conventional single-wire nonresonant dipole — the use of a single coaxial feed line. The usual feed method is to use a coaxial feed line and a choke balun at the feed point as described in the chapter **Transmission Line System Techniques**.

The separation between the dipoles for the various frequencies does not seem to be especially critical. One set of wires can be suspended from the next larger set, using insulating spreaders (of the type used for feed line spreaders) to give a separation of a few inches. Users of this antenna often run some of the dipoles at right angles to each other to help reduce interaction. Some operators use inverted V-mounted dipoles as guy wires for the mast that supports the antenna system. The top (and longest) dipole must support the weight of the rest of the antenna plus the feed line, so use heavy wire (copper-clad steel is the strongest) for the top antenna.

While the separation between dipoles does not seem to be especially critical to final performance, it does affect the amount of interaction between them that makes tuning each dipole difficult. A method of construction and tuning reported by Don Butler, N4UJW (**hamuniverse.com/multidipole.html**) is shown is Figure 12.16B. For dipoles in the 2-18 MHz range, separating the dipoles at the feed point by at least 5½ inches vertically and at the ends by 38 inches results in a final length closer than ±2% of a single dipole.

An interesting method of construction used successfully by Louis Richard, ON4UF, is shown in **Figure 12.21**. The antenna has four dipoles (for 7, 14, 21 and 28 MHz) con-

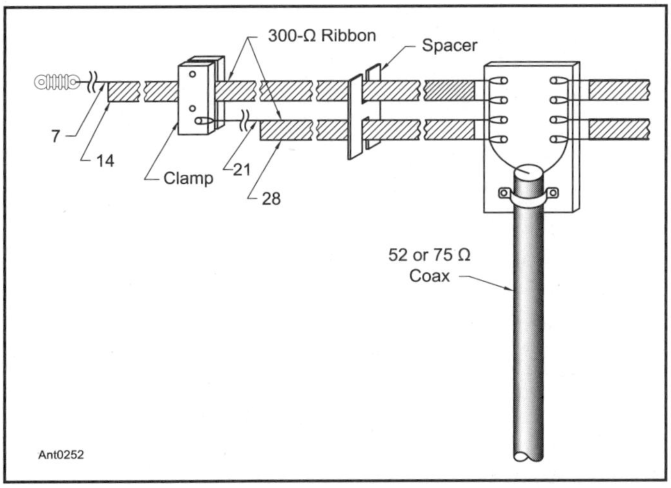

Figure 12.21 — Sketch showing how the twin-lead multiple-dipole antenna system is assembled. The excess wire and insulation are stripped away.

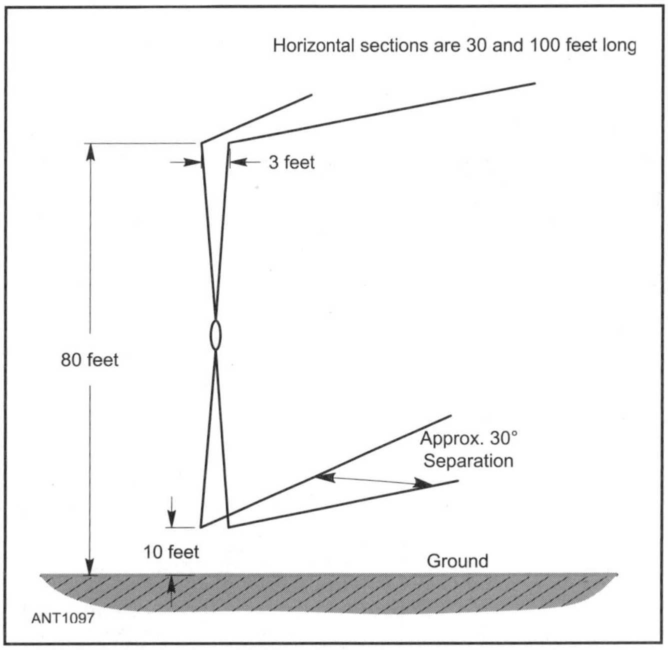

Figure 12.22 — The Double-L antenna by K2KQ is a pair of vertical dipoles with their ends bent to be parallel to the ground. The bottom horizontal wires should be at least 10 feet above ground. For single-band operation, install only a single dipole. The antenna works well as either a single-band or dual-band antenna.

structed from 300-Ω twin lead. A single length of twin lead makes two dipoles. Thus, two lengths, as shown in the sketch, serve to make dipoles for four bands. Be sure to use twin lead with copper-clad steel conductors because all of the weight, including that of the feed line, must be supported by the uppermost wire (450-Ω window line could also be used).

Two pieces of twin lead are first cut to a length suitable for the two halves of the longest dipole. Then one of the conductors in each piece is cut to proper length for the next band higher in frequency. The excess wire and insulation is stripped away. A second pair of lengths is prepared in the same manner, except that the lengths are appropriate for the next two higher frequency bands. (Note the potential for interaction between higher and lower-frequency dipoles that may alter the tuning of previously adjusted dipoles.)

A piece of thick plastic sheet (plexiglass, polycarbonate, or high-density polyethylene) drilled with holes for anchoring each wire serves as the central insulator. The shorter pair of dipoles is suspended the width of the ribbon below the longer pair by clamps also made of poly sheet. Intermediate spacers are made by sawing slots in pieces of plastic sheet so they will fit the ribbon snugly.

The multiple-dipole principle can also be applied to vertical antennas. Parallel or fanned λ/4 elements of wire or tubing can be driven against ground or tuned radials from a common feed point.

PROJECT: DOUBLE-L ANTENNA

The Double-L antenna by Don Toman, K2KQ is a variation of the multi-wire dipole. (**www.yccc.org/Articles/double_l.htm**) Shown in **Figure 12.22**, the antenna is basically a vertical dipole with its ends bent to run horizontally over ground. It can be constructed as a single antenna for one band or a second dipole can be added to use the antenna on two bands.

Construction is not critical. The bottom wires should be at least 10 feet above ground and no radial system is required. If you do construct the dual-band version, the vertical wires are connected together at the feed point and separated by about 3 feet where they bend to become horizontal. The two horizontal sections are separated by about 30°. If the antennas are supported by a metal tower, the vertical section should be at least 3 feet from the tower.

The antenna is inherently unbalanced and may be tuned by removing or adding wire to the lower legs without dramatically affecting performance or feed point impedance. The dimensions given result in an SWR minimum near 1.83 MHz and 3.75 MHz.

12.2 Trap Dipoles

By using tuned circuits of appropriate design strategically placed in a dipole, the antenna can be made to show what is essentially fundamental resonance at a number of different frequencies. The general principle is illustrated by **Figure 12.23**. The tuned circuits are also referred to as "traps" and so an antenna that uses tuned circuits to change its electrical configuration at different frequencies is called a "trap antenna" or a "trapped antenna."

Even though a trap antenna arrangement is a simple one, an explanation of how a trap antenna works can be elusive. For some designs, traps are resonated in our amateur bands, and for others (especially commercially made antennas) the traps are resonant far outside any amateur band.

A trap in an antenna system can perform either of two functions, depending on whether or not it is resonant at the operating frequency. A familiar case is where the trap is parallel-resonant in an amateur band. For the moment, let us assume that dimension A in Figure 12.23 is 32 feet and that each L/C combination is resonant in the 7-MHz band. Because of its parallel resonance, the trap presents a high impedance at that point in the antenna system. The electrical effect at 7 MHz is that the trap behaves as an open circuit. It serves to separate the outside ends, the B sections, from the inner sections of the antenna. The result is easy to visualize — we now have an antenna system that is resonant in the 7-MHz band. Each 33-foot section (labeled A in the drawing) represents λ/4 with the trap acting as an open circuit. We therefore have a full-size 7-MHz antenna.

The second function of a trap, obtained when the frequency of operation is *not* the resonant frequency of the trap, is one of electrical loading. If the operating frequency is below the trap's resonant frequency, the trap behaves as an inductor; if above, as a capacitor. Inductive loading will electrically lengthen the antenna, and capacitive loading will electrically shorten the antenna.

Let's carry our assumption a bit further and try using the antenna we just considered at 3.5 MHz. With the traps resonant in the 7-MHz band, they will behave as inductors when operation takes place at 3.5 MHz, electrically lengthening the antenna. This means that the total length of sections A and B (plus the length of the inductor) may be something less than a physical λ/4 for resonance at 3.5 MHz. Thus, we have a two-band antenna that is shorter than full size on the lower frequency band. But with the electrical loading provided by the traps, the overall electrical length is λ/2. The total antenna length needed for resonance in the 3.5-MHz band will depend on the L/C ratio of the trap elements.

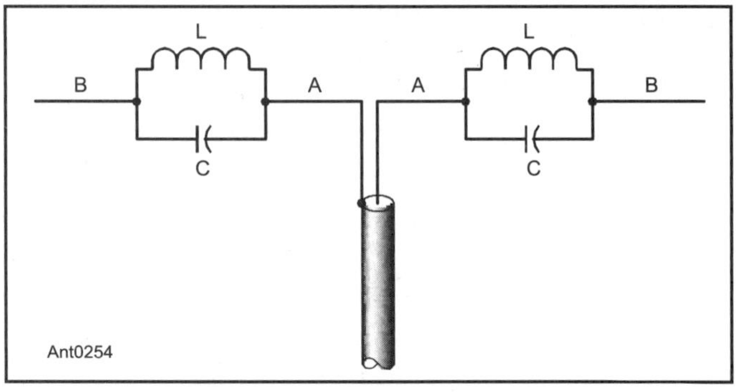

Figure 12.23 — A trap dipole antenna. This antenna may be fed with 50-Ω coaxial line. Depending on the L/C ratio of the trap elements and the lengths chosen for dimensions A and B, the traps may be resonant either in an amateur band or at a frequency far removed from an amateur band for proper two-band antenna operation.

The key to trap operation away from resonance is its L/C ratio, the ratio of the value of L to the value of C. At resonance, however, within practical limitations the L/C ratio is immaterial as far as electrical operation goes. For example, in the antenna we've been discussing, it would make no difference for 7-MHz operation whether the inductor were 1 µH and the capacitor were 500 pF (the reactances would be just below 45 Ω at 7.1 MHz), or whether the inductor were 5 µH and the capacitor 100 pF (reactances of approximately 224 Ω at 7.1 MHz). But the choice of these values will make a significant difference in the antenna size for resonance at 3.5 MHz. In the first case, where the L/C ratio is 2000, the necessary length of section B of the antenna for resonance at 3.75 MHz would be approximately 28.25 feet. In the second case, where the L/C ratio is 50,000, this length need be only 24.0 feet, a difference of more than 15%.

The above example concerns a two-band antenna with trap resonance at one of the two frequencies of operation. On each of the two bands, each half of the dipole operates as an electrical λ/4. However, the same band coverage can be obtained with a trap resonant at, say, 5 MHz, a frequency quite removed from either amateur band. With proper selection of the L/C ratio and the dimensions for A and B, the trap will act to shorten the antenna electrically at 7 MHz and lengthen it electrically at 3.5 MHz. Thus, an antenna that is intermediate in physical length between being full size on 3.5 MHz and full size on 7 MHz can cover both bands, even though the trap is not resonant at either frequency. Again, the antenna operates with electrical λ/4 sections. Note that such nonresonant traps have less RF current flowing in the trap components, and hence trap losses are less than for resonant traps.

Additional traps may be added in an antenna section to cover three or more bands. Or a judicious choice of dimensions and the L/C ratio may permit operation on three or more bands with just a pair of identical traps in the dipole.

An important point to remember about traps is this. If the operating frequency is below that of trap resonance, the trap behaves as an inductor; if above, as a capacitor. The above discussion is based on dipoles that operate electrically as λ/2 antennas. This is not a requirement, however. Elements may be operated as electrical 3⁄2 λ, or even 5⁄2 λ, and still present a reasonable impedance to a coaxial feed line. In trap antennas covering several HF bands, using electrical lengths that are odd multiples of λ/2 is often done at the higher frequencies.

To further aid in understanding trap operation, let's now

choose trap L and C components that each have a reactance of 20 Ω at 7 MHz. Inductive reactance is directly proportional to frequency, and capacitive reactance is inversely proportional. When we shift operation to the 3.5-MHz band, the inductive reactance becomes 10 Ω, and the capacitive reactance becomes 40 Ω. At first thought, it may seem that the trap would become capacitive at 3.5 MHz with a higher capacitive reactance, and that the extra capacitive reactance would make the antenna electrically shorter yet. Fortunately, this is not the case. The inductor and the capacitor are connected in parallel with each other.

$$Z = j\frac{X_L X_C}{X_L + X_C} \qquad (1)$$

where *j* indicates a reactive impedance component, rather than resistive. A positive result indicates inductive reactance, and a negative result indicates capacitive. In this 3.5-MHz case, with 40 Ω of capacitive reactance and 10 Ω of inductive, the equivalent series reactance is 13.3 Ω inductive. This inductive loading lengthens the antenna to an electrical λ/2 overall at 3.5 MHz, assuming the B end sections in Figure 12.23 are of the proper length.

With the above reactance values providing resonance at 7 MHz, X_L equals X_C, and the theoretical series equivalent is infinity. This provides the open-switch effect, disconnecting the antenna ends.

At 14 MHz, where $X_L = 40\ \Omega$ and $X_C = 10\ \Omega$, the resultant series equivalent trap reactance is 13.3 Ω capacitive. If the total physical antenna length is slightly longer than 3⁄2 λ at 14 MHz, this trap reactance at 14 MHz can be used to shorten the antenna to an electrical 3⁄2 λ. In this way, three-band operation is obtained for 3.5, 7, and 14 MHz with just one pair of identical traps. The design of such a system is not straightforward, however, because any chosen L/C ratio for a given total length affects the resonant frequency of the antenna on both the 3.5 and 14-MHz bands.

12.2.1 Trap Losses

Since the tuned circuits have some inherent losses, the efficiency of a trap system depends on the unloaded Q values of the tuned circuits. Low-loss (high-Q) coils should be used, and the capacitor losses likewise should be kept as low as possible. With tuned circuits that are good in this respect — comparable with the low-loss components used in transmitter tank circuits, for example — the reduction in efficiency compared with the efficiency of a simple dipole is small, but tuned circuits of low unloaded Q can lose an appreciable portion of the power supplied to the antenna.

The commentary above applies to traps assembled from conventional components. The important function of a trap that is resonant in an amateur band is to provide a high isolating impedance, and this impedance is directly proportional to Q. Unfortunately, high Q restricts the antenna bandwidth, because the traps provide maximum isolation only at trap resonance.

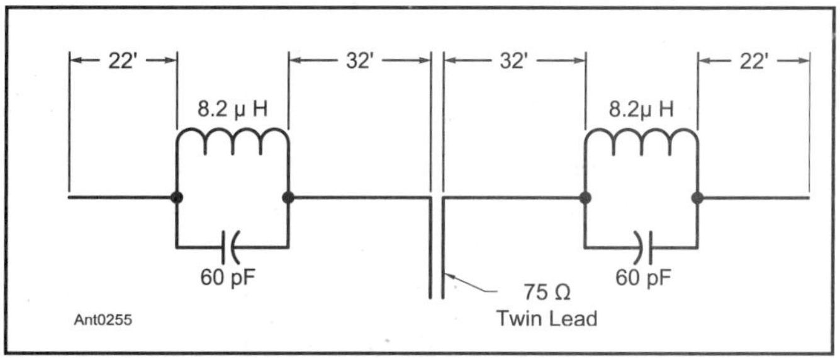

Figure 12.24 — Five-band (3.5, 7, 14, 21, and 28 MHz) trap dipole for operation with 75-Ω feed line at low SWR. The balanced (parallel-conductor) line indicated is desirable, but 75-Ω coax can be substituted with a choke balun at the feed point to maintain symmetry. Dimensions given are for resonance (lowest SWR) at 3.75, 7.2, 14.15, and 29.5 MHz. Resonance is very broad on the 21-MHz band, with SWR less than 2:1 throughout the band.

12.2.2 Five-Band W3DZZ Trap Antenna

C. L. Buchanan, W3DZZ, created one of the first trap antennas for the five pre-1979 WARC amateur bands from 3.5 to 30 MHz. Dimensions are given in **Figure 12.24**. Only one set of traps is used, resonant at 7 MHz to isolate the inner (7-MHz) dipole from the outer sections. This causes the overall system to be resonant in the 3.5-MHz band. On 14, 21 and 28 MHz the antenna works on the capacitive-reactance principle just outlined. With a 75-Ω feed line, the SWR with this antenna is under 2:1 throughout the three highest frequency bands, and the SWR is comparable with that obtained with similarly fed simple dipoles on 3.5 and 7 MHz. (The complete article is included with this book's online material.)

TRAP CONSTRUCTION

Traps frequently are built with concentric coaxial aluminum tubes (usually with plastic tubing in-between them for insulation) for the capacitor, with the coil either self-supporting or wound on a form of larger diameter than the tubular capacitor. The coil is then mounted coaxially with the capacitor to form a unit assembly that can be supported at each end by the antenna wires. In another type of trap devised by William J. Lattin, W4JRW (see References and Bibliography), the coil is supported inside an aluminum tube and the trap capacitor is obtained in the form of capacitance between the coil and the outer tube. This type of trap is inherently weatherproof.

A simpler type of trap can be easily assembled from readily available components. A small transmitting-type ceramic "doorknob" capacitor is used, together with a length of commercially available coil material, these being supported by an ordinary ceramic or plastic antenna strain insulator 4¼ inches long. The circuit constants and antenna dimensions differ slightly from those of Figure 12.24, in order to bring the antenna resonance points closer to the centers of the various phone bands. Construction data are given in **Figure 12.25**. If a 10-turn length of inductor is used, a half turn from each end may be used to slip through the anchor holes in the insulator to act as leads.

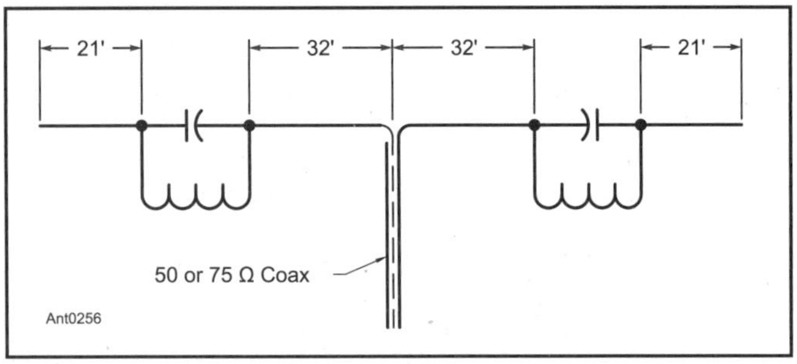

Figure 12.25 — Layout of multiband antenna using traps constructed as shown in Figure 10.26. The capacitors are 100 pF each, transmitting type, 5000-V dc rating (Centralab 850SL-100N). Coils are 9 turns of #12 AWG wire, 2½ inches diameter, 6 turns per inch (B&W 3029) with end turns spread as necessary to resonate the traps to 7.2 MHz. These traps, with the wire dimensions shown, resonate the antenna at approximately the following frequencies on each band: 3.9, 7.25, 14.1, 21.5, and 29.9 MHz (based on measurements by W9YJH).

The components used in these traps are sufficiently weatherproof in themselves so that no additional weatherproofing has been found necessary. However, if it is desired to protect them from the accumulation of snow or ice, a plastic cover can be made by cutting two discs of plastic slightly larger in diameter than the coil, drilling at the center to pass the antenna wires, and cementing a plastic cylinder on the edges of the discs. The cylinder can be made by wrapping two turns or so of 0.02-inch plastic sheet around the discs, if no suitable ready-made tubing is available. Plastic soft-drink bottles or food jars are easily adaptable for use as impromptu trap covers if protected from solar UV.

For low-power use, John Portune, W6NBC, designed a printed-circuit board trap (see References and Bibliography). The trap consists of a pair of rectangular spiral inductors, one on each side of the PCB. The inductor traces mirror each other so they exactly overlap. The two traces and the PCB material make up the trap's capacitance. The trap is tuned by moving a jumper along the inductor.

The common FR-4 glass-epoxy PCB material is somewhat lossy with a dissipation factor of 2% but is adequate for intermittent transmitting at 100 W or less. If polyimide or Teflon (PTFE) are available, those materials are a less-lossy dielectric.

12.2.3 W8NX Multiband, Coax-Trap Dipoles

Over the last 60 or 70 years, amateurs have used many kinds of multiband antennas to cover the traditional HF bands. The availability of the 30, 17 and 12 meter bands has expanded our need for multiband antenna coverage. This section is based on the August 1994 *QST* article "Two New Multiband Trap Dipoles" by Al Buxton, W8NX. This article and two others by the same author are included with this book's online material, providing designs for trap dipoles operating on all of the amateur bands below 30 MHz.

Two different antennas are described here. The first covers the traditional 80, 40, 20, 15, and 10 meter bands, and the second covers 80, 40, 17, and 12 meters. Each uses the same type of W8NX trap — connected for different modes of operation — and a pair of short capacitive stubs to enhance coverage. The W8NX coaxial-cable traps have two different modes: a high- and a low-impedance mode. The inner-conductor windings and shield windings of the traps are connected in series for both modes. However, either the low- or high-impedance point can be used as the trap's output terminal. For low-impedance trap operation, only the center conductor turns of the trap windings are used. For high-impedance operation, all turns are used, in the conventional manner for a trap. The short stubs on each antenna are strategically sized and located to permit more flexibility in adjusting the resonant frequencies of the antenna.

PROJECT: 80, 40, 20, 15, AND 10 METER DIPOLE

Figure 12.26 shows the configuration of the 80, 40, 20, 15, and 10 meter antenna. The radiating elements are made of #14 AWG stranded copper wire. The element lengths are the wire span lengths in feet. These lengths do not include the lengths of the pigtails at the balun, traps and insulators. The 32.3-foot-long inner 40 meter segments are measured from the eyelet of the input balun to the tension-relief hole in the trap coil form. The 4.9-foot segment length is measured from the tension-relief hole in the trap to the 6-foot stub. The 16.1-foot outer-segment span is measured from the stub to the eyelet of the end insulator.

The coaxial-cable traps are wound on PVC pipe coil forms and use the low-impedance output connection. The stubs are 6-foot lengths of ⅛-inch stiffened aluminum or copper rod hanging perpendicular to the radiating elements. The first inch of their length is bent 90° to permit attachment to the radiating elements by large-diameter copper crimp connectors. Ordinary #14 AWG wire may be used for the stubs, but it has a tendency to curl up and may tangle unless weighed down at the end. You should feed the antenna with 75-Ω coaxial cable using a choke balun.

This antenna may be thought of as a modified W3DZZ antenna due to the addition of the capacitive stubs. The length and location of the stub give the antenna designer two extra degrees of freedom to place the resonant frequencies within

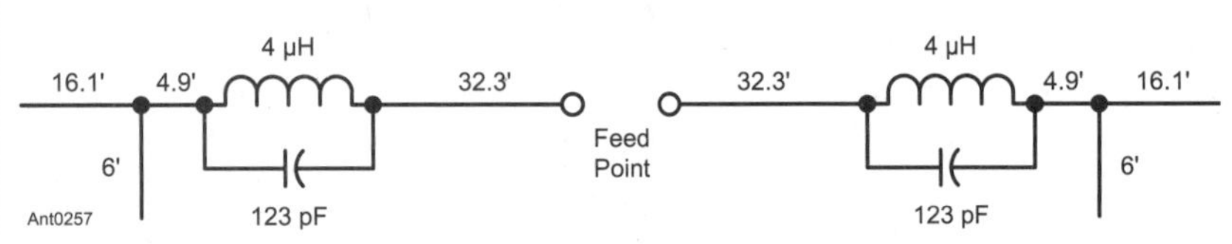

Figure 12.26 — A W8NX multiband dipole for 80, 40, 20, 15 and 10 meters. The values shown (123 pF and 4 µH) for the coaxial-cable traps are for parallel resonance at 7.15 MHz. The low-impedance output of each trap is used for this antenna.

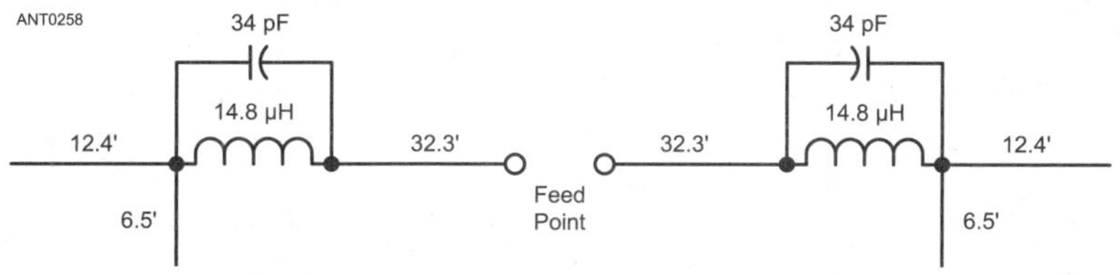

Figure 12.27 — A W8NX multiband dipole for 80, 40, 17, and 12 meters. For this antenna, the high-impedance output is used on each trap. The resonant frequency of the traps is 7.1 MHz.

the amateur bands. This additional flexibility is particularly helpful to bring the 15 and 10 meter resonant frequencies to more desirable locations in these bands. The actual 10 meter resonant frequency of the original W3DZZ antenna is somewhat above 30 MHz, pretty remote from the more desirable low frequency end of 10 meters.

PROJECT: 80, 40, 17, AND 12 METER DIPOLE

Figure 12.27 shows the configuration of the 80, 40, 17, and 12 meter antenna. Notice that the capacitive stubs are attached immediately outboard after the traps and are 6.5 feet long, ½ foot longer than those used in the other antenna. The traps are connected for the high-impedance parallel-resonant output mode. Since only four bands are covered by this antenna, it is easier to fine tune it to precisely the desired frequency on all bands. The 12.4-foot tips can be pruned to a particular 17 meter frequency with little effect on the 12 meter frequency. The stub lengths can be pruned to a particular 12 meter frequency with little effect on the 17 meter frequency. Both such pruning adjustments slightly alter the 80 meter resonant frequency. However, the bandwidths of the antennas are so broad on 17 and 12 meters that little need for such pruning exists. The 40 meter frequency is nearly independent of adjustments to the capacitive stubs and outer radiating tip elements. Like the first antennas, this dipole is fed with a balun and 75-Ω feed line.

Figure 12.28 shows the schematic diagram of the traps. It explains the difference between the low and high-impedance modes of the traps. Notice that the high-impedance terminal is the output configuration used in most conventional trap applications. The low-impedance connection is made across only the inner conductor turns, corresponding to one-half of the total turns of the trap. This mode steps the trap's impedance down to approximately one-fourth of that of the high-impedance level. This is what allows a single trap design to be used for two different multiband antennas.

Figure 12.29 is a drawing of a cross-section of the coax trap shown broadside to the long axis of the trap. Notice that the traps are conventional coaxial-cable traps, except for the added low-impedance output terminal. The traps are 8¾ close-spaced turns of RG-59 (Belden 8241) on a 2⅜-inch-OD PVC pipe (schedule 40 pipe with a 2-inch ID) coil form. The forms are 4⅛ inches long. Trap resonant frequency is very sensitive to the outer diameter of the coil form, so check it carefully. Unfortunately, not all PVC pipe is made with the same wall thickness. The trap frequencies should be checked with a dip meter and general-coverage receiver and adjusted to within 50 kHz of the 7150 kHz resonant frequency before installation. One inch is left over at each end of the coil forms to allow for

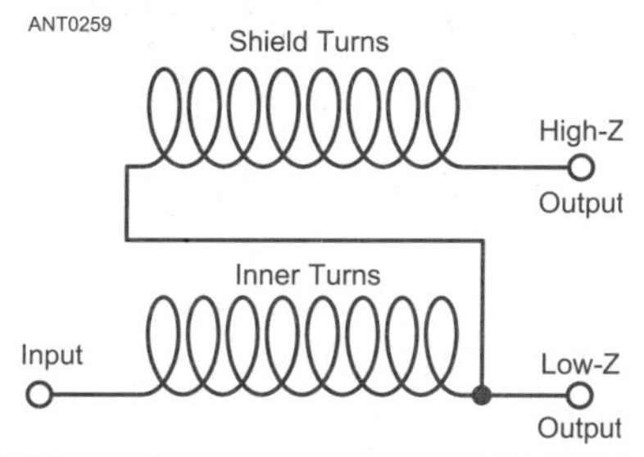

Figure 12.28 — Schematic for the W8NX coaxial-cable trap. RG-59 is wound on a 2⅜-inch OD PVC pipe.

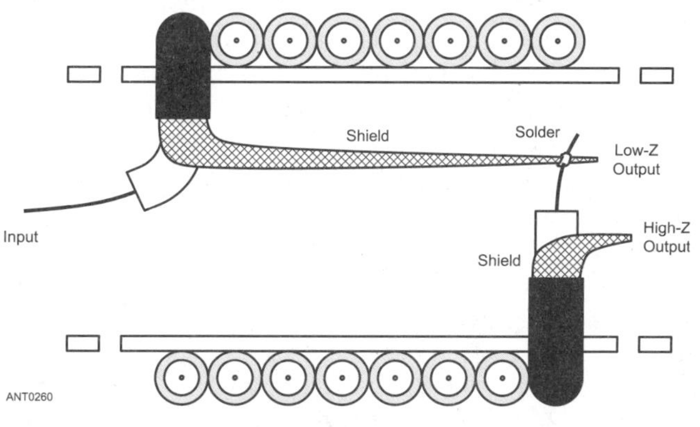

Figure 12.29 — Construction details of the W8NX coaxial-cable trap.

the coax feed-through holes and holes for tension-relief attachment of the antenna radiating elements to the traps. Be sure to seal the ends of the trap coax cable to prevent moisture from entering the coaxial cable. (See the discussion on waterproofing in the **Building Antenna Systems and Towers** chapter.)

Also, be sure that you connect the 32.3-foot wire element at the start of the inner conductor winding of the trap. This avoids detuning the antenna by the stray capacitance of the coaxial-cable shield. The trap output terminal (which has the shield stray capacitance) should be at the outboard side of the trap. Reversing the input and output terminals of the trap will lower the 40 meter frequency by approximately 50 kHz, but there will be negligible effect on the other bands.

Figure 12.30 shows a coaxial-cable trap. Further details of the trap installation are shown in **Figure 12.31**. This drawing applies specifically to the 80, 40, 20, 15, and 10 meter antenna, which uses the low-impedance trap connections.

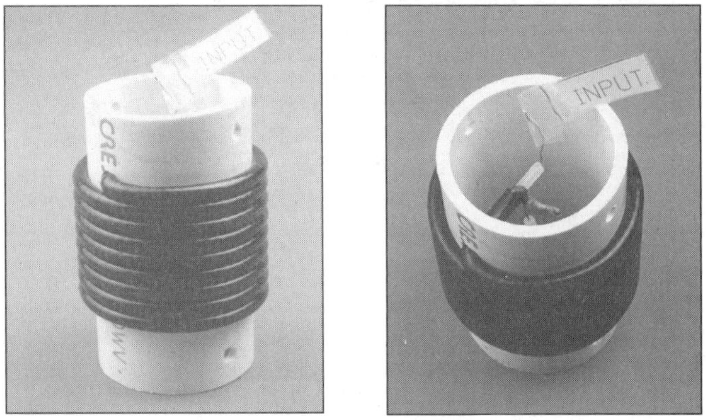

Figure 12.30 — Other views of a W8NX coax-cable trap.

Notice the lengths of the trap pigtails: 3 to 4 inches at each terminal of the trap. If you use a different arrangement, you must modify the span lengths accordingly. All connections can be made using crimp connectors rather than by soldering.

PERFORMANCE

The performance of both antennas has been very satisfactory. W8NX uses the 80, 40, 17, and 12 meter version because it covers 17 and 12 meters. (He has a tribander for 20, 15, and 10 meters.) The radiation pattern on 17 meters is that of a 3/2-wave dipole. On 12 meters, the pattern is that of a 5/2-wave dipole. At his location in Ohio, the antenna runs essentially east and west. It is installed as an inverted V, 40 feet high at the center, with a 120° included angle between the legs. Since the stubs are very short, they radiate little power and make only minor contributions to the radiation patterns. In theory, the pattern has four major lobes on 17 meters, with maxima to the northeast, southeast, southwest and northwest. These provide low-angle radiation into Europe, Africa, South Pacific, Japan and Alaska. A narrow pair of minor broadside lobes provides north and south coverage into Central America, South America and the polar regions.

There are four major lobes on 12 meters, giving nearly end-fire radiation and good low-angle east and west coverage. There are also three pairs of very narrow, nearly broadside, minor lobes on 12 meters, down about 6 dB from the major end-fire lobes. On 80 and 40 meters, the antenna has the usual figure-8 patterns of a half-wave-length dipole.

Both antennas function as electrical half-wave dipoles on 80 and 40 meters with a low SWR. They both function as odd-harmonic current-fed dipoles on their other operating frequencies, with higher, but still acceptable, SWR. The presence of the stubs can either raise or lower the input impedance of the antenna from those of the usual third and fifth harmonic dipoles. Again W8NX recommends that 75-Ω, rather than 50-Ω, feed line be used because of the generally higher input impedances at the harmonic operating frequencies of the antennas.

The SWR curves of both antennas were carefully measured using a 75 to 50-Ω transformer from Palomar Engineers inserted at the junction of the 75-Ω coax feed line and a 50-Ω SWR bridge. The transformer is required for accurate SWR measurement if a 50-Ω SWR bridge is used with a 75-Ω line. Most 50-Ω radios operate satisfactorily with a 75-Ω line, although this requires different tuning and load settings in the final output stage of a vacuum tube amplifier or antenna tuner. The 75 to 50-Ω transformer is used only when making SWR measurements and at low power levels.

Figure 12.32 gives the SWR curves of the 80, 40, 20, 15, and 10 meter antenna. Minimum SWR is nearly 1:1 on 80 meters, 1.5:1 on 40 meters, 1.6:1 on 20 meters, and 1.5:1 on 10 meters. The minimum SWR is slightly below 3:1 on 15 meters. On 15 meters, the stub capacitive reactance combines with the inductive reactance of the outer segment of the antenna to produce a resonant rise that raises the antenna input resistance to about 220 Ω, higher than that of the usual 3/2-wave-length dipole. An antenna tuner may be required on this band to keep a solid-state final output stage happy under these load conditions.

Figure 12.33 shows the SWR curves of the 80, 40, 17, and 12 meter antenna. Notice the excellent 80 meter performance with a nearly unity minimum SWR in the middle of the band. The performance approaches that of a full-size 80 meter wire dipole. The short stubs and the low-inductance traps shorten the antenna somewhat on 80 meters. Also observe the good 17 meter performance, with the SWR being only a little above 2:1 across the band.

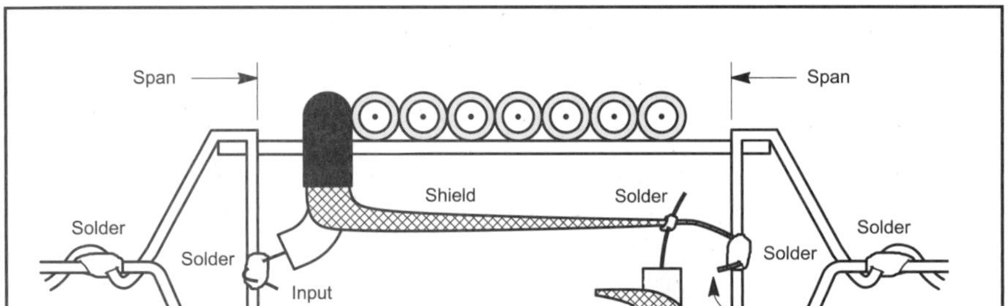

Figure 12.31 — Additional construction details for the W8NX coaxial-cable trap.

But notice the 12 meter SWR curve of this antenna, which shows 4:1 SWR across the band. The antenna input resistance approaches 300 Ω on this band because the capacitive reactance of the stubs combines with the inductive reactance of the outer antenna segments to give resonant rises in impedance. These are reflected back to the input terminals. These stub-induced resonant impedance rises are similar to those on the other antenna on 15 meters, but are even more pronounced.

High SWR in coaxial cables longer than about 100 feet can lead to high feed line losses as shown in the **Transmission**

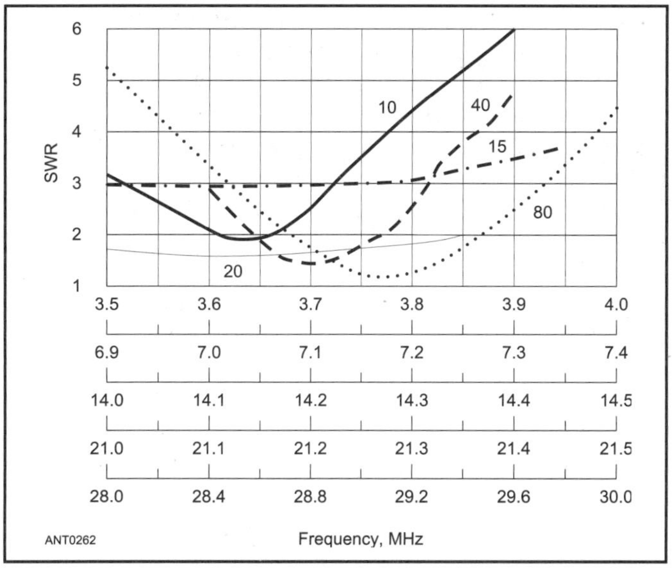

Figure 12.32 — Measured SWR curves for an 80, 40, 20, 15 and 10 meter antenna, installed as an inverted V with 40-ft apex and 120° included angle between legs.

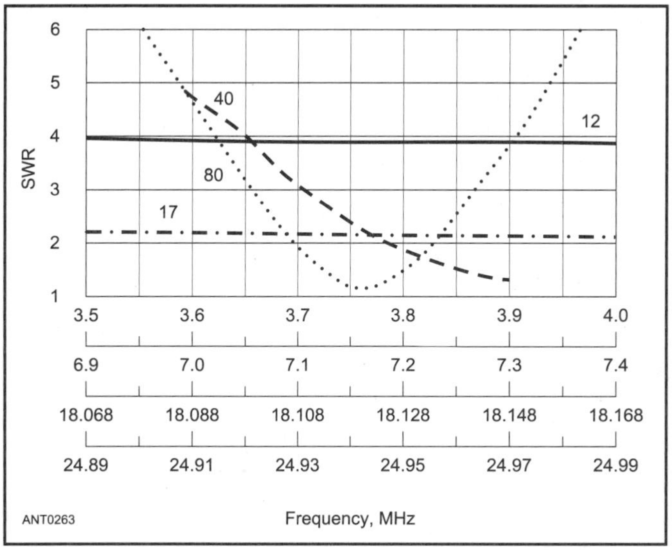

Figure 12.33 — Measured SWR curves for an 80, 40, 17 and 12 meter antenna, installed as an inverted V with 40-ft apex and 120° included angle between legs.

Lines chapter. If you plan on operating this antenna with an SWR of greater than 3:1, make sure the amount of feed line loss is acceptable.

High voltages in the feed line should not cause too much concern. Even if the SWR is as high as 9:1 *no destructively high voltages will exist on the transmission line*. Recall that transmission-line voltages increase as the square root of the SWR in the line. Thus, 1 kW of RF power in 75-Ω line corresponds to 274 V line voltage for a 1:1 SWR. Raising the SWR to 9:1 merely triples the maximum voltage that the line must withstand to 822 V. This voltage is well below the 3700-V rating of RG-11, or the 1700-V rating of RG-59, the two most popular 75-Ω coax lines. Voltage breakdown in the traps is also very unlikely. As will be pointed out later, the operating power levels of these antennas are limited by RF power dissipation in the traps, not trap voltage breakdown or feed line SWR.

TRAP LOSSES AND POWER RATING

Table 12.1 presents the results of trap Q measurements and extrapolation by a two-frequency method to higher frequencies above resonance. W8NX employed a Boonton Q meter for the measurements. Extrapolation to higher-frequency bands assumes that trap resistance losses rise with skin effect according to the square root of frequency, and that trap dielectric loses rise directly with frequency. Systematic measurement errors are not increased by frequency extrapolation. However, random measurement errors increase in magnitude with upward frequency extrapolation. Results are believed to be accurate within 4% on 80 and 40 meters, but only within 10 to 15% at 10 meters. Trap Q is shown at both the high- and low-impedance trap terminals. The Q at the low-impedance output terminals is 15 to 20% lower than the Q at the high-impedance output terminals.

W8NX computer-analyzed trap losses for both antennas in free space. Antenna-input resistances at resonance were first calculated, assuming lossless, infinite-Q traps. They were again calculated using the Q values in Table 12.1. The radiation efficiencies were also converted into equivalent trap losses in decibels. **Table 12.2** summarizes the trap-loss analysis for the 80, 40, 20, 15, and 10 meter antenna and **Table 12.3** for the 80, 40, 17, and 12 meter antenna.

The loss analysis shows radiation efficiencies of 90% or more for both antennas on all bands except for the 80, 40, 20, 15, and 10 meter antenna when used on 40 meters. Here, the radiation efficiency falls to 70.8%. A 1-kW power level at 90% radiation efficiency corresponds to 50 W dissipation per trap. In W8NX's experience, this is the trap's survival limit for extended key-down operation. SSB power levels of 1 kW

Table 10.1
Trap Q

Frequency (MHz)	3.8	7.15	14.18	18.1	21.3	24.9	28.6
High Z out (Ω)	101	124	139	165	73	179	186
Low Z out (Ω)	83	103	125	137	44	149	155

Table 10.2
Trap Loss Analysis: 80, 40, 20, 15, 10 meter Antenna

Frequency (MHz)	3.8	7.15	14.18	21.3	28.6
Radiation Efficiency (%)	96.4	70.8	99.4	99.9	100.0
Trap Losses (dB)	0.16	1.5	0.02	0.01	0.003

Table 10.3
Trap Loss Analysis: 80, 40, 17, 12 meter Antenna

Frequency (MHz)	3.8	7.15	18.1	24.9
Radiation Efficiency (%)	89.5	90.5	99.3	99.8
Trap Losses (dB)	0.5	0.4	0.03	0.006

PEP would dissipate 25 W or less in each trap. This is well within the dissipation capability of the traps.

When the 80, 40, 20, 15, and 10 meter antenna is operated on 40 meters, the radiation efficiency of 70.8% corresponds to a dissipation of 146 W in each trap when 1 kW is delivered to the antenna. This is sure to burn out the traps — even if sustained for only a short time. Thus, the power should be limited to less than 300 W when this antenna is operated on 40 meters under prolonged key-down conditions such as RTTY. A 50% CW duty cycle would correspond to a 600-W power limit for normal 40 meter CW operation. Likewise, a 50% duty cycle for 40 meter SSB corresponds to a 600-W PEP power limit for the antenna.

W8NX knows of no analysis where the burnout wattage rating of traps has been rigorously determined. Operating experience seems to be the best way to determine trap burn-out ratings. In his own experience with these antennas, he's had no traps burn out, even though he operated the 80, 40, 20, 15 and 10 meter antenna on the critical 40 meter band using his AL-80A linear amplifier at the 600-W PEP output level. He did not make a continuous, key-down, CW operating test at full power purposely trying to destroy the traps!

Some hams may suggest using a different type of coaxial cable for the traps. The dc resistance of 40.7 Ω per 1000 feet of RG-59 coax seems rather high. However, W8NX has found no coax other than RG-59 that has the necessary inductance-to-capacitance ratio to create the trap characteristic reactance required for the 80, 40, 20, 15, and 10 meter antenna. Conventional traps with wide-spaced, open-air inductors and appropriate fixed-value capacitors could be substituted for the coax traps, but the convenience, weatherproof configuration and ease of fabrication of coaxial-cable traps is hard to beat.

12.3 The Terminated Folded Dipole

Originally described in the Junee 1949 *QST* by G. L. Countryman, W3HH, the terminated folded dipole (TFD) is often used by amateurs as a wide-band antenna that operates over the HF range without a tuner. (A tilted version is abbreviated as the T2FD. The original article is included with this book's online material.) The antenna, shown in **Figure 12.34** looks like a folded dipole but acts as a traveling wave antenna. It is generally useful over a 5:1 or 6:1 range with its lower frequency, f_L, used to calculate dimensions.

Two common approximations of the antenna dimensions are used:

Wide-Long:
Length (feet) = $300/f_L$ (MHz)
Separation (feet) = $10/f_L$ (MHz) (2)

and

Narrow-Short:
Length (feet) = $200/f_L$ (MHz)
Separation (feet) = $0.5/f_L$ (MHz) (3)

The Narrow-Short configuration is closest to commercial models sold today. Figure 12.34 also provides two sets of

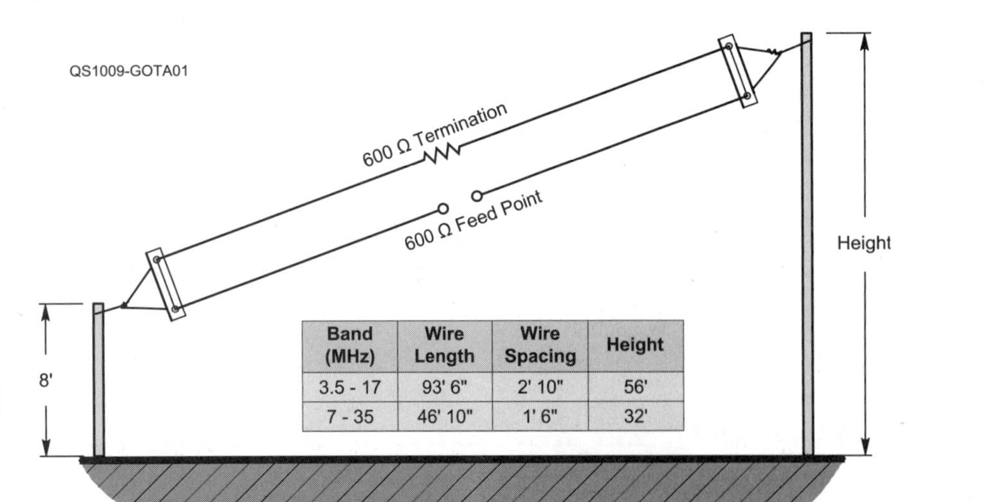

Band (MHz)	Wire Length	Wire Spacing	Height
3.5 - 17	93' 6"	2' 10"	56'
7 - 35	46' 10"	1' 6"	32'

Figure 12.34 — The terminated folded dipole adds a 600-Ω terminating resistor to smooth out the feed point impedance variations over a wide range. The terminating resistor dissipates some of the input power in order to provide a consistent match.

dimensions for different HF frequency ranges.

The termination resistor directly opposite the feed point acts as a *swamping load* to dampen swings in feed point impedance, particularly toward the lower end of the antenna's range. A value of around 600 Ω is required and must be non-inductive. (The resistance value can vary ±10% and still obtain good results.) The resistor can dissipate an appreciable fraction of the applied power and is usually rated at ⅓ of the transmitter output power. If a single resistor is not available, strings of resistors in series-parallel combinations will work.

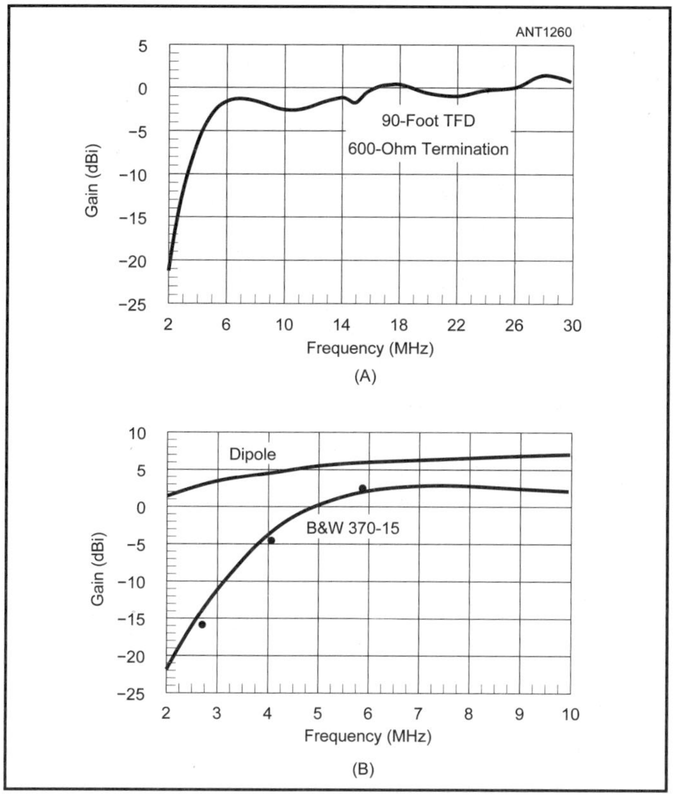

Figure 12.35 — At A, theoretical gain (in dBi) versus frequency for a 90-foot terminated folded dipole (600-Ω termination) in free space. At B, theoretical gain of a dipole and a 90-foot TFD antenna over poor ground (conductivity of 1 mS/m), and measured NVIS gain (the solid circles) for the B&W 370-15 antenna. All antennas were installed horizontally at 30 feet.

The feed point resistance of the antenna requires a 12:1 matching transformer to provide a match to 50-Ω coaxial cable. Designs using an 800-Ω terminating resistor and 16:1 transformer also work well.

The antenna wire chosen must be capable of supporting the antenna's weight, which with the balun and terminating resistor, plus any spacers used to keep the antenna wires separated, can be significantly higher than for a single-wire dipole. A common choice is #12 AWG hard-drawn copper. An analysis by L.B. Cebik, W4RNL, of models of the antennas provides an alternate set of guidelines for the terminating resistor (see References and Bibliography).

As the operating frequency approaches the lower end of the antenna's range, efficiency and gain will begin to fall off rapidly. An analysis of the TFD by Belrose (see References and Bibliography) shows that the antenna's gain is approximately 0 dBi over most of its range (see **Figure 12.35A**) and is significantly less effective than a dipole with a tuner by as much as 10 dB (see Figure 12.35B).

However, at heights of 50 feet or less, the antenna will provide a relatively high-angle, nearly omnidirectional pattern useful for NVIS-like coverage. Tilting the antenna, as described in the original article by W3HH, enhances high-angle radiation even further. This is the TFD's most common application, as a regional coverage antenna over a wide range of HF when only a single, fixed antenna without tuners can be installed, such as for an EOC or station providing temporary public service or disaster relief communications.

12.4 Multiband Vertical Antennas

There are two basic types of vertical antennas; either type can be used in multiband configurations. The first is the ground-mounted vertical and the second, the ground plane. These antennas are described in detail in the chapter **Dipoles and Monopoles**.

The efficiency of any ground-mounted vertical depends a great deal on near-field earth losses. As pointed out in the chapter **Effects of Ground**, these near-field losses can be reduced or eliminated with an adequate radial system. Considerable experimentation has been conducted on this subject by Jerry Sevick, W2FMI, and several important results were obtained. It was determined that a radial system consisting of 40 to 50 radials, 0.2 λ long, would reduce the earth losses to about 2 Ω when a λ/4 radiator was being used. These radials should be on the earth's surface, or if buried, placed not more than an inch or so below ground. Otherwise, the RF current would have to travel through the lossy earth before reaching the radials. In a multiband vertical system, the radials should be 0.2 λ long for the lowest band, that is, 55 feet long for 3.5-MHz operation. Any wire size may be used for the radials. The radials should fan out in a circle, radiating from the base of the antenna. A metal plate, such as a piece of sheet copper, can be used at the center connection. Sevick also contributed a great deal of information about short vertical that are much less than λ/4. This subject is covered in the **General Purpose MF and HF Antennas** chapter.

The other common type of vertical is the ground-plane antenna. Normally, this antenna is mounted above ground with the radials fanning out from the base of the antenna. The vertical portion of the antenna is usually an electrical λ/4, as is each of the radials. In this type of antenna, the system of radials acts somewhat like an RF choke, to prevent RF currents from flowing in the supporting structure, so the number of radials is not as important a factor as it is with a ground-mounted vertical system. From a practical standpoint, the customary number of radials is four or five. In a multiband configuration, λ/4 radials are required for each band of operation with the ground-plane antenna.

This is not so with the ground-mounted vertical antenna, where the ground plane is relied upon to provide an image of the radiating section. Note that even quarter-wave-long radials are greatly detuned by their proximity to ground — radial resonance is not necessary or even possible. In the ground-mounted case, so long as the ground-screen radials are approximately 0.2 λ long at the lowest frequency, the length will be more than adequate for the higher frequency bands.

12.4.1 Full-Size Vertical Antennas

A vertical antenna should not be longer than about ¾ λ at the highest frequency to be used, however, if low-angle radiation is wanted. You can see why from reviewing the radiation patterns for dipoles in the chapter **Dipoles and Monopoles**. As the antenna lengthens, the pattern breaks up into lobes that are at high elevation angles for a vertical antenna. Nevertheless, an antenna that is λ/4 on the lower frequency of operation can still be useful over a 3:1 frequency range or even more if the high-angle radiation can be tolerated. For example, an 80 meter λ/4 vertical around 66 feet high is useful through the 30 meter band and a 25-foot vertical would be useful from about 10 MHz through the 28 MHz band.

The 43-foot vertical monopole shown in **Figure 12.36** has become increasingly popular. While not resonant on any of the amateur bands except 60 meters where it approximates a quarter-wave vertical, with a series inductor at the feed point it has an easily-matched base impedance and is an efficient radiator. In fact, it is very similar to an Extended Double Zepp (EDZ) for 20 meters which is 86 feet long. (See the **Broadside and End-Fire Dipole Arrays** chapter for more information on the EDZ.) If the EDZ is cut in half and placed over a ground plane, it becomes a ⅝ λ vertical for 20 meters and the 43-foot monopole results.

While this antenna is not resonant on 20 meters, and by itself has a high SWR, its base impedance is largely capacitive reactance. Adding a simple inductor in series (about 320 Ω of inductive reactance, or 3.6 μH at 14.15 MHz) will reduce SWR to below 2:1. On other bands, the SWR is more extreme.

There are basically two options to feed the antenna on other bands:

- Install a manual or automatic antenna tuner at the antenna base, with switchable loading coils for some bands if needed, and use that arrangement to provide a match to 50 Ω coax. Then run the matched coax to the station with low losses.
- Accept the mismatch and feed mismatched coax to an antenna tuner at the station end. While this is simple, the high SWR creates extra loss in the feed line as explained in the **Transmission Line System Techniques** chapter.

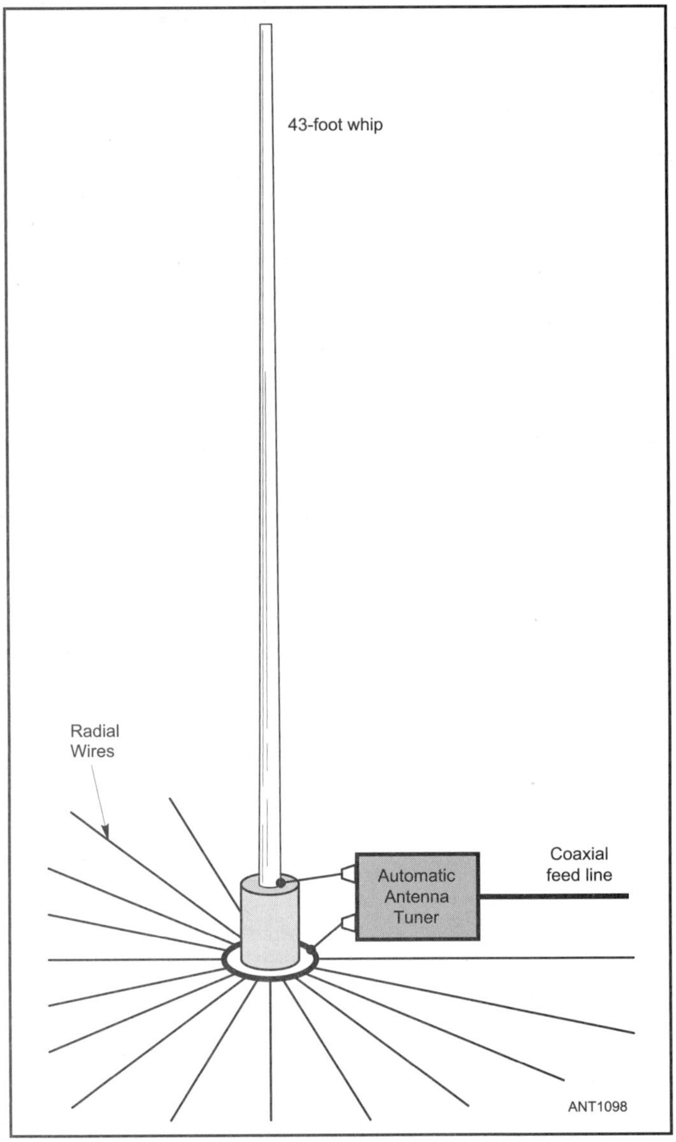

Figure 12.36 — A vertical whip antenna tuned by an automatic antenna tuner at its base is an effective multiband antenna. A 43-foot whip is a popular choice, present-ing reasonable feed point impedances from 80 through 10 meters.

The antenna can also be used on 80 and 160 meters although the high feed point impedance requires special matching techniques. Phil Salas, AD5X, and Steve Masticola, WX2S have designed tunable matching networks for these bands that make the vertical work on all of the HF bands plus the MF band of 160 meters (see References and Bibliography section).

One caveat about this vertical is that on the higher-frequency bands, the radiation pattern breaks up into lobes which radiate a lot of signal at high angles that are not necessarily effective for DX. On 60 through 20 meters, the elevation pattern is similar to that in **Figure 12.37A**. On higher frequency bands, the pattern begins to look more like that in Figure 12.37B. This may be perfectly acceptable, and the antenna's minimal visual impact makes it a feasible choice in many circumstances

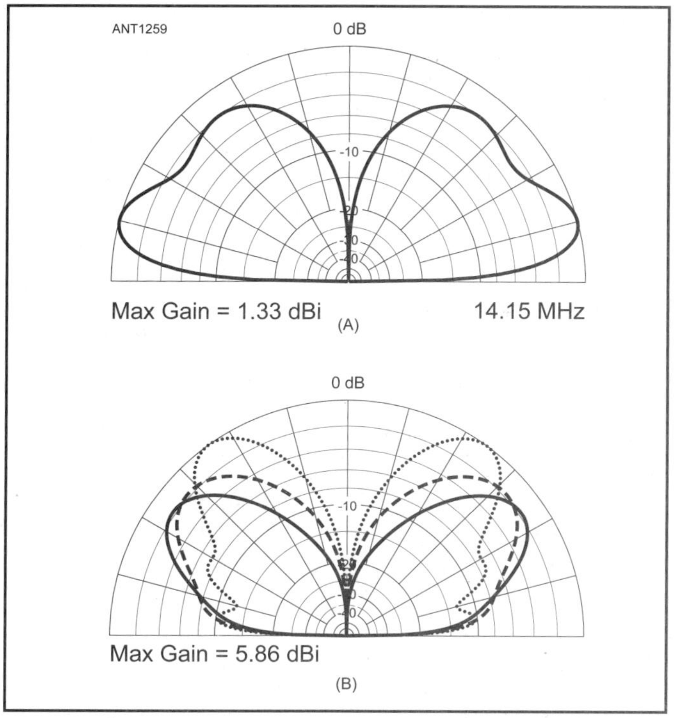

Figure 12.37 — *EZNEC* elevation pattern of ⅝-wavelength ground-mounted monopole on 20 meters over typical ground (A). At B, the same antenna on 17 meters (solid line), 15 meters (dashed line), and 10 meters (dotted line).

where a trap vertical or horizontally polarized antenna might not be possible.

12.4.2 Trap Verticals

The trap principle described in Figure 12.23 for center-fed dipoles also can be used for vertical antennas. There are two principal differences. Only one half of the dipole is used, the ground connection taking the place of the missing half, and the feed point impedance is one half the feed point impedance of a dipole. Thus it is in the vicinity of 30 Ω (plus the ground-connection resistance), so 50 Ω cable should be used since it is the commonly available type that comes closest to matching.

Most amateurs prefer to purchase multiband trap verticals because of the mechanical complexities and requirements to be self-supporting. Commercial multiband trap verticals such as the Hustler 4/5/6BTV series and the Hy-Gain AVQ series have been widely used for many years. The Butternut HF-series is another time-tested design with a different approach than traps to multiband design. All of these antenna families provide effective performance as ground-mounted antennas when used with a good radial system.

Verticals advertised as "ground-independent" are intended to be mounted above ground. Models such as the Cushcraft R9 and R6000 and the Hy-Gain AV-680 are end-fed systems that are electrically longer than λ/4 at the frequency of operation. They have a high feed point impedance that is reduced to 50 Ω with a matching network at the base of the antenna. These are particularly useful antennas for temporary stations and when restrictions prevent the installation of ground systems.

Amateurs interested in designing their own trap vertical may find a dual-band antenna a great way to experiment. Antenna modeling software and trap design utilities can be used to capture the antenna's basic elements. Once built, some adjustment fine-tunes the design. For example, Steve Sutterer, AKØM, started with W1JR's ⅜-wavelength vertical design (see the **General Purpose MF and HF Antennas** chapter) and inserted a 20 meter coax trap about 25 feet up from the feed point. Steve designed the trap using VE6YP's online utility at **www.qsl.net/ve6yp/CoaxTrap.html**. Another 25 feet or so of wire above the trap and an auto-tuner at the base resulted in a 50-foot high, tree-supported, wire vertical that works quite well on 80, 40, 30, and 20 meters. If you choose to build this antenna, its final dimensions will depend on the materials you use and the environment in which it is operated — expect to make some adjustments along the way. This kind of simple antenna project is a great way to learn about antenna and trap design.

12.4.3 A Five-Band Vertical Dipole

This antenna, shown in **Figure 12.38**, is a short version of the 135-foot dipole described earlier, except that it is oriented vertically. The antenna gives good performance on the HF bands from 20 through 10 meters and can be used at a fixed station or for portable or temporary operation. Like the longer horizontal "flat-top," it is fed with 450 Ω window line and a balanced tuner. The antenna can be installed at an angle as long as the feed line is oriented symmetrically at right angles to the antenna. Feed line lengths of more than 33 feet are recommended to help minimize RF common-mode current on the feed line from coupling to the tuner enclosure.

12.4.4 A Fan Vertical

Just like the fan dipole described earlier, the fan vertical shown in **Figure 12.39** connects multiple elements to a single feed point. In the case of the fan vertical, each element is λ/4 long instead of λ/2 and a set of radials is used to provide a

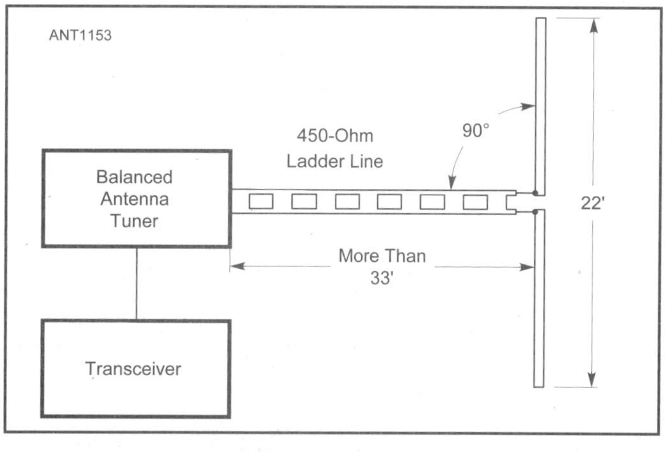

Figure 12.38 — A five-band vertical dipole that covers 20, 17, 15, 12, and 10 meters. The antenna is fed with 450-Ω window line and requires a balanced antenna tuner.

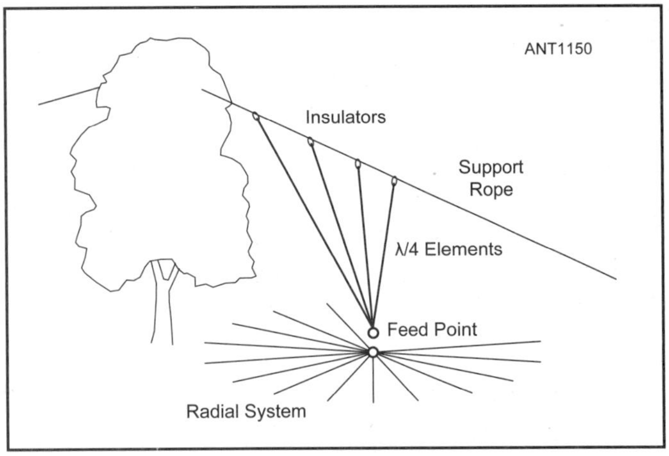

Figure 12.39 — A set of fan verticals. The support must be high enough to clear λ/4 on the lowest frequency band used. At higher frequencies, the shorter elements can be held up with ropes to the support rope as necessary. Radials are required to provide a ground plane.

ground plane. The same cautions apply regarding interaction between the verticals as for the fan dipoles. Getting a full set tuned on each band can be frustrating! Nevertheless, whether used for a fixed station or for portable operation, being able to use multiple bands with a single feed line is very convenient. A similar design was described by Mark Weaver, WB3BJF that places the vertical wires near a tree trunk with radials surrounding the base (see References and Bibliography section).

12.4.5 PROJECT: A Dual-Band Vertical for 80 and 160

When stationed on Guam, Dave Mueller, N2NL, needed a vertical antenna for contesting on 80 and 160 meters but had only one tall vertical support. Based on the idea of a fan vertical, he used homemade open-wire line as a pair of vertical radiators as shown in **Figure 12.40**. The vertical wires are held apart with commercially available spacers for 600 Ω line or homemade spacers of PVC tubing, such as for irrigation systems, that are from 4 to 6 inches long. Initial versions of the antenna used 450 Ω window line as the parallel vertical elements. This proved to be prone to arcing at full power when there was moisture or vegetation on the window line.

If the vertical support is high enough, the 80 meter section can be kept straight. Otherwise, the open-wire line is bent horizontal, and the 80 meter conductor is tuned as an inverted L. To create the 160 meter section, wire is then attached to the end of the other open-wire line conductor and to the point at which open-wire line makes a right-angle bend. These form top loading wires for a 160 meter T or "flat-top" vertical. There can be more than two top-loading wires for the 160 meter section.

At the feed point, both open-wire line conductors are shorted together and connected to the coax center conductor. A set of ground radials is connected to the shield of the coax.

Tune the 80 meter L first, then adjust the length of the 160 meter, top-loading wire connected at its junction with the window line — this results in the least interaction during tuning. You will probably have to adjust each of the conductor lengths to get the resonant point of the system to the desired frequencies.

To raise the base impedance of the vertical, a shunt inductor is connected across the feed point. A vacuum relay or manual switch can be used to change the tap or switch the coil in and out of circuit for different bands.

If you have the equipment, measure the feed point impedance to determine the resistive component. The instructions for hairpin matching in the chapter on **Transmission Line System Techniques** provide a chart of required inductive reactances. Another effective technique is to wind a coil with about 100 Ω of reactance (about 9 μH at 1.8 MHz) and adjust a tap on the coil until a match is obtained.

A similar approach was taken by Kai Jørgensen, OZ4KG, for his dual-band vertical shown in **Figure 12.41**. A single vertical support with a metal bottom section and nonconductive top section is used. Nonconductive sidearms hold the vertical elements. A short extension of the vertical mast holds the 160 meter section, if necessary. He uses a nearby tower as the second support for the 160 meter horizontal section. The tuning process is approximately the same.

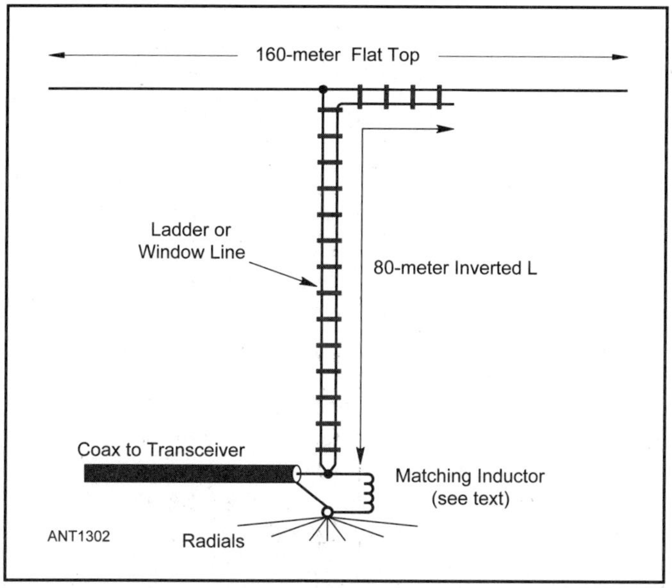

12.40 — Dual-band vertical for 80 and 160 meters as constructed by N2NL/KH2. The 80-meter section forms an inverted L with top loading wires creating a top-loaded 160-meter T vertical.

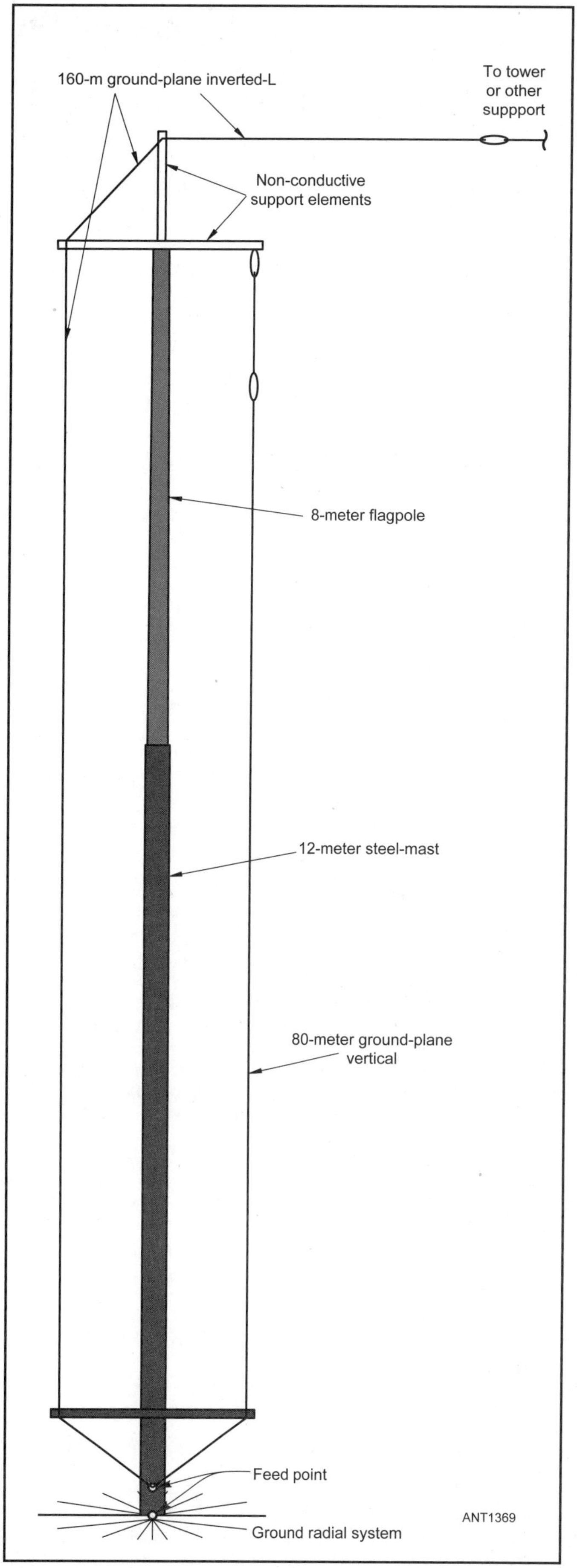

12.41 — Dual-band vertical for 80 and 160 meters as constructed by OZ4KG. The 160-meter section is constructed as an inverted L. A shunt inductor across the base can be used to raise the feed point impedance to 50 ohms as described in the text.

12.5 The Coupled-Resonator Dipole

The *coupled-resonator* or (*C-R*) system is described by Gary Breed, K9AY, in an article in *The ARRL Antenna Compendium, Vol 5*, entitled "The Coupled-Resonator Principle: A Flexible Method for Multiband Antennas." The following is condensed from that article which is included in the online material.

In 1995, *QST* published two antenna designs that use an interesting technique to get multiband coverage in one antenna. Although a C-R antenna requires more complicated construction, its main attraction is in making a multiband antenna that can be built with no compromise in matching or efficiency. Rudy Severns, N6LF, described a wideband 80- and 75 meter dipole using this technique (see the **General Purpose MF and HF Antennas** chapter), and Robert Wilson, AL7KK, showed us how to make a three-band vertical. (See the References and Bibliography.) Both of these antennas achieve multifrequency operation by placing resonant conductors very close to a driven dipole or vertical — with no direct electrical connection.

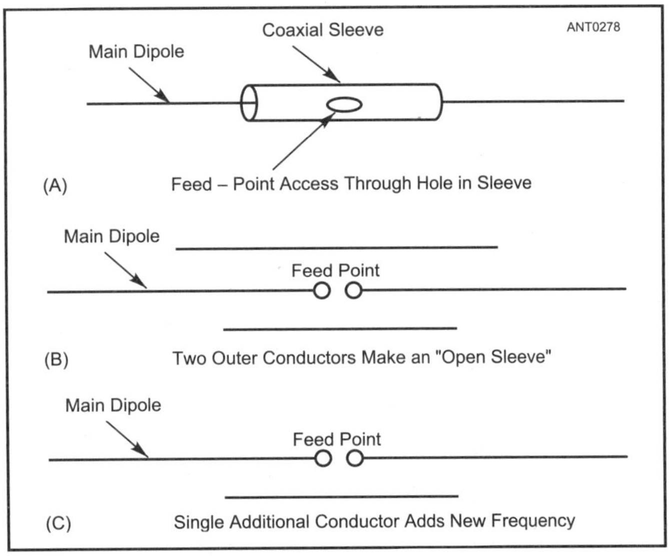

Figure 12.42 — Evolution of coupled-resonator antennas: At A, the *coaxial-sleeve* dipole; at B, the *open-sleeve* dipole; and at C, a *coupled-resonator* dipole, the most universal configuration.

12.5.1 The Coupled-Resonator Principle

The first step toward the C-R antenna was the *coaxial sleeve* antenna (**Figure 12.42**) which covered two frequencies by surrounding a dipole or monopole with a cylindrical tube resonant at the higher of the desired frequencies. It was soon determined that two conductors at the second frequency, placed on either side of the main dipole or monopole, would make a skeleton representation of a cylinder (Figure 12.42B). This is called the *open-sleeve antenna*. Later it was discovered these extra conductors did not need to be added in pairs, and that a single conductor at each frequency could add the extra resonances (Figure 12.42C). This is the method used by several manufacturers to offer multiband Yagis that only require a single feed line or Yagis with broad operating bandwidths.

Figure 12.43 illustrates the general behavior of a C-R antenna. Each figure shows the SWR at the feed point of a dipole, over a range of frequencies. When this dipole is all alone, it will have a very low SWR at its half-wave resonant frequency (Figure 12.43A). Next, if we take another wire or tubing conductor and gradually bring it close to the dipole, we will see a "dip" in the dipole's SWR at the resonant frequency of this new wire. See Figure 12.43B. We are beginning to the see the effects of interaction between the two conductors. As we bring this new conductor closer, we reach a point where the SWR "dip" has become very deep. There is now a good match at both the original dipole's resonant frequency and the frequency of the new conductor, as illustrated in Figure 12.43C.

We can repeat this process for several more conductors at other frequencies to get a dipole with three, four, five, six, or more resonant frequencies. The principle also applies to verticals, so any reference to a dipole can be considered to be valid for a vertical, as well.

The C-R principle is defined this way: *Given a dipole (or vertical) at one frequency and an additional conductor resonant at another frequency, there is an optimum distance between them that results in the resonance of the additional conductor being imposed upon the original dipole, resulting in a low SWR at both resonant frequencies.*

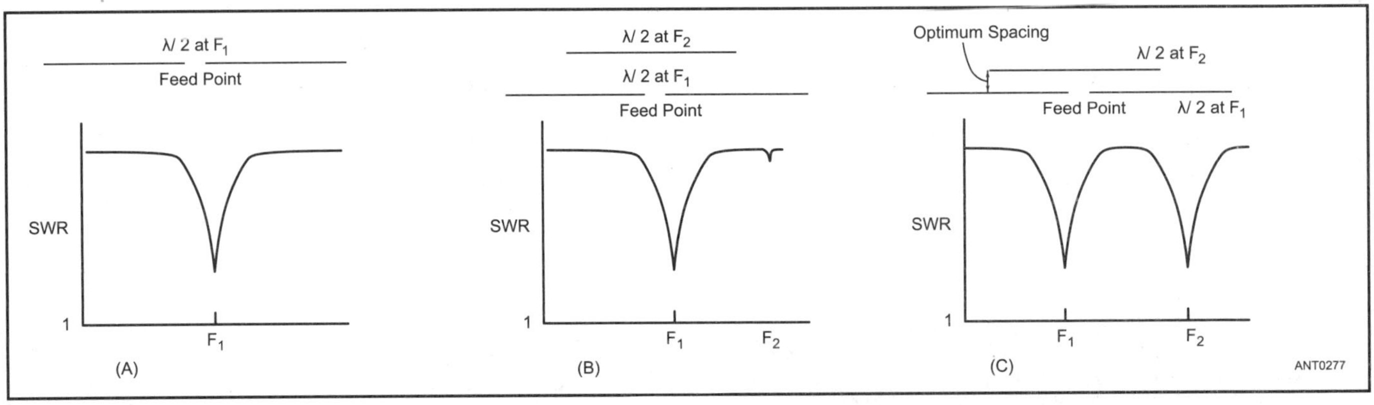

Figure 12.43 — At A, the SWR of a dipole over a wide frequency range. At B, a nearby conductor is just close enough to interact with the dipole. At C, when the second conductor is at the optimum spacing, the combination is matched at both frequencies.

C-R Element Spacing

K9AY's equations in the original article do indeed yield a good "first-cut" value for the spacing between coupled-resonator elements. **Figure 12.A** shows the spacing, in inches, plotted against the ratio of frequencies, for two coupled resonator elements with different diameters, again expressed in inches. This is for an upper frequency of 28.4 MHz. Beyond a frequency ratio of about 1.5:1 (28.4:18.1 MHz), the spacing flattens out to a fixed distance between elements for each element diameter. For example, if ½-inch elements are used at 28.4 and 18.1 MHz, the spacing between the elements is about 3.75 inches.

EZNEC verifies the computations. Note that a large number of segments are necessary for each element when they are closely spaced from each other, and the segments on the elements must be closely aligned with each other. Be sure to run the Average Gain test, as well as Segmentation tests. The modeler should also be aware that if mutually coupled resonators are placed along a horizontal boom (as they would be on multiband Yagis using coupled resonators), the higher-frequency elements will act like retrograde directors, producing some gain (or lack of gain, depending on the azimuth being investigated).

For example, in the *EZNEC* file **K9AY C-R 28-21-14 MHz 1 In.EZ**, using 1-inch diameter elements spaced 6 inches apart, if the 28-MHz element is placed 6 inches behind the 14-MHz driven element (with the 21-MHz element placed 6 inches ahead), on 28 MHz the system will have a F/B of 2.6 dB, favoring the rearward direction. On 21 MHz, the system will exhibit a F/B of 1.6 dB, favoring the forward direction. Of course, there are systems where gain and F/B due to the C-R configuration may be put to good use, such as the multiband Yagis mentioned above. However, if the elements are spaced above/below the 14-MHz driven element there is no distortion of the dipole patterns.

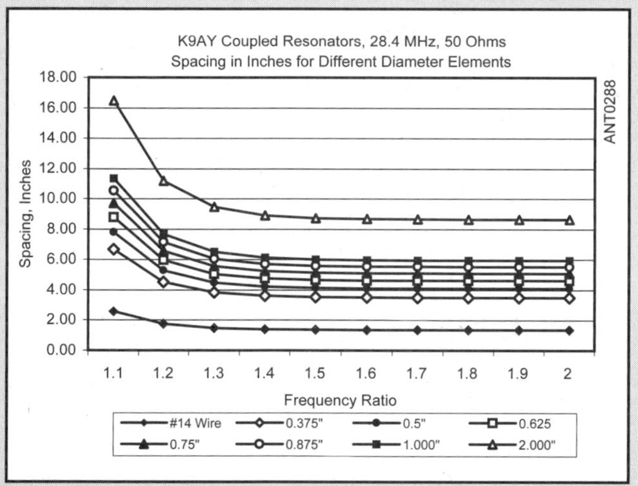

Figure 12.A — Graph of the spacing versus frequency ratio for two Coupled-Resonator elements at 28.4 MHz, for 50 Ω.

12.5.2 Characteristics of C-R Dipoles

There are three primary variables for designing the coupled resonator: (1) the diameter of the conductor, (2) its length, and (3) the position of the resonator relative to the main element. Having the freedom to control these factors gives us the advantage of *flexibility*; we have a wide range of control over the impedance at each added frequency. Another advantage is that the behavior at each frequency is quite *independent*, once the basic design is in place. In other words, making fine-tuning adjustments at one frequency doesn't change the resonance or impedance at the other frequencies. A final advantage is *efficiency*. With conductors close together, and with a resonant target conductor, coupling is very efficient. This avoids traps, stubs, and compensating networks found on other multiband antennas that introduce lossy reactive components.

There are two main disadvantages of C-R antennas. The first is the relative *complexity* of construction. Several conductors are needed, installed with some type of insulating spacers to maintain distance between the conductors. Other multiband antennas have their complexities as well (such as traps that need to be mounted and tuned), but C-R antennas will usually be bulkier. The larger size generally means greater wind load, which is a disadvantage in some circumstances.

The other significant disadvantage is *narrower bandwidth*, particularly at the highest of the operating frequencies. We can partially overcome this problem with large conductors that are naturally broad in bandwidth, and in some cases we might even use an extra conductor to put two resonances in one band. It is interesting to note that the pattern is opposite to that of trap antennas. The C-R bandwidth gets narrower at the highest frequencies of operation, while trap antennas generally have narrowest bandwidth at their lowest frequencies.

There are two special situations that should be noted. First, when the antenna has a resonance near the frequency where the driven dipole is 3/2 λ long (3/4 λ for a vertical), the dipole has a fairly low impedance. The spacing of the C-R element needs to be increased to raise its impedance so that the parallel combination of the main element and C-R element equals the desired impedance (usually 50 Ω). There is also significant antenna current in the part of the main dipole extending beyond the C-R section, contributing to the total radiation pattern. As a result, this particular arrangement radiates as three λ/2 sections in phase and has about 3 dB gain and a narrower directional pattern compared to a dipole (**Figure 12.44**). This might be an advantage for antennas covering bands with a frequency ratio of about three, such as 3.5 and 10.1 MHz, 7 and 21 MHz, or 144 and 430 MHz.

The other special situation is when we want to add a new frequency very close to the resonant frequency of the main dipole. An antenna for 80 and 75 meters would be an example of this. Again, the driven dipole has a fairly low impedance at the new frequency. Add the fact that coupling is very strong between these similar conductors and we find that a wide spacing is required to make the antenna work. A dipole resonant at 3.5 MHz and another wire resonant at 3.8 MHz will need to be 3 or 4 feet apart, while a 3.5 MHz and 7 MHz combination might only need to be spaced 4 or 5 inches.

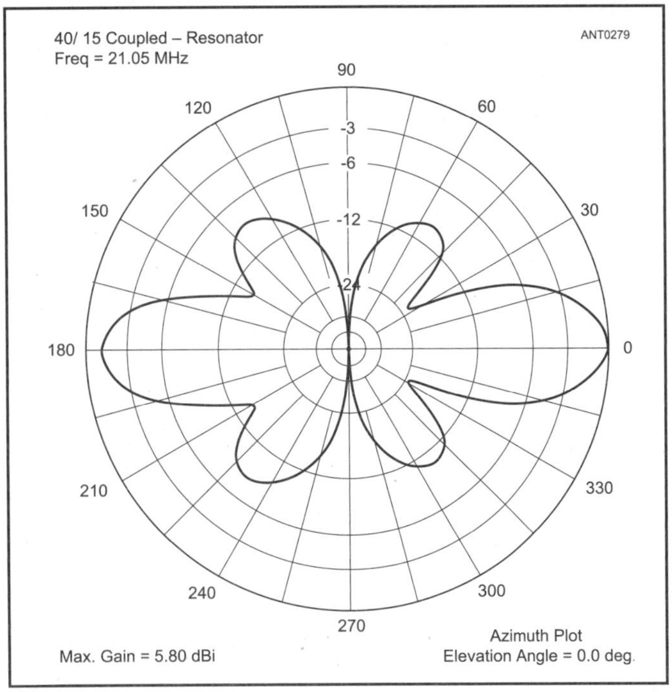

Figure 12.44 — Radiation pattern for the special case of a C-R antenna with the additional resonance at the third harmonic of the main dipole resonant frequency.

Another useful characteristic of C-R antennas is that they are easily and accurately modeled by computer programs based on either *MININEC* or *NEC*, as long as you stay within each program's limitations. With ease of computer modeling, an approximate solution will provide a starting point that can quickly be adjusted for optimum dimensions.

The added resonators influence the lengths of all conductors, due to the capacitance between the conductors. Capacitance causes antennas to look electrically shorter, so each element needs to be about 1% or 2% longer than a simple dipole at the same frequency. As a rule of thumb, use 477/f (in feet) instead of the usual 468/f when calculating dipole length, and 239/f instead of 234/f for a $\lambda/4$ vertical.

PROJECT: A C-R DIPOLE FOR 30/17/12 METERS

The C-R principle is used here to build a dipole covering 30, 17 and 12 meters with a 50 Ω feed point impedance. (See the full article for a 75 Ω feed point impedance.) #12 AWG wire with a diameter of 0.08 inches is used and the main dipole is cut for the 10.1 MHz band. The spacing between the main dipole and the resonators should be 2 inches. Since the antenna will be installed over real ground, these spacing distances may not be exact. Use these initial values into your favorite antenna-modeling program to optimize the dimensions for installation at the height you choose. The predicted spacing is accurate enough to build an antenna with minimum trial-and-error. Like all dipoles, the impedance varies with height above ground, but the 2-inch spacing results in an excellent match on the two additional bands, at heights of more than 25 feet.

The final dimensions of the dipole for 10.1, 18.068 and 24.89 MHz are shown in **Figure 12.45**. These are the final pruned lengths for a straight dipole installed at a height of about 40 feet. If the antenna is installed as an inverted V, you will need each wire to be a bit longer. Pruning this type of antenna is just like a dipole — if it's resonant too low in frequency, it's too long and the appropriate wire needs to be shortened. So, you can cut the wires just a little long to start with and easily prune them to resonance.

A final note: if you want to duplicate this antenna design, remember that the 2-inch spacing is just for #12 AWG wire! The required spacing for a C-R antenna is related to the conductor diameter. This same antenna built with #14 AWG wire needs under 1-1/2-inch spacing, while a 1-inch aluminum-tubing version requires about 7-inch spacing.

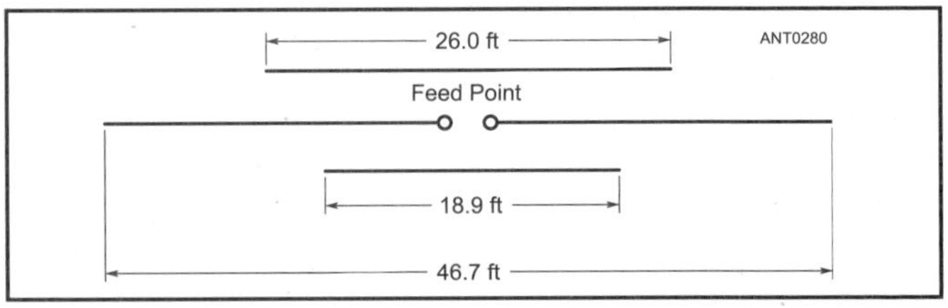

Figure 12.45 — Dimensions of a C-R dipole for the 30-, 17- and 12 meter bands.

12.6 References and Bibliography

Source material and more extended discussion of topics covered in this chapter can be found in the references given below and in the textbooks listed at the end of Chapter 2, **Antenna Fundamentals**.

B. Austin, "An HF Multiband Wire Antenna for Single-Hop Point-to-Point Applications," *Journal of the Institution of Electronics and Radio Engineers*, 57, 1987, pp. 167-173.

W. M. Bell, "A Trap Collinear Antenna," *QST*, Aug. 1963, pp. 30-31.

J. S. Belrose, "The HF Discone Antenna," *QST*, July 1975, pp. 11-14, 56.

J. Belrose and P. Bouliane, "The Off-Center-Fed Dipole Revisited: A Broadband, Multiband Antenna," *QST*, Aug. 1990, pp. 28-34.

J. Belrose, "Technical Correspondence: Terminated Folded Dipole," *QST*, May 1994, pp. 88-89.

H. J. Berg, "Multiband Operation with Paralleled Dipoles," *QST*, July 1956, pp. 42-43.
J. T. Bolljahn and J. V. N. Granger, "Omnidirectional VHF and UHF Antennas," *Antenna Engineering Handbook* , H. Jasik, ed. (New York: McGraw-Hill, 1961) pp. 27-32 through 27-34.
G. A. Breed, "Multi-Frequency Antenna Technique Uses Closely-Coupled Resonators," *RF Design*, November 1994. US Patent 5,489,914, "Method of Constructing Multiple-Frequency Dipole or Monopole Antenna Elements Using Closely-Coupled Resonators," Gary A. Breed, Feb. 6, 1996.
G. H. Brown, "The Phase and Magnitude of Earth Currents Near Radio Transmitting Antennas," *Proc. IRE*, Vol 23, No. 2, Feb. 1935, pp. 168-182.
G. H. Brown, R. F. Lewis and J. Epstein, "Ground Systems as a Factor in Antenna Efficiency," *Proc. IRE*, Vol 25, No. 6, June 1937, pp. 753-787.
C. L. Buchanan, "The Multimatch Antenna System," *QST*, Mar. 1955. pp. 22-23, 155.
L.B. Cebik, "Modeling the T2FD," various online sources.
L.B. Cebik, "Horizontally Oriented, Horizontally Polarized Large Wire Loop Antennas," 29 Mar. 1999, various online sources.
L.B. Cebik, "My Top Five Backyard Multi-Band Wire HF Antennas," various online sources.
S. Davis, "Nested Loop Antennas," *QST*, Mar. 2014, pp. 40-41.
D. DeMaw, "Lightweight Trap Antennas — Some Thoughts," *QST*, June 1983, pp. 15-18.
D. DeMaw, "A Closer Look at Horizontal Loop Antennas," *QST*, May 1990, pp. 28-29, 35.
W. C. Gann, "A Center-Fed 'Zepp' for 80 and 40," *QST*, May 1966, pp. 15-17.
A. Greenberg, "Simple Trap Construction for the Multiband Antenna," *QST*, Oct. 1956, pp. 18-19, 120.
G. L. Hall, "Trap Antennas," Technical Correspondence, *QST*, Nov. 1981, pp. 49-50.
J. Hallas, "Getting On the Air: The Terminated Folded Dipole," *QST*, Sep. 2010, pp. 51-52.
J. Hallas, "The 43 Foot Vertical Monopole — What's the Magic?" *QST*, June 2012, pp. 30-32.
W. Hayward, "Designing Trap Antennas," Technical Correspondence, *QST*, Aug. 1976, p 38.
D. Hollander, "A Big Signal from a Small Lot," *QST*, Apr. 1979, pp. 32-34.
R. H. Johns, "Dual-Frequency Antenna Traps," *QST*, Nov. 1983, pp. 27-30.
W. J. Lattin, "Multiband Antennas Using Decoupling Stubs," *QST*, Dec. 1960, pp. 23-25.
W. J. Lattin, "Antenna Traps of Spiral Delay Line," *QST*, Nov. 1972, pp. 13-15.
R. Littlefield, "A No Compromise Off-Center Fed Dipole for Four Bands" *QST*, June 2008, pp. 32-34.
M. A. Logan, "Coaxial-Cable Traps," Technical Correspondence, *QST*, Aug. 1985, p 43.
J. R. Mathison, "Inexpensive Traps for Wire Antennas," *QST*, Feb. 1977, p. 18.
S. Masticola, "An Improved 160/80-Meter Matching Network for Your 43-Foot Vertical," *QST*, Sep. 2017, pp. 39-41.
L. McCoy, "An Easy-to-Make Coax-Fed Multiband Trap Dipole," *QST*, Dec. 1964, pp. 28-30.
G. E. O'Neil, "Trapping the Mysteries of Trapped Antennas," *Ham Radio*, Oct. 1981, pp. 10-16.
W. I. Orr, editor, "The Low-Frequency Discone," *Radio Handbook*, 14th Edition, (Editors and Engineers, 1956), p. 369.
J. Pawlowski, "Cat Whiskers — The Broadband Multi-Loop Antenna" *QST*, Mar. 2018, pp. 40-44.
J. Portune, "Printed Circuit Board Antenna Traps for HF," *QST*, May 2018, pp. 30-32.
L. Richard, "Parallel Dipoles of 300-Ohm Ribbon," *QST*, Mar. 1957, p. 14.
J. Richter, "Off-Center-Fed Dipoles — Theory and Practice," *HF Dipole Antennas for Amateur Radio* (ARRL, 2019).
P. Salas, "160 and 80 Meter Matching Network for Your 43 Foot Vertical — Part 1 and Part 2," *QST*, Dec. 2009, p. 30-32, and Jan. 2010, pp. 34-35.
W. Sandford, Jr., "A Modest 45-Foot DX Vertical for 160, 80, 40, and 30 Meters," *QST*, Sep. 1981, pp. 27-31. Also see Feedback, Nov. 1981, p. 50.
T. H. Schiller, Force 12, US Patent 5,995,061, "No loss, multi-band, adaptable antenna," Nov. 30, 1999.
H. Scholle and R. Steins, "Eine Doppel-Windom Antenna fur Acht Bander," *cq-DL*, Sep 1983, p. 427. (In English: *QST*, Aug. 1990, pp. 33-34.)
R. Severns, "A Wideband 80 meter Dipole," *QST*, July 1995, pp. 27-29.
M. Seybold, "The Low-Frequency Discone," *CQ*, July 1950, p. 13.
D. P. Shafer, "Four-Band Dipole with Traps," *QST*, Oct. 1958, pp. 38-40.
W. Silver, "Hands-On Radio Experiment 136: The End-Fed Antenna," *QST*, May 2014, pp. 63-64.
R. C. Sommer, "Optimizing Coaxial-Cable Traps," *QST*, Dec. 1984, pp. 37-42.
S. Stearns, "All About the Discone Antenna: Antenna of Mysterious Origin and Superb Broadband Performance," *QEX*, Jan./Feb. 2007, pp. 37-44.
L. Varney, "The G5RV Multiband Antenna . . . Up-to-Date," *The ARRL Antenna Compendium, Vol 1* (Newington: ARRL, 1985), p. 86.
M. Weaver, "A Four-Band "Tree" Vertical," *QST*, Nov. 1995, pp. 69-70.
R. Wilson, "The Offset Multiband Trapless Antenna (OMTA)," *QST*, Oct. 1995, pp. 30-32. Also see Feedback, Dec. 1995, p. 79.
S. Yates, "The End-Fed Half-Wave Antenna," **aa5tb.com/efha.html**.

TABLE OF CONTENTS

Chapter 13 — Online Material

Articles
- "A 10 Meter Moxon Beam" by Allen Baker, KG4JJH
- "A 20 Meter Moxon Antenna" by Larry Banks, W1DYJ
- "Construction of W6NL Moxon on Cushcraft XM240" by Dave Leeson, W6NL
- "Having a Field Day with the Moxon Rectangle" by L.B. Cebik, W4RNL
- "Multimatch Antenna System" by Chester Buchanan, W3DZZ
- "Two Multiband Quad Designs"

Support Files
- "4-Element 48ft Boom OWA Design Information"

Chapter 13
HF Beam Antennas

13.1 Yagi Antennas

Along with the dipole and the quarter-wave vertical, radio amateurs throughout the world make extensive use of the Yagi antenna, more accurately referred to as a Yagi array. Hidetsugu Yagi and Shintaro Uda, two Japanese university professors, invented the Yagi in the 1920s. Uda did much of the developmental work while Yagi introduced the array to the world outside Japan through his writings in English. Although the antenna should properly be called a *Yagi-Uda* array, it is commonly referred to simply as a *Yagi*.

The Yagi is a type of end-fire multielement array as described in the **Phased Arrays** chapter. At the minimum, it consists of a single *driven element* and a single *parasitic element*. These elements are placed parallel to each other on a supporting boom some distance apart. This arrangement is known as a 2-element Yagi. The parasitic element is termed a *reflector* when it is placed behind the driven element, opposite to the direction of maximum radiation, and is called a *director* when it is placed ahead of the driven element. See **Figure 13.1**. In the VHF and UHF spectrum, Yagis employing 30 or more elements are not uncommon, with a single reflector and multiple directors. See the **VHF and UHF Antenna Systems** chapter for details on VHF and UHF Yagis. Large HF arrays may employ 10 or more elements and will be covered in this chapter.

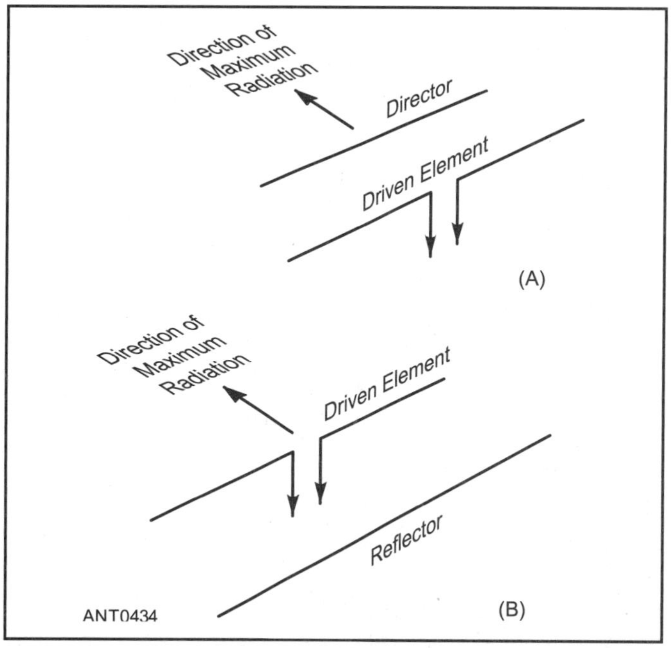

Figure 13.1 — Two-element Yagi systems using a single parasitic element. At A the parasitic element acts as a director, and at B as a reflector. The arrows show the direction in which maximum radiation takes place.

13.1.1 How A Yagi Works — An Overview

The gain and directional pattern of a Yagi array is determined by the relative amplitudes and phases of the currents induced into all the parasitic elements. Unlike directly driven multielement arrays in which the designer must compensate for mutual coupling between elements, proper Yagi operation *relies on* mutual coupling. The current in each parasitic element is determined by its spacing from both the driven element and other parasitic elements, and by the tuning of the element itself. Both length and diameter affect element tuning.

The following discussion is quite over-simplified but serves to illustrate the basic process by which the Yagi antenna creates its radiation pattern. Begin with a dipole driven element resonant at the operating frequency and a single parasitic element configured as a reflector, slightly longer than the driven element. Current in the driven element creates a radiated electromagnetic field (the *direct field*) that induces a current in the parasitic element. That induced current causes a *re-radiated field* just as if the current was caused by a transmitter connected to the element. The re-radiated field combines with the direct field from the driven element to create the antenna's radiation pattern.

Three things determine the phase relationship between the current in the reflector and in the driven element. First, the direct field at the reflector slightly lags the field at the driven element because the direct field must travel from the driven element to the reflector. Second, the induced current is 180° out of phase with the direct field at the location of the reflector. Third, the reflector element is slightly longer than a resonant length and so its self-impedance is inductive, creating additional phase lag in the induced current relative to the direct field.

The combination of phase lags due to the direct field's travel time, the 180° phase inversion for induced current, and the reflector's inductance cause the re-radiated field to partially cancel with the direct field from the driven element along a line from the driven element through the reflector. (Imagine the boom extending beyond the reflector — that is the line being referred to.) This creates the rear null in the Yagi radiation pattern. Similarly, the fields reinforce each other in the opposite direction to the forward direction as shown in Figure 13.1.

The situation is reversed in the case of a director element. The phase lag from travel time and the inversion in the induced current act the same as for the reflector. The director element is slightly shorter than resonant however and so has a capacitive self-impedance, creating a phase lead. The combination results in the fields reinforcing in the forward direction and cancelling in the opposite direction.

Two-element Yagis are useful antennas but even more directivity (gain) can be obtained by adding additional parasitic elements. Additional reflectors are rarely used because field cancellation to the rear of the antenna leaves too little field for them to improve directivity. Thus, multiple directors are used to increase directivity as you will see in practical Yagi designs later in the chapter.

The actual situation is of course far more complicated than this simplistic view of Yagi operation. In a real Yagi, the mutual coupling between all elements must be considered, including between the parasitic elements. This makes determining the optimum spacing and element length for a desired radiation pattern quite involved mathematically and best left to software modeling programs.

13.1.2 Yagi Modeling

For about 50 years amateurs and professionals created Yagi array designs largely by "cut and try" experimental techniques. In the early 1980s, Jim Lawson, W2PV, described in detail for the amateur audience the fundamental mathematics involved in modeling Yagis. His book *Yagi Antenna Design* (now out of print) is highly recommended for serious antenna designers as is his series of articles in *Ham Radio* (see References and Bibliography). The advent of powerful microcomputers and sophisticated computer antenna modeling software in the mid 1980s revolutionized the field of Yagi design for the radio amateur. In a matter of minutes, a computer can try 100,000 or more different combinations of element lengths and spacings to create a Yagi design tailored to meet a particular set of high-performance parameters. To explore this number of combinations experimentally, a human experimenter would take an unimaginable amount of time and dedication and the process would no doubt suffer from considerable measurement errors. With the computer tools available today, an antenna can be designed, constructed and then put up in the air, with little or no tuning or pruning required.

A very popular modeling program is *EZNEC* by Roy Lewallen, W7EL (**eznec.com**). *EZNEC* is well-suited to model Yagi antennas. There are several Yagi antenna models included with this book's online material, and *EZNEC* is discussed in more detail in the **Antenna Modeling** chapter.

THE *YW* MODELING PROGRAM

Included with this book's online material, the *YW* modeling program developed by Dean Straw, N6BV, is designed to evaluate monoband Yagi antennas. (*YW* stands for *Yagi for Windows*.) *YW* results compare very closely with Brian Beezley's *YO* or *YA* programs (no longer sold in the amateur market) and with *NEC*-based programs, such as *EZNEC*, *NEC-Win Plus* or *NEC-4*. *YW* is a special-purpose program, designed strictly for monoband Yagis. It has the advantage of running many times more quickly than general-purpose programs such as *NEC* but it has some attendant limitations.

YW evaluations over ground are done over flat "perfect" ground. Mutual impedances between Yagi elements and the ground are not specifically taken into account in *YW*, so calculations for antennas mounted less than approximately $\lambda/8$ above ground are likely to be inaccurate. Antennas mounted in the presence of other nearby antennas or mounted very low to the ground are the specialties of method-of-moment programs like *EZNEC*. Despite these caveats, *YW* will get you very close to a final design — one where you can simply cut the elements and expect that your Yagi will work as advertised.

13.2 Yagi Performance Parameters

There are three main parameters used to characterize the performance of a particular Yagi — *forward gain*, *pattern* and *drive impedance/SWR*. Another important consideration is mechanical strength. It is very important to recognize that each of the three electrical parameters should be characterized over the frequency band of interest in order to be meaningful. Neither the gain, the SWR nor the pattern measured at a single frequency gives very much insight into the overall performance of a particular Yagi.

Poor designs have even been known to reverse their directionality over a frequency band, while other designs have excessively narrow SWR bandwidths, or gain that peaks excessively in the band. Finally, an antenna's ability to survive the wind and ice conditions expected in one's geographical location is an important consideration in any design. Much of this chapter will be devoted to describing detailed Yagi designs that are optimized for a good balance between gain, pattern and SWR over various amateur bands, and that are designed to survive strong winds and icing.

13.2.1 Yagi Gain

Like any other antenna, the gain of a Yagi must be stated in comparison to some standard of reference. Designers of phased vertical arrays often state gain referenced to a single, isolated vertical element. See the section on "Phased Array Techniques" in the chapter **Phased Arrays**.

Many antenna designers prefer to compare gain to that of an *isotropic radiator in free space*. This is a theoretical antenna that radiates equally well in all directions, and by definition, it has a gain of 0 *dBi* (dB isotropic). Many radio amateurs, however, are comfortable using a dipole as a standard reference antenna, mainly because it is *not* a theoretical antenna.

In free space, a dipole does not radiate equally well in all directions — it has a figure-eight azimuth pattern, with deep nulls off the ends of the wire. In its favored directions, a free-space dipole has 2.15 dB gain compared to the isotropic radiator. You may see the term *dBd*, meaning gain referenced to a dipole in free space. Subtract 2.15 dB from gain in dBi to convert to gain in dBd.

Assume for a moment that we take a dipole out of "free space," and place it one wavelength above the ocean, whose saltwater makes an almost perfect ground. At an elevation angle of 15°, where sea water-reflected radiation adds in phase with direct radiation, the dipole has a gain of about 6 dB, compared to its gain when it was in free space, isolated from any reflections. This and other related effects are addressed in the chapter **Effects of Ground**.

It is perfectly legitimate to say that this dipole has a gain of 6 dBd, although the term "dBd" (meaning "dB dipole") makes it sound as though the dipole somehow has gain over itself! Always remember that gain expressed in dBd (or dBi) refers to the *counterpart antenna in free space*. The gain of the dipole over saltwater in this example can be rated at either 6 dBd (over a dipole in free space), or as 8.15 dBi (over an isotropic radiator in free space). Each frame of reference is valid, as long as it is used consistently and clearly. In this chapter we will often switch between Yagis in free space and Yagis over ground. To prevent any confusion, gains will be stated in dBi.

Yagi free-space gain ranges from about 5 dBi for a small 2-element design to about 20 dBi for a 31-element long-boom UHF design. The length of the boom is the main factor determining the gain a Yagi can deliver. Gain as a function of boom length will be discussed in detail after the sections below defining antenna response patterns and SWR characteristics.

13.2.2 Radiation Pattern Measurements

Figure 13.2 compares the E-plane and H-plane pattern of a 3-element Yagi in free space to those of a dipole and an isotropic radiator. (See the **Antenna Fundamentals** chapter for definitions and conventions associated with measurement of radiation patterns.) These patterns were generated using *NEC-2* modeling software. Figure 13.2A shows that this 3-element Yagi in free space exhibits 7.28 dBi of gain (referenced to isotropic), and has 5.13 dB gain over a free-space dipole. For this particular antenna, the half-power beamwidth is about 66°.

FRONT-TO-BACK RATIO

Again as seen in Figure 13.2A, this antenna's front-to-back ratio is 34 dB comparing response at 180° to that in the forward direction at 0°. (The ratio of the forward response to the averaged response over the entire 180° rearward section is called the *front-to-rear ratio*.) In Figure 13.2A there are two side-lobes, at 120° and at 240°, which are about 24 dB below the peak response at 0°. Since interference can come from any

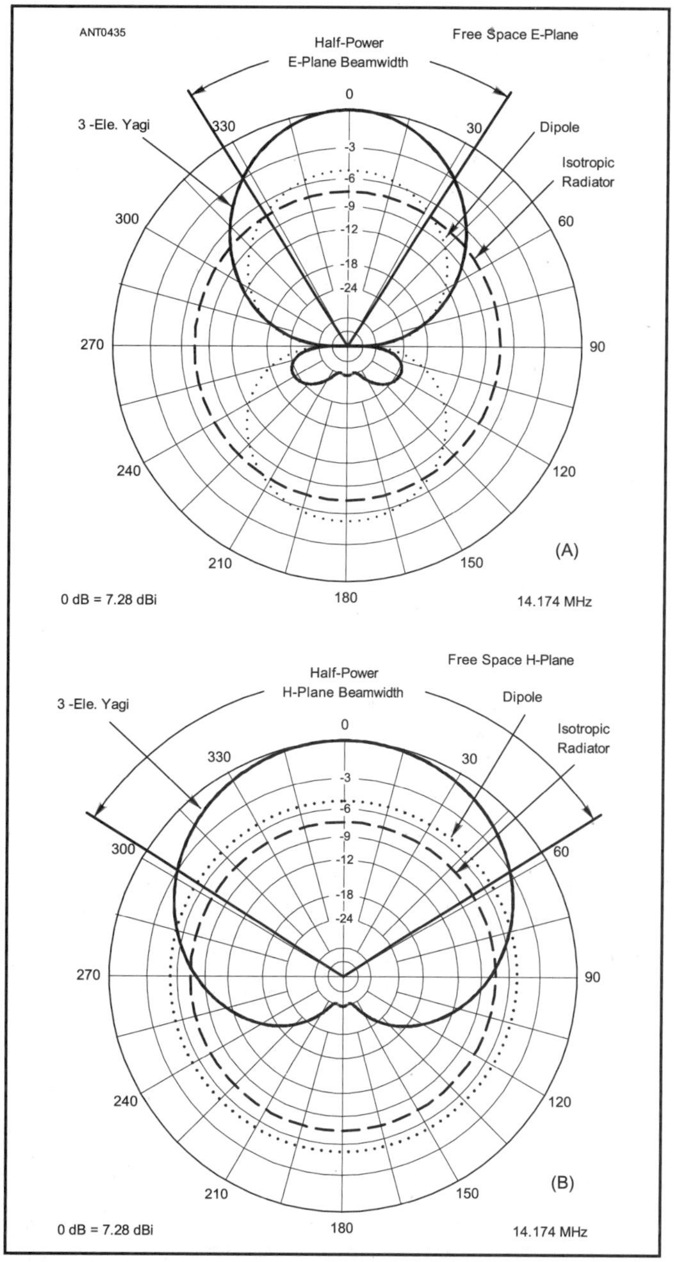

Figure 13.2 — E-Plane (electric field) and H-Plane (magnetic field) response patterns for 3-element 20 meter Yagi in free space. At A the E-Plane pattern for a typical 3-element Yagi is compared with a dipole and an isotropic radiator. At B the H-Plane patterns are compared for the same antennas. The Yagi has an E-Plane half-power beamwidth of 66°, and an H-Plane half-power beamwidth of about 120°. The Yagi has 7.28 dBi (5.13 dBd) of gain. The front-to-back ratio, which compares the response at 0° and at 180°, is about 35 dB for this Yagi. The front-to-rear ratio, which compares the response at 0° to the largest lobe in the rearward 180° arc behind the antenna, is 24 dB, due to the lobes at 120° and 240°.

direction, not only directly off the back of an antenna, these kinds of sidelobes limit the ability to discriminate against rearward signals. The term *worst-case front-to-rear ratio* is used to describe the worst-case rearward lobe in the 180°-wide sector behind the antenna's main lobe. In this case, the worst-case front-to-rear ratio is 24 dB.

In the rest of this chapter the worst-case front-to-rear ratio will be used as a performance parameter, and will be abbreviated as "F/R." For a dipole or an isotropic radiator, Figure 13.2A demonstrates that F/R is 0 dB. Figure 13.2B depicts the H-field response for the same 3-element Yagi in free space, again compared to a dipole and an isotropic radiator in free space. Unlike the E-field pattern, the H-field pattern for a Yagi does not have a null at 90°, directly over the top of the Yagi. For this 3-element design, the H-field half-power beamwidth is approximately 120°.

Figure 13.3 compares the azimuth and elevation patterns for a horizontally polarized 6-element 14-MHz Yagi with a 60-foot boom mounted one wavelength over ground to a dipole at the same height. As with any horizontally polarized antenna, the height above ground is the main factor determining the peaks and nulls in the elevation pattern of each antenna. Figure 13.3A shows the E-field pattern, which has now been labeled as the azimuth pattern. This antenna has a half-power azimuthal beamwidth of about 50°, and at an elevation angle of 12° it exhibits a forward gain of 16.02 dBi, including about 5 dB of ground reflection gain over relatively poor ground, with a dielectric constant of 13 and conductivity of 5 mS/m. In free space this Yagi has a gain of 10.97 dBi.

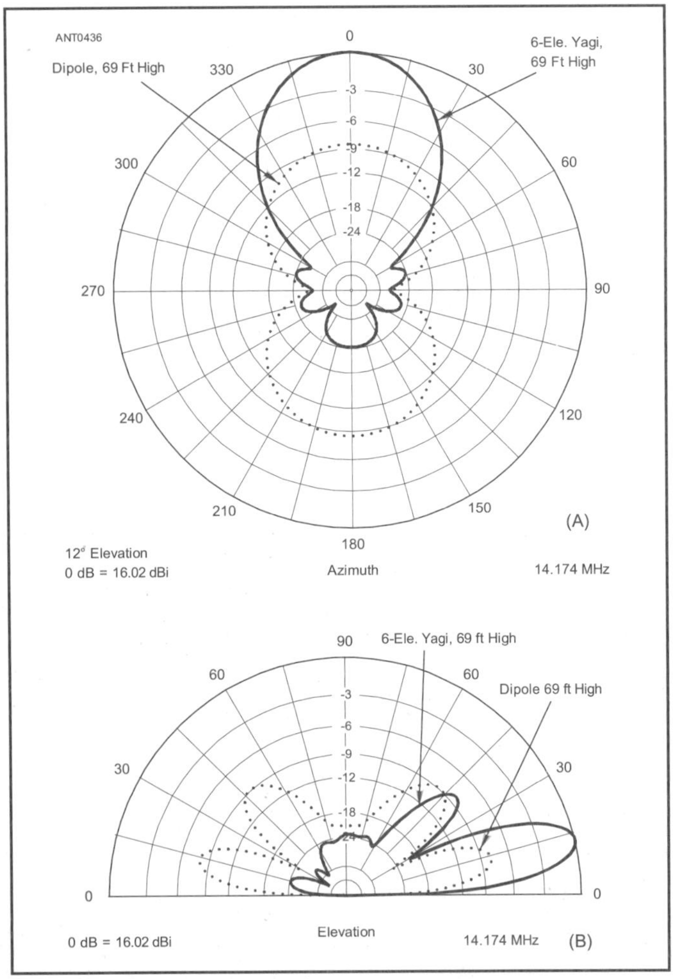

Figure 13.3 — Azimuth pattern for 6-element 20 meter Yagi on 60-foot long boom, mounted 60 feet over ground. At A, the azimuth pattern at 12° elevation angle is shown, compared to a dipole at the same height. Peak gain of the Yagi is 16.04 dBi, or just over 8 dB compared to the dipole. At B, the elevation pattern for the same two antennas is shown. Note that the peak elevation pattern of the Yagi is compressed slightly lower compared to the dipole, even though they are both at the same height over ground. This is most noticeable for the Yagi's second lobe, which peaks at about 40°, while the dipole's second lobe peaks at about 48°. This is due to the greater free-space directionality of the Yagi at higher angles.

The H-field elevation response of the 6-element Yagi has a half-power beamwidth of about 60° in free space, but as shown in Figure 13.3B, the first lobe (centered at 12° in elevation) has a half-power beamwidth of only 13° when the antenna is mounted one wavelength over ground. The dipole at the same height has a very slightly larger first-lobe half-power elevation beamwidth of 14°, since its free-space H-field response is omnidirectional.

Note that the free-space H-field directivity of the Yagi suppresses its second lobe over ground (at an elevation angle of about 40°) to 8 dBi, while the dipole's response at its second lobe peak (at about 48°) is at a level of 9 dBi.

The shape of the azimuthal pattern for a Yagi operated over real ground will change slightly as the Yagi is placed closer and closer to ground. Generally, however, the azimuth pattern doesn't depart significantly from the free-space pattern until the antenna is less than 0.5 λ high. This is just over 17 feet high at 28.4 MHz and just below 35 feet at 14.2 MHz — heights that are not difficult to achieve for most amateurs. Some advanced modeling programs can optimize Yagis at the exact installation height.

13.2.3 Feed Point Impedance and SWR

The impedance at the feed point of the driven element in a Yagi is affected not only by the tuning of the driven element itself, but also by the spacing and tuning of nearby parasitic elements, and to a lesser extent by the presence of ground. In some designs that have been tuned solely for maximum gain, the driven-element impedance can fall to very low levels, sometimes less than 5 Ω. This can lead to excessive losses due to conductor resistance, especially at VHF and UHF. In a Yagi that has been optimized solely for gain, conductor losses are usually compounded by large excursions in impedance levels with relatively small changes in frequency. The SWR can thus change dramatically over a band and can create additional losses in the feed line. **Figure 13.4** illustrates the SWR over the 28 to 28.8 MHz portion of the 10 meter amateur band for a 5-element Yagi on a 24-foot boom, tuned for maximum forward gain at a spot frequency of 28.4 MHz. Its SWR curve is contrasted to that of a Yagi designed for a good compromise of gain, SWR and F/R.

Even professional antenna designers have difficulty accurately measuring forward gain. On the other hand, SWR can easily be measured by professional and amateur alike. Few manufacturers would want to advertise an antenna with the narrow-band SWR curve shown in Figure 13.4!

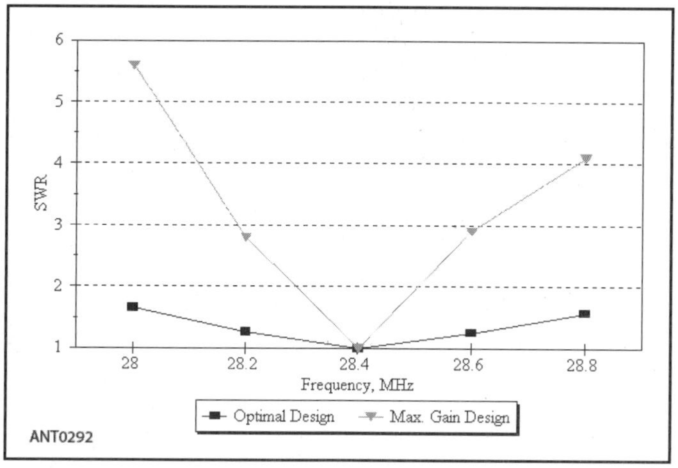

Figure 13.4 — SWR over the 28.0 to 28.8 MHz portion of the 10 meter band for two different 3-element Yagi designs. One is designed strictly for maximum gain, while the second is optimized for F/R pattern and SWR over the frequency band. A Yagi designed only for maximum gain usually suffers from a very narrow SWR bandwidth.

DIRECT FEED YAGIS

By carefully adjusting the position and tuning of the Yagi elements by using modeling software it is possible to create an antenna design for which the feed point impedance is close to 50 Ω and can be fed directly with coaxial cable. The tradeoff is usually a small amount of gain, but such *direct feed* designs are becoming increasingly common with no compromise in performance. Direct feed designs are somewhat more complex mechanically as the driven element must be insulated from the boom. This includes designs that require beta or hairpin matching.

CHOKE BALUNS

The use of a choke or current balun is good practice for any Yagi with a balanced driven element, regardless of feed point impedance. To be sure, many Yagi antennas provide adequate performance without a balun at the feed point but obtaining top performance from the antenna requires decoupling or isolation of the feed line from the antenna. If a choke balun is *not* used, interaction between the outer surface of the coaxial cable shield and the antenna can affect feed point impedance and result in significant common-mode current on the shield. Re-radiation from the common-mode current can fill in the radiation pattern nulls, degrading front-to-side and front-to-back performance. Common-mode current on the feed line can also result in RF-related problems in the station. (See "Common-Mode Transmission Line Currents" in the **Transmission Line System Techniques** chapter.)

13.3 Monoband Yagi Performance Optimization

13.3.1 Yagi Design Goals

The previous section discussing driven-element impedance and SWR hinted at possible design trade-offs among gain, pattern and SWR, especially when each parameter is considered over a frequency band rather than at a spot frequency. Trade-offs in Yagi design parameters can be a matter of personal taste and operating style. For example, one operator might exclusively operate the CW portions of the HF bands, while another might only be interested in the Phone portions. Another operator may want a good pattern in order to discriminate against signals coming from a particular direction; someone else may want the most forward gain possible, and may not care about responses in other directions.

There are only a few variables available to adjust when one is designing a Yagi to meet certain design goals. The variables are:

1) The physical length of the boom
2) The number of elements on the boom
3) The spacing of each element along the boom
4) The tuning of each element
5) The type of matching network used to feed the array.

For elements that are created from telescoping tubing sections, the lengths of individual sections (called the taper schedule) affects antenna performance as well. Taper schedule is usually varied in order to provide mechanical strength and is not considered a primary electrical design variable.

Extensive computer modeling of Yagis indicates that the parameter that must be compromised most to achieve wide bandwidths for front-to-rear ratio and SWR is forward gain. However, not much gain must be sacrificed for good F/R and SWR coverage, especially on long-boom Yagis. Although 10 and 7-MHz Yagis are not rare, the HF bands from 14 to 30 MHz are where Yagis are most often found, mainly due to the mechanical difficulties involved with making sturdy antennas for lower frequencies. The highest HF band, 28.0 to 29.7 MHz, represents the largest percentage bandwidth of the upper HF bands, at almost 6%. It is difficult to try to optimize in one design the main performance parameters of gain, worst-case F/R ratio and SWR over this large a band. Many commercial designs thus split up their 10 meter designs into antennas covering one of two bands: 28.0 to 28.8 MHz, and 28.8 to 29.7 MHz. For the amateur bands below 10 meters, optimal designs that cover the entire band are more easily achieved.

The performance requirements for Yagis used at VHF and UHF are similar to those of HF Yagis but place more emphasis on reduction of side lobes due to the importance of lowering received noise above 30 MHz. In addition, there are differences in feed point matching and considerations of losses are handled differently. These topics are addressed in the **VHF, UHF, and Microwave Antenna Systems** chapter. The remainder of this chapter will focus on HF designs unless specifically noted otherwise.

13.3.2 Gain and Boom Length

As pointed out earlier, the gain of a Yagi is largely a function of the length of the boom. As the boom is made longer, the maximum gain potential rises. For a given boom length, the number of elements populating that boom can be varied, while still maintaining the antenna's gain, provided of course that the elements are tuned properly. In general, putting more elements on a boom gives the designer added flexibility to achieve desired design goals, especially to broaden the response across a frequency band.

Figure 13.5A is an example illustrating gain versus frequency for three different types of 3-element Yagis on 8-foot booms. The three antennas were designed for the lower end of the 10 meter band, 28.0 to 28.8 MHz, based on the following different design goals:

Antenna 1: Maximum mid-band gain, regardless of F/R or SWR across the band

Antenna 2: SWR less than 2:1 over the frequency band; best compromise gain, with no special consideration for F/R over the band.

Antenna 3: "Optimal" case: F/R greater than 20 dB, SWR less than 2:1 over the frequency band; best compromise gain.

Figure 13.5B shows the F/R over the frequency band for these three designs, and Figure 13.5C shows the SWR curves over the frequency band. Antenna 1, the design that strives strictly for maximum gain, has a poor SWR response over the band, as might be expected after the previous section discussing SWR. The SWR is 10:1 at 28.8 MHz and rises to 22:1 at 29 MHz. At 28 MHz, at the low end of the band, the SWR of the maximum-gain design is more than 6:1. Clearly, designing for maximum gain alone produces an unacceptable design in terms of SWR bandwidth. The F/R for Antenna 1 reaches a high point of about 20 dB at the low-frequency end of the band, but falls to only 3 dB at the high-frequency end.

Antenna 2, designed for the best compromise of gain while the SWR across the band is held to less than 2:1, achieves this goal, but at an average gain sacrifice of 0.7 dB compared to the maximum gain case. The F/R for this design is just under 15 dB over the band. This design is fairly typical of many amateur Yagi designs before the advent of computer modeling and optimization programs. SWR can easily be measured, and experimental optimization for forward gain is a fairly straightforward procedure. By contrast, overall pattern optimization is not a trivial thing to achieve experimentally, particularly for antennas with more than four or five elements.

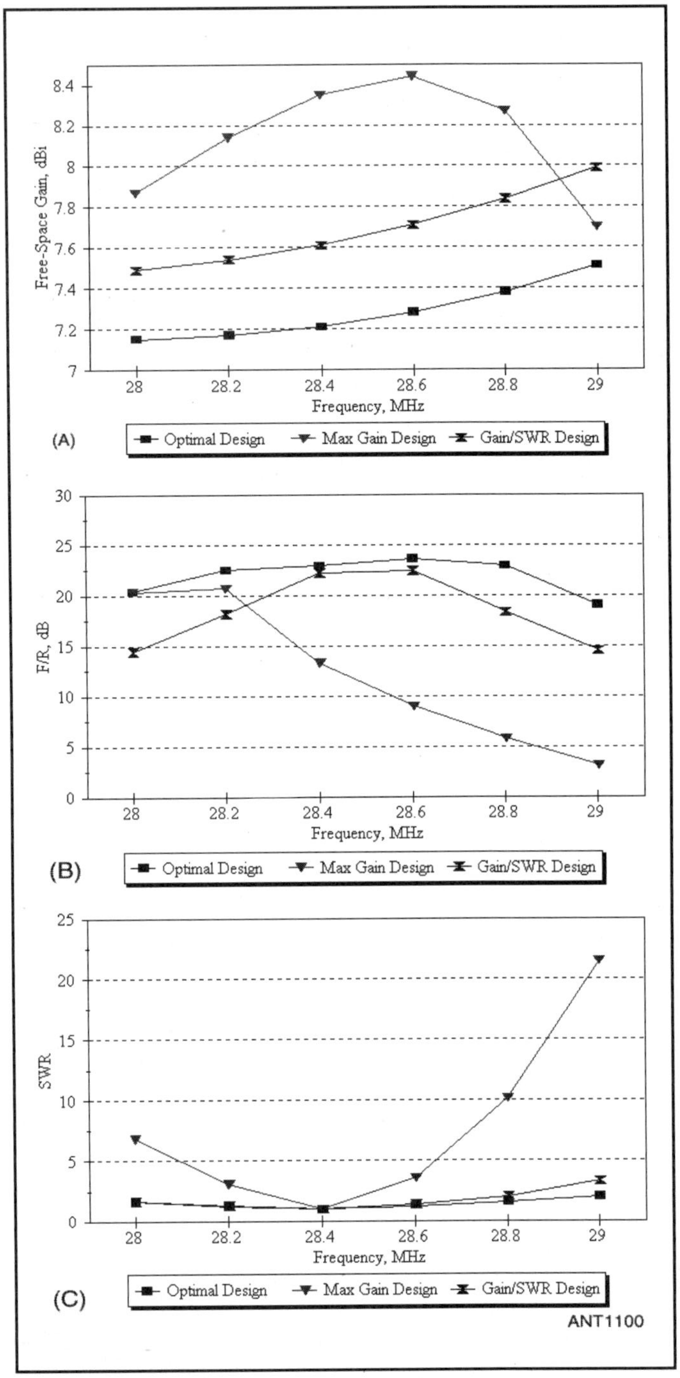

Figure 13.5 — Comparisons of three different 3-element 10 meter Yagi designs using 8-foot booms. At A, gain comparisons are shown. The Yagi designed for the best compromise of gain and SWR sacrifices an average of about 0.5 dB compared to the antenna designed for maximum gain. The Yagi designed for optimal F/R, gain and SWR sacrifices an average of 1.0 dB compared to the maximum-gain case, and about 0.4 dB compared to the compromise gain and SWR case. At B, the front-to-rear ratio is shown for the three different designs. The antenna designed for optimal combination of gain, F/R and SWR maintains a F/R higher than 20 dB across the entire frequency range, while the antenna designed strictly for gain has a F/R of 3 dB at the high end of the band. At C, the three antenna designs are compared for SWR bandwidth. At the high end of the band, the antenna designed strictly for gain has a very high SWR.

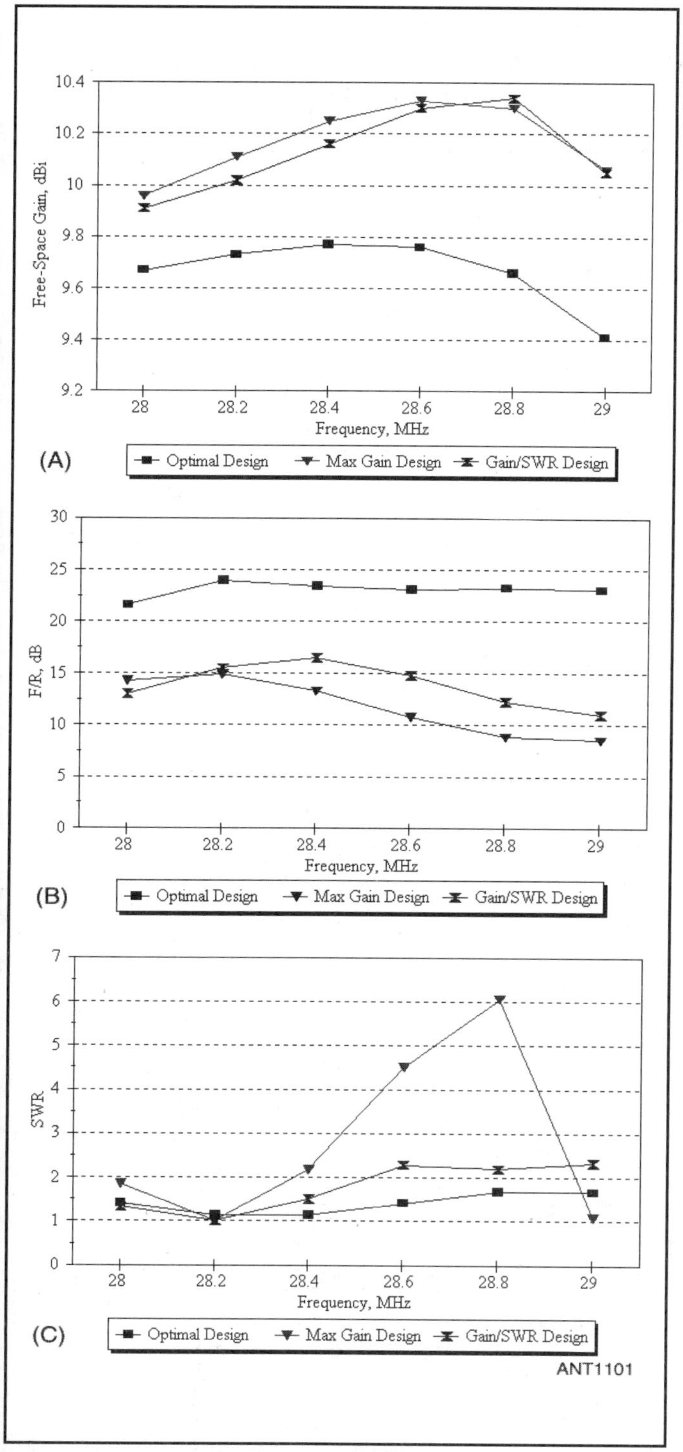

Figure 13.6 — Comparisons of three different designs for 5-element 10 meter Yagis on 20-foot booms. At A, the gain of three different 5-element 10 meter Yagi designs are graphed. The difference in gain between the three antennas narrows because the elements can be stagger-tuned to spread the response out better over the desired frequency band. The average gain reduction for the fully optimized antenna design is about 0.5 dB. At B, the optimal antenna displays better than 22 dB F/R over the band, while the Yagi designed for gain and SWR displays on average 10 dB less F/R throughout the band. At C, the SWR bandwidth is compared for the three Yagis. The antenna designed strictly for forward gain has a poor SWR bandwidth and a high peak SWR of 6:1 at 28.8 MHz.

Antenna 3, designed for an optimum combination of F/R, SWR and gain, compromises forward gain an average of 1.0 dB compared to the maximum gain case, and about 0.4 dB compared to the compromise gain/SWR case. It achieves its design objectives of more than 20 dB F/R over the 28.0 to 28.8 MHz portion of the band, with an SWR less than 2:1 over that range.

Figure 13.6A shows the free-space gain versus frequency for the same three types of designs, but for a bigger 5-element 10 meter Yagi on a 20-foot boom. Figure 13.6B shows the variation in F/R, and Figure 13.6C shows the SWR curves versus frequency. Once again, the design that concentrates solely on maximum gain has a poor SWR curve over the band, reaching just over 6:1 toward the high end of the band. The difference in gain between the maximum gain case and the optimum design case has narrowed for this size of boom to an average of under 0.5 dB. This comes about because the designer has access to more variables in a 5-element design than in a 3-element design, and can stagger-tune the various elements to spread the response out over the whole band.

Figure 13.7A, **B** and **C** show the same three types of designs, but for a 6-element Yagi on a 36-foot boom. The SWR bandwidth of the antenna designed for maximum gain has improved compared to the previous two shorter-boom examples, but the SWR still rises to more than 4:1 at 28.8 MHz, while the F/R ratio is pretty constant over the band, at a mediocre 11 dB average level. While the antenna designed for gain and SWR does hold the SWR below 2:1 over the band, it also has the same mediocre level of F/R performance as does the maximum-gain design.

The optimized 36-foot boom antenna achieves an excellent F/R of more than 22 dB over the whole 28.0 to 28.8 MHz band. Again, the availability of more elements and more space on the 36-foot long boom gives the designer more flexibility in broadbanding the response over the whole band, while sacrificing only 0.3 dB of gain compared to the maximum-gain design.

Figure 13.8A, **B**, and **C** show the same three types of 10 meter designs, but now for a 60-foot boom, populated with eight elements. With eight elements and a very long boom on which to space them out, the antenna designed solely for maximum gain can achieve a much better SWR response across the band, although the SWR does rise to more than 7:1 at the very high end of the band. The SWR remains less than 2:1 from 28.0 to 28.7 MHz, much better than for shorter-boom, maximum-gain designs. The worst-case F/R ratio is never better than 19 dB, however, and remains around 10 dB over much of the band. The antenna designed for the best compromise gain and SWR loses only about 0.1 dB of gain compared to the maximum-gain design, but does little better in terms of F/R across the band.

Contrasted to these two designs, the antenna optimized for F/R, SWR and gain has an outstanding pattern, exhibiting an F/R of more than 24 dB across the entire band, while keeping the SWR below 2:1 from 28.0 to 28.9 MHz. It must sacrifice an average of only 0.4 dB compared to the maximum gain

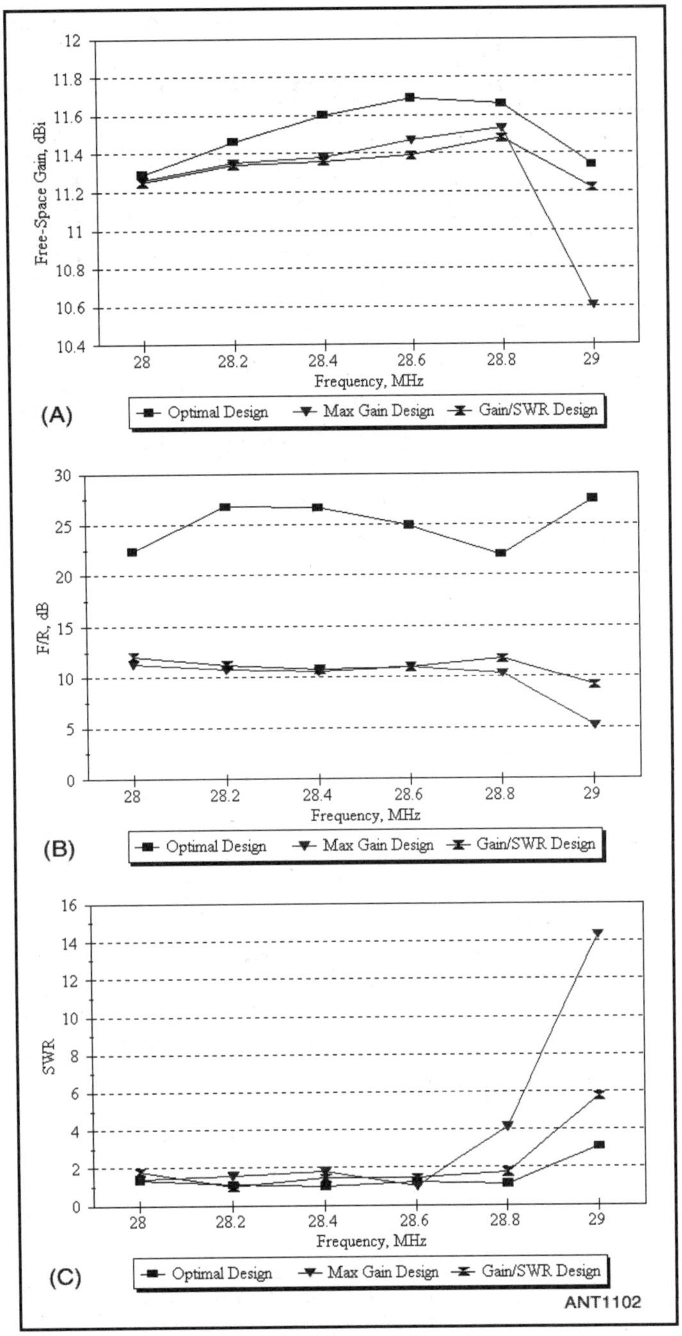

Figure 13.7 — Comparisons of three different 6-element 10 meter Yagi designs on 36-foot booms. At A, gain is shown over the band. With more elements and a longer boom, the tuning can be staggered even more to make the antenna gain more uniform over the band. This narrows the gain differential between the antenna designed strictly for maximum gain and the antenna designed for an optimal combination of F/R, SWR and gain. The average difference in gain is about 0.2 dB throughout the band. At B, the F/R performance over the band is shown for the three antenna designs. The antenna designed for optimal performance maintains an average of almost 15 dB better F/R over the whole band compared to the other designs. At C, the SWR bandwidth is compared. Again, the antenna designed strictly for maximum gain exhibits a high SWR of 4:1 at 28.8 MHz, and rises to more than 14:1 at 29.0 MHz.

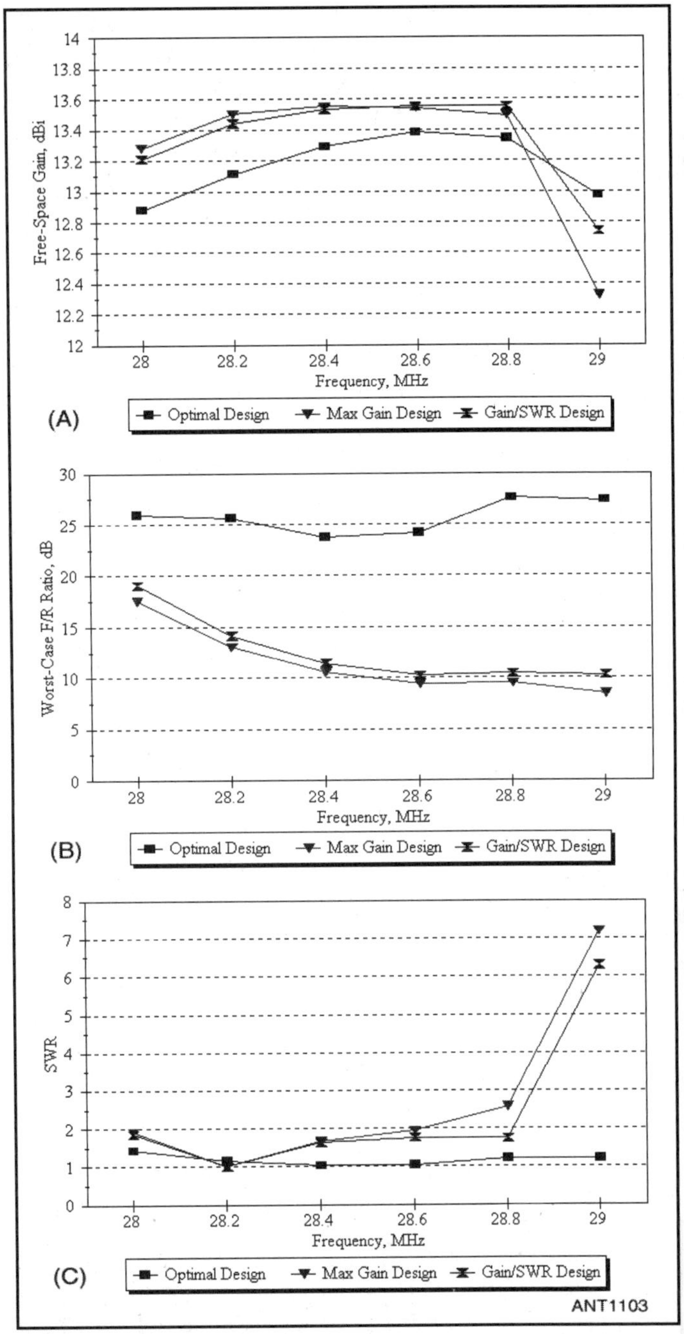

Figure 13.8 — Comparisons of three different 8-element 10 meter Yagi designs using 60-foot booms. At A, gain is shown over the frequency band. With even more freedom to stagger-tune elements and a very long boom on which to place them, the average antenna gain differential over the band is now less than 0.2 dB between the three design cases. At B, an excellent 24 dB F/R for the optimal design is maintained over the whole band, compared to the average of about 12 dB for the other two designs. At C, the SWR differential over the band is narrowed between the three designs, again because there are more variables available to broaden the bandwidth.

design at the low end of the band, and actually has more gain than the maximum gain and gain/SWR designs at the high-frequency end of the band.

The conclusion drawn from these and many other detailed comparisons is that designing strictly for maximum mid-band gain yields an inferior design when the antenna is examined over an entire frequency band, especially in terms of SWR. Designing a Yagi for both gain and SWR will yield antennas that have mediocre rearward patterns, but that lose relatively little gain compared to the maximum gain case, at least for designs with more than three elements.

However, designing a Yagi for an optimal combination of F/R, SWR and gain results in a loss of gain less than 0.5 dB compared to designs designed only for gain and SWR. **Figure 13.9** summarizes the forward gain achieved for the three different design types versus boom length, as expressed in wavelength.

Except for the 2-element designs, the Yagis described in the rest of this chapter have the following design goals over a desired frequency band:

1) Front-to-rear ratio over the frequency band of more than 20 dB

2) SWR over the frequency band less than 2:1

3) Maximum gain consistent with points 1 and 2 above

Just for fun and to illustrate what an imaginative antenna designer can do with modeling software, **Figure 13.10** shows the gain versus boom length for theoretical 20 meter Yagis that have been designed to meet the three design goals above. The 31-element design for 14 MHz would be wondrous to behold. Sadly, it is unlikely that anyone will build one, considering that the boom would be 724 feet long! However, such a design *does* become practical when scaled to 432 MHz. In fact, a K1FO 22-element and a K1FO 31-element Yagi described in the **VHF, UHF, and Microwave Antenna Systems** chapter are the prototypes for the theoretical 14-MHz long-boom designs.

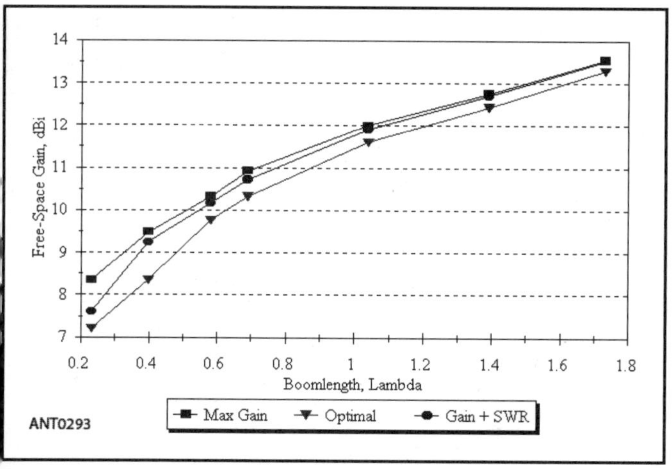

Figure 13.9 — Gain versus boom length for three different 10 meter design goals. The goals are: (1) designed for maximum gain across band, (2) designed for a compromise of gain and SWR, and (3) designed for optimal F/R, SWR and gain across 28.0 to 28.8 MHz portion of 10 meter band. The gain difference is less than 0.5 dB for booms longer than approximately 0.5 λ.

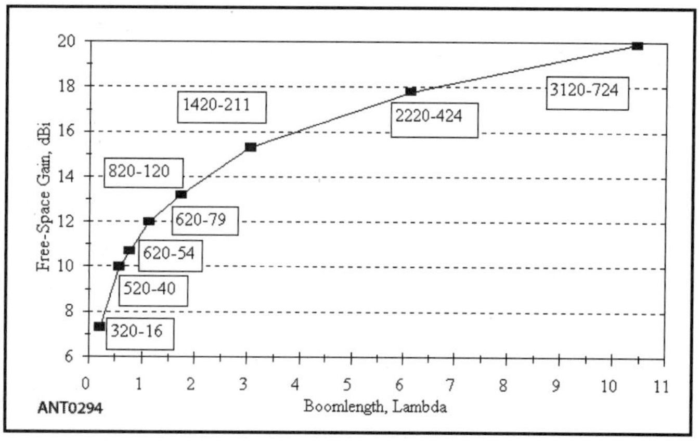

Figure 13.10 — Theoretical gain versus boom length for 20 meter Yagis designed for optimal combination of F/R, SWR and gain across the entire 14.0 to 14.35 MHz band. The theoretical gain approaches 20 dBi for a gigantic 724-foot boom, populated with 31 elements. Such a design on 20 meters is not too practical, of course, but can readily be achieved on a 24-foot boom on 432 MHz.

13.3.3 Optimized Designs and Element Spacing

TWO-ELEMENT YAGIS

Many hams consider a 2-element Yagi to give "the most bang for the buck" among various Yagi designs, particularly for portable operations such as ARRL Field Day. A 2-element Yagi has about 4 dB of gain over a simple dipole (sometimes jokingly called a "one-element Yagi") and gives a modest F/R of about 10 dB to help with rejection of interference on receive. By comparison, going from a 2-element to a 3-element Yagi increases the boom length by about 50% and adds another element, a 50% increase in the number of elements — for a gain increase of about 1 dB and another 10 dB in F/R.

ELEMENT SPACING IN LARGER YAGIS

One of the more interesting results of computer modeling and optimization of high-performance Yagis with four or more elements is that a distinct pattern in the element spacings along the boom shows up consistently. This pattern is relatively independent of boom length, once the boom is longer than about 0.3 λ.

The reflector, driven element and first director of these optimal designs are typically bunched rather closely together, occupying together only about 0.15 to 0.20 λ of the boom. This pattern contrasts sharply with older designs, where the amount of boom taken up by the reflector, driven element and first director was typically more than 0.3 λ. **Figure 13.11** shows the element spacings for an optimized 6-element, 36-foot boom, 10 meter design, compared to a W2PV 6-element design with constant spacing of 0.15 λ between all elements.

A problem arises with such a bunching of elements toward the reflector end of the boom — the wind loading of the antenna is not equal along the boom. Unless properly compensated, such new-generation Yagis will act like wind vanes, punishing and often breaking, the rotators trying to turn, or

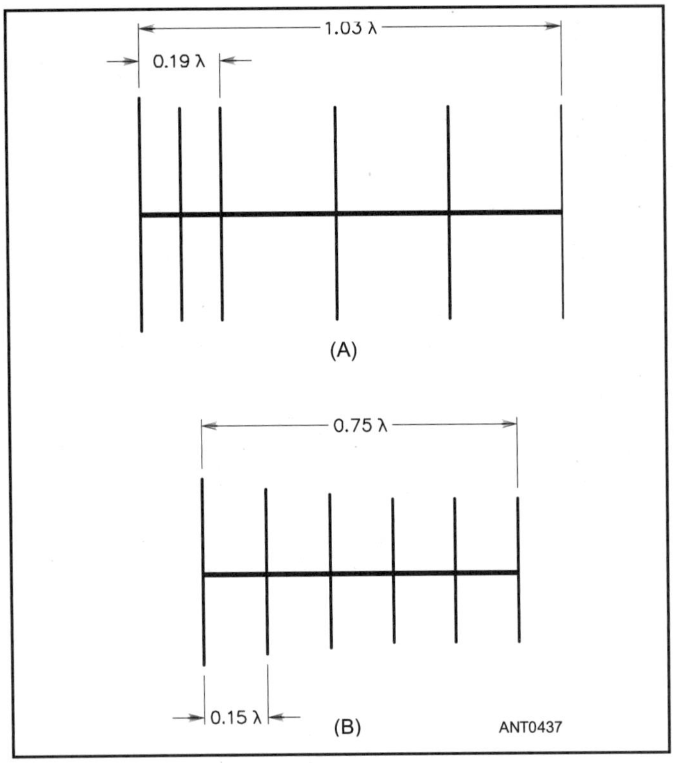

Figure 13.11 — Tapering spacing versus constant element spacing. At A, illustration of how the spacing of the reflector, driven element and first director (over the first 0.19 λ of the boom) of an optimally designed Yagi is bunched together compared to the Yagi at B, which uses constant 0.15 λ spacing between all elements. The optimally designed antenna has more than 22 dB F/R and an SWR less than 1.5:1 over the frequency band 28.0 to 28.8 MHz.

hold, them in the wind. One successful solution to windvaning has been to employ "dummy elements" made of PVC pipe. These non-conducting elements — called *torque compensators* — are placed on the boom close to the last director so the wind load is equalized at the mast-to-boom bracket. Flat plates can also be installed on the boom to oppose the turning force from the elements.

Along with an unbalanced wind load, the weight balance point is likely to be different than wind load balance point. The solution is generally to place a small amount of lead or iron inside one end of the boom in order to balance the antenna weight.

Despite the relatively close spacing of the reflector, driven element and first director, modern optimal Yagi designs are not overly sensitive to small changes in either element length or spacing. In fact, these antennas can be constructed from design tables without excessive concern about close dimensional tolerances. In the HF range up to 30 MHz, building the antennas to the nearest ⅛-inch results in performance remarkably consistent with the computations, without any "tweaking" or fine-tuning when the Yagi is on the tower.

13.3.4 Element Tuning

Element tuning (or *self-impedance*) is a complex function of the effective electrical length of each element and the effective diameter of the element. In turn, the effective length and diameter of each element is related to the taper schedule (if telescoping aluminum tubing is used, the most common method of construction), the length of each telescoping section, the type and size of mounting bracket used to secure the element to or through the boom, and the size of the Yagi boom itself. Note especially that Yagis constructed using wire elements will perform very differently compared to the same antenna constructed with elements made of telescoping aluminum tubing.

The process by which a modern Yagi is designed usually starts out with the selection of the longest boom possible for a given installation. A suitable number of elements of a given taper schedule are then placed on this boom, and the gain, pattern and SWR are calculated over the entire frequency band of interest to the designer. Once an electrical design is chosen, the designer must then ensure the mechanical integrity of the antenna design. This involves verifying the integrity of the boom and each element in the face of the wind and ice loading expected for a particular location. The chapter **Antenna Materials and Construction** discusses the details of tapered telescoping aluminum elements for the upper HF bands. In addition, the ARRL book *Physical Design of Yagi Antennas*, by Dave Leeson, W6NL, describes the mechanical design process for all portions of a Yagi antenna very thoroughly, and is highly recommended for serious Yagi builders. (This book is now out of print.)

13.3.5 The Optimized Wideband Antenna (OWA) Yagi

Jim Breakall, WA3FET, developed the Optimized Wideband Antenna (OWA) in the early 1990s. The goal was to increase the bandwidth of monopole and dipole antennas by placing one or more parasitic elements close to the driven dipole or monopole, in a manner similar to the well-known sleeve antenna. (See the "Coupled Resonator" section of the **Multiband HF Antennas** chapter.) The lengths and spacings of these extra parasitic wires are then optimized to minimize peak SWR in a range of frequencies over a selected band. By applying this method to Yagis, it is possible to cover complete amateur bands with very low SWR, high peak gain, and high front-to-back ratio with this technique.

In a Yagi, the goals of the OWA method are to achieve a direct 50 Ω feed point impedance with a minimum peak SWR, maximize minimum gain, and maximize minimum front-to-back ratio in some three-dimensional angular region of the spatial pattern. The optimization is applied to the entire range of frequency desired. Various weightings can be applied to the goals if one of them is considered more important than the others. For example, lowering peak SWR in a band might be more important than the gain or front-to-back ratio. By tailoring the weighting factors on each of the goals, one can produce many optimum designs with the best characteristics achievable over a desired frequency range band.

The OWA is implemented in a Yagi antenna by placing the first director very close (< .05 λ) in front of the driven element. Because the feed point impedance is raised to 50 Ω, a simple 1:1 current or choke balun is all that is needed to connect a coaxial feed line. This replaces the more complex gamma or beta matching systems used to match the low impedance of most Yagi antennas.

The OWA Yagi design gives up a slight amount of peak forward gain (typically less than 0.5 dB) from a conventional Yagi. This small degradation is impossible to notice in operation and is well worth the simplicity of the direct feed and the excellent bandwidth of the other characteristics.

A set of HF OWA Yagi designs from WA3FET and Justin Johnson, GØKSC, is presented later in this chapter. *EZNEC* models are included for each antenna. Additional discussion of the OWA and other methods of optimization may be found in the "High-Performance Yagi Antennas" section of the **VHF, UHF, and Microwave Antennas** chapter. See the **Antenna Modeling** chapter for a discussion of optimizer software.

13.4 Monoband Yagi Designs

The detailed Yagi design tables that follow are for two taper schedules for HF Yagis covering the 14 through 30-MHz amateur bands. These designs were developed by Dean Straw, N6BV, using the program *YW* (*Yagi for Windows*) that is included with this book's online material. The heavy-duty elements are designed to survive at least 120-mph winds without icing, or 85-mph winds with ¼-inch radial ice. The medium-duty elements are designed to survive winds greater than 80 mph, or 60-mph winds with ¼-inch radial ice.

For 10.1 MHz, the elements shown are capable of surviving 105-mph winds, or 93-mph winds with ¼-inch radial ice. For 7.1 MHz the elements shown can survive 93-mph winds, or 69-mph winds with ¼-inch radial ice. For these two lower frequency bands, the elements and the booms needed are very large and heavy. Mounting, turning and keeping such antennas in the air is not a trivial task.

Each element is mounted above the boom with a heavy rectangular aluminum plate by means of U-bolts with saddles, as shown in **Figure 13.12A**. Stauff clamps (**stauffusa.com**) can also be used to hold the elements on the boom as discussed in the VHF, UHF, and Microwave Antennas chapter. This method of element mounting is rugged and stable, and because the element is mounted away from the boom, the amount of element detuning due to the presence of the boom is minimal. The element dimensions given in each table already take into account any element detuning due to the boom-to-element mounting plate. For each element, the length of the tip determines the tuning, since the inner tubes are fixed in diameter and length.

The element-to-boom mounting plates are modeled as a short section of element equivalent to a cylinder with an effective diameter given for each antenna. These dimensions to simulate the effect of the mounting plate are incorporated in the model files for the *YW* modeling software with this book's online material.

The second column in each design table shows the spacing

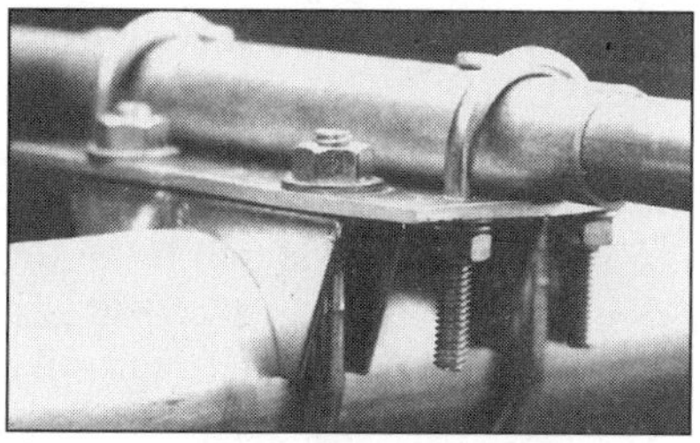

(A)

(B)

Figure 13.12 — Typical construction techniques for an HF Yagi. Photo A shows a typical element-to-boom clamp. U-bolts are used to hold the element to the plate and muffler clamps hold the plate to the boom. Photo B shows a hairpin match on a driven element insulated from a mounting plate that is attached to the boom with muffler clamps and saddles. Outdoor-rated gray PVC conduit sleeves insulate the element from the mounting plate. U-bolts hold the element on the plate. The feed line is connected to the two screws to which the hairpin inductor is attached. Note that the hairpin inductor's center point is attached to the boom at an electrically neutral point. All mounting hardware should be galvanized or stainless steel. The latter requires the use of an anti-seize compound to prevent thread galling.

of each element relative to the next element in line on the boom, starting at the reflector, which itself is defined as being at the 0.000-inch reference point on the boom. The boom for antennas less than 30 feet long can be constructed of 2-inch OD tubing with 0.065-inch wall thickness. Designs larger than 30 feet long should use 3-inch OD heavy-wall tubing for the boom. Because each boom has extra space at each end, the reflector is actually placed 3 inches from the end of the boom. For example, in the 310-08H.YW design (a 10 meter Yagi with 3 elements on an 8-foot boom), the driven element is placed 36 inches ahead of the reflector, and the director is placed 54 inches ahead of the driven element.

The next columns give the lengths for the variable tips for the heavy-duty and then the medium-duty elements. In the example above for the 310-08H.YW Yagi, the heavy-duty reflector tip, made out of ½-inch OD tubing, sticks out 66.750 inches from the ⅝-inch OD tubing. Note that each telescoping piece of tubing overlaps 3 inches inside the piece into which it fits, so the overall length of ⅛-inch OD tubing is 69.750 inches long for the reflector. The medium-duty reflector tip has 71.875 inches protruding from the ⅝-inch OD tube, and is 74.875 inches long overall. As previously stated, the dimensions are not extremely critical, although measurement accuracy to ⅛ inch is desirable.

The last row in each variable tip column shows the length of one-half of the "dummy element" torque compensator used to correct for uneven wind loading along the boom. This compensator is made from 2.5 inches OD PVC water pipe mounted to an element-to-boom plate like those used for each element. The compensator is mounted 12 inches behind the last director, the first director in the case of the 3-element 310-08H.YW antenna. Note that the heavy-duty elements require a correspondingly longer torque compensator than do the medium-duty elements.

HALF ELEMENTS

Each design shows the dimensions for *one-half* of each element, mounted on *one side* of the boom. The other half of each element is symmetrical, mounted on the other side of the boom. The use of a tubing sleeve inside the center portion of the element is recommended, so that the element is not crushed by the mounting U-bolts. Unless otherwise noted, each section of tubing is made of 6061-T6 aluminum tubing, with a 0.058-inch wall thickness. This wall thickness ensures that the next standard size of tubing can telescope with it. Each telescoping section is inserted 3 inches into the larger tubing, and is secured by one of the methods shown in the **Antenna Materials and Construction** chapter, which also includes generic half-element designs rated for specific wind- and ice-loading conditions.

DIRECT FEED HAIRPIN MATCH

Each antenna is designed with a driven-element length appropriate for a hairpin or *beta match* network. The driven element's length may require slight readjustment for best match, particularly if a different matching network is used. *Do not change* either the lengths or the telescoping tubing schedule of the parasitic elements — they have been optimized for best performance and will not be affected by tuning of the driven element!

Note that the center of the hairpin is connected to the boom using a grounding lug. The center of the hairpin inductor is electrically neutral and may be connected to the boom for dc grounding and mechanical stability. The hairpin match requires a choke balun, described in the **Transmission Line System Techniques** chapter. Hairpin dimensions are included at the end of each *YW* file and in the following tables.

Figure 13.12B shows the driven element for a 2-element 17 meter Yagi built by Chuck Hutchinson, K8CH. The aluminum tubing on each side of the boom is 1-inch OD, and the two pieces were mechanically joined together with a ¾-inch OD fiberglass rod insulator. Electrical tape over the insulator protects the fiberglass from UV. Three-inch lengths of 1-inch UV-resistant PVC conduit, split lengthwise, make the grey outer insulators for the driven element. Aluminum plates were obtained commercially. Stainless steel saddle clamps ensure that the elements don't rotate on the boom in heavy winds. Bolts are used to pin the center fiberglass rod to the aluminum tubing, while also providing an electrical connection for the #12 AWG hairpin inductor wire and the feed line coax.

13.4.1 10 Meter Yagis

Figure 13.13 describes the electrical performance of eight optimized 10 meter Yagis with boom lengths between 6 to 60 feet. The end of each boom includes 3 inches of space for the reflector and last-director (or driven element for the 2-element designs) mounting plates. Figure 13.13A shows the free-space gain versus frequency for each antenna; Figure 13.13B shows the front-to-rear ratio, and Figure 13.13C shows the SWR versus frequency. Each antenna with three or more elements was designed to cover the lower half of the 10 meter band from 28.0 to 28.8 MHz, with SWR less than 2:1 and F/R better than 20 dB over that range.

Figure 13.13D shows the taper schedule for two types of 10 meter elements. The heavy-duty design can survive 125-mph winds with no icing, and 88-mph winds with ¼-inch of radial ice. The medium-duty design can handle 96-mph winds with no icing, and 68-mph winds with ¼-inch of radial ice.

The element-to-boom mounting plate for these Yagis is a 0.250-inch thick flat aluminum plate, 4 inches wide by 4 inches long. Each element except for the insulated driven element, is centered on the plate, held by two stainless-steel U-bolts with saddles. Another set of U-bolts with saddles is used to secure the mounting plate to the boom. The mounting plate has an effective diameter of 2.405 inches for the heavy-duty element and 2.310 inches for the medium-duty element. The equivalent length on each side of the boom is 2 inches.

13.4.2 12 Meter Yagis

Figure 13.14 describes the electrical performance of seven optimized 12 meter Yagis with boom lengths between 6 to 54 feet. The end of each boom includes 3 inches of space for the reflector and last director (or driven element) mounting plates. The narrow frequency range of the 12 meter band allows the performance to be optimized easily. Figure 13.14A shows the free-space gain versus frequency for each antenna; Figure 13.14B shows the front-to-rear ratio, and Figure 13.14C shows the SWR versus frequency. Each antenna with three or more elements was designed to cover the narrow 12 meter band from 24.89 to 24.99 MHz, with SWR less than 2:1 and F/R better than 20 dB over that range.

Figure 13.14D shows the taper schedule for two types of 12 meter elements. The heavy-duty design can survive 123-mph winds with no icing, and 87-mph winds with ¼ inch of radial ice. The medium-duty design can handle 85-mph winds with no icing, and 61-mph winds with ¼ inch of radial ice.

The element-to-boom mounting plate for these Yagis is a 0.375 inch thick flat aluminum plate, 5 inches wide by 6 inches long. The mounting plate has an effective diameter of 2.945 inches for the heavy-duty element, and 2.857 inches for the medium-duty element. The equivalent length on each side of the boom is 3 inches. As usual, the torque compensator is mounted 12 inches behind the last director.

13.4.3 15 Meter Yagis

Figure 13.15 describes the electrical performance of eight optimized 15 meter Yagis with boom lengths between 6 feet to a spectacular 80 feet. The end of each boom includes 3 inches of space for the reflector and last director (or driven element) mounting plates. Figure 13.15A shows the free-space gain versus frequency for each antenna; Figure 13.15B shows the worst-case front-to-rear ratio, and Figure 13.15C shows the SWR versus frequency. Each antenna with three or more elements was designed to cover the full 15 meter band from 21.000 to 21.450 MHz, with SWR less than 2:1 and F/R ratio better than 20 dB over that range.

Figure 13.15D shows the taper schedule for two types of 15 meter elements. The heavy-duty design can survive 124-mph winds with no icing, and 90-mph winds with ¼ inch of radial ice. The medium-duty design can handle 86-mph winds with no icing, and 61-mph winds with ¼ inch of radial ice.

The element-to-boom mounting plate for these Yagis is a 0.375-inch thick flat aluminum plate, 5 inches wide by 6 inches long. The mounting plate has an effective diameter of 3.0362 inches for the heavy-duty element, and 2.9447 inches for the medium-duty element. The equivalent length on each side of the boom is 3 inches. As usual, the torque compensator is mounted 12 inches behind the last director.

13.4.4 17 Meter Yagis

Figure 13.16 describes the electrical performance of six optimized 17 meter Yagis with boom lengths between 6 to a heroic 60 feet. As usual, the end of each boom includes 3 inches of space for the reflector and last director (or driven element) mounting plates. Figure 13.16A shows the free-space gain versus frequency for each antenna; Figure 13.16B shows the worst-case front-to-rear ratio, and Figure 13.16C shows the SWR versus frequency. Each antenna with three or more elements was designed to cover the narrow 17 meter band from 18.068 to 18.168 MHz, with SWR less than 2:1 and F/R ratio better than 20 dB over that range.

Figure 13.16D shows the taper schedule for two types of 17 meter elements. The heavy-duty design can survive 123-mph winds with no icing, and 83-mph winds with ¼-inch of radial ice. The medium-duty design can handle 83-mph winds with no icing, and 59-mph winds with ¼ inch of radial ice.

The element-to-boom mounting plate for these Yagis is a 0.375-inch thick flat aluminum plate, 6 inches wide by 8 inches long. The mounting plate has an effective diameter of 3.5122 inches for the heavy-duty element, and 3.3299 inches for the medium-duty element. The equivalent length on each side of the boom is 4 inches. As usual, the torque compensator is mounted 12 inches behind the last director.

13.4.5 20 Meter Yagis

Figure 13.17 describes the electrical performance of eight optimized 20 meter Yagis with boom lengths between 8 to a giant 80 feet. As usual, the end of each boom includes 3 inches of space for the reflector and last director (driven element) mounting plates. Figure 13.17A shows the free-space gain versus frequency for each antenna; Figure 13.17B shows the front-to-rear ratio, and Figure 13.17C shows the SWR versus frequency. Each antenna with three or more elements was designed to cover the complete 20 meter band from 14.000 to 14.350 MHz, with SWR less than 2:1 and F/R ratio better than 20 dB over that range.

Figure 13.17D shows the taper schedule for two types of 20 meter elements. The heavy-duty design can survive 122-mph winds with no icing, and 89-mph winds with ¼ inch of radial ice. The medium-duty design can handle 82-mph winds with no icing, and 60-mph winds with ¼ inch of radial ice.

The element-to-boom mounting plate for these Yagis is a 0.375-inch thick flat aluminum plate, 6 inches wide by 8 inches long. The mounting plate has an effective diameter of 3.7063 inches for the heavy-duty element, and 3.4194 inches for the medium-duty element. The equivalent length on each side of the boom is 4 inches. As usual, the torque compensator is mounted 12 inches behind the last director.

13.4.6 30 Meter Yagis

Figure 13.18 describes the electrical performance of three optimized 30 meter Yagis with boom lengths between 15 to 34 feet. Because of the size and weight of the elements alone for Yagis on this band, only 2-element and 3-element designs are described. The front-to-rear ratio requirement for the 2-element antenna is relaxed to be greater than 10 dB over the band from 10.100 to 10.150 MHz, while that for the 3-element designs is kept at greater than 20 dB over that frequency range.

As usual, the end of each boom includes 3 inches of space for the reflector and last director mounting plates. Figure 13.18A shows the free-space gain versus frequency for each antenna; Figure 13.18B shows the worst-case front-to-rear ratio, and Figure 13.18C shows the SWR versus frequency.

Figure 13.18D shows the taper schedule for the 30 meter elements. Note that the wall thickness of the first two sections of tubing is 0.083 inches, rather than 0.058 inches. This heavy-duty element design can survive 107-mph winds with no icing, and 93-mph winds with ¼ inch of radial ice.

The element-to-boom mounting plate for these Yagis is a 0.500-inch thick flat aluminum plate, 6 inches wide by 24 inches long. The mounting plate has an effective diameter of 4.684 inches. The equivalent length on each side of the boom is 12 inches. These designs require no torque compensator.

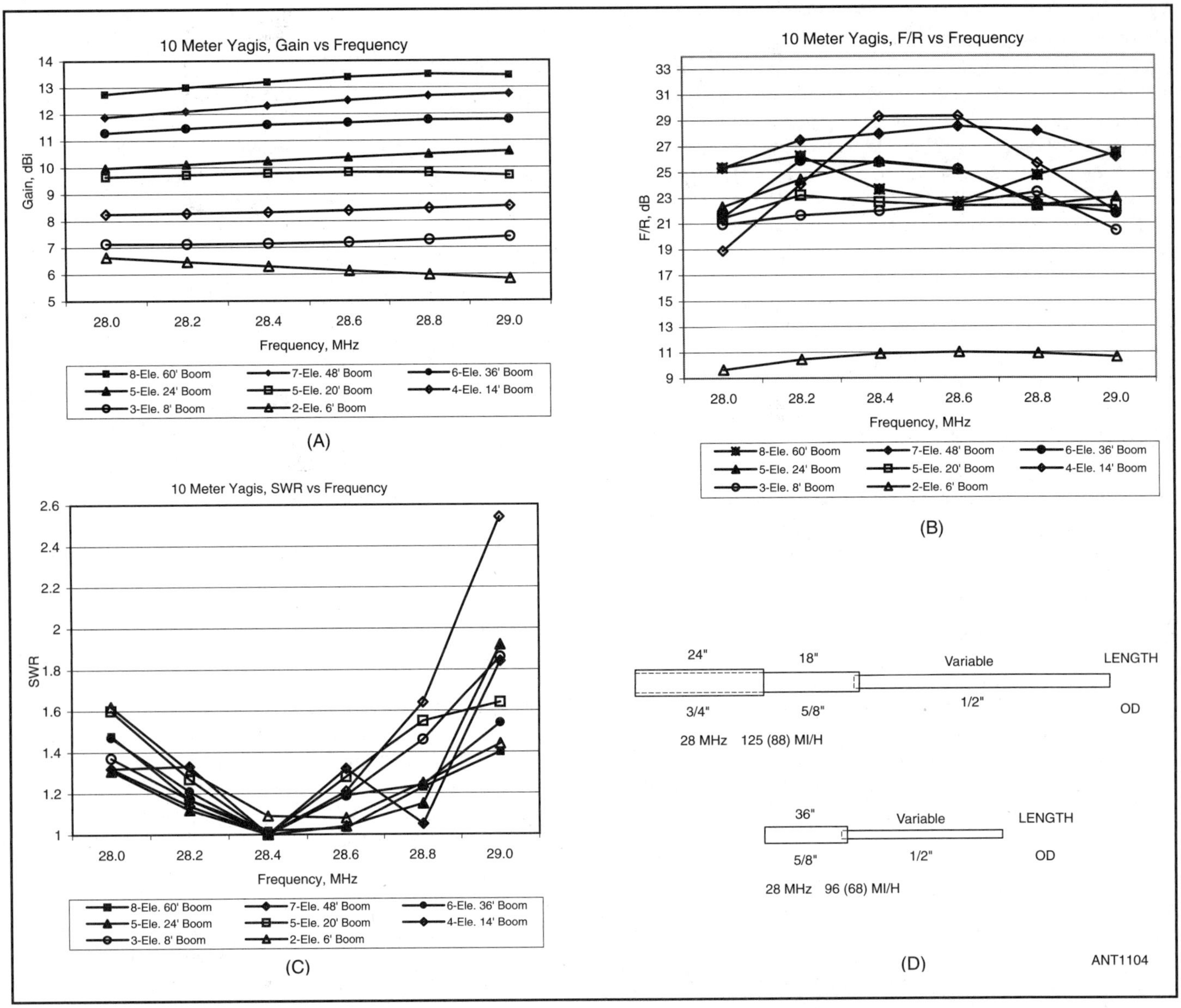

Figure 13.13 — Gain, F/R and SWR performance versus frequency for optimized 10 meter Yagis. At A, gain is shown versus frequency for eight 10 meter Yagis whose booms range from 6 feet to 60 feet long. Except for the 2-element design, these Yagis have been optimized for better than 20 dB F/R and less than 2:1 SWR over the frequency range 28.0 to 28.8 MHz. At B, front-to-rear ratio for these antennas is shown versus frequency, and at C, SWR is shown over the frequency range. At D, the taper schedule is shown for heavy-duty and for medium-duty 10 meter elements. The heavy-duty elements can withstand 125-mph winds without icing, and 88-mph winds with ¼-inch radial ice. The medium-duty elements can survive 96-mph winds without icing, and 68-mph winds with ¼-inch radial ice. The wall thickness for each telescoping section of 6061-T6 aluminum tubing is 0.058 inches, and the overlap at each telescoping junction is 3 inches.

Table 13.1
Optimized 10 Meter Yagi Designs

Two-element 10 meter Yagi, 6 foot boom

Element	*Spacing*	*Heavy-Duty Tip*	*Medium-Duty Tip*
File Name		210-06H.YW	210-06M.YW
Reflector	0.000"	66.000"	71.500"
Driven Element	66.000"	57.625"	63.000"

Three-element 10 meter Yagi, 8 foot boom

Element	*Spacing*	*Heavy-Duty Tip*	*Medium-Duty Tip*
File Name		310-08H.YW	310-08M.YW
Reflector	0.000"	66.750"	71.875"
Driven Element	36.000"	57.625"	62.875"
Director 1	54.000"	53.125"	58.500"
Compensator	12" behind Dir. 1	19.000"	18.125"

Four-element 10 meter Yagi, 14 foot boom

Element	*Spacing*	*Heavy-Duty Tip*	*Medium-Duty Tip*
File Name		410-14H.YW	410-14M.YW
Reflector	0.000"	66.000"	72.000"
Driven Element	36.000"	58.625"	63.875"
Director 1	36.000"	57.000"	62.250"
Director 2	90.000"	47.750"	53.125"
Compensator	12" behind Dir. 2	22.000"	20.500"

Five-element 10 meter Yagi, 24 foot boom

Element	*Spacing, inches*	*Heavy-Duty Tip*	*Medium-Duty Tip*
File Name		510-24H.YW	510-24M.YW
Reflector	0.000"	65.625"	70.750"
Driven Element	36.000"	58.000"	63.250"
Director 1	36.000"	57.125"	62.375"
Director 2	99.000"	55.000"	60.250"
Director 3	111.000"	50.750"	56.125"
Compensator	12" behind Dir. 3	28.750"	26.750"

Six-element 10 meter Yagi, 36 foot boom

Element	*Spacing, inches*	*Heavy-Duty Tip*	*Medium-Duty Tip*
File Name		610-36H.YW	610-36M.YW
Reflector	0.000"	66.500"	71.500"
Driven Element	37.000"	58.500"	64.000"
Director 1	43.000"	57.125"	62.375"
Director 2	98.000"	54.875"	60.125"
Director 3	127.000"	53.875"	59.250"
Director 4	121.000"	49.875"	55.250"
Compensator	12" behind Dir. 4	32.000"	29.750"

Seven-element 10 meter Yagi, 48 foot boom

Element	*Spacing, inches*	*Heavy-Duty Tip*	*Medium-Duty Tip*
File Name		710-48H.YW	710-48M.YW
Reflector	0.000"	65.375"	70.500"
Driven Element	37.000"	59.000"	64.250"
Director 1	37.000"	57.500"	62.750"
Director 2	96.000"	54.875"	60.125"
Director 3	130.000"	52.250"	57.625"
Director 4	154.000"	52.625"	58.000"
Director 5	116.000"	49.875"	55.250"
Compensator	12" behind Dir. 5	35.750"	33.750"

Eight-element 10 meter Yagi, 60 foot boom

Element	*Spacing, inches*	*Heavy-Duty Tip*	*Medium-Duty Tip*
File Name		810-60H.YW	810-60M.YW
Reflector	0.000"	65.000"	70.125"
Driven Element	42.000"	58.000"	63.500"
Director 1	37.000"	57.125"	62.375"
Director 2	87.000"	55.375"	60.625"
Director 3	126.000"	53.250"	58.625"
Director 4	141.000"	51.875"	57.250"
Director 5	157.000"	52.500"	57.875"
Director 6	121.000"	50.125"	55.500"
Compensator	12" behind Dir. 6	59.375"	55.125"

These **10 meter Yagi** designs are optimized for > 20 dB F/R, and SWR < 2:1 over frequency range from 28.000 to 28.800 MHz, for heavy-duty elements (125 mph wind survival) and for medium-duty (96 mph wind survival). For coverage from 28.8 to 29.7 MHz, subtract 2.000 inches from end of each element, but leave element spacings the same as shown here. Only element tip dimensions are shown, and all dimensions are inches. See Fig 13.13D for element telescoping tubing schedule. Torque compensator element is made of 2.5" OD PVC water pipe placed 12 inches behind last director. Dimensions shown for compensators is one-half of total length, centered on boom.

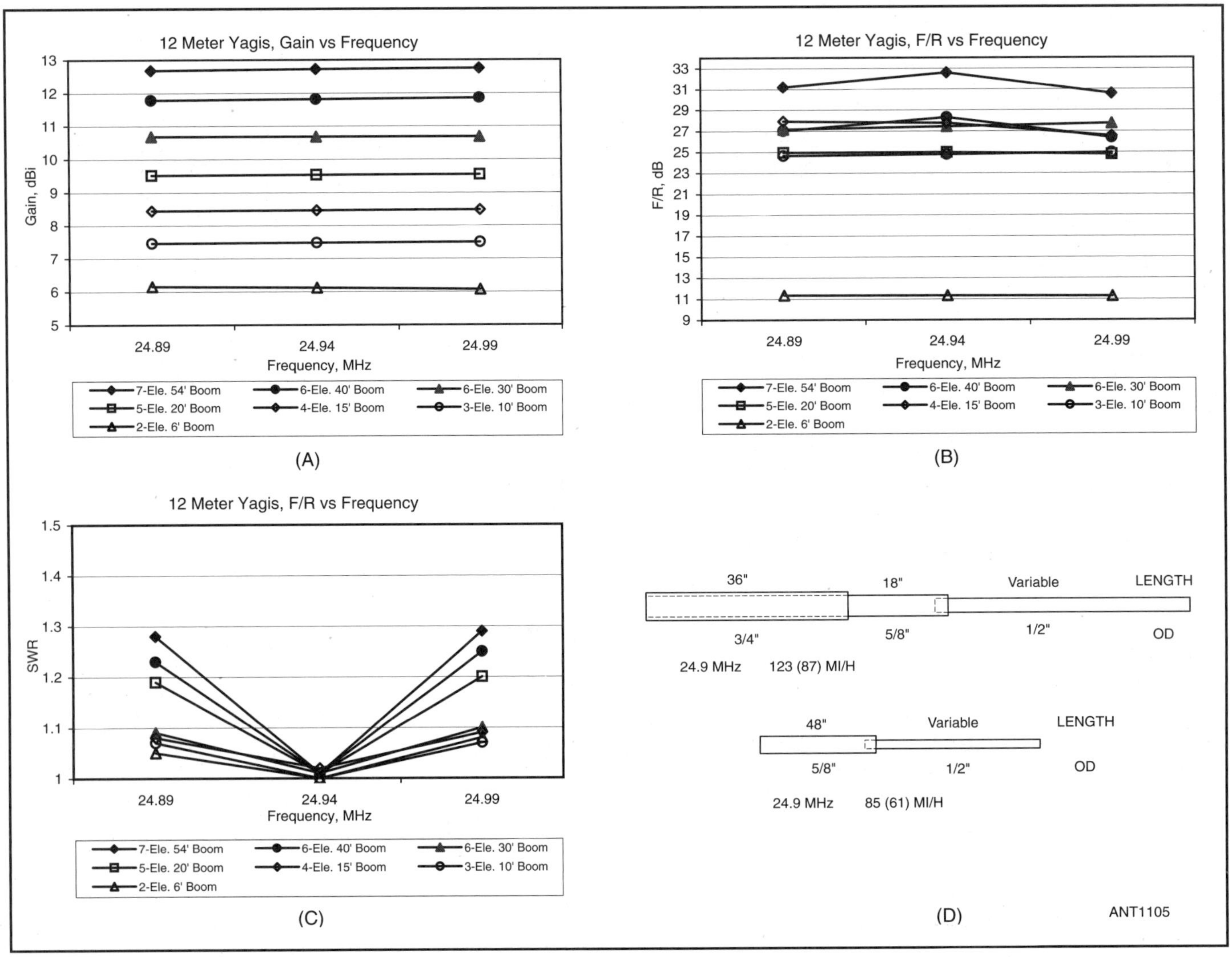

Figure 13.14 — Gain, F/R and SWR performance versus frequency for optimized 12 meter Yagis. At A, gain is shown versus frequency for seven 12 meter Yagis whose booms range from 6 feet to 54 feet long. Except for the 2-element design, these Yagis have been optimized for better than 20 dB F/R and less than 2:1 SWR over the narrow 12 meter band 24.89 to 24.99 MHz. At B, front-to-rear ratio for these antennas is shown versus frequency, and at C, SWR over the frequency range is shown. At D, the taper schedule for heavy-duty and for medium-duty 12 meter elements is shown. The heavy-duty elements can withstand 123-mph winds without icing, and 87-mph winds with ¼-inch radial ice. The medium-duty elements can survive 85-mph winds without icing, and 61-mph winds with ¼-inch radial ice. The wall thickness for each telescoping section of 6061-T6 aluminum tubing is 0.058 inches, and the overlap at each telescoping junction is 3 inches.

Table 13.2

Optimized 12 Meter Yagi Designs

Two-element 12 meter Yagi, 6 foot boom

Element	*Spacing*	*Heavy-Duty Tip*	*Medium-Duty Tip*
File Name		212-06H.YW	212-06M.YW
Reflector	0.000"	67.500"	72.500"
Driven Element	66.000"	59.500"	65.000"

Three-element 12 meter Yagi, 10 foot boom

Element	*Spacing, inches*	*Heavy-Duty Tip*	*Medium-Duty Tip*
File Name		312-10H.YW	312-10M.YW
Reflector	0.000"	69.000"	73.875"
Driven Element	40.000"	60.250"	65.250"
Director 1	74.000"	54.000"	59.125"
Compensator	12" behind Dir. 1	13.625"	12.000"

Four-element 12 meter Yagi, 15 foot boom

Element	*Spacing, inches*	*Heavy-Duty Tip*	*Medium-Duty Tip*
File Name		412-15H.YW	412-15M.YW
Reflector	0.000"	66.875"	71.875"
Driven Element	46.000"	61.000"	66.000"
Director 1	46.000"	58.625"	63.750"
Director 2	82.000"	50.875"	56.125"
Compensator	12" behind Dir. 2	16.375"	14.500"

Five-element 12 meter Yagi, 20 foot boom

Element	*Spacing, inches*	*Heavy-Duty Tip*	*Medium-Duty Tip*
File Name		512-20H.YW	512-20M.YW
Reflector	0.000"	69.750"	74.625"
Driven Element	46.000"	62.250"	67.000"
Director 1	46.000"	60.500"	65.500"
Director 2	48.000"	55.500"	60.625"
Director 3	94.000"	54.625"	59.750"
Compensator	12" behind Dir. 3	22.125"	19.625"

Six-element 12 meter Yagi, 30 foot boom

Element	*Spacing, inches*	*Heavy-Duty Tip*	*Medium-Duty Tip*
File Name		612-30H.YW	612-30M.YW
Reflector	0.000"	68.125"	73.000"
Driven Element	46.000"	61.750"	66.750"
Director 1	46.000"	60.250"	65.250"
Director 2	73.000"	52.375"	57.625"
Director 3	75.000"	57.625"	62.750"
Director 4	114.000"	53.625"	58.750"
Compensator	12" behind Dir. 4	30.000"	26.250"

Six-element 12 meter Yagi, 40 foot boom

Element	*Spacing, inches*	*Heavy-Duty Tip*	*Medium-Duty Tip*
File Name		612-40H.YW	612-40M.YW
Reflector	0.000"	67.000"	71.875"
Driven Element	46.000"	60.125"	65.500"
Director 1	46.000"	57.375"	62.500"
Director 2	91.000"	57.375"	62.500"
Director 3	157.000"	57.000"	62.125"
Director 4	134.000"	54.375"	59.500"
Compensator	12" behind Dir. 4	36.500"	31.625"

Seven-element 12 meter Yagi, 54 foot boom

Element	*Spacing, inches*	*Heavy-Duty Tip*	*Medium-Duty Tip*
File Name		712-54H.YW	712-54M.YW
Reflector	0.000"	68.000"	73.000"
Driven Element	46.000"	60.500"	65.500"
Director 1	46.000"	56.750"	61.875"
Director 2	75.000"	58.000"	63.125"
Director 3	161.000"	55.625"	60.750"
Director 4	174.000"	56.000"	61.125"
Director 5	140.000"	53.125"	58.375"
Compensator	12" behind Dir. 5	43.125"	37.500"

These 12 meter Yagi designs were optimized for > 20 dB F/R, and SWR < 2:1 over frequency range from 24.890 to 24.990 MHz, for heavy-duty elements (123 mph wind survival) and for medium-duty (85 mph wind survival). Only element tip dimensions are shown, and all dimensions are inches. See Fig 13.14D for element telescoping tubing schedule. Torque compensator element is made of 2.5" OD PVC water pipe placed 12" behind last director. Dimensions shown for compensators is one-half of total length, centered on boom.

These **12 meter Yagi** designs were optimized for > 20 dB F/R, and SWR < 2:1 over frequency range from 24.890 to 24.990 MHz, for heavy-duty elements (123 mph wind survival) and for medium-duty (85 mph wind survival). Only element tip dimensions are shown, and all dimensions are inches. See Fig 13.14D for element telescoping tubing schedule. Torque compensator element is made of 2.5" OD PVC water pipe placed 12" behind last director. Dimensions shown for compensators is one-half of total length, centered on boom.

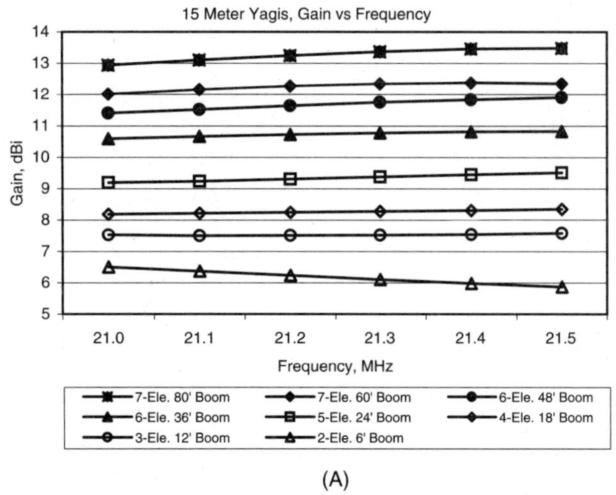

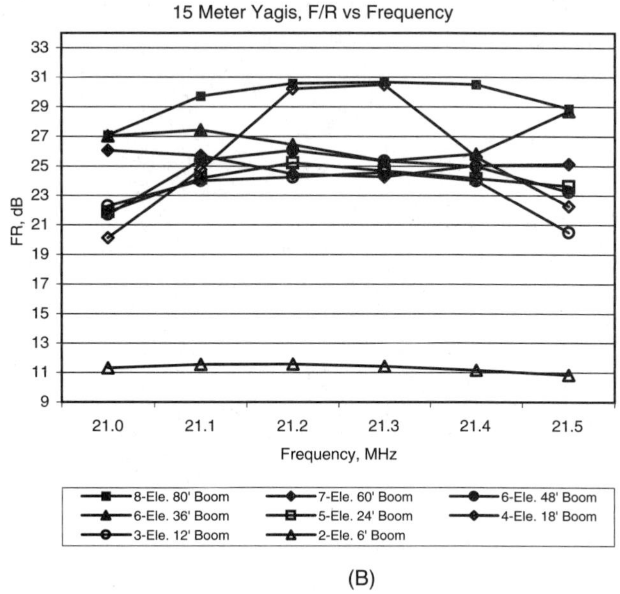

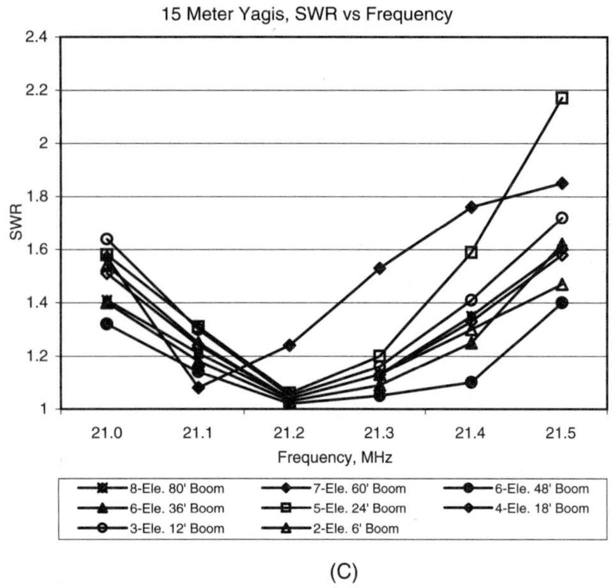

Figure 13.15 — Gain, F/R and SWR performance versus frequency for optimized 15 meter Yagis. At A, gain versus frequency is shown for eight 15 meter Yagis whose booms range from 6 feet to 80 feet long. Except for the 2-element design, these Yagis have been optimized for better than 20 dB F/R and less than 2:1 SWR over the frequency range 21.0 to 21.45 MHz. At B, front-to-rear ratio for these antennas is shown versus frequency, and at C, SWR over the frequency range is shown. At D, the taper schedule for heavy-duty and for medium-duty 15 meter elements is shown. The heavy-duty elements can withstand 124-mph winds without icing, and 90-mph winds with ¼-inch radial ice. The medium-duty elements can survive 86-mph winds without icing, and 61-mph winds with ¼-inch radial ice. The wall thickness for each telescoping section of 6061-T6 aluminum tubing is 0.058 inches, and the overlap at each telescoping junction is 3 inches.

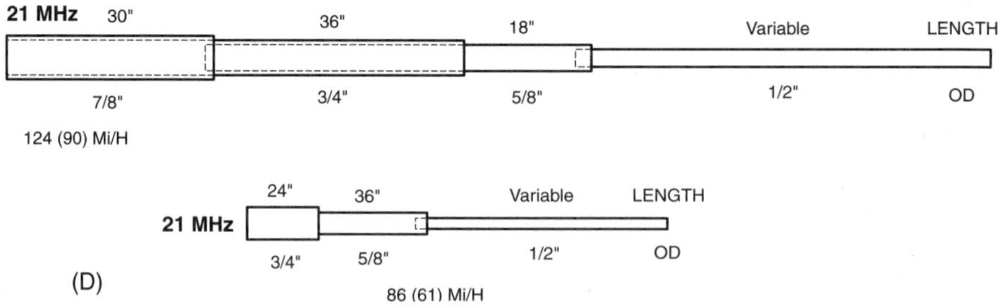

ANT1106

Table 13.3
Optimized 15 Meter Yagi Designs

Two-element 15 meter Yagi, 6 foot boom			
Element	*Spacing*	*Heavy-Duty Tip*	*Medium-Duty Tip*
File Name		215-06H.YW	215-06M.YW
Reflector	0.000"	62.000"	85.000"
Driven Element	66.000"	51.000"	74.000"
Three-element 15 meter Yagi, 12 foot boom			
Element	*Spacing*	*Heavy-Duty Tip*	*Medium-Duty Tip*
File Name		315-12H.YW	315-12M.YW
Reflector	0.000"	62.000"	84.250"
Driven Element	48.000"	51.000"	73.750"
Director 1	92.000"	43.500"	66.750"
Compensator	12" behind Dir. 1	34.750"	37.625"
Four-element 15 meter Yagi, 18 foot boom			
Element	*Spacing*	*Heavy-Duty Tip*	*Medium-Duty Tip*
File Name		415-18H.YW	415-18M.YW
Reflector	0.000"	61.000"	83.500"
Driven Element	56.000"	51.500"	74.500"
Director 1	56.000"	48.000"	71.125"
Director 2	98.000"	36.625"	60.250"
Compensator	12" behind Dir. 2	20.875"	18.625"
Five-element 15 meter Yagi, 24 foot boom			
Element	*Spacing*	*Heavy-Duty Tip*	*Medium-Duty Tip*
File Name		515-24H.YW	515-24M.YW
Reflector	0.000"	62.000"	84.375"
Driven Element	48.000"	52.375"	75.250"
Director 1	48.000"	47.875"	71.000"
Director 2	52.000"	47.000"	70.125"
Director 3	134.000"	41.000"	64.375"
Compensator	12" behind Dir. 3	40.250"	35.125"
Six-element 15 meter Yagi, 36 foot boom			
Element	*Spacing*	*Heavy-Duty Tip*	*Medium-Duty Tip*
File Name		615-36H.YW	615-36M.YW
Reflector	0.000"	61.000"	83.375"
Driven Element	53.000"	52.000"	75.000"
Director 1	56.000"	49.125"	72.125"
Director 2	59.000"	45.125"	68.375"
Director 3	116.000"	47.875"	71.000"
Director 4	142.000"	42.000"	65.375"
Compensator	12" behind Dir. 4	45.500"	39.750"
Seven-element 15 meter Yagi, 48 foot boom			
Element	*Spacing*	*Heavy-Duty Tip*	*Medium-Duty Tip*
File Name		615-48H.YW	615-48M.YW
Reflector	0.000"	62.000"	84.000"
Driven Element	48.000"	52.000"	75.000"
Director 1	48.000"	51.250"	74.125"
Director 2	125.000"	48.000"	71.125"
Director 3	190.000"	45.500"	68.750"
Director 4	161.000"	42.000"	65.375"
Compensator	12" behind Dir. 4	51.500"	45.375"
Seven-element 15 meter Yagi, 60 foot boom			
Element	*Spacing*	*Heavy-Duty Tip*	*Medium-Duty Tip*
File Name		715-60H.YW	715-60M.YW
Reflector	0.000"	59.750"	82.250"
Driven Element	48.000"	52.000"	75.000"
Director 1	48.000"	52.000"	74.875"
Director 2	93.000"	49.500"	72.500"
Director 3	173.000"	44.125"	67.375"
Director 4	197.000"	45.500"	68.750"
Director 5	155.000"	41.750"	65.125"
Compensator	12" behind Dir. 5	58.500"	51.000"
Eight-element 15 meter Yagi, 80 foot boom			
Element	*Spacing*	*Heavy-Duty Tip*	*Medium-Duty Tip*
File Name		815-80H.YW	815-80M.YW
Reflector	0.000"	62.000"	84.000"
Driven Element	56.000"	52.500"	75.500"
Director 1	48.000"	51.500"	74.375"
Director 2	115.000"	48.375"	71.500"
Director 3	164.000"	45.750"	69.000"
Director 4	202.000"	43.125"	66.500"
Director 5	206.000"	44.750"	68.000"
Director 6	163.000"	40.875"	64.250"
Compensator	12" behind Dir. 6	95.000"	83.375"

These **15 meter Yagi** designs are optimized for > 20 dB F/R, and SWR < 2:1 over entire frequency range from 21.000 to 21.450 MHz, for heavy-duty elements (124 mph wind survival) and for medium-duty (86 mph wind survival). Only element tip dimensions are shown. See Fig 13.15D for element telescoping tubing schedule. All dimensions are in inches. Torque compensator element is made of 2.5" OD PVC water pipe placed 12" behind last director, and dimensions shown for compensators is one-half of total length, centered on boom.

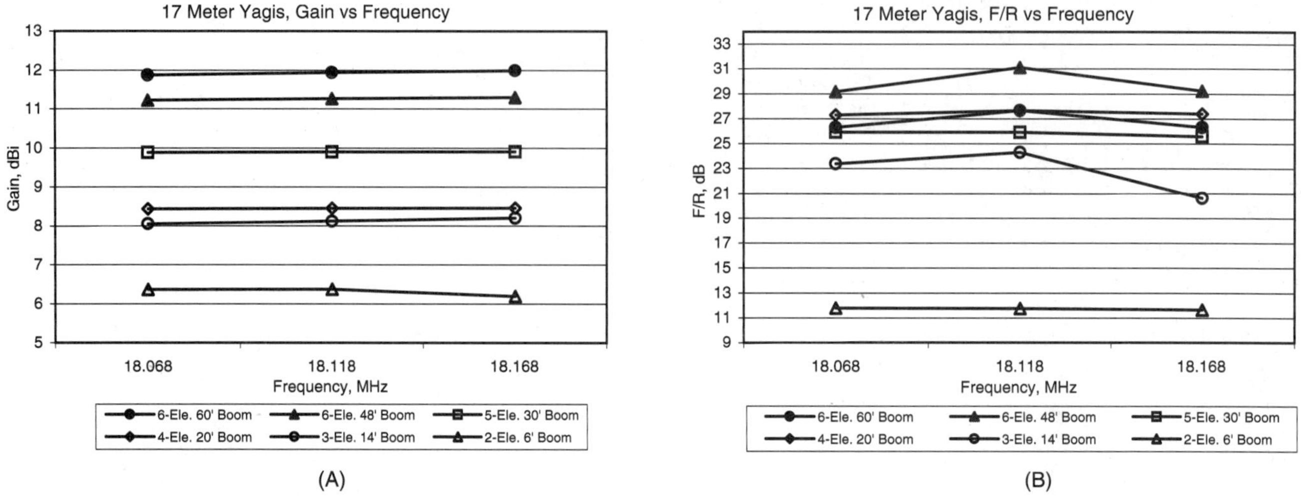

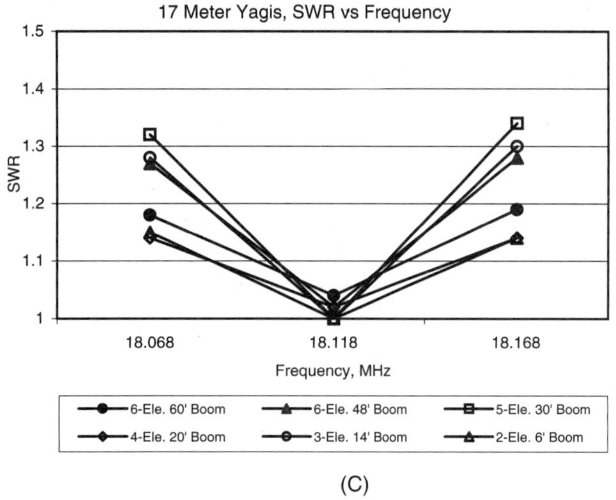

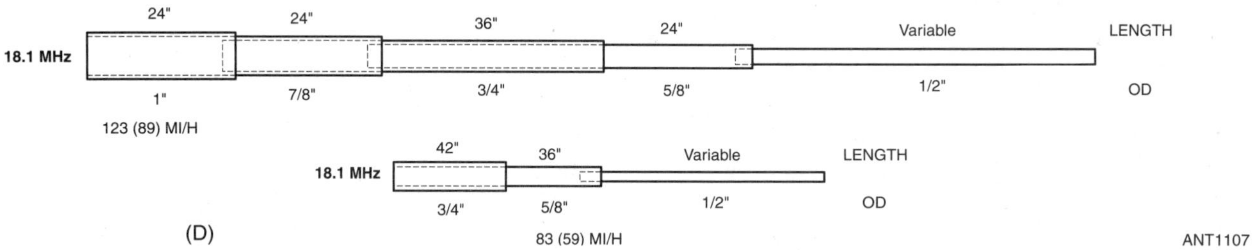

Figure 13.16 — Gain, F/R and SWR performance versus frequency for optimized 17 meter Yagis. At A, gain versus frequency is shown for six 17 meter Yagis whose booms range from 6 feet to 60 feet long. Except for the 2-element design, these Yagis have been optimized for better than 20 dB F/R and less than 2:1 SWR over the narrow 17 meter band 18.068 to 18.168 MHz. At B, front-to-rear ratio for these antennas is shown versus frequency, and at C, SWR over the frequency range is shown. At D, the taper schedule for heavy-duty and for medium-duty 10 meter elements is shown. The heavy-duty elements can withstand 123-mph winds without icing, and 89-mph winds with ¼-inch radial ice. The medium-duty elements can survive 83-mph winds without icing, and 59-mph winds with ¼-inch radial ice. The wall thickness for each telescoping section of 6061-T6 aluminum tubing is 0.058 inches, and the overlap at each telescoping junction is 3 inches.

Table 13.4
Optimized 17 Meter Yagi Designs

Two-element 17 meter Yagi, 6 foot boom			
Element	*Spacing*	*Heavy-Duty Tip*	*Medium-Duty Tip*
File Name		217-06H.YW	217-06M.YW
Reflector	0.000"	61.000"	89.000"
Driven Element	66.000"	48.000"	76.250"
Three-element 17 meter Yagi, 14 foot boom			
Element	*Spacing*	*Heavy-Duty Tip*	*Medium-Duty Tip*
File Name		317-14H.YW	317-14M.YW
Reflector	0.000"	61.500"	91.500"
Driven Element	65.000"	52.000"	79.500"
Director 1	97.000"	46.000"	73.000"
	12" behind Dir. 1	12.625"	10.750"
Four-element 17 meter Yagi, 20 foot boom			
Element	*Spacing*	*Heavy-Duty Tip*	*Medium-Duty Tip*
File Name		417-20H.YW	417-20M.YW
Reflector	0.000"	61.500"	89.500"
Driven Element	48.000"	54.250"	82.625"
Director 1	48.000"	52.625"	81.125"
Director 2	138.000"	40.500"	69.625"
Compensator	12" behind Dir. 2	42.500"	36.250"
Five-element 17 meter Yagi, 30 foot boom			
Element	*Spacing*	*Heavy-Duty Tip*	*Medium-Duty Tip*
File Name		517-30H.YW	517-30M.YW
Reflector	0.000"	61.875"	89.875"
Driven Element	48.000"	52.250"	80.500"
Director 1	52.000"	49.625"	78.250"
Director 2	93.000"	49.875"	78.500"
Director 3	161.000"	43.500"	72.500"
Compensator	12" behind Dir. 3	54.375"	45.875"
Six-element 17 meter Yagi, 48 foot boom			
Element	*Spacing*	*Heavy-Duty Tip*	*Medium-Duty Tip*
File Name		617-48H.YW	617-48M.YW
Reflector	0.000"	63.000"	90.250"
Driven Element	52.000"	52.500"	80.500"
Director 1	51.000"	45.500"	74.375"
Director 2	87.000"	47.875"	76.625"
Director 3	204.000"	47.000"	75.875"
Director 4	176.000"	42.000"	71.125"
Compensator	12" behind Dir. 4	68.250"	57.500"
Six-element 17 meter Yagi, 60 foot boom			
Element	*Spacing*	*Heavy-Duty Tip*	*Medium-Duty Tip*
File Name		617-60H.YW	617-60M.YW
Reflector	0.000"	61.250"	89.250"
Driven Element	54.000"	54.750"	83.125"
Director 1	54.000"	52.250"	80.750"
Director 2	180.000"	46.000"	74.875"
Director 3	235.000"	44.625"	73.625"
Director 4	191.000"	41.500"	70.625"
Compensator	12" behind Dir. 4	62.875"	53.000"

These **17 meter Yagi** designs are optimized for > 20 dB F/R, and SWR < 2:1 over entire frequency range from18.068 to 18.168 MHz, for heavy-duty elements (123 mph wind survival) and for medium-duty (83 mph wind survival). Only element tip dimensions are shown. All dimensions are in inches. Torque compensator element is made of 2.5" OD PVC water pipe placed 12" behind last director, and dimensions shown for compensators is one-half of total length, centered on boom.

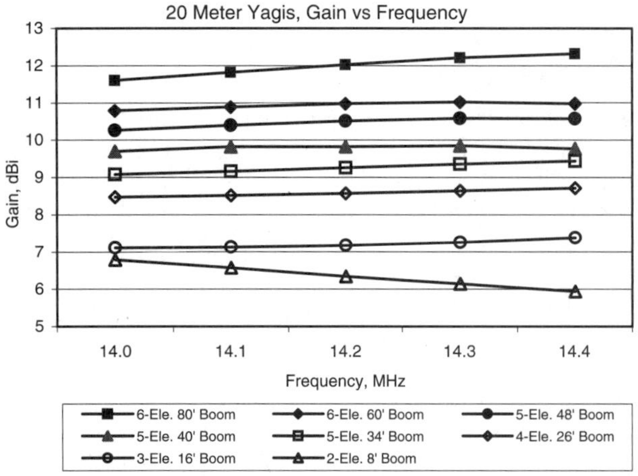

(A)

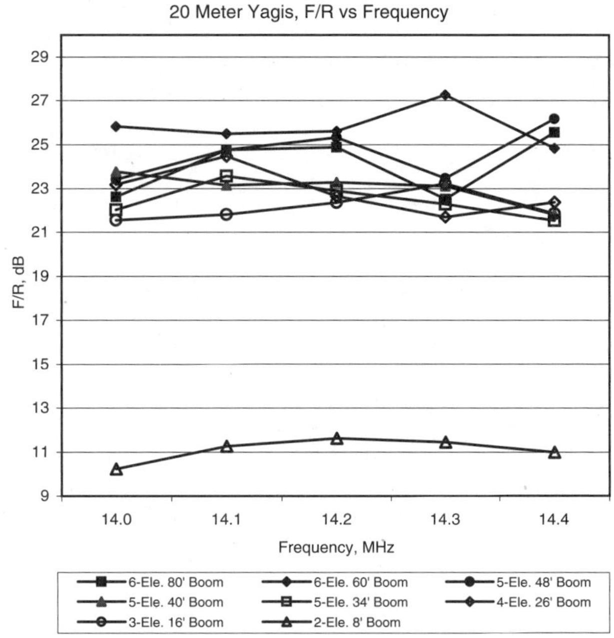

(B)

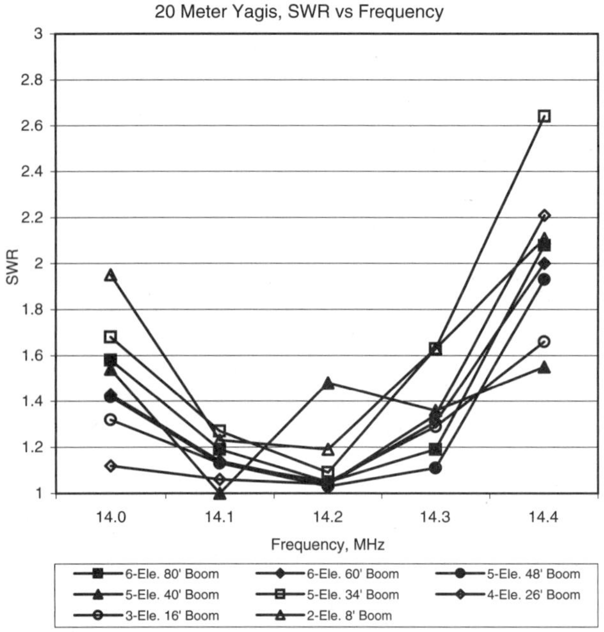

(C)

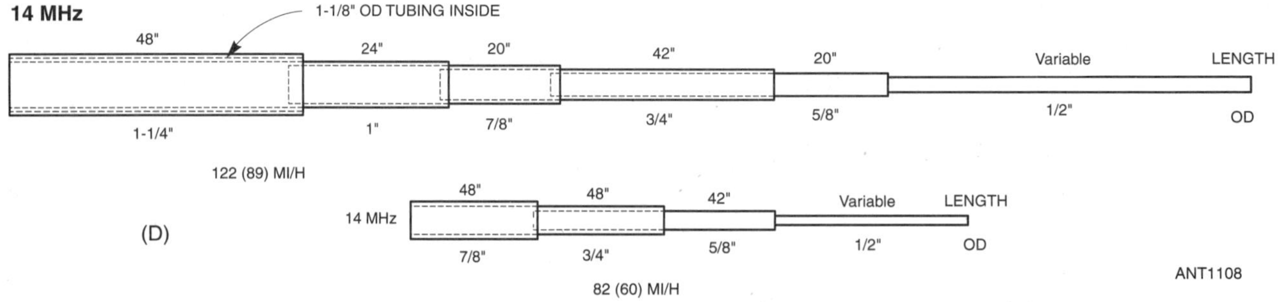

Figure 13.17 — Gain, F/R and SWR performance versus frequency for optimized 20 meter Yagis. At A, gain versus frequency is shown for eight 20 meter Yagis whose booms range from 8 feet to 80 feet long. Except for the 2-element design, these Yagis have been optimized for better than 20 dB F/R and less than 2:1 SWR over the frequency range 14.0 to 14.35 MHz. At B, front-to-rear ratio for these antennas is shown versus frequency, and at C, SWR over the frequency range is shown. At D, the taper schedule for heavy-duty and for medium-duty 20 meter elements is shown. The heavy-duty elements can withstand 122-mph winds without icing, and 89-mph winds with ¼-inch radial ice. The medium-duty elements can survive 82-mph winds without icing, and 60-mph winds with ¼-inch radial ice. The wall thickness for each telescoping section of 6061-T6 aluminum tubing is 0.058 inches, and the overlap at each telescoping junction is 3 inches.

Table 13.5

Optimized 20 Meter Yagi Designs

Two-element 20 meter Yagi, 8 foot boom			
Element	*Spacing*	*Heavy-Duty Tip*	*Medium-Duty Tip*
File Name		220-08H.YW	220-08M.YW
Reflector	0.000"	66.000"	80.000"
Driven Element	90.000"	46.000"	59.000"
Three-element 20 meter Yagi, 16 foot boom			
Element	*Spacing*	*Heavy-Duty Tip*	*Medium-Duty Tip*
File Name		320-16H.YW	320-16M.YW
Reflector	0.000"	69.625"	81.625"
Driven Element	80.000"	51.250"	64.500"
Director 1	106.000"	42.625"	56.375"
Compensator	12" behind Dir. 1	33.375"	38.250"
Four-element 20 meter Yagi, 26 foot boom			
Element	*Spacing*	*Heavy-Duty Tip*	*Medium-Duty Tip*
File Name		420-26H.YW	420-26M.YW
Reflector	0.000"	65.625"	78.000"
Driven Element	72.000"	53.375"	65.375"
Director 1	60.000"	51.750"	63.875"
Director 2	174.000"	38.625"	51.500"
Compensator	12" behind Dir. 2	54.250"	44.250"
Five-element 20 meter Yagi, 34 foot boom			
Element	*Spacing*	*Heavy-Duty Tip*	*Medium-Duty Tip*
File Name		520-34H.YW	520-34M.YW
Reflector	0.000"	68.625"	80.750"
Driven Element	72.000"	52.250"	65.500"
Director 1	71.000"	45.875"	59.375"
Director 2	68.000"	45.875"	59.375"
Director 3	191.000"	37.000"	51.000"
Compensator	12" behind Dir. 3	69.250"	56.250"
Five-element 20 meter Yagi, 40 foot boom			
Element	*Spacing*	*Heavy-Duty Tip*	*Medium-Duty Tip*
File Name		520-40H.YW	520-40M.YW
Reflector	0.000"	68.375"	80.500"
Driven Element	72.000"	53.500"	66.625"
Director 1	72.000"	51.500"	64.625"
Director 2	139.000"	48.375"	61.750"
Director 3	191.000"	38.000"	52.000"
Compensator	12" behind Dir. 3	69.750"	56.750"
Five-element 20 meter Yagi, 48 foot boom			
Element	*Spacing*	*Heavy-Duty Tip*	*Medium-Duty Tip*
File Name		520-48H.YW	520-48M.YW
Reflector	0.000"	66.250"	78.500"
Driven Element	72.000"	53.000"	66.000"
Director 1	88.000"	50.500"	63.750"
Director 2	199.000"	47.375"	60.875"
Director 3	211.000"	39.750"	53.625"
Compensator	12" behind Dir. 3	70.325"	57.325"
Six-element 20 meter Yagi, 60 foot boom			
Element	*Spacing*	*Heavy-Duty Tip*	*Medium-Duty Tip*
File Name		620-60H.YW	620-60M.YW
Reflector	0.000"	67.000"	79.250"
Driven Element	84.000"	51.500"	65.000"
Director 1	91.000"	45.125"	58.750"
Director 2	130.000"	41.375"	55.125"
Director 3	210.000"	46.875"	60.375"
Director 4	199.000"	39.125"	53.000"
Compensator	12" behind Dir. 4	72.875"	59.250"
Six-element 20 meter Yagi, 80 foot boom			
Element	*Spacing*	*Heavy-Duty Tip*	*Medium-Duty Tip*
File Name		620-80H.YW	620-80M.YW
Reflector	0.000"	66.125"	78.375"
Driven Element	72.000"	52.375"	65.500"
Director 1	122.000"	49.125"	62.500"
Director 2	229.000"	44.500"	58.125"
Director 3	291.000"	42.625"	56.375"
Director 4	240.000"	38.750"	52.625"
Compensator	12" behind Dir. 4	78.750"	64.125"

These **20 meter Yagi** designs are optimized for > 20 dB F/R, and SWR < 2:1 over entire frequency range from 14.000 to 14.350 MHz, for heavy-duty elements (122 mph wind survival) and for medium-duty (82 mph wind survival). Only element tip dimensions are shown. See Fig 13.17 for element telescoping tubing schedule. All dimensions are in inches. Torque compensator element is made of 2.5" OD PVC water pipe placed 12" behind last director, and dimensions shown for compensators is one-half of total length, centered on boom.

Table 13.6
Optimized 30 Meter Yagi Designs

Two-element 30 meter Yagi, 15 foot boom

Element	*Spacing*	*Heavy-Duty Tip*
File Name		230-15H.YW
Reflector	0.000"	50.250"
Driven Element	174.000"	14.875"

3-element 30 meter Yagi, 22 foot boom

Element	*Spacing*	*Heavy-Duty Tip*
File Name		330-22H.YW
Reflector	0.000	59.375
Driven Element	135.000	35.000
Director 1	123.000	19.625

Three-element 30 meter Yagi, 34 foot boom

Element	*Spacing*	*Heavy-Duty Tip*
File Name		330-34H.YW
Reflector	0.000"	53.750"
Driven Element	212"	29.000"
Director 1	190"	14.500"

These 30 meter Yagi designs are optimized for > 10 dB F/R, and SWR < 2:1 over entire frequency range from 10.100 to 10.150 MHz for heavy-duty elements (105 mph wind survival). Only element tip dimensions are shown. See Fig 13.18D for element telescoping tubing schedule. All dimensions are in inches. No torque compensator element is required.

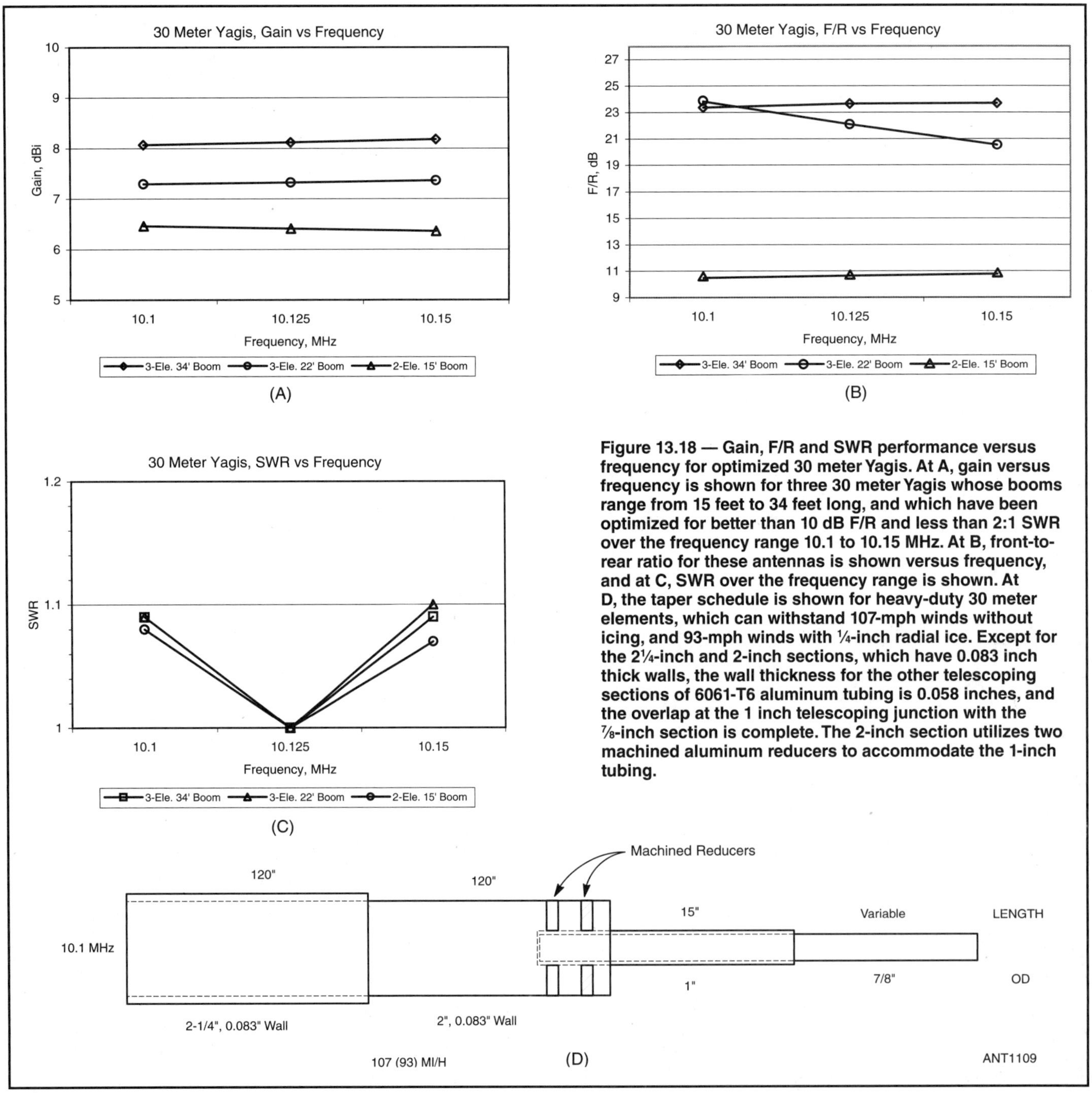

Figure 13.18 — Gain, F/R and SWR performance versus frequency for optimized 30 meter Yagis. At A, gain versus frequency is shown for three 30 meter Yagis whose booms range from 15 feet to 34 feet long, and which have been optimized for better than 10 dB F/R and less than 2:1 SWR over the frequency range 10.1 to 10.15 MHz. At B, front-to-rear ratio for these antennas is shown versus frequency, and at C, SWR over the frequency range is shown. At D, the taper schedule is shown for heavy-duty 30 meter elements, which can withstand 107-mph winds without icing, and 93-mph winds with ¼-inch radial ice. Except for the 2¼-inch and 2-inch sections, which have 0.083 inch thick walls, the wall thickness for the other telescoping sections of 6061-T6 aluminum tubing is 0.058 inches, and the overlap at the 1 inch telescoping junction with the ⅞-inch section is complete. The 2-inch section utilizes two machined aluminum reducers to accommodate the 1-inch tubing.

Table 13.7
Optimized 40 Meter Yagi Designs

Two-element 40 meter Yagi, 20 foot boom

Element	*Spacing*	*Heavy-Duty Tip*
File Name		240-20H.YW
Reflector	0.000"	85.000"
Driven Element	234.000"	35.000"

Three-element 40 meter Yagi, 32 foot boom

Element	*Spacing*	*Heavy-Duty Tip*
File Name		340-32H.YW
Reflector	0.000"	90.750"
Driven Element	196.000"	55.875"
Director 1	182.000"	33.875"

Three-element 40 meter Yagi, 48 foot boom

Element	*Spacing*	*Heavy-Duty Tip*
File Name		340-48H.YW
Reflector	0.000"	81.000"
Driven Element	300.000"	45.000"
Director 1	270.000"	21.000"

These 40 meter Yagi designs are optimized for > 10 dB F/R, and SWR < 2:1 over low-end of frequency range from 7.000 to 7.200 MHz, for heavy-duty elements (95 mph wind survival). Only element tip dimensions are shown. See Fig 13.19D for element telescoping tubing schedule. All dimensions are in inches. No wind torque compensator is required.

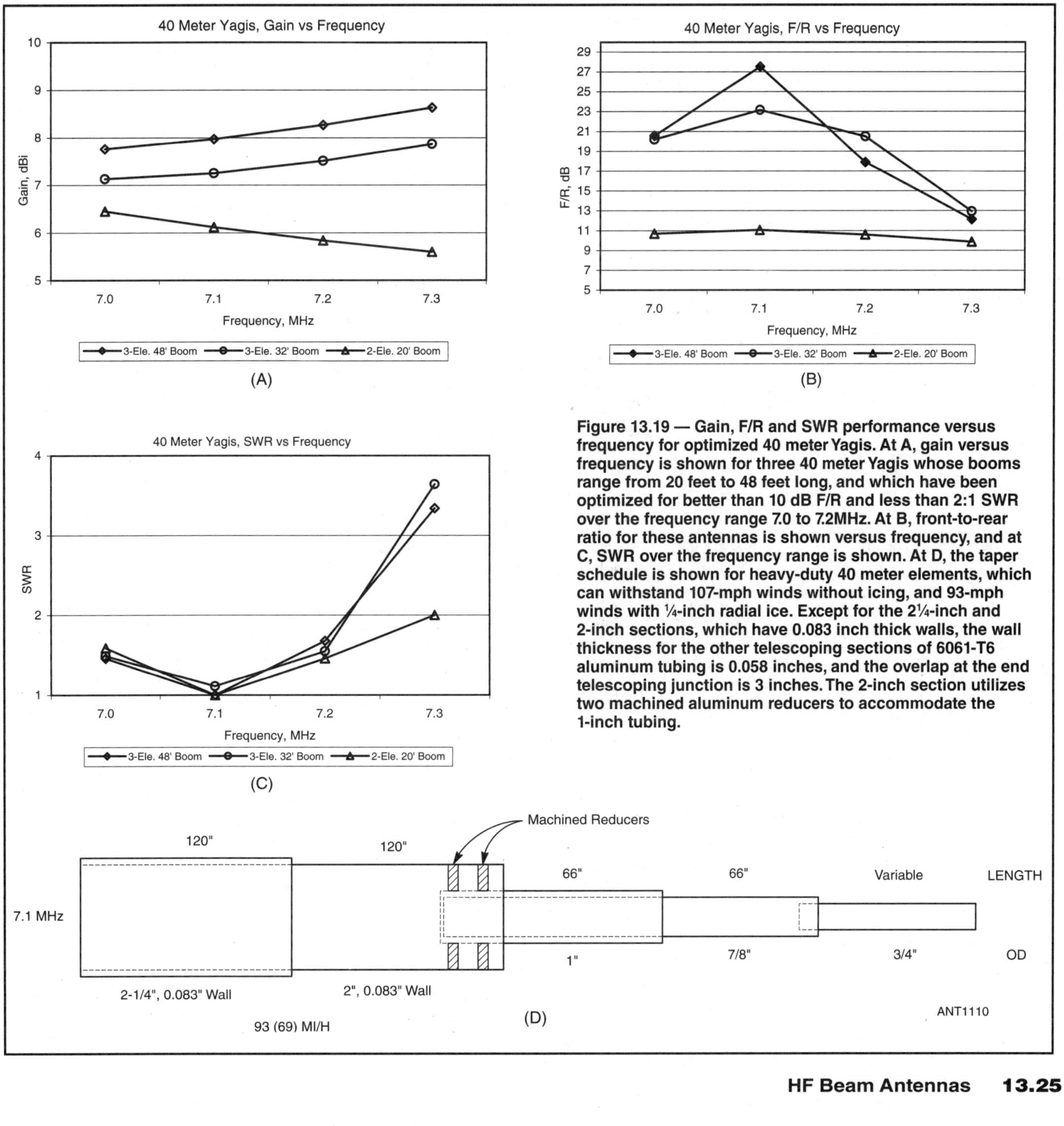

Figure 13.19 — Gain, F/R and SWR performance versus frequency for optimized 40 meter Yagis. At A, gain versus frequency is shown for three 40 meter Yagis whose booms range from 20 feet to 48 feet long, and which have been optimized for better than 10 dB F/R and less than 2:1 SWR over the frequency range 7.0 to 7.2MHz. At B, front-to-rear ratio for these antennas is shown versus frequency, and at C, SWR over the frequency range is shown. At D, the taper schedule is shown for heavy-duty 40 meter elements, which can withstand 107-mph winds without icing, and 93-mph winds with ¼-inch radial ice. Except for the 2¼-inch and 2-inch sections, which have 0.083 inch thick walls, the wall thickness for the other telescoping sections of 6061-T6 aluminum tubing is 0.058 inches, and the overlap at the end telescoping junction is 3 inches. The 2-inch section utilizes two machined aluminum reducers to accommodate the 1-inch tubing.

13.4.7 40 Meter Yagis

Figure 13.19 describes the electrical performance of three optimized 40 meter Yagis with boom lengths between 20 to 48 feet. Like the 30 meter antennas, because of the size and weight of the elements for a 40 meter Yagi, only 2-element and 3-element designs are described. The front-to-rear ratio requirement for the 2-element antenna is relaxed to be greater than 10 dB over the band from 7.000 to 7.300 MHz, while the goal for the 3-element designs is 20 dB over the frequency range of 7.000 to 7.200 MHz. It is exceedingly difficult to hold the F/R greater than 20 dB over the entire =40 meter band without sacrificing excessive gain with a 3-element design.

As usual, the end of each boom includes 3 inches of space for the reflector and last director mounting plates. Figure 13.19A shows the free-space gain versus frequency for each antenna; Figure 13.19B shows the front-to-rear ratio, and Figure 13.19C shows the SWR versus frequency.

Figure 13.19D shows the taper schedule for the 40 meter elements. Note that the wall thickness of the first two sections of tubing is 0.083 inches, rather than 0.058 inches. This element design can survive 93-mph winds with no icing, and 69-mph winds with ¼ inch of radial ice.

The element-to-boom mounting plate for these Yagis is a 0.500-inch thick flat aluminum plate, 6 inches wide by 24 inches long. The mounting plate has an effective diameter of 4.684 inches. The equivalent length on each side of the boom is 12 inches. These designs require no torque compensator.

13.4.8 Modifying Monoband Hy-Gain Yagis

Enterprising amateurs have long used the Hy-Gain "Long John" series of HF monobanders as a source of top-quality aluminum and hardware for customized Yagis. Often-modified older models include the 105BA for 10 meters, the 155BA for 15 meters, and the 204BA and 205BA for 20 meters. Newer Hy-Gain designs, the 105CA, 155CA and 205CA, have been redesigned by computer for better performance.

Hy-Gain antennas have historically had an excellent reputation for superior mechanical design. In the older designs the elements were purposely spaced along the boom to achieve good weight balance at the mast-to-boom bracket, with electrical performance as a secondary goal. Thus, the electrical performance was not necessarily optimal, particularly over an entire amateur band.

Newer Hy-Gain designs are electrically superior to the older ones, but because of the strong concern for weight balance are still not optimal by the definitions used in this chapter. With the addition of wind torque-compensation dummy elements, and with extra lead weights where necessary at the director end of the boom for weight-balance, the electrical performance can be enhanced, using the same proven mechanical parts.

Note that because the HyGain boom-to-element clamps require several inches of boom, the elements can't be mounted exactly at the end of the boom. Since the boom length of 34' is the same as the sum of element spacings, Director 3 cannot have a spacing of 190" and will have to be a few inches less. This does not result in a significant change in the antenna pattern according to designer Dean Straw, N6BV.

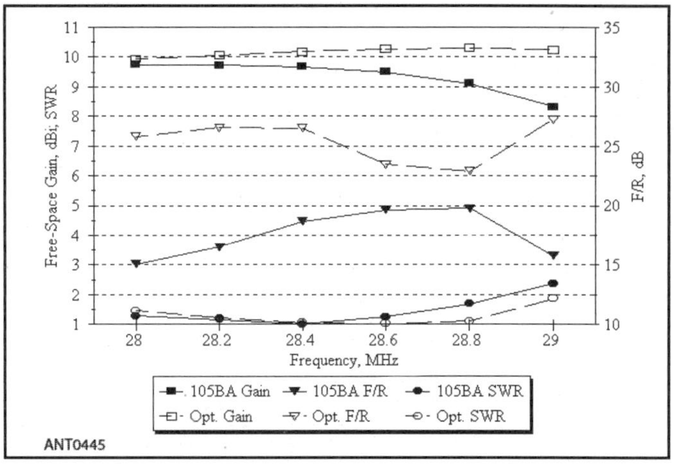

Figure 13.20 — Gain, F/R and SWR over the 28.0 to 28.8 MHz range for original and optimized Yagis using Hy-Gain hardware. Original 105BA design provided excellent weight balance at boom-to-mast bracket, but compromised the electrical performance somewhat because of non-optimum spacing of elements. Optimized design requires wind torque-balancing compensator element, and compensating weight at director end of boom to rebalance weight. The F/R ratio over the frequency range for the optimized design is more than 23 dB. Each element uses the original Hy-Gain taper schedule and element-to-boom clamp, but the length of the tip is changed per Table 13.8.

Table 13.8
Optimized Hy-Gain 20 Meter Yagi Designs

Optimized 204BA, Four-element 20 meter Yagi, 26 foot boom

Element	*Spacing*	*Element Tip*
File Name		BV204CA.YW
Reflector	0.000"	56.000"
Driven Element	85.000"	52.000"
Director 1	72.000"	61.500"
Director 2	149.000"	50.125"

Optimized 205CA, Five-element 20 meter Yagi, 34 foot boom

Element	*Spacing*	*Element Tip*
File Name		BV205CA.YW
Reflector	0.000"	62.625"
Driven Element	72.000"	53.500"
Director 1	72.000"	63.875"
Director 2	74.000"	61.625"
Director 3	190.000"	55.000"
Director 3	190.000"	55.000"

Figure 13.20 shows the computed gain, F/R ratio and SWR for a 24-foot boom, 10 meter optimized Yagi (modified 105BA) using Hy-Gain hardware. **Figure 13.21** shows the same for a 26-foot boom 15 meter Yagi (modified 155BA), and **Figure 13.22** shows the same for a 34-foot boom (modified 205BA) 20 meter Yagi. **Tables 13.8** through **13.10** show dimensions for these designs. The original Hy-Gain taper schedule is used for each element. Only the length of the end tip (and the spacing along the boom) is changed for each element.

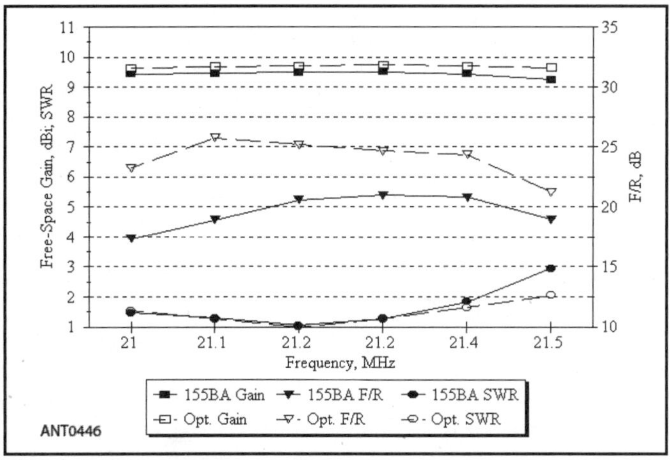

Figure 13.21 — Gain, F/R and SWR over the 21.0 to 21.45 MHz band for original and optimized Yagis using Hy-Gain hardware. Original 155BA design provided excellent weight balance at boom-to-mast bracket, but compromised the electrical performance somewhat because of non-optimum spacing of elements. Optimized design requires wind torque-balancing compensator element, and compensating weight at director end of boom to rebalance weight. The F/R ratio over the frequency range for the optimized design is more than 22 dB. Each element uses the original Hy-Gain taper schedule and element-to-boom clamp, but the length of the tip is changed per Table 13.9.

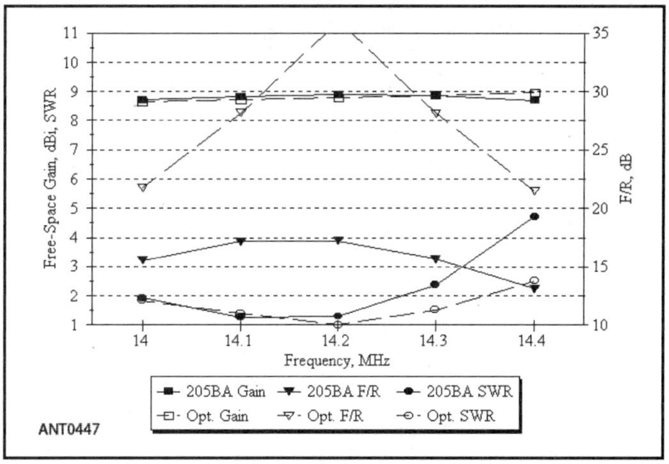

Figure 13.22 — Gain, F/R and SWR over the 14.0 to 14.35 MHz band for original and optimized Yagis using Hy-Gain hardware. Original 205BA design provided good weight balance at boom-to-mast bracket, but compromised the electrical performance because of non-optimum spacing of elements. Optimized design requires wind torque-balancing compensator element, and compensating weight at director end of boom to rebalance weight. The F/R ratio over the frequency range for the optimized design is more than 23 dB, while the original design never went beyond 17 dB of F/R. Each element uses the original Hy-Gain taper schedule and element-to-boom clamp, but the length of the tip is changed per Table 13.10.

Table 13.9
Optimized Hy-Gain 15 Meter Yagi Designs

Optimized 155BA, Five-element 15 meter Yagi, 26 foot boom

Element	*Spacing*	*Element Tip*
File Name		BV155CA.YW
Reflector	0.000"	64.000"
Driven Element	48.000"	65.500"
Director 1	48.000"	63.875"
Director 2	82.750"	61.625"
Director 3	127.250"	55.000"

Table 13.10
Optimized Hy-Gain 10 Meter Yagi Designs

Optimized 105BA, Five-element 10 meter Yagi, 24 foot boom

Element	*Spacing, inches*	*Element Tip*
File Name		BV105CA.YW
Reflector	0.000"	44.250"
Driven Element	40.000"	53.625"
Director 1	40.000"	52.500"
Director 2	89.500"	50.500"
Director 3	112.250"	44.750"

13.4.9 4-Element 40-Meter OWA Yagi

This 4-element 40-meter design on a 48-foot boom was installed at the K3LR contest station and designed by WA3FET and Nathan Miller, NW3Z. The antenna is described in more detail at **naic.edu/~angel/kp4ao/ham/owa.html**.

A simple NEC-4 model shown below was modeled in *EZNEC Pro/4+* (*EZNEC* is now a free program, downloadable at **eznec.com**) and the optimizer program *AutoEZ* (**ac6la.com**) The aluminum tubing elements are not tapered for simplicity, with all elements modeled as constant diameter 1-inch aluminum tubing. The geometry for this Yagi is shown in **Figure 13.23** from *EZNEC*. The element dimensions are shown from

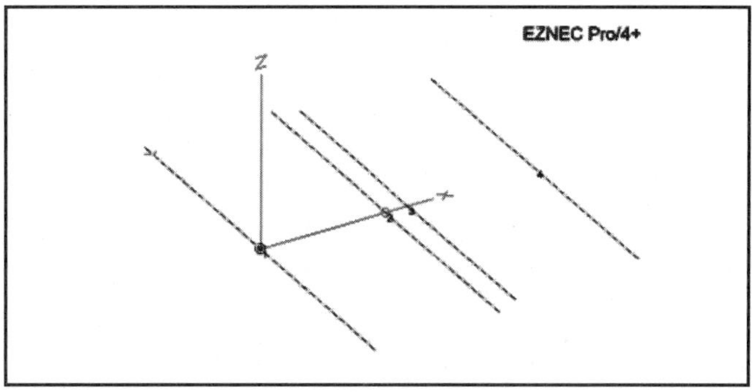

Figure 13.23 — Three-dimensional geometry of the 4-element 40-meter OWA Yagi.

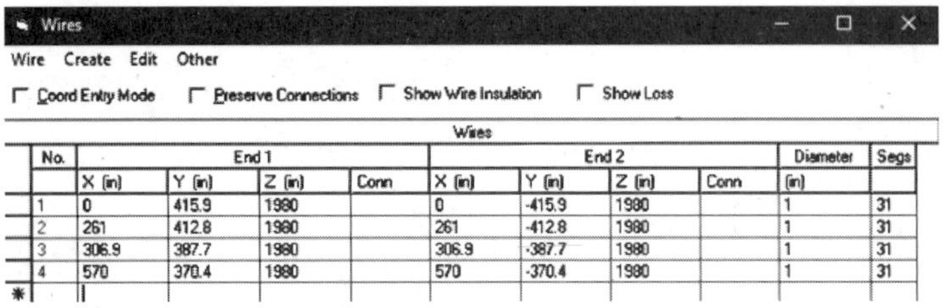

Wires

Wire Create Edit Other

Coord Entry Mode | Preserve Connections | Show Wire Insulation | Show Loss

Wires

No.	End 1				End 2				Diameter	Segs
	X (in)	Y (in)	Z (in)	Conn	X (in)	Y (in)	Z (in)	Conn	(in)	
1	0	415.9	1980		0	-415.9	1980		1	31
2	261	412.8	1980		261	-412.8	1980		1	31
3	306.9	387.7	1980		306.9	-387.7	1980		1	31
4	570	370.4	1980		570	-370.4	1980		1	31

Figure 13.24 — Model dimensions for the 4-element 40-meter OWA Yagi.

Figure 13.25 — SWR performance of the 4-element 40-meter OWA Yagi model across the 40-meter band.

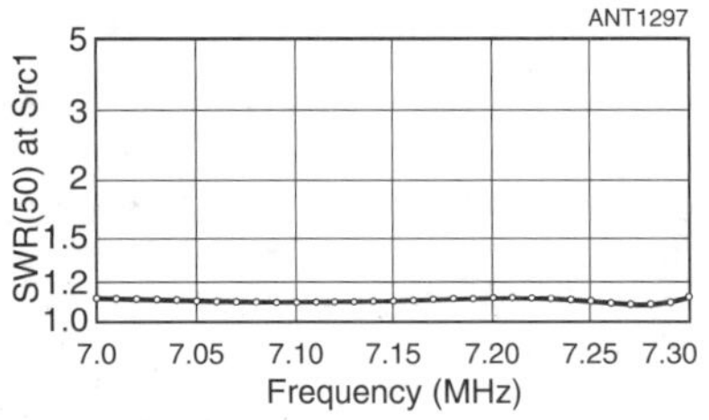

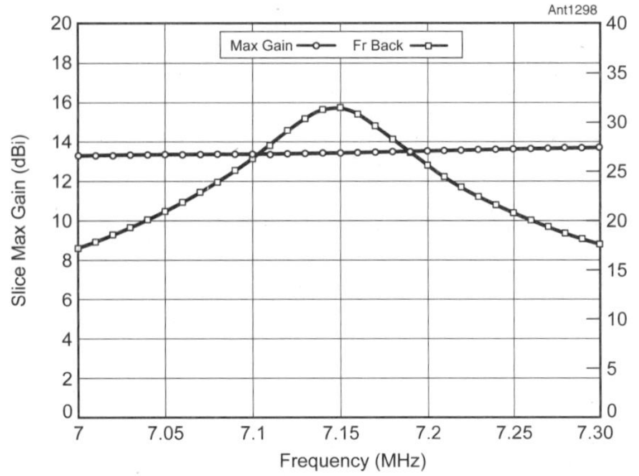

Figure 13.26 — Gain and Front-to-Back ratio (F/B) of the 4-element 40-meter OWA Yagi model across the 40-meter band.

EZNEC in **Figure 13.24** at a height of 165 ft over average ground conditions. (Radiation patterns can be obtained from the *EZNEC* models.)

Figures 13.25 and **26** show the performance of the OWA Yagi. Most important about the performance is the extremely flat SWR (1.1:1) and gain (13 dBi) across the 40-meter band. Front-to-back performance peaks in the center of the band at about 33 dB, dropping to about 18 dB at the band edges. This consistency of performance is very useful as explained in the "High-Performance Yagi Antennas" section of the **VHF, UHF, and Microwave Antennas** chapter.

CONSTRUCTION OF THE ELEMENTS AND BOOM

The elements are made of standard .058-inch wall 6061-T6 aluminum tubing. Where the tubing overlaps, the joint is coated with antioxidant and secured with 4 aluminum pop rivets set 90 degrees apart. See the section on element tapers below.

The parasitic elements are mounted to the boom using four U-bolts and a plate is made of 4 x 0.375-inch-thick aluminum bar stock. Because the OWA Yagi uses direct feed, the driven element must be split in the center and insulated from the boom. To insulate the element, the mounting plate is made of 3/8-inch thick Garolite, a phenolic material. A gap of 2 inches is left in the center of the driven element and a 24-inch section of fiberglass rod of appropriate diameter is inserted inside the tubing to provide strength and to prevent crushing. Two U-bolts are used on each half of the driven element for additional support.

It is important to note that when measuring the driven elements, all measurements are made from the center of the gap, not the beginning of the tubing. For example, if the taper schedule calls for 48 inches of 1.000-inch tubing for the inside part of the element half, the tubing on each side must be cut to 47 inches to leave a 2-inch gap in the center.

48-foot booms must be vertically guyed with a truss from the mast to eliminate most of the sag. Stainless steel eyebolts are installed through the boom. Phillystran, a PVC-protected Kevlar rope, is used for the truss support cables. Because Phillystran is non-conductive, interaction is avoided.

ELEMENT TAPER SCHEDULE

To convert the constant diameter design of the models to a tapered stepped-diameter construction for each element, the following procedures could be used. One could choose an appropriate taper, for example, from the K5GO tapers as described in the **Antenna Materials and Construction** chapter and the online material. One could choose either light duty or heavy duty depending on the wind survivability desired. All the various tapered stepped-diameter sections are given except for the lengths of the tips.

One way to find these tip lengths is to model each constant diameter element by itself and sweep it in frequency to determine its resonance, the frequency where the imaginary part of the impedance will be very close to 0 ohms. Then pick some length for the tapered element tip and run the model with that element at the resonant frequency found from the constant diameter element. Note the value of the imaginary part of the impedance at this resonant frequency for this tapered element with the chosen tip length. Call this tip length L1 and the imaginary impedance value at the resonant frequency X1. Then, pick another tip length not much different than the L1 value, maybe 6 inches longer or shorter or so, appropriately. Call this new tip length L2 and run the model with that element and find the imaginary impedance value at the same resonant frequency and call it X2. Now compute the correct tip length (L) to use by the following formula:

$$L = (L2-L1)/(X2-X1) \times (-X1) + L1$$

You can then check this by running the model's tapered element with this tip length (L) and do a sweep like you did for the constant diameter element. It should resonate very close to the same frequency.

Another method is to use optimization in the *AutoEZ* program with the tip length as a variable and then run it at the resonant frequency and find the tip length that gives the imaginary part of the impedance equal to 0 as a goal. Other programs like *4NEC2* also have optimization built in.

Another method is to use the procedure in the Leeson book *Physical Design of Yagis* that has a spreadsheet method to work with tapered and equivalent constant diameter elements.

13.4.10 HF OWA Yagis

This set of OWA Yagis was designed by Justin Johnson, GØKSC, of Innovantennas (**innovantennas.com**) for the *Antenna Book*. The beams cover 30 meters through 10 meters, with a 9-element, 10-meter beam that will certainly make the most of the band during the peak of Cycle 25.

The set includes the following antennas:

30 Meters: 4 element
20 Meters: 4, 5, and 7 element
17 Meters: 6 element
15 Meters: 6 element
12 Meters: 6 element
10 Meters: 4, 5, 6, 7, 9 element

The elements are made of standard .058-inch wall 6061-T6 aluminum tubing. Where the tubing overlaps, the joint is coated with antioxidant and secured with four aluminum pop rivets set 90 degrees apart. Because the OWA Yagi uses direct feed, the driven element must be split in the center and insulated from the boom using Garolite phenolic or a similar material. Leave a gap of 1 – 2 inches in the center of the driven element and a 12 – 24-inch section of fiberglass rod of appropriate diameter inserted inside the tubing to provide strength and to prevent crushing. Two U-bolts should be used on each half of the driven element for additional support.

When measuring the driven elements, all measurements are made from the center of the gap, not the beginning of the tubing. For example, if the model calls for 48 inches of 1.000-inch tubing for the inside part of the element half, the tubing on each side must be cut to 47 inches to leave a 2-inch gap in the center.

The EZNEC model file for each antenna is provided in the online material. The elements are specified with metric (mm) measurements, but the diameters are chosen to match the inch-based telescoping tubing sizes available in North America. (See the **Antenna Materials and Construction** chapter.) Use 0.049-inch or 0.058-inch wall thickness tubing as specified in Table 10.3. Lengths are given in the model's Wires table in millimeters (0.001-meter) increments. This represents an excellent compromise between accuracy of the model and specifying available materials for North American readers.

13.5 Multiband Yagis

So far, this chapter has discussed monoband Yagis — that is, Yagis designed for a single Amateur Radio frequency band. Because hams have operating privileges on more than one band, multiband coverage has always been very desirable.

INTERLACING ELEMENTS

In the late 1940s, some experimenters tried interlacing Yagi elements for different frequencies on a single boom, mainly to cover the 10 and 20 meter bands (at that time the 15 meter band wasn't yet available to hams). The experimenters discovered that the mutual interactions between different elements tuned to different frequencies are very difficult to handle.

Adjusting a lower-frequency element usually results in interaction with higher-frequency elements near it. In effect, the lower-frequency element acts like a retrograde reflector, throwing off the effectiveness of the higher-frequency directors nearby. Element lengths and the spacing between elements can be changed to improve performance of the higher-frequency Yagi, but the resulting compromise is rarely equal to that of an optimized monoband Yagi. A reasonable compromise for portable operation was developed by VE7CA and is described in the **Portable Antennas** chapter.

TRAP MULTIBANDERS

Multiband Yagis using a single boom can also be made using traps. Traps allow an element to have multiple resonances. The **Multiband HF Antennas** chapter provides details on trap designs. The general function is very similar to trap dipoles in which the traps act as open circuits or reactances that change the electrical length of the element at different frequencies.

Commercial vendors have sold trap antennas to hams since the 1950s and surveys show that after simple wire dipoles and multiband verticals, trap triband Yagis are the most popular antennas in the Amateur Radio Service.

The originator of the trap tribander was Chester Buchanan, W3DZZ, in his March 1955 *QST* article, "The Multimatch Antenna System." On 10 meters this rather unusual tribander used two reflectors (one dedicated and one with traps) and two directors (one dedicated and one with traps). On 20 and 15 meters three of the five elements were active using traps. The W3DZZ tribander employed 12 traps overall, made with heavy wire and concentric tubular capacitors to hold down losses in the traps. Each trap was individually fine-tuned after construction before mounting it on an element.

Another example of a homemade tribander was the 26-foot boom 7-element 20/15/10 meter design described by Bob Myers, W1XT (ex-W1FBY) in the December 1970 *QST*. The W1FBY tribander used only two sets of traps in the driven element, with dedicated reflectors and directors for each frequency band. Again, the traps were quite robust in this design to minimize trap losses, using 7⁄16-inch aluminum tubing for

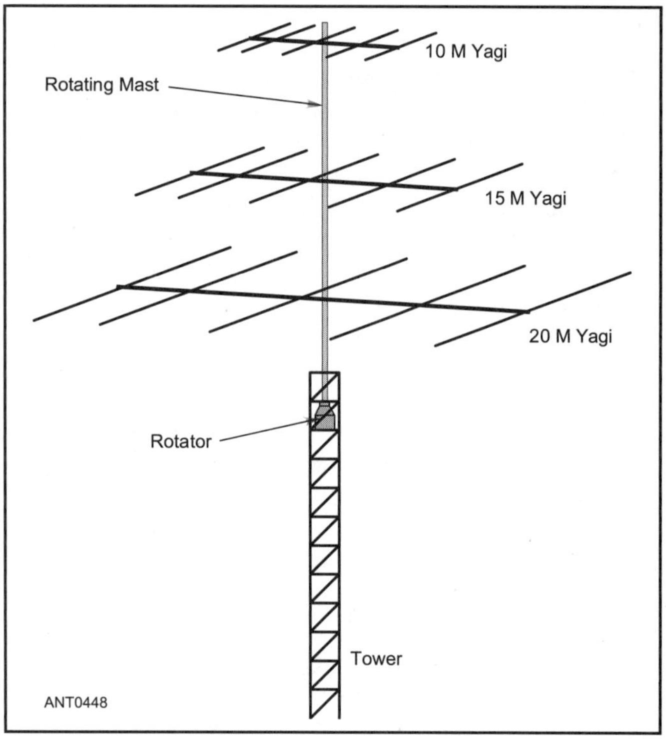

Figure 13.27 — "Christmas Tree" stack of 20/15/10 meter Yagis spaced vertically on a single rotating mast.

the coils and short pieces of RG-8 coax as high-voltage tuning capacitors.

Relatively few hams actually build tribanders for themselves, mainly because of the mechanical complexity and the close tolerances required for such antennas. The traps themselves must be constructed quite accurately for reproducible results, and they must be carefully weatherproofed for long life in rain, snow, and often polluted or corrosive atmospheres.

Traps, like any lumped-constant circuit, have some amount of loss which can be minimized with careful design. The primary compromise incurred in a trap multiband Yagi is the fixed element spacing on all bands. The usual tribander design is optimized for the middle band while the spacing is a bit too long for the highest band and a bit too short for the lowest band. Nevertheless, trap tribanders provide good performance in a compact package.

CHRISTMAS TREE STACKS

Another possible method for achieving multiband coverage using monoband Yagis is to stack them in a "Christmas tree" arrangement as in **Figure 13.27**. For an installation covering 20, 15 and 10 meters, you could mount the 20 meter monobander on the rotating mast just at the top of the tower. Then perhaps 9 feet above that you would mount the 15 meter monobander, followed by the 10 meter monoband Yagi 7 feet

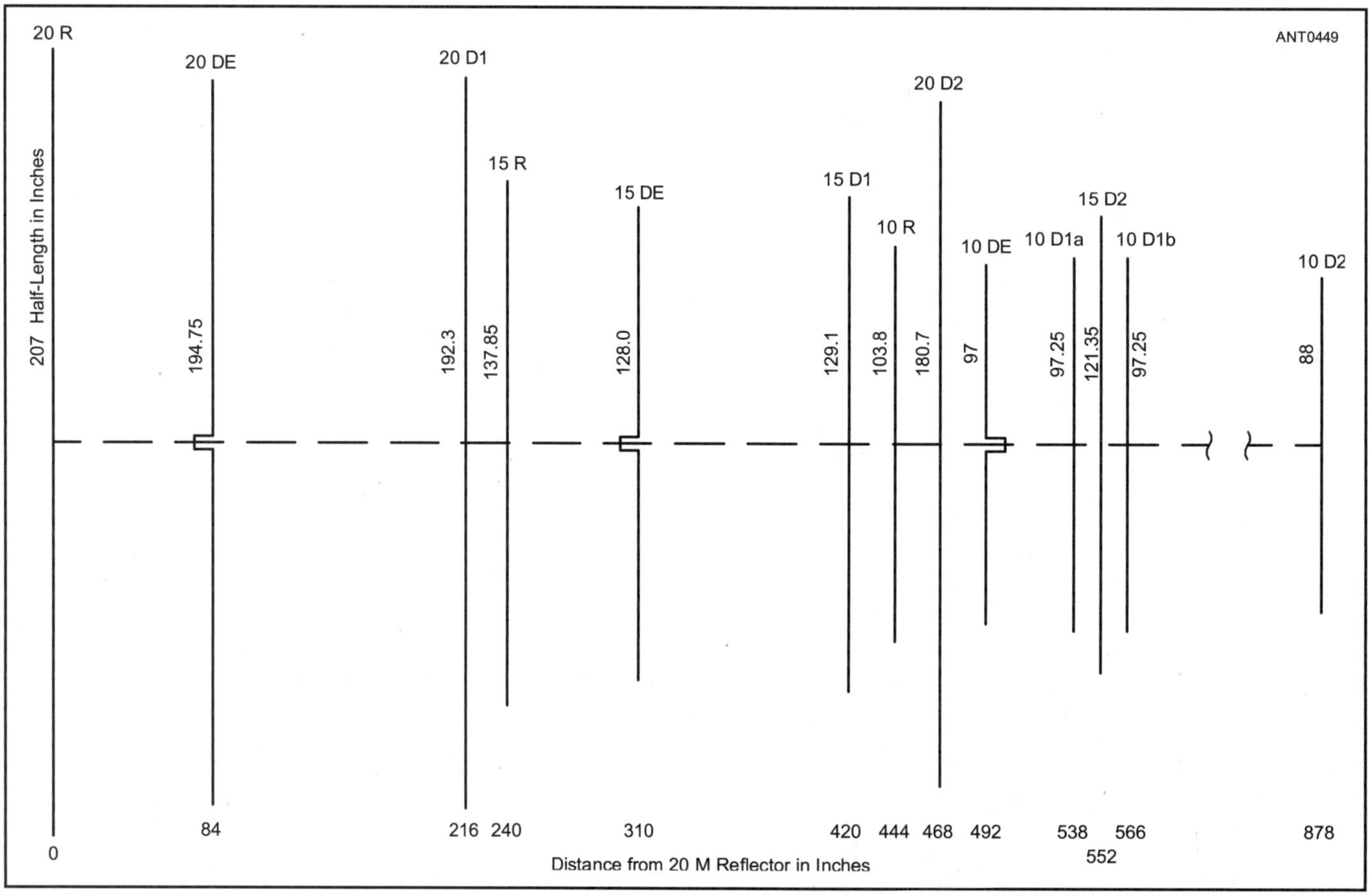

Figure 13.28 — Dimensions of K5RR's trap-less tribander using "forward stagger" and open-sleeve techniques to manage interaction between elements for different frequencies.

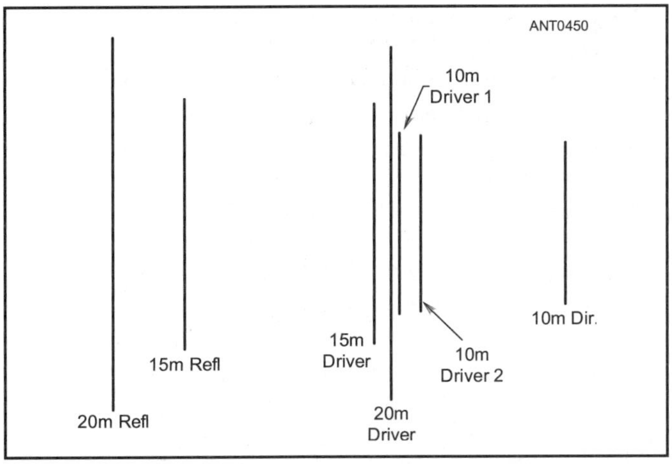

Figure 13.29 — Layout of Force 12 C-3 multiband Yagi. Note that the 10 meter (driver/director) portion of the antenna is "forward staggered" ahead of the 15 meter (reflector/driver) portion, which in turn is placed ahead of the 20 meter (reflector/driver) portion. The antenna is fed at the 20 meter driver, which couples parasitically to the 15 meter driver and the two 10 meter drivers.

further up on the mast. Another configuration would be to place the 10 meter Yagi in between the lower 20 meter and upper 15 meter Yagis. Whatever the arrangement, the antenna in the middle of such a Christmas-tree always suffers the most interaction from the lowest-frequency Yagi.

Dave Leeson, W6NL, mentions that the 10 meter Yagi in a closely stacked Christmas Tree (15 meters at the top, 10 meters in the middle, and 20 meters at the bottom of the rotating mast) loses "substantial gain" because of serious interaction with the 20 meter antenna. (N6BV and K1VR calculated that the free-space gain in the W6NL stack drops to 5 dBi, compared to about 9 dBi with no surrounding antennas.) Monobanders are *definitely not* universally superior to tribanders in multiband installations.

FORWARD STAGGERING

Some hams have built multiband Yagis on a common boom, using a technique called *forward staggering*. This means that that most (or all) of the higher-frequency elements are placed in front of any lower-frequency elements — in other words, most of the elements are not interlaced. Richard Fenwick, K5RR, described his triband Yagi design in the September 1996 *QEX* magazine. This uses forward-stagger and open-sleeve design techniques and was optimized using several sophisticated modeling programs.

Fenwick's tribander used a 57-foot, 3-inch OD boom to hold 4 elements on 20 meters, 4 elements on 15 meters and 5 elements on 10 meters. **Figure 13.28** shows the element placement for the K5RR tribander. Most hams, of course, don't have the real-estate or the large rotator needed to turn such a large, but elegant solution to the interaction problem!

FORCE 12 C-3 "MULTI-MONOBAND" TRIBAND YAGI

Antenna manufacturer Force 12 used forward-stagger layouts and patented combinations of open- and closed-sleeve drive techniques extensively in their product line of multiband antennas, which they called "multi-monoband Yagis." **Figure 13.29** shows the layout for the popular Force 12 C-3 triband Yagi. The C-3 uses no traps, thereby avoiding any losses due to traps. The C-3 consists of three 2-element Yagis on an 18-foot boom, using full-sized elements designed to withstand high winds. (There is a pair of 10 meter driven elements for coverage of the full band.) Similar multiband Yagis with various band combinations are available from several manufacturers.

The C-3 feed system employs open-sleeves, where the 20 meter driver element is fed with coax through a common-mode current balun and parasitically couples to the closely spaced 15 meter driver and the two 10 meter driven elements to yield a feed point impedances close to 50 Ω on all three bands. Open-sleeve dipoles are discussed in the **Multiband HF Antennas** chapter.

Note the use of the forward-stagger technique in the C-3, especially on 10 meters. To reduce interaction with the lower-frequency elements behind it, the 10 meter portion of the C-3 is mounted on the boom ahead of all the lower-frequency elements with the main 10 meter parasitic element (#7) acting as a director. The lower-frequency elements behind the 10 meter section act as retrograde reflectors, gaining some improvement of the gain and pattern compared to a monoband 2-element Yagi. A simplified *EZNEC* model of the C-3 is included with this book's online material.

On 15 meters, the main parasitic element (#2) is a dedicated reflector, but the other elements ahead on the boom act like retrograde directors to improve the gain and pattern somewhat over a typical 2-element Yagi with a reflector. On 20 meters, the C-3 is a 2-element Yagi with a dedicated reflector (#1) at the back end of the boom.

The exact implementation of any Yagi, of course, depends on the way the elements are constructed using telescoping aluminum tubing. The C-3 type of design is no exception.

13.6 Shortening Yagi Elements

Almost any technique that can be used to reduce the physical length of a dipole can also be used to shorten the physical length of a Yagi element. The tradeoffs are additional mechanical complexity and reduced performance with respect to forward gain and SWR bandwidth. As with shortened dipoles and monopoles, placement of the loading structures is critical to obtaining good performance and careful modeling is required. (Caution should be used in modeling wires that are very close to each other, junctions of large-diameter conductors, and other complex mechanical arrangements.)

LINEAR LOADING

The most common size-reducing technique is *linear loading* and it can be applied to Yagis as well as dipoles and verticals. An example of linear loading for a dipole was presented in the **General Purpose MF and HF Antennas** chapter. A very similar example of linear loading for a 2-element 20 meter Yagi can be found in a June 1976 *QST* article by Cole Collings, WØYNF.

Linear loading essentially consists of folding the antenna into a zig-zag pattern. Each back-and-forth folded segment radiates very little because the field from each of the folded conductors partially cancels that of the adjacent conductors. Nevertheless, the folding does extend the electrical length of the antenna. The effective length of the folded antenna is somewhat longer than if the section remained unfolded.

The Hy-Gain 402BA 2-element 40 meter Yagi was a popular linearly-loaded antenna with 46-foot elements. A full-size element on 40 meters is approximately 65 feet long, so linear loading provided a substantial reduction in size.

END LOADING AND INDUCTOR LOADING

The technique of adding capacitance hats near the end of an antenna to lower its resonant frequency is most often encountered in vertical ground-plane antennas for the lower HF bands. The technique can also be put to good use on HF Yagis as seen in the Cushcraft (**cushcraftamateur.com**) MA5B mini-beam for 20/17/15/12/10 meters. The capacitance hats on this multiband Yagi play a major role in reducing the longest element to a bit over 17 feet long — just over λ/4 on 20 meters.

The elements of the MA5B also use traps and that also helps reduce length by inserting inductance into the element below the trap's resonant frequency. The Cushcraft XM240 2-element 40 meter Yagi also uses a combination of capacitance hats and coils to reduce element size.

Inductors on large Yagis for 75/80 meters are used similarly to base loading in verticals. The same general concerns apply with the inductance and placement of the coil, as well as losses in the coil.

13.7 The Moxon Rectangle

L.B. Cebik, W4RNL, has written extensively about the *Moxon rectangle*, an antenna invented by Les Moxon, G6XN, derived from a design by VK2ABQ (see "References and Bibliography" section). The Moxon rectangle beam takes less space horizontally than a conventional 2-element Yagi design, yet it offers nearly the same amount of gain and a superior front-to-back ratio. And as an additional benefit, the drive-point impedance is close to 50 Ω, so that it doesn't need a matching section. For example, rather than a "wingspan" of 17 feet for the reflector in a conventional 2-element 10-meter Yagi, the Moxon rectangle is 13 feet wide, a saving of almost 25%. The closed rectangular mechanical assembly gives some rigidity to the design, keeping it stable in the wind.

The Moxon rectangle functions by virtue of the mutual coupling between parallel element segments and the coupling between the facing element tips, which are both bent toward each other. See **Figure 13.30** which shows the general outline for W4RNL's 10-meter aluminum Moxon rectangle from his June 2000 *QST* article (see "References and Bibliography"). The tips of the elements are kept a fixed distance from each

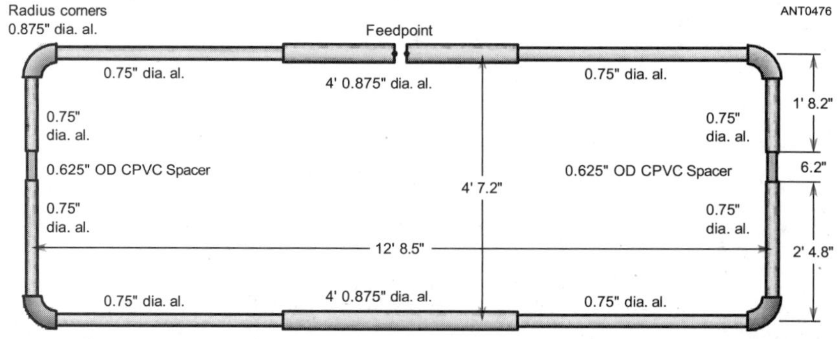

Figure 13.30 — General outline of the 10 meter aluminum Moxon rectangle, showing tubing dimensions.

other by PVC spacers. If any dimension of the Moxon rectangle is critical, it is the separation of the tips. For VHF/UHF Moxons made from solid wire, be sure to secure the tips so they remain aligned across the gap and a fixed distance apart.

The calculations for designing a Moxon have been implemented as software calculator in *MoxGen* by Dan Maguire, AC6LA, which is available at no cost, downloadable from **ac6la.com/moxgen1.html**.

Two additional HF Moxon designs, one for 10 meters and another for 20 meters, are included in the online material. See the **Portable Antennas** chapter for Moxon designs on 17 through 10 meters. Moxon rectangles for 6, 2, 1¼ meters, and 70 cm are included in the **VHF, UHF, and Microwave Antennas** chapter. A list of articles about Moxons by W4RNL is online at **on5au.be/content/a10/moxon/moxpage.html**.

13.7.1 Moxon Design Behavior

DESIGN SENSITIVITIES

• Decreasing or increasing the side-to side dimension (A in Figure 1) raises or lowers the maximum F/B and the resonant feed point frequencies by about 40 kHz. For small changes in dimension A, the resonant feed point impedance does not change.

• Increasing or decreasing only the length of the driver tails (dimension B) by one inch lowers or raises the resonant frequency of the driver by about 70 kHz on 10 meters (proportionally less on other bands). The new resonant feed point impedance will be a few ohms lower (for an increase in driver length) than before the change. The frequency of maximum F/B will not change significantly.

• Increasing or decreasing only the length of the reflector tails (dimension D) by one inch lowers or raises the peak F/B frequency by about 70 kHz (again, on 10 meters). The driver's resonant frequency will not significantly change, but the impedance will be higher (for an increase in reflector length) than before the change.

BEHAVIOR WITH HEIGHT

As the antenna height increases, the take-off angle (or the elevation angle of maximum radiation) decreases, as do the vertical and horizontal beamwidths between half-power points. These properties are in line with those of any horizontally polarized array. Hence, the gain increases slightly with antenna height increases.

Moreover, the feed point impedance of the antenna undergoes only small changes with changes in heights. Indeed, the excellent F/B performance at the low height of ⅜λ holds promise for 40-meter and lower-frequency installations. The conclusion is that a Moxon rectangle falls in the class of "well-behaved" antennas, requiring no finicky field adjustments once the basic design is set and tested.

13.7.2 Wire Moxon Rectangles for 40 – 10 Meters

A table of Moxon rectangle dimensions for the HF amateur bands was published in June 2000 *QST* and is repeated here as **Table 13.11**. Dimensions for other bands may be obtained with the *MoxGen* software. All dimensions are labeled in **Figure 13.31**.

A tubing model will have dimensions close to those for a thin wire model. However, it will exhibit a broader SWR bandwidth. The *MoxGen* software accepts both wire gauge and tubing OD for its calculations.

These antennas have feed point impedances between about 56 and 58 Ω, a close match to the standard amateur 50 Ω coaxial cable. Free space gain and front-to-back ratio are consistent for all the models, averaging 5.8 dBi and greater than 32 dB in free space, respectively, at the design frequency, centered in each band.

Table 13.11
Dimensions of Wire Moxon Rectangles for 80 – 10 Meters

All dimensions refer to designations in Figure 1. Dimensions are in feet and apply to #14 AWG bare-wire antennas.

Band	*Frequency (MHz)*	*A*	*B*	*C*	*D*	*E*
80	3.6	99.98	15.47	2.16	18.33	36.96
75	3.9	92.28	14.28	2.00	16.92	33.20
40	7.09*	50.69	7.82	1.15	9.35	18.32
20	14.175	25.30	3.87	0.62	4.70	9.19
15	21.225	16.88	2.56	0.44	3.14	6.14
10	28.3*	12.65	1.90	0.35	2.36	4.61

*Because of bandwidth versus wire-size considerations, 40- and 10-meter design frequencies are below the mid-band points to obtain less than 2:1 50 Ω SWR over as much of the band as possible. See the text for alternative strategies.

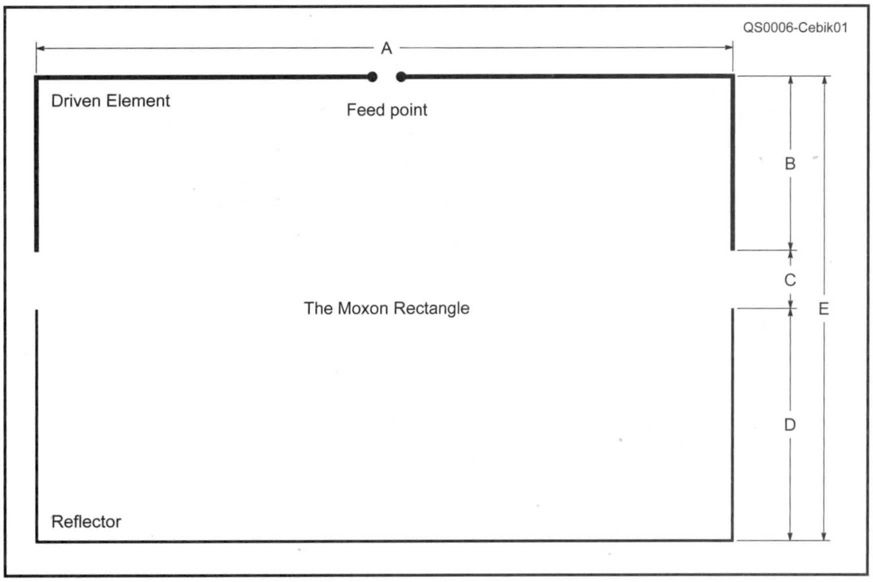

Figure 13.31 — Outline of a Moxon rectangle with various dimensions labeled. Table 13.11 shows the dimensions for various HF amateur bands.

13.7.3 40 Meter Moxon Rectangle

Dave Leeson, W6NL, has modified the Cushcraft XM240 2-element 40-meter Yagi to a Moxon rectangle design shown in **Figure 13.32**. The W6NL Moxon Yagi is a high efficiency design that uses cross elements to provide both loading and the Moxon coupling. The upgrade of the XM240 to the W6NL Moxon consists of replacing the loading coil LCA sections with four new assemblies, each consisting of two new sections and the new T loading element. The remaining parts are original Cushcraft.

The antenna has a gain of more than 10 dBi (including ground reflections) and a high front-to-back ratio (not specified by the designer). As is usual for Moxon designs, the SWR bandwidth is very good — more than 300 kHz with an SWR of less than 1.5:1.

Modifying the XM240 is described in detail in W6NL's design article, "Construction of W6NL Moxon on Cushcraft XM240," included with the online material. The mechanical strength of the antenna is also improved as part of the modification procedure.

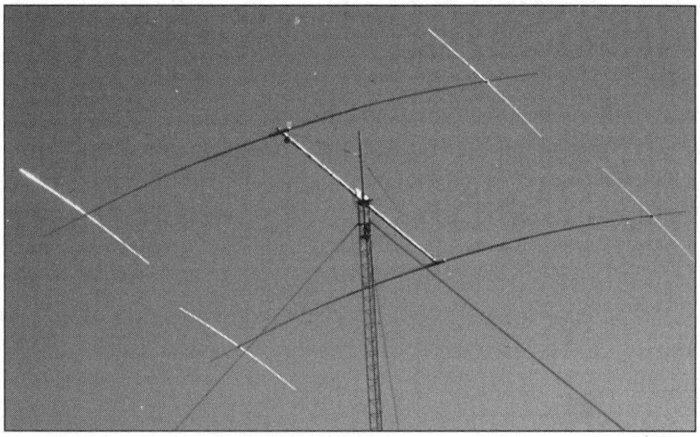

Figure 13.32 — A Cushcraft XM240 2-element 40 meter Yagi is modified by W6NL to become a Moxon Rectangle. The antenna is mechanically strengthened during the modification, as well. (Photo by Dave Leeson, W6NL)

13.8 Quad Antennas

The previous section discussed Yagi arrays as systems of approximately half-wave dipole elements that are coupled together mutually. You can also employ other kinds of elements using the same basic principles of analysis. For example, loops of various types may be combined into directive arrays. A popular type of parasitic array using loops is the *quad antenna*, in which loops having a perimeter of about one wavelength are used in much the same way as half-wave dipole elements in the Yagi antenna.

Clarence Moore, W9LZX, created the quad antenna in the early 1940s while he was at the Missionary Radio Station HCJB in Quito, Ecuador. He developed the quad to combat the effects of corona discharge at high altitudes. The problem at HCJB was that their large Yagi was literally destroying itself by melting its own element tips. This occurred due to the huge balls of corona it generated in the thin atmosphere of the high Andes Mountains. Moore reasoned correctly that closed loop elements would generate less high voltage — and hence less corona — than would the high impedances at the ends of a half-wave dipole element.

Figure 13.33 shows the original version of the two-element quad, with a driven element and a parasitic reflector. The square loops may be mounted either with the corners lying on horizontal and vertical lines, as shown at the left, or with two sides horizontal and two vertical (right). The feed points shown for these two cases will result in horizontal polarization, which is commonly used.

Quad designers may want to look up a copy of Bill Orr, W6SAI's *All About Cubical Quads* (now out of print) for a variety of design notes and ideas. Similarly, R. P. Haviland, W4MB's series of quad-related articles in *Ham Radio* and *QEX* are also worth reading (see References and Bibliography).

13.8.1 Quads Versus Yagis

Since its invention, there has been controversy whether the quad is a better performer than a Yagi. The three main electrical performance parameters of a Yagi are gain, response patterns (front-to-rear ratio, F/R) and feed point impedance/SWR. Proper analysis of a quad also involves checking all these parameters across the entire frequency range over which you

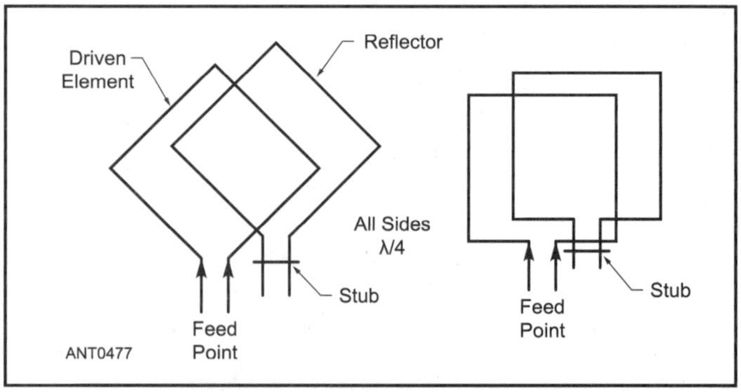

Figure 13.33 — The basic two-element quad antenna, with driven-element loop and reflector loop. The driven loops are electrically one wavelength in circumference (¼ wavelength on a side); the reflectors are slightly longer. Both configurations shown give horizontal polarization. For vertical polarization, the driven element should be fed at one of the side corners in the arrangement at the left, or at the center of a vertical side in the "square" quad at the right.

intend to use it. Both a quad and a Yagi are classified as "parasitic, end-fire arrays." Modern antenna modeling by computer shows that monoband Yagis and quads with the same boom lengths and optimized for the same performance parameters have gains within about 1 dB of each other, with the quad slightly ahead of the Yagi.

Figure 13.34 plots the three parameters of gain, front-to-rear ratio (F/R) and SWR over the 14.0 to 14.35-MHz band for two representative antennas — a monoband three-element quad and a monoband four-element Yagi. Both of these have 26-foot booms and both are optimized for the best compromise of gain, F/R and SWR across the whole band.

While the quad in Figure 13.29 consistently exhibits about 0.5 dB more gain over the whole band, its F/R pattern toward the rear isn't quite as good as the Yagi's over that span of frequencies. This quad attains a maximum F/R of 25 dB at 14.1 MHz, but it falls to 17 dB at the bottom end of the band and 15 dB at the top. On the other hand, the Yagi's F/R stays consistently above 21 dB across the whole 20 meter band. The quad's SWR rises to just under 3:1 at the top end of the band, but stays below 2:1 from 14.0 to almost 14.3 MHz. The Yagi's SWR remains lower than 1.5:1 over the whole band.

The reason the Yagi in Figure 13.29 has more consistent responses for gain, F/R and SWR across the whole 20 meter band is that it has an additional parasitic element, giving two additional variables to play with — that is, the length of that additional element and the spacing of that element from the others on the boom.

Yagi advocates point out that it is easier to add extra elements to a Yagi, given the mechanical complexities of adding another element to a quad. Extra parasitic elements give a designer more flexibility to tailor all performance parameters over a wide frequency range. Quad designers have historically opted to optimize strictly for gain and, as stated before, they can achieve as much as 1 dB more gain than a Yagi with the same length boom. But in so doing, a quad designer typically has to settle for front-to-rear patterns that are peaked over more narrow frequency ranges. The 20 meter quad plots in Figure 13.34 actually represent an even-handed approach, where the gain is compromised slightly to obtain a more consistent pattern and SWR across the whole band.

Figure 13.35 plots gain, F/R and SWR for two 10 meter monoband designs: a five-element quad and a five-element Yagi, both placed on 26-foot booms. The quad now has the same degrees of freedom as the Yagi, and as a consequence the pattern and SWR are more consistent across the range from 28.0 to 28.8 MHz. The quad's F/R remains above about 18.5 dB from 28.0 to 28.8 MHz. Meanwhile, the Yagi maintains an F/R of greater than 22 dB over the same range, but has almost 0.8 dB less gain compared to the quad at the low end of the band, eventually catching up at the high end of the band. The SWR for the quad is just over 2:1 at the bottom of the band, but remains less than 2:1 up to 28.8 MHz. The SWR on the Yagi remains less than 1.6:1 over the whole band.

Figure 13.36 shows the performance parameters for two 15 meter monoband designs: a five-element quad and a five-element Yagi, both on 26-foot booms. The quad is still the leader in gain, but has a less optimal rearward pattern and a somewhat less flat SWR curve than the Yagi. One thing should be noted in Figures 13.34 – 13.36. The F/R pattern on the Yagi is largely determined by the response at the 180° point, directly in back of the frontal lobe. This point is usually referred to when discussing the "front-to-back ratio."

The quad on the other hand has what a sailor might term

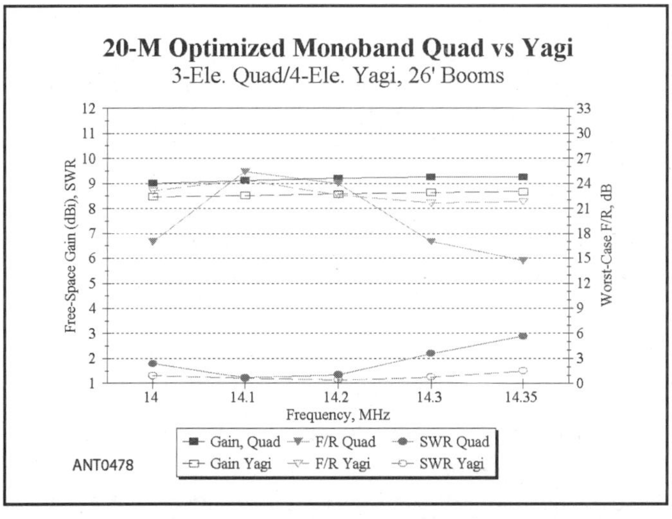

Figure 13.34 — Comparison of gain, F/R and SWR over the 14.0 to 14.35-MHz range for an optimized three-element quad and an optimized three-element Yagi, both on 26-foot booms. The quad exhibits almost 0.5 dB more gain for the same boom length, but doesn't have as good a rearward pattern over the whole frequency range compared to the Yagi. This is evidenced by the F/R curve. The quad's SWR curve is also not quite as flat as the Yagi. The quad's design emphasizes gain more than the other two parameters.

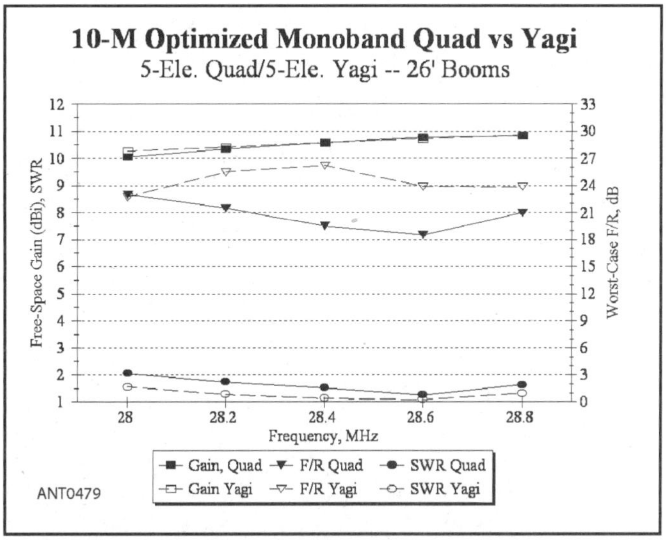

Figure 13.35 — Comparison of gain, F/R and SWR over the 28.0 to 28.8-MHz range for an optimized five-element quad and an optimized five-element Yagi, both on 26-foot booms. The gain advantage of the quad is about 0.25 dB at the low end of the band. The F/R is more peaked in frequency for the quad, however, than the Yagi.

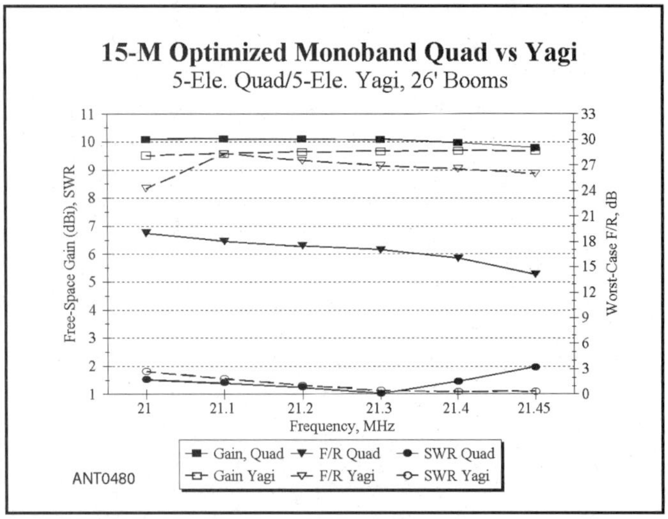

Figure 13.36 — Comparison of gain, F/R and SWR over the 21.0 to 21.45-MHz range for an optimized 5-element quad and optimized 5-element Yagi, both on 26-foot booms. The quad enjoys a gain advantage of about 0.5 dB over most of the band. Its rearward pattern is not as good as the Yagi, which remains higher than 24 dB across the whole range, com-pared to the quad, which remains in the 16-dB average range.

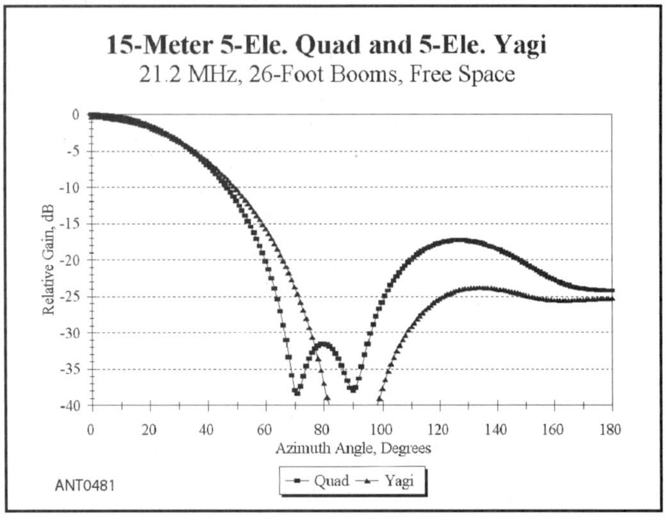

Figure 13.37 — Comparing the pattern of the 15 meter quad and Yagi shown in Figure 13.31. The quad has a slightly narrower frontal beamwidth (it has 0.5 dB more gain than the Yagi), but has higher "rear quartering" sidelobes at about 125° (with a twin sidelobe, not shown, at 235°). These sidelobes limit the worst-case front-to-rear (F/R) to about 17 dB, while the F/B (at 180°, directly at the back of the quad) is more than 24 dB for each antenna.

"quartering lobes" (referring to the direction back toward the "quarterdeck" at the stern of a sailing vessel) in the rearward pattern. These quartering lobes are often worse than the response at 180°, directly in back of the main beam. **Figure 13.37** overlays the free-space E-Field responses of the 15 meter quad and Yagi together. At 21.2 MHz, the quad actually has a front-to-back ratio (F/B) of about 24 dB, excellent in anyone's book. The Yagi at 180° has a F/B of about 25 dB, again excellent.

Table 13.12
Dimensions for Optimized Monoband Quads in Figs 13.34, 13.35 and 13.36, on 26-Foot Booms

	14.2 MHz	*21.2 MHz*	*28.4 MHz*
Reflector	73' 9"	49' 6"	37' 3"
R-DE Spacing	17' 8"	7'	6' 4"
Driven Element	71' 8"	47' 6"	35' 9"
DE-D1 Spacing	8' 3"	5'	5' 6"
Director 1	68' 7"	46' 8"	34' 8"
D1-D2 Spacing	—	6' 8"	6' 9"
Director 2	—	46' 10"	35' 2"
D2-D3 Spacing	—	7' 4"	7' 5"
Director 3	—	45' 8"	34' 2"
Feed method	Direct 50 Ω	Direct 50 Ω	Direct 50 Ω

However, at an azimuth angle of about 125° (and at 235° azimuth on the other side of the main lobe) the quad's "quartering lobe" is down only some 17 dB, setting the worst-case F/R at 17 dB also. As explained in the sections on Yagis, the reason F/R is more important than just the F/B is that on receive, signals can come from any direction, not just from directly behind the main beam.

Table 13.12 lists the dimensions for the three computer-optimized monoband quads shown in Figures 13.34, 13.35, and 13.36.

CUBICAL VERSUS CONCENTRIC QUADS

First — no quad is truly "cubical" in the sense of the distance between the elements being the same as the side of an element. That would place the elements λ/4 apart which is too widely spaced for good performance. The term "cubical quad" generally applies to multiband quads that maintain the same electrical spacing between elements on each band whereas "concentric quad" refers to a set of elements mounted on the spreaders in one plane, concentric to each other. (The two quad antenna projects with this chapter are concentric quads.)

The cubical quad with its consistent electrical spacing has a very slight performance advantage on the higher frequency bands but requires a special spreader mount at the center of the boom to hold the spreaders in the required tilted configuration. In fact, the boom of a true cubical quad is only inches long since the spreaders meet near the center. The cubical quad's spreaders, being both diagonal and tilted, must be a few percent longer than the planar spreaders of the concentric quad.

QUADS VERSUS YAGIS AT LOW HEIGHTS

Another belief held by some quad enthusiasts is that they need not be mounted very high off the ground to give excellent DX performance. Quads are somehow supposed to be greatly superior to a Yagi at the same height above ground. Unfortunately, this is mainly wishful thinking.

Figure 13.38 compares the same two 10 meter antennas as in Figure 13.35, but this time with each one mounted on a 50-foot tower over flat ground, rather than in theoretical free space. The quad does indeed have slightly more gain than a

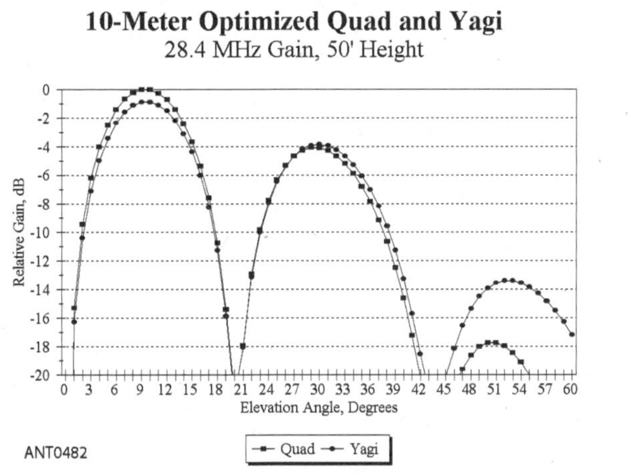

Figure 13.38 — A comparison on 10 meters between an optimized five-element quad and an optimized five-element Yagi, both mounted 50 feet high over flat ground and both employing 26-foot booms. There is no appreciable difference in the peak elevation angle for either antenna. In other words, a quad does not have an appreciable elevation-angle advantage over a Yagi mounted at the same boom height. Note that the quad achieves its slightly higher gain by taking energy from higher-angle lobes and concentrating that energy in the main elevation lobe. This is a process that is similar to what happens with stacked Yagis.

Yagi with the same boom length, as it has in free space. This is evidenced by the very slight compression of the quad's main lobe, but is more obvious when you look at the third lobe, which peaks at about 53° elevation. In effect, the quad squeezes some energy out of its second and third lobes and adds that to the first lobe. However, the difference in gain compared to the Yagi is only 0.8 dB for this particular quad design at a 9° elevation angle. And while it's true that every dB counts, you can also be certain that on the air you wouldn't be able to tell the difference between the two antennas. After all, a 10- to 20-dB variation in the level of signals is pretty common because of fading at HF.

13.8.2 Multiband Quads

On the other hand, one of the valid reasons quads have remained popular over the years is that antenna homebrewers can build multiband quads far more easily than they can construct multiband Yagis. In effect, all you have to do with a quad is add more wire to the existing support arms. It's not quite as simple as that, of course, but the idea of ready expandability for other bands is very appealing to experimenters.

Like the Yagi, the quad does suffer from interactions between wires of different frequencies, but the degree of interaction between bands is usually less for a quad. The higher-frequency bands are the ones that often suffer most from any interaction, for both Yagis and quads. For example, the 10 and 15 meter bands are usually the ones affected most by nearby 20 meter wires in a triband quad, while the 20 meter elements are not affected by the 10 or 15 meter elements.

Modern computer modeling software can help you counteract at least some of the interaction by allowing you to do virtual "retuning" of the quad on the computer screen — rather than clinging precariously to your tower fiddling with wires. However, the programs (such as *NEC-2* or *EZNEC*) that can model three-dimensional wire antennas such as quads typically run far more slowly than those designed for monoband Yagis (such as *YW* included with this book). This makes optimizing rather tedious, but you use the same considerations for tradeoffs between gain, pattern (F/R) and SWR over the operating bandwidth as you do with monoband Yagis.

Two multiband quad design projects are provided in the online material. The first is a large triband 20-, 15-, 10-meter quad built on a 26-foot boom made of 3-inch irrigation tubing. This antenna has three elements on 20 meters, four elements on 15 meters, and five elements on 10 meters.

The second project is a compact two-element triband quad on an 8-foot boom that covers 20, 17, 15, 12, and 10 meters. This antenna uses five concentric wire loops mounted on each of the two sets of spreaders. Either antenna may be constructed in a diamond or square configuration.

While the same basic construction techniques are employed for both multiband quads, the scale of the larger triband antenna makes it a far more ambitious undertaking! The large quad requires a strong tower and a rugged rotator. It also requires a fair amount of real estate in order to raise the quad to the top of the tower without getting tangled in trees or other antennas.

13.8.3 Building A Quad

The parasitic element shown in Figure 13.28 is tuned in much the same way as the parasitic element in a Yagi antenna. That is, the parasitic loop is tuned to a lower frequency than the driven element when the parasitic is to act as a reflector, and to a higher frequency when it is to act as a director. Figure 13.28 shows the parasitic element with an adjustable tuning stub, a convenient method of tuning since the resonant frequency can be changed simply by changing the position of the shorting bar on the stub. In practice, it has been found that the length around the loop should be approximately 3.5% greater than the self-resonant length if the element is a reflector, and about 3.0% shorter than the self-resonant length if the parasitic element is a director. Approximate formulas for the loop lengths in feet are:

$$\text{Driven Element} = \frac{1008}{f_{MHz}}$$

$$\text{Reflector} = \frac{1045}{f_{MHz}}$$

$$\text{Director} = \frac{977}{f_{MHz}}$$

These are valid for quad antennas intended for operation below 30 MHz and using uninsulated #14 AWG stranded copper wire. At VHF, where the ratio of loop circumference to conductor diameter is usually relatively small, the circumference must be increased in comparison to the wavelength. For example, a one-wavelength loop constructed of ¼-inch tubing for 144 MHz should have a circumference about 2% greater than in the above equation for the driven element.

Element spacings on the order of 0.14 to 0.2 free-space wavelengths are generally used. You would employ the smaller spacings for antennas with more than two elements, where the structural support for elements with larger spacings tends to become challenging. The feed point impedances of antennas having element spacings on this order have been found to be in the 40 to 60 Ω range, so the driven element can be fed directly with coaxial cable with only a small mismatch.

For spacings on the order of 0.25 wavelength (physically feasible for two elements, or for several elements at 28 MHz) the impedance more closely approximates the impedance of a driven loop alone — that is, 80 to 100 Ω. The feed methods described in the **Transmission Line System Techniques** chapter can be used, just as in the case of the Yagi.

FEEDING THE MULTIBAND QUAD

There are two approaches to feeding a multiband quad with several driven elements. If the driven elements are all on one set of spreaders the *combined feed* ties all of the elements together at a single feed point. This allows the use of a single feed line but creates a great deal of interaction between harmonically-related elements, reducing gain and F/B dramatically as described by L.B. Cebik in "Feeding the 5-Band Quad" (see References and Bibliography). Using separate feed lines to each driven element results in much less interaction and preserves the quad's performance.

A compromise that allows the use of a single feed line to the station but separate feed lines for each element is to use a remote coax switch such as the Ameritron RCS-4 or RCS-8V (**ameritron.com**). The coax switch can be mounted on the antenna boom or mast and short feed lines run from the switch to the elements. The editor used just such a configuration for a number of years for a five-band 2-element quad with good results.

The impedance of the multiband quad driven elements varies quite a bit from the free-space value of a single loop. Cebik's article mentioned above shows that the feed point impedance varies from close to 50 Ω on 10 meters (the innermost element) to more than 100 Ω on 20 meters (the outermost element). If multiple feed lines are used, quarter-wave matching sections as described in the **Transmission Line System Techniques** chapter can be used to provide an acceptable SWR.

MECHANICAL CONSTRUCTION ISSUES

The most obvious problem related to quad antennas is the ability to build a structurally sound system. If high winds or heavy ice are a normal part of the environment, special precautions are necessary if the antenna is to survive a winter season.

Both multiband quad arrays use fiberglass spreaders. Bamboo is a suitable substitute (if economy is of great importance). However, the additional weight of the bamboo spreaders over fiberglass is an important consideration. A typical 12-foot bamboo pole weighs about 2 pounds; the fiberglass type weighs less than a pound. By multiplying the difference times 8 for a two-element array, times 12 for a three-element antenna, and so on, it quickly becomes apparent that fiberglass is worth the investment if weight is an important factor. Properly treated, bamboo has a useful life of three or four years, while fiberglass life is probably 10 times longer.

One step beyond the conventional fiberglass arm is the pole-vaulting arm. For quads designed to be used on 7 MHz, surplus "rejected" pole-vaulting poles are highly recommended. Their ability to withstand large amounts of bending is very desirable. The cost of these poles is high, and they are difficult to obtain.

Spreader supports (sometimes called *spiders*) described for the multiband quad section below are designed to be less likely to rotate on the boom as a result of wind pressure.

The physical sturdiness of a quad is directly proportional to the quality of the material used and the care with which it is constructed. The size and type of wire selected for use with a quad antenna is important because it will determine the capability of the spreaders to withstand high winds and ice. One of the more common problems confronting the quad owner is that of broken wires. A solid conductor is more apt to break than stranded wire under constant flexing conditions. For this reason, stranded copper wire is recommended. For 14, 21 or 28 MHz operation, #14 or #12 AWG stranded wire is a good choice. Soldering of the stranded wire at points where flexing is likely to occur should be avoided.

You may connect the wires to the spreader arms in many ways. The simplest method is to drill holes through the fiberglass at the appropriate points on the arms and route the wires through the holes. Some amateurs have experienced cracking of fiberglass poles which might be a result of drilling holes through the material. However, this seems to be the exception rather than the rule.

Soldering a wire loop across the spreader is recommended. However, you should take care to prevent solder from flowing to the corner point where flexing could break it. A better method is to clamp a piece of plastic tubing to the spreader with a stainless steel hose clamp and run the wire through the tubing. This allows the wire to slide when the antenna flexes.

Every effort must be placed upon proper construction if you want to have freedom from mechanical problems. Hardware must be secure or vibration created by the wind may cause separation of assemblies. Solder joints should be clamped in place to keep them from flexing, which might fracture a connection point.

While a boom diameter of 2 inches is sufficient for smaller quads using two or even three elements for 14, 21 and 28 MHz, when the boom length reaches 20 feet or longer a 3-inch diameter boom is highly recommended. Wind creates two forces on the boom, vertical and horizontal. The vertical load on the boom can be reduced with a guy-wire truss cable. The horizontal forces on the boom are more difficult to relieve, so 3-inch diameter tubing is desirable.

DIAMOND OR SQUARE?

The question of how to orient the spreader arms has been raised many times over the years. Should you mount the loops in a diamond or a square configuration? Should one set of spreaders be horizontal, giving the loop a diamond shape as shown in Figure 13.33 on the left, or should the wire itself be horizontal to the ground (spreaders mounted diagonally in the fashion of an X) as shown at the right in Figure 13.33? From the electrical point of view, there is not enough difference in performance to worry about.

From the mechanical point of view there is no question which version is better. The diamond quad, with the associated horizontal and vertical spreader arms, is capable of holding an ice load much better than a system where no vertical support exists to hold the wire loops upright. Put another way, the vertical poles of a diamond array, if sufficiently strong, will hold the rest of the system erect. When water droplets are accumulating and forming into ice, it is very reassuring to see water running down the wires to a corner and dripping off, rather than just sitting there on the wires and freezing. The wires of a loop (or several loops, in the case of a multiband antenna) help support the horizontal spreaders under a load of ice. A square quad will droop severely under heavy ice conditions because there is nothing to hold it up straight.

Of course, in climates where icing is not a problem, many amateurs point out that they like the aesthetics of the square configuration. There are thousands of square-configuration quads in temperate areas around the world.

Another consideration will enter into your choice of orientation for a quad. You must mount a diamond quad somewhat higher on the mast or tower than for an equivalent square array, just to keep the bottom spreader away from the tower guys when you rotate the antenna.

GETTING IT UP THE TOWER

The many elements of a Yagi are hard enough to maneuver around guy wires but the quad's three-dimensional structure can make it a challenge to lift to the top of a tower. If the tower is a crank-up or tilt-over, it is much easier to get a quad mounted on the antenna mast. On a fixed, lattice-style tower with guys, you'll have to carefully work the antenna around each set of guy wires as it is raised. The tram technique described in the **Building Antenna Systems and Towers** chapter is highly recommended as the work to rig the tram pays off with a lot easier lift of the antenna over the guys.

Another technique used by builders of the more common concentric quads it to assemble each set of spreaders and elements separately on the ground and lift them one by one to the top of the tower where they are mounted to the boom one after another. This requires some planning but is a lot easier than lifting the entire assembled antenna, particularly when there are guy wires to contend with.

13.9 References and Bibliography

Source material and more extended discussions of the topics covered in this chapter can be found in the references given below and at the end of Chapter 1 **Antenna Fundamentals**, and in the Chapter 8 **Antenna Modeling** References and Bibliograhy section on **Yagi Antenna Optimization**.

13.9.1 Yagi-Uda Arrays

ARRL's Yagi Antenna Classics, ARRL, 2001.

Barton, D., AF6S, "Yagi Gain versus Boom Length," Communications Quarterly, Winter 1994, pp. 94-95.

Bojsen, J., et al., "Maximum Gain of Yagi-Uda Arrays," Electronics Letters, vol. 7, no. 18, Sep. 9, 1971, pp. 531-532.

Chen C., and Cheng, D., "Optimum Element Lengths for Yagi-Uda Arrays," IEEE Transactions on Antennas and Propagation, Vol. 23, No. 1, Jan. 1975, pp. 8 – 15.

Cheng D., and Chen, C., "Optimum Element Spacings for Yagi-Uda Arrays," IEEE Transactions on Antennas and Propagation, Vol. 21, No. 5, Sep. 1973, pp. 615 – 623.

Cheng, D., "Gain Optimization for Yagi-Uda Arrays," IEEE Antennas and Propagation Magazine, Vol. 33, No. 3, June 1991, pp. 42 – 45.

Collinge, C., WØYNF, "Linearly Loaded 20-Meter Beam," *QST*, June 1976, pp. 18 – 19.

Donnelly, D., K2SS, "Yagi vs. Quad," 2-part article, Ham Radio, May 1988, pp. 68 – 71; June 1988, pp. 78 – 83.

Ehrenspeck H., and Poehler, H., "A New Method for Obtaining Maximum Gain from Yagi Antennas," IRE Transactions on Antennas and Propagation, Vol. 7, No. 4, Oct. 1959, pp. 379 – 386.

Fenwick, R., K5RR, "A High-Performance Triband Beam with No Traps," *QEX*, Sep. 1996, pp. 3 – 7.

Greenblum, C., "Notes on the Development of Yagi Arrays," 2-part article, *QST*, Aug. 1956, pp. 11 – 17, 114, 116; Sep. 1956, pp. 23 – 26, 122.

Hoch, G., DL6WU, "Yagi Antennas: Principle of Operation and Optimum Design Criteria," VHF Communications, Vol. 9, No. 3, Autumn 1977, pp. 157 – 166.

Hoch, G., DL6WU, "More Gain with Yagi Antennas," VHF Communications, Vol. 9, No. 4, Winter 1977, pp. 204 – 211.

Hoch, G., DL6WU, "Extremely Long Yagi Antennas," VHF Communications, Vol. 14, No. 3, Autumn 1982, pp. 130 – 138.

Jaffin, S., WB3BGU, "Applied Yagi Antenna Design," 6-part article, Ham Radio, May 1984, pp. 14 – 15, 17 – 20, 23 – 25, 27 – 28; June 1984, pp. 33 – 35, 37 – 38, 41 – 43, 45 – 46; July 1984, pp. 73 – 75, 77 – 79, 81 – 82, 85 – 86; Aug. 1984, pp. 103 – 104, 106 – 110; Sep. 1984, pp. 93 – 99; Oct. 1984, pp. 89 – 92, 95 – 96.

Koontz, F., WA2WVL, "Design and Construction of Wire Yagi Antennas," Communications Quarterly, Winter 1994, pp. 96 – 99.

Lawson, J., W2PV, "Yagi Antenna Design," 9-part article, *Ham Radio*, Jan. 1980, pp. 22 – 27 (errata Sep 1980, p. 67); Feb. 1980, pp. 19 – 27; May 1980, pp. 18 – 26; June 1980, pp. 33 – 40 (errata Oct 1980, p. 78); July 1980, pp. 18 – 31; Sep. 1980, pp. 37 – 45 (errata Feb. 1981, p. 94); Oct. 1980, pp. 29 – 37; Nov. 1980, pp. 22 – 34; Dec. 1980, pp. 30 – 41.

Lawson, J., W2PV, Yagi Antenna Design (ARRL, 1986).

Leeson, D., W6QHS, Physical Design of Yagi Antennas (ARRL, 1992).

Lindsay, J., W0HTH, "Quads and Yagis," *QST*, May 1968, pp. 11 – 19, 150.

Myers, R., W1FBY, "A Wide-Spaced Multielement Tribander," *QST*, Dec. 1970, p. 33 – 38.

Reisert, J., W1JR, "Yagi Facts and Fallacies," Ham Radio, May 1986, pp. 103 – 106, 111 – 113.

Schiller, T., N6BT, Array of Light, Next Generation Antennas, 3rd Edition (2010).

Uda S., and Mushiake, Y., Yagi-Uda Antenna, The Research Institute of Electrical Communication, Tohoku University, Sendai, Japan, 1954.

Welsh, R., N3RW, "Yagi: The Man and His Antenna," *QST*, Oct. 1993, pp. 45 – 47.

Yagi H., and. Uda, S., "Projector of the Sharpest Beam of Electric Waves," Proceedings of the IEE Japan, Vol. 2, No. 2, Feb. 1926, pp. 49 – 52. Reprinted in *QEX*, Mar./Apr. 2021, pp. 10 – 14.

Yagi, H., "Beam Transmission of Ultra Short Waves," Proceedings of the IRE, Vol. 16, No. 6, June 1928, pp. 715 – 740. Reprinted in Proceedings of the IEEE, Vol. 72, No. 5, May 1984, pp. 635 – 652. Also reprinted in Vol. 85, No. 11, Nov. 1997, pp. 1864 – 1874.

13.9.2 Loop Arrays

Bergren, L., WØAIW, "The Multielement Quad," *QST*, May 1963, pp. 11 – 16.

Cebik, L., W4RNL, and Cerreto, R., WA1FXT, "A 2 Element Quad for 17 and 12 Meters Using a Combined Feed," *The ARRL Antenna Compendium, Vol. 8*, pp. 71 – 75 (ARRL, 2010).

Cebik, L., W4RNL, "Feeding the 5-Band Quad," *The ARRL Antenna Compendium, Vol. 7* pp. 146 – 153 (ARRL, 2002),.

Donnelly, D., K2SS, "Yagi vs. Quad," 2-part article, *Ham Radio*, May 1988, pp. 68 – 71; June 1988, pp. 78 – 83.

Handelsman, D., N2DT, "The Rectangular Family of Antennas," 2-part article, *QEX*, Mar./Apr. 2001, pp. 35 – 46; Jan./Feb. 2002, pp. 12 – 22.

Harrison, R., VK2ZTB, "Loop-Yagi Antennas," *Ham Radio*, May 1976, pp. 30 – 32.

Haviland, R., W4MB, "The Quad Antenna," 3-part article, *Ham Radio*, May 1988, pp. 43 – 53; June 1988, pp. 54 – 67; Aug. 1988, pp. 34 – 47.

Haviland, R., W4MB, "The Quad Antenna Revisited," 5-part article, *Communications Quarterly,* Summer 1999, pp. 43 – 73 (errata, Fall 1999, p. 108); Fall 1999, pp. 65 – 85; *QEX,* Nov./Dec. 2000, pp. 10 – 19; Mar./Apr. 2001, pp. 47 – 54; Jan./Feb. 2002, pp. 23 – 32.

Hilliard, D., WØPW, "A 902-MHz Loop Yagi Antenna," *QST*, Nov. 1985, pp. 30 – 32.

Kennicott, J., W4OVO, "Three-Element Quad for 15 – 20 Meters Which Uses Circular Elements," *Ham Radio*, May 1980, pp. 12 – 17.

Landskov, H., W7KAR, "Evolution of a Quad Array," *QST*, Mar. 1977, pp. 32 – 36.

Lindsay, J., "A Parasitic End-Fire Array of Circular Loop Elements," *IEEE Transactions on Antennas and Propagation*, Vol. 8, No. 4, Sep. 1967, pp. 697 – 698.

Lindsay, J., WØHTH, "Quads and Yagis," *QST*, May 1968, pp. 11 – 19, 150.

Mees, D., ON7NQ, "Improving the Cubex Three-Element, Five-Band Quad," *The ARRL Antenna Compendium, Vol. 6*, pp. 119 – 120 (ARRL, 1999).

Orr, W., W6SAI, and Cowan, S., W2LX, *Cubical Quad Antennas*, 3rd Edition. (Radio Amateur Call Book, 1982).

Reynante, J., KD6GLF, "A Five-Element Quad Antenna for 2 Meters," *QST*, Jan. 1995, pp. 67 – 69.

Shen L., and Raffoul, G., "Optimum Design of Yagi Array of Loops," *IEEE Transactions on Antennas and Propagation*, Vol. 22, No. 6, Nov. 1974, pp. 829 – 830.

13.9.3 Moxon Rectangles

Cebik, L., W4RNL, "Having a Field Day With the Moxon Rectangle," *QST*, June 2000, pp. 38 – 42.

Cebik, L., W4RNL, "Modeling and Understanding Small Beams: Part 2 - VK2ABQ squares and Moxon rectangles," *Communications Quarterly*, Spring 1995, pp. 55 – 70.

Cebik, L., W4RNL, "Moxon Rectangles for 6 Meters," (available from online sources)

Cebik, L., W4RNL, "The Moxon Rectangle: A Review," (available from online sources)

Cebik, L., W4RNL, "Wire Moxon Rectangles for 40-10 Meters," (available from online sources)

Moxon, L., G6XN, "Two-element HF Beams," *ham radio*, May 1987, pp. 8 – 32.

TABLE OF CONTENTS

Chapter 14 — Online Material

Articles

- "Station Design for DX, Part I" by Paul Rockwell, W3AFM
- "Station Design for DX, Part II" by Paul Rockwell, W3AFM
- "Station Design for DX, Part III" by Paul Rockwell, W3AFM
- "Station Design for DX, Part IV" by Paul Rockwell, W3AFM
- N6BV and K1VR Stack Feeding and Switching Systems
- "The Effect of Local Terrain on HF Launch Angles" by Dean Straw, N6BV
- "Manually Generating a Terrain Profile"
- "Contest Aerials - Switching Stacked Antennas" by Dave Pruett, K8CC
- Appendix F of FM 11-65: High Frequency Radio Communications, US Army publication

Chapter 14
HF Antenna System Design

This chapter combines information from previous editions into a condensed discussion of HF antenna system design. An amateur just beginning to build an HF station may be more interested in trying out different types of antennas and gaining experience with selecting, building and installing them. Later on, as experience is gained and specific goals are formed, the process of system design becomes important.

No single book can provide a step-by-step procedure for designing antenna systems — there are too many different needs and operating styles. What can be done, however, is to give an overview of the process by which system-level issues are identified and dealt with. Tools such as propagation prediction and antenna modeling software will be discussed from the antenna system perspective. Methods of using antennas to meet certain goals, such as stacking Yagis and using near vertical incidence skywave propagation (NVIS), will be covered.

By thinking about your "antenna farm" as a system — whether a single antenna in a tree or a multiple-tower contest station — you will be able to make better use of your time and materials and have more success on the air.

We will begin with an overview of the system design process and how to approach it. The next step is a section covering the use of propagation prediction tools as a means of assessing the coverage of an antenna system. Then the effects of local terrain on antenna system planning and performance are covered. The final sections address the use of vertical stacks of Yagi antennas to control elevation angle.

14.1 System Design Basics

The most important time spent in putting together an antenna system is the time spent in planning. Later in this chapter the section on Local Terrain will present steps needed to evaluate how your local terrain can affect HF communications. You will need to compare the patterns resulting from your own terrain to the statistically relevant elevation angles needed for coverage of various geographic areas. (The elevation-angle statistics were discussed in the **Radio Wave Propagation** chapter and are located with this book's online material as is the terrain-assessment program *HFTA*.)

The implicit assumptions in using propagation data and terrain analysis are (1) that you know where you want to talk to, and (2) that you'd like the most effective system possible. At the start of such a theoretical analysis, cost is no object. Practical matters, like cost or the desires of your spouse, can come later! After all, you're just checking out all the possibilities. If nothing else, you will use the methodology in this chapter to evaluate any property you are considering buying so that you can build your "dream station."

By using the techniques and tools available in this book, you can rationally and logically plan an antenna system that will be best suited for your own particular conditions. Now, however, you have to get practical. Thinking through and planning the installation can save a lot of time, money and frustration.

One often overlooked part of successful antenna system design (and station design, too) is the keeping of a station notebook. Make sure you save and organize the various computer files and documents associated with your system design. Being able to revisit the steps leading to a decision — successful or unsuccessful — is very important. Important data from measurements or tests should always be clearly labeled and stored so that you can find it later. Think of each page or file as a brick in the grand structure you are building. No one ever regrets having kept good records!

While no one can tell you the exact steps you should take in developing your own master plan, this section, prepared originally by Chuck Hutchinson, K8CH, should help you with some ideas.

14.1.1 Desires and Limitations

Begin planning by spelling out your communications desires and the limitations placed on them. Engineers call these "requirements" and "constraints" — all successful projects begin with clearly understanding and recording them. What bands are you interested in? Who (or where) do you want to talk to? When do you operate? How much time and money are you willing to spend on an antenna system? What physical limitations affect your master plan?

From the answers to the above questions, begin to formulate goals — short, intermediate and long range. Be realistic about those goals. Remember that there are three station effectiveness factors that are under your control. These are: operator skill, equipment in the station, and the antenna system. There is no substitute for developing operating skills. Some tradeoffs are possible between station equipment and antennas. For example, a high-power amplifier can compensate for a less than optimum antenna but only for transmitted signals. By contrast, a better antenna has advantages for receiving as well as for transmitting.

Consider your limitations. Are there regulatory restrictions on antennas in your community? Are there any deed restrictions or covenants that apply to your property? Do other factors (finances, family considerations, other interests, and so forth) limit the type or height of antennas that you can erect? All of these factors must be investigated because they play a major role determining the type of antennas you erect.

Chances are that you won't be able to immediately do all you desire. Think about how you can budget your resources over a period of time. Your resources are your money, your time available to work, materials you may have on hand, friends that are willing to help, etc. One way to budget is to concentrate your initial efforts on a given band or two. If your major interest is in chasing DX, you might want to start with a very good antenna for the 14-MHz band. A simple multiband antenna could initially serve for other frequencies. Later you can add better antennas for those other bands.

14.1.2 Site Planning

A map of your property or proposed antenna site can be of great help as you begin to consider alternative antennas. You'll need to know the size and location of buildings, trees and other major objects in the area. Be sure to note compass directions on your map. Graph or quadrille paper (or a simple CAD program) can be very useful for this purpose. See **Figure 14.1** for an example. It's a good idea to make a few photocopies of your site map so you can mark on the copies as you work on your plans. If you create a master map with CAD software, you can create and save lots of alternatives for comparisons and evaluations.

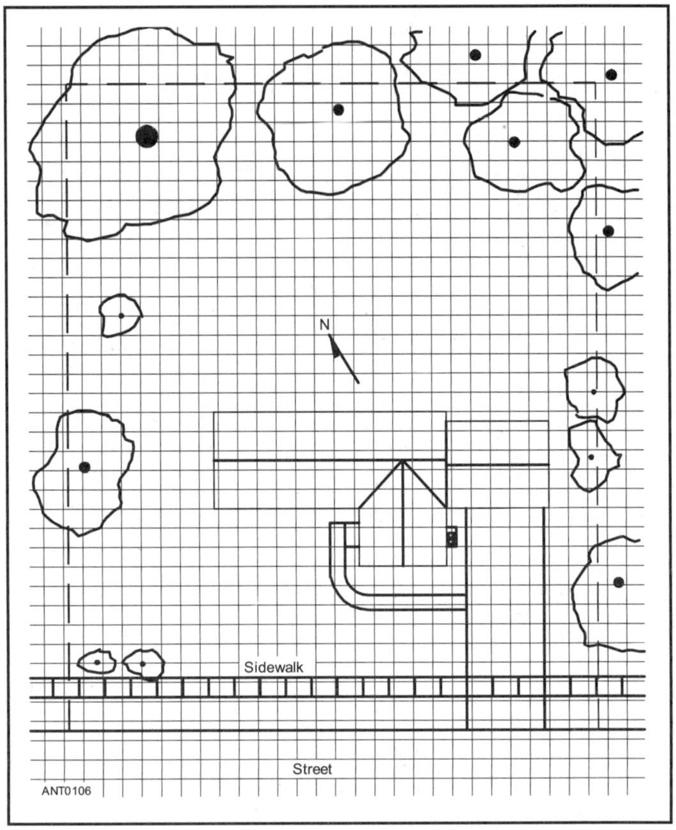

Figure 14.1 — A site map such this one is a useful tool for planning your antenna installation.

Use your map to plan antenna layouts and locations of any supporting towers or masts. If your plan calls for more than one tower or mast, think about using them as supports for wire antennas. As you work on a layout, be sure to think in three dimensions even though the map shows only two.

Be sensitive to your neighbors. A 70-foot guyed tower in the front yard of a house in a residential neighborhood is not a good idea (and probably won't comply with local ordinances!). You probably will want to locate that tower in the back yard.

Be sure to include restrictions and hazards on the map. For example, you may have set-back requirements from property lines for any structure on your property, such as a tower. You may not be allowed to intrude on neighboring "air space" with antenna elements. Power lines and other hazards such as buried utilities should be on your map, as well. It's just as important to identify where antennas can't go as where they can.

As discussed in the **Building Antenna Systems and Towers** chapter, consider access needs when laying out your system. If you will be putting up towers, consider how a backhoe or concrete truck can get to the location of the tower base. You'll need to allow space for towers that fold over or that are tipped up for installation, too.

14.1.3 Initial Analysis

Use the information in this chapter, antenna modeling software and propagation evaluation tools to analyze antenna patterns in both horizontal and vertical planes toward geographic areas of interest. Consider the azimuthal pattern of fixed antennas. You'll want to orient any fixed antennas to favor the directions of greatest interest to you.

Use antenna modeling tools to help you evaluate what type of antenna might be suitable to your own particular style of operating. Do you want a Yagi with a lot of rejection of received signals from the rear? Let's say that terrain analysis shows that you need an antenna at least 50 feet high. Do you really need a steel tower, or would a simple dipole in the trees serve your communication needs just fine? How about a vertical in your backyard? Would that be inconspicuous enough to suit your neighbors and your own family, yet still get you on the air?

If you want to work DX, you'll want antennas that radiate energy at low as well as intermediate angles. An antenna pattern is greatly affected by the presence of ground and by the local topography of the ground. Therefore, be sure to consider what effect ground will have on the antenna pattern at the height you are considering. A 70-foot high antenna is approximately ½, 1, 1½ and 2 wavelengths high on 7, 14, 21 and 28 MHz respectively. Those heights are useful for long-distance communications. The same 70-foot height represents only $\lambda/4$ at 3.5 MHz, however. Most of the radiated energy from a dipole at that height would be concentrated straight up. This condition is not great for long-distance communication, but can still be useful for some DX work and excellent for short-range communications.

Lower antenna heights can be useful for certain types of communications — see the section on NVIS communications later in this chapter, for example. However, for most amateur operation it is generally true that "the higher, the better" as far as communications effectiveness is concerned. This general rule of thumb, of course, should be tempered by an exact analysis of your local terrain. Being located at the top of a steep hill can mean that you can use lower tower heights to achieve good coverage.

There may be cases where it is not possible to install low-frequency dipoles $\lambda/4$ or more above the ground. A vertical antenna with many radials is a good choice for long-distance communications. You may want to install both a dipole and a vertical for the 3.5- or 7-MHz bands. On the 1.8-MHz band, unless extremely tall supports are available, a vertical antenna is likely to be the most useful for DXing. You can then choose the antenna that performs best for a given set of conditions. A low dipole will generally work better for shorter-range communications, while the vertical will generally be the better performer over longer distances.

The performance of ground-mounted monopole antennas depends strongly on a system of ground radials and the characteristics of the ground. Be sure to review this book's chapter **Effects of Ground** when considering how and where to install the antenna. In particular, Al Christman, K3LC's article on maximizing the effectiveness of ground radials for a fixed amount of wire makes good economic sense for practical system builders.

Jim Brown, K9YC, has published an in-depth analysis of vertical versus horizontal antenna selection in his paper "Planning Antenna Systems For the Little Gun Station" at **k9yc.com/publish.htm**. There are also several presentations on antenna selection on his website. Regardless of how you approach the problem, designing an antenna system is always an exercise of balancing constraints against desired performance.

14.1.4 Building a System Plan

At this point, you will enter a repeated sequence of "design-model-adjust" as you evaluate the plan. You should start with modeling and then compare the results to those "desires" you wrote down at the beginning. With each round of modeling and comparison, your antenna system will be improved.

As you refine your system design, you can also build the long-term plan for construction of the antenna system. Chances are that you can divide the actual construction of your system into a series of phases or steps. By keeping the long-range plan in mind you will be able to make better decisions at every step of the way toward achieving your goals!

Say, for example, that you have lots of room and that your long-range plan calls for a pair of towers, one 100-feet high, and the other 70-feet high, to support monoband Yagi antennas. The towers will also support a horizontal 3.5-MHz dipole, for DX work. On your map you've located them so the 80 meter dipole will be broadside to Europe. You decide to build the 70-foot tower with a triband beam and 80 and 40 meter inverted-V dipoles to begin the project.

In your master plan you design the guys, anchors and all hardware for the 70-foot tower to support the load of stacked 4-element 10 and 15 meter monobanders Yagis. So you make sure you buy a heavy-duty rotator and the stout mast needed for the monoband antennas later because you have a long-range plan for them. Thus you avoid having to buy, and then sell, a medium-duty rotator and lighter weight tower materials later on when you upgrade the station. You could have saved money in the long run by putting up a monoband beam for your favorite band, but you decided that for now it is more important to have a beam on 14, 21 and 28 MHz, so you choose a commercial triband Yagi.

The second step of your plan calls for installing the second tower and stacking a 2-element 40 meter and a 4-element 20 meter monoband Yagi on it. You also plan to replace the tribander on the 70-foot tower with stacked 4-element 10 and 15 meter monoband Yagis. Although this is still a "dream system" you can now apply some of the modeling techniques discussed earlier in this chapter to determine the overall system performance.

14.1.5 Modeling Interactions

In this next step of analysis we're going to assume that you have sufficient real estate to separate the 70- and 100-foot towers by 150 feet so that you can easily support an 80 meter dipole between them. We'll also assume that you want the 80 meter dipole to have its maximum response at a

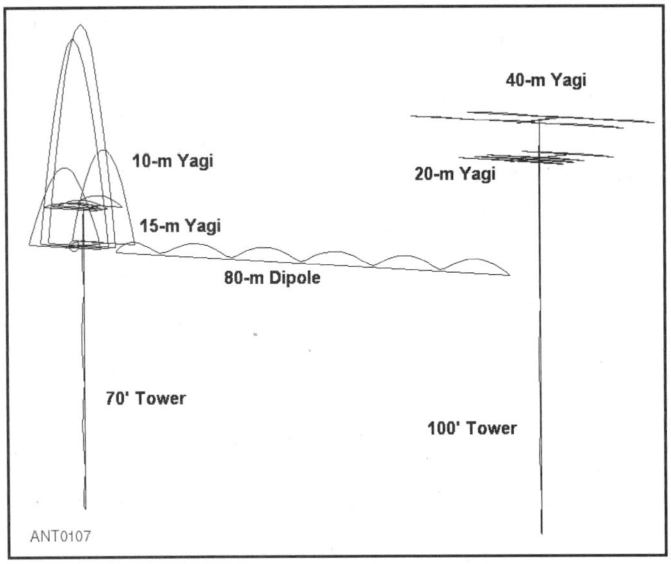

Figure 14.2 — Layout for two-tower antenna system, at 70 and 100 feet high and 150 feet apart. The 70-foot tower has a 4-element 10 meter Yagi at 80 feet on a 10-foot rotating mast and a 4-element 15 meter Yagi at 70 feet. An 80 meter dipole goes from the 70-foot tower to the 100-foot tower, which holds a 2-element 40 meter Yagi at 110 feet and a 4-element 20 meter Yagi at 100 feet. In this figure all the rotatable Yagis are facing the direction of Europe and the currents on the 15 meter Yagi are shown. Note the significant amount of current induced on the nearby 80 meter dipole that will cause a re-radiated signal!

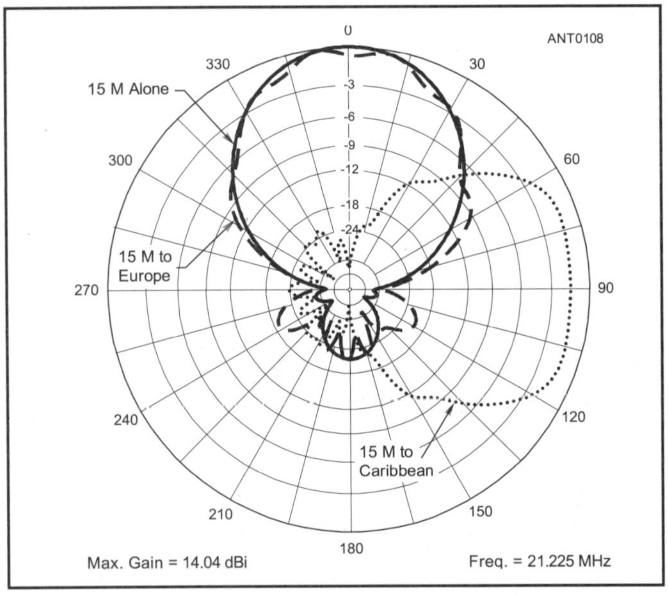

Figure 14.3 — An overlay of azimuth patterns. The solid line is the radiation pattern for the 15 meter Yagi all by itself. The dashed line is the pattern for the 15 meter Yagi, as affected by all the other antennas. The dotted line is the pattern for the 15 meter Yagi when it is pointed toward the Caribbean, with the Yagis on the 100-foot tower pointed toward the 70-foot tower. The peak response of the 15 meter Yagi has dropped by about 1.5 dB.

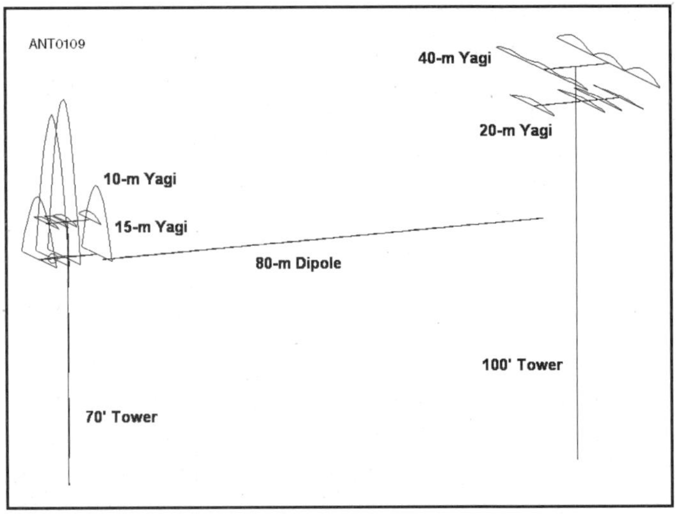

Figure 14.4 — The layout and 15 meter currents when the Yagis on the 100-foot tower are pointed toward the 70-foot tower. The 15 meter Yagi has been rotated to face the direction of the 100-foot tower (toward the Caribbean).

heading of 45° into Europe from your location in Newington, Connecticut. The dipole will also have a lobe facing 225° toward the USA and New Zealand, making it a good antenna for both domestic contacts and DX work. Note that it is important to model interactions for the full system even if you don't plan on building all of it right away. This helps avoid "mid-course corrections."

Let's examine the interactions that occur between the rotatable Yagis for 10, 15, 20 and 40 meters. See **Figure 14.2**, which purposely exaggerates the magnitude of the currents on the 4-element 15 meter Yagi mounted at 70 feet. Here, both sets of Yagis have been rotated so that they are pointing into Europe. There is a small amount of current radiated onto the 10 meter antenna but virtually no current is radiated onto the 40 and 20 meter Yagis. This is good.

However, significant current is picked up by the 80 meter dipole. This undesired current and the subsequent reradiated signal affects the radiation pattern of the 15 meter antenna, as shown in **Figure 14.3**, which overlays the pattern of the 4-element 15 meter Yagi by itself with that of the Yagi interacting with the other antennas. You can see "ripples" in the azimuthal response of the 15 meter Yagi due to the effects of the 80 meter dipole's re-radiation. The magnitude of the ripples is about 1 dB at worst, so they don't seriously affect the forward pattern (into Europe), but the rearward lobes are degraded somewhat, to just below 20 dB.

Figure 14.3 also shows the worst-case situation for the 15 meter Yagi. Here, the 15 and 10 meter stack has been turned clockwise 90°, facing the Caribbean, while the 40 and 20 meter Yagis on the 100-foot tower have been turned counterclockwise 90° (in the direction of Japan) to face the 70-foot tower holding the 10/15 meter Yagis. You can see the layout and the currents in **Figure 14.4**. Now the 40 and 20 meter Yagis re-radiate some 15 meter energy and reduce the maximum gain by about 1.5 dB. Note that in this direction

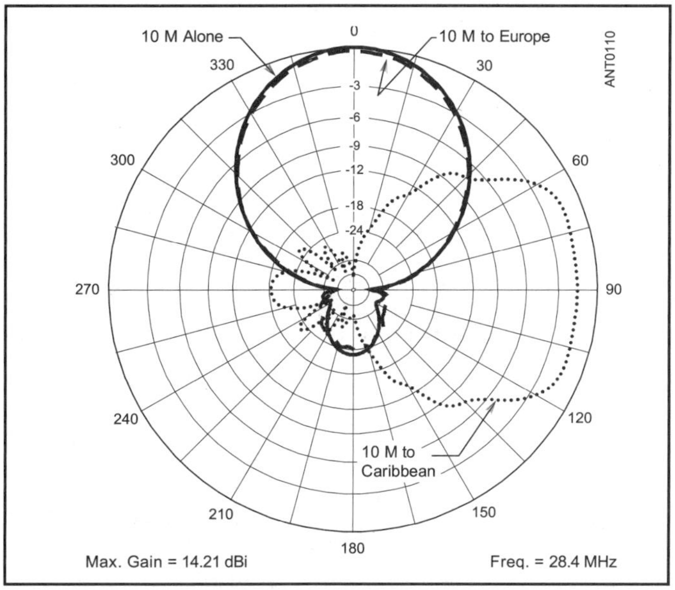

Figure 14.5 — The radiation patterns for the 10 meter Yagi. The solid line is the 10 meter Yagi by itself. The dashed line is for the same Yagi, with all other antenna interactions. The dotted line shows the worst-case pattern, with the stacked Yagis on the 100-foot tower facing the 70-foot tower and the 10 meter Yagi pointed toward the Caribbean. Again, the peak response of the 10 meter Yagi has dropped about 1.5 dB in the worst-case situation.

the 80 meter dipole no longer has 15 meter energy radiated onto it by the 15 meter Yagi.

The shape of the patterns will change depending on whether you specify "current" or "voltage" sources in the models for the other antennas, since this effectively opens up or shorts the feed points at the other antennas so far as 15 meter energy is concerned. In practice, this means that the interaction between antennas will vary somewhat depending on the length of the feed lines going to each antenna and whether each feed line is open-circuited or short-circuited when it is not in use.

You can now see that interactions between various antennas pointing in different directions can be significant in a real-world antenna system. In general, higher frequency antennas are affected by re-radiation from lower-frequency antennas, rather than the other way around. Thus the presence of a 10 or 15 meter stack does not affect the 20 meter Yagi at all.

Modeling can also help determine the minimum stacking distance required between monoband Yagis on the same rotating mast. In this case, stacking the 10 and 15 meter monobanders 10 feet apart holds down interaction between them so that the pattern and gain of the 10 meter Yagi are not impacted adversely. **Figure 14.5** demonstrates this in the European direction, where the patterns for the 10 meter beam by itself looks very clean compared to the same Yagi separated by 10 feet from the 15 meter Yagi below it. The worst-case situation is pointing toward the Caribbean, when the 40 and 20 meter stack is facing the 70-foot tower. This drops the 10 meter gain down about 1.5 dB from maximum, indicating significant interaction is occurring.

In this situation you might find it best to place the 70-foot tower in the direction closest to the Caribbean if this direction is very important to you. Doing so will, however, cause the pattern in the direction of the Far East to be affected on 10 and 15 meters. You have the modeling tools necessary to evaluate various configurations to achieve whatever is most important to you.

14.1.6 Compromises

Because of limitations, most amateurs are never able to build their dream antenna system. This means that some compromises must be made. Do not, under any circumstances, compromise the safety of an antenna installation. Follow the manufacturer's recommendations for tower assembly, installation and accessories. Make sure that all hardware is being used within its ratings. (The ARRL document "Antenna Height and Communications Effectiveness" by Dean Straw, N6BV, and Jerry Hall, K1TD, may be of use in dealing with local regulatory agencies. It is available for downloading at **arrl.org/files/file/antplnr.pdf**.)

Guyed towers are frequently used by radio amateurs because they cost less than more complicated unguyed or freestanding towers with similar ratings. (See the chapter **Building Antenna Systems and Towers** for more information.) Guyed towers are fine for those who can climb or those with a friend who is willing to climb. But you may want to consider an antenna tower that folds over or one that cranks up (and down). Some towers crank up (and down) and fold over too. See **Figure 14.6**. That makes for convenient access to antennas for adjustments and maintenance without climbing. Crank-up towers also offer another advantage. They allow antennas to be lowered during periods of no operation, for aesthetic reasons, or during periods of high winds.

A well-designed monoband Yagi should outperform a multiband Yagi. In a monoband design the best adjustments can be made for gain, front-to-rear ratio (F/R) and matching, but only for a single band. In a multiband design, there are always tradeoffs in these properties for the ability to operate on more than one band. Nevertheless, a multiband antenna has many advantages over two or more single band antennas. A multiband antenna requires less heavy-duty hardware, requires only one feed line, takes up less space and it costs less.

Apartment dwellers face much greater limitations in their choice of antennas. For most, the possibility of a tower is only a dream. (One enterprising ham made arrangements to purchase a top-floor condominium from a developer. The arrangements were made before construction began, and the plans were altered to include a roof-top tower installation.) For apartment and condominium dwellers, the situation is still far from hopeless. The chapters **Stealth and Limited Space Antennas** and **Portable Antennas** present ideas for consideration.

14.1.7 System Design Examples

You can plan according to the preceding sections to put together modest or very large antenna systems. The process may sound intimidating but the hardest part is usually just

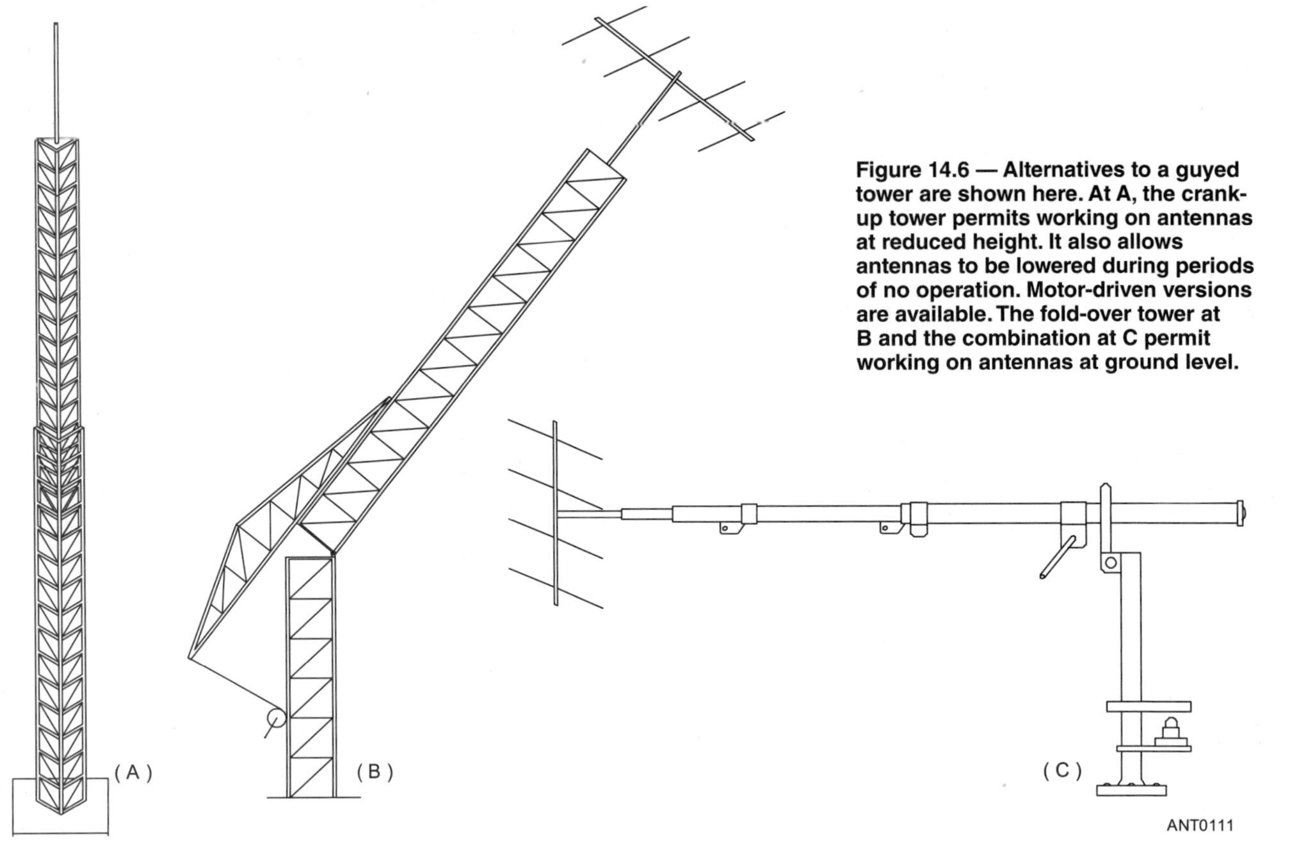

Figure 14.6 — Alternatives to a guyed tower are shown here. At A, the crank-up tower permits working on antennas at reduced height. It also allows antennas to be lowered during periods of no operation. Motor-driven versions are available. The fold-over tower at B and the combination at C permit working on antennas at ground level.

getting started! At this point, some simple examples might be instructive and encourage you to start planning.

ANTENNA SYSTEM EXAMPLE #1

What might a ham put together for antennas to try a little of everything with a modest budget? Let's suppose that the goals are (1) low cost, (2) no tower, (3) coverage of all HF bands and the repeater portion of one VHF band and (4) the possibility of working some DX.

After studying the pages of this book, the station owner decides to first put up a 135-foot center-fed antenna. High trees in the back yard will serve as supports to about 50 feet. This antenna will cover all the HF bands by using a balanced feeder and an antenna tuner. It should be good for DX contacts on 10 MHz and above and will probably work okay for DX contacts on the lower bands. However, the plan calls for a ground-mounted vertical and radial system for 3.5 and 7 MHz to enhance the DX possibilities on those bands. For VHF, a chimney-mounted vertical is included.

ANTENNA SYSTEM EXAMPLE #2

A club station has bigger ambitions. Goals for their station are (1) a good setup for DX on 14, 21 and 28 MHz, (2) moderate cost, (3) one tower, (4) ability to work some DX on 1.8, 3.5 and 7 MHz, and (5) no need to cover the CW portion of the bands.

After considering the options, the club decides to install a 65-foot guyed tower. A large commercial triband Yagi will be mounted on top of the tower. The center of a trap dipole tuned for the phone portion of the 3.5- and 7-MHz bands will be supported by a wooden yardarm installed at the 60-foot level of the tower, with ends drooping down to form an inverted-V. An inverted-L for 1.8 MHz starts near ground level and goes up to a similar yardarm on the opposite side of the tower. The horizontal portion of the inverted L runs away from the tower at right angles to the trap dipole. Later, they will experiment with sloping antennas for 3.5 MHz. If those experiments are not successful, a $\lambda/4$ vertical will be used on that band.

14.1.8 Empirical Testing

Part of system design is "closing the loop" and evaluating the performance of what you have designed. If the performance is as expected, that validates your planning and design approach. If the performance isn't as expected, find out why and use that as a learning experience to improve your skills.

Unfortunately, many amateurs do not know how to evaluate performance scientifically or compare one antenna with another. Typically, they will put up one antenna and try it out on the air to see how it "gets out" in comparison with a previous antenna. This is obviously a very poor evaluation method because there is no way to know if the better or worse reports are caused by changing band conditions, different S meter characteristics or any of several other factors that could influence the reports received.

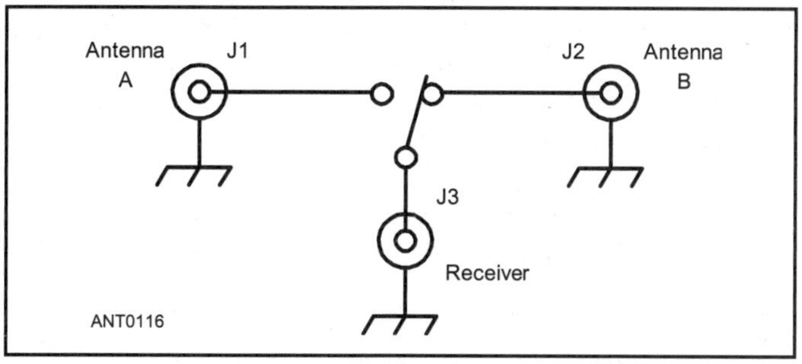

Figure 14.7 — When antennas are compared on fading signals, the time delay involved in disconnecting and reconnecting coaxial cables is too long for accurate measurements. A simple slide switch will do well for switching coaxial lines at HF. The four components can be mounted in a tin can or any small metal box. Leads should be short and direct. J1 through J3 are coaxial connectors.

Many times the difference between two antennas or between two different locations for identical antennas amounts to only a few decibels, a difference that is hard to discern unless instantaneous switching between the two is possible. Those few decibels are not important under strong signal conditions, of course, but when the going gets rough a few dB can make the difference between solid copy and no possibility of real communication.

Very little in the way of test equipment is needed for casual antenna evaluation, other than a communications receiver. You can even do a qualitative comparison by ear, if you can switch antennas instantaneously. Differences of less than 2 dB, however, are still hard to discern. The same is true of S meter readings. Signal strength differences of less than a decibel are usually difficult to see. If you want to measure that last fraction of a decibel, you should use a good ac voltmeter at the receiver audio output (with the AGC turned off).

In order to compare two antennas, switching the coaxial transmission line from one to the other is necessary. No elaborate coaxial switch is needed; even a simple double-throw toggle or slide switch will provide more than 40 dB of isolation at HF. See **Figure 14.7**. Switching by means of manually connecting and disconnecting coaxial lines is not recommended because that takes too long. Fading can cause signal-strength changes during the changeover interval.

Whatever difference shows up in the strength of the received signal will be the difference in performance between the two antennas in the direction of that signal. For this test to be valid, both antennas must have nearly the same feed point impedance, a condition that is reasonably well met if the SWR is below 2:1 on both antennas.

On ionospheric propagated signals (sky wave) there will be constant fading, and for a valid comparison it will be necessary to take an average of the difference between the two antennas. Occasionally, the inferior antenna will deliver a stronger signal to the receiver, but in the long run the law of averages will put the better antenna ahead.

Of course with a ground-wave signal, such as that from a station across town, there will be no fading problems. A ground-wave signal will enable the operator to properly evaluate the antenna under test in the direction of the source. The results will be valid for ionospheric-propagated signals at low elevation angles in that direction. On 28 MHz, all sky-wave signals arrive and leave at low angles. But on the lower bands, particularly 3.5 and 7 MHz, we often use signals propagated at high elevation angles, almost up to the zenith. For these angles a ground-wave test between local stations may not provide a proper evaluation of the antenna, and use of sky wave signals becomes necessary.

14.2 Propagation and Coverage

The section "Elevation Angles for HF Communication" in the **Radio Wave Propagation** chapter is an excellent introduction to the use of propagation prediction software such as *IONCAP* and *VOACAP* to assess the coverage of an HF antenna at different frequencies for a wide range of solar conditions. This book's online material includes a set of elevation angle statistics derived from these tools that you can use when designing your antenna system. This book's previous editor, Dean Straw, N6BV, compiled the statistics to cover more than 240 locations around the world in all 40 CQ zones and covers the five primary HF amateur bands (80 through 10 meters) for 24 hours at six levels of solar activity. These tables show signal strength in S units for easier use by amateurs.

As you plan your antenna system, it is strongly recommended that you become familiar with at least one propagation prediction tool and undertake a study of propagation at your location to the areas of the world with which you want to communicate. Two descriptions of using propagation information are presented as examples of how understanding propagation can inform your antenna system design choices.

14.2.1 Elevation Angles for Low-Band Dxing

In the chapter **Effects of Ground**, the importance was noted of matching the elevation response of your antennas as closely as possible to the range of elevation angles needed for communication with desired geographic areas. **Figure 14.8** shows the statistical 40 meter elevation angles needed over the entire 11-year solar cycle to cover the path from Boston, Massachusetts, to all of Europe. These angles range from 1° (at 9.6% of the time when the 40 meter band is open to Europe) to 28° (at 0.3% of the time). Creating an antenna system that concentrates the radiated energy at these low elevation angles is crucial to work DX on the bands below 10 MHz.

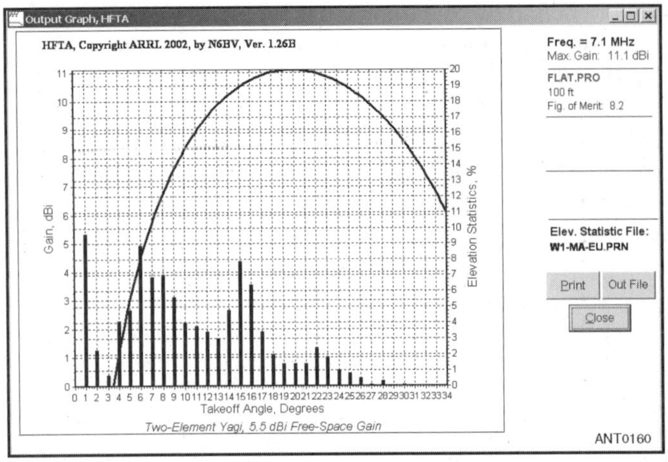

Figure 14.8 — Screen capture from HFTA (HF Terrain Assessment) program showing elevation response for 100-foot high dipole over flat ground on 7.1 MHz, with bar-graph overlay of the statistical elevation angles needed over the whole 11-year solar cycle from New England (Boston) to all of Europe. Even a 100-foot high antenna cannot cover all the necessary angles.

Figure 14.8 also overlays the elevation pattern response of a 100-foot high flattop dipole on the elevation-angle statistics, illustrating that even at this height the coverage is hardly optimum to cover all the necessary elevation angles. While Figure 14.8 is dramatic in its own right, the data can be viewed in another way that emphasizes even more the importance of low elevation angles. **Figure 14.9** plots the *cumulative distribution function*, the total percentage of time 40 meters is open from Boston to Europe, at or below each elevation angle. For example, Figure 14.9 says that 40 meters is open to Europe from Boston 50% of the time at an elevation angle of 9° or less. The band is open 90% of the time at an elevation angle of 19° or less.

Figure 14.10 plots the 40 meter elevation-angle data for six major geographic areas around the world from Boston. In general, the overall range of elevation angles for far-distant locations is smaller, and the angles are lower than for closer-in areas. For example, from Boston to southern Asia (India), 50% of the time the takeoff angles are 4° or less. On the path to Japan from Boston, the takeoff angles is less than or equal to 6° about 70% of the time. These are low angles indeed.

Figure 14.11 shows similar data for the 40 meter band from San Francisco, California, to the rest of the world. The path to southern Africa from the US West Coast is a very long-distance path, open some 65% of the time it is open at angles of 2° or less! The 40 meter path to Japan involves takeoff angles of 10° or less more than 50% of the time. If you are fortunate enough to have a 100-foot high flattop dipole for 40 meters, at a takeoff angle of 10° the response would be down about 3 dB from its peak level at 20°. At an elevation angle of 5° the response would be about 8 dB down from peak. You can see why the California stations located on mountain tops do best on 40 meters for DXing.

Figure 14.12 shows the same percentage-of-time data for the 80 meter band from Boston to the world. Into Europe from Boston, the 80 meter elevation angle is 13° or less more than 50% of the time. Into Japan from Boston, 90% of the time the band is open is at a takeoff angle of 13° or less. (Note that these elevation statistics are computed for "undisturbed" ionospheric conditions. There are times when the incoming

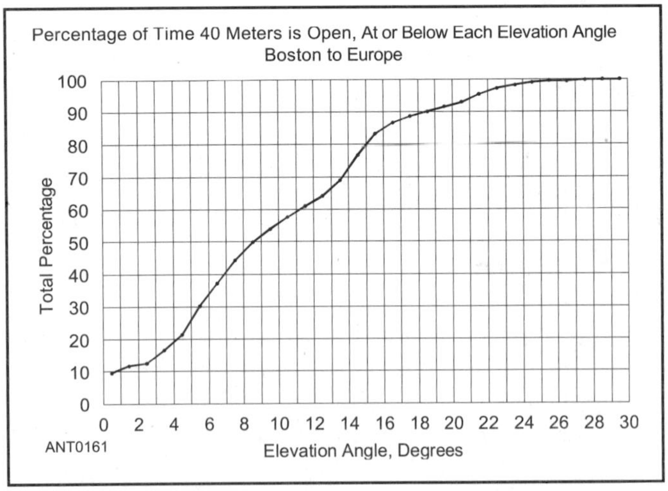

Figure 14.9 — Another way of looking at the elevation statistics from Figure 14.8. This shows the percentage of time the 40 meter band is open, at or below each elevation angle, on the path from Boston to Europe. For example, the band is open 50% of the time at an angle of 9° or lower. It is open 90% of the time at an angle of 19° or lower.

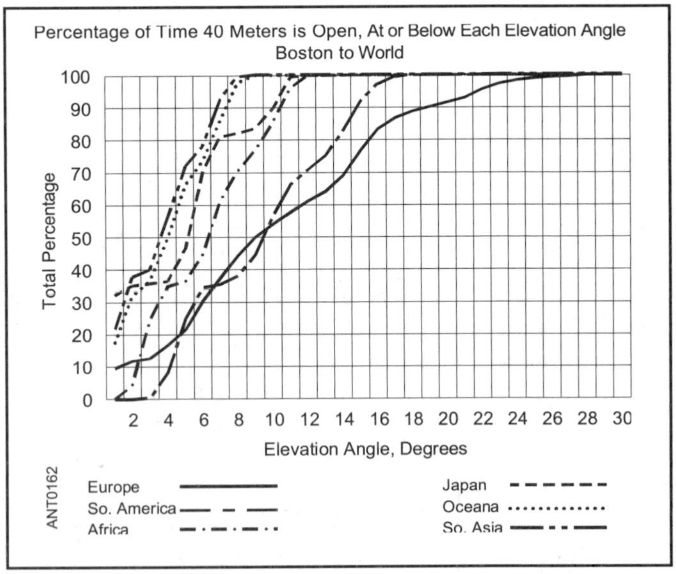

Figure 14.10 — The percentage of time the 40 meter band is open, at or below each elevation angle, for various DX paths from Boston: to Europe, South America, southern Africa, Japan, Oceania and south Asia. The angles are predominantly quite low. For example, on the path from Boston to Japan, 90% of the time when the 40 meter band is open, it is open at elevation angles less than or equal to 10°. Achieving good performance at these low takeoff angles requires very high horizontally polarized antennas, or efficient vertically polarized antennas.

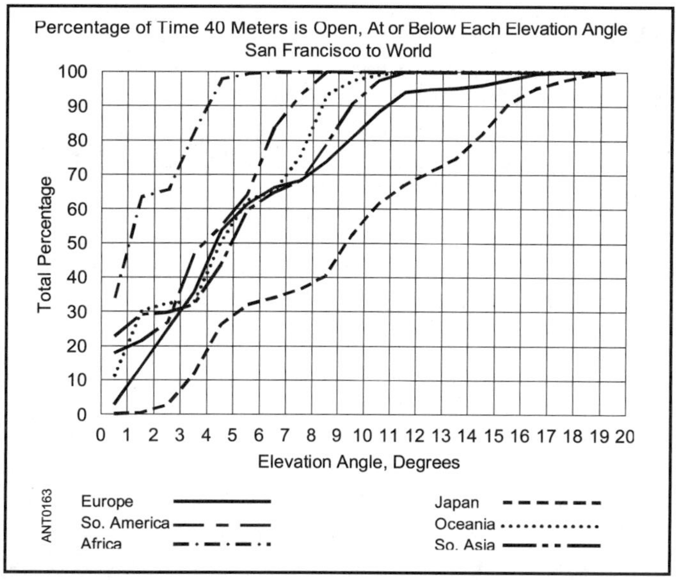

Figure 14.11 — The 40 meter statistics from the West Coast: from San Francisco to the rest of the DX world. Here, 90% of the time the path to Europe is open, it is at takeoff angles less than or equal to 11°. No wonder the hams living on mountain tops do best into Europe from the West Coast.

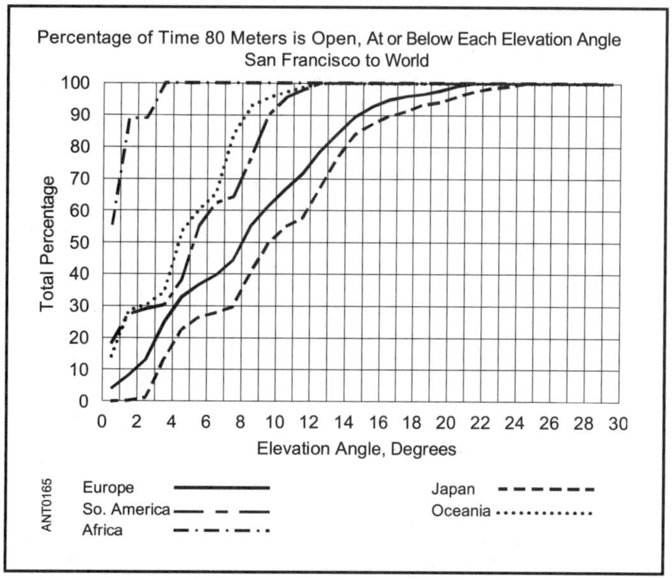

Figure 14.13 — From San Francisco to the rest of the world on 80 meters: 90% of the time on the path to Japan, the takeoff angle is less than or equal to 17°; 50% of the time the angle is less than or equal to 10°; 25% of the time the angle is less than or equal to 6°. A horizontally polarized antenna would have to be 600 feet above flat ground to be optimum at 6°!

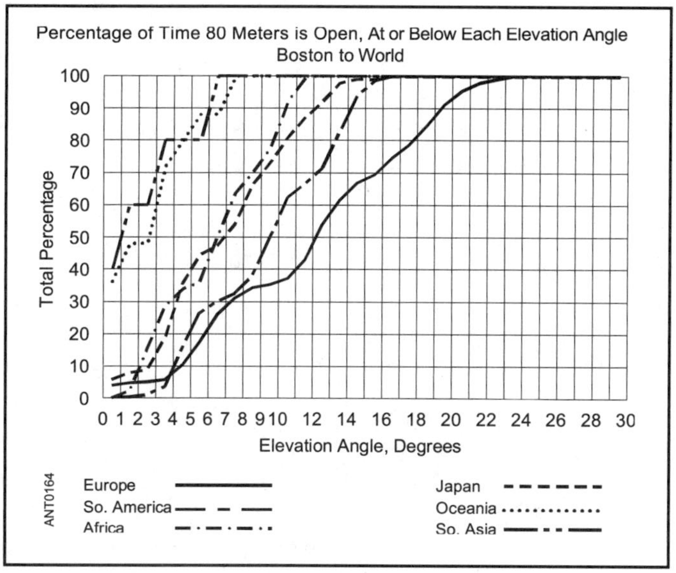

Figure 14.12 — The situation on 80 meters from Boston to the rest of the DX world. Into Europe, 90% of the time the elevation angle is less than or equal to 20°. Into Japan from Boston, 90% of the time the angle is less than or equal to 12°.

angles are affected by geomagnetic storms, and generally speaking the elevation angles rise under these conditions.)

Figure 14.13 shows the 80 meter data from San Francisco to the world. Low elevation angles dominate in this graph and high horizontal antennas would be necessary to optimal coverage. In fact, 50% of the time for all paths, the elevation angle is less than 10°.

14.2.2 NVIS Communication

Not all hams are interested in working stations thousands of miles from them. Traffic handlers and rag chewers may, in fact, only be interested in *nearby* communications — perhaps out to 600 miles from their location. In such cases, the low elevation angles needed for effective DXing may be completely ineffective in providing the required short-range coverage.

For example, a ham in Boston may want to talk with his brother-in-law in Cleveland, Ohio, a path that is just over 550 miles away. Or an operator in Buffalo, New York, may be the net control station (NCS) for a regional net involving the states of New York and New Jersey. She needs to cover distances up to about 300 miles away.

Depending on the time of day, the most appropriate ham frequencies needed for nearby communications are the 40 and 80/75 meter bands, with 160 meters also a possibility during the night hours, particularly during low portions of the sunspot cycle. The elevation angles involved in such nearby distances are usually high, even almost directly overhead for distances beyond ground-wave coverage (which may be as short as a few miles on 40 meters). For example, the distance between the Massachusetts cities of Boston and Worcester is about 40 miles. On 40 meters, 40 miles is beyond ground-wave coverage. So you will need sky-wave signals that use the ionosphere to communicate between these two cities, where the elevation angle is 83° — very nearly straight up.

Hams using vertical antennas for communications with nearby stations may well find that their signals will be below the noise level typical on the lower bands, especially if they aren't running maximum legal power. Such relatively short-range paths involve what is called *NVIS*, "Near Vertical Incidence Skywave," a fancy name for HF communication systems covering nearby geographic areas. The US military discusses NVIS out to about 500 miles, encompassing the territory a brigade might cover. Elevation angles needed to cover distances from 0 to 500 miles range from about 40° to 90°. This also covers the circumstances involved in amateur communications, particularly in emergency situations. (An Internet group on NVIS, **groups.io/g/NVIS**, focuses on antenna designs and techniques for NVIS communication.)

For the interested reader, a study of the use of NVIS communication is presented in "Near Vertical Incidence Skywave Propagation: Elevation Angles and Optimum Antenna Height for Horizontal Dipole Antennas" in the February 2015 issue of *IEEE Antennas and Propagation Magazine* by Ben Witvliet, PA5BW, and other researchers.

The following section is adapted from the article "What's the Deal About NVIS?" by Dean Straw, N6BV, that appeared in the December 2005 *QST*. This article used an example of a hypothetical earthquake in San Francisco to analyze HF emergency communication requirements.

HAM RADIO RESPONSE IN NATURAL DISASTERS

One of San Francisco's somewhat less endearing nicknames is "the city that waits to die." When the *Big Earthquake* does come, you can be assured that all the cell phones and the land-line telephones will be totally jammed, making calling in or out of the San Francisco Bay Area virtually impossible. The same thing occurred in Manhattan on September 11, 2001. The Internet will also be severely affected throughout northern California because of its trunking via the facilities of the telephone network. Commercial electricity will be out in wide areas because power lines will be down.

Table 14.1
Average Elevation Angles for Target Destinations from San Francisco

Location	*Distance (Miles)*	*Average Elevation Angle (Degrees)*
San Jose, CA	43	80
Sacramento, CA	75	78
Fresno, CA	160	63
Reno, NV	185	60
Los Angeles	350	44
San Diego	450	42
Portland, OR	530	30
Denver, CO	950	18
Dallas, TX	1500	8

If the repeaters on the hills around the San Francisco Bay Area haven't been damaged by the shaking itself, there will be some ham VHF/UHF voice coverage in the intermediate area, at least until the backup batteries run down. But connecting to the dysfunctional telephone system will be difficult at best through amateur repeaters.

With little or no telephone coverage, an obvious need for ham radio communications to aid disaster relief would be from San Francisco to Sacramento, the state capital. Sacramento is 75 miles northeast of the Bay Area, well outside VHF/UHF coverage, so amateur HF will be required on this radio circuit. On-the-ground communications directly between emergency personnel (including the armed-forces personnel who will be brought into the rescue and rebuilding effort) will often be difficult on VHF/UHF since San Francisco is a hilly place. So HF will probably be needed even for short distance, operator-to-operator or operator-to-communications center work. Throughout the city, portable HF stations will have to be quickly set up and staffed to provide such communications.

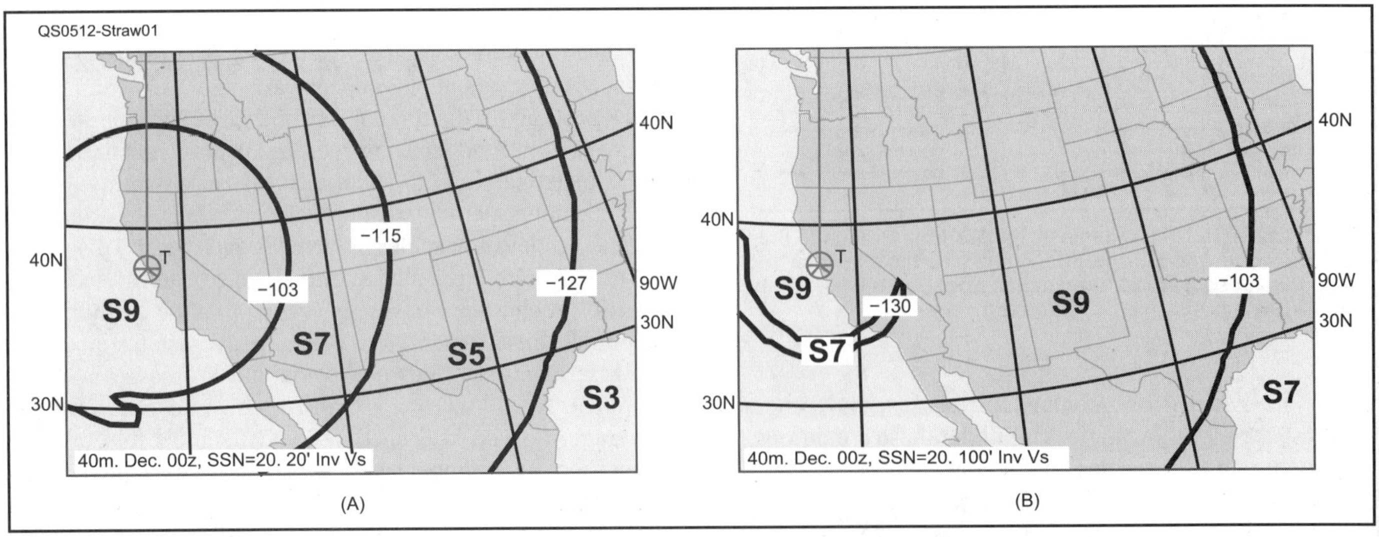

Figure 14.14 — At A, predicted 40 meter geographic coverage plot for a 100 W transmitter in December at 0000 UTC (near sunset), for a SSN (Smoothed Sunspot Number) of 20. The antennas used are 20-foot-high inverted V dipoles. At B, 40 meter coverage for same date and time, but for 100-foot-high flattop dipoles. Most of California is well covered with S9 signals in both cases, but there is more susceptibility in the higher dipole case to thunderstorm crashes coming from outside California, for example from Arizona or even Texas. Such noise can interfere with communications inside California.

Hams half jokingly call short range HF communications on 40 and 80 meters "cloud warming." This is an apt description, because the takeoff angles needed to launch HF signals up into the ionosphere and then down again to a nearby station are almost directly upward. **Table 14.1** lists the distance and takeoff angles from San Francisco to various cities around the western part of the USA. The distance between San Francisco and Sacramento is about 75 miles, and the optimum takeoff angle is about 78°. Launching such a high-angle signal is best done using horizontally polarized antennas mounted relatively close to the ground, such as low dipoles.

GEOGRAPHIC COVERAGE FOR NVIS

Figure 14.14A shows the geographic area coverage around San Francisco for a 100-W station on 7.2 MHz using an inverted V dipole. The center of this antenna is 20 feet above flat ground and the ends are 8 feet high. An actual implementation of such an antenna could be as an 80 meter inverted V, fed in parallel with a 40 meter inverted V dipole at a 90° angle. See **Figure 14.15**. The 8-foot height puts the ends high enough to prevent RF burns to humans (or most animals). The low height of the antenna above ground means that the azimuthal pattern is omnidirectional for high elevation angles.

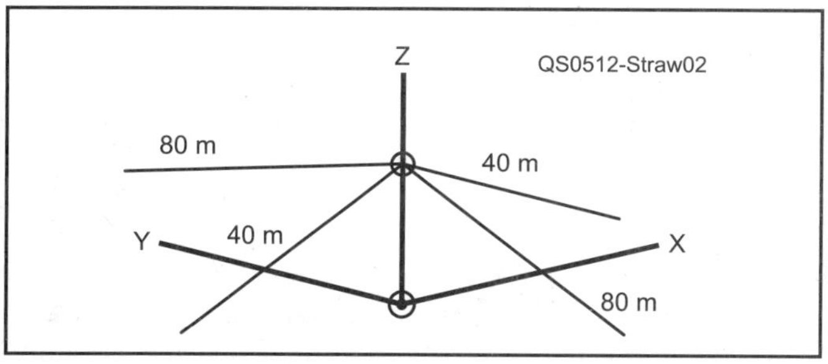

Figure 14.15 — Layout for two band inverted V dipoles for 40 and 80 meters. The two dipoles are fed together at the center and are laid out at right angles to each other to minimize interaction between them. Each end of both dipoles is kept 8 feet above ground for personal safety.

Figure 14.14 was generated using the *VOAAREA* program, part of the *VOACAP* propagation-prediction suite, for the month of December. This was for 0000 UTC, close to sundown, for a low period of solar activity (Smoothed Sunspot Number, SSN of 20). The receiving stations were also assumed to be using identical inverted-V dipoles.

You can see that almost the whole state of California is covered with S9 signals, minus only a thin slice of land near the Mexican border in the southeast portion of the state, where the signal drops to S7. Signals from Texas are predicted to be only S5 or less in strength. Signals (or thunderstorm static) coming from, say, Louisiana would be several S units weaker than signals from central Texas.

Now take a look at Figure 14.14B. Here, the date, time and solar conditions remain the same, but now the antennas are 100-foot high flattop dipoles. California is still blanketed with S9 signals, save for an interesting crescent-shaped slice near Los Angeles, where the signal drops down to S7. Close investigation of this intriguing drop in signal strength reveals that the necessary elevation angle, 44°, from San Francisco to this part of southern California falls in the first null of the 100-foot high antenna's elevation pattern. See **Figure 14.16**, which shows the elevation patterns for five 40 meter antennas at different heights. In the null at a 44° takeoff angle, the 100-foot high dipole is just about equal to a 2-foot high dipole. We'll discuss 2-foot high dipoles in more detail later.

For most of California, the problem with 100-foot high 40 meter antennas is that interfering signals from Texas, Colorado or Washington State will *also* be S9 in San Francisco. So will static crashes coming from thunderstorms all over the West

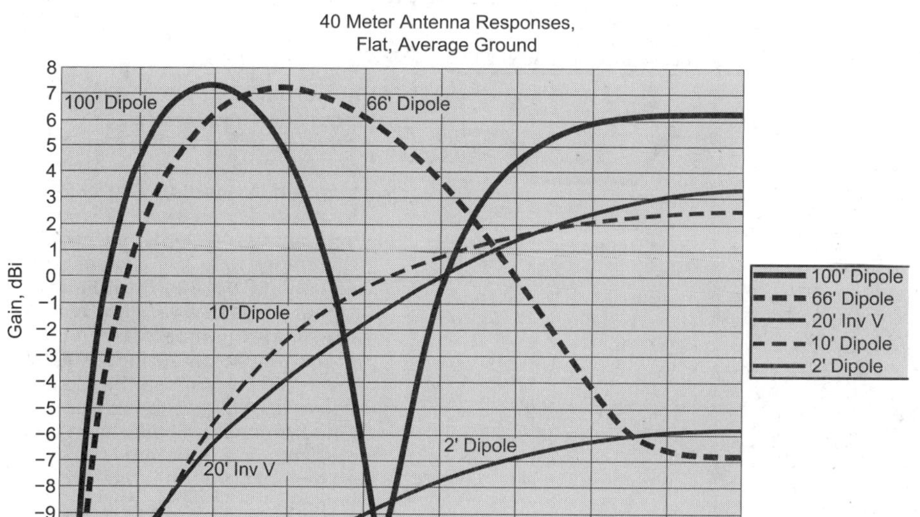

Figure 14.16 — Elevation plots for different 40 meter antennas above flat ground with average ground characteristics (5 mS/m conductivity and dielectric constant of 13). The 10-foot-high flattop dipole and the 20 foot-high inverted V dipoles both have close to the same characteristics. Note that there is a null in the response of the 100-foot-high flattop dipole at a 42° elevation angle. The gain there is roughly that of a 2 foot-high dipole!

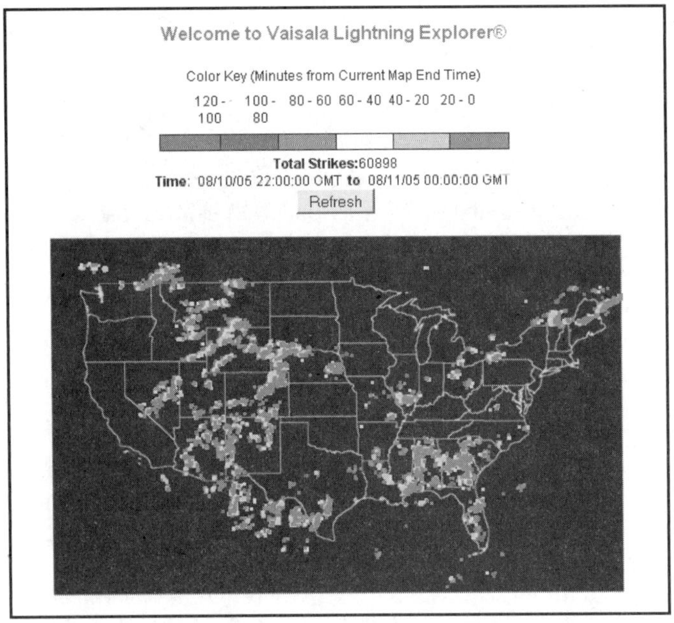

Figure 14.17 — The distribution of lightning strikes across the USA for August 10, 2005 from 2200 to 0000 UTC, in the afternoon California time. There are lots of lightning strikes in the US during the summer — 60,898 of them in this two-hour period! (Courtesy Vaisala Lightning Explorer.)

and much of the Gulf Coast. See **Figure 14.17**, which shows a typical distribution of thunderstorms across the US in the late afternoon, California time, in mid-August. There certainly are a lot of thunderstorms raging around the country in the summer.

The signal-to-noise and signal-to-interference ratios for a 20-foot high inverted V dipole will be superior for medium-range distances, say out to 500 miles from the center, compared to a 100-foot high antenna. The 20-foot high antenna can discriminate against medium-angle thunderstorm noise in the late afternoon coming from the Arizona desert, although it wouldn't help much for thunderstorms in the Sierra Nevada in central Nevada, which are arriving in San Francisco at high angles, along with the desired NVIS signals.

This is the essence of what NVIS means. NVIS exploits the difference in elevation pattern responses of low horizontally polarized antennas compared to higher horizontal antennas, or even verticals. Over the years, many hams have been lead to believe that higher is always better. This is not quite so true for consistent coverage of medium or short distance signals!

If NVIS only involved putting up a low horizontally polarized antenna on 40 meters the story would end here. However, real cloud warming is more complicated. It also involves the intelligent choice of more than just one operating frequency to achieve reliable all day, all-night communications coverage.

Figure 14.18 shows the signal strength predicted using *VOACAP* for the 350-mile path from San Francisco to Los Angeles for the month of December for a period of low solar activity (SSN of 20). The antennas used in this case are 10-foot high dipoles, just for some variety. These act almost like 20-foot high Inverted V dipoles. December at a low SSN was chosen as a worst-case scenario because the *winter solstice* occurs on December 21. This is the day that has the fewest hours of daylight in the year. (Contrast this with the *summer solstice*, on June 21, which has the most hours of daylight in the year.) Note that the upper signal limit in Figure 14.18 is "S10" — a fictitious quantity that allows easier graphing. S10 is equivalent to S9+, or at least S9+10 dB.

The 40 meter curve in Figure 14.18 shows that the MUF (maximum usable frequency) actually drops below the 7.2 MHz amateur band after sunset. The signal becomes quite weak for about 14 hours during the night, from about 0300 to

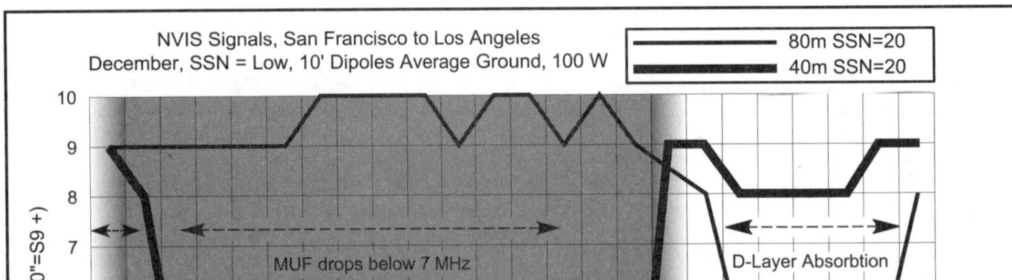

Figure 14.18 — VOACAP calculations for a 350 mile path from San Francisco to Los Angeles, using 10 foot-high flattop dipoles. This plot shows the signal strength in S units ("S10" = S9+10) for a worst-case month/SSN combination — winter solstice, in December, for a low level of solar activity (SSN = 20). The 40 meter signal drops to a very low level during the night because the MUF drops well below 7.2 MHz. The 80 meter signal drops in the afternoon because of D-layer absorption. For 24-hour communications on this path, the rule of thumb is to select 40 meters during the day and 80 meters during the night.

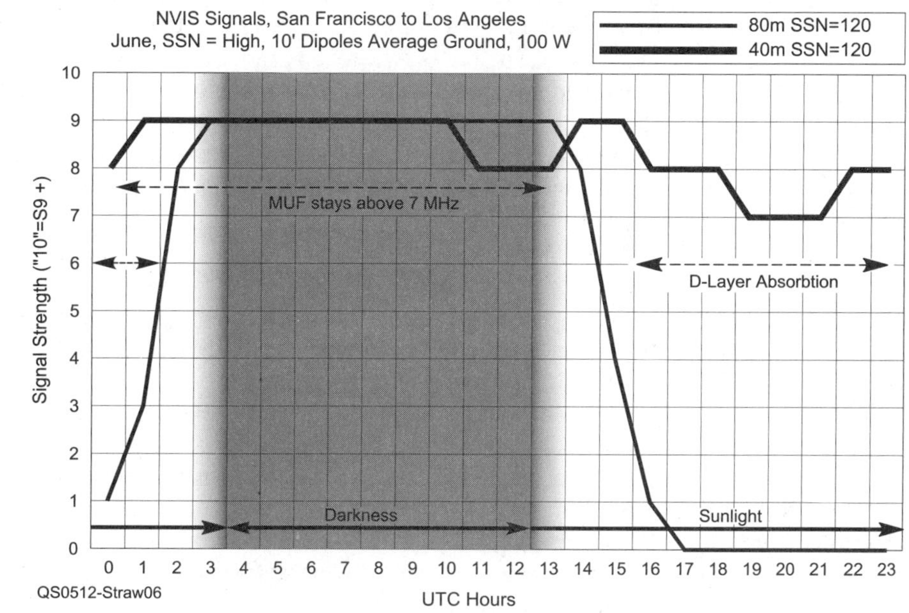

Figure 14.19 — Signal strengths for the San Francisco to Los Angeles path for a worst-case month/SSN combination — summer solstice, in June, for a high level of solar activity (SSN = 120). Now 80 meters drops out more dramatically during the daylight hours, due to increased D-layer absorption. At this high level of solar activity, 40 meters remains open 24 hours with reasonable signal levels. However, the NVIS rule-of-thumb still holds: Use 40 meters during the day; 80 meters at night.

1700 UTC. In a period of low solar activity the 40 meter band thus becomes strictly a *daytime band* on this medium-distance path.

The 80 meter curve in Figure 14.18 shows strong signals after dusk, through the night and up until about an hour after sunrise. After sunrise, 80 meters starts to suffer absorption in the D layer of the ionosphere and hence the signal strength drops. Here, 80 meters is a true *nighttime band.*

Let's see what happens from San Francisco to Los Angeles during a period of high solar activity (SSN of 120) during the summer solstice in June. **Figure 14.19** shows that 40 meters now stays open all hours of the day due to the greater number of hours of sunlight in June and because the ionosphere becomes more highly ionized by higher solar activity. Meanwhile, 80 meters still remains a nighttime band during these conditions on this path.

Now, let's look at a shorter-distance path — our 75-mile emergency communications path from San Francisco to Sacramento. We'll again use June during the summer solstice, at a high level of solar activity (SSN of 120) because this represents another worst-case scenario. **Figure 14.20** shows that 40 meters remains open on this path all day, dropping to a lower signal level just before sunrise. At sunrise, the MUF drops close to 7.2 MHz. 80 meters is still mainly a nighttime band to Sacramento, even though it does yield workable signal levels even during the daylight hours. However, 40 meters is

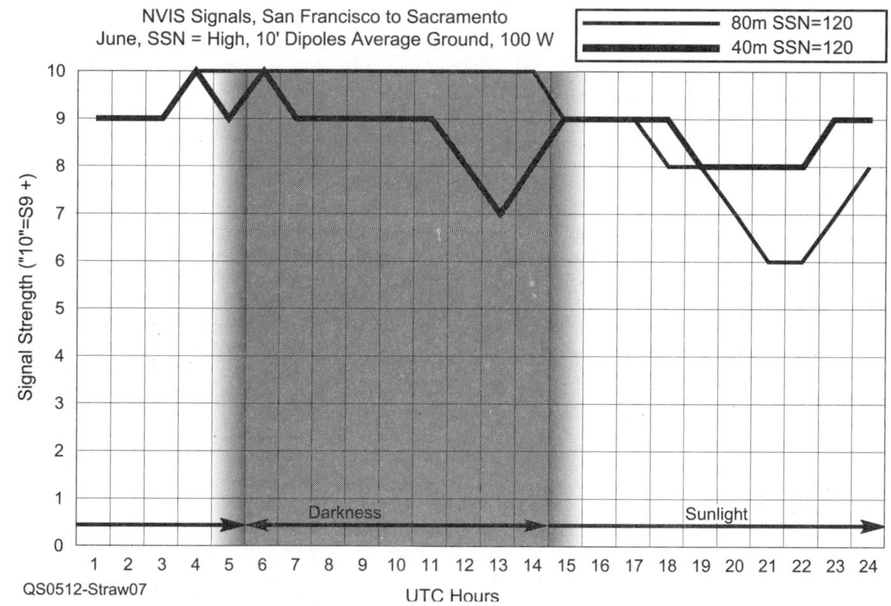

Figure 14.20 — Signal strengths for a 75 mile path — San Francisco to Sacramento. This is for June and SSN = 120. Either band could be used successfully over the full 24-hour period because signal levels are always higher than S6. But the simple NVIS rule-of-thumb still holds: Use 40 meters during the day; 80 meters at night. This simplifies giving instructions to operators unaccustomed to the use of HF.

better from 1200 to 0400 UTC, so 40 would be still the right daytime band for this path during the day.

CHOOSING THE RIGHT NVIS FREQUENCY

You can see that a pattern is developing here for efficient NVIS short/medium-distance communications out to 500 miles:

- You should pick a frequency on 40 meters during the day.
- You should pick a frequency on 80 meters during the night.
- You should choose an antenna that emphasizes moderate to high elevation angles, from 40° to almost directly overhead at 90°.

"What about 60 meters?" you might ask. The characteristics on 60 meters fall in between 40 and 80 meters, although it resembles 40 meters more closely. With characteristics close to that of 40, but with only five channels available and a 50-W power limit, the 60 meter band is currently of low utility for serious NVIS use. (Should the FCC allocate hams more 60 meter frequencies, the band's utility would improve.)

What about 160 meters? For 100-W level radios, even at the worst-case month or during low solar activity, the critical frequency doesn't fall below 3.8 MHz often enough to destroy the ability to communicate, even for short distances. That is a relief, considering that installing a 160 meter half-wave dipole involves a 255-foot length of wire, and it would need to be elevated at least 30 feet in the center. A short loaded vertical such as a 160 meter mobile whip would have poor response at the high elevation angles needed for NVIS. You could probably put a monster 160 meter horizontal dipole up at a permanent location, but hauling such a thing around in the field would not be an easy task.

NVIS STRATEGY

You could pose the question about whether NVIS is an operating *mode* or whether it is actually an operating *strategy*. We maintain that NVIS is a strategy. It involves choosing both appropriate frequencies and then appropriate antennas for those frequencies. Figure 14.20 does show that on short-distance paths, such as between San Francisco and Sacramento, you could stay on 80 meters all day and night. But if you have to give a single rule-of-thumb to operators who are not very experienced at operating HF, we would tell them to operate on the higher frequency band during the day and on the lower frequency band at night.

ANTENNA HEIGHT FOR NVIS

Some NVIS users have advocated placing dipoles only a few feet over ground, something akin to saying, "If low is good for NVIS, then lower must be even better." A very low antenna *might* work in specific instances — for example, covering a small state such as Rhode Island or even just the San Francisco Bay Area. Such an antenna is very unlikely to cover a large region or state, such as California or Texas, especially on 80 meters. **Figure 14.21** shows the computed elevation responses for a number of 80-meter antennas, including a 2-foot-high dipole. Mounting an antenna lower than 10 feet above ground poses safety concerns, as well.

Figure 14.22B shows the 80-meter geographic coverage plot for 2-foot-high flattop dipoles, compared with the plot in Figure 14.22A for 20-foot-high inverted V dipoles on both ends of the path. The 2-foot-high dipoles produce about two S-units less signal across all of California than the 20-foot-high inverted V dipoles, at 0300 UTC in December, with an SSN of 20. Ground losses increase dramatically as the antenna height is reduced.

The differential between California signals and possible interfering signals from, say, New Mexico, is predicted to be four S-units, the same as it is for the higher inverted V dipole at 20 feet. Thus there is no real advantage in terms of signal-to-interference ratio or signal-to-noise ratio (for thunderstorm static crashes) for either height. This is because the shape of all the response curves in Figure 14.21 below 20 feet essentially track each other in parallel.

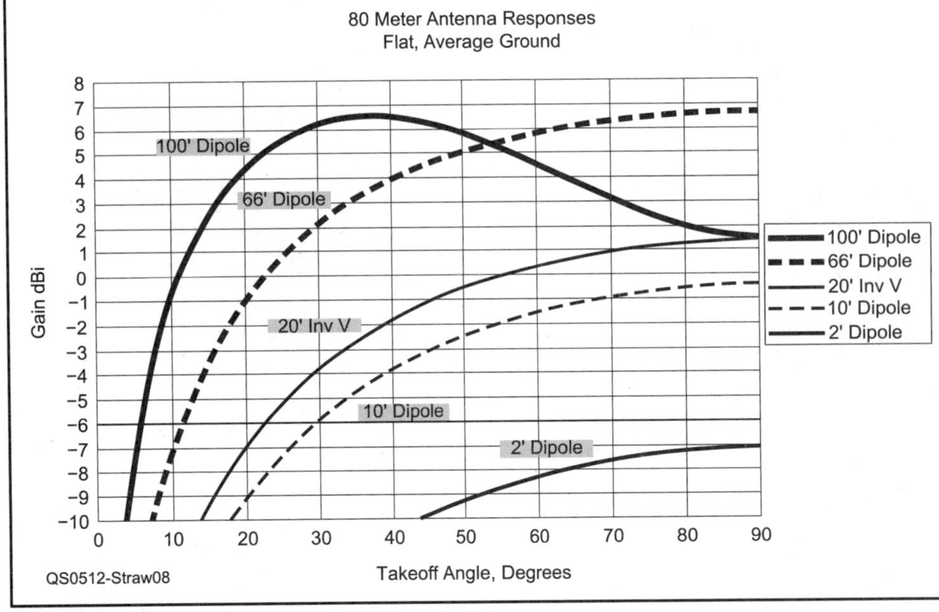

Figure 14.21 — Elevation response patterns for 80-meter antennas over average soil. The shapes track each other rather well, remaining parallel for heights from 2 to 66 feet over flat ground. The 2-foot dipole is substantially down, about 9 dB, from the 20 foot inverted V dipole at all angles.

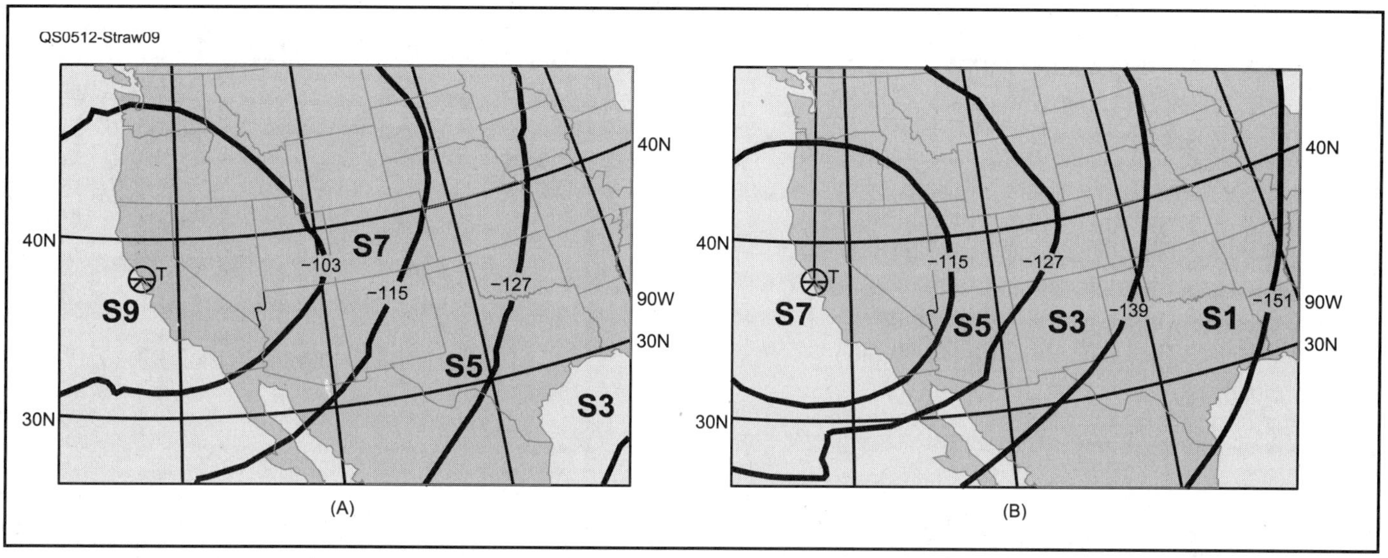

Figure 14.22 — Geographic coverage plots for December, SSN = 20, 0300 UTC. At A, antennas are 20-foot-high inverted V dipoles over Average soil. At B, antennas are 2-foot-high flattop dipoles over Average soil. The response for the 2-foot-high antennas is down about 2 S Units, 8 to 12 dB for a typical communications receiver.

Radials or reflector wires are not recommended, either. The more conductive the ground or ground screen, the less efficient the antenna becomes as the dipole's horizontal E-field component is "shorted out." A feed point impedance that drops below 40 Ω is a clue that efficiency is falling. Not until the reflector is λ/4 or higher above ground does efficiency improve but this is not necessary as the coverage plots demonstrate. This is an edge case with very little practical benefit.

NVIS is a very efficient propagation mode, ideal for low power and portable antennas. If you are in an emergency situation operating on batteries, you can reduce power from 100 W to 10 W with a 20-foot high inverted-V antenna and still maintain the same signal strength as a 2-foot-high dipole at 100 W.

LOW NVIS ANTENNAS AND NOISE

Some advocates of really low antennas have stated that the received noise is much lower than that received from higher antennas, and this therefore leads to better signal-to-noise ratios (SNR). How much this is true depends on the source of the noise. If the noise comes from distant thunderstorms, then the SNR advantage going to a 2-foot antenna from a 20-foot-high one is insignificant, as Figure 14.22 indicates.

If noise is from an arcing insulator on a HV power line half a mile away, that noise will arrive at the antenna as a ground-wave signal. We calculate that the 2-foot antenna receives 4.4 dB less noise by ground-wave than a 20-foot-high inverted V dipole. However, at an incoming elevation angle of 45° — suitable for a signal going from Los Angeles to San Francisco — the signal would be down 7.1 dB on the low dipole compared to the higher antenna. The net loss in SNR for the 2-foot-high dipole is thus 7.1 – 4.4 or 2.7 dB. In this case, the SNR is actually worse on the very low antenna.

Remember that any improvement on SNR from a very low antenna is available only on received signals. This may be useful if an efficient transmit antenna is also available but the losses of such a low antenna make it a poor choice as the only antenna, especially when safety risks are considered.

It's possible that a really low dipole may serve your short-range communication needs just fine. But just as "higher is better" isn't universally true for NVIS (or even longer range) applications, "lower is better" isn't a panacea either.

ELEVATION ANGLES FOR MODERATE DISTANCES ON 75/80 METERS

Figure 14.23 shows the elevation angles statistics for a 75 meter, 550-mile path from Boston to Cleveland, together with overlays of the elevation patterns for several different types of antennas. These elevation statistics cover all parts of the 11-year solar cycle for this path. The responses for the popular G5RV antenna (see the **Multiband Antennas** chapter) are shown for two different heights above flat ground: 50 and 100 feet. An 80 meter half-wave sloper ("full sloper") and an 80 meter ground-plane antenna are also shown. All antenna patterns are for "average ground" constants of 5 mS/m conductivity and a dielectric constant of 13.

At the statistically most significant takeoff angles around 50°, the two horizontally polarized G5RV antennas are about equal. At the second-highest elevation peak near 30°, the 100-foot G5RV has about a 4-dB advantage over its lower counterpart. The full sloper has comparable performance to the 100-foot high G5RV from 1° to about 20° and then gradually rises to its peak at angles higher than 70°. The full sloper is superior to the 50-foot horizontal G5RV at low takeoff elevation angles. The 80 meter ground plane has a deep null directly overhead. At an elevation angle of 70° it is down some 16 dB compared to the 50-foot high horizontal G5RV.

The advantage of antennas suitable for high-angle radiation

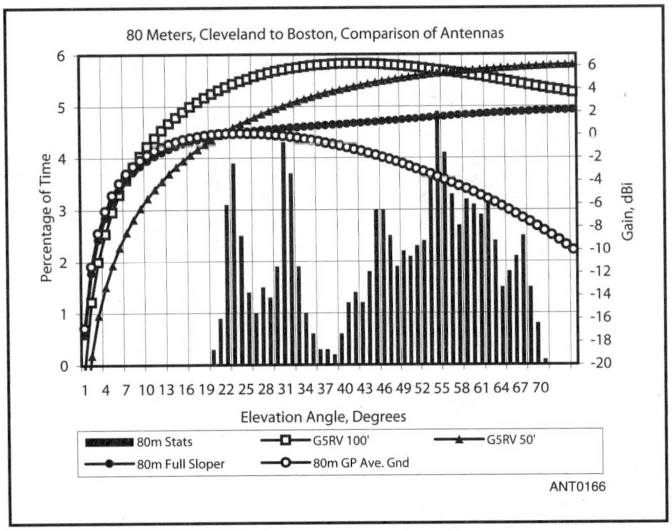

Figure 14.23 — 80/75 meter elevation statistics for all portions of the 11-year solar cycle for the path from Cleveland, Ohio, to Boston, Massachusetts, together with the elevation responses for four different multiband antennas. The 100-foot high horizontally polarized G5RV performs well over the entire range of necessary takeoff elevation angles.

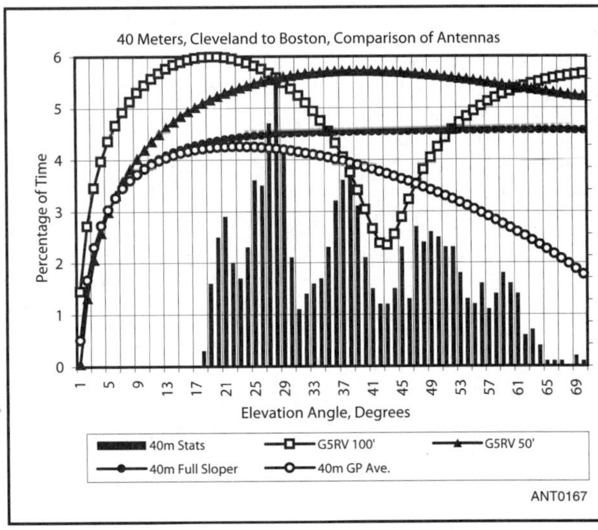

Figure 14.24 — 40 meter elevation statistics for the Cleveland to Boston path, together with elevation patterns for four antennas. Here, the 100-foot high horizontally polarized G5RV would have a null in the middle of the range of elevation angles needed for consistent performance on this path. For multiband use on this path to relatively nearby stations, the 50-foot high horizontal antenna would be a better choice than the 100-foot high antenna.

was vividly demonstrated during a 75 meter QSO one fall evening between N6BV/1 in southern New Hampshire and W1WEF in central Connecticut. This involved a distance of about 100 miles and W1WEF was using his Four-Square vertical array. Although W1WEF's signal was S9 on the Four Square, N6BV/1 suggested an experiment. Instead of connecting the so-called "dump power" connector on his Comtek ACB-4 hybrid phasing coupler to a 50-Ω dummy load (the normal configuration), W1WEF switched the dump power to his 100-foot high 80 meter horizontal dipole. W1WEF's signal came up more than 20 dB! The approximately 100-W of power that would otherwise be "wasted" in the dummy load was converted to useful signal.

ELEVATION ANGLES FOR MODERATE DISTANCES ON 40 METERS

Figure 14.24 shows the situation for the 40 meter band, from Boston to Cleveland, together with the same antennas used for 80 meters in Figure14.23. Note that the 100-foot high horizontally polarized G5RV has about a 16-dB null at an elevation angle of 43°. This doesn't affect things for low elevation angles, but it certainly has a profound effect on signals arriving between about 30° to 60°, especially when compared to the 50-foot high horizontal G5RV. The 40 meter full sloper beats out the high horizontal antenna from about 35° to 50°. And the ground plane is obviously not the antenna of choice for this moderate-range path from Boston to Cleveland, although it is still a good performer on longer-distance paths, with their low takeoff angles.

A 100-foot high multiband dipole is about ⅜-λ high on 75/80 meters. It is an excellent antenna for general-purpose local and DXing operation. But the same dipole used on 40 meters becomes ¾-λ high. At that height, the nulls in its elevation pattern give large holes in coverage for nearby 40 meter contacts. Many operators have found that a 40- to 50-foot high dipole on 40 meters gives them far superior performance for close-in QSOs, when compared to a high dipole, or even a high 2-element 40 meter Yagi.

NVIS SUMMARY

The use of NVIS strategies to cover close-in and intermediate distance communications within about 600 miles involves the intelligent choice of low HF frequencies. As a rule-of-thumb for ham band NVIS, 40 meters is recommended for use during the day; 80 meters during the night.

NVIS involves the choice of antennas suitable for this strategy. Horizontally polarized dual-band 80 and 40 meter flattop dipoles that are mounted higher than about 10 feet high will work adequately for portable operations. Dual-band 80 and 40 meter inverted V dipoles supported 20 feet above the ground at the center can also work well in portable operations.

Single-band 40 meter flattop antennas about 30 feet high and 80 meter flattop antennas about 60 feet high can do a good job for fixed locations.

14.3 Effects of Local Terrain

This section describes a systematic approach to HF antenna system design, based on the following information about your location and available antennas:

1) The effect of local terrain on elevation patterns for horizontally polarized antennas.

2) The range of elevation angles necessary to get from point A to point B

3) The elevation patterns for various types and configurations of antennas

By using this method, you can determine where and how high to install antennas to optimize your transmitted signal during all parts of the 11-year solar cycle. For vertically polarized antennas, see the chapter **Effects of Ground**.

14.3.1 Fresnel Zones and Site Selection

The following material is based on Appendix F of *FM 11-65: High Frequency Radio Communications*, a US Army publication used for planning radio communication links, supplemented by notes and figures from Frank Donovan, W3LPL. The complete text of Appendix F, including planning charts and graphs, is included in the online material and PDF documents or printed copies of the full publication are available online from several sources.

The Fresnel (fruh-NEL) zone is of particular interest in HF and 6-meter operation where ground reflections from the antenna's Fresnel zone can provide up to 6 dB of ground gain. Ground reflections from the antenna's Fresnel zone form the antenna's elevation pattern. This is very important in selecting antenna heights in order to maximize gain at the desired elevation angles. This discussion of Fresnel zones applies to horizontally polarized antennas. A description of reflection of radio waves, both horizontally and vertically polarized, is provided in the **Effects of Ground** chapter.

FRESNEL ZONE DEFINITION

The *first Fresnel zone* is defined as the first area in front of the antenna from which reflected rays reinforce the direct ray at the desired elevation angle. (See the following sections for a discussion of elevation angle selection.)

The geometry of the first Fresnel zone is shown in **Figure 14.25**. The *reference ray* is the reflected ray which is parallel to the direct ray and has a path length λ/2 longer than the direct ray. Since a horizontally polarized ray is phase-shifted by 180 degrees on reflection, the reference ray is in phase with the direct ray.

The ellipse outlined by the dashed line in Figure 14.25 represents the area from which the reflected rays will be less than 180 degrees in phase from the reference ray. All reflections from the elliptical area will reinforce the direct ray. Most of the reinforcement is produced by reflections in the middle half of the first Fresnel zone area where reflections are phase-shifted less than 90 degrees. The exact geometry of the first Fresnel zone is determined by wavelength and elevation angle (θ or A in the figure).

Although second, third, and higher order Fresnel zones surround the first Fresnel zone, they have no relavance to site selection or antenna placement. They are of less importance because at the greater distance from the transmitting antenna, the reflected rays are weaker. In the odd-numbered zones (1, 3, 5, etc.) the reflected ray reinforces the direct ray; in the even-numbered zones (2, 4, 6, etc.) it weakens the direct ray. Unless otherwise specified, all subsequent references to the Fresnel zone will be only to the first Fresnel zone.

Reflections in the Fresnel zone reinforce the direct ray from the antenna, producing up to 6 dB of *ground gain* or *ground reflection gain*. Stated another way, a horizontal λ/2 dipole at least λ/2 high with relatively smooth terrain and few tall

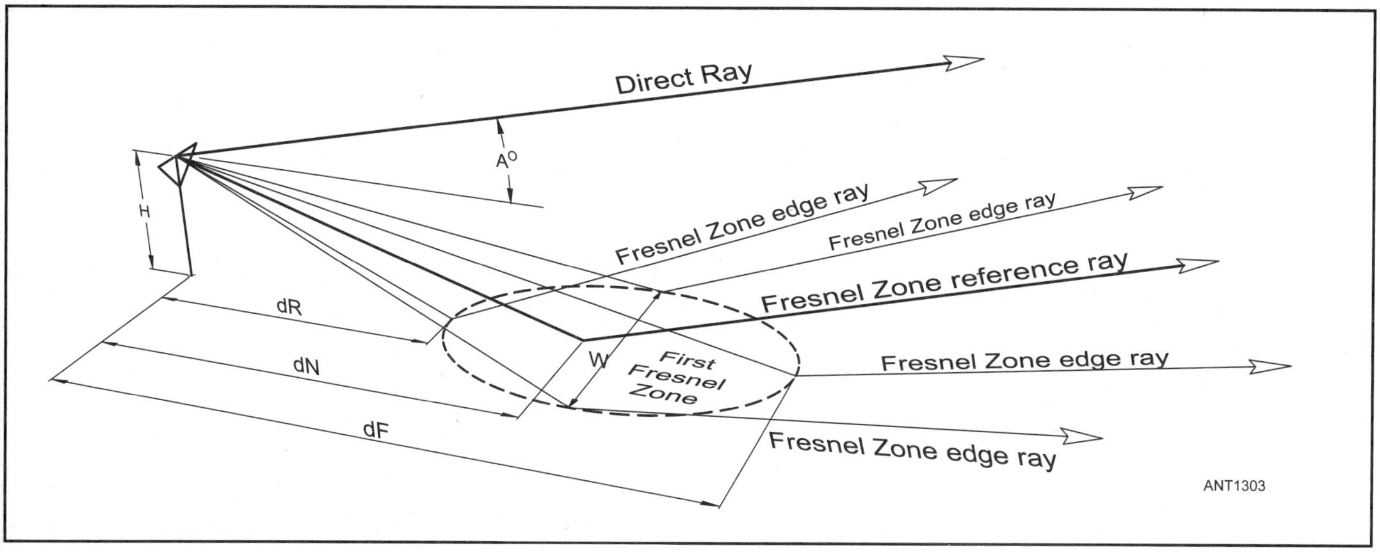

Figure 14.25 — Geometry of the first Fresnel zone. Elevation angle is shown as θ or A in different references.

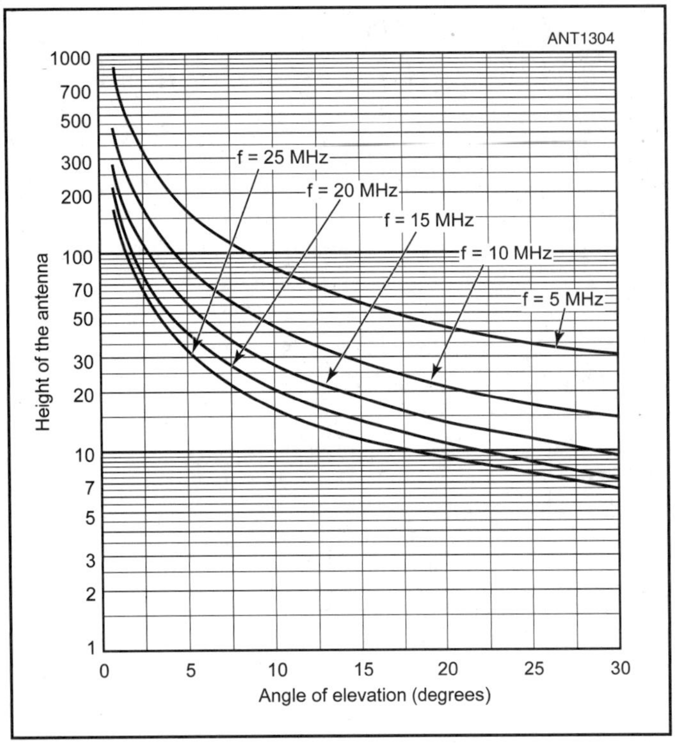

Figure 14.26 — Optimum antenna height as a function of frequency and elevation angle.

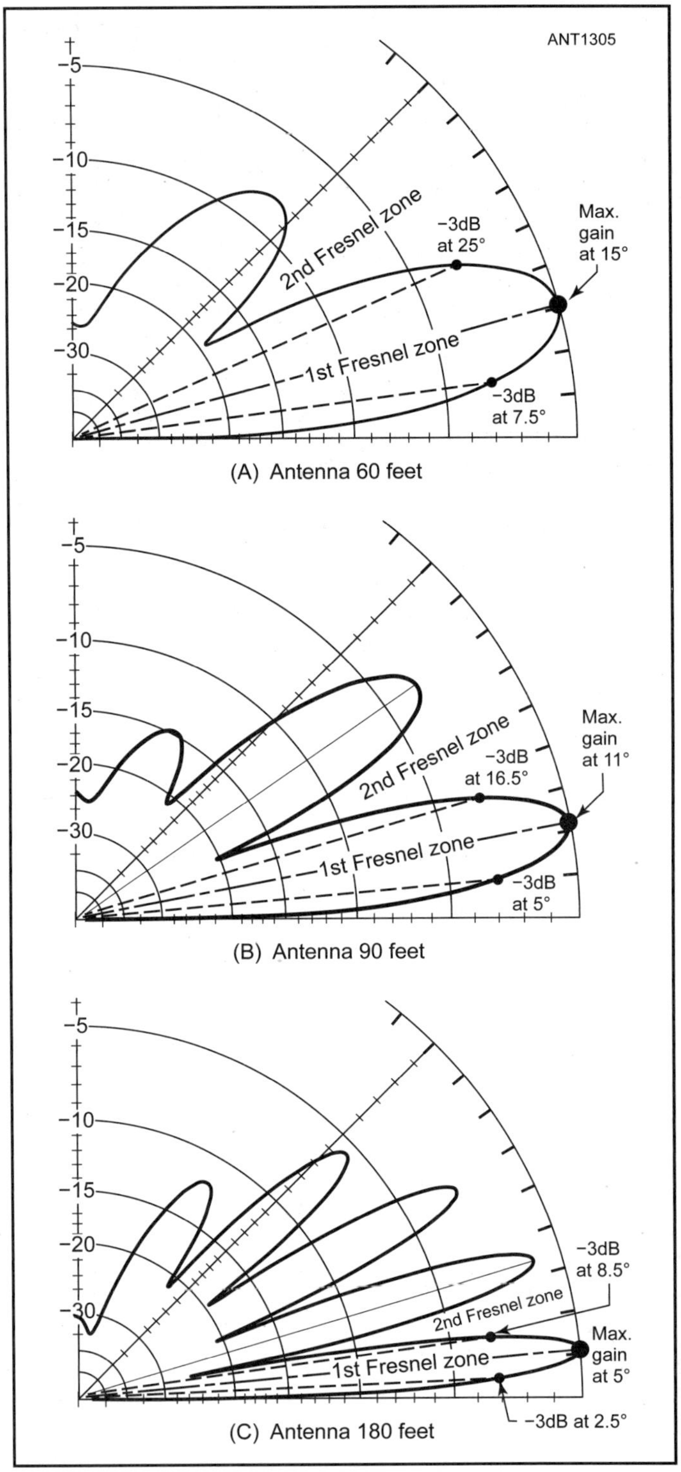

Figure 14.27 — Elevation patterns for a 5-element 20-meter Yagi at the heights in Table 14.2. The effects of the first three Fresnel zones are shown.

buildings or other obstructions in the Fresnel zone has up to 6 dB gain over a horizontal dipole in free space. (This additional ground gain can distort claimed antenna gain figures — be sure it is made clear whether antenna gain is specified in free space or over ground and at what height.)

SITE ANALYSIS AND GUIDELINES

The middle half of the Fresnel zone is where most ground gain is developed. Specifically, the part with less than ±90 degrees of phase shift from the reference ray is part where reflections produce more than 3 dB of ground gain. For that reason, this portion is the most important on HF and 6 meters for:

- Selecting a new permanent or temporary station location
- Selecting a site for DXpedition, Field Day, or mountaintop/hilltop operations
- Placement of a 6 meter EME antenna to optimize ground gain
- Placement of an antenna on or close to sloping terrain to reduce its elevation angle
- Selection of antenna location and height to minimize ground gain degradation from irregular or densely forested terrain or from densely populated locations

Fresnel zone analysis also produces some important guidelines:

- Avoid locations surrounded by tall buildings if possible.
- In a location surrounded by many tall obstructions, install the antenna on a tall tower or on the roof of a tall building if possible.
- Sloping sites are beneficial only if the slope is long enough that most of the middle half of the Fresnel zone is on the sloping terrain and the antenna is installed close to or on the sloping terrain.

ANTENNA HEIGHT AND FRESNEL ZONE LOCATION

After selecting a desired elevation angle, θ, as described in the following sections, the optimum antenna height for a

Table 14.2
Fresnel Zones for Different Elevation Angles on 20 Meters
See Figure 14.25 for parameter definitions; height is over flat, unobstructed ground

Desired elevation angle – A	15°	10°	5°
Optimum antenna height – H (ft)	60	90	180
Fresnel zone near edge – dN (ft)	15	65	330
Fresnel zone maximum width – W (ft)	350	500	1000
Fresnel zone reference point – dR (ft)	200	500	2300
Fresnel zone far edge – dF (ft)	1300	3000	12,000
Fresnel zone reflecting area	8 acres	25 acres	200 acres
Optimum terrain roughness (ft)	<15	<20	<45

specific angle can be determined in terms of angle and frequency by **Figure 14.26** or by calculation as:

$$h = \frac{\lambda}{4\sin\theta}$$

For example, if the desired elevation angle is 10 degrees on 14 MHz, the optimum height for the antenna from Figure 14.26 is 28 m (91 ft). From other charts in Appendix F (see the online material), the reference ray reflects at a distance of about 0.7 km (Figure F-4). In this case, the Fresnel zone begins at about 21 m (Figure F-5), extends to 0.9 km (Figure F-6), and has a width of about 170 m (Figure F-7). The size of and distance to the near edge of the Fresnel zone is greatest for low elevation angles and least for high elevation angles.

Table 14.2 shows distances in Figure 14.25 for a 20-meter Yagi at various heights over flat ground optimized for three elevation angles. As you can see, at low elevation angles, the near edge of the zone moves farther from the antenna (dN) and the zone greatly increases in area and length.

Figures 14.27A-C shows clearly the effect of reflections from the first two Fresnel zones for a 5-element 20-meter Yagi at the heights in Table 14.2. Generally speaking, odd-numbered zones create lobes from the reflected ray as it reinforces the direct ray. Even-numbered zones create nulls as the reflected ray cancels with the direct ray. The greatest effects on elevation pattern come from the first and second Fresnel zones.

The lowest antenna (60 feet) covers 7.5 to 25 degrees with a broad lobe and no null at the high angles for shorter-distance contacts. The middle antenna (90 feet) has its maximum gain between 5 and 16.5 degrees so it will perform well for most DX propagation but would not perform as well as the lower antenna at shorter distances. Finally, the highest antenna (180 feet) has excellent gain at the very low angles needed for above-the-MUF propagation just as the band is opening and after it begins to close. Unfortunately, the 180-foot-high antenna has a deep null at the important 9- to 14-degree elevation angles that are needed for most DX propagation, so this antenna would be too high for most DXing. By comparing the relative gains, you can see that as antenna height increases, there is up to 3 dB improvement at very low angles of 1 to 5 degrees useful for above-the-MUF and very long-distance propagation.

GROUND GAIN DEGRADATION

For reflections from the Fresnel zone to reinforce the direct ray, the area in the zone — especially in the middle half of the zone that produces most of the ground gain — should be as free as possible of large terrain irregularities, large buildings and trees, and other large obstructions typical of a dense urban location. In addition, no reflected ray from any portion of the Fresnel zone should be obstructed. General guidelines for optimizing ground gain from Fresnel zones with irregular terrain and obstructions, especially in the middle half of the Fresnel zone area, are:

- Terrain irregularities: <25% of antenna height, and
- Buildings and trees: <25% of antenna height, and
- Obstructions present in less than 5% of the Fresnel zone area, especially in the middle half of its area.

Large obstructions typical of urban environments will create phase shifts, absorption, reflections, and diffraction can reduce ground gain by several dB or more. Recognizing practical limitations for amateurs, another approach is to acknowledge that irregular terrain and large obstructions in the Fresnel zone will reduce the available amount of ground gain in many urban locations unless the antenna is much higher than the obstructions. When planning a station, you may wish to consider the location of the Fresnel zones for the frequencies, heights, and elevation angles you plan to use.

SLOPING SITES

If the desired elevation angle is less than about 5 degrees, a reasonable antenna height is most conveniently obtained on terrain with a long slope in the direction of the radio path encompassing most of the Fresnel zone. The effect of the slope is an apparent increase in the antenna height, as shown in **Figure 14.28**. When selecting the height of antennas over (or close to) sloping terrain, the angle of the slope, β, can be added to the elevation angle in Figure 14.26 or the equation for height modified:

$$h = \frac{\lambda}{4\sin(\theta + \beta)}$$

For example, if a slope of 5 degrees is available over most of the Fresnel zone, Figure 14.26 shows that the height of the

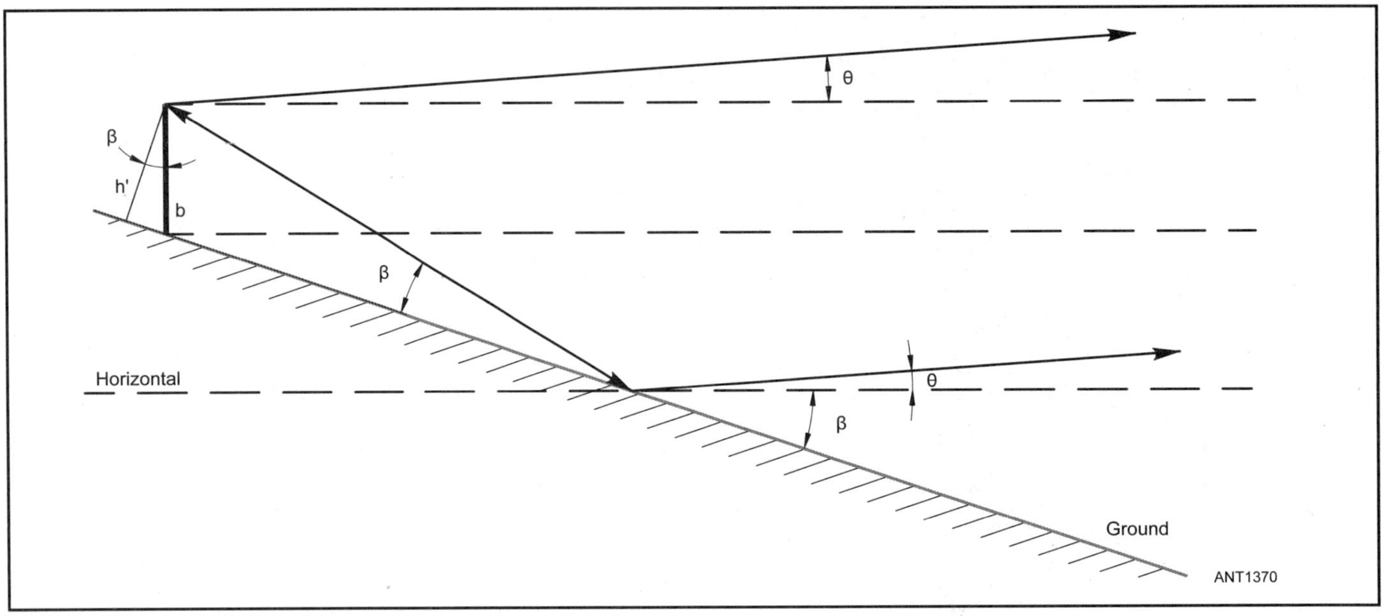

Figure 14.28 — The geometry of reflection for a sloping site.

tower for an elevation angle of 10 degrees is reduced from 28 to 19 meters, about 30 feet lower.

Tower heights may also be reduced by locating the antenna on elevated terrain, like a cliff or some other abrupt drop-off. The terrain above the drop-off should encompass as little of the Fresnel zone as possible and the terrain below the drop-off should encompass as much of the Fresnel zone as possible for the full effect. Remember that the reflected rays should not be obstructed.

14.3.2 Required Range of Elevation Angles

The following material is condensed from the article "The Effect of Local Terrain on HF Launch Angles" by R. Dean Straw, N6BV, in the July 1995 *QEX* magazine and updated in subsequent editions of this book. The terrain analysis program *HFTA* (HF Terrain Assessment) and supporting data files are included with this book's online material. Prior to Straw's article, previous studies of how local terrain affects long-distance operation included four *QST* "How's DX" columns, by Clarke Greene, K1JX, from October 1980 to January 1981. Greene's work was an update of a landmark series of 1966 *QST* articles entitled "Station Design for DX," by Paul Rockwell, W3AFM. (The articles by Straw, Greene, and Rockwell are included with this book's online material.)

To provide the data necessary to design HF antenna systems, ARRL HQ has tabulated the range of elevation angles from all regions of the US to important DX QTHs around the world. This was accomplished using the *IONCAP* computer program. *IONCAP* is considered the standard of comparison for propagation programs by many agencies of the US government, including the Voice of America, Radio Free Europe, and more than 100 foreign governments throughout the world. An online propagation prediction tool, *VOACAP* (**voacap.com**), uses the essential calculating mechanics of *IONCAP* and provides a flexible user interface. See the **Propagation of Radio Waves** chapter for more information about *VOACAP*.

Elevation angle calculations were done for all levels of solar activity, for all months of the year, and for all 24 hours of the day. The results were gathered into databases, from which special custom-written software extracted detailed statistics. The results are included in the online material's HFTA folder. Additional tables of signal strength by location, date, time of day, and solar activity level are also available in the online material's Propagation Prediction Files folder.

Figure 14.29 shows the full range of elevation angles (represented as vertical bars) for the 20-meter path from New England (centered on Newington, Connecticut) to all of Europe. This is for all openings, in all months, over the entire 11-year solar cycle. The most likely elevation angle occurs at 5° for about 13% of the times when the 20-meter band is open to Europe from New England. The band is open from 4° to 6° a total of about 34% of the times the band is open. There is a secondary peak between 10° to 12°, occurring for a total of about 25% of the times the band is open.

Overlaid on Figure 14.29 along with the elevation-angle statistics are the elevation-plane responses for three different horizontally polarized Yagi beams, all over flat ground. The first is mounted 140 feet high, 2 λ in terms of wavelength. The second Yagi is mounted 70 feet high (at 1 λ) and the third is 35 feet (0.5 λ). The 140-foot high antenna has a deep null at 15°, but it also has the highest response (13.4 dBi) of the three at the statistical peak elevation angle of 5°. However, at 12° — where the band is open some 9% of the time — the 140-foot high Yagi is down 4 dB compared to the 70-foot antenna.

The 70-foot high Yagi arguably covers the overall range best, since it has no disastrous nulls in the 1° to 25° range, where most of the action occurs on 20 meters. At 5°, however, its response is only 8.8 dBi, 4.6 dB down from the 140-foot-high antenna at that angle. The 35-foot antenna peaks above 26° in elevation angle and is down some 10.4 dB compared to the 140-foot antenna at 5°. Obviously, no single antenna covers the complete range of elevation angles needed.

Note that the highest Yagi has a strong *second lobe* peaking at 22°. Let's say that you could select between two antennas, one at 140 and one at 70 feet, and that the incoming angle for a particular distant station is 22°. You might be fooled into thinking that the incoming angle is around 6°, favoring the first peak of the higher antenna, when in truth the angle is relatively high. The 70-foot antenna's response would be lower at 22° than the higher one, but only because the 140-foot antenna is operating on its second lobe. (What would clinch a determination of the correct incoming angle — 6° or 22° — would be the response of the 35-foot-high Yagi, which would be close to its peak at 22°, while it would be very far down at 6°.)

Now, we must emphasize that these elevation angles are *statistical entities* — in other words, just because 5° is the "statistically most likely angle" for the 20-meter path from New England to Europe doesn't mean that the band will be open at 11° at any particular hour, on a particular day, in a particular month, in any particular year. In fact, however, experience agrees with the *IONCAP* computations: the 20-meter path to Europe usually opens at a low angle in the New England morning hours, rising to about 11° during the afternoon, when the signals remain strongest throughout the afternoon until the evening in New England.

What would happen if we were to feed all three Yagis at 140, 70, and 35 feet in-phase as a stack? **Figure 14.30** shows this situation, along with a more highly optimized stack at 120, 80 and 40 feet that better covers the overall range of elevation angles from Connecticut to Europe.

Now see **Figure 14.31**, which uses the same 120/80/40-foot stack of 20-meter antennas as in Figure 14.30, but this time from Seattle, Washington, to Europe. For comparison, the response of a single 4-element Yagi at 100 feet over flat ground

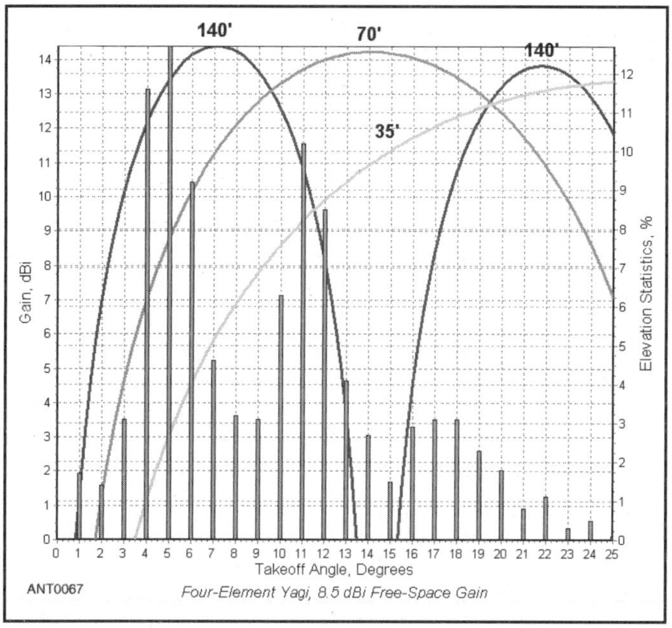

Figure 14.29 — Graph showing 20 meter percentage of all openings from New England to Europe versus elevation angles, together with overlay of elevation patterns over flat ground for three 20 meter antenna systems. The most statistically likely angle at which the band will be open is 5°, although at any particular hour, day, month and year, the actual angle will likely be different. Note the deep null exhibited by the 140-foot high antenna centered at 14°.

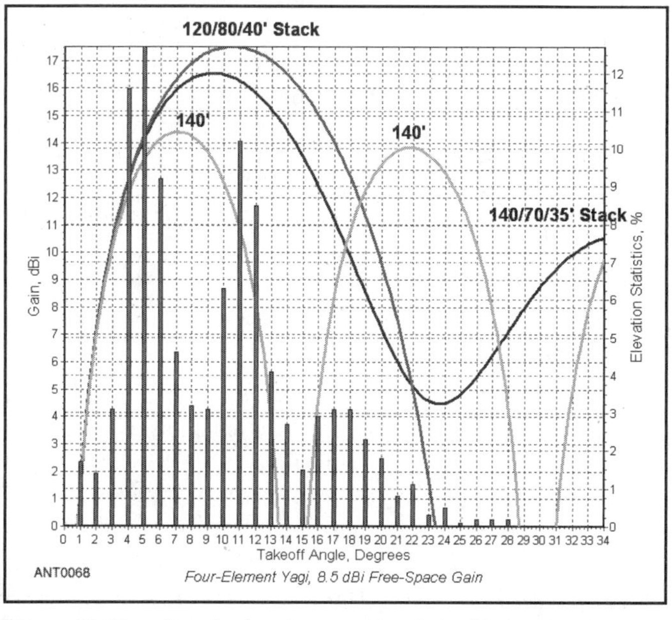

Figure 14.30 — Graph showing results of stacking antennas at different heights on the same tower to cover a wider range of elevation angles, in this case for the path from Connecticut (W1) to all of Europe on 20 meters. The optimized stack at 120/80/40 feet covers the needed range of elevation angles better than the stack at 140/70/35 feet or the single Yagi at 140 feet.

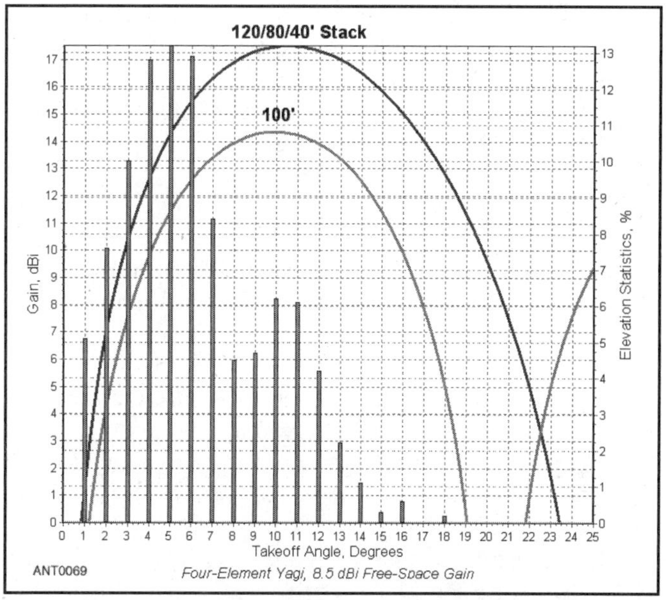

Figure 14.31 — Graph showing 20 meter percentage of all openings, this time from Seattle, WA, to Europe, together with an overlay of elevation patterns over flat ground for two 20 meter antenna systems. The statistically most likely angle on this path is 5°, occurring about 13% of the time when the band is actually open. Higher antennas predominate on this low-angle path.

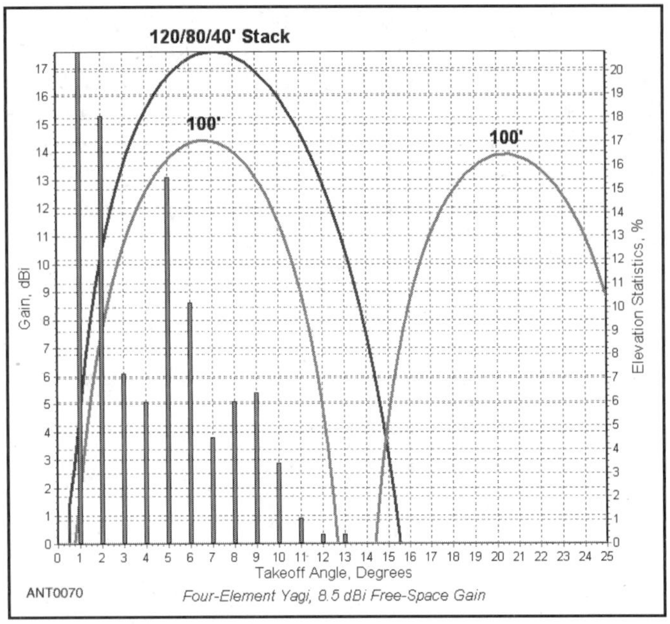

Figure 14.32 — Graph showing 15 meter percentage of all openings from Chicago to Southern Africa, together with overlay of elevation patterns over flat ground for two 15 meter antenna systems. On this long-distance, low-angle path, higher antennas are again most effective.

is also shown in Figure 14.27. Just because 5° is the statistically most prevalent angle (occurring some 13% of the time) from Seattle to Europe on 20 meters, this doesn't mean that the actual angle *at any particular moment in time* might not be 10°, or even 2°. The statistics for W7 to Europe say that 5° is the most likely angle, but 20-meter signals from Europe arrive at angles ranging from 1° to 18°. Note that this range of angles is quite a bit less than from W1 to Europe, which is much closer geographically to Europe than is the Pacific Northwest of the US. If you design an antenna system to cover all possible angles needed to talk to Europe from Seattle (or from Seattle to Europe) on 20 meters, you would need to cover the full range from 1° to 18° equally well.

Similarly, if you wish to cover the full range of elevation angles from Chicago to Southern Africa on 15 meters, you would need to cover 1° to 13°, even though the most statistically likely signals arrive at 1°, for 21% of the time when that the band is open for that path. See **Figure 14.32**.

It is important to recognize that Figures 14.29 through 14.32 are for flat ground. When the antennas are mounted over irregular local terrain, things get much more complicated. First, however, we'll discuss general-purpose antenna modeling programs as they try to model real terrain.

14.3.3 Drawbacks of Computer Models Over Real Terrain

Modern general-purpose antenna modeling programs such as the various forms of *NEC* and *EZNEC* (**eznec.com**) can accurately model almost any type of antenna commonly used by radio amateurs. In addition, there are specialized programs specifically designed to model Yagis efficiently, such as *YW* (Yagi for *Windows*, included with this book's online information). These programs, however, are all unable to model antennas accurately over anything other than *purely flat ground.*

While *NEC* and *EZNEC* can simulate irregular ground terrain, they do so in a decidedly crude manner, employing step-like concentric rings of height around an antenna. Their documentation clearly states that diffraction off these steps is not modeled. Common experience among serious modelers is that the warnings in the manuals are worth heeding.

Although you can analyze and even optimize antenna designs using free-space or flat-ground models, it is *diffraction* that makes the real world a very, very complicated place. This should be clarified — diffraction is hard, even tortuous, to analyze properly, but it makes analysis of real-world results far more believable than a flat-world reflection model does.

14.3.4 Ray-Tracing Over Uneven Local Terrain

THE RAY-TRACING TECHNIQUE

First, let's look at a simple ray-tracing procedure involving only horizontally polarized reflections, with no diffraction. From a specified height on the tower, the radiated signals from an antenna are imagined as straight-line *rays* in 0.25° increments from +35° above the horizon to –35° below the horizon. Each ray is traced over the foreground terrain to see if it hits the ground at any point on its travels in the direction of interest. If it does hit the ground, the ray is reflected following the classical law of reflection. That is, the outgoing angle equals the incoming angle, reflected through the normal to the slope of the surface. Once the rays exit into the ionosphere, the individual contributions are vector-summed to create the overall far-field elevation pattern.

The next step in terrain modeling involves adding *diffractions* as well as reflections. This was initially described in a paper by Jim Breakall, WA3FET, Dick Adler, K3CXZ, Joel Young, and a group of other researchers entitled "The Modeling and Measurement of HF Antenna Skywave Radiation Patterns in Irregular Terrain" in the July 1994 *IEEE Transactions on Antennas and Propagation*. They described in rather general terms the modifications they made to the *NEC-BSC* program. They showed how the addition of a ray-tracing reflection and diffraction model to the simplistic stair-stepped reflection model in regular *NEC* gave far more realistic results. For validation, they compared actual pattern measurements made on a site in Utah (with an overflying helicopter) to the computed patterns made using the modified *NEC* software.

THUMBNAIL HISTORY OF THE UNIFORM THEORY OF DIFFRACTION

It is instructive to look briefly at the history of how *Geometric Optics* (GO) evolved (and still continues to evolve) into the *Uniform Theory of Diffraction* (UTD). The following is summarized from the historical overview in one book found to be particularly useful and comprehensive on the subject of UTD: *Introduction to the Uniform Geometrical Theory of Diffraction*, by McNamara, Pistorius, and Malherbe.

Classical geometric optics has been studied since antiquity

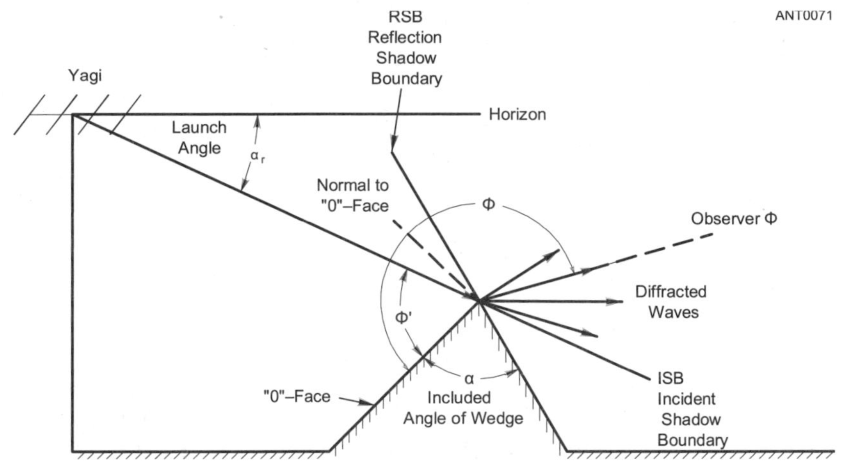

Figure 14.33 — Diagram showing diffraction mechanism of ray launched at angle α_r below the horizon at a diffraction wedge, whose included angle is α. Referenced to the incident face (the o-face as it is called in UTD terminology), the incoming angle is ϕ' (phi prime). The wedge creates an infinite number of diffracted waves. Shown is one whose angle referenced to the o-face is ϕ, the so-called observation angle in UTD terminology.

and deals strictly with geometric shapes. Mathematical analysis of shapes utilizes a methodology that traces the paths of straight-line rays of light. In classical geometric optics, however, there is no mention of three important quantities: phase, intensity and polarization. Indeed, without phase, intensity or polarization, there is no way to deal properly with the phenomenon of *interference*, or its cousin, *diffraction*. These phenomena require theories that deal with *waves* rather than rays.

In the 1600s, a Dutchman named Snell figured out the law of refraction, resulting in *Snell's law*. By the early 1800s, the basic world of classical optics was pretty well described from a mathematical point of view, based on the work of a number of individuals.

Wave theory has also been around for a long time, although not as long as geometry. Workers like Hooke and Grimaldi had recorded their observations of interference and diffraction in the mid-1600s. Huygens had used elements of wave theory in the late 1600s to help explain refraction. By the late 1800s, the work of Lord Rayleigh, Sommerfeld, Fresnel, Maxwell and many others led to the full mathematic characterization of all electromagnetic phenomena, light included. To get a handle on a typical real-world physical situation, a combination of classical ray theory and wave theory was needed.

The breakthrough in the combination of classical geometric optics and wave concepts came from J. B. Keller of Bell Labs in 1953, although he published his work in the early 1960s. In the very simplest of terms, Keller introduced the notion that shooting a ray at a diffraction *wedge* causes wave interference at the tip, with an infinite number of diffracted waves emanating from the diffraction point. Each diffracted wave is considered a point source radiator at the place of generation, the diffraction point. Thereafter, the paths of individual waves can be traced as though they were individual classical optic rays again. What Keller came up with was a reasonable mathematical description of what happens at the tip of the diffraction wedge.

Figure 14.33 is a picture of a simple diffraction wedge, with an incoming ray launched at an angle of a_r, referenced to the horizon, impinging on it. The diffraction wedge here is considered to be perfectly conducting, and hence impenetrable by the ray. The wedge generates an infinite number of diffracted waves, going in all directions not blocked by the wedge itself. The amplitudes and phases of the diffracted waves are determined by the interaction at the wedge tip, and this in turn is governed by the various angles associated with the wedge. Shown in Figure 14.29 are the included angle α of the wedge, the angle ϕ' of the incoming ray (referenced to the incoming surface of the wedge), and the observed angle ϕ of one of the outgoing diffracted waves, also referenced to the wedge surface.

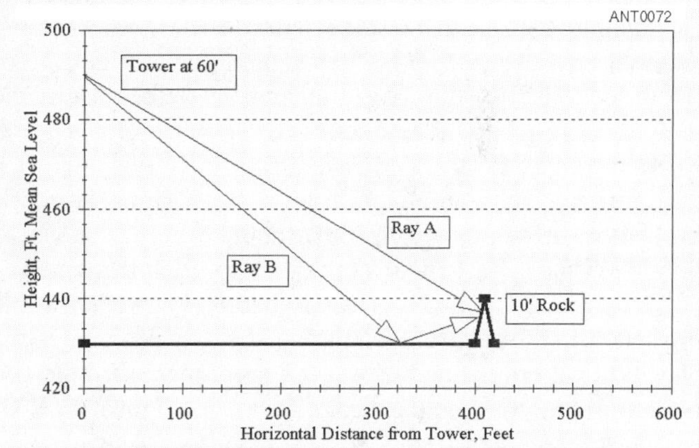

Figure 14.34 — Hypothetical terrain exhibiting so-called "10-foot rock effect." The terrain is flat from the tower base out to 400 feet, where a 10-foot high rock is placed. Note that this forms a diffraction wedge, but that it also blocks direct waves trying to shoot through it to the flat surface beyond, as shown by Ray A. Ray B reflects off the flat surface before it reaches the 10-foot rock, but it is blocked by the rock from proceeding further. A simple Geometric Optics (GO) analysis of this terrain without taking diffraction into account will result in the elevation response shown in Figure 14.31.

The so-called *shadow boundaries* are also shown in Figure 14.33. The *Reflection-Shadow Boundary* (*RSB*) is the angle beyond which no further reflections can take place for a given incoming angle. The *Incident-Shadow Boundary* (*ISB*) is that angle beyond which the wedge's face blocks any incident rays from illuminating the observation point.

Keller derived the amplitude and phase terms by comparing the classical GO solution with the exact mathematical solution calculated by Sommerfeld for a particular case where the boundary conditions were well known — an infinitely long, perfectly conducting wedge illuminated by a plane wave. Simply speaking, whatever was left over had to be diffraction terms. Keller combined these diffraction terms with GO terms to yield the total field everywhere.

Keller's new theory became known as the *Geometric Theory of Diffraction* (abbreviated henceforth as GTD). The beauty of GTD was that in the regions where classical GO predicted zero fields, the GTD "filled in the blanks," so to speak. For example, see **Figure 14.34**, showing the terrain for a hypothetical case, where a 60-foot high 4-element, 15-meter Yagi illuminates a wide, perfectly flat piece of ground. A 10-foot-high rock has been placed 400 feet away from the tower base in the direction of outgoing rays. **Figure 14.35** shows the elevation pattern predicted using reflection-only GO techniques. Due to blockage of the direct wave (A) trying to shoot past the 10-foot-high rock, and due to blockage of (B) reflections from the flat ground in front of the rock by the rock, there is a *hole* in the smooth elevation pattern.

As we know from experience, a single 10-foot-high rock will not have such an effect on a 15-meter signal. Keller's GTD took diffraction effects into account to show that waves do indeed sneak past and over the rock to fill in the pattern.

However, GTD wasn't perfect. Keller's GTD predicts some big spikes in the pattern, even though the overall shape of the elevation pattern is much closer to reality than a simple GO reflection analysis would indicate. The region right at the RSB and ISB shadow boundaries is where problems are found. The GO terms go to zero at these points because of blockage by the wedge, while Keller's diffraction terms tend to go to infinity at these very spots.Nevertheless, despite these nasty problems at the ISB and RSB, the GTD provided a remarkably better solution to diffraction problems than did classical GO.

In the early 1970s, a group at Ohio State University under R. G. Kouyoumjian and P. H. Pathak did some pivotal work to resolve this problem, introducing what amounts to a clever "fudge factor" to compensate for the tendency of the diffraction terms at the shadow boundaries to go to infinity. They introduced what is known as a *transition function*, using a form of Fresnel integral. Most importantly, the Ohio State researchers also created several *FORTRAN* computer programs to compute the amplitude and phase of diffraction components.

The ARRL program derived from this work is called *HFTA*, standing for "HF Terrain Assessment," written by Dean Straw, N6BV. As the name suggests, *HFTA* analyzes the effect of local terrain on HF propagation through the ionosphere. It is designed for horizontally polarized Yagis at various heights, although it will model the effects of a simple flattop dipole also. The accurate appraisal of the effect of terrain on vertically polarized signals is a far more complex problem than for horizontally polarized waves, and *HFTA* doesn't handle verticals. (*HFTA* is included with this book's online material. Note that *HFTA* was originally developed for 32-bit Windows systems. See this book's website for more information on running *HFTA* on 64-bit systems.)

There are a set of help files available with *HFTA* that will be invaluable for first-time users. In addition, the tutorial, "A Beginner's Guide to *HFTA* High Frequency Terrain Assessment" by John White, VA7JW is available online at **orcadxcc.org/content/VA7JW_HFTA_Manual.pdf**.

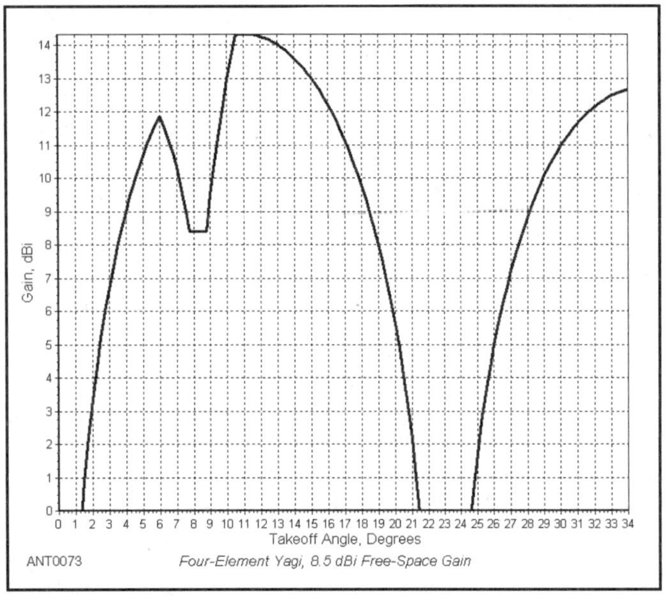

Figure 14.35 — Elevation response for rays launched at terrain in Figure 14.30 from a height of 60 feet using a 4-element Yagi. This was computed using a simple Geometrical Optics (GO) reflection-only analysis. Note the hole in the response between 6° to 10° in elevation. It is not reasonable for a 10-foot high rock to create such a disturbance at 21 MHz!

14.3.5 Simulation Examples

We want to focus first on some simple results, to show that the computations do make some sense by presenting some simulations over simple terrains.

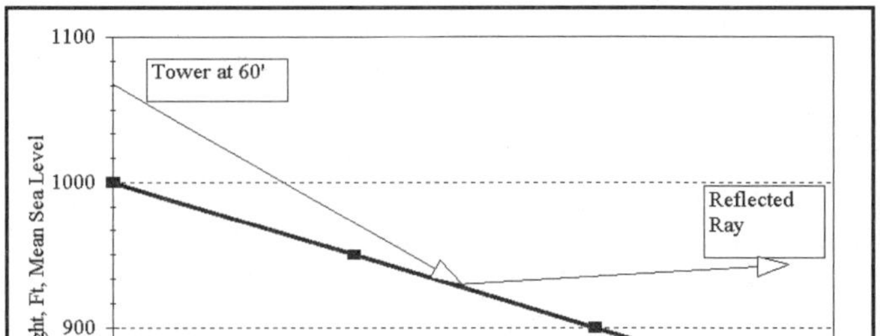

Figure 14.36 — A long, gentle downward-sloping terrain. This terrain has no explicit diffraction points and can be analyzed using simple GO reflection techniques.

SIMPLE TERRAIN EXAMPLES

Look at the simple case shown in **Figure 14.36**, where a very long, continuous down-slope from the tower base is shown. Note that the scales used for the X- and Y-axes are different: the Y-axis changes 300 feet in height (from 800 to 1,100 feet), while the X-axis goes from 0 to 3,000 feet. This exaggerates the apparent steepness of the downward slope, which is actually a rather gentle slope, at $\tan^{-1}$ (1,000 – 850) / (3,000 – 0) = –2.86°. In other words, the terrain falls 150 feet in height over a range of 3,000 feet from the base of the tower.

Figure 14.37 shows the computed elevation response for this terrain profile, for a 4-element horizontally polarized Yagi on a 60-foot tower. The response is compared to that of an identical Yagi placed 60 feet above flat ground. Compared to the "flatland" antenna, the hilltop antenna has an elevation response shifted over by almost 3° toward the lower elevation angles. In fact, this shift is directly due to the –2.86° slope of the hill. Reflections off the slope are tilted by the slope. In this situation there is a single diffraction at the bottom of the gentle slope at 3000 feet, where the program assumes that the terrain becomes flat.

Look at **Figure 14.38**, which shows another simple terrain profile, called a "Hill-Valley" scenario. Here, the 60-foot-high

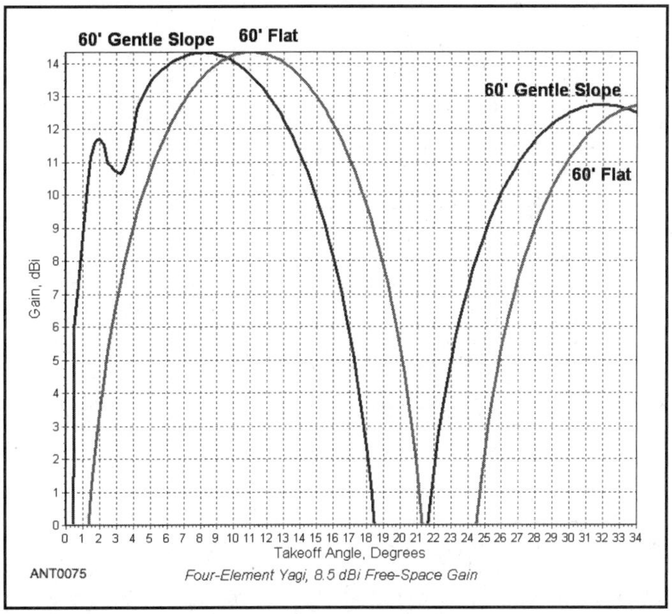

Figure 14.37 — Elevation response for terrain shown in Figure 14.32, using a 4-element 15 meter Yagi, 60 feet high. Note that the shape of the response is essentially shifted toward the left, toward lower elevation angles, by the angle of the sloping ground. For reference, the response for an identical Yagi placed over flat ground is also shown.

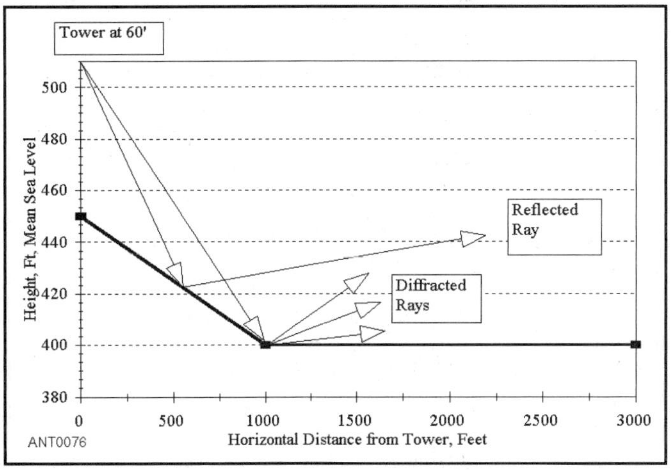

Figure 14.38 — "Hill-Valley" terrain, with reflected and diffracted rays.

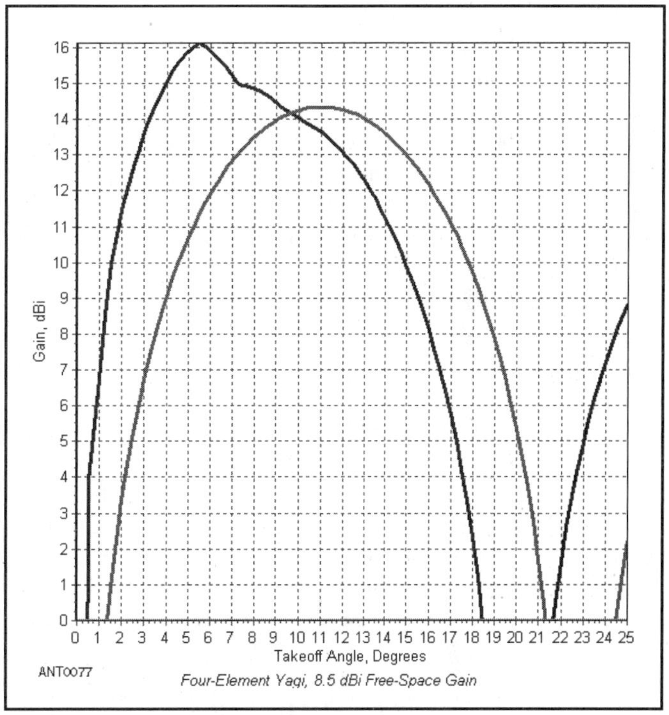

Figure 14.39 — Elevation response computed by HFTA program for single 4-element 15 meter Yagi at 60 feet above "Hill-Valley" terrain shown in Figure 14.34. Note that the slope has caused the response in general to be shifted toward lower elevation angles. At 5° elevation, the diffraction components add up to increase the gain slightly above the amount a GO-only analysis would indicate.

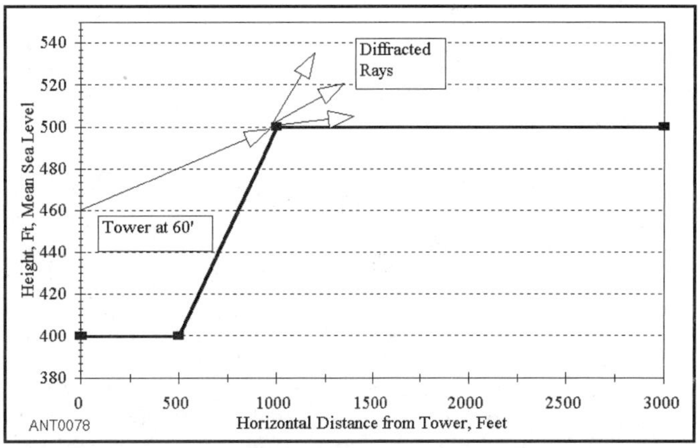

Figure 14.40 — "Hill-Ahead" terrain, shown with diffracted rays created by illumination of the edge of the plateau at the top of the hill.

tower stands on the edge of a gentle hill overlooking a long valley. Once again, the slope of the hill is exaggerated by the different X and Y-axes. **Figure 14.39** shows the computed elevation response at 21.2 MHz for a 4-element Yagi on a 60-foot-high tower at the edge of the slope.

Once again, the pattern is overlaid with that of an identical 60-foot-high Yagi over flat ground. Compared to the flatland antenna, the hilltop antenna's response above 9° in elevation is shifted by almost 3° toward the lower elevation angles. Again, this is due to reflections off the downward slope. From 1° to 9°, the hilltop pattern is enhanced even more compared to the flatland antenna, this time by diffraction occurring at the bottom of the hill.

Now let's see what happens when there is a hill ahead in the direction of interest. **Figure 14.40** depicts such a situation, labeled "Hill-Ahead." Here, at a height of 400 feet above mean sea level, the land is flat in front of the tower, out to a distance 500 feet, where the hill begins. The hill then rises 100 feet over the range 500 to 1000 feet away from the tower base. After that, the terrain is a plateau, at a constant elevation of 500 feet.

Figure 14.41 shows the computed elevation pattern for a 4-element 21-MHz Yagi 60-feet high on the tower, compared again with an overlay for an identical 60-foot high antenna over flat ground. The hill blocks low-angle waves directly radiated from the antenna from 0° to 2.3°. In addition, waves that would normally be reflected from the ground, and that would normally add in phase from about 2.3° to 12°, are blocked by the hill also. Thus, the signal at 8° is down almost 5 dB from the signal over flat ground, all due to the effect of the hill. Diffracted waves start kicking in once the direct wave rises enough above the horizon to illuminate the top edge of the hill. These diffracted waves tend to augment elevation angles above about 12°, which reflected waves can't reach.

Is there is any hope for someone in such a poor locationfor DXing? **Figure 14.42** shows the elevation response for a truly heroic solution. This involves a stack of four 4-element Yagis, mounted at 120, 90, 60 and 30 feet on the tower. Now, the total gain at low angles is just about comparable to that from a single 4-element Yagi mounted over flat ground.

At 5° elevation, four diffraction components add up (there are zero reflection components) to achieve the far-field pattern. This seems reasonable, because each of the four antennas is illuminating the diffraction point separately and we know that

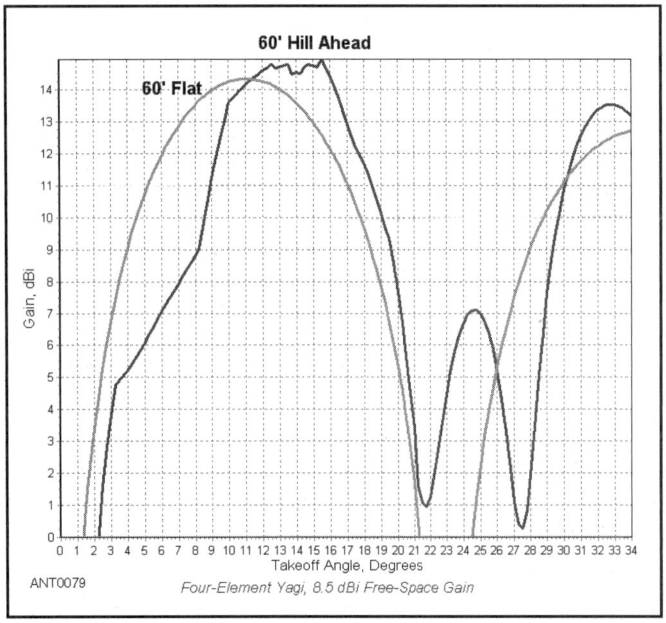

Figure 14.41 — Elevation response computed by HFTA for "Hill-Ahead" terrain shown in Figure 14.36. Now the hill blocks direct rays and also precludes possibility of any constructive reflections. Above 10°, diffraction components add up together with direct rays to create the response shown.

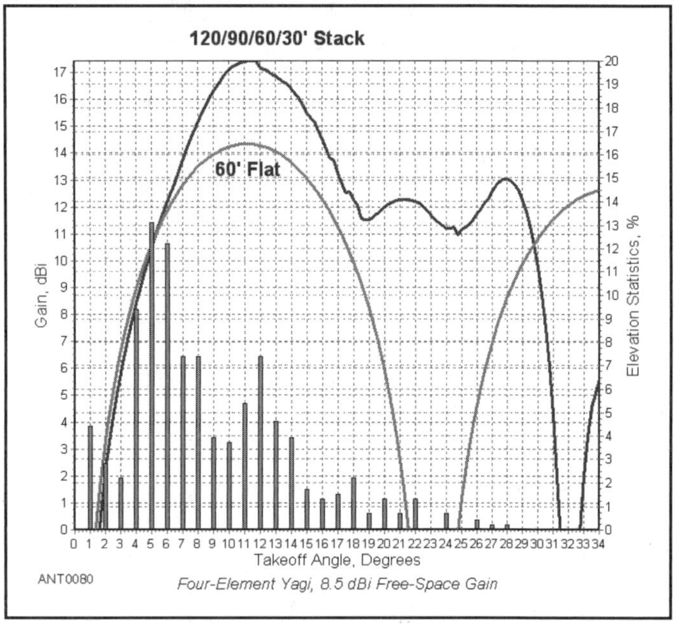

Figure 14.42 — Elevation response of "heroic effort" to surmount the difficulties imposed by hill in Figure 14.36. This effort involves a stack of four 4-element Yagis in a stack starting at 120 feet and spaced at 30-foot increments on the tower. The response is roughly equivalent to a single 4-element Yagi at 60 feet above flat ground, hence the characterization as being a "heroic effort." The elevation-angle statistics from New England to Europe are overlaid on the graph for reference.

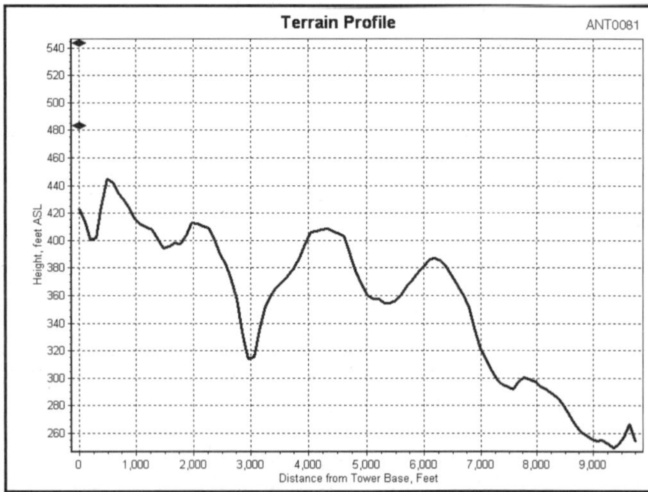

Figure 14.43 — Terrain of N6BV/1 in Windham, NH, toward Japan. HFTA identifies 52 different points where diffraction can occur.

none of the four antennas can *see over* the hill directly to produce a reflection at a low launch angle.

At an elevation angle of 5°, 15 meter signals arrive from Europe to New England about 13% of the total time when the band is actually open. We can look at this another way. For about two-thirds of the times when the band is open on this path, the incoming angle is between 3° to 12°. For about one-third of the time, signals arrive above 10°, where the "heroic" four-stack is really beginning to come into its own.

COMPLEX TERRAIN EXAMPLE

The results for simple terrains look reasonable; let's try a more complicated real-world situation. **Figure 14.43** shows the terrain from N6BV's previous New Hampshire location toward Japan. The terrain was complex, with 52 different points *HFTA* identifies as diffraction points. **Figure 14.44** shows a labeled *HFTA* output for three different types of antennas on 20 meters: a stack at 120 and 60 feet, the 120-foot antenna by itself, and then a 120/60-foot stack over flat ground, for reference. The elevation-angle statistics for New England to Japan are overlaid on the graph also, making for a very complicated looking picture. (Note that output from *HFTA* is in color which is easier to interpret.)

Comparison of the same 120/60-foot stacks over irregular terrain and flat ground is useful to show where the terrain itself is affecting the elevation response. The flatland stack has more gain in the region of 3° to 7° than the same stack over the actual terrain toward Japan. On the other hand, the actual terrain boosts signals in the range of 8° to about 12°. This demonstrates the conservation of energy — you may gain a stronger signal at certain elevation angles, but you will lose gain at others. In this case, the N6BV always felt the station was

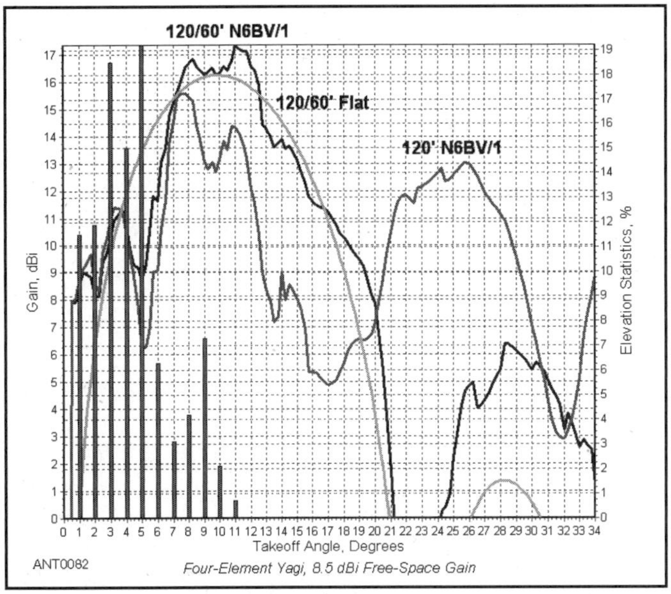

Figure 14.44 — Elevation responses computed by HFTA for N6BV/1 terrain shown in Figure 14.39, for a stack of two 4-element 20 meter Yagis at 120 and 60 feet, together with the response for a single Yagi at 120 feet and a 120/60-foot stack over flat ground for reference. Due to the response, many diffraction and reflection components is quite complicated!

"weak" toward Japan on 20 meters, because the dominant angles are lower on that path.

Examination of the detailed data output from *HFTA* shows that at an elevation angle of 5°, there are 6159 diffraction components. There are many, many signals bouncing around off the terrain on their trip to Japan! Note that because of blockage of some parts of the terrain, the 60-foot-high Yagi cannot illuminate all the diffraction points, while the higher 120-foot Yagi is able to see these diffraction points.

It is fascinating to reflect on the thought that received signals coming down from the ionosphere to the receiver are having encounters with the terrain, but from the opposite direction. It's not surprising, given these kinds of interactions, it is likely that transmitting and receiving are not exactly reciprocal.

The 120/60-foot stack in Figure 14.40 achieves its peak gain of 17.3 dBi at 11° elevation, where it is about 3 dB stronger than the single Yagi at 120 feet. It maintains this 3-dB advantage over most of the range of incoming signals from Japan. This difference in performance between the stack and each antenna by itself was observed many times on the air. Much of the time when comparisons are being made, however, the small differences in signal are difficult to measure meaningfully, especially when the fading varies signals by 20 dB or so during a typical QSO. It should be noted that the stack usually exhibited less fading compared to each antenna by itself.

14.3.6 Using *HFTA*

GENERATING *HFTA* TERRAIN DATA USING THE K6TU DATA SERVICE

Stu Phillips, K6TU, has greatly simplified the process of obtaining terrain elevation data for use with *HFTA* by automating the complete process on **k6tu.net**. The service is available at no charge to any registered user. The signup process is free and gives ongoing access to both the propagation prediction services of the site as well as generating terrain profiles. K6TU obtained the complete 0.5 Tbyte database from USGS. For non-US locations, the service uses the Advanced Spaceborne Thermal Emission and Reflection Radiometer (ASTER: **asterweb.jpl.nasa.gov**) dataset covering 83N to 83S. (The manual method of generating terrain profiles is described in the online material. The MicroDEM-based method is no longer available. Should changes to the **k6tu.net** site be made that affect this section, information will be available on this book's website.)

From the home page's Main Menu (at upper left), select HF TERRAIN ANALYSIS. A page describing the *HFTA* program and how to use the terrain profile service will be displayed. When you are ready to generate a profile for your location, click NEW in the NAVIGATION menu at the upper right then TERRAIN PROFILE REQUEST on the next page displayed (see **Figure 14.45**). The data entry form will be displayed with the following information to be supplied by you:

•Name of the profile, such as "W1AW Tower 1 Terrain Profile"

•Latitude to six decimal degrees (41.714511)

•Longitude to six decimal degrees (–72.727325)

The easiest way to obtain the location lat/lon to this preci-

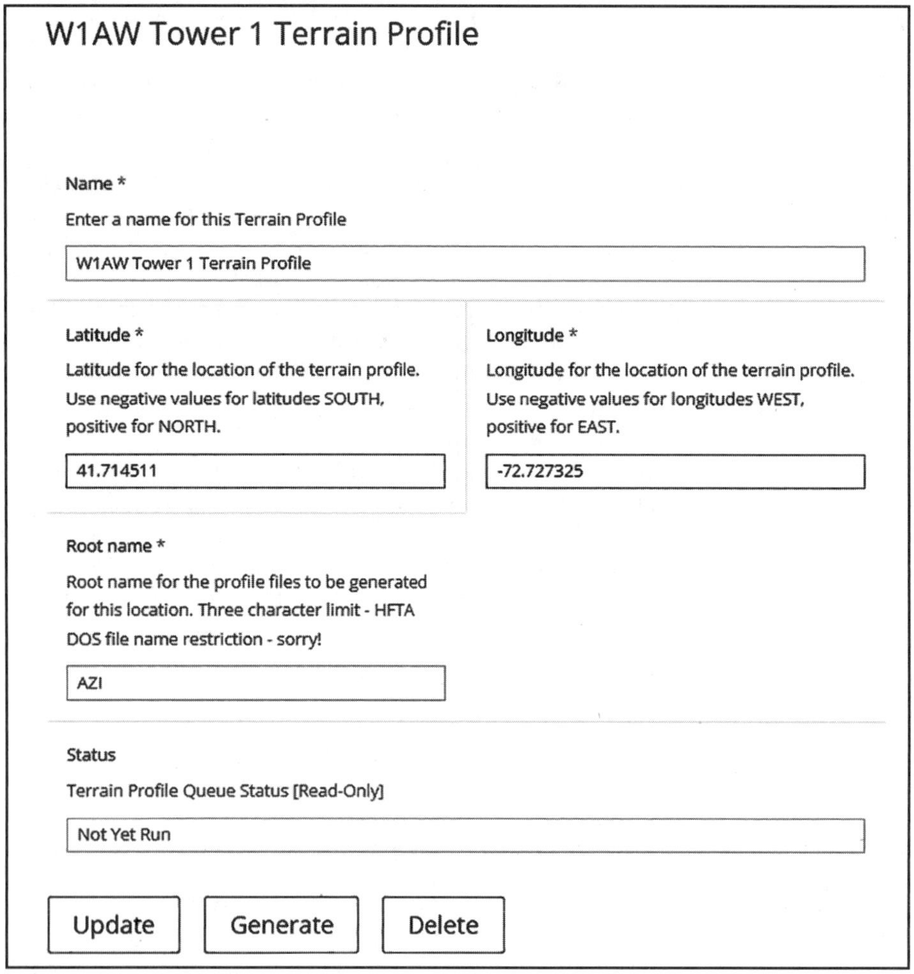

Figure 14.45 — The location data entry screen for one of the towers at W1AW. Google Maps (maps.google.com) was used to obtain the lat/lon location data as described in the text.

sion is to use an online mapping service, such as *Google Maps* (**maps.google.com**). Select the satellite view, then zoom in to the highest resolution view of your location. Click once on the exact location for which you want to generate a profile. The lat/lon position of the location will display in a small label window. Click on the lat/lon data values and a window will appear at the upper left giving lat/lon in both decimal and degree-minute-second format. Highlight and copy the latitude value then paste it into the data service's **LATITUDE** window. Do the same for the longitude value.

With both the latitude and longitude data entered, click **SAVE** and then **GENERATE PROFILE** on the following page. The service will send you an email when the file of data has been created containing the URL where the file is saved. After you click on the URL and see the [filename}.zip file, right-click on the filename and save it. You can then extract all 360 PRO files into a folder with the name you selected for the profile. To use the data, follow the instructions that come with *HFTA* and select the appropriate PRO file.

K6TU also offers a downloadable *HFTASweep* program to generate gain tables. (It has been tested on Windows 7, 8, and 10.) As explained on the K6TU website, *HFTASweep* is a program "wrapper" for *HFTA*. It runs *HFTA* 90 times programmatically to model a *horizontally* polarized antenna over the actual terrain around your location. The program captures the results for each azimuth direction (four different azimuth angles at a time) and at the end, builds a *VOACAP* type 13 antenna model as a file called **antenna.13**.

The **antenna.13** file is a table containing the gain of the modeled antenna over actual terrain for all radiation angles from 0 to 34 degrees (the maximum radiation angle considered by *HFTA*). The remaining elevation angles simply contain the same data for angles 35 through 90 as the data shows for 34 degrees. This is a reasonable simplification for horizontal antennas that are at least a quarter wavelength or more above ground level.

ALGORITHM FOR RAY-TRACING THE TERRAIN

Once a terrain profile is created, there are a number of mechanisms that *HFTA* takes into account as a ray travels over that terrain:

1) Classical ray reflection, with Fresnel ground coefficients.

2) Direct diffraction, where a diffraction point is illuminated directly by an antenna, with no intervening terrain features blocking the direct illumination.

3) When a diffracted ray is subsequently reflected off the terrain.

4) When a reflected ray encounters a diffraction point and causes another series of diffracted rays to be generated.

5) When a diffracted ray hits another diffraction point, generating another whole series of diffractions.

Certain unusual, bowl-shaped terrain profiles, with sheer vertical faces, can conceivably cause signals to reflect or diffract in a backward direction, only to be reflected back in the forward direction by the sheer-walled terrain to the rear. *HFTA* does not accommodate these interactions, mainly because to do so would increase the computation time too much. It only evaluates terrain in the forward direction along one azimuth of interest.

Figure 14.46 shows a portion of an *HFTA* screen capture in the direction toward Europe from the N6BV/1 location in New Hampshire on 21.2 MHz. It compares the results for a 90/60/30-foot stack of TH7DX tribanders to the same stack over flat land, and to a single antenna at 70 feet over flat ground. The 70-foot single antenna represents a pretty typical station on 15 meters. The terrain produces excellent gain at lower elevation angles compared to the same stack over flat

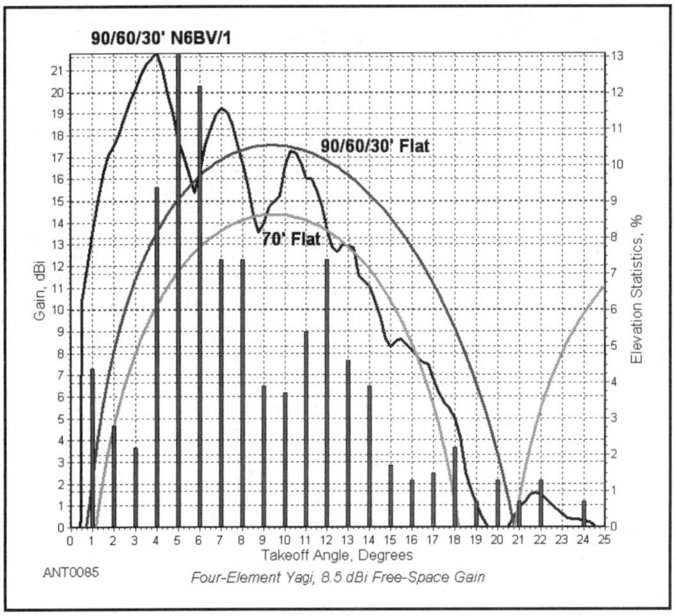

Figure 14.46 — The 21-MHz elevation response for a stack of three TH7DX Yagis mounted on a single tower at 90/60/30 feet, at the N6BV/1 QTH for a 45° azimuth toward Europe. The terrain focuses the energy at low elevation angles compared to the same stack over flat ground. This illustrates once again the conservation of energy: Energy squeezed down into low elevation angles is stolen from other, higher, angles.

ground. The stack is very close to or superior to the single 70-foot-high Yagi at all useful elevation angles. Terrain can indeed exhibit a profound effect on the launch of signals into the ionosphere — for good or for bad.

HFTA'S INTERNAL ANTENNA MODEL

The operator selects the antenna used inside *HFTA* to be anything from a dipole to an 8-element Yagi. The default assumes a simple cosine-squared mathematic response, equivalent to a 4-element Yagi in free space. *HFTA* traces rays only in the forward direction from the tower along the azimuth of interest. This keeps the algorithms reasonably simple and saves computing time.

HFTA considers each antenna in a stack as a separate *point source*. The simulation begins to fall apart if a traveling wave type of antenna like a rhombic is used, particularly if the terrain changes under the antenna — that is, the ground is not flat under the entire antenna. For a typical Yagi, even a long-boom one, the point-source assumption is reasonable. The internal antenna model also assumes that the Yagi is horizontally polarized. *HFTA* does not do vertically polarized antennas, as discussed previously. The documentation for *HFTA* also cautions the user to work with practical spacings between stacked Yagis — 0.5 λ or more because *HFTA* doesn't explicitly model mutual coupling between Yagis in a stack.

HFTA compares well with the measurements for horizontal antennas described earlier by Jim Breakall, WA3FET, using a helicopter in Utah. Breakall's measurements were done with a 15-foot-high horizontal dipole.

MORE ABOUT *HFTA* FREQUENCY COVERAGE

HFTA can be used on frequencies higher than the HF bands, although the graphical resolution is only 0.25°. The patterns above about 100 MHz thus look rather grainy. The UTD is a *high-frequency-asymptotic* solution, so in theory the results become more realistic as the frequency is raised. Keep in mind too that *HFTA* is designed to model launch angles for skywave propagation modes, including E- and F-layer, and even sporadic-E. Since by definition the ionospheric launch angles include only those above the horizon, direct line-of-sight UHF modes involving negative launch angles are not considered in *HFTA*.

See the HFTA.PDF documentation file for further details on the operation of *HFTA*. This file, as well as sample terrain profiles for some *big-gun* stations, is included with this book's online material.

14.4 Stacking Yagis and Switching Systems

The preceding sections illustrate the importance of controlling the elevation angle of an antenna's radiation pattern at HF. In addition, the wide variations also illustrate that a single antenna, no matter how much gain it produces, at a single height is often inadequate in maintaining effective communications over the desired path. For example, during a DX band opening on the upper HF bands the initial signals usually appear at very low elevation angles. Later, as the opening strengthens and spreads, signals at higher elevation angles are the strongest. Finally, as the band closes to that area, signals will again be the strongest at low angles. Being able to select the right elevation angle at the right time is important to sustained success in DXing or contest operation.

In HF amateur stations, the most common arrangement to control elevation angle is a vertical stack of identical Yagis on a single tower. This arrangement is commonly called a *vertical stack*. At VHF and UHF, amateurs sometimes employ collinear stacks, where identical Yagis are stacked side-by-side at the same height. This arrangement is called a *horizontal stack*, and is not usually found at HF, because of the severe mechanical difficulties involved with large, rotatable side-by-side arrays. In addition, whereas on HF a primary goal is being able to control the elevation angle of the radiation pattern for the optimum ionospheric path, on VHF and UHF it is more important to narrow the azimuthal width of the array's main lobe and minimize side lobes to improve the signal-to-noise ratio of very weak signals on both ends of the path.

Figure 14.47 illustrates the two different stacking arrangements. In either case, the individual Yagis making up the stack are generally fed in phase. There are times, however, when individual antennas in a stacked array are purposely fed out of phase in order to emphasize a particular elevation pattern.

Let's look at the reasons hams stack Yagis:

- For more gain
- For a wider elevation footprint in a target geographical area
- For azimuthal diversity — two or more directions at once
- For less fading
- For less precipitation static

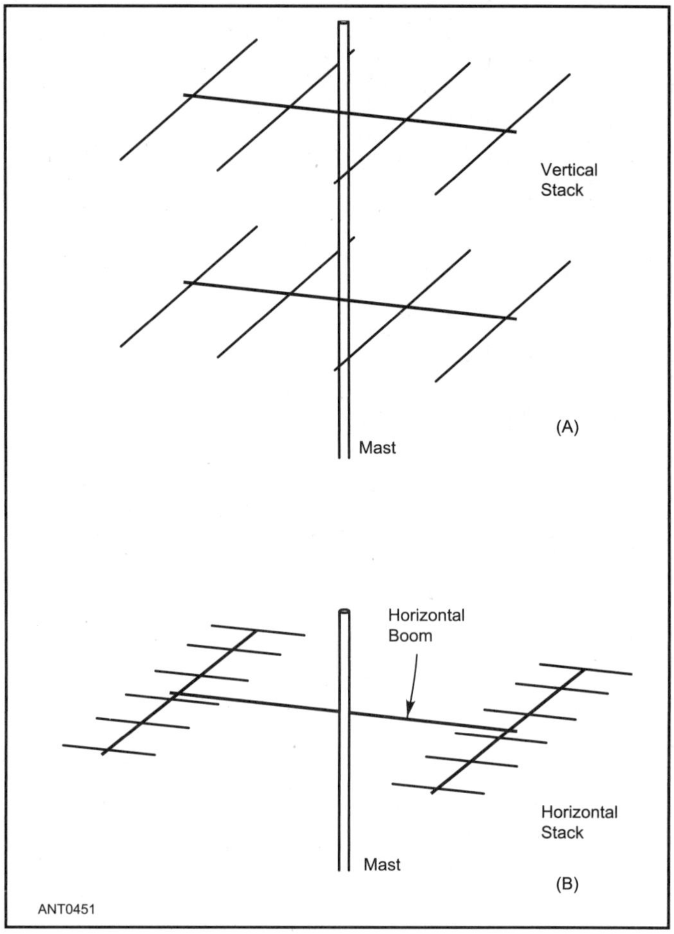

Figure 14.47 — Stacking arrangements. At A, two Yagis are stacked vertically (broadside) on the same mast. At B, two Yagis are stacked horizontally (collinear) side-by-side. At HF the vertical stack is more common because of mechanical difficulties involved with large HF antennas stacked side-by-side, whereas at VHF and UHF the horizontal stack is common.

14.4.1 Stacks and Gain

Figure 14.48 compares the elevation responses for three antenna systems of 4-element 15 meter Yagis. The response for the single Yagi at a height of 120 feet peaks at an elevation of about 5°, with a second peak at 17° and a third at 29°. When operated by itself, the 60-foot high Yagi has its first peak at about 11° and its second peak beyond 34°.

The basic principle of a vertically stacked HF array is that it takes energy from higher-angle lobes and concentrates that energy into the main elevation lobe. The main lobe of the 120/60-foot stack peaks about 7° and is about 2 dB stronger than either the 60- or 120-foot antenna by itself. The shape of the left-hand side of the stack's main lobe is determined mainly by the 120-foot antenna's response. The right-hand side of the stack's main lobe is "stretched" rightward (toward higher angles) mainly by the 60-foot Yagi, while the shape follows

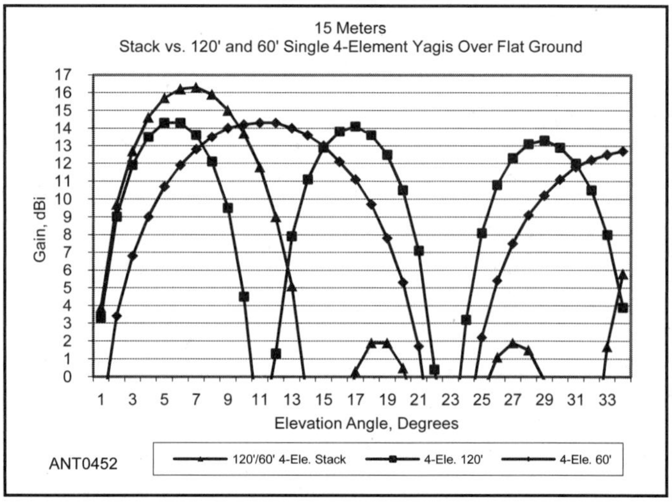

Figure 14.48 — Comparison of elevation patterns on 15 meters for a stack of 4-element Yagis at 120 and 60 feet and individual Yagis at those two heights. The shape of the stack's response is determined mainly by that of the top antenna.

the curve of the 120-foot Yagi.

Look at the second and third lobes of the stack, which appear about 18° and 27°. These are about 14 dB down from the stack's peak gain, showing that energy has indeed been extracted from them. By contrast, look at the levels of the second and third lobes for the individual Yagis at 60 and 120 feet. These higher-angle lobes are almost as strong as the first lobes.

The stack squeezes higher-angle energy into its main elevation lobe, while maintaining the frontal lobe azimuth pattern of a single Yagi. This is the reason why many state-of-the-art contest stations are stacking arrays of relatively short-boom antennas, rather than stacking long-boom, higher-gain Yagis. A long-boom HF Yagi narrows the azimuthal pattern (and the elevation pattern too), making pointing the antenna more critical and making it more difficult to spread a signal over a wide azimuthal area, such as all of Europe and Asiatic Russia at one time.

14.4.2 Stacks and Wide Elevation Coverage

Detailed studies using sophisticated computer models of the ionosphere have revealed that coverage of a wide range of elevation angles is necessary to ensure consistent DX or contest coverage on the HF bands. These studies have been conducted over all phases of the 11-year solar cycle, and for numerous transmitting and receiving QTHs throughout the world.

The chapter **Radio Wave Propagation** covers these studies in more detail, and this book's online material includes a huge number of elevation-angle statistical tables for locations all around the world. The *HFTA* (HF Terrain Assessment) program included with this book's online material can not only compute antenna elevation patterns over irregular local terrain, but it can compare them directly to the elevation-angle statistics for a particular target geographic area.

A 10 METER EXAMPLE

Figure 14.49 shows the 10 meter elevation-angle statistics for the New England path from Boston, Massachusetts, to all of the continent of Europe. The statistics are overlaid with the computed elevation response for three individual 4-element Yagis, at three heights: 90, 60 and 30 feet above flat ground. In terms of wavelength, these heights are 2.60 λ, 1.73 λ and 0.86 λ high.

You can see that the 90-foot high Yagi covers the lower elevation angles best, but it has a large null in its response centered at about 11°. This null puts a big hole in the coverage for some 22% of all the times the 10 meter band is open to Europe. At those angles where the 90-foot Yagi exhibits a null, the 60-foot Yagi would be effective, and so would the 30-foot Yagi. If that is the only antenna you have, the 90-foot high Yagi would be too high for good coverage of Europe from New England.

The peak statistical elevation angle into Europe is 5°, and this occurs about 11% of all the times the 10 meter band is open to Europe from Boston. At an elevation of 5° the 30-foot high Yagi would be down almost 7 dB compared to the 90-foot high Yagi, but at 11° the 90-foot Yagi would be more than 22 dB down from the 30-foot Yagi. There is no single height at which

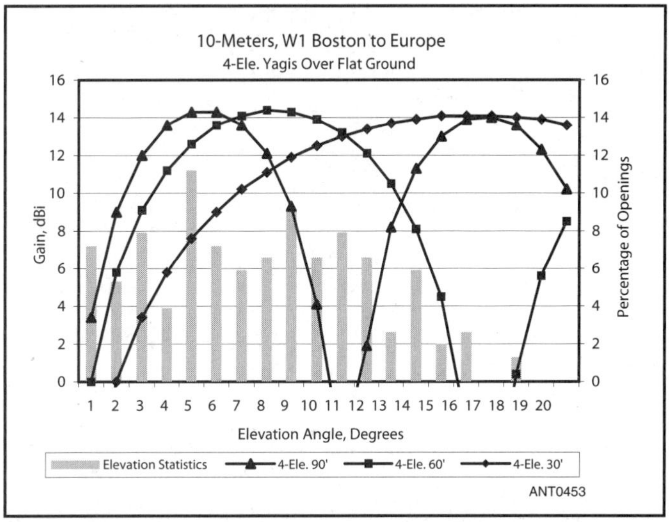

Figure 14.49 — Comparison of elevation patterns and elevation-angle statistics for individual 10-meter TH7DX tribanders mounted over flat ground aiming from New England to Europe. No single antenna can cover the wide range of angles needed — from 1° to 18°.

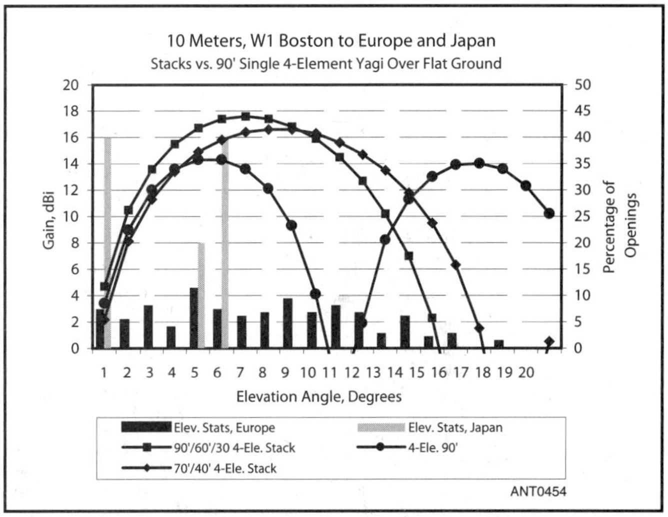

Figure 14.50 — Combinations of 4-element Yagis over flat ground. The elevation-angle statistics into Japan from New England (Boston) are represented by the black vertical bars, while the grey vertical bars represent the elevation-angle statistics to Europe. The 90/60/30-foot stack has the best elevation footprint into Japan, although the 70/40-foot stack performs well also.

one Yagi can optimally cover all the necessary elevation angles, especially to a large geographic area such as Europe — although the 60-foot high antenna is arguably the best compromise for a single height. To cover all the possibilities to Europe, however, you need a 10 meter antenna system that can cover equally well the entire range of elevation angles from 1° to 18°.

Figure 14.50 compares elevation-angle statistics for two 10 meter paths from New England to Europe and to Japan. The elevation angles needed for communications with the Far East are very low. Overlaid on Figure 14.47 for comparison are the elevation responses over flat ground for three different antenna systems, using identical 4-element Yagis:

- Three Yagis, stacked at 90, 60 and 30 feet
- Two Yagis, stacked at 70 and 40 feet
- One Yagi at 90 feet.

The best coverage of all the necessary angles on 10 meters to Europe is with the stack of three Yagis at 90/60/30 feet. The two-Yagi stack at 70 and 40 feet comes in a close second to Europe, and for elevation angles higher than about 9° the 70/40-foot stack is actually superior to the 90/60/30-foot stack.

Both of the stacks illustrated here give a wider *elevation footprint* than any single antenna, so that all the angles can be covered automatically without having to switch from higher to lower antennas manually. This is perhaps the major benefit of using stacks, but not the only one, as we'll see.

To Japan, the necessary range of elevation angles is considerably smaller than that needed to a larger geographic target area like Europe. The 90/60/30-foot stack is still best on the basis of having higher gain at low angles, although the two-Yagi stack at 70 and 40 feet is a good choice too. Note that the single 90-foot high Yagi's performance is very close to the 70/40-foot stack of two Yagis at low angles, but the two-Yagi stack is superior to the single 90-foot antenna for angles higher than about 5° on 10 meters.

A 15 METER EXAMPLE

The situation is similar on 15 meters from New England to Europe. On 15 meters, the range of angles needed to fully cover Europe is 1° to 28°. This large range of angles makes covering all the angles even more challenging. Ken Wolff, K1EA, a devoted contest operator and the author of the famous *CT* contest logging program, put it very clearly when he wrote in the bulletin for the Yankee Clipper Contest Club:

"Suppose you have 15 meter Yagis at 120 feet and 60 feet, but can feed only one at a time. A 15 meter beam at 120 feet has its first maximum at roughly 5° and the first minimum at 10°. The Yagi at 60 feet has a maximum at 10° and a minimum at 2°. At daybreak, the band is just opening, signals are arriving at 3° or less and the high Yagi outperforms the low one by 5-10 dB. Late in the morning, western Europeans are arriving at angles of 10° or more, while UA6 is still arriving at 4-5°. Western Europe can be 20-30 dB louder on the low antenna than the high! What to do? Stack 'em!"

Figure 14.51 illustrates K1EA's scenario, showing the elevation statistics to Europe from Massachusetts and the elevation responses for a 120- and a 60-foot high, 4-element Yagi, both over flat ground, together with the response for both antennas operated as a vertical stack. The half-power beamwidth of the stack's main lobe is 6.9°, while that for the 120-foot antenna by itself is 5.5° and that for the 60-foot antenna by itself is 11.1°. The half-power beamwidth numbers by themselves can be deceiving, mainly because the stack starts out with a higher gain. A more meaningful observation is that the stack has equal to or more gain than either of the two individual antennas from 1° to about 10°.

Is such a stack of 15 meter Yagis at 120 and 60 feet optimal for the New England to Europe path? No, it isn't, as we'll explore later, but the stack is clearly better than either antenna by itself for the scenario K1EA outlined above.

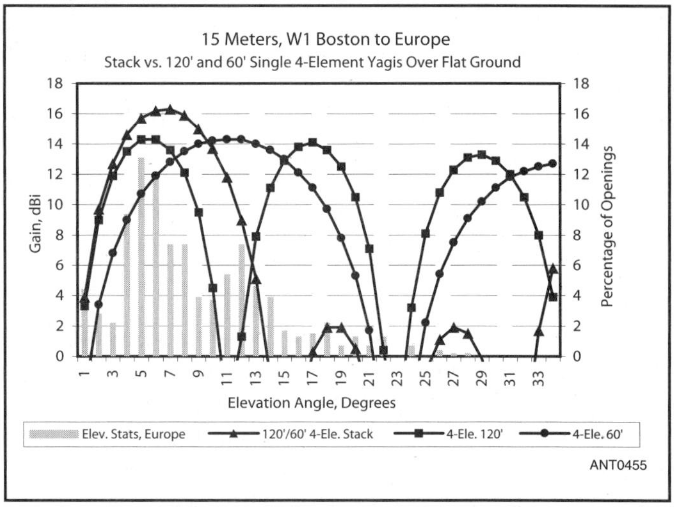

Figure 14.51 — Comparison of elevation patterns for K1EA's illustration about 15 meter Yagis mounted over flat ground, with elevation-angle statistics to Europe added. The stack at 120 and 60 feet yields a better footprint over the range of 3° to 11° at its half-power points, better than either antenna by itself.

A 20 METER EXAMPLE

Take a look now at **Figure 14.52**, which overlays elevation-angle statistics for Europe (gray vertical bars) and Japan (black vertical bars) from Boston on 20 meters, plus the elevation responses for four different sets of antennas mounted over flat ground. Just for emphasis, the highest antenna is a 200-foot high 4-element Yagi. It is clearly too high for complete coverage of all the needed angles into Europe. A number of New

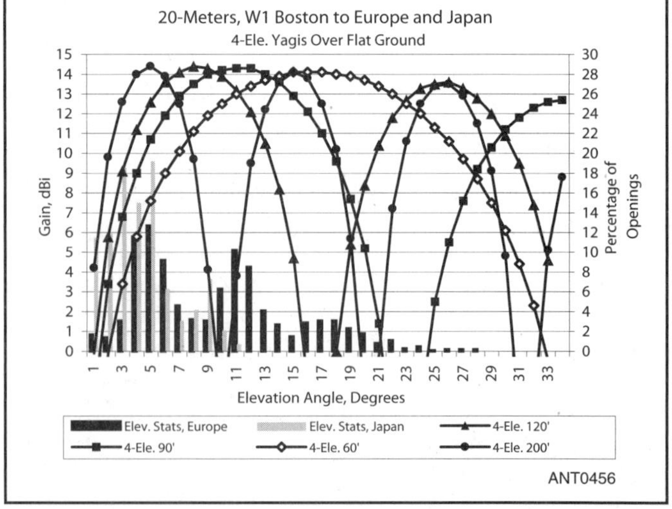

Figure 14.52 — Comparison of elevation patterns for individual 20 meter Yagis over flat ground, compared with the range of elevation angles needed on this band from New England to Europe (gray bars) and to Japan (black bars). For fun, the response of a 200-foot high Yagi is included — this antenna is far too high to cover the needed range of angles to Europe because of its deep nulls at critical angles, like 10°. But the 200 footer is great into Japan!

England operators have verified that this is true — a really high Yagi will open the 20 meter band to Europe in the morning and may shut it down in the afternoon, but during the middle of the day the high antenna gets soundly beaten by lower antennas.

To Japan, however, from New England the range of angles needed narrows considerably on 20 meters, from 1° to only 11°. For these angles, the 200-foot Yagi is the best antenna to work Japan from New England on 20 meters.

This is true provided that the antenna is aiming out over flat ground. The actual, generally irregular, terrain in various directions can profoundly modify the takeoff angles favored by an antenna system, particularly on steep hills. There will be more discussion on this important topic later on.

14.4.3 Eliminating Deep Nulls

Now, let's look closely at some other 20 meter antennas in Figure 14.46, the ones at 120 and 60 feet. At an elevation angle of 8° the difference in elevation response between the 60- and 120-foot high Yagis is just over 3 dB. Can you really notice a change of 3 dB on the air? Signals on the HF bands often rise and fall quickly due to fading, so differences of 2 or 3 dB are difficult to discern. Consequently, the difference between a Yagi at 120 feet and one at 60 feet may be difficult to detect at elevation angles covered well by both antennas. But a *deep null* in the elevation response is very noticeable.

When Dean Straw, N6BV, put up his 120-foot tower in Windham, New Hampshire, his first operational antenna was a 5-element triband Yagi, with 3 elements on 40 and 4 elements on both 20 and 15 meters. Just as the sun was going down on a late August day Straw finished connecting the feed line in the station. The antenna seemed to be playing like it should, with a good SWR curve and a good pattern when it was rotated. So N6BV/1 called a nearby friend, John Dorr, K1AR, on the telephone and asked him to get on the air to make some signal comparisons on 20 meters into Europe.

Straw was shocked that every European they worked that evening said his signal was several S units weaker than K1AR's. Dorr was using a 4-element 20 meter monobander at 90 feet, which at first glance should have been comparable to Straw's 4-element antenna at 120 feet. But N6BV really shouldn't have been so shocked — in New England, the elevation angles from Europe late in the day on 20 meters are almost always higher than 11°, and that is true for the entire solar cycle.

The N6BV/1 station was located on a small hill, while K1AR was located on flat terrain toward Europe. The elevation response for N6BV/1's 120-foot high Yagi fell right into a deep null at 11°. This was later confirmed many times in the following eight years that the N6BV/1 station was operational. During the early morning opening on 20 meters into Europe, the top antenna was always very close to or equal to the stack of three TH7DX tribanders at 90/60/30 feet on the same tower. But in the afternoon the top antenna was *always* decidedly worse than the stack, so much so that Straw often wondered whether something had gone wrong with the top antenna!

So what's the moral to this short tale? It's simple: *The gain you can achieve, while useful, is not so important as the deep nulls you can avoid by using a stack.*

14.4.4 Stacking Distances Between Yagis

So far, we've examined stacks as a means of achieving more gain over an individual Yagi, while also matching the antenna system's response to the range of elevation angles needed for particular propagation paths. Most importantly, we seek to avoid nulls in the elevation response. Earlier we asked whether a 120/60-foot stack was optimal for the path from New England to Europe on 15 meters. Let's examine how the stacking distance between individual antennas affects the performance of a stack.

Figure 14.53 shows overlays of various combinations of 15 meter Yagis. Just for reference, a plot for a single 60-foot high Yagi is also included. Let's start by looking at the most widely spaced stack in the group: the 120/30-foot stack. Here, the spacing is obviously too large, since the second lobe is actually stronger than the first lobe. In terms of wavelength, the 90-foot spacing between antennas in this stack is 1.94 λ, a large spacing indeed.

There is a great deal of folklore and superstition among amateurs about stacking distances for HF arrays. For years, high-performance stacked Yagi arrays have been used for weak-signal DXing on the VHF and UHF bands. The most extreme example of weak-signal work is EME work (Earth-Moon-Earth, also called *moonbounce*) because of the huge path losses incurred on the way to and from the Moon. The most successful arrays used for moonbounce have low sidelobe levels and very narrow frontal lobes that give huge amounts of gain. The low sidelobes help minimize received noise, since the receive levels for signals that do manage to bounce off the Moon and return to Earth are exceedingly weak.

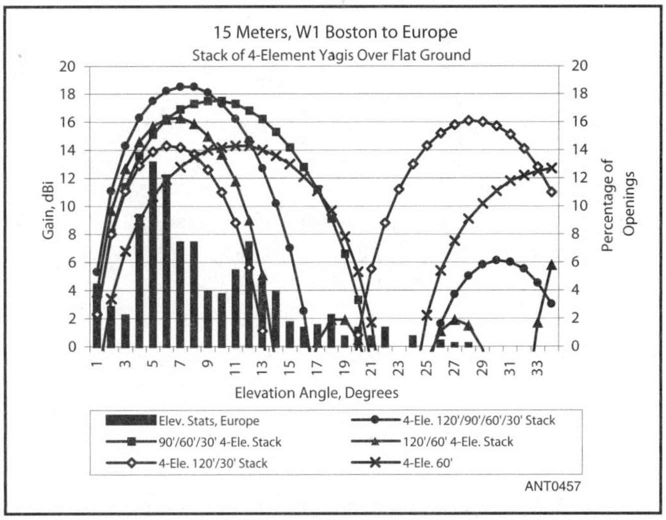

Figure 14.53 — Various stacks toward Europe from New England for 15 meters. The stack at 120 and 30 feet is clearly suboptimal, since the second lobe is higher than the first lobe. The 120/60-foot stack is better in this regard, but is still not as good a performer as the 90/60/30-foot stack. It's debatable whether going to four Yagis in the 120/90/60/30-foot stack is a good idea because it drops below the performance of the 90/60/30-foot stack at about 10° in elevation. The exact distance between practical HF Yagis is not critical to obtain the benefits of stacking. For a stack of tribanders at 90, 60 and 30 feet, the distance in wavelengths between individual antennas is 0.87 λ at 28.5 MHz, 0.65 λ at 21.2 MHz, and 0.43 λ at 14.2 MHz.

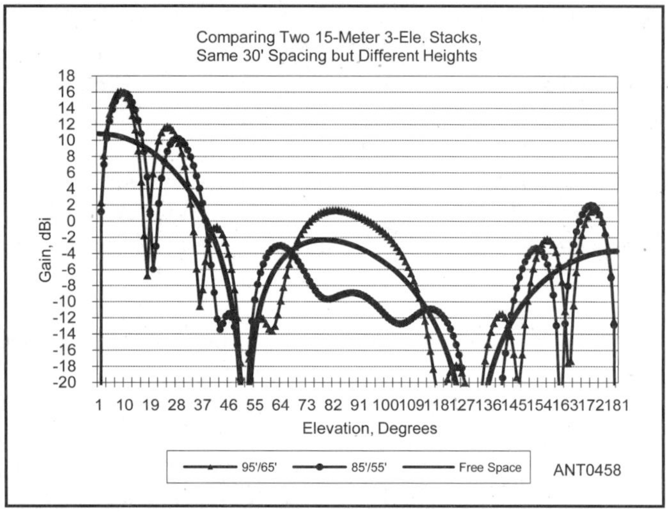

Figure 14.54 — Rectangular plot comparing two 15-meter stacks of 3-element Yagis — each antenna is spaced 30 feet from its partner, but at different heights. The lobes are a complicated function of the antenna height, not the spacing, since that remains constant.

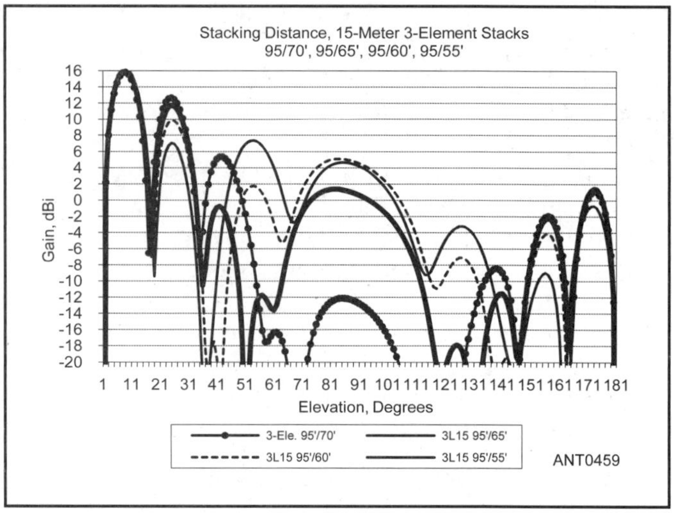

Figure 14.55 — Four spacing scenarios for two 3-element 15-meter Yagis. Things get very complicated. The optimal spacing in terms of stacking gain is 30 feet, which is 0.65 λ. The near-overhead lobes turn out to be ugly looking, but unimportant for skywave propagation.

But HF operation is different from moonbounce in that rigorously trying to minimize high-angle lobes is far less crucial at HF, where we've already shown that the main goal is to achieve gain over a wide elevation-plane footprint without any disastrous nulls in the pattern. The gain gradually increases as spacing in terms of wavelength is increased between individual Yagis in a stack, and then decreases slowly once the spacing is greater than about 1.0 λ. The difference in gain between spacings of 0.5 λ to 1.0 λ for a stack of typical HF Yagis amounts to only a fraction of a decibel. Stacking distances on the order of 0.6 λ to 0.75 λ give best gain commensurate with good patterns.

While the stack at 120/60 feet in Figure 14.50 doesn't have the second-lobe-stronger problem the 120/30-foot stack has — 60 feet between antennas is 1.29 λ, again outside the normal range of HF stack spacings. As a consequence, the 120/60-foot stack doesn't cover the range of elevation angles as well as it could, and is inferior to both the 90/60/30-foot stack and the 120/90/60/30-foot stack. The 120/60-foot two-Yagi stack needs at least one more antenna placed in-between to spread out the elevation-range coverage and to provide more gain.

It could be debated, but the 90/60/30-foot stack seems optimal for coverage of all the angles into Europe from New England on 15 meters. Note that the 30-foot spacing between Yagis is 0.65 λ on 21.2 MHz, right in the middle of the range of typical stack spacings.

SWITCHING OUT YAGIS IN THE STACK

Still, the extra gain that is available at low elevation angles from a 120/90/60/30-foot high, four-Yagi stack in Figure 14.50 is alluring. For those statistically possible, but less likely, occasions when the elevation angle is higher than about 12°, it would be advantageous to switch out the top 120-foot Yagi and operate with only the lower three Yagis in a stack. (This also allows the top antenna to be rotated in another direction, an aspect we'll explore later.) There are even times when the incoming angles are really high and when the top two antennas might be switched out to create a 60/30-foot stack. Later in this chapter we'll present circuitry for such stack switching.

STACKING DISTANCE AND LOBES AT HF

Let's look a little more closely at how a stack achieves gain and a wide elevation footprint. **Figure 14.54** shows a rectangular X-Y graph of the elevation response from 0° to 180° for two 3-element 15 meter Yagis (with 12-foot booms) spaced 30 feet apart (0.65 λ at 21.2 MHz), but mounted at two different heights: 95/65 and 85/55 feet. The rectangular plot gives more resolution than is possible on a polar plot. Note that the heights shown represent typical stacking heights on 15 meters — there's nothing magic about these choices. The free-space H-Plane pattern for the 30-foot spaced stack is also shown for reference.

The worst-case overhead elevation lobe, which ranges from about 60° to 120° in elevation (±30° from straight overhead at 90°), is about 14.7 dB down for the 95/65-foot stack. The overhead lobe peaks broadly at an elevation angle of about 82°. The overhead lobe for the lower 85/55-foot stack occurs at an elevation of about 64°, where it is 19 dB down.

The F/B for both 3-element sets of heights is about 15 dB, well down from the excellent 32 dB F/B for each Yagi by itself. The degradation of F/B is mainly due to mutual coupling to its neighbor in the stack.

The ground-reflection pattern in effect "modulates" the free-space pattern of the individual Yagi, but in a complex and not always intuitive manner. This is quite evident for the 85/55-foot stack at near-overhead angles. In this region things become complicated indeed, because the fourth and fifth lobes due to ground reflections are interacting with the free-space pattern of the stack.

Because the spacing remains constant at 30 feet for these pairs of antennas, however, the main determinant for the upper-elevation angle lobes is the distance of the horizontally polarized antennas above the ground, not the spacing between them.

Changing the Stack Spacing

Figure 14.55 demonstrates just how complicated things get for four different spacing scenarios. Here, the lower Yagi in the stack is moved down in 5-foot increments from the 95/70 feet level, to 95/65, 95/60 and 95/55 feet. The closest spacing, 25 feet in the 95/70-foot stack, yields nominally the "cleanest" pattern in the overhead region from 60° to 120°. The worst-case overhead lobe for the 95/70-foot stack is down 28 dB from peak. The F/B is again about 15 dB.

The worst case overhead lobe for the widest spacing, 40 feet in the 95/55-foot stack, is about 11 dB down from peak. The F/B has increased marginally, but is still only about 16 dB. It is difficult to pinpoint directly whether the spacing or the height above ground is the major determinant for the various lobe amplitudes for the 3-element stack. We'll soon look closely at whether the overhead lobe is important or not for HF work.

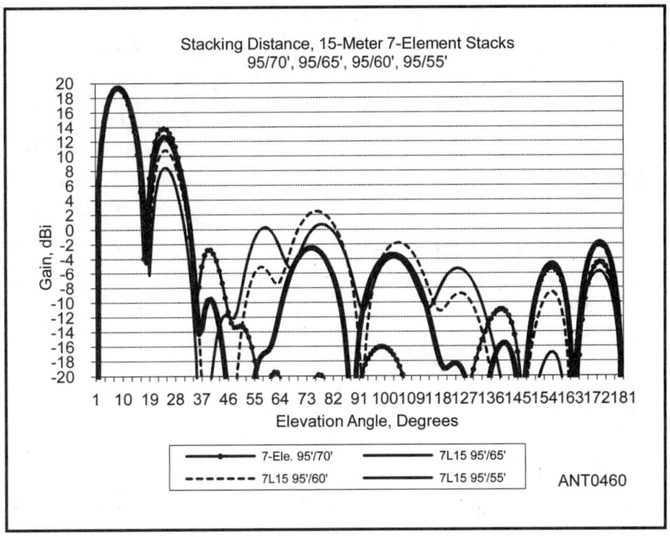

Figure 14.56 — Four spacing scenarios for two large 7-element 15-meter Yagis (on 64-foot booms). Again, a 0.65-λ spacing (30 feet) provides the most stacking gain.

Longer Boom Length and Stack Spacing

Figure 14.56 shows the same type overlay of elevation plots, but this time for two 7-element 15 meter Yagis on gigantic 64-foot booms. These Yagis are also spaced 30 feet apart (0.65 λ at 21.2 MHz), mounted at the same four sets of heights in Figure 14.52. As you'd expect, the free-space elevation pattern for a stacked pair of 7-element Yagis on 64-foot booms is narrower than that for a stacked pair of 3-element Yagis on 12-foot booms. The intrinsic F/B of the longer Yagi is also better than the F/B of the shorter antenna. As a result, all lobes beyond the main lobe of the stacked 7-element pair are lower for both sets of heights than their 3-element counterparts. The worst-case overhead lobe for the 7-element 95/65-foot pair is about 22 dB down at 76° and the F/B at 172° is greater than 21 dB for all four sets of heights.

Table 14.3 summarizes the main performance characteristics for four sets of stacked Yagis. The first entry for each boom length is for the Yagi by itself at a height of 95 feet. Stacked configurations are next listed in order of gain. The column labeled "Worst lobe, dB re Peak" is the amplitude of the second lobe due to ground reflections, and the elevation angle of that second lobe is listed as well.

Besides the 3- and 7-element designs discussed above, we've also added 4- and 5-element designs in Table 14.2.

Table 14.3
Example, Spacing Between 15-Meter Yagis

Antenna	*Peak Gain (dBi)*	*Worst Lobe (dB re Peak)*	*Worst Lobe Angle (°)*	*F/B (dB)*	*Overhead Lobe (dB re Peak)*
3 Elements, 12 Foot Boom					
By itself 95'	13.2	–0.9	21	28.8	–17.5
95'/65' (Δ 30')	16.08	–4.5	25	14.9	–14.7
95'/60' (Δ 35')	16.01	–6.2	24	15.1	–10.9
95'/70' (Δ 25')	15.81	–3.2	24	14.8	–28
95'/55' (Δ 40')	15.71	–8.7	24	16.4	–11
95'/75' (Δ 20')	15.34	–2.3	23	16.3	–17.2
4 Elements, 18 Foot Boom					
By itself 95'	13.92	–1	21	28.3	–20.4
95'/65' (Δ 30')	16.63	–4.5	23	18.5	–17.3
95'/60' (Δ 35')	16.6	–6.2	24	18.2	–13.1
95'/55' (Δ 40')	16.36	–8.7	24	19.8	–13.2
95'/70' (Δ 25')	16.36	–3.3	24	20.4	–31.8
95'/75' (Δ 20')	15.92	–2.5	23	25.9	–19
5 Elements, 23 Foot Boom					
By itself 95'	14.26	–1.1	21	27.9	–22.3
95'/65' (Δ 30')	16.86	–4.6	24	20.8	–19
95'/60' (Δ 35')	16.86	–6.3	24	20.7	–14.4
95'/55' (Δ 40')	16.67	–8.8	24	23.5	–14.4
95'/70' (Δ 25')	16.59	–3.4	24	24.9	–34.4
95'/75' (Δ 20')	16.18	–2.6	23	34.3	–20.2
7 Elements, 64 Foot Boom					
By itself 95'	17.93	–2.2	21	28.9	–17.1
95'/65' (Δ 30')	19.39	–6.9	24.3	21.4	–21.9
95'/60' (Δ 35')	19.38	–8.6	24	21.4	–16.9
95'/55' (Δ 40')	19.29	–10.9	24	25.0	–18.6
95'/70' (Δ 25')	19.26	–5.5	23	24	–35.3
95'/75' (Δ 20')	19.08	–4.6	23	27	–23.4

Over the range of stacking distances between 20 and 40 feet on 15 meters (0.43 λ to 0.86 λ), the peak gain for the 3-element stacks changes less than 0.75 dB, with the 30-foot spacing exhibiting the highest gain. The differences between peak gains versus stacking distance become smaller as the boom length increases. For example, for the 64-foot boom Yagi, the gain varies 19.39 – 19.08 = 0.31 dB for stack spacings from 20 to 40 feet.

In other words, changing the spacing from 20 to 40 feet (0.43 λ to 0.86 λ) doesn't change the gain significantly for boom lengths from 12 to 64 feet (0.26 λ to 1.38 λ). From the point of view of gain, the vertical spacing between individual antennas in an HF stack is not critical.

The worst-case lobes (generally speaking, the second lobe due to ground reflections) are highest for a Yagi operated by itself. After all, a single Yagi doesn't benefit from the redistribution of energy from higher-angle lobes into the main lobe that a stack gives. Thus, the 3-element, 12-foot boom Yagi by itself at 95 feet would have a second lobe at 21° that is only 0.9 dB down from the main lobe, while the stack of two such antennas with 30-foot (0.65 λ) spacing at 95/65 feet would have a second lobe down 4.5 dB. As the spacing between antennas in a vertical stack increases, the second lobe is suppressed more, up to 8.7 dB with 40-foot (0.86 λ) spacing.

Since the free-space elevation pattern for a 3-element Yagi is wider than that for a 7-element Yagi, the second lobe due to ground reflection will be somewhat reduced. This is true for all longer-boom antennas operating by themselves over ground. Used in stacks, the second lobe's amplitude will vary depending on spacing between antennas, but they range only about 6 dB.

The front-to-back ratio will also tend to increase with longer boom lengths on a properly designed Yagi. Table 14.2 shows that the F/B is somewhat better for closer spacings between antennas in a stack, a rather non-intuitive result, considering that the mutual coupling should be greater for closer antennas. For example, the 5-element Yagi stack with 20-foot spacing has a exceptional F/B of 34.3 dB, compared to a F/B of 21.4 dB with the 30-foot spacing distance that gives nominally the most gain. High values of F/B, however, rarely hold over a wide frequency range because of the very critical phasing relationships necessary to get a deep null, so the difference between 34.3 and 21.4 dB would rarely be noticeable in practice.

The near-overhead lobe structure (between 60° to 120° in elevation) tends also to be lower for smaller stack spacings — for all boom lengths — peaking in this example at a spacing of 25 feet for the boom lengths considered here. Since the peak gain actually occurs with smaller spacing between Yagis in this 7-element stack, even relatively large and messy looking overhead lobes are not subtracting from the stacking gain. In the next section we'll now examine whether this overhead lobe is important or not.

STACKING DISTANCES FOR MULTIBAND YAGIS

By definition, a stack of multiband Yagis (such as a "tribander" covering 20/15/10 meters) has a constant vertical spacing between antennas in terms of feet or meters, but not in terms of wavelength. Tribanders are no different than monobanders in terms of optimal spacing between individual antennas. Again, the difference in gain between spacings of 0.5 λ and 1.0 λ for a stack of triband Yagis amounts to only a fraction of a decibel. Furthermore, the main practical constraint that limits choice of stacking distances between any kind of Yagis, multiband or monoband, is the spacing between guy wire sets on the tower itself.

SUMMARY — STACKING DISTANCES

In short, let us summarize that there is nothing magical about stacking distances for practical HF Yagis — a good rule-of-thumb is a stacking distance of 0.65 λ. This is 23 feet on 10 meters, 30 feet on 15 meters and 45 feet on 20 meters for monoband stacks. Practically speaking, however, you've only got limited places where you can mount antennas on the tower — mainly where guy wires allow you to place them. This is especially applicable if you wish to rotate lower antennas on the tower, where you must clear the guys from above the antenna.

14.4.5 Radiation Outside the Main Lobe

THE IMPORTANCE OF HIGHER-ANGLE LOBES

We've already shown that the exact spacing between HF Yagis is not critical for stacking gain. Further, the heights (and hence spacing) of the individual Yagis in a stack interact in a complicated fashion to determine higher-angle lobes.

Let's examine the relevance of such higher-angle lobes for stacked HF Yagis, this time in terms of interference reduction on receive. As the **Radio Wave Propagation** chapter points out, few DX signals arrive at elevation angles greater than about 30°. In fact, DX signals only propagate at elevation angles in the range from 1° to 30° on all the bands where operators might reasonably expect to stack Yagis — nominally from 7 to 29.7 MHz.

You should remember that the definition of the *critical frequency* for HF propagation is the highest frequency for which a wave launched directly overhead at 90° elevation is reflected back down to Earth, rather than being lost into outer space. The maximum critical frequency for extremely high levels of solar flux is about 15 MHz. In other words, high overhead angles do not propagate signals on the upper HF bands.

However, some domestic signals do arrive at relatively high elevation angles. Let's look at some scenarios where higher angles might be encountered and how the elevation patterns of typical HF stacks affect these signals. Let's examine a situation where a medium-range interfering station is on the same heading as a more distant target station.

We'll examine a typical scenario involving stations in Atlanta, Boston and Paris. The heading from Atlanta to Paris is 49°, the same heading as Atlanta to Boston. In other words, the Atlanta station would have to transmit over (and listen through) a Boston station for communication with Paris. The distance between Atlanta and Boston is about 940 miles, while the distance from Atlanta to Paris is about 4350 miles. Ground wave signals obviously cannot travel either of these distances at 21 MHz (ground wave coverage is less than about 10 miles at this frequency), and so the propagation between Atlanta to Boston and Atlanta to Paris will be entirely by means of the ionosphere.

Let's evaluate the situation on 15 meters in the month of October. We'll assume a smoothed sunspot number (SSN) of 100 and that each station puts 1500 W of power into theoretical isotropic antennas that have +10 dBi of gain at all elevation and azimuth angles. [We use such theoretical isotropic antennas because they make it easier to work in *VOACAP*. We will factor in real-world stacks later.] *VOACAP* predicts that the signal from Boston will be S9 + 8 dB in Atlanta at 1400 UTC, arriving at an elevation angle of 21.3° on a single F2 hop. This elevation angle is higher than commonly encountered angles for DX signals, but it is still far away from near-overhead angles.

The signal from Paris into Atlanta is predicted to be about S6 for the same theoretical isotropic antennas, at an incoming elevation angle of 6.4° on three F2 hops. The S6 level validates the rule-of-thumb that each extra hop loses approximately 10 dB of signal strength, assuming that each S unit is about 4 dB, typical for modern receivers.

Now look at **Figure 14.57**, which shows the response for a stack of 3-element Yagis at 90/60/30 feet over flat ground, along with the response for a similar stack of 7-element Yagis. Again, we'll assume that all three stations are using such 3-element 90/60/30-foot stacks. The stations in Atlanta and Boston point their stacks into Europe and the Parisian station points his stack toward the USA. The gain of the Atlanta array at 6.4° into Paris will be about 16 dBi, or 6 dB more than the isotropic array with its +10 dBi of gain selected for use in *VOACAP*. Similarly, the French station's transmitted signal will enjoy a 6 dB gain advantage over the isotropic array used in the *VOACAP* calculation, and thus the French signal into Atlanta will now be S6 + 12 dB, or about S9.

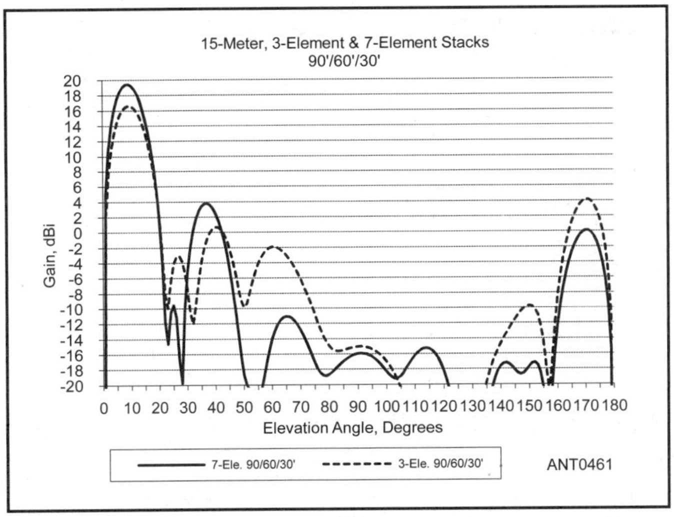

Figure 14.57 — Stacks of three 3-element and 7-element Yagis on 15 meters at 90/60/30 feet heights. The F/B for the 7-element stack is superior to the 3-element stack mainly because the F/B is intrinsically better for the long-boom design.

By comparison, the interfering signal from Boston into Atlanta will be reduced by the rearward pattern of his array, which will launch a signal at 180° – 21.3° = 158.7° in elevation at the single F2 mode from Boston to Atlanta. From Fig-ure 14.51, the Boston station's gain at this rearward elevation is going to drop from the isotropic's +10 dBi of gain down to –11 dBi, a drop of 21 dB. The signal into the Atlanta receiver will also be reduced by the pattern of the Atlanta array on receive, which has a gain of about 0 dBi at 21.3°, compared to the isotropic's +10 dBi gain at 6.4°, a net drop of 10 dB.

Thus, the Boston station's signal will drop by about 21 + 10 = 31 dB, bringing the interfering signal from Boston, which would be S9 + 8 dB for isotropic antennas, down to about S3 due to the combined effects of the arrays. This is a very significant reduction in interference. But you will note that the reduction has nothing to do with the near-overhead lobes, dealing as it does with the trailing edge of the main lobe and the F/B lobe.

Higher Elevation Angles

Now let's evaluate a station that is even closer to Boston, say a station in Philadelphia. The heading from Philadelphia to Paris is 53° and the distance is 3220 miles. On the same day in October as above, *VOACAP* predicts a signal strength of S8 from Paris to Philadelphia, at a 2.7° elevation angle on two F2 hops. Again, the *VOACAP* computations assume isotropic antennas with +10 dBi gain at all three stations. The gain of the 3-element stacks at both ends of the circuit at 2.7° is also about +10 dBi, so the signal level from Paris to Philadelphia would be S8 with the 3-element stacks.

Now *VOACAP* computes the elevation angle from Philadelphia to Boston as 56.3°, on one F_2 hop launched at an azimuth of 53°, well within the azimuthal beamwidth of the stack. *VOACAP* says the predicted signal strength for isotropic antennas with +10 dBi of gain is less than S1!

What's happening here? Boston and Philadelphia are within the "skip" region on 21 MHz and signals are skipping right over Boston from Philadelphia (and vice versa). Actual signals would be much weaker than they would be with theoretical isotropic antennas because of the actual patterns of the transmitting and receiving stacks. At an elevation angle of 56.3° the receiving stack would have a gain of –10 dBi, while at an elevation of 180° – 56.3° = 123.7° the transmitting stack would be down to –10 dBi as well. The net reduction for the stacks compared to isotropics with +10 dBi gain each would be 40 dB, putting the interfering signal well into the receiver noise.

You can safely say that near-overhead angles don't enter into the picture, simply because signals at intermediate distances are in the ionospheric skip zone and interfering signals are very weak in that zone already.

Even in situations where having a poor front-to-back ratio might be beneficial — because it alerts stations tuning across your signal that you are occupying that frequency — the ionosphere doesn't cooperate for intermediate-distance signals that are in the skip zone. Often two stations may be on the same frequency without either knowing that the other is there.

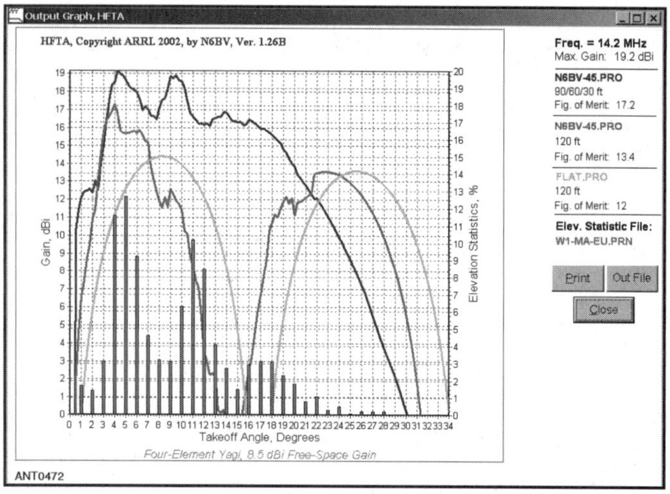

Figure 14.58 — *HFTA* screen shot showing how complicated things become when real-world irregular terrain is analyzed. This is the 20-meter elevation pattern for the N6BV/1 station location in Windham, NH, for the 90/60/30-foot stack of triband TH7DX Yagis and a 4-element Yagi at 120 feet on the same tower. For comparison, the response of a 120-foot Yagi over flat ground is also included.

Ground Wave and Stacks

What happens, you might wonder, for ground-wave signals? Let's look at a situation where the interfering station is in the same direction as the desired target, but is only 5 miles away. Unfortunately, his signal is S9 + 50 dB. Even reducing the level by 30 dB, a huge number, is still going to make his signal 20 dB stronger than signals from your desired target location! There is not much you can do about ground-wave signals and fretting about optimizing stack heights to discriminate against local signals is generally futile.

14.4.6 Real-World Terrain and Stacks

So far, the stacking examples shown have been for flat ground. Things can become a lot more complicated when you deal with real-world irregular terrains! **Figure 14.58** shows the *HFTA*-computed 20 meter elevation responses toward Europe (at an azimuth of 45°) for three antennas at the N6BV/1 location in Windham, New Hampshire. Overlaid as a bar graph are the elevation-angle statistics for the path to all of Europe from New England (Massachusetts). The stack at 90/60/30 feet clearly covers all the angles needed best at 14 MHz. The N6BV/1 120-foot Yagi has a severe null in the region from about 7° to about 20°, with the deepest part of that null occurring at about 13° and is roughly comparable to the 90/60/30-foot stack between 2° to 7°.

In practice, the 120-foot Yagi was indeed comparable to the stack during morning openings to Europe on 20 meters, when the elevation angles are typically about 5°. In the New England afternoon, when the elevation angles typically rise to about 11°, the 120-foot Yagi was always distinctly inferior to the stack.

For reference, the response of a single 120-foot high Yagi over flat ground is also shown. Note that the N6BV 120-foot high Yagi has about 3 dB more gain at a 5° takeoff angle than does its flatland counterpart. This additional gain is due to the focusing effects of the local terrain, which had about a 3° downward slope toward Europe.

Figure 14.59 shows the *HFTA*-computed 15 meter elevation responses toward Europe for the 90/60/30-foot stack at 90/60/30 feet at N6BV/1, compared to the same 120-foot high Yagi and a 90/60/30-foot stack, but this time over flat ground. Again, the N6BV/1 terrain toward Europe has a significant effect on the gain of the stack compared to that of an identical stack over flat ground. In fact, the peak gain of 20.1 dBi at a 4° elevation angle is close to moon-bounce levels.

OPTIMIZING OVER LOCAL TERRAIN

There are only a small number of possibilities to optimize an installation over local terrain:

- Change the antenna height(s) above ground.
- Stack two (or more) Yagis.
- Change the spacing between stacked Yagis.
- Move the tower back from a cliff (or a hill).
- BIP/BOP (Both In Phase/Both Out of Phase).

The *HFTA* program included with this book's online material can be used, together with topographic data available on the Internet, to evaluate all these options.

It is sometimes very surprising to compare elevation responses for different towers located at various points on the same property, particularly when that property is located in the mountains. **Figure 14.60** shows the computed elevation responses for three 100-foot high 14-MHz Yagis over three terrains toward Europe: from the North tower at K1KI's location in West Suffield, Connecticut, from the South tower at K1KI, and over flat ground. The elevation response from the South tower follows that over flat ground well, while the response from the North tower is quite a bit stronger at low elevation angles — about 1.5 dB on average, as the Figure of Merit shows from *HFTA*.

Figure 14.61 shows the reason why this happens — the

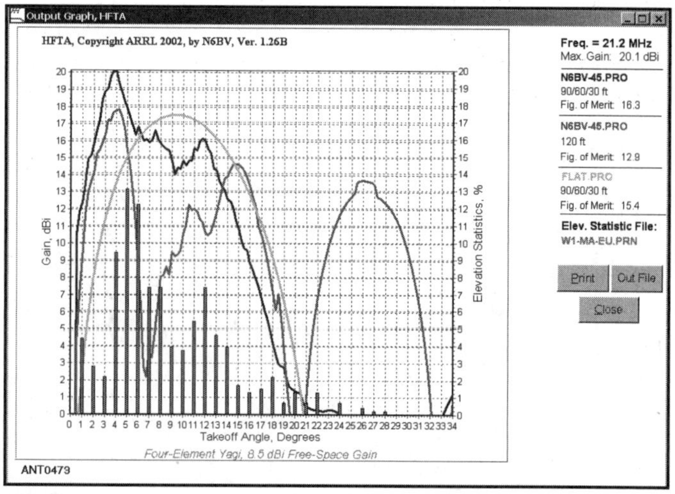

Figure 14.59 — *HFTA* screen shot showing the 15-meter elevation pattern for the N6BV/1 station location in Windham, NH, for the 90/60/30-foot stack of triband TH7DX Yagis and a 4-element Yagi at 120 feet on the same tower. For comparison, the response of a 120-foot Yagi over flat ground is also included.

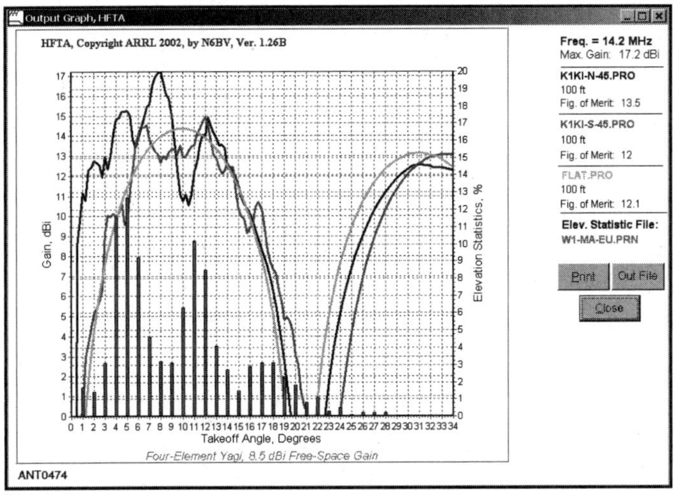

Figure 14.60 — *HFTA* screen shot showing the 20-meter elevation pattern for K1KI's North and South towers, with 100-foot high 4-element Yagis pointing into Europe at an azimuth of 45°. The responses are surprisingly different for two towers separated by only 600 feet.

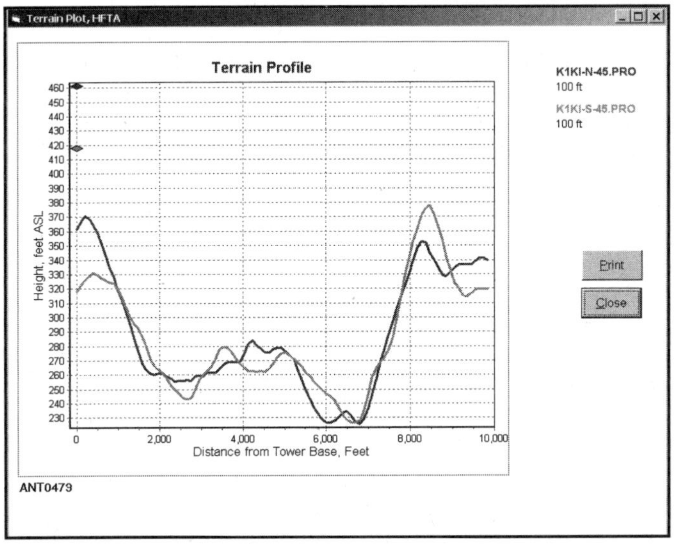

Figure 14.61 — K1KI's terrain profiles for the North and South towers at an azimuth of 45° into Europe.

terrain from the North tower slopes down quickly toward Europe, while the terrain from the South tower goes out almost 900 feet before starting to fall off. These two towers are about 600 feet apart.

14.4.7 Stacking Tribanders

Enterprising amateurs have built stacked tribander arrays even with full recognition that they are compromise antennas when compared one-on-one against monoband Yagis. Bob Mitchell, N5RM, is a prominent example, with his so-called "TH28DX" array of four TH7DX tribanders on a 145-foot-high rotating tower. Mitchell employed a rather complex system of relay-selected tuned networks to choose either the upper stacked pair, the lower stacked pair or all four antennas in stack. Others in Texas have also had good results with their tribander stacks. Contester Tom Owens, K7RI, has very successfully used a pair of stacked KT-36XA tribanders for years.

A major reason why tribanders were used is that over the years both amateurs have had good results using TH6DXX or TH7DX antennas. They are ruggedly built, mechanically and electrically, and their 24-foot long booms are long enough to produce significant gain, despite trap-loss compromises. Trap losses estimated at approximately 0.5 dB are not high enough to be of serious concern. A long-boom tribander like the TH6DXX or TH7DX also has enough space to employ elements dedicated to different bands, so the compromises in element spacing usually found on short-boom 3 or 4-element tribanders can be avoided.

Another factor in favor of tribanders is the serious interaction that can result from stacking monoband antennas closely together on one mast in a Christmas Tree configuration. N6BV's worst experience was with the ambitious 10 through 40 meter Christmas Tree at W6OWQ in the early 1980s. This installation used a Tri-Ex SkyNeedle tubular crank-up tower with a rotating 10-foot-long heavy-wall mast. The antenna suffering the greatest degradation was the 5-element 15 meter Yagi, sandwiched 5 feet below the 5-element 10 meter Yagi at the top of the mast, and 5 feet above the full-sized 3-element 40 meter Yagi, which also had five 20 meter elements interlaced on its 50-foot boom.

The front-to-back ratio on 15 meters was at best about 12 dB, down from the 25+ dB measured with the bottom 40/20 meter Yagi removed. No amount of fiddling with element spacing, element tuning or even orientation of the 15 meter boom with respect to the other booms (at 90° or 180°, for example) improved its performance. Further, the 20 meter elements had to be lengthened by almost a foot *on each end of each element* in order to compensate for the effect of the interlaced 40 meter elements. It was a lucky thing that the tower was a motorized crank-up, because it went up and down hundreds of times as various experiments were attempted!

Interaction due to close proximity to other antennas in a short Christmas Tree can definitely destroy carefully optimized patterns of individual Yagis. Nowadays, such interaction can be modeled using a computer program such as *EZNEC* or *NEC*. A gain reduction of as much as 2 to 3 dB can easily result due to close vertical spacing of monobanders, compared to the gain of a single monoband antenna mounted in the clear. Curiously enough, at times such a reduction in gain can be found even when the front-to-back ratio is not drastically degraded, or when the front-to-back occasionally is actually *improved*.

If you plan on stacking monoband Yagis — for example, putting 15 and 20 meter Yagis on a single tower, do make sure you model the system to see if any interactions occur. You may be quite surprised.

Finally, triband antennas make for less mechanical complexity than do an equivalent number of monobanders. There were five Yagis on the N6BV/1 tower, yielding gain from 40 to 10 meters, as opposed to using 12 or 13 monobanders on the tower.

SIMPLE TRIBANDER STACKS

All this discussion of large stacks of many antennas is simply out of the question for most amateurs. However, many hams already have a tribander on top of a moderately tall tower, typically at a height of about 70 feet. It is not terribly difficult to add another, identical tribander at about the 40-foot level on such a tower. The second tribander can be pointed in a fixed direction of particular interest (such as Europe or Japan), or it can be rotated around the tower on a side mount or a Ring Rotor. If guy wires get in the way of rotation, the antenna can usually be arranged so that it is fixed in a single direction.

Insulate the guy wires at intervals to ensure that they don't shroud the lower antenna electrically. A simple feed system consists of equal-length runs of surplus ½-inch 75-Ω hardline (or more expensive 50-Ω hardline, if you are really obsessed by SWR) from the station up the tower to each antenna. Each tribander is connected to its respective hardline feeder by means of an equal length of flexible coaxial cable, with a ferrite choke balun, so that the antenna can be rotated.

Down in the station, the two runs of hardline can simply be switched in and out of parallel to select the upper antenna only, the lower antenna only, or the two antennas as a stack. See **Figure 14.62**. Any impedance differences can be handled as stated previously, simply by retuning the linear amplifier, or by means of the internal antenna tuner (included in most modern transceivers) when the transceiver is run barefoot. The extra performance experienced in such a system will be far greater than the extra decibel or two that modeling calculates.

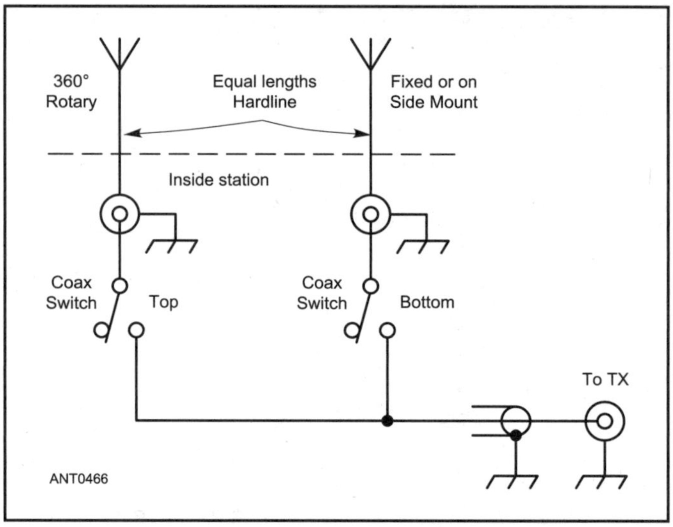

Figure 14.62 — Simple feed system for 70/40-foot stack of tribanders. Each tribander is fed with equal lengths of 0.5-inch 75-Ω Hardline cables (with equal lengths of flexible coax at the antenna to allow rotation), and can be selected singly or in parallel at the operator's position in the station. Again, no special provision is made in this system to equal SWR for any of the combinations.

14.4.8 Stacking Dissimilar Yagis

So far we have been discussing vertical stacks of identical Yagis. Less commonly, hams have successfully stacked dissimilar Yagis. For example, consider a case where two 5-element 10 meter Yagis are placed 46 and 25 feet above flat ground, with a 7-element 10 meter Yagi at 68 feet on the same tower. See **Figure 14.63**, which is a schematic of the layout for this stack. Note that the driven element for the top 7-element Yagi is well behind the vertical plane of the driven elements for the two 5-element Yagis. This offset distance must be compensated for with a phase shift in the drive system for the top Yagi.

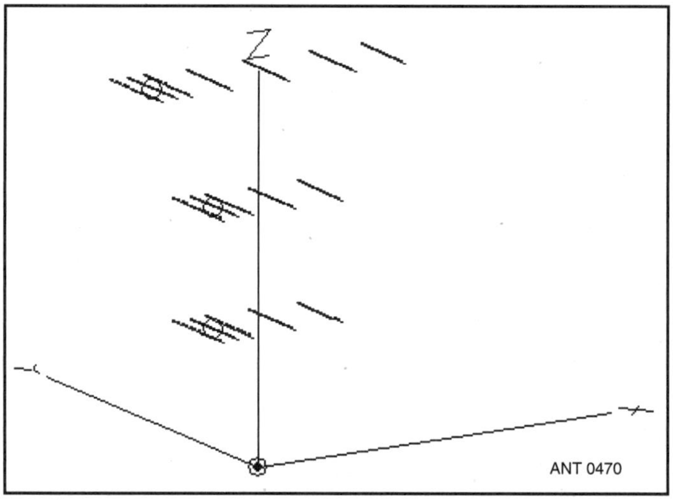

Figure 14.63 — Stacking dissimilar Yagis. In this case a 7-element 10-meter Yagi is stacked over two 5-element Yagis. Note the displacement of the 7-element Yagi's driven element compared to the position of the two 5-element Yagis. This leads to an undesired phase shift for the higher antenna.

Figure 14.64 shows the elevation-pattern responses for uncompensated (equal-length feed lines) and the compensated (additional 150° of phase shift to top Yagi) stacks. These patterns were computed using *EZNEC*. Not only is about 1.7 dB of maximum gain lost, but the peak elevation angle is shifted upward by 11° from the optimal takeoff angle of 8° — where some 10 dB of gain is also lost. Without compensation, this is a severe distortion of the stack's elevation pattern.

For RG-213 coax, the extra length needed to provide an additional 150° of phase shift = 150°/360° λ = 0.417 λ = 9.53 feet at 28.4 MHz. This was computed using the program *TLW* (*Transmission Line for Windows*) included with this book's online material.

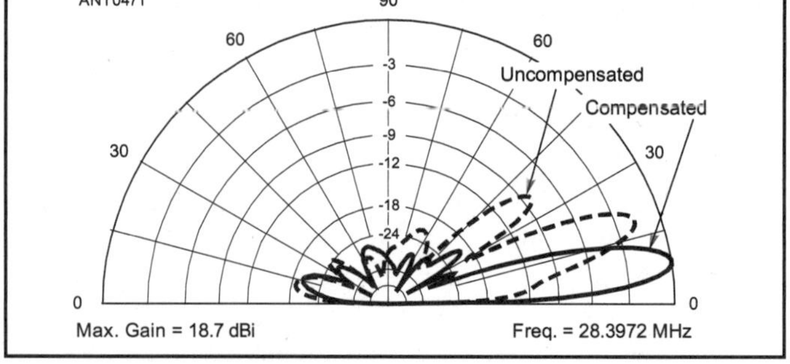

Figure 14.64 — Comparison of elevation responses for 7/5/5-element 10-meter stacks, with and without compensation for driven-element offset.

It is not always possible to compensate for dissimilar Yagis in a stack with a simple length of extra coax, so you should

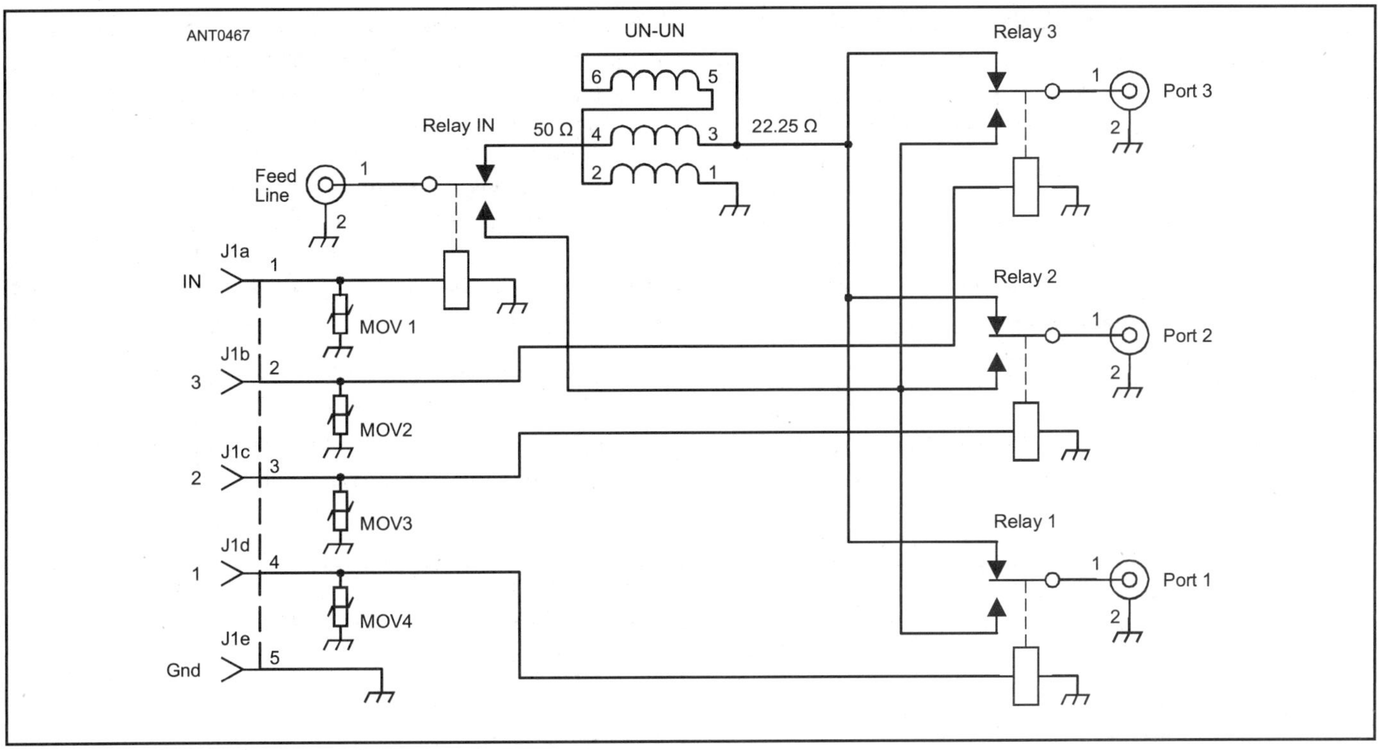

Figure 14.65 — Schematic of WXØB's StackMatch 2000 switchbox, which uses a broadband transmission line transformer using trifilar #12 enamel-insulated wires. (*Courtesy Array Solutions.*)

be sure to model such combinations to make sure that they work properly. A safe alternative, of course, is to stack only identical Yagis, feeding all of them with equal lengths of coax to ensure in-phase operation.

TRANSFORMER-BASED STACKMATCH SWITCHING

Earlier we mentioned how useful it would be to switch various antennas in or out of a stack, depending on the elevation angles that need to be emphasized at that moment. Jay Terleski, WXØB, of Array Solutions (**arraysolutions.com**) has designed switchable matching systems, called *StackMatches*, for stacks of monoband or multiband Yagis. This has become the standard method of switching for stacks of Yagi antennas, whether monoband or triband. (A description of two other systems used by N6BV/1 and K1VR is included with this book's online material.)

The StackMatch uses a 50-Ω to 22.25-Ω broadband transmission-line transformer to match combinations of up to three Yagis in a stack. See **Figure 14.65** for a schematic of the StackMatch. For selection of any 50-Ω Yagi by itself, no matching transformer is needed and Relay IN routes RF directly to the common bus going to Relay 1, 2 and 3. For selection of two Yagis together the parallel impedance is 50/2 = 25 Ω and Relay IN routes RF to the matching transformer. The SWR is 25/22.25 = 1.1:1. For three Yagis used together, the parallel impedance is 50/3 = 16.67 Ω, and the SWR is 22.25/16.67 = 1.3:1.

The broadband transformer consists of four trifilar turns of #12 enamel-insulated wire wound on a Ferrite Corporation FT-240 2.4-inch OD core made of #61 material (μ = 125). WXØB uses 10-A relays enclosed in plastic cases to do the RF switching, selected by a control box at the operating position. (10-A relays can theoretically handle 10 A^2 × 50 Ω = 5000 W.) **Figure 14.66** shows a photo of the transmission-line transformer and StackMaster PCB.

The control/indicator box uses a diode matrix to switch various combinations of antennas in/out of the stack. Three LEDs lined up vertically on the front panel indicate which antennas in a stack are selected.

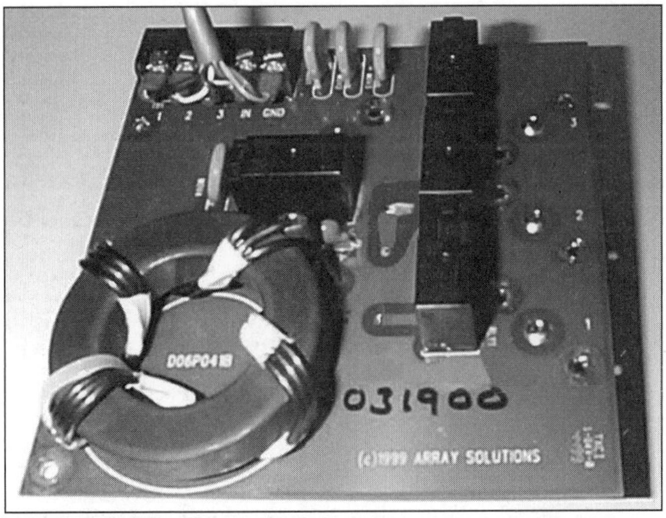

Figure 14.66 — Inside view of StackMatch. (*Photo courtesy Array Solutions.*)

SWITCHING TWO-ANTENNA ARRAYS

The following techniques of switching a pair of antennas was summarized by Dave Pruett, K8CC, in the article "Switching Stacked Antennas" in the January 1988 issue of the *National Contest Journal* (included in the online material).

Most two-element stacks require that each antenna receive equal power at the same phase angle as the other antennas. This is described as "both-in-phase" or "BIP" in a previous section along with being able to drive "both-out-of-phase" or "BOP." When either antenna is selected singly, we must ensure that all the power goes to that antenna alone.

The system in **Figure 14.67** uses a non-shorting switch that may be located in the station or in a remote switching box, as long as equal lengths of feed line are used for each antenna. The system in Figure 14.67B is a variation of the system that uses a shorting switch (unused positions grounded). If a λ/4 length of 50 Ω line is installed between the switch and the main feed line, then 75 Ω antennas and feed lines can be used.

The system in **Figure 14.68** provides A/B/BIP/BOP configurations using a non-shorting switch. A version using a shorting switch is shown in Figure 14.68B. Both systems in Figure 14.68 can be used with either 50 Ω or 75 Ω feed line with no changes. Note the use of the 3λ/4 line at the lower left to invert the phase of one antenna while not changing its feed point impedance.

SWITCHING THREE- OR FOUR-ANTENNA ARRAYS

As a complement to the transformer-based system of Array Solutions, the approach taken by Ken Wolff, K1EA, to stack switching uses only lengths of feed line. Ken relied on the

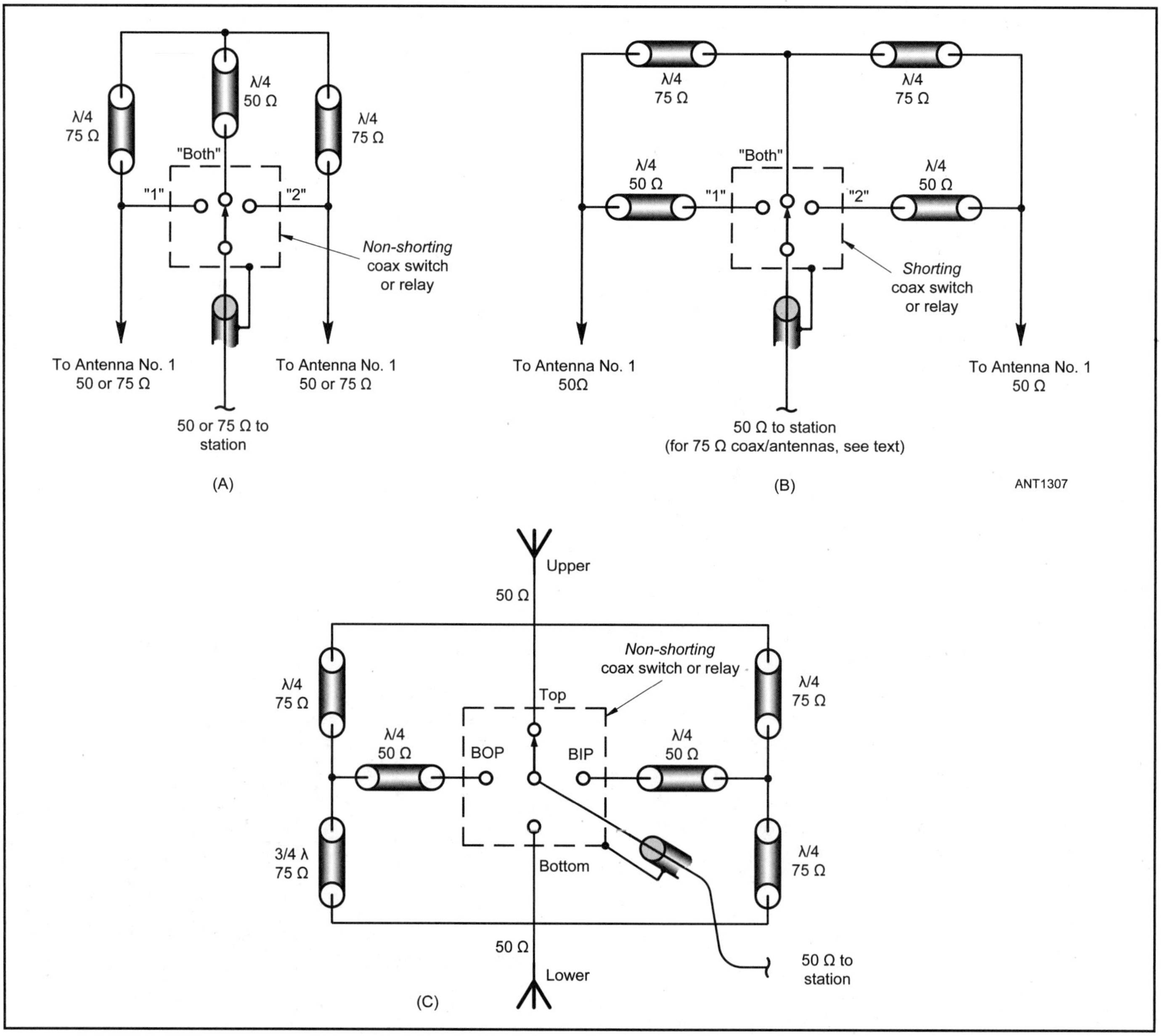

Figure 14.67— An A/Both/B switching system using a non-shorting switch (A) and a shorting switch (B) with unused positions grounded.

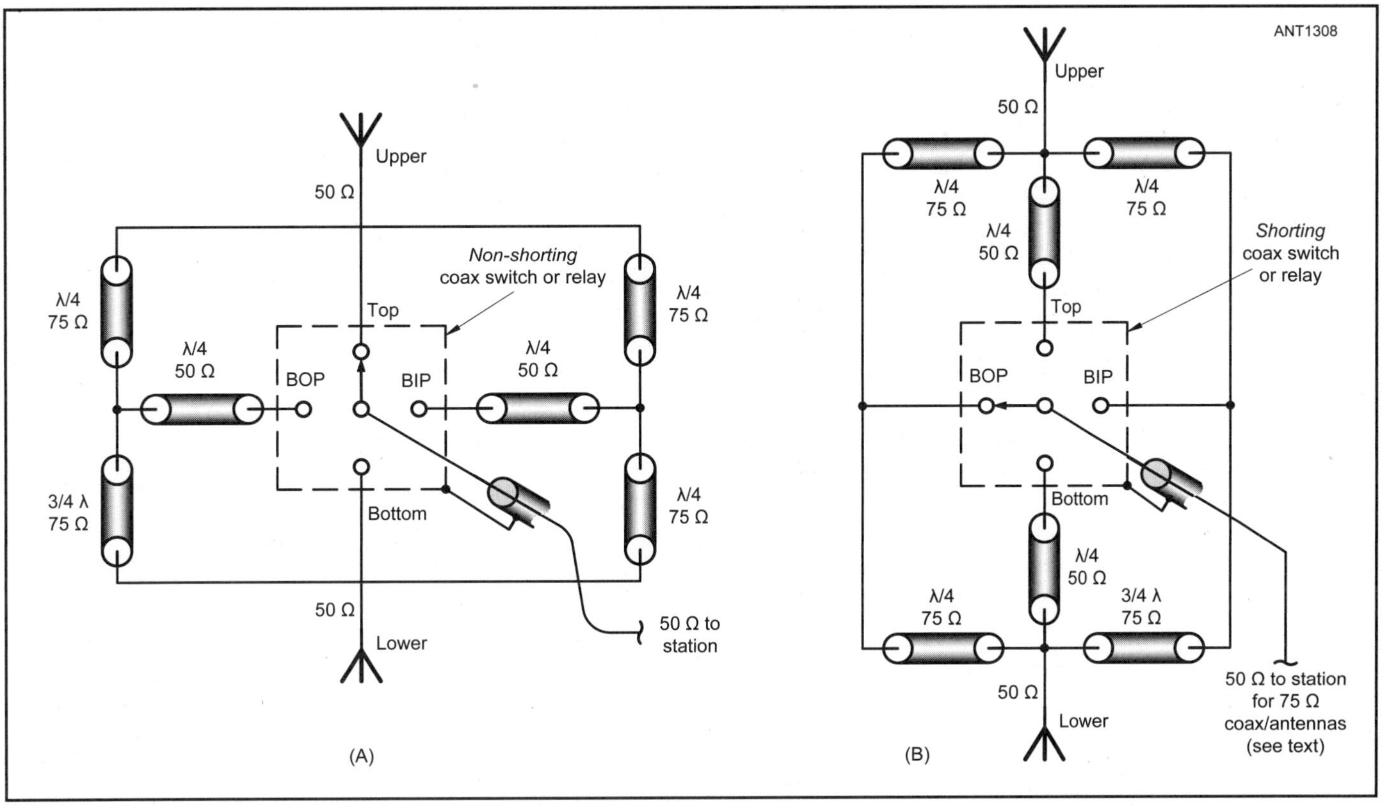

Figure 14.68 — An A/B/BIP/BOP system using a non-shorting switch (A) and a shorting switch (B) with unused positions grounded.

current-forcing effect of a ¼-λ length of feed line and of combining parallel lengths of feed line for impedance transformation. Whether or not you replicate this system, the technique — originally published in the December 1989 issue of the Yankee Clipper Contest Club's *Scuttlebutt* newsletter — can be applied in many different antenna system designs.

The basic approach is to connect all of the antennas in parallel, creating a low impedance, then transforming that impedance to 50 Ω with a ¼-λ transmission line transformer. (See "Quarter-Wave Transformers" in the chapter on **Transmission Line System Techniques**.) Coaxial cables of different impedances (50 and 70 Ω) are combined in parallel to produce low-impedance feed lines. For example, two 50-Ω lines in parallel create a 25-Ω line, and a 50-Ω line in parallel with a 70-Ω line creates a 29.3-Ω cable. For the impedance transformer section, the formula is:

$$Z_{NEW} = (Z_{LINE} \times Z_{LINE}) / Z_{OLD}$$

Some examples of this technique are shown in **Figure 14.69**. For example, the system with three 50-Ω antennas in parallel labeled "Top only or all three" (Figure 14.64B) creates a 50/3 = 16.7 Ω impedance. The 70-Ω and 50-Ω lines in parallel create a 29.2-Ω line. Using the transformation formula, the resulting impedance is (29.2 × 29.2) / 16.7 = 51 Ω which is a very close match to 50 Ω. Similarly, the four-stack (Figure 14.64A) antennas in parallel are 12.5 Ω and the two 50-Ω lines in parallel create a 25-Ω line, resulting in 50 Ω exactly.

Similarly, six 50-Ω antennas in parallel (8.33 Ω) are matched with 20.6-Ω line made of two 70-Ω lines in parallel with one 50-Ω line, creating (20.6 × 20.6) / 8.33 = 50.9 Ω impedance. Five 50-Ω antennas in parallel create 10 Ω which is matched by a 23.3-Ω section of three 70-Ω lines in parallel, creating (23.3 × 23.3) / 10 = 54.4 Ω. The possibilities are endless: switch the antennas to be driven so they are in parallel, then switch in combinations of 50 and 70-Ω lines to match the resulting impedance to 50 Ω.

14.4.9 Miscellaneous Topics

STACKS AND FADING

The following is derived from an article by Fred Hopengarten, K1VR, and Dean Straw, N6BV, in a February 1994 *QST* article. Using stacked Hy-Gain TH7DXs or TH6DXXs at their respective stations, they have solicited a number of reports from stations, mainly in Europe, to compare various combinations of antennas in stacks and as single antennas. The peak gain of the stack is usually just a little bit higher than that for the best of the single antennas, which is not surprising. Even a large stack has no more than about 6 dB of gain over a single Yagi at a height favoring the prevailing elevation angle. Fading on the European path can easily be 20 dB or more, so it is very confusing to try to make definitive comparisons. They have noticed over many tests that the stacks are much less susceptible to fading compared to single Yagis. Even within the confines of a typical SSB

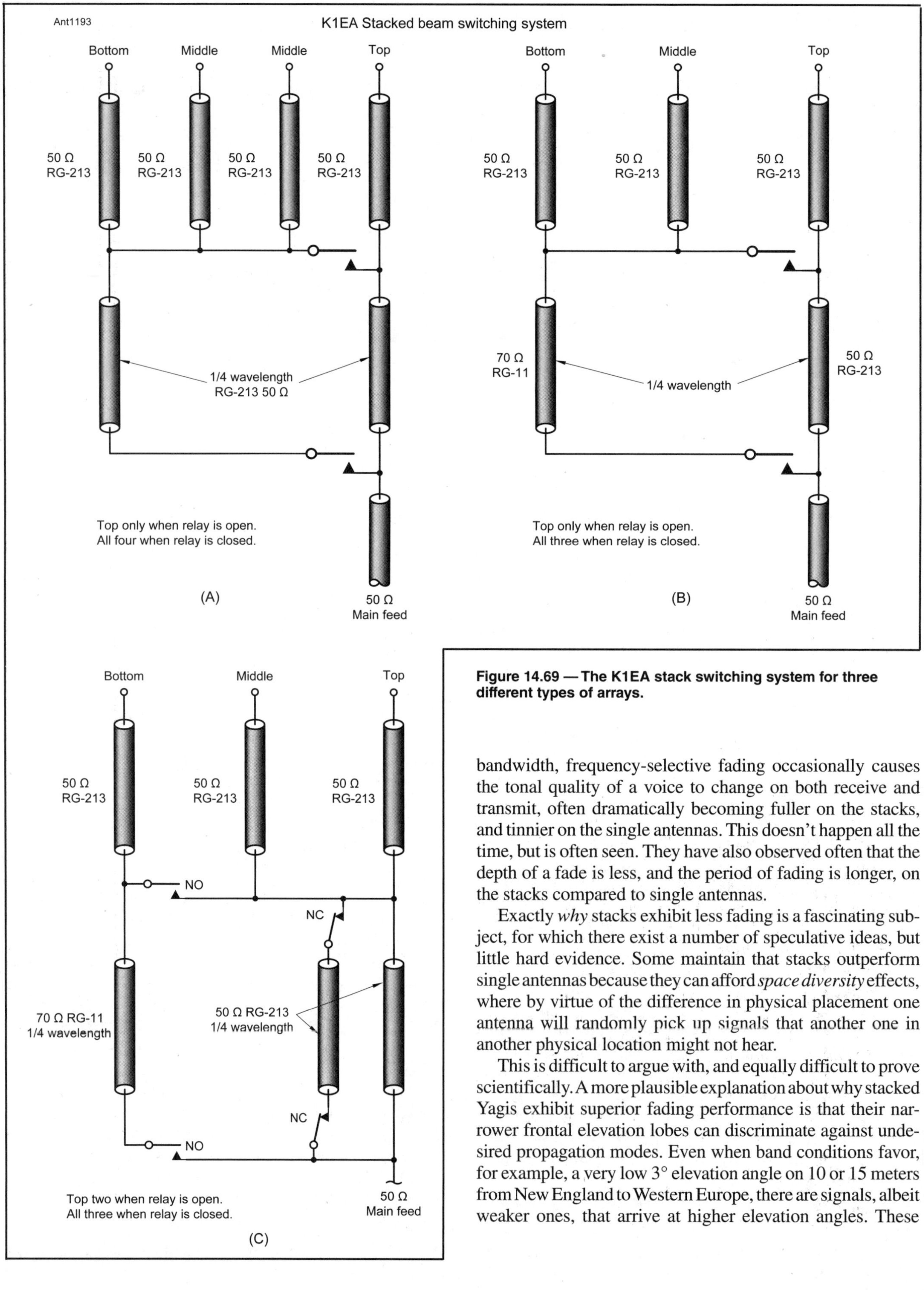

Figure 14.69 — The K1EA stack switching system for three different types of arrays.

bandwidth, frequency-selective fading occasionally causes the tonal quality of a voice to change on both receive and transmit, often dramatically becoming fuller on the stacks, and tinnier on the single antennas. This doesn't happen all the time, but is often seen. They have also observed often that the depth of a fade is less, and the period of fading is longer, on the stacks compared to single antennas.

Exactly *why* stacks exhibit less fading is a fascinating subject, for which there exist a number of speculative ideas, but little hard evidence. Some maintain that stacks outperform single antennas because they can afford *space diversity* effects, where by virtue of the difference in physical placement one antenna will randomly pick up signals that another one in another physical location might not hear.

This is difficult to argue with, and equally difficult to prove scientifically. A more plausible explanation about why stacked Yagis exhibit superior fading performance is that their narrower frontal elevation lobes can discriminate against undesired propagation modes. Even when band conditions favor, for example, a very low 3° elevation angle on 10 or 15 meters from New England to Western Europe, there are signals, albeit weaker ones, that arrive at higher elevation angles. These higher-angle signals have traveled longer distances on their journey through the ionosphere, and thus their signal levels and their phase angles are different from the signals traversing the primary propagation mode. When combined with the dominant mode, the net effect is that there is both destructive and constructive fading. If the elevation response of a stacked antenna can discriminate against signals arriving at higher elevation angles, then in theory the fading will be reduced. Suffice it to say: In practice, stacks do reduce fading.

STACKS AND PRECIPITATION STATIC

The top antenna in a stack is often much more affected by rain or snow precipitation static than is the lower antenna. N6BV and K1VR have observed this phenomenon, where signals on the lower antenna by itself are perfectly readable, while S9+ rain static is rendering reception impossible on the higher antenna or on the stack. This means that the ability to select individual antennas in a stack can sometimes be extremely important for reasons unrelated to elevation angle.

STACKS AND AZIMUTHAL DIVERSITY

Azimuthal diversity is a term coined to describe the situation where one of the antennas in a stack is purposely pointed in a direction different from the main direction of the stack. During most of the time in a DX contest from the East Coast, the lower antennas in a stack are pointed into Europe, while the top antenna is often rotated toward the Caribbean or Japan. In a stack of three identical Yagis, the first-order effect of pointing one antenna in a different direction is that one-third of the transmitter power is diverted from the main target area. This means that the peak gain is reduced by 1.8 dB, not a very large amount considering that signals are often 10 to 20 dB over S9 anyway when the band is open from New England to Europe.

Figure 14.70 shows the 3D pattern of a pair of 4-element Yagis fed in-phase at 95 and 65 feet, but where the lower antenna has been rotated 180° to fire in the –X direction. The backwards lobe peaks at a higher elevation angle because the antenna doing the radiating in this direction is lower on the tower. The forward lobe peaks at a lower angle because its main radiator is higher.

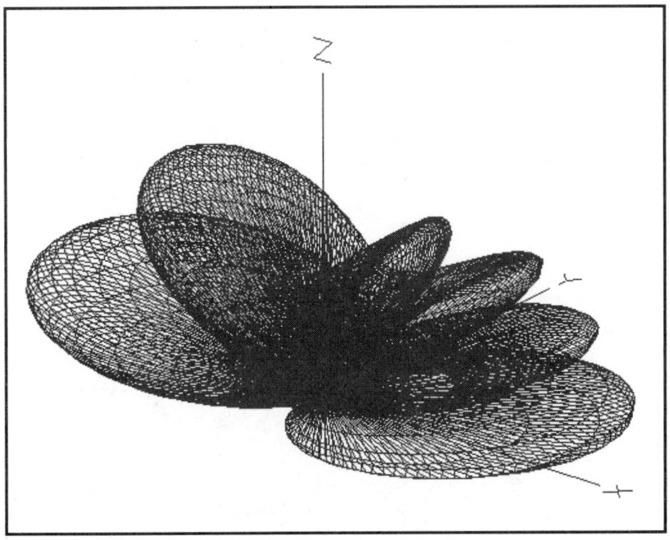

Figure 14.70 — 3D representation of the pattern for two 4-element 15-meter Yagis, with the top antenna at 95 and the bottom at 65 feet, but pointed in the opposite direction.

"BIP/BOP" OPERATION

The contraction "BIP" means "both in-phase," while "BOP" means "both out-of-phase." BIP/BOP refer to stacks containing two Yagis, although the term is commonly used for stacks containing more than two Yagis. In theory, feeding a stack with the antennas out-of-phase will shift the elevation response higher than in-phase feeding.

Figure 14.71 shows a rectangular plot comparing BIP/BOP operation of two 3-element 15 meter Yagis at heights of 2 λ and 1 λ (93 and 46 feet) over flat ground. The BOP pattern is the higher-angle lobe and the two lobes cross over about 14°. The maximum amplitude of the BOP stack's gain is about ½ dB less than the BIP pair. For reference, the pattern of a single 46-foot high Yagi is overlaid on the pattern for the stacks.

The most common method for feeding one Yagi 180° out-of-phase is to include an extra electrical half wavelength of feed line coax going to one of the antennas. This method obviously works on a single frequency band and thus is not applicable to stacks of multiband Yagis, such as tribanders. For such multiband stacks, feeding only the lower antenna(s) — by switching out higher antenna(s) in the stack — is a practical method for achieving better coverage at medium or high elevation angles.

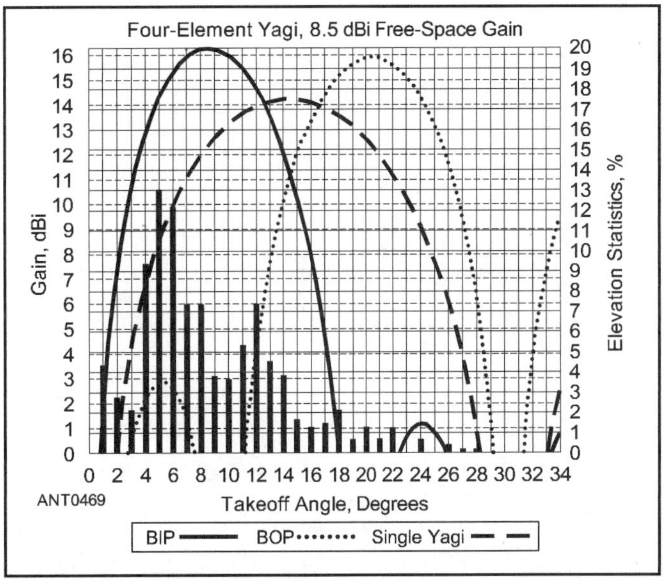

Figure 14.71 — *HFTA* screen shot of "BIP/BOP" operation of two 4-element 15-meter Yagis at 93 and 46 feet above flat ground. The elevation response in BOP (both out-of-phase, dotted line) operation is shifted higher, peaking at about 21°, compared to the BIP (both in-phase, solid line) operation where the peak is at 8°. The dashed line is response of single Yagi at 46 feet.

14.4.10 Stack Antenna Switching Systems

This section applies to schemes to switch between two, three, or four antennas. Quarter-wave sections of 50 or 75 Ω transmission lines can be used to create specific phase relationships between antennas and match the resulting impedances to 50 Ω feed line from the transmitter. These are single-band systems because of the requirement for specific electrical lengths of feed line. Another method describes the use of a transformer to perform the impedance match for different combinations of antennas. This system does not depend on tuned sections of line and can be used on multiple bands with the caveat that the antennas should have very similar impedances on all frequencies to be supported.

The switching systems based on tuned lengths of transmission line (λ/4 or λ/2) use the current-forcing and impedance transformation effects of a λ/4 length of transmission line (see the **Transmission Lines** chapter). These systems rely on four basic rules:

1) The antennas are physically identical.

2) The feed line is identical for all tuned sections — use cable off the same spool or tune each length individually using an antenna or vector impedance analyzer.

3) Use choke baluns to block common-mode RF current on the outside of the feed line shield. (See the **Transmission Line System Techniques** chapter.)

4) Preserve the electrical length of the feed system in any configuration to eliminate the need for amplifier retuning after switching.

The final rule is perhaps the most important for a well-behaved switching system. If λ/4 lines are used for impedance matching and current forcing, the resulting changes to the reactance presented at the antenna feed point will change the loading of the transmitter as the system's configuration is switched. It is also possible to preserve overall path electrical length by using additional λ/2 lines.

Builders of stack switching systems recommend that switching relays should be mounted in a relay box mounted on the tower at the array center. This reduces feed line cost, and the shorter feed lines make it easier to achieve the desired control of antenna phasing. This requires relays that can operate outside, and a control cable is needed.

The switch boxes should be metal, die-cast aluminum is recommended, with the coax connectors and relays mounted on the cover. Relays should be hermetically sealed with very low inductance in the switched path and have contacts rated at 20 A or higher. The low inductance is necessary to preserve phase on the 28 MHz and higher-frequency bands. For internal connections, using ¾-inch flat-weave tinned braid ¼-inch above the surface of the box cover creates an approximately 50 Ω transmission line to reduce phase changes due to inductive wiring. This technique is usable up to 2 meters.

While these methods were originally designed to switch between Yagi antennas in a vertical stack, a similar approach may be used with any type of antenna system in which either/both configurations are used. For example, power could be divided between wire beams or dipoles. The mutual impedances between the antennas should be negligible to ensure correct power splitting and impedance matching. Mixing different types of antennas may lead to unexpected variations in antenna pattern and impedance at different frequencies.

There have been numerous articles on stacked array switching in the *National Contest Journal* in which several useful variations are presented beyond the systems presented here.

14.5 References and Bibliography

Source material and more extended discussions of the topics covered in this chapter can be found in the references given below and at the end of Chapter 4, Propagation of Radio Waves, and Chapter 23, Transmission Lines and Connectors.

Breakall, J., WA3FET, et al, "The Modeling and Measurement of HF Antenna Skywave Radiation Patterns in Irregular Terrain," IEEE Transactions on Antennas and Propagation, July 1994.

Brown, J., K9YC, "Planning Antenna Systems For the Little Gun Station," **k9yc.com/publish.htm**.

"FM 11-65: High Frequency Radio Communications," US Army.

Hopengarten, F., K1VR, Straw, D., N6BV, "Stacking Tribanders: A Super Station – Sorta," *QST* Feb. 1994.

Pistorius, C., McNamara, D., Malherbe, J., *Introduction to the Uniform Geometrical Theory of Diffraction* (Artech House, 1990).

Pruett, D., K8CC, "Switching Stacked Antennas," *National Contest Journal*, Jan. 1988.

Straw, D., N6BV, and Hall, J., K1TD, "Antenna Height and Communications Effectiveness," **arrl.org/files/file/antplnr.pdf**.

Straw, D, N6BV, "What's the Deal About NVIS?" *QST*, Dec. 2005.

Straw, D., N6BV, "The Effect of Local Terrain on HF Launch Angles," *QEX*, July 1995.

Witvliet, B., PA5BW, et al, "Near Vertical Incidence Skywave Propagation: Elevation Angles and Optimum Antenna Height for Horizontal Dipole Antennas," *IEEE Antennas and Propagation Magazine*, Feb. 2014.

White, J., VA7JW, "A Beginner's Guide to HFTA High Frequency Terrain Assessment," **orcadxcc.org/content/VA7JW_HFTA_Manual.pdf**.

Wolff, K., K1EA, "All About Switching," Yankee Clipper Contest Club, *Scuttlebutt*, Dec. 1989.

Notes

Notes

Notes

Notes

Notes

Notes

Notes

Notes

Notes

Notes

Notes

The ARRL
Antenna Book

FOR RADIO COMMUNICATIONS

ARRL The National Association for Amateur Radio®

Twenty-Fifth Edition

- **Volume 1: Antenna Fundamentals — Chapters 1 – 5**
- **Volume 2: Basic Antenna Types and MF and HF Antennas — Chapters 6 – 14**
- **Volume 3: VHF, UHF, and Microwave Antennas and Specialty Applications — Chapters 15 – 24**
- **Volume 4: Building and Maintaining Antenna Systems and Index — Chapters 25 – 29**

Editor
Ward Silver, NØAX

Production Editor
Mark Derks, KC1RVQ

Production
Chris Dunham
Jodi Morin, KA1JPA
David F. Pingree, N1NAS

Editorial Assistant
Maty Weinberg, KB1EIB

Contributors to the Twenty-Fifth Edition
Alan Applegate, KØBG
Jim Breakall, WA3FET
Jim Brown, K9YC
Wyatt Dirks, ACØRA
Frank Donovan, W3LPL
James Duffey, KK6MC
Peter East
Doug Grant, K1DG
Rus Healy, K2UA
Matt Heere, N3NWV
Wolfgang Herrmann
Chuck Higgins
Martin Huml, OK1FUA
Dale Hunt, WB6BYU
Justin Johnson, GØKSC
Al Katz, K2UYH
Hal Kennedy, N4GG
Pete Kobak, KØBAK
Steve Kostro, N2CEI
Sean Kutzko, KX9X
Greg Lapin, N9GL
Dan Layne, ADØCY
Roy Lewellan, W7EL
Fred Maas, KT5X
Richard Marsden
Eduard Mol
Dave Mueller, N2NL
Mario Natali, IØNAA
Ron Ochu, KOØZ
Dave Olean, K1WHS
Steve Olney, VK2XV
Charles Osborne, K4CSO
Jonathan Pettingale
Alex Pettit, KK4VB
John Portune, W6NBC
Bruce Randall, NT4RT
Whitman Reeve
Rick Rosen, K1DS
Rich Russell, ACØUB
Charles Scharlau, NZØI
Tom Schiller, N6BT
Rudy Severns, N6LF
Lowell Silver, WW7SAR
Andrea Slack, K2EZ
Jim Smith, KKØU
John Stanley, K4ERO
Steve Stearns, K6OIK
Patrick Stoddard, WD9EWK
Chip Sufitchi, N2YO
Dave Typinkski, AJ4CO
Ray Uberecken, AAØL
Paul Wade, W1GHZ
Bill Wortman, N6MW
Ken Wyatt, KX7KW

Printed in the USA

ISBN: 978-1-62595-175-5 Hardcover
978-1-62595-176-2 Softcover
978-1-62595-183-0 Four-Volume Set

Kindle eBook Editions
ISBN: 978-1-62595-177-9 — Volume 1
ISBN: 978-1-62595-179-3 — Volume 2
ISBN: 978-1-62595-180-9 — Volume 3
ISBN: 978-1-62595-182-3 — Volume 4

Twenty-Fifth Edition, Second Printing

We strive to produce books without errors. Sometimes mistakes do occur, however. When we become aware of problems in our books (other than obvious typographical errors), we post corrections on the ARRL website. If you think you have found an error, please check **www.arrl.org/arrl-antenna-book-reference** for corrections. If you don't find a correction there, please let us know by sending an email to **pubsfdbk@arrl.org**.

Table of Contents

A detailed Table of Contents is included at the beginning of each chapter.
See the **Project Index** for a complete list of projects.

VOLUME 2

Basic Antenna Types and MF and HF Antennas

VOLUME 3

VHF, UHF, and Microwave Antennas and Specialty Applications

VOLUME 4

Building and Maintaining Antenna Systems and Index

TABLE OF CONTENTS

Chapter 25 — Online Material

Articles

Chapter 25

Antenna Materials and Construction

This chapter contains information on materials and techniques amateurs use to construct antennas. Included is a discussion of useful material types that are readily available at reasonable cost, and tips on working with and using these materials.

The National Electric Code (NEC) from the National Fire Protection Association contains a section on amateur stations in which several recommendations are made concerning minimum size of antenna wire and the manner of bringing the transmission line into the station. The code in itself does not have the force of law, but it is frequently made a part of local building regulations, which are enforceable. The provisions of the code may also be written into, or referred to, in fire and liability insurance documents. See the chapter **Building Antenna Systems and Towers** for more information on applying the NEC to your station's antenna system.

Although antennas are relatively simple structures, they can constitute a potential hazard unless properly constructed. Antennas and supporting ropes or wires should *never* be run under or over public utility (telephone or power) lines. Stay well clear of utility lines when erecting antennas and give yourself plenty of safety margins. Amateurs have lost their lives by failing to observe these precautions.

Basically, any conductive material can be used as the radiating element of an antenna. Almost any insulating material can be used as an antenna insulator. An antenna system must also include some means to support those conductors and maintain their relative positions — the boom for a Yagi antenna, for example. The materials used for antenna construction are limited mainly by physical considerations (required strength and resistance to outdoor exposure) and by the availability of materials. Don't be afraid to experiment with radiating materials and insulators.

The two types of material most often used for antenna conductors are wire and tubing. Wire antennas are generally simple and therefore easier to construct, although arrays of multiple wire elements can become rather complex. When tubing is required, aluminum tubing is used most often because of its light weight, reasonable cost and strength. Aluminum tubing is discussed in a subsequent section of this chapter.

An excellent resource for selecting and using fasteners of all type is *Carroll Smith's Nuts, Bolts, Fasteners and Plumbing Handbook* which is widely available and applies to many facets of antenna and equipment construction.

25.1 Wire for Antenna Systems

25.1.1 Wire Types

Solid copper wire is used for most wire antennas although the use of stranded wire is common. Solid wire is less flexible than stranded wire, but it is available "hard-drawn," which offers good tensile strength and negligible stretch. Special stranded wire with a larger-than-usual number of fine strands (such as Flex-Weave) is available for building antennas. It withstands vibration and bending in the wind better than common stranded wire and better than solid wire. Galvanized steel and aluminum wire are generally not used for antennas because of higher electrical resistance than copper. Galvanized wire also has a strong tendency to rust and making good electrical connections to aluminum wire is difficult — it cannot be soldered directly without special solder fluxes.

Solid wire is also available with and without enamel coating. Enamel coating resists oxidation and corrosion, but bare wire is far more common. Solid wire is also available with a variety of different insulating coatings, including plastics, rubber, and PVC. Unless specifically rated for outdoor use however, wire insulation, including enamel, tends to break down when exposed to the UV in sunlight. Insulation also lowers the velocity factor of wire by a few percent (see the **Transmission Lines** chapter) making it electrically longer than its physical length — this will lower the resonant frequency of an antenna compared to one made of bare wire of equivalent diameter. In addition, insulation increases wind loading without increasing strength. If enameled or insulated wire is used, care should be taken to not nick the wire when removing the coating for an electrical connection. Wire will break at a nick when flexed repeatedly, such as by wind.

"Soft-drawn" or annealed copper wire is easy to handle and obtain. Common THHN-insulated "house wire" is soft-drawn. Unfortunately, soft-drawn wire stretches considerably under load. Soft-drawn wire should only be used in applications where there will be little or no tension, or where some change in length can be tolerated. For example, the length of a horizontal antenna fed at the center with open-wire line is not critical, although a change in length may require some readjustment of an impedance matching unit. Similarly, if the wire stretches significantly, it can be re-trimmed to the desired length. Repeated cycles of stretching followed by trimming and re-tensioning will result in loss of strength and possibly in mechanical failure.

Copper can be applied to a steel core for added tensile strength in two ways. One is through plating, which results in a relatively thin layer of copper on the outside of the steel core. The other is through a process called "cladding" and the resulting copper outer layer is much thicker than plated. Copper-clad steel is manufactured to ASTM-B-452 standards with 40% conductivity in two tempers: soft and hard drawn. The thickness of the copper layer is most important when the wire is used for RF purposes so that the RF energy does not penetrate through the copper layer and into the high loss steel core. Plating is rarely sufficiently thick in relation to the required skin depth for the frequency of operation to maintain the highest efficiency. Another type of wire with a steel core is aluminum clad steel, often called "Alum-O-Weld." This wire is manufactured to ASTM B415 standards with a conductivity of 14% to 40% IACS (relative to annealed copper). It is primarily marketed as stranded for overhead power lines and guy cables. Occasionally it can be found as a single wire (i.e. #12) and is useful for antennas, although the steel core (hard drawn) can cause it to be somewhat brittle and difficult to safely handle. Copper clad steel shares this characteristic.

"Hard-drawn" copper wire and CCS (copper-clad steel, usually sold as the trademarked product Copperweld) wire are more difficult to handle because of their mechanical stiffness and, in the case of CCS, the tendency to have "memory" when unrolled. These types of wire are ideal for applications where high strength for a given weight is required and/or significant stretch cannot be tolerated. Care should be exercised to make sure kinks do not develop in hard-drawn and CCS wire — the wire will have a far greater tendency to break at a kink. The "memory" or tendency of CCS wire to coil up can be reduced by suspending it a few feet above ground for a few days before final use. The wire should not be recoiled before it is installed. The electrical quality of CCS wire varies considerably. A conductivity class of 30% or higher is desirable, meaning the wire has 30% of the conductivity of copper wire of the same diameter but for RF applications at HF it will have close to 100% conductivity due to skin effect.

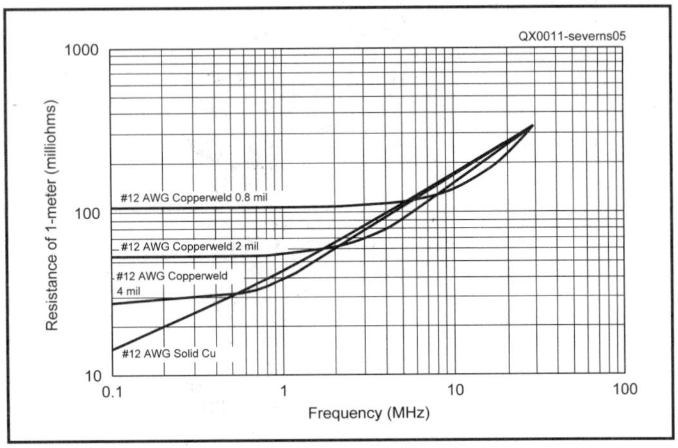

Figure 25.1 — Resistance comparison of 1-meter lengths of #12 AWG solid copper, #12 AWG Copperweld with 4-mil cladding, and an equivalent wire for 19- strand #26 AWG Copperweld with 0.8-mil cladding from 1 to 30 MHz. Derived by N6LF from FEM modeling.

At MF and LF, such as 160 meters, 630 meters, and 2200 meters, CCS wire may have unacceptably high losses due to insufficient thickness of the copper layer. (See the **Transmission Lines** chapter's section on "Choosing and Installing Feed Lines" for the effect on feed line loss.) **Figure 25.1** compares the resistance of copper-clad and solid copper wires at 100 kHz and higher frequencies. As a general guideline, for resistance losses to be equivalent to solid copper, the cladding should be approximately 3 times the skin depth which is:

1.80 MHz (160 meters) —49 μm
475 kHz (630 meters) — 95 μm
136.5 kHz (2200 meters) — 177 μm

Assume the cladding thickness is the typical 10 percent of the overall conductor diameter at 30% conductivity. To have equivalent resistance to solid copper at 1.80 MHz the overall wire diameter must be 490 × 3 = 1470 μm or 0.058 inches and #14 AWG is the smallest wire gauge that is thick enough. Similarly, the smallest wire gauge is #8 AWG at 475 kHz. Less cladding such as for the 20% conductivity CCS will result in increased losses, so at 630 and 2200 meters solid copper conductors are recommended. N6LF also notes that stranded CCS is inferior to solid CCS below 40 meters.

Copper cladding can be damaged by abrasion (typically at insulators) or sharp bends. Plastic insulators of sufficient strength are preferable to ceramic insulators when using CCS; they are soft in comparison and less likely to degrade the copper cladding over time. Induced defects in copper cladding eventually result in mechanical failure due to rusting of the steel core. Breaks in the copper cladding also form high resistance points to RF and will heat considerably when running high power. Heat accelerates oxidation (rusting).

Concerns about the effect of insulation on copper wire for ground radials are addressed by N6LF in his *QEX* article "Insulated Wire and Antennas" (available in the online material), a study of using insulated wire. There are esthetic concerns — the outermost transparent sheath of THHN electrical wire degrades with exposure to UV and is shed by the wire. N6LF also considers mechanical issues — adding insulation increases the weight but also causes ground surface or buried radials to last longer than bare wire. Overall, the effects of the insulation are insignificant with the exception being when used in sparse radial systems. Those effects disappear for more than 12 elevated radials and 16 to 20 radials on or below ground level.

The velocity factor (VF) of bare copper wire is approximately 0.95 compared to the free-space speed of light. Plastic insulation causes additional slowing of a few percent. The exact value varies with the type and thickness of the insulation, but the most common materials (PVC, polyethylene, Teflon) have a VF of 0.95–0.98. Thus, copper wire insulated with one of these materials would have an overall VF from 0.95 × 0.95 = 0.90 to 0.95 × 0.98 = 0.93. If the electrical length of insulated wire is critical to your design, it is recommended to measure the wire's VF or allow for a few percent of adjustment.

Table 25.1
Copper-Wire Table

Wire Size AWG (B&S)	*Dia in Mils[1]*	*Dia in mm*	*Turns per Linear Inch Enamel*	*Feet per Pound Bare*	*Ohms per 1000 ft 25°C[3]*	*Cont.-duty current[2,3] Single Wire in Open Air*
1	289.3	7.348	—	3.947	0.1264	—
2	257.6	6.544	—	4.977	0.1593	—
3	229.4	5.827	—	6.276	0.2009	—
4	204.3	5.189	—	7.914	0.2533	—
5	181.9	4.621	—	9.980	0.3195	—
6	162.0	4.115	—	12.58	0.4028	—
7	144.3	3.665	—	15.87	0.5080	—
8	128.5	3.264	7.6	20.01	0.6405	73
9	114.4	2.906	8.6	25.23	0.8077	—
10	101.9	2.588	9.6	31.82	1.018	55
11	90.7	2.305	10.7	40.12	1.284	—
12	80.8	2.053	12.0	50.59	1.619	41
13	72.0	1.828	13.5	63.80	2.042	—
14	64.1	1.628	15.0	80.44	2.575	32
15	57.1	1.450	16.8	101.4	3.247	—
16	50.8	1.291	18.9	127.9	4.094	22
17	45.3	1.150	21.2	161.3	5.163	—
18	40.3	1.024	23.6	203.4	6.510	16
19	35.9	0.912	26.4	256.5	8.210	—
20	32.0	0.812	29.4	323.4	10.35	11
21	28.5	0.723	33.1	407.8	13.05	—
22	25.3	0.644	37.0	514.2	16.46	—
23	22.6	0.573	41.3	648.4	20.76	—
24	20.1	0.511	46.3	817.7	26.17	—
25	17.9	0.455	51.7	1031	33.00	—
26	15.9	0.405	58.0	1300	41.62	—
27	14.2	0.361	64.9	1639	52.48	—
28	12.6	0.321	72.7	2067	66.17	—
29	11.3	0.286	81.6	2607	83.44	—
30	10.0	0.255	90.5	3287	105.2	—
31	8.9	0.227	101	4145	132.7	—
32	8.0	0.202	113	5227	167.3	—
33	7.1	0.180	127	6591	211.0	—
34	6.3	0.160	143	8310	266.0	—
35	5.6	0.143	158	10480	335	—
36	5.0	0.127	175	13210	423	—
37	4.5	0.113	198	16660	533	—
38	4.0	0.101	224	21010	673	—
39	3.5	0.090	248	26500	848	—
40	3.1	0.080	282	33410	1070	—

[1] A mil is 0.001 inch.
[2] Max wire temp of 212° F and max ambient temp of 135° F.
[3] Ratings are for dc measurements and currents without skin effect.

USING STRAP OR BRAIDED CONDUCTORS

Communications systems expert Frank Donovan, W3LPL, contributed these guidelines for the use of strap (solid metal) or braided conductors. Wide, flat, copper strap is the standard for grounding and bonding when flexibility isn't required — at least 10 mm wide, wider is better. Thin, round conductors are always inferior. Properly designed grounding and bonding systems use wide, flat, copper strap everywhere except for the short lengths where flexibility is required for the last foot or two of connections to moveable equipment.

Braided conductors should not be used for grounding (connections to a facility's grounding electrodes as discussed in the chapter on **Building Antenna Systems and Towers**) and should be as short as reasonably practical. Unnecessarily long pigtails should always be avoided. Terminals should be installed on both ends of the braid conductors, firmly and securely fastened to the equipment and conductors at each end of the braid. (Use a crimp terminal specified for use with braid.)

Tinned, tightly woven, wide (at least 10 mm, preferably much wider), flat copper braid is an excellent bonding conductor provided it is not corroded. Skin effect may force RF to jump across each wire crossover, but there are very many tight connections in parallel.

Braid removed from coaxial cable is an exceptionally poor choice for a bonding conductor because it is round, relatively small diameter, loosely meshed and usually not tinned. It works very well inside coaxial cable because the jacket maintains tight connections in the braid mesh and protects the copper from water and other corrosion agents.

Braid of any kind should never be used outdoors, especially on antennas and towers where lightning currents might be present. The standard is wide, flat, copper strap or #2 AWG solid copper wire where lightning currents might be present. Seven or nineteen strand copper wire can be used where flexibility is required.

25.1.2 Wire Size and Tension

Many factors influence the choice of wire type and size (gage or gauge). Important considerations include the length of the unsupported span, the amount of sag that can be tolerated,

Table 25.2
Stressed Antenna Wire

American Wire Gauge	***Recommended Tension[1] (pounds)*** *Copper-clad steel[2]*	*Hard-drawn copper*	***Weight (pounds per 1000 feet)*** *Copper-clad steel[2]*	*Hard-drawn copper*
4	495	214	115.8	126.0
6	310	130	72.9	79.5
8	195	84	45.5	50.0
10	120	52	28.8	31.4
12	75	32	18.1	19.8
14	50	20	11.4	12.4
16	31	13	7.1	7.8
18	19	8	4.5	4.9
20	12	5	2.8	3.1

[1]Approximately one-tenth the guaranteed breaking strength. Might be increased 50% if end supports are firm and there is no danger of ice loading.
[2] Copperweld, 40% copper

the stability of the supports under wind pressure, the amount of wind and ice loading anticipated and whether a transmission line will be suspended from the span. Some sag is desirable. Removing most or all sag requires additional unnecessary tension and increases the likelihood of failure. **Table 25.1** shows the wire diameter, current-carrying capacity and resistance of various sizes of copper wire. **Table 25.2** shows the recommended maximum working tension of hard-drawn and CCS wire of various sizes. The recommended working tension is approximately 10% of the minimum guaranteed breaking strength of the wire. Together with a calculation of span sag, these two tables can be used to select the appropriate wire size for an antenna.

The National Electrical Code (NEC) specifies minimum conductor sizes for different span-length wire antennas. For hard-drawn copper wire, the NEC specifies #14 AWG wire for open (unsupported) spans less than 150 feet, and #10 AWG for longer spans. CCS, bronze or other high-strength conductors may be #14 AWG for spans less than 150 feet and #12 AWG for longer runs. Lead-in conductors (for open-wire transmission line) should be at least as large as those specified for antennas.

The RF resistance of copper wire increases as the size of the wire decreases. In most common wire antenna designs; however, the antenna's radiation resistance will be much higher than the wire's RF resistance and the efficiency of the antenna will be adequate. Wire sizes as small as #30 AWG, or even smaller, have been used successfully in the construction of "invisible" antennas in areas where more conventional antennas cannot be erected. In most cases, the selection of wire for an antenna will be based primarily on the mechanical properties of the wire, since the suspension of wire from elevated supports places the wire in tension.

CALCULATING WIRE SAG

The following section is based on a *QST* "Technical Correspondence" item by Darrell Emerson, AA7FV, in the March 2014 issue of *QST.* Given the horizontal distance between two antenna masts supporting the wire, the weight per foot of the wire and the tension in the wire, it is possible to predict the wire sag at the lowest point of the wire, halfway between the supports (**Figure 25.2**). Previous editions of the *Antenna Book* used a nomograph based on the original article "Predicting Sag in Long Wire Antennas" in the January 1966 issue of *QST*, by John J. Elengo, Jr, W1DQ.

The equation describing the catenary, the curve of a rope or chain held horizontally between two supports, was first derived in 1691 by Johann Bernoulli and others. The equation is now found in many engineering and mathematical textbooks.

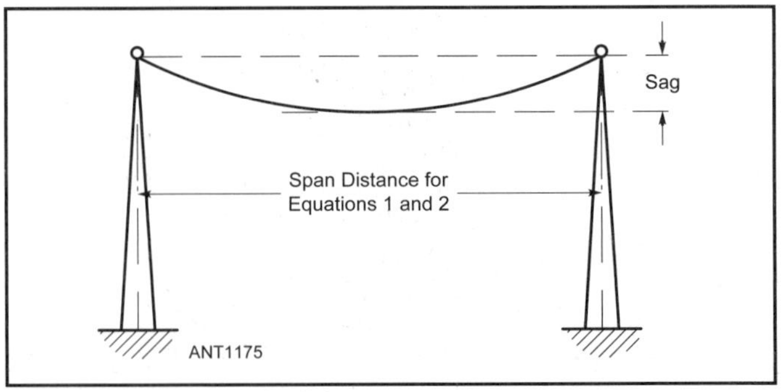

Figure 25.2 — This drawing applies to Equations 1 and 2 for calculating sag of wire antennas.

Wiring Techniques

Working with antenna wire, cables, and terminals requires heavier tools and different techniques from ordinary electronic wiring. Manufacturers of tools and materials often supply "how to" tutorials on their websites and others can be found on YouTube and other internet video sites. A comprehensive course by CED Engineering on wiring techniques can be found online at **cedengineering.com/upload/Wiring%20Techniques.pdf.**

One form of the equation is:

$$\text{Wire sag} = \frac{T}{w}\left[\cosh\left(w\frac{S}{2T}\right) - 1\right]$$

where

cosh = the hyperbolic cosine

T = the tension in the wire in pounds

w = the weight of the wire in pounds per foot

S = the span of the wire, here defined as the *total* horizontal distance in feet between the two supports of the wire.

(There has been some confusion in previous publications about whether S represents half the distance or the total distance between supports. Here, it represents the total distance.) There is a much simpler approximation that is valid in all cases likely to be of interest to the radio amateur.

$$\text{Wire sag} = \frac{wS^2}{8T}$$

This equation is also given in Edmund Laport's *Radio Antenna Engineering*, in his chapter on "Wire Stringing." (For sag between two supports of unequal height, see the "Miscellaneous Data" chapter of *Reference Data for Engineers* or **electricalengineeringinfo.com/2015/01/what-is-sag-tension-in-electrical-transmission-lines.html**)

An example calculation:

w = 0.011 (pounds per foot, 11 pounds per 1000 ft)

S = 420 feet (span, being the total distance between supports.

T = 50 pounds (wire tension).

Substituting w, S and T into the full equation, the computed result for sag is 4.860 feet. Using the much more convenient approximation, the result is 4.861 feet. The simpler formula is certainly adequate.

If the calculated sag is greater than allowable, it may be reduced by any one or a combination of the following:

1) Providing additional supports, thereby decreasing the span

2) Increasing the tension in the wire

3) Decreasing the size (gage or gauge) of the wire

These calculations do not take into account the weight of a feed line supported by the antenna wire.

25.1.3 Wire Splicing and Connections

Wire antennas should preferably be made with unbroken lengths of wire. In instances where this is not feasible, wire sections should be spliced as shown in **Figure 25.3**. Any insulation should be removed for about 6 inches from the end of each section (take care not to nick the wire). Enamel may be removed by scraping with a knife or rubbing with sandpaper until the copper underneath is bright. The turns of wire should be brought up tight around the standing part of the wire by twisting with broad-nose pliers.

The crevices formed by the wire should be completely filled by using solder that does not contain an acid-core flux.

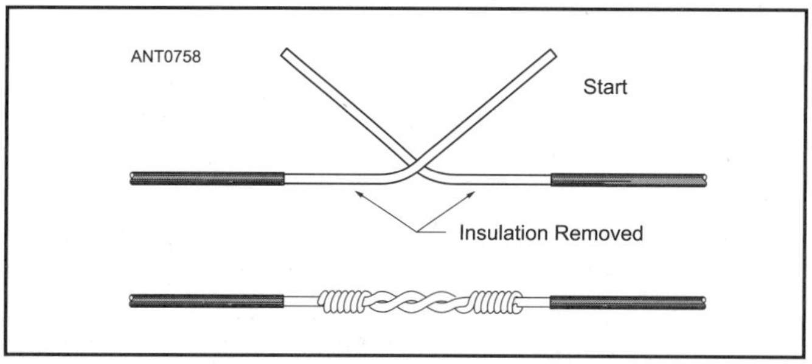

Figure 25.3 — Correct method of splicing antenna wire. Solder should be flowed into the wraps after the connection is completed. After cooling, the joint should be sprayed with acrylic to prevent oxidation and corrosion.

A soldering iron or gun may not be sufficient for heavy wire or in cold temperatures; use a propane or butane torch instead. The joint should be heated sufficiently so the solder flows freely into the joint when the source of heat is removed momentarily. After the joint has cooled completely, it should be wiped clean with a cloth and then sprayed generously with acrylic to prevent oxidation.

25.1.4 Radial Systems

See the chapter **Effects of Ground** for complete information on the requirements for ground radial systems, including references showing how to get the best results from a specific amount of wire.

Bare copper wire is the least expensive for radials with #18 or #20 AWG, the smallest size likely to last due to mechanical abuse. Smaller wire will function electrically and is a good choice for temporary or portable installations. Good prices can be obtained by buying wire in bulk directly from a distributor. Copper will also withstand corrosive soils much better than aluminum.

When attaching radials to the antenna ground system, try to avoid direct contact with the ground due to corrosion from minerals in the ground. Tin-lead solder should not be used in connections in contact with the ground. If soldering is required, such as to a ground ring or plate, use silver-bearing plumbing solder or brazing rods. High-temperature MAPP gas torches are available at plumbing supply stores.

Radials need not be in direct contact with soil except for appearances. Shallow burial slits in a lawn can be made with an edging tool or a wire burial plow. See the previous cautions about avoiding direct contact between soil and solder joints.

Radials can also be placed on closely cropped grass and the grass will grow over and around them in a few weeks, allowing them to lie directly on top of the soil. To hold radials firmly to the ground, landscaping staples can be used, or iron rebar tie wire can be cut into 6-to-10-inch pieces and bent double.

25.2 Antenna Insulators

Insulation is particularly important at the outer end or ends of wire antennas, since these points are always at a comparatively high RF potential. If an antenna is to be installed indoors, it should not touch any building materials. Electric fence insulators are inexpensive and work well for this purpose. Much greater care should be given to the selection of proper insulators when the antenna is located outside where it is exposed to wet weather. Antenna insulators should be made of material that will not absorb moisture. The best insulators for antenna use are made of glass or glazed porcelain although plastic insulators are widely available and suitable for most antennas.

The length of an insulator relative to its surface area is indicative of its comparative voltage stand-off and RF leakage abilities. A long thin insulator will have less leakage than a short thick insulator. Some antenna insulators are deeply ribbed to increase the surface leakage path without increasing the physical length of the insulator. Shorter insulators can be used at low-potential points, such as at the center of a dipole. If such an antenna is to be fed with open-wire line and used on several bands however, the center insulator should be the same as those used at the ends, because high RF potential may exist across the center insulator on some bands.

INSULATOR STRESS

As with the antenna wire, the insulator must have sufficient physical strength to carry the mechanical load of the antenna without danger of breaking. Elastic line ("bungee cord" or "shock cord") or woven fishing line can provide both the end-insulator and support functions at antenna ends, subject to their ability to carry mechanical load. They are often used in antennas of the "invisible" type mentioned in the **Stealth and Limited Space Antennas** and **Portable Antennas** chapters. Abrasion between a woven line and a wire loop will cut through the line fairly quickly unless a fishing swivel or similar metal attachment point is used.

Use of high power approaching and up to the US legal limit of 1500 W may cause sufficient leakage current to melt woven or monofilament line directly connected to a wire loop at the end of a dipole or similar antenna. Bungee cord and similar black products can contain conductive materials and are not suitable in applications where high current and/or high voltage are present in the antenna system. Depending on the antenna type (i.e., dipole, magnetic loop), the voltage and current can vary greatly for the same transmitter output power. A suitable antenna insulator as explained below must be used for the intended antenna system.

For low-power operation with short antennas not subject to appreciable stress, almost any small plastic, glass, or glazed-porcelain insulator will do. Homemade insulators of plastic rod or sheet are usually satisfactory. Many plastics rated for outdoor use make good insulators — this includes Lucite (polycarbonate), Delrin, plexiglass, and even the high-density polyethylene (HDPE) used in cutting boards. More care is required in the selection of insulators for longer spans and higher transmitter power.

For a given material, the breaking tension of an insulator will be proportional to its cross-sectional area. It should be remembered that the wire hole at the end of the insulator decreases the effective cross-sectional area. For this reason, insulators designed to carry heavy strains are fitted with heavy metal end caps, the eyes being formed in the metal cap, rather than in the insulating material itself.

The following stress ratings of ceramic antenna insulators are typical:

- ⅝-inch square by 4 inches long — 400 pounds
- 1 inch diameter by 7 or 12 inches long — 800 pounds
- 1½ inches diameter by 8, 12 or 20 inches long, with special metal end caps — 5000 pounds

These are rated breaking tensions. The actual working tensions should be limited to not more than 25% of the breaking rating. Plastic insulators have significantly lower tension ratings.

The antenna wire should be attached to the insulators as shown in **Figure 25.4A**. Care should be taken to avoid sharp angular bends in the wire when it is looped through the insulator eye. The loop should be generous enough in size that it will not bind the end of the insulator tightly. If the length of the antenna is critical, the length should be measured to the outward end of the loop, where it passes through the eye of the insulator. (See the note below about the loop area affecting the antenna's electrical length.)

Soldering should be done as described earlier for the wire splice. If CCS wire is used, care should be taken to ensure insulator holes and edges are smooth. Any roughness at contact points between the wire and the insulator will cause the

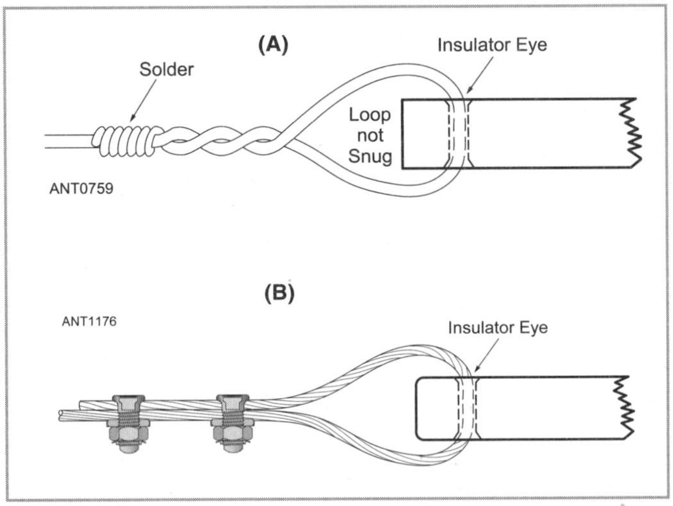

Figure 25.4 — The insulator at A uses a conventional twisted attachment which is soldered with a non-acid flux solder and can be sprayed with outdoor acrylic for protection. At B, split bolts may be used to provide a secure connection between heavy wires, such as at an insulator or splice. Using a pair of split bolts provides a more mechanically secure connection. For both types of connection, do not make the wire loop too snug to allow the wire loop to move in the insulator and flex.

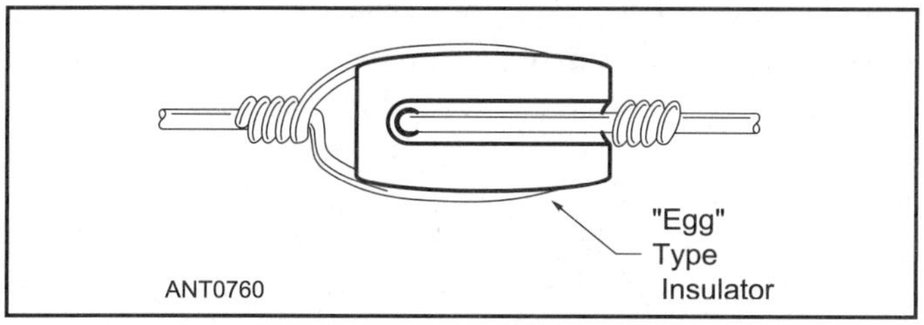

Figure 25.5 — Conventional manner of fastening wire to a strain insulator. This method decreases the leakage path and increases capacitance, as discussed in the text.

copper to be abraded away over time, exposing the wire's steel core and eventually leading to mechanical failure from rust. Assuming they are of sufficient size to handle the mechanical load, plastic insulators are a good choice for use with CCS wire.

An alternative to soldering antenna wires is the use of "split bolts" to clamp wires together as in Figure 25.4B. This allows wire connections to be adjusted, such as when trimming a dipole length. If the proper-size split bolt is used, the connection is secure even for large conductors. Using a pair of split bolts provides a more secure attachment and less stress on the wire than a single bolt. Split bolts are available in solid copper, tinned copper, and aluminum. They can be used for both similar and dissimilar metals and some are designated for direct burial applications (e.g., attaching to rebar or buried ground wires).

Note that the large area of the loop through the insulator adds capacitance to the antenna. The larger the insulator loop, the more capacitance is created, and the greater its effect in lowering the resonant frequency of the antenna. This effect increases with operating frequency. When building a wire antenna, attach the insulators temporarily (without soldering) and adjust the resonant frequency of the antenna before soldering the insulator loop.

STRAIN INSULATORS

Strain or "egg" insulators have their holes at right angles since they are designed to be connected as shown in **Figure 25.5**. This arrangement places the insulating material in compression rather than tension. An insulator connected this way can withstand very high mechanical load.

The principal attribute of strain insulators is that the wire will not fall or fail to carry load if the insulator breaks, since the two loops are interlocked. Insulator failure may go unnoticed however — strain insulators should be visually checked periodically. Because the wires are wrapped around each other, the leakage path is shorter than it would be otherwise and both leakage and capacitive end effects are higher compared to insulators where the wires are not interlinked. For this reason, strain insulators are typically confined to applications such as breaking up resonances in guy wires, where there is high mechanical load and where RF insulation is of minor importance.

Strain insulators are suitable for use at low-potential points on an antenna, such as at the center of a dipole. They may also be used at the ends of antennas used for low power operation.

BOOM-TO-ELEMENT INSULATORS

Black insulators often incorporate carbon or other materials to provide UV resistance. These materials can increase the insulator loss and electrically shorten the section of element next to it, particularly at VHF and UHF. If such insulators are used, every element on the antenna may need to be lengthened a small amount to achieve best performance as modeled. Polypropylene, polyamide (without carbon), and UV-resistant Plexiglass (used for outside signage) are good materials to use. Generally, if the insulator is not black, the impact on the antenna performance is likely to be small if any.

FEED POINT INSULATORS

Often referred to as "center insulators," the insulators used at the feed point of a wire antenna often have special features that help attach and support feed lines. A "dog bone" style insulator as in **Figure 25.6A** is the most common. To attach a coaxial feed line using this style of insulator, the cable's shield and center conductor are separated into "pigtails" that are soldered to the wire at each eye. The cable can be supported by looping it over the insulator and securing it with tape or a

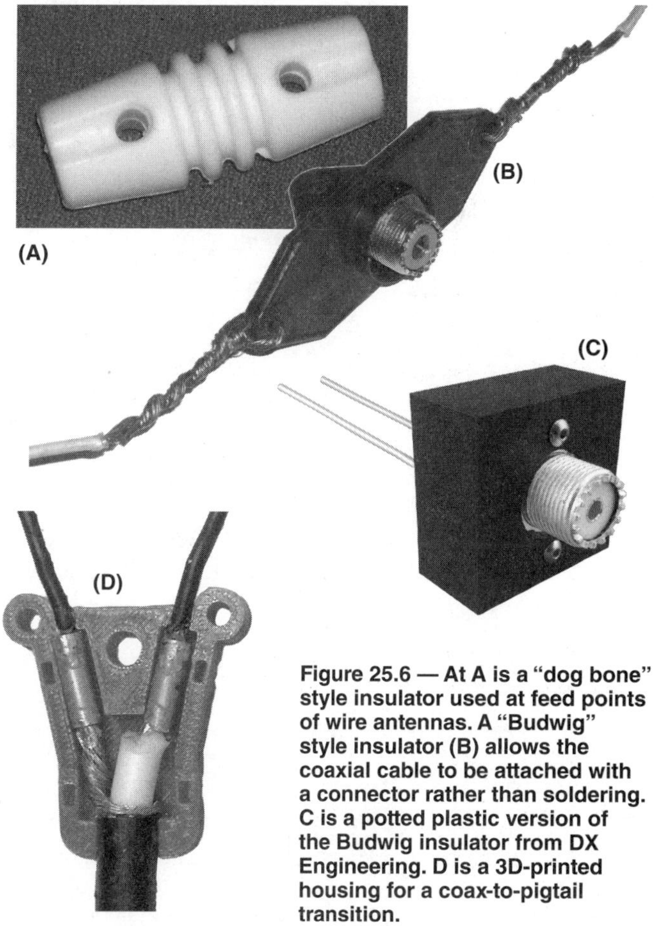

Figure 25.6 — At A is a "dog bone" style insulator used at feed points of wire antennas. A "Budwig" style insulator (B) allows the coaxial cable to be attached with a connector rather than soldering. C is a potted plastic version of the Budwig insulator from DX Engineering. D is a 3D-printed housing for a coax-to-pigtail transition.

clamp. Note that the length of the separated shield and center conductor count as part of the antenna length — that may be significant at higher frequencies. The cable must be carefully waterproofed with a coating such as silicone sealant or Liquid Electrical Tape to prevent water from being wicked into the cable by the exposed shield.

To avoid having to solder or clamp coax pigtails to the antenna, the "Budwig" style of insulator in Figure 25.6B includes an SO-239 connector so that the coaxial cable can be attached with a PL-259 connector instead. The PL-259 and exposed portion of the SO-239 connectors in this case should be waterproofed. This type of center insulator can be made from a PVC pipe cap or other plumbing fittings as shown later in this chapter. The DX Engineering DXE-FPC-SO239 is an updated style of feed point insulator seen in Figure 25.6C. W6NBC has designed a 3D-printed connection block shown in Figure 25.6D for coax terminated in pigtail leads without a connector. (See the online material for details.) If one of these insulators is used, remember that the connection between the connector and the antenna element contributes to the antenna length. This may be significant at VHF and higher frequencies.

Whenever coaxial cable is hanging vertically from a connection, the coax needs to be secured so that the weight of the coax is not pulling on the connector. Wind is another consideration and can add force to the weight on the connection(s) from being blown around by the wind. Securing the coax applies to all connectors. Various types of cable grips and clamps are available — use one that will not cut into or pinch the cable over time.

Figure 25.7 shows a feed point insulator intended for use with parallel-wire feed line. The dog bone style of insulator may be used but cannot support the feed line in the same way as for coaxial cable. Parallel-wire line cannot be looped back on itself with the conductors close together. If left unsupported, the conductors of the feed line continually flex and bend in the wind which causes them to break. The tee-style of insulator in the figure captures the parallel-wire feed line and provides mechanical support, greatly reducing breakage.

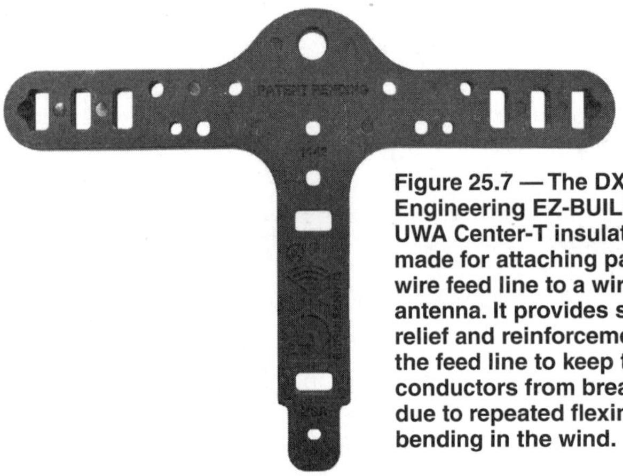

Figure 25.7 — The DX Engineering EZ-BUILD UWA Center-T insulator is made for attaching parallel-wire feed line to a wire antenna. It provides strain relief and reinforcement to the feed line to keep the conductors from breaking due to repeated flexing and bending in the wind.

INSULATORS FOR PARALLEL-WIRE ANTENNAS

Figure 25.8A shows the sketch of an insulator designed to be used at the ends of a folded dipole or a multiple dipole made of parallel conductor line. It should be made approximately as shown, out of insulating material about ¼ inch thick. The advantage of this arrangement is that the strain of the antenna is shared by the conductors and the plastic webbing of the line, which adds considerable strength. After soldering, the screw should be sprayed with acrylic.

Figure 25.8B shows a similar arrangement for suspending one dipole from another in a stagger-tuned dipole system. If better insulation is desired, these insulators can be wired to a conventional insulator.

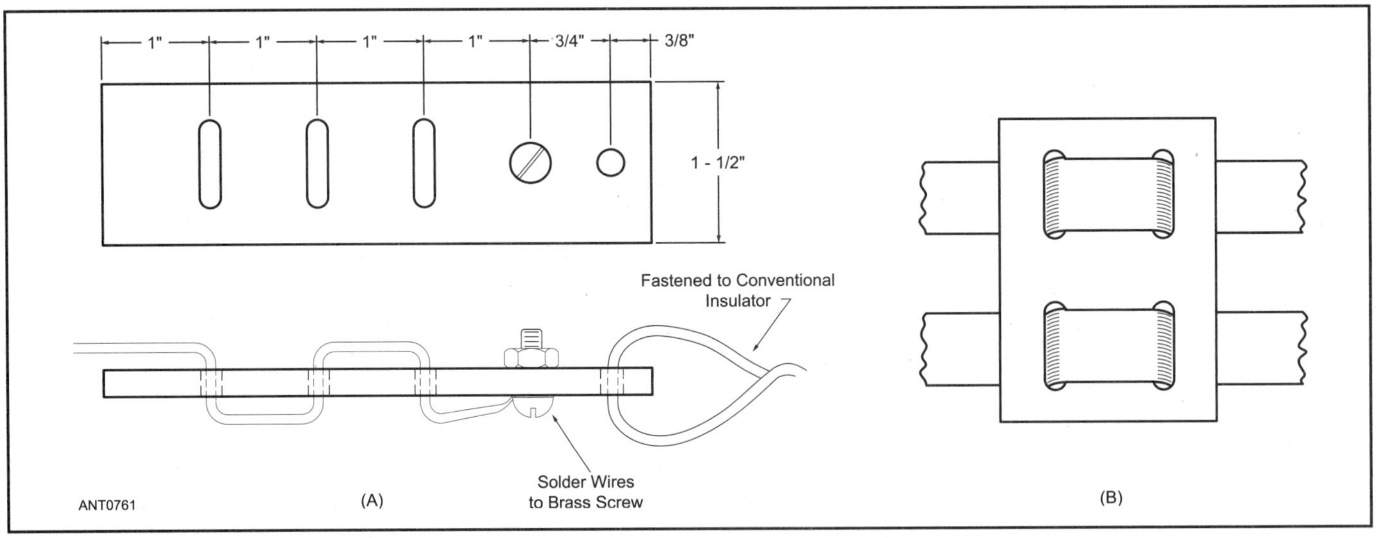

Figure 25.8 — At A, an insulator for the ends of folded dipoles, or multiple dipoles made of parallel-wire line. At B, a method of suspending one ribbon dipole from another in a multiband dipole system.

25.3 Antennas of Aluminum Tubing

Aluminum is a non-toxic, malleable, ductile metal with a density approximately 35% that of iron and 30% that of copper. Aluminum can be polished to a high brightness, and it will retain this polish in dry air. In the presence of oxygen, aluminum forms an oxide coating (Al_2O_3) that protects the metal from further corrosion. Direct contact between aluminum and certain metals (particularly ferrous metals such as iron or steel) in an outdoor environment can bring about galvanic corrosion of aluminum and its alloys. Some protective coating such as Noalox or Penetrox should be applied to any point of contact between dissimilar metals. (See the section on Hardware, Lubricants, and Compounds in this chapter and the section on Corrosion in the chapter **Building Antenna Systems and Towers**.)

Aluminum alloys can be used to build amateur antennas, towers and supports. Light weight and high conductivity make aluminum ideal for these applications. Alloying typically lowers conductivity, but significantly increases tensile strength. Aluminum is typically alloyed with metals such as manganese, silicon, copper, magnesium and zinc. Cold rolling can be employed to further increase the strength.

A four-digit system is used to identify aluminum alloys, such as 6061. Aluminum alloys starting with a 6 contain dimagnesium silicide (Mg_2Si). The second digit indicates modifications of the original alloy or impurity limits. The last two digits designate different aluminum alloys within the category indicated by the first digit.

In the 6000-series, the 6061 and 6063 alloys are commonly used for antenna applications. Both types have good resistance to corrosion, medium strength and are widely available. A further designation like T6 denotes thermal treatment (heat tempering).

6061-T6 aluminum has relatively high strength and good workability. It is highly resistant to corrosion and will bend without taking a "set." Comparative specifications show 6063-T832 tensile strength (PSI) 42,000 and yield strength 39,000. 6061-T6 is tensile strength (PSI) 45,000 and yield strength 40,000.

Aluminum tube is found either as "drawn" tube, or "extruded" and the difference can be a major consideration due to the tolerances for each manufacturing process. The tolerances for extruded tube (most often 6061-T6, not 6063-T832) are quite "loose." Extruded tube wall thickness variations might not allow it telescope as expected every time, as will drawn tube. Extruded tube is also not necessarily always round, as drawn tube will be. Instead, extruded can be slightly oval and still be within tolerance.

In recent years 6063-T832 drawn aluminum tubing has become an attractive alternative to 6061-T6, given its good mechanical properties and comparatively low cost. Often found in commercial antennas, this alloy's low cost is derived from ubiquitous use in household items including aluminum folding chairs. More information on the available aluminum alloys can be found in **Table 25.3**.

Table 25.3
Aluminum Numbers and Alloy Types for Amateur Use

Common Alloy Numbers

Type	*Characteristics*
2024	Good formability, high strength
5052	Excellent surface finish, excellent corrosion resistance, normally not heat treatable for high strength
6061	Good machinability, good weldability
6063	Good machinability, good weldability
7075	Good formability, high strength

Common Tempers

Type	*Characteristics*
T0	Special soft condition
T3	Hard
T6	Hardest, possibly brittle
TXXX	Three digit tempers — usually specialized high strength heat treatments, similar to T6

General Uses

Type	*Uses*
2024-T3	Chassis boxes, antennas, anything that will be bent or flexed repeatedly
7075-T3	High strength, corrosion resistant
6061-T6	Tubing and pipe; angle channel and bar stock
6063-T832	Tubing and pipe; angle channel and bar stock

25.3.1 Selecting Aluminum Tubing

Table 25.4 shows the standard sizes of aluminum tubing that are stocked by most aluminum suppliers or distributors in thc United States and Canada. Note that all tubing comes in 12-foot lengths (local metal suppliers sometimes stock 6- and 8-foot lengths) and larger-diameter sizes may be available in lengths up to 24 feet. Note also that 6063-T832 drawn tube of any diameter will fit snugly into the next larger size, if the larger size has a 0.058-inch wall thickness. For example, ⅝-inch tubing has an outside diameter of 0.625 inch. This will fit into ¾-inch tubing with a 0.058-inch wall, which has an inside diameter of 0.634 inch. A clearance of 0.009 inch is just right for a slip fit or for slotting the tubing and then using hose clamps. Always get the next larger size and specify a 0.058-inch wall to obtain the 0.009-inch clearance. If the tubing is extruded, be careful to test its ability to telescope before purchasing. For extended telescoping. 0.049-inch wall tubing is a better choice. **Table 25.5** provides a cross-reference between common metric and Imperial unit sizes of round tubing.

A little figuring with **Table 25.6** will give you all the information you need to build a beam, including what the antenna will weigh.

25.3.2 Sources of Aluminum Tubing and Mounting Materials

Aluminum tubing can be purchased new and is available from local metal suppliers and some amateur radio dealers. Materials for attaching pieces of tubing, such as boom-to-mast or boom-to-element brackets can be purchased from ham radio antenna manufacturers either as separate products or as replacement parts for a commercial antenna. Stauff clamps (**us.stauff.com**) are a relatively new product for amateurs; they are used in industrial applications to support pipe and tubing. The polyamide version is preferred.

Formed element-to-boom brackets can be used, as well as flat plate or bar with U-bolts over the boom. The type of mounting of the element to the boom affects the element length, whether it is attached directly to the boom, or insulated from the boom (e.g. with PVC tube). The mounting plate needs to be considered in the computer model. A mounting plate that is 4" × 8" has an equivalent diameter of approximately 2.5" and 4" in length for each element half. A mounting plate that is 3" × 6" has an equivalent diameter of about 1.8" and a length of 3" for each element half. The mounting plate equivalent will be the first section in a model of the half-element. Don't overlook sources for used tubing, such as a local metal scrap yard.

Surplus metal dealers often have odd lengths and sizes of aluminum tubing. Understand the difference between "tubing" which is extruded and rated for structural use and "pipe" and tubing for household products which are not as strong. Tubing often has an alloy number printed on it.

Some items to look for include irrigation tubing (fairly common in rural areas and nick-named "hand move"), tent poles for small antennas, tubing and fittings from scrapped antennas, and aluminum angle, rod, and round or square extrusions stock. Occasionally, aluminum tower sections can be found in scrap yards. Garage sales are also good sources of used tubing, and every hamfest flea market has a broken or bent antenna or two that can be "parted out." By being a good scavenger, you can build up a "bone yard" of materials for antenna construction.

Irrigation pipe may be usable as a boom or for longer element sections, although it may require extra support from a truss. Tent poles range in length from 2½ to 4 feet, are usually tapered and can be split on the larger end and mated with the smaller end of another pole of the same diameter. A small stainless-steel hose clamp can be used to fasten the poles at this junction. These make fine elements for small beams, particularly at VHF, and even booms for VHF and UHF Yagis.

For vertical antennas, consumer items such as window-washing and painter's poles can sometimes be used. These are not made of structural strength tubing but are often suitable and are low cost. For larger low-band verticals, irrigation pipe is often used.

Table 25.4
Aluminum Tubing Sizes
6061-T6 (61S-T6) Round Aluminum Tube In 12-Foot Lengths

Tubing Diameter	*Wall Thickness Inches*	*Wall Thickness Stubs Ga.*	*ID, Inches*	*Approximate Weight Pounds Per Foot*	*Approximate Weight Pounds Per Length*
3⁄16 in.	0.035	(#20)	0.117	0.019	0.228
(0.1875 in.)	0.049	(#18)	0.089	0.025	0.330
1⁄4 in.	0.035	(#20)	0.180	0.027	0.324
(0.25 in.)	0.049	(#18)	0.152	0.036	0.432
	0.058	(#17)	0.134	0.041	0.492
5⁄16 in.	0.035	(#20)	0.242	0.036	0.432
(0.3125 in.)	0.049	(#18)	0.214	0.047	0.564
	0.058	(#17)	0.196	0.055	0.660
3⁄8 in.	0.035	(#20)	0.305	0.043	0.516
(0.375 in.)	0.049	(#18)	0.277	0.060	0.720
	0.058	(#17)	0.259	0.068	0.816
	0.065	(#16)	0.245	0.074	0.888
7⁄16 in.	0.035	(#20)	0.367	0.051	0.612
(0.4375 in.)	0.049	(#18)	0.339	0.070	0.840
	0.065	(#16)	0.307	0.089	1.068
1⁄2 in.	0.028	(#22)	0.444	0.049	0.588
(0.5 in.)	0.035	(#20)	0.430	0.059	0.708
	0.049	(#18)	0.402	0.082	0.984
	0.058	(#17)	0.384	0.095	1.040
	0.065	(#16)	0.370	0.107	1.284
5⁄8 in.	0.028	(#22)	0.569	0.061	0.732
(0.625 in.)	0.035	(#20)	0.555	0.075	0.900
	0.049	(#18)	0.527	0.106	1.272
	0.058	(#17)	0.509	0.121	1.452
	0.065	(#16)	0.495	0.137	1.644
3⁄4 in.	0.035	(#20)	0.680	0.091	1.092
(0.75 in.)	0.049	(#18)	0.652	0.125	1.500
	0.058	(#17)	0.634	0.148	1.776
	0.065	(#16)	0.620	0.160	1.920
	0.083	(#14)	0.584	0.204	2.448
7⁄8 in.	0.035	(#20)	0.805	0.108	1.308
(0.875 in.)	0.049	(#18)	0.777	0.151	1.810
	0.058	(#17)	0.759	0.175	2.100
	0.065	(#16)	0.745	0.199	2.399
1 in.	0.035	(#20)	0.930	0.123	1.476
	0.049	(#18)	0.902	0.170	2.040
	0.058	(#17)	0.884	0.202	2.424
	0.065	(#16)	0.870	0.220	2.640
	0.083	(#14)	0.834	0.281	3.372

25.3.3 Construction with Aluminum Tubing

Although there is endless variation in the type of antennas designed and built with aluminum tubing, Yagis are by far the most common. Yagi antennas can be successfully built using rules-of-thumb for element and boom material and sizing. Some of these approaches and a set of example element designs are provided in the following paragraphs. *YagiStress*, a commercially available software program developed and supported by Kurt Andress, K7NV can be used to accurately calculate the loads and survivability of Yagi designs. Designers and builders of large Yagi antennas are well advised to use modeling software such as *YagiStress* to ensure survivability of the antenna while at the same not using more material than

Tubing Diameter	Wall Thickness Inches	Wall Thickness Stubs Ga.	ID, Inches	Approximate Weight Pounds Per Foot	Approximate Weight Pounds Per Length
1⅛ in.	0.035	(#20)	1.055	0.139	1.668
(1.125 in.)	0.058	(#17)	1.009	0.228	2.736
1¼ in.	0.035	(#20)	1.180	0.155	1.860
(1.25 in.)	0.049	(#18)	1.152	0.210	2.520
	0.058	(#17)	1.134	0.256	3.072
	0.065	(#16)	1.120	0.284	3.408
	0.083	(#14)	1.084	0.357	4.284
1⅜ in.	0.035	(#20)	1.305	0.173	2.076
(1.375 in.)	0.058	(#17)	1.259	0.282	3.384
1½ in.	0.035	(#20)	1.430	0.180	2.160
(1.5 in.)	0.049	(#18)	1.402	0.260	3.120
	0.058	(#17)	1.384	0.309	3.708
	0.065	(#16)	1.370	0.344	4.128
	0.083	(#14)	1.334	0.434	5.208
	*0.125	⅛ in.	1.250	0.630	7.416
	*0.250	¼ in.	1.000	1.150	14.832
1⅝ in.	0.035	(#20)	1.555	0.206	2.472
(1.625 in.)	0.058	(#17)	1.509	0.336	4.032
1¾ in.	0.058	(#17)	1.634	0.363	4.356
(1.75 in.)	0.083	(#14)	1.584	0.510	6.120
1⅞ in.	0.058	(#17)	1.759	0.389	4.668
(1.875 in.)					
2 in.	0.049	(#18)	1.902	0.350	4.200
	0.065	(#16)	1.870	0.450	5.400
	0.083	(#14)	1.834	0.590	7.080
	*0.125	⅛ in.	1.750	0.870	9.960
	*0.250	¼ in.	1.500	1.620	19.920
2¼ in.	0.049	(#18)	2.152	0.398	4.776
(2.25 in.)	0.065	(#16)	2.120	0.520	6.240
	0.083	(#14)	2.084	0.660	7.920
2½ in.	0.065	(#16)	2.370	0.587	7.044
(2.5 in.)	0.083	(#14)	2.334	0.740	8.880
	*0.125	⅛ in.	2.250	1.100	12.720
	*0.250	¼ in.	2.000	2.080	25.440
3 in.	0.065	(#16)	2.870	0.710	8.520
	*0.125	⅛ in.	2.700	1.330	15.600
	*0.250	¼ in.	2.500	2.540	31.200

*These sizes are extruded. All other sizes are drawn tubes.

Table 25.6
Aluminum Alloy Strength

Tubing	*Tensile Strength PSI min*	*Yield Strength PSI min*
6005A-T61 extruded	38,000	35,000
6061-T6	45,000	40,000
6061-T8 drawn	45,000	40,000
6063-T832 drawn	42,000	39,000
6063-T6 extruded	35,000	31,000
6063-T52 extruded	27,000	21,000
6082-T6*	42,100	36,300
AW-6060-T66**	31,180	23,200
AW-6005-T6**	36,260	29,000

*<5 mm wall
**<3 mm wall
Strengths shown are approximate. Consult manufacturer data sheets for specifications of materials chosen.

Table 25.5
Metric and Imperial Tubing Sizes

Diameter	*Wall Thickness (mm)*	*Weight (kg/m)*
10 mm	0.7	0.055
6 mm	0.8	0.035
8 mm	0.8	0.049
10 mm	0.8	0.062
1/2"		
12.7 mm	0.8	0.081
19 mm	0.9	0.138
6 mm	1	0.042
8 mm	1	0.059
3/8"		
9.5 mm	1	0.072
11 mm	1	0.085
12 mm	1	0.093
1/2"		
12.7 mm	1	0.099
14 mm	1	0.110
16 mm	1	0.127
19 mm	1	0.153
22 mm	1	0.178
25 mm	1	0.204
28 mm	1	0.229
32 mm	1	0.263
10 mm	1.1	0.083
19 mm	1.1	0.167
17 mm	1.2	0.161
19 mm	1.2	0.181
22 mm	1.2	0.212
3/8"		
9.5 mm	1.3	0.090
11 mm	1.3	0.107
16 mm	1.5	0.184
19 mm	1.5	0.223
25 mm	1.5	0.299
28 mm	1.5	0.337
32 mm	1.5	0.388
35 mm	1.5	0.426
38 mm	1.5	0.464
45 mm	1.5	0.553
50 mm	1.5	0.617
60 mm	1.5	0.744
40 mm	1.7	0.552
20 mm	2	0.305
25 mm	2	0.390
30 mm	2	0.475
40 mm	2	0.645
50 mm	2	0.814
50 mm	2	0.814
70 mm	2	1.154
90 mm	2	1.493
114 mm	2	1.900
4½"		
114.3 mm	2	1.905
30 mm	2.5	0.583
60 mm	2.5	1.219
45 mm	3	1.069
60 mm	3	1.450
70 mm	3	1.705
35 mm	4	1.052
90 mm	5	3.605

required to achieve desired mechanical performance. *YagiStress* was used to calculate the wind-speed ratings of the half-element designs in this chapter and is based on the TIA-222-C "Structural Standard for Antenna Supporting Structures and Antennas." Antenna mechanical design spreadsheets from *Physical Design of Yagi Antennas* by David Leeson, W6NL (see References and Bibliography), are available from **realhamradio.com/Download.htm** (the URL is case-sensitive) and have been updated to TIA-222-F. (Support and availability of *YagiStress* is suspended as of December 2022 but it may be available again at some future date. TIA-222-H is the current revision of the wind speed standard.)

Antennas for frequencies of 14 MHz and above are usually made to be rotated. Rotatable antennas require materials that are strong, lightweight, and easy to obtain. Material selection is dependent on many factors, with weather conditions typically being the most demanding requirement. High winds alone may not cause as much damage to an antenna as does ice loading. Ice in combination with high wind is typically the worst-case condition.

If your location is adjacent to sloping ground, such as a hilltop, be aware of up-lifting winds, as they can cause serious damage by literally lifting the elements and boom, then dropping them. In such locations, large booms and long elements are often trussed on the bottom side, similarly to the typical truss system on the "top side" of the Yagi.

As explained in Section 25.2.1, elements and booms can be made from telescoping tubing to provide the necessary total length. This is referred to as *tapering*. The boom diameter for a rotatable Yagi or quad should be selected to provide required structural strength and to stably support the elements. The appropriate tubing diameter for a boom depends on many factors. Among them are element weight, element length, number of elements and environmental loads, including static loads such as ice and dynamic loads, principally from wind gusts. Tubing of 1-1/4-inch diameter can easily support a three-element 28-MHz antenna and marginally a two-element 21-MHz antenna. A 2-inch diameter boom will be adequate for larger 28-MHz antennas or for harsh weather conditions and for antennas up to three elements on 14 MHz or four elements on 21 MHz. It is not recommended that 2-inch diameter booms be made any longer than 24 feet unless additional support is added to carry both vertical and horizontal loads. Suitable reinforcement for a long 2-inch boom can consist of a truss or a truss and lateral support, as shown in **Figure 25.9**.

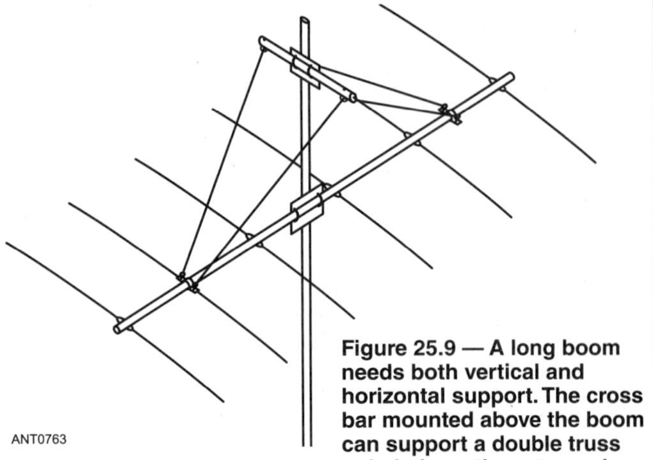

Figure 25.9 — A long boom needs both vertical and horizontal support. The cross bar mounted above the boom can support a double truss to help keep the antenna in position.

A guideline from long-time manufacturers is to decide which component of a Yagi antenna should be the first to fail and which component should be the absolute component "left standing." The boom is the final choice. It is far better to lose element parts, even full elements, rather than have a boom failure, which means the entire Yagi comes down. Repairing a Yagi with a broken boom is much more difficult than fixing elements.

For boom lengths more than 24 feet, 3-inch diameter thin-wall (i.e., 0.065") material is usually required. Thicker wall material, such as 0.125" × 2", 2.25", and 2.5" can be telescoped for a strong boom in this length range. The larger boom diameters provide considerable mechanical stability as well as large clamping surface area for boom-to-element hardware. Clamping surface area is particularly important if heavy icing is anticipated and helps prevent rotation of elements around the axis of the boom.

Pinning an element to the boom with a bolt or, preferably, a swaged, hardened pin, can eliminate this possibility, but the hole introduces a stress riser that can materially reduce the strength of the boom. Element rotation about the boom axis can be minimized by mounting elements under the boom rather than on top. Pinned elements will work loose over time, especially in windy locations and elongate the pinning holes in both the element and the boom. This is a progressive condition resulting in elements that can be so loose-fitting to the boom that their rotational positions change frequently. Although this condition typically does not adversely affect the electrical performance of a Yagi, the mechanical strength of the members involved degrades as the holes elongate. A Yagi with elements at various angles is unsightly as well, although it will continue to perform electrically.

A 3-inch diameter boom with a wall thickness of 0.065 inch is satisfactory for antennas up to about a five-element, 14-MHz array that is spaced on a 40-foot-long boom. A truss is recommended for any boom longer than 24 feet.

Figure 25.10 — Light-duty half-element designs for Yagi antennas. The other side of the element is identical and the center section should be a single piece twice as long as the length shown here for the largest diameter section. Tubing with 0.116-inch wall thickness consists of doubled 0.058-inch wall sections of the same length. Tubing with 0.125-inch or 0.250-inch wall thickness is 6061-T6 alloy, all other tubing is 6063-T832. Doubler (DB) sections consist of a length of tubing inserted completely into the next larger segment, flush with the inner end of that larger segment. Included with this book's online material is a text file file "K5GO Half-Element Designs" gives complete specifications for each half-element along with survivability ratings for ½ inch and 1 inch of radial ice loading.

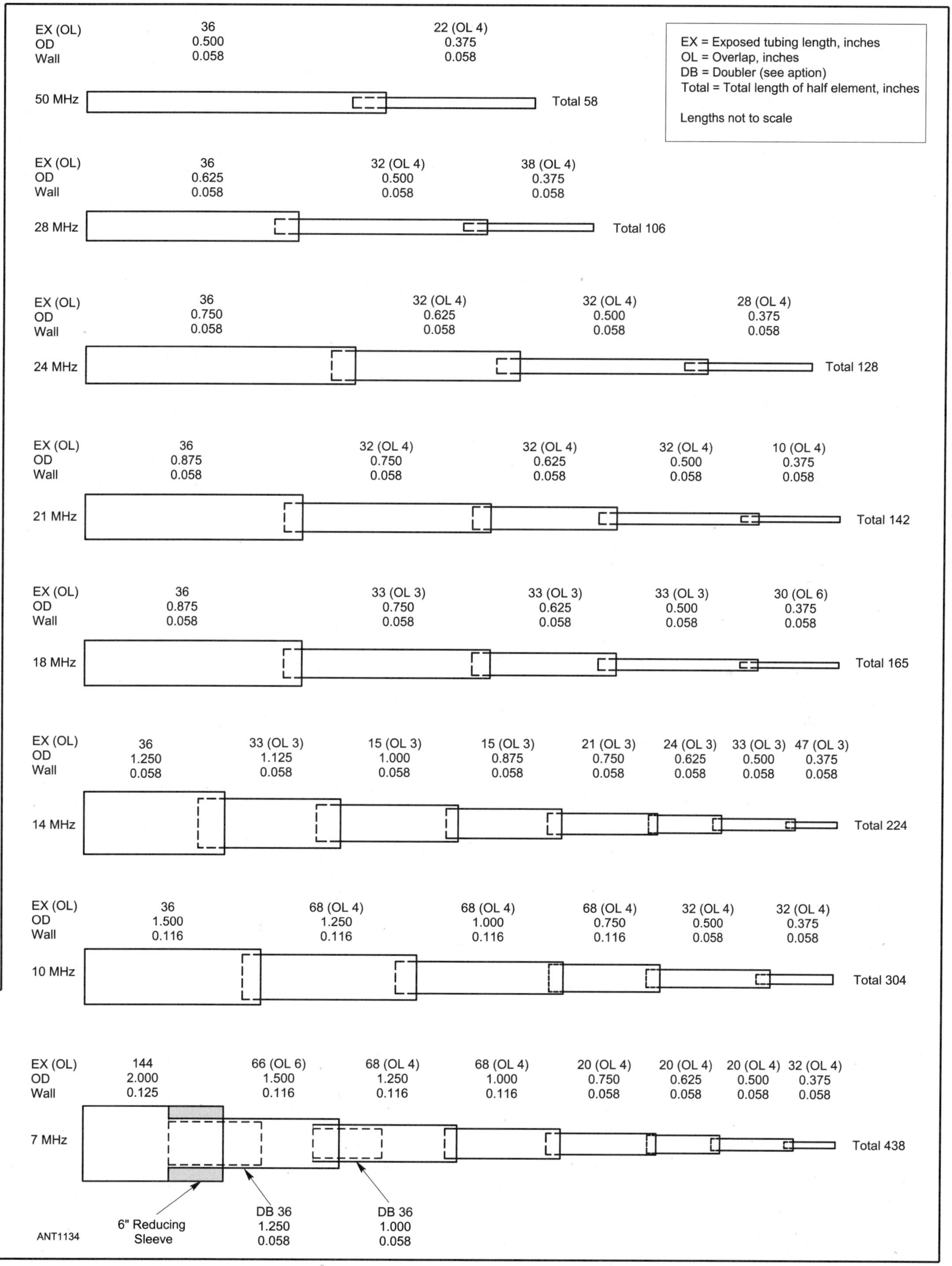
EX = Exposed tubing length, inches
OL = Overlap, inches
DB = Doubler (see aption)
Total = Total length of half element, inches
Lengths not to scale
EX (OL)
OD
Wall
50 MHz
36
0.500
0.058
22 (OL 4)
0.375
0.058
Total 58
28 MHz
36
0.625
0.058
32 (OL 4)
0.500
0.058
38 (OL 4)
0.375
0.058
Total 106
24 MHz
36
0.750
0.058
32 (OL 4)
0.625
0.058
32 (OL 4)
0.500
0.058
28 (OL 4)
0.375
0.058
Total 128
21 MHz
36
0.875
0.058
32 (OL 4)
0.750
0.058
32 (OL 4)
0.625
0.058
32 (OL 4)
0.500
0.058
10 (OL 4)
0.375
0.058
Total 142
18 MHz
36
0.875
0.058
33 (OL 3)
0.750
0.058
33 (OL 3)
0.625
0.058
33 (OL 3)
0.500
0.058
30 (OL 6)
0.375
0.058
Total 165
14 MHz
36
1.250
0.058
33 (OL 3)
1.125
0.058
15 (OL 3)
1.000
0.058
15 (OL 3)
0.875
0.058
21 (OL 3)
0.750
0.058
24 (OL 3)
0.625
0.058
33 (OL 3)
0.500
0.058
47 (OL 3)
0.375
0.058
Total 224
10 MHz
36
1.500
0.116
68 (OL 4)
1.250
0.116
68 (OL 4)
1.000
0.116
68 (OL 4)
0.750
0.116
32 (OL 4)
0.500
0.058
32 (OL 4)
0.375
0.058
Total 304
7 MHz
144
2.000
0.125
66 (OL 6)
1.500
0.116
68 (OL 4)
1.250
0.116
68 (OL 4)
1.000
0.116
20 (OL 4)
0.750
0.058
20 (OL 4)
0.625
0.058
20 (OL 4)
0.500
0.058
32 (OL 4)
0.375
0.058
Total 438
6" Reducing Sleeve
DB 36
1.250
0.058
DB 36
1.000
0.058
ANT1134

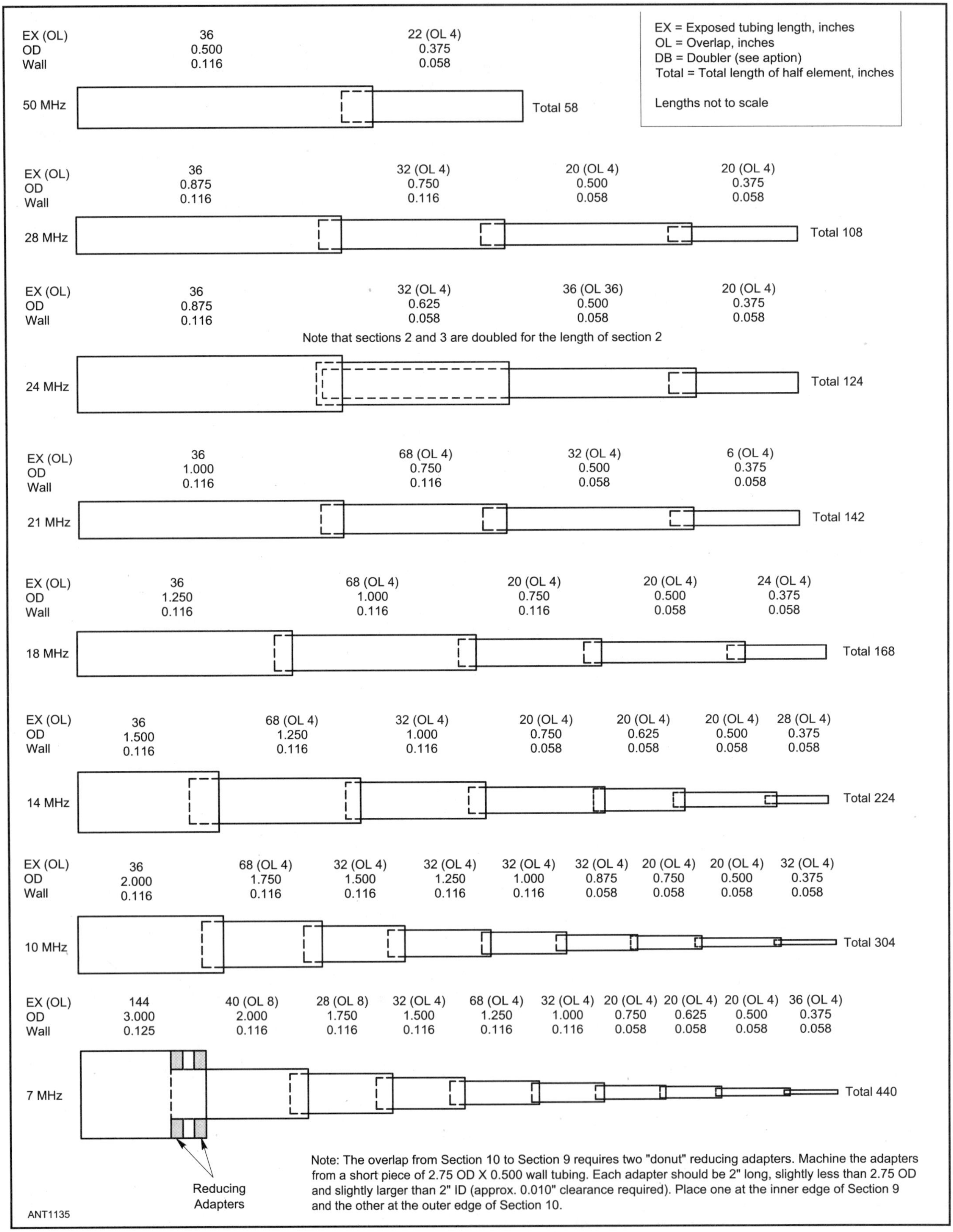

Figure 25.11 — A heavy-duty schedule of Yagi half-element segments. See the Figure 25.10 caption for details. Table 25.4 gives details of aluminum tubing sizes.

ELEMENT ASSEMBLY

Figure 25.10 shows tapered Yagi element designs contributed by Stan Stockton, K5GO, that will survive winds in excess of 80 mi/h. With a ½-inch thickness of radial ice, these designs will withstand winds from 45 to 77 mi/h. Ice increases the surface area subject to wind loading but does not increase the strength of the element. More rugged designs are shown in **Figure 25.11**. With no ice loading, these elements will survive in 118 to 172-mi/h winds, and in winds from 78 to 92 mi/h with ½ inch of radial ice. Deviations from the designs provided require analysis with a program such as *YagiStress* to ensure survivability in the environmental conditions of interest. Except for the very largest 40 meter elements, all required tubing lengths are 6 feet or shorter that can be shipped by parcel services. The file "K5GO Half-Element Designs" showing all element segment lengths, overlaps, tubing specifications and more information on ice loading is included with this book's online material.

Figures 25.10 and 25.11 show only half elements. When the element is assembled, the largest size tubing for each element should be double the length shown in the drawing, with its center being the point of attachment to the boom. These designs are somewhat conservative, in that they are self-resonant slightly below the frequency indicated for each design. Telescoping the outside end sections to shorter lengths for resonance will increase the survival wind speeds. Conversely, lengthening the outside end sections will reduce the survival wind speeds. (See the References and Bibliography listing for David Leeson, W6NL [ex-W6QHS], at the end of this chapter.)

Additional element specifications for HF Yagis built in very high production quantities (>150,000 elements in the field) by Force 12, Inc. and Next Generation Antennas can be found in Array of Light, 3rd and 4th Editions by Tom Schiller, N6BT (**nextgenerationantennas.com**). These elements use riveted construction and were designed in consultation with Kurt Andress, K7NV (*Yagi Stress*); the element-to-boom mounts designed with assistance from Dave Leeson, W6NL (*Physical Design of Yagi Antennas).*

Figure 25.12 shows several methods of fastening antenna element sections together. The slot and hose-clamp method shown in Figure 25.12A works well for joints that require adjustment. Generally, one adjustable joint per element half is sufficient to tune an antenna. Stainless-steel hose clamps work well and are inexpensive. Some do not have stainless steel screws, however. This can be checked with a magnet. Table 25.6 shows available hose-clamp sizes. Wherever tubing sections overlap, a small amount of anti-oxidation compound should be used. This prevents aluminum oxide from forming between the tubing surfaces that can create a high impedance electrical connection and/or mechanically "freeze" the joint.

Over time in the weather, with the elements flexing millions of times, compression clamps can become loose, and the section(s) can move. A simple added safety measure is to wrap top-quality electrical tape (e.g., *Scotch 33+)* over the compression clamp and joint and the last wrap should not have tension (stretch) in the tape.

Figures 25.12B, 12C and 12D show possible fastening methods for joints that do not require adjustment. At B, machine screws and nuts hold the elements in place. At C, sheet metal screws are used. At D, rivets secure the tubing. If the antenna is to be assembled permanently, rivets are the best choice. Once in place they are permanent, although they can be drilled out if necessary. They will not work free, regardless of vibration or wind, if properly installed and seated.

The proper rivets are called "closed-end" rivets are specifically designed for vibration (i.e. used in aircraft). These are usually not found at the typical hardware store but are available at many commercial supply companies. They are a "blind rivet", as there is no back piece and the nickname for them is "pop rivet." **Figure 25.13E** shows this type of rivet and how it secures a tubing junction.

The rivets come with an aluminum body in varying lengths for securing two or three tubing sections. The small ⅛-inch body is sufficient for element sections up to 1⅛" inch diameter and can be pulled by hand. Larger sizes (i.e. 3⁄16") are can be pulled with an inexpensive pneumatic tool. In all cases, use the smallest nozzle for the rivet tool to secure the rivets.

These rivets are available with a choice of mandrels, the

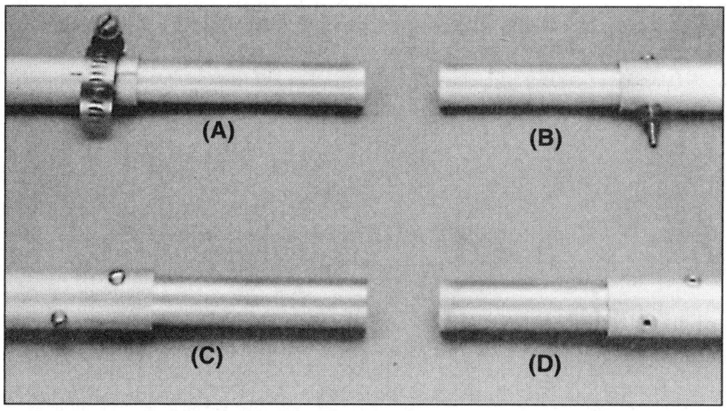

Figure 25.12 — Methods of connecting telescoping tubing sections to build beam elements. See text for a discussion of each method.

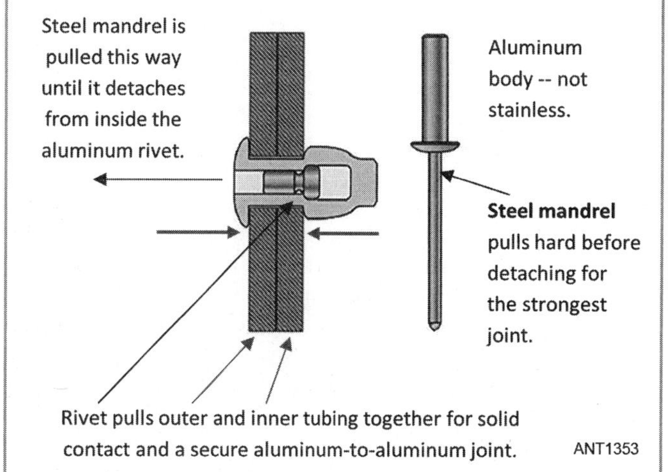

Figure 25.13 — Pop-rivet construction and use.

"nail" that gets pulled to seat it in place. The mandrel is attached internally to the aluminum body and "pops" off when the rivet is seated. Steel mandrels seat the rivets more firmly than aluminum mandrels and in either case, the aluminum body of the rivet makes a secure connection to the tubing. There is no danger of dissimilar-metal corrosion with aluminum rivets and aluminum antenna elements. If the antenna is to be disassembled and moved periodically, either B or C will work. If machine screws are used, however, take all possible precautions to keep the nuts from vibrating free. Use nylon-insert (Nylock) nuts or lock washers and a thread-locking compound.

Very strong elements can be made by using a double thickness of tubing, made by telescoping one size inside another for a portion of, or for the total length. This is usually done at the center of an element where more strength is desired at the boom support point, as in the 14-MHz element in Figure 25.11. Other materials can be used as well, such as wood dowels, fiberglass rods, etc.

Antennas for 50 MHz need not have elements larger than ½-inch diameter, although up to 1 inch is used occasionally. At 144 and 222 MHz the elements are usually ⅛ to ¼ inch in diameter. For 432 MHz, elements as small as 1/16-inch diameter work well if made of stiff rod. Aluminum welding rod of 3/32 to ⅛-inch diameter is fine for 432-MHz arrays, and ⅛ inch or larger is good for the 222-MHz band. Aluminum rod or hard-drawn wire works well at 144 MHz.

Tubing and rod sizes recommended in the paragraph above are usable with most formula dimensions for VHF/UHF antennas. Larger diameter material reduces Q and increases bandwidth; smaller diameter material raises element and overall antenna Q and reduces bandwidth. Much smaller diameters than those recommended will require longer elements, particularly for antennas for 50-MHz and above.

ELEMENT TAPER AND ELECTRICAL LENGTH

The builder should be aware of one important aspect of telescoping or tapered elements. When the element diameter tapers, as shown in Figures 25.10 and 25.11, the electrical length is not the same as it would be for a constant diameter element of the same total length. Length corrections for tapered elements are discussed in the chapter on **HF Beam Antennas**.

25.3.4 Notes on Aluminum Tubing Construction

CONNECTING ELEMENTS TO THE BOOM

Per theory, there is no RF voltage at the center of a parasitic element and insulation is not required at the boom-to-element interface for elements centered on the boom. Driven elements may or may not be electrically connected to the boom depending on the feed system employed. In practice, parasitic elements are usually directly connected to the boom both mechanically and electrically for designs from HF through lower UHF. At upper UHF, grounded elements are subject to detuning because the element-to-boom contact no longer acts as a point but rather as a complex shape of significant area. At HF, unanticipated and unwanted resonances can occur in center-grounded elements. Highly conservative HF designs and many UHF designs insulate all elements from the boom, typically using Garolite at HF and suitable materials such as Teflon at UHF and above.

EFFECT OF BOOM ON ELEMENT LENGTH

Metal booms have a small "shortening effect" on elements that run through them. With materials sizes commonly employed, this is not more than one percent of the element length and may not be noticeable. It is just perceptible with ½-inch tubing booms used on 432 MHz, for example. At VHF and UHF, standard design-formula lengths can be used as given and driven element matching can be adjusted at the desired operating frequency. The center frequency of an all-metal array will tend to be 0.5 to 1 percent higher than a similar system built with insulated elements.

ELEMENT VIBRATION

Metal antenna elements have high mechanical Q and if the tubing sections are too long, they will tend to vibrate, especially in light winds. The result over time will be an element fracture. If the design in use is prone to these vibrations, one way to dampen them is by placing a piece of polypropylene rope or similar material line inside the element throughout its entire length. Choice of damping line material is not critical — the line will not be exposed to the sun's UV. The line will mildew or rot however if something like inexpensive clothesline is used. Cap or tape the end of the element to secure the damping line, leaving a small hole or gap for drainage on each end. If mechanical requirements dictate (a U-bolt going through the center of the element, for instance), the line may be cut into separate pieces for each element half.

The best solution, of course, is to use an element design that does not vibrate, which is one that generally is constructed using shorter segments. This results in more mass towards the element center and a more constant tapering wind load towards the tip. As the element flexes in the wind, the element presents less wind load and sheds the wind vortex.

END CAPS

Plastic end caps are often used in an attempt to seal tubing from water. Unfortunately, due to condensation and leakage, the opposite often results and the tubing fills with water. This can lead to unpleasant surprises when an antenna is tilted for lowering or maintenance! Leaving the element open is the preferred choice but if wind noise is a problem or birds or squirrels attempt to store food in the tubing, a plug of porous plastic mesh or stainless steel wool will keep out larger items while allowing water to drain and dampening noise from the wind blowing across the element.

WORK SUPPORT FIXTURE

Most tubing antennas for 50 MHz and lower frequencies are Yagis with all of the elements in the same plane. Working on these antennas is much easier if the boom is held at close to chest height. A portable workbench (aka, "Workmate") holds the boom securely and prevents it from rotating due to imbalance during construction. A portable PA speaker stand will also support a heavy load at adjustable heights.

25.4 Other Materials for Antenna Construction

25.4.1 Wood and Bamboo

Wood is very useful in antenna work. It is available in a great variety of types and sizes. Rug poles of wood or bamboo make fine booms. Bamboo is quite satisfactory for spreaders in quad antennas.

Round wood stock (doweling) is found in many hardware stores in sizes suitable for small arrays. Wood is good for the framework of multi-bay arrays for the higher bands, as it keeps down the amount of metal in the active area of the array. Square or rectangular boom and frame materials can be cut to order in most lumber yards if they are not available from the racks in suitable sizes.

Wood used for antenna construction should be well seasoned and free of knots or damage. Available materials vary, depending on local sources. Your lumber dealer can help you better than anyone else in choosing suitable materials. Joining wood members at right angles can be done with gusset plates, as shown in **Figure 25.14**. These can be made of thin outdoor-grade plywood. Construction with round material can be handled in ways similar to those used with metal components, such as with U-bolts.

In the early days of radio, hardwood was used as insulating material for antennas, such as at the center and ends of dipoles, or for the center insulator of a driven element made of tubing. Wood dowels cut to length were the most common approach. To drive out moisture and prevent the subsequent absorption of moisture into the wood, it was treated before use by boiling in paraffin. Of course, today's technology has produced superior materials for insulators in terms of both strength and insulating qualities. However, the technique is worth consideration in an emergency or if low cost is a prime requirement. "Baking" the wood in an oven for a short period at 200° F should drive out any moisture. Then treatment as described in the next paragraph should prevent moisture absorption. The use of wood insulators should be avoided at high-voltage points if high power is being used.

All wood or bamboo used in outdoor installations should be protected from the weather with varnish or paint. A good grade of marine spar varnish or UV-stable polyurethane varnish will offer protection for years in mild climates, and one or more seasons in harsh climates. Epoxy-based paints also offer good protection. Bamboo can also be protected by wrapping it with electrical tape. Spray varnish is sometimes applied after wrapping with tape and will provide excellent longevity.

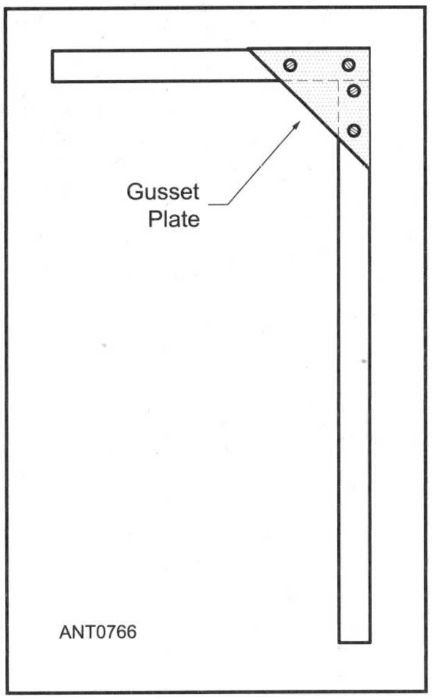

Figure 25.14 — Wood members can be joined at right angles using gusset plates.

25.4.2 Plastics

Plastic tubing and rods of various sizes are available from many building-supply stores. The uses for the available plastic materials are limited only by your imagination. PVC pipe and electrical conduit is quite useful for antenna construction at VHF and UHF. For permanent antennas, be sure the plastic will withstand UV exposure or paint it.

Plastic plumbing and irrigation fittings can also be used to enclose baluns and as the center insulator or end insulators of a dipole, as shown in **Figure 25.15**. The same fittings and adapters can be used to create a portable antenna that is assembled using friction fits between pipe and fittings.

Plastic or Teflon rod can be used as the core of antenna loading coils, including for mobile antennas (**Figure 25.16**), but the material for this use should be selected carefully. Some plastics, particularly PVC, become warm in the presence of a strong RF field. This can result in the core deforming or even catching fire. Where high RF fields are anticipated, fiberglass or Teflon solid rod, or open polycarbonate cylinders are recommended. Home goods stores frequently carry inexpensive drinking glasses made of polycarbonate, in a variety of sizes. These make excellent coil forms for high-power RF applications.

25.4.3 Fiberglass

Fiberglass is lightweight, withstands harsh weather well, and has excellent insulating qualities. Fiberglass rod and tubing are excellent for the nonconductive structure of an antenna. The strongest fiberglass rod is *pultruded*, which orients the fibers lengthwise, as compared to G10 laminate's cross-hatch

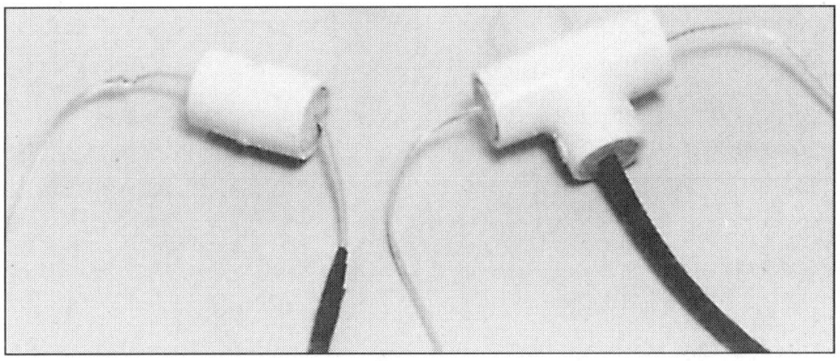

Figure 25.15 — Plastic plumbing parts can be used as antenna center and end insulators.

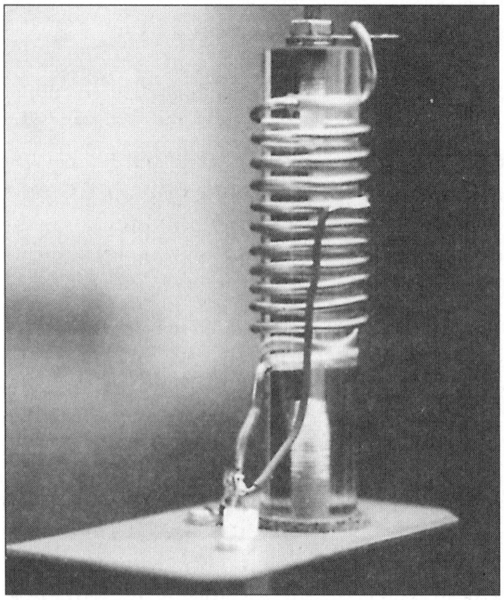

Figure 25.16 — A mobile-antenna loading coil wound on a polystyrene rod.

construction. Fiberglass poles are the preferred material for spreaders for quad antennas, for example. Fiberglass rod or tubing can be used as the boom for VHF and UHF antennas. Extendable fiberglass poles have become very popular as supports for portable wire antennas. The SteppIR family of tunable Yagi, dipole, and vertical antennas use fiberglass tubes with flexible, low-loss, composite metal tape inside as the elements.

Fiberglass should be painted or coated to protect it from exposure to UV when used outdoors. UV breaks down the resin holding the glass fibers together and the surface begins to shed fibers, leading to cracks and water ingress.

Whenever working with fiberglass materials — sawing, cutting, sanding, drilling — gloves and eye protection against loose fiber fragments should be used. If heavy dust is being generated, a dust mask should be worn.

A disadvantage of hollow fiberglass poles is that they may be crushed rather easily. Fracturing occurs at the point where the pole is crushed, causing it to lose its strength. A crushed pole is next to worthless. Some amateurs have repaired crushed poles with fiberglass cloth and epoxy, but the original strength is nearly impossible to regain. Inserting a wooden dowel into the tubing provides additional crush resistance.

25.5 Hardware, Lubricants, and Compounds

The material in this section is an overview of several common topics of interest to antenna system builders. Many more topics are covered in detail by the Workshop Chronicles columns of Don Daso, K4ZA, in the *National Contest Journal* (see the References section). **Table 25.7** is a cross-reference between the closest equivalent standard sizes used in the metric and Imperial unit systems.

25.5.1 Hardware

Antennas should be assembled with good quality hardware intended for outdoor use. Stainless steel is a good choice for long life. Rust will quickly attack plated steel hardware, making nuts difficult, if not impossible, to remove. Stainless-steel nylon-insert (Nylock) nuts are a good choice for antennas since they are subjected to vibration from the wind.

Good quality hardware is expensive, but over time is less expensive and much less frustrating than poor quality "equivalents." Antennas built of high-quality hardware need to be taken down and refurbished much less often. When the time does come to repair or modify an antenna, rusty hardware, particularly at the top of a tower, will seem in retrospect to have been a very poor investment.

If stainless-steel muffler or saddle clamps and hose-clamps are not available, steel hardware can be plated or painted with a good zinc-chromate primer and a one or more finish coats. Rust inhibiting paints are also good protection. Stainless lock washers and nuts can be used with plated U-bolts, and they usually can be removed later, even with rusty threads on the U-bolts. When using stainless-steel hardware, use an anti-seize compound on the threads to prevent the threads from jamming due to galling of the thread surfaces.

Larger antennas and tower fixtures are often assembled with bolts that are rated to have specific strengths and characteristics. The ratings are identified by *grade markings* on the bolt head. A complete list of the markings developed by ASTM, SAE, and ISO standards organizations is available from American Fastener at **americanfastener.com/astm-sae-and-iso-grade-markings-for-steel-fasteners**. It is important to use hardware with the required ratings to prevent failure from overloaded hardware.

GRADES OF BOLTS

Bolt material strength is determined by the alloy and processing method (cold working or heat treating, for example). The two most important properties are *tensile strength* and *yield strength.* Tensile strength refers to the stress at which the material breaks. Yield strength is the stress level at which the material permanently deforms or yields. The load a fastener can carry is calculated by multiplying the material strength by the nominal cross-sectional area of the thread.

A bolt's material strength is designated by *grade*. These grades are indicated by markings that may be stamped directly on the bolt head. A list of grades and strengths is presented in **Table 25.8** which is a guide for sizes of bolts that are com-

Table 25.7
Metric and Imperial Standard Sizes

Nominal Metric Size (mm)	*Closest Imperial Size (inches)*	*Closest Imperial Size Converted to Metric (mm)*
3.3	1/8	3.2
4 – 5	3/16	4.8
6	1/4	6.4
8	5/16	7.9
10	3/8	9.5
12	1/2	12.7
15 – 16	5/8	15.9
20	3/4	19.1
22	7/8	22.2
25	1	25.4
28	1 1/8	28.6
30 – 32	1 1/4	31.7
35	1 3/8	34.9
38	1 1/2	38.1
40	1 5/8	41.3
45	1 3/4	44.5
48	1 7/8	47.6
50	2	50.8
55	2 1/8	54.0
58	2 1/4	57.2
60 – 65	2 1/2	63.5
70	2 3/4	69.9
75	3	76.2
80 – 85	3 1/4	82.6
90	3 1/2	88.9
100	4	101.6
110 – 120	4 1/2	114.3
125 – 130	5	127.0
140	5 1/2	139.7
150	6	152.4
175 – 180	7	177.8
200 – 205	8	203.2

monly found in tower and antenna construction. **Table 25.9** presents guidelines for tightening torque from the standards organization, ASTM International.

STAINLESS STEEL

Steel is an alloy, made from iron with trace amounts of carbon and other elements. Stainless alloys require adding chromium, manganese, molybdenum, nickel, phosphorous, silicon, and/or sulfur. Depending on the mix and ratio of these, various "grades" of stainless steel are produced, and used to make fasteners. The most common stainless steel is called "18-8," which simply means approximately 18 percent chromium and 8 percent nickel. Other common grades of stainless steel for fasteners are 304 and 316, both derived from the 18-8 mix.

Because stainless has little carbon, it cannot be heat treated, which translates into reduced strength. (This explains why, if you want to pin something — say a mast or beam — you should not use stainless, but a much harder fastener, such as a grade 8 bolt.) Stainless bolts have reasonable tensile strength, but reduced yield strength, compared to "regular" steel hardware.

The reason for using stainless steel hardware is to avoid corrosion, not for strength. Yet the reduced strength compared to carbon steel means should be considered for the application; ignoring this basic fact could mean disaster.

Stainless steel is also subject to *galling*, often called cold-welding, which happens when the surfaces of fasteners are put under pressure when tightening. The mating threads rub and break down their surface coatings, which results in friction, causing them to seize together. You're forced to simply twist the bolt off. The potential for galling can be limited by lubricating the threads with an anti-seize compound. Another option is to switch grades—using a 304 nut on 316 threads, for example, so the surfaces have a sufficiently different chemistry to reduce galling.

Table 25.8
A bolt's material strength is designated by grades, sometimes stamped on the bolt head.

Grade	*Marks on Head*	*Material*	*Tensile Strength N/mm²*	*Tensile Strength PSI*	*Yield Strength N/mm²*	*Yield Strength PSI*
2	None	Steel	510	7 4,000	393	57,000
5	3	Steel	827	120,000	634	92,000
8	6	Alloy Steel	1030	150,000	896	130,000
SHCS	None	Alloy Steel	1240	180,000	965	140,000
18-8	None	302 Stainless	690	100,000	448	65,000
316	None	316 Stainless	690	100,000	448	65,000

Table 25.9
These guidelines come from the American Society for Testing and Materials (now called ASTM International), which develops voluntary consensus standards. Some 12,000 such standards are in use today worldwide.

Bolt Size (in)	*TPI*	*Proof Load (lbs)*	*Clamp Load (lbs)*	*Tightening Torque (ft/lbs) Lubed*	*Galv*	*Plain*
1/4	20	1145	859	2	4	4
5/16	18	1886	1415	4	9	7
3/8	16	2790	2093	7	16	13
7/16	14	3827	2870	10	26	21
1/2	13	5108	3831	16	40	32

GALVANIZED STEEL

Galvanized steel generally has a longer life than plated steel, but this depends on the thickness of the galvanizing coat. In harsh climates rust will usually develop on galvanized fittings in a few years. For the ultimate in long-term protection, galvanized steel should be further protected with zinc-chromate primer and then paint or enamel before exposing it to the weather. Cold-galvanizing spray is useful in repairing damage to galvanized surfaces and preventing rust. It is available in-home goods stores.

25.5.2 Lubricants and Compounds

Lubricants are designed to:

• Minimize friction at the points of contact in bearings.

• Protect the polished bearing surfaces from becoming corroded.

• Dissipate heat.

• Remove or prevent foreign particles from getting into a bearing.

Table 25.10 lists a few common lubricants and associated compounds used with amateur antennas and towers.

There are two basic types of lubricants — oil and grease. Each has advantages and limitations. Being liquid, oil can lubricate all surfaces and dissipate heat more readily. It retains its characteristics over a broad range of temperatures, making it ideal for high-speed and high-temperature jobs. Oil levels can usually be carefully controlled and monitored.

Grease is thicker and can seal a bearing better than oil. It can easily be confined inside a bearing, allowing you to "pre-lube" fittings, for instance, as it clings to surfaces better than oil. As such, it lasts longer than oil. Grease consists of three components; a base oil, thickeners, and additives.

Thickeners are what give grease its density — its ability to stay in place — and also play a role in lubrication. The thickener allows the oil to "come to the surface" as needed. Additives help address specific problems in lubrication. Molybdenum disulfide (referred to as "moly") is the most common. Graphite and zinc are also commonly used. Tackifiers make grease sticky, meaning the grease will stay in place. Rust-inhibitors are additives that inhibit oxidation.

THRUST BEARINGS

The typical Rohn bearing is designed to run dry (unlubricated). The speed is so slow that little heat will be generated. When motionless, the load is distributed symmetrically on the ball and the race. When a tangential load causes the ball to roll, the material in the race will "bulge" in front of the ball, while "flattening out" behind it. Since not enough heat is generated from sliding friction in a typical thrust bearing, metal pickup or welding does not occur. But the aluminum race can show evidence of wear. Since it's a softer metal than the ball bearings, it will deform first, sometimes to the point of seizure. Lubrication will not solve that problem so if the manufacturer instructions do not require lubrication, don't add it.

THREAD LUBRICANTS

A minimal amount of lubricant is used to lubricate bolt threads. To derive the full benefit of the lubricant, the nut and bolt should be hand-tightened, drawing the nut down the entire length of the threads, coating them with a fine film of lubricant. You do not need lots of lubricant, just enough to coat the threads with a fine film.

ANTI-OXIDATION COMPOUNDS

Noalox and Penetrox are anti-oxidation compounds. Noalox contains suspended zinc particles which penetrate and cut aluminum oxidation, making it excellent for aluminum-to-aluminum connections (i.e., tubing). Penetrox can also be used for aluminum-to-copper connections. These are not lubricants but can keep telescoping tubing joints movable by preventing corrosion and oxidation.

WIRE ROPE LUBRICANT

The most common use of wire rope in amateur stations is for crank-up masts and towers. As the rope moves, the strands of wire in the rope move against each other, creating friction and wear. In addition, there can be corrosion at clamps and oxidation or rusting of the wire surfaces. A cable or wire rope lubricant is used to reduce friction and provide corrosion protection both inside the rope and on its outside surface where it moves through pulleys. The mast or tower manufacturer may make specific recommendations for a type of lubricant and a schedule for application which should be followed, particularly under warranty. If no guidance is available, select a penetrating type of lubricant that is sprayed on to the rope. The solvent carrying the lubricant then evaporates, coating the wire strands. A coating type of lubricant is generally not recommended since it will collect dirt, grit, and other contaminants.

Table 25.10
Lubricants and Other Compounds

Trade name	*Type*	*Application*
NoAlOx, OxGard	Anti-oxidation,	Aluminum-aluminum joints
Penetrox	Anti-oxidation	Aluminum-aluminum and aluminum-copper joints
Jet-Lube SS-30	Anti-oxidation, corrosion-inhibitor	Aluminum-aluminum, aluminum-copper, and copper-copper joints
Anti-seize		Thread lubricant for stainless steel hardware
WD-40	Penetrating oil	Light lubrication, water displacement
PB Blaster, Kroil	Penetrating oil	Loosening rusted or frozen/seized parts
Lubriplate grease	Aluminum-complex	Tools and bearings
White Lithium grease		Water-resistant, tower assembly, rotators
Dielectric grease		Protect electrical connections
Locktite Threadlock		Prevents loosening, various strengths

25.6 References and Bibliography

Source material and more extended discussion of topics covered in this chapter can be found in the following references:

ARRL's *Wire Antenna Classics*, (Newington, CT: ARRL)

More Wire Antenna Classics, (Newington, CT: ARRL)

ARRL's Yagi Antenna Classics, (Newington, CT: ARRL)

Reference Data for Radio Engineers: Radio, Electronics, Computer, and Communications (Boston, MA: Newnes, an Imprint of Elsevier, 2001)

Daso, D., K4ZA, *Antenna Towers for Radio Amateurs* (Newington, CT: ARRL, 2010)

Daso, D., K4ZA, "Workshop Chronicles," *NCJ*, Mar. 2005 and subsequent issues.

Elengo, J., K1AFR, "Predicting Sag in Long Wire Antennas," *QST*, Jan. 1966, pp. 57 –58.

Laport, E., *Radio Antenna Engineering*, (McGraw-Hill, 1952), Chapter 4, "Wire Stringing," p. 466.

Leeson, D., W6NL, *Physical Design of Yagi Antennas* (Newington, CT: ARRL). [out of print]

Leeson, D., W6NL, "Joint Design for Yagi Booms," *QEX*, June 1993, pp. 6 – 8.

Leeson, D., W6NL, "Strengthening the Cushcraft 40-2CD," *QST*, Nov. 1991, pp. 36 – 42.

Morris, S., K7LXC, *Up the Tower* (Champion Radio Products, 2009)

Weber, R., K5IU, "More on Strengthening Yagi Elements," *QST*, Oct. 1992, pp. 65 – 66.

Weber, R., K5IU, "Structural Evaluation of Yagi Elements," *Ham Radio*, Dec. 1988, pp. 29 – 46.

Schiller, T., N6BT, Array of Light 3rd Edition (Next Generation Antennas, 2010)

Schiller, T., N6BT, Array of Light 4rd Edition (Next Generation Antennas, 2022)

Smith, C., *Carroll Smith's Nuts, Bolts, Fasteners and Plumbing Handbook* (Motorbooks, 1990).

TABLE OF CONTENTS

Chapter 26
Safe Practices

This chapter covers three aspects of antenna system safety: work practices and techniques, electrical grounding and bonding, and managing RF exposure. Amateur radio activities are not inherently hazardous, but like many things in modern life, it pays to be informed. Safety begins with our attitude. Make it a habit to plan work carefully. Don't be the one to say, "I didn't think it could happen to me."

Having a good attitude about safety is not enough, however. We must be knowledgeable about common safety guidelines and follow them faithfully. Safety guidelines cannot possibly cover all situations, but if we approach each task with a measure of common sense, we should be able to work safely. If you are contemplating either a simple or significant tower or antenna project, read the articles and references listed in the References and Bibliography

Consider hiring professional assistance if you are not comfortable with doing antenna or tower work yourself. Tower services with amateur experience are listed on various amateur Product Review websites such as **eham.net**. Be sure to get references and contact recent customers for reviews. Like any service you select to do work around your home or business, the service should be licensed, bonded, and insured in case of accidents or other problems.

Many of the safety products mentioned in this chapter are available from numerous *QST* advertisers. They should be able to supply everything you need in order to do the job safely and correctly, resulting in years of trouble-free service at the least possible risk. Learning and practicing the right way to do things will save you time, money, and worries.

OSHA and Tower Work

OSHA is the federal Occupation Safety and Health Agency (**osha.gov**) that sets minimum safety standards for workers. Each state has an agency that is responsible for enforcing the OSHA regulations in that state. In addition, your state agency may have stricter regulations than OSHA; OSHA regulations are just the minimum requirements.

If you are getting paid or paying someone to do tower work, you or they must comply with the federal and state regulations. If you are simply working on your own system, or someone else's without pay, then you don't fall under the OSHA/state laws. But you should still observe them! You should use only OSHA/state approved safety equipment and follow the regulations applicable to your activity. By doing this, you'll be giving yourself a large and acceptable safety margin while working.

26.1 Antenna Safety and Safety Equipment

An antenna at height is a critical component of an effective amateur radio station. Hams have a long history of finding creative ways to get their antennas to the necessary height. We don't always think of safety first, but regardless of whether the chosen antenna support is as simple as a tree, or as sophisticated as a tall, purpose-built tower, every antenna project can present many potential risks to life and limb. In this section, Jim Idelson, K1IR, guides us through the basics of safely working on antenna systems and the use of safety equipment.

The topic of antenna and tower safety is by no means an academic pursuit. In 2020, there were three recorded fatalities related to amateur radio tower work. There were another three in 2019. Additionally, we know that there have been many additional serious injuries. The causes of these tragic incidents were all identifiable in advance and avoidable. With careful attention and care, no ham needs to lose his or her life or sustain serious injury, when working with an antenna.

Working aloft is dangerous, pure and simple. Whether it's 20 or 200 feet tall, climbing towers is dangerous. You're always at risk. Fear of heights is one of the most common phobias, affecting an estimated 3 to 5 percent of the population.

Due to the dangers inherent in such work, working safely is paramount. That translates into having and using the proper equipment to ensure one does not get hurt. Today, that means a proper harness, proper clothing and tools, along with some training or instruction. This focus on safety and equipment is critical.

In this section, we first introduce some important ways of

thinking and working that are likely to make for a safe day of antenna work. Next, we focus on some of the specific scenarios and hazards most commonly encountered when working with antennas at height. Some of the more frequently encountered hazards include:

- Electrical hazards
- Structural failure of antenna supports.
- Hazards associated with lifting and lowering equipment
- Falls from substantial height
- Challenging weather
- Falling objects
- Equipment failure

26.1.1 Thinking About Safety: a Big-Picture View

Safety isn't something you can just set-and-forget. It's not just buy-and-be-done. It's a way of thinking; it's pervasive; it's ingrained in everything you do. Done right, safety practices result in long-term benefits.

Every antenna project has a lifetime that comprises several important phases. Those phases include Planning, Construction, Maintenance, and Removal. Let's look at each of these phases and the safety practices can be incorporated into each one.

THE PLANNING PHASE

Every antenna project starts with some sort of *Planning*. Whether it's a simple wire antenna or a much bigger installation involving a tower and large rotatable antennas, you'll need to make a range of important decisions before you begin to build.

Design is perhaps the most critical part of the planning phase. In your design, you will consider questions about the location of the antenna system, the size, weight and strength of the pieces and parts that will be employed, and the strength of all the supporting components. Factors like climate and weather, desired lifetime, maintenance requirements, budgets and local regulations will all factor into your antenna system design. Getting the design right is particularly important because it sets the stage for what will happen throughout the full lifetime of the system. A great design is likely to result in a system that will be effective, reliable, and safe over the course of its entire lifetime.

Conservatism is the watchword for a safe project. Maximum safety can best be ensured by creating the path forward from a conservative perspective. Being conservative in terms of goals and specifications will not only increase safety margins for the project; it will reduce the effort and cost associated with the later phases of construction, maintenance, and removal.

THE CONSTRUCTION PHASE

Construction is the time when the antenna or antenna support structure is erected, usually a day — or several — of concentrated effort by a team of hams. Construction is one of the two phases in the life of an antenna or tower during which risks of injury are at their peak. Some of those involved might have experience while others are new to the process of antenna and tower construction. Fueled by excitement and anticipation of the moment when the project will be ready to test, the build team is often motivated to get the job done quickly. With pressure to complete any job in the least possible amount of time comes the temptation to cut corners. The best of planning is easily compromised when team members decide to skip steps, work with insufficient skills or tools, or omit critical procedures. This is a moment when everyone must be vigilant about keeping safe practices front-and-center.

THE MAINTENANCE PHASE

Once the new antenna or tower has been fully installed and tested, its service life begins. Every owner/operator of an antenna system hopes for a long lifetime of trouble-free operation. But wear-and-tear is a fact of life for all mechanical and electrical devices. Some forms of degradation are predictable, like loosening of nuts and bolts, corrosion, wear on ropes exposed to friction. These can and should be addressed with preventative maintenance steps. Annual checkups on hardware are a must. Replacing antenna support ropes before they fail — every other year is reasonable — will prevent much bigger problems and will help to avoid the need for much bigger and riskier repair projects.

THE REMOVAL PHASE

The second very risky phase in the life of an antenna project — particularly with respect to a tower — is the removal phase. A variety of potential risks present themselves during the take-down of an existing tower. Let's consider two of the most common safety risks related to removal of a tower.

Unknown Construction Materials and Methods: It is not uncommon for hams to assist in removal of a tower or other antenna structure when a ham is moving or is otherwise no longer able to use it. After asking the first few questions, it is very common to find that the original construction details were not documented. Is there concrete in the base and anchors of a guyed tower? How much? How is the roof tower attached to the roof? What material was used for the mast? Can it be identified simply by visual inspection? When the answers to these questions are unknown, workers face significant risks. If there isn't sufficient information to make a safety determination, there is only one correct plan of action — walk away and call in the professionals.

Old Installations: When an installation has been in place for many years, the risks multiply. Corrosion is almost certain to have invaded the hardware. Frequently, the corrosion is invisible; even a trained professional may have difficulty detecting it. The opportunity to take on a removal project in exchange for the salvage value of a tower and antennas is a temptation that should be avoided. If you are not highly trained and experienced in working with tower and antenna installations — walk away and call in the professionals.

26.1.2 Build in Safety from the Start

Working at height is a potentially dangerous activity. As you think about climbing a roof or a tree or a tower, it is important to frame every decision you make in the context of safety. Continually asking the question, "Is this the safest way?" should become second nature. In the workplace, this is known as developing a safety mindset. Mature organizations and teams build safety considerations into every project they undertake, fully recognizing the costs in time and resources that come with keeping people and property safe. The world of amateur radio is no different. Payback on investments in safety are returned over the long term, manifest in reduced rates of injury and fatalities. Developing a reputation as a safe pursuit also helps to foster the permissive and trusting environment, we need to pursue the experimental aspects of amateur radio.

In this section, we will focus much of our thoughts on tower climbing, but many of the concepts are transferable to other forms of work aloft.

TAKE ADVANTAGE OF STANDARDS

Commercial and government standards organizations provide a great deal of guidance in the area of safety equipment and technique. In the United States, OSHA, the Federal Occupational Safety and Health Administration, collects and analyzes real-world safety data. The agency then works with industry in the creation and publication of high-level regulations for workplace safety. These rules include minimum standards for equipment and procedures. ANSI (American National Standards Institute) is an independent non-profit organization that encourages further development of detailed requirements for safety in many areas, including tower safety. The Telecommunication Industry Association (TIA) produces the most widely accepted standards package covering tower safety — ANSI/TIA-222-H (publ. 2017). Although amateurs working on their own non-commercial projects without compensation are not bound by these rules and recommendations, you'll be much better off following them. There are many suppliers of excellent safety equipment that meets and exceeds the industry standards. If you choose compliant equipment and learn safe techniques, you will be on the path to a safe and successful project.

BEFORE YOU BEGIN: PREPARATION

As you contemplate any new project, you will need to scope the entire job.

- What are your goals? How will you define success?
- What strategies could you use to achieve the goals?

Consider a job that's diagnostic in nature. You believe there is a problem with a coax cable getting hung up on some tower hardware. The goal is to observe the coax as you turn the antenna rotor through 360 degrees of rotation. The traditional approach would be to send a climber up the tower for direct, close-up observation. But new technology can offer faster, easier, and safer options. Consider observing with binoculars from the safety of the ground or an upper floor in a nearby building. Or send a drone right to the location in question and capture recorded video through the full rotation.

Another concept that can greatly reduce risks and make a job go more smoothly is to break the work down into well-defined modules. Successful completion of one module should be required before proceeding to the next. When you break the work down this way, you gain the ability to manage your time more precisely and you have a built-in series of go/no-go decision points.

When you decide that climbing is the right approach, take stock of all elements of the job.

- Skills and experience
- Parts and supplies
- Safety equipment
- Tools
- People
- Time
- Conditions (weather, physical impediments, safety status)

One of the most important aspects of safety that you must assess is whether you have the knowledge and awareness to do a job safely and efficiently. You must have the mental ability to climb and work at altitude while constantly rethinking all connections, techniques, and safety factors. Safely climbing and working on towers is 90% mental. Mental preparedness is something that must be learned. This is an occasion where there is no substitute for experience. Establishing and maintaining the right mental attitude is the hardest step for most people. But when safety gets the focus it deserves, tower work can be done safely, and accidents are relatively rare.

PRE-WORK SAFETY INSPECTION

Before climbing any tower, perform a careful inspection of the base, anchors, and guys. A pair of binoculars allows inspection of the tower and guys above eye level. There should be no visible rust on the base section, especially where it contacts concrete. Look for splits from water freezing in the tower legs. If any of these are apparent, do not climb until further checks and any repairs are done. A small hole (1⁄8 inch typically) may be drilled in the bottom of a tower leg without compromising strength to drain any standing water. Guys should be correctly tensioned and all guy attachments and anchors secure and free of rust.

If the tower is a crank-up model, lower it completely until the lift cable is slack. Tower sections can bind and appear to be lowered completely but slip from a climber's weight, causing serious injury. Before climbing, block all tower sections with a 2×4 or heavier material so no downward motion is possible. Never assume a crank-up tower is fully nested. For tilt-over towers or tower-lowering fixtures, perform a careful inspection of the winch, cable, and fixture before raising or lowering the tower. When the tower is near horizontal, the tension in the cable can be thousands of pounds. Make sure no one is near the tower as it is raised or lowered and wear protective gear when operating the winch.

26.1.3 Climbing Safely

100% TIE-OFF: A RULE TO LIVE BY

Since its inception in the 1970s, OSHA has focused on workplace safety. In the 1980s, OSHA introduced the *100%*

tie-off standard requiring a worker's body belt to be secured by at least two lanyards. This protection factor became essential for employees who executed tasks at heights. Fortunately, the body belt soon evolved into a full harness, which further protected workers from spinal and internal organ injuries.

OSHA's fall protection standards even cover fall protection for residential construction projects. Personal fall arrest systems must bring a worker to a complete and controlled stop, restrict workers from falling more than 6 feet, and have the strength necessary to endure twice the impact energy of the worker falling 6 feet.

Why is this type of system necessary? 100% tie-off is part of your plan to 'expect the unexpected'. While you can be in top physical condition, equipped with the most modern tools and trained in the latest techniques, unexpected situations still arise. When everything is going well, there is no reason to expect to suddenly 'just let go' when perched on the side of the tower. But workers can and do experience sudden health emergencies. A heart attack or a fainting episode due to dehydration can render a worker unconscious. Unavoidable distractions can occur, as well. What if you were to accidentally disturb a bees' nest and suddenly get stung while in your work position? Any of these situations could cause you to lose consciousness or the attention needed to maintain your footing. In moments like these, your 100% tie-off will save your life.

Fall protection is continuing to evolve in today's workplace, as employers and regulators seek new ways to protect workers. OSHA has ongoing research on how to identify hazards and is also improving current regulations. As a ham, if you know and take to heart the concept of 100% tie-off, you will be well on your way to making your tower work as safe as possible.

Working on a Crank-up Tower

One of the advantages of a crank-up is the ability to bring the antennas down to rooftop level or close to ground level where it's easier for the station owner to work on it. For this convenience, the price that you pay is the added mechanical complexity and cost of the crank-up apparatus. They can cost two to three times the cost of a guyed tower of the same height.

Another limitation is the fact that a crank-up cannot be climbed safely once it is extended. Do not climb a crank-up tower unless it is totally nested and locked in the lowered position! Again, all of the weight of the system is on the cable and pulley systems and if something breaks or comes loose, your toes and fingers are in the path of the tower sections as they fall! If the tower is jammed and won't come down, don't climb it to fix it. Get a boom truck or crane in to lift you up to work on it. Better yet, get professional help.

It is possible to climb a crank-up if you can lock it into place. One method is to use 3- to 4-foot-long pieces of 2×4 or pipes. Another is to place a U-bolt on at least one leg under each section. Insert them at the bottom of each section through the bracing and they'll catch each section before it can move down very far. You can also gently lower the tower until it rests on the safety pieces, thus jamming them into place and eliminating any tower movement at all.

It will still be quite difficult to climb the nested lattice-work sections. Having small feet is an advantage here, as silly as that may sound. It may even be difficult to find a suitably rated attachment point for your lanyard. Again, the safest and simplest method of working on a crank-up tower is to utilize a manlift.

26.1.4 Working Safely

It's not a race! Take your time and climb safely. If you become tired or uncomfortable, stop and rest with your lanyards securely fastened to the tower. If you feel unsafe at any time for any reason — stop and return to a safe position or configuration!

SAFETY CLIMB SYSTEMS

Most commercial towers have a safety climb system, typically a 3/8-inch steel cable that runs from the top to the bottom of the tower. The climber attaches to a special trolley with a cable from the climber's harness. The trolley will slide up freely but clamps the safety cable if weight is put on it, thus preventing you from sliding down the cable and tower. Such installations are rare on amateur towers but are worth considering.

THE MENTAL ASPECT

One of the most important aspects of safety is having the knowledge and awareness that will enable you to perform work safely and efficiently. You must have the mental ability to climb and work at altitude while constantly rethinking all connections, techniques, and safety factors. Climbing and working on towers safely is 90% mental. Mental preparedness is something that must be learned. This is an instance when there's simply no substitute for experience.

When it comes to tower climbing, only a small percentage of people are willing to climb and work aloft. The biggest obstacle is making the mental adjustment. Properly installed towers are inherently safe and accidents are relatively rare. The only thing stopping most people is their own mind and attitude.

Would you have any trouble standing on a 24 × 24-inch piece of plywood on the ground? Probably not! But could you stand on that same four-square-foot platform 100 feet in the air? The only difference is in your mind. It's easier said than done, but you must make the mental adjustment if you are going to do any tower work.

An important lesson learned from mountain climbing that is directly applicable to tower climbing is that when you climb, you have four points of attachment and security — two hands and two feet. When tower climbing, move only one point at a time. That leaves you with three points of contact on the tower and a wide margin of safety if you ever need it. This is in addition to having your fall-arrest lanyard connected at all times.

Another recommended technique to excel at the mental game is to always do everything the same way every time.

Principles of Working Safely

The following safety tenets are founded on three fundamental principles: Do it safely or not at all; There's always time to do it right; and if it's worth doing, do it better.

1. Never load or operate structures or equipment outside the design limits. Be careful with tools, ropes, pulleys, and other equipment that can cause injury or damage if they fail due to overload. Use the right stuff!

2. Always move to a safe, controlled condition and seek assistance when a situation is not understood. This is particularly important when working on towers and antennas. If something doesn't look right or isn't going according to plan, return to a safe state and figure out what to do.

3. Always operate with the safety mechanisms engaged. If a safety mechanism prevents you from doing something, either the task is unsafe, or you may not be using the right equipment.

4. Always follow safe work practices and procedures. Make a plan before you start and don't do something you know is unsafe.

5. Act to stop unsafe practices. The team's safety depends on every team member. Do not hesitate to stop work if you see it is unsafe. Don't be afraid to speak up or ask for help! Regroup and do it right.

6. Clarify and understand procedures before proceeding. This is particularly important when working with a crew. Be sure everyone understands the procedure and how to communicate.

7. Involve people with expertise and firsthand knowledge in decisions and planning. Ask for advice and guidance from experienced hams when planning a task with which you are unfamiliar.

That is, always wear your positioning lanyard on the same D-ring and always connect it in the same way. Always look at your belt D-ring while clipping in with your safety strap. This way you'll always confirm that you're belted in. Don't rely on the "clicking sound," or assume you're belted in and clipped on. Always look — always!

CHECK YOUR SAFETY EQUIPMENT

You should also check your safety equipment every time before you use it. Inspect it for any nicks or cuts to your belt and safety strap. Professional tower workers are required to check their safety equipment every day; follow their example.

INCLEMENT WEATHER

Tower work is the easiest when the weather is nice, and the sun is shining. Unfortunately, that doesn't always coincide with your construction schedule or repair priority. Don't hesitate to call off your project and wait for the right weather. If you're not sure if the weather is good enough, it probably isn't.

For raising tower sections or antennas, a relatively windless day is preferred. Professional climbers usually do their trickiest lifts first thing in the morning when the chance of wind is the least. Don't push on in marginal conditions; you may wind up doing more harm than good. Obviously, you don't ever want to climb during a lightning storm.

As far as rain goes, unless it's coming in horizontally it's more of a nuisance. For ham towers, you'll always be belted in, and you won't be walking across any rain-slicked surfaces, so working in the rain is possible. Just dress in good rain gear and you'll be able to still get some work done. Some ham towers have been painted, and their surfaces become slick when wet, so exercise extreme caution if that's what you're working on. And while the workers aloft may find they can continue working, the ground crew (or the client) suffers more, getting wet while waiting for something to happen. Again, don't hesitate to wait for better conditions.

ELECTRICAL SAFETY

Electrocution due to metal antenna or tower parts touching power lines is the biggest cause of tower related electrical injuries. *Be very careful if you're anywhere near power lines.*

Even without touching a power line, you can still be electrocuted while working on a tower, which is a large, grounded conductor. A major cause of tower injuries and deaths is electrocution. While there usually isn't much 120 V ac circuitry on amateur towers, care should still be taken around ac power. Use battery-powered equipment, if possible, both for convenience and safety. If you do use ac extension cords, make sure they're plugged into a GFCI (ground fault circuit interrupter) for your protection. Power tools operated from ac should be double-insulated. Part of your pre-work safety meeting should be pointing out where the circuit breaker box is in case someone has to turn off the power.

SAFETY TIPS FOR TOWER WORK

- Don't climb with anything in your hands; attach it to your harness if you must climb with it or have your ground crew send it up in a bucket after you're in position.
- Don't put any hardware in your mouth; you could swallow it or choke on it.
- Remove any rings and/or neck chains; they can get hooked on things.
- Be on the lookout for bees, wasps, and their nests; there aren't too many bigger surprises when you're climbing a tower. If you do get stung, apply a meat tenderizing powder containing the enzyme papain, such as Adolph's Meat Tenderizer, directly on the sting moistened with a little water or saliva. The enzyme neutralizes the venom and reduces the pain within a minute or two. Keep a bottle in your tower tool kit. If you are allergic to stings, always have an Epi-Pen or the equivalent with you. Benadryl is another recommended medication for stings.
- Don't climb when tired; that's when most accidents occur.
- Don't try to lift anything by yourself; one person on a tower has very little leverage or strength. Let the ground crew use their strength; save your strength for when you really need it, or you'll quickly run out of arm strength.
- If something doesn't work one way, rerig, and then try again.

26.1.5 Safety Equipment

PERSONAL FALL-ARREST SYSTEM (PFAS)

To implement 100% tie-off, you need the right equipment. OSHA specifies the equipment that must be used. Minimum requirements for *Personal Protective Equipment (PPE)* are given in OSHA Standard 1910.140 - Personal fall protection systems (**osha.gov/laws-regs/regulations/standardnumber/1910/1910.140**).

The most important pieces of safety equipment are the *personal fall arrest system (PFAS)* you wear and the accompanying lanyards that attach to it (**Figure 26.1**). The body harness portion of the PFAS, known as a *fall-arrest harness (FAH)*, has leg loops and suspenders to help spread the fall forces over more of your body and can hold you in a natural position with your arms and legs hanging below you where you're able to breathe normally.

Two or more lanyards are used. The first is the critically-important *dual fall-arrest lanyard.* This lanyard attaches to a D-ring at the base of your neck, between your shoulder blades. Two lanyards extend from this point and attach to the tower above your work position. The dual lanyard design allows you to always have one leg of the lanyard secured to the tower. This is the most flexible and cost-effective way to implement your 100%-attach climbing strategy. The simplest lanyards of this type use 6-foot ropes or ballistic nylon straps. They are inexpensive, but don't offer any shock absorption. Shock absorbing varieties are much preferred. These lanyards typically have bar-tacked stitches that pull apart under force and decelerate you as you fall (see **Figure 26.2**). Should you ever have a fall, and the shock-absorbing component is activated, the lanyard is no longer considered safe. It should be disposed of and replaced before the next climb.

The second lanyard is the *positioning lanyard* shown in **Figure 26.3**. That is, it holds you in a working position and attaches to the D-rings at your waist. Positioning lanyards can be adjustable or fixed and are made from materials such as nylon rope, steel chain or special synthetic materials. An adjustable positioning lanyard will adjust to fit various tower diameters, while a fixed-length lanyard is often either too long or too short for the job. The rope type is the least expensive version. It is important to remember that positioning lanyards are not considered part of the safety system. They provide a level of comfort and convenience during long periods on an antenna support structure, but they do not provide protection in the event of a fall.

Hams should note that leather climbing equipment was

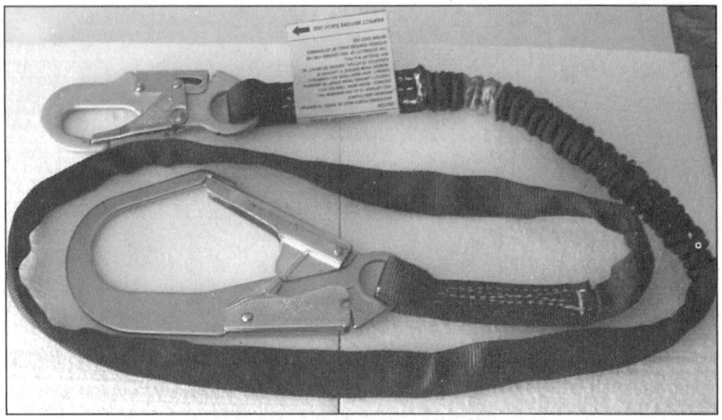

Figure 26.2 — A shock absorbing fall arrest lanyard. Portions of the nylon webbing are sewn together and pull apart under the weight of a climber falling on it, thus decelerating the fall.

(A) (B) (C)

Figure 26.1 — KI4TZ wearing a Personal Fall Arrest system (A). With shock-absorbing lanyard, positioning lanyard, and a safety hook, he's ready to work aloft. Long pants, hard hat, steel shank boots, and eye protection are included. Shock-absorbing lanyard is attached to the Fall Arrest Harness's rear D-ring (B). The waist-level pouch (C) allows easy access to required tools.

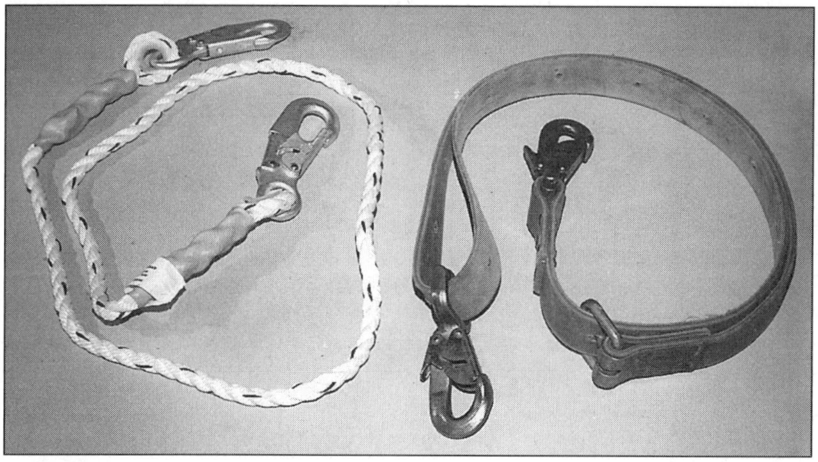

Figure 26.3 — A fixed-length rope positioning lanyard on the left and a versatile Klein adjustable lanyard on the right. They both use double-locking snap-hooks.

outlawed some years ago by OSHA — so don't use it. This includes the old-fashioned climbing belt that utility pole linemen used for years but which offers no fall protection capability. If you drop down while wearing a safety belt, your body weight can cause it to shift up from your waist to your ribcage where it could immobilize your diaphragm and suffocate you. On the other hand, you can use a safety belt for positioning when it's worn in conjunction with a compliant FAH. In fact, most modern fall arrest harnesses also incorporate a waist belt with positioning lanyard attachment rings for a single, convenient, comfortable climber safety package.

But having the right gear doesn't ensure it is in good working order. Take a page from the OSHA regulations. Professional tower workers are required to check their safety equipment every day. Make it your habit to check your safety equipment before you use it — every time. Inspect for any nicks or cuts to your harness and lanyards immediately prior to each use.

Mountain Climbing Harnesses

Mountain climbing harnesses are not an acceptable substitute for a tower-climbing fall arrest harness. While less expensive, they are designed to be used only with climbing ropes and hardware, not with tower tools or equipment. The first problem with using a mountain climbing harness is that most require you to tie the harness directly to a rope or to a carabiner and most hams are not skilled at tying climbing knots properly. You could use a locking carabiner as an attachment point, but it is another piece of hardware that could fail or open up at the wrong time.

Second, there are no D-rings for attaching any sort of positioning lanyard; you can only connect one carabiner to the loops in the front. The nylon loop on the front of the climbing harness is only designed to position the leg loops and is intended to be used only with a climbing rope or carabiner, not the metal snaps of your lanyard that are frequently snapped on and off. They also don't have any provisions for convenient attachment of tool or bolt bags.

The final problems are that a mountain climbing harness may be designed for a force of only 1000 pounds while OSHA-approved fall-arrest gear must be designed for 5000 pounds of strength. Also, the mountain climbing harness has no fall-arrest capability. Although the main advantage of a mountain climbing harness is low cost, its limitations prevent it from being recommended for tower work.

One caveat . . . if you are in the market for new climbing gear, you should not overlook the purpose-built systems offered by manufacturers that have traditionally served the recreation climbing market, but now have professional safety products in their portfolio. Some of their products are outstanding — meeting the highest safety standards while taking advantage of the latest in light-weight, comfortable technical fabrics.

Do not even consider using harnesses not designed for climbing work, such as lightweight tethers or belts for hunting or other recreational uses. Please: use only the tools designed specifically for the job!

BOOTS

Boots should be leather with a steel or fiberglass shank. Diagonal bracing on Rohn 25G tower is only 5/16-inch rod — spending all day standing on that small step will take a toll on your feet. The stiff shank will support your weight and protect your feet; tennis shoes will not. Leather boots are mandatory on towers such as Rohn BX that have sharp X-cross braces, plus your feet are always on a slant and that combination can be quite painful and hard on your feet.

HARD HATS

The hard hat is highly recommended. Just make sure hard hats are OSHA-approved and that you and your crew wear them. As you'll be looking up and down a lot while wearing your hard hat, a chin strap is essential to keep it from falling off. Look for the ANSI or OSHA label on the hard hat; that should be the minimum safety compliance for your helmet.

SAFETY GOGGLES

Approved safety goggles should be worn to prevent eye injury. Look for ANSI or OSHA approval.

GLOVES

If you do a lot of tower work, your hands will take a beating — gloves are essential. Keep several spare pairs for ground crew members who show up without them. Cotton gloves are fine for gardening, but not for tower work; they don't provide enough friction for climbing or working with a haul rope. Leather gloves are the only kind to use; either full leather or leather-palmed are fine.

The softer the gloves the more useful they'll be. Stiff leather construction gloves are fine for the ground crew, but the pigskin and other soft leathers allow you to thread a nut or do just about any other delicate job without removing (and possibly dropping) your gloves.

Framers or mechanics gloves (with the fingertips removed) keep your hands warm with enough fingertip exposed to perform fine work. Some types also have straps for hanging the gloves on the tower while you are working.

SAFETY EQUIPMENT SUPPLIERS

Chances are that you've got a safety equipment store in your area, but your best bet is to search the Internet for what you need since tower climbing equipment is not very common. Manufacturers such as Klein, Petzl, DBI-Sala and others all provide OSHA-approved safety equipment. These are more expensive products, but they're preferred by professionals who wear them and work in them all day. These companies will have many other useful accessories and rigging supplies, such as rope, canvas buckets, tool pouches and other hardware.

26.1.6 Tower Climbing Best Practices

Just having the right gear does not ensure that it will save your life; you must use it correctly. The most important rule for using a full body harness and dual fall arrest lanyard is to follow these 100% tie-off steps as you climb. This is illustrated in **Figure 26.4**.

• Attach one lanyard hook above you as you start your ascent. In the case of most ham towers, attach to a tower leg, as the leg provides the strongest anchor point. A tower rung is often not strong enough to sustain the downward force of a fall as specified in the OSHA requirements.

• As you climb, attach the second lanyard hook to a tower leg a few feet above the first.

• Disconnect the lower hook and continue to climb.

• As you climb, alternately place the hooks such that you are always securely connected by one of lanyard ends and try to keep your attachment point above you as much as possible. This will limit the maximum distance you can fall at any moment.

• When you reach your work position, attach both lanyards above you for redundant fall protection while you do the job.

• Reverse these steps as you descend.

Note that the sequence described above does not call for use of a positioning lanyard. That is because, while a positioning lanyard is helpful in providing comfort while climbing or while working, it does not meet the stringent requirements for fall arrest. So, you can — and should — use positioning

(A) (B) (C)

Figure 26.4 — Proper 100% tie-off technique (see text). Lanyard attachment at the work position is shown in A. Note that the fall-arrest lanyard is attached above the climber. (It is attached to a leg, not a rung.) Getting ready to climb with both fall-arrest lanyards attached is shown at B. Moving one of the fall arrest lanyards while maintaining tie-off is shown at C. (Jim Idelson, K1IR, photos.)

The National Electrical Code (NEC)

The National Electrical Code (a.k.a. — "the Code") is a comprehensive document that details safety requirements for all types of electrical installations. In addition to setting safety standards for house wiring and grounding, the Code also contains a section on Radio and Television Equipment — Article 810. Sections C and D specifically cover "Amateur Transmitting and Receiving Stations." Highlights of the section concerning Amateur Radio stations follow. If you are interested in learning more about electrical safety, you may purchase a copy of *The National Electrical Code* or *The National Electrical Code Handbook*, edited by Peter Schram, from the National Fire Protection Association, Batterymarch Park, Quincy, MA 02269. Both are available at libraries, as well.

Antenna installations are covered in some detail in the Code. It specifies minimum conductor sizes for different length wire antennas. For hard-drawn copper wire, the Code specifies #14 AWG wire for open (unsupported) spans less than 150 feet, and #10 AWG for longer spans. Copper-clad steel, bronze or other high-strength conductors may be #14 AWG for spans less than 150 feet and #12 AWG wire for longer runs. Lead-in conductors (for open-wire transmission line) should be at least as large as those specified for antennas.

The Code also says that antenna and lead-in conductors attached to buildings must be firmly mounted at least 3 inches clear of the surface of the building on nonabsorbent insulators. The only exception to this minimum distance is when the lead-in conductors are enclosed in a "permanently and effectively grounded" metallic shield, such as a conduit. The exception also applies to coaxial cable.

According to the Code, lead-in conductors (except those covered by the exception) must enter a building through a rigid, noncombustible, nonabsorbent insulating tube or bushing, through an opening provided for the purpose allowing a clearance of at least 2 inches or through a drilled windowpane. All lead-in conductors to transmitting equipment must be arranged so that accidental contact is difficult.

Transmitting stations are required to have a means of draining static charges from the antenna system. An antenna discharge unit (lightning arrestor) must be installed on each lead-in conductor (except where the lead-in is protected by a continuous metallic shield that is permanently and effectively grounded, or the antenna is permanently and effectively grounded). An acceptable alternative to lightning arrestor installation is a switch that connects the lead-in to ground when the transmitter is not in use.

Grounding conductors are described in detail in the Code. Grounding conductors may be made from copper, aluminum, copper-clad steel, bronze, or similar erosion-resistant material. Insulation is not required. The "protective grounding conductor" (main conductor running to the ground rod) must be as large as the antenna lead-in, but not smaller than #10 AWG. The "operating grounding conductor" (to bond equipment chassis together) must be at least #14 AWG. Grounding conductors must be adequately supported and arranged so they are not easily damaged. They must run in as straight a line as practical between the mast or discharge unit and the ground rod.

lanyards to make your work easier, but don't depend on them for fall protection.

Many climbers have developed ways of thinking about movement on a tower that force safer behavior. One such model is the concept of having four points of attachment and security — two hands and two feet. When climbing, move only one point at a time. That leaves you with three points of contact and a wide margin of safety if you ever need it. This is in addition to having your fall-arrest lanyard connected at all times. When climbing with a pair of lanyard hooks, make a habit of thinking "On Then Off" so that the disconnected hook is attached *before* taking the connected hook off the tower.

Another recommended technique is to always do everything the same way every time. That is, always wear your positioning lanyard on the same D-ring and always connect it in the same way. Always look at your belt D-ring while clipping in with your safety strap. This way you'll always confirm that you're securely belted in. Always look!

26.1.7 Insurance

It is important that you have insurance that covers any potential liabilities (someone getting hurt on the tower, damage caused by tower failure, and so on) as well as the physical equipment itself.

The ARRL also offers the Ham Radio Equipment Insurance Plan (**arrlinsurance.com**). Your mobile and home station equipment is covered on an all-risk form which includes fire, lightning, theft, collision, and other accidents and natural hazards. Loss or damage to antennas, towers or rotators is covered. Review the policy's coverage at the insurance plan's website for complete information.

Ray Fallen, ND8L, an agent for State Farm Insurance for over 20 years, wrote an informative article on insurance in the February 2009 issue of *QST*, titled "Homeowners Insurance and Your Antenna System," which is included with this book's online material. Ray also wrote Chapter 12 of *Antenna Towers for Radio Amateurs*, which is focused on insurance.

26.2 Grounding and Lightning Protection

The subject of grounding and lightning protection is very broad, covering everything from low-frequency ac safety through RF electromagnetics. As such, a thorough treatment is well beyond the scope of this section although certain important points and concepts can be introduced and discussed. This section covers the requirements for safety grounding of the antenna system components and discusses ground or earth connections used for lightning protection.

Providing "cookbook" solutions for grounding and lightning protection is unrealistic because every station and antenna system is different. Local codes, soils, and lightning environment all vary from location to location. Thus, the goal is to provide general guidance, define common terms, and identify authoritative sources of information so that the underlying principles can be applied to a specific station's needs.

It is recommended that station builders study the references and articles cited in this section, including those provided with this book's online material, when designing and constructing their own station. The References and Bibliography list several well-known and respected industry standards for lightning protection of military and commercial communications and broadcast facilities. While a professional-quality installation may be out of reach for amateurs, the references provide guidance for how to best use the limited resources available.

This section also includes material from the **Safe Practices** chapter of *The ARRL Handbook* and other ARRL publications such as *Grounding and Bonding for the Radio Amateur* (see References and Bibliography). This section includes material updated by Jim Lux, W6RMK, and Ward Silver, NØAX. The ARRL's Technical Information Service web page on Safety (**arrl.org/safety**) also contains lists of useful references and guidelines for installing protective systems in your station.

26.2.1 Station Grounding

PERMITTING

Most locations impose some kind of permit process on significant outdoor building projects, such as towers. Along with mechanical and civil engineering concerns for whether the structure is sound and appropriately sited, electrical requirements for grounding and lightning protection are also included.

Find out what building codes apply in your area and have someone explain the regulations about antenna installation and safety. For more help, look in your telephone directory or online for professional engineers, electricians, and contractors. Accommodating local code requirements at the beginning of your project is usually a lot less expensive than trying to bring a non-compliant system up to code later.

Your local electrical supply houses and distributors are good sources for both references to contractors and materials.

AC SAFETY GROUNDING

The National Electrical Code (NEC) specifies standard methods for dealing with electrical shock and fire hazards. Article 250 deals with grounding and bonding. Article 810 deals with radio and television equipment. Your local code may refer to these sections or may impose different or more stringent requirements. In either case, the local code has authority over your installation. The NEC is published by the National Fire Protection Agency (NFPA) — **nfpa.org**.

Reducing the detailed language of the NEC to simple terms, you must connect all exposed metal from ac-powered equipment to a central, common ground. All equipment electrically connected to an ac-powered device should have a permanent safety connection, even if the equipment is unpowered, such as an antenna tuner or audio switch. This includes masts and towers, along with any other outdoor equipment.

This connection is usually referred to as the *ac safety ground* and the connection is made through the "third wire" of your ac power wiring; the bare or "green" wire. If a short circuit develops between the ac wiring and the enclosure of a piece of equipment, the resulting fault current in the ground connection trips a circuit breaker in the hot conductor. Leakage current is mostly due to capacitance between the phase (hot) ac conductor and the equipment chassis or enclosure. This includes capacitors used for ac line filtering and stray capacitances such as between a power transformer's windings and its metallic frame, which is usually bonded to the equipment enclosure and thus to the equipment's ground connection.

Ground-Fault Circuit Interrupter (GFCI) breakers go one step further and monitor the balance of current on the hot and neutral lines of an ac circuit. If an imbalance is detected, it is assumed that the missing current is flowing on the equipment chassis or enclosure, where it can present a shock hazard and the GFCI trips to remove power.

Since the ac safety ground is concerned with currents at the power line frequency and its first few harmonics, the length of the connection is relatively unimportant. Similarly, resistance in the ac safety ground path or imbalances of a few ohms between circuits doesn't matter much from the perspective of safety. (Imbalances might be significant for signal-level connections.) The important thing is that hazardous current takes a path that doesn't go through a human being.

BONDING

Bonding refers to connecting equipment enclosures and earth connections (ground rods) together, so they are at the same electrical potential, even if there is a short-circuit in the ac power connection or a lightning strike. The goal of bonding is to minimize the voltage between exposed metal surfaces to reduce shock hazards. In the amateur station, bonding also minimizes RF current flow between pieces of equipment that can cause improper or degraded operation.

Equipment can be bonded together regardless of whether there is an earth connection. Bonding connections should be made with conductors that are heavy enough to be mechanically secure and that have minimal resistance.

The minimum recommended grounding conductor for bonding earth connections together is #6 AWG stranded wire. Copper strap (or flashing) should be a minimum of 1.5 inches wide and 0.05 inch thick. In the station, bonding equipment enclosures together can be done with #12 or #14 AWG wire or flat-woven, tinned copper, braided strap. For lowest impedance, the ratio of length to width of the braid should be about five or less. Mounting holes should be installed at each end of the braid, not solder lugs. Use grounding fittings and blocks for connecting the bonding wires and strap together. Use stainless steel washers or shim when connecting conductors of dissimilar metals.

Do not use braided strap outdoors or where exposed to moisture because the individual strands oxidize over time, greatly reducing the effectiveness of braid at RF. Use bare copper for buried ground wires. (There are some exceptions; seek an expert's advice if your soil is corrosive.) Braid removed from coaxial cable should not be used as an RF bonding conductor. Without the protection and compression of the jacket, the weave of round braid loosens, and the individual wires will oxidize.

Exposed runs above ground that are subject to physical damage may require additional protection (such as a conduit) to meet code requirements. Wire size depends on the application but never use anything smaller than #6 AWG for bonding conductors. Local lightning-protection experts, electricians, or building inspectors can recommend sizes for each application.

STATIC DISSIPATION

Wind and precipitation can develop a large static charge on any ungrounded antenna or tower, up to thousands of volts. If not discharged to ground, the charge will eventually build up and arc to a convenient grounded point. This can obviously lead to equipment damage. There are several methods of dissipating static charge without affecting the antenna system. The first two suggestions are primarily useful at HF.

- High-value, non-inductive resistors of 10 kΩ or more connected from ungrounded feed line conductors to ground allow small currents from static electricity to be discharged safely. The resistors should be rated for at least 1 kV to avoid arcing from high RF power levels and induced voltage surges from nearby lightning strikes. Metal-oxide resistors of ½-W or higher power rating can withstand short pulse overloads and are inexpensive.
- RF chokes below their self-resonant frequency provide a dc path to ground for static while having sufficiently high reactance to have a minimal effect on impedance. Note that wire-wound chokes use fine wire that can act as a fuse and become an open circuit from a lightning-induced transient.
- DC-grounded antenna designs use shunt inductors to ground at the feed point or have driven elements connected to a supporting structure, such as a gamma-matched Yagi. These antennas dissipate static charges continually.
- Quarter-wave shorted stubs (see the **Transmission Line System Techniques** chapter) act as an open-circuit at the design frequency while providing a direct path to ground for static electricity. This technique only works on a single band and can't be used on multiband antennas.

26.2.2 Lightning Protection

Effective lightning protection system design is a complex topic. There are a variety of system tradeoffs that must be made and that determine the type and amount of protection needed. Hams can easily follow some general guidelines that will protect their stations against high-voltage events that are induced by nearby lightning strikes or that arrive via utility lines. The basic techniques are thoroughly explored in the three-part 2002 series of *QST* articles "Lightning Protection for the Amateur Station" by Ron Block, KB2UYT (now NR2B), that are included with this book's online material and at **arrl.org/lightning-protection**. Another useful document on lightning protection is available from the IEEE, "How to Protect Your House and Its Contents from Lightning" (see the References and Bibliography).

You may also wish to use the services of a local lightning protection expert to advise you on specific techniques appropriate for your area. Companies that sell lightning-protection products may offer considerable help to apply their products to specific installations. The References and Bibliography section of this chapter contains a partial list of PolyPhaser's publications (**polyphaser.com**).

TOWER AND FEED LINE GROUNDING

Because a tower is usually the highest metal object on the property, it is the most likely strike target. Proper tower grounding is essential to lightning protection. The goal is to establish short multiple paths to the earth so that the strike energy is divided and dissipated.

Connect each tower leg and each fan of metal guy wires to a separate ground rod. Space the rods at least 6 feet apart. Bond the leg ground rods together with #6 AWG or larger copper bonding conductor (form a ring around the tower base, see **Figure 26.5**). Connect a continuous bonding conductor

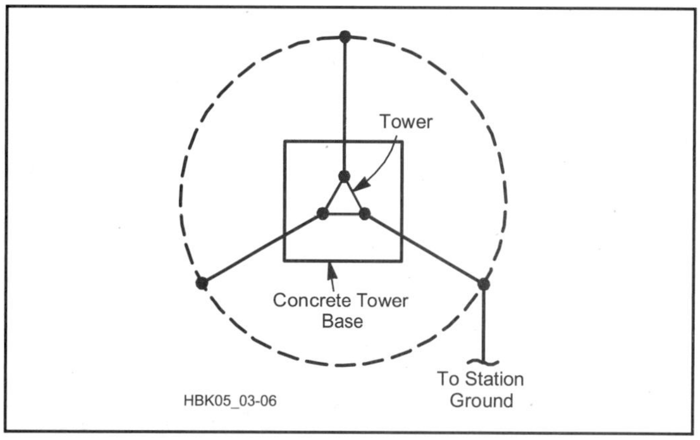

Figure 26.5 — Schematic of a properly grounded tower. A bonding conductor connects each tower leg to a ground rod and a buried (1 foot deep) bare, tinned copper ring (dashed line), which is also connected to the station ground and then to the ac safety ground. Locate ground rods on the ring, as close as possible to their respective tower legs. All connectors should be compatible with the tower and conductor materials to prevent corrosion. See text for conductor sizes and details of lightning and voltage transient protection.

Welded Earth Connections

The best way to ensure that a buried ground rod connection remains effective is to weld it. Conventional welding may be used but a far easier method is to use exothermic welding in which a fast-burning fuel melts the rod and connecting wire together, resulting in a mechanically and electrically secure connection. Often referred to as "one-shots," the best known products in this line are the CADWELD products by Erico (**erico.com**) and the Harger Uni-Shot (**harger.com**).

between the tower ring ground and the station's cable entrance panel.

Make all connections with fittings approved for grounding applications. Do not use lead-tin solder for these connections — it will be destroyed in the heat of a lightning strike. If the fitting is to be buried, use welded connections (see the sidebar) or clamps that are specifically rated for direct burial.

Because galvanized steel (which has a zinc coating) reacts with copper when combined with moisture, use stainless steel hardware between the galvanized metal and the copper grounding materials. Alternatively, Burndy GC-18A and Panduit both supply "dual-rated pipe ground clamps" to connect aluminum and copper or galvanized surfaces and copper. They accept tubing or pipe up to 2" in diameter.

To provide an alternate path for strike energy to entering your home or station via the feed line, ground the feed line outside the home. Ground the coax shield to the tower at the antenna and the base to keep the tower and line at the same potential. Several companies offer grounding blocks that make this job easy to do properly.

BONDING EARTH CONNECTIONS

Significant voltages can be created between separated earth connections, especially during lightning surges. The solution is to bond all external earth connections together, creating a *perimeter ground* as shown in **Figure 26.6**. The perimeter ground may extend completely around the building but if that is not practical, extend the ground system as far as you can with ground rods every 10 feet or so.

The ground rods and buried bonding conductor provide a low impedance path for the lightning's charge outside the building and help keep equipment at close to the same voltage at low frequencies. Both are important — keep as much of the lightning energy as possible outside while minimizing large voltage differences and current surges that damage equipment inside.

The *service entrance ground* connection from the ac power distribution panel to a ground rod establishes a local earth connection for lightning protection of ac-powered equipment and appliances. All other earth connections should be bonded to the service entrance ground.

All earth connection bonding requires at least #6 AWG copper wire. This includes lightning protection conductors, electrical service, telephone, antenna system grounds, and underground metal pipes. Any ground rods used for lightning

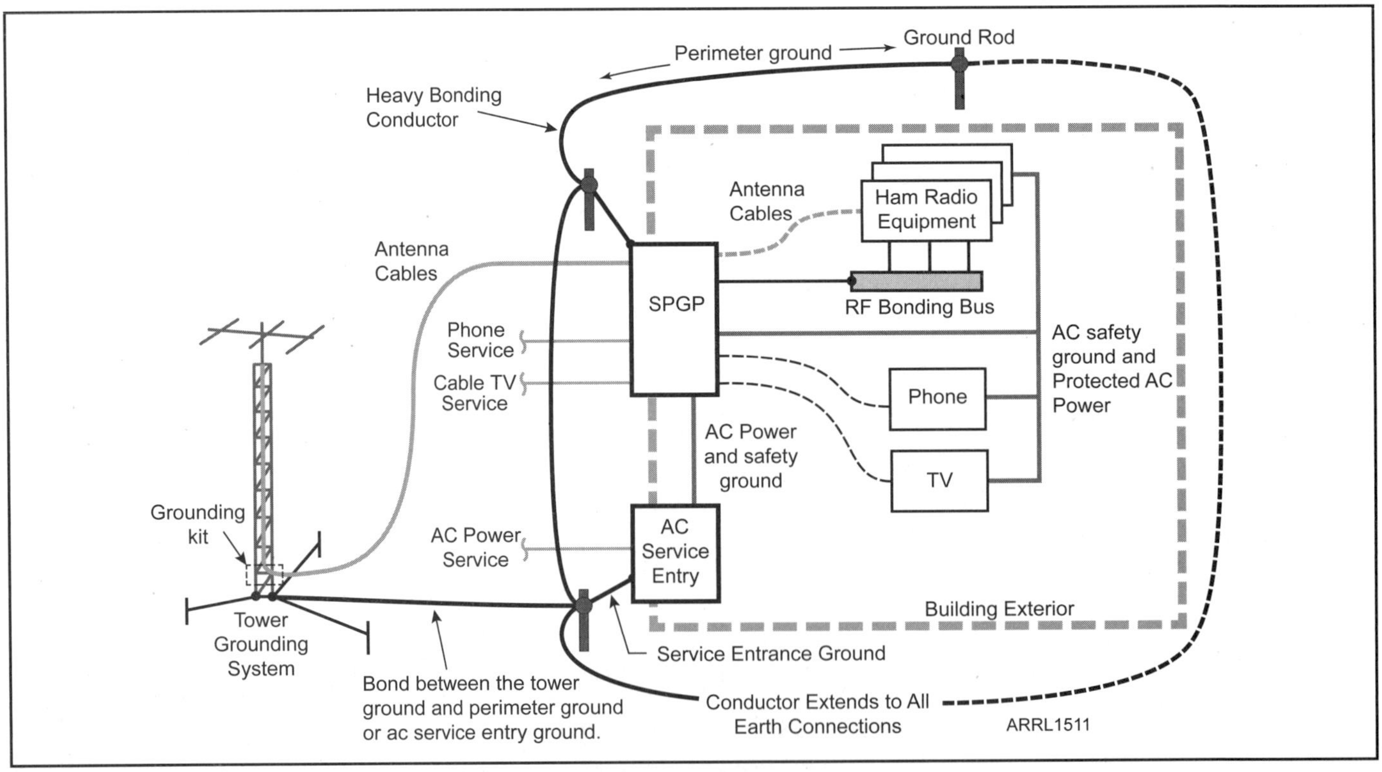

Figure 26.6 — A lightning protection system attempts to dissipate charge outside the home or station. All earth connections are bonded together to provide a low-impedance path for lightning, and to minimize voltage differences and current flow between equipment and circuits.

protection or entrance panel grounding should be spaced at least 6 feet from each other and the electrical service or other utility grounds and then bonded to the ac service entrance ground as required by the NEC and local codes.

UFER GROUND

If you live in a dry or rocky area where ground rods are difficult to install or ineffective, a Ufer ground or "concrete-encased electrode" is often used. The Ufer ground relies on the conductivity of concrete to contact the soil over the large area of a foundation slab or tower footing. Reinforcing bar (rebar) inside the concrete is bonded to building steel or a tower with a heavy bonding conductor.

More information on Ufer grounds is available in NEC section 250.52, the PolyPhaser book *Lightning Protection and Grounding,* and in *Lightning Protection & Grounding Solutions for Communication Sites* (see References and Bibliography).

CABLE ENTRANCE PANEL

The basic concept for lightning protection is to make sure that all the radio and other equipment is bonded so that it "rises and falls together" in the presence of a transient voltage. It's not so important that the station be at "ground" potential, but, rather, that everything is at the *same* potential. For fast rise-time transients, such as the individual strokes that make up a lightning strike, even a short wire has enough inductance that the voltage drop along the wire is significant, so whether you are on the ground floor, or the 10th floor of a building, your station is "far" from Earth potential.

The easiest way to ensure that everything is at the same potential is to tie all the signals to a common reference. In large facilities, this reference would be provided by a grid of large diameter cables under the floor, or by wide copper bars, or even a solid metal floor. A more practical approach for smaller facilities, like an amateur station, is to have a common connection point for all the cables. Often referred to as an *entrance panel* or *single-point ground panel* (SPGP), the object is to provide a common potential reference for shields and signal connections.

The easiest way to create an entrance panel is to install a large metal enclosure or a metal panel as a bulkhead and grounding block in the exterior wall. The panel should be bonded to the lightning dissipation ground with a short wide conductor, and, like all grounds, bonded to the electrical system's ac service entry ground. This is illustrated in Figure 26.6 which shows the various grounding systems at a typical home station. (W8JI's website at **w8ji.com/station_ground.htm** shows examples of an entrance panel system and wiring practices.)

If multiple entry panels or lightning arrestors are used for different services, such as telephone or cable TV, connect those panels or lightning arrestors to the outside bonding conductor that connects all the ground rods together.

Every conductor that enters the structure, including antenna system control lines, should have its own lightning arrestor or surge suppressor on an entrance panel. Suppressors are available from a number of amateur equipment vendors, as well as the usual electrical equipment suppliers such as Square-D, Graybar, and so forth. (See Parts 1 and 2 of the *QST* articles by Block for illustrations.)

Mount all lightning arrestors, protective devices, switches, and relay disconnects on the outside facing wall of the bulkhead. The enclosure or panel should be installed in a way that if lightning currents cause a component to fail, the molten metal and flaming debris do not start a fire. To avoid creating another path to ground around the panel, route the feed lines, rotator control cables, and so on at least six feet away from other nearby grounded metal objects.

LIGHTNING ARRESTORS AND SURGE SUPPRESSORS

Feed line lightning arrestors (known as *antenna discharge units* in the NEC) are available for both coaxial cable and parallel-conductor or balanced line. Most of the balanced line arrestors use a simple spark gap arrangement, but a balanced line impulse suppressor is available from several vendors.

DC-blocking arrestors for coaxial cable have a fixed frequency range. They present a high-impedance to lightning energy below 1 MHz while offering a low impedance to higher-frequency RF.

DC-continuous arrestors (gas tubes and spark gaps) can be used over a wider frequency range than those that block dc. Where the coax carries supply voltages to remote devices (such as a mast-mounted preamp or remote coax switch), dc-continuous arrestors must be used. **Figure 26.7** shows examples of typical arrestors for coaxial cable.

UPPER FLOOR STATIONS

A common situation is for an amateur station to be on the floor of a building well above ground level. For these stations, **Figure 26.8A** shows two common methods of routing feed lines and control cables to an upper floor station, either from ground level or above ground directly to the station. In both cases, an SPGP must be created at the station as close as practical to the cable entry point with its own grounding conductor to the external ground electrode. All lightning protectors must be mounted on the SPGP so they will fire at the

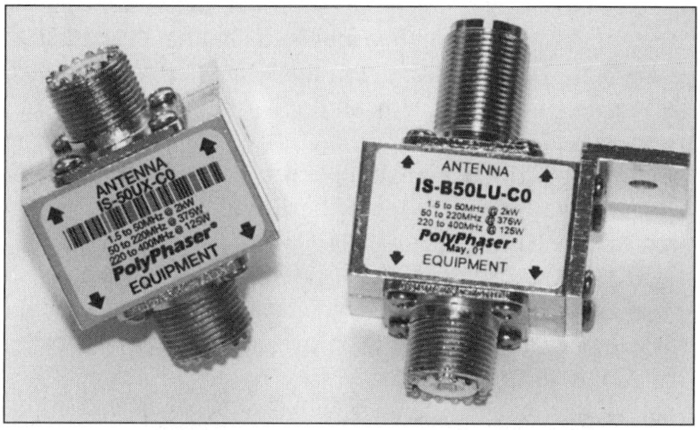

Figure 26.7 — Typical coaxial lightning arrestors from PolyPhaser. These are mounted on a cable entrance panel that is connected to a ground rod.

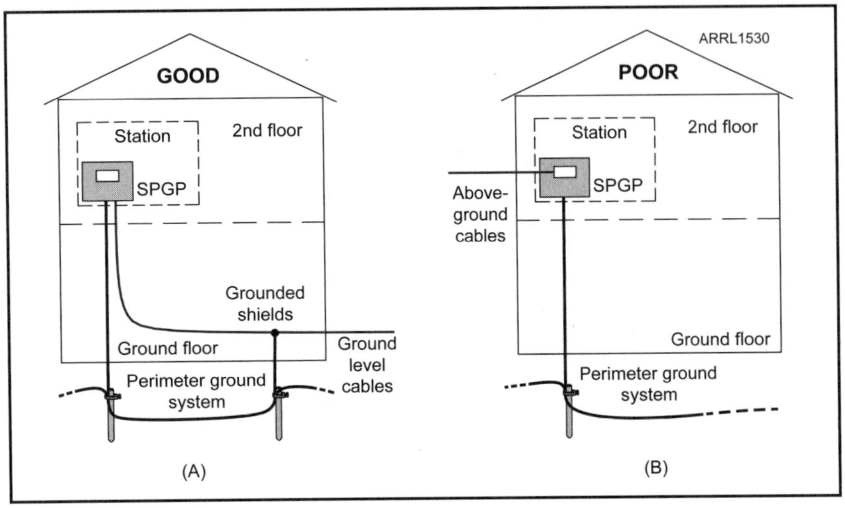

Figure 26.8 — Protecting an upper-floor station against lightning is a challenge. Construct a single-point ground panel (SPGP, see text) entrance panel at the station level. Run a large strap or metal pipe to the perimeter ground system. The best option is to ground the shields of the feed lines at ground level before running the feed lines to the entry panel.

same time. Even if lightning protectors are added elsewhere outside the building, an SPGP at the station is still required. If a separate feed-through panel is used, it must be bonded to the SPGP. The SPGP must have its own grounding conductor to the external ground conductor to the external ground electrode. All cables must be run outside the building until they reach the entry point where the SPGP is located. Your first choice should be running above-ground cables to ground level before they reach the station entry point.

Running antenna feed lines and other cables directly into an upper-floor station as in Figure 26.8B creates a very difficult lightning protection challenge. An entrance panel on the upper floor will help equalize voltage differences between the cables but unless building steel framework is available as a low-impedance earth connection, the long bonding conductor to the perimeter ground will result in significant voltage differences with respect to other ground-referenced wiring such as ac power. In such cases, it is best to disconnect all cables at the entrance panel when not in use or when storms are in the area.

PROJECT: ANTENNA FEED LINE PROTECTION UNIT

This project was originally published as "Antenna Feed Line Control Box" by Phil Salas, AD5X, in the August 2014 issue of *QST*. The complete article, including a parts list, all construction details, design formulas, and performance measurements, is included with this book's online material. Commercial versions of this project are sold as "antenna disconnectors" by Inrad (**vibroplex.com**) and Paradan Radio (**paradanradio.com**) — see the Product Review of these items in the April 2019 issue of *QST*.

Regardless of whether you build the device yourself or purchase it, the connection to your station's ground system is very important. The connection should be direct, low-impedance, and short, using heavy wire or strap. This will help reduce the voltage difference between the protector and any other piece of station equipment also connected to the ground system. The commercial versions completely disconnect both conductors of the feed line from the radio. See this chapter's section on "Grounding and Lightning Protection" for more information about lightning protection.

Design

While the design in **Figure 26.9** is not a substitute for a fully grounded entrance panel with lightning arrestors, it can make a good backup protection system, particularly for stations on upper floors of buildings or with marginal lightning protection grounding. The intent is to deal with voltage pulses caused by nearby lightning strikes and static buildup on antennas, both of which can damage equipment in the station.

Static buildup (often termed precipitation static) on antennas can reach thousands of volts. This can be prevented with bleeder resistors connected from the coax center conductor to ground.

As most of the energy in a lightning strike is concentrated well below 500 kHz, the high reactance of the blocking capacitors will attenuate much of this low frequency energy. This high reactance also allows the voltage to spike. Therefore, gas discharge tubes will fire for impulse voltages greater than about 800 V, shunting that energy to ground.

Normally the blocking capacitors will simply serve to pass RF current from the unit's inputs to the ANT outputs. Capacitors passing heavy current are subject to heating and will dissipate power P_d proportional to the product of the equivalent series resistance ESR and the square of the RF current I: $P_d = I^2 \times ESR$. This power dissipation can be minimized by choosing capacitors with a low dissipation factor DF. Because $ESR = DF \times X_c$ where X_c is the capacitor's reactance, the power dissipated by the capacitor can now be expressed as $P_d = I^2 \times DF \times X_c$. Stating X_c in terms of the frequency f and capacitance C yields $P_d = (I^2 \times DF) / (2\pi \times f \times C)$.

Capacitor heating can be further reduced by paralleling capacitors, which reduces the current through each capacitor. Assuming the aggregate current remains the same, paralleling two equal-value capacitors will reduce the current through each by a factor of 2. Because the power dissipated is proportional to the square of the current, the net effect will be to reduce the individual capacitor power dissipation by a factor of 4.

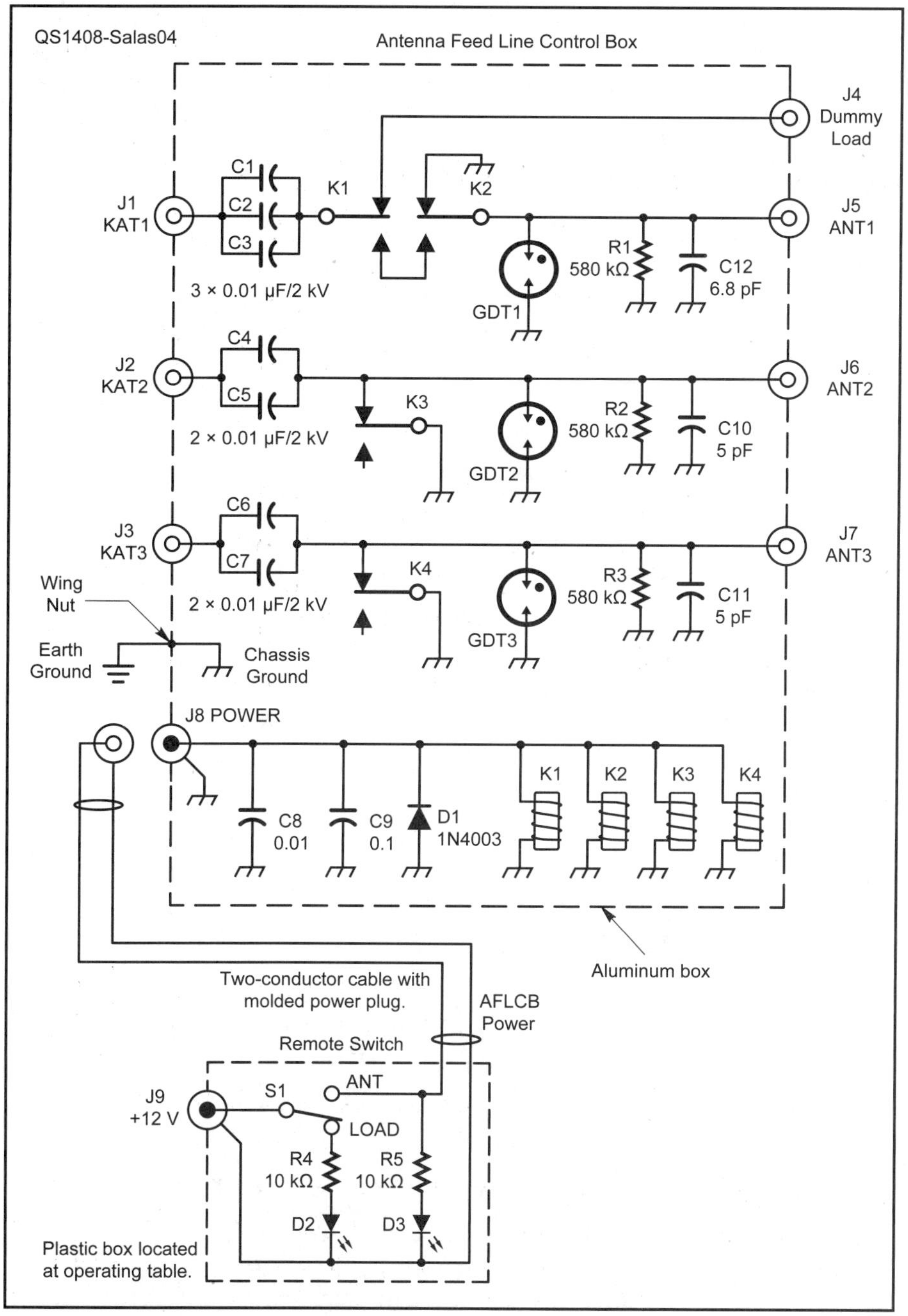

Figure 26.9 — Schematic diagram and parts list for the antenna feed line protection box.

C1 – C7 — 0.01 μF, 2 kV capacitor (Mouser 594-S103M69Z5UP63K7R).
C8 — 0.01 μF, 100 V ceramic disc capacitor (Mouser 140-100Z5-103Z-RC).
C9 — 0.1 μF, 50 V ceramic disc capacitor (Mouser 140-50U5-104M-RC).
C10, C11 — 5 pF, 1 kV capacitor (Mouser 75-561R10TCCV50).
C12 — 6.8 pF, 1 kV capacitor (Mouser 75-561R10TCCV68).
D1 — 1N4003 diode (Mouser 512-1N4003).
D2, D3 — Green LED (Mouser 941-C5SMFGJSCV14Q7S1).
GDT1 – GSD3 — Gas discharge tube 800 V (Mouser 652-2095-80-BLF).
R1 – R3 — 580 kΩ 1/2 W 3.5 kV resistor (Mouser 594-HVR3700005903FR5).
R4 – R5 — 10 kΩ, 1/4 W resistor (Mouser 66-CMF1/41002FLFTR).
J1 – J7 — SO-239 connector (Mouser 601-25-7350).
J8, J9 — DC jack 2.1 × 5.5 mm (Mouser 163-1060-EX).
K1 – K4 — SPDT power relay (Mouser 655-RTB14012F).
S1 — SPDT toggle switch (Mouser 108-0009-EVX).
DC power cable (with compatible plug for J8) — 2.1 × 5.5 mm × 3 feet (Mouser 172-4204).
Plastic box — 1.38 × 1.38 × 0.79 inches (Mouser 546-1551MBK).
Aluminum box — 4.3 × 3.3 × 1.6 inches (Mouser 563-CU-5471).

Schematic

When remote switch S1 is set to LOAD, power is removed from SPDT relays K1 – K4. The disposition of the relay contacts shown in the schematic is for the power off condition, which is also the state of the unit when the station is off the air. In the power off state, relays K2 – K4 connect signals ANT1 – ANT3 (J5 – J7) to chassis ground, which in turn is connected to the station single point ground via a wing nut connection on the outside of the aluminum box. This prevents static charge buildup on the antennas and provides a path to ground for pulses due to nearby lightning strikes. Additionally, when S1 is set to LOAD, relay K1 connects KAT1 to DUMMY LOAD through SO-239 connectors J1 and J4 respectively.

When remote switch S1 is set to ANTS, +12 V is supplied to the unit through POWER input J8, which in turn is connected to the coils of relays K1 – K4. Diode D1 suppresses the kickback voltage from the relay coils when power is removed and capacitors C8 and C9 bypass any RF voltage to ground.

In the power on state, the ground shunt connections made by relays K2 – K4 of signals ANT1 – ANT3 are removed along with the KAT1 connection to DUMMY LOAD through K1. Then, KAT1 is connected to ANT1 via K1 and K2 and blocking capacitors C1 – C3. The author paralleled three capacitors on this port because this is his only 160-meter connection.

The paths through the unit are similar, so we'll take the path from KAT2 (J2) to ANT2 (J6) as a representative case. Capacitors C4 and C5 are wired in parallel to share RF current in order to minimize heating due to their equivalent series resistance. In the power on state, since ANT1 – ANT3 are no longer grounded, the unit is vulnerable to pulses from nearby lightning strikes and to static charge buildup on the antennas. In the event of a nearby lightning strike, the high reactance of C4 and C5 will block much of the energy from the pulse. And if the voltage spike should reach 800 V, gas discharge tube GDT2 will fire, and the energy will be shunted to ground. While operating, static charge buildup is prevented by bleeder resistor R2 connected from the signal line to chassis ground.

The small-value capacitor C10

compensates for the impedance bump due to the unit's internal wiring. Without the capacitor the SWR of the unit degrades to about 1.3:1 on 6 meters when the unit is perfectly terminated. With the capacitor, the SWR is less than 1.03:1 (36 dB return loss) on 6 meters. Of course, these capacitors are not really necessary as the protection box follows any antenna tuner. However, any impedance bump can degrade a less-than perfect feed line SWR enough to make the difference between bypassing a tuner and the need for a tuner to be in-line, especially at the higher frequencies.

Note that the voltage ratings of the components in this path are higher than typical: R2 is rated at 3.5 kV, where an ordinary resistor is only rated for a few hundred volts; blocking capacitors C4 and C5 are rated at 2 kV and C10 is rated at 1 kV. The relays, while inexpensive, work well through 6 meters, providing 1 kV RMS of isolation between contacts, 5 kV RMS of isolation between the coil and the contacts, and contacts rated at 12 A continuous current.

Figure 26.10 — View of parts placement inside the aluminum box housing the unit. The wing nut at the upper left connects the chassis ground of the unit to the station single-point ground at the cable entrance panel.

Construction

Figure 26.10 shows a photo of the completed unit built into a cast aluminum box that has a mounting bracket that permits it to be rigidly mounted in place. The relays (with pins up) are hot-glued to the aluminum box. A step-drill or ⅝-inch Greenlee punch works well for cutting the SO-239 connector holes. Note that there is also a ground stud consisting of a #8 stainless steel screw, nut, lock washer, and wing nut for connecting to your station's single-point ground.

26.3 RF Safety

(This section was written by Gregory Lapin, PhD, PE, N9GL, chair of the ARRL RF Safety Committee. Additional contributions were made by Ric Tell, K5UJU, Kai Siwiak, KE4PT, Peter Zollman, G4DSE, and Ian White, GM3SEK. The ARRL's RF Safety Committee reviewed this content. RF Safety is also addressed in the **Safe Practices** chapter of the *ARRL Handbook* that was updated with the 100th edition and applies to references here.) Additional material is available on the ARRL's RF Exposure web page at **arrl.org/rf-exposure**.

26.3.1 Safe Use of RF Energy

As licensed radio amateurs, we all need to ensure that our transmissions do not expose any people beyond the levels that have been deemed to be acceptably safe by regulatory agencies. Standards organizations have reviewed and analyzed scientific research that has been performed for over 60 years and derived safe levels of exposure for people. Regulatory agencies, such as the FCC in the United States, have developed rules that require all operators of RF transmitters to maintain human exposure levels below specified limits. (See the sidebar "Standards, Science, and the Community.")

An RF safety program should be developed for each amateur radio station. The program begins with an analysis of potential exposure levels around the station. If exposure of people greater than the regulatory limits is possible, the program then determines what forms of mitigation will be necessary to prevent any person's exposure from exceeding the regulatory limits. Although not required by the FCC, documentation of the RF safety program should be filed away so that it can be referred to if there is ever a question about RF exposure from that station.

Standards, Studies, and the Community

You or your neighbors are likely to have questions about where standards for RF Exposure come from or how science evaluates the effects of RF on people. There are two papers on the **arrl.org/rf-exposure** web page and in this book's online material to help understand these questions: "RF Safety Standard Development" and "Types of Scientific Studies." To help you explain to your neighbors that amateur radio is safe because you operate your equipment safely, the discussion "How to Interact with a Concerned Neighbor" can make the conversation easier for everyone. Finally, there is much information online with more coming out every day. "Interpreting the News About RF Exposure 'Discoveries'" will help you deal with information that might be out of context or simply untrue. These documents will be helpful as you navigate questions about RF and amateur radio transmissions.

The ARRL RF Safety Committee

The ARRL maintains an RF Safety Committee that is composed of engineering, scientific and medical experts in the many aspects of the study of RF safety. The RFSC serves as a resource to the ARRL Board of Directors, the ARRL Laboratory, and to the amateur community. It monitors new scientific research and many of its members participate in the scientific committees that develop safety standards for RF exposure. The RFSC participates in generating the RF safety questions for FCC amateur question pools and works with the FCC in developing its environmental regulations. For more information about the RFSC, see **arrl.org/arrl-rf-safety-committee**.

Part of RF safety is recognizing that non-ham acquaintances and family members may have concerns about RF exposure. Material in the *ARRL Handbook* will help you explain what hazards exist and what steps you have taken to ensure that you are operating safely. Some people may have concerns that are difficult to address. For such cases, the ARRL provides "How to Interact with a Concerned Neighbor," which offers some guidance in responding. In addition, the internet is a frequent source of information that is taken out of context, or simply false, which can also raise unwarranted concerns. The paper "Interpreting the News About RF Exposure 'Discoveries'" may be helpful when such issues are brought to you. Both papers are available at the ARRL's RF Exposure website, **arrl.org/rf-exposure**.

26.3.2 What Exposure Limits are Based On

Past health studies have been summarized by standards bodies that then identified levels of exposure considered to be safe. The two most recognized standards bodies are the IEEE International Committee on Electromagnetic Safety, ICES, and the International Commission on Non-Ionizing Radiation Protection, ICNIRP. The FCC has based its exposure regulations on both the IEEE C95.1-1991 standard and recommendations from the National Council on Radiation Protection (NCRP), a scientific group chartered by the U.S. Congress, in their NCRP Report #86. Other countries have based their exposure regulations on the ICNIRP guidelines. When using exposure tools, it is important to be aware of which standards have been used to make the calculations. To read more about the history of electromagnetic energy exposure standards development see the RF Safety Standard Development on the ARRL's RF Exposure web page (**arrl.org/rf-exposure**).

26.3.3 FCC RF Exposure Requirements

The FCC is given the responsibility under the National Environmental Policy Act of 1969 to control the operations of transmitters that it regulates to prevent adverse effects on the environment. In 1996, this duty was expanded to prevent the signals from all transmitters, including those of the Amateur Radio Service, from causing excessive human exposure. Since the FCC is not a health and safety agency, it has relied on accepted RF safety standards and advice from the FDA, NIOSH, and EPA to develop its rules.

MEASURES OF HUMAN EXPOSURE

The FCC RF exposure limits are based on an older IEEE standard, IEEE Std. C95.1-1991, which was later adopted as an American National Standard ANSI/IEEE C95.1-1992. The latest revision of IEEE standard is IEEE Std C95.1-2019, *IEEE Standard for Safety Levels with Respect to Human Exposure to Electric, Magnetic, and Electromagnetic Fields, 0 Hz to 300 GHz*. While the older IEEE standard used *specific absorption rate* (SAR), which is essentially a measure of the amount of RF energy the human body absorbs from a given source, as a basic dosimetric parameter in the formulation of RF exposure limits (Ed. — *a* dosimeter *is an instrument that measures exposure to doses of radiation, e.g., radio waves, X-rays, etc.*), the most recent IEEE standard uses *Dosimetric Reference Limits*, or DRLs, to specify the maximum rate at which RF energy may be safely absorbed within the body. (DRL is equivalent to Basic Restriction, or BR, used in the ICNIRP standard.) A difference introduced in the latest IEEE standard is that SAR is used as the DRL only from 100 kHz to 6 GHz. Incident power density is used as the DRL above 6 GHz, since at the higher frequencies, energy does not penetrate the body but is mostly absorbed at the surface. At those frequencies, the most sensitive organ that could be exposed is the cornea of the eye.

Exposure limits are divided into two classes: Occupational (or Restricted or Controlled), for those people who are aware of their exposure and have been trained how to control it, and General Population (or Unrestricted or Uncontrolled), for everyone else. The FCC considers radio amateurs and the other people living in their homes to be in the occupational category. To make that classification valid, radio amateurs are expected to educate the people living in their homes about the potential for RF exposure caused by operation of their stations and how to reduce exposure if they so desire. The household members should also be made aware of when the station is transmitting.

The basic measure of exposure is the SAR, measured in watts per kilogram of body mass. The SAR limit is a fixed value over much of the frequency spectrum and only varies with the part of the body that is exposed. The DRLs are lowest for whole body averaged exposure (0.08 W/kg for general population and 0.4 W/kg for the occupational population). Higher DRLs apply to localized exposure, which is averaged over a specified mass of tissue. The limits for localized exposure to the head and torso are 2 W/kg (in the C95.1-2019 standard, or 1.6 W/kg in the FCC regulations) for general population and 10 W/kg (in the C95.1-2019 standard or 8 W/kg in the FCC regulations) for the occupational population, averaged over any 10 grams of tissue in the C95.1-2019 standard or over any 1 gram of tissue in the FCC regulations. For the extremities (defined as the limbs and the external portions of the ears, in the C95.1-2019 standard or as the hands, wrists, feet, and ankles in the FCC regulations) the localized exposure

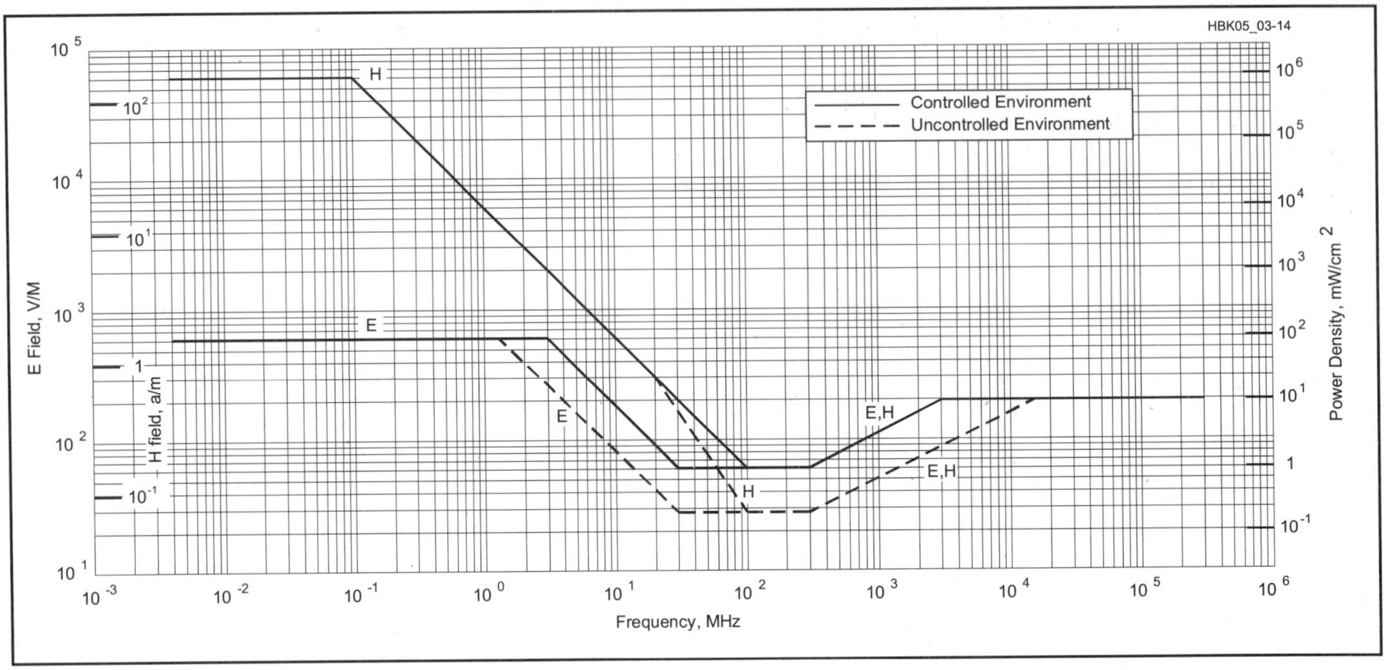

Figure 26.11 — MPE limits from ANSI/IEEE C95.1-1992 for the General Population (Uncontrolled Environment).

limits are 4 W/kg for general population and 20 W/kg for the occupational population, averaged over any 10 grams of tissue.

It is very difficult to measure or model SAR. To measure SAR, one would have to place a probe directly in the tissue at many locations and monitor the rate of absorbed electromagnetic energy over time. Modeling SAR is also complicated since the varying shapes of anatomical structures and their electromagnetic properties must be modeled over the entire body. While these tasks are not impossible, they are complex enough that few amateur licensees would have the ability or means to determine if exposure from their stations exceeds the FCC SAR limits.

To overcome this problem, the C95.1-2019 standard also provides *Exposure Reference Levels*, or ERLs (also known as Reference Levels in the ICNIRP guidelines), each of which relates a SAR limit to corresponding quantities that can be more easily measured or modeled. Unlike SAR limits, however, ERLs do vary with frequency, largely because of the different resonances that arise from the dimensions of body parts relative to the wavelengths.

Although most things affecting radio amateurs are unchanged between the C95.1-2019 standard and the FCC rules, some of the terminology is different. Most obviously, ERL is the new name for the FCC's *Maximum Permissible Exposure* (MPE) term. Another difference that is important to radio amateurs is that the FCC now requires any body part that is exposed within 20 cm of the source to be evaluated with the SAR. A potentially confusing difference between FCC regulations and the current standard is that the FCC references localized exposure to 1 gram of tissue in the head and torso, while the C95.1-2019 standard bases its DRLs on absorption in 10 grams of tissue.

In **Figure 26.11**, which represents MPEs recommended by the IEEE C95.1-1991 standard, the resonance effect is clear. The average human body and its larger structures are most resonant in the VHF range, and because of this the MPEs in that frequency range are the lowest (most restrictive). The entire VHF range, from 30 to 300 MHz, is assigned the most restrictive MPEs in order to encompass the different body sizes in the population, including children, as well as smaller resonant structures in the body, and differences between exposure in free space, and when standing on ground. Where the difference between wavelength and the size of body structures is greater, both at VLF, LF, low HF frequencies, and at microwave and higher frequencies, absorption is less and the MPEs are higher (less restrictive).

At lower frequencies the interaction between electromagnetic energy and the human body differs for the electric and magnetic components of the energy. To address this, in the frequency range up to 300 MHz there are different MPEs for the electric field strength (E, in V/m) and the magnetic field strength (H, in A/m), and both limits must be met. Above 300 MHz the MPE limits are expressed only by the power density (S, in mW/cm^2). When measuring or modeling power density at frequencies below 300 MHz one must be careful to work in terms of "plane-wave equivalent power density." Plane-wave conditions are a defining feature of the far-field, but this assumption is not valid in the near-field. Instead, the power density must be derived from both E and H at any given location, using a vector calculation that takes account of the direction and phase of each component. Because of this complexity, a common, but conservative, practice is to ensure that both the MPEs for E and H are complied with separately.

The graph in Figure 26.11 can also be expressed as a table in **Table 26.1**. The FCC has published their MPE limits in that form in their rules, in 47 CFR §1.1310(e). Unlike the plots

Table 26.1
Limits for Maximum Permissible Exposure (MPE)

Limits for Occupational/Controlled Exposure

Frequency Range (MHz)	*Electric Field Strength (V/m)*	*Magnetic Field Strength (A/m)*	*Power Density (mW/cm2)*	*Averaging Time (minutes)*
0.3-3.0	614	1.63	(100)*	6
3.0-30	1842/f	4.89/f	(900/f2)*	6
30-300	61.4	0.163	1.0	6
300-1500	—	—	f/300	6
1500-100,000	—	—	5	6

f = frequency in MHz
* = Plane-wave equivalent power density (see Notes 1 and 2 in Table 22.5).

Limits for General Population/Uncontrolled Exposure

Frequency Range (MHz)	*Electric Field Strength (V/m)*	*Magnetic Field Strength (A/m)*	*Power Density (mW/cm2)*	*Averaging Time (minutes)*
0.3-1.34	614	1.63	(100)*	30
1.34-30	824/f	2.19/f	(180/f2)*	30
30-300	27.5	0.073	0.2	30
300-1500	—	—	f/1500	30
1500-100,000	—	—	1.0	30

f = frequency in MHz
* = Plane-wave equivalent power density (see Notes 1 and 2).
Note 1: This means the equivalent far-field strength that would have the E or H-field component calculated or measured. It does not apply well in the near field of an antenna. The equivalent far-field power density can be found in the near or far field regions from the relationships:
Pd = |Etotal|2 / 3770 mW/cm^2 or from $P_d = |H_{total}|^2 \times 37.7$ mW/cm^2.
Note 2: $|E_{total}|^2 = |E_x|^2 + |E_y|^2 + |E_z|^2$, and $|H_{total}|^2 = |H_x|^2 + |H_y|^2 + |H_z|^2$

in Figure 26.11, power density is also shown in the table for frequencies below 30 MHz, under the condition that it be plane-wave equivalent power density. However, as illustrated in examples in this section, depending on the antenna type, either E or H can determine the minimum compliance distance below 300 MHz, so both must be evaluated.

Localized Exposure

The MPEs in the FCC rules presume uniform whole-body exposure. This may or may not be the case for typical amateur radio situations. In some circumstances, exposure is mainly limited to a portion of the body. Handheld radios are an important example of this. When you hold a handheld radio in front of your face, most of the exposure from its antenna is to your head and hand. The FCC uses 20 cm (about 8 inches) between the antenna and any portion of a person's body to distinguish between whether whole body MPE can be used for assessing compliance with their rules, or if a localized SAR determination must be made. For any antenna that is used less than 20 cm from a person, the only acceptable form of exposure determination is SAR.

Since SAR determination is beyond the capabilities of most radio amateurs, it is expected that manufacturers of handheld radios will perform that test and provide the results to the amateur, who will be able to use those test results to show compliance with the FCC exposure requirements. This requirement became official on May 3, 2021, and any handheld radios manufactured before that time are assumed to meet the exposure requirements even without SAR testing by the manufacturer. In support of this, a survey was made of the FCC equipment database for commercial handheld radios at frequencies just above the amateur 2-meter band with similar power output and manufacturer-provided flexible "rubber duck" antennas. This limited survey confirmed that the SAR measurements were below the FCC SAR limits, implying that the same could be presumed for comparable devices in the amateur market. Similar results were seen for devices that operated at frequencies just above the amateur 70 cm band (see the reference for Tell).

26.3.4 Averaging Factors Affecting Exposure

The MPEs are based on eliminating the whole-body heating effects caused by RF exposure. Both spatial averaging and time averaging affect actual whole-body exposure. In addition, the duty cycles of modulation type and operating pattern affect time-averaged exposure levels. Although these considerations add complexity to the evaluation process, doing so creates a more accurate determination that may change what seems to be an exposure exceeding the MPE to one that is actually in compliance.

SPATIAL AVERAGING

MPEs are expressed as spatial averages over the dimensions of the human body. This could become important in some circumstances where combining direct and reflected fields results in the observed field strength changing significantly with height. For comparison with the MPE, the spatially averaged E or H value is the square root of the average of the sum of the squares of the field strength samples along the averaging dimension. The FCC defines the vertical spatial averaging dimension to be 2 meters. In lieu of spatial averaging, the FCC also accepts the maximum field strength determined over the same dimension.

TIME AVERAGING

When exposed to RF, the body takes some time to heat up. The MPEs therefore apply to time-averaged exposures over a designated time period of either 6 minutes for the occupational population or 30 minutes for the general population. This process results in controlling the average level of thermal load imposed on the body. The time averaging permits short exposure to very high-power signals if followed by a period of very low exposure. This has the same heating effect on the body as a continuous lower exposure, as long as the average does not exceed the MPE over the designated averaging time.

MODULATION DUTY CYCLE

Many forms of modulation cause the peak power of the transmitted signal to vary over time. In addition, the human voice has variations that may be described in terms of peaks and valleys of voice amplitude and the natural pauses between syllables and words. Digital communication modes also have predictable spaces between transmitted symbols.

The *modulation duty cycle* averages out any short-term variations in power level created by the chosen mode of modulation. The modulation duty cycle of a constant-level unmodulated carrier is 100% as are frequency and phase modulation, which involve no changes in amplitude. The same is true for digital modes based on frequency or phase modulation.

By definition, all forms of amplitude modulation involve variations of amplitude with time. Notable examples include on/off keying of a carrier, AM, and SSB forms of voice modulation. In all these cases the modulation duty cycle is less than 100%, resulting in a reduction of the average RF exposure.

The combined effects of these types of variations lead to an average signal power that can result in lower exposure than would occur with an unmodulated signal. Some common averaging duty cycles for typical amateur radio modulation types are shown in **Table 26.2**.

Table 26.2
Common Exposure Averaging Duty Cycles for Common Modulation Types

Modulation Type	*Averaging Duty Cycle*
SSB (no speech processing)	20%
SSB (heavy speech processing)	50%
CW	40%
AM	50%
FM	100%
FSK/RTTY	100%
AFSK SSB	100%

OPERATIONAL DUTY CYCLE

During the mandated exposure averaging times of 6 minutes or 30 minutes, almost all amateur communication switches between transmit and receive several times. Since exposure occurs only during the transmit portion of the conversation, the transmit-receive cycle represents another reduction factor to the overall time-averaged value of exposure. A typical conversation between two radio amateurs would have each station transmitting for about 50% of the time. While contesting, the transmit duty cycle could either be lower, with short calls and longer times listening or higher with frequent CQ calls and short pauses for a reply. Each amateur should make a best estimate of the duty cycle of transmission for their form of operating.

An operational duty cycle is meaningful only if transmissions are significantly shorter than the averaging period. The mandated averaging times of 6 or 30 minutes represent 'rolling' periods, meaning that one should always evaluate the worst case. For instance, if an amateur transmits continuously for 5 minutes and then listens for 5 minutes, the worst-case duty cycle within a rolling 6-minute averaging period of occupational exposure, is 5 minutes out of every 6 minutes, or 83%. However, for this same transmit-receive timing in calculations for exposure of the general population, with a 30-minute averaging period, transmission takes up 15 minutes out of the 30-minute averaging period and the resulting duty cycle is 50%.

In 'freestyle' operation where much depends on human behavior, determining specific values of operational duty cycle can be difficult and the amateur should make exposure calculations based on the expected worst-case timing. However, the exact opposite is true of many modern digital modes because the operational duty cycle is fixed by the protocol used and can thus be applied with full certainty.

ACTUAL EXPOSURE AFTER AVERAGING

The exposure that is calculated for a person standing near an antenna can be lowered by factoring in both the modulation duty cycle and the operational duty cycle. For instance, imagine that you have calculated that a person in the general population could be exposed to the equivalent of 1.0 mW/cm^2, a power density that exceeds the FCC MPE of 0.2 mW/cm^2 by a large margin, when standing in a certain location near your 10 meter dipole. You may then apply time averaging. If you normally operate with uncompressed SSB and converse with a friend where you each talk for one minute and listen for one minute, Table 26.2 indicates that your modulation duty cycle is 20% (factor of 0.2) and your operational duty cycle will 50% (factor of 0.5). The actual averaged exposure of the person in question would then be:

$$1.0 \text{ mW/cm}^2 \times 0.2 \times 0.5 = 0.1 \text{ mW/cm}^2$$

Therefore, this location will meet the FCC MPE for the general population provided the station is operated as described. However, if you were to switch your modulation from SSB to FM, the modulation duty cycle would increase from 20% to 100% (factor of 1.0). With all other operating parameters being the same, the averaged exposure of the person in question would then become:

$$1.0 \text{ mW/cm}^2 \times 1.0 \times 0.5 = 0.5 \text{ mW/cm}^2$$

which is more than double the FCC MPE for the general population.

From a practical point, however, exposure evaluations are based on several assumptions, particularly about operating behavior. If a station is so close to the MPE limits that it depends on specific operating conditions be compliant, then the amateur should consider other ways to make sure that the station is always below the MPEs. For instance, the assumption of the 50% operational duty cycle may be incorrect the first time the amateur has a long story to tell. The section on RF Exposure Mitigation below may be useful to give the station some margin to allow for variability of the operating assumptions.

26.3.5 Responsibilities of the Radio Amateur

The FCC requires every amateur radio licensee to abide by its exposure rules. Even when you are not required to perform a complicated analysis, it is still your responsibility to make sure that no person is ever exposed above the MPEs listed in the FCC rules.

There are several ways to evaluate the exposure from your amateur radio station. For the purposes of exposure analysis, it is not necessary to determine the exact levels to which people might be exposed. It is only necessary to affirm that any exposure **will be less than** the FCC MPE limits. Any assessment method that provides credible confirmation of that point is all that you need to demonstrate compliance. Unless you want to explore your station's exposure levels further, you may use the simplest assessment method that serves your purpose.

The simplest forms of evaluation are designed to be highly conservative so that any operation passing the evaluation is highly unlikely to ever cause overexposure to any person near the antennas. However, this conservatism may sometimes indicate non-compliance in cases where a more exact analysis could show that no overexposure would occur. For that reason, whenever a simple and conservative calculation indicates non-compliance, that should be interpreted as a signal to analyze the situation in greater depth. For instance, the FCC provides simple, conservative calculations to determine if a situation may be classified as exempt from more detailed analysis. A favorable result indicates compliance, but a negative result does not automatically indicate non-compliance. It only indicates that further analysis may be needed.

Most amateur stations will require multiple analyses to account for differences in frequency bands, antennas, operating modes, power levels, and any other factor that could affect the final outcome. It is not necessary to use the same analysis methods for each set of operating parameters and, as stated earlier, the simplest method that shows compliance with the FCC MPEs is generally the best to use.

The FCC does not require that amateurs submit the results of their exposure assessments to the Commission. However, it is wise to document the assessments for your own records. If there is ever a question about overexposure from your station, you will then be able to provide documentation to show why that is not the case.

CHOOSING AN ANALYSIS METHOD

The two most common analysis methods, measurement and computer modeling, are described in detail in the *ARRL Handbook*. Both measurements and modeling are time consuming and can be complex. Why go to that kind of trouble when simple online calculators exist, such as the one located on the ARRL website? (**arrl.org/rf-exposure-calculator**)

Online calculators are usually based on far-field calculations and specified to their stated exposure limit. Even in the near-field these can be accurate for some *simple* antennas, such as half wave dipoles in free space (the proof of this can be found in the reference for Jordan and Balmain). However, it is important to be aware that all simplified online calculators can give incorrect results in more complex situations. These include:

- Compact antennas with dimensions much smaller than one-half wavelength
- Potentially all antennas that are less than one-tenth wavelength above ground
- Highly directional antennas

Directional antennas are an important case where simplified online calculators will give incorrect results. Such calculators take account of antenna gain by calculating the compliance distance for the main beam of the antenna, but they then apply that same compliance distance in every other direction as well, which is conservative but also physically impossible. Even for a rotatable antenna, where the main beam could be pointing in any horizontal direction, there are still locations at which the antenna is never pointed, such as directly below or above the antenna, leading to false indications of noncompliance in those locations. For example, if the online calculator determines that the compliance distance for a Yagi antenna is 10 meters and the antenna is mounted on a 10-meter high tower, then its results cannot be used to show compliance for a person standing at the base of the tower. Similar inaccuracies exist for roof mounted towers and people inside a house. The only correct conclusion from such an analysis is that a simplified calculator is not applicable to that case.

A very important note is that online calculators must not be used for an antenna operated within 20 cm of the body surface, such as handheld transceivers. The FCC requires a localized SAR assessment of exposure for that operating condition.

PERFORMING AN EXPOSURE ASSESSMENT

Before choosing what type of exposure assessment is needed, there is certain information that must be gathered, which will be used with any assessment. When using modeling techniques, the measure of power that is supplied to the antenna is needed. For this, you need to know the output power of the transmitter and the loss in the feed line. The simplest way to calculate this is to first convert transmitter output power to dBW (dB with respect to a watt, e.g., 100 watts = 20 dBW). Next subtract the feed line loss, which is specified by coaxial cable manufacturers in dB/100 feet, or dB/100 meters, at various frequencies. Choose the closest frequency that is reported by the manufacturer and factor in the length of the coaxial cable. Then subtract cable loss in dB from the transmitter power output in dBW.

If you have a power amplifier that you use occasionally, try applying that high power output in the analysis. If the assessment suggests that exposure may not comply with the MPE, you can analyze with and without the amplifier separately. Clearly, if analysis with the higher transmit power indicates no potential exposure problems, then it is not necessary to repeat the analysis at a lower power level.

Knowledge of the minimum distances that people can be from the antenna is necessary for assessing compliance. It is advantageous to distinguish between exposure to people in the occupational population and exposure to the general population. This can be important for some stations since the FCC MPE values for the occupational population are significantly less stringent. The FCC has designated licensed radio amateurs and the members of their households as members of the occupational population, so it may be necessary to perform two analyses for each antenna on each band, with the distance from the antenna to the nearest people in the occupational population used in one calculation and the distance from the antenna to the nearest people in the general population used in the other calculation.

DECIDING TO MEASURE OR MODEL

Measuring RF fields, particularly in the near field, is not easy to perform accurately. Calibrated equipment is necessary, including both the antenna and the measuring device. In the near field, both E and H must be measured, and the measurement antenna orientation also must be varied to account for differences in polarization. To further complicate matters, objects in the field can also perturb the measurements. Such objects include the person making the measurements and the ground.

Usually, the most accurate measurement is accomplished with a small non-resonant field probe rather than large, full-sized antennas. This is because of the need to evaluate field levels at many points in space to characterize the field distribution. It would be difficult or impossible to orient such a resonant antenna properly, particularly at HF, to perform the necessary measurements. All measurement equipment must be calibrated for its response to determine actual E or H strength. Such calibration is challenging and explains why professional field measurement equipment is usually expensive.

Since measuring RF fields requires expensive equipment and exacting technique in order to obtain accurate results, most radio amateurs are likely to use computer modeling to perform their exposure assessments. Computer software that models emissions from antennas is inexpensive and readily available to most radio amateurs. Often these calculations are used to predict antenna characteristics such as gain and directionality. Such results are only valid in the far-field of the antenna, where the RF signal becomes a plane wave. This is not often particularly useful in determining human exposure since levels of concern occur much closer to the antenna in most cases. Most of the popular antenna modeling software programs also contain calculations for determining near-field E and H strengths.

Near-field calculations in most NEC-based antenna modeling software are typically presented as arrays of numbers. How these are used depends heavily on the frequencies being modeled. One commonality is that the end goal of such modeling is to identify locations in the vicinity of an antenna that exceed the FCC MPE values and to determine whether any person has access to those locations. Keep in mind that such locations might include a neighbor's home or property.

SINGLE POINT ANALYSIS VS. SPATIAL AVERAGING

Generally, the easiest form of analysis is to determine field strengths at individual points in the space around the antenna. To find the locations at which the field is highest, the spatial resolution of the result array must be sufficiently high. With finer and finer spatial resolution, especially if you analyze in 2 or 3 dimensions, the number of analysis points grows quickly and can exceed the capabilities of the modeling software or the computer memory. As long as no point in the space in which a person may be found exceeds the MPE limits, this form of analysis is sufficient to ensure compliance with exposure regulations. However, the actual exposure is classified as the average over the entire body. Thus, even if there is a hot spot that exceeds MPE limits, it is still possible that the whole-body average exposure is below the limits. Under those conditions it is necessary to perform a spatial average of all points in space that can be occupied by the body at one time.

For FCC compliance, the spatial average is taken over a 2-meter vertical line that starts where the person is standing. Spatial averages of both E and H provide a complete assessment of potential RF field exposure.

PRE-ASSESSED EQUIPMENT CONFIGURATIONS

For many antennas, detailed computer modeling of their near-field strengths under a wide range of installation conditions and power levels can be used to create extensive databases of *pre-assessed equipment configurations* (PAECs). Individual amateurs can then use these results to estimate exposure levels and potential exclusion zones for their own stations. The Radio Society of Great Britain has published a large collection of such assessments that can serve as an interesting analysis of different antenna types and locations, and they have been accepted as valid exposure assessments by the British regulator, OFCOM (**rsgb.org/emf**). However, radio amateurs in the United States should be cautioned that the RSGB analysis generates compliance distances based on the ICNIRP 2020 guidelines and not the FCC MPEs, which can be different.

GROUND EFFECTS

Particularly in the near-field, ground reflections play a major part in determining exposure. The ground below an antenna can have a significant effect on the near-field exposure patterns. It is for this reason that the FCC exemption formulae utilize a perfect ground power reflection factor of 4.00, to be highly

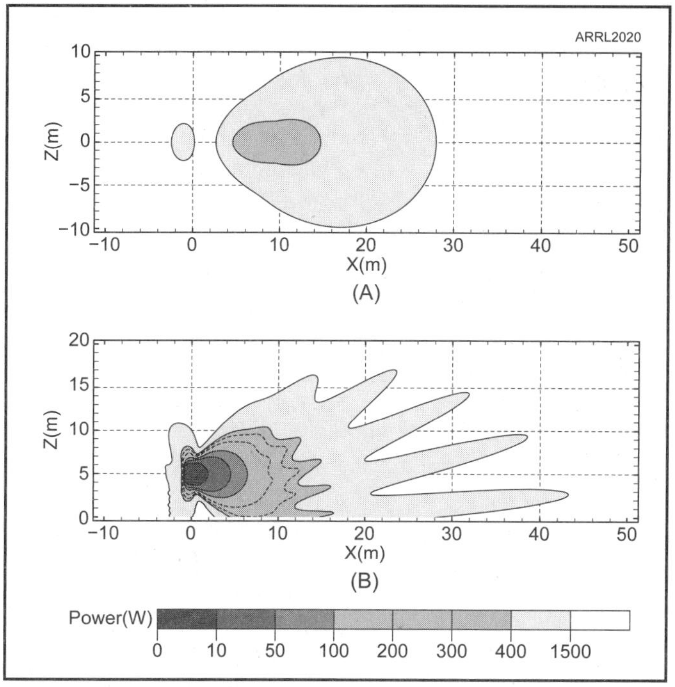

Figure 26.12 — The exclusion zones with fields spatially averaged over 2 meters vertically. The antenna is a 2 meter 5-element Yagi antenna mounted 6 meters high and aligned along the X-axis. Analysis is performed at 144.2 MHz with horizontal polarization and a Clay-type ground. (A) Exclusion zones (in the X-Y plane) spatially averaged from the ground to 2 meters height. The contours represent areas on the ground where no one may be located during their averaging time for different transmit power levels, as indicated in the legend. (B) Exclusion zones in the air along the main beam of the antenna (in the XZ plane) with spatial averaging for 2 meters surrounding every calculated point. The contours are exclusion zones for different power levels, as indicated in the legend, and represent points in space in which no one may be located during their averaging time, such as people on upper floors of neighboring buildings. This analysis was done with the assumptions of 100% duty cycles and no feedline loss. Exclusion zones may get smaller if those assumptions are accounted for. This analysis was provided by Peter Zollman, G4DSE.

conservative and ensure that no station that qualifies for the exemption could possibly overexpose persons at distances greater than those specified in the calculations. Online far-field calculators, such as the ARRL calculator, also factor in ground power reflection, but use a more common factor of 2.56.

Reflections from the ground result in complex interference patterns that can lead to regions of high exposure in unexpected locations. An example of this effect is illustrated in **Figure 26.12b**.

TYPICAL NEAR-FIELD ANALYSIS

In the near field of an antenna, ground reflection can have a strong effect on the pattern, which may spell difficulty for defining a high exposure region. Ideally, a 3-dimensional analysis of all locations around an antenna would be performed, with identification of all locations in space at which the power density exceeds the FCC MPE. This will not be practical for most radio amateurs, since it requires a very large amount of computer memory and relatively long computation times. Instead, modeling portions of space around the antenna can identify areas at which the MPE levels could be exceeded. A common name for such areas is *exclusion zones*, since these are locations at which the presence of a person must be excluded.

One way to limit the number of calculations is to make use of known symmetries around an antenna. If the near-field pattern is known to be symmetric, such as with the vertical plane passing through the center of a horizontal antenna as the plane of symmetry, then it is only necessary to model the near-field in half of the total space irradiated by the antenna and then to mirror the field strength patterns to the other half across that plane.

Another important consideration for near-field analysis is to accurately model the power delivered to the antenna. Although some modeling software permits the input power to the antenna to be defined, other modeling software requires the input of a voltage or current at the feed point and then models the power delivered to the antenna from that input. The relationship is not easily calculable beforehand because the exact antenna feed point impedance is not known until it is accurately determined by the modeling software. A two-pass process permits accurate input power to be modeled. In the first pass, an arbitrary voltage source is applied to the antenna. The results of this pass of the model produces the actual antenna input power based on the modeled input voltage. Using that result, the modeled input voltage can be adjusted as:

$$\text{Desired Input Voltage} = \text{Initial Input Voltage} \times \sqrt{(\text{Desired Power} / \text{Initial Modeled Power})}$$

After this calculation, the Initial Input Voltage is replaced by the Desired Input Voltage in the modeling parameters and then subsequent modeling results will be based on the Desired Power being applied to the antenna. Note that any change of the antenna design or height is likely to change the Initial Modeled Power and the two-pass process must be repeated.

EXAMPLE EXCLUSION ZONES

As an example of determining appropriate exclusion zones, both the RF field on the ground below the antenna as well as the field in nearby buildings must be considered. For example, refer to the plot in Figure 26.12a. It shows the exclusion zones for a person in the general population category standing on the ground below a 5-element, 2-meter beam mounted about three wavelengths above the ground (6 meters). It is calculated at a plane parallel to the ground surrounding the antenna. A typical point analysis might be at a height of 2 meters, or at the top of the head of a tall person. This height is chosen on the presumption that this is as close to the antenna as any portion of an average person can get (a shorter person would be farther away from the antenna and thus would be assumed to have a lower exposure). Exclusion zones for various power levels are indicated by shading of the near-field pattern. Ground effects could result in lower heights with higher field strengths in some locations. A resolution on compliance in this case could be to model several heights between the ground

Analysis of an End-Fed Wire Antenna

The wide availability of automatic antenna tuners (a.k.a., autotuners) enables the use of random-length or resonant end-fed half-wave wire antennas as described in the **General Purpose MF and HF Antennas** chapter. If fed directly without a feed line, the end of the antenna is connected to the tuner's single-wire terminal or the center pin of a coaxial connector. Tuners allow such wires to be loaded on many frequency bands, greatly simplifying multi-band operation, particularly in difficult situations such as portable operation or when stealth antennas are required.

Be aware that such an antenna is prone to creating high field strengths near the operating position. The end of the wire connected to the tuner is part of the antenna and can radiate a strong electric field, E, magnetic field, H, or both, that greatly surpass exposure limits for the occupational population, such as the amateur operator. During portable operation in public areas, such as parks, the exclusion zones are even larger for any non-amateur that is observing the operation, or just happens to be walking past. The operator should determine the exclusion zones for the general population and then make sure that no one is permitted closer than that to the antenna when transmitting.

A near-field analysis was conducted for a second-floor home station with a 20-meter-long wire connected to the autotuner, strung out of the window to a nearby tree. The transmitter generated 100 watts and, for the sake of comparison, transmitted with 100% operational and modulation duty cycles. Analyses were performed at 3.8 MHz, 7.1 MHz, 10.1 MHz, 14.1 MHz, 18.1 MHz, 21.2 MHz, 24.9 MHz, and 28.5 MHz using *NEC-5* with near-field sample spacing of 0.1 meters. **Figure 26.A** shows the region in which E exceeds the FCC occupational MPE at 14.1 MHz. Analysis of H showed smaller compliance distances than for E with this antenna on all frequencies studied. The shapes of the exclusion zones were similar on all bands with only the dimensions changing. For comparison, the ARRL online calculator was used to determine the minimum compliance distance at each frequency from the end of the antenna for an occupational person (the amateur radio operator at the operating position). The comparative results were instructive.

On some bands, the online calculator overestimated the compliance distance to the operating position (i.e., the calculated compliance distance was too high). These were all the bands for which the wire length was greater than or equal to one wavelength (i.e., 20 through 10 meters). On the 40-meter band, where the wire length was approximately equal to one-half wavelength, the calculator and the model were in close agreement. On the 80-meter band, where the wire length was one-quarter wavelength, the online calculator greatly underestimated the compliance distance determined by the model. In every case, however, the occupational compliance distance from the feed point on the tuner was a substantial distance, the greatest being 1.4 meters (4.6 feet) on both the 80- and 30-meter bands and the smallest being 0.3 meters (1 foot) on the 40-meter band. The average was 0.75 meters (2.5 feet) on the 17 through 10-meter bands. For most of these bands, the operator would be sitting in the exclusion zone while transmitting. If more than 100 watts is used or other people who are in the general population are observing the operator, such as at a Field Day site or in a park setting, special care should be taken with this type of antenna.

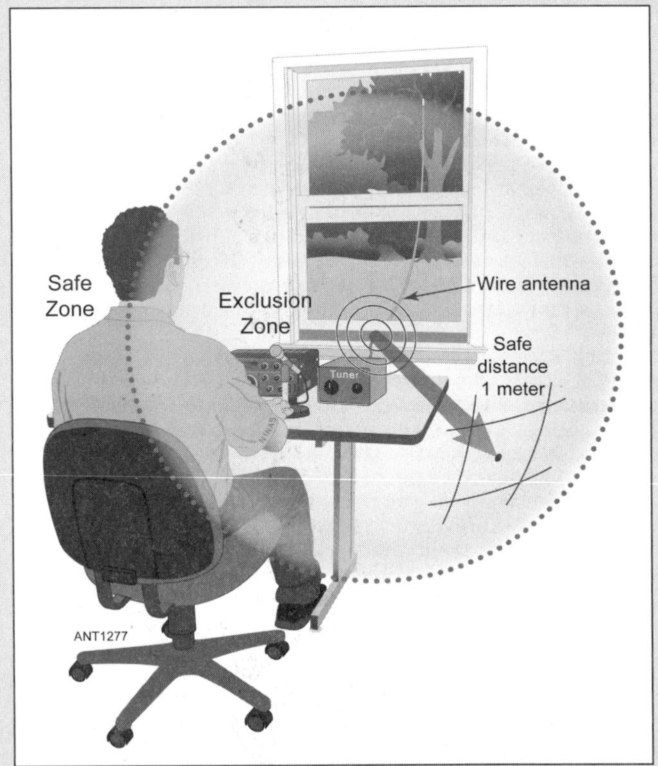

Figure 26.A — The exclusion zones for the occupational population in the space surrounding the feed point of an end-fed random length wire antenna connected to the back of a transceiver with an autotuner installed. Contours are based on E values and represent the regions that a person in that exposure group, such as radio amateurs and the members of their household, must stay away from the feed point. This analysis was done with the assumptions of 100% duty cycles and no feed line loss. Exclusion zones may get smaller if those assumptions are accounted for.

Analysis of a Small HF Loop Antenna

Amateurs who operate a portable station or in space-limited locations, such as apartments, have found the small HF loop antenna to be convenient and efficient. Articles describing this type of antenna have been published in *QST* and *QEX* (see Siwiak and Quick, *QEX* July/Aug. 2018 and Siwiak and Quick, *QST* Sep. 2018). These small loops can have circumferences that are far less than 1/2 wavelength and be tuned over a wide range of frequencies with feed point impedance close to 50 Ω.

Particularly in space-limited locations, this antenna may have to be located relatively close to people. In this type of antenna, the limiting compliance distance can be caused by either E or H, so both must be modeled at all frequency bands used. Even though the far-field pattern shows a null in the direction perpendicular to the plane containing the loop, that null does not exist in the near-field. Thus, at some frequencies and power levels the limiting compliance distance may be in that direction.

Modeling the near-field with *NEC-5* and identifying the locations near the antenna that exceed the FCC MPE limits for E and H, exposure patterns such as in **Figure 26.B** are seen. For a loop with a circumference of 2.8 meters mounted 1 meter above the ground, analyses were performed at 3.8 MHz, 7.1 MHz, 10.1 MHz, 14.1 MHz, 18.1 MHz, 21.2 MHz, 24.9 MHz, and 28.5 MHz using *NEC-5* with near-field sample spacing of 0.1 meters. The transmitter generated 100 watts and, for the sake of comparison, transmitted with 100% operational and modulation duty cycles. For a person in the general population standing on the ground near the antenna, it is H that most often governs the compliance distance. The highest field strengths around the antenna where a person may be standing were in the 20-meter band, modeled at 14.1 MHz. The uncontrolled compliance distance based on the E model is 1.7 meters (5.6 feet) and based on H is 2.2 meters (7.2 feet).

The occupational compliance distances are smaller, as expected. The model calculated an E controlled compliance distance of 1.2 meters (3.9 feet) and an H controlled compliance distance of 1.6 meters (5.2 feet). Clearly, in this case, H constrains compliance with the FCC MPEs rather than E.

The online calculator does not perform badly for this antenna. The antenna gain of the small HF loop is 4 dBi broadside to the loop. Based on that gain at 14.1 MHz with 100 watts transmitted, the online calculator gives compliance distances of 2.4 meters (7.8 feet) for the uncontrolled environment and 1.1 meters (3.5 feet) for the controlled environment. The model shows that this compliance distance is not the same in all directions so, if it is important, an elliptical exclusion zone can be drawn around the antenna, rather than the circular zone that the calculator implies.

When space is constrained, the operator must make sure that the operating position is far enough away from the antenna. Of more concern is when this antenna is used in portable operation, such as in a park or at Field Day. With members of the general population at the location, barriers must be placed to ensure that no one gets closer than 7.2 feet from the antenna, or even more distant if higher power is transmitted. Even better would be to raise the antenna so that compliance is achieved everywhere at ground level.

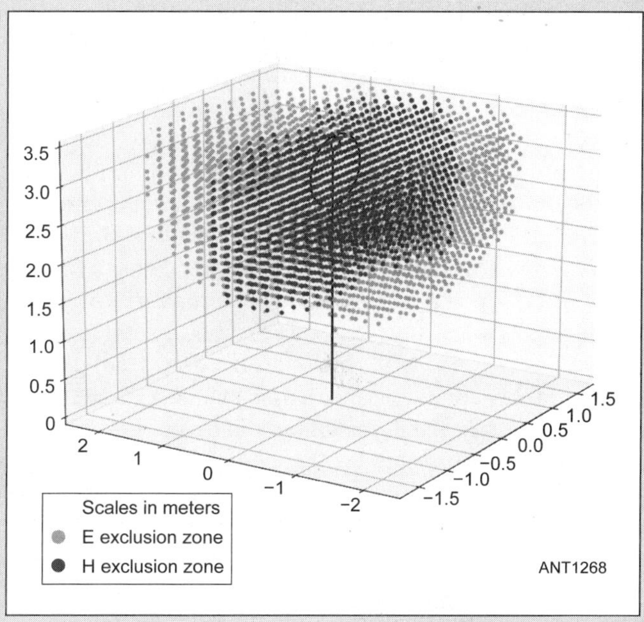

Figure 26.B — The exclusion zones for the general population in the space surrounding a small HF loop antenna. No person in the general population group may enter one of these regions and the H exclusion zone is clearly larger than that of E for this type of antenna. This analysis was done with the assumptions of 100% duty cycles and no feed line loss. Exclusion zones may get smaller if those assumptions are accounted for. This analysis was provided by Kai Siwiak, KE4PT.

and the 2-meter height and then select the largest calculated power density at each location. A more elegant solution to the question of ground effects, that also more accurately determines the whole-body exposure of a standing person, is to calculate the spatially averaged power density between the ground and the 2-meter height. Figure 26.12a was produced this way.

Another useful near-field analysis pattern is a plot in a vertical plane that is parallel to the main beam of the antenna, as in Figure 26.12b. For various power levels it shows the exclusion zones in the air for a person standing at various heights. This type of plot is useful in determining if the exposure of a person on an any floor of a neighboring building will comply with the FCC MPE limits. Reflections from ground result in a highly complex pattern, that, in this example, looks like the fingers of a hand, particularly at higher powers. The exclusion zones are computed by spatially averaging E and H at points along a vertical, 2-meter line above every point and determining the radiated power that would result in a field strength equal to the most restrictive MPE. This accurately accounts for any part of a person's body potentially occupying any of the exclusion zones.

Other examples of antennas that have complicated near-field exposure patterns and thus require modeling to accurately determine compliance distances are described in two sidebars in this section: the "Analysis of an End-Fed Wire Antenna" and the "Small HF Loop Antenna." Both antennas can have unexpectedly large near-fields that may not be accurately represented by more simple forms of analysis, such as online calculators.

26.3.6 RF Exposure Mitigation

It is not the purpose of the FCC human exposure regulations to keep you off the air. The goal is to prevent overexposure of anyone while you are operating. If your calculations indicate that there are locations where people may be present and where field strengths can exceed the FCC MPE (i.e., overexposure is possible), you must perform some type of mitigation to prevent that occurrence. There are many ways that you can mitigate an overexposure situation, some of which may actually improve the performance of your station.

In cases where an exclusion zone exists for a person standing on the ground under the antenna, prevention of such an exclusion zone may be as simple as raising the antenna by a short distance. For antennas that have sections mounted close to the ground, such as the ends of an inverted-V, raising those ends by even a meter could convert a noncompliant station into one meeting the FCC MPEs.

POSITIVE ACCESS CONTROL

The FCC requires *positive access control* (PAC) to prevent people from accessing areas in which they might be overexposed relative to the general population MPE. PAC is a general term that refers to an active measure that keeps people from entering potential overexposure areas. PAC can be as simple as a locked door to a rooftop on which antennas are located. A common way to achieve PAC is to place effective fencing around the areas where access must be controlled.

Wherever PAC is used, the FCC requires that signs be posted to warn about the possibility of overexposure. Signs must be large enough to be read from a distance from which no overexposure can occur. The content of the sign is specified in the FCC rules and, for areas in which the FCC general population MPE may be exceeded, it must have the word "NOTICE" in white letters on a blue banner along with a description of the hazard and your contact information. If exposure to RF fields within the controlled area has the potential to exceed the FCC occupational population MPE, then the sign must have the word "CAUTION" in black letters on a yellow banner along with similar text as the "NOTICE" sign.

OPERATING MODIFICATIONS

Two conditions must be met to cause exposure: the station must be transmitting and there must be a person present in a potential overexposure area. PAC is a means of preventing the latter. However, control of transmissions can also satisfy the requirement that no person be exposed to electromagnetic energy exceeding the MPEs.

The FCC writes its regulations to cover all forms of radio transmission that they regulate. Some of these include broadcast transmitters and cellular base stations. These both differ from amateur radio in two ways: they transmit during all hours of every day of the year and, most importantly, they are usually unattended. In contrast, amateur radio transmissions are generally intermittent, and the amateur radio operator is usually present at the transmitter while it is operating.

Amateurs transmitting with high power into high-gain arrays, such as often used with EME operation, need to pay careful attention not only to where the main beam might intersect adjacent areas where people could be located but also the region immediately beneath the array where high field strengths can occur.

Amateur radio stations that are remotely controlled require special considerations when identifying areas in which individuals might be exposed above the FCC MPEs. For remote operations, the best approach is to design the transmitting site to be inherently compliant with the MPEs without need for real time monitoring of activity within the area.

If a person enters a potential overexposure area and the amateur radio operator then stops operating, no hazard exists, and the station remains compliant with the FCC rules on exposure. Similarly, the amateur radio operator may determine that a lower transmit power removes the hazard, and then decreases the power any time a person is seen to enter the overexposure area. Nonetheless, because the field could exceed the MPE, if the transmitter is not turned off, posting a sign in an appropriate area that alerts individuals to the fact that exposure might exceed the MPE will be expected by the FCC. This scenario could also apply to the situation of an amateur radio operation in a public setting where a person could enter a high field region near the antenna, but the control operator can cease operation upon seeing a person encroaching on the area. The best advice in such cases is to eliminate the potential for overexposure in the first place by moving the antenna further away from places that people can access, which also eliminates the need for posting of RF safety signs.

26.3.7 Research and Guidelines for RF Safety

The preceding material presented an introduction to the reasons for developing an RF Safety program for your station and some basic procedures to help determine those areas around your station in which a person might be exposed to your transmitted signals beyond the limits that the FCC has set. To perform more exact exposure analyses, which may be necessary if your station has marginal exposure results from the more common methods, you can look to the following references:

FCC OET BULLETIN 65

The FCC has provided guidance for complying with their rules on human exposure to electromagnetic energy in their OET Bulletin 65. OET Bulletin 65 Supplement B, which was principally written by members of the ARRL RF Safety Committee, provides more specialized information that applies to the operation of an amateur radio station. These documents are available for free download from the FCC website: **fcc.gov/general/oet-bulletins-line**.

RF EXPOSURE AND YOU

The ARRL published a book in 1998 entitled *RF Exposure and You*, which delves into most aspects of RF exposure and how to comply with FCC regulations. Note that this book was published prior to the FCC rule changes that went into effect on May 3, 2021 so parts of the text and the compliance distances from some of the tables are no longer applicable. See the ARRL's RF Exposure website (**arrl.org/rf-exposure**) for updated information.

STANDARDS, SCIENTIFIC REPORTS, AND STUDIES

- IEEE C95.1-2019: IEEE Standard for Safety Levels with Respect to Human Exposure to Electric, Magnetic, and Electromagnetic Fields, 0 Hz to 300 GHz. Note: The IEEE C95.1-1991 standard was adopted by the American National Standards Institute, ANSI, in 1992, and is what the FCC has used to develop its exposure regulations. The 2019 version of the standard is more up-to-date with respect to the current research, but the parts upon which the FCC exposure regulations are based have few changes from the 1991 version. IEEE RF safety standards are available at no cost through a grant from the United States Department of Defense. To download IEEE C95.1-2019 and any of the associated standards in the C95 family, go to **ieeexplore.ieee.org/browse/standards/get-program/page/series?id=82**
- NCRP Report #86: Biological Effects and Exposure Criteria for Radiofrequency Electromagnetic Fields, National Council on Radiation Protection, 1986.
- ICNIRP, Guidelines for limiting exposure to electromagnetic fields (100 kHz to 300 GHz). Health Phys 118(5), pp 483-524, 2020. Available at **icnirp.org/cms/upload/publications/ICNIRPrfgdl2020.pdf**
- Tell, R., K5UJU, "Amateur portable radios (handheld transceivers): exposure considerations based on SAR," *QEX*, July/Aug. 2021, pp. 11 – 15.
- Jordan, E.C. and Balmain, K., *Electromagnetic waves and Radiating Systems, 2nd Ed.*, Prentiss-Hall, 1968, pp. 333 – 338.
- Siwiak, K., KE4PT, and Quick, R., W4RQ, "Small Gap-resonated HF Loop Antenna Fed by Secondary Loop," *QEX*, July/Aug., 2018, pp. 12 – 17.
- Siwiak, K., KE4PT, and Quick, R., W4RQ, "Small Gap-resonated HF Loop Antenna," *QST*, Sept., 2018, pp. 30 – 33.

26.4 References and Bibliography

Source material and more extended discussions of the topics covered in this chapter can be found in the references listed below.

Block, W., "Lightning Protection for the Amateur Radio Station," Parts 1 – 3, *QST*, June 2002, pp. 56 – 59; July 2002, pp. 48 – 52; Aug. 2002, pp. 53 – 55.

Brock-Fisher, T., "Is Your Tower Still Safe?" *QST*, Oct., 2010, p 43-47.

The Complete Guide to Wiring, 6th edition, Cool Springs Press, 2014.

Daso, D., K4ZA, *Antenna Towers for Radio Amateurs* (ARRL, 2010).

Daso, D., K4ZA, "Workshop Chronicles," columns in *National Contest Journal.*

Frydeniund, M., *Lightning Protection for People and Property* (Van Nostrand Reinhold, 1993).

Hart, W., and Malone, E., *Lightning and Lightning Protection* (Interference Control Technologies, Inc, 1988).

Morris, S., *Up the Tower* (Champion Radio Products, 2009).

PolyPhaser, see **polyphaser.com** and look under the Resources link for White Papers on lightning protection.

Rand, K., Lightning Protection & Grounding Solutions for Communication Sites, Polyphaser Corporation, 2000.

Silver, W., NØAX, *Grounding and Bonding for the Radio Amateur*, Second edition (ARRL, 2021).

Smith, C., *Carroll Smith's Nuts, Bolts, Fasteners and Plumbing Handbook* (Motorbooks, 1990).

Ufer, H., "Investigation and Testing of Footing Type Grounding Electrodes for Electrical Installations," IEEE Transactions Paper #63-1505, Power Apparatus Systems Vol. 83, pp. 1042 – 1048, Oct. 1964.

"How to Protect Your House and Its Contents Against Lightning," IEEE, 2005.
Lightning Protection for Engineers (National Lightning Safety Institute [NLSI], 2004).
Lightning Protection for Radio Transmitter Stations (Nautel Corporation, 1998).
Military Handbook MIL-HDBK-419A, Vol. 1 (Basic Theory) and 2 (Applications), Grounding,
Bonding and Shielding for Electronic Equipments and Facilities, 1987 (**tscm.com/MIL-HDBK-419A.PDF**).
National Electrical Code, NFPA 70, National Fire Protection Association (**nfpa.org**).
Standard for the Installation of Lightning Protection Systems, NFPA 780, National Fire Protection Association (**nfpa.org**).
Standards and Guidelines for Communications Sites, document 68P81089E50-B, Motorola, Inc., 2005. Also referred to as "R56" in the communications industry as it is administered by the R56 committee at Motorola.

TABLE OF CONTENTS

Chapter 27 — Online Material

Articles

- "A One Person, Safe, Portable and Easy to Erect Antenna Mast" by Bob Dixon, W8ERD
- "Antenna Feed Line Control Box" by Phil Salas, AD5X
- "Homeowners Insurance and Your Antenna System" by Ray Fallen, ND8L
- "Installing Yagis in Trees" by Steve Morris, K7LXC
- "Is Your Tower Still Safe?" by Tony Brock-Fisher, K1KP
- Ladder Mast and PVRC Mount
- "Lightning Protection for the Amateur Station, Parts 1, 2 and 3" by Ron Block, KB2UYT
- "Removing and Refurbishing Towers" by Steve Morris, K7LXC
- Rotator Specifications
- "The Care and Feeding of an Amateur's Favorite Antenna Support — The Tree" by Doug Brede, W3AS
- "The Tower Shield" by Baker Springfield, W4HYY and Richard Ely, WA4VHM

Chapter 27

Building Antenna Systems and Towers

Getting an antenna in the air and keeping it there requires decisions and challenges. For example, what kind of antenna support? How do you build it? What tools and techniques should be used? There are many other questions. Begin by reviewing the **Safe Practices** chapter for information on safety and safety equipment for working on antennas and towers. That chapter also includes important information on grounding and bonding for antenna systems, as well as RF safety around antennas and in your station and home.

If you are contemplating either a simple or significant tower or antenna project, read the articles and references listed in the References and Bibliography and attend or view presentations by experienced individuals, such as the "Amateur Radio Tower Safety" series of videos on the DX Engineering channel at **youtube.com/@DXEngineering**.

Consider hiring professional assistance if you are not comfortable with doing such work yourself. Tower services with amateur experience are listed on various amateur Product Review websites such as **eham.net**. Be sure to get references and contact recent customers for reviews. Like any service you select to do work around your home or business, the service should be licensed, bonded, and insured in case of accidents or other problems.

Many of the tools and equipment mentioned in this chapter are available from numerous *QST* advertisers. Hardware and materials are widely available for all project sizes. There's no need to use an under-rated item for such important jobs. You should be able to supply everything you need to do the job safely and correctly, resulting in years of trouble-free service at the least possible risk. This chapter also includes text and numerous graphics from publications by Don Daso, K4ZA, and Steve Morris, K7LXC, listed in the References and Bibliography.

27.1 Trees and Masts

27.1.1 Trees

Trees were among the first antenna supports and have been used successfully by many amateurs over the years. If you're in an area with suitable trees — congratulations! They're free (compared to towers) and generally unregulated for use as antenna supports. Trees make good temporary antenna supports and with care can support an antenna for many years — even a large one. When attaching an antenna to a tree, it's important to traumatize the tree as little as possible. This will ensure a strong, enduring attachment.

Although it's relatively easy to get a wire up into a tree, it's certainly more difficult to keep it there for the long term. Tree-mounted antennas require more maintenance, but their height and low cost more than make up for the added work. (Although uncommon, even Yagi antennas have been installed in trees using the techniques in the short article "Installing Yagis in Trees" included with this book's online material.)

CRITICAL SAFETY TIP — STAY FAR AWAY FROM POWER LINES

Antennas, whether constructed of wire or aluminum tubing, are metallic and must be kept a substantial distance from power lines. A wire dipole coming in accidental contact with a power line is a lethal hazard. When you choose trees as your antenna supports, it is important to avoid using trees that are in close proximity to any power lines. Without expert knowledge, you should assume that any wires on utility poles and any wires entering your home at a service entry point are dangerous and must be avoided by a substantial distance.

USING A LINE LAUNCHER OR DRONE

In this method, you use some sort of line launcher from the ground by which a lightweight line (usually fishing line of a few pounds capacity) with a weight on the end of it is propelled over a branch high up in the tree. Hopefully, the weight drops to the ground, and you use the small line to pull up a bigger line with your antenna attached. Such launchers include slingshots, compressed air cannons, fishing rod and reel, bow and arrow, and even tennis ball throwing aids.

Keep people out of the fall zone around the tree since there will be falling weights, plus lines and antennas at some point. Safety glasses and gloves are always a good idea for these kinds of activities.

Somewhat slower but a lot more controllable, a multi-rotor drone can lift light lines over the highest trees, although not through them. A light line is carried over the tree by the drone, and then the light line is used to pull a heavier line over the tree, just as when using a launcher. The process is described in a pair of *QST* articles: See the entries for an article by Kam Sirageldin, N3KS (better known as TI5W and TI7W), and another by Michael Shandblatt, W3MAS, and Joe Warwick, KB3ZED in the References and Bibliography.

ATTACHING AN ANCHOR

A stouter method of securing a rope in a tree is to climb the tree to install an anchor. For light antenna loads, such as the end of a dipole, a threaded eye-screw is the method of choice. (Use welded or forged (closed end) eye-screws and bolts to prevent them from opening up under load.) Just drill a hole into the tree about 1/16 inch smaller than the screw diameter, then twist in the screw as shown in **Figure 27.1**. Be certain you use a cadmium-plated eye-screw threaded for use in wood. A screw thread length of two or three inches should secure most antennas. Allow 1/2 inch of space or more between the trunk and the eye; this allows for outward growth of the tree with time.

For stouter antennas, such as multi-element wire beams, a different method for securing wires to trees is recommended (see **Figure 27.2**). This procedure involves using an eyebolt longer than the tree diameter, drilling completely through the tree, and securing the eyebolt on each side of the tree with flat washers and nuts. Drilling a hole through a tree causes much less trauma to the tree than wrapping something around it. Much of the core of a tree is dead tissue, used mainly for physical support.

Although there will be some wounding of the tree at the site of a bolt or screw, such trauma will be far less than that which occurs from wrapping a wire around the trunk. Wrapping a line around a tree's branch or trunk strangles the veins in the sapwood the same way a noose around your neck would strangle you. It's important not to wrap anything around the trunk.

You can find a professional tree climber/arborist in the *Yellow Pages* or perhaps use the services of a talented friend to install a pulley and rope system in your trees of choice. A 3/8 or 1/2 inch eyebolt screwed into the tree as described above with a pulley attached is the best method. Use a threaded chain link or "cold shut" (a type of chain attachment link) to attach the pulley to the eye.

A non-swiveling pulley is preferred because the lay of the rope can cause the pulley to turn, twisting the rope and possibly jamming the pulley. Use only all-metal pulleys for permanent installations, preferably ones made from stainless or galvanized materials. Plastic parts will either break or be damaged by UV (ultraviolet) radiation from sunlight. Check your local hardware stores for inexpensive stainless-steel pulleys.

Pulleys intended for marine use are excellent candidates for tree-mounted antennas, but they are expensive. Years of trouble-free service offset this expense, however. Harken and Schaeffer are two brands offering quality products.

Keep in mind that with pulleys and haul ropes there should be minimal clearance between the sheave (the wheel in the pulley) and the pulley body; the rope or halyard should be larger in diameter than that clearance so it can't get jammed

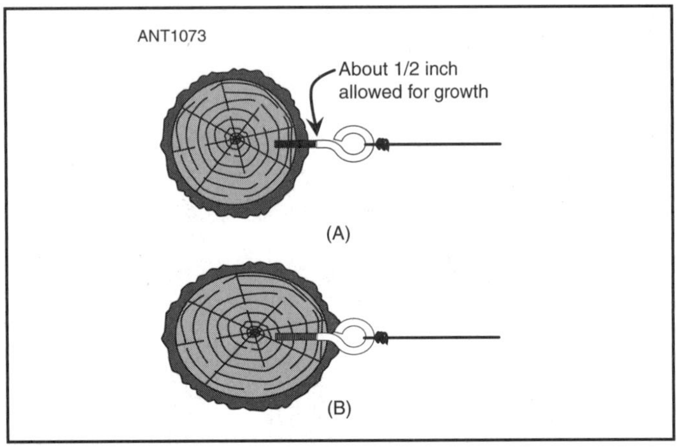

Figure 27.1 — The best way to secure a wire to a tree is with an eye-screw threaded into the wood (A). As the tree grows and expands, however, the eye-screw will become embedded (B) and must be removed and replaced.

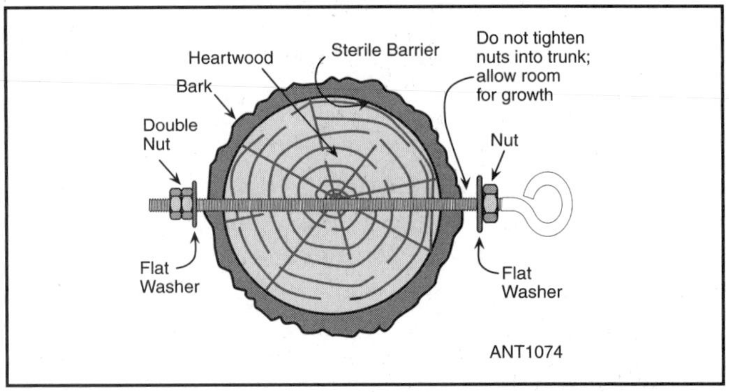

Figure 27.2 — For heavy antenna loads, an eyebolt passed through the trunk or limb will support more weight than an eye-screw. Allow about 1/2 inch of play between the bolt and trunk or limb. Don't tighten the bolt completely; this allows for tree growth.

between the sheave and the pulley — a major annoyance. The best type of rope for this purpose is black Dacron UV-resistant line because it doesn't deteriorate when used outdoors. (See the section "Ropes and Rope Care" later in this chapter for more information.)

Have the climber go up the tree to the desired location, screw in the eye-screw, then attach the pulley. It's all but certain that the tree will have to be pruned to clear a decent window through which the line can travel. It's better to over-prune since new growth will invariably grow back in just a few years. Small branches are incredibly strong and resilient and can cause major problems in any tree project or installation.

Having brought the line along, the climber will put the line through the *back* side of the newly installed pulley (closest side to the tree), attach a weight to the end of the line, and throw it out in the direction that your wire antenna will take. The wire antenna must clear all branches to successfully install your antenna. It's very difficult to install an inverted V in a tree from the ground because it's just about impossible to get both sides of the antenna through the branches. A climber can help by throwing each leg of the antenna through the branches separately.

When the end of the line reaches the ground, remove the weight, then tie the ends of the line together, making the rope into a loop (see **Figure 27.3**). This is because in almost all cases it is the antenna that breaks, not the rope. Without the loop, when the antenna breaks the end of the support rope will be at the top of the pulley and you'll have to send someone up to retrieve it. If you have a loop system, all you have to do is pull the line down and reattach the antenna. Tie an overhand knot loop to form the antenna attachment point (this is usually an insulator) where the rope ends are tied together and you're ready to start hoisting.

If the loop does break and you can still reach it, you can run the replacement through the pulley with a careful splice that still fits through the pulley. One method is to butt the broken and replacement ropes together and wrap the joint with a high-quality flexible tape such as Scotch 33+ on either side of the splice. Thin paracord can be wrapped around the broken rope a few times and wrapped with tape, then used to pull up a replacement rope. Whatever method you choose, pull up the replacement rope or cord slowly and patiently to keep it from getting jammed or pulling the splice apart. Good luck!

In a strong wind that will get the trees swaying, you'll want to have a method that allows the trees to move without breaking the antenna. You can attach a weight of some sort (cement block, plastic milk container filled with water, a bucket with rocks, etc.) to the rope or place tension on it with an elastic cord or strap.

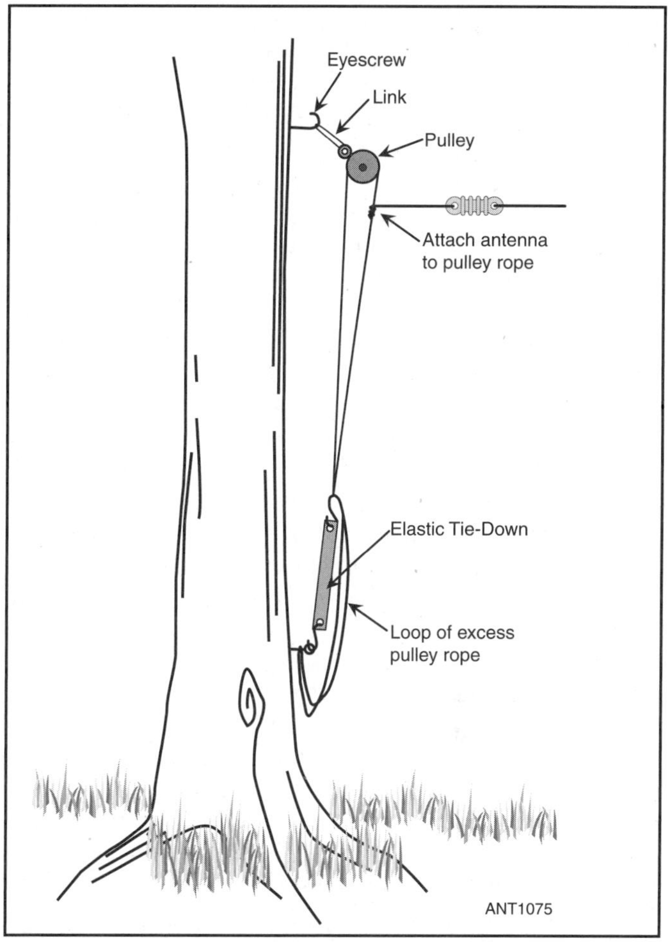

Figure 27.3 — By using a pulley, raising and lowering the antenna for repairs can be done without the need to climb the tree. Elastic cords, straps, or weights can be used to apply tension to the antenna. Loop the excess pulley rope to a second eye-screw in case the cord or strap fails.

TREE CLIMBING

If you're going to climb the tree yourself, you'll need sturdy boots, hard hat, a safety belt with two lanyards and possibly tree climbing spurs. You'll need the two lanyards to leapfrog your belts around branches so that you'll be belted to the tree 100% of the time.

Newer tree climbing techniques don't use steel climbing spikes as it's deemed harmful to the tree. The latest methods use a line thrown or shot over a branch and then the rope is used to climb the tree using rope ascenders like the ones used by mountain climbers and cavers. You don't even touch the trunk of the tree using this technique. The same techniques used to get a line over a branch for an antenna can be used to position a tree-climbing line.

Tree climbing has even become a recreational activity similar to the way that rock climbing has. There are clubs and resources available, and you can find them online. Be sure to check out the equipment and techniques and you can learn to utilize such tools yourself.

27.1.2 Ground-Mounted Masts and Poles

TV AND PUSH-UP MASTS

Stacking TV mast is available in 5- and 10-foot lengths, 1-¼ inches diameter, in both steel and aluminum. These sections are swaged or crimped at one end to permit sections to be joined together. This type of mast is usually mounted on a chimney or some sort of house-mounted bracket and is not intended to be permanently guyed. This mast is suitable for

VHF/UHF verticals and small beams and holding up light wire antennas for HF.

Galvanized steel push-up masts such as Rohn H30 or H40 are intended primarily for TV antennas and wireless Internet antennas. The masts may be obtained with three, four or five 10-foot sections and come complete with guying rings and a means of locking the sections in place after they have been extended. These masts are inherently more suitable for guyed mast installations than the non-telescoping type because the diameters of the sections increase toward the bottom of the mast. For instance, the top section of a 50-foot mast is 1-¼ inches diameter, and the bottom section is 2-½ inches diameter. The mast can be mounted on the ground or on a roof.

While tricky to install (each 10-foot segment must be guyed separately while pushing the mast up section by section), they can provide years of reliable service if not overloaded with anything larger than small VHF/UHF beams and verticals or HF wire antennas. If you are unfamiliar with push-up masts, a local TV antenna installer can perform the actual installation quickly and properly. They cannot be climbed and must be lowered to work on the antennas. Do not attempt to "walk up" these masts when extended.

Push-up masts are available from numerous sources, but the shipping cost often exceeds the cost of the mast. These masts can be ordered online (search for antenna, pushup and mast), through hardware stores and from TV antenna installers.

AB-155 MULTI-SECTION MASTS

The AB-155 is the prototype for multi-section mast kits with aluminum or fiberglass sections. Kits are available with and without a base tripod. Most kits have 4-foot sections which can be stacked safely up to about 40 feet. If a tripod is available, the mast can be assembled like a push-up mast with each new section being added at the bottom with guys attached as the mast grows. Without a tripod base, the mast will have to be assembled lying down and walked up.

The masts are not heavy enough to hold more than very small HF and mid-sized 6-meter beams. They are well-suited to hold wire antennas and VHF beams and verticals. If there is a significant side load, such as holding one end of a dipole, an additional back-guy may be needed, or the guys should be arranged so that one set of guys is aligned to take the extra load.

A critical part sometimes missing from the kits or that may be sold separately are the guy attachment rings. In the original design, the rings slip over the top of each section and rest on collars where the section attaches to the next lowest section. Rings from different types of kits may not fit properly and some kits use a clamp-on ring. If you purchase a used kit, be sure all of the rings are included.

The US Army manual for AB-155, including instructions for putting it up and taking it down safely, is widely available online. The kit also includes a parts list, so be sure all of the necessary items are present before purchasing the kit or starting to put it together. If you are purchasing a used kit, do a careful inspection of all sections, especially where the sections fit together, to be sure the section is not cracked or split.

AB-577 MASTS

The AB-577 "rocket launcher" mast is an all-aluminum crank-up mast kit that was sold as military surplus. It is designed to be field deployed by one or more people and does not require a prepared surface or foundation of any kind. The complete kit in **Figure 27.4** consists of a base section (the "launcher"), eight tube sections, guy wires and all hardware and tools to assemble the 50-foot mast. The standard AB-577 system, with three sets of guys, will support a modest triband Yagi at 45 feet (see **Figure 27.5**).

The supply from military surplus dealers has largely been exhausted but kits are available from other hams. Before purchasing a used kit, be sure that all the required pieces are present (a parts list is included with the kits), and no pieces are damaged beyond normal wear-and-tear.

A total height of 75 feet is possible with the addition of the MK-806 extension kit. It's useful for any application requiring a temporary or permanent tower such as lighting, surveillance, emergency communications, or RF survey work. The quick erection time also makes the mast very useful in neighborhoods where a permanently installed tower is not allowed.

The system consists of several short sections of aluminum

Figure 27.4 — An AB-577 temporary tower system in its transport case. (Photo by Alan Biocca, WB6ZQZ)

Figure 27.5 — Installation of surplus AB-577 tower with tribander at 45 feet at K7NV. (Photo by Kurt Andress, K7NV)

tubing, with special end clamp connectors used to join sections. These can be erected from the base fixture, which has a crank-up type winch-driven elevator platform. The tubing sections are installed in the base fixture and connected to the section above it with an over-center locking *Marmon*-style clamp. Then, the elevator platform is raised with the winch and the new tube is locked in place, high on the base fixture. Then the elevator is lowered to accept the next section. While the tower is extended, the supporting guys are adjusted via the unique *snubber* assemblies at the anchor connection. One person can erect this system, even in windy conditions, when special care is given to keeping the guys properly adjusted during each extension. There are several videos online showing how the system goes up and the original US Army manual is also widely available.

FIBERGLASS POLES

Telescoping fiberglass poles have become widely available in recent years. While they are too light to support rotatable antennas, they are popular as supports for wire antennas. Primarily intended for portable use, if you decide to use one in a permanent installation make sure the surface is coated to resist UV from sunlight or paint it. There is more information on these poles in the **Portable Antennas** chapter.

WOODEN POLES

A seldom-used but sturdy alternative is to use a wooden utility pole. They vary from new ones to used poles that have been pulled from service by utility companies. Make some inquiries to find out the availability and installed cost in your area. You'll need to add pole steps to climb it and will have to fabricate your own antenna mounting hardware. Pole steps are available from suppliers such as MacLean Power Systems. Utility poles are very sturdy, require no guys, and might satisfy your use and budget. Do not use a pole supporting power lines.

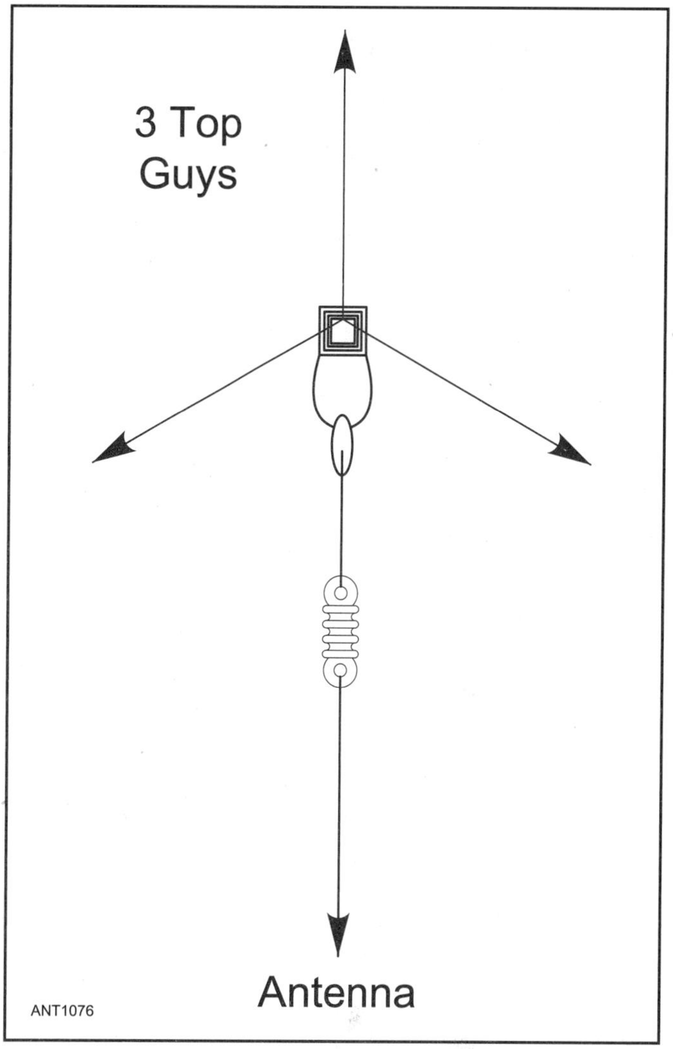

Figure 27.6 — For a mast supporting a wire antenna, the guys should be equally spaced at 120 degrees around the mast with one guy directly in line with the antenna.

27.1.3 Mast Guying

Three guy wires in each set are usually adequate for a mast. These should be spaced equally (every 120 degrees) around the mast. The required number of sets of guys depends on the height of the mast, its stiffness, and the required antenna tension if supporting a wire antenna in one end. A 30-foot-high mast usually requires two sets of guys, and a 50-foot mast needs at least three sets. If supporting the end of a wire antenna, one guy of the top set should be anchored to a point directly opposite the antenna. The other two guys of the same set should be spaced 120° with respect to the first, as shown in **Figure 27.6**.

Generally, the top guys should be anchored at distances from the base of the mast at least 60% of the mast height. The separation of the guy anchors from the mast determines the guy loads and the vertical load compressing the mast. At an anchor distance of 60% of the mast height, the load on the guy wire opposite the wire antenna is approximately twice the antenna tension. The compression in the mast will be 1.66 times the antenna tension. For 80% of the mast height, the guy tension will be 1.6 times larger than the antenna load and the mast compression will be 1.25 times larger.

The largest available and practical anchor spacing should be used. Additional compression on the mast caused by closer anchor spacing increases the tendency of the mast to buckle. Buckling occurs when the compression on the unsupported spans between guys become too great for the unsupported length. The section then bows out laterally and will usually fold over, collapsing the mast. Additional sets of guys reduce the tendency for the mast to buckle under the compression by decreasing the unsupported span lengths and stabilizing the mast, keeping it straight where it best withstands compression.

A natural phenomenon, called *vortex shedding*, can occur when the wind passes over the sections of a guyed mast. For every section size, shape, and length, there is a wind speed that can cause the sections to oscillate mechanically. When all the sections of an antenna support mast are close to the same size and length, it is possible for all of the mast sections to vibrate simultaneously between the guys. To reduce the potential for this, you can place the guys at locations along the mast that will result in different span lengths. This creates different mechanical resonant frequencies for each span, eliminating the possibility of all sections oscillating at the same time.

When determining the guy locations along the mast to treat this problem, you also need to consider the mast buckling requirements. Since compression of the mast is greatest in the bottom span and the least in the top span, the guys should be placed to make the bottom span the shortest and the top span the longest. A general guide for determining the different span lengths is to make the unguyed lengths increase by 10 to 20% with increasing height.

GUY MATERIAL

When used within their safe load ratings, you may use any of the ropes listed later in the chapter for mast guys. Nonmetallic materials have the advantage that there is no need to break them up into sections to avoid unwanted resonant interactions, also discussed later in the chapter. All of these materials are subject to *stretching*, however, which causes mechanical problems in permanent installations. At rated working loads, dry manila rope stretches about 5%, while nylon rope stretches about 20%. Usually, after a period of wind load and wet/dry cycles, the lines will become fairly stable and require less frequent adjustment.

Solid galvanized steel wire is also widely used for light-duty guying. This wire has approximately twice the load ratings of similar sizes of copper-clad wire, but it is more susceptible to corrosion. Stranded galvanized wire sold for guying TV masts is also suitable for light-duty applications but is also susceptible to corrosion. It is prudent to inspect the guys every six months for signs of deterioration or damage. **Figure 27.7** shows how to attach guy wire to strain insulators.

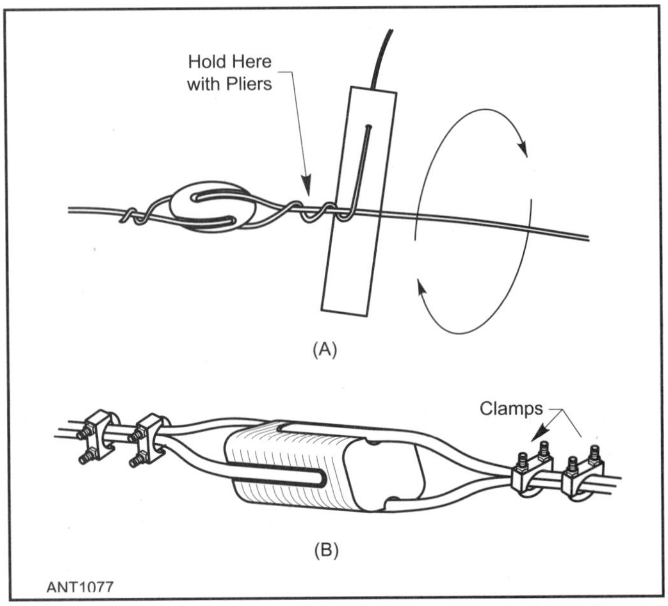

Figure 27.7 — Attaching light-duty guy wires to strain insulators. At (A) a simple lever is used to twist solid wire and at (B) standard Crosby clamps are used for light-duty stranded wire.

GUY ANCHORS

Figures 27.8 and **27.9** show two different kinds of guy anchors. In Figure 27.8 one or more pipes are driven into the ground at right angles to the guy wire. If a single pipe proves to be inadequate, another pipe can be added in tandem, as shown, and connected with a galvanized steel cable. Heavy-gauge galvanized pipe is preferred for corrosion resistance. Steel fence posts may be used in the same manner. Figure 27.9 shows a *dead-man* type of anchor. The buried anchor may consist of one or more pipes 5 or 6 feet long, or scrap automobile parts, such as bumpers or wheels. The anchors should be buried 3 or 4 feet in the ground. The cable connecting the dead-man to the guys should be galvanized wire rope, such as EHS guy cable. You should coat the buried part of the cable with roofing tar to well above the ground and thoroughly dry it prior to burial to enhance resistance to corrosion.

Heavy auger-type anchors that screw into the ground are also used and are commonly used by utilities to anchor power poles. These anchors are usually heavier than required for guying a mast, although they may be more convenient to install. You should conduct annual inspections of the anchors by digging several inches below grade around the anchor to inspect for corrosion.

Trees and buildings may also be used as guy anchors if they are located appropriately. Care should be exercised, however, to make sure that the tree is of adequate size and that any fastening to a building can be made sufficiently secure. See the section above on using trees as antenna supports regarding anchoring to trees.

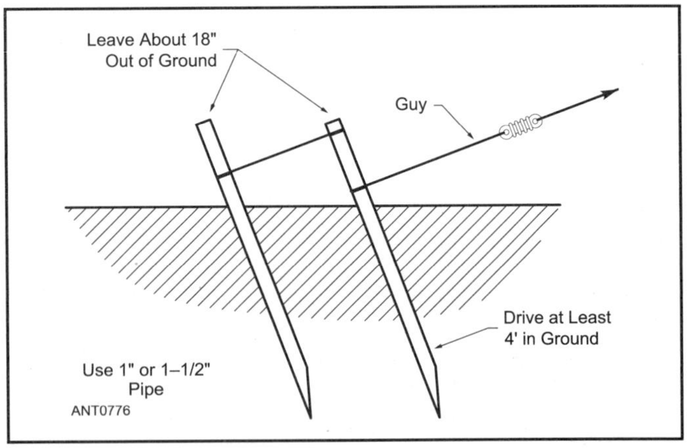

Figure 27.8 — Driven guy anchors. One pipe is usually sufficient for a small mast. For added strength, a second pipe may be added as shown.

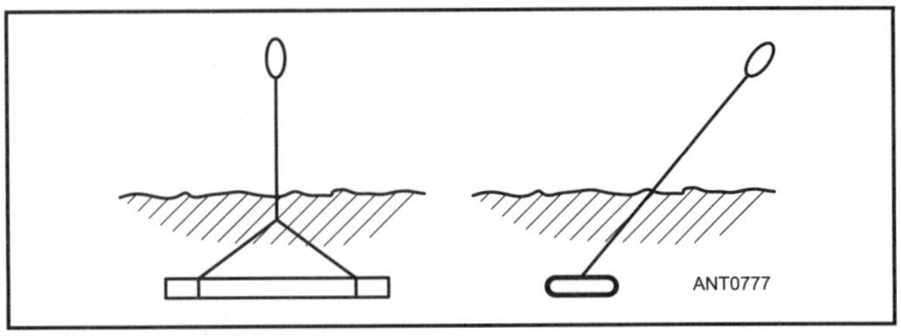

Figure 27.9 — Buried "dead-man" guy anchors (see text).

Another popular type of anchor is a steel pipe or beam set in concrete with the guy attachment a few feet above ground. This makes maintenance around the guys much easier and lessens the chance of someone walking or running into the guys. This type of anchor places heavy bending loads on the pipe or beam and should be designed by an engineer, including the footing since the anchor will transfer the tensioning load to the soil in the direction of the tower.

GUY TENSION

Many troubles encountered in mast guying are a result of pulling the guy wires too tight. Guy-wire tension should never be more than necessary to correct for obvious bowing or movement under wind pressure. Approximately 10% to 15% of the working load is sufficient. In most cases, achieving the necessary tension does not require the use of turnbuckles, with the possible exception of the guy opposite a wire antenna. If any great difficulty is experienced in eliminating bowing from the mast, the guy tension should be reduced, or additional sets of guys are required. The mast should be checked periodically, especially after strong winds, to ensure the guys and anchors have not stretched or moved, allowing the mast to bend away from the required straight alignment.

In the case of rope guys, use of a "trucker's hitch" (see the section on knots) will provide much more tension than can be obtained by just pulling on the rope since it has a 2:1 mechanical advantage.

27.2 Towers

What ham doesn't dream of having a tower? It's a natural yearning that grows out of every ham's on-the-air experience. Regardless of one's individual pursuits within amateur radio, superior results can be achieved with antennas that are high and in the clear. Operations at the longest wavelengths, where tall vertical antennas generate the most desirable radiation patterns, the short-wave bands from 40 to 10 meters, where range is often determined by careful selection of antenna heights from 20 to 200 feet, and the VHF and UHF range, where performance is best when antennas reach above nearby trees and buildings for line-of-sight to the horizon, all present the station builder with the challenge of how to get antennas to the right height.

Towers have always been the structure-of-choice to support amateur antennas. Why? The options offered by commercial tower manufacturers allow each individual tower installation to be purpose-built to suit the needs of the station owner. A wide variety of requirements can be satisfied with selection of the right tower type and proper construction.

A tower is the best answer to a reliable, permanent antenna support structure and they basically come in two types — self-supporting and guyed. Beginning with the small, roof-mounted "four-footed tripod" models, amateurs use towers up to and beyond 200-foot tall broadcast-size structures. This section is an overview of various common types of towers with some of their key characteristics.

Lattice towers consist of two kinds of *members*: *legs* (often called siderails) and diagonal and horizontal *braces*. Members can either be round, such as used with Rohn G-series (**rohnnet.com**), or 90- or 60-degree angled metal. Round-member towers are the most common for amateur towers. The *tower face* is that outward facing area between the legs with the braces between them. Free-standing and guyed lattice towers are built of pre-assembled *sections*, usually 8 to 10 feet long, stacked on top of each other to reach some desired height. Lattice towers are constructed from steel or aluminum, with steel the most common for guyed towers. *Tubular* towers are constructed from telescoping sections of steel tubing. These are referred to as monopoles; some models rotate and can carry significant loads.

> **Learn About Fasteners**
>
> An excellent resource for selecting and using fasteners of all type is *Carroll Smith's Nuts, Bolts, Fasteners and Plumbing Handbook*. Written originally from a background in auto racing, it is widely available and applies to every part of this chapter.

27.2.1 Roof-Mounted Towers

The self-supporting roof-mounted tower is a modest way to support small to mid-sized antennas. This might be your first foray into a tower and directional antenna and a roof-mounted tower offers an inexpensive way to get started. Glen Martin Engineering (**glenmartin.com**) offers several models of four-leg aluminum towers and is a representative source of roof-mounted towers ranging in height from 4.5 to 26 feet. **Figure 27.10** shows a typical installation. Follow the manufacturer's recommendations for installation and grounding.

A roof-mounted tower is attached to the roof with anchor bolts that extend completely through the roof. Do not use lag bolts into the roof trusses. Use a 2×4 or 2×6 across the trusses inside the attic for a backing plate and attach the anchor bolts to them as in **Figure 27.11**. Another similar board can be placed on top of the roof to further distribute the load. Any wood exposed to the weather should be pressure-treated or coated with roofing tar. Roofing tar is also used to seal around the mounting bolt holes to prevent leaks (see **Figure 27.12**).

Roof-mounted antennas and structures provide practical, easy solutions to getting a directional antenna in the air, but roof mounting can be dangerous work. State and federal safety

Figure 27.10 — A roof mounted tower will give your antennas a chance to get up in the air with a minimal impact and cost. (Photo by Redd Swindells, AI2N)

27.12 — Lengths of 2 × 6 on the roof act as footing for the tower legs and are coated by roofing tar to protect against weathering and prevent leaks. (Photo by Jane Wolfert)

Figure 27.11 — The strengthened anchoring for the roofmounted tower. Bolts run through the roof and through the anchor plate (2 × 6) between joists. (Photo by Jane Wolfert)

laws require fall-arrest equipment and it's highly recommended that you use it as well. A fall-arrest harness (FAH) should be attached to an anchor point on the peak of the roof. Most of these small roof-mounted supports cannot be climbed but require a ladder support against them to provide access to the antenna.

27.2.2 Self-Supporting Towers

Self-supporting towers have a smaller footprint but are generally more expensive to install. Significantly more concrete is required for the base of a free-standing tower and the amount of steel or aluminum (which ultimately determines the cost of a tower) is higher. The advantage of a self-supporting tower is that no guy wires are required. This appeals to hams without enough room for the necessary guying system and the cleaner "look" sometimes helps with aesthetic concerns.

Because there are no guys to keep them standing, self-supporting towers depend on bending strength and a large concrete base. The base is generally required to have a volume of at least five to six cubic yards, requiring significant digging and preparation. The weight of the base keeps the tower system's center of gravity low or below ground level, minimizing the overturning force from wind. The soil around the base must be solid enough to withstand the pressure from overturning forces on the tower system. If you have any questions about your ability to properly construct the base, consult a professional engineer or hire a concrete contractor.

FREESTANDING TOWERS

Towers specifically designed and installed for TV antennas are at the low end of suitability for typical HF beams. TV antenna towers have a maximum height of 40 to 60 feet. The most common are the Rohn AX, BX and HDBX series and a tubular-legged type similar to but lighter than Rohn 25G. Universal Manufacturing (**universaltowers.com**) offers similar towers made of aluminum.

The common BX-series towers sketched in **Figure 27.13** are made from stamped steel with X-bracing of the legs. The X braces are not connected to each other, and the most common failure point is between the braces. Also, the rotator and top plates are made from sheet metal and can crack from wind-induced metal fatigue. For small triband HF beams and VHF arrays they are fine but be careful of overloading towers using the smaller stacking sections. These towers should be limited to antennas with boom lengths less than 10 feet since they have minimal resistance to torsion (twisting).

For larger antenna arrays, heavier towers are available that are designed for broadcast and commercial applications. These look much the same as the "TV antenna" towers but are made of heavier material and have heavier and stiffer bracing. The most common models of these towers are the Rohn SSV, Trylon (**trylon.com**), Universal, and AN Wireless (**anwireless.com**). While significantly more expensive, they can handle very large loads, including high winds and icing conditions.

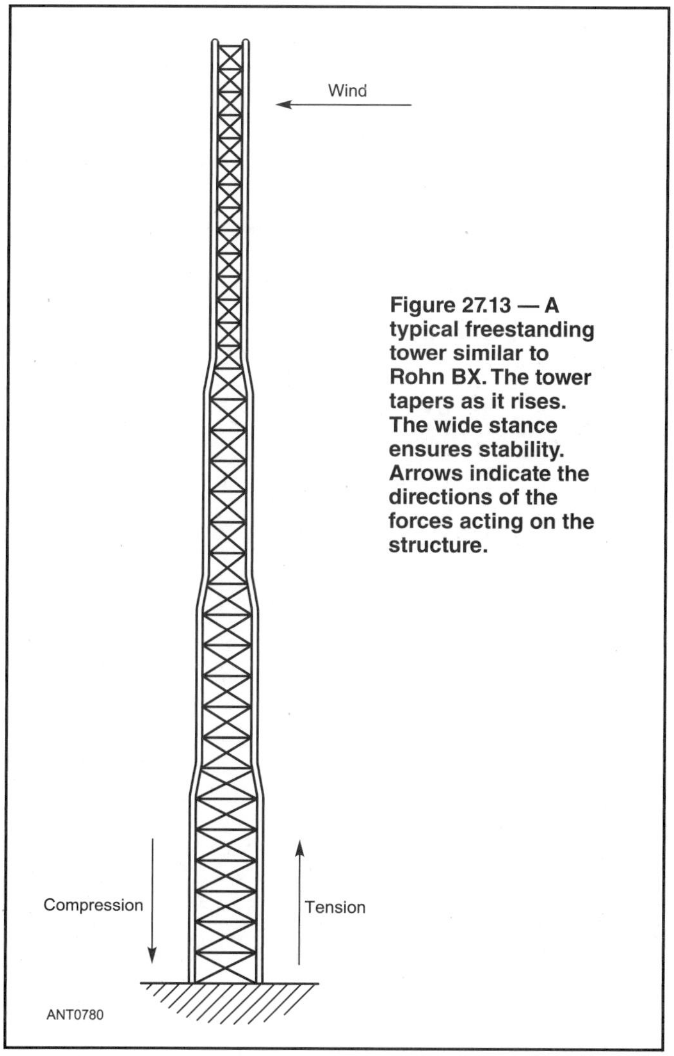

Figure 27.13 — A typical freestanding tower similar to Rohn BX. The tower tapers as it rises. The wide stance ensures stability. Arrows indicate the directions of the forces acting on the structure.

CRANK-UP TOWERS

Crank-ups are a popular type of self-supporting tower. These towers use either a motorized or manual system of cables and pulleys to extend or retract the tower. They are the most expensive tower for the height due to more materials and hardware, but satisfy many hams with limited space, or those who dislike guy wires. When cranked down, a telescoping tower can maintain a low-profile system, out of sight of the neighbors and family.

Tubular crank-ups are generally limited to a single antenna since the rotator is mounted on a plate at the top of the tower without any additional bracing. This limits the size of the antenna and how far above the rotator it can be mounted. Lattice crank-ups generally have the same top structure as a guyed tower and can support much larger antenna and mast combinations.

US Tower (**ustower.com**) dominates the market for crank-ups, manufacturing good products and offering good customer support. Both lattice- and tubular-type crank-up towers are available as shown in **Figure 27.14**.

Do not use guys with normal crank-up towers (those that have no locking devices between sections)! The increased tower compression will be carried by the hoisting cable, which will eventually cause it to fail.

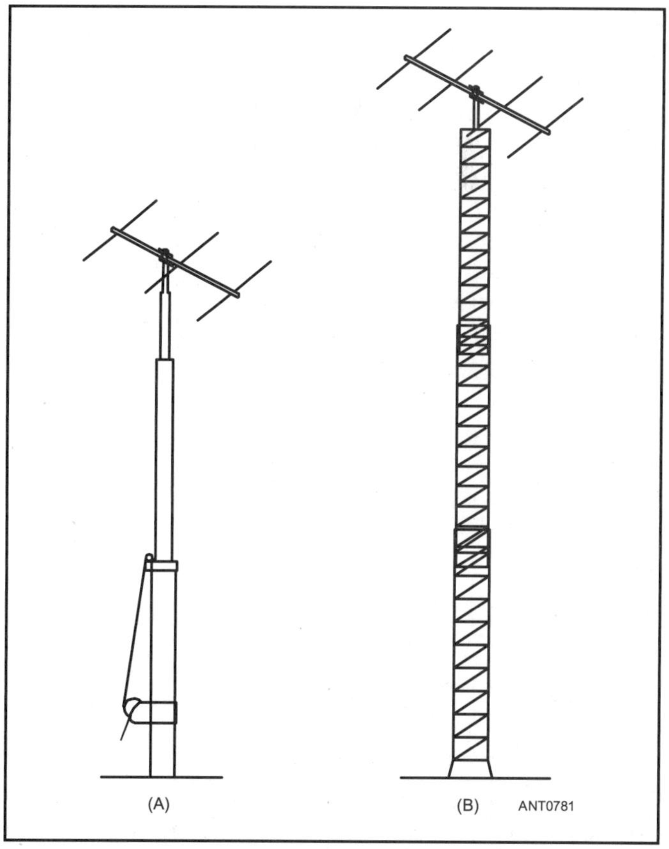

Figure 27.14 — Two examples of crank-up towers. At (A) a tubular style and at (B) a lattice style.

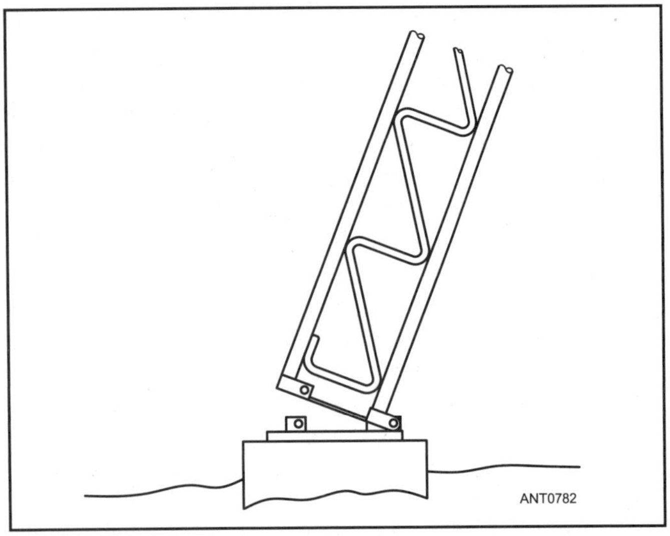

Figure 27.15 — Fold-over or tilting base. There are several different kinds of hinged sections permitting different types of installation. Great care should be exercised when raising or lowering a tilting tower.

TILT-OVER TOWERS

Some free-standing towers have another convenience feature — a hinged section that permits the owner to fold over all or a portion of the tower. The primary benefit is in allowing antenna work to be done close to ground level, without the necessity of removing the antenna and lowering it for service. **Figure 27.15** shows a hinged base used with stacked, guyed tower sections. Many crank-up towers come with optional tilt-over base fixtures that are equipped with a winch and cable system for tilting the fully nested tower between horizontal and vertical positions.

The hinged section can also be designed for portions of the tower above the base. These are usually referred to as *guyed tilt-over towers*, where a conventional guyed tower can be tilted over for installing and servicing antennas.

Misuse of hinged sections during tower erection is a dangerously common practice among radio amateurs. Unfortunately, these episodes can end in accidents. If you do not have a good grasp of the fundamentals of physics, it might be wise to avoid hinged towers or to consult an expert if there are any questions about safely installing and using such a tower. It is often far easier (and safer) to erect a regular guyed tower or self-supporting tower with gin pole and climbing belt than it is to try to walk up an unwieldy hinged tower.

27.2.3 Guyed Towers

Guyed tubular-leg lattice towers are strong, reliable, relatively easy to erect, and have a wide array of accessories designed and intended for ham use. They are usually less expensive to install but need a big footprint for the guying system. Since the typical recommended guy anchor distance from the tower is 80% of the height, the guys for a 100-foot tower need to be anchored 80 feet away from the tower. A set of three guys spaced 120° around the tower is repeated every 30 to 40 feet up the tower (see **Figure 27.16**).

The most widely used guyed towers for amateur applications are the Rohn 25G and 45G. They are well constructed, are hot-dipped galvanized, and have enough accessory items for any use. These towers have tubular legs and Z-bracing rods welded to the legs. The Rohn product catalog (available for downloading from the company website) provides calculations for rated wind load at various heights and all base and guying requirements.

Rohn 25G has a face width of 12 inches and a 10-foot section weighs 40 pounds. A gin pole and a ground crew is the recommended way to install these towers. A practical height limit of 190 feet at 90 MPH wind speed provides 7.8 square feet of antenna load capacity. A 100-foot tower yields 9.1 square

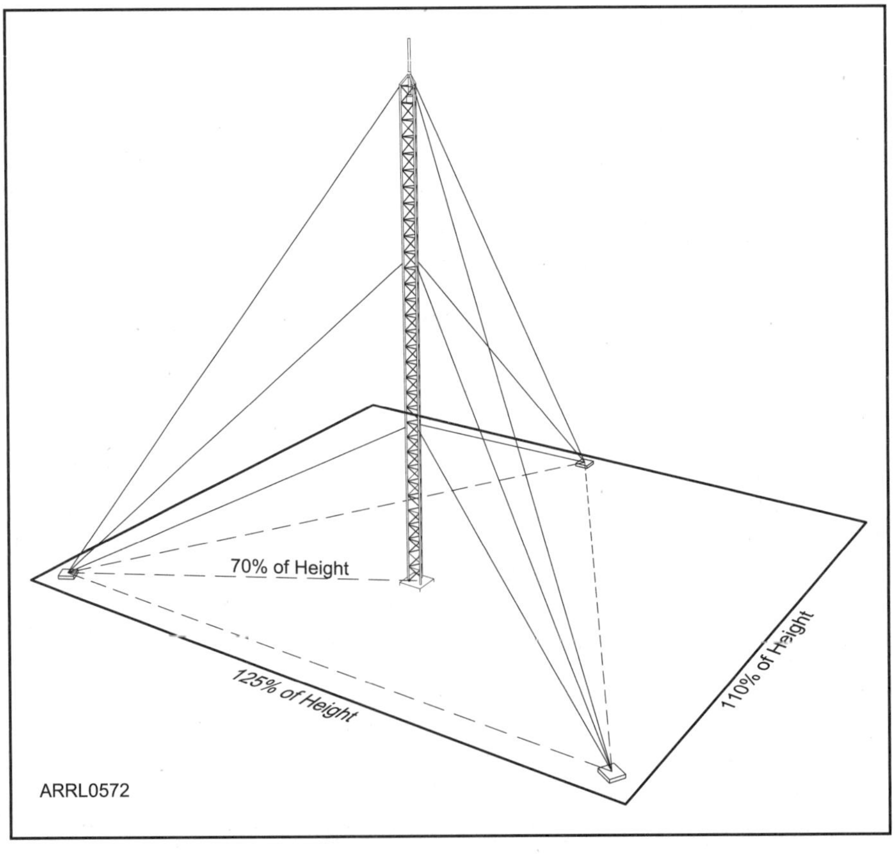

Figure 27.16 — Guyed towers are built from stackable sections that are usually identical except for the base that attaches to the concrete and the top section that supports the rotator and mast. Guy requirements are specified by the manufacturer but spacing the anchors 70 to 80% of the tower height is typical. As seen in the drawing, the tower and guy wires do take up quite a bit of space.

feet of antenna capacity, enough for a small stack of monoband Yagis or a high-performance triband beam. An experienced crew can erect up to a hundred feet a day of this popular tower.

Rohn 45G is 18 inches across the face and a 10-foot section weighs 70 pounds. This robust tower is rated up to 240 feet in 90 MPH winds, with a wind load rating of 16.3 square feet. At a height of 100 feet the wind loading is 21.5 square feet.

Rohn 55G is also 18 inches across the face, weighs 100 pounds per section, and can be installed up to 300 feet with 90 MPH wind ratings. It has a gross capacity of 17.4 square feet in that maximum configuration. The standard Rohn gin pole is not rated for 55G because of its weight.

Tubular-leg towers can also be supported by attaching them to a building with a *house bracket*. The manufacturer will specify how far above a bracket the tower can be extended without guys.

27.3 Engineering the Tower Project

Engineering in this sense means to plan, construct, or manage the practical applications of your tower project. The engineering should be done *before* you begin digging, pouring, or constructing!

The process you go through to design your system may begin with site selection ("What is a suitable location for a tower?") or it may start with selecting a tower ("What tower can I put up on this site?"). Everyone's circumstances are different.

It's not unusual to repeat the process of planning and tower selection for several iterations as you work through the various types of towers and associated costs and constraints. The important thing is to work through the various interacting issues until you are satisfied you have addressed all of them.

It is often very helpful to the novice tower installer to visit other local amateurs who have installed towers. Look over their installations and ask questions. Ask about local permitting processes and requirements. If possible, have a few local experienced amateurs look over your design plans — before you commit yourself. They may be able to offer a great deal of help. If someone in your area is planning to install a tower and antenna system, be sure to offer your assistance. There is no substitute for experience when it comes to tower work and your experience there may prove invaluable to you later.

27.3.1 Site Planning and Permitting

LOCAL ORDINANCES AND CC&RS

Most antenna support structures fall under local building regulations as well as neighborhood restrictions. Many housing developments have Homeowner's Associations (HOAs) as well as Covenants, Conditions and Restrictions (CC&Rs) that may have a direct bearing on whether a tower or similar structure can be erected at all, and if allowed, how it must be constructed for proper safety. This is a broad topic with many facets. Detailed background on these topics is provided in *Antenna Zoning for the Radio Amateur* by Fred Hopengarten, K1VR, an attorney with extensive experience in towers and zoning. You may also want to contact one of the ARRL Field Organization's Volunteer Counsels.

The FCC's PRB-1 memorandum specifies that local regulations must make "reasonable accommodation" for Amateur Radio antennas and support structures but does not pre-empt local regulations including CC&Rs. For more information on PRB-1, consult the information at **arrl.org/prb-1**.

Even without neighborhood restrictions, a building permit is likely to be required. With the proliferation of cellular and other commercial wireless devices and their attendant RF sites, many local governments now require that all antenna support structures meet local building codes. Building permit applications may also require Professional Engineer (PE) calculations and stamp (certification). The ARRL Field Organization's Volunteer Consulting Engineer program may be useful with the engineering side of your project. It is important to understand that these building regulations are the cumulative results of decades of experience – and that compliance with them will result in a structure that is both stronger and safer for you and your neighbors. Working closely with the local building department is often a source of useful knowledge and tips for a better installation. It also gives you permits and signoffs that can provide a level of liability protection in the event of a future incident.

SAFETY

You must consider the safety aspects of your installation. For example, a tower should not be installed in a location where it could fall onto a neighbor's property. Imagine what would happen if your tower or antenna fell — where would it be likely to land? What could it hit on the way down? You may not be able to mitigate every possible outcome but thinking about it before construction may lead to a better plan.

The antenna must be located in such a position that it cannot possibly come in contact with power lines, either during normal operation or if the structure should fall. Consider the proximity of power lines to the tower. Safety rules dictate that all parts of a tower and antenna must remain at least 10 feet from power lines when being erected or after being installed. This is the smallest separation you should consider, and a greater safety margin is strongly recommended.

MANUFACTURERS GUIDELINES

As you explore the various options in the world of towers, you will find that every manufacturer provides a) clear instructions on how their products should be installed, and b) speci-

The 10 Most Common Tower Building Mistakes

1. Not following the manufacturer's specifications

Commercially manufactured towers must comply with current standards for wind loading and structural integrity. Engineers design the towers and make the calculations to make them safe. If you don't follow their specifications at a minimum, the tower will not take the stresses and loads to which it is subjected. In other words, it'll probably fail.

2. Overloading

This is the most common reason for amateur tower failure. You must not exceed the wind load rating. This is even more important for self-supporting and crank-up towers. While you might get away with exceeding the ratings due to built-in design margins, it's never a good idea to overload any part of the tower system. When in doubt, err on the conservative side — you won't regret it.

3. Underestimating wind forces

Wind pressure on a tower and antenna system can be tremendous. Unless you've been on a tower during a windstorm to feel the pressure and the forces, it's difficult to appreciate how significant they are. Increases in wind pressure are not linear; wind loading goes up with the *cube* of wind speed. An increase of 10 MPH in wind speed can increase the wind force by almost 50% in some cases as shown in Table 27.A.

4. Not building for the wind speed rating in your county

While many counties and even whole states in the US are only rated for 70 MPH winds (the minimum rating), many other counties have ratings much higher. For example, Dade and Broward counties in Florida have ratings of 140 MPH. Find out what the wind speed rating is for your county or your specific location and use that as the *minimum* wind speed design parameter for your tower and antenna system. Your county building department can provide wind speed information.

5. Using the wrong mast for the job

This is an all-too-common failure. Stacks of medium to large HF beams can put huge stresses on your mast. Pipe may be fine for small installations where you don't have much wind speed or loading or when there is only one antenna at the top of the tower. Structural tubing is carbon alloy steel rated for strength and is the preferred material.

6. Not tensioning the guy wires properly

Proper guy wire tension is a critical part of a tower's ability to handle wind stresses. Having the wrong tension can be like driving your car with over or under-inflated tires; it is potentially dangerous and is not the proper specification from the manufacturer. Having too little tension can result in wind slamming of the tower and guys as the tower is blown back and forth. Too much tension puts excess preload on the guys and lowers the safety margin significantly.

Around 90% of ham towers use 3⁄16 inch EHS steel guy wires. Guy wire tension is typically 10% of the breaking strength — in the case of 3⁄16 inch EHS that would be 400 pounds. The only inexpensive and accurate way to measure this is to use a Loos Tension Gauge, such as the Loos PT-2 for 3⁄16 and 1⁄4 inch wire rope sizes.

7. Not having a proper ground system

A good ground system is necessary not only for lightning protection but will also protect your equipment, your home and your life. Proper grounding is discussed in the **Safe Practices** chapter of this book and the *ARRL Handbook*.

8. Not doing an annual inspection

Your tower and antennas are undergoing a slow, but constant process of deterioration. The best way to find and fix small problems before they become big problems and potential calamities is by doing an annual inspection.

Look at everything and push and pull on the hardware. You also want to put a wrench on 10% or more of the tower nuts to check for tightness as well as all of the nuts on accessories like antennas, mounts, U-bolts, etc.

9. Not fitting the tower sections on the ground

Tower sections, new or used, may not fit together easily. It's much easier to correct alignment problems on the ground than up on the tower during construction. A handy tool for getting tower sections together (or apart) is a tower jack that combines a leg aligner along with a lever for pulling sections together or pushing them apart.

10. Using the wrong hardware

To slow the process of deterioration, use only hardware that minimizes corrosion. Galvanized or stainless steel materials are the only ones that will survive outdoor use reliably. (See the section on "Corrosion" in this chapter.)

Substituting the wrong hardware can also lead to failure, for example using general hardware store bolts for tower legs when the manufacturer calls for a specific SAE grade. Using hardware totally unsuited for the task is common, i.e., installing the wrong type of 'screw-in' anchor or anchor rods; use of non-closed-eye eyebolts (use only welded or forged ones); use of the wrong guy material (EHS only!); and more.

Table 27.A
Wind Speed and Pressure

Mean Velocity	*Wind Pressure*
50.0 MPH	10.0 PSF
60.0 MPH	14.4 PSF
70.7 MPH	20.0 PSF
86.6 MPH	30.0 PSF
100.0 MPH	40.0 PSF
111.8 MPH	50.0 PSF
122.5 MPH	60.0 PSF

fications of the minimum performance that should be expected when installed per the manufacturer's instructions. This information is developed by design engineers and is usually signed off by a registered professional engineer to certify that the design is in alignment with standard building codes, such as the International Building Code (IBC). The IBC is the base building code used by most jurisdictions in the United States.

The *Prime Directive* is – "DO what the manufacturer says There is a corollary, as well – "DON'T do what the manufacturer doesn't say to do." Professional engineers have analyzed and calculated the proper specifications and conditions for tower structures and their environment. Taking shortcuts or making different decisions will result in a less reliable installation. It is also important to note here that every installation is unique. With the help of your own professional engineer, you may find that even the manufacturer's recommendations are not conservative enough for your particular situation and needs.

AREA AND ACCESS

For a guyed tower, there must be sufficient space for proper guying. The guy anchors should be between 70% and 80% of the tower height in distance from the base of the tower on level ground — sloping terrain may require larger areas (see Figure 27.16).

Erecting the tower and installing the antennas will require some space. Is there enough room to lay out a tower on a tilt-over base and how should the hinged base be oriented? Think about where antennas will be assembled and how they'll be hoisted to the top of the tower. Where will any necessary equipment need to be positioned and how will it get there?

Another part of choosing a tower site has to do with arranging for access. That is, access for base excavation and access for concrete. If you aren't sure, ask a local contractor to evaluate your site and make suggestions. They may spot something important that you've overlooked.

27.3.2 Selecting a Tower

The selection of a tower, its height, and the type of antennas and rotator is probably one of the more complex issues faced by station builders. All aspects of the tower, antenna, and rotator system are interrelated, and you should consider the overall system before making any decisions regarding specific system components.

Selecting a tower must be based on your requirements for what the tower must support, along with considerations such as total budget, permit restrictions, aesthetics, and the specifics of where you intend to install the tower. You should also consider the climate and your ability to maintain a tower. You may already know what general type of tower you want — self-supporting or guyed, lattice or tubular, and so on. Or you may have to select a tower based on the constraints of the available site or other factors.

One of the first things you need to determine in the tower selection process is the type of specification required by the local authorities, if any. Then, you must determine the *Basic Wind Speed* appropriate for the site. The Basic Wind Speed used in most specifications is the average wind speed for one mile of wind passing across the structure. It will be a lower value than the peak readings from an anemometer (wind gauge) installed at the site. For example, a Basic Wind Speed of 70 mph could have a maximum value of 80 mph and a minimum of 60 mph, equally distributed during the passage of the mile of wind. Basic wind speeds can be found in tables or maps contained in the appropriate specifications. Often, the basic wind speed used for the location may be obtained from the local permit authority.

Many building regulations base their specifications for maximum wind speed on TIA-222, "Structural Standard for Antenna Supporting Structures and Antennas." (TIA-222-H is the latest revision as of mid-2023.) Your county building department can provide wind speed information. Remember — these are the minimums, and some building departments use a slightly higher figure for issuing building permits.

Add up the total square feet of antenna area (commercial antennas include area in the antenna specifications) you plan on installing. Compare that combination of wind area and your maximum wind speed rating to manufacturer specifications for the specific models of acceptable towers.

Most tower manufacturers provide catalogues or data packages that represent engineered tower configurations. These are provided as a convenience for users to help determine the most suitable tower configurations. The most commonly used design specifications for towers are the previously mentioned TIA-222 and the UBC (Uniform Building Code). These specifications define how the tower, antenna and guy loads are determined and applied to the system and establish general design criteria for the analysis of the tower. Local authorities often require the review and approval of the installation by a state licensed Professional Engineer (P.E.) to obtain building permits. All local authorities in the United States do not subscribe to the same design standards, so often the manufacturers' general-purpose engineering is not applicable.

DETERMINING TOWER LOAD

Most manufacturers rate their towers in terms of the maximum allowable antenna load that can safely be carried at a specific wind speed. Ensuring that the specific antennas you plan to install meet the tower's design criteria, however, may not always be a straightforward task.

For most towers, the manufacturer assumes that the allowable antenna load is a horizontal force applied at the top of the tower. The allowable load represents a defined amount of exposed antenna area, at a specified wind velocity. Most tower manufacturers rate the load in terms of *Flat Projected Area* (FPA). This is simply the equivalent area of a flat rectangular surface at right angles to the wind. The FPA is not related to the actual shape of the antenna itself, only its rectangular projected area. Some manufacturers provide separate FPAs for antennas made from cylindrical sections and those made from rectangular sections.

In the realm of antenna manufacturers, however, you may encounter another wind load rating called the *Effective Projected Area* (EPA). This attempts to take into account the actual shape of antenna elements. The problem is that there is no agreed-upon standard for the conversion from EPA to load numbers. Different manufacturers may use different conversion factors.

Since most tower manufacturers have provided FPA figures for their towers — allowing us in effect to ignore design-specification details — it would be easiest for us to work only with FPA values for our antennas. This would be fine, if indeed we had good FPA figures for the specific antennas we plan to use! Unfortunately, FPAs are rarely specified for commercially built amateur antennas. Instead, most antenna manufacturers provide effective areas in their specification sheets. You may need to contact the antenna manufacturer directly for the FPA antenna area or for the antenna dimensions so that you can do your own FPA calculations as discussed in **Appendix A** of this chapter.

27.3.3 Designing the Guys

The configuration shown in **Figure 27.17A** is taken from an older (1983) Unarco-Rohn catalog. This configuration has the top set of guys placed at the top of the tower with the lower set halfway up the tower. This configuration is best for most amateur installations, which usually have the antennas mounted on a rotatable mast extending out the top of the tower — thereby placing the maximum lateral loads when the wind blows at the top of the tower (and the bottom of the rotating mast). This configuration can limit the ability to easily tram and install antennas on the mast, or at the tower top, but will work fine for one's first tower installation.

The configuration shown in Figure 27.17B is from the current Rohn catalog (Catalog 3). It shows 5 feet of unsupported tower extending above the top guy set. The lower guy set is approximately halfway between the top guys and the base. The newer configurations are tailored for commercial users who populate the top region of the tower with fixed arrays and/or dishes. The installation in Figure 27.17B cannot safely withstand the same amount of horizontal top load as can the configuration shown in Figure 27.17A, simply because the guys start farther down from the top of the tower.

An overhead view of a guyed tower is given in Figure 27.17C. Common practice is to use equal angular spacings of 120° between guy wires. If you must deviate from this spacing, the engineering staff of the tower manufacturer or a civil engineer should be contacted for advice.

Amateurs should understand that most catalogs show generic examples of tower configurations that work within the cited design specifications. They are by no means the only solution for any specific tower/antenna configuration. You can usually substantially change the load capability of any given tower by varying the size and number of guys. Station builders are encouraged to utilize the services of professional engineers to get the most out of their guyed towers.

To calculate the length of guy wire needed, you'll need to know the height of the guy attachment on the tower (H) and the distance to the guy anchor (D). The length of guy wire (L) is then:

$$L = \sqrt{H^2 + D^2}$$

Then add extra length on each end to account for attaching the guy wire to insulators or turnbuckles. If you break up the guy wire with insulators, you'll need extra guy wire for each side of the insulator. It may be easier to build the guy on the ground, measuring the distance between each insulator as it is attached and cutting the guy wire to the exact length required. Build from the tower end and leave some extra for attaching to the guy anchor.

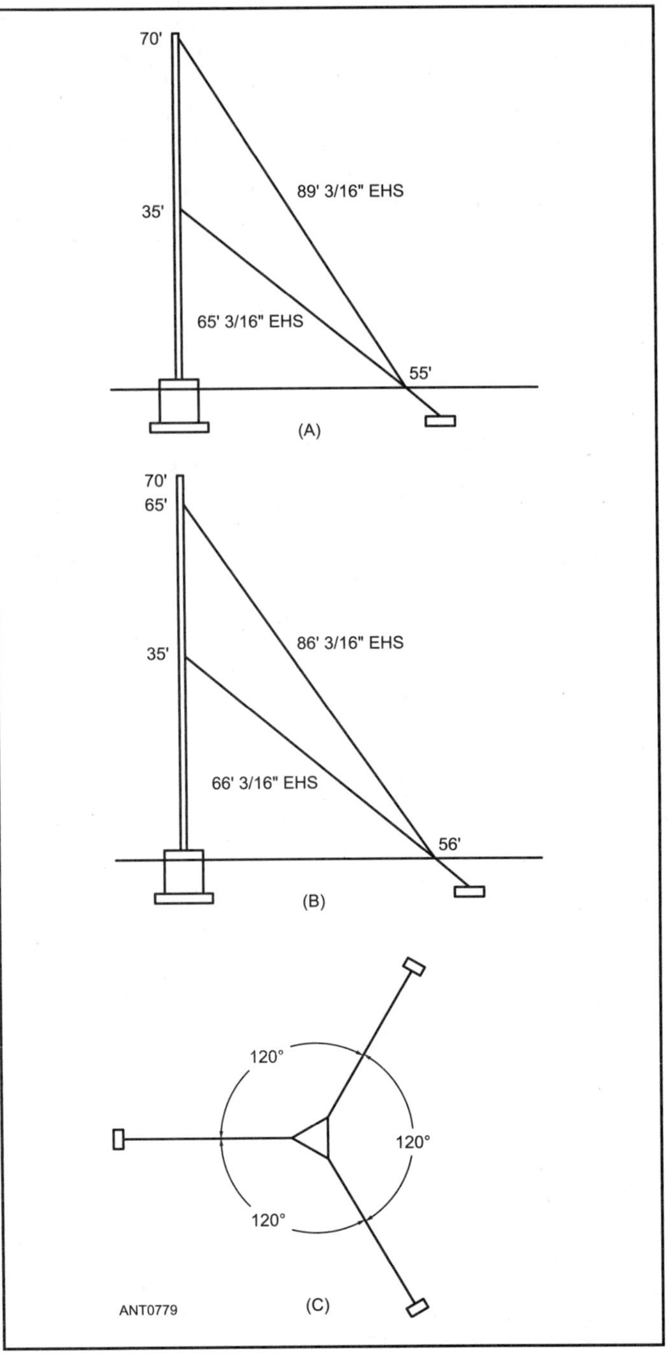

Figure 27.17 — The proper method of installation of a guyed tower. At (A) is the method recommended for most amateur installations. At (B), the method shown in current Rohn catalogs that places considerable stresses on the top section of the tower when large antennas are mounted above the tower (see text). (C) shows the recommended orientation of guy wires, symmetrically spaced around the tower.

GUY ANCHORS

The most common type of guy anchor is the buried "dead man" concrete anchor shown in **Figure 27.18A**. The Rohn catalog contains drawings and specifications for anchors suitable for typical 25G, 45G, and 55G towers. Rohn makes all of the necessary hardware to attach guys to buried anchors.

Screw in earth anchors shown in Figure 27.18B are widely used in commercial tower work for anchoring utility poles and other similar jobs. Smaller anchors are not suited for typical tower installations. Larger anchors that require hydraulic drivers are available from Rohn, AB Chance, and other commercial vendors. Expanding earth anchors are installed in a hole, hammered to secure the anchor fingers in place, and the hole backfilled.

The guide *Encyclopedia of Anchoring* is available online as a four-part PDF document at **hubbellcdn.com/literature/CENT-CP-ENCYCLOPEDIA-OF-ANCHORING-CAT-EN_09_2022_Web.pdf**. Selecting the proper anchor is highly dependent on the soil properties at the tower. If you are unsure of how to install this type of anchor or how to assess soil properties, consult a structural or soils engineer.

Other types of anchors, such as elevated anchors made of steel pipe or I-beams may be more suitable for your circumstances. Guys can also be attached to the structural elements of a building — do not rely on expansion bolts or lugs in masonry. When using an alternative to buried anchors, consult an engineer to be sure your guys will stand up to the load your tower will present in local conditions.

27.3.4 Designing the Base

Several common types of tower bases are shown in **Figure 27.19**. Amateurs use all types of bases. A tower base is its foundation — what supports and carries the weight of the tower itself, along with force directed down the legs from wind and guying tension. Hams often use a variation of the composite base in which part of a tower section, or a "base burial" section is buried in the concrete. Hams also use the pier pin base. A pin in the center of the concrete pier protrudes through a hole in the center of the base plate and holds it in place. The base plate is not bolted down or anchored to the concrete so that some side-to-side and twisting movement is allowed. The pivoted and tapered bases are only used on the largest towers.

Tower manufacturers can provide customers with detailed plans for properly constructing tower bases. **Figure 27.20** is an example of one such plan. This plan calls for a hole that is 3.5 × 3.5 × 6 feet deep. Steel reinforcement bars are tied together to form a cage and placed in the hole.

A strong wooden form is constructed around the top of the hole. The hole and the wooden form are filled with concrete so that the resultant block will be 4 inches above grade. Before it hardens, the anchor bolts are embedded in the concrete, and aligned with the plywood template. The template serves to

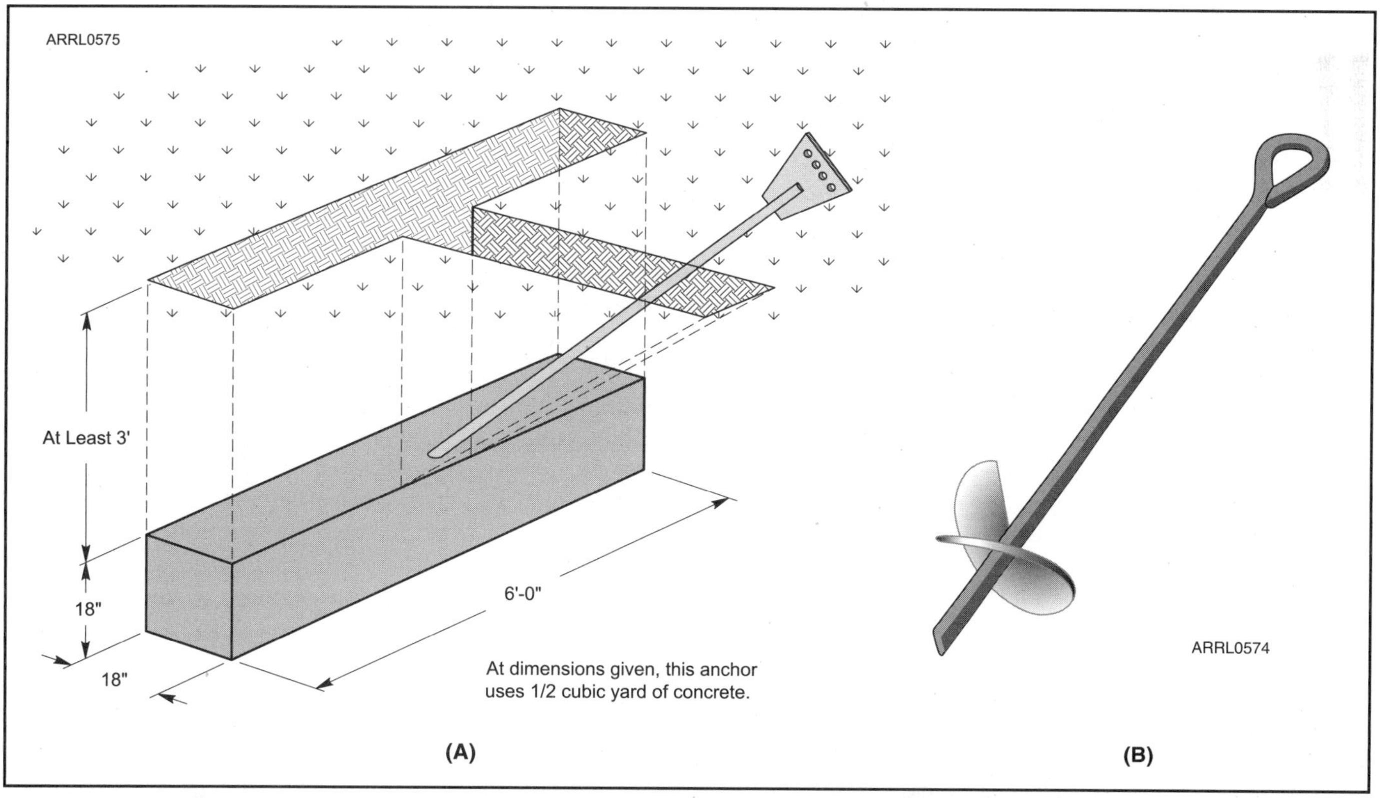

Figure 27.18 — At (A) a "dead man" anchor for guyed towers is basically a block of concrete buried in the ground. Block dimensions are specified by the manufacturer or calculated by an engineer and will vary depending on the height of the tower and wind and ice loading requirements. Various anchor rods are available, depending on the length and strength needed. The typical screw-in anchor is shown at B. Pullout strength is determined by soil characteristics, anchor rod length, and helical screw diameter.

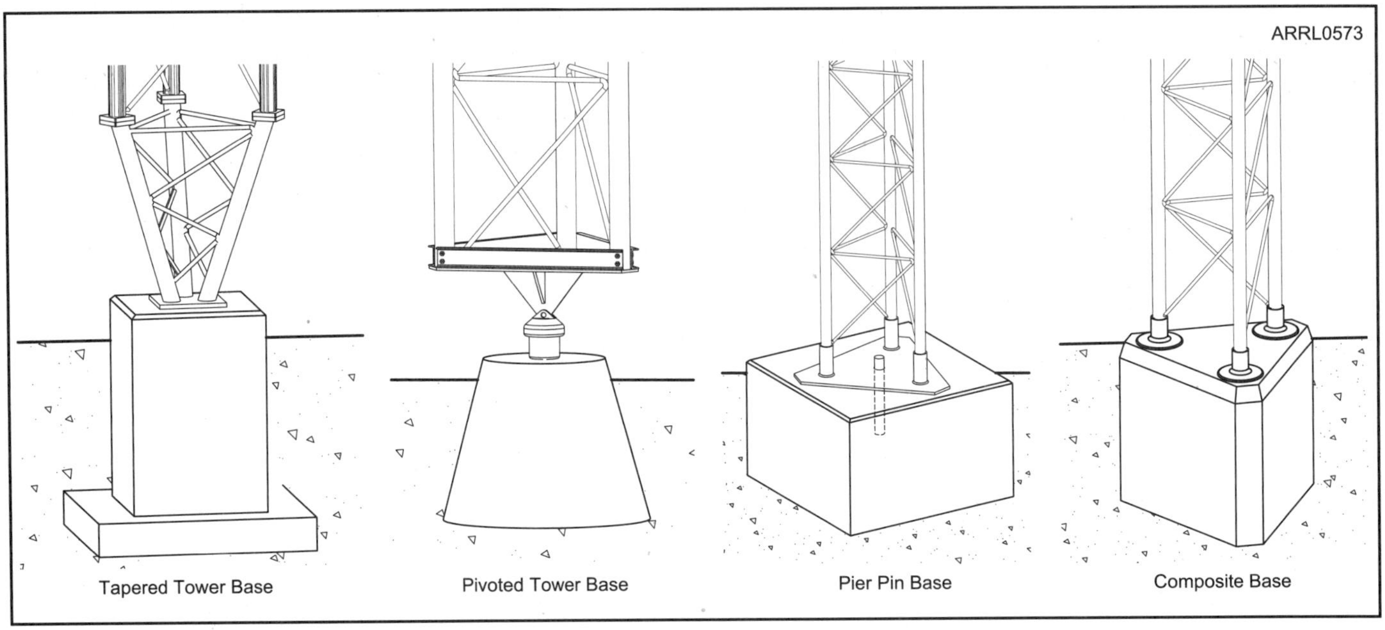

Figure 27.19 — Tower bases can take different forms. Most ham towers use the composite base (or a variation on that with a buried base section) or the pier pin base.

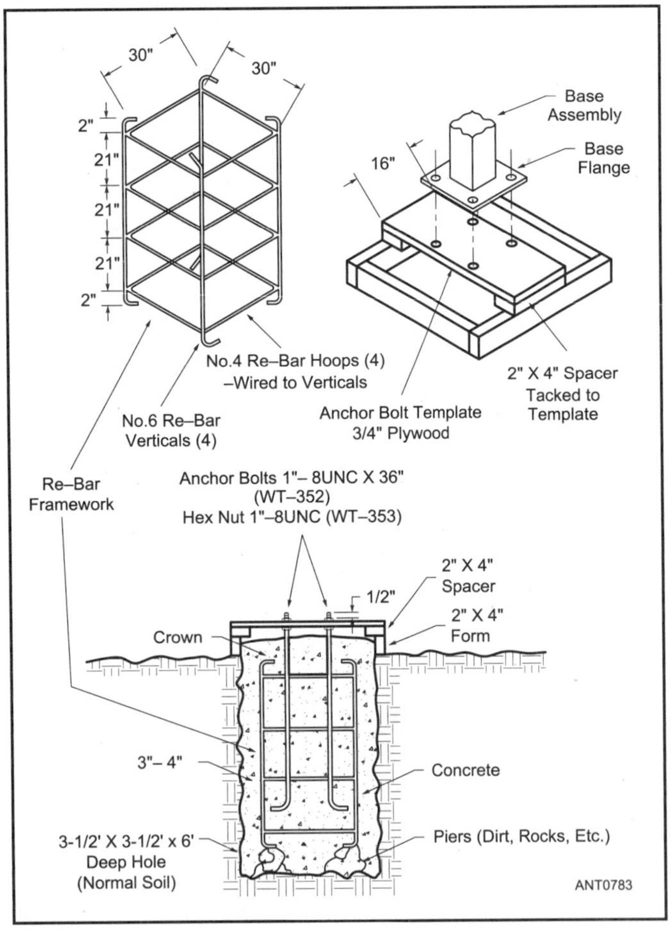

Figure 27.20 — Plans for installing concrete base for a 70-foot tubular crank-up tower. Although the instructions vary from tower to tower, this is representative of the type of concrete base specified by most manufacturers.

align the anchor bolts to properly mate with the tower itself. Once the concrete has cured, the tower base is installed on the anchor bolts and the base connection is adjusted to bring the tower into vertical alignment.

For a tower that bolts to a flat base plate mounted to the footing bolts (as shown in Figure 27.20), you can bolt the first tower section on the base plate to ensure that the base is level and properly aligned. Use temporary guys or wooden braces to hold things exactly vertical while the concrete cures. (The use of such temporary guys also works well when you place the first tower section in the base hole and plumb it vertically before pouring in the concrete.) Manufacturers can provide specific, detailed instructions for the proper mounting procedure. **Figure 27.21** shows a slightly different design for a tower base.

The one assumption so far is that *normal* soil is predominant in the area in which the tower is to be installed. Normal soil is a mixture of clay, loam, sand and small rocks. More conservative design parameters for the tower base should be adopted (usually, using more concrete) if the soil is sandy, swampy, or extremely rocky. If there are any doubts about the soil, the local agricultural extension office can usually provide specific technical information about the soil in a given area. When this information is in hand, contact the engineering department of the tower manufacturer or a civil engineer for specific recommendations to compensate for any special soil characteristics.

PIER-PIN BASES

An important phenomenon in a guyed tower is stretching of the guy cables. All guys stretch under load and when the wind blows the elongated guys allow the tower to lean over somewhat. If the tower base is buried in the concrete footing — as is commonly done in amateur installations — the bending stress at the tower base can become a significant factor. Towers that have been installed with tapered pier-pin bases

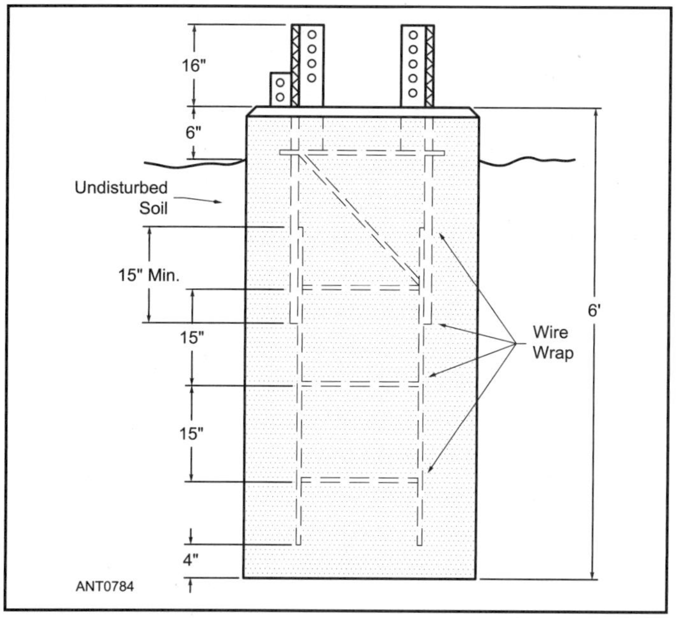

Figure 27.21 — Another example of a concrete base for a 70-foot lattice crank-up tower.

much more freely absorb tower leaning, and they are far less sensitive to guy-elongation problems.

The tapered pier-pin tower installation is not without some drawbacks. These installations often require torque-arm guy brackets or six-guy torque-arm assemblies to control tower rotation due to antenna torque. They also require temporary guys when they are being installed or removed to hold the base steady until the permanent guys are mounted or after the permanent guys have been removed. Some climbers also don't like the flexing when they start to climb these types of towers.

On the positive side, pier-pin base towers have all structural members above the concrete footing, eliminating concerns about hidden corrosion that can occur with buried towers. Most decisions regarding the type of base installation are made according to the preference of the tower builder/maintainer. While either type of base configuration can be successfully used, you would be wise to do the stress calculations (or have a professional engineer do them) to ensure safety, particularly when large antenna loads are contemplated and particularly if guys that can easily stretch are used, such as Phillystran guys.

27.3.5 Designing the Antenna Mast

The antenna mast is the pipe or tubing that extends from the top of the rotator through the top of the tower. Wind loading on the mast can be significant for large antenna systems or for antennas mounted well above the top of the tower. This requires careful selection of the mast material and is an important part of completing your tower system design. **Table 27.1** gives yield strengths for various mast materials. For all but the smallest systems, do not depend on unknown materials for this critical component!

There are two types of round material used for masts — pipe and structural tubing. Pipe is commonly water pipe or conduit and has extremely limited value. Pipe is designed to carry liquids and is not rated for bending strength. While pipe may have a *yield strength* of 30,000 psi (pounds-per-square-inch), that will only accommodate small loads and wind speeds. Another problem is that the OD (outside diameter) of pipe is 1.9 inches which is smaller than the 2.0 inch ham hardware standard. Conduit should not be used as antenna mast at all except for very small antennas.

Tubing on the other hand does come in 2.0 inch sizes and is rated for strength. There are many different materials and manufacturing processes for tubing that may be used for a mast. Yield strengths range from 25,000 psi to nearly 100,000 psi. Knowing the minimum yield strength of the material used for a mast is an important part of determining if it will be safe.

When evaluating a mast with multiple antennas attached to it, special care should be given to finding the worst-case condition (wind direction) for the system. What may appear to be the worst load case, by virtue of the combined flat projected antenna areas, may not always be the exposure that

Table 27.1
Yield Strengths of Mast Materials

Material Specification	*Yield Strength (lb/in.^{2})*
Drawn aluminum tube	
6063-T5	15,000
6063-T832	35,000
6061-T6	35,000
6063-T835	40,000
2024-T3	42,000
Aluminum pipe	
6063-T6	25,000
6061-T6	35,000
Extruded alum. tube	
7075-T6	70,000
Aluminum sheet and plate	
3003-H14	17,000
5052-H32	22,000
6061-T6	35,000
Structural steel	
A36	33,000
Carbon steel, cold drawn	
1016	50,000
1022	58,000
1027	70,000
1041	87,000
1144	90,000
Alloy steel	
2330 cold drawn	119,000
4130 cold worked	75,000
4340 1550 °F quench	162,000
1000 °F temper Stainless steel	
AISI 405 cold worked	70,000
AISI 440C heat-treated	275,000

(From Physical Design of Yagi Antennas by David B. Leeson, W6NL)

creates the largest mast bending moment. Masts with multiple stacked antennas should always be examined to find the exposure that produces the largest mast bending moment. The antenna flat projected areas at 0° and 90° azimuths are particularly useful for this evaluation.

A manual procedure for determining the mast bending stress is available in **Appendix B** of this chapter. There are also several online calculators such as at **dxengineering.com/mastloadestimator**. All you need to know is the wind load of the antenna and how far above the tower you plan on mounting it. If you have any doubts about the strength requirements for your antenna mast, consult a professional installer or engineer.

The often-asked question, "How much mast should be inside the tower?" is certainly important. A good rule-of-thumb is to have 1/3 of the total mast length inside the tower. When selecting the length of the mast, allow for four feet or more of mast extending above the top of the highest antenna on the tower. This extra mast can then be used as a gin pole/pulley attachment point for other antenna or tower work.

27.4 Tools and Equipment

27.4.1 The Tower Toolbox

Any job anywhere is easier and safer if you've got the right tools and tower work is no exception. If you are a weekend mechanic or handyman, you've probably already got most of what you need; all you need to do is add a few specialized items and you're good to go. If, on the other hand, all you have is a hammer, pair of pliers and a screwdriver, you'll need to make a trip or two to the tool store before you can really do anything. Once you have them, you'll be all set whenever any of your friends need help on their tower, too. Have the right tools and be prepared; you'll never go wrong.

Most amateur tower and antenna work can be done with a minimum of hand tools. Nut sizes of 7⁄16, 1⁄2 and 9⁄16 inch are the most common wrench sizes needed. **Table 27.2** lists the tools necessary for building and working on a typical ham tower. Your club may have a gin pole or guy wire tension gauge for members to borrow or you may be able to rent one.

You'll be working outside on towers so it's easy to lose tools when working aloft. Plan on having to locate a tool in grassy or brushy areas. Painting your tower tools with high-visibility paint makes finding them easy and identifies them as yours if they get mixed in with those of other workers.

After you finish a job outside, clean the tools and if there was any precipitation, wipe the tools down with a rag that has been sprayed with some WD-40. This will prevent water from sitting on the tool and starting a rust spot. Threads (adjustable wrenches) and ratchets should get a little extra attention.

When you do climb, take a roll of electrical tape and a knife with you every time. You never know when you'll notice something that needs taping to be secured or wrapped up. The knife (such as a multi-purpose tool or Swiss Army-style) can come in handy, as well.

27.4.2 Specialized Tower Tools

COME-ALONGS

A come-along or hand cable winch, is very useful for pulling tower sections together, tightening tramlines and tensioning guy wires. You'll probably find more uses for it. Cheap ones cost $15-20 and are fine for occasional use. The best ones for tower work have spring-loaded safety latches over the end of the hooks, and very rugged (not stamped) ratchets and pawls.

Table 27.2
Essential tools

1	set of combination wrenches: 7⁄16, 1⁄2 and 9⁄16 inch
1	set of sockets 3⁄8 inch drive
1 each	deep sockets: 7⁄16, 1⁄2, 9⁄16 inch
1 each	screwdrivers (blade and Phillips)
2	adjustable pliers
1	locking pliers (aka "Vise-grips")
1	diagonal cutter
1	razor blade utility knife
2	pulleys
1	drift pin or centering punch (for lining up tower sections)
1	hammer (attach some line for hanging on the tower)
3 each	adjustable wrenches — small, medium, and large
1	bubble level
6	carabiners
6	one-inch nylon webbing slings — 2 feet long
250 ft	rope (or more — this is enough for working on a 100 ft tower)
1	canvas bucket (for parts hauling and storage)
1	Loos PT-2 Tension Gauge
1 set	nutdrivers
1 (or more)	come-along or hand cable winch
1 (or more)	cable grips
1	circular saw with aggregate blade or hand grinder (for cutting metal, including guywires)
1	tag line (1⁄4 inch is fine — you chose the size and length)
1	cordless 1⁄2 inch drill, with assorted bits and socket driver, 18 V recommended
1	set drill bits including step-drill, e.g. Uni-Bit
1	antenna analyzer
1	gin pole
1	soldering gun and solder
1	lighter for sealing cut ropes

CABLE GRIPS

The cable grip in **Figure 27.22** complements the come-along to tighten guy wires. It is a spring-loaded device that slides up the guy wire but clamps down when you put tension on it. Klein is the primary supplier of cable grips and they come in lots of sizes and designs for use with various materials. For amateur use, the Klein 1613-40 is for 3/16 and 1/4 inch EHS guy material — used on the majority of amateur towers. If you have three come-alongs and cable grips you can put initial tension on a full set of three guy wires at the same time.

STEEL CUTTER

A portable (6.5-inch) circular saw with a steel-cutting aggregate blade will work to cut EHS guy wire, well, as will a 4-1/2 inch hand grinder with 1/8-inch steel-cutting blades. Always use safety goggles when cutting metal!

GIN POLE

The purpose of a gin pole (see **Figure 27.23**) is to provide a support point high above the top of the tower for lifting and positioning an object. This allows the necessary work to be done on the object without whoever is doing the work having to support its weight at the same time. The Rohn gin pole (Rohn Erection Fixture EF2545) is rated for sections of Rohn 25G and 45G and comes with clamps to secure it to a leg of the assembled sections. Towers made of angled legs require a special gin pole — contact the manufacturer.

Typical gin pole loads are tower sections (10 feet long) and masts (6 to 22 feet long). Pick up these loads just above their balance point so they will hang naturally in the correct upright position for installation. The Rohn gin pole is 12 feet long, just right for lifting a 10-foot tower section. For 20-foot masts, a 12-foot gin pole is marginal because there is barely 10 feet of working length available from the gin pole. A large mast will probably exceed the rating for the Rohn gin pole (rated for 70 pounds).

Large, heavy-duty masts require special handling; consult an experienced tower worker for instructions on installing large masts. One technique many tower pros use is to install the mast inside the tower at ground level, build the rest of the tower, and haul the mast up through the center of the tower. This greatly reduces risk and complexity by eliminating the need to maneuver the mast over the top of the tower while working high above the ground.

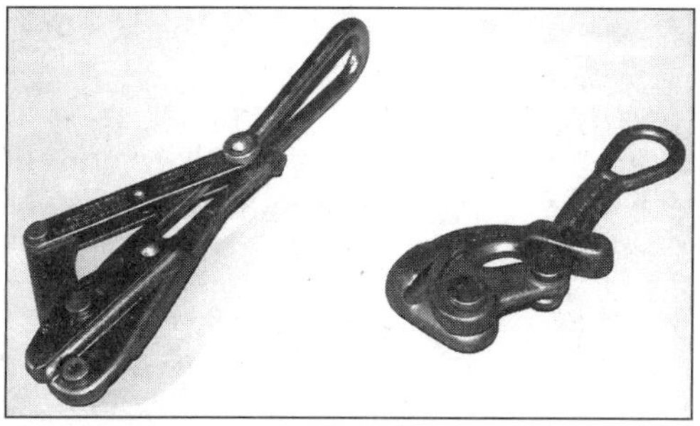

Figure 27.22 — Klein Chicago cable grip on the left and Klein Haven's grip on the right.

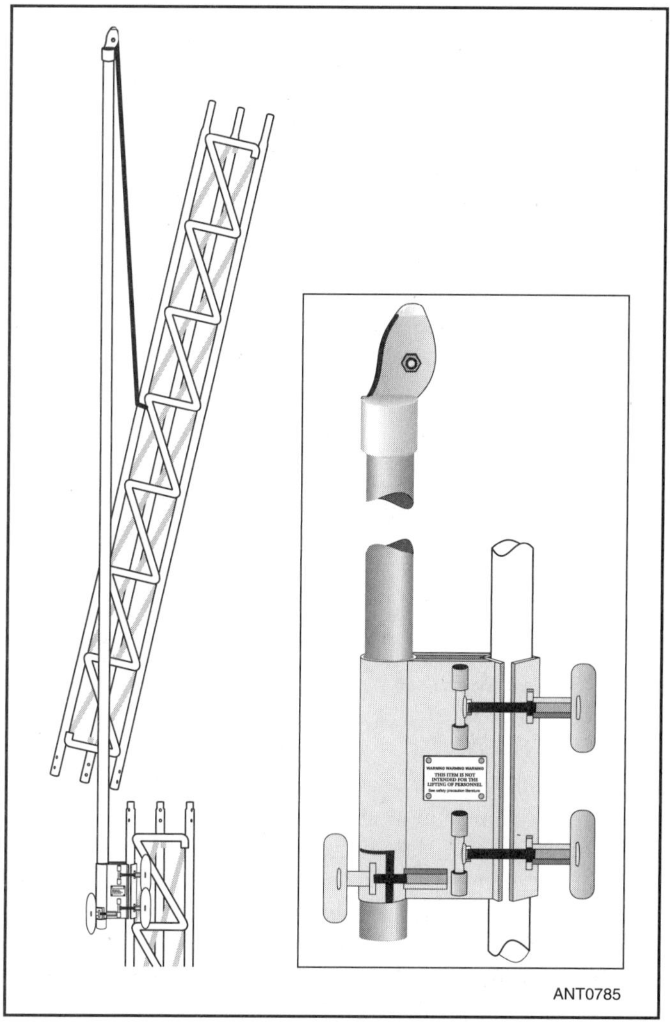

Figure 27.23 — Rohn "Erection Fixture" EF2545 also known commonly as a "gin pole."

CARABINERS

Carabiners are steel or aluminum snap-links with spring loaded gates as seen in **Figure 27.24A** and **27.24B**; they are invaluable for dozens of tower work tasks. A carabiner at the end of your haul rope can be attached to virtually anything that needs to be raised or lowered. A carabiner can be a third hand on the tower; you can clip a carabiner to almost anything with a rung or diagonal brace. You can instantly hang a pulley from a tower rung. Lightweight, they can be clipped on to your climbing harness for easy access. Experienced tower workers may carry twelve to fifteen carabiners on typical jobs. They typically cost $6 to $10 and will last for years with little or no maintenance. If the gate no longer opens and closes smoothly, the carabiner should be discarded.

A word of caution: mountain climbing carabiners are considered to be for private use and not OSHA-approved. Current ratings for mountain climbing carabiners are typically in the 6 to 10 kN (1350-2250 pounds of force) range with the gate open and 18-25 kN (4050-5625 pounds) with the gate closed. A typical rating for an OSHA-approved commercial carabiner

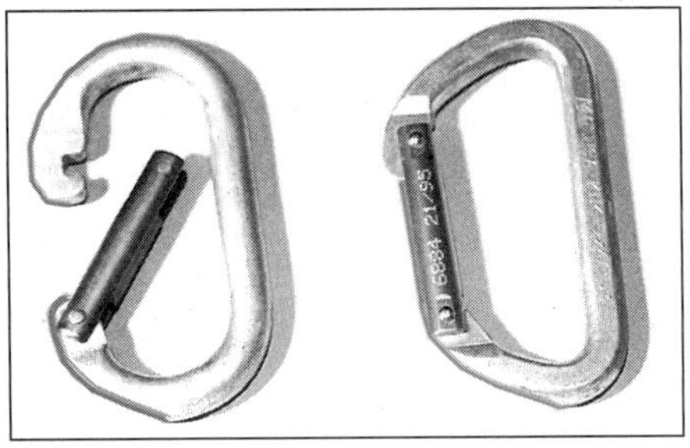

(A)

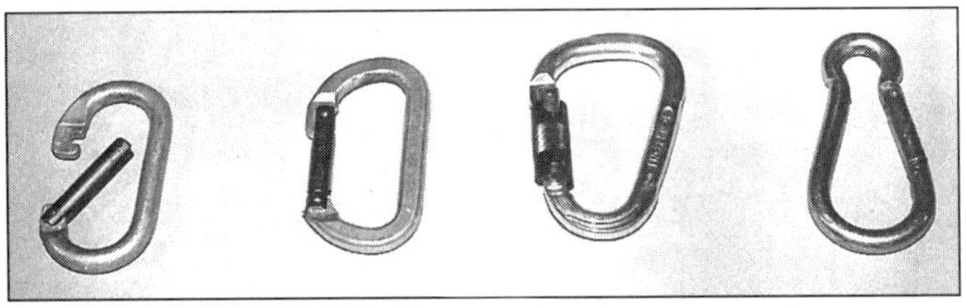

(B)

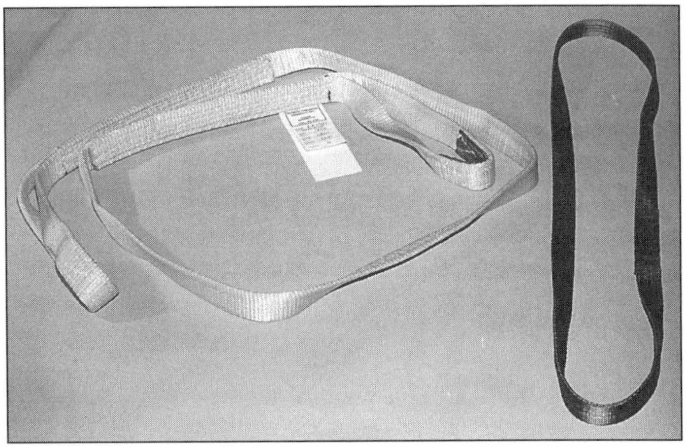

(C)

Figure 27.24 — (A) Oval mountain climbing type carabiners are ideal for tower workloads and attachments. The gates are spring loaded — the open gate is shown for illustration. (B) An open aluminum oval carabiner; a closed oval carabiner; an aluminum locking carabiner; a steel snap link. (C) A heavy duty nylon sling of the left for big jobs and a lighter-duty loop sling on the right for everything else. Make sure the carabiners and slings are rated for lifting heavy loads.

— called a *safety hook* — is 40kN (9000 pounds). If you don't feel that mountain-climbing carabiners are adequately rated, safety hooks are available from safety equipment vendors.

Larger carabiners are available with locking gates; these will give you an added degree of safety, particularly if you are using them for your own protection or if you just want to be doubly safe. They're only a couple of dollars more than the standard, non-locking types.

Big carabiners are used for rescue work and other applications where a wider gate opening is needed. These are sometimes called *gorilla hooks* or *rebar hooks* and are used for larger tower rungs (Rohn BX, for example) and larger loads. OSHA-compliant devices are offered by safety equipment vendors.

USING CARABINERS

Here are some common ways that carabiners are used on tower projects:

1) Attach a sling to a guy anchor rod as an attachment point for the come-along when pulling guy wires.

2) Clip a carabiner onto a rung at the bottom of the tower then attach the haul rope snatch block pulley to it. This will change the direction of the haul rope from vertical to horizontal, making it much easier to pull. It also allows the ground crew workers to watch the load as it goes up or down the tower (without having to strain to look upward) and it removes them from the fall danger zone at the bottom of the tower.

3) Dedicate a sling and carabiner to the gin pole for easy lifting as the tower is assembled.

4) Put a loop through a frequently used tool, then clip it to your belt with a carabiner.

5) Always have a carabiner clipped into the bowline at the end of your haul rope and tag line for quick load attachment.

6) Clip a carabiner into the U-bolt on your rotator to haul it up.

SLINGS

A loop sling is made from one-inch nylon tubular webbing as seen in Figure 27.24C. Mountain-climbing slings are a continuous loop of webbing. A configuration with a sewn loop at each end is also useful. Slings can be wrapped around large or irregularly shaped objects and attached to a rope or tower member with a carabiner. Slings have around the same breaking strength as carabiners (approximately 4000 pounds, or 18.1 kN force) and are very handy for amateur applications and loads. Wrapping one around a tower rung or leg provides a convenient place to hang tools, parts, or a pulley. Like carabiners, slings are not OSHA-approved but they're used for mountain climbing protection. OSHA-approved slings are available from a safety equipment vendor.

LIFTING LOADS WITH SLINGS

Slings are typically used in one of three rigging configurations shown in **Figure 27.25**:

1) *Straight pull* — A simple direct vertical attachment such as for a tower section. Run the sling around a tower member and clip both ends in a carabiner for lifting.

2) *Choker* — Wrap the sling around the load one or more times, insuring that you pull the loop through itself on each wrap, cinch it tight, clip it into a carabiner and pull it up. The more tension you put on the sling, the tighter it gets. Chokers are the best way to lift a mast.

A choker will work in many other cases where you have an irregular load to haul, not just masts. Using a sling as a choker reduces the lifting capacity of the sling by as much as 30% though. A U-clamp can be installed above the sling when

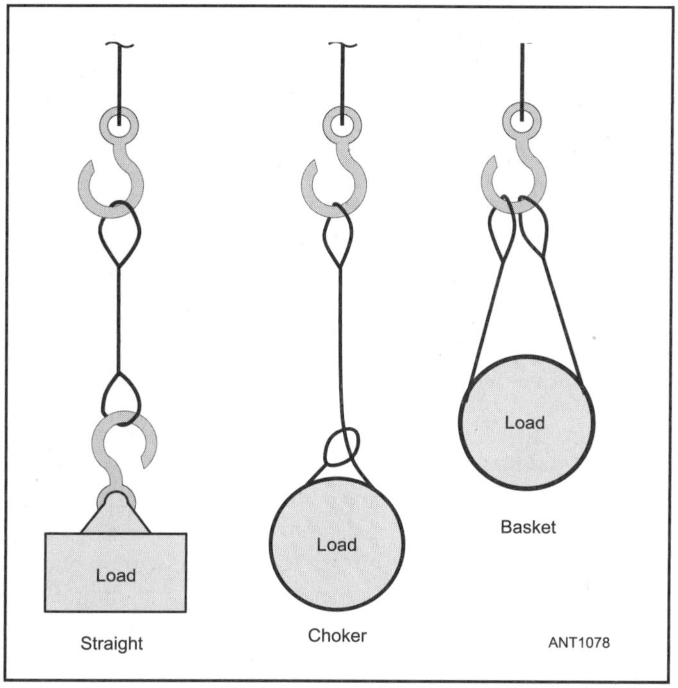

Figure 27.25 — Three basic lifting hitches used with slings and ropes.

raising a mast for redundant protection if desired.

3) *Basket* — Basket hitches distribute a load equally between the two legs of a sling. The greater the angle between the two legs, the smaller the capacity of the sling.

27.4.3 Using a Gin Pole

This section was condensed from the ARRL book *Simple and Fun Antennas for Hams*. We're going to assume in the following discussion that you are using a Rohn EF2545 to install sections of Rohn 45G, which weigh about 70 pounds.

The main working part of the gin pole is the pulley mounted at the top of the 12-foot long heavy-wall aluminum tubing. This pulley has a haul rope going down to the ground crew through the center of the aluminum tube.

An adjustable, sliding clamp toward the bottom of the aluminum tubing is clamped to the tower using a swinging L-bracket-type clamp with two clamping bolts. These have T-bar handles that can be tightened by hand. In fact, this gin pole can be moved and deployed without any tools. The clamp is positioned just below the top braces of the tower section onto which the next tower section is to be installed. Once clamped to the top of the tower, you would loosen the T-bar handle that tightens the clamp against the sliding aluminum tube and slide the tubing up to its maximum extent.

In practice, the following steps are taken as each 10-foot section of tower is installed, one-by-one. We're assuming here that the gin pole starts out on the ground, with at least one person harnessed safely at the top of the tower. We're also assuming that the haul rope has been threaded through the aluminum tube and the top pulley, with a carabiner to prevent it from falling back down the tube.

Here's a rope tip — If the wind is blowing it may be difficult to lower the end of the rope to the ground. Attach a weight to the end of the rope; a wrench works well. If added weight isn't enough, use a carabiner to clip the free end of the haul rope around the other side of the rope. The carabiner will guide the free end back down the haul rope without blowing around.

1) The clamp holding the aluminum tubing is loosened so that the pulley on the tube can be lowered to where it is just above the bottom clamp. Then the T-bar handle for the tube clamp is tightened.

2) The climber lowers a work rope for the ground crew to tie to the gin-pole pull rope. (This work rope has been looped through a temporary pulley clipped to the top of the tower. It is also used to pull up tools and other materials.) The ground crew then pulls the gin pole up to the climber, using the work rope. Once the gin-pole head reaches the top of the tower, the climber clamps the gin pole clamp securely to the top of the tower. The tag line is then removed from the gin pole.

3) The T-bar handle for the tube clamp is loosened, and the aluminum tube is extended to its maximum height, as shown in Figure 27.23. Make sure the free end of the haul rope cannot slip through the top pulley, or else you'll have to lower the gin pole and go through this step again. In other words, the climber keeps the "business end" of the gin pole rope while raising the pole.

4) The free end of the haul rope is then dropped to the ground, often using a weight to keep the rope from waving about. (See the rope tip above.)

5) The ground crew then attaches the free end of the rope *above* the balance point of the tower section. For Rohn 25G or 45G there are eight horizontal cross braces per section. The crew should attach the rope to the fifth horizontal brace from the bottom. Please remember that the tower section should hang with its bottom down so that it is properly oriented when it reaches the top of the tower.

6) Once the bottom of the tower section has been lifted to just above the top of the legs of the bottom tower section, the tower crew can guide the section down onto the top of the three legs, while calling out to the ground crew instructions about *slowly lowering* the new section down onto the legs. See **Figure 27.26**, which illustrates guiding the new section of tower onto the previous section's legs.

7) Once the new tower section has been guided down onto the male ends, the pinning bolts are inserted and tightened with nuts. Note that Rohn uses two different sized bolts on 25G and 45G sections, with the larger diameter bolt on the bottom.

8) Finally, reposition the gin pole for the next section of tower. The T-bar at the clamp is loosened, the tube is dropped down to the level of the clamp, and the climber walks the gin pole up to the top of the section just installed and clamps it there, ready to pull up the next tower section.

Figure 27.26 — Tower worker lowering a new section onto the top of the assembled stack of sections. The gin pole attached to the left leg is bearing the weight as the tower worker gives verbal instructions to the ground crew pulling on the haul rope. (Mike Hammer, N2VR, photo)

27.4.4 Ropes and Rope Care

If you are going to do tower and antenna work, you'll be using ropes. The most common uses are for haul rope, tag lines or work rope and temporary guys. A *halyard* is a rope used for hoisting. A tip from experienced tower climbers is to use the largest rope that is reasonable for the job. That is the easiest to grip and will be the kindest to your hands.

MANILA

Manila is the best known natural fiber rope. Manila must be handled and stored with care as any dampness will cause it to rot and damage its effectiveness and safety.

POLYPROPYLENE

Polypropylene makes lightweight, strong ropes that float on water, are rot-proof and are unaffected by water, oil, gasoline and most chemicals. Polypropylene rope is relatively stiff and doesn't take a knot well.

NYLON

Nylon is the strongest fiber rope commercially available. Due to its elasticity, nylon ropes can absorb sudden shock loads that would break ropes of other fibers. Nylon is particularly recommended for antennas using trees as supports. A disadvantage of new nylon rope is that it stretches by a significant percentage.

Nylon has very good resistance to abrasion and will last four to five times longer than natural fiber ropes. Nylon ropes are rot-proof and are not damaged by oils, gasoline, grease, marine growth or most chemicals.

DACRON

Dacron rope comes in three sizes (3/32, 3/16 and 5/16 inch) and is UV resistant. This is an excellent candidate for any rope used permanently outside such as for wire antenna halyards.

ROPE LAY

All rope is twisted or laid; and nearly all laid rope is *three-strand* construction, typically what you'll find at your local hardware store. Another type of rope is known as *braid-on-braid*, or *kernmantle*. This rope has a laid core covered with a braided jacket to produce a strong, easy-handling rope. In most instances, braid-on-braid rope is stronger than twisted rope of the same material and diameter. It is available in various synthetic fibers. Marine supply stores and mountain climbing stores carry a large variety of braid-on-braid types as well as a variety of types and sizes.

WHICH ROPE TO USE

The best rope for holding up wire antennas with spans up to 150 or 200 feet is ¼-inch nylon rope. Nylon is somewhat more expensive than ordinary rope of the same size, but it weathers much better. UV-resistant Dacron rope is also popular. After an installation with any new rope, it will be necessary to repeatedly take up the slack created by stretching. This process will continue over a period of several weeks, at which time most of the stretching will have taken place. Even a year after installation, however, some slack may still arise from stretching.

For ropes to be used on tower work, first decide which size will suit your needs based on working load. Most amateur loads are less than 100 pounds and very rarely do they exceed 250 pounds. A haul rope having a working load between 100 and 250 pounds will handle just about anything. **Table 27.3** summarizes the sizes and working loads for different types of rope.

Second, choose the type and material of your rope. Polypropylene rope is stiffer and more difficult to knot than nylon. Nylon and braid-on-braid ropes are softer and will take a knot very easily. The softer ropes also coil more easily and are more resistant to kinking.

Finally, choose the length that will be most useful for you. If you double the height of your tower and add 25%, you'll have plenty. A 100-foot tower requires $(100 \times 2) + (100 \times 2 \times 0.25) = 200 + 50 = 250$ feet.

Price varies from less than $20 for 600 feet of ¼ inch polypropylene rope to more than $100 for 165 feet of high

Table 27.3
Rope Sizes and Safe Working Load Ratings in Pounds

3-strand twisted line

Dia. (in.)	Manila	Nylon	Dacron	Polypropylene
1/4	120	180	180	210
3/8	215	405	405	455
1/2	420	700	700	710
5/8	700	1140	1100	1050

Double braided line

Dia. (in.)	Nylon	Dacron
1/4	420	350
3/8	960	750
1/2	1630	1400
5/8	2800	2400

quality kernmantle climbing rope. Have a variety of ropes on-hand for different uses. All are doubled-braided construction. Lengths vary from 100 feet to 600 feet. Each rope is carried and stored in simple plastic tubs, which are labeled appropriately, including dates of purchase. If you simply feed the rope loosely into its container, it will pull back out without kinking or knotting.

Make certain that the rope ends will not unravel. Most supply stores will cut the length with a hot knife; that will do the best job of sealing the ends. You can do it at home by simply melting the ends with a lighter. An alternative is to tightly wrap a few layers of electrical tape or heat-shrink tubing around the ends. Be sure to tape the ends of all your ropes to protect them.

ROPE CARE

Inspect your rope periodically and replace it if there is any visible serious abrasion or damage. Many climbers store their rope loosely in tubs or baskets and do not coil it, which introduces twists. Others prefer spools and the portable garden hose spools that wind up rope will keep it straight and tangle-free, as well.

Here are some additional tips for using ropes:

1) Be certain your rope size is adequate for the job; don't use a rope that is too small.

2) Dry your rope before storing it. Natural fiber (Manila) ropes will mildew and rot if stored wet. You can put nylon ropes in the clothes dryer on low heat if they are really soaked.

3) Don't store ropes in direct sunlight; UV deterioration will significantly weaken them.

4) Cut out and discard any badly worn or abraded portions of a rope; better to have two shorter ropes you can trust than one long one that is suspect.

5) Keep your rope clean. Don't drag it through the mud, or over a rough or gritty surface. Try not to even step on your ropes, especially on wet or muddy soils.

6) Watch for kinks; they can cause permanent damage and weakening.

7) Protect ropes from all chemicals such as acids, oils, gasoline, paints, solvents, etc.

8) Avoid sudden strains; shock loading or jerking may cause failure.

9) Avoid overloading. A safe working load for a rope is 10-20% of its breaking strength.

10) Avoid abrasion. If the rope must run over a tower leg or any surface with a sharp edge, protect it with a layer or two of canvas or other such material.

11) Avoid bending a rope around corners or at sharp angles.

27.4.5 Knots

You can do about 98% of your tower and antenna work with only three knots — and you already know one of them. Remember that any knot will decrease the breaking strength of the rope — usually 40% or more. Choose and use the correct rope and knots for the job, and you should have no problems. Knots not listed here, and additional knot-tying know-how can be found online at Animated Knots (**animatedknots.com**) and Real Knots (**realknots.com/knots**). **Figure 27.27** shows several common knots.

OVERHAND KNOT

Start with an overhand loop, then passing the end under and up through the loop and then tightening. To form an overhand loop in the middle of a rope, double the rope for about two feet and tie an overhand knot with the doubled rope.

BOWLINE

The bowline forms a loop that will not slip, or jam yet unties easily. It is used for hoisting, joining two ropes, and fastening a rope to a ring or carabiner. To tie it, form a small loop in the rope. Run the end up through the loop, behind the standing part, then back down through the loop. Pull tight. Practice this one until you can make it almost automatically.

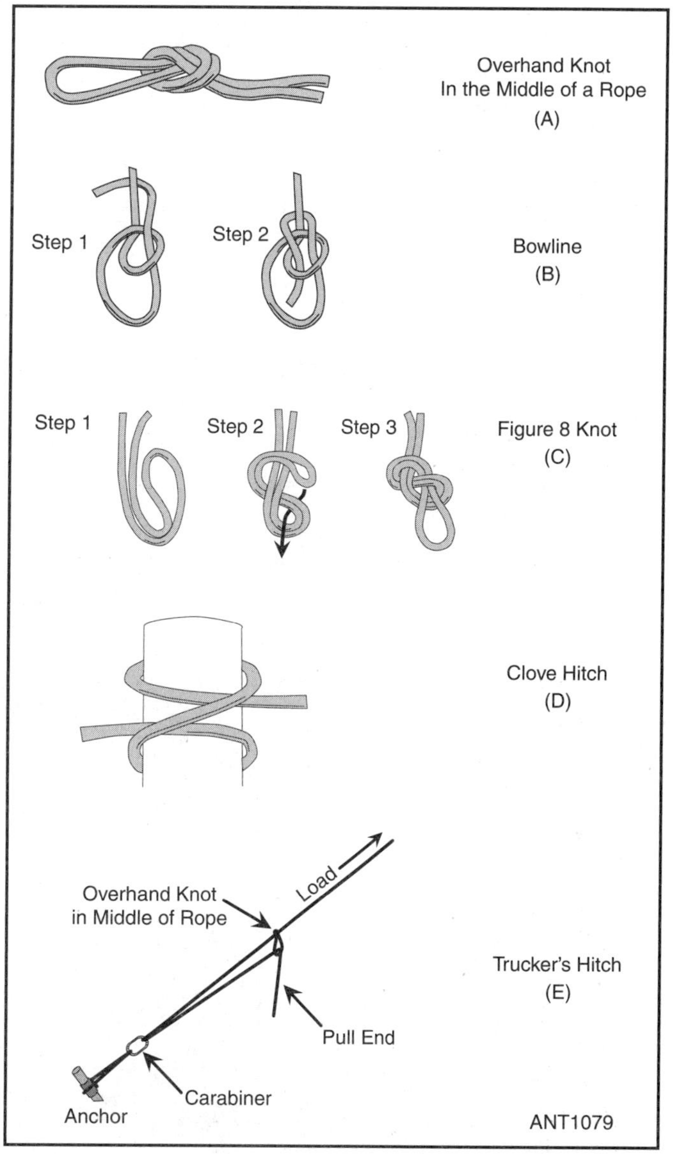

Figure 27.27 — Common knots used in tower and antenna work.

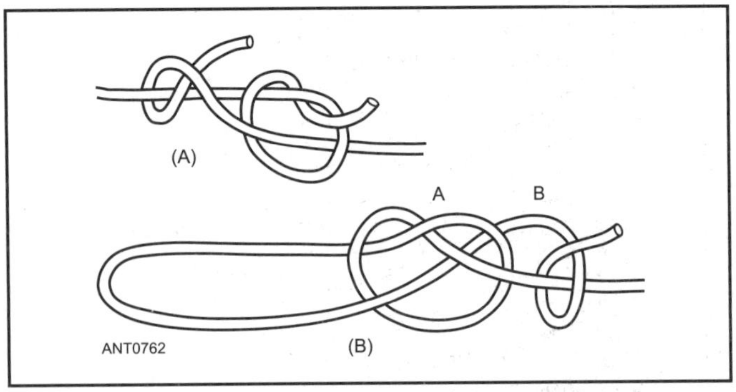

Figure 27.28 — This is one type of knot that will hold with slick types of line. Avoid these types of lines for lifting or safety uses. Shown at A, the knot for splicing two ends. B shows the use of a similar knot in forming a loop, as might be needed for attaching an insulator to a halyard. Knot A is first formed loosely 10 or 12 in. from the end of the rope; then the end is passed through the eye of the insulator and knot A. Knot B is then formed and both knots pulled tight. (Courtesy Richard Carruthers, K7HDB)

FIGURE EIGHT

Simpler than the bowline, a figure-eight knot may be used in most situations in place of a bowline. It is tied like a doubled overhand except that the rope is twisted an additional half-turn before the knot is pulled through the loop. It is one of the few knots that can be easily untied after holding a severe impact load, such as a falling tower section. Its only disadvantage for tower work is that it is a physically larger knot, and it takes a bit more rope than a bowline.

CLOVE HITCH

The clove hitch can be invaluable when you're working with round objects, and it can be put on or around almost any object very quickly.

TRUCKERS' HITCH

The trucker's hitch allows you to tighten the rope as much as you can without a come-along. Tie an overhand loop (see above) toward the load end, run the end of the rope through a carabiner or shackle at a convenient anchor point, pass the end through the loop and then pull to tighten the rope. This technique gives twice the mechanical advantage of pulling on the single rope.

PLASTIC LINE

For types of plastic line that are too slick to hold common knots well, **Figure 27.28** shows a more suitable knot. Needless to say, these lines should probably not be used for lifting loads or holding climbers.

27.4.6 Pulleys

Pulleys are used constantly in tower and antenna projects. One should always be placed at the top of the tower for a haul rope to bring up materials. Steel pulleys costing $25-35 are found in many hardware stores or rigging shops but are heavy. Lightweight nylon pulleys are used by utility company line crews for tower work. Lightweight (aluminum) pulleys designed for "rescue" work are also available. These can be placed on the line at any point and have load ratings comparable to carabiners as well. Wood-sheathed pulleys used in "block and tackle" devices and for sail hoisting should work well for very heavy loads.

Two important things to consider when shopping for pulleys are sheave size and sheave clearance. A sheave is the pulley wheel with a groove in it. A two-inch diameter sheave is the minimum size to use, and larger sizes are better. Use a jam-proof pulley with minimal clearance between the sheave and the pulley body. If there is any way for your haul rope or cable to jump the pulley and get jammed, it almost certainly will.

A *snatch-block* is a pulley with a body that opens up so that it can be placed directly anywhere on a rope without needing one end of the rope to be free. This is useful when the rope is under tension (see **Figure 27.29**).

For supporting wire antennas, avoid small, galvanized pulleys designed for awnings and clothesline pulleys. Use heavier and stronger pulleys intended for outdoor and marine installations with good-quality bearings.

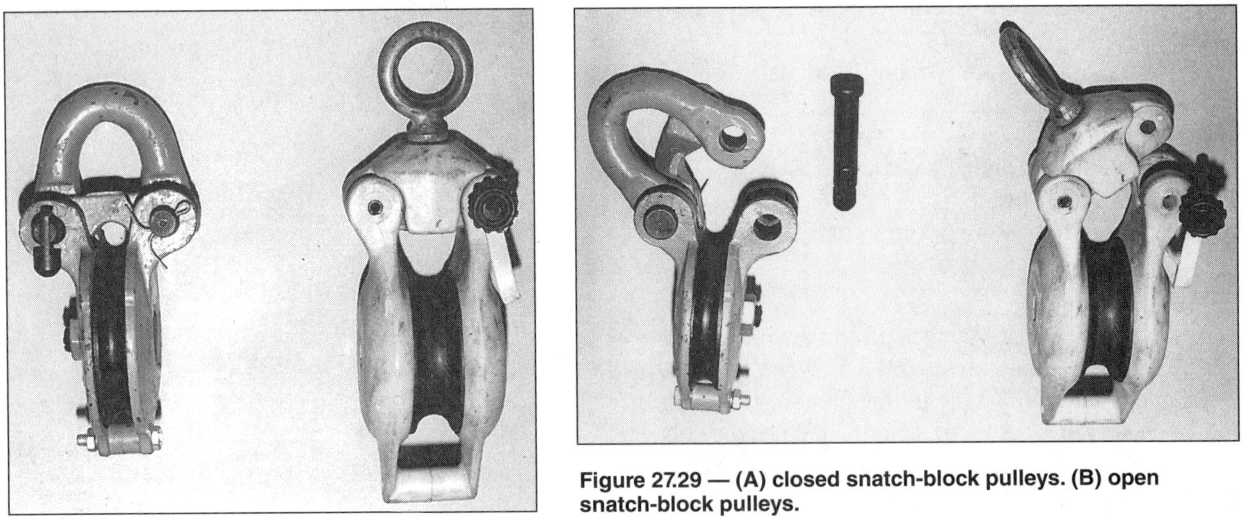

Figure 27.29 — (A) closed snatch-block pulleys. (B) open snatch-block pulleys.

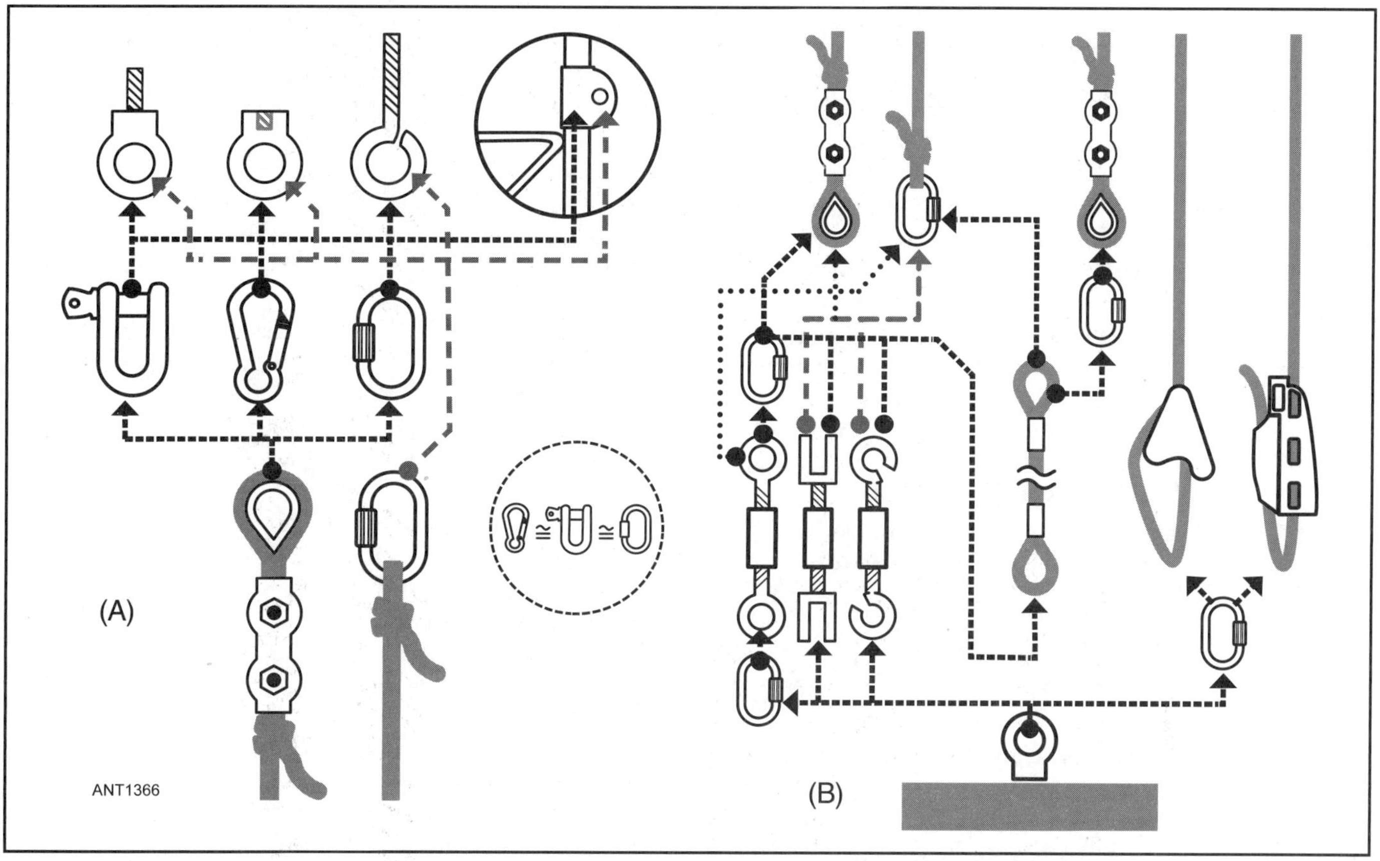

Figure 27.30 — Rope hardware for various tasks. This figure was provided courtesy of Mastrant (mastrant.com).

An important consideration for pulleys to be constantly exposed to the weather is corrosion resistance. Use a good-quality pulley made entirely of alloys and materials that do not corrode readily. Galvanized pulleys will quickly rust. Marine pulleys have good weather-resisting qualities since they are usually made of bronze, but they are comparatively expensive, and the smaller pulleys are not designed to carry heavy loads.

27.4.7 Rope Hardware

The drawing in **Figure 27.30**, from rope manufacturer Mastrant (**mastrant.com**), shows the various hardware elements that are used with rope. Figure 27.30A shows the rope hardware used for terminating and coupling. An attachment point on a tower is shown as an example. Figure 27.30B shows the hardware for attaching rope to tensioning devices like turnbuckles and adjustable cam-locking clips.

27.5 Tower Construction

Now that you have done all the planning and purchased the materials, it's time to start "growing" your tower. We'll start at the bottom!

27.5.1 The Prime Directive

When it comes to tower construction, you are strongly advised to always observe the *Prime Directive* (originally from K7LXC); that is, to "DO what the manufacturer says." Similarly, "DON'T do what the manufacturer doesn't say to do." Follow the specifications for materials, concrete and wind load and you'll minimize the chances for failure, small or large. Follow the directions for assembling equipment and using tools and supplies. Professional engineers have designed every aspect of these systems for safe, long-term and reliable use and it's in your best interests to follow their specifications and directions. Pretty straightforward and simple to follow advice. There are many situations where nothing matches or is covered by the directions. In cases like these, either carefully devise a plan or seek professional advice.

27.5.2 Base Excavation and Rebar

To avoid underground utility lines, please don't dig without calling one of the utility locator services. There are several websites such as **call811.com** that can help, or you can call your local utility for assistance. Avoid expensive and embarrassing surprises. It may even be illegal in your area to begin digging without determining the location of buried utilities!

Hand-digging the hole for a large self-supporting tower base entails a lot of work! Excavating the necessary hole can be done quickly and effectively by a professional contractor. You can also rent excavating equipment and do the job yourself. No matter how you dig the hole, extreme caution should be used when someone is in the hole due to the risk of wall collapse. Many building regulations make it illegal to be in a hole or trench more than 4 feet deep without shoring up the sides of the hole. If you're doing the work yourself, never work alone in a hole that is deeper than your waist.

BUILDING A REBAR CAGE

Once the hole is dug, you'll be installing the reinforcing bar, or rebar. The tower manufacturer will provide a recommended design for the rebar "cage" in the concrete base. **Figure 27.31** shows a typical completed cage.

Rebar is sized in eighths of an inch. For example, #4 rebar is 4⁄8 of an inch, or ½ inch, and #6 rebar is 6⁄8, or ¾ inch. Rebar vendors will cut and bend the rebar to your order which is a lot easier than buying long lengths of it at your local hardware store and trying to cut it yourself.

You can either build the rebar cage on the ground or in the hole. You'll need a backhoe or other piece of equipment to lift the completed cage up and lower it into the hole. Building the cage in the hole is harder since room to work is really restricted. Remember to shore up the hole and don't work by yourself.

To tie the rebar together to form the cage use bailing/tie wire at each joint. Take about 2 feet of bailing wire and bend it in half. Wrap the tie wire through one of the Xs of the joint twice. Next wrap it twice through the other axis of the joint, bring the ends together and wrap them together several times. Use a large pair of pliers, or a wire tie tool, to twist it until snug. To stiffen the cage, add an X cross brace, using two pieces of rebar across each face.

Figure 27.31 — The rebar cage for KX8D's tower base. (Duane Durflinger, KX8D, photo)

Guy anchors are easy to deal with since they're smaller, take less concrete and you don't have to move as much soil. The easiest way to locate the anchors is to temporarily put up a tower section at the desired location and then sight through each face across the opposite leg — that'll give you the angle. Then run your measuring tape out the appropriate distance to the anchor location. A more accurate way of measuring is to use a transit, which will ensure each guy anchor is spaced exactly 120 degrees apart. Suitable transits can be rented quite reasonably.

Once the rebar cage has been placed in the hole, a wooden form surrounding the top of the base hole provides for a neater appearance and raises the top of the base above ground a few inches. This allows water to run off the base and not pool around the legs or bolts of the tower.

INSTALLING THE BASE SECTION

If you're installing a guyed tower such as Rohn 25G or 45G with tubular legs, be sure to put 4 inches or so of crushed stone at the bottom for drainage and set the legs of the base in the gravel. Water will condense in the legs, and if there's no place for the water to drain out, it will build up and split the leg when it freezes.

Place the base section, if used, in the hole without touching the rebar cage and use wooden braces to hold it precisely vertical. Alternately, you can join one of the tower sections to the base section and use temporary guys to hold it up. For bigger tower bases, it's sometimes convenient to attach the leg(s) to the rebar cage with tie wire. A properly constructed rebar cage will be strong enough to support it and you can stand on it if needed.

If anchor bolts are being used, a piece of plywood with the proper hole pattern can be used to hold the bolts in position while the concrete is being poured.

27.5.3 Concrete for Bases

The tower manufacturer will specify the type of concrete required for the base and your building permit may also impose some requirements. The strength specification is generally 2500 to 4000 PSI for tower bases and a slump (a measure of the concrete's workability) of 4. Consult an engineer if you are unfamiliar with ordering or working with concrete. The Wikipedia entry for concrete (**en.wikipedia.org/wiki/Concrete**) provides a great deal of good information.

You can mix the concrete yourself by using bags of premixed concrete and a powered mixer. It takes about 45 80-pound bags of concrete mix to make one cubic yard of concrete so for large bases ordering ready-mix concrete is more practical. The delivery truck will need to be relatively close to the hole (within 10 to 15 feet) to be able to position the delivery chute properly. If the truck cannot get close enough to the hole, you'll have to move the concrete yourself.

To avoid moving tons of concrete (a yard of concrete weighs about 4000 pounds!) in long runs to the hole with a wheelbarrow, use a concrete line pumper — a truck-mounted pump that uses 3-4-inch hoses laid on the ground for concrete distribution. They're not that expensive and can pump up to 400 feet. There are big hydraulic boom pumpers that can work over obstacles such as buildings and fences, and much greater distances, but they are more expensive to hire. In either case, using professional equipment makes the job of moving tons of concrete much easier.

Concrete takes a long time to cure to its rated strength — at least three weeks until it reaches 90% of its rated strength. The concrete supplier can give you complete instructions on how long to wait and whether the concrete needs to be kept damp during the curing period or any other special treatment. It's hard to sit and wait a month before beginning work on the tower but be safe and don't put any load on the base until it is ready to support it. Your building permit may require inspection of the base before tower work can begin.

27.5.4 Working with Guy Wires

Guy wires are the heart of a reliable guyed tower system. Almost any tall amateur tower is going to be guyed. Rohn 25G, 45G and 55G are the most common towers used by amateurs and they all need to be guyed. Before you begin building the tower, familiarize yourself with guy wires and the associated equipment, hardware and techniques. Practice until you are confident of being able to handle guy wires correctly.

GUY WIRE GRADES

Steel guy wire comes in several different grades. Rohn specifications call for EHS (Extra High Strength) cable exclusively. As you can see from **Table 27.4**, this is the strongest steel cable available.

GUY WIRE TERMINATIONS

The three most common methods of terminating guy wires are to use Crosby cable clamps, swaged or crimped pressed fittings, or "preforms" (preformed guy grips). With the advent of the preformed guy grips, Crosby clamp and swage fitting use has declined dramatically.

THIMBLES

At an attachment point, guy wire is put through a thimble or insulator and doubled back for clamping (this is called a

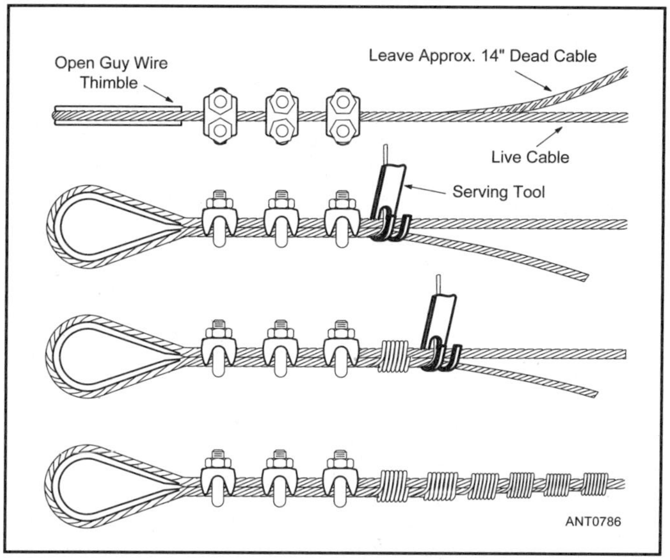

Figure 27.32 — Traditional method for securing the end of a guy wire. This technique is becoming increasingly uncommon as Preforms replace cable clamps.

turnback) as shown in **Figure 27.32**. A thimble is used to prevent the wire from breaking because of a sharp bend at the point of intersection. Conventional wisdom strongly recommends the use of thimbles that are at least two wire sizes larger than the cable to provide a more gradual wire bend radius.

Don't confuse rope thimbles with 'heavy duty' thimbles for use with guy wire. Heavy duty thimbles are thick and virtually impossible to bend. They are also usually galvanized (and should be for use in tower building.) Rope thimbles are very thin steel, often stainless, and easy to open and bend to force over a round member like a turnbuckle eye. The much thicker heavy duty thimble is not easy to bend or deform by hand.

Wrapping the wire around the thimble results in two parallel guy wires. The wire that bears the tension of the guy wire forces is called the "live" end and the short piece that is turned back is called the "dead" end. It's "dead" because it is not load-bearing.

CABLE CLAMPS

The least expensive and most common cable fittings are Crosby clamps consisting of two parts; the U-bolt and the saddle. Always use three clamps per turnback and make certain that the saddle is on the live (load-bearing) side of the guy wire. The saddle portion provides the majority of the holding capacity of the clamp and goes on the "live" side of the cable. To remember the correct method, use the saying, "U goes on the dead end" or "Don't saddle a dead horse." In other words, don't put the saddle on the dead side of the turnback. A clamp mounted backwards loses 40% of the holding capacity of a properly installed clamp.

At one time, the guy wire end strands were unwoven and wrapped around the guy wire itself. This process was known as "serving," and was best done using a special tool. Not only is this procedure quite difficult with EHS, but the wrapped

Table 27.4
Guy Wire Specifications

Typical 3⁄16 inch steel guy wire breaking strengths	
Common Grade	1540 pounds
Utility Grade	2400 pounds
Siemens-Martin Grade	2550 pounds
High Strength Grade	2850 pounds
Stainless Steel Aircraft	3700 pounds
Extra High Strength Grade	3990 pounds
Phillystran HPTG4000	4000 pounds
EHS guy wire sizes and breaking strengths	
3⁄16 inch	3990 pounds
1⁄4 inch	6650 pounds
5⁄16 inch	11,200 pounds
3⁄8 inch	15,400 pounds

bundle will also trap water running down the cable, which will accelerate the rusting process. With the near-universal adoption of preforms, this practice is no longer used very much.

SWAGED FITTINGS

Swaged fittings produce a strong, clean connection. If you don't like the look of lots of cable clamps, swaging may be for you. The most common swages are *Nicopress* fittings shown in **Figure 27.33**. While the fittings themselves are relatively inexpensive, you have to buy or rent a Nicopress tool to crimp them onto the guy wire. Once they're crimped on, they can't be removed.

PREFORMED GUY GRIPS

Preformed guy grips (or *Big-Grip Dead-Ends* from Preformed Line Products — **preformed.com**) are the easiest to use and the most expensive (see **Figure 27.34**). You simply curl them onto the end of the guy wire to produce a permanent termination. Preformed cable grips have virtually replaced cable clamps for power, telephone, and communications companies. Factory specs say that you can remove and reapply the grips twice. If removal is necessary after a guy grip has been installed for a period greater than three months, it must be replaced. If you can't find them locally, they are available from several *QST* advertisers.

Preforms are color-coded for guy wire sizes, as follows:

⅛ inch — blue
³⁄₁₆ inch — red
¼ inch — yellow
⁹⁄₃₂ inch — blue
⁵⁄₁₆ inch — black
⅜ inch — orange

Use only the correct size preforms for the guy wire you are using. Guy wire and related hardware, including cable clamps and preforms, are designed for a certain number of strands in the wire rope, and for a specific lay for each cable size. Do not mix different hardware. Note that preformed grips have two sets of crossover marks. The set closest to the loop is for normal guy wire attachment. The set farthest from the loop is for when the guy wire goes through an insulator.

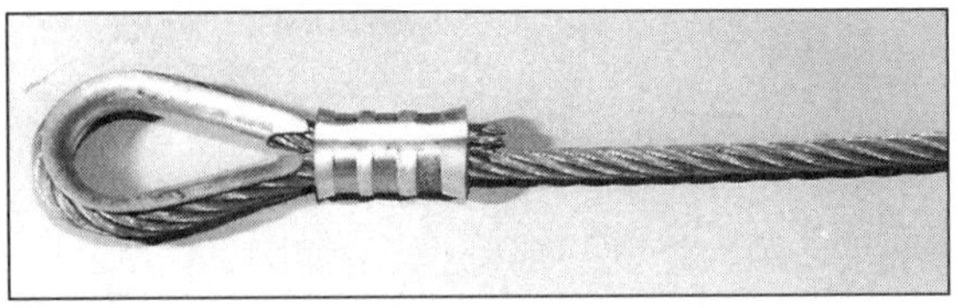

Figure 27.33 — Swaged guy wire end using Nicopress fitting.

INSTALLING PREFORMS

Preformed guy grips are precision devices, designed to be installed by hand; do not use any tools to install them. They should be installed only in conjunction with heavy-duty wire rope thimbles.

1) Insert a heavy-duty thimble into the eye of the preform, then through the attaching hardware (shackle, etc.).

2) Wrap the first leg (either one) around the guy wire with two complete wraps. Simply wrap them around the guy wire. Line up the crossover marks, then wrap the second leg with two complete wraps, ending opposite the first leg.

3) Complete the installation by either simultaneously wrapping both legs (keeping the legs opposite each other) or alternating between the legs a couple of wraps at a time. Bending the EHS guy wire as you wrap the preform leg around it will make it easier to attach.

4) Finish the short leg first, then the long leg.

5) Seat the ends of the legs by hand or use a flat blade screwdriver under the end of the strands. For Phillystran guy wire you may need to separate the strands to finish the ends of the Preform.

6) Attach a black tie-wrap or end sleeve around the grip at the end to secure it.

CUTTING GUY CABLE

Many different methods have been used over the years to cut guy cable. These days, EHS (extra high strength) guy wire is the standard and special cutters are needed to cut this hard wire. Always wear safety goggles when working with guy wires. There can be lots of metal chips floating around when you cut them, or the guy wire can easily whip around and hit you in the face or other body parts.

To cut the guy wire, rent or borrow a bolt cutter. Be certain it will cut EHS, not just soft metal. Another method is to use a circular saw or hand grinder with a metal cutting aggregate blade. These blades are less than $4 at your neighborhood hardware store and will cut pipe mast material as well. Use electrical tape not only to mark where you want to cut, but also to prevent the guy wire from unraveling after it's cut.

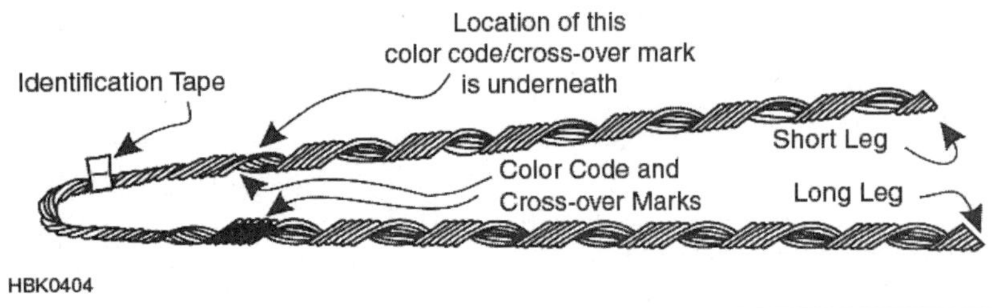

Figure 27.34 — Preformed guy wire "dead end" grip. The grip is wound around the guy wire and holds the load through friction.

Table 27.5
Guy Cable Comparisons

Cable	*Nominal Dia. (inches)*	*Breaking Strength (lbs)*	*Weight (lbs/100 ft)*	*Elongation (inches/100 ft)*	*Elongation (%)*
3⁄16 inch 1×7 EHS	0.188	3990	7.3	6.77	0.56%
1⁄4 inch 1×7 EHS	0.250	6700	12.1	3.81	0.32%
HPTG6700	0.220	6700	3.1	13.20	1.10%
HPTG8000	0.290	8000	3.5	8.90	0.74%
5⁄16 inch 1×7 EHS	0.313	11200	20.5	2.44	0.20%
HPTG11200	0.320	11200	5.5	5.45	0.45%
3⁄8 inch fiberglass rod	0.375	13000	9.7	5.43	0.45%

EHS steel cable information is taken from ASTM A475 (2022), the industry standard specification for steel wire rope.

The HPTG listings are for Phillystran aramid cables, and are based on the manufacturers' data sheets. The elongation (stretch) values are for 100 feet of cable with a 3000-pound load.

PHILLYSTRAN

Introduced in 1973, Phillystran offers the strength of EHS guy wire with the added advantage that it is nonconducting and electrically transparent to RF. It consists of a polyurethane resin-impregnated aramid fiber rope with a thick extruded jacket of specially formulated polyurethane. Its non-conductivity makes it ideal for tower systems where some antennas will be under or close to guy wires. Guy wire interaction with stacks and wire antennas will be eliminated by using Phillystran. **Table 27.5** compares EHS, Phillystran and fiberglass rod guying material.

Recommended Phillystran installation calls for at least 10 to 25 feet of steel cable from the end of the Phillystran to the anchor. This prevents damage from vandalism, accidents and ground fires that can weaken Phillystran and cause a tower failure.

PHILLYSTRAN CABLE GRIPS

Preformed Line Products now manufactures Phillystran-compatible Preformed guy grips. These are different from those used with ¼ or ⅜ inch EHS with a different lay (twist) to match the characteristics of Phillystran. The grips for Phillystran cannot be interchanged with the grips for EHS.

The guy grips are installed generally the same way, except that you must keep some tension on the *Phillystran* while installing them, and you may have to split the strands on the end of the Preform in order to finish wrapping them on. This is because the *Phillystran* is very flexible, particularly when compared to EHS. Other than that, they're installed just like the Preforms for steel guys.

ATTACHING GUY WIRES TO THE TOWER

Figure 27.35 shows two different methods for attaching guy wires to towers. At Figure 27.35A, the guy wire is simply looped around the tower leg and terminated in the usual manner. At Figure 27.35B, a *guy bracket*, with *torque arms* has been added. Even if the torque arms are not required, it is preferred to use the guy bracket to distribute the load from the tower/guy connection to all three tower legs, instead of just one. The torque bracket is more effective at resisting torsion on the tower than simply wrapping the preform or EHS around the tower leg. Rohn offers another guy attachment bracket, called a Torque Arm Assembly (sometimes called a "star guy" bracket), which allows six guys to be connected between the bracket and anchors. This is by far the best method of stabilizing a tower against high torque loads and is recommended for installations with large antennas.

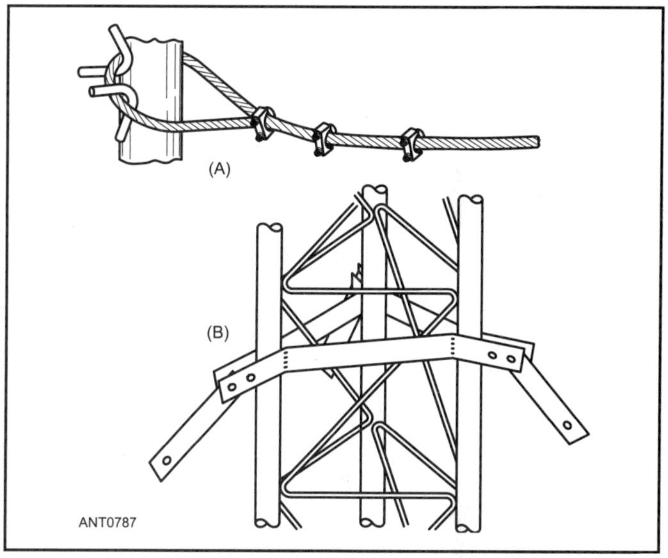

Figure 27.35 — Two methods of attaching guy wires to tower. See text for discussion.

ATTACHING GUYS TO ANCHORS

Turnbuckles and associated hardware are used to attach guy wires to anchors and to provide a convenient method for adjusting tension. **Figure 27.36A** shows a turnbuckle with a single guy wire attached to the eye of the anchor. Turnbuckles are usually fitted with either two eyes, or one eye and one jaw. The eyes are the oval ends, while the jaws are U-shaped with a bolt through each tip. Figure 27.36B shows two turnbuckles attached to the eye of an anchor. The procedure for installation is to remove the bolt from the jaw, pass the jaw over the eye of the anchor and reinstall the bolt through the jaw, through the eye of the anchor and through the other side of the jaw. Only *forged-eye* turnbuckles should be used in which the eye

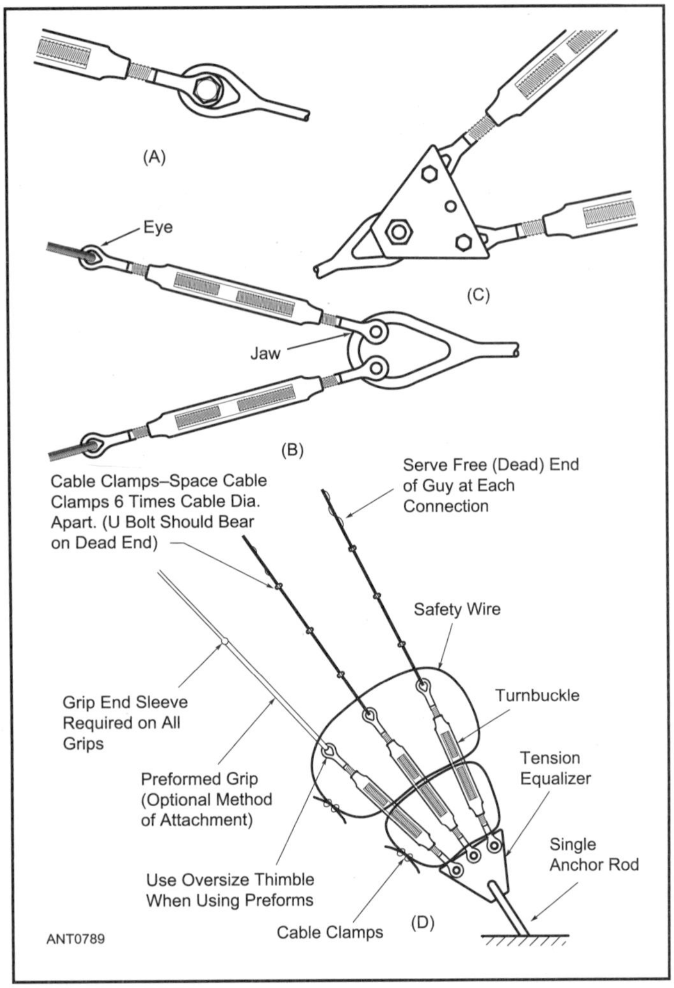

Figure 27.36 — Variety of means available for attaching guy wires and turnbuckles to anchors.

loop is a continuous piece of metal and not formed into a loop by bending. Formed loops can open up and break or slip under heavy loading.

If two or more guy wires are attached to one anchor, *equalizer plates* should be installed (Figure 27.36C). In addition to providing a convenient point to attach the turnbuckles, the plates pivot slightly to equalize the various guy loads and produce a single load applied to the anchor. Once the installation is complete, a safety wire should be passed through the turnbuckles in a figure-eight fashion to prevent the turnbuckles from turning and getting out of adjustment (Figure 27.36D).

PULLING AND TENSIONING GUY WIRES

Once the guys are cut to their appropriate lengths and are attached to the tower, you need to pull them so you can attach them to the turnbuckle at the guy anchor. One method is to pull them by hand with a moderate amount of force (100-200 pounds of pre-tension will stabilize the tower under construction) and then secure them to the anchor. This will deflect the tower slightly but will put some initial tension on them. Another method shown in **Figure 27.37** is to use a come-along and cable grip. Place a nylon sling around the guy anchor for attachment of the other end of the come-along.

Most manufacturers require the final tension of the guy wire to be 10% of its breaking strength. That amount of tension is necessary to eliminate looseness in the cable caused by the spiral wire construction and to eliminate excessive dynamic guy and tower motion under wind loading. For 3/16-inch EHS that amount of tension would be approximately 400 pounds.

How do you know when you've got the right amount of tension? A calibrated dynamometer can be used, but such tools are quite expensive. The Loos Tension Gauge in **Figure 27.38** is an accurate, inexpensive device for measuring guy tension. While originally designed for accurate, repeatable tuning of a sailboat's standing rigging (which typically uses 7X19 SS cable), the accuracy of measurement is quite good enough for ham radio towers, using EHS. It works by measuring the deflection of the guy wire and does not need to be inserted into the guy wire. (If you are using Phillystran, measure tension in the recommended section of steel EHS between the Phillystran and the ground anchor.)

The guy is gripped with a cable grip, which is connected to the anchor below the eye (or equalizer plate) with a block and tackle arrangement (see Figure 27.37) or a come-along. Then the turnbuckle is adjusted to take up the load, the cable grip is released, and the final guy tension is adjusted and checked.

Regardless of how you measure guy tension, the important thing is to tighten all the guys, so tension is approximately the

Figure 27.37 — To tighten guy wires a nylon sling (see at the lower right of the photo) is attached to the guy anchor and a come-along. The come-along is then hooked to a Klein cable grip on the guy wire. The come-along is then tightened until the required guy tension is achieved as measured with a Loos tension gauge or dynamometer. The guy wire can then be attached to the guy anchor and the Klein grip released. (Courtesy of Dale Boggs, K7MJ)

same within each set of guys. Having all tensions equal avoids pulling the tower out of alignment. When you adjust the guys at each level, you should check the tower for vertical alignment and straightness. This is often done with a transit from two ground points located 90° from each other. After plumbing the tower up to the first set of guys, another method is to look up the face of the tower from the bottom. It will be obvious if the sections above the guys are out of plumb.

Resonance in Guy Wires

If steel guy wires are resonant at or near the operating frequency, they can receive and reradiate RF energy. By behaving as parasitic elements, the guy wires may alter and thereby distort the radiation pattern of a nearby antenna. For low frequencies, where a dipole or other simple antenna is used, this is generally of little or no consequence. But at higher frequencies, where a unidirectional antenna is installed, it is desirable to avoid pattern distortion if at all possible. The symptoms of re-radiating guy wires are usually a lower front-to-back ratio and a lower front-to-side ratio than the antenna is capable of producing. The gain of the antenna and the feed point impedance will usually not be significantly affected, although sometimes changes in SWR can be noted as the antenna is rotated. (Of course, other conductors in the vicinity of the antenna can also produce these same symptoms.)

The amount of re-radiation from a guy wire depends on two factors — its resonant frequency, and the degree of coupling to the antenna. Resonant guy wires near the antenna will have a greater effect on performance than those that are farther away. Therefore, the upper portion of the top level of guy wires should warrant the most attention with horizontally polarized arrays. The lower guy wires are usually closer to horizontal than the top level, but by virtue of their increased distance from the antenna, are not coupled as tightly to the antenna.

To avoid resonance, the guys should be broken up by means of egg or strain insulators.

Figure 27.A shows wire lengths that fall within 10% of 1/2-λ resonance (or a multiple of 1/2 λ) over all the HF amateur bands.

Unfortunately, no single length greater than about 14 feet avoids resonance in all bands. If you operate just a few bands, you can locate greater lengths from the chart that will avoid resonance. For example, if you operate only the 14, 21 and 24 MHz bands, guy wire lengths of 27 feet or 51 feet would be suitable, along with any length less than 16 feet.

Of course, you could neutralize the whole problem by using Phillystran at some expense. One way to minimize the cost is to use Phillystran on only the top or top two sets of guys. Further, it's not necessary to use Phillystran all the way down to the anchor. Even using Phillystran for the top 50% will reap benefits.

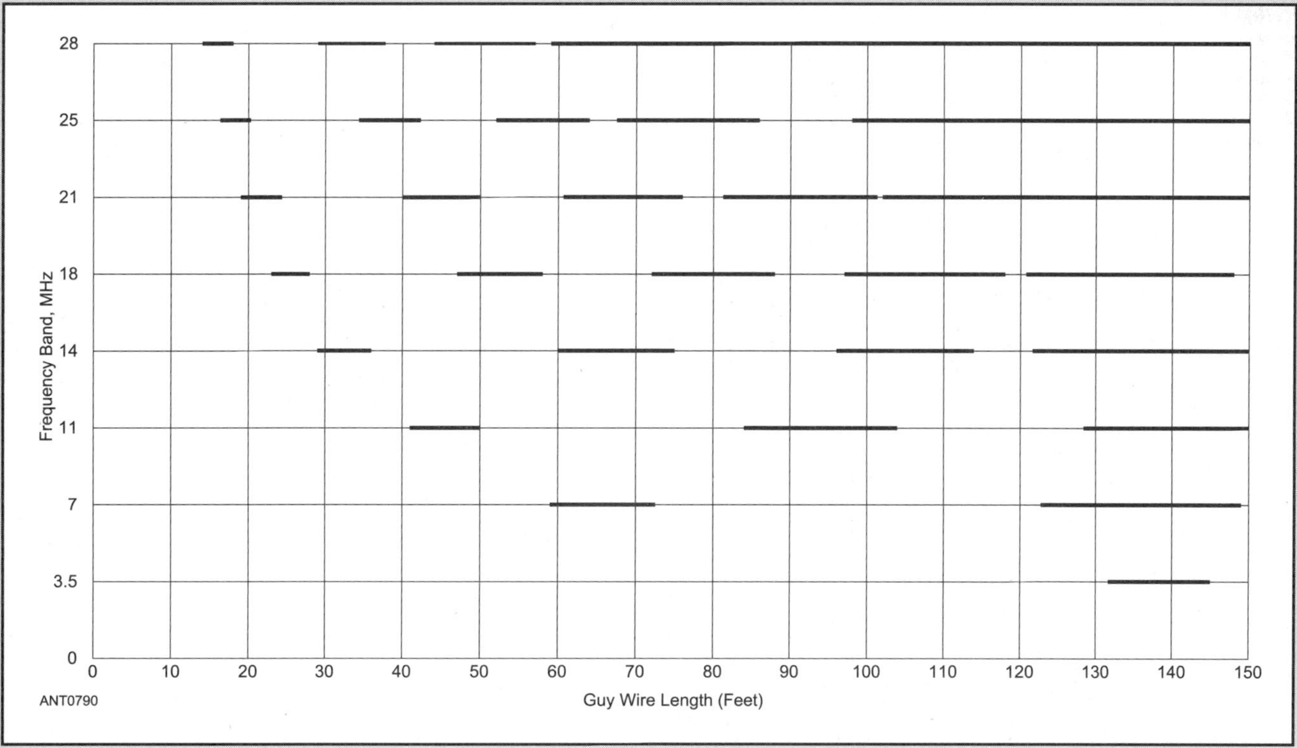

Figure 27.A — The black bars indicate ungrounded guy wire lengths to avoid for the eight HF amateur bands. This chart is based on resonance within 10% of any frequency in the band. Grounded wires will exhibit resonance at odd multiples of a quarter wavelength. (Chart by Jerry Hall, K1TD)

Figure 27.38 — Loos PT-2 guy wire tension gauge.

SAFETY WIRING TURNBUCKLES

The final step in installing guy wires is to safety wire them as shown in **Figure 27.39**. This keeps turnbuckles from loosening from normal vibration and discourages vandalism. Use some pieces of leftover guy cable and loop it through the anchor shackle and the turnbuckles, securing the ends with a cable clamp.

Figure 27.39 — A length of guy cable is used to assure that the turnbuckles do not loosen after they are tightened. This procedure is an absolute requirement in guyed tower systems, as shown by Jodi Morin, KA1JPA (left) and Helen Dalton, KB1HLF.

27.5.5 Before Working on a Tower

THE WORK CREW

For small antenna jobs, two people (one on the tower and one on the ground) are usually enough. Even erecting 25G tower (40 pounds per section) can be accomplished with two people but a third person to handle the tag line is very handy. For 45G, it takes two people on the haul rope as these sections weigh 70 pounds each and a section with guy brackets is close to 100 pounds.

Commercial riggers commonly use some type of winch or windlass to haul up heavy loads. For working with large antennas, such as 40-meter beams, two people on the tower along with one or two tag line handlers, plus two to four on the haul rope, makes for a large crew.

Regardless of size, take care of your crew! Roll out the red carpet for them. They're giving up their time to help you and they deserve it. Make an effort to provide lots of water or iced tea and by all means feed everybody a nice lunch. *No alcohol* until after the work is done!

PRE-WORK MEETING

On project day, the first thing you should do is have a session with the entire crew and go over what is going to be accomplished and the order and manner in which it is going to be done. Cover all safety issues, commands and equipment related to the job. Identify any hazards in the work area, such as power lines. Explain any specialized equipment or tools, including carabiners and slings, come-alongs, hoisting grips and so on. If a come-along or other special tool is going to be used, be sure that someone on the tower and ground crews knows how to use it properly.

One of the most important jobs for the ground crew is to act as a spotter and take care of the safety of the tower crew and the whole team. Point out where a phone is and any phone numbers that may be needed in an emergency. Also discuss and understand what to do in an emergency situation. For minor emergencies, knowing where the closest medical facility is will be valuable. Since just about everyone will have a mobile phone, calling 911 won't present any problems for bigger emergencies. Not many emergency services professionals have been trained for high-angle rescue such as lowering someone off of a tower, so you're probably going to be on your own at least initially. Search and rescue crews are used to working with ropes and other hardware for extrications, so hopefully your 911 operator will be able to put you in touch with them. Physical trauma can set in quickly even with a fall-arrest harness so fast action is vital.

Let your ground crew know not to be standing around the bottom of the tower unless they must specifically be there. This is the danger zone for dropped tools and hardware which reaches high speeds and can bounce a long way from the tower.

Rule #1. The tower crew is in charge. The ground crew should do what the tower crew tells them and not do what the tower crew doesn't tell them. Being on the ground crew is usually pretty boring, but they should not do anything that would have any impact up on the tower. With very few exceptions, the ground crew shouldn't do anything unless directed to. If they are not sure about something, ask the tower crew.

Rule #2. When the ground crew is talking to the tower crew, look up and talk in a loud, concise voice. Although it may be still and quiet on the ground, the ambient noise level on the tower is always significantly higher 50 feet or more in the air and you have major communication obstacles. VHF/UHF handhelds, FRS handhelds, or VOX-operated 47 MHz headsets all work. Make sure that all have fully charged batteries before work begins.

Rule #3. Really communicate. Insist that the ground crew keeps the tower crew really informed. If something is lowered to the ground, the ground crew should tell the tower crew that it's "on the ground." If the tower crew is waiting for the ground crew to do something, they should keep the tower crew informed about status. This prevents the "everybody waiting on everybody" problem.

COMMANDS

Make certain that everyone understands each of the commands — whether they are the ones given here as examples or your own preferred set — and that they all use the same ones. All of the following commands refer to the "load" (antenna, tower section, etc.) and are applied to the "haul rope" (the line to which the load is attached). There are also several common hand signals to use. Simple ones for up, down and stop can be useful, particularly in high-noise situations. Make certain everyone knows what they are. For example:

• "Tension" tells the ground crew to put tension on the line, to take up any slack. Once you have some tension, move the load with "Up" or "Down" commands. Add "Slow" for slower lifting and lowering.

• "Slack" means giving the load some slack.

• "All slack" means the ground crew may gradually and gently release their grip on the load.

• "Stop" is obvious and "Stand by" indicates that they should maintain their assignment while awaiting the next command. Again, the tower crew is in charge; don't do anything without their instruction.

If something drops or falls, alert the ground crew immediately. Yell "Look out below!" or "Headache!" so that they can get out of the way of the wayward bolt, nut or tool. Their hardhats only provide minimal protection against this occurrence. Dropped items are not only dangerous, but it also means sloppy work. Take your time and concentrate on not dropping anything.

THE TOWER CREW

If you're on the tower crew, you should know what you're doing or be working with someone who does. If you are working on a standing tower, before climbing walk around it and make a thorough visual inspection. Look at the base for cracked or rusted legs or missing hardware. Go out to the anchors to check the turnbuckles, clamps and other hardware. Look for bee or wasp nests. Never assume that any tower is safe to climb — always inspect it thoroughly before you take that first step.

Before beginning any maneuver, discuss how you're going to do it and the sequence that will be used. This way, everyone will understand the process and will hopefully do the right thing at the right time. This is particularly important if you're up there with someone you've never worked with before. Sometimes you both assume that the other person is going to do something obvious that needs doing and then neither of you does it — this can be dangerous. Go over everything. This trains an inexperienced person and makes it easier the next time you work together.

Keep your tools either in your bucket or tool bags on your belt or tied to the tower. Try to avoid putting anything on a flat surface such as the rotator plate or thrust bearing plate; they can roll off.

Avoid using ac-powered tools on the tower. Battery powered tools are safer; you can buy, borrow or rent them. If you must use ac-powered tools, make certain they are insulated and that the extension cords are suitable. Zip cord extensions are dangerous. Make certain the ground crew knows where to disconnect the extension cords and where the breaker box is located.

MANAGING THE WORK TO BE DONE

Break everything down into bite-sized pieces and only do one step at a time. Trying to combine two or more steps in the same task is asking for trouble. For example, don't try to bring up the guy wires already attached to the tower section; bring them up after the section and guy brackets are installed. Trying to combine too many steps often results in doing things twice, along with undoing what you've already done. You'll be more efficient and safer by doing things one step at a time.

PREPPING THE MATERIALS

Many tasks are a lot easier to do on the ground than up on the tower. Take the time to prepare all of the materials before hoisting them so that the tower crew's job is as simple as possible.

For a tubular-legged lattice tower, there are several things you can do to make the job easier while the tower sections are still on the ground. First, there will be excess galvanizing from the hot-dip process in many of the leg bolt holes on new tower sections. This will prevent a bolt from going through the hole. Except as a last resort, do not drill out the holes as it exposes the steel. Use a drift pin or taper punch and a hammer to enlarge the hole only enough for a bolt to clear the hole. Next, check the inside of each lower leg for that same excess galvanizing and remove it carefully with a round file. These steps are much easier to do on the ground than in the air.

Check that the sections fit together. It's not uncommon that one leg won't line up, particularly with new sections. Using a piece of pipe or another tower section as a lever, gently bend the out-of-line leg until it slips on properly. This also is easier to do on the ground. Lay out the sections in order of fit and mark one pair of the mating legs with tape or a felt marker to ensure they can be assembled the same way they were on the ground. Be sure to send the sections up the tower in the same order they were checked for fit.

Put a thin coating of white lithium or molybdenum disulfide

grease around the *inside* of each *lower* leg of a tubular-legged tower section. Not only will they slide on more easily during installation, but the grease will help minimize corrosion and oxidation between the sections and make later removal easier. Skipping this step at assembly makes disassembly harder to the point where a jack may be required to get the sections apart! If the tower is to be conductive, for example if it's to be used as a vertical antenna, then use a conductive antioxidant compound such as Noalox or SS-30 instead.

Inspect all of the remaining metal pieces and parts for blocked holes, damaged threads, bent braces or arms — anything that will make it hard to assemble on the tower. These are much easier to repair on the ground. Use a file to round the sharp edges of all plates, steps, brackets or arms to keep them from damaging you.

If more than one antenna is to be installed on the mast, measure and mark where each antenna will be mounted and where the thrust bearing, if any, will be.

Hook the rotator up to its control box and test its operation. Turn the rotator until it indicates North, or another known direction, then get it ready for hoisting. It doesn't really matter which way the rotator is physically installed in the tower as the antennas can be oriented when mounting them on the mast. By knowing the rotator's indicated direction, the antennas can be aligned properly without stopping work.

If you are assembling an antenna for the project, let it sit overnight then retighten all of the nuts and bolts the next day. The hardware will have temperature cycled from warm to cold and warm again and some of the hardware will have loosened up during that temperature-induced expansion and contraction. This is the time to be sure it's all tight!

Antenna element hose clamps invariably catch on anything — wire antennas, guy wires, cables, and so on. To minimize this annoying characteristic, wrap the hose clamp with a layer or two of electrical tape while it's still on the ground. This is another good reason to use rivets instead of hose clamps for beam assembly.

If your rotator doesn't have a connector for the control cable, you can add one using vehicle trailer connectors — they come in a flat 4-wire polarized configuration. Get two sets and install a pigtail to the rotator terminated in one male and one female connector. Do the opposite at the end of your control cable to the station. This will ensure that the cable is always connected correctly. Another system is to use a European Molex connector to connect a short jumper from the rotator to the longer control cable run to the station. Weatherproof the connector by inserting it into a short piece of 2-inch capped PVC tubing taped to a tower leg. Being able to access the rotator wiring at the rotator, without having to look underneath or open up a connector, can be extremely time-saving during troubleshooting.)

When putting something together "temporarily," always install it as though you won't be coming back; "temporary" sometimes means it will be up and used for years!

Tower Climbing Shield

A tower can be legally classified as an "attractive nuisance" that could cause injuries and/or lawsuits. You should take some precautions to ensure that "unauthorized climbers" can't get hurt on your tower.

Generally, the attractive-nuisance doctrine applies to your responsibility to trespassers on your property. (The law is much stricter with regard to your responsibility to an invited guest.) You should expect your tower to attract children, whether they are already technically trespassing or whether the tower itself lures them onto your property. A tower is dangerous to children, especially because of their inability to appreciate danger. (What child could resist trying to climb a tower once they see one?) Because of this danger, you have a legal duty to exercise reasonable care to eliminate the danger or otherwise protect children against the perils of the attraction.

An article describing such a tower shield by Baker Springfield, W4HYY and Richard Ely, WA4VHM was published in the September 1976 *QST*. It has been added to this book's online material, including construction diagrams. Installing it should eliminate the worry.

HARDWARE PREP

Make sure all hardware is stainless or galvanized hardware. Avoid plated hardware.

Never place a load on eyebolts with eyes that can open. Only eyebolts that are forged or welded closed should be use on tower or antenna projects if they will be carrying a load.

Always take extra hardware, nuts and bolts up the tower in case you run short or drop something. If you don't have them, you'll invariably need them.

27.5.6 Assembling the Tower

Route the haul rope through a snatch block at the bottom of the tower to transfer the hoisting effort from pulling down to pulling horizontally. Pulling vertically is all arm strength plus the ground crew is exposed to falling objects. Pull horizontally by putting the rope around your hips and walking backward to hoist the load. This uses larger muscle groups and hoisting will be much easier. In addition, those hoisting can watch the load while staying out of the danger zone. Of course, working with ropes under load near the tower means wearing leather gloves, and a hard hat.

To hold onto a haul rope, put it around your hips then bring the tail or dead end in front of you. *Do not* tie the rope around your waist — this is potentially dangerous. Aiming the tail end in the same direction as the load rope, grasp both ropes with one or both hands. This is the best way to hold or brake a rope load. Don't depend on just using your hands; it's not as reliable and your hands and arms will quickly tire. With the rope secured around you, it can be held comfortably for quite some time.

SECTION STACKING

After rigging the gin pole as in Figure 27.23 (the haul rope goes up the middle of the pipe, across the pulley, then down to the load), attach the leg bracket to the top of the top section, *below* the top brace. Make certain that it's secure before you push the gin pole mast up to the extended position where it'll be ready for the lift. (This is a good procedure to rehearse on the first section above the base with the ground crew watching.) Place a suitably sized piece of plywood or OSB right at the tower base, on which to "stage" or set the sections. This prevents the legs from getting dirt in them. (If you're installing used sections, make sure each siderail is clear of any debris before hauling them aloft.)

The tower section should be rigged so that it hangs more or less vertically; the heavier the section, the more important this becomes. Here are a couple of rigging options: Put a sling around a leg at about 3/4 of the way up the section (double-check for proper top and bottom orientation) to establish the pick-point. Attach the haul rope just above the section midpoint and, on the command of the tower crew, start to pull on the haul rope. Or a carabiner can be clipped to the third horizontal brace down from the section top, and a second carabiner (simply clipped on to the haul rope) can be attached to the topmost horizontal brace.

As soon as it clears the top of the tower, the tower crew should yell "stop," then "down slow" when ready to have the section lowered onto the top of the tower.

If leg alignment problems are discovered (see item 9 in the sidebar on tower building mistakes), use a come-along around the bottom of the whole section and tighten it up to pull the legs together. (This is common practice when putting up angle-legged self-supporting towers.) If you don't have a come-along, a ratchet-operated truck strap may suffice. One person on a tower doesn't have a whole lot of leverage and if the sections line up but won't slide down, use a come-along or TowerJack and pull the section down into place.

Stack an appropriate number of sections (typically up to the next guy point), then bring up the first set of guys and attach them. Your ground crew can use their cable grips and come-alongs to put the initial tension on them, then attach them to the anchors. You'll be able to tell them which ones to tighten and which ones to loosen by using a level on the leg to plumb the tower. Once that's done, repeat the same steps until all the sections are in place and guyed.

It's important to plumb the first set of sections including the first set of guys. Once that segment is plumb, you can look up the face of the tower to see if everything above it lines up — it'll be pretty obvious. If it doesn't, just adjust the come-alongs or turnbuckles to get it straight.

INSTALLING THE MAST

For small and medium masts, use the gin pole to bring it up using a sling as a choker. The choker should be above the balance point so that the mast goes up vertically. Lower the mast into the tower from the top.

Large and heavy masts more than 20 feet long are bigger than the usual gin pole can control properly. When building the tower, put the mast inside the bottom sections. Remove any rotator shelves or other obstructions so the gin pole can lift the mast up through the tower. Once the mast is captured at the top of the tower, it can be raised gradually. The books referenced at the beginning of this chapter provide more details.

Once the mast extends slightly above the top of the tower and is captured by the thrust bearing or a clamp acting as a collar, install the top antenna on the mast if more than one is to be installed. Once the first antenna is installed, pull the mast up with a come-along to where the next antenna is to be installed. Repeat the sequence until all the antennas are installed.

One "trick" to making this vertical lift easier is to replace one of the thrust bearing mounting bolts with a galvanized forged eye or eyebolt. This provides an attachment point for your come-along, making it as close to vertical as possible. This helps ensure the mast comes up as plumb as possible. The eye can, of course, be left in place.

INSTALLING THE ROTATOR

Clip a carabiner and haul rope onto a U-bolt on the rotator clamp or lift it with a bucket. Bring up the rotator, install it under the mast, and lower the mast into the rotator clamp. To haul up the control cable, tie an overhand knot in the cable and snap that into the haul rope carabiner.

To minimize the possibility of binding in the rotator/mast/thrust bearing system, work down when doing the final tightening. Do the thrust bearing first, the rotator mast clamps next, the rotator shelf bolts, and the rotator base bolts last.

When connecting the multiconductor control cable, one way to keep wire colors straight and consistent is to use the resistor color code: black, brown, red, orange, yellow, green, blue, violet, grey and white.

ROTATION LOOPS

The rotation loop should be made with the rotator oriented north (typically its midpoint here in North America), ensuring the loop itself is centered. There are two ways to make a rotation loop for your cables. One way is to tape all the cables coming down the mast into a bundle, leaving an extra 4-5 feet of slack before securing the bundle to a tower leg. The bundle will have some rigidity that will help keep it out of harm's way. Make sure that it doesn't snag on anything as the system rotates and you'll be good to go. If you have a flat-topped tower, wind the cable around the mast 2-3 times in a diameter smaller than the top plate so that the coil lays on the flat surface.

27.6 Raising and Lowering Antennas

While small antennas can simply be pulled up directly on a haul rope, working with HF beams requires some technique. If done properly, the actual work of getting the antenna into position can be executed quite easily with only one person at the top of the tower. The ground crew does all the lifting by using a large pulley attached to the antenna mast or a gin pole with its pulley a foot or two above the point at which the antenna is to be mounted. Because raising an antenna often requires the load to be pulled away from the tower — either to avoid guy wires or as part of a tram or V-track system — this places significant bending forces on a mast or gin pole and may bend them if the pulley is too far above where they are attached to the tower.

The advice and suggestions in this section also apply to removing antennas by following the procedures in reverse. An antenna should probably be removed the same way it was installed. If installing it required a crane, it will most likely have to be removed by a crane.

27.6.1 Avoiding Guy Wires

Guy wires often obstruct the antenna's path to the top of the tower. One method of avoiding them is to tie a tag line to the middle of the boom and to a middle element for leverage (but within reach of the tower crew). The ground crews then pull the antenna out away from the guys as the antenna is raised. With this method, some crew members are pulling up the antenna to raise it while others are pulling down and out to keep the beam clear of the guys. Obviously, the opposing crews must act in coordination to avoid damaging the antenna.

A second method is to tie the haul rope to the center of the antenna. A crew member, wearing a climbing harness, walks the antenna up the tower as the ground crew raises it. Because the haul rope is tied at the balance point, the tower climber can rotate the elements around the guys. A tag line can be tied to the bottom end of the boom so that a ground crew member can help move the antenna around the guys. The tag line must be removed while the antenna is still vertical.

The third method is characterized as the "wig-wag" system by Tom Schiller, N6BT in his book *Array of Light*. It is particularly useful when other antennas are already installed on the antenna mast. The new antenna is lifted until it is immediately below the lowest antenna on the mast and rotated so that its elements are parallel to the boom of the installed antenna. The new antenna's elements are then tipped up to clear the elements of the installed antenna so that the new antenna is rotated around the installed antenna until its boom is above that of the installed antenna. Once above the installed antenna, the new antenna can be lifted so that its elements can be rotated to clear those of the installed antenna and then returned to horizontal and lowered for installation. This can be difficult if more than one antenna is already installed on the antenna mast.

27.6.2 Using a Trolley or Tram System

Sometimes one of the top guys can provide a track to support the antenna as it is pulled upward. Insulators in the guys, however, may obstruct the movement of the antenna. A better track made with rope is an alternative. One end of the rope is secured outside the guy anchors. The other end is passed over the top of the tower and back down to an anchor near the first anchor. So arranged, the rope forms a narrow V-track strung outside the guy wires. Once the V-track is secured, the antenna may simply be pulled up, resting on the track. This is known as the trolley or V-track system. It is not an easy method to use. It requires two trolley cables spread apart some distance with the same tension on each line or the beam will tip. Plus, there is a lot of added friction to the system from the weight of the antenna on the lines.

A much easier system is the tram line in which one line runs from the top of the tower to an anchor on the ground and the antenna is slung below the tramline. The system is illustrated in **Figure 27.40** and a video showing a 10-meter beam being lifted with a tram is available at **youtube.com/@DXEngineering**; search for "tramming".

Install one long (6-foot) sling on each side of the center mounting point of the antenna with two or three wraps around the boom; then bring them together to form a truss with the pick point directly above where the boom plate will attach to the mast.

This assures that the antenna is balanced and will arrive in the correct mounting position. Using two slings on the boom enables you to hoist the beam while it remains horizontal. Even if the antenna is mechanically off balance, you can adjust the slings so that it will remain basically horizontal.

You'll need three pulleys, a haul rope, a length of wire rope for the tram line, an anchor on the ground and miscellaneous slings and carabiners. Suspend the antenna below the tram wire. Small diameter aircraft cable or wire rope such as ⅛ inch or $^3/_{16}$ inch ($^3/_{16}$-inch EHS works well) is sufficient to take the static load of just about any amateur antenna.

To set up the tram line, first secure a sling choker on the antenna mast about three feet above the place where the antenna will be mounted. Use two or three wraps and bring the choker through itself as described earlier.

Clip a carabiner or shackle to the tail of the sling. Then clip a large pulley into the carabiner. Bring up one end of the tram line and clip it into the same carabiner. Obviously, it's easiest if the tram line can "bisect" the guys, which provides the most clearance for the antenna coming up. This may require using some sort of temporary anchor point. A good rule of thumb for tramming is to try and have 1.5 times the tower height of working room; having a shallow angle makes tram work easier.

Secure the other end of the tram line to an anchor. You can use a tree, a fence post, a car, a stake driven into the ground or any other convenient strong point. Use a come-along and

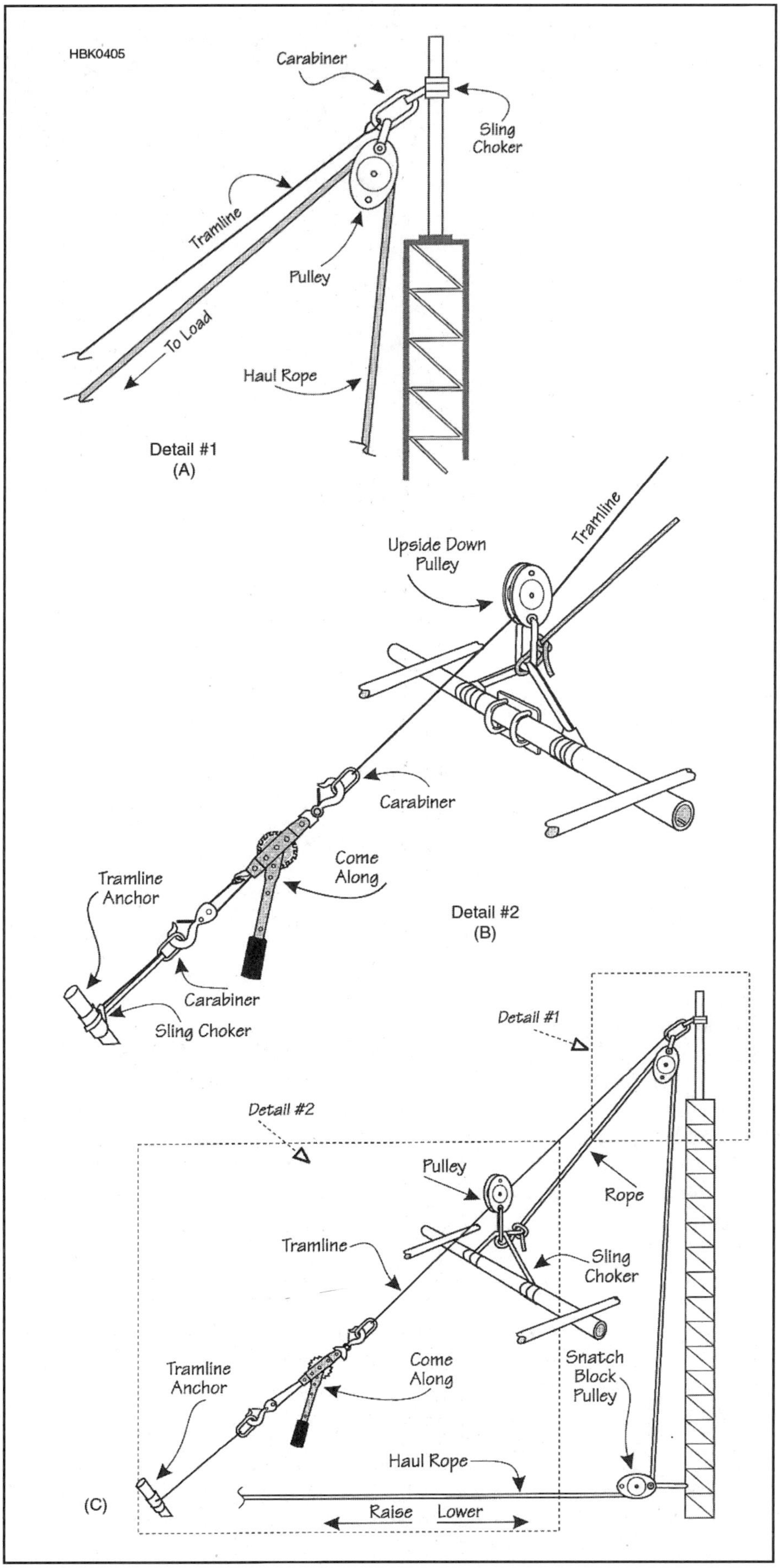

Figure 27.40 — A schematic drawing of the tram line system. At A, rigging the top of the tower for tramming antennas. Note the use of a sling and carabiner. The Hoist Clamp tram line clamp is available from KF7P Metalworks at kf7p.com/KF7P/Hoist_Clamp.html. (B) Rigging the anchor of the tramline. A come-along is used to tension the tramline. (C) The tram system for getting antennas up and down. Run the antenna part way up the tramline for testing before installation. It just takes a couple of minutes to run an antenna up or down once the tramline is rigged.

cable grip to tighten the cable until most of the slack is taken up. Do not over-tighten; you could damage your mast. If the sling on the mast is now high enough to create a significant bending force on your mast (more than four or five feet), back guy it in the opposite direction with another wire line or rope that is anchored to a convenient spot.

Run the haul rope through the back of the antenna mast pulley and out the front in the direction of the ground anchor. Lower the end of the haul rope directly to the ground or tie it to a carabiner clipped onto the tram line and let it slide down the wire. Figure 27.40A shows how the system should look at the top of the tower.

You may have to drop one or two of the top guy wires or any wire antennas closest to the antenna tram path if the antenna is going to be installed close to the top of the tower. Guys should be detached at ground level.

On the ground, attach the tram pulley (**Figure 27.41**) to the tramline. Turn the pulley upside-down (the antenna will be suspended under the tramline) then clip in the load end of the haul rope. Lift the two slings forming the antenna truss to the tram pulley and clip it in. The boom should be at a right angle to the tram line (elements parallel to the line) with the boom-to-mast bracket pointed toward the mast ready to accept U-bolts.

At this point, the haul rope should be attached to the tramline pulley. It goes up through the pulley on the mast, then down the tower to the ground. The third pulley is used at the bottom of the tower to change the direction of the haul rope from vertical to horizontal. At this point the system should look like Figure 27.40.

With antennas that have the elements mounted above the boom the antenna will attempt to flip over or "turn turtle." Minimize that tendency by tying the slings with opposite wraps (one around the boom in one direction, the other wrapped in the other direction).

Another method that helps counteract this unwanted tendency is to use a "tiller" as shown in **Figure 27.42**. It's a 4-foot long or so piece of angle iron or aluminum that is U-bolted to the boom on one end and has a small U-bolt on the front that captures the tramline and acts as a guide. Alternately, the front of the tiller can be attached to the haul rope or lift slings. The tiller holds the antenna boom in a relatively fixed position, thus preventing it from turning over. Once the antenna has been lifted off the ground, the boom can be rotated in the tiller's U-bolt so that the element halves facing the tower are raised to clear the guy wires.

Next, attach any tag lines. Use a small line such as ¼ inch polypropylene because it's light and stiff enough to resist hanging up on any clamps or hardware sticking out on the elements. Tie one end to the boom at a convenient spot that the tower crew can reach to untie it. Wrap the tag line around an adjacent element two or three times. You can add one or two wraps of electrical tape to hold it in place on the element to keep the fulcrum out on the element and away from the boom. The tag line will pull easily through the tape when you're done. If the tag line does hang up on a guy wire, lower the antenna to free it then tram it back up again.

When it's time to launch the antenna, have the ground crew pull the haul rope while another person helps the antenna off the ground. Once the antenna is launched, crew members holding the tag lines can guide it as it goes up. (For big beams, such as 40-meter antennas, more than one person may be required to maneuver it up and on to the tram line and get the lift started.)

Use the tag lines to pull the element halves pointing away from the tower down so that they'll clear the guys. You'll be pulling against the haul rope so don't pull too hard on the tag line. The tag lines can also be used to move the boom so that the antenna will be in the proper mast-mounting orientation.

The tower crew can guide the antenna when it gets close to the tower. Once the antenna has cleared all obstacles and if everything was rigged correctly, the antenna should come right up to the mast. (If using the tiller, once the beam arrives at the mast, you'll find it's now in the way. Having a come-along or a second haul rope you can transfer the load to will allow you to move it out of the way. A second pair of hands can help greatly at this point, too.)

Another advantage is that while on the tram line you can run any on-the-air tests you'd like. Just attach a run of coax before you lift the antenna. To make any adjustments, lower the antenna, make the changes and pull it up again. Make measurements with the boom 90° to the tram line if possible (elements parallel to the tramline).

To take antennas down, rig everything the same way, then lower the antenna down the tram line. Be certain as it comes up from the ground, the haul rope goes behind the boom before it goes through the mast pulley.

Figure 27.41 — A sturdy all-metal pulley suitable for tramming. (Don Daso, K4ZA, photo)

Figure 27.42 — Photo of the tram suspension system used by K7NV. Note that the tiller is attached to the lift sling in this system. The boom of the antenna can be rotated in the U-bolt holding it to the tiller to adjust the tilt of the elements to clear guy wires. (Kurt Andress, K7NV, photo)

27.6.3 Building Antennas on the Tower

A fourth method is to build the antenna on the tower and then swing it into position. Building the Yagi on the tower works particularly well for Yagis mounted partway up the tower, as you might do in a stacked array. The technique works best when the vertical spacing between the guys is greater than the length of the Yagi boom.

Figure 27.43 illustrates the steps involved. A haul rope through a gin-pole or tower-mounted pulley is secured to the boom at the final balance point and the ground crew raises the boom in a vertical position up the tower. A tie rope is used to temporarily secure the upper end of the boom to keep it stable while the boom is being raised. The tower person removes the tie-rope once the boom is raised to the right level and has been temporarily secured to the tower.

The elements are then brought up one at a time and mounted to the boom. It helps if you have a 2- or 3-foot-long spotting mast temporarily attached to the boom to form a 90° frame of reference. This allows the ground crew to spot from below

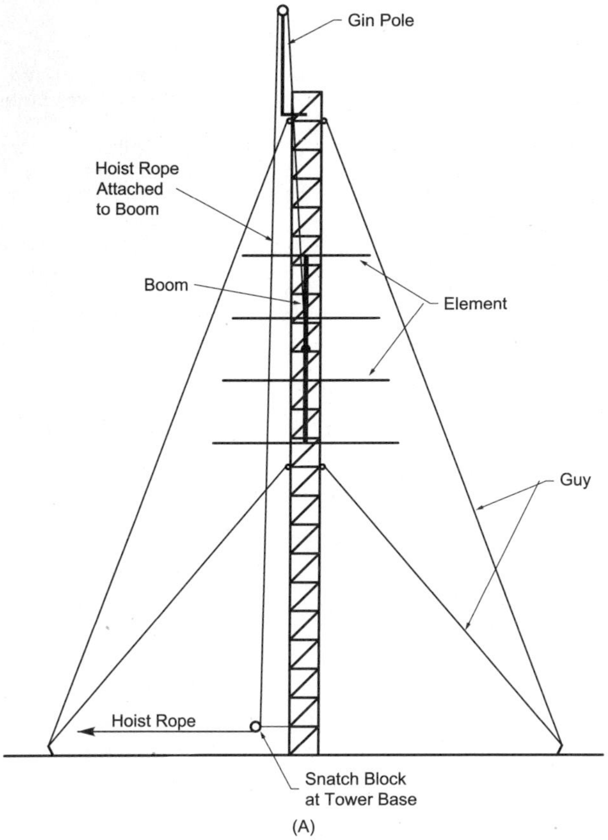

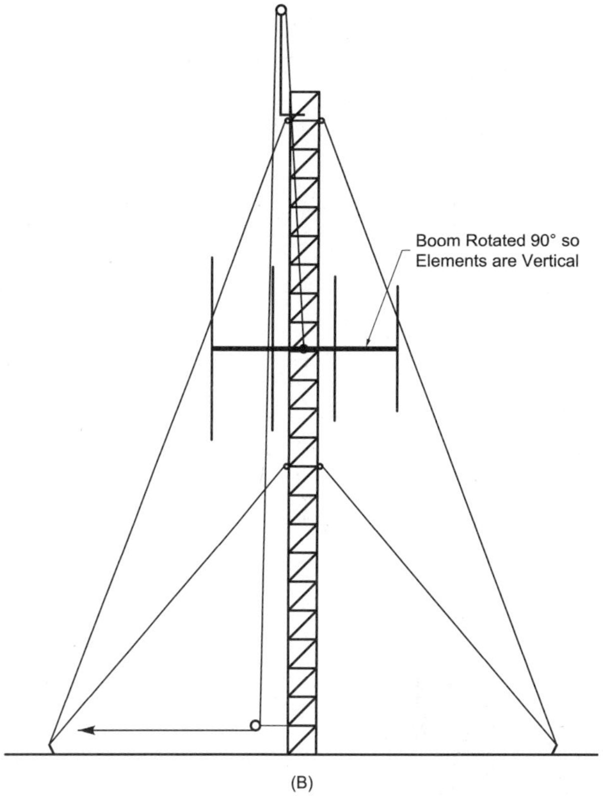

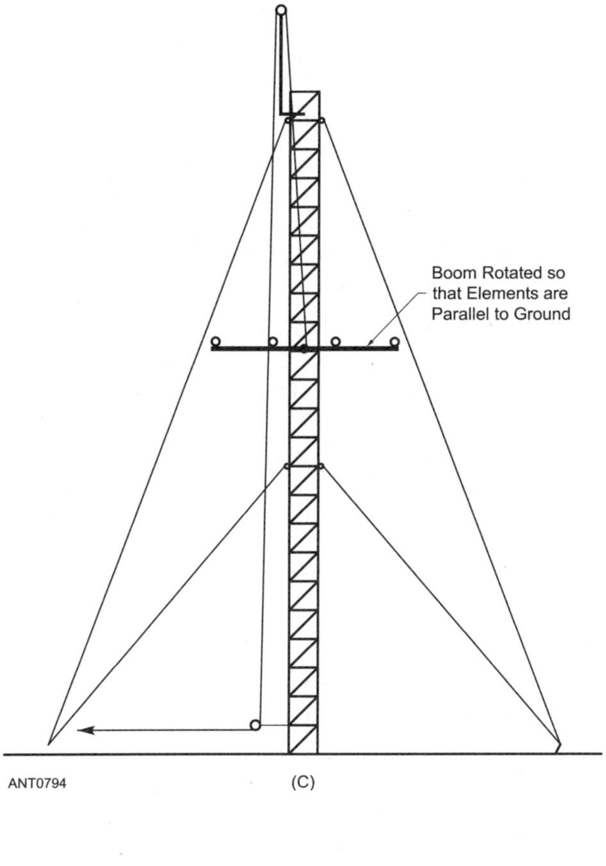

Figure 27.43 — Building a Yagi partway down the tower. At A, the boom is lashed temporarily to the tower and elements are added, starting at the bottom. At B, the temporary rope securing the boom to the tower is removed and the boom is rotated 90° so that the elements are vertical. At C, the boom is rotated another 90°, "weaving through" guy wires, if necessary, until the elements are parallel with the ground, whereupon the boom is secured to the tower.

so that the elements are all lined up in the same plane. After all the elements are mounted and aligned properly, the temporary rope securing the boom to the tower is released, suspending the antenna on the haul rope. The tower person then rotates the boom 90° so that the elements are vertical. Next, the elements are rotated 90° into the tower so that they are parallel to the ground. The ground crew then moves the completed antenna up or down using the haul rope to the final point where it is mounted to the tower.

A modification of this technique also works for building a medium-sized Yagi on the top of the tower. This technique will work if the length of the gin pole at maximum safe extension is long enough (see **Figure 27.44**). (This method is much easier if the top guy set is not exactly at the tower top, as previously mentioned. The extra working space helps immeasurably!)

As usual, the gin pole haul rope is attached to the balance point of the boom and the boom is pulled up the tower in the vertical position, using a rope to temporarily tie the haul rope to the top end of the boom for stability. The boom is temporarily secured to the tower with slings or a short length of rope in the vertical position so that the top end is just higher than the top of the tower. In order to clear the gin pole when the

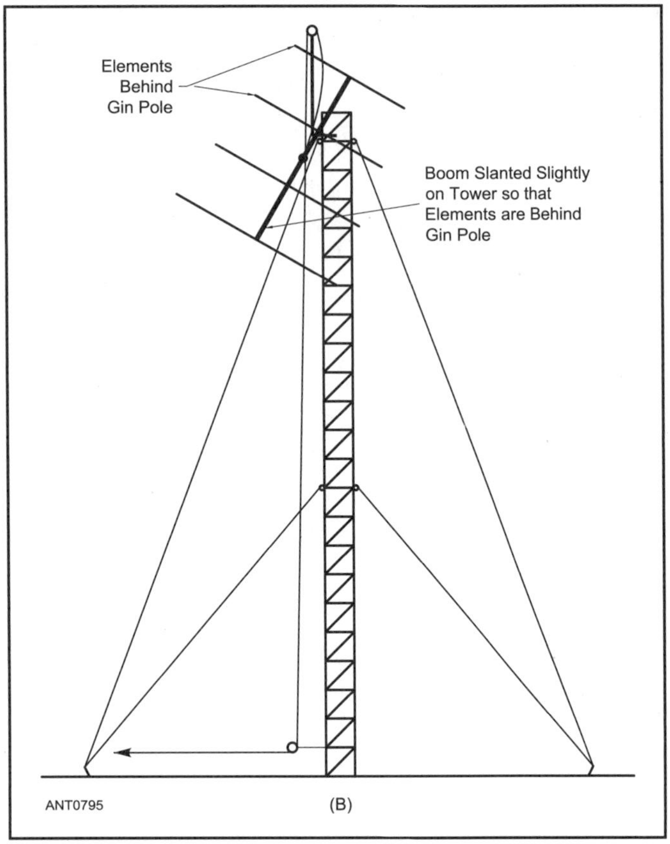

Figure 27.44 — Building a Yagi at the top of the tower. The length of the gin pole must be longer than 1/2 the boom so that the boom can be hoisted upwards to the place where it is mounted to the mast. Usually, the boom is initially lashed to the tower slanted slightly from vertical so that the top element ends up behind the gin pole. The elements are mounted at the bottom end of the boom first to provide stability. Then the element at the top of the boom is mounted and the boom is moved upwards using the gin-pole hoist rope so that the next-to-top element may be mounted, again behind the gin pole. This process is repeated until all elements are mounted (save possibly the middle element if it can be reached easily from the tower once the beam has been mounted to the mast). Then the boom is tilted to the final position, weaving the elements to clear guy wires if necessary.

elements are mounted and the boom is raised higher to mount the next element, you must tilt the boom slightly so that the element mounted to the top end of the boom will be behind the mast. This is very important!

The elements are first mounted to the bottom side of the boom to provide weight down below for stability. Then the top-most element is mounted to the boom. The tower person removes the temporary rope securing the boom to the tower and the ground crew uses the haul rope to move the mast vertically upwards to the point where the next element from the top can be mounted. Once all the elements are mounted and aligned in the same plane (the center element closest to the mast-to-boom bracket can be left off until the boom is in place), the temporary securing rope is removed. The boom is now swung so that the elements can be maneuvered to clear the top guy wires. Once the elements are horizontal the boom is secured to the mast and the center element is mounted if necessary.

A special boom-to-mast mounting plate that supports both building and working on antennas at the top of the tower was designed by members of the Potomac Valley Radio Club. It is described in a short article ("The PVRC Mount") included with this book's online material.

27.7 Notes on Cables and Connectors

The following sections contain information that applies to the construction of antenna and tower systems. For more information on the characteristics of coaxial cable and RF connectors, including cable selection, see the **Transmission Lines** chapter.

A general tip for handling cable and wire — when removing it from a reel or spool, unroll it so that it lies straight and flat along the ground. Pulling it straight up and off the end of the coil twists the cable, leading to strength-reducing kinks and endless aggravation as you attempt to get it untwisted and untangled.

27.7.1 Coaxial Cable

BENDING RADIUS

Bending coax is acceptable as long as the radius of the bend is larger than the specified minimum bending radius. For example, a common minimum bending radius specification for RG-8 is 4 inches (8 times the cable diameter). (See **Table 27.6** for minimum radii for common cable types.) Coax with more rigid shield materials will have a larger bending radius. Bending the coax tighter than the minimum bending radius can cause impedance "bumps" in the line by distorting the geometry of the conductors. It can also cause the center conductor to migrate through the plastic insulation and eventually short to the outer shield.

BURYING COAX

There are several reasons why you might choose to go to all the work of burying your coax. One is that *direct burial* cable is virtually free from storm and UV damage, and usually has lower maintenance cost than cable that is out in the open. Another reason might be aesthetics; a buried cable will be acceptable in almost all communities. Also, being underground reduces common-mode feed line current on the outside of the shield, helping to reduce interstation interference and RFI.

Although any cable can be buried, a cable that is specifically designed for direct burial will have a longer life. The best cable to use is one that has a high-density polyethylene jacket because it is both nonporous and will take a relatively high amount of compressive loads. "Flooded" direct burial cables contain an additional moisture barrier of non-conductive grease under the jacket; this allows the material to leak out, thus "healing" small jacket penetrations. (These can be messy to work with when installing connectors.)

Here are some direct burial tips:

1) Because the outer jacket is the cable's first line of defense, any steps that can be taken to protect it will go a long way toward maintaining the internal quality of the cable.

2) Bury the cable in sand or finely pulverized dirt, without sharp stones, cinders or rubble. If the soil in the trench does not meet these requirements, tamp four to six inches of sand into the trench, lay the cable and tamp another six to eleven inches of sand above it. A pressure-treated board placed in the trench above the sand prior to backfilling will provide some protection against subsequent damage that could be caused by digging or driving stakes.

3) Lay the cable in the trench with some slack. A tightly stretched cable is more likely to be damaged as the fill material is tamped.

4) Examine the cable as it is being installed to be sure the jacket has not been damaged during storage or by being dragged across sharp edges.

5) You may want to consider burying it in plastic pipe or conduit as in **Figure 27.45A** and **B**. Be careful to drill holes in the bottom of the pipe at all low spots so that any moisture can drain out. While PVC pipe provides a mechanical barrier, water incursion is practically guaranteed — you can't keep it out. It will leak in directly or condense from moisture in the

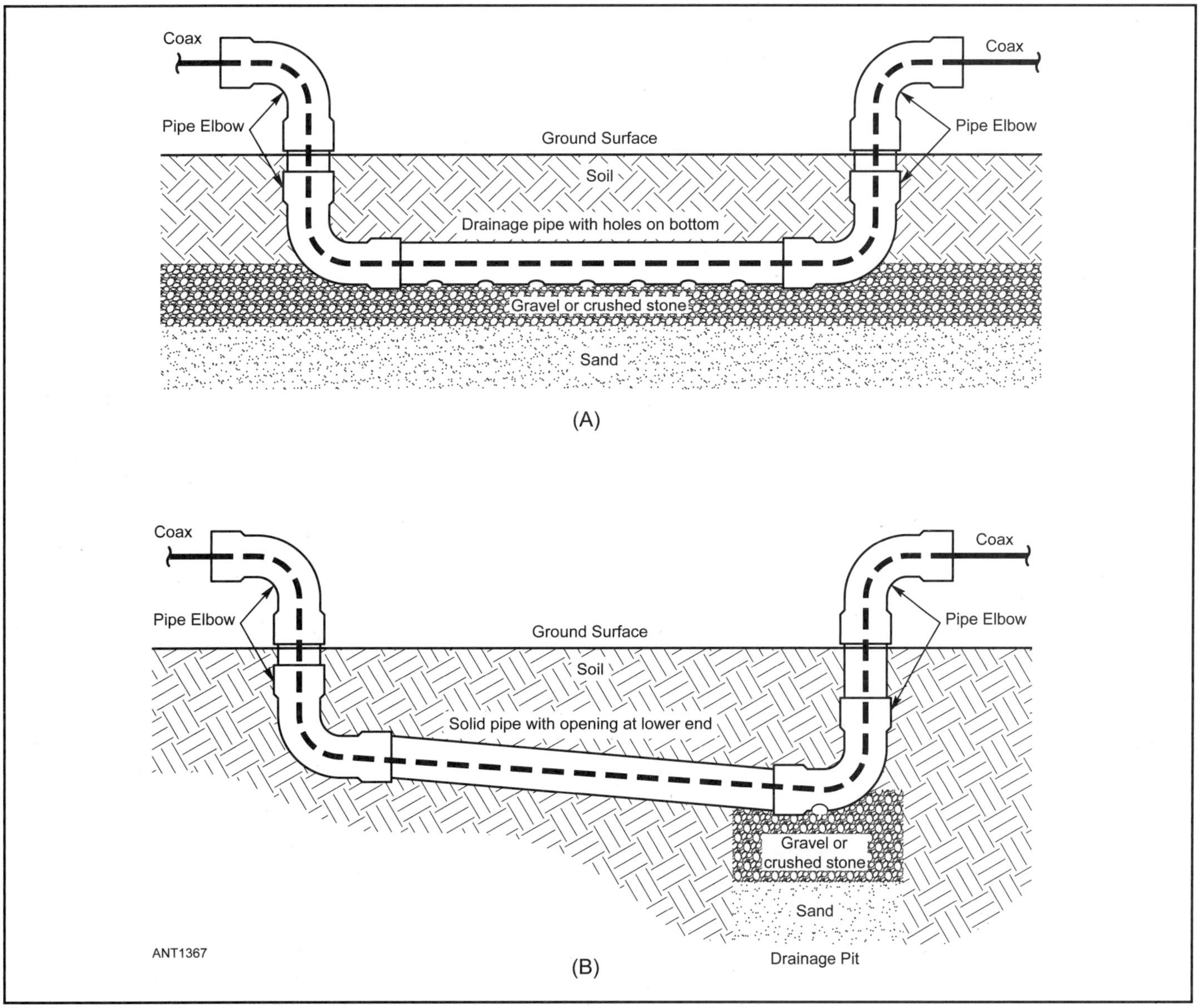

Figure 27.45 – To reduce water accumulation in buried conduit, use drainage pipe with holes on the bottom over gravel or sand (A). Another solution is to use a pitched conduit with one or more low points that drain into drainage pits (B).

Table 27.6
Physical Properties of Selected Transmission Line Types*

Cable Type (Belden p/n)	*Weight/1000 feet (pounds)*	*Max. Tension (pounds)*	*Minimum Bend Radius (inches)*
RG-213U (8267)	104	184	4.0
RG-8U (8237)	104	1909	4.5
RG-8U Foam (9913)	97	300	6.0
RG-8U Foam (8214)	106	230	4.0
RG-8X Foam (9258)	35	75	2.4
RG-58 (7807)	22	25.4	1.9

belden.com

air. Use the perforated type so that any water will just drain out harmlessly.

6) It is important that direct burial is below the frost line to avoid damage by the expansion and contraction of the earth during freezing and thawing of the soil and any water surrounding the buried cables.

COAX JUMPERS

With many beam antennas, the feed point is out of reach from the tower and should be connected to a jumper just long enough to reach from the feed point to the antenna mast. That way, the feed line connection and waterproofing can be done at the most convenient location. If you ever have to remove the antenna in the future, you can just disconnect the jumper and lower the antenna.

COAX "PIGTAILS"

Most manufacturers use some type of feed point system that accepts a PL-259 or N connector. Some antennas require you to split the coax and attach the shield and center conductor to machine screws on the driven element. The exposed end of the coax is very difficult to seal; indeed, it's nearly impossible. Water will wick down the outer shield and into your station unless you take great pains to weatherproof it. Coating the entire pigtail and attachment terminals with Liquid Electrical Tape or some other conformal sealant is a good approach, although UV will degrade such coatings over time. Another approach for HF beams is to use a "Budwig HQ-1" style insulator with the integral SO-239 and wires for connecting to the terminals (See the **Antenna Materials and Construction** chapter for alternatives.) As always, follow the manufacturer's directions.

CHEWING AND CHAFING

The plastic jacket of coax and insulation of balanced feed lines is susceptible to mechanical damage which exposes or even cuts into the inner conductors. On the ground, rodents and other animals may chew on the cable. Flooded cables (most commonly RG-6 TV cable variations) are filled with a viscous sealing compound to keep out water if the jacket is compromised. This also discourages chewing but the jacket is still compromised. There are also various sprays and lotions to make the cables taste bad. Ask a local electrician or installer what works in your area. Steel mesh sleeves (similar to cable grips) and split-loom tubing can also protect the cables where they are exposed.

Chafing can occur where a moving cable contacts a fixed surface, such as a tower leg or guy wire. This eventually wears through the plastic jacket and exposes the shield. Chafing is common where cable is wrapped around a mast or tower sleeve to form a rotator loop. The best solution is to secure the cable in a way that holds it away from surfaces it will rub against from antenna rotation or in the wind. Another solution is to place a protective sleeve or shield on guy wires or tower legs to keep them from contacting the cables.

Everyone develops their own preferences for securing cables to a mast or boom. What often works is to set the antenna system for mid-rotation and secure the cable well away from chafing surfaces. Then rotate the system 180 degrees in each direction to check for the cable coming in contact with or tightening against surfaces that will wear on the jacket.

27.7.2 Control Cables

In addition to coaxial cables, most towers will have some sort of control cable for rotators, antenna switches or other accessories. The manufacturer should provide the size that is necessary and again, you should follow their specifications.

In the case of rotator cables, some rotators are sensitive to voltage drop so bigger sizes should be used. For really long runs, some amateurs use THHN house wire or UF-Romex, (with the motor start capacitor installed at the rotator) from the local hardware store to get reasonably-priced bigger wire. Only the motor and solenoid (if used) conductors typically require the larger wire.

27.7.3 Weatherproofing RF Connectors

The primary purpose of weatherproofing is to keep moisture and contaminants out of your coaxial cable connections. Whether it is rain or condensation, water in a connector can put you off the air.

Properly sealed connector joints will be very effective and reliable in maintaining electrical and mechanical integrity. Here's how to do it as illustrated in **Figure 27.46**:

1) Install the connector correctly on the end of the coax.

2) Use pliers when attaching a PL-259 to a SO-239 or PL-258 barrel connector. Hand-tightened connections are not tight enough! Do not crimp or deform the connector.

3) Apply two wraps of premium electrical tape such as Scotch 33+ or 88.

4) Apply a layer of vapor-wrap material. Vapor-wrap is a butyl rubber material that comes in rolls or sheets and does

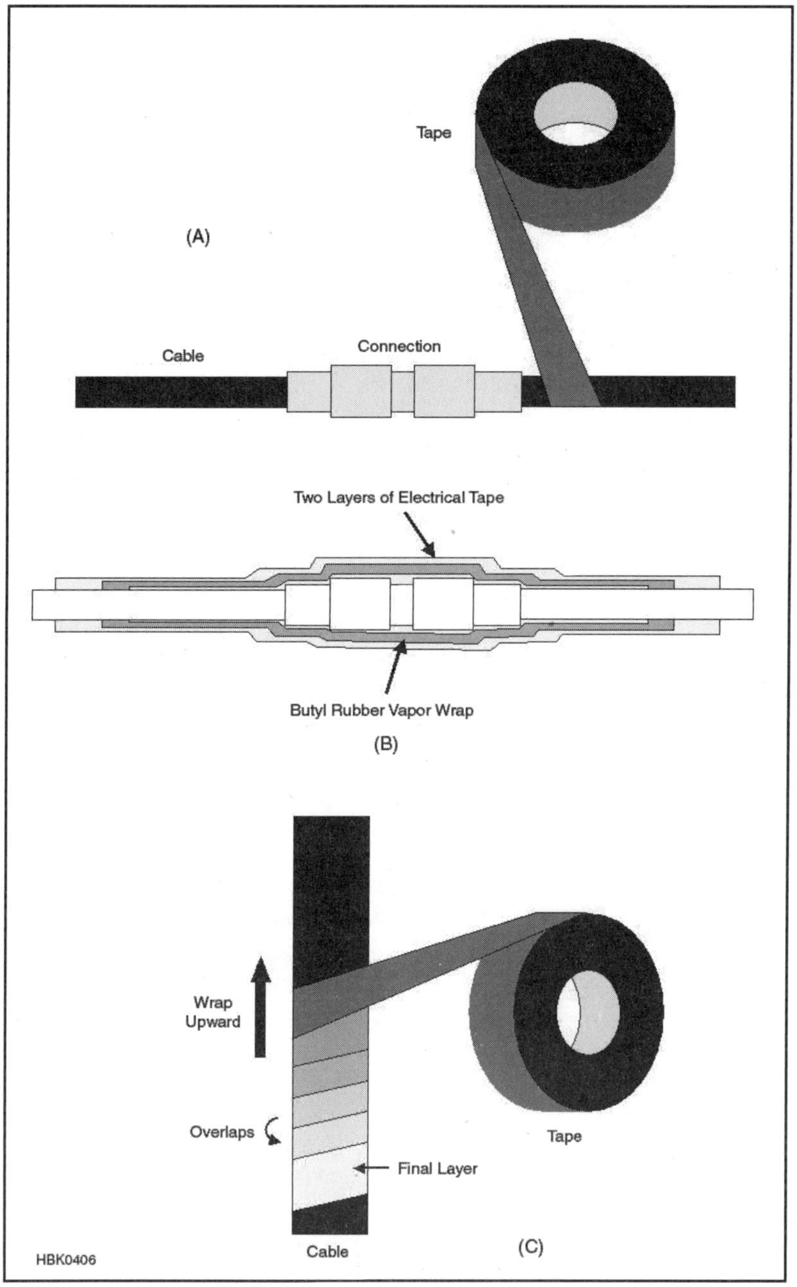

Figure 27.46 — Waterproofing a connector in three steps. At A, cover the connectors with two layers of good-quality electrical tape. B shows a layer of butyl rubber vapor wrap between the two layers of electrical tape. C shows how to wrap tape on a vertical cable so that the tape sheds water away from the connection. (Drawing (C) reprinted courtesy of *Circuitbuilding for Dummies,* Wiley Press)

an excellent job of isolating the joint from the elements. A commercial vapor-wrap, such as from Andrew Solutions or Decibel Products won't stick to connectors and comes off easily. Scotch 130C Linerless Rubber Splicing Tape also works well. By putting one or two wraps of tape over the joint first, your connector will be protected from the vapor-wrap, and it will look as good as new if you ever have to take it apart. To remove, simply take your razor knife, slice down the joint and peel off the weatherproofing.

Putty-type "coax seals" are *not* recommended as the surface can crack and dry out with age. If applied directly to a connector, the connector will become unusable as the inner putty forms a sticky mess. If you want to use putty-type sealants, wrap the connector with a layer of tape first.

5) Apply two or three layers of tape over the vapor wrap.

6) When your coax joint will be vertical, always apply the final layer of tape in an upward direction. This way the tape will overlap in such a way that water will not be conducted into the tape layers (like the shingles on your house). Tape wrapped downward will form little pockets that will trap the rain and conduct it right into your joint.

7) Do not stretch the tape when applying the final few inches. If you put it on under tension, it will eventually "flag," meaning it will come loose and blow around in the wind.

8) Paint the whole joint with a UV-resistant protective layer such as clear acrylic spray paint. That joint should never fail.

9) Install a UV-rated tie wrap above the weatherproofed connector, leaving the free end loose to guide any water running down the coax away from the connector.

SHRINK-FIT TUBING

A recent product for coax joints is shrink-fit tubing impregnated with hot-melt glue along the inside. As you apply heat to the shrink-fit tubing, it shrinks while the glue melts and oozes inside between the fitting and the tubing. It not only keeps the tubing from slipping, but it also fills in the voids in the joint and provides an additional seal. It's an expensive alternative (approximately $1 per inch) but is very simple to use and remove if necessary.

SILICONE SEALANTS

Do not use silicone sealant that gives off acetic acid (a vinegary smell) and absorbs water when curing. Acid and water will migrate into the connection causing problems later. Use only aquarium-type sealants or Dow-Corning 3145 for reliable connections. Be aware that once cured, silicone sealants are very hard to remove from connectors — practically impossible.

Giving off that vinegar-like odor means the silicone is "out-gassing," and tests have shown that once the curing period has passed, the silicone does not corrode or otherwise damage materials.

27.7.4 Tape and Ties

Every amateur installation has many feet of electrical tape used outdoors in a variety of applications. The "3 rolls for $1.00" bargain specials are not recommended for demanding outdoor use, particularly for weatherproofing. Scotch Super 88 is the recommended standard for waterproofing connectors. Besides being conformable to 0 °F (–18 °C), it will perform continuously in ambient temperatures of up to 220 °F (105 °C) and it is UV-resistant. The data sheet says it provides "moisture-tight electrical protection" and it retails in the $4 to $5 range per roll. Another Scotch tape, Super 33+, is another "premium grade, all-weather vinyl insulating tape" with many of the same properties and specs as the Super 88. The only difference is that Super 88 is slightly thicker than Super 33+ (10 mils for 88 vs 7 mils for 33+). Both tapes are easily applied at low temperatures and will even stick to a wet aluminum antenna boom.

Another specialized tape is the Scotch 130C Linerless Rubber Splicing Tape. This is a fairly thick (30 mils vs 7 mils for Super 33+) tape intended for high-voltage splices and is moisture-sealing. 3M makes many products for demanding electrical use — these are just a few of them. You may have your own favorite.

Cable ties or tie-wraps are locking plastic fasteners intended to bundle cables and secure them to brackets and other supports, such as tower legs. They come in a variety of lengths, strengths, and materials at the local hardware store. For outside work, do not use white or translucent tie-wraps; they'll deteriorate quickly from UV exposure, often in less than a year. Black, UV-resistant tie-wraps are better, but they still eventually break down. A wrap of electrical tape will protect the tie.

A tie-wrap also makes a simple drip loop for coax and control cables. Attach a medium-size tie to the cable just before it enters the building with the tie's free end pointing down.

27.7.5 Cable Support

VERTICAL RUNS

When a long run of feed line or control cable is suspended vertically, it is important to provide some strain relief so that the load is not all carried by a connector or terminal strip. Commercial cable support arms such as from KF7P Metalwerks (**kf7p.com**) are available to hold cables away from a crank-up tower or take the load of a long vertical cable run. Another popular solution is the family of *Kellems grips* that are widely available from electrical distributors such as Graybar (**graybar.com**). The smaller grips can be used with larger coaxial cables.

Using electrical tape or wire ties to secure cables to tower legs or other vertical supports will work but makes replacement more difficult since the tape or tie will have to be cut away. An alternative is to use pieces of solid #12 or #14 AWG wire from house wiring cable, twisted around the tower leg and the cables. Such twist-ties can be reused many times.

When running cables down a tower, place them where they will not be stepped on, snag a boot or belt, or compromise your grip on the tower. If possible, run them inside the tower. This also reduces RF picked up from the antennas.

HORIZONTAL RUNS

All practical runs over horizontal spans are actually in the shape of a *catenary*, the shape that a cable takes on if its weight is supported at each end. The distance between horizontal and the lowest part of the cable is called *sag*. Table 27.6 shows typical weights per 1,000 feet, maximum tension, and minimum bend radius for common cables.

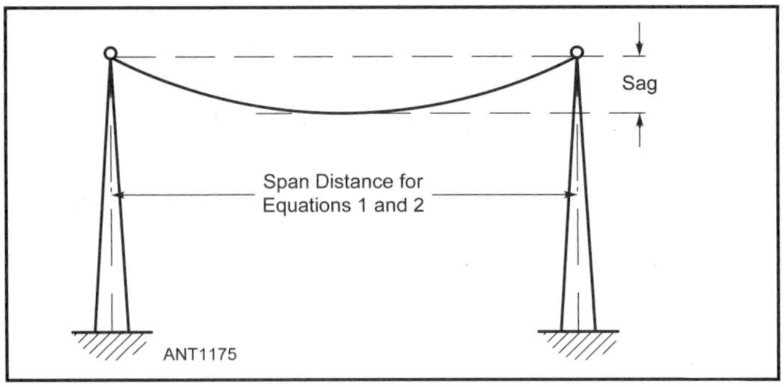

Figure 27.47 — These dimensions are used for calculating sag of wire antennas. (See text.)

Suspending a horizontal run of cables above ground is best done with a *catenary wire* or *messenger wire* anchored to supports on each end. This technique is also referred to as *aerial cable*. The suspended wire supports the weight of the electrical cables, such as feed lines and control cables. The cables can be lashed to the supporting wire (typical of CATV systems) or allowed to move in the supporting hardware.

The tools and materials are very similar to guy wires, using turnbuckles, thimbles, and Crosby clips to secure the catenary ends. The techniques and hardware used are also similar to installing overhead residential power drops and suspended lighting systems.

To determine the amount of tension in a cable or supporting wire needed for a particular amount of sag, use the dimensions in **Figure 27.47** and the formulas in the Wire Size and Tension section of the **Antenna Materials and Construction** chapter.

For amateur installations, ⅛-inch galvanized, or stainless-steel wire rope will be sufficient for typical run lengths. Commercial installers support the cables with J-hooks and bridle rings so they can move, and cables can be removed or added to the bundle. As on tower legs, #12 or #14 AWG solid wire can also be used as reusable ties.

27.7.6 Building Entry

There are already penetrations through building wall provided for utility power, telephone or data service drop lines, and coaxial cable for CATV or satellite TV service, as well as non-electrical facilities such as water, gas, sewage and heating flues. If you are the owner of the property, you have the option of observing the techniques used by those services and doing something similar for radio antenna connecting transmission lines, although some techniques may be much easier to accomplish before the dwelling is completed. Entry points for cables may be required to meet local building codes, as well.

BE SURE YOU KNOW WHAT'S BETWEEN OUTSIDE AND INSIDE

Water supply, drains, electrical service, and gas lines may

be present in exterior walls. One option to avoid them is to make all such penetrations into unfinished space, such as the wall above a basement foundation. Coming in near ground level also has potential benefit in terms of your lightning protection plan. If that is not an option and you can't be sure you don't have service lines in the wall, you may be able to drill through a solid section next to a door or window frame. A PL-259 will require a ¾-inch hole and a 12-inch long (minimum) boring bit will be required.

COMBINING GROUPS OF LINES

For groups of lines, some kind of entrance facility is likely a better idea. For example, interior and exterior electrical boxes on each side of the exterior sheathing and attached on the inside to a stud may make a professional-looking arrangement that is easy to add to just by popping off the covers on both sides. If you're not up to the task, having an hour of discussion time with a contractor or electrician may pay for itself in the long run. This entry panel is likely to be part of your lightning protection system, as well.

WINDOW ENTRANCES

The most popular method for getting cables through windows is to use a *pass-through* or *feed-through panel* such as the MFJ-4602 (**mfjenterprises.com/products/mfj-4602**). These fit under a storm window and are adjustable in length for a solid fit. Weatherproofing foam tape can be installed to seal the gap between the panel and the window. Air-conditioner seals can be used to seal around where the cables go under the inner window and to restore the upper seal between windows. Single-pane window glass can also be replaced with piece of ⅛-inch clear polycarbonate sheet which is easily drilled and will support bulkhead and feedthrough connectors.

Metal connectors through a panel may produce condensation if the inside air is humid, such as in a laundry room or basement. The resulting water drops may drip on to a windowsill or down a wall. A towel under the connectors will help catch the water but should be changed frequently. Regularly check the coax connectors for corrosion or oxidation from the condensation.

Since the window will be partially raised, the latch will be disabled. Use a piece of wood along one or both vertical sides to prevent the window from being opened. Alternately, a small angle bracket can be installed in the window track.

DRIP LOOPS

With any cable entrance arrangement, it is critical that a drip loop be provided. This is easy to do; just make sure that the cable has a lower point before it enters the entrance arrangement. Otherwise, rainwater will follow the cable down and run into the wall or the feed-through connectors. The drip loop allows a lower path from which the water will drip harmlessly onto the ground. A cable tie with the tail left hanging down at the cable's low point or where it attaches to a feed-though connector will also guide water away from the connectors.

BALANCED FEED LINES

All the same techniques can be applied to balanced lines, with just a few caveats. Balanced lines need to avoid being within a few inches of metal ducting, pipes, or wiring. Also, lossy material that will absorb energy from the fields around the line must be avoided. Perhaps most critical is inadvertent coupling to and from signal lines, including telephone, network, or alarm systems. If it is not possible to avoid these situations, it is possible to employ a section of shielded balanced line made from coax as shown in the **Transmission Lines and Connectors** chapter. While this section will likely have a different characteristic impedance than the balanced line, as long as the section is very short in terms of wavelength, it will have a minimal effect on performance.

27.8 Rotators

The rotator is an important component of directional antenna systems, turning the antenna to any direction with a repeatable accuracy of a few degrees. Once in position, the rotator must hold the antenna in place against the wind. The rotator must do this while supporting the weight of the mast and all of the antennas. **Figure 27.48** is a photo of a popular type of rotator with its control box (a Hy-Gain Ham-V rotator with a DCU-1 controller).

Rotators consist of a base assembly mounted to a fixed mast or tower and a rotating assembly atop it with a clamp in which the antenna support mast is held. **Figure 27.49** shows several common ways of attaching rotators to towers, masts, and antennas.

The turning motor and gear train, brake, position indicator, and limit switches are installed in or on the base. The rotating assembly sits on a bearing race resting on the base assembly. A ring gear is the most common method of transferring the motor's rotation to the rotating housing although worm gears are also used.

27.8.1 Rotator Ratings

Rotators are expected to work for many years over a very wide range of temperatures while exposed to the elements, with little or no maintenance. To achieve those performance goals, the rotator's specifications and installation requirements must be respected.

There are three primary rotator ratings: *wind load*, *braking ability*, and *turning torque*. *Effective moment* is the product of antenna weight and turning radius. It is a specification for maximum antenna system size used by the Hy-Gain family of rotators. Many rotators also specify a maximum *vertical load* in pounds or kilograms. **Table 27.7** lists manufacturers for rotators intended for fixed-station installations and **Table 27.8** contains the primary specifications for common rotators.

Wind load is specified both with the rotator mounted inside a tower and with the rotator outside a tower mounted on a mast. When mounted inside a tower section, the tower holds the mast in place straight above the rotator, often using a thrust bearing to hold the mast in place. This eliminates any sideways load on the rotating assembly relative to the base. Wind load is usually specified as a maximum antenna area in square feet, and antenna manufacturers specify the area of their beams for this reason. In the US, wind load calculations are based on the standard EIA/TIA-222-H. In Europe, EN 1991-1-4 is the current standard, corresponding to the German standard DIN 1055-4.

Braking ability is the maximum twisting force the rotator can withstand. This force is primarily by the wind. The rotator's braking action is provided by either a solenoid-controlled wedge or bar inside the housing, or by a worm-gear drive that does not allow backward rotation of the mast under load. Turning torque is the maximum amount of torque the rotator can produce to turn the antennas. Both braking ability and turning torque are given in inch-pounds.

Effective moment is the product of antenna weight in pounds (or kilograms) and turning radius in feet (or meters). Heavy antennas and bigger antennas are harder to turn and to hold in place against the wind, requiring a higher effective moment rating for the rotator.

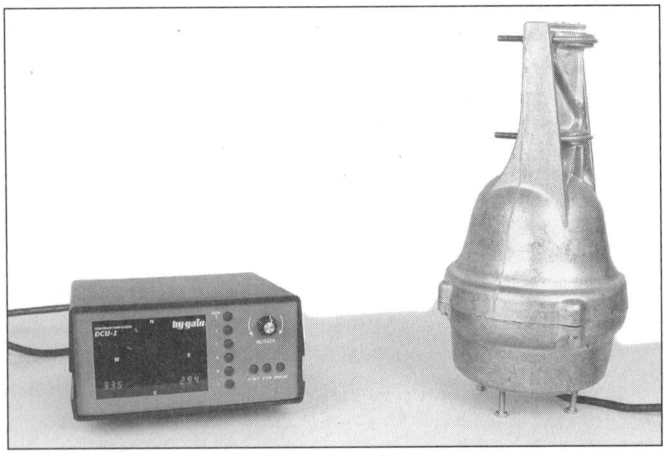

Figure 27.48 — A Ham-V rotator (right) with a DCU-1 digital control unit.

Table 27.7
Rotator Manufacturers (Fixed Station)

AlfaSpid	**alfaradio.ca**
Channel Master	**channelmaster.com**
DX Engineering	**dxengineering.com**
Hy-Gain	**hy-gain.com**
M2	**m2inc.com**
TIC	**ticgen.com**
Yaesu	**yaesu.com**

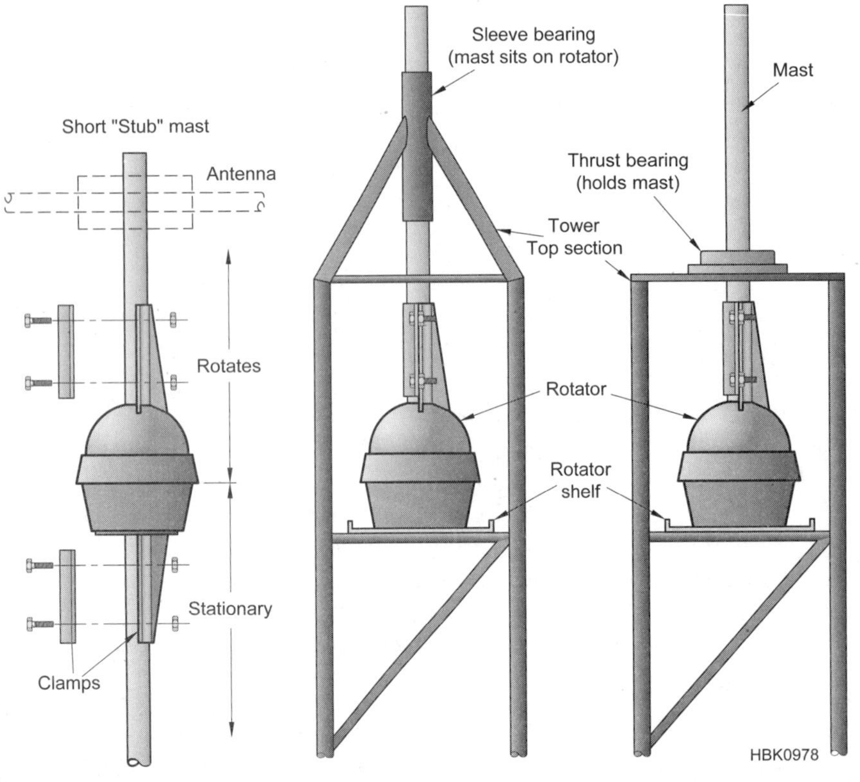

Figure 27.49 — Rotators can be mounted directly on masts or inside lattice-style towers. Rotators can also be mounted directly on the top of towers (not shown).

Table 27.8
Common Rotator Specifications

Manufacturer	*Model*	*Wind Load (in tower, sq ft)*	*Wind Load (outside tower, sq ft)*	*Turning Torque (in-lb)*	*Braking Ability (in-lb)*	*Effective Moment (ft-lb)*	*Brake Type*	*Notes*
AlfaSpid	RAK1			1400	>14,000		Worm gear	Heavy-duty
Channel Master	9521HD			100				Light-duty
DX Engineering	RT4500HD	35	n/a	4400	18,500		Gear reduction	Heavy-duty
Hy-Gain	T2X	20	10	1000	9000	3400	Wedge	Heavy-duty
	AR-40	3	1.5	350	450	300	Disc	Light-duty
	CD-45II	8.5	5	600	800	1200	Disc	Medium-duty
	HAM-IV	15	7.5	800	5000	2800	Wedge	Medium-duty
	HDR-300A	25	n/a	5000	7500		Solenoid lock	Heavy-duty
M2	OR2800	35	n/a	3200	17,000		Worm drive	Heavy-duty
TIC	1022D			6500	6500		Gear reduction	1 motor, heavy-duty
	1032D			7881	7530		Gear reduction	1 motor, heavy-duty
Yaesu	G-450A	10.8	5.4	516	2604		Gear reduction	Light-duty
	G-800DXA	21.5	8	955	3472		Mech and elec	Medium-duty
	G-1000DXA	23.7	8	955	5207		Mech and elec	Medium-duty
	G-2800DXA	32.3	10.8	2170	21700		Mech and elec	Heavy-duty
Az-El Rotators								
Yaesu	G-5500	10.8	10.8	528	3468		Gear reduction	
AlfaSpid	RAS-1	30		1400	14,000		Worm gear	At 12 V
	RAS-2			5000	24,000		Worm gear	At 12 V
	REAL-1			1400	14,000		Worm gear	At 12 V

It is important not to overload a rotator. If you live in a location that is prone to high winds, persistent winds, or large gusts, include a safety factor when selecting a rotator. Persistent twisting from winds can wear out a rotator's brake wedge or housing indentations that hold the brake in place. This can cause the brake to slip or jam. Rotators are not inexpensive, and a failed rotator brake can allow an antenna to "freewheel," damaging the feed line as well. A thrust bearing above the rotator can be used to hold the weight of a large antenna array, leaving only the turning load on the rotator.

"Rotator" or "Rotor"?

The piece of equipment installed on the tower that makes the antennas turn is a "rotator." A "rotor" is the rotary part of a motor or vehicle. For example, the blades of a helicopter form its rotor and the spinning shaft and armature of an electric motor form its rotor. The rotator includes the entire machine, both the stationary and moving parts, making the antenna system turn. Amateurs use both words, rotator and rotor, somewhat interchangeably, regardless.

27.8.2 Types of Rotators

Amateur rotators range from very light-duty models intended for TV antennas all the way to reconditioned "prop-pitch" rotators designed originally to control the pitch or angle of aircraft propeller blades. There are also models for portable and temporary use.

Light-duty antenna rotators are suitable for small VHF and UHF Yagis or log-periodics. They should not be used with HF antennas, large microwave dishes, or antennas with a significant wind load.

Medium-duty rotators can handle a single, mid-sized HF tribander or log-periodic. A small VHF/UHF Yagi can be stacked with the HF beam. These rotators are also good choices for a stack of VHF/UHF antennas. Dish antennas should be evaluated to be sure they won't overload these rotators.

Heavy-duty rotators are able to handle the biggest amateur HF Yagis, including stacks of two or three antennas. Instead of a solenoid brake, some of these rotators use a gear train or worm-drive to provide the braking action. Be sure the tower and supporting hardware is rated to handle the antenna and mast load. These rotators are somewhat larger than the more common medium-duty models and may be difficult to install in smaller lattice-style towers.

Ring rotators, or *orbital ring rotators*, are a special type of rotator installed outside the tower, attached to its legs. The antenna is carried by a motorized cradle that moves around the tower on a circular, toothed track that acts as a drive gear. Ring rotators are generally used at larger stations that use large, stacked HF Yagis.

It is also possible to simply rotate the entire tower with a variety of antennas mounted directly on the tower. The guy wires are attached to bearing rings, allowing the tower to turn inside them.

27.8.3 Rotator Installation

Rotators are usually installed inside a lattice tower section. The top section is usually where the rotator is located although it can be placed anywhere in the tower that can be reached by a mast.

A sleeve or thrust bearing at the top of the section holds the mast so that it is vertical and stabilized against sideways torque. The sleeve is part of the tower's top section. A thrust bearing is mounted on a bearing plate at the top of the tower. The rotator is mounted on a rotator shelf that sits inside the tower and is sold as an accessory by the tower manufacturer. (See **Figure 27.50**.) Smaller rotators usually come with a mast clamp so they can be mounted on a pipe or similar mast with the antenna directly above them.

The common Hy-Gain/CDE rotators (Ham-IV, T2X, and similar models) are designed to accept masts with a maximum of 2-1⁄16-inch OD. Smaller masts, such as 2-inch OD tubing or smaller water pipes will be off-center from the tower section holding the rotator. This can lead to binding in the top sleeve or bearing. The mast can be centered by placing strips of aluminum or brass shim between the mast and the portion of the clamp that is part of the bell housing.

Some crank-up tubular towers have the ability to place the rotator at the base of the tower. The tower is supported on a bearing and the rotator turn the entire tower. This is more convenient for working with the rotator but incurs extra expense.

THRUST BEARINGS

A thrust bearing is mounted at the top of the tower and the antenna mast passes through it to the rotator. The thrust bearing clamps the mast and supports the weight of the antenna system, leaving the rotator to handle the torque load without carrying any weight. Unless directed to do so by the manufacturer, do not put grease in the bearing race. Thrust bearings are designed to run "dry" as they are turned very slowly. Adding grease will capture dirt and debris which cause wear.

Except in unusual circumstances, this is unnecessary since rotator bearings are designed to operate properly at full vertical load. A substitute for an actual thrust bearing is a bushing of heavy plastic or wood that supports a collar clamped to the antenna mast.

Do not use machine shop type pillow-blocks for thrust bearings. They are not intended to be exposed to the weather and will quickly rust solid. Use only outdoor grade (Rohn, galvanized, and so on) thrust bearings. Use stainless steel bolts in the thrust bearing that center the mast so they will not rust.

The thrust bearing can be used to support the weight of the mast and antennas when the rotator is removed. A muffler clamp or U-both on the mast at the thrust bearing will hold even a heavy mast temporarily. Be sure to hold the mast vertically in the tower to keep it from tipping after it becomes top-heavy with no rotator at the bottom to hold it. **Figure 27.51** shows an example of clamps that secure the mast in a tower.

CLAMP HARDWARE AND MOUNTING BOLTS

Use galvanized or stainless-steel U-bolts and mounting bolts to install the rotator and mast clamp. Do not use plated hardware as it will rust and become very difficult to remove. If stainless-steel hardware is used, apply a small amount of

What is an Armstrong Rotator?

The term "armstrong" refers to anything turned or lifted manually, requiring a "strong arm." There are many examples of hams using gears or cranks to turn antennas but the most common is a rope tied to the antenna's boom and pulled from ground level. The antenna's mast turns freely in the tower, or the antenna mount turns on the mast. This is a common temporary solution during antenna system repair or when operating during Field Day and portable.

Figure 27.50 — The accessory shelf in a Rohn 25G top section is pre-drilled with a standard bolt layout to fit the popular Ham-M/1/2/3/4/V series of rotator base plates. This rotator uses a terminal strip for connections although rotators are available with waterproof connectors. After-market kits for converting rotators with terminal strips to waterproof connectors are also available.

Figure 27.51 — Examples of clamps that hold a mast centered in a tower. The clamp shown is the DX Engineering Mastlock (dxengineering.com/parts/dxe-mastlock) or a clamp can be homemade.

anti-seize compound to the threads before tightening to prevent galling. Split-O lock washers should be used on all bolts.

Note that muffler clamps are sized by pipe ID while tubing clamps (masts should be structural tubing) are sized by the tubing OD. A 1-3/4-inch muffler clamp will fit 2-inch tubing with a slight interference fit. A 1-7/8-inch clamp will have a very slight gap on 2-inch tubing.

Because the rotator experiences considerable and repeated twisting, it is possible for the mast clamp or mounting bolts to work loose over time. Heavy-duty aftermarket mast clamps are available for some rotators. If a rotator has worked loose on its rotator shelf or mounting plate, it can be seen from the ground through binoculars in most cases — when the rotator begins to turn, it will turn in the tower before lodging against a strut or leg or a bolt hits the side of a mounting hole. The antenna will begin to turn after that. This must be repaired quickly to avoid damage to the rotator, rotator shelf, or rotator control lines. Make inspection of rotator mounting part of your regular station maintenance.

ROTATOR OR ROTATION LOOP

Leave enough slack in all antenna feed lines for the rotator to turn completely in either direction plus another half-turn in case the mast slips during a high-wind situation. The feed lines can be left hanging below the antenna boom or wrapped around the top sleeve or mast once or twice like a spring. Have someone turn the antenna system completely through its travel with someone on the tower to watch the feed lines as the mast turns. Be sure the feed lines are not chafing against a metal surface that can wear a hole in the jacket or outer insulation, allowing water to get in. In extreme cases, a feed line or control cable can snag on something and break completely. (See "Assembling the Tower" earlier in this chapter.)

27.8.4 Rotator Control

TURNING CONTROL

There are two steps in controlling turning: releasing the brake, if any, and energizing a motor to turn in the desired direction.

An electrically controlled brake consists of a heavy-duty solenoid and a spring-loaded brake wedge or bar that fits into indentations inside the rotating assembly. The rotator's braking torque is determined by how securely the brake is held by the indentations or, if worm gears are used, by the resistance to the gears turning backward under load. To turn the rotator, the solenoid is energized, pulling the brake out of the indentations. Energizing the solenoid is usually the largest current draw of the rotator.

The most common rotator motor is a 2-phase ac motor with a starting capacitor. The capacitor is switched between phases to control direction of rotation. Because of gear reduction, the motor can be fairly small and does not draw much current. Unless blocked by an obstruction, rotators turn a full 360 degrees. Limit switches open at the ends of rotation, removing power from the motor at the extreme ends of travel to prevent feed line damage.

One Control for All

Each rotator family (Ham-IV, Yaesu, M², and so on) comes with a custom control unit for turning and position display. There are also aftermarket control units that operate with any of the common rotators. The most widely used are the Green Heron controllers **(greenheronengineering.com)** and the EA4TX interfaces **(ea4tx.com)**. Both can control most available models of rotators, allowing you to standardize in the station and customize on the tower.

ROTATOR BRAKE CONTROL

When rotation is complete, the solenoid is de-energized and the brake re-engages the indentations, holding the mast in place. Over time, the indentations or brake can wear out, allowing the rotating housing to slip under heavy loads. Wear is accelerated by de-energizing the solenoid while the mast is still turning, causing the brake to impact the sides of the indentations.

To reduce the impact of sudden stops, some controllers allow the mast to stop turning before de-energizing the solenoid. Typical delays are about 5 seconds. Retrofit delay modules are available for the Hy-Gain family of rotators. (See the RBD-5 from Hy-Gain, and rotator repair shops also have modules.) You can make your own brake delay module from the circuit in **Figure 27.52**. It is installed inside the control unit and can be fabricated on perf-board. Whenever power is applied to the rotator motors, C1 is charged and the relay contacts then close, keeping power applied to the brake and motor transformer. When power is removed from the motors, C1 discharges through the relay coil for a few seconds before the relay de-energizes, removing power from the transformer and releasing the brake solenoid. Depending on what relay you select, you may have to experiment with the value of C1

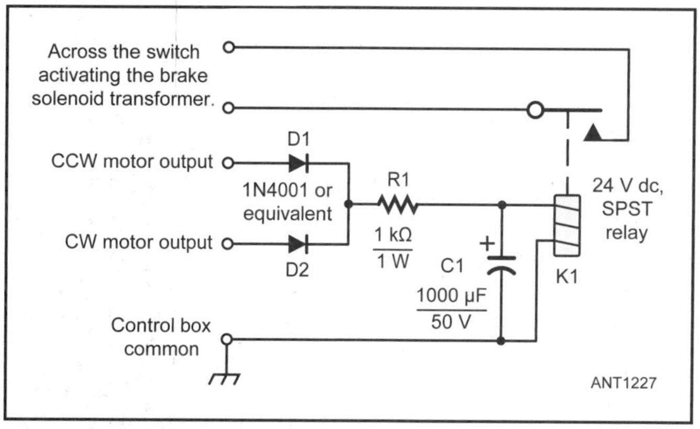

Figure 27.52 — Brake delay circuit for Hy-Gain HAM-series and Tailtwister T2X rotators. D1 and D2 charge C1 through R1, closing the contacts of K1 whenever the motor is energized. The contacts are wired in parallel with the control unit brake and motor power transformer, holding the solenoid on until the relay is no longer energized. When the motor is turned off, C1 discharges through the coil of K1 and power is removed from the brake solenoid when the relay opens.

to get the delay you want.

If your rotator does not have a brake delay, practice keeping the brake energized for a few seconds after you release the turning controls to allow the antennas to stop moving first.

POSITION INDICATION

There are two basic types of position indication — resistance and pulse counting. The most common circuit is a potentiometer ("pot") contained in the rotator's housing and turned in sync with the rotator motor. Current through the pot (typically only a few milliamps) drives an analog meter in the control unit calibrated in degrees. Rotators using pulse counters use a switch to generate pulses that the control unit counts to calculate the number of degrees from one end of travel.

Most rotators in North America are configured as "North center," meaning they can turn an antenna from pointing directly south at one limit, through north at mid-travel, and all the way to south again at the opposite limit. North is the center position on the meter displaying the antenna's direction. (South center meter scales are an option for most rotators.)

To calibrate a pot-indicator rotator's direction, assuming north center, first adjust any meter calibration controls to mid-scale. Then move the rotator to its mid-travel orientation. Loosen the antenna mast clamp and rotate the mast until the antennas point directly north and re-tighten the clamp. (Using a compass or landmark is sufficient resolution for most amateur antennas.) When storing or testing a rotator, make a practice of leaving it set to mid-travel for ease of position calibration and mark or tag it for later reference. For a pulse-count rotator, the manufacturer's manual will provide the necessary directions.

27.8.5 Rotator Wiring

The ARRL appreciates being granted permission by the Hy-Gain Company to reproduce the wiring diagram of the control unit for its widely used Ham-IV/Ham-IVX and Tailtwister T2X rotators as a convenience for readers. (See **Figure 27.53**.)

The connection from the control unit to the rotator requires a multi-conductor control cable (no shield necessary). Most rotators require either 6- or 8-conductor cable. Solenoid brake circuits need heavier wire due to the higher current. If wire is used that is too small, the extra resistance may cause enough voltage drop to result in erratic brake operation or slow turning. Check the manufacturer's recommendation for minimum wire size, which depends on the length of the cable. For the popular Ham-IV series and the Tailtwister T2X rotators, minimum recommended wire sizes are:

- Up to 125 feet: #18 AWG (brake solenoid), #20 AWG (all others)
- 125 to 200 feet: #16 AWG (brake solenoid), #18 AWG (all others)
- 200 to 300 feet: #14 AWG (brake solenoid), #16 AWG (all others)

Rotator Software Control

Recently-designed rotator control units have RS-232 or USB interfaces which are used as COM ports by PC software. There are several different protocols, the Yaesu protocol being the most common. The Rotor-EZ and ERC interfaces can be added to most rotators that don't have a software interface.

Green Heron makes a control unit that can work with almost any common rotator and can be controlled by software, as well. The Green Heron product line includes a number of rotator control accessories and consoles.

Logging software often supports several different protocols and standalone software packages and utilities are also available - Internet searches for "antenna rotator software control" will find many programs.

Going beyond PC interfaces to legacy controllers, popular microcontrollers are being employed to create "smart" rotator systems. For an example, see the SatNOGS system described in the **Antennas for Space Communications** chapter. VK1DSH describes his Arduino-based rotator controller in the July 2016 issue of *Amateur Radio*, the Wireless Institute of Australia's magazine.

Other rotators have similar requirements — consult the manufacturer's instructions.

At the rotator, connections can be made directly to a terminal strip under the unit or to a weatherproofed connector at the end of a short "pigtail." A nearby junction box where the pigtail and cable are attached can also be used. Retrofit kits to replace terminal strips are available from several vendors. The Ham-IV family suggests an 8-pin Cinch-Jones connector with wiring and color code as shown in **Table 27.9**.

For terminal strip connections, use a weatherproofing grease found at electrical and automotive stores. Run the connecting cable down and away from the rotator to guide water away from it. It is highly recommended to use a consistent color code at the rotator and the control unit.

Lightning protection is also recommended for rotator control cables. Several vendors make 8-line lightning protectors for rotators. The protector must be well-grounded and should be installed at a common entry point for all antenna system cables. See the **Safe Practices** chapter for more about grounding and lightning protection practices.

Troubleshooting a rotator electrically can be done from the ground through resistance checks. Table 27.9 shows the nominal resistance values for the Ham-IV family of rotators. When making resistance checks, the control unit must be disconnected. Include the resistance of the cable or test leads when making resistance checks.

For the Hy-Gain family of rotators, if the rotator has been turned to the limit of rotation and can't be turned in the other direction, it is possible to bypass the limit switches at the control box. Jumpering pins 4-6 and 2-5 bypass the limit switches so you can turn the rotator off the limits. This should only be done temporarily.

Table 27.9

Ham-IV/IVX Connector and Cable Wiring

Pin numbers, colors, and resistance values from Ham-IV/IVX Instruction Manual

Pin	Color	Circuit
1	Black	Solenoid and common
2	White	Solenoid
3	Green	Position pot element, + end
4	Blue	Motor winding 1
5	Orange	Right limit switch
6	Yellow	Left limit switch
7	Brown	Position pot element, – end
8	Red	Motor winding 2

Resistance Checks — Read Between Terminals		
1-2	Brake solenoid	0.75 Ω + cable or leads
1-8	1/2 Motor winding	2.5 Ω + cable or leads
1-4	1/2 Motor winding	2.5 Ω+ cable or leads
1-6	1/2 Motor + switch	2.5 Ω + cable or leads
1-5	1/2 Motor + switch	2.5 Ω + cable or leads
8-4	Entire motor	5 Ω + cable or leads
8-5	Right limit switch	0 Ω + cable or leads
4-6	Left limit switch	0 Ω + cable or leads
3-7	Entire pot element	500 Ω
3-1	Pot wiper to element end 1	0 to 500 Ω
7-1	Pot wiper to element end 2	0 to 500 Ω

Note: readings 3-1 and 7-1 should add to 3-7 reading

Keeping Rotators Rotating

Rotators are designed for years of trouble-free service as long as they are not overloaded and subjected to high winds, tightly secured to the mast and tower, and water kept away from the control cable connections and out of the motors and gears. It's your job as the station owner to check on them now and then with regular inspections and make sure the controller and wiring are working properly. Regular resistance and voltage checks at the controller are a good way to look for developing electrical problems.

If your rotator does need repair, there are several service shops that will fix up an old rotator or provide a rebuilt unit. You can also do the job yourself if you are careful and have access to replacement parts — check before opening up the housing! You can find instructions and videos online for most of the common rotators. The *CQ VHF* article "CDE/Hy-Gain Rotors, How to Keep 'em Turning, Parts 1 and 2," by W9FX in the Winter and Spring 2009 issues goes into some detail with photographs of key steps and assemblies.

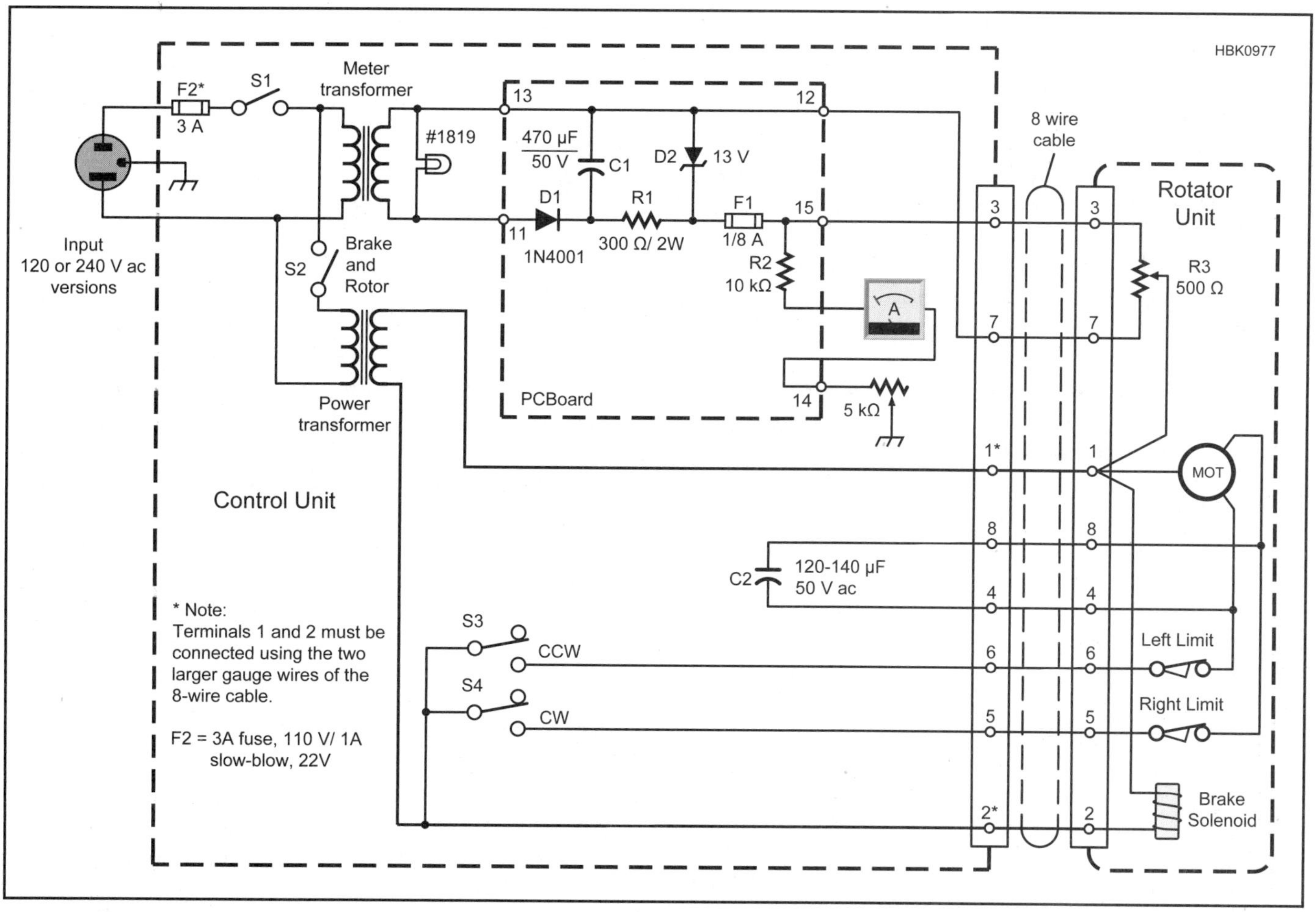

Figure 27.53 — The Hy-Gain Ham-IV/IVX control unit schematic. This control unit will also work with T2X rotators. (Circuit provided courtesy of Hy-Gain, Inc.)

27.9 Corrosion

Corrosion is one of the biggest problems in tower and antenna installations. Knowing more about it will help you to use appropriate materials and stay away from problematic combinations. For detailed information on corrosion, visit the website of Corrosion Source (**corrosionsource.com**), where there are a number of free reports and other downloadable documents on corrosion.

Any metal by itself will eventually oxidize due to exposure to the oxygen in the atmosphere. The aluminum in our antennas combines with oxygen to create the powdery aluminum oxide you find when you take an antenna apart, while oxidation of steel (which is iron) produces the rust that you want to avoid.

When two metals with the right properties are in contact in the presence of an electrolyte, *bimetallic corrosion* takes place. It's the same chemical process that takes place in batteries. Specifically, ions from one metal (called the *anodic* metal) flow across the joint or junction to the other metal (called the *cathodic* metal). In bimetallic joints, the more anodic metal is the one that loses material.

The electrolyte is typically some kind of salt or other compound (such as zinc) dissolved in water making the solution conductive. Rain (particularly acid rain), mist, or condensation are sufficient for bimetallic corrosion to begin.

Galvanically incompatible metals are combinations of metals that readily corrode when in contact because of their ranking on the galvanic chart. The farther apart the metals are in the table, the faster they will corrode when in contact. When you must use different materials, it is best to use metals that are close together in **Table 27.10**. You can see that on a zinc galvanized tower, aluminum and mild steel are the most compatible. If you use materials such as copper and brass when installing your tower ground system on a galvanized tower, you can see that you will have problems with corrosion almost immediately.

One technique for avoiding corrosion on towers is to use an intermediate corrosion-resistant material between two otherwise incompatible metals. For example, to connect a copper ground conductor to a galvanized tower, use a stainless steel washer or shim between the copper and zinc galvanizing and stainless steel hardware to hold them together. Special ground clamps are also available from tower manufacturers for connecting copper ground wires to galvanized towers.

Another technique is to use *sacrificial anodes* that give up material to prevent corrosion of the main structure. A complete discussion of this technique is beyond the scope of this chapter but the *QST* article by Tony Brock-Fisher, K1KP, "Is Your Tower Still Safe?" covers the topic well (see the References and Bibliography or this book's online material).

Table 27.10
Relative Galvanic Series in Sea Water

MORE ANODIC
- Magnesium
- Zinc
- Galvanized steel
- Aluminum
- Mild Steel
- Iron
- 50-50 lead/tin solder
- Stainless Steel
- Tin
- Nickel (active)
- Brass
- Aluminum-bronze
- Copper
- Nickel (passive)
- Silver
- Gold

MORE CATHODIC

27.9.1 Antioxidants

Various compounds are available for combating corrosion. These are *antioxidants* and most commonly-used metals such as copper, aluminum, and steel have several products designed specifically for each of them.

For aluminum antennas, most manufacturers provide a packet of antioxidant with their products. Retarding oxidation is not only a good electrical idea but the compound also functions as an anti-seize coating, aiding you in taking the antenna apart at a later date.

Antioxidants are sometimes incorrectly called "conductive pastes or greases." In general, these antioxidant compounds are comprised of a carrier material with metallic chips in suspension. It is these conductive chips, not the carrier, that give the compound its conductive properties. What happens is that the particles will pierce the layer of oxidation while preventing corrosion by isolating the joint from the air. The compound that comes with Butternut antennas, *Butter-It's-Not*, uses copper dust in a molybdenum suspension while the paste supplied by M^2 Antennas uses copper and graphite flakes in a petroleum base. There are other commercial products available for copper joints which should be used on ground systems. Just be certain to use the right one for the job. **Table 27.11** lists several compounds and their manufacturers. In addition to using antioxidants on towers and antennas, they should be used in ground system joints as well as in marine environments.

27.9.2 Rust

Steel towers and hardware will rust unless steps are taken to prevent it. In the case of towers, use galvanized steel or aluminum. Hardware, including U-bolts, nuts, bolts and other fasteners should either be made out of stainless steel (SS) or be galvanized. Because the galvanizing process deposits a thin coating of zinc on the hardware, you can't interchange SS and galvanized nuts and bolts.

Surface rust is rust that is either deposited when you have water from a rusted piece of hardware run down a surface such as a tower leg or active rust that hasn't yet penetrated the layer of galvanizing. Neither condition is serious, but you should repair those spots during your annual inspection. Use

Table 27.11
Antioxidant Compounds

Product	*Manufacturer*	*Use with*
OX-GARD	GB Electrical — gardnerbender.com	Aluminum-aluminum, aluminum-copper
NOALOX	Ideal Industries, Inc — idealind.com	Aluminum-aluminum
NO-OX-ID "A-SPECIAL"	Sanchem, Inc. — sanchem.com	Steel rust preventative
Penetrox	Burndy — hubbell.com/burndy	Aluminum-aluminum, aluminum-copper
DE-OX	ILSCO Corporation — ilsco.com	Aluminum-aluminum, aluminum-copper

a wire brush to scrub off the rust and then spray the spot with a cold-galvanizing paint. *Cold galvanizing* paint is available at almost any spray paint rack. Check the contents to make sure that it contains zinc. The LPS Company (**lpslabs.com**) makes a very good cold-galvanizing spray that is relatively expensive but adheres very well (LPS Cold Galvanize Corrosion Inhibitor 00516).

27.10 General Maintenance

Having invested time and money installing your dream antenna and tower system, you'll need periodic preventive maintenance (PM) and inspections. The key is to catch anything before it becomes a problem.

If you've followed the directives and steps described in this chapter you've already taken the most important steps in ensuring the safety and reliability of your tower and antenna system. Following the manufacturer's specifications, using the right hardware, using antioxidants and following conservative designs are the true keys to success.

27.10.1 Annual Inspection

An annual inspection is a critical part of your PM program. Most commercial companies do it religiously; many insurance companies require it as a condition of their coverage. An annual inspection entails examining everything in the tower and antenna system, including the ground system, concrete anchors and footings, and the tower structure. In addition to annual inspections, all installations should be inspected after ice storms or windstorms that exceed 60 mph.

You should correct any problems you discover in your inspection. If you're not sure about the seriousness of something you've found, talk to a knowledgeable friend or contact the manufacturer for advice. When you do a tower inspection, you should have enough supplies to redo several coax connectors joints if necessary, as well as a note pad and pencil to write down any discrepancies that may require further action. You'll be able to take care of most problems on the spot, along with knowing what else you may need to finish the repairs. Push and pull on antennas and appurtenances (anything that's attached to the tower) to see if anything is loose. Something might look okay but pushing on it might reveal loose hardware or some other problem.

You should get in the habit of doing a quick visual check every time you climb the tower. Carry a wire brush, a can of cold-galvanizing spray, a roll of electrical tape, and a utility knife to perform small repairs along the way. A station notebook of relevant information found during inspection, along with exceptions and repairs is a handy reference item. The information that follows is based on commercial and TIA-222 tower inspection standards.

TOWER STRUCTURE

1) Check for damaged or faulty tower legs and braces. With welded towers such as Rohn 25G and 45G, the members cannot be replaced without replacing the whole section; minor bends or damage that do not alter the structural integrity can usually be tolerated.

2) Check all welds for integrity.

3) Examine the condition of the finish and any corrosion. Look for rust patches; use a wire brush and cold galvanizing paint to repair it.

4) In addition to visually checking any bolted connections, you should put a wrench to at least 10% of them to check for tightness. Any loose nuts or bolts should be retightened. Also look for missing hardware and replace it immediately.

TOWER ALIGNMENT

1) The tower should be checked for plumb. A guyed tower is allowed a maximum deviation of one part in 400, or three inches per 100 feet. While a transit is the best way to check tower alignment, an electronic level will give you 0.1° accuracy, or a bubble level will indicate relative plumb. Even simpler is a long piece of string with a weight on the end, held an arm's length away from the tower; sight the string along the tower leg for a very quick and fairly accurate indication of tower plumb. For self-supporting towers, the allowed deviation allowed is 1 part in 250 or 4.8 inches in 100 feet.

Even if the tower is perfectly plumb, perform all of the guy wire and hardware checks that follow to be sure nothing is loose or about to fail.

2) Check the guy wires and guy insulators, using binoculars

for the ones that aren't close to the ground or the tower.

3) Examine all guy wire and guy wire hardware including preformed grips, turnbuckles, clamps, and shackles for damage. Make sure that all turnbuckle safety ties are intact. Make sure all bolts and nuts are tight.

4) Check guy wire tension with an instrument or another technique.

5) Examine the tower base and guy anchors. Look for any cracking of the concrete. Also look for evidence of movement in the soil of the anchor rods or base. Look for rust and/or corrosion. Excavate a buried anchor rod for twelve inches to inspect for hidden corrosion — some sources recommend inspecting anchor rods all the way to the concrete anchor.

ANTENNAS, CABLES, AND APPURTENANCES

1) Inspect antenna, boom-to-mast bracket and boom truss hardware for loose or missing hardware. Test nuts for tightness.

2) Look at each feed point joint and coax cable joint for compromised weatherproofing.

3) Check all cables for abrasion, binding and attachment.

4) Examine all appurtenances for missing hardware or corrosion.

ROTATOR

1) Check that all mounting bolts are tight and that they are not slipping in the rotator shelf or plate.

2) Check that the rotator mast clamp is securely holding the mast.

GROUNDING SYSTEM

Do a visual inspection of the grounding system. Inspect all clamp connections. Redo any connections that are corroded or loose.

27.10.2 Crank-Up Maintenance

Crank-up towers are complex mechanical contrivances. While some are hand cranked, many have a motor, gearbox, cables, pulleys and limit switches — all of which should be carefully inspected twice a year.

The electric motors and gearbox are generally bulletproof, and the only inspections are to check the oil level in the gearbox, the condition of the drive belt or chain (some sort of conditioner is helpful for each), and the operation of the cable drum (there are probably some Zerk grease fittings that need attention).

Pulleys are sometimes custom made by the manufacturer so you may not be able to run down to the local store and buy one. Some sheaves are made by the manufacturer and then an off-the-shelf bearing is inserted in the middle. This one you probably can replace.

Pulleys need to turn and not bind so a good thing to do is to watch the pulleys if they're exposed enough while the tower is being raised or lowered and see if there are any problems. (A simple dot or line drawn on the sheave itself will quickly show, even at a distance, whether it's turning or not.)

CRANK-UP CABLES

Crank-up towers depend almost entirely on their cables to operate reliably and safely. Exercise the cables by running the tower up and down a couple of times a month and don't always leave the tower in the same spot all the time, for example at the limit switches. Over time the cable can take a set if it's always at the same place so leaving it at different places spreads the wear over much more of the cable length.

The cables should be lubricated at least annually; twice a year would be even better. Do not use heavy grease or motor oil which will just attract grime and particles. Use a cable lubricant such as PreLube 6 and be sure to check for damage while you're doing the lube job. If you see any of the following, the cables should be replaced:

1) Damage in which a cable is significantly kinked or flattened.

2) Rust. This means serious rust, not surface rust that can be easily wiped or scraped off.

3) Excessive broken strands. Most crank-ups use 7×19 galvanized cable which means it has 133 strands in it. You're allowed to have six total broken strands and three in the same bundle before replacing the cable.

27.10.3 Rotator Maintenance

Most rotator problems are first noticed as misalignment of the antennas with regard to where the control box indicator says they are pointing. With a light duty rotator, this happens frequently when the wind blows the antenna to a different heading. With no brake, the force of the wind can move the gear train and motor of the rotator, while the indicator remains fixed. Such rotator systems have a mechanical stop to prevent continuous rotation during operation, and provision is usually included to realign the indicator against the mechanical stop from inside the station. During installation, the antenna must be oriented correctly for the mechanical stop position, which is usually North.

In larger rotator systems with an adequate brake, indicator misalignment is caused by mechanical slippage in the antenna boom-to-mast hardware. Many texts suggest that the boom be pinned to the mast with a heavy-duty bolt and the rotator be similarly pinned to the mast. There is a trade-off: If there is sufficient wind to cause slippage in the couplings without pins, with pins the wind could break a rotator housing or transmission parts. The slippage will act as a clutch release, which may prevent serious damage to the rotator. On the other hand, you might not like to climb the tower and realign the system after each heavy windstorm.

27.10.4 When Something Fails

Failures to your installation can come in many forms, but wind is generally the common denominator. Rust, metal fatigue and overloading aren't usually a problem until the wind starts to blow. Other causes of failure could be lightning strikes, ice, vandalism or accidents.

ASSESS THE DAMAGE

The first thing to do is a visual inspection. Using binoculars, if possible, take a look at everything from the ground to see if anything is bent or broken. If something is swinging in the wind, that's a major problem. If there is obvious damage, try to determine if it is in danger of falling. If so, evacuate the endangered area immediately and alert local emergency services. This is especially true if it looks as though it could fall on power lines, sidewalks or roadways. If you have damage that isn't an imminent danger to life or property, keep an eye on it until the storm is over to ensure that it doesn't get worse. If you have the opportunity, take some snapshots or video of the damage for documentation.

PREVENT FURTHER DAMAGE

Your next task is to take prudent steps to prevent further damage, both to your property and to the property of others. This is not only common sense but also a requirement of the insurance company. You want to avoid or minimize the possibility of liability lawsuits for personal injury or the property damage of others. Tie anything off you can but *do not* attempt to climb the tower!

FILE AN INSURANCE CLAIM

After the storm is over, call your homeowner's or renter's insurance agent and notify them of the loss. Do it orally first, then follow up with a letter. The insurance company may require a "Proof of Loss." They'll give you a claim number that you'll need to use in all written and verbal communications. Start a file with all your documentation, plus the other paperwork that you'll start accumulating. (See the article on insurance by Ray Fallen, ND8L, included with this book's online material.)

Keep notes of every conversation with your insurance agent or claims adjuster with dates and times; you may have to refer to them in the future. At this point, you may want to write down all pertinent facts surrounding the loss for reference also. Send copies of your photos with your loss letter.

ESTIMATE OF REPAIRS

You'll make things very easy for your claims adjuster if you include an estimate of repair along with your letter and photos. The adjuster has probably never handled a radio tower loss before and will appreciate your help in getting a quote. Contact your local commercial rigger or antenna installation company and they'll give you the quote.

Insurance companies will want professional workers to perform professional repairs to your loss; they expect to pay the going rate and they expect licensed contractors to do the work. Be sure that your estimate for tower repair covers *all* of the work including: dismantling damaged parts, hauling away damaged parts and disposal, clean-up, labor for reinstallation including assembly of antennas, labor for reinstallation of tower, replacing all damaged materials including hardware, cables, rotators, and other items.

Don't be surprised if the estimate comes in quite a bit higher than you expect. Not only are you paying professionals to do all of the work, but a damaged tower or antenna system can be hazardous, and a crane or other piece of equipment may be needed to remove it safely.

STAY IN YOUR COMFORT ZONE

Needless to say, don't consider getting involved in the removal and repair of the damage unless you feel comfortable with it. If there is *any* doubt at all in your mind, either get the professionals in or bring in a piece of equipment such as a crane or boom truck. If anything is at a precarious or dangerous angle, don't touch it — send for the professionals!

27.11 References and Bibliography

Source material and more extended discussions of the topics covered in this chapter can be found in the references listed below.

Encyclopedia of Anchoring, Hubbell Power Systems, **hubbellcdn.com/literature/CENT-CP-ENCYCLOPEDIA-OF-ANCHORING-CAT-EN_09_2022_Web.pdf**

Rohn Product Catalog 3, **www.rohnnet.com/rohn-products-catalog**

Brede, G., "The Care and Feeding of the Amateur's Favorite Antenna Support — The Tree," *QST*, Sep. 1989, pp. 26 – 28, 40.

Daso, D., *Antenna Towers for Radio Amateurs* (ARRL, 2010).

Daso, D., "Workshop Chronicles," columns in *National Contest Journal.*

Hoerner, S., "Fluid Dynamic Drag," (Hoerner Fluid Dynamics, 1993), pp. 1 – 10.

Hutchinson, C., Straw, R., *Simple and Fun Antennas for Hams* (ARRL, 2002).

Keown M., and Lamb, L., "A Simple Technique for Tower-Section Separation," *QST*, Sep. 1979, pp. 37 – 38.

Morris, S., *Up the Tower* (Champion Radio Products, 2009).

Phillabaum, S., "Installation Techniques for Medium and Large Yagis," *QST*, June 1979, pp. 23 – 25.

Pioveson, B., "CDE/Hy-Gain Rotors, How to Keep 'em Turning, Parts 1 and 2," *CQ VHF*, Winter 2009, pp. 38 – 46, and Spring 2009, pp. 18 – 26.

Richards, C., "Mechanical Engineering in Radar and Communications" (Van Nostrand Reinhold Co., 1969), pp. 162 – 165.

Schiller, T., *Array of Light*, Fourth edition, **n6bt.com**.

Shanblatt, M., and Warwick, J., "Hanging a Treetop Antenna with a Drone," *QST*, Mar. 2018, pp. 67 – 68.

Sirageldin, K., "Using a Drone for Antenna Installation at TI5W," *QST*, Nov. 2017, pp. 85 – 86.

Smith, C., *Carroll Smith's Nuts, Bolts, Fasteners and Plumbing Handbook* (Motorbooks, 1990).

Weber, D., "Determination of Yagi Wind Loads Using the Cross-Flow Principle," *Communications Quarterly*, Spring 1993.

White, B., White, E., and White, J., "Assembling Big Antennas on Fixed Towers," *QST*, Mar. 1982, pp. 28 – 29.

Young., W., "Roark's Formulas for Stress & Strain" (McGraw-Hill Co., 1989), pp. 67, 96.

Structural Standards for Steel Antenna Towers and Antenna Supporting Structures, TIA Standard TIA-222-G, Telecommunications Industry Association, Aug 2005 (**tiaonline.org**). May be purchased from IHS/Global Engineering Documents, 15 Inverness Way East, Englewood, CO, 80112-5704, 1-800-854-7179 (**global.ihs.com**).

Appendix A

Determining Antenna Areas and Wind Load

The method for determining the flat projected area (FPA) of an antenna is quite simple. We'll use a Yagi antenna as an example. There are two worst-case areas that should be considered here. The first is the FPA of all the elements when the wind blows in the direction along the boom; that is, at right angles to the elements. The second FPA for a Yagi is when the wind is at right angles to the boom. One of these two orientations produces the worst-case exposed antenna area —all other wind angles present lower exposed areas. The idea is to take the highest of the FPAs for these two wind directions and call that the FPA of the antenna structure. See **Figure 27.54A.**

The element FPA is calculated by multiplying each element's dimension of length by its diameter and then summing the FPAs for all elements. The boom's FPA is computed by multiplying the boom's length by its diameter.

The reason for considering two potential peak-load orientations becomes clear when different frequency antennas are stacked on a mast or tower. Some antennas produce peak loads when the elements are broadside to the wind. This is typical of low-frequency Yagis, where the elements are very long lengths of aluminum tubing. On the other hand, the boom can dominate the surface area computations in higher-frequency Yagis.

The fundamentals responsible for the need to examine both potential FPAs for Yagis relates to how wind flows over a structure and develops loads. Called *The Cross-Flow Principle*, this was introduced to the communications industry by Dick Weber, K5IU, in 1993. The principle is based on the fact that the loads created by wind flowing across an antenna member only produce forces that are normal to (or perpendicular to) the major axis of the member. The resultant and component load calculations for this method are shown in Figure 27.54A.

For a Yagi, this means that wind forces on the elements act in-line with the boom, while forces on the boom act in-line with the elements. Figure 27.54B shows a force diagram for a typical Yagi. Figure 27.54C shows the FPA for a Yagi rotated through 90° of azimuth.

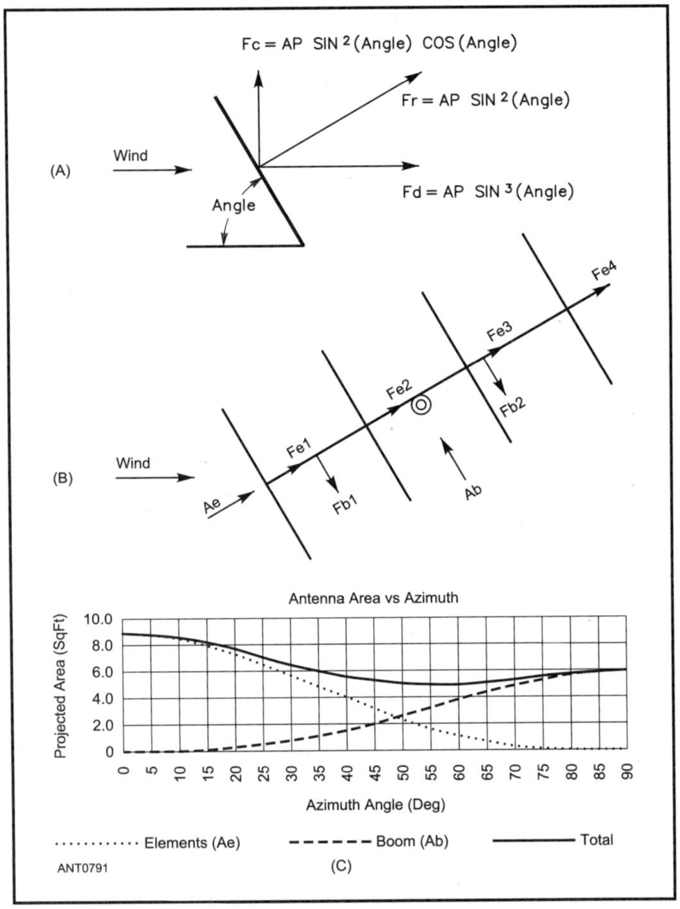

Figure 27.54 — Description of how loads are developed on a Yagi. At A, F_r is the resultant force from the wind load on a generalized member. F_d is the load acting downwind (drag) that creates the load on the tower. F_c is the lateral component of the wind load. The term A is the flat projected area (FPA), which is the broadside area normal to the wind. The term P is the wind pressure. At B, A_e is the total element area, while Ab is the total boom area. All the loads due to the wind act normal to the antenna sections—the force on element #1 (F_{e1}) acts along the axis of the boom, for example. At C, a plot of the effective FPA as a function of the azimuthal wind direction for a Yagi, ignoring drag coefficients. The Yagi in this example has 9.0 square feet of element FPA and 6.0 square feet of boom FPA. The worst-case FPAs occur with the beam pointed in the wind and with the boom broadside to the wind. To determine the actual tower loading, the actual drag coefficients and wind pressures must be used.

Antenna Placement on the Mast/Tower

Another important consideration is where the antenna(s) will be placed on the tower. As mentioned before, most generic tower specifications assume that the entire antenna load is applied at the top of the tower. Most amateur installations have a tubular mast extending above the tower top, turned by a rotator mounted down inside the tower. Multiple Yagi antennas are often placed on the mast above the tower top, and you must make sure that both the tower and the mast can withstand the wind forces on the antennas.

For freestanding towers, you can determine how a proposed antenna configuration compares to the tower manufacturer's rating by using an *Equivalent Moment* method. The method computes the bending moment generated at the base of the tower by wind loads on the tower's rated antenna area located right at the top of the tower and compares that to the case when the antenna is mounted on a mast sticking out of the top of the tower.

The exact value of wind pressure is not important, so long as it is the same for both comparisons. The wind load on the tower itself can be ignored because it is the same in both comparisons and the drag coefficients for the antennas can also be ignored if all calculations are performed using flat projected antenna areas, as we've recommended previously.

Keep in mind that this approach does not calculate *actual* loads and moments relevant to any specific tower design standard, but it does allow equivalent comparisons when the wind pressure is constant and all the antenna areas are of the same type. An example is in order.

Figure 27.55A shows a generic tower configuration, with a concentrated antenna load at the top of the tower. We'll assume that the tower manufacturer rates this tower at 20 square feet of flat projected antenna area. Figure 27.55B shows a typical amateur installation with a rotating mast and an antenna mounted 7 feet above the top of the tower. To make the calculations easy, we select a wind pressure of 1 pound per square foot (1 psf). This makes the tower base moment calculation for Figure 27.55A:

Antenna load = 20 feet^2 × 1 psf = 20 pounds

Base moment = 70 feet × 20 pounds = 1400 foot-pounds.

This is the target value for the comparison. An equivalent configuration would produce the same base moment. For the configuration in Figure 27.55B, we assume a tubular 2-inch diameter mast that is 20 feet long, mounted 5 feet down inside the tower. Note that the lattice structure of the tower allows the wind to "see" the whole length of the mast and that we can consider the force distributed along the mast as being a single force concentrated at the mast's center. The flat projected area of the mast by itself, without the antenna, is:

Mast area = 20 feet × 2 inches / 12 inches/foot = 3.33 square feet

The center of the mast is located at a height of 75 feet. Using the same 1-psf-wind load, the base bending moment due to the mast alone is:

Base moment (due to mast) = 3.33 feet^2 × 1 psf × 75 feet = 249.75 foot-pounds

Including the mast in the configuration reduces the allowable antenna load. The remaining target base moment left for the antenna is found by subtracting the moment due to the mast from the original target value:

New base target moment = 1400 – 249.75 foot-pounds = 1150.25 foot-pounds.

The antenna in Figure 27.55B is located at a height of 77 feet. To obtain the allowable antenna area at this elevation we divide the new base target moment by the antenna height, yielding an allowable antenna load of:

1150.25 foot-pounds / 77 feet = 14.94 pounds.

Since we chose a wind load of 1 psf, the allowable antenna FPA has been reduced to 14.94 square feet from 20 square feet. If the projected area of the antenna we are planning to

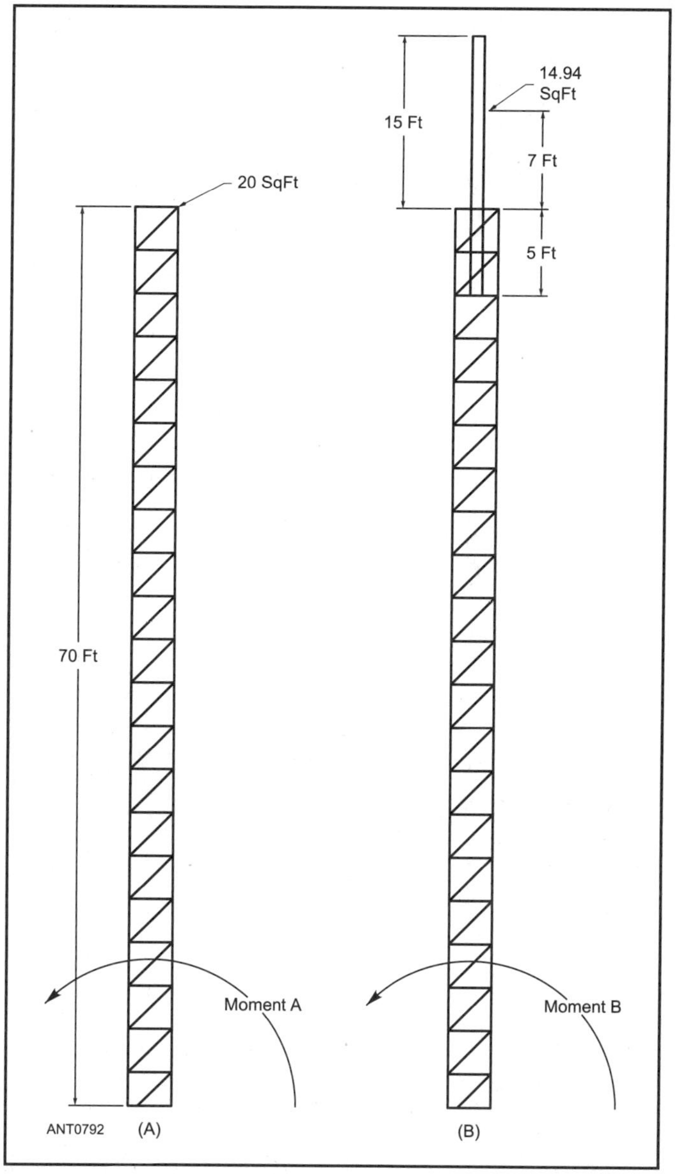

Figure 27.55 — At A, a 70-foot tower rated for 20 square feet of antenna load at the top. At B, the same tower with a 2-inch OD, 20-foot-long mast, with an antenna mounted 7 feet above the top of the tower. Both configurations produce the same tower load.

mount in the new configuration is less than or equal to this value, we have satisfied the requirements of the original design. You can use this equivalent-moment method to evaluate different configurations, even ones involving multiple antennas on the mast or situations with additional antennas placed along the tower below the tower top.

For guyed towers, the analyses become much more rigorous to solve. Because the guys and their behaviors are such a significant portion of the tower support mechanism, these designs can become very sensitive to antenna load placements. A general rule of thumb for guyed towers is never to exceed the original tower-top load rating, regardless of distributed loads along its length. Once you redistribute the antenna load placements along a guyed tower, you should do a fresh analysis, just to be sure.

You can run evaluations using the above method for antennas placed on the mast above a guyed tower top. The use of the Equivalent-Moment method for antennas mounted below the top of a guyed tower, however, can become quite suspect, since many generic tower designs have their intermediate guys sized for zero antenna loads lower down the tower. The proper approach in this case is to have a qualified mechanical engineer check the configuration, to see if guy placement and strength is adequate for the additional antennas down the tower.

Mounting the mast and antenna as shown in Figure 27.55B increases tower loads in the region of the mast. You should investigate these loads to ensure that the tower bracing in that area is sufficient.

Appendix B

Calculating the Required Mast Strength

When you mount antennas on a mast above the tower top, you should examine the bending loads on the mast to ensure that it will be strong enough. This section explains how to perform mast stress calculations for a single sustained wind speed. This procedure does not include height, exposure and gust-response factors found in most tower design standards.

Here are some fundamental formulas and values used to calculate the bending stress in a mast mounted in the top of a tower. The basic formula for wind pressure is:

$$P = 0.00256\ V^2 \quad (1)$$

where

P is the wind pressure is in pounds per square foot (psf)

V = wind speed in miles per hour (mph)

This assumes an air density for standard temperature and atmospheric pressure at sea level. The wind speed is not the Basic Wind Speed discussed in other sections of this chapter. It is simply a steady state (static) wind velocity.

The formula for calculating the force created by the wind on a structure is:

$$F = P \times A \times C_d \quad (2)$$

where

P = the wind pressure from Eq 1

A = the flat projected area of the structure (square feet)

C_d = drag coefficient for the shape of the structure's members.

The commonly accepted *drag coefficient* for long cylindrical members like the tubing used for the mast and antenna is 1.20. The coefficient for a flat plate is 2.0.

The formula used to find the *bending stress* in a simple beam like our mast is:

$$\sigma = (M \times c) / I \quad (3)$$

where

σ = the stress in pounds per square inch (psi)

M = *bending moment* at the base of the mast (inch-pounds)

c = ½ of the mast outside diameter (inches)

I = *moment of inertia* of the mast section (inches^4)

In this equation you must make sure that all values are in the same units. To arrive at the mast stress in pounds per square inch (psi), the other values need to be in inches and pounds also. The equation used to find the moment of inertia for the round tubing mast section is:

$$I = \pi/4\ (R^4 - r^4) \quad (4)$$

where

I = Moment of Inertia of the section (inches^4)

R = Radius of tube outside diameter (inches)

r = Radius of tube inside diameter (inches)

This value describes the distribution of material about the mast *centroid*, which determines how it behaves under load. The equation used to compute the *bending moment* at the base of the mast (where it is supported by the tower) is:

$$M = (F_M \times L_M) + (F_A \times L_M) \quad (5)$$

where

F_M = wind force from the mast (pounds)

L_M = Distance from tower top to center of mast (inches)

F_A = Wind force from the antenna (pounds)

L_A = Distance from tower top to antenna attachment (inches)

L_M is the distance to the center of the portion of the mast extending above the tower top. Additional antennas can be added to this formula by including their F × L. In the installation shown in Figure 27.55B, a wind speed of 90 mph, and a mast that is 2 inches OD, with a 0.250-inch wall thickness, the steps for calculating the mast stress are:

1) Calculate the wind pressure for 90 mph, from Eq 1:

$P = .00256\ V^2 = .00256 \times (90)^2 = 20.736$ psf

2) Determine the flat projected area of the mast. The portion of the mast above the tower is 15 feet long and has an outside diameter of 2 inches, which is 2/12 feet.

Mast FPA, $A_M = 15$ feet × (2 inches / 12 inches/feet) = 2.50 square feet.

3) Calculate the wind load on the mast, from Eq 2:

Mast Force, $F_M = P \times A \times C_d = 20.736$ psf × 2.50 feet^2 × 1.20 = 62.21 pounds

4) Calculate the wind load on the antenna: From Eq 2:

Antenna Force, $F_A = P \times A \times C_d = 20.736$ psf × 14.94 feet^2 × 1.20 = 371.76 pounds

5) Calculate the mast *Bending Moment*, from Eq 5:

$M = (F_M \times L_M) + (F_A \times L_A)$ = (62.21 pounds × 90 inches) + (371.76 pounds × 84 inches) = 36827 inch-pounds

where

$L_M = 7.5$ feet × 12 inches/foot = 90 inches

$L_A = 7.0$ feet × 12 inches/foot = 84 inches.

6) Calculate the mast *Moment of Inertia*, from Eq 4:

$$I = \frac{\pi}{4}\left(R^4 - r^4\right) = \frac{\pi}{4}\left(1.0^4 - 0.75^4\right) = 0.5369 \text{ inches}^4$$

where, for a 2.0-inch OD and 0.250-inch wall thickness tube, R= 1.0 and r = 0.75.

7) Calculate the mast *Bending Stress*, from Eq 3:

$$\sigma = \frac{M \times c}{I} = \frac{36{,}827 \text{ inch-pounds} \times 1.0 \text{ inches}}{0.5369} = 68{,}592 \text{ psi}$$

If the yield strength of the mast material is greater than the calculated bending stress, the mast is considered safe for this configuration and wind speed. If the calculated stress is higher than the mast yield strength, a stronger alloy, or a larger mast, or one with a thicker wall is required.

When evaluating a mast with multiple antennas attached to it, special care should be given to finding the worst-case condition (wind direction) for the system. What may appear to be the worst load case, by virtue of the combined flat projected antenna areas, may not always be the exposure that creates the largest mast bending moment. Masts with multiple stacked antennas should always be examined to find the exposure that produces the largest mast bending moment. The antenna flat projected areas at 0° and 90° azimuths are particularly useful for this evaluation.

One way of reducing the net wind torque on a mast holding multiple antennas is to mount antennas on opposite sides of the mast. This alternate mounting scheme causes the wind torque from each antenna to cancel at least partially, reducing the total torque experienced by the mast.

TABLE OF CONTENTS

Chapter 28 — Online Material

Articles

- "A Reflectometer for Twin-Lead" by Fred Brown, W6HPH
- "An Inexpensive VHF Directional Coupler" and "A Calorimeter for VHF and UHF Power Measurements"
- "Antenna Analyzer Pet Tricks" by Paul Wade, W1GHZ
- "Build a Super-Simple SWR Indicator" by Tony Brock-Fisher, K1KP
- "Improving and Using R-X Noise Bridges" by John Grebenkemper, KI6WX
- "Microwavelengths — Directional Couplers" by Paul Wade, W1GHZ
- "On Tuning, Matching and Measuring Antenna Systems Using a Hand Held SWR Analyzer" by John Belrose, VE2CV
- RF Power Meter (Kaune) support files
- "QRP Person's VSWR Indicator" by Doug DeMaw, W1FB
- "Smith Chart Calculations"
- "SWR Analyzer Tips, Tricks, and Techniques" by George Badger, W6TC, et al
- "Technical Correspondence — A High-Power RF Sampler" by Tom Thompson WØIVJ (plus "More on a High-Power RF Sampler" by Thompson, two files)
- "The Noise Bridge" by Jack Althouse, K6NY
- "Time Domain Reflectometry"
- "The Gadget — An SWR Analyzer Add-On" by Fred Hauff, W3NZ
- "The No Fibbin RF Field Strength Meter" by John Noakes, VE7NI "
- "The SWR Analyzer and Transmission Lines" by Peter Schuch, WB2UAQ
- "The Tandem Match — An Accurate Directional Wattmeter" by John Grebenkemper, KA3BLO
- "Using Single-Frequency Antenna Analyzers"
- "A Pulse Generator For Making TDR Measurements," by Larry Lamano, WAØQZY
- "Calculating Transmission Line Loss with SWR Readings," by John Stanley, K4ERO
- "Return Loss Explained, and Why We Should Care," by John Stanley, K4ERO

Chapter 28

Antenna and Transmission Line Measurements

The principal quantities measured on transmission lines are current or voltage, including phase. From these measurements, forward and reflected power and standing wave ratio (SWR) may be obtained. For antennas, the primary measurement of interest to amateurs is field strength in order to determine an antenna's radiation pattern or to compare relative antenna performance. It is important to note that for most practical purposes, a relative measurement is sufficient. An uncalibrated indicator that shows when the largest possible amount of power is being put into the line is just as useful, in most cases, as an instrument that measures the power accurately. It is seldom necessary to know the actual number of watts going into the line unless the overall efficiency of the system is being investigated. An instrument that shows when the SWR is close to 1:1 is all you need for most impedance-matching adjustments.

Absolute quantitative measurements of amplitude or time (phase) become increasingly difficult at frequencies above a few MHz with numerous sources of error becoming more and more significant. Quantitative measurements of reasonable accuracy demand good design and careful construction in the measuring instruments. They also require intelligent use of the equipment, including knowledge not only of its limitations but also of stray effects in the instrument and also of the test configuration that often lead to false results. Until you know the complete conditions of the measurements, a certain amount of skepticism regarding numerical data resulting from amateur measurements with simple equipment is justified.

Accurate measurement of SWR, for example, is necessary only in studies of antenna characteristics such as bandwidth, or for the design of some types of matching systems, such as a stub match. If such measurements are required, surplus lab equipment of very high quality is commonly available although usually not "in calibration."

On the other hand, purely qualitative or relative measurements, such as comparing one antenna to another, before-and-after, or max/min adjustments are easy to make and quite useful. This chapter presents methods and devices for making these measurements.

28.1 Line Current and Voltage

A current or voltage indicator that can be used with coaxial line is a useful piece of equipment. It need not be elaborate or expensive. Its principal function is to show when the maximum power is being taken from the transmitter for any given set of line conditions (length, SWR, etc). This will occur when you adjust the transmitter output circuits for maximum current or voltage into the transmission line. Although a final-amplifier plate or collector current meter is frequently used for this purpose, it is not always a reliable indicator. In many cases, particularly with a screen-grid tube in the final stage, minimum loaded plate current does not occur simultaneously with maximum power output.

28.1.1 PROJECT: Coaxial RF Voltmeter

You can combine a germanium or Schottky diode in conjunction with a low-range milliammeter and a few resistors to form an RF voltmeter suitable for connecting across the two conductors of a coaxial line, as shown in **Figure 28.1**. It

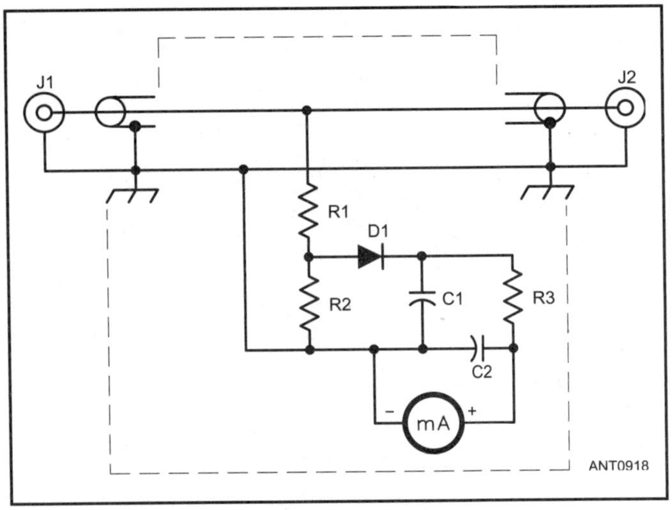

Figure 28.1 — RF voltmeter for coaxial line.
C1, C2 — 0.005- or 0.01-μF disc ceramic.
D1 — 1N34A germanium or 1N5817 Schottky diode.
J1, J2 — Coaxial fittings, chassis-mounting type.
M1 — 0-1 milliammeter (more sensitive meter may be used if desired; see text).
R1 — 6.8 kΩ, metal-oxide, 1 W for each 100 W of RF power.
R2 — 680 Ω, ½ or 1 W carbon-film or metal-oxide.
R3 — 10 kΩ, ½ W (see text).

consists of a voltage divider, R1-R2, having a total resistance about 100 times the Z_0 of the line (so the power consumed will be negligible) with a diode rectifier and milliammeter connected across part of the divider to read relative RF voltage. The purpose of R3 is to make the meter readings directly proportional to the applied voltage, as nearly as possible, by swamping the resistance of D1, since the diode resistance will vary with the amplitude of the current through the diode.

You may construct the voltmeter in a small metal box, indicated by the dashed line in the drawing, and fitted with coax receptacles. R1 and R2 should be noninductive resistors such as carbon-film or metal-oxide types. The power rating for R1 should be 1 W for each 100 W of carrier power in the matched line; separate 1- or 2-W resistors should be used to make up the total power rating required, to the total resistance as given. Any type of resistor can be used for R3; the total resistance should be such that about 10 V dc will be developed across it at full scale. For example, a 0-1 milliammeter would require 10 kΩ, a 0-500 microammeter would take 20 kΩ, and so on. R1 may be a variable resistor so the sensitivity can be adjusted for various power levels.

If more than one resistor is used for R1, the units should be arranged end-to-end with very short leads. R1 and R2 should be kept ½ inch or more from metal surfaces parallel to the body of the resistor. These precautions must be observed if consistent measurements are to be obtained above a few MHz. Stray capacitance and stray coupling limit the accuracy at higher frequencies but do not affect the utility of the instrument for comparative measurements.

CALIBRATION

You may calibrate the meter for RF voltage at low frequencies by comparison with a standard such as an RF ammeter, wattmeter or oscilloscope. If a wattmeter is used, the line must be well matched so the impedance at the point of measurement is equal to the actual Z_0 of the line. The power can be calculated as $E = \sqrt{P Z_0}$. By making voltage measurements at a number of different power levels, you can obtain enough points to draw a calibration curve for your particular setup. Be advised that stray effects and nonlinearities inherent in this simple circuit make a true calibration questionable above a few MHz.

28.1.2 PROJECT: RF Current Meters

The following project was designed by Tom Rauch, W8JI (**w8ji.com/building_a_current_meter.htm**). The circuit of **Figure 28.2** is based on a current transformer (T1) consisting of a T157-2 powdered-iron toroid core with a 20-turn winding. The meter is used with the current-carrying wire or antenna inserted through the middle of the core as a one-turn primary

When 1 A is flowing in the single-turn primary, the secondary current will be 50 mA (equal to primary current divided

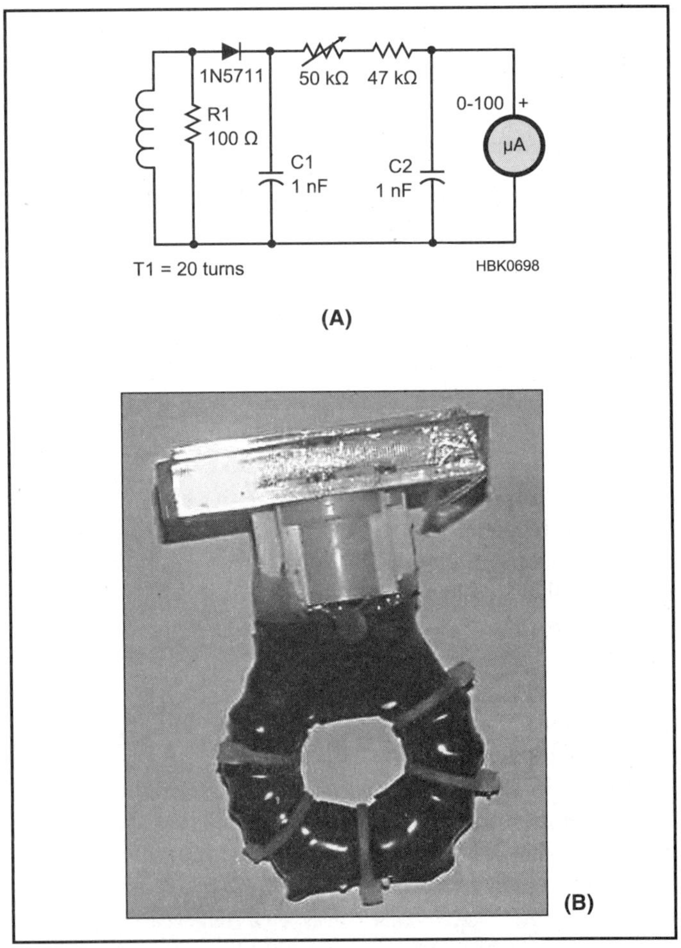

Figure 28.2 — The schematic of the RF current probe (A) and assembly of the RF current probe (B). Use an all-plastic meter and mount the circuits and toroid directly on the back of the meter case.

by the turns ratio of 20:1). R1 across the transformer flattens the frequency response and limits the output voltage. The RF voltage is then detected and filtered by D1 (a low-threshold Schottky diode for minimum voltage drop) and C1. The adjustable sum of R2 and R3 allow for full-scale (FS) calibration of the 100 μA meter. C2 provides additional filtering. The toroid core and all circuitry are glued to the back of the meter case with only R2 exposed — a screwdriver-adjustable calibration pot.

It is important to minimize stray capacitance by using a meter with all-plastic construction except for the electrical parts. The meter in Figure 28.2B has an all-plastic case including the meter scale. The meter movement and all metallic areas are small. The lack of large metallic components minimizes stray capacitance from the proximity of the meter. Low stray capacitance ensures the instrument has the least possible effect on the circuit being tested.

A value of 100 Ω for R1 gave the flattest response from 1.8 to 30 MHz. With 50 mA of secondary current, the voltage across R1 is 0.05 × 100 = 5 V_{RMS}. The peak voltage is then 1.414 × 5 = 7.1 V. At full current, power dissipation in R1 = 50 mA × 5 V_{RMS} = 0.25 W so a ½-W or larger resistor should be used.

The meter used here was a 10,000 Ω/V model so for full-scale deflection from a primary current of 1 A producing a secondary voltage of ~7 V, the sum of R2 and R3 must be set to 7 × 10,000 = 70 kΩ. The low-current meter combined with high detected voltage improves detector linearity.

Calibration of the meter can be performed by using a calibrated power meter and a test fixture consisting of two RF connectors with a short piece of wire between them and through the transformer core. With 50 W applied to a 50-Ω load, the wire will be carrying 1 A of current. Full-scale accuracy is not required in comparison measurements, since the meter references against itself, but linearity within a few percent is important.

This transformer-based meter is much more reliable and linear than thermocouple RF ammeters and perturbs systems much less. Stray capacitance added to the system being tested is very small because of the proximity of the meter and the compact wiring area. Compared to actually connecting a meter with its associated lead lengths and capacitance in line with the load, the advantages of a transformer-coupled meter become apparent.

CLAMP-ON RF CURRENT PROBE

Sometimes it is not practical to disconnect a wire in order to sense RF current on it, such as a power cord or speaker wire. In such cases a clamp-on probe can be used as described in the February 1999 *QST* article by Steve Sparks, N5SV, and shown in **Figure 28.3**. The core is a split-core type — any common material will suffice (type 31, 75, 61, 43, etc) for HF use. If the enclosure is hand-held size, the instrument can be used as a handy detector and "sniffer" for RFI troubleshooting. Because the split core does not close completely and consistently every time, this is not a precision instrument but is effective for relative comparison.

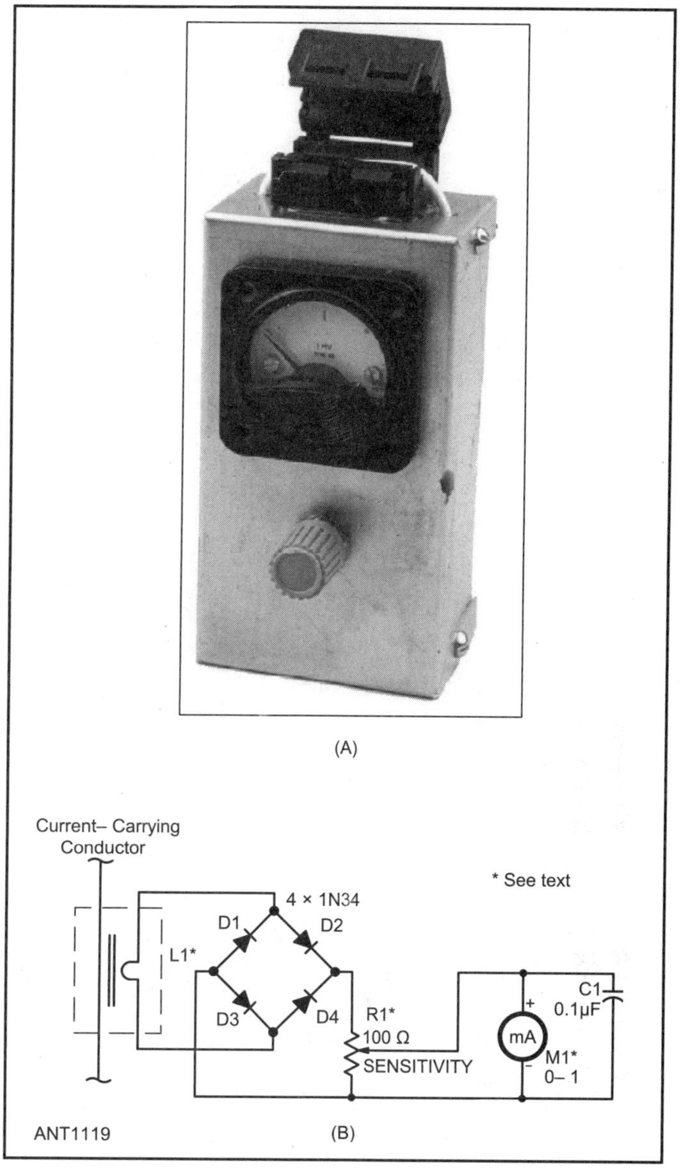

Figure 28.3 — The RF current probe at A is small enough to be hand-carried as an effective RF "sniffer." At B is the schematic of the RF current probe. Use a metal enclosure for the probe.
C1 — 0.1 μF disc ceramic.
D1-D4 — 1N34A germanium or 1N5817 Schottky diodes.
L1 — Single turn of #14 AWG wire through a snap-on ferrite split-core, type 31, 43, 61, 73, or 75 material will work. Glue core to top of metal enclosure.
M1 — 1 mA analog meter. Substitute lower full-scale current for higher sensitivity.
R1 — 100 to 500 Ω panel-mount potentiometer.

28.1.3 RF Ammeters

An RF ammeter is a good way to gauge output power. You can mount an RF ammeter in any convenient location at the input end of the transmission line, the principal precaution being that the capacitance to ground, chassis and nearby conductors should be low. A Bakelite-case instrument can be mounted on a metal panel without introducing enough shunt capacitance to ground to cause serious error up to 30 MHz. When installing a metal-case instrument on a metal panel, you should mount it on a separate sheet of insulating material so that there is ⅛ inch or more separation between the edge of the case and the metal.

A 2-inch instrument can be mounted in a 2 × 4 × 4-inch metal box, as shown in **Figure 28.4**. This is a convenient arrangement for use with coaxial line. Installed this way, a good quality RF ammeter will measure current with an accuracy that is entirely adequate for calculating power in the line. As discussed above in connection with calibrating RF voltmeters, the line must be closely matched by its load so the actual impedance is resistive and equal to Z_0. The scales of such instruments are cramped at the low end, however, which limits the range of power that can be measured by a single meter. The useful current range is about 3 to 1, corresponding to a power range of about 9 to 1.

Figure 28.4 — A convenient method of mounting an RF ammeter for use in a coaxial line. This is a metal-case instrument mounted on a thin bakelite panel. The cutout in the metal clears the edge of the meter by about ⅛ inch.

New RF ammeters are expensive and even surplus pricing can vary widely between $10 and $100 in today's market. AM radio stations are the main users of new units. The FCC defines the output power of AM stations based on the RF current in the antenna, so new RF ammeters are made mainly for that market. They are quite accurate, and their prices reflect that!

The good news is that used RF ammeters are often available. Surplus component and equipment dealers have been a consistent source of RF ammeters. Ham flea markets are also worth trying.

Before buying a used RF ammeter, check to be sure it is actually an RF ammeter — it is common to find meters labeled "RF Amps" that are simple current meters intended to be used with an external RF current sensing unit.

RF AMMETER SUBSTITUTES

Don't despair if you can't find a used RF ammeter. It's possible to construct your own. Both hot-wire and thermocouple units can be homemade. Pilot lamps in series with antenna wires, or coupled to them in various ways, can indicate antenna current or even forward and reflected power. (See the References and Bibliography entrics for Sutter and Wright.)

Another approach is to use a small low-voltage lamp as the heat/light element and use a photodetector driving a meter as an indicator. (Your eyes and judgment can serve as the indicating part of the instrument.) A feed line balance checker could be as simple as a couple of lamps with the right current rating and the lowest voltage rating available. You should be able to tell fairly well by eye which bulb is brighter or if they are about equal. You can calibrate a lamp-based RF ammeter with 60-Hz or dc power.

The optical approach is often taken in QRP portable equipment where an LED is used to replace a meter in an SWR bridge as described by Phil Salas, AD5X, in his Z-Match antenna tuner described in the **Transmission Line System Techniques** chapter.

As another alternative, you can build an RF ammeter that uses a dc meter to indicate rectified RF from a current transformer that you clamp over a transmission line wire as described by Zack Lau, W1VT (see References and Bibliography).

28.2 SWR Measurements

On parallel-conductor lines it is possible to measure the standing wave ratio by moving a current (or voltage) indicator along the line, noting the maximum and minimum values of current (or voltage) and then computing the SWR from these measured values. This cannot be done with coaxial line since it is not possible to make measurements of this type inside the cable. The technique is, in fact, seldom used with open lines because it is not only inconvenient but sometimes impossible to reach all parts of the line conductors. Also, the method is subject to considerable error from antenna currents flowing on the line.

Present-day SWR measurements made by amateurs practically always use some form of *directional coupler* or RF-bridge circuit. The indicator circuits themselves are fundamentally simple, but they require considerable care in construction to ensure accurate measurements. The requirements for indicators used only for the adjustment of impedance-matching circuits, rather than actual SWR measurement, are not so stringent, and you can easily make an instrument for this purpose.

28.2.1 Bridge Circuits

Two commonly used bridge circuits are shown in **Figure 28.5**. The bridges consist essentially of two voltage dividers in parallel, with an RF voltmeter connected between the junctions of each pair of *arms*, as the individual elements are called. When the equations shown to the right of each circuit are satisfied there is no potential difference between the two junctions, and the RF voltmeter indicates zero voltage. The bridge is then said to be in *balance*.

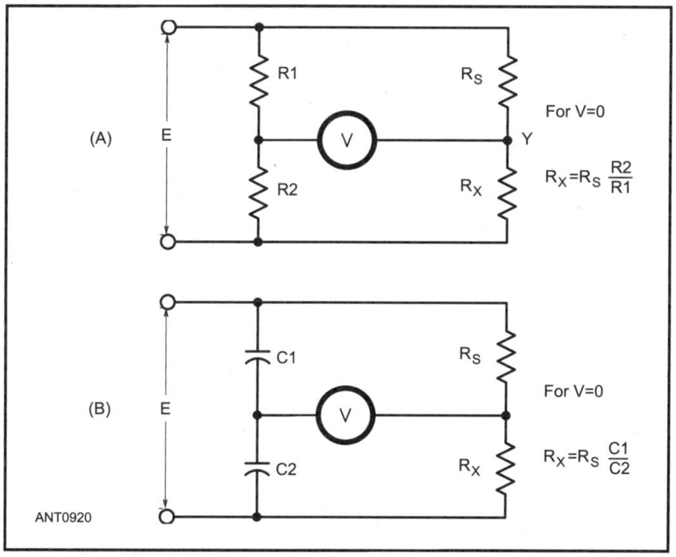

Figure 28.5 — Bridge circuits suitable for SWR measurement. At A, Wheatstone type using resistance arms. At B, capacitance-resistance bridge ("Micromatch"). Conditions for balance are independent of frequency in both types. The voltmeter must be an RF voltmeter.

Taking Figure 28.5A as an illustration, if R1 = R2, half the applied voltage, E, will appear across each resistor. Then if $R_S = R_X$, ½ E will appear across each of these resistors and the RF voltmeter reading will be zero. Remember that a matched transmission line has essentially a purely resistive input impedance. Suppose that the input terminals of such a line are substituted for R_X. Then if R_S is a resistor equal to the Z_0 of the line, the bridge will be balanced.

If the line is not perfectly matched, its input impedance will not equal Z_0 and hence will not equal R_S, since you chose the latter to be equal to Z_0. There will then be a difference in potential between points X and Y, and the RF voltmeter will show a reading. Such a bridge therefore can be used to show the presence of standing waves on the line, because the line input impedance will be equal to Z_0 only when there are no standing waves.

Considering the nature of the incident and reflected components of voltage that make up the actual voltage at the input terminals of the line, as discussed in the **Transmission Lines** chapter, it should be clear that when $R_S = Z_0$, the bridge is always in balance for the incident component. Thus the RF voltmeter does not respond to the incident component at any time but reads only the reflected component. The incident component can be measured across either R1 or R2, if they are equal resistances. The standing wave ratio is then

$$SWR = \frac{E1 + E2}{E1 - E2} \qquad (1)$$

where E1 is the incident voltage and E2 is the reflected voltage. It is often simpler to normalize the voltages by expressing E2 as a fraction of E1, in which case the formula becomes

$$SWR = \frac{1 + k}{1 - k} \qquad (2)$$

where k = E2/E1.

The operation of the circuit in Figure 28.5B is essentially the same, although this circuit has arms containing reactance as well as resistance.

It is not necessary that R1 = R2 in Figure 28.5A; the bridge can be balanced, in theory, with any ratio of these two resistances provided R_S is changed accordingly. In practice, however, the accuracy is highest when the two are equal; this circuit is most commonly used.

A number of types of bridge circuits appear in **Figure 28.6**, many of which have been used in amateur products or amateur construction projects. The bridge at E is most often used in common low-cost SWR meters. (See the References and Bibliography entry for Silver for a description of how these SWR meters work.) All circuits except that at G can have the ground returns of the generator and load at a common potential. At G, the generator and detector ground returns are at a common potential. You may interchange the positions of the detector and transmitter (or generator) in the bridge, and this may be advantageous in some applications.

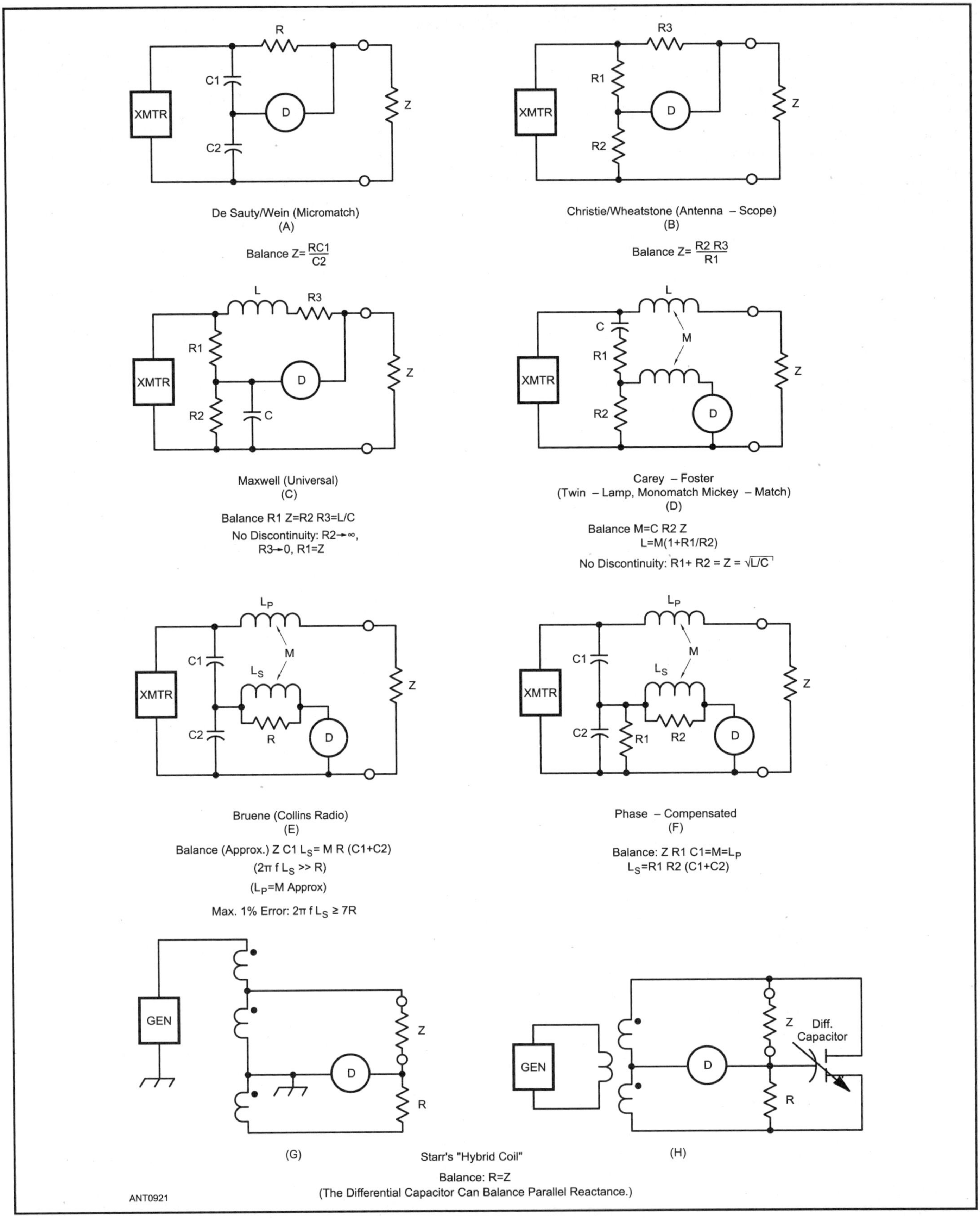

Figure 28.6 — Various types of SWR bridge indicator circuits and commonly known names of bridge circuits or devices in which they have been used. Detectors (D) are usually semiconductor diodes with meters, isolated with RF chokes and capacitors. However, the detector may be a radio receiver. In each circuit, Z represents the load being measured. (*This information provided by David Geiser, WA2ANU*)

The bridges shown at D, E, F and H may have one terminal of the generator, detector and load common. Bridges at A, B, E, F, G and H have constant sensitivity over a wide frequency range. Bridges at B, C, D and H may be designed to show no discontinuity (impedance bump) with a matched line, as shown in the drawing. Discontinuities with A, E and F may be small.

Bridges are usually most sensitive when the detector bridges the midpoint of the generator voltage, as in G or H, or in B when each resistor equals the load impedance. Sensitivity also increases when the currents in each leg are equal.

28.2.2 PROJECT: SWR Resistance Bridge

The basic bridge configuration shown in Figure 28.5B may be home constructed and is reasonably accurate for SWR measurement at HF. A practical circuit for such a bridge is given in **Figure 28.7A** and a representative layout is shown in Figure 28.7B. Properly built, a bridge of this design can be used for measurement of SWRs up to about 15:1 with good accuracy. Resistance bridges cannot be left in the transmission during regular operation due to the power dissipation limits of the resistors. This bridge should be used for test purposes only.

You must also observe these important construction points:

1) Keep leads in the RF circuit short, to reduce stray inductance.

2) Mount resistors two or three times their body diameter away from metal parts, to reduce stray capacitance.

3) Place the RF components so there is as little inductive and capacitive coupling as possible between the bridge arms.

In the instrument shown in Figure 28.7B, the input and line connectors, J1 and J2, are mounted fairly close together so the standard resistor, R_S, can be supported with short leads directly between the center terminals of the connectors. R2 is mounted at right angles to R_S, and a shield partition is used between these two components and the others.

The two 47-kΩ resistors, R5 and R6 in Figure 28.7A, are voltmeter multipliers for the 0-100 microammeter used as an indicator. This is sufficient resistance to make the voltmeter approximately linear (that is, the meter reading is directly proportional to the RF voltage) and no voltage calibration curve is needed. D1 is the rectifier for the reflected voltage and D2 is for the incident voltage. Because of manufacturing variations in resistors and diodes, the readings may differ slightly with two multipliers of the same nominal resistance value, so a correction resistor, R3, is included in the circuit. You should select its value so that the meter reading is the same with S1 in either position, when RF is applied to the bridge with the line connection open. In the instrument shown, a value of 1000 Ω was required in series with the multiplier for reflected voltage; in other cases different values probably would be needed and R3 might have to be put in series with the multiplier for the incident voltage. You can determine this by experiment.

The value used for R1 and R2 is not critical, but you should match the two resistors within 1% or 2% if possible. Keep the resistance of R_S as close as possible to the actual Z_0 of the line you use (generally 50 or 75 Ω). Select the resistor by

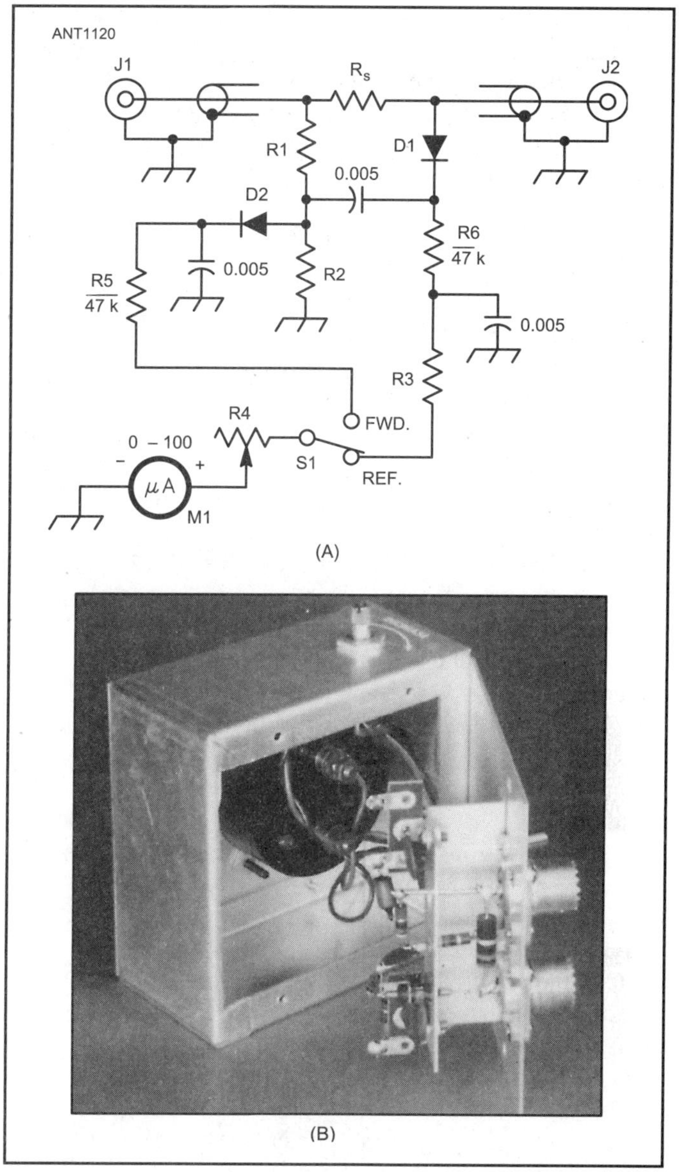

Figure 28.7 — At A, schematic of the resistance bridge for SWR measurement. Capacitors are disc ceramic. Resistors are ½-W composition except as noted below.

D1, D2 — 1N34A germanium or 1N5817 Schottky diodes.
J1, J2 — Coaxial connectors, chassis-mounting type.
M1 — 0-100 dc microammeter.
R1, R2 — 47 Ω, ½-W carbon-film or metal-oxide (see text).
R3 — See text.
R4 — 50-kΩ volume control.
R_S — Resistance equal to line Z_0 (½ or 1 W composition).
S1 — SPDT toggle.

At B, a 2 × 4 × 4-inch aluminum box is used to house this SWR bridge. The variable resistor, R4, is mounted on the side. The bridge components are mounted on one side plate of the box and a subchassis formed from a piece of aluminum. The input connector is at the top in this view. R_S is connected directly between the two center posts of the connectors. R2 is visible behind it and perpendicular to it. One terminal of D1 projects through a hole in the chassis so the lead can be connected to J2. R1 is mounted vertically to the left of the chassis in this view, with D2 connected between the junction of R1-R2 and a tie point.

actual measurement with an accurate resistance bridge, if you have one available.

R4 is for adjusting the incident-voltage reading to full scale in the measurement procedure described below. Its use is not essential, but it offers a convenient alternative to exact adjustment of the RF input voltage.

TESTING

Measure R1, R2 and R_S with a reliable digital ohmmeter or resistance bridge after completing the wiring. This will ensure that their values have not changed from the heat of soldering. Disconnect one side of the microammeter and leave the input and output terminals of the unit open during such measurements to avoid stray shunt paths through the rectifiers.

Check the two voltmeter circuits as described above, applying enough RF (about 10 V peak, 1 V RMS) to the input terminals to give a full scale reading with R4 set for maximum deflection and with the line terminals open. If necessary, try different values for R3 until the reading is the same with S1 in either position.

With J2 open, adjust the RF input voltage and R4 for full-scale reading with S1 in the incident-voltage position. Then switch S1 to the reflected-voltage position. The reading should remain at full scale. Next, short-circuit J2 by touching a screwdriver between the center terminal and the frame of the connector to make a low-inductance short. Switch S1 to the incident-voltage position and readjust R4 for full scale, if necessary. Then throw S1 to the reflected-voltage position, keeping J2 shorted, and the reading should be full scale as before. If the readings differ, R1 and R2 are not the same value, or there is stray coupling between the arms of the bridge. You must read the reflected voltage at full scale with J2 either open or shorted, when the incident voltage is set to full scale in each case, to make accurate SWR measurements.

The circuit should pass these tests at all frequencies at which it is to be used. It is sufficient to test at the lowest and highest frequencies, usually 1.8 or 3.5 and 28 or 50 MHz. If R1 and R2 are poorly matched but the bridge construction is otherwise good, discrepancies in the readings will be substantially the same at all frequencies. A difference in behavior at the low and high ends of the frequency range can be attributed to stray coupling between bridge arms, or stray inductance or capacitance in the arms.

To check the bridge for balance, apply RF and adjust R4 for full scale with J2 open. Then connect a resistor identical with R_S (the resistance should match within 1% or 2%) to the line terminals, using the shortest possible leads. It is convenient to mount the test resistor inside a cable connector (PL-259), a method of mounting that also minimizes lead inductance. When you connect the test resistor the reflected-voltage reading should drop to zero. The incident voltage should be reset to full scale by means of R4, if necessary. The reflected reading should be zero at any frequency in the range to be used. If a good null is obtained at low frequencies but some residual current shows at the high end, the trouble may be the inductance of the test resistor leads, although it may also be caused by stray coupling between the arms of the bridge itself.

If there is a constant low (but not zero) reading at all frequencies the problem is poor matching of the resistance values. Both effects can be present simultaneously. You should make sure you obtain a good null at all frequencies before using your bridge.

BRIDGE OPERATION

You must limit the RF power input to a bridge of this type to 2 W at most, because of the power-dissipation ratings of the resistors. If the transmitter has no provision for reducing power output to this value a simple power-absorber circuit can be made up, as shown in **Figure 28.8**. Lamp DS1 changes resistance as it heats up — from a few ohms when cold to more than 100 Ω at full power. This increasing resistance tends to maintain constant current through the resistor over a fairly wide power range, so the voltage drop across the resistor also tends to be constant. This voltage is applied to the bridge, and with the constants given is in the right range for resistance-type bridges.

To make a measurement, connect the unknown load to J2 and apply sufficient RF voltage to J1 to give a full-scale incident-voltage reading. Use R4 to set the indicator to exactly full scale. Then throw S1 to the reflected voltage position and note the meter reading. The SWR is then found by using these readings in Eq 1.

For example, if the full-scale calibration of the dc instrument is 100 μA and the reading with S2 in the reflected-voltage position is 40 μA, the SWR is

$$\text{SWR} = \frac{100+40}{100-40} = \frac{140}{60} = 2.33:1$$

Instead of calculating the SWR value, you could use the voltage curve in **Figure 28.9**. In this example the ratio of reflected to forward voltage is 40/100 = 0.4, and the SWR value is about 2.3:1.

You may calibrate the meter scale in any arbitrary units, so long as the scale has equal divisions. It is the ratios of the voltages, and not the actual values, that determine the SWR.

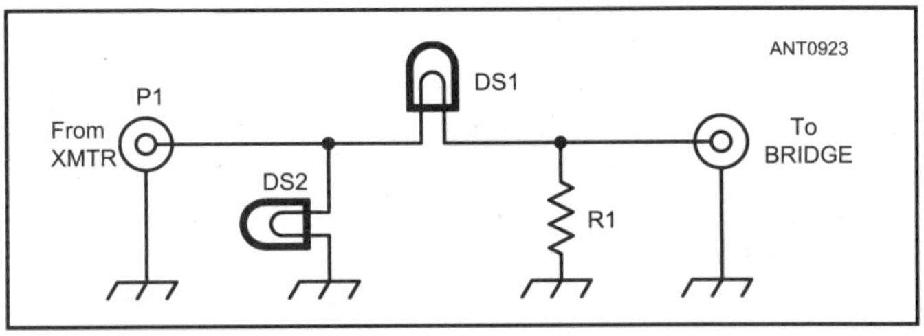

Figure 28.8 — Power-absorber circuit for use with resistance-type SWR bridges when the transmitter has no special provisions for power reduction. For RF powers up to 50 W, DS1 is a 120-V 40-W incandescent lamp and DS2 is not used. For higher powers, use sufficient additional lamp capacity at DS2 to load the transmitter to about normal output; for example, for 250-W output DS2 may consist of two 100-W lamps in parallel. R1 is made from three 1-W 68-Ω resistors connected in parallel. P1 and P2 are cable-mounting coaxial connectors. Leads in the circuit formed by the lamps and R1 should be kept short, but convenient lengths of cable may be used between this assembly and the connectors.

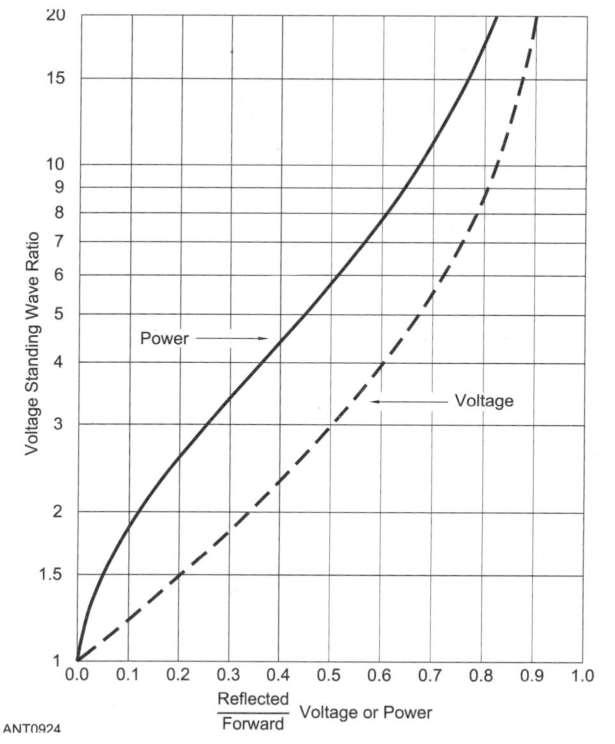

Figure 28.9 — Chart for finding voltage standing wave ratio when the ratio of reflected-to-forward voltage or reflected-to-forward power is known.

28.2.3 Avoiding Errors in SWR Measurements

The principal causes of inaccuracies within the bridge are differences in the resistances of R1 and R2, stray inductance and capacitance in the bridge arms, and stray coupling between arms. If the checkout procedure described above is followed carefully, the bridge in Figure 28.6 should be sufficiently accurate for practical use. The accuracy is highest for low standing wave ratios because of the nature of the SWR calculation; at high SWR the divisor in the equation above represents the difference between two nearly equal quantities, so a small error in voltage measurement may mean a considerable difference in the calculated SWR.

Detector nonlinearity is another source of error. A diode peak detector is approximately linear if the load impedance is high enough, and the signal is much greater than the diode forward-conduction voltage, but it will still have significant nonlinearity at the low end of the scale.

The standard resistor R_S must equal the actual Z_0 of the line. The actual Z_0 of a sample of line may differ by a few percent from the nominal figure because of manufacturing variations, but this has to be tolerated. In the 50- to 75-Ω range, the RF resistance of a noninductive resistor of ½- or 1-W rating is essentially identical with its dc resistance at VHF and below.

COMMON-MODE CURRENTS

As explained in the **Transmission Line System Techniques** chapter, there are two ways in which unwanted *common-mode* currents can flow on the outside of a coaxial line — currents induced onto the line because of its spatial relationship to the antenna and currents that result from the direct connection between the coax outer conductor and (usually) one side of the antenna. Such currents can cause significant SWR measurements error and SWR that changes with line length but for different reasons.

Induced current usually will not be troublesome if the bridge and the transmitter (or other source of RF power for operating the bridge) are shielded so that any RF currents flowing on the outside of the line cannot find their way into the bridge. This point can be checked by inserting an additional section of line (⅛ to ¼ electrical wavelength preferably) of the same Z_0. The SWR indicated by the bridge should not change except for a slight decrease because of the additional line loss. If there is a marked change, you may need better shielding.

Common-mode currents can also flow on the outside of coaxial transmission lines if the outside surface of the shield is connected directly to one side of the antenna. Even if the antenna itself is balanced, this "extra" conductor will unbalance the system and common-mode current will flow on the outside of the line. In such cases, the SWR will vary with line length, even though the bridge and transmitter are well-shielded and the shielding is maintained throughout the system by the use of coaxial fittings. Often, merely moving the transmission line around will cause the indicated SWR to change. This is because the outside of the coax has become part of the antenna system by being connected to the antenna at the feed point. The outside shield of the line thus constitutes a load, along with the desired load represented by the antenna itself. The SWR on the line then is determined by the composite load of the antenna and the outside of the coax. Since changing the line length (or position) changes one component of this composite load, the SWR changes too. This is an undesirable condition since the line is usually operating at a higher SWR than it should — and would if the common-mode current on the outside of the coax were eliminated.

The remedy for both situations is generally to use a common-mode choke balun as described in the **Transmission Line System Techniques** chapter or to detune the outside of the line by proper choice of length so that it presents a high impedance at the frequency of operation. Note that this is not a *measurement error*, since what the instrument reads is the actual SWR on the line.

SPURIOUS FREQUENCIES

Off-frequency components in the RF voltage applied to the bridge may cause considerable error. The principal components of this type are harmonics and low-frequency subharmonics that may be fed through the final stage of the transmitter driving the bridge. The antenna is almost always a fairly selective circuit, and even though the system may be operating with a very low SWR at the desired frequency, it is almost always mismatched at harmonic and subharmonic frequencies.

If such spurious frequencies are applied to the bridge in appreciable amplitude, the SWR indication will be very much in error. In particular, it may not be possible to obtain a null on the bridge with any set of adjustments of the matching circuit. The only remedy is to filter out the unwanted components by increasing the selectivity of the circuits between the transmitter final amplifier and the bridge.

28.2.4 PROJECT: Reflectometers

A reflectometer consists of a coupled pair of transmission-line impedance bridges (see Figure 28.6) operated back-to-back so that the generator and load are reversed in one bridge. The resulting imbalance between bridges is displayed on a meter with a scale calibrated so as to convert the imbalance to SWR. Various simple reflectometers have been described from time to time in *QST* and in *The ARRL Handbook* (see References and Bibliography).

Bridges of this type are usually frequency-sensitive — that is, the meter response increases with increasing frequency for the same applied voltage so that a CAL (calibration) potentiometer is required to set the meter sensitivity for each use.

Because most of these designs are frequency sensitive, it is difficult to calibrate them accurately for power measurement. Similarly, without a guaranteed power calibration, they cannot make accurate quantitative measurements of SWR but their low cost and suitability for use at moderate power levels, combined with the ability to show accurately when a matching circuit has been properly adjusted, make them a worthwhile addition to the amateur station.

A pair of typical reflectometer designs are given in **Figures 28.10** and **28.11**. (Both original *QST* articles are included with this book's online material.) The classic circuit by DeMaw in Figure 28.10 is a very useful circuit at low power levels. It can be scaled up to be used at higher power levels by reducing the number of turns on the toroid primary and increasing the voltage ratings of the voltage sensing capacitors, C1 and C2. (The referenced article by Bruene should be consulted, as well.) The design by Brown in Figure 28.11 is for use with 300-Ω twin-lead and may be used with parallel-wire feed lines of other characteristic impedances such as 450-Ω window line by changing R1 and R2 to match the impedance of the line.

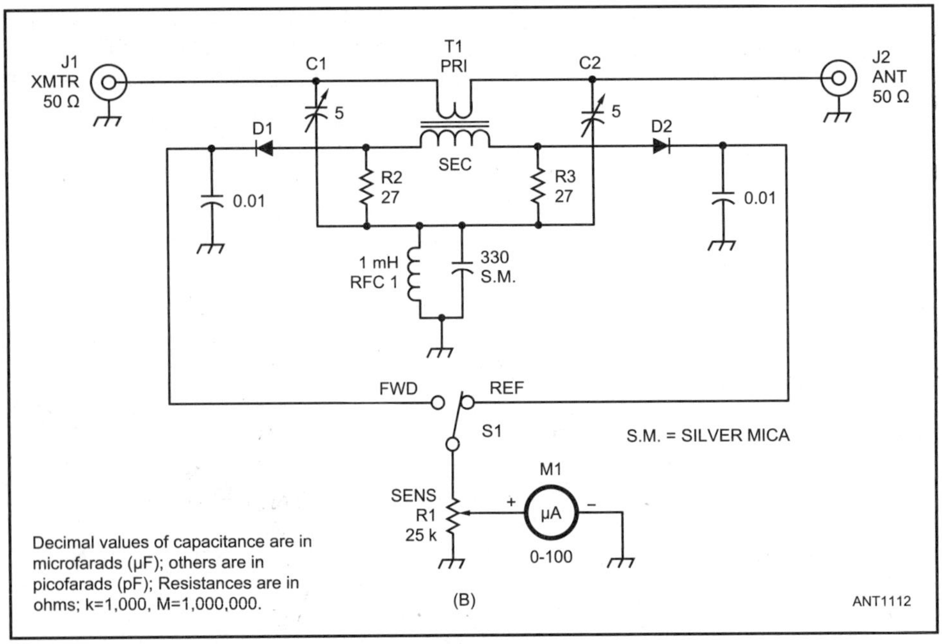

Figure 28.10 — Schematic diagram of the QRP VSWR indicator. Fixed-value capacitors are disc ceramic except those marked with S.M., which are silvered-mica. R2 and R3 and ¼-W carbon-film or metal-film units.
C1, C2 — Miniature PC board mount air trimmer.
D1, D2 — Silicon switching diode, 1N4148 type, matched for equivalent forward resistance by using an ohmmeter.
J1, J2 — RF connector receptacle (phono jack, BNC, UHF).
M1 — Miniature 50- or 100-μA dc meter.
R1 — Linear-taper miniature control, 25 kΩ.
RFC1 — Miniature 1-mH RF choke.
S1 — Miniature SPDT slide or toggle switch.
T1 — Toroidal transformer. Secondary: 60 turns #30 AWG enameled wire on a T68-2 powdered-iron core. Primary is two turns over secondary winding (see text).

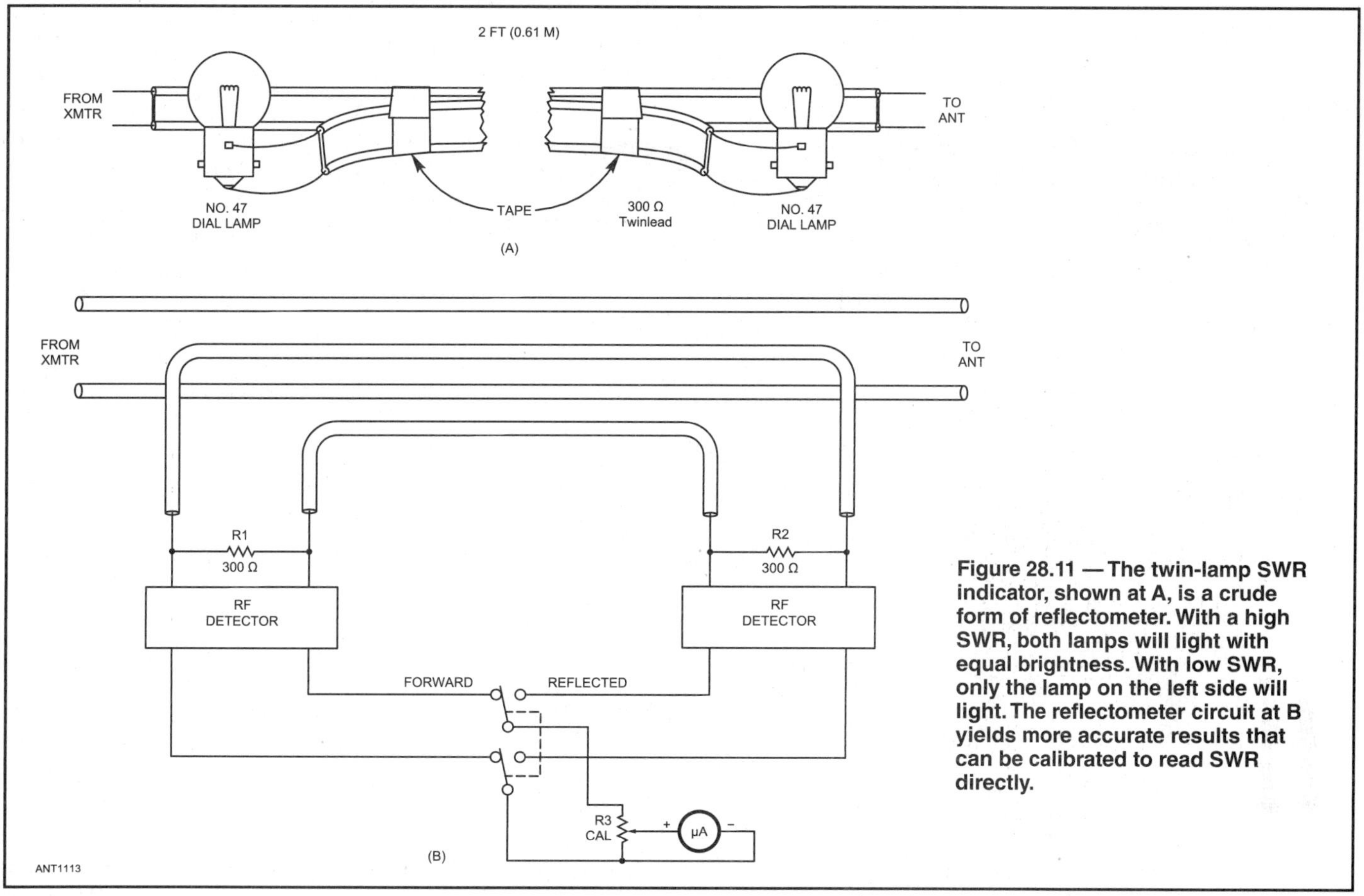

Figure 28.11 — The twin-lamp SWR indicator, shown at A, is a crude form of reflectometer. With a high SWR, both lamps will light with equal brightness. With low SWR, only the lamp on the left side will light. The reflectometer circuit at B yields more accurate results that can be calibrated to read SWR directly.

28.3 RF Power Measurement

The standard commercial instruments amateurs use to measure RF power are the various models of the Thruline *directional wattmeters* from Bird Technologies (**birdrf.com**) such as the popular Model 43. The meter consists of a section of transmission into which is inserted a selectable power-sensing element, popularly referred to as a "slug." The transmission line in the wattmeter is designed to have an element inserted without disrupting normal power flow through the meter.

The element consists of a pickup loop and terminating resistor that form a *directional coupler* — a circuit that couples to a transmission line and extracts a small amount of power flowing in one direction. (See the References and Bibliography entry by Wade for a tutorial on directional couplers, also included with this book's online material.) The construction of the Bird transmission line and sensing element are shown as Figure 3 in the Model 43 operating manual, available for download at **birdrf.com** (look for the Model 43 product information). The element can be rotated so that the directional coupler picks up either forward or reflected power.

The energy from the directional coupler in the element then passes through a rectifying diode and filter capacitor that form an RF detector as described earlier in this chapter. The output of the RF detector then drives a meter that is calibrated in watts. The standard series of elements covers from 2 to 1000 MHz and from 5 W to 5000 W full-scale. A variety of specialty elements are also available.

For a close look at the construction of the Bird Thruline wattmeter and a typical power sensing element, see the Repeater Builders website article, "Photo Tour of a Bird Wattmeter Element," by Robert Meister, WA1MIK at **repeater-builder.com/projects/bird-element-tour/bird-element-tour.html**. A number of excellent white papers and application notes on the use of these ubiquitous instruments are available on the Bird Technologies website under "Resources and Tools."

28.3.1 PROJECT: Directional Power/SWR Meter

The following project is an overview of the January 2011 *QST* article by Bill Kaune, W7IEQ, "A Modern Directional Power/SWR Meter." The complete article including firmware and printed circuit board artwork is included with this book's online material.

The primary use for this unit is to monitor the output power and tuning of a transceiver. The author's station configuration is shown in **Figure 28.12**. RF power generated by the transmitter is routed via RG-8 coaxial cable through a directional coupler to an antenna tuner, which is connected to the antenna

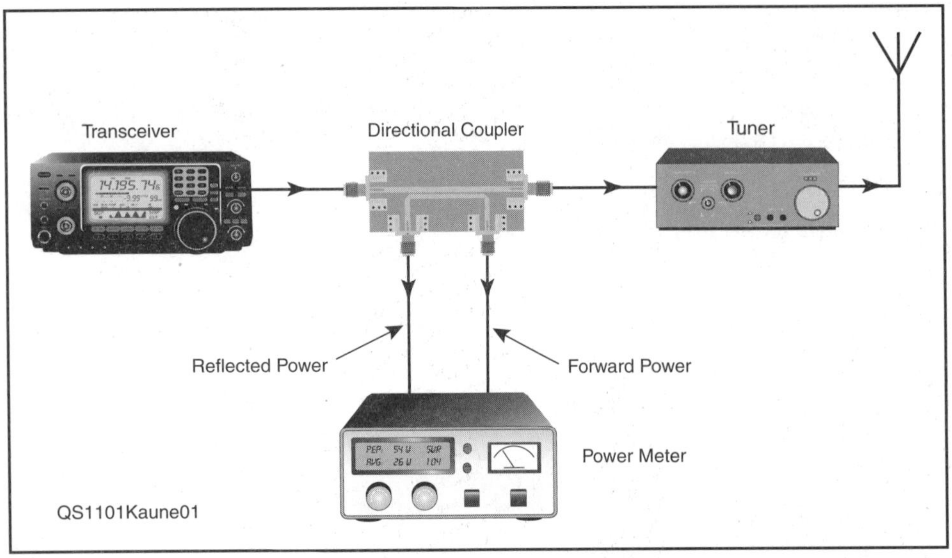

Figure 28.12 — W7IEQ station setup, including the power meter being described here.

The forward and reflected power samples coupled are reduced by a factor of $1/N^2$, where N = 31 is the number of turns of wire on each toroid. Thus the forward and reflected power samples are reduced by about 30 dB. For example, if a transceiver were delivering a power of 100 W to a pure 50 Ω load, the forward power sample from the directional coupler would be about 0.1 W (20 dBm).

The directivity of a directional coupler is defined as the ratio of the forward power sample divided by the reflected power sample when the coupler is terminated in 50 Ω. In this coupler, the directivity measured using an inexpensive network analyzer is at least 35 dB at 3.5 MHz and 28 dB at 30 MHz.

with RG-8. The directional coupler contains circuits that sample the RF power flowing from the transmitter to the tuner (the forward power) and the RF power reflected back from the tuner to the transmitter (the reflected power). These samples are sent via RG-58 cable to the two input channels of the power meter. This project includes the directional coupler and the power meter. Enough detail is provided in the full article so that an amateur can duplicate the device or modify the design.

DIRECTIONAL COUPLER

The directional coupler is based on the unit described in "The Tandem Match" by John Grebenkemper, KI6WX in the January 1987 issue of *QST* and also included with this book's online material. A pair of FT-82-67 toroids with 31 turns of #26 AWG magnet wire over lengths of RG-8 form the basis of the directional coupler shown in **Figure 28.13**.

POWER/SWR METER — CIRCUIT DESCRIPTION

Figure 28.14 shows a front panel view of the power meter. An LCD displays the measured peak (PEP) and average (AEP) envelope powers as well as the standing wave ratio (SWR). The power meter calculates either the peak and average envelope power traveling from the transceiver to load (the forward power) or the peak and average envelope powers actually delivered to the load (the forward power minus reflected power). The average envelope power (AEP) represents an average of the forward or load powers over an averaging period of either 1.6 or 4.8 seconds.

A 1 mA-movement analog meter on the front panel facilitates antenna tuning. This meter continuously displays the quantity 1 – 1/SWR, where SWR is the standing wave ratio on the line. Thus, an SWR of 1.0 corresponds to a meter reading of 0 — no deflection of the meter. An SWR of 2 results

Figure 28.13 — Completed directional coupler.

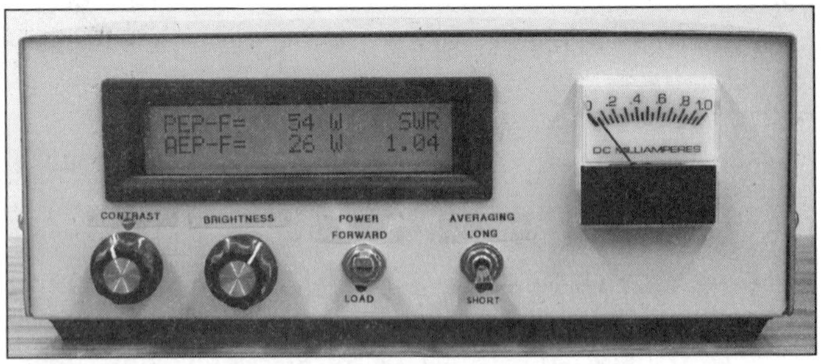

Figure 28.14 — Front panel of power meter. The LCD shows the peak envelope power (PEP), the average envelope power (AEP) and the SWR. The two knobs control the contrast and back lighting of the LCD. One toggle switch determines whether forward or load powers are displayed. A second switch sets the averaging time for the AEP calculation. The meter displays SWR and is used for tuning purposes.

in a 50% deflection of the meter, while an SWR of 5 produces an 80% deflection of the meter.

The forward and reflected power samples from the directional coupler are applied to a pair of Analog Devices AD8307 logarithmic detectors. External 20 dB attenuators (Mini-Circuits HAT-20) reduce the signals from the directional coupler to levels compatible with the AD8307. As noted earlier, the directional coupler has an internal attenuation of about 30 dB, so the total attenuation in each channel is about 50 dB. Thus, a radio operating at a power level of 1 kW (60 dBm) will result in an input to the forward power channel of about 10 dBm. (The schematic diagram and parts list of the power meter are included with this book's online material.) The detectors are configured such that the time constant of their output follows the modulation envelope of the RF signal.

LF398 sample-and-hold ICs stabilize the voltages from the forward and reflected power logarithmic detectors. In this way both voltages can be sampled at the exact same time and held for subsequent analog-to-digital conversion and calculation of power and SWR by the PIC16F876A microprocessor (**microchip.com**). The processor also includes a pulse-width-modulated (PWM) output used to drive the analog SWR meter on the front panel.

28.3.2 PROJECT: High-Power RF Samplers

If one wants to measure characteristics of a transmitter or high-power amplifier, some means of reducing the power of the device to 10 or 20 dBm must be used. The most straightforward way to do this is to use a 30 or 40 dB attenuator capable of handling the high power. A 30 dB attenuator will reduce a 100 W transmitter to 20 dBm. A 40 dB attenuator will reduce a 1 kW amplifier to 20 dBm. If further attenuation is needed, a simple precision attenuator may be used after the signal has been reduced to the 20 dBm level.

The problem with high-power attenuators is that they are expensive to buy or build because the front end of the attenuator must handle the output power of the transmitter or amplifier. If one already has a dummy load, an RF sampler may be used to produce a replica of the signal at a reduced power level. The sampler described here was originally presented by Tom Thompson, WØIVJ, in *QST*'s May 2011 "Technical Correspondence" column. (The original article with construction information plus some supplemental information is included with this book's online material.)

A transformer sampler passes a single conductor (usually the insulated center conductor from a piece of coaxial cable) from the transmitter or amplifier to the dummy load through a toroidal inductor forming a transformer with a single turn primary. The secondary of the transformer is connected to a resistor network and then to the test equipment as shown in **Figure 28.15** The source, whether a transmitter or amplifier, is assumed to be a pure voltage source in series with a 50-Ω resistor. This most likely is not exactly the case but is sufficient for analysis.

If a current, I, flows into the dummy load, then a current, I / N flows in the secondary of the transformer, where N is the number of turns on the secondary. Figure 28.15 also shows the equivalent circuit, substituting a current source for the transformer. 40 dB is selected for the attenuation and 15 turns for the secondary of the transformer. If R_{SHUNT} = 15 Ω, and R_{SERIES} = 35 Ω, then the voltage across a 50-Ω load resistor, R_{SAMPLE}, is 1/100 of the voltage across the dummy load, which is 40 dB of attenuation.

Reflecting this resistor combination back through the transformer yields 0.06 Ω in series with the 50-Ω dummy load impedance. This is an insignificant change. Furthermore, reflecting 100 Ω from the primary to the secondary places 22.5 kΩ in parallel with R_{SHUNT}, which does not significantly affect its value. The test equipment sees a 50-Ω load looking back into the sampler. Even at low frequencies, where the reactance of the secondary winding is lower than 15 Ω, the impedance looking back into the sample port remains close to 50 Ω.

The samplers described here use an FT37-61 ferrite core followed by two resistors as described above. The through-line SWR is good up to 200 MHz, the SWR is fair looking into the sampled port, and the useful bandwidth extends from 0.5 MHz to about 100 MHz. If you are interested in an accurate representation of the third harmonic of your HF transmitter or amplifier, it is important for the sampler to give accurate attenuation into the VHF range.

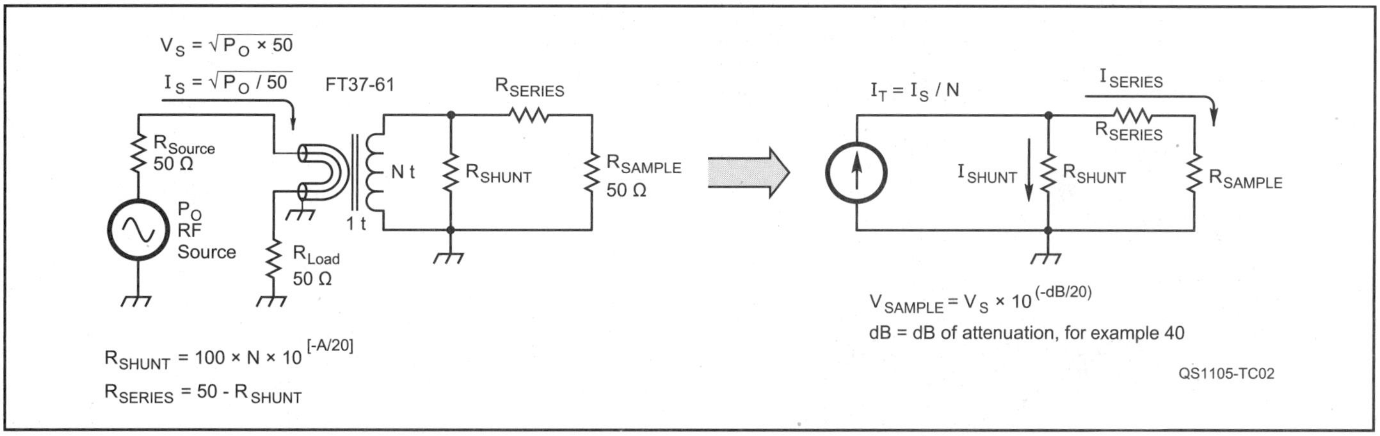

Figure 28.15 — RF sampler circuit diagram and equivalent circuit showing calculations.

Figure 28.16 — RF sampler using box construction.

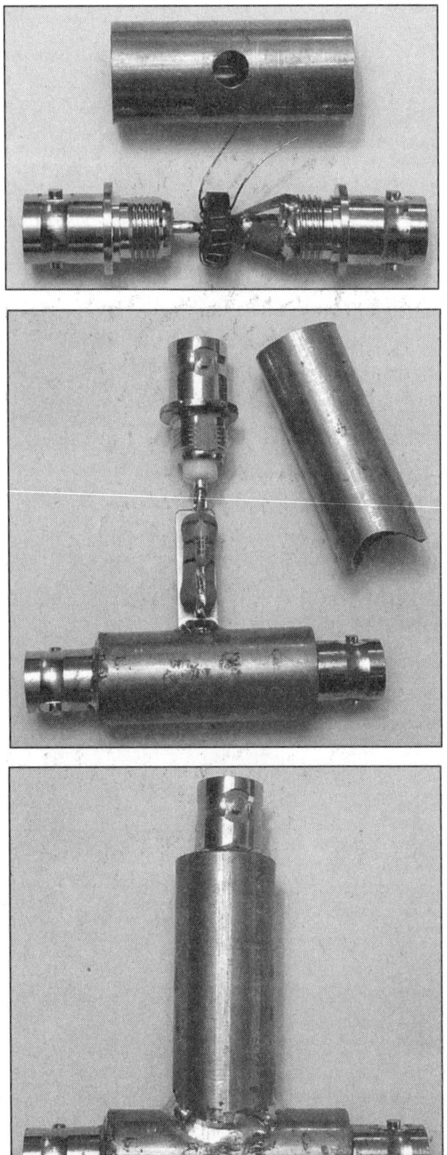

Figure 28.17 — RF sampler using tube construction.

Figure 28.16 shows a photo of a sampler built into a 1.3 × 1.3 × 1 inch (inside dimensions) box constructed from single-sided circuit board material. The through-line connection is made with a short piece of UT-141 semi-rigid coax with the shield grounded only on one side to provide electrostatic shielding between the toroid and the center conductor of the coax. (Do not ground both ends of the shield or a shorted turn is created.) R_{SHUNT} is hidden under the toroid, and R_{SERIES} is shown connected to the sample port. This construction technique looks like a short piece of 200-Ω transmission line in the through-line which affects the SWR at higher frequencies. This can be corrected by compensating with two 3 pF capacitors connected to the through-line input and output connectors as shown in the photo. The through-line SWR was reduced from 1.43:1 to 1.09:1 at 180 MHz by adding the capacitors. This compensation, however, causes the attenuation to differ at high frequencies depending on the direction of the through-line connection. A sampler constructed using the box technique is useable from below 1 MHz through 30 MHz.

Figure 28.17 shows a different approach using 9/16 inch diameter, 0.014 inch wall thickness, hobby brass tubing. This lowers the impedance of the through-line so that no compensation is needed. The through-line SWR for the tube sampler is 1.08:1 at 180 MHz which is as good as the box sampler and the sensitivity to through-line direction is reduced. Although the high frequency attenuation is not as good as the box sampler, the construction technique provides a more consistent result. A sampler constructed using the tube technique should be usable through 200 MHz.

28.3.3 PROJECT: An Inexpensive VHF Directional Coupler

Precision inline metering devices capable of reading forward and reflected power over a wide range of frequencies are very useful in amateur VHF and UHF work, but their rather high cost puts them out of the reach of many VHF enthusiasts. The device shown in **Figures 28.18** through **28.20** is an inexpensive adaptation of their basic principles. It can be made for the cost of a meter, a few small parts, and bits of copper pipe and fittings that can be found in the plumbing section at many hardware stores.

The sampler consists of a short section of handmade coaxial line, in this instance, of 50-Ω impedance, with a reversible probe coupled to it. A small pickup loop built into the probe is terminated with a resistor at one end and a diode at the other. The resistor matches the impedance of the loop, not the impedance of the line section. Energy picked up by the loop is rectified by the diode, and the resultant current is fed to a meter equipped with a calibration control.

The principal metal parts of the device are a brass plumbing T, a pipe cap, short pieces of 3/4-inch ID and 5/16-inch OD copper pipe, and two coaxial fittings. Other available tubing combinations for 50-Ω line may be usable. The ratio of outer conductor ID to inner conductor OD should be 2.4/1. (The complete article with more detail is included with this book's online material.)

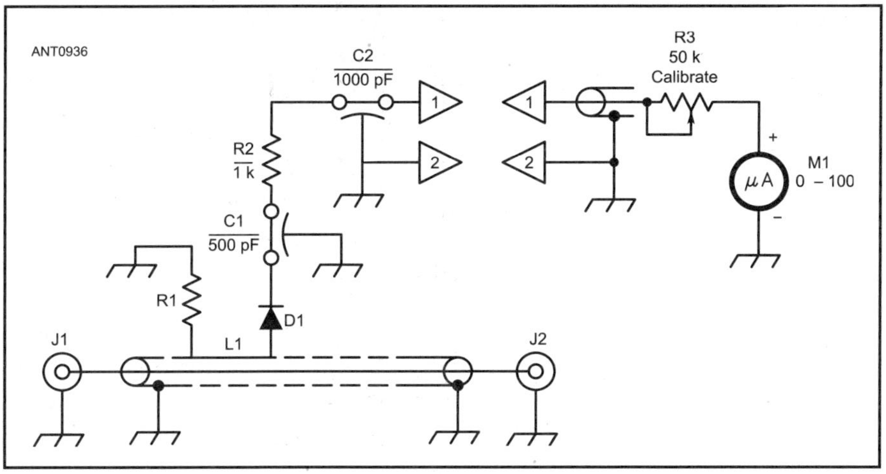

Figure 28.18 — Circuit diagram for the line sampler.
C1 — 500-pF feedthrough capacitor, solder-in type.
C2 — 1000-pF feedthrough capacitor, threaded type.
D1 — 1N34A germanium or 1N5817 Schottky diode.
J1, J2 — Coaxial connector, type N (UG-58A).
L1 — Pickup loop, copper strap 1-inch long × 3⁄16-inch wide. Bend into "C" shape with flat portion 5⁄8-inch long.
M1 — 0-100 μA meter.
R1 — 82 to 100 Ω, carbon-film or metal film.
R3 — 50-kΩ composition control, linear taper.

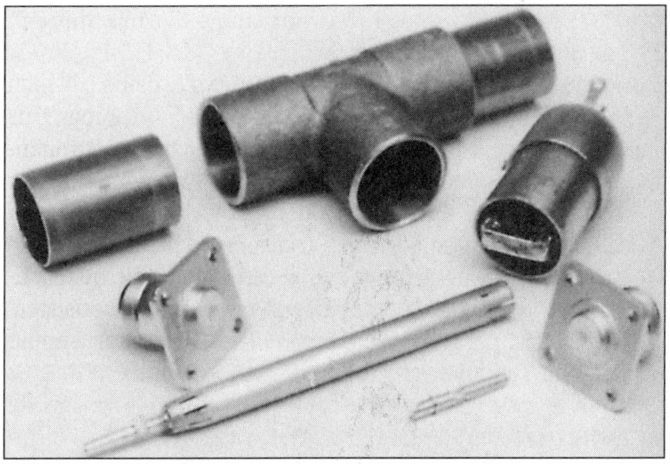

Figure 28.19 — Major components of the line sampler. The brass T and two end sections are at the upper left in this picture. A completed probe assembly is at the right. The N connectors have their center pins removed. The pins are shown with one inserted in the left end of the inner conductor and the other lying in the right foreground.

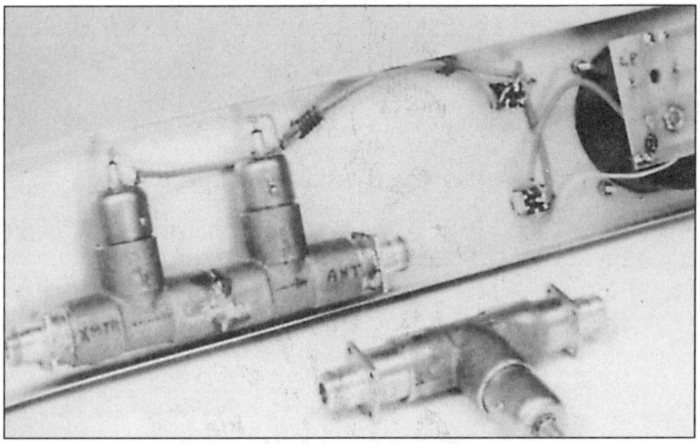

Figure 28.20 — Two versions of the line sampler. The single unit described in detail here is in the foreground. Two sections in a single assembly provide for monitoring forward and reflected power without probe reversal.

The sampler is very useful for many jobs even if it is not accurately calibrated, although it is desirable to calibrate it against a wattmeter of known accuracy as described in the complete article.

28.3.4 PROJECT: RF Step Attenuator

A good RF step attenuator is one of the key pieces of equipment that belongs on your workbench. The attenuator in this project offers good performance yet can be built with a few basic tools. The attenuator is designed for use in 50-Ω systems, provides a total attenuation of 71 dB in 1-dB steps, offers respectable accuracy and insertion loss through 225 MHz and can be used at 450 MHz as shown in **Table 28.1**. This material was originally published as "An RF Step Attenuator" by Denton Bramwell, K7OWJ, in the June 1995 *QST*.

The attenuator consists of 10 resistive pi (π) attenuator sections such as the one in **Figure 28.21**. Each section consists of a DPDT slide switch and three ¼-W, 1%-tolerance metal-film resistors. The complete unit contains single 1, 2, 3 and 5-dB sections, and six 10-dB sections. **Table 28.2** lists the resistor values required for each section.

The enclosure is made of brass sheet stock, readily available at hardware and hobby stores. By selecting the right stock, you can avoid having to bend the metal and need only perform a minimum of cutting.

Table 28.1
Step Attenuator Performance at 148, 225, and 250 MHz
Measurements made in the ARRL Laboratory

Attenuator set for Maximum attenuation (71 dB)		*Attenuator set for minimum attenuation (0 dB)*	
Frequency (MHz)	*Attenuation (dB)*	*Frequency (MHz)*	*Attenuation (dB)*
148	72.33	148	0.4
225	73.17	225	0.4
450	75.83	450	0.84

Note: Laboratory-specified measurement tolerance of ±1 dB

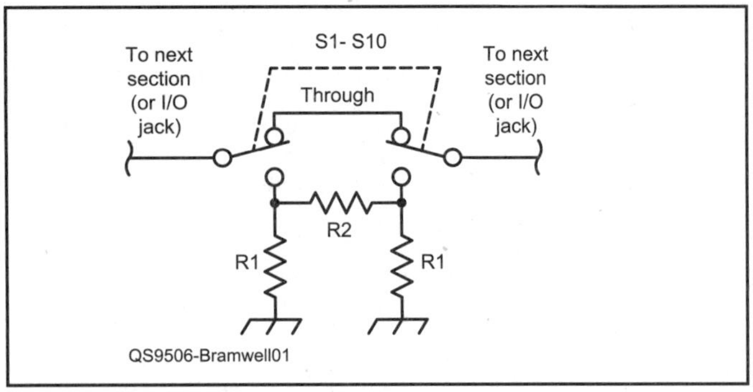

Figure 28.21 — Schematic of one section of the attenuator. All resistors are ¼-W, 1%-tolerance metal-film units. See Table 28.2 for the resistor values required for each attenuator section. There are six 10 dB sections and one each of 1, 2, 3, and 5 dB.

Table 28.2
Closest 1%-Tolerance Resistor Values

Attenuation (dB)	*R1 (Ω)*	*R2 (Ω)*
1.00	866.00	5.60
2.00	436.00	11.50
3.00	294.00	17.40
5.00	178.00	30.10
10.00	94.30	71.50

CONSTRUCTION

The enclosure can be built using only a nibbling tool, drill press, metal shears, and a soldering gun or heavy soldering iron. (Use a regular soldering iron on the switches and resistors.) One method of cutting the small pieces of rectangular tubing to length is to use a drill press equipped with a small abrasive cutoff wheel.

Brass is easy to work and solder. For the enclosure, you'll need two precut 2 × 12 × 0.025-inch sheets and two 1 × 12 × 0.025-inch sheets. The 2-inch-wide stock is used for the front and back panels; the 1-inch-wide stock is used for the ends and sides. For the internal wiring, you need a piece of 5⁄32 × 5⁄16-inch rectangular tubing, a ¼ × 0.032-inch strip, and a few small pieces of 0.005-inch-thick stock to provide interstage shields and form the 50-Ω transmission lines that run from the BNC connectors to the switches at the ends of the step attenuator.

For the front panel, nibble or shear a piece of 2-inch-wide brass to a length of about 9½ inches. Space the switches from each other so that a piece of the rectangular brass tubing lies flat and snugly between them. See **Figure 28.22**. Drill holes for the #4-40 mounting screws and nibble or punch rectangular holes for the bodies of the slide switches.

Before mounting any parts, solder in place one of the 1-inch-wide chassis side pieces to make the assembly more rigid. Solder the side piece to the edge of the top plate that faces the "through" side of the switches; this makes later assembly easier (see **Figure 28.23**). Although the BNC input and output connectors are shown mounted on the top (front) panel, better lead dress and high-frequency performance may result from mounting the connectors at the ends of the enclosure.

DPDT slide switches designed for sub-panel mounting often have mounting holes tapped for #4-40 screws. Enlarge the holes to allow a #4-40 screw to slide through. Before mounting the switches, make the "through" switch connection (see Figure 28.21) by bending the two lugs at one end of each switch toward each other and soldering the lugs together or solder a small strip of brass between the lugs and clip off the lug ends. Mount the switches above the front panel, using 5⁄32-inch-high by 7⁄32-inch-OD spacers. Use the same size spacer on the inside. On the inside, the spacer creates a small post that helps reduce capacitive coupling from one side of the attenuator to the other. The spacers position the switch so that the 50-Ω stripline can be formed later.

The trick to getting acceptable insertion loss in the "through" position is to make the attenuator look as much as possible like 50-Ω coax. That's where the rectangular tubing and the ¼ × 0.032-inch brass strip come into the picture (see Figure 28.22); they form a 50-Ω stripline. (See the **Transmission Lines** chapter for information on stripline.)

Cut pieces of the rectangular tubing about ¾-inch long, and sweat solder them to the front panel between each of the slide switches. Next, cut lengths of the ¼-inch strip long enough to conveniently reach from switch to switch, then cut one more piece. Drill 1⁄16-inch holes near both ends of all but one of the ¼-inch strips. The undrilled piece is used as a temporary spacer, so make sure it is flat and deburred.

Lay the temporary spacer on top of the rectangular tubing between the first two switches, then drop one of the drilled ¼-inch pieces over it, with the center switch lugs through the 1⁄16-inch holes. Before soldering, check the strip to make sure there's sufficient clearance between the ¼-inch strip and the switch lugs; trim the corners if necessary. Use a screwdriver blade to hold the strip flat and solder the lugs to the strip. Remove the temporary spacer. Repeat this procedure for all switch sections. This creates a 50-Ω stripline running the length of the attenuator.

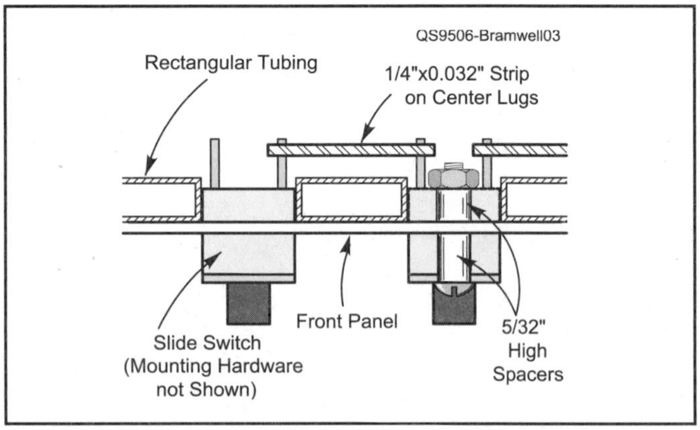

Figure 28.22 — Key to obtaining acceptable insertion loss in the "through" position is to make the whole device look as much as possible like 50-Ω coax. The rectangular tubing and the ¼ × 0.032-inch brass strip between the switch sections form a 50-Ω stripline.

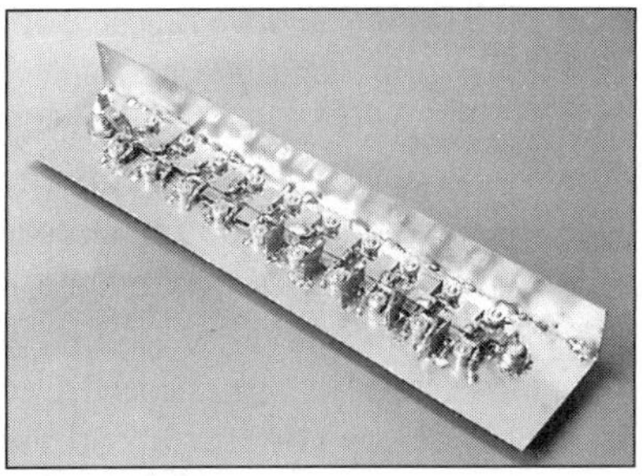

Figure 28.23 — Solder one of the 1-inch-wide chassis side pieces in place to make the assembly more rigid during construction. Solder the side piece to the edge of the top plate that faces the "through" side of the switches; this makes the rest of the assembly easier.

Figure 28.24 — The attenuator before final mechanical assembly. The ¼-inch strips are spaced 0.033 inch apart to form a 50-Ω connection from the BNC connector to the stripline. There are ½-inch square shields between 10-dB sections. The square shields have a notch in one corner to accommodate the end of the rectangular tubing.

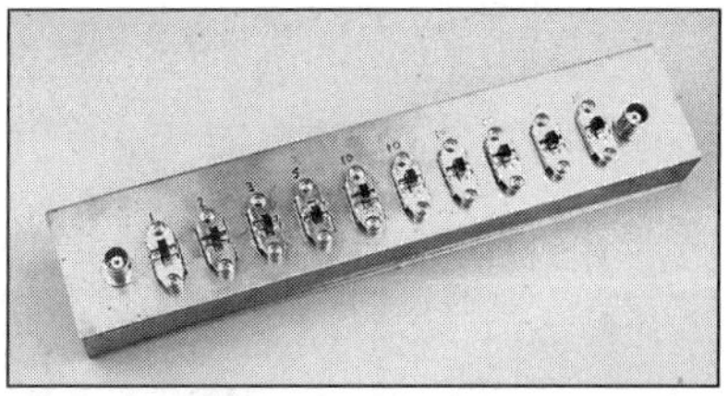

Figure 28.25 — The completed step attenuator in the enclosure of brass sheet. The BNC connectors may be mounted on the front panel at the end of the switches or on the end panels.

Next, solder in place the three 1%-tolerance resistors of each section, keeping the leads as short as possible. Use a generous blob of solder on ground leads to make the lead less inductive. Install a ½-inch-square brass shield between each 10-dB section to ensure that signals don't couple around the sections at higher frequencies.

Use parallel ¼-inch strips of 0.005-inch-thick brass spaced 0.033 inch apart to form 50-Ω feed lines from the BNC connectors to the switch contacts at each end of the stripline as shown in **Figure 28.23**. (Use the undrilled piece of 0.032-inch-thick brass to insure the proper line spacing.) The attenuator with all switches and shields in place is shown ready for final mechanical assembly in **Figure 28.24**

Finally, solder in place the remaining enclosure side, cut and solder the end pieces, and solder brass #4-40 nuts to the inside walls of the case to hold the rear (or bottom) panel. Drill and attach the rear panel and round off the sharp corners to prevent scratching or cutting anyone or anything. Add stick-on feet and labels and your step attenuator of **Figure 28.25** is ready for use.

Remember that the unit is built with ¼-W resistors, so it can't dissipate a lot of power. Remember, too, that for the attenuation to be accurate, the input to the attenuator must be a 50-Ω source and the output must be terminated in a 50-Ω load.

28.4 Field Strength Meters

Few amateur stations, fixed or mobile, are without need of a *field-strength meter* (*FSM*). An instrument of this type serves many useful purposes during antenna experiments and adjustments. In its simplest form, the field strength meter is simply a diode detector and a sensitive meter with a potentiometer wired as a resistive divider to act as a sensitivity control as in **Figure 28.26**. This type of meter is commonly and inexpensively available both new and used. (See the References and Bibliography and this book's online material for the *QST* article describing how to build this simple FSM.)

When work is to be done from many wavelengths away from the antenna, however, such a simple instrument lacks the necessary sensitivity. Further, such a device has a serious fault because its linearity leaves much to be desired and it is very wideband so that the measurement may be upset from any other strong nearby transmitter, such as an AM broadcast station. Thus, a more capable instrument is needed.

28.4.1 PROJECT: Portable Field Strength Meter

The field-strength meter described here takes care of the problems associated with a simple FSM. Additionally, it is small, measuring only 4 × 5 × 8 inches. The power supply consists of two 9-V batteries. Sensitivity can be set for practically any amount desired. However, from a usefulness standpoint, the circuit should not be too sensitive or it will respond to unwanted signals. This unit also has excellent linearity with regard to field strength. (The field strength of a received signal varies inversely with the distance from the source, all other things being equal.) The frequency range includes all amateur bands from 3.5 through 148 MHz, with band-switched circuits, thus avoiding the use of plug-in inductors. All in all, it is a quite useful instrument. The information in this section is based on a January 1973 *QST* article by Lew McCoy, W1ICP (see the References and Bibliography).

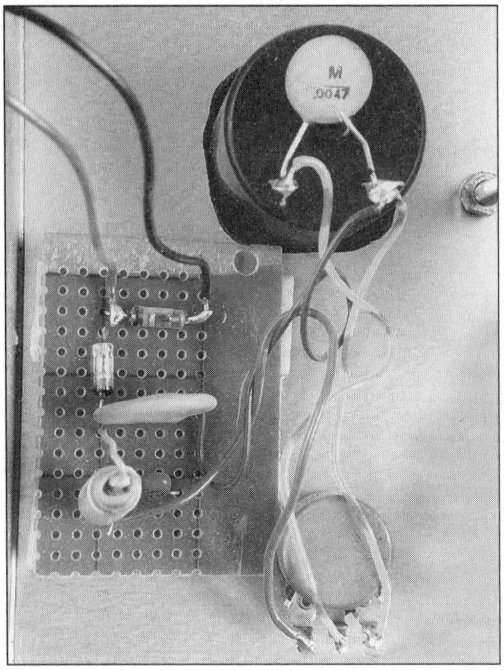

(A)

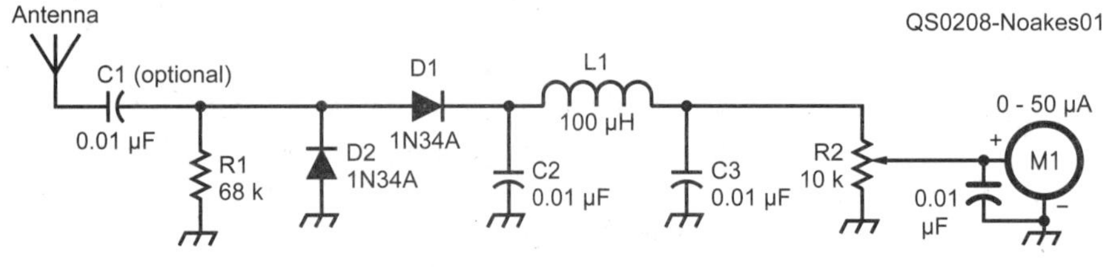

(B)

Figure 28.26 — With the sensitivity control at mid-range, the simple FSM shown at A can easily detect a 1-W, 2 meter signal. B gives the schematic and parts list for the FSM. A metal enclosure is mandatory.

C1-C3 — 0.01 µF disc ceramic capacitor.

D1, D2 — 1N34A (germanium) or 1N5817 (Schottky) diode.

L1 — 100 µH inductor.

M1 — 50 µA analog meter.

R1 — 68 kΩ, ¼-W carbon-film or metal-film resistor.

R2 — 10 kΩ linear or audio taper potentiometer.

Antenna — BNC chassis-mount receptacle.

The unit is pictured in **Figures 28.27** and **28.28**, and the schematic diagram is shown in **Figure 28.29**. A type 741 op-amp IC is the heart of the unit. (Any general-purpose op-amp can be substituted for the 741.) The antenna is connected to J1, and a tuned circuit is used ahead of a diode detector. The rectified signal is coupled as dc and amplified in the op amp. Sensitivity of the op amp is controlled by inserting resistors R3 through R6 in the circuit by means of S2.

With the circuit shown, and in its most sensitive setting, M1 will detect a signal from the antenna on the order of 100 µV. Linearity is poor for approximately the first ⅕ of the meter range, but then is almost straight-line from there to full-scale deflection. The reason for the poor linearity at the start of the readings is because of nonlinearity of the diodes at the point of first conduction. However, if gain measurements are being made this is of no real importance, as accurate gain measurements can be made in the linear portion of the readings.

The 741 op amp requires both a positive and a negative voltage source. This is obtained by connecting two 9-V batteries in series and grounding the center. One other feature of the instru-

Figure 28.27 —The linear field strength meter. The control at the upper left is for C1 and the one to the right for C2. At the lower left is the band switch, and to its right the sensitivity switch. The zero-set control for M1 is located directly below the meter.

Figure 28.28 — Inside view of the field-strength meter. At the upper right is C1 and to the left, C2. The dark leads from the circuit board to the front panel are the shielded leads described in the text under Construction Notes.

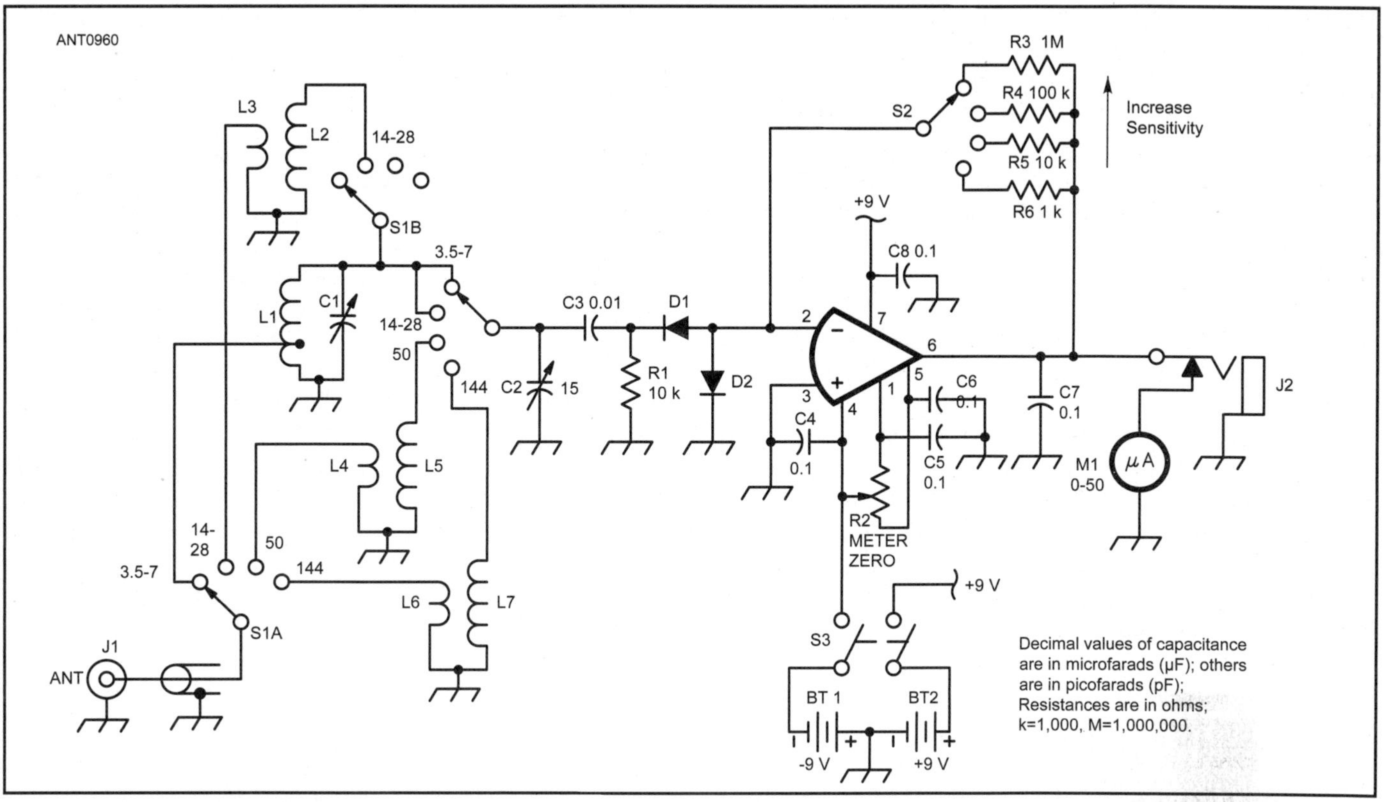

Figure 28.29 — Circuit diagram of the linear field strength meter. All resistors are ¼- or ½-W carbon-film or metal-film types.

C1 — 140 pF variable.
C2 — 15-pF variable
D1, D2 — 1N4148 or equiv.
L1 — 34 turns #24 AWG enameled wire wound on an T-68-2 core, tapped 4 turns from ground end.
L2 — 12 turns #24 AWG enameled wire wound on T-68-2 core.
L3 — 2 turns #24 AWG enameled wire wound at ground end of L2.
L4 — 1 turn #26 AWG enameled wire wound at ground end of L5.
L5 — 12 turns #26 AWG enameled wire wound on T-25-12 core.
L6 — 1 turn #26 AWG enameled wire wound at ground end of L7.
L7 — 1 turn #18 AWG enameled wire wound on T-25-12 core.
M1 — 50 or 100 μA dc.
R2 — 10-kΩ control, linear taper.
S1 — Rotary switch, 3 poles, 5 positions, 3 sections.
S2 — Rotary switch, 1 pole, 4 positions.
S3 — DPST toggle.
U1 — Type 741 op amp or equivalent. Pin numbers shown are for a 14-pin package.

ment is that it can be used remotely by connecting an external meter at J2. This is handy if you want to adjust an antenna and observe the results without having to leave the antenna site.

L1 is the 3.5/7 MHz coil and is tuned by C1. The coil is wound on a toroid form. For 14, 21 or 28 MHz, L2 is switched in parallel with L1 to cover the three bands. L5 and C2 cover approximately 40 to 60 MHz, and L7 and C2 from 130 MHz to approximately 180 MHz. The two VHF coils are also wound on toroid forms.

CONSTRUCTION NOTES

The majority of the components may be mounted on an etched circuit board. A shielded lead should be used between pin 4 of the IC and S2. The same is true for the leads from R3 through R6 to the switch. Otherwise, parasitic oscillations may occur in the IC because of its very high gain.

In order for the unit to cover the 144-MHz band, L6 and L7 should be mounted directly across the appropriate terminals of S1, rather than on a circuit board. The extra lead length adds too much stray capacitance to the circuit. It isn't necessary to use toroid forms for the 50- and 144 MHz coils. They were used in the version described here simply because they were available. You may substitute air-wound coils of the appropriate inductance.

CALIBRATION

The field strength meter can be used as is for a relative-reading device. A linear indicator scale will serve admirably. However, it will be a much more useful instrument for antenna work if it is calibrated in decibels, enabling the user to check relative gain and front-to-back ratios. If you have access to a calibrated signal generator, connect it to the field-strength meter and use different signal levels fed to the device to make a calibration chart. Convert signal-generator voltage ratios to decibels by using the equation

$$dB = 20 \log (V1/V2) \quad (3)$$

where V1/V2 is the ratio of the two voltages and log is the common logarithm (base 10).

Let's assume that M1 is calibrated evenly from 0 to 10.

Next, assume we set the signal generator to provide a reading of 1 on M1, and that the generator is feeding a 100 µV signal into the instrument. Now we increase the generator output to 200 µV, giving us a voltage ratio of 2:1. Also let's assume M1 reads 5 with the 200 µV input. From the equation above, we find that the voltage ratio of 2 equals 6.02 dB between 1 and 5 on the meter scale. M1 can be calibrated more accurately between 1 and 5 on its scale by adjusting the generator and figuring the ratio. For example, a ratio of 126 µV to 100 µV is 1.26, corresponding to 2.0 dB. By using this method, all of the settings of S2 can be calibrated. In the instrument shown here, the most sensitive setting of S2 with R3, 1 MΩ, provides a range of approximately 6 dB for M1. Keep in mind that the meter scale for each setting of S1 must be calibrated similarly for each band. The degree of coupling of the tuned circuits for the different bands will vary, so each band must be calibrated separately.

Another method for calibrating the instrument is using a transmitter and measuring its output power with an RF wattmeter. In this case we are dealing with power rather than voltage ratios, so this equation applies:

$$dB = 10 \log (P1/P2) \quad (4)$$

where P1/P2 is the power ratio.

With most transmitters the power output can be varied, so calibration of the test instrument is rather easy. Attach a pickup antenna to the field-strength meter (a short wire a foot or so long will do) and position the device in the transmitter antenna field. Let's assume we set the transmitter output for 10 W and get a reading on M1. We note the reading and then increase the output to 20 W, a power ratio of 2. Note the reading on M1 and then use Eq 4. A power ratio of 2 is 3.01 dB. By using this method the instrument can be calibrated on all bands and ranges.

With the tuned circuits and coupling links specified in Figure 28.29, this instrument has an average range on the various bands of 6 dB for the two most sensitive positions of S2, and 15 dB and 30 dB for the next two successive ranges. The 30-dB scale is handy for making front-to-back antenna measurements without having to switch S2.

28.5 Impedance, Antenna, and Network Analyzers

Another type of analyzer measures parameters of components, antennas, and circuits over a frequency range. These analyzers are often used to measure impedance and are sometimes referred to as *impedance analyzers*. Analyzers optimized for use with antenna systems are called *antenna analyzers*. More capable analyzers are used to measure gain, filter response, isolation, and other properties of antennas, circuits, and systems that are referred to by the general name of *networks*. These analyzers are called *network analyzers*. The material on vector analyzers was originally updated for the *ARRL Handbook* by Jim Brown, K9YC.

28.5.1 Network Analyzer Basics

Over the last decade, a new generation of relatively low-cost analyzers has emerged, both as free-standing handheld instruments, and as instruments that interface to a computer via a USB port and require software running on the computer to complete the measurement system. Some of the free-standing instruments can operate under computer control and with the computer software greatly expanding their capabilities.

Connections to the analyzer are made at a *port* which is a pair of terminals that serve as an input or output of a network. The port terminals can be anything from alligator clips to a coaxial connector. A network can have more than one port. A dummy load is a *single-port network*, for example, and an amplifier or filter or transmission line are examples of a *two-port network* with an input and output.

A *scalar analyzer* measures only the magnitude of a parameter (a scalar is a value expressed as a single number). Both oscilloscopes and spectrum analyzers are scalar instruments. Scalar parameters include SWR, attenuation, or power. A *vector analyzer* measures the magnitude of two parameters (typically voltages) and the phase difference between them. That combination creates a vector with a magnitude and angle. The measurements are repeated over a range of frequencies. (More information about vectors, scalars, and phasors is available in the **Radio Mathematics** supplement on the ARRL website. Enter "radio mathematics" into the Website Search window at **arrl.org**.)

Vector analyzers make measurements of the *scattering parameters (S-parameters)* of a *single-port* or *two-port network*. They allow any circuit or component to be treated as a "black box" with the S-parameters describing its behavior in terms of amplitude and phase.

The scattering parameters for a two-port network, are:

S_{11}, the reflection coefficient of the input port;

S_{21}, the voltage gain from input to output, also known as Forward Transmission or Forward Gain;

S_{12}, the voltage gain from output to input, also known as Reverse Transmission;

S_{22}, the reflection coefficient of the output port.

Scattering parameters and ports are explained in the following sections S-Parameters and Return Loss.

28.5.2 Antenna Analyzers

Because measurements of antenna systems, including feed lines, are so common, a class of analyzers has been designed with features specifically for that task. The popular name for these instruments is *antenna analyzers*, whether they use analog or digital circuitry.

In many products available today, the measurements and graphs are displayed on a built-in screen about the size of those on mobile phones. The real-time graphing functions are especially helpful when making tuning adjustments in the field. Data is stored internally and can be downloaded to an external PC for processing and storage. (See the sidebar on Touchstone data files)

Antenna analyzers suitable for amateur use are available from several manufacturers — search for "antenna analyzers" on the internet to find them. When shopping for an antenna analyzer, pay careful attention to the capabilities and limitations. Several basic designs are available with tradeoffs between performance and cost.

Be aware that some inexpensive or older units measure only the SWR or impedance magnitude while others measure both the resistive and reactive parts of the impedance. Some of these units may give the magnitude of the reactance but not the sign, requiring the operator to change frequency a small amount and watch the change in impedance magnitude to determine the sign and thus the type of the impedance, inductive or capacitive.

These analyzers have many more uses than impedance and SWR measurements. Paul Wade, W1GHZ, has contributed the paper "Antenna Analyzer Pet Tricks" in this book's online material. It includes how-to guides for a number of useful antenna system measurements.

Some users have reported difficulties in obtaining accurate impedance measurements on low-band antennas when there is a nearby AM broadcast station. This is due to the wideband detector responding to the incoming signal from the AM station. Some manufacturers offer external high-pass broadcast-reject filters to allow the analyzers to be used in the presence of these strong signals. The filter can affect measurements near the broadcast band, particularly in the 3.5 MHz and lower-frequency bands. Check with the filter manufacturer about the limitations of using filters with the analyzer.

SCALAR ANTENNA ANALYZERS

The simplest way to measure the frequency response of a coax-fed antenna to show its resonant frequency or frequency of minimum SWR is to use a standing wave ratio meter. By measuring SWR, the meter shows how well-matched the load or the antenna's feed point impedance is to the characteristic impedance of the feed line. The frequency of the measurement is controlled by changing the frequency of the transmitter. Because SWR is a single numeric value, the combination of a transmitter and SWR meter creates a simple scalar antenna analyzer.

SWR meters are usually calibrated for 50 Ω since most transmitters are designed to drive a 50-Ω load. (Some can be switched to operate in a 75-Ω system.) An SWR meter by itself, however, cannot provide any information about the values of resistance and reactance that make up the impedance.

The first self-contained antenna analyzers replaced the transmitter with a manually tuned low-power signal source and added a detector and analog meter in a single piece of portable equipment. These simple scalar analyzers display SWR and/or impedance magnitude, but not separate values of resistance and reactance. While their capabilities are less sophisticated than today's more advanced equipment, they are still widely used for simple adjustments and troubleshooting antennas.

Today's scalar antenna analyzers add more signal processing capabilities and use phase measurements to calculate both resistance and reactance, along with SWR. Some units have a manually tuned signal source while others make swept frequency measurements under the control of a microprocessor. **Figure 28.30** shows the block diagram of a typical scalar

Determining Reactance Type

Some antenna analyzers can show the positive or negative sign of the reactance. Less expensive units may only show a reactance magnitude. To determine whether reactance is inductive or capacitive on a unit that doesn't indicate the sign, make a slight increase in frequency. If the reactance increases, it is inductive, and if reactance decreases, it is capacitive. This is a helpful trick when adjusting antennas with an analyzer that doesn't indicate reactance sign.

Protecting Antenna Analyzers

Portable antenna analyzers are sensitive instruments with low-power components at their input. They can be damaged by static electricity if care is not exercised. Antennas can collect a significant static charge from rain or wind. Be sure to short the conductors together and ground the transmission line for a few seconds before connecting it to an analyzer to reduce the risk of damage to the sensitive components on the input.

Static buildup can be minimized by providing a dc path between center conductor and the grounded shield for the antenna being measured, offering some protection both for the analyzer and for station equipment. The dc path can be a shunt inductor (typically an RF choke) or a high-value resistor (10 kΩ or more). A very useful protector can be made by soldering the resistor between the center pin and shell of a PL-259 and inserting the PL-259 into a UHF tee adapter that is connected to the analyzer. If the discharge resistor is to be a permanent part of the antenna, it must have a power rating suitable for the transmitter power. Metal oxide resistors of a few watts dissipation are a good choice.

Antennas with non-insulated elements mounted directly on a grounded boom or support are already protected against static. Insulated elements need protection against static. Measure resistance between the feed line center conductor and shield to be sure of whether there is a dc path to drain static.

Using the analyzer when another transmitter is active can also cause damage if enough signal is picked up by the antenna under test. Some analyzers include an isolation relay that protects the input when a measurement is not in progress.

Noise Bridges

Noise bridges (see **Figure 28.A**) were a very common test instrument until the modern antenna analyzer became widespread. The noise bridge includes an adjustable bridge circuit to which a wide-band noise generator is connected as the source and a conventional receiver is attached to the detector port. The receiver is tuned to the desired frequency and the resistance and reactance controls are adjusted for minimum noise in the receiver. If the receiver has a panadapter display, the null frequency can be seen on the screen, speeding the adjustment. The *QST* article "The Noise Bridge" by Jack Althouse, K6NY, gives a more complete description of how a noise bridge works and how it is used. The article is included in the online material.

Figure 28.A — A noise bridge is an alternative to the antenna analyzer, pairing a noise source with an external receiver as the detector. The bridge is adjusted for minimum noise in the receiver and values for R and X are read from calibrated dials.

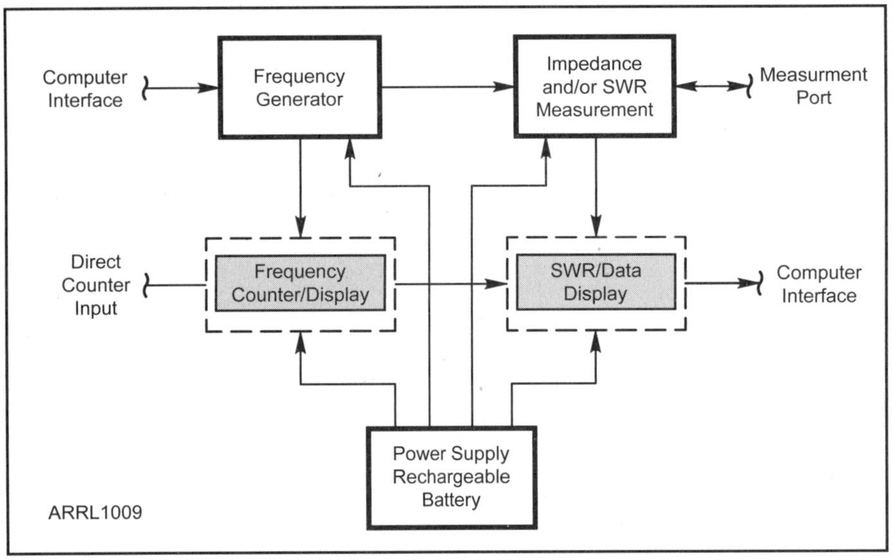

Figure 28.30 — Block diagram showing the elements of typical antenna analyzers. The items shown with dashed lines may not be part of every model.

antenna analyzer. **Figure 28.31** shows examples of both types of scalar antenna analyzers.

Figure 28.32 shows a relatively simple circuit that uses diodes for detecting voltages corresponding to voltage and current at the external load. This inexpensive circuit is useful but at low signal levels the diodes introduce some non-linearity and temperature drift which may be an issue. This type of diode detector responds to signals over a wide frequency range and there may be stray pickup from nearby broadcasting stations that makes the measurement results inaccurate.

VECTOR ANTENNA ANALYZERS

Today, the most common antenna analyzer is a *vector impedance analyzer* (VIA) that measures both the magnitude and the phase angle of impedance at a single-port across a range of frequencies. The impedance is then displayed numerically ($R + jX$), as a phasor ($|Z|\angle\theta$), or some computed value, such as SWR. (Magnitude is denoted by vertical bars, | |, and the angle by ∠.) The measurements can also be displayed on a graph and many instruments can transfer the measurements to a computer for further processing and storage. A VIA can measure the impedance of a single-port network such as an antenna, the input port of an antenna tuner, or a two-terminal component like a resistor or capacitor.

The impedance data can be used to calculate several parameters for the antenna system. Using the specified value for the system reference impedance, which can be any value (it doesn't have to be 50 Ω), For example, the reflection coefficient (ρ or rho) can be calculated as:

$$|\rho| = (Z_L - Z_0) / (Z_L + Z_0)$$

where Z_L is the measured impedance of the load and Z_0 is the specified impedance of the transmission line, which can be any value. Z_L is a complex number; therefore, ρ is, in general, a complex number with a magnitude between zero and one. Rho is approximately equal to zero when the line is matched to the antenna because there is no reflection, all the transmitter power is absorbed by the antenna. When the antenna is poorly matched to the line, ρ is larger and it approaches 1 when the mismatch is large.

$$SWR = (1 + |\rho|) / (1 - |\rho|)$$

Note that SWR only depends on the magnitude of ρ denoted by the vertical bars as $|\rho|$ so it is not a complex number.

28.5.3 Vector Network Analyzers

A *vector network analyzer* (VNA) is an instrument that can measure both the magnitude and phase of the gain and return loss of a two-port network. The VNA's signal generator has a 50-Ω impedance that drives the network port and detector circuits. The generator can drive an input port so that voltages are measured at the output port, or a switch can reverse the signal flow. The first configuration allows the VNA to measure S_{11} and S_{21}. The signal flow is reversed to measure S_{12} and S_{22}.

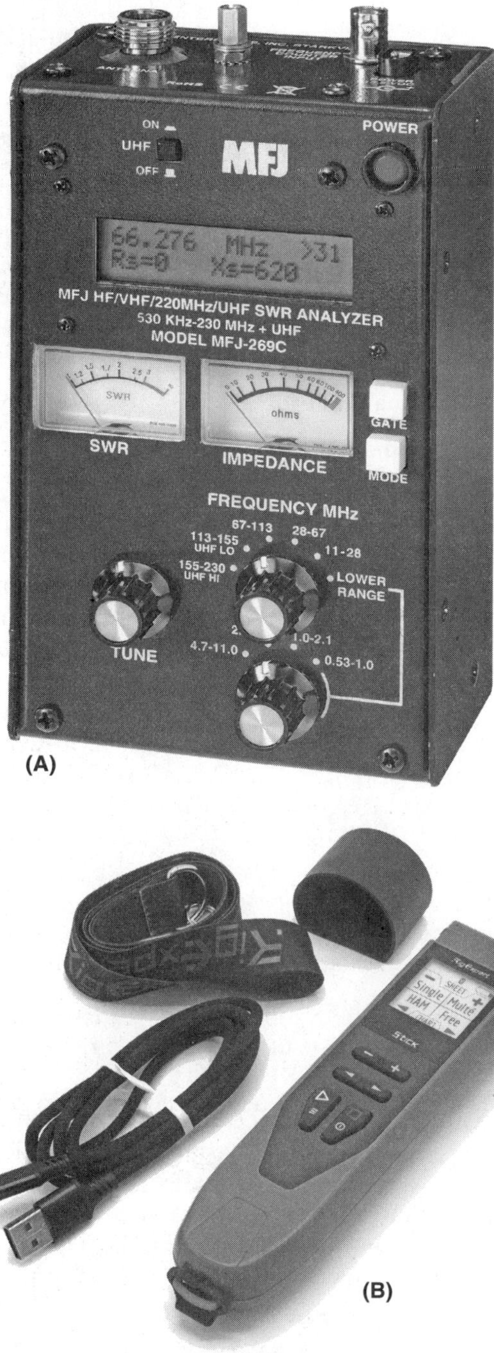

(A)

(B)

Figure 28.31 — The MFJ-259 (A) and -269 series of analyzers revolutionized antenna testing. Rig Expert (B) goes a step further with swept-frequency measurements and a graphical display on the STICK 230.

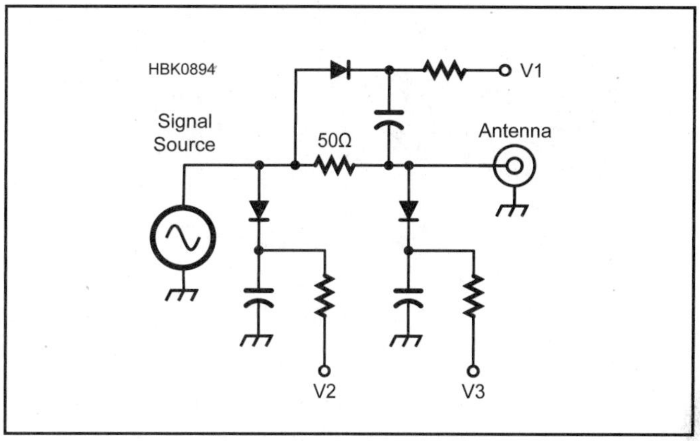

Figure 28.32 — A relatively simple diode detector circuit for measuring impedance that gives adequate results for many applications.

Usually, the VNA or its host software provides the capability to plot the S parameters on a Smith chart for easy evaluation. (See the online material for the **Transmission Lines** chapter for a tutorial on the Smith chart.) S_{11} and S_{21} are often available in a magnitude versus frequency graph, as well.

A VIA has similar construction to a VNA but only includes the 50-Ω generator and bridge circuitry to measure S_{11}, so it can make only single-port measurements such as impedance, SWR, etc. From S_{11}, the analyzer can compute the complex impedance (resistance separately from reactance, including the sign of the reactance to tell us whether it's inductive or capacitive). Most antenna analyzers are VIAs. A two-port VNA can also be configured to do single-port swept frequency measurements. **Figure 28.33** shows a pair of portable VIAs used for antenna measurements. **Figure 28.34** shows a two-port VNA with a coverage of 50 kHz to 4 GHz that is intended for use on the test bench with a host PC.

A complete discussion of how to use a VNA is beyond the scope of this chapter but the user's manual for a VNA will explain the technique of calibrating that particular VNA and using it to make measurements. The second step will be the use of the measurements by computer software to display and transform the measurements into the desired parameters.

In addition, there are many online tutorials and application notes (see the References and Bibliography entries for Agilent) and the text *Microwave Electronics* by Pozar also goes into some detail about what the various measurements are and how they are made.

28.5.4 Calibration and Measurement Plane

The accuracy of a VNA depends on the performance of the directional couplers, detectors, and other circuitry. The coupler directivity is a key contributor to measurement errors. For example, if the directivity is 20 dB, then even a perfect 50-Ω

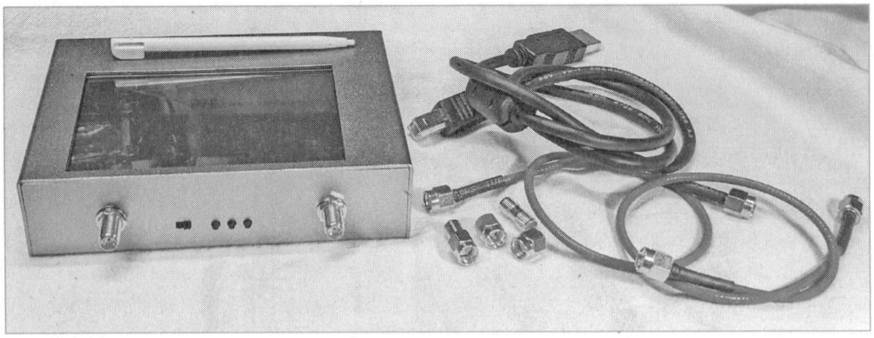

Figure 28.33 — The HCXQS Group NanoVNA covers 50 kHz to 4 GHz and is a complete, two-port vector network analyzer.

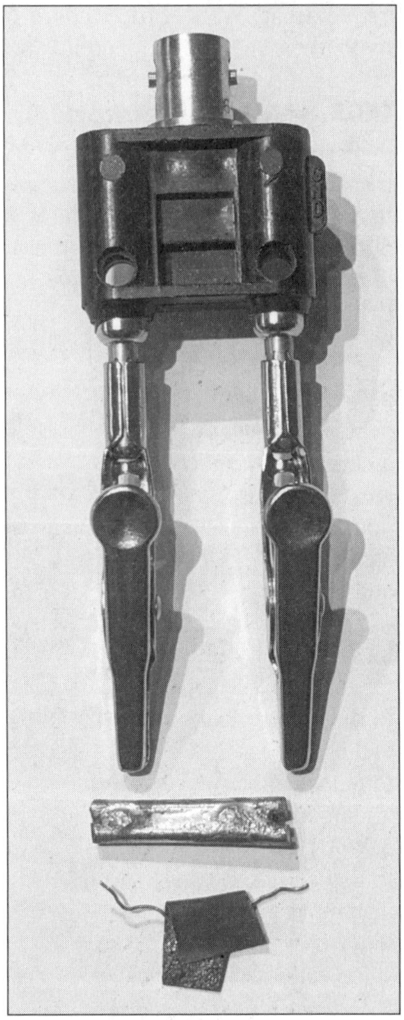

Figure 28.34 — A simple vector impedance analyzer (VIA) test fixture for measuring component impedance into the VHF range. The fixture is a BNC-to-banana plug adaptor with alligator clips on the plugs. Below the clips are the short circuit calibration load and a non-inductive 50 Ω resistor to perform the SOL calibration. (Photo courtesy of Jim Brown, K9YC)

load will show 20 dB of return loss, equivalent to a 1.22:1 SWR. Fortunately, it is possible to eliminate most errors by measuring several terminations of known impedance and calculating a table of correction factors that are applied to the raw measurements. The mathematics involved is quite complex, but it is all handled by the instrument software and is invisible to the user. A thorough discussion of VNA error-correction techniques can be found in Agilent Technologies application note 5965-7709. (**keysight.com**)

The point at which the measurements are to be performed is called the *measurement plane* or *reference plane*. That point can be at the measurement connector of a VIA, at a test fixture connected to the analyzer, or at the ends of cables. (The end of the cable can even be a long distance from the analyzer at the top of a tower!) Calibration consists of taking measurements of known impedances. The analyzer makes the measurements across the desired frequency range and then applies mathematical corrections at each frequency so that the net result is the exact known values. This set of measurements is saved as

Choosing a Computer-Controlled VNA

The most capable and useful VNAs are controlled by a computer. They convert the measurement to base-band audio, which is fed to the computer via an integrated USB interface. Software on the computer controls the analyzer, performs computations to turn the audio into a measured result, and plots the results. These are things to look for:

• It should be a precision instrument, solidly built and with a high-quality bridge, so that it can be reliably calibrated.

• A frequency range of 100 kHz to 600 MHz is a good starting point; 10 kHz to 1.5 GHz is even better.

• The dynamic range of the instrument, including the USB interface, places an upper limit on the range of impedance values (or gain/loss response) that can be displayed in a single measurement. A 12-bit interface (about 88 dB dynamic range) is sufficient for most amateur measurements with a 16-bit interface (about 90 dB dynamic range) desirable for precision work.

• It should be able to make and display measurements at both linear and logarithmic scales; linear sweeps are best for narrow sweeps (a ham band); log sweeps are important for wide sweeps (showing a filter's out-of-band rejection or a choke's impedance and over a wide range).

• An analyzer that can be powered from the USB port simplifies setup both in the field and in the station.

The analyzer's software should include:

• The capability to store and recall calibration data for multiple measurement fixtures or configurations.

• Support the use of correction factors to account for various test fixtures and calibration standards results in greater accuracy for these special setups.

• Perform Time Delay Reflectometry (TDR) from a wide-frequency sweep of a transmission line's S11 (Return Loss). The TDR function should find the line's electrical length and any discontinuities (splices or faults). Cursors should show position in feet or meters with multiple windowing functions for the most useful view of the response.

a *calibration file* that can be stored and retrieved whenever that measurement plane is used in the future. Most analyzers can save multiple configuration files for different measurements. This process is very quick and easy, and the final results of any measurement are much more accurate than from a simple analyzer that does not have a calibration procedure.

One of the sources of error removed by the calibration is any error due to the length of the coaxial cables that connect to the device under test. For that reason, the measurement plane should always be at the end of the cables that are to be used for the test.

Calibration includes the effects of length for any coax between the analyzer and the measurement plane; substituting a different piece of coax requires another calibration. Adding a connector adapter to connect an antenna to the plane that has been calibrated introduces an error that is small enough not to matter for most antenna measurements, but can be significant on the higher HF bands, and much more so at VHF and UHF.

Calibration for S_{21} and S_{12} adds additional calibration measurements with the analyzer output and inputs connected together by identical cables for connecting the VNA output to the Device Under Test (DUT) and the output of the DUT to the VNA input, with the two cables mated through a double female adapter (aka a barrel adaptor).

For antenna measurements, the measurement plane might have to move when one or more connector adapters are attached to the same cable. Since the analyzer is then connected at a different spot on the line and has different geometry, it needs its own calibration.

SOLT CALIBRATION

The most popular technique is called a Short-Open-Load-Through (SOLT) calibration because it uses four calibration standards, a short-circuit (0 Ω), an open-circuit (∞ Ω), a 50-Ω load, and a through (direct) connection between the two ports. If the analyzer is a single-port VIA, there is no through connection and only the SOL calibration is performed. A set of standards with those three impedances over a very wide frequency range are provided with the analyzer or are available as a calibration test kit.

At the location in the antenna system where the measurements are to be made, place the analyzer in its calibration mode. Then follow the instructions to attach the calibration standards one at a time. This calibration procedure can be performed at the end of the feed line to allow measuring the actual driving point impedance of an antenna. The transmission line is disconnected at the antenna and the calibration loads are attached instead of the antenna. This shifts the measurement plane to the antenna itself which is handy when designing matching networks. As mentioned previously, calibration data files can be saved and recalled, so it is not necessary to perform the calibration every time the analyzer's calibration plane is moved.

Most antenna analyzers have a single-ended (unbalanced) output with a coax connector. For making measurements on a balanced transmission line, such as window line or ladder line, a common-mode choke can be added between the analyzer's RF connector and the input to the transmission line as in **Figure 28.35.** The calibration loads are attached to the output side of this choke. This moves the calibration plane to the end of the balanced transmission line.

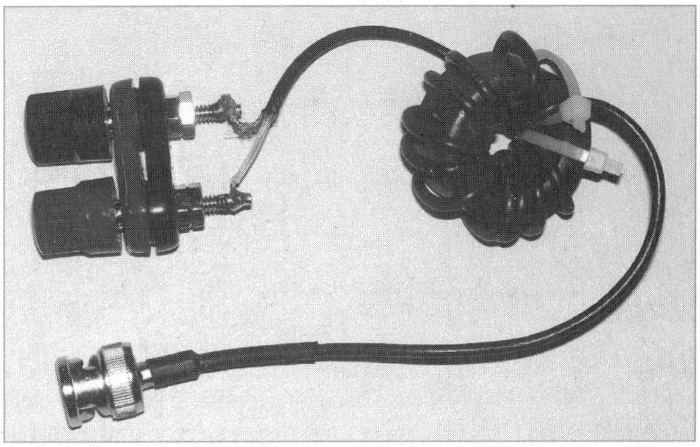

Figure 28.35 — A simple choke balun made by winding multiple turns of coax through a ferrite toroid, creating a high-impedance on the shield's outer surface. The choke is made of 12 turns of RG-193 miniature coax through a 2.4-inch OD, Type 31 core (Fair-Rite 2631803802 or Amidon FT-240-31) and is effective from 1.8 through 30 MHz.

28.5.5 Component Measurement

Many analyzers can also measure component values. Figure 28.34 shows a simple component test fixture for a VIA consisting of a BNC-to-banana plug adaptor with alligator clips slipped on to the plugs. When calibrated, this fixture can make reliable measurements well into the VHF range. For this test fixture, the measurement plane is at the tips of the alligator clips. The simple fixture like that in the figure is typically used at the end of a convenient length of coaxial cable (RG-400 is a good choice) for S_{11} measurement of chokes and other components.

Under the clips is shown the Short test load (#10 bare copper wire folded back on itself and coated with solder to minimize both resistance and inductance) and a large-format 1% 49.9-Ω chip resistor with short leads soldered to it of the minimum length that allows them to attach to the clips.

Inductor values can be measured over a wide frequency range to see if it is resonant within the frequency range where it will be used. **Figure 28.36** shows a graph generated from swept-frequency impedance data collected by an antenna analyzer of the type in Figure 28.33. The figure shows the impedance of an air-core inductor with a nominal inductance of 7.4 μH at low frequencies. (The traces are labeled with the measurement they represent.) The self-resonant frequency is 45.4 MHz. This resonance occurs because of the coil's inter-turn capacitance.

Below the resonant frequency the inductance has a positive reactance, and the coil presents its expected value, 7.4 μH. As the test frequency approaches the self-resonant frequency, the

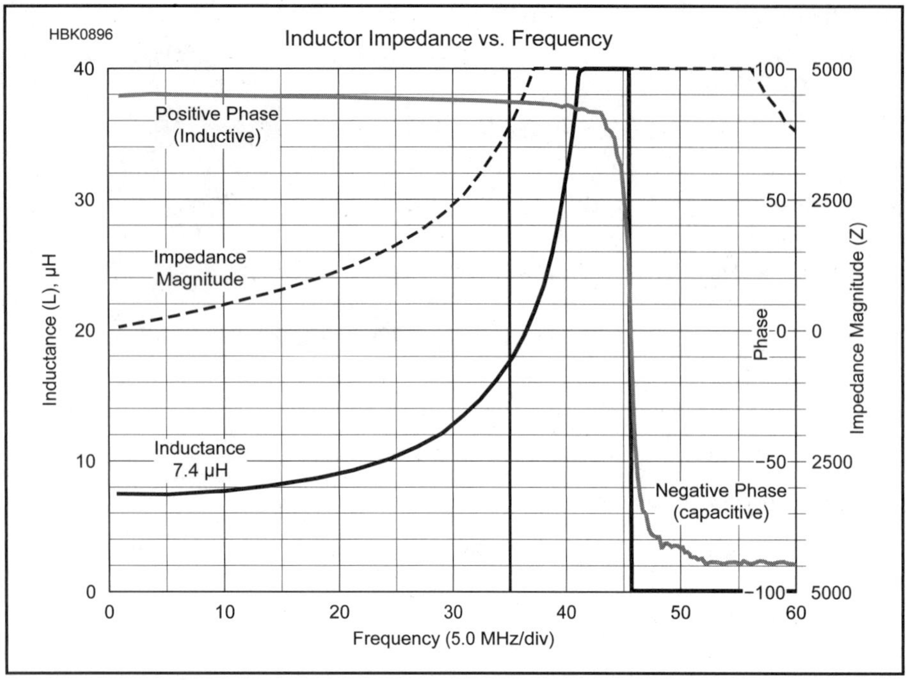

Figure 28.36 — Example of an antenna analyzer (AIM-4170 and companion software) being used to measure an air-core inductor's behavior. The inductor has a nominal value of 7.4 µH and a self-resonant frequency of 45.4 MHz.

inductive reactance actually increases, and the inductance appears to be larger. Above the self-resonant frequency, the component acts like a capacitor, as indicated by the negative phase angle.

The analyzer is also very handy for determining the material of a toroid core. Cores of different ferrite or powdered iron mixes cannot be told apart by their physical appearance, but you can separate them by comparison with a core of known material. Small inductances of one or two turns can be tested side by side to differentiate the cores of different mixes. A single turn through the center of a core or bead will create enough inductance for an accurate measurement on a good analyzer. Be wary of confusing mixes that are designed for EMI suppression with mixes intended for inductive applications. The permeability of the different mixes has resistive and reactive components that dominate over different frequency ranges. See the discussion of ferrites in the **Transmission Line System Techniques** chapter for an explanation. Comparison to cores of known material is much more reliable than a simple calculation of µ (mu — permeability) or A_L, the inductance index.

Capacitors are usually closer to the ideal component than inductors, but they do have some inductance in their leads and eventually at some high frequency they too become self-resonant. This self-resonance should be checked for capacitors that will be used in the VHF/UHF range. Above the self-resonant frequency, a capacitor looks like an inductor.

Resistors have an effective capacitance in parallel with them as well as inductance in their leads, so they are not ideal over a wide frequency range. Physically large power resistors used for dummy loads have larger parasitic components. (Thin-film power resistors in TO-220 packages are available with significantly lower reactance.) Tubular metal and carbon film resistors are often trimmed with a laser to create a spiral track in the deposited film, creating inductance. If the resistor is to be used in an RF circuit, it is prudent to verify its effective frequency range.

28.5.6 Analyzer Data Processing

Firmware built into vector analyzers, and/or software running on an associated computer, can display impedances computed from S_{11} in many forms. Some of the more common are: 1) Smith Chart; 2) Series Resistance (R_S) and Reactance (X_S); 3) Parallel Resistance (R_P) and Reactance (X_P); the magnitude of the Impedance (Z_{MAG} or $|Z|$); 4) Inductance (L) and parallel capacitance (C_P); 5) Capacitance (C) and series resistance (R_S). Depending on firmware and/or software, the frequency sweep can be made, and data presented linearly or logarithmically.

Using built-in firmware or associated computer software, most modern analyzers can perform *time delay reflectometry* (TDR) to determine the length of a transmission line, to look for splices or faults along the line, even find the impedance of each of multiple sections connected in series. (See the discussion of TDR later in this chapter)

Most modern vector analyzers and/or associated computer software can export a measurement for use by design and analysis software like *SimSmith* (**ae6ty.com/smith_charts.html**) and *ZPlots* (**ac6la.com/zplots1.html**) to study feed line loss and design matching networks. (See the sidebar on Touchstone format) These two freeware packages can transform an impedance measurement made anywhere along a feed line to the impedance at the antenna feed point if the length and parameters of the feed line are known. *ZPlots* also offers some design capability for common amateur applications. *SimSmith* allows the design of far more complex two-port networks and transmission line systems.

28.5.7 S-Parameters

In the chapter **Transmission Lines**, the reflection coefficient rho (ρ) is defined as the ratio of the reflected voltage (V_r) to the incident voltage (V_i):

$$\rho = \frac{V_r}{V_i} \tag{5}$$

If we know the load impedance (Z_L) and the transmission line impedance (Z_0) we can calculate ρ from:

$$\rho = \frac{Z_L - Z_0}{Z_L + Z_0} \tag{6}$$

Touchstone Data Files

The most useful analyzers can store sets of measurements in a file for review later and transfer to a PC for further processing and display. The standard format for swept impedance data (frequency and impedance) is called Touchstone **(en.wikipedia.org/wiki/Touchstone_file)**. Files in the Touchstone format can be read and processed by a wide variety of software, including as an input to design or simulation software. Touchstone files are plain text with a line for each data point. A single line header defines the sub-format of the data. The sub-format most widely supported by hardware and software for impedance data is s1p in which the parameter is complex S11, and each line consists of Frequency, Real S11, and Imaginary S11, with spaces as the separator.

Keep in mind that ρ is a complex number (a vector), which we represent by either amplitude and phase (|Z|, θ) or by real and imaginary parts ($R \pm j X$). The two representations are equivalent. Neglecting phase, from |ρ| (the magnitude of ρ) we can then calculate SWR. That's very handy, but here we want to do something different. If we have an instrument that measures ρ and we know Z_0 then we can determine Z_L from:

$$Z_L = Z_0 \left(\frac{1+\rho}{1-\rho} \right) \tag{7}$$

Measuring ρ is one of the things that VNAs do very well. With a VNA, the measurement can be made at one end of a long transmission line with the load at the other end. The effect of the line can be calibrated out, as mentioned above, so that we are in effect measuring right at the load. Note that the symbol Γ (gamma) is also used to represent the reflection coefficient. The two symbols may be used interchangeably.

This approach can be used directly to measure the impedance and resonant frequency of a single antenna element. By open and short circuiting elements in an array of antenna elements we can determine the mutual as well as self-impedances for, and between, all the elements. We can also use this approach to measure component values, inductor Q, etc.

This is an example of a *one-port* measurement; that is, a load at the end of a transmission line. However, to get the most out of a VNA, you need to generalize the above procedure. This is where S-parameters come into play.

VNAs usually have at least two RF connections: the transmit port (T) and the receive port (R). Professional units may have more RF connections. The T output provides a signal from a 50-Ω source and the R port is a detector with a 50-Ω input impedance. Basically, we have a transmitter and a receiver. The transmit port uses a directional coupler to provide measurements of the forward and reflected signals at that output. The receive port measures the signal transmitted through the network.

Using incident and reflected voltages, the two-port network representation is now changed, as shown in **Figure 28.37**, where:

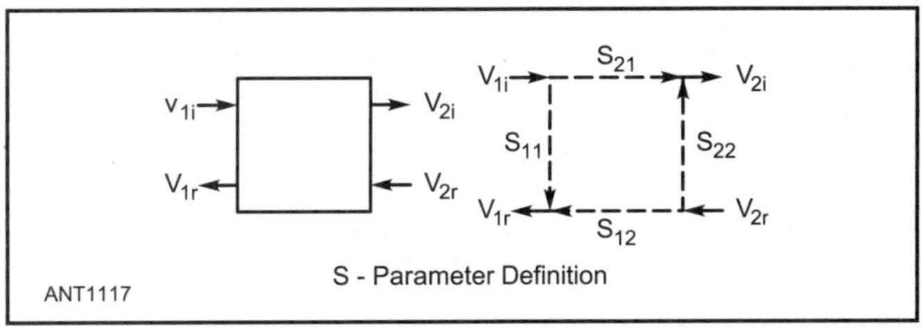

Figure 28.37 — Two-port network with incident and reflected waves.

V_{1i} = incident voltage at port 1
V_{1r} = reflected voltage at port 1
V_{2i} = incident voltage at port 2
V_{2r} = reflected voltage at port 2

We can write an expression in terms of the incident and reflected voltages:

$$\begin{aligned} b_1 &= S_{11}a_1 + S_{12}a_2 \\ b_2 &= S_{21}a_1 + S_{22}a_2 \end{aligned} \tag{8}$$

where:

$$\begin{aligned} a_1 &= \frac{V_{1i}}{\sqrt{Z_0}} = b_1 = \frac{V_{1r}}{\sqrt{Z_0}} \\ a_2 &= \frac{V_{2i}}{\sqrt{Z_0}} = b_2 = \frac{V_{2r}}{\sqrt{Z_0}} \end{aligned} \tag{9}$$

We see that a_n and b_n are simply the incident and reflected voltages at the two ports divided by ($\sqrt{Z_0}$). Because this is a linear network, $S_{21} = S_{12}$.

What are the S_{ij} quantities? These are called the *S-parameters*, which are defined (≡) by:

$$\begin{aligned} S_{11} &\equiv \left.\frac{b_1}{a_1}\right|_{a_2=0} = \left.\frac{V_{1r}}{V_{1i}}\right|_{V_{2i}=0} \\ S_{21} &\equiv \left.\frac{b_2}{a_1}\right|_{a_2=0} = \left.\frac{V_{2r}}{V_{1i}}\right|_{V_{2i}=0} \\ S_{12} &\equiv \left.\frac{b_1}{a_2}\right|_{a_1=0} = \left.\frac{V_{1r}}{V_{2i}}\right|_{V_{1i}=0} \\ S_{22} &\equiv \left.\frac{b_2}{a_2}\right|_{a_1=0} = \left.\frac{V_{2r}}{V_{2i}}\right|_{V_{1i}=0} \end{aligned} \tag{10}$$

Note that the S_{ij} parameters are all ratios of reflected and incident voltages, and they are usually complex numbers. The condition that $a_2 = 0 = V_{2i}$ is the same as saying that port 2 is terminated in a load equal to Z_0 and the network is excited at port 1. This means there is no reflection from the load on port 2, which makes $V_{2i} = 0$. Similarly, if we terminate port 1 with Z_0 and excite port 2, then $V_{1i} = 0 = a_1$.

If we compare Eq 5 to the first line of Eq 10 we see that S11 = ρ_1, the reflection coefficient at port 1. We can now restate Eq 7 in terms of S_{11}:

$$Z = Z_0 \left(\frac{1 + S_{11}}{1 - S_{11}} \right) \tag{11}$$

where Z is the impedance looking into port 1 with port 2 terminated in Z_0. In the case where port 2 does not exist — that is, you are measuring a single impedance (for example, measuring an impedance at port 1 with port 2 open-circuited) or a component, then Z is simply the impedance at that port. Since S_{11} is a standard measurement for VNAs you can calculate Z using Eq 11. In many cases the VNA software will do this calculation for you automatically. You can also measure an impedance at port 2 with port 1 open and determine Z_{22}.

S_{21} represents the ratio of the signal coming out of port 2 (V_{2r}) to the input signal on port 1 (V_{1i}) and is another standard VNA measurement. S_{21} is a measurement of the signal transmission between the ports through the network with port 2 terminated in Z_0; forward gain in most applications.

A full-feature VNA will measure all the S_{ij} parameters at once, but most of the lower-cost units of interest to amateurs are what we call reflection-transmission test sets. What this means is that they only measure S_{11} and S_{21}. To obtain S_{22} and S_{12} we have to interchange the test cables at the ports and run the measurements again. Normally the software will accommodate this as a second entry, and we end up with the full set of S_{ij} parameters.

If we do run a full set of S_{ij} parameters then we can transform these to Z_{ij} using the following expressions, assuming that $S_{21} = S_{12}$:

$$Z_{11} = \frac{(1+S_{11})(1-S_{22})+S_{12}^{\;2}}{(1-S_{11})(1-S_{22})-S_{12}^{\;2}}$$

$$Z_{22} = \frac{(1-S_{11})(1+S_{22})+S_{12}^{\;2}}{(1-S_{11})(1-S_{22})-S_{12}^{\;2}} \tag{12}$$

$$Z_{12} = \frac{2S_{12}}{(1-S_{11})(1-S_{22})-S_{12}^{\;2}}$$

28.5.8 Return Loss

Return loss (RL) is another term for S_{11}, the ratio of the reflected voltage to the incident voltage, usually expressed in dB.

$$RL = -20\log\left(\frac{V_{1i}}{V_{1r}}\right) = -10\log\left(\frac{P_{1i}}{P_{1r}}\right)$$

RL is measured as S_{11} by a VNA.

The name stems from measuring how much voltage is returned from a transmission line wave encountering a termination or impedance discontinuity. If the line is terminated in its characteristic impedance, the entire wave is absorbed, and none is returned so the "loss" at the reflection is total, and RL is infinite. If the line is open- or short-circuited, the entire wave is returned to the source and RL is 0. Note that RL is a positive quantity range from 0 (no transfer of wave energy at the termination) to infinite (all wave energy transferred to the termination). Negative RL would describe a voltage gain. (The article "Return Loss Explained, and Why We Should Care" by John Stanley, K4ERO, is provided in the online material for additional explanation.)

To convert from return loss to SWR, the following formulas are used:

$$|\rho| = 10^{-\frac{RL}{20}} \quad \text{and} \quad SWR = \frac{1+|\rho|}{1-|\rho|}$$

So, for example, a return loss of 20 dB is a reflection coefficient of 0.1, and an SWR of 1.22. A return loss of 10 dB is a reflection coefficient of 0.316, and an SWR of 1.92.

One of the most common measurements made is the standing wave ratio of an antenna. A low SWR means that the antenna input impedance is close to that of the measuring reference impedance. These measurements are from the July/August 2004 *QEX* article by McDermott and Ireland (see References and Bibliography) and show the magnitude of the return loss versus frequency for a KT34XA triband Yagi antenna at the end of 300 feet of hardline cable.

The resonance points are clearly visible. **Figure 28.38** shows the return loss of the antenna swept from 1 MHz to 50 MHz. The 20-meter, 15-meter, and 10-meter band resonances are easily seen. (RL increases toward the bottom of the chart.) **Figure 28.39** shows a close-up of the return loss from 13.5 to 14.5 MHz. **Figure 28.40** shows this same close-up on a Smith chart.

28.5.9 Antenna Array Measurements

The process of building and properly tuning a phased array often involves making a number of different measurements to achieve a desired level of performance, as was pointed out

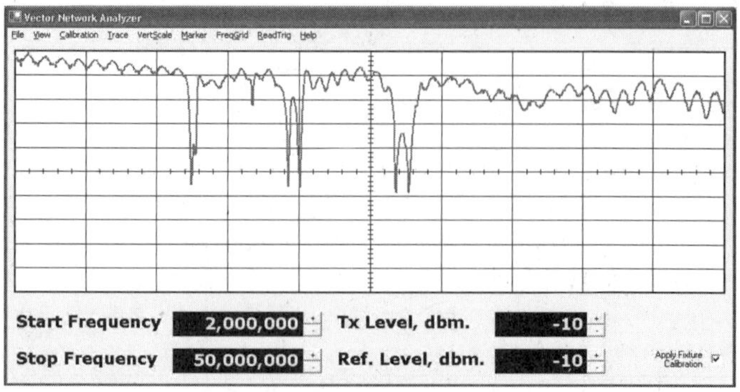

Figure 28.38 — Return loss of KT34XA antenna through 300 feet of hardline. Vertical scale is 5 dB/div. The three resonances at 20, 15, and 10 meters are clearly visible.

in the **Phased Arrays** chapter. This section is adapted from material written by Rudy Severns, N6LF.

After erecting an array, we would like to measure the resonant frequency of each element, the self-impedances of each element and the mutual impedances between the elements. We will also want to know these impedances over the whole operating band to help design a feed network. When building the feed network, we may need to check the values and Qs of the network elements and we will want to determine the electrical lengths of transmission lines.

Final tuning of the array requires that the relative current amplitudes and phases in each element be measured and adjusted, if necessary. We also will want to determine the SWR at the feed point. Doing all of this even moderately well can require quite a bit of equipment, some of which is heavy and requires ac line power. This can be a nuisance in the field, especially if the weather is not cooperating.

In a completed array with its feed network, the network can be excited by the VNA at the feed point and the relative current amplitudes and phases at each element can be measured over a frequency band. Then, adjustments can be made as needed. When the final values for the current amplitudes and phases are known, these values can be put back into an array model in a program like *EZNEC* to determine the pattern of the array across the whole frequency band. A multielement array actually behaves as a multiport network, so using a VNA is a natural solution to the measurement problem.

HF arrays are also large in terms of wavelength. The techniques for measuring forward and reverse powers work well even at 160 meters. For example, even though the array elements may be 100 feet apart, you can place your instruments in a central location and run cables out to each element. The effect of the cables from the VNA to the elements can be absorbed in the initial calibration procedure so the measurements read out at the VNA are effectively those at each element. In other words, the measurement reference points can be placed electrically at the base of the element, regardless of the physical location of the instrumentation and the interconnecting cables.

The discussion of S-parameters in the previous section can be viewed as measuring the characteristics of a 2-element array (see **Figure 28.41**) if one element is attached to the end of the transmission lines at each port as in **Figure 28.42**.

In the case of such an array, S_{21} represents the signal transmission due to the coupling between the elements. i.e., the signal coupled to element 2 as a result of a signal applied to element 1. Transmission lines are assumed to have $Z_0 = 50\ \Omega$ (or the characteristic impedance of the overall system) and may be of any length required by the size of the array.

S-parameters can be determined for an array with any number of elements. In an n-port S-parameter measurement, all ports are terminated in Z_0 at the same time. Measurements are made between one set of ports at a time and repeated until all pairs of ports are measured.

To illustrate the principles of using a VNA we will use a simple 2-element array like that shown in Figure 28.41. To design a feed network to drive this array we need to know the input impedance of each element (Z_1 and Z_2) as a function of the drive currents (I_1 and I_2). The input impedances will depend on the self-impedance of each element, the coupling between them (the mutual impedance) and the drive currents in each element. To manage this problem, we can represent a 2-element array as a

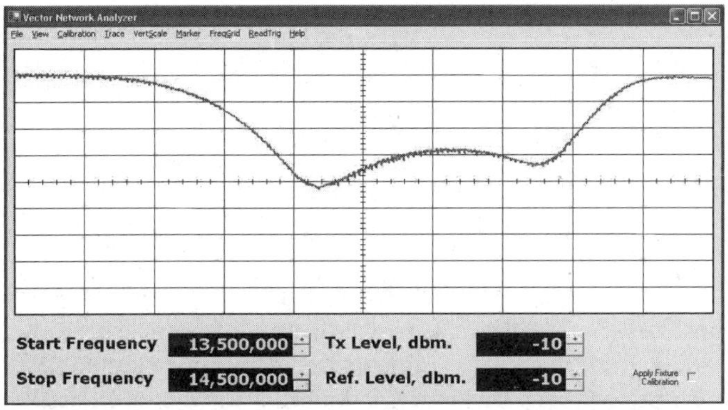

Figure 28.39 — Return loss of Figure 28.38 from 13.5 MHz to 14.5 MHz. A 26 dB return loss (best case at 13.94 MHz) is an SWR of 1.105 (at the in-station end of the feed line). Vertical scale is 5 dB/div.

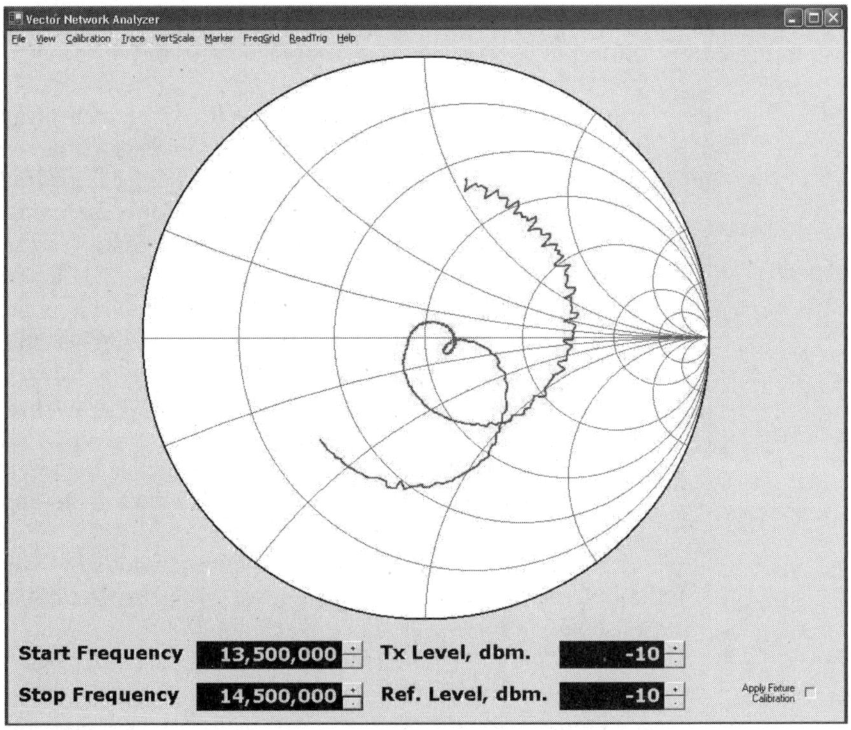

Figure 28.40 — Return loss of Figure 28.39 on a Smith Chart.

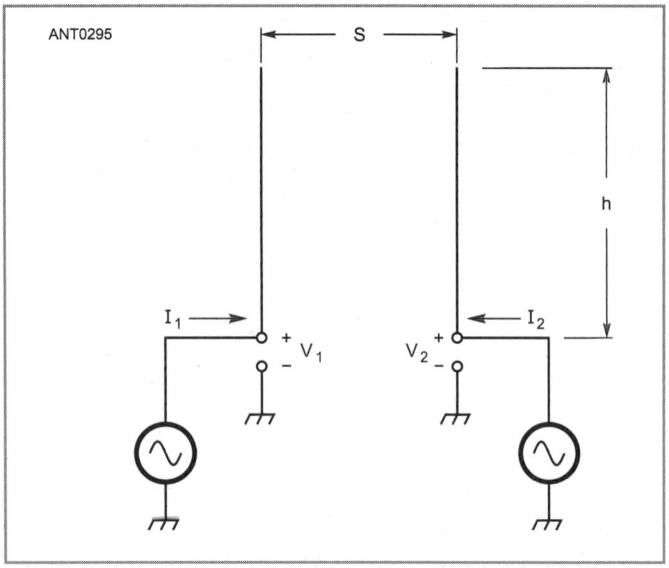

Figure 28.41 — A 2-element array, where h is the element height and S is the spacing between the elements.

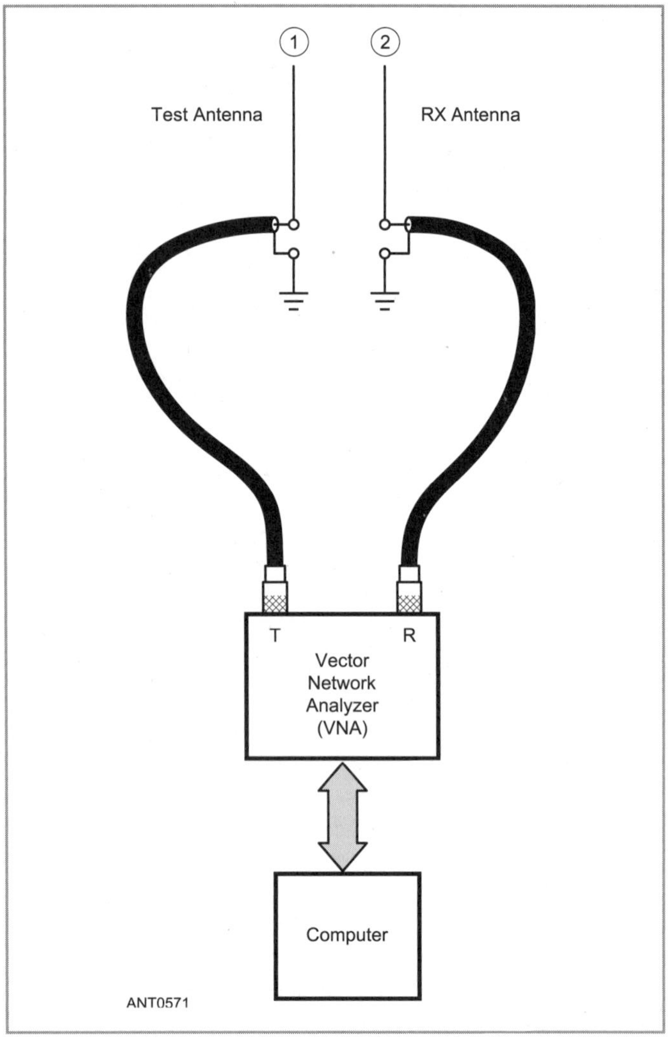

Figure 28.42 — Test setup to measure a 2-element array using a VNA.

two-port network, as shown in **Figure 28.43**. And we can relate the port voltages, currents, and impedances with Eq 25:

$$V_1 = Z_{11}I_1 + Z_{12}I_2$$
$$V_2 = Z_{21}I_1 + Z_{22}I_2 \qquad (13)$$

Normally we know I_1 and I_2 from the design of the array, but we need to determine the resulting element impedances. That's the challenge. Fortunately, an array is a linear network, so $Z_{12} = Z_{21}$, which means we need only determine three variables: the self-impedances Z_{11} and Z_{22} and the mutual impedance, Z_{12}.

Once we know Z_{11}, Z_{12}, Z_{22} and are given I_1 and I_2, we can determine the feed point impedances at each element from:

$$Z_1 = Z_{11} + \left(\frac{I_2}{I_1}\right) Z_{12}$$
$$Z_2 = Z_{22} + \left(\frac{I_1}{I_2}\right) Z_{12} \qquad (14)$$

This is the conventional approach. However, there are some problems here. We have to be able to accurately measure either voltages and currents or impedances in multiple elements that may be separated by large fractions of a wavelength. In addition, accurate measurements of current, voltage and impedance become increasingly more difficult as we go up in frequency.

It turns out that we can get the information more easily by measuring incident and reflected voltages at the ports and from those measurements determine the feed point impedances. A VNA is an instrument for measuring these voltages. It turns out to be easier to measure the ratios of two voltages rather than their absolute values.

The measurement setup using a VNA for a 2-element array is shown in Figure 28.42. A good way to illustrate the use of a VNA for array measurements is to work through an example with a real array. **Figure 28.44** is a picture of a 2-element 20-meter phased array built by Mark Perrin, N7MQ.

Each element is λ/4 (self-resonant at 14.150 MHz) and spaced λ/4 (17 feet 5 inches). In the ideal case, both elements would have the same current amplitude with a 90° phase difference. This gives the cardioid pattern shown in the **Phased Arrays** chapter. There are many schemes for correctly feeding such an array. The one used in this example uses two different 75-Ω transmission lines (one λ/4 and the other λ/2, electrically), as described by Roy Lewallen, W7EL and in Orr and Cowan (see References and Bibliography).

The first task is to resonate the elements individually. With the VNA set to measure S_{11} phase, we will get a graph like that shown in **Figure 28.45**.

At the λ/4 resonant frequency (f_r) we will see a sharp phase transition as we go from -180° to +180°. This is typical of any series resonant circuit. The length of each element is adjusted until the desired f_r is achieved. This is a very sensitive mea-

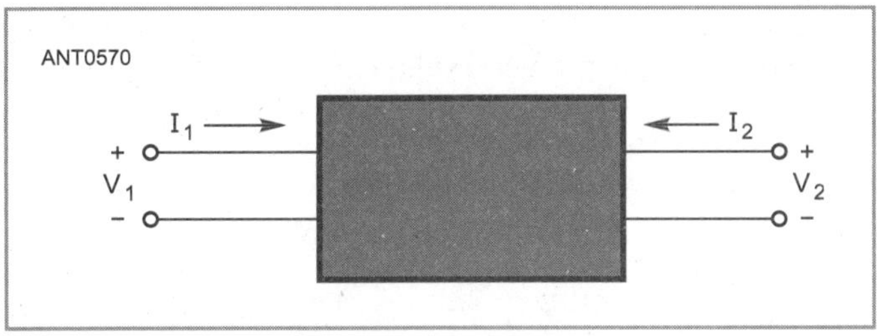

Figure 28.43 — Two-port representation of currents and voltages in the 2-element array in Figure 28.41.

Figure 28.44 — 2-element 20-meter phased array. (Photo courtesy Mark Perrin, N7MQ)

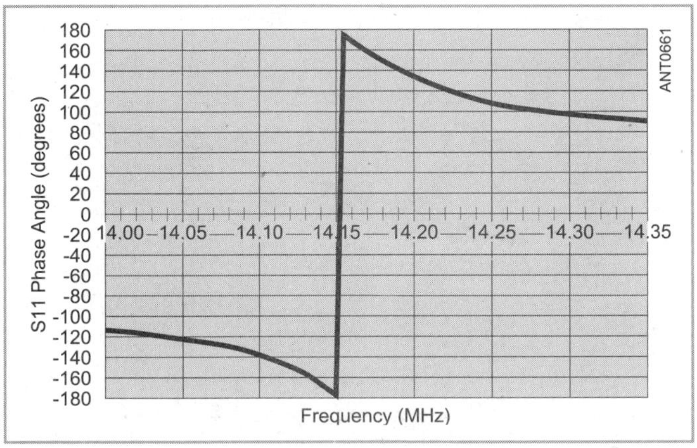

Figure 28.45 — S11 phase plot for an individual element.

surement. You can see the shift in f_r due to the wind blowing, the length of the element changing as it heats up in the sun or any interactions between the feed line and the antenna as you move the feed line around. In fact, this is a very good point in the process to make sure everything is mechanically stable and free of unexpected couplings. Usually, you will find it necessary place choke baluns on each element to reduce stray coupling.

The next step is to determine the self (Z_{11} and Z_{22}) and mutual (Z_{12}) impedances from which the actual driving point impedances present when the array is excited can be determined. There are two ways to go.

First, we can simply use the VNA as an impedance bridge — i.e., make two S_{11} measurements at one element, first with the other element open (Z_{11} or Z_{22}) and then with it shorted (Z_1 or Z_2). We can convert the S_{11} measurements to impedances using Eq 11. The value for Z_{12} can be obtained from:

$$Z_{12} = \pm\sqrt{Z_{11}(Z_{11} - Z_1)}$$
$$Z_{12} = \pm\sqrt{Z_{22}(Z_{22} - Z_2)} \qquad (15)$$

The second approach is to do a full two-port S-parameter measurement (S_{11}, S_{21}, S_{12} and S_{22}) and derive the impedances using Eq 11. Both approaches will work but the second approach has the advantage that the ± ambiguity in Eq 15 is eliminated.

For this example, the impedance values from the measurements at 14.150 MHz, turn out to be:

$$Z_{11} = 51.4 + j0.35$$

$$Z_{22} = 50.3 + j0.299 \qquad (16)$$

$$Z_{12} = 15.06 + j19.26$$

With these values we can now determine the feed point impedances from:

$$Z_1' = Z_{11} + \frac{I_2}{I_1} Z_{12}$$
$$Z_2' = Z_{22} + \frac{I_1}{I_2} Z_{12} \qquad (17)$$
$$\frac{I_1}{I_2} = -j$$

Note that $-j$ represents the 90° phase shift between the currents. Substituting the values from Eq 16 into Eq 17:

$$Z_1' = 32.09 - j14.7 \qquad (18)$$

$$Z2' = 69.61 + j15.32$$

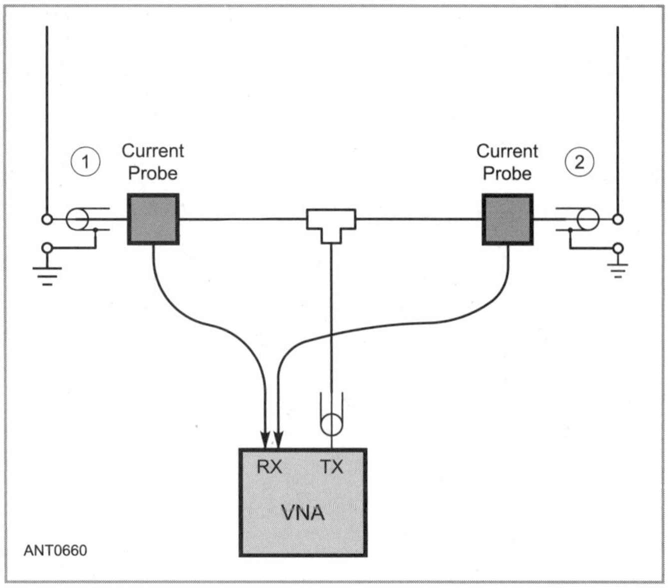

Figure 28.46 — Current phase and amplitude ratio test setup.

With these impedances in hand, we can now design the feed network. In this particular example however, we have decided to use the λ/4 and λ/2 cables as described by Lewallen and accept the results. So, we now proceed to cut and trim the two cables to length.

Again, there are two ways to go. First we can determine the frequency at which each cable is λ/4 long. At this point the input impedance of the cable will be equivalent to a series-resonant circuit, and we can simply measure the phase of S_{11} as we did earlier for f_r and get a plot like that shown in Figure 28.45. In this example the λ/4 resonant frequencies of the two cables are 7.075 MHz and 14.150 MHz.

The second approach would be to measure S_{21} for each cable at 14.150 MHz. The phase shift in S_{21} tells you how long the cable is, in degrees, at a given frequency. Because there is a small variation in cable characteristics with frequency (dispersion) this approach is slightly more accurate since it is done at the desired operating frequency. But this is not very large effect at HF.

This brings us down to the final measurements, which are to check that the relative current amplitudes and phases between the two elements are correct. We can then determine the feed point SWR. The phase and amplitude ratios are made using the S_{12} capability of the VNA and the test setup shown in **Figure 28.46**.

The VNA transmit port is connected to the normal feed point. A current sensor (see the **Phased Arrays** chapter for a discussion of current sensors) is inserted at the base of element 1 and the output of the sensor is returned to the detector or receive port of the VNA. A calibration run is then made to normalize this path. That makes it the reference.

Next, the current sensor is shifted to element 2. The amplitude and phase plots for S_{12} obtained at this point will be the desired relative phase shift and amplitude ratio between the currents in the array when driven at the normal feed point. **Figures 28.47** and **28.48** show the behavior of the example array over the 20-meter band. Note that the amplitude ratio has been converted from dB. We can now use these values in a *EZNEC* model of the array to determine the actual radiation pattern.

The W7EL feed scheme is not perfect, but it has a definite advantage of simplicity. If better performance is desired, we can use the values of Z_1' and Z_2' determined earlier to design and fabricate a new feed network and then proceed to evaluate its performance in the same way.

The final measurement is to connect the transmit port of the VNA to the feed point and measure S_{11}. From this we can calculate the SWR:

$$SWR = \frac{1+|S_{11}|}{1-|S_{11}|}$$

In this example, the return loss, $|S_{11}|$, is about 19 dB over the entire 20-meter band. This corresponds to SWR= 1.25:1.

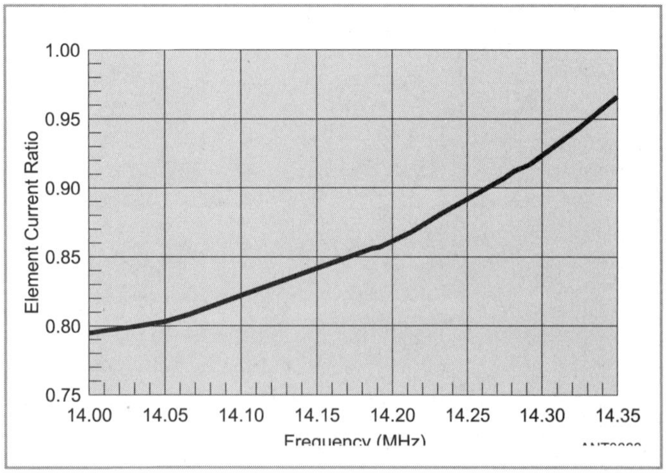

Figure 28.47 — Measured element current ratio over the 20-meter band.

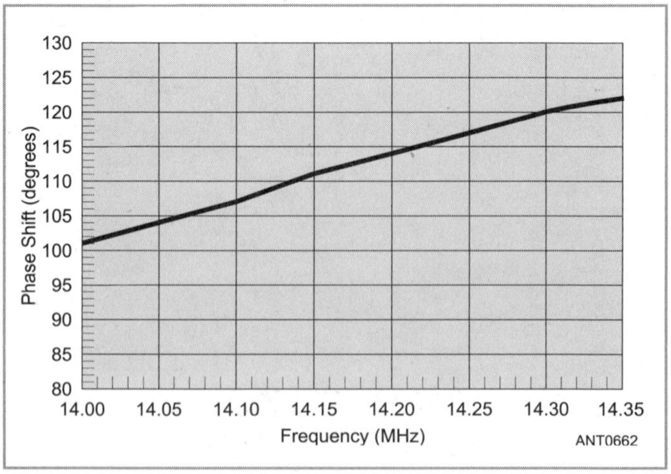

Figure 28.48 — Measured relative current phase shift over the 20-meter band.

28.6 Time-Domain Reflectometry

Time domain reflectometry (TDR) shows what happens to a short, abrupt pulse as it travels through a transmission line. The pulse is reflected by any changes in impedance, such as an open or short (complete reflection) or a change in the line's characteristic impedance (partial reflection). The resulting series of pulses and reflections is displayed as a sequence in time, thus the name of the technique.

In an ideal transmission line terminated by its characteristic impedance, Z_0, the pulse will travel to the far end and be dissipated in the termination, so the trace will be a perfectly flat line. But at any point along the line where the impedance changes (called a *discontinuity*) some of the pulse's energy will be reflected back toward the line's input. The reflected component of the pulse creates an artifact (visually, a "bump") on the otherwise straight line.

The sequence of pulses and their reflections is the *impulse response* of the line. An *impulse* is basically a very short pulse that begins and ends before the system can respond and stabilize. A mathematical impulse is an infinitely narrow pulse made up of all frequencies from zero to infinity.

While it is not possible to create an ideal (perfect) impulse, a very fast rising edge of a longer pulse is a good enough approximation to measure the line's impulse response and the same information can be measured. The longer pulse is called a *step function*. The ideal step function is an infinitely fast change from one level to another, after which it remains at that level. The response of the line to the longer pulse is called the *step response*. Like the impulse, the infinitely fast change in level also contains all frequencies.

A *time domain reflectometer* is the instrument that generates the pulse and displays the results. A TDR displays amplitude (voltage) on the vertical axis and time on the horizontal axis. The position of each artifact along the TDR trace corresponds to the distance from the transmission line input to the discontinuity that produced it. Large discontinuities occur when a line is open-circuited or short-circuited, or when it is connected to an antenna. (Most antennas are matched to the line at their operating frequency, but at other frequencies they are not. Since the pulse contains all frequencies, an antenna is a large discontinuity.) Small discontinuities occur at splices or when a line is damaged.

The delay between the input pulse's rising edge and the artifact is the round-trip time in the line from the TDR to the discontinuity. If the line's velocity of propagation (VF) is known, the physical distance from the input to the discontinuity can be calculated. The shape of each reflection can sometimes provide a clue as to the nature of the discontinuity.

In this sense, the TDR is very much like a radar display in which a pulse is transmitted (shown at the center on a radar display) and any echoes from discontinuities in the air (i.e. targets) reflect some of the pulse back toward the transmitter. The farther away the target, the longer it takes for the pulse to travel to the target and back to the receiver. A radar screen shows echoes from all directions. The TDR only shows echoes from one direction, along the line. The larger the echo, the larger the target or discontinuity.

The usefulness of TDR is not limited to RF systems. Some of the first uses of TDR were for finding faults in cabling systems carrying all sorts of signals, including ordinary telephone lines, and it is still widely used for that purpose.

When performing TDR on a coaxial cable, remove protective components such as lightning arrestors that may add artifacts to or distort the TDR measurements. Non-linear components such as clamp diodes or small light bulbs may also affect the measurement.

28.6.1 Direct Method TDR

There are two common TDR implementations, with variations of both. In the "direct method," which is the oldest and simplest, the line is driven by a pulse. This can be a single pulse or it a train of pulses like a square wave.

The pulse and all of the reflections are displayed on an oscilloscope trace triggered by the pulse's rising edge. The rise time of the pulse must be much shorter than the round-trip time for the impulse to travel to and from the discontinuity. **Figure 28.49** shows an example of a direct method TDR with the pulse generator and cable attached to the oscilloscope which is displaying the pulse.

Although a digital scope can capture a single pulse and its reflections, making the pulse repetitive means that multiple responses can be averaged to improve the signal to noise ratio. The signal to noise ratio of a system excited by a single pulse can be rather limited. Repeated pulses also sustain the trace so the operator to see it if the scope is an analog model. The repetition rate of the pulse must be much slower than it takes for the impulse to make a complete round trip time through the line, however. This insures the line's response dies out completely before the next pulse excites the line again. See the reference article by King about this type of TDR. An up-to-date article by Larry Lamano, WAØQZY, describing a pulse generator suitable for TDR measurements is included in the online material. Diagrams showing typical TDR traces for various conditions are included in the article "Time Domain Reflectometry" in the online material.

Modern antenna analyzers can use the direct method, as well. For example, the AIM family of analyzers, designed by W5BIG and sold by Array Solutions (**arraysolutions.com**), excite the line with a step function waveform having a very fast rise time. This implementation provides a display of the impedance at every point on the line — it can, for example, show the relative impedance of cables having different Z_0, as well as the position of discontinuities. Figure 28.37 shows a TDR display from an AIM analyzer connected to a length of RG-58 (50 Ω cable), a short length of RG-59 (75 Ω), and a 100 Ω load. The discontinuities at the cable and load transitions are clearly shown. Software converts the raw data from the pulse amplitudes into impedance and the time is converted into distance along the line.

The TDR function can be used to determine if the line has been degraded, for example, by water leaking into the coax or

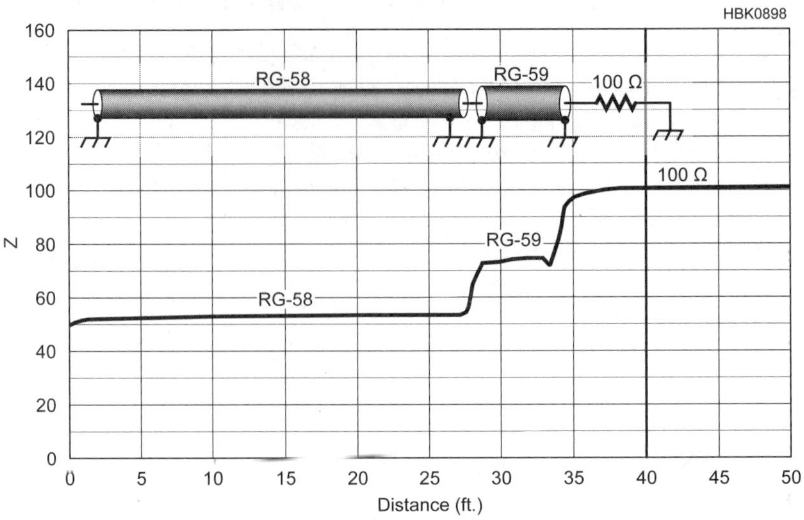

Figure 28.49 — Example of a direct method TDR (see text) connected to a length of RG-58 coaxial cable (50 Ω impedance) followed by a short length of 75 Ω RG-59 and a 100 Ω load. The discontinuities are clearly shown as the reflections add to the input pulse.

if the line has been shorted or cut somewhere between the transmitter and the antenna. Damage or defects can be located within a few inches and this reduces the effort required to repair the line. Defective connectors can also be indicated by short glitches in the trace corresponding to the location of the connectors.

28.6.2 Transform Method TDR

The other common implementation could be described as the "transform method." Instead of determining the impulse response of the cable with a pulse, the excitation is a sine wave swept over a range of frequencies and the analyzer captures the *frequency response*. An *inverse Fourier transform* (IFT) is performed on that frequency response, producing the time-domain response. (See the *ARRL Handbook* for more information on Fourier transforms.)

Frequency and time are the inverse of each other; the complete frequency response of a system contains its time response and the response to an ideal impulse contains the frequency response. A Fourier transform of the time response provides the frequency response, an inverse Fourier transform of the frequency response provides the time response.

Some antenna analyzers and vector network analyzers use this method. (Several well-suited for amateur use can be found by an internet search for "vector antenna analyzer" or "vector network analyzer.") One example is the hand-held SARK-110 Vector Impedance Antenna Analyzer (**sark110.com**). The sine wave exciting the cable need not be a continuous sweep — rather, it can be stepped over a wide range of frequencies and the frequency response is computed from that data. Sweep range, spacing between data points, and the settling time at each data point are set by the user. The process of using a VNA for TDR is also described with examples in the article "Using a VNA Like a Time-Domain Reflectometer" by Roger Denker, in the Sep 2022 issue of *Microwaves & RF*.

An IFT produces spurious artifacts which must be removed by applying a mathematical windowing function to the transformed data. Several mathematically different windowing functions are commonly used, and which of the windows provides the most useful display depends on the shape of the impulse response.

The frequency content of the excitation strongly influences the degree of detail that the measurement can reveal. When the excitation is an impulse, a very fast rise time reveals greater detail. When the excitation is a swept sine wave, a wider frequency range reveals the greatest detail. Currently available analyzers can sweep from 1 kHz to more than 1 GHz.

When performing TDR, a sweep from 5 to 500 MHz (or from 500 MHz to 1 GHz) will clearly show detail that would be missed with a sweep to only 100 MHz, while a 5 MHz to 1 GHz sweep may provide too much detail (or show discontinuities that don't matter below 50 MHz). Beginning the sweep in the HF range avoids smearing of the data due to the variation of VF with frequency (see section below on Effect of Velocity Factor).

EXAMPLES OF USING THE TRANSFORM METHOD

Figure 28.50 is the impulse response of the feed line for a 30 meter half-wave dipole at a height of 100 feet, computed from a sweep over the range of 50 – 500 MHz. Marker 1 shows the effect of lightning protectors at the station entry bulkhead. Marker 2 is a coax splice (two PL-259s and a PL-258 double-receptacle). Markers 3, 4, and 5 are coax defects. Marker 6 is the antenna feed point. Marker 7 is the end of the antenna (displayed distances are for the feed line, so for the antenna are divided by 0.795). Marker 8 is unexplained, but most TDR sweeps show multiple reflections after the antenna.

Figure 28.51 is the step response (similar to that from the direct method shown in Figure 28.37) computed from the same sweep as that used for Figure 28.38. The data revealed (and a visual inspection confirmed) that the cable inside the station and from the bulkhead to the coax splice is 50 Ω, but that the cable from there to the antenna is 75 Ω.

TDR measures time. To convert that measurement to physical

distance, we must provide the velocity factor. The 75 Ω cable is Belden 8213, this sample of which has a measured VF of 0.795 at VHF. The 50 Ω cable has a measured VF of 0.8425 at VHF. In the setup screen for this TDR measurement, VF was set at 0.795, so distance measurements will be correct for the 75 Ω cable, but wrong for the 50 Ω cable. Computed results could be made correct for the 50 Ω cable by changing VF in that setup screen, or by leaving VF at 0.795 and applying a correction factor of (0.8425/0.795) to the dimensions of the 50 Ω cable.

28.6.3 The Effect of Velocity Factor

The velocity of propagation in any transmission line is not constant — it varies with frequency. At the lowest audio frequencies, the velocity factor (VF) is quite a bit lower than the published specification, rising rapidly throughout the audio spectrum, continuing to increase through the radio spectrum until it reaches a nearly constant value in the VHF range. It is this constant value that is computed by simple equations for VF that don't take frequency into account.

For most cables, VF at 2 MHz is typically 1-2% slower than this constant value. In other words, an actual transmission line is 1-2% percent longer electrically than using the "nominal" (VHF) value that the simplified equation for VF predicts. Because of this change, the physical length of a stub for the lower frequency bands is 1-2 percent shorter than predicted from the specified value.

When using software to transform antenna measurements made in the station to the actual impedance at the feed point, the variability of VF must be applied to data for antennas for the lower bands (14 MHz and below). AC6LA's free *Zplots* shareware (based on *Excel* spreadsheets, see **ac6la.com/zplots.html**) computes and plots VF, Z_0, and attenuation versus frequency from measurements of a known length of a transmission made with the far end open and with the far end shorted. *ZPlots* can accept data in the Touchstone format discussed earlier.

Figure 28.52 is a plot of VF and attenuation computed by *ZPlots* from such measurements on a 176-foot length of RG-11 cable with a #14 AWG solid copper center conductor, a foam dielectric, and a copper braid shield. This behavior and general curve shape are typical of all transmission lines as predicted by fundamental transmission line equations. Exact values for each line will differ based on their physical dimensions and their dielectric.

It should also be noted that Z_0 also varies with frequency and below VHF is complex – that is, not a pure resistance, and is slightly capacitive. The *TLW* program (*Transmission Lines for Windows* by N6BV), included with this book's online material, provides Z_0 data for most commonly used cables and can plot voltage and current along the line. These plots clearly show standing waves on a line at 2 MHz when the termination is only resistive. The mismatch is small and the effect on attenuation is insignificant, but it clearly shows up in carefully made measurements of long cable lengths over a range of frequencies as a small ripple in attenuation values.

HBK0899

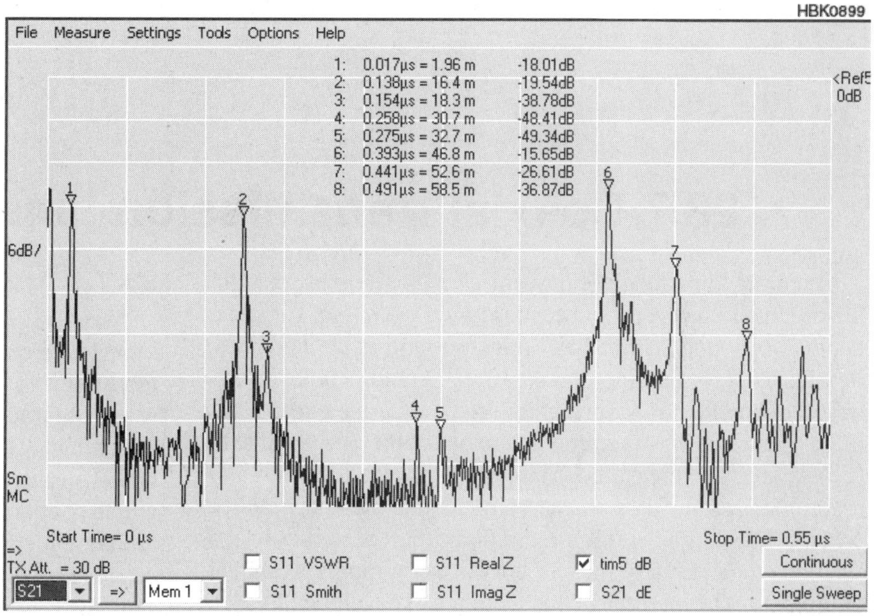

Figure 28.50 — The impulse response of a feed line attached to a 30 meter dipole. The system is swept from 50 to 500 MHz. See text for an explanation of the markers.

HBK0900

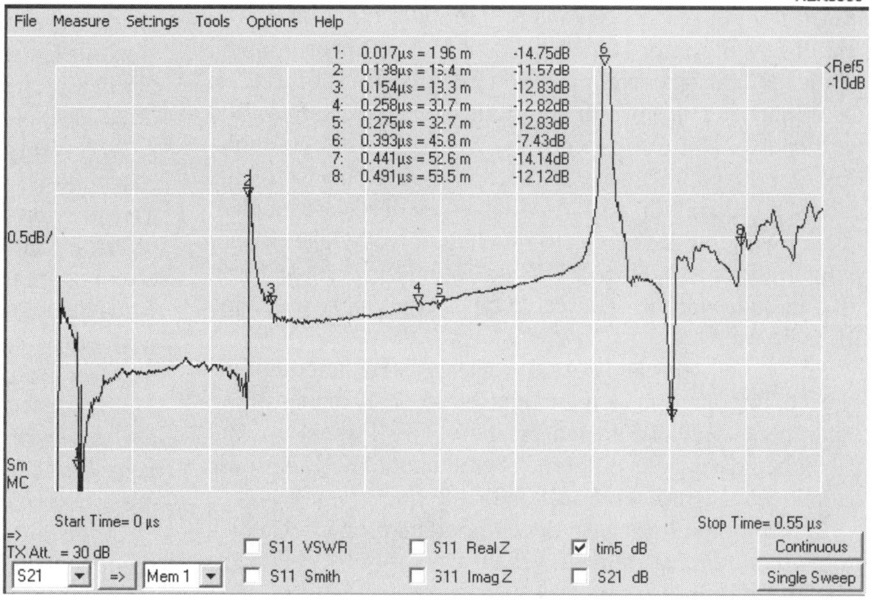

Figure 28.51 — The step response similar to the direct method for the same antenna system in Figure 28.50. See text for an explanation of the graph.

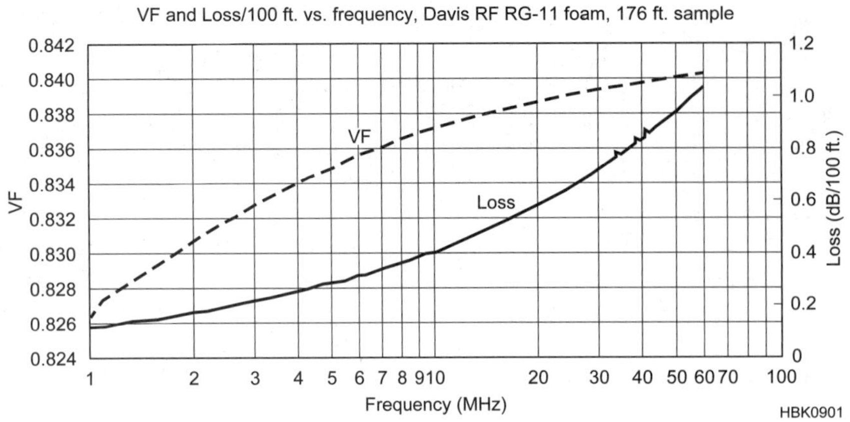

Figure 28.52 — VF and characteristic impedance for an RG-11 cable computed by *ZPlots* software from a swept-frequency measurement. See text.

28.7 Coaxial Transmission Line Measurements

Accurate knowledge of transmission line lengths, characteristic impedance, and sometimes loss is necessary for some applications such as impedance transformation stubs and lines, or lines used for feeding phased arrays. Manufacturer or distributor specifications very often aren't accurate enough for these applications, so the characteristics of the actual cable to be used must be measured. Many of the methods commonly recommended for these measurements fall short in practice when good accuracy is required. This section describes tested methods to accurately make them with commonly available equipment. The material in this section was developed by Roy Lewallen, W7EL.

Measurements of parallel conductor lines are potentially subject to major errors due to common mode currents, balun imperfections, dispersion, and proximity to other objects in addition to the error sources present in coax measurement. While they can be overcome to some degree in specific cases, the frequency range of valid results is likely to be narrow, and verification of the results probably impossible without sophisticated test equipment. So, this section applies only to coaxial cables. It's also assumed that the cables are of at least standard quality with low-loss dielectric such as polyethylene (PE) or polytetrafluoroethylene (PTFE or Teflon®) in some form. This includes all the coax cable types commonly marketed for radio use.

These methods can be used at least to VHF, although the effect of minor errors increases and the accuracy of measuring equipment typically decreases above HF. Typical coaxial cables begin behaving differently below about 1 MHz, so these techniques may not work well in that frequency range.

Transmission lines have three important properties which might require measurement: velocity factor (VF), characteristic impedance (Z_0), and loss.

28.7.1 Are Measurements Necessary

Many applications of transmission lines depend also on knowledge of other factors such as antenna or array element feed point impedance, and these are often not known with great accuracy. Also, techniques for designing impedance matching sections and phased array feed systems are usually based on the assumption that transmission lines are lossless which of course they are not. So even if the line characteristics are known exactly, the system is unlikely to work exactly as predicted. Nevertheless, results are much more likely to be satisfactory if the transmission line characteristics are known with reasonable accuracy.

The level of accuracy required depends on the application and on the level of acceptable deviation from the system design characteristics. For example, quite small changes in transmission line length can cause a substantial reduction in null depth of a phased array pattern. However, even relatively large errors often have only minor effects on gain and main lobe shape. A 2:1 SWR is acceptable for the impedance seen by most HF transmitters, so considerable error might be tolerated in a transmission line matching system. But in general, the effect of errors particularly in transmission line length (due to VF error) is greater when the SWR on the line is high, that is, when the line is terminated with an impedance greatly different from its Z_0. The effect of errors also multiply when longer lines are used, for example $3\lambda/4$ instead of $\lambda/4$ or λ instead of $\lambda/2$. Those factors also narrow the frequency range over which a system will perform satisfactorily. The methods described here should produce VF and Z_0 accuracy in most cases on the order of 2 – 3% at HF with good measurement equipment.

Precision measurements of a large number of solid and foam dielectric coaxial cables have shown the following:

1. Velocity factor and characteristic impedance of most coaxial cables changes with frequency. This occurs primarily below 3 – 5 MHz. The change from the middle of the HF range to 1 MHz is typically on the order of 2 – 3%, increasing more below 1 MHz. So, measurement is recommended for any transmission lines or stubs to be used below 3.5 MHz, and measuring at or near the frequency of use is particularly important in those cases. The remaining comments apply to 3.5 MHz and above.
2. When specific RG cable types such as RG-58/U are mentioned here, they refer only to cables made to the mil-spec standard. Ones marked with just that designation are usually in that category if made by a reputable manufacturer. Cables advertised or labeled as "RG-##/U type" can be just about anything, including having foamed insulation or different center conductor or shield. Note that the popular "RG-8x" is not mil-spec and has no other universal standard.
3. Coaxial cables having solid PE insulation reliably have a velocity factor close to the universally specified 0.66. Measurements consistently showed 0.65 to be closer, but with very little difference from one type or brand of solid PE insulated coax to another. So, you can generally skip velocity factor measurement for this category of cable. Solid PTFE dielectric cables are almost always high quality and reliably have a velocity factor close to 0.695. See item 5 for comments on foam dielectric cable.
4. RG/58-U (see item 2 above) has s specified Z_0 of 53.5Ω, but measurements of a few samples showed them to be close to 57 Ω. RG-8/U, RG-58A/U, and RG-58C/U have a specified Z_0 of 50 Ω and consistently measured 49-50 Ω. So the specified Z_0 of the latter three types could probably be used without measurement. Measurements of some samples of name brand RG-59/U (specified 75 Ω Z_0) showed a Z_0 of 70.6 Ω. So Z_0 measurement of 75 Ω cables is recommended.
5. All parameters of coaxial cables having foam dielectric vary a great deal from brand to brand and lot to lot, and even within a single piece of cable. These should always be measured.

28.7.2 Required Equipment

The methods described here require an accurate means to measure the magnitude of impedance (|Z|, referred to here just as Z). Examples are an antenna analyzer, scalar or vector network analyzer, or impedance bridge. If only R and X are available, Z can be computed as

$$Z = \sqrt{R^2 + X^2}$$

Most of the impedances to be measured will have X much less than R, so in most cases. Equipment should be calibrated with known standards before measurements if possible.

A resistive load such as a low power dummy load is required for some measurements. A 50-Ω load can be used for characterizing lines of characteristic impedances other than 50 Ω. The value of the load generally isn't critical, but it must remain constant over a range of measurement frequencies, so its impedance can't have any significant reactive component that changes with frequency.

There must be an accurate way to measure the physical length of cables before and after cutting to length.

Loss measurement is best done directly with a power meter and a dummy load with resistance equal to the cable's Z_0. It can also be done with SWR measurements, but accuracy suffers at losses less than a dB or so and greater than a few dB as described in the Loss section below.

Methods to short and open circuit one end of a cable are required for many measurements. These are critical to accuracy, so be sure to follow the recommendations below.

28.7.3 Short and Open Circuits

To open circuit a cable end, cut it flush or, if it has a connector, leave it unconnected and not close to other objects.

A short circuit requires more care. A jumper wire (or "pigtail") is unacceptable for two reasons: It adds reactance to the connection, and it provides a path for current along the outside of the cable. Either can have a profound effect on results. There are two ways to make an effective short circuit.

1. If the cable end doesn't have a connector, strip the cable as necessary to twist the shield over the bare center conductor. The connection can temporarily be secured with an electrical wire nut or some kind of clamping device, or more permanently by soldering.
2. If the cable has a connector, construct a mating short-circuited connector. For an SO-239 (UHF) or BNC, the best way is to poke a small hole in a piece of copper foil or braid, insert the center connector pin through the hole, and solder the center conductor to the foil. Then trim the foil until it's larger than the insulator but smaller than the connector flange and solder it to the inner edge of the flange all the way around so there's no path for current from the inside of the shield to the outside. The result is shown in **Figure 28.53** for UHF and B for BNC connectors. Most other solderable metals should work adequately well. A satisfactory alternative is to solder four or more very short and evenly spaced wires between the center pin and the inner edge of the flange, shown for a UHF connector in Figure 28.53C.

28.7.4 Electrical Length and Velocity Factor

You can skip this section if you're already familiar with these terms but it's essential to become familiar with them if not.

The *velocity factor* (VF) of a transmission line is the speed of the waves inside the line expressed as a decimal fraction or percent of the speed of light in free space. This quantity is always less than 100%, or less than one when expressed as a fraction. The *electrical length* of a transmission line is the delay of a wave in the line, expressed as degrees, wavelengths, or length units such as feet or meters. These are correctly preceded by "electrical" as in "electrical degrees" to avoid confusion with the corresponding physical or free-space quantities. Because the wave is travelling more slowly in the line than in free space, the electrical length of a line is always

Figure 28.53 — A photograph of low-impedance short circuits for (left to right) UHF and BNC connectors and an alternative method for UHF connectors.

greater than the physical length:

$$\text{Electrical Length} = \frac{\text{Physical Length}}{\text{Velocity Factor}}$$

If any two of these quantities are known, the third can be calculated.

For example, start *EZNEC* and, with any model, change the frequency to 7.1 MHz and see in the window just below the frequency that the wavelength is 138.53 feet or 42.2243 meters. This is the free space wavelength at that frequency. If we have a transmission line of that physical length which has a velocity factor of 0.66, its electrical length would be 209.89 feet or 63.976 meters. so it would have a delay (when impedance matched) of 1.515 wavelengths, or about 545 degrees. If we wanted the matched cable to have a delay of exactly one wavelength (360°), we need to cut it to 0.66 times the free space wavelength, or about 91.43 feet or 27.87 meters. A piece of the line half this physical length would have an electrical length of λ/2, and so forth.

MEASURING TRANSMISSION LINE ELECTRICAL LENGTH

The electrical lengths of cables having solid polyethylene (PE) dielectric such as RG-8, 11, 58, and 59/(A-C)U can usually be determined for frequencies of a few MHz or more simply by multiplying the physical length by a velocity factor of 0.65. The universally published value of 0.66 will probably be close enough for frequencies of a few MHz or higher, although careful measurements of 15 cable samples puts the actual value closer to 0.65 in the mid HF range. The dielectric constant of PE, which determines the velocity factor, doesn't vary much from cable to cable, and probably not perceptively along the length of a cable.

However, measurements also show a consistent decrease in velocity factor at lower frequencies, typically dropping about 1 – 1.5 percent from 10 to 1 MHz (see the **Transmission Lines** chapter). The phenomenon is due to the change in internal inductance (per unit length) of the conductors as the skin depth decreases to less than many times the wire diameter or shield thickness. The internal inductance is negligible at frequencies where the diameter or thickness is many times the skin depth. But at low frequencies it increases to become a greater part of the total inductance at lower frequencies. Because of the smaller conductors, the effect occurs at higher frequencies with smaller cables like RG-58 than with RG-8. For this reason, VF measurements should be made at approximately the frequency at which the cables will be used.

Foam dielectric coaxial cables such as the popular RG-8x, LMR type, and many promoted as being extra flexible have much more variation in velocity factor – probably due to varying foam density — than solid dielectric cable, and manufacturers' specifications are often very different from what measurements show. So, they require measurement for even moderate accuracy.

The remaining part of this section explains how to make the measurements if it's necessary.

With equipment generally available to amateurs it's possible and practical to directly find the frequency at which the electrical length of a line is some multiple of λ/4. Any other electrical length at any other frequency can be calculated from that. This is done by short or open circuiting the far end of the line and looking for the frequency where the magnitude of the input impedance is minimum or where the reactance (X) is zero and the impedance is low. If this happens with an open circuited line, the electrical length is an odd number of quarter wavelengths. If it happens with a short-circuited line, the electrical length is an even number of quarter wavelengths. Various factors such as line loss, imperfect open or short circuit terminations, instrument imperfections, and other factors limit the accuracy of these measurements, but they'll generally be adequate if carefully done.

The best tool is an antenna analyzer or vector network analyzer both of which are commonly available for a modest cost. However, a scalar network analyzer or noise or resistance bridge can also be used. Simple resistance bridges are described in the *ARRL Handbook's* **Test Equipment and Measurements** chapter.

For the best accuracy you should:

- Measure lengths at frequencies where the input impedance is near zero rather than very high. This is easier for the instrument to accurately determine the desired condition. To produce this condition, open circuit the far end of the line and measure the frequency where the line is an odd number of electrical quarter wavelengths, or short circuit and measure where it's an even number of electrical quarter wavelengths.
- Don't use jumper wires for connections! This not only adds reactance but also provides a path for common

mode current along the outside of the line which can have a dramatic effect on measurements. A good short circuit can be made without a connector or jumpers as described next.

• Terminate the line with as good a short or open circuit as possible. (See Short and Open Circuits above.) Stray shunt capacitance of an open circuit or stray series inductance or resistance of a short circuit will affect the measurement.

• Find a λ/4 multiple length that's close to the frequency at which the line will be used. Line characteristics including velocity factor vary slightly with frequency.

Here's a step-by-step procedure and example:

A cardioid array feed system was designed in the "Simplest" (Transmission Line Only) Feed System subsection of the **Phased Array** chapter. The *Arrayfeed1* calculation results are shown in Fig. 7.21. At the lower right, the 1st SOLUTION wires are shown as being 17.315 and 39.291 feet when using coax having a velocity factor of 0.66 which is standard for solid polyethylene (PE) dielectric. 75-Ω line is needed, so RG-59/U or RG-11/U could be used. Alternatively, foam PE dielectric cable could be used but would be longer – for example, one with 0.75 velocity factor would be 0.75 / 0.66 times as long. But we'll use solid PE dielectric cable for this example.

As explained earlier, a velocity factor of 0.65 or 0.66 might be close enough that you can simply cut the cables to the lengths shown in the figure since it isn't subject to a major cause of variation in foam dielectric cables, namely the foam density. There is, however, some variation due to other causes so measurement is still a good idea. Here's how measurement of the example transmission lines can be done.

1. Find the closest number of electrical quarter wavelengths in the uncut cable is, at the frequency where the transmission line will be used. For our example, The Physical Lengths section of Fig. 7.21 shows the length of an electrical quarter wavelength at 7.15 MHz to be about 22.7 feet, assuming a velocity factor of 0.66. Let's suppose that you have a 100-foot length of coax to cut the transmission lines from. This length is just over four electrical quarter wavelengths (about 91 feet), so the length measurement will be done at the frequency at which the uncut cable is one wavelength.
2. If neither end of the uncut cable has a connector, install one on one end. Connect that end to your measurement device (antenna analyzer, scalar or vector network analyzer, impedance bridge, or other device).
3. Open or short circuit the far end as appropriate – short when measuring an even number of quarter wavelengths and open when measuring an odd number. The one wavelength of our example uncut cable requires a short circuit. See the "best accuracy" list above for methods.
4. Find the resonant frequency where the resistance or magnitude of the impedance is low (typically about an ohm or less) and either the reactance (X) is zero or the magnitude of the impedance (Z) is minimum. We'll call this the "measured frequency" in the following discussion. The example cable would be a bit longer than an electrical wavelength at 7.15 MHz if the cable velocity factor is 0.66, so the measured frequency will probably be a little lower than 7.15 MHz.
5. Calculate the physical lengths of the feed system transmission lines. Although bypassing calculation of the velocity factor would save a step, including it provides a check on the calculations by seeing that it's close to the estimated velocity factor. It also can then be used for quickly making other transmission lines from the same uncut cable, or as a starting point for measuring similar uncut cables.
   A. Find the free space wavelength at the measured frequency. A quick and easy way is to start *EZNEC*, click the FREQUENCY line, and enter the measured frequency. The wavelength is shown in the line below. You can change units by clicking on the UNITS line. For the example, if the measured electrical one wavelength frequency was 6.4 MHz, the free space wavelength is 153.683 feet.
   B. Calculate the length of uncut cable for one electrical wavelength at the measured frequency. In our example, the measurement was done at the one electrical wavelength resonant point, so one electrical wavelength of the cable is simply 100 feet. If, for example, the measurement had been made at ¾ λ, the length for one electrical wavelength would be 4/3× the uncut cable length.
   C. Calculate the velocity factor:
   Velocity factor = electrical wavelength / free space wavelength (both at the same frequency, 6.4 MHz in our example) which for our example = 100 ft / 153.683 ft ≈ 0.651
   D. Find the length of a free space wavelength at the frequency where the transmission lines will be used (as in item a. above except at the operating frequency), then use the velocity factor to find the physical length of a wavelength in the uncut cable. In our example, 7.15 MHz is the frequency of use, and the free space wavelength is 137.562 feet. One physical wavelength of the uncut cable = the velocity factor x the free space wavelength = 89.55 feet. at 7.15 MHz. You can use this velocity factor to find the physical wavelength at other frequencies but be aware that the velocity factor does change slightly with frequency, particularly below about 3 MHz.
   E. Calculate the physical lengths of the feed system transmission lines. If you're using *Arrayfeed1* and have the current solution shown, simply change the VELOCITY FACTOR to the calculated value (0.651 for the example) and read the resulting transmission line physical lengths in the area below (17.078 and 38.756 feet for the example first solution). Or you can calculate the physical lengths. For our example, the Fig. 7.21 1st SOLUTION lengths are 68.65 and 155.80 electrical degrees. One 89.55 ft physical wavelength of cable is 360° so 68.65 electrical degrees = 68.65/360 x 89.55 ft = 17.08 ft. Similarly, the 155.80 electrical degree Line 2 is 155.80/360 x 89.55 ft = 38.76 ft.

28.7.5 Characteristic Impedance (Z_0)

The characteristic impedance (Z_0) of high quality 50 Ω coaxial cables *should* be between 48 and 52 Ω. However, the variation is often much greater, particularly with smaller diameter cables and those with foam dielectric. Measurements of 15 samples of various 50 Ω coax cable types showed Z_0 ranging from 48 to 57 Ω. This variation can be due to small differences in center conductor or dielectric diameter, centering of the center conductor, or dielectric constant of the insulator. It changes with frequency, most notably below about 5 MHz. A decrease of around 3 percent is commonly observed from 5 to 1 MHz, and the increase becomes greater and more rapidly changing below that. Some of this is due to skin effect penetration depth but other factors such as changes in the insulator dielectric constant with frequency are probably involved. So Z_0 should be measured for any application where its value is important. And like velocity factor, it should be measured near the frequency of use and with the actual cable the transmission lines will be cut from.

If you've modeled the array including feed system (highly recommended), you can change the model's transmission line Z_0 to see the effect on the pattern and judge the importance of making careful measurements. The following procedures have been found to produce accurate and repeatable results.

ABOUT Z_0

Any real transmission line has some loss, and this causes its Z_0 to be a complex value, that is, to have reactance as well as resistance. For the types and lengths of cables used for phased arrays, the reactive part of Z_0 is small. The measurements that follow use only the magnitude of the impedances and result in a single real value for Z_0. This is adequately accurate for the purpose, and appropriate for use in array transmission line calculations which generally don't include the effect of cable loss.

METHOD 1: OPEN/SHORT

This method requires measuring the impedance when the far end of the line is shorted or open. Use the same technique for shorting as described in the section dealing with physical length measurement.

Best accuracy is obtained when the measurement is done where the test line is in the vicinity of an odd number of eighth wavelengths long, and the frequency is in the neighborhood of the frequency where the transmission line will be used.

Tune the test instrument (e.g., antenna analyzer, network analyzer, Z bridge) to the frequency where the line will be used. Then tune up or down until you find the nearest frequency where the magnitude of the impedance ($|Z|$, written simply as Z in the following) is about the same with the line open or shorted at the other end. If your analyzer separately reports the reactance, this will be very close to the frequency where the reactance has the same magnitude but a different sign with the line open or shorted. If you can conveniently locate the frequency where Z is the same for both end conditions, that value of Z is Z_0. If you want to reduce the number of times you have to change between an open and short at the far end, just get to where one value is no more than two or three times the other and record Z_{short} and Z_{open}. Then calculate

$$Z_0 = \sqrt{Z_{short} \times Z_{open}}$$

METHOD 2: DUMMY LOAD

This requires measuring the magnitude of the impedance with the far end of the line terminated with a dummy load. The only requirement for the dummy load is that it must be nearly a pure resistance and not change value appreciably at the two frequencies where measurements will be made. A 50-Ω load can be used to measure the Z_0 of either 50 or 75-Ω cables. The far end of the line must connect to the dummy load with a connector and not pigtails or any other kind of contrived connection. Best accuracy is obtained when the measurement is done where the test frequencies are in the neighborhood of the frequency where the transmission line will be used.

Tune the test instrument (e.g., antenna analyzer, network analyzer, Z bridge) to the frequency where the line will be used. Then watch the value of the magnitude of the impedance ($|Z|$, written simply as Z in the following), and tune around that frequency looking for the closest maximum and minimum values of Z. If possible, find a minimum on one side and a maximum on the other. But if the cable length doesn't allow that, look for adjacent minimum and maximum values on one side or the other. If this isn't possible, you'll need to use Method 1 above. If you were able to get the maximum and minimum Z values, calculate

$$Z_0 = \sqrt{Z_{max} \times Z_{min}}$$

28.7.6 Transmission Line Loss

Although transmission line loss is ignored in many, or most, calculations for such things as impedance matching sections and phased array feed systems, even a small amount of loss can sometimes have a profound effect on system operation. Adding the effect of loss to calculations greatly increases their complexity, frequently requiring hyperbolic trigonometry with complex arguments in place of simple trigonometry with real arguments. But many antenna and transmission line programs can take into account the effects of loss and do the necessary complex calculations for accurate results. So, a good approach for designing any system including transmission lines is to model the system as completely as possible. If the underlying design doesn't include transmission line loss, leave it out of the initial model to verify operation as intended. Then add it and readjust the other components of the model if necessary to compensate or to retain satisfactory operation. Of course, to do this you'll need to know the transmission line loss with reasonable accuracy.

Published figures for transmission line loss are, in the author's experience, universally optimistic. So, you can pretty safely use them as a conservative worst case. For higher accuracy, though, you'll need to measure the loss. The two methods to be described use commonly available equipment, but the accuracy of the results is of course ultimately limited by that of the equipment. These methods measure the *matched loss*, that is, the loss when the line is terminated with its characteristic impedance. The actual loss will increase when terminated with other impedances, but the matched loss is the value needed for most calculations and for use in modeling.

ABOUT LINE LOSS

In the MF range and above, the fields in a coaxial cable having an effective shield is almost completely confined to the region between the center conductor and shield, so only that region affects the loss. There are two major contributors of loss in that region. The first, sometimes called "skin effect loss", is resistive I^2R loss in the outer portion of the center conductor and inner portion of the shield. This increases with frequency because of the decreasing skin depth, resulting in the dB loss being approximately proportional to the square root of frequency. The second is dielectric loss in the material between conductors. This loss, in dB, increases in direct proportion to frequency. Below roughly 1 GHz, dielectric loss in PE or PTFE is negligible compared to skin effect loss so the latter dominates. Above that range, the faster-increasing dielectric loss becomes dominant – a log-log graph of dB loss vs frequency for cables with non-plated conductors has a slope close to ½ up to about that frequency, gradually transitioning to a slope of one as frequency increases. Both effects are present at all frequencies but the dielectric loss, very low in the HF range and below but increasing more rapidly than skin effect loss, eventually becomes greater.

So why does foam dielectric cable have less loss than solid dielectric cable at HF, if the loss of even solid dielectric is negligible in that region? The answer is that foam dielectric has a lower dielectric constant, requiring a larger center conductor for the same Z_0. And that decreases the skin effect loss which dominates.

Other factors affect loss including shield roughness and wire plating. At the lower end of the MF range and below, the relationship between inductance and resistance per unit length and the effects of increasing skin depth can cause further changes. For an excellent discussion of these and other phenomena see Johnson and Graham, listed in the References and Bibliography.

METHOD 1: DIRECT MEASUREMENT

The simplest and potentially most accurate method of loss measurement is with an accurate power meter. Terminate the line with a dummy load having the same impedance as the line's Z_0 and apply power which is well within the range of the power meter. The loss in dB is simply

$$Loss(dB) = 10\log\left(\frac{P_{in}}{P_{out}}\right)$$

where P_{out} is the power measured at the termination and P_{in} is the power measured at the line input.

METHOD 2: INPUT SWR OF OPEN OR SHORTED LINE

Short or open circuit the far end of the line (see Short and Open Circuits above) and measure the SWR at the input, then calculate the loss.

$$Loss(dB) = 10\log\left(\frac{SWR+1}{SWR-1}\right)$$

Tests with both a high quality network analyzer and antenna analyzer gave good results with 50 Ω cable when the measured SWR was between about 1.2 and 7 (loss between about 1.3 and 10 dB). Beyond this range, small changes in SWR result in very large percentage changes in calculated loss and dramatically reduce the accuracy of the calculated result. An alternative method based on a color chart was developed by John Stanley, K4ERO, and is provided with the online material in the article "Calculating Transmission Line Loss with SWR Readings."

K4ERO and Steve Stearns, K6OIK have noted that under some circumstances and at low frequencies, the shorted and open measurements may not match. In this case, either use an electrical quarter-wavelength of line for the measurements or average the short and open-circuit values to get the correct value. This measurement error is described in the article "Precautions When Using the Return Loss Method of Measuring Coax Loss," by K4ERO in the January 2022 issue of *QEX.*

28.8 Antenna Field Measurements

Of all the measurements made in Amateur Radio systems, perhaps the most difficult and least understood are various measurements of the radiated field from antennas. For example, it is relatively easy to measure the frequency and CW power output of a transmitter, the response of a filter, or the gain of an amplifier. These are all what might be called *bench measurements* because, when performed properly, all the factors that influence the accuracy and success of the measurement are under control. In making antenna measurements, however, the "bench" is probably your backyard. In other words, the environment surrounding the antenna can affect the results of the measurement.

Control of the environment is not at all as simple as it was for the bench measurement, because now the work area may be rather spacious. This section describes antenna measurement techniques that are closely allied to those used in an antenna measuring event or contest. With these procedures you can make measurements successfully and with meaningful results. These techniques should provide a better understanding of the measurement problems, resulting in a more accurate and less difficult task. The information in this section was provided by Dick Turrin, W2IMU, and was originally published in the November 1974 *QST*. The conventionsused by amateurs to plot radiation patterns and antenna measurements are covered in the **Antenna Fundamentals** chapter.

Antenna Testing with Automated Receiver Networks

Automated receiver networks provide a new tool for antenna testing. It's not the same as making direct measurements of field strength with a single instrument — it's a distributed measurement made by receiving stations worldwide. You can then retrieve their measurements and use them to compare antennas.

WSPR is the ultra-low-power beacon mode for propagation testing that is part of the *WSJT-X* software suite. It's a slow mode with transmit cycles taking two minutes. Since propagation will change during that time, both antennas must be transmitting at the same time on different (but close) frequencies. Over time, the receivers build up a data set, available at wsprnet.org, that gives you a picture of how the antennas are performing. The process of using WSPR to evaluate antennas is described in a paper by Larry Plummer, W6LVP, at **w6lvp.com/wp-content/uploads/2017/01/White-Paper-WSPR-for-antenna-testing.pdf**. Randy Hall, W7AGE, illustrates the process with a video on his YouTube channel at **youtube.com/@K7AGE**: search for the video "Antenna Testing Using WSPR."

You can also use one of the other *WSJT-X* modes, such as FT8 or FT4. The transmit cycles for these modes are 15 and 7.5 seconds, respectively, so there is much less variation during the transmission than during a WSPR cycle. Sites like PSKReporter (pskreporter.info) store reception and decoding reports from all stations that heard your transmission.

An even faster method is to use short CW transmissions that are decoded and measured by stations around the world using the *CW Skimmer* or *RTTY Skimmer* software by Alex Shovkoplyas, VE3NEA. (**dxatlas.com/CwSkimmer**) They measure your station's signal-to-noise ratio and report it to the servers at the RBN (**reversebeacon.net**). The usual method for antenna comparison is to make a couple of short CQs including your call sign and watch for your signal report to be posted. Then call CQ again using a different call sign — perhaps your club's — and watch for that signal to be reported. Or you can make two transmissions on different but closely spaced frequencies with different call signs. There are many possible methods. Then you download the data or use the comparison tools of the RBN. Larry Gauthier, K8UT, describes the process at **k8ut.com/wpfb-file/rbn-as-an-antenna-performance-tool-final-pdf**.

28.8.1 Field Measurement Basics

An antenna is simply a transducer or coupler between a suitable feed line and the environment surrounding it. In addition to the efficient transfer of power from feed line to environment, an antenna at VHF or UHF is most frequently required to concentrate the radiated power into a particular region of the environment.

To be consistent while comparing different antennas, you must standardize the environment surrounding the antenna. Ideally, you want to make measurements with the measured antenna so far removed from any objects causing environmental effects that it is literally in outer space — a very impractical situation. The purpose of the measurement techniques is therefore to simulate, under practical conditions, a *controlled environment*. At VHF and UHF, and with practical-size antennas, the environment can be controlled so that successful and accurate measurements can be made in a reasonable amount of space.

The electrical characteristics of an antenna that are most desirable to obtain by direct measurement are: (1) gain (relative to an isotropic source, which by definition has a gain of unity); (2) space-radiation pattern; (3) feed point impedance (mismatch) and (4) polarization.

POLARIZATION

In general the polarization can be assumed from the geometry of the radiating elements. That is to say, if the antenna consists of a number of linear elements (straight lengths of rod or wire that are resonant and connected to the feed point) the polarization of the electric field will be linear and polarized parallel to the elements. If the elements are not consistently parallel with each other, then the polarization cannot easily be assumed. The following techniques are directed to antennas having polarization that is essentially linear (in one plane), although the method can be extended to include all forms of elliptic (or mixed) polarization.

FEED POINT MISMATCH

The feed point mismatch, although affected to some degree by the immediate environment of the antenna, does not affect the gain or radiation characteristics of an antenna. If the immediate environment of the antenna does not affect the feed point impedance, then any mismatch intrinsic to the antenna tuning reflects a portion of the incident power back to the source. In a receiving antenna this reflected power is reradiated back into the environment, and can be lost entirely.

In a transmitting antenna, the reflected power travels back down the feed line to the transmitter, where it changes the load impedance presented to that transmitter. The amplifier output controls are customarily altered during the normal tuning procedure to obtain maximum power transfer to the antenna. You can still use a mismatched antenna to its full gain potential, provided the mismatch is not so severe as to cause heating losses in the system, especially the feed line and matching devices. (See also the discussion of additional loss caused by SWR in the **Transmission Lines** chapter.)

Similarly, a mismatched receiving antenna may be matched into the receiver front end for maximum power transfer. In any case you should clearly keep in mind that the feed point mismatch does not affect the radiation characteristics of an antenna. It can only affect the system efficiency when heating losses are considered.

Why then do we include feed point mismatch as part of the antenna characteristics? The reason is that for efficient system performance, most antennas are resonant transducers and present a reasonable match over a relatively narrow frequency range. It is therefore desirable to design an antenna, whether it be a simple dipole or an array of Yagis, such that the final single feed point impedance is essentially resistive and matched to the feed line. Furthermore, in order to make accurate, absolute gain measurements, it is vital that the antenna under test accept all the power from a matched-source generator, or that the reflected power caused by the mismatch be measured and a suitable error correction for heating losses be included in the gain calculations. Heating losses may be determined from information contained in the **Transmission Lines** chapter.

While on the subject of feed point impedance, mention should be made of the use of baluns in antennas. A balun is simply a device that permits a lossless transition between a balanced system feed line or antenna and an unbalanced feed line or system. If the feed point of an antenna is symmetric, such as with a dipole, and it is desired to feed this antenna with an unbalanced feed line such as coax, you should provide a balun between the line and the feed point. Without the balun, current will be allowed to flow on the outside of the coax. The current on the outside of the feed line will cause radiation, and thus the feed line will become part of the antenna radiation system. In the case of beam antennas, where it is desired to concentrate the radiated energy is a specific direction, this extra radiation from the feed line will be detrimental, causing distortion of the expected antenna pattern. See the **Transmission Line System Techniques** for additional details on this problem.

28.8.2 Test Site Set-Up and Evaluation

Since an antenna is a reciprocal device, measurements of gain and radiation patterns can be made with the test antenna used either as a transmitting or as a receiving antenna. In general and for practical reasons, the test antenna is used in the receiving mode, and the source or transmitting antenna is located at a specified fixed remote site and unattended. In other words the source antenna, energized by a suitable transmitter, is simply required to illuminate or flood the receiving site in a controlled and constant manner.

As mentioned earlier, antenna measurements ideally should be made under free-space conditions. A further restriction is that the illumination from the source antenna be a *plane wave* over the effective aperture (capture area) of the test antenna. A plane wave by definition is one in which the magnitude and phase of the fields are uniform, and in the test-antenna situation, *uniform over the effective area plane of the test antenna.* Since it is the nature of all radiation to expand in a spherical manner at great distance from the source, it would seem to be most desirable to locate the source antenna as far from the test site as possible. However, since for practical reasons the test site and source location will have to be near the Earth and not in outer space, the environment must include the effects of the ground surface and other obstacles in the vicinity of both antennas. These effects almost always dictate that the test range (spacing between source and test antennas) be as short as possible consistent with maintaining a nearly error-free plane wave illuminating the test *aperture*.

A nearly error-free plane wave can be specified as one in which the phase and amplitude, from center to edge of the illuminating field over the test aperture, do not deviate by more than about 30° and 1 dB, respectively. These conditions will result in a gain-measurement error of no more than a few percent less than the true gain. Based on the 30° phase error alone, it can be shown that the minimum range distance is approximately

$$S_{min} = 2\frac{D^2}{\lambda} \qquad (19)$$

where D is the largest aperture dimension and λ is the free-space wavelength in the same units as D. The phase error over the aperture D for this condition is 1⁄16 wavelength.

Since aperture size and gain are related by

$$\text{Gain} = \frac{4\pi A_c}{\lambda^2} \qquad (20)$$

where A_e is the effective aperture area, the dimension D may be obtained for simple aperture configurations. For a square aperture

$$D^2 = G\frac{\lambda^2}{4\pi} \qquad (21)$$

that results in a minimum range distance for a square aperture of

$$S_{min} = G\frac{\lambda}{2\pi} \qquad (22)$$

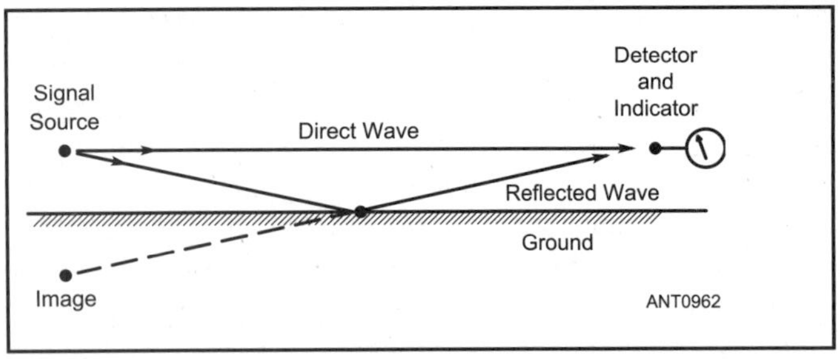

Figure 28.54 — On an antenna test range, energy reaching the receiving equipment may arrive after being reflected from the surface of the ground, as well as by the direct path. The two waves may tend to cancel each other, or may reinforce one another, depending on their phase relationship at the receiving point.

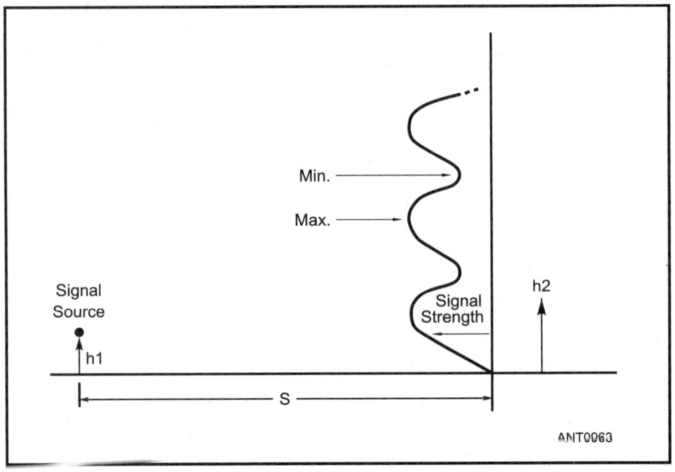

Figure 28.55 — The vertical profile, or plot of signal strength versus test-antenna height, for a fixed height of the signal source above ground and at a fixed distance. See text for definitions of symbols.

and for a circular aperture of

$$S_{min} = G\frac{2\lambda}{\pi^2} \qquad (23)$$

For apertures with a physical area that is not well defined or is much larger in one dimension that in other directions, such as a long thin array for maximum directivity in one plane, it is advisable to use the maximum estimate of D from either the expected gain or physical aperture dimensions.

Up to this point in the range development, only the conditions for minimum range length, S_{min}, have been established, as though the ground surface were not present. This minimum S is therefore a necessary condition even under free-space environment. The presence of the ground further complicates the range selection, not in the determination of S but in the exact location of the source and test antennas above the Earth.

It is always advisable to select a range whose intervening terrain is essentially flat, clear of obstructions, and of uniform surface conditions, such as all grass or all pavement. The extent of the range is determined by the illumination of the source antenna, usually a Yagi, whose gain is no greater than the lowest gain antenna to be measured. For gain measurements the range consists essentially of the region in the beam of the test antenna. For radiation-pattern measurements, the range is considerably larger and consists of all that area illuminated by the source antenna, especially around and behind the test site. Ideally you should choose a site where the test-antenna location is near the center of a large open area and the source antenna is located near the edge where most of the obstacles (trees, poles, fences, etc.) lie.

The primary effect of the range surface is that some of the energy from the source antenna will be reflected into the test antenna, while other energy will arrive on a direct line-of-sight path. This is illustrated in **Figure 28.54**. The use of a flat, uniform ground surface assures that there will be essentially a mirror reflection, even though the reflected energy may be slightly weakened (absorbed) by the surface material (ground). In order to perform an analysis you should realize that horizontally polarized waves undergo a 180° phase reversal upon reflection from the Earth. The resulting illumination amplitude at any point in the test aperture is the vector sum of the electric fields arriving from the two directions, the direct path and the reflected path.

If a perfect mirror reflection is assumed from the ground (it is nearly that for practical ground conditions at VHF/UHF) and the source antenna is isotropic, radiating equally in all directions, then a simple geometric analysis of the two path lengths will show that at various points in the vertical plane at the test-antenna site the waves will combine in different phase relationships. At some points the arriving waves will be in phase, and at other points they will be 180° out of phase. Since the field amplitudes are nearly equal, the resulting phase change caused by path length difference will produce an amplitude variation in the vertical test site direction similar to a standing wave, as shown in **Figure 28.55**.

The simplified formula relating the location of h2 for maximum and minimum values of the two-path summation in terms of h1 and S is

$$h2 = n\frac{\lambda}{4} \times \frac{S}{h1} \qquad (24)$$

with n = 0, 2, 4, . . . for minimums and n = 1, 3, 5, . . . for maximums, and S is much larger than either h1 or h2.

The significance of this simple ground reflection formula is that it permits you to determine the approximate location of the source antenna to achieve a nearly plane-wave amplitude distribution *in the vertical direction* over a particular test *aperture size*. It should be clear from examination of the height formula that as h1 is decreased, the vertical distribution pattern of signal at the test site, h2, expands. Also note that the signal level for h2 equal to zero is always zero on the ground regardless of the height of h1.

The objective in using the height formula then is, given an effective antenna aperture to be illuminated from which a minimum S (range length) is determined and a suitable range site chosen, to find a value for h1 (source antenna height). The required value is such that the first maximum of vertical distribution at the test site, h2, is at a practical distance above the ground, and at the same time the signal amplitude over the aperture in the vertical direction does not vary more than about 1 dB. This last condition is not absolutely necessary but is closely related to the particular antenna under test.

In practice these formulas are useful only to initialize the range setup. A final check of the vertical distribution at the test site must be made by direct measurement. This measurement should be conducted with a small low-gain but unidirectional probe antenna such as a corner reflector or 2-element Yagi that you move along a vertical line over the intended aperture site. Care should be exercised to minimize the effects of local environment around the probe antenna and that the beam of the probe be directed at the source antenna at all times for maximum signal. A simple dipole is undesirable as a probe antenna because it is susceptible to local environmental effects.

The most practical way to instrument the vertical distribution measurement is to construct some kind of vertical track, preferably of wood, with a sliding carriage or platform that may be used to support and move the probe antenna. It is assumed of course that a stable source transmitter and calibrated receiver or detector are available so variations of the order of ½ dB can be clearly distinguished.

Once you conduct these initial range measurements successfully, the range is now ready to accommodate any aperture size less in vertical extent than the largest for which S_{min} and the vertical field distribution were selected. Place the test antenna with the center of its aperture at the height h2 where maximum signal was found. Tilt the test antenna so that its main beam is pointed in the direction of the source antenna. The final tilt is found by observing the receiver output for maximum signal. This last process must be done empirically since the apparent location of the source is somewhere between the actual source and its image, below the ground.

An example will illustrate the procedure. Assume that we wish to measure a 7-foot diameter parabolic reflector antenna at 1296 MHz ($\lambda = 0.75$ foot). The minimum range distance, S_{min}, can be readily computed from the formula for a circular aperture.

$$S_{min} = 2\frac{D^2}{\lambda} = 2 \times \frac{49}{0.75} = 131 \text{ feet}$$

Now a suitable site is selected based on the qualitative discussion given before.

Next determine the source height, h1. The procedure is to choose a height h1 such that the first minimum above ground (n = 2 in formula) is at least two or three times the aperture size, or about 20 feet.

$$h1 = n\frac{\lambda S}{4\,h2} = 2 \times \frac{0.75}{4} \times \frac{131}{20} = 2.5 \text{ feet}$$

Place the source antenna at this height and probe the vertical distribution over the 7-foot aperture location, which will be about 10 feet off the ground.

$$h2 = n\frac{\lambda S}{4\,h1} = 1 \times \frac{0.75}{4} \times \frac{131}{2.5} = 9.8 \text{ feet}$$

Plot the measured profile of vertical signal level versus height. From this plot, empirically determine whether the 7-foot aperture can be fitted in this profile such that the 1-dB variation is not exceeded. If the variation exceeds 1 dB over the 7-foot aperture, the source antenna should be lowered and h2 raised. Small changes in h1 can quickly alter the distribution at the test site. **Figure 28.56** illustrates the points of the previous discussion.

The same set-up procedure applies for either horizontal or vertical linear polarization. However, it is advisable to check by direct measurement at the site for each polarization to be sure that the vertical distribution is satisfactory. Distribution probing in the horizontal plane is unnecessary as little or no variation in amplitude should be found, since the reflection geometry is constant. Because of this, antennas with apertures that are long and thin, such as a stacked collinear vertical, should be measured with the long dimension parallel to the ground.

A particularly difficult range problem occurs in measurements of antennas that have depth as well as cross-sectional aperture area. Long end-fire antennas such as long Yagis, rhombics, V-beams, or arrays of these antennas, radiate as volumetric arrays and it is therefore even more essential that the illuminating field from the source antenna be reasonably uniform in depth as well as plane wave in cross section. For measuring these types of antennas it is advisable to make several vertical profile measurements that cover the depth of the array. A qualitative check on the integrity of the illumination for long end-fire antennas can be made by moving the array or antenna axially (forward and backward) and noting the change in received signal level. If the signal level varies less than 1 or 2 dB for an axial movement of several wavelengths then the field can be considered satisfactory for most demands on accuracy. Large variations indicate that the illuminating field is badly distorted over the array depth and subsequent measurements are questionable. It is interesting to note in connection with gain measurements that any illuminating field distortion will always result in measurements that are lower than true values.

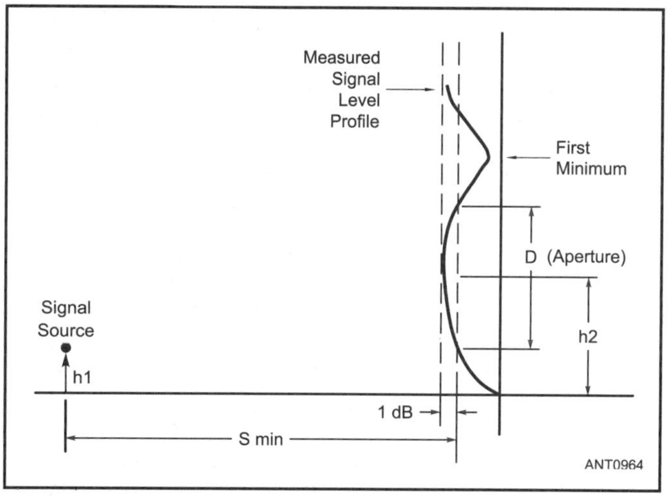

Figure 28.56 — Sample plot of a measured vertical profile.

28.8.3 Absolute Gain Measurement

Having established a suitable range, the measurement of gain relative to an isotropic (point source) radiator is almost always accomplished by direct comparison with a calibrated standard-gain antenna. That is, the signal level with the test antenna in its optimum location is noted. Then you remove the test antenna and place the standard-gain antenna with its aperture at the center of location where the test antenna was located. Measure the difference in signal level between the standard and the test antennas and add to or subtract from the gain of the standard-gain antenna to obtain the absolute gain of the test antenna. Here, *absolute* means with respect to a point source with a gain of unity, by definition. The reason for using this reference rather than a dipole, for instance, is that it is more useful and convenient for system engineering. We assume that both standard and test antennas have been carefully matched to the appropriate impedance and an accurately calibrated and matched detecting device is being used.

A standard-gain antenna may be any type of unidirectional, preferably planar-aperture, antenna, which has been calibrated either by direct measurement or in special cases by accurate construction according to computed dimensions. A standard-gain antenna for VHF and low-UHF bands has been suggested by Richard F. H. Yang (see References and Bibliography). Shown in **Figure 28.57**, it consists of two in-phase dipoles ½ λ apart and backed up with a ground plane 1 λ square. (It is recommended that the builder cut the dipoles close to their free-space length and trim to resonance.)

In Yang's original design, the stub at the center is a balun formed by cutting two longitudinal slots of ⅛-inch width, diametrically opposite, on a ¼-λ section of ⅞-inch rigid 50-Ω coax. An alternative method of feeding is to feed RG-8 or RG-213 coax through slotted ¾-inch copper tubing with a ⅞-inch OD. (Due to variations in ID/OD for stock copper tubing, either bring a section of the rigid coax or take careful measurements with a caliper to check for fit between the coax and the balun tubing.)

Be sure to leave the outer jacket on the coax to insulate it from the copper-tubing balun section. When constructed accurately to scale for the frequency of interest, this type of standard will have an absolute gain of 9.85 dBi (7.7 dBd gain over a dipole in free space) with an accuracy of ± 0.25 dB. (The balun is described in detail in the original article.)

At 1296 MHz it may be more practical to build a reference horn out of sheet metal as described by Paul Wade, W1GHZ on his website (**w1ghz.org**). The waveguide section can also be made out of sheet metal.

28.8.4 Radiation Pattern Measurements

Of all antenna measurements, the radiation pattern is the most demanding in measurement and the most difficult to interpret. Any antenna radiates to some degree in all directions into the space surrounding it. Therefore, the radiation pattern of an antenna is a three-dimensional representation of the magnitude, phase and polarization. In general, and in practical cases for amateur radio communications, the polarization is

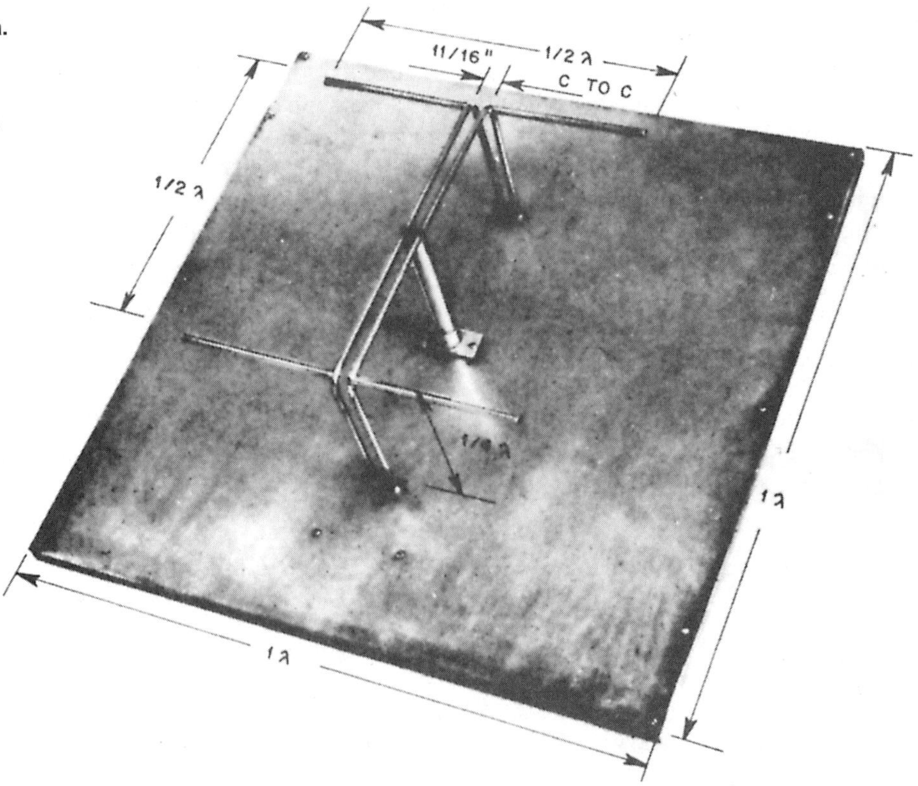

Figure 28.57 — Standard-gain antenna. When accurately constructed for the desired frequency, this antenna will exhibit a gain of 7.7 dB over a dipole radiator, plus or minus 0.25 dB. In this model, constructed for 432 MHz, the elements are ⅜-inch diameter tubing. The phasing and support lines are of 5⁄16-inch diameter tubing or rod.

well defined and only the magnitude of radiation is important.

Furthermore, in many of these cases the radiation in one particular plane is of primary interest, usually the plane corresponding to that of the Earth's surface, regardless of polarization. Because of the nature of the range setup, measurement of radiation pattern can be successfully made only in a plane nearly parallel to the Earth's surface. With beam antennas it is advisable and usually sufficient to take two radiation pattern measurements, one in the polarization plane and one at right angles to the plane of polarization. These radiation patterns are referred to in antenna literature as the principal E-plane and H-plane patterns, respectively. *E-plane* means parallel to the electric field that is the polarization plane and *H-plane* means parallel to the magnetic field in free space. The electric field and magnetic field are always perpendicular to each other in a plane wave as it propagates through space.

When the antenna is located over real Earth, the terms *azimuth* and *elevation* planes are commonly used, since the frame of reference is the Earth itself, rather than the electric and magnetic fields in free space. For a horizontally polarized antenna such as a Yagi mounted with its elements parallel to the ground, the azimuth plane is the E-plane and the elevation plane is the H-plane.

The technique to obtain these patterns is simple in procedure but requires more equipment and patience than does making a gain measurement. First, a suitable mount is required that can be rotated in the azimuth plane (horizontal) with some degree of accuracy in terms of azimuth-angle positioning. Second, a signal-level indicator calibrated over at least a 20-dB dynamic range with a readout resolution of at least 2 dB is required. A dynamic range of up to about 40 dB would be desirable but does not add greatly to the measurement significance.

With this much equipment, the procedure is to locate first the area of maximum radiation from the beam antenna by carefully adjusting the azimuth and elevation positioning. These settings are then arbitrarily assigned an azimuth angle of zero degrees and a signal level of zero decibels. Next, without changing the elevation setting (tilt of the rotating axis), the antenna is carefully rotated in azimuth in small steps that permit signal-level readout of 2 or 3 dB per step. These points of signal level corresponding with an azimuth angle are recorded and plotted on polar coordinate paper. A sample of the results is shown on ARRL coordinate paper in **Figure 28.58**. (See the **Antenna Fundamentals** chapter for more information on coordinate scales.)

On the sample radiation pattern the measured points are marked with an X and a continuous line is drawn in, since the pattern is a continuous curve. Radiation patterns should preferably be plotted on a logarithmic radial scale, rather than a voltage or power scale. The reason is that the log scale approximates the response of the ear to signals in the audio range. Also many receivers have AGC systems that are somewhat logarithmic in response; therefore the log scale is more representative of actual system operation.

Having completed a set of radiation-pattern measurements, one is prompted to ask, "Of what use are they?" The primary answer is as a diagnostic tool to determine if the antenna is

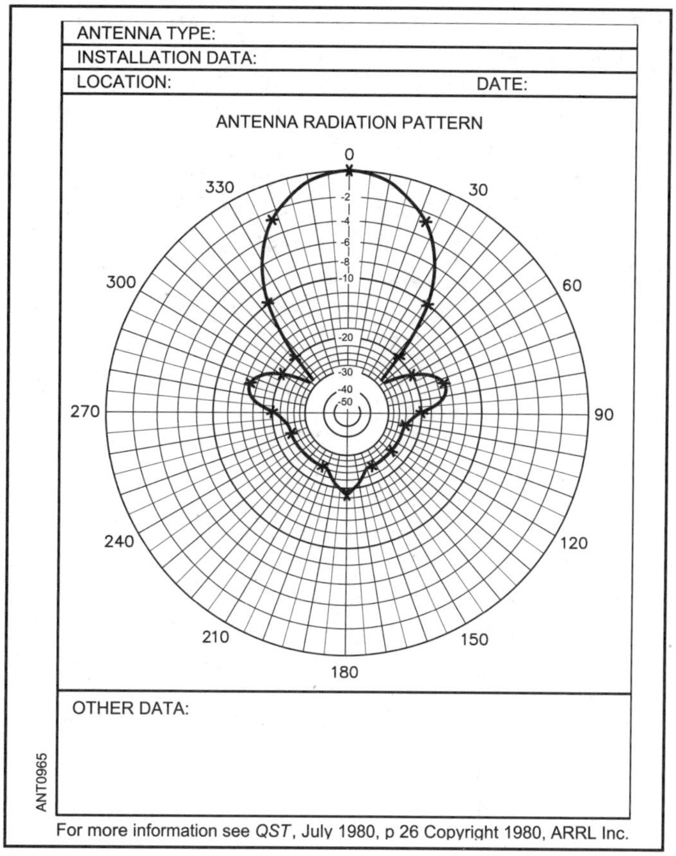

Figure 28.58 — Sample plot of a measured radiation pattern, using techniques described in the text.

functioning as it was intended to. A second answer is to know how the antenna will discriminate against interfering signals from various directions.

Consider now the diagnostic use of the radiation patterns. If the radiation beam is well defined, then there is an approximate formula relating the antenna gain to the measured half-power beamwidth of the E- and H-plane radiation patterns. The half-power beamwidth is indicated on the polar plot where the radiation level falls to 3 dB below the main beam 0-dB reference on either side. The formula is

$$\text{Gain (isotropic)} \cong \frac{41,253}{\theta_E \phi_H} \tag{25}$$

where θ_E and ϕ_H are the half-power beamwidths in degrees of the E- and H-plane patterns, respectively. This equation assumes a lossless antenna system, where any side-lobes are well suppressed. (To obtain gain in dBi, take the log of isotropic gain and multiply by 10.)

To illustrate the use of this equation, assume that we have a Yagi antenna with a boom length of two wavelengths. From known relations (described in the **HF Beam Antennas** chapter) the expected free-space gain of a Yagi with a boom length of 2 λ is about 13 dBi; its gain, G, equals 20. Using the above relationship, the product of $\theta_E \times \phi_H \approx 2062$ square degrees. Since a Yagi produces a nearly symmetric beam shape in cross section, $\theta_E = \phi_H = 45°$. Now if the measured values of θ_E and ϕ_H are much larger than 45°, then the gain will be much lower than the expected 13 dBi.

As another example, suppose that the same antenna (a 2-wavelength-boom Yagi) gives a measured gain of 9 dBi but the radiation pattern half power beamwidths are approximately 45°. This situation indicates that although the radiation patterns seem to be correct, the low gain shows inefficiency somewhere in the antenna, such as lossy materials or poor connections.

Large broadside collinear antennas can be checked for excessive phasing-line losses by comparing the gain computed from the radiation patterns with the direct-measured gain. It seems paradoxical, but it is indeed possible to build a large array with a very narrow beamwidth indicating high gain, but actually having very low gain because of losses in the feed distribution system.

In general, and for most VHF/UHF Amateur Radio communications, gain is the primary attribute of an antenna. However, radiation in directions other than the main beam, called *sidelobe radiation*, should be examined by measurement of radiation patterns for effects such as nonsymmetry on either side of the main beam or excessive magnitude of sidelobes. (Any sidelobe that is less than 10 dB below the main beam reference level of 0 dB should be considered excessive.) These effects are usually attributable to incorrect phasing of the radiating elements or radiation from other parts of the antenna that was not intended, such as the support structure or feed line.

The interpretation of radiation patterns is intimately related to the particular type of antenna under measurement. Reference data should be consulted for the antenna type of interest, to verify that the measured results are in agreement with expected results.

To summarize the use of pattern measurements, if a beam antenna is first checked for gain (the easier measurement to make) and it is as expected, then pattern measurements may be academic. However, if the gain is lower than expected it is advisable to make pattern measurements to help determine the possible causes for low gain.

Regarding radiation pattern measurements, remember that the results measured under proper range facilities will not necessarily be the same as observed for the same antenna at a home-station installation. The reasons may be obvious now in view of the preceding information on the range setup, ground reflections, and the vertical-field distribution profiles. For long paths over rough terrain where many large obstacles may exist, the effects of ground reflection tend to become diffused, although they still can cause unexpected results. For these reasons it is usually unjust to compare VHF/UHF antennas over long paths.

28.9 References and Bibliography

Source material and more extended discussion of topics covered in this chapter can be found in the references given below.

Agilent Technologies, "Understanding the Fundamental Principles of Vector Network Analysis," Application Note AN 1287-1. Other application notes with further information on network analyzers are AN 1287-2 through AN 1287-10 and AN 1287-12, and also 5965-7917E, and 5990-5446EN, available from **keysight.com** (enter the application note number in the search box).

J. Althouse, "The Noise Bridge," *QST*, Sep. 1992, pp. 75-78.

G. Badger, et al, "SWR Analyzer Tips, Tricks and Techniques: SWR Analyzer Hints," *QST*, Sep. 1996, p. 36-40.

T. Baier, "A Small, Simple USB-Powered Vector Network Analyzer Covering 1 kHz to 1.3 GHz," *QEX*, Jan./Feb. 2009, pp. 32-36.

T. Baier, "A Simple S-Parameter Test Set for the VNWA2 Vector Network Analyzer," *QEX*, May/June 2009, pp. 29-32.

A. Bailey, "The Antenna Lab, Parts 1 and 2," *Radio Communication*, Aug. and Sep. 1983.

J. Belrose, "On Tuning, Matching and Measuring Antenna System Impedance Using a Hand-Held SWR Analyzer," *QST*, Sep. 2006, pp. 56-68.

L. Blake, *Transmission Lines and Waveguides* (New York: John Wiley & Sons, 1969), pp. 244-251.

J. H. Bowen, "A Calorimeter for VHF and UHF Power Measurements," *QST*, Dec. 1975, pp. 11-13.

D. Bramwell, "An RF Step Attenuator," *QST*, June 1995, pp. 33-34.

T. Brock-Fisher, "Build a Super-Simple SWR Indicator," *QST*, June 1995, pp. 40-41.

F. Brown, "A Reflectometer for Twin-Lead," *QST*, Oct. 1980, p. 15-17.

W. Bruene, "An Inside Picture of Directional Wattmeters," *QST*, Apr. 1959, pp. 24-28.

CCIR Recommendation 368, Documents of the CCIR XII Plenary assembly, ITU, Geneva, 1967.

J. Carr, "Find Fault with Your Coax," *73*, Oct. 1984, pp. 10-14.

S. Cooper, "A Compensated, Modular RF Voltmeter," *QEX*, Mar./Apr. 2001, pp. 26-34.

P. Danzer, "A Simple Transformer to Measure Your Antenna Current," *QST*, Sep. 2009, p. 35.

D. DeMaw, "In-Line RF Power Metering," *QST*, Dec. 1969, pp. 11-16.

D. DeMaw, "A QRP Person's VSWR Indicator," *QST*, Aug. 1982, p. 45.

D. Fayman, "A Simple Computing SWR Meter," *QST*, July 1973, pp. 23-33.

J. Gibbons and H. Horn, "A Circuit With Logarithmic Response Over Nine Decades," *IEEE Transactions on Circuit Theory*, Vol CT-11, No. 3, Sep. 1964, pp. 378-384.

J. Grebenkemper, "Calibrating Diode Detectors," *QEX*, Aug. 1990, pp. 3-8.

J. Grebenkemper, "The Tandem Match — An Accurate Directional Wattmeter," *QST*, Jan. 1987, pp. 18-26. Also see corrections in Technical Correspondence, *QST*, Jan. 1988, p. 49 and "An Updated Tandem Match" in Technical Correspondence, *QST*, July 1993, p. 50.

J. Grebenkemper, "Improving and Using R-X Noise Bridges," *QST*, Aug. 1989, pp. 27-32, 52; Feedback, *QST*, Jan. 1990, p. 27.

J. Hallas, *Understanding Your Antenna Analyzer,* ARRL, 2013.

E. Hare, "A Current Probe for the RF-Survey Meter," *QST*, Aug. 2000, p. 43.

F. Hauff, "The Gadget — an SWR Analyzer Add-On," *QST*, Oct. 1996, pp. 33-35.

W. Kaune, "A Modern Directional Power/SWR Meter," *QST*, Jan. 2011, pp. 39-43.

T. King, "A Practical Time-Domain Reflectometer," *QST*, May 1989, pp. 22-24.

M. Knitter, "The DG5MK IV Meter — An Accurate Antenna Analyzer," *QEX*, May/June 2017, pp. 3-14.

Z. Lau, "A Relative RF Ammeter for Open-Wire Lines," *QST*, Oct 1988, pp. 15-17, 20.

Z. Lau and C. Hutchinson, "Improving the HW-9 Transceiver," *QST*, Apr. 1988, pp. 26-29.

V. G. Leenerts, "Automatic VSWR and Power Meter," *Ham Radio*, May 1980, pp. 34-43.

J. Lenk, *Handbook of Oscilloscopes* (Englewood Cliffs, NJ: Prentice-Hall, 1982), pp. 288-292.

R. Lewallen, "Notes on Phased Verticals," Technical Correspondence, *QST*, Aug. 1979, pp. 42-43.

I. Lindell, E. Alanen, K. Mannerslo, "Exact Image Method for Impedance Computation of Antennas Above the Ground," *IEEE Trans. On Antennas and Propagation*, AP-33, Sep. 1985.

R. Littlefield, "A Wide-Range RF-Survey Meter," *QST*, Aug. 2000, pp. 42-44.

L. McCoy, "A Linear Field-Strength Meter," *QST*, Jan. 1973, pp. 18-20, 35.

T. McDermott and K. Ireland, "A Low-Cost 100 MHz Vector Network Analyzer with USB Interface," *QEX*, July/Aug. 2004, pp. 3-13.

T. McMullen, "The Line Sampler, an RF Power Monitor for VHF and UHF," *QST*, Apr. 1972, pp. 21-23, 25.

M. W. Maxwell, *Reflections* (Newington: ARRL, 1990), p. 20-3. [Out of print.]

C. Michaels, "Determining Line Lengths," Technical Correspondence, *QST*, Sep. 1985, pp. 43-44.

J. Noakes, "The "No Fibbin" RF Field Strength Meter," *QST*, Aug. 2002, pp. 28-29.

Orr and Cowan, *Vertical Antennas*, Radio Amateur Call Book, 1986, pp. 148-150.

P. Ostapchuk, "A Rugged, Compact Attenuator," *QST*, May 1998, pp. 41-43. Also see Technical Correspondence, *QST*, Dec. 1998, p. 64.

H. Perras, "Broadband Power-Tracking VSWR Bridge," *Ham Radio*, Aug. 1979, pp. 72-75.

D. Pozar, *Microwave Engineering* (New York: John Wiley & Sons, 2004).

S. Ramo, J. Whinnery and T. Van Duzer, *Fields and Waves in Communication Electronics* (New York: John Wiley & Sons, 1967), Chap 1.

Reference Data for Radio Engineers, 5th edition (Indianapolis: Howard W. Sams, 1968), Chapter 28.

W. Sabin, "The Lumped-Element Directional Coupler," *QEX*, Mar. 1995, pp. 3-11.

P. Salas, "A Compact 100-W Z-Match Antenna Tuner," *QST*, Jan. 2003, p. 28-30.

P. N. Saveskie, *Radio Propagation Handbook* (Blue Ridge Summit, PA: TAB Books, 1960).

P. Schuch, "The SWR Analyzer and Transmission Lines," *QST*, July 1997, p. 68.

J. Sevick, "Short Ground-Radial Systems for Short Verticals," *QST*, Apr. 1978, pp. 30-33.

J. Sevick, "Measuring Soil Conductivity," *QST*, Mar. 1981, pp. 38-39.

W. Silver, "Hands-On Radio: Experiment #52 — SWR Meters," *QST*, May 2007, pp. 57-58.

R. Skelton, Ron, "Measuring HF Balun Performance," *QEX*, Nov./Dec. 2010, pp. 39-41.

S. Sparks, "An RF Current Probe for Amateur Use," *QST*, Feb 1999, p. 34.

W. Spaulding, "A Broadband Two-Port S-Parameter Test Set," *Hewlett-Packard Journal*, Nov 1984.

F. Sutter, "What, No Meters?," *QST*, Oct. 1938, pp. 49-50.

D. Turrin, "Antenna Performance Measurements," *QST*, Nov. 1974, pp. 35-41.

F. Van Zant, "High-Power Operation with the Tandem Match Directional Coupler," Technical Correspondence, *QST*, July 1989, pp. 42-43.

P. Wade, "Directional Couplers," Microwavelengths, *QST*, Jan. 2007, pp. 87-88.

P. Wade, "Microwave System Test," Microwavelengths, *QST*, Aug. 2010, pp. 96-97.

C. Wright, "The Twin-Lamp," *QST*, Oct. 1947, pp. 22-23, 110, 112.

R. F. H. Yang, "A Proposed Gain Standard for VHF Antennas," *IEEE Transactions on Antennas and Propagation*, Nov. 1966.

TABLE OF CONTENTS

Chapter 29

Antenna System Troubleshooting

There is not a single amateur who, at some time, has not introduced an error into the antenna system either during installation or use. Similarly, of course, nothing remains faultless forever and those are the subjects of this chapter — finding the errors and faults. The first section of the chapter is directed at the beginner, providing a structured process to hunt for and find the problem. It is adapted from the Wireless Institute of Australia *Amateur Radio* magazine's excellent series of "Foundation Corner" articles for new hams. It was originally written by Ted Thrift, VK2ARA, and Ross Pittard, VK3CE. The second section of the chapter is more detailed and assumes more technical background on the part of the reader. It is adapted from material written by Tom Schiller, N6BT, as part of his book *Array of Light, 3rd Edition* (**n6bt.com**).

The goal of this chapter is not to provide an exhaustive procedure that can be followed "cookbook-style" to troubleshoot any antenna or antenna system. There are just too many variables and configurations for that to be possible. Rather, this chapter suggest systematic approaches and general guidelines to apply in order to find problems. Once problems are identified, the solution is usually obvious and even trivial.

Anyone with experience in maintaining or building systems of more than a few parts — whether related to amateur radio or not — will recognize the value of a systematic approach to troubleshooting. The underlying lesson in this material is that carefully analyzing a problem with a step-by-step approach pays off in effective troubleshooting, saving time and expense. This is true for antennas, transceivers, computer systems — any sort of technology. Whether the reader is just getting on the air or has a lifetime of experience, there is something for everyone in this chapter.

The material on troubleshooting and repair of antenna tuners was contributed by Matt Kastigar, WØMJ. Antenna tuners operate at the same power levels as the final amplifier, so they are subjected to the same stresses plus higher SWR levels. This material helps you find problems and repair the affected components.

The final section of this chapter is more about maintenance than troubleshooting but the two are so closely linked that the information will be helpful. It is another adaptation from the WIA *Amateur Radio* "Foundation Corner" columns, this one written by Ross Pittard, VK3CE, and Geoff Emery, VK4ZPP.

29.1 Antenna System Troubleshooting for Beginners

So you can no longer hear anything and you think your antenna system is faulty. It is very likely that it is, or at least some part of it is faulty. To repair the fault, we first have to find it. To do this we have to treat your antenna system in exactly the same way as fault finding inside a radio. After all, it is an electrical circuit and if not completely correct, will not work in the way that you expect. The process described in the rest of this section can be adapted to most simple antenna systems similar to that shown in **Figure 29.1**.

Start with an inventory of the antenna system. Any of these can be the cause of your problem:

- The support poles and ropes
- The antenna insulators
- The antenna elements
- The feed point or balun
- The feed line
- The entry point to the station
- The jumper cable to the radio

A *jumper cable* (also called a *patch cable*) is a short piece of coaxial cable with RF connectors on each end. It is used to connect two pieces of equipment together. The discussion below assumes that you have a coaxial feed line to the antenna.

Determine the characteristics of the antenna you are troubleshooting:

- Is it a balanced half-wave dipole?
- Is it an off-center-fed (OCF) dipole?
- Is it a multiband antenna, for example, a G5RV?
- What is the primary band it is designed for?

Consider the characteristics of the radio as well:

- Does it have a built in antenna tuner or do you use an add-on antenna tuner?
- Can you transmit a carrier signal on any band?
- Can you adjust the power level of the carrier?

During the following sequence of testing, be alert for mistaken or loose connections, loose or disconnected power and control cables, wires touching each other that shouldn't be, and so forth. Your main system components may be just fine but not connected properly. This is *very* common!

If you haven't started one yet, this is a great time to start your "station notebook" in which you record how your station is built. This is where you write down test results, color codes of control cables, modifications to equipment, dates of installation, etc. This information can be a huge time-saver in the future when you are troubleshooting or designing an addition to the station. A spiral-bound or composition book of graph paper is the best option, but a loose-leaf binder works well, too. Remember to put the date on each page as you make an entry. It's easy to take time-stamped photos, too.

29.1.1 Before Testing

If your radio has a built in auto tuner it has by now attempted to match your antenna system, faults and all. You may have also tried other bands to see if you can get "something" to work. To find the fault we must test the system *on the primary band for which it was designed.* Keep this in mind when you start testing. If your testing indicates that the tuner may be at fault, see the section on Antenna Tuner Troubleshooting and Repair later in this chapter.

TEST EQUIPMENT

In addition to your radio, you will need at least the following items.

- A suitable power/SWR meter.
- A volt-ohm meter to check continuity of cables and wires.
- A suitable 50-Ω dummy load.
- At least two tested and known-good 50-Ω jumper cables.

29.1.2 Initial Testing

This is to ensure that both your radio and your test equipment are working correctly.

1) Remove the antenna coax and connect your test jumper cable.

2) Connect the other end of the jumper cable to your power/SWR meter.

3) Connect your dummy load to the power/SWR meter.

4) Set the power range on the meter to a high scale to prevent overload.

5) Set the radio to the antenna's primary band.

6) Set your radio to CW, AM, or FM.

7) Adjust output power to minimum.

8) Press (PTT) and adjust output power to (say) 5 – 10 W.

9) Check that the power indication in the radio and the power/SWR meter are similar.

10) If your radio passes this test, activate whatever tuner you use, either the transceiver's internal tuner or an external tuner. Then repeat steps 8 and 9 after the internal tuner has adjusted to match the 50 Ω load or you have adjusted the manual tuner to match the 50 Ω load. If the internal or external tuner experience difficulty tuning to the 50 Ω load, proceed to

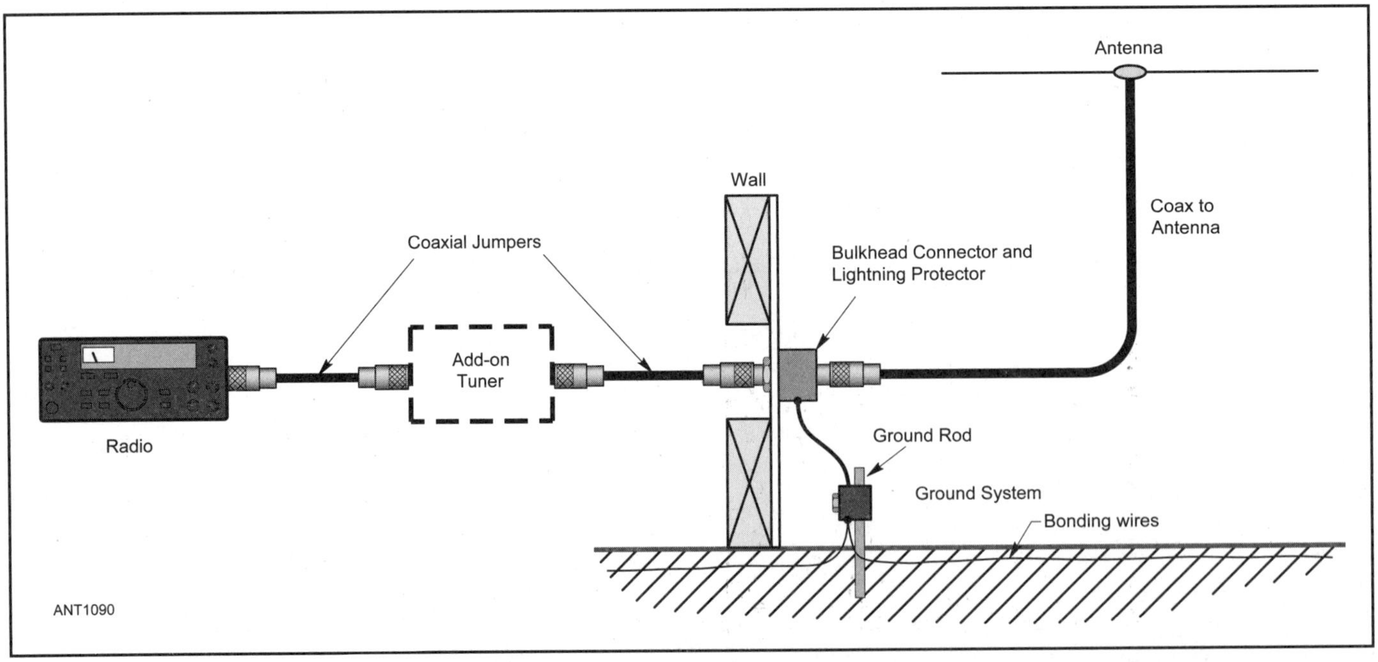

Figure 29.1 — A typical simple antenna system. If the transceiver does not have a built-in auto tuner, an external add-on tuner may be included in the system. It is good practice for the antenna cable to enter the station through a connector on a grounded wall panel. This bulkhead connector is often a lightning arrestor, as well. Coax jumpers connect the various pieces of equipment together.

The Station Notebook

An often neglected but very important part of putting together your station is properly documenting your work. Having the station well documented is an invaluable aid when tracking down a problem. Rather than having to search your memory for information on what you did a long time ago, you'll have the facts on hand.

Each time you install a new antenna, measure the SWR at different points in the band and make a table or plot a curve. Record the frequency of minimum SWR. If you have an analyzer that can measure complex impedance, save the data as a file for later reference, including the date of measurement. Note whether the measurements were made at the antenna, through a length of cable, near the ground, in final position, and so forth. Later, if you suspect a problem, you'll be able to look in your records and compare the antenna with its original performance.

Recording station connections is also important. When adding an accessory, a new antenna, or reconfiguring your existing system, take a photo or make a drawing noting what was changed and why. Even a quick sketch saved as a photo can be invaluable.

the section Antenna Tuner Troubleshooting and Repair.

You have now set a benchmark with known output into a 50-Ω load. *This is an important step.* Do not change any settings on your radio until all tests are completed and the faults fixed.

29.1.3 Antenna System Testing

YOUR SECOND TEST

Here is where we start to eliminate possible causes of your problems. Start by simplifying your antenna system. Remove any extra equipment (switches, filters, etc) between the radio and the antenna, reducing your antenna system to a single connection similar to that in Figure 29.1.

It is likely that you have some kind of receptacle or bulkhead connector (such as a UG-363 adaptor or Amphenol 83-1F) where your antenna coax enters the station. From there you have a jumper cable to your radio. We test this next.

1) Remove the test jumper cable from the radio to the power/SWR meter.

2) Connect your normal jumper cable from the radio to the power/SWR meter.

3) Press (PTT) and observe the power reading. It should be exactly the same as step 8, above. If not, your jumper cable is faulty or not suitable.

TEST AND FIX

First, perform a continuity check of the inner and outer conductors of the jumper cable. Then check the cable insulation — there should be no continuity from the inner to outer conductor. Check that each PL-259 plug is correctly soldered and the pins fit firmly in the SO-239 receptacles. Look for markings on the jacket of the cable to ensure that it is a 50-Ω cable. *If you find a fault and fix it, retest steps 1-3 above.*

YOUR THIRD TEST

Here is another elimination step. It is very common to have bulkhead connectors that are also lightning arrestors. These are not totally fool-proof and can fail due to a lightning hit or moisture. Even non-arrestor connectors fail from moisture or other reasons. We do need to test this connector. If you do not have any connectors between the antenna and your transceiver, skip this test and proceed to the fourth test below.

1) Disconnect the coax to the antenna from the bulkhead connector.

2) Using your "now tested OK" jumper cable, perform a dc test (continuity test) on the connector using the following steps.

3) Connect the jumper cable to the connector on the inside.

4) Test the insulation from inner to outer conductor using a high resistance scale. If the connector is also a lightning arrestor, test the inner conductor to earth ground (should be an open circuit) then test the outer conductor to earth ground (should be a short circuit or very low resistance).

5) The easiest way to test continuity of the connector is to connect your 50-Ω dummy load to the outside of the connector. Look for 50 Ω from the inner to outer conductor.

6) Using two jumper cables and the power/SWR meter, apply power from your transmitter to the dummy load on the outside of the connector.

7) Power should be the same as when you tested your jumper cable.

8) SWR must be no higher than 1.1:1 or the connector is faulty at RF.

BEFORE THE FOURTH TEST

When are we going to test the antenna? Very soon but since it does not work we need to perform a visual inspection. Assuming you have some kind of wire antenna, you need to lower it and in the process, inspect and ensure that:

- On the insulators at each end, there is no possibility of contact between the antenna wire and the supporting wire/ropes.
- If there are any splices in the wire elements, they are well crimped or soldered.
- At the center insulator, there is no possibility of contact between the element wires.
- At the balun or coax connection the element connections are soldered or firmly connected.
  - If it is a center-fed dipole it should be a 1:1 choke balun.
  - If it is an OCF dipole, it should be a 4:1 or 6:1 balun.

Cut away the waterproofing around the coax termination and inspect for water damage. If the connector is discolored or corroded it will need to be cleaned if not replaced and the cable checked as well.

Similar steps apply if you have a Yagi or vertical antenna.

YOUR FOURTH TEST

Now we are going to carefully test the main antenna coax cable *and* its connectors. First some dc tests, then we can RF test.

1) With the coax disconnected from the antenna *and* bulk-

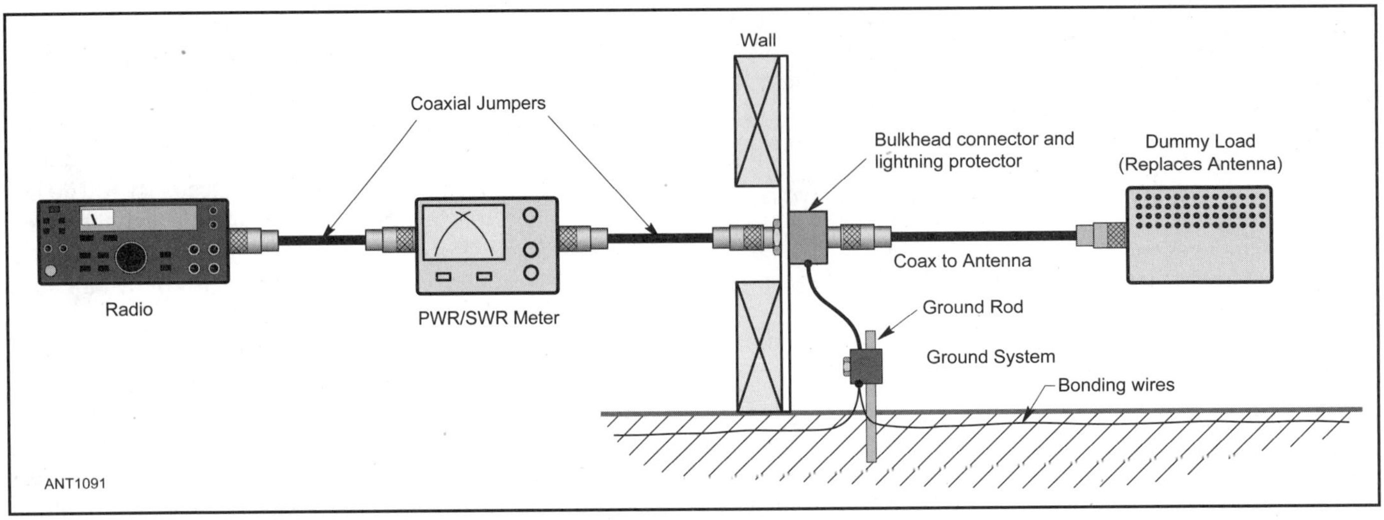

Figure 29.2 — Test setup to check SWR with the antenna replaced by a dummy load.

head connector (or radio), test continuity overall of the inner conductor, then the outer conductor. Test the insulation from the inner to outer conductor on the highest scale.

2) Connect the 50-Ω dummy load to the antenna end of the main coax. At the radio end, measure resistance from the inner to outer conductor. You should see close to 50 Ω.

3) Reconnect the bulkhead connector or radio end of the main coax. You should now have connected in sequence; radio, jumper, power/SWR meter, jumper, bulkhead/wall connector, main coax and dummy load as in **Figure 29.2**.

4) Press PTT and note the power reading: It should be very close to your preset 5 or 10 W. Check SWR; it should be no higher than 1.1:1. Be very wary of seeing no reflected power at all. This could mean that the coax is so lossy that reflected power is unreadable. One more test will determine this.

5) Relocate the power/SWR meter from the station to the antenna end of your main coax but put it where it can be seen. The sequence is now: radio, jumper, bulkhead/wall connector, main coax, power/SWR meter, dummy load as in Figure 29.3.

6) Press PTT and note power reading: it should be at least 75% of your preset 5 or 10 W. If much less, the coax is lossy and should be replaced.

7) Check SWR and it should be no higher than 1.1:1.

8) If you do replace the main coax, repeat all of steps 1 to 7 above.

We are nearly done. Reapply the waterproofing to the connection of coax to balun, or at least some temporary tape. (If it now works you will get so busy you may forget to finish it all!) Pull your antenna back up into position, taking care not to put *any* stress on the coax cable. We are going to test the SWR on the primary band *without* the help of the tuner in the radio.

THE FINAL TEST

Initially we are going to test without the tuner engaged, so we can see how well the antenna is working on the main band

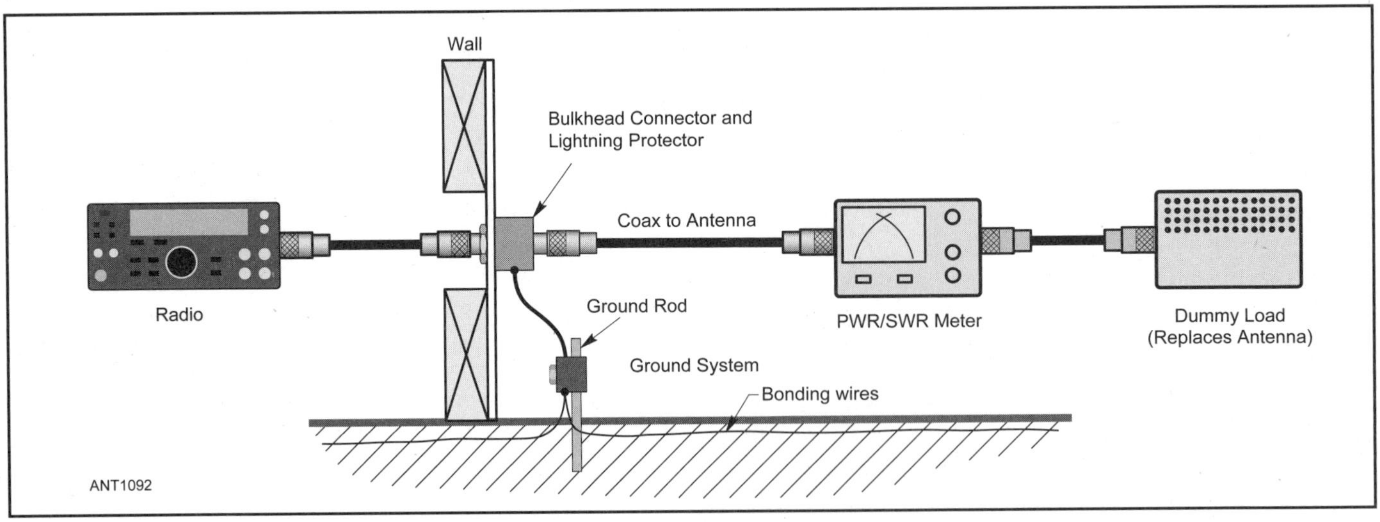

Figure 29.3 — Test setup to check if the main coax is lossy.

that it was designed for. It is only on this band that we can make any adjustments to the length of the wire elements. Before we start adjusting we need to know which direction to go, so we will test the high end, middle and low end of the band.

Connect the power/SWR meter between the radio and the bulkhead connector or between the radio and the coax to the antenna. *Remember that we are now going to be testing "on air" so we need to consider others and ask if the frequency is in use.* (You can also use an SWR analyzer as described in the **Antenna and Transmission Line Measurements** chapter.)

Assuming that the main band is 40 meters, tune the radio to, say, 7250 kHz (near the top of the band) and find a quiet spot. Check/ask if the frequency is in use. If it is not, announce, "[Your call] testing."

- Set carrier power to a low value such as 10 W.
- Set SWR meter calibrate control to near maximum and increase power just enough to be able to set the calibration reading to full-scale.
- Return the meter to read SWR and write down the reading.

Now tune the radio to the middle of the band and find a quiet spot. Repeat the test procedure using an appropriate mode. Repeat near the bottom of the band.

Compare the three SWR readings and decide if the antenna is long (SWR too high at the high end of the band) or short (SWR too high at the low end of the band), or if no adjustment is required. Note that if all SWR readings are 1.5:1 or lower, very little will be gained by adjusting the length. If SWR is uniformly high everywhere in the band, the antenna itself is at fault.

If the antenna SWR measurements are acceptable, it is now time to let the radio and auto tuner operate. If you use an add-on antenna tuner, you'll need to make sure that the jumper between it and the radio is good as described previously, then reinstall it between the radio and antenna. The following paragraph assumes the auto tuner is internal to the radio.

Remove the power/SWR meter from the antenna feed line so that the antenna is connected directly to the radio through the bulkhead connector and jumper. Engage the auto tuner and let it set up on a convenient frequency. Set the output power control to about 75-80% of maximum, then find a clear frequency as before and initiate the auto tuner operation as instructed for your radio. Since you have confirmed your antenna system presents a reasonable SWR to the radio, your tuner should operate normally and you can resume operating! If the tuner does not operate properly, there may be excessive RF current on the feed line's outer surface. Add a choke balun at the output of the radio or antenna tuner and try again. (See the **Transmission Line System Techniques** chapter.) If the tuner still doesn't work, you may have a defective tuner.

If the antenna SWR measurements indicate an antenna fault, the exact troubleshooting sequence will depend on the type of antenna. Remember to write everything down in case you need to contact the manufacturer or ask for other help. Start with a visual inspection of the entire antenna looking for loose or corroded elements and joints. Perform a continuity check of all coils, clamps, and capacitors. Wiggle the various pieces while making measurements to look for intermittent connections. If nothing is obviously wrong, try disassembling the antenna, cleaning the various metal-to-metal surfaces using a nonferrous, nonabrasive synthetic cleaning pad such as a Scotch-Brite pad, then reassemble (checking for proper dimensions and orientation of parts) and test. If this fails to restore normal operation, you should contact the manufacturer's customer service department or ask for help from your local club.

29.2 Guidelines for Antenna System Troubleshooting

The antenna is an electrical device implemented via a mechanical construction; therefore, if it is built properly, it should "work" (especially for production units). There are five general categories of problems:

- Test measurements
- Mechanical
- Proximity
- Feed system
- Misunderstandings

Guidelines for dealing with and approaching each type of problem are presented in the following sections. They will be used in subsequent sections in different ways to address different types of problems. Think of them as a kind of toolbox for troubleshooting. Many of them assume you are testing some type of Yagi or other beam antenna but the general guidelines apply to all types of antennas

It is important to remember this simple rule for adjustments and troubleshooting: Do the simplest and easiest adjustment or correction *first*, and only *one* at a time.

When making on-air comparisons, select signals that are at the "margin" and not pushing your receiver well over S9 where it can be difficult to measure differences of a few dB. Terrain has a lot to do with performance as well. If you are comparing with large stations, keep in mind that station location was probably selected carefully and the antennas were placed exactly where they should be for optimum performance on the property.

Remember the Law of Conservation of Energy: Energy can neither be created, nor destroyed. Therefore, the *sum* of all the energies in a system (an antenna system) is a constant. From the perspective of transmitting, we start with so many watts and the energy will go somewhere, either emitted from the antenna and on its way to the destination, or dissipated as heat due to loss.

If you increase your antenna efficiency, you will expand your performance envelope, and thus be able to hear *and* work more stations, providing more enjoyment from radio. If you increase only your transmit power, you will expand your "transmit envelope," but you won't be able to hear any better!

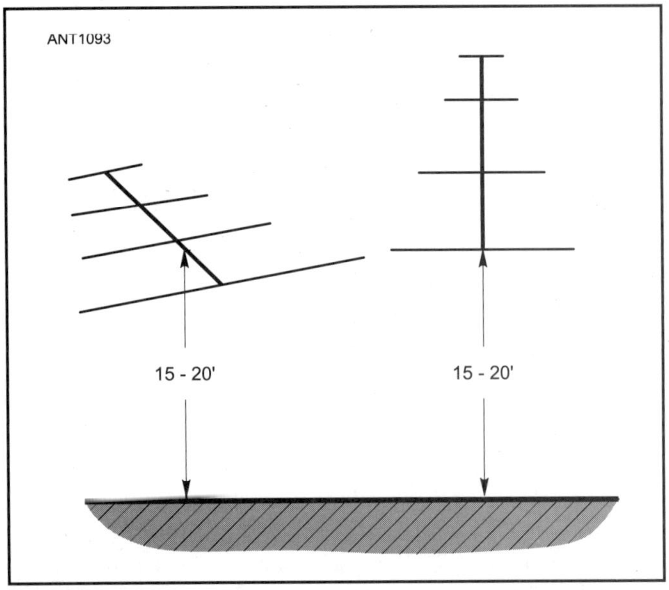

Figure 29.4 — When testing a Yagi or quad antenna, make sure it is at least 15 to 20 feet above ground. If oriented vertically, the reflector should be the closest to ground. Performance will still change as the antenna is raised.

29.2.1 Test Measurements

A. Test the antenna at a minimum height of 15 – 20 feet. (See **Figure 29.4**) This will move the antenna far enough away from the ground (which acts to add capacitance to the antenna) and enable meaningful measurements. Use sawhorses *only* for construction purposes.

- A minimum height of 15-20 feet above ground does not mean 5 feet above a 10-15 foot high roof, it means above ground with nothing in between.
- Antenna resonant frequency will shift upward as it is raised.
- Feed point impedance will change with a change in height and this applies to both horizontal and vertical antennas.
- Some antennas are more sensitive to proximity to ground than others.
- Some antennas are more sensitive to nearby conductive objects (i.e. other antennas) than others.

B. Aiming the antenna upward with the reflector on the ground might coincide with some measurements on rare occasions, but there are no guarantees with this method. The reflector is literally touching a large capacitor (earth) and the driven element is very close, too. Raise the antenna at least 15-20 feet off the ground.

C. When using a hand-held SWR analyzer you are looking for the dip in SWR, not where the impedance or resistance meter indicates 50 Ω. ("Dip" = frequency of lowest SWR value, or lowest swing on the meter.) On the MFJ-259/269 series, the left-hand meter (SWR) is the one you want to watch, not the right-hand meter (IMPEDANCE).

D. Check for nearby broadcast transmitters. The small amount of power used by the hand-held metering devices is no match for several thousand watts. The front-ends of the devices are broadband and will receive this out-of-band broadcast energy and "assume" it is reflected energy. This will manifest itself as the meter never showing a low SWR — sometimes as low as 1.3:1 or higher than 5:1 — all the while the antenna is actually matched properly. The broadcast transmitter will change its power and direction at sunset/sunrise, making daytime and nighttime measurements different. If the signal is from an AM transmitter, you may see the meters move with the programming audio amplitude.

E. Does the SWR and frequency of lowest dip change when the coax length is changed? If so, the balun might be faulty, as in not isolating the load from the coax feed line. Additionally. with an added length of coax and its associated small (hopefully small) amount of loss:

- The value of SWR is expected to be lower with the additional coax and,
- The width of the SWR curve is expected to be wider with the additional coax *when measured at the transmitter end of the coax.*

F. Be sure you are watching for the right dip, as some antennas can have a secondary resonance (another "dip"). It is quite possible to see a Yagi reflector's resonant frequency, or some other dip caused by interaction with adjacent antennas.

29.2.2 Mechanical

A. Are the dimensions correct? Production units should match the documentation (within reason). When using tubing elements, measure each *exposed* element section during assembly and the element *half-length* (the total length of each half of the element) after assembly. Measuring the entire length is sometimes tricky depending on the center attachment to the boom on Yagis, as the element can bow, or the tape might not lie flat along the tubing sections. Self-designed units might have a taper error.

B. Making the average taper diameter larger will make the equivalent electrical element longer. This makes the antenna act as if the physical element is too long.

C. Making the average taper diameter smaller will make the equivalent electrical element shorter. This makes the antenna act as if the physical element is too short.

D. If the element is a mono-taper (tubing element is the same for the entire length), larger diameter elements will be physically shorter than smaller diameter mono-taper elements to give the same electrical performance at the same frequency.

E. The type of mounting of the element to the boom affects the element length, whether it is attached directly to the boom, or insulated from the boom. Incorrect mounting/mounting plate allocation will upset the antenna tuning:

- A mounting plate 4 × 8 inches has an equivalent diameter of approximately 2.5 inches and 4 inches in length for each element half.
- A mounting plate that is 3 × 6 inches has an equivalent diameter of about 1.8 inches and a length of 3 inches for each element half.
- The mounting plate equivalent will be the first section in a model of the element half.

F. In a Yagi, if the elements are designed to be touching, are the elements touching the boom in the correct locations?

G. In a Yagi, if the elements are designed to be insulated, are the elements insulated from the boom in the correct locations?

H. The center of hairpin matching devices (i.e. on a Yagi) can be grounded to the boom.

I. The boom is "neutral," but it is still a conductor! The center of a dipole element is also "neutral" and can be touched while tuning without affecting the reading. With a hairpin match, the center of the hairpin can also be touched while tuning and touching the whole hairpin might not affect the readings much at all.

J. Tests have shown that in installations with several Yagis on a common mast, insulating the elements from the boom can reduce interaction between the individual antennas.

K. Sufficient spacing between Yagis on a common mast is critical to not lose gain and F/B. Even 10 foot spacing between a 20 meter monoband Yagi and a 15 meter Yagi can significantly reduce the gain on 15 (sometimes by 50%), plus almost completely eliminating the F/B on 15.

L. The higher frequency Yagi in a common stack is the one that will be affected by the lower frequency Yagi(s). If a stack of 20, 15 and 10 meter Yagis (20 being the lowest on the mast — which is the correct stacking sequence), the 15 will be affected by the 20, the 10 will be affected by the 15 and possibly also by the 20.

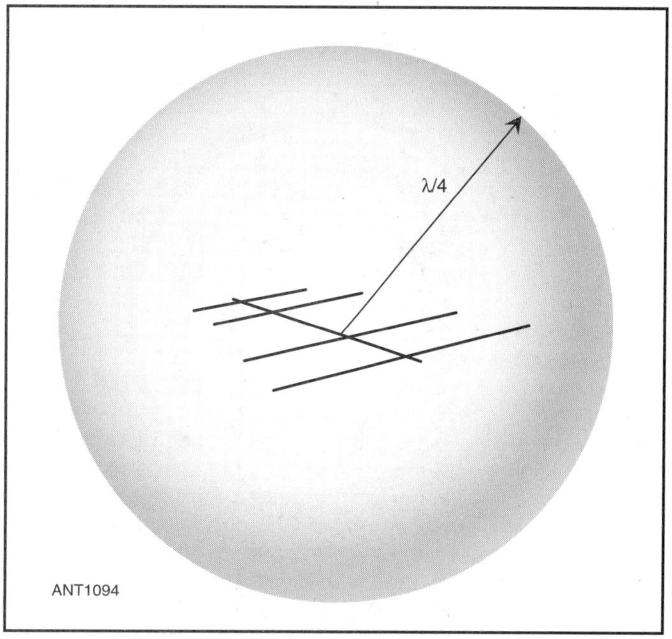

Figure 29.5 — Any conductive material within λ/4 that has a medium to low impedance at the frequency of operation has the potential of interacting with the antenna.

29.2.3 Proximity

A. What else is nearby (roof, wires, guy lines, gutters)? If it can conduct at all, it can and probably will couple to the antenna!

B. Does the SWR change when the antenna is rotated? If so, this indicates interaction. Note that in some combinations of antennas, there can be destructive interaction even if the SWR does not change. Computer models can be useful here.

C. What is within ¼ wavelength of the antenna? Imagine a sphere (like a big ball) with the antenna in question at the center of the sphere, with the following as a radius, depending on frequency. Think in three dimensions like a sphere — up and down and front and rear as in **Figure 29.5**.

160 meters = 140 foot *radius* for ¼ wavelength
80 meters = 70 foot *radius*
40 meters = 35 foot *radius*
20 meters = 18 foot *radius*
15 meters = 12 foot *radius*
10 meters = 9 foot *radius*

D. Interaction occurs whether or not you are transmitting on the adjacent antennas. When receiving, it simply is not as apparent as when transmitting.

E. Wire antennas under a Yagi can easily affect it. This includes inverted V dipoles for the low bands and multiband dipoles. The wire antennas are typically for lower frequency band(s) and will not be affected by the Yagi(s), as the Yagis are used for the higher bands.

F. Are the higher frequency antennas (Yagis) above the lower frequency ones in the stack? Is there adequate distance between the various antennas? Remember, anything within ¼ wavelength in any direction is a potential problem. Careful modeling might not necessarily indicate the interaction in found in the actual installation. Cross polarization between VHF antennas and HF Yagis on the same mast is OK.

G. An 80 meter rotatable dipole should be parallel to nearby Yagi boom(s) so that it is essentially transparent. Other antennas that could interfere *might be able* to be positioned at right angles. Orienting the 80 meter dipole parallel to the boom also makes the installation more neutral in the wind. Most Yagis have more wind load from the elements than from the boom which tends to make them "hunt" in the wind. Adding area to the boom can, therefore, help the installation be more manageable in the wind.

29.2.4 Feed System

The feed system includes:

- the feed line
- switching mechanisms
- pigtails from the feed point on the antenna to the main feed line or switch
- all feed lines inside the radio room

The feed system is the *entire connection* between the radio and the feed point of the antenna.

A. Is the feed line (coax) known to be good? (Start with the easiest first.) Is there water in the coax? This can give strange readings, even frequency-dependent ones. If there is any question, swap the feed line for a known good one and test again.

B. Are the connectors installed properly? Has a connector been stressed (pulled)? Is the rotation loop done properly to not stress the coax? Is it an old existing loop or a new one? Usually it's alright if new. Type N connectors (especially the

Is It Always a Connector?

No, but after many sessions of troubleshooting antenna systems, it can seem like that. Between the transmitter output and the antenna terminals are often many connections — to accessories, switches, tuners, lightning protectors, matching networks, more sections of cable — and each represents a potential point of failure. (See the sidebar "Coax Connectors — Not as Simple as They Appear" in the chapter on **Transmission Lines**.)

Splices, particularly PL-258 "barrel" connectors, seem to have a poor track record because there are many poor-quality PL-258s in circulation at bargain prices. Center contacts can loosen from heating at high power or be too loose to begin with. Too few mating indentations or slightly "off" thread size can lead to intermittent connections that change with flexing or temperature.

Water is the mortal enemy of RF connectors. Learn how to waterproof a connection correctly and with the right materials. (See the **Building Antenna Systems and Towers** chapter.) It takes a lot less time to do the job right than it does to locate the bad connection and re-do it, often requiring replacing the connectors, as well. If possible, route cables so water flows away from connections and protect connections from direct rain, snow, and ice.

Minimize problems by starting with quality connectors and making sure you install them properly, whether crimp or solder. Make sure jumpers in the station are constructed properly, too. If a connector or jumper seems suspicious, rebuild or discard it. Remember that "temporary" often turns into "permanent!" Take care of your connectors and they will take care of you.

older type) are prone to having the center conductor pull out due to the weight of the coax pulling down on the connector. Connectors are easy to do, using the right technique.

C. Is there a "barrel" connector (a PL-258 dual-SO-239 adapter) in the feed line *anywhere*? (See the sidebar Is It "Always a Connector?") Has a new or different barrel been inserted? These are a common failure point, even with new barrels. The failures range from micro-bridges across the face of the barrel shorting out the center and shield, to resistance between the two ends of the barrel. Have the new barrels been tested in a known feed system? Always test them before installing. Use only quality RF adapters as these are common system failure points.

D. Is the coax intact and not frayed such that the shield can come into contact with anything? This can cause intermittent problems as the coax shield touches the tower, such as on rotation loops and coax on telescoping towers.

E. Is the tuner OFF on the radio? This is often overlooked when adding a new antenna. If you are using an external antenna tuner, remove it or set it to the "Bypass" position and see if the problem remains.

F. Are there any new devices in the line? It might be a good idea to remove everything but the essential items when troubleshooting.

G. Is there a remote antenna switch? Swap to another port.

H. Are there band-pass or low-pass filters in the line? Filters can become defective, causing strange SWR readings.

29.2.5 Using Portable Analyzers

With the new ultra-portable antenna analyzers, we can have sophisticated test instruments with us at all times, whether on the ground or at the top of a tower. A low-power QRP dummy load can also go with us in the tool bag. That makes it a lot easier to perform "backward" tests toward the station from the antenna, speeding up the search for a problem in the feed line system somewhere.

Learn how to use the TDR (time-domain reflectometry) function of your analyzer. (See the **Antenna and Transmission Line Measurements** chapter.) Significant impedance changes are easy to spot with a fairly accurate idea of where they are located along the feed line. Along with short- and open-circuits, TDR can also indicate types of mismatches that give a clue to what might be causing the problem. TDR measurements are another good set of data to record for the station records.

Another useful function of a vector analyzer is the ability to move its calibration plane to the antenna terminals. This is done by performing an "short-open-load" (SOL) calibration at the antenna. By applying the set of mathematical corrections from an SOL calibration, the actual antenna impedance can be determined by the analyzer from the other end of the feed line.

29.2.6 Misunderstandings

The antenna can be working properly but there may be a misunderstanding of the anticipated readings versus the actual readings. There can also be discrepancies between the observed "performance" (i.e. F/B ratio) and the specification(s). Having an open mind here is a great asset and will aid in understanding and resolving the situation. "Open mind" means no preconceived ideas or bias, which is sometimes difficult. Remember that we are working toward a solution to improve performance. Common misunderstandings:

A. "A low SWR means the antenna has gain." No, it only means it is matched to the feed line. Remember that a dummy load also has low SWR.

B. "A high SWR means the antenna does not have gain." No, it only means it is not matched or is fed improperly.

C. "An SWR that does not go to 1:1 is a serious problem." No, as long as your tuner can tune it, use it. Reflected power is not totally lost. As long as the feed line loss is acceptable, SWR does not need to be 1:1. (See the **Transmission Lines** chapter for more information on matched line loss with varying SWR.)

D. "My antenna has a great pattern, so it must have a lot of gain." No, these two antenna aspects are not necessarily related. The Beverage receiving antenna has an excellent pattern, but its gain is about –20 dBd.

E. "Once the antenna is up, it will stay there forever." An antenna is an electrical device implemented via a mechanical structure. Mechanical devices require periodic maintenance, just like your car.

29.3 Analyzing an Antenna Problem

Having a specific sequence of steps to take for a systematic resolution of installation questions will make the process easier with less frustration. It will also provide a learning environment and future projects will run smoother and be enjoyable, as the prospect of a higher performance envelope is anticipated!

The following typical debugging sequence is divided into five parts. Each one uses the guidelines to address a specific aspect of the resolution process. Not all the steps will be used each time a new antenna is installed; however, reading through them will be beneficial.

The length of this material and the steps noted should not deter anyone from reading — installation difficulties are usually simple to resolve.

29.3.1 Part 1 — SWR

A. The usual reason for debugging is that the SWR is not as expected. This is the only measurement that can be reliably made by the majority of people.

- If the SWR is showing high values (4:1 or higher), do not attempt any adjustment of the antenna before *first* certifying the feed system is correct.
- High values like this are so far away from the expected values that they essentially eliminate the antenna from being the current problem.

B. Remove all devices in the feed system to eliminate possible components with problems, such as low pass filters (especially if 10 meter SWR readings are not as expected); therefore, we want to work as directly as possible with the antenna in a good location.

C. Isolate the feed system as the first step.

- Place a 50-Ω dummy load at the antenna end of the coax feed line.
- Measure the SWR of the coax feed line at the transmitter end (dummy load at the other).
- If you measure anything other than a low SWR (1.2:1 or less), the coax should be changed and/or;
  - If you see a significant drop in power through the coax (use a wattmeter), the coax should be replaced.
  - If the coax is good, proceed to the next steps.

D. Is the antenna at a reasonable height above ground?

- 15-20 feet above ground and roof.
- If not, place it as high as possible and watch for proximity effects.

E. Does the SWR change as it is rotated?

- What else is on the mast?
- Are there any wire antennas nearby?
- What it is rotating above or below to cause a change in measurement?

F. Are the element lengths correct?

G. Are the elements in the proper location?

H. If a hand-held test unit is used, is there a broadcast transmitter within several miles? This is very important on 160, 80/75 or 40 meter antennas.

29.3.2 Part 2 — Feed System and Antenna Assembly

A. Stay calm.

B. Do the easiest thing(s) first.

C. If a simple change was made (i.e. moved an element a few inches), the problem is most often in the feed system.

D. Swap the coax, even if it takes some effort.

E. Try to remove the parts of the antenna feed system one at a time to isolate the culprit.

F. Be sure to track the correct dip in SWR readings.

G. On production antennas, most problems are identified by checking for:

- Element length and tuning (and location on the boom, but extremely rare).
- A local broadcast transmitter affecting the readings — use your transmitter and its SWR meter.
- Antenna mounted properly; clear of proximity issues, including conductive guy wires.
- Correct feed line and matching system adjusted properly.

29.3.3 Part 3 — Keeping Records

A. Keep sequential notes of each step taken.

- A legal pad or notebook is excellent — number the steps and each page.

Goes Away With Power

This very common symptom has several common causes and here are three common scenarios that may help you find the root of your problem:

Intermittent contacts — Usually noticed on receive, an antenna seems "dead" until a short transmission (one "dit" on CW is usually enough) brings it back to life. This is often caused by corroded, dirty, or worn relay contacts. Relays are commonly found in T-R switches, automatic antenna tuners, antenna switches, and array direction controllers. Start by cleaning relay contacts and visually checking for pitting or corrosion. Sometimes tapping on the relay will restore contact temporarily, locating the problem.

Abrupt change in SWR — Above some power threshold, SWR increasing abruptly is usually caused by arcing. Arcing can occur during transmissions, or it can be caused by a carbon deposit left by a previous arc or lightning surge. Water, dirt, or even insect debris in an enclosure or connector can result in an arc at high power. Perform visual inspections of exposed points at which high voltage may be present. Check RF connectors for signs of arcing from the center pin to the shell.

Worn or defective connector — When high power is applied, SWR gradually increases and may eventually stay high all the time. This is often caused by a connector's center pin losing firm contact with the mating receptacle. Heat causes the receptacle to expand, reducing contact force. A visual inspection may show discoloration. Water in the connector can also cause this symptom. The usual test is simply to replace the suspect connector.

- Write down what was done and then write down the observed result.
- Underline the amount of a change and use a + or – sign, or say "longer" or "shorter."

B. If you make a change, do only one item at a time.

- If you change more than one item, you will not know what caused the observed change.
- If nothing appeared to change and more than one item was changed, it is possible that the items changed countered the effect(s) of each other.
- Changing more than one item at a time makes it *impossible* to track.

C. Write on your notepad the initial observation(s) and the conditions, such as height and proximity.

- This increases your situational awareness and it will provide a documented starting point.

29.3.4 Part 4 — Home-Made Antennas

A. For noncommercial, home-brew, or "one-off" antennas:

- Follow same procedures for production units.
- Element tapering needs to be verified.
- Element mountings might not be properly accounted for (i.e. insulating boots when using old Hy-Gain mounts for a new design)
- Matching techniques might not be working as expected. Check directly at the feed point without the matching device in place.
  - A hairpin will step up the impedance and might just move it to the high side of 50 Ω, making adjustment (down) to 50 Ω impossible.
  - If the design is a "forward stagger" type, the forward Yagi needs to be shorted across the feed point (i.e. hairpin); otherwise, the driver will have an open or shorted coaxial stub (the pigtail feeding it) attached across it.

B. Keeping a station notebook, with as much detail as possible using the same note-taking procedure as described earlier is invaluable.

29.3.5 Part 5 — On-Air Observations

A. F/B is less than expected.

- Antenna height affects F/B, so does the angle being used. Refer to typical plots to acquaint yourself with these issues.
- F/B specification might be too ambitious. Some specifications are given as peak values, available only across a narrow frequency range (if not tuned properly, might be out of band).
- How much to expect?
  - A 2-element full-size parasitic Yagi will be around 12-16 dB and A 2-element shortened, loaded Yagi can be >20 dB if tuned properly.
  - A 3-element full-size Yagi "naturally" wants to be around 20 dB.
- Stacked Yagis on one mast (for example, 20-15-10 meters) can greatly affect the F/B.
- Rotator clamps not secure, mast slipping.
- Antenna attachment might not be secure, even with the clamps tight, and the antenna is slipping on the mast (typical with hard steel masts).

B. How much gain (redistribution of the constant energy) to expect? **Table 29.1** lists real-life, reasonable, verifiable figures for full-size 20 meter antennas, with gain specified as dB compared to a full size dipole at the same height, same location:

Table 29.1
Expected Performance Values of Full-Size Antennas

Gain (dBd)	*Antenna Type*	*Full Size Boom Length (20 meter antenna)*
0	Dipole	Reference*
4.5	2 element	10' boom
5.5	3 element	20' boom
6.5	4 element	30' boom
7.5	5/6 element	42' boom
8.5	7 element	60' boom
9.5	8 element	80' boom
10.5	9 element	105' boom
12.5	12 element	175' boom
14.5	20 element	330' boom

*The Reference is a dipole is in the exact same location as the Yagi, such as over ground)

These figures are increased by 2.14 dB when comparing to the isotropic source (4.5 dB + 2.14 = 6.64 dBi); and if ground reflection gain is also included (i.e. at 1 wavelength above ground), add another 5.8 dB. Using both of these, the 5.5 dB figure for a 3-element Yagi becomes 13.44 dBi. Whichever the case, the reference must be specified; otherwise, you know nothing about the antenna.

C. Does not seem to be competitive or crack pileups.

- Not aimed in the right direction (being off 30 degrees can be a lot).
- Gain specification in error or loss in the antenna system.
- Coax, switches, antenna tuning, antenna components, radial system.
- Could it be a problem with the operator?

29.3.6 Troubleshooting High SWR in Yagi Antennas

This section applies mainly to Yagi antennas; however, it should be useful for other types as well. There are additional sections for other types of antennas.

One thought to keep in mind when tuning a Yagi is that in a Yagi, the primary purpose of the driver (driven element) is to excite the array. The driver tuning has very little to do with the gain and pattern, although the spacing between the driver and adjacent elements is quite important to the Yagi design. In a 2-element Yagi, the driver location does impact the gain and pattern, because it sets the boom length; however, the parasitic element (either a reflector or director) is the primary "controller."

To locate the problem of a high SWR, we need a scope of reasonableness. We need to keep our mind open to locate where the problem is and where the problem is not. Let us say we have just put up a commercially produced Yagi antenna and it has a high SWR at the radio end of the coax. The first thing that comes to mind is that it is the antenna. Perhaps it is, but we need to follow a plan.

Let us say our radio, with the example new antenna, is showing an SWR of 3:1. If the antenna is the culprit, it means the feed line is seeing a load (antenna feed point) that is not close to the characteristic impedance of the coax. In simple form, if the coax is 50 Ω, a 3:1 SWR means the antenna feed point impedance is either 150 Ω or 17 Ω. The feed point impedance of a Yagi can be as low as 17 Ω, but not as high as 150 Ω; therefore, we need to make a choice and we choose that the feed point is 17 Ω. We need to consider there might be an impedance transformation device at the feed point, but this device most likely will never transform the feed point to as high as 150 Ω for an expected 50-Ω feed.

There are times when the feed point will be intentionally transformed high. Band-specific 4:1 coaxial baluns can reduce even harmonics. These baluns require the Yagi feed point to be four times the impedance of the coax, or 200 Ω. This is accomplished using a hairpin across the Yagi feed point. The circuit is: 50-Ω coax up the tower, mast and out to the driver, through the 4:1 coaxial balun and attached to the feed point, which also has the hairpin across the feed point. Back to our example with a 3:1 SWR.

Impedance transformation devices (matching circuits) are used to step up the feed point impedance to match the feed line (some matching circuits step down the impedance, but almost all used in Yagi antennas step up the feed point impedance). If the Yagi feed point impedance is 17 Ω, a hairpin (inductive reactance across the feed point) can be used to increase the feed point to match the 50-Ω feed line. So, if the Yagi really has 17-Ω feed point impedance and there is a hairpin-matching device, one might want to be sure the hairpin is properly attached and adjusted (if there is an adjustment). An important piece of information is to know the untransformed feed point impedance.

Yagis can be designed with a very low feed point impedance, but most production antennas are not. In our example, let us assume the feed point impedance is in the range of 35 Ω. Now what do we do?

The Yagi, without a hairpin, would have a SWR of about 1.4:1, derived by dividing the characteristic impedance of the coax by the feed point: 50 / 35 = 1.4. Therefore, if we remove the impedance transformation device and measure the driven element directly (at a reasonable test height above ground), we should see an SWR of about 1.4:1. If we do, we can eliminate the Yagi as the cause of the 3:1 SWR we saw at the radio end of the coax feed line. This means the problem is elsewhere, so we can go to the common section below.

29.3.7 Troubleshooting High SWR in Non-Yagi Antennas

Full-size dipole antennas will not have a matching device, because their impedance is usually between 45 and 90 Ω, depending on the shape and height above ground. The lower values will be for dipoles that are inverted Vs with the ends not far from the ground. If you have a high SWR on one of these antennas, it is almost sure to be in the feed system.

You should check all the components from the antenna to the radio. This includes the balun (rare to have a problem here), connectors, coax and all equipment in the line, such as SWR meter, antenna switch, etc.

There can be several reasons for a higher than expected SWR. What "a higher than expected SWR" means is that the SWR exceeds the specification by a large margin, such as 2:1, when a 1.3:1 is expected. A difference of a few tenths should not be a serious concern. We are addressing a more major difference. Let us continue the example above for a purchased antenna. If the antenna is one that has been in production, then it is reasonable to expect the antenna to meet the specifications. If the antenna does not meet the specifications, then try the following step first, then move to the longer list below:

Remove the driver element from the array (Yagi), place it on a wooden stepladder. Measure the element to be certain it is built properly, with the correct dimensions. Check the SWR. If the SWR remains at 3:1, the problem is not in the antenna. It must be in the delivery system, because the driver is a dipole and it will not be 3:1 under any circumstances. A dipole's feed point impedance can be between 40 to 90 Ω, depending on its height above ground, which translates into an SWR of not more than 1.8:1. If the SWR is noticeably higher, then the delivery system is suspect and must be checked. This consists of the radio, amplifier, tuner(s), antenna selector(s), all metering equipment (SWR/power meter), all coax lines and connections.

The usual questions to move through the process are:

1) Is the tuner in the line? Many times, a tuner has been left in with settings that cause the rig's SWR indicator to read improperly.

2) Is the SWR/power meter battery powered? A low battery can cause erratic readings.

3) How sure are you that the coax connections are properly made? How good are the solder joints?

4) Is there water in the coax or connectors?

5) Is there an RF choke or balun being used to decouple the coax feed line from the antenna?

If the problem is still not solved, here is the longer list:

1) Remember that an antenna is an electrical design implemented in a mechanical structure; therefore, be sure all joints are mechanically secure, making a solid connection.

2) Remember that an antenna is simply an airborne conductor. There is no "magic"!

3) Be sure the new antenna is at a reasonable test height. Having a Yagi antenna a few feet above ground, such as on sawhorses, will not provide much useful information. An antenna covering 20 through 10 meters should be in the clear about 12 feet above ground, preferably higher. A 40 meter dipole or Yagi can be effectively tested down to 15 feet, but will probably shift upward in frequency as the antenna is raised to its final height. The ground contributes a very large amount of capacitance!

4) Aiming a Yagi upward at the sky with the reflector laying on the ground is also not a good idea. The reflector is very closely coupled to the ground and it will not be properly tuned. If the reflector can be raised several feet above the ground (a quarter wave is perfect), the entire antenna can be accurately

tuned. As discussed earlier, raising the antenna 15 – 20 feet above ground is the best practice.

5) Check the dimensions of the assembled antenna to ensure they are reasonably close to the drawings. Unless a particular design is extremely sensitive, a difference of an inch on 20 meter elements should not cause a serious SWR change.

6) Check for other antennas within proximity to the new antenna. Antennas that are related to the new one are of particular interest. A few guidelines are provided in **Table 29.2**. "New Antenna" refers to the one just put up and assumes it is a horizontal design."Watch These" means other antennas that can be influencing the new one. "Coupling Distance" is the distance that the new antenna can effectively couple through the air to another antenna. The coupling distance implies distance to the closest point between the antennas.

7) If the new antenna is a horizontal type, a vertical antenna within near proximity will not usually cause any problems. The anticipated isolation between horizontal and vertical polarization is 20 dB and this is sufficient to isolate the antennas from causing harmful interference, or sufficient influence as to cause a high SWR on either antenna.

8) Antenna analyzers make antenna adjustments and testing quite easy. These instruments provide a direct SWR readout, along with the frequency. At least one can display a graph of the SWR curve. If one of the new SWR meters is being utilized, caution should be observed. These instruments send out a low level signal to the antenna under test then sense the reflected power, computing the return loss and the resulting SWR at that point on the feed line. The SWR indicated is the SWR at that point on the line, not necessarily at the actual feed point.

The difficulty using these instruments arises when there is RF energy in the area from sources other than from the antenna under test. Some are so sensitive that a SWR reading of 1:1 on the instrument is not possible. The stray RF energy does not have to be near the frequency being tested, as the front end of the instruments are basically untuned and, therefore, very wide.

During the day, AM broadcast stations can dramatically influence these instruments. At dusk, most AM stations reduce their power and redirect the antenna patterns; however, the redirected energy might just now be in the direction of the antenna under test, whereas during the day the energy was directed somewhere else and was not noticed.

AM broadcast harmonics can be a problem up through 40 meters; maybe higher harmonics are multiples of the operating (fundamental) frequency, such as 1200 kHz (AM band) with harmonics at 2400 kHz and 3600 kHz. Although the harmonics are greatly reduced by the filters at the AM transmitter, reducing a harmonic from a 50 kW transmitter so that it is insignificant compared to the reflected power from a 5 mW signal from one of these SWR devices is a tough order.

Any of these instruments can be used as long as one is aware of this possible problem. Using a transmitter at a few watts will be the most accurate, as the energy from other sources will be substantially less than that of the transmitter.

9) If the antenna is physically correct and the antenna is commercially made, then the problem must be due to other conductors or antennas within close proximity, the match, or feed system. The feed system includes the balun or RF choke and the feed line. The feed line has connectors. Some feed lines purchased pre-made use connectors that are only crimped. Some amateurs believe they should be soldered.

10) A balun should have its leads as short as possible, usually about 2½ inches. An RF choke balun should be wound on a cylinder (solenoidal) to be most effective. (See the **Transmission Line System Techniques** chapter.)

11) The split portion attached to the feed point needs to be waterproofed so that water will not wick down inside the coax. If there are guy wires, they should be broken up with insulators into nonresonant lengths, or use nonconductive guy cable.

12) If the coax feed line is old, it is possibly contaminated. The contamination comes from the jacket of the coax contaminating the interior dielectric.

13) The coax might also have water inside. This can happen when the end is split (i.e. such as for an RF choke or pigtail connection) and the water wicks up the braid and goes inside the coax cable. Water can also find a path on the inner dielectric and flow inside the coax cable. Air-dielectric coax is especially sensitive and vulnerable to water. It has even been suggested that water can condense inside coaxial cable with an air dielectric. An air-dielectric cable is one that uses air, rather than a solid material for the space between the center conductor and the shield.

The above information pertains to antennas you design and build as well, except that the actual SWR specification will probably not be readily known but it will be anticipated. Keep in mind that a vast majority of Yagi designs have an impedance of less than 50 Ω at the feed point. It is rare to find any that are even in the 40+ Ω range. Some go as low as 10 Ω. This means the SWR can be as high as 5:1 without a matching circuit.

29.3.8 Yagi Feed Point Impedance Notes

Most Yagi designs are in the high teens to mid 20-Ω range and the unmatched SWR will be a maximum of 2.5:1 for a 20-Ω feed point (assuming no reactance). The matching systems usually utilized (hairpin, gamma, T) are step-up circuits, which means the matching circuit raises feed point impedance. The feed point can be transformed to values above 50 Ω, even as high at 200 Ω if a 4:1 balun is desired to be used (50-Ω coax × 4 = 200 Ω).

Please note that transformed impedance is not the same as "native" impedance, meaning untransformed. An antenna that has a native impedance of 10 Ω will have the same

Table 29.2
Potential Interactions

New Antenna	*Watch These*	*Coupling Distance (ft)*
40 meters	80, 160 meters	35
20 meters	20, 40, 80 meters	18
15 meters	20, 40 meters	12
10 meters	15, 20, 80 meters	10

current flowing in it when the feed point is transformed to 50 Ω to match the coax feed line's characteristic impedance of 50 Ω.

A "dual-driven" 2-element driver design with crossed-over feed straps between the driver elements is a way of transforming the feed point impedance to a higher value, such as 200 Ω. The "native" feed point impedance of that antenna will be much lower, possibly even way below 50 Ω.

29.4 Antenna Tuner Troubleshooting and Repair

Antenna tuners are usually well designed and built with adequately rated parts. Most will last a lifetime but even QRP power levels can result in arcing if the tuner is trying to match a very high impedance or accidentally disconnected from an antenna. *Hot switching* can cause damage when contacts in the tuner are moved while transmitting. Lightning transients and excessive power beyond the tuner's ratings cause damage. Frequently adjusted components and contacts can wear out over time or just get dirty. Any of these can cause problems with normal operation.

For more information on the different types of circuits used in tuners, see the section on Impedance Matching Networks in the chapter on **Transmission Line System Techniques**. Remember that "tuner" is the most common term, but ATU, transmatch, antenna coupler, impedance matching unit, matchbox, and other terms also refer to the same piece of equipment.

If you think your tuner is misbehaving, begin the process of troubleshooting according to the section "Guidelines for Antenna System Troubleshooting." Make sure the problem really is in the tuner! Once you're sure the tuner has a problem, give it a good checkout to get an idea of what you're looking for. Keep notes as you test since these will provide clues to what the problem may be and also point to tests you should perform after repairs to be sure you've fixed the problem.

Whether the tuner is internal to your transceiver or an external unit, start with a close and detailed visual inspection. Are the connectors clean? Do any look like there has been an arc or are they discolored from heat or corrosion? If there are multiple antenna connectors, does the unit perform differently when the antenna is connected to different connectors? Are there any loose parts inside that rattle when the unit is tipped or shaken? Smell the unit, too — if anything smells burned, you may have a damaged component inside. These are also excellent inspections to make before buying a used tuner.

Operate the tuner into a dummy load at low and high power. Does it behave the same way at both power levels? Does it operate erratically? Does wiggling any of the feed lines to or from the tuner cause SWR to change at the transmitter output? Operate the tuner into a known-good antenna and feed line, then repeat the tests.

T network tuners have a reputation of being able to "tune up into themselves" meaning they can match even an open circuit because they have such a wide tuning range. This results in very high voltages and currents inside the tuner, creating all sorts of problems.

Once you've put the tuner through its paces, you'll have a better idea of what problems you're facing. It is also worth considering that there may be more than one problem. When you find something wrong and fix it, repeat your tests to see if the problem is still there or has changed.

29.4.1 Connectors

A "problem" can be chased for hours only to find a connector that is partially seated or loose and is now intermittent. Start by carefully checking them and any connecting cables.

The SO-239 UHF receptacle center socket usually has four fingers that compress around an inserted pin. Check to make sure all of the fingers are still present. Over time or with mechanical stress, a frequently used connector can develop a loose fit between the fingers and pin. You may be able to use a jeweler's screwdriver to bend the finger back toward the center but eventually this will break off the finger. If the fingers are missing or broken, your only option is to replace the entire connector.

Corrosion usually appears as a dark film of tarnish. Silver-plated connectors sometimes develop tarnish that is conductive. Nickel-plated connectors should be clean and bright. Do not use a drill or sandpaper to scrape the socket or connector surfaces — plating will be removed. Use a cotton swab dipped in a contact cleaning solution and insert a test connector several times. Corrosion from water often appears light-colored and crusty. This type of corrosion can be removed with a brush (on the outside of the connector) or swab. Corrosion that can't be removed or that pits the connector surface requires that the connector be replaced.

Replacing the SO-239 entails de-soldering the center pin and the removal of four (sometimes only two) 4-40 screws, nuts, and lock washers. Be sure to use lock washers when reassembling using internal- or external-tooth types so they can make a solid contact to the enclosure. Do not use the split-O lock washers which do not ensure good enclosure contact. If lock washers are not present, consider adding them. If rivets were used, you can substitute screws or re-rivet the connector, making sure the mating metal surfaces are clean.

Internal cables are often smaller TMP connectors that are friction fit into board-mounted jacks. Once installed, these cables are rarely changed or moved and may become intermittent due to oxidation. The cable connector can come loose due to vibration or shock. If the cable is under tension (not recommended) it may pull loose or put stress on the jack.

29.4.2 Manual Antenna Tuners

In a manual tuner, there is generally at least one variable inductor (whether a switch selects taps or continuously adjustable) and one variable capacitor. Usually, there are two variable capacitors with the inductor, forming a T network. If there is a power or SWR meter, switches will select power ranges and forward or reflected power. Some tuners have more than one output antenna connector. Some also have a balun to feed a balanced transmission line and that is selected by a switch, as well. This section will focus on the switches and adjustable inductor and capacitor(s).

SWITCHES

In the presence of high voltage or if a switch is changed while transmitting, arcing of switch contacts is common. Eventually, the arcing results in pitting or burning away of the fixed or sliding contact metal as shown for the *rotary wafer switch* in **Figure 29.6**. Until the contact becomes completely open it will operate more and more erratically. A corroding contact will act the same way. The symptom of failing switches is intermittent high SWR to the transceiver, whether the contact is in the matching circuit or in an antenna select switch. Another possible symptom is SWR that increases during long transmissions as the switch contacts heat up.

Moving the switch to a different position and back again can eventually clear away enough oxide or burned material to make contact again. As the contacts degrade, it becomes more and more difficult to maintain consistent operation and eventually the contact will fail completely. If contact cleaner spray is used be sure it does not leave a residue that can break down from high voltage between and around the contacts.

Most rotary switches have "open" contacts, meaning they are not sealed in a housing, so they can usually be repaired or at least cleaned. Toggle, slide, and rocker switches are not reparable and must be replaced. The only option for smaller rotary switches may be replacement; this is switch-specific.

The contacts of larger rotary switches can be filed, polished, and restored. Note that any sharp points from arcing can make contact distances just slightly smaller or reduce breakdown voltage and arcs can re-occur. Be sure to make all air-insulating gaps as smooth as possible by minimizing the size and sharpness of these points.

For a custom switch for which replacement parts are unavailable, repair is sometimes the only option. Remove the switch and drill-out the rivets holding the contacts. If you can find unused contacts on the same switch, or good contacts on another similar switch, remove a set of contacts and mount them on the switch being repaired with #2 hardware (bolt, washer, and nut) in the needed location. (See **Figure 29.7**.) Do not overtighten the fasteners, as ceramic and older Bakelite or plastics are brittle. If the insulating wafer cracks, it can be repaired with a very small dab of cyanoacrylate. Ceramic can be repaired with white glue or epoxy.

INDUCTORS

Most tuners use fixed-value *air-wound* inductors as in **Figure 29.8** with turns spaced for cooling and higher Q. A switch selects a tap on the coil to change inductance. Impact (tuner was dropped) can cause coil failure by making the plastic or ceramic coil form or formers (strips which holds the turns in place) crack or split. The usual symptom of a broken air-wound coil is a sudden change in coil value that requires a different setting be used for an antenna that hasn't changed.

Figure 29.6 — The switch on the left shows the effects of destructive arcing compared to the new switch at the right. Even heavy-duty switches used in amplifiers and high-power tuners can arc under overload and high-SWR conditions.

Figure 29.7 — A repaired wafer switch section using #2 hardware and contacts removed from another switch. See text for repair instructions.

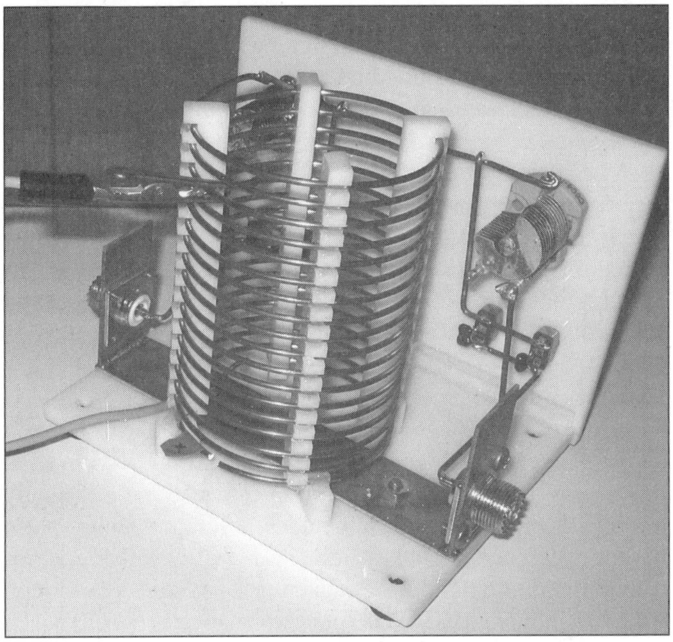

Figure 29.8 — An air-wound coil showing the plastic formers (white strips) that hold the wire turns. Note the large wire gauge and spacing between the turns. In this homebrew tuner, an alligator clip makes the adjustable contact. Commercial tuners use a multi-position switch to select different values of inductance.

Arcing can also occur between turns if they move closer together.

If the coil can be bent back into shape (as uniformly as possible), the former can be glued together (literally) — plastic with modeling cement, ceramic with white glue or epoxy. Plastic forms sometimes crack with age and heat, as well.

If necessary, one (or two) turns can be unwound to replace a connecting lead, then replace the missing section with the proper number of turns and gauge of wire. An "extension" of the coil form made of plastic or wood can be added as a replacement or reinforcement. Glue it so the coil diameter stays as uniform and consistent as possible. If you need to remove and replace many turns, consider replacing the coil.

Toroid core inductors are used in low-power tuners, and the cores can be damaged by impact or by overheating. The symptom of a cracked or broken toroid is usually a large change in inductance that prevents achieving a match.

If the core is just cracked with clean surfaces that fit snugly together, you may be able to glue it back together with a very small amount of glue along the outside edges of the crack. The individual pieces must be fit as closely and tightly as possible. The inductance will be reduced but might still be sufficient to perform its former function — replacement is a better choice if practical. If a core is crushed or pieces are missing, it will have to be replaced.

If you do not have an original part number or specification for the core, assess any paint color and the core's size. Then, carefully remove the coil from its mounting (typically a PC board) and unwind the coil, counting the turns and noting the placement of turns. Taking a photo of the coil in place before it is removed can be very helpful in duplicating the winding. Procure another core of the same material and as close to the same size as possible, then rewind the coil. If the wire is not damaged, i.e. the enamel is still consistently covering the wire and it is not kinked, re-use it; if not, replace with as close to the same size and type of wire (enameled magnet wire, Teflon-covered, solid or stranded). It can be tedious work, but take your time and spread the turns around the core in the same pattern as the original. Replace the coil and test.

Roller inductors as seen in **Figure 29.9** present their own unique challenges. A roller or wiper contacts the coil turns as either it or the roller is moved to increase or decrease inductance. Roller inductors typically develop problems either from corrosion of the coil or moving contact or from overheating at one or more positions on the coil. The usual symptoms of a failing roller inductor are erratic SWR when adjusting inductance or when using high power as the resistance of the moving contact changes.

Corrosion is usually fixable with silver polish or a non-metallic scuffing pad, such as Scotch-Brite. If a crust of corrosion has formed, an old toothbrush and contact cleaner can be used. Do not use steel bristle brushes as they will scratch or remove any plating. A brass bristle brush may be used, but

Figure 29.9 — The roller inductor coil turns while the contact at the top rolls along the wire, moving back and forth along the shaft. The contact is held against the wire by a spring.

gently. The roller contact can also corrode and be cleaned in a similar way. If contact cleaner spray is used, be sure it does not leave a residue that can break down from high voltage between the turns. A polish compound should be flushed from the coil with clean water or alcohol and the coil carefully dried.

If the coil or movable contact has developed a burned spot from prolonged high-power use in one position, repair may be difficult or impossible. Try cleaning the burned spot with a brass bristle brush or toothbrush and contact cleaner. If the contact point is pitted, it is likely to develop additional burning after repair. If replacing the inductor is not an option, try adding a small amount of inductance in series with the roller inductor so that operation takes place at a different location on the inductor. Otherwise, you will probably have to replace the inductor entirely

The spring holding the roller contact against the coil can weaken due to metal fatigue and age. This is often the cause of erratic SWR as the contact moves. Remove the roller for cleaning and add a slight bend to the spring to restore operation; be cautious as over-bending may snap the spring. Test with a continuity checker while running through the length of the coil. If a bad spot with weak or erratic contact is found, adjust the roller to make consistent, solid contact. If the roller rides on a shaft, clean the shaft as well.

CAPACITORS

Air-variable capacitors are almost always present, with or without fixed capacitors that can be switched in for wider ranges or for different bands. Arcing from excessive voltage is most likely to occur on the edges of the capacitor plates, usually on the rounded edges as shown in **Figure 29.10**. Arcing can be from plate to plate or from plates to wiring or chassis hardware. Once an arc develops, prolonged arcing can eat away a capacitor plate (or switch contact), causing ragged edges and points that enable arcs to form at even lower voltages.

The symptom of arcing is proper matching at low power and a sudden jump in SWR as power is increased. You may also be able hear or even smell the arc. At the first sign of arcing, remove power and try to determine what is causing the high voltage in the tuner. High SWR from disconnected or incorrect antennas and bad cables are possible causes, so make sure those aren't the cause before continuing to transmit through the tuner and damaging it.

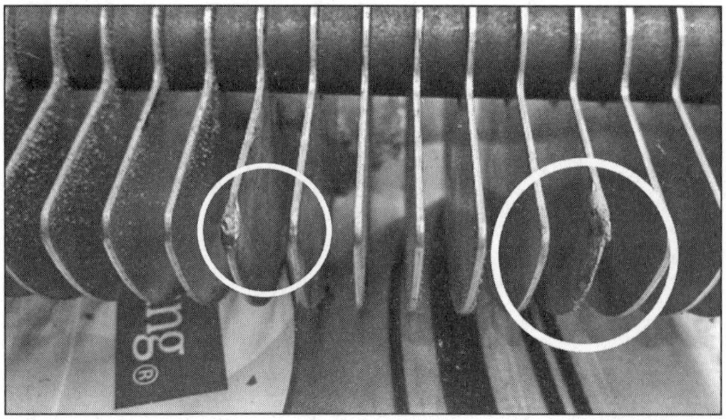

Figure 29.10 — The air-variable capacitor plates in the white circles show the effect of persistent arcing. The edges of the plates are pitted and ragged. The plate on the left is discolored from heat.

When there has been an arc, it is fairly obvious. File down the arced section and smooth to a rounded shape. The amount of capacitance lost to this repair is minimal, versus the cost of repairing the capacitor. Check the bearings, where the shaft of the rotor contacts the frame — clean them with alcohol or safe solvent and add a very slight amount of lubricant. Also check the stator (stationary plates) contacts for connections broken due to fatigue or physical damage.

Fixed transmitting capacitors ("door knob" or vacuum capacitors) can fail, typically at the contacts; test with an LCR meter to verify. The only option is replacement. Carefully remove the screws from the contacts or clamps and replace the entire component. If the original value is unavailable, series or parallel combinations of capacitors can be used although the extra inductance may affect performance above 20 MHz.

BALUNS

Some tuners also include a balun for use of balanced feed lines. The symptom of a balun problem is similar to toroid inductor failures — a sudden change in settings or SWR. If the balun is suspect, measure the winding resistances or count turns. The impedance ratio of a transformer is the square of the turns ratio. If the balun is sealed, estimate the turns ratio as the ratio of the input and output winding resistances. See the previous section on Inductors for repair information. Most failures are from overheating the core, so if the windings are okay, the core is suspect; verify with replacement. If the windings fail, replace with Teflon-covered wire using the same number of (uniformly wound) turns.

29.4.3 Automatic Antenna Tuners

Most of the same components in a manual tuner are present in an automatic tuner along with sensing, selection and indicating circuits. **Figure 29.11** shows the inside of a typical model. The most important new moving parts are relays. Automatic tuners are a bit more difficult to diagnose, but by breaking them down into functional parts, troubleshooting is straightforward.

Typically, automatic tuners have fixed-value capacitors and inductors in an L network. A phase sensor and voltage detectors are read by a microprocessor, which switches in L and C based on an algorithm to arrive at the best SWR match. The inductors and capacitors are switched in or out by relays. In "zero-power" tuners (tuners that do not draw much, if any current once tuning is complete) latching relays are used. These remain open or closed once power is removed.

There are three usual symptoms for automatic tuner problems. The tuner may not be able to achieve a match at all and display some kind of error condition. The tuner can "hunt" without stopping for a match according to its control program — you'll hear the relays clicking or buzzing without stopping. Or the tuner will suddenly start re-tuning at higher power

Figure 29.11 — Typical construction of an automatic tuner (manufactured by LDG) using toroid inductors and fixed-value capacitors. Controlled by a microcontroller, relays (the dark blocks) switch in different values and configurations of L and C to effect the match.

levels during transmission because a component or relay is failing.

If the tuner cannot find a match or keeps hunting to find a match, the issue is typically a bad relay contact. The controller thinks it has the correct L or C value switched-in, but the contact is not being made. A relay driver line can also be open. Also check for a bad interconnect cable between the external antenna connector and the main board.

The most common issue is relay failure. Relays have a coil that is energized to move an *armature* holding the contacts. The usual relay acts as a DPDT switch with a set of NC (*normally closed*) and NO (*normally open*) contacts. To test the coils, first measure the coil resistance. Then energize each relay one at a time and measure the resistance across the contacts with and without power. A coil failure (usually an open-circuit) requires relay replacement.

If the coil is good and you can get to the relay contacts, a strip of bond paper and cleaning solution can sometimes clean the contacts of corrosion or oxide. Carefully remove the cover using a hobby knife along the bottom edge, not poking too far into the relay — you do not want to cut the coil. In miniature relays in particular, be careful to avoid bending the armature while cleaning. Put some cleaning solution on the paper strip, slide the strip between one set of contacts, hold them together, then move the strip back and forth. If there is a dark deposit on the paper, keep sliding it back and forth until no additional deposit is seen. Repeat with the other set of contacts. If the contacts are severely pitted or there is large buildup of material on the contact from arcing, the relay should be replaced. Filing the contacts should be avoided because it removes any plating from the contact surfaces.

If the relays are good, they may not be being actuated by the controller. Check the printed-circuit board traces for burned or open traces. Then work your way back to the driver transistors and test them; microcontrollers do not drive relays directly. Typically, a transistor is used to energize the relay coils. There should be a suppressor diode across each coil, as well. Replace components as necessary.

TUNER CONTROLLER

The microcontroller controls the relays based on its inputs, usually a pair of voltages connected to an analog-to-digital conversion pin. The sensing circuits are made up of a sensing coil, diodes and possibly an operational amplifier. (See the **Antenna and Transmission Line Measurements** chapter material on SWR bridges.) The usual symptom of failed sensing circuits is no forward or reflected power reading, even when the tuner is bypassed. (Most tuners will act as power meters when in the bypass configuration.) Test the sensing coil with an ohmmeter, then the diodes (a common failure especially in a high RF environment). If all check good, replace the op amp that conditions the signals for the microcontroller input.

The last and least likely to fail part is the microcontroller. Testing this component requires a high-impedance oscilloscope and probe, a schematic, and basic logic of how the tuner should work. While trying to tune a dummy load at very low power, verify the inputs (from the sensors) and the outputs (to the driver transistors); as the inputs change, the outputs should also change to adapt. You should hear the relays chatter as the tuner attempts to lower the SWR.

INTERNAL ANTENNA TUNERS

Internal antenna tuners are automatic antenna tuners (as described above), mounted inside a transceiver. They are typically small assemblies, since most transceivers today are of the 100 W class. This complicates troubleshooting only slightly, since the components are smaller. It also makes it slightly easier since internal tuners typically will only correct a mismatch of 3:1 — reducing the number of inductors and capacitors to switch. Use the same techniques to troubleshoot, but be careful of other circuits in the transceiver and be sure than any connecting cables are securely seated. Check the printed-circuit board mounting screws, as they can work themselves loose due to thermal cycling, becoming faulty grounds.

29.4.4 Maintenance and Operation

Think of your antenna tuner as the output stage of an amplifier — it contains many of the same type of components, operates at the same power levels, and needs the same care to operate reliably.

In large manual tuners, keep air-variable capacitors, exposed inductors, and switches clean. Dirt and dust can form paths for arcing and cause heat buildup. Remove dust with a gentle brush and vacuum. If the tuner has been in a smoking environment, smoke deposits can accumulate — clean with a brush first, then apply cleaning solution using a brush and thoroughly dry before operation. Lightly re-oil any bearings in air-variable capacitors after using cleaning solution. Don't put oil on a roller inductor shaft or contact. Check for arcing, broken or heated connections and wires; repair or replace as needed. Use a high-quality cleaning solution on switch contacts, but not to excess and remove any residue. Check connectors and the cables in and out of the tuner.

Check grounding and bonding. The cabinet of an (external) tuner should be bonded to the transceiver, and to any RF bonding bus or plane in the station. If common-mode RF current is upsetting SWR measurements, use ferrite beads or a toroid to block it.

PROPER OPERATION

It sounds backwards, but "tuning a tuner" is the best way to avoid interfering with other stations and is easier on the transceiver, too. There are many antenna analyzer products out today that will put out a very small signal and measure the health of your antenna — they will work in much the same way with the tuner. Connect the analyzer to the input of the tuner (the transmitter port), adjust the analyzer for the desired band / frequency, and tune the tuner for minimum SWR. The result is very low output (reduced interference) and a tuner that is very close to optimal.

If you do not have an antenna analyzer, turn down the transmitter output and use very low power to tune the tuner while your keyer to send a series of 30 WPM dits (followed by your call sign). This is a lower duty cycle than continuous carrier, but will do the same job.

Once you determine the optimal setting for your tuner, write the control settings down by band and frequency — you then have a starting-point for tuning on that band or frequency. Cut an index card to fit over the TUNE and LOAD controls and mark the locations for each band. This way, you are not damaging the front of your tuner. The less time you spend tuning, the less stress you place on your tuner.

29.5 Refurbishing Aluminum Antennas

Whether passed on by another amateur, recovered from the local recycling shop, grabbed as a bargain at a swap meet or just needing to do maintenance, the average amateur often has to bring up to scratch antennas that are the worse for wear.

Two of the most detrimental factors to aluminum are the results of electrolysis caused by poor choice of connectors and the chemistry of the air. Salt near the coast or industrial/automotive particulates can, when mixed with the normal moisture content, eat away at the shiny aluminum. If allowed to progress far enough, the mechanical integrity of the structure is impaired beyond simple repair.

The first procedure is to inspect the antenna. Look for the dreaded white oxide powder around connectors and joints. This points the way to the areas that need particular attention. Next is to try and remove the connector hardware which may be seized beyond recovery. This is particularly the case where steel plated with cadmium or zinc/galvanized hardware has been used. Before struggling with wrenches and screwdrivers, spray the area with penetrating oil such as Kroil or other modern preparations. These are more effective than some of the older preparations such as WD40 and CRC-556.

If the items release, you have had a win. If not then you have to find a suitable method of removal. Sometimes, heating the area with a blowtorch may cause sufficient expansion for the frozen joint to be loosened. Clamps may be cut free using a cutting wheel on a high speed grinding tool — before cutting into the underlying aluminum, try leverage with a small bladed screwdriver and hopefully you will be able to break the metal along the cut without bruising or deformation of the aluminum. Even an old fashioned hacksaw with a fine toothed blade might be suitable in making the cut.

Metal threads that are frozen because of corrosion can be a great frustration. This can be made more difficult if they pass through plastic insulators, as trying to grind the heads off will melt the plastic. A method that has been found helpful is to drill though the head of the metal thread with a sharp drill slightly smaller in diameter than the shank. The hole only needs to be slightly deeper than the depth of the head. Then use a drill slightly smaller than the diameter of the head to remove the head. This method generates less heat from friction than most other methods and is particularly easy to use on PoziDriv or Phillips head hardware as the drill is automatically centered.

Having disassembled the antenna, it is necessary to further inspect its condition and repair and/or treat areas that are damaged. Areas of oxidation need to be removed by abrasion. This can generally be done using a kitchen plastic scouring pad such as Scotch-Brite, if the oxidation is superficial. The advantage of using the plastic pad is that it does not impregnate the surfaces with metal particles of dissimilar metal which will only cause further corrosion later on.

If the pitting is deep, it may be necessary to remove the damaged area and insert a suitable sleeve just to restore mechanical strength. Pitted areas can sometimes be cleaned and an internal sleeve of PVC or similar used but remember to ensure balance if the element or boom section is undamaged on the opposite end. Remember that crystallization occurs in aluminum subjected to constant vibration, a lesson learned from the aircraft industry but obvious in aluminum antennas mounted in windy sites.

If the metal has to be cut, it must be joined to be electrically continuous and particularly at VHF and UHF, the outside diameter must be maintained to keep the tuning characteristics within specification. For this reason, internal sleeves are usually preferable with use of aluminum pop rivets that have aluminum mandrels. Some bargain rivets use steel mandrels and in the right conditions you will have a loose fastening, a nonconductive joint and a noisy antenna.

Once the various components have been cleaned and mended they are ready for reassembly. Replace the hardware with stainless steel and use nylon insert (Nylok) nuts that remain tight without deforming the tubing. Worm drive stainless steel hose clamps are used but not the ones with plain steel worm drives. Boom clamps using U-bolts are expensive items and a wire brush can be used on the threads to remove any rust, followed with a light spray of aluminum-based paint and replacing the washers and nuts with bright steel ones which are then also painted. If possible, after assembly, a further coat of paint is applied to keep the moisture from these components.

Remember that UV light causes many wire jackets to degrade and so any pigtails, whether insulated or not, benefit from having heat shrink tubing applied.

All swaged joints should be cleaned to bright metal on the mating surfaces, remembering the RF skin effect. Use a thin coating of anti-oxidation compound at all metal-to-metal joints as described in the section on Corrosion in the chapter **Building Antenna Systems and Towers**.

On the exterior, if there are concerns of moisture ingress, clean the surface of any contaminants and apply neutral-cure silicone sealant or cover with butyl rubber self-vulcanizing tape. See the section on Waterproofing in the chapter **Building Antenna Systems and Towers**. Do not be tempted to use hot melt glue on external applications as it deteriorates rapidly from UV radiation.

Although there are warnings about painting antennas, particularly where there is evidence of pitting or scratching on the surface a light spray of aluminum-based paint provides added protection against additional damage. The point is that you are not painting a rusty hulk and brushing paint on thickly but lightly coating the surfaces and paint runs will not occur to cause insulation of parts of your antenna.

It is probably wise to have a progressive program of inspection and maintenance of all antenna systems. Birds find our structures good perches, wind can bend things and moisture which is trapped can all cause damage. At least every couple of years is a good program. Look after your antennas and they will serve you well and long.

Advertisers Index

Advertising Department Staff

Janet Rocco, W1JLR, Advertising Relationship Manager

800-243-7768

Direct Line: 860-594-0203 ■ Fax: 860-594-4285 ■ email: ads@arrl.org ■ Web: www.arrl.org/ads

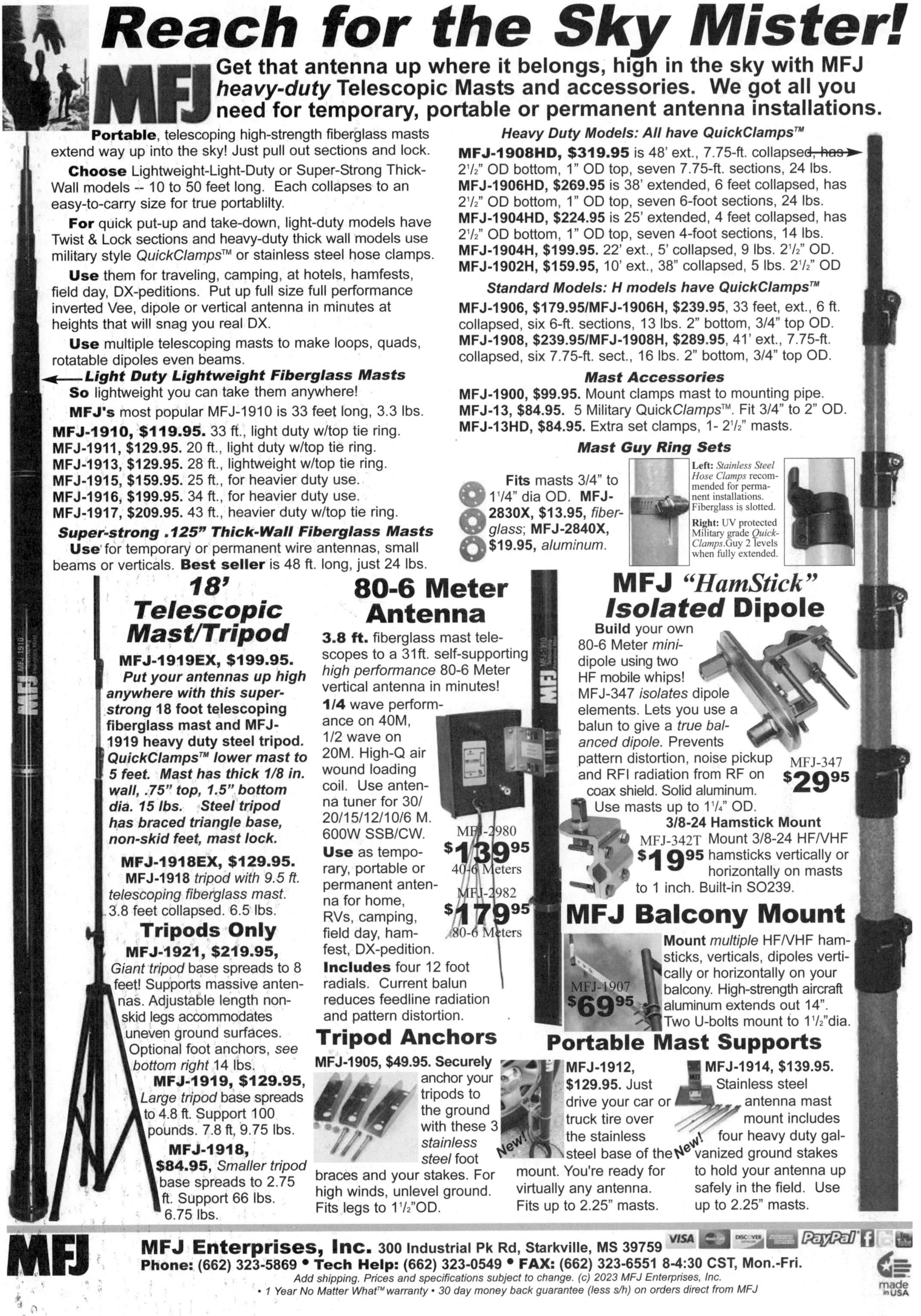
Reach for the Sky Mister!
MFJ
Get that antenna up where it belongs, high in the sky with MFJ heavy-duty Telescopic Masts and accessories. We got all you need for temporary, portable or permanent antenna installations.
Portable, telescoping high-strength fiberglass masts extend way up into the sky! Just pull out sections and lock.
Choose Lightweight-Light-Duty or Super-Strong Thick-Wall models -- 10 to 50 feet long. Each collapses to an easy-to-carry size for true portablilty.
For quick put-up and take-down, light-duty models have Twist & Lock sections and heavy-duty thick wall models use military style QuickClamps™ or stainless steel hose clamps.
Use them for traveling, camping, at hotels, hamfests, field day, DX-peditions. Put up full size full performance inverted Vee, dipole or vertical antenna in minutes at heights that will snag you real DX.
Use multiple telescoping masts to make loops, quads, rotatable dipoles even beams.
Light Duty Lightweight Fiberglass Masts
So lightweight you can take them anywhere!
MFJ's most popular MFJ-1910 is 33 feet long, 3.3 lbs.
MFJ-1910, $119.95. 33 ft., light duty w/top tie ring.
MFJ-1911, $129.95. 20 ft., light duty w/top tie ring.
MFJ-1913, $129.95. 28 ft., lightweight w/top tie ring.
MFJ-1915, $159.95. 25 ft., for heavier duty use.
MFJ-1916, $199.95. 34 ft., for heavier duty use.
MFJ-1917, $209.95. 43 ft., heavier duty w/top tie ring.
Super-strong .125" Thick-Wall Fiberglass Masts
Use for temporary or permanent wire antennas, small beams or verticals. Best seller is 48 ft. long, just 24 lbs.
Heavy Duty Models: All have QuickClamps™
MFJ-1908HD, $319.95 is 48' ext., 7.75-ft. collapsed, has 2 1/2" OD bottom, 1" OD top, seven 7.75-ft. sections, 24 lbs.
MFJ-1906HD, $269.95 is 38' extended, 6 feet collapsed, has 2 1/2" OD bottom, 1" OD top, seven 6-foot sections, 24 lbs.
MFJ-1904HD, $224.95 is 25' extended, 4 feet collapsed, has 2 1/2" OD bottom, 1" OD top, seven 4-foot sections, 14 lbs.
MFJ-1904H, $199.95. 22' ext., 5' collapsed, 9 lbs. 2 1/2" OD.
MFJ-1902H, $159.95, 10' ext., 38" collapsed, 5 lbs. 2 1/2" OD
Standard Models: H models have QuickClamps™
MFJ-1906, $179.95/MFJ-1906H, $239.95, 33 feet, ext., 6 ft. collapsed, six 6-ft. sections, 13 lbs. 2" bottom, 3/4" top OD.
MFJ-1908, $239.95/MFJ-1908H, $289.95, 41' ext., 7.75-ft. collapsed, six 7.75-ft. sect., 16 lbs. 2" bottom, 3/4" top OD.
Mast Accessories
MFJ-1900, $99.95. Mount clamps mast to mounting pipe.
MFJ-13, $84.95. 5 Military QuickClamps™. Fit 3/4" to 2" OD.
MFJ-13HD, $84.95. Extra set clamps, 1- 2 1/2" masts.
Mast Guy Ring Sets
Fits masts 3/4" to 1 1/4" dia OD. MFJ-2830X, $13.95, fiberglass; MFJ-2840X, $19.95, aluminum.
Left: Stainless Steel Hose Clamps recommended for permanent installations. Fiberglass is slotted.
Right: UV protected Military grade Quick-Clamps. Guy 2 levels when fully extended.
18' Telescopic Mast/Tripod
MFJ-1919EX, $199.95.
Put your antennas up high anywhere with this super-strong 18 foot telescoping fiberglass mast and MFJ-1919 heavy duty steel tripod. QuickClamps™ lower mast to 5 feet. Mast has thick 1/8 in. wall, .75" top, 1.5" bottom dia. 15 lbs. Steel tripod has braced triangle base, non-skid feet, mast lock.
MFJ-1918EX, $129.95.
MFJ-1918 tripod with 9.5 ft. telescoping fiberglass mast. 3.8 feet collapsed. 6.5 lbs.
Tripods Only
MFJ-1921, $219.95, Giant tripod base spreads to 8 feet! Supports massive antennas. Adjustable length non-skid legs accommodates uneven ground surfaces. Optional foot anchors, see bottom right 14 lbs.
MFJ-1919, $129.95, Large tripod base spreads to 4.8 ft. Support 100 pounds. 7.8 ft, 9.75 lbs.
MFJ-1918, $84.95, Smaller tripod base spreads to 2.75 ft. Support 66 lbs. 6.75 lbs.
80-6 Meter Antenna
3.8 ft. fiberglass mast telescopes to a 31ft. self-supporting high performance 80-6 Meter vertical antenna in minutes!
1/4 wave performance on 40M, 1/2 wave on 20M. High-Q air wound loading coil. Use antenna tuner for 30/20/15/12/10/6 M. 600W SSB/CW.
Use as temporary, portable or permanent antenna for home, RVs, camping, field day, hamfest, DX-pedition.
Includes four 12 foot radials. Current balun reduces feedline radiation and pattern distortion.
MFJ-2980
$139.95
40-6 Meters
MFJ-2982
$179.95
80-6 Meters
Tripod Anchors
MFJ-1905, $49.95. Securely anchor your tripods to the ground with these 3 stainless steel foot braces and your stakes. For high winds, unlevel ground. Fits legs to 1 1/2"OD.
MFJ "HamStick" Isolated Dipole
Build your own 80-6 Meter mini-dipole using two HF mobile whips! MFJ-347 isolates dipole elements. Lets you use a balun to give a true balanced dipole. Prevents pattern distortion, noise pickup and RFI radiation from RF on coax shield. Solid aluminum. Use masts up to 1 1/4" OD.
MFJ-347
$29.95
3/8-24 Hamstick Mount
MFJ-342T
$19.95
Mount 3/8-24 HF/VHF hamsticks vertically or horizontally on masts to 1 inch. Built-in SO239.
MFJ Balcony Mount
Mount multiple HF/VHF hamsticks, verticals, dipoles vertically or horizontally on your balcony. High-strength aircraft aluminum extends out 14". Two U-bolts mount to 1 1/2"dia.
MFJ-1907
$69.95
Portable Mast Supports
New!
MFJ-1912, $129.95. Just drive your car or truck tire over the stainless steel base of the mount. You're ready for virtually any antenna. Fits up to 2.25" masts.
New!
MFJ-1914, $139.95. Stainless steel antenna mast mount includes four heavy duty galvanized ground stakes to hold your antenna up safely in the field. Use up to 2.25" masts.
MFJ
MFJ Enterprises, Inc. 300 Industrial Pk Rd, Starkville, MS 39759
Phone: (662) 323-5869 • Tech Help: (662) 323-0549 • FAX: (662) 323-6551 8-4:30 CST, Mon.-Fri.
Add shipping. Prices and specifications subject to change. (c) 2023 MFJ Enterprises, Inc.
• 1 Year No Matter What™ warranty • 30 day money back guarantee (less s/h) on orders direct from MFJ
VISA
DISCOVER
PayPal
made in USA

Antenna Helpers
MFJ-259C 530 KHz to 230 MHz
World's most popular analyzer
World's most popular antenna analyzer is new and improved. Now covers 530 KHz to 230 MHz!. Super easy-to-use -- Read antenna SWR, complex impedance, return loss, reflection coefficient. Determine velocity factor, coax cable loss in dB, length of coax and distance to short or open in feet. Read inductance in uH, capacitance in pF. Large two line LCD screen and dual meters clearly display your info. Built-in frequency counter, signal generator, charger circuit, battery saver, low battery warning...
MFJ-259C $299.95
MFJ-269C 530 KHz - 230 MHz
plus 415-470 MHz, 12-bit A/D
MFJ-269C is a super MFJ-259C that adds 415-470 MHz and 12-bit A/D converter that gives you much better accuracy. Complex Impedance Analyzer reads series parallel equivalents and magnitude and phase. CoaxCalculator™ gives line length from electrical degrees and vice-versa for any frequency, Velocity Factor, coax loss in dB. Use any characteristic impedance 10-600 Ohms for more accuracy when using different coax. Has LCD log SWR bargraph, N-connector.
MFJ-269C $399.95
MFJ-226 1 - 230 MHz
VNA Antenna Analyzer
MFJ VNA Antenna Analyzer covers 1 to 230 MHz, 1Hz resolution. •Frequency sweep plots: SWR, Impedance, Resistance, Reactance, Phase Angle, Complex Return Loss, Smith Chart •Sign of reactance positively identifies inductive or capacitive reactance •Amazing accuracy with OSL (Open-Short-Load) calibration - - calibrate through feedline/test cable at different frequencies and store in memory. Measure directly or through feedline. Capture screens and download to PC.
MFJ-226 $319.95
MFJ 300W AutoTuner
Select ultra-wide range 6-3200 Ohms/150 Watts or wide-range 6-1600 Ohms with the world's only dual matching range automatic antenna tuner. 1.8-30 MHz, 4:1 current balun, lighted cross-needle meter plus digital backlit LCD SWR/Wattmeter, 20,000 virtual antenna memories, remote port, rig interface.
MFJ-993B $269.95
MFJ 1.5kW AutoTuner
Full 1500 Watts SSB/CW. Matches 12-1600 Ohms, 1.8-30 MHz. Digital & analog SWR/Wattmeter. Ultra-fast, safe automatic tuning. Antenna switch for two coax/long wire. Automatic amplifier bypass protects your amplifier. Field upgradeable software. Radio interface. MFJ-912, $69.95 external balun.
MFJ-998 $699.95
MFJ Remote AutoTuner
Tune at the antenna! Operating your antenna tuner remotely greatly reduces losses and gives high efficiencies from long coax runs and high SWR. MFJ-998RT has a wide tuning range, 12-1600 Ohms, 1.5 kW, 1.8-30 MHz. Weather protected ABS plastic housing.
MFJ-998RT $819.95
MFJ Manual Tuner
Ultra-wide range precision tuning! More hams use MFJ-949s than any other antenna tuner in the world! Does it all! 1.8-30 MHz, 300 Watts, full size peak/average lighted Cross-Needle SWR/Wattmeter, 8 position antenna switch, dummy load. Highly efficient air inductor, 1kV air-variable tuning capacitors. QRM-Free PreTune™, scratch proof Lexan front panel.
MFJ-949E $189.95
MFJ RFI Isolator
Transform 200 Ohms to 50 Ohms. SO-239. Transmission line transformer construction, low permeability ferrite cores. Stops RFI traveling on the outside of your coaxial shield into your transceiver.
MFJ-915 $29.95
Hi-Performance Ferrite Bead Balun
Like MFJ-913 but handles legal limit. Sturdy cube shaped heavy duty weather housing, Compact 3½Wx3½Hx2½D inches.
MFJ-918 $24.95
MFJ 300W True 4:1 Current Balun
High-permeability ferrite beads on high-quality RG-303 Teflon(R) coax. 1:1, 2 inch diameter by 6 inches in length. 1.5 kW 1.8-30 MHz. Center insulator.
MFJ-913 $29.95
MFJ Lightning Surge Protector
Safeguard equipment. Shunts to 5000A peak impulse current harmlessly to ground. SWR< 1.1:1, < 0.1 dB loss. Use to 1000 MHz, 400W PEP.
MFJ-270 $19.95
MFJ 2-position Antenna Switch
Center ground, automatic grounding of unused positions, lightning surge protection, 2.5 kW PEP, works to over 500 MHz. MFJ-1704, $79.95, four positions.
2 pos.
MFJ-1702C $39.95
Ameritron RCS-4 Remote Coax Switch
Use 1 coax for 4 antennas. No control cable needed. SWR < 1.25, 1.5 - 60 MHz. Use to 100 MHz. 2.5 kW PEP, 1.5 kW average.
RCS-4 $179.95
Ameritron RCS-8V Remote Coax Switch
Replace 5 coax with one! High isolation, < .1 dB loss, 1.2 SWR@ 250 MHz. Use to 450 MHz. 1kW@ 150 MHz. 5kW PEP < 30 MHz, 2.5 kW average.
RCS-8V $199.95
Hy-gain HAM-IV Antenna Rotator
For medium arrays up to 15 sq. ft. wind load. 5-second brake delay, Test/Calibrate function. 2 1/16" mast. Digital control available.
HAM-IV $649.95
Hy-gain AR-500 Antenna Rotator
For VHF/UHF, 6-Meters, TV, FM, magnetic loops, other lightweight antennas. Digital remote control, selectable 110/220 VAC.
AR-500 $159.95
Stainless Steel Telescoping Whips
MFJ-1979, $64.95. 17 ft. ext., 27" collapsed. 3/8-24 base. MFJ-1977, $44.95. 12' ext/26". More lengths available.
Dipole Kit
MFJ-2774K, $64.95. Every thing you need to make your own G5RV, doublet, dipole and others.
Flexible 7-Strand, #14, Solid-Copper Antenna Wire.
Super-strong for very long antennas. No sag!
MFJ-18G100, $29.95. 100 ft.
MFJ-18G250, $64.95. 250 ft.
6-Pack Glazed Ceramic Insulators
MFJ-16C06, $4.99.
Molded High-Strength Center Insulator/SO-239
MFJ-16B01, $19.95
450 Ohm Ladder Line
#18 copper weld wire.
MFJ-18H050, $24.95. 50 ft.
MFJ-18H100, $39.95. 100 ft.
MFJ-18H250, $99.95. 250 ft.
Dealer/Catalog/Manuals
Visit: http://www.mfjenterprises.com or call toll-free 800-647-1800
• 1 Year No Matter What™ warranty • 30 day money back guarantee (less s/h) on orders direct from MFJ
MFJ
MFJ ENTERPRISES, INC.
300 Industrial Pk Rd, Starkville, MS 39759 PH: (662) 323-5869
Tech Help: (662) 323-0549
FAX:(662)323-6551 8-4:30 CST, Mon.-Fri. Add shipping.
Prices and specifications subject to change. (c) 2019 MFJ Enterprises, Inc.
www.mfjenterprises.com

ICOM®
ID-52
IC-705
IC-V3500
ID-5100
IC-7300
IC-7610
IC-9700
HAM RADIO OUTLET®
WORLDWIDE DISTRIBUTION
SALES & SERVICE
SINCE
1971
EXCELLENCE
HAMRADIO.COM
ANAHEIM, CA: 800-854-6046 | ATLANTA, GA: 800-444-7927 | DENVER, CO: 800-444-9476
MILWAUKEE, WI: 800-558-0411 | NEW CASTLE, DE: 800-644-4476 | PHOENIX, AZ: 800-559-7388
PLANO, TX: 877-455-8750 | PORTLAND, OR: 800-765-4267 | SACRAMENTO, CA: 877-892-1745
SALEM, NH: 800-444-0047 | WINTER SPRINGS, FL: 800-327-1917 | WOODBRIDGE, VA: 800-444-4799
Providing excellence in sales, service, and support since 1971.
Find us on facebook
FOLLOW US ON twitter
YouTube

YAESU
The radio
FTDX101MP
FTM-500DR
FTDX10
FT5DR
FT-710
HAM RADIO OUTLET
WORLDWIDE DISTRIBUTION
SALES & SERVICE SINCE 1971 EXCELLENCE
HAMRADIO.COM
ANAHEIM, CA: 800-854-6046 | ATLANTA, GA: 800-444-7927 | DENVER, CO: 800-444-9476
MILWAUKEE, WI: 800-558-0411 | NEW CASTLE, DE: 800-644-4476 | PHOENIX, AZ: 800-559-7388
PLANO, TX: 877-455-8750 | PORTLAND, OR: 800-765-4267 | SACRAMENTO, CA: 877-892-1745
SALEM, NH: 800-444-0047 | WINTER SPRINGS, FL: 800-327-1917 | WOODBRIDGE, VA: 800-444-4799
Providing excellence in sales, service, and support since 1971.
Find us on facebook
FOLLOW US ON twitter
YouTube

Clean up your noisy audio with a..
..bhi DSP noise canceling product!
bhi
ParaPro EQ20 Audio DSP Range
- Improved audio for those with hearing loss
- Two separate mono inputs or one stereo input
- Use with passive speakers or headphones
- Basic EQ units EQ20, EQ20B (use with Dual In-Line, Compact In-Line or In-Line Module)
- DSP noise canceling versions EQ20-DSP, EQ20B-DSP with added Bluetooth on input
QST Dec 2019 review "easy-to-use device that improves the audio clarity of amateur signals"
Parametric Equaliser Response
ParaPro EQ20-DSP
Modular Audio Power Amplifier with DSP Noise Filtering and Parametric Equaliser
In-Line Module
In-Line Module
Amplified DSP Noise Canceling Module
Audio Input
Speaker
Headphones
10V to 16V DC
Dual In-Line
Dual In-Line
Noise Eliminating Module
Great review in Jan '23 QST!
- 5W amplified DSP noise canceling In-Line module
- 8 filter levels 8 to 40dB - Use in-line with a loudspeaker
- Audio bypass feature - 3.5mm mono inputs and outputs
- Headphone socket - Audio input overload feature
Fully featured flexible dual channel DSP noise canceling unit
- 8 Filter levels 9 to 40dB
- 3.5mm mono or stereo inputs
- Line level input/output
- 7 watts mono speaker output
- Headphone socket
- Easy to use controls
Compact In-Line
Compact In-Line Noise Eliminating Module
DSP Noise Cancelling
www.bhi-ltd.com
- Portable DSP noise canceling unit
- Simple controls
- Use in-line with speakers or headphones
- Line/speaker level inputs
- Use with AA batteries or 12V DC supply
New
NEDSP1962-KBD 5W amplified DSP noise canceling speaker retrofit pcb module now available!
DESKTOP MKII
10W amplified DSP noise canceling speaker
- Simple controls
- 8 filter levels
- Separate line level and speaker level audio inputs
- 12V DC power
- Suitable for all radios incl' SDR, Elecraft and FlexRadio products
NES10-2MK4
- 5W DSP Noise Canceling Speaker - 8 filter levels
- Compact speaker for mobile or base station
- Three position switch for all functions
- Headphone socket
- Adjustable metal bracket
NES10-2 MK4
Check out our range of audio accessories
DX ENGINEERING
1-800-777-0703
www.bhi-ltd.com
GigaParts
1-256-428-4644
EA&O
VISA
PayPal
MasterCard

Index

Editor's note: Except for commonly used phrases and abbreviations, topics are indexed by their noun names. Many topics are also cross-indexed.

The Letters "ff" after a page number indicate coverage of the indexed topic on succeeding pages.

A separate **Project Index** and **Author Index** follow the main index.

B

C

D

E

F

G

H

I

N

O

P

Q

R

S

T

U

V

W

Y

Z

Project Index

A

D

Author Index